HANDBOOK OF PHYSIOLOGY

Section 4: Environmental Physiology

HANDBOOK OF PHYSIOLOGY

A critical, comprehensive presentation of physiological knowledge and concepts

Section 4: Environmental Physiology
Volume II

Edited by

MELVIN J. FREGLY
Department of Physiology
University of Florida

CLARK M. BLATTEIS
Department of Physiology and Biophysics
University of Tennessee, Memphis

New York   Oxford
Published for the American Physiological Society
by Oxford University Press
1996

Oxford University Press

Oxford   New York
Athens   Auckland   Bangkok   Bombay
Calcutta   Cape Town   Dar es Salaam   Delhi
Florence   Hong Kong   Istanbul   Karachi
Kuala Lumpur   Madras   Madrid   Melbourne
Mexico City   Nairobi   Paris   Singapore
Taipei   Tokyo   Toronto

and associated companies in
Berlin   Ibadan

Published for the American Physiological Society
by Oxford University Press, Inc.,
198 Madison Avenue, New York, New York 10016

Oxford is a registered trademark of Oxford University Press

Library of Congress Cataloging-in-Publication Data
Environmental physiology
edited by Melvin J. Fregly and Clark M. Blatteis.
p.   cm.—(Handbook of physiology; sect. 4)
Includes bibliographical references and index.
ISBN 0-19-507492-0 (set)
1. Ecophysiology—Handbooks, manuals, etc.
2. Bioclimatology—Handbooks, manuals, etc.
3. Man—Influence of environment—Handbooks, manuals, etc.
4. Adaptation (Physiology)—Handbooks, manuals, etc.
I. Fregly, Melvin J.   II. Blatteis, Clark M.
III. American Physiological Society (1987–)
IV. Series: Handbook of physiology (Bethesda, Md.); sect. 4.
[DNLM: 1. Adaptation, Physiological.   2. Environment.   3. Physiology.
QT 104 H236 1977 sect.4]   QP6.H25   1977 sect. 4   [QP82]   599'.01 s—dc20   [599.05]
DNLM/DLC   for Library of Congress   94-31439

9 8 7 6 5 4 3 2 1

Printed in the United States of America
on acid-free paper

Preface

AT THIS WRITING, it has been 31 years since the predecessor of this *Handbook* was published. The research described in that volume has become classical, and it continues to stand as an essential reference to the field up to that time. Since the earlier edition, various areas of investigation in environmental physiology, then in their infancy, have progressed apace. It is these newer areas that are emphasized in the present volume.

As a result of this expansion of knowledge, we thought it more useful and practical to organize the present volume on the basis of environments, rather than on that of species and organ systems as in the earlier *Handbook*. The present *Handbook* developed as follows. At the invitation of the Publications Committee of the American Physiological Society, we initially prepared a detailed outline of the contents of the proposed volume, which was submitted for review to a number of peers. After incorporation of their many excellent suggestions, the revised outline was reviewed in turn by the Publications Committee, the Handbook Committee, a subcommittee of the Publications Committee, and Oxford University Press. After incorporating their suggestions and receiving final approval from these groups, we identified colleagues with the requisite expertise in the environments covered in the final outline and invited them to become associate editors of their particular environmental topics. We also asked them to suggest expert contributors for the topics to be covered, and those individuals were then formally invited to write the relevant chapters.

The editing process eventually involved scientific review of each submitted chapter within the section by the associate editor and several members of that section, particularly those whose topics could overlap, to identify gaps in coverage and to prevent undue duplication. After this internal review and first revision, the chapters were forwarded to us. We then sent them for external review by at least two peers in the field. After this scientific review, and a subsequent editorial review by us, the chapter was returned to the author(s) for final revision.

In spite of the rigor of this process, we did not attempt to impose a style on the individual authors. Therefore, stylistic differences will be apparent among some of the chapters. Moreover, some readers may perceive that certain chapters are more subjective than others or have emphasized a more controversial point of view. We have recognized and permitted this so as not to stifle controversy and to attest to the currency of the material in this *Handbook*.

Notwithstanding all of this, the reader may find certain gaps and redundancies. Some of these are intentional. Thus we have deliberately included in this volume only those factors in the external environment to which physiological acclimation has been experimentally documented. On this basis, we have included homeostatic reactions to both acute and chronic environmental exposures (that is, not just long-term exposure, as in the previous *Handbook*) and emphasized the mechanisms regulating the physiological responses to these factors. We have excluded reference to potentially toxic environments, for example, pollutants, food additives, drugs, etc., on the grounds that there is, as yet, little substantive evidence of acclimation to them. In this respect, we have also chosen specifically not to consider topics in biometeorology, exercise, aging, pathology and therapeutics, except where they are particularly pertinent to another topic. Furthermore, some of these, for example, aging and exercise, are covered in other sections in the *Handbook* series.

We express our sincere gratitude for the cooperation and support of our associate editors and contributors. Without their commitment, dedication, and enthusiasm we could not have completed this enormous project within the relatively short period of two years. Sadly, two primary authors, Professor Dr. Kurt Brück and Professor A. P. Gagge, died during the first year of the project. In each case, the chapters were completed by their co-authors, Peter Hinckle and Richard R. Gonzalez, respectively.

We are also very grateful to the nearly 150 reviewers who generously criticized the original outline and, later, reviewed the submitted manuscripts for scientific accuracy and appropriateness for inclusion in this volume. Their reviews were most helpful, both to us and to the authors, and in all cases, their comments were well received by all concerned.

It should be evident to our readers that expertise in the areas covered in this volume exists throughout the world. We find no more appropriate way to end this preface, therefore, than to quote from D. B. Dill's preface to the volume of 31 years ago: "The friendly exchanges we have had with our fellow physiologists around the world has been rewarding. It has demonstrated once again that a kindred spirit exists among scientists, that there is only one world of science."

Gainesville, Fla.   M. J. F.
Memphis, Tenn.   C. M. B.
November 1994

Contents

Volume II

4. Diving Mammals

Volume I

HANDBOOK OF PHYSIOLOGY

Section 4: Environmental Physiology

34. Radiation in microgravity

GREGORY A. NELSON | *Space Biological Sciences Group, Jet Propulsion Laboratory, California Institute of Technology, Pasadena, California*

CHAPTER CONTENTS

SPACEFLIGHT RESULTS in unavoidable exposure of astronauts or experimental organisms to the complex space radiation environment while simultaneously subjecting the individuals to gravity unloading and other stresses such as vibroacoustic stimuli. A natural question is whether these independent environmental factors interact in significant ways to compromise the health of spaceflight crews or to significantly modify the outcome of biological experiments focused on an independent variable. Access to time in space and the ability to control acceleration levels of spacecraft have been limited, so important experimental data are limited. Despite these constraints, there are data to suggest that exposure to microgravity may modify organisms' responses to radiation exposure, and there are physiological changes during spaceflight that could plausibly explain such altered responses.

This chapter will first explore the nature of the space radiation environment, particularly exposure to charged particle ionizing radiation, and then review results of selected experiments that have investigated modification of radiation responses by other variables, especially microgravity. Emphasis will be given to findings from animal experiments where the interaction between microgravity and radiation is most directly applicable to human physiology, and the discussion's organization will be by experimental protocol.

CHARACTERISTICS OF THE SPACE RADIATION ENVIRONMENT

Composition and Distribution

The radiation environment of space is extremely complex. Electromagnetic radiation from energetic gamma rays and X rays through ultraviolet and visible light to microwaves and radio waves all permeate the solar system and are dominated overwhelmingly by solar activity. Of importance for human health under normal shielding conditions inside spacesuits and spacecraft are the penetrating X rays and gamma rays emitted directly by the sun or resulting from the interactions of charged particles with spacecraft hardware *(Bremsstrahlung)*. Superimposed on the electromagnetic spectrum are a myriad of subatomic particles such as pions, muons, electrons, and neutrons derived from solar and galactic sources or from interactions of nuclei with the Earth's upper atmosphere and spacecraft materials. Finally, and most importantly, there are large numbers of atomic nuclei of all elements from hydrogen through uranium which travel up to relativistic velocities (31). These nuclei are derived from galactic sources and the sun and, with the exception of protons (H nuclei), are collectively referred to as *cosmic rays* or *high atomic number [Z] energetic (HZE) particles*. Stellar synthesis can only support the production of elements through iron so that the fluence (incident particles per unit area) of elements above $Z = 26$ drops dramatically and represents the galactic cosmic ray (GCR) contribution derived from stellar explosions (novae and supernovae) modified by magnetic fields. The most abundant charged particles are protons (85% of fluence), followed by helium nuclei or alpha particles (13% of fluence); nuclei of higher atomic number make up the remaining 2% of the fluence. But these latter nuclei contribute nearly 50% of the absorbed dose from particles, with iron constituting the single greatest contribution to absorbed dose (energy absorbed per unit mass). This is measured in Gray (Gy), where 1 Gy = 1 Joule/kg; older terminology used rad, where 1 rad = 10^{-2} Gy or 100 erg/g (33).

The fluence of GCR is isotropic except for shielding by the earth for spacecraft in low earth orbit. The flux (fluence per unit time) is generally constant except for long-term variations due to the solar cycle. The fluence of solar particles is less constant and is significantly modified by the magnetic fields of the sun and earth. During solar flares, the fluence may increase dramatically over hours or a few days and tends to localize in "flux tubes" that may or may not intersect the path of the earth. Solar flares consist primarily of relatively low energy protons that fail to penetrate the earth's magnetosphere at low latitudes, due to magnetic focussing effects that divert them back into space. All of these particles penetrate to low altitudes above the earth's magnetic poles and down to regions of about 50° of latitude, as the particles can follow open field lines (31).

Generally speaking, low earth orbital spacecraft receive little dose from cosmic rays, due to their localization in low inclination orbits. Most U.S. spacecraft orbit at inclinations of 28.5°, but some U.S. and most Russian spacecraft have used 50° to 65° inclinations. Of all manned missions, only the Apollo lunar flights have left the earth's magnetic envelope for exposure to the "free space" cosmic ray environment.

The major contribution of radiation to manned spacecraft comes from protons and electrons confined to the trapped radiation belts (Van Allen belts) which occupy toroidal regions of space from altitudes above 500 km to several earth radii. They are distorted to lower altitudes at polar latitudes where they are said to form "horns," according to the appearance of their fluence density in meridional planar sections. Due to asymmetries in the earth's magnetic field, there exists a region over the South Atlantic Ocean called the *South Atlantic Anomaly* (SAA) where the magnetic field is slightly weaker, which allows particles to penetrate to lower altitudes than at other locations around the earth. Thus for orbits at any altitude, the fluence of trapped protons is maximum during traversal of the SAA. The SAA has been the main exposure source for most manned missions and is expected to be the major source of radiation for the international space station (32).

PROPERTIES OF IONIZING AND NONIONIZING RADIATION

Nonionizing electromagnetic radiation excites vibrational or rotational motions in molecules or promotes electronic transitions that, in turn, lead to chemical reactions. These interactions do not usually lead to strand breaks in DNA, but cause specific classes of lesions and elicit repair responses from living cells. The effect of ultraviolet light in causing pyrimidine dimers in DNA is perhaps the best-known example (9). By contrast, ionizing radiation extracts electrons from atoms in a relatively nonspecific pattern and leads to the production of highly reactive chemical species with unpaired electrons, that is, free radicals. The most abundant biological molecule is water whose major ionization product is the hydroxyl radical. Such radicals may diffuse to nearby nucleic acids or proteins where they can cause or lead to strand breaks or the formation of adducts that are difficult to repair. Macromolecules may also be ionized directly without water as an intermediate. The presence of oxygen can accentuate free-radical damage (33).

X rays and gamma rays are low *linear energy transfer (LET)* forms of radiation, which means they lose (transfer to the target) only a small amount of energy per unit of track length (LET = $-dE/dx$ expressed in $keV/\mu m$). For these radiation species small clusters of one to three ionization events are formed every micron or so. The large average distance between ionizations makes DNA strand breakage relatively inefficient. However, as the dose builds, the distribution of ionization events throughout the target volume becomes more uniform; the mean distance between events decreases such that macromolecule strand break probability rises (19). An important consequence of this uniform pattern of dose deposition is that all regions of a biological target are exposed equally, and the large distance between lesions at low doses allows repair mechanisms to keep pace with damage formation (9).

Unique Properties of Charged Particle Radiation

Charged particles are capable of interacting with electrons as well as with nuclei of target atoms. Most interactions are with electrons, but fragmentation of primary and target nuclei is significant for thick targets (human bodies and spacecraft walls), and results in the creation of multiple secondary particles. The consequence is that particles produce dense linear tracks of ionization with little change in direction. The LET is relatively high and follows the Bethe-Bloch relation as a function of range in the target. This means that $-dE/dx$ varies as the primary particle's charge squared divided by its velocity squared ($-dE/dx \propto z^2/v^2$). This behavior is manifested as a so-called Bragg peak in which relative ionization vs. track position rises sharply at the end of a particle's range. Radially, energy loss varies as the inverse square of the off-track distance, which is related to the range of scattered electrons or delta rays. However, delta rays display structured energy deposition themselves, leading to "spurs" and "blobs" of ionization emanating from primary tracks which "thin down" at the limit of their range (19, 34).

There are several consequences of this structured energy deposition. First, concentrated doses are localized to small target regions surrounded by unaffected regions. For example, cells traversed by a heavy ion may sustain the equivalent of many hundreds of rads while neighboring cells are unaffected. Second, the mean distance between ionizations along the track is small, thus multiple radicals or ionizations may be created on the scale of a few DNA base pairs, leading to efficient strand breakage. This is true whether the macroscopic dose is large or small. Third, the concepts of dose and dose rate break down when single particle, single target (cell) interactions are considered; the fundamental deposition event is from a particle moving at nearly the speed of light so that the time associated with traversal of a $3\ \mu$ cell nucleus is about 10^{-14} sec. Fourth, direct ionization of targets becomes more important than that for photons acting on water, and the relative role of active oxygen species in mediating damage is reduced vis-a-vis low LET radiation (19).

Risk Assessment

Risk assessments for charged particles have generally been based on the assumption that it is possible to identify a dose of low LET radiation, such as X rays or gamma rays, which produces the same biological effect as exposure to a dose of charged particles. This has the convenience of allowing risks to be interpreted in the context of familiar hazards, such as medical diagnostic procedures. When effects unique to high LET radiation species are not present, the relative biological effectiveness (RBE) is often calculated: RBE = (dose of standard radiation) ÷ (dose of test radiation) at the same frequency of biological effect. The RBE for a particular endpoint (mutation, tumor incidence, survival) may vary substantially with dose and dose rate, and it generally varies two- to threefold with LET; that is, $RBE_{endpoint} = RBE_{endpoint}$ (dose, dose rate, LET) so inferences must be carefully considered. Infinite RBE values would occur if an endpoint were only observed after high LET radiation exposure. The concept of *dose equivalence* is used to express the enhanced effectiveness of high LET radiation. Dose equivalent [Sv (100 rem)] = absorbed dose [Gy (100 rad)] × RBE. When many endpoints and experimental systems are involved in estimating risk for particular radiation species, a combined RBE value is

usually desired. This value, called the *quality factor,* is established by a panel of experts and is based on the body of published experimental observations. The National Council on Radiation Protection and Measurements (NCRP) is the organization in the United States usually charged with establishing risk estimates and exposure limit recommendations, and it establishes quality factors for its own use. The NCRP has determined that of all the potential health risks associated with exposure to space radiation, radiation-induced cancer is the dominant risk component for long-term space missions (10). Cataractogenesis, mutation, damage to gametes, behavioral deficits, and other problems, rank far behind.

For charged particle–dominated radiation environments, a more useful approach than dose equivalent vs. response for describing the kinetics of damage induction by radiation would be the use of fluence vs. response or probability of response per unit fluence (action cross section, σ) relationships for particles of particular structures and energies. The dose is related to the fluence in the following way: Dose (Gy) = $1.6 \times 10^{-9} \times LET$ ($keV/\mu m$) × fluence (cm^{-2}). This method would allow the prediction of biological effects from mixed populations of particles in a natural environment by weighting the probabilities according to the abundance of the particles of a given LET (or charge state and energy range). High LET-unique endpoints would not be a factor and, for space missions, dose rate would not be an issue because of relatively low flux. This approach is currently under consideration by NASA (10).

Numerous biological effects have been examined resulting from charged particle exposure, and these vary significantly with the details of track structure. Ground-based studies are carried out at particle accelerators (for example, Lawrence Berkeley Laboratory's Bevalac) and provide data from biological systems as diverse as mammals, invertebrates, cultured cells, and microorganisms. Cell survival, cell transformation, tumorigenesis, cataractogenesis, mutation, chromosome aberration, gene expression, life shortening, developmental defects, and virus inactivation are the biological models most frequently used (19). Some of these well-characterized laboratory systems have been examined in the spaceflight environment, but alternate species, media, and incubation conditions have been developed to accommodate restrictions of spaceflight experimentation (for example, pocket mouse and stick insect).

RADIOBIOLOGICAL EXPERIMENTS IN SPACE

The various radiobiological responses of organisms have been studied extensively in the laboratory and the results applied to human health issues. However, there

TABLE 34.1. *Summary of Representative Radiobiology Experiments in Space*

Mission(s) [Experiments]	Organism (Common Name)	Endpoint(s) Measured	Exposure Mode[a]	Flight Acceleration Control Yes/No	μGravity Interaction ± 0[b]
Cosmos 1887 Apollos Spacelabs Others	Brine shrimp	Hatching, development	Natural	No	0
Spacelab D1 IML-1 Cosmos 1887	Stick insect	Hatching, development, growth	Natural with ray tracing[c]	Yes & No	+[d]
IML-1	Nematode	Mutation, development	Natural	Yes & No	0
Apollo 16	Microorganisms	Sporulation, viability	Natural with ray tracing[c]	No	0
Apollo 17	Flour beetle	Development		No	0
[Biostack experiment]	Stick insect	Development		No	0
Apollo 17	Pocket mouse	Brain and eye histology	Natural with ray tracing[c]	No	0
Biocosmos 110	Microorganisms	Phage induction, mutation	Natural	No	0
	Spiderwort	Chromosome aberrations	± Radioprotectant[e]	No	0
	Dog	Physiology		No	0[f]
Cosmos 936[g]	Rat	Retina histology	Natural	Yes	0
Cosmos 782[g]	Rat	Retina histology	Natural	No	0
Gemini III	Human leukocytes	Chromosome aberrations	Artificial ^{32}P-β	No	0
Gemini XI	Human leukocytes	Chromosome aberrations	Artificial ^{32}P-β	No	0
	Bread mold	Viability, mutation		No	0
Biosatellite II	Bacteria	Phage induction	Artificial ^{85}Sr-γ	No	−
	Bread mold	Mutation		No	0
	Parasitic wasp	Mutation, fertility, recombination		No	0
	Fruit fly	Somatic chromosome aberrations		No	0
		Mutation			+
		Recombination			+
		Nondisjunction			0
	Flour beetle	Development		No	0
	Spiderwort	Pollen abortion		No	+
		Stamen hair stunting			+
		Stamen hair mutation			−
Cosmos 605[h]	Rats	Testis mass & spermatogenesis	Natural	No	0
		Haematopoiesis in marrow & spleen			0
		Colony formation in allografts			0
		Muscle mass			0
		General organ morphology			0
Cosmos 690[h]	Rat	Testis mass & spermatogenesis	Artificial ^{137}Cs-γ	No	0

TABLE 34.1. *(continued)*

Mission(s) [Experiments]	Organism (Common Name)	Endpoint(s) Measured	Exposure Mode[a]	Flight Acceleration Control Yes/No	μGravity Interaction ± 0[b]
		Haematopoiesis in marrow & spleen			+[i]
		Colony formation in allografts			−[i]
		Muscle mass			+[k]
		General organ morphology			0
Cosmos 368	H₂ bacteria	Viability	Artificial preflight gamma rays	No	0
	Yeast	Viability			0
	Lettuce & chick pea seeds	Chromosome aberration			0
IML-1	Yeast	Viability in DNA repair conditional mutant	Artificial preflight X rays	No	+[l]

[a]Natural refers to cosmic ray and proton environment of particular mission as attenuated by spacecraft materials. Artificial refers to radioisotope or X-ray exposure superposed on natural environment.

[b]Interaction of radiation and microgravity. +, − 0 refer to enhancement of effect, antagonism of effect or no interaction, respectively.

[c]Ray tracing indicates that individual cosmic ray tracks were correlated with biological targets.

[d]Microgravity and cosmic ray strikes synergistically inhibited development in a stage specific fashion.

[e]An aminothiol was administered to some specimens.

[f]Details of dog physiology are not reported but general condition was described as "good". Interaction was not assessable.

[g]Cosmos 782 and 936 are best interpreted as paired experiments.

[h]Cosmos 605 and 690 are best interpreted as paired experiments.

[i]Microgravity inhibited recovery of radiation-induced pathology.

[j]A slight enhancement of stem cell recovery normally follows irradiation and in the table context is an antagonistic effect.

[k]Radiation inhibited recovery of microgravity-induced pathology.

[l]Microgravity inhibited repair and recovery of permissively grown cells but had no effect on restrictively grown cells.

is no satisfactory way to simulate the complex, low dose rate, omnidirectional, multicomponent environment of space with its complicating feature of microgravity. Therefore, exploratory investigations and experiments to verify ground-based predictions for well-characterized model systems must be done in space.

Some of the earliest space experiments were radiobiological. These were of simple design and typically consisted of inert biological objects not likely to be gravity sensitive. Most used natural radiation exposures and compared data from the microgravity satellite sample with that from matched ground controls. Later, in-flight radioisotope sources were used to enhance the frequency of the biological endpoint to detect negative or positive influences from "space flight factors." More recent experiments have also utilized onboard centrifuges as acceleration controls to examine combined radiation and microgravity as controlled independent variables. The sophistication of the more recent investigations has increased as the supporting flight hardware has improved. Table 34.1 summarizes the most conclusive spaceflight experiments which address the issue of radiation effects in microgravity. The details are discussed below in *Natural Radiation Exposures* and *Artificial Radiation Exposures.*

Exploratory Studies

A number of unmanned and manned spacecraft launched by the United States and the Soviet Union from the late 1950s through the 1970s carried biological specimens consisting primarily of bacteria, bacteriophages, fungal spores, microalgae, plant seeds, arthropod cysts or eggs, and some mammalian cultured cells. Most satellites were placed in low earth orbits for several hours to 2 weeks allowing exposure to natural cosmic rays and trapped proton belts. Controls consisted of matched samples incubated in duplicate flight hardware subjected to vibration and acceleration profiles that simulated rocket launches. In these experiments the natural space radiation induced mutations and chromosome aberrations and effectively disrupted the normal development of invertebrate and plant embryos (24, 32).

Results from exploratory experiments in this genre have been reviewed by Horneck (13) and Antipov (1). They include findings from the Discoverer, NERV, Sputnik, Cosmos, Vostok, and Voshkod flights. The biological specimens included: T4, T7, ECHO 1, and PR8 viruses; *Escherichia, Clostridium, Bacillus,* and *Pseudomonas* bacteria; *Chlorella* algae; *Neurospora* fungus; *Arabidopsis, Nicotiana, Lactuca, Oryza, Crepis, Zea,* and *Tradescantia* plant seeds and microspores; and *Artemia, Tribolium,* and *Drosophila* arthropods.

In this 1950–1970 time period, access to particle accelerators (cyclotrons, linacs, synchrotrons) capable of generating heavy ions in the same energy ranges as cosmic rays was also very limited, so baseline dose vs. response relationships (except for X rays) were not well known for many systems. Vibration, acceleration, and noise, which could modify radiobiological responses, were not common clinical or scientific variables. They became operational issues in military aviation and manned space flight. Thus a database of these and other uncommon environmental effects, which could potentially modify radiobiological responses, had to be developed de novo for each experiment (2).

Natural Radiation Exposures

For experiments utilizing the natural radiation environment to provide exposure, acquisition of large databases were always limited by the volume, mass, and power constraints of life-supporting flight payload hardware, because biological sample sizes had to be large to detect rare genetic and cellular events. Correlations of biological responses with mission duration (proportional to dose) have large variability. These constraints persist.

A second generation of studies arose in which individual cosmic rays were correlated with radiobiological endpoints in space and in which in-flight acceleration controls were provided. The Biostack, Biobloc, and Exobloc series of experiments used ray-tracing techniques to localize cosmic ray tracks to immobilized spores, seeds, and eggs sandwiched between nuclear track detector materials, such as nuclear emulsions, cellulose nitrate, and polycarbonate plastic. In this way, the rare cosmic rays (typically at 4–10/cm^2· day fluence for $Z \geq 6$) could be identified first, then followed by analysis of "hit" biological samples without wasting time and resources on "unhit" specimens. Details of experiments of this type with the brine shrimp, *Artemia,* the stick insect, *Carausius,* the nematode, *Caenorhabditis,* the pocket mouse, *Perognathus,* and the rat, *Rattus,* are presented in the sections following.

Experiments Using the Brine Shrimp, *Artemia salina.* Perhaps the most popular organism for space radiobiological studies is the brine shrimp, *Artemia salina.* This organism can exist as a desiccated gastrula cyst for years inside an eggshell, thereby requiring very modest life support provisions. Upon return to saline solution the shrimp recover and resume development. Following experimental irradiation or exposure to space, rates of hatching, developmental progression, and incidence of developmental anomalies have been recorded. The

results from Cosmos 1887 and ten previous spaceflight experiments have been reviewed by Gaubin (11). Rates of inactivation and anomaly scale with dose but no obvious influence of microgravity on radiation-induced lesions in *Artemia* have been reported. This may be due to the fact that the cysts were essentially dry and the shrimps' structures were rigid and gravity resistant. Therefore, regulatory and repair processes based on soluble enzymes and other biochemical mediators were probably not active.

Experiments Using the Stick Insect, *Carausius morosus.* The stick insect, *Carausius morosus,* has been used on several spaceflights to investigate the effects of gravity and charged particle radiation on development. These insects undergo incomplete metamorphosis in eggs within 75 to 105 days at 18°–22°C. This property permits eggs to be integrated into parallel layers of nuclear track detectors such that ray-tracing methods can be used to identify the eggs hit by HZE particles. This hardware arrangement was integrated into Biostack experiments on several missions, most notably Spacelab D1 and Biocosmos 1887. In each of these missions, eggs in five developmental stages experienced 7–12.6 days of microgravity and were separated into groups hit or unhit by HZE particles during flight and into an unhit ground control. On the D1 mission some eggs were also incubated in-flight in a 1G centrifuge. Hatching rate from eggs, growth rate, and frequency of developmental anomalies were measured. The anomalies were primarily fused, tilted, or stunted segments of antennae, abdomens, and legs (13, 14, 20, 27, 28).

The results from the D1 mission indicated that both microgravity and HZE particle strikes disrupted development in a stage-specific fashion. Hatching rate was reduced and the frequency of anomalies was enhanced. Results from the in-flight centrifuge control insects indicated that microgravity and HZE effects were synergistic, suggesting a role for gravity in the ability of embryos to tolerate or repair damage by HZE particles, or vice versa (27). The sample sizes were limited on D1, so a second experiment on Cosmos 1887 (without the centrifuge) was performed to improve the statistical base, especially for sensitive-stage embryos; approximately 4,300 eggs were flown. Hatching rate was reduced significantly by microgravity exposure (81% control vs. 50% flight for stage II embryos) and a slight additional impairment (additive, not synergistic) was associated with HZE hits. However, developmental anomalies were substantially increased in hit eggs, especially in stage II embryos where unhit flight samples showed approximately a 2% rate of anomaly and flight-hit samples showed a 17% rate of anomaly (28). This compares with ground control rates of less than 0.15% on both D1 and Cosmos 1887 flights. These findings are provocative with respect to the apparent interaction of gravity and radiation on sensitive, stage-specific, developmental processes. They raise the possibility that microgravity and radiation may interact at the level of repair processes, but no specific mechanisms have been proposed.

Experiments Using the Nematode Caenorhabditis elegans. To isolate cosmic ray–induced mutations, the microscopic nematode, *Caenorhabditis elegans,* was flown in 1992 on shuttle STS-42 as part of the Biorack investigations on the International Microgravity Laboratory #1, Spacelab mission (22, 23). Approximately 60 autosomal recessive lethals in a 350 gene region, and 12 mutations in the muscle protein gene, *unc-22,* were isolated from animals held in a dormant larval state or growing in logarithmic culture. Dormant animals were incubated in a laminated assembly next to CR-39 plastic nuclear track detectors so that a ray-tracing procedure could be used to correlate specific cosmic rays with specific mutants. An extensive series of ground experiments using accelerated charged particles provided an interpretive framework for these mutants whose structural properties were analyzed at the molecular level. Unique or characteristic features of authentic cosmic ray–induced mutants were identified (22). The experiment design did not permit an explicit test of gravity effects on mutation kinetics because of the low fluence of particles in space, the small responding fraction of animals, and the requirement of low temperature to immobilize animals for ray tracing which precluded onboard centrifugation. However, in complementary experiments, development and chromosome behavior (segregation and recombination) were measured in rapidly growing cultures as a function of gravity level and found no obvious dependence on gravity (22). The main findings from this experiment were the correlation of molecular structures with authentic cosmic ray hits and the observation of normal chromosome mechanics and development for two successive generations in an animal exposed to microgravity.

Apollo Lunar Flight Experiments with the Pocket Mouse, Perognathus, and Associated Investigations. Experimental animals carried in Apollo 16 and 17 spacecraft were unique in that, along with Zond 5 and 6 lunar flyby missions, they were exposed to radiation fields outside the earth's geomagnetic envelope, and they completely traversed the trapped proton and electron belts. The fluence of particles on these missions included the full energy range for GCR. In earlier Apollo missions crew members observed light flashes that were caused by passage of cosmic ray particles through their dark-adapted retinas (18). The possibility of behavioral decrements and destruction of nervous tissue, including photorecep-

tors, increased interest in HZE effects. Severe constraints on mass, volume, and power limited radiobiological experiments to passive tests (18).

The Apollo Biostack I and II experiments provided correlation of cosmic ray tracks with inactivation of several biological objects. The experiments assessed the efficacy of specific ions and their proximity to bacterial spores, plant seeds and radiculae, brine shrimp cysts, eggs of the flour beetle, *Tribolium,* and the stick insect, *Carausius.* The three animal systems proved to be most sensitive when assayed for inhibition of hatching and the occurrence of developmental abnormalities. These tests, and another experiment (MEED) that provided exposure of organisms to the natural solar light spectrum through quartz windows, did not address interactions of other spaceflight factors (18).

One mammalian experiment was included on the 13 day Apollo 17 flight. In this Biocore study, five pocket mice *(Perognathus longimembris)* were fitted with plastic nuclear track detectors under their scalps to provide a comprehensive map of cosmic rays passing through brain and eye tissues and to assess histologically any damage; in addition, a comprehensive study of all body tissues complemented the main objective (12). Animals were housed separately in cylindrical containers positioned in a relatively heavily shielded location in the command module; four of the five mice survived the flight. When dosimetry and the elaborate stereotactic measurements were completed, 71 cosmic rays of known LET ($Z \geq 6$) were traced into head structures and the direction of deceleration was established for 39 of the tracks. Only five particles intersected retinal tissue of the four mice. Histological examination of serial sections failed to detect lesions in eye or brain tissue which correlated with cosmic ray tracks. The absence of microlesions (34) does not prove that cells were not damaged, but it argues against catastrophic destruction of columns of CNS cells from ions of intermediate atomic number. These Biocore results are illustrative of the difficulties in assessing biological effects of naturally occurring cosmic rays.

Biocosmos 110 Experiments with Dogs and Microorganisms. One early space biology experiment which set the stage for mammalian studies was the 22 day Cosmos 110 mission launched in 1966 using an unmanned 2.3m diameter, spherical Voskhod spacecraft that flew at 51.9° inclination in a 187 × 904 km orbit (2, 17). The satellite contained two terrier dogs (Veterok and Ugolek) kept in separate compartments within the cabin, which were maintained with the human life support system. The animals were largely immobilized, were fed by gastrostomy tubes, and were monitored extensively with physiological sensors. Veterok was given a radiation protectant drug intravenously during the mission, and two additional dogs were treated similarly and simultaneously on the ground as controls. The dogs were recovered in good condition but experienced significant calcium loss and had impaired movement for 8–10 days. Radiobiological endpoint data in the dogs was not available, but the measured dose was 10.5 cGy (approximately 0.5 cGy/day), which was due to protons that the spacecraft encountered as it transitted the lower Van Allen belt at apogee.

Accompanying the dogs were a variety of other organisms, including a lysogenic strain of *E. coli* K12 (lambda), used with and without an aminothiol radioprotectant, plant seeds, *Tradescantia* microspores, and an intact plant. Phage induction was significantly enhanced in flight and suppressed by the radioprotectant. *Tradescantia* microspores showed a variety of chromosome aberrations and mitotic disturbances while mutation in *Chlorella* was not different from ground controls (2).

Biocosmos 782 and 936 Experiments with Rats.

Interest in effects of cosmic rays on mammalian central nervous systems and eyes led to investigations with rats aboard the Soviet biosatellites Cosmos 936 and 782.

Cosmos 936. Cosmos 936 was launched 3 August 1977 for 18.5 days in a 224 × 429 km orbit at 62.8° inclination and for a 90.7 min period. Thirty specific pathogen-free male Wistar rats, fed on a paste diet, were flown. Twenty were exposed to microgravity and ten were incubated in an onboard 1.05G centrifuge. Five rats from each group were used for radiobiological studies and were sacrificed after 25 days of recovery on the ground. Parallel ground control rats were utilized with a 4 day lag time. They were subjected to acoustic noise, mechanical vibration, and acceleration profiles simulating those during launch and recovery. For example, launch noise was 110 dB for 10 min with 50–70 Hz vibration at 0.4 mm amplitude, followed by acceleration for 10 min up to 4G for a 7 min plateau period. Reentry and recovery involved a 5 min acceleration up to 6G for a 3 min plateau, followed by a 10 ms pulse at 50G (16, 29).

Because of the postulated existence of microlesions in the highly organized retinal tissues, the rats' eyes were fixed and examined with light and electron microscopy for linear tracks of damage in the retinas. This examination was prompted by results of accelerated particle experiments at the Bevalac, using neon ions on C57 Black mice and pocket mice, which identified microscopic lesions in cells of the eye and nervous system. HZE particle fluences of 1.75 particles/cm^2·day (Z ≥ 3 and range in lexan ≥ 180 μm) were measured on Cosmos 936 rats using lexan and nitrocellulose track detectors. Total doses of 4.24–5.23 mGy were measured using thermoluminescent detectors (TLDs). Neutron

fluences, stopping protons, and nuclear decay stars, were also quantified using activation foils and nuclear emulsions.

Necrotic retinal cells were swollen and had dense nuclei; membrane debris in widely scattered regions of the outer retinal layer was observed, especially in the rods. Occasional phagocytic activity in pigmented epithelial cells was seen. Macrophages were not present near damaged areas, but the rats were not sacrificed until 25 days after landing. Findings from flight-centrifuged and microgravity-exposed rats were similar with respect to retinal lesions. Necrotic lesions were not observed in the ground control material.

Cosmos 782. Similar necrotic cells were observed in rats flown at 0G on the 25 November 1975 Cosmos 782 mission (19.5 days in a 226 × 405 km orbit at 62.8° inclination) with slightly higher fluence of HZE particles (4.05 particles/cm^2·day). One difference on this mission was that a subset of animals was sacrificed several hours after recovery and in these animals macrophages were seen near lesions. This was interpreted as evidence of passage of HZE particles through retinal tissues with the formation of microlesions; there was no evidence of a gravity interaction (16).

Artificial Radiation Exposures

It is clear from the preceding section that deleterious radiation effects from natural exposures are very rare and mitigate against interpretation of gravity interactions for practical reasons. A different strategy in performing such experiments is to orbit samples in the trapped proton belts to provide exposures of 1–5 Gy, but this would require development of a dedicated unmanned spacecraft. Alternatively, onboard sources of radiation from radioisotopes or irradiation just before or after flight could be used to determine whether flight conditions, especially microgravity, modify radiobiological responses. The latter strategies have been applied in manned and unmanned spaceflights and are discussed in this section.

The S4 Experiments on Gemini III and XI Using Blood Cells and Fungi.

An important early experiment to address radiobiological responses in human cells was conducted in 1965 and 1966 on Gemini spacecraft. The S4 experiment was flown twice to detect possible interactions of microgravity and radiation on human nucleated blood cells (4, 5). On the second mission, bread mold spores were also included to provide independent confirmation of genetic effects (6, 8).

The Gemini III mission (23 March 1965) consisted of only three orbits (0.20 days, 32.5° inclination, perigee of 161 km and apogee of 224 km) and received 0.20–0.45 mGy radiation, measured with thermoluminescent detectors (TLD), depending upon location. The 2.97 day Gemini XI mission (12 September 1966) flew in an elliptical orbit (29° inclination, perigee of 159 km, and apogee of 298 km) except for two orbits with an apogee over Australia of 1370 km and a perigee of 298 km for a rendezvous and docking maneuver that took the spacecraft briefly into the lower Van Allen proton belt. The total TLD dose was 0.23–0.39 mGy, depending upon location.

Fresh blood samples were placed into 3 mm thick glass chambers that astronauts could activate by moving them to an irradiation area in the shielded flight hardware where they were exposed to a series of doses of 0.7 MeV beta particles from ^{32}P; exposures ranged from 0 to 1.74 Gy. Single and multiple break chromosome aberrations were measured in leukocytes held in the G_1 phase of their cell cycle. Survival and mutations at the ad-3A and ad-3B loci were measured in Neurospora conidial spores in suspension, or held on millipore filters and irradiated like the leukocytes, but at exposures of up to 144 Gy. All results were compared with simultaneous ground control responses. No differences between flight and ground samples were detected for mutation or survival in Neurospora irradiated on filters, or in leukocytes examined for multibreak aberrations. Differences between flight and ground samples for single break leukocyte aberrations were observed on Gemini III but were not seen again on Gemini XI. It was suggested that sampling error may explain the earlier results. Differences in survival levels and mutation rates for suspended fungal spores that were metabolically active, unlike inactive filtered spores, could be explained by anoxia. Thus no obvious radiation + gravity interaction was detected.

Biosatellite II. Microorganisms, Plants and Insects Irradiated in Orbit with Gamma Rays.

On September 7, 1967, the unmanned Biosatellite II was launched into a 190 mi circular orbit at 28.5° inclination for 45 h, after which it was recovered (one day earlier than planned) in midair over the Pacific ocean, and its biological payload, consisting of microorganisms, plants, and insects, was sent to Hawaii for analysis. Several of the radiobiological experiments provided known doses of gamma rays to the specimens in orbit, prior to launch, and to matched ground controls to detect the effect of gravity on radiation-induced genetic and developmental lesions. The radiation source was ^{85}Sr in a tungsten holder that could be opened and closed by ground command. The source had an activity of approximately 1.2 Ci and produced 0.513 MeV gamma rays measured by thermoluminescent detectors. The findings provided evidence for a modifying effect of gravity on the radiation effects. Environmental control was acceptable except for formaldehyde vapor (0.8 to 2 ppm), which

exceeded the design specification of 1 ppm and could conceivably have contributed to mutagenesis, especially in the Drosophila experiment.

Experiment P-1135 investigated the induction of P22 virus production from a lysogenic strain of Salmonella typhimurium BS-5 (P22)/P22, and lambda phage from E. coli C-600 (lambda)/lambda [Mattoni, et al., in ref. (30)]. Because of early reentry of the satellite, only the Salmonella experiment could be performed. Small aliquots of 120 cells/ml in broth were placed near the ^{85}Sr source and received 0–16.30 Gy exposures. After flight, differences in cell density and growth rate were detected between ground and flight samples, and a slightly increased resistance to gamma rays in the flight samples was observed. The critical measurement of induced phage production per viable cell immediately after recovery was also measured. A 38% to 42% decrease in yield in the flight samples relative to ground controls which was significant ($P < 0.02$) for doses of 2.65 and 6.45 Gy and at the $P < 0.06$ level for 16.30 Gy. Thus exposure to microgravity appears to show an antagonistic effect on radiation induction of phage.

Experiment P-1037 investigated mutagenesis in heterokaryon conidiospores of the bread mold, Neurospora crassa [de Serres, et al., in ref. (30)]. Mutation at the ad-3A and ad-3B loci were measured from stationary culture cells filtered onto nitrocellulose filters under conditions where they were metabolically inactive. The conidiospores received exposures from 3.40 to 36.00 Gy in space. No significant differences between flight and ground control cells were observed for point mutations or for chromosome deletions. These results agree with those of the previous S4 experiment on Gemini XI for cells held on filters. This is in contrast to the metabolically active cell suspension cultures on Gemini where an antagonistic effect of microgravity on mutagenesis was measured.

Experiment P-1079 examined mutagenesis and genetic recombination in the parasitic wasp Habrobrachon juglandis Ashmead = Bracon hebetor (Say) as a function of gamma ray dose [Von Borstel, et al., in ref. (30)]. Young wasps were placed in culture chambers where they received 20.00 Gy just before launch or inflight doses of 0.07–24.25 Gy. (Brine shrimp cysts and yeast cells were also flown in conjunction with this experiment but showed no differences between flight and ground samples for development and recombination, respectively.) Analysis of sperm from X0 males after recovery revealed no significant difference between flight and ground samples for dominant lethality, recessive lethality, or partial sterility; but there was a slight increase in fertilizing ability. A higher spontaneous rate of recessive lethality was observed, but it was reproduced by vibration on the ground. Oocytes from XX females in the first stage of meiotic metaphase in flight showed a reduction in postfertilization viability from 0.89–0.93 to 0.49–0.56, whereas other stages of oogenesis revealed no differences except for a modest increase of fertility. This oocyte effect correlated with its position in the spacecraft, but not with radiation dose or microgravity, suggesting that other stress such as vibration may have been involved. No effect on recombination in females was observed between the *lemon, honey,* and *cantaloupe* genes on chromosome I. Two additional observations suggested increased longevity of flight females, and disorientation of males leading to decreased male mating behavior in flight, which were not affected by radiation. Thus little or no interaction between microgravity and radiation was observed with respect to production of inherited genetic changes. Only a slight enhancement of sperm and egg fertility was detected.

Experiment P-1159 used *Drosophila melanogaster* larvae, irradiated with 8.32 Gy of gamma rays, to investigate chromosomal alterations and mutation in germ cells and somatic tissues [Oster, in ref. (30)]. One strain of flies of a specific genotype was used to detect alterations in karyotype in cerebral ganglion cells of adults after being irradiated as larvae (P_0 generation). Other strains, including Ring-X and multiply marked flies enabled the detection of point mutations at the *dumpy* locus, recombination, alterations in sex ratio (X-linked lethality), translocations, and nondisjunction or loss of chromosomes following breakage in the F_1 through F_6 generations. The frequency of sex-linked lethals ($0.71 \pm 0.19\%$, $N = 1961$ vs. $0.35 \pm 0.14\%$, $N = 1724$) and recombination between X and Y chromosomes ($0.55 \pm 0.17\%$, $N = 2005$ vs. $0.29 \pm 0.09\%$, $N = 3412$) was increased in flight samples when compared with ground controls. A rare finding was the appearance of seven somatic translocations in the flight flies both with and without radiation. Normally such events are very rare. This surprised the authors who further noted that these translocations were not correlated with radiation dose.

Experiment P-1160 used both pupae and adult *Drosophila melanogaster* [Browning, in ref. (30)]. Four types of flies of complex genotype were irradiated with 40.00 Gy just before launch or with 14.32 Gy in orbit. First, females were mated just prior to flight with males of a complementary genotype so that a homogeneous population of target sperm could be used to identify recessive lethal mutations and translocations (in the F_2 generation) and nondisjunction of the Y chromosome (in the F_1 generation). Second, males were used for matings postflight to detect induced recombination in sperm (normally absent), as well as X chromosome lethals and translocations. Third, young males were irradiated preflight and allowed to mate during flight with complementary virgin females to detect alterations in rejoining of broken chromosomes. Finally, late third instar larvae

and prepupae were irradiated in flight, and males were bred to females to detect X-linked recessive lethals.

Recessive lethal mutations in sperm were observed at a slightly elevated frequency ($P = 0.05$) in flight samples over ground controls, but the difference was matched by rates observed in ground controls which recycled internal satellite gases. This implicated an effect from formaldehyde and glutaraldehyde fixatives used in other experiments. No significant mutation differences were observed at five visible loci (*dp, bw, st, y,* and p^p), or the loss of dominant Y chromosome markers y^+ and *B*. Several translocations were detected from pupal samples, and a slight elevation in recessive lethal mutation was seen in preirradiated males when compared to pooled controls, but control responses suggested that vibration may have played a role. Taken together, these observations did not indicate significant interaction between microgravity and radiation for modifying chromosome behavior or structure.

Experiment P-1039 utilized the flour beetle, *Tribolium confusum* Duval, to study a particular radiation-induced syndrome of abnormal development [Buckhold, et al., in ref. (30)]. Damage to pupal cells in a well localized region caused a misproliferation and deformation of spikes on the membranous wings of the beetle, which prevented closure of the overlying elytra, a feature easily measured. Seven hundred twenty pupae were flown, and they received either no dose or were preirradiated with 13.50 Gy of 180 keV X rays to bring them into the appropriate dose range for the flight. They were placed into the spacecraft as 19–27-h-old pupae; half would be shielded and half would be irradiated in orbit with an additional 7.55 or 9.69 Gy of gamma rays. The flight irradiation significantly ($P < 0.025$) enhanced the incidence of wing abnormalities of flown beetles ($44.8 \pm 3.2\%$) over ground controls ($29.9 \pm 3.0\%$) for pupae receiving approximately 23 Gy. However, after postflight vibration tests and repetitions of ground control procedures, it was concluded that variations in circadian rhythms and vivarium lot differences, but not vibration effects, probably accounted for these results. Thus it appears again that microgravity and radiation do not interact.

Experiment P-1123 used the radiation sensitive Spiderwort plant, *Tradescantia* clone 02, to measure inactivation of pollen, microspore death, spindle defects, somatic mutation in heterozygous petals and stamen hairs from blue to pink or to colorless, as well as stunting of stamen hairs [Sparrow, et al., in ref. (30)]. Thirty-two young plants bearing several flowers each were obtained from axillary cuttings and rooted. They were placed in nutrient tubes, irradiated with 2.23 Gy in orbit, and examined postflight. *Tradescantia* is very sensitive to radiation and showed high frequencies of radiation-induced changes. Pollen abortion occurred at

66% in irradiated flight samples vs. 48% in ground controls. Unirradiated flight and ground samples had a spontaneous pollen abortion incidence of 37% to 39% (this high rate is normal for clone 02). Stunting of stamen hairs occurred at 26.6% for flight-irradiated plants vs. 12.9% for ground controls; unirradiated plants had 10.1% to 10.5% spontaneous abortion rates. These responses, as well as altered nuclei and microspore death, showed clear synergism between radiation and microgravity. By contrast, hair color mutation was clearly antagonized in microgravity with flight irradiated samples showing a 4.4% incidence when compared with a ground control rate of 7.3%. Unirradiated samples had spontaneous mutation rates of 0.2% to 0.3%.

Cosmos 605 and Cosmos 690 Matched Flights with Rats ± Gamma Irradiation.

The most direct mammalian experiment to date which investigated the potential interaction of radiation and microgravity was performed with male Wistar rats on two Soviet satellites, Cosmos 605 and Cosmos 690 (17, 25). The rats were individually housed in cylindrical wire mesh cages 20 cm long by 10 cm in diameter and were fed pelleted food (carrots and beets). Cosmos 605 was launched on 31 October 1973 (21.5 days, 62.8° inclination, 214 km perigee and 424 km apogee) and orbited 27 rats that served as a microgravity control with only low-dose radiation exposure from the natural environment. Cosmos 690 was launched on 22 October 1974 and orbited 15 rats (20.5 days, a similar orbit of 62.9° inclination, 223 km perigee and 389 km apogee). It included a 320 ± 38 Curie ^{137}Cs 0.661 MeV gamma radiation source which provided an 8 Gy exposure over 24 h beginning on flight day 10. Filtering of the source, contained in a tungsten sphere with a collimation cone opening, provided a uniform exposure ± 10%. Simultaneous ground control studies using rats from the same vivarium lots were conducted for both missions. Most analyses were histological or biometric, but some enzymatic and transplantation studies were performed. Most organ systems were studied for alterations due to microgravity or radiation exposure. It was concluded that radiation and microgravity combined led to decreased recovery rates for various reversible alterations induced by microgravity, but there were no significant effects on the development of radiation-induced lesions. The results for selected organ systems are summarized as follows:

1. *Testes*. Examinations of testes from both Cosmos flights and ground control rats demonstrated no effects of microgravity. The pattern of spermatogenesis, frequency of intermediate cell types for gametes, and the histology of seminiferous tubule components was unaffected. Testes weight decrease in irradiated animals was the most dramatic effect, but it was unaffected by microgravity.

2. *Hematopoietic System*. Bone marrow and lymph organs of Cosmos 605 rats showed "a slight inhibition of erythropoiesis in bone marrow and spleen, significant involution of the thymus, marked hypoplasia of lymph tissue of the spleen and, to a smaller extent, lymph nodes" (25) due mostly to reduction of lymphocyte populations. The same responses were affected by irradiation. Histological examinations of bone marrow of Cosmos 690 flight and simulation rats on

… the second post-experimental day (12th post-radiation day) revealed well developed aplasia with discrete foci of hemopoiesis. The pattern of hemopoietic changes was typical of radiation induced bone marrow lesions. However, an exposure of animals to weightlessness influenced the after effect; on the 27th postflight day (37th post-radiation day) the recovery of the hemopoietic tissue was delayed compared with that in ground-based simulation rats. Thus, bone marrow showed radiation-induced changes, and weightlessness affected the course of reparative processes in bone marrow. (25).

Hematopoietic potential was also examined by transplantation of marrow cells to spleens and marrow of lethally irradiated recipients (colony formation assays). Cosmos 690 and control rats showed a reduction to 13%–17% of normal potential for the colony-forming unit (CFU)-spleen test, and to 40%–50% for the CFU-marrow test as expected for the 8 Gy exposure. In spite of the diminished number of stem cells in bone marrow of flight rats, their differentiation pattern did not change. The Cosmos 690 animals showed a trend toward enhanced erythroid potencies of stem cells transplanted to marrow, which may be explained by the more rapid recovery of the erythroid precursor population known to follow irradiation. Thus the qualitative aspects of the recovery may be slightly affected by the combination of microgravity and radiation.

Central Nervous System. There were no effects from any of the treatments except some "morphological signs of enhanced functional activity" in the hypothalamus and pituitary of Cosmos 605 animals which was absent in Cosmos 690 rats (25), suggesting a small modifying influence of radiation.

Heart, Liver and Kidneys. Cosmos 690 specimens exhibited changes in four cardiac enzyme levels consistent with an alteration in carbohydrate metabolism and stimulation of lipid utilization pathways. Some lipid accumulation was noted in the liver, along with some polymorphism in hepatocyte nuclei. Cosmos 605 rats showed no such changes, and the kidneys were unaffected in animals from both flights.

Muscle. Muscle mass loss in hind limbs was a feature of all flight animals, and lactate dehydrogenase isozyme levels showed variations. "In-flight irradiation aggravated weightlessness-induced changes in skeletal muscles and led to a delay in reparative processes and incomplete structural restoration of muscles" post flight (25). This effect was localized in muscles showing microgravity-induced pathologies and is probably a consequence of slowed connective tissue resorption during recovery, which, in turn, delays myofibril regeneration.

Thus responses of tissues sensitive to microgravity, radiation, or both, were obtained. From these responses one unifying concept emerged: *recovery from a pathological condition induced by either radiation or microgravity may be delayed by the presence of the other environmental factor.* Data from hematopoietic tissue are probably the most significant.

Cosmos 368 Experiments with Microorganisms Irradiated on the Ground.

The Cosmos 368 satellite was launched 8 October 1970, for 6.0 days in a 411×211 km orbit at 65° inclination. A variety of microorganisms and plant seeds were studied for any potential interaction between radiation and microgravity (32). Samples were irradiated with a series of doses of gamma rays either immediately before or immediately after flight and results were compared with those from ground controls. Doses of up to 1,600 Gy were employed. Diploid and haploid yeast in suspension or on agar showed no obvious effects of microgravity with respect to colony formation or growth after one to four generations. Similarly, the hydrogen bacteria *Hydrogenomonas eutrophus* (now classified as *Alcaligenes*) in suspension showed no perturbations with respect to plating. Air-dried chick pea and lettuce seeds were scored for chromosome aberrations and meiotic defects (anaphase bridges) with no obvious interference from microgravity. Only a slight difference in rootlet growth and catalase levels could be ascribed to spaceflight factors.

STS-42 Experiment on the Repair-Deficient Yeast.

The design of this Biorack experiment was to preirradiate yeast cells with a series of known radiation doses, incubate them for 1 wk with and without gravity to allow repair, and observe recovery from radiation-induced damage from measurements of colony-forming ability (26). *Saccharomyces cerevisiae* cells in stationary phase were filtered onto a monolayer at $5 \times 10^6/cm^2$ and held on supporting agar blocks. They were irradiated preflight with X rays at five doses from 0 to 1.40 Gy and held at 4°C for transport, launch, and recovery. In flight, the yeast were placed in Biorack incubators at 22° and 36°C for 7 days. Matched cultures were incubated in the ground control (1G) Biorack. The strain used bore a temperature-sensitive allele of the radiation-sensitive mutation *(rad-54-3)* which is defective in repair of DNA double strand breaks at the restrictive temperature (36°C), but has normal repair at the permissive temperature (22°C). Measurements of colony-forming ability vs. dose showed that the survival fraction of flight samples incubated at the permissive temperature was approximately twofold lower than ground samples, and this ratio was independent of dose. Restrictively grown cells were insensitive to gravity and did not repair their X ray–induced damage; these cells served as a control for unexpected effects of spaceflight.

The conclusion was that the repair pathway for radiation-induced DNA double strand breaks is sensitive to gravity levels. These results are important for two reasons. First, a biochemical pathway was identified in a eukaryote that is affected by gravity; this could help to identify a mechanism at the molecular level. Second, they show that microgravity effects may be manifested at the level of single independent cells under conditions where external environmental effects (for example, convection-mediated mass transport of oxygen or nutrients) were minimized.

POSSIBLE MECHANISMS FOR MICROGRAVITY AND RADIATION INTERACTION

There is no substantive influence of gravity on the physical deposition of energy in biological targets. This process is over in less than a picosecond for heavy charged particles, and the relative magnitudes of electrostatic or nuclear forces to gravity are enormous. Once ionizations are produced in a target, however, chemical reactions (for example, free radical attacks) ensue whose rates are limited by diffusion and convective mixing (19). It is these chemical processes that are susceptible to gravitational influences. Though intracellular convection may be overwhelmed by specific transport mechanisms and the cytoskeleton, convective mixing in suspended cells is critical for the exchange of oxygen, carbon dioxide, and nutrients that regulate their metabolism.

Damage to biological macromolecules can be repaired by enzyme complexes whose expression can be induced by radiation exposure, hyperthermia, anoxia, or exposure to heavy metals. Examples are the so-called SOS response (coordinated *recA*- and *lexA*-mediated gene expression phenomenon) in bacteria and the heat shock (stress protein induction) responses in most organisms (9, 19, 21). The repair often requires ATP and is therefore tied to the metabolic state of the affected cells.

Direct control of repair gene expression is also likely. Results from a recent MASER sounding rocket experiment with human epithelial cells indicated that microgravity could alter the expression of protooncogenes *c-fos* and *c-jun*, which act to regulate cell proliferation and differentiation (7). Cogoli and colleagues (see chapter by Gmünder and Cogoli, this *Handbook*) have also shown that differentiation and proliferation of immune cells are rapidly modulated by exposure to microgravity. Many of the enzymes involved in the synthesis of new DNA during cell proliferation have repair roles as well (9, 19), so repair capacity may well be coupled to these functions. The findings of Pross et al. (26) may represent a first step toward identifying the nature of gravity's action on a DNA repair pathway for double strand breaks in yeast. Some recent or pending spaceflight experiments using the ESA Biorack will also directly measure the influence of gravity on radiation-induced genetic damage. G. Horneck and collaborators irradiated frozen bacteria and human cells and allowed them to thaw under matched microgravity and 1xG conditions in space on the International Microgravity Laboratory #2 mission. Bacteria were analyzed for colony-forming ability while human cells were refrozen for electrophoretic analysis of DNA strand breaks (results not available as of this writing). G. Nelson and collaborators will expose *C. elegans* nematodes to beta particles under matched microgravity and 1xG conditions on the Shuttle–Mir mission #3. The nematodes will be analyzed for chromosome aberrations and mutations. These investigations will complement and extend the observations from previous missions and will take advantage of the in-flight acceleration controls available with the Biorack facility.

A variety of fluid redistribution effects and hormonal responses occur in microgravity (see chapters 29 and 36, this *Handbook*) which may, in turn, influence cellular damage induction and repair systems directly, or control the state of oxygenation and hydration of tissues. Another indirect effect of microgravity may occur through modification of circadian rhythms. Radiosensitivity of intestinal crypt cells and bone marrow cells follows a circadian rhythm (15, 33). These responses are thought to reflect entrainment of cell cycles with diurnal rhythms; radio sensitivity correlates with the cell cycle and the corresponding state of DNA, which is less protected by proteins from free radical attack during replication. Thus a variety of direct and indirect effects of gravity unloading could, in theory, modify cellular radiosensitivity and repair.

SUMMARY AND CONCLUSIONS

A variety of experimental approaches has been used to address the effects of radiation in space. Many investigators have attempted to capture rare cosmic ray interactions in situ, whereas others have employed standardized sources of low LET gamma rays and electrons (beta particles) and have varied the gravity levels systematically. All research teams have had to adjust to the frustrating constraints of spaceflight experimentation, which impose limits on protocols that would be unacceptable to laboratory workers on the ground. In spite of these constraints, the efficacy of space radiation to induce genetic and developmental lesions has been demonstrated clearly. Further, there are statistically significant and reproducible differences in the incidence or severity of radiation-induced lesions as a function of gravity.

Several unifying trends emerge. (*1*) The magnitude of microgravity effects on radiobiological endpoints is small and almost never exceeds a two-fold difference either higher or lower than controls. (*2*) Metabolically active or developing systems are much more likely to be affected than inactive systems. The process of development may amplify small differences in embryonic tissues into large observable differences in the morphology of older organisms. (*3*) The direction and magnitude of radiation or microgravity effects is specific to the biological feature or response measured (endpoint), even when two endpoints are measured in the same sample; for example, reduction of mutation and enhancement of pollen abortion in *Tradescantia*. (*4*) Radiation and microgravity effects are stage specific in developing cellular systems. Thus nonspecific effects on all cells are not seen at all times. (*5*) The genetic lesions that seem most likely to appear are those exhibiting chromosomal breakage. This in turn suggests a role for cytoskeletal elements in conferring microgravity sensitivity as these protein complexes are required to move and align chromosomes and enzymes that ligate broken strands. Many of these processes require ATP; inappropriate shunting of energy into different cellular compartments, as a consequence of gravity unloading, could also explain some of the phenomena. To understand the mechanisms of these interactions, more carefully controlled experiments are needed that have sufficiently large statistical samples and precisely defined endpoints.

This manuscript was prepared by the Jet Propulsion Laboratory, California Institute of Technology, under a contract with the National Aeronautics and Space Administration and its Space Radiation Health Program.

REFERENCES

1. Antipov, V. V. Biological studies aboard the spacecraft "Vostok" and "Voskhod." In: *Problems of Space Biology,* edited by N. M. Sisakyan. Moscow: Nauka, 1967, vol. VI, p. 67–83. (NASA TT-F-528)
2. Antipov, V. V., B. I. Davydov, V. V. Verigo, and Yu. M. Svirezhev. Combined effect of flight factors. In: *Foundations of Space*

Biology and Medicine, edited by M. Calvin and O. G. Gazenko. Washington, D.C.: National Aeronautics and Space Administration, 1975, p. 639–667.

3. Antipov, V. V., N. L. Delone, M. D. Nikitin, G. P. Parfyonov, and P. P. Saxonov. Some results of the radiobiological studies performed on Cosmos-110 biosatellite. *Life Sciences and Space Research.* 7: 207–209, 1969.

4. Bender, M. A., P. C. Gooch, and S. Kondo. The Gemini-3 S-4 spaceflight-radiation interaction experiment. *Radiat. Res.* 31: 91–111, 1967.

5. Bender, M. A., P. C. Gooch, and S. Kondo. The Gemini XI S-4 spaceflight-radiation interaction experiment: The human blood experiment. *Radiat. Res.* 34: 228–238, 1968.

6. Bender, M. A., F. J. De Serres, P. C. Gooch, I. R. Miller, D. B. Smith, and S. Kondo. Radiation and zero-gravity effects on human leukocytes and Neurospora crassa. In: *The Gemini Program Biomedical Science Experiments Summary.* Washington, D.C.: National Aeronautics and Space Administration, 1971, p. 205–235. (TM-X-58074)

7. De Groot, R. P., P. J. Rijken, J. Den Hertog, J. Boonstra, A. J. Verkleij, S. W. De Laat, and W. Kruijer. Microgravity decreases *c-fos* induction and serum response element activity. *J. Cell Sci.* 97: 33–38, 1990.

8. De Serres, F. J., I. R. Miller, D. B. Smith, S. Kondo, and M. A. Bender. The Gemini-XI S-4 spaceflight radiation interaction experiment II. Analysis of survival levels and forward-mutation frequencies in *Neurospora crassa. Radiat. Res.* 39: 436–444, 1969.

9. Friedberg, E. C. *DNA Repair.* New York: Freeman, 1985, pp. 1–77, 406–445, 479–497.

10. Fry, R. J. M. Radiation protection guidelines for space missions. In: *Terrestrial Space Radiation and Its Biological Effects.* edited by P. D. McCormack, C. E. Swenberg, and H. Bücker. New York: Plenum, 1988, NATO ASI Series A, vol. 154, p. 715–728.

11. Gaubin, Y., M. Delpoux, B. Pianezzi, G. Gasset, C. Heilmann, and H. Planel. Investigations of the effects of cosmic rays on *Artemia* cysts and tobacco seeds: Results of Exobloc II experiment, flown aboard Biocosmos 1887. *Nucl. Tracks Radiat. Meas.* 17: 133–143, 1990.

12. Haymaker, W., B. C. Look, D. L. Winter, E. V. Benton, and M. R. Cruty. The effects of cosmic particle radiation on pocket mice aboard Apollo XVII: I. Project BIOCORE (M212), a biological cosmic ray experiment: Procedures, summary, and conclusions. *Aviat. Space Environ. Med.* 46: 467–481, 1975. [See also p. 482–654 for dosimetry, engineering, and biology.]

13. Horneck, G. Cosmic ray HZE particle effects in biological systems: Results of experiments in space. In: *Terrestrial Space Radiation and Its Biological Effects.* edited by P. D. McCormack, C. E. Swenberg, and H. Bücker. New York: Plenum, 1988, NATO ASI Series A, vol. 154, p. 129–152.

14. Horneck, G. Impact of spaceflight environment on radiation response. In: *Terrestrial Space Radiation and Its Biological Effects.* edited by P. D. McCormack, C. E. Swenberg, and H. Bücker. New York: Plenum, 1988, NATO ASI Series A, vol. 154, p. 707–714.

15. Ijiri, K., and C. S. Potten. Circadian rhythms in the sensitivity to radiation in mouse intestinal epithelium. *Int. J. Radiat. Biol.* 53: 717–727, 1988.

16. Il'yin, Ye. A., and G. P. Parvenov. (1980) Biological studies on the Kosmos biosatellites. Moscow: Nauka, 1979, p. 1–210. (NASA TM-75769)

17. Johnson, N. L. *Handbook of Soviet Manned Space Flight.* American Astronautical Society Science and Technology Series. San Diego: Univelt, 1980, vol. 48, p. 63–71.

18. Johnston, R. S., L. F. Deitlein, and C. A. Berry. Biomedical results of Apollo. Washington, D.C.: National Aeronautics and Space Administration, 1975, p. 105–113, 343–403. (NASA SP-368)

19. Kiefer, J. *Biological Radiation Effects.* New York: Springer-Verlag, 1990, p. 1–87, 104–120, 137–156.

20. Longdon, N. and V. David (editors). *Biorack on Spacelab D1.* Noordwijk: European Space Agency, 1988, pp. 3–26, 135–145. (ESA SP-1091)

21. Neidhardt, F. C., R. A. Vanbogelen, and V. Vaughn. The genetics and regulation of heat-shock proteins. *Ann. Rev. Genet.* 18: 295–329, 1984.

22. Nelson, G., W. Schubert, G. Kazarians, G. Richards, E. V. Benton, E. R. Benton, and R. P. Henke. Radiation effects in nematodes. Results from IML-1 experiments. *Adv. Space Res.* 14(10): 87–91, 1994.

23. Nelson, G. A., W. W. Schubert, and T. M. Marshall. Radiobiological studies with the nematode *Caenorhabditis elegans.* Genetic and developmental effects of high LET radiation. *Nucl. Tracks Radiat. Meas.* 20: 227–232, 1992.

24. Parfenov, G. P. Genetic investigations in outer space. *Cosmic Res.* 5: 121–133, 1967.

25. Portugalov, V. V., E. A. Savina, A. S. Kaplansky, V. I. Yokoleva, G. N. Durnova, A. S. Pankova, V. N. Shvets, E. I. Alekseyev, and P. I. Katunyan. Discussion of the combined effect of weightlessness and ionizing radiation on the mammalian body: morphological data. *Aviat. Space Environ. Med.* 48: 33–36, 1977.

26. Pross, H. D., M. Kost, and J. Kiefer. Repair of radiation induced genetic damage under microgravity. In: *Proceedings of 5th European Symposium on Life Sciences Research in Space.* Noordwijk: European Space Agency, 1994 p. 193–196. (ESA SP-366)

27. Reitz, G., H. Bücker, R. Facius, G. Horneck, E. H. Graul, H. Berger, W. Rüther, W. Heinrich, R. Beaujean, W. Enge, A. M. Altapov, I. A. Ushakov, Yu. A. Zachvatkin, and D. A. M. Mesland. Influence of cosmic radiation and/or microgravity on the development of *Carausius morosus. Adv. Space Res.* 9: 161–173, 1989.

28. Reitz, G., H. Bücker, W. Rüther, E. H. Graul, R. Beaujean, W. Enge, W. Heinrich, D. A. M. Mesland, A. M. Altapov, I. A. Ushakov, and Yu. A. Zachvatkin. Effects on ontogenesis of *Carausius morosus* hit by cosmic heavy ions. *Nucl. Tracks Radiat. Meas.* 17: 145–153, 1990.

29. Rosenzweig, S. A., and K. A. Souza (editors). *Final Reports of U.S. Experiments Flown on the Soviet Satellite Cosmos 936.* Moffett Field, CA: National Aeronautics and Space Administration, 1979, pp. 3–59, 185–273. (NASA TM-78526)

30. Saunders, J. F. *The Experiments of Biosatellite II.* Washington, D.C: National Aeronautics and Space Administration, 1971, pp. 1–351. (NASA SP-204)

31. Stassinopoulos, E. G. The Earth's trapped and transient space radiation environment. In: *Terrestrial Space Radiation and Its Biological Effects.* edited by P. D. McCormack, C. E. Swenberg, and H. Bücker. New York: Plenum, 1988, NATO ASI Series A, vol. 154, p. 5–35.

32. Tobias, C. A., and Yu. G. Grigor'yev. Ionizing radiation. In: *Foundations of Space Biology and Medicine,* edited by M. Calvin and O. G. Gazenko. Washington, D.C.: National Aeronautical and Space Administration, 1975, p. 473–531.

33. Tobias, C. A., and P. Todd. *Space Radiation Biology and Related Topics.* New York: Academic, 1974, p. 1–100, 115–141, 313–475.

34. Todd, P., and J. T. Walker. The microlesion concept in HZE particle dosimetry. *Adv. Space Res.* 4(10): 187–197, 1984.

35. Effect of spaceflight on lymphocyte function and immunity

FELIX K. GMÜNDER | Basler & Hoffmann Consulting Engineers, Zurich, Switzerland

AUGUSTO COGOLI | Swiss Federal Institute of Technology, Space Biology Group, Zurich, Switzerland

CHAPTER CONTENTS

CHANGES IN IMMUNOLOGICAL FUNCTIONS have been observed in astronauts after missions of short and long duration. The first report was from the Soviet immunologist, Konstantinova, and her coworkers (55), who found that lymphocyte responsiveness to mitogens was remarkably reduced after landing. Lymphocyte responsiveness to mitogens was determined by means of a clinical laboratory test that mimics the events initiated following an infection in lymphocytes isolated from peripheral blood. The Soviet finding was later confirmed by U.S. investigators among Skylab astronauts (51) and most Shuttle flight crew members (78, 80). Despite these observations, astronauts and cosmonauts have generally maintained good health both during and after their flights, except for minor problems with respiratory tract infections and viral gastroenteritis on the Apollo flights and some inflammatory complications in Soviet orbital station crew members (48, 53).

The first attempts to explain the reduction of lymphocyte responsiveness during and after spaceflight included responses to microgravity and radiation. Experimental evidence from ground-based studies in our laboratory pointed to a direct effect of gravity in lymphocytes (21, 22). These studies were conducted at the beginning of an extensive series of experiments in which we examined the effect of microgravity on the in vitro responsiveness of human lymphocytes (6, 16, 17, 20, 23). Others suggested (78, 80) that the stress of spaceflight may affect the immune system. Growing evidence of a link between the neuroendocrine system and the immune system (8) indicated that the depressed lymphocyte response to mitogen in astronauts (in vivo) may have been caused by nervous control of endocrine glands and subsequent mediation of immunocompetent cells via an increased production of stress-related hormones and peptides.

In this chapter we will discuss microgravity effects on lymphocyte reactivity observed in vitro as well as in vivo studies. The notion of "in vivo" assays of lymphocyte responsiveness, however, is inaccurate since this assay is performed ex vivo in cell culture flasks after drawing blood from a subject. Here we will show that lymphocytes drawn from "stressed" astronauts are primed by stress-related hormones and peptides leading to depression of lymphocyte responsiveness to mitogens during and after flight. This type of investigation will be referred to as in vivo. By contrast, for investigations in which lymphocytes are isolated from peripheral blood of unstressed subjects to be studied in space, in the centrifuge, or in the clinostat, we will use the term in vitro. An interesting issue is how in vitro studies relate, if at all, to the in vivo situation.

Elsewhere we have discussed in detail the technology, instruments, and results of experiments with lymphocytes and other single cells in space, on the ground, in clinostats, and in centrifuges (15, 18, 36). In this chapter we will focus on the effect of spaceflight on the human immune system.

In addition to a description of in vitro studies performed in space and on the ground, we will comment on in vivo results. The impact of the neuroendocrine system on immunity under stress situations will be presented and, finally, we will compare stress models and the stress of space flight with respect to lymphocyte function and immunity.

LYMPHOCYTES

In healthy people the total white blood cell count comprises 19% to 48% lymphocytes. *Lymphocytes* are the key cells responsible for the adaptive, specific immune response. Lymphocytes as a population are heterogeneous, being comprised of B cells and T cells that can be further grouped into subpopulations by identifying cell surface marker proteins with monoclonal antibodies. B lymphocytes produce antigen-specific antibodies that form part of the humoral immunity. T cells are composed of a variety of subpopulations, including helper/inducer and suppressor/cytotoxic T cells. The function of T lymphocytes is often referred to as *cell-mediated immunity.*

In human studies, lymphocytes are collected from peripheral blood, but the major portion of the lymphocyte population in mammals, about 95%–98%, resides in tissues such as spleen, lymph nodes, gut mucosa, skin, and bone marrow. Thus in the peripheral blood, only a small fraction of the total population is present. However, there is a constant and large scale recirculation taking place between tissue and blood. Most of the lymphocytes (70%–80%) engaged in recirculation are T cells and a given T cell recirculates once each day in man. Thus lymphocytes obtained from peripheral blood are a fair representation of the whole-body lymphocyte reservoir. In animals, lymphocytes can be isolated from tissues like the spleen or lymphatics. Lymphocytes isolated from body fluids or tissue are nondividing, resting cells, transiently locked in the G_0-phase of the cell cycle, although ample nutrients are available.

In the body, patrolling T lymphocytes enter the G_1 phase of the cell cycle following antigen presentation by *antigen-presenting cells* (macrophages, monocytes), a process that is restricted by the major histocompatibility complex II (MHC II). Binding of antigen and MHC II ligands to the T-cell receptor initiates a sequence of transmembrane and intracellular signals resulting in a maturation process of the lymphocyte, which is often referred to as lymphocyte responsiveness, proliferation, activation, or *blastogenesis*. In the body an antigen triggers only a very restricted number of lymphocytes because the T-cell receptor will recognize a given antigen very specifically. This binding of antigens to the T-

cell receptor is often illustrated with the key/lock analogue.

In the laboratory, plant mitogens such as pokeweed mitogen (PWM), phytohemagglutinin (PHA), and concanavalin A (con A) have been used as stimulants to study the process of blastogenesis, the cell cycle, and metabolism of activated lymphocytes. Activation by mitogens is polyclonal, that is, it involves either all T or B lymphocytes. One molecule of con A, for instance, has four binding sites for α-glucosides. Thus the interaction between con A and glycoproteins on the cell membrane brings about inter- and intracellular clustering of the ligands carrying α-glucosides. The best method to measure the rate of lymphocyte responsiveness following addition of a mitogen is by incorporation of radiolabelled thymidine (^{3}H-thymidine) into DNA (DNA synthesis) or of uridine (^{3}H-uridine) into RNA (protein synthesis) (1). Although both methods have their specific problems (1), such as the optimal concentration to obtain a zero-order reaction and correlation with cell number or biomass, they represent the best methods of estimating lymphocyte responsiveness. The results, expressed as the amount of radiolabelled substrate taken up by the cells, are measured as counts per minute (cpm). The cpm values of a given sample are often referred to a laboratory standard value and expressed as the relative proliferation index (RPI) (41).

DNA synthesis is confined to the S phase of the cell cycle. In lymphocytes it reaches its maximum between 60 and 72 h after exposure to mitogens. The rate of DNA synthesis, measured as incorporation of ^{3}H-thymidine (usually a 2 h pulse after 72 h of incubation), correlates with the proliferation rate of the cells. The increase in RNA synthesis begins shortly (within 1 h) after contact with the mitogen and can be measured with 1 h pulses of ^{3}H-uridine after 20 h of cultivation. The rate of RNA synthesis gives an overall view of the activation of the lymphocytes. Both methods were used to assess human lymphocyte responsiveness following spaceflight.

GRAVITY AND RADIATION EFFECTS ON LYMPHOCYTES

In Vitro Gravity Experiments

Our investigations toward understanding the effects of spaceflight on lymphocytes comprised ground-based experiments at high G in the centrifuge (between 2 and 20 G) as well as at simulated low G in the clinostat, experiments in space on Spacelab 1, D-1, and SLS-1 missions, and in sounding rockets (6, 16, 20, 23, 38).

This series of studies was of a basic biological nature

rather than specific immunological research, for we wanted to study the effect of gravity on cell behavior. In fact, lymphocyte activation with mitogens in vitro is a widely used model for studying the mechanisms of mammalian cell differentiation. Thus the results of these in vitro studies should not be used to explain the changes in lymphocyte responsiveness in astronauts following spaceflight.

Lymphocytes in culture are sensitive to changes in the gravitational environment (Fig. 35.1). Lymphocyte responsiveness to con A is reduced by more than 90% in microgravity in space and by 50%–70% in simulated low G in the fast rotating clinostat. Conversely, responsiveness is enhanced by 20%–40% in the centrifuge at 10G in isolated lymphocytes and by 150% in *whole-blood* lymphocytes, respectively. Several hypotheses have been formulated and discussed to explain these effects (15).

In experiments with isolated lymphocytes in space, the most important question is whether direct or indirect effects of gravity are affecting lymphocyte activation. Gravity may interfere directly with cellular structures and organelles (for example, the nucleus, nucleolus, cytoskeleton) or indirectly by altering the physicochemical environment of the cell (for example, cell-cell contacts, cell-substratum adhesion, fluid convection). For instance, in a study investigating the effect of cell adhesion (of lymphocytes and/or accessory cells like macrophages) to a substratum, a significant correlation between cell adhesion and lymphocyte responsiveness was found (38, 39). The results showed that cells need to anchor and spread prior to achieving an optimal proliferation response. It was concluded that

decreased cell adhesion could contribute to the depressed in vitro lymphocyte responsiveness found in microgravity. The results of an experiment on this subject carried out on SLS-1 supports this view (14a, 16a).

With respect to in vitro lymphocyte function in space, Hungarian and Soviet investigators (5, 76) studied interferon-α production of lymphocytes stimulated by specific inducers on the Soviet space station Salyut-6. These results were later confirmed by the same team (5, 76). They found a four- to eightfold increase in interferon-α production in the space samples as compared to ground controls, and they concluded that lymphocytes are sensitive to factors induced by spaceflight.

In summary, there is substantial evidence that space flight causes—directly or indirectly—metabolic alterations in lymphocytes. Microgravity can be used as a tool to better understand the complex events leading to lymphocyte activation.

In Vitro Radiation Experiments

Besides stress and microgravity, cosmic radiation may also contribute to depressed lymphocyte reactivity in astronauts. In low orbits (for example, 450 km at the equator) the daily dose equivalent of radiation may be as high as 0.1 to 1 mSv. A reduction of lymphocyte responsiveness of 30% was noted in cells exposed to 0.5 Gy X rays prior to the addition of con A (21). However, X rays are not a good representation of low-orbit radiation. Thus in a stratospheric balloon flight, the effect of cosmic radiation on lymphocyte responsiveness was examined (85, 86). Stratospheric cosmic radiation (the spectrum of which cannot be reproduced in a lab-

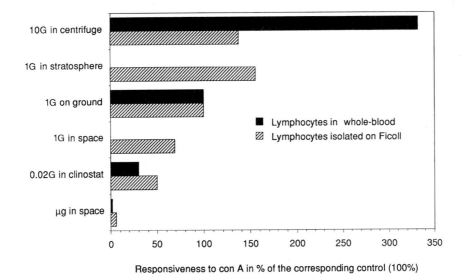

FIG. 35.1. Summary of gravitational effects on the activation of Ficoll-purified lymphocytes *(black columns)* and whole-blood lymphocytes *(hatched columns)*. The results are expressed as a percentage of the corresponding 1G control on Earth.

oratory) is almost identical to that in an Earth orbit, whereas gravity is virtually 1G. In this 1 day experiment, an increased lymphocyte responsiveness was recorded when compared to 1G on the ground (Fig. 35.1). Gualde and Goodwin (45) found no stimulation of T-cell responsiveness to con A and PHA following exposure to low-dose irradiation, however, radiation doses of 0.1 to 0.5 Gy enhanced proliferation of the cytotoxic/suppressor subpopulation. Likewise, low-dose irradiation (5–20 Gy) resulted in an increased human natural killer cell activity (11). Comparable in vivo experiments in mice showed a similar stimulative

effect of low doses of X rays (0.025–0.075 Gy) on the plaque-forming cell reaction of the spleen and on the reactivity of thymocytes to interleukin-1 (61). Compared to an unexposed population, Hiroshima and Nagasaki victims exposed to less than 0.5 Gy showed a higher mitogenic response of lymphocytes to PHA, enhanced natural killer activity, and increased interferon-γ levels in culture supernatants of activated lymphocytes (10). All of these results provide evidence that the immune system might be stimulated by low-level ionizing radiation. This tentative conclusion supports the theory of hormesis where low doses of ionizing radi-

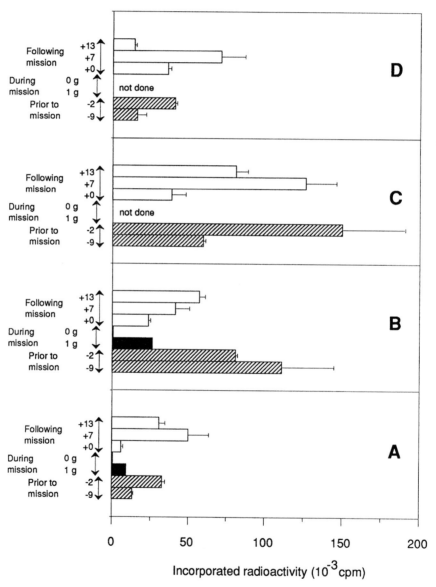

FIG. 35.2. Results of the blood experiment on the D-1 mission. The effect of spaceflight on lymphocyte responsiveness in 4 crew members (named *A*, *B*, *C*, and *D*) was examined in whole-blood cultures. Measurements were made 9 and 2 days prior to the mission, on the 3rd day of the mission, immediately after landing, and 7 and 13 days following the mission. In-flight measurements in subjects *C* and *D* were not possible. The flight samples were incubated at 0G and 1G in the static rack and the gravity-simulating centrifuge, respectively.

ation may result in increased longevity, increased growth and fertility of both plant or animal organisms, and a reduction in cancer frequency (63, 73). Results from balloon experiments may help to identify radiation effects; however, a 1G centrifuge on a spacecraft is the experimental control of choice.

Taken together, the present data suggest that cosmic radiation does not depress lymphocyte responsiveness per se.

NEUROENDOCRINE CONTROL OF LYMPHOCYTE FUNCTION

Another interesting issue is the relation between the in vitro study results addressed above and the depressed lymphocyte responsiveness to mitogens found in ex vivo lymphocyte cultures from astronauts during and after spaceflights (16, 51, 55, 56, 78, 80). As an example, lymphocyte responsiveness in four subjects prior to, during, and after the D-1 mission is shown in Figure 35.2 (16). In the three subjects (A, B, and C) with a decreased lymphocyte responsiveness noted immediately following landing (day +0) as compared to preflight data (day −9 and −2), a recovery within 7 days was noted (day +7).

In subjects A and B, lymphocyte responsiveness was also determined for the first time during flight. The space sample was split into two aliquots: one was kept under microgravity conditions, the other in the centrifuge onboard under 1G conditions. Two interesting results have been recorded: During flight, lymphocyte responsiveness decreased as compared to baseline data (days −9, −2, +7, and +13). Lymphocytes kept under microgravity (0G) responded even less to the mitogenic challenge than those at 1G. This difference in responsiveness between the microgravity and 1G space samples may reflect the prerequisite for lymphocytes to anchor and spread (on the bottom of the tissue culture vessel) prior to achieving an optimal proliferation response as discussed in In Vitro Gravity Experiments, above (38, 39). The 1G in-flight data can be compared with pre- and postflight values.

Such an impairment of lymphocyte function could lead to lowered immunity in astronauts during and after flight. As pointed out earlier, in vivo experiments are fundamentally different from the in vitro experiments, although the effect appears to be the same, namely, reduced lymphocyte responsiveness in space. In fact, the role of stress in mediating the function of lymphocytes in vivo was recognized by Taylor and Dardano (78), and Taylor et al. (80). Cogoli and Tschopp (19) have cautioned that one should not extrapolate from data obtained from in vitro experiments the depressed ex vivo lymphocyte responsiveness in astronauts observed after spaceflight. It is more likely that the multiple stres-

sors associated with spaceflight contribute to the depressed immunity observed in astronauts.

The first researchers to propose a link between the neuroendocrine and the immune system were Besedowsky and Sorkin (8). This hypothesis was based on the existence of afferent–efferent pathways between immune and neuroendocrine structures and was later confirmed by the findings of several other investigators. However, it was already known that in humans, the systemic application of corticosteroids depresses lymphocyte numbers in peripheral blood (43). The mitogenic response of lymphocytes to con A is significantly reduced following in vivo administration of corticosteroids (31). Catecholamines (epinephrine) inhibit lymphocyte proliferation by interacting with β-adrenergic receptors (26). Somewhat more surprising than the general neuroendocrine control over the immune system was the finding that immunocompetent cells produce a variety of hormones and peptides that affect the central nervous system (9).

Taken together, there is strong evidence that the neuroendocrine and immune systems form a regulatory loop (Fig. 35.3) that operates via peptide hormones and transmitters (9).

Regarding the loss of calcium and bone structure during prolonged spaceflight, which may be similar to osteoporosis, a link between bone metabolism and the immune system in cosmonauts has been proposed (53). Whereas bone formation depends on osteoblasts, bone resorption is initiated by osteoclasts. Osteoclasts origi-

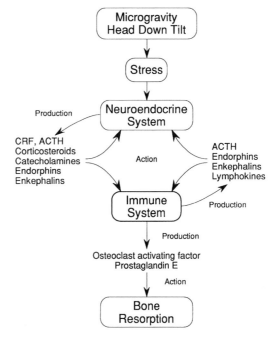

FIG. 35.3. Regulatory loop between the neuroendocrine system and the immune system, and the influence on bone resorption mechanisms.

nate from hemopoietic stem cells like macrophages and are released from the bone marrow. During long-duration head-down tilt (HDT) bed rest of 370 days, an increase in the production of bone resorption–enhancing factors was noted in supernatants of peripheral mononuclear blood cell cultures (53, 60). Such an increase in osteoclast activating factor (OAF) production coincided with a steady increase in lymphocyte responsiveness to PHA during HDT. It is not clear, however, how this increase in lymphocyte responsiveness observed during HDT compares to the reduction in responsiveness observed during and after spaceflight.

STRESS MODELS AND SELECTED ASPECTS OF IMMUNE FUNCTION

The influence of psychological and physical stress, aging, extreme environmental conditions, and nutrition on many immunological parameters has been reviewed (28). Here we present the most interesting findings and relate them to results obtained during and after spaceflight. This approach will help us to understand how spaceflight may alter immunity in crew members, and how this might affect the performance of humans during prolonged spaceflight on space stations and during deep space missions. Since stress-related hormones may influence many immunological parameters we will also discuss changes in the levels of these hormones.

Psychological Stress

It is difficult to evaluate the degree of psychological stress perceived by an individual. There is, however, clear evidence that psychological stress has a depressing effect on immunological parameters. Immunoglobulin A (IgA) secretion was measured in the saliva of 64 dental students five times during an initial low-stress period, three times during high-stress periods, and once during a final low-stress period (50). Secretion of IgA during the high-stress periods was significantly lower compared with that during the preceding and succeeding low-stress periods. Dorian and colleagues studied a large group of medical and psychiatry students on three occasions (28, 29): 2 mo and 2 wk before an important examination, which was considered the peak stress, and 2 wk after the exam. Subjects which felt highly stressed and aroused showed low levels of natural killer cell activity and interleukin-2 blood levels; and lymphocyte responsiveness to con A and PWM was increased after the exam, compared with the two samples obtained prior to the examinations. This group also developed more respiratory infections in the six weeks following the examinations. Furthermore, Cohen et al. (24) exposed 394 healthy subjects to respiratory viruses and found that psychological stress was related in a dose-response manner to the incidence of infection.

Physical Exercise

Short- and long-term physical exercise has dramatic effects on many immunological parameters (71). Physical stress in general profoundly affects the levels of stress-related hormones. It leads to a marked increase in the levels of cortisol, epinephrine, and norepinephrine (12, 32). Leukocytosis after short-term exercise and long-term exercise training has been known for a long time (34). It is explained by a shift in cells into the circulating pool, since differential counts show that the ratio between segmented and nonsegmented cells does not change. During and following exercise lasting more than 24 h, white blood cell counts return to baseline levels (33). The occurrence of a lymphopenia after prolonged physical exercise stress is likely due to the delayed increase in plasma cortisol levels. In general, during and immediately following exercise, the proportion of T cells decreases, but the relative size of the B-cell population does not change. Such relative decreases in the T-cell population is due to a decrease in the T-helper cell subpopulation, which can be explained by the selective, depressing effect of corticosteroids on the number of recirculating T cells (31). The increase in numbers and activity of natural killer cells during exercise is paramount and consistent in all studies.

Lymphocyte responsiveness to mitogens is usually depressed significantly during and following muscular work [Fig. 35.4; (37, 41, 49, 82)].

It appears that changes in white blood cell counts, in the size of leukocytes, lymphocytes, and lymphocyte subpopulations, and the reduction in the mitogenic response of lymphocytes—all of which are observed after short bouts of exercise—are indeed related to a rise in stress-related hormones. In longer lasting exercise the rising level of cortisol may enhance this depression in lymphocyte function. In fact, the administration of catecholamines and cortisol leads to a marked reduction in lymphocyte responsiveness.

Isolation in a Closed Environment

Antarctic research station environments are, in many aspects, very similar to the environment in a spacecraft or on a space station. In both environments, small groups of people live in relative isolation and confinement. Except for the normal protozoan or microbial and viral flora that could cause an opportunistic infection, no infectious agents are present or introduced. This pathogen-free environment could reduce morbidity, but, in isolation, the reduced challenge by new pathogenic or infectious organisms may lower the

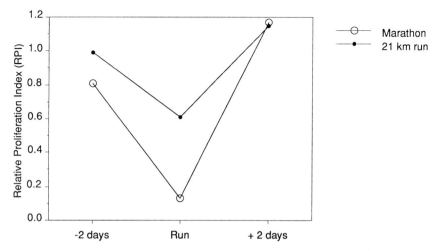

FIG. 35.4. Effect of a marathon run (42 km) and a 21 km run (the median of 8 and 16 test subjects is depicted, respectively) on lymphocyte responsiveness to con A expressed as the relative proliferation index (RPI). The blood draws 2 days prior to and after the run were made following a resting period of 30 min. The reference range of normal, healthy subjects as determined in our laboratory is 0.2 to 1.8 RPI.

responsiveness of the immune system. If this were the case, the confrontation with unfamiliar pathogenic organisms, introduced by visitors or encountered upon the return of astronauts to Earth, might have serious consequences. In addition, small groups of people living in closed environments may be liable to psychosocial stress situations. Cosman and Brandt-Rauf (25) presented an overview of immunological findings in Antarctica, but the data are not consistent. For instance, a decrease in the number of neutrophils and a relative lymphocytosis were noted. The results concerning levels of immunoglobulins are conflicting; there were decreases, increases, or no change. No difference in susceptibility to ambient cold between winter-over groups and newcomers was reported, indicating that any changes in immunological parameters are not clinically important. Nevertheless, due to the lack of data and inconsistent protocols, experiments with small groups living in isolation deserve more intensive study.

Head-down Tilt (HDT) Bed Rest

The physiological effects of HDT bed rest, including those on the immune system, are reviewed by Greenleaf, Fortney, and Schneider in a chapter of this *Handbook*. Therefore, we discuss here only the results of a study carried out by our team in collaboration with the Institute of Aerospace Medicine of the DLR (Deutsche Forschungsanstalt für Luft–Raunfehrt) in Cologne, Germany (2, 35).

In this medium-duration HDT (10 days), the mitogenic response to con A was reduced significantly in all six subjects before and on the last day of HDT, and on the first day of recovery (Fig. 35.5). These changes in

lymphocyte responsiveness correlated well with changes in blood levels of cortisol and catecholamines, and in the numbers of β-adrenoreceptors (35, 64). In stark contrast to lymphocyte responsiveness, delayed-type skin hypersensitivity (DTH) was not affected (Fig. 35.5). No changes in total or differential white blood cell counts were noted. Lymphocyte subpopulations did not change, except for the number of natural killer cells that dropped transiently when the subjects changed from HDT to the normal ambulatory position. Blood levels of the humoral parameters IgA, IgG, and IgM, and concentrations of α_1-microglobulin, α_1-antitrypsin, and c1-inhibitor, were not affected during HDT (42).

ANIMAL MODELS

In animals, and in rodents in particular, the effects of spaceflight, simulated weightlessness, and high G on some immunological parameters have been tested in the pre-Shuttle era (3). On the whole, spaceflight, simulated weightlessness-hypokinesia, and HDT had little impact on immunological parameters in animals in these early studies, the only exception being an enhanced splenocyte (lymphocyte) responsiveness to PHA and con A after landing.

Spaceflight (20 days) significantly increased plasma levels of catecholamines in flight animals compared to those in ground animals, but no difference was found between animals exposed to microgravity conditions and animals kept on a 1G gravity control centrifuge in space (65, 66). Macho and colleagues concluded that the increase in epinephrine and noradrenaline was not due to microgravity, but to other factors associated with

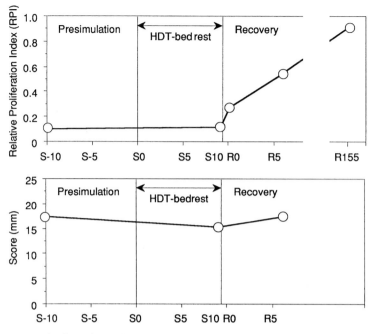

FIG. 35.5. *Top panel*: Effect of HDT-bed rest on the mitogenic responsiveness of lymphocytes, expressed as the relative proliferation index (RPI), and *(bottom panel)* on delayed-type skin hypersensitivity (the mean score of healthy men in Europe is 20 mm). The median of six subjects is shown. HDT consisted of a presimulation period (10 days), followed by the HDT period lasting 10 days. The subjects were also monitored during recovery. The reference range as determined in our laboratory is 0.2–1.8 RPI. The blood draws were performed on the following days: *S-10*, 10 days prior to the beginning of the HDT; *S10*, 10th day of HDT; *R0, R6*, and *R155*, 1, 7, and 156 days after the subjects finished HDT, respectively.

spaceflight and landing. In contrast, plasma corticosteroid levels were significantly increased 1.5- and twofold in animals exposed to microgravity as compared to ground control animals and those subjected to 1G during the mission, respectively. This increase suggests that microgravity was a chronic but mild stress for the rats.

On Biokosmos I (Cosmos 605; 21.5 days, 1973), the flight rats had a decrease in size of lymph nodes, thymus, and spleen, compared to ground control animals (30). The reduced size of these organs was transient and returned to normal 27 days after landing. On Biokosmos III (Cosmos 782; 19.5 days, 1975), Mandel and Balish (67) studied the behavior of cell-mediated defense to *Listeria monocytogenes* and lymphocyte proliferation following stimulation by *L. monocytogenes* antigens, con A, PHA, and PPD. Interestingly, the overall level of the mitogenic responsiveness of lymphocytes from flight rats to these antigens and mitogens was generally higher than in lymphocytes from the three control groups.

Gould and colleagues (44) studied the effect of spaceflight on the production of interferon-γ and interleukin-3 by rat spleen cells on the Spacelab-3 mission. Spleen cells were isolated from flight animals and ground controls 12 h after landing and challenged with

con A, and the concentration of lymphokines was measured in the culture supernatant. Interferon-γ production was clearly inhibited in splenocytes of all flight rats, but interleukin-3 production was normal compared to that in ground-based control animals. It is an open question whether the long transport from the landing site to the laboratory (12 h) was stressful for the animals and led to the observed decrease in interferon-γ production.

The first researchers to study the effect of antiorthostatic suspension (69) on the immune system in rats were Caren et al. (13). Antiorthostatic suspension is used as an HDT model for rats. No significant differences were found between suspended animals and harnessed controls in their normal position with respect to serum immunoglobulin levels, spleen and thymus weight, relative white blood cell counts, and capability to produce anti-sheep red blood cell antibodies.

Sonnenfeld and colleagues (75) studied the effect of suspension on interferon production in rats and mice. With antiorthostatic suspension, interferon-α and -β levels were markedly reduced, but the authors suggested that this reduction was not due solely to the stress of suspension since in orthostatically suspended mice, no effect on interferon production was noted. When antiorthostatically suspended animals returned to normal caging they recovered the ability to produce interferon. Antiorthostatic suspension of mice resulted in a decreased resistance to infection.

In rats, short- and long-term exposure to high gravity provokes a marked response of the hypothalamic–pituitary–adrenal system (72), but Scibetta et al. (74) found no effect on the size of the spleen, thymus, and adrenal glands, or on peripheral leukocyte counts. Likewise, antibody production following immunization with sheep red blood cells was not significantly affected by 3.1G for 9 days. In addition, no effect of deceleration, that is, the change from high gravity conditions (3.1G for 28 days) to normal 1G conditions, was found with respect to the parameters tested mentioned above (4, 74). In contrast, we (70) found that prolonged high-gravity conditions (3.5G for 1 yr) induced a three- to four-fold increase in the responsiveness of splenocytes to con A as compared to that in control animals.

In conclusion, space experiments and ground-based experiments (suspension and high G) show that stress-related hormones and immune systems of rats are sensitive to the G environment. The results suggest that in rats, the function of the immune system is enhanced by the space environment.

SPACE FLIGHT AND SELECTED ASPECTS OF THE IMMUNE FUNCTION

Results from immunological investigations from postflight analysis of earlier missions have been reviewed (7, 14, 19, 56, 77). Here the most relevant findings are summarized.

Stress-related Hormones

Plasma and urinary cortisol levels increased remarkably (4%–32% and up to 190%, respectively) in all crew members aboard three Skylab missions throughout flight (59). This finding was later confirmed by crew members of Shuttle flights (58). During and after spaceflight, blood and urinary levels of catecholamines varied according to the activity of the sympathoadrenal system. Generally, during spaceflight, the catecholamine levels in the blood and urine were reduced compared with those during ground conditions. However, specific physical activities, space adaptation syndrome, launch, and landing, led to intermittent catecholamine surges. In particular, the stress of landing and readaptation to normal gravity led to a heightened response of the sympathoadrenal axis, resulting in increased plasma levels of catecholamines (57, 58, 81). Note that most studies on spaceflight effects on human immunological parameters utilized postflight blood (0–8 days after landing).

The present lack of in-flight data underscores the need to clarify the behavior of human immunity during microgravity.

Immunoglobulins

No significant changes in IgG and IgM blood levels were found after the Apollo flights; the same was reported for IgG, IgA, IgM, IgD, and IgE after the three Skylab missions (51). An increase in IgA, IgG, and IgM serum concentrations was observed after the 49 day flight on the Soviet space station Salyut 5 (47). This increase was explained as a response to the secretion of autoantibodies against degradation products from the atrophy of skeletal muscles that occurred during flight. Other minor immunoglobulin fluctuations were also reported (46, 47, 51, 52, 55). Serum immunoglobulins G, M, A, D, and E concentrations of four crewmen of Spacelab-1 in 1983 also showed no significant changes in flight and postflight (84).

These results do not necessarily exclude an impairment of B-lymphocyte function in space. In fact, only measurement of the specific response to an antigenic challenge in vivo will give information on the efficiency of B-lymphocyte function. Due to ethical reasons, such experiments may be carried out with laboratory animals but not with humans.

Lysozyme

Lysozyme, a protein with bactericidal activity, is localized in saliva and, to a lesser extent, in blood plasma. Compared to preflight levels, salivary lysozyme levels postflight were significantly lower in cosmonauts after 49 days on Salyut 5 (46) and 96 days on Salyut 6 (53). Slightly elevated serum lysozyme levels were found in the Salyut 5 cosmonauts (53) and in two Skylab II astronauts (51), but no changes were observed in the serum of all other Skylab crew members.

Complement

Increased blood levels of C_3 were found after the Apollo flights (52), after Skylab II a decrease was found as well as no change after the other Skylab III and IV missions (51); no change after a 2 day Soyuz flight, and increased levels after 16, 18, and 49 day Salyut flights were found (46, 47). C_4 blood levels were unchanged after the three Skylab missions (51) and after 2, 16, and 18 day Soviet flights, and levels increased after the 49 day Salyut 5 flight (46, 47).

On the whole, the observed fluctuations in blood levels of humoral parameters were not significant and showed little variation within their physiological range.

White Blood Cell Counts

A dramatic increase in total white blood cell (WBC) counts was always observed immediately after landing (16, 51, 78, 80). The number of circulating neutrophils increased (though the ratio between segmented and nonsegmented cells remained unchanged), while lymphocyte counts did not change significantly, or decreased. For instance, Taylor et al. (80) reported an increase of 102% (mean of 41 subjects, range +348% to −21%) in leukocyte counts and a decrease in lymphocyte counts of −13% in 41 subjects (range −56% to +87%) immediately after several shuttle missions.

An increase in leukocyte numbers and constant or slightly decreased lymphocyte counts is the typical response to stress lasting up to several hours (34). It indicates that the stress of landing may contribute substantially to the changes in the immune status recorded in postflight analyses.

Lymphocyte Subpopulations

Changes in the numbers of lymphocyte subpopulations postflight appeared less dramatic and occurred within normal physiological limits. Taylor et al. (80) found no significant changes in subpopulations from 11 crew members after two Space shuttle flights. No changes were noted in the numbers of pan T cells and size of suppressor subset; only a slight increase in size of the helper subset and a slight decrease in the numbers of B lymphocytes occurred. In contrast to these data after long-term flights (75–185 days) a decrease in the numbers of T cells was found in most of the cosmonauts after landing (83). This finding was confirmed by Konstantinova (53, 54), who also noted a decrease in the number of T cells in eight of ten cosmonauts and a smaller size of the B-cell subset in two of four cosmonauts.

Natural Killer Cell Activity

Natural killer (NK) cell activity was measured in blood samples from several Soviet cosmonauts by Konstantinova and her co-investigators (5, 56, 76). In the NK cytotoxicity test, ^{3}H-uridine labeled target cells (human myeloleukemia K-562 cells) were exposed to NK test cells. The amount of radioactivity released after pancreatic RNase treatment was determined and indicated NK-cell activity.

Cytotoxic activity of NK cells was measured prior to (30 days), and 1, 7, and 16–76 days after flight in 35 cosmonauts from short-duration flights (7–10 days), and in 22 cosmonauts from long-duration flights (112–366 days), respectively. The results varied considerably between individual subjects.

Following short spaceflights, in 15 of 35 cosmonauts there were slight decreases in cytotoxic activity which lasted only a few days and returned to baseline levels within 7 days after landing.

NK activity decreased in 11 of 13 subjects after spaceflights lasting 112–175 days. Following long-duration flights (211–366 days), NK activity in six of nine cosmonauts also decreased. Despite the individual differences, Konstantinova and colleagues (56) were able to identify three types of responses: (1) three of 12 cosmonauts who spent between 112 and 175 days in space showed a remarkable decrease in the cytotoxic index (60%–90% lower than preflight) on the first day after landing. Recovery to preflight level, however, occurred within 7 days. (2) five of the 22 cosmonauts who spent at least 112 days in space showed no alteration in cytotoxic activity immediately after flight but did show a significant decrease during the following week. Recovery to baseline level was slow. (3) four cosmonauts displayed low NK-activity immediately after landing, followed by an extended recovery time of 1–2 mo.

These spaceflight results are in agreement with HDT data (39), but they conflict with the observed increase in NK counts and activity following physical effort (37).

Lymphocyte Function

Lymphokines. *Lymphokines* are secreted mainly by T lymphocytes upon antigenic or mitogenic activation. An important class of lymphokines are the *interleukins (IL)*. While IL-1 and IL-6 can be produced by activated macrophages as well as by activated B lymphocytes, IL-2, IL-3, IL-4, IL-5, and tumor necrosis factor are specific T-cell lymphokines. Another important T-lymphocyte product is *interferon-γ*. All of these substances play an important role in the immune response. Since most of the analytical procedures used to determine their blood levels were developed in the last 2 to 5 years, their mechanisms of action are not fully understood. Little is known on how spaceflight may affect lymphokine production; the few data available concern IL-2, and interferon-α and -γ.

Interleukin-2. IL-2 biosynthesis by lymphocytes of cosmonauts was analyzed after induction in vitro with PHA (56, 68). Two aspects were considered: (1) biological activity of IL-2 was evaluated by measuring its ability to induce proliferation of IL-2–dependent cells; and (2) the amount of IL-2 secreted was determined with monoclonal antibodies to a human recombinant IL-2 using an ELISA test. While biological activity in 12 of 13 cosmonauts (in space between 65 and 366 days) dropped significantly after flight (five subjects showed a drop of 50%), the amount of IL-2 produced was higher after flight in eight cosmonauts. It appears that an inactive form of IL-2 was secreted.

Interferon. Lymphocytes from cosmonauts selected from two 7 day and two 9 day flights, respectively, were induced to produce interferon-α in vitro. In two subjects the level dropped by 75% one day after landing; in one subject it returned to preflight baseline within 6 days, in the other it remained low. In two other cosmonauts no change was observed, but both had extremely low baseline levels (5, 76). When interferon-γ biosynthesis was tested after short flights, there was a tendency toward lower postflight levels (56).

Lymphocyte Responsiveness

Apollo program (52). Twenty-one astronauts participated in seven missions (three of which went to the moon) lasting between 6 and 12 days. No effect on lymphocyte activation by PHA was observed after returning to Earth.

Skylab missions (51). The three missions lasted 28, 59, and 84 days. The lymphocytes of nine astronauts were exposed to PHA, and RNA and DNA synthesis was measured. A remarkable depression of RNA synthesis (up to 90%) was detected after flight in eight astronauts. The effect was less evident after the longest mission (84 days), indicating a possible adaptation to flight conditions. Depression of DNA synthesis was also observed in six astronauts; however, the effect was much less evident on DNA than on RNA synthesis. All parameters recovered to baseline values within 7–21 days after landing.

Apollo-Soyuz flight (27). DNA synthesis was reduced in the three astronauts tested after the 9 day flight.

Shuttle missions (78, 80). T-lymphocyte activation by PHA was determined as the rate of DNA synthesis in 41 astronauts flying on 11 missions lasting between 2 and 8 days. Depression of activation was detected in 36 subjects. The average depression for 41 subjects was 25.7%.

Spacelab D-1 (6, 16). We tested lymphocyte activation (DNA-synthesis) by con A on whole-blood cultures on four Spacelab astronauts. For the first time a 1G reference centrifuge on board provided measurement in orbit of samples drawn on the third day in flight. Three astronauts showed depression immediately after landing and two of them showed it also in flight (20%–50%) (Fig. 35.2). Baseline values were reached within 7 days after landing. The experiment has been repeated on the recent SLS-1 mission (Cogoli, Cogoli-Greuter, Bechler, and Criswell, unpublished results).

Soviet cosmonauts (53, 55, 56). The results from Soviet cosmonauts are particularly important in view of the long duration (up to 366 days) of their flights. Kon-

stantinova (53) summarized the data obtained from 50 cosmonauts by activating T lymphocytes with PHA and measuring RNA- and DNA-labeling. The mean responsiveness to PHA (in terms of RNA synthesis) was decreased by 20% in 16 cosmonauts who spent between 122 and 175 days in space, and by 14% in a group of nine cosmonauts who were in orbit between 211 and 366 days. In contrast to these findings the mean response was not decreased in 25 crewmen who flew for 7–10 days. Less evident average changes were detected in all groups when DNA labeling was measured. For those cosmonauts showing depression of lymphocyte responsiveness, recovery occurred within 3–7 days after the 7–10 day flights, whereas recovery took 2–4 wk in the crewmen from longer flights.

Lymphocyte responsiveness following flight was determined by using con A or PHA as mitogens and by measuring the incorporation of ³H-uridine in RNA or ³H-thymidine in DNA (Table 35.1). Depression of lymphocyte responsiveness was found in 72 of 129 (56%) crewmen.

Skin Tests. Cellular immunity can be evaluated with skin tests in which an antigen is applied subcutaneously. The extent of the reaction can be determined by measuring the diameter of the induration on the skin 48 h after application (DTH). Skin tests have been conducted on several Shuttle astronauts (79) and during prolonged spaceflight onboard the orbital space station MIR (40). In three of five cosmonauts tested a decrease below the warning level of DTH was noted during flight (2 crewmen) or following landing (1 cosmonaut). In-flight reductions recovered to normal levels after landing. In one of these two cosmonauts the reduction in DTH was associated with stressful extravehicular activities. Our results are in line with earlier findings of an impaired

TABLE 35.1. *Lymphocyte Responsiveness Following Spaceflight (Listed in Order of Mission Duration)*

Mission	Mission Duration (days)	Total Number of Crew Members/Number with Reduced Response	Average Percentage Depression as Compared to Baseline Levels
Shuttle missions	2–8	41/36	−75%
Apollo	6–12	21/0	0
Spacelab D 1	7	4/3	−30%
Soyuz	7–10	25/0	0
Spacelab SLS 1	10	4/0	0
Skylab	28–84	9/8	−70%
Salyut	122–175	16/16	−20%
Salyut and MIR	211–366	9/9	−14%

lymphocyte function during and after spaceflight using lymphocyte responsiveness to mitogens.

CONCLUSIONS

Our major conclusion is that lymphocyte function is depressed in in vitro cultures under microgravity conditions, and in vivo, in crew members during and following their missions. However, the in vitro and in vivo mechanisms leading to the observed decreases of lymphocyte responsiveness are different. In vitro changes in the physicochemical environment of the cell, and a direct effect of microgravity on cell shape and function, cell-cell and cell-substratum adhesion, and other factors yet to be identified, may contribute to the observed gravity effect. This in vitro mechanism is not well understood. Yet it is very likely that the various stressors of spaceflight lead to the depression of lymphocyte reactivity observed in vivo. It is probable that increased levels of stress-related hormones and peptides in the blood modulate lymphocyte function via membrane receptors and transmembrane and intracellular signal transduction.

Stress models like running also show very similar results to those of spaceflight; but care has to be taken in their interpretation. For instance, long-distance running on the one hand induces a typical stress-related change in T-cell function. On the other hand, prolonged strenuous exercise also activates tissue repair mechanisms and responses directed against damaged muscle tissue as well as the release of endogenous pyrogens and bacterial endotoxins. Microgravity models like HDT provoke responses of the immune system most similar to those resulting from spaceflight.

The present body of knowledge suggests that in the majority of crew members the suspected impairment of immunity may be due to a combination of spaceflight stressors, including microgravity. It is still an open question whether the observed immunological changes will have serious consequences regarding the health of astronauts on long-duration missions on space stations and deep space missions. Testing lymphocyte function by means of responsiveness to mitogens is becoming obsolete. It has been questioned whether this test reveals any clinically meaningful results. Measuring T-lymphocyte reaction in vivo by means of delayed-type skin hypersensitivity is superior for detecting existing impairments in T-cell function under clinical situations, and a high correlation with susceptibility to infection has been established. Delayed-type skin hypersensitivity measurements are presently used on short-term shuttle missions and on long-term missions on MIR. It will be important to continue to measure changes in blood levels of stress-related hormones and peptides during and

following spaceflight, and relate these measurements to changes in immunological parameters. Characteristics of lymphokine and growth factor production by immunocompetent cells, growth factor receptor density and regulation on lymphocyte membranes, and transmembrane and intracellular signal transduction, will become increasingly important in future space immunology investigations.

To the best of our knowledge, astronauts and cosmonauts generally maintained good health both during and following their missions. Although this observation may be somewhat reassuring, possible changes in immunological parameters of astronauts clearly need attention to ensure safe long-duration missions.

SUMMARY

The human lymphocyte is a model cell to study microgravity effects in vitro at the cellular level and one of the key cells for modulating immunity in vivo. In in vitro experiments, lymphocyte responsiveness is reduced under microgravity conditions due to a single or combined action of the following factors: physicochemical environment of the cell, direct effect of microgravity on cell shape and function, cell-cell and cell-substratum adhesion, and other still unidentified factors. In vivo guidence suggests that lymphocyte function is depressed in most crew members by stress-related hormones and immunomodulating lymphokines.

Responses to HDT bed rest are used as analogs to responses to microgravity and provoke changes in immunological parameters strikingly similar to those

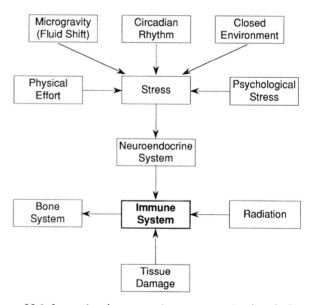

FIG. 35.6. Interactions between various stressors, circadian rhythms, the bone system, and the immune system.

observed during and after spaceflight. Indeed, changes in the levels of cortisol, epinephrine, norepinephrine, leukocyte and lymphocyte counts, numbers of natural killer cells (which parallel natural killer activity based on a per cell activity), and lymphocyte function, all show the same behavior during and after both spaceflight and HDT.

In order to give a synopsis of the effects discussed here, the interactions of the immune system with the most important environmental factors of manned spaceflight are summarized in Figure 35.6.

The authors' work was supported by the Board of the ETH Zürich, the Swiss National Science Foundation (Grants no. 3.382-0.82, 3.338-0.86, and 31-25181.88), the European Space Agency, the PRODEX programme of ESA, Contraves A. G., and Bio-Strath, Inc., Zürich. We especially appreciate the support of all present and former members of the Space Biology Group at ETH-Zürich, including Birgitt Bechler, Marianne Cogoli, Pia Fuchs-Bislin, Juliet Lee, Giovanna Lorenzi, Alex Tschopp, and Myriam Valluchi-Morf.

REFERENCES

1. Adams, R. P. L. Cell culture for biochemists. In: *Laboratory Techniques in Biochemistry and Molecular Biology,* edited by T. S. Work and R. H. Burdon. New York: Elsevier, 1980, p. 181–192.

2. Baisch, F., A. Gaffney, J. M. Karemaker, PH. Arbeille, R. Gerzer, and G. Blomqvist. HDT'88—an international collaborative effort. *Acta Physiol. Scand.* 144 S604: 1–12, 1992.

3. Barone, R. P., and L. D. Caren. The immune system: effects of hypergravity and hypogravity. *Aviat. Space Environ. Med.* 55: 1063–1068, 1984.

4. Barone, R. P., L. D. Caren, and J. Oyama. Effects of deceleration on the humoral antibody response in rats. *Aviat. Space Environ. Med.* 56: 690–694, 1985.

5. Bátkai, L., M. Tálas, I. Stöger, K. Nagy, L. Hiros, I. Konstantinova, M. Rykova, I. Mozgovaya, O. Guseva, and V. Kozharinov. *In vitro* interferon production by human lymphocytes during spaceflight. *Physiologist* 31: S50–S51, 1988.

6. Bechler, B., Cogoli, A., and D. Mesland. Lymphozyten sind schwerkraftempfindlich. *Naturwissenschaften* 73: 400–403, 1986.

7. Beisel, W. R., and J. M. Talbot. The effects of space flight on immunocompetence. *Immunol. Today* 8: 197–200, 1987.

8. Besedowsky, H., and E. Sorkin. Network of immune-neuroendocrine interactions. *Clin. Exp. Immunol.* 27: 1–12, 1977.

9. Blalock, J. E. A molecular basis for bidirectional communication between the immune and neuroendocrine systems. *Physiol. Rev.* 69: 1–32, 1989.

10. Bloom, E. T., M. Akiyama, Y. Kusunoki, and T. Makinodan. Delayed effects of low-dose radiation on cellular immunity in atomic bomb survivors residing in the United States. *Health Phys.* 52: 585–591, 1987.

11. Brovall, C., and B. Schacter. Radiation sensitivity of human natural killer cell activity: Control by X-linked genes. *J. Immunol.* 126: 2236–2239, 1981.

12. Bunt, J. C. Hormonal alterations due to exercise. *Sports Med.* 3: 331–345, 1986.

13. Caren, L. D., A. D. Mandel, and J. A. Nunes. Effect of simulated weightlessness on the immune system in rats. *Aviat. Space Environ. Med.* 51: 251–255, 1980.

14. Cogoli, A. Hematological and immunological changes during space flight. *Acta Astronautica* 8: 995–1002, 1981.

14a. Cogoli, A. The effect of hypogravity and hypergravity on cells of the immune system. *J. Leukocyte Biol.* 54: 253–268, 1993.

15. Cogoli, A., and F. K. Gmünder. Gravity effects on single cells: Techniques, findings, and theory. In: *Advances in Space Biology and Medicine,* edited by S. L. Bonting. Greenwich: JAI Press, 1991, p. 183–248.

16. Cogoli, A., B. Bechler, O. Müller, and E. Hunzinger. Effect of microgravity on lymphocyte activation. In: *Biorack on Spacelab D1,* edited by N. Longdon and V. David. Noordwijk: European Space Agency, ESTEC, 1988, p. 89–100. (ESA SP-1091).

16a. Cogoli, A., B. Bechler, M. Cogoli-Greuter, S. B. Criswell, H. Joller, P. Joller, E. Hunzinger, and O. Müller. Mitogenic signal transduction in T lymphocytes in microgravity. *J. Leukocyte Biol.* 53: 569–575, 1993.

17. Cogoli, A., M. Cogoli, B. Bechler, G. Lorenzi, and F. K. Gmünder. Microgravity and mammalian cells. In: *Microgravity and Mammalian Cells,* edited by T. Duc Guyenne. Noordwijk: European Space Agency, ESTEC, 1989, p. 11–16. (ESA SP-1123).

18. Cogoli, A., and A. Tschopp. Biotechnology in space laboratories. *Adv. Biochem. Eng.* 22: 1–50, 1982.

19. Cogoli, A., and A. Tschopp. Lymphocyte reactivity during spaceflight. *Immunol. Today* 6: 1–4, 1985.

20. Cogoli, A., A. Tschopp, and P. Fuchs-Bislin. Cell sensitivity to gravity. *Science* 225: 228–230, 1984.

21. Cogoli, A., M. Valluchi, J. Reck, M. Müller, W. Briegleb, I. Cordt, and C. Michel. Human lymphocyte activation is depressed at low-g and enhanced at high-g. *Physiologist* 22: S-29–S30, 1979.

22. Cogoli, A., M. Valluchi-Morf, M. Mueller, and W. Briegleb. Effect of hypogravity on human lymphocyte activation. *Aviat. Space Environ. Med.* 51: 29–34, 1980.

23. Cogoli, M., B. Bechler, A. Cogoli, N. Arena, S. Barni, P. Pippia, G. Sechi, N. Valora, and R. Monti. Lymphocytes on sounding rockets. In: *Proc. 4ᵗʰ Eur. Space Agency Symp. Life Sci. Res. in Space,* edited by V. David. Noordwijk: European Space Agency, ESTEC, 1990, p. 229–234. (ESA SP-307)

24. Cohen, S., D. A. J. Tyrell, and A. P. Smith. Psychological stress and susceptibility to the common cold. *N. Engl. J. Med.* 325: 606–612, 1991.

25. Cosman, B. C., and P. W. Brandt-Rauf. Infectious disease in Antarctica and its relation to aerospace medicine: a review. *Aviat. Space Environ. Med.* 58: 174–179, 1987.

26. Crary, B., M. Borysenko, D. C. Sutherland, I. Kutz, J. Z. Borysenko, and H. Benson. Decrease in mitogen responsiveness of mononuclear cells from peripheral blood after epinephrine administration in humans. *J. Immunol.* 130: 694–697, 1983.

27. Criswell, B. S. Cellular immune response. In: *Biospex, Biological Space Experiments,* edited by M. Anderson, J. A. Rummel, and S. Deutsch. Washington D. C.: National Aeronautics and Space Administration, 1979, p. 14. (NASA TM 58217)

28. Dorian, B., and P. E. Garfinkel. Stress, immunity and illness—a review. *Psychol. Med.* 17: 393–407, 1987.

29. Dorian, B., P. Garfinkel, G. Brown, A. Shore, D. Gladman, and E. Keystone. Aberrations in lymphocyte subpopulations and function during psychological stress. *Clin. Exp. Immunol.* 50: 132–138, 1982.

30. Durnova, G. N., A. S. Kaplansky, and V. V. Portugalev. Effect of a 22-day space flight on the lymphoid organs of rats. *Aviat. Space Environ. Med.* 47: 588–591, 1976.

31. Fauci, A. S., and D. C. Dale. The effect of in vivo hydrocortisone on subpopulations of human lymphocytes. *J. Clin. Invest.* 53: 240–246, 1974.

32. Galbo, H. *Hormonal and Metabolic Adaptation to Exercise.* Stuttgart: Georg Thieme, 1983.

33. Galun, E., R. Burstein, E. Assia, I. Tur-Kaspa, J. Rosenblum, and Y. Epstein. Changes of white blood cell count during prolonged exercise. *Int. J. Sports Med.* 8: 253–255, 1987.

34. Garrey, W. E., and W. R. Bryan. Variations in white blood cell counts. *Physiol. Rev.* 15: 597–638, 1935.

35. Gmünder, F. K., B. Bechler, M. Cogoli, P. W. Joller, W. H. Ziegler, J. Müller, and A. Cogoli. Effect of bedrest and head down tilt (10 days) on lymphocyte reactivity. *Acta Physiol. Scand.* 144: 131–141, 1992.

36. Gmünder, F. K., and A. Cogoli. Cultivation of single cells in space. *Appl. Micrograv. Technol.* 1: 115–122, 1988.

37. Gmünder, F. K., P. W. Joller, H. I. Joller-Jemelka, B. Bechler, M. Cogoli, W. H. Ziegler, J. Müller, R. E. Aeppli, and A. Cogoli. Effect of a herbal yeast food supplement and running on immunological parameters. *Br. J. Sports Med.* 24: 103–112, 1990a.

38. Gmünder, F. K., M. Kiess, G. Sonnenfeld, J. Lee, and A. Cogoli. A ground-based model to study effects of weightlessness on lymphocytes. *Biol. Cell* 70: 33–38, 1990.

39. Gmünder, F. K., M. Kiess, G. Sonnenfeld, J. Lee and A. Cogoli. Reduced lymphocyte activation in space: role of cell-substratum interactions. *Adv. Space Res.* 12: (1)55–(1)61, 1992.

40. Gmünder, F. K., I. Konstantinova, A. Cogoli, A. Lesnyak, W. Bogomolov, and A. W. Grachov. Cellular immunity in cosmonauts during long duration spaceflight on board the orbital MIR station. *Aviat. Space Environ. Med.* 65: 419–423, 1994.

41. Gmünder, F. K., G. Lorenzi, B. Bechler, P. Joller, J. Müller, W. H. Ziegler, and A. Cogoli. Effect of long-term physical exercise on lymphocyte reactivity: Similarity to space flight reactions. *Aviat. Space Environ. Med.* 59: 146–151, 1988.

42. Gmünder, F. K., H. Maass, U. Order, B. Bechler, M. Cogoli, P. Joller, J. Transmontano, W. Ziegler, J. Müller, and A. Cogoli. Effect of head down tilt (HDT) on lymphocyte reactivity, β-adrenoceptors and stress-associated hormones. In: *Abstracts of the XXXlth IUPS-Meeting in Helsinki,* edited by L. Hirvonen, J. Tmisjärvi, S. Niiranen and J. Leppäluoto. Oulu: Oy Liitto, 1989, p. 106.

43. Gordon, A. S. Some aspects of hormonal influences upon the leukocytes. *Ann. N. Y. Acad. Sci.* 59: 907–927, 1955.

44. Gould, C. L., M. Lyte, J. Williams, A. D. Mandel, and G. Sonnenfeld. Inhibited interferon-γ but normal interleukin-3 production from rats flown on the space shuttle. *Aviat. Space Environ. Med.* 58: 983–986, 1987.

45. Gualde, N., and J. S. Goodwin. Effect of irradiation on human T-cell proliferation: low dose irradiation stimulates mitogen-induced proliferation and function of the suppressor/cytotoxic T-cell subset. *Cell. Immunol.* 84: 439–445, 1984.

46. Guseva, Y. V., and R. Y. Tashpulatov. Effect of 49-day space flight on parameters of immunological reactivity and protein composition of blood in the crew of Salyut 5. *Space Biol. Aerospace Med.* 13: 1–7, 1979.

47. Guseva, Y. V., and R. Y. Tashpulatov. Effect of flights differing in duration on protein composition of cosmonauts' blood. *Space Biol. Aerospace Med.* 14: 15–20, 1980.

48. Hawkins, W. R., and J. F. Zieglschmid. Clinical aspects of crew health. In: *Biomedical Results of Apollo,* edited by R. S. Johnston, L. F. Dietlein, and C. A. Berry. Washington, D.C.: National Aeronautics and Space Administration, 1975, p. 43–81. (NASA SP-368).

49. Hedfors, E., G. Holm, M. Ivansen, and J. Wahren. Physiological variation of blood lymphocyte reactivity: T-cell subsets, immunoglobulin production, and mixed-lymphocyte reactivity. *Clin. Immunol. Immunopathol.* 27: 9–14, 1983.

50. Jemmott, J. B., M. Borysenko, R. Chapman, J. Z. Borysenko, D. C. McClelland, and D. Meyer. Academic stress, power motivation, and decrease in secretion rate of salivary secretory immunoglobulin A. *Lancet* 1: 1400–1402, 1983.

51. Kimzey, S. L. Hematology and immunology studies. In: *Biomedical Results from Skylab,* edited by R. S. Johnston and L. F. Dietlein. Washington D.C.: National Aeronautics and Space Administration, 1977, p. 249–282. (NASA SP-377).

52. Kimzey, S. L., C. L. Fischer, P. C. Johnson, S. E. Ritzmann, and C. E. Mengel. Hematology and immunology studies. In: *Biomedical Results of Apollo,* edited by R. S. Johnston, L. F. Dietlein, and C. A. Berry. Washington D.C.: National Aeronautics and Space Administration, 1975, p. 197–226. (NASA SP-368)

53. Konstantinova, I. The immune system under extreme conditions: space immunology. In: Problems of Space Biology. Moscow: Nauka, 1988, vol. 59, p. 191–209.

54. Konstantinova, I. V. Immunological research on Salyut-6 prime crews. In: *Results of Medical Research Performed on the Salyut-6-Soyuz Space Station Complex,* edited by N. N. Gurovskiy. Moscow: Nauka, p. 114–124, 1986.

55. Konstantinova, I. V., Y. N. Antropova, V. I. Legenkov, and V. D. Zazhirey. Study of reactivity of blood lymphoid cells in crew members of the Soyuz-6, Soyuz-7 and Soyuz-8 spaceships before and after flight. *Space Biol. Aerospace Med.* 7: 45–55, 1973.

56. Konstantinova I. V., G. Sonnenfeld, A. T. Lesnyak, L. Schaffar, A. Mandel, M. P. Rykova, E. N. Antropova, and B. Ferrua. Cellular immunity and lymphokine production during spaceflights. *Physiologist* 34: S52–S56, 1991.

57. Kvetnansky, R., N. A. Davydova, V. B. Noskov, M. Vicas, I. A. Popova, A. C. Usakov, L. Macho, and A. I. Grigoriev. Plasma and urine catecholamine levels in cosmonauts during long-term stay on space station Salijut-7. *IAF/IAA-86-383*: 1–7, 1986.

58. Leach, C. S. Medical results from STS 1-4: Analysis of body fluids. *Aviat. Space Environ. Med.* 54: S50–S54, 1983.

59. Leach, C. S., and P. C. Rambaut. Biochemical responses of the Skylab crewmen: an overview. In: *Biomedical Results from Skylab,* edited by R. S. Johnston and L. F. Dietlein. Washington D.C.: National Space and Aeronautics Administration, 1977, p. 204–216. (NASA SP-377)

60. Lesniak, A. T., I. V. Konstantinova, N. V. Bodjikov, and P. N. Uchakin. Immunocompetent cells producing humoral mediators of bone tissue. Mineral metabolism during space flight simulation. *Physiologist* 32: S53–S56, 1989.

61. Liu, S. Z., W. H. Liu, and J. B. Sun. Radiation hormesis: Its expression in the immune system. *Health Physics* 52: 579–583, 1987.

62. Lorenzi, G., P. Fuchs-Bislin, and A. Cogoli,. Effects of hypergravity on "whole-blood" cultures of human lymphocytes. *Aviat. Space Environ. Med.* 67: 1131–1135, 1986.

63. Luckey, T. D. *Hormesis with Ionizing Radiation.* Boca Raton: CRC Press, 1980.

64. Maass, H., J. Transmontano, and F. Baisch. Response of adrenergic receptors to 10 days head down tilt bedrest. *Acta Physiol. Scand.* 144: 66–68, 1992.

65. Macho, L., M. Fickova, S. Zorad, R. A. Tigranian, and L. Serova. Effect of space flight and hypokinesia on plasma hormone levels and lipid metabolism in rats. *Physiologist* 31: S61–S62, 1988.

66. Macho, L., R. Kvetnansky, T. Torda, J. Culman, and R. A. Tigranian. Is prolonged space flight of rats a stressful stimulus for the sympatho-adrenal system? In: *Catecholamines and Stress: Recent Advances. Developments in Neurosciences,* edited by E. Usdin, R. Kvetnansky, and I. J. Kopin. New York: Elsevier, 1980, vol. 8, p. 399–409.

67. Mandel, A. D., and E. Balish. Effect of space flight on cell-mediated immunity. *Aviat. Space Environ. Med.* 48: 1051–1057, 1977.

68. Manié, S., I. Konstantinova, J. P. Breittmayer, B. Ferrua, and L. Schaffar. Effects of long duration spaceflight on human T lym-

phocyte and monocyte activity. *Aviat. Space Environ. Med.* 62: 1153–1158, 1991.

69. Morey, E. R. Spaceflight and bone turnover: correlation with a new rat model of weightlessness. *BioScience* 29: 168–172, 1979.

70. Mueller, O., E. Hunzinger, A. Cogoli, B. Bechler, J. Lee, J. Moore, and J. Duke. Increased mitogenic response in lymphocytes from chronically centrifuged mice. In: *Proc. 4th Eur. Space Agency Symp. Life Sci. Res. in Space,* edited by V. David. Noordwijk: European Space Agency, ESTEC, 1990, p. 301–305. (ESA SP-307)

71. Nieman, D. C., and S. L. Nehlsen-Cannarella. Effects of endurance exercise on the immune response. In: *Endurnace in Sport,* edited by R. J. Shephard and P.-O. Åstrand. London: Blackwell, 1992, p. 487–504.

72. Oyama, J. Effects of altered gravitational fields on laboratory animals. In: *Progress in Animal Biometeorology,* edited by S. W. Tromp, J. J. Bouma, and H. D. Johnson. Amsterdam: C. B. Swets and Zeitlinger, vol. 1, 1976, p. 793–808.

73. Sagan, L. A. What is hormesis and why haven't we heard about it before? *Health Phys.* 52: 521–525, 1987.

74. Scibetta, S. M., L. D. Caren, and J. Oyama. The unresponsiveness of the immune system of the rat to hypergravity. *Aviat. Space Environ. Med.* 55: 1004–1009, 1984.

75. Sonnenfeld, G., C. L. Gould, J. Williams, and A. D. Mandel. Inhibited interferon production after space flight. *Acta Microbiol. Hung.* 35: 411–416, 1988.

76. Tálas, M., L. Bátkai, I. Stöger, K. Nagy, L. Hiros, I. Konstantinova, M. Rykova, I. Mozgovaya, O. Guseva, and V. Kozharinov. Results of space experiment program "Interferon." *Acta Microbiol. Hung.* 30: 53–61, 1983.

77. Taylor, G. R. Cell anomalies associated with spaceflight conditions. *Adv. Exp. Med. Biol.* 225: 259–271, 1987.

78. Taylor, G. R., and J. R. Dardano. Human cellular immune responsiveness following space flight. *Aviat. Space Environ. Med.* 54: S55–S59, 1983.

79. Taylor, G. R., and R. P. Janney. In vivo testing confirms a blunting of the human cell-mediated immune mechanisms during space flight. *J. Leukocyte Biol.* 51: 129–132, 1992.

80. Taylor, G. R., L. S. Neale, and J. R. Dardano. Immunological analyses of U.S. space shuttle crewmembers. *Aviat. Space Environ. Med.* 57: 213–217, 1986.

81. Tigranian, R. A., R. Kvetnansky, N. F. Kaita, N. A. Davydova, E. A. Pavlova, and L. I. Voronin. Effect of space flight stress factors on the activity of the sympatic adrenomedullary and the pituitary-adrenocortical system. In: *Catecholamines and Stress: Recent Advances. Developments in Neurosciences,* edited by E. Usdin, R. Kvetnansky, and I. J. Kopin. New York: Elsevier, 1980, vol. 8, p. 409–460.

82. Tvede, N., B. K. Pedersen, F. R. Hansen, T. Bendix, L. D. Christensen, H. Galbo, and J. Halkjær-Kristensen. Effect of physical exercise on blood mononuclear cell subpopulations and in vitro proliferative responses. *Scand. J. Immunol.* 29: 383–389, 1989.

83. Vorobyev, Y. I., O. G. Gazenko, A. M. Genin, N. N. Gurovskiy, A. D. Egorov, and Y. G. Nefedov. Main results of medical studies on Salyut-6-Soyuz program. *Space Biol. Med.* 18: 22–25, 1984.

84. Voss Jr., E. W. Prolonged weightlessness and humoral immunity. *Science* 225: 214–215, 1984.

85. Wiese, C., B. Bechler, F. K. Gmünder, G. Lorenzi, and A. Cogoli. Life science experiments on stratospheric balloons and sounding rockets. In: *Proc. 8th Eur. Space Agency Symp. European Rocket and Balloon Programmes and Related Research,* edited by T. D. Guyenne and J. J. Hunt. Noordwijk: European Space Agency, ESTEC, 1987, p. 371–376. (ESA SP-270)

86. Wiese, C., B. Bechler, G. Lorenzi, and A. Cogoli. Cultures of erythroleukemic cells (K-562) on a stratospheric balloon flight. In: *Terrestrial Space Radiation and Its Biological Effects,* edited by P. D. McCormack, C. E. Swenberg, and H. Bücker. New York: Plenum, 1988, p. 337–343.

36. Exercise and adaptation to microgravity environments

VICTOR A. CONVERTINO

Physiology Research Branch, Clinical Sciences Division, Brooks Air Force Base, Texas

CHAPTER CONTENTS

EXERCISE IS A PHYSICAL CHALLENGE that requires acute systemic responses to sustain the physiological demand of muscular work in normal terrestrial gravity. With regular repeated stimulation from acute exercise in the terrestrial environment, humans can adapt by increasing their physiologic reserves to meet the demands of daily physical activities. In contrast, microgravity environments require a new adaptation level imposed by reduction in hydrostatic pressure gradients within the cardiovascular system, less weight load upon the muscles and bones, and possibly a lower energy requirement and/or output. Travel into space where gravitational forces are lower than on earth has provided opportunities to examine human responses to exercise as a unique way of assessing the influence of gravity on the development and adaptation of our normal physical and physiologic functions.

Exercise is widely used to evaluate physiological adaptation to microgravity. However, limitations in conducting many physiological experiments utilizing exercise in spaceflight have made the interpretation of results difficult; for example, tests are not standardized, there is little control over baseline data, and data are collected from only two or three crew members. These limitations have mandated an emphasis on the development of ground-based experimental programs.

Confinement to bed rest is a valuable analog for assessing the effects of microgravity by inducing similar physical and physiological changes to those that occur when humans are exposed to actual spaceflight. The 6° head-down bed rest model is one of the most effective analogs for spaceflight microgravity [(101); Table 36.1]. Despite the extreme variability in experimental conditions and mission activities associated with spaceflight, the quantitative as well as qualitative comparisons with this bed rest analog are striking. On this premise, it is the purpose of this chapter to present and integrate the results from both spaceflight and ground-based experiments in an attempt to assess the interactions between exercise adaptation and adaptation to microgravity.

EXERCISE PERFORMANCE AFTER ADAPTATION TO MICROGRAVITY

Maximal Oxygen Uptake

Maximal oxygen uptake ($\dot{V}O_{2max}$) provides one integrated index of the systemic changes in cardiorespiratory and muscular functions induced by adaptation to microgravity. Unfortunately, there are no results from documented spaceflight experiments in which $\dot{V}O_{2max}$ has been measured after return to earth, especially in the absence of remedial procedures applied during the space mission. Understanding the isolated impact of microgravity on $\dot{V}O_{2max}$ and maximal work capacity has therefore depended upon ground-based investigations.

Reduced $\dot{V}O_{2max}$ is a common finding from bed rest studies conducted without application of remedial countermeasures (Table 36.2). From five independent studies, 10 days of bed rest reduced supine $\dot{V}O_{2max}$ by

TABLE 36.1. *Comparison of Changes Reported during Rest and Exercise Following Spaceflight and 6° Head-Down Bed Rest (HDBR)*

Physiological Variable	Reference	Microgravity Analog	Days of Exposure	N	Condition	%Δ
Body weight	Thornton et al. (163)	Spaceflight	28	3	Rest	3% ↓
Body weight	Convertino et al. (37)	6° HDBR	30	8	Rest	4% ↓
Venous pressure	Kirsch et al. (108)	Spaceflight	7	2	Rest	58% ↓
Venous pressure	Convertino (31)	6° HDBR	7	11	Rest	32% ↓
Baroreflex	Fritsch et al. (66)	Spaceflight	4–5	16	Rest	28% ↓
Baroreflex	Convertino et al. (35)	6° HDBR	12	11	Rest	31% ↓
Plasma volume	Fischer et al. (63)	Spaceflight	4	2	Rest	9% ↓
Plasma volume	Convertino et al. (35)	6° HDBR	3	11	Rest	12% ↓
Leg volume	Thornton et al. (163)	Spaceflight	28	3	Rest	10% ↓
Leg volume	Convertino et al. (37)	6° HDBR	30	8	Rest	10% ↓
Knee flexors	Thornton and Rummel (164)	Spaceflight	28	3	Maximum torque	8% ↓
Knee flexors	Convertino (29)	6° HDBR	30	8	Maximum torque	10% ↓
Knee extensors	Thornton and Rummel (164)	Spaceflight	28	3	Maximum torque	21% ↓
Knee extensors	Convertino (29)	6° HDBR	30	8	Maximum torque	21% ↓
Ankle flexors	Grigoryeva and Kozlovskaya (81)	Spaceflight	7	12	Maximum torque	25% ↓
Ankle flexors	Kozlovskaya et al. (109)	6° HDBR	14	18	Maximum torque	9% ↓
Ankle extensors	Grigoryeva and Kozlovskaya (81)	Spaceflight	7	12	Maximum torque	22% ↓
Ankle extensors	Kozlovskaya et al. (109)	6° HDBR	14	18	Maximum torque	16% ↓
Oxygen uptake	Kakurin et al. (101)	Spaceflight	5	7	700 kgm/min	4% ↑
Oxygen uptake	Convertino et al. (34)	6° HDBR	7	5	600 kgm/min	5% ↑
Heart rate	Kakurin et al. (101)	Spaceflight	5	7	700 kgm/min	13% ↑
Heart rate	Convertino et al. (34)	6° HDBR	7	5	600 kgm/min	12% ↑
Oxygen pulse	Kakurin et al. (101)	Spaceflight	5	7	700 kgm/min	9% ↓
Oxygen pulse	Convertino et al. (34)	6° HDBR	7	5	600 kgm/min	8% ↓
Stroke volume	Atkov et al. (1)	Spaceflight	237	2	765 kgm/min	31% ↓
Stroke volume	Hung et al. (89)	0° HDBR	10	12	835 kgm/min	28% ↓
Ejection fraction	Atkov et al. (1)	Spaceflight	237	2	765 kgm/min	13% ↑
Ejection fraction	Hung et al. (89)	0° HDBR	10	12	825 kgm/min	20% ↑
Heart rate	Bungo et al. (15)	Spaceflight	7	9	Standing	47% ↑
Heart rate	Convertino et al. (35)	6° HDBR	30	10	Standing	55% ↑

5.2% to 8.7%, with an average reduction of 7.1% ± 0.5% (25, 42, 43, 45, 173). As the duration of bed rest increased, $\dot{V}O_{2max}$ declined so that after 20 days $\dot{V}O_{2max}$ decreased by 12.9% (97) and after 26 days $\dot{V}O_{2max}$ decreased by 19.5% (153). Subjects exposed to 28 days of bed rest decreased $\dot{V}O_{2max}$ by 18.3% (73) and by 18.9% (124), and 30 days of confinement in a space cabin simulator reduced $\dot{V}O_{2max}$ by 20.4% (113). Decrements in mean $\dot{V}O_{2max}$ of 26.4% and 34.6% have been reported following 20 and 30 days of bed rest, respectively (105, 141). Data collected from 214 men and women in 19 independent studies support the notion that reduction of $\dot{V}O_{2max}$ is a natural adaptation to microgravity environments in the absence of preventive or corrective countermeasures (Table 36.2).

Exercise Endurance

Both spaceflight and ground-based experiments designed to determine the effects of adaptation to microgravity on exercise endurance are limited. During the early part of a 96 day flight aboard the Soviet Salyut-6 space station, poorer endurance of a standardized exercise test (125 watts [W] for 5 min) was reported in the two crew members (70). This response was presumed to be caused by physical deconditioning which was manifested by increased heart rate, elevated arterial pressure, reduced stroke volume to lesser work intensity, and an inability to complete the 5 min exercise bout on the 24th day of flight.

Results from numerous ground-based investigations have provided quantitative evidence of reduced endurance by demonstrating that less time was required to reach volitional exhaustion during graded exercise after as compared to before bed rest (25, 42, 43, 45, 49, 87, 92, 105, 106, 113, 127, 154). However, demonstration of the inability of subjects to complete prolonged submaximal work tasks following microgravity is limited since all experiments had exercise durations of no more than 5–10 min.

Oxygen Uptake and Mechanical Efficiency during Submaximal Exercise

Oxygen uptake during 5 min of exercise in spaceflight was slightly but consistently less in all nine astronauts

TABLE 36.2. *Mean Changes in Maximal Oxygen Uptake during Bed Rest without Remedial Procedures* *

References	Days	Bed Rest Model	N	Subjects Age	Sex	Exercise Test	$\dot{V}_{O_{2max}}$, l/min Pre	Post	%Δ
Friman (65)	7	0° BR	22	25 ± 1	M	CE UP	3.30	3.11	−5.8
White et al. (173)	10	0° BR	3	21–26	M	TM UP	3.66	3.47	−5.2
Convertino et al. (43)	10	0° BR	12	45–55	M	CE SUP	2.03	1.91	−5.9
Convertino et al. (43)	10	0° BR	12	45–55	M	CE SUP	2.15	1.81	−15.8
Convertino et al. (42)	10	0° BR	15	45–65	M	CE SUP	2.74	2.52	−8.0
Convertino et al. (42)	10	0° BR	17	45–65	F	CE SUP	1.61	1.49	−7.5
Convertino et al. (45)	10	−6° BR	10	36–51	M	CE SUP	2.42	2.25	−7.0
Convertino (25)	10	−6° BR	10	36–48	M	CE SUP	2.52	2.30	−8.7
Georgievskiy et al. (69)	13	0° BR	4	22–25	M	CE SUP	3.14	2.87	−8.6
Lamb et al. (113)	14	0° BR	8	24–34	M	TM UP	2.43	2.26	−7.0
Convertino et al. (49)	14	0° BR	15	19–23	M	CE SUP	3.52	3.20	−9.1
Stremel et al. (154)	14	0° BR	7	19–22	M	CE SUP	3.83	3.36	−12.3
Chase et al. (19)	15	0° BR	4	21–24	M	CE UP	3.14	3.13	−0.3
Convertino et al. (48)	15	0° BR	4	20–26	M	CE UP	3.86	3.32	−14.0
Convertino et al. (49)	17	0° BR	8	23–34	F	CE SUP	2.06	1.86	−9.7
Saltin et al. (141)	20	0° BR	5	19–21	M	TM UP	3.30	2.43	−26.4
Kakurin et al. (97)	20	0° BR	4	22–24	M	CE UP	3.10	2.70	−12.9
Stevens et al. (153)	26	0° BR	22	18–23	M	TM UP	2.58	2.04	−20.9
Taylor et al. (157)	28	0° BR	2		M		3.85	3.16	−17.8
Meehan et al. (124)	28	0° BR	14	19–24	M	CE UP	3.75	3.04	−18.9
Greenleaf et al. (73)	28	−6° BR	5	32–42	M	CE SUP	3.27	2.60	−20.5
Lamb et al. (113)	30	0° BR	8	17–24	M	TM UP	2.66	2.28	−14.3
Katkovskiy et al. (105)	30	−6° BR	3	24–29	M	CE UP	3.09	2.02	−34.6
Mean							2.96	2.57	−13.2

*BR, bed rest; CE, cycle ergometer; TM, treadmill; UP, upright position; SUP, supine position.

of the U.S. Skylab missions at the same absolute work rate on the cycle ergometer (125, 126, 138). Consistent with this finding are the metabolic responses during supine cycle exercise at 100 W for 7 min performed preflight and 3 to 7 days after 30, 64, and 96 days of spaceflight (Table 36.3), demonstrating that postflight exercise $\dot{V}_{O_2}$ lagged appreciably behind the preflight levels (4, 5). Also, greater $\dot{V}_{O_2}$ during recovery from exercise was reported in cosmonauts after these and other flights (5, 170) and in the three Skylab-4 astronauts (126). These observations verified those from ground-based experiments that demonstrated consistently lower average $\dot{V}_{O_2}$ at equal submaximal work levels after bed rest (43, 49, 154). One explanation for reduced submaximal $\dot{V}_{O_2}$ at equal work output might be increased mechanical efficiency, although this seems unlikely (see below). An alternative hypothesis might be that adaptation to microgravity involves a change in the time constant for the $\dot{V}_{O_2}$ to reach an equilibrium. If the rate change for $\dot{V}_{O_2}$ during the transient phase of exercise were lengthened with microgravity adaptation, then the measured $\dot{V}_{O_2}$ at 3–5 min of exercise may not reach steady state and $\dot{V}_{O_2}$ would be lower compared to pre-exposure levels without a change in mechanical efficiency. This notion is supported by slower $\dot{V}_{O_2}$ kinetics

during the transient phase of exercise after bed rest (Fig. 36.1) (41, 149), and by greater recovery oxygen uptake following exercise in spaceflight (103, 126) and bed rest (41). These changes in $\dot{V}_{O_2}$ dynamics during and after exercise following exposure to microgravity may reflect a greater requirement for anaerobic metabolism during the transient phase of exercise to provide for adequate energy demand. This notion is further supported by higher blood lactate, ventilation, and respiratory exchange ratios following ground-based simulations of microgravity (42, 141, 174).

When exercise metabolism reaches steady state, the energy requirement during post–bed rest cycle ergometry at a constant work output is equal to the pre–bed rest level (3, 34, 39, 40, 46, 141, 174), suggesting that mechanical efficiency does not change. These ground-based results have been verified by spaceflight data. A standardized exercise bout at an intensity of 150 W (2.15 kcal·min⁻¹) was performed on a cycle ergometer by Skylab astronauts before and during spaceflight (126). The average energy expenditure required to perform this work was 10.2 ± .1 kcal·min⁻¹ preflight compared to 9.3 ± .1 kcal·min⁻¹ in flight. Therefore, the mechanical efficiency of performing cycle ergometer exercise during flight (23%) was not appreciably differ-

TABLE 36.3. *Cardiac and Hemodynamic Responses to a Standardized Exercise (600 kgm/min for 3 min) Before and After Spaceflight†*

Variables	30 Day Spaceflight		63 Day Spaceflight		96 Day Spaceflight	
	Rest	Exercise	Rest	Exercise	Rest	Exercise
Heart rate, bpm						
Preflight	60	112	69	120	67	112
Postflight	67*	131*	76*	136*	73*	119*
Oxygen uptake, ml/min						
Preflight	270	1538	268	1469	246	1326
Postflight	283*	1498*	293*	1428*	259*	1272*
Systolic blood pressure, mm Hg						
Preflight	129	138	125	132	137	155
Postflight	140*	158*	127	143*	130	173*
Diastolic blood pressure, mm Hg						
Preflight	71	80	73	76	73	93
Postflight	73	77	70	73	78	85
Stroke volume, ml						
Preflight	98	142	92	133	87	149
Postflight	86*	96*	85*	116*	72*	131*
Cardiac output, l/min						
Preflight	5.9	15.9	6.3	16	5.7	16.6
Postflight	5.8	12.6*	6.4	15.7*	5.2*	15.6*
Rate pressure product, bpm-mm Hg						
Preflight	7,740	15,456	8,625	15,840	9,179	17,360
Postflight	9,380*	20,698*	9,652*	19,448*	9,490*	20,587*

†Values are mean ±SE (N = 2). Asterisk indicates that all subjects changed in the same direction compared to preflight values. Thirty and 63 day data, taken 3 days postflight, are derived from Beregovkin et al. (5), and 96 day data, taken 7 days postflight, are derived from Beregovkin et al. (4). [Table modified from Convertino (26) with permission.]

ent from 1G (21%). Average data for daily exercise levels, energy expenditures, and mechanical efficiency during cycle ergometry from three Skylab-4 astronauts (126) and from two cosmonauts on the 175 day Soviet Salyut-6 mission [146] are presented in Table 36.4. The Soviet data are consistent with the Skylab findings. The $\dot{V}O_2$ predicted during cycle exercise in flight (160 W) was 2.25 liters·min⁻¹ compared to the actual measured

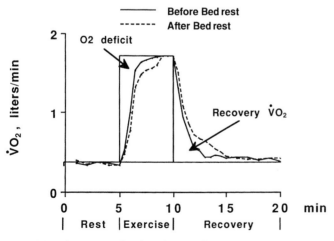

FIG. 36.1. Oxygen uptake ($\dot{V}O_2$) kinetics during constant-load exercise (115 W) before *(solid line)* and after *(broken line)* bed rest. [Modified from Convertino et al. (41) with permission.]

$\dot{V}O_2$ of 2.31 liters·min⁻¹ (approximately 20% mechanical efficiency). Since the preflight and in-flight basal metabolic rates are similar (126), gross efficiency of performing exercise on a mechanically stabilized device such as the cycle ergometer is unaltered by microgravity.

The energy cost of locomotion in microgravity may be much higher when the body cannot be stabilized by postural muscles as it is in terrestrial gravity. This hypothesis was tested by comparing the average work rate and metabolic cost of exercise on the cycle to that on the treadmill during spaceflight (146). Exercise was performed on the treadmill at a speed of 120 m·min⁻¹ (about 4.5 mph) and a system of bungee cords provided stabilization of the subject on the treadmill; the force on the long axis of the body was equivalent to 50 kg (68, 168). Under these conditions, the energy expenditure of treadmill walking/running in Earth gravity would be approximately 5.7 kcal·min⁻¹. However, the measured energy expenditure for treadmill exercise was 7.4 kcal·min⁻¹. This results in a reduction in mechanical efficiency from a predicted 20% in 1G to an actual 15% in microgravity (Table 36.4). These findings were corroborated by the responses to cycle and treadmill exercise during a 1 yr space mission conducted on the Soviet Mir space station. During each in-flight exercise session, a cosmonaut exercise for about 30 min on a treadmill at an average speed of 125 m·min⁻¹ (combined walking and running) and for 26.5 min on a cycle ergometer at

TABLE 36.4. *Estimated Energy Expenditures and Mechanical Efficiencies during Exercise on 175 Day (N = 2)*
and 84 Day (N = 3) Spaceflights*

Variables	175 Day Flight Cycle Ergometer	175 Day Flight Treadmill	84 Day Flight Cycle Ergometer
Average work rate	975 kgm/min	120 m/min	918 kgm/min
Average rate of energy expenditure	11.6 kcal/min	7.4 kcal/min	9.4 kcal/min
Actual oxygen uptake	2.31 liters/min	1.50 liters/min	1.87 liters/min
Predicted oxygen uptake	2.25 liters/min	1.10 liters/min	2.00 liters/min
Estimated mechanical efficiency	20%	15%	23%

*Values represent daily averages. 175 day data are derived from Siminov and Kasyan (146) and 84 day data are derived from Michel et al. (126). [Table is modified from Convertino (26) with permission.]

130 W. Based on prediction equations from 1G, this subject would have required an energy expenditure of approximately 570 kcal during treadmill exercise and 500 kcal during cycling. Thus the total predicted energy expenditure per exercise session would be 1070 kcal (20% mechanical efficiency). However, an average energy cost of 1460 kcal was actually reported, indicating a reduction in mechanical efficiency of the total exercise session to 15% (26). If we assume that the mechanical efficiency of cycle ergometry was not altered from 1G as previously indicated (126, 146), then it can be estimated that 960 kcal were required to perform treadmill exercise rather than the predicted 570 kcal, a reduction in mechanical efficiency from 20% to 11%. It therefore appears that some mechanical advantage of using gravity during exercise with the lower extremities is lost in a microgravity environment. These data should be considered in the design of exercise countermeasures (see below).

Time Course of $\dot{V}O_{2max}$ Change in Microgravity

Duration of exposure to microgravity influences the change in $\dot{V}O_{2max}$. Data from 19 independent investigations suggest that the relative (percent) reduction in $\dot{V}O_{2max}$ is a function of the duration of exposure to microgravity (Fig. 36.2). However, when serial measurements were conducted in the same subjects over 20–30 days of bed rest, $\dot{V}O_{2max}$ decreased rapidly during the initial 3 to 7 days and then showed a more gradual decline (Fig. 36.3). In addition, 6 h of water immersion produced a 10% reduction in $\dot{V}O_{2max}$ in untrained subjects (151) similar to that observed following 10–17 days of bed rest (Table 36.2). These data suggest that the physiological mechanisms associated with the reduction in systemic aerobic capacity involve both fast and slow adapting systems.

Age and Gender

Small sample size, particularly with female astronauts, and a relatively homogeneous age group have limited

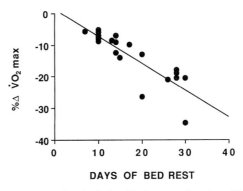

FIG. 36.2. Cross-sectional relationship between duration of bed rest and percent change (%Δ) in $\dot{V}O_{2max}$. Compilation of data from 19 independent investigations (see Table 36.2). The linear regression of best fit is %Δ $\dot{V}O_{2max}$ = −0.85 [Days] + 1.4, r = 0.730.

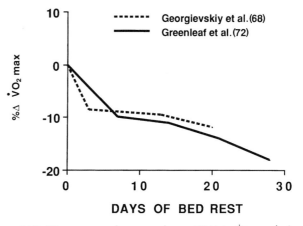

FIG. 36.3. Time course of percent change (%Δ) in $\dot{V}O_{2max}$ during adaptation to simulated microgravity (bed rest) from two independent studies (69, 73).

the use of spaceflight data to evaluate age and gender factors on the exercise response to microgravity. Two studies were designed to investigate responses to exercise before and after exposure to bed rest between men and women of varying age groups. Changes in $\dot{V}O_{2max}$ of young men (19–23 yr) and women (23–34 yr) following 14 and 17 days of continuous bed rest, respectively, were compared in one study (49). In a subsequent series of experiments (41), $\dot{V}O_{2max}$ of middle-aged men and women (45–65 yr) were determined before and after 10 days of bed rest. Despite the significant difference in baseline $\dot{V}O_{2max}$ between the men and women and young and middle-aged subjects, the relative (percent) changes in $\dot{V}O_{2max}$ resulting from exposure to simulated microgravity were similar; $\dot{V}O_{2max}$, measured by supine cycle ergometry, decreased by 9% in the young men, 10% in the young women, 8% in the middle-aged men, and 8% in the middle-aged women. These changes in $\dot{V}O_{2max}$ are similar to those reported from previous bed rest studies in young males (69, 114, 173) and middle-aged men (43). The corresponding changes in maximal work rate and maximal exercise duration, both indices of functional work capacity, were also similar among the groups. Despite significant differences in the absolute $\dot{V}O_{2max}$ between men and women of varying ages, the degree of adaptation to bed rest is relatively equal across age and gender. However, the slope of the regression line between initial $\dot{V}O_{2max}$ and post–bed rest change in $\dot{V}O_{2max}$ was significantly steeper in the younger subjects than in the older subjects (42), suggesting that the rate of absolute reduction in aerobic power is greater in younger subjects. The less marked reduction in $\dot{V}O_{2max}$ in absolute terms for older subjects must be partly explained by an effect of aging; however, middle-aged persons appear to have the same relative ability as younger subjects to reduce their aerobic power in adaptation to microgravity.

Another factor to consider in the comparison of men and women in regard to their exercise performance is the possible effect of the menstrual cycle. During a bed rest study with young female subjects (49), all but two had their menstrual periods before or after the bed rest periods and there appeared to be no consistent effect of menstruation on the $\dot{V}O_{2max}$ response. However, the effect of adaptation to microgravity on menses and exercise performance cannot be assessed adequately from the present data.

Level of Aerobic Fitness

A large $\dot{V}O_{2max}$ reserve may be associated with greater loss of that reserve during adaptation to microgravity compared to that lost in a more sedentary individual with small $\dot{V}O_{2max}$ reserve (141). This notion was supported by the early observation that a 22% reduction in $\dot{V}O_{2max}$ was measured following 28 days of bed rest in a subject with initial $\dot{V}O_{2max} = 4.15$ liters·min^{-1}, compared to a 13% reduction observed in a subject with initial $\dot{V}O_{2max} = 3.54$ liters·min^{-1} (157). This finding was further corroborated by the observation that two subjects with greater aerobic capacity showed

greater relative loss of their $\dot{V}O_{2max}$ than three less fit subjects after 20 days of bed rest (141). When four enduranced-trained runners underwent 6 h of water immersion, their $\dot{V}O_{2max}$ was reduced by 19% compared to only 10% in four untrained subjects (151). Although some analyses have not supported this relationship (76, 77), general findings are consistent with the concept that individuals with high aerobic capacity have greater reduction in $\dot{V}O_{2max}$ during adaptation to microgravity than less fit individuals.

In a study where the variable influence of exercise in the upright posture was controlled with the use of supine cycle ergometry (42), 15 young (19–23 yr) and 15 middle-aged (45–65 yr) men performed similar graded $\dot{V}O_{2max}$ tests before and after bed rest. In both groups, there were significant negative correlations (-0.78 and -0.84, respectively) between the initial pre–bed rest $\dot{V}O_{2max}$ and the percent change in $\dot{V}O_{2max}$. These findings have been corroborated by a subsequent study (44) in which the change in $\dot{V}O_{2max}$ was measured before and after 10 days of bed rest in ten moderately fit subjects ($\dot{V}O_{2max} = 48.5 \pm 1.9$ ml·kg^{-1}·min^{-1}) compared to ten less fit subjects ($\dot{V}O_{2max} = 38.2 \pm 1.8$ ml·kg^{-1}·min^{-1}). The percent reduction in blood volume and $\dot{V}O_{2max}$ was 16% in highly fit subjects compared to 6% in less fit subjects (Fig. 36.4). These results suggest that the relative rate of reduction in aerobic capacity is associated with the initial $\dot{V}O_{2max}$ reserve, and that levels of physical conditioning may influence the degree of adaptation that occurs with prolonged exposure to microgravity. Although several mechanisms may contribute to these results, it is clear that the greater reduction in $\dot{V}O_{2max}$ in fit individuals may be attributed to the association between their larger decreases in blood volume and left ventricular filling (44, 143).

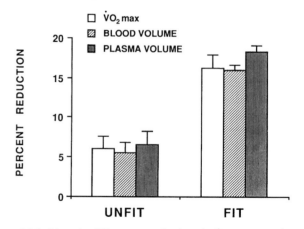

FIG. 36.4. Mean ($\pm$ SE) percent reductions in $\dot{V}O_{2max}$ *(open bars)*, blood volume *(hashed bars)*, and plasma volume *(shaded bars)* after 10 days of bed rest in fit and unfit subjects. [Data from Convertino et al. (44)].

Muscle Function: Strength and Fatigability

The ability to develop and maintain forces with dynamic muscle actions is required for nominal exercise performance. Since skeletal muscles provide the force for moving the body and external objects against Earth's gravity, the absence of gravity removes a major stimulus to maintain normal strength and endurance in microgravity environments. Although this may not be detrimental to exercise performance in microgravity, it could significantly limit one's work output upon return to terrestrial gravity. Indeed, distances of long and high jumps were reduced by an average of 11% and 14%, respectively, after 63 days of spaceflight and were associated with about 10% reduction in force characteristics (20). General loss of strength following spaceflight has been measured in postural muscles of the back (98, 99), knee flexors and extensors (164), elbow flexors and extensors (164), and ankle flexors and extensors (81).

The average decrease in angle-specific peak torque across speeds of concentric muscle actions in 11 subjects was 21% for knee extensors and 10% for knee flexors of the dominant leg following 30 days of 6° head-down bed rest (13, 29, 61). Average body weight loss was 2.5 kg (−3%) and average leg volume was reduced by 10% (37). These changes in body weight, leg volume, and muscle function following bed rest compare favorably with the average changes in body weights, leg volumes, and concentric peak torque development in the same muscle groups reported in the three Skylab astronauts (26, 164) who were in space for a similar duration of 28 days (Table 36.1). The general conclusion is that microgravity reduces muscle strength, primarily in the legs, and the magnitude of strength loss is directly associated with body weight and leg volume reduction, and inversely related to volume (duration × intensity × frequency) of exercise performed in flight (164).

Force–velocity relationships of the ankle flexors (anterior tibialis) and extensors (calf muscles) were measured using isokinetic dynamometry before and after short (7 days) and long (110–237 days) space missions (81). Static and dynamic strength of these muscle groups were reduced during both durations of exposure. The change in the in vivo torque-velocity relationship of bed rest subjects (Fig. 36.5B) demonstrated a reduction in force development across all speeds of limb movement during concentric muscle actions and was qualitatively similar to those changes reported in ankle flexor and extensor muscles following spaceflight (Fig. 36.5A). These bed rest data were the first to demonstrate strength loss for eccentric as well as concentric muscle actions and that the reduction of angle-specific peak torque was not significantly influenced by the type or speed of muscle action.

Data from 17 independent ground-based experiments

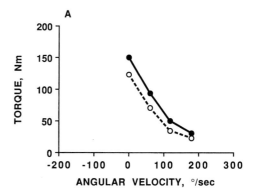

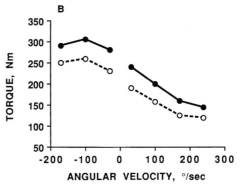

FIG. 36.5. Average torque-velocity relationships before *(closed circles and solid lines)* and after *(open circles and broken lines)* exposure of the calf muscles of 12 subjects to 7 days of spaceflight *(Panel A)* and the knee extensors of 7 subjects to 30 days of 6° head-down bed rest *(Panel B)*. [Modified from Grigoryeva and Kozlovskaya (81) *(Panel A)* and Dudley et al. (61) *(Panel B)* with permission.]

regarding changes in strength of muscle groups during bed rest (Table 36.5) demonstrate that mean reductions in handgrip (−8% ± 3%) and arm strength (−6% ± 5%) were only about one-third as great as the strength losses in the trunk (−22% ± 10%) and leg muscle groups (−20% ± 3%) during exposure periods ranging from 1 to 120 days. These findings have been verified by spaceflight data where the greatest loss of muscular function occurs in the lower extremities with little change in the upper extremities (26, 164). The greater use of arms than legs in microgravity, and the possibility that arm muscles generating relatively small forces on earth are less affected by the absence of gravity than are antigravity muscles of the lower extremities that generate large forces in 1G, could account for smaller strength loss postflight in muscles of the upper as compared to the lower extremities.

Although measurement of strength before and after exposure to bed rest has involved various techniques and produced limited data, some tentative conclusions can be made. Analysis of cross-sectional data suggests that strength in both small-muscle (arm, forearm, hand) and large-muscle (trunk, thigh, leg) groups does not substantially decrease until after 7 to 14 days of bed rest

TABLE 36.5. *Mean Percent Changes in Maximal Strength of Various Muscle Groups after Bed Rest Without Remedial Procedures*

References	Days of Bed Rest	N	Muscle Group	%Δ Strength
Hargens et al. (83)	1	5	Ankle extensors	4
Friden et al. (64)	7	14	Handgrip	−5
			Knee extensors	−5
			Ankle extensors	−7
Trimble and Lessard (167)	7	8	Handgrip	0
Kozlovskaya et al. (109)	14	18	Ankle flexors	−9
			Ankle extensors	−16
Greenleaf et al. (79)	14	7	Handgrip	0
Taylor et al. (157)	21	6	Handgrip	−3
			back	−8
Meehan et al. (124)	28	14	Elbow flexors	9
			Abdomen	5
Dudley et al. (61)	30	11	Knee flexors	−10
			Knee extensors	−20
Gogia et al. (72)	35	15	Elbow flexors	−7
			Elbow extensors	2
			Ankle flexors	−8
			Ankle extensors	−25
			Knee flexors	−8
			Knee extensors	−19
LeBlanc et al. (119)	35	15	Ankle flexors	−10
			Ankle extensors	−26
Birkhead et al. (8)	42	4	Elbow flexors	−5
Deitrick et al. (58)	42–49	4	Handgrip	0
			Elbow flexors	−9
			Ankle flexors	−13
			Ankle extensors	−21
Kakurin et al. (100)	62	3	Back	−19
Yeremin et al. (177)	70	1	Elbow flexors	−28
			Back	−39
			Abdomen	−48
			Knee extensors	−36
			Ankle flexors	−57
			Ankle extensors	−37
Panov and Lobzin (131)	11	1	Handgrip	0
	22		Handgrip	−2
	36		Handgrip	−8
	44		Handgrip	−12
	64		Handgrip	−27
Krupina et al. (110)	95	10	Handgrip	−27
Grigoryeva and Kozlovskaya (81)	120	14	Ankle flexors	−39
			Ankle extensors	−34

(Fig. 36.6). Regression of cross-sectional changes in maximal muscle force development over 120 days of bedrest suggests loss of strength in all muscle groups at a rate of 0.4% per day (Table 36.5). Weight unloading is the apparent prime cause for these losses since minimal degrees of resistance applied to the lower extremities by cycle ergometer exercise training during bed rest resulted in essentially no detectable loss of strength in all measured muscle groups (75, 79, 100, 177).

Data regarding the fatigability characteristics of muscle groups in humans following adaptation to microgravity are limited. Reduction in maximal muscle force development following bed rest is associated with decreased muscle electrical activity and increased fatigability (111). Using spectral power analysis from electromyography recordings, increased fatigability in the gastrocnemius muscles following flight was evidenced by a shift toward lower frequencies in response to maintaining a tension of 50% of maximum voluntary contraction for 1 min (112). Thus increased muscle fatigability has been verified in both actual and ground-based simulations of microgravity.

Effects of Return to Terrestrial Gravity

Reduction of $\dot{V}O_{2max}$ and exercise capacity may be partly a result of an inability of cardiovascular mechanisms to adequately compensate for the orthostatic

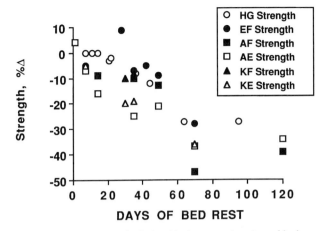

FIG. 36.6. Cross-sectional relationship between duration of bed rest and percent change (%Δ) in strength of handgrip *(HG)*, elbow flexors *(EF)*, ankle flexors *(AF)*, ankle extensors *(AE)*, knee flexors *(KF)*, and knee extensors *(KE)*. Compilation of data from 17 independent investigations (see Table 36.5).

challenge of resuming the upright position following adaptation to microgravity. Exercise in the upright posture is much more dependent upon venous return from the legs and on the Frank-Starling mechanism to augment stroke volume than is exercise in the supine posture (7, 132). Subsequently, mechanisms for control of heart rate and stroke volume would become especially sensitive to venous pooling and to underfilling of the heart during upright exercise following changes in vascular volume and compliance (51). If postural challenges in gravity contribute to reduction in $\dot{V}O_{2max}$, then changes in the cardiorespiratory response to exercise performed in the upright posture following exposure to microgravity should be greater than those changes during supine exercise, assuming responses to supine exercise are similar to those in microgravity.

Following 6 days of spaceflight, selected physiological responses to 75 W on a supine and upright (sitting position) cycle ergometer were compared to preflight responses (170). Exercise heart rate was 12% higher in supine and 17% higher in upright cycling compared to preflight responses. These results are similar to those reported from ground-based studies (23, 43), suggesting that higher supine heart rate after flight probably represented the effect of reduced blood volume, while the additional tachycardia in the upright posture probably represented the added influence of the orthostatic effect of 1G on end-diastolic filling volume and stroke volume. Further, 84 days of spaceflight reduced average upright exercise stroke volume by 24 ml in the three astronauts while supine stroke volume was decreased by only 5 ml (126). These data support the notion that elevated heart rate during terrestrial exercise following spaceflight is a compensatory response to reduced stroke volume, a response exaggerated in the upright posture compared to the supine position.

The average decrease in $\dot{V}O_{2max}$ after bed rest reported from eight independent studies was 9% in the supine posture compared to an average 18% reduction during upright exercise testing (22, 23). In another study, supine and upright $\dot{V}O_{2max}$ following 10 days of bed rest was measured in the same 12 subjects (43). After bed rest, $\dot{V}O_{2max}$ declined by 17% during upright exercise compared to only 7% during supine exercise. The larger reduction in the rate of oxygen transport and utilization in upright as compared to supine exercise following bed rest is associated with a greater reduction in stroke volume and cardiac output (89, 141), with higher heart rate and heart rate–blood pressure product (34, 39, 43, 89), and with larger recruitment of anaerobic energy supply, as suggested by lower submaximal and maximal oxygen uptake associated with greater submaximal oxygen deficit, oxygen debt, $\dot{V}O_2$ time constant, and respiratory quotient (34, 39, 40, 41, 174). It appears that the combined reduction of both regular physical activity and venous system fluid shifts induced by the absence of gravity contribute nearly equally to the reduction in $\dot{V}O_{2max}$ following adaptation to microgravity.

PHYSIOLOGICAL CHANGES ASSOCIATED WITH REDUCED EXERCISE CAPACITY

Pulmonary Function

There is little evidence to support the notion that changes in mean minute ventilation or ventilation–perfusion matching in microgravity might limit gas exchange and adversely affect aerobic processes required to support the metabolic demands of exercise (i.e., $\dot{V}O_{2max}$). Vital capacity, residual volume, total lung capacity, tidal volume, alveolar ventilation, forced expiratory volume, maximum voluntary ventilation, and closing volume are not altered following spaceflight (129, 144). These observations are corroborated by ground-based experiments demonstrating that exposure to bed rest did not change total lung capacity, forced vital capacity, residual lung volume, or 1 s forced expiratory capacity, while diffusing capacity of the lung during exercise showed a tendency to decrease (8, 74, 141). Pulmonary function during exercise in spaceflight is well within the normal limits of in-flight exercise (101). Although reduced maximal minute ventilation volume has been reported following bed rest (48, 141), it is generally unchanged or increased following bed rest despite reductions in $\dot{V}O_{2max}$ (41, 43, 45, 49, 65, 154). In fact, pulmonary efficiency during exercise in spaceflight, measured as ventilation volume during exercise at an oxygen uptake of 2.0 liters·min⁻¹, was essentially unaltered compared to preflight (126). Average ventilation volume during 75% $\dot{V}O_{2max}$ was reported as 82.4 ± 9.9

liters·min^{-1} preflight compared to 83.3 ± 10.8 liters·min^{-1} in flight (103, 126), and was significantly elevated during equal submaximal work rates and $\dot{V}O_2$ following bed rest (34, 45, 48, 174). These data suggest that changes in pulmonary function probably do not limit gas exchange during exercise in microgravity or upon return to terrestrial gravity.

Blood Volume

One of the most consistent and rapid adaptations to microgravity is the reduction in blood volume. Approximately 300 ml of spaceflight-induced hypovolemia comes from the plasma compartment, decreasing from 3.35 liters preflight to 3.05 liters (−8.4% ± 3.0%) following 28 days of spaceflight (94, 118). Plasma volume measurements made after missions of shorter duration (4–11 days) demonstrated similar relative reductions in plasma volume by an average of 8.6% ± 1.3% (range −4% to −16%) (63, 116). The average reduction in plasma volume was greater after longer durations of spaceflight (13% and 16% for 59 and 84 days, respectively) (93). Results from bed rest studies verify the reductions in plasma volume both in magnitude and time course (Fig. 36.7). Taken together, spaceflight and ground-based data suggest an early reduction in blood volume due to the contraction of plasma, followed by a gradual stabilization sometime between 30 and 60 days of microgravity exposure.

Reduction in circulating plasma volume appears to contribute directly to the limitation of hemodynamic responses required to support adequate blood flow and metabolism during moderate to heavy exercise. Reductions in $\dot{V}O_{2max}$ of 7% in ten subjects after 10 days of

bed rest (45) and 14% in four individuals who underwent 14 days of bed rest (48) were associated with mean plasma volume losses of 11% and 17%, respectively. Plasma volume reductions of 11% in 15 men and 13% in 8 women following bed rest were associated with decreased $\dot{V}O_{2max}$ of 9% and 10%, respectively (49). Following 14 days of bed rest, seven young male subjects had a plasma volume loss of 15% with a $\dot{V}O_{2max}$ decrease of 12% (154). In comparison, longer exposures to bed rest of 26 days (153) and 28 days (124) resulted in larger decreases of 21% and 28% in plasma volume and were associated with greater $\dot{V}O_{2max}$ reductions of 20% and 19%, respectively. When the percent change in plasma volume following bed rest is compared to the percent change in $\dot{V}O_{2max}$ from a compilation of cross-sectional data (Fig. 36.8A), a significant relationship exists. This cross-sectional relationship has been reproduced by a subsequent longitudinal experiment (Fig. 36.8B). The high correlation coefficient between the relative changes in plasma volume and $\dot{V}O_{2max}$ suggests that microgravity-induced hypovolemia probably contributes to the reduction in the capacity of the body to transport and utilize oxygen during exercise.

Hypovolemia can contribute to reduced ventricular filling and cardiac output during exercise since it is associated with decreases in mean venous pressure, venous return, and stroke volume (134, 161). The apparent effect on ventricular filling is supported by a close relationship between reduced plasma volume and elevations in maximal heart rate and reductions in stroke volume, cardiac output, and lower $\dot{V}O_{2max}$ following bed rest (11, 48, 141). A rapid 17% to 21% reduction in oxygen uptake during exercise at 160 bpm within 15 days of spaceflight and its restoration within 24–36 h following return to earth suggest that changes in vascular volume contribute to reduced $\dot{V}O_{2max}$ following exposure to microgravity (137, 139, 140). A causal relationship is supported by the observation that reduction in $\dot{V}O_{2max}$ after bed rest has been minimal when hypovolemia was ameliorated by exercise training (Fig. 36.9). However, reversal of hypovolemia alone without exercise does not prevent the decrease in $\dot{V}O_{2max}$ observed following bed rest (152). Further, plasma volume was reduced by nearly 16% in three astronauts after 84 days of spaceflight while in-flight $\dot{V}O_{2max}$ was maintained with extensive exercise programs (126). Clearly, reduced vascular volume is only one mechanism that contributes to impaired cardiovascular function and loss of aerobic capacity during exercise in adaptation to microgravity.

Lower plasma volume might also contribute to reduced $\dot{V}O_{2max}$ by reducing the dilution capacity of the blood for the same amount of acidodic metabolites. This hypothesis is supported by the observations that blood lactate concentration increased during exercise of

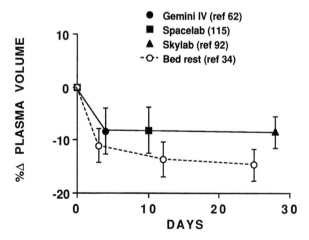

FIG. 36.7. Comparison of time courses of percent change (%Δ) in plasma volume during adaptation to actual spaceflight (closed symbols and solid line) and bed rest (open circles and broken line). Spaceflight data from Gemini IV (63), Spacelab (116) and Skylab IV (93). Bed rest data from Convertino et al. (35).

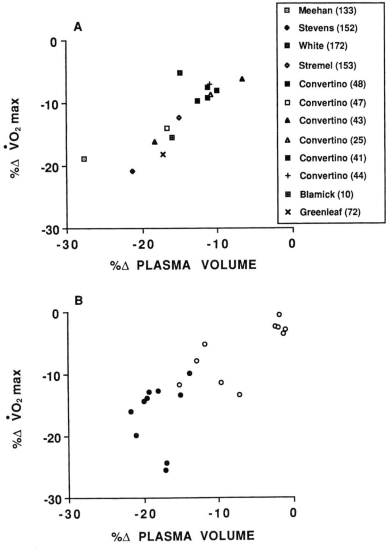

FIG. 36.8. Relationships between percent changes (%Δ) in plasma volume and $\dot{V}O_{2max}$ after adaptation to microgravity (bed rest). *Panel A* represents a cross-sectional compilation of data from 12 independent investigations (%Δ $\dot{V}O_{2max}$ = 0.82 [%Δ *PV*] + 0.3, *r* = 0.841). *Panel B* is generated from Convertino et al. (44) and represents longitudinal data from 10 fit *(closed circles)* and 10 unfit *(open circles)* subjects (%Δ $\dot{V}O_{2max}$ = 0.76 [%Δ *PV*] − 1.7, *r* = 0.787).

equal intensity following 10 days of bed rest that induced 17% hypovolemia, despite no change in total circulating lactate (174). Ventilation volume, expired carbon dioxide, and respiratory exchange ratio during exercise were also higher after bed rest, indicating greater respiratory compensation for acidosis. Therefore, lower plasma volume may contribute to earlier onset of metabolic acidosis, a notion consistent with the significant correlation coefficient between the change in anaerobic threshold and plasma volumes (0.80) following bed rest (45).

Another possible explanation for the effect of vascular hypovolemia on the reduction in $\dot{V}O_{2max}$ may be a lower oxygen-carrying capacity of the blood resulting

from reduced red cell mass. Indeed, a consistent influence of microgravity has been a reduction in circulating hemoglobin and red cell mass (116, 120, 121). Space missions of 7–14 days in duration demonstrated that most crew members reduced their red blood cell mass by an average of 6% to 10% (63, 116) and continued to 14% and 12% after 28 and 59 days, respectively (94). The magnitude of hemoglobin loss was 13%, 26%, 23%, and 24% at 16, 30, 63, and 96 days of spaceflight, respectively (120, 168). These data suggest an early reduction in red cell mass and hemoglobin that reached a plateau after 30 days of flight (93, 116, 121). This represents an approximate 25% reduction in oxygen-carrying capacity of the blood, which could con-

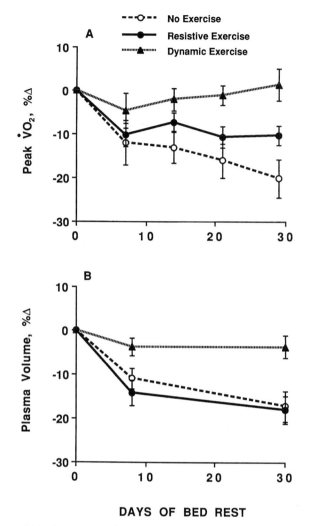

FIG. 36.9. Comparison of time courses of percent changes (%Δ) in peak V̇O₂ and plasma volume during bed rest in subjects who underwent no exercise *(open circles and broken line),* resistive exercise *(closed circles and solid line),* and dynamic cycling exercise *(closed triangles and hashed line).* [Data modified from Greenleaf et al. (73, 80)].

tribute to limited $\dot{V}O_{2max}$ during and following short and prolonged spaceflights. However, maintenance of in-flight $\dot{V}O_{2max}$ occurred in astronauts after 84 days of spaceflight with high levels of in-flight exercise in the presence of 7% reduction in red cell mass (126).

Some ground-based studies have verified spaceflight data by demonstrating that red cell mass can decrease from 5% to 25% during prolonged bed rest (39, 48, 113, 124, 173). However, the relationship between changes in red cell mass and changes in $\dot{V}O_{2max}$ is poor since no change in red cell mass following short-duration bed rest exposure has been reported despite significant reductions in $\dot{V}O_{2max}$ (25, 42, 45, 46, 153, 154, 174). It therefore appears likely that the primary effect of microgravity-induced hypovolemia on reducing

$\dot{V}O_{2max}$ is due to the contraction of the plasma volume, with a lesser contribution coming from reduced red cell mass.

Cardiovascular Function

Impaired $\dot{V}O_{2max}$ involves changes in central hemodynamics, as manifested by elevated heart rate and reductions in stroke and cardiac volume (1, 14, 18, 23, 26, 57, 69, 126, 130, 141, 176) during exercise after spaceflight and bed rest. During exercise at 75% $\dot{V}O_{2max}$, mean heart rate increased from 103 bpm preflight to 114 bpm after 84 days of spaceflight in response to a reduction in stroke volume from 110 ml preflight to 97 ml postflight (126). Elevated exercise heart rates and systolic pressure resulted in an average increase of 25% in their product, an indication of significant elevation in myocardial oxygen demand after returning to 1G (4). In a ground-based experiment, no change in cardiac output during maximal exercise was reported following 20 days of bed rest, despite a reduced $\dot{V}O_{2max}$, because decreased maximal stroke volume was compensated by elevated maximal heart rate (69). In a subsequent ground-based study, relatively constant cardiac output was reported during constant submaximal work rates after bed rest, despite reduced exercise stroke volume (130). It is clear that elevated heart rate can provide a compensatory mechanism by which cardiac output can be maintained during exercise, despite reductions in venous return and stroke volume following adaptation to microgravity. Thus in these experiments, reduced $\dot{V}O_{2max}$ associated with adaptation to microgravity did not appear to be due to reduced cardiac capacity, but instead resulted from some aspect of oxygen transport and utilization by the tissues.

In contrast to the studies cited above (69, 130), lower cardiac output during exercise has been consistently reported during spaceflight (1, 60, 168, 171, 176) and bed rest (89, 107, 140). After flight, stroke volume was reduced by an average of 20% and cardiac output fell an average of 10%, despite an average elevated exercise heart rate of 17 bpm during supine exercise at 100 W performed 3–7 days after 30, 63, and 96 days of exposure to microgravity (Table 36.3). An average 26% decrease in treadmill $\dot{V}O_{2max}$ of five young men following 20 days of bed rest resulted from similar reductions in maximal cardiac output from 20.0 to 14.8 l·min⁻¹ (26%) and maximal stroke volume from 104 to 74 ml (29%) with essentially no change in maximal heart rate or arteriovenous O_2 difference (141). Reduced stroke volume during submaximal exercise after spaceflight (4, 5) and bed rest (107, 141) in both supine and upright postures support the notion that poor postural adaptation, impaired venous return, and/or hypovolemia may not completely account for reductions in cardiac output and $\dot{V}O_{2max}$. Reduced supine exercise stroke volume and an 11% decline in fluoroscopically measured resting heart volume after bed rest advanced the hypothesis that impaired myocardial performance during exercise may be limited by a nonspecific deterioration in ventricular function resulting from cardiac muscle atrophy with adaptation to microgravity (11).

There is little evidence from spaceflight data to support the idea that deterioration of myocardial structure and function occurs and contributes to exercise impairment as an adaptation to microgravity. Cardiac responses during spaceflight demonstrate that there is little effect of microgravity on resting heart rate, stroke volume, and cardiac output (26, 55, 56). However, increased pulse wave propagation velocity at rest has been interpreted as an index of greater left ventricular contractility (55, 82, 136, 169). Using lower-body negative pressure, end-diastolic–volume-stroke volume relationships were generated on astronauts after 84 days in space and demonstrated no change in the slope of the Frank-Starling curve (85). These data suggest that cardiac function at rest is not compromised with adaptation to microgravity. However, cardiovascular responses to exercise during and after spaceflight are characterized by progressive reductions in stroke volume and cardiac output despite dramatic elevations in heart rate in comparison to preflight tests (1, 60, 168, 171, 176). Despite decreased exercise cardiac output, increased myocardial contractility was suggested by greater velocity of propagation of the pulse wave over the aorta (176).

Equilibrium-gated cardiac blood pool scintigraphy demonstrated that myocardial function, measured by left ventricular ejection fraction, was augmented in 12 subjects during graded supine exercise (85 and 135 W) following 10 days of bed rest, despite a significant reduction in $\dot{V}O_{2max}$ (89). The augmentation of both left ventricular ejection fraction and heart rate during submaximal and maximal exercise may represent compensatory mechanisms to ameliorate the reduction in cardiac output, despite lower cardiac end-diastolic volume (preload) and stroke volume after bed rest (Fig. 36.10). These results suggest that reduction in exercise cardiac output during adaptation to microgravity can be accounted for by lower ventricular filling and end-diastolic volume rather than by reduced myocardial function. It is possible that 10 days' exposure to microgravity may not be long enough to induce myocardial atrophy and dysfunction as proposed in longer-duration investigations (11, 141). However, increased ejection fraction and absence of myocardial dysfunction during exercise in bed rest (89) was verified by echocardiographic data collected during rest and exercise on a 237 day spaceflight (1). During flight, both subjects demonstrated 30% lower stroke volume in response to

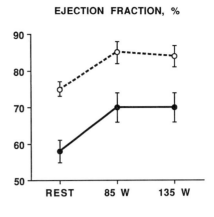

EJECTION FRACTION, %

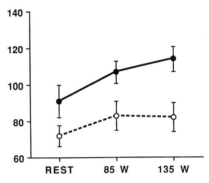

STROKE VOLUME, ml

FIG. 36.10. Mean (± SE) left ventricular ejection fraction and stroke volume during rest and graded exercise before *(closed circles and solid lines)* and after *(open circles and broken lines)* 10 days of bed rest. [Modified from Hung et al. (89) with permission.]

graded exercise (125 and 175 W) compared to preflight responses (Fig. 36.11). Average heart rates were 12% and 17% higher at 125 and 175 W, respectively, compared to preflight rates, but cardiac output remained lower. Left ventricular–filling volume (end-diastolic volume) was significantly depressed during rest and did not increase with exercise. Diminution of stroke volume was prevented by greater myocardial contraction, as indicated by lesser end-systolic reserve and elevated ejection fraction (Fig. 36.11). These flight data were similar to those of ground-based data (Fig. 36.10) and do not support the notion that deterioration of myocardial function plays the leading role in the decrease in left ventricular stroke volumes during exercise following adaptation to microgravity.

Data from spaceflight and ground-based experiments support the notion that there is no apparent cardiac dysfunction during exercise as a consequence of microgravity exposure (16, 58, 60, 169). Maintained or improved cardiac performance during exercise, as indicated by increased ejection fractions and aortic pulse wave propagation velocities, refutes the notion that cardiac "deconditioning" occurs and limits the capacity to

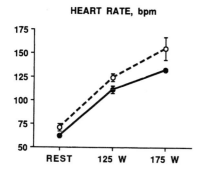

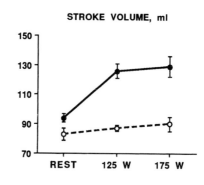

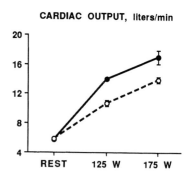

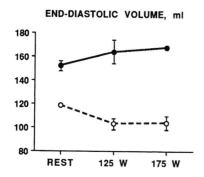

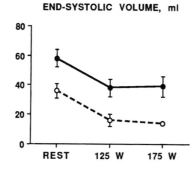

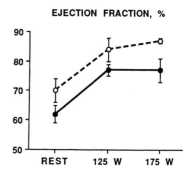

FIG. 36.11. Mean (± SE) cardiac responses of two cosmonauts during rest and at 125 W and 175 W of exercise on a cycle ergometer before *(closed circles and solid lines)* and during *(open circles and broken lines)* a 237 day space mission. [Modified from Atkov et al. (1) with permission.]

perform physical exercise. It appears more likely that reduced stroke volume, cardiac output, and $\dot{V}O_{2max}$ can be primarily attributed to the reduction in circulating plasma volume and its effect on venous return, ventricular filling, and cardiovascular hemodynamics, rather than to an impairment of cardiac function.

Morphological Changes in Skeletal Muscle

Mechanisms involved in the loss of muscle strength with exposure to microgravity are unclear. Loss of load-bearing input to muscle proprioception has been viewed as a primary factor in the development of motor disturbances, particularly during brief exposure to microgravity. The possible contribution of impaired neuromotor mechanisms to the limitation of force development with exposure to microgravity was suggested by high correlations (0.81–0.86) between the reduction in muscle rigidity and force development in antigravity muscles (109). Since this relationship was observed after only 7 days of exposure to microgravity, it is presumed that neuromotor disturbances, rather than muscle atrophy, can be a significant factor in the diminished muscle force development.

Muscle atrophy resulting from unloading and relative disuse has been proposed as a primary contributing factor to the cause of postflight loss in muscle strength (81, 163). Reduction in limb size during spaceflight has been well documented and used as an indication of muscle atrophy (21, 95, 163). However, changes in histochemical and biochemical characteristics of different skeletal muscles and of different fiber types as they relate to reductions in force output and as affected by the type and speed of muscle action have not been measured after spaceflight.

Muscle biopsies, computed tomography, and anthropometric measurements of the thigh and calf were conducted on 11 subjects before and after 30 days of 6° head-down bedrest to determine the effects of simulated microgravity on muscle morphology (13, 37, 38, 86). Following bed rest, there was significant muscle atrophy as evidenced by reduction in the cross-sectional area (CSA) of the total leg muscle compartment (38) and decreased CSA of muscle fibers (29, 86). There were significant decreases in CSA of both fast-twitch and slow-twitch muscle fibers of the vastus lateralis, but only fast-twitch fibers in the soleus following bed rest. In both muscle groups, the relative reduction in muscle fiber CSA tended to be greater in fast-twitch (18%) as compared to slow-twitch fibers (11%) (86). The average decline in peak force production of 21% for the knee extensor group was larger than average reductions in cross-sectional areas of the total thigh muscle compartment (8%) and fiber size of the vastus lateralis (14%). Since maximal force capability is a function of muscle cross-sectional area, factors other than atrophy may contribute to impaired muscle function following bed rest. Various ultrastructural abnormalities such as disorganized myofibrils, cellular edema, irregular Z bodies, and fiber necrosis (disrupted fiber membranes), as indicated by disrupted sarcolemma, abnormal mitochondria, disrupted striation patterns, and mitochondria located in the intercellular spaces, were observed from electron microscope analyses (86). It is therefore possible that in addition to muscle atrophy, ultrastructural changes and other factors associated with the control of the contractile apparatus may contribute to the loss of muscle force generation during long-term exposure to microgravity.

Cellular Metabolism of Skeletal Muscle

In addition to changes in central hemodynamics, reduced $\dot{V}_{O_{2max}}$ and increased muscle fatigability induced by microgravity may reflect increased metabolic acidosis resulting from a reduced capacity to support oxidative metabolism in skeletal muscle. An elevated acidic state in exercising muscle fibers is supported by respiratory compensation with elevations in systemic expired carbon dioxide, ventilation volume, and respiratory exchange ratio (125), and increased recovery oxygen uptake associated with fatigue (103, 126) during exercise of equal intensities in spaceflight as compared to before flight. Similar changes in respiratory gas exchange responses have been reported in subjects exposed to bed rest (34, 39, 41, 43, 141, 174). Anaerobic threshold occurred at a lower absolute (from 93 to 65 W) and relative (from 52% to 42% of $\dot{V}_{O_{2max}}$) steady-state work rate after bed rest (45). The anaerobic threshold data suggest that greater blood lactate accumulation and metabolic acidosis occur at given energy requirements (172), and the data are consistent with observations that measured blood lactate levels during steady-state exercise following bed rest were greater than pre–bed rest levels at the same work rates (141, 174). Taken together, greater respiratory compensation and blood lactate during steady-state exercise suggest that the rate of oxidation of pyruvate within the muscle may decrease, and that a greater proportion of energy demand must rely on anaerobic bioenergetic pathways at the muscle cellular level with adaptation to microgravity.

Reduction in resting muscle blood flow during bed rest (10, 37) may reflect an inability to provide adequate oxygen supply to the working tissue. This is associated with a reduced capillary-to-fiber ratio in muscle fibers of high oxidative capacity (86, 142), suggesting less capacity for delivery and diffusion of oxygen and substrates to the working muscle. Following 30 days of 6° head-down bed rest, there were no changes in activities of lactate dehydrogenase or phosphofructokinase in soleus and vastus lateralis muscles, suggesting that microgravity does not appear to compromise glycolytic metabolism. However, the activities of two enzymes associated with aerobic metabolic pathways, citrate synthase and β-hydroxyacyl-CoA dehydrogenase, were reduced in both muscle groups (86). These data indicate that in addition to compromised capability to generate force, the effect of prolonged muscle unloading and reduced energy requirements associated with exposure to microgravity may reduce the capacity of oxygen delivery to and utilization by muscle fibers at the cellular level. These adaptations may be underlying mechanisms that contribute to the reduction in exercise endurance and $\dot{V}_{O_{2max}}$ following bed rest (23, 26, 87, 92, 106).

Reduced substrate availability to working muscles may also contribute as a mechanism for limiting exercise endurance. Although blood glucose levels were not altered (2) or were slightly elevated during spaceflight (165), plasma insulin was reduced (115). Since peripheral glucose uptake is dramatically reduced following bed rest (59, 77), decreased insulin-receptor sensitivity in muscles may impair the availability and utilization of circulating glucose during exercise following adaptation to microgravity. Paradoxically, a greater proportion of the energy transfer during exercise following spaceflight (103, 125, 126) and bed rest (45, 141, 174) may be dependent upon carbohydrate metabolism. Greater utilization of muscle glycogen stores might be required to sustain exercise with less utilization of free fatty acids, particularly during the performance of moderate to heavy intensities for long duration. Total circulating free fatty acids during exercise are reduced after bed rest (174), perhaps reflecting impaired mobilization as a result of microgravity adaptation. Thus reduction in glucose uptake by and availability of circulating free fatty acids to skeletal muscle may be underlying mechanisms that contribute to decreased exercise performance after adaptation to microgravity.

Venous and Muscle Compliance

Restriction of venous return during exercise can contribute to the reduction in maximal cardiac output and $\dot{V}_{O_{2max}}$ observed following microgravity (1, 11, 14, 23, 26, 69, 130, 141). Increased venous compliance of the lower extremities reported during spaceflight (26, 162, 175) and bed rest (37, 38) can compound the effect of hypovolemia on impairment of venous return during exercise by providing a greater capacity to pool blood under the same hydrostatic pressure (88, 95). This notion is supported by a greater reduction in $\dot{V}O_{2max}$ after bed rest during increased blood pooling (17, 51), and by a lower venous return and stroke volume during exercise in spaceflight that have been associated with increased venous compliance in the legs (175).

The mechanism(s) of the increased venous compliance with microgravity is related to the reduction in size and tone of the muscle compartment (37, 38, 104). Calf compliance is increased early in microgravity (37, 162), suggesting that interstitial fluid efflux from the leg may contribute to this reduction. Muscle atrophy and reduction in the muscle compartment following bed rest are correlated with increased venous compliance (38). When the muscle compartment is reduced, compliance increases (37, 38); when the muscle compartment loss is attenuated with routine muscular activity during bed rest, the compliance changes are eliminated (62). These data indicate that loss of muscle protein and water that occur during unloading of forces characteristic to microgravity environments are important factors contributing to increased venous compliance of the lower extremities, which in turn could lead to the impairment of cardiac output and $\dot{V}O_{2max}$.

Changes in Thermoregulation

Mean daily values for sweat rate in nine astronauts during an average of 1 h of daily exercise decreased from 1,750 ml preflight to 1,560 ml during flight (117). These results suggest that microgravity decreased sweat losses during exercise and possibly reduced insensible skin losses as well. This may be related to the reduction in total body water that averaged from 47.5 liters before flight to 46.3 liters after flight (118). In addition, the microgravity environment apparently promotes the formation of an observed sweat film on the skin surface during exercise by reducing convective flow and sweat drippage. This apparently results in high levels of sustained skin wetness that acts to suppress sweating (117). This effect is unique to spaceflight and cannot be produced by bed rest.

Body heat storage is increased during exercise of equal work intensities after as compared to before bed rest (92). Excessive elevation in rectal temperature above ambulatory control levels in seven men during 70 min of submaximal supine cycle ergometry (~45% $\dot{V}O_{2max}$) at 22°C ambient temperature after 14 days of bed rest indicated a reduced capacity to dissipate heat during adaptation to microgravity (78). No significant differences in total body sweat production were observed, suggesting that there was an inhibition of sweating from the same core temperature stimulus.

Thus limited data from flight and ground-based models indicate that weightlessness causes some impairment in thermoregulation during exercise, which could limit work performance of extended duration.

RECOVERY AFTER RETURN TO THE ONE-GRAVITY ENVIRONMENT

Effects of Ambulation and Exercise during Recovery

Hemodynamic responses during graded submaximal exercise were examined before and periodically throughout 30 days of recovery from 28, 59, and 84 day spaceflight missions (14). During exercise, astronauts exhibited approximately 30% reduction in cardiac output and 50% lower stroke volume. These changes were accompanied by elevations in heart rate and systemic peripheral resistance. Hemodynamic alterations returned to preflight baseline levels within 30 days of recovery (Fig. 36.12). The systematic evaluation of the use of exercise programs during this recovery process is not possible since there is no evidence of formal exercise programs used by astronauts following return from space missions. Evaluation has therefore been dependent upon information gained from ground-based experiments.

There is little convincing evidence that exercise training of moderate intensity enhances recovery of exercise responses from the adaptation to microgravity. Average baseline $\dot{V}O_{2max}$ of 2.77 liters·min^{-1} in four men who underwent 42 days of bed rest remained depressed by 13% 18 days after bed rest, despite the implementation of a recovery exercise program consisting of daily cycle ergometry for 30 min at approximately 55% $\dot{V}O_{2max}$

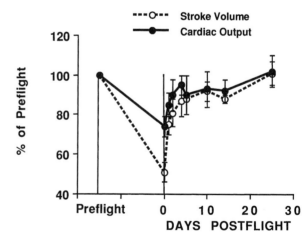

FIG. 36.12. Time course of post-spaceflight recovery of stroke volume *(open circles and broken lines)* and cardiac output *(closed circles and solid line)*for all Skylab missions. Values are mean ± SE expressed as percent of preflight values. [Modified from Buderer et al. (14) with permission.]

(8). The bed rest study of Saltin and co-workers (141) is often cited as evidence favoring the use of exercise programs as an effective technique for enhancing the recovery from the deleterious effects on exercise performance following bed rest. In the three habitually sedentary subjects, $\dot{V}O_{2max}$ levels, which were reduced by 26% following bed rest, were restored within 10–14 days of recovery from bed rest and continued to increase by 36% above pre–bed rest levels at 60 days of recovery. However, the two habitually active subjects required 30–40 days of physical activity to restore $\dot{V}O_{2max}$ values to pre–bed rest levels. These data suggest that exercise training following adaptation to microgravity may be instrumental in enhancing the recovery rate of $\dot{V}O_{2max}$ and physical working capacity, particularly in more sedentary individuals. However, a limitation to this interpretation is that the study was not designed to distinguish between the effects of exercise conditioning and those of resumption of usual activities in the normal upright posture; all five subjects underwent exercise training.

In a study designed to address this issue (22, 52), $\dot{V}O_{2max}$ and other responses to exercise were compared after 10 days of bed rest in six subjects who performed prescribed physical exercise daily for 60 days (exercise group) and six subjects who simply resumed their customary ambulatory activities (control group). Despite a significantly greater increase in the exercise group $\dot{V}O_{2max}$ at 60 days compared to the control group, the $\dot{V}O_{2max}$ and physical work capacity in both groups returned to pre–bed rest levels after 30 days of bedrest, and this was accompanied by significant and similar increases in resting left ventricular–end-diastolic and stroke volumes in both groups. It was suggested that simple resumption of usual physical activities after bed rest was as effective as formal exercise conditioning in restoring the functional capacity to baseline levels. These results are further supported by data demonstrating that a randomized trial of in-hospital exercise conditioning did not increase treadmill performance (147) and that baseline $\dot{V}O_{2max}$ was restored after 14 days of recovery from a 10 day bed rest (46) and after 30 days of recovery from 7 days of bed rest (65) in subjects who merely resumed controlled ambulatory activities with no daily exercise. Since the fitness profile of astronauts is representative of the subjects in these latter studies (6), the available data from both spaceflight and ground-based experiments suggest that resuming normal ambulatory activities in terrestrial gravity appears adequate for the restoration of exercise capacity. However, average baseline $\dot{V}O_{2max}$ of 3.09 liters·min^{-1} in three men who underwent 30 days of bed rest remained depressed by 23% 9 days after bed rest (105), suggesting that longer durations of microgravity exposure may require longer recovery periods. Therefore, the available data do not rule out the possibility that formal exercise training of high intensity may accelerate the recovery process, especially following longer exposure.

Repeated Exposures to Microgravity

Although baseline exercise tolerance and respective cardiovascular and metabolic responses appear to be restored within 7 to 30 days of regular ambulatory activities following spaceflight (5, 14), repeated exposure to microgravity may induce a larger decrement in exercise performance as a result of an additive effect. More than 100 people have flown in space at least two times, but there are few data available from actual space missions to describe the effects of repeated exposure in microgravity on exercise performance during and after flight. Cosmonaut V. V. Ryumin made his 185 day flight 6 mo after a previous 175 day flight. In both missions it was reported that his work capacity remained essentially identical, indicating a complete recovery to his normal state during the intervening 6 mo period (168).

Ground-based experiments have provided the most conclusive data regarding recovery from repeated exposure to microgravity environments. Minute ventilation, $\dot{V}O_2$, and heart rate during submaximal and maximal exercise returned to baseline levels after three repeated 14 day bed rest exposures separated by 3 wk recovery periods which included daily cycle ergometer exercise for 60 min at 50% $\dot{V}O_{2max}$ (154). Similarly, elevated heart rate during 40, 80, and 120 W of cycle ergometer exercise and lower $\dot{V}O_2$ at a heart rate of 160 bpm following each of three 10 day bed rest periods separated by 3 wk periods of normal daily activities, persisted after 1 wk of recovery but returned to pre–bed rest baseline values by the second or third week of recovery (18). Repeated 10 day periods of 6° head-down bed rest, separated by 14 days of recovery consisting of regular daily activities without formal exercise, decreased $\dot{V}O_{2max}$, anaerobic threshold, and oxygen pulse, while submaximal and maximal heart rates, diastolic pressures, and rate–pressure products increased (46). Although submaximal and maximal $\dot{V}O_2$ values returned to pre–bed rest baseline levels following 14 days of recovery, heart rates, diastolic pressures, and rate–pressure products remained significantly elevated following recovery from the second bed rest exposure. It can therefore be concluded that 2 wk of minimal ambulatory activity appear adequate for complete recovery of mechanisms associated with exercise metabolism, but there may be an accumulative effect on cardiovascular mechanisms following a second exposure. Little is known about the time required for recovery of cardiovascular and muscle function from longer periods of exposure to microgravity.

EXERCISE AS A COUNTERMEASURE TO MICROGRAVITY ADAPTATION

$\dot{V}O_2$ and Hemodynamic Responses to Exercise

Periodic exposures to gravity and exercise have been performed by subjects during bed rest in an effort to identify effective and plausible countermeasures against the physiological adaptations that might limit exercise performance during and after exposure to microgravity. When 12 subjects, who performed 2 h treadmill exercise at an intensity of 15%–30% of their $\dot{V}O_{2max}$ twice daily, at one-half gravity during 14 days of combined water immersion and bed rest confinement, were compared to a control group that did not undergo exercise during bedrest, the decline in exercise endurance and plasma volume was twofold less in the exercise group (87). Quiet sitting for 8 h daily during 30 days of bed rest produced a 6% reduction in $\dot{V}O_{2max}$ (9), which was significantly smaller than the 18%–20% reduction expected during complete bed rest of similar duration (Table 36.2). However, daily sitting exercise for 1 h at 100 W (~55% $\dot{V}O_{2max}$) during 24 days of complete bed rest maintained $\dot{V}O_{2max}$ at pre-exposure levels (9). These data clearly indicate that the effects of physiologic adaptations to microgravity on exercise performance are greatly dependent upon less gravity stimuli as well as on a smaller requirement of physical activity. It is therefore reasonable to test exercise as a possible method of reducing or eliminating the effects of adaptation to microgravity.

With the increase in spaceflight duration, extensive exercise during spaceflight has been used routinely as a countermeasure to ameliorate specific adaptations to microgravity associated with impaired $\dot{V}O_2$ and hemodynamic responses during exercise. In the Soviet space program, both a cycle ergometer and a space treadmill equipped with a pulling harness were used during each exercise session (168, 169). There were normally two sessions of exercise daily, one in the morning and one in the evening, each session consisting of an average of 40 min on the cycle ergometer and 34 min on the treadmill (145). Cycle exercise consisted of an average 160 W work rate ($\dot{V}O_2 = 2.3$ liters·min^{-1}) and treadmill exercise consisted of an average locomotion speed of 120 m·min^{-1} (4.5 mph) with about one-third the distance spent walking and two-thirds spent running (average $\dot{V}O_2 = 1.5$ liters·min^{-1}). The average volume of exercise increased in the second half of flight as compared to the first half. Based on exercise heart rate responses and oxygen uptakes, it can be concluded that this exercise represents a training regimen of approximately 50%–75% $\dot{V}O_{2max}$ performed for 2.5 h daily, 6 days per wk for up to 6 mo of spaceflight (145). This

activity has been effective in nearly eliminating cardiovascular and musculoskeletal adaptations to microgravity (145, 166). These exercise regimens applied during 2 mo of spaceflight maintained normal heart rate and stroke volume responses to 70 W ($\dot{V}O_2 = 1.2$ liters/min) at preflight levels (168). After physical deconditioning was induced during the initial 25 days of a 96 day spaceflight, due to the interruption of in-flight exercise, the return to normally scheduled exercise regimens improved the hemodynamic responses, and exercise endurance returned to preflight levels by day 70 and stayed there throughout the remainder of the flight (70). These observations indicate the importance of regular physical exercise for the maintenance of $\dot{V}O_{2max}$ during prolonged exposure to microgravity.

Similar changes in the magnitude of cardiac, hemodynamic, and metabolic responses during exercise, despite differences in the duration of exposure lasting from 30 to 96 days (Table 36.3), reflect the improvement in protective effects resulting from more extensive exercise training programs implemeted during longer flights (4). This observation is best supported by the results of the three U.S. Skylab missions. On the 28 day mission, only a cycle ergometer was used for in-flight exercise, and daily exercise was performed at approximately 70 W ($\dot{V}O_2 = 1.2$ liter·min^{-1}) for an average of 30 min (164). Heart rate was greater and cardiac output less during postflight exercise compared to preflight responses, suggesting a reduced cardiovascular reserve (125). The average amount of work performed on the cycle ergometer was more than doubled on the subsequent 59 day flight so that total daily exercise was about 1 h in duration. With the significant increase in volume and mode of exercise during the second mission, postflight cardiovascular responses to dynamic exercise were similar to that of the 28 day mission (125, 164). On the 84 day mission, a Teflon-coated treadmill was added to the exercise arsenal used on the previous two flights and total daily exercise time was increased to about 1.5 h (164). Following 84 days of spaceflight, the reduction in exercise stroke volume and cardiac output were less than that of the previous two missions, and $\dot{V}O_{2max}$ of all three astronauts actually increased after 79 to 83 days of exposure by an average of about 8% (Fig. 36.13). These observations indicate the possibility of effective prevention of physical deconditioning using exercise during long-term spaceflights.

Generally, the effects of exercise employed during ground-based experiments have corroborated spaceflight data that exercise can ameliorate physiologic adaptations associated with imparied $\dot{V}O_{2max}$ and hemodynamic responses during exercise (Table 36.6). However, ground-based experiments with control groups receiving no exercise during exposure have pro-

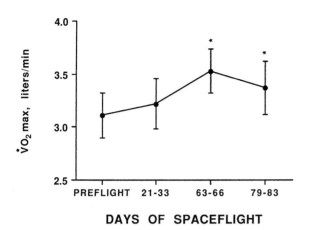

FIG. 36.13. Maximal oxygen uptakes ($\dot{V}O_{2max}$) of three astronauts before and during an 84 day space mission. *Circles* and *lines* represent means ± SE and *asterisks* indicate that all subjects changed in the same direction. [Data plotted from Michel et al. (126)].

cise following 20 days of bed rest was less in subjects who underwent cycle ergometry during bed rest than in nonexercise control subjects (91, 92). A 12% reduction in $\dot{V}O_{2max}$ in nonexercise control subjects was reduced to 5% and 9% when static (21% maximal isometric knee extension torque) and dynamic (cycle ergometry at 68% $\dot{V}O_{2max}$) leg exercises were performed 60 min daily during 14 days of bed rest (154). With bed rest of longer duration (30 days), nonexercised control subjects and resistive (peak isokinetic torque at 100°/sc) exercisers decreased their $\dot{V}O_{2max}$ by 21% and 10%, respectively (73, 75). However, in the latter study, cycle ergometry with a graded protocol desiged to elicit 60%–90% $\dot{V}O_{2max}$ maintained post–bed rest $\dot{V}O_{2max}$ and plasma volume. Exercise eliciting 100% maximal effort performed only once for 15 min at the end of bed rest eliminated the reduction in $\dot{V}O_{2max}$ (25). However, when the intensity of exercise performed during 28–30 days of bed rest was less than 50% $\dot{V}O_{2max}$ (105, 127), the reported reduction in $\dot{V}O_{2max}$ of 19% to 22% was as large as the decrease reported in nonexercised control subjects after the same duration of exposure (73, 75). The accumulated ground-based experiments demonstrate that the average reduction in $\dot{V}O_{2max}$ is only about 4% (Table 36.6) as compared to 13% when no exercise is provided (Table 36.2). These data support the con-

vided a better definition of the effects of exposure duration and intensity of exercise training regimens on aerobic capacity. $\dot{V}O_{2max}$ was increased by 7% with cycle ergometer training and by 16% when the training consisted of supine trampoline for 30 min daily as compared to no change without exercise during 15 days of bed rest (19). The oxygen deficit incurred during exer-

TABLE 36.6. *Effect of Exercise Training during Bed Rest on Maximal Oxygen Uptake**

| | Bed Rest | | Subjects | | | | $\dot{V}O_{2max}$, liters/min | | | | Training Schedule | Training |
References	days	model	N	Age	Sex	Exercise Test	Pre	Post	%Δ	Min/Day	Intensity	Mode
Convertino (25)	10	−6° BR	10	36–48	M	TM UP	3.25	3.21	−1.2	~15 (last day)	100% $\dot{V}O_{2max}$	CE SUP
Stremel et al. (154)	14	0° BR	7	19–22	M	CE SUP	3.80	3.45	−9.2	60	68% $\dot{V}O_{2max}$	CE SUP
Stremel et al. (154)	14	0° BR	7	19–22	M	CE SUP	3.77	3.59	−4.8	60	21% peak torque	RES SUP
Chase et al. (19)	15	0° CR	4	22–26	M	CE UP	3.19	3.42	7.2	30	71% $\dot{V}O_{2max}$	CE SUP
Chase et al. (19)	15	0° CR	4	21–24	M	CE UP	2.96	3.42	15.5	30	76% $\dot{V}O_{2max}$	TRAMP SUP
Rodahl et al. (135)	24	0° BR	2	19–20	M	CE UP	3.00	2.80	−6.7	60	50% $\dot{V}O_{2max}$	CE SUP
Rodahl et al. (135)	24	0° BR	2	18–19	M	CE UP	2.80	2.80	0.0	60	50% $\dot{V}O_{2max}$	CE UP
Miller et al. (127)	28	0° BR	6	18–21	M	TM UP	2.91	2.28	−21.6	60	30% $\dot{V}O_{2max}$	CE SUP
Chase et al. (19)	30	0° CR	4	21–25	M	CE UP	3.17	2.92	−7.9	45	75% $\dot{V}O_{2max}$	TRAMP SUP
Chase et al. (19)	30	0° CR	4	21–22	M	CE UP	3.51	3.19	−9.1	15	75% $\dot{V}O_{2max}$	TRAMP SUP
Greenleaf et al. (73)	30	−6° BR	7	32–42	M	CE SUP	3.13	3.14	1.4	60	90% $\dot{V}O_{2max}$	CE SUP
Greenleaf et al. (73)	30	−6° BR	7	32–42	M	CE SUP	3.24	2.90	−10.2	60	100% peak torque	ISOK SUP
Mean							3.23	3.09	−4.3			

*BR, bedrest; CR, chairrest; CE, cycle ergometer; TM, treadmill; UP, upright position; SUP, supine position.

tention that exercise can be affective in ameliorating a reduction in $\dot{V}O_{2max}$ with adaptation to microgravity of short or long duration when the intensity is greater than 50% $\dot{V}O_{2max}$ and is designed to optimally challenge aerobic metabolic demands.

Effects on Muscle Structure and Function

Despite animal data that show preferential slow-twitch muscle atrophy to microgravity (123, 133), data from human biopsies argue that resistive rather than endurance exercise could be a most appropriate countermeasure against muscle atrophy and dysfunction during adaptation to the unloading characteristic of microgravity environments. The medical report of two cosmonauts of a 16 day orbital mission showed insignificant reductions in muscle tone and limb circumference; strength of the muscles did not change (82). These protective results were associated with in-flight use of a resistance pulling device and wearing spring-loaded suits that provided consistent axial-load resistance of up to 50% of body mass to the musculoskeletal system of arms, legs, and torso during waking hours (8–12 h/day). However, it was not possible to quantify the specificity and amount of exercise performed by these crew members during flight, and the lack of change in muscle size and function may have been merely a result of the short exposure to microgravity.

During the longer U.S. Skylab program, arm and leg strength decreased by 15% to 20% on a 28 day spaceflight mission despite the use of extensive cycle ergometer exercise (164). After this flight, a device that provided concentric resistive isokinetic exercise for the arms and trunk was added to a subsequent 59 day flight and it preserved arm strength (164). However, despite the significant increase in volume and mode of exercise during the 59 day mission, postflight loss of leg strength was similar to that of the 28 day mission (125, 164). On a subsequent 84 day mission, a teflon-coated plate treadmill was added to the exercise arsenal used on the previous 28 and 59 day flights and total daily exercise time was increased (164). The tethering harness of the treadmill device was especially used for jumping and toe rises (126). Following flight, the reduction in body weight, leg volume, and leg strength was not eliminated but was less than half that of the previous two missions (124, 161). Interpretation of these results is difficult because differences in responses may reflect large variability from small sample sizes. However, these observations may indicate the possibility of effective prevention of a reduction in muscle size and function by applying resistive exercise to specific muscle groups during long-term spaceflights.

Limited data are available from ground-based experiments to support the qualitative observations from spaceflight that the use of exercise can ameliorate a reduction in muscle function. A 19% reduction in strength of the back muscle group after 62 days of bed rest was reduced to about 8% when cycle exercise was applied (100), and the decrease of 28% and 36%–57% of the arm and leg muscle groups, respectively, following 70 days of bed rest, was essentially eliminated with the use of cycle or treadmill exercise (177). Exercise during bed rest can maintain total body protein (155), which may be associated with retention of muscle tone and strength (100, 106). It is interesting that dynamic cycle exercise, which provides some resistive stimuli, attenuates reduction of muscle function in groundbased studies, but has been largely ineffective when used during spaceflight (164).

Peak isokinetic torque at 100°/sec was assessed before and after 30 days of 6° head-down bed rest in a group of control subjects who did not receive daily exercise, compared with a group that received 60 min/day, 5 days/wk of cycle exercise at 60%–90% $\dot{V}O_{2max}$ and a group who underwent resistive exercise consisting of peak isokinetic torque at 100°/sec of the knee flexors and extensors (73, 75). The most intriguing finding of this study was that there were no reductions in strength of these muscles in any of the groups, including the control group. This finding may be explained by the experimental protocol that required all groups to undergo strength tests (peak torque development) every 7 days. In essence, even the control group underwent periodic exposure to resistive exercise, which appeared to ameliorate a 20% reduction in strength of knee extensors expected for this duration of exposure to microgravity (Table 36.5). These results may have important implications by indicating that function of large muscle groups may be maintained during long-duration exposure to microgravity with relatively infrequent exposure to peak force development.

Orthostatic Hypotension after Microgravity Exposure

Predisposition toward low blood pressure (orthostatic hypotension) and syncopal symptoms during standing, head-up tilt, or lower body negative pressure is well documented in subjects who have been exposed to periods of actual or simulated microgravity (12, 15, 28, 35, 67, 80, 88, 90, 96, 143, 150). Orthostatic hypotension induced by microgravity is associated with increased venous compliance of the lower extremities (38, 88, 122), reduced plasma and blood volume (12, 26, 28, 30, 35, 44, 80, 143), decreased left ventricular–end-diastolic volume with consequent lowering of stroke volume and cardiac output (31, 143), lower resting venous pressure (31, 108), and attenuated cardiac baroreflex response (28, 33, 35, 68, 128). This reduced capacity of the cardiovascular system to adequately respond and maintain blood pressure during the challenge of standing in Earth's gravity is accompanied by an apparent deconditioning effect associated with physical inactivity and reduced $\dot{V}O_{2max}$. Interestingly, subjects with higher initial aerobic capacity demonstate greater reductions in $\dot{V}O_{2max}$ (44, 141, 150), blood volume (44), and orthostatic tolerance (24, 150) following exposure to bed rest, compared with less fit subjects. The repeated observation that exposure to bed rest induces greater reduction in $\dot{V}O_{2max}$ and a higher incidence of syncope in athletic subjects compared to their sedentary counterparts (24, 143, 150) supports the notion that regular exercise training may provide protection against cardiovascular adaptations to microgravity that underly orthostatic hypotension. The absence of regular physical activity and consequent deconditioning may be one factor that can contribute to reduced effectiveness of blood pressure control during orthostatic challenges following adaptation to microgravity. This raises the intriguing possibility that regular repeated muscular activity designed to protect aerobic capacity during exposure to microgravity may provide protection of the blood pressure control system and orthostatic integrity.

It is well documented that endurance exercise training in ambulatory subjects increases $\dot{V}O_{2max}$ and blood volume (27) and venous pressure (47). Since hypovolemia and reduced central venous pressure occur during adaptation to microgravity and are associated with orthostatic hypotension following spaceflight (15, 26, 88) and bed rest (12, 30, 35, 90, 143), it seems reasonable to propose that endurance exercise training might prove an effective countermeasure against orthostatic hypotension. However, data from both spaceflight and ground-based experiments do not necessarily support this hypothesis. The effectiveness of a daily cycle ergometer exercise program at ~65% $\dot{V}O_{2max}$ in maintaining physical fitness in crews was indicated by an average 8% increase in $\dot{V}O_{2max}$ in three astronauts who completed 84 days of spaceflight (Fig. 36.13). Despite this extensive in-flight exercise training, astronauts experienced 16% plasma volume reduction and increased venous compliance which were associated with orthostatic instability postflight (26, 96). The effects of exercise training during 30 days of bed rest on orthostatic hypotension were assessed in three groups of subjects who had undergone no exercise, endurance exercise, or resistive exercise for two 30 min periods daily for 5 days per wk during bed rest (80). Endurance exercise training maintained $\dot{V}O_{2max}$ and plasma volume at pre–bed rest levels while the other groups experienced significant reductions in these parameters. However, tolerance of head-up tilt was significantly reduced in all groups with no difference between them. Since blood volume was maintained in the endurance-trained subjects, these data provide evidence that physiological mechanisms other

than hypovolemia must be important in the development of orthostatic hypotension following adaptation to microgravity. Results from spaceflight and ground-based studies suggest that repeated endurance exercise designed to defend aerobic capacity may not provide the appropriate stimuli to the mechanism that underlies orthostatic instability.

Attenuated baroreflex control of heart rate occurs during spaceflight (66, 128) and bed rest (26, 28, 30, 35, 36) independent of hypovolemia (28, 30, 35, 161). In addition, greater impairment of baroreflex function and orthostatic hypotension with lower heart rate elevation during standing after bed rest was reported in subjects who experienced syncopal symptoms compared to subjects who tolerated standing with no noticeable difficulty (35). It can be concluded that microgravity leads to substantial and progressive development of baroreflex malfunction, which is significantly related to the occurrence of hypotension and syncope during standing. Development of effective exercise countermeasures for orthostatic hypotension through in-flight physical training might have to include exercise regimens that can specifically increase cardiac baroreflex responses prior to reentry.

Ineffectiveness of using extensive aerobic exercise of moderate intensity and long duration to counteract orthostatic instability following adaptation to microgravity may represent the failure to identify and employ a specific stimulus profile that reverses or attenuates baroreflex impairment, in addition to other factors associated with orthostatic hypotension. Exercise prescriptions used during long-duration spaceflight have been primarily made up of conventional, daily repeated, dynamic exercise on a cycle or treadmill ergometer for durations of 30 min to 2.5 h daily (26). Ground-based experiments have used more conventional exercise training protocols comprised of repeated daily regimens of submaximal intensity and long duration (Table 36.6). Unfortunately, dynamic and resistive training regimens in the earth gravity environment employing repeated daily work sessions have failed to produce chronic changes in the sensitivity of cardiac baroreflex responses (50, 145, 156). In light of the observation that orthostatic intolerance following exposure to microgravity is associated with some impairment of cardiac baroreflex responses, it is not surprising that exercise training that does not alter baroreflex responsiveness in terrestrial gravity fails to provide protection against orthostatic instability following adaptation to microgravity.

There is evidence that use of graded exercise designed to elicit maximal effort can acutely restore various cardiovascular and metabolic capacities attenuated by microgravity. Such exercise leads to substantial and prolonged increase in the sensitivity of cardiac baroreflex responses through at least 24 h of recovery in 1G (32) and during bed rest (36). Bed rest and physical inactivity reduces insulin receptor sensitivity in humans (59, 84); one bout of maximal exercise following 10 days of detraining returned insulin sensitivity to normal levels [84]. One bout of maximal exercise at the end of 10 days of bed rest restored $\dot{V}O_{2max}$, heart rate, blood pressures, rate–pressure product, oxygen pulse, endurance time and orthostatic stability on a treadmill to pre–bed rest levels within 2 h of ambulation (25). Expansion of plasma volume by 10% has also been induced within 24 h by maximal graded exercise (71). A single bout of maximal exercise has reversed fainting episodes following acute exposure to simulated microgravity (150), an effect which might be related to acute blood volume expansion and increased baroreflex sensitivity. The use of one bout of maximal exercise within 24 h of reentry may represent a tenable countermeasure against the development of postflight orthostatic hypotension. The results from these studies provide a physiological basis for future ground-based and spaceflight testing of acute maximal exercise as a protective measure against postflight orthostatic instability.

Considerations of Future Exercise Prescriptions for Microgravity

The assessment of work rates and metabolic costs of inflight exercise indicates that emphasis has been placed on aerobic conditioning with exercise of greater duration and moderate intensity. Unfortunately, the resulting 2 h or more of daily exercise is extremely costly to the operational work day and life support requirements. For instance, the average daily metabolic cost of 1,450 kcal for exercise during Soviet missions represents about half of the total 3,150 kcal intake on these missions (26, 168). If the combination of exercise intensity, duration, and frequency could be optimized to reduce total exercise volume in half, this could save approximately 225,000 kcal over a 6 mo mission, or enough to feed one astronaut for an additional 75 days (26). Finally, long laborious exercise regimens may cause difficulty in compliance; in fact the Soviets reported that some crew members resisted exercising during their mission (26). Since productive working time and life support (oxygen, water, food) are at a premium during spaceflight, it is paramount to design efficient exercise prescriptions.

Significant reductions in enzyme activities of oxidative pathways and capillarity in muscles of the lower extremities following bed rest suggest the need for exercise during spaceflight that will provide a stimulus to aerobic metabolic processes. When such exercise has been employed during spaceflight, aerobic power has been maintained or increased (126), but orthostatic hypotension has persisted (53, 54, 88, 95, 102, 168,

169). There is sufficient evidence from groundbased experiments to suggest that the use of graded exercise designed to elicit $\dot{V}O_{2max}$ can acutely reverse reductions in $\dot{V}O_{2max}$ (25), exercise endurance time (25), plasma volume (71), sensitivity of cardiac baroreflexes (30–32, 148), and fainting episodes (148) following exposure to microgravity. If effectively applied every 7 to 10 days or just prior to return to terrestrial gravity environments, the use of less frequent and more intense exercise as a possible countermeasure against the loss of cardiovascular and metabolic capacities during exercise and development of postflight orthostatic hypotension could be maximally cost-effective by enhancing crew health maintenance and postflight recovery while minimizing in-flight use of work time, food, water, and oxygen for exercise activities (26).

Another consideration for the development of future exercise countermeasure programs is the use of a specific, resistive mode designed to eliminate or minimize muscle atrophy and reduced muscle strength, rather than more traditional endurance exercise that has been employed in spaceflight exercise programs (26). It seems reasonable that preservation of muscle structure and function in microgravity would depend upon replacing muscle actions and forces that are characteristic of the normal terrestrial gravity environment. The consistent reduction in strength of the lower extremities following spaceflight (81, 109, 164), despite the use of extensive dynamic exercise, suggests that greater resistances are developed by normal ambulatory activities in gravity and are therefore required to preserve muscle function. Specifically, use of eccentric muscle actions, in addition to concentric muscle actions during resistive exercise training in spaceflight, may provide several important advantages. In weightlessness, nearly all actions require muscle shortening (concentric actions), whereas muscle lengthening (eccentric actions) is virtually eliminated. In contrast to microgravity environments, eccentric muscle actions are routine on earth and may be critical to the maintenance of muscle size and function since greater force development can be achieved, compared with concentric actions (see Fig. 36.5B). Resistive training on earth, which involves performance of eccentric and concentric muscle actions, induces two to three times the increase in strength as training with only concentric actions (29, 158, 159). In addition, the incorporation of eccentric actions in resistive training does not appreciably increase the energy cost of exercise, despite greater force development (29), and elicits minimal fatigue compared to concentric actions (160). Eccentric muscle actions also provide the additional advantage of optimum force development independent of the speed of limb movement, while slow limb speeds are required for development of optimum force using concentric actions (Fig. 36.5B). Based on results of a 30 day bed rest study (73), the use of isokinetic concentric resistive exercise as a spaceflight countermeasure has been proposed (75). Restriction of resistive exercise to concentric actions only and a single velocity of limb movement may not provide the needed protection of eccentric force development through varying limb speeds necessary for adequate muscular function upon return to the gravitational environment of earth. Eccentric resistive muscle actions during spaceflight could allow astronauts to exercise at varying velocities through their range of limb movements that are individually comfortable, and perhaps provide speeds of motion more representative of those experienced on earth during ambulatory activities without compromising optimum force development. Advantages of incorporating eccentric resistive training as a countermeasure during spaceflight include lower energy cost, less muscle fatigue, greater force development, and greater strength gain than those developed by concentric actions. Implementation of an ergometer designed to generate active, dynamic resistance of varying velocity, through a wide range of motion similar to that experienced by large muscle groups of the lower extremities during ambulation in terrestrial gravity, might provide a greater preventative measure against muscle atrophy and loss of strength during spaceflight at less time and energy cost to the crew.

SUMMARY

Comparisons of exercise responses during and after exposure to 6° head-down bed rest with those responses to spaceflight support the notion that this ground-based model provides an appropriate analog for studying functional human adaptations that occur in microgravity. On this basis, several conclusions can be drawn about adaptation to microgravity and exercise based on certain consistent findings from both spaceflight and ground-based experiments.

The response to dynamic exercise after adaptation to microgravity includes a reduction in physical endurance, although mechanical efficiency appears unaffected. An impaired capacity for systemic oxygen transport and utilization is manifested by significant reductions in oxygen uptake during the trasient phase of submaximal work as well as lower $\dot{V}O_{2max}$. The magnitude of decrease in $\dot{V}O_{2max}$ is related to duration of exposure and baseline level of aerobic fitness independent of age or gender. Reduced exercise endurance and lower $\dot{V}O_{2max}$ following bed rest are associated with various physiological adaptations, including hypovolemia, reduced systemic and muscle blood flows, and lower aerobic enzyme activities, as well as increased venous compliance and higher heart rate. Impaired cardiac output during exercise is primarily the result of reduced

blood volume rather than myocardial dysfunction. Reduced heat dissipation and lower circulating substrate availability may contribute to a limitation of work performance in tasks of long duration. Adaptation to microgravity reduces the force-generating capacity and fatigability of skeletal muscles. The diminution of muscle function is dependent upon the duration of exposure, is more pronounced for large muscle groups of the lower extremities, and is associated with muscle fiber atrophy and ultrastructural disruptions. The recovery of physiological adaptations associated with aerobic challenge may be complete by 2 to 4 wk, depending on the duration of microgravity exposure. The recovery of muscle structure and function is less defined.

Reduced physical and physiological capacities for exercise performance can be restored or maintained by specific corrective or preventive measures that include regular muscular activity during exposure to microgravity. The effective use of exercise as a countermeasure to prevent diminished exercise performance associated with adaptation to microgravity should include an understanding of changes in physiological mechanisms involved and an ability to reverse these changes. While conventional endurance exercise training involving daily workouts of 1–2 h duration during exposure to microgravity has proven effective in ameliorating the reduction in exercise endurance and $\dot{V}O_{2max}$, it has failed to improve post exposure orthostatic hypotension. Single bouts of intense exercise designed to elicit maximal challenge to aerobic metabolism have been successful in reversing numerous changes in physiologic factors associated with reduced exercise performance and postflight orthostatic hypotension, and may be useful in the development of a corrective countermeasure. The use of dynamic resistive exercise throughout exposure to microgravity should be given consideration for the development of a preventive rather than corrective measure against muscle adaptations. The advantages of incorporating dynamic resistive exercise into training programs, with a strong emphasis on the use of eccentric muscle actions in addition to concentric actions, may prove most effective as a protective measure against muscle atrophy and dysfunction caused by long-duration exposure to microgravity environments.

REFERENCES

1. Atkov, O. Y., V. S. Bednenko, and G. A. Fomina. Ultrasound techniques in space medicine. *Aviat. Space Environ. Med.* 58 (Suppl. 9): A69–A73, 1987.
2. Balakhovskiy I. S., and T. A. Orlova. Dynamics of cosmonauts' blood biochemistry during space missions. *Kosm. Biol. Aviakosm. Med.* 12(6): 3–8, 1978.
3. Bassey, E. J., T. Bennett, A. T. Birmingham, P. D. Fentem, D. Fitton, and R. Goldsmith. Effects of surgical operation and bed

rest on cardiovascular responses to exercise in hospital patients. *Cardiovasc. Res.* 7: 588–592, 1973.

4. Beregovkin, A. V., A. S. Vodolazov, V. S. Georgiyevskiy, L.I. Kakurin, V. V. Kalinichenko, N. V. Korelin, V. M. Mikhaylov, and V.V. Shchigolev. Cardiorespiratory system reactions of cosmonauts to exercise following long-term missions aboard the Salyut-6 orbital station. *Kosm. Biol. Aviakosm. Med.* 14(4): 8–11, 1980.

5. Beregovkin, A. V., A. S. Vodolazov, V. S. Georgiyevskiy, V.V. Kalinichenko, N. V. Korelin, V. M. Mikhaylov, Y. D. Pometov, V. V. Shchigolev, and B. S. Katkovskiy. Reactions of the cardiorespiratory system to a dosed physical load in cosmonauts after 30- and 63-day flights in the Salyut-4 orbital station. *Kosm. Biol. Aviakosm. Med.* 10(5): 24–29, 1976.

6. Berry, C. A., W. G. Squires, and A. S. Jackson. Fitness variables and the lipid profile in United States astronauts. *Aviat. Space Environ. Med.* 51: 1222–1226, 1980.

7. Bevegard, S., A. Holmgren, and B. Jonsson. The effect of body position on the circulation at rest and during exercise with special reference to the influence on the stroke volume. *Acta Physiol. Scand.* 49: 279–298, 1960.

8. Birkhead, N. C., J. J. Blizzard, J. W. Daly, G. J. Haupt, B. Issekutz, R. N. Myers, and K. Rodahl. *Cardiodynamic and Metabolic Effects of Prolonged Bed Rest.* Wright-Patterson Air Force Base, OH: Aerosp. Med. Res. Lab., 1963. (AMRL-TDR-63-37)

9. Birkhead, N. C., J. J. Blizzard, J. W. Daly, G. J. Haupt, B. Issekutz, R. N. Myers, and K. Rodahl. *Cardiodynamic and Metabolic Effects of Prolonged Bed Rest with Daily Recumbent or Sitting Exercise and with Sitting Inactivity.* Wright-Patterson Air Force Base, OH: Aerosp. Med. Res. Lab., 1964. (AMRL-TDR-64-61)

10. Blamick, C. A., D. J. Goldwater, and V. A. Convertino. Leg vascular responsiveness during acute orthostasis following simulated weightlessness. *Aviat. Space Environ. Med.* 59: 40–43, 1988.

11. Blomqvist, C. G., J. H. Mitchell, and B. Saltin. Effects of bed rest on the oxygen transport system. In: *Hypogravic and Hypodynamic Environments,* edited by R. H. Murry and M. McCally. Washington, DC: National Aeronautics and Space Administration, 1971, p. 171–176. (NASA SP-269)

12. Blomqvist, C. G., and H. L. Stone. Cardiovascular adjustments to gravitational stress. In: *Handbook of Physiology, Peripheral Circulation and Organ Blood Flow,* edited by J. T. Shepherd and F. M. Abboud. Bethesda, MD: Am. Physiol. Soc., 1983, sec. 2, vol. 3, part 2, p. 1027–1063.

13. Buchanan, P., and V. A. Convertino. A study of the effects of prolonged microgravity on the musculature of the lower extremities in man: an introduction. *Aviat. Space Environ. Med.* 60: 649–652, 1989.

14. Buderer, M. C., J. A. Rummel, E. L. Michael, D. C. Maulden, and C. F. Sawin. Exercise cardiac output following Skylab missions: the second manned Skylab mission. *Aviat. Space Environ. Med.* 47: 365–372, 1976.

15. Bungo, M. W., J. B. Charles, and P. C. Johnson. Cardiovascular deconditioning during space flight and the use of saline as a countermeasure to orthostatic intolerance. *Aviat. Space Environ. Med.* 56: 985–990, 1985.

16. Butusov, A. A., V. R. Lyamin, A. A. Lebedev, A. P. Polyakova, I. B. Svistunov, V. A. Tishler, and A. P. Shulenin. Results of routine medical monitoring of cosmonauts during flight on the Soyuz-9 ship. *Kosm. Biol. Med.* 4(6): 35–39, 1970.

17. Buyanov, P. V., A. V. Beregovkin, and N. V. Pisarenko. Prevention of the adverse effect of hypkinesia on the human cardiovascular system. *Kosm. Biol. Med.* 1(1): 95–99, 1967.

18. Cardus, D. Effects of 10 days recumbency on the response to the bicycle ergometer test. *Aerosp. Med.* 37: 993–999, 1966.

19. Chase, G. A., C. Grave, and L. B. Rowell. Independence of changes in functional and performance capacities attending prolonged bed rest. *Aerosp. Med.* 37: 1232–1238, 1966.

20. Chekirda, I. F., and A. V. Yeremin. Dynamics of cyclic and acyclic locomotion of the Soyuz-18 crew after a 63-day space mission. *Kosm. Biol. Aviakosm. Med.* 11(4): 9–13, 1977.

21. Cherepakhin, M. A., and V. I. Pervushin. Space flight effect on the neuromuscular system of cosmonauts. *Kosm. Biol. Med.* 4(6): 46–49, 1970.

22. Convertino, V.A. Effect of orthostatic stress on exercise performance after bed rest: relation to inhospital rehabilitation. *J. Card. Rehabil.* 3: 660–663, 1983.

23. Convertino, V. A. Exercise responses after inactivity. In: *Inactivity: Physiological Effects,* edited by H. Sandler and J. Vernikos-Danellis. Orlando, FL: Academic, 1986, p. 149–191.

24. Convertino, V. A. Aerobic fitness, endurance training and orthostatic intolerance. *Exerc. Sports Sci. Rev.* 15: 223–259, 1987.

25. Convertino, V. A. Potential benefits of maximal exercise just prior to return from weightlessness. *Aviat. Space Environ. Med.* 58: 568–572, 1987.

26. Convertino, V. A. Physiological adaptations to weightlessness: effects on exercise and work performance. *Exerc. Sport Sci. Rev.* 18: 119–165, 1990.

27. Convertino, V. A. Blood volume: its adaptation to endurance training. *Med. Sci. Sports Exerc.* 12: 1338+–+1348, 1991.

28. Convertino, V. A. Carotid-cardiac baroreflex: relation with orthostatic hypotension following simulated microgravity and implications for development of countermeasures. *Acta Astronautica* 23: 9–17, 1991.

29. Convertino, V. A. Neuromuscular aspects in development of exercise countermeasures. *Physiologist* 34 (Suppl.): S125–S128, 1991.

30. Convertino, V. A. Effects of exercise and inactivity on intravascular volume and cardiovascular control mechanisms. *Acta Astronautica* 27: 123–129, 1992.

31. Convertino, V. A. Adaptation of baroreflexes and orthostatic hypotension. In: *Vascular Medicine,* edited by H. Boccalon. Amsterdam: Elsevier Sci. Pub., 1993, p. 573–577.

32. Convertino, V. A., and W. C. Adams. Enhanced vagal baroreflex response during 24 hours after acute exercise. *Am. J. Physiol.* 260 (*Regulatory Integrative Comp. Physiol.* 31): R570–R575, 1991.

33. Convertino, V. A., W. C. Adams, J. D. Shea, C. A. Thompson, and G. W. Hoffler. Impairment of the carotid-cardiac vagal baroreflex in wheelchair-dependent quadriplegics. *Am. J. Physiol.* 260 (*Regulatory Integrative Comp. Physiol.* 31): R576–R580, 1991.

34. Convertino, V. A., R. Bisson, R. Bates, D. Goldwater, and H. Sandler. Effects of antiorthostatic bedrest on the cardiorespiratory responses to exercise. *Aviat. Space Environ. Med.* 52: 251–255, 1981.

35. Convertino, V. A., D. F. Doerr, D. L. Eckberg, J. M. Fritsch, and J. Vernikos-Danellis. Head-down bedrest impairs vagal baroreflex responses and provokes orthostatic hypotension. *J. Appl. Physiol.: Respir. Environ. Exerc. Physiol.* 68: 1458–1464, 1990.

36. Convertino, V. A., D. F. Doerr, A. Guell, and J. F. Marini. Effects of acute exercise on attenuated vagal baroreflex function during bedrest. *Aviat. Space Environ. Med.* 63: 999–1003, 1992.

37. Convertino, V. A., D. F. Doerr, K. L. Mathes, S. L. Stein, and P. Buchanan. Changes in volume, muscle compartment, and compliance of the lower extremities in man following 30 days of exposure to simulated microgravity. *Aviat. Space Environ. Med.* 60: 653–658, 1989.

38. Convertino, V. A., D. F. Doerr, and S. F. Stein. Changes in size and compliance of the calf following 30 days of simulated micro-

gravity. *J. Appl. Physiol.: Respir. Environ. Exerc. Physiol.* 66: 1509–1512, 1989.

39. Convertino, V. A., D. J. Goldwater, and H. Sandler. Effect of orthostatic stress on exercise performance after bedrest. *Aviat. Space Environ. Med.* 53: 652–657, 1982.

40. Convertino, V. A., D. J. Goldwater, and H. Sandler. Oxygen uptake kinetics of constant-load work: upright vs supine exercise. *Aviat. Space Environ. Med.* 55: 501–506, 1984.

41. Convertino, V. A., D. J. Goldwater, and H. Sandler. VO_2 kinetics of constant-load exercise following bedrest-induced deconditioning. *J. Appl. Physiol.: Respir. Environ. Exerc. Physiol.* 57: 1545–1550, 1984.

42. Convertino, V. A., D. J. Goldwater, and H. Sandler. Bedrest-induced peak VO_2 reduction associated with age, gender and aerobic capacity. *Aviat. Space Environ. Med.* 57: 17–22, 1986.

43. Convertino, V. A., J. Hung, D. J. Goldwater, and R. F. DeBusk. Cardiovascular responses to exercise in middle-aged men following ten days of bed rest. *Circulation* 65: 134–140, 1982.

44. Convertino, V. A., G. M. Karst, S. M. Kinzer, D. A. Williams, and D. J. Goldwater. Exercise capacity following simulated weightlessness in trained and nontrained subjects (abstract). *Aviat. Space Environ. Med.* 56: 489, 1985.

45. Convertino, V. A., G. M. Karst, C. R. Kirby, and D. J. Goldwater. Effect of simulated weightlessness on exercise-induced anaerobic threshold. *Aviat. Space Environ. Med.* 57: 325–331, 1986.

46. Convertino, V. A., C. R. Kirby, G. M. Karst, and D. J. Goldwater. Responses to muscular exercise following repeated simulated weightlessness. *Aviat. Space Environ. Med.* 56: 540–546, 1985.

47. Convertino, V. A., G. W. Mack, and E. R. Nadel. Elevated venous pressure: a consequence of exercise training-induced hypervolemia? *Am. J. Physiol.* 260 (*Regulatory Integrative Comp. Physiol.* 31): R273–R277, 1991.

48. Convertino, V. A., H. Sandler, P. Webb, and J. F. Annis. Induced venous pooling and cardiorespiratory responses to exercise after bedrest. *J. Appl. Physiol.: Respir. Environ. Exerc. Physiol.* 52: 1343–1348, 1982.

49. Convertino, V. A., R. W. Stremel, E. M. Bernauer, and J. E. Greenleaf. Cardiorespiratory responses to exercise after bed rest in men and women. *Acta Astronautica* 4: 895–905, 1977.

50. Convertino, V. A., C. A. Thompson, D. L. Eckberg, J. M. Fritsch, G. W. Mack, and E. R. Nadel. Baroreflex responses and LBNP tolerance following exercise training. *Physiologist* 33: S40–S41, 1990.

51. Cooper, K. H., and J. W. Ord. Physical effects of seated and supine exercise with and without subatmospheric pressure applied to the lower body. *Aerosp. Med.* 39: 481–484, 1968.

52. DeBusk, R. F., V. A. Convertino, J. Hung, and D. Goldwater. Exercise conditioning in middle-aged men after 10 days of bed rest. *Circulation* 68: 245–250, 1983.

53. Degtyarev, V. A., V. G. Doroshev, T. V. Batenchuk-Tusko, Z. A. Kirillova, N. A. Lapshina, S. I. Ponamarev, and V. N. Ragozin. Studies of circulation during LBNP test aboard Salyut-4 orbital station. *Kosm. Biol. Aviakosm. Med.* 11(3): 26–31, 1977.

54. Degtyarev, V. A., V. G. Doroshev, N. D. Kalmykova, Z. A. Kirillova, Y. A. Kukushkin, and N. A. Lapshina. Results of examination of the crew of the Salyut space station in a functional test with creation of negative pressure on the lower half of the body. *Kosm. Biol. Aviakosm. Med.* 8(3): 47–52, 1974.

55. Degtyarev, V. A., V. G. Doroshev, N. D. Kalmykova, Z. A. Kirillova, and N. A. Lapshina. Dynamics of circulatory indices in the crew of the Salyut orbital station during an examination under rest conditions. *Kosm. Biol. Aviakosm. Med.* 8(2): 34–42, 1974.

56. Degtyarev, V. A., V. G. Doroshev, N. D. Kalmykova, Z. A. Kirillova, N. A. Lapshina, A. A. Lepskiy, and V. N. Rabozin. Studies of hemodynamics and phase structure of cardiac cycle in the crew of Salyut-4. *Kosm. Biol. Aviakosm. Med.* 12(6): 9–14, 1978.

57. Degtyarev, V. A., V. G. Doroshev, N. D. Kalmykova, Y. A. Kukushkin, Z. A. Kirillova, N. A. Lapshina, I. I. Popov, V. N. Ragozin, and V. I. Stepantsov. Dynamics of circulatory parameters of the crew of the Salyut space station in functional test with physical load. *Kosm. Biol. Aviakosm. Med.* 12(3): 15–20, 1978.

58. Deitrick, J. E., G. D. Whedon, E. Shorr, V. Toscani, and V. B. Davis. Effects of immobilization upon various metabolic and physiologic functions of normal men. *Am. J. Med.* 4: 3–35, 1948.

59. Dolkas, C., and J. Greenleaf. Insulin and glucose responses during bed rest with isotonic and isometric exercise. *J. Appl. Physiol.* 43: 1033–1038, 1977.

60. Doroshev, V. G., T. V. Batenchuk-Tusko, N. A. Lapshina, Y. A. Kukushkin, N. D. Kalmykova, and V. N. Ragozin. Changes in hemodynamics and phasic structure of the cardiac cycle in the crew on the second expedition of Salyut-4. *Kosm. Biol. Aviakosm. Med.* 11(2): 26–31, 1977.

61. Dudley, G. A., M. R. Duvoisin, V. A. Convertino, and P. Buchanan. Alterations of the *in vivo* torque-velocity relationship of human skeletal muscle following 30 days exposure to simulated microgravity. *Aviat. Space Environ. Med.* 60: 659–663, 1989.

62. Duvoisin, M. R., V. A. Convertino, P. Buchanan, P. D. Gollnick, and G. A. Dudley. Characteristics and preliminary observations of the influence of electromyostimulation on the size and function of human skeletal muscle during 30 days of simulated microgravity. *Aviat. Space Environ. Med.* 60: 671–678, 1989.

63. Fischer, C. L., P. C. Johnson, and C. A. Berry. Red blood cell and plasma volume changes in manned spaceflight. *J. A. M. A.* 200: 579–583, 1967.

64. Friden, J., P. N. Sfakianos, and A. R. Hargens. Muscle soreness and intramuscular fluid pressure: comparison between eccentric and concentric load. *J. Appl. Physiol.: Respir. Environ. Exerc. Physiol.* 61: 2175–2179, 1986.

65. Friman, G. Effect of clinical bedrest for seven days on physical performance. *Acta Med. Scand.* 205: 389–393, 1979.

66. Fritsch, J. M., J. B. Charles, B. S. Bennett, M. M. Jones, and D. L. Eckberg. Short-duration spaceflight impairs human carotid baroreceptor-cardiac reflex responses. *J. Appl. Physiol.: Respir. Environ. Exerc. Physiol.* 73: 664–671, 1992.

67. Gazenko, O. G., A. M. Genin, and A. D. Yegorov. Summary of medical investigations in the USSR manned space missions. *Acta Astronautica* 8: 907–917, 1981.

68. Gazenko, O. G., N. N. Gurovsky, A. M. Genin, I. I. Bryanov, A. V. Eryomin, and A. D. Egorov. Results of medical investigations carried out on board the Salyut orbital stations. *Life Sci. Space Res.* 14: 145–152, 1976.

69. Georgiyevskiy, V. S., L. I. Kakurin, B. S. Katkovskii, and Y. A. Senkevich. Maximum oxygen consumption and functional state of the circulation in simulated zero gravity. In: *The Oxygen Regime of the Organism and its Regulation,* edited by N. V. Lauer and A. Z. Kilchinskaya, Kiev: Naukova Dumka, 1966, p. 181–184.

70. Georgiyevskiy, V. S., N. A. Lapshina, L. Y. Andriyako, L. V. Umnova, V. G. Doroshev, I. V. Alferova, V. N. Ragozin, and Y. A. Kobzev. Circulation in exercising crew members of the first main expedition aboard Salyut-6. *Kosm. Biol. Aviakosm. Med.* 14(3): 15–18, 1980.

71. Gillen, C. M., R. Lee, G. W. Mack, C. M. Tomaselli, T. Nishiyasu, and E. R. Nadel. Plasma volume expansion in humans

after a single intense exercise protocol. *J. Appl. Physiol.: Respir. Environ. Exerc. Physiol.* 71: 1914–1920, 1991.

72. Gogia, P. P., V. S. Schneider, A. D. LeBlanc, J. Krebs, C. Kasson, and C. Pientok. Bed rest effect on extremity muscle torque in healthy men. *Arch. Phys. Med. Rehabil.* 69: 1030–1032, 1988.

73. Greenleaf, J. E., E. M. Bernauer, A. C. Ertl, T. S. Trowbridge, and C. E. Wade. Work capacity during 30-days of bed rest with isotonic and isokinetic exercise training. *J. Appl. Physiol.: Respir. Environ. Exerc. Physiol.* 67: 1820–1826, 1989.

74. Greenleaf, J. E., E. M. Bernauer, L. T. Juhos, H. L. Young, R. W. Staley, and W. van Beaumont. Fluid and electrolyte shifts during bed rest with isometric and isotonic exercise. *J. Appl. Physiol.* 42: 59–66, 1977.

75. Greenleaf, J. E., R. Bulbulian, E. M. Bernauer, W. L. Haskell, and T. Moore. Exercise-training protocols for astronauts in microgravity. *J. Appl. Physiol.: Respir. Environ. Exerc. Physiol.* 67: 2191–2204, 1989.

76. Greenleaf, J. E., and S. Kozlowski. Physiological consequences of reduced physical activity during bed rest. *Exerc. Sport Sci. Rev.* 10: 83–119, 1982.

77. Greenleaf, J. E., and S. Kozlowski. Reduction in peak oxygen uptake after prolonged bed rest. *Med. Sci. Sports Exerc.* 14: 477–480, 1982.

78. Greenleaf, J. E., and R. D. Reese. Exercise thermoregulation after 14 days of bed rest. *J. Appl. Physiol.: Respir. Environ. Exerc. Physiol.* 48: 72–78, 1980.

79. Greenleaf, J. E., W. van Beaumont, V. A. Convertino, and J. C. Starr. Handgrip and general muscular strength and endurance during prolonged bedrest with isometric and isotonic leg exercise training. *Aviat. Space Environ. Med.* 54: 696–700, 1983.

80. Greenleaf, J. E., C. E. Wade, and G. Leftheriotis. Orthostatic responses following 30-day bed rest deconditioning with isotonic and isokinetic exercise training. *Aviat. Space Environ. Med.* 60: 537–542, 1989.

81. Grigoryeva, L. S., and I. B. Kozlovskaya. Effect of weightlessness and hypokinesia on velocity and strength properties of human muscles. *Kosm. Biol. Aviakosm. Med.* 21(1): 27–30, 1987.

82. Gurovskiy, N. N., A. V. Yeremin, O. G. Gazenko, A. D. Yegorov, I. I. Bryanov, and A. M. Genin. Medical investigations during flights of the spaceships Soyuz-12, Soyuz-13, Soyuz-14 and the Salyut-3 orbital station. *Kosm. Biol. Aviakosm. Med.* 9(2): 48–54, 1975.

83. Hargens, A. R., C. M. Tipton, P. D. Gollnick, S. J. Mubarak, B. J. Tucker, and W. H. Akeson. Fluid shifts and muscle function in humans during acute simulated weightlessness. *J. Appl. Physiol.: Respir. Environ. Exerc. Physiol.* 54: 1003–1009, 1983.

84. Heath, G. W., J. R. Gavin, J. M. Hinderliter, J. M. Hagberg, S. A. Bloomfield, and J. O. Holloszy. Effects of exercise and lack of exercise on glucose tolerance and insulin sensitivity. *J. Appl. Physiol.: Respir. Environ. Exerc. Physiol.* 54: 512–517, 1983.

85. Henry, W. L., S. E. Epstein, J. M. Griffith, R. E. Goldstein, and D. R. Redwood. Effect of prolonged space flight on cardiac function and dimensions. In: *Biomedical Results from Skylab,* edited by R. S. Johnston and L. F. Dietlein. Washington, DC: National Aeronautics and Space Administration, 1977, p. 366–371. (NASA SP-377)

86. Hikida, R. S., P. D. Gollnick, G. A. Dudley, V. A. Convertino, and P. Buchanan. Structural and metabolic characteristics of human skeletal muscle following 30 days of simulated microgravity. *Aviat. Space Environ. Med.* 60: 664–670, 1989.

87. Hoche, J., and A. Graybiel. Value of exercise at one-half earth gravity in preventing the deconditioning effects of simulated weightlessness. *Aerosp. Med.* 45: 386–392, 1974.

88. Hoffler, G. W. Cardiovascular studies of U.S. space crews: an overview and perspective. In: *Cardiovascular Flow Dynamics and Measurements,* edited by N. H. C. Hwang and N. A. Normann. Baltimore: University Park Press, 1977, pp. 335–363.

89. Hung, J., D. Goldwater, V. A. Convertino, J. H. McKillop, M. L. Goris, and R. F. DeBusk. Mechanisms for decreased exercise capacity following bedrest in normal middle-aged men. *Am. J. Cardiol.* 51: 344–348, 1983.

90. Hyatt, K. H., and D. A. West. Reversal of bedrest-induced orthostatic intolerance by lower body negative pressure and saline. *Aviat. Space Environ. Med.* 48: 120–124, 1977.

91. Iseyev, L. R., and B. S. Katkovskiy. Unidirectional changes in the human oxygen balance caused by bed confinement and restriction to an isolation chamber. *Kosm. Biol. Med.* 2(4): 117–124, 1968.

92. Iseyev, L. R., and Y. G. Nefedov. Human tolerance to physical stress during four months isolation in a closed space. *Kosm. Biol. Med.* 2(1): 60–65, 1968.

93. Johnson, P. C., T. B. Driscoll, and A. D. LeBlanc. Blood volume changes. In: *Biomedical Results from Skylab,* edited by R. S. Johnston and L. F. Dietlein. Washington, DC: National Aeronautics and Space Administration, 1977, p. 235–241. (NASA SP-377)

94. Johnson, P. C., S. L. Kimzey, and T. B. Driscoll. Postmission plasma volume and red-cell mass changes in the crews of the first two Skylab missions. *Acta Astronautica* 2: 311–317, 1975.

95. Johnson, R. L., G. W. Hoffler, A. Nicogossian, and S. A. Bergman. Skylab experiment M-092: results of the first manned mission. *Acta Astronautica* 2: 265–296, 1975.

96. Johnson, R. L., G. W. Hoffler, A. Nicogossian, S. A. Bergman, and M. M. Jackson. Lower body negative pressure: third manned Skylab mission. In: *Biomedical Results from Skylab,* edited by R. S. Johnson and L. F. Dietlein. Washington, DC: National Aeronautics and Space Administration, 1977, p. 284–312. (NASA SP-377)

97. Kakurin, L. I., R. M. Akhrem-Adhremovich, Y. V. Vanyushina, R. A. Varbaronov, V. S. Georgiyevskii, B. S. Kotkovskiy, A. R. Kotovskaya, N. M. Mukharlyamov, N. Y. Panferova, Y. T. Pushkar, Y. A. Senkevich, S. F. Simpura, M. A. Cherpakhin, and P. G. Shamrov. The influence of restricted muscular activity on man's endurance of physical stress, accelerations and orthostatics. In: *Soviet Conference on Space Biology and Medicine,* Moscow, 1966, p. 110–117.

98. Kakurin, L. I., M. A. Cherepakhin, and V. I. Pervushin. Effect of spaceflight factors on human muscle tone. *Kosm. Biol. Med.* 5(2): 63–68, 1971.

99. Kakurin, L. I., M. A. Cherepakhin, and V. I. Pervushin. Effect of brief space flights on the human neuromuscular system. *Kosm. Biol. Med.* 5(6): 53–56, 1971.

100. Kakurin, L. I., B. S. Kamkovskiy, V. S. Giorgiyevskiy, Yu. N. Purakhan, M. A. Cherenikhin, B. M. Mikhalylov, B. N. Pimukhov, and Ye. N. Buryikov. Functional disturbances during hypokinesia in man. *Vopr. Kurotol. Fizioter. Lech. Fizich. Kult.* 35: 19–24, 1970.

101. Kakurin, L. I., V. I. Lobachik, V. M. Mikhailov, and Yu. A. Senkevich. Antiorthostatic hypokinesia as a method of weightlessness simulation. *Aviat. Space Environ. Med.* 47: 1083–1086, 1976.

102. Kalinichenko, V. V. Dynamics of orthostatic stability of cosmonauts following 2 to 63-day missions. *Kosm. Biol. Aviakosm. Med.* 11(3): 31–37, 1977.

103. Kasyan, I. I., and G. F. Makarov. External respiration, gas exchange and energy expenditures of man in weightlessness. *Kosm. Biol. Aviakosm. Med.* 18(6): 4–9, 1984.

104. Katkov, V. Y., and L. I. Kakurin. The role of skeletal muscle tone in regulation of orthostatic circulation. *Kosm. Biol. Aviakosm. Med.* 12(1): 75–78, 1978.

105. Katkovskiy, B. S., G. V. Machinskiy, P. S. Toman, V. I. Dani-

lova, and B. F. Demida. Man's physical performance after thirty-day hypokinesia with countermeasures. *Kosm. Biol. Med.* 8(4): 43–47, 1974.

106. Katkovskiy, B. S., O. A. Pilysvskiy, and G. I. Smirnova. Effect of long-term hypokinesia on human tolerance to physical stress. *Kosm. Biol. Med.* 3(2): 49–54, 1969.

107. Katkovskiy, B. S., and Y. D. Pometov. Cardiac output during physical exercises following real and simulated space flight. In: *Symposium on Gravitational Physiology.* Berlin: Akademie-Verlag, 1976, p. 301–305.

108. Kirsch, K. A., L. Rocker, O. H. Gauer, and R. Krause. Venous pressure in man during weightlessness. *Science* 225: 218–219, 1984.

109. Kozlovskaya, I. B., L. S. Grigoryeva, and G. I. Gevlich. Comparative analysis of effects of weightlessness and its models on velocity and strength properties and tone of human skeletal muscles. *Kosm. Biol. Aviakosm. Med.* 18(6): 22–26, 1984.

110. Krupina, T. N., and A. Y. Tizul. Changes in the nervous system during a 120-day clinostatic hypokinesia and the prophylaxis of hypokinetic disorders. *Z. Neuropatol. Psikhiatr.* 71: 1611–1617, 1971.

111. LaFevers, E. V., C. R. Booher, W. N. Crozier, and J. Donaldson. Effects of 28 days of bedrest immobilization on the responses of skeletal muscle to isometric stress. In: *JSC/Methodist Hospital 28-day Bedrest Study. Vol. II,* edited by P. C. Johnson and C. Mitchell. Houston, TX: Lyndon B. Johnson Space Center, 1977. (NASA 9-14578)

112. LaFevers, E. V., A. E. Nicogossian, and W. N. Hursta. *Electromyographic Analysis of Skeletal Muscle Changes Arising from 9 Days of Weightlessness in the Apollo-Soyuz Space Mission.* Washington, DC: National Aeronautics and Space Administration, 1976. (NASA TM X-58177)

113. Lamb, L. E., R. L. Johnson, P. M. Stevens, and B. E. Welch. Cardiovascular deconditioning from space cabin simulator confinement. *Aerosp. Med.* 35: 420–428, 1964.

114. Lamb, L. E., P. M. Stevens, and R. L. Johnson. Hypokinesia secondary to chair rest from 4 to 10 days. *Aerosp. Med.* 36: 755–763, 1965.

115. Leach, C. S., S. I. Altchuler, and N. M. Cintron-Trevino. The endocrine and metabolic responses to space flight. *Med. Sci. Sports Exerc.* 15: 432–440, 1983.

116. Leach, C. S., and P. C. Johnson. Influence of spaceflight on erythrokinetics in man. *Science* 225: 216–218, 1984.

117. Leach, C. S., J. I. Leonard, P. C. Rambaut, and P. C. Johnson. Evaporative water loss in man in a gravity-free environment. *J. Appl. Physiol.: Respir. Environ. Exerc. Physiol.* 45: 430–436, 1978.

118. Leach, C. S., and P. C. Rambaut. Endocrine responses in long-duration manned space flight. *Acta Astronautica* 2: 115–127, 1975.

119. LeBlanc, A., P. Gorgio, V. Schneider, J. Krebs, E. Schonfeld, and H. Evans. Calf muscle area and strength changes after five weeks of horizontal bed rest. *Am. J. Sports Med.* 16: 624–629, 1988.

120. Legenkov, V. I., I. S. Balskhovskiy, A. V. Beregovkin, Z. S. Moshkalo, and G. V. Sorokina. Changes in composition of the peripheral blood during 18- and 24-day space flights. *Kosm. Biol. Med.* 7(1): 39–45, 1973.

121. Legenkov, V. I., R. K. Kiselev, V. I. Gudim, and G. P. Moskaleva. Changes in peripheral blood of crew members of the Salyut-4 orbital station. *Kosm. Biol. Aviakosm. Med.* 11(6): 3–12, 1977.

122. Luft, U. C., L. G. Myhre, J. A. Loeppky, and M. D. Venters. A study of factors affecting tolerance of gravitational stress simulated by lower body negative pressure. In: *Research Report on Specialized Physiology Studies in Support of Manned Space*

Flight. Albuquerque, NM: Lovelace Foundation, 1976. (NASA 9-14472)

123. Martin, T. P., V. R. Edgerton, and R. E. Grindeland. Influence of spaceflight on rat skeletal muscle. *J. Appl. Physiol.: Respir. Environ. Exerc. Physiol.* 65: 2318–2325, 1988.

124. Meehan, J. P., J. P. Henry, S. Brunjes, and H. de Vries. *Investigation to Determine the Effects of Long-term Bed Rest on G-tolerance and on Psychomotor Performance.* Los Angeles, CA: Dept. of Physiology, University of Southern California, 1966. (NASA-CR-62073)

125. Michel, E. L., J. A. Rummel, and C. F. Sawin. Skylab experiment M-171 "metabolic activity"—results of the first manned mission. *Acta Astronautica* 2: 351–365, 1975.

126. Michel, E. L., J. A. Rummel, C. F. Sawin, M. C. Buderer, and J. D. Lem. Results of Skylab medical experiment M171—metabolic activity. In: *Biomedical Results from Skylab,* edited by R. S. Johnson and L. F. Dietlein. Washington, DC: National Aeronautics and Space Administration, 1977, p. 372–287. (NASA SP-377)

127. Miller, P. B., R. L. Johnson, and L. E. Lamb. Effects of moderate physical exercise during four weeks of bed rest on circulatory functions in man. *Aerosp. Med.* 36: 1077–1082, 1965.

128. Nicogossian, A. E., J. B. Charles, M. W. Bungo, and C. S. Leach-Huntoon. Cardiovascular function in space flight. *Acta Astronautica* 24: 323–328, 1991.

129. Nicogossian, A. E., C. F. Sawin, and P. J. Bartelloni. Results of pulmonary function tests. In: *The Apollo-Soyuz Test Project Medical Report,* edited by A. E. Nicogossian. Washington, DC: National Aeronautics and Space Administration, 1977, p. 25–28. (NASA SP-411)

130. Orlee, H. D., B. Corbin, G. Dugger, and C. Smith. *An Evaluation of the Effects of Bed Rest, Sleep Deprivation and Discontinuance of Training on the Physical Fitness of Highly Trained Young Men.* Searcy, AR: Harding College, 1973. (NASA-CR-134044)

131. Panov, A. G., and V. S. Lobzin. Some neurological problems in space medicine. *Kosm. Biol. Med.* 2: 59–67, 1968.

132. Poliner, L. R., G. J. Dehmer, S. E. Lewis, R. W. Parkey, C. G. Blomqvist, and J. T. Willerson. Left ventricular performance in normal subjects: a comparison of the responses to exercise in the upright and supine positions. *Circulation* 62: 528–534, 1980.

133. Riley, D. A., S. Ellis, G. R. Slocum, T. Satyanarayana, J. L. W. Bain, and F. R. Sedlak. Hypogravity-induced atrophy of rat soleus and extensor digitorum longus muscles. *Muscle Nerve* 10: 560–568, 1987.

134. Robinson, B. F., S. E. Epstein, R. L. Kahler, and E. Braunwald. Circulatory effects of acute expansion of blood. *Circ. Res.* 19: 26–32, 1966.

135. Rodahl, K., N. C. Birkhead, J. J. Blizzard, B. Issekutz, Jr., and E. D. R. Pruett. Physiological changes during prolonged bed rest. In: *Nutrition and Physical Activity,* edited by G. Blix. Uppsala: Almqvist and Wiksells, 1967, p. 107–113.

136. Rudnyy, N. M., O. G. Gazenko, S. A. Gozulov, I. D. Pestov, P. V. Vasilyev, A. V. Yeremin, V. A. Degtyarev, I. S. Balakhovskiy, R. M. Bayevskiy, and G. D. Syrykh. Main results of medical research conducted during the flight of two crews on the Salyut-5 orbital station. *Kosm. Biol. Aviakosm. Med.* 11(5): 33–41, 1977.

137. Rummel, J. A., E. L. Michel, and C. A. Berry. Physiological responses to exercise after space flight—Apollo 7 to Apollo 11. *Aviat. Space Environ. Med.* 44: 235–238, 1973.

138. Rummel, J. A., E. L. Michel, C. F. Sawin, and M. C. Buderer. Medical experiment M-171: Results from the second manned skylab mission. *Aviat. Space Environ. Med.* 47: 1056–1060, 1976.

139. Rummel, J. A., C. F. Sawin, M. C. Buderer, D. G. Mauldin, and E. L. Michel. Physiological responses to exercise after space flight—Apollo 14 through Apollo 17. *Aviat. Space Environ. Med.* 46: 679–683, 1975.

140. Rummel, J. A., C. F. Sawin, and E. L. Michel. Exercise response. In: *Biomedical Results of Apollo,* edited by R. S. Johnson, L. F. Dietlein and C. A. Berry. Washington, DC: National Aeronautics and Space Administration, 1975, p. 265–275. (NASA SP-368)

141. Saltin, B., G. Blomqvist, J. H. Mitchell, R. L. Johnson, K. Wildenthal, and C. B. Chapman. Response to exercise after bed rest and after training. *Circulation* 38 (Suppl. 7): 1–78, 1968.

142. Saltin, B., and L. B. Rowell. Functional adaptations to physical activity and inactivity. *Federation Proc.* 39: 1506–1513, 1980.

143. Sandler, H. Cardiovascular effects of inactivity. In: *Inactivity: Physiological Effects,* edited by H. Sandler and J. Vernikos. Orlando, FL: Academic, 1986, p. 1–9.

144. Sawin, C. F., A. E. Nicogossian, A. P. Schachter, J. A. Rummel, and E. L. Michel. Pulmonary function evaluation during and following Skylab space flights. In: *Biomedical Results from Skylab,* edited by R. S. Johnson, L. F. Dietlein and C. A. Berry. Washington, DC: National Aeronautics and Space Administration, 1977, p. 388–394. (NASA SP-368)

145. Seals, D. R., and P. B. Chase. Influence of physical training on heart rate variability and baroreflex circulatory control. *J. Appl. Physiol.: Respir. Environ. Exerc. Physiol.* 66: 1886–1895, 1989.

146. Siminov, P. V., and I. I. Kasyan. *Physiological Investigations in Weightlessness.* Moscow: Medicine Publishers, 1983.

147. Sivarajan, E. S., R. A. Bruce, M. J. Almes, B. Green, L. Belanger, B. D. Lindskog, K. M. Newton, and L. W. Mansfield. In-hospital exercise after myocardial infarction does not improve treadmill performance. *N. Engl. J. Med.* 305: 357–362, 1981.

148. Somers, V. K., J. Conway, M. LeWinter, and P. Sleight. The role of baroreflex sensitivity in post-exercise hypotension. *J. Hypertens.* 3: S129–S130, 1985.

149. Stegemann, J., D. Essfeld, and U. Hoffman. Effects of a 7-day head-down tilt (−6°) on the dynamics of oxygen uptake and heart rate adjustment in upright exercise. *Aviat. Space Environ. Med.* 56: 410–414, 1985.

150. Stegemann, J., U. Meier, W. Skipka, W. Hartlieb, B. Hemme, and U. Tibes. Effects of multi-hour immersion with intermittent exercise on urinary excretion and tilt table tolerance in athletes and nonathletes. *Aviat. Space Environ. Med.* 46: 26–29, 1975.

151. Stegemann, J., H.-D. Von Framing, and M. Schiefeling. The effect of a six-hour immersion in thermodifferent water on circulatory control and work capacity in trained and untrained subjects. *Pflugers Arch.* 312: 129–138, 1969.

152. Stevens, P. M., P. B. Miller, C. A. Gilbert, T. N. Lynch, R. L. Johnson, and L. E. Lamb. Influence of long-term lower body negative pressure on the circulatory function of man during prolonged bed rest. *Aerosp. Med.* 37: 357–367, 1966.

153. Stevens, P. M., P. B. Miller, and T. N. Lynch. Effects of lower body negative pressure on physiologic changes due to four weeks of hypoxic bed rest. *Aerosp. Med.* 37: 466–474, 1966.

154. Stremel, R. W., V. A. Convertino, E. M. Bernauer, and J. E. Greenleaf. Cardiorespiratory deconditioning with static and dynamic leg exercise during bed rest. *J. Appl. Physiol.* 41: 905–909, 1976.

155. Syzrantsev, Y. K. Effect of hypodynamia on nitrogen metabolism and importance of graded physical exercises for maintenance of the nitrogen balance. In: *Problems of Space Biology,* edited by V. N. Chernigovskiy. Moscow: Nauka Press, 1967, vol. 7, p. 317–322.

156. Tatro, D. L., G. A. Dudley, and V. A. Convertino. Carotid-cardiac baroreflex response and LBNP tolerance following resistance training. *Med. Sci. Sports Exerc.* 24: 789–796, 1992.

157. Taylor, H. L., A. Henschel, J. Brozek, and A. Keys. Effects of bed rest on cardiovascular function and work performance. *J. Appl. Physiol.* 2: 223–239, 1949.

158. Tesch, P. A., P. Buchanan, and G. A. Dudley. An approach to counteracting long-term microgravity-induced muscle atrophy. *Physiologist* 33: S77–S79, 1990.

159. Tesch, P. A., and E. B. Colliander. Effects of eccentric and concentric resistance training on muscular strength (abstract). *Med. Sci. Sports Exer.* 21 (Suppl.): S88, 1989.

160. Tesch, P. A., G. A. Dudley, M. R. Duvoisin, B. M. Hather, and R. T. Harris. Force and EMG signal patterns during repeated bouts of concentric or eccentric muscle actions. *Acta Physiol. Scand.* 138: 263–271, 1990.

161. Thompson, C. A., D. L. Tatro, D. A. Ludwig, and V. A. Convertino. Baroreflex responses to acute changes in blood volume in humans. *Am. J. Physiol.* 259 (*Regulatory Integrative Comp. Physiol.* 30): R792–R798, 1990.

162. Thornton, W. E., and G. W. Hoffler. Hemodynamic studies of the legs under weightlessness. In: *Biomedical Results from Skylab,* edited by R. S. Johnson and L. F. Dietlein. Washington, DC: National Aeronautics and Space Administration, 1977, p. 324–329. (NASA SP-377)

163. Thornton, W. E., G. W. Hoffler, and J. A. Rummel. Anthropometric changes and fluid shifts. In: *Biomedical Results from Skylab,* edited by R. S. Johnson and L. F. Dietlein. Washington, DC: National Aeronautics and Space Administration, 1977, p. 330–338. (NASA SP-377)

164. Thornton, W. E., and J. A. Rummel. Muscular deconditioning and its prevention in space flight. In: *Biomedical Results from Skylab,* edited by R. S. Johnson and L. F. Dietlein. Washington, DC: National Aeronautics and Space Administration, 1977, p. 191–197. (NASA SP-377)

165. Tigranyan, R. A., I. A. Popova, M. I. Belyakova, N. F. Kalita, Y. G. Tuzova, L. B. Sochilina, and N. A. Davydova. Results of metabolic studies on the crew of the second expedition of the Salyut-4 orbital station. *Kosm. Biol. Aviakosm. Med.* 11(2): 48–53, 1977.

166. Tishler, V. A., A. V. Yeremin, V. I. Stepantsov, and I. I. Funtova. Evaluation of physical work capacity of cosmonauts aboard Salyut-6 station. *Kosm. Biol. Aviakosm. Med.* 20(3): 31–35, 1986.

167. Trimble, R. W., and C. S. Lessard. *Performance Decrement as a Function of Seven Days of Bedrest.* Brooks Air Force Base, TX: School Aerosp. Med., 1970. (SAM-TR-70-56)

168. Vorobyov, E. I., O. G. Gazenko, A. M. Genin, and A. D. Egorov. Medical results of Salyut-6 manned space flights. *Aviat. Space Environ. Med.* 54: S31–S40, 1983.

169. Vorobyev, Y. I., O. G. Gazenko, N. N. Gurovskiy, Y. G. Nefedov, B. B. Yegorov, R. M. Bayevskiy, I. I. Bryanov, A. M. Genin, V. A. Degtyarev, A. D. Yegorov, A. V. Yeremin, and I. D. Pestov. Preliminary results of medical investigations carried out during flight of the second expedition of the Soyuz-4 orbital station. *Kosm. Biol. Aviakosm. Med.* 10(5): 3–18, 1976.

170. Vorobyev, Y. I., O. G. Gazenko, N. N. Gurovskiy, Y. G. Nefedov, B. B. Yegorov, I. I. Spitsa, Y. N. Biryukov, I. I. Bryanov, A. V. Yeremin, and A. D. Yegorov. Experimental Soyuz-Apollo flight. Preliminary results of biomedical investigations carried out during flight of the Soyuz-19 ship. *Kosm. Biol. Aviakosm. Med.* 10(1): 15–22, 1976.

171. Vorobyev, Y. I., O. G. Gazenko, Y. B. Shulzhenko, A. I. Grigoryev, A. S. Barer, A. D. Yegorov, and A. I. Skiba. Preliminary results of medical investigations during 5-month spaceflight aboard Salyut-7–Soyuz-T orbital complex. *Kosm. Biol. Aviakosm. Med.* 20(2): 27–34, 1986.

172. Wasserman, K., B. J. Whipp, S. N. Koyal, and W. L. Beaver. Anaerobic threshold and respiratory gas exchange during exercise. *J. Appl. Physiol.* 35: 236–243, 1973.

173. White, P. D., J. W. Nyberg, and W. J. White. A comparative study of the physiological effects of immersion and recombency. In: *Proc. 2nd Ann. Biomed. Res. Conf. Houston, TX,* 1966, p. 117–166.

174. Williams, D. A., and V. A. Convertino. Circulating lactate and FFA during exercise: Effect of reduction in plasma volume following simulated microgravity. *Aviat. Space Environ. Med.* 59: 1042–1046, 1988.

175. Yegorov, A. D., and O. G. Itsekhovskiy. Study of cardiovascular system during long-term spaceflights. *Kosm. Biol. Aviakosm. Med.* 17(5): 4–6, 1983.

176. Yegorov, A. D., O. G. Itsekhovskiy, A. P. Polyakova, V. F. Turchaninova, I. V. Alferova, V. G. Savelyeva, M. V. Domracheva, T. V. Batenchuk-Tusko, V. G. Doroshev, and Y. A. Kobzev. Results of studies of hemodynamics and phase structure of the cardiac cycle during functional test with graded exercise during 140-day flight aboard the Salyut-6 station. *Kosm. Biol. Aviakosm. Med.* 15(3): 18–22, 1981.

177. Yeremin, A. V., V. V. Bazhanaov, V. L. Marishchuk, V. I. Stepantsov, and T. T. Dzhamgarov. Physical conditioning for man under conditions of prolonged hypodynamia. In: *Problems of Space Biology,* edited by A. M. Genin and P. A. Sorokin. Moscow: Nauka, 1969, p. 192–199.

37. Renal, endocrine, and hemodynamic effects of water immersion in humans

MURRAY EPSTEIN | Nephrology Section, Veterans Affairs Medical Center, and Division of Nephrology, University of Miami School of Medicine, Miami, Florida

CHAPTER CONTENTS

WATER IMMERSION is one of the oldest therapeutic methods. Knowledge or speculation about the therapeutic qualities of water immersion dates back to the earliest days of humanity (8). It is ironic that the recent widespread interest in water immersion as an investigative tool received its impetus not from centuries of hydrotherapeutic practice, but from the modern space program. Reports of orbital manned spaceflights indicate that astronauts undergo a striking natriuresis and diuresis, thought to be a consequence of the cephalad redistribution of body fluids that takes place in a gravity-free environment (34, 43). Since the redistribution of blood volume induced by water immersion parallels that of weightlessness, the water immersion model has been utilized as a means of investigating both normal physiology and deranged volume homeostasis on earth (11).

Water immersion to the neck (NI) has long been known to produce a marked diuresis (8, 34). Several lines of evidence (1, 4, 8, 11, 34, 43) suggest that this effect is mediated by a redistribution of blood volume with a relative increase in central blood volume. The past two decades have witnessed the characterization of many of the hemodynamic alterations of immersion, and the delineation of the myriad effects of immersion on renal function and hormonal change (8, 9, 11). Studies of the "efferent" limb of the immersion model have demonstrated that NI produces a marked natriuresis, kaliuresis, and diuresis, suppression of the renin-aldosterone system, and augmentation of renal vasodilatory prostanoids.

CHARACTERIZATION OF THE "AFFERENT" LIMB OF THE IMMERSION MODEL

Although NI has long been postulated to produce a redistribution of blood volume with a relative increase in central blood volume (8, 33, 34, 39), only in the past two decades have there been data from humans subjects to substantiate this postulate. After the initiation of head-out immersion, there is an *acute* increase in central blood volume of 700 ml with a concomitant increase in central venous pressure from 3 to 15 mm Hg (1, 8, 43). Mean cardiac output increases by 32% and mean stroke volume by 35% (1, 11, 43). Both right atrial and pulmonary arterial transmural pressure gradients increase, while systemic vascular resistance decreases (1, 7, 11, 43). Begin and colleagues (4) determined central hemodynamics serially during a 4 h immersion period utilizing an acetylene rebreathing method. This study confirmed the 25%–36% increment in cardiac index previously noted to occur acutely during immersion (1) and demonstrated that this increment was sustained throughout the period of study. Subsequently, we compared the relative central hemodynamic responses of water immersion to those of standard saline infusion (40). With the identical noninvasive rebreathing method cited earlier (4), the increment in cardiac output induced by head-out water immersion has been shown to be similar to that documented during the extracellular fluid

(ECF) volume expansion attained by acute saline administration (2 liters/120 min) equivalent to 3% body weight (40).

CHARACTERIZATION OF THE "EFFERENT" LIMB OF THE IMMERSION MODEL

Renal Water Handling

Chronologically, the initial emphasis on characterizing the renal effects of immersion was directed toward documentation of changes in renal water handling (3). Within the past three decades, Behn et al. (5) and Epstein et al. (16, 17) have succeeded in characterizing the magnitude and composition of the diuresis. Both groups of investigators observed differing diuretic responses depending on the state of hydration of the study subjects. Although the increase in urine flow (V) induced by immersion was attributable to both an increase in free-water clearance (C_{H_2O}) and osmolal clearance (C_{osm}), the major determinant of the increase in V was attributable to an increase in C_{H_2O} to 4.3 ml/min (16, 18). In contrast to studies of hydrated subjects

(16, 18), the immersion-induced increase in V in the fluid-restricted subjects occurred solely as a function of an increase in C_{osm}, with free-water reabsorption ($T^c_{H_2O}$) remaining constant throughout immersion (26). These results indicate that the magnitude and composition of the diuresis occur as a function of the state of hydration of the subject.

Mechanisms of the Diuresis

Several mechanisms have been documented to contribute to the diuresis attending immersion. These include: (1) suppression of antidiuretic hormone (ADH), (2) increased delivery of filtrate to the diluting site, (3) an increase in the release of endogenous renal prostaglandins, (4) augmentation of atrial natriuretic peptide (ANF), and (5) a decrease in sympathetic nervous system activity. Figure 37.1 summarizes some of the neural, hormonal, and hemodynamic factors that may participate in mediating the diuresis of immersion.

Renal Sodium Handling

We characterized the natriuretic response during various sodium intakes and various depths of immersion

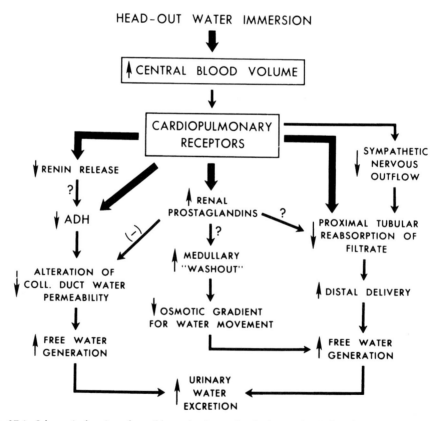

FIG. 37.1. Schematic drawing of possible mechanisms whereby immersion-induced central hypervolemia induces diuresis. *Heavy arrows* indicate pathways for which evidence is available; (—) signifies inhibitory action. [From Epstein (11) with permission.]

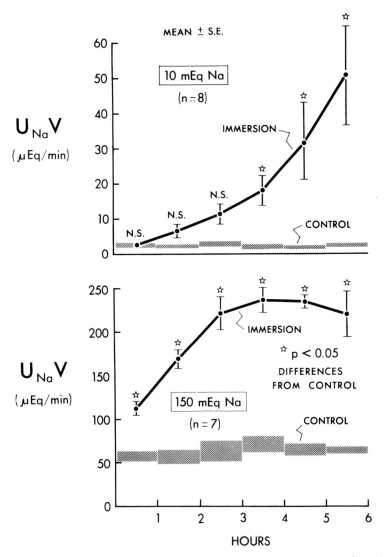

FIG. 37.2. Comparison of effects of immersion on rate of sodium excretion ($U_{Na}V$) in subjects in balance on low-sodium *(top)* and high-sodium *(bottom)* diets. *Shaded areas* represent mean ± SE for control studies. A significant increase in $U_{Na}V$ occurs within the initial hour in sodium-replete subjects, but is delayed to the 4th hour in subjects ingesting a sodium-restricted diet. [Data from sodium-restricted subjects from Epstein and Saruta (32) and for sodium-replete subjects from Epstein et al. (16); figure from Epstein (11) with permission.]

(15, 16, 17, 18, 25, 26, 32). Our studies (32) of mildly sodium-depleted subjects (dietary intake of 10 mEq/day) disclosed that the absolute increase in sodium excretion was less than 7 mEq/6h, reflecting the constraints imposed by the sodium-depleted and volume-contracted state of the subjects. In a subsequent study (16) carried out with a sodium intake approximating more that of the normal diet (150 mEq/day), sodium-replete normal subjects demonstrated an earlier (hour 1 vs. hour 4) and more profound (72 mEq/6h vs. 7 mEq/6h) natriuresis than during sodium depletion (Fig. 37.2). Water immersion to the waist (25) did not induce a natriuresis in either sodium-depleted or sodium-replete subjects, presumably because a lessened pressure gradient induced less central hypervolemia.

Dissociation of Natriuresis from Diuresis

Although immersion is usually associated with both diuresis and natriuresis, the two events do not necessarily always occur together. Thus overnight fluid restriction abolished the diuresis of immersion without attenuating the natriuresis (26). Similarly, the administration of aqueous vasopressin to immersed hydrated normal subjects undergoing immersion abolished the diuresis while the natriuresis remained intact (14). The differences in the temporal profile of the diuresis and natriuresis of immersion merit reemphasis. The diuresis of immersion is usually manifest by hour 1 or 2. In contrast, the natriuresis is progressive and usually peaks by hour 3 or 4. Taken together, these observations suggest the presence of separate mechanisms for diuretic and natriuretic responses.

Mechanisms of the Natriuresis

The demonstration of a highly significant increase in the fractional excretion of sodium during immersion indicates that the natriuresis is attributable primarily to an increased tubular rejection of sodium rather than to alterations in the filtered sodium load. Several lines of evidence suggest that decreased sodium reabsorption occurs at multiple sites in the nephron. A progressive kaliuresis during immersion (16, 18) suggests that the natriuresis of immersion is multifactorial and is mediated in part by an increased rejection of sodium proximal to the diluting site, and is due to an additional component secondary to a decline in circulating aldosterone. This hypothesis is supported by our data indicating that sodium excretion was enhanced when free-water clearance was augmented. This suggests an increase in sodium delivery to the diluting site (16, 18).

The mechanisms mediating the natriuresis are multi-factorial and include aldosterone suppression, augmentation in ANF, a possible humoral natriuretic factor (OLF), an augmentation of renal prostaglandins, and a decrease in sympathetic nervous activity (Fig. 37.3). Clearly, suppression of aldosterone contributes importantly to the natriuresis of NI (18). Enhanced release of endogenous renal prostaglandins also contributes to natriuresis, at least in volume-contracted subjects (21, 22). Increasing evidence suggests that an atrial natriuretic factor (23) and possibly a human ouabain-like compound play a part in natriuresis, as does suppression of sympathetic nervous system activity (36). Finally, the roles of alterations in transcapillary Starling forces and intrarenal blood flow distribution remain to be evaluated.

Renin-Angiotensin-Aldosterone System

Several lines of evidence have suggested the possibility that water immersion would suppress the renin-angiotensin-aldosterone system. In 1960 Bartter and Gann (2) demonstrated that constriction of the supradiaphragmatic inferior vena cava consistently increased aldosterone secretion, presumably by producing relative volume depletion above the constriction. Subsequent studies have demonstrated a parallel increase in plasma renin activity (PRA) in the dog with inferior vena cava

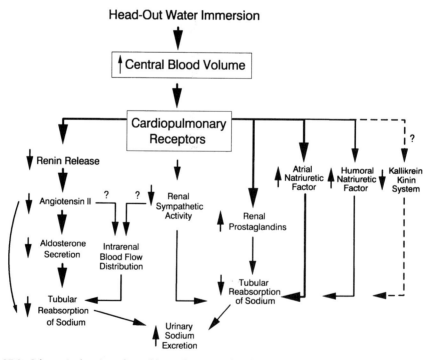

FIG. 37.3. Schematic drawing of possible mechanisms whereby immersion-induced central hypervolemia induces natriuresis. *Heavy arrows* indicate pathways for which evidence is available. Diverse hemodynamic, renal, and hormonal effectors act in concert to promote the natriuresis. [From Epstein (11) with permission.]

constriction. Since immersion to the neck produces an opposite hemodynamic redistribution, characterized by an increase in intrathoracic volume, one would anticipate a suppression of both PRA and aldosterone.

We have systematically characterized the changes of the renin-angiotensin-aldosterone axis during water immersion and have confirmed this formulation (27, 30, 32). Blood was collected serially at 30 min intervals to determine PRA and plasma aldosterone (PA) (27). Immersion resulted in a progressive suppression of PRA within the first 30 min and a significant suppression of PA by 60 min of immersion (27). By 210 min, respective maximal suppression of PRA and PA was 38% and 34% of the prestudy value. PRA and PA returned to prestudy values as soon as 30 min after cessation of immersion.

As detailed previously (8), activation of left atrial and cardiopulmonary receptors with a resultant decrease in sympathetic nerve traffic to the kidney contributes significantly to the suppression of renin release.

Arginine Vasopressin (AVP) or Antidiuretic Hormone

Although ADH suppression during water immersion has been documented only since the early 1980s, considerable earlier evidence had suggested the possibility of such a change. Many investigators reported an increase in solute free–water clearance during head-out water immersion (5, 8, 11, 16, 18, 34). In addition, the administration of vasopressin abolished the diuresis of immersion (14, 34).

In 1975 we utilized a sensitive and specific radioimmunoassay for urinary ADH to document the effects of immersion on urinary ADH excretion (26). Immersion resulted in a progressive decrease in ADH excretion from 80 ± 7 to 37 ± 61 μU/min. Furthermore, cessation of immersion was associated with a marked rebound, with ADH excretion increasing from 37 to 177 μU/min during the recovery hour (26).

The availability of a precise and highly reproducible radioimmunoassay for plasma AVP prompted us to characterize the effects of immersion-induced, acute iso-osmotic volume expansion on plasma AVP in normal human subjects (29). Normal subjects were studied after 14 h of dehydration on two occasions: control and during 4 h of NI. Blood was obtained every 30 min for AVP. Although AVP did not change during the control period, throughout NI it showed prompt and sustained suppression ($P < 0.05$ vs. control). There were no concomitant changes in plasma osmolality. These data support the concept that acute iso-osmotic central volume expansion in humans results in a suppression of plasma AVP.

Renal Prostaglandins

Additional studies demonstrated a profound effect of NI on endogenous prostaglandin synthesis. Studies in sodium-replete subjects disclosed that immersion is associated with a progressive increment in renal prostaglandin E (PGE) excretion, reaching a peak by hour 2 of immersion (21). PGE excretion returned to prestudy levels during the hour postimmersion. Subsequent studies conducted after the administration of indomethacin (50 mg q6h $\times$ 5) disclosed that this cyclooxygenase inhibitor attenuated but did not prevent the immersion-induced increment in PGE (21). In other words, pretreatment with indomethacin decreased basal PGE excretion by more than 50% and lessened the excretion of PGE during subsequent immersion. A similar pattern was noted when five of the subjects treated with indomethacin were restudied after dietary sodium restriction (21).

Although indomethacin produced similar levels of PGE excretion in both the sodium-replete and sodium-depleted groups (21), it did not significantly alter cumulative sodium excretion in the sodium-replete group. In contrast, it virtually abolished the natriuretic response in the sodium-depleted subjects. Such observations lend support to the formulation that renal prostaglandins constitute determinants of renal sodium handling under conditions of diminished effective volume (20).

In contrast to previous reports suggesting a parallelism between renal prostaglandin levels and the renin-aldosterone axis, immersion resulted in a dissociation of these two hormonal systems, with a suppression of PRA while PGE excretion was enhanced (22).

Atrial Natriuretic Peptides (ANF)

Atrial natriuretic peptides (ANF) are found in secretory granules of the atria, and immunoreactive atrial natriuretic factor (irANF) has been detected in animal and human plasma (41, 42). This hormone exerts potent natriuretic and renal hemodynamic effects. The available evidence suggests that augmentation of atrial volume and/or stretch can lead to increased plasma irANF (11, 41, 42). Hence it seemed possible that augmentation of ANF release may contribute to the effects of cardiopulmonary blood volume expansion on the kidney. We studied the effect of central volume expansion induced by water immersion to the neck on the kinetics of ANF release, and the relationship of ANF to renal excretory function (23) in thirteen normal, sodium-replete subjects (Fig. 37.4). Immersion resulted in a prompt and marked increase in plasma irANF from 7.8 ± 1.8 to 19.4 ± 3.8 fmol/ml. These levels fell to 6.3 ± 1.4 fmol/ml after 60 min recovery and were associated

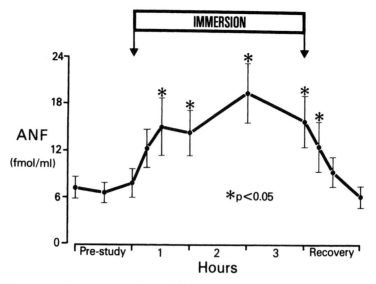

FIG. 37.4. Effect of water immersion on plasma ANP levels in 13 normal subjects. Within 30 min, immersion induced a marked increase in ANP that was sustained throughout immersion. Recovery was associated with a prompt return to the prestudy level. Results are mean ± SE. *$P < 0.05$ or more compared with the level at the end of the prestudy period. [From Epstein et al. (23) with permission.]

with reversible increases in both urine flow rate and in sodium excretion, and with decreases in both PRA ($-66\% \pm 3\%$) and in PA ($-57\% \pm 6\%$). These findings support the postulate that ANF constitutes one of the effectors of the natriuresis of immersion, and presumably of volume homeostasis in normal humans.

Humoral Natriuretic Factor

Evidence has been adduced suggesting that NI stimulates the release of a circulating natriuretic factor that may lead to the encountered natriuresis. This bioassayable factor presumably is identified with an ouabain-like factor (OLF) that cross-reacts with a digitalis receptor (6, 37). We therefore undertook to determine whether the natriuresis of NI is associated with increased activity of a natriuretic factor (12). Urine collected during both seated control and NI studies was fractionated, and the fractions were tested in the rat assay preparation using animals with a single remnant kidney. With the control fractions there was no significant change in sodium excretion. However, the fractions from the NI study resulted in significant increments in both absolute ($U_{Na}V$) and fractional excretion of sodium (FE_{Na}) (12).

Recently Hamlyn and colleagues (37) have purified and structurally identified by mass spectroscopy an endogenous substance from human plasma that binds with high affinity to the ouabain receptor and is indistinguishable from the cardenolide ouabain. We await studies to deliniate the effects of immersion on OLF responsivness.

Catecholamines and Dopa-Dopamine Systems

Because stimulation of left atrial and/or cardiopulmonary receptors in experimental animals results in reduced autonomic nervous system activity, it might be anticipated that maneuvers such as water immersion that augment central blood volume might decrease plasma catecholamine levels. In addition, such alterations might participate in the encountered changes in renal function. Although there have been two previous attempts to examine the response of catecholamines during water immersion (35, 45), methodological considerations and divergent observations have precluded firm conclusions regarding the effect of immersion on catecholamines. We therefore designed a study utilizing more updated methodology to evaluate possible changes in the activities of the sympathetic nervous and dopa-dopamine systems that could contribute to the diuretic and natriuretic response of immersion. We measured plasma and urinary concentrations of norepinephrine, its intraneuronal metabolite dihydroxyphenylglycol (DHPG), epinephrine (EPI), dopa, and dopamine in normal subjects during water immersion (36). The urinary norepinephrine excretion was suppressed consistently during immersion. The urinary excretory rates of EPI and DHPG were also decreased. Urinary excretion of dopa, dopamine, and DHPG, a neuronal metabolite of norepinephrine, changed in a triphasic pattern, with decreased excretion during the first hour of immersion ($P < 0.01$), small but consistent increases during the next two h, and decreased excretion, to below baseline, during recovery ($P < 0.01$ for dopa and dopamine). Our findings suggest that the neurohormonal contribution to the natriuretic response during central hypervolemia is multifactorial and includes persistent sympathoadrenal suppression and a late increase in dopa-dopamine activity.

CONSIDERATIONS FOR SELECTING WATER IMMERSION FOR STUDIES OF VOLUME HOMEOSTASIS

The delineation of the water immersion model has facilitated its application to investigations of renal function and hormonal responsiveness in humans both disease-free and in diverse disease states including edematous disorders (10, 11, 19, 24, 31, 38, 44). Traditional attempts to assess the effects of volume alterations and hypervolemia have utilized rapid volume expansion with exogenous solutions, including saline, mannitol, and albumin. As we have detailed previously, however, several of these maneuvers have a number of drawbacks

TABLE 37.1. *Salient Features of the Model of Head-out Water Immersion in Humans*

1) Immersion produces a prompt redistribution of circulating blood with a relative central hypervolemia

2) Cardiac output is increased by 25–33% and central blood volume by ~700 ml

3) Alterations in central hemodynamics are sustained throughout a 4-h immersion period and are promptly reversible following cessation of immersion

4) Immersion-induced central hypervolemia is associated with a profound and progressive natriuresis and diuresis. These alterations are promptly reversible following cessation of immersion

5) Central hemodynamic and renal effects of immersion are equal in magnitude to those induced by acute saline administration (2 l saline/2 h)

6) Alterations in renal sodium, potassium, and water handling in sodium-replete state generally occur with a concomitant increase in renal plasma flow but in absence of changes in glomerular filtration rate

7) Immersion is associated with a prompt and profound ($\sim$ ⅔) suppression of plasma renin activity and plasma aldosterone. Cessation of immersion is associated with a prompt return of both plasma renin activity and plasma aldosterone to prestudy levels

8) Immersion induces a prompt, marked, and sustained augmentation of atrial natriuretic factor. Cessation of immersion is associated with a prompt return of atrial natriuretic factor to prestudy levels

9) Immersion induces an augmentation of renal prostaglandins as assessed by an increase in urinary prostaglandin E and 6-keto-prostaglandin $F_{1\alpha}$ excretion

10) Above alterations in renal function, renin-aldosterone system, and atrial natriuretic responsiveness occur in the absence of changes in plasma composition

[From Epstein (11) with permission.]

(10, 11). For example, saline infusion nonspecifically increases the volume of all fluid compartments and induces concomitant alterations in plasma composition that precludes definitive statements regarding the etiological role of alterations in plasma volume.

In contrast to the more traditional attempts to achieve extracellular volume expansion, water immersion has several attributes that commend its use. As summarized in Table 37.1, the volume stimulus of immersion is promptly reversible after cessation of immersion, in contrast to the relatively sustained hypervolemia that follows saline administration and thus constitutes an important attribute in minimizing any risk to the study patients. In contrast to saline administration, the volume stimulus of immersion occurs in the absence of changes in plasma composition (11, 18). In addition, the central hypervolemia of water immersion is caused partly by hydrostatic compression of peripheral veins and a consequent decline of venous capacitance, whereas the central hypervolemia of saline administration is due to an elevation of mean circulatory filling pressure as a consequence of increased vascular volume (11).

STUDIES OF DISORDERS CHARACTERIZED BY DERANGED VOLUME HOMEOSTASIS

As detailed in a recent review (11), the immersion model has been successfully utilized as an investigative tool for studying abnormal sodium and water homeostasis in patients with decompensated cirrhosis (10, 19, 28), nephrotic syndrome (38, 44), and, to a lesser extent, essential hypertension (13, 24).

Aside from its utility in investigating the pathogenesis of hypertension, the water immersion model has also been successfully utilized in studies of antihypertensive agents in this patient population. A recent example is the application of water immersion to characterize the natriuretic properties of calcium antagonists (13).

CONCLUSIONS

Several laboratories have succeeded in delineating the circulatory, renal, and endocrine changes induced by water immersion in humans. These studies have demonstrated that immersion in the seated posture results in a redistribution of blood volume with a relative central hypervolemia. Consequently, profound alterations in fluid and electrolyte homeostasis ensue, including a marked natriuresis, kaliuresis, and diuresis as well as a suppression of the renin-aldosterone system and ADH release. Concomitantly, renal prostaglandin and ANF release is stimulated.

Although a delineation of the hormonal and renal responses to immersion is of importance in understanding the normal physiology of volume regulation, it must be underscored that the utility of this model transcends this immediate application. For example, characterization of the effects of immersion has facilitated studies of the pathophysiology of clinical states associated with deranged volume homeostasis, such as advanced liver disease. Thus water immersion has been used successfully to delineate the determinants of sodium and water retention in patients with decompensated cirrhosis (10, 19, 28) and nephrotic syndrome (38, 44). Similarly, the immersion model has been utilized to assess the renin-aldosterone responsiveness of patients with secondary hyperaldosteronism (19), and of anephric patients (31). Finally, the numerous similarities between the effects of water immersion and those of manned spaceflight on the renal and cardiovascular systems commend the use of water immersion as an experimental analogue of weightlessness (11, 34, 43).

REFERENCES

1. Arborelius, N. Jr., U. I. Balldin, B. Lilja, and C. E. G. Lundgren. Hemodynamic changes in man during immersion with the head above water. *Aerospace Med.* 43: 592–598, 1972.
2. Bartter, F. C., and D. S. Gann. On the hemodynamic regulation of the secretion of aldosterone. *Circulation* 21: 1016–1023, 1960.
3. Bazett, H. C., S. Thurlow, C. Corwell, and W. Stewart. Studies on the effects of baths on man. II. The diuresis caused by warm baths together with some observations on urinary tides. *Am. J. Physiol.: Respir. Environ. Exerc. Physiol.* 70: 430–452, 1924.
4. Begin, R., M. Epstein, M. A. Sackner, R. Levinson, R. Dougherty, and D. Duncan. Effects of water immersion to the neck on pulmonary circulation and tissue volume in man. *J. Appl. Physiol.* 40: 293–299, 1976.
5. Behn, C., O. H. Gauer, K. Kirsch, and P. Eckert. Effects of sustained intrathoracic vascular distension on body fluid distribution and renal excretion in man. *Pflugers Arch.* 313: 123–135, 1969.
6. Buckalew, V. M. Natriuretic hormone. In: *The Kidney in Liver Disease* (3rd ed.), edited by M. Epstein. Baltimore: Williams and Wilkins, 1988, p. 417–428.
7. Echt, M., L. Lange, and O. H. Gauer. Changes of peripheral venous tone and central transmural venous pressure during immersion in a thermo-neutral bath. *Pflugers Arch.* 352: 211–217, 1974.
8. Epstein, M. Renal effects of head-out water immersion in man-implications for an understanding of volume homeostasis. *Physiol. Rev.* 58: 529–581, 1978.
9. Epstein, M. Studies of volume homeostasis in man utilizing the model of head-out water immersion. *Nephron* 22: 9–19, 1978.
10. Epstein, M. Renal sodium handling in cirrhosis. In: *The Kidney In Liver Disease* (3rd ed.), edited by M. Epstein. Baltimore: Williams and Wilkins, 1988, p. 3–30.
11. Epstein, M. Renal effects of head-out water immersion in humans: a 15 year update. *Physiol. Rev.* 72: 563–621, 1992.
12. Epstein, M., N. S. Bricker, and J. J. Bourgoignie. The presence of a natriuretic factor in urine of normal men undergoing water immersion. *Kidney Int.* 13: 152–158, 1978.
13. Epstein, M., and A. G. De Micheli. Natriuretic effects of calcium antagonists. In: *Calcium Antagonists In Clinical Medicine,* edited by Epstein M. Philadelphia: Hanley and Belfus, 1992, p. 349–366.
14. Epstein, M., A. G. DeNunzio, and R. D. Loutzenhiser. Effects of vasopressin administration on diuresis of water immersion in normal human. *J. Appl. Physiol.: Respir. Environ. Exerc. Physiol.* 51: 1384–1387, 1981.
15. Epstein, M., A. G. DeNunzio, and M. Ramachandran. Characterization of the renal response to prolonged immersion in normal man. Implications for an understanding of the circulatory adaptation to manned spaceflight. *J. Appl. Physiol.: Respir. Environ. Exerc. Physiol.* 49: 184–188, 1980.
16. Epstein, M., D. Duncan, and L. M. Fishman. Characterization of the natriuresis caused in normal man by immersion in water. *Clin. Sci.* 43: 275–287, 1972.
17. Epstein, M., D. C. Duncan, and B. Meek. The role of posture in the natriuresis of water immersion in normal man. *Proc. Soc. Exp. Biol. Med.* 142: 124–127, 1973.
18. Epstein, M., J. L. Katsikas, and D. C. Duncan. Role of mineralocorticoids in the natriuresis of water immersion in normal man. *Circ. Res.* 32: 228–236, 1973.
19. Epstein, M., R. Levinson, J. Sancho, E. Haber, and R. Re. Characterization of the renin-aldosterone system in decompensated cirrhosis. *Circ. Res.* 41: 818–829, 1977.
20. Epstein, M., and M. Lifschitz. Volume status as a determinant of the influence of renal PGE on renal function. *Nephron* 25: 157–159, 1980.
21. Epstein, M., M. Lifschitz, D. S. Hoffman, and J. H. Stein. Relationship between renal prostaglandin E and renal sodium handling during water immersion in normal man. *Circ. Res.* 45: 71–80, 1979.
22. Epstein, M., M. Lifschitz, R. Re, and E. Haber. Dissociation of renin-aldosterone and renal prostaglandin E during volume expansion induced by immersion in normal man. *Clin. Sci.* 59: 55–62, 1980.
23. Epstein, M., R. D. Loutzenhiser, E. Friedland, R. M. Aceto, M. J. F. Camargo, and S. A. Atlas. Relationship of increased plasma ANF and renal sodium handling during immersion-induced central hypervolemia in normal humans. *J. Clin. Invest.* 79: 738–745, 1987.
24. Epstein, M., R. Loutzenhiser, and R. Levinson. Spectrum of deranged sodium homeostasis in essential hypertension. *Hypertension* 8: 422–432, 1986.
25. Epstein, M., M. Miller, and N. S. Schneider. Depth of immersion as a determination of the natriuresis of water immersion. *Proc. Soc. Exp. Biol. Med.* 146: 562–566, 1974.
26. Epstein, M., D. S. Pins, and M. Miller. Suppression of ADH during water immersion in normal man. *J. Appl. Physiol.* 38: 1038–1044, 1975.
27. Epstein, M., D. S. Pins, J. Sancho, and E. Haber. Suppression of plasma renin and plasma aldosterone during water immersion in normal man. *J. Clin. Endocrinol. Metab.* 41: 618–625, 1975.
28. Epstein, M., D. S. Pins, N. Schneider, and R. Levinson. Determinants of deranged sodium and water homeostasis in decompensated cirrhosis. *J. Lab. Clin. Med.* 87: 822–839, 1976.
29. Epstein, M., S. Preston, and R. E. Weitzman. Iso-osmotic central blood volume expansion suppresses plasma arginine vasopressin in normal man. *J. Clin. Endocrinol. Metab.* 52: 256–262, 1981.
30. Epstein, M., R. Re, S. Preston, and E. Haber. Comparison of suppressive effects of water immersion and saline administration on renin-aldosterone in normal man. *J. Clin. Endocrinol. Metab.* 49: 358–363, 1979.
31. Epstein, M., J. Sancho, G. Perez, E. Haber, R. Re, and R. Loutzenhiser. Volume as a determinant of plasma aldosterone in anephric man. *J. Clin. Endocrinol. Metab.* 46: 309–316, 1978.

32. Epstein, M., and T. Saruta. Effect of water immersion on renin-aldosterone and renal sodium handling in normal man. *J. Appl. Physiol.* 31: 368–374, 1971.

33. Gauer, O. H. Mechanoreceptors in the intrathoracic circulation and plasma volume control. In: *The Kidney in Liver Disease* (1st ed.), edited by M. Epstein. New York: Elsevier, 1978, p. 3–17.

34. Gauer, O. H., J. P. Henry, and C. Behn. The regulation of extracellular fluid volume. *Annu. Rev. Physiol.* 32: 547–595, 1970.

35. Goodall, McC., M. McCally, and D. E. Graveline. Urinary adrenaline and noradrenaline response to simulated weightless state. *Am. J. Physiol.* 206: 431–436, 1964.

36. Grossman, E., D. S. Goldstein, A. Hoffman, I. R. Wacks, and M. Epstein. The effects of water immersion on the sympathoadrenal and dopa-dopamine systems in humans. *Am. J. Physiol.* 262 (*Regulatory Integrative Comp. Physiol.* 33): R993–R999, 1992.

37. Hamlyn, J. M., M. P. Blaustein, S. Bova, D. W. DuCharme, D. W. Harris, F. Mandel, W. R. Mathews, and J. H. Ludens. Identification and characterization of a ouabain-like compound from human plasma. *Proc. Natl. Acad. Sci.* 88: 6259–6263, 1991.

38. Krishna, G. G., and G. M. Danovitch. Effects of water immersion on renal function in the nephrotic syndrome. *Kidney Int.* 21: 393–401, 1982.

39. Lange, L., S. Lange, M. Echt, and O. H. Gauer. Heart volume in relation to body posture and immersion in a thermo-neutral bath. A roentgenometric study. *Pflugers Arch.* 352: 219–226, 1974.

40. Levinson, R., M. Epstein, M. A. Sackner, and R. Begin. Comparison of the effects of water immersion and saline infusion on central haemodynamics in man. *Clin. Sci. Mol. Med.* 52: 343–350, 1977.

41. Maack, T., M. J. F. Camargo, H. D. Kleinert, J. N. Laragh, and S. A. Atlas. Atrial natriuretic factor: structure end function properties. *Kidney Int.* 27: 607–615, 1985.

42. Needleman, P., S. P. Adams, B. R. Cole, M. G. Currie, D. M. Geller, M. L. Michener, C. B. Saper, D. Schwartz, and D. G. Standaert. Atriopeptins as cardiac hormones. *Hypertension* 7: 469–482, 1985.

43. Norsk, P., and M. Epstein. Manned space flight and the kidney. *Am. J. Nephrol.* 11: 81–97, 1991.

44. Peterson, C., B. Madsen, A. Perlman, A. Y. M. Chan, and B. D. Myers. Atrial natriuretic peptide and the renal response to hypervolemia in nephrotic man. *Kidney Int.* 34: 825–831, 1988.

45. Skipka, W. K., A. Deck, and D. Bonning. Effect of physical fitness on vanillylmandelic acid excretion during immersion. *Eur. J. Appl. Physiol.* 35: 271–276, 1976.

38. Head-out water immersion: animal studies

JOHN A. KRASNEY | *Department of Physiology, School of Medicine and Biomedical Sciences, State University of New York at Buffalo, Buffalo, New York*

CHAPTER CONTENTS

HEAD-OUT WATER IMMERSION (WI) is a simple, noninvasive maneuver that has proved to be a useful tool for investigating mechanisms involved in the regulation of blood volume. Head-out water immersion has been used as therapy since Roman times and its purported benefits have provided the basis for treatments given at the great spas of Europe and America. The practice of sitting in baths for social, philosophical, psychological, or hygienic purposes is an institution in Japan and, more recently, the use of hot tubs and Jacuzzi baths has become a major recreational industry in the United States. As a result of experimental animal and human studies carried out over the past two or more decades, initiated primarily by Gauer and Henry (48), there is an emerging body of factual information that has begun to provide important insights into the nature of the profound physiological alterations which follow the simple act of sitting in a bath of warm water.

Several recent reviews of WI have been published (36, 84, 87, 111); the chapters in this volume dealing with the subject focus on adjustments elicited by thermoneutral WI. This chapter will describe the results of studies carried out in animals, including primates, while the companion chapter by Epstein in this *Handbook* (38) deals with responses to WI in humans. Human responses will be referred to in this chapter only for comparative purposes.

During WI, terrestrial animals, including humans, return to the aquatic environment from whence they evolved. As a consequence of the buoyancy effect imparted by the water environment, a microgravity state supervenes. This has led to the use of WI as a model for the study of space physiology (15, 48, 65). In addition, it is obvious that an analysis of the physiology of WI is relevant to understanding the physiology of diving (84, 140). Hong and colleagues (140) have demonstrated that Korean women divers (Ama) spend, on the average, a total time of 188 min in water per diving day. Of this total time, they spend only 52 min or 26.7% of the time actually diving. Thus 136 min are spent in the situation of head-out immersion.

The basic thesis to be advanced in this chapter is that the physiologic responses observed during WI are related to the activation of certain processes which serve to regulate critical physiological variables. At first glance, these physiologic responses might be attributed simply to the removal of the influence of gravity. While this is correct in part, it must be recognized that there is a unique set of stimuli associated with WI that is not present in the true weightless state or space environment. In considering the effects of inducing microgravity, it might also be thought that since bipeds, humans, or subhuman primates face greater antigravity challenges in the terrestrial movement, their responses to WI would be expected to be more profound than those occurring in quadrupeds. While this assumption is reasonable, it may not be entirely correct. Detailed comparisons of WI vs. the weightless state, and bipeds vs. quadrupeds, will be presented later. While WI responses are elicited by both the removal of the effects of gravity and the application of differential hydrostatic pressure, it is useful to view the physiologic regulations occurring in WI as being directed toward the control of systemic and regional tissue oxygen delivery.

THE GAUER-HENRY HYPOTHESIS

Based on a hypothesis first made by Hartshorne in 1847 (66), Peters (116) proposed in 1935, that the body possesses volume receptors which perceive the "fullness of the blood stream." In 1951, however, Gauer and Henry (48) provided the first direct experimental evidence that distention of one of the cardiac chambers is associated with a reflex diuresis. Using an anesthetized open-chest dog preparation, they were able to demonstrate that obstructing the mitral valve via inflation of a balloon led to a diuresis that was characterized primarily by an increase in free water clearance (C_{H_2O}). Subsequently they were able to show that the selective elevation of left atrial pressure by the balloon maneuver caused a reduction of arginine vasopressin (AVP) secretion. The latter response could be abolished by bilateral section of the cervical vagus nerves (48). This demonstration of the diuresis caused by left atrial distention led to the hypothesis that blood volume is regulated by "intrathoracic stretch receptors." The essential point of the work of Gauer and Henry is that they established the existence of a "cardiac–renal link." As shown in Figure 38.1 the initial version of the Gauer-Henry hypothesis proposed that an increase in central blood volume leads to elevated cardiac filling pressure which activates intra-

thoracic, or more specifically, atrial stretch receptors. These atrial "volume receptors" in turn reflexly suppress the secretion of volume regulatory hormones (AVP) and a water diuresis ensues. This concept was later amplified by further studies carried out by Gauer and Henry (48) as well as by the work of others.

RATIO OF BLOOD FLOW TO METABOLISM IN WATER IMMERSION

During WI, there is an increase in the central cardiopulmonary blood volume (1, 6, 92). This augments the cardiac output in a sustained fashion via the Frank-Starling length-tension mechanism (84, 86, 96) and, in certain species such as the dog, by neural mechanisms as well. Since, by definition, systemic O_2 consumption does not change during thermoneutral WI, an unusual situation quickly develops whereby the systemic O_2 delivery and blood flow apparently exceed the O_2 requirements of the systemic tissues (Fig. 38.2). A basic principle of blood flow regulation is that flow is rather precisely regulated in accord with the metabolic demands of the tissues. This is termed "autoregulation of blood flow" (63). Figure 38.2 indicates that the autoregulation relationship is modified during WI such that for any metabolic rate there is a higher level of flow. The idea that WI leads to a relative "luxury perfusion" at least in some tissues is compatible with the data of Hajduczok et al. (63) in the dog and the data of Christie et al. (18) obtained from humans exercising in the water. After

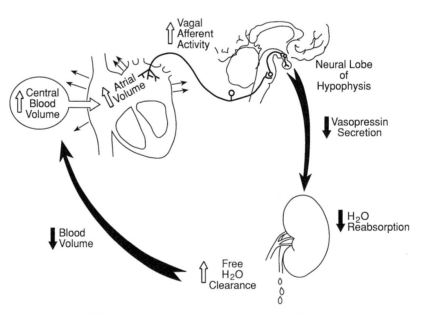

FIG. 38.1. Early version of the Gauer-Henry hypothesis. *Open arrows* indicate an increase or stimulation of the variable, *darkened arrows* indicate a decrease or inhibition of the variable.

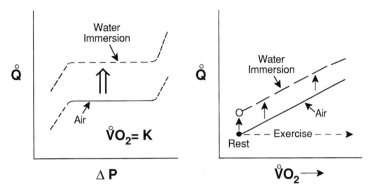

FIG. 38.2. Water immersion increases the ratio of blood flow ($\dot{Q}$) to metabolism ($\dot{V}O_2$) at rest *(left panel)* where $\dot{V}O_2$ is constant. Autoregulation of blood flow occurs, but the level at which systemic flow is held constant over a range of perfusion pressures is shifted upward. As exercise raises the $\dot{V}O_2$ level *(right panel)*, the level of systemic flow is elevated for any $\dot{V}O_2$ level during water immersion. [From Christie et al. (18) with permission.]

some time in the water, usually several hours, the cardiac output, and presumably some regional blood flows, return toward preimmersion levels (21, 115). This would imply that adjustments are activated to bring the blood flow down to a rate that is commensurate with the local metabolic demands. This compensation is of interest in light of the "whole body autoregulation" hypothesis of Guyton and Coleman (60) which predicts that a sustained elevation of cardiac output should lead to a rise in total peripheral resistance, an elevation of arterial pressure, and a return of cardiac output to control levels. This does not occur in WI, even after long periods of immersion.

TRANSCAPILLARY FLUID SHIFT IN WATER IMMERSION

The original view advanced by Gauer and Henry (48), as well as by other investigators (3, 4), proposed that the increase in salt and fluid output by the kidneys in WI ultimately leads to a true correction of the total plasma volume with a decline in the intrathoracic volume back to the preimmersion levels. While this view is to some extent correct, it must be modified to take into account the fact that a major transcapillary fluid shift occurs during WI which acts to elevate the plasma volume (Fig. 38.3). Thus when an animal stands in water, the external hydrostatic pressure decreases the capacity of the venous compartment and shifts blood toward the chest. However, the volume added to the plasma compartment by the fluid shift augments the elevation of central venous pressure. The primary role of the kidney in WI is to minimize the increase in plasma volume which occurs secondary to the fluid shift and otherwise would be quite massive (101, 102, 106, 107).

The original Gauer-Henry hypothesis that postulates

a net reduction of plasma volume in WI thus must be modified to take into account the continuous autotransfusion which occurs in WI (106).

WATER IMMERSION AS A MODEL FOR HYPERVOLEMIA

Methods for Eliciting Hypervolemia

In experimental animals, an increase in blood volume or plasma volume can be induced readily by infusion of volume expanding fluids into the vascular compartment. However, actual volume expansion by infusion (VE) is an invasive procedure which is likely to modify the composition of the plasma as well as produce hypervolemia throughout the circulatory compartment. Accordingly, other methods have been developed to elicit a redistribution of blood volume into the thorax such that the cardiovascular mechanoreceptors can be engaged without being invasive and can avoid other complications associated with VE.

Negative pressure breathing has been used in both humans and animals to elevate thoracic blood volume (48, 67). While this procedure usually elicits a diuresis, the diuresis is typically transient and is not sustained for any length of time. The simple assumption of the supine posture leads to a rise in thoracic blood volume in humans and is associated with a diuresis (21, 115). When quadrupeds such as the dog move from the standing position to the lying down position, there is an increase in ventricular volume and cardiac output (132). Presumably this stimulus would lead to a diuresis and possibly a natriuresis, but no specific studies have been performed to address this question. Head-down tilt has become a method of interest for studying the effects of central redistribution of blood volume in humans (115).

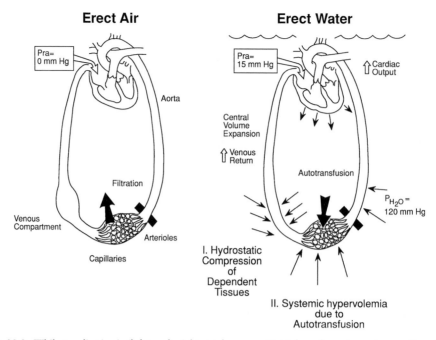

FIG. 38.3. While standing in air *(left panel)*, right atrial pressure (P$_{ra}$) is low, dependent veins are distended, and net filtration occurs in limb capillaries. During water immersion *(right panel)*, central volume expansion occurs with increased Pra and cardiac output because *(I)* there is hydrostatic compression of dependent tissues and *(II)* capillary reabsorption or an autotransfusion occurs in dependent limbs.

This maneuver has not been studied specifically in experimental animals. However, several studies have analyzed cardiovascular and renal responses to head-up tilt in anesthetized and conscious dogs and anesthetized monkeys and have been able to demonstrate that an antidiuresis and antinatriuresis occurs (105, 120, 123). Therefore, it seems likely that head-down tilt would result in diuresis and natriuresis in the dog.

The application of graded levels of lower-body positive pressure has been shown to be an effective method for causing central volume expansion in both humans and animals (78, 131). Churchill et al. (19) have demonstrated that lower-body positive pressure is capable of eliciting a sustained diuresis and natriuresis (24 h) in the awake monkey.

As pointed out above, hypogravity occurs in space in the true weightless condition and in WI. In comparing these two circumstances it must be remembered that WI leads to a shift of fluid into the plasma compartment, resulting in an overall hypervolemic state in addition to the central hypervolemia that results from the application of graded levels of hydrostatic pressure to lower portions of the body. In the conscious dog, graded levels of thermoneutral WI cause graded elevations of peripheral and central venous pressures as well as tissue pressures (106). In comparison, the precise alterations of plasma volume occurring in hypogravity are uncertain. As shown in Table 38.1, it seems likely that the removal

of gravity would result in a rise in central blood volume and the consequent renal response would lead to a true correction, or reduction, of plasma volume. It is not known whether a transcapillary fluid shift occurs in hypogravity. In WI there would be altered inputs to the central nervous system (CNS) from somatic and vestibulo-cerebellar inputs because of the presence of differential hydrostatic pressure, tissue compression, and buoyancy effects. Presumably the input patterns from these receptor systems would also be altered in hypogravity, but it is likely that these patterns differ from those occurring in WI. The effects of these maneuvers on the lung-chest wall system differ also. In WI, negative pressure breathing occurs in both animals (110, 121) and humans (25, 67). The effect of water pressure on the chest wall elicits a pattern of loading on the respiratory muscles resembling an elastic load. This is associated with increased blood flow to the respiratory muscles and increased work of breathing (63, 87, 88).

Volume expansion by infusion causes a central volume expansion, as does hypogravity, head-down tilt, and WI. There is an overall volume expansion as in WI, but this may not be the case in hypogravity and the effect may only be temporary in head-down tilt (Table 38.1). Moreover, fluid probably shifts out of the plasma into the tissues in VE, depending on the nature of the volume expander. It is unlikely that central inputs from peripheral somatic or vestibulo-cerebellar receptors are

TABLE 38.1. *Comparison of Water Immersion, Volume Expansion, Head-down Tilt, and Microgravity*

Response	Water Immersion	Volume Expansion	Head-down Tilt	Microgravity
Intrathoracic hypervolemia	Yes	Yes	Yes	Yes
Systemic hypervolemia	Yes	Yes	Yes	Unknown
Transcapillary fluid shift; direction	Into plasma, sustained	Out of plasma	Into plasma, transient?	Unknown, out of plasma?
Somatic receptor input	Differential pressure, tissue compression, buoyancy	Unchanged	Unloaded	Unloaded
Vestibulo-cerebellar input	Altered by buoyancy	Unaltered	Altered by head-down position	Altered by hypogravity
Lower-limb tissue volume	Decreases	Increases	Decreases	Decreases, "bird-legs"
Pulmonary afferent inputs	Negative-pressure breathing, elastic loading; increased lung vascular volume	Increased lung vascular volume	Increased lung vascular volume, altered loading of respiratory muscles	Increased lung vascular volume, altered loading of respiratory muscles

altered by VE. Lower limb volume increases in VE, while it decreases in WI, head-down tilt, and hypogravity. Lastly, while pulmonary vascular volume rises in WI, VE, head-down tilt, and hypogravity, the actual loading of the lung-chest wall system is quite different amongst the four maneuvers. Other important differences become apparent if the four maneuvers are compared. For example, the nature of the peripheral fluid shift would be quite different in head-down tilt, as compared to the other situations.

Since WI is noninvasive and there is no direct alteration of plasma volume, Gauer and Henry (48) considered WI to be the "investigational tool of choice" for studying responses to VE. While WI has advantages as an experimental method, it cannot be strictly compared with true VE, hypogravity, or head-down tilt. While all of these maneuvers result in central volume expansion, there are unique characteristics associated with each maneuver which must be kept in mind. Epstein (36, 43) has indicated that the magnitude of the diuresis and natriuresis caused by WI is equivalent to that produced by infusion of 2 l of 0.9% NaCl solution. Moreover, both WI and VE with saline cause a rise in the extracellular fluid volume. However, in saline VE fluid moves out of the plasma into the extravascular compartment, while in WI fluid moves from the extravascular compartment into the plasma compartment (102). In addition, it has been shown in the monkey that the onset, rate, and time course of the immersion diuresis and natriuresis differ considerably from those elicited by VE (122).

Methodological Considerations for Animal Studies

By definition, thermoneutral WI requires that there be no change of core temperature or systemic O_2 consumption during the period of WI. In humans the thermoneutral temperature is considered to be approximately 34° or 35°C (24, 36, 130, 151) for immersions of 3–6 h. In initial studies of WI using the anesthetized dog, it was assumed that the thermoneutral temperature of the dog is the same as that of man (141, 153). However, this has not been shown to be the case, as the thermoneutral temperature of the anesthetized dog proved to be much warmer, or 38°C (88). The thermoneutral temperature of the awake dog is slightly cooler, 37°C (62, 86). The conscious sheep also has a thermoneutral temperature of 37°C (personal observation). Awake dogs begin shivering after about 20 min in water with a temperature of 34°C (86, 87). A number of studies have examined the response of the anesthetized (118, 121, 122) and conscious monkey (8) to WI and have assumed that the thermoneutral temperature of the monkey is the same as that of the human. However, there have been no studies which have specifically addressed this question. It is important to recognize that immersions at temperatures other than thermoneutral will elicit deviations from the usual stereotyped responses. For example, cool water will cause peripheral vasoconstriction and may decrease cardiac output, while warm water immersions increase cardiac output (17, 24, 82, 113, 130, 151).

During experiments involving measurement of cardiovascular hemodynamic pressures in animals, an important consideration relates to the location of the hydrostatic indifference point (HIP) (131). In humans the HIP is considered to be located at the midthoracic level at approximately the level of the right atrium and cardiovascular pressures are referred to this level (20). However, it is likely that the HIP shifts during WI and the location of the HIP during WI in humans and animals remains to be determined. In several studies it has been assumed that the HIP shifts to the level of the water surface. In WI to the midcervical level, hemodynamic pressures have been referenced to the water surface (29, 79, 80). This is problematic when immersions to lesser depths are carried out; for example, during immersions of dogs to the leg or mid-chest level, the location of the HIP undoubtedly shifts but the location is uncertain. In some animal studies, the pressure measurements are referred to the level of the right atrium, or two-thirds the distance from the sternum toward the vertebral column (61, 62, 102). This has led to some marked discrepancies in the estimated level of tissue pressure in WI. In an attempt to deal with this problem a pressure reference catheter has been inserted into the midchest via the esophagus (1), or a balloon catheter has been positioned in the pleural space (62). These catheters allow for estimation of the transmural pressure across the heart and great vessels in WI, while the direct pressure measurement from the pleural catheter allows for a quantitative estimate of the hydrostatic compression of the chest wall. To correctly interpret hemodynamic pressure data in WI studies the position of the reference pressure catheter must be identified. As will be discussed (see Transcapillary Fluid Shift), central venous pressure is coupled in a linear manner to both the water pressure and tissue pressure (83, 106, 131). The transmural pressures which distend the heart and great vessels define the level of cardiovascular stretch receptor stimulation.

Animal models have been developed for the study of WI because they allow for invasive manipulations that yield experimental data which in turn cannot be obtained in humans. Our laboratory has developed and standardized the conscious, trained dog model of WI, while Peterson and Benjamin have studied the anesthetized and awake monkey models in WI and volume expansion (117). More recently our laboratory has emphasized the development of a conscious sheep model of WI, which has proved to be quite useful. In addition, pigs and rabbits have been immersed. The cardiovascular response of the pig in WI is quite similar to that of the human, but unfortunately it is difficult to collect urine from pigs. Awake rabbits respond to WI with a diuresis and natriuresis. However, the blood volume of the rabbit is too small to allow for the repeated blood sampling required for hormonal analyses, and urine collections are difficult as well.

If anesthetized animals are studied, then training is not an issue. However, conscious animals generally require training. If dogs are chosen as the model, then they must be selected for temperament and 1–2 wk of training are usually required. On the other hand we have found that sheep require very little training; usually they will stand quietly in the water on the first or second immersion exposure. Pigs are quite similar in this regard in that they appear to enjoy the water. Rabbits usually sit quietly in the water as well. In terms of personal experience with rabbits, it is difficult to determine without specific studies as to whether or not immersion is a stress for this species. Dogs and sheep become rather somnolent in thermoneutral water; in fact, it may be a challenge to keep them awake. As might be anticipated, monkeys require special attention and consideration when studied in the awake state (117). Finally, the rat has been utilized for immersion experiments (77).

All of the mammals studied in WI have been terrestrial. It is pertinent to consider whether or not animals that are aquatic or semiaquatic in nature would respond to WI in a manner similar to terrestrial mammals. It may be that specific adaptations are present in aquatic mammals which would preclude the development of the expected diuresis and natriuresis. However, studies directed to answering this question have not been performed. Humans engaged in aquatic training appear to adapt to WI (21, 115).

The human in the upright position differs physiologically from quadrupeds. The human has a larger column of blood to be displaced centrally. Moreover, systemic vascular compliances differ among species (48, 61, 83, 131). Therefore, the potential for WI to displace blood centrally as well as the ability of the heart and great vessels to stretch should be a consideration. Gilmore (49) has emphasized that there may be major differences between bipeds and quadrupeds in terms of neurohormonal control of blood volume regulation. This point has been reinforced by Peterson (117). Gilmore has argued that the potency of the cardiac receptors to influence the kidney has diminished in bipeds, owing to the upright posture, while the importance of the arterial high pressure baroreceptors in controlling the kidney has increased (49). The level of neural discharge from left atrial receptors for a given level of left atrial pressure is diminished in the primate compared to the dog (49, 117, 153). However, this issue requires further experimental study. These comparisons were made in anesthetized dogs and monkeys, usually after acute surgical preparation of the animal. In addition, the dog typically shows a Bainbridge type cardiac acceleration during volume expansion and WI (16, 62, 64, 86), whereas other subprimate species such as the sheep, pig, and rabbit do not show acceleration of the heart rate, but rather a cardiac slowing, as does the human (96). Indeed, well-trained dogs do not show a cardiac acceleration during WI (85, 140). One might wonder also whether the monkey is as much of a biped as it is presumed to be (19, 49, 78, 87, 117). Discussion of these issues will be presented in greater detail in a later section (see Cardiovascular Receptors: Circulatory, Renal, and Hormonal Influences).

The relative influence of the animal's hydration state on the renal functional response to WI is not clear. Hydration in the form of fluid supplementation has been used to support the normovolemic state in WI because it has been assumed that plasma volume would decline (3, 4, 48). However, as described above (see Transcapillary Fluid Shift in Water Immersion), recent evidence indicates that plasma volume increases due to a transcapillary fluid shift in WI. Thus fluid supplementation would be expected to add to the hypervolemia already caused by WI itself. There have been considerable differences in the prehydration and hydration protocols reported for various WI studies (36, 140). Hydration protocols have involved a single bolus of water ingested before WI (21), hourly boluses of water given during WI (57–59), replacement of all blood and urine losses with saline (62, 86, 140), and maximal water diuresis (126). Unfortunately, the rationale for using a particular hydration protocol has usually not been entirely clear, other than to minimize a predicted decline of plasma volume.

In addition to variations in hydration protocols, WI investigations have involved other variations in protocol. Conditions for standardized human protocols have been described by Epstein and Norsk (36, 112). Animal studies have used varying water bath temperatures without regard for thermoneutrality under anesthetized conditions during spontaneous breathing (88); after acute closure of a thoracotomy (29) or with the thorax intact (88, 118, 122); with the animal on a respirator (110, 121); totally immersed, or head-out (62, 110, 121); in the erect (88, 141) or quadruped position (29, 62, 86); and conscious (62, 86). These varying experimental settings would be expected to have marked effects on the experimental results. The use of positive pressure ventilation would interfere with any central shift of blood into the chest (61). Recent surgery, including opening and closing the thorax, would be expected to raise sympathetic activity and circulating levels of catecholamines, alter local prostaglandin levels, and change levels of important volume regulatory hormones such as atrial natriuretic peptide, vasopressin, and renin-angiotensin II-aldosterone (146). The use of the erect or sitting posture would be expected to induce marked venous pooling in species that are normally quadruped ("the crucifixion response") (87, 88, 131). Thus a greater volume of pooled blood would be pushed centrally during WI (88). Since evidence indicates in humans that pulmonary closing volume increases during WI (124), some degree of arterial desaturation might occur in immersed animals. Therefore, arterial blood gases and pH should be determined, if possible. If awake animals are studied, thorough periods of training are required in order to have a relaxed, unstressed animal in the water.

WI is an innocuous procedure when carried out with awake dogs or other awake species, providing adequate training is instituted. Our practice has been to select dogs that are temperamentally suitable and then we train them to undergo WI for 1 h/day for 1–2 wk prior to the actual study. After this the dogs show no agitation upon entering the water and plasma catecholamines are unchanged. Heart rate may accelerate, but thoroughly trained dogs may not display cardiac acceleration (140) or a rise in mean arterial pressure (85). The sling frame assembly we use is depicted in Figure 38.4. We have used this assembly for dogs, sheep, and pigs. Rabbits can simply be placed in a plastic container. Monkeys can be positioned in restraining chairs that can then be lowered into the water or placed in tanks that are then filled (49, 117, 118, 122).

The time period or duration of the immersion exposure depends on whether anesthesia is used, or, if awake, the level of training and temperament of the animals species studied. Generally, dogs and sheep are studied for periods of 100 min, but they can be studied in the water for many hours. Long-term WI studies in animals are not available, but long-term immersions are important for determining whether time-dependent circulatory, renal, or neurohormonal adjustments are activated (21).

CARDIOVASCULAR RECEPTORS: CIRCULATORY, RENAL, AND HORMONAL INFLUENCES

Cardiac Receptors

These stretch or mechanoreceptors are located in the heart and are divided into receptors having myelinated afferent fibers and those having nonmyelinated afferent fibers. The physiological responses of the receptors having myelinated vagal afferents have been reasonably well studied, as have receptors with nonmyelinated vagal afferents. However, the receptors having nonmyelinated afferents which course centrally in sympathetic pathways are not well understood (14).

Receptors located in the atria give rise to myelinated vagal afferents. These receptors are located in unencapsulated nerve endings and respond to increases in atrial volume. It is probable that the adequate stimulus for these receptors is atrial wall tension (14, 49). In the dog, a group of type B atrial receptors is localized to the pulmonary vein–left atrial junction although these receptors are distributed throughout the atrial walls. An increase in atrial volume increases the activity in the afferent vagal fibers (Fig. 38.1).

In the dog, activation of these type B vagal afferents elicits cardiac acceleration (64, 75, 152), a reduction in vasopressin secretion (76, 95, 135), a depression of renal sympathetic activity (14, 142), and a depression of renin secretion (142, 143). The combined neural and hormonal responses lead to a diuresis and natriuresis

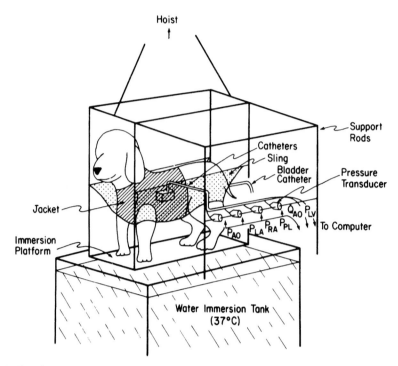

FIG. 38.4. Sling frame assembly for the study of conscious animals during water immersion. P_{AO} = aortic pressure; P_{LA}, Pra = left and right atrial pressures, respectively; P_{PL} = pleural pressure; Q_{AO} = aortic blood flow (electromagnetic flow transducer); P_{LV} = left ventricular pressure (solid state pressure transducer). [Reprinted with permission of the American Physiological Society; from Hajduczok et al. (62).]

(Fig. 38.5, I). In species other than the dog, atrial stretch is less likely to accelerate the heart rate (Bainbridge response), in fact, the heart rate may tend to slow (96).

The experimental conditions under which this reflex is elicited can have profound effects on the nature of the renal response. Table 38.2 compares renal responses elicited by atrial stretch using differing methods in anesthetized vs. awake dogs. In anesthetized dogs, atrial volume expansion leads to a water diuresis, whereas awake dogs show a natriuresis with a smaller increment in free water clearance. The relative importance of vagal afferent pathways appears to be more important in mediating the renal response to volume expansion in conscious monkeys compared to anesthetized monkeys (13, 117, 147). Table 38.2 emphasizes the importance of studying awake animals whenever possible and it indicates that critical components of the atrial–renal volume reflex in the dog are modified by anesthesia.

In addition to myelinated vagal afferents, most of the cardiac ventricular afferent input courses to the central nervous system by way of nonmyelinated C fiber afferents. These C fibers can be chemically activated by veratrine alkaloids and they are probably normally engaged by mechanical stimulation (14). The reflex response has been labeled the *Bezold-Jarisch reflex* and it consists of cardiac slowing, depression of arterial blood pressure, and cessation of breathing (9, 14). In addition, ventricular C fiber stimulation causes inhibition of vasopressin secretion and renal sympathetic nerve activity (142–144). The latter responses could certainly contribute to the diuresis and natriuresis of immersion. Severe bradycardia, hypotension, and apnea, however, are not normally observed during WI. Therefore, the role of ventricular C fiber afferents in the WI response requires clarification.

As will be discussed (see Cardiovascular Responses to Water Immersion), cardiac nerves play a critical role in the cardiovascular and renal responses to immersion. However, the relative contribution of each of the three receptor types (vagal myelinated, vagal nonmyelinated, sympathetic nonmyelinated) in bringing about these responses is uncertain.

Arterial Baroreceptors

The arterial baroreceptors located in the carotid sinus and aortic arch reflexogenic zones regulate the secretion of vasopressin and renin as well as cardiac performance and the level of systemic vascular resistance. An elevation of carotid sinus pressure decreases vasopressin and renin secretion and renal sympathetic nerve activity (135, 136, 142–144).

The arterial baroreceptors are activated during WI. Mean arterial pressure increases in the dog during WI

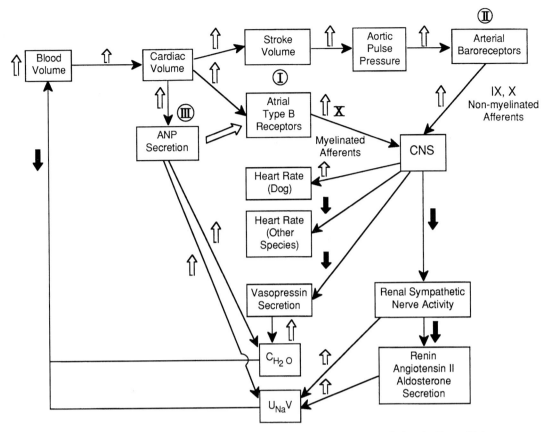

FIG. 38.5. Expanded version of the Gauer-Henry hypothesis. *Arrow symbols* as in Figure 38.1.

TABLE 38.2. *Effects of Anesthesia on Renal Responses to Atrial Stretch*

Anesthetized Dogs	Awake Dogs
Ledsome et al. (94)	Lydtin and Hamilton (98)
Ledsome and Linden (93)	Kaczmarczyk et al. (73)
Kappagoda et al. (76)	Fater et al. (46)
De Torrente et al. (30)	Goetz et al. (56)
Proznitz and Dibona (125)	
Ledsome et al. (95)	
Diuresis: +68%–700%	Diuresis: +64%–500%
Mean: +258%	Mean +314%
Natriuresis: 0%–182%	Natriuresis: 217%–1000%
Mean: +42%	Mean: +500%

(62) but it tends to remain the same in humans, monkeys, and sheep. The arterial pulse pressure widens, due to a rise in cardiac stroke volume in animals that do not display cardiac acceleration in immersion (non-Bainbridge species). Therefore, the arterial baroreceptors are likely to be mechanically loaded via increases in pulsatile and possibly mean arterial pressure (Fig. 38.5, *II*). In the dog the mean arterial pressure is elevated due to a rise in venous return and cardiac output which shifts

blood from the venous compartment into the arterial compartment (62). However, in other species the rise in cardiac output is offset by a decline in total peripheral resistance so that mean arterial pressure does not change.

In the awake dog, there is a rise in heart rate along with the rise in arterial pressure during WI. Therefore, there is a resetting of the arterial baroreflex such that a higher heart rate for any level of mean arterial pressure occurs (152). In addition there is a decrease in the average gain or sensitivity of the baroreflex, accompanied by an increase in the range of pressures over which the baroreflex controls heart rate (Fig. 38.6). Increases in cardiac filling pressure do not raise stroke volume because this "Bainbridge" cardiac accelerator response reduces ventricular filling time (16). However, this response is not consistent in that extensive and rigorous training tends to minimize the heart rate increase in immersion (85, 140). In addition the ventricular inotropic state increases, as judged by an increase in ventricular dP/dt max (62). After cardiac denervation, the awake dog increases its cardiac output to precisely the same level by increasing stroke volume (62). Presumably, in animals which do not display a vigorous heart rate response, the rise in cardiac output is achieved

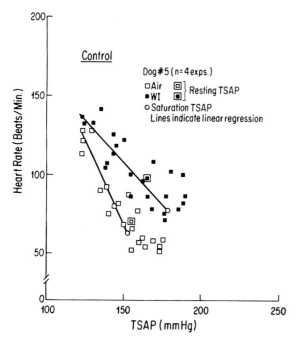

FIG. 38.6. Resetting and change of sensitivity of heart rate limb of arterial baroreflex in an awake dog during water immersion. TSAP = transmural systolic arterial pressure; Saturation TSAP = point at which heart rate stopped changing as TSAP was elevated. Threshold TSAP could not be determined in the standing dog. [Reprinted with permission of the American Physiological Society, from Yoshino et al. (152).]

largely via a rise in ventricular stroke volume due to the length–tension mechanism (6, 96).

With respect to inputs that could account for baroreceptor resetting in WI, it is important to recall that both somatic and vestibulo-cerebellar inputs are altered by WI. The potential effect of these inputs on hormonal and sympathetic control of the kidney remain to be determined. Returning to the issue raised by Gilmore (49), which considers the mechanisms mediating the renal response to volume expansion as differing in primates, it would appear that this issue is far from being resolved (117). More awake-animal studies are needed to answer this question, both in primates as well as in quadrupeds, such as sheep, that do not display a prominent Bainbridge response. It seems that the conscious monkey responds more like the awake dog to volume expansion and that vagal mechanisms mediating volume expansion responses are quite powerful in the awake monkey (138, 147). Although evidence in support of an atriorenal reflex in the monkey is not strong (117), and vagal type B afferent activity from atrial stretch is not striking in the monkey, the central processing of this information is not certain (87). Cornish et al. in 1984 proposed that high-pressure arterial baroreceptors play an important role in controlling blood volume in awake primates (22). The surprising results

of this study were that the diuresis and natriuresis to a 20% volume expansion were potentiated following chronic sinoaortic denervation (SAD). The authors concluded that high-pressure baroreceptors actually inhibit the renal responses to volume expansion. A similar result has been reported by Sit et al. (138) in awake dogs. Based on these results, Gilmore et al. (50) proposed that the inhibitory sinoaortic reflex mechanism might explain why the diuresis and natriuresis to volume expansion are potentiated in anesthetized monkeys, as pentobarbital may inhibit the sinoaortic reflexes.

Regarding the role played by cardiac receptors in the diuresis and natriuresis of volume expansion, Fater et al. (46) reported that total extrinsic denervation of the heart abolishes the renal and cardiac accelerator responses to left atrial balloon inflation in the conscious dog. However, contrary to the notion that atrial afferent mechanisms mediate the response to volume expansion, an increase in blood volume via dextran infusion still caused a potent diuresis and natriuresis. Thus while cardiac reflexes may contribute to the renal response to volume expansion, they are not essential for the response to occur. Central venous pressure and mean arterial pressure rose by similar amounts in the intact and cardiac-denervated dogs of Fater et al. (46). Hence while cardiac denervation removed a reflex response to cardiac stretch, the high-pressure baroreceptors were still present and could have mediated the response. Indeed, it is likely that the cardiac output rose in the intact and denervated dogs in response to the rise in cardiac filling pressure. In the intact dogs the heart rate rose during volume expansion, while in the cardiac denervated dogs the heart rate did not change during volume expansion. Thus with a similar elevation in mean arterial pressure, and most likely, a rise in arterial pulse pressure in the cardiac-denervated dogs due to an expected increase in stroke volume, the loading pattern on arterial baroreceptors would be expected to elicit a more intense baroreceptor discharge. This interpretation cannot be validated, however, since Fater et al. (46) did not present cardiac output or pulse-pressure data. Nevertheless, it seems possible that the high-pressure baroreceptors elicited the renal response in the cardiac-denervated dogs (73). As will be developed below, there is evidence favoring similar mechanisms operating during WI in the awake dog.

CARDIOVASCULAR RESPONSES TO WATER IMMERSION

Central Hemodynamics

In humans, WI to the midcervical level leads to an immediate increase in the cardiac output ranging from

32% to 62%, depending on the method of measuring cardiac output and on the experimental protocol (1, 6, 15, 96). Farhi and Linarsson (45) studied subjects suspended in a harness with their legs hanging free in the air and then during WI. The elevation of cardiac output was graded according to the immersion depth, and the more pronounced elevation of cardiac output in this study (+62%) may be attributed to the fact that the legs were hanging in a dependent position with more blood pooling in the air. Thus during WI more blood was available to be shifted centrally.

In general, mean arterial pressure may rise in humans during WI, but central transmural cardiac filling pressures rise by a similar amount so that the arterial–venous pressure gradient is unchanged. As cardiac output rises, systemic vascular conductance increases (1). The rise in cardiac output can be sustained for several hours via an elevated stroke volume, as heart rate usually slows (96). The heart rate response to WI is dependent on water temperature (82, 96).

The rise in cardiac output occurring in animals is similar to that usually reported to occur in humans—on the order of a 25%–30% increase above the air control outputs in conscious dogs (62) and sheep (personal observation). There is little information on the cardiac output response to WI in either anesthetized or awake monkeys, however, as central venous pressure has usually been reported to rise during WI it is likely that cardiac output increases in WI in monkeys, assuming that the cardiac length–tension mechanism is operative (117). The same is true for the rat (77).

In the conscious dog, the rise in cardiac output occurs in concert with a rise in heart rate and ventricular inotropic state with no change in stroke volume (62). Mean arterial pressure rises with no change in arterial pulse pressure or total peripheral resistance. Following total extrinsic cardiac denervation after the method of Randall et al. (128), with the additional modification of Fater et al. for dissecting around the pulmonary veins (46), the chronotropic and inotropic responses to WI are abolished in the awake dog; yet the cardiac output rises in an identical manner due to an increase in stroke volume. The heart rate response to inflation of a balloon in the left atrium is abolished (62). The mean arterial pressure still increases after cardiac denervation, but now the arterial pulse pressure increases as well due to the rise in cardiac stroke volume. Thus the total peripheral resistance continues to be unchanged during WI after cardiac denervation. As pointed out earlier, the cardiac limb of the baroreflex is reset during WI (Fig. 38.6). However, it is also clear that the limb of the baroreflex which controls total peripheral resistance is also reset during WI and continues to be reset even after cardiac denervation. This is because a rise in arterial pressure on the order of 25 mm Hg above control levels should reflexly reduce total peripheral resistance if the baroreflex were operating normally. Since total peripheral resistance is unchanged, the rise in arterial pressure is due to the compressing effect of WI, leading to a shift in blood volume from the venous side of the circulation into the arterial compartment.

In awake sheep and monkeys there is little change in heart rate during WI, therefore increases in cardiac output are probably mediated by the length–tension mechanism. Moreover, arterial pressure does not rise as much as it does in the dog, and therefore, total peripheral resistance seems more likely to decline. With a more prominent stroke volume response it would seem that the WI response of sheep and monkeys more closely resembles that of the human. As mentioned previously, the potent cardioacclerator and hypertensive responses to WI in the intact dog can be minimized with extensive training (88). Thus the striking baroreceptor resetting "Bainbridge" response of the dog seems to involve a behavioral reaction to WI which can be modulated with training. The response is inconsistent (64, 140). One might wonder whether this baroreflex resetting simply reflects a stress response. This seems unlikely, as minimally trained dogs displaying the rise in heart rate are not agitated; they remain quiet in the water and plasma catecholamines are unchanged. The dog may be a unique species with respect to the Bainbridge phenomenon. It would be of interest to determine whether the Bainbridge response of the awake dog to volume expansion could be modulated by training as well (16).

In contrast to volume expansion and the weightless state, WI has unique effects on the lung–chest wall system. The increase in central blood volume is associated with a rise in pulmonary vascular volume and pulmonary artery pressure (6). The elevation of blood volume in the lungs displaces air and leads to a reduction in the vital capacity (25, 67). If the water level is above the chest then the hydrostatic compression will mechanically load the chest wall and render it more difficult to expand the thorax. This leads to the development of negative pressure breathing (67). The greater inspiratory effort is usually not perceived as dyspnea. However, if the head is submerged and breathing is achieved via a long tube to the surface, pulmonary damage may result. In both anesthetized and awake dogs, WI leads to sustained increases in blood flows to the intercostal muscles and the diaphragm, reflecting the increased work of breathing (63, 88). The greater inspiratory effort further serves to elevate blood into the chest and increase intrathoracic blood volume.

The pattern of breathing during WI resembles that which is associated with elastic loading of the chest wall (67, 88). The conscious dog generally responds to this loading with a tachypnea (62, 63). Although under thermoneutral conditions systemic O_2 consumption does not change, the arterial P_{O_2} tends to decline. However, the arterial blood remains well saturated with O_2 and it is unlikely that the arterial chemoreceptors are engaged in WI. In addition, the arterial P_{CO_2} declines and arterial pH increases in both awake and anesthetized dogs, with a more pronounced decrease in P_{O_2} in the anesthetized dog (88). Prefaut and colleagues (124) observed mild hypoxemia in humans and attributed the response to a decline in pulmonary closing volume. This effect may be offset in the human by more improved perfusion of the apical portions of the lung and better matching of ventilation–perfusion relationships (1). However, it is unlikely that this occurs in quadrupeds during WI because these species probably lack a zone I in the lung (28).

The rise in pulmonary vascular volume and pressure is important to keep in mind if the potential therapeutic implications of WI are considered. This effect would be of concern if patients having congestive heart failure undergo WI.

Both the neural control of breathing and neural control of circulation are probably influenced to a significant extent by elastic chest wall–lung loading in WI due to activation of mechanoreceptors located in the chest wall and the lungs.

In terms of the cardiovascular, hormonal, and renal responses to WI, the general emphasis has focussed on the potential contribution to be derived from low-pressure cardiac receptors and high-pressure arterial baroreceptors. Future studies should be directed toward sorting out the relative importance of these two reflexogenic regions in the physiological regulations of WI. In addition, the relative influence of the various subsets of these receptor groups, such as vagal myelinated vs. vagal C fiber nonmyelinated afferents, requires clarification.

WI exerts differential loading effects on a variety of somatic proprioceptors and mechanoreceptors in addition to those located in the chest wall and diaphragm. The buoyancy effect of WI as well as the reduced tonus in major muscle groups would be expected to markedly alter the nature of the mechanoreceptor input pattern influencing the CNS. Strong evidence in support of cerebellar involvement in the WI response will be discussed below (see Regional Vascular Responses to Water Immersion). This structure can exert powerful effects on the cardiovascular system (33) as well as on somatic neuromuscular systems.

Transcapillary Fluid Shift

As described earlier, Gauer and Henry (48) initially proposed that the physiological adjustment to WI leads to a decline in plasma volume. Bazett reported significant reductions in plasma volume in humans based on measurements of hematocrit in dehydrated subjects (3, 4). Gauer and Henry found similar results (48). However, it has not been reported consistently that plasma volume declines during WI. In fact, most reports indicate either no change or an elevation of plasma volume, particularly if the subjects or animals are studied under hydrated or volume-repleted conditions (7, 57–59, 101).

In the late 1940's, von Diringshofen postulated that a transcapillary fluid shift occurs in WI such that extravascular fluid enters the plasma compartment. This fluid shift was predicted on the basis of theoretical analyses of arterial, venous, and capillary hydrostatic pressure during immersion (149).

The first experimental validation of the von Diringshofen hypothesis was provided by Davis and Dubois (29). They studied anesthetized, acutely instrumented, splenectomized, and bilaterally nephrectomized dogs during WI. Although the water temperature was not reported, it was found that the immersion diuresis in dogs with intact kidneys was not correlated with left atrial transmural pressure, nor was it altered by bilateral vagotomy. In the nephrectomized dogs, there was an increase in plasma volume, as indicated by declines of hematocrit and plasma protein concentration. Since these investigators also found a decline in osmolality, they postulated that fluid entered the plasma compartment during WI, which was hypotonic in composition. Although they did not measure circulating levels of vasopressin, they further postulated that the immersion diuresis was mediated by plasma hypotonicity, leading to a suppression of vasopressin release. Sodium excretion was not reported in this study.

In a later study in humans, Khosla and Dubois (79, 80) also found a hypotonic hemodilution during WI and estimated that plasma volume rises on the order of 6%, based on hematocrit and Evans blue determinations. They suggested that their data were compatible with the idea that intracellular water enters the vascular compartment. Of interest is the observation that K^+ and certain amino acids apparently entered the circulation. Khosla and Dubois (80) used a wick catheter to estimate interstitial fluid pressure and reported that the tissue pressure decreased by 2.1 cm H_2O during WI. This apparent decline in tissue pressure may have been related to the fact that the zero reference pressure was the surface of the water.

Miki and colleagues (101) developed a method for the continuous measurement of blood volume during WI in conscious dogs using ^{51}Cr–labeled red cells and a conductivity cell interposed in an extracorporeal circuit. The dogs were chronically splenectomized and studied in the nonvolume repletion state after overnight food and water restriction. Guyton porous capsules had been implanted chronically in the lower forelimb and upper hind limb to allow for estimation of interstitial fluid pressure. Prior to the study the dogs were mildly pre-hydrated to 2% of their body weight using a solution of 0.45% NaCl.

Figure 38.7 indicates that WI was associated with an immediate increase in plasma volume which attained a peak value about 35 min after the onset of immersion. Thereafter, the plasma volume leveled off, as the increase in urine flow began at this time. The peak elevation of plasma volume amounted to +7% above the preimmersion levels. Upon emersion, after 100 min in the water (37°C), plasma volume declined very rapidly. Plasma osmolality did not change, suggesting that the fluid shift was isotonic. Figure 38.7 also indicates that tissue pressure rose during WI as did calculated capillary hydrostatic pressure, but the rise in tissue pressure (lower forelimb) exceeded the rise in capillary pressure such that net reabsorption occurred, as reflected by the decline in plasma oncotic pressure. In contrast to the study by Khosla and Dubois (80), the reference level for the pressures measured in the Miki study was set at the level of the right atrium both out of and in the water.

Later Miki et al. (107) obtained further evidence that the fluid movement into the vascular compartment involves a transcapillary fluid shift by demonstrating a hemodilution during WI in the conscious dog without

an increase in lymph flow in the thoracic duct. Figure 38.8 indicates that graded WI in turn caused graded increments in forelimb tissue pressure measured by porous capsule and wick catheter, as well as in cephalic vein pressure measured close to the tissue pressure measurement site in the awake dog (106). However, the cephalic vein pressure, which reflects capillary hydrostatic pressure, does not rise when the dogs are only immersed to the level of the limbs. Cephalic vein pressure rises only after immersion to the midchest level, and then increases progressively with further immersion levels. Thus a pressure gradient is established between tissue and capillary in WI which favors capillary reabsorption. The failure of cephalic vein pressure to rise during the first WI level may be related to the central venous compliance. During the deeper levels of WI there is a linear coupling of tissue pressure, cephalic venous pressure, and central venous pressure (106). Thus, central venous pressure may be taken to reflect pressure in the peripheral veins during WI.

The control of extracellular fluid volume (ECFV) during WI was studied by Miki et al. (102) using instrumented dogs that had been chronically splenectomized. On the day of the experiment, the dogs were anesthetized and nephrectomized bilaterally and studied while anesthetized. ECFV was measured continuously using an extracorporeal circuit with ^{125}I-iothalamate which behaves similar to inulin as an indicator of the ECFV. Plasma volume was measured using ^{51}Cr–labeled eryth-

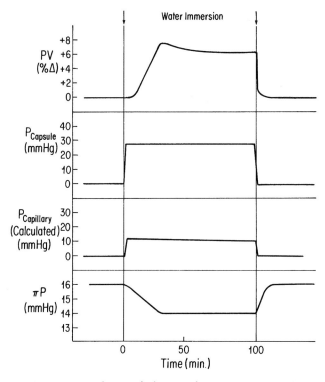

FIG. 38.7. Percent change of plasma volume (PV), pressure in an implanted Guyton capsule in subcutaneous tissue of dog forelimb ($P_{capsule}$), calculated capillary hydrostatic pressure ($P_{capillary}$) and plasma oncotic pressure (πP) in an awake animal undergoing water immersion at 37°C. [Reprinted with permission from the Best Publishing Company; from Krasney (84).]

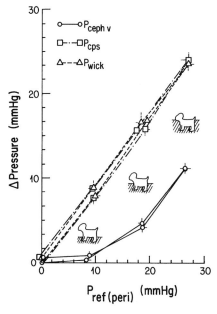

FIG. 38.8. Changes in cephalic vein pressure ($P_{ceph}v$); implanted Guyton capsule pressure (P_{cps}); and wick catheter pressure (P_{wick}) relative to the external hydrostatic reference pressure (P_{ref}) during graded immersion in awake dogs. [Reprinted with permission from the American Physiological Society; from Miki et al. (101).]

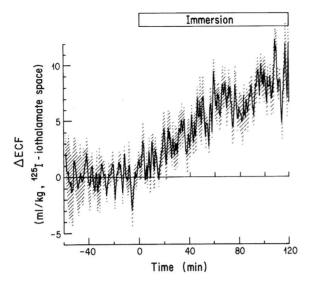

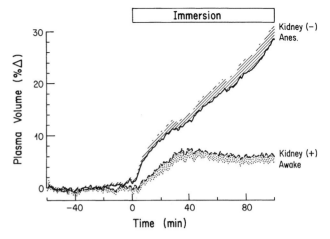

FIG. 38.9. Increase in extracellular fluid volume (△ ECF) during water immersion commencing at time 0 in an anesthetized, splenectomized, bilaterally nephrectomized dog. ECF volume was estimated by the ^{125}I-iothalamate space. [Reprinted by permission of the American Physiological Society, from Miki et al. (102).]

FIG. 38.10. Plasma volume responses to immersion in anesthetized, nephrectomized dogs and in dogs with intact kidneys. The role of the kidney is to minimize hypervolemia during immersion.

rocytes. Figure 38.9 indicates that ECFV increases by 4% during WI, while plasma volume rises in these nephrectomized animals by 33% after 100 min of WI (Fig. 38.10). Extending the WI period to 120 min resulted in a +40% increase in plasma volume. The interstitial fluid volume was estimated by subtracting the ^{51}Cr space from the ^{125}I space and this calculation indicated that the interstitial volume did not change very much. Thus it was estimated that nearly 80% of the shifted volume comes out of the cellular compartment. This view is supported by the observation that plasma K^+ rose in these animals, indicating the entry of a K^+ rich fluid into the vascular compartment. Davis and Dubois (29) also made the latter observation and indicated that the concentrations of certain amino acids in the plasma increase as well.

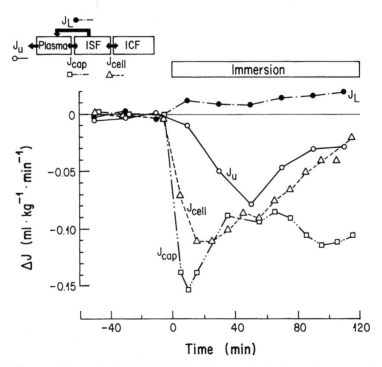

FIG. 38.11. There are three major fluid shifts (△ J) which occur during immersion; across the capillary (J_{cap}), across the cell wall (J_{cell}), and across the kidney (J_u). The fourth fluid shift in the lymphatics (J_L) is minor. ISF = interstitial fluid; ICF = intracellular fluid.

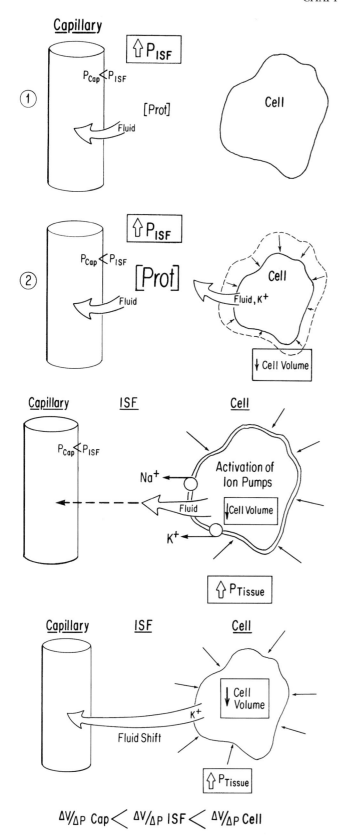

These studies indicate clearly that the role of the kidney in the WI response is to minimize the increase in plasma volume that occurs consequent to the transvascular fluid shift during WI. Figure 38.10 compares plasma volume responses to WI in anesthetized, nephrectomized dogs vs. awake dogs with intact kidneys. In 100 min, 80% of the possible plasma volume increase is eliminated by the kidneys. The total volume of urine and salt eliminated during a WI of several hours duration is far greater than could ever be derived from the plasma compartment alone. Thus it is apparent that the intracellular compartment represents a reservoir of fluid that is readily mobilized during WI. It should be pointed out that, if the original supposition of Gauer and Henry were correct, that is, that WI elicited a decline in the plasma volume, then each time a terrestrial animal entered an aquatic environment, hypovolemia or a circulatory deficit would be expected. Usually this does not occur. One might speculate that in those instances where plasma volume has been reported to decline, the subjects may have been initially dehydrated and nonrepleted and that the WI period was quite prolonged. However, even after very long immersions, as in 6–12 h, the plasma volume as measured by Evans blue remains elevated, suggesting that the intracellular reservoir has not been exhausted (71a).

The magnitude of the fluid shift depends upon hydration conditions during the experiment as well as on the initial hydration conditions. The magnitude of the hemodilution is greater in conscious dogs studied with volume repletion as compared to when they are studied in the absence of volume repletion (140). Moreover, the fluid shift is subject to modulation by hormones. Atrial natriuretic peptide (ANP) has been shown to provoke a shift of fluid out of the plasma compartment (133). This response may be mediated by precapillary vasodilation elevating capillary pressure, or via an increase in capillary permeability (69). Infusions of ANP during WI in doses that approximate physiological plasma concentrations block the transcapillary fluid shift during WI (85).

Thus there are three fluid shifts which occur during WI: (*1*) across the cell membrane; (*2*) across the capillary; and (*3*) across the kidney (Fig. 38.11). The precise mechanism that is responsible for the fluid shift across the cell membrane and ultimately the continuous autotransfusion, is uncertain. As the fluid entering the

FIG. 38.12. *a*: If protein fails to move into the plasma compartment during immersion, the rise of interstitial protein concentration could cause cell water and perhaps ions to leave the cell osmotically. *b*: Graded hydrostatic compression could activate cell membrane mechanoreceptors and membrane pumps which would extrude ions and water from the cells (133A). *c*: Compliance differences between cell, interstitial fluid, and capillary could cause fluid to shift from cell to plasma in immersion. [Reprinted with permission from the Best Publishing Company; from Krasney (84).]

plasma compartment appears to be isotonic (87), or hypotonic, according to Dubois (29, 79, 80), one possibility is that a fluid movement of this type would leave behind proteins in the interstitial compartment. This would cause an elevation of interstitial fluid protein concentration which in turn should pull fluid out of the cell osmotically (Fig. 38.12A). Indeed, limited measurements of fluid from the implanted capsules indicate that the protein concentration rises.

Another possibility is that graded tissue hydrostatic compression activates cell membrane pumps that are normally associated with regulation of cell volume (Fig. 38.12B). If cell volume is increased, then membrane pumps are turned on to move ions and fluid out of the cell to return cell volume to its regulated level. If cells perceive hydrostatic compression as increased cell volume, then this might be the adequate stimulus to move water, K+, and amino acids out of the cell.

Lastly, it is possible that differences in compliance existing between the cell, interstitial compartment, and the capillary, may be responsible for the fluid movement (83, 84). An elevated hydrostatic pressure could decrease cell volume more than the volume of the interstitial compartment (Fig. 38.12C). Thus interstitial pressure would be raised with only a limited increment in interstitial volume and capillary reabsorption would occur. These mechanisms could act individually or

together to move fluid out of the cells during WI. However, such a mechanism must account for the movement of ions as well. It is clear that additional studies involving both animal and mechanical models are required to obtain accurate insights into the nature of this fundamental mechanism. The effects of this immersion–fluid shift mechanism may be applicable in a number of circumstances, for example, the hyperbaric diuresis (68).

To summarize, Figure 38.13 depicts the interrelationships between cardiac output and venous return that occur during equilibrium conditions in WI using the theoretical analysis described by Guyton et al. (61). The graded extravascular hydrostatic compression acts to diminish the capacity of the venous compartment to hold blood. Thus there is a rise in mean circulatory filling pressure, which may be interpreted as being due to a decrease in the unstressed volume (61) of the venous compartment. Actually, measured venous compliance may reflexly increase in WI (35). The elevation of mean circulatory filling pressure causes a rise in venous return and cardiac output increases through the Frank-Starling mechanism (Fig. 38.13, point B); the exception to this is the intact dog in which the rise in cardiac output is neurogenically mediated. As WI continues, the autotransfusion leads to systemic hypervolemia which acts to further elevate mean circulatory filling pressure and cardiac output (Fig. 38.13, point C). Therefore, sys-

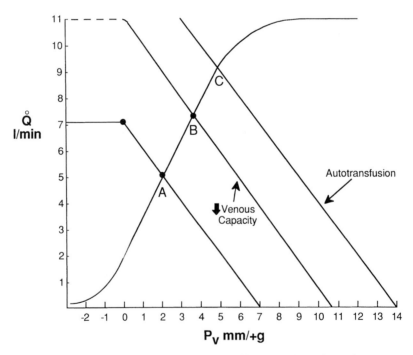

FIG. 38.13. As predicted by Guyton's analysis (61), the equilibrium condition for cardiac output (Q) and venous pressure (P_v) would move from the control state (A) to point B due to decreased venous capacity resulting from hydrostatic compression, and further to point C because of hypervolemia from autotransfusion. The net result is a higher cardiac output at a higher central venous pressure as predicted by the Frank-Starling relationship.

temic O_2 delivery is elevated while systemic O_2 consumption is unchanged. The kidney responds with an increase in fluid output which in turn acts to minimize the increased systemic blood flow.

REGIONAL VASCULAR RESPONSES TO WATER IMMERSION

As described above (see Cardiovascular Responses to Water Immersion), the elevated cardiac output of immersion may or may not be accompanied by an increased arterial pressure, depending upon the species and the extent of training (85, 140). The mechanism for elevation of the cardiac output is related generally to an elevation of cardiac preload, thus it may be expected that cardiac volume work rises in WI. In addition, an elevation of either mean or systolic arterial pressure will raise the ventricular afterload and lead to an increase in cardiac pressure work. Therefore there is a sustained elevation of coronary blood flow during WI [Fig. 38.14; (63, 88)]. Although coronary blood flow is elevated in WI in the awake dog, the transmural distribution of coronary flow is unchanged.

As described previously, there are sustained elevations of blood flow to the intercostal muscles and diaphragm during WI that are related to the increased work of breathing [Fig. 38.14; (63, 88)]. Although the core temperature by definition is unchanged during thermoneutral WI, the transition from room air to the warm water heats up the skin and adjacent insulative tissues. Thus in WI blood flow to the skin and subcutaneous fat increases on the order of several hundred percent. This increased surface blood flow is probably related to superficial heating and to local vasorelaxant responses or to altered neural vasomotor tonus (63).

Thus several of the regional circulatory adjustments in WI can be easily accounted for on the basis of increased local metabolic demand (heart, respiratory muscles) or thermal responses (skin, subcutaneous fat).

The basis for other regional flow adjustments in WI is, however, less certain. Both human and animal studies indicate that there is no change in renal blood flow or glomerular filtration rate. In terms of the other abdominal viscera, there are early increases in blood flow to the pancreas, spleen, hepatic arterial vascular bed, and all segments of the gastrointestinal tract. Since the later occurs, it seems likely that portal venous flow increases as well. These flow increases observed in the awake dog (Fig. 38.14) appear to develop in proportion to the increase in cardiac output. The visceral flow increases are time-dependent in nature, however, as after about 30 min of WI, the visceral flows decline to preimmersion levels and the increased cardiac output is redirected into nonrespiratory skeletal muscles (Fig. 38.14). Early in

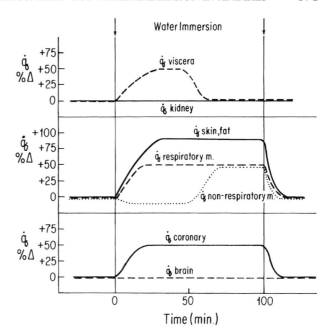

FIG. 38.14. Percent changes (%Δ) in the distribution of regional blood flows (q̇) measured by radiolabeled microspheres during immersion in conscious dogs. Flow increases to certain tissues (skin, fat, respiratory muscles) can be accounted for while flow increases to abdominal viscera and skeletal muscle are not readily explained. [Reprinted with permission from the Best Publishing Company; from Krasney (84).]

WI there were no changes in flow to nonrespiratory skeletal muscles.

Therefore at the onset of WI in the dog, since arterial pressure rises, there is vasoconstriction in nonrespiratory skeletal muscle. Later in WI the nonrespiratory muscles then dilate to receive the increased cardiac output that is rerouted from the viscera. The early vasoconstriction in skeletal muscle may be related to increased sympathetic activity related to arterial baroreflex resetting (152) or vascular autoregulation. The later flow increases could be due to deep muscle warming or to a reduction in muscle tonus, with the vasodilation acting to "steal" flow away from the visceral circuits. On the other hand, later in WI (Fig. 38.14) there is visceral vasoconstriction, the cause of which is also uncertain; it is almost as if a vasodilator factor were being washed out. In the human, Balldin et al. (2) have shown an early increase in limb muscle blood flow in WI as measured by 133Xe clearance. These regional hyperemias in WI seem to be unrelated to local metabolic activity, with the exception of the heart, nonrespiratory skeletal muscle, and skin. If this is correct, these regional hyperemias could be useful in several therapeutic situations, such as treatment of decompression sickness, therapy for muscle injury, and as a nonpharmacological method for increasing drug delivery to certain organs, for example, the liver.

It seems likely that regional vascular responses could play a role in the transvascular fluid shift that occurs in WI. Clearly the early muscle vasoconstriction could diminish capillary hydrostatic pressure, which, in the presence of elevated tissue pressure (99, 106), could lead to elevated capillary reabsorption of fluid. The same is true of the later visceral vasoconstriction. Thus it may be that the origin of the fluid early in WI is from skeletal muscle, and later from the abdominal viscera. This possibility raises an important question as to whether the autotransfusion may be temperature sensitive. Since the fluid shift depends upon the relation of capillary hydrostatic pressure to tissue pressure and the capillary pressure is regulated by vasomotor tone, it might be that more fluid would shift in colder water where there should be more vasoconstriction in skin and skeletal muscle (113), while less fluid would shift in warm water where there is more vasodilation and higher capillary pressures.

The cerebral flow adjustments to WI are of interest. While arterial pressure tends to rise in WI, depending on the species, there is no change in overall cerebral blood flow. This implies that cerebral vasoconstriction might occur, which could be due to autoregulation or to the mild hypocapnia that develops. The latter cerebral constrictor effect is opposed however, by the mild hypoxia that ensues (63, 88). There is evidence that intracranial pressure increases in WI are secondary to the increased pleural and downstream venous pressure. This effect would tend to reduce cerebral perfusion pressure. While total cerebral flow does not change in WI, there are consistent and sustained increases in blood flow to the cerebellum during WI in the awake dog (63). If local flow reflects cerebral metabolism, then there is increased cerebellar metabolism that may be responding to the altered mechanical inputs due to the buoyancy effect of WI. In addition, the cerebellum influences arterial pressure regulation (33).

Several studies have proposed that cardiac receptor afferents can influence the distribution of regional blood flow. However, in awake dogs, following total extrinsic denervation of the heart, the regional systemic blood flow responses were the same as those occurring during WI in a group of sham-denervated dogs. Therefore the regional circulatory adjustment to WI is not dependent on reflex influences from cardiac receptors (63).

Krogh (131) originally proposed a model for the regulation of venous return and cardiac output which depended on the distribution of regional blood flow between vascular beds having a large compliance, such as the splanchnic circuit (slow time constant), vs. vascular beds that have a low compliance, such as skeletal muscle (fast time constant). Thus with blood flow going mainly to the splanchnic circuit, there would be a relative pooling of blood in the compliant veins and venous

return would be less as compared to a situation where flow was largely distributed to a fast time–constant bed, such as muscle. Accordingly, this model would predict that cardiac output should be less early in WI when major flow increases occur in the visceral compartment. However, later when the flow pattern moves away from the viscera into skeletal muscle the cardiac output should rise. Unfortunately, despite the shift in flow from the slow time–constant bed into the fast time–constant bed in the awake dog, the cardiac output remained at a stable, elevated level throughout WI. Therefore this study does not support the Krogh model (131).

It may be concluded that there are striking increases in the ratio of systemic flow to metabolism and the ratios of flow to metabolism in abdominal viscera and skeletal muscles during WI based on awake dog data (63, 88) and limited human studies (2). These responses are not well understood, but the regional hyperemias may be useful for therapeutic purposes. In addition, there is a sustained elevation of blood flow to the cerebellum in WI.

THE RENAL RESPONSE TO WATER IMMERSION

The character of the renal response to WI depends upon the hydration state of the subject or animal. Thermoneutral WI in a hydrated human typically leads to diuresis, natriuresis, kaliuresis, and an increase in free water clearance (36). These responses develop after 20–40 min, and the period for which they are sustained depends upon the fluid repletion protocol. In nonreplete subjects these responses persist for 2–4 h, while in replete subjects responses are sustained for the duration of WI (36). Behn and colleagues (7) found that the diuresis in humans was associated with an elevation of free water clearance, while in dehydrated subjects a smaller diuresis was associated with a rise in osmolal clearance.

The level of physical training influences the character of WI diuresis. Claybaugh and colleagues (21) found that diuresis and natriuresis were significantly reduced in trained swimmers and runners when compared to sedentary control subjects, despite larger and more persistent increases in cardiac output.

The first renal study of WI in the anesthetized dog was reported by Stahl in 1965 (141). Under pentobarbital anesthesia, dogs were moved from the supine position to the vertical erect position which decreased cardiac output, renal blood flow, renal venous tissue tension (as obtained via a renal vein wedge catheter), urine flow ($\dot{V}$), and sodium ($U_{Na}\dot{V}$) and potassium excretion ($U_K\dot{V}$), while arterial pressure remained unchanged. Immersing the upright animal in water (the water temperature was not indicated) led to elevations of $\dot{V}$, $U_{Na}\dot{V}$, and $U_K\dot{V}$ above the air supine levels. Inulin clearance (glomerular filtration rate, [GFR]), renal blood flow, and renal tissue tension also increased above the supine air values. Stahl (141) postulated that the diuresis and natriuresis of WI were due to increases in total renal and medullary blood flow, filtered load, and renal tissue pressure. These renal responses persisted in a group of animals pretreated with desoxycorticosterone acetate (DOCA) and vasopressin that received an infusion of 9-alpha-fluorohydrocortisone during the experiment. This suggests that the diuresis and natriuresis were not due to suppression of vasopressin or aldosterone as predicted by the original Gauer-Henry hypothesis.

Myers and Godley (110) studied chloralose-anesthetized dogs immersed totally in the lateral decubitus position in 33.5°C water. The dogs breathed spontaneously through an endotracheal tube open to room air through a sealed opening in the wall of the tank. Cardiac output and heart rate rose with only a small increase in arterial pressure. $\dot{V}$, $U_{Na}\dot{V}$, fractional sodium excretion (FENa) and osmolal clearance (C_{osm}) rose significantly during WI. However, $U_K\dot{V}$ and C_{H_2O} decreased significantly while renal plasma flow (CPAH) and hematocrit did not change. These data were very similar to results obtained from the same group (52) using only negative pressure breathing. The failure to obtain an increase in C_{H_2O} is not consistent with the original Gauer-Henry hypothesis. Myers and Godley (110) suggested that the renal response to WI was mediated by a rise in medullary blood flow, as did Stahl (141).

Zucker and Gilmore (154) studied pentobarbital-anesthetized dogs in the vertical position in air and in WI at 34°C. They found that WI caused increases in $\dot{V}$, $U_{Na}\dot{V}$, $U_K\dot{V}$, Cosm, C_{H_2O}, Creatinine clearance (C_{cr}), and C_{PAH}. Unfortunately, the cardiac outputs were rather low (0.77 liter/min) and heart rates rather high (198 beats/min) in the vertical air controls. Hematocrit decreased from 45% to 41.9% from air to WI. These responses were not influenced by bilateral vagotomy, or by DOCA and vasopressin pretreatment. Central venous pressure rose during WI but arterial pressure was unchanged. These investigators concluded that the renal response to WI was primarily hemodynamic in nature, as C_{PAH} and C_{cr} rose in WI in agreement with the increase in renal cortical blood flow measured by radiolabeled microspheres in anesthetized, vertically immersed dogs (88).

As described previously (see Transcapillary Fluid Shift), Davis and Dubois (29) showed that WI of anesthetized, acutely instrumented dogs in the quadruped position causes a water diuresis with no change in sodium excretion. The diuresis was unrelated to changes in left atrial transmural pressure or to whether the vagi were intact. These investigators suggested that a hypotonic shift of fluid into the plasma compartment occurs, and they postulated that this caused a decline in vasopressin levels and led to diuresis.

Thus anesthetized dog studies generally report increases in $\dot{V}$, $U_{Na}\dot{V}$, C_{osm}, $U_K\dot{V}$ in association with increased renal blood flow, and GFR. Davis and Dubois (29) postulated that the increase in C_{H_2O} was related to hypotonic hemodilution of the plasma.

Gilmore and Zucker (51) reported results from pentobarbital-anesthetized monkeys studied in the vertical sitting position in air and in WI (35°C). WI caused increases in central venous pressure and arterial pressure with no change in heart rate, suggesting baroreflex resetting. There were increases in $\dot{V}$, C_{osm}, C_{H_2O}, $U_{Na}\dot{V}$, FE_{Na}, and $U_K\dot{V}$, with no changes in C_{PAH} or GFR. Similar responses were observed after vagotomy, but GFR and C_{PAH} rose significantly. These responses were not influenced by pretreatment with DOCA and vasopressin. The authors speculated that the renal response was due to abdominal compression, leading to a redistribution of renal blood flow such that medullary flow increased in agreement with Stahl (141) and Myers and Godley (110).

Later Peterson and colleagues (121) studied anesthetized monkeys during total WI in the recumbent position to determine whether a rise in abdominal pressure was related to the renal response. No renal response or cardiovascular response was observed while central venous and femoral venous pressures rose by equivalent amounts. These monkeys were on positive pressure respiration and the uniform application of hydrostatic pressure may have precluded any translocation of blood into the chest. The authors concluded that abdominal compression plays little role in the renal response.

Subsequently, Peterson et al. (122) compared the magnitude and time course of the renal responses to WI vs. volume expansion in anesthetized monkeys. Both WI and volume expansion resulted in increases in $\dot{V}$, $U_{Na}\dot{V}$, and FE_{Na}, with $\dot{V}$ and $U_{Na}\dot{V}$ being increased significantly after 10 min of WI, but these values were increased significantly only after 30 min of volume expansion. The increases in C_{osm} and C_{H_2O} were also more rapid in WI than in volume expansion. Neither C_{PAH} nor C_{cr} increased in WI, but both were increased after volume expansion. Since the renal responses occurred more rapidly in WI than with volume expansion, the mechanisms mediating the responses may differ.

As pointed out previously in Table 38.2, renal responses to atrial stretch by several methods differ considerably when results from anesthetized, acutely instrumented dogs are compared to those from conscious, chronically prepared dogs.

In general the studies utilizing anesthetized dogs have reported that the renal plasma flow, as estimated by C_{PAH}, and glomerular filtration rate (as estimated by either creatinine clearance or inulin clearance), both increase. This seems to be particularly true if the anesthetized dogs are immersed in the vertical position (88, 141, 154). Thus there is renal vasoconstriction when the dogs are moved from the horizontal position in air to the vertical position. WI then acts to restore hemodynamics to the supine level. The renal functional response in this situation has been attributed to an increase in filtered load. By comparison the immersed anesthetized monkey does not show a change in renal hemodynamics, although C_{PAH} and GFR rise during volume expansion (122).

The awake human shows no change in renal hemodynamics during WI (36, 42). Similar observations have been made in conscious dogs by measuring C_{PAH} (85, 140) and by injecting radiolabeled microspheres (63). In addition, the distribution of renal cortical flow from outer to inner layers in unchanged in WI in the awake dog (63). As estimated by creatinine clearance GFR also does not appear to change. Arterial pressure can rise substantially during WI in the awake dog (62, 63). Therefore the constant renal blood flow indicates renal vasoconstriction which may involve autoregulation. On the other hand, arterial pressure does not increase as much in well-trained dogs (85, 140), and a rise in pressure does not occur in humans (1), awake sheep, or awake monkeys (8, 117, 119). Thus it may be concluded that the natriuretic and diuretic responses of WI are not dependent upon a rise in arterial pressure (47, 81) or changes in filtered load in conscious animals.

The renal responses of WI in awake animals are related to tubular responses. Sondeen et al. (140) utilized the lithium clearance (C_{Li}) method to estimate changes in proximal tubular reabsorption during WI in the awake dog. The C_{Li} method assumes that lithium is handled only by the proximal tubule. In mildly hydrated, volume-repleted dogs, WI increased urine flow ($\dot{V}$) fourfold and fractional excretion of Na (FE_{Na}) approximately threefold. This was associated with a decrease in proximal tubule fractional sodium reabsorption from 0.82% to 0.69% and a decrease in distal fractional sodium reabsorption from 0.96% to 0.88%. By comparison, in mildly hydrated, nonvolume-replete dogs, there were smaller increments of $\dot{V}$ and FE_{Na} and no change in proximal fractional sodium reabsorption. However, distal fractional sodium reabsorption decreased from 0.97% to 0.93%. Thus distal tubular and proximal tubular sodium reabsorption are modified in WI, depending upon the degree of volume repletion. Of interest is the fact that these differing tubule response patterns occurred despite similar changes in arterial and left atrial pressures in the replete vs. nonreplete experiments.

Osmolal clearance increases during WI in both humans and conscious animals. This response is depen-dent upon the degree of hydration (7) and is mainly related to the increase in $U_{Na}\dot{V}$. An increase in potassium excretion ($U_K\dot{V}$) occurs also in humans but this response is less consistent in awake animals. A rise in $U_K\dot{V}$ is more likely to occur in WI in anesthetized dogs and monkeys (122, 141). In awake dogs the increase in $U_K\dot{V}$ is more transient and variable (62, 85, 140). It seems likely that a rise in $U_K\dot{V}$ would occur in WI since a K^+ rich fluid enters the plasma via the fluid shift (102). The $U_K\dot{V}$ response itself seems to be related primarily to increased washout of fluid from the distal tubule.

A rise in C_{H_2O} may be expected during WI in humans, depending on the level of hydration (7, 36). As pointed out earlier, a rise in C_{H_2O} as a consequence of cardiac distention is more prominent in anesthetized than awake animals. Hajduczok and colleagues (62) found that C_{H_2O} was unchanged during WI in volume-replete, intact, awake dogs, and that the diuresis was mainly due to a rise in $U_{Na}\dot{V}$ and C_{osm}. Miki et al. (101) found, however, that the diuresis of WI was primarily associated with a rise in C_{H_2O} in nonreplete, splenectomized, conscious dogs while $U_{Na}\dot{V}$ was unchanged. In an attempt to determine whether the hydration state was the cause of this discrepancy, Sondeen et al. (140) studied WI responses in both replete and nonreplete, intact dogs. This study demonstrated that the WI diuresis was related to increases in both C_{osm} and C_{H_2O} and the responses were graded according to the level of hydration (Fig. 38.15). Thus by exclusion it was proposed that the presence or absence of the spleen might be responsible for the variable C_{H_2O} response in the Hajduczok et al. (62) vs. Miki et al. (101) experiments. However, Sondeen and colleagues (139) demonstrated that renal responses to WI are similar in both intact and splenectomized, awake dogs. In a later experiment, Miki et al. (105) indicated that the diuresis of WI was mainly associated with a rise in $U_{Na}\dot{V}$ and C_{osm} in conscious, intact, nonreplete dogs. C_{H_2O} was unchanged.

HORMONAL RESPONSES TO WATER IMMERSION

Methodological Considerations

Striking adjustments of plasma levels of volume regulatory hormones can occur in WI depending upon the experimental situation. A key methodological concern in attempting to determine the role of certain volume regulatory hormones in WI relates to the size of the species selected for study. It is clear that smaller animals such as the rat and the rabbit display cardiovascular and renal responses to WI. However, if the time course of hormonal responses is to be studied, then repeated blood samples are required. In general, larger animals are studied during WI because their blood volumes can support the repeated blood sampling protocol.

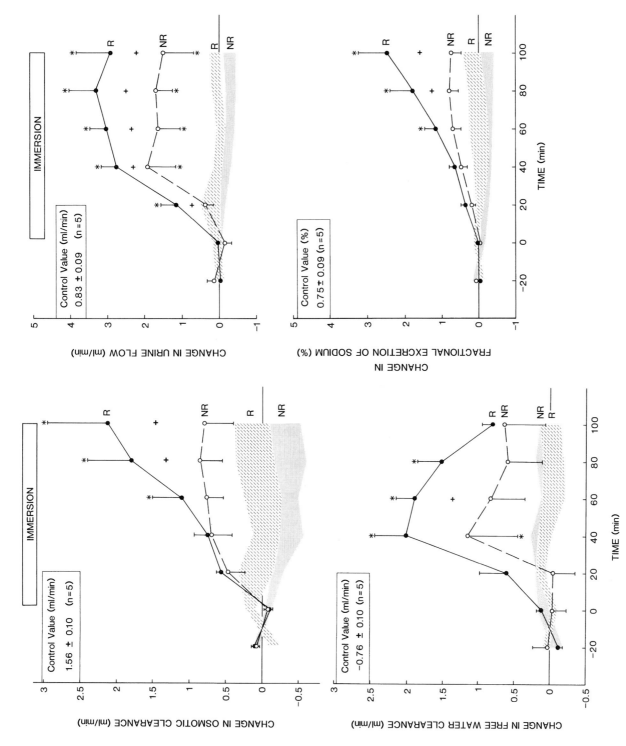

FIG. 38.15. Renal responses to water immersion in volume replete (R) and nonreplete (NR) conscious dogs. *Shaded* and *dashed areas* represent responses during air-timed control studies.

In addition, the investigator must be aware of how hormones are transported in the species chosen for study. For example, there is significant binding of aldosterone in sheep blood such that even pre-WI levels of aldosterone can be quite low (27). If this is the case, then preanalysis extraction procedures may be required.

Lastly, the animal must be tractable during WI. Some animals will display signs of stress and agitation upon the initial immersion which may influence catecholamines, sympathetic activity, and endocrine responses. Therefore, some initial training may be required and the immersion tank should be housed in a quiet room where extraneous influences can be ruled out. Clearly, the thermoneutral temperature must be determined accurately for the species under study. It is also necessary to regulate the room temperature to which the animal is exposed prior to immersion. If the appropriate precautions are attended to, the animal should stand quietly in air, undergo a smooth transition into the water, and stand quietly during WI. In dogs the major disturbance is related to "wet dog shakes" upon emersion.

Vasopressin

Davis and Dubois (29) on the basis of studies in anesthetized, acutely instrumented dogs found that a dilute fluid shifted into the plasma compartment rendering the plasma hypotonic. They postulated that plasma hypotonicity reduced the secretion of vasopressin, leading to the rise of $\dot{V}$ in WI. They failed to observe a rise in $U_{Na}\dot{V}$ in their dogs, which incidentally, were splenectomized. Unfortunately, the idea that the plasma becomes hypotonic in WI has not been confirmed by other studies of immersed humans or animals.

In the absence of changes in plasma osmolality, the primary stimulus for control of vasopressin secretion in WI involves the cardiovascular mechanoreceptors located in the heart and in the carotid sinus and aortic arch. Cardiac stretch may load atrial and ventricular mechanoreceptors and increase activity in vagal myelinated and nonmyelinated afferents. In addition, increases in mean and pulsatile arterial pressure may load the high-pressure arterial baroreceptors. The increased activity in the cardiac and arterial mechanoreceptor afferents can reflexly inhibit vasopressin secretion.

Norsk and Epstein (111) have indicated that a reduction in plasma vasopressin levels is responsible for the rise in C_{H_2O} which can be demonstrated in humans undergoing WI. However, a major problem in interpreting these studies is that vasopressin levels are quite low in hydrated subjects and it is quite difficult to detect reductions in circulating vasopressin under these conditions. There is a great deal of uncertainty as to whether such small reductions can be significant phys-

iologically. Thus vasopressin undergoes trivial suppression, if levels change at all, in hydrated humans displaying an immersion water diuresis (57–59, 89). By contrast, the levels of plasma vasopressin may be quite high in dehydrated subjects prior to WI and these levels decline in a major way, yet no increase in C_{H_2O} is observed. Urine remains in the hypertonic range (7, 148).

A clear correlation between plasma vasopressin levels and the appearance of a C_{H_2O} response in WI has been difficult to establish. The evidence which supports a role for vasopressin in the WI response is from experiments which indicate that the C_{H_2O} response is abolished when subjects are pretreated with vasopressin (39). Bie et al. (11) have shown in conscious dogs that the kidney is quite sensitive to very small changes in vasopressin levels. Moreover, changes in other volume regulatory hormones or mediators such as ANP or prostaglandins can interfere with the actions of vasopressin at the collecting duct (87).

In terms of mechanoreceptor input, it has been shown that vasopressin levels in humans do not always decline when central venous pressure is increased (11). A better correlation has been obtained between changes in mean arterial pressure and arterial pulse pressure and plasma vasopressin levels.

Although Gauer and Henry (48) originally demonstrated that atrial distention causes a reflex decrease in vasopressin secretion in anesthetized dogs, it has been difficult to detect changes in plasma vasopressin levels during WI in awake dogs even though left atrial pressure rises markedly. In intact volume replete awake dogs, Hajduczok (62) found no change in either plasma vasopressin or C_{H_2O}. Sondeen et al. (140) found graded increments in C_{H_2O} during WI in conscious dogs studied during nonvolume repletion and volume repletion. However, plasma vasopressin was unchanged in either case. In contrast, Hajduczok et al. (62) found that vasopressin declined and flow to the hypophysis declined in conscious dogs following total extrinsic denervation of the heart. In these dogs, the rise in $\dot{V}$ was identical to that occurring in intact dogs, except it was associated entirely with a rise in C_{H_2O}. Thus in the dog a good correlation between C_{H_2O} and vasopressin levels is only observed after cardiac denervation suggesting that the arterial baroreceptors may play a role in vasopressin suppression. One might interpret this result to indicate that cardiac receptors act to elevate vasopressin levels in WI (109, 150). However, this explanation is contradicted by the large number of studies which indicate that cardiac receptors inhibit vasopressin secretion (72, 76, 95, 135, 142). On the other hand, through baroreceptor resetting or other means, cardiac receptors could inhibit an arterial baroreflex-mediated depression of vasopressin secretion (89). Since heart rate increases along with mean arterial pressure in the intact dog during WI, the loading pattern on the baroreceptors is different. After cardiac denervation, an identical rise in cardiac output occurs which is associated with a rise in stroke volume and arterial pulse pressure. Thus a reflex inhibition of vasopressin secretion by the arterial baroreflexes could occur or a mechanical effect could be elicited at the sella turcica by the rise in pulse pressure which could be inhibitory (127).

In line with the above reasoning, a depression of vasopressin secretion occurs in the anesthetized monkey during both WI and volume expansion and a depression of vasopressin secretion may occur in the awake human during WI. In terms of the monkey response to WI, Benjamin et al. (8) studied WI responses in conscious, ketamine-anesthetized, and pentobarbital-anesthetized monkeys. Vasopressin levels were elevated in air in the pentobarbital group (56 pg/ml) and in the ketamine group (14 pg/ml), compared to the conscious group (2 pg/ml). Plasma vasopressin declined during WI in both anesthetized groups, but not in the awake group of monkeys. Both anesthetized groups of monkeys had diuretic responses which were about 50% of the response elicited by WI in the conscious group. In contrast, the immersion natriuresis was blunted in the ketamine group (3 µEq/min to 14 µEq/min) as compared to the pentobarbital group (7 µEq/min to 40 µEq/min) and the conscious group (9 µEq/min to 42 µEq/min). Thus the awake monkey shows little alteration in plasma vasopressin during WI, a response similar to that of the awake dog. The effect of varying hydration conditions on the vasopressin response in the monkey remains to be determined.

In primates, both humans and monkeys, there is a rise in cardiac output during WI which is not associated with a rise in heart rate as in the dog, but rather a rise in stroke volume and arterial pulse pressure. Thus the mechanical loading profile presented to the arterial baroreceptors in the primate during WI may differ from that occurring in the dog. Nevertheless, the plasma vasopressin responses to WI appear to be similar in dogs, monkeys, and humans in the conscious state. By exclusion, the canine cardiac denervation experiments (62) and the notion that the primate has attenuated cardiac receptor neurocirculatory control suggest that the arterial baroreceptors may be the primary determinant of vasopressin secretion.

In humans the type and level of athletic training can influence the vasopressin response to WI. Claybaugh et al. (21) found that WI decreases urinary vasopressin excretion significantly in mildly hydrated sedentary subjects but no changes were noted in swimmers, and decreases occurred only in the second hour of WI in trained runners. The WI diuresis and natriuresis were also blunted significantly in the trained subjects despite larger and more sustained increases in cardiac output. It is unclear whether physical training can modulate the WI cardiovascular–hormonal–renal response in experimental animals (5).

Renin-Angiotensin II-Aldosterone System

In humans plasma renin activity and plasma aldosterone levels decline in a more consistent manner than plasma vasopressin and the responses appear to be less related to the degree of hydration (26, 36, 41, 44). However, the diuretic response usually is present within 40 min of WI and is too rapid in onset to be related to effects of aldosterone. In humans the diuresis of WI is not influenced by pretreatment with DOCA but the natriuresis is blunted, implying that the $U_{Na}\dot{V}$ response may be partly related to the suppression of aldosterone (36, 44, 49). As mentioned previously, the kaluretic response in humans is probably related either to natriuresis or to distal tubular washout, rather than to aldosterone suppression.

Data on the nature of the renin-angiotensin II-aldosterone response to WI in the monkey are lacking. However, pretreatment of anesthetized monkeys with DOCA does not influence diuresis or natriuresis (51). The rapid onset of the immersion diuresis in monkeys also is not compatible with an effect derived from aldosterone suppression (122).

Hajduczok and colleagues (62) found that plasma renin activity (PRA) tended to be decreased when conscious, intact dogs were immersed. Significant decreases in PRA were observed when conscious, cardiac denervated dogs were immersed. Since arterial pressure rose in both groups of dogs, the decrease in PRA may be related to the elevation of arterial pressure. However, intact cardiac afferents are not essential for suppression of PRA. Subsequently, Sondeen et al. (140) found identical significant decreases in PRA during WI in conscious dogs studies either in the volume-replete or nonreplete state. Since the $\dot{V}$ and $U_{Na}\dot{V}$ responses were significantly attenuated in the nonreplete experiments, the results indicate that the renal responses are poorly correlated with PRA levels in WI. Angiotensin II and aldosterone levels were not measured in this study so it is unclear as to what extent the PRA response reflects angiotensin II or aldosterone responses. Krasney et al. (85) found that PRA levels were not correlated with $\dot{V}$ or $U_{Na}\dot{V}$ in conscious dogs studied in air. However, in WI, PRA declined significantly and a significant negative correlation between PRA and $\dot{V}$ and $U_{Na}\dot{V}$ developed. Although this negative correlation was significant, it is unclear whether the relationship is more than simply coincidental.

To summarize, the data from human and conscious dog studies indicate that PRA, and probably aldosterone, decline during WI. The onset of the $\dot{V}$ and $U_{Na}\dot{V}$ responses is too fast to be related to an influence from aldosterone. The $U_{Na}\dot{V}$ response, but not the $\dot{V}$ response, is attenuated somewhat in humans, but not in anesthetized monkeys, by pretreatment with DOCA. In awake dogs, PRA is suppressed in WI, although it is difficult to establish a functional relationship between the PRA response and the renal responses. Claybaugh et al. (21) found that there were similar PRA and aldosterone responses to WI in sedentary and trained subjects but there were depressed $\dot{V}$ and $U_{Na}\dot{V}$ responses in the trained group. The precise modulatory influence of the renin-angiotensin II-aldosterone system in WI may be clarified by measurements of angiotensin II levels and differentiating responses which occur early in WI from those occurring later in WI.

Atrial Natriuretic Peptide (ANP)

Distention of the atria has been shown to release ANP into the circulation. ANP has been demonstrated to provoke diuresis, natriuresis, vascular relaxation, and transcapillary fluid shifts. It has been demonstrated that WI elevated plasma levels of ANP in humans (53, 115), dogs (103), and rats (77). Although the elevation of ANP in the plasma is quite rapid at the onset of WI and a number of aspects of the WI response have been attributed to ANP (Fig. 38.5, III) (37), its specific role in the WI response remains unclear.

The relative time courses of the ANP response and the renal response to WI differ. While plasma ANP levels rise and level off rapidly, it takes about 40 min for the immersion natriuresis and diuresis to become evident (103). It is not certain whether the WI responses can actually be elicited by the levels of circulating ANP that are attained during WI. Goetz et al. (55, 56) found in awake dogs that inflating a balloon in the left atrium caused elevations in ANP, $\dot{V}$, and $U_{Na}\dot{V}$. After cardiac denervation, left atrial balloon inflation still elevated plasma ANP, but the $\dot{V}$ and $U_{Na}\dot{V}$ responses were abolished.

If ANP contributes to the renal response of WI, it may not do so immediately. Bie et al. (12) infused small doses of ANP which did not immediately elicit diuretic and natriuretic responses. However, after 45 min of constant infusion, these small doses of ANP were associated with significant increments of $\dot{V}$ and $U_{Na}\dot{V}$. The renal effects of ANP have usually been shown to be associated with doses that cause increases in renal blood flow and GFR. As already pointed out, human and conscious animal studies indicate that the renal response to WI is not associated with any changes in renal hemodynamics. In this regard, Bie et al. (12) found that renal tubular responses occurred after small doses of ANP in the absence of changes in renal hemodynamics.

Miki et al. (103) found that WI in conscious dogs elicits an elevation of plasma ANP levels. However, the pattern of the $\dot{V}$ and $U_{Na}\dot{V}$ responses was poorly correlated with the pattern of the ANP response. Sondeen et al. (140) showed that in awake dogs, the patterns and magnitudes of the ANP elevations were identical when the dogs were immersed under volume-replete and nonreplete conditions, with the nonreplete dogs displaying attenuated $\dot{V}$ and $U_{Na}\dot{V}$ responses.

Although a direct causal relation between the plasma levels of ANP that are attained during WI and the time course and amplitude of the $\dot{V}$ and $U_{Na}\dot{V}$ responses is not readily apparent, ANP could conceivably interact with other volume regulatory hormones so as to contribute to the renal response. ANP has been shown to interfere with aldosterone secretion, inhibit the hydroosmotic actions of vasopressin, and inhibit the release of renin from the kidneys (53).

Metzler and Ramsay demonstrated that the renal responses to doses of ANP that elicited physiological increments of plasma ANP levels are potentiated significantly when conscious dogs were volume-expanded with hypertonic saline (100). Accordingly, Krasney et al. (85) infused ANP in doses of 5 and 25 ng/kg/min in awake dogs in air and during volume-replete WI. Plasma levels of ANP, vasopressin, and renin activity were measured. The results indicated that the diuretic and natriuretic responses to a given plasma level of ANP are significantly potentiated during WI. These results are in accord with the hypertonic expansion study of Metzler and Ramsay (100) and they underscore the fact that the renal response to ANP is dependent upon the hydration state or level of volume expansion. The results of Krasney et al. (85) indicate that there is a true increase in the renal sensitivity to ANP in that the slopes of the lines relating the $\dot{V}$ and $U_{Na}\dot{V}$ responses to plasma ANP level were elevated. However, the intercepts of the relationships were unchanged, indicating that a "resetting" did not occur.

These results demonstrate that the renal functional response to ANP cannot be predicted from the plasma level of ANP because the sensitivity of the kidney to a given level of ANP depends upon the hydration state. Thus the discrepancies and lack of good correlation between plasma ANP level and the renal responses observed in several previous studies (53, 108) may be accounted for on the basis of the altered renal sensitivity to ANP in WI under differing hydration states. The levels of ANP occurring during WI may therefore contribute to the renal functional responses (Fig. 38.5, III).

The mechanisms involved in the enhanced sensitivity of the kidney to ANP in WI or hypertonic saline expansion are uncertain. Although ANP has been shown to inhibit renin secretion, Krasney et al. (85) found that the decreases in PRA in WI were identical, regardless of the plasma level of ANP. In addition, the observation that vasopressin levels are unchanged in the awake dog during WI was confirmed and vasopressin levels were not influenced by the plasma level of ANP. As will be discussed below (see Adrenergic System in Water Immersion), there is strong evidence that indicates that changes in renal sympathetic nerve activity play an important role in the WI response. In this regard, Thoren et al. (145), Imaizumi et al. (71), and Schultz et al. (134) have demonstrated that ANP can inhibit renal sympathetic nerve traffic via an influence on cardiac vagal afferents (Fig. 38.5, *III*). These studies may account partly for the observations of Goetz et al. (55, 56) mentioned previously which indicate that the $\dot{V}$ and $U_{Na}\dot{V}$ responses to atrial balloon inflation are abolished after cardiac denervation, despite similar elevations of plasma ANP.

An interesting effect of ANP is related to its actions in bringing about reductions in plasma volume and elevations of hematocrit (133). This observation indicates that ANP can provoke a transcapillary fluid shift out of the plasma compartment. The reduction in cardiac output by ANP and its hypotensive effects have been attributed to the fluid shift (133). The mechanism of the ANP fluid shift is not clear, although it may involve selective venous contraction and/or arteriolar vasodilation, leading to elevated capillary pressure (133) and/or increased capillary permeability (69).

Krasney et al. (85) found that infusion of 25 ng/kg/min of ANP which elevated plasma ANP levels to those within the physiological range caused a hemoconcentration in conscious dogs standing in air. WI evoked a transcapillary shift of fluid into the plasma, as indicated by a significant reduction in the hematocrit. When ANP was infused in doses of 5 and 25 ng/kg/min during WI, it was observed that ANP prevented the hemodilution in a graded fashion such that at the dose of 25 ng/kg/min, the inward fluid shift was completely prevented. Thus ANP at physiological levels can act to modulate the transcapillary fluid shift during WI. This interactive inhibitory effect may be related to the ability of ANP to elevate capillary hydrostatic pressure. Normally tissue pressure rises more than estimated capillary hydrostatic pressure rises during WI and net capillary reabsorption occurs (Figs. 38.7, 38.8). An elevation of capillary hydrostatic pressure by ANP could prevent the increased capillary reabsorption.

Renal Prostaglandins

The urinary excretion of prostaglandin E increases during WI in both hydrated and dehydrated humans (36). In addition, pretreatment with indomethacin attenuates the increase in UNaV during WI in humans (49). Therefore prostaglandins may act to modulate the renal

responses to WI. This effect may be via the inhibitory effect of prostaglandins on the hydroosmotic actions of vasopressin in the kidney. There is a paucity of information as to the role of prostaglandins during WI in experimental animals.

Adrenergic System in Water Immersion

It has been difficult to establish a strong correlational relationship between the responses of volume regulatory hormones and the diuretic and natriuretic responses of WI. By exclusion, several investigations have sought to determine whether adrenergic mechanisms might be involved in the primary mediation of the renal response to WI.

In terms of circulating catecholamines, Epstein et al. (40) have demonstrated that plasma levels of catecholamines are depressed significantly in humans during thermoneutral WI. More recently, Norsk et al. (112) demonstrated that WI in humans led to a rise in arterial mean and pulse pressure which was associated with a decrease in plasma norepinephrine levels by 50% from pre-WI values. These authors postulated that the decline in circulating norepinephrine concentration could account entirely for the diuresis and natriuresis. By comparison, Hajduczok et al. (62, 63) reported that circulating catecholamine levels are unchanged in conscious dogs during WI.

The kidney receives an elaborate sympathetic innervation. Dibona and others (31, 32, 125) have indicated that renal sympathetic neural activity (RSNA) can elicit important changes in kidney function in the absence of renal hemodynamic responses. Atrial distention can bring about striking reductions in RSNA which, as has been discussed, is associated with increases in $\dot{V}$ and $U_{Na}\dot{V}$ (31, 32, 142, 143). Stimulation of RSNA at intensities which do not alter renal blood flow or GFR elicits antinatriuretic effects. Suppression of RSNA brings about natriuresis (31, 32). Stimulation of cardiac vagal myelinated afferents as well as cardiac C fiber nonmyelinated afferents elicit a reflex inhibition of RSNA (14, 142, 143). Similar effects can be derived from mechanical loading of the high-pressure arterial baroreceptors (31, 32, 144). The relative potency of cardiac vs. sinoaortic mechanoreceptors in the reflex suppression of RSNA is controversial, particularly when primates are compared with subprimate species (117). However, specific studies to address this issue have not been performed in conscious animals. The influence of RSNA, and presumably circulating norepinephrine, on sodium excretion is mediated via tubular alpha receptors (31, 32).

Hajduczok et al. (62) studied renal and hormonal responses to WI in awake, trained dogs that had been either sham operated (S) or subjected to total extrinsic denervation of the heart (CD) by the intrapericardial approach of Randall et al. (128). Inflation of an atrial balloon in the S dogs caused an increase in $\dot{V}$ and $U_{Na}\dot{V}$. This response did not occur in the CD dogs. In the S dogs studied under volume replete conditions, WI caused diuresis, natriuresis, and increased C_{osm} (Fig. 38.16). These responses were detected 20 min after entry into the water. C_{H_2O} and UKV were largely unchanged. As pointed out earlier, plasma vasopressin levels were unchanged. By comparison, the CD dogs showed an identical increase in $\dot{V}$ to that observed in the S dogs. However, the natriuresis was abolished and the response was entirely related to a rise in C_{H_2O}; the increase in C_{osm} was abolished also. This study indicated clearly that, in the awake dog, influences from cardiac afferents are crucial for natriuresis to occur. However, redundant systems are present in that a water diuresis of identical amplitude and time course occurred in the CD dogs (Fig. 38.16). The rise in C_{H_2O} was correlated with the appearance of a significant depression of plasma vasopressin.

It has been proposed that a rise in arterial pressure can cause natriuresis and diuresis (47, 81) and that these responses may be related to a rise in intrarenal pressure (141). In the Hajduczok study (62) the rise in mean arterial pressure occurred in both the S and CD dogs, yet the natriuresis was abolished in the CD dogs. Therefore the WI natriuresis is not due to arterial pressure effects. In addition, as plasma ANP rises during WI in intact dogs (101)—and Goetz et al. (56) have shown that atrial distention in CD dogs elevates plasma ANP levels—it is probable that plasma ANP levels rose during WI in the CD dogs. However, as Goetz et al. (56) found with atrial balloon inflation, this rise in ANP is not sufficient to elicit natriuretic effects in the absence of cardiac nerves.

Thus the CD studies indicate that mechanical loading of cardiac receptors in WI is responsible for a reflex natriuresis. This response is not associated with major alterations of plasma vasopressin in hydrated humans, monkeys and dogs, but rather it is likely related to reflex suppression of renal adrenergic systems. The decline in PRA usually observed in WI may be easily related to a reflex decline in RSNA.

Miki et al. (104) were able to record RSNA in conscious, instrumented dogs during WI and they provided the first direct evidence that RSNA is suppressed during WI. The magnitude of the depression of RSNA is about 50% below pre-WI levels. The renal response in these dogs consisted of a diuresis and natriuresis which was not associated with any change in C_{H_2O}. These same dogs were then subjected to chronic bilateral denervation of the kidneys. After renal denervation, the diuresis and natriuresis responses to WI were abolished.

These studies in conscious dogs indicate strongly that

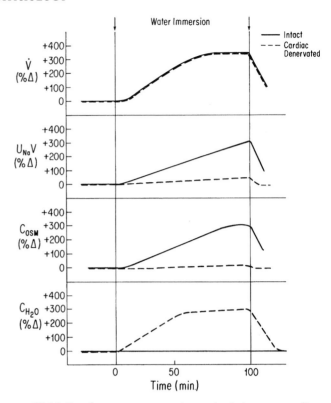

FIG. 38.16. Renal responses to water immersion in intact vs. cardiac-denervated (CD) dogs. $\dot{V}$ = urine flow, $U_{Na}\dot{V}$ = sodium excretion, C_{osm} = osmotic clearance, C_{H_2O} = free water clearance. An identical increase in $\dot{V}$ occurred in the CD dogs, but it was caused by a rise in C_{H_2O} instead of by the rise in C_{osm} and $U_{Na}\dot{V}$ observed in intact dogs. [Reprinted with permission from the Best Publishing Company; from Krasney (84).]

the primary mechanism responsible for the natriuresis of WI is a reflex suppression of RSNA due to mechanical loading of cardiac receptors. Since CD abolishes the natriuresis, it would seem that the important afferent input is from the heart, confirming the original view of Gauer and Henry and earlier investigators. However, reflex activity from sinoaortic receptors can also inhibit RSNA, and specific studies to evaluate the role of the sinoaortic input in WI have not been performed.

While the renal denervation experiments of Miki et al. (104) point to a primary role for the renal sympathetics in the WI natriuretic response, these data must be interpreted with caution. It is likely that these dogs experienced a renal denervation diuresis (120) and thus their pre-WI extracellular fluid volumes may have been different. Second, renal denervation is rather traumatic and the data prior to WI indicate that GFR was reduced, although not significantly so. Finally, it would seem likely that the levels of certain volume regulatory hormones, most likely ANP, were increasing or decreasing during WI after denervation of the kidneys, and it is surprising that the kidney did not respond to this change. If the decline of PRA in WI is due to reductions in RSNA, then PRA levels probably did not change after renal denervation. Further studies are required to sort these issues out.

In addition to the potential for circulating levels of norepinephrine to decline in WI and the clear evidence for reflex suppression of RSNA, there also is evidence for involvement of another adrenergic mediator, dopamine, in the renal response to WI. Administration of dopamine antagonists blocks the natriuretic response to volume expansion in awake monkeys (70) and humans (91). In addition, dopamine antagonists attenuate the natriuretic response to WI in humans (23). While dopamine appears to be important in mediating the natriuresis of WI, it is difficult to reconcile these observations with the powerful attenuating effect of renal denervation.

Circadian rhythms can exert a profound influence on the renal response to WI. Krishna and Danovitch (90) were the first to demonstrate that the renal response to WI is attenuated in humans if the study is done at night. Subsequently, Shiraki et al. (137) reported similar results with the additional observation that the suppression of PRA, vasopressin, and aldosterone observed at night was the same as that observed during the day. The elevations in cardiac output were similar during the day and night as well. Later Miki et al. (108) showed in humans that the elevation of plasma ANP levels during WI was the same at night as during the day, despite significant inhibition of the diuretic and natriuretic response.

In terms of animal studies, the response to WI has not been studied at night. However, Kass et al. (78) showed that renal responses to lower body positive pressure in monkeys were suppressed at night. In contrast, Goetz et al. (54) were unable to demonstrate nocturnal suppression of renal responses to volume expansion in awake dogs. Although further studies are required in other species, it may be that the circadian effect on renal function exists only in primates.

In terms of the relative effects of ANP, vasopressin, and renin-angiotensin II-aldosterone vs. reflex control of adrenergic systems in WI, the data available at present point to a primary role for reflex suppression of RSNA, with the possible suppression of circulating norepinephrine and the release of dopamine as being primary stimuli for the natriuresis of WI. Certain studies indicate that redundant mechanisms, such as suppression of vasopressin, can be brought into play to promote the elimination of fluid in WI if the primary mechanism is blocked. However, by and large it would appear that the role of the volume regulatory hormones in WI is to modulate the primary renal response evinced through suppression of sympathetic activity. Indeed, the circadian influence in primates whereby the hormonal responses becomes less coupled to the renal responses may involve a circadian inhibition of the cardiac renal sympathetic reflex via the central nervous system. The increased kidney sensitivity to ANP in WI may involve actions of ANP in sensitizing cardiac afferents (145) or a central effect on baroreceptor reflex loops.

CONCLUSIONS

Head-out water immersion has been used for therapeutic purposes for centuries and it would seem that an understanding of the complex integrative physiological response to this simple maneuver is beginning to develop. The response to thermoneutral WI involves a sustained elevation of venous return and cardiac output in relation to a constant systemic O_2 consumption. Part of the elevated systemic O_2 delivery is directed to tissues that experience increased metabolic demand, such as the heart and respiratory muscles, or have thermal requirements, such as skin and subcutaneous fat. However, there are time-dependent elevations in flow and O_2 delivery to the viscera and skeletal muscle which involve a mismatching of flow and metabolic demand.

It seems reasonable to postulate that the physiologic response in WI is geared toward readjusting blood flow to metabolism. The end-organ for this readjustment is the kidney, which responds to cardiac stretch caused not only by the displacement of blood from dependent tissues into the thorax, but also by a sustained transcapillary shift of fluid from the cellular compartment into the plasma (Figs. 38.10, 38.17).

Figure 38.18 summarizes the circulatory, hormonal, and renal responses to thermoneutral WI. The response pattern appears to be quantitatively and qualitatively similar in the several species studied thus far in the

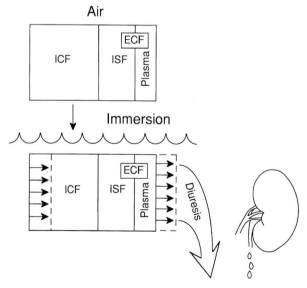

FIG. 38.17. The role of the kidney in water immersion is to minimize the hypervolemia due to the shift of fluid out of the cell compartment.

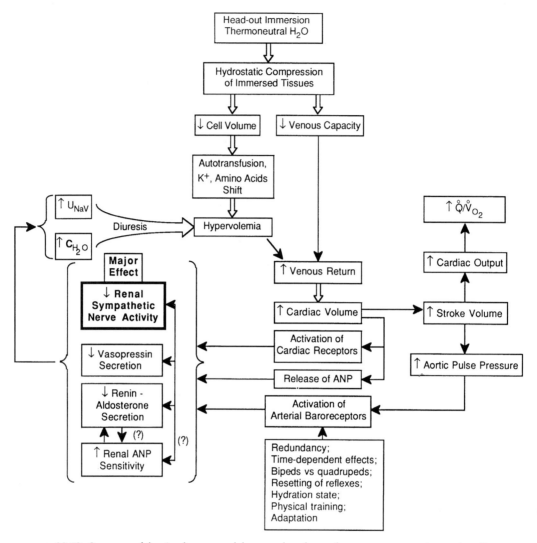

FIG. 38.18. Summary of the circulatory, renal, hormonal, and neural responses to water immersion. Central volume expansion and the rise of cardiac output is caused by compression of dependent veins and by autotransfusion. The rise in cardiac output exceeds metabolic demand at rest and during exercise. Activation of cardiovascular mechanoreceptors and increased ANP secretion sets into motion neural and hormonal mechanisms which promote salt and fluid loss to minimize the hypervolemia and theoretically readjust systemic $\dot{Q}$ to the prevailing $\dot{V}O_2$. The most powerful efferent mechanism affecting the kidney appears to be the sympathetic nerves. The associated hormonal responses appear to modulate the primary renal response. A large number of factors can modulate the basic renal neurohumoral response to immersion.

quadruped position and in the human. An exception appears to be the conscious dog, which typically responds with a Bainbridge type of cardiac acceleration and a rise in arterial pressure. Otherwise the dog is similar to other species. It may be that these atypical canine responses can be "trained away."

Figure 38.17 illustrates that the response of the kidney acts to minimize the hypervolemia in WI that otherwise would be quite massive. This renal compensation can be quite long in duration, especially when WI periods of 6–12 h are studied. Eventually cardiac filling pressure and output return to pre-WI levels, despite con-

tinuance of the immersion (21, 115). This would be expected to be associated with a significant reduction in plasma volume and loss of K^+ and water from the cellular compartment (102).

The fact that a long time period is required to readjust the elevated thoracic and plasma volumes is perhaps fortuitous. If the plasma volume were corrected rapidly and the autotransfusion did not occur, then each time humans or terrestrial animals entered the water, they would emerge with a plasma volume deficit which would have functional consequences. The capacity for autotransfusion is significant (97).

The physiologic adaptation to WI has important applications for species immersed for extended periods. Extended head-out immersion in salt water would be expected to elicit more profound central shifts of blood and transcapillary fluid shifts because of the greater density of salt water. To the extent that WI is applicable to weightlessness in space, prolonged central volume expansion due to microgravity may lead to subnormal plasma volume, which could compromise the circulation upon reexposure to gravity.

In terms of therapeutic applications, WI appears to be especially useful for the treatment of fluid retention disorders such as hypertension of pregnancy (34) and cirrhosis of the liver or the nephrotic syndrome (10, 37, 129). In the latter case, it has been found that certain patients respond to WI while others do not (10, 37). As an investigational tool WI will be useful in determining why some patients do not respond to the condition by increasing salt and water output. It might have utility in treating congestive heart failure. However, the potential of WI to stimulate salt and fluid loss must be weighed against the possible detrimental influence of elevated cardiac preload.

This chapter has focussed on thermoneutral WI. Future studies should be directed toward understanding the effects of deviations of water temperature in either direction from the thermoneutral zone.

The research from the author's laboratory described in this chapter was supported by Program project Grant PO1-HL-28542 from the National Heart, Lung, and Blood Institute and U.S. Army Health Sciences Command. The author extends his appreciation to the numerous co-investigators referred to in the reference list and without whom this work would not have been accomplished, and to Ms. Susan McMahon for preparing the manuscript.

REFERENCES

1. Arborelius, M. J., U. I. Balldin, B. Lilja, and C. E. G. Lundgren. Hemodynamic changes in man during immersion with head above water. *Aerosp. Med.* 43: 592–598, 1972.
2. Balldin, U. I., C. E. G. Lundgren, J. Lundvall, and S. Mellander. Changes in the elimination of 133-xenon from the anterior tibial muscle in man induced by immersion in water and shifts in body position. *Aerosp. Med.* 42: 489–493, 1971.
3. Bazett, H. C. Studies on the effects of baths on man. I. Relationship between the effects produced and the temperature of the bath. *Am. J. Physiol.* 70: 412–429, 1924.
4. Bazett, H. C., S. Thurlow, C. Corwell, and W. Stewart. Studies on the effect of baths on man. II. The diuresis caused by warm baths, together with some observations on urinary tides. *Am. J. Physiol.* 70: 430–452, 1924.
5. Bedford, T. G., and C. M. Tipton. Exercise training and the arterial baroreflex. *J. Appl. Physiol.: Respir. Environ. Exerc. Physiol.* 63: 1926–1932, 1987.
6. Begin, R., M. Epstein, M. A. Sackner, R. Levinson, R. Dougherty, and D. Duncan. Effects of water immersion to the neck on pulmonary circulation and tissue volume in man. *J. Appl. Physiol.* 40: 293–299, 1976.
7. Behn, C., O. H. Gauer, K. Kirsch, and P. Eckert. Effects of sustained intrathoracic vascular distension on body fluid distribution and renal excretion in man. *Pflugers Arch.* 313: 123–135, 1969.
8. Benjamin, B. A., L. C. Keil, M. S. Shapiro, M. A. Kirschenbaum, N. S. Bricker, and H. Sandler. Physiologic response to water immersion in the Rhesus monkey. *Physiologist* 26: A-11, 1983.
9. Bezold, A. Von, L. Hirt. Über die physiologischen Wirkungen des essigsauren Veratrins. *Unter Physiol. Lab. Wurzburg* 1: 75–156, 1867.
10. Bichet, D. G., V. J. Van Patten, and R. W. Schrier. Potential role of increased sympathetic activity in impaired sodium and water excretion in cirrhosis. *N. Engl. J. Med.* 307: 1552–1557, 1982.
11. Bie, P., M. Mumksdorf, and J. Warburg. Renal effects of overhydration during vasopressin infusion in conscious dogs. *Am. J. Physiol.* 247 (*Renal Fluid Electrolyte Physiol.* 18): F103–F109, 1984.
12. Bie, P., B. C. Wang, R. J. Leadley, Jr., and K. L. Goetz. Hemodynamic and renal effects of low-dose infusions of atrial peptide in conscious dogs. *Am. J. Physiol.* 254 (*Regulatory Integrative Comp. Physiol.* 25): R161–R169, 1988.
13. Billman, G. E., M. J. Keyl, D. T. Dickey, D. C. Kem, L. C. Keil, and H. L. Stone. Hormonal and renal response to plasma volume expansion in the primate *Macaca Mulatta*. *Am. J. Physiol.* 244 (*Heart Circ. Physiol.* 13): H201–H205, 1983.
14. Bishop, V. S., A. Malliani, and P. Thoren. Cardiac mechanoreceptors. In: *Handbook of Physiology. The Cardiovascular System*, edited by J. T. Shepherd and F. M. Abboud. Bethesda, MD: Am. Physiol. Soc., 1983 sect. 2, vol. III, chapt. 15, p. 497–556.
15. Blomqvist, C. G., and H. L. Stone. Cardiovascular adjustments to gravitational stress. In: *Handbook of Physiology. The Cardiovascular System*, edited by J. T. Shepherd and F. M. Abboud. Bethesda, MD: Am. Physiol. Soc., 1983, sect. 2, vol. III, chapt. 28, p. 1025–1063.
16. Boettcher, D. H., S. F. Vatner, G. R. Heyndrickx, E. Braunwald. Extent of utilization of the Frank-Starling mechanism in conscious dogs. *Am. J. Physiol.* 234 (*Heart Circ. Physiol.* 3): H338–H345, 1978.
17. Choukroun, M. L., and P. Varene. Adjustments in oxygen transport during head-out immersion in water at different temperatures. *J. Appl. Physiol.: Respir. Environ. Exerc. Physiol.* 68: 1475–1480, 1990.
18. Christie, J. L., L. M. Sheldahl, F. E. Tristani, L. S. Wann, K. B. Sagar, S. G. Levandoski, M. J. Ptacin, K. A. Sobocinski, and R. D. Morris. Cardiovascular regulation during head-out water immersion exercise. *J. Appl. Physiol.: Respir. Environ. Exerc. Physiol.* 69: 657–664, 1990.
19. Churchill, S. E., D. M. Pollock, M. E. Natale, and M. C. MooreEde. Renal response to 7 days of lower body positive pressure in squirrel monkeys. *Am. J. Physiol.* 260 (*Regulatory Integrative Comp. Physiol.* 29): R724–R732, 1991.
20. Clark, J. H., D. R. Hooker, and L. H. Weed. The hydrostatic factor in venous pressure measurements. *Am. J. Physiol.* 109: 166–177, 1934.
21. Claybaugh, J. R., D. R. Pendergast, J. E. Davis, C. Akiba, M. Pazik, and S. K. Hong. The effect of training on hormonal and urinary responses to supine posture and immersion. *J. Appl. Physiol.: Respir. Environ. Exerc. Physiol.* 61: 7–15, 1986.
22. Cornish, K. G., T. McCulloch, and J. P. Gilmore. Sinoaortic baro-receptors and the control of blood volume in the non-

human primate. *Am. J. Physiol.* 247 (*Renal Fluid Electrolyte Physiol.* 16): F539–F542, 1984.

23. Coruzzi, P., A. Biaggi, L. Musicri, C. Ravanetti, P. F. Vescovi, and A. Novarini. Dopamine blockade and natriuresis during water immersion in normal man. *Clin. Sci.* 70: 523–526, 1986.

24. Craig, A. B., Jr., and M. Dvorak. Thermal regulation during water immersion. *J. Appl. Physiol.* 21: 1577–1585, 1966.

25. Craig, A. B., Jr., and M. Dvorak. Expiratory reserve volume and vital capacity of the lungs during immersion in water. *J. Appl. Physiol.* 38: 5–9, 1975.

26. Crane, M. G., and J. J. Harris. Suppression of plasma aldosterone by partial immersion. *Metabolism* 23: 359–368, 1974.

27. Curran-Everett, D. C., J. R. Claybaugh, S. K. Hong, and J. A. Krasney. Hormonal and electrolyte responses of conscious sheep to 96 hours of normobaric hypoxia. *Am. J. Physiol.* 255 (*Regulatory Integrative Comp. Physiol.* 24): R274–R283, 1988.

28. Curran-Everett, D. C., K. McAndrews, and J. A. Krasney. Regional pulmonary blood flow responses to 96 hours of hypoxia in conscious sheep. *J. Appl. Physiol.: Respir. Environ. Exerc. Physiol.* 61: 2136–2143, 1986.

29. Davis, J. T., and A. B. DuBois. Immersion diuresis in dogs. *J. Appl. Physiol.: Respir. Environ. Exercise Physiol.* 42: 915–922, 1977.

30. De Torrente, A., G. L. Robertson, K. M. McDonald, and R. W. Schrier. Mechanism of diuretic response to increased left atrial pressure in anesthetized dog. *Kidney Int.* 8: 355–361, 1975.

31. Dibona, G. F. The functions of renal nerves. *Rev. Physiol. Biochem. Pharmacol.* 94: 75–181, 1982.

32. Dibona, G. F. Neural regulation of renal tubular sodium reabsorption and renin secretion. *Federation Proc.* 44: 2816–2822, 1985.

33. Doba, N., and D. J. Reis. Cerebellum: role in reflex cardiovascular adjustment to posture. *Brain Res.* 39: 495–500, 1972.

34. Doniec-Ulman, I., F. Ko Kox, G. Wambach, and M. Drab. Water immersion-induced endocrine alterations in women with EPH gestosis. *Clin. Nephrol.* 28: 51–55, 1987.

35. Echt, M., L. Lange, and O. H. Gauer. Changes of peripheral venous tone and central transmural venous pressure during immersion in a thermo-neutral bath. *Pflugers Arch.* 352: 211–217, 1974.

36. Epstein, M. Renal effects of head-out water immersion in man: implications for an understanding of volume homeostasis. *Physiol. Rev.* 58: 529–581, 1978.

37. Epstein, M. Renal sodium handing in liver disease. In: *The Kidney in Liver Disease* (3rd ed.), edited by M. Epstein. Baltimore, MD: Williams and Wilkins, 1988, chapt. 1, p. 3–29.

38. Epstein, M. Renal, endocrine, and hemodynamic effects of water immersion in humans. In: *Handbook of Physiology. Environmental Physiology*, edited by M. J. Fregly. New York: Oxford University Press, chapt. 37 (in press), 1995.

39. Epstein, M., A. G. DeNunzio, and R. D. Loutzenhiser. Effects of vasopressin administration on diuresis of water immersion in normal humans. *J. Appl. Physiol.: Respir. Environ. Exerc. Physiol.* 51: 1384–1387, 1981.

40. Epstein, M., G. Johnson, and A. G. DeNunzio. Effects of water immersion on plasma catecholamines in normal humans. *J. Appl. Physiol.: Respir. Environ. Exerc. Physiol.* 54: 244–248, 1983.

41. Epstein, M., J. L. Katsikas, and D. C. Duncan. Role of mineralocorticoids in the natriuresis of water immersion in normal man. *Circ. Res.* 32: 228–236, 1973.

42. Epstein, M., R. Levinson, and R. Loutzenhiser. Effects of water immersion on renal hemodynamics in normal man. *J. Appl. Physiol.* 41: 230–233, 1976.

43. Epstein, M., D. S. Pins, R. Arrington, A. G. DeNunzio, and R. Engstrom. Comparison of water immersion and saline infusion as a means of inducing volume expansion in man. *J. Appl. Physiol.* 39: 60–70, 1975.

44. Epstein, M., and T. Saruta. Effects of water immersion on renin-aldosterone and renal sodium handling in normal man. *J. Appl. Physiol.* 31: 369–374, 1971.

45. Farhi, L. E., and D. Linnarsson. Cardiopulmonary readjustments during graded immersion in water at 35°C. *Respir. Physiol.* 30: 35–50, 1977.

46. Fater, D. C., H. D. Schultz, W. D. Sundet, J. S. Mapes, and K. L. Goetz. Effects of left atrial stretch in cardiac-denervated and intact conscious dogs. *Am. J. Physiol.* 242 (*Heart Circ. Physiol.* 11): H1056–H1064, 1982.

47. Firth, J. D., A. E. G. Raine, and J. G. G. Ledingham. The mechanism of pressure natriuresis. *J. Hypertens.* 8: 97–103, 1990.

48. Gauer, O. H., and J. P. Henry. Neurohumoral control of plasma volume. In: *Cardiovascular Physiology II*, edited by A. C. Guyton and A. W. Cowley. Baltimore, MD: University Park Press, vol. 9, p. 145–189, 1976.

49. Gilmore, J. P. Neural control of extracellular volume in the human and non-human primate. In: *Handbook of Physiology. The Cardiovascular System*, edited by J. T. Shepherd and F. M. Abboud. Bethesda, MD: Am. Physiol. Soc., 1983, sect. 2, vol. III, p. 885–915.

50. Gilmore, J. P., K. G. Cornish, and M. Barazanji. Pentobarbital potentiates natriuretic response to acute volume expansion in monkeys. *Am. J. Physiol.* 254 (*Regulatory Integrative Comp. Physiol.* 23): R727–R734, 1988.

51. Gilmore, J. P., and I. H. Zucker. Contribution of vagal pathways to the renal responses to head-out immersion in the non-human primate. *Circ. Res.* 42: 263–267, 1978.

52. Godley, J. A., J. Waryers, and D. A. Rosenbaum. Cardiovascular and renal function during continuous negative pressure breathing in dogs. *J. Appl. Physiol.* 22: 568–572, 1967.

53. Goetz, K. L. Physiology and pathophysiology of atrial peptides. *Am. J. Physiol.* 254 (*Endocrinol. Metab.* 17): E1–E15, 1988.

54. Goetz, K. L., and B. C. Wang. Canine renal responses to atrial stretch or intravenous saline are not attenuated at night. *Am. J. Physiol.* 250 (*Regulatory Integrative Comp. Physiol.* 19): R638–645, 1986.

55. Goetz, K. L., B. C. Wang, P. Bie, R. J. Lendley, Jr., and P. S. Geer. Natriuresis during atrial distention and a concurrent decline in atriopeptin. *Am. J. Physiol.* 255 (*Regulatory Integrative Comp. Physiol.* 24): R259–R267, 1988.

56. Goetz, K. L., B. C. Wang, P. G. Geer, R. J. Leadley, Jr., and H. W. Reinhardt. Atrial stretch increases sodium excretion independently of release of atrial peptides. *Am. J. Physiol.* 250 (*Regulatory Integrative Comp. Physiol.* 19): R946–R950, 1986.

57. Greenleaf, J. E., J. T. Morese, P. R. Baines, J. Silver, and L. C. Keil. Hypervolemia and plasma vasopressin response during water immersion in man. *J. Appl. Physiol.: Respir. Environ. Exerc. Physiol.* 55: 1688–1693, 1983.

58. Greenleaf, J. E., E. Schvartz, and L. C. Keil. Hemodilution, vasopressin suppression, and diuresis during water immersion in man. *Aviat. Space Environ. Med.* 52: 329–336, 1981.

59. Greenleaf, J. E., E. Schvartz, S. Kravik, and L. C. Keil. Fluid shifts and endocrine responses during chair rest and water immersion in man. *J. Appl. Physiol.: Respir. Environ. Exerc. Physiol.* 48: 79–88, 1980.

60. Guyton, A. C., and T. G. Coleman. Quantitative analysis of the pathophysiology of hypertension. *Circ. Res.* 24: 1–12, 1980.

61. Guyton, A. C., C. E. Jones, and T. G. Coleman. *Circulatory Physiology: Cardiac Output and its Regulation.* Philadelphia, PA: Saunders, 1973.

62. Hajduczok, G., S. K. Hong, J. R. Claybaugh, and J. A. Krasney. The role of cardiac nerves in the hemodynamic and renal responses to head-out water immersion in conscious dogs. *Am. J. Physiol.* 253 (*Regulatory Integrative Comp. Physiol.* 22): R242–R253, 1987.

63. Hajduczok, G., K. Miki, J. R. Claybaugh, S. K. Hong, and J. A. Krasney. Regional circulatory responses to head-out water immersion in conscious dogs. *Am. J. Physiol.* 253 (*Regulatory Integrative Comp. Physiol.* 22): R254–R263, 1987.

64. Hakumaki, M. O. K. Seventy years of the Bainbridge Reflex. *Acta Physiol. Scand.* 130: 177–185, 1987.

65. Hargens, A. R. Introduction and historical perspectives. In: *Tissue Fluid Pressure and Composition,* edited by A. R. Hargens. Baltimore, MD: Williams and Wilkins, 1987, p. 1–9.

66. Hartshorne, H. *Water versus Hydro Therapy or an Essay on Water and its True Relations to Medicine.* Philadelphia, PA: Lloyd P. Smith, p. 28, 1847.

67. Hong, S. K., P. Ceretelli, J. C. Cruz, and H. Rahn. Mechanisms of respiration during submersion in water. *J. Appl. Physiol.* 27: 535–538, 1969.

68. Hong, S. K., and J. R. Claybaugh. Hormonal and renal responses to hyperbaria. In: *Hormonal Regulation of Fluid and Electrolytes: Environmental Effects.* New York, NY: Plenum, 1989, chapt. 4, p. 117–142.

69. Huxley, V. H., V. L. Tucker, K. M. Verburg, and R. H. Freeman. Increased capillary hydraulic conductivity induced by atrial natriuretic peptide. *Circ. Res.* 60: 304–307, 1987.

70. Iaffaldano, R., J. Eye, J. P. Gilmore, and K. G. Cornish. The effects of dopamine blockade on the renal response to volume expansion in the primate. *Federation Proc.* 44: 1751, 1985.

71. Imaizumi, T., A. Takeshita, H. Higoshi, and M. Nakamura. ANP alters reflex control of lumbar and renal sympathetic nerve activity and heart rate. *Am. J. Physiol.* 253 (*Heart, Circ. Physiol.* 22): H1136–H1140, 1987.

71a. Johansen, L. B., N. Foldager, C. Stadeager, M. S. Kristensen, P. Bie, J. Warberg, M. Kamegai, and P. Norsk. Plasma volume, fluid shifts, and renal responses in humans during 12 h of head-out water immersion. *J. Appl. Physiol.* 73: 539–544, 1992.

72. Kaczmarczyk, G. A., W. Christe, R. Mohnhaupt, and H. W. Reinhardt. An attempt to quantitate the contribution of antidiuretic hormone to the diuresis of left atrial distension in conscious dogs. *Pflugers Arch.* 396: 101–105, 1983.

73. Kaczmarczyk, G. A., A. Krake, R. Eisele, R. Mohnhaupt, M. I. M. Noble, B. Singer, J. Stubbs, and H. W. Reinhardt. The role of cardiac nerves in the regulation of sodium excretion in conscious dogs. *Pflugers Arch.* 390: 125–130, 1981.

74. Kaczmarczyk, G. A., V. Unger, R. Mohnhaupt, and H. W. Reinhardt. Left atrial distension and intrarenal blood flow distribution in conscious dogs. *Pflugers Arch.* 390: 44–48, 1981.

75. Kappagoda, C. T., R. J. Linden, and H. M. Snow. A reflex increase in heart rate from distension of the junction between the superior vena cava and the right atrium. *J. Physiol.* 220: 177–197, 1972.

76. Kappagoda, C. T., R. J. Linden, H. M. Snow, and E. M. Whitaker. Left atrial receptors and the antidiuretic hormone. *J. Physiol.* 237: 663–683, 1974.

77. Katsube, N., D. Schwartz, and P. Needleman. Release of atriopeptin in the rate by vasoconstrictors or water immersion correlates with changes in right atrial pressure. *Biochem. Biophys. Res. Commun.* 133: 937–944, 1985.

78. Kass, D. A., F. M. Sulzman, C. A. Fuller, and M. C. Moore-Ede. Renal responses to central vascular expansion are suppressed at night in conscious primates. *Am. J. Physiol.* 239 (*Renal Fluid Electrolyte Physiol.* 8): F343–F351, 1980.

79. Khosla, S. S., and A. B. Dubois. Fluid shifts during initial phase of immersion diuresis in man. *J. Appl. Physiol.: Respir. Environ. Exerc. Physiol.* 46: 703–708, 1979.

80. Khosla, S. S., and A. B. Dubois. Osmoregulation and interstitial fluid pressure changes in humans during water immersion. *J. Appl. Physiol.: Respir. Environ. Exerc. Physiol.* 51: 686–692, 1981.

81. Khraibi, A. A., and F. G. Knox. Renal interstitial hydrostatic pressure during pressure natriuresis in hypertension. *Am. J. Physiol.* 255 (*Regulatory Integrative Comp. Physiol.* 24): R756–R759, 1988.

82. Knight, D. R. and S. M. Horvath. Urinary responses to cold temperature during water immersion. *Am. J. Physiol.* 248 (*Regulatory Integrative Comp. Physiol.* 17): R560–R566, 1985.

83. Koubenec, H. J., W. D. Risch, and O. H. Gauer. Effective compliance of the total vascular system of man sitting in air and immersed in a bath. *Pflugers Arch.* 355 (Suppl. 1): R24, 1975.

84. Krasney, J. A. Physiological responses to head-out water immersion. In: *Man in the Sea,* edited by Y. C. Lin and K. K. Shida. San Pedro, CA: Best Publishing, 1990, chapt. 1, p. 1–32.

85. Krasney, J. A., M. Carroll, E. Krasney, J. Iwamoto, J. R. Claybaugh, and S. K. Hong. Renal, hormonal, and fluid shift responses to ANP during head-out water immersion in awake dogs. *Am. J. Physiol.* 261 (*Regulatory Integrative Comp. Physiol.* 30): R188–R197, 1991.

86. Krasney, J. A., G. Hajduczok, C. Akiba, B. W. McDonald, D. R. Pendergast, and S. K. Hong. Cardiovascular and renal responses to head-out water immersion in canine model. *Undersea Biomed. Res.* 11: 169–183, 1984.

87. Krasney, J. A., G. Hajduczok, K. Miki, J. R. Claybaugh, J. L. Sondeen, D. R. Pendergast, and S. K. Hong. Head-out water immersion: a critical evaluation of the Gauer-Henry hypothesis. In: *Hormonal Regulation of Fluid and Electrolytes,* edited by J. R. Claybaugh and C. E. Wade. New York: Plenum, 1989, chapt. 5, p. 147–185.

88. Krasney, J. A., D. R. Pendergast, E. Powell, B. W. McDonald, and J. L. Plewes. Regional circulatory responses to head-out water immersion in the anesthetized dog. *J. Appl. Physiol.: Respir. Environ. Exerc. Physiol.* 53: 1625–1633, 1982.

89. Kravik, S. E., L. C. Keil, J. E. Silver, N. Wong, W. A. Spaul, and J. E. Greenleaf. Immersion diuresis without the expected suppression of vasopressin. *J. Appl. Physiol.: Respir. Environ. Exerc. Physiol.* 57: 123–128, 1984.

90. Krishna, G. G., and G. M. Danovitch. Renal response to central volume expansion in humans is attenuated at night. *Am. J. Physiol.* 244 (*Regulatory Integrative Comp. Physiol.* 13): R481–R486, 1983.

91. Krishna, G. G., G. M. Danovitch, F. W. J. Beck, and J. R. Sowers. Dopaminergic mediation of the natriuretic response to volume expansion. *J. Lab. Clin. Med.* 105: 214–220, 1983.

92. Lange, L., S. Lange, M. Echt, and O. H. Gauer. Heart volume in relation to body posture and immersion in a thermo-neutral bath. *Pflugers Arch.* 352: 219–226, 1974.

93. Ledsome, J. R., and R. J. Linden. The role of left atrial receptors in the diuretic response to left atrial distension. *J. Physiol.* 198: 487–503, 1968.

94. Ledsome, J. R., R. J. Linden, and W. J. O'Connor. The mechanisms by which distension of the left atrium produces diuresis in anesthetized dogs. *J. Physiol.* 159: 87–100, 1961.

95. Ledsome, J. R., J. Ngsee, and N. Wilson. Plasma vasopressin concentrations in the anesthetized dog before, during, and after atrial distension. *J. Physiol.* 338: 413–421, 1983.

96. Lin, Y. C. Circulatory functions during immersion and breath-hold dives in humans. *Undersea Biomed. Res.* 11: 123–138, 1984.

97. Lundvall, J., and T. Lanne. Large capacity in man for effective plasma volume control in hypovoloemia via fluid transfer from tissue to blood. *Acta Physiol. Scand.* 137: 513–520, 1989.

98. Lydtin, H., and W. F. Hamilton. Effect of acute changes in left atrial pressure on urine flow in unanesthetized dogs. *Am. J. Physiol.* 207: 530–536, 1964.

99. Mellander, S. Comparative studies on the adrenergic neuro-hormonal control of resistance and capacitance blood vessels in the cat. *Acta Physiol. Scand.* 50 (Suppl.): 176, 1960.

100. Metzler, C. J., and D. J. Ramsay. Atrial peptide potentiates renal responses to volume expansion in conscious dogs. *Am. J. Physiol.* 256 (*Regulatory Integrative Comp. Physiol.* 25): R284–R289, 1989.

101. Miki, K., G. Hajduczok, S. K. Hong, and J. A. Krasney. Plasma volume changes during head-out water immersion in the conscious dog. *Am. J. Physiol.* 251 (*Regulatory Integrative Comp. Physiol.* 20): R582–R590, 1986.

102. Miki, K., G. Hajduczok, S. K. Hong, and J. A. Krasney. Extracellular fluid and plasma volumes during water immersion in nephrectomized dogs. *Am. J. Physiol.* 252 (*Regulatory Integrative Comp. Physiol.* 21): R972–R978, 1987.

103. Miki, K., G. Hajduczok, M. K. Klocke, J. A. Kraney, S. K. Hong, and A. J. Debold. Atrial natriuretic factor and renal function during head-out water immersion in conscious dogs. *Am. J. Physiol.* 251 (*Regulatory Integrative Comp. Physiol.* 20): R1000–R1008, 1986.

104. Miki, K., Y. Hayashida, S. Sagawa, and K. Shiraki. Renal sympathetic nerve activity and natriuresis during water immersion in conscious dogs. *Am. J. Physiol.* 256 (*Regulatory Integrative Comp. Physiol.* 25): R299–R305, 1989.

105. Miki, K., Y. Hayashida, F. Tajima, J. Iwamoto, and K. Shiraki. Renal sympathetic nerve activity and renal responses during head-up tilt in conscious dogs. *Am. J. Physiol.* 257 (*Regulatory Integrative Comp. Physiol.* 26): R358–R364, 1989.

106. Miki, K., M. R. Klocke, S. K. Hong, and J. A. Krasney. Interstitial and intravascular pressures in conscious dogs during head-out water immersion. *Am. J. Physiol.* 256 (*Regulatory Integrative Comp. Physiol.* 26): R358–R364, 1989.

107. Miki, K., M. Pazik, E. Krasney, S. K. Hong, and J. A. Krasney. Thoracic duct lymph flow during head-out water immersion in conscious dogs. *Am. J. Physiol.* 252 (*Regulatory Integrative Comp. Physiol.* 21): R782–R785, 1987.

108. Miki, K., K. Shiraki, S. Sagawa, A. J. Debold, and S. K. Hong. Atrial natriuretic factor during head-out immersion at night. *Am. J. Physiol.* 254 (*Regulatory Integrative Comp. Physiol.* 23): R235–R241, 1988.

109. Mills, E., and S. C. Wang. Liberation of antidiuretic hormone: location of ascending pathways. *Am. J. Physiol.* 207: 1399–1404, 1964.

110. Myers, J. W., and J. A. Godley. Cardiovascular and renal function during total body water immersion of dogs. *J. Appl. Physiol.* 22: 573–579, 1967.

111. Norsk, P., and M. Epstein. Effects of water immersion on arginine vasopressin release in humans. *J. Appl. Physiol.: Respir. Environ. Exerc. Physiol.* 64: 1–10, 1988.

112. Norsk, P., F. Bonde Peterson, and N. J. Christensen. Catecholamines, circulation and the kidney during head-out water immersion in humans. *J. Appl. Physiol.: Respir. Environ. Exerc. Physiol.* 69: 479–484, 1990.

113. Park, Y. S., D. R. Pendergast, and D. W. Rennie. Decrease in body insulation with exercise in cool water. *Undersea Biomed. Res.* 11: 159–168, 1984.

114. Pendergast, D. R., A. J. Debold, M. Pazik, and S. K. Hong. Effect of head-out immersion on plasma atrial natriuretic factor in man. *Proc. Soc. Exp. Biol. Med.* 184: 429, 435, 1987.

115. Pendergast, D. R., A. J. Olszowka, M. A. Rokitka, and L. E. Farhi. Gravitational force and the cardiovascular system. In: *Comparative Physiology of Environmental Adaptations,* edited by P. Dejours. Basel: Karger, 1986, p. 15–26.

116. Peters, J. P. *Body Water: The Exchange of Fluids in Man.* Springfield, IL: Thomas, p. 287, 1935.

117. Peterson, T. V. Cardiac reflexes and control of renal function in primates. In: *Reflex Control of the Circulation,* edited by I. H. Zucker and J. P. Gilmore. Boca Raton: CRC Press, 1991, p. 313–358.

118. Peterson, T. V., B. A. Benjamin, and N. L. Hurst. Effect of vagotomy and thoracic sympathectomy on responses of the monkey to water immersion. *J. Appl. Physiol.: Respir. Environ. Exerc. Physiol.* 63: 2476–2481, 1987.

119. Peterson, T. V., B. A. Benjamin, and N. L. Hurst. Renal nerves and renal responses to volume expansion in conscious monkeys. *Am. J. Physiol.* 255 (*Regulatory Integrative Comp. Physiol.* 24): R388–R394, 1988.

120. Peterson, T. V., N. L. Chase, and D. K. Gray. Head-up tilt in the non-human primate. Effects of renal denervation. *Renal Physiol. Biochem.* 7: 265–270, 1984.

121. Peterson, T. V., J. P. Gilmore, and I. H. Zucker. Renal responses of the recumbent nonhuman primate to total body immersion. *Proc. Soc. Exp. Biol. Med.* 161: 260–265, 1979.

122. Peterson, T. V., J. P. Gilmore, and I. H. Zucker. Initial renal responses of nonhuman primate to immersion and intravascular volume expansion. *J. Appl. Physiol.: Respir. Environ. Exerc. Physiol.* 48: 243–248, 1980.

123. Peterson, T. V., N. L. Hurst, and J. A. Richardson. Renal nerves and renal responses to head-up tilt in dogs. *Am. J. Physiol.* 252 (*Regulatory Integrative Comp. Physiol.* 21): R979–R986, 1987.

124. Prefaut, C., F. Dubois, C. Roussos, R. Amaral-Marques, P. T. Macklem, and F. Ruff. Influence of immersion to the neck in water on airway closure and distribution of perfusion in man. *Respir. Physiol.* 37: 313–323, 1979.

125. Prosnitz, E. H., and G. F. Dibona. Effect of decreased renal sympathetic nerve activity on renal tubular sodium reabsorption. *Am. J. Physiol.* 235 (*Renal Fluid Electrolyte Physiol.* 4): F557–F563, 1978.

126. Rabelink, T. J., H. A. Koomans, W. H. Boer, H. J. Van Riju, and E. J. Dorhout Mees. Lithium clearance increases in water-immersion-induced natriuresis in humans. *J. Appl. Physiol.: Respir. Environ. Exerc. Physiol.* 66: 1744–1748, 1989.

127. Rabischong, P., C. Clay, J. Vignaud, and R. Polierae. Approche hemohynamique de la signification fonctionelle due sinus cavernuex. *Neurochirurgie* 18, n-7: 613–622, 1972.

128. Randall, W. C., M. P. Kaye, J. X. Thomas, Jr., and M. J. Barber. Intrapericardial denervation of the heart. *J. Surg. Res.* 29: 101–109, 1980.

129. Rascher, W., T. Tulassay, H. W. Seyberth, H. Humbert, U. Lang, and K. Scharer. Diuretic and hormonal responses to head-out water immersion in nephrotic syndrome. *J. Pediatr.* 109: 609–614, 1986.

130. Rennie, D. W., P. D. Prampero, and P. Ceretelli. Effects of water immersion on cardiac output, heart rate and stroke volume of man at rest and during exercise. *Med. Dello Sport* 24: 223–228, 1971.

131. Rowell, L. B. *Human Circulation Regulation during Physical Stress.* New York: Oxford University Press, 1986.

132. Rushmer, R. F. *Cardiovascular Dynamics* (3rd ed.). Philadelphia, PA: Saunders, 1970, p. 238.

133. Rutlen, D. L., G. Christensen, K. G. Helgesen, and A. Ilebekk.

Influence of atrial natriuretic factor on intravascular volume displacement in pigs. *Am. J. Physiol.* 259 (*Heart Circ. Physiol.* 28): H1595–H1600, 1990.

133A. Sachs, F. Mechanical transduction in biological systems. *CRC Crit. Rev. Biomed. Eng.* 16: 141–169, 1988.

134. Schultz, H. D., D. G. Gardner, C. F. Deschepper, H. M. Coleridge, and J. C. G. Coleridge. Vagal c-fiber blockade abolishes sympathetic inhibition by atrial natriuretic factor. *Am. J. Physiol.* 255 (*Regulatory Integrative Comp. Physiol.* 24): R6–R13, 1988.

135. Share, L. Effects of carotid occlusion and left atrial distention on plasma vasopressin titer. *Am. J. Physiol.* 208: 219–223, 1965.

136. Share, L., M. N. Levy. Carotid sinus pulse pressure, a determinant of plasma antidiuretic hormone concentration. *Am. J. Physiol.* 211: 721–724, 1966.

137. Shiraki, K., N. Konda, S. Sagawa, J. R. Claybaugh, and S. K. Hong. Cardiorenal-endocrine responses to head-out immersion at night. *J. Appl. Physiol.: Respir. Environ. Exerc. Physiol.* 60: 176–183, 1986.

138. Sit, S. P., H. Morita, and S. F. Vatner. Responses of renal hemodynamics and functions to acute volume expansion in the conscious dog. *Circ. Res.* 54: 185–194, 1984.

139. Sondeen, J. L., J. R. Claybaugh, S. K. Hong, and J. A. Krasney. Splenectomy does not alter the natriuretic response to head-out water immersion. *Physiologist* 34 (4): 260, 1991.

140. Sondeen, J. L., S. K. Hong, J. R. Claybaugh, and J. A. Krasney. Effect of hydration state on renal responses to head-out water immersion in conscious dogs. *Undersea Biomed. Res.* 17: 395–411, 1990.

141. Stahl, W. M. Renal hemodynamics: The effect of gravity on sodium and water excretion. *Aerospace Med.* 36: 917–922, 1965.

142. Thames, M. D. Contribution of cardiopulmonary baroreceptors to the control of the kidney. *Federation Proc.* 37: 1209–1213, 1978.

143. Thames, M. D., M. Jarecki, and D. E. Donald. Neural control of renin secretion in anesthetized dogs: Interactions of cardiopulmonary and carotid baroreceptors. *Circ. Res.* 42: 237–245, 1978.

144. Thames, M. D., Miller, B. D., and F. M. Abboud. Baroreflex regulation of renal nerve activity during volume expansion. *Am. J. Physiol.* 243 (*Heart Circ. Physiol.* 12) H810–H814, 1982.

145. Thoren, P., A. L. Mark, D. A. Morgan, T. P. O'Neill, P. Needleman, and M. J. Brody. Activation of vagal depressor reflexes by atriopeptins inhibits renal sympathetic nerve activity. *Am. J. Physiol.* 251 (*Heart Circ. Physiol.* 20): H1252–H1259, 1986.

146. Vatner, S. F., and E. Braunwald. Cardiovascular control mechanisms in the conscious state. *N. Engl. J. Med.* 293: 970–976, 1975.

147. Vatner, S. F., W. T. Manders, and D. R. Knight. Vagally mediated regulation of renal function in conscious primates. *Am. J. Physiol.* 250 (*Heart Circ. Physiol.* 19): H546–H549, 1986.

148. Von Ameln, H., M. Laniado, L. Rocker, and K. A. Kirsch. Effects of dehydration on the vasopressin response to immersion. *J. Appl. Physiol.: Respir. Environ. Exerc. Physiol.* 58: 114–120, 1985.

149. Von Diringshofen, H. Die Wirkungen des hydrostatischen Druckes des Wasserbades auf den Blutdruck in den Kapillaren und die Bindegewebsentwasserung. *7. Kreislaufforsch.* 37: 382–390, 1948.

150. Wang, B. C., G. Flora-Ginter, R. J. Leadley, Jr., and K. L. Goetz. Ventricular receptors stimulate vasopressin release during hemorrhage. *Am. J. Physiol.* 254 (*Regulatory Integrative Comp. Physiol.* 23): R204–R211, 1988.

151. Weston, C. F., J. P. O'Hare, J. M. Evans, and R. J. M. Corroll. Haemodynamic changes in man during immersion in water at different temperatures. *Clin. Sci. Loud.* 73: 613–615, 1987.

152. Yoshino, H., D. C. Curran-Everett, S. K. Hong, and J. A. Krasney. Altered heart rate-arterial pressure relation during head-out water immersion in the conscious dog. *Am. J. Physiol.* 254 (*Regulatory Integrative Comp. Physiol.* 23) R595–R601, 1988.

153. Zucker, I. H., and J. P. Gilmore. Responsiveness of type B atrial receptors in the monkey. *Brain Res.* 95: 159–165, 1975.

154. Zucker, I. H., and J. P. Gilmore. Contribution of peripheral pooling to the renal response to water immersion in the dog. *J. Appl. Physiol.: Respir. Environ. Exerc. Physiol.* 45: 786–790, 1978.

39. The physiology of bed rest

SUZANNE M. FORTNEY
VICTOR S. SCHNEIDER
JOHN E. GREENLEAF

Medical Sciences Division, NASA-Johnson Space Center, Houston, Texas

Life Science Division, NASA-Ames Research Center, Moffett Field, California

    Early period (1855–1929)
    Intermediate period (1930–1959)
Cardiopulmonary System
    Initial hemodynamic responses
    Fluid–electrolyte responses
        Water and electrolytes: early changes
        Water and electrolytes: chronic changes
        Head-down tilt vs. horizontal position
        Red cell mass: time course
        Red cell mass: physical conditioning
        Red cell mass: mechanisms for decrease
        Plasma volume: time course
        Plasma volume: mechanisms for hypovolemia
        Plasma volume: countermeasures
        Total body water
        Extracellular water
    Cardiac function
        Time course
        Cardiac size
        Cardiac morphology and mass
        Cardiac output
        Cardiac contractility
        Heart rate and electrical activity
    Cardiovascular autonomic regulation
        Acute responses
        Chronic responses
        Recovery
        Arterial baroreflex function
        Cardiopulmonary baroreflex function
        Chemoreceptor function
    Venous compliance
        Time course
        Mechanism
    Pulmonary function
    Recovery
        Orthostasis
        Exercise responses and capacity
    Summary of cardiopulmonary adaptation to bed rest
Musculoskeletal System
    Bone and calcium metabolism
    Normal metabolism
        Calcium
        Phosphorus and other minerals
        Bone matrix (collagen)
        Bone densitometry
    Bone loss and muscle atrophy
    Body composition
    Conclusions

Metabolism and Thermoregulation
    Energy metabolism
        Basal metabolic rate
        Submaximal exercise metabolism
        Peak exercise metabolism
    Thermal regulation
Immune Cellular and Humoral Parameters
    Bed rest <30 days
    Bed rest >30 days
Psychophysiological Factors
    Group interaction
    Response stages
    Performance
    Visual and auditory responses
        Visual parameters
        Auditory parameters
    Sleep
    Posture: equilibrium, balance, and gait

PROLONGED REST IN BED has been utilized by physicians and other health-care workers to immobilize and confine patients for rehabilitation and restoration of health since time immemorial. The sitting or horizontal position is sought by the body to relieve the strain of the upright or vertical postures, for example during syncopal situations, bone fractures, muscle injuries, fatigue, and probably also to reduce energy expenditure.

Most health-care personnel are aware that adaptive responses occurring during bed rest proceed concomitantly with the healing process; signs and symptoms associated with the former should be differentiated from those of the latter. Not all illnesses and infirmities benefit from prolonged bed rest. Considerations in prescribing bed rest for patients—including duration, body position, mode and duration of exercise, light–dark cycles, temperature, and humidity—have not been investigated adequately.

More recently, adaptive physiological responses have been measured in normal, healthy subjects in the horizontal or slightly head-down postures during prolonged bed rest as analogs for the adaptive responses of astronauts exposed to the microgravity environment of outer space with encouraging results (see Chapter 36 in this *Handbook*). The recumbent position and prolonged continuous bed rest result in loss of most hydrostatic pressures, virtual elimination of longitudinal compression on the spine and long bones of the lower extremity, reduced muscular force on bones, somewhat reduced total energy utilization, perhaps some nutritional and gastrointestinal changes from a new diet, and psychosocial changes when the patients or subjects move from their previous ambulatory environment to the hospital or metabolic unit setting and routine. Modified afferent stimuli from these varying stressors induce adaptive responses of the body to the new environmental conditions (102). In the major sections of this chapter—cardiopulmonary, musculoskeletal, metabolic, immune, and psychophysiological—the responses to these stimuli and their integration are discussed.

HISTORY OF BED-REST RESEARCH

Before the American and Russian space programs in the 1960s gave significant impetus to bed-rest research, a few studies were conducted on bed-rested patients and on normal, healthy subjects. The following brief review of those studies will be divided into the early period (1855–1929) and the intermediate period (1930–1959). Findings from the modern period (1960 to the present) will be covered in the remainder of this chapter. Four annotated bibliographies have summarized results from bed-rest studies through 1988 (131, 138, 247, 279).

Before starting the review of early period studies, we should state that ancient physicians, for example, Hippocrates, knew that prolonged rest resulted in physiological and functional deterioration (47):

> For if the whole body is rested much more than usual, there is no immediate increase in strength. In fact, should a long period of inactivity be followed by a sudden return to exercise there will be an obvious deterioration. The same is true of each separate part of the body. The feet and limbs would suffer in the same way if they were unaccustomed to exercise, or were exercised suddenly after a period of rest. The same is true of the teeth and of the eyes.

Early Period (1855–1929)

The earliest study appears to be that of Beigel (20) where urinary urea was measured in four healthy subjects during periods of movement and short periods of rest. Mean daily urea excretion (movement period) was 52 g compared with 46 g when the subjects rested on a sofa. From a clinical study of circadian rhythms of respiratory nitrogen and carbon dioxide, and body temperature, Johansson (180) concluded that rhythmicity of these three variables was due primarily to the degree

of muscular activity of the subjects and generally independent of the time of day. Shaffer (332) observed one man during 2 days in bed, followed by 4 days in bed with a few hours each day sitting in a chair, and found that 24 h urinary nitrogen was 4.8 g and 4.4 g, respectively; urinary 24 h sulfur excretion was 0.44 g and 0.42 g, respectively. He concluded that various levels of muscular activity within physiological limits had no effect on protein metabolism. Campbell and Webster (44) also studied nitrogen metabolism on one 28-yr-old normal, healthy male laboratory assistant during 5 days of complete bed rest, during 6.5 h/day of his normal, ambulatory routine, and during the 6.5 h/day routine plus 5 h/day of heavy (225 kg-m/min) cycle ergometer exercise. This man ate a normal daily diet (ad libitum) during the study. Urinary total nitrogen, and creatinine, urea, uric acid, and amino acid excretions were greater at night than during the day. Administration of 35%–40% oxygen for 37 h over a period of 58 h failed to affect the urine composition. Nocturnal diuresis associated with the horizontal body position had not been mentioned in any study thus far. In the first study where both daily caloric and fluid intakes were controlled, Cuthbertson (70) studied a group of nearly healthy patients (5 men, 19–40 yr and 2 women, 19–37 yr) who had loose cartilage in their knee joints. They had a 5-day ambulatory dietary equilibration period and were confined to bed for 6–40 days. The subjects' data were analyzed as case studies. He found increased urinary excretions of sulfur and nitrogen (reflecting muscle protein loss) and calcium and phosphorus (reflecting bone loss) by the second day of bed rest, and gradually decreasing basal oxygen uptake during the bed-rest period.

Thus in this early period before 1930 these investigators initiated research on problems still being studied: effects of horizontal, sitting, and upright postures combined without and with body movement and intensive exercise on protein metabolism, calcium-phosphate metabolism, circadian rhythms, and basal and respiratory metabolism.

Intermediate Period (1930–1959)

There was little, if any, bed-rest research conducted in the 1930s, but one notable event should be mentioned. Mr. Jonnie Richardson, at 16 yr of age in 1932 was put to bed by his mother for imbibing spirits at a ballgame, and he has remained there ever since (7). In 1982 his physician diagnosed heart trouble and an inoperable goiter: his frame was "lily-white with limbs as thin as the legs of a ladder-back chair." He had trouble holding a book up to read and his eyesight was failing. In 1988 he was still alive and doing relatively well (Greenleaf, personal communication, 1988).

World War II seemed to simulate interest in bed rest

and bed-rest research. Harrison (160) chaired a symposium on "The Abuse of Rest in the Treatment of Disease" in a section of Experimental Medicine and Therapeutics at the annual meeting of the American Medical Association in Chicago. Harrison's paper addressed cardiovascular (heart) disease and the other speakers covered obstetrics (85), surgery (296), use of sedatives, arterial compression, and countermeasures such as deep breathing and other systemic muscular exercise (80), orthopedic surgery (110), and psychiatry (258). This symposium possibly marked the beginning of the general practice of early ambulation for hospitalized patients. In 1947 Asher (12) published two poems involving bed rest which succinctly describe the signs, symptoms, and prognosis:

Look at a patient lying long in bed,
what a pathetic picture he makes!
The blood clotting in his veins,
the scybala stacking up in his colon,
the flesh rotting from his seat,
the urine leaking from his distended bladder,
and the spirit evaporating from his soul!

Teach us to live that we may dread
unnecessary time in bed,
Get people up and we may save
our patients from an early grave.

There was a continuation of studies concerning basal metabolism, metabolic end products of muscle atrophy, and bone demineralization (66, 73, 161, 165, 330, 331, 345, 383). Research began on carbohydrate metabolism and the effects of inactivity (33) and inactivity plus exercise training (246) on oral glucose intolerance engendered by exposure to prolonged bed rest.

Much more emphasis was placed on the use of countermeasures to ameliorate bed-rest deconditioning (loss of physical fitness), and general adaptation: for example, exercise and exercise training before, during, and after bed rest to maintain and restore ambulatory physical fitness (200, 353) and plasma and blood volumes (340, 352, 386). Deitrick and colleagues (73) were the first to utilize lower extremity bivalve casts during bed rest to restrict movement of test subjects; to investigate concomitant orthostatic intolerance on the tilt-table; and to employ whole subject head-to-foot (+Gz to −Gz) oscillations as a countermeasure. The oscillation treatment resulted in significant reductions in orthostatic intolerance and in urinary excretion of calcium and creatinine which were associated with attenuated loss of bone, muscle mass, and strength. Tenney (355) performed the first investigation concerning effects of cephalic fluid shifts during 30 min periods of recumbancy on thoracic fluid volume changes and respiratory

functions. Included in his subject population were two legless men. He concluded that the lower extremities contribute blood to the thorax during recumbency without decreasing total lung volume.

CARDIOPULMONARY SYSTEM

Initial Hemodynamic Responses

The cardiopulmonary adaptation to bed rest begins with initial hemodynamic responses to the greatly attenuated foot-to-head hydrostatic pressure gradient. During water immersion and when moving from an upright to a supine position, approximately 700–900 ml of blood redistribute from the lower body into the central circulation (8, 309). Most of this fluid is diverted to the heart and lungs, increasing pulmonary blood flow by 20%–30%, while smaller amounts are distributed to vasculature of the arms and head. Increase in cardiac filling augments stroke volume (29) and stimulates cardiopulmonary baroreceptors, resulting in slight decreases in heart rate and peripheral vascular resistance. Despite regional changes in vascular pressures, mean arterial blood pressure is maintained (159). These early hemodynamic changes initiate most of the body fluid and cardiovascular responses thereafter.

Fluid–Electrolyte Responses

Water and Electrolytes: Early Changes. Diuresis occurs within the first few hours of bed rest, reaches a maximum (45% increase) after 4–8 h (151), and then urinary flow returns to the pre–bed rest level by 24 h (144, 151). There were accompanying significant losses of sodium (+49%) and potassium (+136%) during the first 24 h of bed rest that resulted in decreased plasma sodium concentration during the first 3 days. The increased urinary water and electrolyte losses occurred without thirst or increased fluid intake. This pattern of change in body water and electrolytes has been observed repeatedly (71, 107, 369). Other relevant reviews discuss changes in body fluids and electrolytes during bed rest (106, 122–124, 134, 146, 369).

A rapid fall in circulating levels of plasma renin activity (PRA), plasma vasopressin (PVP), and aldosterone (ALD) during the first 4–8 h of recumbency (Fig. 39.1), and a rise in atrial natriuretic peptide (ANP) may be responsible for the diuresis and loss of sodium during the first few days of bed rest (71, 107, 108, 281). Increased urinary excretion of potassium begins at once, becomes significant after the first wk of bed rest, and continues thereafter (52, 93, 146). There is a continuing isotonic loss of plasma volume during this time interval

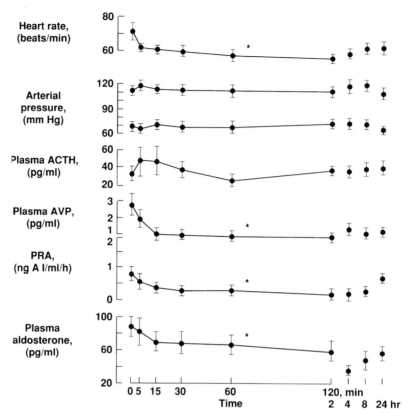

FIG. 39.1. Mean (± SE) heart rate, arterial pressure, and hormonal responses in eight men during 24 h of 6° recumbency. *P < 0.05 from time 0. [From Dallman et al. (71) with permission.]

while plasma sodium concentration is unchanged (6, 34, 49, 52, 93, 130, 146, 256, 281, 322, 399).

Water and Electrolytes: Chronic Changes. Compared to pre–bed rest, a state of decreased total body water is maintained during bed rest with periodic episodes of diuresis (144, 146, 219, 284, 398). A 350 ml/day diuresis was maintained during the first 36 days of a 120 day bed-rest study (284). By day 53 the diuresis had ceased and fluid balance became positive, but the diuresis had begun again on day 83. The decrease in diuresis by day 53 was preceded by a period of sodium retention and elevated aldosterone, which led to plasma vasopressin suppression and recurrence of the diuresis.

During prolonged bed rest there is a progressive loss of total body potassium and slight lowering of serum potassium concentration (146, 171, 178). The slight decline in plasma potassium concentration after 60 days of bed rest was associated with an increased breakdown of tissue proteins (178); for each gram of nitrogen excreted, 3 mEq of potassium were lost. Thus an increase in muscle-wasting was associated with a decline in plasma potassium concentration and with increased urinary loss. Hypokalemia may affect cardiac stability (171, 209).

Uncoupling of the normally tight relationship between PRA and ALD responses may occur during bed rest, even in the absence of changes in the levels of adrenocorticotropic hormone (ACTH), electrolytes, and other factors known to alter this relationship (71, 318, 369). Vernikos (369) observed, in several bed rest studies, that PRA was restored and increased with continued bed rest, while ALD levels remained depressed or were unchanged. Similar uncoupling responses of PRA and ALD occur in water-immersed subjects who were hemorrhaged prior to immersion (139). Dallman et al. (71) demonstrated that the normal adrenal response to stimuli that regulate ALD (inhibition following volume-induced changes in PRA and stimulation following ACTH infusion) was maintained during 14 days of bed rest. Renal sensitivity to ALD was also unchanged during bed rest. Sandler (318) suggested that the uncoupled PRA–ALD relationship may be due to the stimulation of arterial and/or cardiopulmonary baroreceptors and associated renal changes, or to increased ANP secretion. Hypovolemia may play a role, since the uncoupling occurs in both bed-rested and immersed subjects.

Head-down Tilt vs. Horizontal Position. Compared with bed rest in the horizontal position, subjects undergoing slight (−4° to −8°) head-down bed rest exhibit changes in body fluid and electrolyte parameters that appear to be greater and occur at a somewhat accelerated rate (5, 6, 146, 148, 186, 189, 220, 221, 282). Noskov et al. (282) compared the body fluid responses of two groups of five men during 7 day horizontal and 6° head-down bed rest, respectively. Both groups had diuresis and natriuresis resulting in net fluid loss; however, the natriuresis was 1.5 times greater in the head-down group and the diuresis and natriuresis developed faster and more intensely (Table 39.1). The head-down group lost 1.6 kg while the horizontal group had no significant change in body weight during bed rest (Fig. 39.2). Thus the magnitude of the foot-to-head hydrostatic gradient appears to be a prime determinant of total fluid losses during bed rest.

Red Cell Mass: Time Course.

Kimzey et al. (204) compiled results from bed-rest studies from 2 to 35 days and developed an equation to predict red cell mass (RCM) loss:

$$\% \text{ change } RCM = 0.89 + 0.24 \times (BR \text{ days})$$

They described a linear loss of RCM during at least the first mo of bed rest. Others (42, 77, 168, 351) observed the linear decline in RCM for at least 60 days of bed rest which reached equilibrium approximately 10% below the pre–bed rest level. However, the actual rate of loss and equilibrium level may be influenced by the angle of the bed since loss of RCM after 30 days in the 6° head-down position was comparable to that after 100 days in the horizontal position (205).

Loss of RCM continues beyond 60 days of bed rest, although at a slower rate (206, 240). Kiselev et al. (206) and Smirnova et al. (337) conducted studies of 50 and 120 days, respectively, in which supplements of vitamin B_{12} and folic acid were used to stimulate hemopoiesis; there was partial restoration of hemoglobin mass during bed rest and faster recovery of RCM after bed rest.

The decline in RCM continues for some time after

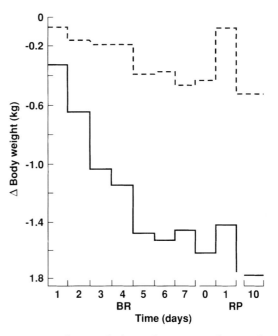

FIG. 39.2. Mean change in body weight in 10 men during 7 days of bed rest (BR) and 10 days of the recovery period (RP) in the horizontal (*dashed line*, N = 5) and 6° head-down (*solid line*, N = 5) positions. [From Noskov et al. (282) with permission.]

reambulation (90, 91, 204, 264, 352). After 14 and 28 day bed-rest periods, RCM continued to decrease during reambulation such that 2 wk after 14 days of bed rest the RCM level equalled that found after 28 days of bed rest (204).

Red Cell Mass: Physical Conditioning.

A decline in red cell mass becomes apparent after approximately 2 wk of bed rest (16, 42, 84, 93, 204–206, 221, 228, 265, 313, 344). For example, RCM decreased by 6.2% after a 20 day period of inactive-horizontal bed rest and was gradually restored during the 55 day reambulation and aerobic exercise training period (313).

TABLE 39.1. *Mean (± SE) Daily Excretion of Urinary Parameters during Baseline, and First Days of Bed Rest and Recovery in Group 1 and Group 2*†

Parameter	Baseline (N = 10)	1st Day of Bed Rest		1st Day of Recovery	
		Group 1 (N = 5)	Group 2 (N = 5)	Group 1	Group 2
Diuresis, ml/kg	11.3 ± 0.5	10.3 ± 0.5	15.3 ± 2.7*	10.6 ± 0.8	10.8 ± 1.3**
Sodium, meq	137 ± 3	175 ± 14*	239 ± 20*	105 ± 8**	108 ± 15**
Potassium, meq	49 ± 1	41 ± 5	49 ± 5	63 ± 6	53 ± 15
Calcium, meq	9.2 ± 0.4	8.3 ± 1.0	10.9 ± 2.0*	8.9 ± 0.3	11.2 ± 0.7
Magnesium, meq	8.5 ± 0.2	8.1 ± 0.3	9.2 ± 0.4	9.3 ± 1.0	11.9 ± 1.6
Chloride, meq	138 ± 4	153 ± 10	200 ± 24*	142 ± 12	147 ± 9**
Sodium/potassium	1.05 ± 0.04	1.8 ± 0.3*	2.0 ± 0.2*	0.64 ± 0.02**	0.83 ± 0.08**
Creatinine, mmol	19.4 ± 0.3	17.7 ± 1.8	22.1 ± 3.5*	18.6 ± 1.8	22.1 ± 2.7

†Group 1 = 5 men, horizontal bed rest; group 2 = 5 men, 6° head-down bed rest. [Adapted from Noskov et al. (282) with permission.]
*$P < 0.05$ from baseline; **$P < 0.05$ from bed rest.

Physical conditioning in ambulatory persons has variable effects on RCM, depending on the mode and intensity of the exercise. Prolonged aerobic conditioning has been reported to increase both plasma volume and red cell mass (207); although eight wk of high-intensity cycle exercise training expanded plasma volume, it had no significant effect on RCM (121). On the other hand, Weight et al. (381), in discussing athletes' anemia which has been observed occasionally even in nonimpact sports, suggested that increased erythrocyte destruction may be due to intravascular hemolysis accompanying increased circulatory rate, increased body temperature, compression of erythrocytes by muscular activity, elevated levels of catecholamines, or acute exercise acidosis.

Aerobic exercise training during bed rest usually appears to have a protective effect on RCM (16, 130, 144, 145, 205). Even moderate to intensive levels of supine cycle exercise can prevent the loss of RCM observed during 14 day (130) and 30 day (144, 145) periods of bed rest. However, after 28 days of bed rest, six subjects who performed a program of progressively increasing cycle exercise intensity had greater decreases in RCM (539 ml) than 12 subjects who did not exercise (180 ml) during bed rest (265). The increased exercise intensity during the last 11 days of bed rest may have contributed to increased destruction of red blood cells. Burkovskaya et al. (42) also noted a greater loss of RCM in bed-rested subjects who performed intensive exercise.

Red Cell Mass: Mechanisms for Decrease. The etiology for the decline in RCM is most likely the result of multiple factors which act with varying intensity and occur during different phases of bed rest (Table 39.2). During the first months of bed rest the decline in RCM is due mainly to inhibition of red cell production (16, 42, 205, 228, 351). One likely factor is the increased hematocrit due to more rapid loss of plasma volume than loss of RCM. Renal and extrarenal receptors regulating synthesis of erythropoietin—a hormone that stimulates red cell production—are stimulated by a decrease in the oxygen-carrying capacity of blood which depends on blood hemoglobin concentration. Therefore, hemoconcentration, which is most marked during the first few days of bed rest, may result in reduced erythropoietin production. However, Dunn et al. (84) reported no significant change in erythropoietin levels (measured with bioassay) in six men after 28 days of horizontal bed rest and in crewmembers during the 10 day Spacelab 1 mission. When a more sensitive radioimmunoassay technique became available, archived samples from the Spacelab 1 crewmembers were retested and a significant decrease in erythropoietin was found (230). Lancaster (228) also measured erythropoietin (mouse bioassay)

TABLE 39.2. *Possible Mechanisms for Decline in Red Cell Mass during Bed Rest*

Reduction or Suppression of Red Cell Production Due to

Loss of plasma volume and increased hematocrit

Hormone-like inhibitors of erythropoiesis (oubain-like factor, marrow-inhibiting lymphokines, inhibition, interferon)

Erythrocytic chalones (tissue-specific products of differentiated cells that selectively inhibit early cells of the same lineage)

Inadequate caloric or protein intake

Increased plasma phosphorus

Altered iron, folic acid, Vitamin B_{12}, or protein uptake or metabolism

Decreased tissue oxygen demand

Bone demineralization and altered calcium balance

Metabolic disturbances resulting from muscle atrophy and negative nitrogen balance

Increased Destruction or Loss of Circulating Red Cells Due to

Microhemorrhages in dependent regions

Capillary oozing of red cells in dependent regions

Sequestration of Red Cells Due to

Changes in red cell morphology

Altered blood flow distribution, especially in the spleen

[Adapted from Talbot and Fisher (351) with permission.]

and found a decreasing trend after 30 days of bed rest and an increasing trend following reambulation. Although these changes were nonsignificant, the strong trend toward decreasing erythropoietin, together with the decreases in total circulating reticulocytes and reticulocyte percentages, suggests that erthropoiesis decreases during bed rest.

Other factors contributing to reduced RCM could be P_{O_2}, renal blood flow, and circulating hormones or hormone-like inhibitors that may be released during bed rest and affect key cellular processes, such as cell membrane transport, resulting in decreased erythropoiesis (351). Balakhovksiy et al. (16) isolated a tissue-specific inhibitor of cell proliferation, which they termed "erythrocytic chalone," from the blood of subjects kept at strict bed rest for 8 days. Decrease in red cell formation may be associated with changes in body metabolism, which include inadequate caloric or protein intakes, increased plasma phosphorus concentration producing a shift in the hemoglobin P_{50}-causing red cells to give up bound oxygen (182), and lack of intake or mobilization of body stores of iron, folic acid, and vitamin B_{12}. Findings from the model of Dunn et al. (83), which used animal and human data, suggest that the primary cause of erythroid suppression during spaceflight is related to changes in body weight. They found positive multivariate correlations between RCM loss and changes in caloric intake, exercise performed, and lean body mass, with lean body mass being the major factor. It is probable that RCM loss during spaceflight or bed rest is a response to decreased tissue oxygen demand. The positive effect of dynamic aerobic exercise training (with its higher oxygen uptake) on maintenance of RCM during bed rest supports this theory (144), while isokinetic exercise training (with it's lower oxygen uptake) does not attenuate RCM loss during 30 days of bed rest.

Other possible factors for RCM reduction during bed rest include bone demineralization and negative calcium balance, which may modify the milieu of the hemopoietic marrow to alter its response to erythropoietin (16, 351). There is no evidence of gross hemolysis or shortened red cell life span during bed rest (228).

Sequestered red cells from the redistribution of blood and body fluids may contribute to reduced RCM during bed rest (351). An increase in splenic blood flow may result in splenomegaly with entrapment of red blood cells not normally taken up by the spleen. Splenic trapping may be enhanced because of potential changes in red cell shape such as those observed during spaceflight (202, 203). This red cell deformation may have been caused by many of the same factors that occur during bed rest (351), such as rapid decrease in plasma volume, altered calcium metabolism (79), or changes in plasma or cellular constituents (lipids, electrolytes, osmols, ATP, 2-3 DPG). Whether similar changes in red cell deformation occur during bed rest is not known.

In conclusion, although decreased red cell mass has been observed consistently after bed rest, the mechanisms are not understood. An excellent summary of possible mechanisms and suggestions for future research can be found in a report by Talbot and Fisher (351).

Plasma Volume: Time Course. Contrary to the slow and progressive decline in RCM during bed rest, the decreasing plasma volume (PV) response begins immediately upon assuming the recumbent position. An initial transient 6%–7% increase in PV (hemodilution hypervolemia) occurs during the first 2 h of recumbency (108, 386), and a 14% hemodilution hypervolemia occurs during the first hour with the 10° head-down position (108). This acute hypervolemia, due probably to attenuated filtration from leg capillaries, is followed by a decrease in PV of approximately 4% (125 ml) after 6 h, and 5%–10% (150–300 ml) after 24 h of horizontal rest (95, 124, 281, 376). The PV contraction is progressive during the first 3–6 days; then the rate of loss decreases and gradually approaches a new steady state (90, 93, 374, 375). Greenleaf et al. (130) compiled data from several studies and derived the following exponential equation (Fig. 39.3) to predict change in PV as a function of days of horizontal bed rest:

$$PV\ loss\ (ml) = [BR\ days/-0.011]$$
$$- 0.0013 \times (BR\ days).$$

The decline in PV may reach approximately 21% (700 ml) after 80 days of bed rest; that is, the volume shifted headward during head-out water immersion (8).

The initial hemodilution during bed rest is a result of the decrease in hydrostatic forces and capillary filtration responses to change in posture. The headward movement of approximately 700–900 ml of fluid from the lower to the upper body changes transcapillary hydrostatic pressures to favor fluid reabsorption and less filtration in interstitial tissues of the lower body and fluid filtration in tissues of the upper body (156). For exam-

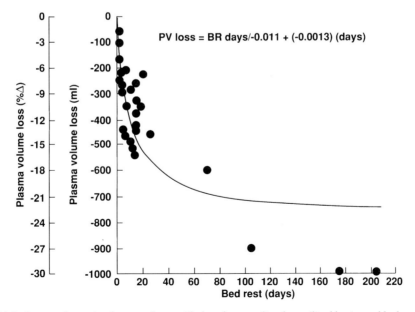

FIG. 39.3. Percent change in plasma volume with data from studies that utilized horizontal bed rest with no remedial procedures. [From Greenleaf et al. (130) with permission.]

ple, the mean capillary pressure in the toes of a standing person is approximately 90 mm Hg, 30 mm Hg in the horizontal position, and 20 mm Hg in the 6° head-down position (158, 238). With a sudden change from the upright to the supine position, vascular hydrostatic pressures decrease immediately as venous blood is displaced from the lower to the upper body, while tissue interstitial fluid pressures equilibrate more slowly. Upon reclining, the initial transcapillary pressure balance is offset toward lower intravascular pressures with a net gradient for fluid reabsorption in the lower body. The opposite gradients for fluid transfer occur in the upper body favoring net fluid filtration. Therefore, depending on the angle of tilt and the relative proportion of the body experiencing a decrease in capillary hydrostatic pressure, changing to a horizontal (bed-rest) position results in an overall net fluid reabsorption and transient hemodilution.

With continued recumbency this initial PV expansion, combined with headward displacement of body fluids, results in increased cardiac filling and a transient (6 h) rise in central venous pressure (31, 95, 281). This increased cardiac filling is postulated to stimulate reflex neural and hormonal responses which result in a twofold increase in urine output and increased sodium, potassium, and osmolar excretions during the first 6 h of bed rest (95, 129, 157). Within 20 to 24 h, as the central hypervolemia is reduced, the diuresis also decreases. Diuresis may not be evident during bed rest if fluid intake is not controlled (282). Voluntary fluid intake can decrease after the first 4 h of 5° head-down recumbency resulting in negative fluid balance without diuresis (157). The mechanism for this hypodipsia is unknown, but may be related to thoracic or intracranial hypervolemia (126) since plasma osmolality remains essentially constant or may be increased by 5–7 mosmol/kg during prolonged bed rest (124, 130, 144).

Plasma Volume: Mechanisms for Hypovolemia. Several hypotheses have been proposed to explain the diuresis and hypovolemia during bed rest. This diuretic response initially was attributed to the Gauer-Henry reflex (97, 98), in which an increase in central blood volume elevates central venous pressure stimulating atrial and blood vessel mechanoreceptors; stimuli travel in vagus nerves to the thalamic region with resulting inhibition of hypothalamic vasopressin release (75). This reflex is accompanied by inhibition of the renin–angiotensin–aldosterone system (315, 397) and decreased renal sympathetic nerve activity (192). A consequence of decreased renal sympathetic nerve stimulation would be increased salt and water excretion either through alterations in renal hemodynamics or by a direct neural effect on renal tubular sodium reabsorption (297). The

overall effect of the Gauer-Henry reflex would be diuresis and natriuresis.

Although the Gauer-Henry reflex has been established in nonhuman primates, it has been difficult to document the hormonal changes required to validate their theory in humans (6, 118, 149, 376). Plasma renin activity (PRA), vasopressin (PVP), and aldosterone (PALD) concentrations decrease somewhat during the first few hours of recumbency (71, 95, 157, 173, 281). However, it is difficult to ascribe the entire diuretic response during bed rest to only the Gauer-Henry reflex since the changes in these hormones have been relatively small. For example, the vasopressin concentration of 1.4 pg/ml, measured in subjects in the supine position, was well within the normal range and decreased to about 0.5 pg/ml following 4 h of 5° head-down bed rest (281). In another study, control PVP was about 2.8 pg/ml during the standing baseline condition, and decreased to about 1.0 pg/ml after 60 min of 6° head-down bed rest (71). Such small changes in vasopressin may contribute to instigating diuresis, but are probably not large enough to account for the continuing diuretic response during bed rest.

Other evidence that questions the Gauer-Henry reflex as the sole mechanism for diuresis after central volume expansion comes from studies of nonhuman primates (111, 112). Gilmore et al. (112) demonstrated that dorsal rhizotomy (C6-T7), bilateral vagotomy, and sinoaortic denervation, which should abolish input from both cardiopulmonary and sinoaortic baroreceptors, failed to prevent diuresis in response to a 15% plasma volume expansion. These findings suggest that, in primates, neural reflexes from the cardiopulmonary region are not essential for instigating diuresis in response to central cardiac volume expansion. Gilmore concluded that "in the primate the volumetric control of salt and water homeostasis has shifted from low-pressure to primarily high-pressure receptors. This shift may have resulted from the evolution to an upright or semiupright position" (111). In addition, cardiac-denervated patients had heart rate, stroke volume, PV, PRA, and PVP responses during acute 6° head-down bed rest similar to those of normal healthy subjects (65). Thus cardiac volume receptor stimulation is not the only mechanism regulating fluid volume during acute volume shifts in humans, but changes in ANP, urodilatin, arterial baroreflexes, and intracranial regulatory systems may also be involved.

A more recent theory to explain the initial PV loss during bed rest involves secretion of ANP. Atriopeptides are found in cells of heart atria and pulmonary arteries. They are released in response to mechanical stretch caused by increased pressure or volume. Genest and Cantin (101) suggest that the diuretic and natriuretic actions of ANP are due mainly to its vasorelaxing effect on renal arteries and arterioles, and to an increase in glomerular filtration rate and filtration fraction resulting in increased excretion of filtered sodium. Atriopeptide concentrations increase transiently during the first 30–60 min of head-down recumbency (3, 106, 107, 149, 252). Gharib et al. (106) reported a twofold increase in plasma ANP during the first 30 min of 9° head-down recumbency, and Allen et al. (3) found that plasma ANP rose from 8.1 to 11.4 pg/ml after 60 min of 10° head-down recumbency. When ANP compounds were administered to humans in pharmacological doses they ellicited marked natriuresis and increased glomerular filtration rate (3, 101, 114). However, it is unclear whether ANP has a significant effect on fluid–electrolyte parameters in the physiological concentrations present at the onset of recumbency. Goetz (114) has concluded that ANP in physiological concentrations contributes minimally to natriuresis and diuresis in humans. On the other hand, Morice et al. (267) and Cottier et al. (68) have reported increases in urinary volume and sodium excretion in response to low-dose infusions of ANP synthetic analogues that increased plasma ANP-like immunoreactivity by only two- to fourfold. Although increases in both plasma ANP and urinary sodium excretion occurred during head-down bed rest (3), there was considerable variation (2.8–21.8 pg/ml) in the plasma ANP response and lack of significant correlation between plasma ANP and sodium excretion. Allen et al. (3) concluded that their findings were "not compatible with a direct cause and effect relationship between plasma ANP and sodium excretion during head-down tilt". Other roles for ANP might be to decrease cardiac output, cardiac filling pressure, and arterial blood pressure; to promote fluid shifts from plasma to the interstitial space; to inhibit the sympathetic nervous system; to reduce the rate of release of PVP and PALD; and to inhibit voluntary salt and water intake. Brain ANP also may reduce the rate of release of vasopressin (114) which may act as a dipsogen (126).

Natochin et al. (274) suggested that changes in fluid and electrolyte balance after prolonged bed rest and spaceflight may be the result of "inadequate" responses of the kidney to fluid-regulating hormones or other regulatory substances. For example, postflight blood levels of PVP were significantly elevated from preflight levels (Fig. 39.4), yet urine osmotic concentrations were reduced for a given level of diuresis which may indicate decreased renal sensitivity to vasopressin. There appeared to be a dissociation between PVP levels and urinary free water clearance postflight. They also observed depressed renal response to vasopressin (dDAVP) administration in subjects after exposure of up to 370 days of head-down bed rest. In ambulatory control subjects, administration of dDAVP resulted in an elevation of urinary osmolality to 1,088 mosmol/kg,

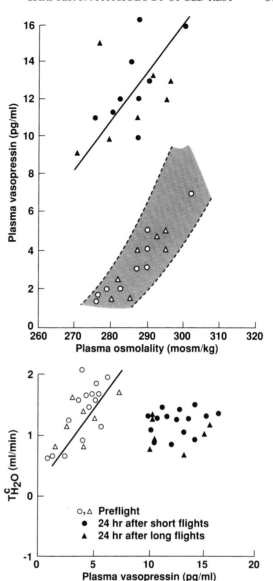

FIG. 39.4. Regression of plasma vasopressin on plasma osmolality *(upper panel)*, and free water clearance $(T^c_{H_2O})$ on plasma vasopressin *(lower panel)* in cosmonauts before and after short- and long-term spaceflights. *Shaded area* indicates normal limits. [From Natochin et al. (274) with permission.]

while after bed rest, urine osmolality increased to only 868 mosmol/kg. Several possible mechanisms for such a decrease in renal sensitivity to vasopressin include decreased intracellular potassium, altered formation of second messengers in response to vasopressin receptor stimulation, altered prostaglandin synthesis, altered blood calcium, or elevated catecholamine levels (274).

An important mechanism for PV regulation is the contribution of plasma proteins. Since each gram of plasma protein binds osmotically approximately 12–15 ml of water (305), factors which reduce vascular protein content will also reduce plasma volume. Total protein content decreases during bed rest while protein concentrations remain stable (91, 144, 367). Van Beaumont et al. (367) observed proportional changes but not equal relative increases in hematocrit and total protein content during bed rest. They hypothesized that part of the hypovolemia during bed rest may be caused by diminished return of protein from the lymphatic system in response to decreased circulation from reduced physical activity. Adjustments in both plasma and interstitial fluid compartment volumes during bed rest may occur in response to changes in protein production, distribution, or degradation.

Plasma Volume: Countermeasures. With prolonged bed rest without remedial procedures (countermeasures), the decline in PV is sustained. Smirnova (337) measured fluid balance and body fluid compartments before and after 120 days of 5° head-down bed rest without and with remedial procedures that included exercise training. Frequent exercise training (undefined) during bed rest tended to exert a hypervolemic effect; compared with pre–bed rest levels, PV was expanded by 3% in the exercise group and decreased by 3%–8% in the nonexercise group during bed rest. Others have also found that vigorous lower body (130, 144, 145, 240, 337) or upper body (364) aerobic exercise training with adequate fluid replacement attenuated the decrease in PV during bed rest. For example, Greenleaf et al. (145)

found that 1 h/day of a supine, aerobic exercise training program maintained PV within 3.7% of the pre–bed rest level, compared to a significant 17.2% loss of PV in a nonexercise group and a significant 18% loss of PV in subjects who performed isokinetic exercise training for 1 h/day.

Other remedial procedures that attenuate or restore hypovolemia during bed rest include periodic application of occlusive cuffs on the arms and legs (263, 342), prolonged exposure to lower-body negative pressure (89, 150, 175, 291, 343, 344), supplemental intake of water and salt (147, 148), intake of water and salt with an antidiuretic agent (adiuretine) (148), or mineralocorticoid administration (34, 170, 342).

The overall PV response to bed rest in young and middle-aged females is similar to that of males (90, 140, 323). However, it has been reported that older women (55–65 yr) may sustain a greater loss of PV than older men and also have more severe orthostatic responses after bed rest (120). Fortney et al. (90) suggested that the PV response of young women to bed rest may be altered by hormonal action associated with menstrual function.

Total Body Water. There is a slow, progressive decrease in total body water (TBW), measured with serial isotope techniques, during 28 days of horizontal bed rest (93). Rapid water loss during the first 2 days was mainly

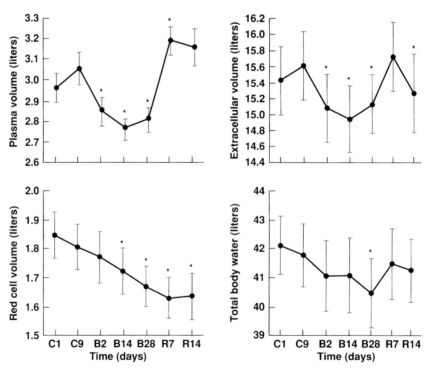

FIG. 39.5. Mean (± SE) plasma, extracellular, red cell, and total body water volumes in 10 men before (*C1, C9*), during (*B2, B14, B28*), and after (*R7, R14*) 28 days of horizontal bed rest. *P < 0.05 from C9 value. [From Fortney et al. (93) with permission.]

extracellular fluid (209 ml of plasma and 324 ml of interstitial water), while the majority of the deficit during the remainder of bed rest was intracellular fluid (Fig. 39.5). The total decrease in TBW was 1,316 ml, with 476 ml extracellular fluid and 840 ml intracellular fluid. The decline in intracellular water was attributed to equilibration with the extracellular fluid and to the decrease in lean body mass. After 48 days of 4° head-down bed rest, Krotov et al. (221) found TBW reduced by 2,840 ml.

There are few data regarding fluid-compartment volumes with bed rest beyond 28 days. Lobachik et al. (240) observed a continued decline in TBW during 120 days of 5° head-down bed rest. Physical exercise training attenuated the decline in TBW by 25%–50% on bed rest day 60, but the efficacy of the countermeasure diminished thereafter.

Extracellular Water. There has been some disagreement about the response of the extracellular fluid (ECF) volume during bed rest. Some investigators (124, 256) have reported that, after 5–14 days of horizontal bed rest, the initial decline in ECF was attenuated and it returned to the pre–bed rest level. With prolonged bed rest, Greenleaf (124) postulated that ECF was restored by expansion of the interstitial fluid volume, which compensates for the reduction in PV. The exact mechanism of this compensatory mechanism is unknown, but may involve bed rest–induced changes in hydrostatic and colloid osmotic pressures, or peripheral-vascular action of ANP. Others (93, 170, 179, 374, 385) have reported continuing decrease in ECF throughout at least the first 14 days of bed rest.

Cardiac Function

Time Course. Change in cardiac function during bed rest can be described in three stages (318). Stage one occurs immediately upon reclining and, involves mainly cardiovascular responses to the headward fluid shift. The enlarged central blood volume causes a transient increase in cardiac filling pressure (31, 95, 118, 242, 243, 281) and stroke volume in accordance with changes predicted by Starling's Law of the heart. Cardiac output may (35, 95, 243, 387) or may not (31, 95, 242, 281) increase briefly, depending on whether the increase in central filling also provoked compensatory bradycardia and peripheral vasodilation. The immediate changes in arterial pressures during movement from a horizontal to head-down position are apparently passive, with transient increases in dependent cranial areas and decreases in arteries in the elevated lower extremities (387). Mean arterial pressure at heart level does not change during 5° to 10° head-down angles (95, 118, 242, 243, 281). The second (hemodynamic) stage

occurs over the next 24–48 h as neurohumoral reflex responses reduce the central volume overload. During this phase, stroke volume and cardiac output decrease slightly in response to the diuresis-induced hypovolemia. The third stage occurs if bed rest continues without remedial measures. Cardiac output and stroke volume continue to decrease and eventually stabilize below the pre–bed rest level. This reduced cardiac output is consistent with the resultant decreased oxygen demand, loss of active muscle mass, decreased blood volume, and decreased circulation to some vascular beds. It is also consistent with Gauer's thesis that the normal operating condition for the human cardiovascular system is at a level commensurate with demands of the upright posture, rather than the supine posture. Thus during bed rest, the cardiovascular system adapts to maintain baseline levels similar to those in the upright position (32), that is, a homeostatic response.

Cardiac Size. During the first 24 h of bed rest there appears to be an increase in cardiac chamber volume, as estimated by the left ventricular end diastolic diameter (31, 281), which may return to pre–bed rest levels during the first 24 h (294). Cardiac size may not increase further with movement from the horizontal to the head-down position (294). These findings accompany unchanged central venous pressures reported when a subject moves between the horizontal and head-down positions (318). Rushmer (310) reported that left ventricular dimensions were maximal when dogs were horizontal, when compared to the 30° head-down position or after intravenous infusion of blood. Sandler (318) concluded that the variable reports of changes in cardiac volumes during the first few hours of horizontal or head-down recumbency may be related to variation in compliance of the upper body vasculature. Nixon et al. (281) attributed small changes in central venous pressures and marked differences in cardiac volumes during 24 h of 5° head-down bed rest to shift of blood from systemic veins to pulmonary veins. Increases in pulmonary artery pressure, and increases in right ventricular work and contractility also support pulmonary influx of large quantities of blood during the first few hours of head-down recumbency (195).

With prolonged bed rest, cardiac size diminishes (15, 216, 313, 353). Saltin et al. (313) noted an 11% decrease in heart size estimated from biplane radiographs after 3 wk of horizontal bed rest. After 70–73 days of bed rest, Krasnykh (216) found a 13%–18% decrease in X ray–derived heart size in three standing subjects who had not recovered to the pre–bed rest level after 20 days of reambulation. Decreased heart size in the supine and upright positions may be due to the decrease in PV. Since PV is restored within a few days after bed rest, the continued decrease in heart size may have been due to peripheral displacement of fluids upon standing.

Cardiac Morphology and Mass. Degenerative changes in myocardial morphology and biochemistry have been found in animals after whole-body immobilization (213, 308, 316, 318). After 20 days of immobilization, cardiac mass in rats had decreased by 23% while total body mass decreased by 54%; after 100 days left ventricular mass had decreased by 20% and right ventricular mass by 22%. When converted to dry ventricular weight, to account for differences in tissue fluid–electrolyte changes, left ventricular weight was still reduced by 24% and the right ventricle by 26% (213, 318). Measurements of left ventricular protein turnover rates also support the change in myocardiac mass during bed rest. Left ventricular incorporation of 35-sulfur labelled methionine tended to decrease over the first 2–5 days of immobilization, and was significantly decreased by 27%, 30%, and 60% after 15, 30, and 100 days, respectively (213). After 30 and 100 days of immobilization, total myocardial protein had decreased by 9% and 18%, respectively. After 14 days of immobilization there was an increase in number and decrease in size of rat cardiac mitochondria; on days 45–60 the size and number appeared normal, and by day 120 the size and number of mitochondria were greater than in control nonrestrained animals (308). On the other hand, there were significant decreases in the volume density of quadriceps femoris muscle mitochondria in dogs confined for 2 and 5 mo that were restored to preconfinement levels with exercise retraining during the ambulatory recovery period (275). Other degenerative cardiac changes reported in rats (298, 308), rabbits (372), and monkeys (36, 191, 326) after 2–6 mo of whole-body immobilization have included accumulation of fat droplets, muscle fiber degeneration, and an increase in connective tissue content. The myocardium from immobilized monkeys had elevated levels of hydroxyproline, supporting electron and light microscopic findings of increased fibrous tissue (36). Lysosomal enzyme activity (free and total) was also elevated in both ventricles, suggesting an increase in number of lysozymes, and increased protein degradation or reduced protein synthesis was indicated by decreases in ribonucleic acid. Such morphological findings have not been documented in humans undergoing bed rest.

Cardiac Output. After a possible transient increase in cardiac output associated with the headward fluid shifts (35, 95, 243, 387), cardiac output and stroke volume have been reported to either return to pre–bed rest levels (343, 353), to decrease (188, 197, 198, 313, 317, 401), or to remain increased (187, 194, 199, 213, 295). These variable responses may be related to the different methods employed, which have included cardiac catheterization using Fick or indicator dilution methods (100, 170, 194, 256, 313), echocardiography (317, 321), various external techniques including apexcardiography (174, 359, 371), mechanocardiography (105, 359, 334), and CO_2 or acetylene rebreathing (35, 95, 199, 281, 295, 401). Catheterization and echocardiography are the more reliable methods and results from these techniques have generally indicated either no change or a slight decrease in cardiac output and stroke volume during extended bed rest.

Cardiac Contractility. During the initial hours of recumbency there is no significant change in cardiac contractile function as measured with echocardiography (29, 242, 281). Within the first 24 h of 5° head-down bed rest, Nixon et al. (281) found a small increase in left ventricular ejection fraction, no change in heart rate or arterial pressures, and no change in the velocity of circumferential fiber shortening, in spite of marked increases in cardiac stroke volume. These data are consistent with an effect of a marked increase in preload on cardiac function, with little or no change in ventricular performance (32).

However, under more drastic conditions of acute central volume overload, where seven men were abruptly tilted from a 70° head-up position to a 30° head-down position for 60 min, Gazenko et al. (100) have reported a transient decrease in cardiac contractility ascribed to a short-term imbalance between the coronary circulation and oxygen requirements of the myocardium. After the first 10–15 min of head-down tilt, left ventricular contractility indices had returned to the pre-tilt level.

The effects of prolonged bed rest on cardiac contractile function are unclear. Cardiac stroke volume was depressed during both supine and upright exercise after 20 days of horizontal bed rest (313). Saltin et al. (313) assumed that decreased cardiac filling would be abolished in the supine position, and that the smaller supine stroke volume may be an indication of depressed intrinsic myocardial function. However, more recent data have questioned their assumption because cardiac filling gradually declines during bed rest, even in the supine position, and may contribute to a smaller stroke volume (31, 281). Goldwater et al. (119) found no significant alteration in cardiac function curves derived from echocardiographic measurements during LBNP before and after 10 days of bed rest. Hyatt et al. (173) concluded that the delayed recovery of maximum oxygen uptake in their subjects after 28 days of horizontal bed rest, in the presence of unaltered pulmonary function and restoration of blood volume, implicated a loss of myocardial function. Later, Hyatt et al. (174) concluded, from systolic time interval and apexcardiography data, that significant decrements in myocardial contractile state were evident by the second wk of bed rest and persisted during 2 wk of recovery. There was no significant correlation between changes in contractility indices and change in PV. They attributed the nonsignificant changes in ejection fraction observed in previous bed-rest studies to technical difficulties in finding the same echo window during repeated testing. Hung et al. (167), on the other hand, found an increase in ejection fraction using equilibrium-gated blood pool scanning at rest and during supine and upright exercise after 10 days of bed rest; they concluded that the post+–+bed rest increase in myocardial systolic performance offset a diminished left ventricular volume and thus minimized the reduction in exercise performance. Georgiyevskiy et al. (105) found no change in myocardial contractility (measured with mechanocardiography) in three men after 62 days of horizontal bed rest. Katkov et al. (195, 196) reported decreased right ventricular contractility after 5 days of 4.5° head-down bed rest from direct pressure measurements obtained from catheters placed into the pulmonary artery, right ventricle, and a radial or brachial artery. Data from a subsequent study (193) indicated no significant change in left ventricular contractile state (max dp/dt, max dp/dt/P, V max) after 5–6 days of 15° head-down bed rest.

Thus there is no consistent pattern of myocardial function during extended bed rest without remedial procedures; it may increase, decrease, or remain unchanged.

Heart Rate and Electrical Activity. Heart rate might be expected to vary during bed rest when considering the influence of changes in blood volume, body fluid distribution, and autonomic nervous function. The initial response to recumbency is a decrease in heart rate.

In most prolonged bed-rest studies there has been progressive increase in resting heart rate (2, 21, 27, 30, 105, 187, 263, 264, 365). Taylor et al. (353) were the first to observe that heart rate increased by about 0.5 bpm with each day of bed rest, and this rate of increase generally continued for 30 days. In bed-rest studies up to 10 days, resting heart rate increased by 12–32 bpm, 30 day studies by 26 bpm, 62 day studies by 25 bpm, and in 70–120 day periods heart rate increased by 1–5 bpm for each ensuing week (318). While basal heart rate increased progressively during 30 days of 6° head-down bed rest in subjects with daily isokinetic and no exercise training, heart rate was unchanged in subjects undergoing daily dynamic (isotonic) exercise training (128).

The mechanism for the progressive increase in heart rate with extended bed rest may be related to ongoing adjustments in autonomic nervous function. During the first few days of bed rest the increase in central blood volume activates cardiopulmonary baroreceptors, resulting in relative withdrawal of sympathetic and increase in parasympathetic function. However, as bed rest continues, the sympathetic/parasympathetic balance gradually reverses towards increasing sympathetic and decreasing parasympathetic responses.

Goldberger et al. (116) hypothesized that a reduction in variability and narrowing of the heart rate power spectrum response to LBNP found after 7–10 days of bed rest is a sign of autonomic dysfunction, causing decreased ability to respond to a gravitational stress. Decreased amplitude of the power spectral frequency peak at 0.25 Hz after 4 days of 6° head-down bed recumbency (Fig. 39.6) corresponded to the vagally mediated respiratory heart rate variation and suggests reduction in parasympathetic function (117).

Change in electrocardiogram (ECG) responses measured from standard limb leads in subjects during 4–10 wk of bed rest could result from changes in heart position, relative inhibition of conduction, impaired myocardial perfusion, or altered repolarization (2, 21, 73, 210, 313, 365, 378). Korolev (209) reported that T_{v1}-wave amplitudes increased, T_{v6} amplitudes decreased, and U-waves of increasing amplitude were detected in pericardial leads and became significant by the second or third wk of bed rest. T-wave changes were interpreted as repolarization disturbances possibly due to altered blood and tissue levels of potassium or calcium. Increasing P-wave amplitudes in leads 2, 3 and aVF during bed rest have been interpreted as resulting from reduced pulmonary or cardiac filling (210), and a variety of other ECG changes have been attributed to "metabolic changes in the myocardium" (2, 73). Changes in the ECG during passive tilt after 70 days of horizontal bed rest suggested increased myocardial ischemia (depression of S-T segment and inversion of T-waves) or decreased cardiac filling (increased amplitude of P and R waves and changes in the T wave) (210). Korolev (210) thought that ischemia of subendocardial layers of the myocardium and the frontal wall of the left ventricle might arise as a result of "a well-expressed decrease in blood inflow into the heart; this results in a decrease in stroke volume and a filling of the coronary vessels . . . . Rotation of the heart clockwise about its longitudinal axis, forward movement of the apex cordis, tachycardia, and irritation of the sympathetic nerve (which in turn results in an increase in oxygen consumption by the myocardium) also to some degree affect change in the ECG T-waves. All these factors can cause both a relative and absolute cardiac inadequacy." Voskresenskiy et al. (378) considered that the adverse ECG responses during bed rest were not due to changes in water–electrolyte balance since variations in plasma potassium, sodium, or calcium concentrations were too small to induce cardiac abnormalities. They hypothesized that the ECG changes reflected autonomic changes, such as decreased parasympathetic and

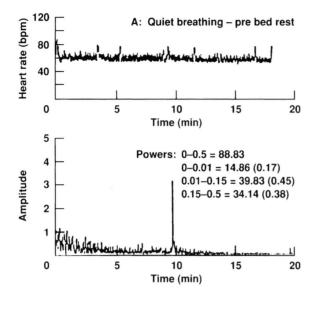

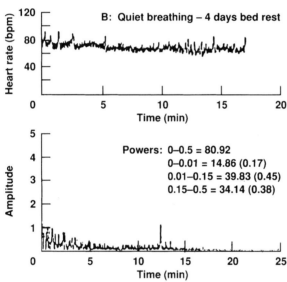

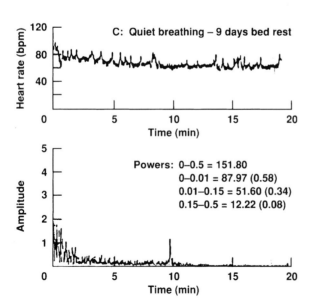

increased adrenergic actions on the heart. Individuals with healthy hearts but unstable autonomic nervous systems often produce altered S-T segments and flat or negative T-deflections. However, interpretations of ECG changes after bed rest must be made with caution since cardiac volume decreases with prolonged bed rest (313, 321, 323). A change in cardiac volume or its position within the thorax could induce "abnormal" ECG changes similar to those discussed above (278).

Cardiovascular Autonomic Regulation

Acute Responses. The acute autonomic regulatory response when subjects move from an upright to a supine or head-down position is overall decreased sympathetic nervous function as indicated by decrease in the levels of plasma and urinary catecholamines (67, 108, 109, 243), decreased renin secretion (108, 243), decreased total systemic vascular resistance (95, 243), and increased venous compliance and arm and leg blood flows (287). Upon moving from a sitting to horizontal position, there was a 20% reduction in plasma norepinephrine and no change in epinephrine or dopamine; after moving from a sitting to 10° head-down position, there was a similar (20%) decrease in norepinephrine, a decrease in epinephrine, and again no change in dopamine (108). Thus the initial catecholamine response to recumbency depends upon the control body position (upright or supine) and the subsequent body angle.

Chronic Responses. The effect of chronic recumbency on autonomic nervous system function is still unresolved. During prolonged bed rest without remedial procedures most authors have reported a slow progressive increase in resting heart rate (2, 21, 27, 30, 73, 105, 187, 263, 264, 352, 353, 365); increase (105, 273), decrease (328), or no change (73, 109, 128, 334) in systolic and mean blood pressures; decreased heart rate variability (116); and either no change (58, 290) or decreases (52, 67, 72, 109, 229, 323, 327, 328) in blood or urinary catecholamine levels. The decline in resting blood catecholamine levels may indicate decreasing sympathetic activity due to reduced energy output and fewer cardiovascular adjustments during bed rest (67, 109). One exception to the decline in catecholamines occurred in eight older men (55–65 yr) after a 5 day bed-rest period where blood pressure, plasma norepinephrine, and cortisol increased significantly. It remains to be determined whether there is a true age-related difference in autonomic response to bed rest (273).

FIG. 39.6. Heart rate time series and spectra from one healthy woman before (A) and during (B, day 4; C, day 9) bed rest. [From Goldberger and Rigney (117) with permission.]

Cholinergic activity and sympathetic nervous function, evaluated from changes in blood catecholamines and their metabolic products in urine, were assessed during 120 days of 4.5° head-down bed rest (72). Resting epinephrine and acetylcholine levels were elevated; blood concentrations of norepinephrine, dopamine, and normetanephrine decreased progressively, and synthesis of norepinephrine and dopamine also decreased, suggesting that sympathetic function decreases and parasympathetic function increases during prolonged bed rest. The threefold increase in respiratory arrhythmia during the first 3 wk of bed rest also suggests increased parasympathetic function (22). These findings are in contrast to the heart rate evidence suggesting depressed parasympathetic function during bed rest (117).

Recovery. One hypothesis accounting for orthostatic intolerance observed immediately after bed rest is impaired sympathetic reflex responses (369), but the evidence is inconclusive. In response to exercise or orthostatic stresses during or just after bed rest, a greater increase in heart rate (172, 354), unaltered vasoconstrictor (95) and potentiated venoconstrictor responses (88), and either a normal (17, 52, 58, 362) or attenuated (369) release of norepinephrine have been reported. Change in autonomic nervous function during bed rest may occur at the central level (central vasomotor centers), and/or at a peripheral level (altered peripheral nerve function, or altered receptor number, sensitivity, or second messenger response). It is possible that basal nervous regulation is altered during bed rest due to a central nervous "resetting" in favor of a greater sympathetic/parasympathetic balance; while at the peripheral level the cardiovascular reflexes, which mediate heart rate and vasoconstrictor responses to abrupt, transient changes in arterial blood pressure (58, 86, 87) or cardiopulmonary volume (94), are impaired. Such impairment of cardiovascular reflexes may result from inappropriate vagal or sympathetic reflex responses, inadequate neurotransmitter release, or down-regulation via decreased number or sensitivity of cardiac or vascular receptors.

Autonomic receptor function and change in receptor sensitivity were determined by measuring cardiovascular effector responses to drug infusion (23, 52, 256, 327, 328, 349). Alpha-adrenergic function was determined with infused graded doses of norepinephrine (0.005–0.160 μg·kg^{-1}·min^{-1}) and angiotensin II (0.5–16.0 μg·kg^{-1}·min^{-1}) into six healthy men; the elevation of mean arterial pressure following infusion was unchanged after 2–3 wk bed rest (52). The infusion doses of norepinephrine or angiotensin II required to produce a given increase in forearm vascular resistance and venous tone were not altered significantly (Fig. 39.7), but the vasoconstrictor and venoconstrictor actions of tyramine were attenuated after 12 days of bed rest (327, 328). The unaltered vascular response to norepinephrine and angiotensin II indicated no change in postjunctional receptor function, while the attenuated action of tyramine suggested decreased synthesis or storage of norepinephrine in prejunction nerve endings. On the other hand, after 14 and 28 days of immobilization, the vascular response of rhesus monkeys to bolus injections of norepinephrine was attenuated, suggesting either a decrease in the number or sensitivity of postjunctional alpha-adrenergic receptors, or a decrease in the resting diameter of the peripheral vessels (23, 76). Thus there is some evidence for attenuation of α-adrenergic function during prolonged bed rest. Impaired α-adrenergic function would be expected to impair arterial vasoconstrictor and splanchnic and cutaneous venoconstrictor reflex responses.

Activation of β-adrenergic neurons results in increased heart rate and contractility (β_1), arterial and venous vasodilation (β_2), and increased renin secretion (β_2). Some have suggested there is increased β-adrenergic activity during bed rest from trends toward increasing levels of plasma renin activity during tilt and infusion of isoproterenol (256), and limited improvement in post–bed rest orthostatic responses after β-adrenergic blockade (256, 319, 369).

Arterial Baroreflex Function. The carotid sinus and aortic arch arterial baroreceptors act primarily to adjust heart rate and peripheral vascular tone to counter abrupt changes in arterial blood pressure (58, 87). Therefore, one outcome of impaired arterial baroreflex function would be altered heart rate and blood pressure responses to orthostasis.

Impairment of the carotid sinus–cardiac baroreflex response has been reported in 6° head-down bed-rest studies (58, 86, 87) where carotid sinus function was assessed by altering pressure within a neck chamber designed to selectively stimulate the carotid sinus baroreceptors. A heart rate R-R interval vs. carotid systolic distending pressure (neck cuff pressure) response curve was used to characterize the carotid sinus baroreflex control of heart rate. There were significant reductions in the R-R interval range and in the maximum slope of the baroreflex response (58). These changes became evident after 12 days of bed rest and persisted for up to 5 days after the 30 day bed rest period. The greatest reductions in maximum slope were found in subjects who had the greatest post–bed rest orthostatic intolerance (Fig. 39.8). Eckberg and Fritsch (86) also found a reduction in the R-R interval response and a trend towards a decrease in maximum slope of the baroreflex response after 10 days of bed rest. They speculated that the mechanism for the impaired cardiac baroreflex mechanism during bed rest may be "autonomic neural

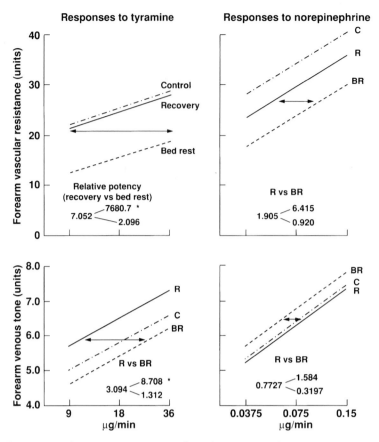

FIG. 39.7. Forearm vascular resistance *(upper panels)* and venous tone *(lower panels)* during infusions of tyramine and norepinephrine in four men during control (C, day 8), horizontal bed rest (BR, day 12), and ambulatory recovery (R, day 6) periods. The dose–response curves were constructed from data from all three (C, BR, R) periods. Relative potency values compare dose of drug needed to produce given levels of resistance or venous tone. For example, a dose of tyramine 7.052 times larger than that in the recovery period was required to result in a certain level of resistance. The upper and lower limits of the dose are 7,680.7 and 2.096, respectively. * Indicates upper vs. lower limit differences are significant ($P < 0.05$). [From Schmid et al. (327) with permission.]

plasticity"; that is, neural adaptation to prolonged alteration of cardiovascular sensory input possibly induced through changes in arterial and venous pressures, changes in input from cardiopulmonary baroreceptors, and loss of the normal pressure gradient between carotid and aortic pressures.

These findings suggest significant impairment in heart rate response to carotid sinus stimulation after bed rest. However, few investigators have questioned whether peripheral vascular responses to arterial baroreceptor stimulation are impaired by bed rest. Leg peripheral vasoconstrictor response to lower-body negative pressure (LBNP), estimated from change in arterial pulse volume measured by impedance, was unaltered following 10 days of 6° head-down bed rest, suggesting unimpaired baroreceptor-induced arterial vasomotor function (28).

One unanswered question is whether attenuation of arterial baroreflex function during bed rest is related to

altered input from cardiopulmonary baroreceptors. Acute increases (isotonic solution ingestion, leg elevation) or decreases (Furosemide, LBNP) in central blood volume, which produced variations in central venous pressure from 1.1 to 9.0 mm Hg, had no significant effect on the carotid–cardiac baroreflex response in humans (350, 357). Acute increases in central blood volume induced by head-down tilt also did not alter arterial baroreceptor function assessed by R-R interval vs. systolic blood pressure plots following bolus injections of phenylephrine (243, 350). On the other hand, Billman et al. (24) reported that heart rate response to the rise in systolic arterial pressure induced by bolus injections of 4.0 µg/kg phenylephrine into rhesus monkeys was attenuated after 90° head-down tilt and after 20% blood volume expansion.

Limited data suggest that chronic changes in central blood volume may alter arterial baroreflex function. In another study with rhesus monkeys, Billman et al. (23)

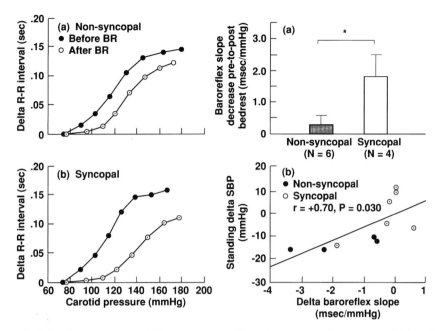

FIG. 39.8. *Left column:* mean carotid baroreceptor–cardiac reflex responses in nonsyncopal (N = 6) and syncopal (N = 4) men before and after 30 days of 6° head-down bed rest. *Right column:* mean (± SE) decreases in baroreflex slopes (pre– vs. post–bed rest) in the nonsyncopal and syncopal men; *$P < 0.05$ (*a*); and regression of change in standing systolic blood pressure on change in maximum baroreflex slope (N = 10) in *b*. [From Convertino et al. (58) with permission.]

found that the heart rate/systolic pressure response to a bolus phenylephrine infusion was attenuated after 7, 14, and 28 days of immobilization. Thus it is possible that the attenuated carotid sinus response reported after 12 days of bed rest in humans (58) may have been due in part to prolonged stimulation of the cardiopulmonary receptors.

Cardiopulmonary Baroreflex Function. Another unresolved question is the importance of possible changes in cardiopulmonary baroreflex function during bed rest. Acute increases in central blood volume induced by ingestion of isotonic fluid, infusion of an isooncotic protein solution, or movement to the 6° head-down position attenuate cardiopulmonary-induced peripheral vasoconstrictor responses during application of LBNP (248, 357). Mack et al. (248) found a significant inverse relationship between the gain of forearm peripheral vascular resistance and the level of blood volume. Thus the larger the blood or central cardiac volume, the smaller the gain of cardiopulmonary baroreceptor function. Pannier et al. (287), on the other hand, found no significant change in the forearm vasoconstrictor response to mild levels of LBNP (<20 mm Hg) after 24 h of 5° head-down bed rest. Because their baseline data were collected with the subjects in the supine position, the magnitude of the changes in central blood volumes between the supine and 5° head-down positions might

not have been sufficient to attenuate the cardiopulmonary reflex vasoconstrictor response.

During prolonged bed rest, cardiac volumes declined below their pre–bed rest supine levels. Thus recent findings suggest that the cardiopulmonary baroreflex setpoint is reset during bed rest to maintain normal upright central circulatory function in subjects maintained in the supine or head-down position (94). Hemodynamic responses to hyperhydration, induced by infusion of isotonic saline (22 ml·kg⁻¹), were compared in six men before and after 6 days of head-down bed rest. By day 6, both resting blood volume and cardiac output had decreased by 13%. Similar hemodynamic changes occurred in response to the volume expansion before and during bed rest. The resistance to volume expansion and rapid return to the preinfusion level during bed rest led to the conclusion that "a new set-point or operating point for intravascular volume control had been established" (94).

Chemoreceptor Function. Heart rate and vascular tone are also regulated by chemoreceptors. Two primary types of cardiac chemoreceptors are arterial receptors located in the carotid and aortic bodies which respond to changes in blood gases and pH, and diffuse cardiopulmonary vagal nerve endings which respond to blood-borne substances such as veratrum alkaloids, phenyl diguanide, and capsaicin (55). Both types of chemoreceptors produce profound cardioinhibition including bradycardia, decreased contractility, and possibly coronary vasodilation. However, these direct cardiac effects are often masked by cardioacceleration which accompanies the increased ventilation also provoked by chemoreceptor stimulation. Arterial chemoreceptor stimulation causes peripheral vasoconstriction, whereas vagal cardiopulmonary receptor stimulation induces peripheral vasodilation. There is a growing interest in investigating the chemoreceptor response of afferent fibers from the heart and great vessels that travel over sympathetic pathways to the spinal cord. These fibers are also stimulated by various blood-borne substances (veratridine, serotonin, bradykinin, and cyanide) and produce cardioacceleration and peripheral vasoconstriction. They may be stimulated during myocardial ischemia though local release of prostaglandins and bradykinin, which cause reflex increases in heart rate and contractility, and blood pressure (55).

The effect of an acute change from upright to the supine position on the chemical control of breathing has been evaluated for CO_2 (303, 382) and for hypoxia (336). Despite differences in breathing patterns and relative hypoventilation in the supine position, no significant differences in response to inhaled CO_2 or to hypoxia were evident.

The effect of prolonged bed rest on arterial chemoreceptor function has not been critically evaluated, but ventilatory reaction to inhaled CO_2 was apparently diminished because of depressed sensitivity of the respiratory center during 120 days of bed rest (377). The most profound effect on chemoreceptor function may occur immediately after bed rest. The carotid bodies have an exceptionally high metabolic rate and perhaps the highest blood flow of any tissue in the body. Blood vessels surrounding carotid bodies are richly innervated with sympathetic fibers and even a slight restriction of blood perfusion may result in stagnant hypoxia and chemoreceptor stimulation (163). Upon standing after bed rest, the potentiated sympathetic vasoconstrictor response might reduce carotid body perfusion and temporarily increase input from the arterial chemoreceptors.

Venous Compliance

Including the liver and spleen, approximately 70% of the circulating blood volume resides in the venous system (309). The activation state of venous capacitance vessels (tone) determines the distribution of cardiac output. Changes in venous tone also determine the level of postcapillary resistance and thus assist in regulating fluid movement across capillaries. Veins also function to regulate heat exchange between the body core and skin, and from skin to the ambient environment. There are three general types of veins in the body which differ

in morphology and function. Splanchnic veins, rich in smooth muscle and sympathetic innervation, are responsive to stimulation from arterial and cardiopulmonary baroreceptors (154). Like splanchnic veins, cutaneous veins are richly innervated but are not as sensitive to baroreceptor stimulation; they respond more to deep respiration and to emotional and thermal stimuli. Muscle veins contain relatively little smooth muscle, and the deep veins in particular have practically no sympathetic innervation (250). Muscle veins, relatively insensitive to arterial baroreceptor stimuli, are regulated almost exclusively by external pressure provided by the surrounding skeletal muscle. During standing, approximately 80%–90% of the venous volume in the legs is located in the deep muscle veins (41). Voluntary contraction of lower extremity muscles can eject approximately 30% of the previously dislocated blood headward (245), resulting in 15%–20% increases in stroke volume and cardiac output (302).

An immediate vascular response to assumption of the horizontal body position is transient vaso- and venodilation. Within the first 30 min of head-down tilt, forearm blood flow increases and forearm venous tone and vascular resistance decrease (243). Pannier et al. (287) found transient increases in both forearm and leg blood flows and decreased venous tone in both limbs during the first 24 h after 5° head-down tilt. Some investigators (95, 243) have reported transient decrease in total systemic vascular resistance at the onset of head-down tilt, while others (242, 361) found no significant change.

Loss of venous "tone" (reduced lower limb compliance) has been suggested as a prominant contributing mechanism to post–bed rest orthostatic intolerance (27, 40, 59, 244, 259, 325). During exposure to 50 mm Hg LBNP after 120 days of bed rest, the volume of blood shifted to the legs was 20% greater than that during pre–bed rest (Fig. 39.9) and was accompanied by marked increases (by 9%–10%) in blood outflow from the head and chest (325). Blood flow to the stomach during LBNP decreased by 3% before bed rest and increased by 3% after bed rest. These changes in blood flow during LBNP were ascribed to changes in venous tone, that is, increased compliance of lower limb vessels.

Time Course. After 4 h of 6° head-down recumbency, Butler et al. (43) found marked orthostatic intolerance in eight men in spite of no significant change in plasma volume and enhanced increase in total peripheral resistance. Only 30–60 min of recumbency are required to elicit the full response of the veins to head-down tilt. They suggested that venous return may have been compromised because of relaxation of venous smooth muscle, which then decreased baseline tone of the leg veins. In the upright position, leg veins are exposed to continuous hydrostatic pressure, activating a local sympa-

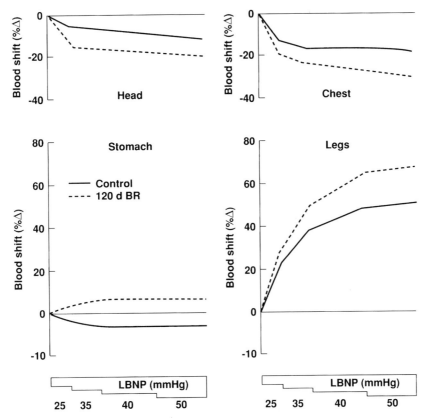

FIG. 39.9. Percent changes in blood volume shifts (^{133}Indium) with lower-body negative pressure (*LBNP*) during control and after 120 days of head-down bed rest. [From Savilov et al. (325) with permission.]

thetic reflex mechanism which increases arterial smooth muscle tone and, in turn, results in decreased muscle blood flow (162). Increased venous stretch also activates afferent neurons that synapse in the spinal cord and stimulate efferent motor neurons to increase local skeletal muscle tone (358). During head-down bed rest, venous pooling is virtually eliminated in the lower limbs, resulting in less local sympathetic nerve stimulation and a decline in local arterial vasoconstrictor and skeletal muscle tone.

Some investigators have reported increases in leg compliance during bed rest. Use of thigh occlusion (40, 59, 244) to induce changes in calf venous pooling have resulted in increased leg compliance after 4–20 days of head-down bed rest. Changes in leg compliance were assumed to be due, in part, to changes in venous compliance. On the other hand, others (119, 257, 259) have found no significant change in leg compliance after 4–14 days of bed rest.

Mechanism. Several factors could account for a reduction in lower limb venous tone during bed rest, including local relaxation of vascular smooth muscle tone (43), decreased perivenous tissue pressures, muscle atrophy–induced reduction of skeletal muscle tone (4, 40,

59, 244), attenuated reactivity to circulating vasoconstrictive substances, reduced cardiopulmonary or arterial baroreceptor sensitivity, or depressed autonomic nervous function. A predominant hypothesis for the increased venous distensibility after bed rest is the loss of mechanical support of the deep muscle veins from reduction of skeletal muscle tone and mass (41).

Pulmonary Function

Factors affecting pulmonary function when a person moves to a supine position include decreased lung volumes and increased airway resistance resulting from increases in intrathoracic and pulmonary blood volumes, upward shifts of abdominal contents and the diaphragm, and direct compression of airways by blood pooling in the lungs (266, 380). During the initial minute of recumbency, headward movement of blood stimulates cardiopulmonary baroreceptors, resulting in reduced peripheral vascular resistance and decreased circulatory catecholamine concentrations. This reduction in sympathetic activity may be accompanied by increased vagal tone (1), which could also contribute to acute increased airway resistance. On the other hand, some positive effects of recumbency include decreased physiological dead space (304) and improved ventilation–perfusion matching and lung diffusion capacity (281). Each of these positive factors should result in increased alveolar–capillary surface area and gas diffusion across the surface of the lungs, resulting in increased arterial P_{O_2} (P_{aO_2}) and a reduced alveolar-arterial O_2 difference (A–a)DO_2.

The (A–a)DO_2 in the supine position was lower than in the standing position (304), but forced vital capacity and peak expiratory flow rate were unchanged in ten men acutely tilted from the horizontal to a 5° head-down position (78). Nikolayenko et al. (280) found a 3.5% decrease in total lung capacity (TLC) when subjects were tilted from the horizontal to a 15° head-down position. TLC decreased by 11% after 7 h of head-down recumbency, remained at that level for 3 days, and then began to recover, reaching 6% below the pre–bed rest level after 7 days. The initial decline in TLC resulted from decreases in both residual volume and vital capacity; after the 7th h of recumbency, vital capacity began to recover while residual volume remained depressed.

With prolonged bed rest progressive changes in pulmonary function restore lung volumes but reduce gas exchange. Compared to supine pre–bed rest values, total lung capacity and forced vital capacity increased progressively during 11–12 days of horizontal bed rest (19), possibly compensating for the initial decline in lung volume when a subject becomes recumbent from an upright position. There was a tendency for diffusing capacity to decrease. When blood volume was reduced in ambulatory subjects with diuretics to a similar degree as during bed rest, Beckett et al. (19) concluded that the increase in lung volumes during bed rest was not due solely to the concurrent 21% decrease in plasma volume, but may have resulted from subtle changes in respiratory mechanics; for example, a shift in the anatomical relationship of the inspiratory muscles to the chest wall which improved their mechanical advantage.

Other investigators have reported no significant changes in pulmonary volumes during prolonged bed rest (30, 73, 261, 313). Deitrick et al. (73) found no significant changes in vital capacity, maximum ventilation capacity, and breath-hold time in men during 6 and 7 wk of horizontal bed rest with lower-body casting. During 70 days of horizontal bed rest, Mikhasev et al. (261) reported decreased resting ventilation, and no substantial changes in TLC or vital capacity. When measured upon first awakening with no subject movement, basal metabolism is unchanged during 30 days of bed rest (125).

Cardus (45) reported that arterial P_{aO_2} decreased from 103 mm Hg in the supine position before to 94 mm Hg after 10 days of bed rest in seven men with increased (A–a)CO_2 gradient. The decline in P_{aO_2} could have been due to pulmonary circulatory stasis or atelectasis of some parts of the lung which decreased the diffusion of gases across the alveolar capillary membrane. The decline in P_{aO_2} levels during prolonged head-down bed rest may also be due to expiratory closure of airways within the normal tidal volume, causing regional changes in alveolar ventilation and ventilation-perfusion matching which in turn results in larger venous admixture and lower arterial oxygen saturation (280). However, after several months of bed rest, adaptive changes may occur which restore pulmonary gas exchange. During the first half of 120 days of bed rest, P_{aCO_2} increased significantly (by 4 mm Hg) and P_{aO_2} decreased, suggesting impaired gas exchange (377). During the second half of bed rest arterial carbon dioxide tension decreased to below pre–bed rest levels, while arterial oxygen tension improved somewhat from 82 to 87 mm Hg. These findings were attributed to the altered gas exchange characteristics in response to changes in distribution of blood flow to lungs. On the other hand, Mikhasev et al. (261) reported that arterial oxygen saturation (oximetry) remained at 96%–98% throughout 70 days of bed rest.

Ventilatory volumes during exercise after 20 days of horizontal bed rest showed only minor variations, although the breathing pattern was modified (30). Respiratory rate was higher, tidal volume was lower, and maximal ventilation was reduced from 129 liters/min before to 99 liters/min after bed rest. This decrease in maximum ventilation was directly proportional to the reduction in maximum work capacity ($\dot{V}_{O_2}$ peak). Maximum respiratory frequency increased from 43 to 50 breaths/min, and maximal exercise tidal volume decreased from 2.9 liters/min (50% of vital capacity) to 2.1 liters/min (34% of vital capacity). The ventilatory coefficient ($\dot{V}e/\dot{V}_{O_2}$) during maximal exercise was unchanged.

Recovery

Orthostasis. Physiologic responses to head-up tilt, passive standing, +Gz acceleration, and LBNP are altered after short exposures to bed rest (43, 104, 132, 141, 255, 373). Vogt (373) found reduced head-up tilt tolerance (time to presyncope), greater increases in heart rate, and a faster decline in blood pressure after 12 h of recumbency. After 2 wk of bed rest the increase in heart rate upon standing was almost double the increase seen before bed rest, and the fall in stroke volume and cardiac output were about twice as great (159). These higher heart rates and lower stroke volumes for a given level of orthostatic stress are assumed to be indicative of a lower orthostatic tolerance (159, 172), but this hypothesis has not been proven satisfactorily.

Few investigators have actually measured orthostatic tolerance following bed rest. All subjects of Miller et al.

(264) exhibited significantly elevated heart rates during a 30 min 90° head-up tilt after bed rest, but only 42% actually exhibited presyncopal signs or symptoms. Fortney et al. (92) determined tolerance, assessed as the level of LBNP tolerated without symptoms of syncope, after 13 days of 6° head-down bed rest. Although all 10 subjects had elevated heart rates during LBNP after bed rest, only five had reduction in the level of LBNP pressure tolerated. Thus elevated heart rate is an appropriate compensatory response to gravitational stress, but it may not necessarily indicate reduced tolerance.

Multiple factors influence orthostatic function after bed rest, including decreased blood volume, increased capillary filtration, increased lower body compliance, decreased baroreceptor sensitivity, and altered autonomic function. The prime prerequisite for reduced gravitational tolerance following bed rest is a sustained change to the horizontal or head-down body position. Birkhead et al. (25) found that control responses to head-up tilting were maintained for three of four subjects allowed to sit quietly for 8 h/day during 24 days of horizontal bed rest; exercise training of sufficient intensity to maintain the maximal level of oxygen uptake ($\dot{V}_{O_{2max}}$) during bed rest failed to prevent impairment of those head-up tilt responses (25, 145).

Countermeasures that have maintained (restored) plasma volume during bed rest have not consistently maintained gravitational tolerance (92, 145, 342, 349), and there is no significant correlation between the hypovolemia and gravitational responses or tolerance (43, 52, 92, 145, 172, 181, 320, 348). Intravenous saline infusion, sufficient to restore central venous pressure to pre–bed rest levels, failed to restore orthostatic responses (31, 32); normalization of central venous pressure required an infusion volume approximately twice the blood volume loss. On the other hand, several investigators have reported improvement of orthostatic responses during bed rest in which plasma volume was restored (34, 147, 175, 343, 344). Although it is likely that bed rest–induced hypovolemia contributes to decrements in gravitational responses, other factors must also make significant contributions to the reduction in tolerance.

Dynamic (isotonic) exercise training during bed rest at a level sufficient to attenuate the loss of physical work capacity offers little or no protection against gravitational intolerance (25, 26, 48, 145, 320, 364), but there are contrary findings (104, 122, 187, 285). If inactivity accompanies bed rest, cardiovascular changes that contribute to loss of aeobic capacity (313, 314) most likely add to the hydrostatic-induced deconditioning effects, as each probably contributes to the loss of orthostatic function after bed rest. Most studies, where exercise training facilitated gravitational responses or tolerance, involved very long periods of bed rest. Resistive or iso-

metric exercise training programs designed to maintain lower-body skeletal muscle tone and mass during 14 days of bed rest can attenuate post–bed rest +Gz acceleration intolerance (132). Thus exercise training may improve gravitational function by maintaining or reestablishing lower-body muscle mass during bed-rest periods of sufficient length for such responses to be significant.

Exercise Responses and Capacity. Many cardiovascular adaptive responses during bed rest have significant impact on exercise capacity. Aerobic exercise capacity is defined as the maximal level of oxygen uptake ($\dot{V}_{O_{2max}}$) attained during exhaustive dynamic (isotonic) exercise that requires a large active muscle mass involving changes in the cardiac output ($\dot{Q}$) and arterial–venous (a–v) oxygen difference:

$$\dot{V}_{O_{2max}} = (\dot{Q}_{max}) \times max(a\text{–}v)_{O_2} \text{ difference.}$$

Most factors which enhance $\dot{V}_{O_{2max}}$ are affected negatively by bed rest (127); for example, hypovolemia, decreased red cell mass, decreased strength, muscle atrophy, and decreased arterial baroreceptor sensitivity (Fig. 39.10). On the other hand, factors such as increased beta$_1$-adrenergic sensitivity and increased ventilation–perfusion matching, might enhance $\dot{V}_{O_{2max}}$. Often, after bed rest the maximal oxygen uptake, as indicated by a plateau of the oxygen uptake curve with a further increase in exercise intensity, cannot be attained. Thus maximal aerobic capacity after bed rest is often reported as "$\dot{V}_{O_{2peak}}$," the highest oxygen uptake measured under existing circumstances, where the limiting factors may have been extreme muscle fatigue or substrate availability (133) rather than the cardiorespiratory limit.

After bed rest, at a given oxygen uptake, cardiovascular responses are exaggerated: there are increased heart rates, $\dot{V}_e/\dot{V}_{O_2}$ levels, blood lactate concentrations, diastolic blood pressures, and respiratory exchange ratios (30, 57, 60–64, 73, 133, 313, 349, 353, 389); attenuated increases in stroke volume resulting in a reduced maximal cardiac output (30, 172, 313, 349); reduced exercise endurance; and unchanged systolic blood pressure. Left ventricular ejection fraction is either increased (167) or unchanged (349) during exercise after bed rest.

$\dot{V}_{O_2}$ kinetics are slower in response to change in load during upright exercises after bed rest (61, 341). Before bed rest a "steady-state" level of oxygen consumption was attained in 3–5 min; after bed rest attaining the steady-state level took longer than 5 min. Once steady state had been reached, the level of oxygen consumption for a given exercise intensity was the same as the pre–bed rest value (57, 60, 313, 353).

Reduced $\dot{V}_{O_2peak}$ occurs after virtually all bed rest

$$\dot{V}O_{2max} = \dot{Q}_{max} \times (a{-}v)O_2 \text{ Difference}_{max}$$

HR	SV	Arterial O_2	Venous O_2
↓ Arterial baroreceptor sensitivity	↓ Blood volume	↓ RCM	↓ Muscle mass
↑ Symp/PS balance	↓ α-adrenergic VC	↓ $\dot{V}_{Emax}$	↓ Muscle blood flow
	↑ Leg compliance	↑ $\dot{V}A/\dot{Q}$	↓ Capillary density
	↑ Splanchnic compliance	↑ Pulmonary diffusion	↑ Tissue diffusion distance
	↓ Skeletal muscle tone		↓ O_2 extraction
	↓ Skeletal muscle mass		↓ Substrate delivery
	↓ Baroreceptor VC responses		↓ Substrate metabolism
	↑ Beta-adrenergic vasodilation		↓ Energy stores
			↓ Myoglobia
			↓ Mitochondria

FIG. 39.10. Adaptive responses of prolonged bed rest that influence maximal oxygen uptake.

studies conducted without remedial exercise training, and the percentage decline is similar in men and women (62, 323). Decrements in $\dot{V}O_2$peak occur rapidly during the first 3–6 days of bed rest, and then more gradually reaching approximately 22% by 30 days of bed rest. After 120 days of 4.5° head-down bed rest, subjects who did not exercise had a 37% decrease in $\dot{V}O_2$peak as measured during upright exercise on a treadmill (260). Changes in responses during submaximal exercise and the decline in $\dot{V}O_2$peak after bed rest were less pronounced when the exercise test was performed in the supine as compared to upright position (60). After 10 days of bed rest the decline in $\dot{V}O_2$peak in healthy, middle-aged men averaged 15% during upright exercise, compared to a 6% decline during supine exercise (63, 167). This suggests that the additional orthostatic stress during upright exercise adversely influences $\dot{V}O_2$peak levels.

Part of the decline in aerobic exercise capacity during bed rest is due to a "deconditioning" effect that periodic exercise training should attenuate or restore. The minimal intensity, duration, and the most appropriate mode of exercise training required to maintain $\dot{V}O_2$peak during bed rest have not been defined. Supine exercise appears to be less effective than upright exercise in preventing loss of exercise capacity during bed rest (25, 26). Birkhead et al. (25) reported that a daily 1 h upright (sitting) exercise session at about 55% of $\dot{V}O_2$peak was sufficient to maintain $\dot{V}O_2$peak during 24 days of horizontal bed rest, while 1 hr of comparable supine exercise was not as effective. These results suggest that reduction of the foot-to-head hydrostatic gradient is a critical factor contributing to deterioration of exercise tolerance during bed rest, and high exercise intensity may be required as an effective countermeasure. In general, only supine exercise-training programs employing intensities greater than 50% of the pre–bed rest $\dot{V}O_2$peak result in attenuating the decline in aerobic capacity. Greenleaf et al. (128) employed an alternating dynamic (isotonic) exercise-training program (5 min warm-up at 40% $\dot{V}O_2$peak followed by alternating 2 min intervals between 40% and 90% $\dot{V}O_2$peak) for two 30 min periods/day for 5 days/wk which maintained supine $\dot{V}O_2$peak during 30 days of 6° head-down bed rest. A similar but more complex alternating exercise program also maintained $\dot{V}O_2$peak during 49 days of bed rest (188). Reduced baroreceptor responsiveness, as assessed by heart rate and blood pressure responses during phenylephrine infusion, and reduced upright (sitting) peak exercise capacity both occurred after 3 wk of bed rest in nonexercised control subjects; neither occurred in subjects who performed sitting exercise training during bed rest (349). In addition, daily dobutamine (a synthetic catecholamine) infusions resulted in maintenance of exercise capacity during bed rest, suggesting that maintenance of sympathetic responsiveness may attenuate or eliminate the decline in exercise capacity. A single bout of supine peak exercise eliminated the reduction in upright $\dot{V}O_2$peak after 10 days of bed rest, which was attributed to acute restoration of plasma volume and cardiovascular reflexes by exercise stimuli (56).

Mechanism(s) for a decline in aerobic exercise capacity during prolonged bed rest involve changes in cardiovascular function, reduced active muscle mass, altered neuromuscular function, and perhaps impaired substrate utilization and availability. Orthostatic factors also contribute to the decline in $\dot{V}O_2$peak, especially in upright sitting exercise (60, 63, 135). During periods of bed rest of less than 10 days, important factors that contribute to reduced $\dot{V}O_{2peak}$ involve decreased cardiac filling exacerbated by reduced plasma volume. As bed rest continues, oxygen delivery to exercising tissues probably decreases as red cell mass declines at about 2 wk. Further reduction in venous return occurs, due to greater pooling of blood in the legs from decreased skeletal muscle tone and mass (at about 2 wk), and possibly to less forceful cardiac and baroreflex vasoconstrictor responses (after 10 days). Peripheral vasoconstrictor responses might also be compromised from depletion of presynaptic norepinephrine and enhanced beta-adrenergic function. Greater reliance on anaerobic metabolism may occur because of reduced muscle blood flow, altered muscle morphology, impaired cellular metabolism from dehydration, and reduced oxidative metabolism. Exercise responses and endurance may be further impaired due to less effective thermoregulation (88, 136).

Summary of Cardiopulmonary Adaptation to Bed Rest

The term *cardiovascular deconditioning* was coined by Keys (200) to describe deleterious effects of bed rest on cardiovascular function: mainly the decreases in exercise and orthostatic tolerance when assuming the upright posture immediately after bed rest. Another perspective is to view these so-called deleterious deconditioning changes as positive adaptive responses to the bed-rest environment.

Figure 39.11 illustrates several proposed mechanisms for the acute and long-term effects of bed-rest adaptation on cardiovascular function. The initial stimuli which trigger cardiovascular changes during bed rest are reduction in hydrostatic pressure and the ensuing cephalad redistribution of body fluids. The resulting increased thoracic fluid volume stimulates cardiopulmonary and arterial baroreceptors, which trigger neural and hormonal reflex responses, resulting in activation of the cardiovascular depressor reflex and fluid and electrolyte secretion in an attempt to restore cardiac volumes to their normal supine level and, with continued bed rest exposure, to their normal ambulatory level. Less understood are possible effects of increased intracranial pressure, from the headward fluid shifts, on cardiovascular and body fluid regulation. Decreased lower body perfusion results in local reduction in tissue interstitial pressures.

Adaptive responses occurring with prolonged bed rest include possible resetting of cardiopulmonary baroreceptors to accommodate the reduced blood volume (94), possible decreased sensitivity of arterial baroreceptors to change in mean arterial pressure (58, 87), and the probable resetting of sympathetic/parasympathetic

autonomic balance (72, 318, 354). Peripheral vascular compliance may increase (43, 325, 357), due to both local vascular changes and reduced reactivity to neural and humoral stimuli (23, 327). Long-term exposure to bed rest may result in extensively altered afferent sensory processing such that a virtual restructuring of reflex neural regulation of the circulation occurs (99).

With prolonged bed rest, especially with significantly reduced physical activity, concurrent changes in functioning of other organ systems may also affect cardiovascular function. Lack of mechanical stress or deformation on the skeletal system from reduced intensity of muscular contractions results in negative calcium and phosphorus balances, as described in the following section. Altered plasma concentrations of phosphorus and especially calcium can alter cardiac, smooth muscle, and nervous system excitability. Bed rest deconditioning is usually accompanied by altered emotional and psychological stimuli (214) which may contribute to changes in baseline autonomic function, such as reduced α-adrenergic and enhanced β-adrenergic activity, and a reduced or more sluggish response to stressors, which could also affect resting and stress-induced cardiovascular responses.

MUSCULOSKELETAL SYSTEM

Bone and Calcium Metabolism

As people age their level of activity usually decreases. This may occur because of a combination of changes related to diet, nature of their work, recreation habits, and the development of acute or chronic illnesses. Relative inactivity induced by prolonged bed rest has been used to investigate musculoskeletal changes from the unloading that may simulate those during travel in microgravity.

Normal Metabolism

The normal, healthy adult is generally considered to be in calcium balance; for example, calcium ingested is equal to the calcium lost. Approximately 400 mg of calcium are reabsorbed from bone each day, while the same amount is utilized for mineralization of new bone. The range of daily calcium ingestion in the adult ranges from 400 to 1,200 mg (277). Assuming an average calcium intake of 1,000 mg/day, approximately 30% (300 mg) is absorbed, mainly in the upper intestinal tract, under the influence of the hormone calcitriol (compounds having vitamin D activity); 140 mg of absorbed calcium is secreted into the intestine both directly and through the biliary tract of the liver, but 30% of that (42 mg) is reabsorbed, resulting in about 200 mg of

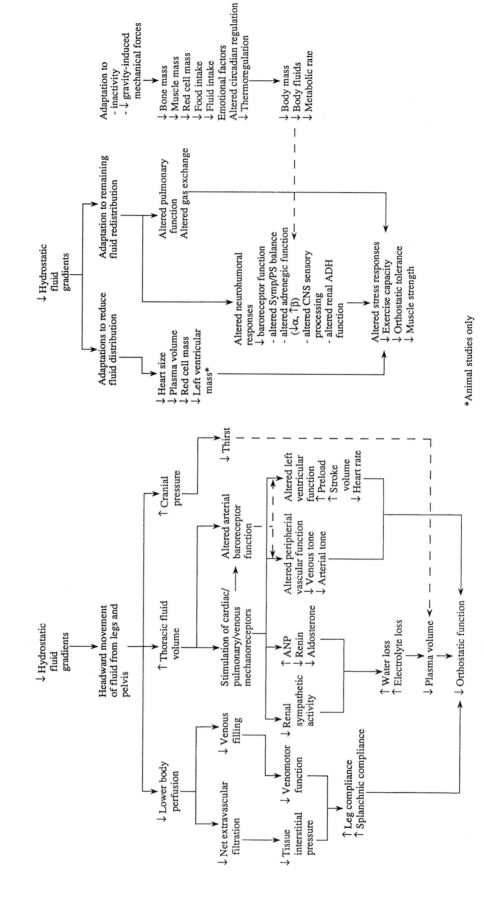

FIG. 39.11. Interaction of physiological responses to deconditioning: spaceflight, bed rest, or water immersion. [Revised from Sandler (318) with permission.]

calcium absorbed into the circulation. The remaining 800 mg are lost in the urine and feces. Parathyroid hormone, which stimulates renal cyclic AMP, is necessary for normal metabolism (turnover) of bone. The kidney filters 6 g of calcium/day. Under the influence of parathyroid hormone 5.8 g are reabsorbed, leaving 200 mg to be excreted in the urine. Thus, about 600 mg/day are lost in the feces. Parathyroid hormone also induces synthesis of calcitriol by the kidney.

The predominate metabolic activity of the adult skeleton is bone turnover or remodeling. Bone remodeling occurs in two phases; a resorption period which lasts 4–6 wk, and a formation period which lasts for 8–12 wk. Since bone resorption and formation periods are coupled, resorption must preceed formation. Osteoclasts, which resorb bone, are derived from hematopoietic mononuclear stems cells in the bone marrow. The colony-forming unit in the marrow for the granulocyte macrophage is activated by a combination of permissive systemic hormones (parathyroid hormone and calcitriol) and local factors (including prostaglandins and cytokines) to develop mature bone-resorbing cells (270). Osteoblasts, which form bone, are differentiated from mesenchymal stem cells (184, 301). Mature osteoblasts produce predominantly type 1 collagen which is secreted into the resorption cavity produced by the osteoclast. The collagen is then mineralized with crystals of hydroxyapatite $[Ca_{10}(PO_4)_6(OH)_2]$. This mineralization may occur either spontaneously or with the aid of the osteoblast. Once mineralization begins, the osteoblast produces an unknown substance which controls the rate and quantity of mineral produced.

Calcium. Subjects in a 17 wk bed-rest study (81) remained in calcium balance during the ambulatory control phase, but during bed rest there were increases in both urinary and fecal calcium excretion, causing a negative balance. Calcium is also lost in shedded skin and sweat; if only urine and fecal calcium are measured, the measured calcium balance will be incorrect. The average change in calcium balance, including skin and sweat losses of 20 mg/day, after 17 wk of bed rest was 179 ± 15 mg/day; the cumulative calcium loss was 21,301 mg, or approximately 1.7% of total body calcium stores (81). After the first mo of bed rest the rate of calcium loss is about 0.5%/mo (329).

Urinary calcium daily output is constant during the ambulatory control period when measured in individuals consuming a fixed diet while living in a metabolic ward. Urinary output is proportional to calcium ingested (236), and is modified by the quantity of meat intake (249). Urinary calcium increases daily over the first 1–2 wk of bed rest, continues to increase more slowly over the following 2–3 wk, and plateaus in the 5th or 6th wk of bed rest (81, 329). The initial rise in

urinary calcium may accompany increase in urinary sodium over the first 24–72 h of bed rest. However, after urine sodium decreases toward normal levels, urinary calcium continues to increase. Deitrick et al. (73) showed that urinary calcium did not increase until the third bed rest day when the predominant saluresis should have subsided.

There is no change in net calcium absorption from the gastrointestinal tract over the first 5 wk of bed rest. Calcium absorption declines during the 6th to 8th wk and then plateaus for the remainder of bed rest in spite of slower gastrointestinal transit time (9). Calcium absorption, measured by calcium[47] kinetics, was unchanged at 5 wk of bed rest, but calcium secretion increased at both 5 and 18 wk of bed rest (241).

Urinary calcium, however, reflects not only bone turnover rate but also dietary calcium intake. As bone turnover increases, there is a rise in serum calcium concentration which suppresses parathyroid hormone secretion which, in turn, suppresses calcitriol levels. Bone turnover is dependent on skeletal maturity, that is, whether longitudinal growth is still occurring (modeling phase), or whether there is normal activity and repair (remodeling) in the adult. Therefore, people with growing bones with high turnover would be expected to excrete higher levels of calcium during bed rest, as has been reported in patients with fractures and paralysis (73, 165, 329, 392). The increase in urinary calcium during bed rest depends on the degree of pre–bed rest trauma or the level of inactivity, and the initial bone turnover rate of the subject. Thus nerve damage–induced paralysis results in the greatest urinary calcium loss, and bed rest with "normal activity" in the horizontal plane the least (329, 392). Intermediate losses occur in patients who were immobilized from trauma (165) or in subjects total-body casted [Fig. 39.12; (73)].

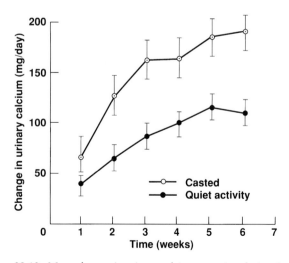

FIG. 39.12. Mean changes in urinary calcium excretion during 6 wk of bed rest in control (quiet activity) and body-casted subjects.

Fecal calcium may decrease during the first 2 wk of bed rest, suggesting increased gastrointestinal absorption. Thereafter, fecal calcium increases for 6–8 wk and remains at that level throughout 36 wk of bed rest (81). Fecal calcium is derived from dietary intake and from secretion into the intestines. Secreted calcium is derived from dietary calcium and from skeletal turnover. The mechanisms responsible for control of secreted calcium are not fully defined, but resorption is under the same control as dietary calcium ingestion. Nonhumoral factors which may decrease calcium absorption include decreased gastric acidity, increased gastrointestinal motility, and increased intake of fiber and insoluble calcium salts. Decrease in parathyroid hormone concentration and subsequent suppression of calcitriol will decrease calcium absorption. The increase in fecal calcium content during bed rest reflects a decrease in absorption of calcium (both ingested and secreted), and may also occur secondary to a predominately increased calcium secretion.

Kidney stones can occur in up to 25% of patients immobilized with trauma or paralysis (73, 292). Urine is normally supersaturated with stone-forming elements (283) and prolonged bed rest may accentuate development of kidney stones; not only because of the increased urinary calcium and phosphorus levels, but also because of increased concentrations of uric acid and oxalate perhaps due to hypovolemia, and decreased concentrations of stone inhibitors, for example, citrate. Hwang et al. (169) reported increased urinary pH, an 8% increase in uric acid concentration, and increased oxalate excretion during prolonged bed rest; but urinary volume and ammonium and sulfate concentrations were unchanged. In the normal individual at bed rest, there is a clinically significant risk of renal stone formation.

Phosphorus and Other Minerals. Phosphorus, the other major component of the bone crystal, is lost during bed rest in direct proportion to the loss of calcium (81, 166, 241, 329). Cutaneous phosphorus loss was undetectable (81). Mean urinary phosphorus rose during bed rest by 94 ± 6 mg/day, and phosphorus balance decreased by 108 ± 10 mg/day.

Bed-rested subjects also exhibited negative magnesium balance, and zinc balance decreased by 44.1 mg after 5 wk of bed rest; copper balance was unchanged (218).

Increases in urinary phosphorus and in trace minerals like magnesium and zinc that are stored in bone, rise in proportion to the amount of bone loss (73, 81, 169, 218, 329). In like manner, bone collagen fractions, measured as urinary hydroxyproline and hydroxylysine, increase in molar ratio to the rise in urinary calcium during bed rest (9). The elevation of mineral excretion in urine continues until a new steady state for bone turn-

over occurs. Urinary calcium does not return to ambulatory baseline levels until reambulation, since healthy subjects bed-rested for 36 wk showed elevated urinary calcium and phosphorus throughout bed rest (329). In paraplegics, urinary mineral excretion may remain elevated for 6 to 18 mo (384).

Bone Matrix (Collagen). Bone atrophy reflects a loss of both bone mineral and bone matrix, and is the initial stage of bone remodeling. Bone is composed of a collagen matrix, secreted by osteoblasts, in which hydroxyapatite crystallizes to make mature bone. In steady-state bone turnover, loss of urinary hydroxyproline is thought to reflect the loss of bone collagen. The hydroxyproline content of collagen occurs only by hydroxylation of proline in established collagen fibrils, and is not incorporated directly into the polypeptide chains of collagen during cellular protein synthesis (307). Thus collagen proline, not exogenous hydroxyproline, is the precursor of hydroxyproline in collagen degradation products. If diet is controlled rigidly in metabolic or bed-rest studies, change in urinary hydroxyproline should eminate from collagen degradation. Collagen is found not only in bone, but also in skin, tendons, ligaments, joint cartilage, intervertebral discs, and basement membranes; but during bed rest bone should be the major contributor for increased collagen turnover. Urinary hydroxyproline increases proportionally with the increase in urinary calcium (Fig. 39.13), indicating that the increased urinary calcium does not come from dietary calcium but from increased bone loss.

Blood variables. Serum total and ionized calcium, and phosphorus levels (Fig. 39.14) do not change substantially during bed rest (9, 81, 329); however, clinical hypercalcemia has been reported (10, 346). Hypercal-

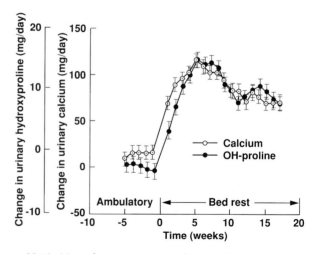

FIG. 39.13. Mean changes in urinary calcium and hydroxyproline in the ambulatory control period and during 17 wk of bed rest.

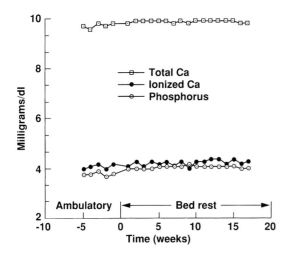

FIG. 39.14. Mean serum total calcium, ionized calcium, and phosphorus concentrations in the ambulatory control period and during 17 wk of bed rest. N = 30.

cemic individuals may be at greater risk for developing kidney stones, or for induction of metastatic calcification in eye lens or in the vascular bed. Increased serum calcium concentration will cause a more rapid decline in parathyroid hormone and calcitriol levels (10, 346).

Bone Densitometry. Bone densitometry measurement precision and accuracy have improved over the last several years with the use of dual energy from X rays to determine the proportion of bone, soft tissue, and fat in the body. Measurement of bone density of the whole skeleton, regional areas of the skeleton from the whole-body scan, and site-specific scans (spine, hip) can be performed quickly and with relatively low radiation exposure (231, 253, 254). Data obtained from these scans indicate either total calcium or bone mineral density. The latter is not true density, but density area, since the skeletal scans are performed in only two dimensions. Measurement of whole-body calcium can be used to determine calcium balance, eliminating the need for costly dietary, urine, and fecal calcium analyses.

Whole-body densitometry and calcium data from six subjects, measured repeatedly during 17 wk of horizontal bed rest and 6 mo of ambulatory recovery [Table 39.3; (234)], indicate that total body calcium changes were almost identical to those in the balance studies reported above. Mean changes in total calcium balance and total bone density from pre– to post–bed rest were −1.7% ± 0.8% and −1.4% ± 0.8%, respectively. Skeletal calcium is redistributed during bed rest (234) such that the greatest loss occurs in the calcaneus, followed by the hip. No significant change occurs in the upper extremities, but calcium content of the head apparently increases. It is unknown if calcium content increases in the soft tissues, in the cranium, or both. The mechanism

for this calcium shift has not been determined; it may be related to structural differences of the skull and scalp with their rich blood supplies (when compared to the long bones), or to the headward shift of fluid and decreased blood and tissue fluid pressures or flows in the head with assumption of the horizontal body position during bed rest.

The rate of skeletal recovery after bed rest is more important than the rate of loss during bed rest. Even though the subjects had completed 6 mo of normal activity after bed rest, total body calcium had not returned to pre–bed rest levels (Table 31.3). There appeared to be little or no recovery of bone mineral in the spine and hip, but there was nearly complete recovery in the calcaneus. Skeletal recovery probably continues beyond 6 mo; however, the residual deficit may never be replaced. This possible long-term bone deficit may mimic bone loss with aging; that is, aging bone loss could be due to disuse from decreasing activity. Perhaps intensive strength training will be necessary for rebuilding residual bone loss after prolonged bed rest.

Bone Loss and Muscle Atrophy

The precise relationship between muscle size and strength and bone size and strength is unclear. In general, the size of bone is proportional to the force applied to it, either directly through impact or muscle contraction (363) as when comparing normal sedentary individuals with athletes in training (269, 333). This is apparent in tennis players, whose dominant arm has both greater muscle and bone mass compared to the nondominant arm (183); and in partially paralyzed people, whose muscle and bone atrophy occurs in the non-innervated area compared to the normally functioning area (96).

Statistically significant lower leg muscle area and strength decrements occur after 5 wk of bed rest (115, 232) in plantarflexor area by −12.5%, and in strength by −26.8%; while dorsiflexor area (−4.0%) and

TABLE 39.3. *Mean (± SD) Percent Changes from Ambulatory Control of Bone Mineral from Total Body and Other Sites After 17 wk Bed Rest and Following 6 mo Reambulation*

Site	Bed Rest	Reambulation
Total body calcium	−1.4 ± 0.5*	−1.4 ± 0.8
Head	+3.2 ± 0.8*	+4.2 ± 1.0
Proximal radius	+1.0 ± 0.5	0.0 ± 0.8
Lumbar spine	−3.9 ± 0.7*	−3.2 ± 0.8
Femoral neck	−3.7 ± 0.8*	−3.6 ± 1.6
Femoral trochanter	−4.6 ± 0.8*	−3.4 ± 1.3
Calcaneus	−10.4 ± 1.7*	−1.8 ± 2.3*

*$P < 0.05$

strength (-0.8%) were unchanged. The plantarflexor muscles, predominately the soleus and gastrocnemius, insert into the calcaneus via the calcaneal tendon. The greatest bone loss during bed rest occurs in the calcaneus [Table 39.3, (234)] because of a marked decrease in the force (tone and contraction) of plantarflexor muscles through the calcaneal tendon, or change in force on the calcaneus directly from standing, walking, or running. There is loss of lean body mass, measured by dual photon absorptiometry (DPA), during 17 wk of bed rest (233). Most of the decrease was in the legs, with minimal changes in the trunk and arms. Although the absolute loss in lean mass in the upper leg was twice that in the lower leg, both had the same percent loss. Back and leg muscles, measured by magnetic resonance imaging, showed significant atrophy in the early weeks of bed rest. After 17 wk of bed rest, muscle volume was unchanged in the psoas, but exhibited losses of 9% in the intrinsic lower back group, 16%–18% in the quadricep and hamstring groups, 21% in the ankle flexors, and 30% in the ankle extensors. Muscle strength, measured with isokinetic dynamometry, followed the decreases in muscle volume or lean body mass.

Negative nitrogen balance during bed rest suggested that lean body mass is lost continuously throughout bed rest (73). Urinary 3-methylhistidine, a degradation product of muscle metabolism, was not significantly elevated during bed rest, suggesting that it is not a reliable marker of muscle atrophy. Bed rest–induced decreases in lean body mass and nitrogen balance return toward control levels during reambulation; after 8 wk, muscle volume and muscle strength had returned to pre–bed rest values, and nitrogen balance increased to 70% of control. Since measurements of changes in lean body mass and nitrogen balance gave similar results, they appear to be more reliable indicators of muscle atrophy.

Body Composition

In studies where caloric intake was maintained at ambulatory control levels throughout bed rest, body weight was unchanged (73, 81, 129, 144, 155, 166, 217, 218, 234, 235, 241, 329). If body weight is unchanged and lean body (muscle) mass is lost (115, 129, 232), percentage of fat mass must increase. Change in body composition was determined with hydrostatic weighing, potassium[40] counting, skinfold measurements, and nitrogen balance on six subjects whose mean body wt was $75.0 \pm$ SD 5.5 kg before, and 75.0 ± 4.9 kg after 5 wk of bed rest (217). Reductions in lean body mass, measured with potassium[40] and nitrogen balance, were 2.25 ± 1.88 kg ($P < 0.05$) and 1.82 ± 0.33 kg ($P < 0.05$), respectively; thus total body fat increased by 2.70 ± 1.78 kg ($P < 0.05$) and 2.27 ± 1.00 kg ($P < 0.05$), respectively. Nonsignificant trends toward reduction in

lean body mass and increase in body fat were found with hydrostatic weighing and skinfold measurements.

The expected decrease in activity in these bed-rested subjects would have resulted in a weight gain of about 4.3 kg. Since fecal energy (by bomb calorimetry) was not different, the lack of expected weight and fat gain was not explained by change in energy absorption. Since caloric (energy) intake was essentially constant, the mechanism responsible for maintaining body weight during bed rest was a comparable energy output; that is, energy utilization by unrestrained, healthy subjects during prolonged bed rest is similar to that of moderately active ambulatory people.

Conclusions

The mechanism for bone atrophy during bed rest is unknown. Changes in physiological responses to parathyroid hormone, calcitriol, and calcitonin activity probably occur in response to decreased forces on bones during bed rest, and therefore are permissive rather than causal. Recent identification of autocrine and paracrine factors including the prostaglandins, cytokines, and other growth factors (301), have suggested that the mechanism of bone loss involves a change in mechanical stress which, in turn, activates local factors to induce subsequent bone loss. The mechanical changes may be associated with a decrease or absence of external pressure transmitted to the bone directly, reduction or absence of muscle tension on the bone, and changes in blood pressure or blood flow within or through the bone.

ENERGY METABOLISM AND THERMOREGULATION

Energy Metabolism

One major response to prolonged sedentary bed rest is reduction of total daily energy production (metabolism) which contributes to the deconditioning syndrome. Some factors influencing energy metabolism and energy balance during bed rest are caloric content and composition of the diet; degree of physical restraint (confinement); level of exercise metabolism, which contributes to efficiency of oxygen delivery to tissues; reduction of hydrostatic pressure which, via probable changes in blood flow and vascular dynamics, might modify oxidative metabolism (133); and the level of psychological stress, which could alter digestive and gastrointestinal function, thus altering food and fluid preferences and intake patterns (137, 237).

Basal Metabolic Rate (BMR).   This is defined as the lowest level of metabolism in a living human, and is measured upon waking under quiet conditions before the subject arises. The stability of the BMR during bed rest is questionable. Some have reported that BMR decreases by 2%–22% during 7–70 day periods of sedentary bed rest (70, 73, 187, 198, 261, 353, 366), while others (39, 340, 383) found no change over 14–60 days of bed rest. If the subject's movement is restricted, decreased metabolism may occur. Deitrick et al. (73) reported a 6.9% reduction in BMR in four subjects resting horizontally in bivalved plaster casts. Since 6.9% of 0.25 liter/min is 0.02 liter/m, within the error of measurement, this should be considered as unchanged BMR. Most data on BMR were obtained from sedentary subjects. Exercise training was performed in two (198, 261) of the six studies where BMR was reported decreased, and in two (39, 348) of three studies with unchanged BMR. Stremel et al. (348) found no change in BMR (range 0.23–0.27 liter/min), measured under well-controlled conditions, in sedentary or exercising subjects during 14 days of horizontal bed rest. In the six studies where sedentary subjects' BMR was reported decreased, diet was controlled in three; BMR varied from −3% to −22%, similar to that reported when diet was not controlled. Thus the degree of dietary control or amount of exercise training appear to have no significant effect on the basal metabolic rate during prolonged bed rest.

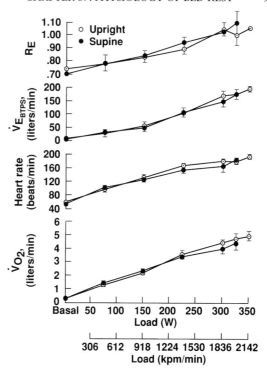

FIG. 39.15. Mean (± SE) cardiorespiratory responses in five men during submaximal and peak loads in sitting and supine positions. Data at 350 W from only two subjects. [From Greenleaf et al. (143) with permission.]

Submaximal Exercise Metabolism

Submaximal exercise covers the range from the resting level to 99% of the maximal oxygen uptake, so any meaningful discussion must specify the exercise load. Submaximal oxygen uptake has been reported to be either unchanged (18, 57, 60, 61, 64, 313) or decreased (63, 64, 177, 348); there appear to be no reports of increased metabolism. These inconsistent findings could have been the results of altered cardiorespiratory function in the various postures (standing, sitting, supine) used for testing, depending upon exercise intensity. In ambulatory subjects there were no differences in the linear increases of respiratory exchange ratio, ventilation, heart rate, or oxygen uptake during continuous, submaximal exercise in the sitting or supine positions on a cycle ergometer to about 225 W (Fig. 39.15). At higher submaximal and maximal loads these responses exhibit greater variability (142, 143).

After 14–17 days of sedentary (no exercise) bed rest, submaximal $\dot{V}O_2$, during leg exercise in the supine position at 55%–60% of peak $\dot{V}O_2$, was unchanged in 15 men but significantly lower by 12% in eight women (64). The lower oxygen consumption in the women was attributed to their lower blood O_2–carrying capacity associated with lower hematocrit and red blood cell concentration. With similar pre– and post–bed rest (10 days) submaximal metabolic rates, post–bed rest heart rates and respiratory exchange ratios were increased sig-

nificantly in both the men and women (62), indicative of greater stress after bed rest. Essentially similar results were found in seven men subjected to 14 days of horizontal bed rest without and with two modes (static and dynamic) of leg exercise training (Fig. 39.16). Below 200 W, oxygen uptakes were similar before and after bed rest; but above 200 W to the peak level, heart rates were increased significantly with all three regimens, and oxygen uptakes were decreased significantly only with the dynamic and no exercise regimens. Static exercise training maintained oxygen uptake but not exercise heart rate.

These accentuated metabolic and cardiorespiratory responses after bed rest could be caused by increased glycolysis leading to metabolic acidosis from enhanced blood lactate (306, 313), decreased oxygen delivery from possibly compromised heart function (313) and decreased peripheral blood flow, skeletal muscle atrophy with reduced capillarization, and loss of plasma volume and erythrocyte mass (31, 46, 129, 264).

Peak Exercise Metabolism

Aerobic exercise capacity ($\dot{V}O_2$ peak) can be reduced by 31% in normal, sedentary bed-rested subjects (128), and probably further in prolonged bedridden patients. Coupled with the potential increase of $\dot{V}O_2$ peak with exercise training of 44% in normal, ambulatory subjects (164), the range of induced

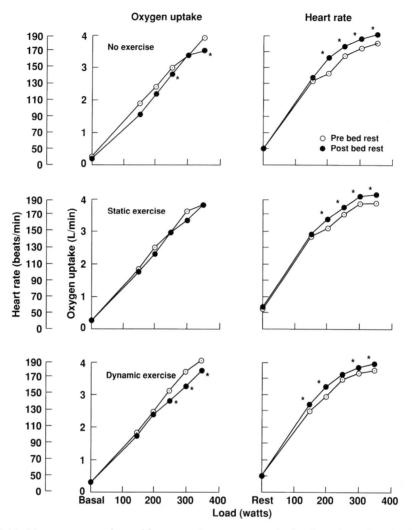

FIG. 39.16. Mean oxygen uptakes and heart rates in seven men under basal, resting, submaximal, and peak exercise loads for three regimens during 14 days of horizontal bed rest. *$P < 0.05$ from corresponding pre–bed rest value. [From Stremel et al. (348) with permission.]

change in aerobic capacity approaches 80%. A summary equation relating percent change in peak oxygen uptake during sedentary bed-rest periods less than 30 days is:

$$\% \ \Delta \dot{V}o_2 \ peak = -0.82 \times (BR \ day) \\ + 0.10; \ (r = -0.77, P < 0.02)$$

Thus peak $\dot{V}o_2$ (liters/min) decreases by 0.8%/day of rest (125). This is probably not a linear curve because peak $\dot{V}o_2$ would reach zero (100% loss) after 122 days.

Thermal Regulation

Dynamic exercise training in ambulatory subjects in thermoneutral and hot environments (heat acclimation) results in adaptive cardiorespiratory and thermoregulatory responses, including reduced resting and exercis-

ing heart rate, blood pressure, and rate of rise of core temperature; and increased endurance, sweating, and muscle and skin blood flows. There appear to be no significant changes in basal or resting core temperatures in ambulatory, exercise-trained subjects.

In reverse fashion, some body temperature and thermoregulatory responses may be altered during bed-rest deconditioning. Basal oral temperatures (T_{or}) were unchanged during 14 day horizontal (130, 141) and 30 day (128) head-down bed-rest periods, but basal auditory canal temperature (T_{ac}) decreased by 0.05°–0.7°C ($P < 0.05$) after 56 days of horizontal bed rest (391). Exercise training during 30 days of 6° head-down bed rest had no effect on mean ($\pm$ SE) basal T_{or} in five nonexercising men (36.31° $\pm$ 0.02°C), in seven men who had performed heavy dynamic leg exercise training (36.33° $\pm$ 0.01°C), or in seven other men who utilized heavy isokinetic leg exercise training (36.30° $\pm$ 0.02°C) for 1 h/d (128). Resting core temperatures after 12–14 days of bed rest were unchanged in men (136) or increased in women (88), while "excessive" increases in rectal [Fig. 39.17; (136)] and esophageal temperatures (88) during moderate submaximal exercise have been attributed mainly to reduced tissue conductance heat loss resulting from enhanced peripheral vasoconstriction associated with probable reduced sweating and evaporative heat loss, but independent of the bed rest–induced reduction in plasma volume. Dogs confined closely for 8 wk in cages, where they could stand but could not turn around, had no change in resting rectal temperature (T_{re}) (276, 293).

Body temperature circadian rhythm occurs during bed rest (211, 239, 285, 370, 391). Changes in T_{or} rhythms occurred in three men during 7 days of 8° head-down bed rest (239). Oral temperature increased by 0.22°C at the minimal point at night and decreased by 0.20°C at the maximal daily point; that is, a flattening of the temperature oscillations. After 56 days of horizontal bed rest Winget et al. (391) found no change in the circadian wave form, but a 0.5°–0.7°C depression of morning T_{ac} in eight men that did not return to pre–bed rest levels. Daily range of T_{ac} was about 1.0°C. There was a tendency for T_{ac} rhythms to become desynchronized during bed rest, where the nonexercise group

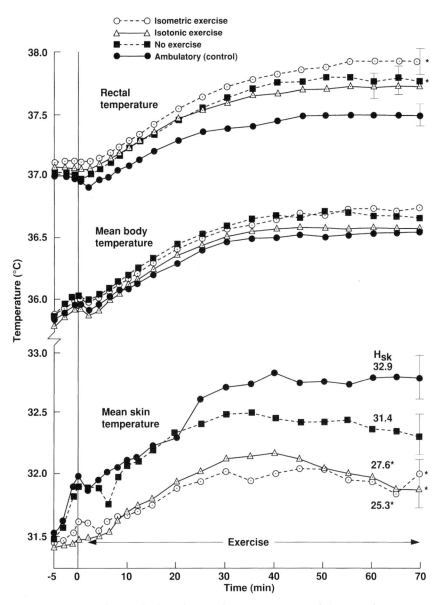

FIG. 39.17. Average rectal, mean body and mean skin temperatures, and tissue conductances (H_{sk}) in seven men during supine submaximal (43%–48% $\dot{V}_{O2}$ peak) leg exercise during ambulatory control, and after 14 days of horizontal bed rest with no exercise, and isotonic (dynamic) and isometric exercise training. [From Greenleaf and Reese (136) with permission.]

had less stable circadian oscillations than the exercise groups (isotonic and isometric exercise at about 800 kcal/day). The desynchronosis occurred in spite of well-regulated light–dark cycles, ambient temperature and humidity, meal times and composition, and exercise level (370).

There are reports of altered thermal sensations during bed rest (224, 285, 286, 353, 360). Nonexercising subjects felt cooler than normal, especially in the legs, after 2 mo of bed rest (353). The time for basal T_{re}, elevated by immersing an arm into 45°C water for 30 min, to return to control level was increased ($P < 0.05$) from 52 min ($\delta T_{re} = 0.7°C$) on the first day to about 80 min ($\delta t_{re} = 0.2°$–$0.3°C$) on day 90 of bed rest (360). thus more time was required to reduce a much smaller heat load. krupina and tizul's (224) nine subjects reported a feeling of heat and a rush of blood to the head during the first 15–20 min of being tilted 4°–6° head down during a 49 day bed-rest period. this sensation decreased after 1.5 h. most subjects reported cold feet by the end of the second day, and this was their most prominant complaint after the sixth day. the 3°–7°C drop in distal limb skin temperatures was attributed to unstable autonomic regulation, although hypovolemia would also contribute to reduced peripheral blood flow. the skin temperature threshold for sensing locally applied heat was unchanged during 120 days of bed rest, but the threshold for sensing cold decreased by 1.1°C (285). in addition, after bed rest the core temperature had to increase beyond 37.5° or decrease below 36.0°C ($\delta = 1.5°C$) before subjects reported discomfort. thus there was an increase in the thermal comfort zone during bed rest adaptation. one subject even reported feeling comfortable with a core temperature of 38.8°C, a pulse rate of 102 beats/min, and profuse sweating. in hot environments, normal ambulatory people usually feel more comfortable when exercising moderately than resting. tizul (360) reported similar thermal sensations and decreases in skin temperature in 10 subjects during 120 days of bed rest. but after the second month there was equalization of zonal differences in skin temperatures with elevation on the distal extremities resulting in less difference between oral–caudal and proximal–distal locations. this peripheral temperature equalization could be part of the adaptation process. fortney et al. (91) and greenleaf and reese (136) have suggested that the bed-rest–induced hypovolemia may result in reduced sweating sensitivity or increased core temperature threshold for sweating. after 14 days of bed rest, seven men had a greater rise in rectal temperature during exercise (compared with ambulatory responses) but similar rates of sweating, hence decreased sweating sensitivity after bed rest (136). Fortney et al. (91) also found neither significant change in total body sweat rates nor change in core temperature thresholds for

sweating or sweating sensitivity in 11 women after 13 days of bed rest. On the other hand, Williams and Reese (388) observed that after 14 days of bed rest in resting men exposed to 41°C ambient temperature, the time to onset of calf sweating (sensitivity) was increased significantly and occurred at a lower mean skin temperature when compared with pre–bed rest control responses. Thus, in spite of bed rest–induced body hypohydration and hypovolemia, which would tend to reduce sweating and sweating sensitivity, the latter was increased in men after bed-rest deconditioning, but not in women. Sweating responses after deconditioning may be specific to the mode of heat-load induction, as in an external ambient heat load vs. an exercise-induced internal heat load.

IMMUNE CELLULAR AND HUMORAL PARAMETERS

Bed Rest < 30 Days

The general conclusion from bed-rest studies of less than 8 days is that bed-rest deconditioning without remedial countermeasures results in unchanged or depressed immune function. The unchanged parameters were usually humoral: immunoglobulins (IgA, IgG, IgM) from mixed venous blood (226, 227, 268)—especially IgG and IgM (226, 268), as well as bactericidal activity of blood serum. Mukhina et al. (268), on the other hand, reported decreased IgA and IgG in blood coming from the brain, liver, and lower limbs. Kut'kova et al. (226) found no change in the number of β lymphocytes.

In spite of no changes in the various lymphocyte counts, lymphocyte responsiveness to Concanavalin A (Con A) was decreased in six men before, during, and at the end of 10 days of −6° head-down bed rest (113). Plasma cortisol levels were elevated before and during bed rest, suggesting that the early reduced lymphocyte responsiveness was the result of psychological stress. Of the many additional immunological variables measured, only natural killer cells decreased transiently at the end of bed rest. Since the subjects' stress levels seemed to abate during the last few days of bed rest, it was suggested that the reduced lymphocyte responsiveness was a response to the stress of cephalic fluid shifts and subsequent reduction in total body water. But hypovolemia reaches near equilibrium by day 4 of bed rest (130), so stress from fluid shifts should have passed, but complete adaptation to the hydration may not have occurred.

There were decreases (significance unspecified) in the phagocytic activity of neutrophils (PAN), T lymphocytes, and the lysozyme titer in saliva (LTS) (215, 226, 227). Increases in spontaneous agglutination of leukocytes, activity of beta-lysines, and increased skin microflora (215, 226, 227) also indicate decreased immune function.

On the other hand, Criswell and Kimzey (69) investigated two groups of six men, with one group bed-rested for 14 and the other for 28 days, and found no change in total lymphocytes, T cells, β cells, or in non-reactive lymphocytes (total minus T + β cells). But they reported significant ($P < 0.05$) *increases* in lymphocyte responsiveness to phytohemagglutinin (PHA), poke-weed mitogen (PWM), and Con A after 28 days of bed rest.

Bed Rest > 30 Days

There have been at least five Russian bed-rest studies from 60 to 182 days where immune parameters were measured. The two earlier studies of 62 days (185, 262) were followed by three of 70–105 days (53, 54, 338, 339); all utilized exercise training in some subjects during bed rest. In nonexercised subjects there were

decreases (significance level unspecified) in blood pro-perdin (a component of the alternative complement pathway), LTS, and skin bactericidal activity (185, 262), as well as decreased PAN (53, 54, 185, 262, 338). Exercise training during bed rest attenuated these adverse responses and restored skin bactericidal activity, leukocyte phagocytic activity, and PAN to control levels. Some subjects developed inflammatory ailments, which included increased pathogenic activity of staphylococcus.

In a more recent 120 day study without exercise training, there were no changes in T cells and natural killer cell activity (NKA) (37), nor in T lymphocytes and T-helper cells (208); but an increase in T-suppressor cells and decreases in β lymphocytes and PHA activity occurred (208). Natural killer cell activity was un-changed during bed rest in seven subjects (Fig. 39.18,

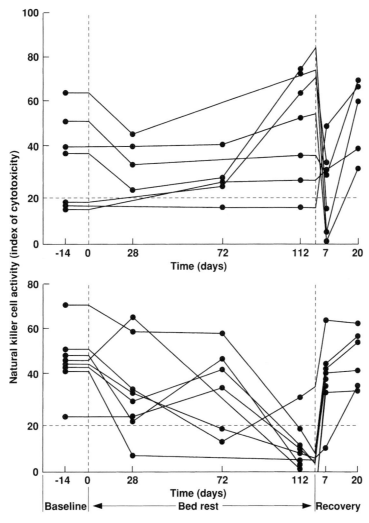

FIG. 39.18. Natural killer (NK) cell activity during 120 days of 4.5° head-down bed rest in six male nonresponders *(upper panel)*, and in nine male responders *(lower panel)*. The index of cytotoxicity was determined in vitro from response of NK cells on ³H-uridine-labeled human K-562 leukocytes. [Redrawn from Konstantinova (208) with permission.]

upper panel) and decreased in eight others (Fig. 39.18, *lower panel*). The depressed NKA was not fully expressed until 72 days, but before 112 days of bed rest. With two exceptions, the nonresponders exhibited acute depression of NKA during the first week of recovery, while all those with decreased NKA during bed rest exhibited acute increases during recovery (Fig. 39.18). Similar divergent recovery NKA responses occurred in eight women after 8 days of bed rest (208). Nonspecific Con A–induced suppressor activity of lymphocytes was unchanged throughout the 120 day bed-rest period. Teplinskaya (356) also reported increased sensitivity to staphylococcus and streptococcus allergens in five of 21 subjects after 60–90 days of bed rest that coincided with decreased PHA reactivity.

Cellular immunity factors were measured in 16 patients (11 with systemic osteoporosis; five with osteoporosis and osteomalacia) and compared with those of nine subjects from the 120 day bed-rest study (37). The ^{3}H-thymidine reactivity was moderately decreased in both groups; T-lymphocyte concentration was unchanged in the bed-rest subjects but elevated to 70%–92% in the patients, while NKA was unchanged or depressed in the subjects and high in the patients. Thus the similarity of immune response in osteoporotic patients and those undergoing prolonged bed rest was not apparent in this study.

It is clear that all investigators who have reported unchanged or decreased immune functions have found them to be distributed over the various immune parameters. The only increased immune function was lymphocyte responsiveness reported by Criswell and Kimzey (69). A clearer description of immune cellular and humoral parameter responses after bed rest is needed with investigation of the mechanisms involved.

PSYCHOPHYSIOLOGICAL FACTORS

Test subjects, and especially patients, subjected to prolonged bed-rest experience restricted mobility, moderate confinement, varying degrees of isolation, and altered dietary intake. Both subjects and patients usually experience back pain and headache (223); patients may have additional pain from their illness or infirmity, and both groups often report symptoms of gastrointestinal distress. In addition, there are reduced proprioceptor and kinesthetic stimuli in response to change from the intermittent, upright, ambulatory posture to the continuous, horizontal bed-rested posture (223, 311, 312, 400). Restriction of body movement with moderate confinement, combined with perceptual deprivation, can induce impaired intellectual and perceptual motor task performance in otherwise normal ambulatory people

(400). With all these sensory perturbations it is surprising that bed-rested people do not exhibit more aberrant behavior.

Group Interaction

Sorokin et al. (338) have described a subject interaction phenomenon referred to as migration of authority. "In it, the liveliest, most cheerful and mobile personality among the subjects of each group acquires seniority at the beginning of hypodynamia. At about the middle of hypodynamia, the intragroup hierarchy is readjusted, with this subject turning up at the bottom of the ladder. 'Seniority' is transferred to the quietest, best-balanced subject, and he retains it to the end of the experiment." This group hierarchy readjustment appears to be an attempt to minimize external irritants that probably influence subjects' attitudes and their reactions to some tests and, consequently, test results. A variant of this migration of authority was observed in the 1971 and 1972 bed-rest studies at Ames Research Center. The emerged leader of the 1971 study subject group applied as a subject for the 1972 study. During its pre–bed rest control period he found that he was not the leader, so he resigned.

Response Stages

Healthy subjects bed-rested to 120 days have been reported to suffer from asthenia, defined as loss of strength and energy; decreased motivation and concentration; and, with greater fatigability, sleep impairment, and increased sensitivity to physical and psychological stressors (11, 69, 251, 272, 300, 339, 379, 390). The following is a composite time sequence of psychophysiological responses during absolute (no exercise) bed rest (11, 212, 223, 251, 288):

Stage 1 (1–2 days): euphoria, much talking, a positive mood, increased cortical (EEG) excitability, good performance, some uncertainty, sound sleep, constipation.

Stage 2 (3–6 days): physical discomfort, various pains, headache, restlessness, desire to get up, difficulty sleeping, increased movement during sleep.

Stage 3 (7–20 days): pains disappeared, mood instability, outbursts, group incompatibility, periods of depression, poor sleep, appearance of abnormal glucose tolerance response, increased vestibular sensitivity, increased skin temperature in hands and feet.

Stage 4 (21–35 days): beginning of asthenic symptoms, increased irritability, intolerance, monotony, rapid fatigability, difficulty lying still, decreased thermal comfort, shallow sleep, more dreaming, performance stabilization, decrease in muscular strength, increasing joint pain.

Stage 5 (36–60 days): increased irritability, interpersonal conflicts, touchiness, capriciousness, sleep disturbances, and aggravation; increased tendon and periosteal reflexes particularly in the legs, increased plantar reflex times, reduced muscular tone particularly in the legs and eyelid; and finger tremor when arms are extended.

Stage 6 (61–105 days): very talkative, further decrease in muscular tone, decreased circumference at shoulder, forearm, and knee; reduced strength in forearm muscles, sharp increase in proprioceptive reflexes, drowsiness, decreased cortical excitability, increased sleep disorders, occasional Babinski reflex.

Stage 7 (106–365 days): increased intensity of tremor in fingers and toes, increasing cortical excitability.

Stage 8 (recovery): very excited, uninhibited, gay, repeat bed-rest subjects were less responsive, general physical weakness, poor exercise tolerance, gross disturbance of coordination, posture, and gait; reduced orthostatic tolerance, pain in soles of feet, subjects cannot stand for 2+–+3 min without support, difficulty walking for 3 days (some subjects need a cane for 8–10 days), leg pain for 3–7 days, increased vestibular sensitivity for 7–10 days; most neurological changes returned to normal by 60 days of recovery.

Performance

Performance of most tracking tasks decreased during absolute bed rest (176, 312, 347, 353, 395, 396). Taylor et al. (353) were probably the first to measure coordination and speed of hand movements after bed rest and found a small decrement in pattern tracing; speed of small hand movements, of medium arm and hand movements, and gross body and arm movements were unchanged after 21 days of bed rest where the six men were allowed to get up for 10 min/day. The bed-rest period was preceded by 6 wk of physical training which may have contributed to the maintenance of limb speed during bed rest. Coordinated visual tracking resulted in 47% errors after 10 days of absolute bed rest, compared with 17% errors before bed rest (395). Error ratios for a similar tracking test were 5.4 after 100 days of absolute bed rest, 2.6 with exercise training during bed rest, compared with 1.2 during ambulatory control; the greatest reductions in the error ratios were on bed-rest days 40 and 80 (396). Simple hand–eye reaction time and hand steadiness (Mercury test) were decreased significantly in eight men after 35 days of bed rest, but responses to more complex tasks, such as a multidimensional pursuit test (manipulation of rudder pedals, stick, and throttle), short-term visual memory, auditory coding, and arithmetic, were unchanged (312). But in that study the subjects stood immediately after bed rest to perform those tests. In a subsequent similar 5 wk bed-

rest study, there were no significant changes in the Mercury test results when the subjects practiced the test during bed rest and took the test in the supine position after bed rest (311). Similar decreased performance on a multidimensional pursuit test was attributed to lack of practice during the 14 days of bed rest (347). Storm and Giannetta (347) also found no decrements in complex tracking performance, hand–eye coordination, or problem solving ability during bed rest with a 30 torr (47%) CO_2 environment. But performance of exercise during bed rest has been reported to attenuate decrements in visual tracking (396). On the other hand, exposure of bed-rested subjects to −30 mm Hg of lower-body negative pressure for 3 h twice/day for 30 days resulted in gross deterioration in a performance test of continuous counting, finding numbers, production of text from memory, and memorizing words (176). Visual memory was more stable than auditory memory, and restoration of performance required 10 days. If LBNP was applied only during the last 5 days of bed rest, there was only a minor decrease in performance that was restored by the second recovery day (176).

Results from a study of the effect of daily isotonic (dynamic) and isokinetic leg exercise training during 30 days of bed rest (74) indicated that proficiency on 10 performance tests increased significantly in the nonexercise control and in both exercise groups, and there were no differences between groups (Fig. 39.19). The isotonic exercise group's mood responses were distinguished from those of the other two groups by significant relative decreases in motivation and concentration, a significant relative increase in sleep quality, and a decrease in psychological tension (Fig. 39.20). The former group's decrements were attributed to slight overtraining from the intense isotonic exercise training. Because the nonexercise group also increased its performance and mood responses, it was concluded that decreased stimuli from reduction of muscular exercise are not the primary cause of reduced psychological performance during prolonged bed rest. Also, optimal selection of the subjects and positive subject–staff interaction and cooperation probably contributed to the positive performance and mood results.

Visual and Auditory Responses

Visual Parameters. Sixteen men were studied over 70 days of horizontal bed rest (82). In Series I, four men underwent absolute bed rest where they were not allowed to turn over on their stomachs; Series II–V involved bed rest for 12 men together with various combinations of countermeasures including pharmaceutical, cardiovascular, and nervous system stimulants: amphetamine, securinine, and caffeine (II); treadmill exercise (III); combined cycle ergometer exercise and thigh

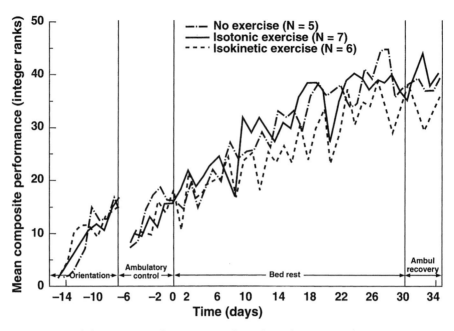

FIG. 39.19. Mean daily composite values in 18 men for eight performance tests during orientation (10 trials), ambulatory–control, bed rest, and ambulatory–recovery periods for the three exercise groups. [From Greenleaf (127) with permission.]

occlusion training (IV); and treadmill exercise with thigh occlusion training (V); while remaining in the horizontal body position (338).

In Series I subjects, the pre–bed rest visual acuity averaged 1.25, peripheral vision extended to 62.3°, intraocular pressure was 20 mm Hg, the near point of clear vision was a 8.5 cm, the blind spot area was 75 cm² (measured at 1.0 m with a 0.5 cm white object), and recovery time of visual acuity to a bright flash was 27 s. The ocular fundus was normal, the optic disk was pale pink with sharp boundaries, and blood vessels were of

normal size and shape. After 45 days of bed rest, the visual acuity had decreased by 21%, peripheral vision decreased by 11°, intraocular pressure decreased by 3 mm Hg, and the near point was extended by 3.5 cm. After 67 days of bed rest, the peripheral vision decreased to 15° and the near point was extended to 12.5 cm. There were no changes in field or color perception during bed rest. On the second day of ambulatory recovery visual acuity remained slightly depressed by 21%, the blind spot had enlarged by 38%, and visual acuity recovery had increased by 120%. On the 20th recovery

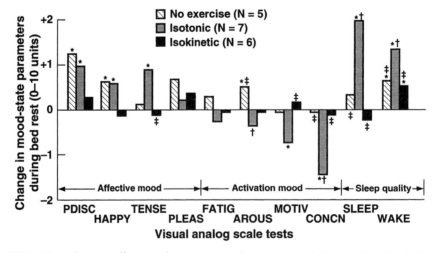

FIG. 39.20. Mean change in affective and activation mood parameters and sleep quality scales during bed rest in the three exercise groups. Zero (y-axis) indicates no change: positive is improvement and negative is deterioration. *$P < 0.05$ from zero slope (linear regression), †$P < 0.05$ from no-exercise group, and ‡$P < 0.05$ from isotonic exercise group. [From DeRoshia and Greenleaf (74) with permission.]

day visual acuity had not returned to normal, the blind spot was still enlarged by 19%, and visual acuity recovery was still increased by 50%. Despite these apparently adverse visual responses, there was no perceptible influence on visual psychomotor performance.

Auditory Parameters. Raised auditory tone thresholds (indicating poorer hearing) from 15–35 db in the 2–8 kHz range occurred during the first wk of 62 day (394) and 49 day (368) horizontal bed-rest studies. The increased thresholds gradually declined and had generally returned to control levels by the end of bed rest. There was complete restoration of thresholds in all subjects after 2 wk of recovery. Similarly increased thresholds by 20–25 db, primarily at the higher frequencies, were also reported by Parin et al. (289) in 10 subjects undergoing 120 days of bed rest, and in nine men during 30 days of bed rest (393). Bone and air conduction auditory responses were not different (13, 368).

Hearing threshold changes appear to respond more quickly during head-down bed rest. Thresholds improved slightly for most frequencies on the first day, followed by deterioration on day 3 and returned to control levels by the end (day 7) of bed rest (14). The 4, 6, and 8 kHz frequencies were the most stable during bed rest and recovery.

The faster response intervals for decline and recovery of auditory thresholds with head-down compared with horizontal bed rest suggests cephalic fluid shifts may play an important role. Diuresis and hypovolemia occur during the first few days of bed rest and body water balance reaches near equilibrium in 3–4 wk, when the increased auditory thresholds decrease towards normal levels. The greater hydrostatic pressure resulting from the shift from the horizontal to 6° head-down position increases stasis in the intracranial veins and increases intracranial pressure (ICP) by about 350 Pa (2.6 mm Hg) (38). From tympanic membrane displacement to a 1000 Hz, 110 db auditory stimulus, Murthy et al. (271) calculated that ICP increased by about 3.0 mm Hg from the horizontal to 6° head-down position; the increase in ICP from sitting to 6° was 17 mm Hg (Fig. 39.21). It is not clear how increased intracranial-perilymphatic pressure would selectively raise high frequency thresholds.

The effect of exercise training performed during bed rest on auditory function is not clear. Initially, Yakovleva and Matsnev (394) reported that graduated exercise on a veloergometer (protocol unspecified) had no effect on the increased auditory thresholds during 60 days of horizontal bed rest or during recovery. From results of a later study (30 days of 4° bed rest), Yakovleva et al. (393) concluded that performance of treadmill exercise (1 h/day), plus periodic application of lower-body negative pressure (2.5 h/day at −36 to −44 mm Hg), and electrical stimulation of leg and trunk

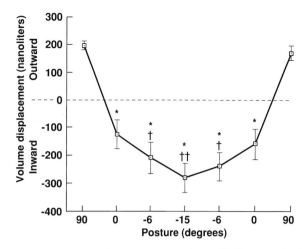

FIG. 39.21. Mean (± SE) tympanic membrane displacement in six subjects in various postures: 90° is sitting, 0° is horizontal, and −6° and −15° are head-down tilt. *$P < 0.05$ from 90°, ††$P < 0.05$ from 0°, and †$P < 0.05$ from −6°. Contraction of the stapedius muscle moves the tympanic membrane (via the oval window and ossicles) inward or outward, reflecting increased or decreased, respectively, perilymphatic pressure which reflects intracranial pressure. [Redrawn from Murthy et al. (271) with permission.]

muscles during bed rest relieved symptoms of head fullness and nasal discomfort. On the other hand, Vasil'yev and Diskalenko (368) found no effect of bone vibration or LBNP on the increased auditory thresholds. One might expect that the hypovolemic effect of dynamic leg exercise would reduce the intracranial pressure and relieve head congestion and discomfort. Apparently this occurs during exercise in microgravity (A. Bean, personal communication).

Sleep

The quantity and quality of sleep is definitely altered in normal, healthy, bed-rested subjects and, as with normal ambulatory people, is a measure of their general emotional state (74, 272, 311, 312). Subjects generally sleep soundly the first few days of bed rest, and then the quality and quantity start to deteriorate; more slowly when exercise is performed. By the second wk some subjects sleep normally, but in others increasing time is needed to fall asleep, thereby decreasing the duration of sleep and inducing a greater incidence of dreaming. After 3 wk, essentially every subject sleeps poorly; sleep comes slowly and less frequently, it is more fitful with unpleasant dreams, and emotional problems increase. These signs and symptoms generally continue to the end of bed rest when normal sleep patterns begin to appear in anticipation of reambulation. The normal electroencephalogram (EEG) response before bed rest consists of predominantly alpha-rhythm with a frequency of 10–12 oscillations/s and amplitudes of 30–200 uV. By the fourth day of bed rest some subjects exhibit slow waves (3–6 oscillations/s), which predominated in all subjects by the end of the second wk; that is, the alpha waves appeared again in some subjects after 50 days of bed rest. Normal EEG rhythms had not returned by the third day of ambulatory recovery. Performance of 1,200 kcal/day of exercise during bed rest retarded onset of the slow waves and tended to normalize the pattern. If the motor (exercise) outlet for psychic energy is reduced, perhaps it is shifted to dreaming as an outlet.

Ryback et al. (311, 312) also found that both exercise (600 kcal/day) and nonexercise groups spent a larger proportion of their sleep in the deep (slow-wave delta) mode during bed rest, with a greater tendency toward deep sleep in the nonexercise group. The number of reported dreams and rapid eye movements (REM) increased for both groups during bed rest, and then decreased toward normal levels in the recovery period. Reduction in delta-wave (non-REM) sleep has been attributed to a physiological reparative process which could be associated with muscular atrophy in both exercised and nonexercised subjects during bed rest. The muscular atrophy would stimulate muscle restoration, and the catabolic effect of exercise would also stimulate muscle repair. Thus the muscle repair process would reduce delta-wave (non-REM) and stimulate alpha-wave (REM) sleep. The level of human growth hormone has been positively correlated with the occurrence of deep (delta-wave) sleep (324). Perhaps this hormone participates in the repair process; that is, the muscle atrophy in the nonexercise group could require more repair, hence more deep sleep.

A more recent study (74) involved the effect of intensive, daily, isokinetic and isotonic-dynamic leg exercise training during 30 days of 6° head-down bed rest on sleep quality (trouble falling asleep [*SLEEP*]) and number of waking episodes (*WAKE*; Fig. 39.20). There was significantly greater trouble falling asleep and more waking episodes in the isotonic exercise group when compared with the isokinetic and nonexercise groups, which exhibited similar responses. The poorer motivation and concentration in the isotonic group would logically be associated with the poorer sleep and waking responses. These deteriorated activation mood and sleep quality responses were attributed to slight overtraining stimuli from the intensive (twice daily) isotonic exercise training regimen.

Posture: Equilibrium, Balance, and Gait

These static (equilibrium) and dynamic (gait) body control functions are usually measured immediately upon reambulation after prolonged bed rest. Some bed-rest deconditioning factors influencing these functions are the modified vestibular state, decreased muscular strength, reduced thresholds for pain, dehydration, altered autonomic nervous system function resulting from cephalic fluid shifts, reduced hydrostatic pressure from assuming the horizontal posture, and reduced exercise-induced stimuli with lower energy utilization. As mentioned previously, upon arising after rest, the subjects are weak and exhibit gross disturbance of coordination, balance, and gait. They are often unable to stand without support, sometimes feel faint, and report pain in the legs—especially in the feet.

Lower-body negative pressure tolerance or orthostatic tolerance to head-up tilting are often measured before the subjects arise. Adaptive physiological responses to these tests would tend to ameliorate the bed-rest deconditioning responses, and may influence the subsequent recovery process.

Taylor et al. (353) were probably the first to measure body ataxia after prolonged bed rest. The range of body sway during 2 min was determined in six men standing with their eyes closed (Fig. 39.22). Body sway increased from a control level of 33 cm to 47 cm on the first day of recovery, and decreased to control levels by day 5. The Graybiel-Fregly postural equilibrium test was given to 10 men after 10 days of horizontal bed rest: the total number of walking steps decreased by 9%, standing time (eyes open) decreased by 30%, and standing time (eyes closed) decreased by 17% (385).

Subsequently, intensive evaluation of equilibrium and gait parameters were measured in 62 day (222, 225, 300), 70 day (102, 152, 190, 201, 335), and 120 day (223, 289, 299) bed-rest studies. The results were generally similar: upon the first day of recovery the subjects exhibited an acute decrease of standing stability (amplitude of body center gravity oscillation increased two- to fivefold), stabilograms showed distinct high-frequency oscillations (1–12 Hz) corresponding to the frequency of physiological tremor, and stability deteriorated when the subjects raised their arms. Frequency and amplitude of the center of gravity with eyes closed were always

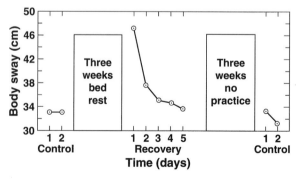

FIG. 39.22. Mean body sway in six men during pre–bed rest control and for 5 days after 3 wk of horizontal bed rest. Lack of practice for 3 wk had no effect on post–bed rest responses. [Redrawn from Taylor et al. (353) with permission.]

greater in the sagittal plane after bed rest, but were greater in the frontal plane before bed rest. Kotovskaya et al. (212) found that body instability (oscillation frequency) increased significantly during the first few days and remained relatively stable throughout 182 days of bed rest (Fig. 39.23, upper half). Body oscillation amplitude increased more slowly and reached equilibrium at about 30–40 days of bed rest (Fig. 39.23, lower half). The greater frequency and amplitude changes occurred with eyes closed with circular head movements on a hard platform, and the lesser responses with the eyes open on a hard platform emphasizing the importance of vision for maintaining body stability during bed-rest deconditioning.

The basic kinematics and dynamics of walking are retained after bed rest, but there are considerable variations in step length, step width, and stride length (335); and decreased amplitudes of joint motion, greater body instability, and lack of smoothness in ankle joint motion (152). Walking with eyes closed compounded these responses and shortened stride length, and one-leg stand times were reduced.

In general, those subjects who performed exercise-training regimens during the 62 day (51, 300), 70 day (103, 152), and 120 day (299) bed-rest studies had attenuated loss of muscle tone and attenuated deleterious equilibrium and gait responses. Cinematographical analysis confirmed that walking and running exercise training during 30 days of bed rest improved gait coordination after bed rest, while electrical stimulation of muscles during bed rest did not (50). Exercise during bed rest has never been reported to return equilibrium and gait responses to normal. Amplitude of arm tremor fluctuations increased by about 100% during the first wk of recumbency, and declined somewhat thereafter over 62 days of bed rest (Fig. 39.24). But this amplitude was unchanged in the group exercising at 1,200 kcal/day. Gurfinkel et al. (152) and Skrypnik (335) also reported that subjects who performed cycle ergometer exercise had more difficulty walking during reambulation than those who performed treadmill exercise during bed rest. But the ergometer subjects had faster recovery of step-length during the first mo. Haines (153) measured 11 body balance functions including floor line walk, rail balance, and Romberg tests in seven men subjected to daily isotonic leg exercise, isometric leg exercise, and no exercise. Results indicated that: (1) 14 days without body balance testing did not affect subsequent balance performance; (2) neither isotonic nor isometric exercise training prevented the deteriorating effects of bed rest on body balance measures, suggesting that other sensory, proprioceptive, and kinesthetic neural information may be altered during deconditioning; (3) the most sensitive balance test (the one that produced the greatest difference pre– vs. post–bed rest) was rail

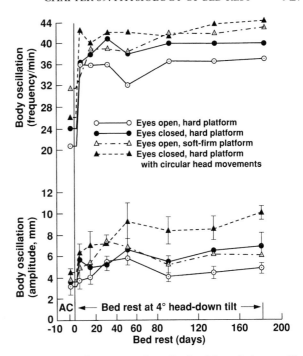

FIG. 39.23. Mean frequency and amplitude of frontal plane oscillations of six men (35–40 yr) before (ambulatory control, AC) and during 182 days of 4° head-down bed rest. Variability unspecified. [Redrawn from Kotovskaya et al. (212) with permission.]

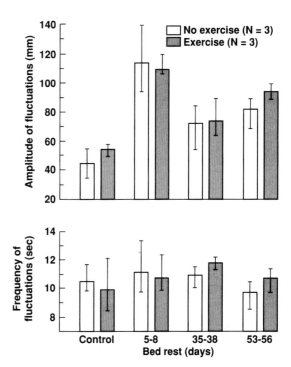

FIG. 39.24. Mean (± range) arm tremor during bed rest. Exercise group utilized leg cycle ergometer and arm resistance exercise at 1,200 kcal/day. [Redrawn from Purakhin and Petukhov (300) with permission.]

walking with eyes open; and (4) balance skills returned to normal by the third day of recovery.

These psychophysiological parameters must interact to maintain homeostasis during adaptation to prolonged bed rest, thus making it difficult to determine which altered stimuli induced by the horizontal posture contribute most to this adaptive response. Reduced foot-to-head hydrostatic pressure in the recumbent posture engenders a headward fluid shift, stimulating diuresis and reducing total body water (dehydration); attenuated exertion contributes to the posture-induced hypovolemia and reduces muscle kinesthesis and strength; and loss of longitudinal pressure unloads bones and also influences kinesthetic stimuli. The combination of increased intracranial pressure accompaning prolonged reduction in total body water may induce novel neuromuscular stimuli. Increased body oscillation and raised auditory thresholds which occur within the first few days of recumbency are perhaps responding to the early fluid shifts. More data are needed on the time course of changes in psychophysiological parameters during bed rest; pre– and post–bed rest measurements are no longer sufficient. Confusion will continue until experiments are designed to control basic parameters, for example, type and intensity of exercise, hydrostatic pressure, and degree of bone loading, while measuring the well-defined psychophysiological responses.

The authors thank Ms. Mary Jackson, Angela Hardesty, and Farzaneh Ghiasvand for their excellent literature search and manuscript preparation. Research reported in this chapter was supported by NASA Tasks 199-18-12-07, 199-26-11-01, 199-14-11-13, and NIH Grants RR-00350 and RR-02558.

REFERENCES

1. Abboud, F. M. Integration of reflex responses in the control of blood pressure and vascular resistance. *Am. J. Cardiol.* 44: 903–911, 1979.
2. Aleksandrov, A. N., and A. K. Kochetov. Effect of 30-day hypokinesia in combination with LBNP training on some indices of the functional state of the cardiovascular system at rest. *Kosm. Biol. Aviakosm. Med.* 8: 71–72, 1974.
3. Allen, M. J., V. T. Ang, and D. Bennett. A comparison of head-down tilt with low-dose infusion of atrial natriuretic peptide in man. *J. Physiol. (Lond.)* 410: 341–350, 1989.
4. Amberson, W. R. Physiologic adjustments to the standing posture. *Bull. Sch. Med. Univ. Maryland* 27: 127–145, 1943.
5. Annat, G., A. Güell, G. Gauquelin, M. Vincent, J. L. Bascands, G. Geelen, A. Sassolas, and C. Gharib. Plasma renin activity during 5-hour antiorthostatic hypodynamia. *Physiologist* 27: S49–S50, 1984.
6. Annat, G., A. Güell, G. Gauquelin, M. Vincent, M. H. Mayet, C. A. Bizollon, J. J. Legros, J. M. Pottier, and C. Gharib. Plasma vasopressin, neurophysin, renin and aldosterone during a 4-day head-down bed rest with and without exercise. *Eur. J. Appl. Physiol.* 55: 59–63, 1986.
7. Anonymous, Associated Press. Puzzling illness keeps man in bed 50 years. *San Jose Mercury News*, May 26, 1982. p. 12A.
8. Arborelius, M., Jr., U. I. Balldin, B. Lilja, and C. E. G. Lundgren. Hemodynamic changes in man during immersion with the head above water. *Aerospace Med.* 43: 592–598, 1972.
9. Arnaud, S. B., V. S. Schneider, and E. Morey-Holton. Effects of inactivity on bone and calcium metabolism. In: *Inactivity: Physiological Effects*, edited by H. Sandler and J. Vernikos. Orlando: Academic, 1986, p. 49–76.
10. Arnaud, S. B., D. J. Sherrard, N. Maloney, R. T. Whalen, and P. Fung. Effects of 1-week head down tilt bed rest on bone formation and the calcium endocrine system. *Aviat. Space Environ. Med.* 63: 14–20, 1992.
11. Artishuk, V. N., A. N. Litsov, V. P. Stupnitskiy, and Yu. V. Yakushkov. Effect of 30-day hypokinesia on the dynamics of higher nervous activity and sleep of an operator. *Kosm. Biol. Aviakosm. Med.* 8: 75–79, 1974.
12. Asher, R. A. J. The dangers of going to bed. *Br. Med. J.* 4: 967–968, 1947.
13. Aust, G., H. Denz, and F. Baisch. Inner ear characteristics during 7 day antiorthostatic bedrest (6° head down tilt). in: *Proc. 2nd Eur. Symp. Life Sci. Res. in Space, Porz Wahn, Germany, 1984.* Noordwijk: ESA Scientific and Technical Publications Branch, p. 251–255. (ESA SP-212, 1984)
14. Aust, G., A. Putzka, and F. Baisch. Effects of head down tilt (HDT) fluid volume shift on cerebral sensory responses. In: *Proc. 35th Congr. Int. Astronautical Federation, Lausanne, Switzerland, 1984.* New York: American Institute of Aeronautics and Astronautics, p. 1–5. (IAF-84-191)
15. Baisch, F., and L. Beck. Left heart ventricular function during a 7 day 0-G simulation (6° head down tilt). In: *Proc. 2nd Eur. Symp. Life Sci. Res. in Space, Porz Wahn, Germany, 1984.* Noordwijk: ESA Scientific and Technical Publications Branch, p. 125–132. (ESA SP-212, 1984)
16. Balakhovskiy, I. S., V. I. Legen'kov, and R. K. Kiselev. Changes in hemoglobin mass during real and simulated space flights. *Kosm. Biol. Aviakosm. Med.* 14: 14–20, 1980.
17. Bascands, J. L., G. Gauquelin, G. Annat, J. P. Pequignot, C. Gharib, and A. Güell. Effect of muscular exercise during 4 days simulated weightlessness on orthostatic tolerance. *Physiologist* 27: S63–S64, 1984.
18. Bassey, E. J., T. Bennett, A. T. Birmingham, P. H. Fentem, D. Fitton, and R. Goldsmith. Effects of surgical operation and bed rest on cardiovascular responses to exercise in hospital patients. *Cardiovasc. Res.* 7: 588–592, 1973.
19. Beckett, W. S., N. B. Vroman, D. Nigro, S. Thompson-Gorman, J. E. Wilkerson, and S. M. Fortney. Effect of prolonged bed rest on lung volume in normal individuals. *J. Appl. Physiol.* 61: 919–925, 1986.
20. Beigel. *Verh. Kaiserl. Leopold Carol. Akad. Naturforsch.* 25: 477, 1855.
21. Beregovkin, A. V., P. V. Buyanov, A. V. Galkin, N. V. Pisarenko, and Ye. Ye. Sheludyakov. Results of investigations of the cardiovascular system during the after-effect of 70-day hypodynamia. In: *Problems of Space Biology. Prolonged Limitation of Mobility and its Influence on the Human Organism*, edited by A. M. Genin and P. A. Sorokin. Moscow: Nauka, 1969, vol. 13, p. 221–227. (NASA TT F-639, 1970)
22. Beregovkin, A. V., and V. V. Kalinichenko. Reactions of the cardiovascular system during 30-day simulation of weightlessness by means of antiorthostatic hypokinesia. *Kosm. Biol. Aviakosm. Med.* 8: 72–77, 1974.
23. Billman, G. E., D. T. Dickey, H. Sandler, and H. L. Stone. Effects of horizontal body casting on the baroreceptor reflex control of heart rate. *J. Appl. Physiol.: Respir. Environ. Exerc. Physiol.* 52: 1552–1556, 1982.

24. Billman, G. E., D. T. Dickey, K. K. Teoh, and H. L. Stone. Effects of central venous blood volume shifts on arterial baro-reflex control of heart rate. *Am. J. Physiol.* 241 (*Heart Circ. Physiol.* 10): H571–H575, 1981.

25. Birkhead, N. C., J. J. Blizzard, J. W. Daly, G. J. Haupt, B. Issek-utz, Jr., R. N. Myers, and K. Rodahl. *Cardiodynamic and Metabolic Effects of Prolonged Bed Rest with Daily Recumbent or Sitting Exercise and with Sitting Inactivity.* Wright-Patterson Air Force Base, Ohio: AMRL-Technical Report 61, 1964.

26. Birkhead, N. C., J. J. Blizzard, B. Issekutz, Jr., and K. Rodahl. *Effect of Exercise, Standing, Negative Trunk and Positive Skeletal Pressure on Bed Rest–induced Orthostasis and Hypercalciuria.* Wright-Patterson Air Force Base, Ohio: AMRL Technical Report 6, 1966.

27. Birkhead, N. C., G. J. Haupt, and R. N. Myers. Effect of prolonged bedrest on cardiodynamics. *Am. J. Med. Sci.* 245: 118–119, 1963.

28. Blamick, C. A., D. J. Goldwater, and V. A. Convertino. Leg vascular responsiveness during acute orthostasis following simulated weightlessness. *Aviat. Space Environ. Med.* 59: 40–43, 1988.

29. Blomqvist, C. G., F. A. Gaffney, and J. V. Nixon. Cardiovascular responses to head-down tilt in young and middle-aged men. *Physiologist* 26: S81–S82, 1983.

30. Blomqvist, C. G., J. H. Mitchell, and B. Saltin. Effects of bed rest on the oxygen transport system. In: *Hypogravic and Hypodynamic Environments,* edited by R. H. Murray and M. McCally. Washington, DC: NASA Special Publication 269, 1971, p. 171–185.

31. Blomqvist, C. G., J. V. Nixon, R. L. Johnson, Jr., and J. H. Mitchell. Early cardiovascular adaptation to zero gravity simulated by head-down tilt. *Acta Astronautica* 7: 543–553, 1980.

32. Blomqvist, C. G., and H. L. Stone. Cardiovascular adjustments to gravitational stress. In: *Handbook of Physiology. The Cardiovascular System,* edited by J. T. Shepherd and F. M. Abboud. Bethesda, MD: Am. Physiol. Soc., 1983 Sect. 2, vol. III, pt. 28, chapt. 2, p. 1025–1063.

33. Blotner, H. Effect of prolonged physical inactivity on tolerance of sugar. *Arch. Intern. Med.* 75: 39–44, 1945.

34. Bohnn, B. J., K. H. Hyatt, L. G. Kamenetsky, B. E. Calder, and W. M. Smith. Prevention of bedrest induced orthostatism by 9-alpha-fluorohydrocortisone. *Aerospace Med.* 41: 495–499, 1970.

35. Bonde-Petersen, F., Y. Suzuki, and T. Sadamoto. Cardiovascular effects of simulated zero-gravity in humans. *Acta Astronautica* 10: 657–661, 1983.

36. Bourne, G. H., M. N. Golarz de Bourne, and H. M. McClure. Orbiting primate experiment. In: *The Use of Non-Human Primates in Space,* edited by R. Simmonds and G. H. Bourne. Washington, DC: NASA Conference Publication 005, 1977, p. 51–82.

37. Bozhikov, N. V., M. P. Rykova, Ye. N. Antropova, and A. T. Lesnyak. Quantitative and functional parameters of T-lymphocytes and activity of normal killers in patients suffering from systemic osteoporosis and subjects undergoing 120 days of hypokinesia with head-down tilt. In: *Space Biology and Aerospace Medicine,* edited by O. G. Gazenko. Moscow: Nauka, 1986, p. 333–334.

38. Bradley, K. C. Cerebrospinal fluid pressure. *J. Neurol. Neurosurg. Psychiatry* 33: 387–397, 1970.

39. Brannon, E. W., C. A. Rockwood, Jr., and P. Potts. The influence of specific exercises in the prevention of debilitating musculoskeletal disorders; implication in physiological conditioning for prolonged weightlessness. *Aerospace Med.* 34: 900–906, 1963.

40. Buckey, J. C., L. D. Lane, G. Plath, F. A. Gaffney, F. Baisch, and C. G. Blomqvist. Effects of head down tilt for 10 days on the compliance of the lower limb. *Physiologist* 33: S167–S168, 1990.

41. Buckey, J. C., R. M. Peshock, and C. G. Blomqvist. Deep venous contribution to hydrostatic blood volume change in the human leg. *Am. J. Cardiol.* 62: 449–453, 1988.

42. Burkovskaya, T. Ye., A. V. Ilyukhin, V. I. Lobachik, and V. V. Zhidkov. Erythrocyte balance during 182-day hypokinesia. *Kosm. Biol. Aviakosm. Med.* 14: 50–54, 1980.

43. Butler, G. C., H. C. Xing, D. R. Northey, and R. L. Hughson. Reduced orthostatic tolerance following 4 h head-down tilt. *Eur. J. Appl. Physiol.* 62: 26–30, 1991.

44. Campbell, J. A., and T. A. Webster. LXXX. Day and night urine during complete rest, laboratory routine, light muscular work and oxygen administration. *Biochem. J.* 15: 660–664, 1921.

45. Cardus, D. O$_2$ alveolar-arterial tension difference after 10 days recumbency in man. *J. Appl. Physiol.* 23: 934–937, 1967.

46. Celsing, F., J. Nystrom, P. Pihlstedt, B. Werner, and B. Ekblom. Effect of long-term anemia and retransfusion on central circulation during exercise. *J. Appl. Physiol.* 61: 1358–1362, 1986.

47. Chadwick, J., and W. N. Mann. *The Medical Works of Hippocrates.* Oxford: Blackwell, 1950, p. 140.

48. Chase, G. A., C. Grave, and L. B. Rowell. Independence of changes in functional and performance capacities attending prolonged bedrest. *Aerospace Med.* 37: 1232–1238, 1966.

49. Chavarri, M., A. Ganguly, J. A. Luetscher, and P. G. Zager. Effect of bedrest on circadian rhythms of plasma renin, aldosterone, and cortisol. *Aviat. Space Environ. Med.* 48: 633–636, 1977.

50. Chekirda, I. F., A. V. Yeremin, V. I. Stepantsov, and I. P. Borisenko. Characteristics of human gait after 30-day hypokinesia. *Kosm. Biol. Aviakosm. Med.* 8: 48–51, 1974.

51. Cherepakhin, M. A. Effect of prolonged bedrest on muscle tone and proprioceptive reflexes in man. *Kosm. Biol. Aviakosm. Med.* 2: 43–47, 1968.

52. Chobanian, A. V., R. D. Lille, A. Tercyak, and P. Blevins. The metabolic and hemodynamic effects of prolonged bed rest in normal subjects. *Circulation* 49: 551–559, 1974.

53. Chukhlovin, B. A., and S. A. Burov. Resistance to infection under conditions of hypodynamia. In: *Problems of Space Biology. Prolonged Limitation of Mobility and its Influence on the Human Organism,* edited by A. M. Genin and P. A. Sorokin. Moscow: Nauka, 1969, vol. 13, p. 116–122. (NASA TT F-639, 1970)

54. Chukhlovin, B. A., P. B. Ostroumov, and S. P. Ivanova. Development of staphylococcal infection in human subjects under the influence of some spaceflight factors. *Kosm. Biol. Med.* 5: 61–65, 1971.

55. Coleridge, J. C. G., and H. M. Coleridge. Chemoreflex regulation of the heart. In: *Handbook of Physiology. The Cardiovascular System. The Heart,* edited by R. M. Berne and N. Sperelakis. Bethesda MD: Am. Physiol. Soc., 1979, sect. 2, vol. I, chapt. 18, p. 653–676.

56. Convertino, V. A. Potential benefits of maximal exercise just prior to return from weightlessness. *Aviat. Space Environ. Med.* 58: 568–572, 1987.

57. Convertino, V. A., R. Bisson, R. Bates, D. Goldwater, and H. Sandler. Effects of antiorthostatic bedrest on the cardiorespiratory responses to exercise. *Aviat. Space Environ. Med.* 52: 251–255, 1981.

58. Convertino, V. A., D. F. Doerr, D. L. Eckberg, J. M. Fritsch, and J. Vernikos-Danellis. Head-down bed rest impairs vagal baroreflex responses and provokes orthostatic hypotension. *J. Appl. Physiol.* 68: 1458–1464, 1990.

59. Convertino, V. A., D. F. Doerr, and S. L. Stein. Changes in size and compliance of the calf after 30 days of simulated microgravity. *J. Appl. Physiol.* 66: 1509–1512, 1989.

60. Convertino, V. A., D. J. Goldwater, and H. Sandler. Effect of orthostatic stress on exercise performance after bedrest. *Aviat. Space Environ. Med.* 53: 652–657, 1982.

61. Convertino, V. A., D. J. Goldwater, and H. Sandler. V̇O₂ kinetics of constant-load exercise following bed rest–induced deconditioning. *J. Appl. Physiol.: Respir Environ. Exerc. Physiol.* 57: 1545–1550, 1984.

62. Convertino, V. A., D. J. Goldwater, and H. Sandler. Bedrest-induced peak V̇O₂ reduction associated with age, gender, and aerobic capacity. *Aviat. Space Environ. Med.* 57: 17–22, 1986.

63. Convertino, V. A., J. Hung, D. Goldwater, and R. F. DeBusk. Cardiovascular responses to exercise in middle-aged men after 10 days of bedrest. *Circulation* 65: 134–140, 1982.

64. Convertino, V. A., R. W. Stremel, E. M. Bernauer, and J. E. Greenleaf. Cardiorespiratory responses to exercise after bed rest in men and women. *Acta Astronautica* 4: 895–905, 1977.

65. Convertino, V. A., C. A. Thompson, B. A. Benjamin, L. C. Keil, W. M. Savin, E. P. Gordon, W. L. Haskell, J. S. Schroeder, and H. Sandler. Haemodynamic and ADH responses to central blood volume shifts in cardiac-denervated humans. *Clin. Physiol.* 10: 55–67, 1990.

66. Cordonnier, J. J., and B. S. Talbot. The effect of the ingestion of sodium-acid phosphate on urinary calcium in recumbency. *J. Urol.* 60: 316–320, 1948.

67. Cottet-Emard, J. M., J. M. Pequignot, A. Güell, L. Peyrin, and C. Gharib. Catecholamines during short- and long-term head-down bedrest. *Physiologist* 33: S69–S72, 1990.

68. Cottier, C., L. Matter, P. Weidmann, S. Shaw, and M. P. Gnädinger. Renal response to low-dose infusion of atrial natriuretic peptide in normal man. *Kidney Int.* 25: (Suppl. 25): S72–S78, 1988.

69. Criswell, B. S., and S. L. Kimzey. The influence of bedrest on the lymphocytic response of man. Alexandria, VA: *Preprints, Aerospace Med. Assoc.,* 1977, p. 191–192.

70. Cuthbertson, D. P. CXLV. The influence of prolonged muscular rest on metabolism. *Biochem. J.* 23: 1328–1345, 1929.

71. Dallman, M. F., J. Vernikos, L. C. Keil, D. O'Hara, and V. Convertino. Hormonal, fluid and electrolyte responses to 6° anti-orthostatic bed rest in healthy male subjects. In: *Stress: The Role of Catecholamines and Other Neurotransmitters,* edited by E. Usdin, R. Kvetnansky, and J. Axelrod. London: Gordon and Breach, 1984, vol. II, p. 1057–1077.

72. Davydova, N. A., S. K. Shishkina, N. V. Korneyeva, Ye. V. Suprunova, and A. S. Ushakov. Biochemical aspects of some neurohumoral system functions during long-term antiorthostatic hypokinesia. *Kosm. Biol. Aviakosm. Med.* 20: 91–95, 1986.

73. Deitrick, J. E., G. D. Whedon, and E. Shorr. Effects of immobilization upon various metabolic and physiologic functions of normal men. *Am. J. Med.* 4: 3–36, 1948.

74. DeRoshia, C. W., and J. E. Greenleaf. Performance and mood-state parameters during 30-day 6° head-down bed rest with exercise training. *Aviat. Space Environ. Med.* 64: 522–527, 1993.

75. DeTorrente, A., G. L. Robertson, K. M. McDonald, and R. W. Schrier. Mechanism of diuretic response to increased left atrial pressure in the anesthetized dog. *Kidney Int.* 8: 355–361, 1975.

76. Dickey, D. T., K. K. Teoh, H. Sandler, and H. L. Stone. Changes in blood volume and response to vaso-active drugs in horizontally casted primates. *Physiologist* 22: S27–S28, 1979.

77. Dietlein, L. F., and R. S. Johnston. U.S. manned space flight: the first twenty years. A biomedical status report. *Acta Astronautica* 8: 893–906, 1981.

78. Dikshit, M. B., and J. M. Patrick. Vital capacity and airflow measured from partial flow-volume curves during 5° head-down tilt. *Aviat. Space Environ. Med.* 58: 343–346, 1987.

79. Dintenfass, L. Speculations on depletion of the red cell mass in astronauts, and on space sickness. *Clin. Hemorheology* 6: 435–437, 1986.

80. Dock, W. The evil sequelae of complete bedrest. *J.A.M.A.* 125: 1083–1085, 1944.

81. Donaldson, C. L., S. B. Hulley, J. M. Vogel, R. S. Hattner, J. H. Bayers, and D. E. McMillan. Effect of prolonged bed rest on bone mineral. *Metabolism* 19: 1071–1094, 1970.

82. Drozdova, N. T., and O. N. Nesterenko. State of the visual analyzer during hypodynamia. In: *Problems of Space Biology. Prolonged Limitation of Mobility and its Influence on the Human Organism,* edited by A. M. Genin and P. A. Sorokin. Moscow: Nauka, 1969, vol. 13, p. 189–191. (NASA TT F-639, 1970)

83. Dunn, C. D. R., P. C. Johnson, and J. I. Leonard. Erythropoietic effects of spaceflight re-evaluated. *Physiologist* 24: S5–S6, 1981.

84. Dunn, C. D. R., R. D. Lange, S. L. Kimzey, P. C. Johnson, and C. S. Leach. Serum erythropoietin titers during prolonged bedrest: relevance to the "anaemia" of space flight. *Eur. J. Appl. Physiol.* 52: 178–182, 1984.

85. Eastman, N. J. The abuse of rest in obstetrics. *J.A.M.A.* 125: 1077–1079, 1944.

86. Eckberg, D. L., and J. M. Fritsch. Carotid baroreceptor cardiac-vagal reflex responses during 10 days of head-down tilt. *Physiologist* 33: S177, 1990.

87. Eckberg, D. L., and J. M. Fritsch. Human autonomic responses to actual and simulated weightlessness. *J. Clin. Pharmacol.* 31: 951–955, 1991.

88. Fortney, S. M. Thermoregulatory adaptations to inactivity. In: *Adaptive Physiology to Stressful Environments,* edited by S. Samueloff and M. K. Yousef. Boca Raton, FL: CRC Press, 1987, p. 75–83.

89. Fortney, S. M. Development of lower body negative pressure as a countermeasure for orthostatic intolerance. *J. Clin. Pharmacol.* 31: 888–892, 1991.

90. Fortney, S. M., W. S. Beckett, A. J. Carpenter, J. Davis, H. Drew, N. D. LaFrance, J. A. Rock, C. G. Tankersley, and N. B. Vroman. Changes in plasma volume during bed rest: effects of menstrual cycle and estrogen administration. *J. Appl. Physiol.* 65: 525–533, 1988.

91. Fortney, S. M., W. B. Beckett, N. B. Vroman, J. Davis, J. Rock, A. Kimbell, N. LaFrance, and H. Drew. *Bedrest in Healthy Women: Effects of Menstrual Function and Oral Contraceptives.* Washington, DC: NASA Contractor Report 171946, 1986.

92. Fortney, S. M., L. Dussack, T. Rehbein, M. Wood, and L. Steinmann. Effect of prolonged LBNP and saline ingestion on plasma volume and orthostatic responses during bed rest. In: *Proc. First Joint NASA Cardiopulmonary Workshop,* edited by S. Fortney and A. R. Hargens. Houston, TX: NASA Conference Publication 10068, 1991, p. 61–70.

93. Fortney, S. M., K. H. Hyatt, J. E. Davis, and J. M. Vogel. Changes in body fluid compartments during a 28-day bed rest. *Aviat. Space Environ. Med.* 62: 97–104, 1991.

94. Gaffney, F. A., J. C. Buckey, L. D. Lane, A. Hillebrecht, H. Schulz, M. Meyer, F. Baisch, L. Beck, M. Heer, H. Maass, P. H. Arbeille, F. Patat, and C. G. Blomqvist. The effects of a 10-day period of head-down tilt on the cardiovascular responses to intravenous saline loading. *Physiologist* 33: S171–S172, 1990.

95. Gaffney, F. A., J. V. Nixon, E. S. Karlsson, W. Campbell, A. B. C. Dowdey, and C. G. Blomqvist. Cardiovascular deconditioning produced by 20 hours of bedrest with head-down tilt (−5°) in middle-aged healthy men. *Am. J. Cardiol.* 56: 634–638, 1985.

96. Garland, D. E., C. A. Stewart, R. H. Adkins, S. S. Hu, C. Rosen, F. J. Liotta, and D. A. Weinstein. Osteoporosis after spinal cord injury. *J. Orthop. Res.* 10: 371–378, 1992.

97. Gauer, O. H., and J. P. Henry. Circulatory basis of fluid volume control. *Physiol. Rev.* 43: 423–481, 1963.

98. Gauer, O. H., J. P. Henry, and C. Behn. The regulation of extracellular volume. *Ann. Rev. Physiol.* 32: 547–595, 1970.

99. Gazenko, O. G., A. I. Grigor'yev, and A. D. Yegorov. Classification and periodicity of adaptive responses of humans on long-term space flight. In: *Mechanisms Underlying the Development of Stress: Stress Adaptation and Functional Disorders,* edited by F. I. Furduy, S. Kh. Kaydarliu, Ye. I. Shrirdy, A. I. Nadvodnyuk, and L. M. Mamalyga. Kishinev, Moldavia: Shtiintsa, 1987, p. 33–52.

100. Gazenko, O. G., V. I. Shumakov, L. I. Kakurin, V. E. Katkov, V. V. Chestukhin, V. M. Mikhailov, A. Z. Troshin, and V. N. Nesvetov. Central circulation and metabolism of the healthy man during postural exposures and arm exercise in the head-down position. *Aviat. Space Environ. Med.* 51: 113–120, 1980.

101. Genest, J., and M. Cantin. Regulation of body fluid volume: the atrial natriuretic factor. *Newsletter Int. Physiol. Soc.* 1: 3–5, 1986.

102. Genin, A. M., and P. A. Sorokin. Prolonged limitation of mobility as a model of the influence of weightlessness on the human organism. In: *Problems of Space Biology. Prolonged Limitation of Mobility and its Influence on the Human Organism,* edited by A. M. Genin and P. A. Sorokin. Moscow: Nauka, 1969, vol. 13, p. 9–16. (NASA TT F-639, 1970)

103. Genin, A. M., P. A. Sorokin, G. I. Gurvich, T. T. Dzhamgarov, A. G. Panov, I. I. Ivanov, and I. D. Pestov. Basic results from studies of the influence of 70-day hypodynamia on the human organism. In: *Problems of Space Biology. Prolonged Limitation of Mobility and its Influence on the Human Organism,* edited by A. M. Genin and P. A. Sorokin. Moscow: Nauka, 1969, vol. 13, p. 248–253. (NASA TT F-639, 1970)

104. Georgiyevskiy, V. S., V. A. Gornago, L. Ya. Divina, N. D. Kalmykova, V. M. Mikhaylov, V. I. Plakhatnyuk, Yu. D. Pometov, V. V. Smyshlyayeva, N. D. Vikharev, and B. S. Katkovskiy. Orthostatic stability in an experiment with 30-day hypodynamia. *Kosm. Biol. Aviakosm. Med.* 7: 61–68, 1973.

105. Georgiyevskiy, V. S., and V. M. Mikhaylov. Effect of hypokinesia on human circulation. *Kosm. Biol. Med.* 2: 48–51, 1968.

106. Gharib, C., G. Gauquelin, G. Geelen, M. Cantin, J. Gutkovska, J. L. Mauroux, and A. Güell. Volume regulating hormones (renin, aldosterone, vasopressin, and natriuretic factor) during simulated weightlessness. *Physiologist* 28: S30–S33, 1985.

107. Gharib, C., G. Gauquelin, G. Geelen, M. Vincent, F. Ghaemmaghami, Ch. Grange, M. Cantin, J. Gutkovska, and A. Güell. Levels of plasma atrial natriuretic factor (alpha hANF) during acute simulated weightlessness. In: *Proc. 2nd Int. Conf. Space Physiol., Toulouse, France, 1985.* Noordwijk: ESA Publications Division, p. 173–176. (ESA SP-237, 1986)

108. Gharib, C., G. Gauquelin, J. M. Pequignot, G. Geelen, C.-A. Bizollon, and A. Güell. Early hormonal effects of head-down tilt (−10°) in humans. *Aviat. Space Environ. Med.* 59: 624–629, 1988.

109. Gharib, C., A. Maillet, G. Gauquelin, A-M. Allevard, A. Güell, R. Cartier, and P. Arbeille. Results of a 4-week head-down tilt with and without LBNP countermeasure: 1. Volume regulating hormones. *Aviat. Space Environ. Med.* 63: 3–8, 1992.

110. Ghormley, R. K. The abuse of rest in bed in orthopedic surgery. *J.A.M.A.* 125: 1085–1087, 1944.

111. Gilmore, J. P. Neural control of extracellular volume in the human and nonhuman primate. In: *Handbook of Physiology. The Cardiovascular System. The Peripheral Circulation,* edited by J. T. Shepherd and F. M. Abboud. Bethesda MD: Am. Physiol. Soc., 1983, sect. 2, vol. III, pt. 2, chapt. 24, p. 885–915.

112. Gilmore, J. P., T. V. Peterson, and I. H. Zucker. Neither dorsal root nor baroreceptor afferents are necessary for eliciting the renal responses to acute intravascular volume expansion in the primate macaca fascicularis. *Circ. Res.* 45: 95–99, 1979.

113. Gmünder, F. K., F. Baisch, B. Bechler, A. Cogoli, M. Cogoli, P. W. Joller, H. Maass, J. Müller, and W. H. Ziegler. Effect of head-down tilt bedrest (10 days) on lymphocyte reactivity. *Acta Physiol. Scand.* 144 (S604): 131–141, 1992.

114. Goetz, K. L. Physiology and pathophysiology of atrial peptides. *Am. J. Physiol.* 254 (*Endocrinol. Metab.* 17): E1–E15, 1988.

115. Gogia, P., V. S. Schneider, A. D. LeBlanc, J. Krebs, C. Kasson, and C. Pientok. Bed rest effect on extremity muscle torque in healthy men. *Arch. Phys. Med. Rehabil.* 69: 1030–1032, 1988.

116. Goldberger, A. L., D. Goldwater, and V. Bhargava. Atropine unmasks bed-rest effect: a spectral analysis of cardiac interbeat intervals. *J. Appl. Physiol.* 61: 1843–1848, 1986.

117. Goldberger, A. L., and D. R. Rigney. Cardiovascular dynamics during space sickness and deconditioning. In: *Proc. First Joint NASA Cardiopulmonary Workshop,* edited by S. Fortney and A. R. Hargens. Houston, TX: NASA Contractor Report 10068, 1991, p. 155–163.

118. Goldsmith, S. R., G. S. Francis, and J. N. Cohn. Effect of head-down tilt on basal plasma norepinephrine and renin activity in humans. *J. Appl. Physiol.* 59: 1068–1071, 1985.

119. Goldwater, D., L. Montgomery, G. W. Hoffler, H. Sandler, and R. Popp. Echocardiographic and peripheral vascular responses of men (ages 46 to 55) to lower body negative pressure (LBNP) following 10 days of bed rest. Alexandria, VA: *Preprints, Aerospace Med. Assoc.,* 1979, p. 51–52.

120. Goldwater, D. J., and H. Sandler. Orthostatic and acceleration tolerance in 55 to 65 year old men and women after weightlessness simulation. Alexandria, VA: *Preprints, Aerospace Med. Assoc.,* 1982, p. 202–203.

121. Green, H. J., J. R. Sutton, G. Coates, M. Ali, and S. Jones. Response of red cell and plasma volume to prolonged training in humans. *J. Appl. Physiol.* 70: 1810–1815, 1991.

122. Greenleaf, J. E. Physiological responses to prolonged bed rest and fluid immersion in humans. *J. Appl. Physiol.: Respir. Environ. Exerc. Physiol.* 57: 619–633, 1984.

123. Greenleaf, J. E. Physiology of fluid and electrolyte responses during inactivity: water immersion and bed rest. *Med. Sci. Sports Exerc.* 16: 20–25, 1984.

124. Greenleaf, J. E. Hormonal regulation of fluid and electrolytes during prolonged bed rest: implications for microgravity. In: *Hormonal Regulation of Fluid and Electrolytes: Environmental Effects,* edited by J. R. Claybaugh and C. E. Wade. New York: Plenum, 1989, p. 215–232.

125. Greenleaf, J. E. Energy and thermal regulation during bed rest and spaceflight. *J. Appl. Physiol.* 67: 507–516, 1989.

126. Greenleaf, J. E. Problem: thirst, drinking behavior, and involuntary dehydration. *Med. Sci. Sports Exerc.* 24: 645–656, 1992.

127. Greenleaf, J. E. (editor). *Exercise Countermeasures for Bed-rest Deconditioning (1986): Final Report.* Moffett Field, CA: NASA Technical Memorandum 103987, 1993.

128. Greenleaf, J. E., E. M. Bernauer, A. C. Ertl, T. S. Trowbridge, and C. E. Wade. Work capacity during 30 days of bed rest with isotonic and isokinetic exercise training. *J. Appl. Physiol.* 67: 1820–1826, 1989.

129. Greenleaf, J. E., E. M. Bernauer, L. T. Juhos, H. L. Young, J. T. Morse, and R. W. Staley. Effects of exercise on fluid exchange and body composition in man during 14-day bed rest. *J. Appl. Physiol.: Respir. Environ. Exerc. Physiol.* 43: 126–132, 1977.

130. Greenleaf, J. E., E. M. Bernauer, H. L. Young, J. T. Morse, R. W. Staley, L. T. Juhos, and W. Van Beaumont. Fluid and electrolyte shifts during bed rest with isometric and isotonic exercise. *J. Appl. Physiol.: Respir. Environ. Exerc. Physiol.* 42: 59–66, 1977.

131. Greenleaf, J. E., C. J. Greenleaf, D. Van Derveer, and K. J. Dor-

chak. *Adaptation to Prolonged Bedrest in Man: A Compendium of Research*. Moffett Field, CA: NASA Technical Memorandum X-3307, 1976.

132. Greenleaf, J. E., R. F. Haines, E. M. Bernauer, J. T. Morse, H. Sandler, R. Armbruster, L. Sagan, and W. Van Beaumont. +Gz tolerance in man after 14-day bedrest periods with isometric and isotonic exercise conditioning. *Aviat. Space Environ. Med.* 46: 671–678, 1975.

133. Greenleaf, J. E., L. T. Juhos, and H. L. Young. Plasma lactic dehydrogenase activities in men during bed rest with exercise training. *Aviat. Space Environ. Med.* 56: 193–198, 1985.

134. Greenleaf, J. E., and S. Kozlowski. Physiological consequences of reduced physical activity during bed rest. In: *Exercise and Sport Sciences Reviews*, edited by R. L. Terjung. Philadelphia: Franklin Institute Press, 1982, vol. 10, p. 84–119.

135. Greenleaf, J. E., and S. Kozlowski. Reduction in peak oxygen uptake after prolonged bed rest. *Med. Sci. Sports Exerc.* 14: 477–480, 1982.

136. Greenleaf, J. E., and R. D. Reese. Exercise thermoregulation after 14 days of bed rest. *J. Appl. Physiol.: Respir. Environ. Exerc. Physiol.* 48: 72–78, 1980.

137. Greenleaf, J. E., and F. Sargent II. Voluntary dehydration in man. *J. Appl. Physiol.* 20: 719–724, 1965.

138. Greenleaf, J. E., L. Silverstein, J. Bliss, V. Langenheim, H. Rossow, and C. Chao. *Physiological Responses to Prolonged Bed Rest and Fluid Immersion in Man: A Compendium of Research (1974–1980)*. Moffett Field, CA: NASA Technical Memorandum 81324, 1982.

139. Greenleaf, J. E., K. Simanonok, E. M. Bernauer, C. E. Wade, and L. C. Keil. *Effect of Hemorrhage on Cardiac Output, Vasopressin, Aldosterone, and Diuresis During Immersion in Men*. Moffett Field, CA: NASA Technical Memorandum 103949, 1992.

140. Greenleaf, J. E., H. O. Stinnett, G. L. Davis, J. Kollias, and E. M. Bernauer. Fluid and electrolyte shifts in women during +Gz acceleration after 15 days' bed rest. *J. Appl. Physiol.: Respir. Environ. Exerc. Physiol.* 42: 67–73, 1977.

141. Greenleaf, J. E., W. Van Beaumont, E. M. Bernauer, R. F. Haines, H. Sandler, R. W. Staley, H. L. Young, and J. W. Yusken. Effects of rehydration on +Gz tolerance after 14 days bedrest. *Aerospace Med.* 44: 715–722, 1973.

142. Greenleaf, J. E., W. Van Beaumont, P. J. Brock, J. T. Morse, and G. R. Mangseth. Plasma volume and electrolyte shifts with heavy exercise in sitting and supine positions. *Am. J. Physiol.* 236 (*Regulatory Integrative Comp. Physiol.* 5): R206–R214, 1979.

143. Greenleaf, J. E., A. L. Van Kessel, W. Ruff, D. H. Card, and M. Rapport. Exercise temperature regulation in man in the upright and supine positions. *Med. Sci. Sports* 3: 175–182, 1971.

144. Greenleaf, J. E., J. Vernikos, C. E. Wade, and P. R. Barnes. Effect of leg exercise training on vascular volumes during 30 days of 6° head-down bed rest. *J. Appl. Physiol.* 72: 1887–1894, 1992.

145. Greenleaf, J. E., C. E. Wade, and G. Leftheriotis. Orthostatic responses following 30-day bed rest deconditioning with isotonic and isokinetic exercise training. *Aviat. Space Environ. Med.* 60: 537–542, 1989.

146. Grigor'yev, A. I., B. R. Dorokhova, G. I. Kozyrevskaya, Yu. V. Namochin, G. S. Arzamazov, and V. B. Noskov. Water-salt metabolism and the functional state of the kidneys during bedrest of varying duration. *Fiziol. Cheloveka* 5: 660–669, 1975.

147. Grigor'yev, A. I., B. S. Katkovskiy, A. A. Savilov, V. S. Georgiyevskiy, B. R. Dorokhova, and V. M. Mikhaylov. Effects of hyperhydration on human endurance of orthostatic and LBNP tests. *Kosm. Biol. Aviakosm. Med.* 12: 20–24, 1978.

148. Grigoriev, A. I., B. L. Lichardus, V. I. Lobachik, N. Mihailowsky, V. V. Zhidkov, and Yu. V. Sukhanov. Regulation of man's hydration status during gravity-induced blood redistribution. *Physiologist* 26: S28–S29, 1983.

149. Grundy, D., K. Reid, F. J. McArdle, B. H. Brown, D. C. Barber, C. F. Deacon, and I. W. Henderson. Trans-thoracic fluid shifts and endocrine responses to 6° head-down tilt. *Aviat. Space Environ. Med.* 62: 923–929, 1991.

150. Güell, A., L. Braak, A. P. LeTraon, and C. Gharib. Cardiovascular adaptation during simulated microgravity: Lower body negative pressure to counter orthostatic hypotension. *Aviat. Space Environ. Med.* 62: 331–335, 1991.

151. Güell, A., Ph. Dupui, G. Fanjaud, A. Bes, J. P. Moatti, and C. Gharib. Hydroelectrolytic and hormonal modifications related to prolonged bedrest in antiorthostatic position. *Acta Astronautica* 9: 589–592, 1982.

152. Gurfinkel, V. S., Ye. I. Pal'tsev, A. G. Feldman, and A. M. El'ner. Changes in certain human motor functions after prolonged hypodynamia. In: *Problems of Space Biology, Prolonged Limitation of Mobility and its Influence on the Human Organism*, edited by A. M. Genin and P. A. Sorokin. Moscow: Nauka, 1969, vol. 13, p. 148–161. (NASA TT F-639, 1970)

153. Haines, R. F. Effect of bed rest and exercise on body balance. *J. Appl. Physiol.* 36: 323–327, 1974.

154. Hainsworth, R. The importance of vascular capacitance in cardiovascular control. *Newsletter Int. Physiol. Soc.* 5: 250–254, 1990.

155. Hantman, D. A., J. M. Vogel, C. L. Donaldson, R. Friedman, R. S. Goldsmith, and S. B. Hulley. Attempts to prevent disuse osteoporosis by treatment with calcitonin, longitudinal compression and supplementary calcium and phosphate. *J. Clin. Endocrinol. Metab.* 36: 845–858, 1973.

156. Hargens, A. R. Fluid shifts in vascular and extravascular spaces during and after simulated weightlessness. *Med. Sci. Sports Exerc.* 15: 421–427, 1983.

157. Hargens, A. R., C. M. Tipton, P. D. Gollnick, S. J. Mubarak, B. J. Tucker, and W. H. Akeson. Fluid shifts and muscle function in humans during acute simulated weightlessness. *J. Appl. Physiol.: Respir. Environ. Exerc. Physiol.* 54: 1003–1009, 1983.

158. Hargens, A. R., D. E. Watenpaugh, and G. A. Breit. Control of circulatory function in altered gravitational fields. *Physiologist* 35: S80–S83, 1992.

159. Harper, C. M., and Y. M. Lyles. Physiology and complications of bed rest. *J. Am. Geriatr. Soc.* 36: 1047–1054, 1988.

160. Harrison, T. R. Abuse of rest as a therapeutic measure for patients with cardiovascular disease. *J.A.M.A.* 125: 1075–1077, 1944.

161. Heilskov, N. C. S., and F. Schónheyder. Creatinuria due to immobilization in bed. *Acta Med. Scand.* 151: 51–56, 1955.

162. Henriksen, O., and P. Sejrsen. Local reflex in microcirculation in human skeletal muscle. *Acta Physiol. Scand.* 99: 19–26, 1977.

163. Heymans, C., and E. Neil. Histology, embryology, anatomy and blood supply of the carotid and aortic chemoreceptor areas. In: *Reflexogenic Areas of the Cardiovascular System*, edited by C. Heymans and E. Neil. Boston: Little, Brown, 1958, p. 114–130.

164. Hickson, R. C., H. Bomze, and J. O. Holloszy. Linear increase in aerobic power induced by a strenuous program of endurance exercise. *J. Appl. Physiol.: Respir. Environ. Exerc. Physiol.* 42: 372–376, 1977.

165. Howard, J. E., W. Parson, and R. S. Bigham, Jr. Studies on patients convalescent from fracture. III. The urinary excretion of calcium and phosphorus. *Bull. Johns Hopkins Hosp.* 77: 291–313, 1945.

166. Hulley, S. B., J. M. Vogel, C. L. Donaldson, J. H. Bayers, R. J. Friedman, and S. N. Rosen. The effect of supplemental oral phosphate on the bone mineral changes during prolonged bedrest. *J. Clin. Invest.* 50: 2506–2518, 1971.

167. Hung, J., D. Goldwater, V. A. Convertino, J. H. McKillop, M.

L. Goris, and R. F. DeBusk. Mechanisms for decreased exercise capacity after bed rest in normal middle-aged men. *Am. J. Cardiol.* 51: 344–348, 1983.

168. Huntoon, C. L., P. C. Johnson, and N. M. Cintron. Hematology, immunology, endocrinology, and biochemistry. In: *Space Physiology and Medicine,* edited by A. Nicogossian, C. Huntoon, and S. Pool. Philadelphia: Lea & Febiger, 1989, p. 222–226.

169. Hwang, I. S., K. Hill, V. Schneider, and C. Y. C. Pak. Effect of prolonged bedrest on the propensity for renal stone formation. *J. Clin. Endocrinol. Metab.* 66: 109–112, 1988.

170. Hyatt, K. H. Hemodynamic and body fluid alterations induced by bedrest. In: *Hypodynamic and Hypogravic Environments,* edited by R. H. Murray and M. McCally. Washington, DC: NASA Special Publication 269, 1971, p. 187–209.

171. Hyatt, K. H., P. C. Johnson, G. W. Hoffler, P. C. Rambaut, J. A. Rummel, S. B. Hulley, J. M. Vogel, C. Huntoon, and C. P. Spears. Effect of potassium depletion in normal males: an Apollo 15 simulation. *Aviat. Space Environ. Med.* 46: 11–15, 1975.

172. Hyatt, K. H., L. G. Kamenetsky, and W. M. Smith. Extravascular dehydration as an etiologic factor in post-recumbency orthostatism. *Aerospace Med.* 40: 644–650, 1969.

173. Hyatt, K. H., W. M. Smith, J. M. Vogel, R. W. Sullivan, W. R. Vetter, B. E. Calder, B. J. Bohnn, and V. M. Haughton. *A Study of the Role of Extravascular Dehydration in the Production of Cardiovascular Deconditioning by Simulated Weightlessness (Bedrest).* Washington, DC: NASA Contractor Report 114808 (pt. 1); 114809 (pt. 2), 1970.

174. Hyatt, K. H., R. W. Sullivan, W. R. Spears, and W. R. Vetter. *A Study of Ventricular Contractility and Other Parameters Possibly Related to Vasodepressor Syncope.* Washington, DC: NASA Contractor Report 128968, 1973.

175. Hyatt, K. H., and D. A. West. Reversal of bedrest-induced orthostatic intolerance by lower body negative pressure and saline. *Aviat. Space Environ. Med.* 48: 120–124, 1977.

176. Ioseliani, K. K. Man's mental performance under conditions of prolonged hypokinesia with use of lower body negative pressure. *Kosm. Biol. Aviakosm. Med.* 8: 86–87, 1974.

177. Iseyev, L. R., and B. S. Katkovskiy. Unidirectional changes in the human oxygen balance caused by bed confinement and restriction to an isolation chamber. *Kosm. Biol. Med.* 2: 67–72, 1968.

178. Ivanov, I. I., B. F. Korovkin, and N. P. Mikhaleva. Investigation of certain biochemical blood serum indicators during prolonged hypodynamia. In: *Problems of Space Biology. Prolonged Limitation of Mobility and its Influence on the Human Organism,* edited by A. M. Genin and P. A. Sorokin. Moscow: Nauka, 1969, vol. 13, p. 100–107. (NASA TT F-639, 1970)

179. Jacobson, L. B., K. H. Hyatt, and H. Sandler. Effects of simulated weightlessness on responses of untrained men to +Gz acceleration. *J. Appl. Physiol.* 36: 745–752, 1974.

180. Johansson, J. E. Daily fluctuations of metabolism and body temperatures in sober condition and complete muscular rest. *Scand. Arch. Physiol.* 8: 85–142, 1898.

181. Johnson, P. C. Fluid volumes changes induced by spaceflight. *Acta Astronautica* 6: 1335–1341, 1979.

182. Johnson, P. C. The erythropoietic effects of weightlessness. In: *Current Concepts in Erythropoiesis,* edited by C. D. R. Dunn. New York: Wiley, 1983, p. 279–300.

183. Jones, H. H., J. D. Priest, W. C. Hayes, C. C. Tichenor, and D. A. Nagel. Humeral hypertrophy in response to exercise. *J. Bone Joint Surg.* 59: 204–208, 1977.

184. Judge, D., V. Schneider, and A. LeBlanc. Disuse osteoporosis: histomorphometry in bed rest. *J. Bone Miner. Res.* 4: S238, 1989.

185. Kakurin, L. I. Effect of long-term hypokinesia on the human body and the hypokinetic component of weightlessness. *Kosm. Biol. Med.* 2: 59–63, 1968.

186. Kakurin, L. I. Simulation of the physiological effects of weightlessness. Moscow: USSR Academy of Sciences, 1976. (NASA TT F-17285)

187. Kakurin, L. I., B. S. Katkovskiy, V. S. Georgiyevskiy, Yu. N. Purakhin, M. A. Cherepakhin, V. M. Mikhaylov, B. N. Petukhov, and Ye. N. Biryukov. Functional disturbances during hypokinesia in man. *Vopr. Kurotol. Fizioter. Lech. Fiz. Kult.* 35: 19–24, 1970.

188. Kakurin, L. I., B. S. Katkovskiy, V. A. Tishler, G. I. Kozyrevskaya, V. S. Shashkov, V. S. Georgiyevskiy, A. I. Grigor'yev, V. M. Mikhaylov, O. D. Anashkin, G. V. Machinskiy, A. Savilov, and Ye. P. Tikhomirov. Substantiation of a set of preventative measures referable to the objectives of missions in the Salyut orbital station. *Kosm. Biol. Aviakosm. Med.* 12: 20–26, 1978.

189. Kakurin, L. I., V. I. Lobachik, V. M. Mikhailov, and Yu. A. Senkevich. Antiorthostatic hypokinesia as a method of weightlessness simulation. *Aviat. Space Environ. Med.* 47: 1083–1086, 1976.

190. Kalin, G. S., and V. G. Terent'yev. State of nervous-system functions during aftereffects of hypodynamia. In: *Problems of Space Biology. Prolonged Limitation of Mobility and its Influence on the Human Organism,* edited by A. M. Genin, and P. A. Sorokin. Moscow: Nauka, 1969, vol. 13, p. 214–220. (NASA TT F-639, 1970)

191. Kaplanskiy, A. S., Ye. A. Savina, P. B. Kazakova, I. P. Khoroshilova-Maslova, G. M. Kharin, V. I. Yakovleva, G. I. Plakhuta-Plakutina, G. N. Durnova, Ye. I. Il'ina-Kakuyeva, Ye. I. Alekseyev, A. S. Pankova, V. N. Shvets, and T. Ye. Burkovskaya. Morphological study of antiorthostatic hypokinesia in monkeys. *Kosm. Biol. Aviakosm. Med.* 19: 53–60, 1985.

192. Karim, F., C. Kidd, C. M. Malpus, and P. E. Penna. The effects of stimulation of the left atrial receptors on sympathetic efferent nerve activity. *J. Physiol. (Lond.)* 227: 243–260, 1972.

193. Katkov, V. E., V. V. Chestukhin, and L. I. Kakurin. Coronary circulation of the healthy man exposed to tilt tests, LBNP, and head-down tilt. *Aviat. Space Environ. Med.* 56: 741–747, 1985.

194. Katkov, V. E., V. V. Chestukhin, E. M. Nikolayenko, S. V. Gvozdev, V. V. Rumyantsev. T. M. Guseynova, and I. A. Yegorova. Central circulation in the healthy man during 7-day head-down hypokinesia. *Kosm. Biol. Aviakosm. Med.* 16: 45–51, 1982.

195. Katkov, V. E., V. V. Chestukhin, E. M. Nikolayenko, V. V. Rumyantsev, and S. V. Gvozdev. Central circulation of a normal man during 7-day head-down tilt and decompression of various body parts. *Aviat. Space Environ. Med.* 54 (Suppl. 1): S24–S30, 1983.

196. Katkov, V. E., V. V. Chestukhin, O. Kh. Zybin, S. S. Sukhotskiy, S. V. Abrosimov, and V. N. Utkin. The effect of short-term antiorthostatic hypokinesia on central and intracardiac hemodymamics and metabolism of a healthy person. *Kardiologiya* 12: 69–75, 1978. (NASA TM-76525, 1981)

197. Katkov, V. E., L. I. Kakurin, V. V. Chestukhin, and K. Kirsch. Central circulation during exposure to 7-day microgravity (head-down tilt, immersion, space flight). *Physiologist* 30: S36–S41, 1987.

198. Katkovskiy, B. S. Human basal metabolism during prolonged bedrest. *Kosm. Biol. Med.* 1: 67–71, 1967.

199. Katkovskiy, B. S., and Yu. D. Pometov. Change in cardiac ejection under the influence of 15-day bed confinement. *Kosm. Biol. Aviakosm. Med.* 5: 69–74, 1971.

200. Keys, A. Deconditioning and reconditioning in convalescence. *Surg. Clin. North Am.* 25: 442–454, 1945.

201. Khilov, K. L., A. Ye. Kurashvili, and V. P. Rudenko. Influence of prolonged hypodynamia on the state of the vestibular ana-

lyzer. In: *Problems of Space Biology. Prolonged Limitation of Mobility and its Influence on the Human Organism*, edited by A. M. Genin, and P. A. Sorokin. Moscow: Nauka, 1969, vol. 13, p. 182–188. (NASA TT F-639, 1970)

202. Kimzey, S. L. The effects of extended spaceflight on hematologic and immunologic systems. *J. Am. Med. Wom. Assoc.* 30: 218–232, 1975.

203. Kimzey, S. L. A review of hematology studies associated with space flight. *Biorheology* 16: 13–21, 1979.

204. Kimzey, S. L., J. I. Leonard, and P. C. Johnson. A mathematical and experimental simulation of the hematological response to weightlessness. *Acta Astronautica* 6: 1289–1303, 1979.

205. Kiselev, R. K., I. S. Balakhovskiy, and O. A. Virovets. Change in hemoglobin mass during prolonged hypokinesia. *Kosm. Biol. Aviakosm. Med.* 9: 80–84, 1975.

206. Kiselev, R. K., A. M. Chayka, and V. I. Legenkov. Effect of coamide and folicobalamin on erythropoiesis under normal living conditions and during antiorthostatic hypokinesia. *Kosm. Biol. Aviakosm. Med.* 20: 48–53, 1986.

207. Kjellberg, S. R., U. Rudhe, and T. Sjöstrand. Increase of the amount of hemoglobin and blood volume in connection with physical training. *Acta Physiol. Scand.* 19: 146–151, 1949.

208. Konstantinova, I. V. Space flight factors and the human immune system: Hypokinesia. In: *The Immune System Under Extreme Conditions: Space Immunology* (Problems of Space Biology No. 59). Moscow: Nauka, 1988, p. 125–146.

209. Korolev, B. A. Changes in myocardial repolarization in healthy persons during restriction of motor activity. *Kosm. Biol. Aviakosm. Med.* 2: 81–85, 1968.

210. Korolev, B. A. Pattern of changes of electrocardiograms and cardiac contraction phases during orthostatic tests after long-term hypokinesia. *Kosm. Biol. Aviakosm. Med.* 3: 67–71, 1969.

211. Koroleva-Munts, V. M. Circadian rhythm of physiological functions in clinostatic hypokinesia. *Fiziol. Zh. SSSR* 60: 1145–1149, 1974.

212. Kotovskaya, A. R., L. N. Gavrilova, and R. R. Galle. Effect of hypokinesia in head-down position on man's equilibrium function. *Kosm. Biol. Aviakosm. Med.* 15: 26–29, 1981.

213. Kovalenko, Ye. A., and N. N. Gurovskiy. Circulatory system change during hypokinesia. In: *Hypokinesia*. Moscow: Meditsina, 1980, p. 107–208. (NASA TM 76395)

214. Kovalenko, Ye. A., and I. I. Kasyan. On the pathogenesis of weightlessness. *Space Med.* 8: 9–18, 1989.

215. Kozar, M. I. Effect of Spaceflight Factors on Indices of Natural Antibacterial Body Resistance (in Russian). Moscow: 1966. Doctoral dissertation.

216. Krasnykh, I. G. Influence of prolonged hypodynamia on heart size and the functional state of the myocardium. *Probl. Kosm. Biol.* 13: 65–71, 1969.

217. Krebs, J. M., V. S. Schneider, H. Evans, M. C. Kuo, and A. D. LeBlanc. Energy absorption, lean body mass, and total body fat changes during 5 weeks of continuous bed rest. *Aviat. Space Environ. Med.* 61: 314–318, 1990.

218. Krebs, J. M., V. S. Schneider, and A. D. LeBlanc. Zinc, copper, and nitrogen balances during bed rest and fluoride supplementation in healthy adult males. *Am. J. Clin. Nutr.* 47: 509–514, 1988.

219. Krotov, V. P. Water metabolism regulating mechanisms in hypokinesia. *Patol. Fiziol. Eksp. Ter.* 10: 15–18, 1980. (NASA TM 76309, 1980)

220. Krotov, V. P., and L. L. Romanovskaya. Effects exerted on water metabolism by body position relative to the gravitational vector. *Kosm. Biol. Aviakosm. Med.* 10: 37–41, 1976.

221. Krotov, V. P., A. A. Totov, Ye. A. Kovalenko, V. V. Bogomolov, L. L. Stazhadze, and V. P. Masenko. Changes in fluid metabolism during prolonged hypokinesia with the body in antiorthostatic position. *Kosm. Biol. Aviakosm. Med.* 11: 32–37, 1977.

222. Krupina, T. N., and A. Ya. Tizul. The significance of prolonged clinostatic hypodynamia in the clinical picture of nervous diseases. *Zh. Nevropatol. Psikhiatr.* 68: 1008–1014, 1968.

223. Krupina, T. N., and A. Ya. Tizul. Changes in the nervous system during a 120-day clinostatic hypokinesia and the prophylaxis of hypokinesic disorders. *Zh. Nevropatol. Psikhiatr.* 71: 1611–1617, 1971.

224. Krupina, T. N., and A. Ya. Tizul. Clinical aspects of changes in the nervous system in the course of 49-day antiorthostatic hypokinesia. *Kosm. Biol. Aviakosm. Med.* 11: 26–31, 1977.

225. Krupina, T. N., A. Ya. Tizul, N. M. Boglevskaya, B. P. Baranova, E. I. Matsnev, and Ye. A. Chertovskikh. Functional changes in the nervous system and functioning of certain analyzers in response to the combined effect of hypokinesia and radial acceleration. *Kosm. Biol. Med.* 1: 61–66, 1967.

226. Kut'kova, O. N., Ye. I. Kuznets, E. V. Yakovleva, G. A. Shalnova, A. F. Bobrov, P. T. Yastrebov, A. D. Nevinnaya, and B. A. Utekhin. Changes in immunological protection factors in humans undergoing simulated weightlessness. In: *Symposium Dedicated to K. E. Tsiolkovskiy*, edited by V. B. Malinin, F. P. Kosmolinsky, and Ye. I. Kuznets. Kaluga: 1983–1984. p. 40–45.

227. Kuznets, Ye. I., O. N. Kut'kova, E. V. Yakovleva, G. A. Shal'nova, I. I. Malkiman, and P. T. Yastrebov. Selection of parameters indicative of human immune status under conditions simulating space flight factors. In: *Symposium Dedicated to K. E. Tsiolkovskiy*, edited by V. B. Malinin, F. P. Kosmolinskiy, and Ye. I. Kuznets. Kaluga: 1988. p. 101–105.

228. Lancaster, M. C. Hematologic aspects of bed rest. In: *Hypogravic and Hypodynamic Environments*, edited by R. H. Murray and M. McCally. Washington, DC: NASA Special Publication 269, 1971, p. 299–307.

229. Leach, C. S., S. B. Hulley, P. C. Rambaut, and L. F. Dietlein. The effect of bedrest on adrenal function. *Space Life Sci.* 4: 415–423, 1973.

230. Leach, C. S., P. C. Johnson, and N. M. Cintron. The endocrine system in space flight. *Acta Astronautica* 17: 161–166, 1988.

231. LeBlanc, A. D., H. J. Evans, C. Marsh, V. Schneider, P. C. Johnson, and S. G. Jhingran. Precision of dual photon absorptiometry measurements. *J. Nucl. Med.* 27: 1362–1365, 1986.

232. LeBlanc, A., P. Gogia, V. Schneider, J. Krebs, E. Schonfeld, and H. Evans. Calf muscle area and strength changes after five weeks of horizontal bed rest. *Am. J. Sports Med.* 16: 624–629, 1988.

233. LeBlanc, A. D., V. S. Schneider, H. J. Evans, C. Pientok, R. Rowe, and E. Spector. Regional changes in muscle mass following 17 weeks of bed rest. *J. Appl. Physiol.* 73: 2172–2178, 1992.

234. LeBlanc, A. D., V. S. Schneider, H. J. Evans, D. A. Engelbreston, and J. M. Krebs. Bone mineral loss and recovery after 17 weeks of bed rest. *J. Bone Miner. Res.* 5: 843–850, 1990.

235. LeBlanc, A., V. Schneider, J. Krebs, H. Evans, S. Jhignran, and P. Johnson. Spinal bone mineral after 5 weeks of bed rest. *Calcif. Tissue Int.* 41: 259–261, 1987.

236. Lemann, J. The urinary excretion of calcium, magnesium and phosphorus. In: *Primer on the Metabolic Bone Diseases and Disorders of Mineral Metabolism*, edited by M. J. Favus. Richmond: William Byrd, 1990, p. 36–39.

237. Lepkovsky, S. The appetite factor. In: *Nutrition in Space and Related Waste Problems*, edited by T. C. Helvey. Washington, DC: NASA Special Publication 70, 1964, p. 191–194.

238. Levick, J. R., and C. C. Michel. The effects of position and skin temperature on the capillary pressures in the fingers and toes. *J. Physiol. (Lond.).* 274: 97–109, 1978.

239. Lkhagva, L. Circadian rhythm of human body temperature in antiorthostatic position. *Kosm. Biol. Aviakosm. Med.* 14: 59–61, 1980.

240. Lobachik, V. I., V. V. Zhidkov, and S. V. Abrosimov. Body fluid

status during a 120-day period of hypokinesis with head-down tilt. *Kosm. Biol. Aviakosm. Med.* 23: 57–61, 1989.

241. Lockwood, D. R., J. M. Vogel, V. S. Schneider, and S. B. Hulley. Effect of the diphosphonate EHDP on bone mineral metabolism during prolonged bed rest. *J. Clin. Endocrinol. Metab.* 41: 533–541, 1975.

242. Löllgen, H., U. Gebhardt, J. Beier, J. Hordinsky, H. Borger, V. Sarrasch, and K. E. Klein. Central hemodynamics during zero gravity simulated by head-down bedrest. *Aviat. Space Environ. Med.* 55: 887–892, 1984.

243. London, G. M., J. A. Levenson, M. E. Safar, A. C. Simon, A. P. Guerin, and D. Payen. Hemodynamic effects of head-down tilt in normal subjects and sustained hypertensive patients. *Am. J. Physiol.* 245 (*Heart Circ. Physiol.* 14): H194–H202, 1983.

244. Louisy, F., C. Gaudin, J. M. Oppert, A. Güell, and C. Y. Guezennec. Haemodynamics of leg veins during a 30 days–6 degrees head-down bedrest with and without lower body negative pressure. *Eur. J. Appl. Physiol.* 61: 349–355, 1990.

245. Ludbrook, J. The musculovenous pumps of the human lower limb. *Am. Heart J.* 71: 635–641, 1966.

246. Lutwak, L., and G. D. Whedon. The effect of physical conditioning on glucose tolerance. *Clin. Res.* 7: 143–144, 1959.

247. Luu, P., V. Ortiz, P. R. Barnes, and J. E. Greenleaf. *Physiological Responses to Prolonged Bed Rest in Humans: A Compendium of Research (1981–1988).* Moffett Field, CA: NASA Technical Memorandum 102249, 1990.

248. Mack, G. W., B. M. Quigley, T. Nishiyasu, X. Shi, and E. R. Nadel. Cardiopulmonary baroreflex control of forearm vascular resistance after acute blood volume expansion. *Aviat. Space Environ. Med.* 62: 938–943, 1991.

249. Margen, S., J. Y. Chu, N. A. Kaufmann, and D. H. Calloway. Studies in calcium metabolism. I. The calciuretic effect of dietary protein. *Am. J. Clin. Nutr.* 27: 584–589, 1974.

250. Marshall, J. M. The venous vessels within skeletal muscle. *Newsletter Int. Physiol. Soc.* 6: 11–15, 1991.

251. Maslov, I. A. Mental states during prolonged hypokinesia. *Zh. Nevropatol. Psikhiatr.* 68: 1031–1034, 1968. (NASA TT F-15, 585, 1974)

252. Maurice, M., B. Roussel, H. Mehier, G. Gauquelin, and C. Gharib. Relationship between hormones and brain water content measured by 1H magnetic resonance spectroscopy during simulated weightlessness in man. *Physiologist* 33: S104–S105, 1990.

253. Mazess, R. B., and H. S. Barden. Measurement of bone by dual-photon absorptiometry (DPA) and dual-energy X-ray absorptiometry (DEXA). *Ann. Chir. Gynaecol.* 77: 197–203, 1988.

254. Mazess, R., B. Collick, J. Trempe, and J. Hanson. Performance evaluation of a dual-energy x-ray bone densitometer. *Calcif. Tissue Int.* 44: 228–232, 1989.

255. McCally, M., T. E. Piemme, and R. H. Murray. Tilt table responses of human subjects following application of lower body negative pressure. *Aerospace Med.* 37: 1247–1249, 1966.

256. Melada, G. A., R. H. Goldman, J. A. Luetscher, and P. G. Zager. Hemodynamics, renal function, plasma renin, and aldosterone in man after 5 to 14 days of bedrest. *Aviat. Space Environ. Med.* 46: 1049–1055, 1975.

257. Melchior, F. M., and S. M. Fortney. Orthostatic intolerance during a 13-day bed rest does not result from increased leg compliance. *J. Appl. Physiol.* 74: 286–292, 1993.

258. Menninger, K. The abuse of rest in psychiatry. *J.A.M.A.* 125: 1087–1090, 1944.

259. Menninger, R. P., R. C. Mains, F. W. Zechman, and T. A. Piemme. Effect of two weeks bed rest on venous pooling in the lower limbs. *Aerospace Med.* 40: 1323–1326, 1969.

260. Mikhailov, V. M., G. V. Machinskiy, V. P. Buzulina, V. S. Geogriyevskiy, E. N. Nechayeva, and S. G. Kryutchenko. Tolerance for provocative tests under conditions of a 1-year exposure to

hypokinesia with head-down tilt. *Kosm. Biol. Aviakosm. Med.* 23: 54–56, 1989.

261. Mikhasev, M. I., V. I. Sokolkov, and M. A. Tikhonov. Certain peculiarities of external respiration and gas exchange during prolonged hypodynamia. In: *Problems of Space Biology. Prolonged Limitation of Mobility and its Influence on the Human Organism,* edited by A. M. Genin and P. A. Sorokin. Moscow: Nauka, 1969, vol. 13, p. 72–78. (NASA TT F-639, 1970)

262. Mikhaylovskiy, G. P., N. N. Dobronravova, M. I. Kozar, M. M. Korotayev, N. I. Tsiganova, V. M. Shilov, and I. Ya. Yakovleva. Variation in overall body tolerance during a 62-day exposure to hypokinesia and acceleration. *Kosm. Biol. Med.* 1: 66–70, 1967.

263. Miller, P. B., B. O. Hartman, R. L. Johnson, and L. E. Lamb. Modification of the effects of two weeks of bedrest upon circulatory functions in man. *Aerospace Med.* 35: 931–939, 1964.

264. Miller, P. B., R. L. Johnson, and L. E. Lamb. Effects of four weeks of absolute bed rest on circulatory functions in man. *Aerospace Med.* 35: 1194–1200, 1964.

265. Miller, P. B., R. L. Johnson, and L. E. Lamb. Effects of moderate physical exercise during four weeks of bed rest on circulatory functions in man. *Aerospace Med.* 36: 1077–1082, 1965.

266. Moreno, F., and H. A. Lyons. Effect of body posture on lung volumes. *J. Appl. Physiol.* 16: 27–29, 1961.

267. Morice, A., J. Pepke-Zaba, E. Loysen, R. Lapworth, M. Ashby, T. Higenbottam, and M. Brown. Low dose infusion of atrial natriuretic peptide causes salt and water excretion in normal man. *Clin. Sci.* 74: 359–363, 1988.

268. Mukhina, N. N., V. V. Chestukhin, V. Ya. Katkov, and A. P. Karpov. Effect of brief antiorthostatic hypokinesia on blood immunoglobulin content. *Kosm. Biol. Aviakosm. Med.* 14: 74–75, 1980.

269. Müller, E. A. Influence of training and of inactivity on muscle strength. *Arch. Phys. Med.* 51: 449–462, 1970.

270. Mundy, G. Bone resorbing cells. In: *Primer on the Metabolic Bone Diseases and Disorders of Mineral Metabolism,* edited by M. J. Favus. Richmond: William Byrd, 1990, p. 18–22.

271. Murthy, G., R. J. Marchbanks, D. E. Watenpaugh, J.-U. Meyer, N. Eliashberg, and A. R. Hargens. Increased intracranial pressure in humans during simulated microgravity. *Physiologist* 35: S184–S185, 1992.

272. Myasnikov, V. I. Characteristics of the sleep of men in simulated space flights. *Aviat. Space Environ. Med.* 46: 401–408, 1975.

273. Natelson, B. H., C. DeRoshia, and B. E. Levin. Physiological effects of bed rest. *Lancet* 1: 51, 1982.

274. Natochin, Yu. V., A. I. Grigoriev, V. B. Noskov, R. G. Parnova, Yu. V. Sukhanov, D. L. Firsov, and E. I. Shakhmatova. Mechanism of postflight decline in osmotic concentration of urine in cosmonauts. *Aviat. Space Environ. Med.* 62: 1037–1043, 1991.

275. Nazar, K., J. E. Greenleaf, D. Philpott, E. Pohoska, K. Olszewska, and H. Kaciuba-Uscilko. Muscle mitochondrial density after exhaustive exercise in dogs: prolonged restricted activity and retraining. *Aviat. Space Environ. Med.* 64: 306–313, 1993.

276. Nazar, K., J. E. Greenleaf, E. Pohoska, E. Turlejska, H. Kaciuba-Uscilko, and S. Kozlowski. Exercise performance, core temperature, and metabolism after prolonged restricted activity and retraining in dogs. *Aviat. Space Environ. Med.* 63: 684–688, 1992.

277. Neer, R. M. Calcium and inorganic phosphate homeostasis. In: *Endocrinology,* edited by L. J. DeGroot, New York: Saunders, 1989, vol. 2, p. 927–953.

278. Nelson, C. V., P. W. Rand, E. T. Angelakos, and P. G. Hugenholtz. Effect of intracardiac blood on the spatial vectorcardiogram. I. Results in the dog. *Circ. Res.* 31: 95–104, 1972.

279. Nicogossian, A. E. T., A. A. Whyte, H. Sandler, C. S. Leach, and P. C. Rambaut. *Chronological Summaries of United States,*

European, and Soviet Bedrest Studies. Washington, DC: NASA, 1979.

280. Nikolayenko, E. M., V. Ye. Katkov, S. V. Gvozdev, V. V. Chestukhin, M. I. Volkova, and M. I. Berkovskaya. Respiratory tract "closing volume" and structure of total lung capacity during seven-day hypokinesia in head-down position. *Kosm. Biol. Aviakosm. Med.* 17: 39–43, 1983.

281. Nixon, J. V., R. G. Murray, C. Bryant, R. L. Johnson, Jr., J. H. Mitchell, O. B. Holland, C. Gomez-Sanchez, P. Vergne-Marini, and C. G. Blomqvist. Early cardiovascular adaptation to simulated zero gravity. *J. Appl. Physiol.: Respir. Environ. Exerc. Physiol.* 46: 541–548, 1979.

282. Noskov, V. B., G. I. Kozyrevskaya, B. V. Morukov, Ye. M. Artamasova, and L. A. Rustam'yan. Body position during hypokinesia, and fluid-electrolyte metabolism. *Kosm. Biol. Aviakosm. Med.* 19: 31–34, 1985.

283. Pak, C. Y. C., C. Skurla, and J. Harvey. Graphic display of urinary risk factors for renal stone formation. *J. Urol.* 134: 867–870, 1985.

284. Pak, Z. P., G. I. Kozyrevskaya, Yu. S. Koloskova, A. I. Grigor'yev, Yu. Ye. Bezumova, and Ye. N. Biryukov. Peculiarities of water-mineral metabolism during 120-day hypokinesia. *Kosm. Biol. Aviakosm. Med.* 7: 56–59, 1973.

285. Panferova, N. Ye. Cardiovascular system during hypokinesia of different duration and degree of expression. *Kosm. Biol. Aviakosm. Med.* 10: 15–20, 1976.

286. Panferova, N. Ye. Heat regulation under prolonged limitation of muscular activity. *Fiziol. Cheloveka* 4: 835–839, 1978.

287. Pannier, B. M., P. J. Lacolley, C. Gharib, G. M. London, J. L. Cuche, J. L. Duchier, B. I. Levy, and M. E. Safar. Twenty-four hours of bed rest with head-down tilt: venous and arteriolar changes in limbs. *Am. J. Physiol.* 260 (*Heart Circ. Physiol.* 29): H1043–H1050, 1991.

288. Panov, A. G., and V. S. Lobzin. Some neurological problems in space medicine. *Kosm. Biol. Med.* 2: 59–67, 1968.

289. Parin, V. V., T. N. Krupina, G. P. Mikhaylovskiy, and A. Ya. Tizul. Principal changes in the healthy human body after a 120-day bed confinement. *Kosm. Biol. Med.* 4: 59–64, 1970.

290. Pequinot, J. M., A. Güell, G. Gauquelin, E. Jarsaillon, G. Annat, A. Bes, L. Peyrin, and C. Gharib. Epinephrine, norepinephrine, and dopamine during a 4-day head-down bed rest. *J. Appl. Physiol.* 58: 157–163, 1985.

291. Pestov, I. D., and B. F. Asyamolov. Negative pressure on the lower part of the body as a method for preventing shifts associated with change in hydrostatic blood pressure. *Kosm. Biol. Aviakosm. Med.* 6: 59–64, 1972.

292. Plum, F., and M. F. Dunning. The effect of therapeutic mobilization on hypercalciuria following acute poliomyelitis. *Arch. Intern. Med.* 101: 528–536, 1958.

293. Pohoska, E. The effect of restriction of physical activity on adaptation to prolonged exercise in dogs. *Acta Physiol. Pol.* 27: 199–202, 1976.

294. Polese, A., D. Goldwater, L. London, D. Yuster, and H. Sandler. Resting cardiovascular effects of horizontal (0°) and head-down (−6°) bed rest (BR) on normal men. Alexandria, VA: *Preprints, Aerospace Med. Assoc.* 1980, p. 24–25.

295. Pometov, Yu. D., and B. S. Katkovskiy. Variations in cardiac output and gas exchange at rest during hypokinesia. *Kosm. Biol. Aviakosm. Med.* 6: 39–46, 1972.

296. Powers, J. H. The abuse of bed rest as a therapeutic measure in surgery. *J.A.M.A.* 125: 1079–1083, 1944.

297. Prosnitz, E. H., and G. F. DiBona. Effect of decreased renal sympathetic nerve activity on renal tubular sodium reabsorption. *Am. J. Physiol.* 235 (*Renal Fluid Electrolyte Physiol.* 4): F557–F563, 1978.

298. Pruss, G. M., and V. I. Kuznetsov. Contractile function of the myocardium during hypodynamia. *Kosm. Biol. Aviaksom. Med.* 8: 45–49, 1974.

299. Purakhin, Yu. N., L. I. Kakurin, V. S. Georgiyevskiy, B. N. Petukhov, and V. M. Mikhaylov. Regulation of vertical posture after flight on the 'SOYUZ-6'-'SOYUZ-8' ships and 120-day hypokinesia. *Kosm. Biol. Med.* 6: 47–53, 1972.

300. Purakhin, Yu. N., and B. N. Petukhov. Neurological changes in healthy subjects induced by two-month hypokinesia. *Kosm. Biol. Med.* 2: 51–56, 1968.

301. Puzas, J. The osteoblast. In: *Primer on the Metabolic Bone Diseases and Disorders of Mineral Metabolism,* edited by M. J. Favus. Richmond: William Byrd, 1990, p. 11–15.

302. Rattan, S. N., R. M. Glaser, F. J. Servedio, and S. R. Collins. Skeletal muscle pumping via voluntary and electrical induced contractions. *Physiologist* 28: 363, 1985.

303. Rigg, J. R. A., A. S. Rebuck, and E. J. M. Campbell. Effect of posture on the ventilatory response to CO_2. *J. Appl. Physiol.* 37: 487–490, 1974.

304. Riley, R. L., S. Permutt, S. Said, M. Godfrey, T. O. Cheng, J. B. L. Howell, and R. H. Shepard. Effect of posture on pulmonary dead space in man. *J. Appl. Physiol.* 14: 339–344, 1959.

305. Röcker, L., K. Kirsch, J. Wicke, and H. Stoboy. Role of proteins in the regulation of plasma volume during heat stress and exercise. *Isr. J. Med. Sci.* 12: 840–843, 1976.

306. Rodahl, K., N. C. Birkhead, J. J. Blizzard, B. Issekutz, Jr., and E. D. R. Pruett. Physiological changes during prolonged bed rest. In: *Nutrition and Physical Activity,* edited by G. Blix. Uppsala: Almqvist & Wiksells, 1967, p. 107–113.

307. Rodwell, V. W. Biosynthesis of the nutritionally non-essential amino acids. In: *Harper's Biochemistry,* edited by R. K. Murray, D. K. Granner, P. A. Mayes, and V. W. Rodwell. Norwalk: Appleton and Lange, 1990, p. 269.

308. Romanov, V. S. Quantitative evaluation of ultrastructural changes in the rat myocardium during prolonged hypokinesia. *Kosm. Biol. Aviakosm. Med.* 10: 50–54, 1976.

309. Rowell, L. B. The venous system. In: *Human Circulation Regulation During Physical Stress.* New York: Oxford University Press, 1986, p. 44–77.

310. Rushmer, R. F. Postural effects on the baselines of ventricular performance. *Circulation* 20: 897–905, 1959.

311. Ryback, R. S., O. F. Lewis, and C. S. Lessard. Psychobiologic effects of prolonged bed rest (weightless) in young, healthy volunteers (study II). *Aerospace Med.* 42: 529–535, 1971.

312. Ryback, R. S., R. W. Trimble, O. F. Lewis, and C. L. Jennings. Psychobiologic effects of prolonged weightlessness (bed rest) in young healthy volunteers. *Aerospace Med.* 42: 408–415, 1971.

313. Saltin, B., G. Blomqvist, J. H. Mitchell, R. L. Johnson, Jr., K. Wildenthal, and C. B. Chapman. Response to exercise after bed rest and after training. A longitudinal study of adaptive changes in oxygen transport and body composition. *Circulation* 38 (Suppl. 7): VII-1–VII-78, 1968.

314. Saltin, B., and L. B. Rowell. Functional adaptations to physical activity and inactivity. *Federation Proc.* 39: 1506–1513, 1980.

315. Sanchez, R. A., E. J. Marco, C. Oliveri, F. J. Otero, O. Degrossi, L. I. Moledo, and S. Julius. Role of cardiopulmonary mechanoreceptors in the postural regulation of renin. *Am. J. Cardiol.* 59: 881–886, 1987.

316. Sandler, H. Effects of bed rest and weightlessness on the heart. In: *Hearts and Heart-Like Organs,* edited by G. H. Bourne. New York: Academic, 1980, vol. 2, p. 435–524.

317. Sandler, H. Cardiovascular effects of inactivity. In: *Inactivity: Physiological Effects,* edited by H. Sandler and J. Vernikos. New York: Academic, 1986, p. 11–47.

318. Sandler, H. *Cardiovascular Effects of Weightlessness and Ground-based Simulation.* Washington, DC: NASA Technical Memorandum 88314, 1988.

319. Sandler, H., D. J. Goldwater, R. L. Popp, L. Spaccavento, and D. C. Harrison. Beta blockade in the compensation for bed-rest cardiovascular deconditioning: Physiologic and pharmacologic observations. *Am. J. Cardiol.* 55: 114D–119D, 1985.

320. Sandler, H., R. L. Popp, and D. C. Harrison. The hemodynamic effects of repeated bed rest exposure. *Aviat. Space Environ. Med.* 59: 1047–1054, 1988.

321. Sandler, H., R. Popp, and E. P. McCutcheon. Echocardiographic studies of bed rest induced changes during LBNP. Alexandria, VA: *Preprints, Aerospace Med. Assoc.,* 1977, p. 242–243.

322. Sandler, H., P. Webb, J. Annis, N. Pace, B. W. Grunbaum, D. Dolkas, and B. Newsom. Evaluation of a reverse gradient garment for prevention of bed-rest deconditioning. *Aviat. Space Environ. Med.* 54: 191–201, 1983.

323. Sandler, H., and D. L. Winter. *Physiological Responses of Women to Simulated Weightlessness.* Washington, DC: NASA Special Publication 430, 1978.

324. Sassin, J. F., D. C. Parker, J. W. Mace, R. W. Gotlin, L. C. Johnson, and L. G. Rossman. Human growth hormone release: relation to slow-wave sleep and sleep-walking cycles. *Science* 165: 513–515, 1969.

325. Savilov, A. A., V. I. Lobachik, and A. M. Babin. Cardiovascular function of man exposed to LBNP tests. *Physiologist* 33: S128–S132, 1990.

326. Savina, E. A., A. S. Kaplansky, V. N. Shvets, and G. S. Belkaniya. Antiorthostatic hypokinesia in monkeys (experimental morphological study). *Physiologist* 26: S76–S77, 1983.

327. Schmid, P. G., M. McCally, T. E. Piemme, and J. A. Shaver. Effects of bed rest on forearm vascular responses to tyramine and norepinephrine. In: *Hypogravic and Hypodynamic Environments,* edited by R. H. Murray and M. McCally. Washington, DC: NASA Special Publication 269, 1971, p. 211–223.

328. Schmid, P. G., J. A. Shaver, M. McCally, J. J. Bensy, L. G. Pawlson, and T. E. Piemme. Effects of two weeks of bed rest on forearm venous responses to norepinephrine and tyramine. Alexandria, VA: *Preprints, Aerospace Med. Assoc.,* 1968, p. 104.

329. Schneider, V. S., and J. McDonald. Skeletal calcium homeostasis and countermeasures to prevent disuse osteoporosis. *Calcif. Tissue Int.* 36: S151–S154, 1984.

330. Schónheyder, F., and P. J. Christensen. The mechanism of creatinuria during immobilization in bed. *Scand. J. Clin. Lab. Invest.* 9: 107–108, 1957.

331. Schónheyder, F., N. S. C. Heilskov, and K. Olesen. Isotopic studies on the mechanism of negative nitrogen balance produced by immobilization. *Scand. J. Clin. Lab. Invest.* 6: 178–188, 1954.

332. Shaffer, P. A. Diminished muscular activity and protein metabolism. *Am. J. Physiol.* 22: 445–455, 1908.

333. Shaver, L. G. The relationship between maximum isometric strength and relative isotonic endurance of athletes with various degrees of strength. *J. Sports Med. Phys. Fitness* 13: 231–237, 1973.

334. Simonenko, V. V. Hemodynamic changes during prolonged hypokinesia according to mechanocardiographic data. In: *Problems of Space Biology. Prolonged Limitation of Mobility and its Influence on the Human Organism,* edited by A. M. Genin, and P. A. Sorokin. Moscow: Nauka, 1969, vol. 13, p. 42–49. (NASA TT F-639, 1970)

335. Skrypnik, V. G. Changes in the biochemical peculiarities of walking under the influence of hypodynamia according to ichnographic data. In: *Problems of Space Biology. Prolonged Limitation of Mobility and its Influence on the Human Organism,* edited by A. M. Genin, and P. A. Sorokin. Moscow: Nauka, 1969, vol. 13, p. 161–170. (NASA TT F-639, 1970)

336. Slutsky, A. S., R. G. Goldstein, and A. S. Rebuck. The effect of posture on the ventilatory response to hypoxia. *Can. Anaesth. Soc. J.* 27: 445–448, 1980.

337. Smirnova, T. M., G. O. Kozyrevskaya, V. I. Lobachik, V. V. Zhidkov, and S. V. Abrosimov. Individual distinctions of fluid-electrolyte metabolism during hypokinesia with head-down tilt for 120 days, and efficacy of preventive agents. *Kosm. Biol. Aviakosm. Med.* 20: 21–24, 1986.

338. Sorokin, P. A., A. M. Genin, M. I. Tishchenko, P. V. Vasil'yev, R. I. Gismatulin, and I. D. Pestov. Organizational and methodological principles for the conduct of prolonged hypodynamia researches. In: *Problems of Space Biology. Prolonged Limitation of Mobility and its Influence on the Human Organism,* edited by A. M. Genin and P. A. Sorokin. Moscow: Nauka, 1969, vol. 13, p. 8–14. (NASA TT F-639, 1970)

339. Sorokin, P. A., V. V. Simonenko, and B. A. Korolev. Clinical observations in prolonged hypodynamia. *Problems of Space Biology. Prolonged Limitation of Mobility and its Influence on the Human Organism,* edited by A. M. Genin and P. A. Sorokin. Moscow: Nauka, 1969, vol. 13, p. 24–34. (NASA TT F-639, 1970)

340. Spealman, C. R., E. W. Bixby, J. L. Wiley, and M. Newton. Influence of hemorrhage, albumin infusion, bedrest, and exposure to cold on performance in the heat. *J. Appl. Physiol.* 1: 242–253, 1948.

341. Stegemann, J., D. Essfeld, and U. Hoffman. Effects of a 7-day head-down tilt (−6°) on the dynamics of oxygen uptake and heart rate adjustment in upright exercise. *Aviat. Space Environ. Med.* 56: 410–414, 1985.

342. Stevens, P. M., T. N. Lynch, R. L. Johnson, and L. E. Lamb. Effects of 9-alphaflurohydrocortisone and venous occlusive cuffs on orthostatic deconditioning of prolonged bed rest. *Aerospace Med.* 37: 1049–1056, 1966.

343. Stevens, P. M., P. B. Miller, C. A. Gilbert, T. N. Lynch, R. L. Johnson, and L. E. Lamb. Influence of long-term lower body negative pressure on the circulatory function of man during prolonged bed rest. *Aerospace Med.* 37: 357–367, 1966.

344. Stevens, P. M., P. B. Miller, T. N. Lynch, C. A. Gilbert, R. L. Johnson, and L. E. Lamb. Effects of lower body negative pressure on physiologic changes due to four weeks of hypoxic bed rest. *Aerospace Med.* 37: 466–474, 1966.

345. Stevenson, F. H. The osteoporosis of immobilization in recumbency. *J. Bone Joint Surg.* 34B: 256–265, 1952.

346. Stewart, A. F., M. Adler, C. M. Byers, G. V. Segre, and A. E. Broadus. Calcium homeostasis in immobilization: an example of resorptive hypercalciuria. *N. Engl. J. Med.* 306: 1136–1140, 1982.

347. Storm, W. F., and C. L. Giannetta. Effects of hypercapnia and bedrest on psychomotor performance. *Aerospace Med.* 45: 431–433, 1974.

348. Stremel, R. W., V. A. Convertino, E. M. Bernauer, and J. E. Greenleaf. Cardiorespiratory deconditioning with static and dynamic leg exercise during bed rest. *J. Appl. Physiol.* 41: 905–909, 1976.

349. Sullivan, M. J., P. F. Brinkley, D. V. Unverferth, J. H. Ren, H. Boudoulas, T. M. Bashore, A. J. Merola, and C. V. Leier. Prevention of bedrest-induced physical deconditioning by daily dobutamine infusions. *J. Clin. Invest.* 76: 1632–1642, 1985.

350. Takeshita, A., A. L. Mark, D. L. Eckberg, and F. M. Abboud. Effect of central venous pressure on arterial baroreflex control of heart rate. *Am. J. Physiol.* 236: (*Heart Circ. Physiol.* 5) H42–H47, 1979.

351. Talbot, J. M., and K. D. Fisher. Influence of space flight on red blood cells. *Federation Proc.* 45: 2285–2290, 1986.

352. Taylor, H. L., L. Erickson, A. Henschel, and A. Keys. The effect of bedrest on the blood volume of normal young men. *Am. J. Physiol.* 44: 227–232, 1945.

353. Taylor, H. L., A. Henschel, J. Brozek, and A. Keys. Effects of bedrest on cardiovascular function and work performance. *J. Appl. Physiol.* 2: 223–239, 1949.

354. Ten Harkel, A. D. J., F. Baisch, L. Beck, and J. M. Karemaker. The autonomic nervous system in blood pressure regulation during 10 days 6° head down tilt. *Physiologist* 33: S178–S179, 1990.

355. Tenney, S. M. Fluid volume redistribution and thoracic volume changes during recumbency. *J. Appl. Physiol.* 14: 129–132, 1959.

356. Teplinskaya, G. P. The effects of space flight factors on the functional activity of T-lymphocytes responsible for delayed hypersensitivity. In: *Space Biology and Aerospace Medicine,* edited by O. G. Gazenko. Moscow: Nauka, 1986, p. 259.

357. Thompson, C. A., D. L. Tatro, D. A. Ludwig, and V. A. Convertino. Baroreflex responses to acute changes in blood volume in humans. *Am. J. Physiol.* 259 (*Regulatory Integrative Comp. Physiol.* 28): R792–R798, 1990.

358. Thompson, F. J., and B. J. Yates. Venous afferent elicited skeletal muscle pumping: a new orthostatic venopressor mechanism. *Physiologist* 26: S74–S75, 1983.

359. Tishchenko, M. I., B. A. Korolev, V. A. Degtyarev, and B. F. Asyamolov. Phase changes in the cardiac cycle during prolonged hypodynamia according to polycardiographic and kinetocardiographic data. In: *Problems of Space Biology. Prolonged Limitation of Mobility and its Influence on the Human Organism,* edited by A. M. Genin and P. A. Sorokin. Moscow: Nauka, 1969, vol. 13, p. 59–64. (NASA TT F-639, 1970)

360. Tizul, A. Ya. The function of thermoregulation in protracted limitation of motor activity (hypokinesia). *Zh. Neuropatol. Psikhiatr.* 73: 1791–1794, 1973. (NASA TT F-15, 566, 1974)

361. Tomaselli, C. M., R. A. Kenney, M. A. B. Frey, and G. W. Hoffler. Cardiovascular dynamics during the initial period of head-down tilt. *Aviat. Space Environ. Med.* 58: 3–8, 1987.

362. Torphy, D. E. Effects of short-term bedrest and water immersion on plasma volume and catecholamine response to tilting. *Aerospace Med.* 37: 383–387, 1966.

363. Treharne, R. W. Review of Wolff++++++'s law and its proposed means of operation. *Orthop. Rev.* 10: 35–47, 1981.

364. Triebwasser, J. H., A. F. Fasola, A. Stewart, and M. C. Lancaster. The effect of exercise on the preservation of orthostatic tolerance during prolonged immobilization. Alexandria, VA: *Preprints, Aerospace Med. Assoc.,* 1970, p. 65–66.

365. Turbasov, V. D. Effect of prolonged antiorthostatic position on cardiac bioelectrical activity according to EKG tracings from corrected orthogonal leads. *Kosm. Biol. Aviakosm. Med.* 14: 54–59, 1980.

366. Udalov, Yu. R., R. V. Kudrova, M. I. Kuznetsov, P. O. Lobzin, V. A. Petrovykh, I. G. Popov, I. A. Romanova, Yu. K. Syzrantsev, A. M. Terilovskiy, L. N. Rogatina, and N. A. Chelnokova. Effect of qualitative differences in diet on metabolism in hypodynamia. *Probl. Kosm. Biol.* 7: 348–354, 1969.

367. Van Beaumont, W., J. E. Greenleaf, and L. Juhos. Disproportional changes in hematocrit, plasma volume, and proteins during exercise and bed rest. *J. Appl. Physiol.* 33: 55–61, 1972.

368. Vasil'yev, A. I., and V. V. Diskalenko. Effect of prolonged hypokinesis on hearing. *Voyenno Med. Zh.* 7: 76–77, 1977.

369. Vernikos, J. Metabolic and endocrine changes. In: *Inactivity: Physiological Effects,* edited by H. Sandler and J. Vernikos. New York: Academic, 1986, p. 99–121.

370. Vernikos-Danellis, J., C. M. Winget, C. S. Leach, and P. C. Rambaut. *Circadian, Endocrine, and Metabolic Effects of Prolonged Bedrest: Two 56-day Bedrest Studies.* Washington, DC: NASA Technical Memorandum X-3051, 1974.

371. Vetter, W. R., R. W. Sullivan, and K. H. Hyatt. Deterioration of left ventricular function: A consequence of simulated weight-lessness. Alexandria, VA: *Preprints, Aerospace Med. Assoc.,* 1971, p. 56–57.

372. Vikhert, A. M., V. I. Metelitsa, V. D. Baranova, and I. Ye. Galakhov. Morphological and biochemical changes in rabbits subjected to considerable limitation of mobility. *Kardiologiya* 12: 143–146, 1972.

373. Vogt, F. B. Tilt table and plasma volume changes with short term deconditioning experiments. *Aerospace Med.* 38: 564–568, 1967.

374. Vogt, F. B., and P. C. Johnson. Plasma volume and extracellular fluid volume changes associated with 10 days bed recumbency. *Aerospace Med.* 38: 21–25, 1967.

375. Vogt, F. B., P. B. Mack, P. C. Johnson, and L. Wade, Jr. Tilt table response and blood volume changes associated with fourteen days of recumbency. *Aerospace Med.* 38: 43–48, 1967.

376. Volicer, L., R. Jean-Charles, and A. V. Chobanian. Effects of head-down tilt on fluid and electrolyte balance. *Aviat. Space Environ. Med.* 47: 1065–1068, 1976.

377. Vorob'yev, V. Ye., V. R. Abdrakhmanov, A. P. Golikov, L. L. Stazhadze, I. B. Goncharov, I. V. Kovachevich, S. G. Voronina, and A. V. Vabishchevich. Effect of 120-day antiorthostatic bedrest on gas exchange and pulmonary circulation in man. *Kosm. Biol. Aviakosm. Med.* 18: 23–26, 1984.

378. Voskresenskiy, A. D., B. A. Korolev, and M. D. Ventsel. Changes in electrocardiogram and statistical structure of cardiac rhythm in the course of confinement to bed. In: *Problems of Space Biology. Prolonged Limitatin of Mobility and its Influence on the Human Organism,* edited by A. M. Genin and P. A. Sorokin. Moscow: Nauka, 1969, vol. 13, p. 35–41. (NASA TT F-639, 1970)

379. Voskresenskiy, A. D., B. B. Yegorov, I. D. Pestov, S. M. Belyashin, V. M. Tolstov, and I. S. Lezhin. Organization of the experiments and overall condition of the subjects. *Kosm. Biol. Med.* 6: 28–32, 1972.

380. Wade, O. L., and J. C. Gilson. Effect of posture on diaphragmatic movement and vital capacity in normal subjects with a note on spirometry as an aid in determining radiological chest volumes. *Thorax* 6: 103–126, 1951.

381. Weight, L. M., M. J. Byrne, and P. Jacobs. Haemolytic effects of exercise. *Clin. Sci.* 81: 147–152, 1991.

382. Weissman, C., B. Abraham, J. Askanazi, J. Milic-Emili, A. I. Hyman, and J. M. Kinney. Effect of posture on the ventilatory response to CO$_2$. *J. Appl. Physiol.: Respir. Environ. Exerc. Physiol.* 53: 761–765, 1982.

383. Whedon, G. D., J. E. Deitrick, and E. Shorr. Modification of the effects of immobilization upon metabolic and physiologic functions of normal men by the use of an oscillating bed. *Am. J. Med.* 6: 684–711, 1949.

384. Whedon, G. D., and E. Shorr. Metabolic studies in paralytic acute anterior poliomyelitis. I. Alterations in nitrogen and creatine metabolism. *J. Clin. Invest.* 36: 942–965, 1957.

385. White, P. D., J. W. Nyberg, L. M. Finney, and W. J. White. *A Comparative Study of the Physiological Effects of Immersion and Bedrest.* Santa Monica, CA: Douglas Aircraft Co. Report DAC-59226, 1966.

386. Widdowson, E. M., and R. A. McCance. The effect of rest in bed on plasma volume: as indicated by haemoglobin and haematocrit levels. *Lancet* 1: 539–540, 1950.

387. Wilkins, R. W., S. E. Bradley, and C. K. Friedland. Acute circulatory effects of head-down position (negative G) in normal man, with a note on some measures designed to relieve cranial congestion in this position. *J. Clin. Invest.* 29: 940–949, 1950.

388. Williams, B. A., and R. D. Reese. Effect of bed rest on thermoregulation. Alexandria, VA: *Preprints, Aerospace Med. Assoc.,* 1972, p. 140–141.

389. Williams, D. A., and V. A. Convertino. Circulating lactate and

FFA during exercise: effect of reduction in plasma volume following exposure to simulated microgravity. *Aviat. Space Environ. Med.* 59: 1042–1046, 1988.

390. Winget, C. M., C. W. DeRoshia. Psychosocial and chronophysiological effects of inactivity and immobilization. In: *Inactivity: Physiological Effects,* edited by H. Sandler and J. Vernikos. New York: Academic, 1986, p. 123–147.

391. Winget, C. M., J. Vernikos-Danellis, S. E. Cronin, C. S. Leach, P. C. Rambaut, and P. B. Mack. Circadian rhythm asynchrony in man during hypokinesis. *J. Appl. Physiol.* 33: 640–643, 1972.

392. Wyse, D. M., and C. J. Pattee. Effect of oscillating bed and tilt table on calcium, phosphorus and nitrogen metabolism in paraplegia. *Am. J. Med.* 17: 645–661, 1954.

393. Yakovleva, I. Ya., V. P. Baranova, L. N. Kornilova, M. V. Nefedova, E. V. Lapayev, and S. R. Raskatova. Study of reactions of human otorhinolaryngological organs during hypokinesia. *Kosm. Biol. Med.* 6: 49–54, 1972.

394. Yakovleva, I. Ya., and E. I. Matsnev. Functional state of the human auditory analyzer in an experiment with two-month hypokinesia. *Kosm. Biol. Med.* 1: 66–70, 1967.

395. Zav'yalov, Ye. S., and S. G. Mel'nik. The scanning activity of a man operator exposed to space flight factors. *Kosm. Biol. Med.* 1: 57–62, 1967.

396. Zav'yalov, Ye. S., S. G. Mel'nik, G. Ya. Chuganov, and A. A. Vorona. Performance of operators during prolonged bed confinement. *Kosm. Biol. Med.* 4: 61–65, 1970.

397. Zehr, J. E., J. A. Hasbargen, and K. D. Kurz. Reflex suppression of renin secretion during distention of cardiopulmonary receptors in dogs. *Circ. Res.* 38: 232–239, 1976.

398. Zorbas, Y. G. Water-mineral metabolism wavelike changes during 180-day hypokinesia. Alexandria, VA: *Preprints, Aerospace Med. Assoc.,* 1980, p. 68–69.

399. Zubeck, J. P. Urinary excretion of adrenaline and nor-adrenaline during prolonged immobilization. *J. Abnorm. Psychol.* 73: 223–225, 1968.

400. Zubeck, J. P., and M. MacNeill. Effects of immobilization: behavioral and EEG changes. *Can. J. Psychol.* 20: 316–336, 1966.

401. Zvonarev, G. P. Dynamics of minute volume during prolonged hypokinesia as estimated by the acetylene method. *Kosm. Biol. Aviakosm. Med.* 5: 50–53, 1971.

2 | HYPERGRAVITY

40. Adaptation to acceleration environments

RUSSELL R. BURTON | Armstrong Laboratory, Brooks Air Force Base, Texas

ARTHUR H. SMITH | University of California, Davis, Davis, California

CHAPTER CONTENTS

INCREASED ACCELERATION ENVIRONMENTS, which may be sustained from a few seconds to as long as several years (an animal's lifetime), are all produced for research purposes with centrifuges. Such environments are created by human technology and in that respect are quite different from the other environments that occur naturally and are discussed in this *Handbook*.

Acceleration environments hold special interest for two groups of physiologists: one is concerned with the study of biological effects of the earth's gravity (g), and the other with the physiologic effects of inertial fields (G) from acceleration developed by aircraft maneuvering or by rocket propulsion.

Since acceleration acting on a mass produces an inertial force that cannot be distinguished from the attracting force of gravity (as determined by Mach and later by Einstein in their theories of equivalence), it is useful in studying the physiologic effects of earth gravity. Since most of the effects of gravity are long-term, these studies take the form of long-duration experiments (several weeks to years). The G intensities of gravitational studies are inversely related to body mass: 5–6G for animals of 50–100 gm size; 3G for 1–2 kg animals; and is probably 1.5G maximal for humans. This type of long-duration, low-G environment is well tolerated because animals are able to physiologically adapt to its inertial forces. Tolerance to this environment is best determined from slowly changing physiologic parameters such as food requirements, body composition, muscle mass, work capacities, blood constituents, and reproduction. Data are mainly obtained at 1G after the increased-G exposure and compared with 1G (never accelerated) control values. These experiments, using laboratory animals that have physiologically adapted to the increased-G environment, address that adaptive process.

The other acceleration environments of interest are those generated by high-performance aircraft and rockets. These extremely dynamic environments of much higher G levels are too stressful physiologically to be tolerated continuously; thus, exposures are limited to a maximum of only a few minutes. Exposure to high-G environments does not involve adaptation, but rather rapidly occurring physiologic accommodation using existing homeostatic processes. However, some physiologic adaptation to this high-G environment can develop with repeated exposures or by cross-adaptation from other physiologically related treatments. Since this environment is encountered in high-performance flight, results are directly applicable to aircrew. Therefore, humans are usually the test subjects for these studies, but animals are used in the more demanding experiments, particularly in those involving extensive invasive procedures.

PHYSICS OF ACCELERATION

Several physical laws of acceleration apply equally in physical as well as biologic systems. The differences in responses to sustained G and chronic G exposures arise from biological processes.

Interaction of Forces and Mass

The interactions of forces and matter were largely described by Galileo in 1638 (59) and later summarized by Newton in 1687 into three laws of motion (114). The *first law of motion* deals with the resistance of an object to change in its state of motion (or rest) when acted upon by an external force—it establishes the concept of inertia. The *second law of motion* relates the progressive change in positions (acceleration, a: $cm \cdot sec^{-2}$) of a body (m, gm mass) when acted upon by an external force (F, dynes): F = ma. However, as the body is restrained from motion in an acceleration field, an inertial force (G) occurs (weight). In space (microgravity), mass, not weight, is the proper descriptor for material magnitude. The *third law of motion*, "for every action, there is an equal and opposite reaction," establishes the inertial force (G) that develops in opposition to acceleration.

Acceleration of Circular Motion (Physics of the Centrifuge)

Circular motion at a uniform rate, as used in centrifugation studies, is unusual in mechanics in that it involves the continual application of force upon an object without changing its rate of motion or its kinetic energy. An understanding of the physical basis of circular motion began with the investigations of Huygens in 1658 (78), a contemporary of both Galileo and Newton. Huygens derived an equation relating the acceleration force (a) of an object in circular motion with its velocity (V, cm/s) and radius of rotation (r, cm): $a = V^2/r$ (Fig. 40.1).

Acceleration force (a) and inertial force (G) are commonly evaluated in terms of the Earth's gravitational constant (g, 980 dynes): $a/g = v^2/rg = G$. The ratio a/g, or for inertial forces G/g, has been designated G, a dimensionless quantity evaluating acceleration fields as multiples of Earth gravity, g. It is also the weight/mass

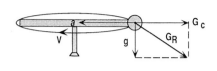

FIG. 40.1. Diagram showing physics of acceleration. r, radius of rotation (cm); V, rate of motion (cm/s; velocity); a, acceleration (dynes; centripetal) force; G_c inertial (dynes; centrifugal) force; g, Earth's gravitational constant (980 dynes); G_R, resultant G vector.

$$G_R = (G^2_c + g^2)^{1/2} \text{ (i.e., } g^2 = 1)$$

ratio (by definition, 1.00 under conditions of Earth's gravity); the alteration of the weight/mass ratio is the physical basis for acceleration physiology.

Since gravity is always present during centrifugation, the net or resultant inertial force (G_R) is a combination of the centrifugal (G_C) and gravitational (g) forces calculated with the Pythagorean Theorem (Fig. 40.1). At lower increased G_C levels, the force of gravity is significant and therefore must be included in accurately determining the level of G exposure; that is, at $2G_C$, g increases the net G by over 10%, but at $5G_C$, not including g results in only a 2% error.

Acceleration without Inducing Weight

The weightless or microgravity environment can be produced on or near earth using three different methods. Since Earth's gravity is pervasive and cannot be altered, its physiologic effects can only be counteracted when a = g as with freefall, or when the "a" vector is equal and opposite to the "g" as in parabolic flight, or orbiting the earth (Fig. 40.2). Interestingly, orbiting flight generally does not produce true weightlessness because of a small atmospheric-induced drag. Hence a is slightly less than g so that a net gravitational effect of approximately g^{-5} creates an environment known as *microgravity*.

Acceleration of Aircraft

Heavier-than-air flight involves rapid motion essential for generating lift. Changes in flight direction (aircraft maneuvering) generate acceleration. Since modern, high-performance fighter aircraft have greater thrust with less weight, maneuverability can routinely generate inertial forces up to 9G for sustained durations of several minutes.

Acceleration of Rockets

Rocket-powered space flight accelerates and decelerates its occupants during launch and landing. In the NASA

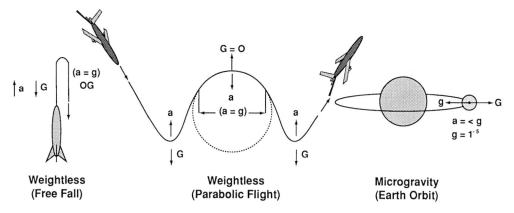

FIG. 40.2. Methods of providing a weightless or microgravity environment.

Mercury Program, these acceleration levels were very high (8–9G). However, since these high-G levels are not necessary with the Space Shuttle, liftoff does not exceed 3G; during reentry, inertial forces are no greater than 2G, but last for 20 to 30 min.

RESEARCH CENTRIFUGES

Although all centrifuges share common physical principles, their operations and design elements vary considerably to accommodate special requirements for specific applications.

Sustained-G Centrifuges

Modern, high-performance, "human-use" centrifuges used in sustained-G experimentation have a center spindle connected to the drive system about which it rotates. Extending from the center is a rigid arm; at its distal end is an enclosed compartment (gondola) that houses the restrained subject. Those research centrifuges presently in operation have arms ranging from about 4.6 m (15 ft) to 15.2 m (50 ft).

The two major performance characteristics of centrifuges are rate of onset of G and maximum level of G. The onset rate is usually expressed as the rate of change per second of the G vector within the gondola (for example, 6 G/s onset rate). High-G onset capability is important for simulating the very high–G onset rates of high-performance aircraft (that is, > 10 G/s); these high onset rates have significant physiologic effects.

There are two methods used for gondola angular control. The simplest is passive or inertial control in which the gondola is mounted on bearings and rotates freely in the roll axis. The other method is active control utilizing drive motors with computer controls to solve the acceleration vector equations so that the G_R vector (Fig. 40.1) is closely maintained. This control system virtu-

ally eliminates the annoying effects on the subject's vestibular system of tangential G forces.

The two centrifuge characteristics, G-level and G-onset rates, are induced by closed- or open-loop methods. In closed-loop control, the subject operates the centrifuge, usually using an aircraft-type control stick. Open-loop systems provide control by other personnel independent from the subject.

Since physiologic data are acquired during G exposure, the gondola is a physiology laboratory. Some have highly sophisticated, state-of-the-art electronics equipment. Data can be obtained from the subject in real time using high-quality slip rings; that is, a "continuous" connection between sensor, detector, and recorder.

Animal restraint platforms or gondolas for research are usually located on the end opposite the human-use gondola.

Chronic-G Centrifuges

Centrifuges used for chronic G research are designed exclusively for animal research with lower operational requirements and are much simpler to operate than high-performance sustained-G machines. An important component of this type of centrifuge is the structure of the cages (animal "gondolas") that contain the animals during G exposures. Their design is usually specific for a particular animal type that provides adequate ambulation and addresses husbandry concerns during the long-duration G exposures. Usually, physiologic data are not collected during G exposure. Designs and operational characteristics for several of these centrifuges have been reported (15, 84, 147, 148, 173).

ACCELERATION NOMENCLATURE

A standardized, internationally recognized method of identifying the position of the animal or human subject within the acceleration field has been developed because physiologic responses to increased G can be significantly influenced by body orientation. Both body restraint and orientation, in specific G fields, can be extremely stressful, particularly for animals.

A major physiologic effect of increased G is on the distribution of blood within the body. Increased intravascular hydrostatic pressure (P_H) develops that is a function of blood-specific density, G level, and column height:

$$P_H = hdG \ldots \ldots \ldots \quad (1)$$

where

P_H = hydrostatic pressure
h = height of the fluid (blood) column
d = specific density of blood
G = the inertial force

To relate column height to the direction of G exposure, nomenclature has been developed around the three major axes of the body that uses z, y, and x symbols: vertical (z); lateral or right/left (y); and horizontal or supine/prone (x). Body orientation along these axes is identified by plus or minus (Table 40.1).

The aircraft pilot and space shuttle astronaut upon reentry are seated upright with $+G_z$ acting upon the body. In this position, long vascular columns are parallel to the G vector; that is, h of Eq. 1 for these columns is at its maximum. Since intravascular P_H directly affects blood flow, relatively moderate levels of sustained $+G_z$ can have profound effects on blood distribution when compared with responses in the $\pm G_x$ or $\pm G_y$ positions where shorter intravascular columns relative to the G vector are in effect.

Quadrupeds as experimental subjects must be restrained in an upright bipedal position when they are exposed to $+G_z$ if studies are to simulate human exposure. Even though this position is not "natural," physiologic responses to $+G_z$ in certain quadrupeds (as in miniature swine) can be quite similar to those of humans (21).

Other body orientations are used less frequently since there are fewer areas of application. However, during the 1960s in the NASA Mercury era when space travel required high G during liftoff and reentry, many studies were conducted where astronauts were oriented in the supine ($+G_x$) position.

Long-duration gravitational studies generally use quadrupeds. These animals are usually naturally oriented ($-G_x$) and are allowed complete mobility in the gondola during centrifugation.

SUSTAINED G

Humans and animals have adapted genetically to terrestrial gravity. Gravity is the physical stimulus that initiates the homeostasis of many physiologic mechanisms. Those mechanisms responsive to gravity can be identified with increased-G exposures, thus forming the basis for animal and human tolerances to sustained increased G. Interestingly, although g-induced homeostasis has evolved under the force of only Earth's gravity, it has a remarkable physiologic reserve capacity—the ability to rapidly accommodate to several multiples of G.

In addition to these physiologic accommodations, greater increases in tolerances can be produced with (1) repeated sustained G exposures, and (2) cross-adaptation. Cross-adaptation produces useful adaptates for one type of environment from exposure to another more easily tolerated.

High Sustained G (HSG)

As animals and humans are exposed to higher G levels, specific life-supporting physiologic functions fail, identifying their natural G-tolerance limits. The G environment, above this natural G-tolerance, is called *high sustained G (HSG)*, where failing physiologic functions must be sustained with anti-G support if tolerance is to be achieved. These anti-G methods are quite effective; they were developed for operational use by military aircrew flying high-performance aircraft. The most important methods have a unique physiologic basis that directly affects arterial blood pressure: (1) the anti-G straining maneuver (AGSM), and (2) continuous, assisted positive pressure breathing during G (PBG). Another important anti-G protection method is the anti-G suit (or trouser) which applies external body pressure preserving venous return.

TABLE 40.1. *Nomenclature of Body Position as Related to the G Vector*

Acceleration Force	Physiological Response	Inertial Force
Forward $+a_x$	Transverse, A-P* G, supine G, chest to back	$+G_x$
Backward $-a_x$	Transverse, P-A G†, prone G, back to chest	$-G_x$
Headward $+a_z$	Positive G‡, toward feet	$+G_z$
Footward $-a_z$	Negative G, toward head	$-G_z$
To left $+a_y$	Right lateral G	$+G_y$
To right $-a_y$	Left lateral G	$-G_y$

*A-P: anterior-posterior; P-A: posterior-anterior.   †Natural posture for quadrupeds relative to gravity.   ‡Natural posture for bipeds relative to gravity.

TABLE 40.2. *Accommodation to Exercise During 3G**

Condition	Pa† (mm Hg)	HR (bpm)	CO (liters/m)	TPR (units)	SV (ml)
1G (rest)	99 ± 3‡	64 ± 3	7.1 ± 0.6	0.86 ± 0.07	114 ± 13
3G (rest)	126 ± 3	116 ± 7	6.3 ± 0.4	1.23 ± 0.10	57 ± 7
1G (exercise)	115 ± 2	130 ± 2	16.7 ± 1.0	0.42 ± 0.02	129 ± 9
3G (exercise)	139 ± 3‡	154 ± 3	15.8 ± 1.6	0.56 ± 0.07	103 ± 12

*Table from (11) with permission.   †Pa = arterial pressure atheart level; TPR = total peripheral resistance in units = Pa ÷ CO, CO in ml/s; SV = stroke volume.   ‡n = 6; mean ± S.D.

Cardiovascular Accommodation to +G_z

The application of sustained G activates a combination of several cardiovascular functions developed genetically in response to more "natural" types of stressors: (*1*) exercise, (*2*) orthostasis, (*3*) hemorrhage, (*4*) hypoxia, and perhaps even (*5*) fear. As Claude Bernard in 1865 noted, the external environment can affect an animal only by changing its internal environment (9). In that regard, it is well known that internal changes can be similar in response to different external environments. Of these natural stressors, cardiovascular responses to acute hemorrhage and orthostasis are probably most characteristic of sustained G. Briefly, major cardiovascular responses to these stressors are initiated by a decrease in arterial pressure (Pa) that is primarily sensed by the body's baroceptors. Reflexly, through the central and autonomic nervous systems, heart rate and cardiac contractility increase, followed by a vasoconstriction increasing total peripheral resistance. An increase in heart rate, albeit with a reduced stroke volume and adequate venous return, maintains sufficient cardiac output (CO) to support eye-level Pa and cerebral blood flow (Table 40.2, 117). Although this response is rapid enough to compensate for acute hemorrhage and usually orthostasis, it can be too slow to be effective for a significant rapid increase in G; for example, one that occurs within 1 s. It is for this reason, during the first few seconds of G exposure, that physi-ologic accommodative capacity is at its lowest (Fig. 40.3).

Arterial Pressure and Cardiac Output.

The limiting physiologic function to increased G is the absence of Pa at the head, reducing blood flow to the brain to a critically low level. The problem with maintaining head-level Pa is the opposing increased P_H with G (Eq. 1). So compelling is this relationship that G tolerances of animals and humans can be grouped according to their ability to maintain heart level Pa during increased G. Humans, most nonhuman primates, and the miniature swine (the only quadruped) are capable of maintaining the CO during increased G necessary for the heart to generate Pa of 100–115 mm Hg found at 1G. Consequently this group has G tolerances limited solely by P_H of the height of the vascular column from the heart to the head. This relationship has been described mathematically (23) and validated experimentally (Fig. 40.4):

$$G = Pa \cdot d/h \dots\dots\dots \quad (2)$$

where

G = G-level tolerance
Pa = heart-generated pressure of 100–120 mm Hg
d = specific density of blood related to Hg (approximately 1/13.6), and
h = eye-heart vertical distance in mm

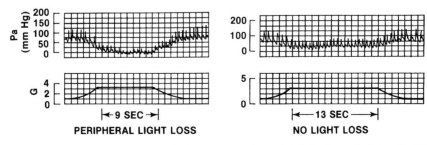

FIG. 40.3. Eye-level Pa analog recording of a human exposed to +3.2 G_z for 9 s on the left and exhibiting insufficient cardiovascular recovery. Grayout (peripheral light loss) occurred that would have been quickly followed with loss of consciousness if the centrifuge had not been stopped. The recording of +3 G_z on the right shows adequate cardiovascular recovery.

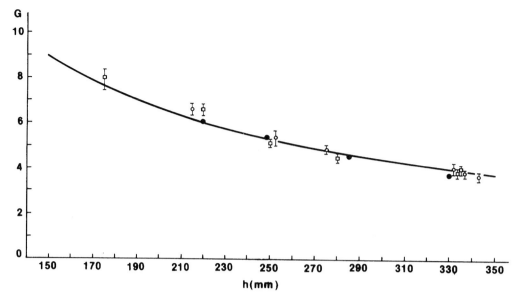

FIG. 40.4. Measured G-level tolerances of the human compared with a curve derived from Eq. 2 (23).

In other animals who are incapable of maintaining adequate CO during increased G exposure, Pa found at 1G at heart level cannot be sustained during increased G. Consequently G tolerances cannot be calculated using the Pa they develop at 1G. However, this equation can be used in predicting G tolerance for this group of animals if the Pa at any G level at heart level is known.

The CO responses of the human and of a typical quadruped (dog) to increasing G are compared showing the increased vulnerability of the dog to G_z:

Human (102):

$$CO = 7.5–7.7G \ldots \ldots \quad (3)$$

where

CO = % change from 1G controls
G = $+G_z$

Dog (74):

$$CO = 16.7–17.6G \ldots \ldots \quad (4)$$

where

CO = % change from 1G controls
G = $+G_z$

Similarly, because of this effect of G on CO, a reduction in Pa occurs that was quantified during 4G sustained exposures and compared with the human, nonhuman primate, and dog. Pertzoff and Britton (126), by integrating areas of carotid Pa deficiency with those of G exposures, were able to demonstrate superior cardiovascular homeostasis in bipeds over the dog (Fig. 40.5).

Light-loss phenomenon. A direct consequence of significant reduction in Pa and blood flow at head level is a phenomenon known as *blackout* (loss of vision) that is most peculiar to the acceleration environment. When Pa is reduced to low levels (50 mm Hg) at head level, vision begins to dim (grayout), narrowing the visual field until it disappears (blackout) with a Pa of 20 mm Hg. Yet with this 20 mm Hg of Pa, blood flow to the brain continues; that is, blood flow ceases in the retina at higher Pa since it is opposed by the resistance of an introcular pressure of approximately 20 mm Hg. Blackout is used by aircrew and research physiologists as a measure of G-level tolerance.

Anti-G straining maneuver. By applying the anti-G straining maneuver, the Pa can be raised by as much as 100 mm Hg. (The AGSM is sometimes referred to as the *Valsalva, L-1,* or *M-1 maneuvers*.) In performing the AGSM, the glottis is closed voluntarily and a forced exhalation attempted. At 3–4 s intervals the glottis is opened and a rapid (<1 s duration) breath is taken. This momentary interruption of the Valsalva effort supports venous return. The AGSM has been used by pilots to increase their tolerance to HSG since World War II (183). The benefit of this maneuver is an immediate increase in the pulmonary pressure, approximately doubling Pa at heart level. An increase in Pa from the AGSM is additive to the Pa of 100–120 mm Hg generated by the heart (Fig. 40.6).

Positive pressure breathing. Continuous, assisted positive pressure breathing during G exposure can increase *pulmonary pressure* up to 70 mm Hg in support of the AGSM. The energy requirements for breathing against positive pressure are reduced with chest counterpressure using an inflatable vest, hence the name *assisted PBG.* The elevated pulmonary pressure raises Pa using mech-

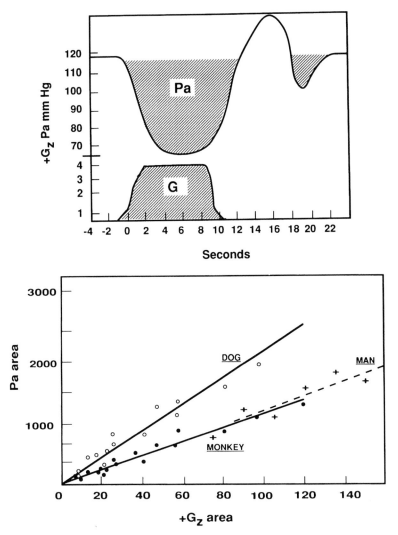

FIG. 40.5. The reduction in carotid arterial pressure *(Pa)* during +G$_z$ exposure is integrated, correlated with the +G$_z$ area, and compared for the dog, nonhuman primate, and the human [Modified from Pertzoff and Britton (126).]

anisms similar to those of the AGSM. Since assisted PBG increases Pa by as much as 50 mm Hg with minimal physical effort, the AGSM is required to generate only an additional 50 mm Hg for the Pa requirement of 225 mm Hg at 9G. Consequently, PBG reduces the fatigue caused by the AGSM and significantly extends G-duration tolerance (18, 139).

Venous Return and Cardiac Output. Direct consequences of inadequate venous return are rapid decreases in CO and Pa (Fig. 40.3). Venous return below the heart is opposed by P$_H$ with increasing G; therefore, because of the compliant nature of veins, blood supply to the heart is diminished and, in most animals, limits +G$_z$ tolerance. Greenfield (66) demonstrated its importance in supporting Pa during sustained G by exposing cats with and without water immersion. Cats immersed in water up to the apex of their hearts tolerated +15–20

G$_z$ (as predicted for the heart-head vertical distance of the cat, Eq. 2). However, without immersion the cat's tolerance was only 3–4 G. The role of venoconstriction in supporting venous return during G exposure was examined in dogs by Salzman and Leverett (136). They found increased venous return from venoconstriction was directly correlated to partial recovery of Pa following its decrease at the onset of 3G.

Venous pressure gradients supporting venous return have been measured in conscious miniature swine during sustained exposures to +G$_z$ (Fig. 40.7). A positive venous gradient always exists between the abdominal and superior vena cavas within the G-tolerance limits of the swine. During inspiration, the diaphragm descends, compressing the hepatic and splanchnic vessels; with subatmospheric pulmonary pressure, venous return is rapidly increased in support of CO (20).

The pumping action from skeletal muscle contrac-

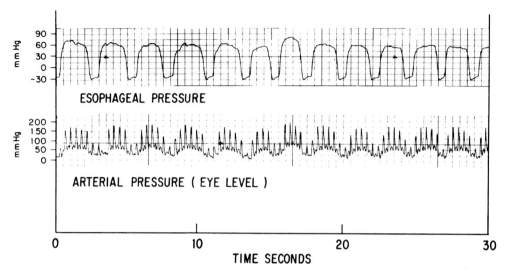

FIG. 40.6. Direct eye-level arterial blood pressure and esophageal pressure (a measure of pulmonary pressure) changes during $+G_z$ of a subject while performing an AGSM. Mean Pa falls to near zero during the inspiratory phase.

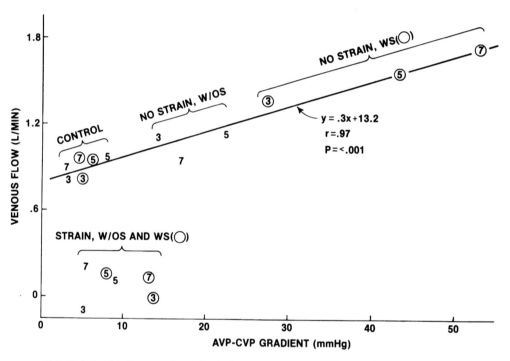

FIG. 40.7. Relationship between the abdominal venous pressure *(AVP)*–central venous pressure *(CVP)* gradient, and venous flow (L/min) found in swine before (control) and during the 3, 5, and 7 $+G_z$ levels, with and without AGSM. Data are shown with *(WS)* and without *(W/OS)* G-suit inflation. *Circled numbers* are data at that $+G_z$ level WS. Regression equation includes control plus no strain data. [From Burns et al. (20) with permission.]

tions of the legs supports venous return below heart level. This muscle pump is particularly active while exercising the legs; vascular dilation reflexly occurs to support an increase in venous return and CO. On the other hand, vascular constriction is the usual accommodative response to increased G at rest. This interesting com-

bination of stressors was studied by Bjurstedt et al. (11). Males (n = 6) were exercised on a cycle ergometer at 50% of their maximum aerobic capacity for 6 min at 3G. Physiologic data obtained from this study are compared for 1G and 3G at rest and with exercise in Table 40.2. Interestingly, the response is one of physiologic compromise that develops nearly the same CO at 3G exercise as found at 1G exercise. But clearly the basis for exercise:G homeostasis is improved venous return with leg muscle pumping. This adaptive strategy that supports venous return at 3G exercise depends less on increased vascular resistance most useful at 3G rest.

Fluid Homeostasis. During HSG exposure, redistribution of blood volume below the heart occurs rapidly; increasing P_H causes a *relative* reduction in blood volume. The kinetics of this fluid shift in the abdominal (pelvic) and thigh regions of the human is shown in Figure 40.8 (96). During HSG exposures, not all fluid redistribution is passive. The body actively compensates for a decrease in CO by redistributing blood from organs of the splanchnic area to the more important organs of the heart and brain (99).

During exposure to HSG, significant reductions in *absolute* plasma volume occur as it moves into extravascular spaces. In the human, 1 min at 6G results in an 11% reduction in plasma volume returning to pre-G levels in about 30 min at 1G (111). Repeated exposures to a 10G peak exposure resulted in a 16.5% reduction in blood volume that returned to pre-G levels after about 20 min at 1G (22). Reduced plasma volume prior to HSG exposure significantly reduces G-duration tolerance (2, 68, 116), and G-level tolerance (142).

Fluid homeostasis during exposures to <3G or in the absence of a significant heart-eye vertical gradient at higher G levels is more dependent upon loss of absolute plasma volume. During long-duration exposures, if adequate plasma volumes are not maintained, Pa will fall and limit G tolerance (41, 129, 163).

Fluid homeostasis during all HSG exposures must contend with: (1) a rapid reduction in relative blood volume as it is redistributed below the heart within the compliant venous system, and (2) an absolute reduction in blood volume as plasma fluids move extravascularly. During low-level, long-duration G, such as during Shuttle reentry, absolute blood volume deficits primarily disrupt cardiovascular homeostasis.

Anti-G suit. The anti-G suit, developed during World War II to support fluid homeostasis, increased tolerance of fighter pilots by 1 to 1.5G for short durations (183). The anti-G suit fits the body snugly, and abdominal and leg bladders are automatically pressurized with air regulated by an anti-G valve during increased G. The inflated abdominal bladder, by pushing into the stomach region below the ribs and diaphragm, compresses the splanchnic vessels and elevates the heart toward the head, thus reducing h of Eq. 2 by 3 cm (133). The inflated leg bladders exert exterior pressure on the major muscles of the calf and thigh. The suit, by increasing vascular resistance, prevents the redistribution of blood volume below the heart and by doing so, supports venous return (Figs. 40.7, 40.8). Suit pressure alone, without the subject performing an AGSM, increases G-level tolerance to 1–1.5 G, half of which comes from elevating the heart. Anti-G suits similar to this design are still used today.

Modern anti-G suits utilize more extensive leg coverage that more effectively sustains fluid homeostasis during HSG (96). Anti-G suits with extended leg coverage, but without an abdominal bladder, are most effective in maintaining circulating blood volumes for long-duration G exposures of <3G (42, 94, 163).

Heart Rate and Rhythm. During sustained G, heart rate is increased rectilinearly with levels of G rising to a maximum rate of approximately 170 bpm at about 7G. On

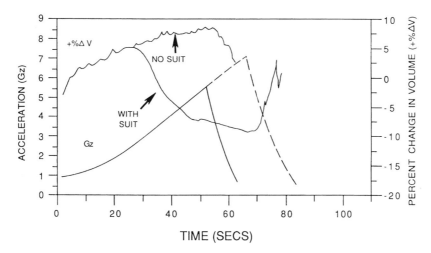

FIG. 40.8. Relative (%) volume changes in the pelvis (abdominal region), thigh, and calf of the human during onset of G with and without the anti-G suit. Volume increases are represented with changes in electrical resistance using an impedance plethysmograph.

rare occasions, 200 bpm is exceeded. A psychologically induced sympathetic response can cause a significant increase in heart rate in subjects on the centrifuge immediately *prior* to a high-G exposure. This anticipatory response is directly proportional to the expected level of G exposure of up to 9G (30). The increase in heart rate is reduced by 62% with propranolol (beta blocker) at 3G without affecting Pa (11).

Ventricular ectopy, including short, self-limited runs of ventricular tachycardia, is frequently associated with HSG exposure. These disturbances of physiologic origin are probably due to exceptionally high sympathetic tone (176). Certain individuals have had paradoxical bradycardia during $+G_z$ stress, but the exact nature of this phenomenon is unknown (141).

Coronary Circulation. During tolerated levels of G exposures, coronary perfusion pressures remain greater than 100 mm Hg. Consequently endocardial/epicardial blood flow ratios are above 1:1, preventing myocardial ischemia. As shown in the swine, heart muscle perfusion at 7G requires 300–500 ml/min/100 g of tissue which equates to 50%–75% of maximum coronary blood flow (98).

Cerebral Blood Circulation. Failure of cerebral blood circulation is the limiting factor to G tolerance. Cerebral circulation in conscious miniature swine at 7G, although well tolerated, is significantly reduced (100). In humans, blood flow of the middle cerebral artery during G onset is reduced by 49% during a 4G exposure with a much greater reduction of 85% in eye-level Pa (117).

Ossard and colleagues (117) reported a reduction in head-level Pa of 110% during onset with brain blood flow reduced by only 61% at 5G; that is, blood flow to the brain can continue even though head-level Pa is zero. This interesting phenomenon occurs because the incompressibility of the skull maintains a constant brain volume with reductions in both blood and cerebral-spinal fluid volumes during increased G. Consequently, blood vessels in the brain remain open as these pressures reach zero. A "siphon effect" is caused by the subatmospheric jugular vein pressure, which can reach −60 mm Hg, such that a constant arteriovenous differential of approximately 50 mm Hg is maintained (Fig. 40.9). However, for this siphon effect to continue, Pa must be sufficiently high to keep the carotid vessels open up to the base of the skull, but also the jugular veins must remain open below the skull (1, 72).

The physics and structure of this siphon effect is one of a "closed" circuit, with ascending and descending units counterbalancing the energy expenditures required of blood flow in the gravitational/acceleration environment (75). This closed loop conservation of energy probably benefits heart level Pa and cardiac output requirements of the majority of the circulating system during sustained G.

Following sustained-G exposures, particularly after an episode of unconsciousness, cerebral hyperemia doubles relative blood flow at 1G for 30 s (64, 175).

Neurologic Accommodation to $+G_z$

G-induced loss of consciousness will occur if there is a critical reduction in blood flow to the brain sustained for 5 s (25). This rapid occurrence of loss of consciousness, the mechanism of which remains unknown, has been considered an adaptive mechanism for terrestrial animals to reduced brain blood flow (177). Loss of con-

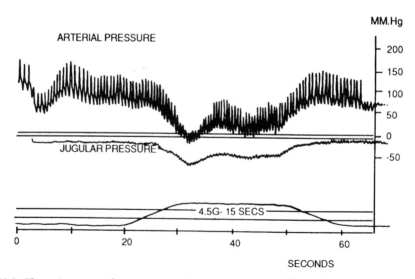

FIG. 40.9. The maintenance of a constant arterio–venous pressure differential is shown as the arterial pressure becomes subatmospheric at head level. [Adapted from Henry et al. (72) with permission.]

sciousness which helps blood return to the brain is particularly effective in bipeds as a protective mechanism against gravity with loss of blood flow to the brain. Bipeds collapse with loss of consciousness and assume a horizontal ($\pm G_x$) posture, which immediately eliminates P_H. Loss of consciousness also reduces oxygen consumption by the brain and the potential for brain injury from prolonged anoxia.

Interestingly, repeated exposure to G-induced loss of consciousness over a period of several months in baboons showed more rapid recovery, indicating that physiologic adaptations to its recovery had occurred. Likewise, humans who have experienced G-induced loss of consciousness recover more rapidly than those who haven't.

Lung and Respiration Accommodation to $+G_z$

All levels of sustained-G exposures affect lung ventilation ($\dot{V}$) and blood perfusion ($\dot{Q}$) that changes ($\dot{V}/\dot{Q}$) ratios. Even at 1G the $\dot{V}/\dot{Q}$ ratio is not uniform over the entire lung; that is, there is more blood perfusion at the base of the lung than at the apex, while the reverse holds true for ventilation: "functional regions" within the lung develop. These ventilation–perfusion relationships are exacerbated with increased G, making the lung less efficient. Pulmonary arterial blood can flow through the base of the lung without gas exchange, as alveolae are closed during the entire breathing cycle—a condition commonly known as *shunting*. Ventilation occurs in the upper lung, but without gas exchange in the absence of blood perfusion. The physiologic effects of these ventilation–perfusion changes in the lung are major increases in alveolar pressure: arterial pressure ($P_A:Pa$) gradients resulting in a reduction in PaO_2, but interestingly without any change in $PaCO_2$, all closely correlated with increasing G (Fig. 40.10).

Alveolae at the base of the lung during HSG collapse from the P_H of the blood of the lung, developing a condition called *compression atelectasis* (30). Another form of acceleration-induced atelectasis is caused by breathing oxygen-enriched gas (>60% oxygen) and wearing an inflated anti-G suit (71). The increased oxygen content in alveolae without ventilation in the lower part of the lung results in *absorption atelectasis*, or collapsing alveolae. It is not known if the lung can adapt to these functional changes with repeated G exposures, as in less reduction in SaO_2 during sustained-G exposures. An excellent review of the effects of G on respiration is available (63).

In spite of this significant reduction in the efficiency of gas exchange, hypoxemia does not limit tolerance to sustained $+G_z$ and consequently does not appear to affect any adaptive process of $+G_z$ tolerance.

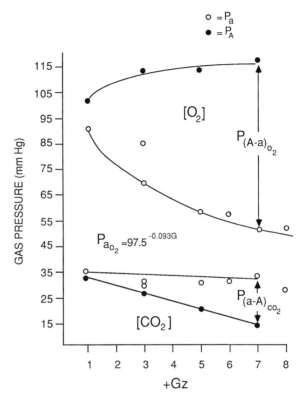

FIG. 40.10. Oxygen and CO_2 pressure gradients between alveoli (P_A) and Pulmonary blood (P_a) as a function of increasing G. [Data from Burton et al. (30).]

Physiological Accommodation to $-G_z$, $\pm G_x$ and $+G_y$

Negative Acceleration ($-G_z$). Negative G_z commonly occurs during civilian aerobatics, where specific maneuvers require outside loops. With maximum levels of -6 G_z following $+7$ G_z at 6 G/s, changes in heart rate of 100 bpm within a few seconds are common. One person had a bradycardia of 32 bpm at -3.8 G_z with individual heart rates commonly going from 175 bpm to 40 bpm within 5 s during $\pm G_z$ cycles. The effect of various levels of sustained $-G_z$ on heart rate in the human is shown in Figure 40.11. However, $-G_z$ induced bradycardia can be less severe in acrobatic pilots than expected because of a form of adaptation called the *batman syndrome* (13). Indeed, repeated exposure to -1 G_z (head-down position in a 1G environment) 3 times/wk for 8 wk promote improved $-G_z$ tolerance, indicating that some adaptation to $-G_z$ had occurred (186).

A concern regarding $-G_z$ exposures is brain capillary rupture resulting in cerebral hemorrhage. Pressure differences between cerebral spinal fluid (CSF) and systemic blood pressures of 100 mm Hg are sufficient to rupture small vessels. Since intravascular pressures at the level of the brain will increase by 25 mm Hg per unit of $-G_z$, rupturing pressures could be reached at -3 to -3.5 G_z. However, Rushmer et al. (134) demonstrated

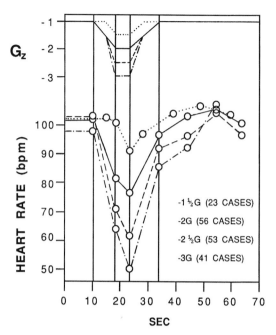

FIG. 40.11. The effect of various levels of $-G_z$ on group mean heart rates. [Modified from Ryan et al. (135).]

that, during centrifugation of cats, increases in CSF pressure were nearly identical to increases in the cerebral venous pressure, thereby affording adequate protection (Fig. 40.12). However, increases in brain P_a were 1.3 times greater than those in the veins, hence counterpressure protection of the brain arteriole was less complete than for the veins. Still, a pressure differential between vessel and tissue of less than 100 mm Hg was consistently maintained. Their findings from studies using cats, which have a small h (Eq. 2), were confirmed up to -8.9 G_z by Beckman (6), using goats that have a much larger h. He also showed that counterpressure of the CSF would prevent brain hemorrhage at venous pressure levels—up to 210 mm Hg.

Prone/Supine Acceleration ($\pm G_x$). Since h is essentially eliminated during assumption of the prone ($-G_x$) or supine ($+G_x$) body positions, physical effects of increased G on P_H of the cardiovascular system are insignificant. Wood et al. (184) reported an increase in CO of 11% in humans at $+5$ G_x, due primarily to a 35% increase in heart rate (HR); mean P_a was increased by 17%. Smedal et al. (145) found a 25% increase in systolic P_a at $+8$ G_x. Negative G_x (prone position) cardiovascular effects were generally similar to those of $+G_x$, but tended to be more variable among individuals

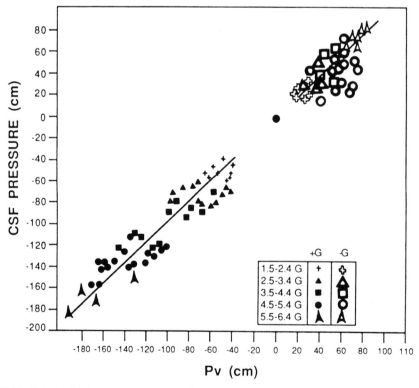

FIG. 40.12. Relationship between cerebrospinal fluid *(CSF)* pressure and venous pressure *(Pv)* (cm) during positive and negative radial accelerations of varying magnitudes. Increased pressures occurred under negative G. [Modified from Rushmer et al. (134).]

(145). Cardiac rhythm disturbances were considerably more common with $+G_x$ than with $-G_x$ (145). Significant cardiovascular accommodation to the $\pm G_x$ environment apparently does not occur.

Lateral G has a greater effect on the cardiovascular system than $\pm G_x$ exposure. A 30% *reduction* in Pa occurred during $+4\ G_y$ exposure of 1 min (145).

Respiration effects of $+G_x$ are quite different from $-G_x$. At $+6\ G_x$ vital capacity (VC) is reduced by 75% over 1G values, whereas at $-6\ G_x$, 85% of 1G VC remains. Alveolar ventilation ($\dot{V}_A$) is reduced to about 50% of 1G values at $+8$ G, but is increased to 150% of control values at $-8\ G_x$. At $+14\ G_x\ \dot{V}_A$ nears 0 (146). Differences in these respiratory functions between $\pm G_x$ occur because during $-G_x$, the lung is not restricted by the shape of the diaphragm and the spine as it is during $+G_x$. This restriction of lung movement significantly reduces gas exchange so there is a much greater reduction in S_aO_2 at $+G_x$ as compared to at $-G_x$, as shown in Figure 40.13, (146). Since cardiovascular changes are not limiting for $\pm G_z$, $\pm G_x$ exposures can be maintained for long durations with due regard for aerobic limitations.

Acceleration absorption atelectasis occurs at $+5.6$ to $+6.4\ G_x$ while breathing 100% oxygen, but without wearing the anti-G suit as required with $+G_z$ because of the restriction of the diaphragm and spine (63).

G Tolerances

The biological effectiveness of natural (genetic) adaptation to gravity can be readily quantified by increasing G exposure until it becomes physiologically limiting. This capacity is called *G tolerance* which has G-level and G-duration components. As noted earlier, tolerance varies significantly with the direction of G vector relative to body orientation and the homeostatic capacity of the species.

Positive G. Animal G-level tolerances generally are parabolic with G duration for all species, similar to the rat survival/mortality curve shown in Figure 40.14 (45). However, considerable variability exists within individual species, and there is even greater variability between animal species; for example, lethal 6G durations are 3 min for rabbits, 6 min for mice, 10 min for rats, and 11 min for chickens (15, 159). The cardiovascular system is primarily the physiologically limiting parameter, but when animal exposure continues beyond cardiovascular limits, neurologic functions become limiting.

G-level tolerance (using light loss and loss of consciousness criteria) for relaxed humans is well defined with a characteristic parabolic curve qualitatively similar to the rat tolerance curve (compare Figures 40.14 and 40.15). Tolerances of subjects exerting any muscular tensing will be higher and less reproducible than those determined on relaxed subjects. The parabolic shape of the relaxed tolerance curve identifies an early component of short duration independent of G level that has a neurologic reserve basis (7). The other component of this curve based on G level depends on cardiovascular homeostasis. Relatively higher G-level tolerances occur with slower onsets of G because of compensatory mechanisms in the body initiated by bar-

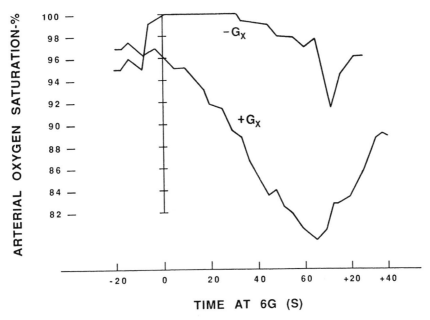

FIG. 40.13. Arterial oxygen saturations measured in subjects during exposure to $\pm 6\ G_x$ for 1 min. [Modified from Smedal et al. (146).]

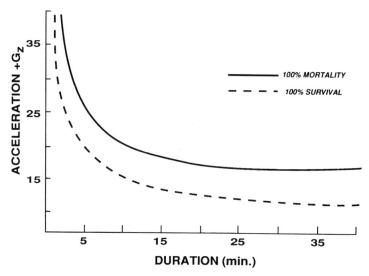

FIG. 40.14. Tolerance of rats exposed to positive +G_z acceleration. Curves delineate 100% survival and 100% mortality. [Modified from Cranmore (45).]

oceptors primarily located in the carotid sinus. Tolerance measurements are useful in quantifying changes in adaptation to G.

G-level tolerance for humans depends on eye-level Pa. Therefore an increase in pulmonary pressure with the AGSM elevates Pa (by as much as 100 mm Hg). When the AGSM is combined with the anti-G suit, G-level tolerances can be increased by 5G so that 9G can be sustained for 45 s (125) and 8G for 60 s (30).

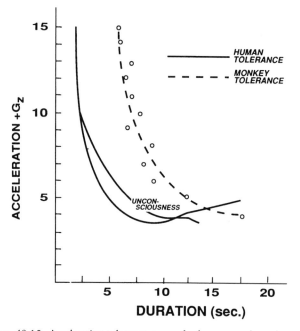

FIG. 40.15. Acceleration tolerance curves for humans and monkeys. Unconsciousness end points were used for both species. [Modified from Kydd and Stoll (97).]

G-duration tolerance is usually measured as the time (in seconds) that subjects can tolerate sustained-G levels until they become fatigued, and although reproducible for the individual, it varies greatly among subjects (33). G tolerance at the higher G-levels depends on the subject's ability to perform an effective AGSM involving muscular strength and anaerobic capacity (26). Above 5G, duration tolerances are longer than anticipated from lower-G data. Thus it appears that different homeostatic processes are in effect; that is, different energy pools are involved (Fig. 40.16).

Negative G. Discomfort, associated with head/face soft tissue edema, limits human *voluntary* tolerance to $-G_z$. Although no systematic tolerance determinations have been made in humans, pilots tolerate -6 G_z routinely during aerobatics (13). A military pilot, exposed to -9 G_z for approximately 30 s during an aircraft accident, survived without permanent medical sequelae. Uncomfortable though it may be, human tolerance to $-G_z$ is probably quite high because of CSF pressures that protect the cerebral vascular system (Fig. 40.12).

Prone/supine G. Since the horizontal body position in the G field significantly reduces h, the cardiovascular system is not limiting for $\pm G_x$ tolerance. However, the difficulty of inspiration, particularly for $+G_x$, becomes important. Difficulty and fatigue of breathing, and hypoxemia from reduced tidal volumes and increased $\dot{V}/\dot{Q}$ inequities, limit $\pm G_x$ tolerance in the human to about ± 15 G_x. Extended exposures to $\pm G_x$ result in complaints of increased difficulty in breathing, stressful discomfort, and soft tissue edema (swelling) in the throat. In one study +12 G_x was tolerated for 60 s, +10 G_x for

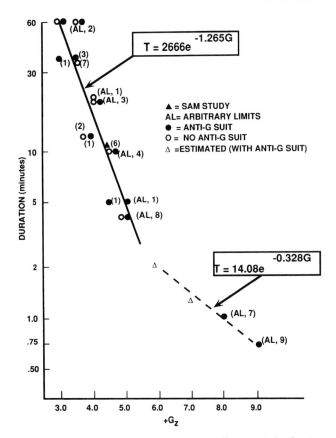

FIG. 40.16. Human tolerance times (min) at different $+G_z$ levels using fatigue as the end point. Number of subjects is in parentheses. [Modified from Burton (26); data from Miller et al. (109).]

2 min, +8 G_x for 2.5 min, +6 G_x for 4.5 min, +4 G_x for 8 min, and +3 G_x for 15 min. Similar tolerances were found for $-G_x$ exposures (3, 4).

Factors That Alter G Tolerances. Tolerances to sustained $+G_z$ can be altered using drugs or inhaled gases that affect Pa. Those factors that stimulate the cardiovascular system will usually increase $+G_z$ tolerances, as in people with high resting Pa or the use of sympathomimetic drugs (176). A high pre-G heart rate was indicative of survival of chickens to HSG (159). Inspired CO_2 concentration of 7.6% increased Pa by 30 mmHg, subsequently increasing G-level tolerance by 1G (51, 95). Hypoglycemia lowers tolerance, and alcohol reduces performance exponentially with G (29). Interestingly, even though propranolol reduced heart rate during G, it did not affect G tolerance since it did not affect Pa (11).

Other factors affect $+G_z$ tolerance. Increased environmental temperature and dehydration significantly reduce HSG duration tolerances (2, 116). Three percent body dehydration causes a 40% reduction in HSG duration tolerance. Smith and colleagues (159) found that body size and age were inversely related to HSG survival tolerance in domestic fowl. Immediately following a 30 s exposure of −1.0, −1.4, or −1.8 G_z, tolerances to $+G_z$ were reduced from 4.1G to 2.5G, a reduction of about 40% (101).

Adaptation to Sustained G

Physiological Stress Response. To develop physiological adaptation, a stress response must be initiated by a stressor for some minimum period (138). Sustained G is such a stressor. Increased levels of serum cortisol, a stress response, occur in humans exposed to short periods of $+G_z$. The magnitude of this increase is a direct function of the level of a 1 min exposure to $+G_z$ and heart-eye vertical distance, yet it is modified with the anti-G suit (110, 111). Since the inflated anti-G suit diminishes the physiologic stress response more than expected from the reduction in h with abdominal bladder inflation, the suit's support of venous return must be a major factor of the stress reduction.

Single and repeated acute, acceleration exposures are associated with heterophilia and lymphopenia, indicating the occurrence of stress (45, 46, 158). Hematological changes of this nature result from a stress response via increased adreno-cortical activity under hypothalamic control. Also, Cranmore and Ratcliffe (46) found enlarged adrenal glands in rats and guinea pigs, but interestingly found no change in pituitary weights following repeated daily accelerations. Yet adaptive responses to acute acceleration can be abolished in rats by hypophysectomy (131) and adrenalectomy (115).

Recovery from sustained-G exposure. The duration of stimulation must be sufficient to develop adaptates to altered environments. This period is measured by the time it takes for physiologic parameters stimulated during G to return to control G-levels.

Much of the recovery from exposure to sustained G in humans and in animals is relatively rapid. After 20 min, cardiovascular, respiratory, sympathetic, and some metabolism parameters have generally returned to pre-G levels. However, subjective fatigue takes several hours for complete recovery. Blood lactate levels are still elevated two to four times pre-G levels 20 min after a fatiguing bout of acceleration (22, 167). Stress indicators such as increased serum cortisol levels required 2 h to return to pre-G levels for subjects not wearing anti-G suits after a 1 min 6G exposure. The use of the anti-G suit cut the recovery time in half, presumably reducing the level of the initial stress response (110).

Belay and colleagues (8) exposed rabbits and dogs to +9 G_x or +12 G_x acceleration until physiological failure had occurred. At 15–30 min postacceleration when the animals exhibited control electrocardiogram (ECG) and respiratory indices, they measured physiological responses to caffeine and strychnine. At this time adverse changes of the animal's responses to these analeptics were still evident, indicating that neurologic recovery was not complete. Other neurologic criteria have been measured by Korneyeva and Ushakov (88), who found that it took 5 days for acetylcholine content in rabbit blood to reach control levels after -8 G_x acceleration exposures of 7–10 min duration.

Therefore, even though sustained-G exposures are short (frequently less than 1 min), some accommodative processes may be stimulated for several hours or perhaps even days. Thus it appears that sustained G can be effective in initiating physiologic adaptation processes from stimulating the stress response.

Repeated Exposures to Sustained G. Repeated exposures produced cumulative adaptive or debilitative effects that involve changes in tolerances and alterations in anatomic structure and physiologic function.

Adaptive effects. Repeated acceleration exposures enhance acceleration tolerance in rats (46, 47, 57, 73, 132), guinea pigs (46), dogs (70, 164), and humans (70, 181). Britton and colleagues (15), who repeatedly exposed rats to increasing $+G_z$ over a period of several days, reported a 50% reduction in mortality compared with previously nonaccelerated controls.

Burns and co-workers (19) observed that repeated exposure of miniature swine to $+G_z$ over periods up to 6 min greatly enhanced their acceleration tolerance. This adaptation, indicated by lower maximum heart rate for equivalent G levels, decreased catecholamines both during and following the acceleration, and decreased the incidence of cardiac pathology.

In humans, repeated daily exposures for 2 wk to slow-onset G enhanced cardiovascular compensatory mechanisms, effectively increasing low $+G_z$ tolerances by an average of 1G (181). Spence and co-workers (162) found that weekly G-exposure was necessary to maintain maximum $+G_z$ duration tolerance to variable high G exposures. Also, Boutellier et al. (14) reported, "only subjects who were regularly exposed to $+3$ G_z were able to tolerate a half-hour exposure to that acceleration."

Frazer and Reeves (57) found that to elicit adaptive responses to sustained G, the G level of the exposure must approximate the final testing acceleration conditions. High-G exposures of $+12$ G_z enhanced tolerances when tested to $+20$ G_z, whereas acceleration training exposures of $+2$ G_z reduced tolerance levels below 1G controls (Fig. 40.17).

After repeated accelerations, Kotovskaya (90) observed an increase in RNA content of various tissues which represented increased protein synthesis—an adaptive change in dogs and monkeys. This response agrees well with the work of Stepantsov and Yeremin (164) who found several "morphohistochemical changes of the adaptation type" in dogs following an adaptive regimen of repeated acute G_x accelerations. These changes included myocardial hypertrophy, increased energy reserves and oxidative enzymes in several organs, plus several vascular changes. Subcutaneous petechiaisis, small pinpoint hemorrhages under the skin, occurs frequently during $+G_z$ exposure, especially in the lower part of the body (called "high-G measles"). This condition is much less common in pilots frequently flying at high-G and in centrifuge subjects regularly exposed to HSG.

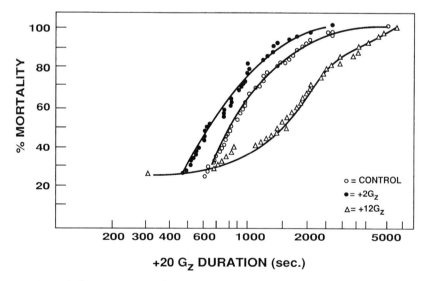

FIG. 40.17. Survival of rats at $+20$ G_z with prior repeated acceleration exposures of $+2$ G_z *(closed circles)* and $+12$ G_z *(open triangles)* and controls with no prior acceleration exposure. [Modified from Frazer and Reeves (57).]

Britton et al. (15) reported that after 90 s exposures daily in dogs over a period of 200 days, 30% fewer physiologic G-induced responses occurred. Parin and co-workers (124), using gastric motility as an indicator of general physiological response to sustained acceleration exposure in dogs, found "marked adaptation" occurred following repeated accelerations at 4–6 day intervals.

Regarding $+G_x$ tolerances of humans, Stepantsov and Yeremin (164) reported an average tolerance increase of 2.6G from repeated G_x exposures that remained in effect for 6 mo following the end of training.

Debilitative effects. Interestingly, in spite of acknowledged adaptive benefits, repeated acceleration exposures can also produce pathologic effects as well as reduced survival rates.

Murray and colleagues (112) exposed dogs twice a week to hourly periods of low intensity $+G_z$ for a total of 30 treatments that were eventually lethal to 67% of the animals. Similar findings were reported for baboons (108). Plus 12 G_z applied for 30 min/day for 30 days, was fatal to 25% of the exposed rats (57).

Smith and co-workers (157) repeatedly exposed domestic fowel to moderate G levels. Three groups were developed depending on their duration of survival: (1) susceptible that died on the first day; (2) moderately tolerant, and; (3) highly resistant. Those animals that died, frequently had subendocardial hemorrhage. Their level of physiologic stress, as indicated by relative lymphopenia, was an inverse prediction of survival. This study showed that homeostasis is extremely variable even with the same breed, and moderate G stress is debilitative with inadequate periods of recovery from the stress of G. Homeostatic mechanisms were not adequately responsive to the rapidity of the application of the stressors.

Barr (5) found that frequently repeated acceleration increased the rate of arterial oxygen desaturation during centrifugation, and decreased its recovery; both effects are typical of acceleration-induced atelectasis. Repeated centrifugation of monkeys produced electroencephalogram (EEG) waves of larger amplitude than anticipated which persisted for longer durations (16). Britton et al. (15) repeatedly exposed female rats to 2 min of $+14\ G_z$ acceleration for 1 wk which "markedly inhibited" estrus activity.

A one-time exposure to levels as high as 18G for 1 min reduced growth for 3 wk in domestic fowl even though they showed only minor debilitation symptoms immediately following G. This response was related to possible mechanical injury of the hypothalamus (158).

Plasma enzyme concentrations, indicative of tissue damage, have been found in monkeys after repeated centrifugation (44), and Burton (22) reported increases of muscle tissue enzymes in humans following repeated HSG exposures over a period of 30 min.

Anatomic vascular changes in rats and rabbits exposed to repeated high-G for several months were found by Prives (132). These alterations were considered pathologies since adaptation to sustained G, prior to these high-G exposures, significantly reduced their occurrence. Silvette and Britton (144) reported "significant pathological changes" in the kidneys of rats which had repeated daily acceleration exposures for 3 wk. Kotova and Savina (89) found increased eye pathology in rabbits in conjunction with repeated acceleration exposures.

Subendocardial hemorrhage of the right or left ventricle are commonly found in animals exposed to $\pm G_z$. This pathology has been attributed to high sympathetic activity in conjunction with a low ventricular volume as a result of insufficient venous return (61). However, the importance of the reduction in ventricular blood volume is not clear. Burton and MacKenzie (31) commonly found left ventricular subendocardial hemorrhage in swine following G exposures well within their G tolerance levels; that is, reduced ventricular blood volume was clearly not a factor. But as discussed in the previous section on adaptive effects, miniature swine were able to adapt to repeated HSG exposures over a 6 mo period, no longer exhibiting these heart lesions (19).

Cats and monkeys that were protected with water immersion of the lower body to assure adequate venous return, after long and repeated exposures to high G, eventually exhibited electrocardiographic changes indicative of right myocardial failure. Indeed, upon postmortem examinations of the cats, myocardial hemorrhage of the right ventricle was evident (66, 79). Similarly, water immersion "protection" was lethal to dogs during a single exposure to a moderate, sustained G level although their hearts were not examined for pathology (182). Propranolol administration prevented, and atropine exacerbated, heart hemorrhage in swine exposed to HSG, indicating the essential role of the sympathetics in inducing this pathology. MacKenzie and Burton (104) found myocardiopathy of the left ventricles due to HSG associated with subendocardial hemorrhage. These heart pathologies do not appear to affect heart function and the hemorrhage is resolved within 14 days in the absence of more G exposures (31).

Cross Adaptation.

Hypoxia. Adaptation to reduced barometric pressure and extremely high intensity $+G_z$ exposures that measure survival tolerances in animals may have a similar hypoxic basis. If such is the case, cross-adaptation between these two treatments would be anticipated. Indeed, increased survival tolerance to $+G_z$ was achieved in rats and monkeys after acclimatization to hypoxia (15, 80). In the primate study, the tolerance criterion was extinction of the EEG for 30 sec, that is probably a measure of the metabolic (oxygen) reserve of the brain with a hypoxic basis, and not one of cardiovascular function, the usual measure of $+G_z$ tolerance that does not have a hypoxic basis.

Several other investigators (91–93, 128, 169, 171) found that high altitude residence (2,000–4,200 m) or periodic hypoxia exposure in a low pressure chamber increased $+G_x$ tolerance by 1.75 ± 1.4 G. Similar findings were reported for mice, rats, and guinea pigs. This anticipated cross-adaptive effect, involving erythropoietic responses, was retained for approximately 3 wk.

Nonspecific stressors. Cross-resistance between acute acceleration exposure and bacterial infection was found in mice at $+25$ G_x, and in rats at $+20$ G_z with endotoxin injections, but there was no effect on hamsters at $+17$ G_x (77, 165). This substantial increase in G tolerance was attributed to a nonspecific resistance perhaps involving stimulation of the reticuloendothelial system.

Physical conditioning. Physical conditioning programs in humans that increase major muscle group strength and anaerobic exercise capacity have been shown to increase $+G_z$ duration tolerance. On the other hand, aerobic training that increased maximal oxygen uptake (Vo_{2max}) had no effect (53, 54). This training effect, however, is not primarily one of physiologic adaptation to $+G_z$ per se, but to an increased capacity to perform the AGSM.

Adaptation to gravitational G. Adaptation to chronic G increases sustained-G tolerance. Chronic exposure of animals for several months to low intensity fields (or 2.5G) greatly enhances tolerance (2.5-fold) to acute exposures of greater intensity, 6G [Table 40.3; (160)]. This tolerance was retained for at least 6 mo residence at Earth gravity.

Also, the development of adaptates from chronic G exposure useful in the sustained-G environment has

been reported by Duling (52). Four-week exposures of hamsters to 4G produced (1) a minor decrease (-10%) in femoral venous pressure; (2) a minor increase (15%, statistically significant) in femoral arterial pressure; (3) a significant decrease (-20%) in basal peripheral resistance—a function of vascular geometry and blood viscosity; (4) a marked two-fold increase in myogenic resistance in support of peripheral vasoconstriction; (5) a similar increase in the functional properties of the baroreflexes; and (6) a repressed function of the chemoreflexes. These changes are cardiovascular responses useful in tolerating a sustained G environment.

Natural Bases for Sustained-G Homeostasis

Sustained G is not a naturally occurring change of the environment. Therefore, to tolerate sustained G, homeostatic processes were developed that are useful in response to naturally occurring stressors, principally those with similar physiologic responses. There are two such stressors occurring naturally that probably provide the basis for sustained G homeostasis: (1) orthostasis, and (2) acute hemorrhage.

Orthostasis. Although orthostasis and increased $+G_z$ each increase intravascular P_H and the cardiovascular responses of each are similar, individual tolerances for each environment are not always directly correlated. Dlusskaya and Khomenko (48) examined this relationship between orthostasis and relaxed $+G_z$ tolerances using tilt and water loading tests with tolerances of $+G_z$ in 37 men, and were able to group the results accordingly: (1) 18 men with both good orthostatic tolerances (OT) and G tolerances (GT); (2) five with good OT and poor GT; (3) eight with good GT and poor OT; and (4) six with both poor OT and GT. Klein et al. (87) in an earlier study came to a similar, but more profound conclusion that orthostatic and relaxed $+G_z$ tolerances are "almost completely independent qualities of the human body."

The reasons for these different tolerances in some people of similar stressors are not completely clear, but perhaps a principal reason is the different distribution of blood volume at the outset of these two conditions. During tests for orthostatic tolerance, such as the tilt table, the body must rapidly respond to a greater change in blood volume redistribution than occurs with a seated person exposed to sustained G. When a person arises from a reclined position, where the majority of blood volume is at heart level, as much as 1 L of blood moves rapidly *below* the heart. The body responds to either condition to maintain blood flow to the head using similar physiologic homeostatic mechanisms, but each requires different levels of physiologic responses. Specifically, orthostasis is far more dependent upon

TABLE 40.3. *Mean HSG Tolerance of Chronically Accelerated Domestic Fowl and Their Age Controls**

	Chronically Accelerated	Control Group	P
Sample size (n)	(5)	(16)	—
Age (days)	289	289 ± 38	—
Body mass (kg)	2.69 ± 0.19†	3.16 ± 0.34*	<0.001
Tolerance (min. at $+6$ G_z)	30.0 ± 21.7	8.41 ± 5.22	<0.05
Initial heart rate (bpm)	294 ± 17	296 ± 43	n.s.
Acceleration change in HR (%)	$+11.2 \pm 10.7$	-4.56 ± 20.1	<0.05

*Table from (160) with permission.   †$\pm$ S.D.

venous system homeostasis than sustained G. Since individual tolerances are not always correlated between orthostasis and $+G_z$, perhaps homeostasis mechanisms developed for another natural stressor are also useful for sustained G. The most likely other stressor is acute hemorrhage. Unfortunately, no cross-adaptation studies involving acutehemorrhage and $+G_z$ exposure have been conducted.

Acute Hemorrhage. Homeostasis of acute hemorrhage is probably one of the oldest physiologic accommodative mechanisms that exists. In fact, it may have provided the basis for orthostatic homeostasis.

Evans and Boyes (55) studied the effects of acute hemorrhage upon several cardiovascular parameters in three dogs. There is a striking similarity of the cardiovascular responses via the sympathetic nervous system to $+G_z$ compared to those for hemorrhage. The percent reduction in CO which was calculated from their data was directly and rectilinearly related to the percent blood loss:

$$CO = 5.7 + 1.57 \text{ B.L.} \ldots \quad (5)$$

where

CO = % reduction in cardiac output compared with pre-hemorrhage values; and
B.L. = % blood loss of total blood volume

Changes in CO were calculated relative to blood loss using Eq. (5) and compared with CO of $+G_z$ using Eq. (4) (Table 40.4). Similar reductions in CO parallel animal tolerances for the two treatments with similar survival durations for dogs at $+4\ G_z$, or 40% blood loss through hemorrhage. Also, as with exposures to $+G_z$, Pa and venous return are reduced and total peripheral resistance increased, changes that are similar to those of $+G_z$ as shown in Table 40.2. It is reasonable, therefore, to believe that some of the homeostatic processes used to tolerate sustained G are those developed in response to acute hemorrhage. Unfortunately, no cross-adaptation studies involving acute hemorrhage and $+G_z$ exposure have been conducted.

TABLE 40.4. *Comparison of $+G_z$ and Acute Hemorrhage in Dogs Necessary to Produce Similar Change in CO as Determined Using Equations (4) and (5)* *

Cardiac Output (%)	Acceleration $(+G_z)$	Hemorrhage (% blood loss)
−18.5	2	−15.4
−36.1	3	−26.6
−53.7	4	−37.8
−71.3	5	−49.0

*See text.

CHRONIC G

The first suggestion for use of chronic centrifugation to simulate changes in Earth gravity was made by an English nutritionist, E. C. Dodds, in 1950 (49). He proposed the existence of "weight sensing" receptors as regulators of food intake and proposed centrifugation as an experimental procedure for their study. The first results of protracted centrifugation experiments were reported 3 years later by Matthews who was interested in the effect of leg loading upon the characteristics of postural muscles (106). After a year of 3G exposure, muscles of the centrifuged rats had a greatly increased (decerebrate) extensor tonus. Since Matthews's initial work, several chronic acceleration programs have indicated that gravitational fields exert a broad influence upon the form and function of animals (118, 156, 185).

Adaptation Process

The influence of exposure time upon the development of acceleration effects is complex. The first few days of centrifugation is stressful for animals (138). Signs and symptoms include immobility, anorexia, impaired thermoregulation, altered glucose and fat metabolism, and occasional neurological abnormalities (34, 36, 56, 58, 76, 120, 150). Liver glycogen deposition is increased several fold in rats during the first few hours of centrifugation and it is eliminated by adrenalectomy and hypophysectomy (119). When the centrifuged animals are returned to earth gravity, stress indicators disappear in a few hours and the animals resume normal behavior. Postmortem examination of animals who have died from this treatment does not reveal any characteristic lesions, either grossly or microscopically, other than those indicative of physiologic stress, for example, small testis and enlarged adrenals (34). No "deacceleration stress" has been reported, but microgravity-adapted animals become stressed upon return to Earth gravity (62).

When animals are maintained in a gravitational field for weeks or months, the signs of acceleration stress will ameliorate and the animals will become adapted, thriving in the new, previously toxic, environment (36).

Adapting animals to increased gravitational force has been accomplished two ways: (1) continuous exposure to a gradually increasing G field by 0.5G increments until the desired G-level is reached; or (2) daily exposures for a given period to the desired G-level. Adaptation to 3G (a very stressful environment for 2 kg animals) is optimal with 4 h exposures to 3G each day (38). This exposure developed a significant stress (as indicated by a relative lymphopenia) which was resolved before the next day's exposure. Longer exposure to 3G (for example, 8 h) did not allow for the resolution of the stress before the next centrifugation so the stress became progressively more severe. Shorter exposure periods (for example, 1 h) simply were not stressful and therefore did not stimulate the adaptive processes. Once animals are adapted to 2G, they retain tolerance to this field while living at 1G for as long as 7 mo (152).

If animals adapted to acceleration fields are reproduced serially, after 5 or 6 generations strains develop with a significantly increased capacity for adaptation to acceleration (149), indicating that the capacity for chronic acceleration adaptation is highly heritable.

Structural and Functional Changes

The load-bearing (musculoskeletal) systems of animals respond readily to changes in loading. This load-bearing phenomenon occurs in animals even under conditions of Earth gravity (39, 127, 137) in which load-bearing capacity increases proportionally to the square of some dimension of body size, but the load increases proportionally to the cube of that dimension of body size. Consequently, in a series of terrestrial animals of increasing size, there is a selective increase in muscle and skeleton sizes—about 5% each per kg increase in body mass (121).

Similar changes in the musculoskeletal system are induced by altered gravitational fields. Small laboratory mammals (0.15 kg–3.8 kg) adapted to a 2G environment (123) exhibit an 18% increase in skeletal mass (as indicated by body calcium content) and a 5% loss of muscle mass (as indicated by body creatine content). However, much of the loss in muscle mass is selectively flexor muscle, which does not have an antigravity func-

tion (28). For example, in domestic fowl at Earth gravity a paired leg extensor/flexor muscle mass ratio is 0.85, and this ratio increases to 1.7 at 2G (Fig. 40.18). The greater extensor mass in these high-G adapted fowl produced an increased exercise capacity. Running times to exhaustion of animals adapted to 1.75G was approximately three times those of Earth-gravity controls (35). In 4 wk exposures of hamsters to 4G (40), there was a 37% increase in tetanic contraction strength and a greater resistance to fatigue in the isolated gastrocnemius muscle. These gravitationally induced changes in the size and function of load-bearing elements are consistent with the principle of "functional demand" (65, 170). But the effect of increased gravitational G on intrafusal muscle fibers of the muscle spindle in domestic fowl has been inconclusive (105).

As discussed previously, blood volume and circulatory characteristics are affected by alterations of the ambient acceleration field. Hydrostatic pressures in an increased acceleration field and the compliant nature of the venous vasculature lead to a major and rapid displacement of the blood along the G vector. The resulting decreased thoracic blood volume and other G-induced physiological responses stimulate additional intake of fluid, and reduce its loss, leading to an increase in plasma volume (60). This chronic increase in plasma volume has been observed uniformly in chronically centrifuged animals (37, 52). Adaptation of the circulation to an increased G-field also leads to quantitative changes in cardiovascular reflexes that regulate circulation (52). Regional bloodflow to the eyes, kidneys, and total skeletal muscle was significantly increased in 2G adapted domestic fowl (174). Generally, these adap-

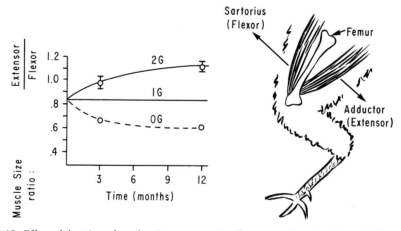

FIG. 40.18. Effect of duration of acceleration on extensive:flexor muscle ratios. Rates of change (with time [t] in months) in the mass ratios of adductor (E) and sartorius (F) muscles have the kinetics:

$$\overline{E}:\overline{F}_{(2G)} = 1.17 - 0.33e^{-0.16t}$$
$$\overline{E}:\overline{F}_{(0G)} = 0.62 + 0.22e^{-0.50t}$$

Data for 0G represents intercept values from observations at several fields at the exposure times indicated. [From Burton et al. (28).]

tations in circulatory function are appropriate for the survival of animals living in an increased G environment.

Energy metabolism also increases in animals resident in fields greater than earth gravity reflecting the increased work requirement for postural maintenance and locomotion. Feed balance measurements with ambulant domestic fowl and rabbits (all about 2 kg body mass) conducted at several field strengths indicated that the maintenance energy requirement increased an average of 24 kcal/kg/G/day (82, 153, 154). At Earth gravity, mature animals of diverse species have a basal energy metabolism (M) proportional to the 0.75 power of body mass (B): $M = kB^{0.75}$ (12). However, in a group of small mammals of diverse size in a 2G environment, M was proportional to the 0.8 power of body mass (123). These observations are consistent with Kleiber's hypothesis (85, 86) that M is a compound function, consisting of a thermoregulatory component proportional to $B^{0.5}$ (107), and an antigravity component proportional to $B,^{1.0}$ which combined are $B^{0.75}$.

Body composition is modified in animals adapted to hyperdynamic environments, particularly the reduction of body fat which has been observed in mice (83), rats (130), rabbits, (82, 154) and chickens (156). This defatting can be substantial; for example, hens at Earth gravity may contain about 30% body fat, but in a 3G field, body fat may be reduced to 2%–3%. This loss of fat appears to be the result of metabolic shifts. Glucose utilization is enhanced and glucose conversion to lipids is reduced to provide increased energy requirements of domestic fowl adapted to 2.5G (56). Certainly this gravitational defatting is not the result of inability to acquire food or of metabolic deficiency. Chickens at 3G have a reduced body mass (20% less than their gravity controls) which can be reduced further by a superimposed fast. However, upon realimentation, this loss of body mass is regained even more rapidly than in Earth gravity controls (151). Thus loss of body mass (and fat) in a high-G environment is a regulated process. Surprisingly, this gravitational defatting has not been observed in monkeys (121, 155), and no rationale is apparent for this interspecies difference.

Vestibular function of domestic fowl exhibits interesting forms of adaptation to chronic acceleration. In some birds, a return from the increased G-field to Earth gravity was accompanied by abnormal postural responses; a hyperflexion or hyper extension of the neck, or a form of paralysis with one leg extended and the other flexed (34). These postural defects were transient and usually disappeared after 8–12 h at Earth gravity. Rotatory stimulation with a Barany device adapted for chickens (179) failed to elicit a nystagmic response, indicating some structural or functional

change in the vestibular apparatus of these chickens. Breeding experiments indicated that the capacity for these postural defects is highly heritable. Another sort of vestibular response was encountered in New Hampshire chicks when returned to Earth gravity after 1 wk exposure to 1.5G. At Earth gravity, 25% of these chicks promptly put their heads between their feet and somersaulted repeatedly. If the somersaulting chicks were returned to 1.5G, they immediately resumed a normal posture and locomotion. An analogous response has been reported in humans, 25% of the humans subjected to weightlessness (in aircraft performing the parabolic maneuver) reporting a sensation of inversion. This response suggests that the somersaulting chicks, like humans, may feel inverted (150). Interestingly, similar experiments with rabbits or rats did not result in any abnormal postural responses when animals were returned to earth gravity. Possibly, these responses in fowl are related to their bipedal posture, or to the unusually large labyrinths characteristic of birds.

Measurement of the vestibular responses of centrifuged birds (when removed to Earth gravity) indicates the centrifugation may preserve their vestibular sensitivity to rotatory stimulation. Ordinarily, animals respond to repeated rotatory stimuli by a habituation, decreasing the threshold stimulus for production of nystagmus, the duration of an induced nystagmus and an increase in the stimulus necessary to produce the maximum nystagmus (179). However, birds maintained at 1.5 or 2G in between rotatory stimulation did not exhibit the characteristics of habituation (180). Also, after the period of rotatory stimulation, birds maintained at Earth gravity returned to the pretreatment state only very slowly, whereas centrifuged birds returned to the initial condition in a few weeks. It is difficult to rationalize this influence of centrifugation upon sensitivity of vestibular function. At 2G, the rotation rate is about the same as that used for vestibular stimulation, 192°/sec (179). But in the centrifuge there is a 12' radius of rotation, and with one degree of freedom there may be no sensation of rotation. It appears that the increased gravitational loading of the vestibular apparatus may affect its functional characteristics that is a condition of physiologic adaptation to G.

Adaptation to Hypo- and Hypergravity Fields

The relation between the biological effects of gravitational fields of lesser and greater strength than Earth gravity is unresolved. The concept that such a relationship does exist has been called the *continuity principle*. In gravitational experiments, the independent variable is the acceleration field and the biological responses are dependent on its magnitude. Since acceleration fields are continuous above and below earth gravity, the biological responses should a priori exhibit a similar continuity (Fig. 40.19). But some exceptions can be anticipated, particularly at low field strengths. Most biological processes will probably require some minimum field strength, that is, a threshold, to stimulate gravitational effects.

The load-bearing musculoskeletal system is particularly responsive to changes in loading, and the reduction in size of the unloaded musculoskeletal system, called *disuse atrophy* or *deconditioning,* is an example of the principle of functional demand (65). Reduction in skeletal muscle masses have been reported in rats after 7 days of orbital microgravity exposure (62, 113) and in astronauts after longer periods in space (168, 178). Similar decreases in muscle mass and strength have been reported in bed rest subjects (168). However, in either bed rest or spaceflight these adverse changes can be prevented, at least partially, by physical exercise (43, 69). Consequently, it appears that there is an exercise component and a gravitational component for maintenance of muscle mass (28, Fig. 40.18).

The skeleton also undergoes atrophic changes when it is unloaded, either in bed rest (50) or in spaceflight (161). In rats, microgravity exposure leads to reduction in the mass of load-bearing bones after 7 days, but not in non–load bearing bones (172). The losses are principally of mineral content which reduces the mechanical properties of bones (62, 166). Skeletal loading in hyper-

dynamic environments also increases the skeletal mineral content (122). Therefore, in the skeleton as in muscle, there appears to be at least a qualitative continuity of hypo-hypergravitational effects.

Chronic exposure to microgravity or bed rest (recumbency) decreases the blood volume primarily dependent upon P_H that will change in strict proportion to the intensity of the gravitational field. Circulatory function and particularly the effectiveness of the baroreflex are reduced in bed rest (17, 140) and in spaceflight (81). Since these decreases in red blood cell mass and plasma volume are the reverse of those occurring in hyperdynamic fields, circulatory function appears to be qualitatively continuous with the gravitational field intensity (37).

Changes in energy metabolism also appear to be at least qualitatively related to gravitational field strengths above and below Earth gravity. Bed rest rapidly reduces energy metabolism by about 34%, which represents the energetic cost of normal ambulation at 1G, and over a few days the basal (fasting) metabolism is reduced by about 8% (10, 67). Centrifugation increases maintenance requirements by 24 kcal/kg/G/day, irrespective of body size (153). Thus the gravitational influence on energy metabolism appears to be continuous, directly related to G fields above and below Earth gravity.

Utilization of Gravitational G in Space

Loss of adaptation to G in microgravity is a concern particularly for extended missions, such as a round-trip to Mars. Scheduled periodic exposure of space travelers to sustained increased-G using a centrifuge has been proposed as a countermeasure (24, 27, 32). Since G level and G duration appear to be physiologically interactive, it has been suggested that increasing G level above 1G might reduce the time required to prevent or ameliorate deconditioning, for example, 180 min of 1G vs. only 30 min of 3G. Controlled research findings have given support to this hypothesis. Ludwig and colleagues (103) found that 4 h of daily 1G exposure prevented orthostatic (tilting) intolerance after 4 days of continuous bed rest. Shulzhenko and Vil-Viliams (143) reported that during 3 days of immersion, daily exposures of 1.9G using a short-radius centrifuge was more effective than the same duration at 1.2G in maintaining 3G tolerance.

The ability to transport gravity into space using a personal centrifuge to prevent physiological deconditioning appears to be an opportunity that nature has provided, one which will one day benefit space travelers.

CONCLUSION

Physiologic adaptation occurs in response to long-duration exposure to a wide range of acceleration fields. This

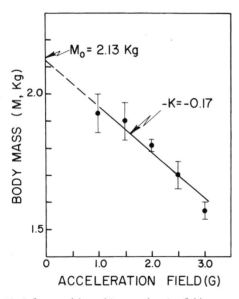

FIG. 40.19. Influence of the ambient acceleration field on mature body mass of chickens. Mature body mass (M) of chronically accelerated chickens decreases rectilinearly with increasing field strength G:

$$M = M_0 - kG$$

where M is mature body mass (kg), G is a field of G-strength, M_0 is weightlessness; i.e., $G = 0$; and, $-k$ is the proportionality coefficient. For the indicated observations (151), the constants in the above equation have the values: $M_0 = 2.13$ kg and $-k = -0.17$; with a correlation coefficient $r = 0.63$ ($P < 0.01$). [From Smith and Burton (151).]

adaptation follows the classic *general adaptation syndrome* described by Selye (138). Involved in adaptation to chronic G are major physiologic functions and structures that include body mass, muscle mass (extensor:flexor differential), body composition, food intake, bone mass, and many others. It has been proposed that adaptive responses developed in an increased gravitational field are related to adaptive responses in a weightless (microgravity) environment. This hypothesis is based on several chronic acceleration studies (62, 113, 122, 172) including the premise that gravity-dependent homeostasis is a continuous function of the intensity of the acceleration (or gravitation) environment beginning with microgravity, extending through Earth gravity into hypergravitational fields. Application of this hypothesis would prevent microgravity deconditioning with periodic centrifuge exposures of astronauts to hypergravity in space.

Animals and humans accommodate sustained G using homeostatic processes previously developed in response to other stressors, probably orthostasis and acute hemorrhage. All have a cardiovascular basis and use the same accommodative processes. Intravascular hydrostatic pressure (P_H) that increases during $+G_z$ is the physical stressor; it limits animal and human tolerance by directly reducing eye-level arterial pressure, and by impeding venous return. Yet the human homeostasis involving $+G_z$ is simply remarkable in that exposures to fields more than three times Earth's gravity (1G) are tolerated using established accommodative physiologic processes. Although lung function is significantly restricted by this environment, it does not appear to be limiting to $+G_z$. Adaptation to sustained G occurs with repeated exposures or by adapting to other types of stressors known to stimulate common adaptates. However, if repeated sustained-G exposures are too stressful, then debilitation instead of adaptation will occur.

Physiologic limitations to a range of G levels in the human is summarized in Table 40.5. Human tolerances to $+G_Z$ have 4 ranges of G levels with differing (*1*) durations and (*2*) limiting physiologic functions. From <1 to 1.5G, physiologic homeostasis appears to be limited by the aging process. Humans have tolerated 1.5G continuously for 1 wk without significant physiologic consequences. The transitional G-level range between 1.5G and 3.5G has not been systematically studied, but one human exposure at 2G for 24 h was terminated voluntarily, but clearly demonstrated that a significant fluid inbalance had occurred and homeostasis through physiologic adaptation was not happening (41). Three G for 1 h in seated humans was without symptoms, but 3.5G was limiting with symptoms of fatigue at 30 min (Fig. 40.16). Above 4.5 to 5.0G, a seated person's physiologic tolerance is limited by cardiovascular insufficiency (Fig. 40.4).

TABLE 40.5. *Ranges of Human Tolerance to $+G_z$ without G-Protection Methods*

$+G_z$ Levels (Range)	Duration	Limiting Physiologic Function*
<1–1.5G	Lifetime	Aging
>1.5 G <3.5G	Hours	Fluid imbalance
>3.5 G	Minutes	Fatigue
>5 G	Seconds	Cardiovascular insufficiency

*Homeostasis is not obtainable.

REFERENCES

1. Akesson, S. On the blood supply to the brain during acceleration. *Acta Physiol. Scand.* 15: 237–244, 1948.
2. Balldin, U. I., A. Sporrong, and P. A. Tesch. Dehydration and G tolerance, psychomotor performance and muscle function. *Aviat. Space Environ. Med.* 55: 467, 1984.
3. Ballinger, E. R., and C. H. Dempsey. The effect of prolonged acceleration in the human body in the prone and supine positions. Wright Air Development Center, 1952. (WADC Report No. 52-250)
4. Barer, A. S., G. A. Golov, V. B. Zubavin, Ye. I. Sorokina, and Ye. Tikhomirov. Oxygen balance of the body during extended accelerations. *Aerosp. Med.* 39: 253–257, 1968.
5. Barr, P. O. Hypoxemia in man induced by prolonged acceleration. *Acta Physiol. Scand.* 54: 128–137, 1962.
6. Beckman, E. L. Protection afforded the cerebrovascular system by the cerebrospinal fluid under the stress of negative G. *J. Aviat. Med.* 20: 430–438, 1949.
7. Beckman, E. L., T. D. Duane, J. E. Ziegler, and H. N. Hunter. Some observations on human tolerance to acceleration stress: Phase IV. Human tolerance to high positive G applied at a rate of 5 to 10 G per second. *J. Aviat. Med.* 25: 50–66, 1954.
8. Belay, V. Ye., P. V. Vasil'yev, G. D. Glod, and M. I. Bryuzgina. Reactivity of animals to caffeine and strychnine in the period of after-effect to transverse accelerations. *Kosm. Biol. Med.* 1: 47–53, 1967.
9. Bernard, C. *An Introduction to the Study of Experimental Medicine*. New York: Dover, 1957.
10. Bernauer, E. M., and W. C. Adams. The effect of a nine day recumbency with and without exercise on the redistribution of body fluids and electrolytes, renal function and metabolism. 1968. NASA Hdq., Wash., D.C. (NASA CR-73664)
11. Bjurstedt, H., G. Rosenhamer, and G. Tyden. Acceleration stress and effects of propranolol on cardiovascular responses. *Acta Physiol. Scand.* 90: 491–500, 1974.
12. Blaxter, R. *Energy Metabolism in Animals and Man*. Cambridge: Cambridge Univ. Press, 1989.
13. Bloodwell, R. D., and J. E. Whinnery. Acceleration exposure during competitive civilian aerobatics. Alexandria, VA: *Preprints, Ann. Meeting Aerosp. Med. Assoc.*, p. 167–168, 1982.
14. Boutellier, U., R. Arceli, and L. E. Farhi. Ventilation and CO_2 responses during $+G_z$ acceleration. *Resp. Physiol.* 62: 141–151, 1985.
15. Britton, S. W., E. L. Corey, and G. A. Stewart. Effects of high acceleration forces and their alleviation. *Am. J. Physiol.* 146: 35–51, 1946.
16. Britton, S. W., V. A. Pertzoff, C. R. French, and R. F. Kline. Circulatory and cerebral change and protective aids during

exposure to acceleratory forces. *Am. J. Physiol.* 150: 7–26, 1947.

17. Bungo, M. W., J. B. Charles, and P. C. Johnson. Cardiovascular deconditioning during spaceflight and the use of saline as a countermeasure. *Aviat. Space Environ. Med.* 56: 985–90, 1985.

18. Burns, J. W., and U. I. Balldin. Assisted positive-pressure breathing for augmentation of acceleration tolerance time. *Aviat. Space Environ. Med.* 59: 225–233, 1988.

19. Burns, J. W., M. H. Laughlin, W. M. Witt, J. T. Young, and V. P. Ellis. Pathophysiologic effects of acceleration stress in the miniature swine. *Aviat. Space Environ. Med.* 54:881–893, 1983.

20. Burns, J. W., M. J. Parnell, and R. R. Burton. Hemodynamics of miniature swine during $+G_z$ stress with and without anti-G support. *J. Appl. Physiol.: Respir. Environ. Exerc. Physiol.* 60: 1628–1637, 1986.

21. Burton, R. R. Positive ($+G_z$) acceleration tolerance of the miniature swine: Application as a human analog. *Aerosp. Med.* 44: 294–298, 1973.

22. Burton, R. R. Human response to repeated high sustained G simulated aerial combat maneuvers. *Aviat. Space Environ. Med.* 51: 1185–1192, 1980.

23. Burton, R. R. A conceptual model for predicting pilot group G tolerance for tactical fighter aircraft. *Aviat. Space Environ. Med.* 57: 733–744, 1986.

24. Burton, R. R. A human-use centrifuge for space stations: Proposed ground-base studies. *Aviat. Space Environ. Med.* 59: 579–582, 1988.

25. Burton, R. R. G-induced loss of consciousness: definition, history, current status. *Aviat. Space Environ. Med.* 59: 2–5, 1988.

26. Burton, R. R. Human physiologic limitations to G in high-performance aircraft. In: *Environmental Physiology*, edited by L. E. Farhi, and C. V. Paganelli. New York: Springer-Verlag, chapt. 10, p. 123–137, 1989.

27. Burton, R. R. Periodic acceleration stimulation in space. In: *19th Intersoc. Conf. Environ. Systems, Soc. Auto. Engr.*, 1989. San Diego, CA, SAE Tech. Paper Series #891434 Warrendale, PA.

28. Burton, R. R., E. L. Besch, S. J. Sluka and A. H. Smith. Differential effect of chronic acceleration upon skeletal muscles. *J. Appl. Physiol.* 23: 80–84, 1967.

29. Burton, R. R., and J. L. Jaggars. Influence of ethyl alcohol ingestion on a target task during sustained $+G_z$ centrifugation. *Aerosp. Med.* 45: 290–296, 1976.

30. Burton, R. R., S. D. Leverett, Jr., and E. D. Michaelson. Man at high sustained $+G_z$. *Aerosp. Med.* 45: 1115–1136, 1974.

31. Burton, R. R., and W. F. MacKenzie. Heart pathology associated with exposures to high sustained $+G_z$. J.C.A.P. *Aviat. Space Environ. Med.* 46: 1251–1253, 1975.

32. Burton, R. R., and L. J. Meeker. Physiologic validation of a short-arm centrifuge for space application. *Aviat. Space Environ. Med.* 63: 476–481, 1992.

33. Burton, R. R., and R. M. Shaffstall. Human tolerance to aerial combat maneuvers. *Aviat. Space Environ. Med.* 51: 641–648, 1980.

34. Burton, R. R., and A. H. Smith. Chronic acceleration sickness. *Aerosp. Med.* 36: 39–44, 1965.

35. Burton, R. R., and A. H. Smith. Muscle size, gravity and work capacity. Presented at the *XVI Int. Congr. Aviat. Space Med. Lisbon, Portugal, 1967.*

36. Burton, R. R., and A. H. Smith. Criteria for physiological stress produced by increased chronic acceleration. *Proc. Soc. Exp. Biol. Med.* 128: 608–611, 1968.

37. Burton, R. R., and A. H. Smith. Hematological findings associated with chronic acceleration. *Space Life Sci.* 1: 501–513, 1969.

38. Burton, R. R., and A. H. Smith. Stress and adaptation responses to repeated acute acceleration. *J. Appl. Physiol.* 222: 1505–1509, 1972.

39. Calder, W. A., III. *Size, Function and Life History.* Cambridge: Harvard Univ. Press, 1984.

40. Canonica, P. G. Effects of prolonged hypergravity stress on the myogenic properties of the gastrocnemius muscle. Columbia: Univ. of South Carolina, 1966. Masters dissertation.

41. Clark, C. C. Observations of a human experiencing 2G for 24 hrs. Presented at the 31st *Ann. Meeting Aerosp. Med. Assoc.*, 1960.

42. Clark, C. C., I.D.R. Gardiner, A. K. McIntyre, and H. Jorgenson. The effect of positive acceleration on fluid loss from blood to tissue spaces in human subjects on the centrifuge. *Federation Proc.* 5: 17, 1946.

43. Convertino, V. A. Neuromuscular aspects in development of exercise countermeasures. *Physiologist* 34 (Suppl. 1): S125–128, 1991.

44. Cope, F. W., and B. D. Polis. Change in plasma transaminase activity of Rhesus monkeys after exposure to vibration, acceleration, heat or hypoxia. *Aerosp. Med.* 30: 90–96, 1959.

45. Cranmore, D. Behavior, mortality and gross pathology of rats under acceleration stress. *Aerosp. Med.* 27: 131–140, 1956.

46. Cranmore, D., and H. L. Ratcliffe. A study of adaptation to acceleration with rats and guinea pigs as test animals. 1956. Avia. Med. Accel. Lab., U.S. Naval Air Dev. Ctr., Johnsville, PA, p. 1–16. (N.A.D.C.-MA-5602)

47. DeMarco, A. O., and I. Geller. Effects of acceleration forces on timing behavior in the white rat. *Aerosp. Med.* 35: 30–32, 1964.

48. Dlusskaya, I. G., and M. N. Khomenko. Responses of subjects differing in tolerances of $+G_z$ to tilt tests and water loading tests. *Kosm. Biol. Med.* 19: 22–27, 1985.

49. Dodds, E. C. Contribution to discussion on obesity. *Proc. R. Soc. Med.* 43: 342–344, 1950.

50. Donaldson, C. L., S. B. Hulley, J. M. Vogel, R. S. Hattner, J. H. Bayers, and D. E. McMillan. Effect of prolonged bed rest on bone mineral. *Metabolism* 19: 1071–1084, 1970.

51. Dripps, R. O., and J. H. Comroe. The respiratory and circulatory response of normal man to inhalation of 7.6 and 10.4 percent CO_2 with a comparison of the maximal ventilation produced by severe muscular exercise, inhalation of CO_2, and maximal voluntary hyperventilation. *Am. J. Physiol.* 149: 43–51, 1947.

52. Duling, B. R. Effects of chronic centrifugation at 3 G's on cardiovascular reflexes in the rat. *Am. J. Physiol.* 213: 466–472, 1967.

53. Epperson, W. L., R. R. Burton, and E. M. Bernauer. The influence of differential physical conditioning regimens on simulated aerial combat maneuvering tolerance. *Aviat. Space Environ. Med.* 53: 1091–1097, 1982.

54. Epperson, W. L., R. R. Burton, and E. M. Bernauer. The effectiveness of specific weight training regimens on simulated aerial combat maneuvering to G tolerance. *Aviat. Space Environ. Med.* 56: 534–539, 1985.

55. Evans, D., and H. W. Boyes. The effects of rapid massive haemorrhage and retransfusion on various cardiovascular parameters in the dog. *Can. Anaesth. Soc. J.* 16: 385–394, 1969.

56. Evans, J. W., and J. M. Boda. Glucose metabolism and chronic acceleration. *Am. J. Physiol.* 219: 893–896, 1970.

57. Frazer, J. W., and E. Reeves. Adaptation to positive acceleration. 1958. Avia. Med. Accel. Lab., U.S. Naval Air Dev. Ctr., Johnsville, PA, p. 1–14. (N.A.D.C.-MA-5818)

58. Fuller, C. A., J. M. Horowitz and B. A. Howitz. Effects of acceleration on the thermoregulatory response of unanesthetized rats. *J. Appl. Physiol.* 42: 74–77, 1977.

59. Galileo, G. *Discourses and Demonstrations Concerning Two New Sciences* [Leyden: Elsevirii, 1638]. Translated by H. Crew and A. DeSalvio. New York: Macmillan, 1934; trans. by S. Drake. Madison: Univ. Wisconsin Press, 1974.

60. Gauer, O. H., and J. P. Henry. Circulatory basis of fluid volume control. *Physiol. Rev.* 43: 423–481, 1963.

61. Gauer, O. H., and J. P. Henry. Negative ($-G_z$) acceleration in relation to arterial oxygen saturation, subendocardial hemorrhage and venous pressure in the forehead. *Aerosp. Med.* 35: 533–545, 1964.

62. Gazenko, O. G., E. A. Ilyin, E. A. Savina, L. V. Serova, A. S. Kaplansky, I. A. Popova, V. S. Oganov, K. V. Smirnov and I. V. Konstantinova. Study of the initial period of adaptation to microgravity in the rat experiment on board Cosmos-1967. *Physiologist* 30 (Suppl. 1): S53–S55, 1987.

63. Glaister, D. H. *The Effect of Gravity and Acceleration in the Lung.* Slough, England: Technivision Services, AGARDograph 113, 1970.

64. Glaister, D. H. Current and emerging techniques in G-LOC detection: Noninvasive monitoring of cerebral microcirculation using near infrared. *Aviat. Space Environ. Med.* 59: 23–28, 1988.

65. Goss, R. J. *The Physiology of Growth.* New York: Academic, 1978.

66. Greenfield, A.D.M. Effect of acceleration on cats with and without water immersion. *J. Physiol.* 104: 5–6, 1945.

67. Greenleaf, J. E. Energy and thermal regulation during bed rest and space-flight. *J. Appl. Physiol. Respir. Environ. Exerc. Physiol.* 67: 507–516, 1989.

68. Greenleaf, J. E. Importance of fluid homeostasis for optimal adaptation to exercise and environmental stress: acceleration. In: *Perspectives in Exercise and Sports Medicine. Fluid Homeostasis During Exercise,* edited by C. V. Gisolfi and D. R. Lamb. Dubuque, Indiana: Benchmark, vol. 3, chapt. 9, p. 307–346, 1990.

69. Greenleaf, J. E., R. Bulbulian, E. M. Bernauer, W. L. Haskell, and T. Moore. Exercise training protocols for astronauts in microgravity. *J. Appl. Physiol. Respir. Environ. Exerc. Physiol.* 67: 2191–204, 1989.

70. Hallenbeck, G. A. The response of normal dogs to prolonged exposure to centrifuge force. 1944. Nat. Res. Council, Div. Med. Sci., from: Accel. Lab., Mayo Aero Med. Unit, Rochester, Minn. p. 1–5. (N.R.C., 279)

71. Haswell, M. S., W. A. Tacker, U. I. Balldin, and R. R. Burton. Influences of inspired oxygen concentration on acceleration atelectasis. *Aviat. Space Environ. Med.* 57: 432–437, 1986.

72. Henry, J. P., O. H. Gauer, S. S. Kety, and K. Kramer. Factors maintaining cerebral circulation during gravitational stress. *J. Clin. Invest.* 30: 292–300, 1951.

73. Herrick, R. M., J. L. Myers, and R. E. Burke. Discriminative behavior following repeated exposure to negative acceleration. 1957. Avia. Med. Accel. Lab. U.S. Naval Air Dev. Ctr., Johnsville, PA, p. 1–9. (N.A.D.C.-MA-5716)

74. Hershgold, E. J. and S. H. Steiner. Cardiovascular changes during acceleration stress in dogs. *J. Appl. Physiol.* 15: 1065–1068, 1960.

75. Hicks, J. W. and H. S. Badeer. Gravity and the circulation "open" vs. "closed" systems. *Am. J. Physiol.* 262: *(Regulatory Integrative Comp. Physiol. 33)* R725–732, 1992.

76. Horowitz, J. M., and B. A. Horwitz. Thermoregulatory responses of unanesthetized rats exposed to gravitational fields of 1 to 4 G. COSPAR: *Life Sci. Space Res.* Ed. R. Holmquist and A. C. Stickland. Pergamon Press, Oxford and New York. XVI: 77–82, 1978.

77. Horwitz, K. B., R. J. Ball, and J. P. Schmidt. Resistance to infection of mice and hamsters following short-term acceleration stress. *Aerosp. Med.* 40: 1248–1251, 1969.

78. Huygens, C. *The Pendulum Clock or Geometrical Demonstration Concerning the Motion of Pendula as Applied to Clocks.* [Hague, 1658]. Translated by R. J. Blackwell. Ames: Iowa State Univ. Press, p. 182, 1986.

79. Jasper, H. H., and A. J. Cipriani. Physiological studies on animals subject ed to positive G. *J. Physiol.* 104: 6–7, 1945.

80. Jasper, H., A. Cipriani, and E. Lotspeich. Physiological studies on the effects of positive acceleration in cats and monkeys. 1942. Report to Subcomit. Accel. of Comit. Avia. Med., Nat. Res. Council Wash., D.C. Proj. A.M. 14. (N.R.C., C-2225)

81. Johnson, R. L., G. W. Hoffler, A. E. Nicogossian, S. A. Bergman, Jr., and M. M. Jackson. Lower body negative pressure: Third manned skylab mission. In: *Biomedical Results from Skylab,* edited by R. S. Johnson and L. F. Dietlein. (NASA SP337) Washington DC: National Aeronautics and Space Administration, chapt. 29, p. 284–312, 1977.

82. Katovich, M. J., and A. H. Smith. Body mass, composition and food intake in rabbits during altered acceleration fields. *J. Appl. Physiol. Respir. Environ. Exerc. Physiol.* 45: 51–55, 1978.

83. Keil, L. C. Changes in growth and body composition of mice exposed to chronic centrifugation. *Growth* 33: 176–180, 1969.

84. Kelly, C. F., A. H. Smith, and C. M. Winget. An animal centrifuge for prolonged operation. *J. Appl. Physiol.* 15: 753–757, 1960.

85. Kleiber, M. *Fire of Life.* New York: Wiley, 1961.

86. Kleiber, M. Further consideration of the relation between metabolic rate and body size. In: *Energy Metabolism of Farm Animals,* edited by K. L. Blaxter, J. Kielanowski, and G. Thorbek. Newcastle on Tyne: Oriel, p. 505–511, 1969.

87. Klein, K. E., H. Bruner, D. Jovy, L. Vogt, and H. M. Wegman. Influence of stature and physical fitness in tilt-table and acceleration tolerance. *Aerosp. Med.* 40:293–297, 1969.

88. Korneyeva, N. V., and A. S. Ushakov. Effect of transverse accelerations on the acetylcholine content and cholinesterase activity of the blood of experimental animals. *Kosm. Biol. Med.* 1: 34–38, 1967.

89. Kotova, E. S. and Ye. A. Savina. Clinical and morphological characteristics of hemodynamic peculiarities in the eye vascular system of test animals exposed to accelerations. *Kosm. Biol. Med.* 2: 38–43, 1968.

90. Kotovskaya, A. R. Certain problems resulting from effects of acceleration during space flight (effects of cumulation and adaptation). Presented at the 17th Mt. Astron. Cong., 9–15 Oct., 1966, Madrid, Spain. N 67-26624, NASA TT-F10412, p. 1–18.

91. Kotovskaya, A. R., R. A. Vartbaronov, F. V. Babchinskiy, and S. F. Simpura. The effect of adaptation to hypoxia under pressure-chamber conditions on tolerance to transverse G-loads. *Problems of Space Biol.* 8: 41–49, 1969. (NASA TT F-580)

92. Kotovskaya, A. R., R. A. Vartbaronov, and S. F. Simpura. Physiological reactions of man to the effect of transverse G-loads after adaptation to high altitude conditions. *Problems of Space Biol.* 8: 17–41, 1969. (NASA, TT F-580)

93. Kotovskaya, A. R., P. V. Vasil'yev, R. A. Vartivaronov, and S. F. Simpura. The effect of preliminary acclimatization in mountains on man's endurance of transverse G-loads. *Problems of Space Biol.* 8: 7–16, 1969. (NASA TT F-180)

94. Kravik, S. E., L. C. Keil, G. Greelen, C. E. Wade, P. R. Barnes, W. A. Spaul, C. A. Elder, and J. E. Greenleaf. Effect on antigravity suit inflation on cardiovascular, PRA and PVP response in humans. *J. Appl. Physiol.: Respir. Environ. Exerc. Physiol.* 61: 766–774, 1986.

95. Krutz, R. W. Effects of elevated CO_2 breathing mixture on $+G_z$ tolerance. *USAFSAM AFOSR Res. Rev.,* Oct., 1974.

96. Krutz, R. W., R. R. Burton, and E. M. Forster. Physiologic correlates of protection afforded by anti-G suits. *Aviat. Space Environ. Med.* 61: 106–111, 1990.

97. Kydd, G. H. and A. M. Stoll. G tolerance in primates I. Unconsciousness end point. *J. Aviat. Med.* 29: 413–421, 1958.

98. Laughlin, M. H. The effects of $+G_z$ on the coronary circulation: A review. *Aviat. Space Environ. Med.* 57: 5–16, 1986.

99. Laughlin, M. H., J. W. Burns, and M. J. Parnell. Regional distribution of cardiac output in unanesthetized baboons during +G$_z$ stress with and without an anti-G suit. *Aviat. Space Environ. Med.* 53: 133–141, 1982.

100. Laughlin, M. H., W. M. Witt, and R. N. Whittaker. Regional cerebral blood flow in conscious miniature swine during high sustained +G$_z$ acceleration stress. *Aviat. Space Environ. Med.* 50: 1129–1133, 1979.

101. Lehr, A-K., A.R.J. Prior, G. Langewouters, B. Ullrich, H. Leipner, S. Zollner, P. Lindner, H. Pongratz, H. A. Dieterich, and K. Theisen. Previous exposure to negative G$_z$ reduces relaxed +G$_z$ tolerance. *Aviat. Space Environ. Med.* 63: 405 (Abst. 119), 1992.

102. Lindberg, E. F., W. F. Sutterer, H. W. Marshall, R. N. Headley, and E. H. Wood. Measurement of cardiac output during headward acceleration using the dye-dilution technique. *Aerosp. Med.* 31: 817–834, 1960.

103. Ludwig, D. A., J. Vernikos, M. R. Duvoisin, and J. L. Stinn. The efficacy of periodic +1G$_z$ exposure in the prevention of bedrest induced orthostatic intolerance. *Avia. Space Environ. Med.* 63:449(Abst. 635), 1992.

104. MacKenzie, W. F., and R. R. Burton. Ventricular pathology in swine at high sustained +G$_z$. *AGARD Conf. Proc.* 189: A2-1–A2-3, 1976.

105. Maier, A., E. Eldred, and R. R. Burton. Effects of long-term increased gravitational load on intrafusal fibers on the avian muscle spindle. *Brain Res.* 112: 180–182, 1976.

106. Matthews, B.H.C. Adaptation to centrifuge acceleration. *J. Physiol.* 122: 31p, 1953.

107. McNab, B. K. Body weights and the energetics of temperature regulation. *J. Exp. Biol.* 53: 329–348, 1970.

108. Menninger, R. P., R. H. Murray, and F. R. Robinson. Repeated, prolonged low intensity +G$_z$ exposures: Anatomical studies in baboons. *Aerosp. Med.* 38: 337–339, 1967.

109. Miller, H., M. B. Riley, S. Bondurant, and E. P. Hiatt. The duration of tolerance to positive acceleration. *J. Aviat. Med.* 30: 360–366, 1959.

110. Mills, F. J. The endocrinology of stress. *Aviat. Space Environ. Med.* 56: 642–650, 1985.

111. Mills, F. J., and V. Marks. Human endocrine responses to acceleration stress. *Aviat. Space Environ. Med.* 53: 537–540, 1982.

112. Murray, R. H., J. Prine, and R. P. Menninger. Repeated, prolonged low-intensity +G$_z$ exposures. Anatomical studies in dogs. *Aerosp. Med.* 36: 972–976, 1965.

113. Musacchia, X. J., J. M. Steffen, R. D. Fell, and M. J. Dumbrowski. Comparative morphometry of fibers and capillaries in soleus following weightlessness (SL-3) and suspension. *Physiologist* 31 (Suppl. 1): S28–S29, 1988.

114. Newton, I. *Philosophia Naturalis Principia Mathematica*. [London: Royal Academy, 1687.] Translated by F. Cajori. Berkeley/Los Angeles: Univ. California Press, p. 680, 1934.

115. North, W. C., and J. A. Wells. Modification by previous trauma and temperature of tolerance to tumbling shock in rats. *Federation Proc.*, 11: 380, 1952.

116. Nunneley, S. A., and R. F. Stribley. Heat and acute dehydration effects on acceleration responses in man. *J. Appl. Physiol.: Respir. Environ. Exerc. Physiol.* 47: 197–200, 1979.

117. Ossard, G., J. M. Clere, M. Kerguelen, F. Melchior, A. Roncin, and J. Seylaz. Response of human cerebral blood flow to +G$_z$ accelerations. *J. Appl. Physiol.* 76: 2114–2118, 1994.

118. Oyama, J. Effects of altered gravitational fields on laboratory animals. In: *Progress in Biometerology*, edited by: S. W. Trump. J. J. Bouma, and H. D. Johnson. Amsterdam: Swets and Zeitlinger, vol. I, chapt. 9, 1976.

119. Oyama, J., and W. T. Platt. Metabolic alterations in rats exposed to acute acceleration stress. *Endocrinology* 76: 203–204, 1965.

120. Oyama, J., W. T. Platt, and V. B. Holland. Deep-body temperature changes in rats exposed to chronic centrifugation. *Am. J. Physiol.* 221: 1271–1277, 1971.

121. Pace, N., D. F. Rahlmann, A. M. Kodama, and A. H. Smith. Changes in the body composition of monkeys during long-term exposure to high acceleration fields. *COSPAR: Life Sci. Space Res.*, edited by R. Holmquist and A. C. Stickland. Pergamon Press, Oxford and NewYork. XVI: 71–76, 1978.

122. Pace, N., D. F. Rahlmann, and A. H. Smith. Skeletal mass change as a function of gravitational loading. *Physiologist* 28 (Suppl. 6): S17–S20, 1985.

123. Pace, N., and A. H. Smith. Scaling of metabolic rate on body mass in small animals at 2.0G. *Physiologist* 26 (Suppl. 6): S125–S126, 1983.

124. Parin, V. V., R. M. Baevskii, and M. D. Emelyanov. *Monographs on Space Physiology*. Moscow: Meditsina, 1968.

125. Parkhurst, M. J., S. D. Leverett, Jr., and S. J. Shubrooks, Jr. Human tolerance to high sustained +G$_z$ acceleration. *Aerosp. Med.* 43: 708–712, 1972.

126. Pertzoff, V. A., and S. W. Britton. Force and time elements in circulating changes under acceleration: carotid arterial pressure deficiency area. *Am. J. Physiol.* 512: 492–498, 1948.

127. Peters, R. H. *The Ecological Implications of Body Size*. Cambridge: Cambridge Univ. Press, 1983.

128. Petruklin, V. G. The effect of preliminary interrupted stay in a rarefied atmosphere on the tolerance of rats to transverse G-loads. *Problems of Space Biol.* 8: 112–118, 1969. (NASA TTF-580)

129. Piemme, T. E., A. S. Hyde, M. McCally, and G. Potos, Jr. Human tolerance to +G$_z$ 100 percent gradient spin. *Aerosp. Med.* 37: 16–21, 1966.

130. Pitts, G. C., L. S. Bull, and J. Oyama. Effect of chronic centrifugation on body composition in the rat. *Am. J. Physiol.* 223: 1044–48, 1972.

131. Polis, B. D., A. Zella, and J. D. Hardy. Hormonal factors in the resistance to acceleration stress. *Aerosp. Med.* 28: 214, 1957.

132. Prives, M. D. The effect of G-force on the structure of the vascular system. *Arkh. Anat., Gistol. i Embriol.* (Moscow), 45: 3–13, 1963.

133. Rushmer, R. F. A roentgenographic study of the effects of a pneumatic anti-blackout suit on the hydrostatic columns in man exposed to positive radial acceleration. *Am. J. Physiol.* 151: 459–468, 1947.

134. Rushmer, R. F., E. L. Beckman, and D. Lee. Protection of the cerebral circulation by the cerebrospinal fluid under the influence of radial acceleration. *Am. J. Physiol.* 151: 355–365, 1947.

135. Ryan, E. A., W. K. Kerr, and W. R. Franks. Some physiologic findings on normal men subjected to negative G. *J. Aviat. Med.* 21: 173–194, 1950.

136. Salzman, E. W., and S. D. Leverett, Jr. Peripheral venoconstriction during acceleration and orthostasis. *Circul. Res.* 4: 540–545, 1956.

137. Schmidt-Nielsen, K. *Scaling*, Cambridge: Cambridge Univ. Press, 1984.

138. Selye, H. *Stress*. Montreal: Acta, 1950.

139. Shaffstall, R. M., and R. R. Burton. Evaluation of assisted positive pressure breathing on +G$_z$ tolerance. *Aviat. Space Environ. Med.* 50: 820–824, 1979.

140. Shen, X., Y. Sun, Q. Xiang, J. Meng, L. Xu, X. Yan, H. Yu, and X. Zhuang. The study of baroreceptor reflex function before and after bed rest. *Physiologist* 31 (Suppl. 1): S22–S23, 1988.

141. Shubrooks, S. J., Jr. Changes in cardiac rhythm during sustained high levels of positive (+G$_z$) acceleration. *Aerosp. Med.* 43: 1200–1206, 1972.

142. Shubrooks, S. J., Jr. Relationship of sodium deprivation to +G$_z$ acceleration tolerance. *Aerosp. Med.* 43: 954–956, 1972.

143. Shulzhenko, E. B., and I. F. Vil-Viliams. Short radius centrifuge as a method in long term space flights. *Physiologist* 35: (Suppl.): S122–S125, 1992.

144. Silvette, H. and S. W. Britton. Acceleratory effects on renal functions. *Am. J. Physiol.* 155: 195–202, 1948.

145. Smedal, H. A., G. R. Holden, and J. R. Smith. Cardiovascular responses to transversely applied acceleration. *Aerosp. Med.* 34: 749–752, 1963.

146. Smedal, H. A., T. A. Rogers, T. D. Duane, G. R. Holden, and J. R. Smith. The physiological limitations of performance during acceleration. *Aerosp. Med.* 34: 48–55, 1963.

147. Smith, A. H. Principles of gravitational biology. In: *Foundations of Space Biology and Medicine,* edited by M. Calvin and O. G. Gazenko. Washington, D.C.: U.S. Government Printing Office, USA (NASA) and USSR (Acad. Sci.), vol. II, book I, chapt. 4, p. 129–162, 1975.

148. Smith, A. H. The study of high-g effects in animals. *Physiologist* 22 (Suppl. 6): 7–10, 1979.

149. Smith, A. H. Enhancement of chronic acceleration tolerance by selection. *Physiologist* 26 (Suppl. 6): S85–S86, 1982.

150. Smith, A. H. Some genetic aspects of adaptation to hyper-dynamic environments. *Physiologist* 36 (Suppl. 1) S-28–S-30, 1993.

151. Smith, A. H., and R. R. Burton. The influence of the ambient accelerative force on mature body size. *Growth* 31: 317–29, 1967.

152. Smith, A. H., and R. R. Burton. Gravitational adaptation of animals. *Physiologist* 23 (Suppl.): 113–114, 1980.

153. Smith, A. H., R. R. Burton, and C. F. Kelly. Influence of gravity on the maintenance feed requirements of chickens. *J. Nutr.* 101: 13–24, 1971.

154. Smith, A. H., and M. J. Katovich. Gravitational influences upon the maintenance requirements of rabbits. *COSPAR Life Sci. Space Res.* Eds. R. Holmquist and A. C. Strickland, Pergamon Press, Oxford and New York. XV: 257–61, 1977.

155. Smith, A. H., D. F. Rahlman, A. M. Kodama, and N. Pace. Metabolic responses of monkeys to increased gravitational fields. *COSPAR Life Sci. Space Res.* Ed. P.H.A. Sneath, Akademie-Verlag, Berlin. XII: 129–32, 1974.

156. Smith, A. H., O. Sanchez P., and R. R. Burton. Gravitational effects on body composition in birds. *COSPAR Life Sci. Space Res.* Ed. P. H. A. Sneath, Akademie-Verlag, Berlin. XIII: 21–27, 1975.

157. Smith, A. H., W. L. Spangler, R. R. Burton, and E. A. Rhode. Responses of domestic fowl to repeated +G$_z$ acceleration. *Aviat. Space Environ. Med.* 50: 1134–1138, 1979.

158. Smith, A. H., W. L. Spangler, B. Carlisle, and G. Kinder. Effects of brief exposure of domestic fowl to very intense acceleration fields. *Aviat. Space Environ. Med.* 50: 126–133, 1979.

159. Smith, A. H., W. L. Spangler, J. M. Goldberg, and E. A. Rhode. Tolerance of domestic fowl to high sustained +G$_z$. *Aviat. Space Environ. Med.* 50: 120–125, 1979.

160. Smith, A. H., W. L. Spangler, E. A. Rhode, and R. R. Burton. Influence of the ambient acceleration field upon acute acceleration tolerance in chickens. *COSPAR Life Sci. Space Rev.* Ed. R. Holmquist, Pergamon Press, Oxford and New York. XVII: 235–239, 1979.

161. Smith, M. C., Jr., P. C. Rambaut, J. M. Vogel, and M. W. Whittle. Bone mineral measurement-experiment MO78. In: *Biomedical Results from Skylab,* edited by R. S. Johnson and L. F. Dietlein. Washington, DC: National Aeronautics and Space Administration, chapt. 20, p. 183–190, 1977. (NASA SP-337)

162. Spence, D. W., M. J. Parnell, and R. R. Burton. Abdominal muscle conditioning as a means of increasing tolerance to +G$_z$ stress. *Preprints, Ann. Meeting Aerosp. Med. Assoc.,* p. 148–149, 1981.

163. Stegmann, B. J., R. W. Krutz, R. R. Burton, C. F. Sawin. Comparisons of three anti-G suit configurations during long duration, low onset, +G$_z$. *Avia. Space Environ. Med.* 63: 209, 1992.

164. Stepantsov, V. I., and A. V. Yeremin. Basic principles for development of acceleration training schedules. *Kosm. Biol. Med.* Moscow, 1969.

165. Stiehm, E. R. Acceleration protection by means of stimulation of reticuloendothelial system. *J. Appl. Physiol.* 17: 293–298, 1962.

166. Stupakov, G. P. Biomechanical changes of bone structure: Changes following real and simulated weightlessness. *Physiologist* 31(Suppl. 1): S4–S7, 1988.

167. Tamir, A., R. R. Burton, and E. M. Forster. Optimum sampling times for maximum blood lactate levels after exposures to sustained +G$_z$. *Aviat. Space Environ. Med.* 59: 54–56, 1988.

168. Thornton, W. E., and J. A. Rummel. Muscular deconditioning and its prevention in spaceflight. In: *Biomedical Results from Skylab,* edited by R. S. Johnson and L. F. Dietline. (NASA SP337) Washington, DC: National Aeronautics and Space Administration, chapt. 21, p. 191–197, 1977.

169. Uglova, N. M. The effect of prolonged stay under conditions of lowered barometric pressure on tolerance of G-loads. *Problems of Space Biol.* 8: 105–109, 1969. (NASA TTF-58)

170. Uglova, N. M., B. Z. Zaripov, and I. A. Mamatakhunov. The regularities of relationships between structure and function under different functional loads (homeostasis and homeomorphosus). *Physiologist* 26 (Suppl. 6): S53–S54, 1983.

171. Vasil'yev, P. V., and N. M. Uglova. The effect of adaptation to conditions of a changed gaseous environment on tolerance to G-loads. *Problems of Space Biol.* 8: 119–133, 1969. (NASA TTF-580)

172. Vico, L., D. Chappard, A. V., Balukin, V. E. Novikov, and C. Alexandre. Effects of 7-day spaceflight on weight-bearing and non-weight-bearing bones in rats (COSMOS 1667). *Physiologist* 30(Suppl. 1): S45–S46, 1987.

173. Walters, G. R., C. C. Wunder, and L. Smith. Multi-field centrifuge for life-long exposure of small mammals. *J. Appl. Physiol.* 15: 307–308, 1960.

174. Weidner, W. J., L. F. Hoffman, and S. D. Clark. Regional blood flow in the domestic fowl immediately following chronic acceleration. *Aviat. Space Environ. Med.* 53: 666–669, 1982.

175. Werchan, P. M. Physiologic bases of G-induced loss of consciousness (G-LOC). *Aviat. Space Environ. Med.* 62: 612–614, 1991.

176. Whinnery, J. E. +G$_z$ tolerance correlation with clinical parameters. *Aviat. Space Environ. Med.* 50: 736–741, 1979.

177. Whinnery, J. E. The G-LOC syndrome. 1990. Naval Air Dev. Ctr., Warminster, PA, p. 1–9. (NADC-91042-60)

178. Whittle, M. W., R. Herron, and J. Cuzzi. Biostereometric analysis of body form. In: *Biomedical Results from Skylab,* edited by R. S. Johnson and L. F. Dietlein. Washington, DC: National Aeronautics and Space Administration, chapt. 22, p. 198–202, 1977. (NASA SP337)

179. Winget, C. M. and A. H. Smith. Quantitative measurement of labyrinthine function in the fowl by nystagmography. *J. Appl. Physiol.* 17: 712–718, 1962.

180. Winget, C. M., A. H. Smith, and C. F. Kelly. Effects of chronic acceleration on induced nystagmus in the fowl. *J. Appl. Physiol.* 17: 709–711, 1962.

181. Wojtkowiak, M. Human centrifuge training of men with lowered +G$_z$ acceleration tolerance. *Physiologist* 34 (Suppl.): S80–S81, 1991.

182. Wood, E. H. Potential hazards of high anti-G$_z$ suit protection. *Aviat. Space Environ. Med.* 63: 1024–1026, 1992.

183. Wood, E. H., E. H. Lambert, E. J. Baldes, and C. F. Code. Effects of acceleration in relation to avaiation. *Federation Proc.* 5: 327–344, 1946.

184. Wood, E. H., W. F. Sutterer, H. W. Marshall, E. F. Lindberg, and R. N. Headley. Effect of headward and forward acceleration on the cardiovascular system. Wright-Patterson Air Force Base, OH, p. 1– 48, 1961. Wright Air Dev. Div (WADD) (WADD-TR-60-634)

185. Wunder, C. C., and L. O. Lutherer. Influence of chronic exposure to increased gravity upon growth and form of animals. *Int. Rev. Gen. Exp. Zool.* 1: 344–416, 1964.

186. Zavadovskiy, A. F., M. M. Korotayev, S. V. Kopanev, I. A. Plyasovabakunina, and Y. N. Vavakin. The effect of active training in a head-down position on tolerance of cranial fluid shift. *Kosm. Biol. Med.* 19: 83–85, 1985.

IV | THE HYPERBARIC ENVIRONMENT

Associate Editor   S. K. Hong

1 | UNDERWATER PHYSIOLOGY OF MAN

41. Hyperbaria/diving: introduction

JAMES VOROSMARTI | *Rockville, Maryland*

HISTORY OF DIVING

HUMANS HAVE ALWAYS ATTEMPTED to become familiar with the underwater world, whether out of curiosity or in the search for food or items of commerce. Until the sixteenth century, however, the only method of exploration available was simple breath-hold diving, which is probably still the most common method of diving in the world, considering all the recreational uses of masks, fins, and snorkels, as well as the breath-hold diving still done in many areas for seafood and other items. Breath-hold divers are limited by the time and depth they are able to spend under water, which is dictated by breath-hold time and lung capacity. Another disadvantage is the possibility of loss of consciousness under water caused by hyperventilation before the dive and a strong conscious effort on the part of the diver to hold his breath. The advantages are that this type of diving requires little or no apparatus and avoids problems encountered with a pressurized system, such as decompression sickness.

The earliest record of divers appears to be Herodotus's account (about 460 B.C.) of Scyllus, a diver employed by Xerxes to recover treasure from wrecked Persian ships (8). In 268 B.C. the activities of the ama of Japan were chronicled (15). There is no record of further advances in human incursions under water between the first century A.D. and the Middle Ages. In the fifteenth and sixteenth centuries, many unusual designs appeared to allow humans to work under water. There is no record that any of these were built or used, and indeed, many would have been inimical to the health and lives of those who did try them. A favorite design was that of a breathing pipe with one end on the surface and the other end terminating in a mouthpiece or mouth covering, a hood, or a hood and jacket combination. Leonardo da Vinci, as may be expected, had several designs for underwater apparatus. The lack of physiological knowledge of dead space, negative pressure breathing, the depletion of oxygen, and the increase in carbon dioxide in a closed atmosphere is apparent in all of these designs. We do not know whether these shortcomings were ever manifested by actual use, but we do know that increasing attention was paid to diving bells.

The first recorded diving bell was used in an attempt to find and raise Caligula's galleys in the lake of Nemi in 1531 (8) with a bell built by Guglielmo Lorena. Other bells were used for salvage work or for entertainment through the sixteenth and seventeenth centuries. A famous bell was designed by Phipps (1680), who salvaged 200,000 pounds sterling in treasure from a wreck in one operation (8). In 1690, the astronomer Sir Edmund Halley constructed the first bell with a supply of fresh air, though he was not the first to design such a bell. Lead-lined barrels could be filled with air at the surface, bunged, and then lowered to the bell. A diver led a hose to the bell when the barrel was below the level of the bell and a bottom bung opened, allowing the water pressure to force the air into the bell. On one occasion, Halley and four colleagues spent an hour and a half at a depth of nine or ten fathoms without any ill effects (10). Halley was also the first to describe an ear squeeze:

> The only inconvenience that attends it (pressurization) is found in the ears, within which there are cavities opening only outwards, and that by pores so small so as not to give admission even to the air itself, unless they be dilated and distended by considerable force. Hence on the first descent of the bell, a pressure begins to be felt on each ear, which by degrees grows painful, like as if a quill were thrust through the hole of the ear.

In 1788 Thomas Smeaton devised the first practical pump to provide compressed air from the surface to either a bell or a diver, or later to a caisson (16). The engineering of a practical air compressor allowed divers to reach greater depths and to remain longer but also exposed them to the new dangers of decompression sickness, nitrogen narcosis, cerebral gas embolism, and oxygen toxicity.

The late eighteenth and early nineteenth centuries saw even more inventive ideas for a diver's dress, but the only successful one, which served as a prototype for

975

other systems, was developed by Augustus Siebe. His first equipment was an "open dress," which consisted of a helmet attached to a jacket extending to the waist and open at the bottom so that the air pumped to the diver could escape. This was based on the first diving helmet invented by John Deane in 1830. The major problem was a considerable risk of flooding the jacket and helmet if the diver bent over too far. The next major advance, the closed diving dress, was also designed by Siebe (1837) and first demonstrated on the wreck of the Royal George (11). A helmet/suit combination was the practical diving dress available for the next century and is still in constant use in much of the world for salvage diving and underwater construction in relatively shallow depths.

A harbinger of the future in diving equipment appeared in 1872—the Rouyquayrol–Denarouze self-contained apparatus (8)—which consisted of a closed suit attached to a hood with face mask. Air was supplied through a demand regulator from a compressed air reservoir on the diver's back. This was not a practical type of diving dress because of the bulkiness of the equipment and the low air volume. The first successful self-contained breathing apparatus was designed in 1878 by Henry Fluess in conjunction with the Siebe, Gorman Company for working in poisonous atmospheres as well as under water (8). The system used pure oxygen with a carbon dioxide absorbent system and was a great success in several flooded collieries and in dewatering the flooded Severn Tunnel. No instances of oxygen poisoning were reported during the Severn Tunnel operation, though we now know that there was a distinct risk because the breathing gas contained oxygen at the partial pressure of 2.2 atmospheres (atm).

The first organized and scientific approach to the problem of decompression from high pressure was conducted by J. S. Haldane at the beginning of the twentieth century. The British Admiralty, realizing that the capability to perform underwater operations would provide a military advantage in the future, provided funding to Haldane for scientific investigation to provide regulations for the safe conduct of diving. From this work came the first decompression tables (5). The ideas embodied in this work are still important to the formulation of decompression schedules.

The next major advance in diving, the use of helium, appeared later in the century. The first person to suggest the use of helium (or hydrogen) was Elihu Thomson, who theorized that the lower solubility of helium in fat and its greater diffusivity would increase a diver's time on the bottom and decrease decompression time (8). Experiments in the United States by Sayers et al. (17) and Behnke and Yarborough (3) showed that there was no advantage in decompression time, though deeper

work was allowable. However, the narcosis encountered by divers using air at great depths [first described by Junod in 1835 (12)] was prevented by using helium. The first practical operational use of helium diving was during the salvage in 1939 of the U. S. S. Squalus, a submarine which sank off Portsmouth, New Hampshire, during sea trials.

The first practical self-contained diving equipment was invented by Captain Yves Le Prieur in the 1920s (2). The defect with this system was that there was not a demand regulator so that air flowed continuously, greatly limiting the underwater duration of the apparatus. In 1943 Cousteau and Gagnan invented the "aqua lung" with a demand valve which allowed the flow of air only when the diver inhaled (2). An advance based on the designs of Fluess and of Cousteau and Gagnan uses sophisticated gas equipment to mix inert gases and oxygen. These systems, which can be of the semiclosed or closed type, conserve breathing gas and allow the diver to remain undetected for long periods of time. In the semiclosed type, some but not all of each expired breath is lost to the water, while the remainder is recirculated through a carbon dioxide removal system and returned to the inhalation side of the equipment. In the closed type, the diver breathes either pure oxygen (which significantly limits depth because of oxygen toxicity) or an inert gas–oxygen mix, which is constantly recirculated with carbon dioxide removal and oxygen replenishment. In a new sophisticated system, oxygen partial pressure can be kept at a constant level by electronic sensors and valves regardless of the depth of the dive.

The last major breakthrough in diving came in the late 1950s and early 1960s with the introduction of practical saturation diving. The term *saturation* means that a diver stays at one depth long enough for the gas dissolved in tissue to be equilibrated with the partial pressures of gas in the ambient environment. The advantage is that no matter how long a diver stays at a certain pressure, his decompression time will remain the same. This provides a great advantage for both economy and safety by reducing the number of times a diver must be exposed to a potentially dangerous decompression. This was first realized as a possibility by Boycott et al. (5), but the idea was never developed until the need for deeper, longer dives became evident. The first intentional saturation dive was done by End and Nohl in 1938, when they stayed at a depth of 101 ft for 27 h breathing air (2). In 1958 Workman et al. (18) did the first experiments using animals to provide the basis for further human work. The first open sea saturation dive was done by Stenuit of the Link group; he remained at 200 feet in a small chamber for 24 h (2). The loss of the U. S. S. Thresher in 1962 provided a great impetus for developing a deep diving capability, not only for submarine escape, rescue, and salvage but also for exploitation of the continental shelf. The U. S. Navy SEALAB experiments led the way for the development of this technique. This capability is now present in many countries, with centers for deep diving research in England, France, Germany, Italy, Norway, the People's Republic of China, and several countries of the Former Soviet Union. Other countries have plans for facilities. Saturation diving has become standard in the diving industry and has also been used to staff underwater habitats in which divers live and work. This type of diving has introduced a new hazard, the High Pressure Nervous Syndrome. It has also spurred research in the use of trimix breathing gases (the use of different inert gases at different times during decompression) and has renewed interest in using hydrogen as the main inert gas component in deep diving. Hydrogen was first investigated for diving by Case and Haldane in 1941 (7) as a substitute for helium, which was available from only two countries and expensive for large-scale operations. Hydrogen, however, was easily obtainable and cheap. Unfortunately, the problems of handling hydrogen and oxygen mixes were not easy. In the late 1960s research was resumed, and hydrogen has since been used in several deep diving experiments in France. One unusual result of saturation diving research has been the establishment of an underwater hotel, where one can stay for up to 22 h at a depth of five fathoms dining, sleeping, making underwater excursions, or watching the fish through large viewing ports.

PRESSURE UNITS

Pressure is defined as force acting on a unit area and is expressed mathematically as

$$Pressure = Force/Area \text{ or } P = F/A.$$

The basic unit of pressure in the International System of Units (SI) is the pascal: one pascal equals one newton per square meter. There are many units used to denote pressure and anyone working with pressure must be familiar with them all. Even though SI units are now required or preferred in scientific publications, the older literature uses various units and in many cases the gauges or other systems of measuring pressure in the field are not yet calibrated in SI units. In addition, gauge pressure must be differentiated from absolute pressure. *Gauge pressure* refers to the pressure shown on a gauge which is calibrated to zero at 1 atm pressure. *Absolute pressure* refers to the total pressure, including atmospheric pressure.

TABLE 41.1 *Units Commonly Used to Express Pressure*

1 Pa = 1N/m²*	1 bar = 10.00 msw
1 atm = 1.013247 bar	1 MPa = 10.000 bar
1 atm = 101.3247 kPa	1 psi = 6,894.76 Pa*
1 atm = 14.6959 psi	1 psi = 51.7151 torr
1 atm = 760.00 torr*	1 psi = 2.251 fsw
1 atm = 33.08 fsw	1 torr = 133.322 Pa*
1 atm = 10.13 msw	1 fsw = 3.063 kPa
1 bar = 100.000 kPa	1 fsw = 22.98 torr
1 bar = 100,000 Pa*	1 msw = 10.000 kPa‡,§
1 bar = 14.50377 psi	
1 bar = 750.064 torr	1 msw = 1.450 psi
1 bar = 32.646 fsw†,§	1 msw = 75.01 torr

*Signifies a primary definition from which other equalities were derived.

†Primary definition for fsw; assumes a density for seawater of 1.02480 at 4°C (the value often used for depth gauge calibration).

‡Primary definition for msw; assumes a density for seawater of 1.0972 at 4°C.

§Primary definitions for fsw and msw are arbitrary since the pressure below a column of seawater depends on the density of the water, which varies widely from point to point in the ocean. These two definitions are consistent with each other if a density correction is applied.

Pa = pascal, N = newton, psi = pounds per square inch, msw = meters of seawater, fsw = feet of seawater.

From *Undersea and Hyperbaric Research,* published by the Undersea and Hyperbaric Medical Society.

Table 41.1 lists units commonly used to express pressure, with equivalents as an aid to those who are not familiar with these units.

MAXIMAL DEPTH OF DIVING FOR HUMANS

As late as the 1950s it was considered dangerous to dive deeper than 150 ft, and diving to 300 ft, which was possible with helium, was undertaken only if absolutely required. The advent of saturation diving has made it possible to dive much deeper (686 m is the deepest dive at the time of this writing), and the limits for diving are not known with any certainty. During the history of diving, limits which have been met have always been overcome by changes in technique or engineering advances. However, we may be approaching the physiological limits of diving. The obvious barriers to very deep diving are gas density and inert gas effects/hydrostatic pressure. Increased gas density has a direct effect on respiration by increasing both intrinsic and extrinsic resistance to gas flow and, therefore, to the work a diver can do at depth. Lanphier and Camporesi (14) extrapolated measured maximum voluntary ventilations to show that moderate work should be possible to depths of around 900 m with appropriate helium–oxygen mixtures. Other studies which have used dense inert gases at pressure to simulate gas density to depths of 1,515 m indicate that divers may be able to do useful work to between 900 and 1,515 m (1, 13). The use of hydrogen as an inert gas in deep diving may extend that depth. A problem is that these estimates ignore the external resistance imposed by the breathing apparatus. They are also based on a small number of subjects, and we know that some individuals will experience severe subjective dyspnea at shallower depths. In any case, at some depth a limit will be reached where ventilation is too low to support any activity; whether this is between 900 and 1,515 m or deeper is still a question.

High Pressure Nervous syndrome (HPNS) occurs in humans beginning at depths of about 180 m and varies between individuals as to depth of onset and type and severity of symptoms. Experimentation with animals has shown that a high enough pressure will result in convulsions and death and that anesthetics can reverse this condition or delay its onset to greater depths (6). Human dives using tri-mix (N_2–He–O_2) have shown that the addition of nitrogen is useful in preventing symptoms (4). The level of nitrogen is crucial: too much produces symptoms of nitrogen narcosis and too little will not alleviate the adverse nervous symptoms. In addition, because of individual variation, the amount may be different for different divers. Whether nitrogen prevents HPNS or only masks the symptoms is of some concern. Unless some type of intervention to prevent HPNS is found, the limit to routine helium–oxygen diving will remain in the region of 350 m; for highly selected divers, however, the limit may be 450–500 m. Perhaps the future will provide a drug that will obviate HPNS.

The ultimate limit to high ambient pressure will probably be the effects of pressure on chemical events in the body. We already have evidence that pressures close to those at the deepest oceanic depths found on earth can cause protein coagulation, enzyme inactivation, and disintegration of red blood cells (9). Whether these effects will occur in humans at lower pressures remains to be seen, but HPNS appears to be early evidence that they may become apparent at lesser depths than might be expected.

The question for the future is whether humans will be as fortunate at finding solutions to problems imposed by high pressure as they have been in the past.

REFERENCES

1. Anthonisen, N. R., M. E. Bradley, J. Vorosmarti, and P. G. Linaweaver. Mechanics of breathing with helium–oxygen and neon–oxygen mixtures in deep saturation diving. In: *Proc. Fourth Symp. Underwater Physiology*, edited by C. J. Lambertsen. New York: Academic, 1971, p. 339–345.
2. Bachrach, A. J. A short history of man in the sea. In: *The Physiology and Medicine of Diving* (3rd ed.), edited by P. B. Bennett and D. H. Elliott. San Pedro, CA: Best, 1982, p. 1–14.
3. Behnke, A. R., and O. D. Yarborough. Physiologic studies of helium. *U.S. Naval Med. Bull.* 36: 542–558, 1938.
4. Bennett, P. B., G. D. Blenkarn, J. Roby, and D. Youngblood. Suppression of the high pressure nervous syndrome in human deep dives by He–N_2–O_2. *Undersea Biomed. Res.* 1: 221–237, 1974.
5. Boycott, A. E., G.C.C. Damant, and J. S. Haldane. The prevention of decompression sickness. *J. Hyg. (Lond.)* 8: 342–443, 1908.
6. Brauer, R. W., S. M. Goldman, R. W. Beaver, and M. E. Sheehan. N_2, H_2 and N_2O antagonism of high pressure neurological syndrome in mice. *Undersea Biomed. Res.* 1: 59–72, 1974.
7. Case, E. M., and J.B.S. Haldane. Human physiology under high pressure. *J. Hyg.* 41: 225–249, 1941.
8. Davis, R. H. *Deep Diving and Submarine Operation* (7th ed.). Chessington, UK: Siebe, Gorman, 1962.
9. Fenn, W. O. Possible role of hydrostatic pressure in diving. In: *Proc. Third Underwater Physiology Symp.* edited by C. J. Lambertsen. Baltimore, MD: Williams and Wilkins, 1967, p. 395–403.
10. Halley, E. The art of living underwater. *Phil. Trans. R. Soc. Lond.* 29: 492–499, 1917.
11. Colonel Pasley's operations at Spithead. Hampshire, England: *Hampshire Telegraph*, 21 September 1840.
12. Junod, T. Recherches sur les effets physiologiques et therapeutiques de la compression et de rarefaction de l'air, taut sur le corps que les membres isoles. *Ann. Gen. Med.* 9: 157, 1835.
13. Lambertsen, C. J., R. Gelfand, R. Peterson, R. Strauss, W. B. Wright, J. G. Dickson, C. Puglia, and R. W. Hamilton, Jr. Human tolerance to He, Ne, and N_2 at respiratory gas densities equivalent to He–O_2 breathing at depths of 1200, 2000, 3000, 4000, and 5000 feet of sea water (Predictive Studies III). *Aviat. Space Environ. Med.* 48: 843–855, 1977.
14. Lanphier, E., and E. M. Camporesi. Respiration and exercise. In: *The Physiology and Medicine of Diving*, edited by P. B. Bennett and D. H. Elliott. San Pedro, CA: Best, 1982, p. 99–156.
15. Nukada, M. Historical development of the ama's diving activities. In: *Physiology of Breath-Hold Diving and the ama of Japan*, edited by H. Rahn. Washington, DC: *Natl. Acad. Sci–Natl. Res. Council*, 1965, p. 25–40.
16. O'Neill, W. Breathing gas delivery. In: *A Pictorial History of Diving*, edited by A. J. Bachrach, B. W. Desiderata, and M. M. Matzen. San Pedro, CA: Best, 1988, p. 68–73.
17. Sayers, R. R., W. P. Yant, and J. H. Hildebrand. Possibilities in the use of helium–oxygen mixtures as a mitigation of caisson disease. U. S. Bureau of Mines, Rep. Invest. 2670, 1925.
18. Workman, R. D., G. F. Bond, and W. F. Mazzone. Prolonged exposure of animals to pressurized normal and synthetic atmospheres. U. S. Naval Med. Res. Lab. Rep. 374, 1962.

42. Hyperbaria: breath-hold diving

YU-CHONG LIN | Department of Physiology, University of Hawaii, John A. Burns School of Medicine, Honolulu, Hawaii

SUK KI HONG | Department of Physiology, State University of New York at Buffalo, School of Medicine and Dentistry, Buffalo, New York

CHAPTER CONTENTS

BREATH-HOLD, FREE, AND SKIN DIVING probably developed from retrieving articles lost in water or harvesting edibles under water. Records were left after breath-hold diving had become important economically. There are archeological finds, dating back to 4,500 B.C., of objects fashioned from pearl shells that must have been obtained by divers (7). Skin divers performed underwater salvage and underwater warfare (20, 23, 104) before compressed air supply systems became available. Breath-hold diving has been practiced as a profession for some 2,000 years and continues today in Japan and Korea.

PROFESSIONAL BREATH-HOLD DIVERS

Brief History

Teruoka (123) kindled scientific interest in breath-hold diving in 1932 when he reported a detailed investigation of diving operations, diving patterns, equipment, seasonal variations, and gas exchange in Japanese breath-hold divers, the ama. *Ama*, meaning sea-woman in Japanese, is an accepted term for breath-hold diving professionals of either gender and there are various synonyms in Japan (92). Scientific inquiries into breath-hold diving culminated in a symposium in Tokyo in 1965, which paid appropriate tribute to Teruoka's accomplishment (107).

Besides having to contend with ocean conditions and biological hazards, professional breath-hold divers have physiological constraints that severely limit diving time and depth. The history of breath-hold diving among the Japanese and Korean ama has shown that rather than fighting these limitations divers have focused on prolonging bottom time well within normal physiological limits. They have done so by such means as shortening descent and ascent times by assists, thus reducing oxygen consumption during descent, ascent, or both; increasing work efficiency with hand tools, fins, float, and boat; improving underwater vision (92); optimizing surface intervals (47, 50); and using thermal protection (96, 116). Distinct features of diving procedures evolved from these developments. An unassisted ama, called *cachido* (or kachido), descends and ascends by his or her own power; but an assisted ama, *funado*, receives assistance during descent, ascent, or both. The latest development is that an ama pulls up the counterweight by himself or herself after getting back on the boat, thus eliminating the need for the assistance of a traditional funado.

Geographical Distribution

Recreational breath-hold diving, such as underwater swimming, spear fishing, and underwater hockey, can be found wherever bodies of water exist. However, the number of professions that use breath-hold diving, such as harvesters of pearl shells and sponges, has declined drastically. Only the ama of Japan and Korea are still active in breath-hold diving professions. Besides the Japanese and Korean divers, professional breath-hold divers once congregated at the Torres Strait for shells, in the Tuamotu Archipelago for pearls, and in the Aegean Sea for sponges.

The population of ama in Korea has declined steadily since the 1940s, when some 27,000 divers were working off the southern shores. By 1956, they numbered 22,000 and by 1965 they had dropped to 16,000. This prompted a warning that breath-hold diving as a profession might be extinct by the turn of the century (41). In recent years, however, the number of professional breath-hold divers in Korea and Japan has remained steady at around 20,000 (97), but this is substantially less than the 1967 estimate of 30,000 (50). In Korea, most divers are women, but in Japan about 65% are men. Sensible practice of conservation principles and lucrative economic rewards have contributed to the break in the predicted decline.

Equipment

Besides the equipment described by Teruoka (123), modern divers wear wet suits and fins. Ama wear weight belts to counter the buoyancy of the wet suit and carry a counterweight of about 13 kg for passive and rapid descent. The development of goggles is noteworthy; a pressure-uncompensated goggle severely limits diving depth. The diminishing gas volume in the goggle during a dive creates a negative pressure and causes conjunctival bleeding. Development of goggles with a reduced total volume limits the total gas volume change and consequently decreases the displacement of tissue around the eyes (103). Pressure-compensated goggles originated around the 1790s, according to Teruoka (123).

Work Patterns

For thermally unprotected divers, water temperature dictates the duration of a dive shift; typically about 60 min in summer and less than 30 min in winter (47, 51, 130). Divers cease diving activity when core temperature reaches 35°C. They often resume diving for one or more shifts following rewarming on a boat or on land (59, 60). Water temperature is of lesser concern for thermally protected divers (60, 116). Typically, wet-suited divers stay in water for 2 h or more in a single dive shift.

According to Kang et al. (60), water current and fatigue are the major determinants of duration of a dive shift.

Irimoto (55) studied the daily routine of 22 male abalone divers in the Boso Peninsula (Chiba Prefecture, Japan). Divers' activity in the sea occurred around noon, for 1–2 h, but they spent considerably longer hours on activities on land and at the port. They dive 25–35 times during a shift for about 60 s each dive and collect as many as seven abalones in one shift.

Diving Patterns

Most of the details on diving patterns are derived from research on Korean and Japanese ama. In a 3 year project, Hong et al. (47) investigated the natural diving patterns of Korean and Japanese ama, men and women in wet suits, by using dive data recorders. Earlier, Park et al. (96) clocked a bottom time of 12.3 s/min work shift or 30.8 s/min dive in Japanese and Korean ama diving to a 5 m depth and 5.7 s/min work shift or 16.7 s/min dive for dives to a 10 m depth. These results showed no major differences in the diving patterns reported earlier (51, 123, 130). Scholander et al. (115) reported an average diving time of 53 s (range, 23–91 s) in 21 pearl divers but no detailed diving pattern (Table 42.1).

The funado's diving patterns differ substantially from the cachido's. Japanese men who dove with assistance worked longer per shift (305 min) with fewer dives (23 times), which were longer and deeper (68.5 s/dive and 10 m as opposed to 26.8–37.0 s/dive and 4–7 m for cachido). The Japanese male ama descended more rapidly than the cachido and so had greater bottom time per dive (45 s vs. 13.2–18.3 s for cachido) (Table 42.1).

Energy Expenditure

The ama's traditional cotton suit incurs a cumulative net heat loss of 390 kcal per dive shift (or approximately 10 kcal/min) in the summer and 560 kcal (or about 30 kcal/min) in the winter (58, 59). Cold stress is apparent, even in the summer, to the degree that water temperature dictates the duration of a work shift as noted above. However, an ama tolerates a net heat loss of about 1,000 kcal in a dive shift regardless of the season (longer shift in summer and shorter shift in winter) (50).

The oxygen debt incurred during a dive is surprisingly invariant, ranging from 800 to 1,200 ml, regardless of the form of assists, cachido and partially or fully assisted funado, but the depth reached is greater in funado. A fully assisted funado can reach 20 m by conserving energy during both descent and ascent (130).

Since the late 1970s, the wet suit has become widely available, significantly alleviating cold stress. The net heat loss during a dive shift in a wet-suited ama was

TABLE 42.1. *Natural Diving Pattern of Wet-Suited Korean and Japanese Ama*

Parameter	Cachido				Funado
	1989 Summer 11 KF	1990 Summer 6 KF	1991 Winter 9 KF	1989 Summer 4 JM	1990 Summer 9 JM
Dive shift, min	188	179	170	201	305
Surface time, min	136	125	113	138	279
Diving time, min	52	54	57	63	26
No. of dives	115	129	129	109	23
Single dive					
Total time, s	28.9	26.8	28.2	37.0	68.5
Descent time, s	7.9	7.3	9.6	9.8	10.0
Ascent time, s	6.7	6.0	5.4	8.9	13.5
Bottom time, s	14.3	13.5	13.2	18.3	45.0
Bottom time					
Total, min/day	27.4	29.0	28.4	32.0	17.2
% of dive shift	14.6	16.2	16.7	15.9	5.7
% of diving min	52.7	53.7	49.8	50.8	66.3
Average depth, m	3.7	3.6	3.6	6.9	9.7

Mean values from Hong et al. (53).
KF, JF, and JM are Korean female, Japanese female, and Japanese male, respectively.

estimated to be 260 kcal in summer and 370 kcal in winter (60).

DURATION

Diving Time

Although exceptionally long diving times have been recorded (16, 115, 123), diving times are usually less than 60 s for professional breath-hold divers. Breath-hold times in excess of 3 min have been observed in competition for depth records and longer times in some experimental conditions and in survival from an apparently accidental drowning (77).

Time Course of a Breath-Hold Dive

Breath-holding involves (1) voluntary inhibition of respiratory muscular activity, with the glottis closed and the intrathoracic pressure stable and slightly above the ambient; (2) the onset of involuntary inspiratory activity with the glottis closed, causing subatmospheric and cyclic intrathoracic pressures, which intensifies as breath-holding continues; and (3) opening of the airway. This sequence corresponds, respectively, to the event markings *A*, *B*, and *C* in Figure 42.1. At *B*, involuntary ventilatory activity (diaphragmatic contraction, Mueller or Valsalva maneuver) begins while inhibition of the glottal opening is still possible. The involuntary ventilatory activity signals the physiological breaking point. A variety of terms have been used to describe the onset of the physiological breaking point, such as air hunger, desire to breathe, want of oxygen, end of the easy-going phase, and diaphragm contraction. The major determinant of the physiological breaking point is the arterial blood P_{CO_2} (1, 21, 81, 95) and is fairly constant for an individual in a given condition. Professional breath-hold divers heed this physiological sign and terminate the dive promptly, typically within 30 s (see Diving Patterns, above). However, it is possible to prolong the dive if necessary.

Breaking Point

At *C*, the conventional breaking point, or simply the breaking point, is reached. The duration between *B* and *C* varies greatly from subject to subject and even within a given individual.

When the physiological breaking point is reached, the intensity and frequency of the involuntary ventilatory activity rises progressively until the breath-hold is eventually terminated (81, 126). However, the perception of the respiratory neuromuscular output, on which the breaking point depends, can be influenced by a variety of unrelated factors which contribute to the large variation in breath-hold time. Breath-hold times ranging from 20 s (113) to 270 s (49) have been reported after an identical command: "exhale maximally, without prior hyperventilation, followed by maximal inhalation and hold the breath as long as possible." Swallowing, respiratory movement with glottis closed, squeezing a rubber ball, willingness, and psychological state are known to influence breath-hold time, albeit unpredictably at times. The involvement of subjective factors renders rigid concepts of chemical and mechanical thresholds inadequate when interpreting the breaking point (36, 126). Nevertheless, chemical and mechanical

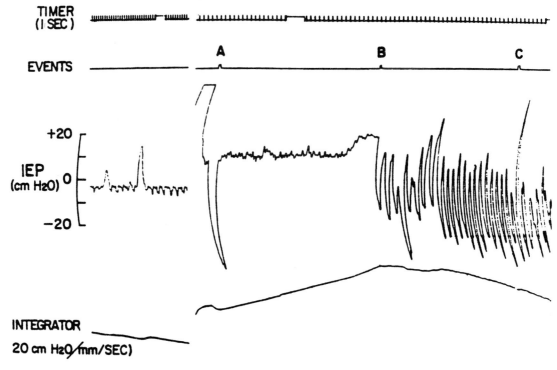

FIG. 42.1. Time course of breath-hold. Record shows intraesophageal pressure (IEP) during a course of breath-hold. Events are *A*, beginning of breath-hold; *B*, onset of involuntary ventilatory activity; and *C*, termination of breath-hold. Period between *A* and *B* was 39 s and between *B* and *C* 31 s. Subject was exercising at 167 kg-m/min while breath-holding. A similar result was observed during breath-hold at rest but with a longer breath-hold time. Modified from Lin et al. (81).

thresholds constitute the major determinants of breath-hold time. This can be quantified by investigating changes in lung volume and alveolar gas composition. Chemical and mechanical factors influencing breath-hold time have been reviewed extensively (36, 70, 88, 126). A listing of major factors follows.

Oxygen Supply–Demand Relationship. The oxygen supply–demand relationship was summarized by Klocke and Rahn (61):

$$\text{BH time} \atop \text{with air} = \frac{TLC \times \text{F}_{\text{A}}\text{O}_2}{\dot{\text{V}}_{\text{O}_2}} \times \frac{P_B - 47}{863} \tag{1}$$

$$\text{BH time} \atop \text{with O}_2 = \frac{VC}{\dot{\text{V}}_{\text{O}_2}} \times \frac{P_B - 47}{863} \tag{2}$$

where breath-hold (*BH*) time is expressed in min; *VC* and *TLC* represent the vital capacity and total lung capacity, respectively, in liters BTPS (body temperature, pressure, saturation with water vapor); $\dot{\text{V}}_{\text{O}_2}$ (oxygen consumption) in ml STPD (standard temperature and pressure, dry)/min; $\text{F}_{\text{A}}\text{O}_2$ the fractional alveolar concentration of oxygen; $P_B - 47$ the barometric pressure less the water vapor pressure at 37°C; and the constant 863 consolidates physical condition of gas volumes and units of expression. With the assumption that residual

volume (*RV*) limits the shrinkage of lung volume, *TLC* is used for breath-hold with air, because $TLC \times \text{F}_{\text{A}}\text{O}_2$ exceeds $RV \times \text{F}_{\text{A}}\text{O}_2$ regardless of the duration of breath-hold, and in breath-hold with O₂ only *VC* is usable. Estimating the O₂ supply from the lung alone, a breath-hold time of 4 min with air and 16 min with O₂ is clearly

TABLE 42.2. *Published Records of Breath-Hold Time*

Conditions	Breath-Hold Time (min.sec)	Subject	Reference
Air	4.30	Subject SKH	50
	4.00	Student	51
Air, HV*	4.11	Subject PM	29
	5.04	Subject EM	29
	5.09	Subject RM	29
O₂	6.29	Pouliquen	36a
O₂, HV	13.00	Subject SH	61
	13.48	Subject MT	61
	14.00	Subject HR	61
	15.13	Student	113
	20.05	Frechette	88
O₂, HV-W†	13.43	Foster	36b

*Hyperventilation.
†Hyperventilation and submergence in water.

possible, and in fact, these values have been observed in resting subjects under laboratory conditions. Table 42.2 summarizes published maximal breath-hold times (77).

Initial Lung Volume and O_2 Volume.

Equations 1 and 2 predict a proportional increase in breath-hold with lung volume, at a given FAO_2 and P_B. Mithoefer (87) presented breath-hold time as a linear function of lung volume based on 279 observations from nine sources. This linear relationship predicts a breath-hold time of 78 s with TLC (air). This relationship and numerous other published breath-hold times can be termed "comfortable" because they fall short of maximal breath-hold time (74). Obviously, the initial amount of available O_2 dictates breath-hold time. At a given lung volume, an increased P_B, FAO_2, or both extends breath-hold time, and a decreased P_B, FAO_2, or both shortens it. Therefore, hyperbaria prolongs breath-hold time (2, 39) and hypobaria shortens it (106, 108, 113). The ability to hold breath improves somewhat with acclimation to altitude (106).

Oxygen Consumption.

Equations 1 and 2 predict an inverse relationship between breath-hold time and the rate of O_2 usage. The long breath-hold time demonstrated in resting subjects (Table 42.2) cannot be obtained in exercising subjects, even with prior hyperventilation. The inverse relationship between $\dot{V}O_2$ and breath-hold time was verified in subjects breath-holding with TLC and functional reserve capacity (FRC) (81).

Hyperventilation.

Hyperventilation is the most effective means of delaying the breaking point; it lowers $PaCO_2$ and raises PaO_2, increasing the volume of O_2 at the onset of a breath-hold. Klocke and Rahn (61) reported astonishingly long breath-hold times in seven subjects after hyperventilation and breath-hold with O_2, increasing from 3.1–8.5 min to 6–14 min (see also Table 42.2).

Temperature.

Exposure to cold elevates $\dot{V}O_2$ and consequently lowers breath-hold time (37, 121).

Mechanical Factors.

Lung volume affects breath-hold time in many ways: (1) it determines the initial O_2 volume (see Initial Lung Volume and O_2 Volume, above), (2) the shrinking lung volume during a dive diminishes tolerance to hypercapnia as well as hypoxia (87), and (3) the cessation of breathing movements constitutes a source of inspiratory stimuli (36, 126). Subjects tolerate a higher degree of hypoxia and hypercapnia by rebreathing (31, 34, 36), by interrupting breath-holding with a single or a few breaths without improving alveolar gas composition (63, 126), or by the Mueller or Valsalva maneuver (6).

Psychological Factors.

Psychological factors have long been recognized as major determinants of breaking point and undoubtedly contribute to the wide range of reported breath-hold times. Even squeezing a rubber ball, which does not involve the respiratory system, is effective in prolonging breath-hold time (6). Since psychological factors play such an important role and since it is unreasonable to assume the constancy of these factors within or between individuals, it is not possible to predict breath-hold time accurately from alveolar gas data.

DEPTH LIMIT

Does the depth at which the lung volume is reduced (or compressed) to the residual volume set the maximal depth of a breath-hold dive? For instance, if a diver with a residual volume of 1.2 l descends when his lungs contain a TLC of 6 l, Boyle's law dictates that the volume will be reduced to the residual volume at a depth of 40 m [or 131 ft, 5 atm absolute (ATA)]. In other words, the TLC/residual volume ratio (6.0 l/1.2 l = 5.0, ~40 m depth) would determine the maximal theoretical depth limit. It is thus predicted that breath-hold divers descending with TLC would be expected to be safe if they dive to a depth somewhat less than that predicted by the TLC/residual volume ratio. However, professional breath-hold divers of Korea and Japan appear to dive to a depth that is considerably shallower than the theoretical maximum. Korean women who do unassisted breath-hold dives have an average residual volume of 1.14 l and an average TLC of 4.58 l (or TLC/residual volume = 4.58/1.14 = 4.0, ~30 m depth) (117) and usually dive to a depth of 5–10 m (47, 51). Japanese men who do assisted breath-hold dives usually do not dive to a depth beyond 20 m (123). However, the Tuamotu male breath-hold divers are known to dive to a depth of about 110–120 ft, or 33–36 m (19), though it is not known if Tuamotu divers are diving beyond the depth limit dictated by the TLC/residual volume ratio due to an absence of lung volume data.

The concept of the theoretical maximal depth for breath-hold diving based on the TLC/residual volume ratio is inconsistent with depth records set by various divers. As shown in Figure 42.2, the depths reached by various record-setting divers have been increasing almost linearly since the early 1950s (85). The record as of 1983 was held by Jacques Mayol, who dove to 105 m (86), far greater than the maximal depth predicted by the TLC/residual volume ratio (determined in the laboratory) of 28 m (TLC/residual volume = 7.22 liters/1.88 liters = 3.8, ~28 m depth). In 1989, Mayol's record was broken by Francisco Ferrera of Cuba and Angela Bandini of Italy, who dove to 112 m and 107 m, respectively (32, 127). Furthermore, Ferrera broke

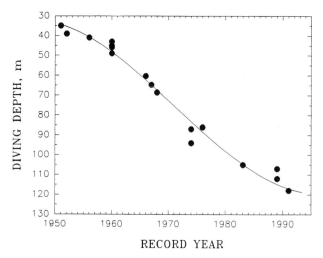

FIG. 42.2. Record depths of breath-hold diving. Maximal depth records for breath-hold diving before 1967 obtained from Craig (17). Modified from Hong (45) and Lin (77).

his own record in July 1991 when he reached 117 m (27). The large difference between the predicted and the actual depths of dives cannot be easily explained; either TLC increases or residual volume decreases during breath-hold diving. In this regard, the finding of Schaefer et al. (111) that thoracic blood volume (estimated by impedance pneumograph) is increased by nearly 1 l during open-sea dives to 40 m is important. If the thoracic cage is rigid during breath-hold diving, it may be logical to assume that an increase in intrathoracic blood volume could replace an equivalent volume of residual gas in the lung, resulting in a corresponding decrease in residual volume. It is technically difficult, if not impossible, to prove the above notion, but it might explain how Mayol reached a depth of 105 m. If we assume that Mayol's residual volume decreased from 1.88 to 0.63 l (a reduction of 1.25 l) during diving, then he would have had a TLC/residual volume ratio of 7.22/0.63 (~11.5), equivalent to a depth of 105 m. Another line of indirect evidence indicates that residual volume may be decreased during diving. If residual volume does not change during diving, one should not be able to dive following a full expiration to residual volume. However, Craig (17) observed that a subject with a residual volume of 2 l, who was asked to dive following a full expiration, was able to dive to a depth of 4.5 m without developing a significant difference between the ambient pressure at that depth and the intrathoracic pressure. Craig interpreted these results to mean that the subject's residual volume must have been compressed from 2.0 l at the surface to 1.4 liters during diving.

Moreli and Data (90) studied breath-hold divers with underwater, radiological apparatus. They took chest X-ray pictures at the surface (standing in air before diving)

and again after the divers reached a maximal depth of 20 m. They found: (1) a narrowing of the intercostal spaces; (2) an elevation of the diaphragmatic dome; (3) a marked contraction of the lung volume; (4) an engorgement of the pulmonary vasculature; and (5) an increase of cardiac transverse diameter. These findings do not by any means demonstrate an actual decrease in residual volume but are consistent with the notion that an increase in intrathoracic blood could displace pulmonary gas during breath-hold diving.

New depth records for breath-hold diving continue to be set despite the controversies as to how such depths can be reached physiologically. Lin (78) proposed that depth records are determined by breath-holding time, based on his concept that the probability of frank alveolar hemorrhage during breath-hold diving is low because during rapid compression: (1) absolute lung volume reduction is buffered by chestward displacement of blood from the periphery and by shape and size changes of the rib cage and the diaphragm, and (2) wall stress failure in pulmonary capillaries is very high, of the same order of magnitude as the aorta (8×10^5 dyne/cm^2 or 8×10^4 N/m^2) (124).

Obviously, actual measurements of the changes in residual volume associated with breath-hold diving are desirable to answer this physiologically interesting phenomenon.

ALVEOLAR GAS EXCHANGE

When a person in an air environment does not breathe, the P_{O_2} of the blood and the alveolar gas decrease continuously and the P_{CO_2} increases. However, the basic pattern of alveolar gas exchange during breath-hold diving is quite different because of the changes in ambient pressure, that is, hydrostatic compression during descent followed by decompression during ascent. Hong et al. (51) were able to collect alveolar gas samples from three Korean female divers immediately before descent, to an 8 m depth, and immediately after surfacing, using a Lanphier–Morin underwater alveolar gas sampler (see ref. 51 for details).

Alveolar gas at the very beginning of a dive (or immediately before descent) is composed of 4% CO_2, 17% O_2, and 79% N_2. This is slightly different from a usual composition in that P_{CO_2} is about 11 torr lower and P_{O_2} is 20 torr higher than the corresponding normal values (Table 42.3). These small deviations from normal are most likely due to a mild hyperventilation, which hydrostatic compression of the chest induces a decrease in lung volume, which in turn results in depth-dependent increases in partial pressures of O_2, CO_2, and N_2. If there were no alveolar gas exchange, the partial pres-

TABLE 42.3. *Alveolar Gas Composition (%) and Pressures (Torr) Immediately Before Descent, on the Bottom (8 m) and Immediately After Returning to the Water Surface*

	O_2		CO_2		N_2	
	%	Torr	%	Torr	%	Torr
Normal resting*	14.0	100	5.6	40	80.0	563
Before descent†	16.7	120	4.0	29	79.3	567
On the bottom† (8 m depth)	11.1	149	3.2	42	85.7	1,143
Return to surface†	5.9	41	5.9	42	88.2	631

*Data of Song et al. (117).
†Data of Hong et al. (51).

sure of these gases would have nearly doubled at an 8 m depth and the concentration (in %) of these gases should have remained unchanged. The gas pressures of the mixed venous blood can be expected to remain unchanged during the first 20 s (circulation time) of compression. Therefore, one would expect diffusion of all three gases from the alveolus to the blood. As a result, the magnitude of increase in the alveolar O_2 and CO_2 gas pressures at an 8 m depth is less than twice the corresponding values observed before descent, and the alveolar concentrations of O_2 and CO_2 are actually lower than the corresponding values before descent. Note also in Table 42.3 that the alveolar P_{N_2} at an 8 m depth was nearly twice as high as before descent and that the alveolar concentration of N_2 was only 6% higher than before descent. This indicates that diffusion of N_2 from the alveolus to the blood is very limited, despite the presence of a significant diffusion gradient for N_2. This is not surprising if we consider that the rate of diffusion of N_2 from the alveolus to the blood is slow compared to that of O_2 or CO_2. In fact, the alveolar concentration of N_2 progressively increases during descent as a result of far more rapid removal of O_2 and CO_2 during compression (descent).

It is important to recognize that the alveolar P_{O_2} is high during descent and at depth, creating no problems with O_2 supply, though the diver is not breathing. However, a considerable amount of CO_2 accumulates in the body during descent and at depth. In fact, the normal direction of CO_2 diffusion (blood to alveolus) is reversed during descent because of the reversal of the diffusion gradient. As a result, P_{CO_2} of blood (and of alveolar gas) is increased by 40% (from 29 torr before descent to 42 torr at an 8 m depth). Perhaps it is fortunate that the P_{CO_2} of blood increases markedly while the diver stays on the bottom because it is this high P_{CO_2} that signals the respiratory center to return to the surface. This mechanism is particularly important since there is no hypoxic stimulus due to the high P_{O_2} during descent and at depth.

Once the diver starts toward the surface, the lung will re-expand rapidly due to decompression, resulting in progressive decreases in the alveolar P_{O_2} and in the diffusion gradient for O_2. Consequently, the alveolar P_{O_2} upon reaching the surface is as low as 41 torr. This is equal to the mixed venous blood P_{O_2} so that there is no O_2 diffusion gradient between the alveolus and the blood. The diver is in a critical state of hypoxia, and, in fact, the majority of diving accidents occur during this ascending phase (16, 105). In some instances, especially when the diver stays on the bottom performing severe exercise, the normal direction of O_2 transfer could be completely reversed during ascent (67, 94). The CO_2 retained in the blood during descent (and on the bottom) now leaves the blood for the lungs as alveolar P_{CO_2} decreases continuously during ascent. A small amount of N_2 that entered the circulation and tissue during descent will also leave slowly according to the reversed diffusion gradient.

The amount of N_2 accumulated in the body during breath-hold diving is small, so the incidence of decompression sickness among professional breath-hold divers is extremely low. However, if breath-hold divers engage in deep, long dives with high frequency, sufficient amounts of N_2 can be accumulated to induce a decompression sickness (65, 98) (see BLOOD N_2 AND O_2 PROFILE, below).

The end tidal P_{O_2} value (~40 torr) was virtually the same as that of mixed venous blood P_{O_2} even during a short (40 s) and shallow (7–12 m) dive performed by unassisted Korean female divers. Lanphier and Rahn (67) not only reproduced the above pattern of alveolar gas exchange in simulated breath-hold diving experiments but also obtained evidence that with longer (80 s) and deeper (10 m) dives, at a moderate workload ($\dot{V}_{O_2}$ = 590 ml/min), O_2 is being extracted from the blood into the alveolar space at the end of the dive, coinciding with impairment of consciousness. Record-setting divers dive to a depth greater than 100 m (11 ATA). Olszowka and Rahn (93) conducted a computer analysis of long (220 s) 100 m breath-hold dives and again showed that during the last phase of lung expansion (that is, during ascent) alveolar P_{O_2} falls rapidly. On occasion it may fall below the mixed venous blood level, thereby inducing O_2 reversal.

One major determinant of alveolar P_{O_2} at the end of a dive is the total energy cost (or O_2 cost). Only estimated values are available in the literature. Depending on the assumptions made for various factors involved in the act of diving, total energy cost has been estimated to be 0.8–1.4 l O_2/min (18, 99, 131) or 0.5–1.0 l O_2/min (28). These represent only about 20%–30% of an individual's $\dot{V}_{O_{2max}}$ and seem to support the hypothesis that there is an O_2-conserving diving reflex in humans adapted to diving. However, the existence of such a reflex is still obscure and more solid data are needed.

At the end of a breath-hold dive, alveolar P_{O_2} is lower than in a normal breath-hold of the same duration at the surface. This can be attributed to fast removal of alveolar O_2 during descent and on the bottom. Theoretically, it is possible that the direction of CO_2 transfer could be reversed during simple breath-holding in air but the amount of CO_2 transferred from the alveolus to the blood during simple breath-holding is much less than that during a breath-hold dive.

BLOOD N_2 AND O_2 PROFILE

Venous Blood N_2 Tension During Repetitive Breath-Hold Diving

A small amount of N_2 is expected to diffuse from the alveolus into the circulation during descent (and on the bottom) due to compression of the chest, which raises alveolar P_{N_2} (see ALVEOLAR GAS EXCHANGE, above). Although it diffuses much more slowly than O_2 and CO_2, a sufficient amount of N_2 could be taken up by the body tissue if a breath-hold diver goes to deep and too often (65). Under such conditions, Paulev (98) experienced what appeared to be a decompression sickness. Despite concerns about N_2 accumulation, the blood level of P_{N_2} has never been measured in working divers. Radermacher et al. (102) measured the kinetics of P_{N_2} in the brachial vein blood in Korean female divers at various intervals during a natural diving work shift (150 min) at the surface (average diving depth of 4–6 m, with duration of 30–40 s). The blood P_{N_2} rapidly reached a plateau of 640 torr after five dives ($t_{1/2}$ of 6.2 min) and declined exponentially to baseline levels ($t_{1/2} = 35.7$ min) after the shift. These results indicate that the absolute P_{N_2} level in working Korean female divers is only modestly higher than the normal level. This is most likely due to the shallow depth of the dives, and the results are consistent with the absence of decompression

sickness among these divers. "Silent" bubbles were reported by Spencer and Okino (119) in a Japanese ama "after a 51 min period of 30 consecutive open sea dives to 15 m; these divers spent 12 sec each for descent and ascent and the dives averaged 53.3 sec," with an average surface interval between successive dives (surface time) of 59 s. Nashimoto (91) also reported grade 1 bubble signals in one out of 33 ama who had returned home after 5 h of repetitive dives. Cross (19) reported the "Taravana" syndrome in breath-hold male pearl divers of Tuamotu Archipelego in French Polynesia. These divers make 40–60 dives per day to a depth of 30–40 m or more, each dive lasting approximately 2 min with 3–4 min of surface interval for recovery. They dive for about 6 h a day during the diving season, and 10%–30% of the divers are known to develop what they call taravana (*tara*, to fall; *vana*, crazily) by the end of the day. Taravana symptoms include vertigo, nausea, partial or complete paralysis, temporary unconsciousness, and in extreme cases death (19). These neurological symptoms are similar to those described by Paulev (see ALVEOLAR GAS EXCHANGE, above) and suggest that some of the Taravana symptoms may be identical to decompression sickness.

Arterial Blood O_2 and CO_2 Tensions During a Breath-Hold Dive

During descent the alveolar O_2 and CO_2 pressures increase (see ALVEOLAR GAS EXCHANGE, above), so a considerable amount of O_2 and CO_2 could diffuse from the alveolus into the blood (28, 65). There is a possibility that alveolar gas pressures (see Table 42.3) may not accurately reflect arterial gas pressures. Qvist et al. (101) collected blood from the radial artery of ama during head-out immersion immediately prior to diving, at depth, during continued breath-holding upon return to the surface, and during air breathing after breaking the breath-hold (recovery). Samples were also collected on land while the ama rested in a sitting position, then immediately prior to breaking voluntary vital capacity breath-holds in air. As shown in Table 42.4, arterial blood P_{O_2} increased from 99 torr at surface before a dive to 141 torr (compression hyperoxia) at a 5 m depth, and P_{CO_2} increased from 42 to 49 torr. As a result of this increase in P_{CO_2} (compression hypercapnia), arterial blood pH decreased from 7.42 before the dive to 7.37 at a 5 m depth. In other words, compression-induced hyperoxia and hypercapnia were demonstrated, as expected, during the course of breath-hold diving. During the early recovery phase, which begins immediately after return to the surface following a breath-hold dive, arterial blood P_{O_2} decreased precipitously from 141 torr at a 5 m depth to 68 torr. However, arterial blood P_{CO_2} remained virtually the same at depth

TABLE 42.4. *Arterial Blood Gas and pH Profile During Breath-Hold Diving (Mean ± SD)*

	PaO_2 (Torr)	$PaCO_2$ (Torr)	pH_a	SaO_2 (%)
Ocean Dives				
Predive (0 s)	100 ± 10	42 ± 2	7.42 ± 0.02	98.3 ± 0.4
Bottom (24 s)	141 ± 2*	47 ± 2*	7.37 ± 0.01*	99 ± 7
Surface breath-hold	63 ± 14*	50 ± 5*	7.35 ± 0.02*	87 ± 8*
Recovery				
0–20 s	68 ± 8	47 ± 9	7.37 ± 0.04	89 ± 3*
>20 s	106 ± 15	41 ± 4	7.41 ± 0.02	98 ± 1
Laboratory Breath-Hold in Air				
Control	110 ± 4	40 ± 2	7.41 ± 0.03	98 ± 0.4
Breath-hold (43 s)	86 ± 13*	44 ± 2*	7.37 ± 0.01*	96 ± 2

*$P < 0.05$ compared to predive value.
SaO_2 = arterial O_2 saturation.
Data from Qvist et al. (101).

and immediately after return to the surface. In one long dive (45 s), the lowest arterial blood PO_2 was 33 torr, and the highest arterial blood PCO_2 of 59 torr was observed upon surfacing. Radial arterial blood PO_2 remained below baseline values for the first 20 s after breaking the breath-hold and returned to the normal (predive) baseline level after approximately 20 s of air breathing. These findings reemphasize the development of a critical hypoxia during the ascending phase (see ALVEOLAR GAS EXCHANGE, above). In dives of usual working duration (30–40 s), arterial blood gas pressure profile remained within normal ranges (101).

Continuous oximetry during breath-holding by Korean female breath-hold divers also indicated that these divers do not subject themselves to critical hypoxia during routine work shifts (120). However, as shown in Table 42.4, the radial arterial O_2 saturation (SaO_2) at depth was maintained at normal predive levels (that is, 98%) due to compression. Although SaO_2 remained greater than 90% for up to 45 s during the subsequent breath-hold, desaturation occurred during longer periods of breath-holding. The lowest SaO_2 (60%) was recorded after an 84 s breath-hold, corresponding to PaO_2 of 33 torr. The SaO_2 remained low and ranged from 76.5 to 98.1 for the first 20 s of air breathing after breaking the breath-hold. These results on radial arterial blood confirm and extend the results of Hong et al. (51), who sampled alveolar gas of Korean ama during unassisted diving.

Qvist et al. (101) observed that arterial hematocrit increased slightly during the dive, consistent with earlier findings (54) and that the spleen contracts in professional breath-hold divers as an adaptation to repetitive breath-hold diving (see ADAPTATION, below). Qvist et al. (101) also found that whole-blood 2, 3-diphospho-

glycerate (DPG) levels increased in the ama, which could induce a minor decrease of hemoglobin O_2 affinity.

Olszowka and Rahn (94) described changes in gas stores during repeated breath-hold dives with a computer program. This simulated the diving pattern of Japanese funado (assisted) divers who perform some 30 dives to a depth of 20 m during a regular work shift of 1 h/day. As reported by previous investigators, an alveolar PO_2 value as low as 30 torr was computed, in agreement with the value measured in the field (see ALVEOLAR GAS EXCHANGE, above). Most likely, the low alveolar PO_2 values frequently observed at the end of dives lead to ascent blackout and probably account for many reported breath-hold diving accidents. Olszowka and Rahn (94) predicted that the alveolar PO_2 of about 30 torr at the end of a breath-hold dive is below the mixed venous PO_2 value, representing a temporary reversal of O_2 flow. The simulation also showed that the blood O_2 stores remain essentially unaltered, while the O_2 needs during breath-hold diving are supplied by the lung and myoglobin stores. To make use of the myoglobin stores, the muscle PO_2 during the dive has to fall to less than 2 torr. This can only be achieved by reducing cardiac output during the dive to 6.5 liters/min and raising it to 9.5 liters/min during recovery. Intense peripheral vasoconstriction (including a marked reduction of muscle blood flow) would be required to lower muscle PO_2 below the P_{50} level of myoglobin (~5 torr). As discussed under DIVING RESPONSES, below, breath-hold diving is always associated with intense peripheral vasoconstriction and is consistent with the utilization of O_2 stored in the muscle. In fact, diving animals such as seals have a higher concentration of myoglobin in the muscle compared to nondiving animals. However, it is not known whether professional breath-hold divers, such as Korean and Japanese ama, have as an adaptation a higher myoglobin concentration in their muscles.

DANGER OF HYPERVENTILATION

Intense hyperventilation dramatically lengthens breath-hold time, particularly when oxygen is used in the process (see Table 42.2) but is a double-edged sword. Diving after excessive hyperventilation delays the desire to breathe at depth, and blackouts often occur without warning during ascent due to the extreme hypoxia on re-expansion of the lungs (see ALVEOLAR GAS EXCHANGE, above). This "shallow water blackout" (it could appear at any depth) occurs approximately 7,000 times a year in the United States, despite detailed scientific findings attributing the blackout to hyperventilation before diving (14, 51, 67, 105) as well as repeated published warnings (15, 16, 42, 50). The peril of excessive hyperventilation, which can lead to a loss of consciousness and drowning, should be emphasized in all swimming and diving programs.

Lanphier (66) summarized the following factors as leading to blackout: (1) excessive hyperventilation, (2) unusual depth, (3) failure to heed air hunger (the physiological breaking point), (4) maneuvers to forestall involuntary inspiratory activities, (5) delays during ascent, and (6) carelessness upon surfacing. Although these sound like common sense, they have sound physiological bases.

DIVING RESPONSES

Breath-hold diving consists of repeated cycles of water immersion up to the neck, breath-holding and submersion, and underwater exercise. With a few exceptions, breath-hold divers encounter modest pressure, which, nevertheless, plays an important role (see ALVEOLAR GAS EXCHANGE, above).

Head-Out Water Immersion

Water immersion up to the neck produces a predive state with prominent changes in cardiovascular, respiratory, and renal functions. In addition, cold stress is inevitable in natural diving unless protection by thermal insulation is employed. In brief, the submerged tissues encounter (1) hydrostatic compression that reduces venous capacitance and displaces abdominal contents chestward; (2) an immediate environment which has a density resembling human tissues, including the blood, effectively rendering them weightless (or neutrally buoyant); (3) a negative transthoracic pressure of about −16

cm H_2O that also promotes cephalad redistribution of blood; and (4) a transcapillary fluid shift that raises blood volume. Blood volume in intrathoracic vasculature, including the heart, increases as a consequence. The distended cardiovascular structures in the thoracic region are known (1) to activate cardiac mechanoreceptors that, via the vagal afferents, force the hypothalamus to perceive them as hypervolemia, though total blood volume remains constant; (2) to encroach on the pulmonary air space and to alter respiratory mechanics; and (3) to enhance ventricular diastolic filling. Thus the functional expressions of water immersion are diuresis, restricted ventilation, and sustained increase in cardiac stroke output. These subjects have been reviewed extensively (25, 26, 33, 35, 42, 72).

Hypothermia, which follows a prolonged stay in water with temperatures below 34°C, further complicates this physiological condition. This predive state exists in natural diving but is not present in most laboratory experiments. Therefore, one should keep in mind these predive differences when interpreting results obtained in the laboratory vs. in the field.

Cardiovascular Responses

With the exception of heart rate and blood pressure, cardiovascular responses are rarely mentioned. Hemodynamic data during breath-hold diving are derived mostly from laboratory work, though, as we have noted, one should interpret laboratory data with respect to differences in the predive state. Cardiovascular responses to breath-hold in humans have been compared and contrasted to those of diving mammals (3, 10, 23, 70, 73, 75, 114).

Diving Bradycardia. Although diving bradycardia (apneic bradycardia or breath-hold bradycardia) was recognized early in birds and aquatic mammals, a systematic study of this phenomenon did not take place until the 1950s (128, 129). Scholander et al. (115) convincingly demonstrated diving bradycardia in pearl divers in the Torres Strait. This was followed by Irving's study (56) on alligator wrestlers in Florida. Studies flourished thereafter, in the field and in the laboratory (53, 109).

In the laboratory, breath-hold bradycardia can be induced by a simple breath-hold or by face immersion with breath-hold. The subject can be seated or in a prone position in a comfortable thermal ambient, consuming O_2 at a resting rate. Although the onset of bradycardia is immediate, heart rate gradually reaches a minimum (71, 82). The maximal response of heart rate in humans can be influenced by a variety of factors during the intervening time, including lung volume, alveolar P_{O_2}, P_{CO_2}, immersion of the face, temperature, intrathoracic pressure, metabolic states, mental condition, breath-hold duration, and preexisting pathological states affecting the chemosensitivity and autonomic nervous functions (70, 118). In addition, factors such as age, gender, physical conditioning, and underwater experience are said to affect the magnitude of diving bradycardia (43, 84).

In the field, diving involves exercise in addition to these factors. Figure 42.3 summarizes the effect of pre-breath-hold heart rate, exercise, and face immersion temperature on breath-hold bradycardia. Each point represents the group mean from 25 reports, except where marked with an *a* to denote the response of a single subject. Face immersion in cold water (<15°C) potentiates bradycardia both at rest and during exercise. In general, at rest heart rate decreases from the pre-breath-hold value by approximately 15% during breath-hold in air, 20% during face immersion in warm water, and 30% during face immersion in cold water.

Figure 42.3 shows clearly that exercise potentiates breath-hold bradycardia. Jung and Stolle (57) demonstrated that a swimmer's heart rate levels off at 55 beats/min during a 50 m underwater swim. Such activity requires between 5 and 10 times the resting $\dot{V}O_2$. Heart rate would be 180 beats/min if breathing were allowed. As shown in Figure 42.3, the decrease may be as much as 90% of the pre-breath-hold heart rate during exercise. The lowest heart rate, 5.6 beats/min, during face immersion in cold water was recorded from a resting subject (4). This point fits on the line of a 90% reduction (lower left-hand corner in Figure 42.3). A review revealed 16 published reports showing heart rate during breath-hold of less than 25 beats/min (77). In 1991, Ferrigno et al. (29) showed that heart rate in three elite divers decreased rapidly to 20–24 beats/min during breath-hold dives at sea to 65 and 45 m.

A potent bradycardia is found when exercise occurs in cold water. Vagus nerves mediate breath-hold bradycardia as in other diving mammals and nondiving vertebrates (8, 30, 38). The sympathetic branch of the autonomic nervous system usually plays a minor role, but an elevated background level of sympathetic activity enhances cardiodepressor effects on the heart (68). This "accentuated antagonism" is present during breath-hold diving because the circulating catecholamine level is elevated during cold exposure and exercise.

Apnea, hypoxia, and hypercapnia. Hyperoxia attenuates the breath-hold bradycardiac response, inferring that hypoxia potentiates it (22, 89). Less clear, however, are the effects of hypoxia, hypercapnia, and breath-holding individually. Lin and colleagues performed a series of five experiments which permitted an estimation of the individual effects in young male subjects (82). The series included BH-1, continuous breath-hold with air; BH-2, rebreathing at 15 s intervals during breath-hold with air; BH-3, same as BH-2 but CO_2 build-up was prevented; BH-4, rebreathing during breath-hold with O_2; and BH-5, rebreathing through a CO_2 remover during breath-hold with O_2. BH-1 represents the sum of respiratory arrests, hypoxia, and hypercapnia. The difference in heart rate responses between BH-1 and BH-2 is the effect of respiratory arrest. The difference between BH-4 and BH-5 represents the effect of hypercapnia. The hypoxic effect is obtained by subtracting the effect of apnea and hypercapnia from the total bradycardiac response. From these experiments it was concluded that out of a 31% reduction from the pre-breath-hold heart rate, apnea and hypoxia accounted for 19% and 18%, respectively, thus making up a 37% reduction. Hypercapnia reverses bradycardia, accounting for the other 6%.

Cardiac arrhythmia. Significant cardiac arrhythmias are often found in healthy subjects during common respiratory maneuvers such as deep inspiration, prolonged inspiration, breath-holding, and release of breath-hold (64). Furthermore, arrhythmias are prominent in cold-water exposure. Therefore, it should not be surprising to find cardiac arrhythmias in breath-hold diving. High frequency arrhythmias associated with immersion are reported in natural diving, particularly during the winter months (53, 109, 115).

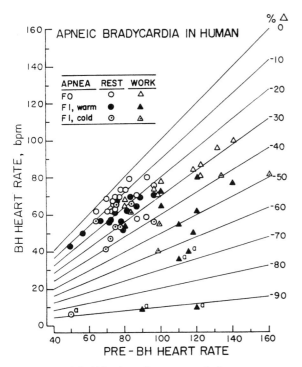

FIG. 42.3. Breath-hold bradycardiac response in humans at rest and during exercise. FI and FO are breath-hold with and without face immersion in water, respectively. Reproduced from Lin (77) with permission.

Cardiac Output. During breath-holding, there are two major differences between humans and other diving mammals. First, the cardiac output in humans changes little (48, 70, 83), while in marine mammals cardiac output decreases greatly in proportion to bradycardia. Second, breath-holding produces an elevated blood pressure in humans (38, 48, 70) but not in marine mammals. Blood pressure in dogs responds similarly to that of marine mammals during voluntary snout immersion (80). Blood pressure in rats responds similarly to that in humans, but cardiac output changes in rats resemble those in marine mammals (69, 79). Bjertnaes et al. (9) reported in 1984 a substantial decrease in cardiac output in breath-hold exercising subjects, in keeping with an enhanced breath-hold bradycardiac response during exercise (Figure 42.3).

Arterial Blood Pressure and Vascular Resistance. Breath-holding induces an elevation of blood pressure in humans, with little change in cardiac output, indicating peripheral vasoconstriction. Calculations based on available data show that humans raise total peripheral resistance (TPR) by 26%–53% (48, 125). In comparison, diving species elevate TPR during a dive by as much as 12 times predive values (11). These contrasting features in circulatory adjustments explain in part why humans are not good breath-hold divers.

Blood Flow and Oxygen Conservation. Limb blood flow decreases moderately in humans during breath-holding with face immersion; forearm or calf flow decreases by 7%–50% (12, 22, 70). A review by Elsner and Gooden (23) confirms that vasoconstriction occurs but that it is unable to shut down limb blood flow completely during breath-holding, which renders O_2 conservation ineffective in humans. Reduction in blood flow to aerobically active tissues will not result in conservation of O_2, since the arterio-to-venous difference in O_2 content widens in the face of reduced blood flow (48, 70).

ADAPTATION

Breath-hold divers are exposed to certain physical and physiological stresses, including the hydrostatic compression of the chest during descent followed by reexpansion of the chest during ascent, hypercapnia and attendant changes in acid-base balance, hypoxia, physical exercise, and body heat loss. Divers who engage in breath-hold diving for many years seem to develop specific physical and physiological adaptations to the above stresses (44) that appear to improve their diving ability.

Lung Volume and Maximal Respiratory Pressure

Studies of U.S. Navy male divers (13, 110), Korean female divers (117), and Japanese ama divers (both male and female) (116, 122) clearly show that vital capacity is significantly greater (by about 700 ml) in all diver groups as compared to paired controls. This larger vital capacity was associated with larger inspiratory capacity, except in one study on Japanese male (Tsushima) divers in whom expiratory reserve volume, rather than inspiratory capacity, was significantly greater than in control subjects (116). The maximal inspiratory pressure is significantly greater than normal in Korean female divers (117), indicating that inspiratory muscle force at a given lung volume is greater.

Longitudinal studies conducted on U.S. Navy divers showed that, after 1 yr of diving training, inspiratory capacity, vital capacity, and TLC are elevated significantly (13). Similar longitudinal studies carried out by Bachman and Horvath (5) before and after 4 months of either swimming or wrestling training showed a significant increase in inspiratory and vital capacities in swimmers but not in wrestlers. These studies strongly suggest that higher inspiratory and vital capacities observed in divers cannot be attributed to daily physical workload in the air environment per se but to aquatic activity.

Breath-hold divers float at the surface between two successive dives, engaging in head-out immersion, in which they are subjected to negative-pressure breathing equivalent to −16 cm H_2O (46). The muscles work harder to inspire against this hydrostatic pressure gradient across the chest wall, and, hence, the inspiratory muscles are developed to a greater extent than the expiratory muscles. This probably accounts for significant increases in the inspiratory capacity and the maximal inspiratory pressure observed in divers.

The maximal depth of a dive is primarily determined by the TLC/residual volume ratio (see DEPTH LIMIT, above). Professional breath-hold divers have higher TLC than controls, but what happens to the residual volume? Bachman and Horvath (5) found a significant decrease (260 ml) in residual volume after 4 months of swimming training, strongly suggesting that residual volume is also subject to adaptation to diving. Kobayashi et al. (62) found a greater "body wall pliability" in Japanese ama divers, which may, at least in part, account for the unexplained lowering of residual volume.

CO_2 Adaptation

A considerable amount of CO_2 accumulates in the blood and tissues during the course of a breath-hold dive (see ALVEOLAR GAS EXCHANGE, above). In response to repetitive exposures to hypercapnia, most breath-hold divers (that is, Korean and Japanese ama and U.S. Navy divers) show a reduced sensitivity of the ventilatory system to hypercapnia (110, 117). Moreover, the high tolerance to CO_2 observed in U.S. Navy divers is reversed after a 3 month layoff (110). According to Honda et al. (40), a reduced CO_2 sensitivity is not demonstrable in Japanese cachido (unassisted) divers who dive to shallow depths (about 5 m) only during warm seasons. Adaptation to CO_2 appears to be associated with a reduced adrenergic and stress response to CO_2 (39).

Adaptation to Hypoxia

Alveolar P_{O_2} is kept high during descent and at depth because of hydrostatic compression of the chest, but it decreases rapidly during ascent because of rapid decompression, resulting in a severe (or fatal) hypoxia (see ALVEOLAR GAS EXCHANGE, above). In response to repeated exposure to hypoxia, the ventilatory response to low O_2 is considerably attenuated in both escape-training tank instructors (110) and Japanese funado (assisted) divers engaged in deeper and longer dives (40). In cachido divers of Japan and Korea, this hypoxic adaptation is not demonstrable (117). Schaefer (110) also reported that the lower ventilatory response to low O_2 breathing observed in the escape-training tank instructors is associated with the formation of a large O_2 debt. This suggests that tissue oxidation may decrease in divers during hypoxia.

Other Adaptations

The "diving bradycardia" induced by apneic face immersion in cold water appears to be increased above normal in both Hawaiian male divers (49) and Korean female divers (B. S. Kang, unpublished data). Korean female divers engaged in cold-water diving without wet suits until 1977 were subjected to severe cold-water stress. In response to repeated exposure to cold all year round, they developed many different types of adaptation. However, this subject is beyond the scope of this chapter. For further information readers are referred to Hong et al. (52).

As described in the chapter by Lahiri, many diving animals seem to have mechanisms that conserve O_2 during breath-hold diving. Although these animals are most often engaged in short, shallow, aerobic dives, they occasionally perform very deep, long dives in which extreme bradycardia, peripheral vasoconstriction, marked reduction in cardiac output, and constant blood pressure are observed. As a result, O_2 consumption

decreases markedly, prolonging dive time. Whether such O_2-conserving mechanisms exist in humans is not known. Although human divers manifest cardiovascular responses to breath-hold diving that are qualitatively similar to those observed in diving animals, the magnitude of change is much less marked (see DIVING RESPONSES, above). In human nondivers, total peripheral resistance is always high, but during breath-holding (with or without face immersion), cardiac output either remains unchanged, decreases, or increases. In other words, the rise in systemic vascular resistance is not closely adjusted to the reduction in cardiac output, thus reducing the O_2-saving potential because of the increase in cardiac afterload (see ref. 42). Recently, Ferretti et al. (28) estimated that the power output in elite professional breath-hold divers during breath-hold dives ranged from 513 to 929 ml O_2/min, corresponding to about 20%–30% of $\dot{V}_{O_2max}$. On the basis of these and other data (that is, lactate and muscle blood flow), the investigators proposed that O_2-conserving reflexes exist in humans adapted to breath-hold diving. It would be of great interest to test this hypothesis on Korean and Japanese ama divers.

Qvist et al. (100) observed a consistent increase of arterial hemoglobin concentration during breath-hold diving in the Weddell seal. They suggested that an extension of the sympathetic outflow of the diving reflex may have caused profound contractions of the seal's very large spleen. Although they have no direct evidence that this splenic contraction takes place, this type of contraction could inject large quantities of red blood cells saturated with oxygen into the seal's central circulation during diving, prolonging diving time. Hurford et al. (54) ultrasonically measured the actual splenic size in Korean ama divers before and immediately after the daily 3 h workshift and found that the splenic volume decreased by 19.5%, while hematocrit increased by 10.5%. No such changes were observed in Japanese male scuba divers, who do not routinely engage in breath-hold diving. Splenic contraction in response to breath-hold diving may represent yet another manifestation of adaptation.

REFERENCES

1. Agostoni, E. Diaphragm activity during breath holding: factors related to its onset. *J. Appl. Physiol.* 18: 30–36, 1963.
2. Alvis, H. J. Breath-hold breaking point at various increased pressures. *U.S. Navy Med. Res. Lab. Rep.* 177, 10: 110–120, 1951.
3. Andersen, H. T. Physiological adaptations in diving vertebrates. *Physiol. Rev.* 46: 212–243, 1966.
4. Arnold, R. W. Extremes in human breath-hold, facial immersion bradycardia. *Undersea Biomed. Res.* 12: 183–190, 1985.
5. Bachman, J. C., and S. M. Horvath. Pulmonary function

changes which accompany athletic conditioning program. *Res. Q.* 30: 235–239, 1969.

6. Bartlett, D., Jr. Effects of Valsalva and Mueller maneuvers on breath-holding time. *J. Appl. Physiol.: Respir. Environ. Exerc. Physiol.* 42: 717–721, 1977.

7. Beebe, W. *Half Mile Down.* New York: Duel, Sloan and Pearce, 1934.

8. Berk, J. L., and M. N. Levy. Profound reflex bradycardia produced by transient hypoxia or hypercapnia in man. *Eur. Surg. Res.* 9: 75–84, 1977.

9. Bjertnaes, L., A. Hauge, J. Kjekshus, and E. Soyland. Cardiovascular responses to face immersion and apnea during steady state muscle exercise. *Acta Physiol. Scand.* 120: 605–612, 1984.

10. Blix, A. S., and B. Folkow. Cardiovascular adjustments to diving in mammals and birds. In: *Handbook of Physiology. Peripheral Circulation and Organ Blood Flow,* edited by J. T. Shepherd and F. M. Abboud. Bethesda, MD: Am. Physiol. Soc., 1983, sect. 2, vol. 3, p. 917–945.

11. Blix, A. S., J. K. Kjekshus, I. Enge, and A. Bergan. Myocardial blood flow in the diving seal. *Acta Physiol. Scand.* 96: 277–280, 1976.

12. Brick, I. Circulatory responses to immersing the face in water. *J. Appl. Physiol.* 21: 33–36, 1966.

13. Carey, C. R., K. E. Schaefer, and H. J. Alvis. Effect of skin diving on lung volume. *J. Appl. Physiol.* 8: 519–523, 1956.

14. Craig, A. B. Cause of loss of consciousness during underwater swimming. *J. Appl. Physiol.* 16: 583–586, 1961.

15. Craig, A. B. Underwater swimming and loss of consciousness. *JAMA* 176: 255–258, 1961.

16. Craig, A. B. Summary of 58 cases of loss of consciousness during underwater swimming and diving. *Med. Sci. Sports* 8: 171–175, 1967.

17. Craig, A. B. Depth limits of breath-hold diving (an example of fennology). *Respir. Physiol.* 5:14–22, 1968.

18. Craig, A. B., and W. L. Medd. Oxygen consumption and carbon dioxide production during breath-hold diving. *J. Appl. Physiol.* 24: 190–202, 1968.

19. Cross, E. R. Taravana. Diving syndrome in the Tuamotu diver. In: *Physiology of Breath-Hold Diving and the Ama of Japan,* edited by H. Rahn and T. Yokoyama. Washington, DC: NAS–NRC Publ. 1341, 1965, p. 207–219.

20. Davis, R. W. Deep diving and underwater rescue. *J. R. Soc. Arts* 82: 1032–1047, 1934.

21. Douglas, C. G., and J. S. Haldane. The causes of periodic or Cheyne-Stokes breathing. *J. Physiol.* 38: 401–419, 1909.

22. Elsner, R., D. L. Franklin, R. L. Van Citters, and D. W. Kenney. Cardiovascular defense against asphyxia. *Science* 153: 941–949, 1966.

23. Elsner, R., and B. Gooden. *Diving and Asphyxia.* Cambridge: Cambridge University Press, 1983, p. 60–73.

24. Elsner, R. W., B. A. Gooden, and S. M. Robinson. Arterial blood gas changes and the diving response in man. *Aust. J. Exp. Biol. Med. Sci.* 49: 435–444, 1971.

25. Epstein, M. Cardiovascular and renal effects of head-out water immersion in man. *Circ. Res.* 39: 619–627, 1976.

26. Epstein, M. Renal effect of head-out water immersion in man: implications for an understanding of volume homeostasis. *Physiol. Rev.* 58: 529–581, 1978.

27. Ferrera, F. Personal experience at deep breath-hold diving [Abstract]. In: *The Physiology of Deep Breath-Hold Diving,* edited by P. D. Data. Chieti, Italy: 1991.

28. Ferretti, G., M. Costa, M. Ferrigno, B. Grassi, C. Marconi, C. E. G. Lundgren, and P. Cerretteli. Alveolar gas composition and exchange during deep breath-hold diving and dry breath-

holds in elite divers. *J. Appl. Physiol.: Respir. Environ. Exerc. Physiol.* 70: 794–802, June 28–30, 1991.

29. Ferrigno, M., B. Grassi, G. Ferretti, M. Costa, C. Marconi, P. Cerretelli, and C. Lundgren. Electrocardiogram during deep breath-hold dives by elite divers. *Undersea Biomed. Res.* 18: 81–91, 1991.

30. Finley, J. P., J. F. Bonet, and M. B. Waxman. Autonomic pathways responsible for bradycardia on facial immersion. *J. Appl. Physiol.: Respir. Environ. Exerc. Physiol.* 47: 1218–1222, 1979.

31. Fowler, W. S. Breaking point of breath holding. *J. Appl. Physiol.* 6: 539–545, 1954.

32. Gamba, R. New world records of breath-hold diving depth, Francisco Ferrera, 112 m; and Angela Bandini, 107 m (in Japanese). *Marine Diving* Jan. (suppl.): 6–9, 1990.

33. Gauer, O. H., and J. P. Henry. Neurohormonal control of plasma volume. In: *International Review of Physiology, Cardiovascular Physiology II,* edited by A. C. Guyton and A. W. Cowley. Baltimore, MD: University Park, 1976, p. 145–190.

34. Gautier, H., R. Lefrancois, and P. Pasquis. Breath holding and rebreathing at low and high altitude. *Respir. Physiol.* 23: 201–207, 1975.

35. Greenleaf, J. E. Physiological responses to prolonged bed rest and fluid immersion in humans. *J. Appl. Physiol.: Respir. Environ. Exerc. Physiol.* 57: 619–633, 1984.

36. Godfrey, S., and E. J. M. Campbell. The control of breath holding. *Respir. Physiol.* 5: 385–400, 1968.

36a. *Guinness Book of World Records,* edited by N. McWhirter, D. A. Boehm, S. Topping, and C. Smith. New York, NY: Sterling Publ. Co., 1984, p. 30.

36b. *Guinness Book of World Records,* edited by N. McWhirter, S. Greenberg, D. A. Boehm, and S. Topping. New York, NY: Sterling Publ. Co., 1981, p. 45.

37. Hayward, J. S., C. Hay, B. R. Mathews, C. H. Overweel, and D. D. Radford. Temperature effect on the human dive response in relation to cold water near-drowning. *J. Appl. Physiol.: Respir. Environ. Exerc. Physiol.* 56: 202–206, 1984.

38. Heistad, D. D., F. M. Abboud, and J. W. Eckstein. Vasoconstrictor response to simulated diving in man. *J. Appl. Physiol.* 25: 542–549, 1968.

39. Hesser, C. M. Breath holding under high pressure. In: *Physiology of Breath-Hold Diving and the Ama of Japan,* edited by H. Rahn and T. Yokoyama. Washington, DC: NAS–NSC Publ. 1341, 1965, p. 165–181.

40. Honda, Y., Y. Masuda, F. Hayashi, and A. Yoshida. Differences in ventilatory responses to hypoxia and hypercapnia between assisted (funado) and non-assisted (cachido) breath-hold divers. In: *Hyperbaric Medicine and Underwater Physiology,* edited by K. Shiraki and S. Matsuoka. Kitakyushu, Japan: Univ. Occup. Environ. Hlth., 1983, p. 45–58.

41. Hong, S. K. Hae-nyo, the diving women of Korea. In: *Physiology of Breath-Hold Diving and the Ama of Japan,* edited by H. Rahn and T. Yokoyama. Washington, DC: NAS–NRC Publ. 1341, 1965, p. 99–111.

42. Hong, S. K. The physiology of breath-hold diving. In: *Diving Medicine,* edited by R. H. Strauss. New York: Grune and Stratton, 1976, p. 269–286.

43. Hong, S. K. Breath-hold bradycardia in man: an overview. In: *The Physiology of Breath-Hold Diving,* edited by C. E. G. Lundgren and M. Ferrigno. Bethesda, MD: Undersea Hyperbaric Med. Soc., 1987, p. 158–173.

44. Hong, S. K. Physical and physiological adaptations to breath-hold diving in humans: a review. In: *Underwater and Hyperbaric Physiology, IX,* edited by A. A. Bove, A. J. Bachrach, and L. J. Greenbaum, Jr. Bethesda, MD: Undersea Hyperbaric Med. Soc., 1987, p. 57–65.

45. Hong, S. K. Diving physiology—man. In: *Comparative Pulmonary Physiology—Current Concepts*, edited by S. C. Wood. New York: Marcel Dekker, 1989, p. 787–802.

46. Hong, S. K., P. Cerretelli, J. C. Cruz, and H. Rahn. Mechanics of respiration during submersion in water. *J. Appl. Physiol.* 27: 535–538, 1969.

47. Hong, S. K., J. Henderson, A. Olszowka, W. E. Hurford, K. J. Falke, J. Qvist, P. Radermacher, K. Shiraki, M. Mohri, H. Takeuchi, W. J. Zapol, D. W. Ahn, J. K. Choi, and Y. S. Park. Daily diving pattern of Korean and Japanese breath-hold divers (ama). *Undersea Biomed. Res.* 18: 433–443, 1991.

48. Hong, S. K., Y. C. Lin, D. A. Lally, B. J. B. Yim, N. Kominami, P. W. Hong, and T. O. Moore. Alveolar gas exchanges and cardiovascular functions during breath-holding with air. *J. Appl. Physiol.* 30: 540–547, 1971.

49. Hong, S. K., T. O. Moore, G. Seto, H. K. Park, W. R. Hiatt, and E. M. Bernauer. Lung volumes and apneic bradycardia in divers. *J. Appl. Physiol.* 29: 172–176, 1970.

50. Hong, S. K., and H. Rahn. The diving women of Korea and Japan. *Sci. Am.* 216: 34–43, 1967.

51. Hong, S. K., H. Rahn, D. H. Kang, S. H. Song, and B. S. Kang. Diving pattern, lung volumes, and alveolar gas of the Korean diving women (ama). *J. Appl. Physiol.* 18: 457–465, 1963.

52. Hong, S. K., D. W. Rennie, and Y. S. Park. Cold acclimatization and deacclimatization in Korean women divers. *Exerc. Sports Sci. Rev.* 14: 231–268, 1986.

53. Hong, S. K., S. H. Song, P. K. Kim, and C. S. Suh. Seasonal observations on the cardiac rhythm during diving in the Korean ama. *J. Appl. Physiol.* 23: 18–22, 1967.

54. Hurford, W. E., S. K. Hong, Y. S. Park, D. W. Ahn, K. Shiraki, M. Mohri, and W. M. Zapol. Splenic contraction during breath-hold diving in the Korean ama. *J. Appl. Physiol.: Respir. Environ. Exerc. Physiol.* 69: 932–936, 1990.

55. Irimoto, T. Daily space use patterns of male breath-hold abalone divers. *J. Hum. Ergol. (Tokyo)* 2:59–74, 1973.

56. Irving, L. Bradycardia in human divers. *J. Appl. Physiol.* 18: 489–491, 1963.

57. Jung, K., and W. Stolle. Behavior of heart rate and incidence of arrhythmia in swimming and diving. *Biotelem. Pat. Monit.* 8: 228–239, 1981.

58. Kang, B. S., S. H. Song, C. S. Suh, and S. K. Hong. Changes in body temperature and basal metabolic rate of the ama. *J. Appl. Physiol.* 18: 483–488, 1963.

59. Kang, D. H., P. K. Kim, B. S. Kang, S. H. Song, and S. K. Hong. Energy metabolism and body temperature in the ama. *J. Appl. Physiol.* 20: 46–50, 1965.

60. Kang, D. H., Y. S. Park, Y. D. Park, I. S. Lee, D. S. Yoen, S. H. Lee, D. W. Rennie, and S. K. Hong. Energetics of wet-suit diving in Korean women breath-hold divers. *J. Appl. Physiol.: Respir. Environ. Exerc. Physiol.* 54: 1702–1707, 1983.

61. Klocke, F. J., and H. Rahn. Breath-holding after breathing of oxygen. *J. Appl. Physiol.* 14: 689–693, 1959.

62. Kobayashi, S., T. Ogawa, C. Adachi, F. Ishikawa, and K. Takahashi. Maximal respiratory pressure and pliability of the body wall of the Japanese ama. *Acta Med. Biol.* 18: 249–260, 1971.

63. Kobayashi, S., and C. Sasaki. Breaking point of breath holding and tolerance time in rebreathing. *Jpn. J. Physiol.* 17: 43–56, 1967.

64. Lamb, L. E., G. Dermksian, and C. A. Sarnoff. Significant cardiac arrhythmia induced by common respiratory maneuvers. *Am. J. Cardiol.* 2: 563–571, 1958.

65. Lanphier, E. Application of decompression tables to repeated breath-hold dives. In: *Physiology of Breath-Hold Diving and the Ama of Japan*, edited by H. Rahn and T. Yokoyama. Washington, DC: NAS–NRC Publ. 1341, 1965, p. 227–236.

66. Lanphier, E. H. Breath-hold and ascent blackout. In: *The Physiology of Breath-Hold Diving*, edited by C. E. G. Lundgren and M. Ferrigno. Bethesda, MD: Undersea Hyperbaric Med. Soc., 1987, p. 32–42.

67. Lanphier, E. H., and H. Rahn. Alveolar gas exchange during breath-hold diving. *J. Appl. Physiol.* 18: 471–477, 1963.

68. Levy, M. N. Cardiac sympathetic–parasympathetic interactions. *Federation Proc.* 43: 2598–2602, 1984.

69. Lin, Y. C. Autonomic nervous control of cardiovascular response during diving in the rat. *Am. J. Physiol.* 227: 601–605, 1974.

70. Lin, Y. C. Breath-hold diving in terrestrial mammals. In: *Exercise and Sport Sciences Reviews*, edited by R. L. Terjung. Philadelphia: Franklin, 1982, *vol. 10*, p. 270–307.

71. Lin, Y. C. Cardiopulmonary physiology of nondiving mammals during breath-hold dives. In: *Hyperbaric Medicine and Underwater Physiology*, edited by K. Shiraki and S. Matsuoka. Kitakyushu, Japan: Univ. Occup. Environ. Hlth., 1983, p. 25–35.

72. Lin, Y. C. Circulatory functions during immersion and breath-hold dives in humans. *Undersea Biomed. Res.* 11: 123–138, 1984.

73. Lin, Y. C. Breath-hold diving: human imitation of aquatic mammals. In: *Diving in Mammals and Man*, edited by A. O. Brubakk, J. W. Kanwisher, and G. Sundnes. Trondheim, Norway: Tapir, 1986, p. 81–89.

74. Lin, Y. C. Effect of O_2 and CO_2 on breath-hold breaking point. In: *The Physiology of Breath-Hold Diving*, edited by C. E. G. Lundgren and M. Ferrigno. Bethesda, MD: Undersea Hyperbaric Med. Soc., 1987, p. 75–86.

75. Lin, Y. C. Human imitation of marine mammals and its clinical significance. In: *Underwater and Hyperbaric Physiology IX*, edited by A. A. Bove, A. J. Bachrach, and L. J. Greenbaum. Bethesda, MD: Undersea Hyperbaric Med. Soc., 1987, p. 29–45.

76. Lin, Y. C. Applied physiology of diving. *Sports Med.* 5: 41–56, 1988.

77. Lin, Y. C. Physiological limitations of humans as breath-hold divers. In: *Man in the Sea*, edited by Y. C. Lin and K. K. Shida, 1990, *vol. II*, p. 33–56.

78. Lin, Y. C. Physiological limitations of diving depth in humans [Abstracts]. In: *Proceedings of the Workshop of Deep Breath-Hold Diving*, edited by P. G. Data, Chieti, Italy: June 28–30, 1991.

79. Lin, Y. C., and D. G. Baker. Cardiac output and its distribution during diving in the rat. *Am. J. Physiol.* 228: 733–737, 1975.

80. Lin, Y. C., E. L. Carlson, E. P. McCutcheon, and H. Sandler. Cardiovascular functions during voluntary apnea in dogs. *Am. J. Physiol.* 245 (*Regulatory Integrated Comp. Physiol.* 16): R143–R150, 1983.

81. Lin, Y. C., D. A. Lally, T. O. Moore, and S. K. Hong. Physiological and conventional breath-hold breaking points. *J. Appl. Physiol.* 37: 291–296, 1974.

82. Lin, Y. C., K. K. Shida, and S. K. Hong. Effect of hypercapnia, hypoxia, and rebreathing on heart rate response to apnea. *J. Appl. Physiol.: Respir. Environ. Exerc. Physiol.* 54: 166–171, 1983.

83. Lin, Y. C., K. K. Shida, and S. K. Hong. Effect of hypercapnia, hypoxia, and rebreathing on circulatory response to apnea. *J. Appl. Physiol.: Respir. Environ. Exerc. Physiol.* 54: 172–177, 1983.

84. Manley, L. Apnoeic heart rate responses in humans, a review. *Sports Med.* 9: 286–310, 1990.

85. Mayol, J. Jacques Mayol—apnea a meno cento. *Frateli Fabbri Editori* Jan., 1–96, 1976.

86. Missiroli, di F. M. Mayol a 105 m. *Mondo Sommerso* N271: 32–37, 1983–1984.

87. Mithoefer, J. C. Breath holding. In: *Handbook of Physiology. Respiration,* edited by W. O. Fenn and H. Rahn. Washington, DC: Am. Physiol. Soc., 1965, vol. II, p. 1011–1025.

88. Mithoefer, J. C. The breaking point of breath holding. In: *Physiology of Breath-Hold Diving and the Ama of Japan,* edited by H. Rahn and T. Yokoyama. Washington, DC: NAS–NRC Publ. 1341, 1965, p. 195–205.

89. Moore, T. O., R. Elsner, Y. C. Lin, D. A. Lally, and S. K. Hong. Effects of alveolar P_{O_2} and P_{CO_2} on apneic bradycardia in man. *J. Appl. Physiol.* 34: 795–798, 1973.

90. Morreli, L., and P. G. Data. Thoracic radiological changes during deep breath-hold diving [Abstract]. In: *Proceedings of the Workshop on the Physiology of Deep Breath-Hold Diving,* edited by P. G. Data. Chieti, Italy: June 28–30, 1991.

91. Nashimoto, I. Intravascular bubbles following repeated breath-hold dives (in Japanese). *Jpn. J. Hyg.* 31: 439, 1976.

92. Nukada, M. Historical development of the ama's diving activities. In: *Physiology of Breath-Hold Diving and the Ama of Japan,* edited by H. Rahn and T. Yokoyama. Washington, DC: NAS–NRC Publ. 1341, 1965, p. 25–41.

93. Olszowka, A. J., and H. Rahn. Breath-hold diving. In: *Extreme Environments: Coping Strategies of Animals and Man,* edited by J. R. Sutton, C. S. Houston, and G. Cowles. New York: Praeger, 1987, p. 417–428.

94. Olszowka, A. J., and H. Rahn. Gas store changes during repetitive breath-hold diving. In: *Man in Stressful Environments— Diving, Hyper- and Hypobaric Physiology,* edited by K. Shiraki and M. K. Yousef. Springfield, IL: Thomas, 1987, p. 41–56.

95. Otis, A. B., H. Rahn, and W. O. Fenn. Alveolar gas exchanges during breath hold. *Am. J. Physiol.* 152: 674–686, 1948.

96. Park, Y. S., H. Rahn, I. S. Lee, S. I. Lee, D. H. Kang, S. Y. Hong, and S. K. Hong. Patterns of wet suit diving in Korean women breath-hold divers. *Undersea Biomed. Res.* 10: 203–215, 1983.

97. Park, Y. S., K. Shiraki, and S. K. Hong. Energetics of breath-hold diving in Korean and Japanese professional divers. In: *Man in the Sea,* edited by Y. C. Lin and K. K. Shida. San Pedro, CA: Best, 1990, vol. II, p. 75–87.

98. Paulev, P. Decompression sickness following repeated breath-hold dives. *J. Appl. Physiol.* 20: 1028–1031, 1965.

99. Pendergast, D. R. Energetics of breath-hold diving. In: *The Physiology of Breath-Hold Diving,* edited by C. E. G. Lundgren and M. Ferrigno. Bethesda, MD: Undersea Hyperbaric Med. Soc., 1987, p. 135–147.

100. Qvist, J., R. D. Hill, R. C. Schneider, K. J. Falke, G. C. Leggins, M. Guppy, R. L. Elliot, P. W. Hochachka, and W. M. Zapol. Hemoglobin concentrations and blood gas tensions of free-diving Weddell seals. *J. Appl. Physiol.: Respir. Environ. Exerc. Physiol.* 61: 1560–1569, 1986.

101. Qvist, J., W. E. Hurford, P. Radermacher, G. Guyton, K. J. Falke, Y. S. Park, D. W. Ahn, S. K. Hong, K. Stanek, and W. M. Zapol. Arterial blood gas tensions during breath-hold diving in the Korean ama. *FASEB J.* 5: A1127, 1991.

102. Radermacher, P., K. J. Falke, Y. S. Park, D. W. Ahn, J. Qvist, S. K. Hong, and W. M. Zapol. Nitrogen tensions in brachial vein blood of Korean divers (ama). *FASEB J.* 5: A1125, 1991.

103. Rahn, H. The physiological stresses of the ama. In: *Physiology of Breath-Hold Diving and the Ama of Japan,* edited by H. Rahn and T. Yokoyama. Washington, DC: NAS–NRC Publ. 1341, 1965, p. 113–137.

104. Rahn, H. Breath-hold diving: a brief history. In: *The Physiology of Breath-Hold Diving,* edited by C. E. G. Lundgren and M. Ferrigno. Bethesda, MD: Undersea Hyperbaric Med. Soc., 1987, p. 1–3.

105. Rahn, H. Breath-hold diving: alveolar O_2 and blackout. In: *Underwater and Hyperbaric Physiology IX,* edited by A. A.

Bove, A. J. Bachrach, and L. J. Greenbaum. Bethesda, MD: Undersea Hyperbaric Med. Soc., 1987, p. 3–15.

106. Rahn, H., H. T. Bahyson, J. F. Muxworthy, and J. M. Hagen. Adaptation to high altitude: changes in breath-holding time. *J. Appl. Physiol.* 6: 154–157, 1953.

107. Rahn, H., and T. Yokoyama. *Physiology of Breath-Hold Diving and the AMA of Japan.* Washington, DC: NAS–NRC Publ. 1341, 1965.

108. Robard, S. Effect of oxygen, altitude and exercise on breath-holding time. *Am. J. Physiol.* 150: 148–152, 1947.

109. Sasamoto, H. The electrocardiogram pattern of the diving ama. In: *Physiology of Breath-Hold Diving and the Ama of Japan,* edited by H. Rahn and T. Yokoyama. Washington, DC: NAS–NRC Publ. 1341, 1965, p. 271–280.

110. Schaefer, K. E. Adaptation to breath-hold diving. In: *Physiology of Breath-Hold Diving and the Ama of Japan,* edited by H. Rahn and T. Yokoyama. Washington, DC: NAS–NRC Publ. 1341, 1965, p. 237–252.

111. Schaefer, K. E., R. D. Allison, J. H. Dougherty, Jr., C. R. Carey, R. Walker, F. Yost, and D. Parker. Pulmonary and circulatory adjustments determining the limits of depth in breath-hold diving. *Science* 162: 1020–1023, 1968.

112. Schneider, E. C. Observation on holding the breath. *Am. J. Physiol.* 94: 464–470, 1930.

113. Schneider, E. C. Respiration at high altitude. *Yale J. Biol. Med.* 4: 537–550, 1932.

114. Scholander, P. F. Physiological adaptation to diving in animals and man. The Harvey Lectures 57: 93–110, 1961–1962.

115. Scholander, P. F., H. T. Hammel, H. LeMessurier, E. Hemmingsen, and W. Garey. Circulatory adjustment in pearl divers. *J. Appl. Physiol.* 17: 184–190, 1962.

116. Shiraki, K., N. Konda, S. Sagawa, Y. S. Park, T. Komatsu, and S. K. Hong. Diving pattern of Tsushima male breath-hold divers (Katsugi). *Undersea Biomed Res.* 12: 439–452, 1985.

117. Song, S. H., D. H. Kang, B. S. Kang, and S. K. Hong. Lung volumes and ventilatory responses to high CO_2 and low O_2 in the ama. *J. Appl. Physiol.* 18: 466–470, 1963.

118. Song, S. H., W. K. Lee, Y. A. Chung, and S. K. Hong. Mechanism of apneic bradycardia in man. *J. Appl. Physiol.* 27: 323–327, 1969.

119. Spencer, M. P., and H. Okino. Venous gas emboli following repeated breath-hold dives. *Federation Proc.* 31: 355, 1972.

120. Stanek, K., G. P. Guyton, W. E. Hurford, Y. S. Park, D. W. Ahn, J. Qvist, K. J. Falke, S. K. Hong, H. Kobayashi, K. Kobayashi, and W. M. Zapol. Continuous pulse oximetry in the breath-hold diving women of Korea and Japan. *FASEB J.* 5: A1127, 1991.

121. Sterba, J. A., and C. E. G. Lundgren. Diving bradycardia and breath-holding time in man. *Undersea Biomed. Res.* 12: 139–150, 1985.

122. Tatai, K., and K. Tatai. Anthropometric studies on the Japanese ama. In: *Physiology of Breath-Hold Diving and the Ama of Japan,* edited by H. Rahn and T. Yokoyama. Washington, DC: NAS–NRC Publ. 1341, 1965, p. 71–83.

123. Teruoka, G. Die Ama und ihre Arbeit. *Arbeitsphysiol.* 5: 239–251, 1932.

124. West, J. B., K. Tsukimoto, O. Mathieu-Costello, and R. Prediletto. Stress failure in pulmonary capillaries. *J. Appl. Physiol.: Respir. Environ. Exerc. Physiol.* 70: 1731–1742, 1991.

125. Whayne, T. F., N. T. Y. Smith, E. I. Eger, II, R. K. Stoeling, and C. E. Whitcher. Reflex cardiovascular responses to simulated diving. *Angiologia* 23: 500–508, 1972.

126. Whitelaw, W. A., B. McBride, J. Arnar, and K. Corbet. Respiratory neuromuscular output during breath-holding. *J. Appl. Physiol.: Respir. Environ. Exerc. Physiol.* 50: 435–443, 1981.

127. Woods, V. Angela Bandini. *Vogue* 180: 228–230, 1990.

128. Wyss, V. Electrocardiogram of apneic subjects during immersion in water at various depths. *Boll. Soc. Ital. Biol. Sper.* 32: 503–506, 1956.

129. Wyss, V. Swimming under water in apnea and the nature of the electrocardiogram. *Boll. Soc. Ital. Biol. Sper.* 32: 506–509, 1956.

130. Yokoyama, T. Energy expenditure by the diving ama in Japan. In: *Underwater and Hyperbaric Physiology IX,* edited by A. A. Bove, A. J. Bachrach, and L. J. Greenbaum, Jr. Bethesda, MD: Undersea Hyperbaric Med. Soc., 1987, p. 17–28.

131. Yokoyama, T., and S. Iwasaki. Ecology of the Japanese ama. In: *Human Adaptability,* edited by H. Yoshimura and S. Kobayashi. Tokyo: University of Tokyo, 1975, vol. 3, p. 199–209.

2 | GAS PHYSIOLOGY IN DIVING

43. Gas physiology in diving

CLAES E. G. LUNDGREN

ANDREA HARABIN

PETER B. BENNETT

HUGH D. VAN LIEW

EDWARD D. THALMANN

Center for Research and Education in Special Environments, State University of New York at Buffalo, Buffalo, New York
Naval Medical Research Institute, Bethesda, Maryland
Hyperbaric Center and Department of Anesthesiology, Duke University Medical Center, Durham, North Carolina
Department of Physiology, State University of New York at Buffalo, Buffalo, New York
Department of Environmental and Occupational Medicine and F. G. Hall Hypo/Hyperbaric Center, Duke University Medical Center, Durham, North Carolina

CHAPTER CONTENTS

FOR STEADY-STATE GAS EXCHANGE under water, a diver must rely on breathing gear to obtain gas in sufficient amounts at an adequate pressure and composition. The section that follows will mainly address the first two requirements, while the last will be dealt with later in the chapter.

BREATHING UNDER WATER: VENTILATORY NEEDS*

Effects of Gas Compression

The amount of breathing gas needed (per unit time) is directly proportional to water depth, that is, the ambient pressure. This relationship is primarily dictated by the necessity of adequate carbon dioxide elimination. As far as oxygen supply goes, the most common breathing gas in diving, namely air, provides for an abundance of oxygen at depth. For example, the oxygen content of 1.0 liters [standard temperature and pressure, dry (STPD)] of air at the surface is 0.21 liters, while in 1.0 liters of dry air at 33 ft of depth [10 m, about 2 atmospheres (atm)], that is, at twice the pressure, it is 0.42 liters (STPD), at 66 ft (20 m or 3 atm) it is 0.63 liters (STPD), and so on. [The exact relationship between depth and pressure depends on the density of the water (cf. pressure conversion table in the chapter by Vorosmarti in this *Handbook*).]

With CO_2 elimination, however, the increased pressure of breathing gas at depth does not offer an advantage. The physiological regulation of breathing is primarily geared to maintaining an acceptable alveolar CO_2 pressure ($P_{A}CO_2$). It follows that when the pressure of the respired gas, for instance, is twice the normal atmospheric pressure (at 33 ft or 10 m), the alveolar CO_2 fraction ($F_{A}CO_2$) must be reduced by half to maintain the same $P_{A}CO_2$. The $F_{A}CO_2$ is proportional to $\dot{V}CO_2/\dot{V}_A$, where $\dot{V}CO_2$ is the amount of CO_2 eliminated and $\dot{V}_A$ is the alveolar ventilation. Since $\dot{V}CO_2$,

---

*This section was written by Claes E. G. Lundgren.

expressed in liters BTPS (body temperature, ambient pressure, saturated with water vapor), is reduced by about half (disregarding a small bias due to constant water vapor pressure) with the doubled ambient pressure, the necessary 50% reduction of $F_{A}CO_2$ will be achieved as long as $\dot{V}_A$ (expressed in liters BTPS) remains the same. In a more generalized and well-known form:

$$\dot{V}_A = \frac{863 \times \dot{V}_{CO_2}}{P_{A}CO_2} \text{ or } P_{A}CO_2 = \frac{863 \times \dot{V}_{CO_2}}{\dot{V}_A}$$

where $\dot{V}_{CO_2}$ is expressed in l STPD and $P_{A}CO_2$ in mm Hg and 863 is a correction factor for the BTPS–STPD transformation. As can be seen in this expression, $P_{A}CO_2$ depends on metabolism and ventilation and is independent of depth.

The high gas density encountered in diving limits the ventilation that can be achieved. This is the case whether high density is caused primarily by pressure or by gas composition (110). The deterioration in maximal voluntary ventilation (MVV) with increasing depth and different gas compositions is shown in Figure 43.1. Relative to MVV at 1.0 atmosphere absolute (ATA) MVV at depth has been given as

$$MVV_{depth} = MVV_0 \times \rho^{-k}$$

where ρ is the gas density, k is a best-fit regression exponent (usually between 0 and 0.56), and MVV_0 is recorded at $\rho = 1.0$ (102). Maximal voluntary ventilation has been used as a predictor of ventilatory performance during maximal exertion. Thus exercise ventilation is expected to be a fraction, smaller than 1.0, of MVV. However, in the diving environment this relationship is variable. It appears that MVV at high gas densities is not limited by respiratory muscle performance; some studies have found ventilation levels during heavy exercise at (simulated) depth which were almost as high as the 15 s MVV measured at the same pressure (97, 116, 182) or even exceeded it (147). However, some reports claim that the relationship between exercise ventilation and MVV is the same as at the surface (5, 54, 61). The differences in observations may be due to dissimilar methodology. The idea that exercise ventilation approaches MVV at depth is supported by the notion that pulmonary gas flow may be limited by dynamic airway collapse, which would be enhanced by high gas density (155, 182). Maximal expiratory flow (at lung volumes larger than 25% of vital capacity) appears to be proportional to the -0.4 to -0.45 power of gas density (83, 84, 182). As effort-independent flow develops at depth, respiratory muscle performance will not set a limit on MVV or exercise ventilation.

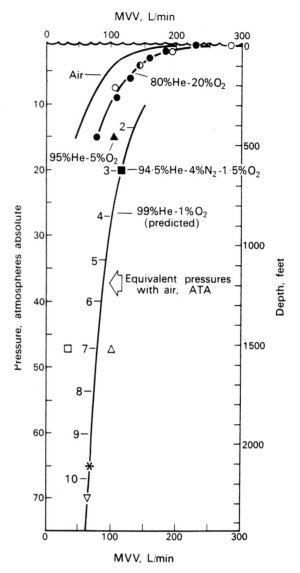

FIG. 43.1. Actual and predicted relationships between pressure (depth) and MVV while breathing air or various helium–oxygen mixtures. Numbers on curve to the right refer to air pressure in atm that were predicted to allow the same MVV achievable with helium–oxygen mixtures at the pressures given on the ordinate. Reproduced with permission from Lanphier and Camporesi (102).

CO₂ Elimination

Even with ample supply of breathing gas, CO_2 accumulation may be a problem in diving. The tendency for hypoventilation in divers has been linked both to factors in the environment and to traits unique to divers as a group. Some of these factors are illustrated in Figure 43.2. The first column illustrates that some divers breathing air, even at 1.0 ATA, showed a tendency for CO_2 accumulation. When pure oxygen was inhaled at 1.8 ATA there was a further increase in $P_{A}CO_2$, shown in the second column. This might have been a combined

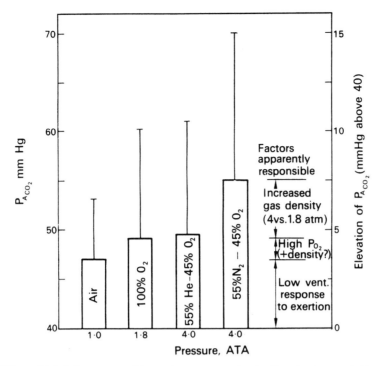

FIG. 43.2. Effects of inhaled gas composition and pressure on alveolar P_{CO_2} levels recorded by Lanphier at U. S. Navy Experimental Diving Unit. Bars indicate mean values; lines above bars show highest individual values. Total gas pressure at bottom of graph; gas composition within bars. Oxygen P_{O_2} was the same in the second, third, and fourth bars. Comparison of bars allows some conclusions (see text) about *Factors apparently responsible* for observed hypercapnia as tabulated in the figure. Reproduced with permission from Lanphier and Camporesi (102).

effect of oxygen per se and density. Pressure in itself had little effect when increased by adding He "on top" of the 1.8 ATA of O_2 pressure, as seen in the third column. With P_{O_2} still at 1.8 ATA and the addition of N_2, there was a further increase in P_{ACO_2}, depicted in the fourth column. The ventilatory depression caused by N_2 is apparently related to its density and not its narcotic action (107).

The phenomenon of CO_2 retention in divers has attracted interest since the 1950s. In an excellent review Lanphier and Camporesi (102) noted that individuals who showed a tendency for CO_2 accumulation were common among divers but were also found occasionally in the general population of athletic young subjects. Several studies have found that scuba (self-contained underwater breathing apparatus) divers breathing air usually show a tendency for increased steady-state alveolar CO_2 pressure (P_{ACO_2}) when exercising at both normal and increased ambient pressures (102). The question of what makes divers more prone to CO_2 accumulation than nondivers has been addressed in a few studies. Divers, ex-divers, and nondivers were monitored by Kerem et al. (92) during oxygen breathing at rest and during exercise at 1 atm. During moderate exer-

cise, pronounced hypoventilation and hypercapnia were observed in the divers and ex-divers alike, the end-tidal CO_2 pressures (P_{ETCO_2}) in both groups exceeding by 7–8 mm Hg (0.93–1.06 kPa) the normocapnic level exhibited by nondivers under the same conditions. When the subjects also performed breath-holds after the exercise, end-expiratory P_{CO_2} was 15 mm Hg (2.0 kPa) higher in divers than in nondivers. Superimposed inspired CO_2 loads caused similar elevations in P_{ETCO_2} in divers and nondivers. It was concluded that during oxygen breathing and exercise CO_2 retention in divers cannot be accounted for solely as a conditioned breathing behavior but may be due partly to a reduced central responsiveness to CO_2. Furthermore, CO_2 retention was apparently not dependent on current diving activity (92). However, the authors felt that the results of their study did not provide a method for reliable screening of diver candidates to identify potential CO_2 retainers.

In additional studies of respiratory control in divers Sherman et al. (136) recorded CO_2 sensitivity in terms of ventilatory response and occlusion pressure ($P_{0.1}$) during CO_2 challenge (rebreathing technique). Scuba divers of varying experience and diving activity were compared with nondivers. All the divers' values were in the lower range of nondiver control values, and about one-third of the divers were below the normal range (mean $\pm$ 2 SD) for CO_2 sensitivity. The authors concluded that their divers were a subgroup of normal healthy subjects with either an inherent or an acquired relatively low CO_2 response (136).

A striking feature of the divers' hypercapnia is that it may go unnoticed at depth while reaching levels that may incapacitate the diver (118, 169). The CO_2 effect is additive to nitrogen narcosis (82).

THE NEED FOR PRESSURE BALANCE

Adequate pressure is the requirement on gas supply to the diver that is the most difficult to satisfy. This quickly becomes evident if one dives while using the simplest of underwater breathing devices, the snorkel. An artificial extension of the diver's own airways, the snorkel typically consists of a tube 15–20 mm in diameter fitted with a mouthpiece at one end and curved to allow the other, open end to protrude above the water when the diver/swimmer assumes a prone posture.

The disadvantage of the snorkel is that (disregarding small pressure fluctuations due to breathing) the alveolar air communicating with the atmosphere via the tube remains at 1 atm of pressure regardless of the depth to which the diver descends. Thus increasing water depth will create a transmural pressure difference across the diver's respiratory organs so that the diver is exposed to negative pressure breathing (also known as negative static lung loading or SLL). The ability of human inspiratory muscles to overcome such pressure differences is quite limited: the maximal (static) inspiratory pressure is on the order of 100 cm H_2O (10 kPa) (130). Hence, the snorkel should only be used for surface swimming; trying to get any deeper rapidly becomes exhausting or even dangerous.

In contrast to the snorkel, other types of breathing gear are typically designed to provide breathing gas at the appropriate pressure, namely the pressure of the depth at which the diver is located. In the classical "hard hat" diving suit, air is pumped through a hose down to the diver (Fig. 43.3). The pressure of the respired air is automatically balanced to the water pressure at chest level since the upper part of the flexible suit is kept inflated. Excess air is bled off through a valve on the helmet that the diver can adjust. This equipment works well but carries the potential risk of so-called squeeze accidents. Should the air hose become severed at some location above the diver and, at the same time, there is a leak in the check valve at the air inlet to the helmet, the gear is transformed to a snorkel outfit in the sense that the gas pressure in the helmet and in the diver's lungs will fall to the pressure existing at the upper, sev-

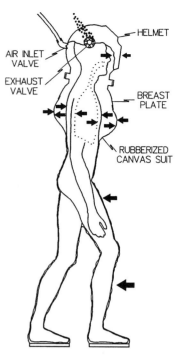

FIG. 43.3. Gas supply in a heavy (hard hat) diving suit. Size of the arrows illustrates the relative magnitude of water and air pressure. Air pressure in lungs is the same as air pressure in "bubble" surrounding the chest, which is determined by water pressure at chest level. Further details in text.

ered end of the hose. Not only does this make inspiration impossible even at very modest hose lengths but severe mechanical injury, known as whole-body squeeze, may result. If the hose length above the diver is, for example, 100 ft (pressure difference about 3 atm or 306 kPa), the force pushing the diver's body into the helmet would amount to about 11.8 kN (1,200 kgf, 2,646 lbf) assuming a 400 cm^2 large opening to the helmet. A squeeze mechanism may also develop if the diver suffers a sufficiently deep fall under water and the ambient pressure increase is not balanced rapidly enough by the flow of air into the suit.

The scuba breathing regulator (Fig. 43.4) consists essentially of a small tin from which the diver inhales through a tube. One side of the tin is formed by a rubber diaphragm, one side of which faces the water while the other side operates a compressed-air valve in the tin so that the gas pressure in the tin always equals the water pressure on the diaphragm. Since the regulator typically is connected to a mouthpiece or a breathing mask, the pressure of the inhaled air is the same as the water pressure at the mouth level. Depending on the diver's posture under water, this pressure does not necessarily match the water pressure on the chest (Fig. 43.5A); that is, a situation of static lung loading may exist. Other types of scuba, such as closed and semiclosed rebreathing gear with so-called counterlungs or breathing bags,

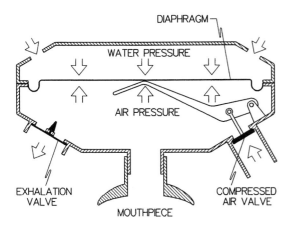

FIG. 43.4. Schematic of scuba breathing regulator. Breathing air is provided at the same pressure as water pressure acting on the membrane. A pressure differential (higher outside than inside), whether caused by inspiration or descent, will move the membrane inward and provide for inflow of compressed air.

may also expose the diver to static lung loads (Fig. 43.5B).

Respiratory Problems

For this discussion, the concept of the pressure centroid of the chest is useful. This is a location at which water pressure is a representative mean of the hydrostatic pressures at all other locations on the chest. From a practical point of view, the vertical location of the centroid relative to the chest may be found by putting a length of hose between a diver's breathing regulator and a mouthpiece and adjusting the regulator's position relative to the submerged diver's chest so that the expiratory reserve volume becomes normal. In the upright posture, the pressure centroid is in a plane about 14 cm below the sternal notch, and in the prone posture it is 7 cm above (more shallow than) the sternal angle. However,

there is not complete consensus as to whether a diver's breathing gas should be offered exactly at the centroid pressure (109a). Nonetheless, SLL by a mere -10 to -20 cm H_2O (-1 to -2 kPa) was connected with dyspnea in air-breathing subjects performing exercise in simulated dives to 190 ft (58 m, 6.76 atm) (85, 147). Under the same conditions, a SLL of 10 cm H_2O (1 kPa) was the least conducive to dyspnea. Remarkably, there were no differences in PETCO$_2$, with different SLL ranging from 20 to -30 cm H_2O (2 to -3 kPa) (85, 147). Dyspnea was uniformly described as difficulty to inspire and the same was the case with helium-breathing subjects at 43.4 atm (1,400 ft, 430 m) (54) and 49.5 ATA (1,600 ft, 490 m) (138). Inspiratory dyspnea is likely to be relieved by positive pressure breathing, which tends to aid inspiration. Such a beneficial effect has also been confirmed by Thalmann and Piantadosi (146).

In exercising subjects just below the surface, negative SLL was associated with less expiratory flow limitation during the prone posture than when upright (51). This contrasts with observations in subjects exposed to negative SLL during submersion at 190 ft (57 m, 6.76 atm) who suffered more dyspnea in the prone than in the upright posture (85, 147). This was proposed to be due to the lower pressure on the extrathoracic airways in the erect posture, a notion that gains credence from the observation by Flook and Fraser (63) of dynamic compression of the extrathoracic trachea during forced inspirations by subjects exposed to high gas densities at depth.

Cardiovascular Aspects

Cardiovascular injury or malfunction appears to be the main danger during snorkel breathing. Even the shallow immersion attained when sitting upright in water to the neck has profound circulatory effects, indicating that

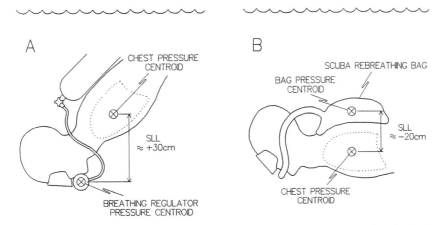

FIG. 43.5. Scuba diver with breathing regulator at mouth, experiencing positive static lung loading in the head-down position (A). Scuba diver with counterlung (breathing bag) on back is exposed to negative SLL in the prone position (B).

"deep" snorkel breathing may induce dangerously high cardiac preloads. Thus Arborelius et al. (6) recorded a redistribution of about 700 ml of blood into the chest in subjects undergoing head-out immersion. This was associated with extra systoles and an increase in mean right atrial pressure of 13 mm Hg (1.73 kPa). One subject exhibited right atrial pressure peaks of about 30 mm Hg (3.99 kPa) and several subjects showed pressure oscillations similar to the so-called W-pattern, which is typically seen in constrictive pericarditis with right heart engorgement. The deeper submersion made possible by a snorkel could probably lead to even more pronounced intrathoracic blood pooling and strain on the heart. One aspect of this would be an increase in the cardiac afterload on the heart as the pressure of the external water column is transmitted to the systemic circulation (100). In subjects lying recumbent at a depth of about 100 cm (3.3 ft) under water while snorkel breathing, arterial pressure increased slightly, exceeding the pressure of the water column. Lanari et al. (100) warned that if such a submersion were to be deep or long lasting, there might be a risk of pulmonary edema or rupture of bronchial vessels (the only systemic vessels unsupported by the external water column). The potential danger of the long snorkel is underscored by Stigler's 1911 report (140). He used submersion and snorkel breathing to determine the strength of the inspiratory muscles. Experimenting on himself he tried in vain for a few seconds to inhale through the tube while submerged with his chest at a depth of 2 m. Immediately after surfacing he suffered severe prostration and "delirium cordis" (atrial fibrillation) and was diagnosed with acute heart dilatation.

An increase in cardiac output may occur secondary to the redistribution of blood into the chest that is brought about by immersion and negative SLL. This increase has been given as about 30% (6, 108) or even 60% (62). The difference between these numbers appears to be due to differences in orthostatic load in the nonimmersed control situations. The hyperkinetic circulation induced by immersion may potentially impact on the morbidity of decompression sickness. Immersion increases the rate of whole-body nitrogen elimination during oxygen breathing (14) as well as radioactive xenon elimination from a muscle deposit (13). Immersion in warm water (37°C) has been employed during nitrogen washout in humans to reduce the incidence of decompression sickness during subsequent experimental decompression (11). However, this may to some extent have been a temperature effect (109).

Ventilation–Perfusion Matching

The effects of negative SLL during immersion extend to ventilation–perfusion matching. Functional residual capacity (FRC) is reduced due to a smaller expiratory reserve volume. The smaller FRC is conducive to airway closure, which may be enhanced by intrathoracic blood pooling (7). Two mechanisms have been proposed for this blood pooling effect: vascular engorgement of airway mucosa and increased lung weight (46). The tendency for airway closure is confirmed by an increased closing volume during immersion (32, 46). When the breathing medium is oxygen, this airway closure may cause absorption atelectases (12, 47). The above-mentioned mechanisms and relative hypoxia secondary to airway closure in dependent lung regions tend to shift ventilation as well as perfusion toward the apices of the lungs of erect immersed subjects (7, 128).

The influence of the changes in ventilation–perfusion distribution on gas exchange has been variously described as a drop in arterial P_{O_2} (43), no change (7), or slight reductions or increases by as much as 15 mm Hg (2 kPa) in the alveolar–arterial oxygen pressure difference ($P_{A-a_{O_2}}$) (129). The variation in $P_{A-a_{O_2}}$ depends on whether the subjects breathe so as to have an overlap between tidal volume and closing volume. With increasing age and obesity in an immersed subject, an increasing portion of the airways remains closed during the breath. The resulting shunting of blood flow (in terms of O_2 uptake) causes a drop in Pa_{O_2} (129). By contrast, in the absence of this overlap (large breaths, young individuals) between the tidal volume and the closing volume, oxygen exchange during immersion is dominated by improved diffusing capacity due to expansion of the pulmonary capillary bed (129) and/or hyperventilation relative to oxygen consumption (50). These changes may explain observations of increased Pa_{O_2} during immersion.

EXTERNAL BREATHING IMPEDIMENT

Dyspnea and CO_2 Exchange

In addition to the possible SLL, breathing gear also introduces more or less external breathing resistance (that is, flow resistance). Hence the question, what is an acceptable level of breathing resistance and what are the consequences of exceeding this level? Warkander et al. (170) exposed subjects in simulated, immersed dives at 1.45 atm (15 ft, 4.5 m, 147 kPa) and 6.8 atm (190 ft, 57 m, 675 kPa) to external breathing resistance while monitoring their respiratory performance and obtaining dyspnea scores. Three levels of resistance were used: the highest level was 8–12 cm H_2O (0.8–1.2 kPa)/liters/s at flow rates of 2–3 liters/s. With this resistance, the subjects who were breathing air and exercising at 60% of their maximal oxygen uptake generated mouth-pressure fluctuations of ±25 cm H_2O (2.5 kPa). A moderate 85% of the highest resistance and a low 65% were also used. The minimal (control) resistance was 2 cm H_2O

(0.2 kPa)/liters/s at 3 liters/s. In the six subjects who were studied, external breathing resistance induced varying degrees of hypoventilation and CO_2 retention as well as varying degrees of dyspnea, the two changes tending to be mutually exclusive and typical of individual subjects. The results of 96 experiments are shown in Figure 43.6, where high levels of end-tidal P_{CO_2}, despite some scatter, tend to be associated with the least dyspnea. By contrast, when the subject was able to maintain a relatively low end-tidal P_{CO_2}, this occurred at the cost of suffering more pronounced dyspnea. Both CO_2 retention and dyspnea increased with higher breathing resistance. Detailed deliberations about acceptable levels of breathing resistance are offered by

Warkander et al. (170) but are beyond the scope of this discussion. Briefly, considerations of safety and practicality inspired the authors to propose that the external resistance of divers' breathing gear (for air breathing at pressures between 1 and 6.8 atm or 101 and 690 kPa) should not impose respiratory work in excess of 1.5–2.0 J/liter in the ventilation range of 30 to 75 l (BTPS)/min. In addition to imposing flow resistance, breathing gear may be the source of elastic and inertial loading. The importance of such loads (of a realistic magnitude) for the diver's well-being and performance has not been determined.

BAROTRAUMA OF THE LUNG

As the ambient pressure decreases during ascent, the expanding lung air normally escapes through the airways. However, if the diver performs a breath-hold preventing excess air volume from escaping, the lung may become overdistended to the point of barotrauma, that is, disruption of lung tissue. Another injury mechanism involves air trapping in a small section of the lung. This may be due to a variety of abnormalities (for a review, see ref. 70), some of which may be hard to detect in medical screening of diver candidates. One possible abnormality is lower-than-normal static lung compliance with unevenly distributed lung elastance, as proposed by Colebatch et al. (44) who compared divers who had suffered pulmonary barotrauma with matched controls.

In barotrauma the air that penetrates the tissues may cause mediastinal and subcutaneous emphysema, pneumothorax, and/or arterial gas embolism. The incidence of pulmonary barotrauma in divers has been given at one per 19,800 dives (105) and the relative incidence of mediastinal emphysema relative to arterial embolism was reported to be 23/117 (104). Arterial gas embolism is potentially fatal, and prompt recompression treatment, the only remedy when symptoms from the central nervous system (CNS) develop, is usually very effective.

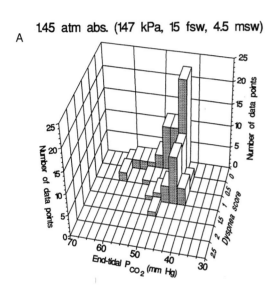

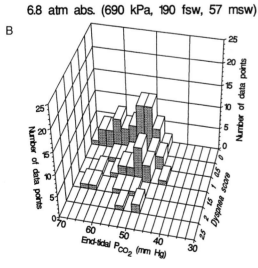

FIG. 43.6. Results of end-tidal P_{CO_2} recordings and dyspnea scoring (0 = no dyspnea, 2.5 very severe dyspnea) in subjects performing exercise under water at 147 kPa (15 ft, 4.5 m) (*A*) and 690 kPa (190 ft, 57 m) (*B*). Low dyspnea scores are associated with high CO_2 levels and vice versa. Reproduced with permission from Warkander et al. (170).

OXYGEN TOXICITY*

Oxygen breathing offers important physiological and operational advantages for diving. Because O_2 is metabolized, it probably does not contribute to decompression sickness (171). A high P_{O_2} also hastens decompression by increasing the gradient for inert gas removal from saturated tissues; thus recompression with O_2 is the treatment of choice for decompression sickness (153). Finally, in some military operations where

*This section was written by Andrea Harabin.

exhaled bubbles and decompression obligation cannot be tolerated, 100% O_2 is employed in a closed-circuit scuba.

The advantages of high P_{O_2} must be weighed against the inevitable toxicity resulting from prolonged exposure. The lung is a primary target, probably because it is exposed to the highest P_{O_2} of all vital organs. At 1 bar, most animals can survive for only about 3 days in 100% O_2. A P_{O_2} of 0.4–0.5 bar is generally considered safe for indefinite exposures.

At P_{O_2} greater than 3 bar in dry exposures and around 1.6 in immersed exposures, pulmonary toxicity occurs, but effects on the CNS limit exposures. Convulsions often occur. Both pulmonary and CNS symptoms of toxicity occur sooner at higher P_{O_2} and worsen with length of exposure.

This section will describe the pulmonary and CNS effects of O_2 breathing on humans, the limits of human tolerance to O_2 breathing, the modulating influences on development of toxicity, and finally, mechanisms.

Physiological Response to O_2 Breathing

The physiological response to O_2 breathing has been studied in humans breathing O_2 up to 3.5 bar (40, 55, 98, 99, 179). The effects are small and transient and appear more marked at higher P_{O_2}. They may be summarized as follows: there is an immediate increase in resting ventilation, which often results in a small, but significant decrease in Pa_{CO_2}. Small decrements in heart rate and cerebral blood flow and increased total peripheral resistance are also common. Retinal vessels constrict (132), and animal studies show that there may be vasoconstriction in other peripheral beds (especially muscle, liver, and kidney) (151, 166).

Some of these acute responses can be explained by effects on hemoglobin. Because arterial saturation is nearly complete on normal air breathing, the amount of O_2 in physical solution increases as PI_{O_2} is increased. At about 3 bar, theoretically, enough O_2 could be in physical solution that venous hemoglobin would remain fully saturated ($S\bar{v}_{O_2}$). $S\bar{v}_{O_2}$ has been shown to be elevated in humans breathing O_2 at this depth (98, 179), though not as much as predicted, probably because of pulmonary arterial–venous shunts and blood flow decrements. This observed decrease in unsaturated hemoglobin represents sufficient loss of buffering capacity to increase local brain hydrogen ion concentration and to stimulate ventilation. The small decrease in cerebral blood flow may also result from arterial hypocapnia rather than a direct constrictor effect of O_2 (99).

The response to exercise appears to be altered in 100% O_2. Lambertsen et al. (99) showed that Pa_{CO_2} was increased in subjects exercising in 2 bar of O_2 over

that attained with the same work load in air at the surface or under pressure.

Toxic Effects: Pulmonary

After about 6–14 h of exposure to 100% O_2 at 1 bar, humans begin to feel symptoms of pulmonary toxicity including tracheobronchial irritation, substernal pain, a burning sensation on inspiration, and dyspnea. These changes occur with shorter latency as P_{O_2} is raised and become more severe as exposure is lengthened. Many aspects of pulmonary function have been monitored in humans breathing O_2 for 6–74 h at pressures ranging from 0.8 to 3 bar and, in additional studies where exposure to P_{O_2} was only mildly elevated (PI_{O_2} from 0.23 to 0.47 bar), for up to 17 days (39, 74). Decrements in static and dynamic lung volumes and flows and diffusion characteristics have been observed. Decrements in vital capacity (% Δ VC) have been most widely studied and proposed as a predictive index for monitoring the onset, rate of development, and degree of severity of pulmonary toxicity (41). A quantitative analysis of the data (74) showed a relationship between duration and P_{O_2} of exposure such that % Δ VC can be calculated using the formula:

$$\% \ \Delta \ VC = -0.011 \cdot (P_{O_2} - 0.5) \cdot time$$

where P_{O_2} is in bar and time is in min. Average decrements from 4% to 10% have been considered tolerable for therapeutic O_2 usage. Wide variability in individual responses limits the usefulness of Δ VC as an index. This analysis also showed that there were measurable decrements in vital capacity even at a P_{O_2} of 0.4 bar. Results not included in this analysis demonstrate that pulmonary toxicity occurs in humans exposed to O_2 at 3 bar (39). Of practical consequence in scuba diving is the observation by Baer et al. (10) that immersion and O_2 breathing for only 0.5–2 h causes much larger reductions in vital capacity (>20%). All documented decrements in pulmonary function have been reversible. Recovery may be immediate or may take as long as several months.

Animals given long O_2 exposures at 1 bar demonstrate severe pulmonary histopathology (41, 45). First, endothelium is destroyed, platelets and then neutrophils infiltrate, and epithelial cells are damaged. Ultimately, atelectasis and interstitial and alveolar edema form. The processes are similar in all animal species, with mainly the time course varying. Primates may be more tolerant than other species; young animals are more tolerant than older ones. Inflammatory cells appear to augment rather than to cause this damage, though this notion is still somewhat controversial. Exposure to an FI_{O_2} as low as 60% O_2 at 1 bar for 7 days caused biochemical and physiological injury to the lungs of rats (45).

Although pulmonary pathology suggested that there might be progressive impairment in gas exchange, leading to severe terminal hypoxia and hypercapnia, studies in unanesthetized animals (73, 112) do not support this idea. Gas exchange, cardiac output, and pulmonary and arterial blood pressures are remarkably well maintained until a precipitous terminal response. Severe hypoxia is not a feature of the terminal response; in fact PaO_2 decreases by only 100–200 torr. Terminal acidosis develops, which may be respiratory, metabolic, or mixed, depending on the species. Impairment of respiratory control and chemoreflexes occurs following only long exposures (95, 150).

Toxic Effects: Central Nervous System

In scuba diving, as opposed to saturation exposures, oxygen breathing is generally limited by CNS effects rather than the pulmonary ones described above. Convulsion or loss of consciousness may occur, and common minor symptoms include nausea, twitching, paresthesias, and alterations in vision, hearing, and mood. Donald (52, 53) provided the first thorough investigation of the limits of human tolerance to hyperbaric O_2 (HBO). He demonstrated variability in the response to HBO: different subjects tolerate different amounts of time at a given PO_2 before developing symptoms of toxicity. Furthermore, he showed that individual subjects tolerated different amounts of time on different days. He constructed a dose–response relationship between PO_2 and onset of symptoms: earlier onset occurred at greater depths. Finally, he showed that simply immersing a subject decreased the average time and depth for toxicity and that exercise further augmented the risk of CNS O_2 toxicity in an immersed subject. The mechanisms responsible for exacerbation by immersion and exercise are still not well understood, though both might increase cerebral blood flow and O_2 delivery.

Because of the military's continuing interest in the use of 100% O_2 for treatments and certain underwater swimming applications, additional studies of human tolerance have been performed. A series of over 600 exposures was completed by Butler and Thalmann (36) at depths ranging from 1.6 to 2.5 bar for up to 4 h. Extended depth–time O_2 exposure limits based on these studies have been adopted (152). The same array of symptoms was noted as in Donald's studies, but toxic symptoms developed sooner and at more shallow depths in Donald's studies than in Butler and Thalmann's (36). It appears that the investigators of both studies were careful about achieving an FIO_2 near 1.0 and keeping CO_2 low. Carbon dioxide is known to exacerbate O_2 toxicity (41) and can accumulate if a closed-circuit breathing apparatus malfunctions. One difference between the two studies was the type of exer-

cise. Donald's subjects were performing vigorous arm exercise and Butler and Thalmann's were pedaling a bicycle (leg) ergometer. Submaximal arm exercise results in higher O_2 consumption, heart rate, ventilation, and blood pressure than leg exercise at the same power output (163).

As with pulmonary O_2 toxicity, it would be useful to have a noninvasive index of toxicity for the design of HBO treatments and usage. Less serious symptoms preceded convulsions only about half the time in the studies cited above. When a subject felt as if a convulsion was about to occur, removal of O_2 did not always prevent it. Lambertsen and co-workers (96) have studied cardiovascular, pulmonary, and brain functions in subjects exposed to 1.5 to 3.0 bar of O_2. With the exception of the pulmonary effects (39) and a progressive narrowing and recovery of visual field and electroretinogram, these human studies agree with animal studies in suggesting that subtle changes in physiological function are not detectable early enough in HBO exposures to predict more severe manifestations of toxicity.

Modulation of Toxicity

Intermittency. The most effective means of preventing or at least postponing O_2 toxicity is by interrupting exposure with short periods of air breathing. Intermittent exposure effectively slows pulmonary toxicity at 2 bar in humans (81) and postpones the development of convulsions and death in animals (75). There does not appear to be any optimal schedule for these interruptions; basically, shorter O_2 times and longer air times decrease the risk of toxicity. Thus other considerations must guide the choice of schedule.

Inert Gases. In humans, development of pulmonary O_2 toxicity appears to be unaffected by the presence of inert gas (55). Addition of 3 bar of helium or nitrogen to 5 bar O_2 shortened the latency to CNS toxicity in rats (31). The effect of inert gases in human CNS toxicity is not known, though this is a question of some interest for diving.

Mechanism of O_2 Toxicity

The biochemistry of O_2 toxicity has been well reviewed (68, 77, 88) and will be summarized here. Following in vitro exposures, enzymes (particularly sulfhydryl-containing ones) are inactivated, cellular and mitochondrial metabolisms are depressed, and lipids are peroxidized. Regulation of the inhibitory neurotransmitter γ-aminobutyric acid (GABA) is affected by HBO exposure in a manner consistent with a role in O_2-induced convulsions. Evidence suggests a primary role for O_2-derived free radicals. Progressive univalent reduction of molecular O_2 forms reactive intermediates: superoxide, peroxide, and hydroxyl radicals. It is now known that radical formation is a part of normal metabolism and that there is a system of defenses in place.

The role of free radicals in pulmonary and CNS O_2 toxicity has been suggested by two lines of study: (1) when O_2 levels are increased, increased production of partially reduced O_2 species, oxidized glutathione, lipid peroxides, and hydrogen peroxide has been shown in vitro and in intact in vivo systems; (2) manipulation of the enzyme system that metabolizes the partially reduced forms of molecular O_2 (superoxide dismutase, catalase, and glutathione peroxidase) affects O_2 tolerance. Animals with inherently high resistance to pulmonary O_2 toxicity have elevated lung levels of these enzymes. Young animals, shown to be more tolerant to O_2 toxicity, have an enhanced ability to induce these antioxidant enzymes. Treatments which increased brain enzyme levels postponed CNS toxicity in rats; conversely, depletion of glutathione enhanced toxicity (178) and short HBO exposures elevated lung enzymes (72). Efforts to protect animals by supplementing α-tocopherol, a natural antioxidant important in membrane maintenance, have generally failed, though it has been shown consistently that animals deficient in tocopherol develop more lipid peroxidation and O_2 toxicity.

NITROGEN NARCOSIS*

Signs and Symptoms

Exposure to compressed air at pressures above 4 bar (100 ft) induces, in humans and animals, signs and symptoms of intoxication or narcosis (23, 64). The effects are similar to those caused by alcoholic intoxication, hypoxia, or the early stages of anesthesia. Behnke et al. (16) summarized the narcosis as euphoria, retardation of higher mental processes, and impaired neuromuscular coordination. Memory is impaired and concentration is difficult, with perceptual narrowing. Intellectual functions are affected to a greater degree than manual functions.

The signs and symptoms show wide interpersonal variability and become increasingly worse the greater the depth or pressure. At 300 ft (91 m) the effects are severe and the diver may be incapable of any useful work. At still greater depths, bizarre signs and symptoms, such as catalepsy or aphasia and potential loss of consciousness, result (2).

Reduction of pressure results in an immediate return to normal, with perhaps an amnesia of what had occurred while narcotized.

*This section was written by Peter B. Bennett.

A number of factors will potentiate the severity of the narcosis, including hard work under water (2, 3), alcohol (89), apprehension and anxiety (49), and any conditions which result in a higher retained carbon dioxide tension (82).

Causes and Mechanisms

Inert Gas Theory. There are two basic theories accounting for compressed air intoxication: the inert gas theory and the carbon dioxide theory. The cause is now widely recognized, as first suggested by Behnke et al. (16, 17), to be due to the raised nitrogen partial pressure. The comparative narcotic potencies of inert gases in the rare gas series (for example, helium, neon, argon, krypton, and xenon) as well as nitrogen have been related to lipid solubility in accordance with the Meyer-Overton theory (115, 122) and with other physical constants (37). Thus argon is about twice as potent as nitrogen and helium 4.3 times less potent or only weakly narcotic. Substitution of helium–oxygen for nitrogen–oxygen in the gases breathed at pressure results in elimination of narcosis (17).

Carbon-Dioxide Theory. From time to time an alternative theory has been advanced, first attributed to Bean (15) and later supported by others (135, 154), that compressed air intoxication was due to carbon dioxide narcosis. This, it was suggested, resulted from the impairment of ventilation due to the increased density of the breathing gases and the relatively fast compressions to depth resulting in carbon dioxide retention.

However, arterial blood gas measurements in humans exposed to air or oxygen–helium at 286 ft (90 m) while performing psychomotor tests indicated a significant psychometric decrement only with air but with no significant increase in P_{aCO_2} (25).

Electrophysiological Mechanisms. Marshall and Fenn (111) showed that after 260 min exposures of frogs to nitrogen at 530 ft (162 m) or argon at 300 ft (91 m) it was possible to reversibly block reflex preparations but that helium had no effect. To cause a block in propagation along isolated peripheral nerve, extreme pressures, such as 314–344 bar (310–340 ATS) of argon, were required (37). These and other studies (8, 67, 103) pointed to the site of inert gas and nitrogen narcosis as central synapses, possibly at the ascending reticular activating system of the brain stem and cortex. This was further confirmed (20) by averaged auditory evoked potential responses (AER) from the cat cortex and reticular activating system at 373 ft (114 m). Additional studies with auditory and visual evoked potentials in divers reported a correlation between decrements in evoked potentials and psychometric performance while breathing compressed air at increased pressure but no effects with helium or neon (24, 94, 133). Ackles and Fowler (1), using visual evoked responses (VER) and AER with performance tests, compared argon with nitrogen exposure in divers. However, they found no differential effect of the two gases, though there was significantly more narcosis present with argon. The reasons for this discrepancy remain unclear.

Cellular and Membrane Mechanisms.

As with general anesthesia there have been many theories about mechanisms of narcosis at the molecular level. Mullins's (119) "critical volume of occupation" concept, involving occlusion of cell pores by inert gas molecules, was adopted by Miller et al. (117) in their "critical volume" hypothesis. This maintains that narcosis is the result of expansion of a hydrophobic region of neuronal synaptic membranes to a certain critical volume (0.4% above normal) as a result of adsorption of narcotic molecules (106).

As a result of criticisms (30, 66, 71), this theory was modified by Halsey et al. (71) to a "multi-site expansion" hypothesis, suggesting that more than one molecular site with differing physical properties may be involved. Further, they suggested that pressure reversal of the narcosis may not necessarily involve the reverse effect (constriction) at the same site. Later work by Bennett et al. (29) reported a fall in surface tension of a phospholipid monolayer, indicating a lateral expansion of membranes in the presence of high pressures of nitrogen, in accordance with Clements and Wilson (42). However, with helium there was the opposite effect of an increase in surface tension. The relevance of such changes to synaptic propagation has yet to be determined, but it is probable that it will involve the Ca^{2+}-mediated release of neurotransmitters at the presynaptic site.

Pharmacological Studies.

Prevention of nitrogen narcosis has commonly relied on substitution of nitrogen by helium or nitrogen–oxygen mixtures with less nitrogen and more oxygen. The use of pharmacological agents has received less interest (21). However, Bennett (18) compared eleven drugs on nitrogen narcosis and oxygen toxicity in rats and reported that a single drug could be effective in controlling both. Frenquel (alpha-4-piperydyl benzhydrol hydrochloride) was found to be an effective agent in both humans and rats (19). In addition, cationic detergents, such as stearylamine and cetyltrimethyl ammonium bromide, significantly prevented the depression of evoked potentials in rats caused by nitrogen narcosis (26). It was postulated that this was due to stabilization of neuronal membranes with prevention of an associated increased permeability of ions, which has also been invoked in the mechanism of nar-

cosis (27). Further, surface active agents, such as lithium and α-tocopherol, were found to be equally effective in preventing the loss of rat righting response due to narcosis (22, 28).

Adaptation

There is consensus that many individuals experience an adaptation to nitrogen narcosis with time at depth or repeated exposures. However, there is no objective support, suggesting that it is merely the result of psychological habituation (38, 93, 101). There have been attempts to lower the degree of narcosis on the basis of adaptation by saturating divers at 60 ft (19 m) and making excursions to 200, 250, or 300 ft (63, 78, 91 m). However, objective data again did not provide sufficient evidence for such an adaptation to nitrogen other than subjective sensations of improvement in some divers.

PHYSICS OF BUBBLE FORMATION*

The simple beauty of a bubble is belied by the complexity of describing the physics of its formation (59, 134). This section is restricted to the specific case of growth and shrinkage of gas bubbles in animal tissues, as an adjunct to understanding and controlling decompression sickness (see ref. 35).

It is generally accepted that gas bubbles that form in supersaturated body fluids are the fundamental cause of decompression sickness, but manifestations of the malady may be due to secondary effects of the bubbles on the sufferer's physiological or biochemical systems. Supersaturation occurs when a person goes from one environmental pressure to a lesser pressure, as when a diver returns from depth or an aviator or astronaut is subjected to altitude. Eventually excess gas in the tissues is carried by blood to the lungs for escape to the exterior. "Wash-in" and "wash-out" are usually assumed to approach equilibrium in an exponential manner, rapid at first with an ever slower tail, and relatively fast in tissues which have high perfusion with blood and in which solubility of inert gas is low. It was formerly hoped that decompression tables (instructions for ascent) could avoid bubbling by arranging rates of ascent that would allow the excess gas to wash out while still in dissolved form. Application to diving practice has indicated that this idea is overly simplistic; for example, techniques to detect gas emboli in veins have revealed that bubbles are often present when there are no manifestations of decompression sickness (139).

Supersaturation by itself does not necessarily cause extensive bubbling. When a can of carbonated beverage

---

*This section was written by Hugh D. Van Liew.

is opened, one expects a few bubbles to appear in the liquid while it is drunk from the can or poured into a glass. However, if the container is jarred by being dropped on the floor just before it is opened, the beverage in the opened container bubbles profusely. What happened when the can was jarred? A reasonable explanation is that many gaseous "nuclei" became distributed throughout the liquid. When the fluid was decompressed by opening the can, gas diffused from the supersaturated liquid into the many nuclei. The nuclei may have come from gas quantities that were entrapped by splashing within the unopened can, from small bubbles which broke loose from gas quantities that were sequestered somehow at the walls of the container, from nongaseous "incipient nuclei" free in the liquid that were potentiated by the impact, or from de novo generation by the impact. The carbonated beverage may be a model for events that happen to a much lesser degree in the body fluids of human beings and animals; there is no evidence that massive bubbling can be initiated by jarring the body, but exercise before decompression has been reported to increase decompression sickness incidence in humans (35) and movement of joints causes bubbles to form in certain animals (113).

How much gas can come from a given amount of tissue when bubbles form? Solubility of nitrogen in watery tissue is 0.00014 $ml_{gas}/(ml_{tissue}$ kPa$)^{-1}$. Upon decompression from 200 to 100 kPa [33 feet of seawater (FSW) to surface] or other 2/1 decompressions, the formation of bubbles may clear the excess gas from a volume of tissue or blood that is roughly 100 times the volume of the liberated gas, so the bubbles are about 1% of the volume of the tissue for a 2/1 decompression, or 4% for a 5/1 decompression.

Diffusion Equations

The two phenomena that are most decisive in the formation of persistent bubbles are diffusion of gas molecules and the action of surface tension. Equation 1 can be used to characterize gas diffusing through membranes and the exchanges of nitrogen-containing subcutaneous pockets in air-breathing rats (157).

$$dV_{N_2}/dt = (KA/L)(P_{tis}N_2 - P_{bub}N_2) \qquad (1)$$

In the equation, dV_{N_2}/dt is rate of nitrogen gas passing across the boundary, A is area of the exchange surface, L is path length for diffusion, K is a permeation coefficient related to the solubility and diffusivity of nitrogen in the boundary layer material, $P_{bub}N_2$ is partial pressure of nitrogen inside the bubble, and $P_{tis}N_2$ is partial pressure in the general surroundings of the bubble beyond the distance L, that is, beyond the range of the diffusion gradient. If partial pressure of nitrogen is greater outside

than inside, nitrogen will enter, causing bubble growth; the bigger the difference, the faster the rate of entrance.

A bubble cannot be made up of only one gas, except perhaps fleetingly. The main constituent is the inert gas in the person's breathing mixture, nitrogen in a person breathing air or helium in a person breathing helium–oxygen, but the metabolic gases (oxygen and carbon dioxide) are also represented. In what follows, it is assumed that air is being breathed, but the principles also apply to other breathing mixtures. Because each gas in the bubble is governed by an equation of the form of equation 1, the partial pressures of O_2, CO_2, and N_2 come as close as possible to their counterparts in the surroundings. After any change in conditions, such as compression, decompression, or change of breathing gas, there is a transient phase in which the various gases jockey for position to bring their respective $(P_{tis}-P_{bub})$ differences to a minimum, thus bringing the bubble to a new steady state of approximately constant composition (157).

Equation 1 is a relatively simple example of many possibilities for diffusion equations. All will contain terms related to the geometry of the gas phase and its surroundings and to the gas concentration gradient. Equation 1 is not adequate to describe small bubbles because of the implication that there is a fixed boundary layer or unstirred shell. The more complex equation 2 is appropriate for estimating diffusion in small spherical bubbles in animal tissues, which are close to the steady state of gas concentrations and at constant pressure (158). The dV/dt term has been replaced by a term related to the in situ size of the bubble; dR/dt is rate of change of bubble radius with time. Equation 2 reduces to equation 1 when applied to large bubbles and is very similar to equations used for bubbles in vitro (59).

$$dR/dt = K'(\lambda + 1/R)(P_{tis}N_2 - P_{bub}N_2)/P_{bub}N_2 \qquad (2)$$

K' is a permeation coefficient which is slightly different from K in equation 1 and λ is blood perfusion in tissue around the bubble.

Surface Tension

Action of surface tension can be understood in the context of equation 2 by noting that surface forces exert a pressure on the gases inside a spherical bubble according to the equation of Laplace: $P = 2\gamma/R$, where P is pressure inside a sphere, γ is surface tension, and R is radius. The total pressure inside the bubble, P_{tot}, is higher than ambient pressure, P_B:

$$P_{tot} = P_B + 2\gamma/R \qquad (3)$$

The high P_{tot} elevates the partial pressure of inert gas inside the bubble ($P_{bub}N_2$ in equation 2). A simplistic interpretation of the Laplace equation is that free bubbles cannot begin to form because surface tension will immediately shrink them to extinction. However, bubbles are found in the body after decompression. One way around this logical problem has been to posit that the body contains gas nuclei, bubble formation centers, or gas nucleation processes; according to this idea, the body contains small gaseous entities which either persist chronically or are generated occasionally.

The simplest idea of a nucleus is that it is a small gas quantity in the body which could grow into a bubble if the appropriate conditions prevail. Investigators have postulated that stabilized nuclei are free in body fluids (184), in crevices within the tissues (for example, ref. 149), or at boundaries between dissimilar tissues (35). Possibly some sort of structure which is ordinarily nongaseous plays a role in bubble formation. Tensile strength of pure water is enormous (formation of vapor bubbles is said to require 140,000 kPa [186], leading to the belief that bubbles are unlikely to form de novo in the body and supporting the idea that there are stable nuclei. This may be a misconception, however, because bubbles in the body are not pure water vapor. Free energy concepts suggest the finite possibility that bubbles may form occasionally when molecules of dissolved inert gas aggregate, perhaps near hydrophobic surfaces (125, 173).

After decompression, the action of surface tension involves a positive feedback loop, illustrated for an air-breathing person in Figure 43.7. Consider a small bubble or nucleus which has a high P_{tot} in accordance with equation 3. If it is in a supersaturated tissue or blood and the $P_{tis}N_2$ becomes slightly higher than $P_{bub}N_2$ (box 1 of Fig. 43.7), a small amount of N_2 will diffuse (box 2), according to equation 2. This increases the radius slightly (box 3); pressure due to surface tension (box 4) falls, as predicted by equation 3 and total pressure falls (box 5), with a consequent fall of P_{N_2} (box 6) and an

increase of the diffusion gradient (box 1), thus continuing the self-propagating feedback until the radius is large enough that surface tension pressure (box 4) no longer contributes significantly to the total pressure in the bubble.

Other phenomena can act on the feedback loop. For example, some sort of physical influence could temporarily increase the radius of the gas phase (box 3) or metabolically produced CO_2 could enter the gas phase; this would lower P_{N_2} inside (box 6) as well as increase the radius (box 3). The loop works in reverse during bubble absorption (envision the dark arrows reversed); when the bubble is small enough for surface tension to increase pressure inside, the gas is absorbed more and more rapidly.

Growth and Decay of a Preformed Bubble

Figure 43.8 illustrates the time courses of factors which affect a preexisting bubble in a person who ascends from 2 atm to normal pressure (200 to 100 kPa). The figure was generated on a personal computer with a numerical integration technique which uses a system of equations, including equations 2 and 3, to account for the interactions of many of the influences which cause a bubble to grow or shrink (156).

The traces of four variables in Figure 43.8 will be described in descending order from where they start at the top left of the figure. Consider first the broken trace for ambient pressure (P_B). At 5 min, ambient pressure decreases from 200 to 100 kPa at a rate of 200 kPa/min, close to the standard ascent rate for U.S. Navy diving tables (60 FSW/min = 184 kPa/min); on the time scale of the figure, the fall of P_B appears to be almost instantaneous. Partial pressure of N_2 inside the bubble falls synchronously with ambient pressure (see the $P_{bub}N_2$ trace). The sum of partial pressures inside a bub-

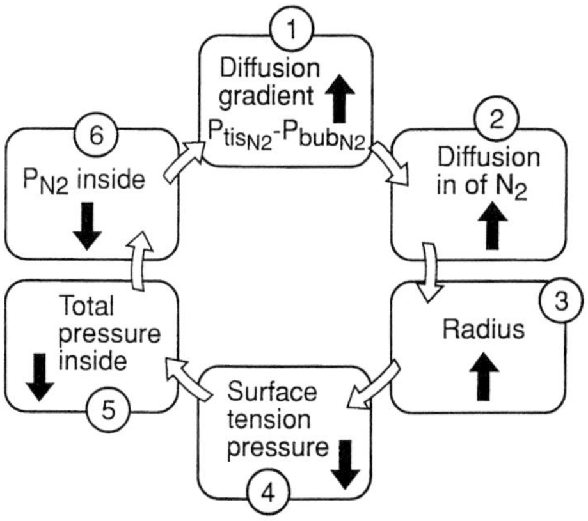

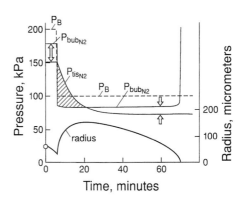

FIG. 43.7. Feedback loop due to interaction between surface tension and gas diffusion. Action of the loop causes explosive growth of small bubbles. Reproduced from Van Liew (156), with permission.

FIG. 43.8. Simulation of the time course of variables when a preexisting bubble in the body is subjected to decompression. It is assumed that surface tension and barometric pressure are the only physical forces acting on the bubble. Redrawn from Van Liew (156), with permission.

ble this large is assumed to be equal to ambient pressure; the $P_{bub}N_2$ trace is a little lower than the P_B trace because the bubble also contains water vapor plus partial pressures of O_2 and CO_2 that are very near tissue levels. The $P_{bub}N_2$ trace is above the $P_{tis}N_2$ trace at the start. In living animal tissues, oxygen consumption causes a deficit of oxygen partial pressure, which is incompletely balanced by production of CO_2. Because O_2 and CO_2 permeate very rapidly, their tissue levels tend to set the partial pressure of inert gas in a bubble to a level that is greater than in the surrounding tissues and blood (87, 157). This is known as inherent unsaturation or the oxygen window. Magnitude of the oxygen window is indicated in Figure 43.8 by the open arrows before and after decompression.

The hatching in the figure shows what can be called a "crossover area," where partial pressure of N_2 in the tissue ($P_{tis}N_2$) is temporarily higher than partial pressure of N_2 in the bubble; tissue N_2 must await wash-out by the circulation. Consider the trace for radius of the bubble (arbitrarily started at 60 μm). At first the bubble shrinks to 30 μm due to the oxygen window. If the ascent had been slightly later, the bubble would have ceased to exist. When ascent occurs, the radius increases immediately because of gas expansion by Boyle's law. This doubles the volume and increases the radius from 30 to 38 μm, but the crossover of N_2 causes the bubble to grow by inward diffusion of N_2 until radius becomes 150 μm. The bubble shrinks again when tissue PN_2 becomes lower than bubble PN_2 (after 20 min). At 70 min, radius is small, so surface tension causes a precipitous rise in $P_{bub}N_2$, seen as an almost vertical trace at the right of the figure.

For illustrative purposes, Figure 43.8 was drawn for a tissue having rapid wash-out of dissolved N_2. A tissue with slow wash-out would have a much slower decline of $P_{tis}N_2$ than the 5 min halftime for wash-out in Figure 43.8, the hatched bubble-growth area would tend to

extend as a rectangle from the time of decompression toward the right side of the figure, and the bubble would grow for a much longer time than shown. Note also that it is assumed that wash-in of N_2 into this particular tissue had been completed before decompression; slower tissues might not be saturated with N_2.

When a person breathes pure oxygen, the tissues and blood become denitrogenated, so absorption rate is maximal. This emphasizes the value of oxygen breathing, which can cause bubbles to shrink over 10 times more rapidly than with air breathing (157). For treating decompression sickness, there is a choice of making bubbles small by compression and thereby perhaps causing immediate relief of symptoms, of administering oxygen to hasten absorption of the bubbles, or of combining the two kinds of treatment.

Nucleus to Bubble

Figure 43.9 illustrates the generation of one stable, large bubble from a very small bubble or nucleus for the same decompression as in Figure 43.8. Before ascent, a large bubble, in which the effect of surface tension is negligible, would have an internal PN_2 of about 180 kPa (short broken line segment at the extreme left of the figure), whereas the particular bubble size chosen for this case (1.58 μm radius) would have $P_{bub}N_2$ of 240 kPa, as shown; the difference is caused by the high surface tension pressure of the small bubble. For this simulation, it is necessary to forbid the small bubble to shrink; a real bubble would be extinguished rapidly, as seen in Figure 43.8. At the arrow, a crossover area, which is too small to be seen, initiates action of the feedback loop shown in Figure 43.7: surface tension pressure decreases and PN_2 inside the bubble falls (from the level of the arrow to 90 kPa). Almost immediately, the total pressure in the gas phase (box 5 in Fig. 43.7) is well below that in the tissue, so the bubble grows rapidly.

When initial bubble radius was set at 1.57 μm, just one one-hundredth of a micrometer smaller than in Figure 43.9, a crossover did not occur and the bubble did not grow. After ascent, the PN_2 in the small bubble was about 150 kPa, less than the 240 kPa it was before ascent but more than the 90 kPa of a large bubble. There may be a range of sizes of gaseous nuclei in tissues (141) or of sizes generated by nucleation processes; if so, larger nuclei will be expected to grow into stable bubbles during decompression and smaller ones will not.

DECOMPRESSION*

Decompression sickness (or caisson disease) is a multisystem disease that occurs when an individual is sub-

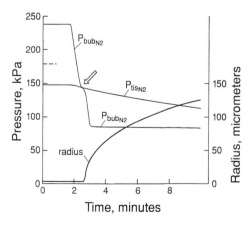

FIG. 43.9. Simulation of the consequences of decompression for a very small bubble or "nucleus." Redrawn from Van Liew (156), with permission.

*This section was written by Edward D. Thalmann.

jected to a too rapid reduction in ambient pressure. The exact mechanisms leading to symptoms of decompression sickness are unknown, but the preponderance of studies point to expanding gas bubbles as the cause (87). The previous section, PHYSICS OF BUBBLE FORMATION, addressed the gas-phase growth phenomenon. This section provides an overview of decompression sickness and the development of procedures to avoid the disease. There is a considerable volume of literature on this subject, and an in-depth review would consume considerably more space than allotted here. The referenced literature is extensive and current, and with this section as a guide, the interested reader may delve as deeply into the subject as he or she wishes.

Those at risk of decompression sickness include aviators or astronauts, who may suddenly undergo an ambient pressure reduction, as well as diving or compressed air caisson workers, who, after spending time at increased ambient pressures, undergo a pressure reduction during the return to atmospheric pressure. During the pressure reduction that occurs with ascent, a point will be reached where dissolved gas tension in body tissue exceeds ambient pressure, a condition known as supersaturation. This supersaturation will increase the gradient for gas elimination by the circulation, as well as the gradient for driving dissolved gas into any gas phases that may exist in the tissue, causing them to grow. The presumed sequence of events was covered in the preceding section. These initial processes are thought to be purely physical–chemical, but once supersaturation-driven gas-phase growth begins, cellular mechanisms are probably triggered that eventually lead to symptoms of decompression sickness (joint pain, lethargy, paresthesia, paresis) (58, 70). One could ascend slowly enough so that gas wash-out keeps up with the ambient pressure reduction and supersaturation or excessive gas-phase growth does not occur, but these ascent rates are so slow as to make them impractical. Experience has shown that faster rates are possible and that some degree of supersaturation may be safely tolerated. The goal of decompression procedures is to provide ascent (or decompression) profiles that avoid symptoms of decompression sickness while minimizing the time it takes to return to the surface after a pressure exposure.

As an occupational disease, decompression sickness was an insignificant problem until Triger perfected the compressed air caisson in 1839 (165) and Seibe perfected the closed diving dress in 1837 (9). This resulted in a surge of exposures to increased ambient pressure. The first physiological descriptions were reported by Trouessart in 1845 (137), and the pathological effects were described in 1854 by Pol and Watelle (126). Paul Bert is credited with consolidating these observations into a description of the clinical syndrome of decompression sickness in 1878 (165). Keays (91) firmly estab-

lished the value of recompressing divers (that is, returning them to increased ambient pressure) as a treatment for decompression sickness.

Initially, decompression profiles were determined by the personal experience and judgment of those in charge of decompressing caisson workers or divers. In 1906 the British Royal Navy commissioned Haldane to come up with a systematic approach for decompressing divers, resulting in the first set of standardized decompression tables (33). Later decompression tables were computed using mathematical concepts based on those developed by Haldane (80, 142–144, 183) as well as newer approaches (123, 159, 161, 174, 180).

It must be stressed that the presumed sequence of events leading to decompression sickness is a supposition; no direct observations linking cause and effect have been made. Gas-phase formation in tissue was first observed in the eye of a pit viper exposed to pressure reduction in a vacuum chamber by Robert Boyle in 1670 (34). More recent observations have shown gas-phase formation to have occurred postmortem in spinal cords of animals suffering signs of decompression sickness (65). These, coupled with observations of intravenous gas phases detected ultrasonically in humans following dives (127), have confirmed earlier studies pointing to gas-phase formation as the likely initiating event in decompression sickness.

Calculation of decompression tables has centered around the development of mathematical models which generally have two parts: gas kinetics and ascent criteria. Gas kinetics is a euphemism for a time-dependent function that describes how one accumulates and dissipates a decompression "dose." In decompression models, this dose is usually the calculated amount of gas that is dissolved in body tissue at any time during a dive. Ascent criteria describe the conditions under which a certain decompression dose will lead to symptoms of decompression sickness. The above section, PHYSICS OF BUBBLE FORMATION, gives a set of equations that describe how bubbles grow and decay with changes in ambient pressure. The dose would be the amount of dissolved gas present before bubble growth occurs. A decompression model incorporating this concept may use as its ascent criteria a function that relates bubble volume during ascent to the likelihood of decompression sickness occurring (161).

The fundamental issue in the physiological approach to modeling is whether gas-phase formation can be avoided altogether or whether gas nuclei are always present and their growth can only be contained. The former approach was used in early models that described gas kinetics as perfusion-limited, using exponential kinetics (183), or as diffusion-limited, describing diffusion into an infinite slab of tissue (79, 80). The dose was tissue-gas tension and ascent criteria were constraints on the tissue-gas tension to some maximal value, usually depth-dependent. The presumption was that as long as this maximal supersaturation was never exceeded, gas formation would not occur and decompression sickness would be avoided. Another approach assumed that gas nuclei were always present and that growth of the gas phase always took place, even at modest levels of supersaturation, which resulted in a significant asymmetry between gas uptake and elimination (142). Ascent criteria were a set of constraints on total gas phase volume that, if exceeded, would result in decompression sickness.

The idea of a preexisting gas phase was fostered by observations in shrimp (48, 60) and liquid-breathing rats (160) that a large but brief increase in ambient pressure (spike) just before decompression reduced the occurrence of observable bubbles or symptoms of decompression sickness. Since these animals had no gas reservoirs (liquid-breathing rats had their lungs filled with and were completely immersed in a fluorocarbon liquid with a high oxygen solubility), the predecompression pressure spike was purely hydrostatic and no increase in tissue-gas tension occurred. However, it was postulated that during the spike, preexisting gas nuclei were rapidly compressed and forced into solution and were then unavailable for growth during the subsequent decompression, thus reducing the number of symptoms. These observations led to more modern models that used sophisticated mathematics to describe bubble growth or gas nuclei formation and elimination (159, 161, 180, 181). One of the problems with postulating preexisting gas nuclei is that spherical bubbles will eventually collapse on their own as they are forced back into solution because of surface tension effects. However, various mechanisms for stabilizing gas nuclei in tissue have been proposed (149).

The goal of the physiological approach to decompression modeling of developing a model from "first principle" has not been realized. The lack of consensus on the exact events leading from tissue supersaturation to symptoms of decompression sickness has resulted in the wide variety of approaches used in developing decompression models. In fact, all current models have several adjustable parameters for which no a priori values can be assigned. Trials of many human exposures have to be conducted and model parameters adjusted until safe profiles are computed based on the observed incidence of decompression sickness. Little statistical confidence can be gained from this approach since each profile tested has to be considered independently and the expected incidence of decompression sickness computed from the binomial distribution based only on the number of replicated exposures for each profile. Time and expense have restricted the numbers of such exposures to as few as four and as many as 30–40, resulting in substantial confidence bounds even when no decompression sickness occurred (145). While this approach has produced useful decompression procedures, the risk of decompression sickness may vary widely from table to table.

Both of these shortcomings have been addressed in the probabilistic approach to decompression sickness. This approach uses the principle of failure time analysis (90) or survival analysis (57) to pool nonreplicated dive data from a variety of sources to allow procedures with higher levels of statistical confidence to be developed (174). A time-dependent risk function is postulated that presumably describes the rate at which decompression sickness will occur at any time. In its initial implementation this risk function is related to the amount of supersaturation present in tissue at any time. The tenets of survival analysis provide a formal mathematical relationship between this risk function and the probability of decompression sickness occurring. Unknown parameter values in the risk function (for example, exponential time constants, thresholds, weighing factors, etc.) are determined by the method of maximum likelihood to fit the risk function to a data base of dive profiles in which the outcome (decompression sickness or not) is known for each diver (174, 176) (for a review of the technique applied to diving, see ref. 161).

Using the probabilistic approach, not only has there been initial success in actually estimating the probability of decompression sickness from a particular dive profile (172, 177) but estimates of the most likely times for symptom occurrence now seem possible (176). The technique allows computation of decompression profiles to a specified level of risk. Another feature of the survival analysis approach is that objective measures of the comparative success of different decompression models in predicting risk are possible (175), which has resulted in demonstrating the superiority of asymmetrical over symmetrical gas kinetics in describing the risk of decompression sickness (123).

While decompression models do not necessarily reflect actual physiological processes at the cellular level, essential features of the more successful models (that is, symmetrical or asymmetrical gas uptake or elimination, predicted time distribution of instantaneous risk) may give clues to underlying physiological mechanism (that is, how gas nuclei may be stabilized, whether gas kinetics is perfusion- or diffusion-limited).

Closing in on the specific physiological process involved in decompression sickness is similar to the way in which physicists closed in on atomic structure. Basic research describing individual mechanisms, such as inert gas exchange (121) on the one side and more macroscopic phenomenological descriptions on the other, may ultimately converge on an accurate, quantitative description of the physiological interactions involved. While driven by commercial or military diving needs, research on decompression sickness has application to understanding the basic mechanisms of gas exchange at the macroscopic and the microscopic levels, and concepts can be applied equally well to describing the kinetics involved in tissue uptake and elimination of anesthetic gases and toxic inhalants.

True adaptation of humans to decompression has probably not occurred, but observations in caisson workers that the incidence of decompression sickness decreased with increased numbers of exposures (4, 124, 162, 164) have raised the possibility that some sort of acclimatization does take place. If exposures were stopped for several days, the incidence again increased during the first exposures after returning to work (69, 78, 114, 131).

In divers, apparent acclimatization to decompression sickness has been reported anecdotally for years, but hard evidence remains elusive (56). Most of the documented evidence in diving has come from helium–oxygen breathing gas decompression model validation trials (78, 148) supporting the concept of "working up" a diver, that is, conducting several mild pressure exposures to decrease the risk of decompression sickness on a subsequent severe exposure. One possible mechanism for acclimatization is that frequent exposures gradually deplete the number of gas nuclei in the body that could grow and cause symptoms (76, 185). Another mechanism that has been proposed involves changes in passive tissue relaxation (86). Still another explanation comes from the hypothesis that symptoms of decompression sickness result from gas-phase activation of the complement system (167) and that repeated exposures cause gradual depletion of complement, resulting in a lower symptom occurrence (168). Whether or not acclimatization occurs has been mainly of academic interest and has not been widely investigated as a practical means of avoiding decompression sickness. However, manned trials in which decompression procedures are being tested are careful to consider the possibility of acclimatization (145) and allow a minimum of several days to elapse between exposures. The reason is that if acclimatization does occur, decompression tables might be found safe based on manned trials where there are frequent pressure exposures, then a higher incidence of decompression sickness may occur when later used by individuals not having had recent pressure exposures.

REFERENCES

1. Ackles, K. N., and B. Fowler. Cortical evoked response and inert gas narcosis in man. *Aerospace Med.* 41: 1184, 1971.
2. Adolfson, J. Deterioration of mental and motor functions in hyperbaric air. *Scand. J. Psychol.* 6: 26–31, 1965.
3. Adolfson, J., and A. Muren. Air breathing at 13 atmospheres. Psychological and physiological observations. *Sartryck Forsvarsmed.* 1: 31–37, 1965.
4. Aldrich, C. J. Compressed-air illness, caisson disease. *Int. Clin.* 10: 73–88, 1900.
5. Anthonisen, N. R., G. Utz, M. H. Kruger, and J. S. Urbanetti. Exercise tolerance at 4 and 6 ATA. *Undersea Biomed. Res.* 3: 95–102, 1976.
6. Arborelius, M., Jr., U. I. Balldin, B. Lilja, and C. E. G. Lundgren. Hemodynamic changes in man during immersion with the head above water. *Aerospace Med.* 43: 592–598, 1972.
7. Arborelius, M., Jr., U. I. Balldin, B. Lilja, and C. E. G. Lundgren. Regional lung function in man during immersion with the head above water. *Aerospace Med.* 43: 701–707, 1972.
8. Arduini, A., and M. G. Arduini. Effect of drugs and metabolic alterations on brain stem arousal mechanism. *J. Pharmacol.* 110: 76–85, 1954.
9. Bachrach, A. J. A short history of man in the sea. In: *The Physiology of Diving and Compressed Air Work* (3rd ed.), edited by P. B. Bennett and D. H. Elliott. London: Bailliere Tindall, 1982, p. 1–14.
10. Baer, R., G. O. Dahlbäck, and V. I. Balldin. Pulmonary mechanics and atelectasis during immersion in O_2-breathing subjects. *Undersea Biomed. Res.* 14: 229–240, 1987.
11. Balldin, U. I. The preventive effect of denitrogenation during warm water immersion on decompression sickness in man. Proc. First Annual Scientific Meeting Eur. Undersea Biomed. Soc., Stockholm 1973. *Försvarsmedicin* 9: 239–243, 1973.
12. Balldin, U. I., G. O. Dahlbäck, and C. E. G. Lundgren. Changes in vital capacity produced by oxygen breathing during immersion with the head above water. *Aerospace Med.* 42: 384–387, 1971.
13. Balldin, U. I., C. E. G. Lundgren, J. Lundwall, and S. Mellander. Changes in the elimination of ^{133}xenon from the anterior tibial muscle in man induced by immersion in water and by shifts in body position. *Aerospace Med.* 42: 489–493, 1971.
14. Balldin, U. I., and C. E. G. Lundgren. Effects of immersion with the head above water on tissue nitrogen elimination in man. *Aerospace Med.* 42: 1101–1108, 1972.
15. Bean, J. W. Tensional changes of alveolar gas in reactions to rapid compression and decompression and question of nitrogen narcosis. *Am. J. Physiol.* 161: 417–425, 1950.
16. Behnke, A. R., R. M. Thomson, and E. P. Motley. The psychologic effects from breathing air at 4 atmospheres pressure. *Am. J. Physiol.* 112: 554–558, 1935.
17. Behnke, A. R., and O. K. Yarbrough. Respiratory resistance, and oil–water solubility and mental effects of argon compared with helium and nitrogen. *Am. J. Physiol.* 126: 409–415, 1939.
18. Bennett, P. B. Comparison of the effects of drugs on nitrogen and oxygen toxicity in rats. *Life Sci.* 12: 721–727, 1962.
19. Bennett, P. B. Prevention in rats of the narcosis produced by inert gases at high pressures. *Am. J. Physiol.* 305: 1013–1018, 1963.
20. Bennett, P. B. The effects of high pressures of inert gases on auditory evoked potentials in cat cortex and reticular formation. *Electroencephalogr. Clin. Neurophysiol.* 17: 388–397, 1964.
21. Bennett, P. B. Review of protective pharmacological agents in diving. *Aerospace Med.* 43: 184–192, 1972.
22. Bennett, P. B. Pharmacological effects of inert gases and hydrogen. In: *Underwater Physiology*, edited by C. J. Lambertsen. Bethesda, MD: Am. Soc. Exp. Biol., 1975.
23. Bennett, P. B. Inert gas narcosis. In: *The Physiology and Medicine of Diving* (4th ed.), edited by P. B. Bennett and D. H. Elliott. London: Saunders, 1993.
24. Bennett, P. B., K. N. Ackles, and V. J. Cripps. Effects of hyperbaric nitrogen and oxygen on auditory evoked responses in man. *Aerospace Med.* 40: 521–525, 1969.

25. Bennett, P. B., and G. D. Blenkarn. Arterial blood gases in man during inert gas narcosis. *J. Appl. Physiol.* 36: 45–48, 1974.

26. Bennett, P. B., and A. N. Dossett. Mechanisms and prevention of inert gas narcosis and anesthesia. *Nature* 228: 1317–1318, 1970.

27. Bennett, P. B., and A. J. Hayward. Electroencephalographic and other changes induced by high partial pressures of nitrogen. *Electroencephalogr. Clin. Neurophysiol.* 13: 91–98, 1961.

28. Bennett, P. B., B. Leventhal, R. Coggin, and L. Racanska. Lithium effects: protection against nitrogen narcosis potentiation of HPNS. *Undersea Biomed Res.* 7: 11–16, 1980.

29. Bennett, P. B., D. Papahadjopoulos, and A. D. Bangham. The effect of raised pressures of inert gases on phospholipid model membranes. *Life Sci.* 6: 2527–2533, 1967.

30. Bennett, P. B., S. Simon, and Y. Katz. High pressures of inert gases and anesthesia mechanisms. In: *Molecular Mechanisms of Anesthesia,* edited by B. R. Fink. New York: Raven, 1975, vol. 1, p. 367–403.

31. Bitterman, N., A. Laor, and Y. Melamed. CNS O_2 toxicity in O_2-inert gas mixtures. *Undersea Biomed. Res.* 14: 477–483, 1987.

32. Bondi, K. R., J. Murray Young, R. M. Bennett, and M. E. Bradley. Closing volumes in man immersed to the neck in water. *J. Appl. Physiol.* 40: 736–740, 1976.

33. Boycott, A. E., G. C. C. Damant, and J. S. Haldane. The prevention of compressed air illness. *J. Hyg.* 8: 342–443, 1908.

34. Boyle, R. Continuation of the observations concerning respiration. *Phil. Trans.* 5: 2035–2056, 1670.

35. Brubakk, A. O., B. B. Hemmingsen, and G. Sundnes (Eds). *Supersaturation and Bubble Formation in Fluids and Organisms.* Trondheim, Norway: Tapir, 1989.

36. Butler, F. K., and E. D. Thalmann. CNS O_2 toxicity in closed circuit scuba divers II. *Undersea Biomed. Res.* 13: 193–223, 1986.

37. Carpenter, F. G. Anesthetic action of inert gases on the central nervous system in mice. *Am. J. Physiol.* 172: 471–474, 1953.

38. Case, E. M., and J. B. S. Haldane. Human physiology under high pressure. *J. Hyg.* 41: 225–249, 1941.

39. Clark, J. M., R. M. Jackson, C. J. Lambertsen, R. Gelfand, W. D. B. Hiller, and M. Unger. Pulmonary function in men after O_2 breathing at 3 ATA for 3.5 h. *J. Appl. Physiol.* 71: 878–885, 1991.

40. Clark, J. M., and C. J. Lambertsen. Rate of development of pulmonary O_2 toxicity in man during O_2 breathing at 2 ATA. *J. Appl. Physiol.* 30: 739–752, 1971.

41. Clark, J. M., and C. J. Lambertsen. Pulmonary oxygen toxicity: a review. *Pharmacol. Rev.* 23: 37–133, 1971.

42. Clements, J. A., and K. M. Wilson. The affinity of narcotic agents for interfacial films. *Proc. Natl. Acad. Sci. USA* 48: 1008–1014, 1962.

43. Cohen, R., W. H. Bell, H. A. Saltzman, and J. A. Kylstra. Alveolar–arterial oxygen pressure difference in man immersed up to the neck in water. *J. Appl. Physiol.* 30: 720–723, 1971.

44. Colebatch, H. J. H., M. M. Smith, and C. K. Y. Ng. Increased elastic recoil as a determinant of pulmonary barotrauma in divers. *Respir. Physiol.* 26: 55–64, 1976.

45. Crapo, J. D. Morphologic changes in pulmonary oxygen toxicity. *Annu. Rev. Physiol.* 48: 721–731, 1986.

46. Dahlbäck, G. O. Lung Mechanics During Immersion in Water. Lund, Sweden: Univ. of Lund, 1978, Thesis.

47. Dahlbäck, G. O., and U. I. Balldin. Positive-pressure oxygen breathing and pulmonary atalectasis during immersion. *Undersea Biomed. Res.* 10: 39–44, 1983.

48. Daniels, S., K. C. Eastaugh, W. D. M. Paton, and E. B. Smith. Micronuclei and bubble formation: a quantitative study using the common shrimp, crangon cragnon. In: *Underwater Physiology VIII, Proc. Eighth Symp. Underwater Physiol.* edited by

A. J. Bachrach and M. M. Matzen. Bethesda, MD: Undersea Med. Soc., 1984, p. 147–157.

49. Davis, F. M., J. P. Osborne, A. D. Baddeley, and I. M. F. Graham. Diver performance: nitrogen narcosis and anxiety. *Aerospace Med.* 43: 1079–1082, 1972.

50. Derion, T., H. J. B. Guy, K. Tsukimoto, W. Schaffartzik, R. Prediletto, D. C. Poole, D. R. Knight, and P. D. Wagner. Ventilation–perfusion relationships in the lung during head-out water immersion. *J. Appl. Physiol.* 72: 64–72, 1992.

51. Derion, T., W. G. Reddan, and E. H. Lanphier. Effects of body position and static lung loading during immersion on end-expiratory lung volume and peak expiratory flow [Abstract]. *Undersea Biomed. Res.* 15: 69, 1988.

52. Donald, K. W. O_2 poisoning in man, part I. *Br. Med. J.* 1: 667–672, 1947.

53. Donald, K. W. O_2 poisoning in man, part II. *Br. Med. J.* 1: 712–717, 1947.

54. Dwyer, J., H. A. Saltzman, and R. O'Bryan. Maximal physical work capacity of man at 43.4 ATA. *Undersea Biomed. Res.* 4: 359–372, 1977.

55. Eckenhof, R. G., J. H. Dougherty, Jr., A. A. Messier, S. F. Osborne, and J. W. Parker. Progression of and recovery from pulmonary O_2 toxicity in humans exposed to 5 ATA air. *Aviat. Space Environ. Med.* 58: 658–667, 1987.

56. Eckenhoff, R. G., and J. S. Hughes. Acclimatization to decompression stress. In: *Underwater Physiology VIII, Proc. Eighth Symp. Underwater Physiol.,* edited by A. J. Bachrach and M. M. Matzen. Bethesda, MD: Undersea Med. Soc., 1984, p. 93–100.

57. Elandt-Johnson, R. C., and N. O. Johnson. *Survival Models and Data Analysis.* New York: Wiley, 1980.

58. Elliott, D. H., and R. E. Moon. Manifestations of the Decompression Disorders. In: *The Physiology and Medicine of Diving* (4th ed.), edited by P. B. Bennett and D. H. Elliott. London: W. B. Saunders, 1993, p. 481–505.

59. Epstein, P. S., and M. S. Plesset. On the stability of gas bubbles in liquid–gas solutions. *J. Chem. Physics* 18: 1505–1509, 1950.

60. Evans, A., and D. N. Walder. Significance of gas micronuclei in the aetiology of decompression sickness. *Nature* 222: 251–252, 1969.

61. Fagraeus, L., and D. Linnarsson. Maximal voluntary and exercise ventilation at high ambient air pressures. *Försvarsmedicin* 9: 275–278, 1973.

62. Farhi, L. E., and D. Linnarsson. Cardiopulmonary readjustments during graded immersion in water at 35°C. *Respir. Physiol.* 30: 35–50, 1977.

63. Flook, V., and I. M. Fraser. Inspiratory flow limitation in divers. *Undersea Biomed. Res.* 16: 305–311, 1989.

64. Fowler, A., K. N. Ackles, and G. Porlier. Effects of inert gas narcosis on behavior—a critical review. *Undersea Biomed. Res.* 12: 369–402, 1985.

65. Francis, T. J. R., J. L. Griffin, L. D. Homer, G. H. Pezeshkpour, A. J. Dutka, and E. T. Flynn. Bubble-indiced dysfunction in acute spinal cord decompression sickness. *J. Appl. Physiol.* 68: 1368–1375, 1990.

66. Franks, N. P., and W. R. Lieb. Where do narcotics act? *Nature* 274: 339–342, 1978.

67. French, J. D., M. Verzeano, and H. W. Magoun. A neural basis of the anesthetic state. *AMA Arch. Neurol.* 69: 519–529, 1953.

68. Fridovich, I., and B. Freeman. Antioxidant defenses in the lung. *Annu. Rev. Physiol.* 48: 693–702, 1986.

69. Golding, F. C., P. Griffiths, H. V. Hempleman, W. D. M. Paton, and D. N. Walder. Decompression sickness during construction of the Dartford tunnel. *Br. J. Indust. Med.* 17: 167–180, 1960.

70. Hallenbeck, J. M., and J. C. Andersen. Pathogenesis of the decompression disorders. In: *The Physiology and Medicine of*

Diving (3rd ed.), edited by P. B. Bennett and D. H. Elliott. London, Bailliere Tindall, 1982, p. 435–460.

71. Halsey, M. J., B. W. Wardley-Smith, and C. J. Green. Pressure reversal of general anesthesia—a multi-site expansion hypothesis. *Br. J. Anaesthesia* 50: 1091–1097, 1978.

72. Harabin, A. L., J. C. Braisted, and E. T. Flynn. Response of antioxidant enzymes to intermittent and continuous hyperbaric O₂. *J. Appl. Physiol.: Respir. Environ. Exerc. Physiol.* 69: 328–335, 1990.

73. Harabin, A. L., L. D. Homer, and M. E. Bradley. Pulmonary O₂ toxicity in awake dogs: metabolic and physiological effects. *J. Appl. Physiol.: Respir. Environ. Exerc. Physiol.* 57: 1480–1488, 1984.

74. Harabin, A. L., L. D. Homer, P. K. Weathersby, and E. T. Flynn. An analysis of decrements in vital capacity as an index of pulmonary O₂ toxicity. *J. Appl. Physiol.: Respir. Environ. Exerc. Physiol.* 63: 1130–1135, 1987.

75. Harabin, A. L., S. S. Survanshi, P. K. Weathersby, J. R. Hays, and L. D. Homer. The modulation of O₂ toxicity by intermittent exposure. *Tox. Appl. Pharmacol.* 93: 298–311, 1988.

76. Harvey, E. N. Physical factors in bubble formation. In: *Decompression Sickness*, edited by J. F. Fulton. Philadelphia: Saunders, 1951, p. 90–114.

77. Haugaard, N. Cellular mechanisms of O₂ toxicity. *Physiol. Rev.* 48: 311–373, 1968.

78. Hempleman, H. V. Decompression procedures for deep, open sea operations. In: *Underwater Physiology III*, Baltimore: Williams and Wilkins, 1967, p. 255–266.

79. Hempleman, H. V. British decompression theory and practice. In: *The Physiology and Medicine of Diving and Compressed Air Work* (1st ed.), edited by P. B. Bennett and D. H. Elliott. London: Bailliere, Tindall and Cassell, 1969, p. 291–318.

80. Hempleman, H. V. History of the evolution of decompression procedures. In: *The Physiology of Diving and Compressed Air Work* (3rd ed.), edited by P. B. Bennett and D. H. Elliott. London: Bailliere Tindall, 1982, p. 319–351.

81. Hendricks, P. L., D. A. Hall, W. L. Hunter, and P. J. Haley. Extension of pulmonary O₂ tolerance in man at 2 ATA by intermittent O₂ exposure. *J. Appl. Physiol.* 42: 593–599, 1977.

82. Hesser, C. M., J. Adolfson, and L. Fagraeus. Role of CO₂ in compressed-air narcosis. *Aerospace Med.* 42: 163–168, 1971.

83. Hesser, C. V., F. Lind, and B. Faijerson. Effects of exercise and raised air pressures on maximal voluntary ventilation. In: *Proc. Eur. Underwater Biomed. Soc. 5th Ann. Scientific Meeting*, edited by J. Grimstad. Bergen, Norway: Eur. Undersea Biomed. Soc., 1979, p. 203–212.

84. Hickey, D. D., C. E. G. Lundgren, and A. J. Pasche. Influence of exercise on maximal voluntary ventilation and forced expiratory flow at depth. *Undersea Biomed. Res.* 10: 241–254, 1983.

85. Hickey, D. D., W. T. Norfleet, A. J. Påsche, and C. E. G. Lundgren. Respiratory function in the upright working diver at 6.8 ATA (190 fsw). *Undersea Biomed. Res.* 14: 241–262, 1987.

86. Hills, B. A. Acclimatization to decompression sickness: a study of passive relaxation in several tissues. *Clin. Sci.* 37: 109–124, 1969.

87. Hills, B. A. *Decompression Sickness, Vol I: The Biophysical Basis of Prevention and Treatment*. New York: Wiley, 1977, p. 310–315.

88. Jamieson, D., B. Chance, E. Cadenas, and A. Boveris. The relation of free radical production to hyperoxia. *Annu. Rev. Physiol.* 48: 703–719, 1986.

89. Jones, A. W., R. D. Jennings, J. Adolfson, and C. M. Hesser. Combined effects of ethanol and hyperbaric air on body sway and heart rate in man. *Undersea Biomed. Res.* 6: 15–25, 1979.

90. Kalbfleish, J., and R. L. Prentice. *The Statistical Analysis of Failure Time Data*. New York: Wiley, 1980.

91. Keays, F. L. *Compressed Air Illness*. Ithaca, NY: Cornell University Medical College, 1909, vol. 2, p. 1–55.

92. Kerem, D., Y. Melamed, and A. Moran. Alveolar PCO₂ during rest and exercise in divers and non-divers breathing O₂ at 1 ATA. *Undersea Biomed. Res.* 7: 17–26, 1980.

93. Kiesling, R. J., and C. H. Maag. Performance impairment as a function of nitrogen narcosis. *J. Appl. Psychol.* 46: 91–95, 1962.

94. Kinney, J. S., and C. L. McKay. The visual evoked response as a measure of nitrogen narcosis in Navy divers. Report 664, New London: U.S. Naval Submarine Medical Center, 1971.

95. Lahiri, S., E. Mulligan, S. Andronikou, M. Shirahata, and A. Mokashi. Carotid body chemosensory function in prolonged normobaric hyperoxia in the cat. *J. Appl. Physiol.: Respir. Environ. Exerc. Physiol.* 62: 1924–1931, 1987.

96. Lambertsen, C. J., J. M. Clark, R. Gelfand, J. B. Pisarello, W. H. Cobbs, J. E. Bevilacqua, D. M. Schwartz, D. J. Montaban, C. S. Leach, P. C. Johnson, and D. E. Fletcher. Definition of tolerance to continuous hyperoxia in man. In: *Ninth Int. Symp. Underwater Hyperbaric Physiol.*, edited by A. A. Bove, A. J. Bachrach, and L. J. Greenbaum, Jr. Bethesda, MD: Undersea Hyperbaric Med. Soc., 1987, p. 717–735.

97. Lambertsen, C. J., R. Gelfand, R. Peterson, R. Strauss, W. B. Wright, J. G. Dickson, Jr., C. Puglia, and R. W. Hamilton, Jr. Human tolerance to He, Ne and N₂ at respiratory gas densities equivalent to He–O₂ breathing at depths to 1200, 2000, 3000, 4000, and 5000 feet of sea water (Predictive Studies III). *Aviat. Space Environ. Med.* 48: 843–855, 1977.

98. Lambertsen, C. J., R. H. Kough, D. Y. Cooper, G. L. Emmel, H. Y. H. Loeschke, and C. F. Schmidt. O₂ toxicity. Effects in man of O₂ inhalation at 1 and 3.5 atmospheres upon blood gas transport, cerebral circulation and cerebral metabolism. *J. Appl. Physiol.* 5: 471–486, 1953.

99. Lambertsen, C. J., S. G. Owen, H. Wendel, M. W. Stroud, A. A. Lurrie, W. Lochner, and G. F. Clark. Respiratory and cerebral circulatory control during exercise at .21 and 2.0 atm inspired PO₂. *J. Appl. Physiol.* 14: 966–982, 1959.

100. Lanari, A., A. Lambertini, F. L. Zubiaur, and B. Bromberger Barnea. Las modificaciones de la presion intratoracica, arterial sistemica, y venosa periferica durante la sumersion. *Medicina (B. Aires)* 20: 159–163, 1960.

101. Lanphier, E. H. Influences of increased ambient pressures upon alveolar ventilation. In: *Proc. 2nd Symp. Underwater Physiol.*, edited by C. J. Lambertsen and L. J. Greenbaum. Washington, DC: *Natl. Acad. Sci. Natl. Res. Council*, 1963.

102. Lanphier, E. H., and E. M. Camporesi. Respiration and exercise. In: *The Physiology and Medicine of Diving*, edited by P. B. Bennett and D. H. Elliott. San Pedro, CA: Best, 1982, p. 99–156.

103. Larrabee, M. G., and J. M. Posternak. Selective action of anesthetics on synapses and axons in mammalian sympathetic ganglia. *J. Neurophysiol.* 15: 91–114, 1952.

104. Leitch, D. R., and R. D. Green. Pulmonary barotrauma in divers and the treatment of cerebral arterial gas embolism. *Aviat. Space Environ. Med.* 57: 931–938, 1986.

105. Leitch, D. R., and R. D. Green. Recurrent pulmonary barotrauma. *Aviat. Space Environ. Med.* 57: 1039–1043, 1986.

106. Lever, M. J., K. M. Miller, W. D. M. Paton, and E. B. Smith. Pressure reversal of anaesthesia. *Nature* 231: 371–386, 1971.

107. Linnarson, D., and C. M. Hesser. Dissociated ventilatory and central respiratory responses to CO₂ at raised N₂ pressure. *J. Appl. Physiol.: Respir. Environ. Exerc. Physiol.* 45: 756–761, 1978.

108. Löllgen, H., G. von Nieding, and R. Horres. Respiratory and hemodynamic adjustment during head-out water immersion. *Int. J. Sports Med.* 1: 25–29, 1980.

109. Lundgren, C. E. G. Discussion. In: *The Physiological Basis of*

Decompression Sickness, edited by R. D. Vann. Bethesda, MD: Undersea Hyperbaric Med. Soc., 1989, p. 69.

109a. Lundgren, C. E. G., and D. E. Warkander, eds. Discussion. In: *Physiological and Human Engineering Aspects of Underwater Breathing Apparatus.* Bethesda, MD: Undersea and Hyperbaric Medical Society, Inc., 1989, p. 213–221.

110. Maio, D. A., and L. E. Farhi. Effect of gas density on mechanics of breathing. *J. Appl. Physiol.* 23: 687–693, 1967.

111. Marshall, J. M., and W. O. Fenn. The narcotic effects of nitrogen and argon on the central nervous system of frogs. *Am. J. Physiol.* 163: 733, 1950.

112. Matalon, S., M. S. Nesarajah, and L. E. Farhi. Pulmonary and circulatory changes in conscious sheep exposed to 100% O_2 at 1 ATA. *J. Appl. Physiol.: Respir. Environ. Exerc. Physiol.* 53: 110–116, 1982.

113. McDonough, P. M., and E. A. Hemmingsen. Bubble formation in crabs induced by limb motions after decompression. *J. Appl. Physiol.: Respir. Environ. Exerc. Physiol.* 57: 117–122, 1984.

114. McWhorter, J. E. The etiological factors of compressed-air illness. The gaseous content of tunnels: the occurrence of the disease in workers. *Am. J. Med. Sci.* 139: 373–383, 1910.

115. Meyer, K. H., and H. Hopff. Narcosis by inert gases under pressure. *Hoppe-Seyler's Z. Physiol. Chem.* 126: 288–298, 1923.

116. Miller, J. N., O. D. Wangensteen, and E. H. Lanphier. Respiratory limitations to work at depth. In: *Proc. Third Int. Conf. Hyperbaric Underwater Physiol.,* edited by X. Fructus. Paris, Doin, 1972, p. 118–123.

117. Miller, K. W., W. D. M. Paton, R. Smith, and E. M. Smith. The pressure reversal of general anaesthesia and the critical volume hypothesis. *Mol. Pharmacol.* 9: 131–143, 1973.

118. Morrison, J. B., J. T. Florio, and W. S. Butt. Observations after loss of consciousness under water. *Undersea Biomed. Res.* 5: 179–187, 1978.

119. Mullins, L. J. Some physical mechanisms in narcosis. *Chem. Rev.* 54: 289–323, 1954.

120. Nemiroff, M. J., G. R. Saltz, and J. G. Weg. Survival after cold-water near-drowning: the protective effect of the diving reflex. *Am. Rev. Respir. Dis.* 115: 145, 1977.

121. Novotny, J. A., D. L. Mayers, Y. F. J. Parsons, S. S. Survanshi, P. K. Weathersby, and L. D. Homer. Xenon kinetics in muscle are not explained by a model of parallel perfusion-limited compartments. *J. Appl. Physiol.: Respir. Environ. Exerc. Physiol.* 68: 876–890, 1990.

122. Overton, E. *Studien Über die Narkose.* Jena Fischer, 1901.

123. Parker, E. C., S. S. Survanshi, P. K. Weathersby, and E. D. Thalmann. Statistically based decompression tables VIII: linear exponential kinetics. NMRI Technical Report 92-73. Bethesda, MD: Naval Medical Research Institute, 1992.

124. Paton, W. D. M., and D. N. Walder. Compressed air illness. An investigation during the construction of the Tyne tunnel 1948–50. Medical Research Council Special Report 281. London: Her Majesty's Stationary Office, 1954.

125. Piccard, J. Aero-emphysema and the birth of gas bubbles. *Proc. Staff Meetings Mayo Clin.* 16: 700–704, 1941.

126. Pol, B., and T. J. J. Watelle. Memoire sur les effects de la compression de l'air applique au creusement des puits a hoville. *Ann. d'Hyg. Pub. Med. Legate* 2: 241–279, 1854.

127. Powell, M. R., M. P. Spencer, and O. T. Von Ramm. Ultrasonic surveillance of decompression. In: *The Physiology and Medicine of Diving* (3rd ed.), edited by P. B. Bennett and D. H. Elliot. London: Bailliere Tindall, 1982, p. 404–434.

128. Prefaut, C., F. Dubois, C. Roussos, R. Amaral-Marques, P. T. Macklem, and F. Ruff. Influence of immersion to the neck in water on airway closure and distribution of perfusion in man. *Respir. Physiol.* 37: 313–323, 1979.

129. Prefaut, C., M. Ramonatxo, R. Boyer, and G. Chardon. Human gas exchange during water immersion. *Respir. Physiol.* 34: 307–318, 1978.

130. Ringqvist, T. The ventilatory capacity in healthy subjects. *Scand. J. Clin. Lab. Invest.* 18: 1–179, 1966.

131. Rubenstein, C. J. Role of decompression conditioning in the incidence of decompression sickness in deep diving. NEDU Report 12-68. Panama City, FL: Navy Experimental Diving Unit, 1968.

132. Saltzman, H. A., L. Hart, H. O. Sieker, and E. J. Duffy. Retinal vascular response to hyperbaric oxygenation. *J. Am. Med. Soc.* 191: 114–116, 1975.

133. Schreiner, H. R., R. W. Hamilton, and T. D. Langley. Neon: an attractive new commercial diving gas. In: *Proc. Offshore Technology Conf. Houston, May 1–3, 1972.*

134. Scriven, L. E. On the dynamics of gas phase growth. *Chem. Eng. Sci.* 10: 1–13, 1959.

135. Seusing, J., and H. Drube. The importance of hypercapnia in depth intoxication. *Klin. Wschr.* 38: 1088–1090, 1960.

136. Sherman, D., E. Eilender, A. Shefer, and D. Kerem. Ventilatory and occlusion-pressure responses to hypercapnia in divers and non-divers. *Undersea Biomed. Res.* 7: 61–74, 1980.

137. Smith, A. H. The Effects of High Atmospheric Pressure Including the Caisson Disease. New York: Eagle Print, 1873. Prize Essay of the Alumnae Association of the College of Physicians and Surgeons New York.

138. Spaur, W. H., L. W. Rahmond, M. M. Knott, J. C. Crothers, W. R. Braithwaite, E. D. Thalmann, and D. F. Uddin. Dyspnea in divers at 49.5 ATA: mechanical, not chemical in origin. *Undersea Biomed. Res.* 4: 183–198, 1977.

139. Spencer, M. P., and H. F. Clarke. Precordial monitoring of pulmonary gas embolism and decompression bubbles. *Aerospace Med.* 43: 762–767, 1972.

140. Stigler, R. Die Kraft unserer Inspirations Muskulatur. *Pflugers Arch.* 139: 234–254, 1911.

141. Strauss, R. H. Bubble formation in gelatin: implications for the prevention of decompression sickness. *Undersea Biomed. Res.* 1: 169–174, 1974.

142. Thalmann, E. D. Phase II testing of decompression algorithms for use in the U.S. Navy Underwater Decompression computer. NEDU Report 1-84. Panama City, FL: Navy Experimental Diving Unit, 1984.

143. Thalmann, E. D. Air-N_2O_2 decompression computer algorithm development. NEDU Report 8-85. Panama City, FL: Navy Experimental Diving Unit, 1985.

144. Thalmann, E. D. Development of a decompression algorithm for constant 0.7 ATA oxygen partial pressure in helium diving. NEDU Report 1-85. Panama City, FL: Navy Experimental Diving Unit, 1985.

145. Thalmann, E. D. USN experience in decompression table validation. In: *34th UHMS Workshop, Validation of Decompression Tables,* edited by H. R. Schreiner and R. W. Hamilton. Bethesda, MD: Undersea Hyperbaric Med Soc., Pub 74, 1989, p. 1–88.

146. Thalmann, E. D., and C. A. Piantadosi. Submerged exercise at pressure up to 55.55 ATA [Abstract]. *Undersea Biomed. Res.* 8: 25, 1981.

147. Thalmann, E. D., D. K. Sponholtz, and C. E. G. Lundgren. Effects of immersion and static lung loading on submerged exercise at depth. *Undersea Biomed. Res.* 6: 259–290, 1979.

148. Thalmann, E. D., J. L. Zumrick, H. J. C. Schwartz, and F. K. Butler. Accommodation to decompression sickness in HeO$_2$ divers. *Undersea Biomed. Res.* 11(suppl. 1): 6, 1984.

149. Tikuisis, P. Modeling the observations of in vivo bubble formation with hydrophobic crevices. *Undersea Biomed. Res.* 13: 165–180, 1986.

150. Torbati, D., A. Mokashi, and S. Lahiri. Effects of acute hyperbaric oxygenation on respiratory control in cats. *J. Appl. Physiol.: Respir. Environ. Exerc. Physiol.* 67: 2351–2356, 1989.

151. Torbati, D., D. Parolla, and S. Lavy. Organ blood flow, cardiac output, arterial blood pressure, and vascular resistance in rats exposed to various oxygen pressures. *Aviat. Space Environ. Med.* 50: 256–263, 1979.

152. U. S. Department of Navy. *U. S. Navy Diving Manual* (rev. 2), Washington, DC: Naval Sea Systems Command, 1987, vol. 2. (NAVSEA 0994-LP-001-9020).

153. U. S. Navy. *U. S. Navy Diving Manual* (rev. 3), Washington, DC: Naval Sea Systems Command, 1991, vol. 2 (NAVSEA 0994-LP-001-9020).

154. Vail, E. G. Hyperbaric respiratory mechanics. *Aerospace Med.* 42: 536–546, 1971.

155. Van Liew, H. D. Mechanical and physical factors in lung function during work in dense environments. *Undersea Biomed. Res.* 10: 255–264, 1983.

156. Van Liew, H. D. Simulation of the dynamics of decompression sickness bubbles and the generation of new bubbles. *Undersea Biomed. Res.* 18: 333–345, 1991.

157. Van Liew, H. D., B. Bishop, P. Walder-D, and H. Rahn. Effects of compression on composition and absorption of tissue gas pockets. *J. Appl. Physiol.* 20: 927–933, 1965.

158. Van Liew, H. D., and M. P. Hlastala. Influence of bubble size and blood perfusion on absorption of gas bubbles in tissues. *Respir. Physiol.* 7: 111–121, 1969.

159. Vann, R. D. Decompression theory and application. In: *The Physiology of Diving and Compressed Air Work* (3rd ed.), edited by P. B. Bennett and D. H. Elliot. London: Bailliere Tindall, 1982, p. 352–382.

160. Vann, R. D., J. Grimstad, and C. H. Nielsen. Evidence for gas nuclei in decompressed rats. *Undersea Biomed. Res.* 7: 107–112, 1980.

161. Vann, R. D., and E. D. Thalmann. Decompression physiology and practice. In: *The Physiology of Diving and Compressed Air Work* (4th ed), edited by P. B. Bennett and D. H. Elliott. London: Bailliere Tindall, 1993.

162. Van Rensselaer, H. The pathology of the caisson disease. *Trans. Med. Soc. N.Y.* 408–444, 1891.

163. Vokac, Z., H. Bell, E. Bautz-Holter, and K. Rodahl. O_2 uptake/heart rate relationship in leg and arm exercise, sitting and standing. *J. Appl. Physiol.* 39: 54–59, 1975.

164. Walder, D. N. Adaptation to decompression sickness in caisson work. In: *Proc. Third Int. Biometerology Congr. Oxford*, 1968, p. 350–359.

165. Walder, D. N. The compressed air environment. In: *The Physiology of Diving and Compressed Air Work* (3rd ed.), edited by P. B. Bennett and D. H. Elliot. London: Bailliere Tindall, 1982, p. 15–30.

166. Walker, B. R., A. A. Attallah, J. B. Lee, S. K. Hong, B. K. Mookerjee, L. Share, and J. A. Krasney. Antidiuresis and inhibition of PGE_2 excretion by hyperoxia in the conscious dog. *Undersea Biomed. Res.* 7: 113–126, 1980.

167. Ward, C. A., D. McCullough, and W. D. Fraser. Relation between complement activation and susceptibility to decompression sickness. *J. Appl. Physiol.: Respir. Environ. Exerc. Physiol.* 62: 1160–1166, 1987.

168. Ward, C. A., D. McCullough, D. Yee, D. Stanga, and W. D. Raser. Complement activation involvement in decompression sickness of rabbits. *Undersea Biomed. Res.* 17: 51–66, 1990.

169. Warkander, D. E., W. T. Norfleet, G. K. Nagasawa, and C. E. G. Lundgren. CO_2 retention with minimal symptoms but severe dysfunction during wet simulated dives to 6.8 atm abs. *Undersea Biomed. Res.* 17: 515–523, 1990.

170. Warkander, D. E., W. T. Norfleet, G. K. Nagasawa, and C. E. G. Lundgren. Physiologically and subjectively acceptable breathing resistance in divers' breathing gear. *Undersea Biomed. Res.* 19: 427–445, 1992.

171. Weathersby, P. K., B. L. Hart, E. T. Flynn, and W. F. Walker. Role of O_2 in the production of human decompression sickness. *J. Appl. Physiol.: Respir. Environ. Exerc. Physiol.* 63: 2380–2387, 1987.

172. Weathersby, P. K., J. R. Hayes, S. S. Survanshi, L. D. Homer, B. L. Hart, E. T. Flynn, and M. E. Bradley. Statistically based decompression tables. II. Equal risk air diving decompression schedules. NMRI Technical Report 85-17. Bethesda, MD: Naval Medical Research Institute Technical, 1985.

173. Weathersby, P. K., L. D. Homer, and E. T. Flynn. Homogeneous nucleation of gas bubbles in vivo. *J. Appl. Physiol.: Respir. Environ. Exerc. Physiol.* 53: 940–946, 1982.

174. Weathersby, P. K., L. D. Homer, and E. T. Flynn. On the likelihood of decompression sickness. *J. Appl. Physiol.: Respir. Environ. Exerc. Physiol.* 53: 815–825, 1984.

175. Weathersby, P. K., S. S. Survanshi, J. R. Hayes, and M. E. MacCallum. Statistically based decompression tables III: comparative risk using U. S. Navy, British, and Canadian standard air schedules. NMRI Technical Report 86-50. Bethesda, MD: Naval Medical Research Institute, 1986.

176. Weathersby, P. K., S. S. Survanshi, J. R. Hayes, E. Parker, and E. D. Thalmann. Predicting the time of occurrence of decompression sickness. *J. Appl. Physiol.: Respir. Environ. Exerc. Physiol.* 72: 1541–1548, 1992.

177. Weathersby, P. K., S. S. Survanshi, L. D. Homer, B. L. Hart, R. Y. Nishi, E. T. Flynn, and M. E. Bradley. Statistically based decompression tables. I. Analysis of standard air dives: 1950–1970. NMRI Technical Report 85-16. Bethesda, MD: Naval Medical Research Institute, 1985.

178. Weber, C. A., C. A. Duncan, M. J. Lyons, and S. G. Jenkinson. Depletion of tissue glutathione with DEM enhances hyperbaric O_2 toxicity. *Am. J. Physiol. (Lung Cell Mol. Physiol. 2)* 258: L308–L312, 1990.

179. Whalen, R. E., H. A. Saltzman, D. H. Holloway, H. D. McIntosh, H. O. Sieker, and I. W. Brown. Cardiovascular and blood gas response to hyperbaric oxygenation. *Am. J. Cardiol.* 15: 638–646, 1965.

180. Wienke, B. R. Tissue gas exchange models and decompression computations: a review. *Undersea Biomed. Res.* 16: 53–89, 1989.

181. Wienke, B. R. *Basic Decompression Theory and Application.* Flagstaff, AZ: Best, 1991.

182. Wood, L. D. H., and A. C. Bryan. Exercise ventilatory mechanics at increased ambient pressure. *J. Appl. Physiol.: Respir. Environ. Exerc. Physiol.* 44: 231–237, 1978.

183. Workman, R. D., and R. C. Bornmann. Decompression theory: American practice. In: *The Physiology and Medicine of Diving* (2nd ed.), edited by P. B. Bennett and D. H. Elliot. London: Bailliere Tindall, 1975, p. 307–330.

184. Yount, D. E., T. D. Kunkle, J. S. D'Arrigo, F. W. Ingle, C. M. Yeung, and E. L. Beckman. Stabilization of gas cavitation nuclei by surface-active compounds. *Aviat. Space Environ. Med.* 48: 185–191, 1977.

185. Yount, D. E., and R. H. Strauss. Bubble formation in gelatin: a model for decompression sickness. *J. Appl. Physics* 44: 5081–5089, 1976.

186. Zheng, Q., D. J. Durben, G. H. Wolf, and C. A. Angell. Liquids at large negative pressures: water at the homogeneous nucleation limit. *Science* 254: 829–832, 1991.

3 | MIXED-GAS SATURATION DIVING

44. Mixed-gas saturation diving

SUK KI HONG | Department of Physiology, State University of New York at Buffalo, Buffalo, New York

PETER B. BENNETT | Hyperbaric Center and Department of Anesthesiology, Duke University Medical Center, Durham, North Carolina

KEIZO SHIRAKI | Department of Physiology, University of Occupational and Environmental Health, Kitakyushu, Japan

YU-CHONG LIN | Department of Physiology, University of Hawaii, Honolulu, Hawaii

JOHN R. CLAYBAUGH | Physiology Section, Department of Clinical Investigation, Tripler Army Medical Center, Honolulu, Hawaii

CHAPTER CONTENTS

THE MIXED-GAS SATURATION-DIVING TECHNIQUE was developed in the 1950s because of the potential dangers, such as oxygen toxicity and nitrogen narcosis, associated with deep-sea diving with air. To avoid oxygen toxicity, the P_{O_2} of the breathing gas is kept at a level slightly higher than that of air and the nitrogen tension is kept low by compressing with other inert gases such as helium, neon, and hydrogen. Using such a technique, human divers reached a maximal depth of nearly 690 meters seawater (MSW) [approximately 70 atmosphere absolute (ATA)] in a multiday dive. As the depth of the dive increases, many physiological functions begin to be affected. This chapter deals with physiological changes during exposure of human divers to hyperbaric environments. These affect nerve functions, as manifested by changes in the electroencephalogram (EEG) and in psychomotor activity; thermoregulatory functions caused by increased body heat loss; physical work capacity and maximal aerobic power; cardiopulmonary functions, hyperbaric bradycardia, and changes in cardiovascular fitness; and lastly, body fluid balance and the mechanisms of hyperbaric diuresis/natriuresis.

RATIONALE

With the development of air scuba-diving techniques, the duration and depth of dives can be much greater than with breath-hold diving. However, it soon became evident that scuba diving with compressed air also presented many problems, including the possibilities of N_2 narcosis and O_2 toxicity, as well as the long decompression time required for the elimination of N_2 gas taken up during the hyperbaric phase. Thus another approach was necessary to dive deeper and longer than air scuba diving permitted. The mixed-gas saturation-diving technique was conceived in 1957 by U. S. Navy Captain George Bond. In 1962, the Conshelf I and Man in Sea projects exposed human subjects to multiday, open-sea, deep saturation dives. The present depth record of 2,250 FSW (feet seawater, equivalent to 69 ATA) was a dry saturation dive directed by Peter Bennett of Duke University in 1981 (12).

To minimize or eliminate the N_2 narcosis that occurs

when the PN_2 of the breathing gas is high, helium is used as the major inert gas in the breathing mixture. Compared to N_2, He has lower molecular weight (7:1), density (7:1), oil–water solubility (3:1), and narcotic potency (4.3:1). Even with the use of He, however, the high pressure nervous syndrome (HPNS) is known to occur at a depth beyond 500 FSW if compression is rapid. To overcome this problem, Bennett proposed the use of 5%–10% N_2 in deep saturation dives and, in fact, used a combination of He and N_2 as the diluent gas in the record dive to 69 ATA mentioned above. When a gas mixture contains N_2, He, and O_2 it is referred to as "trimix."

Avoidance of O_2 toxicity is relatively easy. A normoxic state is desirable, but breathing gas with PO_2 of 0.20 ATA seems to induce hypoxic signs (this phenomenon is referred to as The Chouteau effect) (37). It is customary to increase the breathing gas PO_2 to 0.3–0.5 ATA. The mechanism of this interesting Chouteau effect has not been critically assessed and requires further study. Another question related to O_2 is the upper safe limit of PO_2 in the gas mix. This requires long-term animal studies.

Decompression time becomes greater the longer the bottom time in any type of dive. For example, if an air scuba diver stays at a depth of 190 FSW for 60 min, a total ascent time of 232 min is required, according to the *U. S. Navy Diving Manual*, with a ratio of bottom time to total diving time of 60/232 = 0.26; that is, bottom time is only 26% of total dive time (Fig. 44.1, left panel). In saturation diving, a state of equilibrium exists between gas in body tissues and ambient gas; body tis-

sues become saturated with ambient gas within 24 h, after which no more inert gas is taken up. Once the body is saturated, decompression (ascent) time is the same, independent of the duration of the dive, and, in fact, the bottom time/total diving time increases as the bottom time increases (see Fig. 44.1, right panel), thereby making the dive more economical as the duration increases. A decompression rate of 100 FSW/24 h [equivalent to ~30 MSW/24 h] is usually used. Suppose a diver stayed 2 days at 200 FSW (~7 ATA). The bottom time/total diving time ratio of 48 h/96 h is 0.5. However, if the same diver stayed 7 days (168 h) at the same 200 FSW, then the total bottom time/total dive time ratio is increased to 168 h/216 h = 0.78, a 55% increase. For commercial diving, it is important to increase the bottom time while decreasing the decompression time.

Another advantage of saturation diving is that the deeper the saturation depth, the deeper divers can make excursion dives, without decompression on ascent to saturation depth (164a). In other words, saturation depth is not necessarily the depth of work, and if one can dive to 100 or so feet deeper than excursion depth, this makes final decompression that much shorter.

The use of He avoided many adverse effects of high pressure N_2, so He has been used as a diluent gas in the majority of saturation dives conducted so far. However, the use of He presents a new problem associated with thermoregulation of divers. Compared to N_2, He has a five- to sixfold higher thermoconductivity and specific heat. Therefore, heat loss from the body in a He environment, especially in hyperbaric conditions, is very rapid; the thermoneutral temperature at which a person

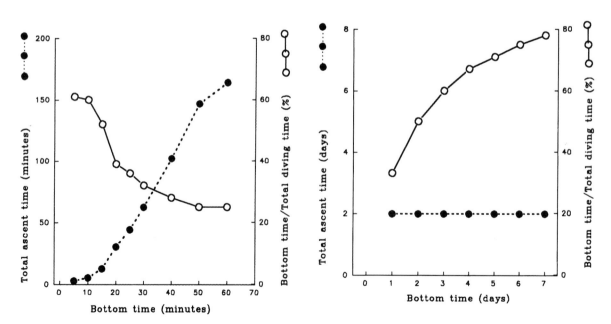

FIG. 44.1. Total ascent time and bottom time/total diving time ratio as a function of bottom time. *Left:* Air scuba dive to 190 ft; *right:* He–O_2 saturation dive to 200 ft.

can maintain body temperature and comfort is typically as high as ~32°C at $\geq$ 30 ATA He–O_2 in saturation dives (167).

As human divers progress deeper, a question arises as to the depth limit for saturation diving. The depth record for human divers of 2,250 FSW (or 69 ATA) was beaten when Rostain et al. (143) carried out a saturation diving experiment with baboons *(Papio papio)*. The investigators used He–N_2–O_2 gas instead of He–O_2 and successfully subjected the animals to a pressure of 109 ATA.

One of the major problems in deep saturation diving is gas density, which increases as a linear function of ambient pressure. As breathing gas density increases, the total resistance of lung airways increases, hindering pulmonary ventilation. The molecular weight and the density of H_2 are lowest among potential inert diluent gases, so the potential of H_2 as a major diluent gas is being critically assessed in several laboratories. French scientists successfully conducted a saturation dive at a simulated depth of 450 MSW (46 ATA) using H_2 (54% of breathing gas, or P_{H_2} = 25 ATA) (61). Compared to He–O_2 diving, the reductions in respiratory resistance and in breathing effort observed during H_2–O_2 diving appeared to be a major factor in improving performance at depths greater than 300 MSW. The use of H_2 also attenuated clinical symptoms of HPNS, presumably by antagonizing the pressure effects. However, H_2 has high thermoconductivity and high specific heat, so rapid, in fact, that heat loss may become a major problem unless the ambient temperature is raised (for example, ~34°C at 31 ATA-dry in a dog) (51). Limited studies conducted so far provide no evidence for a toxic effect of H_2. Nevertheless, Fructus (54) expressed concern that H_2 narcosis appears to be the primary deterrent to the use of H_2 for deep diving.

HIGH PRESSURE NERVOUS SYNDROME

Signs and Symptoms

To prevent nitrogen narcosis when diving to depths greater than 300 ft (91 m), the nonnarcotic helium–oxygen is substituted, as described in the chapter by Lundgren et al. in this *Handbook*. In 1965, however, Bennett unexpectedly observed decrements in motor and intellectual performance in humans in rapid oxygen–helium compressions of 100 ft/min (30 m/min) to 600 ft (183 m) and 800 ft (244 m) (6, 7, 14). These decrements were accompanied by dizziness, nausea, vomiting, and a marked tremor of the hands and trunk that resulted in the condition then called helium tremors. Brauer et al. (31) and Fructus et al. (57) reported additional factors such as dysmetria and an increase in slow-wave theta

(4–6 Hz) electroencephalogram (EEG) activity and a fall in alpha (8–13 Hz) activity with bouts of somnolence termed "microsleep" and increased stages 1 and 2 and decreased stages 3 and 4 of sleep. Similar changes also were reported in animals (30).

Since that time, a significant effort has been made to study and prevent these effects in humans and animals during exposure to very high pressures, now termed the high pressure nervous syndrome, or HPNS (9, 10, 68, 87, 99). This condition becomes worse the greater the pressure or depth over 600 ft (183 m) and the faster the rate of compression. However, there is considerable interindividual variation in susceptibility (9, 26, 27, 87, 139, 140). Further, during early studies of HPNS, it was noted that unlike nitrogen narcosis or compressed air intoxication there is often some adaptation to the initial signs and symptoms on arrival at depth (19). Thus in the initial discovery of HPNS the individual performance and symptoms returned to normal in 1.5 h (6, 14). Brauer et al. (30) reported similar adaptations in animals. However, at depths greater than 1,300 ft (396 m) adaptation is less likely and some individuals may actually become worse with time.

Prevention of HPNS

Bennett (8) analyzed 23 dives to depths greater than 1,000 ft (300 m) between 1965 and 1975 and reviewed a number of methods potentially able to ameliorate the signs and symptoms of HPNS. Among these were selecting the least susceptible divers, choosing a slow exponential compression profile with stages en route to permit adaptation to compression, allowing time for adaptation on arrival at depth before starting work, and adding nitrogen to helium–oxygen to produce the so-called trimix (helium–nitrogen–oxygen).

Diver Selection. Different divers exposed to high helium–oxygen pressures show different severities of HPNS symptoms. For example, for EEG changes, whereas one diver may indicate a 1,000% increase in theta activity, another may show less than 200% or be unaffected (99, 136, 137). This variation in effect is stable.

Many attempts have therefore been made to select divers who might be less affected by high pressures. In most cases this took the form of a rapid compression (100 ft/min) to 600 ft (183 m) with measurements of EEG, psychological performance, and finger tremors, eliminating those with marked changes or decrements.

Rostain et al. (137), using EEG, divided divers into three groups after a rapid (15 min) compression to 590 ft (180 m) and a stay there for 10 min. Group 0 showed no significant increase in theta activity (<10%); group 1 showed increases between 10% and 100% and group 2 an increase in excess of 100%. Comparison of the results with a standard helium–oxygen dive to 450 m (1,476 ft) gave a rank Spearman correlation coefficient of 0.72 ($P = 0.01$), and it was recommended that group 2 should not be exposed to depths greater than 300 m (984 ft).

Nevertheless, such selection tests are not infallible, and it is wise to use such a method and then increase depth exposure slowly so that the first dive is to 200 m (656 ft) then perhaps 300 m (984 ft) followed by 450 m (1476 ft), as used in a large series of German dives with trimix (He–N_2–O_2) as a further preventive of HPNS (10, 18, 19).

Compression Rate. Very slow compression to 1,000 ft (305 m) of about 50 ft/h (12 m/h) in a U. S. Navy dive at Duke University Hyperbaric Center in 1969 (124, 147, 159) indicated no signs or symptoms of HPNS. In the early 1970s British (15, 20, 21, 120) and French (55, 56) researchers in a series of record deep research dives to 1,500 ft (457 m) and 1,706 ft (520 m), respectively, confirmed that slow exponential compression rates with stages permitted divers to reach these extraordinary depths with the presence of some HPNS, though not to an incapacitating degree as in the earlier studies. Tremor first appeared at 1,000–1,150 ft (305–350 m) and was marked by 1,600 ft (480 m). It showed two forms, postural and intentional, with frequencies indicating enhanced resting tremor. There was a decrement in psychometric test abilities and usual increased EEG theta and decreased alpha activities.

In June 1972, the French first exposed divers to 2,001 ft (610 m) for 80 min with a slower compression rate and stages during a dive called Physalie VI. Signs and symptoms of HPNS first appeared at 1,150 ft (350 m). Performance decrements varied from 18% to 48%, and effects were considered less than in the earlier British dive to 1,500 ft (457 m) (58). However, to achieve such depths with divers not incapacitated by HPNS, but with it nevertheless present, would take as long as 8–10 days of compression. Experiments continued throughout the 1970s to investigate the effect of compression rate and different profiles on the incidence of HPNS (59, 163). The data indicated that, while very slow rates of compression will prevent HPNS during compression, the hydrostatic pressure itself may still induce severe incapacitating HPNS at depths over 1,400 ft (427 m). Surprisingly, however, in 1990 a German dive to 600 m with an 11 day compression time produced only mild HPNS in three divers (107).

Trimix Using Nitrogen or Hydrogen. Johnson and Flagler (88) noted that addition of ethyl alcohol to a tank of tadpoles caused them to fall unconscious to the bottom. Application of 150 bar [15 megapascals (MPA)] hydrostatic pressure antagonized the alcohol effects and

the tadpoles resumed swimming normally. Thus was pressure reversal of narcosis first reported.

Bennett et al. (17), while studying the effects of raised pressures of helium, neon, nitrogen, argon, etc. on the surface tension of phospholipid model membranes, noted that the narcotic gases nitrogen and argon induced a fall in surface tension, whereas helium and neon caused a rise. Lever et al. (101) noted in mice a reversal of loss of righting reflex induced by nitrogen of raised pressure by application of helium at higher pressures. Therefore, Bennett et al. (11) utilized these phenomena in the reverse way, that is, addition of a narcotic nitrogen to helium–oxygen to ameliorate or prevent the pressure (helium)-induced HPNS. The trimix of helium–nitrogen–oxygen effectively suppressed the nausea, vomiting, dizziness, and intentional and postural tremors seen with helium–oxygen dives and improved psychometric performance.

French researchers (36, 132) in three dives to 1,000 ft (300 m) compared the addition of 4.5% or 9% nitrogen to helium–oxygen and noted the optimum was 4.5%. In the early 1980s at the Duke Hyperbaric Center an extensive series of studies, called Atlantis, were made with a combination of fast and slow compression rates and either 5% or 10% nitrogen in heliox, while studying the effects of inspired gas density, hydrostatic pressure, and narcosis on various psychological, neurophysiological, respiratory, and circulatory measurements. The dives reached the remarkable depth of 2,250 ft (686 m), with the divers fit and able to function and minimal signs or symptoms of HPNS (12, 13, 16).

This research was followed by an extensive German series of over 18 deep trimix dives with 5% nitrogen in helium–oxygen to depths between 984 and 1,968 ft (300–600 m). Results indicated the presence of little or no HPNS. Divers were fit to work on arrival with no tremors, nausea, vomiting, or undue fatigue (19).

That hydrogen may be useful as a gas for saturation diving was predicted by Brauer and Way (29). Its lower density is more suitable for easier breathing and its narcotic potency makes it an effective additive to helium–oxygen instead of nitrogen (54).

Early experiments using hydrogen with animals were contradictory, varying from control of HPNS (29, 50, 51, 69) to no difference from helium in baboons (138) to death in other animals (116).

However, French workers have carried out significant research on animals and humans to evaluate hydrogen. When used without helium as a hydrogen–oxygen mixture, hydrogen narcosis became apparent in divers at 787 ft (240 m) (35, 54).

Subsequent attempts were concentrated on hydrogen, helium, and oxygen mixtures. Studies compared a mixture of 54%–56% hydrogen (COMEX HYDRA IV dive) in helium–oxygen with previous dives using nitrogen–helium–oxygen (DRET 79/31 and ENTEX dive series) to 1,476 ft (450 m). Results were most encouraging, with few signs or symptoms of HPNS or performance decrement (98, 99). However, EEG theta activity increased, alpha decreased, and sleep disruptions occurred.

In a later study at 1,640 ft (500 m) called HYDRA VI, with the same compression profile and using 49% hydrogen to reach 500 m, similar results occurred. Surprisingly, one diver developed psychosis a few hours after tremors appeared (128).

Psychotic disorders also occurred in two other divers breathing hydrogen–oxygen at 1,000 ft (300 m) in HYDRA IX (135). These findings have placed some caution on the continued use of hydrogen, even though COMEX performed a helium, hydrogen, oxygen open ocean dive to 500 m (49% H_2) with six divers, who made excursions to 520 m and 530 m for 26 h with no debilitating HPNS effects (62).

Mechanisms of HPNS

Neurotransmitters. Neurophysiological studies indicate that HPNS is complex and that many factors may be involved from pressure effects on the peripheral spinal cord or brain systems (24, 47, 89, 90, 142). A common but equally complex suggestion is that the likely cause involved alteration of neurotransmitter mechanisms at the synapse. Thus the epileptic-like seizures in the baboon have been linked to changes in aspartate or gamma-aminobutyric acid (GABA), an inhibitory transmitter (48, 91, 142, 171), as have changes in visual evoked potentials in humans (132) and somatic evoked potentials in baboons (86). Sleep changes have been related to the changes in the noradrenaline–serotonin system (134, 141), tremor changes to an increase in catecholamines (109), and behavioral changes and hallucinatory effects to changes in monoamines (1, 28, 115). The peripheral cholinergic system has been involved from reflex studies in baboons and humans (46, 74, 75, 76, 85, 86).

Further, the narcotic antagonism of HPNS with nitrogen or hydrogen is not simple and involves more than one site of action (70, 72, 144).

Drugs. A number of drugs with specific GABA or aspartate neurotransmitter action have been utilized, including sodium valproate, nepecotic acid, diaminobutyric acid, beta-alanine, and muscimol (22, 126, 133, 142). Most of these drugs raised the threshold for tremor and convulsions in animals.

Similarly, the supposition that cerebral monoamines are involved has resulted in the use of specific drugs, such as reserpine (28) and FLA-63 (92), a dopamine-beta-hydroxylase inhibitor. However, increasing dopamine by L-DOPA (25, 71) did not raise the seizure threshold. McLeod et al. (115) showed opposite effects of pressure on dopamine in different regions of the brain. However, using implanted multifiber carbon electrodes sensitive to dopamine, Abraini and Rostain (1) reported that an increase of extracellular dopamine in the caudate nucleus is pressure dependent.

Clearly the mechanism of HPNS is complex and global in regard to its site of origin and mechanism within the nervous system. Considerably more research will be required to relate these many changes to specific signs and symptoms of HPNS generated by increased pressure.

THERMOREGULATORY FUNCTION

Soon after helium was introduced as a respirable gas to extend the safe depth limits for deep-sea diving, it was recognized as a cause of increased rate of body heat loss, rather than the nitrogen in air (5). A number of dry saturation diving experiments have contributed greatly to our understanding of the physiology of humans in the hyperbaric environment. When humans are confined in a high-pressure helium–oxygen atmosphere, they are far more affected by the environmental temperature than humans living in a normal atmosphere. In this context we have to consider a different heat exchange mode between body and environment and produce a different definition of thermal comfort.

Many observations indicate that hyperbaric gas is thermally an unusual environment. Consider the simple fact that humans in hyperbaric helium environments are comfortable only in rather warm temperatures (above 30°C) and quickly sense small changes in temperature (23, 53, 167).

In hyperbaric chamber exposures to helium–oxygen, it has been possible to provide thermal comfort and stable body temperatures by simply maintaining a high ambient temperature (157). Thus Raymond et al. (129) observed no evidence of cold stress in lightly clad men at rest in a thermally comfortable helium–oxygen atmosphere of 4.3 to 14.6 ATA. However, Webb (166) has presented the view that thermal homeostasis may be achieved in hyperbaric helium–oxygen at the price of an increase in metabolic rate, despite being thermally indifferent in such atmospheres.

The thermal vulnerability of mammals in helium-rich atmospheres, first demonstrated by Leon and Cook (100) has now been confirmed by numerous studies on rats (157) and humans (129, 131, 153, 162), which share a common feature: as ambient pressure is increased, a higher and more precisely adjusted ambient temperature must be provided to insure thermal balance and comfort because of the physical characteristics of the environment.

Physical Characteristics of Helium Environment

The helium–oxygen atmosphere of deep saturation diving presents major problems in maintaining thermal homeostasis. These problems are a consequence of the physical properties of helium, which, though only one-seventh as dense as nitrogen, has a thermal conductivity six times greater and a specific heat at constant pressure five times greater than nitrogen. These physical properties require an environmental temperature of approximately 31°–32°C for human subjects to be in a thermally comfortable zone (167).

The major changes in heat transfer processes caused by a helium–oxygen atmosphere under pressure occur in the heat transfer coefficients for convection and evaporation of sweat. At sea level the convective heat loss by respiration is about 1% of metabolic heat production, but under pressure this term becomes greater.

Comfortable Temperature

Figure 44.2 provides comfort temperatures for a prolonged stay in helium–oxygen atmospheres at various pressures ranging from 1 to 49.5 ATA.

Convective Heat Loss

The body surface is a major pathway for heat loss by convection in a normal air environment, but it becomes the dominant pathway in hyperbaric and helium-rich environments (Fig. 44.3), as observed by several investigators (45, 129, 153, 167). It is possible to discuss convective heat loss from a theoretical standpoint, but the convective heat transfer coefficient (hc) associated with the physical properties of helium-rich environments is not simple to measure. Gagge and Nishi (60) have shown that hc is affected by the physical properties of the gas phase as follows:

$$hc \approx \lambda \left(\frac{\rho}{\eta}\right)^{0.55} (\eta \cdot C_p/\lambda)^{0.33} \text{ for forced}$$
convection, and

$$hc \approx \lambda \left(\frac{\rho}{\eta}\right)^{0.5} (\eta \cdot c_p/\lambda)^{0.25-0.333} \text{ for free}$$
convection,

where λ is thermal conductivity, ρ is density, η is viscosity, and c_p is specific heat at constant pressure. The principal physical properties of gas are listed in Table 44.1. It is possible to determine the ratio of hc for any gas mixture at pressure (%O_2, %He, %N_2, or %H_2) to that for air at 1 ATA. The ρ term is only affected by barometric pressure. According to these equations, the forced hc in a helium-rich, 31 ATA environment (1.3% O_2 balanced with 98.7% He) is 15 times greater than that for air at 1 ATA. Convection is also affected by

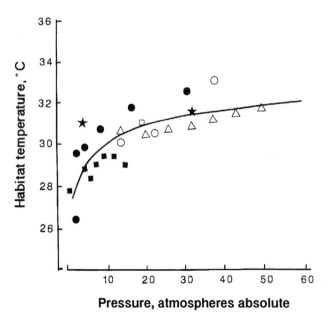

FIG. 44.2. Comfort temperatures for prolonged stay in hyperbaric helium atmosphere. The empirical line is drawn from various observations of saturation dives (●, ref. 162; ■, ref. 129; Δ, ref. 131; ○, ref.; □, ref. 167; ★, refs. 153, 154). Adapted from Webb et al. (167), with permission.

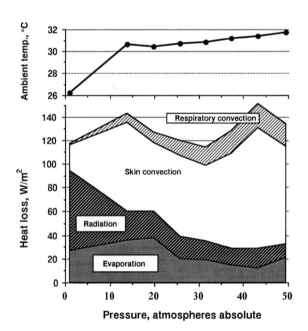

FIG. 44.3. Routes of body heat loss in resting men in helium–oxygen at pressures ranging between 1 and 49.5 ATA. Adapted from Raymond et al. (131), with permission.

TABLE 44.1. *Gas Characteristics at 1 ATA and 27°C (300°K)*

Gas	Density, ρ (kg·m^{-3})	Specific Heat, c_p (kcal·kg^{-1}·K^{-1})	Viscosity, η (poise × 10^{-5})	Thermal Conductivity, λ (cal·cm^{-1}·s^{-1}·K^{-1} × 10^{-5})
Air	1.161	0.2403	18.46	6.288
O$_2$	1.283	0.2198	20.72	6.270
N$_2$	1.123	0.2486	17.82	6.175
He	0.159	1.241	19.96	37.40
H$_2$	0.0823	3.42	8.86	43.7

movement of the ambient gas as well as by activity on the part of the subject. In general, air movement is very complicated in hyperbaric environments because of the intentionally increased ventilation to avoid an accumulation of CO_2 and unequal temperature distribution inside the chamber. As convective heat exchange is the product of hc multiplied by the temperature difference between skin surface and ambient atmosphere, factors which alter skin temperature should be taken into consideration. The increased convection reduces mean skin temperature ($\bar{T}_{sk}$) in hyperbaric helium, even though the environmental temperature is warmer than that during control in air (93). Consequently, the positive skin-to-ambient gas temperature gradient is reduced in helium atmospheres as ambient pressure increases (Fig. 44.4).

Core temperature is not necessarily reduced greatly at pressure, despite the lower values of $\bar{T}_{sk}$. Thus, core-to-skin temperature difference becomes large in helium-rich hyperbaric environments. An increase in the temperature difference may indicate a decreased heat drain from core to periphery via bloodstream, that is, vasoconstriction. There is evidence that the maintenance of

normal overall metabolism and core temperature is attributable to peripheral vasoconstriction. Using partitional calorimetry, Shiraki et al. (153) observed a reduction of the heat transfer coefficient from core to skin and hence vasoconstriction in five subjects who stayed for 4 days in a 4 ATA helium–oxygen environment. Raymond et al. (129) also postulated that skin conductance was reduced by one-fourth at 15 ATA in He–O_2 as compared with air at 1 ATA.

Radiation

Radiant heat exchange between the body and chamber walls can be calculated as follows (Stefan-Boltzmann law):

$$Q = \sigma A_1 (T_1^4 - T_2^4)/[1/\epsilon_1 + A_1/A_2(1/\epsilon_2 - 1)]$$

where Q is radiant heat exchange (W·m^{-2}), σ is the Stefan-Boltzmann constant (5.67 × 10^{-8}·W·m^{-2}·K^{-4}), A_1 is the surface area of the human body (m^2), A_2 is the surface area of the chamber walls (m^2), T_1 is mean skin temperature, T_2 is mean wall temperature in degrees Kelvin (K), ϵ_1 is emissivity of the body surface, and ϵ_2 is emissivity of the chamber wall.

The skin-to-gas thermal gradient in the comfort zone is reduced as the ambient pressure increases (Fig. 44.4), and this contributes to the reduction of radiant heat transfer from body surface to chamber walls (31°–32°C in a helium–oxygen environment; Fig. 44.4). However, if the ambient temperature happens to be lower, as it might be in a personnel transfer chamber, radiant heat loss becomes considerable. There is danger that the high radiant heat exchange between human surface and chamber also may contribute to a rapid hyperthermia if the environmental temperature bounds are breached or when heavy work loads are given to divers in chambers with relatively high temperature.

Because radiant heat exchange depends on emissivity of the chamber walls, whose surface area is usually not large enough to neglect the ratio of the wall surface to the human body surface (A_1/A_2 in the equation), heat exchange should be calculated by accounting for the nature of the wall surface, including paint materials and color. Emissivity of a typical chamber wall painted with

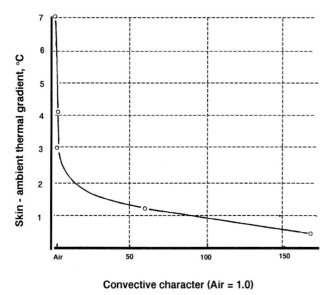

Convective character (Air = 1.0)

FIG. 44.4. Skin-to-ambient gas thermal gradient for mildly active men in various ambient temperatures. Adapted from Webb (165), with permission.

flat dark ivory lacquer is 0.95. However, when the steel walls of a personnel transfer chamber are unstained glossy metal, radiant heat exchange can be reduced considerably, being 0.52–0.56.

When a human is in a large room, A_1/A_2 then becomes small enough to neglect the radiant heat exchange, which can be calculated as in a normal atmosphere:

$$Q = \sigma A_1(T_1^4 - T_2^4)\epsilon_1$$

Evaporation

Evaporative heat transfer in a hyperbaric environment would be reduced because of physically reduced evaporative power in a hyperbaric helium gas environment and also by the pressure per se. As discussed above, increased convection brings about a reduction in mean skin temperature in hyperbaric helium, even though the environmental temperature is warmer than that in a normal atmosphere (129). The cooler skin surface will in turn cause a reduced evaporative heat loss from the skin (32). Indeed, a reduced insensible water loss was observed during a saturation dive at 31 ATA (154), as well as at 4 ATA (153). However, whether the reduction of evaporative water loss in a hyperbaric environment should be attributed to a reduced sweating ability is not yet explored. An experiment conducted at an acute 4 ATA air exposure (152) suggests that the hyperbaric environment may not impair sweating ability. In this experiment, the authors measured a significantly reduced change in resting body mass during 1 h exposure at 38°C on a Potter bed scale (0.5 g sensitivity), whereas local sweating rate measured by a dry-air ventilated capsule was independent of pressure. This experiment indicates that humans sweat normally in the high pressure environment, but evaporation of sweat from the skin surface is reduced below that at 1 ATA because of the attenuated diffusion of water vapor at increased pressure (125). Humidity at the boundary layer on the skin surface has a great influence on evaporation.

Light clothing gives considerable protection against the cold with increasing atmosphere, and in addition to this effect, clothing creates a vapor barrier against evaporative heat loss from the skin surface (60). Thus in a hyperbaric environment normal clothing tends to act impermeable to vapor.

Respiratory Heat Loss

Another potentially critical problem associated with saturation dives is related to respiratory heat loss. At 31 ATA, about 95% of respiratory heat loss is from convection, and the rest is via evaporation via humidification of the inspired gas. At 1 ATA air, however, the convective component accounts for only about 10% of respiratory heat loss (119).

Heat is needed to warm cool and dry inspired air to body temperature; additional heat is needed to evaporate moisture from the lining of the upper airway. As gas density and specific heat increase, the warming of the gas becomes the dominant element. The quantity of respiratory heat loss depends upon gas temperature, density, specific heat, and volume flow rate. The basic expression is

$$H_{resp} = \dot{V}\rho c_p(T_e - T_i) + 0.058\ \dot{V}(W_e - W_i)$$

where H_{resp} is the rate of respiratory heat loss in kcal/min; $\dot{V}$ is respiratory minute volume in 1/min [body temperature, ambient pressure, and saturated with water vapor (BTPS)]; ρ is the density of the gas in g/l; c_p is the specific heat of the gas in kcal/g·°C; T_e and T_i are the temperatures of expired and inspired gas, respectively, in °C; 0.058 is the latent heat of vaporization in kcal/g; W_e and W_i are the water contents of expired and inspired gas, respectively, in g/l. This equation tells us that the product of density and specific heat determines respiratory heat loss per unit quantity of air breathed. In other words, the denser the gas and the higher its specific heat the more heat is lost via ventilation. This problem can be eliminated only by heating the breathing gas (127).

It is evident that respiratory heat loss is a sizable avenue for body heat loss, especially when the environment is cooler than the comfort zone.

Possible Metabolic Effects

Raymond et al. (129) measured resting metabolism by analyzing expired gases in five subjects over a pressure range of 4.3–14.6 ATA. They observed that overall resting metabolism remained close to the control value in helium-rich hyperbaric environments. They also observed no pressure-related changes in rectal temperature.

Similarly, Shiraki et al. (153) failed to observe any elevation of resting metabolic rate in four subjects during the day in a 4 ATA helium-rich environment but found a significant increase in metabolic rate only early in the morning (standard metabolism; equivalent to the basal metabolic rate in a normal environment). There was no change in rectal temperature in all subjects during the entire period of observation. Continuous 24 h monitoring of body temperature over 12 days failed to show any significant changes in rectal temperature during 31 ATA saturation dives (93, 154). Thus even in a helium-rich hyperbaric environment, rectal temperature is well maintained.

What causes the small metabolic increase during sleep

in hyperbaric environments? The most likely explanation is that there is a chronically increased heat production to match a persistently high heat drain. It may be argued that humans really do not expose themselves to cold for long but rather find ways behaviorally to avoid prolonged periods of increased heat loss. Behavioral adjustment of heat dissipation by adding or removing clothing may be easier for humans during the day than at night during sleep.

Another hypothesis involves the increased work of breathing dense gases, a specific pharmacological effect, or some unknown effect of pressure, but the thermal balance argument is quite plausible.

Physical Activity

The importance of physical activity in maintaining thermal balance was suggested by Raymond et al. (131) based on the fact that a higher ambient temperature was necessary for thermal comfort during sleep. An elevated metabolism upon waking (153) may support this notion. In the actual hyperbaric environment, temperature is not always in the comfort range. Under emergency conditions it is not always possible to provide a comfortable temperature in a hyperbaric helium–oxygen environment. It is of prime importance for diving or medical officers to know how quickly body heat will be lost at a given temperature and atmospheric pressure when no special protective equipment is provided. We have no systematic studies of this problem, but there is much anecdotal observation to suggest that life-threatening hypothermia can develop within a few hours if the environmental temperature is below 20°C and pressure is greater than ~10 ATA. Raymond et al. (131) estimated the fall in rectal temperature as a function of time at 49.5 ATA for various temperatures between 0° and 30°C and concluded that unprotected subjects would be rendered helpless due to hypothermia in a few hours at temperatures below ~25°C.

Specific Technical Problems

Because convective heat loss from the body surface as well as from the respiratory tract is unusually high in hyperbaric environments, it is important that whenever saturation dives are undertaken, either in experimental chambers or in the water, enough measurements be taken to permit direct calculation of convective heat loss. This requires measurements of the gas temperature, pressure, viscosity, vapor pressure, velocity, wall temperature, and, for humans, skin temperature, internal temperature, and metabolic rate.

Measurement of oxygen consumption to predict metabolic rate is of prime importance, but there are technical problems to be solved. At increased ambient pressures, determination of fractional concentration of oxygen with sufficient accuracy is difficult because for a constant partial pressure the fractional concentration of gas is inversely proportional to depth. As an example, when the measurement of oxygen consumption is made while breathing air at 1 ATA, the oxygen concentration difference between inspired and expired mixed gas is 3% during light activity. However, at 31 ATA in helium–oxygen (99% He, 1% O_2) the difference would be 0.03%, assuming an unchanged alveolar ventilation. If one assumes an accuracy of $\pm 0.1\%$ (an accuracy typical of commercially available respiratory gas mass spectrometers) in determining the difference between inspired and mixed expired oxygen concentration, it is apparent that the error is unacceptably large. Unless we have oxygen analyzers with accuracy of 0.001%, we cannot expect accurate measurements of O_2 consumption in hyperbaric environments.

WORK CAPACITY

As discussed in the preceding sections, deep-sea saturation divers breathe a gas mixture that is usually far more dense than air at sea level. Even at a modest ambient pressure of 18.6 ATA, breathing He–O_2 with P_{O_2} of 0.3 ATA gas density is nearly four times that of 1 ATA air, which would correspondingly increase the resistance to breathing, thus limiting the physical exercise (or work) capacity of divers. Exercise under hyperbaric conditions has often been thought to be limited by ventilatory insufficiency. Nonetheless, human divers perform moderate physical work even in a dry chamber breathing a gas mixture, which is equivalent to He–O_2 breathing at a pressure of 152 ATA (5,000 FSW) (94). The safe depth limits for human exposure are not yet known, nor are the factors which may set such limits.

Linnarsson and Fagraeus (106) determined the maximal aerobic power ($\dot{V}_{O_2max}$) under high ambient pressure in a dry chamber as a measure of the integrated function of the oxygen transport system in human divers exposed to 1–3 ATA air (with P_{O_2} ranging from 0.2 to 0.6 ATA). They found that ($\dot{V}_{O_2max}$) increased by 8% during exposure to 1.4 ATA air (with P_{O_2} of 0.28 ATA) and that the time to reach exhaustion was prolonged in proportion to the raised oxygen uptake (Table 44.2). These phenomena were attributed to the moderate hyperoxia accompanying the slight rise in ambient pressure of the air. Similar results were reported by Dressendorfer et al. (42), who found a small but significant (3%) increase in $\dot{V}_{O_2max}$ when divers were exposed to a dry ambient pressure of 18.6 ATA with P_{O_2} of 0.3 ATA. The idea that the increase in maximal aerobic power at high pressure is primarily due to the slight hyperoxia is supported by additional findings that

TABLE 44.2. *Maximal Aerobic Power ($\dot{V}O_{2max}$) of Human Divers During Saturation Dives*

Total Pressure (ATA)	P_{O_2} (ATA)	Density Relative to 1 ATA Air	$\dot{V}O_{2max}$ ($1 \cdot min^{-1}$)			References
			(1) 1 ATA Air	(2) At Pressure	(2)/(1) (%)	
1.4 (air)	0.28	1.50	3.55	3.83	108.0	Linnarsson and Fagraeus (106)
3.0 (air)	0.62	3.30	—	—	106.0	Linnarsson and Fagraeus (106)
3.0 (He–O₂)	0.62	1.08	—	—	112.5	Linnarsson and Fagraeus (106)
18.6 (He–O₂)	0.30	3.80	1.59	1.64	103.0	Dressendorfer et al. (42)
18.6 (He–O₂)	0.21	2.80	1.59	1.54	97.0	Dressendorfer et al. (42)
31.0 (He–O₂)	0.40	7.30	3.11	2.71	87.0	Ohta et al. (123)
43.4* (He–O₂)	0.40	6.70	2.59	2.02	78.0	Dwyer et al. (43)
47–66 (He–N₂–O₂)	0.50	9.20–14.60	2.81	2.51	89.0	Salzano et al. (147)

*Wet dive; all others are dry dives.

$\dot{V}O_{2max}$ did not change when the same divers were exposed to the same 18.6 ATA (He–O₂) with normoxia (0.2 ATA P_{O_2}). Also, Dressendorfer et al. (42) noted an increase of 2 min in endurance time during hyperoxic gas breathing at 18.6 ATA, whereas normoxic values at 18.6 ATA were similar to those at 1 ATA air. It thus appears that, over a moderate pressure range, moderate hyperoxia ($P_{O_2} \geq 0.3$ ATA) is responsible for maintenance of normal work capacity. However, as ambient pressure increases breathing gas density should increase and at some point should counteract the beneficial effect of hyperoxia.

Other studies conducted at dry ambient pressures higher than 30 ATA (with $P_{O_2} = 0.3$–0.5 ATA) also support the notion that work capacity decreases at depth. Ohta et al. (123) reported a 13% decrease in $\dot{V}O_{2max}$ in divers exposed to 31 ATA ($P_{O_2} = 0.4$ ATA). However, Lambertsen et al. (94), although they did not specifically determine $\dot{V}O_{2max}$ itself, studied the performance success of an increasingly severe pattern of exercise on a bicycle ergometer at approximately 80% of the subject's maximal work capacity under normal conditions at 1 ATA. Failure to complete this work occurred only under a condition equivalent to performing the heaviest work at a respired gas density equivalent to that expected with helium at a simulated depth of 5,000 FSW (152 ATA) or with hydrogen at 10,000 FSW (304 ATA). However, Lambertsen et al. (94) caution that there are factors other than gas density (for example, hydrostatic pressure, narcotic or other pharmacological effects of inert gases, diffusion limitation),

which introduce truly detrimental effects upon physiological systems and purposeful activity.

In another deep dry saturation dive at 47–66 ATA trimix (He–N₂–O₂) conducted at Duke University, Salzano et al. (146) noted that dyspnea at rest and during exercise was evident in all divers and was predominantly inspiratory in nature. Despite dyspnea, divers were able to perform work requiring a $\dot{V}O_2$ greater than 2 l/min at each depth. Compared with surface measurements, moderate work at depth resulted in alveolar hypoventilation, arterial hypercapnia, higher levels of arterial lactate, and signs of simultaneous respiratory and metabolic acidosis. Interestingly, the increase of ventilation that accompanies the onset of acidemia at the surface was not seen at depth; acidemia at depth was more severe, and its onset occurred at lower work rates than at 1 ATA.

Work tolerance of divers engaged in a wet dive to simulated depths of 1,400 to 1,600 FSW (43.4–49.5 ATA) was also considerably less than predicted on the basis of effects of gas density alone (43, 161). Since these studies were conducted in the wet pot (that is, under water), the low work tolerance may have been caused by submersion or by exposure to a hydrostatic pressure larger than that in the studies from which the predictor was formulated. Regardless of the mechanisms underlying the effects of immersion on work tolerance, it is worth noting that during a wet dive to 1,400 FSW (43.4 ATA He–O₂) $\dot{V}O_{2max}$ decreased by 22%, the biggest pressure-induced reduction in maximal aerobic power reported in the literature (see Table 50.1 in the chapter by Cerretelli and Hoppeler in this *Handbook*). In addition, Dwyer et al. (43) noted a 30% reduction in maximal external work at 43.4 ATA. Dwyer et al. (43) and Spaur et al. (156) also noted severe dyspnea during exercise in the wet pot at both pressures.

Another important finding by Salzano et al. (146) is that there were no large differences in a variety of responses when inspired gas had a density of 7.9 g/l at 47 ATA or a density of 17.1 g/l at 66 ATA. Apparently the impact of the hyperbaric environment on the physiological response to physical work was almost fully manifested at a pressure of 47 ATA with a He–O_2 gas mixture.

Severe dyspnea becomes much more exaggerated during underwater work at high ambient pressure. The exact mechanism underlying this dyspnea is not known. Spaur et al. (156) observed severe dyspnea with normal or low arterial CO_2 and normal oxygen transport, thus excluding hypercapnia and/or hypoxemia as the factors responsible. Structural changes in the lungs were also excluded as a factor in the limiting symptoms of dyspnea. Additional studies should be directed toward the elucidation of the mechanism for the development of dyspnea of divers (especially submerged divers) exposed to high ambient pressure.

During multiday saturation dives, divers are exposed to high ambient pressure with a phase of compression and decompression and an artificial atmosphere, where Po_2 of the breathing gas is usually elevated; moreover, the lungs are exposed to the high density of the breathing gas under pressure, thereby increasing the work of breathing. During decompression, venous gas microemboli may be generated that are subsequently filtered in the pulmonary circulation, where they can induce a pulmonary inflammatory reaction and gas exchange abnormalities (79). Pathological conditions in the lungs induced by these processes can restrict inert gas elimination during decompression and may reduce the efficiency of the lung as a filter for venous gas emboli (34). Thorsen et al. (161) studied pulmonary function and exercise tolerance before and after three saturation dives (He–O_2) to a pressure of 39 ATA with bottom times of 3–13 days and found that $\dot{V}O_{2max}$ decreased by 14% after the dive; there was a significant correlation between decrease in $\dot{V}O_{2max}$ and accumulated bubble load on the pulmonary circulation, indicating a peripheral pulmonary lesion with impaired gas exchange.

Because reduced ventilatory capacity is one of the most important factors that may restrict physical work capacity in hyperbaric environments, H_2 with its low specific density can be expected to give even lower breathing resistance than He, so divers engaged in a H_2–O_2 saturation dive could be expected to show a high work capacity. However, in the first deep saturation HYDRA dive using H_2 (65) (total pressure of 46 ATA with PH_2 of 25 ATA), the minute volume ($\dot{V}_E$) tended to be smaller and end-tidal PCO_2 greater in H_2–O_2 than in He–O_2 dives, interpreted as evidence for alveolar hypoventilation. Moreover, the divers felt subjectively less difficulty in performing a given exercise in H_2–O_2 than in He–O_2. These observations are preliminary in nature, and quantitative data on the physiological parameters of physical exercise, such as the $\dot{V}O_{2max}$ and endurance time, during H_2 dives are not yet available.

CARDIOPULMONARY FUNCTIONS

Since the 1960s, comfortable hyperbaric conditions, compression profiles, and decompression procedures have been systematically worked out to provide physiological as well as psychological requirements that allow humans to live and work under high ambient pressures. The goal has been to enable divers to do useful work at moderate intensities through the refinement of hyperbaric environments, instruments, and tools. The word "useful" is emphasized here because exercise capacity is diminished by respiratory limitations at very high pressure. Divers cannot perform at sea-level $\dot{V}O_{2max}$; even if they could, there would be little physiological reserve remaining.

Chouteau (37) reported that goats exhibited symptoms of hypoxia under pressure while breathing a normoxic He–O_2 mixture. Modest elevation of PIO_2 abolished hypoxia and symptoms. We now know that besides depressing ventilatory capacity, increased gas density may interfere with pulmonary gas exchange by a variety of interacting processes, such as $\dot{V}_A/\dot{Q}$ inhomogeneity, stratified inhomogeneity, intrapulmonary gas mixing, dead space volume, affinity of hemoglobin for O_2, and binary diffusion coefficients. Some of these are predictable, some are not, and some even produce off-setting effects. Lanphier and Camporesi (96) reviewed these factors to provide partial explanations for the Chouteau effect that occurs at depth while breathing normoxic mixtures, particularly during exercise.

In saturation diving, there have been no startling revelations as far as cardiorespiratory function is concerned. Human subjects retain about 90% of the sea-level $\dot{V}O_{2max}$ in a 66 ATA trimix (He–N_2–O_2) condition, where divers breathe a gas mixture 15.5 times the density of sea-level air (146; also see WORK CAPACITY, above). Except for occasional arrhythmias which are no more frequent than at sea level and a transient orthostatic intolerance, cardiovascular function is apparently normal. The circulatory system has been dismissed as the limiting factor for working in hyperbaria. However, breathing high-density gases raises the work of respiration and limits exercise capacity, so the relationships between gas density, adequacy of ventilation, homeostasis of gas exchange, and work of respiration have received the most attention.

Respiratory Functions

Pulmonary Ventilation. Fluid dynamics and lung mechanics predict increased pulmonary resistance and respiratory work with a rise in gas density. Lanphier and Camporesi (96) compiled available data showing the relationship between maximal voluntary ventilation (MVV) and gas density as $MVV = (MVV_o)\rho^k$, where $k = -0.5$ and MVV_o denotes sea-level value. This equation shows that the effect on MVV trails off as gas density (ρ) increases (96). Further data have shown that k values range from -0.37 to -0.45, possibly resulting from the refinement of breathing apparatus and/or better training of divers (for review see ref. 63). A small value of k indicates a smaller decrease of MVV with increasing gas density. In this MVV–density relationship, it is the relative density that matters as far as ventilatory capacity is concerned. It makes no difference whether the density is reached by varying pressure or by using gas mixtures. Lambertsen et al. (94) predict adequate ventilatory performance of humans at 5,000 FSW, or 150 ATA, using a He–O_2 mixture 22 times the density of 1 ATA air, if other physiological factors are not limiting. Theoretically, this performance could be extended using hydrogen.

Maximal voluntary ventilation values provide a rough estimate of the expected $\dot{V}_{Emax}$ and predict sustainable ventilation for useful work at depth. Estimates of $\dot{V}_{Emax}$ range from a pessimistic 50% to an optimistic 80% of MVV at pressure. Performing near MVV at a depth of 5,000 FSW in an equivalent He–O_2 condition, Lambertsen et al. (94) showed that subjects encountered severe ventilatory limitation but were able to accomplish work up to 1,200 kpm/min. At more moderate work levels, ventilatory capacity was sufficient to keep P_{ACO_2} below 50 mm Hg. Morrison and Florio (120) tested two subjects up to 1,500 FSW depth in He–O_2 environments and reported no significant respiratory problems during rest or performance of moderate work. However, as the required level of ventilation approaches MVV at depth, dyspnea appears, especially during work under water (see WORK CAPACITY, above).

Airway Resistance and Work of Respiration. Gas flow is density-dependent in central airways because of a small cross-sectional area, high flow velocity, turbulence, and convective acceleration. Consequently, the work of breathing increases in hyperbaria. Airway resistance during expiration is about two times the inspiratory

resistance at rest. Expiratory flow resistance rose further during voluntary hyperventilation at all levels of gas density (164). Maio and Farhi (108) showed that the nonelastic pressure (ρ_{nonel}) increases with flow rate ($\dot{V}$) and gas density (ρ) in an empirical equation, $\rho_{nonel} = 0.40(\rho + 0.7)\dot{V}^{1.6}$. All of these point to the resistive causes of diminished ventilatory capacity and aerobic power at depth.

In breathing dense gases, as in added airway resistance, it is advantageous for the respiratory system to adapt a pattern of slow breathing with large tidal volume to reduce resistive work and to adapt an increased preinspiratory lung volume to allow for faster expiration with modest expiratory effort. Indeed, divers breathing the chamber gas directly or through scuba have tuned their respiratory activity this way without being aware of it.

Pulmonary Gas Exchange. Besides the increased work of breathing, alveolar gas exchange may also be hampered in dense gas environments because of a decreased diffusion coefficient of gases. On the one hand, a lowered O_2 and CO_2 exchange is expected because of poor diffusive mixing in alveoli at depth, but on the other hand, the axial penetration of dense gases may be enhanced and $\dot{V}_A/\dot{Q}$ distribution improved (38, 170). The end result is that minimum (A − a)P_{O_2} and (a − A)P_{CO_2} exist in mixed-gas saturation diving when P_{IO_2} is kept between 0.3 and 0.4 atmospheres (atm).

Ventilatory Drive. Homeostasis of alveolar gas composition depends on ventilatory responses to mechanical, neural, and chemical stimuli. An inappropriate response leads promptly to hypo- or hypercapnia. Among the causes of relative hypoventilation and CO_2 retention in hyperbaria, increased gas density and reduced CO_2 responsiveness count the most, with hyperoxia playing a minor role (95). Ventilatory response to CO_2 decreases by 21% in a 31 ATA He–O_2 environment (123), and in extreme cases the reduction reaches 80% (64). Alveolar and arterial hypercapnia result from exercise and are made especially worse when working in water.

Cardiovascular Functions

Electrocardiographic monitoring is routine in saturation dives, but it yields few clinically significant changes at rest or during exercise (44, 169). Bradycardia with varied persistency occurs upon pressure exposure (see Hyperbaric Bradycardia, below). Associated with bradycardia are occasional atrioventricular–nodal escape rhythms, isolated premature ventricular beats, and a lengthening of QT intervals. All changes were asymptomatic, however. Bradycardia and the suspicion of retarded ventricular depolarization triggered a series of studies on electrical events in the myocardium showing that very high pressure per se depresses cardiac excitability (for review see ref. 80).

Hyperbaric Bradycardia. Bradycardia occurs at rest (Fig. 44.5) and during exercise (Fig. 44.6) in hyperbaric environments (105). Some studies reported that bradycardia diminishes spontaneously within days of pressure exposure. In most studies, however, bradycardia persists throughout the dive, with a gradual return toward the predive level. Even so, the heart remains responsive to pressure changes, with bradycardia occurring in each pressure excursion (169). Matsuda et al. (112) reported that bradycardia persisted throughout a 7 day exposure at a 7 ATA He–O$_2$ environment, and heart rate remained low despite intentional cold exposure during the fourth and fifth days. Salzano et al. (146) reported an absence of hyperbaric bradycardia at rest, but during exercise a trend for relative bradycardia existed for a given work rate: the higher the density, pressure, and/or P$_{N_2}$ the lower the heart rate.

Two major categories have been identified in hyperbaria-induced bradycardia: oxygen-dependent and non-oxygen-dependent. The involvement of hyperoxia in bradycardia is clear at sea level as well as in hyperbaric conditions, at rest and during exercise, and in both humans and animals (105, 150). Hyperoxia causes vasoconstriction, elevates vagal tone (41, 49), and depresses sympathetic activity (78, 149) via baroreceptors and chemoreceptors.

Hyperbaric bradycardia remains after eliminating

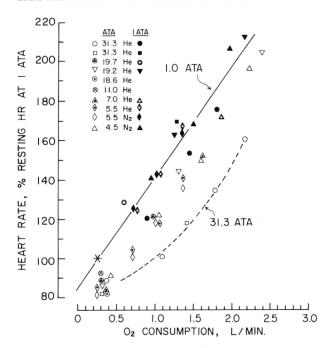

FIG. 44.6. Heart rate during exercise in hyperbaric environments. Numerical values and sources are listed in Table 2 in Lin and Shida (105). Reproduced with permission (105).

hyperoxic effects by experimental designs. Aside from the studies in isolated cardiac tissues, the nonoxygen-dependent effects of gas density, pressure, and inert gases are not easily separable. The relationship between pressure and gas density is such that changes in gas density at a given pressure cannot be achieved without introducing various diluent gases. Flynn et al. (52) studied heart rate in exercising humans at 1.00, 3.27, and 5.45 ATA, with gas density remaining constant at 1.11 g/l, and showed that heart rate falls with increased pressure. Although this result indicates that pressure per se depresses heart rate, in the same study it was shown that increased gas density lowers heart rate, with ambient pressure remaining constant (52). Adding to the confusion, inert gases at high pressure may also produce cardiac depression (104) and change in the β-adrenergic activity (49).

Hemodynamics. In a 1965 review, Schaefer (148) stated: "There is general agreement that the pulse rate decreased under increased pressure, whereas findings on blood pressure vary so much that no definite statement is possible." Unfortunately, not much progress has been achieved since then. Although there have been few measurements, there are indications that arterial blood pressure is apparently normal in hyperbaria. The absence of blood pressure abnormalities contributes to the lack of enthusiasm for monitoring arterial blood pressure. Although not related to prolonged inactivity, transient

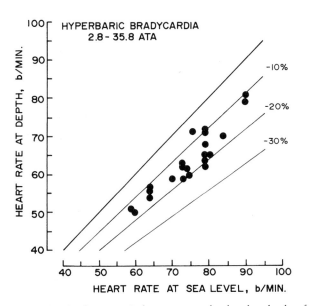

FIG. 44.5. Resting heart rate in humans at sea level and at depths of up to 35.8 ATA. See Table 1 in Lin and Shida (105) for numerical values and references. Reproduced with permission (105).

blood pressure instability and orthostatic intolerance may occur early in a hyperbaric exposure (see Orthostatic Intolerance, below). Elevated arterial Po_2 causes vasoconstriction notably in the retina, brain, kidney, and heart. Consequently, cardiac output falls by 10%–15% in humans breathing 1 atm or greater O_2 (41, 168). In hyperbaria cardiac output is either increased or unchanged compared to the precompression state because an increase in stroke volume offsets hyperbaric bradycardia (3, 110, 123, 155). Nonelastic respiratory pressure increases during inspiration, accentuating the lowering of intrathoracic pressure and causing a pooling of central blood volume. This increases the preload of the heart and the stroke volume. Since the afterload does not change in hyperbaria, an enhanced contractility could also increase stroke volume. A positive inotropic effect of increased pressure has been reported in isolated myocardium (4) and in anesthetized rats (158).

Orthostatic Intolerance. Fainting occurs during 15 min of 90° head-up body tilt in a 31 ATA He–O_2 environment (3, 102) and in a trimix (He–N_2–O_2) at 46 ATA (103); before or after exposure the same test did not produce fainting. Cardiovascular deconditioning typically occurs following prolonged inactivity, but indications of deconditioning appear earlier than can be accounted for by prolonged inactivity. Using the sum of perturbations in the systolic and diastolic blood pressures and heart rate (cardiovascular index of deconditioning or CID), circulatory deconditioning can be documented without the need for syncope as the endpoint. An increased CID compared to preexposure values suggests deconditioning (33). Typically, CID rises within 24 h of exposure, returns to sea-level values, and rises again during prolonged chamber confinement. The initial rise in CID corresponds to the hyperbaric diuresis (see BODY FLUID BALANCE, below), and the late increase in CID reflects deconditioning following prolonged inactivity. Deconditioning is preventable by prudent management of the diver's fluid intake and implementation of an exercise program.

BODY FLUID BALANCE

Profound alterations in the manner in which water and electrolyte balances are achieved occur in humans exposed to hyperbaric environments. This potential problem and area of research interest was recognized first by Hamilton (73), who reported diuresis in subjects performing a dry saturation dive to 20 ATA. Much has been learned regarding the characteristics of this diuretic response, but our understanding of the underlying mechanisms is less clear. This area has been most

comprehensively reviewed by Hong and Claybaugh (81).

Physical Characteristics of the Environment Which Affect Body Fluid Balance

The change in thermoneutral temperature (see review under THERMOREGULATORY FUNCTION, above) was among the first effects recognized as potential contributors to the diuresis associated with hyperbaria. During early hyperbaric experiments, subjects were simultaneously exposed to cold stress, so diuresis was attributed to the well-known effect of cold on urine flow (15). However, when chamber temperatures were made thermoneutral, diuresis still occurred (112) and was sustained during the hyperbaric phase with no accompanying increase in thirst. Although this appeared as a water imbalance, total body water was maintained better than would be calculated from the discrepancy in water intake and urinary excretion (112). From these observations, Hong and Paganelli (see review, ref. 84) hypothesized that insensible water loss was reduced due to reduction of the diffusivity of water vapor in the hyperbaric environment. In Hana Kai II, a He–O_2 saturation dive to 18.6 ATA conducted in 1977, during 17 days at pressure, daily urine flow remained high by about 1 l per day, with water intake remaining relatively constant. No change in total body water was detected, and a reduction of about 35% in insensible water loss was measured (82). Paganelli and Kurata (125) calculated that the diffusion of water vapor would decrease by about 80% from 1 ATA air values when compared to 18.6 ATA in an environment of 98% He and 2% O_2. Thus, theoretically, the hyperbaric environment poses an additional perturbation on body fluid balance, but the theoretical 80% reductions in evaporative water loss are more severe than the 35% reductions of insensible water loss that were actually measured at 18.6 ATA (82). A role of reduced evaporative water loss in hyperbaric diuresis could only be significant if thirst were maintained at normal volumes of water intake. The reduced insensible water loss would then lead to decreased plasma osmolality and subsequently a reduced release of antidiuretic hormone (ADH).

Another physical aspect of the hyperbaric environment which probably affects body fluid balance is the effect of increased gas density on ventilation mechanics, and hence, on intrathoracic pressure. It would be expected, for instance, that high gas density would increase the inspiratory resistance similar to negative pressure breathing. Thus it has been proposed (39, 82) that the negative pressure during inspiration increases intrathoracic blood volume and contributes to the hyperbaric diuresis via the Gauer–Henry reflex. This potential effect of gas density is relevant when the threshold for the hyperbaric diuretic response is examined. Niu et al. (122) reported no diuresis in subjects exposed to 2.5 ATA N_2–O_2 at a density of 3.16 kg·m^{-3}, and no diuresis occurred at 4 ATA in He–O_2 at a density of 2.0 kg·m^{-3} (153); however, at a similar depth (4 ATA) in N_2–O_2 at a density of 5.0 kg·m^{-3} diuresis was observed (2). It appears that gas density plays some role in the diuretic response of hyperbaria.

Characteristics of Hyperbaric Diuresis

In a previous review (81), 14 saturation dive studies were statistically analyzed to determine if urine flow rate and the magnitude of pressure were correlated. The 14 studies were selected because thermoneutral temperatures were achieved and urine flow was measured. There was no correlation between urine flow and pressure. In nine of these studies and in three subsequent studies, urine osmolality was also determined; a summary of these 12 dives is included in Figure 44.7. Although diuresis was present in all of these dives, its magnitude is not correlated with the depth of the dive (Fig. 44.7, panel a). However, a significant negative correlation

exists between urine osmolality and ambient pressure (Fig. 44.7, panel b). Furthermore, there is a significant negative correlation between urine osmolality and urine flow (Fig. 44.7, panel c) but not between urine flow and the daily excretion of osmotic particles (Fig. 44.7, panel d). These data indicate that the majority of hyperbaria-induced diuresis is a result of increased excretion of a dilute urine and is not osmotic in nature. It appears paradoxical that urine osmolality could be negatively correlated with pressure and urine flow, while at the same time pressure and urine flow were not significantly correlated. This unlikely combination of events seems to be the result of a significant negative correlation between the excretion of osmotic particles and pressure (Fig. 44.7, panel e). This latter observation may be due to a reduced increase in urea excretion as pressure becomes greater. A large portion of the osmotic substances excreted at 7 ATA can be attributed to a 50% increase in excretion rate of urea (113), and at 31 and 49.5 ATA a 30% increase is observed (130, 151). Since collecting duct reabsorption of both water and urea are under the influence of ADH, increased excretion of free water or urea suggests the same mechanism. Currently we are

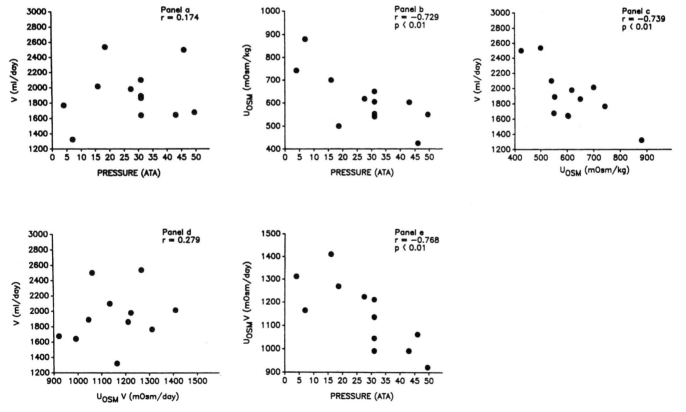

FIG. 44.7. Correlations of data obtained from Alexander et al. (2), Matsuda et al. (113), Hong et al. (82), Neuman et al. (121), Leach et al. (97), Claybaugh et al. (39), Shiraki et al. (151), Raymond et al. (130), Goldinger et al. (66), Sagawa et al. (145), and Miyamoto et al. (117) were used to compile the 12 points. Leach et al. (97) was used for two pressures. Correlation coefficients and level of significance are shown. V = urine flow, U_{OSM} = urine osmolality, $U_{OSM}V$ = excretion of osmotic particles.

without explanation for the decreased excretion of osmotic substances at greater pressures.

In addition to increases in free water and urea excretion rates, however, the early phase of hyperbaric diuresis is frequently accompanied by temporary increases in excretion rates of sodium, potassium, and organic phosphorus in the absence of any change in creatinine excretion (81). These transient negative balances in electrolytes undoubtedly contribute to diuresis, and, although the contribution would appear to be small over several days of hyperbaria, as concluded from Figure 44.1, the resulting body-weight losses of about 1 kg (2, 82, 151) are significant. For instance, a loss of body fluid was accompanied by an increased hematocrit and a calculated reduction in plasma volume of about 10% after 2 days at 31 ATA (151).

Reduced insensible water loss and continued water intake could explain all of the free water diuresis during hyperbaria, provided there is a decrease of plasma osmolality, which would result in a reduction in ADH release in response to osmoreceptor inhibition. Although plasma osmolality is reduced in some saturation dives (151), it is frequently unchanged (82, 130) or increased (2, 39). These observations suggest that mechanisms other than reduced plasma osmolality are causing ADH to be reduced.

Finally, studies have shown hyperbaric diuresis to be primarily nocturnal (39, 145, 151). Urine flow was significantly increased during both daytime and nighttime, but the increase in nighttime urine flow was greater.

This nocturnal diuresis was due to an increase in osmotic clearance, and the primary osmotic substance, when measured, was urea. Negative free water clearance was reduced during both the daytime and the nighttime (151).

Mechanisms of Hyperbaric Diuresis

In most previous saturation-dive studies, urinary excretion of ADH has been reduced (39, 40, 82, 97, 113); exceptions include a transient period during comression and early during the isopressure stage of hyperbaria (82, 130). Among several possibilities, this increase may be related to discomforts of nausea associated with slight episodes of HPNS. Even in these instances, ADH was significantly reduced later in the dives. The reductions of urinary ADH have now been confirmed by measurements of plasma ADH concentration (40, 114, 117). A significant portion of free water diuresis and the increased excretion of urea appear to be due to decreased plasma levels of ADH.

The mechanism causing the reduction in ADH levels is not clear. In two of the studies where plasma ADH has been reported (40, 117), simultaneous plasma samples indicated either a decrease (151) or no change (117) in plasma osmolality. The consistency of reduced urinary excretion of ADH and the inconsistency of the plasma osmolality response strongly suggest mechanisms other than, or in addition to, osmoreceptor inhibition of ADH. As stated above, the possible influence

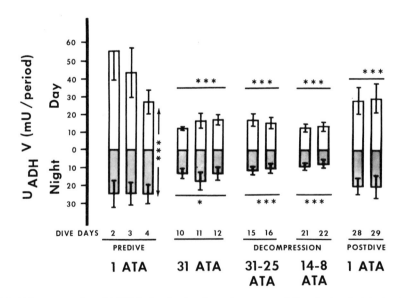

FIG. 44.8. Urinary excretion of ADH during daytime (open bars, upward direction) and nighttime (shaded bars, downward direction) at predive, 31 ATA, decompression, and postdive. Vertical lines on histogram bars represent ± SE. Vertical arrows with *** represent $P < 0.005$ for day vs. night when all three predive days are considered. Horizontal lines indicate differences between that period when compared to corresponding daytime or nighttime values for days 2, 3, and 4 combined at $P < 0.05$ (*), $P < 0.01$ (**), and $P < 0.005$ (***). Reproduced from Claybaugh et al. (40), with permission.

of increased gas density and subsequent increased resistance to breathing was suggested by Hong et al. (82). In theory, negative pressure breathing could increase thoracic blood volume and reduce ADH release by the Gauer–Henry reflex. To test this possibility, Hebden et al. (77) subjected people to negative pressure breathing (-15 cm H_2O for 60 min, inspiratory phase only), once in the seated position and once in the upright position and observed no effect on plasma ADH. Similarly, Tanaka et al. (160) found that even continuous negative pressure breathing (-11 mm Hg for 60 min) had no effect on ADH. The degree of negative pressure breathing produced in these studies exceeds the -3.9 cm H_2O measured in saturation divers at 49.5 ATA by Raymond et al. (130). However, it could be argued that longer exposures to negative pressure breathing may have a different effect than these acute studies.

Studies have shown that the normal increase in plasma ADH concentration in response to passive 90° head-up tilt disappears at 31 ATA (114). In addition (as seen in Fig. 50.8 in the chapter by Cerretelli and Hoppeler in this *Handbook*), the normal circadian pattern of ADH excretion in urine, high during the day and low during the night, disappears at 31 ATA (39, 40). This circadian rhythm is thought to be a result of the upright posture during the daytime and not of a central zeitgeber. Thus both the tilt table and the circadian rhythm studies support a reduced baroreceptor sensitivity of ADH. However, although ADH is decreased most during the daytime, nighttime excretion rates have also been shown to decrease (Fig. 44.8). The question arises, can ADH release be stimulated in people exposed to a hyperbaric environment? A single example has been observed; ADH release can still be evoked by sensations of syncope (114). The decreased ADH excretion coincided with a reduced negative free water clearance during both daytime and nighttime (151).

Reduced baroreceptor sensitivity probably does not involve afferent activity from the baroreceptors because, while ADH responses to upright tilt are eliminated, plasma renin activity is enhanced (114). The enhanced responses of renin would be expected since there is a reduced plasma volume at hyperbaria.

Figure 44.9 shows the plasma levels of various hormones which affect body fluid homeostasis in response to 31 ATA. These responses are typical, though occasional deviations have been reported. Plasma ADH concentration decreased in all studies where it was measured (40, 114, 117). Increases in plasma renin activity and aldosterone, possibly resulting from the decreased plasma volume, would appear to have a role in the kaliuresis which is usually observed. Additionally, an occasional antinatriuresis occurring after several days of saturation at 31 ATA (152) and 49.5 ATA (130) has been attributed to increased aldosterone. Parathyroid hor-

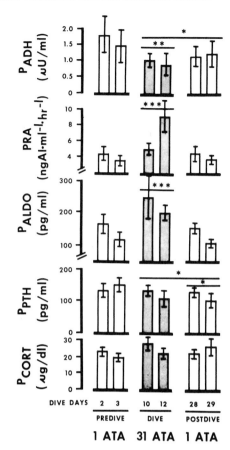

FIG. 44.9. Plasma concentrations of water and electrolyte regulating hormones at predive, 31 ATA *(Dive)*, and postdive. Antidiuretic hormone *(ADH)*, plasma renin activity *(PRA)*, aldosterone *(ALDO)*, parathyroid hormone *(PTH)*, and cortisol *(CORT)* were measured in plasma. Horizontal bars over histogram bars indicate significant differences compared to predive days 2 and 3 at $P < 0.05$ (*), $P < 0.01$ (**), and $P < 0.005$ (***). Reproduced from Claybaugh et al. (40), with permission.

mone has been measured only once (40) in an attempt to explain the occasional increase in plasma Ca^{2+} concentration and the more frequently occurring phosphaturia. Parathyroid hormone was not found to be significantly elevated, only a modest phosphaturia and no increase in plasma Ca^{2+} being observed in that study (151).

To provide information on the mechanism for the natriuresis that occasionally occurs, particularly at nighttime (117, 145), plasma concentrations of atrial natriuretic peptide (ANP) have been reported in three saturation dives. Two studies indicate no response of ANP in subjects exposed to 31 (145) or 46 (118) ATA; one shows a significant increase in ANP at 16 ATA (117). Taken as is, the ANP results indicate an inconsistent response that is not correlated with urinary excretion of sodium. Also, the renin–angiotensin–aldosterone system is usually stimulated and cannot, therefore, be implicated in a natriuresis occurring at hyperbaria. An alternative possibility has been provided by Goldinger et al. (67), who showed an inhibition of active sodium transport in the human erythrocyte by high hydrostatic pressure. They observed a 30% decrease in sodium efflux at 30 ATA that was reversible. Additional experiments have shown that sodium transport across toad skin, a model for human collecting duct, is inhibited by 20% at 50 ATA (83). Taken together, these results are compatible with a reduced renal tubular reabsorption of sodium resulting from inhibition of active sodium transport by high hydrostatic pressure.

Effects of Immersion

Field applications of saturation diving incorporate the two severe environmental challenges of hyperbaria and episodes of water immersion. It is reasonable to expect that the combined effects could produce serious reductions in body fluid. Studies of body fluid regulation and control in open-sea saturation dives are limited by obvious technical difficulties. In one such study reported by Neuman et al. (121), divers were saturated at 850 FSW and did four excursion dives to 950 FSW. An exaggerated diuresis was not observed. The general characteristics of the diuresis, however, were similar to those previously observed in hyperbaria. More detailed studies have been conducted in chambers equipped with a temperature-controlled water tank. During immersion in water at 15°C, diuresis was greater at 11 ATA than at 1 ATA (111). Although this increase in urine flow was evident during immersion, it was greatest after the subject came out of water. This was attributed to the additional cold stress at hyperbaria but represents a realistic situation and points out the need for thermal protection at depth. When water temperature was maintained at 35°C during immersion at 31 ATA, urine flow was not greater than at 1 ATA (110).

Thus immersion diuresis is similar in magnitude during hyperbaria when water temperature is not low. The characteristics of immersion diuresis are similar with respect to free water and osmotic clearance. Urinary excretion of ADH was decreased by immersion, especially in cold water at 1 ATA, but with hyperbaria-induced reductions in basal levels of ADH a significant reduction in response to immersion was not detected at 11 ATA or 31 ATA. However, when ADH levels are low, further slight reductions that, due to assay and sampling variability are not statistically significant, can produce large increases in urine flow. Plasma aldosterone concentration was reduced by similar degrees during immersion at both 1 ATA and 31 ATA (110). It appears that immersion diuresis is similar at 1 ATA and at hyperbaria and is additive to hyperbaric diuresis.

CONCLUSIONS

Hyperbaric environments greatly alter the normal mechanisms whereby human beings regulate water balance, yet water balance is maintained quite well. Certainly one of the leading events which leads to the cascade of physiological adjustments is the reduced insensible water loss at hyperbaria, and fluid intake remains at normal levels instead of being appropriately depressed. Also, ADH is reduced, allowing the body to void the free water and maintain normal plasma osmolality and total body water. These observations are consistent in virtually all studies, and it follows that the deeper the dive and the more severe the reduction in evaporative water loss, the more dilute the urine must become to maintain normal plasma osmolality. Taken together, all of the studies suggest this response. The major issue remaining is to explain the reduction of ADH which often coincides with normal or increased plasma osmolality and usually with a decreased plasma volume. The increased loss of osmotic substances, mostly due to increases in urea excretion and therefore perhaps a response to decreased ADH also, leads to a transient negative water balance in the compression phase and early in the isopressure phase. In addition, the role of high hydrostatic pressure per se on Na^+ transport mechanisms may contribute to natriuresis in early phases of hyperbaria. The resulting body fluid volume deficit appears to persist (130, 151) and probably contributes to the increased plasma renin and aldosterone levels. The characteristic increase in K^+ excretion is likely influenced by elevated aldosterone levels.

The threshold for hyperbaric diuresis would appear to be pressure-dependent and gas density–dependent at approximately 4 ATA and at a gas density above 3 $kg \cdot m^{-3}$. The exact interplay of these two parameters on diuresis deserves further research.

REFERENCES

1. Abraini, J. H., and J. C. Rostain. Pressure induced striated dopamine release correlates with hyperlocomotor activity in rats exposed to high pressure. *J. Appl. Physiol.: Respir. Environ. Exerc. Physiol.* 71: 638–643, 1991.
2. Alexander, W. C., C. S. Leach, C. L. Fischer, C. J. Lambertsen, and P. C. Johnson. Hematological, biochemical, and immunological studies during a 14-day continuous exposure to 5.2% O_2 in N_2 at pressure equivalent to 100 FSW (4 ata). *Aerospace Med.* 44: 850–854, 1973.
3. Arita, H., Y. C. Lin, M. Sudoh, I. Kuwahira, Y. Ohta, H. Saiki, S. Tamaya, and H. Nakayama. Seadragon VI: a 7-day dry saturation dive at 31 ATA. V. Cardiovascular responses to a 90° body tilt. *Undersea Biomed. Res.* 14: 425–436, 1987.
4. Ask, J. A., and I. Tyssebotn. Positive inotropic effect on the rat atrial myocardium compressed to 5, 10, and 30 bar. *Acta Physiol. Scand.* 134: 277–283, 1988.
5. Behnke, A. R., and O. D. Yarborough. Physiologic studies of helium. *U. S. Navy Med. Bull.* 36: 542–548, 1938.

6. Bennett, P. B. *Psychometric Impairment in Men Breathing Oxygen–Helium at Increased Pressures.* London: U. K. Medical Research Council, R.N. Personnel Research Committee, Underwater Physiology Subcommittee Report 251, 1965.

7. Bennett, P. B. Performance impairment in deep diving due to nitrogen, helium, neon and oxygen. In: *Proc. 3rd Symp. Underwater Physiol.,* edited by C. J. Lambertsen. Baltimore: William and Wilkins, 1967, p. 327–340.

8. Bennett, P. B. A strategy for future diving. In: *Proc. 8th UHMS Workshop. The Strategy for Future Diving to Depths Greater than 1000 ft.* Report W.S. 6.15.75. Bethesda: Undersea Med. Soc., 1975, p. 71–86.

9. Bennett, P. B. The high pressure nervous syndrome. In: *The Physiology of Diving and Compressed Air Work,* edited by P. B. Bennett and D. H. Elliott. London: Saunders, 1993, p. 194–237.

10. Bennett, P. B. Physiological limitations to underwater exploration and work. *Comp. Biochem. Physiol.* 93A: 295–300, 1989.

11. Bennett, P. B., G. D. Blenkarn, J. Roby, and D. Youngblood. Suppression of the high pressure nervous syndrome in human deep dives by He–N_2–O_2. *Undersea Biomed. Res.* 1: 221–237, 1974.

12. Bennett, P. B., R. Coggin, and M. McLeod. Effect of compression rate on use of trimix to ameliorate HPNS in man to 686 m (2250 ft). *Undersea Biomed. Res.* 9: 335–351, 1982.

13. Bennett, P. B., R. Coggin, and J. Roby. Control of HPNS in humans during rapid compression with trimix to 650 m (2132 ft). *Undersea Biomed. Res.* 8: 85–100, 1981.

14. Bennett, P. B., and A. N. Dossett. *Undesirable Effects of Oxygen–Helium Breathing at Great Depths.* Medical Research Council, R.N. Personnel Research Committee, Underwater Physiology Subcommittee Report 260, 1967.

15. Bennett, P. B., and S. P. Gray. Changes in human urine and blood chemistry during a simulated oxygen–helium dive to 1500 ft. *Aerospace Med.* 42: 868–874, 1971.

16. Bennett, P. B., and M. McLeod. Probing the limits of human deep diving. In: *Diving and Life at High Pressures,* edited by W. D. M. Paton, D. H. Elliott, and E. B. Smith. London: Royal Soc., 1984, p. 105–117.

17. Bennett, P. B., D. Papahadjopoulos, and A. D. Bangham. The effect of raised pressures of inert gases on phospholipid model membranes. *Life Sci.* 6: 2527–2533, 1967.

18. Bennett, P. B., and H. Schafstall. The value of TRIMIX 5 to control HPNS. In: *Man in the Sea,* edited by Y. C. Lin and K. K. Shida. San Pedro, CA: Best, 1990, p. 101–115.

19. Bennett, P. B., and H. Schafstall. Scope and design of the GUSI International Research Program. *Undersea Biomed. Res.* 19: 231–241, 1992.

20. Bennett, P. B., and E. J. Towse. Performance efficiency of men breathing oxygen–helium at depths between 100 feet and 1500 feet. *Aerospace Med.* 42: 1147–1156, 1971.

21. Bennett, P. B., and E. J. Towse. The High Pressure Nervous syndrome during a simulated oxygen–helium dive to 1500 ft. *Electroencephalogr. Clin. Neurophysiol.* 31: 383–393, 1971.

22. Bichard, A. R., and H. J. Little. Drugs that increase gamma aminobutyric acid transmission protect against the High Pressure Nervous syndrome. *Br. J. Pharmacol.* 76: 447–452, 1982.

23. Bowerd, R. W. Metabolic and thermal responses of man in various He-O_2 air environments. *J. Appl. Physiol.* 23: 561–565, 1967.

24. Bowser-Riley, F. Mechanistic studies on the high pressure neurological syndrome. *Phil. Trans. R. Soc. Lond. B.* 304: 31–41, 1984.

25. Bowser-Riley, F., J. A. Dobbie, W. D. M. Paton, and E. B. Smith. A possible role for monoaminergic inhibition in the high pressure nervous syndrome. *Undersea Biomed. Res.* 9(suppl.): 32–33, 1982.

26. Brauer, R. W., R. W. Beaver, and H. W. Gillen. Correlation studies of individual variation in susceptibility to various components of HPNS in mice. *J. Appl. Physiol.: Respir. Environ. Exerc. Physiol.* 50: 272–278, 1981.

27. Brauer, R. W., R. W. Beaver, C. D. Hogue, B. Ford, S. M. Goldman, and R. T. Venters. Intra and interspecies variability of vertebrate high pressure neurological syndrome. *J. Appl. Physiol.* 37: 844–851, 1974.

28. Brauer, R. W., R. W. Beaver, and M. E. Sheehan. The role of monoamine neurotransmitters in the compression rate dependence of HPNS convulsions. In: *Underwater Physiology VI,* edited by C. W. Shilling and M. W. Beckett. Bethesda: MD: FASEB, 1978, p. 49–59.

29. Brauer, R. W., and R. O. Way. Relative narcotic potencies of hydrogen, helium, nitrogen and their mixtures. *J. Appl. Physiol.* 29: 23–31, 1970.

30. Brauer, R. W., R. O. Way, M. R. Jordan, and D. E. Parrish. Experimental studies on the High Pressure Nervous syndrome in various mammalian species. In: *Proc. 4th Symp. Underwater Physiol.,* edited by C. J. Lambertsen. New York: Academic, 1971, p. 487–500.

31. Brauer, R. W., R. O. Way, and R. A. Perry. Narcotic effects of helium and hydrogen and hyperexcitability phenomena at simulated depths of 1500 to 4000 feet of sea water. In: *Toxicity of Anaesthetics,* edited by B. R. Fink. Baltimore, MD: Williams and Wilkins, 1968, p. 241–255.

32. Brebner, D. F., D. McK. Kerslake, and J. L. Waddell. The effect of atmospheric humidity on the skin temperature and sweat rates of resting man at two ambient temperatures. *J. Physiol. (Lond.)* 144: 299–306, 1958.

33. Bungo, M. W., and P. C. Johnson. Cardiovascular examinations and observations of deconditioning during the space shuttle orbital flight test program. *Aviat. Space Environ. Med.* 54: 1001–1004, 1983.

34. Butler, B. D., and J. Katz. Vascular pressures and passage of gas emboli through the pulmonary circulation. *Undersea Biomed. Res.* 15: 203–209, 1988.

35. Carlioz, M., M. C. Gardette-Chauffour, J. C. Rostain, and B. Gardette. Hydrogen narcosis: psychometric and neurophysiological study. In: *Proc. Xth Congr. Eur. Undersea Biomed. Soc.,* edited by T. Nome, G. Susbielle, M. Comet, M. Jaquin, and R. Sciarli. Marseille: European Undersea Medical Soc. Aberdeen, 1984, p. 97–109.

36. Charpy, J. P., E. Murphy, and C. Lemaire. Performances psychometriques apres compressions rapides a 300 m. *Med. Subaq. Hyperbare* 15: 192–195, 1976.

37. Chouteau, J. Respiratory gas exchange in animals during exposure to extreme ambient pressure. In: *Underwater Physiology IV. Proc. 4th Symp. Underwater Physiol.,* edited by C. J. Lambertsen. New York: Academic, 1971, p. 385–397.

38. Christopherson, S., and M. P. Hlastala. Pulmonary gas exchange during altered density gas breathing. *J. Appl. Physiol.: Respir. Environ. Exerc. Physiol.* 52: 221–225, 1982.

39. Claybaugh, J. R., S. K. Hong, N. Matsui, H. Nakayama, Y. S. Park, and M. Matsuda. Responses of salt- and water-regulating hormones during a saturation dive to 31 ATA (Seadragon IV). *Undersea Biomed. Res.* 11: 65–80, 1984.

40. Claybaugh, J. R., N. Matsui, S. K. Hong, Y. S. Park, H. Nakayama, and K. Shiraki. Seadragon VI. A 7-day dry saturation dive at 31 ATA. III. Alterations in basal and circadian endocrinology. *Undersea Biomed. Res.* 14: 401–411, 1987.

41. Daly, W. J., and S. Bondurant. Effects of oxygen breathing on

the heart rate, blood pressure and cardiac index of normal men resting, with reactive hyperemia, and after atropine. *J. Clin. Invest.* 41: 126–132, 1962.

42. Dressendorfer, R. H., S. K. Hong, J. F. Murlock, J. Pegg, B. Respicio, R. M. Smith, and C. Yelverton. Hana Kai II: a 17 day dry saturation dive at 18.6 ATA. V. Maximal oxygen uptake. *Undersea Biomed. Res.* 4: 283–296, 1977.

43. Dwyer, J., H. A. Saltzman, and R. O'Bryan. Maximal physical-work capacity of man at 43.4 ATA. *Undersea Biomed. Res.* 4: 359–372, 1977.

44. Eckenhoff, R. G., and D. R. Knight. Cardiac arrhythmias and heart rate changes in prolonged hyperbaric air exposure. *Undersea Biomed. Res.* 11: 355–367, 1984.

45. Epperson, W. L., D. G. Quigley, W. C. Robertson, V. S. Behar, and B. E. Welch. Observations on man in an oxygen–helium environment at 380 mmHg total pressure: III. Heat exchange. *Aerospace Med.* 37: 457–462, 1966.

46. Fagni, L., M. Hugon, and J. C. Rostain. Facilitation des potentials evoques comesthesiques (PES) en plongee profonde a saturation. *J. Physiol. Paris* 76: 17A, 1980.

47. Fagni, L., M. Weiss, J. Pellet, and M. Hugon. The possible mechanism of the pressure induced motor disturbances in the cat. *Electroencephalogr. Clin. Neurophysiol.* 53: 590–601, 1982.

48. Fagni, L., F. Zinebi, and M. Hugon. Helium pressure potentiates the N-methyl-D-aspartate and d,L-homocysteate induced decreases of field potentials in the rat hippocampal slice preparation. *Neurosci. Lett.* 81: 285–290, 1987.

49. Fagraeus, L. Cardiorespiratory and metabolic functions during exercise in the hyperbaric environment. *Acta Physiol. Scand.* Suppl. 414: 1–40, 1974.

50. Fife, W. P. The use of nonexplosive mixtures of hydrogen and oxygen for diving. Texas A & M University Hyperbaric Lab Report TAMU-SG-79-201 College Station, 1979.

51. Fife, W. P. The toxic effects of hydrogen–oxygen breathing mixture. In: *Hydrogen as a Diving Gas,* edited by R. W. Brauer. Bethesda, MD: Undersea Hyperbaric Med. Soc., 1987, p. 13–23.

52. Flynn, E. T., T. E. Berghage, and E. F. Coil. Influence of increased ambient pressure and gas density on cardiac rate in man. Washington, DC: U. S. Navy Exp. Diving Unit Report 4-72, 1972.

53. Fox, E. L., H. S. Wiss, R. L. Bartels, and E. P. Hiatt. Thermal responses of man during rest and exercise in a helium oxygen environment. *Arch. Environ. Health* 13: 23–28, 1966.

54. Fructus, X. Hydrogen pressure and HPNS. In: *Hydrogen as a Diving Gas,* edited by R. B. Brauer. Bethesda, MD: Undersea Hyperbaric Med. Soc., 1987, p. 125–140.

55. Fructus, X. R. Down below the great depths. In: *Proc. 3rd Int. Conf. Hyperbaric Underwater Physiol.* Paris: Doin, 1972, p. 13–22.

56. Fructus, X. R., C. Agarate, R. Naquet, and J. C. Rostain. Postponing the High Pressure Nervous syndrome (HPNS) to 1640 feet and beyond. In: *Proc. 7th Symp. Underwater Physiol.,* edited by C. J. Lambertsen. Bethesda, MD: FASEB, 1976, p. 21–33.

57. Fructus, X. R., R. W. Brauer, and R. Naquet. Physiological effects observed in the course of simulated deep chamber dives to a maximum of 36.5 atm in helium–oxygen atm. In: *Proc. 4th Symp. Underwater Physiol.,* edited by C. J. Lambertsen. New York: Academic, 1971, p. 545–550.

58. Fructus, X. R., and J. P. Charpy. Etude psychometrique de 2 sujets lors dune plongee fictive jusqua 52.42 ATA. *Bull. Medsubhyp.* 7: 3–12, 1972.

59. Fructus, X. R., and J. C. Rostain. HPNS: a clinical study of 30 cases. In: *Proc. 6th Symp. Underwater Physiol.,* edited by C.

W. Shilling and M. W. Beckett. Bethesda, MD: FASEB, 1978, p. 1–8.

60. Gagge, A. P., and Y. Nishi. Heat exchange between human skin surface and thermal environment. In: *Handbook of Physiology. Reactions to Environmental Agents.* Bethesda, MD: Am. Physiol. Soc., 1977, sect. 9, p. 69–92.

61. Gardette, B. Human deep hydrogen dives 1983–1985. In: *Hydrogen as a Diving Gas,* edited by R. W. Brauer. Bethesda, MD: Undersea Hyperbaric Med. Soc., 1987, p. 109–118.

62. Gardette, B. Compression procedures for mice and human hydrogen deep diving. COMEX HYDRA program. In: *High Pressure Nervous Syndrome 20 Years Later,* edited by J. C. Rostain, E. Martinez, and C. Lemaire. Marseille: ARAS-SNHP, 1989, p. 217–231.

63. Gelfand, R. Concepts of ventilatory and respiratory gas homeostasis in simulated undersea exposure. In: *Underwater Physiology VIII,* edited by A. J. Bachrach and M. M. Matzen. Bethesda, MD: Undersea Med. Soc., 1984, p. 515–533.

64. Gelfand, R., C. J. Lambertsen, and R. E. Peterson. Human respiratory control at high ambient pressures and inspired gas densities. *J. Appl. Physiol.: Respir. Environ. Exerc. Physiol.* 48: 528–539, 1980.

65. Giry, P., A. Battesti, R. Hyaclinte, and H. Burnet. Ventilatory tolerance to exercise during a hydrogen–helium–oxygen saturation dive (HYDRA V). In: *Hydrogen as a Diving Gas,* edited by R. W. Brauer. Bethesda, MD: Undersea Hyperbaric Med. Soc., 1987, p. 179–198.

66. Goldinger, J. M., S. K. Hong, J. R. Claybaugh, A. K. C. Niu, S. I. Gutman, R. G. Moon, and P. B. Bennett. *Undersea Biomed. Res.* 19: 287–294, 1992.

67. Goldinger, J. M., B. S. Kang, Y. E. Choo, C. V. Paganelli, and S. K. Hong. Effect of hydrostatic pressure on ion transport and metabolism in human erythrocytes. *J. Appl. Physiol.: Respir. Environ. Exerc. Physiol.* 49: 224–231, 1980.

68. Halsey, M. J. Effects of high pressure on the central nervous system. *Physiol. Rev.* 62: 1341–1377, 1982.

69. Halsey, M. J., E. I. Eger, D. W. Kent, and J. P. Warne. High pressure studies of anesthesia. In: *Molecular Mechanisms of Anesthesia (Progress in Anesthesiology),* edited by B. R. Fink. New York: Raven, 1975, p. 353–361.

70. Halsey, M. J., C. J. Green, and B. Wardley-Smith. Renaissance of nonunitary molecular mechanisms of general anesthesia. In: *Molecular Mechanisms of Anesthesia,* edited by B. R. Fink. New York: Raven, 1980, p. 273–283.

71. Halsey, M. J., and B. Wardley-Smith. High pressure neurological syndrome: do anticonvulsants prevent it. *Br. J. Pharmacol.* 72: 502–503, 1981.

72. Halsey, M. J., B. Wardley-Smith, and C. J. Green. The pressure reversal of general anesthesia—a multi site expansion hypothesis. *Br. J. Anaesth.* 50: 1091–1097, 1978.

73. Hamilton, R. W., Jr. Physiological responses at rest and in exercise during saturation at 20 atmospheres of He-O$_2$. In: *Underwater Physiology. Proc. Third Symp. Underwater Physiol.,* edited by C. J. Lambertsen. Baltimore, MD: Williams and Wilkins, 1967, p. 361–374.

74. Harris, D. J. Hyperbaric hyperreflexia: tendon jerk and Hoffman reflexes in man at 43 bar. *Electroencephalogr. Clin. Neurophysiol.* 47: 680–692, 1979.

75. Harris, D. J. Observations on the knee-jerk on oxy-helium at 31 and 43 bar. *Undersea Biomed. Res.* 6: 55–74, 1979.

76. Harris, D. J., and P. B. Bennett. Force and duration of muscle twitch contractions in humans at pressures up to 70 bar. *J. Appl. Physiol.: Respir. Environ. Exerc. Physiol.* 54: 1209–1215, 1983.

77. Hebden, R. A., B. J. Freund, J. R. Claybaugh, W. M. Ichimura, and G. M. Hashiro. Effect of inspiratory-phase negative pres-

sure breathing on urine flow in man. *Undersea Biomed. Res.* 19: 21–29, 1992.

78. Hesse, B., I. L. Kanstrup, N. J. Christensen, T. Ingemann-Hasen, J. F. Hansen, J. Halkjaer-Kristensen, and F. B. Peterson. Reduced norepinephrine responses to dynamic exercise in human subjects. *J. Appl. Physiol.: Respir. Environ. Exerc. Physiol.* 51: 176–178, 1981.

79. Hlastala, M. P., H. T. Robertson, and B. K. Rose. Gas exchange abnormalities produced by venous gas emboli. *Respir. Physiol.* 36: 1–17, 1972.

80. Hogan, P. M. The response of cardiac muscle cells to elevated hydrostatic pressure. In: *Hyperbaric Medicine and Physiology,* edited by Y. C. Lin and A. K. C. Niu. San Pedro, CA: Best, 1988, p. 27–36.

81. Hong, S. K., and J. R. Claybaugh. Hormonal and renal responses to hyperbaria. In: *Hormonal Regulation of Fluid and Electrolytes: Environmental Effects,* edited by J. R. Claybaugh and C. E. Wade. New York: Plenum, 1989, p. 117–146.

82. Hong, S. K., J. R. Claybaugh, V. Frattali, R. Johnson, F. Kurata, M. Matsuda, A. A. McDonough, C. V. Paganelli, R. M. Smith, and P. Webb. Hana Kai II: a 17-day dry saturation dive at 18.6 ATA. III. Body fluid balance. *Undersea Biomed. Res.* 4: 247–265, 1977.

83. Hong, S. K., M. E. Duffey, and J. M. Goldinger. Effect of high hydrostatic pressure on sodium transport across the toad skin. *Undersea Biomed. Res.* 11: 37–47, 1984.

84. Hong, S. K., and C. V. Paganelli. Water exchange in hyperbaria. In: *Physiological Function in Special Environments,* edited by C. V. Paganelli and L. Farhi. New York: Springer-Verlag, 1989, p. 82–94.

85. Hugon, M., L. Fagni, and J. C. Rostain. Cycle d'excitabilité des reflexes monosyneptiques et mechanismes cholinergic. Observations de physiologie hyperbare chez les primates. *J. Physiol. Paris* 76: 19A, 1980.

86. Hugon, M., L. Fagni, J. C. Rostain, and K. Seki. Somatic evoked potentials and reflexes in monkey during saturation dives in dry chamber. In: *Underwater Physiology, VII,* edited by A. J. Bachrach and M. M. Matzen. Bethesda, MD: Undersea Med. Soc., 1981, p. 381–390.

87. Hunter, W. L., and P. B. Bennett. The causes, mechanisms and prevention of the high pressure nervous syndrome. *Undersea Biomed. Res.* 1: 1–28, 1974.

88. Johnson, F. H., and E. A. Flagler. Hydrostatic pressure reversal of narcosis in tadpoles. *Science* 112: 91–92, 1950.

89. Kaufmann, P. G., P. B. Bennett, and J. C. Farmer. Effect of cerebellar ablation on the high pressure nervous syndrome in rats. *Undersea Biomed. Res.* 5: 63–70, 1978.

90. Kaufmann, P. G., C. C. Finley, P. B. Bennett, and J. C. Farmer. Spinal cord seizures elicited by high pressures of helium. *Electroencephalogr. Clin. Neurophysiol.* 47: 31–40, 1979.

91. Kendig, J. J., and Y. Grossman. How can hyperbaric pressure increase central nervous system excitability? In: *Current Perspectives in High Pressure Biology,* edited by H. W. Jannash, R. E. Marquis, and A. M. Zimmerman. New York: Academic, 1987, p. 159–169.

92. Koblin, D. D., H. J. Little, A. R. Green, S. Daniels, E. G. Smith, and W. D. M. Paton. Brain monoamines and the high pressure neurological syndrome. *Neuropharmacology* 19: 1031–1038, 1980.

93. Konda, N., K. Shiraki, H. Takeuchi, H. Nakayama, and S. K. Hong. Seadragon VI: a 7-day dry saturation dive at 31 ATA. IV. Circadian analysis of body temperature and renal functions. *Undersea Biomed. Res.* 14: 413–423, 1987.

94. Lambertsen, C. J., R. Gelfand, R. Peterson, R. Strauss, W. B. Wright, J. G. Dickson, Jr., C. Puglia, and R. W. Hamilton, Jr. Human tolerance to He, Ne, and N_2 at respiratory gas densities

equivalent to He–O_2 breathing at depths to 1200, 2000, 3000, 4000, and 5000 feet of sea water (predictive studies III). *Aviat. Space Environ. Med.* 48: 843–855, 1977.

95. Lanphier, E. H. Influence of increased ambient pressure upon alveolar ventilation. In: *Underwater Physiology II,* edited by C. J. Lambertsen and L. J. Greenbaum, Jr. Washington, DC: Nat. Res. Council, Publ. 1181, 1963, p. 124–133.

96. Lanphier, E. H., and E. M. Camporesi. Respiration and exercise. In: *The Physiology of Medicine of Diving* (3rd ed.), edited by P. B. Bennett and D. H. Elliott. San Pedro, CA: Best, 1982, p. 99–156.

97. Leach, C. S., J. R. M. Cowley, M. T. Troell, J. M. Clark, and C. J. Lambertsen. Biochemical, endocrinological, and hematological studies. In: *Predictive Studies IV: Work Capacity and Physiological Effects in He–O_2 Excursions to Pressures of 400–800–1200 and 1600 Feet of Seawater,* edited by C. J. Lambertsen, R. Gelfand, and J. M. Clark. Univ. Pennsylvania, Philadelphia, Inst. Environ. Med. Report 78-1, 1978, p. 1–59.

98. Lemaire, C. Hydrogen narcosis, nitrogen narcosis and HPNS: a performance study. In: *IX Int. Symp. Underwater Hyperbaric Physiol.,* edited by A. A. Bove, A. J. Bachrach, and L. T. Greenbaum. Bethesda, MD: Undersea Hyperbaric Med. Soc., 1987, p. 579–582.

99. Lemaire, C., and J. C. Rostain. *The High Pressure Nervous Syndrome and Performance.* Marseille: Octares, 1988.

100. Leon, H. A., and S. F. Cook. A mechanism by which helium increases metabolism in small animals. *Am. J. Physiol.* 199: 243–245, 1960.

101. Lever, M. J., K. W. Miller, W. D. M. Paton, W. B. Street, and E. B. Smith. Effects of hydrostatic pressure on mammals. In: *Proc. 6th Underwater Physiol. Symp.,* edited by C. J. Lambertsen. New York: Academic, 1971, p. 101–108.

102. Lin, Y. C. Cardiovascular deconditioning in hyperbaric environments. In: *Man in Stressful Environments. Diving, Hyper- and Hypobaric Physiology,* edited by K. Shiraki and M. K. Yousef. Springfield, IL: Thomas, 1987, p. 72–92.

103. LIn, Y. C., J. R. Claybaugh, J. Holthaus, H. G. Schafstall, and P. B. Bennett. Orthostatic intolerance during GUSI-18 dive, a simulated trimix saturation dive at 46 ATA. *Undersea Biomed. Res.* 18(suppl.): 97–98, 1991.

104. Lin, Y. C., and E. N. Kato. Effects of helium gas on heart rate and oxygen consumption in unanesthetized rats. *Undersea Biomed. Res.* 1: 281–290, 1974.

105. Lin, Y. C. and K. K. Shida. Brief review: Mechanisms of hyperbaric bradycardia. *Chin. J. Physiol.* 31: 1–22, 1988.

106. Linnarsson, D., and L. Fagraeus. Maximal work performances in hyperbaric air. In: *Underwater Physiology V,* edited by C. J. Lambertsen. Bethesda, MD: FASEB, 1976, p. 55–60.

107. Lorenz, J., G. Athanassenas, P. Hampe, K. Muller, G. Plath, and J. Wenzel. Human brainstem auditory evoked potentials (BAEP) in deep saturation diving. In: *Proc. XVII Annual Meeting EUBS.* Heraklion, Greece, Sept.–Oct., 1991, p. 25–48.

108. Maio, D. A., and L. E. Farhi. Effect of gas density on mechanics of breathing. *J. Appl. Physiol.* 23: 687–693, 1967.

109. Barsden, C. A. Functional aspects of S. hydroxtryptamine neurons. Application of electrochemical monitoring in vivo. *Trends Neurosci.* 5: 1–16, 1979.

110. Matsuda, M., S. K. Hong, H. Nakayama, H. Arita, Y. C. Lin, J. R. Claybaugh, R. M. Smith, and C. E. G. Lundgren. Physiological responses to immersion at 31 ATA (Seadragon IV). In: *Underwater Physiology VII, Proc. Seventh Symp. Underwater Physiol.,* edited by A. J. Bachrach and M. M. Matzen. Bethesda, MD: Undersea Med. Soc., 1981, p. 283–296.

111. Matsuda, M., H. Nakayama, H. Arita, J. F. Morlock, J. R.

Claybaugh, R. M. Smith, and S. K. Hong. Physiological responses to head-out immersion in water at 11 ATA. *Undersea Biomed. Res.* 5: 37–52, 1978.

112. Matsuda, M., H. Nakayama, A. Itoh, N. Kirigaya, F. K. Kurata, R. H. Strauss, and S. K. Hong. Physiology of man during a 10-day dry heliox saturation dive (SEATOPIA) to 7 ATA. I. Cardiovascular and thermoregulatory functions. *Undersea Biomed. Res.* 2: 101–118, 1975.

113. Matsuda, M., H. Nakayama, F. K. Kurata, J. R. Claybaugh, and S. K. Hong. Physiology of man during a 10-day dry heliox saturation dive (Seatopia) to 7 ATA. II. Urinary water, electrolytes, ADH, and aldosterone. *Undersea Biomed. Res.* 2: 119–131, 1975.

114. Matsui, N., J. R. Claybaugh, Y. Tamura, H. Seo, Y. Murata, K. Shiraki, H. Nakayama, Y. C. Lin, and S. K. Hong. Seadragon VI. A 7-day dry saturation dive at 31 ATA. VI. Hyperbaria enhances renin but eliminates ADH responses to head-up tilt. *Undersea Biomed. Res.* 14: 437–447, 1987.

115. McLeod, J., P. B. Bennett, and R. L. Cooper. Rat brain catecholamine release at 1, 10, 20 and 100 ATA heliox, nitrox and trimix. *Undersea Biomed. Res.* 15: 211–221, 1988.

116. Michaud, A., J. Parc, L. Barthelemy, J. Lechutton, J. Corriol, J. Chouteau, and F. Lebougher. Premieres données sur une limitation de l'utilisation du melange oxygene–hydrogene pour la plongee profonde à saturation. *C. R. Acad. Sci. III* 269: 497–499, 1969.

117. Miyamoto, N., N. Matsui, I. Inoue, H. Seo, K. Nakabayashi, and H. Oiwa. Hyperbaric diuresis is associated with decreased antidiuretic hormone and decreased atrial natriuretic polypeptide in human divers. *Jpn. J. Physiol.* 41: 85–99, 1991.

118. Moon, R. E., E. M. Comporesi, T. Xuan, J. Holthaus, P. R. Michell, and W. D. Watkins. ANF and diuresis during compression to 450 and 600 MSW. *Undersea Biomed. Res.* 14(suppl.): 43–44, 1987.

119. Moore, T. O., J. F. Morlock, D. A. Lally, and S. K. Hong. Thermal cost of saturation diving: respiratory and whole body heat loss at 16.1 ATA. In: *Underwater Physiology, Proc. Fifth Symp. Underwater Physiol.*, edited by C. J. Lambertsen. Bethesda, MD: FASEB, 1976, p. 741–754.

120. Morrison, J. B., and J. T. Florio. Respiratory function during a simulated dive to 1500 ft. *J. Appl. Physiol.* 30: 724–732, 1971.

121. Neuman, T. S., R. F. Goad, D. Hall, R. M. Smith, J. R. Claybaugh, and S. K. Hong. Urinary excretion of water and electrolytes during open-sea saturation diving to 850 FSW. *Undersea Biomed. Res.* 6: 291–302, 1979.

122. Niu, A. K. C., S. K. Hong, J. R. Claybaugh, J. M. Goldinger, O. Kwon, M. Li, E. Randall, and C. E. G. Lundgren. Absence of diuresis during a 7-day saturation dive at 2.5 ATA N_2–O_2. *Undersea Biomed. Res.* 17: 189–199, 1990.

123. Ohta, Y., H. Arita, H. Nakayama, S. Tamaya, C. E. G. Lundgren, Y. C. Lin, R. M. Smith, R. Morin, L. E. Farhi, and M. Matsuda. Cardiopulmonary functions and maximal aerobic power during a 14-day saturation dive at 31 ATA (Seadragon IV). In: *Underwater Physiology VII*, edited by A. J. Bachrach and M. M. Matzen. Bethesda, MD: Undersea Med. Soc., 1981, p. 209–221.

124. Overfeld, E. M., H. A. Saltzman, J. V. Salzano, and J. A. Kylstra. Respiratory gas exchange in normal men breathing 0.9 oxygen in helium at 31.1 ats. *J. Appl. Physiol.* 27: 471–475, 1969.

125. Paganelli, C. V., and F. Kurata. Diffusion of water vapor in binary and ternary gas mixtures at increased pressure. *Respir. Physiol.* 30: 15–26, 1977.

126. Pearce, P. C., D. Clarke, C. J. Dore, M. J. Halsey, N. P. Luff, and C. J. Maclean. Sodium valproate interactions with the HPNS: EEG and behavioral observations. *Undersea Biomed. Res.* 16: 99–113, 1989.

127. Piantadosi, C. A., and E. D. Thalmann. Thermal responses in humans exposed to cold hyperbaric helium–oxygen. *J. Appl. Physiol.: Respir. Environ. Exerc. Physiol.* 49: 1099–1106, 1980.

128. Raoul, Y., J. L. Meliet, and B. Broussole. Troubles psychiatriques et plongee profonde. *Med. Arm.* 16: 269–270, 1988.

129. Raymond, L. W., W. H. Bell, II, K. R. Bondi, and C. R. Lindberg. Body temperature and metabolism in hyperbaric helium atmospheres. *J. Appl. Physiol.* 24: 678–684, 1968.

130. Raymond, L. W., N. S. Raymond, V. P. Frattali, J. Sode, C. S. Leach, and W. H. Spaur. Is the weight loss of hyperbaric habituation a disorder of osmoregulation? *Aviat. Space Environ. Med.* 51: 397–401, 1980.

131. Raymond, L. W., E. Thalmann, G. Lindgren, H. C. Langworthy, W. H. Spauer, J. Crothers, W. Braithwaite, and T. Berghage. Thermal homeostasis of resting man in helium–oxygen at 1–50 atmospheres absolute. *Undersea Biomed. Res.* 2: 51–67, 1975.

132. Rostain, J. C., and S. Dimov. Potentials evoques visuels et cycle d'excitabilité au cours d'une plongée simulée à 610 m en atmosphere helium–oxygen (physalie VI). *Electroencephalogr. Clin. Neurophysiol.* 41: 287–300, 1976.

133. Rostain, J. C., B. Gardette, M. C. Gardette-Chauffour, and C. Forni. HPNS of baboons during helium–nitrogen–oxygen slow exponential compression. *J. Appl. Physiol.: Respir. Environ. Exerc. Physiol.* 59: 341–350, 1984.

134. Rostain, J. C., M. C. Gardette-Chauffour, J. P. Gourret, and R. Naquet. Sleep disturbances in man during different compression profiles up to 62 bars in helium oxygen mixture. *Electroencephalogr. Clin. Neurophysiol.* 69: 127–135, 1988.

135. Rostain, J. C., M. C. Gardette-Chauffour, and R. Naquet. Studies of neurophysiological effects of hydrogen–oxygen mixture in man up to 30 bars. *Undersea Biomed. Res.* 17: 159, 1990.

136. Rostain, J. C., C. Lemaire, M. C. Gardette-Chauffour, J. Doucet, and R. Naquet. Criteria analysis of selection for deep diving (EEG and performance). In: *Underwater Physiology VII*, edited by A. J. Bachrach and M. M. Matzen. Bethesda, MD: Undersea Med. Soc., 1981, p. 435–443.

137. Rostain, J. C., C. Lemaire, M. C. Gardette-Chauffour, J. Doucet, and R. Naquet. Estimation of human susceptibility to the high pressure nervous syndrome. *J. Appl. Physiol.: Respir. Environ. Exerc. Physiol.* 54: 1063–1070, 1983.

138. Rostain, J. C., and R. Naquet. Resultats preliminaires d'une etude comparative de l'effet des melanges oxygene–helium et oxygene–hydrogene et de hautes pressions sur de babouin Papio papio. In: *Proc. Troisiemes Journées Int. d'Hyperbarie et de Physiol. Subaquatique*, edited by X. Fructus. Marseille: Doin, 1970, p. 44–49.

139. Rostain, J. C., and R. Naquet. Le syndrome nerveux des hautes pressions: characteristiques et evolution en fonction de divers modes de compression. *Rev. EEG Neurophysiol.* 4: 107–124, 1974.

140. Rostain, J. C., and R. Naquet. Human neurophysiological data obtained from two simulated dives to a depth of 610 meters. In: *Underwater Physiology VI*, edited by C. W. Shilling and M. W. Beckett. Bethesda, MD: FASEB, 1978, p. 9–19.

141. Rostain, J. C., G. Regesta, M. C. Gardette-Chauffour, and R. Naquet. Sleep organization in man during long stays in helium–oxygen mixture at 30–40 bars. *Undersea Biomed. Res.* 18: 21–36, 1991.

142. Rostain, J. C., B. Wardley-Smith, C. Forni, and M. J. Hasley. Gamma amino butyric acid and the high pressure neurological syndrome. *Neuropharmacology* 25: 545–554, 1986.

143. Rostain, J. C., B. Wardley-Smith, and M. J. Hasley. Effects of sodium valproate on HPNS in rats: the probable role of GABA. In: *Underwater Physiology VIII*, edited by A. J. Bachrach and M. M. Matzen. Bethesda, MD: Undersea Med. Soc., 1984, p. 601–605.

144. Rowland-James, P., M. W. Wilson, and K. W. Miller. Pharmacological evidence for multiple sites of action of pressure in mice. *Undersea Biomed. Res.* 8: 1–11, 1981.

145. Sagawa, S., J. R. Claybaugh, K. Shiraki, Y. S. Park, M. Mohri, and S. K. Hong. Characteristics of increased urine flow during a dry saturation dive at 31 ATA. *Undersea Biomed. Res.* 17: 13–22, 1990.

146. Salzano, J. V., E. M. Camporesi, B. W. Stolp, and R. E. Moon. Physiological responses to exercise at 47 and 66 ATA. *J. Appl. Physiol.: Respir. Environ. Exerc. Physiol.* 57: 1055–1068, 1984.

147. Salzano, J., D. C. Rausch, and H. A. Saltzman. Cardiorespiratory responses to exercise at simulated seawater depth of 1000 ft. *J. Appl. Physiol.* 28: 34–41, 1970.

148. Schaefer, K. E. Circulatory adaptation to the requirements of life under more than one atmosphere of pressure. In: *Handbook of Physiology Circulation,* edited by W. F. Hamilton. Washington, DC: Am. Physiol. Soc., 1965, sect. 2, p. 1843–1873.

149. Seals, D. R., D. G. Johnson, and R. F. Fregosi. Hyperoxia lowers sympathetic activity at rest but not during exercise in humans. *Am. J. Physiol.* 260 (*Regulatory Integrative Comp. Physiol.* 31): R873–R878, 1991.

150. Shida, K. K., and Y. C. Lin. Contribution of environmental factors in development of hyperbaric bradycardia. *J. Appl. Physiol.: Respir. Environ. Exerc. Physiol.* 50: 731–735, 1981.

151. Shiraki, K., S. K. Hong, Y. S. Park, S. Sagawa, N. Konda, J. R. Claybaugh, H. Takeuchi, N. Matsui, and H. Nakayama. Seadragon VI: a 7-day dry saturation dive at 31 ATA. II. Characteristics of diuresis and nocturia. *Undersea Biomed. Res.* 14: 387–400, 1987.

152. Shiraki, K., N. Konda, S. Sagawa, and K. Miki. Changes in cutaneous circulation at various atmospheric pressures in modifying temperature regulation of man. In: *Thermal Physiology,* edited by R. S. Hales. New York: Raven, 1984, p. 263–266.

153. Shiraki, K., N. Konda, S. Sagawa, H. Nakayama, and M. Matsuda. Body heat balance and urine excretion during a 4-day saturation dive at 4 ATA. *Undersea Biomed. Res.* 9: 321–333, 1982.

154. Shiraki, K., S. Sagawa, N. Konda, H. Nakayama, and M. Matsuda. Hyperbaric diuresis at a thermoneutral 31 ATA He–O₂ environment. *Undersea Biomed. Res.* 11: 341–353, 1984.

155. Smith, R. M., S. K. Hong, R. H. Dressendorfer, J. H. Dwyer, E. Hayashi, and C. Yelverton. Hana Kai II: a 17-day dry saturation dive at 18.6 ATA. IV. Cardiopulmonary functions. *Undersea Biomed. Res.* 4: 267–281, 1977.

156. Spaur, W. H., L. W. Raymond, M. M. Knott, J. C. Crothers, W. R. Braithwaite, E. D. Thalmann, and D. F. Uddin. Dyspnea in divers at 49.5 ATA: mechanical, not chemical in origin. *Undersea Biomed. Res.* 4: 183–198, 1977.

157. Stetzner, L. C., and B. Deboer. Thermal balance in the rat during exposure to helium–oxygen from 1 to 141 atmospheres. *Aerospace Med.* 43: 306–309, 1972.

158. Stuhr, L. E. B., J. A. Ask, and I. Tyssebotn. Cardiovascular changes in anesthetized rats during exposure to 30 bar. *Undersea Biomed. Res.* 17: 383–393, 1990.

159. Summit, J. K., J. S. Kelly, J. M. Herron, and H. A. Saltzman. 1000 foot helium saturation exposure. In: *Proc. 4th Symp. Underwater Physiol.,* edited by C. J. Lambertsen. New York: Academic, 1971, p. 519–527.

160. Tanaka, H., S. Sagawa, K. Miki, F. Tajima, J. R. Claybaugh, and K. Shiraki. Sympathetic nerve activity and urinary responses during continuous negative pressure breathing in humans. *Am. J. Physiol.* 261 (*Regulatory Integrative Comp. Physiol.* 32): R276–R282, 1991.

161. Thorsen, E. T., J. Hjelle, K. Segadal, and A. Golsoik. Exercise tolerance and pulmonary gas exchange after deep saturation dives. *J. Appl. Physiol.: Respir. Environ. Exerc. Physiol.* 68: 1809–1814, 1990.

162. Timbal, J., H. Vieillefond, H. Guenard, and P. Varene. Metabolism and heat losses of resting man in a hyperbaric helium atmosphere. *J. Appl. Physiol.* 36: 444–448, 1974.

163. Torok, Z. The compression strategy in the Alverstoke deep dives series. In: *Man in the Sea,* edited by Y. C. Lin and K. K. Shida. San Pedro, CA: Best, 1990, vol. 1, p. 23–41.

164. Vorosmarti, J., M. E. Bradley, and N. R. Anthonisen. The effects of increased gas density on pulmonary mechanics. *Undersea Biomed. Res.* 2: 1–10, 1975.

164a. Vorosmarti, J., R. de G. Hansen, and E. E. P. Barnard. Further studies in decompression from steady state exposure to 250 meters. In: *Underwater Physiology VI,* edited by C. W. Shilling and M. W. Beckett. Bethesda, MD: FASEB, 1978, p. 438–445.

165. Webb, P. Body heat loss in undersea gaseous environments. *Aerospace Med.* 41: 1282–1288, 1970.

166. Webb, P. The thermal drain of comfortable hyperbaric environments. *Nav. Res. Rev.* 26: 1–7, 1973.

167. Webb, P., S. J. Troutman, Jr., V. Frattalli, R. H. Dressendorfer, J. Dwyer, T. O. Moore, J. F. Morlock, R. M. Smith, Y. Ohta, and S. K. Hong. Hana Kai II: a 17-day dry saturation dive at 18.6. II. Energy balance. *Undersea Biomed. Res.* 4: 221–246, 1977.

168. Whalen, R. E., H. A. Saltzman, D. H. Lolloway, R., H. D. McIntosh, H. O. Sieker, and I. W. Brown, Jr. Cardiovascular and blood gas responses to hyperbaric oxygen. *Am. J. Cardiol.* 15: 638–646, 1965.

169. Wilson, J. M., P. D. Kligfield, G. M. Adams, C. Harvey, and K. E. Schaefer. Human ECG changes during prolonged hyperbaric exposure breathing N₂–O₂ mixtures. *J. Appl. Physiol.: Respir. Environ. Exerc. Physiol.* 42: 614–623, 1977.

170. Wood, L. D. H., A. C. Bryan, S. K. Bau, T. R. Weng, and H. Levison. Effect of increased gas density on pulmonary gas exchange in man. *J. Appl. Physiol.* 41: 206–210, 1976.

171. Zinebi, F., L. Fagni, and M. Hugon. Excitatory and inhibitory amino-acidergic determinants of the pressure induced neuronal hyperexcitability in rat hippocampal slices. *Undersea Biomed. Res.* 17: 487–493, 1990.

4 | DIVING MAMMALS

45. Diving physiology of the Weddell seal

WARREN M. ZAPOL | Department of Anaesthesia, Harvard Medical School and Massachusetts General Hospital, Boston, Massachusetts

SOME CARNIVOROUS MAMMALS over the past 20 million years have evolved sophisticated physiological systems enabling them to forage under water at great depths. Paleontologists believe that dog-like creatures, which were the first mammals to hunt the sea for fish and crustaceans, evolved into the 17 families of phocid or earless seals we now watch sporting about the seashores of every continent on the globe (11, 20). Seal evolution was particularly prolific on the continent of Antarctica; when the Andean chain submerged approximately 20 million years ago and severed the link between the Palmer Peninsula and Tierra del Fuego, the last land-based predators disappeared. Free to adapt to the increasingly icy conditions, glaciation, and pack ice of Antarctica, phocid seals evolved into five distinct species. The Antarctic Weddell seal, *Leptonychotes weddelli,* the species most thoroughly studied, is remarkable in that it can dive longer than 1 h and 10 min to depths of more than 500 m (8). Other seals can also dive for long periods to great depths (12).

To survive the challenge of the extremely cold air that plummets to $-40°C$ in August at McMurdo Station, Antarctica, with winds of 38 m/s, and the less frigid $(-1.9°C)$ sea, Weddell seals developed thick layers of blubber (one-quarter of their body weight) and a large adult size (350–450 kg). Because little blood perfuses their skin and blubber, the seals lose little heat to seawater.

Another hurdle for the Antarctic seal is the profound depth of its prey. The coastal waters of Antarctica reach 250–600 m one-quarter of a mile offshore. These plunges result from the scouring action of glaciers and ice sheets, which scrape sand and debris from the ocean bottom. It is there, within 50 m of the ocean floor, that the Antarctic cod, *Disostichus mawsonii,* slowly swims (4). This 20–30 kg fish is the staple of the Weddell diet. We should not be surprised, then, to find that the Weddell's dives can exceed an hour, twenty times a human's capacity to hold his breath safely, and that the Weddell can withstand pressures at the bottom near 750 psi.

In terrestrial mammals, pressure poses problems to air pockets, such as sinus cavities and middle ear chambers, and to nervous tissue, notably the brain. Pressure also can cause a dangerous uptake of nitrogen gas. The deep diving seal conveniently has evolved without sinuses (8). Its method of coping with nervous tissue pressure is uncertain. Perhaps it is protected by cortisol production, which is discussed later in this chapter, as is the seal's solution to problems of nitrogen uptake.

During a dive, the seal's gas transport system must store and distribute oxygen, remove carbon dioxide, and maintain safe levels of nitrogen. These functions will be treated separately, and in each case laboratory findings will be considered before field discoveries. Let us turn first to oxygen storage and distribution. There are two ways all mammals store large amounts of oxygen: as a gas in the lungs and in cells of two types, circulating red cells, in which oxygen is bound to hemoglobin, and muscle cells, in which oxygen is bound to myoglobin. Seals use their lungs to oxygenate blood at the surface but not at pressures below 40 m, where the lungs collapse to airlessness (5) and the hemoglobin/myoglobin oxygen storage system takes over as the sole oxygen reservoir. These oxygen-binding systems are little influenced by pressure and operate well at great depths. Air stored in the lung is a detriment to the diving seal for another reason: near the surface it provides buoyancy. The seal's solution is to have evolved lungs small for its weight and to exhale before diving, further reducing lung volume.

To meet its underwater storage demands, a seal relies on an enormous blood volume and on high hemoglobin and myoglobin concentrations. Lenfant and co-workers (13) showed that, in contrast to humans with blood comprising 7% of their body weight, the Weddell seal has twice this quantity, 14% or about 60 l of blood in a 450 kg adult seal. In human blood, hemoglobin-filled red cells occupy 35%–45% of the volume. In the Weddell seal, 65% or more of a blood sample taken after surfacing from a dive contains red cells (18). We note that the total quantity, rather than the dissociation, of hemoglobin is special in the seal (13, 19). The Weddell seal also produces large quantities of myoglobin (4.5 g%) to store oxygen locally in the muscles vital for swimming (2).

Scientists have known for over a hundred years that

bradycardia occurs when animals are forced to dive. Bert (1), a French physiologist, reported in 1870 that the pulse rate of ducks slowed markedly when they were submerged. This slowing of the heart rate upon immersion of the face is believed to be universal to all mammalian species. Observed in humans, it has been found to be most profound in those species which habitually dive, such as seals and whales. It is a reaction regarded as part of the diving response; neural impulses received from the face trigger the brain to cease respiratory efforts, slow the heart, and constrict certain arteries (22). The major effect of the reduced heart rate is to decrease cardiac output. Indeed, the Weddell seals studied in our laboratory at McMurdo Station reduced cardiac output from 20 liters/min at rest to 5 l/min during experimental diving and increased it to 60 l/min during recovery periods after the dive (23).

Reasoning that this fourfold reduction in blood flow must mean that some tissues receive less flow during dives at low heart rates, we measured blood flow to the various tissues by injecting radioactive microspheres during the laboratory dive and comparing this with normal blood flow. We found that the brain, retina, spinal cord, and other nervous system tissues remained at normal blood flow levels during diving (see Fig. 45.1); all of these tissues apparently were selected by the seal as vital for underwater navigation and motor control. That blood flow to the brain remained normal underscores the fact that the marine mammal's brain requires oxygen; however, it does not require as much as that of terrestrial mammals, functioning comfortably at oxygen partial pressures at which humans black out (18).

We found that two other tissues continue to receive elevated blood flow fractions during diving: the adrenal and the placenta of pregnant seals (14, 15). The adrenal makes cortisol and the Weddell and other deep diving seals are among those mammals with the highest blood cortisol concentrations (ten to fifteen times human plasma levels) (14). Although we know that cortisol has many important effects, we do not know why the blood levels in deep diving seals are so high. Cortisol may stabilize nervous cells to prevent high pressure nervous syndrome.

Blood flow to the nonpregnant uterus ceases during laboratory simulated diving; using radioactive microspheres we reported that the gravid uterus receives the largest share of blood flowing from the descending aorta during diving (15). Space does not permit a description of seal fetal physiological adaptations (see refs. 7, 15).

With the exception of the central nervous system and the lung's pulmonary arterial circulation, all the seal's other organ systems and tissues are turned off during laboratory diving; their supplying arteries are constricted via a vessel constrictor reflex mediated in the brain (22). This neural reflex, another part of the diving response, is triggered by the sympathetic nervous system, which can also be activated in a less specific manner in fear and flight or fright reactions.

We measured lactic acid in the blood of every seal which surfaced after a forced dive in the laboratory (16).

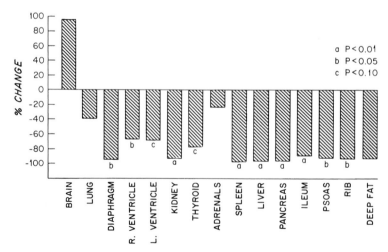

FIG. 45.1. Weddell seal regional blood flow measured by injecting radioactive microspheres during laboratory diving: percent change of organ blood flow, ml · min · g⁻¹, during simulated diving.

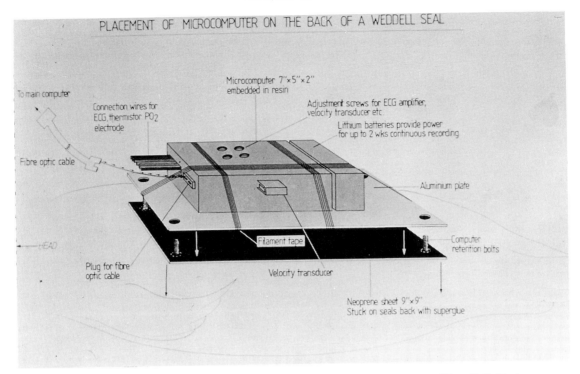

FIG. 45.2. Schematic diagram of diving computer and mounting system (courtesy of Drs. G. C. Liggins and R. Elliott).

High levels of lactic acid lower the pH of arterial blood. Scholander's (22) studies suggested that the seal avoids this danger by restricting metabolism, and thus lactic acid, to the skeletal muscles, which, because they are already constricted, are isolated from vulnerable organs such as the heart and brain until the dive is over. At that time, spilling lactic acid into the blood is harmless; the seal can gradually metabolize it through the liver, lung, and other organs over a period of 10–30 min.

Hoping to test the applicability of our laboratory observations to true field conditions and to fill in gaps in our understanding of the seal's methods of coping with both great pressure and oxygen deprivation, we decided to take advantage of electronic technology for studying the free diving seal. In 1982 and 1983 Hill (7) built a small diving computer, which was battery powered and encapsulated to withstand the 500 m diving depths of the Weddell (see Fig. 45.2). It was fastened to a rubber sheet, which was glued to the Weddell seal's dorsal fur, and could be unfastened without injury to the seal. When the seal moulted during the summer, the rubber sheet was shed. The computer recorded heart rate and depth at predetermined intervals for several days. The computer required a brief high speed fiberoptic linkup, which in 10 s could transfer data from the diving computer to the fish house computer and give the seal computer new instructions. The computer also con-

trolled an electrical blood pump, which sampled arterial blood at specified times and depths and flushed the sampling line with a solution containing heparin and hypertonic saline, thus preventing the tubing from clotting or freezing in the seawater. After collection, the blood sample was pumped into a bag or serial sampling device tethered to the fiber-optic line. The seal was rigged with the computer and released at an isolated hole using the technique of Kooyman et al. (8, 10). A portable fish hut placed over the hole allowed us to have heated access to the seal, which could only surface at this hole (see Fig. 45.3).

Kooyman and co-workers (10) noted that almost

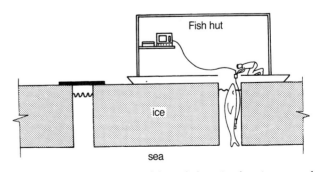

FIG. 45.3. Layout of ice hut and diving hole on ice sheet (courtesy of Dr. R. D. Hill).

TABLE 45.1. *A Comparison of Short and Long Dives*

Short Dives (<20 min)	Long Dives (20 min–1 h)
Feeding	Exploratory, escape
Aerobic	Anaerobic
Short surface recovery time	May require long recovery time to metabolize lactic acid
Variable diving heart rates	Low invariate diving heart rates
Ascent tachycardia, skeletal muscles probably perfused	Skeletal muscles probably not perfused

95% of the Weddell seal's free dives last less than 20 min. During these short feeding dives, they do not perform the anaerobic metabolism we had observed in all laboratory dives (16). Using the microcomputer sampler, Hochachka's group (6, 18) discovered that seals do not release lactic acid into arterial blood during or after these dives. Thus the seals appear to provide their muscles with sufficient oxygen, perhaps by resupplying myoglobin–oxygen stores from oxygenated hemoglobin circulating to muscle tissue. If this is so, their metabolic pattern would closely resemble that of a marathon runner: oxygen consumed by performing muscles is balanced by oxygen delivered by perfused blood to maintain aerobic metabolic efficiency. With no acid to metabolize at the surface, the seal, like the marathon runner, can keep going without long rest periods. We observed that after a few breaths at the surface, the seal reoxygenated its blood and set off on another dive. Le

Boeuf and co-workers (12) made similar observations in monitored northern elephant seals.

A very different picture was obtained when we analyzed lactic acid production in the 5% of seal dives lasting 20–30 min, dives they take when they need to explore distant routes or to escape from predators. Using the diving blood sampler, we learned that these dives were characterized by a profound bradycardia, with little variability of heart rate, and that after, but not during, these long dives they release lactic acid into the bloodstream (6, 18). The seal "decides" early to reduce its heart rate markedly, probably by totally shutting down blood flow to muscles and maximally conserving oxygen for use by the brain and heart, as it did in the laboratory forced dive. Having switched to anaerobic metabolism, the Weddell seal can remain under water for an hour or more (8). A summary of the characteristics of long and short dives is presented in Table 45.1.

Qvist et al. (18) made the remarkable discovery that red cell concentration increased 50% during the first 10–15 min of a dive, going from 35%–40% after resting at the hole several minutes to 60% during the dive (Fig. 45.4). Although oxygen pressures in arterial blood rise with compression and then fall rapidly at the onset of a dive (Fig. 45.5), the oxygen content of arterial blood during these first 10–15 min of long dives does not change, indicating an even balance between the supply and consumption of oxygen during this period (18). Thus oxygenated red cells may be added to circulating blood during the dive and removed during recovery at

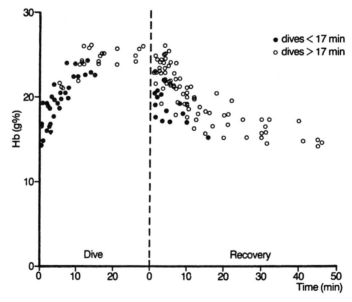

FIG. 45.4. Arterial hemoglobin (*Hb*) changes during diving and after resurfacing. Dives were divided into short (<17 min) and long (>17 min) dives. Serial sampling during long dives showed that hemoglobin stabilized within 10–12 min. Rate of rise of hemoglobin was close to $1 \text{ g} \cdot 100 \text{ ml}^{-1} \cdot \text{min}^{-1}$ during the first 10 min. Rate of decrease during recovery was similar.

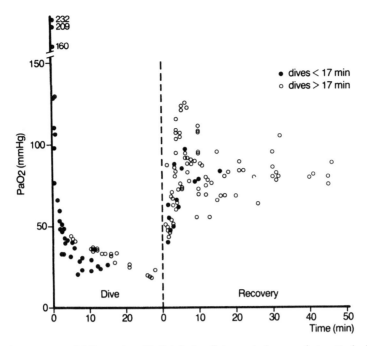

FIG. 45.5. Changes in arterial O_2 tensions (PaO_2) during diving and after resurfacing. Early diving compression hyperoxia is apparent. Lowest PaO_2 recorded was 18.2 torr at the end of a 27 min dive. Similar low PaO_2 values were recorded at the end of short dives (that is, dives < 17 min). Highest postdive PaO_2 values were recorded after dives of long duration. These high values may reflect cooling of the seal.

the surface. Where are these oxygenated red cells coming from? We suspect the spleen for several reasons. Although a poorly understood organ, the spleen of many animals is known to contract under sympathetic nervous stimulation, thereby injecting red cells into circulating blood and raising the red cell concentration. Horses have large spleens and inject their splenic red cells (about 55% of red cell mass) into circulating blood after a few minutes of heavy exercise (17). The Weddell seal spleen is proportionally one of the largest mammalian spleens and is matched (see Table 45.2) only by that of the southern elephant seal, another long diving species. We estimate that the Weddell seal spleen would have to store approximately 60% of the red cells to inject the 20 l of cells necessary to raise hemoglobin concentration by 50% during diving (Fig. 45.6) (18). The Weddell seal also has large sacs in its venous system, which could act as expandable reservoirs to accommodate this large injected volume.

A contractile spleen would be an enormous advantage to the Weddell seal, serving as a source of oxygen during diving and at the same time helping dilute increasing blood CO_2 and nitrogen levels. The adrenergic sympathetic diving reflex could quite plausibly extend to the spleen, causing it to contract and extrude its blood supply only while diving. When the seal dives repeatedly for short periods, surfacing briefly in between, the spleen may provide maximal oxygen supply for aerobic

muscles by remaining contracted: the large volume of circulating red cells could be rapidly reoxygenated in the lung at the surface when cardiac output is high.

The anatomy of the seal's lungs differs from the human in ways which appear to protect it from nitrogen narcosis and decompression sickness. The small and large airways of terrestrial mammals, including humans, are reinforced with horseshoe-shaped cartilages, sufficient to support bronchi but containing soft walls which collapse when subjected to large external pressures. Denison and Kooyman's (3) morphological studies

TABLE 45.2. *Comparison of Autopsy Spleen Weight/ Body Weight and Estimates of Red Cell Storage Capacity of the Spleen*

Species	% Spleen/ Body Weight	Spleen Red Cell Storage Capacity (% Total)
Weddell seal	0.9	60
Southern elephant seal	0.8	—
Harbor seal	0.4	—
Baleen whale	0.02	—
Porpoise	0.02	—
Human	0.25	<10
Sheep	0.2	26
Horse	0.3	54
Dog	0.2	20

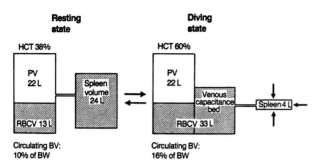

Dynamic spleen function of Weddell seals

FIG. 45.6. One hypothesis explaining the events that lead to a marked increase in circulating hemoglobin concentrations during diving by a 350 kg Weddell seal. Red blood cells (*RBC*) stored in the spleen at near-normal O_2 and CO_2 tensions are released into the portal circulation and subdiaphragmatic capacitance veins and then enter the central circulation via the inferior vena caval sphincter. Effects of RBC release are (*1*) maintenance or increase of aortic arterial O_2 content until reservoir is depleted (usually 10–12 min into long dives), (*2*) lack of buildup of CO_2 until reservoir is emptied (dilution effect), and (*3*) reduction of initially high N_2 tensions (dilution effect). Splenic storage capacity of RBC may amount to 60% of RBC mass, and splenic weight during rest totals at least 7% of body weight; splenic weight is probably greater because splenic venous effluent contains both plasma and RBC (Hct < 100%). *PV* = plasma volume; *RBCV* = RBC volume; *Hct* = hematocrit; *BV* = blood volume; *BW* = body weight.

showed that seals evolved ring-shaped (annular) cartilages in their small and large airways, which served as armored reservoirs for storing air. This anatomical evidence suggested that when the seal descends, gas must be squeezed out of the lung's alveolar regions and stored

within the airway reservoirs from which it cannot dissolve into the blood. Upon ascent, gas in the airways would expand, reinflating the collapsed lung.

Evidence collected in the 1960s supported this view. Photographs taken at depth by Ridgeway et al. (21) of dolphin chests showed a collapsed thorax, and their analysis of gas exhaled at surfacing showed no evidence of higher CO_2 or lower O_2 levels than in air, indicating that reservoir-stored air did not contact blood during the dive. Further, Kooyman et al.'s (9) studies of seals forced to dive in compression chambers suggested that the lungs collapsed when the seal reached a depth of 50–70 m.

Our field studies estimated the depth at which lung collapse occurs by measuring the actual arterial nitrogen pressures reached by free diving seals. These measurements were not easily accomplished. Falke and co-workers (5) performed the van Slyke blood nitrogen extraction procedure in our fishing hut on the McMurdo Sound ice sheet, quantifying the amount of nitrogen in arterial blood samples during a single dive. Serial measurements of blood nitrogen tensions during a dive to 90 m are shown in Figure 45.7. Arterial blood nitrogen pressure (PaN_2), near 550 mm Hg at atmospheric pressure, increases as the seal descends, reaching a peak value of 1,500–2,500 mm Hg at a depth of approximately 40 m. This then is the depth at which the lung collapses, limiting nitrogen uptake in the blood. In 47 measurements obtained while seals dove up to 230 m, the highest arterial PaN_2 was 2,400 mm Hg (5). As the dive progresses beyond 40 m, the blood nitrogen

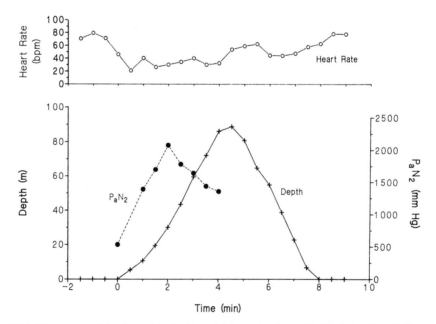

FIG. 45.7. Heart rate and depth combined with serial determinations of PaN_2 during dive. PaN_2 values determined early during a dive when pulmonary gas exchange spaces are collapsing. Each sampling time was 30 s.

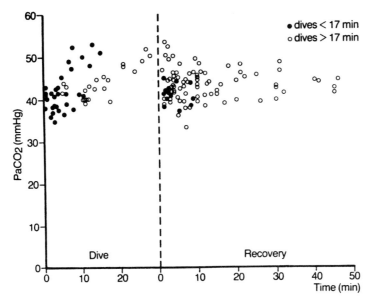

FIG. 45.8. Changes in arterial CO_2 tension ($PaCO_2$) during dive and after resurfacing from short and long dives. $PaCO_2$ in long dives did not increase above resting value (42.5 ± 1.63 torr) until 12–13 min into dive but then rose to levels similar to those measured in short dives. Lack of rise in $PaCO_2$ early in long dives is believed to be caused by addition of red blood cells with a normal CO_2 content. $PaCO_2$ in the recovery period rarely decreased below normal levels, despite marked increase of respiratory rate after resurfacing. Decrease of $PaCO_2$ after resurfacing was slower if the seal went to sleep.

pressures fall, probably for several reasons. First, there is blubber, in which nitrogen is four and one-half times more soluble than in blood or muscle. Second, there are body tissues, which are able to store some of this nitrogen. Third, there is the blood itself, some of which may not have been exposed to nitrogen diffusing into lung capillaries from the alveoli prior to collapse, either because this blood was circulating elsewhere in the body or because it was in the spleen. A mixing of this low PaN_2 blood with the high PaN_2 arterial blood would dilute the nitrogen concentration.

The seal's ability to moderate its blood nitrogen pressures at depth also serves to protect it from decompression sickness. The remaining blood nitrogen could still present problems; however, the seal again reduces its risk. At great depths the tiny gas nuclei around which bubbles form are themselves so thoroughly compressed as to disappear, removing nucleation sites.

Finally, we turn to the seal's methods for regulating CO_2 levels. In the laboratory, blood CO_2 levels rose markedly during the dive, increasing from 48 to 73 mm Hg in 20 min dives (23). In our field studies, CO_2 rose much less, from 39 to 48 mm Hg in dives lasting 20 min (see Fig. 45.8). (18). We believe that the splenic injection of red cells diluted the blood CO_2 of the free diving seals, maintaining near-normal levels. This then adds yet another aspect to the importance of splenic constriction during diving: it moderates the rise of nitrogen and carbon dioxide as well as possibly supplying tissues and organs with oxygen-rich red cells.

Why study the Weddell seal and its adaptations to the hostile climate of the Antarctic Ocean and its shores? Our studies have been motivated by a desire to discover nature's strategies for dealing with extremes of cold, pressure, and breath-holding. What we have learned enables us to marvel at the effectiveness and sophistication of these mammalian evolutionary adaptations.

The author thanks all of his collaborators for helping obtain, record, and analyze the data and Margaret Flynn for secretarial assistance. This work was made possible by grants from the U. S. National Science Foundation, Division of Polar Programs (DPP-9118192), with logistical support in Antarctica provided by the U. S. Navy and Air Force.

REFERENCES

1. Bert, P. *Lecons sur la Physiologie Comparée de la Respiration.* Paris: Balinière, 1870, p. 523–553.
2. Castellini, M. A., and G. N. Somero. Buffering capacity of vertebrate muscle: correlations with potentials for anaerobic function. *J. Comp. Physiol.* 143: 191–198, 1981.
3. Denison, D. M., and G. L. Kooyman. The structure and function of the small airways in pinniped and sea otter lungs. *Respir. Physiol.* 17: 1–10, 1973.
4. Eastman, J. T., and A. L. DeVries. Antarctic fishes. *Sci. Am.* 255: 106–114, 1986.

5. Falke, K. J., R. D. Hill, J. Qvist, R. C. Schneider, M. Guppy, G. C. Liggins, P. W. Hochachka, R. E. Elliott, and W. M. Zapol. Seal lungs collapse during free diving: evidence from arterial nitrogen tensions. *Science* 229: 556–558, 1985.

6. Guppy, M., R. D. Hill, R. C. Schneider, J. Qvist, G. C. Liggins, W. M. Zapol, and P. W. Hochachka. Microcomputer-assisted metabolic studies of voluntary diving of Weddell seals. *Am. J. Physiol.* 250 (*Regulatory Integrative Comp. Physiol.* 19): R175–R187, 1985.

7. Hill, R. D., R. C. Schneider, G. C. Liggins, A. H. Schuette, R. L. Elliott, M. Guppy, P. W. Hochachka, J. Qvist, K. J. Falke, and W. M. Zapol. Heart rate and body temperature during free diving of the Weddell seal. *Am. J. Physiol.* 253 (*Regulatory Integrative Comp. Physiol.* 24): R344–R351, 1987.

8. Kooyman, G. L. *Weddell Seal: Consummate Diver.* Cambridge, UK: Cambridge University Press, 1981.

9. Kooyman, G. L., J. P. Schroeder, D. M. Denison, D. D. Hammond, J. J. Wright, and W. P. Bergman. Blood nitrogen tensions of seals during simulated deep dives. *Am. J. Physiol.* 223: 1016–1020, 1972.

10. Kooyman, G. L., E. A. Wahrenbrock, M. A. Castellini, R. W. Davis, and E. E. Sinnett. Aerobic and anaerobic metabolism during voluntary diving in Weddell seals: evidence of preferred pathways from blood biochemistry and behavior. *J. Comp. Physiol.* 138: 335–346, 1980.

11. Laws, R. M. Seals. In: *Antarctic Ecology,* edited by R. M. Laws. London: Academic, vol. 2, 1984, p. 621–715.

12. Le Boeuf, B. J., D. P. Costa, A. C. Huntley, and S. D. Feldkamp. Continuous, deep diving in female northern elephant seals, *Mirounga angustirostris. Can. J. Zool.* 66: 446–458, 1988.

13. Lenfant, G., R. Elsner, G. L. Kooyman, and D. M. Drabek. Respiratory function of the blood of the adult and fetal Weddell seal—*Leptonychotes weddelli. Am. J. Physiol.* 216: 1595–1597, 1969.

14. Liggins, G. C., J. T. France, R. C. Schneider, B. S. Knox, and W. M. Zapol. Concentrations, metabolic clearance rates, production rates and binding of cortisol in Antarctic phocid seals. *Acta Endocrinol.* (Copenhagen) 129: 356–359, 1993.

15. Liggins, G. C., J. Qvist, P. W. Hochachka, B. J. Murphy, R. K. Creasy, R. C. Schneider, M. T. Snider, and W. M. Zapol. Fetal cardiovascular and metabolic responses to simulated diving in the Weddell seal. *J. Appl. Physiol.: Respir. Environ. Exerc. Physiol.* 49: 424–430, 1980.

16. Murphy, B., W. M. Zapol, and P. W. Hochachka. Metabolic activities of the heart, lung and brain during diving and recovery in the Weddell seal. *J. Appl. Physiol.: Respir. Environ. Exerc. Physiol.* 48: 596–605, 1980.

17. Persson, S. G. B., L. Ekman, G. Lydin, and G. Tufvesson. Circulatory effects of splenectomy in the horse I–IV. II. Effect on plasma volume and total and circulating red cell volume. *Zentralbl. Veterinarmed.* 20: 456–468, 1973.

18. Qvist, J., R. D. Hill, R. C. Schneider, K. J. Falke, G. C. Liggins, M. Guppy, R. L. Elliott, P. W. Hochachka, and W. M. Zapol. Hemoglobin concentrations and blood gas tensions of free-diving Weddell seals. *J. Appl. Physiol.: Respir. Environ. Exerc. Physiol.* 61: 1560–1569, 1986.

19. Qvist, J., R. E. Weber, and W. M. Zapol. Oxygen equilibrium properties of blood and hemoglobin of fetal and adult Weddell seals. *Am. J. Physiol.* 50: 999–1005, 1981.

20. Ridgeway, S. H., and R. J. Harrison. *Handbook of Marine Mammals,* Seals, edited by S. H. Ridgeway and R. J. Harrison. London: Academic, 1981, vol. 2.

21. Ridgeway, S. H., B. L. Scronce, and J. Kanwisher. Respiration and deep diving in the bottlenose porpoise. *Science* 166: 1651–1653, 1969.

22. Scholander, P. F. *Experimental Investigations on the Respiratory Function in Diving Mammals and Birds. Hvalradets Skrifter,* Norske Videnskaps-Akad (Oslo), 1940.

23. Zapol, W. M., G. C. Liggins, R. C. Schneider, J. Qvist, M. T. Snider, R. K. Creasy, and P. W. Hochachka. Regional blood flow during simulated diving in the conscious Weddell seal. *J. Appl. Physiol.: Respir. Environ. Exerc. Physiol.* 47: 968–973, 1979.

V | THE TERRESTRIAL ALTITUDE ENVIRONMENT

Associate Editor   S. Lahiri

46. Evolutionary aspects of atmospheric oxygen and organisms

DANIEL L. GILBERT | *Unit on Reactive Oxygen Species, Biophysics Section, Basic Neurosciences Program, NINDS, National Institutes of Health, Bethesda, Maryland*

CHAPTER CONTENTS

. . . the process of cosmic evolution is indissolubly linked with the fundamental characteristics of the organism . . . logically, in some obscure manner, cosmic and biological evolution are one.

—L. J. Henderson (122)

DETERMINING THE ORIGINS OF ATMOSPHERIC OXYGEN is basic to our understanding of the relationship of oxygen to the biosphere, yet it cannot be fully tested by scientific methods. We need to follow carefully where the scarce available scientific data lead us. The processes that led to the birth of oxygen in our universe will be considered here. Why is it that the most abundant elements in the universe, with the exception of helium, are also the most abundant elements in the biosphere? An attempt will be made to answer this question and relate it to the origin of living organisms. Another question that will be considered is why the Earth's atmosphere has evolved from one that was essentially devoid of molecular oxygen to our present one composed chiefly of molecular oxygen and nitrogen. This question will be related to the evolution of the Earth's biosphere. What are the rates of evolution of the Earth's biosphere and atmosphere? Considering the universe, Weinberg (307)

wrote: ". . . this present universe has evolved from an unspeakably unfamiliar early condition, and faces a future extinction of endless cold or intolerable heat." Of course, this will not happen in the foreseeable future. We cannot stop the evolution of the biosphere and atmosphere from facing eventual extinction, but we humans have the power to change the kinetics of evolution on the short-term scale, speeding this process dramatically forward to the death of the biosphere as we know it. However, we can attempt to slow the process down. Understanding the evolution of the relationship of oxygen to the biosphere should help us choose the proper course of action for the future.

THE ORIGIN OF OXYGEN IN THE UNIVERSE

What was the universe like before oxygen was present? To answer this question, we must consider the origin of our universe itself. According to the "Big Bang" theory, the universe originated from an explosion (211, 271, 307). This event occurred between 13 Gyrs and 18 Gyrs ago (1 Gyr represents 1 Gigayear, or $1 \cdot 10^9$ years; 292), or an estimated average of 15 Gyrs ago. The most distant galaxy known is 8C 1435+63, and the age of this galaxy has been estimated to be 15.8 Gyrs (279a). The calculated expansion age of the universe is less than this value (11a, 212a). It has been suggested that both ages can be made to agree if the expansion rate is decreased from the currently accepted estimates (11a). Although there are some difficulties with this theory (209), there is no better theory to replace it. Quantum theory has also been applied to the universe to explain some discrepancies of the "Big Bang" theory; this approach is termed *quantum cosmology* (119). After the first few minutes, the only nuclei that were present were hydrogen and helium, hydrogen being much more abundant than helium. About 500,000 years later, the temperature of the universe had cooled to 3,000 K; electrons and nuclei formed neutral atoms.

As the universe evolved, primordial clouds condensed into galaxies. It seems that this process took about 3 Gyr in the development of our own galaxy (66, 272).

1059

Stars were formed within the galaxies. A star, such as our Sun, releases energy from the following net nuclear fusion reaction in its core (163):

$$4 \ {}^{1}\text{H} \rightarrow {}^{4}\text{He} \qquad (1)$$

In this fusion process, ${}^{2}\text{H}$ (deuterium) and ${}^{3}\text{He}$ (an isotope of helium) are intermediates; the temperature required for this process is at least $5 \cdot 10^{6}$ K. The ${}^{4}\text{He}$ nucleus is very stable; it is termed an alpha particle. Likewise, the ${}^{1}\text{H}$ particle is termed a proton. The ${}^{4}\text{He}$ nucleus is composed of two protons and two neutrons, for a total of four nucleons. This reaction produces ${}^{2}\text{H}$ and ${}^{3}\text{He}$ as intermediates. After about 10 Gyr, all the hydrogen in the core becomes fused into helium. Then the core temperature will have risen to at least $100 \cdot 10^{6}$ K and the helium in the core will fuse to produce carbon and oxygen as follows:

$$3 \ {}^{4}\text{He} \rightarrow {}^{12}\text{C} \qquad (2)$$

$$ {}^{12}\text{C} + {}^{4}\text{He} \rightarrow {}^{16}\text{O} \qquad (3)$$

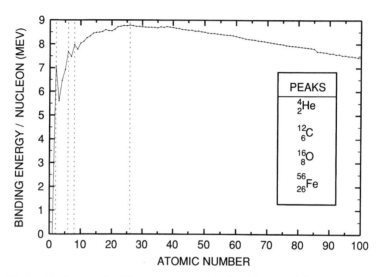

FIG. 46.1. Nuclear binding energies. These values were calculated from the following equation (314):

$$E_{BA} = \frac{Z \cdot H + N \cdot n - M_e}{Z + N},$$

where E_{BA} = binding energy per nucleon; Z = number of protons; N = number of neutrons; M_e = mass excess in MeV; H = mass excess in MeV for the proton; and n = mass excess in MeV for the neutron. The mass excess equals the atomic mass minus the integer atomic mass unit. The atomic mass of ${}^{12}\text{C}$ is defined to equal 12 integer atomic mass units. $H = 8.0714$ MeV and $n = 7.2890$ MeV. For ${}^{16}\text{O}$, the atomic mass equals 15.99491466 and the atomic mass number equals 16. Thus the mass excess equals -0.00508534. The Einstein equation (energy equals mass times the square of the speed of light) enables one to calculate the equivalent energy for this mass excess. One MeV is the equivalent energy of 0.001073535 atomic mass units. Dividing the -0.00508534 mass units by 0.001073535 mass units/MeV gives -4.737 MeV for the mass excess of ${}^{16}\text{O}$ (202). For ${}^{16}\text{O}$, $Z = N = 8$, so by substitution into this equation, the binding energy per nucleon equals 7.9762 MeV.

Values for the mass excesses were taken from McGervey (183). Only one isotope for each element was chosen. These isotopes were generally either the most common or the longest-lived ones for the element. Note that about 99% of naturally occurring hydrogen, helium, carbon, nitrogen, and oxygen is in the form of the isotopes ${}^{1}\text{H}$, ${}^{4}\text{He}$, ${}^{12}\text{C}$, ${}^{14}\text{N}$, and ${}^{16}\text{O}$, respectively. For almost all the elements, it was not significant what isotope was used for representing the element. Note that the atomic number equals the number of protons in the nucleus and that the number of nucleons equals the number of protons plus neutrons. The dotted lines represent peaks in nuclear stability. The first peak on the left is the ${}^{4}\text{He}$ peak, which points out that ${}^{4}\text{He}$ is much more stable than ${}^{1}\text{H}$. This fact accounts for the energy released in reaction (1) (in the text). Similarly, note that ${}^{12}\text{C}$ and ${}^{16}\text{O}$ are also peaks, which accounts for the energy released in reactions (2) and (3). Note that ${}^{56}\text{Fe}$ is the most stable nucleus of all the atoms; this is the reason why the big stars end up with an iron core. Fusion of atoms lighter than iron can result in a release of energy, whereas fission of atoms heavier than iron can also result in a release of energy. According to the shell theory of the nucleus, Mayer (182) has shown that nuclei that contain 2, 8, 20, 28, 50, 82, or 126 protons or neutrons are unusually stable. These nuclear shells are analogous to the atomic valence shells of electrons. Mayer's chart indicates that the only nuclei to contain double magic numbers are ${}^{4}\text{He}$, ${}^{16}\text{O}$, ${}^{40}\text{Ca}$, ${}^{48}\text{Ca}$, and ${}^{208}\text{Pb}$. Of these five nuclei, only ${}^{4}\text{He}$, ${}^{16}\text{O}$, and ${}^{40}\text{Ca}$ have the same number of protons and neutrons. This figure illustrates the stability of ${}^{4}\text{He}$ and ${}^{16}\text{O}$.

Now the star becomes a red giant. When the star is about 14 Gyr old, it ejects its hydrogen-rich outer shell containing ^{12}C and ^{16}O and becomes a white dwarf. Finally, the white dwarf cools into a black dwarf.

If the star is born much larger than our Sun, the core temperatures are increased so that many more nuclear fusion reactions occur. When the temperature is at least $20 \cdot 10^6$ K, ^{12}C, ^{13}N, ^{13}C, ^{14}N, ^{15}O, and ^{15}N are used as catalysts for the production of helium (equation 1). This net reaction is termed the carbon cycle or the CNOH burning process. Then the star evolves into a red supergiant. The lifetime of such a star is much less than that of the smaller low-temperature stars, such as our Sun. As the temperature increases to over $2 \cdot 10^9$ K, fusion reactions continue until the core is essentially iron. These fusion reactions result in energy production and atomic nuclei, whose atomic number is less than or equal to that of iron. When no more fusion reactions can occur in the core, there is a sudden, violent explosion called a *supernova*. The supernova sprays the environment with fused atomic nuclei. Explosive nucleosynthesis of atomic nuclei, whose atomic number is greater than iron, also occurs. The star now degenerates into either a neutron star or, if the star is big enough, a black hole (271). The oxygen that we breathe today is derived from a breakdown of stars.

Indeed, the recent Supernova 1987A showed the detection of highly ionized atoms, such as those of nitrogen, carbon, oxygen, and helium (46). The nuclear binding energies are shown in Figure 46.1. Note that the peaks of stability are helium, carbon, oxygen, and iron. Massive stars first produce helium, then carbon and oxygen, and finally iron. For all nuclei possessing a lower atomic number than iron, fusion of nuclei releases energy; however, for all nuclei possessing a higher atomic number than iron, fission of nuclei releases energy. It seems that the stability of atomic nuclei depends on the number of either protons or neutrons in a nucleus. If these numbers have values of 2, 8, 20, 28, 50, 82, or 126, then the nuclear structure is extremely stable. A theory to explain these so-called magic numbers is the nuclear shell theory (182). The *nuclear shell theory* is analogous to the atomic electronic shell theory, in which chemical stability occurs when the electrons in the outer shells are completed, as in the inert gases. The percent atom abundance of the stable isotopes of some elements are given in Table 46.1. The oxygen atom has eight protons and eight neutrons, thus possessing a double magic number. The only other atoms that possess the same number of protons and neutrons and the double magic numbers are helium and calcium. Hence, the ^{16}O nucleus is extremely stable.

OXYGEN COMPOUNDS IN THE UNIVERSE

Table 46.2 gives the relative cosmic abundances estimated by Greenberg (111). There is evidence suggesting that the ratio of oxygen and the other elements to hydrogen is only 40% of the accepted abundances in our galaxy, The Milky Way (187a). Now we are in a position to understand why hydrogen is the most prevalent atom in the universe and why helium is the next most abundant universal atom (see equation 1). Oxygen is the third most abundant element in the universe (see equation 3). At least part of the reason is the nuclear stability of the oxygen atom. Temperatures have to be exceptionally high to fuse the oxygen atoms to form the bigger nuclei. Another reason for this abundance of oxygen is the fact that it has a small nucleus. Carbon is also relatively abundant due to the stellar fires in the cores of many stars producing carbon nuclei (see equation 2). Nitrogen is an intermediate in the carbon-catalyzed helium production in the big stars. Next in abundance are magnesium, silicon, iron, and sulfur;

TABLE 46.1. *Percent Atom Abundance of Stable Isotopes in the Cosmos*

Element	Isotope	Percent
Hydrogen	^1H	99.998
	^2H (or) ^2D	0.002
Helium	^3He	0.0142
	^4He	99.9858
Carbon	^{12}C	98.89
	^{13}C	1.11
Nitrogen	^{14}N	99.634
	^{15}N	0.366
Oxygen	^{16}O	99.758
	^{17}O	0.038
	^{18}O	0.204
Sulfur	^{32}S	95.02
	^{33}S	0.75
	^{34}S	4.21
	^{36}S	0.017
Calcium	^{40}Ca	96.94
	^{42}Ca	0.647
	^{43}Ca	0.135
	^{44}Ca	2.09
	^{46}Ca	0.0035
	^{48}Ca	0.187
Iron	^{54}Fe	5.8
	^{56}Fe	91.8
	^{57}Fe	2.15
	^{58}Fe	0.29

Values (5) given are the percent of atoms to the total atoms for each element. Note that the predominant isotopes for each element are ^1H, ^4He, ^{12}C, ^{14}N, ^{16}O, ^{32}S, ^{40}Ca, and ^{56}Fe.

magnesium, silicon, and sulfur are also intermediates in the nuclear reactions leading to the end product of iron.

If we consider the chemistry of these five most abundant elements, helium is a member of the noble gases; it is a gas above 2.18 K ($-269°C$), and it has the highest first-ionization energy (2,370 kJ/mol) of any element (268). None of the noble gases forms stable negative ions (168). Some noble gases, such as krypton and xenon, do form a few stable compounds, in spite of the fact that noble gases possess the highest ionization energies in their respective periods (268). We know of no stable compound of helium because of these large energies. For this reason, helium does not play a role in the chemistry and prechemistry of living processes.

Since there is such a predominance of hydrogen, most of the oxygen would be bound to hydrogen in the form of water. Likewise, hydrogen would be bound to carbon in the form of methane and to nitrogen in the form of ammonia.

Tables 46.3, 46.4, and 46.5 list chemicals detected in the dense interstellar clouds of our Milky Way; in addition, molecular hydrogen was found (59). When red

TABLE 46.2. *Percent Atom Abundance of the Most Abundant Atoms in the Universe*

Atom	% Abundance	SEM
H	92.969	0.0
He	6.903	0.311
O	0.068	0.003
C	0.040	0.002
N	0.009	0.0006
All others	0.011	0.0005
Total	100.000	

Calculated from Greenberg (111). Greenberg gave four different sources for these values. We averaged them and calculated the standard error of the mean (SEM) for each atom. Greenberg gave these values as relative to the hydrogen atom. We have assumed that all the others are Mg, Si, Fe, and S, with the remainder of the atoms having a negligible value.

TABLE 46.3. *Oxygen Chemical Species in Interstellar Space*

Symbol	Chemical Species
H_2O	Water
OH	Hydroxyl
CO	Carbon monoxide
C_3O	Tricarbon monoxide
HCO	Formyl
HCO^+	Formyl ion
H_2CO	Formaldehyde
HOCO	Protonated carbon dioxide
HCOOH	Formic acid
CH_2CO	Ketene
CH_3OH	Methyl alcohol
HCC_2HO	Propynal
$HCOOCH_3$	Methyl formate
$(CH_3)_2O$	Dimethyl ether
CH_3CHO	Acetaldehyde
CH_3CH_2OH	Ethyl alcohol
NO	Nitric oxide
HNO	Nitroxyl
HNCO	Isocyanic acid
NH_2CHO	Formamide
SO	Sulfur monoxide
SO_2	Sulfur dioxide
OCS	Carbonyl sulfide
SiO	Silicon monoxide

From Dalgarno (59).

TABLE 46.4. *Carbon Chemical Species in Interstellar Space*

Symbol	Chemical Species
C_2	Carbon
CH	Methylidyne
C_2H	Ethynyl
C_2H_2	Acetylene
C_3H_2	Cyclopropenylidene
C_4H	Butadinyl
C_5H	Pentynylidyne
CH_3CN	Methyl cyanide
CH_3C_2H	Methyl acetylene
CH_3C_4H	Methyl diacetylene
C_6H	Hexatrinyl
CN	Cyanogen
C_3N	Cyanoethynyl
HCN	Hydrogen cyanide
HNC	Hydrogen isocyanide
HC_3N	Cyanoacetylene
H_2NH	Methanimine
CH_2CN	Cyanomethyl radical
NH_2CH	Cyanamide
CH_3NH_2	Methylamine
CH_3C_3N	Methyl cyanoacetylene
CH_2CHCN	Vinyl cyanide
CH_3CH_2CN	Ethyl cyanide
HC_5N	Cyanodiacetylene
HC_7N	Cyanohexatriyne
HC_9N	Cyano-octatetra-yne
$HC_{11}N$	Cyano-decapenta-yne
CS	Carbon monosulfide
C_2S	Dicarbon sulfide
C_3S	Tricarbon sulfide
H_2CS	Thioformaldehyde
CH_3SH	Methyl mercaptan
HCS^+	Thioformyl ion
HCNS	Isothiocyanic acid

From Dalgarno (59).

TABLE 46.5. *Nitrogen and Other Chemical Species in Interstellar Space*

Symbol	Chemical Species
NH_3	Ammonia
N_2H^+	Protonated nitrogen
NS	Nitrogen sulfide
PN	Phosphorus nitride
H_2S	Hydrogen sulfide
SiS	Silicon sulfide
HCl	Hydrogen chloride

From Dalgarno (59).

giant stars become white dwarfs, they eject massive circumstellar envelopes. These envelopes contain many of the same chemicals in Tables 46.3, 46.4, and 46.5 as well as SiC, SiC_2, C_4Si, C_3H, CH_3CN, CP, HSC_2, $HSiC_2$, NaCl, KCl, AlCl, and AlF (204). A major portion of the chemicals in interstellar space are the polycyclic aromatic hydrocarbons, containing as many as 90 atoms (3, 165). The chemical composition of comets has been shown to contain hydrogen, oxygen, carbon, and nitrogen compounds (78, 129, 197). Comets contain significant amounts of water, carbon monoxide, and methyl alcohol (128).

The bulk of the interstellar dust is composed of particles of about 0.1 μm. Greenberg (111) has proposed that ultraviolet (UV) radiation breaks up the molecules on the surface, or mantles, of these particles; these mantles have a thickness of about 0.015 μm. Chemical radicals in these mantles would be stable at a temperature

of about 15 K; however, an explosive mixture can occur due to a sudden rise in temperature of about 10 K caused by grains colliding or by high-energy cosmic radiation. Taking into account the probabilities of these events, Greenberg (111) has proposed that there is a mantle or surface evolution of these particles, as depicted in Figure 46.2. Note that the model starts out with a chemical composition of water, methane, and ammonia. Energy is added and the chemical composition changes to carbon dioxide, carbon monoxide, dioxygen, and dinitrogen. The result is oxidation by a dehydrogenation process. The dust particle becomes a pinpoint of oxidation in a reducing environment composed of mainly hydrogen. Note the similarity of the evolution of this particle to the evolution of the Earth, which we consider now.

EARLY HISTORY OF THE EARTH

Possibly a shock wave caused by an explosion, such as a supernova, compressed an interstellar cloud of dust and gas to form a protoplanetary disk of condensed matter. The condensed matter became little planets, or *planetesimals,* as they rotated around a center. Planetary systems might be common throughout the universe (214). The center eventually became our Sun (163). Our Sun was born about 4.7 Gyr ago, or 4,700 Myr (1 Myr $= 1 \cdot 10^6$ years). It seems that the Sun's initial luminosity was about 70% of its present value and that its luminosity has increased at a more or less constant rate (106). Aggregation of these planetesimals continued

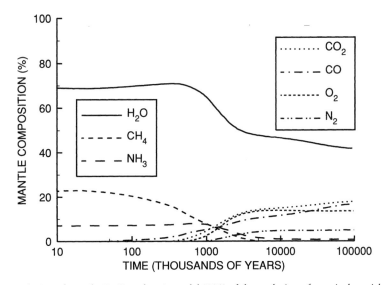

FIG. 46.2. Evolution of mantle. In Greenberg's model (111) of the evolution of a typical particle in a dust cloud, there is a change in the chemical composition over the comparatively short period of time lasting for about $1 \cdot 10^6$ years, or 1 Myr. Note that initial conditions of water, methane, and ammonia are reducing in nature. As the evolution of the mantle progresses, there is a change toward an oxidized mantle, consisting of carbon dioxide, carbon monoxide, and molecular nitrogen. Thus these dust particles become a pinpoint of oxidation in a reducing interstellar dust cloud.

until a proto-Earth was formed, probably in less than 0.1 Gyr (246). The energy released from bombardments of these planetesimals could cause some loss of the planet's protoatmospheres (1). Cameron and Benz (39) proposed that toward the end of this aggregation process, a large planetary body collided with the proto-Earth to produce the Earth–Moon system. This Giant Impact would vaporize some rock that would result in a Giant Blowoff of the vaporized rock and the protoatmosphere of the Earth. If this event occurred about 4.5 Gyr ago, then the newborn Earth would have been very hot (186).

Geological times on Earth begin about this time, as depicted in Table 46.6. However, this excess surface heat would have disappeared within a period of much less than 0.1 Myr (275). Evidence from lunar rocks and lunar craters indicates that the Moon was bombarded with objects such as asteroids, comets, comet dust, and meteorites during the period from about 4,400 to 3,800 Myr ago. Radioisotopes have been used to estimate geological times (75). Presumably, the Earth was undergoing a similar bombardment during this time (143). Many of the prebiotic chemicals existing in outer space

could be brought to the Earth by this bombardment. We will call a chemical "biotic" if it has been synthesized by some living organism. Therefore, by definition, all chemicals before life originated cannot be called biotic.

A method to distinguish between organic prebiotic chemicals and organic abiotic chemicals is to determine the chirality, that is, the mirror image isomeric configuration, of the organic chemical. With but few exceptions, D-sugars and L-amino acids are used by the biosphere. L-sugars are not metabolized (167). Interestingly, an L-amino enzyme has been synthesized (190). A method to distinguish between organic abiotic chemicals and organic biotic chemicals is to measure the stable isotope ratios in the organic chemical. These ratios are often different, depending on their biological or abiotic origin. Table 46.7 gives some natural isotope ratios and an explanation of the usual δ, that is, parts per thousand, method of presentation.

The Murchison meteorite contained nonracemic amino acids; the ratio of D-alanine to L-alanine was 0.85, suggesting that chiral chemicals were formed in the solar system, preceding life on Earth (71). In addi-

TABLE 46.6. *Geological Times*

Eon	Era	Period	Epoch	Beginning Time
Precambrian	Archean	Hadean		4,500 Myr
		Early Archean		3,900 Myr
		Late Archean		2,900 Myr
	Proterozoic	Early Proterozoic		2,500 Myr
		Middle Proterozoic		1,600 Myr
		Late Proterozoic		900 Myr
Phanerozoic	Paleozoic	Cambrian		550 Myr
		Ordovician		500 Myr
		Silurian		440 Myr
		Devonian		410 Myr
		Carboniferous*		360 Myr
		Permian		290 Myr
	Mesozoic	Triassic		250 Myr
		Jurassic		210 Myr
		Cretacous		140 Myr
	Cenozoic	Tertiary	Paleocene	65 Myr
			Eocene	55 Myr
			Oligocene	35 Myr
			Miocene	25 Myr
			Pliocene	5.0 Myr
		Quaternary	Pleistocene	1.8 Myr
			Holocene	0.01 Myr

*Note that sometimes the Carboniferous period is not used. In its place, the Mississippian period, which began 360 Myr ago, and the Pennsylvanian period, which began 320 Myr ago, are used. Enhanced carbon burial occurred during the Carboniferous period, giving rise to major coal deposits (17).

Generally, the timings of the earliest geological events are less certain than the more recent events. Thus the tabulations of these events are usually given as the most recent past event on top. However, we wish to emphasize the evolution of these events as they occurred, so we have placed the oldest event on top. This also makes it easier to compare the table to Table 46.10. Values of this table were generally taken from Ernst (73) and Schopf (251). Myr represents 1 million years ($1 \cdot 10^6$ years).

TABLE 46.7. *Natural Isotopic Ratios of Some Elements*

Element	Isotope Ratio	Value
Hydrogen	$^2H/^1H$	0.000150
Carbon	$^{13}C/^{12}C$	0.0111
Nitrogen	$^{15}N/^{14}N$	0.00371
Oxygen	$^{18}O/^{16}O$	0.00200
Sulfur	$^{34}S/^{32}S$	0.0443

These natural isotopic ratios were calculated from Lide (168). The isotopic abundance of the elements varies to some extent in different compounds. Let us consider the following isotopic exchange reaction:

$$^lX_S + {}^hX_R \rightarrow {}^lX_R + {}^hX_S,$$

where superscripts refer to the mass of the isotope X, subscript S refers to the sample compound or phase, and subscript R refers to the reference compound or phase. The equilibrium constant of this reaction or fractionation factor equals the isotope ratio (r_s) in sample S divided by the isotope ratio (r_r) in reference R. A common method of expressing this fractionation phenomenon is given in the following equation:

$$\delta^hX = 1000 \cdot \left[\frac{r_s}{r_r} - 1\right],$$

where $r_s = {}^hX/{}^lX$ in the sample; $r_r = {}^hX/{}^lX$ in the reference standard; and δ^hX of the sample is measured in parts/thousand or abbreviated as ‰.

It is customary for superscript h to be greater than superscript l. The δ^hX is a measure of the distribution of the heavy isotope (^hX) as compared to the light isotope (^lX) in a sample. Fractionation factors are temperature-dependent.

TABLE 46.8. *Comparison of the Percent Atom Abundance between the Biosphere and Lithosphere*

Element	Biosphere	Lithosphere
Hydrogen	50.4	2.1
Carbon	29.4	0.0
Oxygen	14.1	61.4
Nitrogen	3.8	0.0
Sodium	0.6	2.0
Potassium	0.4	1.2
Calcium	0.3	1.8
Silicon	0.3	21.8
Chlorine	0.2	0.0
Phosphorus	0.2	0.1
Sulfur	0.2	0.1
Magnesium	0.1	1.6
Iron	0.0	1.7
Aluminum	0.0	6.2
Others	0.0	0.0
Total	100.0	100.0

These values were obtained by calculating the parts per mass element averages of land plants, ocean plants, land animals, and ocean animals for the biosphere average and the averages of continental crust and granite shell for the lithosphere (233). These averages were converted to the percent atom abundance for each element. Note the lack of correlation between the percent abundance in the biosphere and lithosphere. The four most abundant elements in the biosphere are also the most abundant elements in the universe, excluding the chemically inert helium (see Table 46.2). Other elements used in the biosphere are manganese, cobalt, nickel, copper, and zinc.

tion, these investigators found that the $\delta^{13}C$ for the amino acids were higher than for biological amino acids, indicating that these acids were not of terrestrial biological origin. The formation of chiral chemicals, such as the L-amino acids and the D-sugars, could have arisen in outer space (26, 26a, 48, 71, 108). Of the four forces (electromagnetic, weak, strong, and gravity), only the weak force is chiral (174a). Calculations using the weak force have been made to show that L-amino acids are slightly more stable than D-amino acids (174a). Thus it seems that prebiotic chemicals originated in outer space. We take the view that during the Hadean period (4,500–3,900 Myr ago), organic precursors were deposited on the Earth by massive bombardments of comets (207, 208), asteroids, meteorites, cosmic dust, etc. from outer space (50). However, about 70% of the organic matter was volatilized when the carbonaceous Murchison meteorite was shocked to a peak pressure of 36 GPa (288). We have already presented evidence of organic chemicals existing in outer space. Consequently, we find that the biosphere is just a droplet of the universe. Table 46.8 illustrates that the biospheric elements are closely correlated with the cosmic elements and not with the lithospheric elements. Table 46.9 summarizes our view on the origin and evolution of life on Earth, which we have divided into seven stages. Stage 1 represents this prebiotic buildup of metastable chemicals in the reducing cosmos.

By definition, the Hadean period (4,500–3,900 Myr ago) contains no existing rock records (73). The following period, the Early Archean period (3,900–2,900 Myr ago), contains the oldest known rocks on Earth, which belong to the Isua Supracrustal Group in western Greenland. These rocks have an age of about 3,800 Myr.

Unless offset by some other factors, the Earth's surface would have been significantly colder than at present due to the decreased luminosity of the Sun at this time. Sagan and Mullen (243) realized that there must be other factors present on the early Earth that would raise its temperature, since the geologic evidence, such as evidence for liquid water, did not support such a cold, frozen Earth. They concluded that at this time there must have been present in the atmosphere one or more greenhouse gases. A *greenhouse gas* is an atmospheric gas that absorbs infrared radiation at about 10 μm and thus inhibits the hot infrared radiation from leaving the Earth; the result is a warming of the Earth's surface (90). They suggested that the very early Earth may have had an abundance of hydrogen, which gradually changed to an ammonia–water vapor atmosphere. The centers of absorption wavelength in micrometers for some greenhouse gases are 15.0 for carbon dioxide; 10.53 for

TABLE 46.9. *Stages of Origin and Evolution of Life on Earth*

Environment	Stage	Reaction	Free Energy (kJ/mol)
Reducing	1	Metastable compound synthesis in outer space	
		$2H_2O + CH_4 \rightarrow CO_2 + 4H_2$	129.8
		$H_2O + CH_4 \rightarrow \frac{1}{6}C_6H_{12}O_6 + 2H_2$	135.6
	2	Energy used by first organisms	
		$4H_2 + CO_2 \rightarrow 2H_2O + CH_4$	−129.8
		$2H_2 + \frac{1}{6}C_6H_{12}O_6 \rightarrow H_2O + CH_4$	−135.6
	3	Biological replenishment of metastable compounds	
		$2H_2O + CH_4 \rightarrow CO_2 + 4H_2$	129.8
		$H_2O + CH_4 \rightarrow \frac{1}{6}C_6H_{12}O_6 + 2H_2$	135.6
Transition	4	Biological replenishment by Photosystem I	
		$2H_2 + CO_2 \rightarrow \frac{1}{6}C_6H_{12}O_6 + H_2O$	5.8
Oxidizing	5	Inorganic atmospheric production of oxygen	
		$2H_2O \rightarrow O_2 + 2H_2$	474.0
	6	Energy used by biological aerobic respiration	
		$O_2 + \frac{1}{6}C_6H_{12}O_6 \rightarrow H_2O + CO_2$	−479.8
	7	Biological replenishment by photosynthesis	
		$H_2O + CO_2 \rightarrow \frac{1}{6}C_6H_{12}O_6 + O_2$	479.8

Free energies were obtained from Latimer (164).

Stage 1 represents the buildup of metastable species in outer space, where the major constituents are hydrogen, helium, water, methane, and ammonia. In such a reducing environment, energetic photons could produce metastable species, such as carbon dioxide and sugar, as illustrated in Figure 46.4. The reactions shown in this table are just representative chemical reactions and show only net reactions. Figure 46.5 illustrates some of the energetics of the nitrogen-metastable species. Other metastable species include chemicals such as the metastable species of carbon and nitrogen, i.e., amino acids. Stage 2 shows some representative net reactions, which the first living organisms on Earth could utilize for energy. The reaction between hydrogen and carbon dioxide is used by the methanogenic archaebacteria for their source of energy. Note that the energy that can be released by the hydrogen reduction of carbon dioxide and of sugar is almost the same. Stage 3 shows the biological replenishment reactions necessary for the continuance of the survival of living organisms. Hydrogen can be photochemically produced from biogenic porphyrins as well as from ferrous ions (181). For the biosphere to survive the exposure of hydrogen, it had to develop antihydrogen defense mechanisms; hydrogen would not only serve as a biological energy source but also tend to reduce the metastable constituents. An evolutionary pressure of hydrogen toxicity resulted in antihydrogen defenses.

Stage 4 shows the transition phase when organisms began using light as an energy source and decreasing the hydrogen gas so that the atmosphere became less reducing. This stage can be designated the nonoxygenic phase. Kasting et al. (146) have pointed out that Photosystem I could use H_2 or H_2S as the hydrogen donor; if H_2S is used, then elemental S is also produced. Stage 5 shows that water vapor in the upper atmosphere would be photolyzed by solar energy; the hydrogen gas would escape from the Earth, resulting in a dehydrogenation of the atmosphere and releasing oxygen gas. Stage 6 shows that the biosphere can use up the metastable compounds by oxidation with oxygen gas and release much more energy by aerobic respiration than was ever made available to the biosphere in the previous stages. Stage 7 shows that photosynthesis can restore the metastable compounds using solar energy.

ammonia; 7.66 for methane; 7.35 for sulfur dioxide; and 4.5, 7.78, and 17.0 for nitrous oxide (N_2O; 135). Hydrogen, water vapor, and ozone are also greenhouse gases. Comparing the greenhouse gas effectiveness of ozone and methane to carbon dioxide on a per mole of gas basis, ozone is about 2000 times and methane is about 25 times more effective than carbon dioxide (90).

PRODUCTION OF THE FIRST LIVING ORGANISMS

In 1952, Urey (291) postulated that the primitive atmosphere at the time of the origin of life was composed of hydrogen, water, methane, and ammonia. In 1953, Miller (188) experimentally showed that amino acids could be synthesized in such an atmosphere. Before Urey's postulation it was realized that the origin of life occurred in the absence of free oxygen in the atmosphere (116, 205). Rubey (236) has pointed out that volcanoes and hot springs release carbon dioxide, con-

cluding that the primitive atmosphere was lost and that carbon dioxide was an important constituent in a secondary atmosphere.

It seems likely that before a living cell could be formed, the metastable chemicals must have been concentrated, and not just in a dilute "prebiotic soup." The lifetime of these metastable chemicals could be increased if they were shielded from electromagnetic radiation, especially in the UV region. Sagan (240, 241) has emphasized the role of UV radiation on early life when there was no atmospheric oxygen. If these chemicals sank to a depth of 10 m of water, then they would be adequately protected from the damaging UV light (242, 263). It is interesting to note that obligate photoautotrophic bacteria in the anoxic regions of the Black Sea can function at greater depths, where there is very little light; these bacteria contain enhanced concentrations of photopigments (60). There are also other methods of biological protection, such as that afforded by the outer layers of microbial communities to the underlying layers of these organisms, natural protective chem-

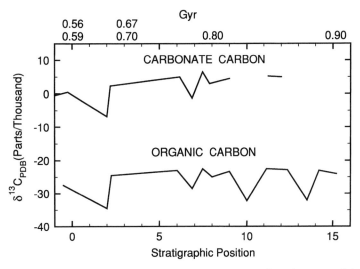

FIG. 46.3. Carbon isotopic values. (See Table 46.7 for explanation.) The reference used for measuring $\delta^{13}C$ was the Peedee belemnite standard (PBS). The $\delta^{13}C$ is a measure of the distribution of heavy carbon (^{13}C) as compared to light carbon (^{12}C) in a sample. Note that inorganic carbonate carbon has relatively more heavy carbon than organic carbon. This is due to the kinetic fixation of carbon by the biosphere. Most of the carbon fixation proceeds by the ribulose 1,5-bisphosphate carboxylase (RuBP, RUBISCO) pathway to phosphoglycerate (PGA), a three-carbon compound as in C3 plants. Of lesser importance is the phosphoenolpyruvate (PEP) pathway to produce oxaloacetate (OAA), a four-carbon compound, as in C4 plants. Carbon fixed by the RUBISCO pathway has $\delta^{13}C$ values between −20‰ and −30‰, whereas carbon fixed by the PEP pathway has $\delta^{13}C$ values between −2‰ and −3‰ (248). Note that the geological samples are obtained at different depths (stratigraphic position), corresponding to the different ages when the samples were deposited. The ages are shown on top and they correspond to the Late Proterozoic, which lasted from about 550 Myr to 900 Myr. Individual data points are not shown, but they clearly showed dips in the $\delta^{13}C$ for both carbonate and organic carbons. Dips are due to an increase of biological activity. If more carbon dioxide is involved in photosynthesis, then the organic carbon would be composed of more light carbon. If photosynthesis is increased, then the heavy carbon would be diluted by biological respiration in both the atmosphere and hydrosphere. This would explain the dips in both records. One consequence of this activity would be an increase in the oxygen concentrations in both the atmosphere and biosphere. Now, as a result of the increased oxygen, there would be an increase of oxygen toxicity. The increased oxygen would produce an evolutionary pressure to combat this poison by producing more natural antioxidant defenses. This would lead to the cyclic characteristics seen in both records. The figure was adapted from Holland (124).

icals against UV irradiation, biological repair mechanisms of UV damage, and negative phototaxis (177).

Concentration of chemicals can occur at phase boundaries, especially at charged surfaces; the activity of the chemical ionic species depends on the surface charge (104). Thus the concentration of a positively charged chemical species is higher than the same species in the bulk solution. Bernal (16) has suggested that clay could act as such a surface and concentrate prebiotic chemicals. Although the details have yet to be established, it seems probable that an ordered selection of prebiotic chemicals and chemical reactions gave rise to the first life-forms on Earth (84, 290). Figure 46.3 illustrates how carbon isotopic values are used to distinguish between carbon of biological origin and of nonbiological origin. Bacteria attached to solid surfaces can behave differently from those that are freely suspended (82). The formation of lipid membranes enclosing complex biochemical systems would have to occur, concentrating these increasingly complex systems before there would be a viable life system. Active transport mechanisms would have to be evolved to maintain osmotic equilibrium across the lipid membranes (95, 175).

In addition, these complex biochemical systems would have to be able to replicate themselves before any biosphere could exist (142). Perhaps clays participated in the replication process (35). Self-replicating molecules have been synthesized (79); such simple molecules may have played an important role in the origin of the first living organisms on Earth (206). Perhaps self-replicating proteins evolved before the onset of ribonucleic acid (RNA). There is evidence suggesting that *prions*, the infectious agents for such diseases as kuru and sheep scrapie, are thought to be self-replicating proteins (89, 220). Prions are extremely stable, both chemically and physically, even to UV light (89); such stability would be needed for survival of the primitive organisms. Self-replication of RNA probably preceded the appearance of desoxyribonucleic acid (DNA) self-replication (55, 304).

Keosian (148) expressed the view that after the synthesis of many complex biochemicals, it was necessary that many series of increasingly complex systems occurred before the appearance of a primitive biosphere. Schueler (257) refers to these increasingly complex systems as levels of integration; they include molecular systems, polymolecular systems, and the cellular (or poly-polymolecular) systems.

ENERGETICS OF THE FIRST LIVING ORGANISMS

Taking the view that many prebiotic chemicals originate in the cosmos, where there is much more hydrogen than any other element, we now consider the energy involved in the chemical reactions that can take place in a hydrogen environment. Hydrogen gas reacts with oxygen gas according to the following net reaction:

$$2H_2 + O_2 \rightarrow 2H_2O \qquad (4)$$

The free energy for this reaction is -474 kJ/mol (164); thus water is a very stable compound. Likewise, the most thermodynamically stable carbon compound in the presence of hydrogen is methane, as shown in Figure

46.4. Figure 46.5 shows that ammonia is the most thermodynamically stable nitrogen compound in the presence of hydrogen.

Energy sources on the primitive Earth include solar radiation; ionizing radiation from radioisotopes ^{40}K, ^{232}Th, ^{235}U, ^{238}U, and ^{244}Pu; heat from volcanoes; electric discharges; and cosmic radiation (215). These energy sources can dehydrogenate the water, methane, and ammonia into oxidized species, some of which are illustrated in Figures 46.4 and 46.5. However, the thermodynamic potential is such that reverse or back reactions release energy. If some of these oxidized species are in outer space, then the chance of a collision is small due to a low density of particles, and thus these oxidized species can exist for some time. Also, if the energy of activation of these oxidized chemicals is high, they can exist for a long time.

It appears that the Earth lost its very reducing atmosphere before life originated. However, we postulate that the first living organisms developed in some "hydrogen niches" on Earth; subsequently, we shall see that hydrogen environments exist even at the present time. Therefore, the first organisms could obtain their energy from the nonbiological, metastable complex chemical systems in a reducing atmosphere, as represented by stage 2 in Table 46.9 For example, the

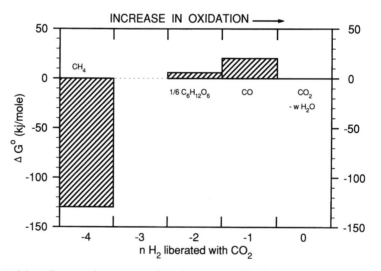

FIG. 46.4. Stability of some carbon compounds in the presence of hydrogen. The ordinate represents the decrease in free energy for the various compounds, according to the following equation (96):

$$\frac{1}{a} C_a H_b O_{a(2-w)} + wH_2O \rightarrow CO_2 + nH_2,$$

where $b = 2a(n - w)$.

The more thermodynamically stable a chemical is in the presence of hydrogen, the lower it is on this diagram. The free energies for Figures 46.4, 46.5, 46.7, and 46.8 were obtained from Latimer (164). The standard free energy for $\frac{1}{6}C_6H_{12}O_6$ is 5.8 kJ/mol, so the free energy for the reaction of $\frac{1}{6}C_6H_{12}O_6$ plus H_2O yielding CO_2 plus $2H_2$ is -5.8 kJ/mol. Similarly, the standard free energy for CH_4 is -129.8 kJ/mol, so the free energy for the reaction of CO_2 plus $4H_2$ yielding CH_4 plus $2H_2O$ is -129.8 kJ/mol. Summing these two reactions gives a value of -135.6 kJ/mol for the net reaction of $\frac{1}{6}C_6H_{12}O_6$ plus $2H_2$ yielding CH_4 plus H_2O.

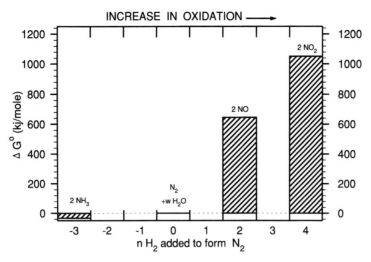

FIG. 46.5. Stability of some nitrogen compounds in the presence of hydrogen. The ordinate represents the decrease in free energy for the various compounds, according to the following equation (96):

$$aH_bN_{2/a}O_{w/a} + nH_2 \rightarrow N_2 + wH_2O,$$

where $b = 2(w - n)/a$.
For an explanation of the type of diagram, see Figure 46.4.

methanogenic archaebacteria obtain their energy from the following reaction (24):

$$CO_2 + 4H_2 \rightarrow CH_4 + 2H_2O \qquad (5)$$

The legend for Figure 46.4 points out that the free energy of this reaction is -129.8 kJ/mol. These organisms obtain their energy from this reaction; no other sources of energy have been found (309). These bacteria are commonly grown in 0.5 to 0.8 bar of H_2; they can survive 3 bar of H_2 (reviewed in 102). Woese et al. (312) have pointed out that these bacteria are unique organisms, which can thrive in the most unusual environments on Earth today (141, 310).

Woese et al. (311) proposed that these organisms be classified in the kingdom Euryarchaeota, which would be part of a new domain, Archaea. They generally use ammonia as their nitrogen source (24). They are strict anaerobes and cannot tolerate any oxygen. The Archaebacteria have been shown to be related to each other, using ribosomal RNA as a molecular chronometer (310). The methanogens contain unusual coenzymes, the thermophiles live at extremely high temperatures, and the halphiles have very high intracellular salt concentrations. Woese (310) has postulated that the first organisms could have been the thermophiles, which derive their energy from the reduction of sulfur; in fact, the optimum temperature for the growth of *Pyrodictium occultum* is 110°C. Sulfur isotope ratios have been interpreted as indicating that bacterial sulfate reduction was present 2.7 Gyr ago (249). Methanogenesis probably existed 2.7–2.9 Gyr ago, when there were extreme depletions of $\delta^{13}C$ from sedimentary organic matter

(121). An abyssal methanogenic archaebacteria *(Methanopyrus kandleri)* has been identified, which grows at an optimum temperature of 98°C using energy derived from reaction 5 (174, 234). These bacteria were found in the hot vents of the Guaymas Basin, Gulf of California; the depth of these vents is 2,000 m. A significant amount of the Earth's heat flux is due to the natural decay of radioactive elements in the Earth. The heat produced by this process must have been much higher during the Hadean and Early Archean periods, when the amount of radioactive elements was much greater (127). There must have been many hot vents during these periods. These organisms also possess a dehydrogenase enzyme that forms H_2; the standard free energy for the reaction that this enzyme catalyzes is $+5.5$ kJ/mol (174). Thus hydrogen formation requires an energy input. Geothermal energy from the hydrothermal vents can act as an energy input analogous to solar energy, which is used for photosynthesis. These hot spots in water could act as a concentrating area, where the conditions were just ripe for the origin of the first living forms. Since UV cannot reach the depths of the oceans, Sagan (242) has proposed that abyssal organisms were the ancestors of the prokaryotes. Thermal models of prebiotic chemistry have already been postulated (85, 126). These organisms would be subjected to a high pressure. Studies of deep-sea microorganisms generally show that their metabolism is slowed, presumably due to high pressure and low temperature (138).

Perhaps the present Archaebacteria are not the first primitive living organisms on Earth, but these organisms or their ancestors could certainly be good candidates for the first living cells on Earth. Even if that statement proves to be false, the biochemical reactions of these organisms could be the remnants of the biochemical reactions present in the first cells. In summary, metastable carbon dioxide combines with hydrogen to produce energy in a back reaction; this energy was used by the first cells for biological maintenance.

For these first living organisms to survive, the biosphere had to regenerate these metastable compounds or nutrients. Stage 3 in Table 46.9 shows that about 130 kJ of energy is required for each mole of methane oxidized. This regeneration, such as the replenishment of hydrogen by the hydrogen-producing bacteria, requires an energy input into the biosphere. Methane-producing bacteria often have symbiotic relationships with hydrogen-producing bacteria. Hydrogen gas produced by the so-called S organism is taken up by the methanogenic bacteria to form methane. These S organisms cannot tolerate an excess of hydrogen gas, such as 0.5 bar of H_2; other examples of hydrogen toxicity have been documented (reviewed in 102). The strict anaerobic archaebacterium *Pyrococcus furiosus* produces hydrogen gas and carbon dioxide, using a fermentative metabolism, growing optimally at 100°C. These organisms reduce sulfur to hydrogen sulfide as an antihydrogen defense mechanism against hydrogen toxicity (8). Hydrogenase is a remarkable enzyme; it enables the organism either to use hydrogen as an energy source or to liberate it; all forms of hydrogenase contain iron and sulfur, and the majority contain nickel (31).

Figure 46.6 represents a life-energy profile in different atmospheres and illustrates that the energy responsible for life is also the energy that can destroy life by uncontrolled back reactions. Living cells in a reducing atmosphere take up hydrogen as an energy source; most of the gas is used by the cell for energy and anabolic synthesis. However, some of the gas destroys the cellular constituents. Antihydrogen defenses necessarily had to be developed for cell survival.

Replenishment of H_2 eventually would have to occur as the H_2 was used for energy by the primitive biosphere. Hydrogen can be formed photochemically from ferrous ions; the reaction has a high quantum yield for generating hydrogen (181). Porphyrins, which are present in all living matter, even in the archaebacteria, can also photochemically produce hydrogen (181).

Hydrogen gas can also be produced biologically using carbon monoxide. The thermophylic anaerobic organism *Clostridium thermoaceticum* utilizes CO_2 and H_2 as a source of carbon for growth; CO enters this mechanism via carbon monoxide dehydrogenase (222, 315). The free energy is −20.2 kJ/mol for this net reaction, which is:

$$H_2O + CO \rightarrow CO_2 + H_2 \qquad (6)$$

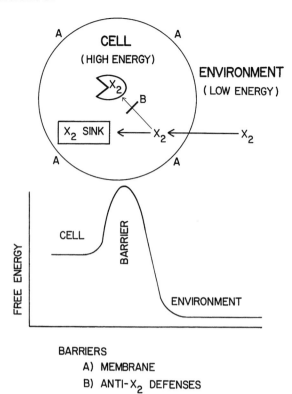

FIG. 46.6. Life-energy profile in different atmospheres. A living cell is characterized by possessing more energy than its external environment. What keeps this energy difference? The cell membrane represented by *A* provides a barrier to the flow of energy out of the cell by means of active transport mechanisms. Gases, such as hydrogen and oxygen, represented by X_2, can cross the membrane barrier by diffusion. X_2 is taken up by the cell to produce biological energy, represented in this diagram by the X_2 *SINK*. Some of the X_2 in the X_2 *SINK* can be used for anabolic synthesis as well. However, some of the X_2 can destroy the cellular constituents. As a consequence, anti-X_2 defenses, represented by *B*, had to be developed by the cell to survive. Thus the first primitive cells had to deal with hydrogen as an energy source and hydrogen toxicity, whereas modern aerobic cells have to deal with oxygen as an energy source and oxygen toxicity.

Carbon monoxide dehydrogenase is an important enzyme in the acetyl-CoA pathway for fixing carbon dioxide (72, 109, 222). The methanogen archae *Methanothrix soehngenii* has been shown to have a carbon monoxide dehydrogenase (139).

A slow metabolism would allow the primitive organisms to decrease the rate of the damaging back reactions and, therefore, would be an antihydrogen defense. The first experimental evidence showing that hydrogen is toxic was published in 1941 (316). The finding of a superoxide dismutase (SOD) gene in the strict anaerobe archaebacterium *Methanobacterium thermoautotropicum* (285) presents some difficulties. SOD is an antioxidant enzyme; it dismutates the superoxide radical anion into dioxygen and hydrogen peroxide (88). What is it doing in this anaerobe? Fridovich (87) has suggested that anaerobes possessing SOD might have developed the need for this enzyme, as they would be exposed temporarily to an oxygen environment in their travel from one anaerobic site to another. Alternatively, perhaps this enzyme had another unknown previous function in this organism, possibly as an antihydrogen enzyme.

In summary, the first organisms could use energy, such as light and heat, to use up the prebiotic organic chemicals, which originated in outer space. Then, other organisms were able to replenish these prebiotic organic chemical nutrients by producing hydrogen. However, the hydrogen, which provided these primitive cells with energy and nutrients, also was toxic to these early life-forms.

WHEN DID THE EARTH'S BIOSPHERE ORIGINATE?

Schidlowski (248) noted that the $\delta^{13}C$ in the organic carbon from the Isua Supracrustal Group was less than that from the carbonate carbon from the same formation and concluded that an early biosphere was possibly operating 3,800 Myr ago. However, this evidence is indirect. Schopf (253) has pointed out that in such old samples, metamorphism due to high temperatures can cause the release of hydrogen from organic matter, resulting in a decreased ratio of hydrogen to carbon. Schopf has also pointed out that $\delta^{13}C$ of the organic matter that has undergone metamorphism is decreased. Evidence obtained from *stromatolites* (columnar or mound-shaped layered rock structures) and microfossils, as well as from determinations of $\delta^{13}C$ values from the Swaziland Supergroup in South Africa and from the Pilbara Supergroup in Western Australia, indicates that life existed there about 3,500 Myr ago (252–254). Hence, the origin of life probably occurred between 3,800 Myr and 4,200 Myr ago. Table 46.10 gives a general timetable of some of the major events that have occurred on Earth.

THE TRANSITION STAGE BETWEEN A REDUCING AND AN OXIDIZING ATMOSPHERE

A biological light process known as *photoreduction* represents the transition between a reducing and an oxidizing atmosphere; this is shown in stage 4 of Table 46.9 (22). Photoreduction has been shown to occur in the anaerobic metabolism in the green algae *Scenedesmus obliquus;* this process requires a hydrogenase that takes up hydrogen gas, and also requires Photosystem I. The reaction requires an extremely small amount of energy. The net effect of this reaction is to reduce the hydrogen gas in the environment; thus dehydrogenation would make the atmosphere less reducing and more oxidizing.

The transition environment can be called nonoxygenic. Some organisms can accomplish this net reaction using heat; the hyperthermophilic archaebacterium *Pyrodictium brockii* grows at an optimum temperature of 105°C and can accomplish the same result (213).

As the atmosphere became more oxidizing, the ammonia in the atmosphere decreased due to the fact that the primitive biosphere was using ammonia as its nitrogen source for synthesizing amino acids, proteins, and other cellular constituents. Probably a more important reason is that dinitrogen is thermodynamically more stable in an oxygen atmosphere than ammonia. In addition, ammonia is photodissociated by UV radiation (160). Some organisms had to develop a means of fixing dinitrogen to ammonia or other forms of fixed nitrogen. Nitrogenase, the nitrogen-fixation enzyme, is inhibited by oxygen; also, the synthesis of nitrogenase is inhibited by oxygen and ammonia (77). The production of hydrogen gas is accompanied by nitrogen fixation. However, the hydrogenase enzyme removes the hydrogen gas (77).

There is some evidence that organisms living almost 3,500 Myr ago were producing oxygen (254). Living organisms are responsible for the origin of the Earth's oxygen in the atmosphere (145). Light is absorbed in plants by Photosystem II, releasing dioxygen from water and producing ATP; electron transfer goes from Photosystem II to Photosystem I to the reductant NADPH (reduced nicotinamide-adenine dinucleotide phosphate; 196) The older Photosystem I and the younger Photosystem II probably evolved separately (45). The 3,800-Myr-old Isua rocks are the oldest rocks showing banded iron formation (BIF; 125). According to Holland (125), iron in these formations is probably due to hydrothermal ocean circulation, possibly at the hot midocean ridges, which was brought close to the ocean surface and precipitated by many factors, including oxidation. Ferrous iron (Fe^{2+}) is soluble, whereas ferric iron (Fe^{3+}) is not soluble. Thus BIF formations can be due to oxidation as well as ocean stratification. No BIFs have occurred in rocks aged between 1,900 and 800 Myr. The appearance of *red beds*, that is, terriginous sedimentary rocks containing red ferric oxide, which is also in BIFs, and analysis of *paleosols*, that is, ancient soil profiles, at about 1,900 Myr ago show that they are highly oxidized. In addition, Holland (125) points out that uranium in the reduced state is not found in rocks younger than 2,000 Myr, concluding from these observations that prior to 2,200 Myr ago, the O_2 pressure was less than or equal to 2 mbar (0.2 kPa, where kPa represents kiloPascal; 1.6 torr), which is roughly the value that can support respiration (45). However, the cellular oxygen levels, where oxygen was produced, would be higher. The minimum O_2 pressure after 1,900 Myr ago was 30 mbar (3 kPa; 24 torr). The finding of 1700-Myr-old fossil sterols from northern Australia indicates that the oxygen pressure was at least 0.2 kPa (2 mbar; 1.6 torr) at this time, which is the minimum oxygen tension needed for the biosynthesis of membrane sterols (237). Photosynthesis by primitive cyanobacteria, sometimes known as blue-green algae, caused this relatively rapid increase in O_2 pressure. Liberated oxygen would then oxidize reduced chemicals such as Fe^{2+}. Oxygen produced by abiotic photodissociation of water would not be enough to increase oxygen in the atmosphere; the conclusion is that the rise in atmospheric oxygen was due to the biological process of photosynthesis (302).

Many cyanobacteria have the ability to fix nitrogen, that is, to produce ammonia from nitrogen gas. The enzyme that catalyzes this fixation is nitrogenase. Since nitrogenase is inhibited by oxygen and since oxygen is produced photosynthetically, cyanobacteria had to develop many different types of antioxidant defense (77).

Nitrification is the biotic process of oxidizing ammonia or nitrite for energy purposes. *Nitrosomonas* bacteria oxidize ammonia, whereas *Nitrobacter* bacteria oxidize nitrite (HNO_2). These aerobic organisms grow slowly because of the small amount of energy derived from these oxidations (see Table 46.5). They grow best at low oxygen pressures, probably due to the decreased production of oxidizing free radicals. In chemistry, the definition of radical has changed from the time when Guyton de Morveau first introduced it in 1787 (103a). Currently, the definition of a free radical is a chemical entity containing an unpaired electron. Generally, an electronic orbital contains two electrons with opposite spins, resulting in a net electron spin of zero. Free radicals are usually very reactive due to the presence of an unpaired electron (103a). For example, *Nitrosocystis oceanus* grow best at the reduced pressure of 21 mbar (16 torr; 219).

Denitrification is the biological process by which nitrogen gas is produced from nitrates. Experiments on denitrification in *Pseudomonas stutzeri* showed that this process requires nitric oxide reductase and proceeds from nitrite (NO_2^-) to nitric oxide (NO) to nitrous oxide (N_2O) to N_2 (30). When *Pseudomonas aeruginosa* was grown aerobically and subsequently transferred to an anaerobic environment, NO and N_2O were produced; when this organism was grown anaerobically, NO was reduced to N_2 via nitrogen dioxide (NO_2; 295). Nitrogen oxides are reactive species of oxygen.

DEVELOPMENT OF THE OXIDIZING ATMOSPHERE

In the upper atmosphere water vapor is photodissociated by UV into hydrogen atoms and hydroxyl free radicals. Hydrogen atoms escape into outer space, since they are sufficiently light to reach the escape velocity of the Earth. Hydroxyl free radicals are extremely reactive and undergo a series of reactions, which eventually produce molecular oxygen. Thus the Earth is becoming oxidized by a process of dehydrogenation. Stage 5 in Table 46.9 represents this nonbiological source of oxygen.

An analogous situation occurs on Europa, the second large moon of Jupiter, which also contains oxygen in its atmosphere. The water on Europa is dissociated by Jupiter's magnetospheric electrons, resulting in the light hydrogen escaping, and the oxygen being retained in the atmosphere. The oxygen pressure is only 10^{-11} bar, or 10 pbar (117a). Since oxygen has such a high potential, there must be many oxygen sinks on the surface of Europa, which prevent any significant increase in the free molecular oxygen. Indeed, the presence of a significant amount of dioxygen on any celestial body in the universe is evidence for a presence of a biosphere. Ganymede, another moon of Jupiter, also contains oxygen, but in ice, instead (131a).

Photodissociation of dioxygen or molecular oxygen produces atomic oxygen, which can then combine with dioxygen in the presence of a third inert particle to form trioxygen or ozone. Since ozone is also photodissociated by UV into atomic oxygen and dioxygen, it becomes a shield for the damaging effects of UV on biological processes (15). However, it seems that the presence of only 0.2 mbar of oxygen in our atmosphere can produce an adequate ozone screen against this potentially damaging effect to the primitive biosphere (297). Water absorbs UV weakly, and the presence of a liquid water column of at least 10 m could be an effective ozone screen (177, 242, 263).

Living organisms could then utilize the energy released by the oxidation of metastable compounds, such as sugar, for biological purposes. Stage 6 in Table 46.9 represents this biological oxidation or respiration. Stage 7 in Table 46.9 represents the reaction producing oxygen by photosynthesis. Note that the solar energy input in photosynthesis is practically the same as the energy required for the photodissociation of water. Thus from this point of view, the carbon in the carbohydrate acts merely like a sponge to absorb the gaseous hydrogen. Henderson (122) wrote in 1913: "hydrogen and oxygen are likely to confer great chemical activity wherever they are, and that they are quite unrivaled in this respect." It has been suggested that the release of energy by the burning of hydrogen gas by oxygen could help solve the "world energy crisis" (112). *Hydrogenomonas* bacteria, a type of pseudomonad, actually can use their energy by this oxidation of hydrogen by oxygen (176, 250). A proposal has been made not to use the genus *Hydrogenomonas* classification but in its place to divide the bacteria into the genus *Alcaligenes* and the genus *Pseudomonas* (61). The transition phase

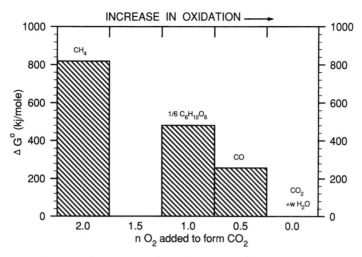

FIG. 46.7. Stability of some carbon compounds in the presence of oxygen. The ordinate represents the decrease in free energy for the various compounds, according to the following equation (96):

$$\frac{1}{a} C_a H_{2aw} O_b + nO_2 \rightarrow CO_2 + wH_2O,$$

where $b = a(w + 2 - 2n)$.

For an explanation of the type of diagram, see Figure 46.4.

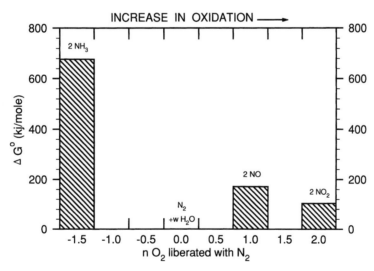

FIG. 46.8. Stability of some nitrogen compounds in the presence of oxygen. The ordinate represents the decrease in free energy for the various compounds, according to the following equation (96):

$$aH_{2w/a}N_{2/a}O_b \rightarrow N_2 + nO_2 + wH_2O,$$

where $b = (2n + w)/a$.

For an explanation of the type of diagram, see Figure 46.4.

reaction of stage 4 illustrates that an extremely small quantity of energy is required for the carbon in carbon dioxide to absorb the hydrogen gas to form a carbohydrate.

The biological reactions available in the oxidizing atmosphere are much larger than was available in the primitive reducing atmosphere. Possibly, during this transition phase, there could be local "oxygen oases" in this anoxic environment (144). Thus Figures 46.7 and 46.8 illustrate that much more energy is associated with the compounds in an oxygen atmosphere. Figure 46.7 shows that this is true for the carbon compounds, as compared to Figure 46.4. Likewise, Figure 46.8 shows that this is also true for the nitrogen compounds, as compared to Figure 46.5. These facts allowed the biosphere to develop into its many different forms of life. However, as illustrated in Figure 46.6, antioxidant mechanisms had to be developed before the biosphere could exist in such a metastable environment.

The toxic effects of oxygen are mediated by free radicals (94). These free radicals are responsible for oxygen toxicity. When oxygen pressure to lung tissue was increased, there was an increase in the superoxide radical anion and hydrogen peroxide (137). An example of this toxicity can be seen in the measurement of survival times of organisms. When the environmental atmosphere was experimentally increased to 1 bar of pure oxygen, female mice survived only 111 ± 1 h (92) and *Paramecium caudatum* cultures survived 148 ± 1 h (93). Increasing the oxygen pressure five times the normal atmospheric sea-level oxygen pressure decreased the survival time of these mice 100-fold. Oxygen is toxic! As Gerschman (91) stated: "A better understanding of the fundamental mechanism involved inclines us to marvel at the continuous and powerful cellular defenses against oxygen rather than to be surprised at its potential destructive action."

Cyanobacteria contain antioxidant defenses, such as carotenoids, tocopherols, SOD, and catalase; however, only some contain ascorbate peroxidase. It seems that ascorbate peroxidase was a later development in the cyanobacteria (191). Other antioxidant defenses include glutathione, glutathione peroxidase, etc. (86, 118, 159). Carotenoids are unusual antioxidants since they are found to protect against photosensitized oxygen toxicity in plants; they can quench the electronically excited oxygen molecule (singlet oxygen) and inhibit lipid peroxidation (41, 159). β-carotene acts as an antioxidant when the oxygen is less than 150 torr, whereas it acts as a prooxidant at higher oxygen pressures (33). SOD can sometimes act as a prooxidant (134). Under certain circumstances, reduced glutathione can damage DNA (105, 189). Flavonoids are polyphenols found in plants; many of them possess antioxidant activity by inhibiting lipid peroxidation (27, 278). However, flavonoids can cause oxidative damage to DNA (244, 278). The antioxidant ascorbate can also possess prooxidant activity (179). Many of these prooxidant activities depend upon the presence of a transition metal, such as iron and copper. Nitric oxide is another oxygen species that under the appropriate circumstances can be either a prooxidant or an antioxidant (235a). Gilbert (97) wrote: "Depending on the oxygen pressure, an antioxidant effect can be pronounced or diminished to such a degree that the effect will be a pro-oxidant one."

C_3 plants comprise 80%–90% of all land plants (225). Photosynthesis in C_3 plants is inhibited 33%–55% in 0.2 bar oxygen, compared to the rate in 0.01–0.03 bar oxygen (317). This inhibition is due to the competition of oxygen and carbon dioxide for the active site on the enzyme ribulose 1,5-bisphosphate carboxylase/oxidase (RuBP; RUBISCO). RUBISCO catalyzes the fixation of carbon dioxide with ribulose 1,5-bisphosphate to produce two molecules of 3-phosphoglyceric acid (PGA), a 3-carbon acid (277). A mutant plant resistant to oxygen toxicity was found to have an increased catalase (318).

C_4 plants evolved later to avoid this oxygen inhibition. In C_4 plant photosynthesis, carbon dioxide is fixed with phosphoenol-pyruvate (PEP) to produce oxaloacetic acid (OAA), a 4-carbon acid; this reaction is catalyzed by the enzyme PEP carboxylase, which has a much greater affinity for carbon dioxide than for oxygen. Thus C_4 plants are better able to perform photosynthesis in the presence of oxygen than C_3 plants, but the C_3 plants can more easily perform photorespiration. C_4 plants grow in the tropics (277). Crabgrass is an example of a C_4 plant (196).

Crassulacean Acid Metabolism (CAM) plants have another way to fix carbon under hot desert conditions. At night, carbon dioxide gets fixed as malic acid, using PEP carboxylase; during the day, the malic acid gets decarboxylated and the carbon dioxide gets refixed, using the RUBISCO pathway (277). Thus plants can be much more resistant to the toxic effects of oxygen than mammals.

Intracellular oxygen concentrations in C_3 plants are about the same as air-equilibrium values; in C_4 plants they can be more than twice air-equilibrium values; and in CAM plants they can be almost four times air-equilibrium values. Extracellular oxygen concentrations in some aquatic plants can be more than four times air-equilibrium values (225). About 1%–4% of the oxygen uptake by mitochondria results in the production of the superoxide radical anion and hydrogen peroxide (28). Probably inorganic ferric ion was used by the biosphere for destroying the potential damaging effect of hydrogen peroxide before the development of enzymes, such as catalase and peroxidases. This ion can catalyze the destruction of hydrogen peroxide; the free energy for this catalyzed reaction is −210 kJ:

$$2H_2O_2 \rightarrow 2H_2O + O_2 \qquad (7)$$

When ferric ion is enclosed in a porphyrin molecule, the resulting complex catalyzes reaction 7 1,000 times faster than bare ferric ion; when porphyrin is associated with a protein, such as catalase, the catalytic efficiency is increased $1 \cdot 10^7$, as compared to the ferric ion (38).

In spite of antioxidant defenses, which are not perfect, oxygen toxicity is a problem for aerobic organisms. These defenses are of six different types.

The first type of defense for a living organism is to possess a low instead of a high metabolic rate. A decrease in the metabolic rate means that there are fewer chemical reactions per given time period; subsequently, there will also be fewer damaging chemical reactions occurring in this time period. Lowering the temperature usually decreases the metabolic rate. However, oxygen toxicity is more of a problem to organisms that have a high temperature and a high metabolic rate. For example, mice, which have a higher metabolic rate than humans, also have more oxidized DNA damage per day (4). Mice pay for this increased metabolic rate by having a shorter lifespan than humans.

The second type of defense is the development of preventative measures. For example, animals have developed circulatory systems with oxygen carriers, such as hemoglobin. The effect is to have a plentiful supply of oxygen to the cells at a lower oxygen concentration than the environment. Thus the lower oxygen concentration means less oxygen toxicity. Another preventative measure is to keep essential cellular transition metals, such as iron and copper, in their proper place at the proper time and away from locations (location meaning dynamic, not static) in the cell, where these metals can catalyze damaging oxidations. Iron is not in a free state but is sequestered by siderophores, transferrin, and ferritin. Mechanisms exist in cells that provide dynamic barriers to dangerous oxidants so that these oxidants are kept in their proper place at the proper time.

The third type of defense is the development of antioxidant chemicals and enzymes against the toxic reactive oxygen species (ROS). ROS refers to free radicals, such as the superoxide radical anion, hydroxyl radical, nitric oxide, nitrogen dioxide, and other reactive oxygen species, such as hydrogen peroxide, ozone, and singlet oxygen. ROS mediate oxygen toxicity. Glutathione, β-carotene (pro-vitamin A), α-tocopherol (vitamin E), uric acid, ascorbic acid (vitamin C), SOD, catalase, and glutathione peroxidase are examples of antioxidant chemicals and enzymes. A mole of urate is about 10 times more effective as an antioxidant than a mole of ascorbate (14).

The fourth type of defense is the development of repair mechanisms for getting the damaged cell or cellular constituent back to its original functioning state. DNA repair mechanisms (4) as well as tissue regeneration processes are examples of this type of defense.

The fifth type of defense includes mechanisms for enhanced protection, such as the induction of antioxidant enzymes as a response against an increased oxidative stress. Such mechanisms are called *regulons* and involve the induction of many genes (64, 83). For example, the bacterial oxyR regulon responds to hydrogen peroxide and induces the production of catalase, and the bacterial soxR regulon responds to the superoxide radical anion and induces the production of SOD. Catalase destroys hydrogen peroxide and SOD removes the superoxide radical anion. Both of these regulons induce other changes as well (64).

The sixth type of defense occurs in multicellular organisms where selected cells die due to programmed cell death or apoptosis. Low concentrations of reactive oxygen species have been implicated as some of the causative agents inducing apoptosis (169, 210, 227, 258). Perhaps apoptosis is a *quality control* mechanism in that mild adverse environments cause the removal of slightly defective cells from the cellular populations and permit only the survival of those cells capable of enduring in adverse environments. Apoptosis can also help dispose of unwanted cells during the development of the organism; thus *survival of the fittest* also occurs in many cell populations in the organism. Apoptosis is initiated by specifically activating receptors on the cell surface, causing a sequence of events which generally results in the fragmentation of nuclear DNA and subsequent death of the cell. Apoptosis occurs in many diverse cell types and in many multicellular organisms (259). DNA fragmentation is used by unicellular organisms to protect against phage infection (155). Bacteria populations use programmed cell death to survive under changing environmental conditions (316a).

These six types of defense can resist the damaging effects of oxygen toxicity, but they cannot prevent them completely. The phenomenon of aging in animals is at least partly due to ROS (4, 58, 91, 120, 239, 280). Gutteridge (114) has pointed out, however, that other variables must also be important.

CHRONOLOGY OF EVENTS IN THE OXYGEN ATMOSPHERE

Before the development of a significant increase in the atmospheric oxygen content, there were probably many niches where anaerobic conditions persisted. Even today, there exist anoxic sites, such as the Black Sea, the world's largest anoxic basin (60). As the cyanobacteria polluted the Earth with toxic oxygen, there were undoubtedly many organisms that did not survive. The very process of generating oxygen by photosynthesis is also inhibited by oxygen. Antioxidant defenses had to be developed by the biosphere to regulate oxygen metabolism. The mineral ores also were sinks for oxygen until they became more oxidized. Cloud (53) points out that these factors explain the delay from the first signs of life on Earth about 3,500 Myr ago to 2,100 Myr ago, when the geological record shows evidence of a significant oxygen content in the atmosphere.

Oxygen toxicity is mediated by ROS, which are composed of species of varying reactivities. The transition metals, such as iron and copper, have been shown to change the relatively unreactive superoxide radical anion into the highly reactive hydroxyl radical. Thus in free solution, these metal ions act as prooxidants (86, 118). However, iron is needed by most cells for enzymatic functions; iron, if not properly controlled, can promote ROS (123). The fact that *Lactobacillus plantarum,* an aerobe, does not require iron provides this organism with an antioxidant defense; this organism does not contain catalase, yet it is viable in 7mM hydrogen peroxide (9). The function of SOD is replaced by Mn^{2+} in this organism (87); the catalase function is replaced by a pseudocatalase, which contains manganese (19, 153).

Most aerobic bacteria need to take up ferric iron, which is very insoluble in water; the calculated Fe^{3+} concentration at a pH of 7 is about $1 \cdot 10^{-15}M$, and the calculated Fe^{2+} concentration at a pH of 7 is about $1M$ (273). However, in aerobic waters, Fe^{2+} is oxidized to Fe^{3+}, which precipitates out of solution. These bacteria transport Fe^{3+} by chelating it to iron carriers, or *siderophores.* Strict anaerobes do not possess siderophores. A well-known siderophore is ferrioxamine B (Desferal), which has been used as an iron chelator (130, 199). Since anaerobes are not in contact with oxygen, they do not have to protect themselves against the prooxidant iron. Today, the iron in our aerobic waters is transported in colloidal solution (265) or is associated with clays (43). Freshly formed ferric hydroxide particles in waters act as scavengers for many different compounds (294). In the surface waters of the open ocean, iron is adsorbed onto dust (aerosol) particles (320). Some unicellular organisms can obtain their iron by producing a reductase which reduces the insoluble Fe^{3+} in the external environment to the soluble Fe^{2+}; the soluble ferrous ion is then transported into the cell (151).

Life changed on Earth when some cells liberated oxygen into the atmosphere. Photosynthetic cells, such as those in plants, pollute the Earth with gaseous molecular oxygen. What is good for one species may be bad for another species. Anaerobes cannot survive in the presence of oxygen; at the same time, aerobes cannot survive in the absence of oxygen. There are also the microaerophiles, which require oxygen but at a lower concentration than is present in air (158). For example, today phosphates are considered bad since they promote phytoplankton growth in lakes (70). This proliferation of algae and plant growth, or *eutrophication,* is a natural aging process of lakes. However, it can cause undesirable effects, such as hypoxic conditions caused by bacterial decay of algae. This hypoxia inhibits fish growth (224).

Atmospheric oxygen increased due to cyanobacteria photosynthetic activity. This increase led to other forms of oxygen in the atmosphere, including the hydroxyl radical (OH·), hydroperoxyl radical (HO₂·), hydrogen peroxide (H_2O_2), and ozone (O_3). Ozone screened out the damaging UV radiation so that living forms could prosper in sunlight close to the air/water interface.

The dissociation constant (pK) of the hydroperoxyl radical in water is 4.8 so that in water at a pH of 7.4 almost all of this weak acid is dissociated into the superoxide radical anion ($O_2^{\cdot-}$). The decay constant of this radical system ($HO_2^{\cdot}/O_2^{\cdot-}$) is maximal at a pH of 4.8. At a pH of 7.4, the decay constant is about 100 times less; at a pH of 6.4, the decay constant is only about 10 times less (21). Since many biological membranes have a negative surface charge, the membrane surface pH is less than the bulk pH, as measured with a pH electrode (104). Therefore, the hydroperoxyl radical may participate in the damaging lipid peroxidation of biological membranes. The hydroperoxyl radical belongs to a class of radicals known as peroxyl radicals ($RO_2^{\cdot}$), which can propagate damaging chain reactions and are part of ROS (200).

Knoll (152) has postulated a sequence of events for the evolution of the biosphere based mainly on subunit ribosomal RNA (rRNA) and the fossil record, suggesting that the archaebacteria and bacteria had a common ancestor and that this common ancestor shared another common ancestor with organisms containing nucleated cells (eukaryotes). The fossil evidence shows that cyanobacteria existed at least as early as 2,000 Myr ago and probably much earlier, perhaps 3,500 Myr ago. Geological evidence also shows that eukaryotes existed 1,700–1,900 Myr ago (152, 252). Eukaryotes also generally contain organelles, such as mitochondria; and the photosynthetic ones also contain plastids, such as chloroplasts (176). Chloroplasts may have risen several times (152). A red alga has been dated to be between 950 and 1,260 Myr old; this is one of the earliest evidences of a multicellular organism (152). Multicellular organisms evolved from unicellular organisms many times (176). Evidence of land plants indicates that they were in existence 460 Myr ago (226). Fossils of primitive animals have been dated to 580 Myr (152).

As photosynthetic production of oxygen by cyanobacteria increased, so did the demand for oxygen increase. The presence of oxygen had three effects on the biosphere. First, the synthesis of cellular constituents was altered. Second, more energy was available to the biosphere. Third, the biosphere was forced to cope with the toxic effects of oxygen (99). The atmospheric oxygen pressure was at least 0.03 bar (24 torr) 1,900 Myr ago, in the middle of the Early Proterozoic period. It seems that the oxygen pressure rose to its present level of 0.2 bar by the beginning of the Paleozoic era, 550 Myr ago (17, 237; see Table 46.10 and Fig. 46.3).

Excluding hydrogen, carbon, oxygen, and nitrogen, the biosphere also requires sodium, potassium, calcium, chlorine, phosphorus, sulfur, magnesium, and iron.

TABLE 46.10. *Timetable of Some Major Events on Earth*

Biosphere Events	Oxygen	Land Events	Time
First living cells	0.0 mbar	Many small continents	4,000 Myr
Cyanobacteria	2 mbar	60% land	2,200 Myr
	15 mbar	BIF decrease	2,100 Myr
	60 mbar	Red bed formation	1,900 Myr
Nucleated cells	55 mbar		1,800 Myr
Multicellular organisms	145 mbar	97% land	1,100 Myr
	210 mbar	Mountain formation	950 Myr
	190 mbar	Mountain formation ends	750 Myr
Animals	210 mbar		580 Myr
Land plants	210 mbar	Gondwana—continent	550 Myr
Vertebrates	210 mbar		490 Myr
Species extinction = 85%	210 mbar		440 Myr
Species extinction = 82%	230 mbar		360 Myr
Vascular plants and insects	235 mbar	Laurentia—continent	350 Myr
Reptiles and conifer plants	295 mbar		290 Myr
Species extinction = 96%	240 mbar		250 Myr
	200 mbar	Pangaea formed	220 Myr
Species extinction = 76%	185 mbar		210 Myr
Mammals and birds	190 mbar		190 Myr
	195 mbar	Splitting of Pangaea	180 Myr
Flowering plants	210 mbar		150 Myr
Dinosaur extinction	215 mbar	South America	65 Myr
Primates	210 mbar		60 Myr
	210 mbar	India collided into Asia	45 Myr
	210 mbar	Australia	35 Myr
Anthropoids	210 mbar		30 Myr
Hominoids	210 mbar		20 Myr

The oxygen pressure in the atmosphere is given in mbar (1,000 mbar = 100 kPa = 750 torr). The data given in this table represent only some general ideas and are not necessarily to be taken as accurate. The information in this table was obtained from the text as well as from other references (17, 53, 136, 176, 300). See Figure 46.3 for possible minor oxygen oscillations.

The age of the Earth is about 4,500 Myr, so the origin of the biosphere probably occurred when the Earth was 500 Myr old. There is evidence that the first glaciation on Earth occurred about 2,500 Myr ago. After this glaciation, the Earth was ice-free until about 900 Myr ago (57). There is an indication that the Earth's temperature was about 5°C warmer about 140 Myr ago (56). Thus probably the temperature on Earth has varied substantially.

It seems that there were two major increases in the Earth's land mass. The first, which occurred between 3,000 and 2,500 Myr ago, increased the Earth's land to almost 60% of the present value. The next occurred between 2,200 and 1,600 Myr ago; the land mass 1,600 Myr ago was 97% of our present value (171).

There were no polar land masses about 500 Myr ago. Gondwana, a supercontinent, moved southward and was composed of South America, Africa, Antarctica, India, and Australia. Laurentia, a northern supercontinent composed of North America and Europe, collided with Gondwana to form one land mass, the supercontinent of Pangaea. Pangaea extended from the northern pole to the southern pole, before it split into Gondwana and Laurentia.

About 550 Myr ago, the Cambrian explosion occurred when invertebrates appeared within a few million years (29, 173, 238). There occurred 65 Myr ago an extinction of 76% of the species, including the dinosaurs; this was one of the five biggest extinctions that have occurred. The biggest extinction occurred at the end of the Permian period, about 250 Myr ago and marked the transition between the end of the Paleozoic era and the beginning of the Mesozoic era. The other three big extinctions occurred 440, 360, and 250 Myr ago (see Table 46.6). Also, smaller extinctions happened about 190, 140, 90, and 35 Myr ago (136). A change in oxygen pressure has been hypothesized to provide a reason for the dinosaur extinction; one theory was that oxygen decreased, causing death due to hypoxia, and another theory was that oxygen increased, causing death due to hyperoxia (see 102). Natural extinctions can be caused by a depletion of nutrients (286). However, the extinction that took place 65 Myr ago when the dinosaurs died appears to be due to one or more extraterrestrial impacts from asteroids or comets (20); the evidences for this conclusion are the appearance of an iridium layer and a charcoal and soot layer in the rocks dated at this time (313) and the finding of a large-diameter structure in the Yucatán part of Mexico also dated from this time (25, 269, 270). The diameter of this structure might be 400 km (269). Perhaps more than one impact occurred at that time, since other sites have been suggested for this impact (25, 228). The charcoal and soot layer is evidence of major wildfires caused by the impacts. It has been estimated that the frequency of impacts caused by bodies having a diameter of at least 5 km is 1 per 10 Myr (184).

It appears that there was an increase of burial of organic carbon and liberation of oxygen 2,000 and 900 Myr ago (65). As a first approximation, we have assumed that the oxygen increased linearly with time between 2,200 and 600 Myr ago and also that the oxygen pressure was 210 mbar 600 Myr ago, the same as it is today. Superimposed on this linear relationship are apparently two oscillations of oxygen increases that occurred during these two periods of mountain formation, when organic carbon was buried (65). When organic carbon gets buried, the liberated oxygen does not necessarily stay in the atmosphere; some of it is able to oxidize other chemicals, such as iron. From 600 Myr ago to the present, we have made use of a model in which there were two other oxygen oscillations (17). However, we have assumed a minimum deviation from an oxygen pressure of 210 mbar; the values were sometimes at the edge of their error zone. If the oxygen pressure was substantially greater than our present value of 210 mbar, then additional antioxidant defense mechanisms would have been evolved to cope with the problem of oxygen toxicity. It is our contention that under such conditions, the present biosphere would be much less sensitive to the toxic effects of oxygen.

However, the nutrients that are usually considered to be of short supply are phosphorus and fixed nitrogen (264). Iron deficiency occurs in Drake Passage waters in the Antarctic, and also in the Gulf of Alaska, that prevents phytoplankton in these waters from blooming (178). The phytoplankton biomass production was noted to be highest in iron-rich waters of the Southern ocean (63a).

Table 46.8 shows that chlorine, phosphorus, and sulfur are concentrated in the biosphere. Both desoxyribonucleic acid and ribonucleic acid contain phosphorus and are involved in biological replication reactions. Adenosinetriphosphate (ATP) also contains phosphorus and is used in biological energy reactions. Sulfur is also an important constituent in antioxidant chemicals, such as glutathione and cysteine. The purple sulfur bacteria *Chromatiaceae* use photosynthesis to produce elemental sulfur (S) and fixed carbon compounds from hydrogen sulfide (H_2S) and carbon dioxide. Elemental sulfur and carbon dioxide can photosynthetically produce sulfate (SO_4^{2-}) and carbohydrate. *Desulfovibrio*, an anaerobe, uses sulfate as an oxidant to reduce lactate to acetate for energy, and in this reaction sulfide (S^{2-}) is produced. Photosynthetic reactions require little energy, and the energy released using sulfate as an oxidant is also very small (249). Thus these organisms release sulfide, which ends up as pyrite (ferrous sulfide, FeS_2, which is insoluble), commonly known as fool's gold.

OXYGEN AS A BIOLOGICAL ENERGY STORE

Oxygen is unique. As previously pointed out (96), it is ideally suitable for serving as a biological energy source for the following reasons. First, oxygen is abundant. It is the third most prevalent atom in the universe, partly due to the fact that ^{16}O is a very stable atom. The only other elements that are more prevalent are hydrogen and the chemically inert helium. Second, oxygen is easily available. It is a gas at temperatures above $-183°C$, and gases can easily distribute themselves in the atmosphere, so they are available all over the surface of a planet. The solubility of oxygen into water is temperature-dependent so that a decrease of temperature increases the solubility. Third, oxygen has a high thermodynamic potential, meaning that much energy can be stored in oxygen. A few other species have a higher thermodynamic potential, such as fluorine, ozone, atomic oxygen, and chlorine. Fourth, oxygen is sluggish in its behavior. If not for this fact, oxygen could not be very useful in serving as a storage form of energy; if it were not sluggish, then the stored energy would be dissipated. Fluorine, ozone, atomic oxygen, and chlorine are much more reactive. Schrödinger (255) wrote about "the marvelous faculty of a living organism . . . [to] . . .

delay[s] the decay into thermodynamical equilibrium (death)." The sluggishness of oxygen is responsible for this facility.

We take the view that the energy of aerobic life is derived from the splitting of the water molecule into oxygen and hydrogen by photosynthesis; subsequently, the hydrogen is concentrated by being fixed to carbon. Solar energy is thus transduced to stored biological energy. In the reverse process of respiration, carbohydrates release hydrogen by dehydrogenation reactions; the released hydrogen finally combines with oxygen in the cytochrome system. The energy derived in the process of respiration is due to the combining of the gaseous oxygen with the hydrogen bound to carbon compounds located at specific sites in living organisms to form water. It seems to us that during the evolution of the Earth, the origin of life was not a chance occurrence, but rather was inevitable. Further, we believe that the evolution of the aerobic biosphere was also a necessary consequence.

During this dehydrogenation process, ATP is produced, which is the actual source of biological energy. Actually, the oxidation of glucose to carbon dioxide is equal to 36 molecules of ATP formed for each molecule of glucose. Using ATP as the final biological energy source is a way that energy can be directed toward a site-specific area for a specialized biological function. Perhaps the precursor for ATP was an inorganic pyrophosphate (217). Oxygen is not necessary at the site where the biological energy is used, such as muscle contraction, active transport, etc. These sites are thus protected against the threat of oxygen toxicity by the use of ATP as a secondary energy source.

OXYGEN COMPOUNDS IN THE ATMOSPHERE

There are two chemical species of carbon oxide in the atmosphere: carbon monoxide and carbon dioxide. Some higher plants produce about 3 Emol of carbon monoxide/Myr (262). It has been estimated that about 100 Emol of carbon monoxide/Myr are produced by natural processes; anthropogenic release is about 35 Emol/Myr (40). Carbon monoxide is barely detectable in human plasma (13 nmol/l with a range of 5–21 nmol/l; 44), where it is known to interfere with the oxygen binding to hemoglobin (267). Carbon monoxide is oxidized to carbon dioxide in the heart and muscle of frogs (52, 80), is released from heme oxidase, activates guanylyl cyclase, and may be a neurotransmitter (293).

Although the increase in fossil-fuel burning is alarming since it produces a significant increase in carbon dioxide, it has only a negligible influence on atmospheric oxygen (98). The oxygen concentration in the atmosphere is very difficult to change in a short geological time; however, ozone, another form of oxygen, is increasing in the troposphere and decreasing in the stratosphere. Table 46.11 gives a very approximate summary of the various layers in the atmosphere as a function of altitude. The gaseous constituents are given in Table 46.12. Although sulfur dioxide and the nitrogen oxides seem small quantitatively, they play a major role in air pollution. Organic chemicals are also found in the Earth's atmosphere (135), as well as in interstellar space (Table 46.13).

Table 46.14 gives some of the reactions in the stratosphere and in the troposphere. The atmospheric reactions are dominated by oxidizing reactions. Ozone is present in both the stratosphere and troposphere. Water is photodissociated in the atmosphere, releasing the hydrogen atom and the hydroxyl radical. Since these radicals are extremely reactive, most of them react with each other to produce water (297). Above 20 km, water dissociation is complete (303). The hydrogen atoms in the outer limits of the atmosphere achieve escape velocity and enter outer space; the rate of this escape is diffusion-limited (297, 303).

Ozone absorbs UV radiation, having absorption wavelengths principally from 210 to 290 nm; molecular oxygen absorbs wavelengths less than 200 nm; nitrogen dioxide absorbs wavelengths between 180 and 410 nm. Stratospheric ozone protects the biosphere from dam-

TABLE 46.11. *Atmospheric Layers*

| Atmospheric Layer | Bottom | | |
	Altitude, km	Temperature, °C	Pressure, torr
Troposphere	0.0	15.0	760.0
Tropopause	11.1	−56.5	167.6
Stratosphere	20.0	−56.5	41.5
Stratopause	47.4	−2.5	0.83
Mesosphere	51.0	−2.5	0.53
Mesopause	86.0	−86.3	0.0028
Thermosphere	91.0	−86.3	0.0012

The layers are defined by temperature breaks as a function of altitude; the suffix "pause" indicates no temperature change. The temperature rises in the thermosphere, and at an altitude of 500 km the kinetic temperature is about 725°C, which is about the exospheric temperature. The values were obtained from the U.S. Standard Atmosphere 1976 (198).

These values are only approximate and are latitude-dependent. At high latitudes, the beginning of the tropopause is at 8 km; at low latitudes, it is at about 18 km. The tropopause forms folds and has breaks. The ozone layer extends from 20 km to 30 km. The light gases hydrogen and helium are the predominant species above 600 km. The upper boundary of outer space is about 1,000 km (135).

The percent mass of the atmosphere for these layers is as follows: 82.3% for the troposphere plus the tropopause; 17.6% for the stratosphere and stratopause; and 0.1% for the mesosphere and the mesopause (303).

TABLE 46.12. *Sea Level Gases in the Dry Gas Atmosphere*

Gas	Pressure mbar	Gas	Pressure μbar
N_2	781	H_2	0.5
O_2	209	CO	0.5
Ar	9.3	N_2O	0.5
CO_2	0.3	Xe	0.09
Ne	0.02	SO_2	0.07
He	0.005	O_3	0.05
CH_4	0.002	NO_2	0.01
Kr	0.001	NO	0.002

The pressures for the quantitatively minor gases vary and are only approximate. Also, NH_3, HNO_3, H_2SO_4, H_2O_2; the free inorganic radicals, OH and HO_2; and organic chemicals may or may not be present at quantitatively insignificant concentrations (81, 135). Nitric oxide (NO) and nitrogen dioxide (NO_2) are collectively called NO_x species, and are usually expressed as NO_2. Natural and anthropogenic sources of nitrogen oxides are about equal in quantity. These nitrogen oxides in the presence of solar radiation and organic compounds (see Table 46.13) give rise to ozone. About 40% of the SO_2 is derived from human activities, such as the burning of sulfur-containing fuels. Volcanoes release SO_2 and smaller amounts of H_2S, but this amounts to only 2% of the total sulfur released into the atmosphere, with 18% from sea spray and 40% from biogenic origins (81).

The nitrogen oxides are oxidized by the OH radical to produce nitric acid; likewise, SO_2 is oxidized by OH to produce sulfuric acid. Such reactions produce acid rain.

TABLE 46.13. *Some Organic Chemicals in the Atmosphere that Are in Interstellar Space*

Symbol	Chemical Species
H_2CO	Formaldehyde
HCOOH	Formic acid
CH_3OH	Methyl alcohol
CH_3CHO	Acetaldehyde
CH_3CH_2OH	Ethyl alcohol
C_2H_2	Acetylene
CH_3C_2H	Methyl acetylene

From Isidorov (135). Only those organic chemicals listed in Isidorov (135) and in Tables 46.3–46.5 are given here. There are surely many others.

aging UV radiation (81). UV radiation is subdivided into three groups depending upon the spectral wavelengths (90). The first group, UV-A, is the weakest energetically and is comprised of 320–400 nm wavelengths; the second group, UV-B, is more energetic and is comprised of 280–320 nm wavelengths; and the third group, UV-C, is the most energetic and is comprised of 200–280 nm wavelengths. Another terminology has also been used for UV radiation: *far UV* for wavelengths of 10–200 nm, *UV* for wavelengths of 200–399 nm, and *near UV* for wavelengths of 300–380 nm. The shorter the wavelength, the more damaging it is to the biosphere. UV-C radiation is strongly absorbed by the

TABLE 46.14. *Some Atmospheric Reactions*

Effect	Stratospheric Reactions	Reaction
O_3 Formation	$O_2 + h\nu \rightarrow 2O$	S1
	$O_2 + O + M \rightarrow O_3 + M$	S2
O_3 Destruction	$X + O_3 \rightarrow XO + O_2$	S3
	$XO + O \rightarrow X + O_2$	S4
	$O_3 + O \rightarrow 2O_2$	S5
O_3 Destruction	$O_3 + h\nu \rightarrow O + O_2$	S6

Effect	Tropospheric Reactions	Reaction
O_3 Formation	$HO_2 + NO \rightarrow OH + NO_2$	T1
	$NO_2 + h\nu \rightarrow NO + O$	T2
	$HO_2 + h\nu \rightarrow OH + O$	T3
	$O_2 + O + M \rightarrow O_3 + M$	T4
O_3 Destruction	$O_3 + NO \rightarrow O_2 + NO_2$	T5
O_3 Destruction	$O_3 + h\nu \rightarrow O + O_2$	T6
OH Formation	$O + H_2O \rightarrow 2OH$	T7
HO_2 Formation	$CO + OH \rightarrow CO_2 + H$	T8
	$H + O_2 + M \rightarrow HO_2 + M$	T9
H_2O_2 Formation	$2OH \rightarrow H_2O_2$	T10
O_3 Destruction	$O_3 + OH \rightarrow HO_2 + O_2$	T11
CH_3 Radical formation	$CH_4 + OH \rightarrow CH_3 + H_2O$	T12

This table gives some of the main reactions occurring in the stratosphere and troposphere. The following references were used in compiling this table: Finlayson-Pitts and Pitts (81), Isidorov (135), Thompson (287), and Campbell (40). In reaction S1, the UV radiation splits the bond between the two oxygen atoms, which results in the formation of an oxygen atom. For simplicity, we will make no distinction between the ground states and the excited states of the oxygen atom and dioxygen. Subsequently, in reaction S2, there is a three-body collision between the oxygen atom, dioxygen, and a catalytic neutral particle (*M*) to form trioxygen, or ozone.

Note that reaction T4 is the same as reaction S2 in the stratosphere. Because of the destruction of O_3 by NO (reaction T5), O_3 cannot exist in the presence of significant amounts of NO at the same time and place (81).

Some of the destruction of ozone is the same as in the stratosphere; reactions T6 and S6 are the same. The resulting atomic oxygen from reaction T6 can react with water to produce the reactive OH radical (reaction T7). The radical OH reacts with CO (reaction T8), O_3 (reaction T11), and CH_4 (reaction T12). CO and CH_4 are the two largest sinks of OH in the troposphere; the CH_4 sink, including feedback processes, is approximately as large as the direct CO sink (172). The hydrogen atom formed in reaction T8 can react with dioxygen to form the radical HO_2, as in the neutral particle catalytic reaction T9. Hydrogen peroxide occurs when two hydroxyl radicals dimerize to each other as in reaction T10.

OH and HO_2 can react with hydrocarbons to produce R and RO_2 radicals; these radicals can then participate in chain reactions that form smog. H_2O_2 also plays a role as an oxidant (287).

atmosphere, and almost none of it penetrates to the Earth's surface. UV-B radiation can cause significant biological damage, and to a much lesser degree so can UV-A radiation (90). Water is dissociated into the hydrogen atom and the hydroxyl radical at UV radiation below 242 nm (303).

Measurements have been made in the Antarctic during the cold winter season, showing a depletion of total column ozone (245). A decrease in the total vertical ozone profile has been observed in the Northern Hemisphere (282, 306). Decreases in stratospheric ozone have been observed in both the Southern and Northern hemispheres, especially during winter and spring (107, 232, 282, 322). This decrease in stratospheric ozone is the so-called ozone hole and has reached 50% in the austral spring in the Antarctic. This deletion is mainly caused by the anthropogenic release of chlorofluorocarbons, which are used as refrigerants (235). Chlorofluorocarbons form the chlorine atom, which can act as chemical X in reaction S3 in Table 46.14 and deplete ozone as depicted in the chain reaction of S3 plus S4 (193). Other chemical free radicals that can react as X in reaction S3 are OH and odd nitrogen species, such as NO. The XO formed in reaction S3 then reacts with atomic oxygen to reform X, as illustrated in reaction S4. This chain reaction is summed in reaction S5. It has been suggested that ClO can react by other means to deplete the stratospheric ozone (192). Polar stratospheric clouds evaporate in the spring and then the ClO can react with NO_2 to form $ClONO_2$; the resulting $ClONO_2$ can be photochemically dissociated into reactive X species. Possibly, these reactions can take place in the warmer lower latitudes in the spring (289). Hydrogen chloride is released from volcanic eruptions into the atmosphere, but it seems that most of the hydrogen chloride is removed in condensed supercooled water.

The argument has been made that volcanic eruptions probably do not provide many chlorine atoms for the removal of ozone (284). However, the June 15, 1991, eruption of Mt. Pinatubo in the Philippines was the largest volcanic eruption of the twentieth century, producing a large aerosol load into the atmosphere. This aerosol load shielded the Earth's surface from solar radiation and temporarily reversed the global warming produced by the anthropogenic release of the "greenhouse gases" (182a.) Particulate matter from the Pinatubo eruption can help form the polar stratospheric clouds and transform inert forms of chlorine, as in HC1, into the more reactive form of chlorine gas. The chlorine gas is then photodissociated into the chlorine atom and removes stratospheric ozone by reaction S3 in Table 46.14. Record low depletions of stratospheric ozone over Antarctica was observed in the years 1991 to 1993 (182a). However, this does not diminish the long-term effect of the anthropogenic release of chlorofluorocarbons on decreasing the stratospheric ozone layer.

Stratospheric ozone is also destroyed by solar energy, as shown in reaction S6. It has also been suggested that energetic solar protons are responsible for ozone depletion; the secondary radiation emanating from these energetic particles would have enough reactivity to oxidize N_2 to NO; X would then represent NO in reaction S3 to deplete O_3, as shown in Table 46.14 (281).

The stratospheric ozone decreases the damaging UV radiation reaching the Earth. In 1990, the ratio of UV-B radiation to total radiation (280–700 nm) increased, which resulted in a 6%–12% decrease in phytoplankton production in Antarctic waters (279).

The lower atmosphere or troposphere contains oxidants such as the hydroperoxyl radical, hydroxyl radical, and hydrogen peroxide (287). UV radiation absorption in clouds produces hydrogen peroxide, as well as other peroxides (76). The mechanism involves ferric iron complexed to organic ligands being photochemically dissociated to ferrous iron and organic radicals (321). Clouds also have a major influence on reducing many of the oxidizing species in the troposphere (166). These oxidants can give rise to tropospheric ozone. Since 1900, the tropospheric ozone has probably increased about 25% (133). Ozone has been shown to be toxic to living organisms (187).

A major source of tropospheric ozone is the release of NO from automobile exhaust, which reacts with HO_2 to form a highly reactive scavenger, the OH radical, as shown in Table 46.14 by reaction T1. NO_2, released from reaction T1, then photochemically forms NO and atomic oxygen, as in reaction T2. Summing T1 and T2 results in T3, which results in the photochemical production of OH and atomic oxygen. Reaction T4 shows that ozone is formed from the resulting oxygen atom combining with dioxygen, with a neutral particle serving as a catalyst. Thus photochemical air pollution caused by exhaust from automobiles produces ozone (81). It is interesting to note that NO is an important regulator of cellular processes (47, 194, 195).

Tropospheric ozone is also produced in the harvesting of sugarcane. A common method for harvesting sugarcane fields is to burn them; this burning produces carbon monoxide and ozone. Harvesting these fields produces not only sugar but also alcohol, which is used as an automobile fuel (149).

Thus humans are adding insult to injury by producing oxidants in the lower atmosphere and by increasing the damaging effects of UV radiation. The inability to keep ozone in its proper place does pose a serious problem for the biosphere.

STABILITY OF THE OXYGEN CYCLE

The Earth's crust is composed of the lithosphere, hydrosphere, atmosphere, and biosphere. Oxygen atoms comprise 58% of all the atoms in the lithosphere, which has been referred to as the oxysphere (102). There are 371,000 Emol of oxygen in the lithosphere. The prefix abbreviation "E" was used by Gilbert (98) to represent Erda (Old High German for Earth), or $1 \cdot 10^{18}$. However, we will use the approved International System of Units (SI), in which the prefix "exa," also abbreviated as E, represents the multiple of $1 \cdot 10^{18}$ (168). The hydrosphere contains 38,500, the atmosphere contains 37.7, and the biosphere contains 0.0812 Emol of oxygen. Of the 37.7 Emol of oxygen in the atmosphere, 37.0 are in the form of O_2, 0.6 are in the form of H_2O, and 0.0619 are in the form of CO_2 (Figs. 46.9 and 46.10). Basically, the atmospheric oxygen produced by photosynthesis is removed by respiration. The net production of oxygen by photosynthetic organisms is 6,500 Emol/Myr, or 0.0065 Emol/year, so the atmospheric oxygen turnover rate due to this production equals 37 Emol divided by 0.0065 Emol/year, or 5,700 years.

Weathering exposes reduced species, such as iron, carbon, and sulfur in their reduced states, so that atmospheric oxygen can react with them. The rate of this process is about 10 Emol/Myr. Some living organisms can use these oxidized species for energy by reducing these oxidants. Much slower rates are due to hydrogen loss from the Earth to outer space (297, 299, 300), hydrogen release from volcanoes, and reduced iron release from the mantle at sites of midocean ridges (299). Note that the rate of fossil-fuel combustion is small; the turnover of the oxygen reservoir is about 1 Myr. However, since the carbon dioxide reservoir is only 0.0619 Emol at the present time, the rate of fossil-fuel combustion on it is about 200 years. We have arbitrarily assigned other removal rates on the atmospheric oxygen reservoir due to decay processes involving the oxidation of methane, sulfur, nitrogen, and iron. Ignoring the anthropogenic influence of using fossil fuels, the errors are too large to decide if the oxygen reservoir is currently increasing, decreasing, or remaining the same. Measurements have been obtained using a new extremely sensitive interferometric technique, which shows that the rate of atmospheric molecular oxygen loss is 670 ± 170 Emol/Myr (147). As a first approximation and ignoring human activities in the twentieth century, we have assumed that this reservoir is in a steady state.

The ^{18}O content of oxygen in the atmosphere is greater than the ^{16}O content of water. This phenomenon is termed the *Dole effect* and has been shown to be due to the enzyme systems involved with photorespiration. The $\delta^{18}O$ of air is 21.2‰; the reference used was the standard mean ocean seawater (SMOS; 18; see Table 46.7 for explanation).

It has been assumed that 600 Myr ago, the oxygen reservoir was about the same as it is at present. Has it remained constant since then? The trend has certainly been increasing from the time when the first aerobic cells appeared. However, even though the trend has been

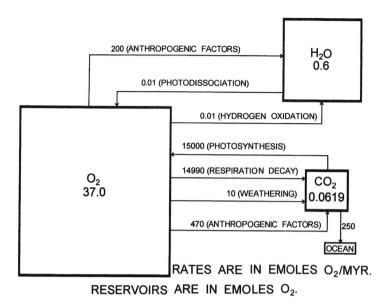

FIG. 46.9. O_2 exchanges in the atmosphere. Most of the values are derived from Gilbert (102). The value for the CO_2 reservoir was calculated from Hall (117). For simplicity, we have rounded off the rates determined by Keeling and Shertz for the anthropogenic factors as well as the rate of CO_2 entering the ocean (147).

The three most abundant oxygen species residing in the atmosphere are molecular oxygen or dioxygen, water vapor, and carbon dioxide. Other important oxygen species are ozone, carbon monoxide, hydrogen peroxide, nitrous oxide (N_2O), sulfur oxides, the free radicals nitric oxide (NO) and nitrogen dioxide (NO_2), the hydroperoxyl radical (HO_2), and the hydroxyl radical (OH). Note that the bulk of the oxygen in the atmosphere is in the form of molecular oxygen or dioxygen. Photosynthetic production of oxygen is the predominant fast process, and the corresponding respiratory uptake rate of oxygen occurs at almost the same rate. Plants that produce oxygen also use oxygen. The plant respiration rate is 8,500 Emol/Myr, so the net rate of photosynthesis is 15,000 Emol/Myr minus 8,500 Emol/Myr, or 6,500 Emol/Myr. Thus the 37 Emol of dioxygen in the atmosphere is replaced by photosynthesis every 5,700 years.

Weathering and erosion processes expose reduced chemicals to the atmosphere and allow these chemicals to be oxidized by atmospheric oxygen. The release of reduced chemicals from volcanoes and midocean ridges due to tectonic activity also has the same effect as weathering on the atmospheric oxygen reservoir; therefore, this activity has been considered as part of weathering (17, 300). However, decay and burial of organic chemicals removes carbon dioxide and increases oxygen (17). Mountain formation can bury organic carbon and lead to a release of oxygen (65). Planting trees also has the same effect as burial processes, while deforestation has the opposite effect. It has been suggested by many authors that planting trees can decrease the buildup of carbon dioxide produced by the burning of fossil fuels (69, 132, 201, 256). Deforestation is occurring today at such a high rate that tree planting barely offsets its effect. Although these processes are of considerable importance to the small carbon dioxide reservoir, they are of little significance to the much larger oxygen reservoir.

Other gases, such as methane and carbon monoxide, also are oxidized by dioxygen. When carbohydrates are burned, one mole of oxygen is consumed for each mole of carbon dioxide released (see stage 6 in Table 46.9). Physiologists have used the respiratory exchange ratio or respiratory quotient (R) as the symbol for the ratio of carbon dioxide released to oxygen consumed by organisms (276, 308). Similarly, we can use the atmospheric exchange of these two gases as also being represented by the symbol R. Thus when carbohydrates are burned, the value of the atmospheric exchange ratio, or atmospheric quotient R, is one. However, when hydrocarbons are burned, the atmospheric R is less than one, as illustrated in the following reaction:

$$2O_2 + CH_4 \rightarrow CO_2 + 2H_2O.$$

In this reaction, the atmospheric R is 0.5. This results in more water being formed. The burning of fossil fuels reduces the atmospheric dioxygen at the rate of 670 Emol/Myr; 470 Emol/Myr of carbon dioxide are produced from this burning and 200 Emol/Myr of water are also produced. The atmospheric R due to anthropogenic sources is equal to 470 Emol/Myr divided by 670 Emol/Myr, or 0.7. About 250 Emol/Myr of carbon dioxide enter the ocean so that the net rate of increase in the atmospheric carbon dioxide reservoir is about 220 Emol/Myr.

What will be the final outcome of these anthropogenic factors on these atmospheric reservoirs? The carbon in the Earth's fossil-fuel reservoir is about 0.9 Emol (102). If we assume that the burning of carbon from this reservoir is the principal anthropogenic factor leading to the increase of atmospheric carbon dioxide and that the rate of this increase is kept constant, then we can answer this question. The fossil-fuel reservoir will be depleted in roughly 2,000 years (0.9 Emol divided by 0.47 Emol/thousand years). The increase of the atmospheric carbon dioxide reservoir will then equal 2,000 years times 0.22 Emol/thousand years, or 0.44 Emol; this reservoir will then equal 0.0619 Emol plus 0.44 Emol, or roughly 0.5 Emol. Dividing the 0.5 Emol by 0.0619 Emol gives a value of 8.1; this means that the present carbon dioxide reservoir will have increased by 810%! The corresponding decrease of the atmospheric oxygen will then equal 2,000 years times 0.67 Emol/thousand years, or 1.34 Emol; this reservoir will then equal 37 Emol minus 1.34 Emol, or roughly 35.7 Emol. Dividing the 35.7 Emol by 37 Emol gives a value of 0.965; this means that the present oxygen reservoir will have decreased by only 3.5%. Hence, the atmospheric carbon dioxide is easier to change than the relatively stable oxygen reservoir.

Walker (298) assumed that atmospheric oxygen was in a steady state, but he emphasized that this was just an assumption. Kump (161) has pointed out that due to the lack of precise knowledge about cycling rates it is not possible to state if the atmospheric dioxygen reservoir is decreasing, increasing, or remaining the same.

1082

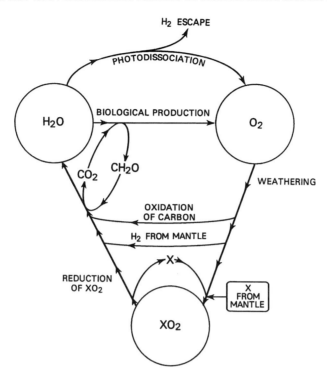

FIG. 46.10. Slow processes in the oxygen cycle. Photosynthesis is the predominant biological process for producing oxygen; this process is reversed by respiration (see Fig. 46.9). The mantle releases gases in volcanoes and solid materials in midocean rifts caused by plate tectonics. Iron is released in the ferrous state in these rifts. X represents reduced chemicals such as iron (300), sulfur (17, 299, 301), and other gases, such as methane and carbon monoxide (300). Reduction of the oxidized X (XO_2) is accomplished by biological activity. Hydrogen is released from volcanoes and is removed from the Earth by photodissociation of water in the upper atmosphere at about equal rates (300).

increasing, feedback mechanisms have always been present so that oscillations in the reservoir were changing over short geological times. For example, consider the effect of a decreased ozone in the atmosphere. There would be an increase in UV radiation reaching photosynthetic cells. As a result, many of these cells would die and less oxygen would be released into the atmosphere. If the change in the ozone concentration was gradual enough, then some cells would evolve protective mechanisms against this increased flux of UV radiation. Then, more oxygen would be released by these adapted cells into the atmosphere. Finally, the atmospheric ozone level would increase as a consequence of more oxygen being released. There are other feedback mechanisms for regulating atmospheric oxygen released by aerobic cells. We know that big changes occurred in the biosphere, atmosphere, hydrosphere, and lithosphere from the time that the first cells appeared up until 600 Myr ago.

Waters next to land masses are areas where living organisms tend to concentrate. When they die, their remains are used for food by other living organisms, but some of these remains get buried into those sediments close to land masses. Plate tectonics cause the sediments to be uplifted in the formation of mountains. Weathering and erosion of these newly formed mountains expose the buried organic compounds to the atmospheric oxygen, which in turn oxidizes them. In addition, reduced gases and reduced minerals released by volcanoes and midocean ridges, also due to plate tectonics, react with atmospheric oxygen. Thus plate tectonic activity decreases the oxygen in the atmosphere, just as the process of respiration does (300). However, the burial of organic carbon increases with tectonic activity associated with mountain formation; this phenomenon releases oxygen. The oxygen is then able to oxidize chemicals such as iron and sulfur, and the excess remains in the atmosphere as dioxygen (65).

There were many small islands on Earth 4,000 Myr ago (see Table 46.10). That means the water–land interface was much larger than it is today, when there are big continents. Due to plate tectonics, these small islands began to coalesce into bigger and bigger islands until continents were formed (300). As a result of bigger land masses and reduced shorelines, there would be less of the rich organic sediments to be brought up by tectonic activity. Hence, there would be less oxidation from weathering processes. The conclusion is that the atmospheric oxygen would rise, assuming that the photosynthetic rate would not change. During the Carboniferous period, which began about 360 Myr ago, the supercontinent of Pangaea was beginning to form. This consolidation of land masses would be a factor contributing to an atmospheric oxygen increase. Since a good part of Pangaea was located in tropical areas, the suggestion has been made that the hot, humid conditions would increase photosynthesis (32).

Another factor was the rise of vascular land plants at this time. Vascular land plants were significant to the atmospheric oxygen because of their lignin content. *Lignin* is the substance in the cell wall of plants that gives it strength. It is a carbohydrate, which is very difficult to degrade biotically or abiotically. Plants from the Carboniferous period probably contained about 40% lignin; modern land plants have a much lower lignin content, amounting to about 20%. Most of the lignin today is degraded by fungi, such as mushrooms. However, during the Carboniferous period, there was little degradation, resulting in organic carbon burial and a corresponding atmospheric oxygen increase. The buried organic carbon eventually became coal (230). Robinson (231) concludes that about 320 Myr ago, while atmospheric oxygen pressures were high, atmospheric carbon dioxide pressures were low.

Both of these scenarios conclude that the oxygen level rose during this time (17). A model has been made by Berner and Canfield (17) to estimate the increase of oxygen in the atmosphere. These authors use some of the same arguments; in addition, they mention the release of oxygen from sulfate reduction to ferrous sulfide.

It has been suggested that fire has been a major force in evolution, especially when there were episodes of higher oxygen pressures relative to the present oxygen pressure (229). Jones and Chaloner (140) proposed that if the oxygen pressure is less than 130 mbar, wildfires cannot be sustained; if the oxygen pressure is above 350 mbar, wildfires would destroy too many plants. An oxygen pressure of 130 mbar occurs at an altitude of 3,800 m (12,467 feet; 198). Therefore, they believe that the maximum oxygen pressure never was higher than 350 mbar. Watson et al. (305) believe that an oxygen pressure of 250 mbar would have been too high for the survival of land vegetation, due to the probability of fires at this oxygen pressure.

It is our contention that if oxygen rose significantly above the present value of 210 mbar, there would have developed additional antioxidant defense mechanisms that would persist even now. We concur that oxygen rose to a maximum of 300 mbar 300 Myr ago. This figure was determined from the lower limit of atmospheric oxygen predicted by the model of Berner and Canfield (17).

Nitrogen in our atmosphere serves as an antioxidant defense and can reduce the risk of fire. If the atmosphere is not diluted with some form of relatively inert gas, such as nitrogen, then the probability of fires and explosions is increased. The first U.S. astronauts survived breathing 100% dioxygen at 0.33 bar for a number of days (37, 62). In addition, the presence of an inert gas is a factor in preventing lung collapse (67, 223).

We know that there is an intimate relationship between the atmosphere and biosphere. However, we take issue with Lovelock's Gaia hypothesis about this relationship (170). Lovelock theorizes that there is a purposeful relationship between these two spheres of the Earth. We object to the model of this relationship as being purposeful instead of being simply an interdependence (101). The Gaia hypothesis uses *teleological* reasoning, that is, reasoning that relates a phenomenon to a purpose. "Mammals have eyes so that they can see" is an example of teleological reasoning. One gets into trouble using this kind of reasoning to explain why mammals have an appendix, which serves no useful purpose. This type of reasoning does not involve mechanisms. That is not to say that one cannot use teleological reasoning for an intuitive approach to a problem. Cannon (42) quoted E. von Bruecke as stating, "Teleology is a lady without whom no biologist can live. Yet he is ashamed to show himself with her in public." Scientists can use a range of nonscientific methods, including teleological reasoning, intuition, and imagination, for initial approaches to a problem. However, they should use a scientific approach to the particular problem, in which purpose plays no role. We agree with Gould (110), who wrote: "Gaia strikes me as a metaphor, not a mechanism. (Metaphors can be liberating and enlightening, but new scientific theories must supply new statements about causality . . .)."

There are negative feedback processes that tend to regulate or stabilize this relationship over the short range. However, there are also some positive feedback processes that can destabilize this relationship. An understanding of these processes is essential in explaining the profound changes in the evolution of both spheres from a reducing primary atmosphere and no biosphere to the present oxidizing atmosphere and the current biosphere.

DUAL EFFECTS OF OXYGEN ON THE BIOSPHERE

Oxygen not only provides energy for the biosphere but also impairs and kills living organisms. Since a killing dose of X irradiation releases a very small amount of energy and since X irradiation biological damage is due to many ROS, it has been postulated that X irradiaton acts like a catalyst for the toxic effects of oxygen (100). Antioxidant defenses developed by the biosphere have allowed the biosphere to exist in the presence of oxygen. During the development of the Earth's oxygen atmosphere, the biosphere continually had to evolve the many different forms of antioxidant defense.

Iron is a necessary component for many enzymes, in spite of the fact that the presence of iron in the free state can catalyze the production of the highly toxic hydroxyl radical. For cells to survive, there must be a minimum concentration of iron ions in the fluids of any organism. Mammals use transferrin for iron transport. Transferrin is a high molecular weight glycoprotein, which can bind two iron atoms in the ferric state. Mammals cannot use the low molecular weight siderophores as iron carriers since they can be lost through glomerular filtration in the kidneys (151). The transferrin receptor protein then transports the ferric iron into the intracellular ferritin. Aconitase, an enzyme involved in the Krebs cycle, contains iron–sulfur clusters and regulates the intracellular iron concentration by a feedback mechanism (150). Ferritin is a large hollow protein, which can sequester about 4,500 atoms in the ferric state (296). The release of iron from ferritin is controlled by aconitase, which senses the free iron (150).

Living organisms have actually used the toxic effects of oxygen for their own survival. For example, oxidation of trees by fires can act as an antioxidant defense. Controlled forest fires are used to aid the survival of forests. The heat of the small controlled fires helps germination of many seeds. In addition, the fires help the trees to develop barks more resistant to bigger and hotter forest fires (23).

Living organisms have also used oxygen toxicity for killing purposes against other living organisms. Multicellular organisms use oxygen not only for energy but also for defense against other organisms that invade their bodies. For example, white blood cells (11), brain microglial cells (54), and other phagocytic cells (118), when stimulated by foreign invaders or unwanted cells, take up an extra amount of oxygen, called an *oxygen burst*, and then use this oxygen to produce reactive oxygen species to destroy such cells. Some invader cells, such as a virulent strain of *Nocaria asteroides*, are not killed by phagocytic cells; this organism secretes SOD into the external environment and possesses a high intracellular level of catalase (12, 13).

In higher plants, mitochondrial respiration can proceed by taking up oxygen; this part is coupled to the production of ATP. However, there is an alternate pathway that is cyanide-resistant and not coupled to ATP production (115). Could this be similar to the oxygen burst observed in some animal cells? Plants are attacked by herbivores; this might be a protective defense against unwanted cells.

Most of the cells in mammals are exposed to oxygen pressures much less than that in the ambient air, where the oxygen pressure is 158 torr (21 kPa or 210 mbar). Oxygen pressure in the lung is about the highest in human body, about 100 torr (13 kPa or 130 mbar); oxygen pressure in the mixed venous blood is about 40 torr (5 kPa or 50 mbar). A strain of *Mycobacterium tuberculosis,* which can live in the lungs, has the capability to withstand this increased oxygen stress by also secreting SOD into the external environment (162).

In addition, there are other cells in the human body that produce ROS. When the sea urchin egg becomes fertilized, the egg undergoes an *oxygen burst,* releasing hydrogen peroxide, which induces oxidative polymerization of the hard fertilization envelope surrounding the egg; this envelope protects the egg from other sperm and noxious chemicals (266).

There are reactive oxygen species producing cells which are intimately connected with initiation and promotion of cancers (34). Activation of oncogenes as well as loss or mutation of tumor suppressor genes seem to be involved in the production of colorectal cancer (6). Mutations can arise from oxidative stress, due to the fact that DNA is subject to oxidative damage in vivo (203, 274). Perhaps colon cancer cells, which are very difficult to kill, have adapted to the increased oxidative state. Colon cells are normally subjected to a higher oxidative stress than most other normal tissues (10). These colon cancer cells, which are highly resistant to oxidative stress, have an elevated level of reduced glutathione (GSH) and glutathione peroxidase, which is indicative of an increased antioxidant defense (185). These particular cancer cells have the capacity to proliferate but do not have the capacity to form a self-propagating, organized system of cells, capable of an independent existence in the absence of a host. Perhaps these cells are being adapted to an increased oxidative stress and with time they can evolve into a new autonomous life-form. It is interesting to note that some human tumor cells release hydrogen peroxide (283).

Protection against apoptosis or programmed cell death can often be harmful to the organism. For example, the protooncogene *bcl-2* protects against oxidative stress inducing apoptosis, thus allowing tumor cells to grow (227, 319). However, production of erythropoietin in the kidney and liver is stimulated by hypoxia (74); this hormone then inhibits apoptosis in the erythroid progenitor cells in the bone marrow, which subsequently become erythrocytes (156, 157). Thus the oxidative state does participate in the control of cell populations.

It has been found that the aquatic plant *Chara fragilis* produces oxygen bubbles, which inhibit the breeding of mosquitoes (180). *Symbiodinium microadriaticum,* an alga, lives intracellularly in the sea anemone *Anthopleura elegantissima*. These two organisms have a symbiotic relationship: the anemone uses the carbon compounds produced photosynthetically by the algae and the algae obtain their metabolites from the anemone. When the anemone is exposed to light, the recorded values of oxygen pressure in gastrodermal tissue are much higher than the 158 torr in the sea-level atmosphere; the highest was 328 torr. In response to this oxygen stress, SOD activity increased significantly (68).

The bombardier beetle of the genus *Brachinus* can use ROS for explosive secretory discharge as an offensive weapon to attack ants (7, 63). Higher aquatic plants, possessing lacunal spaces, generally do not live below a 10–12 m depth; this depth corresponds to a total pressure of 2.0–2.2 bar (equivalent to roughly doubling the sea-level atmospheric pressure) in these spaces. The growth of these plants is inhibited due to the toxicity of oxygen (212). Diving animals, such as turtles, birds, and mammals, also may be subject to high pressures, since the maximum depths some of these animals can reach in their dives is about 1,000 m; these organisms reduce their gas volume at these depths (154). Air pockets at 1,000 m below the surface of the ocean would be about 20 bar. The highest oxygen pressure in living organisms is about 500 bar in swimbladders of fish; yet the other tissues in the fish are sensitive to oxygen toxicity (103). Scheid et al. (247) have noted that one of the main reasons that oxygen is prevented from leaving the swimbladder is that the swimbladder wall is covered by a gas-impermeable layer, which is impregnated with crystals of guanine and some hypoxanthine. Removal of this layer results in an increased gas permeability of about 100 times.

THE FUTURE ATMOSPHERE ON EARTH

Of course, we can only speculate on what the future atmosphere will be. The increase of carbon dioxide in the atmosphere should roughly correspond to a decrease in atmospheric dioxygen. Since there is so much more dioxygen in the atmosphere compared to atmospheric carbon dioxide, the percent change of atmospheric dioxygen is extremely small.

Eventually, the slow dehydrogenation on Earth probably will cause the Earth to lose its water and the resulting free oxygen to become bound to other elements, such as carbon and iron. However, this condition will not occur for about 2.5 Gyr (36). Since Mars probably possesses a very oxidized soil (216) and water (260), it is possible that Mars is in a postlife stage (96, 102).

We believe that other planets in the universe during their evolution could possibly support a biosphere, using oxygen as an energy source. Evolution on such planets would require a prelife reducing atmosphere and sufficient time to produce an oxygen atmosphere by a primitive anaerobic biosphere. Later, the atmosphere would be composed of carbon dioxide and nitrogen, as it is on Venus (131, 218). Ultimately, the star explodes and the solar system is destroyed. Our Sun will last about 5 Gyr more before it enters the red giant phase. There is a possibility that asteroids can demolish a good part or all of a planet's biosphere. Within the past 160 Myr, there have been only seven known craters, with diameters of at least 40 km (113). In 1908, a stony asteroid crashed into the earth (51). In 1991, an asteroid came within 170,000 km of the Earth, which is less than half the distance from the Earth to the Moon (261). Possibly, the Earth contains an asteroid belt (221). However, small asteroids will explode in the upper atmosphere and consequently produce little or no damage on Earth (49). The use of rockets and nuclear explosions has been suggested to deflect and fragment asteroids from crashing into the Earth (2).

SUMMARY AND CONCLUSION

The Earth is a pinpoint of oxidation in a reducing universe. When the Earth was born 4.5 Gyr ago, the atmosphere was a reducing one. Debris from outer space contained the organic chemicals that were the precursors of the biosphere. The primitive biosphere received its energy from the reduction of oxidized carbon compounds, such as carbon dioxide, by hydrogen, forming methane and water. This reduction releases a substantial amount of energy. Energy from hot vents in the ocean and from solar radiation was the source for the biosphere to reverse this reaction by the biological production of hydrogen. There was a slow loss of hydrogen from the Earth's atmosphere and the development of photosynthetic production of oxygen, which made the atmosphere more oxidizing. About 2.0 Gyr ago, the atmosphere contained a significant amount of oxygen. Respiration was now possible, using oxygen as an energy source, with photosynthesis replenishing this oxygen. Since that time, the oxygen in the atmosphere increased to its present value. Much more energy is associated with oxygen burning of the metastable compounds than with hydrogen reacting with these same compounds.

We believe that any biosphere in the universe must contain hydrogen, oxygen, carbon, and nitrogen. In addition, we take the view that the origin of life on Earth was a most probable event (or events) and not an improbable one. Given the same conditions on other planets in the universe, we believe that the occurrence of other biospheres is most probable and that their biochemistry would also be based on these four elements. Also, we believe that, given a sufficient period of time and depending on the temperature and pressure, some living organisms will develop an aerobic form of life.

Questions about the physiology and the intelligence of these organisms can only be speculated upon at this time.

The energy of aerobic life is derived from the marriage of hydrogen and oxygen. It takes solar energy to break this union. Photolysis of water occurs as an abiotic reaction or as a biotic reaction, which we call photosynthesis. Carbon is used to bind hydrogen and prevent its escape as a gas from the biosphere. Nitrogen is required for structure in living cells. Living cells can make use of other elements only in their local environment for specialized functions. In general, biotic reactions are the same as abiotic reactions: the difference is that biotic reactions have faster kinetics.

Oxygen is the best storage form of energy in the universe. However, all energy-storage forms for the biosphere represent a double-edged sword. The energy-storage form not only supplies the energy for life but also destroys living material. Thus oxygen, which is the storage form for aerobic life, will also destroy living material. The price of living is dying.

I thank Drs. J. William Schopf for showing me his papers and books in press and for his helpful discussions in directing me to some pertinent literature; Heinrich D. Holland for sending me his paper in press; J. C. Walker, Cyril Ponnamperuma, and J. A. Berry for sending me their reprints; and Jennifer M. Robinson for sending me her reprints and for inviting me to give a talk at the 1990 Oxygen Workshop, which I could not attend due to unforeseen difficulties: I was planning to present part of this paper at the meeting. The ideas presented in this paper are solely my own and may or may not reflect the ideas of the acknowledged individuals. Finally, I would like to express my gratitude to Dr. Claire Gilbert for her editorial comments and to Raymond L. Gilbert for his assistance.

REFERENCES

1. Ahrens, T. J. Impact erosion of terrestrial planetary atmospheres. *Annu. Rev. Earth Planet. Sci.* 21: 525–555, 1993.

2. Ahrens, T. J., and A. W. Harris. Deflection and fragmentation of near-Earth asteroids. *Nature* 360: 429–433, 1992.

3. Aihara, J. Why aromatic compounds are stable. *Sci. Am.* 266 (No. 3): 62–68, 1992.

4. Ames, B. N., M. K. Shigenaga, and T. M. Hagen. Oxidants, antioxidants, and the degenerative diseases of aging. *Proc. Natl. Acad. Sci. USA* 90:7915–7922, 1993.

5. Anders, E., and M. Ebihara. Solar-system abundances of the elements. *Geochim. Cosmochim. Acta* 46: 2363–2380, 1982.

6. Anderson, M. J., and E. J. Stanbridge. Tumor suppressor genes studied by cell hybridization chromosome transfer. *FASEB J.* 7: 826–833, 1993.

7. Aneshansley, D. J., T. Eisner, J. M. Widom, and B. Widom. Biochemistry at 100°C: explosive secretory discharge of bombardier beetles *(Brachinus)*. *Science* 165: 61–63, 1969.

8. Aono, S., F. O. Bryant, and M. W. Adams. A novel and remarkably thermostable ferredoxin from the hyperthermophilic archaebacterium *Pyrococcus furiosus*. *J. Bacteriol.* 171: 3433–3439, 1989.

9. Archibald, F. S. *Lactobacillus plantarum*, an organism not requiring iron. *FEMS Microbiol. Lett.* 19: 29–32, 1983.

10. Babbs, C. F. Free radicals and the etiology of colon cancer. *Free Radic. Biol. Med.* 8: 191–200, 1990.

11. Babior, B. M., R. S. Kipnes, and J. T. Curnette. Biological defense mechanisms. The production by leukocytes of superoxide, a potential bactericidal agent. *J. Clin. Invest.* 52: 741–744, 1973.

11a. Bartlett, J. G., A. Blanchard, J. Silk, and M. S. Turner. The case for a Hubble constant of 30 km·s^{-1}·Mpc^{-1}. *Science* 267:980–983, 1995.

12. Beaman, L., and B. L. Beaman. The role of oxygen and its derivatives in microbial pathogenesis and host defense. *Annu. Rev. Microbiol.* 38: 27–48, 1984.

13. Beaman, L., and B. L. Beaman. Monoclonal antibodies demonstrate that superoxide dismutase contributes to protection of *Nocardia asteroides* within the intact host. *Infect. Immun.* 58: 3122–3128, 1990.

14. Becker, B. F. Towards the physiological function of uric acid. *Free Radic. Biol. Med.* 14: 615–631, 1993.

15. Berkner, L. V., and L. C. Marshall. The history of growth of oxygen in the earth's atmosphere. In: *The Origin and Evolution of Atmospheres and Oceans*, edited by P. J. Brancazio and A. G. W. Cameron. New York: Wiley, 1964 p. 102–126.

16. Bernal, J. D. The physical basis of life. *Proc. Phys. Soc.* 62 (A): 537–558, 1949.

17. Berner, R. A., and D. E. Canfield. A new model for atmospheric oxygen over Phanerozoic time. *Am. J. Sci.* 289: 333–361, 1989.

18. Berry, J. A. Biosphere, atmosphere, ocean interactions: a plant physiologist's perspective. In: *Primary Productivity and Biogeochemical Cycles in the Sea. Brookhaven Symposium in Biology No. 37*, edited by P. G. Falkowski and A. B. Woodhead. New York: Plenum, 1992, p. 441–454.

19. Beyer, W. F., Jr., and I. Fridovich. Pseudocatalase from *Lactobacillus plantarum*: evidence for a homopentameric structure containing two atoms of manganese per subunit. *Biochemistry* 24: 6460–6467, 1985.

20. Bice, D. M., C. R. Newton, S. McCauley, P. W. Reiners, and C. A. McRoberts. Shocked quartz at the Triassic-Jurassic boundary in Italy. *Science* 255: 443–446, 1992.

21. Bielski, B. H. J., D. E. Cabelli, R. L. Arudi, and A. B. Ross. Reactivity of HO_2/O_2^- radicals in aqueous solution. *J. Phys. Chem. Ref. Data* 14: 1041–1100, 1985.

22. Bishop, N. I., and L. W. Jones. Alternate fates of the photochemical reducing power generated in photosynthesis: hydrogen production and nitrogen fixation. *Curr. Topics Bioenerg.* 8: 3–31, 1978.

23. Biswell, H. H. *Prescribed Burning in California Wildlands Vegetation Management*. Berkeley: Univ. California Press, 1989.

24. Blaut, M., V. Müller, and G. Gottschalk. Energetics of methanogens. In: *The Bacteria. Volume 12. Bacterial Energetics*, edited by T. A. Krulwich. New York: Academic, 1990, p. 505–537.

25. Blum, J. D., C. P. Chamberlain, M. P. Hingston, C. Koeberl, L. E. Marin, B. C. Schuraytz, and V. L. Sharpton. Isotopic comparison of K/T boundary impact glass with melt rock from the Chicxulub and Manson impact structures. *Nature* 364: 325–327, 1993.

26. Bonner, W. A. Terrestrial and extraterrestrial sources of molecular homochirality. *Orig. Life Evol. Biosph.* 21: 407–420, 1992.

26a. Bonner, W. A. Chirality and life. *Orig. Life Evol. Biosph.* 25:175–190, 1995.

27. Bors, W., W. Heller, C. Michel, and M. Saran. Flavonoids as antioxidants: determination of radical-scavenging efficiencies. In: *Methods in Enzymology. Oxygen Radicals in Biological Systems. Oxygen Radicals and Antioxidants,* edited by L.

Packer, and A. N. Glazer. San Diego, CA:Academic, 1990, vol. 186, pt. B, p. 343–355.

28. Boveris, A., and E. Cadenas. Production of superoxide radicals and hydrogen peroxide in mitochondria. In: *Superoxide Dismutase,* edited by L. W. Oberley. Boca Raton, FL: CRC, 1982, vol. II, p. 15–30.

29. Bowring, S. A., J. P. Grotzinger, C. E. Isachsen, A. H. Knoll, S. M. Pelechaty, and P. Kolosov. Calibrating rates of early Cambrian evolution. *Science* 261: 1293–1299, 1993.

30. Braun, C., and W. G. Zumft. Marker exchange of the structural genes for nitric oxide reductase blocks the denitrification pathway of *Pseudomonas stutzeri* at nitric oxide. *J. Biol. Chem.* 266: 22785–22788, 1991.

31. Bryant, F. O., and M. W. Adams. Characterization of hydrogenase from the hyperthermophic archaebacterium, *Pyrococcus furiosus. J. Biol. Chem.* 264: 5070–5079, 1989.

32. Budyko, M. I., A. B. Ronov, and A. L. Yanshin, *History of the Earth's Atmosphere,* translated by S. F. Lemehko and V. G. Yanuta. New York: Springer-Verlag, 1987.

33. Burton, G. W., and K. U. Ingold. β-Carotene: an unusual type of lipid antioxidant. *Science* 224: 569–573, 1984.

34. Byers, V. S., and R. W. Baldwin (Eds). *Immunology of Malignant Diseases.* Boston: MTP, 1987.

35. Cairns-Smith, A. G., and H. Hartman (Eds). *Clay Minerals and the Origin of Life.* New York: Cambridge Univ. Press, 1986.

36. Caldeira, M., and J. F. Kasting. The life span of the biosphere revisited. *Nature* 360: 721–723, 1992.

37. Calloway, D. H. (Ed). *Human Ecology in Space Flight.* New York: New York Academy of Sciences, 1966.

38. Calvin, C. *Chemical Evolution. Molecular Evolution Towards the Origin of Living Systems on the Earth and Elsewhere.* New York: Oxford Univ. Press, 1969.

39. Cameron, A. G. W., and W. Benz. The origin of the moon and the single impact hypothesis IV. *Icarus* 92: 204–216, 1991.

40. Campbell, I. M. *Energy and the Atmosphere. A Physical-Chemical Approach* (2nd ed). New York: Wiley, 1986.

41. Canfield, L. M., J. W. Forage, and J. G. Valenzuela. Carotenoids as cellular antioxidants. *Proc. Soc. Exp. Biol. Med.* 200: 260–265, 1992.

42. Cannon, W. B. *The Way of an Investigator. A Scientist's Experience in Medical Research* (1968 ed.). New York: Hafner, 1945, p. 108.

43. Carroll, D. Role of clay minerals in the transportation of iron. *Geochim. Cosmochim. Acta* 14: 1–28, 1958.

44. Chalmers, A. H. Simple, sensitive measurement of carbon monoxide in plasma. *Clin. Chem.* 37: 1442–1445, 1991.

45. Chapman, D. J., and J. W. Schopf. Biological and biochemical effects of the development of an aerobic environment. In: *Earth's Earliest Biosphere. Its Origin and Evolution,* edited by J. W. Schopf. Princeton, NJ: Princeton Univ. Press, 1983, p. 302–320.

46. Chevalier, R. A. Supernova 1987A at five years of age. *Nature* 355: 691–696, 1992.

47. Chiueh, C. C., D. L. Gilbert, and C. A. Colton (Eds). *The Neurobiology of NO· and ·OH.* Annals NY Acad. Sci. New York: New York Academy of Sciences, 1994, vol. 738.

48. Chyba, C. A. Extraterrestrial amino acids and terrestrial life. *Nature* 348: 113–114, 1990.

49. Chyba, C. F. Explosions of small Spacewatch objects in the Earth's atmosphere. *Nature* 363: 701–703, 1993.

50. Chyba, C., and C. Sagan. Endogenous production, exogenous delivery and impact-shock synthesis of organic molecules: an inventory for the origins of life. *Nature* 355: 125–132, 1992.

51. Chyba, C. F., P. J. Thomas, and K. J. Zahnle. The 1908 Tunguska explosion: atmospheric disruption of a stony asteroid. *Nature* 361: 40–44,1993.

52. Clark, R. T., Jr., J. N. Stannard, and W. O. Fenn. The burning of CO to CO_2 by isolated tissues as shown by the use of radioactive carbon. *Am.J. Physiol.* 161: 40–46, 1950.

53. Cloud, P. Beginnings of a biospheric evolution and their biogeochemical consequences. *Paleobiology* 2: 351–357, 1976.

54. Colton, C. A., and D. L. Gilbert. Microglia, an *in vivo* source of reactive oxygen species in the brain. In: *Advances in Neurology. Neural Regeneration,* edited by F. Seil. New York: Raven, 1993, vol. 59, p. 321–326.

55. Crick, F. Forward. In: *The RNA World. The Nature of Modern RNA Suggests a Prebiotic RNA World,* edited by R. F. Gesteland and J. F. Atkins. Plainview, NY: Cold Spring Harbor Lab., 1993, p. xi–xiv.

56. Crowley, T. J. Past CO_2 changes and tropical sea surface temperatures. *Paleoceanography* 6: 387–394, 1991.

57. Crowley, T. J., and G. R. North. *Paleoclimatology.* New York: Oxford Univ. Press, 1991.

58. Cutler, R. G. Antioxidants and aging. *Am. J. Clin. Nutr.* 53(suppl. 1): 373S–379S, 1991.

59. Dalgarno, A. Interstellar chemistry. In: *Chemistry in Space,* edited by J. M. Greenberg and V. Pironello. Boston: Kluwer Acad. Pub., 1991, p. 71–87.

60. Damsté, J. S. S., S. G. Wakeham, M. E. L. Kohnen, J. M. Hayes, and J.W. de Leeuw. A 6,000-year sedimentary molecular record of chemocline excursions in the Black Sea. *Nature* 362: 827–829, 1993.

61. Davis, D. H., M. Doudoroff, R. Y. Stanier, and M. Mandel. Proposal to reject the genus *Hydrogenomonas*: taxonomic implications. *Int. J. Syst. Bacteriol.* 19: 375–390, 1969.

62. Davis, J. C., and T. K. Hunt (Eds). *Hyperbaric Oxygen Therapy.* Bethesda, MD: Undersea Medical Society, 1977.

63. Dean, J., D. J. Aneshansley, H. E. Edgerton, and T. Eisner. Defensive spray of the bombardier beetle: a biological pulse jet. *Science* 248: 1219–1221, 1990.

63a. de Baar, H. J. W., J. T. M. de Jong, D. C. E. Bakker, B. M. Löscher, C. Veth, U. Bathmann, and V. Semtacek. Importance of iron for planton blooms and carbon dioxide drawdown in the Southern ocean. *Nature* 373: 412–415, 1995.

64. Demple, B. Regulation of bacterial oxidative stress genes. *Annu. Rev. Genet.* 25: 315–337, 1991.

65. Des Marais, D. J., H. Strauss, R. E. Summons, and J. M. Hayes. Carbon isotope evidence for the stepwise oxidation of the Proterozoic environment. *Nature* 359: 605–609, 1992.

66. Dickens, R. J., B. F. W. Croke, R. D. Cannon, and R. A. Bell. Evidence from stellar abundances for a large age difference between two globular clusters. *Nature* 351: 212–214, 1991.

67. DuBois, A. B., T. Turaids, R. E. Mammen, and F. T. Nobrega. Pulmonary atelectasis in subjects breathing oxygen at sea level or at simulated altitude. *J. Appl. Physiol.* 21: 828–836, 1966.

68. Dykens, J. A., and J. M. Shick. Oxygen production by endosymbiotic algae controls superoxide dismutase activity in their animal host. *Nature* 297: 579–580, 1982.

69. Dyson, F. J. Can we control the carbon dioxide in the atmosphere? *Energy* 2: 287–291, 1977.

70. Edmundson, W. T. Phosphorus, nitrogen, and algae in Lake Washington after diversion of sewage. *Science* 169: 690–691, 1970.

71. Engel, M. H., S. A. Macko, and J. A. Silfer. Carbon isotope composition of individual amino acids in the Murchison meteorite. *Nature* 348: 47–49, 1990.

72. Ensign, S. A., and P. W. Ludden. Characterization of the CO oxidation/H_2 evolution system of *Rhodospirillum rubrum.* Role of a 22-kDa iron-sulfur protein in mediating electron transfer between carbon monoxide dehydrogenase and hydrogenase. *J. Biol. Chem.* 266: 18395–18403, 1991.

73. Ernst, W. G. The early Earth and the Archean rock record. In:

Earth's Earliest Biosphere. Its Origin and Evolution, edited by J. W. Schopf. Princeton, NJ: Princeton Univ. Press, 1983, p. 41–52.

74. Fandrey, J., and H. F. Bunn. In vivo and in vitro regulation of erythropoietin mRNA: measurement by competitive polymerase chain reaction. *Blood* 81: 617–623, 1993.

75. Faure, G. *Principles of Isotope Geology* (2nd ed.). New York: Wiley, 1986.

76. Faust, B. C., C. Anastasio, J. M. Allen, and T. Arakaki. Aqueous-phase photochemical formation of peroxides in authentic cloud and fog waters. *Science* 260: 73–75, 1993.

77. Fay, P. Oxygen relations of nitrogen fixation in cyanobacteria. *Microbiol. Rev.* 56: 340–373, 1992.

78. Feldman, P. D. The volatile composition of comets deduced from ultraviolet spectroscopy. In: *Chemistry in Space,* edited by J. M. Greenberg and V. Pirronello. Boston: Kluwer, 1991, p. 339–361.

79. Feng, Q., T. K. Park, and J. Rebek. Crossover reactions between synthetic replicators yield active and inactive recombinants. *Science* 256: 1179–1180, 1992.

80. Fenn, W. O., and D. M. Cobb. The burning of carbon monoxide by heart and skeletal muscle. *Am. J. Physiol.* 102: 393–401, 1932.

81. Finlayson-Pitts, B. J., and J. N. Pitts, Jr. *Atmospheric Chemistry: Fundamentals and Experimental Techniques.* New York: Wiley, 1986.

82. Fletcher, M. The physiological activity of bacteria attached to solid surfaces. *Adv. Microb. Physiol.* 32: 53–85, 1991.

83. Fornace, A. J., Jr. Mammalian genes induced by radiation; activation of genes associated with growth control. *Annu. Rev Genet.* 26: 507–526, 1992.

84. Fox, S. W. Life from an orderly cosmos. *Naturwissenschaften* 67: 576–581, 1980.

85. Fox, S. W., and K. Harada. Synthesis of uracil under conditions of a thermal model of prebiological chemistry. *Science* 133: 1923–1924, 1961.

86. Frank, L. Oxygen toxicity in eukaryotes. In: *Superoxide Dismutase. Pathological States,* edited by L. W. Oberley. Boca Raton, FL: CRC, 1985, vol. III, p. 1–43.

87. Fridovich, I. Oxygen toxicity in prokaryotes: the importance of superoxide dismutase. In: *Superoxide Dismutase.* edited by L. W. Oberley. Boca Raton, FL: CRC, 1982, vol. I, p. 79–88.

88. Fridovich, I. Superoxide dismutases. An adaptation to a paramagnetic gas. *J. Biol. Chem.* 264: 7761–7764, 1989.

89. Gajdusek, D. C. Transmissible and non-transmissible amyloidoses: autocatalytic post-translational conversion of host precursor proteins to β-pleated configurations. *J. Neuroimmunol.* 20: 95–110, 1988.

90. Gates, D. M. *Climate Change and Its Biological Consequences.* Sunderland, MA: Sinauer, 1993.

91. Gerschman, R. Oxygen effects in biological systems. In: *Symp. Spec. Lect. XXI Int. Cong. Physiol. Sci.* Buenos Aires, 1959, p. 222–226.

92. Gerschman, R., D. L. Gilbert, and D. Caccamise. Effect of various substances on survival times of mice exposed to different high oxygen tensions. *Am. J. Physiol.* 192: 563–571, 1958.

93. Gerschman, R., D. L. Gilbert, and J. N. Frost. Sensitivity of *Paramecium caudatum* to high oxygen tensions and its modification by cobalt and manganese ions. *Am. J. Physiol.* 192: 572–576, 1958.

94. Gerschman, R., D. L. Gilbert, S. W. Nye, P. Dwyer, and W. O. Fenn. Oxygen poisoning and x-irradiation: a mechanism in common. *Science* 119: 623–626, 1954.

95. Gilbert, D. L. Relationship between ion distribution and membrane potential during a steady state. *Bull. Math. Biophys.* 22: 323–349, 1960.

96. Gilbert, D. L. Speculation on the relationship between organic and atmospheric evolution. *Perspect. Biol. Med.* 4: 58–71, 1960.

97. Gilbert, D. L. The role of pro-oxidants and antioxidants in oxygen toxicity. *Radiat. Res.* (suppl. 3): 44–53, 1963.

98. Gilbert, D. L. Cosmic and geophysical aspects of the respiratory gases. In: *Handbook of Physiology: Respiration,* edited by W. Fenn and H. Rahn. Washington, DC: Am. Physiol. Soc., 1964, sect. 3, vol. I, p. 153–176.

99. Gilbert, D. L. Atmosphere and oxygen. *Physiologist* 8: 9–34, 1965.

100. Gilbert, D. L. Introduction: Oxygen and life. *Anesthesiology* 37: 100–111, 1972.

101. Gilbert, D. L. Discussion: What controls atmospheric oxygen? *BioSystems* 12: 123–124, 1980.

102. Gilbert, D. L. Significance of oxygen on earth. In: *Oxygen and Living Processes: An Interdisciplinary Approach,* edited by D. L. Gilbert. New York: Springer-Verlag, 1981, p. 73–101.

103. Gilbert, D. L. Oxygen: an overall biological view. In: *Oxygen and Living Processes: An Interdisciplinary Approach,* edited by D. L. Gilbert. New York: Springer-Verlag, 1981, p. 376–392.

103a. Gilbert, D. L. From the breath of life to reactive oxygen species (ROS). In: *Free Radicals—Mechanisms in Neurology and Psychiatry,* edited by J. L. Cadet and J. E. Johnson. New York: Wiley, 1995 (in press).

104. Gilbert, D. L., and G. Ehrenstein. Membrane surface charge. *Curr. Top. Membr. Trans.* 22: 407–421, 1984.

105. Gilbert, D. L., R. Gerschman, J. Cohen, and W. Sherwood. The influence of high oxygen pressures on the viscosity of solutions of sodium desoxyribonucleic acid and of sodium alginate. *J. Am. Chem. Soc.* 79: 5677–5680, 1957.

106. Gilliland, R. L. Solar evolution. *Palaeogeogr. Palaeoclimatol. Palaeoecol. (Global Planet. Change Sect.)* 75: 35–55, 1989.

107. Gleason, J. F., P. K. Bhartia, J. R. Herman, R. McPeters, P. Newman, R. S. Stolarski, L. Flynn, G. Labow, D. Larko, C. Seftor, C. Wellemeyer, W. D. Komhyr, A. J. Miller, and W. Planet. Record low global ozone in 1992. *Science* 260: 523–526, 1993.

108. Goldanskii, V. I., and V. V. Kuzmin. Chirality and cold origin of life. *Nature* 352: 114, 1991.

109. Gorst, C. M., and S. W. Ragsdale. Characterization of the NiFeCO complex of carbon monoxide dehydrogenase as a catalytically competent intermediate in the pathway of acetyl-coenzyme A synthesis. *J. Biol. Chem.* 266: 20687–20693, 1991.

110. Gould, S. J. *Bully for Brontosaurus. Reflections in Natural History.* New York: Norton, 1992, p. 339.

111. Greenberg, J. M. Physical, chemical and optical interactions with interstellar dust. In: *Chemistry in Space,* edited by J. M. Greenberg and V. Pirronello. Boston: Kluwer, 1991, p. 227–261.

112. Gregory, D. P. The hydrogen economy. *Sci. Am.* 228: 13–21, 1973.

113. Grieve, R. A. F. When will enough be enough? *Nature* 363: 670–671, 1993.

114. Gutteridge, J. M. C. Ageing and free radicals. *Med. Lab. Sci.* 49: 313–318, 1992.

115. Guy, R. D., J. A. Berry, M. L. Fogel, and T. C. Hoering. Differential fractionation of oxygen isotopes by cyanide-resistant and cyanide-sensitive respiration in plants. *Planta Med.* 177: 483–491, 1989.

116. Haldane, J. B. S. The origins of life. In: *Adventures in Earth History. 1970,* edited by P. Cloud. San Francisco: Freeman, 1954, p. 377–384.

117. Hall, D. O. Carbon flows in the biosphere: present and future. *J. Geol. Soc. Lond.* 146: 175–181, 1989.

117a. Hall, D. T., D. F. Strobel, P. D. Feldman, M. A. McGrath, and H. A.Weaver. Detection of an oxygen atmosphere on Jupiter's moon Europa. *Nature* 373: 677–679, 1995.

118. Halliwell, B., and J. M. C. Gutteridge. *Free Radicals in Biology and Medicine* (2nd ed.). New York: Oxford Univ. Press, 1989.

119. Halliwell, J. J. Quantum cosmology and the creation of the universe. *Sci. Am.* 265: 76–79, 82–85, 1991.

120. Harman, D. Aging: a theory based on free radical and radiation chemistry. *J. Gerontol.* 11: 298–299, 1956.

121. Hayes, J. M. Geochemical evidence bearing on the origin of aerobiosis, a speculative hypothesis. In: *Earth's Earliest Biosphere. Its Origin and Evolution,* edited by J. W. Schopf. Princeton, NJ: Princeton Univ. Press, 1983, p. 291–301.

122. Henderson, L. J. *The Fitness of the Environment.* Boston: Beacon Press, 1958.

123. Hill, H. A. O. Iron: an element well-fitted for its task? In: *The Biological Chemistry of Iron,* edited by H. B. Dunford, D. Dolphin, K. N. Raymond, and L. Sieker. Boston: Reidel, 1982, p. 3–12.

124. Holland, H. Chemistry and evolution of the Proterozoic ocean. In: *The Proterozoic Biosphere. A Multidisciplinary Study,* edited by J. W. Schopf and C. Klein. New York: Cambridge Univ. Press, 1992, p. 169–172.

125. Holland, H. Early Proterozoic atmospheric change. In: *Early Life on Earth: Nobel Symposium 84,* edited by S. Bengtson. New York: Columbia Univ. Press, 1994, p. 237–244.

126. Holm, N. G. Why are hydrothermal systems proposed as plausible environments for the orign of life? *Orig. Life Evol. Biosph.* 22: 5–14, 1992.

127. Holm, N. G., and R. J. Hennet. Hydrothermal systems: their varieties, dynamics, and suitability for prebiotic chemistry. *Orig. Life Evol. Biosph.* 22: 15–31, 1992.

128. Hoban, S. Serendipitous images of methanol in comet Levy (1900 XX). *Icarus* 104: 149–151, 1993.

129. Huebner, W. F., and D. C. Boice. Comets as a possible source of prebiotic molecules. *Orig. Life Evol. Biosph.* 21: 299–315, 1992.

130. Hughes, M. N., and R. K. Poole. *Metals and Micro-organisms.* New York: Chapman and Hall, 1989.

131. Hunten, D. M. Atmospheric evolution of the terrestrial planets. *Science* 259: 915–920, 1993.

131a. Hunten, D. M. Europa's oxygen atmosphere. *Nature* 373: 654, 1995.

132. Idso, S. B. The aerial fertilization effect of CO_2 and its implications for global carbon cycling and maximum greenhouse warming. *Bull. Am. Meteorol. Soc.* 72: 962–965, 1991.

133. Isaksen, I. S. A., T. Berntsen, and S. Solberg. Estimates of past and future tropospheric ozone changes from changes in human released source gases. In: *Ozone in the Atmosphere,* edited by R. D. Bojkov and P. Fabian. Hampton, VA: A. Deepak, 1989, p. 576–579.

134. Ishii, T., H. Iwahashi, R. Sugata, and R. Kido. Superoxide dismutase enhances the toxicity of 3-hydroxyanthranilic acid to bacteria. *Free Radic. Res. Commun.* 14: 187–194, 1991.

135. Isidorov, V. A. *Organic Chemistry of the Earth's Atmosphere,* translated by E. A. Koroleva. New York: Springer-Verlag, 1990.

136. Jablonski, D. Extinctions: a paleontological perspective. *Science* 253: 754–757, 1991.

137. Jamieson, D., B. Chance, E. Cadenas, and A. Boveris. The relation of free radical production to hyperoxia. *Annu. Rev. Physiol.* 48: 703–719, 1986.

138. Jannasch, H. W., and C. D. Taylor. Deep-sea microbiology. *Annu. Rev. Microbiol.* 38: 487–514, 1984.

139. Jetten, M. S. M., A. J. Pierik, and W. R. Hagen. EPR characterization of a high-spin system to carbon monoxide dehydro-

genase from *Methanothrix soehngenii. Eur. J. Biochem.* 202: 1291–1297, 1991.

140. Jones, T. P., and W. G. Chaloner. Fossil charcoal, its recognition and palaeoatmospheric significance. *Palaeogeogr. Palaeoclimatol. Palaeoecol. (Global Planet. Change Sect.)* 97: 39–50, 1991.

141. Jones, W. J., D. P. Nagle, Jr., and W. B. Whitman. Methanogens and the diversity of archaebacteria. *Microbiol. Rev.* 51: 135–177, 1987.

142. Joyce, G. F. RNA evolution and the origins of life. *Nature* 338: 217–224, 1989.

143. Kasting, J. F. Bolide impacts and the oxidation state of carbon in the earth's early atmosphere. *Orig. Life Evol. Biosph.* 20: 199–231, 1990.

144. Kasting, J. F. Models relating to Proterozoic atmospheric and ocean chemistry. In: *The Proterozoic Biosphere. A Multidisciplinary Study,* edited by J. W. Schopf and C. Klein. New York: Cambridge Univ. Press, 1992, p. 1185–1187.

145. Kasting, J. F. Earth's early atmosphere. *Science* 259: 920–926, 1993.

146. Kasting, J. F., H. D. Holland, and L. R. Kump. Atmospheric evolution: the rise of oxygen. In: *The Proterozoic Biosphere. A Multidisciplinary Study,* edited by J. W. Schopf and C. Klein. New York: Cambridge Univ. Press, 1992, p. 159–163.

147. Keeling, R. F., and S. R. Shertz. Seasonal and interannual variations in atmospheric oxygen and implications for the global carbon cycle. *Nature* 358: 723–727, 1992.

148. Keosian, J. Life's beginnings—origin or evolution? *Orig. Life* 5: 285–293, 1974.

149. Kirchhoff, V. W. J. H., E. V. A. Marinho, P. L. S. Dias, E. B. Pereira, R. Calheiros, R. André, and C. Volpe. Enhancements of CO and O_3 from burnings in sugar cane fields. *J. Atmos. Chem.* 12: 87–102, 1991.

150. Klausner, R. D., and T. A. Rouault. A double life: cytosolic aconitase as a regulatory RNA binding protein. *Mol. Biol. Cell* 4: 1–5, 1993.

151. Klausner, R. D., T. A. Rouault, and J. B. Harford. Regulating the fate of mRNA: the control of cellular iron metabolism. *Cell* 72: 19–28, 1993.

152. Knoll, A. H. The early evolution of eukaryotes: a geological perspective. *Science* 256: 622–627, 1992.

153. Kono, Y., and I. Fridovich. Functional significance of manganese catalase in *Lactobacillus plantarum. J. Bacteriol.* 155: 742–746, 1983.

154. Kooyman, G. L. Pressure and the diver. *Can. J. Zool.* 66: 84–88, 1988.

155. Korona, R., and B. R. Levin. Phage-mediated selection and the evolution and maintenance of restriction-modification. *Evolution* 47: 556–575,1993.

156. Koury, M. J., and M. C. Bondurant. Erythropoietin retards DNA breakdown and prevents programmed death in erythroid progenitor cells. *Science* 248: 378–381, 1990.

157. Koury, M. J., and M. C. Bondurant. Control of red cell production: the roles of programmed cell death (apoptosis) and erythropoietin. *Transfusion* 30: 673–674, 1990.

158. Krieg, N. R., and P. S. Hoffman. Microaerophily and oxygen toxicity. *Annu. Rev. Microbiol.* 40: 107–130, 1986.

159. Krinsky, N. I. Mechanism of action of biological antioxidants. *Proc. Soc. Exp. Biol. Med.* 200: 248–254, 1992.

160. Kuhn, W. R., and S. K. Atreya. Ammonia photolysis and the greenhouse effect in the primordial atmosphere of the Earth. *Icarus* 37: 207–213, 1979.

161. Kump, L. R. Chemical stability of the atmosphere and ocean. *Palaeogeogr. Palaeoclimatol. Palaeoecol. (Global Planet. Change Sect.)* 75: 123–136, 1989.

162. Kusunose, E., K. Ichihara, Y. Noda, and M. Kusunose. Super-

oxide dismutase from *Mycobacterium tuberculosis. J. Biochem.* 80: 1343–1352, 1976.

163. Kutter, G. S. *The Universe and Life. Origins and Evolution.* Boston: Jones and Bartlett, 1987.

164. Latimer, W. M. *The Oxidation States of the Elements and Their Elements in Aqueous Solutions* (2nd ed.). New York: Prentice-Hall, 1952.

165. Léger, A., L. d'Hendecourt, L. Verstraete, and P. Ehrenfreund. Small grains and large aromatic molecules. In: *Chemistry in Space,* edited by J. M. Greenberg and V. Pironello. Boston: Kluwer, 1991, p. 211–225.

166. Lelieveld, J., and P. J. Crutzen. The role of clouds in tropospheric photochemistry. *J. Atmos. Chem.* 12: 229–267, 1991.

167. Levin, G. V., and L. R. Zehner. L-sugars: Lev-O-Cal. In: *Alternative Sweeteners* (2nd ed.), edited by L. O. Nabors and R. C. Gelardi. New York: Dekker, 1991, p. 117–125.

168. Lide, D. R. (Ed.). *CRC Handbook of Chemistry and Physics. A Ready-Reference Book of Chemical and Physical Data. 1992–1993* (73rd ed.). Boca Raton, FL: CRC, 1992.

169. Little, G .H., and A. Flores. Inhibition of programmed cell death by catalase and phenylalanine methyl ester. *Comp. Biochem. Physiol.* 105A: 79–83, 1993.

170. Lovelock, J. *The Ages of Gaia. A Biography of Our Living Earth,* New York: Bantam, 1990.

171. Lowe, D. R. Major events in the geological development of the Precambrian Earth. In: *The Proterozoic Biosphere. A Multidisciplinary Study,* edited by J. W. Schopf and C. Klein. New York: Cambridge Univ. Press, 1992, p. 67–75.

172. Lu, Y., and M. A. K. Khalil. Methane and carbon monoxide in OH chemistry: the effects of feedbacks and reservoirs generated by the reactive products. *Chemosphere* 26: 641–655, 1993.

173. Luria, S. E., S. J. Gould, and S. Singer. *A View of Life.* Menlo Park, CA: Benjamin/Cummings, 1981.

174. Ma, K., C. Zirngibl, D. Linder, K. O. Stetter, and R. K. Thauer. N^5,N^{10}-Methylenetetrahydromethanopterin dehydrogenase (H_2-forming) from the extreme thermophile *Methanopyrus kandleri. Arch. Microbiol.* 156: 43–48, 1991.

174a. Macdermott, A. J. Electroweak enantioselection and the origin of life. *Orig. Life Evol. Biosph.* 25: 191–199, 1995.

175. Maloney, P. C., and T. H. Wilson. The evolution of ion pumps. *BioScience* 35: 43–48, 1985.

176. Margulis, L., and K. V. Schwartz. *Five Kingdoms. An Illustrated Guide to the Phyla of Life on Earth* (2nd ed.). New York: Freeman, 1988.

177. Margulis, L., J. C. G. Walker, and M. Rambler. Reassessment of roles of oxygen and ultraviolet light in Precambrian evolution. *Nature* 264: 620–624, 1976.

178. Martin, J. H., R. M. Gordon, S. E. Fitzwater. Iron in Antarctic waters. *Nature* 345: 156–158, 1990.

179. Maskos, Z., and W. H. Koppemol. Oxyradicals and multivitamin tablets. *Free Radic. Biol. Med.* 11: 609–610, 1991.

180. Matheson, R. The utilization of aquatic plants as aids in mosquito control. *Am. Nat.* 54: 56–86, 1930.

181. Mauzerall, D. The photochemical origins of life and photoreaction of ferrous ion in the Archean oceans. *Orig. Life Evol. Biosph.* 20: 293–302, 1990.

182. Mayer, M. G. The structure of the nucleus. *Sci. Am.* 184(3): 22–26, 1951.

182a. McCormick, M. P., L. W. Thomason, and C. R. Trepte. Atmospheric effects of the Mt Pinatubo eruption. *Nature* 373: 399–404, 1995.

183. McGervey, J. D. *Introduction to Modern Physics.* New York: Academic, 1971.

184. McLaren, D. J., and W. D. Goodfellow. Geological and bio-

logical consequences of giant impacts. *Annu. Rev. Earth Planet. Sci.* 18: 123–171, 1990.

185. Mekhail-Ishak, K., N. Hudson, M. Tsao, and G. Batist. Implications for therapy of drug-metabolizing enzymes in human colon cancer. *Cancer Res.* 49: 4866–4869, 1989.

186. Melosh, H. J. Giant impacts and the thermal state of the early earth. In: *Origin of the Earth,* edited by H. E. Newsom and J. H. Jones. New York: Oxford Univ. Press, 1990, p. 69–83.

187. Menzel, D. B. Ozone: an overview of its toxicity in man and animals. *J. Toxic Environ. Health* 13: 183–204, 1984.

187a. Meyer, D. M., M. Jura, I. Hawkins, and J. A. Cardelli. The abundance of interstellar oxygen toward Orion: evidence for recent infall. *Astrophys. J.* 437: L59–L61, 1994.

188. Miller, S. L. A production of amino acids under possible primitive Earth conditions. *Science* 117: 528–529, 1953.

189. Milne, L., P. Nicotera, S. Orrenius, and M. J. Burkitt. Effects of glutathione and chelating agents on copper-mediated DNA oxidation: pro-oxidant and antioxidant properties of glutathione. *Arch. Biochem. Biophys.* 304: 102–109, 1993.

190. Milton, R. C. d., S. C. F. Milton, and S. B. H. Kent. Total chemical synthesis of a D-enzyme: the enantiomers of HIV-1 protease show demonstration of reciprocal chiral substrate specificity. *Science* 256: 1445–1448, 1992.

191. Miyake, C., F. Michihata, and K. Asada. Scavenging of hydrogen peroxide in prokaryotic and eukaryotic algae: acquisition of ascorbate peroxidase during the evolution of cyanobacteria. *Plant Cell. Physiol.* 32: 33–43, 1991.

192. Molina, M. J. Chemistry of stratospheric ozone depletion. In: *Atmospheric Chemistry. Models and Predictions for Climate and Air Quality,* edited by C. S. Sloane and T. W. Tesche. Chelsea, MI: Lewis, 1991, p. 1–8.

193. Molina, M. J., and F. S. Rowland. Stratospheric sink for chlorofluoromethanes: chlorine atom-catalysed destruction of ozone. *Nature* 249: 810–812, 1974.

194. Moncada, S., and E. A. Higgs. Endogenous nitric oxide: physiology, pathology and clinical relevance. *Eur. J. Clin. Invest.* 21: 361–374, 1991.

195. Moncada, S., R. M. J. Palmer, and E. A. Higgs. Nitric oxide: physiology, pathophysiology, and pharmacology. *Pharmacol. Rev.* 43: 109–142, 1991.

196. Mooney, H. A. Photosynthesis. In: *Plant Ecology,* edited by M. J. Crawley. Boston: Blackwell Scientific, 1986, p. 345–373.

197. Mukhin, L. M., Y. P. Dikov, E. N. Evlanov, M. N. Fomenkova, A. D. Grechinskiy, M. A. Nazarov, O. F. Prilutskiy, T. V. Ruzmaikina, R. Z. Sagdeev, and B. V. Zubkov. Chemical composition of Halley's dust component from the PUMA-2 data. In: *Chemistry in Space,* edited by J. M. Greenberg and V. Pironello. Boston: Kluwer, 1991, p. 399–414.

198. National Oceanic and Atmospheric Administration. *U.S. Standard Atmosphere, 1976.* Washington, DC: National Oceanic and Atmospheric Administration, 1976.

199. Neilands, J. B. Microbial iron transport compounds (siderophores) as chelating agents. In: *Development of Iron Chelators for Clinical Use,* edited by A. E. Martell, W. F. Anderson, and D. G. Badman. New York: Elsevier, 1981, p. 13–31.

200. Neta. P., R. E. Huie, and A. B. Ross. Rate constants for reactions of peroxyl radicals in fluid solutions. *J. Phys. Chem. Ref. Data* 19: 413–513, 1990.

201. Norby, R. J., C. A. Gunderson, S. D. Wullschleger, E. G. O'Neill, and M. K. McCracken. Productivity and compensatory responses of yellow-polar trees in elevated CO_2. *Nature* 357: 322–324, 1992.

202. Ohanian, H. C. *Modern Physics.* Englewood Cliffs, NJ: Prentice-Hall, 1987.

203. Olinski, R., T. Zastawny, J. Budzbon, J. Skokowski, W. Zegar-

sky, and M. Dizdaroglu. DNA base modifications in chromatin of human cancerous tissues. *FEBS Lett.* 309: 193–198, 1992.

204. Omont, A. Circumstellar chemistry. In: *Chemistry in Space,* edited by J. M. Greenberg and V. Pironello. Boston: Kluwer, 1991, p. 171–197.

205. Oparin, A. I. *The Origin of Life,* translated by S. Morgulis. New York: Dover, 1953.

206. Orgel, L. E. Molecular replication. *Nature* 358: 203–209, 1992.

207. Oró, J. Comets and the formation of biochemical compounds on the primitive earth. *Nature* 190: 389–390, 1961.

208. Oró, J., T. Mills, and A. Lazcano. Comets and the formation of biochemical compounds on the primitive earth—a review. *Orig. Life Evol. Biosph.* 21: 267–277, 1992.

209. Pagel, E. J. Beryllium and the Big Bang. *Nature* 354: 267–268, 1991.

210. Parchment, R. E. The implications of a unified theory of programmed cell death, polyamines, oxyradicals and histogenesis in the embryo. *Int. J. Dev. Biol.* 37: 75–83, 1993.

211. Peebles, P. J. E., D. N. Schramm, E. L. Turner, and R. G. Kron. The case for the relativistic hot Big Bang cosmology. *Nature* 352: 769–776, 1991.

212. Peñuelas, J. High oxygen tension inhibits vascular aquatic plant growth in deep waters. *Photosynthetica* 21: 494–502, 1987.

212a. Pierce, M. J., D. L. Welch, R. D. McClure, S. van den Bergh, R. Racine, and P. B. Stetson. The Hubble constant and Virgo cluster distance from observations of Cepheid variables. *Nature* 371: 385–389, 1994.

213. Pihl, T. D., R. N. Schicho, R. M. Kelly, and R. J. Maier. Characterization of hydrogen-uptake activity in the hyperthermophile *Pyrodictium brockii. Proc. Natl. Acad. Sci. USA* 86: 138–141, 1989.

214. Podsiadlowski, P., J. E. Pringle, and M. J. Rees. The origin of the planet orbiting PSR1829–10. *Nature* 352: 783–784, 1991.

215. Ponnamperuma, C., Y. Honda, and R. Navarro-Conzález. Chemical studies on the existence of extraterrestrial life. *J. Br. Interplanet. Soc.* 45: 241–249, 1992.

216. Ponnamperuma, C., A. Shimoyama, M. Yamada, T. Hobo, and R. Pal. Possible surface reaction on Mars: implications for Viking biology results. *Science* 197: 455–457, 1977.

217. Pramanik, A. M., S. Bingsmark, M. Lindahl, H. Baltscheffsky, M. Baltscheffsky, and B. Andersson. Inorganic-pyrophosphate-dependent phosphorylation of spinach thylakoid proteins. *Eur. J. Biochem.* 198: 183–186, 1991.

218. Prinn, G. G., and B. Fegley, Jr. The atmospheres of Venus, Earth, and Mars: a critical comparison. *Annu. Rev. Earth Planet. Sci.* 15: 171–212, 1987.

219. Prosser, J. I. Autotrophic nitrification in bacteria. *Adv. Microb. Physiol.* 30: 125–181, 1989.

220. Prusiner, S. B. Transgenic investigations of prion diseases of humans and animals. *Phil. Trans. R. Soc. Lond. B* 339: 239–254, 1993.

221. Rabinowitz, D. L., T. Gehrels, J. V. Scotti, R. S. McMillan, M. L. Perry, W. Wisniewski, S. M. Larson, E. S. Howell, and E. A. Mueeler. Evidence for a near-Earth asteroid belt. *Nature* 363: 704–706, 1993.

222. Ragsdale, S. Enzymology of the acetyl-CoA pathway of CO_2 fixation. *Crit. Rev. Biochem. Mol. Biol.* 26: 261–300, 1991.

223. Rahn, H., and L. E. Farhi. Gaseous environment and atelectasis. *Federation Proc.* 23: 1035–1041, 1963.

224. Rast, W., and A. O. Ryding. Introduction. In: *The Control of Eutrophication of Lakes and Reservoirs,* edited by S. O. Ryding and W. Rast. Park Ridge, NJ: Parthenon, 1989, p. 1–4.

225. Raven, J. A. Plant responses to high O_2 concentrations: relevance to previous high O_2 episodes. *Palaeogeogr. Palaeocli-matol. Palaeoecol. (Global Planet. Change Sect.)* 97: 19–38, 1991.

226. Richardson, J. B. Origin and evolution of the earliest land plants. In: *Major Events in the History of Life,* edited by J. W. Schopf. Boston: Jones and Bartlett, 1992, p. 95–118.

227. Richter, C. Pro-oxidants and mitochondrial Ca^{2+}: their relationship to apoptosis and oncogenesis. *FEBS Lett.* 325: 104–107, 1993.

228. Robin, E., L. Froget, C. Jéhanno, and R. Rocchia. Evidence for a K/T impact event in the Pacific Ocean. *Nature* 363: 615–617, 1993.

229. Robinson, J. M. Phanerozoic O_2 variation, fire, and terrestrial ecology. *Palaeogeogr. Palaeoclimatol. Palaeoecol. (Global Planet. Change Sect.)* 75: 223–240, 1989.

230. Robinson, J. M. Lignin, land plants, and fungi: biological evolution affecting Phanerozoic oxygen balance. *Geology* 15: 607–610, 1990.

231. Robinson, J. M. Phanerozoic atmospheric reconstructions: a terrestrial perspective. *Palaeogeogr. Palaeoclimatol. Palaeoecol. (Global Planet. Change Sec.)* 97: 51–62, 1991.

232. Rodriguez, J. M. Probing stratospheric ozone. *Science* 261: 1128–1129, 1993.

233. Romankevich, Y. A. Biogeochemical aspects of the earth's living matter, translated from *Geokhim.* 2: 292–306. *Geochemistry Int.* 25: 123–136, 1988.

234. Rospert, S., J. Breitung, K. Ma, B. Schwörer, C. Zirngibl, R. K. Thauer, D. Linder, R. Huber, and K. O. Stetter. Methylcoenzyme M reductase and other enzymes involved in methanogenesis from CO_2 and H_2 in the extreme thermophile *Methanopyrus kandleri. Arch. Microbiol.* 156: 49–55, 1991.

235. Rowland, F. S. Ozone depletion theory. *Science* 261: 1102–1103, 1993.

235a. Rubbo, H., R. Radi, M. Trujillo, R. Telleri, B. Kalyanaraman, S. Barnes, M. Kirk, and B. A. Freeman. Nitric oxide regulation of superoxide and peroxynitrite-dependent lipid peroxidation. Formation of novel nitrogen-containing oxidized lipid derivatives. *J. Biol. Chem.* 269: 26066–26075, 1994.

236. Rubey, W. W. Geologic history of sea water. An attempt to state the problem. *Bull. Geol. Soc. Am.* 62: 1111–1148, 1951.

237. Runnegar, B. Precambrian oxygen levels estimated from the biochemistry and physiology of early eukaryotes. *Palaeogeogr. Palaeoclimatol. Palaeoecol. (Global Planet. Change Sect.)* 97: 97–111, 1991.

238. Runnegar, B. Evolution of the earliest animals. In: *Major Events in the History of Life,* edited by J. W. Schopf. Boston: Jones and Bartlett, 1992, p. 65–93.

239. Rusting, R. L. Why do we age? *Sci. Am.* 267: 130–135, 138–141, 1992.

240. Sagan, C. Radiation and the origin of the gene. *Evolution* 11: 40–55, 1957.

241. Sagan, C. On the origin and planetary distribution of life. *Radiat. Res.* 15: 174–192, 1961.

242. Sagan, C. Ultraviolet selection pressure on the earliest organisms. *J. Theoret. Biol.* 39: 195–200, 1973.

243. Sagan, C., and G. Mullen. Earth and Mars: evolution of atmospheres and surface temperatures. *Science* 177(6): 52–56, 1972.

244. Sahu, S. C., and G. C. Gray. Interactions of flavonoids, trace metals, and oxygen: nuclear DNA damage and lipid peroxidation induced by myricetin. *Cancer Lett.* 70: 73–79, 1993.

245. Sanders, R. W., S. Solomon, M. A. Carroll, and A. L. Schmeltekopf. Ground-based measurements of O_3, NO_2, $OClO$, and BrO during the 1987 Antarctic ozone depletion event. In: *Ozone in the Atmosphere,* edited by R. D. Bojkov and P. Fabian. Hampton, VA: A. Deepak, 1989, p. 65–69.

246. Sasaki, S. The primary solar-type atmosphere surrounding the

accreting earth: H_2O-induced high surface temperature. In: *Origin of the Earth,* edited by H. E. Newsom and J. H. Jones. New York: Oxford Univ. Press, 1990, p. 195–209.

247. Scheid, P., B. Pelster, and H. Kobayashi. Gas exchange in the fish swimbladder. In: *Oxygen Transport to Tissue XII,* edited by J. Piiper, T. K. Goldstick, and M. Meyer. New York: Plenum, 1990, p. 735–742.

248. Schidlowski, M. A 3,800-million-year isotopic record of life from carbon in sedimentary rocks. *Nature* 333: 313–318, 1988.

249. Schidlowski, M., J. M. Hayes, and I. R. Kaplan. Isotopic inferences of ancient biochemistries: carbon, sulfur, hydrogen, and nitrogen. In: *Earth's Earliest Biosphere. Its Origin and Evolution,* edited by J. W. Schopf. Princeton, NJ: Princeton Univ. Press, 1983, p. 149–186.

250. Schlegel, H. G. Physiology and biochemistry of Knallgasbacteria. *Adv. Comp. Physiol. Biochem.* 2: 185–236, 1966.

251. Schopf, J. W. (Ed). *Major Events in the History of Life.* Boston: Jones and Bartlett, 1992.

252. Schopf, J. W. The oldest fossils and what they mean. In: *Major Events in the History of Life,* edited by J. W. Schopf. Boston: Jones and Bartlett, 1992, p. 29–63.

253. Schopf, J. W. The oldest known records of life: Archean stromatolites, microfossils, and organic matter. In: *Early Life on Earth: Nobel Symposium 84,* edited by S. Bengtson. New York: Columbia Univ. Press, 1994, p. 192–206.

254. Schopf, J. W. Microfossils of the Early Archean Apex Chert: new evidence of the antiquity of life. *Science* 260: 640–646, 1993.

255. Schrödinger, E. *What is Life? And Other Scientific Essays.* Garden City, NY: Doubleday, 1956.

256. Schroeder, P., and L. Ladd. Slowing the increase of atmospheric carbon dioxide: a biological approach. *Climatic Change* 19: 283–290, 1991.

257. Schueler, F. W. *Chemobiodynamics and Drug Design.* New York: McGraw-Hill, 1960.

258. Schwartz, J. L., D. Z. Antoniades, and S. Zhao. Molecular and biochemical reprogramming of oncogenesis through the activity of prooxidants and antioxidants. *Ann. NY Acad. Sci* 686: 262–279, 1993.

259. Schwartzman, R. A., and J. A. Cidlowski. Apoptosis: the biochemistry and molecular biology of programmed cell death. *Endoc. Rev.* 14: 133–151, 1993.

260. Scott, D. H., J. W. Rice, Jr., and J. M. Dohm. Martian paleolakes and waterways: exobiological implications. *Orig. Life Evol. Biosph.* 21: 189–198, 1991.

261. Scotti, J. V., D. L. Rabinowitz, and B. G. Marsden. Near miss of the earth by a small asteroid. *Nature* 354: 287–289, 1991.

262. Seiler, W., H. Giehl, and G. Bunse. The influence of plants on atmospheric carbon monoxide and dinitrogen oxide. *Pure Appl. Geophys. (Pageoph)* 116: 439–451, 1978.

263. Seliger, H. H. Environmental photobiology. In: *The Science of Photobiology,* edited by K. C. Smith. New York: Plenum, 1977, p. 143–173.

264. Shaffer, G. A model of biogeochemical cycling of phosphorus, nitrogen, oxygen, and sulphur in the ocean: one step toward a global climate model. *J. Geophys. Res.* 94: 1979–2004, 1989.

265. Shapiro, J. Effect of yellow organic acids on iron and other metals in water. *J. Am. Water Works Assoc.* 56: 1062–1082, 1964.

266. Shapiro, B. M. The control of oxidant stress at fertilization. *Science* 252: 533–536, 1991.

267. Sharma, V. S., D. Bandyopadhyay, M. Berjis, J. Rifkind, and G. Boss. Double-mixing kinetic studies of the reactions of monoliganded species of hemoglobin: $\alpha_2^{(CO)1}\beta_2$ and $\alpha_2\beta_2^{(CO)1}$. *J. Biol. Chem.* 266: 24492–24497, 1991.

268. Sharpe, A. G. *Inorganic Chemistry* (3rd ed). New York and Essex, UK: Wiley and Longman, 1992.

269. Sharpton, V. L., K. Burke, A. Carnago-Zanoguera, S. A. Hall, D. S. Lee, L. E. Marín, G. Suárez-Reynoso, J. M. Queaada-Muñeton, P. D. Spudis, and J. Urrutia-Fucugauchi. Chicxulub multiring impact basin: size and other characteristics derived from gravity analysis. *Science* 261: 1564–1567, 1993.

270. Sharpton, V. L., G. B. Dalrymple, L. E. Marín, G. Ryder, B. C. Schuraytz, and J. Urrutia-Fucugauchi. New links between the Chicxulub impact structure and the Cretaceous/Tertiary boundary. *Nature* 359: 819–821, 1992.

271. Silk, J. *The Big Bang. The Creation and Evolution of the Universe.* San Francisco: Freeman, 1980.

272. Silk, J. Slow-motion galactic birth. *Nature* 351: 191, 1991.

273. Sillén, L. G., and A. E. Martell. *Stability Constants of Metal-Ion Complexes. Special Publication 17.* London: Burlington House, 1964.

274. Simic, M. G. Urinary biomarkers and the rate of DNA damage in carcinogenesis and anticarcinogenesis. *Mutat. Res.* 267: 277–290, 1992.

275. Sleep, N. H., K. J. Zahnle, J. F. Kasting, and H. J. Morowitz. Annihilation of ecosystems by large asteroid impacts on the early Earth. *Nature* 342: 139–142, 1989.

276. Slonim, N. B., and L. H. Hamilton. *Respiratory Physiology* (4th ed.). St. Louis: Mosby, 1981.

277. Smith, B. N. Evolution of C_4 photosynthesis in response to changes in carbon and oxygen concentrations in the atmosphere through time. *BioSystems* 8: 24–32, 1976.

278. Smith, C., B. Halliwell, and O. I. Aruoma. Protection by albumin against the pro-oxidant actions of phenolic dietary components. *Food Chem.Toxicol.* 30: 483–489, 1992.

279. Smith, R. C., B. B. Prézelin, K. S. Baker, R. R. Bidigare, N. P. Boucher, T. Coley, D. Karentz, S. MacIntyre, H. A. Matlick, D. Menzies, M. Ondrusek, Z. Wan, and K. J. Waters. Ozone depletion: ultraviolet radiation and phytoplankton biology in Antarctic waters. *Science* 255: 952–959, 1992.

279a. Spinrad, H., A. Dey, and J. R. Graham. Keck observations of the most distant galaxy: 8C 1435+63 AT $z = 4.25$. *Astrophys. J.* 438: L51–L54, 1995.

280. Stadtman, E. R. Protein oxidation and aging. *Science* 257: 1220–1224, 1992.

281. Stephenson, J. A. E., and M. W. J. Scourfield. Importance of energetic solar protons in ozone depletion. *Nature* 352: 137–139, 1991.

282. Stolarski, R., R. Boijkov, L. Bishop, C. Zerefos, J. Staehelin, and J. Zawodny. Measure trends in stratospheric ozone. *Science* 256: 342–349, 1992.

283. Szatriwski, T. P., and C. F. Nathan. Production of large amounts of hydrogen peroxide by human tumor cells. *Cancer Res.* 51: 794–798, 1991.

284. Tabazadeh, A., and R. P. Turco. Stratospheric chlorine injection by volcanic eruptions: HCl scavenging and implications for ozone. *Science* 260: 1082–1085, 1993.

285. Takao, M., A. Yasui, and A. Oikawa. Unique characteristics of a superoxide dismutase of a strictly anaerobic archaebacterium *Methanobacterium thermoautotrophicum. J. Biol. Chem.* 266: 14151–14152, 1991.

286. Tappan, H. Extinction or survival: selectivity and causes of Phanerozoic crises. In: *Geological Implications of Impacts of Large Asteroids and Comets on the Earth. Geological Special Paper 190,* edited by L. T. Silver and P. Schultz. Boulder, CO: Geological Society of America, 1982, p. 265–276.

287. Thompson, A. M. The oxidizing capacity of the earth's atmosphere: probable past and future. *Science* 256: 1157–1165, 1992.

288. Tingle, T. N., J. A. Tyburczy, T. J. Ahrens, and C. H. Becker.

The fate of organic matter during planetary accretion; preliminary studies of the organic chemistry of experimentally shocked Murchison meteorite. *Orig. Life Evol. Biosph.* 21: 385–397, 1992.

289. Toumi, R., R. L. Jones, and J. A. Pyle. Stratospheric ozone depletion by $ClONO_2$ photolysis. *Nature* 365: 37–39, 1993.

290. Tyagi, S., and C. Ponnamperuma. Nonrandomness in prebiotic peptide synthesis. *J. Mol. Evol.* 30: 391–399, 1990.

291. Urey, H. C. On the early chemical history of the Earth and the origin of life. *Proc. Natl. Acad. Sci. USA* 38: 351–363, 1952.

292. van den Bergh, S. The age and size of the universe. *Science* 258: 421–424, 1992.

293. Verma, A., D. J. Hirsch, C. E. Glatt, G. V. Ronnett, and S. H. Snyder. Carbon monoxide: a putative neural messenger. *Science* 259: 381–384, 1993.

294. von Gunter, U., and W. Schnedier. Primary products of the oxygenation of iron (II) at an oxic-anoxic boundary: nucleation, aggregation, and aging. *J. Coll. Interface Sci.* 145: 127–139, 1991.

295. Voßwinkel, R., I. Neidt, and H. Bothe. The production and utilization of nitric oxide by a new, denitrifying strain of *Pseudomonas aeruginosa*. *Arch. Microbiol.* 156: 62–69, 1991.

296. Waldo, G. S., J. Ling, J. Sanders-Loehr, and E. C. Theil. Formation of an Fe(III)-tyrosinate complex during biomineralization of H-subunit ferritin. *Science* 259: 796–798, 1993.

297. Walker, J. C. G. *Evolution of the Atmosphere*. New York: Macmillan, 1977.

298. Walker, J. C. G. How life affects the atmosphere. *BioScience* 34: 486–491, 1984.

299. Walker, J. C. G. Iron and sulfur in the pre-biologic ocean. *Precambrian Res.* 28: 205–222, 1985.

300. Walker, J. C. G. *Earth History. The Several Ages of the Earth*. Boston: Jones and Bartlett, 1986.

301. Walker, J. C. G. Global geochemical cycles of carbon, sulfur and oxygen. *Marine Geology* 70: 159–174, 1986.

302. Walker, J. C. G., C. Klein, M. Schidlowski, J. W. Schopf, D. J. Stevenson, and M. R. Walter. Environmental evolution of the Archean-Early Proterozoic earth: In: *Earth's Earliest Biosphere: Its Origin and Evolution*, edited by J. W. Schopf. Princeton, NJ: Princeton Univ. Press, 1983, p. 260–291.

303. Warneck, P. *Chemistry of the Natural Atmosphere*. New York: Academic, 1988.

304. Watson, J.D. Prologue. Early speculations and facts about RNA templates. In: *The RNA World. The Nature of Modern RNA Suggests a Prebiotic RNA World*, edited by R. F. Gesteland and J. F. Atkins. Plainview, NY: Cold Spring Harbor Lab., 1993, p. xv–xxiii.

305. Watson, A., J. E. Lovelock, and L. Margulis. Methanogenesis, fires and the regulation of atmospheric oxygen. *BioSystems* 10: 293–298, 1978.

306. Wege, K. Extremely low temperatures in the stratosphere and very low total ozone amount above Northern and Central

Europe during winter 1989. *J. Atmos. Chem.* 12: 381–390, 1991.

307. Weinberg, S. *The First Three Minutes. A Modern View of the Origin of the Universe*. New York: Basic Books, 1977.

308. West, J. B. *Respiratory Physiology—The Essentials* (4th ed.). Baltimore: Williams & Wilkins, 1990, p. 53–54.

309. Whitman, W. B. Methanogenic bacteria. In: *The Bacteria. A Treatise on Structure and Function. Archaebacteria*, edited by C. R. Woese and R. S. Wolfe. New York: Academic, 1985, vol. 8, p. 3–84.

310. Woese, C. R. Bacterial evolution. *Microbiol. Rev.* 51: 221–271, 1987.

311. Woese, C. R., O. Kandler, and M. L. Wheelis. Towards a natural system of organisms: proposal for the domains Archaea, Bacteria, and Eucarya. *Proc. Natl. Acad. Sci. USA* 87: 4576–4579, 1990.

312. Woese, C. R., L. J. Magrum, and G. E. Fox. Archaebacter. *J. Mol. Evol.* 11: 245–252, 1978.

313. Wolbach, W. S., I. Gilmour, and E. Anders. Major wildfires at the Cretaceous/Tertiary boundary. In: *Global Catastrophes in Earth History; An Interdisciplinary Conference on Impacts, Volcanism, and Mass Mortality. Geological Society of America Special Paper 247*, edited by V. L. Sharpton and P. D. Ward. Geological Society of America: Boulder, CO, 1990, p. 391–400.

314. Wong, S. S. M. *Introductory Nuclear Physics*. Englewood Cliffs, NJ: Prentice Hall, 1990.

315. Wood, H. G. Life with CO or CO_2 and H_2 as a source of carbon and energy. *FASEB J.* 5: 156–163, 1991.

316. Wyss, O., and P. W. Wilson. Mechanism of biological nitrogen fixation. VI. Inhibition of *Azotobacter* by hydrogen. *Proc. Natl. Acad. Sci. USA* 27: 162–168, 1941.

316a. Yarmolinsky, M. B. Programmed cell death in bacterial populations. *Science* 267: 836–837, 1995.

317. Zelitch, I. Selection and characterization of tobacco plants with novel O_2-resistant photosynthesis. *Plant Physiol.* 90: 1457–1464, 1989.

318. Zelitch, I. Physiological investigations of a tobacco mutant with O_2-resistant photosynthesis and enhanced catalase activity. *Plant Physiol.* 93: 1521–1524, 1990.

319. Zhong, L. T., T. Sarafian, D. J. Kane, A. C. Charles, S. P. Mah, R. H. Edwards, and D. E. Bredesen. bcl-2 inhibits death of central neural cells induced by multiple agents. *Proc. Natl. Acad. Sci. USA* 90: 4533–4537, 1993.

320. Zhuang, G., and R. A. Duce. The adsorption of dissolved iron on marine aerosol particles in surface waters of the open ocean. *Deep-Sea Res.* Part I 40: 1413–1429, 1993.

321. Zuo, Y., and J. Hoigné. Evidence for photochemical formation of H_2O_2 and oxidation of SO_2 in authentic fog water. *Science* 260: 71–73, 1993.

322. Zurer, P. Ozone hits low levels over Antarctica, U. S. *Chem. Eng. News* Oct. 4: 5, 1993.

47. Tissue capacity for mitochondrial oxidative phosphorylation and its adaptation to stress

WILLIAM L. RUMSEY | Department of Pharmacology, Zeneca Pharmaceuticals, Wilmington, Delaware

DAVID F. WILSON | Department of Biochemistry and Biophysics, Medical School, University of Pennsylvania, Philadelphia, Pennsylvania

CHAPTER CONTENTS

SURVIVAL OF AN ORGANISM requires that each of its essential parts (tissues) is able to maintain the metabolic energy supply necessary for maintaining the dynamic steady state of the structural elements and intracellular environment. In addition, the organism requires energy for food gathering and other functions involving mechanical work. The amount of adenosine triphosphate (ATP) that can be derived from a given weight of foodstuff is much greater for oxidative phosphorylation than for nonoxidative metabolism. Carbohydrate metabolism via glycolysis can provide 2 ATP/glucose (~3 ATP per glucose from glycogen), while oxidative phosphorylation can provide from 36 to 38 ATP/glucose. The difference is even larger for fats and protein. Thus, the high rate of ATP utilization by higher organisms required to support neural activity and mobility can only be supported by mitochondrial oxidative phosphorylation. Only through the greater amount of ATP produced per gram of food can the high level of metabolic activity be obtained without intake of excessive amounts of food. It can be stated that mitochondrial oxidative phosphorylation provides at least 95% of the ATP supply of higher organisms. For diets in which the caloric intake includes fat and protein, 98%–99% or greater of the ATP is supplied by oxidative phosphorylation.

Most of mitochondrial oxidative phosphorylation is based on oxidation of intramitochondrial NADH by molecular oxygen:

$$NADH + H^+ + \tfrac{1}{2}O_2 + 3ADP + 3Pi \rightarrow NAD^+ + H_2O + 3ATP \quad (1)$$

Some reducing equivalents are also provided by reduced flavins, but these represent only a relatively small fraction of the total. The dehydrogenase enzymes that oxidize NADH and most of the reduced flavin for the respiratory chain, NADH dehydrogenase and succinate dehydrogenase, are associated with the mitochondrial inner membrane. These enzymes and the metabolic pathways that produce the NADH and reduced flavin for respiration (such as the tricarboxylic acid cycle, β-oxidation of fatty acids, glutamate dehydrogenase, and the other oxidative enzymes) are within the mitochondrial matrix. Thus, the metabolism that "feeds" the respiratory chain is separated from the cytoplasm by two membranes. The outer mitochondrial membrane, which is permeable to molecules less than about 2,000 daltons, has little effect on the movements of metabolites. The inner membrane is impermeable to uncharged hydrophilic molecules equivalent to more than about 4 carbon sugars and to all hydrophilic charged molecules. Several transport systems in the membrane are designed to selectively permit metabolites to move between the cytosol and the inner mitochondrial matrix. The inner membrane isolates the mitochondrial matrix from the cytoplasm and determines the flow of metabolites, and thereby information, between the compartments. Within this environment the capacity of mitochondria to carry out oxidative phosphorylation is absolutely dependent on: (*i*) transport of oxidizable substrates across the mitochondrial membrane; (*ii*) transport of ATP, ADP, and inorganic phosphate across the mitochondrial membrane; (*iii*) the activity of the dehydrogenases of the catabolic pathways in the mitochondrial matrix; (*iv*) the cellular content of respiratory enzymes; and (*v*) the availability of molecular oxygen at appropriate pressures.

What is needed is a quantitative understanding of the relationships among the factors that regulate oxidative phosphorylation under physiological conditions and the effects of the physiological conditions on this relationship. The adaptive responses of tissue to physiological conditions that induce metabolic stress must occur through metabolic factors that are substantially altered in response to the challenge. These factors then serve as messengers that communicate this information to other metabolic pathways. This chapter first addresses regulation of mitochondrial respiration in vivo and in vitro, focusing on the metabolic factors that determine the rate of respiration and/or are altered during periods of increased ATP demand. We will then examine the effects of adaptation on each of these metabolic determinants, with the premise that adaptation is "successful" only if it leads to increased capacity of those parameters responsible for limiting the oxidative capacity of the tissue. Comparison of the effects of different physiological conditions for stressing oxidative metabolism permits identification of those parameters specific for oxidative metabolism and characterization of the extent of the adaptive process. An effort will be made to identify the metabolic parameters that are responsible for detecting changes in cellular energy levels and for providing the signal(s) for the adaptive responses to environmental challenges to oxidative metabolism.

REGULATION OF THE RATE OF MITOCHONDRIAL OXIDATIVE PHOSPHORYLATION IN VIVO

"Supply" and "Demand" in Determining the Mitochondrial Respiratory Rate

In vivo the rate of mitochondrial respiration is tightly coupled to the rate of utilization of ATP (demand). This coupling is tight enough that in resting skeletal muscle, for example, the respiratory rate of skeletal muscle at rest may be less than 1% of that at maximal work rates. In the bee-flight muscle at rest the rate is only about 0.1% of the value at maximal work. In each case, the resting respiratory rate can be accounted for by ATP utilization for protein synthesis, ion transport, and other "basic maintenance" reactions of the cells. Thus, the rate of oxidative phosphorylation of muscle in vivo is determined by the requirement (demand) for ATP, and there is very little respiration without ATP synthesis. After isolation from the tissue, most mitochondria have a substantial rate of respiration in the absence of net ATP synthesis. It should be noted, however, that this effect is due primarily from the isolation procedure that results in damage to the mitochondria. Thus, conclusions concerning the regulation of mitochondrial oxidative phosphorylation based on data obtained using isolated preparations should be viewed with caution until they have been shown to be consistent with observations in vivo.

"Supply," as it applies to mitochondrial oxidative phosphorylation, relates to the supply of reducing equivalents (NADH, reduced flavin). NADH is continuously being oxidized to NAD^+ and must be re-reduced using substrates transported into the mitochondrial matrix. These substrates are then oxidized by dehydrogenases that couple their oxidation to reduction of NAD^+ to NADH. This supply side of oxidative phosphorylation is highly regulated (for reviews, see refs. 23, 39). Receptors for hormones in the plasma membrane, for example, regulate the intracellular Ca^{2+} concentration and thereby the intramitochondrial Ca^{2+} concentration. Calcium ion concentration is an important activator for many of the catabolic dehydrogenases, in particular those of the tricarboxylic acid cycle. Simply changing from one substrate to another (fats vs. carbohydrates vs. amino acids) results in utilization of different dehydrogenases with different capacities for dehydrogenation. Oxidation of fats by the liver, for example, results in a more reduced state of the intramitochondrial NAD couple than does oxidation of carbohydrates (see, for example, refs. 121, 122). Substrate transport across the mitochondrial membrane may also be regulated, although this has not yet been established. Regulation at any of these levels would be expected to influence the availability of NADH for oxidation by the respiratory chain and thereby the capacity for ATP synthesis.

As noted above, oxidative phosphorylation operates in a balance between supply and demand. Since the respiratory rate of tissue is very low in the absence of ATP utilization, ATP utilization is the primary determinant of the respiratory rate. As ATP is continuously hydrolyzed by the cell, the small intracellular pool of ATP requires that, on the average, it be resynthesized as rapidly as it is used. If the rate of utilization of ATP becomes appreciably greater than the rate of synthesis, the cellular content of ATP falls. The cellular ATP content is sufficient to provide the cellular needs for only a few seconds at normal metabolic rates, and failure to correct the imbalance between utilization and synthesis is pathological. The rate of mitochondrial oxidative phosphorylation is of necessity, therefore, tightly coupled to the rate of ATP utilization. Somehow an excess or deficit in the rate of ATP synthesis must signal the mitochondria to increase or decrease the respiratory rate until the rate of ATP synthesis again equals the rate of utilization. General agreement has not been reached on the mechanism of regulation of mitochondrial respiration (see, for examples, refs. 4, 20, 29, 121, 122). In the present discussion, we will use one of the proposed mechanisms, the one we believe to provide the best available understanding of the overall process, as a framework for presentation. The overall picture is one in which the primary determinant of the mitochondrial respiratory rate is the rate of ATP utilization by the cell. As long as the rate of ATP utilization exceeds the rate of synthesis, the [ATP]/[ADP][Pi] ratio continuously decreases. This decrease must, in turn, activate respiration and therefore increase the rate of ATP synthesis until it is again equal to the rate of utilization. The activity of the dehydrogenases, on the other hand, determines the energy level ([ATP]/[ADP][Pi]) at which oxidative phosphorylation can synthesize ATP at a given respiratory rate.

On the Mechanism of Regulation of Mitochondrial ATP Synthesis In Vivo

In healthy cells, the first two sites of oxidative phosphorylation are near equilibrium (32, 121, 122). The overall equation for the two sites is:

$$NADH + H^+ + 2c^{3+} + 2ADP + 2Pi$$
$$\rightarrow NAD^+ + 2ATP + 2c^{2+} \quad (2)$$

where NADH and NAD^+ are the free concentrations of the intramitochondrial coenzymes; ATP, ADP, and Pi are the concentrations of these metabolites free in the cytosol, and c^{3+} and c^{2+} are the oxidized and reduced forms of cytochrome c. Reduced cytochrome c, the product of reaction 2, is the substrate for cytochrome c oxidase that is responsible for reduction of molecular oxygen to water:

$$4c^{2+} + O_2 + 2ADP + 2Pi + 4H^+$$
$$\rightarrow 4c^{3+} + 2ATP + 2H_2O \quad (3)$$

All of the reducing equivalents for reduction of molecular oxygen to water must pass through cytochrome c, and in normal mitochondria this is their only fate. Thus,

the rate of mitochondrial respiration is equal to the rate of the cytochrome c oxidase reaction. Any factors altering the rate of oxidation of cytochrome c will correspondingly alter the rate of oxidative phosphorylation. Conversely, any change in the rate of respiration must occur through factors that affect the rate of the cytochrome c oxidase reaction (117–120). It has been shown that, in suspensions of well-coupled mitochondria at a given value of [ATP]/[ADP][Pi], the rate of oxygen reduction by cytochrome c oxidase increases markedly with increasing reduction of cytochrome c (120). The level of reduced cytochrome c is, in turn, controlled by reaction 2. The equilibrium constant for reaction 2 is expressed:

$$K_{eq} = [NAD^+]/[NADH] \times (c^{2+}/c^{3+})^2$$
$$\times ([ATP]/[ADP][Pi])^2 \quad (4)$$

At near equilibrium, the level of reduction of cytochrome $c(c^{2+}/c^{3+})$ is determined by the [NADH]/[NAD^+] ratio and [ATP]/[ADP][Pi]:

$$c^{2+}/c^{3+} = ([NADH]/[NAD^+])^{1/2}$$
$$\times [ADP][Pi]/[ATP] \times (1/K_{eq})^{1/2} \quad (5)$$

Thus, at constant [NADH]/[NAD^+] reduction of cytochrome c increases as [ATP]/[ADP][Pi] decreases. Measurements both in preparations of isolated mitochondria and in intact cells have shown that the respiratory rate increases with decreasing energy state ([ATP]/[ADP][Pi]) (see, for example, refs. 45, 86). Equation 5 predicts that the level of reduction can also be increased without changing [ATP]/[ADP][Pi] if the [NADH]/[NAD^+] is increased. The latter may increase due either to regulatory factors that increase the activity of the dehydrogenases or to alterations in oxidizable substrate levels or type. This latter case was first shown using perfused liver and isolated hepatocytes, where changing from oxidizing carbohydrate to oxidizing fatty acids results in an increase in [NADH]/[NAD^+], an increase in respiratory rate, and an *increase* in cellular [ATP]/[ADP] (see, for example, ref. 121). Later studies using phosphorus NMR to study the heart in situ have shown that when the heart is stimulated by norepinephrine the respiratory rate increases up to fourfold with an *increase* in [phosphocreatine]/[Pi] ratio, indicating an increase in cellular energy state (66, 91). In the dog heart, an increase in [NADH]/[NAD^+] ratio was reported based on surface fluorescence measurements, but the increase in pyridine nucleotide reduction was not quantitated (4).

Oxygen Dependence of Mitochondrial Oxidative Phosphorylation In Vivo

Mitochondrial oxidative phosphorylation requires oxygen in the local environment at a pressure sufficient for carrying out ATP synthesis under the prevailing metabolic state of the cell. Oxygen consumption by the mitochondria results in the oxygen pressure being lower at this site than that in the capillaries, and this oxygen pressure difference provides the driving force for net diffusion of oxygen. The oxygen pressure difference between the capillaries and the mitochondria can be divided into two parts: from the capillaries to the interstitial space and from the interstitial space to the mitochondria. The oxygen pressure at the venous end of the capillaries varies somewhat among tissues (for example, 30 to 40 torr in brain cortex). Most of the oxygen pressure difference appears to be between the capillary lumen and the extracellular space. Direct measurements of the oxygen pressure difference between the extracellular space and the mitochondria indicate it is 1 torr or less (17, 22, 24, 33, 74, 93, 95, 117, 124). On the other hand, it has been suggested (see refs. 64, 65) that: (i) the spatial organization of mitochondria within cells is an important determinant of the oxygen dependence of cellular respiration, and (ii) the distribution of mitochondria within some cells is actively controlled as part of a mechanism for determining the oxygen dependence of the cellular respiration. This hypothesis is based on data showing an oxygen pressure difference between the extracellular space and the mitochondria of several torr. Such large diffusion gradients are in marked contrast to the values measured by other investigators (see above). Although mitochondria are often observed to be asymmetrically distributed in cells, it is likely that this is due to factors other than oxygen diffusion gradients. Diffusion of ADP, for example, is much more likely to be limiting than diffusion of oxygen. The ADP flux is higher than that for oxygen (the ADP/O_2 is approximately 6) and the diffusivity of ADP is less than that of oxygen because ADP is larger, binds to many cellular components, and has a net electrical charge. Thus, improvement in function may be achieved by placing the mitochondria near the site of ADP generation or, as in muscle, by inducing the creatine phosphokinase reaction and having creatine facilitate movement of ADP to the mitochondria.

Delivery of oxygen is clearly very important for mitochondrial oxidative phosphorylation in tissue. This delivery is dependent on the capillary oxygen pressure, the intercapillary distance, the oxygen consumption rate, and the diffusivity of oxygen in tissue. The capillary oxygen pressures generally remain nearly constant or decrease as respiration increases. The oxygen diffusivity is established by the composition of most mammalian tissue and has been shown to be near that of physiological saline. Thus, adaptation to altered oxygen consumption or delivery to tissue involves primarily changes in intercapillary distance and/or capillary surface area per unit of tissue volume.

MITOCHONDRIAL ENZYME CONTENT IN CELLS WITH DIFFERENT RATES OF RESPIRATION

The variation of mitochondrial enzyme content with cell type is well recognized. In general, cells with higher average respiratory rates (per gram wet weight) also have a higher cytochrome content (per gram wet weight). For example, the respiratory rates of hepatocytes, cultured kidney cells, and perfused rat heart have been reported to be 1.1, 1.3, and 10.6 μmol O_2/min g wet wt, respectively (30), all measured in a nominally "resting" state. When the amount of cytochrome c was measured as an indicator of the content of mitochondrial respiratory enzymes, these values were 20, 16, and 51 nmol/g wet weight of cells. Thus, the cytochrome content increased almost in proportion to the respiratory rate, such that the turnover number of cytochrome c (the average number of times each cytochrome c is oxidized and reduced per second) is similar for many cell types, ranging from 6/sec to 14/sec. The reason for the differences in respiratory enzyme content could be genetic, e.g., the genetic reading pattern that determines the differentiated cell type could also determine the mitochondrial enzyme content. On the other hand, as cells differentiate this alters their enzyme content and thereby their metabolic pattern and the rate of ATP utilization. The content of mitochondrial enzyme in the cells could, therefore, be the result of adaptation to the metabolic requirements for ATP as opposed to being an inherent property of the cell type. The latter would be consistent with the content of mitochondrial enzymes being coupled to the cellular energy state and/or the intramitochondrial reduction of the NAD couple.

METABOLIC ADAPTATION OF HEART MUSCLE TO ALTERATIONS IN PHYSIOLOGICAL WORK RATE

The heart is different from skeletal muscle in that the basal work rate of the heart is high and continuous throughout the day. Thus, the heart is never truly at rest. Even when an individual is inactive, the heart must continue to operate at about 20%–25% of its maximal capacity. This is in contrast to skeletal muscle, where exercise results in respiratory rates 10- to 100-fold greater than at rest. It is not surprising that the adaptive response of myocardial metabolism to endurance training is generally considered to be negligible (85, see ref. 7 for review). Techniques other than exercise must be used to induce metabolic adaptations in the heart. Thyroid hormone, for example, has marked effects on the basal metabolic rate of animals. It has been shown that levels of mitochondrial dehydrogenases in various tissues have been altered by thyroid hormone treatment

TABLE 47.1. *Content of Mitochondrial Proteins in Hearts from Hypothyroid, Euthyroid, and Hyperthyroid Rats*

Condition	Cytochrome $c + c_1$	Cytochrome $a + a_3$	Mitochondrial Protein	Oxygen Consumption
Hypothyroid	39	23	66	5.5
Euthyroid	56	32	70	8.2
Hyperthyroid	69	42	80	11.3

The data (means) were taken from Nishiki et al. (83). Hearts were first perfused retrogradely at 80 cm H_2O. Cytochrome content is expressed as nmol/g wet wt and oxygen consumption is expressed as mmol O_2/min/g wet wt.

(73). When comparison was made of hearts from hypothyroid (thyroidectomized), euthyroid (sham operated), and hyperthyroid (intraperitoneal injection of 35 µg thyroid hormone/100 g body weight each day for 10–15 days) rats, large changes in their enzyme content were found (83). Following retrograde perfusion at 80 cm H_2O pressure with glucose as the sole substrate, hearts from each group were freeze-clamped for determination of their enzyme content and of their metabolic status. The results are summarized in Table 47.1. The content of mitochondrial respiratory chain enzymes (cytochrome c and cytochrome $a + a_3$) per g wet weight increased by almost twofold from hypothyroid to hyperthyroid, as did the respiratory rate of the perfused hearts. Measurements of the metabolic status during the perfusion indicated the hearts were all operating at the same [ATP]/[ADP][Pi] and [NADH]/[NAD$^+$], consistent with full adaptation of the metabolism to the different work loads imposed by the different thyroid hormone levels in vivo.

METABOLIC ADAPTATIONS OF SKELETAL MUSCLE IN RESPONSE TO ENDURANCE TRAINING AND TO ELECTRICAL STIMULATION

General Observations

Skeletal muscle serves as a useful model for discussion of metabolic adaptation because metabolic activity can increase manyfold as the tissue progresses from the resting state (inactivity) to a working state in which there is sustained hydrolysis of ATP supporting the cycling of the contractile myofilaments. Regularly performed activity of the endurance type or continuous (or intermittent) electrical stimulation results in an enhanced capacity to synthesize ATP in skeletal muscle. The adaptative response of skeletal muscle is so remarkable that it can be used to serve as an auxillary pump to assist failing heart muscle (2). Skeletal muscle blood flow is low at rest and rises in accordance with the metabolic needs of the parenchymal tissue, thereby providing adequate delivery of oxygen and oxidizable fuels (glucose and lipids). Except under extremely high metabolic or pathological conditions, there is normally a good match of blood flow and metabolism. Any impedance to blood flow may, however, temporarily compromise the ability of skeletal muscle to produce ATP. We will also discuss the long-term consequences of partial occlusion of blood flow as well as the changes brought about by reduction of functional demand on oxidative capacity, and how improvements in the latter are obtained with exercise training. The precise mechanism(s) for these types of biochemical adaptation has not been fully elucidated. Some of these adaptations will be described, and an attempt will be made to provide insights regarding the cellular stimuli and the mechanisms by which the biochemical changes are induced.

Effects of Increased Energy Demand on Enzyme Content and Metabolism of Skeletal Muscle

The increase in oxidative capacity that results from conversion of fast (glycolytic)-type to slow (oxidative)-type skeletal muscle, induced by an increase in energy demand via sustained contractile activity, has been well documented (36, 48, 49, 88, 96, 98). Experimental paradigms that utilize exercise to increase energy demand either in humans or animals are subject to less stringent controls than those utilizing electrical stimulation of the muscle through chronically implanted electrodes. On the other hand, whole-animal oxygen consumption and maximal endurance are more relevant when measured in the protocols utilizing exercise for endurance training. For the purposes of this review, however, we shall often group exercise and electrical stimulation protocols together, because the changes induced in the muscles are similar. Brief comments on the protocols are included for clarity.

Among the morphological and ultrastructural changes that take place within the muscle fibers in response to either exercise training or electrical stimulation is an increase of the volume fraction of mitochondria. This change was first suggested (89) to be simply due to diminution of muscle fiber area at a constant mitochondrial volume. It has since been reported, however, that electrical stimulation results in an increase in mitochondrial content and weight per muscle fiber (26). Mitochondrial volume fraction increased markedly after about $1\frac{1}{2}$ weeks of continuous electrical stimulation (10 Hz) of the tibialis anterior of the hind limb as compared to the unstimulated contralateral control leg. The rise in the volume fraction of mitochondria remained elevated for the next 2–7 wk of stimulation. Thereafter, and despite continued stimulation, the volume fraction of mitochondria declined, although to levels still above control values.

The maximal velocities (V_{max}) of various enzymes have been used typically to evaluate changes of oxidative capacity in tissue. An increase in V_{max} is consistent with an elevation in enzyme content (given that turnover number is unchanged). This type of change should therefore be correlated with the observed increase in the volume fraction of mitochondria described above.

Pette and coworkers (90) and other investigators (for review, see refs. 88, 96) have established that long-term electrical stimulation stimulates an increase of maximal velocity of a number of mitochondrial enzymes in skeletal muscle. In a recent and perhaps the most comprehensive investigation of the adaptive response of enzyme content to repetitive electrical stimuli, Henriksson and coworkers (41) examined the time dependence of changes in V_{max} of several enzymes within six metabolic pathways of both the cytoplasmic and mitochondrial compartments. These authors reported that maximal activities (expressed per unit dry weight of muscle) of 12 mitochondrial enzymes measured in the tibialis anterior muscle (mixed fiber type) were normally about 30% to 84% less than those determined in the soleus (primarily oxidative fibers). Electrical stimulation (10 Hz, 24 h/day) of these muscles resulted in levels of V_{max} of the mitochondrial enzymes that were 3- to 14-fold greater than their nonstimulated counterpart. Values were markedly greater than those in the unstimulated soleus muscle. Substantial enhancement of enzyme activity was observed as early as 2 wk after initiation of the stimulus, but values reached peak levels after 3–5 wk of stimulation. With continued stimulation, some of the mitochondrial enzyme activities declined from their peak values. Nonetheless, these values remained greater than those in unstimulated control muscle. For example, after 3 wk of stimulation, V_{max} of citrate synthase and succinate dehydrogenase increased by six- and eightfold (Table 47.2), respectively, but thereafter decreased to about 50% of their respective peak levels. This biphasic response is similar to that described above regarding the time-dependent changes in the volume fraction of mitochondria.

In contrast to findings based on mitochondrial enzymes, key enzymes in the pathways of glycolysis, glycogenolysis, and glycogenesis decline in response to electrical stimulation (41). The maximal activities of creatine kinase, adenylate kinase, and adenylate acid deaminase were also reported to decline. One notable exception is hexokinase. Hexokinase activity increased linearly during the first two weeks of stimulation and then declined in a manner similar to that described above for the mitochondrial enzymes.

It is well recognized that endurance training enhances an individual's capacity to perform aerobic work (see,

TABLE 47.2. *Effects of Long-term Electrical Stimulation on Maximal Activities of Skeletal Muscle Enzymes*

Enzyme (Mitochondrial)	Control Leg	Stimulated Leg	% of Control
Succinate dehydrogenase	1.08	8.3	770
Citrate synthase	2.8	17	607
Malate dehydrogenase	7.7	34	440
3-OH-butyrate dehydrogenase	2.04	51	735
Carnitine acetyltransferase	0.39	2.3	590
Glutamate dehydrogenase (cytosolic)	0.26	1.59	611
Lactate dehydrogenase	442	218	49
Phophofructokinase	31	18	58
Fructose-2,6-bisphosphatase	0.52	0.23	44

The data (means) were taken from Henriksson et al. (41). Enzyme activities are expressed as mole of product formed per kg protein per h. The values were obtained from the tibialis anterior muscle from rabbits after 3 wk of continuous electrical stimulation.

for example, ref. 97), which is, in part, related to the adaptive change brought about in the levels of oxidative enzymes in skeletal muscle cells. Holloszy (47) showed initially that the activities (expressed per gram wet weight) of NADH cytochrome c reductase, succinate dehydrogenase, and cytochrome c oxidase were increased by about twofold in hind limbs of rats subjected to a program of strenuous treadmill running (animals were progressively exposed to running at 31 m/min at an 8-degree grade for 120 min/day, 5 days/wk for a 12 wk period). The concentration of cytochrome c was also increased by twofold, indicating an effective change in content of cytochrome c oxidase. Total mitochondrial protein, however, was increased by only 60%. These findings stimulated a series of investigations by Holloszy and colleagues (for review, see refs. 48, 49). Subsequently, it was reported that some enzymes of the Krebs cycle (50) and of the pathway of long-chain fatty acid transfer into the mitochondrion and β-oxidation (81) also increased by twofold in response to a similar exercise regime. For the Krebs cycle enzymes, the maximal activities of citrate synthase, NAD^+-isocitrate dehydrogenase were increased by a factor of two in muscle homogenates, whereas those of glutamate dehydrogenase, mitochondrial malate dehydrogenase, and α-ketoglutarate dehydrogenase were less affected by endurance training (35%, 50%, and 50%, respectively). It might be expected that the level of α-ketoglutarate dehydrogenase and to some extent that of mitochondrial malate dehydrogenase would change in proportion with the activities of citrate synthase and NAD^+-isocitrate dehydrogenase, since these enzymes are generally considered as flux-determining steps in the Krebs cycle (39). Evaluation of the pathway of β-oxidation revealed that the increase in enzyme activities (palmityl CoA synthetase, carnitine palmytyltransferase, and palmityl CoA dehydrogenase) was coupled to a twofold improvement in the oxidation of the long-chain fatty acids, palmitate, oleate, and linoleate. The content of the mitochondrial F_1ATPase was also shown to increase in proportion to the constituents of the respiratory chain, but the maximal activities of the mitochondrial creatine phosphokinase, cytoplasmic creatine phosphokinase, and adenylate kinase were not similarly affected by exercise (84). Since the activities of the creatine phosphokinases did not change in response to endurance training and not all of the activities of the Krebs cycle enzymes changed equally, these authors suggested that the composition of the mitochondria of trained animals may differ from those of sedentary ones.

Investigations in which biopsies of muscles were obtained from endurance-trained athletes and from more sedentary individuals or from athletes participating in nonendurance events confirmed the findings from animal models (for review, see ref. 98). For example, Gollnick and coworkers (36) reported that succinate dehydrogenase activity was greater by about 49% in the vastus lateralis muscle of endurance-trained men (runners) than in the same muscle of untrained individuals of comparable ages. Moreover, enzyme activity was greatest in muscles actively engaged in the endurance task. By contrast, phosphofructokinase activity was similar in both groups of men, although glycogen content was higher in trained muscle.

In a more recent and comprehensive investigation, Davies and coworkers (21) addressed the issue of mitochondrial composition and whether it may differ in exercised trained animals relative to sedentary ones. Rats were exercised by treadmill running using a protocol similar to that used in the investigations of Holloszy (26.8 m/min at 8.5-degree grade and progressively increased training time to 120 min/day for a total of 10 wk). Exercise training resulted in an increase in whole-body maximal oxygen consumption (VO_{2max}) from 76.6 to 87.7 ml/kg/min, a change of 14%, and an enhanced endurance capacity, a change of 408%. The mitochondrial concentrations of iron-sulfur clusters, cytochromes c ($+c_1$), b, a, and flavoprotein in hind limb muscles increased in accordance with the twofold rise in oxidative capacity. These constituents of the respiratory chain maintained constant proportion to one another. Table 47.3 shows that the activities (expressed as mmol of acceptor reduced/g muscle wet wt/min) of succinate dehydrogenase, NADH dehydrogenase, and choline dehydrogenase were increased by 209%, 171%, and

TABLE 47.3 *Effect of Endurance-Type Exercise Training on Maximal Activities of Mitochondrial Enzymes and Cytochrome Content in Skeletal Muscle*

Enzyme	Normal	Trained	% Increase
Succinate dehydrogenase	11.9	24.9	209
NADH dehydrogenase	71.8	123	171
Choline dehydrogenase	0.31	0.83	268
Cytochrome $c + c_1$	13.5	28.7	211
Cytochrome a	6.4	13.3	208
Cytochrome $c + c_1/a$	2.1	2.1	0

The values (means for 9 rats) were selected from Davies et al. (21). Enzyme activities are expressed as mmol acceptor reduced/min/g wet wt and cytochrome content is expressed as nmol/g wet wt.

268%, respectively. These findings indicate that the specific activities of at least most mitochondrial enzymes is not affected by exercise training. Moreover, the protein:(lipid + protein) ratios of skeletal muscle mitochondria were not significantly altered by training, demonstrating that mitochondrial volumes were not modified. Thus, it was concluded from these studies that mitochondrial content doubled in response to the exercise regime.

It was suggested by Holloszy and colleagues that the changes in mitochondrial enzyme content brought about by exercise were regulated in a coordinated manner and maintained in constant proportion each to the other (81, 84). The coordinated regulation of enzyme content is likely common to both types of conditioning discussed above. In the investigation of Henriksson and coworkers (41), not all of the measured mitochondrial enzymes followed the biphasic pattern of change. In some cases, the mitochondrial enzyme activity increased and remained at that elevated level throughout the rest of the experimental period. Consequently these authors concluded that electrical stimulation may have resulted in mitochondria that were different in composition than those from control muscles. The enzymes of carbohydrate metabolism, however, decreased in parallel, suggesting they were under coordinated control. Chi et al. (16) considered that more than one signal is operative for regulating the changes in protein synthesis and/or degradation since the activity of hexokinase changes in contrast to its companions within the same metabolic pathway. The work of Davies (21), however, strongly supports the notion that the biosynthesis of mitochondrial proteins is subject to integrated regulatory control.

The question then arises: Is the increase in V_{max} of mitochondrial proteins in response to either exercise training or electrical stimulation due primarily to a slowing of the degradation rate or to an enhancement of biosynthesis? In an early study, Booth and Holloszy (8) reasoned that the turnover of a protein can be estimated by following the time course of a change in enzyme activity or protein concentration after withdrawal of the stimulus (such as exercise) that induced the change. These authors showed that the half-times for the increase in enzyme content and the subsequent return to control levels are both about 6–8 days. This suggested that an increased synthetic rate was responsible for greater enzyme activity in trained muscles.

Enhancement of Protein Synthesis

The techniques of molecular biology have been used to examine the biochemical events that result in enhanced oxidative capacity of skeletal muscle in response to increased functional demand (for an additional perspective, see ref. 9). Williams and coworkers provided evidence that increases in functional demand are accompanied by increased gene expression. The concentrations of mitochondrial DNA, mitochondrial ribosomal RNA, and cytochrome b mRNA (a mitochondrial gene product) were shown to be present in muscle in proportion to its oxidative capacity, i.e., cardiac muscle > type I red, oxidative skeletal muscle > type II white, glycolytic skeletal muscle (112). When type II skeletal muscles were subjected to chronic electrical stimulation, the above parameters increased in proportion to the change in oxidative capacity, indicating that expression of mitochondrial genes in skeletal muscle is proportional to their copy number. This proportionality between mitochondrial DNA and oxidative capacity is observed for tissue during normal development and cell differentiation, as well as when oxidative capacity is altered by external stresses. Williams (112) suggested that skeletal muscle cells retain the ability to alter mitochondrial DNA synthesis even though these cells no longer have the capacity to replicate nuclear DNA. Moreover, skeletal muscle cells do not express a greater level of mitochondrial mRNA than mitochondrial rRNA in those cells having a higher capacity for oxidative energy production, in contrast to other types of cells. These findings support the hypothesis that stimulation of oxidative capacity in tissue requires an enhanced rate of mitochondrial DNA synthesis.

Mitochondrial proteins are expressed from both mitochondrial DNA and nuclear DNA. It therefore seems likely that biosynthesis of mitochondrial proteins in response to conditioning be subject to coordinated expression of the applicable genes. Moreover, conditioning, at least by electrical stimulation, evokes differential responses from cytoplasmic and mitochondrial proteins, i.e., the activity of the former decreases whereas the latter increases. Williams and colleagues (112, 114, 115) pointed out that protein synthesis may be regulated by one or more of the various stages of the process. These could include the efficiency of transcription, the rate of RNA processing, stability of the mRNA, the efficiency of translation, the rate of posttranslational modification or transport, and the stability of the protein product.

Cytochrome c oxidase, the terminal component of the electron transport chain, consists of subunits encoded by both nuclear (subunit VIC) and mitochondrial genomes (subunit III). In order to determine whether the two genomic systems were regulated in a coordinated manner, Hood and colleagues (54) examined the time course of changes in the tissue levels of mRNAs encoding these subunits in response to chronic electrical stimulation of the tibialis anterior muscle of the rat. It was shown that both mRNAs increased in parallel over the stimulation period, supporting a role for coordinated regulation. The consequent rise in activity of cytochrome c oxidase corresponded to that in the levels of both mRNAs but only during the initial 14 days of stimulation. Thereafter, the change in enzyme activity exceeded that in the levels of the two mRNA. These authors concluded that the increase in cytochrome oxidase content resulted from both a rise in specific mRNAs and from changes at the translational level.

Williams and coworkers (116) found that 21 days of electrical stimulation produced a fivefold increase of cytochrome b mRNA and a marked decrease in the level of aldolase mRNA. After 5 days of stimulation, however, the decrease in the level of aldolase mRNA was already evident, whereas the concentration of cytochrome b mRNA had not yet increased. Since the changes in V_{max} of citrate synthase and aldolase paralleled the levels of cytochrome b mRNA and aldolase mRNA, the authors concluded that there was no change in the efficiency of translation. Moreover, the similarity between the levels of mitochondrial DNA and RNA transcripts of mitochondrial genes indicated that the increased level of mitochondrial DNA was responsible for the increased concentration of mitochondrial transcripts in response to electrical stimulation. For two nuclear genes encoding mitochondrial proteins (bF_1-ATPase and subunit VIC of cytochrome oxidase), increased levels of mRNA were found in skeletal muscles from electrically stimulated hind limbs (114). The increased levels of mRNA were not equivalent to the increase in mitochondrial volume fraction, maximal activity of mitochondrial marker enzymes, or mRNA transcribed from mitochondrial genes. This indicates that regulation of pretranslational events cannot account completely for the enhancement of mitochondrial oxidative capacity. Other aspects of protein synthesis, such as those enumerated above, must also be regulated in mitochondrial biogenesis.

Daily administration of chloramphenicol, an inhibitor of translation of mitochondrial ribosomes, blocked the increase in activity of cytochrome oxidase induced by chronic electrical stimulation (115). On the other hand, chloramphenicol had no effect on activity-induced changes of enzyme activity for proteins derived from nuclear genes, citrate synthase, and aldolase. Neither did it affect the rise in mRNA transcribed from nuclear genes, in this case for F_1-ATPase and myoglobin, or from mitochondrial genes encoding rRNA for cytochrome *b*. These data suggested that adaptation to enhanced contractile activity does not utilize mitochondrial gene products as pretranslational regulatory effectors of nuclear or mitochondrial gene expression.

It is evident from the discussion above that further work in the area of mitochondrial biogenesis and adaptation is needed. However, the investigations of Williams and coworkers and other investigators (54, 82, 102) show that increases in protein synthesis are responsible for enhancement of oxidative capacity of skeletal muscle, in agreement with the earlier conclusions of Booth and Holloszy (8).

Possible Second Messengers

There are reports that the adaptive response of mitochondrial biogenesis to exercise or electrical stimulation may involve β-adrenergic receptors and/or adenosine 3',5'-cyclic monophosphate (cAMP). The density of β-adrenergic receptors has been correlated with the level of oxidative capacity of skeletal muscle (113). Administration of β-adrenergic antagonists inhibits the exercise-induced increase of mitochondrial enzyme activity (63). By elevating intracellular levels of cAMP in skeletal myotubes, it has been possible to stimulate expression of some mitochondrial proteins (71). Chronic electrical stimulation was also shown to increase the density of β-adrenergic receptors and to increase cAMP levels in skeletal muscle (70). The time course of this increase was similar to that for induction of mitochondrial enzymes. Treatment of animals with propranolol at doses sufficient to lower heart rate and decrease cAMP in skeletal muscle was without effect on the activity-induced increases of cAMP, β-adrenergic receptors, and expression of mRNA products of proteins encoding mitochondrial proteins. The latter results suggested that stimulation of β-adrenergic receptor activity was not a necessary prerequisite for mitochondrial protein synthesis. At this time, the contribution of the β-adrenergic system to the adaptation of skeletal muscle mitochondrial protein synthesis is not clear and remains controversial.

Resistance to Fatigue During Exertion

Endurance exercise training and chronic electrical stimulation improve resistance to fatigue. For example, the time to exhaustion is longer for exercise-trained rats than for sedentary controls (21). Fatigue resistance

results, at least in part, from the increased capacity for oxidative phosphorylation. Magnetic resonance (^{31}P-NMR) studies of electrically conditioned latissimus dorsi muscle in dogs showed that phosphocreatine concentration in conditioned and contralateral control muscles was decreaed in proportion to a given level of electrical stimulus (18). The magnitude of change was, however, less for conditioned muscle than for its contralateral control. When the level of exertion was progressively increased, the tension-time index reached a plateau as the ratio of inorganic phosphate to phosphocreatine (Pi/Pcr) continued to rise in control muscles. On the other hand, conditioned muscles were able to reach values of tension-time index above that for controls with less change of Pi/Pcr at each step of the work regimen. In similar studies, it was reported that the ratio of contractile tension developed to oxygen consumed during stimulation was markedly greater for electrically conditioned muscles during moderate (300 msec on) and intense (800 msec on) protocols, suggesting improved efficiency of the conditioned muscle (1). Thus, the conditioned muscle is capable of performing work at less of an energy cost per unit of completed work (25).

EFFECTS OF CONDITIONING ON THE CAPACITY FOR OXYGEN DELIVERY TO LOCAL REGIONS OF THE MUSCLE

It is well known that exercise training enhances cardiovascular function. From the discussion above, the increase in mitochondrial enzyme content within skeletal muscle brought about by conditioning suggests that these changes must be coupled to enhance oxygen delivery to the mitochondrion in order to optimize the synthesis of ATP. At steady state, the consumption of oxygen must equal its delivery. Thus, not only are improvements in oxygen delivery necessary at the systems level of organization but also within the local domain, specifically within skeletal muscle tissue and the working myocytes. Accordingly, the present discussion will focus on changes in capillarity and cellular myoglobin content, two important determinants of the oxygen delivery capacity. For more extensive reviews on these topics, we suggest reading papers such as those of Hudlicka (55, 56), Wittenberg (123), and Wittenberg and Wittenberg (125, 126).

Changes in Muscle Capillarity

The driving force for oxygen transport from the capillaries to the mitochondria is the difference in oxygen pressure between these points of supply and utilization. The capillary surface area and the intercapillary distances are, therefore, an inherently important feature of oxygen delivery to the mitochondria. The "functional" surface area for exchange is delimited by the total capillary length per unit of tissue volume and the aggregate red cell length, including plasma gaps of less than about 5 μm (53). The capillaries may serve as a potential "bottleneck" for diffusive flux of oxygen (53), most especially in tissue such as skeletal muscle in which the demand for oxygen can rise to very high levels. At rest, the number of open capillaries, i.e., those containing red cells, is low in skeletal muscle but increases markedly during the transition to the working condition (51), a process called *recruitment*. As oxygen consumption is increased above the basal state, the increase in the number of capillaries containing red cells enhances the oxygen transport capacity of the tissue by both increasing the cross-sectional area of the effective capillary bed and decreasing the velocity of red cell transit through capillaries at a given volume of flow through the tissue (53). Slowing of red cell velocity allows more time for unloading oxygen from oxyhemoglobin, affording a greater oxygen extraction by the tissue. Clearly, changes in tissue capillarity can be an important factor affecting tissue oxidative capacity.

Methods used for determination of tissue capillarity have been subject to criticism (for review, see ref. 98). As can be concluded from the discussion above, it may not be sufficient to simply count the number of capillaries supplying tissues; rather, it may be more appropriate to measure the diameter and length of capillaries in order to estimate the surface area available for oxygen exchange. These types of measurements are difficult due to the considerable variation in capillary structure among species and skeletal muscle types. Moreover, the anatomical structures that these microvessels may assume, i.e., a straight length of vessel or a more convoluted, tortuous one winding about the muscle fiber, may further complicate these measurements. Nonetheless, counting the number of capillaries per unit of cross-sectional area or the capillary-to-fiber ratio have provided an estimate of muscle capillarity.

Biopsies obtained from human skeletal muscle have shown that endurance-type exercise training results in an increase in capillary density with little or no change in muscle fiber diameter. Brodal and coworkers (11) biopsied the quadriceps femoris muscle of young men (18–34 years old) who were either sedentary or who had participated in regular physical training (4–7 days per week) for at least 5 years. Using electron microscopy to identify capillaries within fixed sections of muscle, these authors found that capillary diameters (3–5 μm) and fiber diameters (49 μm) were similar in both groups of subjects. As compared to the sedentary individuals, however, the capillary/fiber ratio was 41% greater in the trained men. Moreover, the average number of capillaries around each fiber was 33% higher in trained than in untrained subjects. The latter result indicated

that, in muscle of the endurance-trained person, each capillary was shared by fewer fibers. The quadriceps femoris is composed of a mix of fiber types, including both slow oxidative (red) fibers and fast glycolytic (white) ones. The number of capillaries in a given fiber type was consistent with the content of mitochondria, independent of the individual's level of training. Since fiber size was unchanged in the trained subjects, the increased capillarity indicated a "true" increase in the capillary density.

Conditioning by electrical stimulation results in a decrease of muscle fiber size during the transformation from the fast to slow type (16, 26, 89). Continuous electrical stimulation seems to have a more pronounced effect on fiber size than intermittent stimulation, consistent with the changes in oxidative capacity (89). Noticeable declines in weight of the whole tibialis anterior muscle were observed after only 9 days of continuous stimulation, and after 12 wk of stimulation the weight of this muscle had declined to 44% of its contralateral unstimulated control (26). Chi and coworkers (16) reported that continuous stimulation for a period of 5 wk decreased fiber size from 0.58 ± 0.03 to 0.31 ± 0.06 μg/mm, a decrease of nearly 50%. Hudlicka and Tyler (60) reported that fiber size may undergo a transient increase during the early phase of transformation due to edema that seems to disappear with continued stimulation. These findings were in agreement with those of Chi et al. (16), who also reported an increase in fiber size in the first few weeks of stimulation followed by a significant decrease. The magnitude and the direction of change in fiber size will clearly affect the determination of capillary density.

When skeletal muscles of rabbit were stimulated either intermittently at high frequency (40 Hz, 8 h/day) or continuously (10 Hz), capillary density and the capillary/fiber ratio increased markedly (60). With continuous stimulation, increased capillarity was observed at an earlier time than with intermittent stimulation. In the former case, marked enhancement of capillary density was found after 4 days, whereas in the latter significant changes were not obtained until the tissue had been stimulated intermittently for 7 days. In the extensor digitorum longus muscle, the capillary/fiber ratio was 1.25 ± 0.02 and increased by 49% after 14 days, and was increased further after 28 days of stimulation (total change represented a 66% increase). Capillary density doubled in this muscle after 28 days of stimulation. These changes were preceded by significant increases in succinate dehydrogenase activity. A shift toward a greater number of fast oxidative fibers was observed as early as 4 days of stimulation. In an earlier study, Hudlicka and coworkers (57) reported that the increase in capillary density in the extensor digitorum muscle seems to be due in part to the growth of capillaries supplying the fast glycolytic fibers. Growth of capillaries in the tibialis anterior muscle also seemed to be associated with those supplying fast glycolytic fibers.

It should be noted that an early study on the effects of endurance-type exercise training on capillary density in the quadriceps femoris muscle of humans showed that the number of capillaries per mm^2 of muscle was not significantly different between trained and untrained men: 640 ± 54 vs. 600 ± 24, respectively (42). The trained subjects had been engaged in athletic competition for at least 5 years in events such as marathon or cross-country running or bicycling, and their maximal rates of oxygen uptake were about 42% greater than those of the control subjects. The size of the muscle fibers of the trained group, however, were found to be larger by about 30% than those of the untrained individuals. As a result, the number of muscle cells per unit area of muscle was less but the capillary/fiber ratio was increased by about 38% in the trained group. When the diffusion distance (average half-distance between two capillaries) was measured, no significant differences were evident between the two groups of men. These measurements utilized techniques in which the endothelial basement membrane was stained, thereby showing all of the capillaries (those in both the open or closed state). In part, these results seem inconsistent with those from later work and should be regarded with caution.

A thorough description of the factors responsible for increased capillary growth is beyond the scope of the present discussion (the reader is directed to ref. 56), and only a brief commentary regarding the contribution of metabolites and mechanical effects associated with enhanced flow is included here. Hudlicka and coworkers (61) reported that infusion of rabbits with adenosine or a methylxanthine derivative for 7 days resulted in a significant increase in capillarity in both cardiac and skeletal muscle. It has been reported that adenosine is released from muscle during conditions that alter oxygen consumption and the energy state of the tissue (68, 69). Adenosine at a concentration of 5 μM increased proliferation of cultured coronary microvascular endothelial cells, possibly via a G protein and subsequent stimulation of intracellular levels of adenosine 3′,5′-cyclic monophosphate (80). Metabolic by-products of adenosine, inosine, and hypoxanthine were without effect on cell proliferation. By contrast, Burton and Barclay (13) were unable to show any effect of adenosine (final concentration = 100 μM) on growth of cultured endothelial cells from dog hind limb. In the latter study, however, brief exposure (one h/day for a period of 7 days) to low oxygen pressures (35 torr vs. 65 torr for normoxic controls) stimulated a proliferation (22% above initial levels) of these cells, whereas addition of 5mM lactic acid decreased cell numbers. Addition of lactic acid did not alter osmolarity of the culture media but significantly increased the hydrogen ion concentration. Increasing the partial pressure of CO_2 that the cells were exposed to or addition of 5mM sodium lactate to the culture media were without significant effect on cell proliferation.

Hudlicka and coworkers (58, 61, 62) have concluded that vasodilation resulting from adenosine infusion and not adenosine per se was the stimulus for subsequent capillary growth. A similar conclusion was made from more recent work (59). In the latter investigation, these workers limited the blood supply to the tibialis anterior muscle of rats (the contralateral one served as a control) and, in some cases, these muscles were electrically stimulated. The capillary/fiber ratio was unaffected by flow limitation or flow limitation coupled with electrical stimulation, despite large increases in oxidative capacity resulting from both conditions. These authors suggested that endothelial cell damage resulting from changes in shear stress on capillary walls may provide the necessary stimulus for induction of endothelial proliferation.

Changes in Myoglobin Content

Myoglobin is a heme-containing protein that is found in cardiac, skeletal, and smooth muscles as well as in many plant and bacterial cell types. In general, the concentration of myoglobin in skeletal muscle is proportional to that of cytochrome oxidase (72). Myoglobin concentration appears to be, in part, determined by the distance through which oxygen must diffuse in order for the latter to reach the mitochondria (123). This is borne out by the finding that red skeletal myocytes contain about twice the concentration of myoglobin and have a twofold greater diffusion radius for oxygen than do cardiac myocytes (72). That myoglobin aids the diffusion of oxygen in vitro (101, 123) suggests that it serves the same purpose in muscle cells. By rendering myoglobin unable to bind oxygen with agents such as hydrogen peroxide (19) or sodium nitrite (10), contractile efforts of both skeletal and cardiac muscle, respectively, were shown to be markedly diminished. Although myoglobin may also act as a temporary reservoir of intracellular oxygen (3), a primary function of myoglobin is to facilitate oxygen diffusion and therefore provide uniform oxygenation of those cells in which it is expressed.

Whipple (111) was likely the first investigator to suggest that exercise conditioning would enhance the concentration of myoglobin in skeletal muscle. This hypothesis was supported by the observations that muscles from active, sporting dogs contained greater levels of myoglobin than those from sedentary animals. These types of observations and others made by Lawrie (72) prompted Pattengale and Holloszy (87) to examine the effects of regularly performed treadmill running of progressively increasing intensity on myoglobin concentration in skeletal muscles of rats. The exercise regime resulted in about an 80% increase in the concentration of myoglobin in both the quadriceps and hamstring muscle groups as compared to sedentary controls, whereas no difference was observed for the rectus abdominus muscle. The latter muscle would not be expected to be utilized during exercise. These results have since been confirmed by other investigators using animal models (38, 43, 99), although studies in humans have not been conclusive (108). Myoglobin concentration has also been shown to be increased in muscles receiving chronic electrical stimulation (67). Electrical impulses (10 Hz, 10 h daily) from implanted electrodes delivered for 28 days resulted in about a 150% increase in myoglobin content of the tibialis anterior muscle of rats. The latter change paralleled that of citrate synthase activity. Underwood and Williams (109) reported that myoglobin mRNA increased by about 15-fold following 21 days of electrical stimulation, suggesting that pre-translational mechanisms were important for myoglobin gene expression.

Honig and coworkers (52) have pointed out that the concentration of oxymyoglobin is about 100 times that of free intracellular oxygen during heavy exercise. This difference occurs despite the fact that oxygen diffusivity is about 20-fold greater than that of myoblobin (123). Consequently, the intracellular flux of oxygen is mediated primarily by myoglobin. By increasing the concentration of myoglobin within the muscle cells, carrier-facilitated oxygen flux would be enhanced (31), permitting greater oxygen conductance (52).

EVOLUTIONARY DESIGN OF MUSCLES FOR DIFFERENT WORK LOADS

The capacity for muscles to adapt to altered work loads is apparently limited by genetics and/or physiology in the muscles of individual animals and men. These limits are much less important in the evolutionary design of muscles for different functions, since time is available for alteration of the fiber type and capillary geometry if this is of significant advantage to performance. Mathieu-Costello (76), Mathieu-Costello et al. (77), and Suarez et al. (104) have carried out comprehensive studies of the relationships of: (i) capillary number and surface area, (ii) muscle fiber number and area, and (iii) mitochondrial area and volume in muscles of very different oxidative capacity. The highest known rates of oxidative metabolism in any tissue occur in the flight muscle of bats and hummingbirds (>2.1 ml O_2 per minute per g), and these muscles have been compared with muscles in which the maximal metabolic rates are much lower, such as the bat hind limb and rat soleus. Com-

parisons among these muscles show a close correlation between the content of mitochondria (respiratory capacity) and capillary length per unit of fiber volume. Comparison of hummingbird flight muscle, bat flight muscle, bat hind limb, and rat soleus muscle, for example, gave mitochondrial volumes of 35%, 35%, 16.5%, and 6.1%, respectively, while the relative capillary length per unit of fiber volume was 5.4, 5.4, 2.2 and 1.0, respectively. The fiber cross-sectional areas were 200, 318, 447, and 2,200 μm^2, respectively, consistent with the fiber diameter decreasing with increasing maximal respiratory capacity (see refs. 76, 105). The capillary surface area is directly correlated with the mitochondrial volume of the tissue, and the function of the decreasing fiber diameter is to allow a closer packing of the capillaries, not necessarily decreased diffusion distances. Thus, the capacity to deliver oxygen to tissuue is directly dependent on the capillary surface area per tissue volume, while oxidative capacity is directly dependent on the content of mitochondrial respiratory enzymes. Optimal muscle design requires a match between oxygen supply and demand. The number of capillaries per fiber in muscle is nearly constant, and decrease in fiber diameter permits both increase in capillary surface area and increase in respiratory enzyme content (mitochondrial volume) per unit volume of tissue. It should be emphasized that mitochondrial volume is only an approximate measure of the content of respiratory enzymes, since the amount of respiratory enzyme per unit of mitochondrial protein can vary over two- to threefold. The area of mitochondrial inner membrane is a better morphological indicator of respiratory enzyme content, and, for example, this area is significantly higher per unit volume for mitochondria in hummingbird flight muscle than in rat soleus muscle. A corresponding increase in the capillary surface area per unit volume of mitochondria would be required to obtain an appropriate "match" of oxygen supply and demand. Muscles with very high oxidative capacity also have high myoglobin contents, and this may account for the observation that the required capillary surface area per unit mitochondrial volume does not decrease significantly with decreasing fiber diameter. Evidence has been obtained that the primary decrease in oxygen pressure occurs near the capillary wall and, in muscle, myoglobin-facilitated oxygen diffusion results in rather shallow oxygen gradients within the muscle fiber (33, 35). The adaptive changes in fiber diameter and capillary surface area/fiber volume among muscle types and species are very large, whereas those in individual muscles in response to increased work are small. The adaptive response of developed muscle may be limited to only small alterations in fiber diameter and capillary density by physical and genetic restraints.

EFFECTS OF DECONDITIONING ON THE ACTIVITY OF OXIDATIVE ENZYMES IN MUSCLE

Cessation of conditioning by exercise or an electrical stimulus quickly results in a reduction of oxidative capacity to levels found prior to application of the conditioning stimulus. The activities of mitochondrial oxidative enzymes were shown to decline exponentially after 6 wk of continuous electrical stimulation of the tibialis anterior muscle was discontinued (12). The half-life for the measured enzymes was about 4 to 5 wk and was well correlated to changes in the volume fraction of mitochondria. Susceptibility to fatigue also returned and followed the loss of oxidative capacity. These changes marked the transition of muscle fiber type from conditioned slow oxidative fibers to fast glycolytic ones.

These aspects of deconditioning may be applicable to changes observed during senescence. It is well known that maximal oxygen consumption and the ability to perform aerobic work decreases in sedentary individuals undergoing the normal process of aging. This is in part due to lower levels of oxidative enzymes in skeletal muscle. It has been shown that activities (expressed as per gram of muscle wet weight) of enzymes within the respiratory chain, the Krebs cycle, β-oxidation (40, 94), and glycolysis (5, 94) were decreased in skeletal muscles of older animals as compared to younger ones. These findings are consistent with a similar decline in substrate oxidation rates (94). Since cytochrome oxidase activity fell in parallel with the decline in its heme components and without a significant change in the turnover number, the reduced oxidative capacity was likely due to a loss of mitochondrial content per gram of muscle. Loss of respiratory activity occurred at a time when there was marked decline in muscle mass and muscle protein. By contrast, cardiac muscle, which is continually active, undergoes very little age-related decrement of oxidative capacity (103). Peripheral skeletal muscles of the hind limb (for example, rat gastrocnemius muscle was analyzed in the studies of Rumsey and coworkers [94]) are active intermittently when the animal chooses to move about within the confines of its environment. Voluntary motions of caged laboratory animals decrease as the animals grow older (37, 92). It is likely, therefore, that the age-related loss of oxidative capacity is at least in part due to lack of voluntary locomotion (79). Moreover, these types of changes are consistent with those found with more extreme forms of disuse, such as limb immobilization. Short-term limb immobilization results in skeletal muscle atrophy accompanied by decreased capacity to oxidize carbohydrate and lipid substrates in the affected limb (48). The levels of proteins within the metabolic pathways responsible for energy production seemed to have adapted to the lower requirements of skeletal muscle for ATP synthesis. The latter conclusion is strengthened by the finding that moderate-intensity treadmill walking performed regularly from adulthood to old age improves oxidative capacity of skeletal muscle during senescence (94) in a similar manner to that described above for younger animals. For example, animals that exercised regularly (5 days per wk, 20 min/day at 18 m/min up an 8-degree incline) maintained significantly greater levels of cytochrome oxidase activity and cytochrome content than age-matched sedentary controls. Senescence was associated with some loss of oxidative capacity in the exercised animals as compared to trained younger ones. Oxidative capacity in trained animals measured near the end of their natural lifespan was, however, equivalent to that obtained in sedentary young adult controls. It is clear that exercise enhances mitochondrial oxidations at any age. If, however, voluntary motions are diminished during nonexercise periods, respiratory activity is fated to decline in a manner similar to that seen in more sedentary animals.

EFFECTS OF CHRONIC DIMINISHED SUBSTRATE DELIVERY ON ENERGY METABOLISM IN MUSCLE

Since energy production in most cells is primarily a function of mitochondrial oxidative phosphorylation, continuous delivery of oxygen is essential to the ability to maintain an optimal phosphorylation-state ratio for the performance of contractile activity in skeletal muscle. The ability of mitochondria to synthesize ATP in situ is dependent on oxygen pressures at the cell surface of about 20 to 30 torr (95, 117–120). Below these levels, synthesis of ATP may occur at a lower than normal [ATP]/[ADP][Pi], potentially compromising cellular function. Increasing the number of mitochondrial proteins provides a greater capacity to synthesize ATP at a given oxygen pressure.

Effect of Restricted Blood Flow in Muscle

Several reports indicate that maximal activities of key enzymes in the pathways of β-oxidation, the Krebs cycle, and the electron transport chain are increased in skeletal muscle measured in the legs of patients with peripheral vascular occlusive disease as compared to unaffected individuals (14, 46, 75). For example, the activity of cytochrome oxidase (expressed as μmol of O_2/min/g of protein) was 110 ± 21 in a patient group and 55 ± 6 in normal subjects (15). When oxygen pressures were measured using an oxygen microelectrode inserted into the head of the gastrocnemius muscle of normal and affected individuals at rest, there was no significant difference between the two groups (15). Tissue oxygen pressure is dependent on the level of blood flow, which in these cases will vary according to the degree of occlusion, and the prevailing level of metabolic activity. At rest, blood flow to peripheral skeletal muscle is normally very low due to a low metabolic rate. During exercise, ATP hydrolysis is markedly enhanced and blood flow to muscle rises severalfold in normal persons. As a result of the increased metabolic activity, tissue oxygen pressures decrease. Bylund-Fellenius and coworkers (15) showed that the exercise-induced decrease in skeletal muscle oxygen pressures was, however, more pronounced in the legs of patients with vascular occlusive disease than in those of the nondiseased subjects. Measurements of metabolites showed that the phosphocreatine-to-creatine ratio (Pcr/Cr) was lower in patients than in normal subjects. At rest, the Pcr/Cr ratio was about 1.18 in patients and 1.54 in normal subjects and decreased after exercise to 0.49 and 0.87, respectively. On the other hand, [lactate]/[pyruvate] ratios tended to be lower both at rest and after exercise in the legs of patients than in the normal subjects. It should be noted that the patients were unable to perform the exercise task as well as normal subjects.

Interestingly, in patients who had undergone surgery in order to restore flow to affected musculature, levels of oxidative enzymes in this tissue decreased as compared to values measured prior to surgery (75). Patients who participated in a moderate exercise program but did not receive reconstructive surgery were able to further increase their level of oxidative enzymes. In those patients who received both surgical reconstruction and exercise training, levels of oxidative enzymes remained unchanged after the therapeutic interventions. All three modes of therapy improved walking performance.

The results described above were confirmed and extended further using a rat model of peripheral vascular occlusive disease (27, 28). Ligation of the common iliac artery resulted in a marked reduction, 76%–93%, in blood flow to the affected hind limb during electrical stimulation. This stimulation increased blood flow by 5- to 25-fold in normally perfused tissue. Both hind limbs of the animals were subjected to intermittent electrical stimulation for 6 days. The level of stimulation was low and insufficient to alter oxidative capacity in legs with normal blood flow. As compared to the leg receiving normal blood flow, Table 47.4 shows that maximal activities of oxidative enzymes (citrate synthase and cytochrome oxidase) increased by more than 20% in the soleus and extensor digitorum longus muscles of the conditioned ligated leg (28). When resting metabolite values obtained from the ligated legs were compared to those from nonligated ones, values of [ATP]/[ADP] were decreased while those of [Pcr]/[Cr] and glycogen were increased. Immediately following a

TABLE 47.4. *Effects of Electrical Stimulation on Maximal Activities of Enzymes from Ischemic Skeletal Muscle*

Enzyme	Control Leg	Ligated Leg	% Increase
Soleus Muscle			
Phosphofructokinase	331	353	107
Citrate synthase	184	229	124
Cytochrome oxidase	70	89	127
Extensor digitorum longus muscle			
Phosphofructokinase	1262	1306	103
Citrate synthase	184	222	121
Cytochrome oxidase	55	68	124

The data (means) were taken from Elander et al. (28). Flow in the ligated leg was reduced to 76%–93% of flow in the contralateral control leg. Both legs received electrical stimulation of the sciatic nerves for 6 days (10–20 min/session, 3–4 times/day).

10 min period of stimulation, Elander and colleagues (27) found that the soleus muscle from the adapted (ligated) leg maintained a better [Pcr]/[Cr] ratio (1.7 vs. 1.87 at rest) than that from the nonadapted leg (0.99 vs. 1.33 at rest). [Lactate] and the [lactate]/[pyruvate] ratio increased in both adapted and nonadapted legs as a result of the stimulation period, but the magnitude of change was less in the muscles of the adapted legs. Glycogen was spared in the gastrocnemius muscle of the adapted legs, whereas it was diminished significantly in that from the control legs. The results above indicate that the adaptive response to vascular occlusive disease is very similar to that induced by endurance exercise training or electrical stimulation in normal subjects, and that these effects are potentially additive.

Many diseases, in particular those that affect oxygen exchange within the lungs, influence the level of oxygen delivery to skeletal muscle. From the previous discussion, however, the phenomenon of stress-induced adaptation of tissue oxidative capacity is primarily related to mechanisms that take place within the local environment of the muscle. For this reason, we have omitted in-depth discussion of changes in muscle oxidative capacity in response to conditions that influence delivery of oxygen to the entire organism. On the other hand, some mention of the metabolic changes that result from exercise training at high altitude is warranted (for more in-depth review, see refs. 44, 107, 110).

High-Altitude Induced Adaptation

Ascent to high altitude results in remarkable changes in oxygen flux to tissue. For example, in a study (106) designed to simulate an ascent to the summit of Mount

Everest (8,848 m), arterial oxygen pressure was about 30 mm Hg in six young healthy males (ages 21–31 years). At an inspired oxygen pressure of 43 mm Hg, maximal oxygen uptake was markedly decreased as compared to that obtained at sea level: 1.17 ± 0.08 liters/min vs. 3.98 ± 0.2 liters/min, respectively. The ability of an individual to perform work or exercise is compromised until appropriate acclimation has taken place. Altitude-adapted persons are able to work at higher rates following acclimatization, with less production of muscle and blood lactate. Comparison of Andean natives to either endurance-trained or highly trained power athletes showed that the Andeans demonstrated equivalent calf muscle work rates despite a lower anaerobic capacity than the power-trained athletes (78). Moreover, when calf muscle high-energy phosphates were measured during work using magnetic resonance spectroscopy, the Andeans displayed similar or reduced perturbations of these parameters as compared to the other two groups of trained individuals. Similarly, when seven young males underwent a prolonged exercise challenge at sea level, within 4 wk of arrival at an altitude of 4300 m, and after a 3 wk residence at this altitude, the ATP-to-free-ADP ratio was significantly less following the brief acclimatization period than that on arrival to high altitude (34). These results suggest that the acclimatization process invokes marked changes in the capacity of muscle to oxidize substrate. Terrados et al. (108) showed that citrate synthase activity and myoglobin content of skeletal muscle trained at hypobaric conditions was greater than that trained at normobaric conditions. Moreover, diaphragm muscle of rats endurance-trained at high altitude have a greater glucose phosphorylation potential, as measured by the activity of hexokinase, than their sea-level counterparts (6). During altitude acclimatization, however, the whole body is involved and many of the physiological processes that influence oxygen flux to tissue are likely to undergo change (see, for example, ref. 100). This makes it difficult to draw reliable conclusions concerning the biochemical basis for alterations observed in individual tissues.

SUMMARY

Cellular adaptation to increased utilization of ATP for metabolic work is an orchestrated increase in each of the factors that help increase rates of mitochondrial oxidative phosphorylation. These factors include:

(*i*) Increased supply of intramitochondrial NADH for mitochondrial oxidative phosphorylation. This is accomplished in two parts, the extent of each dependent on the initial metabolic characteristics of the cell:

1. Increased activity of the mitochondrial dehydrogenases associated with the tricarboxylic acid cycle, and β-oxidation.

2. Increased capacity for metabolism of fat and, to some extent, glycogen.

(*ii*) Increased delivery of oxygen to the mitochondria, a process involving improvements in every aspect of oxygen delivery. These include:

1. Increased tissue vascularization with resulting increase in the capacity for delivery of oxygen due to increased tissue blood volume and capillary surface area.

2. In some muscles, the fiber diameter may decrease. This decreases the diffusion distance for oxygen from the extracellular space to the mitochondria and the diffusion distances for ATP and Pi (in muscle, diffusion of ADP is facilitated by creatine).

3. Increased concentration of the intracellular oxygen carrier, myoglobin.

(*iii*) Increased content of mitochondria and respiratory enzymes. The capacity of mitochondrial oxidative phosphorylation to synthesize ATP is directly related to the content of respiratory enzymes, other parameters being the same. The increase can occur in both the volume percent of the tissue that is mitochondrial and the fraction of the mitochondrial protein that is related to the respiratory enzymes.

REFERENCES

1. Acker, M., W. A. Anderson, R. L. Hammond, F. DiMeo Jr., J. McCullum, M. Staum, M. Velchik, W. E. Brown, D. Gale, S. Salmons, and L. W. Stephenson. Oxygen consumption of chronically stimulated skeletal muscle. *J. Thorac. Cardiovasc. Surg.* 94: 702–709, 1987.

2. Acker, M., R. L. Hammond, J. D. Mannion. S. Salmons, and L. W. Stephenson. Skeletal muscle as the potential power source for a cardiovascular pump: Assessment in vivo. *Science* 236: 324–327, 1987.

3. Astrand, I., P. O. Astrand, E. H. Christensen, and R. Hedman. Myoglobin as an oxygen-store in man. *Acta Physiol. Scand.* 48: 454–460, 1960.

4. Balaban, R. S. Regulation of oxidative phosphorylation in the mammalian cell. *Am. J. Physiol.* 258 (*Cell Physiol.* 27): C377–C389, 1990.

5. Bass, A., E. Gutmann, and V. Hanzlikova. Biochemical and histochemical changes in energy supply-enzyme pattern of muscles of the rat during old age. *Gerontologia (Basel).* 21: 31–45, 1975.

6. Bigard, A. X., A. Brunet, B. Serrurier, C. Y. Guezennec, and H. Monod. Effects of endurance training at high altitude on diaphragm muscle properties. *Pflugers Arch.* 422 (3): 239–244, 1992.

7. Blomqvist, C. G., and B. Saltin. Cardiovascular adaptations to physical training. *Annu. Rev. Physiol.* 45: 169–189, 1983.

8. Booth, F. W., and J. O. Holloszy. Cytochrome *c* turnover in rat skeletal muscles. *J. Biol. Chem.* 252 (2): 416–419, 1977.

9. Booth, F. W., and D. B. Thomason. Molecular and cellular adaptation of muscle in response to exercise: Perspectives of various models. *Physiol. Rev.* 71 (2): 541–585, 1991.

10. Braunlin, E. A., G. M. Wahler, C. R. Swayze, R. V. Lucas, and I. J. Fox. Myoglobin facilitated oxygen diffusion maintains mechanical function of mammalian cardiac muscle. *Cardiovasc. Res.* 20: 627–636, 1986.

11. Brodal, P., F. Ingjer, and L. Hermansen. Capillary supply of skeletal muscle fibers in untrained and endurance-trained men. *Am. J. Physiol.* 232 (6) (*Heart Circ. Physiol.* 3): H705–H712, 1977.

12. Brown, J. M. C., J. Henriksson, and S. Salmons. Restoration of fast muscle characteristics following cessation of chronic stimulation: physiological, histochemical and metabolic changes during slow-to-fast transformation. *Proc. R. Soc. Lond. B. Biol. Sci.* 235: 321–346, 1989.

13. Burton, H. W., and J. K. Barclay. Metabolic factors from exercising muscle and the proliferation of endothelial cells. *Med. Sci. Sports Exerc.* 18(4): 390–395, 1986.

14. Bylund, A.-C., J. Hammarsten, J. Holm, and T. Schersten. Enzyme activities in skeletal muscle from patients with peripheral arterial insufficiency. *Eur. J. Clin. Invest.* 6(6): 425–429, 1976.

15. Bylund-Fellenius, A.-C., P. M. Walker, A. Elander, S. Holm, J. Holm, and T. Schersten. Energy metabolism in relation to oxygen partial pressure in human skeletal muscle during exercise. *Biochem. J.* 200: 247–255, 1981.

16. Chi, M. M.-Y., C. S. Hintz, J. Henriksson, S. Salmons, R. P. Hellendahl, J. L. Park, P. M. Nemeth, and O. H. Lowry. Chronic stimulation of mammalian muscle: enzyme changes in individual fibers. *Am. J. Physiol.* 251 (*Cell Physiol.* 20): C633–C642, 1986.

17. Clark, A. Jr., P. A. A. Clark, R. J. Connett, T. E. J. Gayeski, and C. R. Honig. How large is the drop in PO₂ between cytosol and mitochondrion? *Am. J. Physiol.* 252 (*Cell Physiol.* 21): C583–C587, 1987.

18. Clarke III, B. J., M. A. Acker, K. McCully, H. V. Subramanian, R. L. Hammond, S. Salmons, B. Chance, and L. W. Stephenson. In vivo ³¹P-NMR spectroscopy of chronically stimulated canine skeletal muscle. *Am. J. Physiol.* 254 (*Cell Physiol.* 23): C258–C266, 1988.

19. Cole, R. P. Myoglobin function in exercising skeletal muscle. *Science* 216: 523–525, 1982.

20. Connett, R. J., C. R. Honig, T. E. J. Gayeski, and G. A. Brooks. Defining hypoxia: a systems view of VO₂, glycolysis, energetics, and intracellular PO₂. *J. Appl. Physiol.* 68 (3): 833–842, 1990.

21. Davies, K. J. A., L. Packer, and G. Brooks. Exercise bioenergetics following sprint training. *Arch. Biochem. Biophys.* 209: 539–554, 1981.

22. Degn, H., and H. Wohlrab. Measurements of steady-state values of respiration rate and oxidation levels of respiratory pigments at low oxygen tensions. A new technique. *Biochim. Biophys. Acta* 245: 347–355, 1971.

23. Denton, R. M., and J. G. McCormack. Calcium transport by mammalian mitochondria and its role in hormone action. *Am. J. Physiol.* 249 (*Endocrinol. Metab.* 12): E543–E554, 1985.

24. Dione, K. E. Oxygen transport to respiring myocytes. *J. Biol. Chem.* 265: 15400–15402, 1990.

25. Dudley, G. A., P. C. Tullson, and R. I. Terjung. Influence of mitochondrial content on the sensitivity of respiratory control. *J. Biol. Chem.* 262 (19): 9109–9114, 1987.

26. Eisenberg, B. R., and S. Salmons. The reorganization of subcellular structure in muscle undergoing fast-to-slow type transformation. *Cell. Tissue Res.* 220: 449–471, 1981.

27. Elander, A., J.-P. Idstrom, S. Holm, T. Schersten, and A.-C. Bylund-Fellenius. Metabolic adaptation to reduced muscle

blood flow. II. Mechanisms and beneficial effects. *Am. J. Physiol.* 249 (*Endocrinol. Metab.* 12): E70–E76, 1985.

28. Elander, A., J.-P. Idstrom, T. Schersten, and A.-C. Bylund-Fellenius. Metabolic adaptation to reduced muscle blood flow. I. Enzyme and metabolic alterations. *Am. J. Physiol.* 249 (*Endocrinol. Metab.* 12): E63–E69, 1985.

29. Erecińska, M., and D. F. Wilson. Regulation of cellular energy metabolism. *J. Membr. Biol.* 70: 1–14, 1982.

30. Erecińska, M., D. F. Wilson, and K. Nishiki. Homeostatic regulation of cellular energy metabolism: experimental characterization in vivo and fit to a model. *Am. J. Physiol.* 234 (3) (*Cell Physiol.* 3): C82–C89, 1978.

31. Federspiel, W. J. A model study of intracellular oxygen gradients in a myoglobin-containing skeletal muscle fiber. *Biophys. J.* 49: 857–868, 1986.

32. Forman, N. G., and D. F. Wilson. Energetics and stoichiometry of oxidative phosphorylation from NADH to cytochrome *c* in isolated rat liver mitochondria. *J. Biol. Chem.* 257: 12908–12915, 1982.

33. Gayeski, T. E. J., and C. R. Honig. O₂ gradients from the sarcolemma to cell interior in red muscle at maximal VO₂. *Am. J. Physiol.* 251 (*Heart Circ. Physiol.* 20): H789–H799, 1986.

34. Green, H. J., J. R. Sutton, E. E. Wolfe, J. T. Reeves, G. E. Butterfield, and G. A. Brooks. Altitude acclimatization and energy metabolic adaptations in skeletal muscle during exercise. *J. Appl. Physiol.* 73 (6): 2701–2708, 1992.

35. Groebe, K., and G. Thews. Role of geometry and anisotropic diffusion for modelling PO₂ profiles in working red muscle. *Respir. Physiol.* 79: 255–278, 1990.

36. Gollnick, P. D., R. B. Armstrong, C. W. Saubert IV, K. Piehl, and B. Saltin. Enzyme activity and fiber composition in skeletal muscle of untrained and trained men. *J. Appl. Physiol.* 33 (3): 312–319, 1972.

37. Goodrick, C. L., D. K. Ingram, M. A. Reynolds, J. R. Freeman, and N. L. Cider. Differential effects of intermittent feeding and voluntary exercise on body weight and life span in adult rats. *J. Gerontol.* 38: 36–45, 1983.

38. Hagler, L., R. I. Coppes, E. W. Askew, A. L. Hecker, and R. H. Herman. The influence of exercise and diet on myoglobin and metmyoglobin reductase in the rat. *J. Lab. Clin. Med.* 95: 222–230, 1980.

39. Hansford, R. G. Control of mitochondrial substrate oxidation. *Curr. Top. Bioenerg.* 10: 217–283, 1980.

40. Hansford, R. G., and F. Castro. Age-linked changes in the activity of enzymes of the tricarboxylate cycle and lipid oxidation and of carnitine content in muscles in the rat. *Mech. Ageing Dev.* 19: 191–201, 1982.

41. Henriksson, J., M. M.-Y. Chi, C. S. Hintz, D. A. Young, K. K. Kaiser, S. Salmons, and O. H. Lowry. Chronic stimulation of mammalian muscle: changes in enzymes of six metabolic pathways. *Am. J. Physiol.* 251 (*Cell Physiol.* 20): C614–C632, 1986.

42. Hermansen, L., and M. Wachtlova. Capillary density of skeletal muscle in well-trained and untrained men. *J. Appl. Physiol.* 30: (6), 860–863, 1971.

43. Hickson, R. C. Skeletal muscle cytochrome *c* and myoglobin, endurance and frequency of training. *J. Appl. Physiol.* 51: 746–749, 1981.

44. Hochacka, P. W. Muscle enzymatic composition and metabolic regulation in high altitude adapted natives. *Int. J. Sports Med.* 13 (1): S206–S209, 1992.

45. Holian, A., C. S. Owen, and D. F. Wilson. Control of respiration in isolated mitochondria: quantitative evaluation of the dependence of respiratory rates on [ATP], [ADP], and [Pi]. *Arch. Biochem. Biophys.* 181: 164–171, 1977.

46. Holm, J. P. Bjorntorp, and T. Schersten. Metabolic activity in

human skeletal muscle. Effect of peripheral arterial insufficiency. *Eur. J. Clin. Invest.* 2 (5): 321–325, 1972.

47. Holloszy, J. Biochemical adaptations in muscle. Effect of exercise on mitochondrial oxygen uptake and respiratory enzyme activity in skeletal muscle. *J. Biol. Chem.* 242 (9): 2278–2282, 1967.

48. Holloszy, J. O., and F. W. Booth. Biochemical adaptation to endurance exercise in muscle. *Annu. Rev. Physiol.* 38: 273–291, 1976.

49. Holloszy, J. O., and E. F. Coyle. Adaptations of skeletal muscle to endurance exercise and their metabolic consequences. *J. Appl. Physiol.* 56 (4): 831–838, 1984.

50. Holloszy, J. O., L. B. Oscai, I. J. Don, and P. A. Mole. Mitochondrial citric acid cycle and related enzymes: adaptive response to exercise. *Biochem. Biophys. Res. Commun.* 40: 1368–1373, 1970.

51. Honig, C. R. Hypoxia in skeletal muscle at rest and during the transition to steady work. *Microvasc. Res.* 13: 377–398, 1977.

52. Honig, C. R., R. J. Connett, and T. E. Gayeski. O_2 transport and its interaction with metabolism; a systems view of aerobic capacity. *Med. Sci. Sports Exerc.* 24 (1): 47–53, 1992.

53. Honig, C. R., J. L. Frierson, and T. E. J. Gayeski. Anatomical determinants of O_2 flux density at coronary capillaries. *Am. J. Physiol.* 256 (*Heart Circ. Physiol.* 25): H375–H382, 1989.

54. Hood, D. A., R. Zak, and D. Pette. Chronic stimulation of rat skeletal muscle induces coordinate increases in mitochondrial and nuclear mRNAs of cytochrome-*c*-oxidase subunits. *Eur. J. Biochem.* 179: 275–280, 1989.

55. Hudlicka, O. Effect of training on macro- and microcirculatory changes in exercise. *Exerc. Sport Sci. Rev.* 5: 181–231, 1977.

56. Hudlicka, O. Development of microcirculation: capillary growth and adaptation. In: *Handbook of Physiology. Cardiovascular System.* Editors: E. M. Renkin, C. C. Michel and S. R. Geiger. Bethesda, MD: Am. Physiol. Soc., 1984, sect. 2, chapt. 5, p. 165–216.

57. Hudlicka, O., L. Dodd, E. M. Renkin, and S. D. Gray. Early changes in fiber profile and capillary density in long-term stimulated muscles. *Am. J. Physiol.* 243 (*Heart Circ. Physiol.* 12): H528–H535, 1982.

58. Hudlicka, O., D. Pette, and H. Staudte. The relation between blood flow and enzymatic activities in slow and fast muscles during development. *Pflügers Arch.* 343: 341–346, 1973.

59. Hudlicka, O., and S. Price. The role of blood flow and/or muscle hypoxia in capillary growth in chronically stimulated fast muscles. *Pflügers Arch.* 417: 67–72, 1990.

60. Hudlicka, O., and K. R. Tyler. The effect of long-term high-frequency stimulation on capillary density and fibre types in rabbit fast muscles. *J. Physiol.* 353: 435–445, 1984.

61. Hudlicka, O., K. R. Tyler, A. J. A. Wright, and A. M. Ziada. The effect of long-term vasodilation on capillary growth and performance in rabbit heart and skeletal muscle. *J. Physiol.* 334: 49P, 1983.

62. Hudlicka, O., A. J. A. Wright, and A. M. A. R. Ziada. Angiogenesis in the heart and skeletal muscle. *Can. J. Cardiol.* 2 (2): 120–123, 1986.

63. Ji, L. L., D. L. F. Lennon, R. G. Kochan, F. J. Nagle, and H. A. Lardy. Enzymatic adaptation to physical training under β-blockade in the rat. Evidence of a β₂-adrenergic mechanism in skeletal muscle. *J. Clin. Invest.* 78: 771–778, 1986.

64. Jones, D. P. Intracellular diffusion gradients of O_2 and ATP. *Am. J. Physiol.* 250 (*Cell Physiol.* 19): C663–C675, 1986.

65. Jones, D. P., T. Y. Aw, C. Bai, and A. H. Sillau. Regulation of mitochondrial distribution: an adaptive response to changes in oxygen supply. In: *Response and Adaptation to Hypoxia: Organ*

to Organelle, edited by S. Lahiri, N. S. Cherniack, and R. S. Fitzgerald. Oxford: Oxford Univer. Press, 1991, p. 25–35.

66. Katz, L. A., J. A. Swain, M. A. Portman, and R. S. Balaban. Relation between phosphate metabolites and oxygen consumption in heart in vivo. *Am. J. Physiol.* 256 (*Heart Circ. Physiol.* 25): H265–H274, 1989.

67. Kaufmann, M., J.-A. Simoneau, J. H. Veerkamp, and D. Pette. Electrostimulation-induced increases in fatty acid-binding protein and myoglobin in rat fast-twitch muscle and comparison with tissue levels in heart. *FEBS Lett.* 245: 181–184, 1989.

68. Kiviluoma, K. T., K. J. Peuhkurinen, and I. E. Hassinen. Role of cellular energy state and adenosine in the regulation of coronary flow during variation in contraction frequency in an isolated perfused heart. *J. Mol. Cell. Cardiol.* 18: 1133–1142, 1986.

69. Knabb, R. M., S. W. Ely, A. N. Bacchus, R. Rubio, and R. M. Berne. Consistent parallel relationships among myocardial oxygen consumption, coronary blood flow, and pericardial infusate adenosine concentration with various interventions and β-blockade in the dog. *Circ. Res.* 53: 33–41, 1983.

70. Kraus, W. E., T. S. Bernard, and R. S. Williams. Interactions between sustained contractile activity and β-adrenergic receptors in regulation of gene expression in skeletal muscles. *Am. J. Physiol.* 256 (*Cell Physiol.* 25): C506–C514, 1989.

71. Lawrence Jr., J. C., and W. J. Salsgiver. Evidence that levels of malate dehydrogenase and fumarase are increased by cAMP in rat myotubes. *Am. J. Physiol.* 247 (*Cell Physiol.* 16): C33–C38, 1984.

72. Lawrie, R. A. The activity of the cytochrome system in muscle and its relation to myoglobin. *Biochem. J.* 55: 298–305, 1953.

73. Lee, Y. P., and H. Lardy. Influence of thyroid hormone on L-α-glycerophosphate dehydrogenase and other dehydrogenases in various organs of the rat. *J. Biol. Chem.* 240: 1427–1436, 1965.

74. Longmuir, I. S. Respiration rate of rat-liver cells at low oxygen concentrations. *Biochem. J.* 65: 378–382, 1957.

75. Lundgren, F., A.-G. Dahllof, T. Schersten, and A.-C. Bylund-Fellenius. Muscle enzyme adaptations with peripheral arterial insufficiency: spontaneous adaptation, effect of different treatments and consequences on walking performance. *Clin. Sci. (Colch)* 77: 485–493, 1989.

76. Mathieu-Costello, O. Comparative aspects of muscle capillary supply. *Annu. Rev. Physiol.* 55: 503–525, 1993.

77. Mathieu-Costello, O., J. M. Szewczak, R. B. Logemann, and P. J. Agey, Geometry of blood tissue exchange in bat flight muscle compared to bat hindlimb and rat soleus muscle. *Am. J. Physiol.* 262 (*Regulatory Integrative Comp. Physiol.* 31): R955–R965, 1992.

78. Matheson, G. O., P. S. Allen, D. C. Ellinger, C. C. Hanstock, D. Gheorghiu, D. C. McKenzie, C. Stanley, W. S. Parkhouse, and P. W. Hochacka. Skeletal muscle metabolism and work capacity: a 31P-NMR study of Andean natives and lowlanders. *J. Appl. Physiol.* 70 (5): 1963–1976, 1991.

79. McCarter, R. J. M., E. J. Masoro, and B. P. Yu. Rat muscle structure and metabolism in relation to age and food intake. *Am. J. Physiol.* 242 (*Regulatory Integrative Comp. Physiol.* 13): R89–R93, 1982.

80. Meininger, C., and H. J. Granger. Mechanisms leading to adenosine-stimulated proliferation of microvascular endothelial cells. *Am. J. Physiol.* 258 (*Heart Circ. Physiol.* 27): H198–H206, 1990.

81. Mole, P. A., L. B. Oscai, and J. O. Holloszy. Adaptation of muscle to exercise. Increase in levels of palmitoyl CoA synthetase, carnitine palmityltransferase, and palmityl CoA dehydrogenase, and in the capacity to oxidize fatty acids. *J. Clin. Invest.* 50: 2323–2330, 1971.

82. Morrison, P. R., R. B. Biggs, and F. W. Booth. Daily running for 2 wk and mRNAs for cytochrome c and α-actin in rat skeletal muscle. Am. J. Physiol. 257 (Cell Physiol. 26): C936–C939, 1989.

83. Nishiki, K., M. Erecińska, D. F. Wilson, and S. Cooper. Evaluation of oxidative phosphorylation in hearts from euthyroid, hypothyroid, and hyperthyroid rats. Am. J. Physiol. 235 (Cell. Physiol. 4): C212–C219, 1978.

84. Oscai, L. B., P. A. Mole, and J. O. Holloszy. Biochemical adaptations in muscle. II. Response of mitochondrial adenosine triphosphatase, creatine phosphokinase, and adenylate kinase activities in skeletal muscle to exercise. J. Biol. Chem. 246 (22): 6968–6972, 1971.

85. Oscai, L. B., P. A. Mole, and J. O. Holloszy. Effects of exercise on cardiac weight and mitochondria in male and female rats. Am. J. Physiol. 220 (6): 1944–1948, 1971.

86. Owen, C. S., and D. F. Wilson Control of respiration by the mitochondrial phosphorylation potential. Arch. Biochem. Biophys. 161: 581–591, 1974.

87. Pattengale, P., and J. O. Holloszy. Augmentation of skeletal muscle myoglobin by a program of treadmill running. Am. J. Physiol. 213 (3): 783–785, 1967.

88. Pette, D. Activity-induced fast to slow transitions in mammalian muscle. Med. Sci. Sports Exerc. 16 (6): 517–528, 1984.

89. Pette, D., W. Muller, E. Leisner, and G. Vrbova. Time dependent effects on contractile properties, fibre population, myosin light chains and enzymes of energy metabolism in intermittently and continuously stimulated fast twitch muscles of the rabbit. Pflügers Arch. 364: 103–112, 1976.

90. Pette, D., M. E. Smith, H. W. Staudte, and G. Vrbova. Effects of long-term electrical stimulation on some contractile and metabolic characteristics of fast rabbit muscles. Pflügers Arch. 338: 257–272, 1973.

91. Portman, M. A., F. W. Heineman, and R. S. Balaban. Developmental changes in the relation between phosphate metabolites and oxygen consumption in the sheep heart in vivo. J. Clin. Invest. 83: 456–464, 1989.

92. Reaven, E. P., and G. M. Reaven. Structure and function changes in the endocrine pancreas of aging rats with reference to the modulating effects of exercise and caloric restriction. J. Clin. Invest. 68: 75–84, 1981.

93. Robiolio, M., W. L. Rumsey, and D. F. Wilson. Oxygen diffusion and mitochondrial respiration in neuroblastoma cells. Am. J. Physiol. 256 (Cell Physiol. 25): C1207–C1213, 1989.

94. Rumsey, W. L., Z. V. Kendrick, and J W. Starnes. Bioenergetics in the aging Fischer 344 rat: Effects of exercise and food restriction. Exp. Gerontol. 22: 271–287, 1987.

95. Rumsey, W. L., C. Schlosser, E. M. Nuutinen, M. Robiolio, and D. F. Wilson. Cellular energetics and the oxygen dependence of respiration in myocytes isolated from adult rat heart. J. Biol. Chem. 265: 15392–15402, 1990.

96. Salmons, S., and J. Henriksson. The adaptive response of skeletal muscle to increased use. Muscle Nerve 4: 94–105, 1981.

97. Saltin, B., and P.-O. Astrand. Maximal oxygen uptake in athletes. J. Appl. Physiol. 23: 353–358, 1967.

98. Saltin, B., and P. D. Gollnick. Skeletal muscle adaptability: significance for metabolism and performance. In: Handbook of Physiology. Skeletal Muscle. Bethesda, MD: Am. Physiol. Soc., 1983, sect. 10, chapt. 19, p. 555–631.

99. Saunders, D. K., and M. R. Fedde. Physical conditioning: Effect on the myoglobin concentration in skeletal and cardiac muscle of bar-headed geese. Comp. Biochem. Physiol. 100A (2): 349–352, 1991.

100. Scheel, K. W., E. Seavey, J. F. Gaugl, and S. E. Williams. Coronary and myocardial adaptations to high altitude in dogs. Am. J. Physiol. 259 (Heart Circ. Physiol. 28): H1667–H1673, 1990.

101. Scholander, P. F. Oxygen transport through hemoglobin solutions. Science 131: 585–590, 1960.

102. Seedorf, U., E. Leberer, B. J. Kirschbaum, and D. Pette. Neural control of gene expression in skeletal muscle. Effects of chronic stimulation on lactate dehydrogenase isoenzymes and citrate synthase. Biochem. J. 239: 115–120, 1986.

103. Starnes, J. W., and W. L. Rumsey. Cardiac energetics and performance of exercised and food restricted rats during aging. Am. J. Physiol. 254 (Heart Circ. Physiol. 23): H599–H608, 1988.

104. Suarez, R. K., J. R. B. Lighton, G. S. Brown, and O. Mathieu-Costello. Mitochondrial respiration in hummingbird flight muscles. Proc. Natl. Acad. Sci. USA 88: 4870–4873, 1991.

105. Sullivan, S. M., and R. N. Pittman. Relationship between mitochondrial volume density and capillarity in hamster muscles. Am. J. Physiol. 252 (Heart Circ. Physiol. 21): H149–H155, 1987.

106. Sutton, J. R., J. T. Reeves, B. M. Groves, P. D. Wagner, J. K. Alexander, H. N. Huffgren, A. Cymerman, and C. S. Houston. Oxygen transport and cardiovascular function at extreme altitude: lessons from Operation Everest II. Int. J. Sports Med. 13(1): S13–S18, 1992.

107. Terrados, N. Altitude training and muscular metabolism. Int. J. Sports Med. 13 (1): S206–S209, 1992.

108. Terrados, N., E. Jansson, C. Sylven, and L. Kaijser. Is hypoxia a stimulus for synthesis of oxidative enzymes and myoglobin? J. Appl. Physiol. 68 (6): 2369–2372, 1990.

109. Underwood, L. E., and R. S. Williams. Pretranslational regulation of myoglobin gene expression. Am. J. Physiol. 252 (Cell Physiol. 21): C450–C453, 1987.

110. Wagner, P. D. Adaptation of O₂ transport and utilization at altitude in man. Adv. Exp. Med. Biol. 317: 75–94, 1992.

111. Whipple, G. H. The hemoglobin of striated muscle. I. Variations due to age and exercise. Am. J. Physiol. 76: 693–707, 1926.

112. Williams, R. S. Mitochondrial gene expression in mammalian striated muscle. Evidence that variation in gene dosage is the major regulatory event. J. Biol. Chem. 261 (26): 12390–12394, 1986.

113. Williams, R. S., M. G. Caron, and K. Daniel. Skeletal muscle β-adrenergic receptors: variations due to fiber type and training. Am. J. Physiol. 246 (Endocrinol. Metab. 9): E160–E167, 1984.

114. Williams, R. S., M. Garcia-Moll, J. Mellor, S. Salmons, and W. Harlan. Adaptation of skeletal muscle to increased contractile activity. Expression of nuclear genes encoding mitochondrial proteins. J. Biol. Chem. 262 (6): 2764–2767, 1987.

115. Williams, R. S., and W. Harlan. Effects of inhibition of mitochondrial protein synthesis in skeletal muscle. Am. J. Physiol. 253 (Cell Physiol. 22): C866–C871, 1987.

116. Williams, R. S., S. Salmons, E. A. Newsholme, R. E. Kaufman, and J. Mellor. Regulation of nuclear and mitochondrial gene expression by contractile activity in skeletal muscle. J. Biol. Chem. 261 (1): 376–380, 1986.

117. Wilson, D. F., and M. Erecińska. Effect of oxygen pressure on cellular metabolism. Chest 88S: 229S–232S, 1985.

118. Wilson, D. F., M. Erecińska, C. Drown, and I. A. Silver. The oxygen dependence of cellular energy metabolism. Arch. Biochem. Biophys. 195: 485–493, 1979.

119. Wilson, D. F., C. S. Owen, and M. Erecińska. Quantitative dependence of mitochondrial oxidative phosphorylation on oxygen concentration: a mathematical model. Arch. Biochem. Biophys. 195: 494–504, 1979.

120. Wilson, D. F., C. S. Owen, and A. Holian. Control of mitochondrial respiration: a quantitative evaluation of the roles of cytochrome c and oxygen. Arch. Biochem. Biophys. 182: 749–762, 1977.

121. Wilson, D. F., M. Stubbs, N. Oshino, and M. Erecińska. Thermodynamic relationships between the mitochondrial oxidation-

reduction reactions and cellular ATP levels in ascites tumor cells and perfused rat liver. *Biochemistry* 13: 5305–5311, 1974.

122. Wilson, D. F., M. Stubbs, R. L. Veech, M. Erecińska, and H. A. Krebs. Equilibrium relations between the oxidation-reduction reactions and the adenosine triphosphate synthesis in suspensions of isolated liver cells. *Biochem. J.* 140: 57–64, 1974.

123. Wittenberg, J. B. Myoglobin-faciliated oxygen diffusion: role of myoglobin in oxygen entry into muscle. *Physiol. Rev.* 50 (4): 559–636, 1970.

124. Wittenberg, B. A., and J. B. Wittenberg. Oxygen pressure gradients in isolated cardiac myocytes. *J. Biol. Chem.* 260: 6548–6554, 1985.

125. Wittenberg, B. A., and J. B. Wittenberg. Transport of oxygen in muscle. *Annu. Rev. Physiol.* 51: 857–878, 1989.

126. Wittenberg, J. B., and B. A. Wittenberg. Mechanisms of cytoplasmic hemoglobin and myoglobin function. *Annu. Rev. Biophys. Biophys. Chem.* 19: 217–241, 1990.

48. Metabolic defense adaptations to hypobaric hypoxia in man

P. W. HOCHACHKA | *Department of Zoology, University of British Columbia, Vancouver*

CHAPTER CONTENTS

IT IS WELL KNOWN FROM NUMEROUS earlier studies (6, 11, 14, 19, 20, 22, 25, 30, 33–35, 41, 42, 46, 50–54) that natives indigenous to high-altitude environments display a number of functional and structural adaptations that allow them, at least partially, to circumvent the main metabolic problem they face: maintaining an acceptably large scope for sustained aerobic metabolism despite reduced availability of O_2 in the inspired air. As a result of adaptations at various levels of organization, maximal O_2 uptake ($\dot{V}_{O_2max}$) values at altitudes in the range of 3,500–4,500 m are reduced only modestly. In our studies of Quechuas, for example, metabolic rates at about 4,000 m altitude are reduced to 89%–95% of normoxic values (28). This represents a metabolic inhibition of only $\frac{1}{2}$ to $\frac{1}{4}$ of that experienced by most lowlanders under similar hypoxic stress, and implies relatively hypoxia resistant functions of many key tissues and organs over wide dynamic ranges of whole-body metabolism and work. How such adaptations are achieved at the metabolic level in different tissues and organs is not well understood (22) and is the central issue of this chapter. It will be analyzed for three tissues in particular: brain, heart, and skeletal muscle. To put this problem into proper perspective, however, it is essential to begin with the concept of adaptation itself.

ADAPTATION AND ITS INTERPLAY WITH TIME

Adaptation can be defined as the summation of those processes and traits that enhance survival and therefore reproductive success of the individual, of its population, and of its species. The site of action of natural selection is necessarily the individual, but the net effect is expressed at the population level. By definition, then, hypoxia adaptations are those responses that favor survival in spite of potentially O_2-limiting conditions. Generally, adaptations are categorized with respect to time available for the response (24, 26):

(i) Acute responses to hypoxia (as to any kind of external environmental stress) are essentially instantaneous and are designed to defend the system against O_2 limitation. They are utterly dependent on preexisting macromolecular and physiological machinery (because of the acute nature of the stress). What is regulated is the output of the system; there is not enough time to change the nature of the system. Finally, the response can be compensatory, as in the case of the Pasteur effect (activated to make up the energy deficit). Alternatively, the responses can be exploitive (taking advantage of the O_2 limitation to achieve some other biological purpose). For example, the prolonged and deep diving in large seals may take advantage of ischemia- or hypoxia-induced metabolic suppression in order to cover more distance and spend more time submerged on a given breathold than would otherwise be possible (21). Similarly, aquatic turtles take advantage of both a hypoxia- and hypothermia-induced metabolic suppression in lake bottom muds for overwintering submergence where they are safe from much more hypoxia-sensitive predator species (24).

(ii) Acclimatory responses to hypoxia occur over periods of days to weeks and perhaps even longer (14, 37, 40, 53, 54). Like acute responses, these are designed to defend against the stress parameter (O_2 limitation in this case). However, because of the time available for these responses, they clearly need not depend only on preexisting macromolecular machinery; there is ample time for restructuring and reorganizing. Preexisting macromolecular components may be up- or down-regulated; new receptors, new membrane bilayer phospholipids, new membrane proteins (channels, transporters, pumps), new isozymes or new respiratory pigments may be synthesized, allowing the orchestration of responses that are simply unattainable in acute exposure (40). In

short, what is regulated during acclimation is functional organization as well as output (26). Again, the overall strategy of acclimatory responses may be compensatory or exploitative, depending on cell, tissue, and species.

(iii) Long-term or phylogenetic adaptations to hypoxia also obviously include options of restructuring and biochemical reorganization but involve generations of time to become fixed in the species genome. Interestingly, most biomedical scientists refer only to the third of the above three time courses of response as "genetic" adaptation, an error that seems to have arisen initially as a kind of convenient shorthand terminology (22). This is unfortunate, because now it is commonly overlooked by biomedical and biological scientists alike that, at the cell level, all three time courses of adaptation depend on genetically determined molecular machinery. Thus a genetic base or linkage is to be anticipated for acute, acclimatory, and long-term phylogenetic adaptations. Be that as it may, it is in this broad context that we will examine metabolic defense responses to chronic hypoxia in indigenous highlanders. An initial focus on three specific tissues (muscle, heart, and brain) is designed to facilitate trying to mechanistically understand the basis for integrated whole-organism adaptations to hypoxia.

HYPOXIA ADAPTATIONS IN MUSCLE METABOLISM

Because the resting metabolic rate (RMR) of skeletal muscle is low (26), a commonly used source of information about metabolic adaptation to hypoxia in humans involves assessing the relative contributions to energy supply of aerobic vs. anaerobic metabolic pathways during exercise and recovery. The usual assumption is that perturbations in plasma and tissue lactate pools indicate significant mismatch between muscle energy demands and O_2 fluxes (between muscle adenosine triphosphate [ATP] demands and the capacity to supply ATP through aerobic metabolic pathways). In lowlanders during incremental $\dot{V}O_{2max}$ tests, plasma lactate concentrations remain stable for about the first half of the test, but these values begin to rise at about 50% of maximum work; at fatigue; plasma concentrations in excess of 12–15 mM are not uncommon (51), and higher values can be seen in well-trained athletes (17). When such lowlanders exercise under hypoxic conditions, they form more lactate (for any given power output) than in normoxia (51), a metabolic response common to many animals under O_2-limiting conditions (24); it is interpreted as an attempt to make up the energy deficit caused by O_2 insufficiency and is a special expression of the Pasteur effect (24).

The acute metabolic response in native highlanders (Quechuas and Sherpas) differs strikingly from this standard metabolic pattern (20, 27, 51). At all exercise intensities to fatigue, plasma lactate concentrations are significantly lower than in lowlanders. Curves of plasma lactate concentration versus power output in watts are right-shifted so that, at fatigue, plasma lactate values are in the 5–7 mM range, or about half those seen in lowlanders (27). Although the data indicate that O_2 fluxes to working muscles are more closely balanced with ATP demands in highlanders than in lowlanders, they are perplexing for an obvious reason: despite hypobaric hypoxia, *high-altitude natives produce less, rather than more, lactate for a given power output*. This kind of metabolic response is so counterintuitive to workers in the field that it has become known as the lactate paradox (20, 51). First described for Andean natives over 50 years ago (11), this metabolic characteristic in lowlanders is an acclimation; it requires about 10–20 days at altitude to be expressed, and it deacclimates along a similar time course on descent (14, 40, 53, 54). In highland natives, in contrast, the lactate paradox is expressed even 6 wk after descent (27), and, indeed, after 6 wk of reacclimation to high altitude (Hochachka and Stanley, unpublished data). Thus, we conclude that it is a more stable metabolic property than in lowlanders. Because it is an expression of oxidative metabolism that seems relatively fixed and does not deacclimate much, we assume a strong developmental or genetic basis for the adaptation.

Although the adaptive significance of the lactate paradox at high altitude is well recognized (Edwards [11] as long ago as 1936 realized that the excessive production and accumulation of lactate are counterproductive), the molecular mechanisms underlying these unique regulatory interactions between aerobic and anaerobic metabolism have remained a mystery for some 50 years. Recent (37) magnetic resonance spectroscopy (MRS) and muscle enzyme data (28) imply that lactate production in muscles of altitude natives is regulated at two coarse and finely tuned levels. The finly tuned mechanism for defending skeletal muscle in Sherpas and Quechuas against overproduction of lactate (28) acts directly at the level of pyruvate, whose flux to lactate is attenuated by *maintaining high ratios of pyruvate kinase (PK)/lactate dehydrogenase (LDH) and of malate dehydrogenase (MDH)/LDH*. High PK/LDH ratios assure mitochondrial metabolism of a "pyruvate push" from glycolysis, while high MDH/LDH ratios assure that most cytosolic NADH is oxidized by MDH (for transfer of malate to the mitochondria), not by LDH, which would lead to lactate accumulation. As a result of these adjustments and of generally low glycolytic and oxidative enzyme activities, these muscles behave as if anaerobic glycolysis were down-regulated and as if metabolism were simplified, with phosphocreatine (PCr) and oxidative phosphorylation as the two main sources of ATP and with modest maximum flux capacities (indexed by relatively low oxidative enzyme contents).

The coarse control mechanism, on the other hand, acts at the level of pathway control, and in effect serves to temper activation of the glycolytic path during muscle work. The evidence for this concept is derived from MRS monitoring of phosphate metabolites during an incremental exercise protocol to fatigue (using a calf muscle ergometer designed for function in a 1 m bore, 1.5 T magnet). These studies (20, 27, 37) hypothesize that the difference between skeletal muscle metabolism in Quechuas and in lowlanders is analogous to the difference between cardiac and skeletal muscles, but that the difference is less extreme. In cardiac muscle, energy demand and energy supply functions are so closely balanced that large changes in work rate are achieved with minimal or no changes in concentrations of high-energy phosphate metabolites and with minimal or no lactate production (2). In contrast, in working skeletal muscle, change in power output is usually accompanied by changes in high-energy phosphate metabolites and, particularly, in the phosphorylation potential (7, 8, 12, 13, 45). These metabolite signals change less during incremental work in indigenous highlanders than in sedentary or power-trained lowlanders (37), and are widely considered (7, 8, 44) to "drive" oxidative metabolism and, thereby, to account for the large increase in $\dot{V}O_2$ that skeletal muscle can sustain during work. What is often overlooked or underemphasized is that the same metabolite signals that turn on oxidative metabolism serve (kinetically and thermodynamically) to activate glycolysis; the smaller the magnitude of these signals, the less the glycolytic activation (1, 20). That is why aerobically working skeletal muscle in highlanders generates less lactate than in lowlanders (20) and why cardiac muscle, in contrast, produces no lactate at all under these conditions (2).

An interesting test of this model is found in subjects sustaining an unusual myopathy—M4 LDH deficiency (32). In muscles of these individuals, anaerobic glycolysis is down-regulated because of a genetic accident—loss of a 20-base sequence in exon 6 of the gene for M4 LDH (38). The energy-coupling hypothesis of the lactate paradox (20) would predict that the high-energy phosphate profiles during rest-work transitions in muscles of these subjects should be rather similar to those in altitude-adapted natives (where anaerobic glycolysis is down-regulated as an adaptation to chronic hypoxia [28]). Preliminary MRS studies of gastrocnemius during submaximal work appear to be consistent with the concept that, in muscles whose ATP-generating metabolism is dominated by PCr and oxidative phosphorylation, rest-to-work transitions occur with smaller perturbations in "high energy" phosphate metabolites than in

"normal" muscles that express a high-activity glycolytic path (Hochachka and Kanno, unpublished data). Put another way, muscles with a genetically down-regulated anaerobic glycolysis (with M4 LDH-deleted) seem to regulate [adenylate] and [PCr] in manners qualitatively more similar to cardiac muscle than to skeletal muscles retaining an intact, high-capacity anaerobic glycolytic pathway, and this outcome in fact would be predicted from the MRS studies of Quechua muscles with adaptively down-regulated anaerobic glycolysis.

Given such a spectrum of metabolic organization (from the classic situation, as exemplified by the human gastrocnemius [37], to the cardiac muscle situation [2]), these studies suggest that high-altitude *acclimation* in Caucasian lowlanders (14) moves skeletal muscle metabolic regulation and functional organization toward the cardiac end of the spectrum and that *in high-altitude-adapted Quechuas and Sherpas, the mechanism allowing closer ATP demand–ATP supply coupling has moved even further along this adaptation line (20, 27, 28). In altitude* natives, this key regulatory component in the final link of the path of O_2 in vivo either acclimates and deacclimates slowly or not at all, which is why we tentatively conclude that it, too, represents a true, long-term (developmental or genetic) adaptation.

ADAPTATION OF OUTPUT AND OF REGULATION

In mechanistic terms, then, adaptation of muscle metabolism in indigenous highlanders seems to be based on two kinds of changes. The first involves adjustments in enzyme and thus in gene expression affecting

(i) the potential for pyruvate flux to lactate per se,

(ii) the overall flux potentials of both glycolytic and oxidative pathways of energy metabolism (these two processes together are able to account for the low lactate levels found during exercise in $\dot{V}O_{2max}$ tests, and indeed the low $\dot{V}O_{2max}$ values found in highland natives), and

(iii) the efficiency or tightness of coupling between ATP demand and ATP supply pathways (this allows reduced perturbations of the adenylates and PCr during rest → fatigue incremental work protocols).

A second kind of adjustment seems focused on regulation rather than metabolic output capacity per se. Almost nothing is known about this level of regulation of muscle metabolism and organization. From first principles, however, it most likely involves adjustments in the expression of regulatory proteins (and genes) that alter the response time constants during acclimation to change in O_2 availability. It is this kind of adjustment that presumably underpins the strikingly different acute and acclimatory responses in muscle metabolism between native highlanders and lowlanders. The question arises of whether or not analogous adaptations occur in other tissues. Interestingly, the answer seems to be affirmative at least for the heart and brain.

HYPOXIA ADAPTATIONS IN CARDIAC METABOLISM

While it is widely appreciated that the human heart has high O_2 demands, it is often overlooked just how high these are. In a healthy adult at rest (43), the energy demands are equivalent to about 25 μmol ATP $\cdot$ $g^{-1} \cdot min^{-1}$; on a g weight basis, this approaches the ATP turnover rate of leg muscles of a marathon runner in the heat of a race (5). Of course, during the race, the ATP turnover rate of the heart further increases by three- to severalfold, at this point representing one of the highest mass-specific metabolic rates in the human body. Metabolic rates of up to, or even higher than, 100 μmol ATP $\cdot g^{-1} \cdot min^{-1}$ could be sustained by endogenous fuels for only limited time periods—for 1 to 2 min if only anaerobic glycogenolysis were possible, for close to an hour if intracellular glycogen were to be used in a fully aerobic metabolism. So it is not surprising that for its function to be indefinitely sustainable, the heart must be supported by exogenous fuels (usually glucose or free fatty acids [FFA]) and by O_2-based metabolic pathways (3, 10). It is of course this consequent critical need for high O_2 fluxes that makes the heart so sensitive to any kind of O_2 limitation—hypoxia, hypoxemia, or ischemia—and it is this sensitivity that has led to enormous scientific and medical interest in cardiac defense mechanisms against O_2 limitation. As a result, an impressive array of studies have been performed on a variety of animal and organ-level models (2–4, 9, 10), on normal subjects compared to clinically compromised ones (16, 18, 36), and on normoxic subjects compared to acutely hypoxia stressed or to hypoxia-acclimated individuals (39, 43). However, with a few exceptions (43), there has been a notable lack of studies of heart metabolic defense mechanisms in subjects who have become adapted to chronic hypoxia over phylogenetic time. We consider this to be a serious oversight, since it is in these kinds of subjects that we may be able to identify mechanisms selected by nature for defending this most sensitive organ against limiting O_2 availability.

To begin correcting for the dearth of information in this area, we recently turned our attention to a study of regional glucose metabolic rates in hearts of Quechuas, born and living all their lives at high altitudes, as have their ancestors, for many generations. Using positron emission tomography (PET), we found that the heart utilizes glucose at higher rates than in normoxic controls. Measured during whole-body resting conditions and after an overnight fast for metabolic standardization, rates of glucose uptake (MR_{gluc}) in all six interro-

gated regions of interest (ROIs) were nearly twice as high as in Quechuas as in lowlanders studied for comparison; the lowest values for lowlanders were only $\frac{1}{10}$ of the highest MR_{gluc} values for Quechuas (29). We consider that the biochemical significance of this result arises from the fact that *glucose metabolism uses O_2 more efficiently than does FFA metabolism (9, 19, 31, 48, 49). Because of a so-called O_2 wasting effect of FFA, the heart of the Quechua can gain 25%–60% more ATP/mole of O_2 whenever it preferentially uses glucose instead of FFA, the predominant fuel for the heart of lowlanders under normoxic conditions.* Especial importance is attached to this metabolic defense mechanism against hypoxia because it appears to be such a dominant adaptation in a people conditioned over generations to hypobaric conditions.

In principle it might be possible to bring about these adaptations in fuel preferences of the heart with appropriate adjustments in substrate and hormone concentrations (3). Although measurements of glucose, FFA, catecholamines, insulin, and glucagon indicate that all these components occur at concentrations well within the normal range, it is evident that minimally two conditions (appropriate hormonal and appropriate fuel-availability adjustments) underpin the heart's metabolic response to hypobaric hypoxia in Quechuas. The key hormonal conditions favoring increased glucose contribution to heart ATP production are low ratios of insulin/epinephrine; these conditions influence the pictures for both plasma glucose and FFA, but the key parameter seemingly regulated is glucose availability per se, since there is a strong direct relationship between heart glucose metabolic rates and [glucose]; as would be expected, a weaker inverse correlation is observed between glucose metabolic rates and FFA availability (29).

A final question arising from this work on the heart concerns the biological significance of the observed hypoxia defense responses of the Quechua heart. That these adaptations are important are indicated by two kinds of relationships. The first shows that the work of the heart (measured as cardiac output) in resting Quechuas varies as a function of plasma glucose availability, while it varies inversely with FFA. Perhaps even more striking is the second observation that the work of the heart in resting subjects varies directly with the glucose metabolic rate (MR_{gluc}) of the heart (Hochachka, Holden, and McKenzie, unpublished data) supporting by direct empirical test the conclusion that the glucose preference adaptations of the Quechua heart make measurable biological impacts on heart performance (29). Parenthetically, it should be mentioned that no data are available on MR_{gluc} for the heart during maximum exercise protocols, so it is not known if these relationships extend into this range of heart work. Because at high work rates, the heart's preference for lactate rises, as does lactate availability (10), it is probable that these simple relationships do not extend into zones of maximum whole-body exercise.

Because of the enormous implications to clinical research of these data (especially if it becomes possible to manipulate and manage the response to hypoxia), the question of whether or not cardiac metabolism of indigenous highlanders adjusts on shorter time scales is especially interesting. Unfortunately, no information is available on the effects of acute hypoxia on these shifts in fuel preference (40). That the adaptation at least to some degree is plastic, however, is indicated by deacclimation studies, which show some attentuation of glucose metabolic rates in Quechuas (and thus of glucose preference by the heart) after 3 wk at low altitudes (29). Thus, as in the case of muscle, chronic hypoxia adaptation appears to affect both the basic metabolic organization of the heart and its regulation patterns.

HYPOXIA ADAPTATIONS IN BRAIN METABOLISM

The central nervous system in humans, as in most mammals, also displays surprisingly high mass-specific metabolic rates (about 20 μmol ATP $\cdot$ g$^{-1}\cdot$min^{-1}) and is widely considered to be the most hypoxia-sensitive organ in the body. For this reason, and because cerebrovascular disease is still a leading cause of death in modern societies, a very large research interest focuses on mechanisms of defense against O$_2$ limitation in the mammalian brain. Most studies in this area deal with acute (hypoxic or ischemic) exposures and concentrate (i) on mechanism of damage, and (ii) on intervention strategies, usually pharmacological in nature (55). Longer-term acclimation effects on brain hypoxia tolerance are less well understood (15), and the dearth of knowledge on effects of hypoxia adaptation of the human brain over lifetime or generational time is even greater. For practical purposes, the literature here is nonexistent.

While hypoxia defense adaptations over phylogenetic time have not been explored in the human species, they have received intense research interest in animal species that are known to be exceptionally tolerant of extended hypoxia or even total anoxia. These studies establish that the vertebrate brain can be protected for prolonged time periods of hypoxia and indicate profound metabolic adaptations that increase hypoxia tolerance. In ectothermic species and in hypoxia-tolerant (diving) mammals, the pivotal piece in hypoxia defense adaptation is brain hypometabolism with coordinate suppression of energy demand via transmitter-mediated change in metabolism-membrane integration (22 and references therein). Interestingly, from PET measure-

ments of regional brain metabolic rates using the tracer analogue, ^{18}fluorodeoxyglucose (FDG), two striking characteristics emerge (23). First, region-by-region comparisons indicate a general pattern of modestly lower brain-glucose metabolic rates after deacclimation than on arrival. This effect is most notable for the thalamic nuclei, where upon arrival from high altitude, the metabolic rates are slightly lower than normal and then fall after deacclimation.

Secondly, and more significantly, such region-by-region comparisons indicate that glucose metabolic rates of essentially all 26 brain regions analyzed in Andean natives are systematically lower than in lowlanders. In view of the inherent resolution limits of the PET scanning technique and the resulting underestimation of regional metabolism due to partial volume contributions of cerebrospinal fluid, white matter, and gray matter, the measured magnitude of these differences are quite dramatic. For example, average cortical metabolic rates are 15% and 21% lower on arrival and after (3 wk) deacclimation, respectively, than in the control lowlanders. For these reasons we attach especial significance (i) to the general decline in regional metabolic rate upon deacclimation, and (ii) to the generally lower metabolic rates of brain regions in highlanders than in lowlanders.

The latter observations indicate that in man, as in animals, adaptation to chronic hypoxia leads to relative hypometabolism of the brain in order to minimize O$_2$-limitation impacts (21, 23). By comparison, the brain metabolic rate of deep-diving seals is only about $\frac{1}{3}$ to $\frac{1}{2}$ that of man (26), while metabolic rates are even lower in the brain of the most hypoxia-tolerant vertebrate known, the aquatic turtle (24). In these systems, energy-demanding functions of the brain are either down-regulated or are made more efficient, so that energy demand–energy supply coupling is not violated; in either event, the organ is less affected by hypoxia because it can function longer on a given amount of oxygen (22). It is tempting to speculate that chronic hypobaric hypoxia in man leads to a similar adaptation, which would account for brain metabolic rates in highlanders that are generally lower than in lowlanders. But how are we to explain the deacclimation effects?

As a working framework, we assume that, relative to lowland normoxia, chronic exposure to hypobaric hypoxia in man as in other mammals leads to reversible up-regulation of glucose metabolic capacities (due to increased expression of glucose transporters, enzyme machinery, and possibly even capillarity [15]). On exposure to lower altitudes, we assume that any such comparable adaptations in the human brain would be at least partially reversed and thus would lead to a gradual reduction in glucose metabolic rates down to some new setpoint. Such a process could well account for the lower region-by-region brain metabolic rates in high-altitude natives after deacclimation from high altitudes. In qualitative terms, brain metabolism in Quechuas seems to operate between two setpoints determined by normoxia or hypoxia adaptation state, both of which are seemingly lower than in lowlanders.

HYPOXIA ADAPTATION IN MAN: TOWARD UNRAVELING THE INTEGRATED RESPONSE

To date, then, rather detailed analyses are available for metabolic defense adaptations specific to three tissues—muscle, heart, and brain—in man after exposure to high altitude for generations. It is useful to compare and contrast the three tissue-specific responses in the whole-organism framework, in order to begin to piece together the integrated meaning of these tissue-specific metabolic responses to hypoxia.

In muscle, the main defense strategy against chronic hypoxia appears to be a kind of stoichiometric efficiency adaptation: maximizing the percent contribution of aerobic vs. anaerobic ATP-generating pathways. One adaptive advantage clearly arises from the much higher yield of ATP/mole of glucose-catabolized (36 ATP rather than 2 ATP per glucose), and from avoiding complications due to accumulating anaerobic end products.

Down-regulation of anaerobic glycolytic function during muscle exercise is the most telling phenotypic expression of this adaptation. By using this as a kind of marker of adaptation, it becomes clear that all three time courses of hypoxia adaptation are different in highlanders compared to lowlanders. This is consistent with the concept of genetic influence on the molecular and metabolic machinery underpinning all three time courses of the hypoxia defense responses *as well as on their regulation*. The latter may be as fundamental as the former.

In the heart, the situation is perhaps clearest of all, for during chronic exposure to hypoxia the heart defends itself with the most basic of available strategies: maximizing the yield of ATP per mole O_2 consumed. Because of the O_2 wasting effect of free fatty acids, glucose oxidation uses O_2 up to 60% more efficiently than does free fatty acid oxidation. Again, the adaptation is one of up regulating stoichiometric efficiencies. At this time, the up-regulation of percent contribution of glucose oxidation to heart ATP turnover rates (based on low insulin/ephinephrine but high glucose/FFA ratios) remains the only described "phenotypic" expression of this adaptation, although subsequent studies may well find modifications of cardiac lipid or lactate metabolism as equally instructive phenotypic hallmarks of the adaptation.

As with muscle, we can use this identified phenotypic trait as a kind of adaptation marker to probe the acute, acclimatory, and phylogenetic levels of hypoxia adaptation. Although much remains to be done along these lines, indications of differences along all three time courses are already evident. For example, in the heart of the lowlander, an overnight fast can lead to severe suppression of glucose contributions to energy turnover, while in the highlander, cardiac metabolism continues to include the use of glucose even after an overnight fast. Thus, as in muscle, chronic hypoxia adaptation leads to fine-tuning of expression of regulatory machinery (as well as of metabolic output). However, whether or not the heart's response to acute hypoxia differs in the two subject groups is unknown; similarly, even if glucose contribution to energy turnover declines after 3 wk of deacclimation in Quechuas, it is not known if similar acclimation phenomena may occur in lowlanders.

In the brain, the strategies of up-regulating stoichiometric efficiencies (either as in muscle or as in heart) are not available alternatives; the brain in man, as in most mammals, already displays a near-absolute preference for glucose as carbon and energy source, and what is more, fully oxidizes essentially all the glucose it consumes. Thus, improving the ratio of aerobic/anaerobic contributions to energy turnover is not an available option; neither is it possible to further maximize the yield of ATP per O_2 consumed (at least with available biochemical, especially enzymatic, composition). However, it is possible to minimize the impact of chronic O_2 limitation by down-regulating ATP turnover; dropping ATP turnover rates by 20% extends by the same fractional amount the length of time the organ can function on a given amount of O_2, a time-honored hypoxia defense strategy that is almost universally found in the world of hypoxia-tolerant animals.

At this time, the only phenotypic expression of this adaptational strategy is glucose-uptake rate. Again, many further studies remain to be done to extend our understanding of this adaptational strategy. However, it is especially instructive that after a deacclimation period, brain-glucose metabolic rates in altitude Quechuas do not up-regulate toward the nominally higher levels of the lowlander; just the opposite occurs. An obvious implication is that key processes in glucose regulation and metabolic homeostasis in the brain of altitude natives operate differently from those of lowlanders (or between different setpoint than in lowlanders). Whether or not acute or acclimatory responses to hypoxia also differ, however, is unknown at this time.

In all three tissues for which good quantitative data are available, then, it is evident that hypoxia defense adapations involve

(i) adjustments of metabolic output potential per se, and

(ii) adjustments of the regulation system controlling the way metabolic output (and cell work) are used.

In functional terms, the former (adjustments in metabolic output)

(i) balance the relative contributions of aerobic and anaerobic pathways to overall ATP turnover (best represented in muscle),

(ii) balance the relative contributions of carbohydrate and fat to overall ATP synthesis by oxidative phosphorylation (best represented in the heart, but also evident in skeletal muscle), and

(iii) coordinately down-regulate ATP demand, presumably by improved efficiency or by simply slowing down ATP-requiring processes (22), and thus allow similar suppression of ATP-producing pathways to stay in energy balance despite the low ATP-turnover rates (best represented in brain).

In contrast, the latter or regulatory adaptations influence the expression of currently unknown components (genes and proteins) controlling the nature, degree, and timing of acute and acclimatory responses to hypoxia.

EVALUATING TISSUE-SPECIFIC METABOLIC ADAPTATIONS AGAINST HYPOXIA

The final question concerns how big an impact, at the whole-organism level, these tissue-specific responses make. The heart at rest (26) accounts for about 10% of whole-body resting metabolic rate (RMR). By preferentially burning glucose, it maximally gains about a 60% advantage. At normal cardiac O_2 consumption rates, this means the heart's contribution to RMR could drop to 6.25%, or a modest 3%–4% advantage and impact on whole-organism O_2 needs. On the other hand, it extends by a maximum factor of 1.6-fold the length of time the heart can survive on a given amount of oxygen—a considerable advantage whenever O_2 is limiting.

Considering its modest size, the brain, like the heart, makes a surprisingly large contribution to overall whole-body ATP turnover rates: almost 20% of the RMR is normally ascribed to brain metabolism (26). By resetting its basal ATP turnover rates at a hypometabolic level, on average close to 20% below normal, the energy demands of the brain can be reduced to 15% of RMR. As in the case of the heart, this represents a maximum 4% advantage and impact on whole-body O_2 requirements and, as in the heart, this could extend by a factor of about 1.2-fold the length of time the brain can survive on a given amount of oxygen—again a useful advantage whenever O_2 is limiting.

The adaptational response in muscle tends toward maximizing the percent contribution of aerobic versus anaerobic pathways to overall ATP production. Compared to the above, the payoff of this adaptational strategy is enormous—because of the enormous (18-fold) difference between ATP yield per mole of glucose in oxidative phosphorylation versus anaerobic glycolysis. A simple calculation drives home this point. Assume a working metabolic rate of 38 μmol ATP $\cdot$ g^{-1} $\cdot$ min^{-1}, 1 μmol of glucose fully oxidized supplies 36 of the required 38 μmols of ATP, while 1 μmole fermented to lactate supplies the additional 2 μmols ATP; the lactate is then reconverted to glucose (costing 6 ATP) for subsequent complete oxidation. The two-way glucose ↔ lactate flux shows a net cost of 4 ATP, so that when this glucose is subsequently fully oxidized its net gain is only 32, not 36, ATP; relying on glucose fermentation to "top up" or "make up" for any hypoxia-based energetic deficit, then, reduces the efficiency of O_2 utilization to only 88% of theoretical maximum. At rest, skeletal muscle accounts for about 30% of RMR, so a 12% drop in efficiency again has only a 3%–4% impact. However, in heavy exercise, skeletal muscle can account for up to 90% of whole-organism metabolic rate (47); at this rate of O_2 flux, the 12% drop in efficiency represents a metabolism essentially equal to RMR, and thus represents a much larger absolute impact on overall whole-body O_2 management and on hypoxia tolerance. That may be the ultimate selective advantage of muscle glycolytic down-regulation in indigenous highlanders; in effect, the adaptational process seems to be maximizing the contribution of mitochondrial metabolism to energy production in these muscles, while minimizing potential glycolytic contributions to their work. Not being able to rely on a significant glycolytic contribution to meeting high ATP demands may well carry a price of its own: a reduced ceiling on maximum (aerobically supportable) muscle work. Interestingly, rather low $\dot{V}O_{2max}$ values seem to characterize both Sherpas (33) and Quechuas (27), whose muscles also display low mitochondrial volume densities (6, 33) and low activities of mitochondrial enzymes (28).

It could be said that trade-off is the name of the game in evolutionary adaptation of both animals and humans.

Many thanks are extended to my collaborators in Canada, the U.S., Peru, and Nepal. Most of this work was supported by NSERC (Canada).

REFERENCES

1. Arthur, P. G., M. C. Hogan, D. E. Bebout, P. D. Wagner, and P. W. Hochachka. Modelling effects of hypoxia on ATP turnover in exercising muscle. *J. Appl. Physiol.* 73: 737–742, 1992.
2. Balaban, R. S. Regulation of oxidative phosphorylation in the

mammalian cell. *Am. J. Physiol.* 258 (*Cell Physiol.* 27): C377–C389, 1990.

3. Barrett, E. J., R. G. Schwartz, C. K. Francis, and B. L. Zaret. Regulation by insulin of myocardial glucose and fatty acid metabolism in the conscious dog. *J. Clin. Invest.* 74: 1073–1079.

4. Brindle, K. M., M. J. Blackledge, R. A. J. Challis, G. K. Radda. ^{31}P NMR Magnetization transfer measurements of ATP turnover during steady-state isometric muscle contraction in the rat hind limb *in vivo. Biochemistry* 28: 4887–4893, 1989.

5. Callow, M. A., M. Morton, and M. Guppy. Marathon fatigue: the role of plasma free fatty acids, muscle glycogen, and blood glucose. *Eur. J. Appl. Physiol.* 55: 654–661, 1986.

6. Cerretelli, P., B. Kayser, H. Hoppeler, D. Pette. Muscle morphometry and enzymes in acclimatization. In: *Hypoxia—The Adaptations* edited by Sutton, J. R., G. Coates, and J. E. Remmers. Toronto: B. C. Decker, 1990, p. 220–224.

7. Connett, R. J. Analysis of metabolic control: new insights using scaled creatine kinase model. *Am. J. Physiol.* 254 (*Regulatory Integrative Comp. Physiol.* 23): R949–R959, 1988.

8. Connett, R. J., and C. R. Honig. Regulation of $\dot{V}O_{2max}$. Do current biochemical hypotheses fit the *in vivo* data? *Am. J. Physiol.* 256 (*Regulatory Integrative Comp. Physiol.* 25): R898–R906, 1989.

9. Daut, J., and G. Elzinga. Substrate dependence of energy metabolism in isolated guinea pig cardiac muscle: a microcalorimetric study. *J. Physiol.* 413: 379–387, 1989.

10. Drake, A. Substrate utilization in the myocardium. *Basic Res. Cardiol.* 19: 1–11, 1985.

11. Edwards, H. T. Lactic acid in rest and work at high altitude. *Am. J. Physiol.* 116: 367–375, 1936.

12. From, A. H. L., S. D. Zimmer, S. P. Michurski, P. Mohanakrishnan, V. K. Ulstad, W. J. Thomas, and K. Ugurbil. Regulation of oxidative phosphorylation in the intact cell. *Biochemistry* 29: 3733–3743, 1990.

13. Funk, C. I., A. Clark, Jr., R. J. Connett. A simple model of metabolism: applications to work transitions in muscle. *Am. J. Physiol.* 258 (*Cell Physiol.* 27): C995–C1005, 1990.

14. Green, H. J., J. Sutton, P. Young, A. Cymerman, and C. S. Houston. Operation Everest II: muscle energetics during maximal exhaustive exercise. *J. Appl. Physiol.* 66: 142–150, 1989.

15. Harik, S. I., R. E. Behmand, and J. C. Lamanna. Chronic hypobaric hypoxia increases the density of cerebral capillaries and their glucose transporter protein. *J. Cerebr. Blood Flow Metab.* 11: 5496, 1991.

16. Henes, C. G., S. R. Bermann, M. N. Walsh, B. E. Sobel, and E. M. Geltman. Assessment of myocardial oxidative metabolic reserve with positron emission tomography and carbon-II acetate. *J. Nucl. Med.* 30: 1489–1499, 1989.

17. Hermansen, L., and O. Vaga. Lactate disappearance and glycogen synthesis in human muscle after maximal exercise. *Am. J. Physiol.* 233 (*Endocrinol. Metab. Gastrointest. Physiol.* 2): E422–E429, 1977.

18. Hicks, R. J., W. H. Herman, V. Halff, E. Molina, E. R. Wolfe, G. Hutchins, and M. Schwaiger. Quantitative evaluation of regional substrate metabolism in the human heart by positron emission tomography. *J. Am. Coll. Cardiol.* 18: 101–111, 1991.

19. Hochachka, P. W. Exercise limitations at high altitude: the metabolic problem and search for its solution. In: *Circulation, Respiration, and Metabolism,* edited by R. Gilles. Berlin: Springer-Verlag, 1985, p. 240–249.

20. Hochachka, P. W. The lactate paradox: analysis of underlying mechanisms. *Ann. Sports Med.* 4: 184–188, 1988.

21. Hochachka, P. W. Metabolic biochemistry and the making of a mesopelagic mammal. *Experientia* 48: 570–575, 1992.

22. Hochachka, P. W. The Monge legacy: learning from hypoxia adapted animals and man. In: *Hipoxia—Investigaciones Basicas y Clinicas.Homenaje a Carlos Monge Cassinelli,* edited by Leon-Velarde, F., and A. Arregui, Lima: UPCH, 1993, p. 361–374.

23. Hochachka, P. W., C. M. Clark, W. D. Brown, C. Stanley, C. Stone, R. J. Nickles, G. G. Zhu, P. S. Allen, J. E. Holden. The brain at high altitude: Hypometabolism as a defense against chronic hypoxia? *J. Cerebr. Blood Flow Metab.,* 14: 671–679, 1994.

24. Hochachka, P. W., and M. Guppy. *Metabolic Arrest and the Control of Biological Time.* Cambridge, MA: Harvard University Press, 1987, p. 1–227.

25. Hochachka, P. W., T. P. Mommsen, J. H. Jones, and C. R. Taylor. Substrate and O_2 fluxes during rest and exercise in a high-altitude-adapted animal, the llama. *Am. J. Physiol.* 253 (*Regulatory Integrative Comp. Physiol.* 22): R298–R305, 1987.

26. Hochachka, P. W., and G. N. Somero. Biochemical Adaptation. Princeton University Press, Princeton, 1984, p. 1–537.

27. Hochachka, P. W., C. Stanley, G. O. Matheson, D. C. McKenzie, P. S. Allen, and W. S. Parkhouse. Metabolic and work efficiencies during exercise in Andean natives. *J. Appl. Physiol.* 70: 1720–1730, 1991.

28. Hochachka, P. W., C. Stanley, D. C. McKenzie, A. Villena, and C. Monge. Enzyme mechanisms for pyruvate-to-lactate flux attenuation; a study of Sherpas, Quechuas, and Hummingbirds. *Int. J. Sports Med.* 23: S119–S122, 1992.

29. Holden, J. E., W. D. Brown, C. Stone, C. Stanley, R. J. Nickles, and P. W. Hochachka. Enhanced cardiac metabolism of plasma glucose in high altitude natives. *J. Appl. Physiol.,* 1995, in press.

30. Hurtado, A., T. Velasquez, C. Reynafarje, R. Lozano, R. Chavez, H. A. Salazar, B. Reynafarje, C. Sanchez, and J. Munoz. *Mechanisms of Natural Acclimatization. Studies on the Native Resident of Morococha, Peru, at an Altitude of 14,900 Feet.* Randolph AFB, Texas: USAF School of Aviation Medicine, 1956, p. 1–62 (Rep. 56-1).

31. Hutter, J. F., H. M. Piper, and P. G. Spieckermann. Effect of fatty acid oxidation on efficiency of energy production in rat heart. *Am. J. Physiol.* 249 (*Heart Circ. Physiol.* 18): H723–H728, 1985.

32. Kanno, T., K. Sudo, I. Takeuchi, S. Kanda, N. Honda, and K. Oyama. Hereditary deficiency of lactate dehydrogenase M-subunit. *Clin. Chim. Acta* 108: 267–276, 1980.

33. Kayser, B., H. Hoppler, H. Classen, and P. Cerretelli. Muscle ultrastructure and performance capacities of Himalayan Sherpas. *J. Appl. Physiol.* 70: 1938–1942, 1991.

34. Kollias, J., E. R. Buskirk, R. F. Akers, E. K. Prokop, P. T. Baker, and E. Picon-Reatequi. Work capacity of longtime residents and newcomers to altitude. *J. Appl. Physiol.* 24: 792–799, 1968.

35. Little, M. A., and J. M. Hanna. The response of high altitude populations to cold and other stresses. In: *The Biology of High Altitude Peoples,* edited by P. T. Baker. Cambridge, UK: Cambridge Univ. Press, 1978, p. 251–298.

36. Marwick, T. H., W. J. MacIntyre, E. E. Salcedo, R. T. Go, G. Saha, and A. Beachler. Identification of ischemic and hibernating myocardium: feasibility of post-exercise F-18 deoxyglucose positron emission tomography. *Cathet. Cardiovasc. Diagn.* 22: 100–106, 1991.

37. Matheson, G. O., P. S. Allen, D. C. Ellinger, C. C. Hanstock, D. Gheorghiu, D. C. McKenzie, C. Stanley, W. S. Parkhouse, and P. W. Hochachka. Skeletal muscle metabolism and work capacity: a ^{31}P-NMR study of Andean natives and lowlanders. *J. Appl. Physiol.* 70: 1963–1976, 1991.

38. Mayekawa, M., K. Sudo, S. S. Li, and T. Kanno. Genotypic analysis of families with lactate dehydrogenase A (M) deficiency by selective DNA amplification. *Hum. Genet.* 88: 34–38, 1991.

39. McKenzie, C. D., L. S. Goodman, B. Davidson, C. C. Nath, G. O. Matheson, W. S. Parkhouse, P. W. Hochachka, P. S. Allen, C.

Stanely, and W. Ammann. Cardiovascular adaptations in Andean natives after 6 wk of exposure to sea level. *J. Appl. Physiol.* 70: 2650–2655, 1991.

40. McLellan, T., I. Jacobs, and W. Lewis. Acute altitude exposure and altered acid-base status. I. Effects on exercise ventilation and blood lactate responses. *Eur. J. Appl. Physiol.* 57: 435–444, 1988.

41. Monge, C. M. *Acclimatization in the Andes.* Baltimore, MD: John Hopkins University Press, 1948.

42. Monge, C. M., and F. Leon-Velarde. Physiological adaptation to high altitude: Oxygen transport in mammals and birds. *Physiol. Rev.* 71: 1135–1172, 1991.

43. Moret, P. Myocardial metabolism: acute and chronic adaptation to hypoxia. *Med. Sport Sci.* 19: 48–63, 1985.

44. Nioka, S., Z. Argov, G. P. Dobson, R. E. Forster, H. V. Subramanian, R. L. Veech, and B. Chance. Substrate regulation of mitochondrial oxidative phosphorylation in hypercapnic rabbit muscle. *Am. J. Physiol.* 72: 521–528, 1991.

45. Rumsey, W. L., C. Schlosser, E. M. Nuutinen, M. Robiollo, and D. F. Wilson. Cellular energetics and the oxygen dependence of respiration in cardiac myocytes isolated from adult rate. *J. Biol. Chem.* 265: 15392–15402, 1990.

46. Sutton, J. R. The hormonal responses to exercise at sea level and at altitude. In: *Hypoxia, Exercise, and Altitude.* New York: Liss, 1983, p. 325–358 (Proc. 3rd Banff Int. Hypoxia Symp.).

47. Taylor, C. R., R. H. Karas, E. R. Weibel, and H. Hoppeler. Adaptive variation in the mammalian respiratory system in relation to energetic demand. II. Reaching the limits to oxygen flow. *Respir. Physiol.* 69: 7–26, 1987.

48. Van Hardeveld, C. Effects of thyroid hormone on oxygen consumption, heat production, and energy economy. In: *Thyroid Hormone Metabolism,* edited by G. Hennemann. New York: Dekker, 1986, p. 579–608.

49. Vik-Mo, H., and O. D. Mjos. Influence of free fatty acids on myocardial oxygen consumption and ischemic injury. *Am. J. Cardiol.* 48: 361–365, 1981.

50. Way, A. B. Exercise capacity of high altitude Peruvian Quechua Indians migrant to low altitude. *Hum. Biol.* 48: 175–191, 1976.

51. West, J. Lactate during exercise at extreme altitudes. *Federation Proc.* 45: 2953–2957, 1986.

52. Winslow, R. M., and C. C. Monge. *Hypoxia, Polycythemia and Chronic Mountain Sickness.* Baltimore, MD: Johns Hopkins University Press, 1987.

53. Young, A. J., W. J. Evans, A. Cymerman, K. B. Pandolf, J. J. Knapik, and J. T. Maher. Sparing effects of chronic high-altitude exposure on muscle glycogen utilization. *J. Appl. Physiol.* 52: 857–862, 1982.

54. Young, P. M., P. B. Rock, C. S. Fulco, L. A. Trad, V. A. Forte, Jr., and A. Cymerman. Altitude acclimation attenuates plasma ammonia accumulation during submaximal exercise. *J. Appl. Physiol.* 63: 758–764, 1988.

55. Zivin, J. A., and D. W. Choi. Stroke therapy. *Sci. Am.* 265: 56–63, 1991.

49 Hypoxia, erythropoietin gene expression, and erythropoiesis

PETER J. RATCLIFFE | *Institute of Molecular Medicine, John Radcliffe Hospital, Oxford, England*

KAI-UWE ECKARDT | *Physiologisches Institüt der Universität Regensburg, Regensburg, Germany*

CHRISTIAN BAUER | *Physiologisches Institüt der Universität Zürich, Zürich, Switzerland*

ADULT HUMAN BEINGS NORMALLY produce about 2 million erythrocytes per second to compensate for those cells that, after a journey that lasts for 120 days and covers a distance of about 300 kilometers in the circulation, are removed by phagocytes in spleen, liver, and bone marrow. The process by which this continuous regeneration of erythrocytes occurs, termed *erythropoiesis,* comprises multiple steps of division and maturation of erythroid precursor cells in the bone marrow. Normally the rate of erythropoiesis is geared to contribute to the maintenance of a constant oxygen supply to the oxygen-consuming tissues. Blood oxygen supply can be reduced in two ways. First, the oxygen saturation of hemoglobin may decline if ambient oxygen tension decreases; for instance, upon ascent to high altitude or in heart-lung disease. Second, blood cells may be lost, either from bleeding or from pathological reduction of erythrocyte lifespan. Under all these conditions cardiovascular adaptations can be observed, which may partially compensate for the hypoxic condition, but fundamental correction requires an adaptation of erythropoiesis. During severe hypoxic stimulation the normal production rate of red cells can increase up to tenfold (115). The major humoral factor that is responsible both for maintenance of a normal blood erythrocyte count under conditions of constant oxygen availability, and for adaptation of erythropoiesis to changes in oxygen availability, is a glycoprotein named erythropoietin. In adults it is predominantly produced in the kidneys, but for reasons that are still unclear, during fetal life and the early neonatal period, the liver is the major source. The concentration of this hormone in the blood is normally fairly constant, being approximately 5 pmol/l. However, when the oxygen content of blood decreases, as in severe chronic anemias, the erythropoietin concentration can increase up to 1,000-fold above normal.

The existence of erythropoietin was postulated as early as 1906 (30). However, it took the diligence and experimental imagination of later researchers to better define the concept of a hormone, erythropoietin, that is produced in response to hypoxia (69, 85, 116, 152, 219). Major breakthroughs in erythropoietin research were achieved following the purification of a few milligrams of human erythropoietin (189), the subsequent cloning of the gene, and its successful expression in permanent mammalian cell lines (122, 167). These achievements led to the availability of recombinant human erythropoietin, which has been used in clinical therapy with great success, in particular for the treatment of renal anemias (for review, see refs. 3, 75). The availability of recombinant hormone in large quanitities has also greatly facilitated studies of the physiological regulation of erythropoietin, and the development of molecular probes has permitted studies of erythropoietin gene regulation.

While this progress has turned erythropoietin research into a rapidly developing and expanding field, many aspects of the biosynthesis and action of this hormone still remain unknown. However, convincing evidence has been assembled indicating that a major control in the feedback loop that regulates erythropoiesis in an oxygen-dependent fashion occurs through changes in the expression of the erythropoietin gene. In this overview, we will put major emphasis on this latter aspect, and try to outline the current, albeit fragmentary, knowledge of the regulation of the erythropoietin gene, since this appears to be a fitting example of how an adaptive process that is important for tissue oxygenation of the whole organism can be studied at the molecular level. To elucidate the physiological importance of erythropoietin and its regulation we will, however, first provide a brief summary of the functionally important parts of the molecule, its action on erythropoietic target cells, and then describe the principles of its biosynthesis and metabolism under different physiological conditions. A comprehensive survey of these aspects, however, is beyond the scope of this contribution, and for further detailed information the reader is referred to recent reviews and monographs (9, 125, 148).

The Functional Anatomy of the Erythropoietin Molecule

In this section the structure of erythropoietin will be discussed, with particular emphasis on those parts that are thought to be of functional relevance with regard to its interaction with the erythropoietin receptor. The erythropoietin genes from humans (122, 167), old world monkeys (166), and mice (177, 247) have been cloned. The gene encoding human erythropoietin lies on the long arm of chromosome 7, 7q21 (211). It contains four introns and five exons and codes for a 193-amino-acid polypeptide (122, 167). A 27-amino-acid leader sequence at the N-terminal part is cleaved off during secretion of the hormone, in addition to the removal of the carboxy-terminal arginine during posttranslational processing. The mature product is therefore des-Arg 165 erythropoietin, both as the recombinant form and from urinary sources (217).

Sedimentation equilibrium experiments showed erythropoietin to be a single macromolecular component with a molecular weight of 30,400 and a carbohydrate content of 40% (49). The carbohydrate part consists of four oligosaccharide side chains, three of which are N-linked to asparagine residues 24, 38, and 83, and one is O-linked to serine 126 in human and monkey erythropoietin (27, 68, 160, 224). In mouse erythropoietin, serine 126 is replaced by proline, but it is not yet known whether mouse erythropoietin has an alternative O-linked glycosylation site elsewhere.

The carbohydrate moiety may be responsible for the expanded molecular structure of erythropoietin, which can be inferred from a difference between the calculated (20A) and the experimentally observed (32A) Stokes radius (49). It has been postulated that the complex oligosaccharide chains (224, 260) extend into the solvent and hence significantly increase the Stokes radius (49). Such a molecular fence may protect the erythropoietin molecule from unwanted interactions with plasma proteases or neuraminidases. The carbohydrate portion of the molecule carries about 12 sialic acid residues per mole erythropoietin, which renders the molecular very acidic at physiological pH, isoionic point 4.2–4.5 (49, 53, 120). In contrast, both desialiated and nonglycosylated erythropoietin have isoionic points around 9.0 (49, 120).

The in vivo activity of the hormone is strongly dependent on the number of sialic acids (120). Most probably this arises because desialiation exposes galactose residues and leads to a rapid uptake of partially or fully desialiated erythropoietin via galactose receptors in the liver, thereby reducing the half-life of the hormone. In vitro, the position is quite different and partially or fully desialiated erythropoietin shows two- to fourfold greater in vitro activity than the intact form when tested in cultures of bone marrow cells (101, 120). Even when all of the carbohydrates are removed, the "naked" protein still binds to the erythropoietin receptor and is able to sustain growth of erythroid progenitor cells in vitro (269). Nevertheless, deglycosylated derivatives of erythropoietin have a strong tendency to aggregate, and it is only the monomeric form that exhibits a biological activity even in vitro (53, 101). Therefore, the hydrophilic oligosaccharide structures may maintain the conformation of the hydrophobic polypeptide structure even if they are not directly involved in the interaction with cellular receptors for erythropoietin (101, 193). Thus, the sugar chains on the erythropoietin molecule appear to have two major roles: extending the residence time in plasma, and contributing to the structural integrity of the protein component of the molecule.

We now turn to the protein component of erythropoietin and consider experiments directed toward the identification of those regions that interact with the

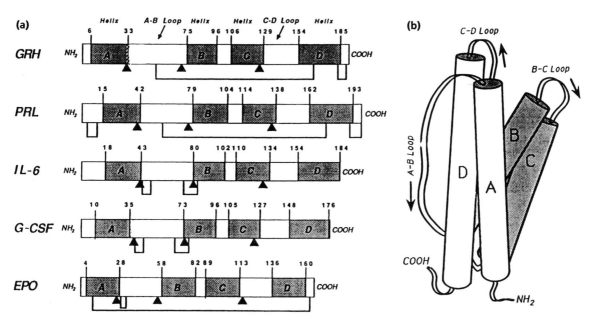

FIG. 49.1. Secondary and tertiary structure prediction for a group of GRH-like cytokines. (a) Location of secondary structural elements in the human sequences for GRH, PRL, IL-6, G-CSF, and EPO. The four α helices from the GRH X-ray structure are labeled A to D; loops between helices are appropriately named. Known disulfide bridge connections are marked in black lines. Gene exon boundaries are indicated beneath the respective protein sequences by black triangles. (b) Drawing of the GRH fold emphasizing the helix bundle core and loop connectivity. The exposed surface of helix D will play an important role in receptor binding (from ref. 11, with permission).

erythropoietin receptor. Earlier experiments have shown that the two disulfide bonds located between residues Cys 7 and Cys 161 and between Cys 29 and 33 (160, 177) are essential for the correct overall folding of the erythropoietin backbone and structural interaction with its receptor (274). More recently, investigators have sought to examine the structure-function relationship of defined amino acid sequences using oligonucleotide-directed mutagenesis. Before discussing the results, some consideration should be given to the molecular model of erythropoietin. The algorithm developed by Chou and Fasman (1978) to predict secondary structures from the amino acid sequence was used to calculate a model of erythropoietin's molecular structure (177). However, the Chou-Fasman rules do not take into account any interaction of the protein part with the carbohydrate moiety in glycoproteins and are therefore of limited predictive value in such sugar-conjugated proteins. More recently, erythropoietin has been predicted to consist of four antiparallel α-helices, with two long and one short loop connection, and to resemble growth hormone (Fig. 49.1). This model is consistent with circular dichroic spectra that indicate a preponderance of helical structures in erythropoietin (49, 160). Furthermore, deletion mutants made within the proposed helical domains resulted in unstable products of erythropoietin with no detectable biologic activity

(22, 34). In contrast, deletion mutants that lie outside the predicted α-helical bundles (for instance Δ111–119 and Δ120–129) lead to mutants of erythropoietin that were stable with full retention of biologic activity (22, 34). These results are in agreement with studies using antibodies directed against conformation-dependent and conformation-independent parts of the protein part of erythropoietin. Only antibodies that recognize conformation-dependent epitopes, probably confined to the α-helical core, inhibited binding of erythropoietin to its receptor (46, 258). However, proof of this model, and whether, as suggested by Bazan (10, 11), helix D is the most critical part of the ligand must await the resolution of the erythropoietin-receptor complex by X-ray crystallography.

THE ERYTHROPOIETIN RECEPTOR

Many investigators have used radiolabeled erythropoietin to demonstrate specific binding to cells derived from erythroid lineages. These comprise virally transformed spleen cells, murine and human erythroid erythroleukemia cells, as well as normal erythroid progenitors both from adult and fetal sources (for reviews, see refs. 47, 148). Scatchard analysis has revealed that ~500–1,000 erythropoietin receptors are present on the surface of purified human erythroid precursor cells that have a dissociation constant of ~200 pM (26, 228). In some cell lines, two classes of erythropoietin receptors have been documented: a high-affinity receptor (K_D~200 pM) and a low-affinity receptor (K_D~600–800 pM). The functional significance of these two receptor classes is not clear. The recent cloning of the murine (46) and human (134, 277) erythropoietin receptor has provided new insights into the structure of the binding domains and erythropoietin signal transduction. As inferred from the sequence of the cDNA, the cloned mouse erythropoietin receptor is a 507-amino-acid molecule with one single membrane-spanning domain between amino acids 250 and 272. The extracytoplasmic N-terminal part of the receptor contains the erythropoietin binding domain. The C-terminal cytoplastmic tail is associated with the signal transduction. It is now recognized that the erythropoietin receptor shares structural homology with several other receptors for hematopoietic growth factors, lymphokines, and growth-hormone-related proteins that all belong to a new and growing superfamily of receptor proteins (10, 11, 35). All these receptors share important morphological and functional characteristics. For instance, just outside the transmembrane domain is a highly conserved protein motif, Trp-Ser-X-Trp-Ser. Site-directed mutagenesis of this motif prevents function of the erythropoietin receptor (35, 279). Erythropoietin receptor molecules with mutations of this motif were not processed correctly, did not bind erythropoietin, and could not activate cells (279).

One important control over expression of the erythropoietin receptor is exerted by the recently isolated nuclear DNA–binding protein GATA-1. GATA-1 not only plays a key role in regulation of the erythropoietin receptor, but also governs the expression of various erythroid-specific genes during the development of blood cells. The name GATA-1 is derived from a consensus motif in the promotor region of a number of genes, which includes that coding for the erythropoietin receptor (36, 113, 285). Although GATA-1 was originally thought to be erythroid-specific (266, 267), it was subsequently found in megakaryocytes and mast cell lineages (171, 222). Comparison of erythropoietin receptor mRNA levels, and assays of the receptor itself in the interleukin-3-dependent mouse cell line 32D, indicates the existence of control mechanisms for erythropoietin receptor expression other than transcriptional activation by GATA-1. These experiments have suggested that another important regulatory step determining the erythroid-specific response to erythropoietin is the efficiency of the translocation of the erythropoietin receptor to the cell surface (183). Recently, evidence of an alternatively spliced and truncated erythropoietin receptor has been reported (192). In these experiments the truncated receptor was found to be more prevalent in earlier erythroid progenitors and to be functionally different. Although the truncated form could transduce a mitogenic signal, it appeared to be less effective in preventing programmed cell death. Alternative splicing of the erythropoietin receptor gene appears therefore to be yet another control mechanism for receptor function.

Many studies have addressed the question of possible second messengers that may or may not be involved in the complex chain of events leading from erythropoietin receptor ligand binding to erythroid differentiation (for review, see ref. 148). Though many of these events are not yet well understood, recent studies have demonstrated very clearly that binding of erythropoietin induces tyrosine phosphorylation of a defined set of proteins (7, 151, 215), including the cytosolic part of the erythropoietin receptor (55, 188). Since no catalytic activity is predicted from sequence analysis of the cloned erythropoietin receptor molecule, these phosphorylation events could be mediated by proteins that have a tyrosine kinase activity and are associated with the erythropoietin receptor (47, 55). It is worth noting in this connection that a clear correlation exists between the ability of erythropoietin to bind and its abilities to support growth and to induce tyrosine phosphorylation (188).

ERYTHROPOIESIS AND ERYTHROID DIFFERENTIATION

Blood cells continuously originate from small numbers of pluripotent stem cells that can make identical copies of themselves; i.e., they continuously renew themselves in the bone marrow. These stem cells generate progenitor cells that are committed to either erythrocytic, megakaryocytic, granulocytic, or monocytic differentiation. Control of these differentiation pathways is determined, in part, through specific growth factors binding to their cognate receptors. The first morphologically recognizable erythroid-committed cell is the proerythroblast, which divides four times and therefore yields 16 reticulocytes that later become the mature red blood cells. The immediate progenitor of the proerythroblasts is named the colony-forming unit-erythroid (CFU-E), which in turn is derived from the burst-forming unit-erythroid (BFU-E), the earliest erythroid progenitor cell type identified so far. BFU-E and CFU-E differ in that the former cell pool requires much larger concentrations of erythropoietin than the CFU-E, which may be related to the fact that the BFU-E have many fewer erythropoietin receptors on their surface than CFU-E (26, 225, 227). The BFU-E can be further divided into a more immature form that is dependent on the pressure of interleukin 3, but does not respond to erythropoietin, and a more mature form that is dependent both on interleukin 3 and erythropoietin (225). Thus, immature BFU-E undergo a well-controlled switch from a dependence on interleukin 3 to a dependence on erythropoietin for viability, with a concomitant expansion of erythropoietin receptors (148). The main target cells for the action of erythropoietin is the CFU-E and their immediate descendants, the proerythroblasts. CFU-E/Proerythroblasts have the highest receptor density of all erythroid progenitor cells, but erythropoietin binding continuously declines with further erythroid differentiation to undetectable levels at the reticulocyte stage (26, 226, 227). Indeed, it has been shown that it is largely the CFU-E compartment that increases with increasing erythropoietin concentrations and decreases when erythropoietin levels fall after transfusing red blood cells (4, 105, 121, 200). In view of the fact that erythropoietin promotes the survival of the CFU-E cells (142), any increase in erythropoietin levels would allow a greater proportion of cells entering the CFU-E stage to survive.

The time from induction of hypoxia-induced erythropoietin production to a significant increase in hematocrit is 60–70 h following anemic hypoxia in mice (147). This time course fits with the 60–70 h needed for CFU-E to differentiate into mature erythrocytes (141, 144), thereby corroborating the notion that erythropoietin-dependent erythropoiesis is mainly regulated at the level of CFU-E. Erythropoietin binding to CFU-E is associated with a number of differentiation events that are characteristic for this maturation pathway. These comprise induction of hemoglobin synthesis, extrusion of the nucleus from erythroblasts, increased expression of transferrin receptors, erythropoietin receptors, and erythrocyte membrane components such as glycophorin, band 3, and band 4.1 (reviewed in ref. 148).

The transcription factor GATA-1 seems to provide a master switch for erythroid differentiation (36; and earlier references given therein). Inactivation of the gene that codes for GATA-1 by targeted disruption results in a lack of mature red blood cells in chimeric transgenic mice (206). If it is true that binding of erythropoietin to its receptor enhances GATA-1 expression in erythroid progenitor cells (36), this protein may also be involved in the prevention of programmed death (apoptosis) in erythroid cells. Apoptosis of CFU-E was shown to be greatly inhibited by erythropoietin, allowing them to develop into red blood cells (142). Such a model predicts that in normal human beings the majority of CFU-E do not survive, but with hypoxia-induced increases of the erythropoietin concentration the cells are rescued and differentiate into mature red blood cells.

MODULATION OF SERUM ERYTHROPOIETIN LEVELS

The discovery of erythropoietin as the first hematopoietic growth factor was a consequence of the fact that it is unique in being produced mainly, if not exclusively, outside the bone marrow and transported to its target cells via the bloodstream. Erythropoietin thus fulfills criteria of a true hormone and, in fact, measurements of erythropoietin concentrations in blood serum provide much of the basis of the current knowledge on erythropoietin regulation (for example, see Fig. 49.2).

Assays for Erythropoietin

Early estimations of erythropoietin in biological fluids were based on the induction of an increase in red cell mass following the injection of serum, plasma, or urine into otherwise untreated rodents. Suppressing endogenous erythropoietin production and erythropoiesis in

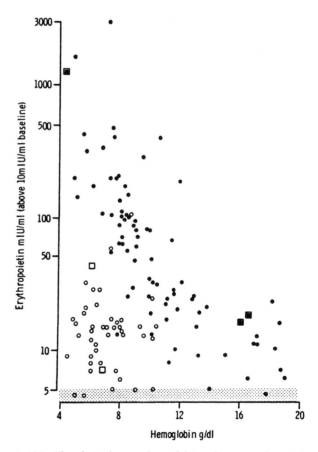

FIG. 49.2. The relation between hemoglobin and serum erythropoietin in patients without renal disease *(closed circles)*, patients with a functioning renal transplant *(closed squares)*, and patients with anemia associated with renal failure managed by dialysis (nephric *(open circles)* and anephric *(open squares)* subjects) (from ref. 41, with permission).

recipient animals by starvation (210), hypophysectomy (92) or polycythemia (42), and using ^{59}Fe incorporation into red blood cells as the response parameter (210) allowed improvement in the sensitivity and accuracy of such "in vivo bioassays" for erythropoietin. Erythropoietin can also be determined by "in vitro bioassays" (54, 149). In these assays a proliferation of erythroid progenitors (180) or induction of hemoglobin synthesis (149) is measured in short-term cultures of hematopoietic cells. However, substances other than erythropoietin may also exert erythropoietic activity or interfere with it, and thus the specificity of these in vitro assays is low. Following the purification of native erythropoietin and expression of the recombinant hormone, immunoassays for the hormone were developed that have nowadays largely replaced the bioassays (40, 67, 95, 96, 218, 246, 278). Although immunological cross-reactants may occasionally invalidate erythropoietin measurements (45, 245), estimates by immunoassay and by bioassay in vivo are generally very similar (67, 95, 96, 218, 246). The sensitivity, however, of immunoassays for erythropoietin is approximately tenfold greater than that of in vivo bioassays and their availability has greatly facilitated the assessment of normal and subnormal as well as elevated serum erythropoietin concentrations.

Serum Erythropoietin Levels and Their Dependence on Oxygenation

Erythropoietin is traditionally quantified as "international units" (IU). One IU was originally defined as the amount of erythropoietic stimulating material that produced an erythropoietic response equivalent to 5 μmol cobaltous chloride in assay rats. Normal serum values for erythropoietin of healthy subjects not living at extreme altitudes are mostly in the range of 5–30 mU/ml when determined by radioimmunoassays (40, 96, 125, 172, 218). Assuming a specific activity of the hormone of 130,000 IU/mg protein, this corresponds to about 1–7 pmol/l. Baseline erythropoietin levels were found to be independent of age beyond two months (60, 114) and also independent of gender (40, 96, 218).

This "baseline" concentration of erythropoietin in serum appears essential to maintain erythropoiesis at a rate that compensates for the physiologic demise of erythrocytes, and thus to maintain the red cell mass. When animals are immunized against erythropoietin they develop a severe and lethal anemia (95, 153). In turn, any variation in oxygen supply of the organism usually leads to opposite changes in serum erythropoietin concentrations, and the ensuing inverse relationship between oxygenation and serum erythropoietin levels provides the basis for an efficient feedback modulation of erythropoiesis. Erythropoietin levels rise irrespective of whether tissue oxygen supply is reduced through a reduction in hemoglobin concentration, as in anemia (40, 74, 96, 130), through a reduction in oxygen transport capacity of hemoglobin, as occurs with carbon monoxide inhalation (62, 88, 129, 229, 259), through a reduction in arterial oxygen saturation of hemoglobin (1, 58, 184, 232), or through increased oxygen affinity of hemoglobin (162). Thus, it appears that it is the amount of oxygen available rather than the number of red cells per se that determines the erythropoietin response, and that control operates through sensing of tissue oxygenation at one or more sites. Conversely, when blood oxygen content is increased, as in polycythemia rubra vera, serum erythropoietin levels are on average decreased (17, 43, 71, 96, 139, 218).

Kinetics and Quantity of Increases in Serum Erythropoietin Levels

When man or rodents are exposed to acute hypoxia, erythropoietin levels remain unchanged for about 1.5 to 2 h and thereafter increase approximately linearly. Less than 15 min of hypoxia are, however, sufficient to trigger a measurable increase in erythropoietin formation (28, 62). The rest of the lag phase appears to reflect the time required for signal transduction, and synthesis and secretion of the hormone. The rate of rise of serum erythropoietin and the peak concentration that is reached after 6 to 24 h in rodents and within 48 h in humans are exponentially related to the degree of hypoxic stimulation. Following more prolonged hypoxia erythropoietin concentrations decline again, and remain only slightly elevated (1, 32, 58, 91, 124, 234). This reduction occurs well before red cell mass is increased, presumably through an adaptation of the oxygen-sensing mechanisms that govern erythropoietin formation (see below). Serum concentrations of erythropoietin that are only moderately elevated or even within the normal range appear to be sufficient, however, to maintain an increased rate of erythropoiesis under conditions of chronic hypoxia (43).

Disturbances in Oxygen-Dependent Control of Serum Erythropoietin Levels

A variety of conditions have been found to modulate or even abrogate the response of erythropoietin levels to reduction in oxygen availability. Thus, the most obvious exception to the exponential relationship between hemoglobin and erythropoietin concentrations occurs in the anemia that accompanies chronic renal failure, where serum erythropoietin levels generally remain within, or only slightly above, the normal range despite marked reductions in hemoglobin concentrations (31, 44, 76, 179) (Fig. 49.2). Although the pathogenesis of renal anemia is considered to be multifactorial (75), the significance of inappropriately low erythropoietin formation has in recent years been clearly confirmed through the successful treatment of uremic patients with recombinant erythropoietin (44, 76).

An important role of pituitary-dependent hormones in erythropoietin formation is seen from the results of hypophysectomy, when basal and hypoxia-induced erythropoietin formation is greatly reduced (112, 158, 202). This influence of the pituitary is apparently mediated through the combined action of a number of hypophyseal hormones or their mediators, such as adrenocorticotropic hormone (204), steroid hormones (204), thyroid hormones (205), growth hormone (182, 203), and insulinlike growth factor I (155), which have all been shown to stimulate the production of erythropoietin in experimental animals.

Following the observation that patients with carbon dioxide retention due to pulmonary insufficiency may fail to develop polycythemia to a degree expected from the extent of their hypoxia (94), several investigators have shown that respiratory and metabolic acidosis suppress hypoxia-induced erythropoietin formation (8, 38, 62, 186, 187). During hypoxia, acidosis may improve tissue oxygenation first by increasing the ventilatory response through chemoreceptor activation, and second by lowering the oxygen affinity of hemoglobin. Although both effects may contribute to the reduction in erythropoietin formation, they do not seem to account for all of the inhibition. First, hypercapnia was found to suppress erythropoietin levels even when increased arterial oxygenation was prevented by artificial ventilation (8). Second, respiratory and metabolic acidosis were also found to reduce erythropoietin levels in anemic animals (186) or animals exposed to carbon monoxide (62).

Several studies have in recent years shown that immunoreactive erythropoietin levels in serum may be inappropriately low in patients with the so-called anemia of chronic disorders, including rheumatoid arthritis (6, 117), AIDS (253), or cancer (185). Such inhibition may be partially mediated by immunomodulatory cytokines. For instance, interleukin-1β inhibits erythropoietin formation in isolated perfused kidneys (128). This group has also shown that interleukin 1 and tumor necrosis factor inhibited Epo production by HepG2 cells (128). In Hep3B cells another group have found that interleukin 1α and β, tumor necrosis factor-α, and transforming growth factor β inhibit the synthesis of erythropoietin, whereas interleukin 6 is synergistic with hypoxia in inducing erythropoietin production (79). This effect on erythropoietin production most probably synergizes with inhibitory effects of cytokines and interferons on the proliferation of erythropoietic precursors in the pathogenesis of the anemia of chronic disease (181).

A number of observations suggest that the size or the functional activity of the erythron might influence serum erythropoietin levels independently of changes in blood oxygen content. Thus, an increase in serum erythropoietin levels may occur after treatment of patients with cytotoxic drugs, even before any fall of hemoglobin levels (18, 208), and, conversely, elevated serum erythropoietin may fall in anemic patients immediately after the administration of iron (19) or vitamin B_{12} (83), before any improvement in oxygen delivery or increase in reticulocytes is apparent.

Clearance and Metabolic Fate of Erythropoietin

Pharmacokinetic studies suggest that erythropoietin is distributed between the plasma and the extravascular space, and average values of distribution volume in humans given an intravenous bolus of erythropoietin have been reported between 30 and 90 ml/kg body weight (254). Plasma clearance of the hormone after intravenous application was in most studies found to be biexponential. Average values of half-life times for the inital distribution phase were found between 0.4 and 2.3 h and average values of the elimination half-life times have been reported in the range of 4 to 12 h (254). This variability may partially be related to differences in the protocols of pharmacokinetic studies and their evaluation, as well as subtle differences in the preparations of recombinant erythropoietin used for kinetic studies. Nevertheless, using the same preparation in a single study, marked variations have been found (44, 254). The mechanisms of erythropoietin clearance are poorly understood. Although the liver is considered a likely site for erythropoietin metabolism, direct evidence for hepatic uptake has only been provided for asialoerythropoietin (93, 255). Whether desialation of the erythropoietin molecule plays a physiological role in its metabolism remains unknown. The kidneys do not seem to play a major role in erythropoietin clearance, since its disappearance rate is not markedly altered in the anephric state (138, 256). The size of the erythron, which might influence erythropoietin clearance rate through erythropoietin consumption by erythropoietic target cells (257), was shown, in fact, not to influence erythropoietin survival in circulation in a quantitatively significant manner (209). Furthermore, experiments in rodents have provided no evidence that the disappearance time of erythropoietin depends on oxygen supply (58). Thus, despite the lack of precise knowledge about the metabolic fate of erythropoietin, it appears that oxygen-dependent variations in serum hormone concentrations directly reflect alterations in erythropoietin production rates. Consequently, most researchers have concentrated on defining the source and the mechanisms underlying control of production.

THE SOURCE OF ERYTHROPOIETIN

Organs Producing Erythropoietin

A dominant role of liver and kidneys in erythropoietin formation was originally deduced from the results of organ ablation studies. These investigations revealed that, in adult animals, the increase in serum erythropoietin levels in response to stimulation is severely blunted after bilateral nephrectomy (123, 232, 273), the remaining erythropoietin formation being abolished when nephrectomy is combined with subtotal hepatectomy (89). Direct support for renal synthesis of erythropoietin arose when biologically active erythropoietin was detected in blood-perfused (85, 152) or in serum-free perfused kidneys (70), and, furthermore, could be demonstrated in the homogenate of hypoxic rat kidneys (90, 126). Finally, cloning of the erythropoietin gene provided proof of de novo synthesis, since with the use of genomic and cDNA probes for erythropoietin, it became possible to demonstrate erythropoietin mRNA in kidneys of hypoxic animals (23, 238).

Apart from kidneys, erythropoietin mRNA was detected in the liver (23, 238), and, more recently, using RNAse protection, low levels have also been found in brain, testis, lung, and spleen (261, 262). Using in situ hybridization, investigators have furthermore postulated the presence of erythropoietin mRNA in bone marrow macrophages (271), but this observation has not been confirmed with other techniques (261, 262). Possibly, the list of organs containing erythropoietin mRNA might extend with further investigation and the use of even more sensitive techniques. It is of interest, however, that in all of the organs in which erythropoietin mRNA has so far been detected, with the exception only of lung, hypoxia was found to increase erythropoietin mRNA levels, indicating that oxygen-sensing mechanisms controlling erythropoietin mRNA levels may be widely distributed (261, 262). In order to judge the physiological importance of erythropoietin mRNA expression in different organs, however, it is essential to consider their quantitative contribution. Such estimations in adult rats under hypoxic stimulation revealed that the liver is clearly the major extrarenal site of erythropoietin mRNA accumulation, whereas even the combined contribution of extrarenal organs apart from liver was quantitatively insignificant (262).

Relative Contribution of Liver and Kidneys to Erythropoietin Formation

Considering liver and kidneys as the physiologically important production sites for erythropoietin, at least under hypoxic conditions, raises the question about their relative contribution, and this appears to vary with age, species, and the severity of hypoxia.

Both indirect evidence from organ ablation studies

(29, 109, 201, 273, 280–282) and quantitative measurements of erythropoietin mRNA in liver and kidneys (66, 145) indicate that the liver is the predominant production site for erythropoietin during fetal life, whereas the kidneys produce most erythropoietin in adults (Fig. 49.3). The time of this "shift" in erythropoietin production, however, differs between species. It may be complete during late gestation, as in mice (143), it may not be initiated before late gestation and only partially achieved at birth, as in sheep (280), or the liver may remain the predominant production site for erythropoietin for up to 2 to 4 wk of postnatal life, as in rats (66, 109, 273). The mechanisms underlying the ontogeny of erythropoietin production have not been resolved. The species-dependent variations suggest, however, that it is not directly related to the changes in hemodynamics and oxygenation, which occur at birth. It has also been shown in rats that the total amount of erythropoietin mRNA in liver does not decrease with age, but increases up to 2 wk of age, after which it remains rather constant, and that the "shift" in the predominance from liver to kidneys primarily results from a marked increase in the amount of renal erythropoietin mRNA during postnatal development (66).

Although in adult rats erythropoietin formation after bilateral nephrectomy is reduced to about 15% (73, 123, 232, 273), this appears to underestimate the hepatic contribution, because hepatic erythropoietin mRNA accumulation was found to be reduced in bilaterally nephrectomized animals (66). In fact, in intact

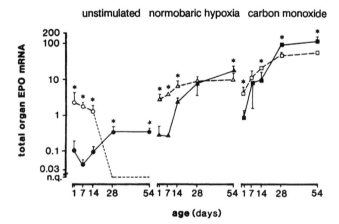

FIG. 49.3. Changes with development in the total amount of erythropoietin mRNA in rat kidney *(closed symbols)* and livers *(open symbols)*. Erythropoietin mRNA was quantified by RNAse protection assays; units are arbitrary and relate values to a standard preparation of erythropoietin mRNA. Before 28 days the liver contains the majority of total-body erythropoietin mRNA. Although renal erythropoietin mRNA increases with development and exceeds the hepatic contribution, the liver still contains approximately 33% of the total erythropoietin mRNA in severe stimulated animals. This contribution is similar whether stimulation is by normobaric hypoxia or exposure to carbon monoxide (from ref. 66, with permission).

adult rats erythropoietin mRNA in the liver amounts to about 30% to 40% of the total under conditions of severe hypoxic stimulation, including normobaric hypoxia, carbon monoxide exposure, and severe acute or prolonged hemorrhage (66, 261, 262). However, under both normobaric hypoxia or hemorrhagic anemia, the hepatic contribution to total erythropoietin mRNA was found to depend on the severity of the stimulus applied and it was less significant, i.e., below 20% of the total, under mild hypoxic stimulation (261). It appears, therefore, that the liver is less sensitive in terms of erythropoietin formation than the kidneys. This might also explain why the liver, despite its large potential for erythropoietin mRNA expression, does not compensate for loss of the renal production site. Tan et al. have found that rats rendered moderately uremic by subtotal nephrectomy became anemic, although in this model, hepatic erythropoietin mRNA formation under severe normobaric hypoxia was just as high as in control animals (262). It appears that at ambient oxygen tensions, the anemia in these animals was not sufficient to trigger an increase in hepatic erythropoietin formation that was adequate to restore red cell mass.

In humans the hepatic capacity for erythropoietin formation has not been quantified, but observations in anemic patients with chronic renal failure also indicate a significant potential for erythropoietin formation, despite an inappropriate erythropoietin response toward their reduction in hemoglobin levels. For instance, marked increases in serum erythropoietin levels were found in uremic patients when intercurrent illness (33, 272) or exposure to high altitude (21) lead to hypoxia.

Cellular Site of Erythropoietin Formation

Despite much effort, the question about the cell types producing erythropoietin in liver and kidneys has only recently been resolved. While the demonstration of putative erythropoietin-immunoreactivity in glomeruli (87) and the finding that cultured mesangial cells produce an in vitro erythropoietic activity (157) have in the past favored a glomerular origin of erythropoietin production in the kidney, this concept had to be revised following the application of nucleic acid hybridization techniques. Using RNA blot analysis, it was found that after partial disaggregation of hypoxic rat kidneys erythropoietin mRNA is not present in the glomerular fraction, but rather in the tubular fraction (238). Subsequently, erythropoietin mRNA was demonstrated on histological kidney sections by in situ hybridization. Although some investigators using this technique concluded that erythropoietin mRNA is localized in tubular cells (170, 174), Koury et al. and Lacombe et al. have shown quite convincingly that hybridization signals arise neither from glomerular nor from tubular cells, but

originate in cells located between tubuli in the interstitium of the renal cortex of mice and rats (145, 159). In accordance with these observations, erythropoietin mRNA was furthermore detected in interstitial cells of human polycystic kidneys (64). Since the peritubular space consists in a major part of peritubular capillaries, it was suggested that endothelial cells may be the production site of erythropoietin (145, 159). Nevertheless, other interstitial cells such as fibroblasts remain possible candidates. In fact, recent experiments using a digoxigenin-labeled riboprobe demonstrate colocalization of erythropoietin mRNA and immunohistochemical staining of a renal interstitial population with antibodies to 5' ectonucleotidase. Although this antigen is present on other cell populations, a fibroblastlike cell population is strongly positive within the renal interstitium (164), and these studies, together with a recent study of marked gene expression in transgenic mice (175), indicate that it is this fibroblast cell population that produces renal erythropoietin (5).

In livers of anemic transgenic mice in situ hybridization revealed the presence of erythropoietin mRNA in two different cell types; while approximately 80% of the cells containing erythropoietin mRNA were hepatocytes, the remaining 20% were cells with nonepithelial morphology, located in or adjacent to the sinusoidal spaces (146). This observation was confirmed by erythropoietin mRNA measurements in liver cells isolated from hypoxic rats and separated into fractions of parenchymal and nonparenchymal cells. Erythropoietin mRNA was present in both fractions, but parenchymal cells appeared to contain the majority (65, 237). Although some studies have suggested that Kupffer cells secrete erythropoietin (108, 109, 199), recent studies using marker gene expression in transgenic mice indicate that the nonepithelial erythropoietin producing cells are Ito cells (173a).

One interesting difference between liver and kidney has emerged from studies of cellular localization of erythropoietin mRNA in situ hybridization. In kidney, cells appeared to express erythropoietin mRNA in a near all-or-none fashion so that the major determinant of total renal erythropoietin mRNA was the number of positive cells (147). In contrast, in liver graded responses were seen in individual cells (146). Though this might indicate an important difference in the mechanism of oxygen sensing it could equally arise if oxygen gradients were much steeper in kidney.

OXYGEN SENSING IN THE CONTROL OF ERYTHROPOIETIN FORMATION

In understanding the mechanism of oxygen-dependent control of erythropoietin formation, it is necessary to consider the sites of oxygen sensing and what determines tissue oxygenation at these sites, as well as the mechanism of signal reception and transduction.

Evidence for Intrarenal Oxygen Sensing

Isolated perfused kidney experiments, which show oxygen-regulated modulation of erythropoietin mRNA levels (216) and production of erythropoietin (198, 231), demonstrate that all the events necessary for detection of hypoxia and production of erythropoietin can operate intrarenally—though not necessarily in the same cell. Although some of these studies demonstrated that production of erythropoietin mRNA and hormone over a 2–3 h period was rather similar to that observed in vivo (216, 231, 262), others have found lower rates of production in isolated kidneys (198). Pagel and colleagues also observed that varying the hematocrit of the perfusate in isolated perfused kidneys had no effect on erythropoietin production (198), and that reduction of renal perfusion in vivo by renal artery constriction (197) is a much less effective stimulus for erythropoietin production than anemia. In view of previous findings showing that hypothalamic stimulation increases plasma erythropoietin levels and that hypophysectomy severely reduces the erythropoietin response to hypoxia (111, 112), they proposed that an additional extrarenal sensing system contributes importantly to control of renal erythropoietin production. Some studies have also reported an influence of bilateral renal nerve section on erythropoietin formation (16, 81), suggesting that an extrarenal influence on erythropoietin formation might be transmitted via the nervous system. This was, however, not confirmed in a recent investigation, where erythropoietin mRNA levels in paired kidneys of unilaterally denervated rats were not found to be different (63). Consequently, it appears that any extrarenal influence, if present, would have to be humoral. However, direct evidence that such an additional extrarenal sensing system contributes importantly to control of renal erythropoietin production has yet to emerge, and it is possible that during constriction of the renal artery, and some hypoxic conditions used in kidney perfusion, the appropriate level of intrarenal hypoxia is not achieved or that inappropriately severe hypoxia actually damages the system.

It has long been appreciated that despite its high blood flow, the kidney contains poorly oxygenated regions (165, 220). Severe tissue hypoxia in the organ principally responsible for erythropoietin production raises the possibility that the tissue specificity of erythropoietin production might be determined directly by local oxygenation. Several lines of evidence demonstrate that this is not entirely the case. First, local ischemia of other organs does not lead to erythrocytosis. Second, as discussed previously, the developmental shift of eryth-

ropoietin production from liver to kidney does not, in most species, correlate with the shift in organ oxygenation associated with birth. Third, transgenic experiments to be discussed later show that 5′ DNA sequence makes a dominant contribution to organ distribution of expression. Nevertheless, tissue hypoxia may make some contribution to the localization of erythropoietin production within organs. In fact, this is more clearly demonstrated in liver than in kidney.

The Role of Local Oxygenation in Hepatic Erythropoietin Formation

The expression of erythropoietin mRNA in pericentral areas of the hepatic lobules in mice bearing a human transgene (146) is consistent with local oxygen gradients determining expression (135). This possibility is also in keeping with observations that several hepatic cell culture models show oxygen-regulated erythropoietin production in vitro. In some of these culture systems, including organ cultures of mouse fetal liver (286), mixed cultures of isolated fetal liver cells (156), and isolated Kupffer cells (199), demonstration of erythropoietin production depended on bioassays, whose specificity was somewhat uncertain. Later, Goldberg et al., who screened a number of cell culture lines for their ability to produce erythropoietin, found that two human hepatoma cell lines, Hep G2 and Hep 3B, produce erythropoietin in an oxygen-dependent fashion (100). These cell lines have been widely used in studies of both signal transduction and gene expression. More recently, we have observed that in primary cultures of hepatocytes from juvenile rats, accumulation of erythropoietin mRNA and erythropoietin production occurs in an oxygen-dependent fashion (65). These observations in hepatoma cells and isolated hepatocytes clearly indicate that erythropoietin production and oxygen sensing can occur in the same cell.

The Role of Local Oxygenation for Renal Erythropoietin Formation

Intrarenal oxygenation is remarkably inhomogenous (163, 235). Regional hypoxia is believed to arise from two considerations: first, a high rate of energy consumption (37, 51, 137, 161), and second, the countercurrent arrangement of arterial and venous vessels that provides the basis for shunt diffusion of blood gases (150, 165, 235). Direct microelectrode measurements have shown that the renal cortical tissue pO_2 is well below that in the renal veins, and even lower pO_2, in the region of about 10 mm Hg, is found in renal medulla (163). It is tempting to speculate that the recruitment of erythropoietin-producing interstitial cells from inner to outer renal cortical tissue with increasing severe anemia

(147) or hypoxia (61) might arise from local changes in renal cortical oxygenation. However, direct measurements of renal tissue oxygen tensions have not been performed in conjunction with studies of erythropoietin production. Furthermore, some aspects of the distribution of erythropoietin-producing cells in the kidney indicate that factors other than local oxygen tension may contribute to the restriction of erythropoietin mRNA expression. Thus, it is clear that the lowest oxygen tensions are found in the renal medulla, which does not appear to produce erythropoietin. Within the cortex, expression of erythropoietin is found primarily in the cortical labyrinth and not the medullary rays (61). Finally, it has not yet been possible to observe regulated erythropoietin production in isolated renal cell populations, so that it is important to be aware that some doubt remains as to whether or not the renal cell population that produces erythropoietin senses tissue oxygenation directly.

One other aspect of renal physiology deserves discussion. In the normal-functioning kidney, transport work accounts for the majority of oxygen consumption (37, 51, 137, 161). Since more than 95% of the glomerular filtrate is reabsorbed, it follows that renal oxygen consumption should be closely determined by the glomerular filtration rate. Thus, parallel changes in renal blood flow and glomerular filtration rate might not disturb the balance of oxygen consumption and supply and might enable the kidney to respond to the change in blood-oxygen availability rising from anemia without confounding effects from changes in renal hemodynamics (72). Some evidence that reduction in tubular transport work might reduce erythropoietin production has been provided by studies of the effect of diuretic agents on erythropoietin production by hypoxic animals (59, 86). Whereas agents acting on distal tubular transport had no effect, acetazolamide reduced erythropoietin production. The action of acetazolamide is primarily on the proximal tubule, and it is of interest in this connection that Koury et al. found that the cells that produce erythropoietin lie adjacent to proximal tubules (145). Observations in isolated perfused kidneys might also support a role for transport work in determination of erythropoietin production. When isolated kidneys are perfused with hyperoncotic bovine serum albumin to block glomerular filtration, levels of erythropoietin mRNA are reduced at equivalent conditions of perfusate oxygen delivery (264). Nevertheless, this reduction cannot be securely interpreted as arising from reduction of transport work, because hyperoncotic perfusion does not, in fact, reduce total renal oxygen consumption greatly. It appears that at least under some circumstances reduction in transport work can be counterbalanced by an increase in oxygen consumption by metabolic reactions such as gluconeogenesis (249). Thus, although it is

appealing that the unusual parallel relationship between renal blood flow and oxygen consumption could be important in understanding how precise control of erythropoietin production and hematocrit is achieved, it is still not at all clear how this operates.

Mechanism of Oxygen Sensing and Signal Transduction

Recognition that the hypoxia-erythropoietin signaling system is critical to the control of erythropoiesis and has led to many attempts to define the nature of the sensor in pharmacological studies. Of importance is the finding that hypoxia cannot be mimicked by the application of inhibitors of mitochondrial respiration. Thus experiments in vivo (194), in hepatoma cells (77), and in isolated perfused kidneys (263) have shown that neither cyanide nor other inhibitors of the respiratory chain stimulate erythropoietin production. Failure of erythropoietin gene activation in these experiments does not arise from nonspecific damage, since in all these systems it is possible to observe stimulation by hypoxia in the presense of toxic concentrations of cyanide. These experiments suggest that the sensing system is distinct from cell-stress responses to nonspecific damage, and make it less likely that the hypoxic sensing system operates through metabolic derangements such as alteration of phosphorylation potential and cellular redox potential, which would be expected to arise from both exposure to hypoxia and application of inhibitors of respiratory chain (25).

Based on the results of experiments performed on hepatoma cells, it has been proposed that the oxygen sensor is a heme protein that can reversibly bind molecular oxygen (98). Goldberg and colleagues noted that not only hypoxia, but also cobalt, nickel, and manganese, could stimulate erythropoietin production in human hepatoma cells and proposed that these ions could substitute for the ferrous ion in a putative heme protein and lock it in the deoxy form. Several experiments were performed that support this hypothesis. Exposure to cobaltous or nickel ions did not summate with hypoxic stimulation, suggesting that they do indeed act through a common pathway. The behavior of carbon monoxide was studied on the basis that it specifically binds ferrous heme proteins. Exposure to carbon monoxide greatly reduced the erythropoietin response to hypoxia but not to cobalt. This response argues against nonspecific toxicity and would be consistent with binding of carbon monoxide to the ferrous heme protein but not cobalt protoporphyrin. Finally, when hepatoma cells were incubated with desferrioxamine and 4,6-dioxoheptanoic acid to reduce heme synthesis, erythropoietin production in response to hypoxia, nickel, and cobalt was also reduced.

Consistent with the view that oxygen sensing involves the operation of a heme protein are the findings of Fandrey and colleagues, who studied the action of pharmacological probes that induce or inhibit the cytochrome p450 system in HepG2 cells (78). Inducing agents such as phenobarbital and 3-methylcholanthrene increased erythropoietin production. Some inhibitory agents—diethyldithiocarbonate and cysteamine chloride, but not metyrapone—reduced erythropoietin production. The authors postulated that the cytochrome p450 system might play a role in the oxygen-sensing system controlling erythropoietin production.

Although oxygen gradients exist within cells (133), there is no compelling reason for siting oxygen sensing in any particular intracellular location. It is thus plausible that oxygen could interact directly with nuclear proteins to induce binding, or that it could modify proteins bound to DNA to induce transcription of the erythropoietin gene. Some evidence that this is not the case comes from application of the protein-synthesis inhibitor cycloheximide. In hepatoma cells cycloheximide blocks induction of erythropoietin gene expression (98). It is difficult to exclude the possibility that some permissive effect is being abrogated by cycloheximide or even that the effect simply arises from toxicity. However, sensitivity of the inducing system to cycloheximide would be compatible with the requirement for new protein synthesis in a cascade of events leading to erythropoietin gene expression.

Many experimenters have sought to implicate classical second messengers in transduction of the oxygen sensing system. These studies have been reviewed in depth elsewhere (84, 125) and will only be outlined here.

In a number of systems, agents that interact with cAMP are reported to influence erythropoietin production. Thus, administration of cAMP or its analogues to whole animals is reported to increase erythropoietin production. Agents such as adenosine (A_2), receptor agonists (270) and β_2 adrenoreceptor agonists (82), and prostanoids (84) that operate through activation of adenyl cyclase can increase erythropoietin levels. Administration of cAMP analogues to a human renal carcinoma cell line increased erythropoietin production under both normoxic and hypoxic conditions (244). However, no response to cAMP was found by using hepatoma cells (98) and isolated hepatocytes (65) under normoxic conditions. In another study of hepatoma cells, it was reported that forskolin and cholera toxin did not increase erythropoietin production under normoxic conditions but augmented the production during hypoxia (270). In recent studies of isolated perfused kidneys, no effect of forskolin on kidneys perfused under both well-oxygenated and hypoxic conditions was observed (231). In another study of isolated perfused kidneys, the authors did not observe any effect of forskolin, dbcAMP, salbutamol, or the A_2 receptor agonist N6-ethyl-carboxamidoadenosine on erythropoietin mRNA levels (265). These studies therefore do not indicate that cAMP directly mediates oxygen-dependent changes in erythropoietin gene expression. Rather, under certain circumstances cAMP may modulate erythropoietin production. It is not clear that this reflects involvement in the physiological response to hypoxia or some non-oxygen-dependent modulating influence on erythropoietin production.

Recently, several groups reported independently that the application of phorbol esters inhibits erythropoietin formation in human hepatoma cells (127, 154) and a renal carcinoma cell line (110). Phorbol esters are known to activate protein kinase C, but the precise role of protein kinase C for EPO formation is unclear, since inhibitors of this enzyme were also found to reduce EPO formation in hepatoma cells (127, 154).

CONTROL OF ERYTHROPOIETIN GENE EXPRESSION

The preceding discussion indicates that regulation of erythropoietin gene expression must reflect the coordination of at least three distinct influences: developmental changes, tissue restriction, and inducing environmental stimuli. Of particular interest is the rapid and very high amplitude change in expression observed in response to hypoxia. One of the attractions in the erythropoietin system for the more general study of gene regulation is that the very large body of existing physiological data permits a more accurate physiological appraisal of effects observed in systems adapted for the study of gene regulation than is often the case.

Modulation of Erythropoietin mRNA Levels

Immunohistochemical studies have not demonstrated large cellular stores of preformed erythropoietin, and the response to hypoxic stimulation is blocked by inhibitors of mRNA and protein synthesis (233). These findings indicate that the large increase in blood levels of the hormone that follow hypoxic stimulation are achieved by large increases in the rate of de novo synthesis. A substantial body of evidence shows that this occurs mainly through modulation of the level of erythropoietin mRNA.

In rats and mice, the increase in erythropoietin mRNA is discernible within 1 h of severe hypoxic exposure, and reaches a level 300–500-fold over baseline by 4–8 h (66, 145). This response is similar in time course and magnitude to the change in hormone levels in the blood. When stimuli of graded severity are applied, a graded increase is observed in erythropoietin mRNA that again closely parallels the increase in hormone production, irrespective of whether stimulation is by hypoxic or anemic hypoxia (261).

Modulation of the steady-state level of mRNA can be achieved through changes in the rate of gene transcription, or in the rate of degradation of the mRNA. Evidence that both mechanisms operate to control erythropoietin mRNA levels has been obtained by comparison of the amplitude of modulation of erythropoietin mRNA with direct measurement of transcriptional rate by nuclear run-on experiments (99). Schuster and colleagues demonstrated increased erythropoietin transcription in nuclei isolated from kidneys of anemic hypoxic and cobalt-treated rats (236). However, they were unable to detect transcription in nuclei from unstimulated animals, so the precise amplitude of the increase could not be determined. Using nuclei from Hep 3B cells, Goldberg and colleagues were able to measure transcription in unstimulated cells grown normoxically, and estimated that the increase in transcription under hypoxia was approximately tenfold (99). This was less than the 50-fold increase in erythropoietin mRNA which these authors observed in the same cells. Although neither measurement it easy to make accurately, the discrepancy is clear and the authors have reasonably argued that these findings would be consistent with oxygen-dependent changes in erythropoietin mRNA stability. To demonstrate this directly, mRNA half-life must be measured under varying conditions of oxygenation. When Hep 3B cells were switched from a hypoxic to a normoxic environment, steady-state erythropoietin mRNA decreased by 50% within 1.5 to 2 h. Because new transcription was not blocked, this result represents a maximum estimate of the half-life in normoxic cells. However, when new transcription was blocked by actinomycin D the half-life of erythropoietin mRNA was surprisingly increased to approximately 8 h and was, in fact, similar in normoxic and hypoxic cells (99). This effect of actinomycin D itself on mRNA stability precluded measurement of the effect of oxygenation on mRNA stability by this method, but is itself of interest. Increased erythropoietin mRNA stability was also observed after application of the protein synthesis inhibitor cycloheximide. Similar observations have been made with respect to a number of other genes (191, 248) and have been interpreted as indicating the existence of a specific ribonuclease that is itself turning over rapidly. Though it is possible that hypoxia might affect erythropoietin mRNA stability by an interaction with such a system, this is not necessarily the case, and although an erythropoietin mRNA-binding activity has been described that is regulated by hypoxia in brain, no regulation was observed in liver and kidney (223) (see later, under *Cis*-acting Sequences in Erythropoietin mRNA).

In summary, while evidence for oxygen-dependent changes in erythropoietin gene transcription is clear, the case for additional regulation of erythropoietin mRNA by oxygen-dependent changes in mRNA stability is not yet completely proven.

Cis-Acting Sequences that Coordinate Gene Expression

Cis-acting regulatory elements are sequences located on the same DNA molecule in the vicinity of the gene that coordinate the DNA–protein, protein–protein, and possibly DNA–DNA interactions, which regulate gene transcription (for review, see refs. 169, 212). In mRNA, cis-acting sequences also interact with proteins to control splicing, transport, degradation, and rate of translation (for review, see refs. 103, 173).

A great deal of research has been devoted toward determination of these cis-acting sequences. Such studies form the basis of our understanding of the anatomy of gene regulation (for review, see refs. 57, 132, 196). They also provide the basis for identification, and purification of the associated DNA-binding proteins that mediate interaction between the signal-transduction pathway and the gene (136, 250). For the cell physiologist there are other interests. Cis-acting sequences can be used to bring selectable reporter genes under the control of the inducing stimulus, permitting the selection of cell lines with mutations in the signaling pathway (131). The existence of similar motifs in control regions of different genes may indicate a common regulatory pathway (221). Where multiple physiological stimuli contribute to regulation of a single gene, functional studies of isolated cis-acting sequences may indicate whether or not signal transduction operates through a final common pathway. Finally, as will be illustrated later, transfection of a cis-acting sequence coupled to a reporter provides a readout that enables the experimenter to test for the existence of the interacting sensor and signal system in different cells.

Sequences lying in the immediate vicinity of the 5' end of the gene determine accurately sited initiation of transcription, and is termed the promoter (24). Elements within the promoter may also contribute to regulation of transcriptional rate. Other cis-acting regulatory sequences may be lying in almost any position in the vicinity of the gene (107, 169). For many of these sequences, termed *enhancers,* recombinant experiments have demonstrated that the DNA will mediate a functional interaction independent of orientation and at variable distances from a promoter. Although these cis-acting elements may operate in distinct ways to control the tissue specificity of gene expression or inducible expression, the same sequence motifs can be found, on occasion, in cis-acting sequences associated with promoter function, inducible enhancer function, and alteration of chromatin structure (80, 230).

Studies of cis-acting sequence fall broadly into two categories: first, structural studies that attempt to demonstrate binding to DNA of proteins—with the implication of a functional role in the regulation of expression; and second, studies in which function is assayed directly by deletion or mutation of the sequence under investigation and expression in transfected cells. In transfection studies, exogenous DNA is introduced into cells by a variety of techniques that permit DNA to cross cell membranes and enter the nucleus (for review, see ref. 102). Most of the transfection studies described in this chapter employ transient transfection assays in which the gene product is assayed within a short period varying from hours to a few days of transfection. Since the exogenous DNA is not fully integrated into chromatin, transcriptional responses will reflect some, but not necessarily all, of the constraints on gene expression that operate in vivo (80) and references therein). In assessing the merits of structural and functional studies, one might imagine that direct demonstration of function would be more secure evidence of a physiological role. This is not necessarily the case, since the simplified systems used to demonstrate function may create nonphysiological rate-limiting steps. For instance, transient transfection may present the nuclear environment with a large number of copies of the test sequence, with consequent distortion of the relative concentration of interacting species. It follows that the best evidence of a physiological role is provided where different types of study all implicate a given sequence.

Sequence Conservation in Erythropoietin Genes

Rates of silent nucleotide substitution are rather constant at around 1% per million years. Genomic sequence from species of known evolutionary divergence can therefore be compared, to determine where enhanced conservation of nucleotide sequence indicates that selective pressure has operated, presumably arising from some functional role. In the case of human and murine erythropoietin genes, divergence took place in the region of 80 million years ago, so that very little homology would be expected in functionless regions. However, significant homologies outside coding sequences are present in several regions (177, 247). Immediately 5' to the gene, in the promoter region, there is 90% homology over 140 bp. The long 5' and 3' untranslated regions also contain regions of significant homology. In the 3' untranslated region there is a region of 80% homology for 120 bp lying 100 bp 3' to the stop codon. Of surprise, however, was the finding that the first intron has an overall homology of 65% and contained regions of up to 50 bp with over 90% homology (177, 247).

Using low-stringency hybridization to search for possible homology between murine and human genes outside the sequenced regions, homology at the 5' end of the gene was located entirely within 400 bp of the start of the gene (247). Hybridization studies at the 3' end of the gene were only performed against the human cDNA, so that highly conserved control sequences that lie 3' to the cDNA were not detected. Aside from this problem, such techniques cannot reliably exclude the existence of control sequences in a region. The conserved DNA binding motif for a particular controlling protein is often short (for instance, 6–8 nucleotides). Even if, as is common, a series of DNA-binding sites are clustered together (56, 283), a particular order is not always necessary for function. In this situation conservation of a cluster of motifs would be impossible to detect by cross-hybridization and may be difficult to see even by computer searching of sequence data. The existence of control elements at a particular site can therefore only be disproven by appropriate functional assays.

Hypersensitive Sites in Erythropoietin Genes

Chromosomal DNA is complexed with histones to form nucleosomes and other higher-order interactions that package DNA in chromatin (for review, see refs. 80, 178) The structure of chromatin is altered at sites of gene expression, and these structural alterations may be probed to show the location of active genes and cis-acting control sequences. Such sites are termed *hypersensitive* because of their pronounced sensitivity to chemical modification or nuclease cleavage by agents such as DNAse I (for review, see ref. 106). The precise chemical basis of these hypersensitive sites is not clear. However, they are thought to reflect access of the nuclease where nucleotides are complexed with functional binding proteins. Such protein binding also creates associated areas of reduced sensitivity, termed *footprints*. Determination of this detail within hypersensitive sites in genomic DNA requires selection or amplification of the DNA sequence of interest (190). However, the approximate position of hypersensitive sites can be determined quite easily by pretreatment of nuclei with DNAse I prior to DNA preparation and analysis by Southern blotting (106). Such studies provide a good guide to the presence of functional sites, although the range of functions is much broader than those associated with control of gene expression, and includes operation of the replication apparatus, as well as transcriptional control (106).

Semenza and colleagues have demonstrated DNAse I hypersensitivity 3' to the human erythropoietin gene (241) (Fig. 49.4d). These sites were initially mapped in a transgenic mouse line carrying six tandemly arranged copies of a 10 kb human erythropoietin transgene. The

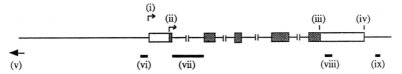

FIG. 49.4a. Schematic diagram of the erythropoietin gene indicating the position of possible cis-acting control regions. The gene consists of five exons *(boxed areas)* and four introns (not drawn to scale). The exonic regions contain long 5' and 3' untranslated region *(open boxes)* lying on either side of coding sequence *(hatched boxes)*. Features indicated on the diagram are (*i*) the transcriptional start site, (*ii*) translation initiation, (*iii*) the stop codon, (*iv*) the poly A addition site, (*v*) the existence of distant sequence lying 5' to the gene, which acts to control tissue specificity of expression and possibly as an inducible element in kidney (Fig. 49.4*b*), (*vi*) the promoter region, (*vii*) the 1st intron, which is unusually highly conserved, (*viii*) the position in the 3' untranslated region of an mRNA protein binding site (Fig. 49.4*c*), (*ix*) the position of a transcriptional enhancer lying 3' to the poly A addition site (Fig. 49.4*d,e*).

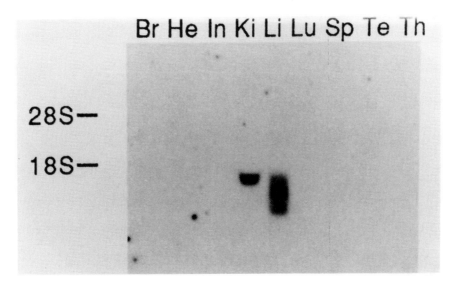

FIG. 49.4b. Northern blot analysis of expression of human erythropoietin mRNA in the organs of an anemic transgenic mouse bearing a 22 kb human erythropoietin transgene. Organs are brain (*Br*), heart (*He*), intestine (*In*), kidney (*Ki*), liver (*Li*), lung (*Lu*), spleen (*Sp*), testes (*Te*), and thymus (*Th*). Expression pattern mimics that of the endogenous gene in that expression is limited to the liver and kidney. Comparison of this expression pattern with that of human erythropoietin transgenes of different lengths in other lines of transgenic mice indicates the existence of a region between 6 and 14 kb 5' to the human erythropoietin gene that is required for regulated expression in kidney (from ref. 240, with permission).

head-to-tail arrangement of the integrated DNA enabled the authors to probe a unique internal restriction fragment of greater intensity, and demonstrate hypersensitivity sites lying just beyond the 3' end of the gene.

In the transgenic mice, the sites were detected in nuclei prepared from the liver but were not detected in nuclei prepared from brain, spleen, and kidney of these animals (241). The authors were careful to point out that while their findings were evidence for a functionally important site in liver, they did not exclude function of that site in kidney. Since this transgene was not expressed in kidney the negative result is difficult to interpret, and if the hypersensitive sites were present only in a small proportion of nuclei from kidney, they might not be detected anyway.

The Promoter of the Erythropoietin Gene

Cis-acting sequences that regulate gene expression are commonly found in the region of the promoter. However, despite recognition that the erythropoietin promoter has a rather unusual structure, surprisingly little is known about its mechanism of action. Mapping of the transcriptional start site was first performed by Shoemaker and Mitsock in mouse kidney (247). Homologous sequence is present in the human gene, and primer extension studies have confirmed that it is used as a site of initiation (39). The highly conserved region lying immediately upstream and downstream from this site presumably corresponds to the promoter. This sequence is extremely GC-rich but does not contain a consensus TATA motif upstream from initiation. For many genes

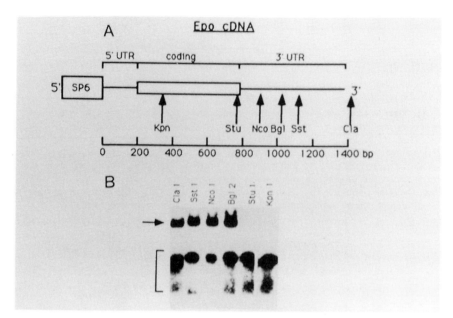

FIG. 49.4c. Demonstration of erythropoietin mRNA protein complexes by band-shift assays. (A) A restriction map of the full-length erythropoietin cDNA. Transcripts running to the indicated restriction sites were generated using SP6 RNA polymerase after linearization of the template at each of those restriction sites. (B) Autoradiograph of band-shift assay using cytosolic lysates from normoxic Hep3B cells. The arrow designates the erythropoietin RNA band-shifted complex, while the bracket represents free RNA. Complex formation is not observed when the transcripts stop at the Stu I or KpnI sites, indicating a binding site in the 3' UTR (from ref. 223).

the TATA motif lying at position −30 is essential for correct initiation (24, 104). However, a significant proportion of genes resemble erythropoietin in not containing this consensus sequence. For some of these genes, mutational studies have indicated that the bases immediately at the site of initiation provide the alignment for correct initiation (13, 251). The erythropoietin initiation site, however, also differs from this consensus sequence, and the mechanism aligning the transcriptional initiation complex for erythropoietin is unknown.

Some uncertainty in analysis of the erythropoietin promoter arises from Semenza et al.'s studies of transcripts from the human erythropoietin gene in transgenic mice (239). Anemia-inducible transcripts were found to arise, not only from the initiation site homologous to that described by Shoemaker and Mitsock (247), but from other sites lying mainly in a region 240–320 bp upstream from that site. Semenza also described a similar pattern of upstream initiation in Hep3B cells and in erythropoietin transcripts from human fetal liver. However, when the human transgene was correctly expressed in mouse kidney, the initiation was predominantly or exclusively from the "downstream" position homologous to that used in normal mouse kidney (240). Our own studies of mouse erythropoietin gene expression demonstrate transcripts arising predominantly from this site in kidney and liver of

normal mice exposed to hypoxia, and in the liver and kidney of hypoxic mice carrying a marked mouse erythropoietin transgene (Tan, Maxwell, and Ratcliffe, unpublished observation). There is therefore good evidence that the highly homologous region extending upstream and perhaps downstream from this site does function as a major promoter in both liver and kidney. The physiological significance of upstream transcriptional starts from the human gene in liver, and whether this represents an important species difference between mouse and man, remains to be established.

Several lines of evidence indicate that promoter sequences in isolation can interact with the oxygen-sensing system. Thus, DNA binding studies have demonstrated factors in nuclear extract of mouse kidney that bind a deoxynucleotide sequence corresponding to bases −61 to −45 relative to the mouse initiation site (15). Enzymatic studies have indicated that this complex consists of both RNA species and proteins. Comparison of nuclear extracts from normal and cobalt-stimulated mice showed that the amount of two of the RNA species binding the DNA was reduced after stimulation. The authors proposed that negative regulation of erythropoietin gene transcription was operating through this ribonuclear protein complex. However, functional studies that corroborate this evidence have not yet been published, and recent transfection studies of the human erythropoietin gene indicate that this sequence is not

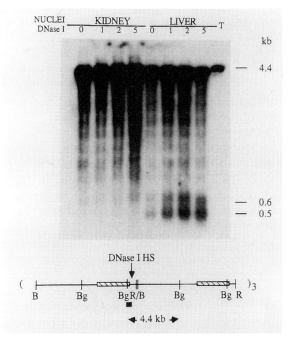

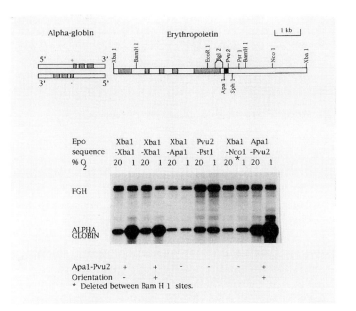

FIG. 49.4d. Demonstration of hypersensitive sites lying 3' to the human erythropoietin gene. Nuclei were prepared from liver and kidney of transgenic mice containing six copies of a 10 kb human transgene lying head to tail (two are shown in the diagram). This created an internal Bgl2 restriction fragment (4.4 kb). When nuclei were treated with the indicated amounts of DNAse I (μg/ml) prior to DNA extraction and Bgl2 digestion, cleavage occurred in the liver nuclei to create the additional 0.6 and 0.5 kb fragments. The position of this site is indicated by a vertical arrow and correlates with the functional demonstration of an enhancer element (from ref. 241, with permission).

FIG. 49.4e. Operation of the erythropoietin 3' enhancer in transiently transfected HepG2 cells. Portions of the mouse erythropoietin gene were coupled to a human α globin gene and tested for their ability to convey hypoxic regulation on the α globin reporter. (a) Position of the linked human α globin and mouse erythropoietin gene in the recombinant test plasmids. (b) Autoradiograph of RNAse protection assay. Alternate lanes show normoxic and hypoxic α globin expression and expression of the control plasmid containing a ferritin–growth homrone fusion gene (FGH). The erythropoietin gene fragments for each test plasmid are indicated above each lane. The presence of the Apa1–Pvu 2 fragment containing the active enhancer sequence is indicated below the autoradiograph, and its orientation with respect to the α globin gene is indicated as + or −. The element conveys oxygen-regulated expression independently of distance or orientation with respect to the α_1-globin promoter (from ref. 213, with permission).

sufficient to convey oxygen-regulated promoter function (20).

In vitro transcription experiments also provide evidence that the isolated promoter region can mediate oxygen-regulated transcription. In these experiments, nuclear extract is added, in vitro, to a DNA template containing the putative promoter, and the quantity of labeled transcript is assessed after polyacrylamide gel electrophoresis (52). Nuclear extract from anemic mouse kidney was found to support transcription from a template containing 0.2 kb of 5' human erythropoietin sequence (268). Interestingly, activity was not observed when 0.4 kb of 5' erythropoietin sequence was used as template, raising the possibility that the extra sequence mediates repressive interactions. This is of interest considering the rather weak operation of 0.4 kb promoter sequence of both mouse and human erythropoietin genes in hepatoma cell lines and the near-absent activity that has been observed in other cell lines. Other studies of in vitro transcription have been performed on nuclear extract from Hep3B cells (39). In each case, increased transcription was observed when nuclei were

obtained from a stimulated source. This indicates that DNA-binding interactions can be preserved in at least a partially functional state through the procedure used to obtain the nuclear extract.

Very recently, transfection studies in Hep3B cells have defined more clearly functional elements mediating oxygen-regulated gene expression in the erythropoietin promoter (20). In these studies spurious transcription, commencing at aberrant initiation sites in the plasmid and extending through the area of interest, was eliminated by insertion of an SV40 polyA addition site 5' to the erythropoietin sequence. The sensitive luciferase reporter gene (5) was then used to assay promoter function. These techniques permitted assessment of both basal and inducible promoter activity, despite the rather weak operation of the erythropoietin promoter. Both basal and inducible activity were fully represented when the promoter was deleted to 117 bp from the CAP site, whereas further deletion to 65 bp from the CAP virtually abolished both basal and inducible activity. Interestingly, sequence similarities were noted between this promoter region and the 3' enhancer, and these sequences appeared to operate synergistically to increase the amplitude of oxygen-dependent transcriptional control.

Enhancer Elements 3' to the Erythropoietin Gene

Transfection studies have demonstrated that important control elements lie 3' to the erythropoietin gene. Beck et al. used transient transfection of a shortened human erythropoietin gene in Hep3B cells (14). By making successive deletions they demonstrated that sequence within a 150 bp restriction fragment lying 120 bp 3' to the human erythropoietin gene appeared to be responsible for mediation of the hypoxia-induced transcription. This element directed oxygen-regulated transcription irrespective of whether the erythropoietin promoter was intact, suggesting that it could operate on cryptic promoter sites in the transfected plasmid DNA. The region defined in these studies corresponded with the 3' portion of a 256 bp region of DNA shown by Semenza and colleagues to operate on the SV40 promoter in transfected Hep3B cells (241). In a similar analysis of the mouse erythropoietin gene, using both α globin and the ferritin gene as reporters, Pugh et al. found the active sequence to be located in an identical position 120 bp 3' to the mouse erythropoietin gene (213) (Fig. 49.4*e*). In experiments similar to those reported by Beck et al. on the human gene (14), Pugh et al. found that when the element was excised from the 3' end of the mouse erythropoietin gene, expression in transfected HepG2 cells was reduced to an undetectable level. When it was replaced 5' to the erythropoietin promoter, oxygen-dependent expression was completely restored (213). In these experiments the transcriptional start was mapped by RNAse protection, demonstrating directly that the 3' element conveyed O_2-dependent transcriptional control both on the promoter itself and on aberrantly initiated transcription. This sequence therefore has the classical features of a eukaryotic transcriptional enhancer in operating independently of orientation and distance on a variety of heterologous promoters.

As expected for a functionally important sequence, a high degree of homology with the human sequence was observed. Though the sequence homology extends over at least 140 bp, the minimal functional element required for enhancer activity in transiently transfected cells is shorter, and most probably varies with the distance from the promoter on which it is operating. Approximately 60–70 bp were necessary when the enhancer element was placed 1.5 kb from an alpha globin promoter (213), whereas a 43 bp sequence was sufficient for enhancer function when placed close to the erythropoietin promoter or close to a herpes simplex virus thymidine kinase promoter (20). The requirement for a more extensive DNA protein complex for operation at greater distances and in chromatin is well established in other systems (80, 230). It is probable that the more extensive sequence homology in this region indicates the need for more complex DNA protein interactions for the operation of this sequence in vivo from its position 4 kb 3' of the erythropoietin gene promoter.

Semenza et al. have used a combination of in vitro DNAse I footprinting and gel retardation assays to demonstrate in vitro the binding of nuclear factors to this region (241). The DNAse I footprinting technique allows the binding sites of proteins to be studied in detail, provided that the binding characteristics can be reproduced in the in vitro system. A similar criterion applies to gel retardation studies in which retardation of the migration of a labeled double-stranded oligonucleotide sequence is observed, and gives information about the nature of both the binding site and the binding protein. DNAse I footprinting using nuclear extracts from anemic and nonanemic mouse livers demonstrated four protected areas in a region extending somewhat beyond the minimal element that is necessary and sufficient for activity in the transient transfection assays. Gel retardation assays were performed with nuclear extracts from both kidney and liver. Each of four oligonucleotides that bound nuclear factors showed a different retardation pattern, indicating multiple binding proteins. Changes were observed between anemic and nonanemic nuclei, but interestingly the patterns observed were similar between liver and kidney, indicating the potential for this enhancer sequence to contribute to the oxygen-dependent regulation of the erythropoietin gene in kidney as well as liver. One difficulty in interpretation of these studies arises from the use of whole organs to prepare nuclear extract when erythropoietin production is confined to a subset of cell within the organ. When Hep3B or HepG2 cells are used to prepare the nuclear extract, a different pattern of DNA-binding activities is observed, with striking DNAse protection seen in a region within the functionally defined element that contains a direct repeat of a steroid/thyroid hormone receptor half-site (20). However, this region alone is insufficient for enhancer function, implying the need for one or more additional DNA–protein interactions as the minimal requirement for generation of the active complex (20, 213). Mutational and DNA-binding studies of the region of the human enhancer lying 5' to this direct repeat sequence have demonstrated at least two further binding sites, one of which appears to bind a hypoxia-inducible nuclear factor termed HIF-1 (243). Mutational studies of the first 48 nucleotides of the mouse Epo enhancer also indicate at least three protein binding sites within this region corresponding to areas defined in studies of the human enhancer (214).

In addition to the enhancer lying 3' to the gene, less powerful enhancer activity has been observed in the 3' untranslated region of the gene (119). A 255 bp portion of the 3' untranslated region of the human gene was found to produce a small increase in transcription of human erythropoietin/growth-hormone fusion genes in Hep3B cells. More important, in that study the authors also reported that a stably transfected cloned Hep3B line containing both a 1,192 bp human erythropoietin 5' sequence and this 255 bp 3' sequence surrounding the growth hormone, produced a 16-fold induction of growth hormone under hypoxic stimulation. Which element is responsible for this effect cannot be determined, but this experiment does clearly demonstrate that high-amplitude changes in erythropoietin gene expression in Hep3B cells are not necessarily dependent on the powerful enhancer lying 3' to the gene in all circumstances.

Operation of Distant Cis-Acting Elements to Control Tissue Specificity of Gene Expression

Introduction of a foreign gene into the germline of mice to create transgenic lines allows the effects of the presence or absence of cis-acting elements to be studied in vivo. This is of particular use in studying developmental and tissue-specific expression. Semenza and colleagues (239, 240, 242) have described human erythropoietin gene expression in transgenic mice containing four different human erythropoietin transgenes; three contained 0.7 kb of 3' sequence, which includes the enhancer element described in detail previously, but differed at the 5' end in containing 0.4 kb, 6 kb, and 14 kb of 5' flanking sequence. The fourth transgene contained 16.5 kb of 5' sequence and 2.2 kb of 3' sequence. The 0.4 kb transgene was widely expressed but was inducible only in liver, the 6 kb transgene was inducibly expressed in liver but was not expressed elsewhere, and the longest two transgenes were expressed in an inducible manner in the liver and kidney (Fig. 49.4b). This pattern led the authors to propose the existence of an element between 0.4 kb and 6 kb that represses expression in most tissues, and the existence of an element between 6 and 14 kb that permits or controls renal gene expression. Even the largest transgene was not expressed in a copy-number-dependent fashion and the hepatic-to-renal ratio of gene expression was large in relation to what is observed in vivo, possibly implying that still more relevant sequence lies outside these areas.

The achievement of erythropoietin tissue-specific gene expression in transgenic mice allows another approach to the identification of the erythropoietin-producing cells. If an identifiable reporter gene is placed behind the erythropoietin promoter, then erythropoietin tissue–specific gene expression of the reporter gene may be produced. The use of a transforming oncogene

as a reporter may induce tumor formation and aid the setting up of cell lines from specific tissues. This has recently been achieved using a 17 kb mouse erythropoietin construct containing 9 kb of 5' erythropoietin sequence fused to the SV40 virus "T" antigen (175). These animals show regulated expression of "T" antigen in the nuclei of an interstitial cell population that does not stain with leukocyte common antigen, the macrophage marker F4/80 or the endothelial marker CD31. Double immunohistochemical labeling with antibodies to 5' ectonucleotidase (175) strongly supports recent in situ hybridization studies (5) that demonstrate colocalization of erythropoietin mRNA with that antigen, indicating that the renal erythropoietin-producing cells are a fibroblastlike population.

Cis-Acting Sequences in Erythropoietin mRNA

The elements discussed in the preceding sections may all contribute to oxygen-dependent control of erythropoietin gene transcription. Experiments described previously indicate that additional controls of erythropoietin mRNA stability may operate. The long 5' and 3' untranslated regions surrounding erythropoietin coding sequence have led to the suggestion that these sequences may mediate such control, though in other genes examples of stability determinants can be found within coding sequence (248). Using binding studies to RNA transcribed in vitro, retarding protein species were found to specifically interact with a 120 bp region lying in a highly conserved portion of the 3' untranslated region (223) (Fig. 49.4c).

Comparison between cytoplasmic extracts from organs of unstimulated and mildly stimulated mice did not show differences in lung, kidney, muscle, and liver but did show increased binding from brain and spleen. These data therefore do not completely resolve the uncertainty as to whether changes in erythropoietin mRNA stability are regulated by oxygen. Although the authors postulate that this erythropoietin mRNA cis element might be present in undescribed transcripts in brain and spleen, it should be noted that these organs do contain a very small amount of erythropoietin mRNA that is increased by hypoxia (261, 262).

The above studies indicate that multiple interactions with cis-acting sequences coordinate erythropoietin gene expression. For a given gene, differentiated mammalian tissues may exhibit differences approaching 10^9 between transcription in "nonexpressing" cells and the "fully active" state (275). To achieve inducibility against this background thermodynamic principles indicate that control should best be achieved by multiple factors, thus permitting multiplication of the effects of modest changes in concentration or energetics. Whether the multiple influences that have been defined for erythropoietin gene expression do interact in this way, or whether there is substantial redundancy, is not yet clear.

RELATIONSHIP TO OTHER ADAPTIVE RESPONSES TO HYPOXIA

Several lines of circumstantial evidence suggest that this system may operate more widely than in control of erythropoietin gene expression itself. Hypoxia is a basic physiological stimulus to which many genes in many organisms show responses (2, 252, 284). Studies in hepatoma cells have suggested the possibility that a hemoprotein is the oxygen sensor (98), and it is therefore of interest that heme and hemoproteins are known to play a central role in oxygen-regulated responses in many lower organisms. For instance, expression of the fixL/J proteins that control the oxygen-sensitive nitrogen-fixation genes of *Rhizobium meliloti* reveals fixL to be a hemoprotein that reversibly binds oxygen (48, 97). FixL and fixJ are homologous with a family of bacterial regulators for which the mode of signal transduction is phosphorylation. However, although fixL was shown to possess kinase activity, the mode of signal transduction following oxygen ligation has not yet been formally defined.

In yeast, heme also appears to be closely involved with transcriptional regulation of oxygen-responsive genes (reviewed in ref. 284). Exogenous heme will restore an aerobic pattern of gene transcription to anaerobically grown cells, whereas strains containing mutants in the heme biosynthetic pathway show a constitutive anaerobic pattern of gene expression (118). Oxygen interacts in a number of steps in heme biosynthesis and degradation, and it is proposed that cellular heme levels act as an intermediate in the mechanism that senses oxygen in yeast cells (284). Heme itself interacts with transcriptional activity by the zinc finger protein HAP-1 (heme-activating protein). Mutational studies indicate the existence of a domain in HAP-1 that masks its own DNA-binding region in the absence of heme. Deletion of this domain results in heme independent constitutive activation by HAP-1 (207). This activates a set of genes involved with aerobic function together with the repressor Rox-1, which acts to repress a set of anaerobic genes (168). Clearly, this scheme is different from the one proposed by Goldberg et al. for the involvement of heme in erythropoietin regulation. However, in neither case do the experimental data precisely specify the model proposed, and it may be that there are some common elements.

Other evidence that a mechanism of oxygen sensing similar to that which controls erythropoietin production might operate widely at least in mammalian cells has recently been obtained from transfection studies. A transcriptionally active complex can form on transiently transfected DNA without full operation of repressive effects associated with the packaging of DNA into chromatin. It is therefore possible to use transient transfection of DNA containing the regulatory sequence to assay for the presence of the interacting oxygen-sensing system in cells that do not make erythropoietin but might operate a similar oxygen-sensing system for other purposes. When this was done (176) it was found that the enhancer sequence did indeed mediate oxygen-dependent transcriptional effects in a wide variety of cell lines that did not make erythropoietin and were derived from organs such as skin, lung, and ovary that are not associated with significant erythropoietin gene expression in vivo (Fig. 49.5). More detailed study of one of these cell lines, the Chinese hamster lung fibroblastoid cell lines a23 (276), showed that in several important respects the physiological features of erythropoietin gene expression were mimicked by the operation of the DNA sequence in this non-erythropoietin-producing cell line. Thus, the enhancer conveyed on an α globin reporter gene responses to both hypoxia and cobaltous ions. The response was not mimicked by cyanide, and was blocked by the protein synthesis inhibitor, cycloheximide, as is also the case for regulation of the endogenous erythropoietin gene in hepatoma cells (98). The implication of these findings is that an oxygen-sensing system, similar or identical to the one involved in erythropoietin regulation, operates much more widely in mammalian cells than the tightly tissue-restricted expression of the erythropoietin gene. The existence of tissue-specific gene repression, superimposed on a widespread sensing and signaling system, is well established for several responses such as the adenylate cyclase system and glucocorticoid responsive genes (12, 195, 221). These systems operate widely but induce particular genes in particular cells, because tissue-specific repressive influences such as arise from chromatin structure allow only a given subset of genes with cAMP or glucocorticoid response elements to be transcribed in a given cell. It may be that a similar hierarchical system operates for the oxygen-sensing system that regulates erythropoietin production, and that the same system serves other functions in non-erythropoietin-producing cells. Many genes are transcribed in an oxygen-dependent manner, and for some, such as platelet-derived growth factor β chain (140), features similar to the erythropoietin response have been noted. However, whether other genes are in fact controlled by the same sensing system as erythropoietin will require further definition of the DNA-binding motifs within the cis-acting control elements, and identification of their cognate DNA-binding proteins. Evidence that certain glycolytic genes do indeed respond to the same oxygen-dependent transcriptional control system as erythropoietin has

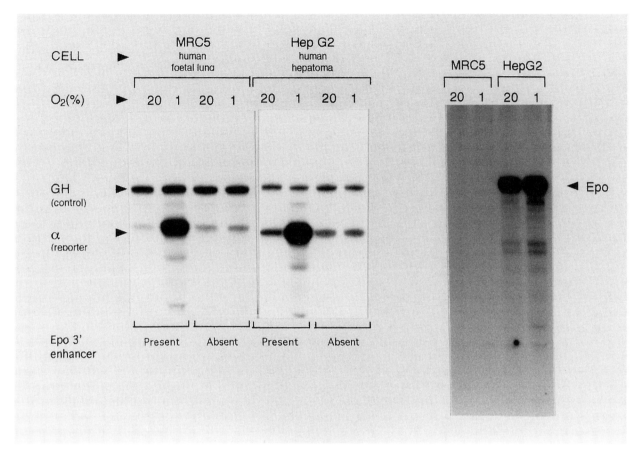

FIG. 49.5. In transfected cells the oxygen-dependent operation of the erythropoietin 3' enhancer is more widespread than expression of the native erythropoietin gene. (a) Autoradiograph of RNAse protection assay showing oxygen-dependent operation of the erythropoietin 3' enhancer in transfected MRC5 and HepG2. Alternate lanes show normoxic and hypoxic expression of the test plasmids containing α globin with or without the erythropoietin 3' enhancer. A ferritin–growth hormone fusion gene was used on the cotransfected control plasmid. A similar increase in transcription of the α globin reporter gene is observed in hypoxic cells of each type when the enhancer is present. (b) Autoradiograph of RNAse protection assay of endogenous erythropoietin gene expression in MRC5 (human lung fibroblast cell line) and HepG2 (human hepatoma cell line). Endogenous erythropoietin gene expression is seen in HepG2 but not MRC5.

recently been obtained by demonstration of functional similarities in the regulation of these genes and the operation of similar cis-acting sequences which bind to the hypoxically inducible nuclear factor HIF-1 (83a).

The authors are grateful to Mrs. E. M. Rose and Miss C. E. Bastable for the preparation of this manuscript.

REFERENCES

1. Abbrecht, P. H., and J. K. Littell. Plasma erythropoietin in men and mice during acclimatization to different altitudes. *J. Appl. Physiol.* 32: 54–58, 1972.
2. Adair, T. H., W. J. Gay, and J.-P. Montani. Growth regulation of the vascular system: evidence for a metabolic hypothesis. *Am. J. Physiol.* 259 (*Regulatory Integrative Comp Physiol.* 28): R393–R404, 1990.
3. Adamson, J. W., and J. W. Eschbach. Treatment of the anemia of chronic renal failure with recombinant human erythropoietin. *Annu. Rev. Med.* 41: 349–360, 1990.
4. Adamson, J. W., B. Torok-Storb, and N. Lin. Analysis of erythropoiesis by erythroid colony formation in culture. *Blood Cells* 4: 89–103, 1978.
5. Bachmann, S., M. LeHir, and K.-U. Eckardt. Colocalization of erythropoietin mRNA and ecto-5'-nucleotidase immunoreactivity in peritubular cells of rat renal cortex indicates that fibroblasts produce erythropoietin. *J. Histochem. Cytochem.* 41: 335–341, 1993.
6. Baer, A. N., E. N. Dessypris, E. Goldwasser, and S. B. Krantz. Blunted erythropoietin response to anaemia in rheumatoid arthritis. *Br. J. Haematol.* 66: 559–564, 1987.
7. Bailey, S. C., R. Spangler and A. J. Sytkowski. Erythropoietin induces cytosolic protein phosphorylation and dephosphorylation in erythroid cells. *J. Biol. Chem.* 226: 24121–24125, 1991.
8. Baker, R., J. R. Zucali, B. J. Baker, and J. Strauss. Erythropoietin and intrarenal oxygenation in hypercapnic versus normocapnic hypoxemia. *Adv. Exp. Med. Biol.* 69: 597–609, 1984.

9. Bauer, C., and A. Kurtz. Oxygen sensing in the kidney and its relation to erythropoietin production. *Annu. Rev. Physiol.* 51: 845–856, 1989.

10. Bazan, J. F. Haemopoietic receptors and helical cytokines. *Immunol. Today* 11: 350–354, 1990.

11. Bazan, J. F. Structural design and molecular evolution of a cytokine receptor superfamily. *Proc. Natl. Acad. Sci. USA* 87: 6934–6938, 1990.

12. Beato, M. Gene regulation by steroid hormones. *Cell* 56: 335–344, 1989.

13. Beaupain, D., J. F. Eleouet and P. H. Romeo. Initiation of transcription of the erythroid promoter of the porphobilinogen deaminase gene is regulated by a *cis*-acting sequence around the cap site. *Nucleic Acids Res.* 18: 6509–6515, 1990.

14. Beck, I., S. Ramirez, R. Weinmann, and J. Caro. Enhancer element at the 3′-flanking region controls transcriptional response to hypoxia in the human erythropoietin gene. *J. Biol. Chem.* 266: 15563–15566, 1991.

15. Beru, N., D. Smith, and E. Goldwasser. Evidence suggesting negative regulation of the erythropoietin gene by ribonucleoprotein. *J. Biol. Chem.* 265: 14100–14104, 1990.

16. Beynon, G. The influence of the autonomic nervous system in the control of erythropoietin secretion in the hypoxic rat. *J. Physiol. (London)* 266: 347–360, 1977.

17. Birgegard, G., O. Miller, J. Caro, and A. J. Erslev. Serum erythropoietin levels by radioimmunoassay in polycythemia. *Scand. J. Haematol.* 29: 161–167, 1982.

18. Birgegard, G., L. Wide, and B. Simonsson. Marked erythropoietin increase before fall in Hb after treatment with cytostatic drugs suggests mechanism other than anaemia for stimulation. *Br. J. Haematol.* 72: 462–466, 1989.

19. Bjarnason, I., P. M. Cotes, S. Knowles, C. Reid, R. Wilkins, and T. J. Peters. Giant lymph node hyperplasia (Castelman's syndrome) of the mesentery. Observations on the associated anemia. *Gastroenterology* 87: 216–223, 1984.

20. Blanchard, K. L., A. M. Acquaviva, D. L. Galson, and H. F. Bunn. Hypoxic induction of the human erythropoietin gene: cooperation between the promoter and enhancer, each of which contains steroid receptor response elements. *Mol. Cell. Biol.* 12: 5373–5385, 1992.

21. Blumberg, A., H. Keller, and H. R. Marti. Effect of altitude on erythropoiesis and oxygen affinity in anaemic patients on maintenance dialysis. *Eur. J. Clin. Invest.* 3: 93–97, 1973.

22. Boissel, J.-P., and H. F. Bunn. Erythropoietin structure-function relationships. In: *The Biology of Hematopoiesis,* edited by N. Dainiak, E. P. Cronkite, R. McCaffrey, and R. D. Shadduck. New York: Wiley-Liss, 1990, p. 227–232.

23. Bondurant, M. C., and M. J. Koury. Anemia induces accumulation of erythropoietin mRNA in the kidney and liver. *Mol. Cell. Biol.* 6: 2731–2733, 1986.

24. Breathnach, R., and P. Chambon. Organization and expression of eukaryotic split genes coding for proteins. *Annu. Rev. Biochem.* 50: 349–383, 1981.

25. Brezis, M., P. Shanley, P. Silva, K. Spokes, S. Lear, F. H. Epstein, and S. Rosen. Disparate mechanisms for hypoxic cell injury in different nephron segments. *J. Clin. Invest.* 76: 1796–1806, 1985.

26. Broudy, V. C., N. Lin, M. Brice, B. Nakamoto, and T. Papayannopoulou. Erythropoietin receptor characteristics on primary human erythroid cells. *Blood* 77: 2583–2590, 1991.

27. Broudy, V. C., J. F. Tait, and J. S. Powell. Recombinant human erythropoietin: purification and analysis of carbohydrate linkage. *Arch. Biochem. Biophys.* 265: 329–336, 1988.

28. Cahan, C., P. L. Hoekje, E. Goldwasser, M. J. Decker, and K. P. Strohl. Assessing the characteristic between length of hypoxic exposure and serum erythropoietin levels. *Am. J. Physiol.* 258 (*Regulatory Integrative Comp. Physiol.* 27): R1016–1021, 1990.

29. Carmena, A. O., D. Howard and F. Stohlman. Regulation of erythropoiesis. XXII. Erythropoietin production in the newborn animal. *Blood* 332: 376–382, 1968.

30. Carnot, P., and C. Deflandre. Sur l'activite hemopoietique du serum au cours de la regeneration du sang. *C.R. Acad. Sci. Paris* 143: 384–386, 1906.

31. Caro, J., S. Brown, O. Miller, T. Murray, and A. J. Erslev. Erythropoietin levels in uremic nephric and anephric patients. *J. Lab. Clin. Med.* 93: 449–458, 1979.

32. Caro, J., and A. J. Erslev. Biologic and immunologic erythropoietin in extracts from hypoxic whole rat kidneys and in their glomerular and tubular fractions. *J. Lab. Clin. Med.* 103: 922–931, 1984.

33. Chandra, M., G. K. Clemons, and M. I. McVicar. Relation of serum erythropoietin levels to renal excretory function. Evidence for lowered set point for erythropoietin production in chronic renal failure. *J. Pediatr.* 113: 1015–1021, 1988.

34. Chern, Y., T. Chung, and A. J. Sytkowski. Structural role of amino acids 99–110 in recombinant human erythropoietin. *Eur. J. Biochem.* 202: 225–229, 1991.

35. Chiba, T., H. Amanuma, and K. Todokoro. Tryptophan residue of TRP-SER-X-TRP-SER motif in extracellular domains of erythropoietin receptor is essential for signal transduction. *Biochem. Biophys. Res. Commun.* 184: 485–490, 1992.

36. Chiba, T., Y. Ikawa, and K. Todokoro. GATA-1 transactivates erythropoietin receptor gene, and erythropoietin receptor-mediated signals enhance GATA-1 gene expression. *Nucleic Acids Res.* 19: 3843–3848, 1991.

37. Cohen, J. J. Relationship between energy requirements for Na$^+$ reabsorption and other renal functions. *Kidney Int.* 29: 32–40, 1986.

38. Cohen, R. A., M. E. Miller, J. F. Garcia, G. Moccia, and E. P. Cronkite. Regulatory mechanism of erythropoietin production: Effects of hypoxemia and hypercarbia. *Exp. Hematol.* 9: 513–521, 1981.

39. Costa-Giomi, P., J. Caro, and R. Weinmann. Enhancement by hypoxia of human erythropoietin gene transcription *in vitro. J. Biol. Chem.* 265: 19185–10188, 1990.

40. Cotes, P. M. Immunoreactive erythropoietin in serum. *Br. J. Haematol.* 50: 427–438, 1982.

41. Cotes, P. M. Physiological studies of erythropoietin in plasma. In: *Erythropoietin,* edited by W. Jelkmann and A. J. Gross. Berlin: Springer-Verlag, 1989, p. 57–79.

42. Cotes, P. M., and D. R. Bangham. Bioassay of erythropoietin in mice made polycythaemic by exposure to air at a reduced pressure. *Nature* 191: 1065–1067, 1961.

43. Cotes, P. M., C. J. Dore, J. A. Liu Yin, M. Lewis, M. Messinezy, T. C. Pearson, and C. Reid. Determination of serum immunoreactive erythropoietin in the investigation of erythrocytosis. *N. Engl. J. Med.* 315: 283–287, 1986.

44. Cotes, P. M., M. J. Pippard, C. D. L. Reid, C. G. Winearls, D. O. Oliver, and J. P. Royston. Characterization of the anaemia of chronic renal failure and the mode of its correction by a preparation of human erythropoietin. *Q. J. Med.* 70: 113–137, 1989.

45. Cotes, P. M., R. C. Tam, P. Reed, and M. Hellebostad. An immunological cross-reactant of erythropoietin in serum which may invalidate EPO radioimmunoassay. *Br. J. Haematol.* 73: 265–268, 1989.

46. D'Andrea, A. D., P. J. Szklut, H. F. Lodish, and E. M. Alderman. Inhibition of receptor binding and neutralization of bioactivity by anti-erythropoietin monoclonal antibodies. *Blood* 75: 874–880, 1990.

47. D'Andrea, A. D., and L. I. Zon. Erythropoietin receptor. *J. Clin. Invest.* 86: 681–687, 1990.

48. David, M., M.-L. Daveran, J. Batut, A. Dedieu, O. Domergue, J. Ghai, C. Hertig, P. Boistard, and D. Kahn. Cascade regulation of nif gene expression in *Rhizobium meliloti. Cell* 54: 671–683, 1988.

49. Davis, J. M., T. Arakawa, T. W. Strickland, and D. A. Yphantis. Characterization of recombinant human erythropoietin produced in Chinese hamster ovary cells. *Biochemistry* 26: 2633–2638, 1987.

50. de Wet, J. R., K. V. Wood, M. De Luca, D. R. Helinski, and S. Subramani. Firefly luciferase gene: structure and expression in mammalian cells. *Mol. Cell. Biol.* 7: 725–737, 1987.

51. Deetjen, P., and K. Kramer. Die Abhangigkeit des O_2-Verbrauches der Niere von der Na-Ruckresorption. *Pflugers Arch.* 273: 636–650, 1961.

52. Dignam, J. D., R. M. Lebovitz, and R. G. Roeder. Accurate transcription initiation by RNA polymerase II in a soluble extract of isolated mammalian nuclei. *Nucleic Acids Res.* 11: 1475–1489, 1983.

53. Dordal, M. S., F. F. Wang, and E. Goldwasser. The role of carbohydrate in erythropoietin action. *Endocrinology* 116: 2293–2299, 1985.

54. Dunn, C. D. R., J. H. Jarvis, and J. M. Greenman. A quantitative bioassay for erythropoietin using mouse fetal liver cells. *Exp. Hematol.* 3: 65–78, 1975.

55. Dusanter-Fourt, I., N. Casadevall, C. Lacombe, O. Muller, C. Billat, S. Fischer, and P. Mayeux. Erythropoietin induces the tyrosine phosphorylation of its own receptor in human erythropoietin-responsive cells. *J. Biol. Chem.* 267: 10670–10675, 1992.

56. Dynan, W. S. Modularity in promoters and enhancers. *Cell* 58: 1–4, 1989.

57. Dynan, W. S., and R. Tjian. Control of eukaryotic messenger RNA synthesis by sequence-specific DNA-binding proteins. *Nature* 316: 774–778, 1985.

58. Eckhardt, K., J. Dittmer, R. Neumann, C. Bauer, and A. Kurtz. Decline of erythropoietin formation at continuous hypoxia is not due to feedback inhibition. *Am. J. Physiol.* 258 (*Renal Fluid Electrolyte Physiol.* 27): F1432–F1437, 1990.

59. Eckardt, K., A. Kurtz, and C. Bauer. Regulation of erythropoietin production is related to proximal tubular function. *Am. J. Physiol.* 258 (*Renal Fluid Electrolyte Physiol.* 27): F942–F947, 1989.

60. Eckardt, K.-U., W. Hartmann, U. Vetter, F. Pohlandt, R. Burghardt and A. Kurtz. Serum immunoreactive erythropoietin of children in health and disease. *Eur. J. Pediatr.* 149: 459–464, 1990.

61. Eckardt, K.-U., S. T. Koury, C. C. Tan, S. J. Schuster, P. J. Ratcliffe, B. Kaissling, and A. Kurtz. Distribution of erythropoietin producing cells in rat kidneys during hypoxic hypoxia. *Kidney Int.* 43: 815–823, 1993.

62. Eckardt, K.-U., A. Kurtz, and C. Bauer. Triggering of erythropoietin formation by hypoxia is inhibited by respiratory and metabolic acidosis. *Am. J. Physiol.* 258 (*Regulatory Integrative Comp. Physiol.* 27): R678–R683, 1990.

63. Eckardt, K.-U., M. Le Hir, C. C. Tan, P. J. Ratcliffe, B. Kaissling, and A. Kurtz. Renal innervation plays no role in oxygen dependent control of erythropoietin mRNA levels. *Am. J. Physiol.* (*Renal Fluid Electrolyte Physiol.* 32): F925–F930, 1992.

64. Eckardt, K.-U., M. Möllmann, R. Neumann, R. Brunkhorst, H.-U. Burger, G. Lonnemann, H. Scholz, G. Keusch, B. Buchholz, U. Frei, C. Bauer, and A. Kurtz. Erythropoietin in polycystic kidneys. *J. Clin. Invest.* 84: 1160–1166, 1989.

65. Eckardt, K.-U., C. W. Pugh, P. J. Ratcliffe, and A. Kurtz. Oxy-gen dependent modulation of erythropoietin mRNA in rat hepatocytes *in vitro. Pflugers Arch.* 1993, in press.

66. Eckardt, K.-U., P. J. Ratcliffe, C. C. Tan, C. Bauer, and A. Kurtz. Age dependent expression of the erythropoietin gene in rat liver and kidneys. *J. Clin. Invest.* 89: 753–760, 1992.

67. Egrie, J. C., P. M. Cotes, J. Lane, R. E. Graines Das, and R. C. Tam. Development of radioimmunoassays for human erythropoietin using recombinant erythropoietin as tracer and immunogen. *J. Immunol. Methods* 99: 235–241, 1987.

68. Egrie, J. C., T. W. Strickland, J. Lane, K. Aoki, A. M. Cohen, R. Smalling, G. Trail, F. K. Lin, J. K. Browne, and D. K. Hines. Characterization and biological effects of recombinant human erythropoietin. *Immunobiology* 172: 213–224, 1986.

69. Erslev, A. J. Humoral regulation of red cell production. *Blood* 8: 349–357, 1953.

70. Erslev, A. J. *In vitro* production of erythropoietin by kidneys perfused with a serum-free solution. *Blood* 44: 77–85, 1974.

71. Erslev, A. J., and J. Caro. Pure erythrocytosis classified according to erythropoietin titers. *Am. J. Med.* 76: 57–61, 1984.

72. Erslev, A. J., J. Caro, and A. Besarab. Why the kidney? *Nephron* 41: 213–216, 1985.

73. Erslev, A. J., J. Caro, E. Kansu, and R. Silver. Renal and extra-renal erythropoietin production in anaemic rats. *Br. J. Haematol.* 45: 65–72, 1980.

74. Erslev, A. J., J. Wilson, and J. Caro. Erythropoietin titers in anemic, nonuremeic patients. *J. Lab. Clin. Med.* 109: 429–433, 1987.

75. Eschbach, J. W. The anemia of chronic renal failure: Pathophysiology and the effects of recombinant erythropoietin. *Kidney Int.* 35: 134–148, 1989.

76. Eschbach, J. W., M. H. Abdulhadi, J. K. Browne, B. G. Delano, M. R. Downing, J. C. Egrie, R. W. Evans, E. A. Friedman, S. E. Graber, and N. R. Haley. Recombination human erythropoietin in anemic patients with end-stage renal disease. *Ann. Intern. Med.* 111: 992–1000, 1989.

77. Fandrey, J., W. Jelkmann, and C. P. Seigers. Control of the production of erythropoietin in hepatoma cell cultures (HepG2). *Funktionsanal Biol. Syst.* 33: 165–177, 1991.

78. Fandrey, J., F. P. Seydel, C.-P. Siegers, and W. Jelkmann. Role of cytochrome P_{450} in the control of the production of erythropoietin. *Life Sci.* 47: 127–134, 1990.

79. Faquin, W. C., T. J. Schneider, and M. A. Goldberg. Effect of inflammatory cytokines on hypoxia-induced erythropoietin production. *Blood* 79: 1987–1994, 1992.

80. Felsenfeld, G. Chromatin as an essential part of the transcriptional mechanism. *Nature* 355: 219, 1992.

81. Fink, G. D., and J. W. Fisher. Erythropoietin production after renal denervation or beta-adrenergic blockage. *Am. J. Physiol.* 230: 508–513, 1976.

82. Fink, G. D., and J. W. Fisher. Stimulation of erythropoiesis by beta adrenergic agonists. I: Characterization of activity in polycythaemic mice. *J. Pharmacol. Exp. Ther.* 202: 192–198, 1977.

83. Finne, P. H., R. Skoglund, and S. Wetterhus. Urinary erythropoietin during initial treatment of pernicious anaemia. *Scand. J. Haematol.* 10: 62–68, 1973.

83a. Firth, J. D., B. L. Ebert, C. W. Pugh, and P. J. Ratcliffe. Oxygen-regulated control elements in the phosphoglycerate kinase 1 and lactate dehydrogenase A genes: Similarities with the erythropoietin 3′ enhancer. *Proc. Natl. Acad. Sci. USA* 91: 6496–6500, 1994.

84. Fisher, J. W. Pharmacologic modulation of erythropoietin production. *Annu. Rev. Pharmacol. Toxicol.* 28: 101–122, 1988.

85. Fisher, J. W., and B. J. Birdwell. The production of an erythropoietic factor by the *in situ* perfused kidney. *Acta Haematol. (Basel)* 26: 224–232, 1961.

86. Fisher, J. W., D. B. Knight, and C. Couch. The influence of

several diuretic drugs on erythropoietin formation. *J. Pharmacol Exp. Ther.* 141: 113–121, 1963.

87. Fisher, J. W., G. Taylor, and D. D. Porteous. Localisation of erythropoietin in glomeruli of sheep kidney by fluorescent antibody technique. *Nature* 205: 611–612, 1965.

88. Fogh, J. A sensitive erythropoietin assay on mice exposed to CO-hypoxia. *Scand. J. Clin. Lab. Invest.* 18: 33–44, 1966.

89. Fried, W. The liver as a source of extrarenal Epo production. *Blood* 40: 671–677, 1972.

90. Fried, W., J. Barone-Varelas, and M. Berman. Detection of high erythropoietin titres in renal extracts of hypoxic rats. *J. Lab. Clin. Med.* 97: 82–96, 1981.

91. Fried, W., C. Johnson, and P. Heller. Observations on regulation of erythropoiesis during prolonged periods of hypoxia. *Blood* 38: 607–616, 1970.

92. Fried, W., L. Plazk, L. O. Jacobson, and E. Goldwasser. Erythropoiesis II: Assay of erythropoietin in hypophysectomized rats. *Proc. Soc. Exp. Biol. Med* 92: 203–207, 1956.

93. Fukuda, M. N., H. Sasaki, L. Lopez, and M. Fukuda. Survival of recombinant epo in the circulation: The role of carbohydrates. *Blood* 73: 84–89, 1989.

94. Gallo, R. C., W. Fraimow, R. T. Cathcart, and A. J. Erslev. Erythropoietic response in chronic pulmonary disease. *Arch. Intern. Med.* 113: 559–568, 1964.

95. Garcia, J. F., and G. K. Clemons. The radioimmunoassay of erythropoietin. In: *Recent Advances in Nuclear Medicine,* edited by J. H. Lawrence and S. Winchell. New York: Grune and Stratton, 1983, p. 19–40.

96. Garcia, J. F., S. N. Ebbe, L. Hollander, H. O. Cutting, M. E. Miller, and E. Cronkite. Radioimmunoassay of erythropoietin: circulating levels in normal and polycythemic human beings. *J. Lab. Clin. Med.* 99: 624–635, 1982.

97. Gilles-Gonzalez, M. A., G. S. Ditta, and D. R. Helinski. A haemoprotein with kinase activity encoded by the oxygen sensor of *Rhizobium meliloti. Nature* 350: 170–172, 1991.

98. Goldberg, M. A., S. P. Dunning, and H. F. Bunn. Regulation of the erythropoietin gene: evidence that the oxygen sensor is a heme protein. *Science* 242: 1412–1415, 1988.

99. Goldberg, M. A., C. C. Gaut, and H. F. Bunn. Epo mRNA levels are governed by both the rate of gene transcription and posttranscriptional events. *Blood* 77: 271–277, 1991.

100. Goldberg, M. A., G. A. Glass, J. M. Cunningham, and H. F. Bunn. The regulated expression of epo by two human hepatoma cell lines. *Proc. Natl. Acad. Sci. USA* 84: 7972–7976, 1987.

101. Goldwasser, E., J. F. Eliason, and D. Sikkema. An assay for epo *in vitro* at the milliunit level. *Endocrinology* 97: 315–323, 1974.

102. Gorman, C. High efficiency gene transfer into mammalian cells. In: *DNA Cloning,* edited by D. M. Glover. Oxford: IRL Press, 1985, p. 143–190.

103. Green, M. R. Pre-mRNA processing and mRNA nuclear export. *Curr. Opin. Cell Biol.* 1: 519–525, 1989.

104. Greenblatt, J. Roles of TFIID in transcriptional initiation by RNA polymerase II. *Cell* 66: 1067–1070, 1991.

105. Gregory, C. J., and A. C. Eaves. Three stages of erythropoietic progenitor cell differentiation distinguished by a number of physical and biologic properties. *Blood* 51: 527–537, 1978.

106. Gross, D. S., and W. T. Garrard. Nuclease hypersensitive sites in chromatin. *Annu. Rev. Biochem.* 57: 159–197, 1988.

107. Grosveld, F., G. B. van Assendelft, D. R. Greaves, and G. Kollias. Position independent, high-level expression of the human β-globin gene in transgenic mice. *Cell* 51: 975–985, 1987.

108. Gruber, D. F., J. R. Zucali, and E. A. Mirand. Identification of erythropoietin producing cells in fetal mouse liver cultures. *Exp. Hematol.* 5: 392–398, 1977.

109. Gruber, D. F., J. R. Zucali, J. Wleklinski, V. LaRussa, and E. A. Mirand. Temporal transition in the site of rat erythropoietin production. *Exp. Hematol.* 5: 399–407, 1977.

110. Hagiwara, M., K. Nagakura, M. Ueno, and J. W. Fuisher. Inhibitory effects of tetradecanoylphorbol acetate and diacylglycerol on erythropoietin production in human renal carcinoma cell cultures. *Exp. Cell Res.* 173: 129–136, 1987.

111. Halvorsen, S. Plasma erythropoietin levels following hypothalamic stimulation in the rabbit. *Scand. J. Clin. Lab. Invest.* 13: 564–575, 1961.

112. Halvorsen, S., B. L. Roh, and J. W. Fisher. Erythropoietin production in nephrectomized and hypophysectomized animals. *Am. J. Physiol.* 215: 349–352, 1968.

113. Heberlein, C., K.-D. Fischer, M. Stoffel, J. Nowock, A. Ford, U. Tessmer, and C. Stocking. The gene for erythropoietin receptor is expressed in multipotential hematopoietic and embryonal stem cells: Evidence for differentiation stage-specific regulation. *Mol. Cell. Biol.* 12: 1815–1826, 1992.

114. Hellebostad, M., P. Haga, and P. M. Cotes. Serum immunoreactive erythropoietin in healthy normal children. *Br. J. Haematol.* 70: 247–250, 1988.

115. Hillman, R. S., and C. A. Finch. Erythropoietin: normal and abnormal. *Semin. Hematol.* 4: 327–336, 1967.

116. Hjort, E. Reticulocyte increase after injection of anemic serum. *Norsk Mag. F. Laegevidensk* 97: 270–277, 1936.

117. Hochberg, M. C., C. M. Arnold, B. B. Hogans, and J. L. Spivak. Serum immunoreactive erythropoietin in rheumatoid arthritis: impaired response to anemia. *Arthritis Rheum.* 31: 1318–1321, 1988.

118. Hodge, M. R., G. Kim. K. Singh, and M. G. Cumsky. Inverse regulation of the yeast *COX5* genes by oxygen and heme. *Mol. Cell. Biol.* 9: 1958–1964, 1989.

119. Imagawa, S., M. A. Goldberg, J. Doweiko, and H. F. Bunn. Regulatory elements of the erythropoieting gene. *Blood* 77: 278–285, 1991.

120. Imai, N., A. Kawamura, M. Higuchi, M. Oh-Eda, T. Orita, T. Kawaguchi, and N. Ochi. Physicochemical and biological comparison of recombinant human erythropoietin with human urinary erythropoietin. *J. Biochem. (Tokyo)* 107: 352–359, 1990.

121. Iscove, N. N. The role of erythropoietin in regulation of population size and cell cycling of early and late erythroid precursors in mouse bone marrow. *Cell Tissue Kinet.* 10: 323–334, 1977.

122. Jacobs, K., C. Shoemaker, R. Rudersdorf, S. D. Neill, R. J. Kaufman, A. Mufson, J. Seehra, S. S. Jones, R. Hewick, E. F. Fritsch, M. Kawakita, T. Shimizu, and T. Miyake. Isolation and characterization of genomic and cDNA clones of human epo. *Nature* 313: 806–810, 1985.

123. Jacobson, L. O., E. Goldwasser, W. Fried, and L. Pizak. Role of the kidney in erythropoiesis. *Nature* 179: 633–634, 1957.

124. Jelkmann, W. Temporal pattern of erythropoietin titers in kidney tissue during hypoxic hypoxia. *Pflügers Arch.* 393: 88–91, 1982.

125. Jelkmann, W. Erythropoietin: Structure, control of production, and function. *Physiol. Rev.* 72: 449–489, 1992.

126. Jelkmann, W., and C. Bauer. Demonstration of high levels of erythropoietin in rat kidneys following hypoxic hypoxia. *Pflügers Arch.* 392: 34–39, 1981.

127. Jelkmann, W., A. Huwiler, J. Fandrey, and J. Pfeilschifter. Inhibition of erythropoietin production by phorbol ester is associated with down regulation of protein kinase C-α isoenzyme in hepatoma cells. *Biochem. Biophys. Res. Commun.* 179: 1441–1448, 1991.

128. Jelkmann, W., H. Pagel, M. Wolff, and J. Fandrey. Monokines inhibiting erythropoietin production in human hepatoma cul-

tures and in isolated perfused rat kidneys. *Life Sci.* 50: 301–308, 1992.

129. Jelkmann, W., and J. Seidl. Dependence of erythropoietin production on blood oxygen affinity and hemoglobin concentration in rats. *Biomed. Biochim. Acta* 46: S304–S308, 1987.

130. Jelkmann, W., and G. Wiedemann. Serum erythropoietin level: relationship to blood hemoglobin concentration and erythrocytic activity of the bone marrow. *Klin. Wochenschr.* 68: 403–407, 1990.

131. John, J., R. McKendry, S. Pellegrini, D. Flavell, I. M. Kerr, and G. R. Stark. Isolation and characterisation of a new mutant human cell line unresponsive to alpha and beta interferons. *Mol. Cell. Biol.* 11: 4189–4195, 1991.

132. Johnson, P. F., and S. L. McKnight. Eukaryotic transcriptional regulatory proteins. *Annu. Rev. Biochem.* 58: 799–839, 1989.

133. Jones, D. P. Intracellular diffusion gradients of O_2 and ATP. *Am. J. Physiol.* 250 (*Cell Physiol.* 19): C663–C675, 1986.

134. Jones, S. S., A. D. D'Andrea, L. L. Haines, and G. G. Wong. Human erythropoietin receptor: Cloning, expression and biological characterization. *Blood* 76: 31–35, 1990.

135. Jungermann, K., and N. Katz. Functional specialization of different hepatocyte populations. *Physiol Rev.* 69: 708–764, 1989.

136. Kadonaga, J. R., and R. Tjian. Affinity purification of sequence-specific DNA binding proteins. *Proc. Natl. Acad. Sci. USA* 83: 5889–5893, 1986.

137. Kiil, F., K. Aukland, and H. E. Refsum. Renal sodium transport and oxygen consumption. *Am. J. Physiol.* 201: 511–516, 1961.

138. Kindler, J., K.-U. Eckardt, K. Jandeleit, A. Kurtz, A. Schreiber, P. Scigalla, and H.-G. Sieberth. Single dose pharmacokinetics of recombinant human erythropoietin (rHuEPO) in patients with various degrees of renal failure. *Nephrol. Dial. Transplant* 4: 345–349, 1989.

139. Koeffler, H. P., and E. Goldwasser. Erythropoietin radioimmunoassay in evaluating patients with polycythemia. *Ann. Intern. Med.* 94: 44–47, 1981.

140. Kourembanas, S., R. L. Hannan, and D. V. Fuller. Oxygen tension regulates the expression of the platelet-derived growth factor-β chain gene in human endothelial cells. *J. Clin. Invest.* 86: 670–674, 1990.

141. Koury, M. J., and M. C. Bondurant. Maintenance by erythropoietin of viability and maturation of murine erythroid precursor cells. *J. Cell. Physiol.* 137: 65–74, 1988.

142. Koury, M. J., and M. C. Bondurant. Erythropoietin retards DNA breakdown and prevents programmed death in erythroid progenitor cells. *Science* 248: 378–381, 1990.

143. Koury, M. J., M. C. Bondurant, S. E. Graber, and S. T. Sawyer. Epo messenger RNA levels in developing mice and transfer of ^{125}I-epo by the placenta. *J. Clin. Invest.* 82: 154–159, 1988.

144. Koury, M. J., S. T. Sawyer, and M. C. Bondurant. Splenic erythroblasts in anemia-inducing Friend disease: a source of cells for studies of erythropoietin mediated differentiation. *J. Cell. Physiol.* 121: 526–532, 1984.

145. Koury, S. T., M. C. Bondurant, and M. J. Koury. Localization of erythropoietin synthesizing cells in murine kidneys by *in situ* hybridization. *Blood* 71: 524–527, 1988.

146. Koury, S. T., M. C. Bondurant, M. J. Koury, and G. L. Semenza. Localization of cells producing erythropoietin in murine liver by *in situ* hybridization. *Blood* 77: 2497–2503, 1991.

147. Koury, S. T., M. J. Koury, M. C. Bondurant, J. Caro, and S. E. Graber. Quantitation of epo-producing cells in kidneys of mice by *in situ* hybridization: correlation with hematocrit, renal epo mRNA and serum epo concentration. *Blood* 74: 645–651, 1989.

148. Krantz, S. B. Erythropoietin. *Blood* 77: 419–434, 1991.

149. Krantz, S. B., O. Gallien-Lartigue, and E. Goldwasser. The effect of epo upon heme synthesis by marrow cells *in vitro*. *J. Biol. Chem.* 238: 4085–4092, 1963.

150. Kriz, W. Structural organization of the renal medulla: comparative and functional aspects. *Am. J. Physiol.* 241 (*Regulatory Integrative Comp. Physiol.* 12): R3–R16, 1981.

151. Kuramochi, S., T. Chiba, H. Amanuma, A. Tojo, and K. Todokoro. Growth signal erythropoietin activates the same tyrosine kinases as interleukin 3, but activates only one tyrosine kinase as differentiation signal. *Biochem. Biophys. Res. Commun.* 181: 1103–1109, 1991.

152. Kuratowska, Z., B. Lewartowski, and E. Michalak. Studies on the production of erythropoietin by isolated perfused organs. *Blood* 18: 527–534, 1961.

153. Kurtz, A., K.-U. Eckardt, R. Neumann, R. Kaissling, M. Le Hir, and C. Bauer. Site of epo formation. *Contr. Nephrol.* 76: 14–23, 1989.

154. Kurtz, A., K.-U. Eckardt, P. J. Ratcliffe, C. W. Pugh, and P. Corvol. Requirement of protein kinase C activity for regulation of erythropoietin production in human hepatoma cells (HepG2). *Am. J. Physiol.* 262 (*Cell Physiol.* 31): C1204–1210, 1992.

155. Kurtz, A., K.-U. Eckardt, L. Tannahill, and C. Bauer. Regulation of epo production. *Contrib. Nephrol.* 66: 1–16, 1988.

156. Kurtz, A., W. Jelkmann, A. Pfuhl, K. Malmstrom, and C. Bauer. Erythropoietin production by fetal mouse liver cells in response to hypoxia and adenylate cyclase stimulation. *Endocrinology* 118: 567–572, 1986.

157. Kurtz, A., W. Jelkmann, F. Sinowatz, and C. Bauer. Renal mesangial cell cultures as a model for study of erythropoietin production. *Proc. Natl. Acad. Sci. USA* 80: 4008–4011, 1983.

158. Kurtz, A., J. Zapf, K.-U. Eckardt, G. K. Clemons, E. R. Froesch, and C. Bauer. Insulin-like growth factor I stimulates erythropoiesis in hypophysectomized rats. *Proc. Natl. Acad. Sci. USA* 85: 7825–7829, 1988.

159. Lacombe, C., C. Chesne, and P. Varlet. Expression of the epo gene in the hypoxic mouse liver. *Blood* 72: 92a, 1988.

160. Lai, P. H., R. Everett, F. F. Wang, T. Arakawa, and E. Goldwasser. Structural characterization of human epo. *J. Biol. Chem.* 261: 3116–3121, 1986.

161. Lassen, N. A., O. Munck, and J. H. Thaysen. Oxygen consumption and sodium reabsorption in the kidney. *Acta Physiol. Scand.* 51: 371–384, 1961.

162. Lechermann, B., and W. Jelkmann. Epo production in normoxic and hypoxic rats with increased blood oxygen affinity. *Respir. Physiol.* 60: 1–8, 1985.

163. Leichtweiss, H.-P., D. W. Lubbers, C. Weiss, N. Baumgartl, and W. Reschke. The oxygen supply of the rat kidney: measurements of intrarenal pO_2. *Pflugers Arch.* 309: 328–349, 1969.

164. Lemley, K. V., and W. Kriz. Anatomy of the renal interstitium. *Kidney Int.* 39: 370–381, 1991.

165. Levy, M. N., and G. Sauceda. Diffusion of oxygen from arterial to venous segments of renal capillaries. *Am. J. Physiol.* 196: 1336–1339, 1959.

166. Lin, F. K., C. H. Lin, P. H. Lai, K. Jeffrey, J. K. Browne, J. C. Egrie, R. Smalling, G. M. Fox, K. K. Chen, M. Castro, and S. Suggs. Monkey erythropoietin gene: cloning, expression and comparison with the human erythropoietin gene. *Gene* 44: 201–209, 1986.

167. Lin, F. K., S. Suggs, C. H. Lin, J. K. Browne, R. Smalling, J. C. Egrie, K. K. Chen, G. M. Foox, F. Martin, Z. Stabinsky, S. M. Badrawi, P. H. Lai and E. Goldwasser. Cloning and expression of the human erythropoietin gene. *Proc. Natl. Acad. Sci. USA* 82: 7580–7584, 1985.

168. Lowry, C. V., and R. S. Zitomer. *ROX1* encodes a heme-

induced repression factor regulating *ANB1* and *CYC7* of *Saccharomyces cerevisiae*. *Mol. Cell. Biol.* 8: 4651–4658, 1988.

169. Maniatis, T., S. Goodbourn, and J. A. Fischer. Regulation of inducible and tissue-specific gene expression. *Science* 236: 1237–1245, 1987.

170. Maples, P. B., D. H. Smith, N. Beru, and E. Goldwasser. Identification of erythropoietin-producing cells in mammalian tissues by *in situ* hybridization. *Blood* 68: 170, 1986.

171. Martin, D. I. K., L. I. Zon, G. Mutter, and S. H. Orkin. Expression of an erythroid transcription factor in megakaryocytic and mast cell lineages. *Nature* 344: 444–447, 1990.

172. Mason-Garcia, M., B. S. Beckman, J. W. Brookins, J. S. Powell, W. Lanham, S. Blaisdell, L. Keay, S.-C. Li, and J. W. Fisher. Development of a new radioimmunoassay for erythropoietin using recombinant erythropoietin. *Kidney Int.* 38: 969–975, 1990.

173. Mattaj, I. W. Splicing stories and poly (A) tales: an update on RNA processing and transport. *Curr. Opin. Cell Biol.* 2: 528–538, 1990.

173a. Maxwell, P. H., D. J. P. Ferguson, M. K. Osmond, C. W. Pugh, A. Heryet, B. G. Doe, M. H. Johnson, and P. J. Ratcliffe. Expression of a homologously recombined erythropoietin-SV40 T antigen fusion gene in mouse liver: Evidence for erythropoietin production by Ito cells. *Blood* 84:1823–1830, 1994.

174. Maxwell, A. P., T. R. J. Lappin, C. F. Johnston, J. M. Bridges, and M. G. McGeown. Erythropoietin production in kidney tubular cells. *Br. J. Haematol.* 74: 535–539, 1990.

175. Maxwell, P. H., C. W. Pugh, M. Osmond, A. Herryet, L. G. Nicholls, B. Doe, D. Ferguson, M. Johnson, and P. J. Ratcliffe. Identification of the renal erythropoietin producing cells using transgenic mice expressing SV40 large T antigen directed by erythropoietin control sequences (abstract) *Kidney Int.* 44: 1149–1162, 1993.

176. Maxwell, P. H., C. W. Pugh, and P. J. Ratcliffe. Inducible operation of the erythropoietin 3′ enhancer in multiple cell lines: evidence for a widespread oxygen sensing system. *Proc. Natl. Acad. Sci. USA* 90: 2423–2427, 1993.

177. McDonald, J. D., F. K. Lin, and E. Goldwasser. Cloning, sequencing and evolutionary analysis of the mouse erythropoietin gene. *Mol. Cell. Biol.* 6: 842–848, 1986.

178. McGhee, J. D., and G. Felsenfeld. Nucleosome structure. *Annu. Rev. Biochem.* 49: 1115–1156, 1980.

179. McGonigle, R. J. S., J. D. Wallin, R. K. Shadduck, and J. W. Fisher. Erythropoietin deficiency and inhibition of erythropoiesis in renal insufficiency. *Kidney Int.* 25: 437–444, 1984.

180. McLeod, D. L., M. M. Shreeve, and A. A. Axelrad. Improved plasma culture system for production of erythrocytic colonies *in vitro*: quantitative assay method for CFU-E. *Blood* 44: 517–534, 1974.

181. Means, R. T., and S. B. Krantz. Progress in understanding the pathogenesis of the anemia of chronic disease. *Blood* 80: 1639–1647, 1992.

182. Meineke, A., and R. C. Crafts. Further observations on the mechanism by which androgens and growth hormone influence erythropoiesis. *Ann. N.Y. Acad. Sci.* 149: 298–307, 1968.

183. Migliaccio, A. R., G. Migliaccio, A. D.'Andrea, M. Baiocchi, S. Crotta, S. Nicolis, S. Ottolenghi, and J. W. Adamson. Response to erythropoietin in erythroid subclones of the factor-dependent cell line 32D is determined by translocation of the erythropoietin receptor to the cell surface. *Proc. Natl. Acad. Sci. USA* 88: 11086–11090, 1991.

184. Milledge, J. S., and P. M. Cotes. Serum erythropoietin in humans at high altitude and its relation to plasma renin. *J. Appl. Physiol.* 59: 360–364, 1985.

185. Miller, C. B., R. J. Jones, S. Piantadosi, M. D. Abeldoff, and J. L. Spivak. Decreased erythropoietin response in patients with the anemia of cancer. *N. Engl. J. Med.* 322: 1689–1692, 1990.

186. Miller, M. E., and D. Howard. Modulation of epo concentrations by manipulation of hypercarbia. *Blood Cells* 5: 389–403, 1979.

187. Miller, M. E., M. Rorth, H. H. Parving, D. Howard, I. Reddington, C. R. Valeri, and F. Stohlman. pH effect on erythropoietin response to hypoxia. *N. Engl. J. Med.* 288: 706–710, 1973.

188. Miura, O., A. D'Andrea, D. Kabat, and J. N. Ihle. Induction of tyrosine phosphorylation by the erythropoietin receptor correlates with mitogenesis. *Mol. Cell. Biol.* 11: 4895–4902, 1991.

189. Miyake, T., K. H. Kung, and E. Goldwasser. Purification of human erythropoietin. *J. Biol. Chem.* 252: 5558–5564, 1977.

190. Mueller, P. R., and B. Wold. In vivo footprinting of a muscle specific enhancer by ligation mediated PCR. *Science* 246: 780–786, 1989.

191. Mullner, E. W., and L. C. Kuhn. A stem-loop structure in the 3′ untranslated region mediates iron-dependent regulation of transferrin receptor mRNA stability in the cytoplasm. *Cell* 53: 815–825, 1988.

192. Nakamura, Y., N. Komatsu, and H. Nakauchi. A truncated erythropoietin receptor that fails to prevent programmed cell death of erythroid cells. *Science* 257: 1138–1141, 1992.

193. Narhi, L. O., T. Arakawa, K. H. Aoki, R. Elmore, M. F. Rohde, T. Boone, and T. W. Strickland. The effect of carbohydrate on the structure and stability of erythropoietin. *J. Biol. Chem.* 266: 23022–23026, 1991.

194. Necas, E., and E. B. Thorling. Unresponsiveness of erythropoietin-producing cells to cyanide. *Am. J. Physiol.* 222: 1187–1190, 1972.

195. O'Malley, B. W. Steroid hormone action in eucaryotic cells. *J. Clin. Invest.* 74: 307–312, 1984.

196. Pabo, C. O., and R. T. Sauer. Transcription factors: Structural families and principles of DNA recognition. *Annu. Rev. Biochem.* 61: 1053–1095, 1992.

197. Pagel, H., W. Jelkmann, and C. Weiss. A comparison of the effects of renal artery constriction and anemia on the production of erythropoietin. *Pflugers Arch.* 413: 62–66, 1988.

198. Pagel, H., W. Jelkmann, and C. Weiss. Isolated serum-free perfused rat kidneys release immunoreactive erythropoietin in response to hypoxia. *Endocrinology* 128: 2633–2638, 1991.

199. Paul, P., S. A. Rothmann, J. T. McMahon, and A. S. Gordon. Epo secretion by isolated rat Kupffer cells. *Exp. Hematol.* 12: 825–830, 1984.

200. Peschle, C., C. Cillo, I. A. Pappaport, M. C. Magli, G. Migliaccio, F. Pizzella, and G. Mastroberardino. Early fluctuations of BFU-E pool size after transfusion or erythropoietin treatment. *Exp. Hematol.* 7: 87–93, 1979.

201. Peschle, C., G. Marone, A. Genovese, C. Cillo, C. Magli, and M. Condorelli. Erythropoietin production by the liver in fetal-neonatal life. *Life Sci.* 17: 1325–1330, 1975.

202. Peschle, C., I. A. Rappaport, M. C. Magli, G. Marone, F. Lettieri, C. Cillo, and A. S. Gordon. Role of the hypophysis in erythropoietin production during hypoxia. *Blood* 51: 1117–1124, 1978.

203. Peschle, C., I. A. Rappaport, G. F. Sasso, A. S. Gordon, and M. Condorelli. Mechanism of growth hormone (GH) action on erythropoiesis. *Endocrinology* 91: 511–517, 1972.

204. Peschle, C., G. F. Sasso, G. Mastroberardino, and M. Condorelli. The mechanism of endocrine influences on erythropoiesis. *J. Lab. Clin. Med.* 78: 20–29, 1971.

205. Peschle, C., E. D. Zanjani, A. S. Gidari, W. D. McLaurin, and A. S. Gordon. Mechanism of thyroxine action on erythropoiesis. *Endocrinology* 89: 609–612, 1971.

206. Pevny, L., M. C. Simon, E. Robertson, W. H. Klein, S. F. Tsai, V. D'Agati, S. H. Orkin, and F. Costantini. Erythroid differ-

entiation in chimaeric mice blocked by a targeted mutation in the gene for transcription factor GATA-1. *Nature* 349: 257–260, 1991.

207. Pfeifer, K., K.-S. Kim, S. Kogan, and L. Guarente. Functional dissection and sequence of yeast HAP1 activator. *Cell* 56: 291–301, 1989.

208. Piroso, E., A. J. Erslev, and J. Caro. Inappropriate increase in erythropoietin titers during chemotherapy. *Am. J. Hematol.* 32: 248–254, 1989.

209. Piroso, E., A. J. Erslev, K. K. Flaharty, and J. Caro. Erythropoieting life span in rats with hypoplastic and hyperplastic bone marrows. *Am. J. Hematol.* 36: 105–110, 1991.

210. Plzak, L., W. Fried, L. O. Jacobson, and W. F. Bethard. Demonstration of stimulation of erythropoiesis by plasma from anemic rats using Fe59. *J. Lab. Clin. Med.* 46: 671–678, 1955.

211. Powell, J. S., K. L. Berkner, R. V. Lebo, and J. W. Adamson. Human erythropoietin gene: high level expression in stably transfected mammalian cells and chromosome localization. *Proc. Natl. Acad. Sci. USA* 83: 6465–6469, 1986.

212. Ptashne, M. and A. A. F. Gann. Activators and targets. *Nature* 346: 329–331, 1990.

213. Pugh, C., C. C. Tan, R. W. Jones, and P. J. Ratcliffe. Functional analysis of an oxygen-related transcriptional enhancer lying 3' to the mouse Epo gene. *Proc. Natl. Acad. Sci. USA* 88: 10553–10557, 1991.

214. Pugh, C. W., B. L. Ebert, O. Ebrahim, and P. J. Ratcliffe. Characterisation of functional domains within the mouse erythropoietin 3' enhancer conveying oxygen-regulated responses in different cell lines. *Biochem. Biophys. Acta* 1217: 297–306, 1994.

215. Quelle, F. W., and D. M. Wojchowski. Proliferative action of erythropoietin is associated with rapid protein tyrosine phosphorylation in responsive B6SUt.EP cells. *J. Biol. Chem.* 266: 609–614, 1991.

216. Ratcliffe, P. J., R. W. Jones, R. E. Phillips, L. G. Nicholls, and J. I. Bell. Oxygen-dependent modulation of erythropoietin mRNA levels in isolated rat kidneys studied by RNAase protection. *J. Exp. Med.* 172: 657–660, 1990.

217. Recny, M. A., H. A. Scoble, and Y. Kim. Structural characterization of natural human urinary and recombinant DNA-derived erythropoietin. *J. Biol. Chem.* 262: 17156–17163, 1987.

218. Rege, A. B., J. Brookins, and J. W. Fisher. A radioimmunoassay for erythropoietin: serum levels in normal human subjects and patients with hemopoietic disorders. *J. Lab. Clin. Med.* 100:829–843, 1982.

219. Reissmann, K. R., T. Nomura, R. W. Gunn, and F. Brosius. Erythropoietic response to anaemia or Epo injection in uraemic rats with or without functioning renal tissue. *Blood* 16: 1411–1422, 1960.

220. Rennie, D. W., R. B. Reeves, and J. R. Pappenheimer. Oxygen tension in urine and its relation to intrarenal blood flow. *Am. J. Physiol.* 195: 120–132, 1958.

221. Roesler, W. J., G. R. Vandenbark, and R. W. Hanson. Cyclic AMP and the induction of eukaryotic gene transcription. *J. Biol. Chem.* 263: 9063–9066, 1988.

222. Romeo, P.-H., M.-H. Prandini, V. Joulin, V. Mignotte, M. Prenant, W. Vainchenker, G. Marguerie, and G. Uzan. Megakaryocytic and erythrocytic lineages share specific transcription factors. *Nature* 344: 447–449, 1990.

223. Rondon, I. J., L. A. MacMillan, B. S. Beckman, M. A. Goldberg, T. Schneider, H. F. Bunn, and J. S. Malter. Hypoxia up-regulates the activity of a novel erythropoietin mRNA binding protein. *J. Biol. Chem.* 266: 16594–16598, 1991.

224. Sasaki, H., B. Bothner, A. Dell, and M. Fukuda. Carbohydrate structure of erythropoietin expressed in Chinese hamster ovary cells by a human erythropoietin cDNA. *J. Biol. Chem.* 262: 12059–12076, 1987.

225. Sawada, K., S. B. Krantz, C.-H. Dai, S. T. Koury, S. T. Horn, A. D. Glick, and C. I. Civin. Purification of human blood burst-forming units-erythroid and demonstration of the evolution of erythropoietin receptors. *J. Cell. Physiol.* 142: 219–230, 1990.

226. Sawada, K., S. B. Krantz, J. S. Kans, E. N. Dessypris, S. Sawyer, A. D. Glick, and C. I. Civin. Purification of human erythroid colony-forming units and demonstration of specific binding of erythropoietin. *J. Clin. Invest.* 80: 357–366, 1987.

227. Sawada, K., S. B. Krantz, S. T. Sawyer, and C. I. Civin. Quantitation of specific binding of erythropoietin to human erythroid colony-forming cells. *J. Cell Physiol.* 137: 337–345, 1988.

228. Sawyer, S. T., S. B. Krantz, and K. Sawada. Receptors for erythropoietin in mouse and human erythroid cells and placenta. *Blood* 74: 103–109, 1989.

229. Scaro, J. L., M. A. Carrera, A. R. A. P. De Tombolesi, and C. Miranda. Carbon monoxide and erythropoietin production in mice. *Acta Physiol. Pharmacol. Ther. Latinoam.* 25: 204–210, 1975.

230. Schatt, M. D., S. Rusconi, and W. Schaffner. A single DNA-binding transcription factor is sufficient for activation from a distant enhancer and/or from a promoter position. *EMBO J.* 9: 481–487, 1990.

231. Scholz, H., H.-J. Schurek, K.-U. Eckardt. A. Kurtz, and C. Bauer. Oxygen-dependent erythropoietin production by the isolated perfused rat kidney. *Pflugers Arch.* 417: 1–6, 1991.

232. Schooley, J. C., and L. J. Mahlmann. Erythropoietin production in the anephric rat. I. Relationship between nephrectomy, time of hypoxic exposure and erythropoietin production. *Blood* 39: 31–38, 1972.

233. Schooley, J. C., and L. J. Mahlmann. Evidence for the de novo synthesis of erythropoietin in hypoxic rats. *Blood* 40: 662–670, 1972.

234. Schooley, J. C., and L. J. Mahlmann. Hypoxia and the initiation of Epo production. *Blood Cells* 1: 429–448, 1975.

235. Schurek, H. J., U. Jost, H. Baumgartl, H. Bertram, and U. Heckmann. Evidence for a preglomerular oxygen diffusion shunt in rat renal cortex. *Am. J. Physiol.* 259 (*Renal Fluid Electrolyte Physiol.* 28): F910–F915, 1990.

236. Schuster, S. J., E. V. Badiavas, P. Costa-Giomi, R. Weinmann, A. J. Erslev, and J. Caro. Stimulation of erythropoietin gene transcription during hypoxia and cobalt exposure. *Blood* 73: 13–16, 1989.

237. Schuster, S. J., S. T. Koury, M. Bohrer, S. Salceda, and J. Caro. Cellular sites of extrarenal and renal erythropoietin production in anemic rats. *Br. J. Haematol.* 81: 153–159, 1992.

238. Schuster, S. J., J. H. Wilson, A. J. Erslev, and J. Caro. Physiologic regulation and tissue localization of renal erythropoietin messenger RNA. *Blood* 70: 316–318, 1987.

239. Semenza, G. L., R. C. Dureza, M. D. Traystman, J. D. Gearhart, and S. E. Antonarakis. Human erythropoietin gene expression in transgenic mice: multiple transcription initiation sites and cis-acting regulatory elements. *Mol. Cell. Biol.* 10: 930–938, 1990.

240. Semenza, G. L., S. T. Loury, M. K. Nejfelt, J. D. Gearhart, and S. E. Antonarakis. Cell type-specific and hypoxia-inducible expression of the human epo gene in transgenic mice. *Proc. Natl. Acad. Sci. USA* 88: 8725–8729, 1991.

241. Semenza, G. L., M. K. Nejfelt, S. M. Chi, and S. E. Antonarakis. Hypoxia-inducible nuclear factors bind to an enhancer element located 3' to the human epo gene. *Proc. Natl. Acad. Sci. USA* 88: 5680–5684, 1991.

242. Semenza, G. L., M. D. Traystman, J. D. Gearhart, and S. E. Antonarakis. Polycythaemia in transgenic mice expressing the

human erythropoietin gene. *Proc. Natl. Acad. Sci. USA* 86: 2301–2305, 1989.

243. Semenza, G. L., and G. L. Wang. A nuclear factor induced by hypoxia via de novo protein synthesis binds to the human erythropoietin gene enhancer at a site required for transcriptional activation. *Mol. Cell. Biol.* 12: 5447–5454, 1992.

244. Sherwood, J. B., E. R. Burns, and D. Shouval. Stimulation of cAMP of erythropoietin secretion by an established human renal carcinoma cell line. *Blood* 69: 1053–1057, 1987.

245. Sherwood, J. B., L. D. Carmichael, and E. Goldwasser. The heterogeneity of circulating human serum epo. *Endocrinology* 122: 1472–1477, 1988.

246. Sherwood, J. B., and E. Goldwasser. A radioimmunoassay for epo. *Blood* 54: 885–893, 1979.

247. Shoemaker, C. B., and L. D. Mitsock. Murine erythropoietin gene: cloning, expression and human gene homology. *Mol. Cell Biol.* 6: 849–858, 1986.

248. Shyu, A.-B., M. E. Greenberg, and J. G. Belasco. The c-fos transcript is targeted for rapid decay by two distinct mRNA degradation pathways. *Genes Dev.* 3: 60–72, 1989.

249. Silva, P., R. Hallac, K. Spokes, and F. H. Epstein. Relationship among gluconeogenesis, QO$_2$ and Na$^+$ transport in the perfused rat kidney. *Am. J. Physiol.* 242 (*Renal Fluid Electrolyte Physiol.* 13): F508–F513, 1982.

250. Singh, H., R. Sen, D. Baltimore, and P. A. Sharp. A nuclear factor that binds to a conserved sequence motif in transcriptional control elements of immunoglobulin genes. *Nature* 319: 154–158, 1986.

251. Smale, S. T., and D. Baltimore. The 'initiator' as a transcription control element. *Cell* 57, 1989.

252. Spiro, S., and J. R. Guest. FNR and its role in oxygen-regulated gene expression in *Escherichia coli*. *FEMS Microbiol. Rev.* 75: 399–428, 1990.

253. Spivak, J. L., D. C. Barnes, E. Fuchs, and T. C. Quinn. Serum immunoreactive erythropoietin in HIV-infected patients. *JAMA* 261: 3104–3107, 1989.

254. Spivak, J. L., and P. M. Cotes. The pharmacokinetics and metabolism of erythropoietin. In: *Erythropoietin—Molecular, Cellular and Clinical Biology*, edited by A. J. Erslev, J. W. Adamson, J. W. Eschbach, and C. G. Winearls. Baltimore and London: The Johns Hopkins University Press, 1991, p. 162–183.

255. Spivak, J. L., and B. B. Hogans. The in vivo metabolism of recombinant human epo in the rat. *Blood* 73: 90–99, 1989.

256. Steinberg, S. E., J. F. Garcia, G. R. Matzke, and J. Mladenovic. Erythropoietin kinetics in rats: generation and clearance. *Blood* 67: 646–649, 1986.

257. Stohlman, F., and G. Brecher. Humoral regulation of erythropoiesis V. Relationship of plasma erythropoietin level to bone marrow activity. *Proc. Soc. Exp. Biol. Med.* 100: 40–43, 1959.

258. Sytkowski, A. J., and K. A. Donahue. Immunochemical studies of human erythropoietin using site-specific anti-peptide antibodies. *J. Biol. Chem.* 262: 1161–1165, 1987.

259. Syvertsen, G. R., and J. A. Harris. Epo production in dogs exposed to high altitude and carbon monoxide. *Am. J. Physiol.* 225: 293–299, 1973.

260. Takeuchi, M., N. Inoue, T. W. Strickland, M. Kubota, M. Wada, R. Shimizu, S. Hoshi, H. Kozutsumi, S. Takasaki, and A. Kobata. Relationship between sugar chain structure and biological activity of recombinant human erythropoietin produced in Chinese hamster ovary cells. *Proc. Natl. Acad. Sci. USA* 86: 7819–7822, 1989.

261. Tan, C. C., K.-U. Eckhardt, J. D. Firth, and P. J. Ratcliffe. Feedback modulation of renal and hepatic erythropoietin mRNA in response to graded anemia and hypoxia. *Am. J. Physiol.* 252 (*Renal Fluid Electrolyte Physiol.* 32): F474–F481, 1992.

262. Tan, C. C., K.-U. Eckhardt, and P. J. Ratcliffe. Organ distribution of erythropoietin messenger RNA in normal and uremic rats. *Kidney Int.* 40: 69–76, 1991.

263. Tan, C. C., and P. J. Ratcliffe. Effects of inhibitors of oxidative phosphorylation on erythropoietin mRNA in isolated perfused rat kidneys. *Am. J. Physiol.* 261 (*Renal Fluid Electrolyte Physiol.* 30): F982–F987, 1991.

264. Tan, C. C., and P. J. Ratcliffe. Oxygen sensing and Epo mRNA production in isolated perfused rat kidneys. In: *Pathophysiology and Pharmacology of Erythropoietin*, edited by H. Pagel, C. Weiss, and W. Jelkmann. Berlin, Heidelberg: Springer-Verlag, 1992a, p. 57–68.

265. Tan, C. C., and P. J. Ratcliffe. Rapid oxygen-dependent changes in Epo mRNA in isolated perfused rat kidneys: evidence against mediation by cAMP. *Kidney Int.* 41: 1581–1587, 1992.

266. Trainor, C. D., T. Evans, G. Felsenfeld, and M. S. Boguski. Structure and evolution of a human erythroid transcription factor. *Nature* 343: 92–96, 1990.

267. Tsai, S.-F., D. I. K. Martin, L. I. Zon, A. D. D'Andrea, G. G. Wong, and S. H. Orkin. Cloning of cDNA for the major DNA-binding protein of the erythroid lineage through expression in mammalian cells. *Nature* 339: 446–451, 1989.

268. Tsuchiya, T., H. Ochiai, S. Imajob-Ohmi, M. Ueda, T. Suda, M. Nakamura, and S. Kanegasaki. *In vitro* reconsitution of an erythropoietin gene transcription system using its 5'-flanking sequence and a nuclear extract from anemic kidney. *Biochem. Biophys. Res. Commun.* 182: 137–143, 1992.

269. Tsuda, E., G. Kawanishi, M. Ueda, S. Masuda, and R. Sasaki. The role of carbohydrate in recombinant human erythropoietin. *Eur. J. Biochem.* 188: 405–411, 1990.

270. Ueno, M., I. Seferynska, B. Beckman, J. Brookins, J. Nakashima, and J. W. Fisher. Enhanced erythropoietin secretion in hepatoblastoma cells in response to hypoxia. *Am. J. Physiol.* 257 (*Cell Physiol.* 26): C743–749, 1989.

271. Vogt, C., S. Pentz, and I. N. Rich. A role for the macrophage in normal hemopoiesis. III. In vitro and in vivo Epo gene expression in macrophages detected in situ hybridisation. *Exp. Hematol.* 17: 391–397, 1989.

272. Walle, A. J., and W. Niedermayer. Are erythropoietin levels in uremic patients on hemodialysis dependent on the kidney disease and the duration of hemodialysis treatment? *Proc. EDTA* 16: 481–486, 1979.

273. Wang, F., and W. Fried. Renal and extrarenal epo production male and female rats of various ages. *J. Lab. Clin. Med.* 79: 181–186, 1972.

274. Wang, F. F., C. K. H. Kung, and E. Goldwasser. Some chemical properties of human epo. *Endocrinology* 116: 2286–2292, 1985.

275. Weintraub, H. Assembly and propagation of repressed and derepressed chromosomal states. *Cell* 42: 705–711, 1985.

276. Westerveld, A., R. P. L. S. Visser, P. M. Khan, and D. Bootsma. Loss of human genetic markers in man-Chinese hamster somatic cell hybrids. *Nature [New Biol.]* 234: 20–24, 1971.

277. Winkelmann, J. C., L. A. Penny, L. L. Deaven, B. G. Forget, and R. B. Jenkins. The gene for the human erythropoietin receptor: analysis of the coding sequence and assignment to chromosome 19p. *Blood* 76: 24–30, 1990.

278. Wognum, A. W., P. M. Lansdorp, A. C. Eaves, and G. Krystal. An enzyme-linked immunosorbent assay for erythropoietin using monoclonal antibodies, tetrameric immune complexes, and substrate amplification. *Blood* 74: 622–629, 1989.

279. Yoshimura, A., T. Zimmers, D. Neumann, G. Longmore, Y. Yoshimura, and H. F. Lodish. Mutations in the Trp-Ser-X-Trp-

Ser motif of the erythropoietin receptor abolish processing, ligand binding and activation of the receptor. *J. Biol. Chem.* 267: 11619–11625, 1992.

280. Zanjani, E. D., J. I. Acensao, P. B. McGlave, M. Banisadre, and R. C. Ash. Studies on the liver to kidney switch of Epo production. *Clin. Invest.* 67: 1183–1188, 1981.

281. Zanjani, E. D., E. N. Peterson, A. S. Gordon, and L. R. Wasserman. Erythropoietin production in the fetus: role of the kidney and maternal anemia. *J. Lab Clin. Med.* 83: 281–287, 1974.

282. Zanjani, E. D., J. Poster, H. Burlington, L. I. Mann, and L. R. Wasserman. Liver as the primary site of epo formation in the fetus. *J. Lab. Clin. Med.* 89: 640–644, 1977.

283. Zenke, M., T. Grundstrom, H. Matthes, M. Wintzerith, C. Schatz, A. Wildeman, and P. Chambon. Multiple sequence motifs are involved in SV40 enhancer function. *EMBO J.* 5: 387–397, 1986.

284. Zitomer, R. S., and C. V. Lowry. Regulation of gene expression by oxygen in *Saccharomyces cerevisiae. Microbiol. Rev.* 56: 1–11, 1992.

285. Zon, L. I., H. Youssoufian, C. Mather, H. F. Lodish, and S. H. Orkin. Activation of the erythropoietin receptor promoter by transcription factor GATA-1. *Proc. Natl. Acad. Sci. USA* 88: 10638–10641, 1991.

286. Zucali, J. R., V. Stevens, and E. A. Mirand. In vitro production of erythropoietin by mouse fetal liver. *Blood* 46: 85–90, 1975.

50. Morphologic and metabolic response to chronic hypoxia: the muscle system

P. CERRETELLI | *Department of Physiology, University of Geneva, Geneva, Switzerland, and Department*
H. HOPPELER | *of Anatomy, University of Bern, Bern, Switzerland*

STUDIES OF HUMANS (partly because of the practical interest in mountaineering at extreme altitudes) have led the way ever since modern scientists started to ask questions about the physiological phenomena dominating life at high altitude, whereas relatively few studies of animals in a natural hypoxic habitat have been carried out. Thus our understanding of adaptive changes at high altitude is based mainly on data from human experimentation (137, 141, 190).

Organisms have different strategies and options to adapt to a hostile environment. *Adaptation* is defined here as change that reduces the physiological strain produced by stressful aspects of the environment (22). Adaptations may be developmentally or genetically fixed and unchangeable *(genotypic adaptations)* or they may evolve during the lifetime of an animal *(phenotypic adaptations* or *acclimatizations)* (95). Acclimatizations may be activated within a fraction of the lifetime of an organism and may subside if the physiological stress is removed. The capacity for acclimatization may be itself a genetically fixed trait of an individual or a species. In reference to hypoxia, *acclimation* is generally used to describe phenotypic adaptations that are introduced by simulated as opposed to real exposure to high altitude (6). A third category of adaptive options comprises regulatory phenomena that can be turned on almost instantaneously. Much of the scientific work on the effects of the hypoxic environment has focused on acute adaptations occurring as a consequence of exposure of unacclimatized humans to hypoxia. In this chapter, the phenotypic adaptations of muscle structure and metabolism will be explored and analyzed in the context of observed changes of muscle function and/or exercise capacity.

HUMAN MUSCULAR PERFORMANCE AT ALTITUDE: HISTORICAL ASPECTS

Physiology

After the pioneering work of Loewy (124) to determine at sea level, both quantitatively and qualitatively, the energy sources for exercise from the analysis of gas exchange, several physiologists tried to measure O_2 consumption and the respiratory quotient at altitude. The initial studies concerned possible changes of resting metabolism as a function of altitude. Up to 3,800 m, resting oxygen consumption ($\dot{V}O_2$) was practically unaltered and the energy expenditure for a given mechanical load was apparently greater on the Mt. Rosa glacier

(3,800 m) and at the Gnifetti hut (3,620 m) than at sea level (125, 173). The investigators, however, could not offer a satisfactory explanation for the latter finding. Peak O_2 uptake values up to about $1.5 \, l \cdot min^{-1}$ were recorded during uphill walking, but maximal oxygen consumption ($\dot{V}O_{2max}$) was apparently not attained. Durig and Zuntz (61) confirmed these findings.

Douglas et al. (60) reinvestigated the problem of walking efficiency at altitude on Pike's Peak (4,300 m) and concluded that "we do not see any reason why in the one case we have investigated the metabolism should have been greater at the high altitude than at sea level." The lower efficiency in altitude walking was attributed to the conditions of the ground, which in the mountains were presumably more similar to a grass track at Oxford than to laboratory conditions. The Douglas et al. paper is extremely interesting also because the authors made an important observation that may be of general interest for today's exercise physiologists: "Our attempts therefore, so far as we went, to obtain indirect evidence of lactic acid production on Pike's Peak gave negative results up to the pace of four miles an hour." This speed corresponds to a $\dot{V}O_2$ of ~1,600 $ml \cdot min^{-1}$, probably ~70%–80% of the subjects' $\dot{V}O_{2max}$ (~$2.1 \, l \cdot min^{-1}$), a workload at which the so-called lactic anaerobic threshold was apparently still not attained. This may be the first reported indication of the possible shift of the lactic threshold toward greater fractions of $\dot{V}O_{2max}$ than at sea level, a consequence of the "lactate paradox" (93).

The first measurements of $\dot{V}O_{2max}$ were probably those by Robinson et al. (166). The first study aimed at measuring the loss of maximal power as a function of acute hypoxia was probably that of Margaria (128), who, working in a hypobaric chamber, found a quasi-monoexponential decrease of power output during all-out exercises on the bicycle ergometer. At 5,000 m, he quantitated the reduction of maximal external power to 62% of the sea-level control value.

During the International Physiological Expedition to the Andes in 1935, Christensen and Forbes (41) were the first to study the effect of chronic hypoxia (up to 5,300 m) on the power of men working on a bicycle ergometer. Their results indicate a progressive decrease of $\dot{V}O_2$ (probably $\dot{V}O_{2max}$, based on heart rates) to 81% at 2,810 m, to 70% at 3,660 m, to 60% at 4,700 m, and down to 49% of sea-level control at 5,340 m (see also ref. 52).

Biochemistry

Reynafarje (163) was, to our knowledge, the first to report biochemical data on human muscle biopsies. These were obtained from nine healthy young male natives of the Peruvian mining city of Cerro de Pasco

(4,400 m above sea level) and the results were compared to those of nine age-matched native controls resident of Lima (50 m above sea level). The explicit purpose of this study was to investigate possible species specificities of altitude adaptations. The important findings were a significantly increased myoglobin concentration and a greater activity of oxidative enzymes in the sartorius muscle of men adapted to high altitude. This landmark paper confirmed results obtained earlier on dogs native to high altitude (105), and some of the data were compatible with the results of a study of Valdivia (184) reporting a great increase of skeletal muscle capillary densities in Andean guinea pigs. The combined results of the above papers can be held as the basis of much of the thinking on how environmental and/or local muscle hypoxia affect, through phenotypic plasticity, the structural and metabolic conditions of skeletal muscle tissue. Hochachka et al. (97) have condensed the salient features of the conventional view on the mechanisms governing high-altitude acclimatization into what they called an interpretive hypothesis. They proposed that the central problem of organisms faced with high-altitude hypoxia was to "maintain an acceptable high scope for aerobic metabolism in the face of the reduced oxygen availability of the atmosphere." According to their hypothesis this goal was achieved by increasing the capacity for oxygen transfer to the tissues and by improving cell metabolism through an increase in mitochondrial enzyme activity and abundance. Thus the increased capillary density, decreased diffusion distance, increased facilitated diffusion capacity, and increased oxidative capacity of muscles chronically exposed to hypoxia observed in the classical papers could all be explained within a coherent conceptual framework. As we will see later in this chapter, this proposed notion of skeletal muscle acclimatization to chronic hypoxia has been partially abandoned as a consequence of human experimentation (see Muscle Structure and Metabolic Markers, below).

ACUTE VS. CHRONIC EXPOSURE: EFFECTS OF ALTITUDE AND EXPOSURE TIME

Acute Exposure

Measurements of $\dot{V}O_{2max}$ in acute hypoxia have been carried out in hypobaric chambers as well as while breathing hypoxic mixtures at sea level. In general, these experiments yield results that can be superimposed on $\dot{V}O_{2max}$ measurements obtained shortly after reaching equivalent altitudes in the mountains (Fig. 50.1).

Åstrand (3) determined in an experiment on himself that upon acute exposure in a pressure chamber $\dot{V}O_{2max}$ decreased from 4.6 to 4.2 (91%) at 3,000 m and to 3.0

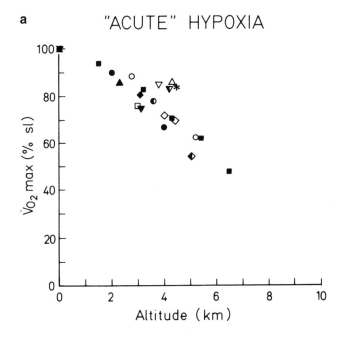

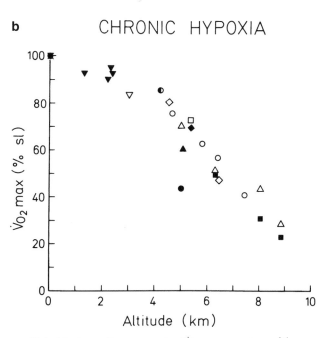

FIG. 50.1. Maximum O_2 consumption ($\dot{V}O_{2max}$; percentage of the sea-level value made equal to 100) as a function of altitude (km) after acute (a, less than 5 days) and chronic (b, more than 3 wk) exposure. Each symbol represents data from a different reference as follows: a: ○, 128; ●, 3; ▽, 55; ▼, 81; □, 54; ■, 36; △, 89; ▲, 65; ◇, 170; ♦, 53; ◐, 34; *, 46; ▼, 96; ◈, 74b. b: ○, 154; ●, 38; ▽, 5; ▼, 4; □, 34; ■, 192; △, 179; ▲, 74b; ◇, 64; ♦, 34; ◐, 96.

(65%) $1 \cdot min^{-1}$ at 4,000 m. Dill et al. (54, 56) found a 44% decrease of $\dot{V}O_{2max}$ during the first week at 3,800 m, with a small but significant improvement over the following 5 wk spent at altitude. A subsequent study carried out at 3,100 m on high school champion runners (53) showed a decrease in $\dot{V}O_{2max}$ of 18% upon arrival, a slight increase during 2 wk exposure, and a 4.2% increase upon return to sea level. These results are, to some extent, at variance with those of Grover and Reeves (80), who found a 4.6% drop in $\dot{V}O_{2max}$ upon return to sea level after 2 wk at 3,100 m, during which $\dot{V}O_{2max}$ was constant at about −20% compared with pre-altitude reference values. Cerretelli et al. (36) found in two subjects breathing hypoxic mixtures [from 0.21 to 0.09 F_{IO_2} (concentration of inspired oxygen) corresponding to altitudes from sea level to 6,500 m] a progressive decrease in $\dot{V}O_{2max}$ down to 47% of the normoxic value. At a simulated equivalent altitude of 5,350 m the drop in $\dot{V}O_{2max}$ was about 40% compared to sea level, somewhat greater than the corresponding value (~−30%) for acclimatized lowlanders at Mt. Everest base camp (5,350 m; see refs. 33, 34). Saltin et al. (170) report a decrease in $\dot{V}O_{2max}$ to 72% of control values when breathing 13% O_2 at sea level (~4,000 m) and to 70% of sea-level control when breathing air at 4,300 m.

Blatteis (21) found in four lowlanders within 12 h following arrival to Morococha (4,540 m) a decrease of 38% of the estimated maximum O_2 uptake ($\Delta\dot{V}O_{2max}$) compared to sea level. Over the following 6 wk of altitude residence $\Delta\dot{V}O_{2max}$ dropped progressively to 18% and absolute $\dot{V}O_{2max}$ reached the average level of a group of sex- and age-matched natives (46 vs. 44 $ml \cdot min^{-1} \cdot kg^{-1}$).

Cymerman et al. (46) found a 16% decrease in $\dot{V}O_{2max}$ at 4,300 m, a reduction that is definitely lower than the average figure found by different investigators.

Chronic Hypoxia

Pugh (154) and Pugh et al. (156) determined $\dot{V}O_{2max}$ at increasing altitudes on members of the Mt. Cho Oyu and Mt. Everest expeditions and of the Silver Hut team. Maximal aerobic power was found to decrease to 57% and 41% of sea-level control value at 6,400 m and 7,440 m, respectively. Pugh (154) estimated that approaching the summit of Everest $\dot{V}O_{2max}$ would drop to about 15% of the sea-level value, just enough to assure the actual consumption required by mountaineers during normal walking. *Gross efficiency,* defined as the thermal equivalent of the rate of work divided by the thermal equivalent of total oxygen consumption, increased with higher climbing rates ranging between 13% and 22% for rates of 400 and 1,000 $kg \cdot m \cdot min^{-1}$, respectively. Pugh et al. (156) showed that the relationship between O_2 uptake and work rate is not affected by altitude up to 7,400 m. $\dot{V}_{O_{2max}}$ was determined by Cerretelli and Margaria (38) on five subjects after 60 days of climbing at altitudes between 3,800 m and 7,500 m. At 5,000 m the decrease was 55% compared to sea-level control values. Cerretelli (34) measured $\dot{V}_{O_{2max}}$ of acclimatized lowlanders (n = 36) at 5,350 m after 8 wk exposure to altitudes up to the summit of Mt. Everest. $\dot{V}_{O_{2max}}$ was 36.8 ± 4.5 (S.D.) ml · kg^{-1} · min^{-1}, that is, ~25%–30% lower than control sea-level values. The energy cost of stepping up and down a 30 cm bench at two different rates was the same as that found before the expedition.

Administration of a normoxic mixture (33, 34) at high altitude (after 6–8 wk exposure) did not restore sea-level $\dot{V}_{O_{2max}}$, which resumed only 92% of its preexpedition level, an observation indirectly made also by Edwards (62), likely a consequence of muscle deterioration (see also Mitochondria, below).

West et al. (192) determined $\dot{V}_{O_{2max}}$ in the course of the American Medical Research expedition to Everest up to 6,300 m. The lowest points were obtained by giving subjects acclimatized to 6,300 m mixtures containing 16% and 14% O_2 to breathe, the latter being equivalent to the altitude of the summit of Mt. Everest. The drop in $\dot{V}_{O_{2max}}$ was 50% at 6,300 m, 69% at 8,050 m, and 77% at the summit (8,848 m). West et al. (192) also extended previous observations by Pugh et al. (156), Lahiri et al. (120), and Cerretelli (34), confirming that the relationship between O_2 consumption and external work is not affected by chronic hypoxia.

During Operation Everest II, $\dot{V}_{O_{2max}}$ was determined in at least five subjects breathing air, with P_{IO_2} varying between 150 m (sea level) and 43 mm Hg (8,848 m) and $\dot{V}_{O_{2max}}$ reduced to a minimum of ~30% of sea-level control (179).

The effect on $\dot{V}_{O_{2max}}$ of a 5 wk sojourn at 5,050 m has been studied in the Pyramid laboratory at Lobuche (Nepal) on a group of 10 lowlanders (74b). The results, shown in Figure 50.2, indicate that after an initial 47% fall in $\dot{V}_{O_{2max}}$, maximal aerobic power undergoes a partial recovery, reductions being 44% and 40%, respectively, after 15 and 35 days. Upon descent to sea level, $\dot{V}_{O_{2max}}$ increased to only 92% of the preexpedition level and kept at this lower level for at least 5 wk. Another interesting preliminary finding by the same authors (Fig. 50.3) is the greater decrease of oxygen hemoglobin saturation (HbO$_2$) with increasing workload in an athlete compared to a sedentary subject at altitude. This was confirmed by further measurements performed by Marzorati et al. (130b) in the same experimental conditions. This result could be explained with a greater diffusion limitation across the alveolocapillary membrane due to a shorter erythrocyte–air contact time in the athlete's lung.

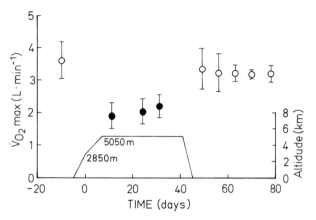

FIG. 50.2. $\dot{V}_{O_{2max}}$ as a function of time before exposure, in the course of altitude exposure, and following return to sea level (solid symbols, altitude data; open symbols, sea-level measurement; mean of 10 subjects ± S.D.) (75).

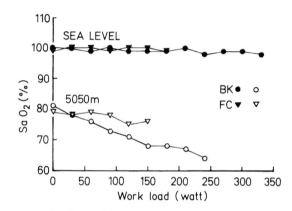

FIG. 50.3. Blood hemoglobin saturation (SaO_2, %) at rest and during exercise at sea level and at 5,050 m in a sedentary (FC) and an athletic (BK) subject (75).

High-Altitude Natives vs. Acclimatized Lowlanders

Hurtado (106) compared groups of native residents of Morococha (4,540 m) and Lima (sea level), Peru, while performing submaximal exercise on a treadmill (132 m · min^{-1} on a +11% gradient). Exhaustion time was found to be longer (59 vs. 34 min) in the altitude than in the sea-level group. Oxygen consumption per kg · m^{-1} external work was moderately lower (2.43 vs. 2.66 ml O_2) at Morococha, corresponding to a net (after subtracting from total $\dot{V}_{O_2}$ the resting value) efficiency of 22.2% and 19.9%, respectively. The author stresses the higher degree of physical efficiency "as one of the most important characteristics of the native residents and possibly the best index of their acclimatization." This difference, if real, cannot be the consequence of a more favorable respiratory quotient, which, in fact, was 0.93 for lowlanders and 0.95 for highlanders.

Saha (168) determined the gross mechanical efficiency of Sherpa Tensing Norgay (one of the Mt. Everest summiters in 1953) during climbing at ~2,000 m altitude at rates varying between 472 and 678 kg · m · min^{-1}. The data were 23% and 20%, respectively. In another study, Das and Saha (48) determined gross mechanical efficiency of 66 Sherpas when walking on a treadmill on a +20% gradient at a speed of 3.2 km · h^{-1}. The mean value was 22.9% ± 2.3.

Elsner et al. (64) determined $\dot{V}_{O_2max}$ in native Morococha (4,540 m) residents when breathing ambient air and hypoxic mixtures equivalent to an altitude of 6,400 m and compared these figures with those of acclimatized lowlanders. $\dot{V}_{O_2max}$ at 4,540 m (40.7 ml · kg^{-1} · min^{-1}) compares well with the data of Pugh (154) and of Cerretelli and Margaria (38). Administration of normoxic mixtures raised $\dot{V}_{O_2max}$ in two subjects by about 30%, whereas a reduction of P_{IO_2} corresponding to an altitude of 6,400 m lowered $\dot{V}_{O_2max}$ to 24 ml · kg^{-1} · min^{-1}, that is, to about 60% of the control-hypoxic value.

Coudert and Paz-Zamora (44) examined highland Bolivian professional and collegiate soccer players (age 23 yr) and found a mean $\dot{V}_{O_2max}$ of 59 ml · kg^{-1} · min^{-1}, a very high value indeed. Considering that measurements were made at 3,700 m, the value found for this group of athletes is higher than the average level for most players of the leading European soccer teams and national selections. Lahiri et al. (120) measured $\dot{V}_{O_2max}$ of four Sherpas and two acclimatized lowlanders at 4,880 m after a 9 wk sojourn. Corresponding values were 45.3 and 33.6 ml · kg^{-1} · min^{-1}, respectively. Net efficiency determined on both Sherpas and lowlanders was ~0.25, whereas maximal blood lactate concentration [La$_b$] was 6.6 and 9.2 mM, respectively. Frisancho et al. (71) determined $\dot{V}_{O_2max}$ at 3,400 m on groups (age 21.5–24 yr) of Peruvian natives, Peruvian lowlanders acclimatized to chronic hypoxia, and Peruvian and U.S. white acclimatized (from 4 months to 4 yr) lowlanders. $\dot{V}_{O_2max}$ was 46.3 ± 5 (S.D.), 46.0 ± 6.3, 38 ± 5.2, and 38.5 ± 5.8 ml O$_2$ · kg^{-1} · min^{-1}, respectively. Cerretelli (34) determined $\dot{V}_{O_2max}$ on 21 Sherpas at Mt. Everest base camp (5,350 m). The average value was 39.7 ± 7.1 (S.D.) ml · kg^{-1} · min^{-1}, slightly but not significantly higher than the average figure for acclimatized lowlanders (36.8 ± 4.5). The energy cost for an identical stepping exercise was the same in Sherpas and lowlanders. Greksa et al. (78) determined $\dot{V}_{O_2max}$ in trained male and female youths (8.8–19.5 yr) native to La Paz, Bolivia (3,700 m) and found average values of 46.9 and 39.3 ml · kg^{-1} · min^{-1}, respectively. These values tended to be 10%–20% lower than those of sea-level athletes (mainly competitive swimmers). Greksa et al. (79) compared at La Paz (~3,600 m) groups of "young" and "old" adolescents born and raised in the same area, categorized by ethnicity (European and

Aymara). $\dot{V}_{O_2max}$ (47.0 ± 6.8 vs. 50.9 ± 6.5) for "old" European and Aymara groups, respectively, did not differ significantly between ethnic groups. For boys not undergoing competitive training these values are rather high compared with homologous data obtained on lowlanders at sea level. Sun et al. (177) measured $\dot{V}_{O_2max}$ on 16 Tibetan lifelong residents of Lhasa (3,658 m) and 20 Han (Chinese) 8 ± 1 yr residents of the same altitude matched for age, height, weight, and lack of exercise training. Values were 51 ± 1 (S.E.) and 46 ± 1 ml · kg^{-1} · min^{-1} for Tibetan and Chinese, respectively. Hochachka et al. (96) determined $\dot{V}_{O_2max}$ on six Quechua Indians at 4,000 m. They found an average $\dot{V}_{O_2max}$ of 43 ± 2 ml · kg^{-1} · min^{-1}, which was raised to 50 ± 2 by normoxic breathing at altitude and to 53 ± 2 ml by 3 wk deacclimatization at sea level. Second-generation Tibetans (n = 20, 17.8 yr old college students born and living at Kathmandu, 1,300 m) had a $\dot{V}_{O_2max}$ of 37 ± 1.1 ml ·kg^{-1} · min^{-1}, the same average value (36.7 ± 1.1) found in a Nepalese control group (n = 20, 20.2 yr) (personal observation, 1995). Favier et al. (65b) determined the effects of normoxic breathing in 50 altitude natives of La Paz (3,600 m) Bolivia. $\dot{V}_{O_2max}$ increased from 2.76 to 2.99 liters · min^{-1} or 45 to 48.7 ml · kg^{-1} · min^{-1}, i.e. by only ~8%.

From the above data, which are far from covering the innumerable measurements carried out at altitude, the following conclusions can be drawn. (1) the decrease in $\dot{V}_{O_2max}$ max as a function of altitude up to 5,000–6,000 m is essentially the same for acute and chronic hypoxia (Fig. 50.1). This is probably only coincidental due to the interplay between increased maximal O$_2$ transport ($\dot{Q}_{max} \times C_{aO_2}$) in the course of acclimatization, resulting from increased hemoglobin concentration ([Hb]) and only moderately decreased $\dot{Q}_{max}$, and the possible reduction and/or deterioration of the active muscle mass, which may result in reduced $\dot{V}_{O_2max}$. The latter phenomenon is reflected in part by the postexposure vs. preexposure reduction of $\dot{V}_{O_2max}$ and by the failure to resume sea-level $\dot{V}_{O_2max}$ when administering normoxic mixtures to acclimatized lowlanders. From the few available data, it would appear that in altitude natives the gain of $\dot{V}_{O_2max}$ upon acute normoxia is percentage-wise less than the loss undergone by acclimatized lowlanders when shifting from normoxia to comparable hypoxic levels. The cause of this difference is unknown. (2) Gross and net mechanical efficiencies of habitual exercise (such as stepping up and down from a bench or walking) are the same in hypoxia as in normoxia. With two exceptions (96, 106), no differences in working efficiency were found between high-altitude natives and acclimatized lowlanders. (3) At altitudes around 4,000 m, it would appear that natives show a tendency toward a slightly higher absolute $\dot{V}_{O_2max}$. However, it may be difficult to make objective comparisons among groups with different physical and socioeconomic histories. (4) High-altitude natives with different ethnic heritages do not show statistically different $\dot{V}O_{2max}$ values.

COMPARING ANIMAL AND HUMAN DATA

As previously noted, the bulk of knowledge on acclimatization to altitude hypoxia stems from human experimentation. A major problem in extrapolating findings obtained on animals exposed to chronic hypoxia to humans consists in the species specificity of the phenotypic plasticity of the respiratory system. Monge and Leon-Velarde (137) have discussed the role of "preadaptation" in this context (24, 25). *Preadaptation* is the genotypic capacity to rapidly invade a stressful physiological environment because existing structural and/or functional traits offer a selective advantage in the new environment. Monge and Leon-Verlarde (137) cite the example of camelids that, because of their natural hypoxia tolerance, suffered less from natural selection than Castilian chickens when brought to the extreme environmental hypoxia of the Andean highlands. When studying the adaptive variation of the respiratory system (181), it was noted that F_{IO_2} could be reduced to as low as 12% in goats without compromising $\dot{V}O_{2max}$, whereas dogs (similar to humans) were quite susceptible to even small changes in F_{IO_2} (111, 150). From this it was concluded that goats would fare much better in a situation of hypoxia. It becomes clear from these examples that the approach of comparative physiology is useful in understanding strategies of hypoxic acclimatization in animals in general but that extrapolation or generalization to other species must be made with much circumspection.

Banchero (6) has provided an excellent review on the cardiovascular response to hypoxia, concentrating mostly on data from animal experimentation. He states that much of the older literature on acclimatization to hypoxia in animals is fraught with the problem that not enough attention has been paid to the fact that hypoxia exposure was often combined with, but not controlled for, cold exposure. In particular, little attention has been paid to the body weight to body surface area ratio and to habitual factors influencing the thermal microclimate to which animals were exposed during studies purportedly aimed exclusively at elucidating aspects of hypoxic acclimatization. With regard to skeletal muscle acclimatization, Banchero (6) concludes that there is little evidence from animal experimentation that simple hypoxia is a sufficient stimulus leading to capillary neoformation. He maintains that to study the phenotypic plasticity of skeletal muscle tissue in animals it is necessary

to combine chronic hypoxia with exercise or cold exposure to further challenge the respiratory system.

In conclusion, it would appear that it is difficult to extend results from animal experimentation to humans exposed to chronic hypoxia. For one thing, there is distinct evidence for species specificity of a malleability of the respiratory system to hypoxia. Moreover, humans exposing themselves to extreme altitudes are often subjected to stress, exhausting physical exercise, poor nutrition, and cold. One or several of these stimuli have either been poorly controlled or were entirely missing in most animal experimentation in chronic hypoxia.

STRUCTURAL AND METABOLIC CHARACTERISTICS OF HIGH-ALTITUDE NATIVES

Body Composition

High-altitude natives are characterized by smaller body mass (50–55 kg vs. 60–75 kg), lower percentage of body fat (5%–8% vs. 11%–15%), and shorter stature (155–162 cm vs. 166–180 cm) than most homologous sea-level populations. It has not been established whether and to what extent these differences are the consequence of genetic, nutritional, and/or environmental (for example, hypoxia, cold, etc.) factors.

According to Hurtado (106), adult men native to Morococha, Peru (4,540 m) have an average body weight of 53.7 $\pm$ 0.6 (S.E.) kg compared to 63 $\pm$ 1 kg for adult men in Lima (sea level). The corresponding height was 159.2 $\pm$ 1.1 vs. 168 $\pm$ 0.9 cm, respectively. The differences between the two groups are probably the consequence of genetic factors. Similar values were obtained by Picòn-Reàtegui et al. (149).

Frisancho and Baker (69) reported for a group of Quechua Indian men living at ~4,000 m average weight and height of 55.9 $\pm$ 6.5 (S.D.) kg and 160 $\pm$ 4.9 cm, respectively. Corresponding values for women were 54 $\pm$ 6.5 kg and 148 $\pm$ 5.2 cm, respectively. Men of the same population living at altitudes above 4,500 m were significantly heavier than their counterparts living at 4,000 m, whereas stature was the same. Quechua men (18–20 yr old) living at Lamas, Peru (1,300 m), were characterized by a height of 158.7 cm, only slightly taller than their high-altitude counterparts (156.7 cm) Parenthetically, a control group of 18–20 yr old Mestizos (a Spanish-speaking, genetically different population that has settled in the same region since the Spanish conquest) had a height of 160.9 cm (70). The body weight of a group of Bolivian natives from La Paz (n = 50, 24.4 yr) was found to be 61.4 $\pm$ 1.1 kg with a height of 167.9 $\pm$ 0.9 cm (65b).

With regard to Sherpas (an eastern Tibetan population that moved to its present habitat about 300–500 years ago), the average body weight of a group aged 35 yr living at ~4,000 m was found to be 54.6 ± 0.65 (S.E.) kg with a stature of 162.2 ± 0.9 cm (190).

Sloan and Masali (175) reported for a group of 18–35 yr old Sherpas born and living at ~2,600 m (Phaplu, Nepal) average body weight and height of 55.9 ± 5.7 (S.D.) kg and 162.7 ± 5.9 cm, respectively. Body weight and stature of Tibetan refugees near Kathmandu, Nepal (~1,000 m) were also measured by Weitz (190). Values were 58.1 ± 1.1 (S.E.) kg and 166.9 ± 1 cm, respectively, that is, somewhat higher than for homogolous populations living at high altitude. Gupta and Basu (83) found, at variance with the results of Weitz (190), significantly higher weight, stature, and body fat values for Sherpas of the upper Khumbu (3,500–4,500 m), Nepal, than for their migrant counterparts at lower altitudes (1,000–1,500 m) of the Darjeeling district, West Bengal. Kayser et al. (115b) determined the body weight of 20 second-generation 17.8 yr old Tibetans born and raised at 1,300 m. The average value (53.4 ± 1.2 kg) was not significantly different from that of a control Nepalese group (20.2 yr, 57.2 ± 2.6 kg).

The average body mass of Ethiopians living at 3,000 m was found to be 56.8 ± 0.8 (S.E.) kg, with a stature of 167.3 ± 0.6 cm compared with 53.6 ± 0.8 kg and 168.8 ± 0.85 cm, respectively, for a genetically identical group living at 1,500 m (91).

From the above data it appears that (1) mountain populations, independent of their specific habitat, have smaller body mass and are shorter of stature than average Caucasians or African comparable groups (Nilo-Hamitics); (2) the differences become negligible if a comparison is made with most Asian ethnic groups; (3) within a given population, altitude per se seems to play a minor role as a determinant of body mass and stature; (4) high-altitude natives have a smaller fraction of body fat (down to 5.7%, see refs. 71, 83, 175) than sea-level residents, a finding that could justify, in part, the greater specific (per kilogram body weight) maximal aerobic power of high-altitude natives compared with sea-level control subjects.

An interesting observation is that Tibetans raised at ~1,200 m, like high-altitude Sherpas (of Tibetan descent) as well as Peruvian Amerindians, are characterized by a reduced growth rate (14, 146). When comparing Tibetan lowlanders with the altitude Sherpa group, the finding is rather surprising considering the different degree of hypoxia and nutritional conditions under which they live. According to Pawson (146), it is plausible to hypothesize for these populations a common, strong genetic component in growth. The difference in size upon maturation could perhaps be the effect of the restriction placed on growth by the hypoxic envi-

ronment. That a reduced growth may be associated with altitude and that the threshold of altitude effects appears to be near 1,500 m was also shown for North American children by Yip et al. (197).

It is beyond the scope of this chapter to describe the characteristics of the child at birth or of childhood growth as well as the various physiological changes reported in altitude natives. These topics have been adequately covered by specialists to whom the reader is referred. In any case, an account of some of these studies is given for the reader's convenience. Biologists have determined the rates of growth of Caucasian (197), Andean (10, 11, 14, 68–71), Himalayan (13, 83, 175, 190), and African (Ethiopian) (91) populations and have made interesting comparisons (70, 71, 146) among them. Other specialized studies have pointed out some differences in the morphology of the chest of altitude natives compared to sea-level residents (12), the relative length of the lower limbs, and the varying distribution of the limb muscles (large calf muscles) related to load carrying.

Physiologists and biochemists have investigated: (1) the hematological characteristics of several highland populations, such as Peruvian and Chilean natives (106, 195, 196), Tibetans (1, 15, 16), and Sherpas (139, 172, 195); (2) arterial blood gases and acid base balance, respiration, and lung function of Andean and Himalayan natives (30, 45, 65b, 84, 107, 118–120, 147, 161, 172b); (3) the cardiovascular function of Peruvian natives at both altitude and sea level (186); and (4) myoglobin content, enzymatic activities, and muscle ultra-structure of altitude natives of the Andes and of Nepal (115, 163). For a discussion of the possibility that altitude natives (Quechua Indians) are characterized by greater efficiency values of muscle contraction, see The Lactic Mechanism, below.

Muscle Structure and Metabolic Markers

Due to ethical and transcultural problems related to obtaining muscle biopsies from native highlanders, little information is currently available on skeletal muscle tissue composition in high altitude populations. In his classical paper, Reynafarje (163) reports that blood hemoglobin content and muscle myoglobin concentration were 18% and 16% greater, respectively, in high-altitude natives compared to lowlanders. Muscle oxidative capacity was estimated by measuring the mitochondrial enzymes cytochrome c reductase and transhydrogenase, which were both significantly elevated by 78% and 77%, respectively. Saltin and Gollnick (169) have questioned these results. They claimed, based on their own work (see refs. 169, 171), that the greater aerobic potential of high-altitude natives was most likely due to more

TABLE 50.1. *Morphometric Data Obtained by Electron Microscopy on Biopsies of M. Vastus Lateralis from Various Groups of Subjects*

		Units	Sherpas (n = 5)	Professional Mountaineers (n = 5)	Mountaineers before Expedition (n = 14)	Mountaineers after Expedition (n = 14)	Lowlanders Untrained (n = 15)	Lowlanders after 6–8 wk Endurance Training (n = 15)	Professional Bicyclists (n = 3)
Fiber size	x ± SE	μm²	3,190 ± 260	3,110 ± 140	4,170 ± 190	3,360 ± 160	3,795 ± 160	3,949 ± 150	6,160 ± 650
Capillary density	x ± SE	mm⁻²	467 ± 11	542 ± 57	483 ± 20	538 ± 24	438 ± 16	506 ± 20	487 ± 10
Capillary to fiber ratio	x ± SE	unitless	1.48 ± 0.11	1.67 ± 0.17	1.99 ± 0.05	1.79 ± 0.10	1.64 ± 0.07	1.99 ± 0.09	3.01 ± 0.38
Vv(mt,f)	x ± SE	%	3.96 ± 0.27	4.95 ± 0.21	5.85 ± 0.32	4.76 ± 0.17	4.72 ± 0.14	6.59 ± 0.26	11.37 ± 0.82
Vv(ms,f)	x ± SE	%	0.43 ± 0.10	0.74 ± 0.11	0.96 ± 0.13	0.52 ± 0.056	0.57 ± 0.05	1.30 ± 0.10	3.49 ± 0.54
Vv(mc,f)	x ± SE	%	3.52 ± 0.19	4.21 ± 0.15	4.88 ± 0.22	4.23 ± 0.14	4.15 ± 0.12	5.29 ± 0.20	7.89 ± 0.67
Vv(li,f)	x ± SE	%	0.45 ± 0.15	0.83 ± 0.27	0.81 ± 0.14	0.92 ± 0.17	0.61 ± 0.09	1.05 ± 0.20	1.40 ± 0.50
Vv(fi,f)	x ± SE	%	80.6 ± 1.3	79.4 ± 1.1	77.8 ± 0.72	80.6 ± 0.64	82.1 ± 0.54	74.2 ± 0.92	60.44 ± 4.41

Vv(mt,f), volume density of total mitochondria; Vv(ms,f), volume density of subsarcolemmal mitochondria; Vv(mc,f), volume density of interfibrillar mitochondria; Vv(li,f), volume density of lipid droplets; Vv(fi,f), volume density of myofibrils.

active living habits. It may be noted that when Reynafarje published his landmark paper it was generally not agreed that endurance training increases muscle aerobic capacity (see ref. 98).

The finding of an elevated myoglobin concentration in skeletal muscle of high-altitude natives by Reynafarje (163) has never been checked. In contrast to rats, whose myoglobin concentration increases with endurance exercise (92, 145), humans are believed to maintain constant myoglobin concentrations even with extreme forms of endurance training (109, 180). This would argue for the fact that the observed difference in myoglobin concentration between highlanders and lowlanders was indeed related to living in a hypoxic environment. Further support comes from a training study carried out at sea level on humans showing that in the same subject myoglobin concentration increased in one leg trained in hypoxia but not in the contralateral leg trained in normoxia (182). Likewise, in dog sternothyroid muscle myoglobin concentration increased by 50% when barometric pressure was lowered from 635 mm Hg to 435 mm Hg for 3 wk (73).

A morphometric evaluation on muscle biopsies from the vastus lateralis of five male Sherpas (age 28 ± 2.8 yr) living at altitudes of 3,000–5,000 m for all of their lives has been conducted (115). Subjects were characterized by a low body mass (55 ± 3.2 kg) and a normal maximal oxygen uptake capacity (48.5 ± 5.4 ml $O_2 \cdot min^{-1} \cdot kg^{-1}$; determined indirectly with a one-point submaximal test procedure at 1,300 m altitude in Kathmandu) when compared to sedentary lowlanders. $\dot{V}O_{2max}$ values reported for these Sherpas are therefore similar to those reported by Hochachka et al. (96) for a group of Quechua Indians. Analysis showed that muscle fiber size was similar to that observed in Caucasian

elite high-altitude climbers and in mountaineers after return from Himalayan expeditions but significantly smaller than that of either trained or untrained lowlanders or mountaineers before expeditions (Table 50.1). Descriptors of muscle tissue capillarity, such as capillary to fiber ratio and capillary density, showed values similar to those observed in elite mountaineers and untrained Caucasians (Table 50.1). Surprisingly, a very low volume density of total mitochondria was found in the Sherpa vastus lateralis muscle (3.96 ± 0.27%, S.E.), in fact lower than in any population analyzed so far. Distributions of interfibrillar (3.52 ± 0.19%) and subsarcolemmal (0.43 ± 0.21%) mitchondria and the content of myofibrils (80.56 ± 1.3%) were similar to those observed in untrained populations. The relative volume of intracellular lipid deposits (0.45 ± 0.15%) was not different from that observed in populations not trained for high endurance. The finding of a low volume density of mitochondria is in contrast to the "interpretative hypothesis" of Hochachka et al. (97), which would imply an exceptionally high mitochondrial content for high altitude–adapted individuals to ensure adequate substrate fluxes and fast readjustment rates at low oxygen concentrations (see also The lactic mechanism, below).

Since Sherpas are characterized by normal values for variables describing oxygen supply to muscle tissue (capillaries) and a low value for the structural variable describing tissue oxidative capacity (mitochondrial density), it is possible that in these subjects mitochondrial oxygen availability may be optimized. There is no plausible explanation for the observation that Sherpas reach the same weight-specific $\dot{V}O_{2max}$ with a lower muscular mitochondrial concentration than trained and untrained Caucasians (115). Not enough histochemical data on Sherpa muscle tissue are available. It would be of great interest to determine the fiber type composition of limb muscles in different native highlanders. Should the fiber pattern be predominantly type I (slow twitch), this could account for the tighter ATP demand–supply coupling apparently found for Quechua Indians (96, 131; see The Lactic Mechanism, below). Preliminary results of measurements of fiber type distribution and mitochondrial density in the vastus lateralis muscle of second-generation Tibetans (T) and of a control group of Nepalese (N), all born and raised at 1,300 m are now available (personal observation, unpublished). Muscle biopsy analysis showed similar fiber type distribution (type I : 57 ± 3.4% S.D. [T] vs 58.6 ± 3.4 [N], type IIa : 22.3 ± 2.9 vs 24.1 ± 3.5%, type IIb : 15.9 ± 2.9 vs. 17.4 ± 1.4%). The Tibetans tended to have smaller average fiber cross sectional areas than the control subjects (3,413 ± 677 [T] vs 3,895 ± 447 μm^2 [N], P = 0.074), but a similar number of capillaries per muscle fiber (1.35 ± 0.23 [T] vs. 1.46 ± 0.23[N]) and per muscle fiber cross sectional area (399 ± 29 [T] vs .382 ± 65 mm^{-2} [N]). Despite the similar specific $\dot{V}O_{2max}$, T had a much lower total mitochondrial volume density (3.99 ± 0.17 [T] vs. 5.51 ± 0.19% [N], P<0.025). Both subsarcolemmal and central fractions of mitochondrial densities were significantly smaller in T compared to N. Lower mitochondrial volume densities in T were accompanied by proportionally lower activities of the mitochondrial enzymes citrate-synthetase (Krebs-cycle) and beta-hydroxy-acyl-dehydrogenase (FFA-oxidation), whereas the activity of the cytoplasmatic enzymes lactate-dehydrogenase and hexokinase (anaerobic glycolysis) were identical in the two groups.

A particularly interesting finding is the identity of the found low mitochondrial volume density of second-generation Tibetans with that of altitude Sherpas (115). It would appear that Tibetans are characterized by an inborn trait that still prevails in second generation lowland offspring. On the basis of these data, the hypothesis made above does not seem to be supported by either ultrastructural or biochemical data. This discrepancy is confirmed by the data of Rosser and Hochachka (167b) on three Quechua Indians.

STRUCTURAL AND PHYSIOLOGICAL ADAPTATIONS TO CHRONIC HYPOXIA

Changes of Body and Muscle Mass and Nutrition

After prolonged altitude exposure, body weight decreases. The members of the Himalayan Scientific and Mountaineering expedition lost between 6.4 and 9 kg after spending ~8 months above 5,000 m (155). Consolazio et al. (43) reported an average drop of 2.7 and

3.8 kg, respectively, in two groups undergoing gradual vs. abrupt exposure after 4 wk at 4,300 m. In a large group of Caucasian lowlanders, an 8–10 wk exposure to altitudes between 5,350 and the summit of Mt. Everest (8.848 m) led to a 7% decrease of body weight (33). A similar reduction (~8%) was reported by West et al. (192) for the 15 members of the American Medical Research expedition after a 5 wk sojourn at or above 5,400 m. During Operation Everest II, the subjects lost in 38 days an average of 7.4 ± 2.2 (S.D.) kg or ~9% of initial body weight (167). Westerterp et al. (193b) have reported ten subjects with a loss of body mass from 65.7 to 60.8 kg after a 21-day sojourn at 6,542 m. The authors' conclusion is that lowlanders with ad libitum access to food cannot attain energy balance in the investigated conditions. The energy deficit appears to be mainly a consequence of malnutrition, whereas the role of malabsorption is limited.

When allowance is made for changes of body weight induced by exercise and/or training, which in lowlanders are mainly at the expense of body fat and water, ~70% of the high altitude–induced loss is attributable to fat-free mass, particularly to muscles (28, 29, 67, 127, 167). The reduction of muscle mass is associated with a marked deterioration of muscle tissue, accumulation of lipofuscin (130), a decreased volume of muscle mitochondria (100), and a reduced activity of several oxidative enzymes (102). Sherpas appear to maintain body weight and muscle mass during residence above 5,400 m (29).

The fall of body weight at high altitude is due to the combination of several factors, which are discussed in the following sections.

Increased Basal Metabolic Rate. Measurements of basal metabolic rate (BMR) in acclimatized lowlanders date back to the end of the nineteenth century when Schumburg and Zuntz (173) showed that resting O_2 uptake was essentially unaffected by hypoxia up to about 4,000 m. More refined measurements in lowlanders who had been at 5,800 m altitude for between 82 and 113 days showed average BMR levels ~10% higher than at sea level (72).

In Andean native miners, Picòn-Reàtegui (148) found a BMR 11% higher than predicted from the formula of Harris and Benedict (90). In three Sherpas mean BMR at 5,800 m was found to be 21% higher than predicted (72). The BMR increase, in general, cannot be justified by the extra O_2 cost of increased ventilation, changes of body composition, or exposure to cold. The increased BMR could be associated with an enhanced thyroid function, as shown by the increase in serum total T_4, free T_4, T_3 and concentration of thyroid-stimulating hormone after 18 days' exposure to 5,400 and 6,300 m (138). At 4,300 m, Hannon and Sudman (87) found a transient increase of BMR, with a peak at 36 h of exposure, followed by a slow return toward normal values, which were resumed after 40 days. After a week at 3,300 m, Nair et al. (142) found a 12% increase of BMR. $\dot{V}o_2$ was back to normal by the second week and even below control by the third week. Changes in BMR seem to follow changes of serum T_3 and T_4 (159). Hochachka et al. (96) found in Quechua Indians serum T_3 and T_4 levels within the normal range for sea-level control individuals.

Thus it appears that BMR, at altitudes above 5,000 m, may be enhanced by a hyperthyroxinemic state.

Primary Decrease of Food Intake due to Hypoxia-Induced Anorexia.

Ad libitum food intake has been shown to be depressed at altitude compared to sea level in the same individual (32). Moreover, if exposure is abrupt, leading to symptoms of acute mountain sickness, energy intake may be very low over several days (85). Also, during climbing food intake tends to be rather low and increases occasionally during rest at the base camp (82, 193). Westerterp et al. (193b) have found that the average daily metabolic expenditures of ten lowlanders sojourning at 6,542 m was 24% higher than energy intake. An important component of altitude anorexia may be lack of palatable food and general discomfort. This has been confirmed by Narici et al. (143) and by Kayser et al. (115d), who failed to find significant changes of body weight, skin-fold thickness, and limb circumference in eight Caucasian males during a 30 day sojourn at the Italian Research Laboratory in Lobuche, Nepal, at 5,050 m. This structure provides outstanding facilities, including a wide choice of palatable foods. At altitudes above 5,000 m, primary anorexia becomes a constant symptom contributing, together with increased BMR and activity level, to an overall energy deficit.

Decrease of Body Water.

Water may be lost by increased evaporation due to altitude hyperventilation, decreased fluid intake resulting from decreased thirst, or changes in water and electrolyte metabolism and associated hormone responses (42).

According to Pugh (155), water loss during climbing could attain very high levels, requiring intakes up to 7 l/day. During a climb of Mt. Everest, a group of mountaineers lost 2.6–4.0 liters/day of water, taking in only 1.8–2.7 liters/day of fluids. The difference was mainly accounted for by the production of ~1 liter/day of metabolic water (193).

In lowlanders, total body water was found to decrease by 2.25 kg after a 6 day sojourn at 4,300 m. In the same subjects, intracellular water was reduced by 3.5 liters, whereas extracellular water increased by 1.3 liters though not significantly (117). These figures, contrary to some previous reports (86), were confirmed by

Hoyt et al. (103), who found in adult males, after 10 days at 4,300 m, a reduction of total body water by 1.7 liters ($H_2^{18}O$ dilution), a nonsignificant increase of the extracellular space (1.4 liters, by NaBr dilution), a decrease (−3 liters) of the intracellular space, and a 20% reduction of plasma volume. Plasma volume was also found to drop 0.7 liter after 30 days' exposure to 4,300 m in young women (86). With regard to high-altitude Andean natives, Picòn-Reàtegui et al. (148) reported intracellular fluid volumes similar to those found at sea level but slightly greater extracellular fluid volumes.

Decreased Absorption of Nutrients from the Gastrointestinal Tract.

Despite anecdotal reports of increased fat in the stool, fat and carbohydrate absorption at high altitude have been found to be normal, at least up to 4,700 m. Rai et al. (158) reported normal digestibility and utilization of dietary fats up to 324 g/day at 3,800 m and to 232 g/day at 4,700 m for an exposure time of over 4 months. Sridharan et al. (176) reported normal D-xylose excretion, an index of unchanged absorption activity for carbohydrates of the upper part of the small intestine, after 22 days at 3,500 m. Also, gastric secretion was found to be unaffected by altitude in both acclimatized lowlanders and altitude natives. Boyer and Blume (29) reported decreased absorption of fat (by 48.5%) and decreased excretion of xylose (by 24.3%) at 6,300 m.

As indicated above, a sizeable fraction of altitude-dependent weight loss is due to a reduction of muscle mass. To test the hypothesis that muscle loss is due to malabsorption of dietary protein, Kayser et al. (113) administered ^{15}N-labeled soya protein to six healthy male subjects in the course of a 3 wk sojourn at 5,000 m. Protein absorption was found to be the same as at sea level (97% vs. 96%). Westerterp et al. (193b) found a gross energy digestibility of 85% in subjects sojourning 21 days at 6,542 m, but based on the low energy content of the feces, tend to exclude significant malabsorption.

Muscle loss could, at least in part, be the result of detraining, especially at extreme altitudes. However, there are indicators that hypoxia per se may negatively influence amino acid metabolism. Rennie et al. (162) have shown that acute hypobaric hypoxia decreases the turnover and uptake of leucine from the forearm muscle compartment. In this respect, it is noteworthy that normobaric hypoxia, as observed in patients with chronic obstructive pulmonary disease, is often accompanied by marked loss of muscle mass due to disturbances in protein metabolism (140). Hypoxia may interfere directly (152) or indirectly by altered hormonal secretion with protein synthesis. However, further research is needed to gain insight into the mechanisms controlling the turnover of body protein stores.

Limiting Factors to Aerobic and Anaerobic Exercise

Factors Limiting $\dot{V}O_{2max}$. The characteristics of aerobic metabolism of exercising, sedentary, and trained subjects have been described in detail in many review articles. Also, the effects of environmental factors on the maximal aerobic performance ($\dot{V}O_{2max}$) are quite well known. In contrast, the factor(s) limiting development of maximal aerobic power by the muscle system is still a matter for discussion. Until a few years ago, the rate-limiting role to $\dot{V}O_{2max}$ was attributed to a single predominant factor, typically convective O_2 flow (maximum cardiac output $\times$ maximum O_2 binding capacity of hemoglobin). An analysis carried out by di Prampero (58) and resumed with a somewhat different approach by Wagner et al. (187) and by Wagner (186b) points out that the limits to $\dot{V}O_{2max}$ are imposed, even though to a different extent, by all factors contributing to O_2 uptake in the lungs, transport to the periphery, and utilization by the mitochondria, that is, gas exchange; O_2 binding by hemoglobin, thoracic, and peripheral circulation; and the oxidative potential of the tissues. Such factors may be looked upon as resistances in series, each contributing to limit the maximal rate of O_2 flow to, and utilization by, the tissues.

The proposed multifactorial model describing the factors limiting $\dot{V}O_{2max}$ is a simplified application of the O_2 conductance equation downstream from the lung (174). Oxygen is assumed to flow under a gradient from the alveoli to the mitochondria overcoming a series of resistances. The latter may be reduced for simplicity to three resistive elements with specific physiological meaning: the first is related to blood convective O_2 transfer; the second, resistance transfer (R_Q), to the peripheral O_2 transfer (R_Q); the third, (R_m), to the overall capacity of mitochondria to consume O_2. The total resistance to O_2 flow (R_T) is given by the sum of the above three elements. For any given PO_2 gradient, the weight of each factor (expressed as a fraction F_Q, F_t, F_m of the total made equal to 1) in limiting $\dot{V}O_{2max}$, may result from the interplay of R_Q, R_t, and R_m, so that $F_Q = R_Q/R_T$; $F_t = R_t/R_T$; $F_m = R_m/R_T$. The fractional limitation of $\dot{V}O_{2max}$ is calculated for various environmental conditions in Table 50.2 (66). It may be seen that in normoxia F_Q is by far the major determinant of $\dot{V}O_{2max}$ (0.7).

TABLE 50.2. *Fractional Limitation to* $\dot{V}O_{2max}$ Downstream from the Lung

Condition	F_Q	F_t	F_m
Normoxia	0.70	0.15	0.15
Acute hypoxia	0.48	0.26	0.26
Chronic hypoxia	0.45	0.20	0.35
Acute normoxia	0.70	0.11	0.19

Peripheral factors become more important in chronic hypoxia, accounting jointly for 0.55 instead of 0.3.

Factors Limiting Anaerobic Performance at Altitude

The alactic mechanism. Few investigations have been devoted to the study of the effects of hypoxia on maximum anaerobic power. On the one hand, di Prampero et al. (59) studied the effects of acute and chronic hypoxia on the maximal power ($\overline{w}$) that can be sustained during a short (<10 s) all-out effort on the bicycle ergometer. On the other hand, in a subsequent study carried out on the members of the 1981 Swiss Lhotse Shar expedition Cerretelli and di Prampero (37) assessed the effects of high-altitude acclimatization on maximal instantaneous or "peak" power ($\hat{w}$), that is, the power developed during the performance of a maximal vertical jump off both feet (26), an exercise lasting only a fraction of a second. The results of these studies are summarized in Table 50.3. It appears that (*1*) acute hypoxia (12% and 14.5% O_2 in N_2) has no detectable effects on $\overline{w}$; (2) chronic hypoxia (3 wk at altitudes of 4,540–5,200 m) does not affect either $\overline{w}$ and $\hat{w}$; (3) chronic hypoxia exceeding 5 wk reduces significantly $\hat{w}$.

A study conducted on six members of the 1986 Swiss Mt. Everest expedition by Ferretti et al. (67) indicates that maximal power determined during a maximal vertical jump off both feet on a force platform by the method of Davies and Rennie (49) upon return after an 8–10 wk sojourn at altitudes exceeding 5,000 m, was slightly but significantly reduced. However, because of the concomitant decrease of the cross-sectional area of

TABLE 50.3. *Effects of Hypoxia on Maximal Anaerobic Power*

Altitude (m)	FIO_2	Exposure	$\overline{w}$ (W·kg^{-1})	n
Sea level	1.0	Acute	11.1 ± 1.2	24
Sea level	0.145	Acute	11.3 ± 1.0	24
Sea level	0.121	Acute	11.2 ± 1.0	24
4,540	0.209	3 wk	11.0 ± 0.9	24
Altitude (m)	FIO_2	Exposure	$\hat{w}$ (W·kg^{-1})	n
400	0.209	Before	21.0 ± 1.9	15
5,200	0.209	3 wk	20.8 ± 3.9	25
5,200	0.209	5 wk	16.0 ± 1.8	15
400	0.209	After	17.4 ± 1.5	20

Maximal mechanical power during a 10 s all-out effort on the cycle ergometer and during the pushing phase of a standing high jump off both feet (average of first five jumps in a row).
$\overline{w}$ = power determined during a 10 s all-out effort on the cycle ergometer. $\hat{w}$ = peak power obtained during the pushing phase of a standing high jump. Mean values ± SD.
n = number of measurements.
From Cerretelli and di Prampero (37).

the extensor muscles of the thigh, maximal power, expressed per unit cross-sectional area of the muscle, was unchanged, which allows us to conclude that the decrease in power depends only on a net loss of muscle mass. Indeed, data obtained at the Pyramid Laboratory (5,050 m) indicate that, in the absence of a significant reduction of body weight and muscle mass, the maximal vertical jumping height, an indirect measurement of maximal power, is not influenced by a 5 wk sojourn at altitude (115 d; 143).

As is well known, at the onset of muscular exercise, O_2 consumption and lactic acid production do not contribute significantly to the energy requirement of the working muscles since their time course of readjustment is rather slow compared with the mechanical events of the contraction. As a consequence, maximal muscular power output is an index of the maximal rate of ~P splitting in muscle (57).

It can be concluded that neither acute nor chronic (up to 5 wk) exposure to hypoxia has any effect on the maximal rate of ~P hydrolysis in muscle, at least up to an altitude of 5,200 m. This finding is compatible with the data of Knuttgen and Saltin (116), who have shown that adenosine triphosphate (ATP) and phosphocreatine (PCr) concentrations are not affected by acute hypoxia, at least up to a simulated altitude of 4,000 m. Raynaud and Durand (160) have shown that the O_2 debt paid within the first minute of recovery after 3 wk at 3,800 m is essentially the same as at sea level. Since this part of the debt can be taken mainly as representative of ~P resynthesis after exercise, these data also suggest that hypoxia has no major effect on this aspect of muscle metabolism. From the data of Table 50.3, it appears, however, that for altitudes above 5,200 m and for prolonged exposures to hypoxia, maximal power may undergo a substantial drop, probably as a consequence of muscle mass reduction and/or deterioration (100).

Maximal anaerobic power has been determined also by performing a force-velocity test and a 30 s Wingate test (mean power) (9). Whereas the first of the above measurements has the same physiological significance as the one described above for the determination of sustained maximal power ($\overline{w}$), the 30 s Wingate test is a measurement of the average power developed by the subject during a period in which not only ~P splitting but also oxidations and anaerobic glycolysis contribute a sizeable amount of energy. Grassi et al. (75) found on six subjects after one month sojourn at 5,050 m a reduction of ~10% of the power developed during the 30 s Wingate test. By contrast, the power developed by the same subjects during a 10 s all-out exercise on the bicycle ergometer was unchanged compared to sea level. Bedu et al. (17), in a group of Bolivians native to a 3,700 m altitude (7–15 yr old), found that the performance of the force-velocity test was not different from that of an age-matched sea-level control group, whereas that of the Wingate test was about 15% lower.

It may be concluded that chronic hypoxia does not affect the capability of carrying out "explosive" efforts. Possible limitations may only be the consequence of a reduction of muscle mass in relation to body weight.

The lactic mechanism. With regard to acute hypoxia, Barcroft (7) was the first to report that two subjects the morning after arrival at Mt. Rosa (~4,500 m) were characterized by blood lactate concentration [La_b] of about 4 mM, that is, greater than normal sea-level control values (~1 mM). Bock et al. (23), however, found little or no increase in a group of subjects after breathing a hypoxic mixture corresponding to an altitude of 6,700 m for 1 h.

The pioneering study of Edwards (62) showed constant resting arterial [La_b] in high-altitude natives and in subjects acclimatized up to 6,140 m. Maximum blood lactate concentration ([La_bmax]) as a consequence of supramaximal exercise decreases considerably as altitude increases. Edwards (62) found that at 6,140 m even exhaustive, repetitive exercise hardly raised [La_b] above 2 mM when one of his subjects (all acclimatized lowlanders) was administered a sea-level equivalent of oxygen to breathe. Measurements carried out by Pugh et al. (156) on members of the British Scientific Himalayan expedition, by Lahiri et al. (120), and by Hansen et al. (88) essentially confirmed Edwards's results.

In the course of various expeditions to the Khumbu Valley of Nepal and to the Andes a sizeable number (n = 41) of acclimatized and unacclimatized natives (Sherpas and Peruvian Indians) and Caucasian lowlanders were investigated at altitude, at sea level, and in a hypobaric chamber (39) to assess one or more of the following variables: (1) resting blood lactate, (2) maximal blood lactate following continuous or intermittent supramaximal workloads until voluntary exhaustion while breathing air or oxygen at ambient pressure, (3) the relationship between blood hydrogen ion concentration [H^+] and [La_b] after supramaximal efforts of varying duration, and (4) the kinetics of lactate washout from blood during recovery following supramaximal loads while breathing air or oxygen. The results obtained at altitude are summarized in Table 50.4, from which is appears that (1) resting lactate is not affected by chronic hypoxia (in both Caucasians and highlanders [La_b] is around 1 mM, as previously shown by Edwards [62]) and (2) [La_bmax] at an altitude of about 5,350 m is reduced to about half the average sea-level control value in all subjects and is essentially not affected by O_2 administration. The latter observation, as indicated in the previous paragraph, had already been made by Edwards (62), who also noticed, after a long sojourn at altitude, limited effects of O_2 breathing on

TABLE 50.4. *Blood (Venous) Lactate Concentration (± S.E.) at Rest and after Exhaustive Exercise at Various Altitudes in Different Ethnic Groups*

Metabolic Conditions	Breathing Conditions	Altitude (m)	Subjects	Lactate (mM)	Reference
Rest	Air	5,050	Caucasians	1.43 ± 0.07	(74b)
		5,350	Caucasians	0.97 ± 0.07	(35)
		5,350	Sherpas	0.92 ± 0.12	(35)
		4,540	Peruvian Indians	1.09 ± 0.04	(39)
	O_2	5,350	Caucasians	0.94 ± 0.09	(35)
Exhaustive exercise	Air	5,050	Caucasians	6.29 ± 0.27	(74b)
		5,350	Caucasians	5.93 ± 0.35	(35)
		5,350	Sherpas	6.52 ± 1.19	(35)
		4,540	Peruvian Indians	5.84 ± 0.10	(39)
	O_2	5,350	Caucasians	6.93 ± 0.53	(35)

both exhaustion time and [La_bmax] and was confirmed by Grassi et al. (74b) on ten subjects who had been sojourning for one month at 5,050 m. Sherpas (n = 3) exercising supramaximally at sea level were able to reach [La_b] up to at least 11 mM, close to Caucasians breathing air at sea level (~14 mM) or in a hypobaric chamber at a simulated altitude of 3,800 m (~13 mM).

Extremely low [La_bmax] levels were also reported by West et al. (192) in lowlanders acclimatized to 6,300 m and by Sutton et al. (179) during Operation Everest II at a pressure of 282 mm Hg corresponding to an altitude of ~7,600 m.

Cerretelli et al. (39) attributed the decrease of the maximal lactic acid capacity at high altitude to the reduction of plasma bicarbonate concentration consequential to the drop of Pa_{CO_2} and speculated that "should Pa_{CO_2} at the summit of Mt. Everest drop to 10 torr, the size of the maximal lactacid O_2 debt would be practically nil."

Cerretelli et al. (39) also carried out measurements of blood pyruvate and found concentrations of this metabolite similar to those expected at sea level both at rest and after exhausting exercise.

The relationships between the increase in blood [H^+] and the corresponding rise in blood [La_b] determined at sea level in Caucasians and at 5,350 m in acclimatized lowlanders and Sherpas appear to be linear (35). The buffer value for the whole body, calculated from the slope of the above line, is reduced to about half at altitude. The highest [H^+] at exhaustion is approximately the same for the two conditions, independent of [La_bmax].

The pattern of lactic acid disappearance from blood during recovery starting from the same absolute concentration level (for example, 6 mM, that is, the maximum found at 5,350 m altitude) may be slightly different in the mountains compared to sea level. In the latter condition the function describing lactic acid washout

during recovery is close to monoexponential, with a half-time of ~15 min (129). In chronic hypoxia a variable delay is observed before [La_b] starts dropping. Such a delay, which occurs after heavy exercise with [La_b] > 10 mM even at sea level, may be, in part, the consequence of the reduced net release of lactic acid by the exercising muscles resulting from acclimatization (18). It is also influenced by the duration of exercise, the length of altitude exposure, and circulatory factors (for example, peripheral vasoconstriction), as shown by the consequences of O_2 administration at altitude, which results in a probably retarded lactic acid washout from the muscles and blood (39). Independent of the initial time lag, the following decay of the [La_b] curve is characterized by similar half-times in normoxia and chronic hypoxia, which seems to indicate that the rates of glucose resynthesis and/or lactate oxidation during recovery are not greatly influenced by altitude. This observation was confirmed by the results of Grassi et al. (75) in six subjects exposed for about one month at 5,050 m.

It has been proposed that the reduced lactic capacity at altitude, also known as "lactate paradox" (93), may have become in high-altitude natives a developmental or genetic adaptation ("perpetual lactate paradox"; 96). This metabolic feature in chronically hypoxic lowlanders is a consequence of acclimatization, since the expression of the so-called lactate paradox is reversible, with a half-time of about 5 days for both the on- and the off-phase (deacclimatization), as shown by Grassi et al. (74). Hochachka et al.'s (96) hypothesis, as well as its metabolic implications with respect to oxidative metabolism, is based on observations made on Quechua Indians examined both at altitude (La Raya, 4,200 m) and at sea level and requires confirmation by more extensive studies on other altitude populations. Moreover, doubts were cast even on the reproducibility of the finding on which the hypothesis is based, that is, the constancy of the lactate paradox when highlanders descend to sea level. This is indeed the prerequisite for admitting a possible but improbable genetic adaptation of the basic energy-yielding metabolic pathways in altitude natives. The following facts appear to be particularly relevant at this time and must be taken into account before this hypothesis becomes acceptable: (1) The Sherpas of Nepal, a much more established and genetically identified high-altitude population than the Quechua Indians, undergo the same adaptive metabolic changes in chronic hypoxia as acclimatized lowlanders. In fact, they do not seem to display the so-called perpetual lactate paradox (see ref. 39). In addition, first-generation Tibetans born at 1,300 m are characterized by the same [La_bmax] (~12 mM) as control subjects (personal observation, unpublished). (2) Sherpas have the same efficiency for aerobic work as sea-level residents and acclimatized lowlanders (34, 120). (3) The muscle mitochondrial volume density of Sherpas is low (3.96 ± 0.27), even lower than that of sedentary sea-level dwellers (~4%–4.5%) (115). This finding, if confirmed for the Quechuas, argues against the possibility of a metabolic organization of skeletal muscle "toward the cardiac end of the spectrum" that would be coherent with the theory (96, 131). Indeed, according to Hochachka (94), a more tight coupling in ATP demand–ATP supply is at the heart of the perpetual lactate paradox found in his Andean subjects. (4) A tighter ATP demand–ATP supply coupling would bring about only a negligible reduction of the muscle PCr stores at the onset of constant-load exercises. This is the case for myocardium (112) but not for the calf muscles of Quechua Indians (see, for example, Fig. 9B in ref. 131). The latter appear in fact to behave, with regard to the kinetics of PCr resynthesis, as sedentary subjects at sea level.

Apart from the hypothesis that the lactate paradox may perpetuate itself and become a genetic feature of high-altitude natives, a possibility not supported by the measurements carried out in second-generation Tibetans (see point (1) above), no clear-cut evidence has been offered for this adaptational change. West (191) and Sutton and Heigenhauser (178) and Cerretelli et al. (37b) have indicated the following possible determinants: (1) Physical training. Training at altitude might hypothetically result in a progressively greater recruitment of slow twitch fibers and thus lower [La_b max] levels. To our knowledge this possibility is not supported by experimental evidence. Moreover, according to Green et al. (76b), the reduction in glycolytic flux with acclimatization cannot be explained by changes in the muscle oxidative potential, muscle fiber size, or muscle fiber capillarization. (2) A change of the rate of lactic acid removal and uptake from or by muscle or other organs. Evidence has been given of an apparently sluggish diffusion of lactic acid from muscle to blood in

chronic hypoxia (18). A diffusion limitation, however, could possibly lead to a change of shape of the [La_b] recovery curve after exercise but, because of the relatively slow kinetics of lactic acid oxidation and/or glucose resynthesis from lactic acid, not to a reduction of the peak [La_b] level. However, Green et al. (77) have shown that in chronic hypoxia (Operation Everest II) lactic acid accumulation after exercise is drastically reduced in both blood and muscle. (3) Enzyme activity changes along the glycolytic pathway. Howald et al. (102) have shown that the activities of hexokinase, phosphofructokinase, lactate dehydrogenase, and glyceraldehyde phosphate dehydrogenase in the vastus lateralis of acclimatized mountaineers 10–12 days after return from a 6–8 wk high altitude (>5,000 m) exposure are essentially unchanged in respect to preexpedition control values. Similar results, including data on glycogen phosphorylase, were obtained after a 15 day exposure to 4,300 m by Young et al. (199) and by Green et al. (76) in the course of Operation Everest II. Rosser and Hochachka (167b) have shown in Quechua Indians an increase of lactate dehydrogenase. Thus it would appear that the main regulatory enzymes of the glycolytic pathway are not affected by chronic hypoxia and cannot be the basis of the lactate paradox. The maximal potential activity of glycolytic enzymes, however, does not provide information concerning the in vivo maximal flux along the glycolytic pathway. The latter, in fact, could also be affected by changes in the velocity constants of rate-limiting enzymes and/or by some form of "upstream" inhibition, possibly attributable to the β-adrenergic modulation of glycogenolysis (see below) and/or to a reduced neuromuscular activation (see below). The maximal glycolytic flux was evaluated indirectly by Grassi et al. (75), and it was found to be reduced at 5,050 m compared to sea level, indicating that a lower maximal glycolytic flux could be responsible, at least in part, for the reduced [La_bmax] at altitude. (4) Substrate availability. Determinations of muscle glycogen stores in chronic hypoxia show only minor reductions (76b, 198) or no reduction at all (77) compared to normoxia. In contrast, an increased dependence on blood glucose has been shown both at rest and during exercise, possibly as a consequence of enhanced hepatic gluconeogenesis, by Brooks et al. (31). Depletion of muscle glycogen stores, therefore, appears an unlikely candidate in determining the reduction of [La_bmax]. (5) Hormones. Epinephrine is known to activate phosphorylase a, leading to enhanced muscle glycogen breakdown. During altitude acclimatization, plasma epinephrine response to exercise, after a transient increase (4 h of exposure to 4,300 m), resumes sea-level control values as shown by measurements carried out after 21 days. Plasma norepinephrine is slightly enhanced compared to control levels in both acute and chronic hypoxia (135). Thus a reduced activation of phosphorylase *a* could only occur via a down-regulation of the β-adrenergic receptors. Richalet et al. (164, 165) have shown, in fact, a hypoxia-induced reduction of the β receptor activity of both lymphocytes and the myocardium, which could restrict the maximum rate of glycolysis and, therefore, maximal lactic capacity in chronic hypoxia. In any case, experiments carried out with β blockers (200) indicate that chronic stimulation of β-adrenergic receptors by elevated levels of circulatory norepinephrine is not the mechanism responsible for the lactate paradox during altitude acclimatization. Moreover, Mazzeo et al. (135b) published results according to which the sympathoadrenal system does not entirely account for the blood lactate changes observed duringexercise at altitude. (6) Reduction of the tissue buffer capacity. The lactate paradox has also been associated with a reduced blood and tissue buffer capacity. This hypothesis was tested by Kayser et al. (114) by bicarbonate (0.3 g/kg of body weight) loading before and at the end of a 35 day sojourn at 5,050 m. Net (observed minus resting) peak [La$_b$] accumulation increased in six subjects from 12.9 to 16.6 m*M* at sea level and only from 6.85 to 7.95 (not significant) m*M* at high altitude. These results indicate that an increase in blood and muscle buffer capacity at high altitude does not allow resuming sea-level peak lactic capacity.

Independent of its mechanism, which is still a matter for investigation, a reduction of lactic acid accumulation in blood and in the tissues, particularly the muscles, may prevent changes of pH that could be harmful for the functioning of the cell. The blockade of anaerobic glycolysis, however, does not seem to occur through a negative feedback via a H$^+$-mediated inactivation of a key enzyme. Modulation of the activity of glycogen phosphorylase *a* by catecholamines could be the mechanism responsible for the lactate paradox, even though we cannot rule out the possibility of a more complicated interaction of some of the factors listed above.

The alternate theory proposed by Hochachka (93) considers the lactate paradox not as the consequence of an impaired energy flow through the glycolytic pathway but rather as an upstream reflection of an improved function of the respiratory chain. Unfortunately, this hypothesis, even though theoretically sound, does not have enough biochemical and/or physiological experimental support. Kayser et al. (115c) have hypothesized that during chronic hypobaric hypoxia the central nervous system may play a primary role in limiting exhaustive exercise and maximal accumulation of lactate in blood. This would confirm the hypothesis put forward by Bigland-Ritchie and Vollestad (20b), according to which a complete neuromuscular activation may not be achieved in chronic hypobaric hypoxia. Under these conditions, inhibitory reflexes elicited from limb or res-

piratory muscles would act centrally, limiting maximal motor drive in relation to their O$_2$ availability. According to data by Grassi et al. (74) at 5,050 m, the reduced [La$_b$max] at altitude is related, at least in part, to the duration of the exercise protocol. Indeed, [La$_b$max] determined at the end of short (30–45 s) supramaximal exhaustive exercise bouts increases during altitude acclimatization, whereas [La$_b$max] determined at the end of longer (5–20 min) exercise decreases in the course of acclimatization (74b). Moreover, [La$_b$max] determined at the end of all-out 10–30 s exercise bouts at altitude does not decrease at all or decreases only slightly (10 s), compared to sea level (75).

MORPHOLOGICAL CHANGES OF SKELETAL MUSCLE TISSUE

Before engaging in the discussion on the findings related to skeletal muscle changes induced by chronic hypoxia in humans, we have to consider some methodological limitations which influence the interpretation of morphometric (and to some extent biochemical) data collected from tissue samples obtained with the classical Bergström needle used in virtually all research on muscle tissue discussed in this review (20). (*1*) The needle biopsy technique permits retrieval of some 50–100 mg of muscle tissue, usually from the vastus lateralis. Structural and biochemical studies generally tacitly assume the observed muscle tissue composition to be representative of muscle tissue as a whole and of the prevailing experimental conditions. Neither of these assumptions has been verified in human experimentation due to ethical considerations and discomfort related to the need of multiple biopsies, which would be necessary for implementing sound sampling strategies (122, 123). (*2*) In most biopsy studies the volume of the muscle from which a biopsy was taken has not been measured. As a consequence, only relative changes of structural or biochemical variables can be reported. As long as muscle mass remains constant over the experimental period, this limitation may be immaterial. However, weight loss with a concomitant loss of muscle mass is a hallmark of chronic exposure to hypoxia, as discussed earlier in some detail (see Changes of Body and Muscle Mass and Nutrition). If a change in muscle mass does occur, it becomes mandatory to report the relative changes of the variables describing tissue characteristics (for example, mitochondrial volume as percent of tissue volume) as well as the absolute changes (for example, mitochondrial volume in milliliters). Absolute changes have been calculated by multiplying the change in muscle tissue volume by the relative change of the tissue variable under observation (100, 126). Because of the difficulty of measuring directly the individual volume of the vastus lateralis muscle, measurement of its cross-sectional area or of the cross-section of all thigh muscles by means of a computed tomography scan or nuclear magnetic resonance imaging at mid-thigh level has been adopted. Absolute values of some tissue components (such as total capillary length or total mitochondrial volume) could be calculated for a slice of vastus lateralis muscle 1 cm thick (100). The implicit assumptions in this calculations are that the observed changes of muscle cross-sectional area are representative of the change of muscle volume and that the biopsy sample analyzed is representative of the whole muscle mass under investigation. (3) During the biopsy procedure and when handling muscle samples prior to fixation, muscle fibers lose their parallel arrangement and get disorganized in many ways (202). Moreover, the state of muscle contraction (that is, sarcomere length) cannot be controlled in muscle biopsies.

The assessment of the state of muscle contraction is considered a prerequisite for a detailed analysis of muscle capillarity (132, 134). In our experience sarcomere length is fairly uniform around 1.95 μm in muscle biopsies fixed for electron microscopy (average of ten biopsies). The situation is further complicated by shrinking artifacts introduced by the hyperosmolar concentration of the fixation media used in the preparation for electron microscopy. Shrinking artifacts may be largely eliminated when tissue is cryofixed for histochemistry, leading to larger values for fiber cross-sectional area and lower values for capillary density. The distortion of muscle architecture in needle biopsy samples imposes limitations on the precision with which fiber size and muscle capillarity can be estimated (202). In contrast, all volume density estimates describing the quantitative ultrastructural composition of muscle fibers are unaffected by shrinking artifacts (63) as well as by the disorganization of the anisotropic structure of muscle tissue (189). The following discussion of the changes of muscle structure with exposure to chronic hypoxia accounts for the technical limitations encountered when performing structural studies on biopsied tissue.

Fiber Size and Fiber Type Distribution

As a consequence of the pronounced atrophy incurred by chronic exposure to hypoxia, virtually all studies report a significant reduction in muscle fiber cross-sectional area with prolonged hypoxic exposure. Hoppeler et al. (100) analyzed muscle biopsies from the vastus lateralis of 14 mountaineers from two expeditions (sojourn of 8 wk at altitudes >5,000 m) and found a reduction of the average muscle fiber cross-sectional area of 20%, concomitant with a reduction of thigh cross-sectional area of 11%. During Operation Everest II, decreases of muscle cross-sectional area of 13% and

15% for thigh and upper arm muscles, respectively, were observed in six subjects (127). Biopsies from vastus lateralis obtained during Operation Everest II showed a significant 25% decrease of fiber cross-sectional area of type II fibers, while type I fibers decreased by 26% (not significant). From the Swedish Mount Everest expedition in 1987 (3 months at 5,200 m), a reduction of fiber cross-sectional area of close to 15% was reported for both vastus lateralis and biceps brachii muscles (188). This study did not report on changes in muscle cross-sectional area. No data are available indicating whether or to what extent the decrease in fiber cross-sectional area is reversible upon return to a normoxic environment. Interestingly, fiber size in Sherpas (115) as well as in world-class high-altitude mountaineers with multiple exposure to chronic severe hypoxia (144) is somewhat smaller than that of lowlanders and in the same range as that seen in mountaineers after an expedition (Table 50.1; Fig. 50.4).

Data of both the Swedish Mt. Everest expedition (188) and Operation Everest II (77) show no indication of fiber type transformation as a consequence of chronic exposure to hypoxia. In both cases the investigated subjects showed similar distributions of slow and fast twitch fibers in both vastus lateralis and biceps brachii muscles. The finding of an unchanged fiber type pattern with hypoxic exposure in humans is in contrast with the results of experiments on rodents exposed to simulated altitudes of 4,000 m for 10 wk (108). This experiment showed significant changes of the fiber distribution pattern in soleus and extensor digitorum muscle in rats. The world-class high-altitude climbers studied by Oelz et al. (144) were found to have over 70% slow twitch fibers in the vastus lateralis. This is similar to long-term endurance athletes and could be the result of endurance activity related to high-altitude mountaineering or of genetic predisposition.

Capillary Supply

Ever since Valdivia (184) reported greater capillary densities in guinea pigs native to the Peruvian mountains, it has been a major contention that chronic hypoxia promotes capillary growth (see ref. 104). It came as a surprise to find that in humans the capillary to fiber ratio (a structural variable scarcely affected by fiber size changes and relatively insensitive to the distortions of muscle architecture introduced by the biopsy procedure) is unchanged or even reduced in both vastus lateralis and biceps brachii after prolonged chronic hypoxia (76, 100, 188). The finding of an unchanged or even decreased capillary to fiber ratio argues strongly against capillary neoformation induced by chronic hypoxia in humans. From this it follows that the increase in capillary density (ranging from 9% to 12% in the vastus lateralis) observed in all structural studies (76, 100, 127,

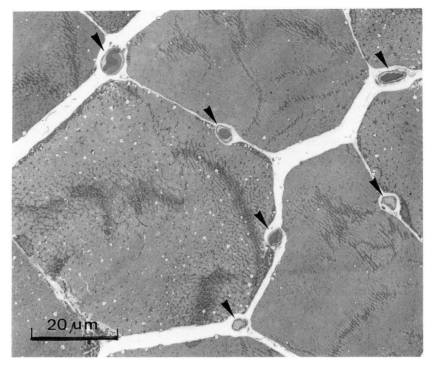

FIG. 50.4. Micrograph of cross-section of portions of muscle fibers in a muscle biopsy of M. vastus lateralis as used for morphometry of capillarity and fiber size (arrows indicate capillaries).

188) must entirely be the consequence of the decrease in fiber size, that is, a consequence of muscle fiber atrophy. Hoppeler et al. (100) have calculated the total capillary length in a slice of vastus lateralis of 1 cm thickness to be on the order of 100 km, which remains unchanged after 8 wk of hypoxic exposure. For this calculation, it was assumed that hypoxia does not change the degree of capillary tortuosity, a finding supported by evidence from rats and mice maintained at a 3,800 m altitude and compared to sea-level controls (133, 151). The latter studies also demonstrated, at least in rodents, that capillary diameter is unchanged with chronic hypoxia exposure.

With regard to capillary supply of human skeletal muscles, there is no evidence of capillary neoformation related to hypoxia (76, 100, 127, 188). However, the same capillary bed subserves a smaller muscle fiber volume. Together with a possible increase in muscle myoglobin concentration, the reduction of the fiber area supplied by one capillary favors oxygen transfer from capillaries to muscle mitochondria.

Mitochondria

Hypoxia has been considered a major stimulus for increasing muscle oxidative capacity (see ref. 97). However, Operation Everest II showed no increase of the activity of the citric acid cycle enzymes citrate synthase and succinate dehydrogenase (SDH) over the first 33

days of progressive decompression to 282 torr in five subjects (76). Rather unexpectedly, a significant decrease of citrate synthase and SDH activity (−14% and −31%, respectively), accompanied by a 33% reduction of hexokinase, occurred during the last 7 days of exposure to 282 torr. It was speculated that the reduction in mitochondrial oxidative capacity might have been, at least in part, a consequence of the drastic reduction of physical activity at this altitude. These results are supported by a study on seven subjects from whom muscle biopsies were obtained before and after the Swiss Mt. Everest expedition (102). Citrate synthase and the respiratory chain enzyme cytochrome oxidase were significantly reduced by 23%. Similar reductions were seen for enzymes related to beta oxidation of fatty acids and utilization of ketone bodies. Furthermore, this study showed strong statistical correlations between the reduction of mitochondrial enzyme activities and the reduction of the volume density of mitochondria determined with morphometric methods in the same subjects. From this it was concluded that the observed changes in mitochondrial enzyme activities were due to a loss of mitochondrial structure rather than to qualitative changes of mitochondrial function. In contrast, morphometric analysis of biopsy samples from five subjects of Operation Everest II showed no significant difference of the volume density of mitochondria between pre- and postexposure biopsies, despite the significant reduction of citrate synthase activity (76, 127). The smaller number of subjects analyzed and a possibly smaller decrease in muscle oxidative capacity observed in this study may have contributed to this result.

If the morphometric data of the seven subjects of the Swiss Lhotse expedition in 1981 and the seven subjects participating in the Swiss Mt. Everest expedition in 1986 are combined, a significant reduction of the volume density of total mitochondria by 19% is observed (100). Both the larger population of interfibrillar (-13%) and the smaller population of subsarcolemmal (-43%) mitochondria were significantly reduced, though not to the same degree. This finding was supported by a topological analysis, further indicating that

mitochondria from the fiber periphery seemed to be more readily lost than mitochondria from the fiber center during hypoxic exposure. Relating mitochondrial volume density to $\dot{V}O_{2max}$, it was observed that data of the mountaineers before expedition closely fitted previously established relationships between these variables (99, 101). After the expedition, the relationship between $\dot{V}O_{2max}$ and mitochondrial volume density was changed. This was due to the fact that, in general, mountaineers with high preexpedition mitochondrial densities lost more mitochondria at high altitude but, despite that, were able to maintain a relatively elevated maximal oxygen uptake capacity.

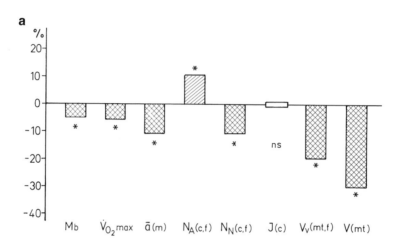

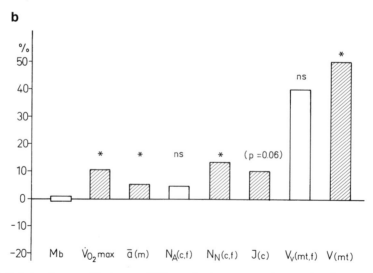

FIG. 50.5. Relative changes of body mass (Mb), maximal oxygen uptake capacity ($\dot{V}O_{2max}$, measured in normoxia after continuous and in hypoxia after discontinuous exposure to severe hypoxia), muscle cross-sectional area [$\bar{a}(m)$], capillary density [$N_A(c,f)$], capillary to fiber ratio [$N_N(c,f)$], total capillary length [$J(c)$], mitochondrial volume density [$V_V(mt,f)$], and total mitochondrial volume [$V(mt)$] with continous (a) and discontinuous (b) exposure to hypoxia. (From Hoppeler and Desplanches, 1992.)

To better understand the hypoxia-induced changes of variables describing muscle metabolism, absolute values for the various tissue components were calculated by multiplying muscle volume by morphometric volume density data. High altitude exposure resulted for the 14 subjects of the combined Swiss expeditions (100) in a 10% reduction of muscle volume. As a consequence, total mitochondrial volume in the vastus lateralis was reduced by as much as 30%. The combined morphometric analysis of muscle structure and ultrastructure indicates that loss of muscle volume may be a major finding with prolonged high altitude adaptation. The loss of muscle bulk can be calculated as being primarily due to a loss of myofibrillar proteins (100). The structural equivalent is a decrease in muscle fiber size observed in virtually all biopsy studies. Both biochemical and structural studies indicate a loss of mitochondrial oxidative capacity on the order of 10%–20% in relative terms and greater when the loss of muscle volume is also considered. As the capillary network is spared from atrophy, the same capillary bed serves a reduced muscle oxidative capacity (Fig. 50.5).

Fiber Damage and Regenerative Events

Ultrastructural analysis of biopsies taken after the Swiss Lhotse and Mt. Everest expeditions indicated the presence of increased amounts of lipofuscin in subsarcolemmal locations of muscle fibers (Fig. 50.6; 100). Lipofuscin is a pigment which accumulates when autophagy in lysosomes is high or when degradation of biomaterial is hindered by chemical alterations such as lipid peroxidation (8, 40). A quantitative analysis indicated that lipofuscin is an exceedingly rare component of muscle tissue, representing only some 2×10^{-4}% of the muscle fiber volume in mountaineers before the expedition (130). This study indicated a significant threefold increase in volume density of lipofuscin over the period of the expedition in the seven members of the Swiss Mt. Everest expedition. Lipofuscin content varied considerably among individuals. Partly this was due to the difficulty of quantitating an exceedingly rare tissue component. However, there seemed to be a trend for those mountaineers with previous extreme altitude climbing experience to have higher preexpedition lipofuscin values.

The same study also showed a significant increase in the volume density of satellite cells. Satellite cells are dormant myogenic stem cells wedged between the basement membrane and the sarcolemma of the muscle fiber (Fig. 50.7). Satellite cells represent the regenerative capacity of skeletal muscle tissue (2). They can go through mitosis and myoblast and myotube formation in case of repair of muscle damage. Like lipofuscin, satellite cells, which are rare in human muscles—accounting for only $1.7 \cdot 10^{-4}$% of the muscle volume (130)—

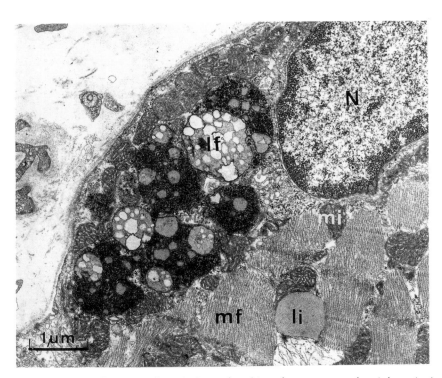

FIG. 50.6. Micrograph of a section of muscle tissue of a subject after exposure to chronic hypoxia. Accumulation of the degradation pigment lipofuscin (*lf*) close to the muscle fiber nucleus (*N*) can be seen (*li*, lipid droplet; *mf*, myofibrils; *mi*, mitochondria).

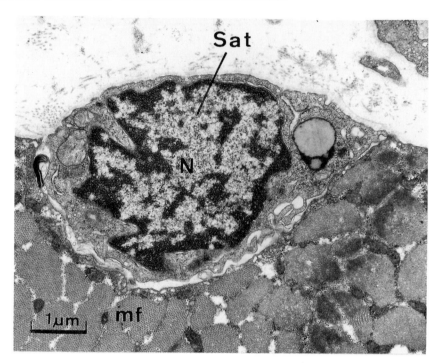

FIG. 50.7. Micrograph of a satellite cell (*Sat*) located under the basement membrane in close apposition to muscle fiber (*N*, nucleus of satellite cell; *mf*, myofibrils in muscle fiber).

underwent a threefold increase in volume density in biopsies obtained after the expedition, whereas the volume fraction of myonuclei remained constant. Due to technical difficulties, the absolute number of satellite cells was not determined. Some myoblast formation could be detected. However, this phenomenon was too rare to be quantitated. The significance of the increase in satellite cell volume with prolonged hypoxia exposure cannot be appreciated while the role of satellite cells during muscle adaptive events, such as training (not involving gross muscle fiber damage), is not understood.

TRAINING AND MUSCLE ADAPTATIONS WITH HYPOXIA

Altitude Training

Altitude training has become rather popular mainly for two reasons: (*1*) the frequent occurrence of athletic competitions at altitudes between 1,500 and 3,700 m and (*2*) the common observation that East African highlanders normally training at altitudes of ~2,000 m perform particularly successfully in endurance competitions when at sea level. Acclimatization and training protocols must necessarily be adopted to prevent symp-

toms of acute hypoxia and to give athletes the opportunity to test and practice the more favorable pace imposed by reduced environmental P_{O_2}. The possibility has been suggested that prolonged training at higher elevations, besides raising blood hemoglobin concentration, may confer special benefits because of an hypothesized altitude–training synergistic effect. The latter would materialize in an enhanced muscle respiratory capacity, the underlying mechanism being a prompting of the oxidative machinery by the hypoxic stimulus.

Whereas the usefulness of training in hypoxia for altitude performances cannot be questioned, the benefits from hypoxic training for competing at sea level have been, and still are, a matter for debate. Several studies aimed at determining the evolution of $\dot{V}_{O_{2max}}$ upon exposure to moderate hypoxia (usually corresponding to ~2,500 m altitude) and sudden return to sea level have been carried out since 1969 (for details, see refs. 27, 121). In view of the lack of standardization of the experimental protocols, the results appear to be rather controversial, the average improvement being small (~5%) and not statistically significant. This appears to be the case particularly for subjects whose absolute baseline $\dot{V}_{O_{2max}}$ value was higher (27), even though there is evidence of an extreme variability of the results even among highly trained subjects (157). A reason for the decrease of $\dot{V}_{O_{2max}}$ found by some authors after altitude exposure could be an overall detraining effect, the cause of which is the obligatory reduction of the absolute workload at which most subjects can train. Indeed, above 1,500–1,800 m, maximal exercise could enhance the normal tendency for athletes toward a decrease of arterial blood O_2 saturation (50), which would necessarily limit maximal aerobic performance. Upon return to sea level, the athlete would have lost his or her habitual normoxic pace and, therefore, would experience a limitation in performance.

Physically active though nonathletic subjects, when brought to higher altitudes, after the initial loss of $\dot{V}O_{2max}$, tend to progressively improve maximal aerobic power. However, upon return to sea level, $\dot{V}O_{2max}$ never reaches prealtitude levels (33, 47, 75) in spite of increased blood [Hb]. This is due mainly to detraining, even though at altitudes above 5,000 m hypoxia-induced muscle deterioration might also be involved (100).

Terrados et al. (183) have studied in competitive cyclists the effects of training at sea level and at simulated altitude (corresponding to 2,300 m) on $\dot{V}O_{2max}$ and maximal work capacity determined in the same conditions. They concluded that work capacity at altitude was increased more by training at altitude (33%) than at sea level (14%), whereas work capacity at sea level was equally improved (33% and 22%, not significant) by the two training conditions. A decrease of muscle phosphofructokinase activity was found after simulated altitude training.

Mizuno et al. (136) studied the effects of 2 wk of altitude (2,100–2,700 m) training in a group of well-trained cross-country skiers. No significcnat changes of $\dot{V}O_{2max}$ were found after hypoxic training, but short-term running time was 17% improved, probably as a result of increased glycolytic capacity. The gastrocnemius muscle maintained a prealtitude capillary supply with a 10% decreased mitochondrial enzyme activity, likely a consequence of detraining. The triceps brachii, by contrast, showed increased capillarization and unchanged mitochondrial enzyme activities as a result of greater activation. Thus the hypothesis that altitude training could be beneficial from the standpoint of muscle respiration does not seem to be supported by the above experimental results. The average very high performance capacity of highlands African endurance athletes is likely to be ascribed to higher specific $\dot{V}O_{2max}$.

No data are available, to our knowledge, on the characteristics of isometric muscle training in chronic hypoxia. This problem is particularly interesting in view of the possible implications from findings showing impaired muscle metabolism in chronic hypoxia. Narici and Kayser (143) found that isometric training of the elbow flexors for 1 month at 5,050 m altitude with the same absolute load increased both cross sectional area of the muscle (+11.3%) and the maximum voluntary contraction (+9.5%). Corresponding figures at sea level were +17.7% and 13.6%, respectively. Thus chronic hypoxia seems to reduce the potential for hypertrophy of human skeletal muscle.

In conclusion, training at altitude is beneficial, if not necessary, for performances to be carried out at altitude, but it does not seem to provide additional benefits compared to equivalent training at sea level for sea-level competitions.

Training in Normobaric Hypoxia

Evidence from studies in which hypoxia was imposed only over the relatively short periods of actual training indicate that local lack of oxygen might in fact be a potent stimulus for the formation of muscle respiratory enzymes (110, 182). In one study (110) eight subjects performed one-legged bicycle exercise in a pressurized chamber enclosing only the lower part of the body. Both legs were trained at the same workload, but one leg was rendered ischemic during training by increasing pressure to 50 mm Hg. The leg trained in ischemic hypoxia increased performance more than the leg trained under normal perfusion conditions. It was noted that there was a larger increase of citrate synthase activity in muscle biopsies obtained from the leg trained in ischemia than from the control leg. Similar results are reported by Terrados et al. (182). These authors had ten subjects doing one-legged bicycle exercise training in a large environmental pressure chamber. Again, both legs were trained at the same workload; however, one leg was trained with the chamber at normobaric, the other leg with the chamber at hypobaric (2,300 m) conditions. Time to fatigue increased significantly more in the leg trained under hypobaric conditions. Concomitantly there was a significantly larger increase in citrate synthase activity in the hypobarically trained leg. Moreover, myoglobin concentration decreased in the normobarically trained leg and increased in the hypobarically trained leg. It was reasoned that "substrate flux, presumed to be similar in both conditions, is a less likely candidate as a stimulus for the enzyme synthesis. Instead, the stimulus seems to be related to the lowered blood oxygen content or tension, creating a more pronounced disturbance of the local energy balance."

Recent results (51) confirm the findings of a larger increase in oxidative enzyme activities with training regimes which use hypoxia as an additional stimulus during the training phase. Five subjects were trained on a bicycle ergometer for 3 wk (2 h/day, 6 days/wk) at a load corresponding to 80% of peak heart rate, measured in hypoxia while breathing 10% O_2 in N_2, equivalent to an elevation of 5,700 m altitude. After a 14 month detraining period, subjects were retrained at the same absolute workloads but in normoxia. Training in normoxia had no effect on any of the investigated variables. Training in hypoxia increased hypoxic (but not normoxic) $\dot{V}O_{2max}$, increased muscle cross-sectional area, interfibrillary mitochondrial volume density, and capillary per fiber ratio by 10%, 42%, and 13%, respectively.

Studies using discontinuous hypoxia as an added stimulus during training open exciting new venues for gaining insight into the nature of the training stimulus. Apart from the fact that energy balance of contracting muscles must be more profoundly altered with hypoxia as an added stimulus, other mechanisms governing muscle tissue malleability must also be considered. A number of signal peptides, such as heat shock transcription factors (19), oxygen-regulated proteins (194), and hypoxia-associated proteins (201) were found to be expressed in cell cultures exposed to hypoxia. It is conceivable that these or similar proteins might influence protein metabolism in muscle cells exposed to hypoxia.

Muscle deterioration as a consequence of continuous exposure to severe hypoxia might further be related to the finding that there is a significant reduction in synthesis of myofibrillar as well as cytoplasmic proteins in skeletal (but not heart) muscle of rats exposed to hypoxia corresponding to altitudes at which mountaineers classically sojourn during expeditions to the highest peaks on earth (153). Clearly, research into phenotypic adaptations of humans to hypoxia will continue to be a major challenge for future multidisciplinary studies of both an applied (for example, clinical) and a fundamental nature.

REFERENCES

1. Adams, W. H., and L. J. Strang. Haemoglobin level in persons of Tibetan ancestry living at high altitude. *Proc. Soc. Exp. Biol. Med.* 149: 1036–1039, 1975.

2. Allbrook, D. B. Skeletal muscle regeneration. *Muscle Nerve* 4: 234–245, 1981.

3. Åstrand, P. O. The respiratory activity in man exposed to prolonged hypoxia. *Acta Physiol. Scand.* 30: 343–368, 1954.

4. Åstrand, P. O., and K. Rodahl. *Textbook of Work Physiology.* New York: McGraw Hill, 1970, p. 573.

5. Balke, B., J. T. Daniels, and J. A. Faulkner. Training for maximum performance at altitude. In: *Exercise at Altitude,* edited by R. Margaria. Amsterdam: Excerpta Medica, 1967, p. 179–186.

6. Banchero, N. Cardiovascular responses to chronic hypoxia. *Annu. Rev. Physiol.* 49: 465–476, 1987.

7. Barcroft, J. The respiratory function of the blood. In *Part I. Lessons from High Altitudes.* Cambridge: Cambridge University Press, 1925.

8. Barden, H. J. The histochemical relationship of neuromelanin and lipofuscin. *J. Neuropathol. Exp. Neurol.* 28: 419–441, 1969.

9. Bar-Or, O. The Wingate anaerobic test. An update on methodology, reliability and validity. *Sports Med.* 4: 381–394, 1987.

10. Beall, C. M. Growth in a population of Tibetan origin at high altitude. *Am. Hum. Biol.* 8: 31–38, 1981.

11. Beall, C. M. Optimal birthweights in Peruvian populations at high and low altitudes. *Am. J. Phys. Anthropol.* 56: 209–216, 1981.

12. Beall, C. M. A comparison of chest morphology in high altitude Asian and Andean populations. *Hum. Biol.* 54: 145–163, 1982.

13. Beall, C. M. Aging and growth at high altitudes in the Himalayas. In: *The People of South Asia: The Biological Anthropology of India, Pakistan and Nepal,* edited by J. R. Lukacs. New York: Plenum, p. 365–385, 1984.

14. Beall, C. M., P. T. Baker, T. S. Baker, and J. D. Haas. The effects of high altitude on adolescent growth in southern Peruvian Amerindians. *Hum. Biol.* 49: 109–124, 1977.

15. Beall, C. M., and M. C. Goldstein. Hemoglobin concentration of pastoral nomads permanently resident at 4850–5450 meters in Tibet. *Am. J. Phys. Anthropol.* 73: 433–438, 1987.

16. Beall, C. M., and A. B. Reichsman. Hemoglobin levels in a Himalayan high altitude population. *Am. J. Phys. Anthropol.* 63: 302–306, 1984.

17. Bedu, M., N. Fellmann, H. Spielvogel, C. Falgairette, E. van Praagh, and J. Coudert. Force-velocity and 30-s Wingate tests in boys at high and low altitudes. *J. Appl. Physiol.* 70: 1031–1037, 1991.

18. Bender, D. R., B. M. Groves, R. E. McCullough, R. G. McCullough, L. Trad, A. J. Young, A. Cymerman, and J. T. Reeves. Decreased exercise muscle lactate release after high altitude acclimatization. *J. Appl. Physiol.* 1456–1462, 1989.

19. Benjamin, I. J., B. Kroger, and R. S. Williams. Activation of the heat shock transcription factor by hypoxia in mammalian cells. *Proc. Natl. Acad. Sci. USA* 87: 6263–6267, 1990.

20. Bergström, J. Muscle electrolytes in man. *Scand. J. Clin. Lab. Invest.* 14 (suppl. 68): 1962.

20b. Bigland-Ritchie, B., and N. K. Vollestad. Hypoxia and fatigue: How are they related? In *Hypoxia. The Tolerable Limits,* edited by J. R. Sutton, C. S. Houston, and G. Coates. Indianapolis, IN: Benchmark Press, 1988, pp. 315–328.

21. Blatteis, C. M. Oxygen uptake and blood lactate in man during mild exercise at altitude. In: *Environmental Stress: Individual Human Adaptations,* edited by L. J. Folinsbee, J. A. Wagner, J. F. Borgia, B. L. Drinkwater, J. A. Gliner, and J. F. Bedi. New York: Academic, 1978, p. 351–371.

22. Bligh, J., and K. G. Johnson. Glossary of terms for thermal physiology. *J. Appl. Physiol.* 35: 941–961, 1973.

23. Bock, A. V., D. B. Dill, and H. T. Edwards. Lactic acid in the blood of resting man. *J. Clin. Invest.* 11: 775–788, 1932.

24. Bock, W. J. Preadapation and multiple evolutionary pathways. *Evolution* 13: 194–211, 1959.

25. Bock, W. J. The definition and recognition of biological adaptation. *Am. Zool.* 20: 217–227, 1980.

26. Bosco, C., P. Luhtanen, and P. V. Komi. A simple method for measurement of mechanical power in jumping. *Eur. J. Appl. Physiol.* 50: 273–282, 1983.

27. Bouissou, P., F. Peronnet, Y. Guezennec, and J. P. Richalet. In: *Performance et Entraînement en Altitude.* Quebec: Aspects Physiologiques et Physiopathologiques, Décarie, 1987, p. 91–101.

28. Boutellier, U., H. Howald, P. E. di Prampero, D. Giezendanner, and P. Cerretelli. Human muscle adaptations to chronic hypoxia. In: *Hypoxia, Exercise and Altitude,* edited by J. R. Sutton, C. S. Houston, and N. Jones. New York: Liss, 1983, p. 273–281.

29. Boyer, S. J., and F. D. Blume. Weight loss and changes in body

composition at high altitude. *J. Appl. Physiol.: Respir. Environ. Exerc. Physiol.* 57: 1580–1585, 1984.

30. Brody, S. B., S. Lahiri, M. Simpser, E. K. Motoyama, and T. Velasquez. Lung elasticity and airway dynamics in Peruvian natives to high altitude. *J. Appl. Physiol: Respir. Environ. Exerc. Physiol.* 42: 245–251, 1977.

31. Brooks, G. A., G. E. Butterfield, R. R. Wolfe, B. M. Groves, R. S. Mazzeo, J. R. Sutton, E. E. Wolfel, and J. T. Reeves. Increased dependence on blood glucose after acclimatization to 4300 m. *J. Appl. Physiol.* 70: 919–927, 1991.

32. Butterfield, G. E. Elements of energy balance at altitude. In: *Hypoxia: The Adaptations,* edited by J. R. Sutton, G. Coates, and J. E. Remmers. Toronto: Decker, 1990, p. 88–93.

33. Cerretelli, P. Limiting factors to oxygen transport on Mount Everest. *J. Appl. Physiol.* 40: 658–667, 1976.

34. Cerretelli, P. Metabolismo ossidativo ed anaerobico nel soggetto acclimatato all'altitudine. *Min. Aerosp.* 67(suppl.): 11–26, 1976.

35. Cerretelli, P. Gas exchange at high altitude. In: *Pulmonary Gas Exchange,* edited by J. B. West. New York: Academic, 1980, vol. II, p. 97–147.

36. Cerretelli, P., U. Bordoni, R. Debijadij, and F. Saracino. Respiratory and circulatory factors affecting the maximal aerobic power in hypoxia. *Arch. Fisiol.* 67: 344–357, 1967.

37. Cerretelli, P., and P. E. di Prampero. Aerobic and anaerobic metabolism during exercise at altitude. In: *Medicine and Sport Science. Muscle Bioenergetics,* edited by E. Jokl and M. Hebbelinck. Basel: Karger, 1985, p. 1–19.

37b Cerretelli, P., B. Grassi, and B. Kayser. Anaerobic metabolism at altitude: recent developments. In *Hypoxia: Investigaciones bàsicas y clinicas,* edited by F. Léon-Velarde and A. Arregui. Lima, Peru: IFEA-UPCH, 1993, pp. 167–179.

38. Cerretelli, P., and R. Margaria. Maximum oxygen consumption at altitude. *Int. Z. Angew. Physiol. einschl. Arbeitsphysiol.* 18: 460–464, 1961.

38b Cerretelli, P., M. Narici, and B. Kayser. Esperienza italiana nello studio del metabolismo muscolare. *Med. Sport.* 47: 391–400, 1994.

39. Cerretelli, P., A. Veicsteinas, and C. Marconi. Anaerobic metabolism at high altitude: the lactacid mechanism. In: *High Altitude Physiology and Medicine,* edited by W. Brendel and R. A. Zink. New York: Springer, 1982, p. 94–102.

40. Chio, K. S., U. Reiss, B. Fletcher, and A. L. Tappel. Peroxidation of subcellular organelles: formation of lipofuscinlike fluorescent pigments. *Science* 166: 1535–1536, 1969.

41. Christensen, E. H., and W. H. Forbes. Sauerstoffaufnahme und respiratorische Funktionen in grossen Höhen. *Skand. Arch. Physiol.* 76: 88–100, 1937.

42. Claybaugh, J. R., D. P. Brooks, and A. Cymerman. Hormonal control of fluid and electrolyte balance at high altitude in normal subjects. In: *Hypoxia and Mountain Medicine,* edited by J. R. Sutton, G. Coates, and C. S. Houston. Burlington, VT: Queen City, 1992, p. 61–72.

43. Consolazio, C. F., L. O. Matoush, H. L. Johnson, and T. A. Daws. Protein and water balances of young adults during prolonged exposure to high altitude (4300 meters). *Am. J. Clin. Nutr.* 21: 154–161, 1968.

44. Coudert, J., and M. Paz-Zamora. Estudio del consumo de oxigeno en La Paz (3700 mts) sobre un grupo de atletas nativos en la altura. *Annu. Inst. Boliv. Biol. Altura (Bolivia),* 1970.

45. Cruz, J. Mechanics of breathing in high altitude and sea level subjects. *Respir. Physiol.* 17: 146–161, 1973.

46. Cymerman, A., K. B. Pandolf, A. J. Young, and J. T. Maher. Energy expenditure during load carriage at high altitude. *J. Appl. Physiol.: Respir. Environ. Exerc. Physiol.* 51: 14–18, 1981.

47. Cymerman, A., J. T. Reeves, J. R. Sutton, P. B. Rock, B. M. Groves, M. K. Malconian, P. M. Young, P. D. Wagner, and C. S. Houston. Operation Everest II: maximal oxygen uptake at extreme altitude. *J. Appl. Physiol.* 66: 2446–2453, 1989.

48. Das, S. K., and H. Saha. The respiratory metabolism of the Sherpas (hill-people) during climbing: a study of sixty-six cases of normal healthy adults. *Ind. J. Med. Res.* 55: 579–583, 1967.

49. Davies, C. T. M., and E. Rennie. Human power output. *Nature* 217: 770–771, 1968.

50. Dempsey, J. A., P. G. Hanson, and K. S. Henderson. Exercise-induced arterial hypoxaemia in healthy human subjects at sea level. *J. Physiol. (Lond.)* 355: 161–175, 1984.

51. Desplanches, D., H. Hoppeler, M. T. Linossier, C. Denis, H. Claassen, D. Dormois, J. R. Lacour, and A. Geyssant. Effects of training in normobaric hypoxia on human muscle ultrastructure. *Pflugers Arch.* 425: 263–267, 1993.

52. Dill, D. B. *Life, Heat and Altitude.* Boston: Harvard University Press, 1938, p. 168.

53. Dill, D. B., and W. C. Adams. Maximal oxygen uptake at sea level and at 3090 m altitude in high school champion runners. *J. Appl. Physiol.* 30: 854–859, 1971.

54. Dill, D. B., L. G. Myhre, D. K. Brown, K. Burrus, and G. Gehlsen. Work capacity in chronic exposures to altitude. *J. Appl. Physiol.* 23: 555–560, 1967.

55. Dill, D. B., L. G. Myhre, E. E. Phillips, Jr., and D. K. Brown. Work capacity in acute exposure to altitude. *J. Appl. Physiol.* 21: 1168–1176, 1966.

56. Dill, D. B., S. Robinson, W. Balke, and J. L. Newton. Work tolerance age and altitude. *J. Appl. Physiol.* 19: 483–488, 1964.

57. di Prampero, P. E. Energetics of muscular exercise, *Rev. Physiol. Biochem. Pharmacol.* 89: 143–122, 1981.

58. di Prampero, P. E. Metabolic and circulatory limitations to $\dot{V}O_{2max}$ at the whole animal level. *J. Exp. Biol.* 115: 319–331, 1985.

59. di Prampero, P. E., P. Mognoni, and A. Veicsteinas. The effects of hypoxia on maximal anaerobic alactic power in man. In: *High Altitude Physiology and Medicine,* edited by W. Brendel and W. R. Zink. New York: Springer, 1982, p. 88–93.

60. Douglas, C. G., J. S. Haldane, Y. Henderson, and E. C. Schneider. Physiological observations made on Pike's Peak, Colorado, with special reference to adaptation to low barometric pressure. *Phil. Trans. R. Soc. Lond.* B 203: 185–318, 1913.

61. Durig, A., and N. Zuntz. Beiträge zur Physiologie des Menschen in Hochgebirge. *Arch. f. A. u. Ph.* (suppl.): 417–456, 1904.

62. Edwards, H. T. Lactic acid in rest and work at high altitude. *Am. J. Physiol.* 116: 367–375, 1936.

63. Eisenberg, B. R. Quantitative ultrastructure of mammalian skeletal muscle. In: *Handbook of Physiology, Skeletal Muscle,* edited by L. D. Peachy and R. H. Adrian. Bethesda, MD: Am. Physiol. Soc. 1983, p. 73–112.

64. Elsner, R. W., A. Bolstad, and C. Forno. Maximum oxygen consumption of Peruvian Indians native to high altitude. In: *The Physiological Effects of High Altitude,* edited by W. H. Weihe. Oxford: Pergamon, 1964, p. 217–223.

65. Faulkner, J. A., J. Kollias, C. B. Favour, E. R. Buskirk, and B. Balke. Maximum aerobic capacity and running performance at altitude. *J. Appl. Physiol.* 24: 685–691, 1968.

65b. Favier, R., H. Spielvogel, D. Desplanches, G. Ferretti, B. Kayser, and H. Hoppeler. Maximal exercise performance in chronic hypoxia and acute normoxia in high-altitude natives. *J. Appl. Physiol.,* 1995 (in press).

66. Ferretti, G. On maximal oxygen consumption in hypoxic humans. *Experientia* 46: 1188–1194, 1990.

67. Ferretti, G., H. Hauser, and P. E. di Prampero. Muscular exer-

cise at high altitude: VII. Maximal muscular power before and after exposure to chronic hypoxia. *Int. J. Sports Med.* 11(suppl. 1): S31–S34, 1990.

68. Frisancho, A. R. Developmental responses to high altitude hypoxia. *Am. J. Phys. Anthropol.* 32: 401–408, 1970.

69. Frisancho, A. R., and P. T. Baker. Altitude and growth: a study of the patterns of physical growth of a high altitude Peruvian Quechua population. *Am. J. Phys. Anthropol.* 32: 279–292, 1970.

70. Frisancho, A. R., K. Guire, W. Babler, G. Borkan, and A. Way. Nutritional influence on childhood development and genetic control of adolescent growth of Quechuas and Mestizos from the Peruvian lowlands. *Am. J. Phys. Anthropol.* 52: 367–375, 1980.

71. Frisancho, A. R., J. Sanchez, D. Pallardel, and L. Yanez. Adaptive significance of small body size under poor socio-economic conditions in southern Peru. *Am. J. Phys. Anthropol.* 39: 255–262, 1973.

72. Gill, M. B., and L. G. C. E. Pugh. Basal metabolism and respiration in men living at 5800 m (19000 ft). *J. Appl. Physiol.* 19: 949–954, 1964.

73. Gimenez, M., R. J. Sanderson, O. K. Reiss, and N. Banchero. Effects of altitude on myoglobin and mitochondrial protein in canine skeletal muscle. *Respiration* 34: 171–176, 1977.

74. Grassi, B., G. Ferretti, B. Kayser, M. Marzorati, A. Colombini, C. Marconi, and P. Cerretelli. Maximal rate of blood lactate accumulation during exercise at altitude in humans. *J. Appl. Physiol.* 1995 (in press).

74b. Grassi, B., B. E. J. Kayser, T. Binzoni, M. Marzorati, M. Bordini, C. Marconi, and P. Cerretelli. Peak blood lactate concentration during altitude acclimatization and deacclimatization in humans. *Pflugers Arch.* 420: R165(A), 1992.

75. Grassi, B., P. Mognoni, M. Marzorati, A. Colombini, S. Mattiotti, E. Caspani, C. Marconi, and P. Cerretelli. Effect of chronic hypoxia on peak capillary lactate after exhaustive exercises of various durations (Abstract). Exp. Biology 1995. Atlanta, Georgia (USA). April 9–13, 1995 *FASEB J.* 1995 (in press).

76. Green, H. J., J. R. Sutton, A. Cymerman, P. M. Young, and C. S. Houston. Operation Everest II: Adaptations in human skeletal muscle. *J. Appl. Physiol.* 66: 2454–2461, 1989.

76b. Green, H. J., J. R. Sutton, E. E. Wolfel, J. T. Reeves, G. E. Butterfield, and G. A. Brooks. Altitude acclimatization and energy metabolic adaptations in skeletal muscle during exercise. *J. Appl. Physiol.* 73:2701–2708, 1992.

77. Green, H. J., J. Sutton, P. Young, A. Cymerman, and C. S. Houston. Operation Everest II: muscle energetics during maximal exhaustive exercise. *J. Appl. Physiol.* 66: 142–150, 1989.

78. Greska, L. P., J. D. Haas, T. L. Leatherman, H. Spielvogel, M. Paz Zamora, L. Paredes Fernandez, and G. Moreno-Black. Maximal aerobic power in trained youths at high altitude. *Ann. Hum. Biol.* 9: 201–209, 1982.

79. Greska, L. P., H. Spielvogel, and L. Paredes Fernandez. Maximal exercise capacity in adolescent European and Amerindian high-altitude natives. *Am. J. Phys. Anthropol.* 67: 209–216, 1985.

80. Grover, R. F., and J. T. Reeves. Exercise performance of athletes at sea level and 3100 meters altitude. In: *The Effects of Altitude on Physical Performance,* edited by R. F. Goddard. Chicago: Athletic Inst., 1967, p. 80–87.

81. Grover, R. F., J. T. Reeves, E. B. Grover, and J. G. Leathers. Exercise performance of athletes at sea level and 3100 m altitude. *Med. Thorac.* 23: 129–143, 1966.

82. Guilland, J. C., and J. Klepping. Nutritional alterations at high altitude in man. *Eur. J. Appl. Physiol.* 54: 517–523, 1985.

83. Gupta, R., and A. Basu. Variations in body dimensions in relation to altitude among the Sherpas of the eastern Himalayas. *Ann. Hum. Biol.* 8: 145–151, 1981.

84. Hackett, P. H., J. T. Reeves, C. D. Reeves, R. F. Grover, and D. B. Rennie. Control of breathing in Sherpas at low and high altitude. *J. Appl. Physiol.: Respir. Environ. Exerc. Physiol.* 49: 374–379, 1980.

85. Hannon, J. P. Nutrition at high altitude. In: *Environmental Physiology: Aging, Heat and Altitude,* edited by S. M. Horvath and M. K. Yousef. Amsterdam: Elsevier, 1980, p. 309–327.

86. Hannon, J. P., K. S. K. Chinn, and J. L. Shields. Effects of acute high-altitude exposure on body fluids. *Federation Proc.* 28: 1178–1184, 1969.

87. Hannon, J. P., and D. M. Sudman. Basal metabolic and cardiovascular function of women during altitude acclimatization. *J. Appl. Physiol.* 34: 471–477, 1973.

88. Hansen, J. E., G. P. Stelter, and J. A. Vogel. Arterial pyruvate, lactate, pH, and PCO_2 during work at sea level and high altitude. *J. Appl. Physiol.* 23: 523–530, 1967.

89. Hansen, J. E., J. A. Vogel, G. P. Stelter, and C. F. Consolazio. Oxygen uptake in man during exhaustive work at sea level and high altitude. *J. Appl. Physiol.* 23: 511–522, 1967.

90. Harris, J. A., and F. G. Benedict. *A Biometric Study of Basal Metabolism in Man,* Washington, DC: Carnegie Inst., 1919, p. 279.

91. Harrison, G. A., C. F. Kuchemann, M. A. S. Moore, A. J. Boyce, T. Baju, A. E. Mourant, M. J. Godber, B. G. Glasgow, A. C. Kopec, D. Tillis, and A. J. Clegg. The effects of altitudinal variation in Ethiopian populations. *Phil. Trans. R. Soc. Lond. B.* 256: 147–182, 1969.

92. Hickson, R. C. Skeletal muscle cytochrome c and myoglobin, endurance, and frequency of training. *J. Appl. Physiol.: Respir. Environ. Exerc. Physiol.* 51: 746–749, 1981.

93. Hochachka, P. W. The lactate paradox: analysis of underlying mechanisms. *Ann. Sports Med.* 4: 184–188, 1988.

94. Hochachka, P. W. Principles of physiological and biochemical adaptation. High-altitude man as a case study. In: *Physiological Adaptation in Vertebrates,* edited by S. E. Wood, R. E. Weber, A. R. Hargens, and R. W. Millard. New York: Dekker, 1992, p. 21–35.

95. Hochachka, P. W., G. O. Matheson, W. S. Parkhouse, J. Sumar-Kalinowski, C. Stanley, C. Monge, D. C. McKenzie, J. Merkt, P. S. F. Man, R. Jones, and P. S. Allen. Path of oxygen from atmosphere to mitochondria in andean natives: adaptable versus constrained components. In: *Hypoxia: The Adaptations,* edited by J. R. Sutton, G. Coates, and J. E. Remmers. Toronto: Decker, 1990, p. 72–87.

96. Hochachka, P. W., C. Stanley, G. O. Matheson, D. C. McKenzie, P. S. Allen, and W. S. Parkhouse. Metabolic and work efficiencies during exercise in Andean natives. *J. Appl. Physiol.* 70: 1720–1730, 1991.

97. Hochachka, P. W., C. Stanley, J. Merkt, and J. Sumar-Kalinowski. Metabolic meaning of elevated levels of oxidative enzymes in high altitude adapted animals: an interpretive hypothesis. *Respir. Physiol.* 52: 303–313, 1983.

98. Holloszy, J. O., L. B. Oscai, I. J. Don, and P. A. Mole, Mitochondrial citric acid cycle and related enzymes: adaptive response to exercise. *Biochem. Biophys. Res. Commun.* 40: 1368–1373, 1970.

99. Hoppeler, H., H. Howald, K. E. Conley, S. L. Lindstedt, H. Claassen, P. Vock, and E. R. Weibel. Endurance training in humans: aerobic capacity and structure of skeletal muscle. *J. Appl. Physiol.* 59: 320–327, 1985.

100. Hoppeler, H., E. Kleinert, C. Schlegel, H. Claassen, H. Howald, and P. Cerretelli. Muscular exercise at high altitude: II. Morphological adaptation of skeletal muscle to chronic hypoxia. *Int. J. Sports Med.* 11(suppl.): S3–S9, 1990.

101. Hoppeler, H., P. Luethi, H. Claassen, E. R. Weibel, and H. Howald. The ultrastructure of the normal human skeletal muscle. A morphometric analysis on untrained men, women, and well-trained orienteers. *Pflugers Arch.* 334: 217–232, 1973.

102. Howald, H., D. Pette, J.-A. Simoneau, A. Uber, H. Hoppeler, and P. Cerretelli. Muscular exercise at high altitude: III. Effects of chronic hypoxia on muscle enzyme activity. *Int. J. Sports Med.* 11(suppl. 1): S10–S14, 1990.

103. Hoyt, R. W., M. J. Durkot, G. H. Kamimori, D. A. Schoeller, and A. Cymerman. Chronic altitude exposure (4300 m) decreases intracellular and total body water in humans [Abstract]. In: *Hypoxia and Mountain Medicine,* edited by J. R. Sutton, G. Coates, and C. S. Houston. Burlington, VT: Queen City, 1992, p. 306.

104. Hudlicka, O. Growth of capillaries in skeletal and cardiac muscle. *Circ. Res.* 50: 451–461, 1982.

105. Hurtado, A. Respiratory adaptation in the Indian natives of the Peruvian Andes: Studies at high altitude. *Am. J. Phys. Anthropol.* 17: 137–165, 1932.

106. Hurtado, A. Animals in high altitudes: resident man. In: *Handbook of Physiology. Adaptation to the Environment,* edited by D. B. Dill and E. F. Adolph. Bethesda, MD: Am. Physiol. Soc., 1964, p. 843–860.

107. Hurtado, A., and H. Aste-Salazar. Arterial blood gases and acid-base balance at sea level and at high altitudes. *J. Appl. Physiol.* 1: 304–325, 1948.

108. Itoh, K., T. Moritani, K. Ishida, C. Hirofuji, S. Taguchi, and M. Itoh. Hypoxia-induced fibre type transformation in rat hindlimb muscles—histochemical and electromechanical changes. *Eur. J. Appl. Physiol.* 60: 331–336, 1990.

109. Jansson, E., C. Sylven, and E. Nordevang. Myoglobin in the quadriceps femoris muscle of competitive cyclists and untrained men. *Acta Physiol. Scand.* 114: 627–629, 1982.

110. Kaijser, L., C. J. Sunberg, O. Eiken, A. Nygren, M. Esbjoernsson, C. Sylven, and E. Jansson. Muscle oxidative capacity and work performance after training under local leg ischemia. *J. Appl. Physiol.* 69: 785–787, 1990.

111. Karas, R. H., C. R. Taylor, J. H. Jones, S. L. Lindstedt, R. B. Reeves, and E. R. Weibel. Adaptive variation in the mammalian respiratory system in relation to energetic demand: VII. Flow of oxygen across the pulmonary gas exchanger. *Respir. Physiol.* 69: 101–116, 1987.

112. Katz, L. A., J. A. Swain, M. A. Portman, and R. S. Balaban. Relation between phosphate metabolites and oxygen consumption of heart in vivo. *Am. J. Physiol.* 256 (*Heart Circ. Physiol.* 27): H265–H274, 1989.

113. Kayser, B., K. Acheson, J. Decombaz, E. Fern, and P. Cerretelli. Protein absorption and energy digestibility at high altitude. *J. Appl. Physiol.* 73: 2425–2431, 1992.

114. Kayser, B., G. Ferretti, B. Grassi, T. Binzoni, and P. Cerretelli. Maximal lactic capacity at altitude: effect of bicarbonate loading. *J. Appl. Physiol.* 75: 1070–1074, 1993.

115. Kayser, B., H. Hoppeler, H. Claassen, and P. Cerretelli. Muscle structure and performance capacity of Himalayan Sherpas. *J. Appl. Physiol.* 70: 1938–1942, 1991.

115b. Kayser, B., C. Marconi, T. Amatya, B. Basnyat, A. Colombini, B. Broers, P. Cerretelli. The metabolic and ventilatory response to exercise in Tibetans born at low altitude. *Respir. Physiol.* 98: 15–26, 1994.

115c. Kayser, B., M. Narici, T. Binzoni, B. Grassi, and P. Cerretelli. Fatigue and exhaustion in chronic hypobaric hypoxia: influence of exercising muscle mass. *J. Appl. Physiol.* 76: 634–640, 1994.

115d. Kayser, B., M. Narici, S. Milesi, B. Grassi, and P. Cerretelli. Body composition and alactic anaerobic performance during a one month stay at altitude. *Int. J. Sports Med.* 14: 244–247, 1993.

116. Knuttgen, H. G. and B. Saltin. Oxygen uptake, muscle high energy phosphates, and lactate in exercise under acute hypoxic conditions in man. *Acta Physiol. Scand.* 87: 368–376, 1973.

117. Krzywicki, H. J., C. F. Consolazio, H. L. Johnson, W. C. Nielsen, Jr., and P. A. Barnhart. Water metabolism in humans during acute high-altitude exposure (4300 m). *J. Appl. Physiol.* 30: 806–809, 1971.

118. Lahiri, S. Respiratory control in Andean and Himalayan high altitude natives. In: *High Altitude and Man,* edited by J. B. West and S. Lahiri. Baltimore, MD: Williams and Wilkins, 1984, p. 147–162.

119. Lahiri, S., and J. S. Milledge. Sherpa physiology. *Nature* 207: 610–612, 1965.

120. Lahiri, S., J. S. Milledge, H. P. Chattopadyay, A. K. Bhattacharyya, and A. K. Sinha. Respiration and heart rate of Sherpa highlanders during exercise. *J. Appl. Physiol.* 23: 545–554, 1967.

121. Levine, B. D., R. C. Roach, and C. S. Houston. Work and training at altitude In: *Hypoxia and Mountain Medicine,* edited by J. R. Sutton, G. Coates, and C. S. Houston. Burlington, VT: Queen City, 1992, p. 192–201.

122. Lexell, J., and C. C. Taylor. Variability in muscle fibre areas in whole human quadriceps muscle: how to reduce sampling errors in biopsy techniques. *Clin. Physiol.* 9: 333–343, 1989.

123. Lexell, J., and C. C. Taylor. A morphometrical comparison of right and left whole human vastus lateralis muscle—how to reduce sampling errors in biopsy techinques. *Clin. Physiol.* 11: 271–276, 1991.

124. Loewy, A. Die Wirkung ermüdender Muskerlarbeit auf den respiratorischen Stoffwechsel. *Arch. Ges. Physiol. Menschen Tiere* 49: 405–422, 1891.

125. Loewy, A. Ueber den Einfluss der verdünnten Luft und des Höhenklimas auf den Menschen. *Pflügers. Arch.* 66: 477–538, 1897.

126. Luethi, J. M., H. Howald, H. Claassen, K. Roesler, P. Vock, and H. Hoppeler. Structural changes in skeletal muscle tissue with heavy-resistance exercise. *Int. J. Sports Med.* 7: 123–127, 1986.

127. MacDougall, J. D., H. J. Green, J. R. Sutton, G. Coates, A. Cymerman, P. Young, and C. S. Houston. Operation Everest-II—structural adaptations in skeletal muscle in response to extreme simulated altitude. *Acta Physiol. Scand.* 142: 421–427, 1991.

128. Margaria, R. Die Arbeitsfähigkeit des Menschen bei vermindertem Luftdruck. *Arbeitphysiol.* 2: 261–272, 1929.

129. Margaria, R., H. T. Edwards, and D. B. Dill. The possible mechanism of contracting and paying the oxygen debt and the role of lactic acid in muscular contraction. *Am. J. Physiol.* 106: 689–714, 1933.

130. Martinelli, M., R. Winterhalder, P. Cerretelli, H. Howald, and H. Hoppeler. Muscle lipofuscin content and satellite cell volume is increased after high altitude exposure in humans. *Experientia* 46: 672–676, 1990.

130b. Marzorati, M., C. Marconi, B. Grassi, A. Colombini, M. Conti, E. Caspani, and P. Cerretelli. VO_{2max} in chronic hypoxia: greater reduction in athletes than in sedentary subjects (Abstract) Exp. Biology 1995, Atlanta, Georgia; April 9–13, 1995. *FASEB J.* 1995 (in press).

131. Matheson, G. O., P. S. Allen, D. C. Ellinger, C. C. Hanstock, D. Gheorghiu, D. C. McKenzie, C. Stanley, W. S. Parkhouse, and P. W. Hochachka. Skeletal muscle metabolism and work capacity: a ^{31}P-NMR study of Andean natives and lowlanders. *J. Appl. Physiol.* 70: 1963–1976, 1991.

132. Mathieu, O., L. M. Cruz-Orive, H. Hoppeler, and E. R. Weibel. Estimating length density and quantifying anisotropy in skeletal muscle capillaries. *J. Microsc.* 131: 131–146, 1983.

133. Mathieu-Costello, O. Muscle capillary tortuosity in high altitude mice depends on sarcomere length. *Respir. Physiol.* 76: 289–302, 1989.

134. Mathieu-Costello, O., H. Hoppeler, and E. R. Weibel. Capillary tortuosity in skeletal muscles of mammals depends on muscle contraction. *J. Appl. Physiol.* 66: 1436–1442, 1989.

135. Mazzeo, R. S., P. R. Bender, G. A. Brooks, G. E. Butterfield, B. M. Groves, J. R. Sutton, E. E. Wolfel, and J. T. Reeves. Arterial catecholamine response during exercise with acute and chronic high-altitude exposure. *Am. J. Physiol.* 261 (*Endocrinol. Metab.* 24): E419–E424, 1991.

135b. Mazzeo, R. S., G. A. Brooks, G. E. Butterfield, A. Cymerman, A. C. Roberts, M. Selland, E. E. Wolfel, and J. T. Reeves. β-adrenergic blockade does not prevent the lactate response to exercise after acclimatization to high altitude. *J. Appl. Physiol.* 76: 615–616, 1994.

136. Mizuno, M., C. Juel, T. Bro-Rasmussen, E. Mygind, E. Schibye, B. Rasmussen, and B. Saltin. Limb skeletal muscle adaptation in athletes after training at altitude. *J. Appl. Physiol.* 68: 496–502, 1990.

137. Monge, C., and F. Leon-Velarde. Physiological adaptation to high altitude—oxygen transport in mammals and birds. *Physiol. Rev.* 71: 1135–1172, 1991.

138. Mordes, J. P., F. D. Blume, S. Boyer, M. R. Zheng, and L. E. Braverman. High-altitude pituitary–thyroid dysfunction on Mount Everest. *N. Engl. J. Med.* 308: 1135–1138, 1983.

139. Morpurgo, G., A. Arese, A. Bosia, G. P. Pescarmona, M. Luzzana, G. Modiano, and S. Krishna Ranjit. Sherpas living permanently at high altitude: a new pattern of adaptation. *Proc. Natl. Acad. Sci. U.S.A.* 73: 747–751, 1976.

140. Morrison, W. L., J. N. A. Gibson, C. Scrimgeour, and M. J. Rennie. Muscle wasting in emphysema. *Clin. Sci.* 75: 415–420, 1988.

141. Mosso, A. *Life of Man in the High Alps.* London: Fisher Unwin, 1898.

142. Nair, C. S., M. S. Malhotra, and P. M. Gopinarth. Effect of altitude and cold acclimatization on the basal metabolism in man. *Aerosp. Med.* 42: 1056–1059, 1971.

143. Narici, M. V. and B. Kayser. Hypertrophic response of human skeletal muscle to strength training in hypoxia and normoxia. *Eur. J. Appl. Physiol.* 70: 1995 (in press).

144. Oelz, O., H. Howald, P. di Prampero, H. Hoppeler, H. Claassen, R. Jenni, A. Buehlmann, G. Ferretti, J.-C. Brueckner, A. Veicsteinas, M. Gussoni, and P. Cerretelli. Physiological profile of world-class high-altitude climbers. *J. Appl. Physiol.* 60: 1734–1742, 1986.

145. Pattengale, P. K., and J. O. Holloszy. Augmentation of skeletal muscle myoglobin by a program of treadmill running. *Am. J. Physiol.* 213: 783–785, 1967.

146. Pawson, I. G. Growth and development in high altitude populations: a review of Ethiopian, Peruvian and Nepalese studies. *Proc. R. Soc. Lond. B* 194: 83–98, 1976.

147. Peñaloza, D., F. Sime, N. Banchero, R. Gamboa, J. Cruz, and E. Marticorena. Pulmonary hypertension in healthy men born and living at high altitudes. *Am. J. Cardiol.* 11: 150–157, 1963.

148. Picòn-Reàtegui, E. Basal metabolic rate and body composition at high altitudes. *J. Appl. Physiol.* 16: 431–434, 1961.

149. Picòn-Reàtegui, E., R. Lozano, and J. Valdivieso. Body composition at sea level and high altitudes. *J. Appl. Physiol.* 16: 589–592, 1961.

150. Piiper, J., P. Cerretelli, F. Cuttica, and F. Mangili. Energetic metabolism and circulation in dogs exercising in hypoxia. *J. Appl. Physiol.* 21: 1143–1149, 1966.

151. Poole, D. C., and O. Mathieu-Costello. Skeletal muscle capillary geometry: adaptation to chronic hypoxia. *Respir. Physiol.* 77: 21–30, 1989.

152. Preedy, V. S., D. M. Smith, and P. H. Sugden. The effects of 6 hours hypoxia on protein synthesis in rat tissues in vivo and in vitro. *Biochem. J.* 228: 179–185, 1985.

153. Preedy, V. R., and P. H. Sugden. The effects of fasting or hypoxia on rates of protein synthesis in vivo in subcellular fractions of rat heart and gastrocnemius muscle. *Biochem. J.* 257: 519–527, 1989.

154. Pugh, L. G. C. E. Muscular exercise on Mount Everest. *J. Physiol. Lond.* 141: 233–261, 1958.

155. Pugh, L. G. C. E. Physiological and medical aspects of the Himalayan scientific and mountaineering expedition, 1960–61. *Br. Med. J.* 2: 621–627, 1962.

156. Pugh, L. G. C. E., M. B. Gill, S. Lahiri, J. S. Milledge, M. P. Ward, and J. B. West. Muscular exercise at great altitude. *J. Appl. Physiol.* 19: 431–440, 1964.

157. Rahkila, P., and H. Rusko. Effect of high altitude training on muscle enzyme activities and physical performance characteristics of cross-country skiers. In: *International Series on Sports Sciences. Exercise and Sport Biology,* edited by P. V. Komi. Champaign, IL: Human Kinetics, 1982, vol. 12, p. 143–151.

158. Rai, R. M., M. S. Malhotra, G. P. Dimri, and T. Sampathkumar. Utilization of different quantities of fat at high altitude. *Am. J. Clin. Nutr.* 28: 242–245, 1975.

159. Rastogi, G. K., M. S. Malhotra, M. C. Srivastava, R. C. Sawhney, G. L. Dua, K. Sridharan, R. S. Hoon, and I. Singh. Study of the pituitary–thyroid functions at high altitude in man. *J. Clin. Endocrinol. Metab.* 44: 447–452, 1977.

160. Raynaud, J., and J. Durand. Oxygen deficit and debt in submaximal exercise at sea level and high altitude. In: *High Altitude Physiology and Medicine,* edited by W. Brendel and R. A. Zink. New York: Springer, 1982, p. 103–106.

161. Remmers, J. E., and J. C. Mithoefer. The carbon monoxide diffusing capacity in permanent residents at high altitudes. *Respir. Physiol.* 6: 233–244, 1969.

162. Rennie, M. J., P. Babij, J. R. Sutton, J. J. Tonkins, W. W. Read, C. Ford, and D. Halliday. Effects of acute hypoxia on forearm leucine metabolism. *Prog. Clin. Biol. Res.* 136: 317–323, 1983.

163. Reynafarje, B. Myoglobin content and enzymatic activity of muscle and altitude adaptation. *J. Appl. Physiol.* 17: 301–305, 1962.

164. Richalet, J.-P., C. Delavier, J.-L. Le Trong, C. Dubray, and A. Keromes. Désensibilisation des béta recepteurs lymphocytaires humains en hypoxie d'altitude (4350 m). *Arch. Int. Physiol. Biochim.* 96: A468, 1988.

165. Richalet, J.-P., P. Larmignat, C. Rathat, A. Keromes, P. Baud, and F. Lhoste. Decreased cardiac response to isoproterenol infusion in acute and chronic hypoxia. *J. Appl. Physiol.* 65: 1957–1961, 1988.

166. Robinson, S., H. T. Edwards, and D. B. Dill. New records in human power. *Science* 85: 409–410, 1937.

167. Rose, M., C. S. Houston, C. S. Fulco, G. Coates, J. R. Sutton, and A. Cymerman. Operation Everest II: nutrition and body composition. *J. Appl. Physiol.* 65: 2545–2551, 1988.

167b. Rosser, B. W. C., and P. W. Hochachka. Metabolic capacity of muscle fibers from high-altitude natives. *Eur. J. Appl. Physiol.* 67: 513–517, 1993.

168. Saha, H. Studies on the oxygen uptake and efficiency of climbing of Tensing Norgay and other subjects. *Q. J. Exp. Physiol.* 43: 295–298, 1958.

169. Saltin, B., and P. D. Gollnick. Skeletal muscle adaptability: significance for metabolism and performance. In: *Handbook of Physiology. Skeletal Muscle,* edited by L. D. Peachy, R. H. Adrian, and S. R. Geiger. Baltimore, MD: Williams and Wilkins, 1983, p. 555–631.

170. Saltin, B., R. F. Grover, C. G. Blomqvist, L. H. Hartley, and

R. L. Johnson, Jr. Maximal oxygen uptake and cardiac ouput after 2 weeks at 4300 m. *J. Appl. Physiol.* 25: 400–409, 1968.

171. Saltin, B., E. Nygaard, and B. Rasmussen. Skeletal muscle adaptation in man following prolonged exposure to high altitude [Abstract]. *Acta Physiol. Scand.* 109: 31A, 1980.

172. Samaja, M., A. Veicsteinas, and P. Cerretelli. Oxygen affinity of blood in altitude Sherpas. *J. Appl. Physiol.: Respir. Environ. Exerc. Physiol.* 47: 337–341, 1979.

172b. Samaja, M., C. Mariani, A. Prestini, and P. Cerretelli. Arterial blood acid-base balance and gas content at altitude: respiratory alkalosis as buffer of blood O_2 transport (Abstract). Exp. Biology 1995, Atlanta, Georgia, April 9–13, 1995. *FASEB J* 1995 (in press).

173. Schumburg, and N. Zuntz. Zur Kenntniss der Einwirkungen des Hochgebirges auf den menschlichen Organismus. *Pflugers Arch.* 63: 461–493, 1896.

174. Shephard, R. J. A non-linear solution of the oxygen conductance equation: applications to performances at sea level and at altitude of 7350 ft. *Int. Zeitschr. Angew. Physiol.* 27: 212–225, 1969.

175. Sloan, A. W., and M. Masali. Anthropometry of Sherpamen. *Ann. Hum. Biol.* 5: 453–458, 1978.

176. Sridharan, K., M. S. Malhotra, T. N. Upadhayay, S. K. Grover, and G. L. Dua. Changes in gastro-intestinal function in humans at an altitude of 3500 m. *Eur. J. Appl. Physiol.* 50: 145–154, 1982.

177. Sun, S. F., T. S. Droma, J. G. Zhang, J. X. Tao, S. Y. Huang, R. G. McCullough, R. E. McCullough, C. S. Reeves, J. T. Reeves, and L. G. Moore. Greater maximal O_2 uptakes and vital capacities in Tibetan than Han residents of Lhasa. *Respir. Physiol.* 79: 151–162, 1990.

178. Sutton, J. R., and G. J. F. Heigenhauser. Lactate at altitude. In: *Hypoxia: The Adaptations,* edited by J. R. Sutton, G. Coates, and J. E. Remmers. Toronto: Dekker, 1990, p. 94–97.

179. Sutton, J. R., J. T. Reeves, P. D. Wagner, B. M. Groves, A. Cymerman, M. K. Malconian, P. B. Rock, P. M. Young, S. D. Walter, and C. S. Houston. Operation Everest II: oxygen transport during exercise at extreme simulated altitude. *J. Appl. Physiol.* 64: 1309–1321, 1988.

180. Svedenhag, J., J. Henriksson, and C. Sylven. Dissociation of training effects on skeletal muscle mitochondrial enzymes and myoglobin in man. *Acta Physiol. Scand.* 117: 213–218, 1983.

181. Taylor, C. R., R. H. Karas, E. R. Weibel, and H. Hoppeler. Adaptive variation in the mammalian respiratory system in relation to energetic demand: II. Reaching the limits to oxygen flow. *Respir. Physiol.* 69: 7–26, 1987.

182. Terrados, N., E. Jansson, C. Sylven, and L. Kaijser. Is hypoxia a stimulus for synthesis of oxidative enzymes and myoglobin? *J. Appl. Physiol.* 68: 2369–2372, 1990.

183. Terrados, N., J. Melichna, C. Sylven, E. Jansson, and L. Kaijser. Effects of training at simulated altitude on performance and muscle metabolic capacity in competitive road cyclists. *Eur. J. Appl. Physiol.* 57: 203–209, 1988.

184. Valdivia, E. Total capillary bed in striated muscle of guinea pigs native to the Peruvian mountains. *Am. J. Physiol.* 194: 585–589, 1958.

185. Viault, F. G. Sur l'augmentation considerable du nombre des globules rouges dans le sang chez les habitants des hauts plateaux de l'Amerique du sud. *C. R. Acad. Sci. III* 111: 917–918, 1890.

186. Vogel, J. A., L. H. Hartley, and J. Cruz. Cardiac output during exercise in altitude natives at sea level and high altitude. *J. Appl. Physiol.* 36: 173–176, 1974.

186b. Wagner, P. D. Algebraic analysis of the determinants of $V_{O_{2max}}$. *Respir. Physiol.* 93: 221–237, 1993.

187. Wagner, P. D., H. Hoppeler, and B. Saltin. Determinants of maximal oxygen uptake. In: *The Lung: Scientific Foundations,* edited by R. G. Crystal and J. B. West. New York: Raven, 1991, p. 1585–1593.

188. Wahlund, E., P. Weng, N.-H. Areskog, and B. Saltin. Swedish Mount Everest Expedition 1987. Rapport fran den svenska Mount Everest-expeditionen. Svensk expeditionstradition att bevara—en anledning att starta projektet. *Lakartidningen* 85: 3161–3169, 1988.

189. Weibel, E. R. *Stereological Methods. Practical Methods for Biological Morphometry.* London: Academic, 1979, vol. I chap. 4, 6.

190. Weitz, C. A. The Effects of Aging and Habitual Activity Pattern on Exercise Performance Among a High Altitude Himalayan Population. University Park: Pennsylvania State Univ., 1973. Ph.D. Thesis.

191. West, J. B. Lactate during exercise at extreme altitude. *Federation Proc.* 45: 2953–2957, 1986.

192. West, J. B., S. J. Boyer, D. J. Graber, P. H. Hackett, K. H. Maret, J. S. Milledge, R. M. Peters, Jr., C. J. Pizzo. M. Samaja, F. H. Sarnquist, R. B. Schoene, and R. M. Winslow. Maximal exercise at extreme altitudes on Mount Everest. *J. Appl. Physiol.: Respir. Environ. Exerc. Physiol.* 55: 688–698, 1983.

193. Westerterp, K., B. Kayser, F. Brouns, J.-P. Herry, and W. Saris. Energy expenditure climbing Mt. Everest [Abstract]. *Int. J. Sports Med.* 13: 87, 1992.

193b. Westerterp, K. R., B. Kayser, L. Wouters, J.-L. Le Trong, and J.-P. Richalet. Energy balance at high altitude of 6,542 m. *J. Appl. Physiol.* 77: 862–866, 1994.

194. Wilson, R. E., and R. M. Sutherland. Enhanced synthesis of specific proteins, RNA, and DNA caused by hypoxia and reoxygenation. *Int. J. Radiat. Oncol. Biol. Phys.* 16: 957–961, 1989.

195. Winslow, R. M., K. W. Chapman, C. G. Gibson, M. Samaja, C. C. Monge, E. Goldwasser, M. Sherpa, F. D. Blume, and R. Santolaya. Different hematologic responses to hypoxia in Sherpas and Quechua Indians. *J. Appl. Physiol.* 66: 1561–1569, 1989.

196. Winslow, R. M., C. C. Monge, N. J. Statham, C. G. Gibson, S. Charache, J. Wittembury, O. Moran, and R. L. Berger. Variability of oxygen affinity of blood human subjects native to high altitude. *J. Appl. Physiol.: Respir. Environ. Exerc. Physiol.* 51: 1411–1416, 1981.

197. Yip, R., N. J. Binkin, and F. L. Trowbridge. Altitude and childhood growth. *J. Pediatr.* 113: 486–489, 1988.

198. Young, A. J., W. J. Evans, A. Cymerman, K. B. Pandolf, J. J. Knapik, and J. T. Maher. Sparing effect of chronic high-altitude exposure on muscle glycogen utilization. *J. Appl. Physiol.: Respir. Environ. Exerc. Physiol.* 52: 857–862, 1982.

199. Young, A. J., W. J. Evans, E. C. Fisher, R. L. Sharp, D. L. Costill, and J. T. Maher. Skeletal muscle metabolism of sea-level natives following short-term high-altitude residence. *Eur. J. Appl. Physiol.* 52: 463–466, 1984.

200. Young, A. J., P. M. Young, R. E. McCullough, L. G. Moore, A. Cymerman, and J. T. Reeves, Effect of beta-adrenergic blockage on plasma lactate concentration during exercise at high altitude. *Eur. J. Appl. Physiol.* 63: 315–322, 1991.

201. Zimmermann, L. H., R. A. Levine, and H. W. Farber. Hypoxia induces a specific set of stress proteins in cultured endothelial cells. *J. Clin. Invest.* 87: 908–914, 1991.

202. Zumstein, A., O. Mathieu, H. Howald, and H. Hoppeler. Morphometric analysis of the capillary supply in skeletal muscles of trained and untrained subjects—its limitations in muscle biopsies. *Pflugers Arch.* 397: 277–283, 1983.

51. Peripheral chemoreceptors and their sensory neurons in chronic states of hypo- and hyperoxygenation

SUKHAMAY LAHIRI

Department of Physiology, University of Pennsylvania, School of Medicine, Philadelphia, Pennsylvania

CHAPTER CONTENTS

RESPONSES AND ADAPTATIONS of peripheral chemoreceptors to chronic hypoxia and hyperoxia are the primary focus of this chapter. Some acute responses are included to the extent that they provide relevant background material. The underlying theme here is that the

family of heme pigments (chromophores) which bind and/or react with O_2 characterizes the chemoreceptor cell function. There is a large number of these pigments in many cells (149), and they are potentially present in carotid body cells. Synthesis and metabolism of these heme pigments are expected to be affected by the oxygen environment. Throughout this chapter hints will appear as to the importance of these heme pigments in O_2 chemoreception and in adaptive responses to O_2 pressure changes, not only in the peripheral chemoreceptors but in some other cells as well. Searches for common mechanisms are in order. Some of the well-known intracellular pigments collectively contribute to metabolic hypothesis of O_2 chemoreception, and the putative membrane pigments form the basis of the membrane hypothesis of O_2 chemoreception. However, cytosolic heme enzyme, for example, guanylate cyclase, is a potential player in the inhibitory control of the chemoreception process. In addition, the chapter will refer to the sensory neurons because of the hypothesis that these cells would show the effects of depolarization initiated either at the innervation of the target cells or directly by hypoxia. The importance of the bidirectional trophic effects between the glomus and sensory ganglion cells is emphasized.

Not unexpectedly, there are controversies in some areas, and I will project my own assessment to stimulate future research. Needless to say, a balanced view of the more established areas will be attempted. However, the view may not be shared by all. The literature reviewed in other chapters in this *Handbook* will not be fully recited here. Instead, those chapters will be cross-referenced. Finally, a complete account of the peripheral chemoreceptors is beyond the scope of this chapter.

OVERVIEW: PERIPHERAL CHEMORECEPTORS AND SENSORY NEURONS

Carotid bodies are small organs located near the carotid sinus at the bifurcation of external and internal carotid arteries, where the baroreceptors are also located. The blood supply usually consists of a short artery or two directly arising from the large artery, in parallel with several other arteries through which blood flows rapidly so that samples of arterial blood from heartbeat to heartbeat are rapidly available to the carotid body. DeCastro and Rubio (32) emphasized the special thin-walled structure of the carotid body artery and its collapsibility at low perfusion pressure, comparable to the carotid sinus. This structure may have functional significance in terms of blood flow and chemosensory response (see below, under *Carotid Body Blood Volume and Flow*), which are innervated by baroreceptors.

The artery gives rise to numerous small blood vessels and large fenestrated capillaries which surround the islands of carotid body cells, and several glomus cells (type I) are tightly enclosed by sustentacular cells (type II). Most of the glomus cells are innervated, and they are the only cells known to contain transmitter granules. The major innervation is derived from the bipolar sensory cells in the petrosal ganglion, and a very minor efferent innervation comes from the pre- and postganglionic cells of the cervical sympathetic nervous system. The latter is mostly directed to the carotid body blood vessels. Bare nerve endings between the clusters of cells are sparsely found (112) and may secrete neuropeptides (83) controlling integrated carotid body function.

Aortic bodies, which are a smaller collection of glomoids, are spread over wide areas of blood vessels near the heart (see ref. 67). The blood supply routes are very different from those of the carotid bodies. Slender and long blood vessels are the common features, which seem to restrain oxygen delivery to the target cells (82; see also refs. 1, 43, 49 for reviews). Accordingly, increased hematocrit during chronic hypoxia may make important differences between aortic and carotid body function (see refs. 43, 100 for reviews). Sensory innervation is derived from the bipolar cells in the nodose ganglion.

There are even smaller specks of glomoid tissues in the lungs (103), thorax, abdomen, and neck (see ref. 112 for review). The responses of these structures to chronic hypoxia are not known but are expected to mimic those of carotid chemoreceptors.

Oxygen sensing is manifested in two forms: acute responses are instantaneous and chronic cellular changes occur over days and weeks. The bases for the two responses are not clearly understood, and it is not known how the two responses are linked. The consensus model of the expression of O_2 chemoreception is that upon initiation vesicular neurotransmitters are released which generate neural discharge. However, nonvesicular neurotransmitters, like nitric oxide and carbon monoxide (142, 164) may play important roles (see below, under *Oxygen and Chemoreceptive Pigments*). Practically all of the observations on chronic

effects come from carotid body studies. Although there are many differences between the responses of the carotid and aortic body chemoreceptors (see 43, 84 for reviews) and their systemic effects (see 49 for review), the basic structural and functional elements are the same. Accordingly, the carotid body can be viewed as a prototype of the peripheral chemoreceptors in terms of adaptation to chronic hypoxemia. However, in terms of adaptation to chronic hypotension and anemia, aortic bodies are expected to differ from carotid bodies because aortic bodies are stimulated by these systemic events (see 100 for review); but there is practically no observation on aortic bodies in chronic states of hypoxia, anemia, polycythemia, hypotension, or hypertension or in any combination of these systemic states.

The sensory neurons which innervate the glomus cells are depolarized as a consequence of their axons at the target cells being depolarized during chemoreception. Accordingly, these sensory cells are expected to show the effects of chronic depolarization in chronic hypoxia. To distinguish between this effect and that due to hypoxia alone, it is necessary to generate data on the sensory ganglion cells without innervation of the target glomus cells. The processes of the sensory cells also synapse with the neurons in the brain stem. The two axonal processes of the same cell are expected to manifest similar constitutive structures and functions—for example, distribution of nitric oxide synthase (NOS), tyrosine hydroxylase, substance P, vasoactive intestinal peptide (VIP), etc. may be similar at the two ends—but little is known at this time.

The responses of the carotid and aortic bodies to oxygen pressure changes are by themselves of little physiological benefit to the organism unless the appropriate reflexes and feedback mechanisms are activated and the functions of the target organs are influenced. These oxygen sensors are the initiating components of the reflex arcs, controlling many functions of the autonomic nervous system. The target organs of the system are numerous, the more well known of which are respiratory muscles. Control of airway and vascular smooth muscles and cardiac muscles is less known. Chemoreflex control of exocrine and endocrine organs and of all other organs under autonomic influence are suspected, but very little is known. Reviews of some of these functions are included in other chapters of this section.

Reviews of adaptations to hypoxia have appeared (see refs. 88, 93, 156 for general reviews).

Oxygen is a potent chemical for both aerobic and anaerobic cells. Oxygen metabolism by aerobic cells generates adenosine triphosphate (ATP), which is essential for the living processes. A small part of metabolic oxygen generates reactive oxygen species (ROS), which may function as second messengers and/or cause structural damage. Accordingly, these cells have also developed defensive antioxidant mechanisms (see, for example, ref. 51). Anaerobic cells do not possess these defenses and do not survive in the presence of high partial pressure of oxygen (PO_2). However, there are cells (for example, macrophages, neutrophils, and neuroglial cells) in the aerobic organism that use ROS to damage the invading bacteria and other cells for survival. However, excessive ROS during prolonged hyperoxia can overwhelm the defensive mechanisms of the organism itself and cause cellular damage. The cells in turn respond with possible damage control, desensitization, and repair. Oxygen chemoreceptor cells, which live by oxygen, may also die by oxygen because of its special sensitivity to ROS.

Although glomus cells are considered to be oxygen sensors, the role of the petrosal ganglion cells and their innervation in the peripheral target organs is not completely ruled out (114a).

OXYGEN CONTINUUM AND OPTIMUM

A very low oxygen pressure in the environment is incompatible with life, and breathing 100% oxygen even at 1 atmosphere is a killer. The deleterious effect of hyperoxia is at least in part due to a blockade of oxidative metabolism (136), creating a cellular hypoxia-like condition. Membrane proteins and ion channels are also affected (72). Accordingly, there is a continuum in the biological effects of oxygen between the two extremes, and between the two low points there is an optimum. This concept of an oxygen continuum applies to all types of cell and system, though one may vary from another. Gilbert (54) emphasizes and illustrates this concept in another chapter of this *Handbook*. Hochachka (66) defined the metabolic defense in hypoxic adaptation, and Lutz (108) described the mechanism for anoxic survival in the vertebrate brain. Since the peripheral chemoreceptors and carotid and aortic bodies are specifically designed to express responses to oxygen pressure changes, the concept of an oxygen continuum will be applied, and the effects of both chronic hypoxia and hyperoxia will be considered.

pH_o–pH_i RELATIONSHIP AND CO_2–H^+ STIMULUS INTERACTION WITH HYPOXIA: ROLE OF CARBONIC ANHYDRASE

A special feature of carotid glomus cells is that their pH_i varies linearly with pH_o, covering the physiological range, independent of CO_2–HCO_3^- (19, 20, 63, 158). It makes sense for chemoreceptor cells to possess this property to elicit linear physiological neural and consequent reflex responses. Incidentally, central chemoreceptor cells are expected to show similar pH_i and pH_o responses as ventilatory responses predict. The persistent linear ventilatory response to hypercapnia after a compensated respiratory alkalosis at high altitude (see ref. 11) indicates that H^+ is the stimulus species for both types of chemoreceptor cell.

Evidence that pH_i determines the chemoreceptor cell response to pH_o came from the studies of several laboratories. A step increase in PCO_2 in vivo and in vitro results in a sharp increase in the chemosensory nerve discharge, followed by adaptation to a steady-state activity. The initial fast peak response is eliminated by the inhibitor of carbonic anhydrase both in vivo and in vitro (13, 69, 70). The steady-state CO_2 response is only slightly reduced, consistent with the fact that equilibrium of the CO_2 reaction is not altered by inhibition of the enzyme function. However, in the face of a reduced rate of intracellular H^+ production during hypercapnia, with intact mechanisms of ionic regulation, pH_i of the chemoreceptor cell may turn alkaline after inhibition of carbonic anhydrase (19, 20, 70).

The observations that isohydric hypercapnia (pH_o unchanged) only transiently stimulated the chemosensory nerve discharge (57) and that it is eliminated by permeable inhibitors of carbonic anhydrase (70) also emphasize the role of intracellular carbonic anhydrase. However, a sustained but small residual effect of CO_2 appears to persist (Fig. 51.1). How closely this residual response follows pH_i of glomus cells is not clear, though Buckler et al. (19) drew a parallel between the two on the basis of Gray's (57) observation of the chemosensory nerve responses. Gray, however, seldom found any overshoot in the responses to hypercapnia, unlike in vivo (13) and in vitro (70) observations. Taken together it is reasonable to conclude that glomus cell carbonic anhydrase plays time-dependent physiological roles through the production of intracellular H^+: (1) most importantly, rapid neural responses to the oscillations of respiratory gases and (2) less importantly, steady-state responses to CO_2 equilibrium in the moving blood, the lack of which could attenuate the stimulus response relationship, as reported by Teppema et al. (147).

At high altitude, carbonic anhydrase inhibitor is recommended as medication against acute mountain sickness (59). One of the effects is a rise in local tissue PCO_2 because transport of the same volume of CO_2 from tissues to blood necessitates a rise in PCO_2. The net effect, however, would depend on the ratio between CO_2 production and tissue perfusion. Accordingly, carotid body tissue PCO_2 may not rise as much as in the brain. Thus the central chemoreceptors are stimulated and ventilation increases after carbonic anhydrase inhibition. Consequently, alveolar and atrial PO_2 rise and PCO_2 decreases, which in turn would decrease carotid body

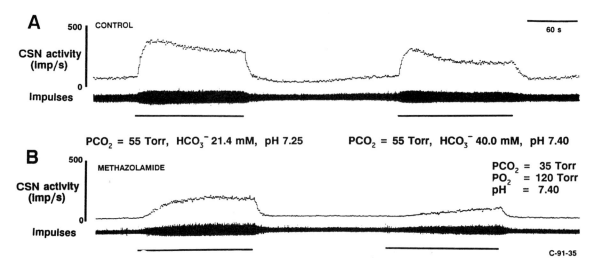

FIG. 51.1. Effects of CO_2 hydration and pH_o on chemosensory discharge of cat carotid body. (*A*) Acidic hypercapnia stimulated discharge with gradual adaptation. The same hypercapnia without changing pH_o showed a similar response pattern but its magnitude was less. This means that acid pH_o contributed to the response, and hypercapnia itself was a stimulus, which at least in part was due to fast intracellular hydration mediated by carbonic anhydrase. (*B*) To examine the idea, the foregoing tests were repeated after treating the carotid body with methazolamide (40 μM). Methazolamide reduced baseline activity, eliminated the fast initial response, and delayed and dimished the magnitude of the responses to both acidic and isohydric hypercapnia. Thus intracellular carbonic anhydrase is critical for fast response to CO_2. Without fast CO_2 hydration the response to isohydric hypercapnia, developed slowly, but acidic hypercapnia produced a faster and larger response, indicating a significant effect of pH_o.

stimulation. pH_i oscillations in the glomus cells would decrease and attenuate breathing periodicity particularly during sleep at high altitudes (144).

The effect of carbonic anhydrase inhibition would also depend in vivo on the first product of decarboxylation. If it is CO_2, hydration would be delayed and tissue P_{CO_2} would rise for the same CO_2 output and increase extra- and intracellular acidity. If it is H^+/HCO_3^-, dehydration would be delayed, lowering tissue P_{CO_2}. The physiological effects of carbonic anhydrase inhibition in the brain appear to suggest that CO_2 is the first product of dehydration.

CO_2–H^+ and Interaction with Hypoxia

Both acute and chronic hypoxia are associated with respiratory alkalosis. Accordingly, adaptation to chronic hypoxia is also adaptation to alkalosis. It is well known that hypoxic chemosensory discharge is attenuated by acute alkalosis and augmented by acidosis (see ref. 90 for review). The role of CO_2–H^+ in hypoxic adaptation is an important consideration. However, studies have found that the effects of alkalosis in chronic hypoxia are difficult to measure in the intact organism because alkalosis is partially compensated with time. Bisgard and his colleagues (11) circumvented the problem by perfusing the carotid bodies alone with normocapnic hypoxia (and variations of it) in unanesthetized goats and measuring ventilation. In a series of papers they

reported that the ventilatory response increased over hours (ventilatory acclimatization), and they attributed this to an increase in the carotid chemosensory response over time. However, neurotransmitter responses were not measured. Carotid chemosensory responses in anesthetized goats (12) also increased over hours.

Another way to assess the effects of alkalosis on hypoxia is to compare the responses to hypoxia alone with those of hypoxia plus appropriate CO_2 in inspired air. We performed these experiments with rats and found that CO_2 actually suppressed the effect of chronic hypoxia on the catecholamine content of rat carotid bodies (dopamine, ng/μg protein: 0.614 ± 0.116 in isocapnic vs. 2.754 ± 0.260 in poikilocapnic hypoxia). These studies are incomplete. Cruz et al. (29) also reported that isocapnic hypoxia did not augment ventilatory acclimatization. Dhillon et al. (34) reported morphometric analysis of the carotid body of chronically hypercapnic and hypoxic rats.

Catecholamine Release and Chemosensory Nerve Discharge

In this context it is highly relevant to note that a correspondence between CO_2–O_2 stimulus interaction in the chemosensory discharge and in the catecholamine release from the carotid body has not been established, though a parallel between the two responses during hypoxia alone has been found (45). In fact, whether

CO_2–H^+ releases carotid body catecholamine is not clear. Fitzgerald and colleagues (48) did not find any acute effect of 70 torr P_{CO_2} in vivo on carotid body catecholamines. The claim of Gonzalez and colleagues (121, 132) that superfusate acidity (up to pH of 6.6 at P_{CO_2} of 140 torr) increased dopamine release from cat and rabbit carotid bodies in vitro is in conflict with the results of Fitzgerald et al. (49). Rigual et al. (132) found a clear correlation between dopamine release and carotid sinus nerve (CSN) discharge rate due to acid stimulus in the cat carotid body, but the correlation is not the same as with hypoxia, though a low pH of 6.6 is adequate to elicit a maximal sensory nerve discharge in the cat carotid body. Dopamine release was only a fraction of that induced by hypoxia and dinitrophenol (DNP). The same group also reported that Ca^{2+}-free solution reduced dopamine release by 95%, whereas CSN discharge was reduced by only 55% due to hypoxia. Accordingly, a systematic correspondence between dopamine release and carotid chemosensory discharge is yet to be established. Thus dopamine alone cannot account for the sensory nerve responses. There are hardly any data on CO_2–O_2 stimulus interactions in dopamine release. The question is important. In chronic hypoxia the carotid body with an enormously increased dopamine content and turnover rate (60, 126) offers an attractive testing model, but observations are scarce.

The dominant view is that dopamine has a feedback inhibitory function (65, 126; see ref. 11 for review). This view militates against the proposal that dopamine is the excitatory neurotransmitter for both hypoxic and acid stimuli (133). The two events, dopamine release and chemosensory nerve discharge, could take place in parallel but independently. This writer takes the view that it is premature to assign a cause and effect relationship between dopamine release and chemosensory nerve discharge. Instantaneous voltametric/amperometric measurement of dopamine release along with chemosensory nerve discharge (21, 37) is likely to be an important tool in resolving this important issue.

It is highly relevant to note that graded tissue hypoxia caused proportionate dopamine release from the brain structure of piglets whether it was hypoxic or ischemic hypoxia.

OXYGEN DELIVERY

Po₂ vs. O₂ Content

Reduction of oxygen delivery by a reduction of P_{O_2} stimulates both carotid and aortic bodies, but an equivalent reduction of O_2 delivery by a decrease of O_2 content alone (anemia and carboxyhemoglobinemia) fails to stimulate carotid chemosensory discharge, though aortic chemoreceptors are stimulated (100). Accord-

ingly, a chronic reduction of O_2 content alone would elicit aortic but not carotid body response. Indeed, carotid bodies do not respond to chronic anemia (see ref. 64 for review), but there is no observation available on aortic chemoreceptors (see Carbon Monoxide/Oxygen, below).

Arterial Perfusion Pressure

Systemic hypotension to the level of 50 torr during normoxia or hypoxia hardly stimulates the carotid chemosensory discharge, whereas aortic chemoreceptors respond vigorously (2,100). Thus chronically hypotensive animals may not show any response (structural, at the catecholamine level, or in chemosensory nerve discharge) in the carotid body but should manifest in the aortic body responses.

The response of aortic bodies to acute stimuli can be explained by the postulate that their blood supply is already limiting and that further reduction lowers tissue P_{O_2}, causing excitation. According to this explanation, O_2 flow to the carotid chemoreceptors has to be higher than to the aortic body.

The collapsible large arterial blood vessels (see refs. 64, 67 for reviews) and numerous small blood vessels through the parenchyma may have something to do with the response to acute stimuli. It is unlikely that low blood flow is compatible with the results, and it is likely that tissue P_{O_2} is greater than the threshold stimulus. Until this threshold is reached by the reduction of O_2 content, carotid chemoreceptors are not stimulated. Indeed, hypoperfusion of carotid bodies in vitro with cell-free solution generally may not stimulate the chemosensory discharge until the perfusion pressure falls below 40 torr (L. Morelli, A. Mokashi, and S. Lahiri, unpublished observations). Accordingly, O_2 delivery by content alone was not sensed by the carotid body. However, microvascular P_{O_2} was maintained around 40 torr at the perfusate P_{O_2} of 125 torr in vitro (133a). Tissue P_{O_2} must have been lower than 40 torr, and yet a compromise of O_2 delivery by content or perfusion pressure did not affect it. Similar results are seen in vivo (99). According to the Fick principle, an autoregulated increased flow may offer an alternative explanation, but its mechanism is unclear. In any event, a lack of response of carotid chemosensory nerve to a decreased O_2 delivery except by P_{O_2} remains a fascinating feature of acute and chronic states.

Carotid Body Blood Volume and Flow

Profuse vascularization of carotid body is well known (112). In contrast, aortic bodies appear to be less vascular (82). Chronic hypoxia with increased erythropoiesis shows significantly increased blood volume in the carotid body (113, 126). The physiological effect of the increased microvascular blood volume is not known, but, if anything, an increased O_2 content in the vicinity of the chemoreceptor cell is expected to keep tissue P_{O_2} higher than it would be otherwise. It is known, however, that experimentally induced anemia does not excite chemosensory discharge (43).

INTEGRATED CAROTID BODY AND GLOMUS CELL

Carotid chemosensory responses are the most reliable index to date of chemoreception (89). The conceptual model here is that chemoreception occurs at the glomus cell level and is expressed in the chemosensory discharge through several intervening steps—cell and/or mitochondrial membrane depolarization, Ca^{2+} mobilization, and neurotransmitter release (86). Accordingly, a parallel between the glomus cell and sensory responses is predicted in chronic hypoxia. However, the bipolar cell which innervates the glomus cells need not be just a relaying station and could modulate the responses through interaction with the target cells.

Results on isolated glomus cell studies are conflicting in the areas of membrane potential, depolarization, and Ca^{2+} mobilization. For example, at a depolarizing holding potential hypoxia decreased K^+ conductance in some (106) but increased it in other (7, 8) studies. In studies by Biscoe and Duchen (7) and Sato et al. (135) $[Ca^{2+}]_i$ increased, though it decreased in the hands of Donnelley and Kholwadwala (38). That extracellular Ca^{2+} is a necessary factor for neurotransmitter release and chemosensory nerve responses to hypoxia is supported by most but not all observations (8, 38, 56, 139). Unless these baseline data are established, understanding the role of ion channels and ionic dynamics in chronic hypoxia and hyperoxia, which change the chemosensory responses, will be limited. Incidentally, several of these studies on glomus cells were made at around 22°C, which totally blocks the chemosensory responses to hypoxia (see ref. 90 for review). Biscoe and Duchen (7) demonstrated a parallel effect in glomus cells. Despite the foregoing uncertainties, glomus cell responses have been found to correspond to chemosensory nerve responses: one area is intracellular pH and sensory responses to hypercapnia (19, 57, 70, 158) and the other area is the effect of metabolic inhibitors on attenuating hypoxic responses of glomus cells and chemosensory nerves of the carotid body (39, 40, 119).

OXYGEN AND CHEMORECEPTIVE PIGMENTS

Biological pigments are associated with the functions of oxygen. Two prominent functions are the transport of O_2 by hemoglobin and myoglobin and the transfer and use of oxygen by enzymes. The terminal cytochrome oxidase is of specific interest in view of the long-standing metabolic hypothesis of O_2 chemoreception. There are many other heme proteins involved in one or the other oxygen function in the cell. These heme proteins presumably evolved together with O_2. The subject is therefore relevant to the question of O_2 sensing and its adaptation in chronic changes in the P_{O_2} environment (93). Alteration of oxygen sensing could be due to concentrations of the heme proteins by environmental O_2 pressures. One well-known example is hemoglobin (5). Carotid body heme pigments in O_2 chemoreception and adaptation require further attention.

Carbon Monoxide/Oxygen

Because CO behaves like and competes with O_2, it impairs the physiological functions involving O_2 (see refs. 26 and 27 for review). Heme iron provides the common linkage for O_2 and CO. This linked function occurs at two levels—transport of O_2 by hemoglobin and myoglobin and utilization and binding of O_2 at the level of subcellular organelles, cytosol, and cell membrane. Because of the extremely high affinity for CO, blockade of O_2 transport by hemoglobin in vivo becomes the limiting factor for survival during CO inhalation; thus studies of the cellular pigments with lower CO affinity are not possible. The organism succumbs to lack of O_2 at P_{CO} of less than 1 torr, at which level mitochondrial metabolic effect is minimal. However, there are certain subtle metabolic effects (18, 23). Also, a small amount of CO is produced endogenously from the breakdown of heme and may control cyclic GMP production by acting on the heme enzyme guanylate cyclase (see ref. 149a). The presence of heme oxygenase II and CO production in the carotid body tissue are, however, not known.

It is well known that CO exposure leads to quiet death without respiratory effort. This is in part due to lack of chemoreflex stimulation (28) and to suppression of the controller (brain) functions (41). It has now been demonstrated that carotid chemoreceptors are not stimulated by carboxyhemoglobinemia up to 40% or more during normoxia, while aortic chemoreceptors respond vigorously (96). The most likely reason is that cellular P_{O_2} is not lowered in the carotid body (99) and that P_{CO} of less than 1 torr in the arterial blood and carotid body tissue does not stimulate the cellular mechanism of O_2 chemoreception. This level of P_{CO}, however, would stimulate cyclic guanosine monophosphate (cGMP) production, which may inhibit chemosensory nerve discharge (149a). Wang et al. (152) reported inhibition of chemosensory nerve discharge by cGMP. The hypothesis that low P_{CO} does not stimulate carotid body chemoreception was tested with chronic CO inhalation (0.35 torr) in the rat at sea level. After 3 wk, hematocrit increased from 40% to 75% but carotid bodies did not respond and glomus cells did not show hypertrophy or hyperplasia (138) nor did they show an increase in catecholamine contents (98), unlike those seen in chronic hypoxia for the same increase in hematocrit (113, 125). Sensory discharge presumably did not increase. It is possible but unlikely that the increased hematocrit counteracted any stimulatory effect at the cellular level. The erythropoietic effect of carboxyhemoglobinemia was presumably due to hypoxia of the erythropoietin-producing tissues (5).

At some concentrations CO could block the hypoxic response of oxygen-sensitive cells (105). Goldberg et al. (55) showed that 70 torr P_{CO} suppressed erythropoietin production due to hypoxia by the hepatic cell line. Lahiri et al. (91) found that CO produced dual effects on carotid chemosensory nerve discharge in vitro; Figure 51.2 illustrates these effects. Transition from hypoxic perfusate (P_{O_2} = 50 torr) to carboxylate perfusate (P_{CO} = 300 torr) with the same hypoxia first inhibited then stimulated activity. The latter was inhibited by illumination of the carotid body with visible bright light (from lamp, 150 W, 15 V through light guide). Inhibition of the hypoxic effect by CO mimicked the effect of O_2 and was presumably due to binding of CO with an O_2 binding site. The subsequent photosensitive stimulation was presumably due to reaction of CO with cyto-chrome c oxidase and blockade of oxidative phosphorylation (79), but with bright light the CO complex also suppressed the hypoxic response. This inhibition by CO means that another process of chemoreception was also taking place. The transient decline of the hypoxic response can be found at a lower P_{CO} (60 torr), which also suppressed the subsequent hypoxic stimulation. It appeared that the inhibition was not clearly photosensitive. The results exhibited dual effects of CO. The photolabile stimulatory effect of high steady-state P_{CO} was demonstrated during normoxia (73). An extension of these observations with a time course of effects has now been made (91). The maximal effect of CO was consistently less than the maximal response to ischemia, supporting the idea that CO suppresses a part of the hypoxic response, but high P_{CO} also elicits a response mimicking the effect of hypoxia. On the one hand, the combined observations provided strong evidence in favor of participation of cytochrome oxidase in O_2 chemoreception. There is no other way than by complexing with intracellular components that CO can elicit the excitatory response. Action spectra of carotid chemosensory response at high P_{CO} matches absorbance spectra at 590 mm and 430 nm. On the other hand, there is also evidence that other CO-binding pigments contribute to O_2 chemoreception. This could be nonrespiratory membrane-bound and cytosolic heme pigments. Cytosolic guanylate cyclase is a potential target for the

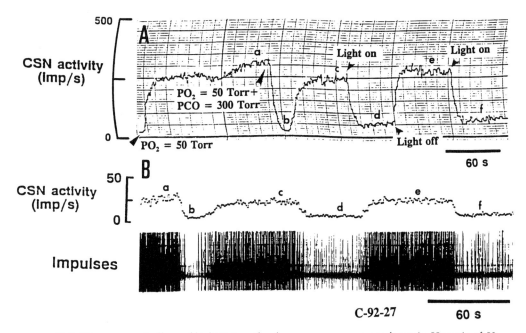

FIG. 51.2. Time course of effects of high P_{CO} on the chemosensory response to hypoxia. Hypoxia of 50 torr stimulated discharge in the absence of bright light. At point a the perfusate was changed to the same hypoxia plus 300 torr P_{CO}. The immediate effect was a depression of the activity (b), followed by excitation. The excitation was eliminated by bright light (c) (photodissociation of the CO complex) even though the same hypoxia was present. Clearly, hypoxia did not cause the usual stimulation in the presence of CO. However, undissociated CO complex elicited chemoreception.

CO effect (149a). Lopez-Lopez and Gonzalez (107) reported that K^+ current suppression by 40 torr P_{O_2} was 70% reversed by 70 torr P_{CO}. It was not clear from the paper whether the glomus cells were exposed to light under the experimental conditions. It would have been interesting to know if the CO effect could be reversed by light. In the context it needs to be pointed out that the maximal decrease in K^+ channel conductance occurred by lowering extracellular P_{O_2} from 150 torr to 90 torr (56). That is, K^+O_2 is hyperoxia-sensitive. Accordingly, the observation at 40 torr P_{O_2} would be the same at 90 torr P_{O_2}. In other words, it is not clear if hypoxia of 40 torr was necessary to manifest the CO effect. Furthermore, the CO effect was inhibitory rather than excitatory and did not negate the metabolic hypothesis of O_2 chemoreception. These aspects have been discussed at length elsewhere (86). In any case, the finding adds to the concept that CO-sensitive hemoprotein in the glomus cell may contribute to O_2 chemoreception. If these CO-binding pigments are responsible for the initiation of O_2 chemoreception, it is likely that chronic hypoxia would augment their concentration and activity. There is indeed evidence that at an early stage of chronic hypoxia the chemosensory response is augmented (12, 97). However, there can be other associated effects—an increased neurotransmitter synthesis, for example. Stea et al. (143) reported that dissociated glomus cells of rat carotid body cultured in chronic hypoxia ($P_{O_2} \simeq 40$ torr) manifested increased Na^+ channel density and hypertrophy. The effects resemble those of cyclic adenosine monophosphate (cAMP), but activation of the cyclase may be secondary to the released neurotransmitters. Cellular acidosis may not be the mediator of the hypoxic effect (63, 71, 131, 158). Microvascular P_{O_2} values in the carotid body (99) and in other tissues (159) are compatible with the metabolic dependence on P_{O_2}. However, the current understanding is not adequate. It is important to test the hypothesis that the CO-binding pigments in the carotid body are key to acute responses and that the pigment effects are enhanced during chronic hypoxia.

Nitric Oxide/Oxygen

Nitric oxide has about 30,000 times greater affinity for hemoglobin than has oxygen and is a potential molecule for reacting with heme proteins. It is an endogenous gas, produced from L-arginine by the enzyme NOS present in many cells, and a transmitter (see refs. 17, 52, 117, 142). The evidence that it is involved in blood vessel relaxation is overwhelming. The signaling pathways are endothelium-dependent in some and dependent on peripheral nerves and central neurons in other. Nitric oxide binds with the iron heme of cytosolic guanylate cyclase, enhancing its catalytic activity and the forma-

tion of cGMP. The latter in turn presumably increases phosphorylation of target protein(s) by cGMP-dependent protein kinase. The endothelium-dependent pathway is stimulated by various agents, like acetylcholine, substance P, ATP, and bradykinin, and requires Ca^{2+}. Nitric oxide synthase (NOS) in the neural innervation of blood vessels and gut can be activated by neural excitation. The resulting nitric oxide diffuses to adjacent smooth muscles, eliciting vascular relaxation. A similar pathway exists in some parts of the brain, where activation of glutamate receptors (N-methyl-D-aspartate [NMDA] subtype) increases cytosolic $[Ca^{2+}]$, which binds to calmodulin, activating NOS. There are several synthetic analogs of L-arginine which reversibly block NOS activity and nitric oxide synthesis and are used as tools to test nitric oxide effects. These effects can be perturbed by blocking the cGMP pathway: cytosolic guanylate cyclase can be inhibited by agents like methylene blue (nonspecific), and cGMP metabolism may be blocked by inhibiting phosphodiesterase activity. With a half-life of 5 s, nitric oxide removal does not require any special mechanism. However, it is produced on demand.

Peripheral chemoreceptors are potentially well equipped for the nitric oxide mechanism: there is a conspicuously large amount of vascular endothelium and smooth muscle with autonomic innervation. Glomus cells are innervated by the processes of the petrosal ganglion cells (112). Here the hypothesis is that vasodilator tone is high due to the L-arginine–nitric oxide mechanism which allows high blood flow. Moderate systemic hypotension, therefore, does not sufficiently reduce microvascular P_{O_2} of carotid body and does not elicit chemosensory responses (99). Another hypothesis is that an increased nitric oxide production upon neural stimulation generates cGMP, which in turn may decrease chemosensory discharge. Permeable cGMP has been reported to decrease chemosensory nerve response to hypoxia (152). It is noteworthy, in the context, that 8-bromo-cGMP potentiated adrenal chromaffin cell secretion due to low doses of nicotine (123). Accordingly, it is possible that nitric oxide would stimulate release of neurotransmitters from activated glomus cells.

In vivo and in vitro studies showed that micromolar doses of L-arginine analogs are excitatory for the carotid chemosensory nerve discharge in the cat (129; S. J. Fiodone, personal communication). We also found similar results (unpublished observations). However, the excitatory response developed slowly over minutes and was eliminated sharply by sodium nitroprusside (< 1 μM). These results suggest that a part of the effect involves vascular responses of the carotid body. Clearly, nitric oxide generated in the carotid body parenchyma would have access to both vessels and carotid body cells and nerves. Consistent with these observations NOS has been localized in the carotid body. Prabhakar et al. (129) described reduced nicotinamide adenine dinucleotide phosphate (NADPH)-diaphorase–positive nerve fibers enveloping glomus cell in the cat carotid body. Some glomus cells were also diaphorase-reactive. Wang et al. (153) found only NOS-immunoreactive axons from the petrosal ganglion innervating rat carotid body cells and blood vessels. We found that nerve fibers were diaphorase- and NOS-positive in rat and cat carotid bodies (58). These nerve endings displayed varicosities and were VIP-positive. Further detailed studies are needed to demonstrate NOS-positive innervation of glomus cells, though some of the petrosal ganglion cells may be NOS-positive. It appears that nitric oxide may be involved in neuropeptide release. It is of interest to note that inhibition of chemosensory discharge by L-arginine alone is not a proof of nitric oxide participation because both L-arginine and D-arginine have been found to inhibit carotid chemosensory nerve discharge in the cat in vitro (D. Bebout, personal communication).

Nitric oxide synthase has now been localized in the carotid body (58, 129, 153). The effects of nitric oxide are presumably mediated through heme-containing guanylate cyclase in smooth muscle and glomus cells, but nitric oxide can bind with many other pigments. The combined effects seem to be inhibitory for the chemosensory nerve discharge. Nitric oxide synthase itself is a heme enzyme. It is not clear if NOS and guanylate cyclase are directly affected by hypoxia.

CHRONIC HYPOXIA VS. HYPEROXIA

Structure

Direct comparisons of the effects of chronic short-term hypoxia and hyperoxia are made in this section. The effects of long-term and life-long hypoxia will be reviewed in a later section (see LIFE-LONG HYPOXIA AND DEVELOPMENTAL ASPECTS). Some of the effects with respect to control of ventilation have been discussed in the chapter by Bisgard and Forster (11) in this *Handbook*.

Hypoxia. A majority of the acute studies have used almost deadly levels of inspired PO_2 (17–35 torr) and have reported a reduction in the number of dense-core vesicles (see ref. 112 for review). With less severe hypoxia (inspired PO_2 of 70 torr) these changes were not perceptible. It is clear that an unusually severe acute hypoxic stimulus is needed to manifest a structural correlate of the biochemical and functional responses.

Chronic hypoxia, however, shows profound changes in the carotid bodies of various species, and there seems

to be correlation between structure and chemical changes and function. Glomus cells show hypertrophy (113, 125) and on occasion hyperplasia (9, 64, 120). What mechanisms control cell division and cell growth are not known. However, the mechanisms probably reside in the carotid body and are not humorally mediated (98, 138). Stea et al. (143) reported hypertrophy of cultured glomus cells of rat carotid body during chronic hypoxia. Figure 51.3 compares ultrastructures of carotid bodies of normoxic and chronically hypoxic (PIO_2 = 70 torr for 28 days) cats. Chronic hypoxia increased volume density of glomus cells. However, the subcellular organelles appeared comparable to those in the control. Dense-core vesicles were not visibly increased, though dopamine and norepinephrine concentrations increased significantly. In the rat chronic hypoxia also did not show as significant an increase in glomus cell granules (113, 125) as catecholamine concentrations (60, 126). Heath and Smith (64) provide more information on dense-core vesicles in other species.

Figure 51.4 compares synaptic nerve endings from two cat carotid bodies. The clear core vesicles in the nerve endings are comparable except that in chronic hypoxia the vesicles appeared to be more organized and focused on the synaptic function. There are centric arrays of vesicles. The dense-core vesicles on the glomus side of the membrane also appeared to focus toward the active site of the membrane. The vesicles in the vicinity of the cell membrane have proliferation of the coated pit, and the membrane itself is thickened on both sides. On the whole, the area gives an appearance of brisk activity. These illustrations are given to indicate potential areas of research to explore and establish the structure–function relationship.

Hyperoxia. Figures 51.5 and 51.6 are electron micrographs of carotid bodies of cats exposed to inspired PO_2 of about 700 torr for 63 h. There are several striking changes. Each glomus cell contained an unusually large amount of membranous organelles—rough endoplasmic reticulum, Golgi apparatus, and tubules. The Golgi appeared to have lost its normal structure close to the nuclear membrane in the centrosome region. The appearance of the dispersed Golgi is sometimes indistinguishable from the endoplasmic reticulum. The mixed Golgi–endoplasmic reticulum compartment along with the abundance of microtubules indicated a significant change in membrane traffic. The Golgi apparatus is broken up as if in preparation for cell division. It is thought that many cellular functions are inhibited at this stage of cell division (80). This could lead to a block in the secretory pathway. The meaning of this membrane trafficking in chronic hyperoxia has to be sought in the fundamentals of the normal processes. The mitochondria,

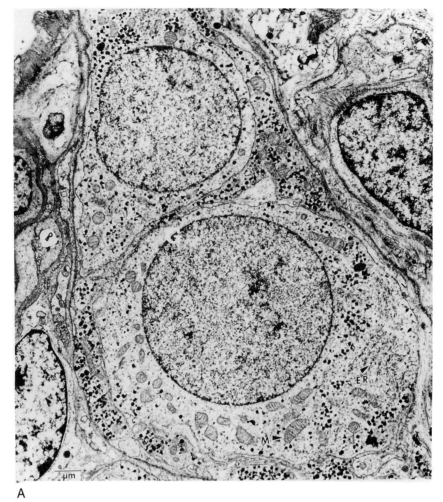

A

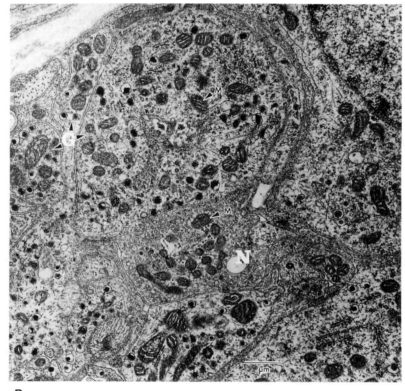

B

particularly in the glomus cells, showed fewer cristae. Dense-core vesicles were in abundance, and many appeared lobular, as if the vesicles were fusing. Figure 51.6 also shows a cross-section of a cilium, having nine microtubule triplets. There are numerous membranous structures around the glomus cells and nerve fibers, giving an appearance of active sustentacular and Schwann cells, performing repair functions from injury. Timed structural studies are needed to understand to what extent these changes consist of physiological responses as opposed to damage control and repair. The roles of superoxide and nitric oxide in toxicity need consideration.

Neurotransmitters and Gene Expression

Hypoxia. Acute hypoxia increases release of carotid body dopamine (see ref. 44 for review), and prolonged hypoxia increases carotid body catecholamine content and concentrations in the rat (60, 126, 150). Chronic hypoxia also increases turnover rates of carotid body dopamine and norepinephrine, including their release (125, 126). Hypoxia is also known to increase tyrosine hydroxylase activity not only in the carotid body but also in the sympathetic ganglia (44). The mechanism and pathways of the effects of acute and chronic hypoxia are not known. The effect of hypoxia could be specific and direct and not mediated through secondary blood-borne messengers released due to the stress of hypoxia. Indeed, it is known that prolonged hypoxia increases tyrosine hydroxylase activity also in the superior cervical ganglion and adrenal medulla (46, 62, 78; see ref. 45 for review). Czyzyk-Krzeska and colleagues (30) demonstrated that the concentration of tyrosine hydroxylase mRNA increased in the carotid body type I cells but not in the cells of the superior cervical ganglion or the adrenal medulla. The response of type I cells occurred in the denervated carotid body of adrenalectomized rats. The observation does not explain the lack of catecholamine responses of the superior cervical ganglion and the adrenal medulla. However, it appears that hypoxia exerts an additional effect on tyrosine hydroxylase gene expression in the carotid body. Dopamine released by hypoxia could have stimulated cAMP (50, 127), which augmented tyrosine hydroxylase activity. Chronic administration of cAMP to cultured glomus cells increased Na^+ channel density (143), suggesting that the effect of chronic hypoxia was indirect. Hyper-

capnic effects on tyrosine hydroxylase activity were found to be negligible.

There are other putative neurotransmitters in the carotid body. The finding of Fidone and colleagues that chronic hypoxia increases neuropeptides in the carotid body is noteworthy as well (S. J. Fidone, personal communication). Neuropeptides are likely to play a role in carotid body function (114, 130). It seems that there occurs an overabundance of the putative neurotransmitters in chronic hypoxia but that the functional changes are less dramatic.

Glomus cells have been found to stain heavily for galanin, a neuropeptide (P. Grimes, S. Lahiri, and R. Stone, unpublished observations) whose function in the carotid body has not been elucidated. However, in other cells it is a strong hyperpolarizing agent and is inhibitory (4). In general, it decreases $[Ca^{2+}]_i$ and inhibits Ca^{2+}-dependent processes. If released during hypoxia, galanin is expected to inhibit chemoreceptor excitation by a feedback system.

Taken together it seems that specific neurotransmitters for the chemosensory nerve responses to hypoxia have not been identified, though acetylcholine, dopamine, and substance P have been proposed (44). Their roles in chronic hypoxia are less clear. There seems to be a network of effects of the neurotransmitter-like substance in the glomus cells and nerves.

Hyperoxia. Both dopamine and norepinephrine are increased in content and concentration in the carotid bodies of cats after exposure to inspired Po_2 of 700 torr for 60 h (115). The reason for the raised level of catecholamines is not clear from these results alone, but their underutilization is a plausible cause because that would explain the significant attenuation of the chemosensory responses to hypoxia following chronic hyperoxic exposure in the cat (95, 116) and rat (104). The effects are dose- and duration-dependent (97, 104, 148).

Chemosensory Nerve Function

Hypoxia. Short-term chronic hypoxia (hours to days) has long been known to cause a gradual increase in ventilation (11). Increases in carotid chemosensory responses to hypoxia (in the goat and cat) could account for a part of the chemoreflex response (3, 10, 12, 31, 128, 151). The mechanism could be due to an increased

FIG. 51.3. (A) Control cat carotid body (150 torr of inspired oxygen pressure). Two glomus cells are in view. ER, endoplasmic reticulum; M, mitochondria. There are numerous dense-core granules and an average number of mitochondria. Bar indicates 1 μm magnification. (B) Carotid body of a chronically hypoxic cat (28 days of 70 torr inspired oxygen pressure). Parts of several glomus (G) cells and a nerve ending (N) are seen. Overabundance of mitochondria (M) in the cells and nerve ending, are apparent. Bar indicates 1 μm magnification.

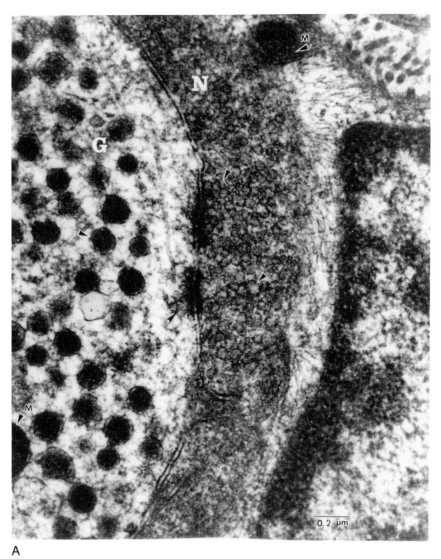

A

FIG. 51.4. (A) Neuroglomus junction in a normal cat carotid body (150 torr of inspired oxygen pressure). Nerve ending (N) on the glomus cell (G) with a cleft between is seen. Clear-core granules in the nerve ending are of various sizes. They are particularly concentrated at the synaptic condensations. Some of the dense-core vesicles in the glomus cell appear to have lost their contents and are less dense. Bar indicates 0.2 μm magnification.

turnover rate of the neurotransmitters (126). Also, glomus cell hypertrophy (126) and an altered relationship between the vesicles in the nerve endings and in the glomus cells may be important (see Fig. 51.4). These structural changes require a longer duration of hypoxia than a few minutes or hours. However, details of the microcircuitry of structure and function have not been worked out.

Chronic hypoxia of greater strength in the cat, according to one report (145), attenuated both chemosensory and ventilatory responses to hypoxia. Catecholamine levels in these carotid bodies were not measured but were unlikely to be very different from those in cats exposed to moderate hypoxia. Another study

(101) showed that efferent inhibition of the hypoxic response was greater than the control. The inhibition seems to be dopaminergic since dopamine receptor blockade also blocked the efferent effect (102). Earlier, hypoxic control of ventilation was found to be depressed in cats chronically exposed to severe hypoxia and was attributed to CSN (146). Endogenous opiates were not the determinant (128). The role of nitric oxide in these peripheral and central inhibitions in chronic hypoxia is not known.

There are numerous observations (157) that human sojourners on high mountains maintain their hypoxic ventilatory drive, as seen in resting alveolar P_{CO_2}. Also, newcomers to high altitudes manifest periodic breath-

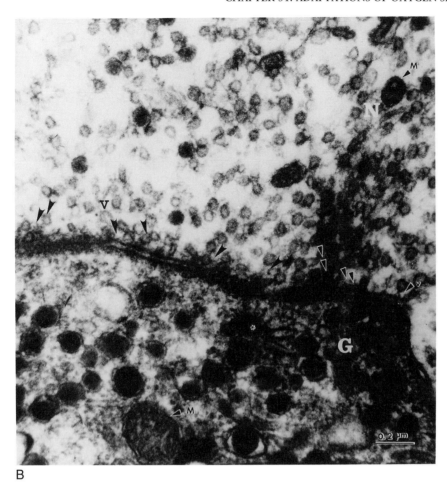

FIG. 51.4 (B) Neuroglomus junction in a carotid body of a chronically hypoxic cat (28 days of 70 torr inspired oxygen pressure). Along the length of the synaptic junction the clear-core vesicles (V) is arranged in concentric rings. Three such arrangements can be identified (arrows). These vesicles are gravitated toward the synapse. The dense-core vesicles in the glomus (G) are also concentrated near one end of the synapse. Bar indicates 0.2 μm magnification.

ing, which in part is dependent on the peripheral chemoreceptor drive. Elimination of hypoxia eliminates periodic breathing. Also, subjects who show blunted hypoxic chemoreflex drive also lack breathing periodicity (92).

Hyperoxia. Chronic normobaric hyperoxia for 60–67 h clearly attenuated carotid chemosensory responses to hypoxia in the cat, leaving the responses to nicotine and hypercapnia intact (94, 95). Since these characteristic responses resembled those of oligomycin, a blocker of mitochondrial ATP synthesis, it was postulated that one of the effects of chronic hyperoxia was on the oxidative metabolism of the chemoreceptor cell (94). Chronic hyperoxia is known to cause selective inactivation of some key mitochondrial enzymes in some cells (74, 136). According to this scheme, PO_2 changes would not influence oxidative metabolism any more during chronic hyperoxia and hence would not initiate the O_2 chemoreception mechanism. However, these cells with an interrupted oxidative pathway are more dependent on glycolytic metabolism for energy supply. Any depletion of this source of energy, for example, by acidosis, should exaggerate the response, as seen in the augmented chemosensory effect of CO_2. However, this may not be the full explanation.

Aortic chemoreceptors were not, however, affected by chronic hyperoxia presumably because the reactive oxygen species were not overwhelming owing to low blood flow and tissue PO_2 (116).

Hyperbaric oxygenation (HBO) also attenuated the chemoreflex ventilatory response to hypoxia (104), but the magnitude of effect was far less. Torbati et al. (148) reported that chemosensory responses to hypoxia were slightly but significantly diminished after HBO, consistent with the structural responses of the carotid body, but these effects were far less than those in chronic hyperoxia (95, 116). It appears that adequate time is

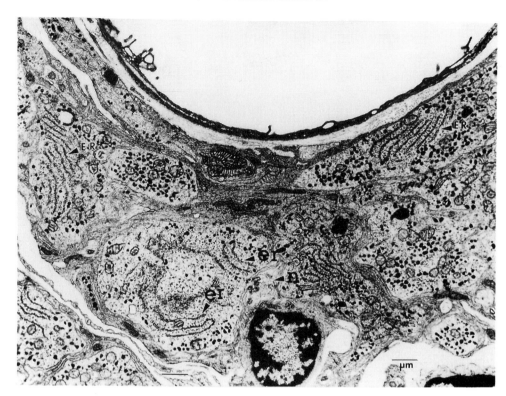

FIG. 51.5. Effects of chronic hyperoxia on cat carotid body ultrastructure (63 h of 700 torr inspired oxygen pressure). Numerous endoplasmic reticulum (*ER/er*) are seen in all glomus cells. Dense-core vesicles are preponderant, and mitochondria in glomus cells manifest low levels of cristae, unlike in another structure near the blood vessel at top. Bar indicates 1 µm magnification.

needed during hyperoxia for the inactivation processes to develop. The working hypothesis is that ROS attack pigment molecules in the mitochondrial and plasma membranes, which are presumably responsible for the initiation of O_2 chemoreception. This leaves the responses to hypercapnia and nicotine intact. In fact, the hypercapnic response is exaggerated presumably because of a metabolic shift from oxidative to glycolytic energy metabolism in the chemoreceptor cells due to ROS (148). Under a suppressed oxidative response, cells were more dependent on glycolytic metabolism for energy, and the well-known depressant effect of acidosis on fructokinase activity and glycolytic metabolism could lead to an apparent exaggerated chemoreception response to hypercapnic acidosis (Pasteur effect). These ideas are consistent with the effects of metabolic inhibitors on carotid and aortic chemosensory responses to hypercapnia (42, 119) but require further study.

CHRONIC CELL DEPOLARIZATION

Depolarization of catecholamine-containing sympathetic ganglia has long been known to increase tyrosine hydroxylase activity (14). Katz and collegues (76, 78) reported that cultured cells of nodose and petrosal ganglia showed a similar trait. Accordingly, continued depolarization of sensory cells during chronic hypoxia could increase tyrosine hydroxylase activity in the sensory cells (164). However, it has been reported that prolonged and chronic hypoxia increase tyrosine hydroxylase activity of both innervated and chronically denervated carotid bodies (30), suggesting that hypoxia may depolarize glomus cells, where enzyme activity is located. There are several reports that hypoxia may suppress K^+ current in isolated glomus cells and presumably cause their depolarization (33, 106, 124). Depolarization results in cellular ionic changes, which could lead to a cascade of events (for example, Ca^{2+} mobilization and catecholamine release) and stimulation of tyrosine hydroxylase activity. More recently, it has been reported that rats which show a blunted ventilatory response to hypoxia also lack of charybdotoxin-sensitive K^+ channels in their glomus cells. This appears to prevent them from depolarizing (and hence triggering Ca^{2+} influx and neurosecretion) during acute hypoxia, although the glomus cells possess O_2-sensitive K^+ channels (162a).

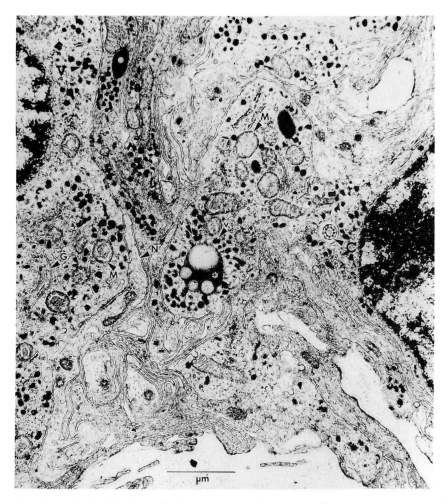

FIG. 51.6. Carotid body of a chronically hyperoxic cat at a higher magnification than Figure 51.5 (63 h of 700 torr inspired oxygen pressure). Membranous structures in and around glomus cells are numerous. A nerve ending (N) on a glomus (G) cell appears normal, but clear-core vesicles are only few. Dense-core vesicles are quite abundant but appear to be fusing in many instances. Mitochondria (M) in glomus cells show fewer cristae. There is a lipofuscin body (*) in a glomus cell. This structure appears to be a repository of some of the dense-core vesicles. Another glomus cell shows a cilium (C). Bar indicates 1 μm magnification.

EFFERENT CONTROL

Petrosal and Nodose Ganglia

Petrosal ganglion cells, which innervate glomus cells, are relevant in the following context: (1) diverse neurotransmitters are synthesized and transported down the axons to the glomus cells (44, 46), (2) polarization, particularly of their vesicles at the synapse, and (3) possible feedback control of the glomus cell function. This means that, although glomus cells may detect and respond to PO_2, bipolar cells may strongly influence structural and possibly functional organization of the glomus cells (53). Three types of sensory nerve have been recognized (81), but it is not clear if they originate from three different types of neuron. Some of these nerves form reciprocal synapses with glomus cells, sug-

gesting both their afferent and efferent functions at the synapse. A significant number of these neurons are dopaminergic (77) and may release dopamine on to the glomus cells. During chronic hypoxia, the dopaminergic property of the carotid body is augmented (60, 126), but little is known about petrosal ganglion neurons, except that dopaminergic efferent activity is augmented during chronic hypoxia (101, 145). Finley and colleagues (46) believe that petrosal ganglion chemosensory cells exert dopaminergic and peptidergic effects on glomus cells.

Nitric oxide is now known to be a potential neurotransmitter in peripheral innervation (117, 142). As noted earlier, NOS has been found in carotid body innervation (58, 129, 153), and block of NOS elicited chemosensory excitation (25, 74, 129, 153). These preliminary results implicated nitric oxide pathways for inhibitory control of carotid body function.

Structural studies also provide morphological substrates for such functions of petrosal ganglion nerve endings on the carotid body, as shown in Figures 51.3 and 51.4. Clear-core vesicles are arranged in concentric rings which make contact at the synapse (Fig. 51.4B). The radial arrangement seems to be guided by microfilaments or microtubules. Dense-core vesicles in glomus cells are concentrated at one point and appear to show processes which reach out to other vesicles and the synaptic junction. There is a message here, but it is not completely understood except that this type of arrangement is more common in chronic hypoxia. The suspected trophic interaction between glomus and petrosal ganglion cells is presumably further enhanced in chronic hypoxia, but very little is known about petrosal ganglia in chronic hypoxia, though it is known that the denervated carotid body in the rat does respond in a way similar to the innervated carotid body. However, a quantitative difference is not known. Much less is known about nodose ganglion and its trophic interaction with aortic bodies.

The functions of the reciprocal synapse in chemical transmission are not understood. However, evidence favors the view that these synapses function as feedback inhibition, presumably through the nitric oxide pathway. Efferent inhibition of the chemosensory discharge seems to increase with the chemosensory discharge (see ref. 122 for review).

Sympathetic Ganglia

Pre- and postganglionic sympathetic supply to the carotid body is a small fraction of the total innervation. Blood vessels receive much of the sympathetic supply (112).

The contribution of the sympathetic nerve in the control of chemosensory nerve responses to acute hypoxia seems to be minimal. Only the vascular effects of chronic hypoxia were blocked by propranolol, indicating noradrenergic effects (126). The physiological effects of these changes are not known. However, it is known that chronic denervation does not significantly alter catecholamine and structural responses to chronic hypoxia.

Cyclic GMP, Vascular Smooth Muscle, Glomus Cells, Petrosal Ganglion Processes, and Innervation of Carotid Body Elements

Cyclic GMP is now known to be an important regulator of vascular smooth muscle contractility. Although very little of its role in carotid and aortic body blood vessels is known, it is worthwhile to recount briefly the results in other vascular tissues. The pulmonary artery under-

goes contraction in hypoxia, which also decreases cGMP concentrations (151a). These changes are attenuated by the blockade of guanylate cyclase activation (22). These responses are unrelated to endothelium-derived relaxing factor (EDRF)/nitric oxide and prostaglandins. The mechanism of hypoxic effect is not known, but oxygen metabolites related to H_2O_2 may be involved (162). Activation of soluble guanylate cyclase coupled to oxygen tension is a plausible pathway for the physiological response. The pathway for cyclase activation seems to differ from that for nitric oxide, but once cGMP is generated the mechanism of effect is the same. A decreased cGMP level due to hypoxia would augment physiological responses—smooth muscle contraction and increased chemosensory discharge, for example—and hyperoxia would decrease cGMP concentration, leading to smooth muscle relaxation and inhibition of chemosensory nerve discharge. Superoxide anion, however, may inhibit cGMP-associated relaxation of bovine pulmonary artery (24).

Atrial natriuretic peptide (ANP) also increases cellular cGMP but by a different mechanism. Plasma membrane—bound guanylate cyclase is activated by ANP for the cGMP response (160).

Endogenous CO, which is formed from heme metabolism and shares some of the properties of nitric oxide, could also generate cGMP and influence biological functions (111, 149a). However, nothing is known about CO production and its function in the carotid body.

Cyclic GMP levels in the whole carotid body have been reported to be increased by nitrovasodilator (152) and ANP (154, 155) and to be decreased by acute hypoxia (152). These responses could be attributed, at least in part, to the vascular smooth muscles in the carotid body, as seen in other blood vessels (137). The same applies to cAMP responses of the whole carotid body (44). However, Fidone and colleagues (44) thought that the physiological chemosensory nerve responses were due to glomus cells because they used superfused carotid bodies apparently without vascular effect. Furthermore, permeable cGMP decreased carotid chemosensory activity. However, an effect on the sensory nerve ending has not been excluded, and studies with isolated glomus cells have not been reported. Innervations of the carotid body vasculature are dominantly NOS-positive (58). Studies with adrenal chromaffin cells showed potentiation of secretion due to low doses of nicotine by 8-bromo-cGMP, nitroprusside, and ANP (123). The mechanism is unclear with respect to the inhibitory effects of nitroprusside and ANP.

Taken together, cGMP seems to be the point of convergence for several physiological functions which are potentially capable of influencing chemosensory discharge. The effect of cGMP is inhibitory, and how it may fit into hypoxic excitation has not emerged.

In the context of acute effects, it is of great physiological interest to know the chronic effects of hypoxia and hyperoxia on cGMP concentration in the carotid body.

COBALT, OTHER TRANSITION METALS, AND Ca^{2+}

Cobalt and nickel, among the transitional metals, are well known to stimulate erythropoietin release and erythropoiesis (47). These divalent ions are also Ca^{2+} channel blockers and, in small concentrations, stimulate carotid chemosensory discharge both in vivo (36) and in vitro (C. Di Giulio, unpublished observations) presumably through surface receptors (141). In high concentration in vitro Co^{2+} and Cd^{2+} (1–2 mM) block hypoxic chemoreception (33, 139). At these high concentrations, however, responses to CO_2 and nicotine are also suppressed (R. Iturriaga and S. Lahiri, unpublished observations). These diverse results are cited to indicate that the effect of Co^{2+} may not be entirely mediated through Ca^{2+} mechanisms. Also, the pattern of Ca^{2+} mobilization in glomus cells by hypoxia is not clearly understood (38–40, 135). If the hypoxic effect is mediated by Ca^{2+} mobilization, its chronic blockade by Co^{2+} may not elicit hypoxia-like effects, but chronic administration of Co^{2+} not only augments erythropoiesis but adds to the hypoxic effect also (75). Chronic Co^{2+} also causes hypertrophy of carotid body glomus cells (35). Thus the cells which are specially designed to respond to hypoxia also manifest some similar responses to agents which interfere with Ca^{2+} function. There are, however, many unknowns.

LIFELONG HYPOXIA AND DEVELOPMENTAL ASPECTS

Life for mammals begins with hypoxia, which presumably offers optimal conditions for controlled embryonic and fetal development. At fetal arterial Po_2 of 25–30 torr and slightly acidic pH at sea level, peripheral chemoreceptors would be maximally active by the adult standard and the consequent reflex drive, particularly respiratory, would be strong but wasteful. Thus although the genetically coded development must take its course during gestation, it has been hypothesized that the hypoxic environment might favor delayed maturity of the peripheral chemoreceptors. In favor of this hypothesis, the blunted chemoreflex response in subjects with congenital heart disease may be cited (for example, 16). Also, kittens born and raised in the hypoxic environment showed a delayed development of the hypoxic chemoreflex drive (61). However, older human subjects studied at high altitude, as well as children, gain full maturity before it may decline later with age (87). This decline in the ventilatory chemoreflex function at high

altitude is well illustrated in Figure 51.7, where alveolar or arterial Pco_2 values are plotted against inspired Po_2. With the decreasing inspired Po_2, alveolar Po_2 and Pco_2 values decreased steeply in new arrivals at altitudes, but the decrease was significantly less in life-long residents, indicating their attenuated peripheral chemoreflex drive (134).

The mechanisms of the effects of continued hypoxia on maturity are not well known, but it is well known that chemosensory responses to both hypoxia and hypercapnia are not fully mature at birth (15, 110, 118). Measuring the metabolic dynamics of dopamine in fetal and newborn rats, Hertzberg (65) suggested that a diminished turnover rate of dopamine and disinhibition were responsible for the rise of chemosensory responses with age. But Donnelly and Doyle (37a) disagreed in that both the dopamine and CSN discharge were low in the newborn rats and they increased with age. Marchal et al. (109, 110) reported that exogenous dopamine exerted both an inhibitory and an excitatory effect on chemosensory responses in newborn kittens; it required more than 2 wk to develop a fully mature response. They also found that dopamine D_2 receptors were already developed in these kittens. Dopamine clearly exerts an important effect on the chemosensory responses to hypoxia. Chronic hypoxia influences both dopamine and norepinephrine dynamics, but the meaning of the interaction between hypoxia and catecholamine dynamics is yet to be clearly understood. Alter-

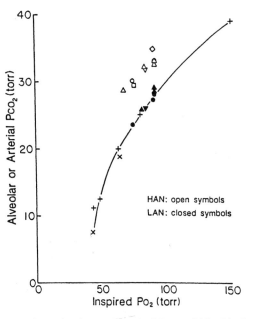

FIG. 51.7. Relationship between inspired Po_2 at high altitudes and alveolar/arterial Pco_2, indicating high respiratory drive in the low-altitude native *(LAN)* and low respiratory drive in the high-altitude native *(HAN)*. (Reprinted with permission from Santolaya et al., *Respir. Physiol.* 77: 253–262, 1989, see ref. 134.)

ation of ionic channel concentrations of the plasma membrane during chronic states of hypoxia is an important consideration for excitable cells. Stea et al. (143) reported a significantly increased concentration of Na^+ channels in the membranes of glomus cells cultured in chronic hypoxia. The meaning of blunting of response to hypoxia in life-long hypoxia in terms of glomus cell ion channels is beginning to be understood (162a). Hypertrophy of glomus cell and carotid body is maintained during life-long hypoxia (64, 85).

PERIPHERAL CHEMORECEPTORS IN CARDIOPULMONARY AND VASCULAR DISEASES

Cardiopulmonary and Vascular Diseases

Cardiopulmonary diseases which result in hypoxemia also lead to structural and functional changes in the carotid body, analogous to those seen in chronic hypoxia. A comprehensive review can be found in Heath and Smith (64). Winslow and Monge (161) discuss many aspects of chronic mountain sickness.

We found that cats with arteriosclerotic disease showed intense carotid chemosensory discharges at arterial P_{O_2} of 112 torr. Raising inspired P_{O_2} to 710 torr only gradually diminished activity (unpublished observations). Structural alterations can be found in sclerotic diseases (64). The implication is that patients with this type of disease will have normally heightened reflex responses, which would have to be taken into account in management.

As discussed earlier, life-long hypoxemia results in underdevelopment of the chemoreflex responses, but a distinction cannot be made between the contributions of carotid and aortic chemoreceptors from the overall reflexes.

Taken together, peripheral chemoreceptors with abnormal functions are not uncommon in many cardiovascular and pulmonary diseases. The abnormal reflex functions need attention in these diseases.

Diseases of the Glomoids

Tumors of carotid and aortic bodies and of other glomera have been reviewed in Heath and Smith (64) and Zak and Lawson (163). These tumors mostly consist of glomus cells. Nerve endings are rare, suggesting that most of these glomus cells are not innervated. Sustentacular cells are relatively few in number. Apparently glomus cells contain normal volumes of dense-core vesicles, which are known to contain putative neurotransmitters. The functional abnormalities have not been documented in these conditions.

A greater incidence of carotid body chemodectoma at high altitudes is found in many reports (64). The disease has also a familial trait. A direct connection between hypoxia and the disease is speculative.

Baroreceptor vs. Peripheral Chemoreceptors

Baroreceptors and carotid and aortic chemoreceptors reside in the same vascular areas, and the sensory neurons are located in the same ganglia. Hypertension stimulates baroreceptors and prevents chemoreceptor excitation. Baroreceptor discharge inhibits sympathetic nerve activity, and carotid chemoreceptor activity excites sympathetic nerve activity (84). The interaction between the two receptor functions in chronic states is an important consideration, particularly in the areas of electrolyte and fluid balance and in renal function, which also brings up the question of ANP. Plasma membrane–bound guanylate cyclase is a receptor for ANP (137). For glomus cells, the effect is inhibitory for the chemosensory response to hypoxia in vitro (154). In chronic states of hypertension the carotid body is known to become larger, consisting particularly of a proliferation of sustentacular cells and an increase in chemosensory activity. These responses and interplay between several factors and integration and adaptation are dealt with in the chapter by Hoyt and Honig (68) in this Handbook.

PERSPECTIVES

Because the peripheral chemoreceptors are responsive to many blood-borne agents, including P_{O_2} and P_{CO_2}–pH, osmolarity, and hormones, the mechanisms by which the responses are initiated and modulated are bound to be multifaceted in both acute and chronic states. As in any other cell, it is necessary to take apart and study the membranes, organelles, and cytosolic components to understand the contribution of each part, but it is also critical to put the parts together for an integrated story. The carotid body, as a prototype of arterial chemoreceptors, is ideal for such studies and for unravelling the principles related to oxygen biology.

It has become increasingly obvious that, as expressed in the sensory discharge, glomus cells are responsive to a wide range of P_{O_2}, unlike adrenal medullary cells, which are of a similar embryological origin (see refs. 8, 43 for reviews). Also, O_2-sensitive K^+ channels of glomus cell membranes may not account for the responses of the chemosensory nerve and dopamine release from glomus cells. Since a single mechanism cannot explain the sensitivity over a wide range of P_{O_2}, a hypothesis has been put forward that a family of O_2-binding pigments, including cytochrome c oxidase, is responsible for the initiation and modification of the responses to hypoxia (86). Carbon monoxide and action spectra of the CO complexes are likely to be useful tools in this study, for example, in Ca^{2+} and pH_i signaling. The cascade of events that follows can be studied independently of the initial events. Adaptations to chronic stimuli marshal the processes which are all integrated at the cellular level.

The effects of sustained cell depolarization need to be separated from those of hypoxia alone. The signaling pathways include an array of proteins [receptors, guanosine triphosphate (GTP)-binding proteins, cyclases, protein kinases, regulatory and target proteins]. There are interactions between the signaling pathways, consisting of feedback regulation. The dynamics of regulation, particularly in chronic hypoxia and hyperoxia, provide stimulating challenges in carotid body research.

The effects of CO_2–HCO_3^- are intimately associated with the effects of O_2, but the bases of the connection are not clearly understood. Studies of specific gene activation and molecular expression necessarily have to be a part of the investigative programs. The field is wide open.

The role of nitric oxide and carbon monoxide as neurotransmitters in peripheral innervation and smooth muscles (117, 142) is eminently applicable to the peripheral chemoreceptors in health and disease. The prospect of integrating the functions of these gaseous molecules, which rapidly diffuse from one site to another, is exciting.

Carotid and aortic bodies are the gateways of information for the rest of the organism to respond to oxygen pressure changes. Without this information, the body is unaware of the danger of low oxygen pressure in arterial blood. The protective reflex responses of the autonomic nervous system, including respiratory and cardiovascular functions, depend on the sensory signals from the carotid and aortic bodies. It is interesting that so few cells can do so much for the rest of the organism.

I thank Valerie Johnson and Wendy Patriquin for their secretarial assistance. Supported in part by the Department of Physiology, University of Pennsylvania, and by Grant HL-43413.

REFERENCES

1. Acker, H. P_{O_2} chemoreception in arterial chemoreceptors. *Annu. Rev. Physiol.* 51: 835–844, 1989.
2. Anand, A., and A. S. Paintal. Oxygen sensing by arterial chemoreceptors. In: *Response and Adaptation to Hypoxia: Organ to organelle,* edited by S. Lahiri, N. S. Cherniack, and R. S. Fitzgerald. New York: Oxford University Press, 1991, p. 81–94.
3. Barnard, P. S., S. Andronikou, M. Pokorski, N. J. Smatresk, and A. Mokashi. Time-dependent effect of hypoxia on carotid body chemosensory function. *J. Appl. Physiol.: Respir. Environ. Exerc. Physiol.* 63: 685–691, 1987.
4. Bartfai, T., G. Fidone, and G. Laugel. Galanin and galanin antagonists: molecular and biochemical perspectives. *TIPS* 13: 312–317, 1992.
5. Bauer, C., P. Ratcliffe, and K.-U. Eckhardt. Hypoxia, erythropoietin gene expression, and erythropoiesis. In: *Handbook of Physiology. Environmental, Physiology* edited by M. J. Fregly and C. M. Blatteis. Bethesda, MD: Am. Physiol. Soc.
6. Bee, D., D. J. Pallot, and G. R. Barer. Division of type 1 and epithelial cells in the hypoxic rat carotid body. *Acta Anat. (Basel)* 126: 226–229, 1986.
7. Biscoe, T. J., and M. R. Duchen. Cellular basis of transduction in carotid chemoreceptors. *Am. J. Physiol.* 258 (*Gastrointest. Liver Physiol.* 21): G271–G278, 1990.
8. Biscoe, T. J., and M. R. Duchen. Monitoring P_{O_2} by the carotid chemoreceptors. *NIPS* 5: 229–233, 1990.
9. Biscoe, T. J., and M. R. Duchen. Responses of type I cells dissociated from the rabbit carotid body to hypoxia. *J. Physiol. (Lond.)* 428: 39–59, 1990.
10. Bisgard, G. E., M. A. Busch, and H. V. Forster. Ventilatory acclimatization to hypoxia is not dependent on cerebral hypocapnic alkalosis. *J. Appl. Physiol.: Respir. Environ. Exerc. Physiol.* 60: 1011–1015, 1986.
11. Bisgard, G. E., and H. Forster. Ventilatory responses to acute and chronic hypoxia. In: *Handbook of Physiology. Environmental, Physiology* edited by M. J. Fregly and C. M. Blatteis. New York: Oxford University Press, for the American Physiological Society, 1995, p. 1207–1239.
12. Bisgard, G. E., A. Nielsen, E. Vidruk, L. Daristotle, M. Engwall, and H. V. Forster. Mechanisms of ventilatory acclimatization to hypoxia in goats. In: *Chemoreceptors and Reflexes in Breathing: Cellular and Molecular Aspects,* edited by S. Lahiri, R. Forster II, R. O. Davies, and A. I. Pack. New York: Oxford University Press, 1989, p. 208–215.
13. Black, A. M., D. I. McCloskey, and R. W. Torrance. The responses of carotid body chemoreceptors in the cat to sudden changes of hypercapnic and hypoxic stimuli. *Respir. Physiol.* 13: 36–49, 1971.
14. Black, I. B., D. M. Chikaraishi, and E. J. Lewis. Trans-synaptic increase in RNA coding for tyrosine hydroxylase in rat sympathetic ganglion. *Brain Res.* 339: 151–153, 1985.
15. Blanco, E. E., G. S. Dawes, M. A. Hanson, and H. B. McCook. The response to hypoxia of arterial chemoreceptors in fetal sheep and newborn lambs. *J. Physiol. (Lond.)* 351: 25–37, 1984.
16. Blesa, M., S. Lahiri, W. Rashkind, and A. P. Fishman. Normalization of the blended ventilatory response to acute hypoxia in congenital cyanotic heart disease. *N. Engl. J. Med.* 296: 237–241, 1977.
17. Bredt, D. S., and S. H. Synder. Nitric oxide as a neuronal molecular messenger. *Neuron* 8: 3–11, 1992.
18. Brown, S. D., and C. D. Piantadosi. In vivo binding of carbon monoxide to cytochrome c oxidase in rat brain. *J. Appl. Physiol.: Respir. Environ. Exerc. Physiol.* 68: 604–610, 1990.
19. Buckler, K. J., R. D. Vaughan-Jones, C. Peers, D. Lagadic-Gossman, and P. C. G. Nye. Effects of extracellular pH, P_{CO_2} and HCO_3^- on intracellular pH in isolated type-1 cells of the neonatal rat carotid body. *J. Physiol. (Lond.)* 444: 703–721, 1991.
20. Buckler, K. J., R. D. Vaughan Jones, C. Peers, and P. C. Nye. Intracellular pH and its regulation in isolated type I carotid body cells of the neonatal rat. *J. Physiol. (Lond.)* 436: 107–129, 1991.
21. Buerk, D., S. Lahiri, D. K. Chugh, and A. Mokashi. Electrochemical detection of rapid dopamine release kinetics during

hypoxia in perfused/superfused cat carotid bodies. *J. Appl. Physiol.* (in press), 1995.

22. Buerk-Wolin, T. M., and M. S. Wolin. H2O2 and cGMP may function as an O2 sensor in the pulmonary artery. *J. Appl. Physiol.: Respir. Environ. Exerc. Physiol.* 66: 107–170, 1989.

23. Chance, B., M. Erechinska, and M. Wagner. Mitochondrial responses to carbon monoxide toxicity. *Ann. N. Y. Acad. Sci.* 174: 193–204, 1970.

24. Cherry, P. D., H. A. Omar, K. A. Farrell, J. Stuart, and M. S. Wolin. Superoxide anion inhibits cGMP-associated bovine pulmonary arterial relaxation. *Am. J. Physiol.* 259 (*Heat Circ. Physiol* 28): H1056–1062, 1990.

25. Chug, D., M. Katayama, A. Mokashi, and S. Lahiri. Nitric oxide–related inhibition of carotid chemosensory nerve activity in the cat. *Respir. Physiol.* 97: 147–156, 1994.

26. Coburn, R. F., E. R. Allen, S. Ayres, D. Bartlett, Jr., S. M. Horvath, L. H. Kuller, V. Laties, L. D. Longo, and E. P. Radford, Jr. *Carbon Monoxide.* Washington, DC: Natl. Acad. Sci., 1977.

27. Coburn, R. F., and H. Forman. Carbon monoxide toxicity. In: *Handbook of Physiology. The Respiratory System,* edited by L. E. Farhi and S. M. Tenney. Bethesda, MD: Am. Physiol. Soc., 1986, vol. IV p. 439–458.

28. Comroe, J. H., Jr., and C. F. Schmidt. The part played by reflexes from the carotid body in the chemical regulation of respiration in the dog. *Am. J. Physiol.* 121: 75–96, 1938.

29. Cruz, J. C., J. T. Reeves, R. F. Grover, J. T. Maher, R. E. McCullough, A. Cymerman, and J. C. Denniston. Ventilatory acclimatization to high altitude is prevented by CO2 breathing. *Respiration* 39: 121–130, 1980.

30. Czyzyk-Krzeska, M. F., D. A. Bayliss, E. E. Lawson, and D. E. Milhorn. Regulation of tyrosine hydroxylase gene expression in the rat carotid body by hypoxia. *J. Neurochem.* 58: 1538–1546, 1992.

31. Data, P. G., C. Di Giulio, A. Mokashi, W.-X. Huang, A. K. Sherpa, D. G. Penney, K. Albertine, and S. Lahiri. Mechanisms and site of effect of chronic erythropoietic stimuli on carotid body. In: *Arterial Chemoreception,* edited by C. Eyzaguirre, S. J. Fidone, R. S. Fitzgerald, S. Lahiri, and D. McDonald. New York: Springer-Verlag, 1990, p. 323–329.

32. DeCastro, F., and M. Rubio. The anatomy and innervation of the blood vessels of the carotid body and the role of chemoreceptive reactions in the autoregulation of the blood flow. In: *Arterial Chemoreceptors,* edited by R. W. Torrance. Oxford: Blackwell, 1968, p. 267–277.

33. Delpiano, M. A., and J. Heschler. Evidence for a Po2 sensitive K+ channel in the type-1 cell of the rabbit carotid body. *FEBS Lett.* 249: 195–198, 1989.

34. Dhillon, D. P., G. R. Barer, and M. Walsh. The enlarged carotid body of the chronically hypoxic and chronically hypercapnic rat: a morphometric analysis. *Q. J. Exp. Physiol.* 69: 301–317, 1984.

35. Di Giulio, C., P. G. Data, and S. Lahiri. Chronic cobalt causes of hypertrophy of glomus cells in the rat carotid body. *Am. J. Physiol.* 261 (*Cell Physiol.* 30): C102–C105, 1991.

36. Di Giulio, C., W.-X. Huang, S. Lahiri, A. Mokashi, and D. B. Buerk. Cobalt stimulates carotid body chemoreceptors. *J. Appl. Physiol.: Respir. Environ. Exerc. Physiol.* 68: 1844–1849, 1990.

37. Donnelley, D. F. Electrochemical detection of catecholamine release from rat carotid body in vitro. *J. Appl. Physiol.: Respir. Environ. Exerc. Physiol.* 74: 2330–2337, 1993.

37a. Donnelly, D. F., and T. P. Doyle. Developmental changes in hypoxia-induced catecholamine release from rat carotid body, in vitro. *J. Physiol. (Lond.)* 475:267–275, 1994.

38. Donnelley, D. F., and K. Kholwadwala. Hypoxia decreases intracellular calcium in adult rat carotid body glomus cells. *J. Neurophysiol.* 67: 1543–1551, 1992.

39. Duchen, M. R., and T. J. Biscoe. Mitochondrial function in type I cells isolated from rabbit arterial chemoreceptors. *J. Physiol. (Lond.)* 450: 13–31, 1992.

40. Duchen, M. R., and T. J. Biscoe. Relative mitochondrial membrane potential and [Ca2+]i in type I cells isolated from the rabbit carotid body. *J. Physiol. (Lond.)* 450: 33–61, 1992.

41. Edelman, N. H., J. E. Melton, and J. A. Neubauer. Central adaptation to hypoxia. In: *Response and Adaptation to Hypoxia: Organ to Organelle,* edited by S. Lahiri, N. S. Cherniack, and R. S. Fitzgerald. New York: Oxford University Press, 1991, p. 235–244.

42. Erhan, B., E. Mulligan, and S. Lahiri. Metabolic regulation of aortic chemoreception responses to CO2. *Neurosci. Lett.* 24: 143–147, 1981.

43. Eyzaguirre, C., R. S. Fitzgerald, S. Lahiri, and P. Zapata. Arterial chemoreceptors. In: *Handbook of Physiology. The Cardiovascular System,* edited by J. T. Shepherd and F. M. Abboud. Washington, DC: Am. Physiol. Soc., 1983, vol. III, p. 557–621.

44. Fidone, S. J., C. Gonzalez, B. Dinger, A. Gomez-Niño, A. Obeso, and K. Yoshizaki. Cellular aspects of peripheral chemoreceptor function. In: *The Lung,* edited by R. G. Crystal, J. B. West, P. J. Barnes, N. S. Cherniack, and E. R. Weibel. New York: Raven, 1991, p. 1319–1332.

45. Fidone, S. J., C. Gonzalez, A. Obeso, A. Gomez-Niño, and B. Durger. Biogenic amine and neuropeptide transmitters in carotid body chemotransmission: experimental findings and perspectives. In: *Hypoxia: The Adaptations,* edited by J. R. Sutton, G. Coates, and J. E. Remmors. Toronto: Decker, 1990, p. 116–126.

46. Finley, J. C. W., J. Polak, and D. M. Katz. Transmitter diversity in carotid body afferent neurons: dopaminergic and peptidergic phenotypes. *Neuroscience* 51: 973–987, 1992.

47. Fisher, J. W. Pharmacologic modulation of erythropoietin production. *Annu. Rev. Pharmacol. Toxicol.* 28: 101–122, 1988.

48. Fitzgerald, R. S., P. Garger, M. C. Hauer, H. Raff, and L. Fechter. Effect of hypoxia and hypercapnia on catecholamine content in cat carotid body. *J. Appl. Physiol.: Respir. Environ. Exerc. Physiol.* 54: 1408–1413, 1983.

49. Fitzgerald, R. S., and S. Lahiri. Reflex responses to chemoreceptor stimulation. In: *Handbook of Physiology. The Respiratory System,* edited by N. S. Cherniack and J. G. Widdicombe. Washington, DC: Am. Physiol. Soc., 1986, p. 313–362.

50. Fitzgerald, R. S., E. M. Rogus, and A. Deghani. Catecholamines and 3′,5′ cyclic AMP in carotid body chemoreception in the cat. In: *Tissue Hypoxia and Ischemia,* edited by M. Reivich, R. F. Coburn, S. Lahiri, and B. Chance. New York: Plenum, 1977, p. 245–260.

51. Fridovitch, I. The biology of oxygen radicals: general concepts. In: *Oxygen Radicals and Tissue Injury,* edited by B. Halliwell. Bethesda, MD: FASEB, 1988, p. 1–3.

52. Furchgot, R. F., W. Martin, P. D. Cherry, D. Jothianandan, and G. Villani. Endothelium-dependent relaxation and cyclic GMP. In: *Vascular Neuroeffector Mechanisms,* edited by J. A. Bevan, T. Godfriend, R. A. Maxwell, J. C. Stodet, and M. Worcel. Amsterdam: Elsevier, p. 105–114.

53. Gallego, R., I. Ivorra, and A. Morales. Effects of central or peripheral anatomy on membrane properties of sensory neurons in the petrosal ganglion of the cat. *J. Physiol. (Lond.)* 391: 39–56, 1987.

54. Gilbert, D. L. Evolutionary aspects of atmospheric oxygen and organisms. In: *Handbook of Physiology. Environmental Physiology,* edited by M. J. Fregly and C. M. Blatteis. New York: Oxford University Press, for the American Physiological Society, 1995, p. 1059–1094.

55. Goldberg, M. A., S. P. Dunning, and H. F. Bunn. Regulation of the erythropoietin gene: evidence that the oxygen sensor is a heme protein. *Science* 242: 1412–1414, 1988.

56. Gonzalez, C., L. Almaraz, A. Obeso, and R. Rigual. Oxygen and acid chemoreception in the carotid body chemoreceptors. *TIPS* 15: 146–153, 1992.

57. Gray, B. A. Response of the perfused carotid body to changes in pH and P_{CO_2}. *Respir. Physiol.* 4: 229–245, 1968.

58. Grimes, P., S. Lahiri, R. Stone, A. Mokashi, and D. Chug. Nitric oxide synthase occurs in neurons and nerve fibers of the cat carotid body. In: *Arterial Chemoreceptors: Cell to System.* New York: Plenum, 1994, p. 221–224.

59. Hackett, P. *Mountain Sickness: Prevention, Recognition and Treatment.* New York: American Alpine Club, 1980.

60. Hanbauer, I., F. Karoum, S. Hellstrom, and S. Lahiri. Effects of long-term hypoxia on the catecholamine content in rat carotid. *Neuroscience* 6: 81–86, 1981.

61. Hanson, M. A., G. J. Eden, J. G. Nighuis, and P. J. Moore. Peripheral chemoreceptors and other oxygen sensors in the fetus and newborn. In: *Chemoreceptors and Reflexes in Breathing: Cellular and Molecular Aspects,* edited by S. Lahiri, R. E. Forster, R. O. Davies, and A. I. Pack. New York: Oxford University Press, 1989, p. 113–120.

62. Hayashi, Y., S. Miwa, K. Lee, K. Koshimura, K. Hamahata, H. Hasegawa, M. Gujiwara, and J. Watanabe. Enhancement of in vivo tyrosine hydroxylation in the rat adrenal gland under hypoxic conditions. *J. Neurochem.* 54: 1115–1121, 1990.

63. He, S.-F., J.-Y. Wei, and C. Eyzaguirre. Effects of relative hypoxia and hypercapnia on intracellular pH and membrane potential of cultured carotid body glomus cells. *Brain Res.* 556: 333–338, 1991.

64. Heath, D., and P. Smith. *Diseases of the Human Carotid Body.* London: Springer-Verlag, 1992.

65. Hertzberg, T. *Postnatal Adaptation of Peripheral Arterial Chemoreceptors.* Stockholm: Nobel Institute for Neurophysiology, 1990.

66. Hochachka, P. W. Metabolic defense adaptations to hypobaric hypoxia in man. In: *Handbook of Physiology. Environmental Physiology,* edited by M. J. Fregly and C. M. Blatteis. New York: Oxford University Press, for the American Physiological Society, 1995, p. 1115–1123.

67. Howe, A., and E. Neil. Arterial chemoreceptors. In: *Handbook of Sensory Physiology: Entereceptors,* edited by E. Neil. Berlin: Springer-Verlag, 1972, p. 47–80.

68. Hoyt, R., and A. Honig. Energy metabolism and fluid balance at high altitude. In: *Handbook of Physiology. Environmental Physiology,* edited by M. J. Fregly and C. M. Blatteis. Bethesda, MD: Am. Physiol. Soc. 1994.

69. Iturriaga, R., and S. Lahiri. Carotid body chemoreception in the absence and presence of CO_2–HCO_3^-. *Brain Res.* 568: 253–260, 1991.

70. Iturriaga, R., A. Mokashi, and S. Lahiri. Dynamics of carotid body responses in the presence and absence of CO_2–HCO_3^-: role of carbonic anhydrase. *J. Appl. Physiol.: Respir. Environ. Exerc. Physiol.* 75: 1587–1594, 1993.

71. Iturriaga, R., W. L. Rumsey, S. Lahiri, D. Spergel, and D. F. Wilson. Intracellular pH and oxygen chemoreception in the cat carotid body in vitro. *J. Appl. Physiol.: Respir. Environ. Exerc. Physiol.* 72: 2259–2266, 1992.

72. Jamieson, D. Oxygen toxicity and reactive oxygen metabolites in mammals. *Free Radical Biol. Med.* 7: 87–108, 1989.

73. Joels, N., and E. Niel. The action of high tensions of carbon monoxide on the carotid chemoreceptors. *Arch. Int. Pharmacodyn. Ther.* 138: 528–534, 1962.

74. Joenje, H., J. J. P. Gille, A. B. Oostra, and P. van de Valk. Some characteristics of hyperoxia-adapted Hela cells. *Lab. Invest.* 52: 420–428, 1985.

75. Katsuoka, Y., B. Beckman, W. J. George, and J. W. Fisher. Increased level of erythropoietin in kidney extracts of rats treated with and hypoxia. *J. Physiol. (Lond.)* 244: 129–133, 1983.

76. Katz, D. M. Molecular mechanisms of carotid body afferent neuron development. In: *Response and Adaptation to Hypoxia: Organ to Organelle,* edited by S. Lahiri, N. S. Cherniack, and R. S. Fitzgerald. New York: Oxford University Press, 1991, p. 133–142.

77. Katz, D. M., and M. J. Erb. Developmental regulation of tyrosine hydroxylase expression in primary sensory neurons of the rat. *Dev. Biol.* 137: 233–242, 1990.

78. Katz, D. M., K. A. Markey, M. Goldstein, and I. B. Black. Expression of catecholaminergic characteristics by primary sensory neurons in the normal adult rat in vivo. *Proc. Natl. Acad. Sci. USA* 80: 3526–3530, 1986.

79. Keilin, D. *The History of Cell Respiration and Cytochrome.* Cambridge: Cambridge University Press, 1970.

80. Klausner, R. D., J. G. Donaldson, J. Lippincott-Schwatz, and A. Brefeldin. Insights in the control of membrane traffic and organelle structure. *J. Cell. Biol.* 116: 1071–1080, 1992.

81. Kummer, W. Three types of neurochemically defined autonomic fibers innervate the carotid baroreceptor and chemoreceptor regions in the guinea-pig. *Anat. Embryol.* 181: 477–489, 1990.

82. Kummer, W., and K. Addicks. Quantitative studies of the aortic bodies of the cat. 1. Point-counting analysis of tissue components. *Acta Anat. (Basel)* 126: 48–53, 1986.

83. Kummer, W., and J.-O. Habeck. Light and electronmicroscopical and immunohistochemical investigation of the innervation of the human carotid body. In: *Neurobiology and Cell Physiology of Chemoreception,* edited by P. G. Data, H. Acker, and S. Lahiri. London: Plenum, 1993, p. 67–71.

84. Lahiri, S. Peripheral chemoreflex. In: *The Lung,* edited by R. G. Crystal, J. B. West, P. J. Barnes, N. S. Cherniack, and E. R. Weibel. New York: Raven, 1991, p. 1333–1340.

85. Lahiri, S. Respiration in chronic hypoxia and hyperoxia: role of peripheral chemoreceptors. In: *Control of Breathing and Its Modelling Perspective,* edited by Y. Honda, Y. Miamoto, K. Kono, and J. G. Widdicombe. New York: Plenum, 1992, p. 433–440.

86. Lahiri, S. Chromophores in O_2 chemoreception: the carotid body model. *NIPS* 9: 162–165, 1994.

87. Lahiri, S., J. S. Brody, T. Velasquez, E. K. Motoyama, M. Simpser, R. G. Delaney, and G. Polgar. Pulmonary adaptation to high altitude: genetic vs. environment. *Nature* 261: 133–135, 1976.

88. Lahiri, S., N. S. Cherniack, and R. S. Fitzgerald (Eds). *Response and Adaptation to Hypoxia: Organ to Organelle.* New York: Oxford University Press, 1991.

89. Lahiri, S., and R. G. Delaney. Stimulus interaction in the responses of carotid body chemoreceptor single afferent fibres. *Respir. Physiol.* 24: 249–266, 1975.

90. Lahiri, S., and R. G. Gelfand. Mechanisms of acute ventilatory responses. In: *Regulation of Breathing,* edited by T. F. Hornbein. New York: Dekker, 1981, p. 773–843.

91. Lahiri, S., R. Iturriaga, and A. Mokashi. CO reveals dual mechanisms of O_2 chemoreception in the cat carotid body. *Respir. Physiol.* 94: 227–240, 1993.

92. Lahiri, S., K. Maret, and M. G. Sherpa. Dependence of high altitude sleep apnea on ventilatory sensitivity to hypoxia. *Respir. Physiol.* 52: 281–301, 1983.

93. Lahiri, S., A. Mokashi, C. Di Giulio, A. K. Sherpa, W.-X. Huang, and P. G. Data. Carotid body adaptation. In: *Hyp-*

oxia—The Adaptations, edited by J. R. Sutton, G. Coates, and J. E. Remmers. Toronto: Dekker, 1990, p. 127–130.

94. Lahiri, S., A. Mokashi, M. Shirahata, and S. Andronikou. Chemical respiratory control in chronically hyperoxic cats. *Respir. Physiol.* 82: 201–216, 1990.

95. Lahiri, S., E. Mulligan, S. Andronikou, M. Shirahata, and A. Mokashi. Carotid body chemosensory function in prolonged normobaric hyperoxia in the cat. *J. Appl. Physiol.: Respir. Environ. Exerc. Physiol.* 62: 1924–1931, 1987.

96. Lahiri, S., E. Mulligan, T. Nishino, A. Mokashi, and R. O. Davies. Relative responses of aortic body and carotid body chemoreceptors to carboxyhemoglobinemia. *J. Appl. Physiol.: Respir. Environ. Exerc. Physiol.* 50: 580–586, 1981.

97. Lahiri, S., E. Mulligan, N. J. Smatresk, P. Barnard, A. Mokashi, D. Torbati, M. Pokorski, R. Zhang, P. G. Data, and K. Albertine. Mechanisms of carotid body responses to chronic low and high oxygen pressures. In: *Chemoreceptors and Reflexes in Breathing: Cellular and Molecular Aspects,* edited by S. Lahiri, R. E. Forster, R. O. Davies, and A. I. Pack. New York: Oxford University Press, 1989, p. 215–227.

98. Lahiri, S., D. G. Penney, A. Mokashi, and K. H. Albertine. Chronic CO inhalation and carotid body catecholamines: testing of hypotheses. *J. Appl. Physiol.: Respir. Environ. Exerc. Physiol.* 67: 239–242, 1989.

99. Lahiri, S., W. L. Rumsey, D. F. Wilson, and R. Iturriaga. Contribution of in vivo microvascular PO_2 in the cat carotid body chemotransduction. *J. Appl. Physiol.: Respir. Environ. Exerc. Physiol.* 75: 1035–1043, 1993.

100. Lahiri, S., N. J. Smatresk, and E. Mulligan. Responses of peripheral chemoreceptors to natural stimuli. In: *Physiology of the Peripheral Arterial Chemoreceptors,* edited by R. G. O'Regan and H. Acker. Amsterdam: Elsevier, 1983, p. 221–256.

101. Lahiri, S., N. J. Smatresk, M. Pokorski, P. Barnard, and A. Mokashi. Efferent inhibition of carotid body chemoreception in chronically hypoxic cats. *Am. J. Physiol.* 245 (*Regulatory Integrative Comp. Physiol.* 14): R678–R683, 1983.

102. Lahiri, S., N. J. Smatresk, M. Pokorski, P. Barnard, A. Mokashi, and K. H. McGregor. Dopaminergic inhibition of carotid body chemoreceptors in chronically hypoxic cats. *Am. J. Physiol.* 247 (*Regulatory Integrative Comp. Physiol.* 17): R24–R28, 1984.

103. Lauweryns, J. M., and A. V. Lommel. Morphometric analysis of hypoxia-induced synaptic activity in intrapulmonary neuroepithelial bodies. *Cell Tissue Res.* 226: 201–214, 1982.

104. Liberzon, I. R., R. Arieli, and D. Kerem. Attenuation of hypoxic ventilation by hypobaric O_2: effects of pressure and exposure time. *J. Appl. Physiol.: Respir. Environ. Exerc. Physiol.* 66: 851–856, 1989.

105. Lloyd, B. B., D. J. C. Cunningham, and R. C. Goode. Depression of hypoxic hyperventilation in man by sudden inspiration of carbon monoxide. In: *Arterial Chemoreceptors,* edited by R. W. Torrance. Oxford: Blackwell, 1968, p. 145–147.

106. Lopez-Barneo, J., J. R. Lopez-Lopez, J. Urena, and C. Gonzalez. Chemotransduction in the carotid body: K^+ current modulated by PO_2 in type I chemoreceptor cells. *Science* 241: 580–582, 1988.

107. Lopez-Lopez, J. R., and C. Gonzalez. Time course of K^+ current inhibition by low oxygen in the chemoreceptor cells of adult rabbit carotid body. *FEBS Lett.* 299: 251–254, 1992.

108. Lutz, P. L. Mechanisms for anoxic survival in the vertebrate brain. *Annu. Rev. Physiol.* 54: 601–618, 1992.

109. Marchal, F., A. Bairam, P. Hauzi, J. P. Crance, C. Di Giulio, P. Vert, and S. Lahiri. Carotid chemoreceptor response to natural stimuli in the newborn kitten. *Respir. Physiol.* 87: 183–193, 1992.

110. Marchal, R., A. Bairam, P. Hauzi, J. M. Hascoet, J. P. Crance, P. Vert, and S. Lahiri. Dual responses of carotid chemosensory afferents to dopamine in the newborn kitten. *Respir. Physiol.* 90: 173–183, 1992.

111. Marks, G. S., J. F. Brien, K. Nakatsu, and B. E. McLaughlin. Does carbon monoxide have a physiological functions? *TIPS* 12: 185–188, 1991.

112. McDonald, D. Peripheral chemoreceptors. In: *Regulation of Breathing,* edited by T. F. Hornbein. New York: Dekker, 1981, p. 105–319.

113. McGregor, K. H., J. Gil, and S. Lahiri. A morphometric study of the carotid body in chronically hypoxic rat. *J. Appl. Physiol.: Respir. Environ. Exerc. Physiol.* 57: 1430–1438, 1984.

114. McQueen, D. S. Pharmacological aspects of putative transmitters in the carotid body. In: *Physiology of the Peripheral Arterial Chemoreceptors,* edited by H. Acker and R. G. O'Regan. Amsterdam: Elsevier, 1983, p. 149–196.

114a. Mitchell, R. A., A. K. Sinha, and D. M. McDonald. Chemoreceptive properties of regenerated endings of the carotid sinus nerve. *Brain Res.* 43: 680–685, 1972.

115. Mokashi, A., C. DiGuilio, L. Morelli, and S. Lahiri. Chronic hyperoxic effects on cat carotid body structure and catecholamines. *Respir. Physiol.* 97: 25—32, 1994.

116. Mokashi, A., and S. Lahiri. Aortic and carotid body chemoreception in prolonged hyperoxia in the cat. *Respir. Physiol.* 86: 233–243, 1991.

117. Moncada, R. M., J. Palmer, and E. A. Higgs. Nitric oxide: physiology, pathophysiology and pharmacology. *Pharmacol. Rev.* 43: 109–142, 1991.

118. Mulligan, E. Discharge properties of carotid bodies: developmental aspects. In: *Development Neurobiology of Breathing,* edited by G. G. Haddard and J. P. Farber. New York: Dekker, p. 321–340.

119. Mulligan, E., S. Lahiri, and B. T. Storey. Carotid body O_2 chemoreception and mitochondrial oxidative phosphorylation. *J. Appl. Physiol.: Respir. Environ. Exerc. Physiol.* 51: 438–446, 1981.

120. Nurse, C. A., and C. Vollmer. Effects of hypoxia on cultured chemoreceptors of the rat carotid body: DNA synthesis and mitotic activity in glomus cells. In: *Neurobiology and Cell Physiology of Chemoreception,* edited by P. G. Data, H. Acker, and S. Lahiri. London: Plenum, 1993, p. 79–84.

121. Obeso, A., A. Rocher, S. Fidone, and C. Gonzalez. The role of dihydrophyridine-sensitive Ca^{2+} channels in stimulus-evoked catecholamine release from chemoreceptor cells of the carotid body. *Neuroscience* 47: 463–472, 1992.

122. O'Regan, R. G., and S. Majcherczyk. Control of peripheral chemoreceptors by efferent nerves. In: *Physiology of the Peripheral Arterial Chemoreceptors,* edited by H. Acker and R. G. O'Regan. Amsterdam: Elsevier, 1983, p. 257–298.

123. O'Sullivan, A. J., and R. D. Burgoyne. Cyclic GMP regulates nicotine-induced secretion from cultured bovine adrenal chromaffin cells: effects of 8-bromo-cyclic GMP, atrial natriuretic peptide and nitroprusside (nitric oxide). *J. Neurochem.* 54: 1805–1908, 1990.

124. Peers, C., and J. O'Donnell. Potassium currents recorded in type I carotid body cells from the neonatal rat and their modulation by chemoexcitatory agents. *Brain Res.* 522: 259–266, 1990.

125. Pequignot, J. M., J. M. Cottet-Emard, Y. Dalmaz, and L. Peyrin. Dopamine and norepinephrine dynamics in rat carotid body during long-term hypoxia. *J. Auton. Nerv. Syst.* 21: 9–14, 1987.

126. Pequignot, J. M., S. Hellström, and T. Hertzberg. Long-term hypoxia and hypercapnia in the cat carotid body: a review. In: *Arterial Chemoreception,* edited by C. Eyzaguirre, S. J. Fidone,

R. S. Fitzgerald, S. Lahiri, and P. Zapata. New York: Springer-Verlag, 1990, p. 100–114.

127. Perez-Garcia, M. T., L. Alvaraz, and C. Gonzalez. Effects of different types of stimulation on cyclic AMP content in the rabbit carotid body: functional significance. *J. Neurochem.* 55: 1287–1293, 1990.

128. Pokorski, M., and S. Lahiri. Endogenous opiates and ventilatory acclimatization to chronic hypoxia in the cat. *Respir. Physiol.* 83: 211–222, 1991.

129. Prabhakar, N. R., G. K. Kumar, C. H. Chang, F. H. Agani, and M. A. Haxhiru. Nitric oxide in the sensory function of the carotid body. *Brain Res.* 625: 10–22, 1993.

130. Prabhakar, N. R., S. C. Landis, G. K. Kumar, D. Mullikin-Kilpatrick, N. S. Cherniack, and S. Leeman. Substance P and neurokinin A in the cat carotid body: localization, exogenous effects and changes in content in response to arterial P_{O_2}. *Brain Res.* 481: 205–214, 1989.

131. Ray, D. K., A. Mokashi, M. Katayama, F. Botré, and S. Lahiri. Lack of parallelism between the effects of hypoxia on the glomus cell pHi and carotid chemosensory response. *Neurosci. Abstr.* 19: 1407, 1993.

132. Rigual, R., J. R. Lopez-Lopez, and C. Gonzalez. Release of dopamine and chemoreceptor discharge induced by low pH and high P_{CO_2} stimulation of the cat carotid body. *J. Physiol. (Lond.)* 433: 519–531, 1991.

133. Roumy, M., C. Armengand, and L.-M. Leitner. Catecholamines in the carotid body. In: *Arterial Chemoreception,* edited by C. Eyzaguirre, S. J. Fidone, R. S. Fitzgerald, S. Lahiri, and D. McDonald. New York: Springer-Verlag, 1990, p. 115–123.

133a. Rumsey, W. L., R. Iturriaga, D. Spergel, S. Lahiri, and D. F. Wilson. Optical measurements of the dependence of chemoreception on oxygen pressure in the cat carotid body. *Am. J. Physiol.* 261 (*Cell Physiol.* 30): C614–C622, 1991.

134. Santolaya, R. B., S. Lahiri, R. T. Alfaro, and R. B. Schoene. Respiration adaptation in the highest inhabitants and highest Sherpa mountaineers. *Respir. Physiol.* 77: 253–262, 1989.

135. Sato, M., K. Ikeda, K. Yoshizaki, and H. Koyano. Response of cytosolic calcium to anoxia and cyanide in cultured glomus cells of newborn rabbit carotid body. *Brain Res.* 551: 527–530, 1991.

136. Schoonen, W. G. E. J., A. H. Wanamarta, J. M. van der Klei-van Moorsel, C. Jakobs, and H. Joenje. Respiration failure and stimulation of glycolysis in Chinese hamster ovary cells exposed to normobaric hyperoxia. *J. Biol. Chem.* 265: 11118–11124, 1990.

137. Schulz, S., M. Chinkers, and D. L. Garbers. The guanylate cyclase/receptor family of proteins. *FASEB J.* 3: 2026–2035, 1989.

138. Sherpa, A. K., K. H. Albertine, D. G. Penney, B. Thompkins, and S. Lahiri. Chronic CO exposure stimulates erythropoiesis but not glomus cell growth. *J. Appl. Physiol.: Respir. Environ. Exerc. Physiol.* 67: 1383–1387, 1989.

139. Shirahata, M., and R. S. Fitzgerald. Dependency of hypoxic chemotransduction in cat carotid body on voltage-gated calcium channels. *J. Appl. Physiol.: Respir. Environ. Exerc. Physiol.* 71: 1062–1069, 1991.

140. Shirahata, M., and R. S. Fitzgerald. The presence of CO_2/HCO_3^- is essential for hypoxic chemotransduction in the in vivo perfused carotid body. *Brain Res.* 545: 297–300, 1991.

141. Smith, J. B., C. D. Dwyer, and L. Smith. Cadmium evokes inositol polyphosphate formation and calcium mobilization. *J. Biol. Chem.* 264: 7115–7118, 1989.

142. Snyder, S. H. Nitric oxide: first in a new class of neurotransmitters? *Science* 257: 494–496, 1992.

143. Stea, A., A. Jackson, and C. A. Nurse. Hypoxia and N^6,O^2-dibutyryladenosine 3'5'-cyclic monophosphate, but not nerve growth factor, induce Na^+ channels and hypertrophy in chromaffin-like arterial chemoreceptors. *Proc. Natl. Acad. Sci. U.S.A.* 89: 9469–9473, 1992.

144. Sutton, J. R., C. S. Houston, A. L. Mansell, M. McFadden, P. Hackett, and A. C. Powles. Effects of acetazolamide on hypoxemia during sleep at high altitude. *N. Engl. J. Med.* 301: 1329–1331, 1979.

145. Tatsumi, K., C. K. Pickett, and J. V. Weil. Attenuated carotid body hypoxic sensitivity after prolonged hypoxic exposure. *J. Appl. Physiol.: Respir. Environ. Exerc. Physiol.* 70: 748–755, 1991.

146. Tenney, S. M., and L. C. Ou. Hypoxic ventilatory response of cats at high altitude: an interpretation of blunting. *Respir. Physiol.* 30: 185–199, 1977.

147. Teppema, L. J., F. Rocheffe, and M. Demedts. Ventilatory effects of acetazolamide in cats during hypoxemia. *J. Appl. Physiol.: Respir. Environ. Exerc. Physiol.* 72: 1717–1723, 1992.

148. Torbati, D., A. K. Sherpa, S. Lahiri, A. Mokashi, K. H. Albertine, and C. Di Giulio. Hyperbaric oxygenation alters carotid body ultrastructure and function. *Respir. Physiol.* 92: 183–196, 1993.

149. Vanderkooi, J. M., M. Erecinska, and I. A. Silver. Oxygen in mammalian tissue: methods of measurement and affinities of various reactions. *Am. J. Physiol.* 260 (*Cell Physiol.* 29): C1131–C1150, 1991.

149a. Verna, A., D. J. Hirsch, C. E. Glatt, G. V. Ronnett, and S. H. Snyder. Carbon monoxide: a putative neural messenger. *Science* 259: 381–384, 1993.

150. Verna, A., A. Scharnel and J.-M. Pequignot. Long-term hypoxia increases the number of norepinephrine-containing glomus cells in the rat carotid body: a correlative immunocytochemical and biochemical study. *J. Auton. Nerve Syst.* 44: 171–177, 1993.

151. Vizek, M., C. K. Pickett, and J. V. Weil. Increased carotid body hypoxic sensitivity during acclimatization to hypobaric hypoxia. *J. Appl. Physiol.: Respir. Environ. Exerc. Physiol.* 63: 2403–2410, 1987.

151a. Voelkel, N. F. Mechanisms of hypoxic pulmonary vasoconstriction. *Am. Rev. Respir. Dis.* 133: 1186–1195, 1986.

152. Wang, W.-J., G.-F. Chen, B. G. Dinger, and S. J. Fidone. Effects of hypoxia on cyclic nucleotide-formation in rabbit carotid body in vitro. *Neurosci. Lett.* 105: 164–168, 1989.

153. Wang, Z.-Z., D. S. Bredt, S. J. Fidone, and L. J. Stensaas. Neurons synthesizing nitric oxide innervate the mammalian carotid body. *J. Comp. Neurol.* 336: 419–432, 1993.

154. Wang, Z.-Z., L. He, L. J. Stensaas, B. G. Dinger, and S. J. Fidone. Localization and in vitro actions of atrial natriuretic peptide in the cat carotid body. *J. Appl. Physiol.: Respir. Environ. Exerc. Physiol.* 70: 942–946, 1991.

155. Wang, Z.-Z., L. J. Stensaas, W.-J. Wang, B. G. Dinger, J. de Vente, and S. J. Fidone. Atrial natriuretic peptide increases cyclic guanosine monophosphate immunoreactivity in the carotid body. *Neuroscience* 49: 479–486, 1992.

156. Weil, J. V. Ventilatory control at high altitude. In: *Handbook of Physiology. The Respiratory System,* edited by N. S. Cherniack and J. G. Widdicombe. Washington, DC: Am. Physiol. Soc., 1986, vol. II, p. 703–727.

157. West, J. B. Life at extreme altitude. In: *Handbook of Physiology. Environmental Physiology,* edited by M. J. Fregly and C. M. Blatteis. Bethesda, MD: Am. Physiol. Soc., 1994.

158. Wilding, T. J., B. Cheng, and A. Roos. pH regulation in adult rat carotid body glomus cells: importance of extracellular pH, sodium and potassium. *J. Gen. Physiol.* 100: 593–608, 1992.

159. Wilson, D. F., W. L. Rumsey, T. J. Green, and J. M. Vanderkooi. The oxygen dependence of mitochondrial oxidative phos-

phorylation measured by a new optical method for measuring oxygen concentration. *J. Biol. Chem.* 263: 2712–2718, 1988.

159a. Wilson, D. F., A. Mokashi, D. Chugh, S. Vinogradov, S. Osanai, and S. Lahiri. The primary oxygen sensor of the cat carotid body is cytochrome a₃ of the mitochondrial respiratory chain. *FEBS Lett.* 351: 370–374, 1994.

160. Winquist, R. J., E. P. Faison, S. A. Waldman, K. Schwartz, F. Murad, and R. M. Rapport. Atrial natriuretic factor elicits an endothelium independent relaxation and activates particulate guanylate cyclase in vascular smooth muscle. *Proc. Natl. Acad. Sci. U.S.A.* 81: 7661–7664, 1984.

161. Winslow, R. M., and C. Monge. *Hypoxia, Polycythemia and Chronic Mountain Sickness.* Baltimore: Johns Hopkins University Press, 1987.

162. Wolin, M. S., and T. M. Burke. Hydrogen peroxide elicits activation of bovine pulmonary arterial soluble guanylate cyclase by a mechanism associated with its metabolism by catalase. *Biochem. Biophys. Res. Commun.* 143: 20–25, 1987.

162a. Wyatt, C. N., C. Wright, D. Bell, and C. Peers. O₂-sensitive K⁺ currents in carotid body chemoreceptor cells from normoxic and chronically hypoxic rats and their roles in hypoxic chemotransduction. *Proc. Natl. Acad. Sci. USA* 92: 295–299, 1995.

163. Zak, F. G., and W. Lawson. *The Paraganglionic Chemoreceptor System.* New York: Springer-Verlag, 1982.

164. Zigmond, R. E., and C. W. Bowers. Influence of nerve activity on the macromolecular content of neurons and their effector organs. *Annu. Re. Physiol.* 43: 673–687, 1981.

52. Ventilatory responses to acute and chronic hypoxia

GERALD E. BISGARD | Department of Comparative Biosciences, School of Veterinary Medicine, University of Wisconsin, Madison, Wisconsin

HUBERT V. FORSTER | Department of Physiology, Medical College of Wisconsin, Milwaukee, Wisconsin

CHAPTER CONTENTS

AN INCREASE IN BREATHING is one of the most rapid physiological changes that occurs in response to environmental hypoxia (Fig. 52.1). This response usually enhances well-being by maintaining lung and blood partial pressure of oxygen (PO_2) at a higher level than would exist without the hyperpnea.

The effect of hypoxia on breathing varies with its duration (Fig. 52.1). For humans an initial phase, *acute hypoxia*, is at least 60 min in duration (74, 88, 114, 164, 308). A second phase, the *short-term* or *acclimatization* phase, is initiated within the first few hours of hypoxia and is sustained for years of hypoxic exposure (74, 114, 308, 332). The third phase, *long-term* acclimatization, is characteristic of breathing in high-altitude natives (74, 114, 211, 212, 330). The first three sections of this chapter are concerned with these three phases. The fourth section summarizes findings on the effects of hypoxia on breathing during sleep. The final section provides a perspective on some of the advantages and disadvantages of these changes in breathing during hypoxia. Throughout, the emphasis is on humans, but information from other species is also included.

VENTILATORY RESPONSES TO ACUTE HYPOXIA

In humans at sea level reducing partial pressure of arterial oxygen (PaO_2) from its normal value of about 95 torr to 60 torr has minimal effect on pulmonary ventilation ($\dot{V}_E$) and $PaCO_2$ (74, 114, 188, 308). However, as PaO_2 is reduced from about 60 to 30 torr there is a progressive increase in $\dot{V}_E$ and a decrease in $PaCO_2$ (74, 114, 188, 308). These changes are evident within seconds after PaO_2 is decreased, and usually hyperpnea and hypocapnia are progressive over the first 2–3 min of hypoxemia. This temporal pattern largely reflects the fact that the stimulus for hyperpnea, PaO_2, does not immediately reach its nadir when partial pressure of inspiratory oxygen (PIO_2) is reduced because of the lung store of O_2. After reaching a peak in 2–3 min, $\dot{V}_E$ begins to gradually decline over 30–60 min of sustained hypoxemia (Fig. 52.1) (88, 89, 107, 133, 135, 164, 186). Nevertheless, $\dot{V}_E$ remains slightly above baseline in most adult humans after 60 min of hypoxemia, but in newborns the secondary decrease may reduce $\dot{V}_E$ to levels below baseline (22, 25, 35–37, 142, 226, 259). This secondary decline has been termed *ventilatory roll-off.*

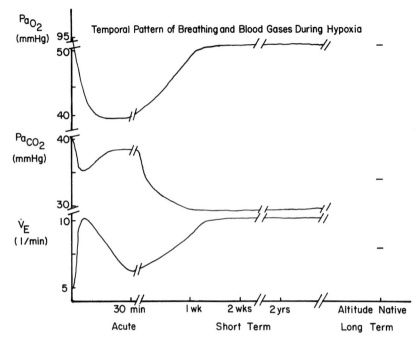

FIG. 52.1. Schematic of the temporal pattern of pulmonary ventilation (V̇E), partial pressure of carbon dioxide in arterial blood (PaCO2), and partial pressure of oxygen in arterial blood (PaO2) during exposure of humans to an altitude of approximately 4,300 m.

Hypoxemic hyperpnea and hyperventilation have also been observed in numerous nonhuman species studied at sea level (74, 114). However, the magnitude of the initial hyperventilation varies, as shown by findings that after 5 min at a PaO2 of 40 torr PaCO2 is reduced by about 2, 6, 8, and 10 torr in goat, human, pony, and rat, respectively (74). The roll-off phenomenon has been documented in many nonhuman species (adult and newborn), but whether it is characteristic across all species is not known (22, 25, 35, 37, 136, 204–206, 226, 232, 235, 273, 369). Finally, in nonhumans it is well known that hypoxia reduces fetal breathing movements (177).

Mechanisms of Ventilatory Changes during Acute Hypoxia

Mechanisms Stimulating Breathing. Much if not all of the hyperpnea during the initial minutes of hypoxemia is mediated by the carotid chemoreceptors (Fig. 52.2) (64, 101, 111, 166, 182, 214, 219). This conclusion is based on findings that as PaO2 is decreased in anesthetized animals, the increase in neural activity of the carotid sinus nerve parallels the increase in V̇E in both the anesthetized and the awake state (214). Moreover, in both states removal of these chemoreceptors attenuates and/or eliminates hyperpnea (18, 26, 28, 32, 112, 113, 344, 378). However, there are at least two other mechanisms which may contribute to hyperpnea. First,

chemoreceptors in the arch of the aorta can be activated by hypoxemia (Fig. 52.2) (58, 64, 165, 218). These might contribute to ventilatory drive in intact animals and/or they may act as "back-up" mechanisms whose importance is increased in the absence of the primary carotid chemoreceptors. This possibility is suggested by findings in cats (340) and ponies (15) that the loss of this chemoreflex for the first few weeks after carotid deafferentation is followed by a gradual, partial recovery of the reflex; this recovery can then be abolished in ponies (15) by denervating the aortic chemoreceptors. Second, mechanisms within the central nervous system (CNS) are capable of eliciting an increase in breathing (Fig. 52.2). Probably the most striking example is the tachypneic hyperpnea during severe hypoxemia (28, 128, 136, 249, 250), which is probably mediated by a suprapontine mechanism related to changes in monoamine metabolism (24, 70, 128). It is not known if this mechanism involves chemoreception in the same manner since it occurs in the periphery. However, some hypothalamic and hippocampal neurons depolarize as PO2 is lowered even in the absence of synaptic input, which suggests that the neurons are directly responding to the hypoxemia (85, 86, 124, 149, 180).

Mechanisms Inhibiting Breathing and Potentially Causing Roll-Off. Several mechanisms have been postulated to account for or contribute to V̇E roll-off. First, there may be a change in pulmonary mechanics such that a con-

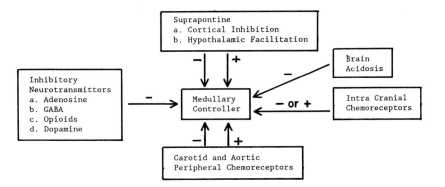

Acute Hyperpnea

1. ↑ Peripheral
   Chemoreceptor
   Stimulation

2. ↑ Suprapontine
   facilitation

Acute roll-off

1. ↑ Respiratory Impedance

2. Peripheral Chemoreceptors
   a) ↓ stimulation
   b) ↑ inhibition

3. ↓ Intracranial Chem.

4. Hypoxic Brain Depression
   a) metabolic impairment
   b) ↑ cortical inhibition
   c) ↑ inhibitory neuro.t.
      adenosine
      GABA
      opioids
      dopamine

5. Brain Acidosis

6. ↓ Metabolic Rate

Short-Term Acclimatization

1. ↑ Peripheral Chemo-
   receptor Hypoxic
   Sensitivity

2. ↑ CNS Facilitation

3. ↑ Intracranial Chemo-
   receptor Stimulation

4. ↓ Respiratory Impedance

Long Term Adaptation

1. ↓ Peripheral Chemo-
   receptor Sensitivity

2. ↓ CNS Facilitation

FIG. 52.2. Schematic of some of the structures involved in ventilatory changes caused by hypoxia *(upper portion)* and a listing of the postulated mechanism for the ventilatory changes during each phase of hypoxia.

stant level of muscle activity results in a progressive decrease in $\dot{V}_E$ (205). This is supported by measured changes in the mechanics of newborn, awake monkeys during hypoxia (204, 205). Moreover, during acute hypoxia in this species, a $\dot{V}_E$ roll-off coexists with sustained elevated airway occlusion pressure and diaphragmatic electromyograms (indexes of respiratory drive) (205, 206). However, Lawson and Long (226) found a roll-off in phrenic activity of piglets during acute hypoxia, and Blanco et al. (22) did not find a change in mechanics accompanying $\dot{V}_E$ roll-off during acute hypoxia in kittens. As a result, factors other than a change in pulmonary mechanics must contribute to $\dot{V}_E$ roll-off.

A second hypothesis is that there is a reduction in carotid chemoreceptor activity after the third minute of hypoxemia. This possibility has been directly tested, and it was found that carotid chemoreceptor activity was sustained at an elevated level throughout 30 min of hypoxemia; thus $\dot{V}_E$ roll-off must be due to other mechanisms (235, 369).

A third hypothesis is that carotid chemoreceptors directly mediate $\dot{V}_E$ roll-off (57). This idea is supported by the data of St. John and Bianchi (348, 349), who found that pharmacological or hypoxic activation of carotid chemoreceptors resulted in a decrease in discharge frequency of many medullary respiratory neurons. Since the responses with pharmacological agents could not be due to a CNS effect, the data raise the possibility that carotid afferents have an inhibitory effect on some medullary neurons. This theory is also supported by findings of a roll-off in breathing frequency during, and a depressed frequency after, electrical stimulation of the carotid sinus nerve (57). Finally, breathing of carotid body–denervated (CBD) animals often does not change during hypoxia (31, 36, 112, 231, 339); thus during acute hypoxia carotid afferents may directly mediate both the initial hyperpnea and the subsequent $\dot{V}_E$ roll-off.

A fourth hypothesis is that there is a reduction in stimulation at the intracranial chemoreceptor due to hypocapnia and/or alkalosis secondary to the initial hyperventilation (74, 114, 332) and/or an increase in cerebral blood flow (92, 179, 196, 276, 331). Unfortunately, the role of this chemoreceptor remains uncertain simply because it is impossible with the available methods to determine the stimulus level at this receptor

(74). The exact location of these receptors (dorsal or ventral medulla or hypothalamus) and the nature of the adequate stimulus (intra- or extracellular H^+ or CO_2) remain highly controversial (71, 72, 85, 86, 105, 175, 246, 268, 374). Nevertheless, available evidence suggests that cerebral interstitial fluid (ISF) pH increases during the first few minutes of hypoxic and carbon monoxide–induced brain hypoxia, but subsequently, due to an increase in lactic acid production by the brain, ISF pH decreases to below normal (30, 104, 178, 271, 272, 276). This lactacidosis corresponds temporally to $\dot{V}_E$ roll-off. Moreover, cerebral intracellular pH of rats decreases during acute hypoxia (262) as does directly measured lumbar cerebrospinal fluid (CSF) pH in awake humans after 1 h of hypoxia (76). The only evidence of alkalosis in the brain is in the cisternal CSF of awake ponies after 1 h of hypoxia (286). Accordingly, although the status of the stimulus level at the intracranial chemoreceptor during acute hypoxia remains speculative, the available evidence suggests an actual increase rather than a decrease in stimulus level during the $\dot{V}_E$ roll-off phase; thus $\dot{V}_E$ roll-off is probably mediated by a mechanism other than stimulus level at the intracranial chemoreceptor.

A fifth potential mechanism for $\dot{V}_E$ roll-off is what has generally been termed *hypoxic brain depression* (HBD). This depression manifests as the reduction in breathing that occurs during (1) hypoxia in the absence of the carotid chemoreceptor (47, 48, 231, 258, 371) and (2) recovery from hypoxia (126, 164, 251, 313, 385). It may also manifest as an increase in breathing with hyperoxia in a CBD animal (112, 113, 127, 317). It is evident most uniformly under anesthetized conditions, but in the awake state it is not always observed with brain hypoxia (269, 271). Nevertheless, available data (discussed below) warrant concluding that the $\dot{V}_E$ roll-off during 1 h of hypoxia is due in part to HBD.

One of the mechanisms proposed to account for HBD is metabolic impairment of medullary respiratory neurons (269). During very severe hypoxia [oxygen saturation (SaO_2) < 45%] it is likely that neuronal dysfunction occurs (type III HBD) (269, 334). However, during less severe hypoxia (type II HBD) $\dot{V}_E$ roll-off and HBD occur, but when these levels of hypoxemia are combined with normal or increased levels of other $\dot{V}_E$ stimuli responses are normal or even enhanced (49, 74, 75, 90, 134, 188, 245, 288, 291). In other words, when PaO_2 is 40–60 torr, $\dot{V}_E$ responses to hypercapnia, exercise, and pharmacological agents are greater than they are when PaO_2 is normal or hyperoxic. This enhanced responsiveness has been termed *positive interaction*, and at least a component of this is mediated at the carotid chemoreceptors (69, 101, 166, 182, 215). The complete mechanism of these interactions has not been elucidated, and its physiological significance remains speculative. Nevertheless, HBD is evident in a subject hypoxic

at rest, but if exercise is initiated, the $\dot{V}_E$ response is actually greater than during normoxic exercise. This indicates that HBD is a specific inhibitory response, and the data are inconsistent with hypoxic depolarization and synaptic transmission failure of medullary respiratory neurons at these levels of hypoxemia (124, 269, 377).

It has been suggested that HBD reflects stimulation by hypoxia of cortical regions, which then inhibit breathing. The best evidence in support of this theory is that electrical stimulation of cortical regions inhibits breathing (4, 108) and that decortication or midcollicular transection enhances the $\dot{V}_E$ response to hypoxia under certain conditions (359–362). Accordingly, it seems that cortical regions might be directly or indirectly responsible for a portion of HBD.

A third proposed mechanism of HBD is that it reflects a decrease in excitatory and/or an increase in inhibitory neurotransmitters or neuromodulators in the brain (230, 269). Indeed, it has been shown that even with mild hypoxemia there is a change in synthesis or release of several excitatory and inhibitory neurotransmitters (70, 280, 392, 396, 397). The net effect of these changes would be hyperpolarization of a sufficient number of respiratory neurons so that firing frequency and therefore ventilation would decrease. The evidence suggests that altered levels of four inhibitory neuromodulators may mediate HBD.

One of these inhibitory neurotransmitters is adenosine. Brain concentration of adenosine correlates inversely with hypoxemia, spontaneous activity of cortical neurons is suppressed with application of adenosine (or an analog), and breathing is depressed with exogenous administration of adenosine into the brain (155, 193, 194, 207, 318, 396). Furthermore, infusion of an adenosine antagonist, aminophylline, has been shown to attenuate the inhibitory influence on breathing of adenosine (96), to attenuate $\dot{V}_E$ roll-off in adult humans (89, 133), and to alleviate HBD in some other preparations (126, 178, 207, 251, 273). However, other studies have found that this antagonist did not alleviate HBD (136); thus it appears that other neurotransmitters must also contribute to HBD under certain conditions.

A second inhibitory neurotransmitter is γ-aminobutyric acid (GABA) (154). Brain concentration of GABA increases progressively as PaO_2 decreases due to both an increase in production and a decrease in reuptake once released (176, 392, 397). The depressant effect of GABA has been shown by medullary topical and iontophoretic application to result in a reduction of breathing and cardiovascular functions (191, 400, 401). Administration of bicuculline, an antagonist of GABA (63), prevents depression of breathing by GABA (400). Finally, in the anesthetized cat HBD resulting from CO inhalation can be prevented by prior infusion of bicuculline (269).

In newborns and adults of some species opioids also probably contribute to HBD (52, 142, 270, 321, 324). Opioids depress breathing, and their levels increase during physiological stresses (321). However, opioids do not measurably increase during acute hypoxia (122). Nevertheless, administration of the opioid antagonist naloxone attenuates HBD in newborn and adult cats and dogs (52, 142, 270, 321). Naloxone does not appear to alter $\dot{V}_E$ roll-off during hypoxia in adult humans (186). As a result, it seems that opioids are contributing to HBD only under certain conditions.

Tatsumi et al. (356) suggest that a CNS dopaminergic mechanism mediates hypoxic ventilatory roll-off in awake cats. The glomus cells of the carotid bodies and regions within the CNS contain dopamine (54, 150, 157, 363, 368), which at these locations acts predominantly as an inhibitory modulator of breathing (16, 24, 56, 338). In cats haloperidol, a peripheral and centrally acting dopaminergic antagonist, will prevent hypoxic roll-off, but domperidone, a peripherally acting dopamine receptor blocker, has no effect (203, 356). These data indicate that a CNS dopamine mechanism mediates roll-off in this species.

It has also been proposed that lactacidosis in the brain contributes to ventilatory roll-off and HBD (272). Indeed, acidosis decreases the excitability of most neurons, including respiratory premotorneurons (5, 181, 200, 234). This might occur through an effect of acidosis on Ca influx into cells and/or intracellular Ca binding (316). An increase in ISF-ionized Ca decreases membrane excitability probably through an effect on the sodium channel protein molecule (5, 123, 174, 297). Acidosis may also cause depression through its well-known effect on enzymatic activity, which could contribute to alterations of excitatory and inhibitory neurotransmitters (399). Indeed, evidence suggests that intracellular acidosis results in an increase in production of the inhibitory neurotransmitter GABA (269). That lactacidosis may cause HBD is suggested by data in the anesthetized cat CO model of HBD, which showed a negative correlation between brain [H$^+$] and phrenic nerve activity (272). Furthermore, in this model preventing lactacidosis during moderate hypoxia by infusing dichloroacetate (DCA) prevents HBD (272). However, infusion of DCA in humans does not alter ventilatory roll-off (131). Nevertheless, the possibility remains that brain lactacidosis may actually be depressing breathing during acute hypoxia in spite of the fact that [H$^+$] at the chemoreceptor is probably increased. This notion is consistent with findings in anesthetized CBD rabbits, where $\dot{V}_E$ was increased during P_{CO_2}-induced medullary extracellular fluid (ECF) acidosis but not when medullary ECF pH was decreased to the same degree by hypoxia-induced lactacidosis (189).

Finally, it has been proposed that $\dot{V}_E$ roll-off is secondary to a decrease in metabolic rate. In several species metabolic rate ($\dot{V}_{O_2}$) decreases during hypoxia (120, 160, 260, 261, 315). Neonates (121, 160, 260, 261) are particularly susceptible to this effect, and in adults the magnitude of this effect is inversely related to body size (120). In other words, rats (278) are very susceptible, whereas ponies (232) show minimal effect of hypoxia on $\dot{V}_{O_2}$. This effect might be another manifestation of HBD, or the reduced $\dot{V}_{O_2}$ might be mediated by mechanisms independent of HBD. Nevertheless, it is clear that, on the one hand, reduced $\dot{V}_{O_2}$ contributes to a $\dot{V}_E$ roll-off in certain circumstances, yet, on the other hand, the temporal pattern and magnitude of roll-off in other conditions indicate that $\dot{V}_{O_2}$ has a minimal role.

In conclusion, acute hypoxia elicits multiple changes that influence breathing. The temporal patterns of these changes are not all the same, and the magnitude of each effect may vary with age, body size, environmental temperature, availability of substrates, etc. These changes are likely to account for the fact that the temporal pattern of $\dot{V}_E$ over an hour of hypoxia is variable between and within species. More importantly, the changes in key ventilatory stimuli proceed in opposite directions. For example, hyperventilation in response to low Pa_{O_2} results in a fall in Pa_{CO_2} and [H$^+$], which inhibits breathing. The balance between these two factors depends upon the individual sensitivities to hypoxia and CO_2–[H$^+$]. Since there are individual variations in sensitivity to these stimuli, the balance will differ between individuals, resulting in variations in $\dot{V}_E$ to acute hypoxia (320). Variation in responses is not restricted to hypoxia and CO_2–[H$^+$]. Clearly, it is the balance between all excitatory and inhibitory influences that determines the breathing responses to hypoxia. In adults the balance favors excitation; in the fetus inhibition is dominant; and in the newborn initially it is excitation, but inhibition often dominates.

VENTILATORY ACCLIMATIZATION TO SHORT-TERM HYPOXIA

It has long been recognized that pulmonary ventilation is increased upon ascent to high altitude. The progressive, time-dependent increase in ventilation that occurs after the first hour of hypoxia is termed *short-term ventilatory acclimatization*.

Historical Perspective

A very interesting historical summary of studies of ventilatory effects of altitude exposure has been published by Kellogg (187). He highlights key studies in the early twentieth century by such pioneers as C. G. Douglas and J. Barcroft and their colleagues, who clearly showed reduced alveolar P_{CO_2} in subjects sojourning at high altitudes. In 1911 John Haldane and C. G. Douglas led

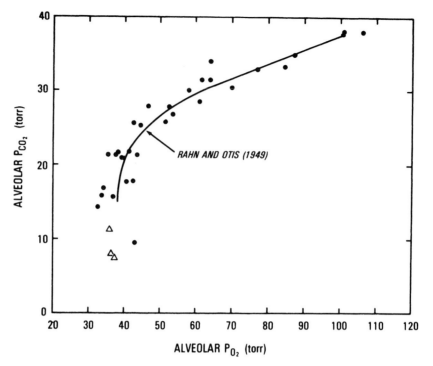

FIG. 52.3. O_2–CO_2 diagram indicating mean values of alveolar $P_{A}CO_2$ after completion of ventilatory acclimatization to various levels of prolonged hypoxia according to Rahn and Otis (308). Triangle values were added by West et al. (381) for extreme altitudes. [From West et al. (391).]

a party to the summit of Pike's Peak and demonstrated the time-dependent fall in alveolar PCO_2 and the continued hyperventilation on return to low altitude, both considered hallmarks of ventilatory acclimatization. Kellogg also summarizes the work of Mabel FitzGerald (109, 110), who carefully measured alveolar PCO_2 and found it to be lowered as a function of altitude in residents of the Colorado mountains.

One of the most cited and most historically significant papers is that of Rahn and Otis (308), who measured alveolar gas composition in human subjects acutely exposed to varying simulated altitudes of up to 22,000 feet in a hypobaric chamber. In addition, they documented the change in alveolar gases over a 3 wk exposure at a simulated altitude of 9,500 feet. They plotted their data along with data obtained by other pioneering investigators who had measured alveolar gases in acclimatized subjects at a variety of altitudes. The alveolar PO_2–PCO_2 plot (the Fenn diagram) derived from these data is a classic in the history of study of ventilatory acclimatization to hypoxia. This plot shows the fall in alveolar PCO_2 and the rise in alveolar PO_2 that occur with completion of acclimatization (Fig. 52.3) (391).

Time Course

Rahn and Otis (308) found that acclimatization was complete after 4 days at an altitude of 9,500 feet. Subsequent studies have suggested that the time course of

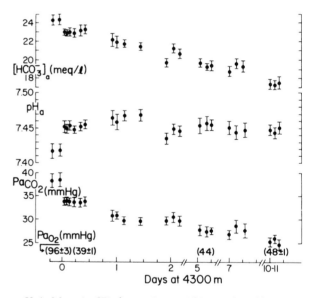

FIG. 52.4. Mean ($\pm$ SE) changes in arterial PO_2 and acid-base status in five human subjects during 10–11 days' sojourn at 4,300 m. [From Forster et al. (116).]

acclimatization in humans is dependent on the altitude of exposure, greater altitudes requiring longer periods of acclimatization. Moderate altitude (4,300 m) requires approximately 10 days (Fig. 52.4) (116), while the highest altitudes (>8,000 m) require more than 30 days (389). For most nonhuman mammals, acclimatization requires a much shorter period of time than for

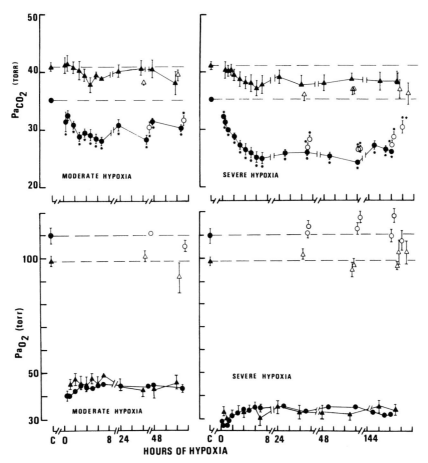

FIG. 52.5. Mean ($\pm$ SE) changes in arterial P_{CO_2} (Pa_{CO_2}) and arterial P_{O_2} (Pa_{O_2}) during exposure of carotid body intact (circles) and carotid body–denervated (triangles) goats to two levels of hypobaric hypoxia (moderate, *left*; severe, *right*). C indicates control normoxia values; open symbols indicate acute return to normoxia. Asterisks indicate significant difference from control normoxia value. [From Smith et al. (339).]

humans (74). For example, ventilatory acclimatization in goats exposed to an altitude of 4,300 m, as determined by measurement of arterial P_{CO_2}, is complete in approximately 4–6 h (112, 339) (Fig. 52.5). At the same altitude ponies exhibit a significant degree of acclimatization after the same duration of hypoxia (32, 113) but with an additional 2 torr fall after 24 h of hypoxia (32). Dogs acclimatize to an altitude of 3,550 m in 3 h (34), whereas rats appear to have a time course similar to humans, with about 2 wk required to acclimatize to 4,300 m (279). We are not aware of time course data for cats exposed to hypoxia, but there is strong evidence of acclimatization after 48 h at a simulated altitude of 4,570 m (370).

Variation

The previously mentioned variability of the response in individuals to acute hypoxia may also affect the magnitude of their short-term acclimatization to hypoxia. This has been illustrated in human subjects in whom

ventilation after acclimatization was significantly correlated with ventilatory response to acute isocapnic hypoxia (169). However, a notable exception to this was a subject who had a low level of hypoxic ventilatory sensitivity but who acclimatized normally (75). Individuals who have a strong ventilatory response to acute hypoxia at sea level are more likely to have a greater capacity to perform work at high altitude, such as climbing to great heights (236, 326).

MECHANISMS OF SHORT-TERM ACCLIMATIZATION TO HYPOXIA

Excellent reviews have been published, which are thorough in their coverage of the state of the art up to the early to mid-1980s (74, 380). We briefly summarize earlier studies and then focus on later discoveries.

The first main mechanism, suggested in the early 1900s, was that plasma lactic acidosis in hypoxia was responsible for stimulation of breathing during acclimatization to prolonged hypoxia; however, when the ability to measure blood pH became available, it became apparent that acclimatization was accompanied by respiratory alkalosis. After the discovery of the arterial chemoreceptors, it was thought that even though the blood was alkaline, the renal lowering of plasma bicarbonate could help to explain the progressive hyperventilation characteristic of acclimatization (187). Dempsey and Forster (74) extensively reviewed the literature on acid-base changes in arterial blood during acclimatization in humans and a variety of animals and showed that there was no uniform positive relationship between arterial [H$^+$] and level of ventilation during acclimatization. Arterial blood remains alkaline during acclimatization as PaCO$_2$ progressively falls (Fig. 52.4). Since PaO$_2$ also rises during acclimatization, it was concluded that "changes in ventilation during short-term hypoxia are the cause, rather than the result, of changes in arterial pH and PaO$_2$" (74).

Role of Medullary Chemoreceptors

Historically, since arterial blood gas stimuli did not seem to provide a stimulus which could explain the time-dependent increase in ventilation via the peripheral chemoreceptors, physiologists looked to central chemoreceptors as a possible source of stimulation during acclimatization. One major theory put forth by Mitchell et al. (254–256) was that the ventilatory response to any acid-base derangement is the sum of stimuli from the H$^+$-sensitive medullary and peripheral chemoreceptors. In these studies the putative central chemoreceptors were believed to be on or near the ventral surface of the medulla and that their H$^+$ stimulus level could be reflected in the [H$^+$] of the bulk CSF. This led to studies by Severinghaus et al. (332), who found normal CSF pH in humans after 2 and 8 days at high altitude. They assumed that CSF was alkaline during acute hypoxia and therefore suggested that acclimatization to hypoxia was due to time-dependent restoration of CSF [H$^+$] to normal. This was thought to be due to active transport of HCO$_3^-$ or H$^+$ and/or an increase in brain production of lactic acid (342, 344). The attractive Severinghaus theory was accepted by many physiologists; however, their enthusiasm was dampened by subsequent studies which strongly indicated that CSF remains alkaline during acclimatization in humans and animals (34, 60, 76, 77, 113, 116, 286, 386).

In contrast to the Mitchell theory, Fencl, Pappenheimer, and their colleagues (105, 292) found evidence to suggest that the intracranial chemoreceptor responded to cerebral ISF pH and not to CSF. Therefore, other investigators sought to determine if [H$^+$] closer to the putative central chemoreceptors could be responsible for increasing ventilation. Davies (67) calculated that

cerebral ISF acidity increased in the environment of the putative central chemoreceptors in hypoxic anesthetized dogs even though bulk CSF remained alkaline. Subsequently, Fencl et al. (104), using the ventriculocisternal perfusion technique in awake goats after 5 days of hypoxia, concluded that cerebral ISF and CSF were in marked ionic disequilibrium during hypoxia and that lactic acidosis occurred at the putative central chemoreceptor, potentially explaining ventilatory acclimatization. However, this hypothesis is not consistent with findings that acclimatization does not occur in CBD animals that theoretically also have cerebral lactacidosis. Moreover, as already summarized (see Mechanisms Inhibiting Breathing and Potentially Causing Roll-Off), evidence suggests that lactic acid production in hypoxia, rather than being a respiratory stimulant, inhibits breathing by increasing the level of the inhibitory neurotransmitter GABA (272). Finally, Musch et al. (262) found during moderate and severe hypoxia in rats that brain tissue and ISF lactacidosis did not correlate with the temporal pattern of ventilatory acclimatization.

Others have proposed that stimulation of the carotid body could produce hypocapnic alkalosis and that it is the combination of hypoxia and alkalosis that induces a localized metabolic acidosis in the CNS, which in turn stimulates central chemoreceptors. Indeed, Eger et al. (93) hypothesized that acclimatization was dependent on hypocapnia and that the time-dependent correction of ISF pH near the central chemoreceptor increased ventilatory drive during acclimatization. Their theory was that maintaining PaCO$_2$ constant during hypoxia should prevent acclimatization. They studied human subjects exposed to 8 h of hypoxia or normoxia with varying levels of maintained PaCO$_2$. Their results showed only partial prevention of acclimatization during normocapnic hypoxia, making a firm conclusion impossible.

In a similar study, Cruz et al. (61) exposed subjects to hypobaric hypoxia in an altitude chamber and attempted to maintain isocapnia by adding 3.77% CO$_2$ to the chamber atmosphere. Unfortunately, the results of the study are not clear because end-tidal CO$_2$ was not obtained in the CO$_2$-supplemented group at 3 h of hypoxia and was elevated 3 torr above control after 27 h of hypoxia, confounding interpretation of the ventilatory response. In addition, the control group failed to exhibit hypocapnic alkalosis on exposure to hypoxia; therefore, they could not document the basic premise that hypocapnic alkalosis is a requisite factor initiating acclimatization. Thus neither study in human subjects could clearly confirm its hypothesis, though the data indicated that acclimatization was attenuated based on the study's criteria for acclimatization. We have examined this hypothesis in awake goats and have clearly demonstrated acclimatization to hypoxia during whole-body isocapnic exposure (97, 319) and during isolated carotid body exposure (13), strongly suggesting that the mechanism for ventilatory acclimatization to short-term hypoxia is not dependent on hypocapnia; rather, hypocapnia appears to be a consequence of acclimatization.

Goldberg et al. (138) utilized ^{31}P-nuclear magnetic resonance to determine changes in brain intracellular pH during acclimatization to hypoxia and during deacclimatization in human subjects. Their objective was to determine if brain intracellular pH decreased in a uniform manner in response to hypoxia, which could suggest acid stimulation of central chemoreceptors. The investigation revealed no change in intracellular pH in a 2.5 cm thick slice through the brain during hypoxic exposure. However, there was a transient decrease in brain pH 15 min after return to normoxia following 7 days of hypoxia and another fall 12 h after return to normoxia. These data do not support a ventilatory stimulation role for brain intracellular acidosis in acclimatization to hypoxia. However, because of the possible localized nature or extracellular position of central chemoreceptors, their role in acclimatization is not ruled out by the findings.

In summary, after an extensive history searching for a role for central chemoreceptors in acclimatization, little convincing evidence remains that acidification and stimulation of central chemoreceptors is responsible for the increased ventilatory drive during acclimatization to short-term hypoxia. However, the entire topic of central chemoreception remains controversial. For example, data indicate that chemoreceptive elements have been located not only near the ventral medullary surface (175, 264–266) but also in other medullary (71, 72, 268) and hypothalamic (374) locations. Until we better understand the structure, location, and function of the central chemoreceptors, their role in acclimatization will not be clear.

Role of Peripheral Chemoreceptors

Carotid Body Denervation Studies. A great deal of attention has been directed toward the role of the peripheral arterial chemoreceptors, especially the carotid bodies, in acclimatization to hypoxia. The primary method of assessing this has been by denervation of the carotid bodies to ascertain whether acclimatization is modified or eliminated. Bouverot and Bureau (26) found that CBD dogs with residual aortic chemoreceptor function exhibited delayed acclimatization to hypoxia. Intact dogs acclimatized within 3 h, as demonstrated by the maximal fall in $PaCO_2$, whereas CBD dogs did not hyperventilate until after 24 h of hypoxia. Carotid and aortic body–denervated ponies exhibited an attenuated acclimatization to hypoxia, as indicated by a fall in $PaCO_2$ of 3 torr after 8 h of hypoxia compared to a 5 torr fall in intact ponies (113). Decreased $PaCO_2$ was not sustained in the denervated ponies beyond 18 h of hypoxia, while it remained present for more than 92 h in the intact ponies.

Similarly, two studies in goats found that acclimatization was clearly attenuated in CBD animals (112, 339). The study by Smith et al. (339) was perhaps the most thorough of any on the subject of acclimatization in intact and CBD goats. These investigators used two arterial levels of hypoxia (44 torr and 33 torr) and carefully measured duplicate and triplicate arterial blood gases. At both PaO_2 levels acclimatization was significantly attenuated in CBD animals (Fig. 52.5). One notable study did not reproduce this effect in goats (347). These investigators found that CBD goats exhibited significant hyperventilation (a reduction in $PaCO_2$ of 10 torr) after 3 days at a barometric pressure of 446 torr. This difference is difficult to reconcile with those of Smith et al. (339) and Forster et al. (113), whose results, however, are supported by studies using the isolated perfused carotid body in awake goats and showing that acclimatization was not induced during systemic hypoxia while the carotid body remained normoxic (387).

Responses to Isolated Carotid Body Hypoxia. To specifically address the issue of acclimatization to hypoxia, we devised the carotid body perfusion model in awake goats. The goat was considered a good model because it exhibits acclimatization within 4 h and has a cerebral circulation which allows separation of carotid body blood flow to the CNS (40). In this model blood gases perfusing the carotid body are controlled by an extracorporeal circuit, while systemic (and CNS) blood gases are determined by inhaled gas and the level of pulmonary ventilation. The first question asked was: Is brain hypoxia required for induction of acclimatization? This was addressed in two studies, one in which systemic arterial blood gases were used to assess acclimatization (Fig. 52.6, left) and one in which ventilation was measured and systemic $PaCO_2$ was maintained constant to prevent hypocapnic alkalosis (13, 40). The results of these studies showed that neither CNS hypoxia nor hypocapnic alkalosis was required to induce acclimatization in the goat.

In another study we wished to determine if a unique stimulus at the carotid body was required to initiate acclimatization. In this case normoxic hypercapnia was used as the stimulus (12). A sufficient level of normoxic hypercapnic stimulus was given to the carotid body to achieve the same acute stimulation of ventilation as with hypoxic stimulation. In this case no time-dependent acclimatization occurred, indicating that the unique stimulus for acclimatization to hypoxia in the goat is hypoxia of the carotid body (Fig. 52.6, right).

These studies were followed by experiments designed to determine if carotid body afferent neural traffic

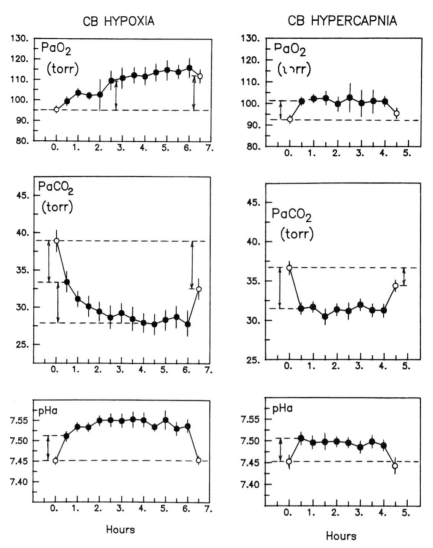

FIG. 52.6. *Left panel:* Mean (± SE) changes in systemic arterial P_{O_2}, P_{CO_2}, and pH in six awake goats during 6 h of isolated perfusion of the carotid body with blood controlled at a P_{O_2} of 40 torr by an extracorporeal circuit. Open symbols indicate normoxic conditions at the carotid body. The time-dependent fall in Pa_{CO_2} from 0.5 h to 4 h is indicative of acclimatization in the absence of brain hypoxia. *Right panel:* Mean (± SE) changes in systemic arterial P_{O_2}, P_{CO_2}, and pH in six awake goats during 4 h of isolated perfusion of the carotid body with blood controlled at a mean normoxic P_{CO_2} of 78 torr by an extracorporeal circuit. Open symbols indicate normocapnic conditions at the carotid body. There was no evidence of acclimatization (no time-dependent fall in Pa_{CO_2}) with hypercapnic stimulation of the carotid body. Arrows indicate significant differences between means ($P < 0.05$). [From Bisgard et al. (12).]

changed in a time-dependent manner in anesthetized goats. It was found that up to 4 h of isocapnic hypoxia produced time-dependent increases in carotid chemoreceptor afferent activity (Fig. 52.7) (275), while a similar period of normoxic hypercapnia stimulation produced no time-dependent change (100). These results are consistent with the findings of carotid body perfusion studies in awake goats and provide evidence that hypoxia can increase afferent activity of the carotid body in a time-dependent manner.

Natural acclimatization is accompanied by a falling Pa_{CO_2} and a rising Pa_{O_2}; however, in the above carotid body perfusion studies in awake goats and in the carotid sinus nerve recording studies steady-state isocapnic hypoxic stimulation to the carotid body ($Pa_{O_2} = 40$ torr) was used in all cases. Steady-state stimuli were used to minimize changes in stimulus levels in carotid body arterial blood and thus to ascertain the responses of the specific effect of hypoxia alone on the carotid body and on the resulting $\dot{V}_E$ changes. Maintaining these stimuli

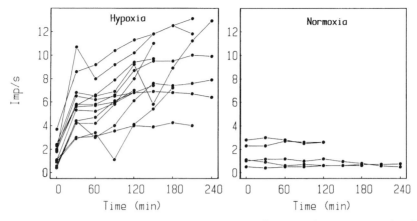

FIG. 52.7. *Left panel:* Data from chloralose-anesthetized goats illustrating the time course of single fiber carotid body afferent discharge frequency up to 4 h of arterial hypoxia (PaO$_2$ = 40 torr, n = 13). *Right panel:* Data from control normoxic goats (n = 5). [From Nielsen et al. (275).]

constant is unlike natural acclimatization but was not expected to change the basic hypoxia-induced mechanisms affecting the carotid body.

Vizek et al. (370) examined ventilatory and carotid chemoreceptor afferent activity in cats after 48 h at a barometric pressure of 440 torr. They found that both carotid body afferent activity and the ventilatory response to acute hypoxia were elevated after hypoxic exposure. In a second study on cats the same laboratory confirmed that these responses were augmented (357), thus supporting the view that increased sensitivity of the carotid chemoreceptors to hypoxia contributes to acclimatization to hypoxia.

Barnard et al. (7) examined carotid chemoreceptor activity for up to 3 h in cats and found no time-dependent increase; however, this may only indicate that the cat undergoes a slower time course of acclimatization than the goat. In the same study after 28 days of hypoxia cats exhibited an increased chemoreceptor afferent response to acute hypoxia. This finding is consistent with increased carotid body hypoxic sensitivity after hypoxic exposure.

Ventilatory Responses to Acute Hypoxia during and after Short-Term Acclimatization. Studies in goats (97, 319) using both poikilocapnic and isocapnic exposure to hypoxia have shown that the ventilatory response to acute hypoxia is augmented upon return to normoxic conditions as compared to the response measured prior to hypoxic exposure. These findings are consistent with an increasing carotid body response to acute hypoxia as a primary mechanism of acclimatization. Similar results were reported by Vizek et al. (370) and Tatsumi et al. (357) in cats, as mentioned above. Rats have also been found to have augmented ventilatory responses to acute

hypoxia during exposure to a simulated altitude of 5,500 m (287).

Most earlier studies in humans indicated little or no change in the response to acute hypoxia during altitude sojourn (see ref. 115 for review). However, subsequent studies in humans have shown that the ventilatory response to acute hypoxia increases during exposure of lowlanders to chronic hypoxia (115, 138, 322, 327, 393). Forster et al. (115) found that the response to hypoxia was increased significantly on days 4 and 45 of exposure to 3,100 m altitude and that it remained elevated on day 7 after return to sea level. White et al. (393) showed that the ventilatory response to acute hypoxia increased progressively with time of exposure to 4,300 m. A similar finding was reported by Sato et al. (322). In the Operation Everest II project six subjects exposed to simulated altitude at barometric pressures of 428 and 305 torr had increased hypoxic ventilatory responses, which were greatly augmented after return to sea level pressure (765 torr) (327).

Goldberg et al. (138) found that all four subjects exposed to a pressure of 447 torr had an increased hypoxic ventilatory response 2–4 days after return to sea level. These studies indicating increased hypoxic sensitivity in human subjects are consistent with the increase in carotid chemoreceptor sensitivity to hypoxia described in animal studies (7, 12–14, 97, 275, 319, 370). However, it is not possible to rule out increased CNS sensitivity to carotid body input in the human studies. The general ventilatory hyperresponsiveness found in human subjects (114) (see Role of the Central Nervous System, below) could reflect primarily a CNS effect, a carotid body effect, or a combination of the two.

Studies in goats, limited to the first 4–6 h of hypoxia, point to increased carotid body sensitivity to hypoxia as a cause for the increased ventilatory drive in acclimatization. The data indicate no evidence for a change in CNS mechanisms being responsible for acclimatization in the goat. One cannot rule out the possibility that some CNS mechanisms could come into play with more prolonged or severe hypoxia. Species differences may also be important. It is possible that CNS mechanisms are of greater importance in some species, such as humans, and that in species such as the goat carotid body mechanisms predominate.

Possible Mechanisms for Increased Carotid Body Sensitivity during Short-Term Acclimatization

The reason(s) for increased sensitivity of the carotid body to hypoxia during chronic hypoxia remains unknown. Potential mechanisms for increased sensitivity of the carotid body include intrinsic influences such as increased neurochemical facilitation, a change in the fundamental chemical transduction mechanism, or depression of a neurochemical inhibitory mechanism. Possible extrinsic mechanisms include influence of circulating substances or changes in efferent neural modulation of the carotid body sensory mechanisms. Most efforts to understand the mechanism for increased sensitivity have been directed toward changes in neuromodulator activity, especially dopamine, a putative inhibitory carotid body amine. There have been no studies aimed at determining if the fundamental transduction process is changed by prolonged hypoxia; therefore, no discussion of it is included in this review. The transduction process remains to be fully elucidated but is currently under intense investigation (139). Similarly, we are unaware of any effort to determine if any circulating substance(s) could change the sensitivity of the peripheral chemoreceptors to hypoxia during acclimatization. While no such substance has been suggested, this possibility cannot be completely eliminated from consideration.

Possible Changes in Intrinsic Neurochemical Modulation of the Carotid Body.

Dopamine is generally considered an important inhibitory neurochemical modulator of carotid body afferent output (16, 20, 87, 229, 402). Carotid body type I cells contain a large quantity of stored dopamine, which is released by acute hypoxic stimulation causing a reduction in dopamine content (150, 156, 280). In rats more prolonged hypoxia (days to weeks) causes a marked increase in carotid body dopamine content (151, 280) and in the rate of dopamine turnover (294).

Studies have been carried out to determine if carotid body dopamine could play a role in short-term acclimatization in goats (19, 97, 99). One hypothesis tested was that effective dopaminergic blockade would greatly accelerate the rate of acclimatization to hypoxia, if acclimatization was due to a time-dependent depletion of carotid body dopamine, that is, loss of dopaminergic inhibition. This was tested by giving goats exposed to hypoxia an effective dopamine D2 antagonist, domperidone, which does not cross the blood–brain barrier (203). The expected result was that there would be a more rapid acclimatization with domperidone; however, the results were equivocal as the rate of acclimatization was hastened but not clearly enough for a firm conclusion (19). Two additional studies did not support a role for dopaminergic mechanisms in the acclimatization of the goat to hypoxia (97, 99). In one case no change in carotid body dopamine content was found after 4 h of hypoxia, suggesting that no time-dependent depletion of dopamine occurred (99). In the other study ventilatory responses to intravenous infusions of dopamine were given before and after acclimatization to hypoxia. Depression of ventilation during these infusions has been demonstrated to be carotid body–mediated in goats (16, 197). There were no changes in the responses to dopamine infusion after acclimatization to hypoxia, suggesting that no change in carotid body dopamine receptor sensitivity occurred (97).

In a study using cats, Tatsumi et al. (357) obtained results contrary to our findings in goats. They found that under control conditions the dopamine antagonist domperidone augmented the ventilatory and carotid sinus chemoreceptor afferent responses to acute hypoxia and hypercapnia. However, after acclimatization these responses were high and could not be augmented further by administration of domperidone, thus, indicating that it is likely that decreased dopaminergic activity in the peripheral chemoreceptors is playing an important role in acclimatization to short-term hypoxia in the cat.

While the cat and goat studies are not in agreement, it appears that there is a possible role for dopaminergic mechanisms in short-term acclimatization. Indeed, the studies in goats, while not supporting a role for dopamine, cannot entirely rule it out.

Another important carotid body catecholamine is norepinephrine. It is abundant in the carotid body (2, 157, 238), in both sympathetic nerve terminals, and in chemosensory type I cells (150, 375). However, its role in carotid body function is not clear. Both excitatory and inhibitory roles for norepinephrine in the carotid body have been described (20, 62, 78, 183, 229). Kou et al. (195) provided convincing evidence for inhibitory α-2 adrenergic receptors in the carotid body of the adult cat. Guanabenz, the α-2 agonist they used, caused an inhibition of carotid chemoreceptor afferent activity. Cao et al. (44), from this same group, reported that the inhibitory response to guanabenz was diminished in cats exposed to 24–36 h of hypoxia. This suggests that an inhibitory modulating effect of adrenergic receptors could be down-regulated during exposure to hypoxia and, therefore, be responsible for increased carotid body sensitivity to hypoxia. The increased hypoxic sensitivity could provide additional drive to breathe in cats exposed to chronic hypoxia. Exogenous norepinephrine has also been shown to be inhibitory in the goat carotid body (301, 376), and a similar study of acclimatization to hypoxia as that carried out by Cao et al. was undertaken (319). In this study intracarotid norepinephrine infusions were given before, during, and after acclimatization in awake goats to determine if there was a loss of the inhibitory effect, as suggested by the study of Cao et al. (44). In the goat infusion study there was no attenuation of the inhibitory effect of norepinephrine after acclimatization to hypoxia, indicating no significant change in carotid body adrenergic receptor sensitivity. Thus the possible α-2 adrenergic carotid body mechanism, which may contribute to acclimatization in cats, does not seem to be present in goats.

In addition to dopamine and norepinephrine, the carotid body contains many other neurochemical agents which could be involved in the hypoxic response. These include adenosine, serotonin, acetylcholine, and many neuroactive peptides such as substance P, neurokinin A, enkephalins, vasoactive intestinal peptide, calcitonin gene-related peptide, atrial natriuretic peptide, and neuropeptide Y (106, 201, 202, 244, 305). None of these agents has been assessed for a possible role in the response of the carotid body to prolonged hypoxia as in acclimatization to short-term hypoxia. Among the more interesting of the peptides in the carotid body is substance P (304–307). Exogenously administered substance P has been reputed to increase carotid body afferent discharge (307), and substance P antagonists are reported to depress the hypoxic response of the carotid body while sparing its response to hypercapnia (304, 306). Therefore, it may be an important link in the hypoxia sensory mechanism of the carotid body. As such, a change in its metabolism or in its receptor function induced by prolonged hypoxia could play a role in the response of the carotid body to prolonged hypoxia.

Possible Changes in Extrinsic Modulation of Hypoxic Sensitivity of the Carotid Body.

The carotid body has sympathetic efferent innervation that is mediated by a few preganglionic neurons and a greater proportion of postganglionic sympathetic neurons in the ganglioglomerular nerve from the superior cervical ganglion (102, 239, 367). These fibers innervate the vasculature of the carotid body as well as some of the type I glomus cells (239, 240). The sympathetic neurons, in addition to

containing norepinephrine, also have been found to contain peptides, including neuropeptide Y (192, 201) and possibly others (277). The influence of sympathetic innervation on carotid body function has been the object of considerable interest. However, carotid body afferent responses to acute hypoxia and hypercapnia are not significantly altered by sympathectomy (237). Thus there seems to be no significant role for sympathetic innervation in the response of the carotid body to acute physiological stimuli. However, the content and turnover of dopamine and norepinephrine in the carotid body are increased during chronic hypoxia (151, 280, 294), and this is at least partially under sympathetic neural control (295). Therefore, one cannot yet rule out the possible involvement of sympathetic efferent control of the carotid body playing a role in modulating carotid body sensitivity during prolonged hypoxia. We are unaware of any studies directly examining the role of sympathetic nervous control of ventilation in short-term acclimatization to hypoxia.

Another efferent influence on the carotid body is reported to exist in the fibers descending from the medulla in the carotid sinus nerve. Efferent inhibitory control of carotid body activity via carotid sinus nerve efferents has been studied extensively (277). The source of these fibers is primarily the rostral part of the nucleus ambiguus (10, 68, 73). The effect of carotid sinus nerve efferent fibers on the carotid body is inhibitory, for if the main part of the carotid sinus nerve is severed while recording afferent activity from a small slip of the nerve, hypoxic discharge frequency increases (267). The inhibition may be mediated by dopamine release (221). The role of the carotid sinus efferents in carotid body activity during chronic hypoxia has been examined in anesthetized cats exposed to hypoxia (10% O_2) for 21–49 days by Lahiri et al. (220, 221). These workers found evidence for increased efferent inhibition in chronically hypoxic cats which may be dopaminergically mediated and for increased sensitivity to hypoxia after sinus nerve section. Their results do not directly indicate the role carotid sinus efferents have but suggest that one possible mechanism of increased sensitivity of the carotid body to hypoxia after prolonged exposure could be a reduction in efferent inhibition.

While efferents may play a role in increased sensitivity of the carotid body to hypoxia, this role is apparently not essential. Nielsen et al. (275) found time-dependent increasing chemoreceptor discharge frequency in anesthetized goats (fig. 52.7) during prolonged hypoxia with the carotid sinus nerve cut (eliminating efferents) for recording. Similarly, Vizek et al. (370) recorded carotid chemoreceptor discharge frequency from intact and proximally sectioned carotid sinus nerves and found that the increased hypoxic sensitivity remained unchanged after efferents were eliminated in hypoxia-acclimatized cats.

Role of the Central Nervous System

Forster and Dempsey (114) postulated that a generalized CNS hyperexcitability occurs in hypoxia that contributes to ventilatory acclimatization to hypoxia. Their hypothesis was based on studies by Tenney and coworkers (289, 361) in cats suggesting that cortical facilitation of ventilation is induced during prolonged hypoxia. Other evidence to support a CNS facilitory contribution include increased ventilatory responses to exercise (75), hypercapnia (74, 77, 115, 188, 286), and intravenous doxapram (113, 117) and spontaneous and visually evoked electroencephalographic (EEG) changes during sojourns at high altitude (118, 325).

It has been postulated that disturbances in CNS monoamine metabolism could result in increased ventilatory facilitation during acclimatization. Olson et al. (280) examined monoamine metabolism in the CNS of rats exposed to 7 days of hypoxia. These investigators found that norepinephrine turnover was unaffected but that dopamine turnover was reduced significantly. Serotonin turnover fell 50% during acute hypoxia but returned to normal. On return to normoxia, both dopamine and serotonin turnover increased above prehypoxia control levels. Thus there were changes in CNS monoamine metabolism during hypoxia; however, the roles for these changes in acclimatization are unclear.

Millhorn et al. (252, 253) showed that a long-term potentiation of CNS ventilatory drive may be induced following strong peripheral chemoreceptor stimulation in cats. This effect was diminished by pretreatment with methysergide, parachlorophenylalanine, or 5,7 dihydroxytryptamine, indicating a serotonergic mechanism. They postulated that acclimatization could be a serotonergically mediated CNS effect initiated by strong hypoxic stimulation of the carotid body. Olson (278) directly tested the serotonin hypothesis in awake rats, finding that acclimatization proceeded normally even after inhibition of serotonin synthesis with parachlorophenylalanine, thus diminishing support for such a hypothesis.

We have found that acclimatization can be induced by isolated hypoxic stimulation of the carotid body in awake goats but not with stimulation of the carotid body by normoxic hypercapnia at a level sufficient to produce the same acute stimulation of ventilation as hypoxia does (12, 40). Therefore, nonspecific (nonhypoxic) stimulation of the carotid body by itself will not induce acclimatization in goats. This finding does not support the Millhorn serotonergic hypothesis for acclimatization.

Does hypoxic CNS depression play a role in accli-

matization to hypoxia? In other words, is there a time-dependent alleviation of the depression that occurs during acute hypoxia (see under Mechanisms Inhibiting Breathing and Potentially Causing Roll-Off), which could explain the progressive increase in ventilation with acclimatization? To directly test this possibility we examined the ventilatory effects of CNS hypoxia utilizing the awake goat carotid body perfusion model (387). Systemic (CNS) hypoxia (Pa_{O_2} 40 torr) was induced by inhalation of hypoxic gas while the carotid body was maintained normoxic, thus avoiding compensatory effects following chronic denervation and allowing normoxic tonic carotid body activity to continue. The intact perfused carotid body also allowed testing of the response to acute carotid body stimulation. Animals were selected which did not have significant aortic body function; therefore, systemic hypoxia with carotid body normoxia could be interpreted as CNS hypoxia. We found that 4 h of CNS hypoxia alone with carotid body normoxia produced mild ventilatory stimulation but no time-dependent increase in ventilation typical of acclimatization. In addition, there was no change in the ventilatory response to carotid body hypoxia after prolonged CNS hypoxia, suggesting that prolonged brain hypoxia did not produce CNS changes which would modify the input from the carotid body. Thus these studies indicate that brain hypoxia per se is not an important factor in acclimatization to hypoxia in the goat. However, these results and those from previous studies in goats and ponies (32, 113, 339) suggest that there may be a small and short-lived CNS contribution to ventilatory drive during the first hours to days of hypoxic exposure.

Ventilation, Metabolic Rate, and Lung Mechanics

Changes in alveolar or arterial P_{CO_2} have been the main variables used to quantify short-term acclimatization to hypoxia, and this is appropriate because these variables accurately reflect the relationship between metabolic rate and alveolar ventilation. In human subjects resting oxygen consumption during prolonged altitude sojourn is normal or only slightly elevated (9, 61, 75, 141, 190, 373). Assuming that volume of CO_2 elimination (V_{CO_2}) follows the same pattern of change, which was the case in studies reported by Klausen et al. (190), then the typical fall in Pa_{CO_2} would be associated with true alveolar hyperventilation. Indeed, an increase in minute ventilation is typically found in human studies of altitude acclimatization (74, 190, 393). A consistent finding is that Pa_{CO_2} falls progressively in goats and ponies during the time course of acclimatization; however, resting minute ventilation follows a less consistent pattern, sometimes lower than would be expected (32, 97, 112, 113). A change in metabolic rate has been examined as a potential cause for the discrepancy between Pa_{CO_2} and ventilation. These measurements have shown variable but modest changes in resting metabolism that cannot fully account for the lower than expected ventilation (17, 112, 113, 168, 347). It is possible that factors other than increased ventilatory drive may sometimes contribute to the fall in Pa_{CO_2} that is highly reproducible during acclimatization. Changes in ventilatory pattern, airway volume, or differential respiratory muscle recruitment, which may change dead space ventilation and/or gas exchange, could be responsible for these changes (32).

Gautier et al. (130) reported changes in lung mechanics in human subjects during acclimatization to short-term hypoxia. These include a significant reduction in airway resistance, mildly decreased lung elastic recoil, and lower than predicted functional residual capacity (FRC). In addition, they found that mouth occlusion pressure ($P_{0.1}$) increased on the first day at 3,547 m and then remained constant each day, while mean inspiratory flow (V_T/T_I) progressively increased (129, 130). On this basis they suggested that a change in lung mechanics, rather than increased neural drive, could be responsible for increased ventilation at high altitude.

DEACCLIMATIZATION FOLLOWING ACCLIMATIZATION TO SHORT-TERM HYPOXIA

Deacclimatization is the hyperventilation that persists and then diminishes in a time-dependent manner to a normal level of ventilation upon return from altitude sojourn. It is well known that hyperventilation (deacclimatization) continues after return to normoxic conditions in acclimatized human subjects and animals (74). Few systematic studies have been carried out on deacclimatization. Perhaps the best data on the topic are those of Dempsey et al. (76), who studied human subjects and ponies during recovery from a simulated altitude of 4,300 m. These workers found that Pa_{CO_2} increased progressively over the first 13 h of normoxia, and in their last measurement, after 24–25 h of normoxia, Pa_{CO_2} was near the normal. However, arterial pH was slightly acidotic relative to prehypoxia values because of slow metabolic compensation for the rising Pa_{CO_2}. Both CSF and arterial acid-base changes were dissociated from the rising Pa_{CO_2} and, therefore, could not explain either the continued hyperventilation or the time-dependent reduction in ventilation during deacclimatization. Thus the cause of the maintained hyperventilation was not revealed.

The general hypothesis has been that the mechanism for deacclimatization is likely to be the same as that for acclimatization. Those workers espousing a change in central chemoreceptor stimulus level as a cause for acclimatization have assumed that the same mechanism is responsible for deacclimatization, that is, that a small rise in P_{CO_2} would increase acidity at the central chemoreceptor and maintain hyperventilation. Again, this mechanism cannot be ruled out until we can isolate the central chemoreceptor and determine its stimulus level in deacclimatization.

In studies indicating that increased peripheral chemoreceptor drive contributes to acclimatization to short-term hypoxia, the possibility that deacclimatization may be a separate mechanism has emerged. This was suggested after noting that hyperventilation continued in acclimatized goats returned to normoxia and subjected to either isolated carotid body hypoxia (13, 40) or whole-body hypoxia (97) when Pa_{CO_2} was allowed to fall naturally (poikilocapnic hypoxia), whereas goats maintained systemically normocapnic during either whole-body or isolated carotid body hypoxia did not hyperventilate on return to normoxic conditions (97, 319). Acclimatization occurs whether maintaining systemic normocapnia or not, while persistent hyperventilation requires the development of hypocapnic alkalosis. Thus it appears that prolonged systemic (CNS) hypocapnic alkalosis is required during acclimatization to produce hyperventilation during deacclimatization. During poikilocapnic acclimatization both central and peripheral mechanisms may be present, but the central mechanism (shift to a lower CO_2 set point) is overwhelmed by the peripheral chemoreceptor drive. When the peripheral chemoreceptor drive is removed, the central drive associated with respiratory alkalosis is manifested. Vizek et al. (370) also suggested that deacclimatization and acclimatization were separate mechanisms based on the presence of continued hyperventilation in acclimatized cats after sinus nerve section. It appears that carotid bodies play a significant role in acclimatization, while CNS mechanisms are responsible for deacclimatization. That carotid bodies do not contribute to hyperventilation on return to normoxia is further supported by finding no increase in normoxic carotid chemoreceptor afferent activity after prolonged hypoxia in anesthetized goats (14).

VENTILATION AND VENTILATORY CONTROL IN LONG-TERM HYPOXIA

Short-term ventilatory acclimatization to hypoxia may last for years in human sojourners, with hyperventilation maintained at as high a level as at the end of the initial period of acclimatization. However, in natives or very long-term residents of altitude hyperventilation is attenuated and Pa_{CO_2} rises. There seems to be no ventilatory control difference between very long-term residents and those adapted from birth at high altitude, thus we refer to these groups together as highlanders. The new level of Pa_{CO_2} is indicative of continued hyperventilation relative to sea-level natives and is intermediate between unacclimatized and short-term-acclimatized subjects when plotted on the Fenn diagram (fig. 52.1) (74, 208, 212, 227, 328, 380). This decrease in ventilation in highlanders is accompanied by a dramatic loss of ventilatory sensitivity to acute hypoxia *(blunting)* (115, 216, 227, 247, 328, 329, 381) and in most studies no change in the response to hypercapnia (55, 115, 247, 329), though a diminished response has been found using transient (343) or nonsteady-state (381) methods, which could reflect the influence of diminished peripheral chemoreceptor function. Similar reductions in responses to transient hyperoxia, two or three breaths of 100% O_2, or to hypoxia, two to five breaths of N_2, in highlanders suggest a reduction in the peripheral chemoreflex (227).

It has been reported that administration of O_2 produces hyperventilation in highlanders compared to lowlanders sojourning at altitude (145, 216, 247). This observation is unexpected since a decrease in ventilation from hyperoxic depression of the peripheral chemoreceptor drive is the typical response. This hyperventilation in highlanders has been interpreted as being possibly due to relief of central hypoxic depression (380). It may also be a direct CNS effect of hyperoxia itself (65). Hyperoxic hyperventilation is commonly observed in CBD animals (65, 249) and after acclimatization to hypoxia (113, 339). Thus a number of possible mechanisms could be compatible with this hyperventilation: lack of peripheral chemosensitivity, relief of central hypoxia, or a direct CNS effect of hyperoxia itself.

Loss of hypoxic sensitivity in highlanders is apparently an acquired characteristic (that is, not inherited) for it appears to be a function of time in residence at altitude. Weil et al. (381) demonstrated a significant time-dependent decreased ventilatory response to acute hypoxia in nonnative subjects residing in Leadville, Colorado, at 3,100 m for 3–39 yr. The mean reduction of this group compared to control was 43%. Native subjects born and residing at 3,100 m for 29–56 yr had a mean response 9.6% of that of sea-level control subjects. A similar trend was found by Forster et al. (115). Hackett et al. (145) produced data from Himalayan Sherpas, many of them mountaineers, indicating that the blunting of hypoxic sensitivity was a function of degree of hypoxia (kilometers above sea level) and time in residence at altitude (km × yr). Children resident of Leadville (3,100 m) have normal responses to acute hypoxia, indicating that birth or early life at altitude does not confer loss of the acute hypoxic response (41). Similar observations were made on children resident in the Peruvian Andes at 3,850 m (212). The blunting of hypoxic chemosensitivity remains for many years upon return to sea level and was once thought to be a per-

manent change (216, 345). However, some patients formerly hypoxemic from right to left congenital cardiac shunts and with depressed hypoxic responses (346) may recover hypoxic sensitivity after correction of the hypoxemia (23, 91). Highlanders resident for long periods at low altitude may also regain some hypoxic sensitivity (211).

Mechanisms of Acclimatization to Long-term Hypoxia

Mechanisms of acclimatization to long-term hypoxia have remained obscure; however, the hypoxic chemoreflex has been the object of the most intense investigation, for the characteristics of the ventilatory changes clearly indicate disorder in the ventilatory response to acute hypoxia. The carotid body itself may be the site of change during chronic hypoxia, not only because of the loss of the hypoxic response but also because it has been well recognized that prolonged hypoxia produces significant morphological and biochemical changes in the carotid body. Enlarged carotid bodies have been found in animals and humans chronically hypoxic from disease or altitude residence (84, 152, 222, 223, 243, 296). While it is controversial whether there is hyperplasia of type I glomus cells in chronically hypoxic animals some investigators (84, 243, 296), agree that the type I cells hypertrophy. There is increased vascular volume (84, 243) as well, and it has been suggested that catecholamine-containing dense-core vesicles in type I cells increase in size and number (296). This would be consistent with the increase in dopamine and norepinephrine catecholamine content occurring in chronically hypoxic rats (151, 280). Unfortunately, morphological and biochemical changes in animals have not correlated with changes in ventilatory control. For example, the well-described changes outlined above in rat carotid bodies have been associated with only small and transient depression of the ventilatory response to hypoxia (6).

Animals born and residing at high altitude have been found to have normal or near normal ventilatory responses to hypoxia (21, 29, 167, 209, 210); therefore, a natural animal model of blunted hypoxic sensitivity does not exist. However, cats placed in hypobaric hypoxia simulating an altitude of 5,500 m for 3–4 wk develop a blunted hypoxic response similar to that seen in human highlanders (354, 360). This model has provided a means to study the phenomenon of blunting, and in the original study by Tenney and Ou (360) it was found that midcollicular decerebration restored hypoxic ventilatory response, suggesting that cortical influences played a role in the blunting phenomenon. This observation may or may not apply to the highlander human who has developed blunting because the cat quickly recovers its ventilatory response after a few days at sea level while the human does not (211, 360). Tatsumi et al. (354) used the Tenney and Ou cat model and showed that both ventilatory and carotid body afferent discharge responses to acute hypoxia were diminished in hypoxia-exposed cats (5,500 m for 3–4 wk). These findings support a direct carotid body role in the blunted acute response to hypoxia. They also found that cutting the carotid sinus nerve increased the carotid body afferent discharge response to hypoxia, suggesting that descending efferents may play a role in inhibition of the hypoxic response. Dopaminergic mechanisms in the carotid body have also been indicated (355). Their data also suggested that CNS translation of carotid body efferent traffic was reduced, indicating the possibility of a CNS role in producing the blunted response. Thus there is good evidence for both a carotid body and a CNS role in producing blunting of the response to severe hypoxia in cats (354, 355, 360).

Pokorski and Lahiri (302, 303) have tested the hypothesis that blunting of the hypoxic response in the chronic cat model was mediated by opiate mechanisms in the carotid body. They showed that the opiate antagonist naloxone failed to inhibit the blunted hypoxic response of the carotid body, suggesting that opiates are not involved in the blunting of the hypoxic response. While no other animal model exhibits the blunting response seen in cats, other animal models show an attenuation of acclimatization after about 1–2 wk at high altitude characterized by a reduction in hyperventilation toward the level expected with exposure to acute hypoxia. This has been observed in dogs (34), cats (360), goats (213, 384), and ponies (286). None of these animals, except cats exposed to an altitude of 5,500 m, exhibits a blunted response to acute hypoxia.

Sheep and yaks native to an altitude of 3,800 m (209) and llamas native to 3,790 m (29) do not exhibit blunted ventilatory response to acute hypoxia, nor do Hereford calves native to 3,100 m (21). Therefore, it does not appear that all of the same mechanisms are operating in these animals as in the cat exposed to 5,500 m or in the human highlander. Nevertheless, the reduction in hyperventilation is similar to what is seen in human highlanders, and elucidation of the mechanism for why short-term acclimatization is not sustained in the various animal models may provide clues for the lower degree of hyperventilation in highlanders. Weinberger et al. (384) tested the hypothesis that opioids may be responsible for modulating long-term ventilatory acclimatization to hypoxia in goats. They concluded that opioids did not play a role in the attenuation of acclimatization after 2 wk at a simulated altitude of 4,300 m. Thus there is evidence that hypoxic acclimatization is attenuated by chronic hypoxia in many species. In cats exposed to great heights and in human highlanders this is also accompanied by a blunted response

to acute hypoxia. The evidence to date seems to point to both CNS and carotid body mechanisms in the blunted hypoxic response. The mechanism of attenuated resting hyperventilation characteristic of many species is not known. It is possible that one or more mechanisms previously mentioned in this paper as a cause for hypoxic ventilatory depression may operate in the chronically hypoxic subject, for example, cerebral lactacidosis or an imbalance toward excessive inhibitory neurochemicals such as GABA or adenosine (see under Mechanisms Inhibiting Breathing and Potentially Causing Roll-Off, above). One could speculate that while carotid body sensitivity to an acute hypoxic stimulation is attenuated (blunted), tonic discharge from the carotid chemoreceptors is elevated, providing the moderate hyperventilation in highlanders.

The consequences of loss of the full degree of hyperventilation in highlanders may be offset by other adaptive mechanisms, such as increased blood oxygen-carrying capacity (increased hematocrit), increased capillary density in muscle relative to muscle fiber mass, increased oxidative capacity of muscle (390), lower lactic acid accumulation in hypoxic exercise in highlanders (lactate paradox) (163), and possibly redistribution (185) or increased density of muscle mitochondria (290), though this is controversial (379). These adaptive changes may assist oxygen transport at moderate altitudes (less than 5,000 m), but at extreme altitudes failure of adaptation mechanisms overcomes their value (see the chapter by West in this *Handbook*).

EFFECTS OF HYPOXIA ON BREATHING DURING SLEEP

Breathing changes during sleep at sea level (298, 299, 312, 341) and to an even greater extent during sleep at high altitude (248, 274, 311, 352, 353). As a result, it is appropriate not only to review the data on breathing during hypoxia but to briefly summarize the characteristics of sleep at sea level, how hypoxia influences sleep per se, and how breathing changes during sleep at sea level.

Sleep at Sea Level

Wakefulness, rapid eye movement (REM) sleep, and nonrapid eye movement (NREM) sleep are defined by brain-wave patterns, muscle tone, and oculomotor activity (143, 298, 309). In young adults sleep generally consists of four or five NREM–REM cycles, each lasting for 90–110 min (309). The natural progression is NREM sleep stage I (light sleep) to NREM stages II, III, and IV followed by REM sleep. The duration of each stage or phase varies slightly over the five cycles, with, for example, REM sleep being up to 30 min longer in cycle 5 than in cycle 1. The NREM stages differ in the degree of EEG synchronization, reflecting a progressive decrease in neuronal activity. Studies have not identified a specific substance or anatomical location underlying NREM sleep, but cholinergic mechanisms in the forebrain and brain stem are probably involved (233, 241, 242). There are two stages of REM sleep, tonic and phasic. Throughout REM, there is a suppression of muscle activity due to active hyperpolarization of spinal cord motoneurons (tonic REM) (233, 241). This atonia is interspersed with phasic rapid eye movements and/or abrupt muscle twitches due to random fluctuation in motoneuron excitability (333). As with NREM, the precise mechanism of REM is not certain; however, a cholinergic pontine mechanism appears to be of critical importance during tonic REM, and the pons, lateral geniculate, and occipital lobe are involved during phasic REM (242, 333). These characteristics of sleep develop gradually over the first 3 months of life in humans, with minimal change thereafter until at least the fifth decade of life (59, 333).

Sleep during Hypoxia

Beginning the first evening that a sea-level resident is exposed to hypoxia producing a PaO_2 of 50 torr or less, the most common observations or complaints are the frequency of arousal, sleeplessness, and insomnia (248, 274, 311, 353). Indeed, in actual sleep studies frequency of arousal is a major change from sea level. In addition, there is a reduction in stages III and IV NREM sleep and a tendency for a reduction in REM sleep (225, 274). These objective measurements suggest that changes in sleep during hypoxia are not as drastic as subjective evaluations indicate. With continued exposure to hypoxia, both subjective and objective observations indicate some return of sleep to the sea-level pattern (311). Nevertheless, changes in sleep between sea level and high altitude in altitude natives are qualitatively similar to those in a sea-level resident during acute hypoxia (212; J. H. Cootes, personal communication). However, frequency of arousal, sleepiness, insomnia, and changes in sleep stage are less in an altitude native relative to a sea-level resident acutely exposed to hypoxia.

Changes in Breathing during Sleep at Sea Level

A hallmark of breathing in adult humans during NREM sleep at sea level is that breathing decreases and there is a 2–4 torr increase in $PaCO_2$ above the awake state (82, 83, 263, 298, 299, 312, 341). This hypoventilation is in part due to loss of the "wakefulness" stimulus for breathing. This stimulus is probably suprapontine in origin, and in its absence during NREM sleep there is a reduction of spontaneous activity of ventral medullary

respiratory neurons (282–285). Another factor contributing to hypoventilation during NREM sleep at sea level is an increase in upper airway resistance. This increase is secondary to a reduction in phasic activation of upper-airway abductor muscles (83, 173, 323, 336). The increased resistance represents an increased ventilatory load, which in the awake state elicits compensatory responses (3, 27, 173, 394, 395). The compensatory response has a rapid (probably mechanoreceptor) and a slow (chemoreceptor) component. Another effect of sleep is loss of the immediate portion of this compensatory response (3, 27, 173, 394, 395). The chemoreceptor component of this reflex remains as indicated by the fact that there is actually increased stimulation of the diaphragm during sleep, which is eliminated by passive hyperventilation to restore $PaCO_2$ to awake levels (83, 159, 337). Evidence that there are indeed a "controller" and a "mechanics" component to the hypoventilation of NREM sleep is found in studies showing that elimination of increased resistance reduces but does not eliminate sleep-induced hypoventilation (83, 159, 314).

Absence of the suprapontine influence on breathing during NREM sleep not only partially accounts for the hypoventilation but also results in breathing being relatively more dependent on the so-called metabolic or automatic control system (66, 83, 298, 299, 335). The CO_2–H^+ chemoreceptors located in the brain stem are a vital component of this system. The importance of this system is indicated by findings that reducing and/or eliminating chemoreceptor activity by hyperventilation in the awake state usually does not cause apnea, but hypocapnia to 1–2 torr below the awake $PaCO_2$ will cause apnea during NREM sleep (66, 81, 158, 335). Since breathing becomes dependent on CO_2–H^+ stimulation, it is not surprising that the whole-body response to inspired CO_2 changes minimally during NREM sleep. The small decrease in CO_2 sensitivity observed by some investigators (83, 298, 299, 383) may reflect the effect of the added ventilatory load and/or the fact that, due to a change in cerebral blood flow, $PaCO_2$ does not provide an adequate index of stimulus level at the intracranial chemoreceptor. Finally PaO_2 decreases only a few torr during sleep and $\dot{V}_E$ sensitivity to hypoxia changes minimally; thus there is no increase in hypoxic stimulation of breathing during NREM sleep at sea level (300).

Another characteristic of NREM sleep, particularly in young and aging humans, is that regular breathing is interspersed with periods of apnea (46, 83, 170, 298, 299). This is more prevalent in males than in females. This periodic breathing is of two types. One is termed *obstructive sleep apnea* because breathing movements continue but due to upper airway blockade there is no flow of air (171). The blockade results from a total collapse of the oropharynx secondary to the absence of abductor muscle contraction alluded to previously (171). The second type is termed *central apnea*, during which breathing movements cease due to failure of the central respiratory-pattern generator. Both of these may represent the extreme influence of the mechanisms, which in most humans manifest as hypoventilation and increased airway resistance during NREM sleep (46, 83, 170, 173). However, multiple factors can induce periodic breathing under different specific conditions (50, 51). Conceivably, a combination of factors (reductions in metabolic rate, upper airway afferents, and suprapontine input) may, as Andrews et al. (1) propose, create a "central rhythm" which manifests as periodic breathing. Indeed it has been hypothesized that "the underlying abnormality in these patients may be the periodicity of the respiratory controller" (171).

The hallmark of breathing during REM sleep (particularly phasic REM) is that it is irregular and tachypneic and exhibits minimal chemoreceptor responsiveness (11, 82, 83, 298, 299). There is also evidence that both pulmonary and alveolar ventilation are above awake levels, resulting in a reduced Pa_{CO_2} and an increased Pa_{O_2} (298, 299). The minimal chemoreceptor responsiveness suggests that breathing during REM sleep functions essentially independently of the metabolic control system (that is, the chemoreceptors). Accordingly, it appears that breathing during REM sleep is under the control of the so-called behavioral mechanisms (282, 284). The anatomical location of these influences is not certain, but there is evidence that pontine mechanisms are involved (283, 333). Breathing during REM sleep fits into the general characteristics of this state, that is, increased, irregular, nonpurposeful activity mediated in part by pontine mechanisms.

Changes in Breathing during Sleep at High Altitude

Sleep at altitude results in a slight increase in Pa_{CO_2} similar to sleep at sea level (11). However, Pa_{CO_2} during sleep at altitude remains well below sea-level values (351, 352). As a result, hypoxia does not appear to change the nature of the basic mechanisms regulating breathing during sleep nor does sleep appear to change the basic mechanism mediating ventilatory acclimatization to hypoxia.

The major change and the hallmark of breathing during sleep at altitude is that breathing becomes more periodic in most individuals (11, 33, 148, 382). This change, which becomes more severe as Pa_{O_2} is decreased (74), has been termed Cheyne-Stokes breathing because it is similar to that found by Cheyne (53) and Stokes (350) in patients at sea level with circulatory problems. The general pattern is for cyclic waxing and waning, particularly of tidal volume, with or without periods of

apnea (Fig. 52.8). The periods of apnea during sleep are often terminated by an arousal with a strong sensation of dyspnea. Several breaths follow and sleep is resumed, but the repetitive nature of these events results in sensations of sleeplessness and chronic dyspnea. This pattern of breathing has been observed during all stages of NREM sleep, but it does not seem to occur during REM sleep (11, 274, 311). Breathing periodicity and sleep disturbances are usually most severe the first few nights of hypoxia, with a diminution as ventilatory acclimatization is completed (11, 31), and there is a relatively stable pattern of breathing in high-altitude natives (217). Nonhuman species, such as the pony, also breathe periodically at high altitude and, in this species, also while in the awake state (fig. 52.8).

A likely contributor to breathing periodicity during hypoxia is an instability of the metabolic or chemical component of the respiratory control system (11, 50, 51, 217, 372). In feedback regulatory control systems, instability can be caused by decreased dampening, increased circulatory delay, and/or increased sensory gain (50, 51, 144). It has been proposed that the latter of these is of most importance during hypoxia. Specifically, Berssenbrugge et al. (11) proposed that increased carotid chemoreceptor gain during acclimatization to hypoxia combined with a hypocapnia-induced apnea threshold underlies periodic breathing. In other words, during hypoxia, carotid chemoreceptors cause hyperventilation, lowering P_{CO_2} or $[H^+]$ at the intracranial chemoreceptors to below threshold levels, resulting in apnea. During apnea, Pa_{O_2} decreases and Pa_{CO_2} increases progressively to the point where, acting synergistically at the carotid chemoreceptors, they elicit another period of hyperventilation. The cycle repeats, resulting in self-sustained periodic oscillations. This proposed schema is consistent with the concept of greater or exclusive dependence during sleep on the metabolic control system (284). It is also consistent with the finding that at sea level passive hyperventilation during sleep causes apnea (66, 81, 158, 335); during hypoxia hyperventilation is active as a result of carotid chemoreceptor stimulation. Moreover, the schema is supported by at least six experimental findings. First, the length of the cycles or periods usually is about 20 s, which is consistent with the instability of chemical feedback control (372). Second, administration of high O_2 and/or high CO_2 gas mixtures attenuates altitude breathing periodicities (11). Third, high-altitude natives with a blunted responsiveness to hypoxia have a relatively stable breathing pattern during sleep (274). Fourth, increasing carotid chemoreceptor sensitivity and inducing hyperventilation by adenosine infusion produces periodic breathing in sleeping humans at sea level (137). Fifth, some (11, 312) but not all (274) studies have found that breathing periodicity does not occur

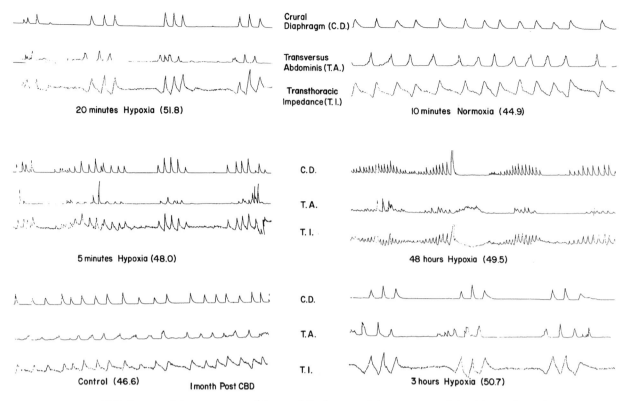

FIG. 52.8. Integrated electromyogram of the crural diaphragm and the transversus abdominis muscles and the transthoracic impedance measurement of pulmonary ventilation in one carotid body–denervated pony during sea-level control, after 5 and 20 min and 3 and 48 h of hypoxia (PaO$_2$ ≈ 45 torr) and after 10 min of return to sea level. Numbers in parentheses are PaCO$_2$ in torr. Note the constant waxing and waning (Cheynes-Stokes) of breathing throughout hypoxia, which was alleviated within 10 min of return to sea level.

during REM sleep, when breathing regulation is thought to be independent of the chemical control system (11, 284). Finally, high-altitude breathing periodicity is reduced by CBD (1, 31).

Breathing instability during hypoxia has also been explained in terms of loss of ventilatory afterdischarge (VAD) (98, 132). This phenomenon refers to the persistent elevation of ventilation that gradually decreases after cessation of a stimulus that elicited hyperpnea (95). It has been hypothesized to be due to a central neural process, most probably through the medium of reverberating neural circuits (95). Acute hypoxia can elicit VAD, but in most studies a concomitant slight decrease in PaCO$_2$ will attenuate the hypoxia-induced VAD (98, 132). It is concievable that the apnea following hypoxic hyperventilation occurs because the afterdischarge stimulus is below threshold. As PaCO$_2$ and [H$^+$] increase during apnea, input to the respiratory control system reaches a critical level, resulting in resumed breathing. In other words, VAD normally stabilizes breathing and prevents apnea after a period of intense stimulation of breathing, but the hypocapnia during hypoxic hyperventilation will not allow VAD to serve this stabilizing function.

Some observations are not readily explained by the two schema outlined above. First, although CBD reduced the incidence of altitude-induced breathing periodicity in ponies, it did not eliminate this phenomenon (Fig. 52.8) (31). As a result, the high-gain carotid chemoreceptors and VAD are not critical for the periodicities. Second, dopamine infusion, which inhibits carotid chemoreceptor activity, does not attenuate periodic breathing in newborn lambs during hypoxia (1). Third, the length of the periods is not constant during hypoxia, and some of the lengths are longer than those generally attributed to chemical instability (31). Fourth, during an altitude sojourn by a lowlander, the incidence and severity of the periodicities seems to decrease over the first few days of hypoxia, which is when V̇$_E$ acclimatization occurs (11, 31). A characteristic of acclimatization is increased gain of the chemoreflex; thus there does not appear to be a positive correlation during this period between chemoreceptor gain and periodicity.

It seems that understanding altitude-induced breathing periodicity will be enhanced as the basic characteristics of the central pattern generator become clearer. Indeed, Feldman (103) has proposed that there are multiple pattern generators; each of which may or may not be influenced by chemical instability. Certainly, the apparent absence of high-altitude breathing periodicity in REM sleep is consistent with the concept of multiple pattern generators (11). Conceivably, with the changing neurotransmitter milieu in the brain during hypoxia there are frequent changes in which pattern generator is dominant. Finally, since multiple factors can create periodic breathing, it might be a combination of factors that underlies high-altitude breathing (1, 11, 31, 83). As indicated earlier, the concept of a CNS mechanism underlying periodic breathing has been proposed (1, 31, 170, 171).

BENEFITS AND SHORTCOMINGS OF BREATHING CHANGES DURING HYPOXIA

Existence in a hypoxic environment can elicit two different adaptive types of response (161, 162, 211, 269). One response is to reduce the utilization of oxygen to prolong survival with the minimum available oxygen. This response is characteristic of (1) diving animals in an environment that does not permit replenishing oxygen (365); (2) the fetus, whose oxygen availability is not totally under its own control (121); and (3) many mammals during at least acute exposure to low environmental oxygen (121, 261). The second response is to change the function of several physiological systems to enhance delivery of oxygen to the tissues. This response is characteristic of adaptation to high-altitude residence and to physiological (exercise) or pathological (congenital heart disease, anemia) conditions that affect delivery of oxygen to the tissues. Generally, it seems the first response is most appropriate for short periods of hypoxia and during conditions such as fetal life when "conservation of energy may be more important than any advantage breathing may impart" (177). Conversely, the second response seems most appropriate for normal function in a chronically hypoxic environment. Indeed, available evidence supports this logic. Moreover, during environmental hypoxia, it appears that the adaptations proceed in a logical sequence that progressively enhances physical and mental performance. Furthermore, it appears that the sequence of changes in breathing during acute and chronic hypoxia follow the "wisdom of the body" doctrine (43).

As summarized, during acute hypoxia at rest breathing is increased minimally until PaO_2 is less than 50 torr, and even at these low levels the peak increase is sustained for only 2–3 min. Indeed, it appears that a prominent response during this period in most mammals is a decrease in oxygen utilization and/or a major influence on breathing of HBD. However, if exercise is required during acute hypoxia, there is an increase in breathing considerably above normal. In other words, if the demands for oxygen are low (rest), breathing does not need to increase to meet the tissue needs unless PaO_2 is very low. However, if the demands are increased, an increase in breathing is required to meet the tissue needs. The positive interaction between hypoxia and exercise, or CO_2, (alluded to earlier), results in an enhanced or appropriate ventilatory response. In other words, the basic mechanism causing HBD is overridden by the strong synergistic influences on breathing. Accordingly, breathing responses during acute hypoxia seem appropriate.

As also summarized, a prominent feature of short-term hypoxia or altitude sojourn is ventilatory acclimatization, which increases PaO_2 and decreases $PaCO_2$ several torr from levels during acute hypoxia (74). At an altitude of 4,300 m this increase in PaO_2 at rest and during exercise is approximately 10 torr (from about 40 to 50 torr), which results in over a 10% increase in SaO_2 and over a 2 vol% increase in CaO_2 (74). The benefits of this increase during exercise are obvious as, for example, maximal work output should increase by the same percentage as CaO_2. Moreover, the importance of this change increases with altitude; successful scaling of Mt. Everest would be impossible without ventilatory acclimatization (388). Indeed, it has been stated that ventilatory acclimatization is the single most important physiological response to chronic hypoxia (390).

Is the 10 torr increase in PaO_2 beneficial or needed at rest during sojourn at 4,300 m? It seems that under basal conditions the benefit of this increase is minimal. However, a normal life-style at altitude that requires above basal physical or mental energy would benefit from acclimatization.

A second prominent feature of short-term hypoxia is periodic breathing during sleep, which with the accompanying apnea is probably the cause of frequent arousals. This change does not seem to be beneficial because it results in sensations of sleeplessness and fatigue. Fortunately, periodic breathing subsides over the first several days at altitude, and indeed the temporal pattern of the decrease somewhat follows ventilatory acclimatization (382). Since PaO_2 is elevated during sleep at altitude from acute hypoxia levels, a major benefit of ventilatory acclimatization might be in attenuating periodic breathing (351, 352). This attenuation would restore sleep toward normal, reduce sensations of sleeplessness and fatigue, and thereby considerably enhance performance at rest and during exercise in the awake state.

The beneficial effects of ventilatory acclimatization might best be exemplified by the effects of acetazolamide on acute mountain sickness (AMS). This condition commences in humans within 4–8 h after ascent to altitude; persists for several hours; and manifests as anorexia, nausea, headache, and vomiting (147, 184, 224). Acetazolamide pretreatment clearly reduces the symptoms of AMS (140, 148, 224, 274, 382). This agent inhibits carbonic anhydrase, resulting in CO_2 retention, an increase in plasma and CSF [H^+], and an enhanced hyperventilation, which increases PaO_2 during awake and sleeping states (38, 42, 119, 125, 148, 353). The precise cause of AMS is not known; it has been proposed to be due to an inadequate ventilatory response to hypoxia, impaired cellular oxidation processes, changes in fluid retention, shifts in fluid between intra- and extracellular compartments, and an increase in cerebral blood flow (38, 94, 146, 311, 398). As previously discussed, there is considerable variation between individuals in the ventilatory response to CO_2 and hypoxia. Conceivably, those who have a small net increased drive to breathe at altitude are more susceptible to AMS. However, it is clear that ventilatory responsiveness is not the sole factor underlying AMS; some individuals with negligible acute ventilatory chemosensitivity do not develop AMS but do exhibit normal ventilatory acclimatization to altitude (75). In any event, it has been proposed that enhanced hyperventilation, PaO_2, and tissue oxygenation with acetazolamide administration are responsible for alleviation of AMS symptoms (38, 94, 310, 398). Accordingly, the effect of acetazolamide on the symptoms of AMS supports the concept that the natural reduction of AMS between 24 and 48 h of hypoxia in humans could be due to the increase in PaO_2 resulting from natural ventilatory acclimatization. Moreover, even sojourners not experiencing clinical AMS feel better in the acclimatized state than during acute hypoxia.

The major shortcomings of ventilatory acclimatization to hypoxia are the increased utilization of oxygen by the respiratory muscles and the constant sensation of dyspnea. At extreme altitudes, the cost of breathing may approximate or exceed the cost of ambulation (364, 388). Breathing is clearly then a limitation to exercise at extreme altitude. Even though the impact of the dyspnea cannot be exactly quantitated, it unquestionably has a major impact on the well-being or life-style of an altitude sojourner. Even at moderate altitudes of 3,000 m, the simple tasks of walking, climbing stairs, or household chores elicit a hyperpnea that is discomforting. As a result, the pace of these tasks is slowed or they are avoided when possible. Clearly then in many sojourners at altitude, ventilatory acclimatization does not provide a life-style comparable in all respects to that at sea level.

As already summarized, high-altitude natives breathe less than sea-level residents sojourning at altitude. In spite of the reduced breathing, PaO_2 is nearly the same in natives and sojourners. The native has the same PaO_2 in spite of a relatively lower PaO_2 because pulmonary diffusing capacity for oxygen has increased as a result of proliferation of alveolar–capillary units (8, 39, 45).

An increase in the surface area for diffusion reduces the need for the relatively high pressure gradient for diffusion created by hyperventilation in the sojourner. In other words, the final outcome of pulmonary adaptation to hypoxia, PaO_2, is equal in natives and sojourners, but it is achieved differently. In essence then, the native has the same benefits (summarized earlier) as the sojourner, but the native does not experience either of the two major shortcomings to the extent experienced by the sojourner.

Adaptation of the native appears desirable over adaptation of the sojourner. Why then does the sojourner not similarly adapt? The probable reason relates to the fact that native adaptation requires structural changes which probably require several years to develop. Furthermore, the lung must have the capability to undergo proliferation. The importance of these factors is demonstrated by the fact that adults who move to altitude permanently as adults will only increase diffusing capacity minimally and never to the level of the native (45, 79, 80). However, sea-level children who move before puberty will over several years increase diffusing capacity to the native level (45). In other words, the structural changes appear minimal for mature lungs, but for a lung that is capable of additional natural growth chronic hypoxia will induce increased proliferation.

Of interest is the association between increased pulmonary diffusing capacity and reduced hyperventilation in long-term altitude residents. There is no readily apparent direct link between the two. Conceivably the permissive factors for both are similar so that one does not normally occur without the other.

In some adults native to high altitude, a condition known as chronic mountain sickness (CMS), or Monge's disease, develops (153, 172, 257). Prominent features of CMS include polycythemia, elevated pulmonary artery pressure, right heart hypertrophy, congestive heart failure, and neuropsychic symptoms. The exact cause of CMS is unknown. It occurs primarily in adult males and in native and nonnative long-term residents of high altitude, such as Himalayan (Tibetan) natives exceptional in their resistance to the condition (293). It is usually associated with a lower PaO_2 and SaO_2 and a higher $PaCO_2$ than are found in healthy natives of altitude. In other words, relative alveolar hypoventilation during eupnea, possibly due to virtual insensitivity to hypoxia and low sensitivity to CO_2, seems to underlie CMS (212). Apparently, insensitivity to hypoxia per se does not cause hypoventilation and CMS, but rather it plays a permissive role when some other extrinsic factor, such as obesity, tends to cause hypoventilation (199). In a sense, the risk of the normal native adaptation to high altitude is that $\dot{V}_A$ is at or near a nadir, and thus a further reduction will in some individuals result in CMS. The importance of $\dot{V}_A$ in CMS is indicated by the fact that CMS is treated clinically by pharmacological respiratory stimulation (198).

To our knowledge, pulmonary adaptation to hypoxia is one of the best-known examples of the "wisdom of the body" doctrine (43). If O_2 demands are low (rest) during acute hypoxia breathing increase is negligible, but if O_2 demands are increased (exercise), mechanisms are present to increase breathing to minimize hypoxemia. During short-term hypoxia, ventilatory acclimatization significantly increases PaO_2 and CaO_2, which greatly enhances physical and mental performance. However, acclimatization comes with the cost of increased O_2 use by respiratory muscles and dyspnea, both of which limit performance. This hyperpnea is therefore not an optimal pulmonary adaptation. The apparent optimum is observed in the high-altitude native who breathes less but as a result of the increased alveolar–capillary surface area for diffusion has the same PaO_2 as the sojourner. This adaptation is not achievable by adults who move to altitude. Adaptation of the sojourner is quickly mobilized and represents the best the pulmonary system can contribute to a normal life-style at altitude. The danger of the relatively lower $\dot{V}_A$ in the native is that they have no reserve when other factors further reduce $\dot{V}_A$, and thus they are more susceptible to CMS. Finally, in another illustration of the wisdom of the body (43), hematological, vascular, and tissue changes further enhance O_2 delivery in a highly efficient manner during long-term hypoxia (228, 281, 358, 366, 379).

REFERENCES

1. Andrews, D. C., P. Johnson, and M. E. Symonds. Metabolic rate and periodic breathing in the developing lamb. *J. Physiol. (Lond.)* 417: 137P, 1989.
2. Armengaud, C., L. M. Leitner, C. H. Malber, M. Roumy, M. Ruckebusch, and J. F. Sutra. Comparison of the monoamine and catabolite content in the cat and rabbit carotid bodies. *Neurosci. Lett.* 85: 153–157, 1988.
3. Badr, M. S., J. B. Skatrud, J. A. Dempsey, and R. L. Begle. Effect of mechanical loading on expiratory muscle and inspiratory activity during NREM sleep. *J. Appl. Physiol.* 68: 1195–1202, 1990.
4. Bakey, P., and W. H. Sweet. Effects on respiration, blood pressure and gastric mobility of stimulation of orbital surface of frontal lobe. *J. Neurophysiol.* 3: 276–281, 1940.
5. Balestrino, M., and G. G. Somjen. Concentration of carbon dioxide, interstitial pH and synaptic transmission in hippocampal formation of the rat. *J. Physiol. (Lond.)* 396: 247–266, 1988.
6. Barer, G. R., C. W. Edwards, and A. I. Jolly. Changes in the carotid body and the ventilatory response to hypoxia in chronically hypoxic rats. *Clin. Sci. Mol. Med.* 50: 311–313, 1976.
7. Barnard, P., S. Andronikou, M. Pokorski, N. Smatresk, A. Mokashi, and S. Lahiri. Time-dependent effect of hypoxia on carotid body chemosensory function. *J. Appl. Physiol.* 63: 685–691, 1987.
8. Bartlett, D., and J. E. Remmers. Effects of high altitude exposure on the lungs of young rats. *Respir. Physiol.* 13: 116–125, 1971.

9. Bender, P. R., B. M. Groves, R. E. McCullough, R. G. McCullough, S. Huang, A. J. Hamilton, P. D. Wagner, A. Cymerman, and C. S. Reeves. Oxygen transport to exercising leg in chronic hypoxia. *J. Appl. Physiol.* 65: 2592–2597, 1988.
10. Berger, A. The distribution of the cat's carotid sinus nerve afferent and the efferent cell bodies using the horseradish peroxidase technique. *Brain Res.* 190: 309–320, 1980.
11. Berssenbrugge, A., J. Dempsey, and J. Skatrud. Hypoxic versus hypocapnic effects on periodic breathing during sleep. In: *High Altitude and Man*, edited by J. B. West and S. Lahiri. Bethesda, MD: Am. Physiol. Soc., 1984, p. 115–127.
12. Bisgard, G. E., M. A. Busch, L. Daristotle, A. Berssenbrugge, and H. V. Forster. Carotid body hypercapnia does not elicit ventilatory acclimatization in goats. *Respir. Physiol.* 65: 113–125, 1986.
13. Bisgard, G. E., M. A. Busch, and H. V. Forster. Ventilatory acclimatization to hypoxia is not dependent on cerebral hypocapnic alkalosis. *J. Appl. Physiol.* 60: 1011–1015, 1986.
14. Bisgard, G. E., M. J. A. Engwall, N. Weizhen, A. M. Nielsen, and E. Vidruk. The effect of prolonged stimulation of afferent activity of the goat carotid body. In: *Chemoreceptors and Chemoreceptor Reflexes,* edited by H. Acker, A. Trzebski, and R. G. O'Regan. New York: Plenum, 1990, p. 165–170.
15. Bisgard, G. E., H. V. Forster, and J. P. Klein. Recovery of peripheral chemoreceptor function after denervation in ponies. *J. Appl. Physiol.: Respir. Environ. Exerc. Physiol.* 49: 964–970, 1980.
16. Bisgard, G. E., H. V. Forster, J. Klein, M. Manohar, and V. A. Bullard. Depression of ventilation by dopamine in goats—effects of carotid body excision. *Respir. Physiol.* 41: 379–392, 1980.
17. Bisgard, G. E., H. V. Forster, J. Mesina, and R. G. Sarazin. Role of the carotid body in hyperpnea of moderate exercis in goats. *J. Appl. Physiol.: Respir. Environ. Exerc. Physiol.* 52: 1216–1222, 1982.
18. Bisgard, G. E., H. V. Forster, J. A. Orr, D. D. Buss, C. A. Rawlings, and B. Rasmussen. Hypoventilation in ponies after carotid body denervation. *J. Appl. Physiol.* 40: 184–190, 1976.
19. Bisgard, G. E., N. A. Kressin, A. M. Nielsen, L. Daristotle, C. A. Smith, and H. V. Forster. Dopamine blockade alters ventilatory acclimatization to hypoxia in goats. *Respir. Physiol.* 69: 245–255, 1987.
20. Bisgard, G. E., R. A. Mitchell, and D. A. Herbert. Effects of dopamine, norepinephrine and 5-hydroxytryptamine on the carotid body of the dog. *Respir. Physiol.* 37: 61–80, 1979.
21. Bisgard, G. E., A. V. Ruiz, R. F. Grover, and J. A. Will. Ventilatory acclimatization to 3,400 meters altitude in the Hereford calf. *Respir. Physiol.* 21: 271–296, 1974.
22. Blanco, C. E., M. A. Hanson, P. Johnson, and H. Rigatto. Breathing pattern of kittens during hypoxia. *J. Appl. Physiol.: Respir. Environ. Exerc. Physiol.* 56: 12–17, 1984.
23. Blesa, M. I., S. Lahiri, D. Phil, W. Rashkind, and A. P. Fishman. Normalization of the blunted ventilatory response to acute hypoxia in congenital cyanotic heart disease. *N. Engl. J. Med.* 296: 237–241, 1977.
24. Bonora, M., and H. Gautier. Influence of dopamine and norepinephrine on the central ventilatory response to hypoxia in conscious cats. *Respir. Physiol.* 71: 11–24, 1988.
25. Bonora, M., D. Marlot, H. Gautier, and B. Duron. Effects of hypoxia on ventilation during postnatal development in conscious kittens. *J. Appl. Physiol.: Respir. Environ. Exerc. Physiol.* 56: 1464–1471, 1984.
26. Bouverot, P., and M. Bureau. Ventilatory acclimatization and CSF acid-base balance in carotid denervated dogs at 3550 m. *Pflugers Arch.* 361: 17–23, 1976.
27. Bowes, G., L. F. Kozar, S. M. Andrey, and E. A. Phillipson. Ventilatory responses to inspiratory flow-resistive loads in

awake and sleeping dogs. *J. Appl. Physiol.: Respir. Environ. Exerc. Physiol.* 54: 1550–1557, 1983.

28. Bowes, G., E. R. Townsend, L. F. Kozar, S. M. Bromley, and E. A. Phillipson. Effect of carotid body denervation on arousal response to hypoxia in sleeping dogs. *J. Appl. Physiol.: Respir. Environ. Exerc. Physiol.* 51: 40–45, 1981.

29. Brooks, J. G., III, and S. M. Tenney. Ventilatory response of llama to hypoxia at sea level and high altitude. *Respir. Physiol.* 5: 269–278, 1968.

30. Brown, D. L., and E. E. Lawson. Brain stem extracellular fluid pH and respiratory drive during hypoxia in newborn pigs. *J. Appl. Physiol.: Respir. Environ. Exerc. Physiol.* 64: 1055–1059, 1988.

31. Brown, D. R., H. V. Forster, A. S. Greene, and T. F. Lowry. Breathing periodicity in intact and carotid chemoreceptor denervated ponies during normoxia and chronic hypoxia. *J. Appl. Physiol.: Respir. Environ. Exerc. Physiol.* 74: 1073–1082, 1993.

32. Brown, D. R., H. V. Forster, T. F. Lowry, M. A. Forster, A. L. Forster, S. M. Gutting, B. K. Erickson, and L. G. Pan. Effect of chronic hypoxia on breathing and EMGs of respiratory muscles in awake ponies. *J. Appl. Physiol.: Respir. Environ. Exerc. Physiol.* 72: 739–747, 1992.

33. Brusil, P. J., T. B. Waggener, R. E. Kranauer, and P. Gulesian, Jr. Methods for identifying respiratory oscillations disclose altitude effects. *J. Appl. Physiol.: Respir. Environ. Exerc. Physiol.* 48: 545–556, 1980.

34. Bureau, M. A., and P. Bouverot. Blood and CSF acid-base changes and rate of ventilatory acclimatization of awake dogs to 3550 m. *Respir. Physiol.* 24: 203–216, 1975.

35. Bureau, M. A., A. Côté, P. W. Blanchard, S. Hobbs, P. Foulon, and D. Dalle. Exponential and diphasic ventilatory response to hypoxia in conscious lambs. *J. Appl. Physiol.: Respir. Environ. Exerc. Physiol.* 61: 836–842, 1986.

36. Bureau, M. A., J. Lomarche, P. Foulon, and D. Dalle. The ventilatory response to hypoxia in the newborn lamb after carotid body denervation. *Respir. Physiol.* 60: 109–119, 1985.

37. Bureau, M. A., R. Zinman, P. Foulon, and R. Begin. Diphasic ventilatory response to hypoxia in newborn lambs. *J. Appl. Physiol.: Respir. Environ. Exerc. Physiol.* 56: 84–90, 1984.

38. Burki, N. K., S. A. Khan, and M. A. Hameed. The effects of acetazolamide on the ventilatory response to high altitude hypoxia. *Chest* 101: 736–741, 1992.

39. Burri, P., and E. Weibel. Environmental oxygen tension and lung growth. *Respir. Physiol.* 11: 247–256, 1971.

40. Busch, M. A., G. E. Bisgard, and H. V. Forster. Ventilatory acclimatization to hypoxia is not dependent on arterial hypoxemia. *J. Appl. Physiol.: Respir. Environ. Exerc. Physiol.* 58: 1874–1880, 1985.

41. Byrne-Quinn, E., I. E. Sodal, and J. V. Weil. Hypoxic and hypercapnic ventilatory drives in children native to high altitude. *J. Appl. Physiol.* 32: 44–46, 1972.

42. Cain, S. M., and J. E. Dunn. Low doses of acetazolamide to aid accommodation of men to altitude. *J. Appl. Physiol.* 21: 1195–1200, 1966.

43. Cannon, W. B. *The Wisdom of the Body.* New York: Norton, 1932.

44. Cao, H., Y. R. Kuo, and N. R. Prabhakar. Absence of chemoreceptor inhibition by alpha-2 adrenergic receptor agonist in cats exposed to low P_{O_2}. *FASEB J.* 5: A1118, 1991.

45. Cerny, F. C., J. A. Dempsey, and W. G. Reddan. Pulmonary gas exchange in nonnative residents of high altitude. *J. Clin. Invest.* 52: 2993–2999, 1973.

46. Chapman, K. R., E. N. Bruce, B. Gothe, and N. S. Cherniack. Possible mechanisms of periodic breathing during sleep. *J. Appl. Physiol.: Respir. Environ. Exerc. Physiol.* 64: 1000–1008, 1988.

47. Chapman, R. W., T. V. Santiago, and N. H. Edelman. Effects

of graded reduction of brain blood flow on ventilation in unanesthetized goats. *J. Appl. Physiol.: Respir. Environ. Exerc. Physiol.* 47: 104–111, 1979.

48. Chapman, R. W., T. V. Santiago, and N. H. Edelman. Brain hypoxia and control of breathing: neuromechanical control. *J. Appl. Physiol.: Respir. Environ. Exerc. Physiol.* 49: 497–505, 1980.

49. Cherniack, N. S., N. H. Edelman, and S. Lahiri. Hypoxia and hypercapnia as respiratory stimulants and depressants. *Respir. Physiol.* 11: 113–126, 1970.

50. Cherniack, N. S., B. Gothe, and K. P. Strohl. Mechanisms for recurrent apneas at altitude. In: *High Altitude and Man*, edited by J. B. West and S. Lahiri. Bethesda, MD: Am. Physiol. Soc., 1984, p. 129–140.

51. Cherniack, N. S., C. von Euler, I. Homma, and F. F. Kao. Experimentally induced Cheyne-Stokes breathing. *Respir. Physiol.* 37: 185–200, 1979.

52. Chernick, V., and R. J. Craig. Naloxone reverses neonatal depression caused by fetal asphyxia. *Science* 216: 1252–1253, 1982.

53. Cheyne, J. A case of apoplexy, in which the fleshy part of the heart was converted into fat. *Dublin Hosp. Rep.* 2: 216, 1818.

54. Chiocchio, S. R., A. M. Biscardi, and J. H. Tramezzani. Catecholamines in the carotid body discharge and ventilation. *J. Appl. Physiol.: Respir. Environ. Exerc. Physiol.* 60: 370–375, 1986.

55. Chiodi, H. Respiratory adaptation to chronic high altitude hypoxia. *J. Appl. Physiol.* 10: 81–87, 1957.

56. Chow, C. M., C. Winder, and D. J. C. Read. Influences of endogenous dopamine on carotid body discharge and ventilation. *J. Appl. Physiol.: Respir. Environ. Exerc. Physiol.* 60: 370–375, 1986.

57. Coles, S. K., F. Havashi, G. S. Mitchell, and D. R. McCrimmon. Carotid sinus nerve stimulation and hypoxia have both facilitatory and depressant effects on phrenic nerve activity in rats. *FASEB J.* 6: 1386, 1992.

58. Comroe, J. H. The location and function of the chemoreceptors of the aorta. *Am. J. Physiol.* 127: 176–190, 1939.

59. Coons, S., and C. Guilleminault. The development of sleep-wake patterns and non-rapid eye movement sleep stages during the first six months of life in normal infants. *Pediatrics* 69: 793–798, 1982.

60. Crawford, R. D., and J. W. Severinghaus. CSF pH and ventilatory acclimatization to altitude. *J. Appl. Physiol.: Respir. Environ. Exerc. Physiol.* 45: 275–283, 1978.

61. Cruz, J. C., J. T. Reeves, R. F. Grover, J. T. Maher, R. E. McCullough, A. Cymerman, and J. C. Denniston. Ventilatory acclimatization to high altitude is prevented by CO_2 breathing. *Respiration* 39: 121–130, 1980.

62. Cunningham, D. J. C., E. N. Hey, and B. B. Lloyd. The effect of noradrenaline infusion on the relation between pulmonary ventilation and alveolar P_{O_2} and P_{CO_2}. *Ann. N. Y. Acad. Sci.* 109: 756–771, 1963.

63. Curtis, D. R., A. W. Duggan, D. Felix, and G. A. R. Johnson. GABA, bicuculline, and central inhibition. *Nature* 226: 1222–1224, 1970.

64. Daly, M. De B., and A. Ungar. Comparison of the reflex responses elicited by stimulation of the separately perfused carotid and aortic body chemoreceptors. *J. Physiol. (Lond.)* 182: 379–403, 1966.

65. Daristotle, L., M. J. A. Engwall, N. Weizhen, and G. Bisgard. Ventilatory effects and interactions with change in Pa_{O_2} in awake goats. *J. Appl. Physiol.: Respir. Environ. Exerc. Physiol.* 71: 1254–1260, 1991.

66. Datta, A. K., S. A. Shea, R. L. Horner, and A. Guz. The influence

of induced hypocapnia and sleep on the endogenous respiratory rhythm in humans. *J. Physiol. (Lond.)* 440: 17–33, 1991.

67. Davies, D. G. Evidence for cerebral extracellular fluid [H+] as a stimulus during acclimatization to hypoxia. *Respir. Physiol.* 32: 167–182, 1978.

68. Davies, R. O., and M. Kalia. Carotid sinus nerve projections to the brain stem in the cat. *Brain Res.* 6: 531–541, 1981.

69. Davies, R. O., and S. Lahiri. Absence of carotid chemoreceptor response during hypoxic exercise in the cat. *Respir. Physiol.* 18: 92–100, 1973.

70. Davis, J. N., and A. Carlsson. Effect of hypoxia on monoamine synthesis, levels and metabolism in rat brain. *J. Neurochem.* 21: 783–790, 1973.

71. Dean, J. B., D. A. Bayliss, J. T. Erickson, W. L. Lawing, and D. E. Millborn. Depolarization and stimulation of neruons in nucleus tractus solitarii by carbon dioxide does not require chemical synaptic input. *Neuroscience* 36: 207–216, 1990.

72. Dean, J. B., W. L. Lawing, and D. E. Millhorn. CO_2 decreases membrane conductance and depolarizes neurons in the nucleus tractus solitarii. *Exp. Brain Res.* 76: 656–661, 1989.

73. DeGroat, W. C., I. Nadelhaft, C. Morgan, and T. Schauble. The central origin of efferent pathways in the carotid sinus nerve of the cat. *Science* 205: 1017–1018, 1979.

74. Dempsey, J. A., and H. V. Forster. Mediation of ventilatory adaptations. *Physiol. Rev.* 62: 262–346, 1982.

75. Dempsey, J. A., H. V. Forster, M. L. Birnbaum, W. G. Reddan, J. Thoden, R. F. Grover, and J. Rankin. Control of exercise hyperpnea under varying durations of exposure to moderate hypoxia. *Respir. Physiol.* 16: 213–231, 1972.

76. Dempsey, J. A., H. V. Forster, G. E. Bisgard, L. W. Chosy, P. G. Hanson, A. L. Kiorpes, and D. A. Pelligrino. Role of cerebrospinal fluid [H+] in ventilatory deacclimatization from chronic hypoxia. *J. Clin. Invest.* 64: 199–205, 1979.

77. Dempsey, J. A., H. V. Forster, and G. A. Dopico. Ventilatory acclimatization to moderate hypoxemia in man. The role of spinal fluid [H+]. *J. Clin. Invest.* 53: 1091–1100, 1974.

78. Dempsey, J. A., N. Gledhill, W. G. Reddan, H. V. Forster, P. G. Hanson, and A. D. Claremont. Pulmonary adaptation to exercise: effects of exercise type and duration, chronic hypoxia and physical training. *Ann. N. Y. Acad. Sci.* 301: 243–261, 1977.

79. Dempsey, J. A., W. G. Reddan, M. L. Birnbaum, H. V. Forster, J. S. Thoden, R. F. Grover, and J. Rankin. Effects of acute through life-long hypoxic exposure on exercise pulmonary gas exchange. *Respir. Physiol.* 13: 62–89, 1971.

80. Dempsey, J. A., W. G. Reddan, G. A. do Pico, F. Cerny, and H. V. Forster. Determinants of acquired changes in pulmonary gas exchange in man via chronic hypoxic exposure. *Prog. Respir. Res.* 9: 180–186, 1975.

81. Dempsey, J. A., and J. B. Skatrud. A sleep-induced apneic threshold and its consequences. *Am. Res. Resp. Dis.* 133: 1163–1170, 1986.

82. Dempsey, J. A., and J. B. Skatrud. Fundamental effects of sleep state on breathing. In: *Current Pulmonology,* edited by D. H. Simmons. Chicago: Year Book, 1988, vol. 9, p. 267–304.

83. Dempsey, J. A., J. B. Skatrud, M. S. Badr, and K. G. Henke. Effects of sleep on the regulation of breathing and respiratory muscle function. In: *The Lung: Scientific Foundations,* edited by R. G. Crystal and J. B. West. New York: Raven, 1991, p. 1615–1629.

84. Dhillon, D. P., G. R. Barer, and M. Walsh. The enlarged carotid body of the chronically hypoxic and chronically hypoxic and hypercapnic rat: a morphometric analysis. *Q. J. Exp. Physiol.* 69: 301–317, 1984.

85. Dillon, G. H., and T. G. Waldrop. In vitro response of caudal hypothalamic neurons to hypoxia and hypercapnia. *Neuroscience* 51: 941–950, 1992.

86. Dillon, G. H., and T. G. Waldrop. Responses of feline caudal hypothalamic cardiorespiratory neurons to hypoxia and hypercapnia. *Exp. Brain Res.* 96: 260–272, 1993.

87. Docherty, R. J., and D. S. McQueen. Inhibitory action of dopamine on cat carotid chemoreceptors. *J. Physiol. (Lond.)* 279: 425–436, 1978.

88. Easton, P. A., and N. R. Anthonisen. Carbon dioxide effects on the ventilatory response to sustained hypoxia. *J. Appl. Physiol.* 64: 1451–1456, 1988.

89. Easton, P. A., and N. R. Anthonisen. Ventilatory response to sustained hypoxia after pretreatment with amino phylline. *J. Appl. Physiol.: Respir. Environ. Exerc. Physiol.* 64: 1445–1450, 1988.

90. Edelman, N. H., P. E. Epstein, S. Lahiri, and N. S. Cherniack. Ventilatory responses to transient hypoxia and hypercapnia in man. *Respir. Physiol.* 17: 302–314, 1973.

91. Edelman, N. H., S. Lahiri, L. Braudo, N. S. Cherniack, and A. P. Fishman. The blunted ventilatory response to hypoxia in cyanotic congenital heart disease. *N. Engl. J. Med.* 282: 405–411, 1970.

92. Edelman, N. H., T. V. Santiago, and J. A. Neubauer. Hypoxia and brain blood flow. In: *High Altitude and Man,* edited by J. B. West and S. Lahiri. Bethesda, MD: Am. Physiol. Soc., 1984, p. 101–113.

93. Eger, E. I., II, R. H. Kellogg, A. H. Mines, M. Lima-Ostos, C. G. Morrill, and D. W. Kent. Influence of CO_2 on ventilatory acclimatization to altitude. *J. Appl. Physiol.* 24: 607–615, 1968.

94. Ehrenreich, D. L., R. A. Burns, R. W. Alman, and J. F. Fazekas. Influence of acetazolamide on cerebral blood flow. *Arch. Neurol.* 5: 227–232, 1961.

95. Eldridge, F. L. Central neural respiratory stimulatory effect of active respiration. *J. Appl. Physiol.* 37: 723–735, 1974.

96. Eldridge, F. L., D. E. Millhorn, and J. P. Kiley. Antagonism by theophylline of respiratory inhibition induced by adenosine. *J. Appl. Physiol.: Respir. Environ. Exerc. Physiol.* 59: 1428–1433, 1985.

97. Engwall, M. J. A., and G. E. Bisgard. Ventilatory responses to chemoreceptor stimulation after hypoxic acclimatization in awake goats. *J. Appl. Physiol.: Respir. Environ. Exerc. Physiol.* 69: 1236–1243, 1990.

98. Engwall, M. J. A., L. Daristotle, W. Z. Niu, J. A. Dempsey, and G. E. Bisgard. Ventilatory afterdischarge in the awake goat. *J. Appl. Physiol.: Respir. Environ. Exerc. Physiol.* 71: 1511–1517, 1991.

99. Engwall, M. J. A., E. B. Olson, Jr., and G. E. Bisgard. Carotid body amine levels in goats exposed to hypoxia and hypercapnia. *Neurosci. Lett.* 107: 221–226, 1989.

100. Engwall, M. J. A., E. H. Vidruk, A. M. Nielsen, and G. E. Bisgard. Response of the goat carotid body to acute and prolonged hypercapnia. *Respir. Physiol.* 74: 335–344, 1988.

101. Eyzaguirre, C., and H. Kayano. Effects of hypoxia, hypercapnia, and pH on the chemoreceptor activity of the carotid body in vitro. *J. Physiol. (Lond.)* 178: 385–409, 1965.

102. Eyzaguirre, C., and P. Zapata. Perspectives in carotid body research. *J. Appl. Physiol.: Respir. Environ. Exerc. Physiol.* 57: 931–957, 1984.

103. Feldman, J. L., Neurophysiology of breathing in mammals. In: *Handbook of Physiology. Intrinsic Regulatory Systems of the Brain,* edited by F. E. Bloom. Washington, DC: Am. Physiol. Soc., 1985, sect. 1, vol. 4, p. 463–524.

104. Fencl, V., R. A. Gabel, and D. Wolfe. Composition of cerebral fluids in goats adapted to high altitude. *J. Appl. Physiol.: Respir. Environ. Exerc. Physiol.* 47: 508–513, 1979.

105. Fencl, V., T. B. Miller, and J. R. Pappenheimer. Studies on the respiratory response to disturbances of acid-base balance, with

deductions concerning the ionic composition of cerebral interstitial fluid. *Am. J. Physiol.* 210: 459–472, 1966.

106. Fidone, S., C. Gonzales, B. Dinger, and L. Stensaas. Transmitter dynamics in the carotid body. In: *Chemoreceptors and Chemoreceptor Reflexes,* edited by H. Acker, A. Trzebski, and R. G. O'Regan. New York: Plenum, 1990, p. 3–14.

107. Filuk, R. B., D. J. Berezanski, and N. R. Anthonisen. Depression of hypoxic ventilatory response in humans by somatostatin. *J. Appl. Physiol.: Respir. Environ. Exerc. Physiol.* 65: 1050–1054, 1988.

108. Fink, B. R., R. Katz, H. Reinhold, and A. Schoolman. Suprapontine mechanisms in regulation of respiration. *Am. J. Physiol.* 202: 217–220, 1962.

109. Fitzgerald, M. P. The changes in the breathing and the blood at various high altitudes. *Phil. Trans. R. Soc. Lond. B* 203: 351–371, 1913.

110. Fitzgerald, M. P. Further observations on the changes in the breathing and the blood at various high altitudes. *Proc. R. Soc. Lond. B* 88: 248–258, 1914.

111. Fitzgerald, R. S., and G. A. Dehghani. Neural responses of the cat carotid and aortic bodies to hypercapnia and hypoxia. *J. Appl. Physiol.: Respir. Environ. Exerc. Physiol.* 52: 596–601, 1982.

112. Forster, H. V., G. E. Bisgard, and J. P. Klein. Effect of peripheral chemoreceptor denervation on acclimatization of goats during hypoxia. *J. Appl. Physiol.: Respir. Environ. Exerc. Physiol.* 50: 392–398, 1981.

113. Forster, H. V., G. E. Bisgard, B. Rasmussen, J. A. Orr, D. D. Buss, and M. Manohar. Ventilatory control in peripheral chemoreceptor-denervated ponies during chronic hypoxemia. *J. Appl. Physiol.* 41: 878–885, 1976.

114. Forster, H. V., and J. A. Dempsey. Ventilatory adaptations. In: *Lung Biology in Health and Disease. Regulation of Breathing,* edited by T. Hornbein. New York: Dekker, 1981, p. 845–904.

115. Forster, H. V., J. A. Dempsey, M. L. Birnbaum, W. G. Reddan, J. Thoden, R. F. Grover, and J. Rankin. Effect of chronic exposure to hypoxia on ventilatory response to CO$_2$ and hypoxia. *J. Appl. Physiol.* 31: 586–592, 1971.

116. Forster, H. V., J. A. Dempsey, and L. W. Chosy. Incomplete compensation of CSF [H$^+$] in man during acclimatization to high altitude (4,300 m). *J. Appl. Physiol.* 38: 1067–1072, 1975.

117. Forster, H. V., J. A. Dempsey, E. Vidruk, and G. A. DoPico. Evidence of altered regulation of ventilation during exposure to hypoxia. *Respir. Physiol.* 20: 379–392, 1974.

118. Forster, H. V., R. J. Soto, J. A. Dempsey, and M. J. Hosko. Effect of sojourn at 4,300 m altitude on electroencephalogram and visual evoked response. *J. Appl. Physiol.* 39: 109–113, 1975.

119. Forwand, S. A., M. Landowne, J. N. Follansbee, and J. E. Hansen. Effect of acetazolamide on acute mountain sickness. *N. Engl. J. Med.* 279: 839–845, 1968.

120. Frappell, P., C. Lanthier, R. V. Baudinette, and J. P. Mortola. Metabolism and ventilation in acute hypoxia: a comparative analysis in small mammalian species. *Am. J. Physiol.* 262 (*Regulatory Integrative Comp. Physiol.* 31): R1040–R1046, 1992.

121. Frappell, P., C. Saiki, and J. P. Mortola. Metabolism during normoxia, hypoxia and recovery in the newborn kitten. *Respir. Physiol.* 86: 115–124, 1991.

122. Freedman, A., A. T. Scardella, N. H. Edelman, and T. V. Santiago. Hypoxia does not increase CSF or plasma β-endorphin activity. *J. Appl. Physiol.: Respir. Environ. Exerc. Physiol.* 64: 966–971, 1988.

123. Fry, C. H., and P. A. Poole-Wilson. Effects of acid-base changes on excitation–contraction coupling in guinea pig and rabbit cardiac ventricular muscles. *J. Physiol. (Lond.)* 313: 141–160, 1981.

124. Fujiwara, N., H. Higashi, K. Shimaji, and M. Yoshemura.

Effects of hypoxia on rat hippocampal neurons in vitro. *J. Physiol. (Lond.)* 384: 131–151, 1987.

125. Galdston, M. Respiratory and renal effects of a carbonic anhydrase inhibitor (diamox) on acid-base balance in normal man and in patients with respiratory acidosis. *Am. J. Med.* 19: 516–532, 1955.

126. Gallman, E. A., and D. E. Millhorn. Two long-lasting central respiratory responses following acute hypoxia in glomectomized cats. *J. Physiol. (Lond.)* 395: 333–347, 1988.

127. Gautier, H., and M. Bonora. Effects of carotid body denervation on respiratory pattern of awake cats. *J. Appl. Physiol.: Respir. Environ. Exerc. Physiol.* 46: 1127–1131, 1979.

128. Gautier, H., and M. Bonora. Possible alterations in brain monoamine metabolism during hypoxia-induced tachypnea in cats. *J. Appl. Physiol.: Respir. Environ. Exerc. Physiol.* 49: 769–777, 1980.

129. Gautier, H., J. Milic-Emili, G. Miserocchi, and N. M. Siafakas. Pattern of breathing and mouth occlusion pressure during acclimatization to high altitude. *Respir. Physiol.* 41: 365–377, 1980.

130. Gautier, H., R. Peslin, A. Grassino, J. Milic-Emili, B. Hannhart, E. Powell, G. Miserocchi, M. Bonora, and J. T. Fischer. Mechanical properties of the lungs during acclimatization to altitude. *J. Appl. Physiol.: Respir. Environ. Exerc. Physiol.* 52: 1407–1415, 1982.

131. Georgopoulos, D., D. Berezanski, and N. R. Anthonisen. Effect of dichloroacetate on ventilatory response to sustained hypoxia in normal adults. *Respir. Physiol.* 82: 115–122, 1990.

132. Georgopoulos, D., Z. Bshouty, M. Younes, and N. R. Anthosnisen. Hypoxic exposure and activation of the afterdischarge mechanism in conscious humans. *J. Appl. Physiol.: Respir. Environ. Exerc. Physiol.* 69: 1159–1164, 1990.

133. Georgopoulos, D., S. G. Holtby, D. Berezanski, and N. R. Anthonisen. Aminophylline effects on ventilatory response to hypoxia and hyperoxia in normal adults. *J. Appl. Physiol.: Respir. Environ. Exerc. Physiol.* 67: 1150–1156, 1989.

134. Georgopoulos, D., S. Walker, and N. R. Anthonisen. Increased chemoreceptor output and ventilatory response to sustained hypoxia. *J. Appl. Physiol.: Respir. Environ. Exerc. Physiol.* 67: 1157–1163, 1989.

135. Georgopoulos, D., S. Walker, and N. R. Anthonisen. Effect of sustained hypoxia on ventilatory response to CO$_2$ in normal adults. *J. Appl. Physiol.: Respir. Environ. Exerc. Physiol.* 68: 891–896, 1990.

136. Gershan, W., H. V. Forster, T. F. Lowry, and M. A. Forster. Central ventilatory depression by hypoxia in goats. *Am. Rev. Respir. Dis.* 145: A673, 1992.

137. Gleeson, K., and C. W. Zwillich. Adenosine infusion and periodic breathing during sleep. *J. Appl. Physiol.: Respir. Environ. Exerc. Physiol.* 72: 1004–1009, 1992.

138. Goldberg, S. V., R. B. Schoene, D. Haynor, B. Trimble, E. R. Swenson, J. B. Morrison, and E. J. Banister. Brain tissue pH and ventilatory acclimatization to high altitude. *J. Appl. Physiol.: Respir. Environ. Exerc. Physiol.* 72: 58–63, 1992.

139. Gonzales, C., L. Almaraz, A. Obeso, and R. Rigual. Oxygen and acid chemoreception in the carotid body chemoreceptors. *Trends Neurosci.* 15: 146–153, 1992.

140. Greene, M. K., A. M. Keer, I. B., McIntosh, and R. J. Prescott. Acetazolamide in prevention of acute mountain sickness; a double-blind controlled cross-over study. *BMJ* 283: 811–813, 1981.

141. Grover, R. F. Basal oxygen uptake of man at high altitude. *J. Appl. Physiol.* 18: 909–912, 1963.

142. Grunstein, M. M., T. A. Hazinski, and M. A. Schlueter. Respiratory control during hypoxia in newborn rabbits: implied action of endorphins. *J. Appl. Physiol.: Respir. Environ. Exerc. Physiol.* 51: 122–130, 1981.

143. Guilleminault, C., and W. C. Dement. General physiology of

sleep. In: *The Lung: Scientific Foundations,* edited by R. G. Crystal and J. B. West. New York: Raven, 1991, p. 1609–1614.

144. Guyton, A. C., J. W. Crowell, and J. W. Moore. Basic oscillating mechanism of Cheyne-Stokes breathing. *Am. J. Physiol.* 187: 395–398, 1956.

145. Hackett, P. H., J. T. Reeves, C. D. Reeves, R. F. Grover, and D. Rennie. Control of breathing in Sherpas at low and high altitude. *J. Appl. Physiol.: Respir. Environ. Exerc. Physiol.* 49: 374–379, 1980.

146. Hackett, P. H., D. Rennie, S. E. Hofmeister, R. F. Grover, E. B. Grover, and J. T. Reeves. Fluid retention and relative hypoventilation in acute mountain sickness. *Respiration* 43: 321–329, 1982.

147. Hackett, P. H., D. Rennie, and H. D. Levine. The incidence, importance, and prophylaxis of acute mountain sickness. *Lancet* 2: 1149–1154, 1976.

148. Hackett, P. H., R. C. Roach, G. L. Harrison, R. B. Schoene, and W. J. Mills. Respiratory stimulants and sleep periodic breathing at high altitude. *Am. Rev. Respir. Dis.* 135: 896–898, 1987.

149. Haddad, G. G., and D. F. Donnelly. O₂ deprivation induces a major depolarization in brain stem neurons in the adult but not in the neonatal rat. *J. Physiol. (Lond.)* 429: 411–428, 1990.

150. Hanbauer, I., and S. Hellstrom. The regulation of dopamine and noradrenaline in rat carotid body and its modification by denervation and hypoxia. *J. Physiol. (Lond.)* 282: 21–34, 1978.

151. Hanbauer, I., F. Karoum, S. Hellstrom, and S. Lahiri. Effects of hypoxia lasting up to one month on the catecholamine content in rat carotid body. *Neuroscience* 6: 81–86, 1981.

152. Heath, D., P. Smith, and R. Jago. Hyperplasia of the carotid bodies. In: *The Peripheral Arterial Chemoreceptors,* edited by D. J. Pallot. London: Croom Helm, 1984, p. 277–281.

153. Heath, D., and D. R. Williams. Monge's disease. In: *Man at High Altitude* (2nd ed.), New York: Churchill Livingstone, 1981, p. 167–179.

154. Hedner, J., T. Hedner, P. Wessberg, and J. Jonason. An analysis of the mechanism by which γ-aminobutyric acid depresses ventilation in the rat. *J. Appl. Physiol.: Respir. Environ. Exerc. Physiol.* 56: 849–856, 1984.

155. Hedner, T., J. Hedner, P. Wessberg, and J. Jonason. Regulation of breathing in the rat: indications for a role of central adenosine mechanisms. *Neurosci. Lett.* 33: 147–151, 1982.

156. Hellstrom, S., I. Hanbauer, and E. Costa. Selective decrease of dopamine content in rat carotid body during exposure to hypoxic conditions. *Brain Res.* 118: 352–355, 1976.

157. Hellstrom, S., and H. Koslow. Biogenic amines in carotid body of adult and infant rats, a gas chromatographic–mass spectrometric assay. *Acta Physiol. Scand.* 93: 540–547, 1975.

158. Henke, K. G., A. Arias, J. B. Skatrud, and J. A. Dempsey. Inhibition of inspiratory muscle activity during slep. *Am. Rev. Respir. Dis.* 138: 8–15, 1988.

159. Henke, K. G., J. A. Dempsey, J. M. Kowitz, S. Badr, and J. B. Skatrud. Effects of sleep-induced increases in upper airway resistance on respiratory muscle activity. *J. Appl. Physiol.: Respir. Environ. Exerc. Physiol.* 69: 617–624, 1990.

160. Hill, J. R. The oxygen consumption of new-born and adult mammals. Its dependence on the oxygen tension in the inspired air and on the environmental temperature. *J. Physiol. (Lond.)* 149: 346–373, 1959.

161. Hochachka, P. W. Assessing metabolic strategies for surviving O₂ lack: role of metabolic arrest coupled with channel arrest. *Mol. Physiol.* 8: 331–350, 1985.

162. Hochachka, P. W. Defense strategies against hypoxia and hypothermia. *Science* 231: 234–241, 1986.

163. Hochachka, P. W., G. O. Matheson, W. S. Parkhouse, J. Sumar-Kalinoski, C. Stanley, C. Monge, D. C. McKenzie, J. Merkt, S. F. P. Man, R. Jones, and P. S. Allen. Path of oxygen from the

atmosphere to mitochondria in Andean natives: adaptable versus constrained components. In: *Hypoxia, The Adaptations,* edited by J. R. Sutton, G. Coates, and J. E. Remmers. Toronto: Dekker, 1990, p. 72–74.

164. Holtby, S. G., D. J. Berezanski, and N. R. Anthonisen. Effect of 100% O₂ on hypoxic eucapnic ventilation. *J. Appl. Physiol.: Respir. Environ. Exerc. Physiol.* 65: 1157–1162, 1988.

165. Hopp, F. A., J. L. Seagard, J. Bajic, and E. J. Zuperku. Respiratory responses to aortic and carotid chemoreceptor activation in the dog. *J. Appl. Physiol.: Respir. Environ. Exerc. Physiol.* 70: 2539–2550, 1991.

166. Hornbein, T. F., Z. J. Griffo, and A. Roos. Quantitation of chemoreceptor activity: interrelation of hypoxia and hypercapnia. *J. Neurophysiol.* 24: 561–568, 1961.

167. Hornbein, T. F., and S. C. Sorensen. Ventilatory response to hypoxia and hypercapnia in cats living at high altitude. *J. Appl. Physiol.: Respir. Environ. Exerc. Physiol.* 27: 834–836, 1969.

168. Hoyt, R. W., M. J. Durkot, V. A. Forte, Jr., L. J. Hubbard, L. A. Trad, and A. Cymerman. Hypobaric hypoxia (380 torr) decreases intracellular and total body water in goats. *J. Appl. Physiol.: Respir. Environ. Exerc. Physiol.* 71: 509–513, 1991.

169. Huang, S. Y., J. K. Alexander, R. Grover, J. T. Maher, R. E. McCullough, R. G. McCullough, L. G. Moore, J. B. Sampson, J. V. Weil, and J. T. Reeves. Hypocapnia and sustained hypoxia blunt ventilation on arrival at high altitude. *J. Appl. Physiol.: Respir. Environ. Exerc. Physiol.* 56: 602–606, 1984.

170. Hudgel, D. W., K. R. Chapman, C. Faulks, and C. Hendricks. Changes in inspiratory muscle electrical activity and upper airway resistance during periodic breathing induced by hypoxia during sleep. *Am. Rev. Respir. Dis.* 135: 899–906, 1987.

171. Hudgel, D. W., and T. Harasick. Fluctuation in timing of upper airway and chest wall inspiratory muscle activity in obstructive sleep apnea. *J. Appl. Physiol.: Respir. Environ. Exerc. Physiol.* 69: 443–450, 1990.

172. Hurtado, A. Some clinical aspects of life at high altitude. *Ann. Intern. Med.* 53: 247, 1960.

173. Iber, C., S. F. Davies, R. C. Chapman, and M. M. Mahowald. A possible mechanism for mixed apnea in obstructive sleep apnea. *Chest* 89: 800–805, 1986.

174. Irisawa, H., and R. Sato. Intra- and extracellular actions of proton on the calcium current of isolated guinea pig ventricular cells. *Circ. Res.* 59: 348–355, 1986.

175. Issa, F. G., and J. E. Remmers. Identification of a subsurface area in the ventral medulla sensitive to local changes in Pco₂. *J. Appl. Physiol.: Respir. Environ. Exerc. Physiol.* 72: 439–446, 1992.

176. Iversen, K., T. Hedner, and P. Lundborg. GABA concentrations and turn-over in neonatal rat brain during asphyxia and recovery. *Acta Physiol. Scand.* 118: 91–94, 1983.

177. Jansen, A. H., and V. Chernick. Fetal breathing and development of control of breathing. *J. Appl. Physiol.: Respir. Environ. Exerc. Physiol.* 70: 1431–1446, 1991.

178. Javaheri, S., L. J. Teppema, and J. A. Evers. Effects of aminophylline on hypoxemia-induced ventilatory depression in the cat. *J. Appl. Physiol.: Respir. Environ. Exerc. Physiol.* 64(9): 1837–1843, 1988.

179. Jensen, J. B., A. D. Wright, N. A. Lassen, M. E. Raichle, and A. R. Bradwell. Cerebral blood flow at altitude. In: *Hypoxia: The Adaptations,* edited by J. R. Sutton, G. Coates, and J. E. Remmers. Toronto: Decker, 1990, p. 296.

180. Jiang, C., and G. G. Haddad. Effect of anoxia on intracellular and extracellular potassium activity in hypoglossal neurons in vitro. *J. Neurophysiol.* 66: 103–111, 1991.

181. Jodkowski, J. S., and J. Lipski. Decreased excitability of respiratory motoneurons during hypercapnia in the acute spinal cat. *Brain Res.* 386: 296–304, 1986.

182. Joels, N., and E. Neil. The influence of anoxia and hypercapnia separately and in combination on chemoreceptor impulse discharge. *J. Physiol. (Lond.)* 155: 45, 1961.

183. Joels, N., and H. White. The contribution of the arterial chemoreceptors to the stimulation of respiration by adrenaline and noradrenaline in the cat. *J. Physiol. (Lond.)* 197: 1–23, 1968.

184. Johnson, T. S., and P. B. Rock. Current concepts: acute mountain sickness. *N. Engl. J. Med.* 319: 841–845, 1988.

185. Jones, D. P., T. W. Aw, C. Bai, and A. H. Sillau. Regulation of mitochondrial distribution: an adaptive response to changed in oxygen supply. In: *Reponses and Adaptation to Hypoxia, Organ to Organelle,* edited by S. Lahiri, N. S. Cherniack, and R. S. Fitzgerald. New York: Oxford University Press, 1991, p. 25–31.

186. Kagawa, S., M. J. Stafford, T. B. Waggener, and J. W. Severinghaus. No effect of naloxone on hypoxia-induced ventilatory depression in adults. *J. Appl. Physiol.: Respir. Environ. Exerc. Physiol.* 52: 1030–1034, 1982.

187. Kellogg, R. H. Altitude acclimatization, a historical introduction emphasizing the regulation of breathing. *Physiologist* 11: 37–57, 1969.

188. Kellogg, R. H. Oxygen and carbon dioxide in the regulation of respiration. *Federation Proc.* 36: 1658–1663, 1977.

189. Kiwull-Schone, H., and P. Kiwull. Hypoxic modulation of central chemosensitivity. In: *Central Neurone Environment,* edited by M. E. Schlafke, H. P. Koepchen, and W. R. See. Berlin: Springer-Verlag, 1983, p. 88–95.

190. Klausen, K., B. Rasmussen, J. Gjellerod, H. Madsen, and E. Petersen. Circulation, metabolism and ventilation during prolonged exposure to carbon monoxide and to high altitude. *Scand. J. Clin. Lab. Invest.* 103: 26–38, 1968.

191. Kneussl, M. P., P. Pappagianopoulos, B. Hoop, and H. Kazemi. Reversible depression of ventilation and cardiovascular function by ventriculo-cisternal perfusion with gamma-aminobutyric acid in dogs. *Am. Rev. Respir. Dis.* 133: 1024–1028, 1986.

192. Kondo, H., H. Kuramoto, and T. Fujita. Neuropeptide tyrosine-like immunoreactive nerve fibers in the carotid body chemoreceptor of rats. *Brain Res.* 372: 353–356, 1986.

193. Koos, B. J., A. Chao, and W. Doany. Adenosine stimulates breathing in fetal sheep with brain stem section. *J. Appl. Physiol.: Respir. Environ. Exerc. Physiol.* 72: 94–99, 1992.

194. Koos, B. J., and K. Matsuda. Fetal breathing, sleep state, and cardiovascular responses to adenosine in sheep. *J. Appl. Physiol.: Respir. Environ. Exerc. Physiol.* 68: 489–495, 1990.

195. Kou, Y. R., P. Ernsberger, P. A. Cragg, N. S. Cherniack, and N. R. Prabhakar. Role of alpha-2 adrenergic receptors in the carotid body response to isocapnic hypoxia. *Respir. Physiol.* 83: 353–364, 1991.

196. Krasney, J. A., J. B. Jensen, and N. A. Lassen. Cerebral blood flow does not adapt to sustained hypoxia. *J. Cereb. Blood Flow Metab.* 10: 759–764, 1990.

197. Kressin, N. A., A. M. Nielsen, R. Laravuso, and G. E. Bisgard. Domperidone induced potentiation of ventilatory responses in awake goats. *Respir. Physiol.* 65: 169–180, 1986.

198. Kryger, M., R. McCollough, D. Collins, C. Scoggin, J. Weil, and R. Grover. Treatment of chronic mountain polycythemia (CMP) with respiratory stimulation. *Am. Rev. Respir. Dis.* 115: 345–352, 1977.

199. Kryger, M., R. McCullough, R. Dockel, D. Collins, J. V. Weil, and R. F. Grover. *Am. Rev. Respir. Dis.* 118: 659–666, 1978.

200. Krynjevic, K., M. Randic, and B. K. Siesjo. Cortical CO_2 tension and neuronal excitability. *J. Physiol. (Lond.)* 176: 105–122, 1965.

201. Kummer, W. Three types of neurochemically defined autonomic fibres innervate the carotid baroreceptor and chemoreceptor regions in the guinea-pig. *Anat. Embryol.* 181: 477–489, 1990.

202. Kummer, W., and J. Habeck. Substance P and calcitonin gene-related peptide-like immunoreactivities in the human carotid body studied at light and electron microscopical level. *Brain Res.* 554: 286–292, 1991.

203. Laduron, P. M., and J. E. Leysen. Domperidone, a specific in vitro dopamine antagonist, devoid of in vivo central dopaminergic activity. *Biochem. Pharmacol.* 28: 2161–2165, 1979.

204. Lafromboise, W. A., R. D. Guthrie, T. A. Standaert, and D. E. Woodrum. Pulmonary mechanics during ventilatory response to hypoxemia in the newborn monkey. *J. Appl. Physiol.: Respir. Environ. Exerc. Physiol.* 55: 1008–1014, 1983.

205. LaFramboise, W. A., T. A. Standaert, D. E. Woodrum, and R. D. Guthrie. Occlusion pressures during the ventilatory response to hypoxemia in the newborn monkey. *J. Appl. Physiol.: Respir. Environ. Exerc. Physiol.* 50: 1169–1174, 1981.

206. LaFromboise, W. A., and D. Woodrum. Elevated diaphragm electromyogram during neonatal hypoxic ventilatory depression. *J. Appl. Physiol.: Respir. Environ. Exerc. Physiol.* 59: 1040–1045, 1986.

207. Lagercrantz, H., Y. Yamamoto, B. B. Fredholm, N. R. Probhakas, and C. von Euler. Adenosine analogues depress ventilation in rabbit neonates. Theophylline stimulation of respiration via adenosine receptors? *Pediatr. Res.* 18: 387–390, 1984.

208. Lahiri, S. Acid base in sherpa altitude residents and lowlanders at 4880 m. *Respir. Physiol.* 2: 323–334, 1967.

209. Lahiri, S. Unattenuated ventilatory hypoxic drive in ovine and bovine species native to high altitude. *J. Appl. Physiol.* 32: 95–102, 1972.

210. Lahiri, S. Ventilatory response to hypoxia in intact cats living at 3850 m. *J. Appl. Physiol.: Respir. Environ. Exerc. Physiol.* 43: 114–120, 1977.

211. Lahiri, S. Adaptive respiratory regulation—lessons from high altitudes. In: *Environmental Physiology: Aging, Heat and Altitude,* edited by S. M. Horvath and M. K. Yousef. Amsterdam: Elsevier, 1980, p. 341–347.

212. Lahiri, S. Respiratory control in Andean and Himalayan high-altitude natives. In: *High Altitude and Man,* edited by J. B. West and S. Lahiri. Bethesda, MD: Am. Physiol. Soc., 1984, p. 147–162.

213. Lahiri, S., N. S. Cherniack, N. H. Edelman, and A. P. Fishman. Regulation of respiration in goat and its adaptation to chronic and life-long hypoxia. *Respir. Physiol.* 12: 388–403, 1971.

214. Lahiri, S., and R. G. Delaney. Relationship between carotid chemoreceptor activity and ventilation in the cat. *Respir. Physiol.* 24: 267–286, 1975.

215. Lahiri, S., and R. G. Delaney. Stimulus interaction in the responses of carotid body chemoreceptor single afferent fibers. *Respir. Physiol.* 24: 249–266, 1975.

216. Lahiri, S., F. F. Kao, T. M. Velasquez, C. Martinez, and W. Pezzia. Irreversible blunted respiratory sensitivity to hypoxia in high altitude natives. *Respir. Physiol.* 6: 360–374, 1969.

217. Lahiri, S., K. H. Maret, M. G. Sherpa, and R. M. Peters, Jr. Sleep and periodic breathing at high altitude: Sherpa natives versus sojourners. In: *High Altitude and Man,* edited by J. B. West and S. Lahiri. Bethesda, MD: Am. Physiol. Soc., 1984, p. 73–90.

218. Lahiri, S., A. Mokashi, E. Mulligan, and T. Nishino. Comparison of aortic and carotic chemoreceptor responses to hypercapnia and hypoxia. *J. Appl. Physiol.: Respir. Environ. Exerc. Physiol.* 51: 55–61, 1981.

219. Lahiri, S., T. Nishino, A. Mokashi, and E. Mulligan. Interaction of dopamine and haloperidol with O_2 and CO_2 chemoreception in carotid body. *J. Appl. Physiol.: Respir. Environ. Exerc. Physiol.* 49: 45–51, 1980.

220. Lahiri, S., N. Smatresk, M. Pokorski, P. Barnard, and A. Mokashi. Efferent inhibition of carotid body chemoreception in chronically hypoxic cats. *Am. J. Physiol.* 245 (*Regulatory Integrative Comp. Physiol.* 16): R678–R683, 1983.

221. Lahiri, S., N. Smatresk, M. Pokorski, P. Barnard, A. Mokashi, and K. H. McGregor. Dopaminergic efferent inhibition of carotid body chemoreceptors in chronically hypoxic cats. *Am. J. Physiol.* 247 (*Regulatory Integrative Comp. Physiol.* 18): R24–R28, 1984.

222. Laidler, P., and J. M. Kay. Ultrastructure of carotid body in rats living at a simulated altitude of 4300 metres. *J. Pathol.* 124: 27–33, 1978.

223. Laidler, P., and J. M. Kay. A quantitative morphological study of the carotid bodies of rats living at a simulated altitude of 4300 metres. *J. Pathol.* 117: 183–191, 1975.

224. Larson, E. B., R. C. Roach, R. B. Schoene, and T. F. Hornbein. Acute mountain sickness and acetazolamide. *JAMA* 248: 328–332, 1982.

225. Laszy, J., and A. Sarkadi. Hypoxia-induced sleep disturbance in rats. *Sleep* 13: 205–217, 1990.

226. Lawson, E. E., and W. A. Long. Central origin of biphasic breathing pattern during hypoxia in newborns. *J. Appl. Physiol.: Respir. Environ. Exerc. Physiol.* 55: 483–488, 1983.

227. Lefrancois, R., H. Gautier, and P. Pasquis. Ventilatory oxygen drive in acute and chronic hypoxia. *Respir. Physiol.* 4: 217–228, 1968.

228. Lenfant, C., J. Torrance, E. English, C. A. Finch, C. Reynafarje, J. Ramos, and J. Faura. Effect of altitude on oxygen binding by hemoglobin and on organic phosphate levels. *J. Clin. Invest.* 47: 2652–2656, 1968.

229. Llados, F., and P. Zapata. Effects of adrenoreceptors stimulating and blocking agents on carotid body chemosensory inhibition. *J. Physiol. (Lond.)* 274: 501–509, 1978.

230. Long, W. A., and E. E. Lawson. Neurotransmitters and biphasic respiratory response to hypoxia. *J. Appl. Physiol.: Respir. Environ. Exerc. Physiol.* 57: 213–222, 1984.

231. Long, W. Q., G. G. Giesbrecht, and N. R. Anthonisen. Ventilatory response to moderate hypoxia in awake chemodenervated cats. *J. Appl. Physiol.: Respir. Environ. Exerc. Physiol.* 74: 805–810, 1993.

232. Lowry, T. F., H. V. Forster, M. Korducki, A. L. Forster, and M. A. Foster. Comparison of ventilatory responses to sustained reduction in arterial O₂ tension vs. content in awake ponies. *J. Appl. Physiol.: Respir. Environ. Exerc. Physiol.* 76: 2147–2153, 1994.

233. Lydic, R. Central regulation of sleep and autonomic physiology. In: *Clinical Physiology of Sleep,* edited by R. Lydic, and J. L. Biebuyck. Bethesda, MD: Am. Physiol. Soc., 1988, p. 1–20.

234. Marshall, K. C., and I. Engberg. The effects of hydrogen ion on spinal neurons. *Can. J. Physiol. Pharmacol.* 58: 650–655, 1980.

235. Martin-Body, R. L., and B. M. Johnston. Central origin of the hypoxic depression of breathing in the newborn. *Respir. Physiol.* 71: 25–32, 1988.

236. Masuyama, S., H. Kimura, T. Sugita, T. Kuriyama, K. Tatsumi, F. Kunimtomo, S. Okita, H. Tojima, Y. Yuguchi, S. Watanabe, and Y. Honda. Control of ventilation in extreme-altitude climbers. *J. Appl. Physiol.: Respir. Environ. Exerc. Physiol.* 61: 500–506, 1986.

237. Matsumoto, S., A. Mokashi, and S. Lahiri. Influence of ganglioglomerular nerve on carotid chemoreceptor activity in the cat. *J. Auton. Nerv. Syst.* 15: 7–20, 1986.

238. McDonald, D. M. Peripheral chemoreceptors. In: *Regulation of Breathing,* edited by T. F. Hornbein. New York: Dekker, 1981, pt. I, p. 105–319.

239. McDonald, D. M., and R. A. Mitchell. The innervation of glomus cells, ganglion cells and blood vessels in the rat carotid body: a quantitative ultrastructural analysis. *J. Neurocytol.* 4: 177–230, 1975.

240. McDonald, D. M., and R. A. Mitchell. The neural pathway involved in "efferent inhibition" of chemoreceptors in the cat carotid body. *J. Comp. Neurol.* 201: 457–476, 1981.

241. McGinty, D. J., R. Drucker-Colin, A. Morrison, and P. O. Parmeggiani (Eds). *Brain Mechanisms of Sleep.* New York: Raven, 1985.

242. McGinty, D. J., and M. B. Sterman. Sleep suppression after basal forebrain lesions in the cat. *Science* 160: 1253–1255, 1968.

243. McGregor, K. H., J. Gil, and S. Lahiri. A morphometric study of the carotid body in chronically hypoxic rats. *J. Appl. Physiol.: Respir. Environ. Exerc. Physiol.* 57: 1430–1438, 1984.

244. McQueen, D. S. Pharmacological aspects of putative transmitters in the carotid body. In: *Physiology of the Peripheral Arterial Chemoreceptors,* edited by H. Acker and R. G. O'Regan. New York: Elsevier, 1983, p. 149–195.

245. Melton, J. E., J. A. Neubauer, and N. H. Edelman. CO₂ sensitivity of cat phrenic neurogram during hypoxic respiratory depression. *J. Appl. Physiol.: Respir. Environ. Exerc. Physiol.* 65: 736–743, 1988.

246. Miles, R. Does low pH stimulate central chemoreceptors located near the ventral medullary surface? *Brain Res.* 271: 349–353, 1983.

247. Milledge, J. S., and S. Lahiri. Respiratory control in lowlanders and Sherpa highlanders at altitude. *Respir. Physiol.* 2: 310–322, 1967.

248. Miller, J. C., and S. M. Horvath. Sleep at altitude. *Aviat. Space Environ. Med.* 48: 615–620, 1977.

249. Miller, M. J., and S. M. Tenney. Hyperoxic hyperventilation in carotid-deafferented cats. *Respir. Physiol.* 23: 23–30, 1975.

250. Miller, M. J., and S. M. Tenney. Hypoxia-induced tachypnea in carotid-deafferented cats. *Respir. Physiol.* 23: 31–39, 1975.

251. Millhorn, D. E., F. L. Eldridge, J. P. Kiley, and T. G. Waldrop. Prolonged inhibition of respiration following acute hypoxia in glomectomized cats. *Respir. Physiol.* 57: 331–340, 1984.

252. Millhorn, D. E., F. L. Eldridge, and T. G. Waldrop. Prolonged stimulation of respiration by a new central neural mechanism. *Respir. Physiol.* 41: 87–103, 1980.

253. Millhorn, D. E., F. L. Eldridge, and T. G. Waldrop. Prolonged stimulation of respiration by endogenous central serotonin. *Respir. Physiol.* 42: 171–188, 1980.

254. Mitchell, R. A. Cerebrospinal fluid and the regulation of respiration. In: *Advances in Respiratory Physiology,* edited by C. G. Caro. Baltimore, MD: Williams and Wilkins, 1966, p. 1–47.

255. Mitchell, R. A., C. T. Carman, J. W. Severinghaus, B. W. Richardson, M. M. Singer, and S. Shnider. Stability of cerebrospinal fluid pH in chronic acid-base disturbances in blood. *J. Appl. Physiol.* 20: 443–452, 1965.

256. Mitchell, R. A., H. H. Loeschcke, J. W. Severinghaus, B. W. Richardson, and W. H. Massion. Regions of respiratory chemosensitivity on the surface of the medulla. *Ann. N. Y. Acad. Sci.* 109: 661–681, 1963.

257. Monge, M. C. Chronic mountain sickness. *Physiol. Rev.* 23: 166, 1943.

258. Morrill, C. G., J. R. Meyer, and J. V. Weil. Hypoxic ventilatory depression in dogs. *J. Appl. Physiol.* 38: 143–146, 1975.

259. Mortola, J. P. Dynamics of breathing in newborn mammals. *Physiol. Rev.* 67: 187–243, 1987.

260. Mortola, J. P., and R. Rezzonico. Metabolic and ventilatory rates in newborn kittens during acute hypoxia. *Respir. Physiol.* 73: 55–68, 1988.

261. Mortola, J. P., R. Rezzonico, and C. Lanthier. Ventilation and oxygen consumption during acute hypoxia in newborn mammals; a comparative analysis. *Respir. Physiol.* 78: 31–43, 1989.

262. Musch, T. I., J. A. Dempsey, C. A. Smith, G. S. Mitchell, and N. T. Bateman. Metabolic acids and [H⁺] regulation in brain tissue during acclimatization to chronic hypoxia. *J. Appl. Physiol.: Respir. Environ. Exerc. Physiol.* 55: 1486–1495, 1983.

263. Naifeh, K. H., and J. Kamiya. The nature of respiratory change associated with sleep onset. *Sleep* 4: 49–59, 1981.

264. Nattie, E. E. Diethyl pyrocarbonate (an imidazole binding substance) inhibits rostral VLM CO_2 sensitivity. *J. Appl. Physiol.: Respir. Environ. Exerc. Physiol.* 61: 843–850, 1986.

265. Nattie, E. E., and L. Aihua. Fluorescence location of RVLM kainate microinjections that alter the control of breathing. *J. Appl. Physiol.: Respir. Environ. Exerc. Physiol.* 68: 1157–1166, 1990.

266. Nattie, E. E., J. W. Mills, L. C. Ou, and W. M. St. John. Kainic acid on the rostral ventrolateral medulla inhibits phrenic output and CO_2 sensitivity. *J. Appl. Physiol.: Respir. Environ. Exerc. Physiol.* 65: 1525–1534, 1988.

267. Neil, E., and R. G. O'Regan. Efferent and afferent impulse activity recorded from few-fibre preparations of otherwise intact sinus and aortic nerves. *J. Physiol. (Lond.)* 215: 33–47, 1971.

268. Neubauer, J. A., S. F. Gonsalves, W. Chou, H. M. Geller, and N. H. Edelman. Chemosensitivity of medullary neurons in explant tissue cultures. *Neuroscience* 45: 701–708, 1991.

269. Neubauer, J. A., J. E. Melton, and N. H. Edelman. Modulation of respiration during brain hypoxia. *J. Appl. Physiol.: Respir. Environ. Exerc. Physiol.* 68: 441–451, 1990.

270. Neubauer, J. A., M. A. Posner, T. V. Santiago, and N. H. Edelman. Naloxone reduces ventilatory depression of brain hypoxia. *J. Appl. Physiol.: Respir. Environ. Exerc. Physiol.* 63: 699–706, 1987.

271. Neubauer, J. A., T. V. Santiago, M. A. Posner, and N. H. Edelman. Ventral medullary pH and ventilatory responses to hyperperfusion and hypoxia. *J. Appl. Physiol.: Respir. Environ. Exerc. Physiol.* 58: 1659–1668, 1985.

272. Neubauer, J. A., A. Simone, and N. H. Edelman. Role of brain lactic acidosis in hypoxic depression of respiration. *J. Appl. Physiol.: Respir. Environ. Exerc. Physiol.* 65: 1324–1331, 1988.

273. Neylon, M., and J. M. Marshall. The role of adenosine in the respiratory and cardiovascular response to systemic hypoxia in the rat. *J. Physiol. (Lond.)* 440: 529–545.

274. Nicholson, A. N., P. A. Smith, B. M. Stone, A. R. Bradwell, and J. H. Coote. Altitude insomnia: studies during an expedition to the Himalayas. *Sleep* 11: 354–361, 1988.

275. Nielsen, A. M., G. E. Bisgard, and E. H. Vidruk. Carotid chemoreceptor activity during acute and sustained hypoxia in goats. *J. Appl. Physiol.: Respir. Environ. Exerc. Physiol.* 65: 1796–1802, 1988.

276. Nolan, W. F., and D. G. Davies. Brain extracellular fluid pH and blood flow during isocapnic and hypocapnic hypoxia. *J. Appl. Physiol.: Respir. Environ. Exerc. Physiol.* 53: 247–252, 1982.

277. O'Regan, R. G., and S. Majcherczyk. Control of peripheral chemoreceptors by efferent nerves. In: *Physiology of the Peripheral Arterial Chemoreceptors*, edited by H. Acker and R. G. O'Regan. Amsterdam: Elsevier, 1983, p. 257–298.

278. Olson, E. B., Jr. Ventilatory adaptation to hypoxia occurs in serotonin-depleted rats. *Respir. Physiol.* 69: 227–235, 1987.

279. Olson, E. B., Jr., and J. Dempsey. Rat as a model for human-like ventilatory adaptation to chronic hypoxia. *J. Appl. Physiol.: Respir. Environ. Exerc. Physiol.* 44: 763–769, 1978.

280. Olson, E. B., Jr., E. H. Vidruk, D. R. McCrimmon, and J. A. Dempsey. Monoamine neurotransmitter metabolism during acclimatization to hypoxia in rats. *Respir. Physiol.* 54: 79–96, 1983.

281. Opitz, E. Increased vascularization of the tissue due to acclimatization to high altitude and its significance for oxygen transport. *Exp. Med Surg.* 9: 389–463, 1951.

282. Orem, J. Medullary respiratory neuron activity: relationship to tonic and phasic REM sleep. *J. Appl. Physiol.: Respir. Environ. Exerc. Physiol.* 48: 54–65, 1980.

283. Orem, J., and W. C. Dement. Neurophysiological substrates of the changes in respiration during sleep. In: *Advances in Sleep Research*, edited by E. D. Weitzman. New York: Spectrum, 1975, vol. 2, p. 1–42.

284. Orem, J., and A. Netick. Behavioral control of breathing in the cat. *Brain Res.* 366: 238–253, 1986.

285. Orem, J., I. Osorio, E. Brooks, and T. Dick. Activity of respiratory neurons during NREM sleep. *J. Neurophysiol.* 54: 1144–1156, 1985.

286. Orr, J. A., G. E. Bisgard, H. V. Forster, D. D. Buss, J. A. Dempsey, and J. A. Will. Cerebrospinal fluid alkalosis during high-altitude sojourn in unanesthetized ponies. *Respir. Physiol.* 25: 23–37, 1975.

287. Ou, L. C., J. Chen, E. Fiore, J. C. Leiter, T. Brinck-Johnsen, G. F. Birchard, G. Clemons, and R. P. Smith. Ventilatory and hematopoetic responses to chronic hypoxia in two rat strains. *J. Appl. Physiol.: Respir. Environ. Exerc. Physiol.* 72: 2354–2363, 1992.

288. Ou, L. C., M. J. Miller, and S. M. Tenney. Hypoxia and carbon dioxide as separate and interactive depressants of ventilation. *Respir. Physiol.* 28: 347–358, 1976.

289. Ou, L. C., W. M. St. John, and S. M. Tenney. The contribution of central mechanisms rostral to the pons in high altitude ventilatory acclimatization. *Respir. Physiol.* 54: 343–351, 1983.

290. Ou, L. C., and S. M. Tenney. Properties of mitochondria from hearts of cattle acclimatized to high altitude. *Respir. Physiol.* 8: 151–159, 1970.

291. Ou, L. C., and S. M. Tenney. The role of brief hypocapnia in the ventilatory response to CO_2 with hypoxia. *Respir. Physiol.* 28: 333–346, 1976.

292. Pappenheimer, J. R. The ionic composition of cerebral extracellular fluid and its relation to control of breathing. *Harvey Lect.* 6: 71–93, 1967.

293. Pei, S. X., X. J. Chen, B. Z. Siren, Y. H. Liu, X. S. Ching, E. M. Harris, I. S. Anand, and P. C. Harris. Chronic mountain sickness in Tibet. *Q. J. Med.* 71: 555–574, 1989.

294. Pequignot, J. M., J. M. Cottet-Emard, Y. Dalmaz, and L. Peyrin. Dopamine and norepinephrine dynamics in rat carotid body during long-term hypoxia. *J. Auton. Nerv. Sys.* 21: 9–14, 1987.

295. Pequignot, J. M., Y. Dalmaz, J. Claustre, J. M. Cottet-Emard, N. Borghini, and L. Peyrin. Preganglionic sympathetic fibers modulate dopamine turnover in rat carotid body during long-term hypoxia. *J. Auton. Nerv. Syst.* 32: 243–250, 1991.

296. Pequignot, J. M., S. Hellstrom, and H. Johannsson. Intact and sympathectomized carotid bodies of long-term hypoxic rats: a morphometric ultrastructural study. *J. Neurocytol.* 13: 481–493, 1984.

297. Philipson, K. D., M. M. Bersohn, and A. Y. Nishimoto. Effects of pH on Na^+–Ca^{++} exchange in canine cardiac sarcolemmal vesicles. *Circ. Res.* 50: 287–293, 1982.

298. Phillipson, E. A. Control of breathing during sleep. *Am. Rev. Respir. Dis.* 118: 909–939, 1978.

299. Phillipson, E. A., and C. E. Sullivan. Respiratory control mechanisms during NREM and REM sleep. In: *Sleep Apnea Syndromes*, edited by C. Guilleminault and W. C. Dement. New York: Liss, 1978, p. 47–64.

300. Phillipson, E. A., C. E. Sullivan, D. J. C. Read, E. Murphy, and L. F. Kozar. Ventilatory and waking responses to hypoxia in sleeping dogs. *J. Appl. Physiol.: Respir. Environ. Exerc. Physiol.* 44: 512–520, 1978.

301. Pizarro, J., M. M. Warner, M. L. Ryan, G. S. Mitchell, and G. E. Bisgard. Intracarotid norepinephrine infusions inhibit ventilation in goats. *Respir. Physiol.* 90: 299–310, 1992.

302. Pokorski, M., and S. Lahiri. Effects of naloxone on carotid body chemoreception and ventilation in the cat. *J. Appl. Physiol.: Respir. Environ. Exerc. Physiol.* 51: 1533–1538, 1981.

303. Pokorski, M., and S. Lahiri. Endogenous opiates and ventilatory acclimatization to chronic hypoxia in the cat. *Respir. Physiol.* 83: 211–222, 1991.

304. Prabhakar, N. R., Y. R. Kou, and M. Runold. Chemoreceptor responses to substance P, physalaemin and eledoisin: evidence for neurokinin-1 receptors in the cat carotid body. *Neurosci. Lett.* 120: 183–186, 1990.

305. Prabhakar, N. R., S. C. Landis, G. K. Kumar, D. Mullikin-Kilpatrick, S. Cherniack, and S. Leeman. Substance P and neurokinin A in the cat carotid body: localization, exogenous effects and changes in content in response to arterial PO_2. *Brain Res.* 481: 205–214, 1989.

306. Prabhakar, N. R., J. Mitra, and N. S. Cherniack. Role of substance P in hypercapnic excitation of carotid chemoreceptors. *J. Appl. Physiol.: Respir. Environ. Exerc. Physiol.* 63: 2418–2425, 1987.

307. Prabhakar, N. R., R. Runold, M. Yamamoto, H. Lagercrantz, and C. von Euler. Effect of substance P antagonist on the hypoxia-induced carotid chemoreceptor activity. *Acta Physiol. Scand.* 121: 301–303, 1984.

308. Rahn, H., and A. B. Otis. Man's respiratory response during and after acclimatization to high altitude. *Am. J. Physiol.* 157: 445–462, 1949.

309. Rechtschaffen, A., and A. Kales. *A Manual of Standardized Terminology Techniques and Criteria for Scoring States of Sleep and Wakefulness in Newborn Infants.* Los Angeles: UCLA Brain Information Service/Brain Research Institute, 1971.

310. Reeves, J. T., L. G. Moore, R. G. McCullough, G. Harrison, B. I. Tranner, A. J. Micco, et al. Headache at high altitude is not related to internal carotid arterial blood velocity. *J. Appl. Physiol.: Respir. Environ. Exerc. Physiol.* 59: 909–915, 1985.

311. Reite, M., D. Jackson, R. L. Cahoon, and J. V. Weil. Sleep physiology at high altitude. *Electroencephalogr. Clin. Neurophysiol.* 38: 463–471, 1975.

312. Remmers, J. E., D. Bartlett, Jr., and M. D. Putnam. Changes in the respiratory cycle associated with sleep. *Respir. Physiol.* 28: 227–238, 1976.

313. Reynolds, W. J., and H. T. Milhorn, Jr. Transient ventilatory response to hypoxia with and without controlled alveolar PCO_2. *J. Appl. Physiol.* 35: 187–196, 1973.

314. Rist, K. E., J. A. Daubenspeck, and J. F. McGovern. Effects of non-REM sleep upon the respiratory drive and the respiratory pump in humans. *Respir. Physiol.* 63: 241–256, 1986.

315. Robinson, K. A., and E. M. Haymes. Metabolic effects of exposure to hypoxia plus cold at rest and during exercise in humans. *J. Appl. Physiol.: Respir. Environ. Exerc. Physiol.* 68: 720–725, 1990.

316. Rose, B., and R. Rick. Intracellular pH and intracellular free Ca and junctional cell–cell coupling. *J. Membr. Biol.* 44: 377–415, 1978.

317. Rosenstein, R., L. E. McCarthy, and H. L. Borison. Slow respiratory stimulant effect of hyperoxia in chemodenervated decerebrate cats. *J. Appl. Physiol.* 39: 767–772, 1975.

318. Runold, M., H. Lagercrantz, and B. B. Fredholm. Ventilatory effect of an adenosine analogue in unanesthetized rabbits during development. *J. Appl. Physiol.: Respir. Environ. Exerc. Physiol.* 61: 255–259, 1986.

319. Ryan, M. L., M. S. Hedrick, J. Pizarro, and G. E. Bisgard. Carotid body noradrenergic sensitivity in ventilatory acclimatization to hypoxia. *Respir. Physiol.* 92: 77–90, 1993.

320. Sahn, S. A., C. W. Zwillich, N. Dick, R. E. McCullough, S. Lakishminarayan, and J. V. Weil. Variability of ventilatory responses to hypoxia and hypercapnia. *J. Appl. Physiol.: Respir. Environ. Exerc. Physiol.* 43: 1019–1025, 1977.

321. Santiago, T. V., and N. H. Edelman. Opioids and breathing. *J. Appl. Physiol.: Respir. Environ. Exerc. Physiol.* 59: 1675–1685, 1985.

322. Sato, M., J. W. Severinghaus, F. L. Powell, F. Xu, and M. J. Spellman, Jr. Augmented hypoxic ventilatory response in men at altitude. *J. Appl. Physiol.: Respir. Environ. Exerc. Physiol.* 73: 101–109, 1992.

323. Sauerland, E. K., and R. M. Harper. The human tongue during sleep: electromyographic activity of the genioglossus muscle. *Exp. Neurol.* 51: 160–170, 1976.

324. Schaeffer, J. I., and G. G. Haddad. Ventilatory response to moderate and severe hypoxia in adult dogs: role of endorphins. *J. Appl. Physiol.: Respir. Environ. Exerc. Physiol.* 65: 1383–1388, 1988.

325. Schmeling, W. T., H. V. Forster, and M. J. Hosko. Effect of sojourn at 3200 m altitude on spinal reflexes in young adult males. *Aviat. Space Environ. Med.* 48: 1039–1045, 1977.

326. Schoene, R. B., S. Lahiri, P. H. Hackett, R. M. Peters, Jr., J. S. Milledge, C. J. Pizzo, F. H. Sarnquist, S. J. Boyer, D. J. Graber, K. H. Maret, and J. B. West. Relationship of hypoxic ventilatory response to exercise performance on Mount Everest. *J. Appl. Physiol.: Respir. Environ. Exerc. Physiol.* 56: 1478–1483, 1984.

327. Schoene, R. B., R. C. Roach, P. H. Hackett, J. R. Sutton, A. Cymerman, and C. S. Houston. Operation Everest II: ventilatory adaptation during gradual decompression to extreme altitude. *Med. Sci. Sports Exerc.* 22: 804–810, 1990.

328. Severinghaus, J. W. Hypoxic respiratory drive and its loss during chronic hypoxia. *Clin. Physiol.* 2: 57–79, 1972.

329. Severinghaus, J. W., C. R. Bainton, and A. Carcelen. Respiratory insensitivity to hypoxia in chronically hypoxic man. *Respir. Physiol.* 1: 308–334, 1966.

330. Severinghaus, J. W., and A. Carcelén. Cerebrospinal fluid in man native to high altitude. *J. Appl. Physiol.* 19: 319–321, 1964.

331. Severinghaus, J. W., H. Chiodi, E. I. Eger, B. Brandstater, and T. F. Hornbein. Cerebral blood flow in man at high altitude: role of cerebrospinal fluid pH in normalization of flow in chronic hypocapnia. *Circ. Res.* 19: 274–282, 1966.

332. Severinghaus, J. W., R. A. Mitchell, B. W. Richardson, and M. M. Singer. Respiratory control at high altitude suggesting active transport regulation of CSF pH. *J. Appl. Physiol.* 18: 1155–1166, 1963.

333. Siegel, J. M. Brain stem mechanisms generating REM sleep. In: *Principles and Practice of Sleep Medicine*, edited by M. Kryger, R. Roth, and W. C. Dement. Philadelphia: Saunders, 1988 p. 104–120.

334. Siesjo, B. K., and L. Nilsson. The influence of arterial hypoxemia upon labile phosphates and upon extracellular and intracellular lactate and pyruvate concentration in the rat brain. *Scand. J. Clin. Lab. Invest.* 27: 83–96, 1971.

335. Skatrud, J. B., and J. A. Dempsey. Interaction of sleep state and chemical stimuli in sustaining rhythmic ventilation. *J. Appl. Physiol.: Respir. Environ. Exerc. Physiol.* 55: 813–822, 1983.

336. Skatrud, J. B., and J. A. Dempsey. Airway resistance and respiratory muscle function in snorers during NREM sleep. *J. Appl. Physiol.: Respir. Environ. Exerc. Physiol.* 59: 328–335, 1985.

337. Skatrud, J. B., J. A. Dempsey, S. Badr, and R. L. Begle. Effects of high airway impedance on CO_2 retention and respiratory muscle activity during NREM sleep. *J. Appl. Physiol.: Respir. Environ. Exerc. Physiol.* 65: 1676–1685, 1988.

338. Smatresk, N. J., M. Pokorski, and S. Lahiri. Opposing effects of dopamine receptor blockade on ventilation and carotid chemoreceptor activity. *J. Appl. Physiol.: Respir. Environ. Exerc. Physiol.* 54: 1567–1573, 1983.

339. Smith, C. A., G. E. Bisgard, A. M. Nielsen, L. Daristotle, N. A. Kressin, H. V. Forster, and J. A. Dempsey. Carotid bodies are required for ventilatory acclimatization to chronic hypoxia. *J.*

Appl. Physiol.: Respir. Environ. Exerc. Physiol. 60: 1003–1010, 1986.

340. Smith, P. G., and E. Mills. Restoration of reflex ventilatory response to hypoxia after removal of carotid bodies in the cat. *Neuroscience* 5: 573–580, 1980.

341. Snyder, F., J. A. Hobson, D. F. Morrison, and F. Goldfrank. Changes in respiration, heart rate and systolic blood pressure in human sleep. *J. Appl. Physiol.* 19: 417–422, 1964.

342. Sorensen, S. C. Ventilatory acclimatization to hypoxia in rabbits after denervation of peripheral chemoreceptors. *J. Appl. Physiol.* 28: 836–839, 1970.

343. Sorensen, S. C., and J. C. Cruz. Ventilatory response to a single breath of CO₂ in O₂ in normal man at sea level and high altitude. *J. Appl. Physiol.* 27: 186–190, 1969.

344. Sorensen, S. C., and A. H. Mines. Ventilatory responses to acute and chronic hypoxia in goats after sinus nerve section. *J. Appl. Physiol.* 28: 832–835, 1970.

345. Sorensen, S. C., and J. W. Severinghaus. Irreversible respiratory insensitivity to acute hypoxia in man born at high altitude. *J. Appl. Physiol.* 25: 217–220, 1968.

346. Sorensen, S. C., and J. W. Severinghaus. Respiratory insensitivity to acute hypoxia persisting after correction of tetralogy of Fallot. *J. Appl. Physiol.* 25: 221–223, 1968.

347. Steinbrook, R. A., J. C. Donovan, R. A. Gabel, D. E. Leith, and V. Fencl. Acclimatization to high altitude in goats with ablated carotid bodies. *J. Appl. Physiol.: Respir. Environ. Exerc. Physiol.* 55: 16–21, 1983.

348. St. John, W. M. Respiratory neuron responses to hypercapnia and carotid chemoreceptor stimulation. *J. Appl. Physiol.: Respir. Environ. Exerc. Physiol.* 51: 816–822, 1981.

349. St. John, W. M., and A. L. Bianchi. Responses of bulbospinal and laryngeal respiratory neurons to hypercapnia and hypoxia. *J. Appl. Physiol.: Respir. Environ. Exerc. Physiol.* 59: 1201–1207, 1985.

350. Stokes, W. *The Diseases of the Heart and Aorta.* Dublin: Hodges and Smith, 1854, p. 302–337.

351. Sutton, J. R., G. W. Gray, C. S. Houston, and A. C. P. Powles. Effects of duration at altitude and acetazolamide on ventilation and oxygenation during sleep. *Sleep* 3: 455–464, 1980.

352. Sutton, J. R., G. W. Gray, C. S. Houston, and A. C. P. Powles. Effects of acclimatization on sleep hypoxemia at altitude. In: *High Altitude and Man,* edited by J. B. West and S. Lahiri. Bethesda, MD: Am. Physiol. Soc., 1984, p. 141–146.

353. Sutton, J. R., C. S. Houston, A. L. Mansell, M. D. McFadden, P. M. Hackett, J. R. A. Rigg, and A. C. P. Powles. Effect of acetazolamide on hypoxemia during sleep at high altitude. *N. Engl. J. Med.* 301: 1329–1331, 1979.

354. Tatsumi, K., C. K. Pickett, and J. V. Weil. Attenuated carotid body hypoxic sensitivity after prolonged hypoxic exposure. *J. Appl. Physiol.: Respir. Environ. Exerc. Physiol.* 70: 748–755, 1991.

355. Tatsumi, K., C. K. Pickett, and J. V. Weil. Effects of a dopamine antagonist on decreased carotid body and ventilatory hypoxic sensitivity in chronically hypoxic cats. *Am. Rev. Respir. Dis.* 143: A189, 1991.

356. Tatsumi, K., C. K. Pickett, and J. V. Weil. Effects of haloperidol and domperidone on ventilatory roll off during sustained hypoxia in cats. *J. Appl. Physiol.: Respir. Environ. Exerc. Physiol.* 72: 1945–1952, 1992.

357. Tatsumi, K., C. K. Pickett, and J. V. Weil. Possible role of dopamine in ventilatory acclimatization to high altitude. *Respir. Physiol.* 99: 63–73, 1995.

358. Tenney, S. M., and L. C. Ou. Physiological evidence for increased tissue capillarity in rats acclimatized to high altitude. *Respir. Physiol.* 8: 137–150, 1970.

359. Tenney, S. M., and L. C. Ou. Ventilatory response of decorticate

and decerebrate cats to hypoxia and CO₂. *Respir. Physiol.* 29: 81–92, 1976.

360. Tenney, S. M., and L. C. Ou. Hypoxic ventilatory response of cats at high altitude: an interpretation of "blunting." *Respir. Physiol.* 30: 185–199, 1977.

361. Tenney, S. M., P. Scotto, L. C. Ou, D. Bartlett, Jr., and J. E. Remmers. Suprapontine influences on hypoxia ventilatory control. In: *High Altitude Physiology: Cardiac and Respiratory Effects,* edited by R. Porter and J. Knight. London: Churchill Livingstone 1971, p. 89–102.

362. Tenney, S. M., and W. M. St. John. Is there localized cerebral cortical influence on hypoxic ventilatory response? *Respir. Physiol.* 41: 227–232, 1980.

363. Thierry, A. M., L. Stinus, G. Blanc, and J. Glowinski. Some evidence for the existence of dopaminergic neurons in the rat cortex. *Brain Res.* 50: 230–234, 1973.

364. Thoden, J. S., J. A. Dempsey, W. G. Reddan, M. L. Birnbaum, H. V. Forster, R. F. Grover, and J. Rankin. Ventilatory work during steady-state response to exercise. *Federation Proc.* 28: 1316–1321, 1969.

365. Ultsch, G. R., and D. C. Jackson. Long-term submergence at 3 degrees C of the turtle, *Chrysemys picta belli* in normoxic and severly hypoxic water. I. Survival, gas exchange, and acid-base status. *J. Exp. Biol.* 96: 11–28, 1982.

366. Valdivia, E. Total capillary bed in striated muscle of guinea pigs native to the Peruvian mountains. *Am. J. Physiol.* 194: 585–589, 1958.

367. Verna, A. Observations on the innervation of the carotid body of the rabbit. In: *The Peripheral Arterial Chemoreceptors,* edited by M. J. Purves. London: Cambridge University Press, 1975, p. 75–97.

368. Versteeg, D. H. G., J. Van Der Gugten, W. DeJong, and M. Palkovits. Regional concentrations of noradrenaline and dopamine in rat brain. *Brain Res.* 113: 563–574, 1976.

369. Vizek, M., C. K. Pickett, and J. V. Weil. Biphasic ventilatory response of adult cats to sustained hypoxia has central origin. *J. Appl. Physiol.: Respir. Environ. Exerc. Physiol.* 63: 1658–1664, 1987.

370. Vizek, M., C. K. Pickett, and J. V. Weil. Increased carotid body hypoxic sensitivity during acclimatization to hypobaric hypoxia. *J. Appl. Physiol.: Respir. Environ. Exerc. Physiol.* 63: 2403–2410, 1987.

371. Wade, J. G., C. P. Larson, Jr., R. F. Hickey, W. K. Ehrenfeld, and J. W. Severinghaus. Effect of carotid endarterectomy on carotid chemoreceptor and baroreceptor function in man. *N. Engl. J. Med.* 282: 823–829, 1970.

372. Waggener, T. B., P. J. Brusil, R. E. Kronauer, R. A. Gabel, and G. F. Inbar. Strength and cycle time of high-altitude ventilatory patterns in unacclimatized humans. *J. Appl. Physiol.: Respir. Environ. Exerc. Physiol.* 56: 576–581, 1984.

373. Wagner, P. D., J. R. Sutton, J. T. Reeves, A. Cymerman, B. M. Groves, and M. K. Malconian. Operation Everest II: pulmonary gas exchange during a simulated ascent of Mt. Everest. *J. Appl. Physiol.: Respir. Environ. Exerc. Physiol.* 63: 2348–2359, 1987.

374. Waldrop, T. G. Posterior hypothalamic modulation of the respiratory response to CO₂. *Pflugers Arch.* 418: 7–13, 1991.

375. Wang, Z. Z., L. J. Stensaas, B. Dinger, and S. J. Fidone. Co-existence of tyrosine hydroxylase and dopamine beta-hydroxylase immunoreactivity in glomus cells of the cat carotid body. *J. Auton. Nerv. Syst.* 32: 259–264, 1991.

376. Warner, M. M., G. S. Mitchell, J. Pizarro, M. L. Ryan, and G. E. Bisgard. Ventilatory responses to norepinephrine infusion in normoxic and hypoxic goats. *FASEB J.* 5: A1119, 1991.

377. Wasicko, M. J., J. E. Melton, J. A. Neubauer, N. Krawciw, and N. H. Edelman. Cervical sympathetic and phrenic nerve

responses to progressive brain hypoxia. *J. Appl. Physiol.: Respir. Environ. Exerc. Physiol.* 68: 53–58, 1990.

378. Watt, J. G., P. R. Dumke, and J. H. Comroe, Jr. Effects of inhalation of 100 percent and 14 percent oxygen upon respiration of unanesthetized dogs before and after chemoreceptor denervation. *Am. J. Physiol.* 138: 610–617, 1943.

379. Weibel, E. R., and H. Hoppeler. Respiratory system adaptation to hypoxia: lung to mitochondria. In: *Response and Adaptation to Hypoxia, Organ to Organelle,* edited by S. Lahiri, N. S. Cherniack, and R. S. Fitzgerald. New York: Oxford University Press, 1991, p. 3–13.

380. Weil, J. V. Ventilatory control at high altitude. In: *Handbook of Physiology. Control of Breathing,* edited by N. S. Cherniack and J. G. Widdicombe. Bethesda, MD: Am. Physiol. Soc., 1986, sect. 3, vol. II, pt. 2, p. 703–727.

381. Weil, J. V., E. Byrne-Quinn, I. E. Sodal, G. F. Filley, and R. F. Grover. Acquired attenuation of chemoreceptor function in chronically hypoxic man at high altitude. *J. Clin. Invest.* 50: 186–195, 1971.

382. Weil, J. V., M. H. Kryger, and C. H. Scoggin. Sleep and breathing at high altitude. In: *Sleep Apnea Syndromes,* edited by C. Guilleminault and W. C. Dement. New York: Liss, 1978, p. 119–136.

383. Weil, J. V., D. P. White, N. J. Douglas, and C. W. Zwillich. Ventilatory control during sleep in normal humans. In: *High Altitude and Man,* edited by J. B. West and S. Lahiri. Bethesda, MD: Am. Physiol. Soc., 1984, p. 91–100.

384. Weinberger, S. E., R. A. Steinbrook, D. B. Carr, V. Fencl, R. A. Gabel, D. E. Leith, J. E. Fisher, R. Harris, and M. Rosenblatt. Endogenous opioids and ventilatory adaptation to prolonged hypoxia in goats. *Life Sci.* 40: 605–613, 1987.

385. Weiskopf, R. B., and R. A. Gabel. Depression of ventilation during hypoxia in man. *J. Appl. Physiol.* 39: 911–915, 1975.

386. Weiskopf, R. B., R. A. Gabel, and V. Fencl. Alkaline shift in lumbar and intracranial CSF in man after 5 days at high altitude. *J. Appl. Physiol.* 41: 93–97, 1976.

387. Weizhen, N., M. J. A. Engwall, L. Daristotle, J. Pizarro, and G. E. Bisgard. Ventilatory effects of prolonged systemic (CNS) hypoxia in awake goats. *Respir. Physiol.* 87: 37–48, 1992.

388. West, J. B. Man on the summit of Mount Everest. In: *High Altitude and Man,* edited by J. B. West and S. Lahiri. Bethesda, MD: Am. Physiol. Soc., 1984, p. 5–17.

389. West, J. B. Rate of ventilatory acclimatization to extreme altitude. *Respir. Physiol.* 74: 323–333, 1988.

390. West, J. B. Acclimatization and adaptation: organ to cell. In: *Responses and Adaptation to Hypoxia, Organ to Organelle,* edited by S. Lahiri, N. S. Cherniack, and R. S. Fitzgerald. Bethesda, MD: Am. Physiol. Soc., 1991, p. 177–190.

391. West, J. B., P. H. Hackett, K. H. Maret, J. S. Milledge, R. M. Peters, Jr., C. J. Pizzo, and R. M. Winslow. Pulmonary gas exchange on the summit of Mt. Everest. *J. Appl. Physiol.: Respir. Environ. Exerc. Physiol.* 55: 678–687, 1983.

392. Weyne, J., F. VanLeuven, and I. Leusen. Brain amino acids in conscious rats in chronic normocapnic and hypocapnic hypoxemia. *Respir. Physiol.* 31: 231–239, 1977.

393. White, D. P., K. Gleeson, C. K. Pickett, A. M. Rannels, A. Cymerman, and J. V. Weil. Altitude acclimatization: influence on periodic breathing and chemoresponsiveness during sleep. *J. Appl. Physiol.: Respir. Environ. Exerc. Physiol.* 63: 401–412, 1987.

394. Wiegand, L., C. Zwillich, and D. White. Sleep and the ventilatory response to resistive loading in normal man. *J. Appl. Physiol.: Respir. Environ. Exerc. Physiol.* 64: 1186–1195, 1988.

395. Wilson, P. A., J. B. Skatrud, and J. A. Dempsey. Effects of slow wave sleep on ventilatory compensation to inspiratory elastic loading. *Respir. Physiol.* 55: 103–120, 1984.

396. Winn, H. R., R. Rubio, and R. M. Berne. Brain adenosine concentration during hypoxia in rats. *Am. J. Physiol.* 241 (*Heart Circ. Physiol.* 10): H235–H242, 1981.

397. Wood, J. D., W. J. Watson, and A. J. Drucker. The effect of hypoxia on brain gamma-aminobutyric acid levels. *J. Neurochem.* 15: 603–608, 1968.

398. Wright, A. D., A. R. Bradwell, J. Jensen, and N. Lassen. Cerebral blood flow in acute mountain sickness and treatment with acetazolamide. *Clin. Sci. (Colch.)* 74(suppl. 18): 1P, 1988.

399. Wu, J.-Y. Purification, characterization, and kinetic studies of GAD and GABA-T from mouse brain. In: *GABA in Nervous System Function,* edited by E. Roberts, T. Chase, and D. Tower. New York: Raven, 1976, p. 7–55.

400. Yamada, K. A., P. Hamosh, and R. A. Gillis. Respiratory depression produced by activation of GABA receptors in hindbrain of cat. *J. Appl. Physiol.: Respir. Environ. Exerc. Physiol.* 51: 1278–1286, 1981.

401. Yamada, K. A., W. P. Norman, P. Hamosh, and R. A. Gillis. Medullary ventral surface GABA receptors affect respiratory and cardiovascular function. *Brain Res.* 248: 71–78, 1982.

402. Zapata, P. Effects of dopamine on carotid chemo- and baroreceptors in vitro. *J. Physiol. (Lond.)* 224: 235–251, 1975.

53. The cardiovascular system at high altitude

MIRSAID M. MIRRAKHIMOV | *Kyrgyz Institute of Cardiology, Bishkek, Kyrgyzstan*
ROBERT M. WINSLOW | *University of California, San Diego, California*

ALTERATIONS IN THE CARDIOVASCULAR SYSTEM are among the most important physiological responses to either acute or chronic hypoxia. Except for patients with cardiovascular, pulmonary, or other diseases, however, significant physiological responses are not detected at altitudes lower than 1,000 m. The vast majority of subjects are unable to adapt to altitudes above 5,000–6,000 m; they develop a deterioration syndrome. During adaptation to hypoxia, physiological responses occur in both the systemic and the pulmonary circulations.

NORMAL CARDIAC OUTPUT

The major adjustment in the O_2 delivery system when tissue demand increases is an increase in cardiac output.

In normal sea-level residents cardiac output can increase by about fourfold; this increase is mediated primarily by increased heart rate and secondarily by increased stroke volume. During exercise, most increased cardiac output is directed to exercising muscle.

Stroke Volume

Starling (156) believed that stroke volume played the central role in determining cardiac output. That conclusion, however, was derived from in vitro experiments with a heart–lung preparation; later work by Rushmer (151) showed the constancy of stroke volume in exercise. It is now known that under severe stress stroke volume does increase somewhat in exercise (65). Furthermore, stroke volume is higher in the supine than in the erect posture (24), and it increases after physical training (152). The mechanism of stroke volume increase is either reduction of afterload (resistance to flow), an increase in preload (venous return to the heart), or an increase in the myocardial contractile state (167).

Heart Rate

The heart rate increase during exercise is under the general control of the autonomic nervous system, either by a decrease in parasympathetic restraint or by a sympathetic stimulation. The latter can occur either by neural stimulation or by an increase in circulating catecholamines. In normal, resting sea-level residents baseline heart rate is about 60 beats per minute (bpm) and can increase to about 200 bpm. The actual value for a person is age-dependent, and the maximum decreases with age.

CARDIAC OUTPUT AT HIGH ALTITUDE

It is well known that cardiac output increases during short-term adaptation to altitude (42, 76, 102, 168). However, resting cardiac output may not show significant changes at altitudes up to 3,500–4,000 m. The increases in cardiac output observed in the early stages of altitude adaptation are mostly due to tachycardia rather than to increases in stroke volume.

During the first days at 3,200 and 3,600 m, significant shortening of both right ventricular (111) and left ventricular (108, 111) ejection times are observed. Although the contractile function of the heart may eventually deteriorate, cardiac output remains increased (12). However, there is evidence that during short-term adaptation to altitude changes in left ventricular systolic function are due to decreased venous return (5, 37), resulting in decreased stroke volume. When mountaineers ascend to various altitudes in the Himalayas left ventricular ejection time shortens. At altitudes of 4,904 m and 5,620 m left ventricular ejection time was longer in native highlanders of Tibet than in Chinese living at moderate altitudes (182). However, echocardiographic indices of ejection fraction increased in these subjects, suggesting some enhancement of left ventricular systolic function. Ejection fraction increases at altitudes above 3,719 m, and at 4,004 m circulation was 1.67 circuits (circ)/s compared to 1.25 circ/s at 2,261 m. Changes in left ventricular systolic and diastolic dimensions also indicated an improvement in myocardial function at high altitude.

The first determination of cardiac output in high-altitude natives was carried out in Morococha, Peru, in 1938 by Rotta and co-workers (cited in ref. 180) using the acetylene technique (42). They found a slight increase in resting cardiac output in the Morococha natives compared with sea-level controls. Dye dilution measurements also showed increased cardiac output in Morococha natives (163). Measurements using the Fick technique in the same population showed no differences in cardiac output. Other conflicting conclusions regarding high-altitude natives have also been published (13, 58). Also, measurements in Cerro de Pasco, Peru (3,400 m), using modern thermodilution techniques gave results higher than those of direct Fick when both techniques were used simultaneously (181). These unsettling results suggest that great care must be taken in the selection of methods and to the assumption of the steady state, implicit in all techniques using the Fick principle.

Geographic differences further complicate the question of cardiac output in high-altitude natives, despite some studies having been done by the same groups of investigators using the same methods. Studies in North American high-altitude residents may not be strictly comparable to those in Andean residents, inasmuch as the subjects are of different sizes. Nevertheless, results suggest that cardiac output is higher in Andean residents, and they compare favorably with expected sea-level values. Another important difference between the natives of the two areas is that hematocrit levels in Andeans are higher, and a satisfactory description of the effect of hematocrit on cardiac output in humans is still incomplete. Detailed measurements were presented in one subject in Cerro de Pasco, whose cardiovascular hemodynamics at rest and during exercise were studied using both thermodilution and direct Fick methods (181). In this subject elevated hematocrit lowered cardiac output, an effect that could be reversed (at least temporarily) by hemodilution.

Heart Rate at High Altitude

Tachycardia is one of the early responses to hypoxia, and the mechanism is very complex. Activation of the sympathetic nervous system (123) and beta-adrenoceptor stimulation are doubtlessly involved in the response. Preliminary administration of beta-blockers (propranolol) can prevent increases in heart rate at high altitude (3,200 m).

Hypocapnia and alkalosis have no role in heart rate changes during hypoxia. In one study heart rate decreased in dogs in which hypoxic hyperventilation was prevented by artificial ventilation and skeletal muscle paralysis simultaneously induced (81). Bradycardia also occurred in hypoxic dogs with denervated lungs (27). Thus hypoxic tachycardia in dogs is apparently mediated by pulmonary stretch receptors, which stimulate a regulatory increase in sympathetic activity.

Similar mechanisms of tachycardia have not been demonstrated in humans. However, endothelin, which can produce both vasoconstriction and enhanced sino-atrial node activity, may mediate hypoxic tachycardia (72). Tachycardia may also result from inhibition of vagal influences through increased vascular chemoreceptor discharge (90). A role in maintaining altitude tachycardia is also played by the state of aortic (19) and sinocarotid chemoreceptors. At 3,450 m, tachycardia persists for 8–15 days (41), and at greater altitudes (6,000–7,000 m) heart rates does not return to baseline levels despite long-term adaptation.

Unlike newcomers, native highlanders usually show normal or decreased heart rates. During chronic hypoxia bradycardia is due more to diminished sympathetic effects than to enhanced parasympathetic influences (28), though the latter are enhanced somewhat (142). Changes in adrenoreceptor sensitivity to sympathetic effects are also important: in certain highlanders subcutaneous administration of atropine does not affect heart rate, and heart rate response to adrenergic stimulation during submaximal exercise is attenuated at 3,800 m (141). Finally, the heart rate response to chronic hypoxia may also be related to direct changes in the functional state of the sinus node.

Effect of Polycythemia on Cardiac Output

Guyton and Richardson (53) studied venous return to the hearts of dogs as a function of right atrial pressure and found a very predictable relationship between the two and that hematocrit had a profound effect on this. That is, in polycythemia a given right atrial pressure results in a lowered venous return. Since cardiac output must equal venous return in the steady state, this means that cardiac output is profoundly reduced by hematocrit.

Castle and Jandl (23) drew attention to the fact that blood volume increases in polycythemia and that it is necessary to consider the combined effect of multiple variables on the control of cardiac output. Very few experimental observations are available to draw a significant parallel between laboratory models and the in vivo effects of polycythemia on cardiac output. Guyton et al. (52) pointed out that polycythemic subjects need not have reduced cardiac output if systemic pressure is increased. This is almost exactly the finding in dogs made acutely polycythemic (143).

In chronic polycythemia blood volume increases, increasing venous return and systemic pressure. If, in addition to this change, peripheral resistance decreases due to vascularization, venous return can be restored to normal. In fact, with cardiac hypertrophy and a lessened resistive load on the heart, the cardiac output curve may actually return to normal. Unfortunately, no measurements of this degree of sophistication have been made in high-altitude natives, and one must look to circumstantial evidence to try to evaluate this model for the relationship between hematocrit, viscosity, and cardiac output (180).

Available data are consistent with the view that the lifelong high-altitude native does not necessarily have increased cardiac output on the basis of chronic hypoxia. In view of the increased pulmonary artery pressures (see PULMONARY CIRCULATION AT ALTITUDE, below), Guyton's predictions for chronic polycythemia may be correct: a high-pressure, low-velocity type of circulation develops with expanded blood volume in systemic, pulmonary, and coronary vascular beds. Furthermore, this type of circulation, while serving as a reservoir of O_2, responds sluggishly to sudden increases in metabolic demand (such as during exercise), lowers partial pressure of oxygen (PO_2) in tissue, and increases the O_2 debt that must be paid on exertion.

Sojourners Exposed to Altitude

Normal lowland subjects taken to a simulated altitude of 4,000 m (barometric pressure 462 torr) do not show a reduced cardiac output within the first hour of exposure (157). However, in lowlanders transported to similar altitudes for longer periods cardiac output falls, reaching a minimum 3 days after arrival; reduced stroke volume is not compensated by tachycardia (64). Pugh (133) showed that cardiac output remains depressed even after several months of high-altitude exposure.

An important difference between sojourners and natives is that the latter tend to be more polycythemic. The heart of the high-altitude native increases in size, muscle mass, and metabolism, and its failure is often considered part of the clinical picture of chronic mountain sickness (Monge's disease). In addition, adjustments of the regional distribution of cardiac output affect all organ systems and their functions. Sometimes the interactions between blood perfusion and organ function are exceedingly complex, as in those of the brain and the lung, and may involve complicated feedback regulatory loops.

PULMONARY CIRCULATION AT ALTITUDE

Role of Hematocrit in Pulmonary Hypertension

Detailed studies of cardiovascular hemodynamics in high-altitude natives were carried out by Peñaloza and co-workers (129) in the Peruvian Andes. They selected ten subjects with chronic mountain sickness and compared them with appropriate sea-level controls. The results showed that subjects with chronic mountain sickness had elevated right ventricular, pulmonary artery, and systemic vascular pressures, thereby suggesting similarities between this clinical entity and brisket disease of cattle (59). Peñaloza and colleagues went on to show that the signs of cor pulmonale (right heart failure) could be reversed by descent to sea level. They distinguished two causes for pulmonary hypertension: hypoxia and polycythemia. Hypoxia is relieved immediately after descent, while the reduction in pulmonary artery pressure is much slower.

Drawing on the anatomical data of Arias-Stella and Kruger (9), Peñaloza et al. (129) concluded that chronic mountain sickness is due primarily to alveolar hypoventilation, which in turn leads to muscularization of the pulmonary vessels and pulmonary hypertension. When these data and conclusions were presented at a symposium in 1971, the discussion pointed out a potential flaw in this hypothesis: removal of blood by phlebotomy was known in some cases to reverse apparent hypoventilation (129). In other words, partial pressure of arterial oxygen (PaO_2) and oxygen saturation (SaO_2) increased, while $PaCO_2$ decreased; this seems to indicate that even if hypoventilation is the primary cause of chronic mountain sickness, it is at least partly reversible by hematocrit reduction.

Role of Hypoxia in Pulmonary Hypertension

Hypoxia appears to be one of the most important factors affecting the diameter of small pulmonary vessels, whose role in changing pulmonary artery pressure was first reported by von Euler and Liljestrand (174). A hypoxia-induced rise in pulmonary artery pressure has been confirmed by many subsequent studies (36, 88, 119). Hurtado (69) found an enlarged proximal segment of the pulmonary artery in native highlanders at 4,540 m, and subsequent direct measurements confirmed elevated pulmonary artery pressure (45, 128, 145).

Hypoxia-induced increases in both local and total pulmonary vascular resistance are considered to be a biologically appropriate adaptive response aimed at optimizing intrapulmonary redistribution of blood, thereby improving the interaction between ventilation and pulmonary blood flow (125). Referring to numerous studies, Harris and Heath (56) state that the decreases in pulmonary blood flow observed during hypoxia indicate vasoconstriction. Experiments performed by Bergofsky and Haas (17) demonstrated that the contribution of pulmonary veins to decreases in pulmonary blood flow during hypoxia is negligible, pulmonary arteries being most important in this respect.

During the first days of adaptation by lowlanders to an altitude of 3,200 m, pulmonary vascular resistance increases, and the changes are even more pronounced at 3,600–4,200 m (112). Measurements of pulmonary artery pressure in natives of high altitude (43, 49, 67, 128) show definite but nonlinear increases with altitude from 3,000 to 4,540 m. Vogel et al. (173) believe that pulmonary hypertension develops in normal subjects only at altitudes over 3,000 m. Extensive electrocardiographic studies (102) found signs of right ventricular hypertrophy in 5% of the apparently healthy highlanders living permanently at 2,050–2,500 m, a fact that can only be explained by the effect of altitude.

Vogel (170) also emphasizes the important role of moderate hypoxia in pulmonary vascular abnormalities. If, for example, altitude-sensitive calves with preliminary unilateral pulmonary artery ligation are taken to an altitude of 1,620 m, they develop progressive pulmonary hypertension with right-heart failure. In children with congenital interventricular septal defect (left-to-right shunts) permanent residence at this moderate altitude is accompanied by significantly higher pulmonary vascular resistance than in their sea-level counterparts. Sleep apnea, commonly occurring at this altitude, may contribute to this process by producing arterial desaturation (95). Desaturation might also result from blunted ventilatory responses, often observed in highlanders (30, 84, 85). However, periodic breathing occurs more frequently in subjects with high ventilatory sensitivity to hypoxia (38). Thus pulmonary hyperten-

sion at relatively low altitudes could occur in subjects sensitive to hypoxia.

The range of pulmonary artery pressure increases is rather great even in highlanders living at the same altitude. In residents of Leadville, Colorado (3,100 m), pulmonary artery pressure is sometimes significantly higher than in residents of La Oroya, Peru (3,750 m) (67, 173). This discrepancy was explained by the difference in development of acclimatization in the subjects studied: people settled in the mountainous regions of Peru about 10,000 years ago and in Leadville only a few hundred years ago. Consequently, it is possible that during very prolonged acclimatization to altitude pulmonary vascular response to hypoxia may become attenuated.

The severity of pulmonary hypertension is sometimes excessive in highlanders who have lost their natural acclimatization (129, 182). Wu et al. (182) studied 30 healthy subjects and 26 patients with chronic mountain sickness and found that mean pulmonary artery pressure, measured by ultrasound, reached 38.9 ± 1.5 mm Hg compared to 24.4 ± 4.5 mm Hg in normal highlanders. Red blood cell count, hemoglobin concentration, hematocrit, and Pa_{CO_2} values proved to be significantly higher, while Pa_{O_2} and Sa_{O_2} values were lower in patients with chronic mountain sickness. Experiments performed by Hill and Ou (61), who studied hypoxic responses in two strains of rat (sensitive and nonsensitive to hypoxia), failed to explain the higher degree of pulmonary hypertension and right ventricular hypertrophy in hypoxia-sensitive rats by differences in vasoconstrictive responses. These authors attributed the higher pulmonary artery pressure values and the more pronounced right ventricular hypertrophy in the sensitive rats to structural alterations in pulmonary vessels.

Significant differences in hypoxic responses were observed in randomized groups of sea-level cattle, which prompted Will et al. (179) to suggest that individual sensitivities differ. Subsequent studies by the same group confirmed this hypothesis for cattle (49) and humans (26).

In another study 35 apparently healthy natives of Tien-shan, China (2,000–2,500 m), were studied at the Institute of Cardiology in Bishkek, Russia (760 m), several days after descent from the mountains (113). Twenty-one of them had less pronounced rises in pulmonary artery pressure (normal responders), while rises were much more pronounced in 14 (hyperresponders) (Table 53.1) after consecutive inhalation of gas mixtures containing 14%, 12%, and 10% O_2 in nitrogen for 10 min each. Despite similar initial hemodynamic profiles in hyper- and normal responders, hypoxic responses in the former were characterized not only by significantly greater increases in pulmonary artery pressure and resistance but also by increases in heart rate and cardiac output (Table 53.1).

TABLE 53.1. *Effect of Hypoxic Gas Mixtures on Hemodynamics in Highlanders (2,000–2,500 m)*

Indices	Group	Normoxia	$F_{IO_2} = 0.14$	$F_{IO_2} = 0.12$	$F_{IO_2} = 0.10$
Heart rate (bpm)	N	82.8 ± 3.8			86.6 ± 6.1
	H	74.6 ± 3.0			84.7 ± 4.0*
PAPs (mm Hg)	N	22.8 ± 2.0	23.8 ± 2.2	28.1 ± 2.0*	31.1 ± 2.1**
	H	26.2 ± 2.0	38.2 ± 3.5	42.3 ± 5.3*	48.3 ± 2.8**
PAPd (mm Hg)	N	8.8 ± 1.4	11.4 ± 1.8*	10.6 ± 2.6*	13.2 ± 1.7*
	H	8.2 ± 1.3	16.6 ± 2.2	22.1 ± 2.9*	23.0 ± 1.8*
PAP (mm Hg)	N	12.8 ± 1.3	15.6 ± 1.7	17.5 ± 1.3	21.1 ± 2.1*
	H	16.2 ± 1.6	27.6 ± 2.3	32.2 ± 3.6	34.8 ± 2.7
Cardiac index (l/min/m²)	N	3.2 ± 0.3			3.1 ± 0.3
	H	3.4 ± 0.3			4.8 ± 0.7*
TPR (dyne/s/cm)	N	99.1 ± 11.1			277.2 ± 39.4*
	N	215.1 ± 34.5			384.6 ± 39.4*

Abbreviations: F_{IO_2}, fraction of inspired oxygen; bpm, beats per minute; PAPs, systolic pulmonary artery pressure; PAPd, diastolic pulmonary artery pressure; PAP, mean pulmonary artery pressure; TPR, total peripheral resistance; N, normal responders (n = 21); H, hyperresponders (n = 14).
*$P < 0.05$.
**$P < 0.01$.

Adrenergic Control of Pulmonary Vasculature

Beta-adrenoceptors (β-AR) on peripheral blood lymphocytes were assessed in subjects with different hypoxic ventilatory responses. Under basal conditions density proved to be significantly higher in normal responders compared to hyperresponders (6.8 ± 3.2 and 3.7 ± 1.4 fmol/10⁶ cells, respectively). During hypoxic tests B_{max} showed a 1.9-fold increase in hyperresponders, whereas changes in normal responders were not significant (Fig. 53.1). At the same time the β-AR dissociation constant (kd) was not significantly different in the two groups. Beta-adrenoceptor-dependent activation of adenylate cyclase caused by isoproterenol was similar in both groups in basal conditions and during hypoxia (Table 53.2).

Hypoxia had no effect on β-AR density but increased activity by 20% in normal responders. As already noted, hypoxia almost doubled the number of β-AR in hyperresponders: specific activity per receptor proved to be almost 1.8 times lower. These findings indicate that almost all de novo exposed β-AR on the cell surface show signs of dissociation of the receptor from the G protein–adenylate cyclase system.

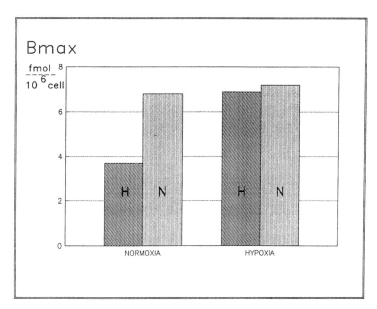

FIG. 53.1. Density of β-adrenoreceptors on peripheral blood lymphocytes of humans who demonstrate normal (N) and increased (H) pulmonary artery pressure rise in response to hypoxia. In normoxic conditions density is higher in normals compared to hyperresponders. When breathing a hypoxic gas mixture ($F_{IO_2} = 0.10$) only hyperresponders increased density.

TABLE 53.2. *Activation of Lymphocyte Adenylate Cyclase by Isoproterenol (10 mM) in Highlanders*

Subjects	Adenylate Cyclase (fmol/10 min/10⁶ cells)	
	Basal	Stimulated
Normal responders		
Normoxia	7.9 ± 0.8	129 ± 8
Hypoxia	7.1 ± 0.8	149 ± 13*
Hyperresponders		
Normoxia	7.4 ± 0.4	138 ± 7
Hypoxia	7.5 ± 1.1	141 ± 15

* $P < 0.05$.

Adenylate cyclase is known to transform neuromediator signals into biological cellular responses (for example, relaxation of vascular smooth muscle cells). The above-mentioned dissociation was confirmed by experiments that showed normal G protein and adenylate cyclase activity under direct stimulation with specific stimulants. Dissociation of β-AR from G protein and adenylate cyclase is one of the major signs of heterologous desensitization (154). Interestingly, in yaks with initially low β-AR density transmission of intracellular signals was more effective during acute hypoxia than in calves (158). Voelkel et al. (169) found that hypoxia and the associated hypernorepinephrinemia decrease β-AR density in the rat heart.

Several studies showed that peripheral blood lymphocytes contain a homogeneous population of β-AR, which to a certain extent reflect their state in the heart (21) and lungs (1). These findings suggest that the state of β-AR in effector cells of the heart and vessels (for example, those in the pulmonary circulation) can be estimated from the response of blood lymphocyte β-AR. It was previously found that β-AR activation is only accompanied by dilatation of pulmonary vessels—the higher the initial degree of pulmonary vasoconstriction, the greater the observed effect (71). This suggests that β-AR activation by hypoxia may be one of the adaptive mechanisms preventing excessive pulmonary vasoconstriction. It seems that there is some initial defect in the β-AR system of hyperresponders that is responsible for the excessive vasoconstrictive effect of hypoxia. In this example it can be seen why excessively enhanced physiological responses can be the cause of pathological states and even disease.

Aldashev et al. (4) studied the state of β-AR and the associated transmission of intracellular signals on mononuclear peripheral blood cells in highlanders (Eastern Pamir, 3,600–4,200 m) suffering from pulmonary hypertension with various degrees of right ventricular hypertrophy. In patients with second- and third-grade hypertrophy β-AR density was 4.5 times lower

than in controls (2.27 ± 0.22 vs. 9.8 ± 1.28 fmol/10⁶ cells). Values (kd) also proved to be 2.5 times slower (0.57 ± 0.14 vs. 1.44 ± 0.18 fmol/10⁶ cells). When adenylate cyclase was stimulated with isoproterenol and other beta-agonists, it exhibited less activation (+33% and +120%, respectively). These findings indicate the presence of β-AR desensitization in patients with altitude pulmonary hypertension and evidence of severe right ventricular hypertrophy caused by altitude hypoxia. This desensitization does not occur in pulmonary hypertension secondary to mitral stenosis.

Exposure of cell cultures (monolayers of embryonal myocardiocytes of chicks) to $Po_2 < 1.5$ torr decreases the number of β-AR, whose affinity for an agonist or antagonist remains unchanged (91). The authors of the study concluded that hypoxia induces the loss of β-AR on the cell surface of intact myocytes with a concomitant blunting of the response of adenylate cyclase to agonists. During reoxygenation the number of receptors and adenylate cyclase activity are restored in a coordinated manner. Data on altered signal transmission through β-ARs on cell surfaces indicate that these alterations are different during acute and chronic hypoxia, though in both cases the amount of receptors decreases. In this connection it should be noted that cardiac response to infusion of sympathetic agonists weakens during human adaptation to high altitude (141). It is recognized that hypoxia causes β-AR and adenosinergic receptor down-regulation and muscarinic receptor up-regulation (142).

Pulmonary Artery Morphology

Hypoxia is an important contributor to pulmonary hypertension. Enlargement of arterial muscular layers of the pulmonary trunk and altered morphology of smooth muscle cells in small pulmonary arteries previously devoid of these cells have been described by a number of authors (9, 56). This muscularization, however, is not accompanied by fibrosis. For example, at high altitude dogs, cats, and cattle show medial thickening of small arteries (less than 150 μm in diameter). Well-adapted llamas (56) and yaks (78, 79) do not demonstrate these changes.

In native highlanders and in hypoxia-sensitive animals pulmonary arteries are characterized by external and internal elastic laminae with an intervening thick medial layer. Longitudinal muscular bundles penetrating the intimal layer are observed both in muscular pulmonary vessels and in arterioles (176). Muscularization of pulmonary arteries with a diameter less than 70 μm is characteristic of hypoxia. Pulmonary veins in permanent high-altitude residents have mildly increased amounts of smooth muscle cells (120, 176).

Although the intensity and sequence of the responses may vary (139), vascular remodeling can be clearly observed as early as the end of the second week of hypoxic exposure, and alterations of smooth muscle cell precursors, smooth muscle cells, fibroblasts, and endothelial cells may be significant in the first few days. Rapid morphological alterations in pulmonary vessels are confirmed by the poor efficiency of O_2 uptake during elevated (twice the initial value) pulmonary artery pressure in volunteers exposed to hypoxia equivalent to that of Mt. Everest (51).

Factors other than hypoxia also mediate the increase in muscle cell pulmonary artery muscle mass: cultured smooth muscle cells do not exhibit hypertrophy or hyperplasia in response to hypoxia in the range of physiological variations (16). Species variation is also not explained by hypoxia alone (46, 50, 127, 137, 177, 178). For example, hypoxia does not induce marked hypertensive responses in yaks, but high sensitivity is characteristic of calves. In addition, elevated pulmonary artery pressure is not completely reversed by O_2 inhalation (67) or by administration of calcium antagonists. The latter agents, as is now known, decrease hypoxic vasoconstriction and pressor responses (98, 121, 122, 136).

During exercise, mean pulmonary artery pressure increases by only 20%–30% from the initial level in highlanders. This is probably due not only to muscularization of the pulmonary arterial bed but also to an increase in the collagenous content of the vessels, a result of stimulation of adventitial fibroblasts and pericytes communicating with endothelial cells. The thickened adventitial layer of pulmonary arteries, with a decrease in the external diameter, is not returned to normal by arterial dilatation (139). Thus initially reversible pulmonary hypertension becomes relatively stable, refractory to correction by oxygen inhalation or pharmacological agents (112).

Connective Tissue Growth: Hypertrophy

Hypoxia-induced morphological alterations affect the activity of cells in large and small arteries, as well as those in the precapillary segment (29, 62, 101). Figure 53.2 illustrates the remodeling of a pulmonary arteriole with a diameter of about 50 μm (77). During chronic hypoxia pulmonary vascular smooth muscle cells produce a substance which strongly stimulates connective tissue growth (99). Hypoxia (5% O_2 in N_2) dramatically enhances synthesis of proteins with various molecular weights, including a 38 kd protein having functional properties similar to cycline (3). Synthesis of this protein is enhanced by increasing Ca^{2+} concentrations in the cell medium. Future studies will show whether this protein (or other molecular mediators) is involved in vascular remodeling.

Other factors also may contribute to the control of pulmonary artery hypertrophy. Administration of cis-4-hydroxy-L-proline (cHyp), an inhibitor of collagen accumulation, significantly decreases collagen content in hypoxic pulmonary vessels. Animals treated with this agent demonstrate reduced pulmonary vascular reactivity to angiotensin II, norepinephrine, and KCl, whereas reactivity to PGF_2d is maintained (164). Administration of cilazopril, an angiotensin-converting enzyme (ACE) inhibitor, prevents medial thickening of the pulmonary artery, though mean pulmonary arterial pressure remains increased and leads to right ventricular hypertrophy (25).

Genetic factors also play a role in pulmonary artery hypertrophy at altitude. Right ventricular hypertrophy in rats with a genetic form of pulmonary emphysema is not associated with morphological changes in the pulmonary arterial bed (94). At rest pulmonary hypertension plays an important role in intrapulmonary blood redistribution, but during exercise it is probably inconsequential because even at sea level exercise increases the uniformity of blood flow through lung apexes (45).

At altitudes of 2,500–4,200 m the frequency of pulmonary hypertension in females is half that in males and is usually characterized by a milder course (107). This may be explained by differences in hormone levels: estradiol attenuates hypoxic vasoconstrictive responses (39).

Elevated blood viscosity as a result of polycythemia is an additional determinant of pulmonary artery hypertrophy (140, 161). Mean hematocrit exceeds 55% in residents of Eastern Pamir (3,600 m above sea level) (104), a figure somewhat higher than in residents of La Oroya (3,790 m) and Morococha (4,540 m) in Peru (70, 180). Chronic mountain sickness is characterized by undue increases in circulating packed red cells in high-altitude natives who exhibit electrocardiographic and vectorcardiographic evidence of severe right ventricular hypertrophy and overt signs of right-heart failure with decreased sensitivity to oxygen of carotid sinus and aortic chemosensitive zones (114, 115, 180). Peñaloza et al. (129) are of the opinion that chronic mountain sickness represents cor pulmonale related to loss of acclimatization to high altitude. Bloodletting improves pulmonary pressure (54, 57, 181).

Mechanism of Pulmonary Vasoconstriction

The mechanisms underlying the development of hypoxic pulmonary hypertension and vasoconstriction are not fully understood, despite intensive studies (35, 118, 175). However, it is clear that vasoconstriction at the precapillary level is impossible without Ca^{2+} involve-

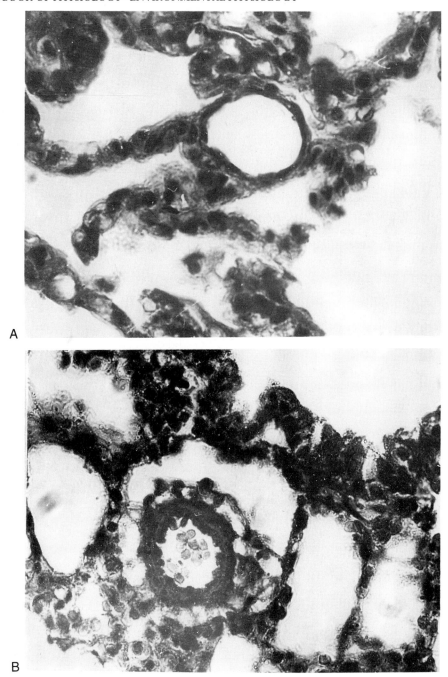

FIG. 53.2. Morphology of pulmonary arterioles (diameter 45 μm). (A) Pulmonary arteriole in a low-altitude (760 m) native. The wall is composed of elastic fibers. (B) Pulmonary arteriole of a native high-lander (3,600 m). Muscular layer is enlarged and situated between internal and external elastic membranes.

ment in smooth muscle contraction. Whether hypoxia can affect calcium translocation in the cell or the latter is effected by mediators remains unresolved. However, vascular reactivity can be enhanced by acidosis and oxidative phosphorylation inhibitors.

Endothelial cells, rich in receptors for several vasoactive substances, including nitric oxide, may play a role (33). Both acetylcholine and bradykinin exert a relaxing action on isolated arteries only when vascular endothelium is intact. Relaxation requires contact between endothelial and smooth muscle cells. Indeed, the internal elastic lamina has numerous fenestrations through which these cells can come into contact with each other. Electron microscopic studies show that, compared to calves, yaks well adapted to high altitude have (Fig. 53.3) considerably larger endothelial cells with larger

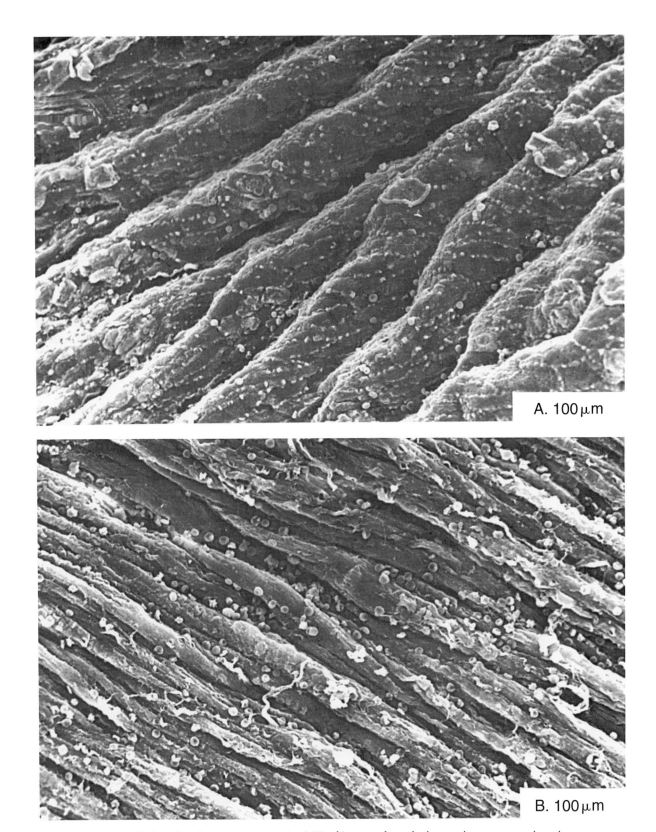

FIG. 53.3. Scanning electron microscopy (× 360) of inner surface of a large pulmonary artery branch (diameter 10,000 μm). (*A*) Endothelial cells of yaks are enlarged with cytoplasmic microprocesses with a cellular structure on the surface. (*B*) Endothelial cells of calves are smaller, and cytoplasmic microprocesses are absent. Blood cells are seen on the endothelial surface.

surface areas, suggesting high functional activity (79). It should be noted that nitric oxide inhibitors and cell-free hemoglobin significantly attenuate hypoxia-induced vasodilatation related to endothelium-derived relaxing factor (EDRF) release. The low P_{O_2} in small arteries and arterioles may be a physiological stimulus for prolonged nitric oxide release (131).

EFFECTS OF HYPOXIA ON THE HEART

Morphology

Hypoxia-induced alterations in cardiac function are reflected in morphological changes. As already discussed, pulmonary vasoconstriction increases the load on the right heart, raising right ventricular systolic pressure. Rats exposed to hypoxia show moderate increases in right ventricular dp/dt_{max} (184) and right ventricular hypertrophy. Humans show right (AQRS) deviation in the early stages of altitude adaptation, which also progresses to right ventricular hypertrophy (106, 130, 144), though regional variation has been noted (146). It is important to note that exercise performance in subjects with evidence of right ventricular hypertrophy is significantly decreased ($P < 0.001$) and that these subjects often develop chronic mountain sickness.

Metabolic Effects

Increased O_2 extraction by tissue is a possible physiological mechanism to adapt to hypoxia (67). This could be mediated by enhancement of the metabolic pathways to meet tissue energy requirements. Well-adapted Tibetans with long-term residence at Lhasa (3,658 m) are able to achieve greater maximum oxygen consumption during muscular exercise owing to increased oxygen transport to working muscles, compared to Hans living at that altitude for 8 ± 1 yr (160). Some permanent residents of high-altitude Peru also have a higher aerobic capacity compared to sojourners (96). Myocardial energy supply may also be related to the increased myocardial content of total lipids, total phospholipids, cholesterol, and sphingomyelin observed in various species (55). High-altitude myocardial capillarization and myoglobin content increase, and aerobic and anaerobic metabolism improve (116). Changes in structure and function of mitochondria lead to the choice of more economical substrates: carbohydrates are preferred to free fatty acids. Despite the observed alterations in myocardial metabolism, working capacity increases (44).

The adapted myocardium apparently becomes more resistant to anoxia (97, 132). Increases in the number and size of intracellular ultrastructures (124, 153, 183) suggest metabolic changes: nucleic acid and protein syn-

thesis are induced, and inhibition of RNA synthesis by actinomycin suppresses oxygen consumption in experimental animals (100). Adaptation to hypoxia apparently persists for some time after return to sea level. After a 50 day adaptation of young volunteers (aged 20–24 yr) to 3,200 m tolerance to severe chamber hypoxia (7,500 m) was increased, and this tolerance persisted for 6 months following descent (89, 103).

Hypoxic training increases the portion of the lactate dehydrogenase (LDH) subunit M involved in meeting energy requirements by glycolysis (7) and increases phosphate potential and lactate production and consumption (116). At the same time, lactate production may decrease during exercise in chronic hypoxia (68). In conscious animals the system of electron transmission in mitochondria is sensitive to the changes in P_{O_2} observed in the range of normoxic variations. The observed decreases in cytochrome oxidase activity in hypoxia (above 5,000 m) may be due to the lower food consumption (semistarvation) observed at high altitude (20). Altitude hypoxia enhances citrate synthase and hydroxyacyl COA dehydrogenase activity in the left ventricular myocardium of mammals (63).

Oxygen delivery to the myocardium may also improve as a consequence of increased concentrations of myoglobin (8). In humans this increase is likely due to hypoxia rather than to conditioning (162). However, rats do not show significant increases in myocardial myoglobin concentration during hypoxia (3,500 m altitude) without simultaneous muscle training.

Microcirculation

Numerous publications suggest that tissue aerobic metabolism is increased due to increased capillary density, primarily in the myocardium (10, 183, 185). Increased capillary density shortens the distance across which oxygen diffuses into the cell (73, 87, 166). Experiments carried out in goose embryos bred under normoxic ($P_{O_2} = 120$ torr) and hypoxic ($P_{O_2} = 94$ torr) conditions showed increased muscular tissue capillarization at high altitude, whereas hypoxia per se did not affect myoglobin concentrations in the myocardium (155).

CORONARY CIRCULATION

It is commonly believed that the incidence of coronary artery disease and systemic hypertension is lower in high-altitude natives than it is in persons at sea level. This is obviously a complex matter because these properties are governed by many factors, including diet, exercise, age, genetic makeup, and smoking habits, just to mention the common ones. Nevertheless, the clinical impression of lower systemic blood pressure in high-altitude natives was substantiated in epidemiological studies in Peru (147). The results of these studies, summarized by Peñaloza et al. (129), were believed to be due to increased vascularization of the capillary beds and diminished peripheral resistance in response to hypoxia. However, Peñaloza et al. pointed out that the difference in blood pressures was only in the systolic value, and diastolic pressures were found by Ruiz and Peñaloza (149) to be slightly higher in high-altitude natives. These authors attributed the increase to polycythemia because they claimed the values significantly decreased after bleeding. Unfortunately, no blood volume measurements are available from these studies.

No case of myocardial infarction was reported in a series of 300 autopsies in Andean natives (134). Moreover, further epidemiological studies by Ruiz and co-workers (148) showed that electrocardiographic diseases were less frequent in high-altitude subjects than in sea-level controls. Because of these provocative facts, Arias-Stella and Topilsky (10) performed postmortem studies of the coronary trees of the hearts of ten natives of Cerro de Pasco; these were compared with those of ten sea-level controls. They found branching of the coronary circulation at several levels and concluded that increased vascularization could explain the decreased incidence of myocardial disease.

In spite of this increased vascularization of the myocardium in high-altitude natives, Moret (117) and Grover et al. (47) showed that coronary blood flow was reduced compared with sea-level controls. Thus it would seem that there may be an increased volume of myocardial blood and, perhaps, of O_2 reserve. Moret (117) also measured coronary blood flow and O_2 transport parameters in four subjects with chronic mountain sickness. The results indicated increased coronary blood flow (probably because of increased left ventricular work), normal O_2 content of coronary sinus blood, and underperfusion of some areas of the myocardium. Moret calculated the O_2 consumption for the hearts of patients with chronic mountain sickness to 9.7 mm/mg of tissue compared with 7 mm/mg for high-altitude controls.

Two hour hypoxemia enhances coronary blood flow (83), and flow increases during the first days at altitude (34, 116). Investigation of the anatomy of the coronary vascular bed in ten native Peruvian highlanders (Cerro de Pasco, 4,380 m) revealed a relatively high density of peripheral coronary artery branches (10). Stereoangiographic examination (22) of the heart in 53 natives of Puno (3,466–4,287 m) and 30 sea-level residents showed prevalence of right-type coronary artery vessel distribution during hypoxia, as well as a consistently increased number of first-order vessels and the presence of an increased number of intercoronary shunts in highlanders. Hypoxia is usually considered the most potent factor in causing vasodilatation (18), and it is presumed that adenosine released by the heart during hypoxia mediates coronary vasodilatation. Intercoronary shunts observed at altitude during physical training require further corroboration by serial coronarographic studies (74).

Right ventricular hypertrophy and pulmonary hypertension alter coronary circulation because of increased myocardial mass and changes in the phasic blood flow circuit in the major vessels of the hypertrophied right ventricle (174). Increases in hypoxic coronary flow are associated with increased cardiac output (60, 171). However, as the duration of adaptation increases, cardiac output returns to normal (102) or even decreases (6), and in calves a 20 day exposure to high altitude results in a significant reduction in coronary blood flow (172). Studies in three healthy subjects performed at sea level and on the tenth day at Leadville (3,100 m) showed decreased coronary blood flow accompanied by increased oxygen extraction (48). The site of adaptation to hypoxia is probably peripheral (that is, at the tissue level), and autoregulatory changes in coronary blood flow are aimed at maintaining stable myocardial oxygen tension.

SYSTEMIC CIRCULATION

Effects of Altitude

In addition to the changes in cardiac output and heart rate that occur at altitude, blood flow distribution is altered due to enhanced sympathetic activity. The outflow of blood from visceral organs increases in the early stage of adaptation (82), and experiments in rats (165) at simulated 4,400 m altitude for 44 days show that the redistribution is persistent if the rats are additionally exposed to acute hypoxia [fraction of inspired oxygen (FIO_2) = 0.11]. Flow to the heart and working muscles is at the expense of the cutaneous vascular bed, kidneys, the splanchnic region, resting muscles, and fatty tissue. Rheoencepholograms, rheopulmonograms, and limb rheograms (80) simultaneously recorded in 26 healthy subjects before ascent to high altitude and during a 25 day stay at 3,200 m revealed increases in cerebral blood flow and in the tone of resistance limb vessels. Decreases in cutaneous blood flow are also characteristic of native highlanders and long-standing sojourners at high altitude (93). Decreases in renal blood flow were observed in both highlanders (15) and sojourners (126), but renal function was not affected.

During short-term altitude adaptation decreases in systolic blood pressure may be observed, but on longer exposure to 4,300 m systemic blood pressure increases in parallel with rises in norepinephrine concentrations (138). Arterial hypertension in highlanders and subjects with long-term adaptation is well documented (102, 115). No clear-cut relation between arterial pressure and age is seen in Eastern Himalayan Sherpas (14).

Atrial Natriuretic Factor

Atrial natriuretic factor (ANF), predominantly synthesized during right atrial stretch, may play a role in the control of blood pressure (40). In one study bicycle exercise, performed at 70%–75% of maximal capacity during inhalation of hypoxic gas mixtures ($FIO_2 = 0.16$), caused greater rises in plasma ANF (45%) and aldosterone (36%) compared to sea-level gas mixtures (86). Thus acute hypoxia appears to enhance the effects of stimuli leading to ANF release during dynamic exercise, and it is possible that ANF is a contributing factor in the dissociation of aldosterone and renin secretion.

In addition to hypoxia, rises in ANF levels seem to be related to rises in pulmonary artery pressure. Hypoxic right ventricular hypertrophy can enhance ANF synthesis (159), and hypoxia can cause extraatrial expression of myocardial ANF synthesis. Exposure of rats to hypoxia ($FIO_2 = 0.10$) for 3 wk was accompanied by a ninefold increase in immunoreactive ANF and 160-fold increases in ANF-transporting mRNA in the hypertrophied right ventricle. Development of right ventricular hypertrophy and the increase in ANF occurred simultaneously and were evident on day 7 of the hypoxic experiments, reaching a plateau by day 14.

Enhancement of ANF synthesis can change arterial pressure. In this respect, decreases in ACE activity in the lungs (75) may also be important. These decreases are apparently not related to the effect of hypoxia but rather to hemodynamic changes in the lung (135).

Venous Tone

Venous tone increases during inhalation of hypoxic gas mixtures ($FIO_2 = 0.115$). During more severe hypoxia ($FIO_2 = 0.07$) active venoconstriction is observed, the latter contributing to the redistribution of blood from the limbs to the great vessels (32). Peripheral venous pressure decreases during the early phases of adaptation to 3,200–3,600 m but is elevated in permanent residents of various altitudes (2,020–4,200 m). This can be partly explained by increases in circulating blood volume and venous tone (102).

Summary of Effects on the Circulatory System

In summary, changes in the circulatory system during hypoxia have been observed not only in the functional state of the heart and vascular system per se but also in the cellular and subcellular structures, leading to long-term adaptation. Early effects of hypoxia include increases in heart rate and cardiac output, as well as blood flow redistribution resulting from regulatory changes in the autonomic nervous system. Following the immediate adaptive changes, physiological indices virtually reach baseline values. If during the transitional period the subject is exposed to an additional stimulus—graded exercise, for example—then the circulatory response may become unstable. It is only after 20–25 days (at 3,200 m) that cardiovascular responses to exercise become relatively stable (103). This stable phase represents adaptation, and during this phase many physiological variables reach values characteristic of native highlanders.

APPLIED ASPECTS

Hypoxic training improves the functional capacity of the cardiovascular system: endurance, resistance to severe chamber hypoxia (equivalent to 7,500 m), and thermal tolerance (2). Adaptation to 3,200 m persists for at least 6 months after descent from altitude. Possibly, hypoxic training could be used to prevent certain diseases (100, 104) since systemic hypertension (11, 31, 150) and ischemic heart disease and its risk factors are less frequent among highlanders than among lowlanders (110).

Hypoxic training in an altitude chamber normalizes the high arterial pressure observed in early essential hypertension (104, 105), and hypoxia inhibits progressive hypertension (66, 92, 104). The beneficial effects are mainly related to changes in the angiotensin–aldosterone system and in ANF synthesis. Negative adaptive changes may also result from hypoxia-induced cardiovascular alterations. In addition to chronic mountain sickness and acute pulmonary and cerebral edema, these include the development of severe pulmonary hypertension and right ventricular hypertrophy, associated with various cardiac arrhythmias (109, 111) and heart failure. Experimental studies in dogs revealed that right ventricular hypertrophy induced by pulmonary artery ligation decreases the sensitivity threshold of the myocardium and shortens the refractory period (107). Natives of high altitude (3,600–4,200 m) should be under systematic medical supervision, and when signs and symptoms of right-heart failure develop, they should be urged to move to lower altitudes.

REFERENCES

1. Aarons, R. D., and P. B. Molinoff. Changes in the density of beta adrenergic receptors in rat lymphocytes, heart and lung after chronic treatment with propranolol. *J. Pharmacol. Exp. Ther.* 221: 439–443, 1982.

2. Agadzhanian, N. A., and M. M. Mirrakhimov. *Organism Resistance.* Moscow: Nauka, 1970, p. 184.

3. Aldashev, A. A., A. Agibetov, Yugai, and A. Shamshiev. Specific proteins synthesized in human lymphocytes during hypoxia. *Am. J. Physiol.* 261: 92–96, 1991.

4. Aldashev, A. A., U. M. Borbugulov, B. A. Davletov, and M. M. Mirrakhimov. Human adrenoceptor system response to the development of high-altitude pulmonary arterial hypertension. *J. Mol. Cell. Cardiol.* 21: 175–179, 1989.

5. Alexander, J. K., and R. F. Grover. Mechanism of reduced cardiac stroke volume at high altitude. *Clin. Cardiol.* 6: 301–303, 1983.

6. Alexander, J. K., L. H. Hartley, M. Modelski, and R. F. Grover. Reduction of stroke volume during exercise in man following ascent to 3100 m altitude. *J. Appl. Physiol.* 23: 849–858, 1967.

7. Anderson, G. L., and R. W. Bullard. Effect of high altitude on lactic dehydrogenase isozymes and anoxic tolerance of the rat myocardium. *Proc. Soc. Exp. Biol. Med.* 138: 441–443, 1971.

8. Antony, A., E. Ackerman, and G. K. Strother. Effects of altitude acclimatization on rat myoglobin: changes in myoglobin content of skeletal and cardiac muscle. *Am. J. Physiol.* 196: 512–519, 1959.

9. Arias-Stella, J., and H. Kruger. Pathology of high altitude pulmonary edema. *Arch. Pathol.* 76: 147, 1963.

10. Arias-Stella, J., and M. Topilsky. Anatomy of the coronary circulation at high altitude. In: *High Altitude Physiology: Cardiac and Respiratory Aspects,* edited by R. Porter and J. Knight. Edinburgh: Churchill Livingstone, 1971, p. 149–154.

11. Baker, P. T. The adaptive fitness of high-altitude population Biological Programme. In: *The Biology of High Altitude Peoples. International Biological Programme 14,* edited by P. T. Baker. Cambridge: Cambridge University Press, 1977, p. 317–350.

12. Balasubramanian, V., O. P. Mathew, S. C. Tiwari, A. Behl, S. C. Sharma, and R. S. Hoon. Alteration in left ventricular function in normal man on exposure to high sltitude (3658 m). *Br. Heart J.* 40: 276–285, 1978.

13. Banchero, N., F. Sime, D. Penaloza, J. Cruz, R. Gamboa, and E. Marticorena. Pulmonary pressure, cardiac output, and arterial oxygen saturation during exercise at high altitude and at sea level. *Circulation* 33: 249–262, 1966.

14. Basu, A., R. Gupta, P. Mitra, A. Dewanji, and A. Sinha. Variations on resting blood pressure among Sherpas of the eastern Himalaya. In: *Proceedings of the Indian Statistical Institute Golden Jubilee, International Conference on Human Genetics and Adaptation,* edited by K. C. Basu. 1982, p. 60–69.

15. Becker, E. L., J. A. Schilling, and R. B. Harvey. Renal function in man acclimatized to high altitude. *J. Appl. Physiol.* 10: 79, 1957.

16. Benitz, W. E., J. D. Coulson, D. S. Lesslor, and M. Bernfield. Hypoxia inhibits proliferation of fetal pulmonary arterial smooth muscle cells in vitro. *Pediatr. Res.* 20: 966–972, 1986.

17. Bergofsky, E. H., and F. Haas. An investigation of the site of pulmonary vascular response to hypoxia. *Bull. Physiopathol. Resp.* 4: 91–103, 1968.

18. Berne, R. M., R. Rubio, B. R. Duling, and V. T. Wied Meier. Effects of acute and chronic hypoxia on coronary blood flow. In: *Hypoxia, High Altitude, and the Heart,* edited by S. Karger. Basel: Munchen, 1970, p. 56–66.

19. Biro, G. P., J. D. Hatcher, and D. B. Jenings. The role of the aortic body chemoreceptors in the cardiac and respiratory responses to acute hypoxia in the anesthetized dog. *Can. J. Physiol. Pharmacol.* 151: 249–259, 1973.

20. Bonverot, P. *Adaptation to Altitude—Hypoxia in Vertebrates.* Berlin: Springer-Verlag, 1985, p. 176.

21. Brodde, O. E., N. Stuna, V. Demuth, et al. Alpha- and beta-adrenoceptors in circulating blood cells of essential hypertensive patients increased receptor density and responsiveness. *Clin. Exp. Hypertens.* 7: A1135- A1150, 1985.

22. Carmelino, M. cited by D. Heath and D. R. Williams. *Man at High Altitude* (2nd ed.), Edinburgh: Churchill Livingstone, 1981, p 347.

23. Castle, W. B., and J. H. Jandl. Blood viscosity and blood volume: opposing influences upon oxygen transport in polycythemia. *Semin. Hematol.* 3: 193–198, 1966.

24. Chapman, C. B., J. N. Fisher, and J. B. Sproule. Behavior of stroke volume at rest and during exercise in human beings. *J. Clin. Invest.* 39: 1208–1213, 1960.

25. Clozel, J. P., C. Saunier, D. Hartemann, and W. Fischli. Effect of cilazapril, a novel angiotensin converting enzyme inhibitor on the structure of pulmonary arteries of rats exposed to chronic hypoxia. *J Cardiovasc. Pharmacol.* 17: 36–40, 1991.

26. Collins, D. D., C. H. Scoggin, C. W. Zwillich, and J. V. Weil. Hereditary aspects of decreased hypoxic response. *J. Clin. Invest.* 62: 105, 1978.

27. Daly, M., and M. J. Scott. The cardiovascular responses to stimulation of carotid body chemoreceptors in the dog. *J. Physiol. (Lond.)* 165: 179–197, 1963.

28. Daniyarov, S. B. *The Results and Perspectives of the Role of Organism Nervous System Adapting to High Altitude.* Frunze: Meditsinskii Institute, 1989, p. 51.

29. Davies, P., F. Maddalo, and L. Reid. Effects of chronic hypoxia on structure and reactivity of rat lung microvessels. *J. Appl. Physiol.: Respir. Environ. Exerc. Physiol.* 58: 795–801, 1985.

30. Dubinina, G. S., R. I. Kovalyova, and M. M. Mirrakhimov. *Physiology and Pathology of the Organism at High Altitude.* Kyrgyz State Medical Institute, 1967, p. 7–11.

31. Dzhailobaev, A. D., B. Y. A. Grinshtein, G. S. Dubinina, and O. N. Narbekov. *Physiology and Pathology of the Organism at High Altitude in Residents of High Altitude in Kyrgyzia.* Kyrgyz State Medical Institute, 1973, p. 155–163.

32. Eckstein, J., and A. W. Horsley. Effects of hypoxia on peripheral venous tone in man. *J. Lab. Clin. Med.* 56: 847–853, 1960.

33. Editorial. EDRF. *Lancet* 11: 137–138, 1987.

34. Feinberg, H., A. Gerola, and L. N. Katz. Effect of hypoxia on cardiac oxygen consumption and coronary flow. *Am. J. Physiol.* 195: 593–600, 1958.

35. Fishman, A. P. *The Pulmonary Circulation: Normal and Abnormal. Mechanisms, Management, and the National Registry.* Philadelphia: University of Pennsylvania Press, 1990.

36. Fishman, A. P., H. W. Fritts, and A. Cournand. Effects of acute hypoxia and exercise on the pulmonary circulation. *Circulation* 5: 263, 1952.

37. Fowles, R. F., and H. N. Hultgren. Left ventricular function at high altitude examined by systolic time intervals and M-mode echocardiography. *Am. J. Cardiol.* 52: 862–866, 1983.

38. Goldenberg, F., J. P. Richalet, I. Ounen, and A. M. Antezana. Sleep apneas and the high altitude newcomer. *Int. J. Sports Med.* 13: S34–S36, 1992.

39. Gordon, J. B., R. C. Wetzel, M. L. McGeady, N. F. Adkinson, Jr., and J. T. Sylvestor. Effects of indomethacin in on estradiol-induced attenuation of hypoxic vasoconstriction in lamb lungs. *J. Appl. Physiol.: Respir. Environ. Exerc. Physiol.* 61: 2116–2121, 1986.

40. Graham, R. M., and J. P. Zisfein. Atrial natriuretic factor: biosynthetic regulation and role in circulatory homeostasis. In: *The heart and cardiovascular system,* edited by H. A. Fozzard, E. Haber, R. B. Jennings, A. M. Katz, and H. E. Morgan. New York: Raven, 1986, p. 1559–1572.

41. Grandjean, E. Physiologi du climat de la montagne. *J. Physiol. Paris* 40: 1A–96A, 1948.

42. Grollman, A. Physiological variations of the cardiac output of man. The effect of high altitude on the cardiac output and its related functions: an account of experiments conducted on the

summit of Pikes' Peak Colorado. *Am. J. Physiol.* 93: 19–40, 1930.

43. Grover, R. F. Comparative physiology of hypoxic pulmonary hypertension. In: *Cardiovascular and Respiratory Effects of Hypoxia.* Basel: Karger, 1966.

44. Grover, R. F. Limitation of aerobic working capacity at high altitude. *Adv. Cardiol.* 5: 11–16, 1970.

45. Grover, R. F. Chronic hypoxic pulmonary hypertension. In: *The Pulmonary Circulation: Normal and Abnormal,* edited by A. P. Fishman. Philadelphia: University of Pennsylvania Press, 1990, p. 283–299.

46. Grover, R. F., J. R. Johnson, R. G. McCullough, R. E. McCullough, S. E. Hotmeister, W. B. Campbell, and R. C. Reynolds. Pulmonary hypertension and pulmonary vascular reactivity in beagles at high altitude. *J. Appl. Physiol.: Respir. Environ. Exerc. Physiol.* 65: 2632–2640, 1988.

47. Grover, R. F., R. Lufchanowski, and J. K. Alexander. *Hypoxia, High Altitude, and the Heart.* Basel: Karger, 1971.

48. Grover, R. F., R. Lufschanowski, and J. K. Alexander. Decreased coronary blood flow in man following ascent to high altitude. *Adv. Cardiol.* 5: 72–79, 1990.

49. Grover, R. F., J. T. Okin, H. R. Overy, A. Treger, and F. H. N. Spracklen. Natural history of pulmonary hypertension in normal adult residents of high altitude. *Circulation* 32: 102, 1965.

50. Grover, R. F., J. H. K. Vogel, K. H. Averill, and S. G. Blount, Jr. Pulmonary hypertension. Individual and species variability relative to vascular reactivity. *Am. Heart J.* 66: 1, 1963.

51. Groves, B. M., J. T. Reeves, J. R. Sutton, P. D. Wagner, A. Cymmerman, M. K. Malconian, P. B. Rock, P. M. Young, and C. S. Houston. Operation Everest II: elevated high-altitude pulmonary resistance unresponsive to oxygen. *J. Appl. Physiol.* 63: 521–530, 1987.

52. Guyton, A. C., C. E. Jones, and T. G. Coleman. *Cardiac Output and its Regulation* (2nd ed.). Philadelphia: Saunders, 1973.

53. Guyton, A. C., and T. Q. Richardson. Effect of hematocrit on venous return. *Circ. Res.* 9: 157–164, 1961.

54. Hakim, T. S., and R. B. Malik. Hypoxic vasoconstriction in blood and plasma perfused lungs. *Physiology* 72: 109–122, 1988.

55. Harris, P. Some observation on the biochemistry of the myocardium at the high altitude. In: *High Altitude Physiology,* edited by R. Porter and J. Knight. Edinburgh: Churchill Livingstone, 1971, p. 125–129.

56. Harris, P., and D. Heath. *The Human Pulmonary Circulation* (2nd ed.). Edinburgh: Churchill Livingstone, 1977.

57. Harrison, B. D., and T. C. Strokes. Secondary polycythaemia: its causes, effects and treatment. *Br. J. Dis. Chest* 76: 313, 1982.

58. Hartley, L. H., J. K. Alexander, M. Modelski, and R. F. Grover. Subnormal cardiac output at rest and during exercise at 3,100 m altitude. *J. Appl. Pysiol.* 23: 839–848, 1967.

59. Hecht, H. H., H. Kuida, R. L. Lange, J. L. Thorne, and A. M. Brown. Brisket disease II. Clinical features and hemodynamic observations in altitude-dependent right heart failure of cattle. *Am. J. Med.* 32: 171–183, 1962.

60. Hellems, H. K., J. W. Ord, F. N. Talmers, and R. C. Christensen. Effects of hypoxia on coronary blood flow and myocardial metabolism in normal human subjects. *Circulation* 16: 893, 1957.

61. Hill, N. S., and L. C. Ou. The role of pulmonary vascular responses to chronic hypoxia in the development of chronic mountain sickness in rats. *Resp. Physiol.* 58: 171–185, 1984.

62. Hislop, A., and L. Reid. Changes in the pulmonary arteries of the rat during recovery from hypoxia-induced pulmonary hypertension. *Br. J. Exp. Pathol.* 58: 653–662, 1977.

63. Hochachka, P. W., C. Stanley, J. Merkt, and J. Sumar-Kalinowski. Metabolic meaning of elevated levels of oxidative

enzymes in high altitude adapted animals: an interpretive hypothesis. *Respir. Physiol.* 52: 303–313, 1983.

64. Hoon, R. S., V. Balasubramanian, O. P. Mathew, S. C. Tiwari, S. C. Sharma, and K. S. Chadha. Effect of high-altitude exposure for 10 days on stroke volume and cardiac output. *J. Appl. Physiol.* 42: 722, 1977.

65. Horwitz, L. D., J. M. Atkins, and S. J. Leshin. Role of the Frank-Starling mechanism in exercise. *Circ. Res.* 31: 868, 1972.

66. Hultgren, H., and R. F. Grover. Circulatory adaptation to high altitude. *Annu. Rev. Med.* 19: 119–152, 1968.

67. Hultgren, H. W., J. Kelly, and H. Miller. Effect of oxygen upon pulmonary circulation in acclimatized man at high altitude. *J. Appl. Physiol.* 20: 239–243, 1965.

68. Hurtado, A. Animals in high altitude: resident man. In: *Handbook of Physiology. Adaptation to the Environment,* edited by D. B. Dill and E. F. Adolph. Washington, DC: Am. Physiol. Soc., 1964, sect. 4, p. 843–860.

69. Hurtado, A. *Rev. Med. Peruana,* 1932. Cited in H. N. Hultgren and R. F. Grover. Circulatory adaptation to high altitude. *Annu. Rev. Med.* 19: 119–152, 1968.

70. Hurtado, A., A. Merino, and E. Delgado. Influence of anoxemia on the hemopoetic activity. *Intern. Med.* 75: 284–324, 1945.

71. Hyman, A. L., and P. J. Kadowitz. Enhancement α and β adrenoceptor responses by elevation in vascular tone in pulmonary circulation. *Am. J. Physiol.* 250 (*Heart Circ. Physiol.* 21): H1109–H1116, 1986.

72. Ishikawa, T., M. Yanagisawa, S. Kimura, K. Goto, and T. Masaki. Positive chronotropic effects of endothelin, a novel endothelium-derived vasoconstrictor peptide. *Pflugers Arch.* 413: 108–110, 1988.

73. Ivanov, K. P., and Y. Y. Kislyakov. The efficacy of main physiological reactions of brain adaptation to hypoxia. *Proc. Acad. Sci. USSR* 233: 997–1000, 1977.

74. Jackson, F. The heart at high altitude. *Br. Heart J.* 30: 291–294, 1968.

75. Jederlinic, P., N. S. Hill, L. C. Ou, and B. L. Fenburg. Lung angiotensin converting enzyme activity in rats with differing susceptibilities to chronic hypoxia. *Thorax* 43: 703–707, 1988.

76. Jungmann, H. Kreislaufwirkungen von Bergfahrten (Oberstdorf-Nebelhorn). *Z. Exp. Med.* 119: 280–285, 1952.

77. Kadyraliev, T. K. Morphological alterations in pulmonary resistant vessels in development of high-altitude pulmonary hypertension. *Arch Pathol* 5: 36–40, 1990.

78. Kadyraliev, T. K. Functional restruction morphology of resistant vessels and pulmonary capillaries during adaptation to high altitude. *Arch. Pathol.* 5: 40–45, 1991.

79. Kadyraliev, T. K., and M. M. Mirrakhimov. Functional morphology of resistant vessels and pulmonary capillaries in species and individual organism adaptation to high altitude. *Bull. Exp. Biol. Med.* 7: 100–104, 1992.

80. Khamzamulin, R. O., M. M. Mirrakhimov, N. N. Brimkulov, and Kalyuzhnyi. Obsidanum effect on cardiovascular system during high-altitude hypoxia. *Kardiologiya* 10: 112–117, 1978.

81. Kontos, H. A., H. P. Mauck, D. W. Richardson, and J. L. Patterson, Jr. Mechanism of circulatory responses to systemic hypoxia in the anesthetized dog. *Am. J. Physiol.* 209: 397–403, 1965.

82. Korner, P. I. Integrative neural cardiovascular control. *Physiol. Rev.* 51: 312–367, 1971.

83. Koyama, T., and K. Nakugawa. The effect of hypoxia on the coronary blood flow reserpinized dogs. *Am. Heart J.* 84: 487–495, 1972.

84. Lahiri, S., and P. G. Data. Chemosensitivity and regulation of ventilation during sleep at high altitudes. *Int. J. Sports Med.* 13(suppl.): S31–S33, 1992.

85. Lahiri, S., N. H. Edelman, N. S. Cherniack, and A. P. Fishman.

Blunted hypoxic drive to ventilation in subjects with life-long hypoxemia. *Federation Proc.* 28: 1289–1295, 1969.

86. Lawrence, D. L., and Y. Shenker. Effect of hypoxic exercise on atrial natriuretic factor and aldosterone regulation. *Am. J. Hypertens.* 4: 341–347, 1991.

87. Lenfant, C., and K. Sullivan. Adaptation to high altitude. *N. Engl. J. Med.* 284: 1298–1309, 1971.

88. Logaras, G. Further studies of the pulmonary arterial blood pressure. *Acta Med. Scand.* 14: 120, 1947.

89. Malyshev, I. U. The Phenomenon of Adaptational Stabilization Structures and the Role of Heat Shock Proteins in it. Moscow:, 1992. Synopsis of MD thesis.

90. Margaria, R., and P. Cerretelli. Physiological aspects of life at extreme altitudes. In: *Biometeorol. Proc. Second Bioclimatol. Congr., Sept. 4–10, London,* edited by S. W. Tromp. 1960. p. 3–25.

91. Marsh, J. D., and K. A. Sweeney. Beta-adrenergic receptor regulation during hypoxia in intact cultured heart cells. *Am. J. Physiol.* 256 (*Heart Circ. Physiol.* 27): H275–H281, 1989.

92. Marticorena, E., L. Ruiz, J. Severino, J. Calvez, and D. Peñaloza. Systemic blood pressure in white men born at sea level: changes after long residence at high altitude. *Am. J. Cardiol.* 23: 364–368, 1969.

93. Martineaud, J. P., J. Durand, J. Coudert, and Seroussi. La circulation cutanée au cours de l'adaptation altitude. *Pflugers Arch.* 310: 264–276, 1969.

94. Martorana, P. A., M. Wilkinson, P. Van Even, and G. Lungarella. Tsk mice with genetic emphysema. Right ventricular hypertrophy occurs without hypertrophy muscular pulmonary arteries or muscularization of arterioles. *Am. Rev. Respir. Dis.* 142: 333–337, 1990.

95. Masuyama, S. H., S. H. Kohchiyama, T. Shinozari, S. H. Okita, F. Kunitotno, H. Tojima, H. Kimura, T. Kuriyama, and Y. Honda. Periodic breathing at high altitude and ventilatory responses to O_2 and CO_2. *J. Physiol. (Lond.)* 39: 523–535, 1989.

96. Mazess, R. B. Exercise performance at high altitude (4000 m) in Peru. In: *High Altitude Adaptation in a Peruvian Community,* 1968, p. 167–185.

97. McGrath, J. J., J. Prochazka, V. Pelouch, and B. Ostadal. Physiological responses of rats to intermittent high-altitude stress: effects of age. *J. Appl. Physiol.* 34: 289–293, 1973.

98. McMurtry, I. F., A. B. Davidson, J. T. Reeves, and R. F. Grover. Inhibition of hypoxic pulmonary vasoconstriction by calcium antagonists in isolated rat lungs. *Circ. Res.* 38: 99, 1976.

99. Mecham, R. P., L. A. Whitehouse, D. S. Wrenn, W. C. Parks, G. L. Griffin, R. M. Senior, E. C. Crouch, K. R. Stenmark, and N. F. Voelkel. Smooth muscle–mediated connective tissue remodelling in pulmonary hypertension. *Science* 237: 423–426, 1987.

100. Meerson, F. Z. Adaptation to high-altitude hypoxia. In: edited by O. G. Gazenko. Moscow: Nauka, 1986, p. 222–251.

101. Meyrick, B., and L. Reid. Endothelial and subintimal changes in rat hilar pulmonary artery during recovery from hypoxia. A quantitative ultrastructural study. *Lab. Invest.* 42: 603–615, 1980.

102. Mirrakhimov, M. M. Cardiovascular system at high altitude. In: Meditsina, 1968, p. 156.

103. Mirrakhimov, M. M. Ventilatory and circulatory adaptation to Tien-Shan and the Pamirs altitude in man. In: *Proc. Union Physiol. Sci.* Delhi: 1974.

104. Mirrakhimov, M. M. Biological and physiological characteristics of the high altitude natives of Tien Shan and the Pamirs. In: *The Biology of High-Altitude Peoples. International Biological Programme 14,* edited by P. T. Baker. Cambridge: Cambridge University Press, 1977, p. 299–315.

105. Mirrakhimov, M. M., A. S. Dzhumagulova, E. Shatemirova, Y. u. M. Ishmatov, and V. A. Zelenshchikova. Hypoxic training during hypertension. *CV World Rep.* 3: 13–219, 1990.

106. Mirrakhimov, M. M., and B. Ya. Grinshtein. Cardiac muscle adaptation to high altitude. In: *Physiology and pathology of cardiovascular system. Proc. Kyrgyz State Medical Institute,* edited by Y. A. M. Snezhko. 1966. p. 71–80.

107. Mirrakhimov, M. M., and T. F. Kalko. Peripheral chemoreceptors and human adaptation to high altitude. *Biomed. Biochem. Acta.* 47: 89–91, 1988.

108. Mirrakhimov, M. M., R. I. Kovalyova, Z. M. Kudaiberdiev, and O. N. Narbekov. The study of the circulation function at the altitudes of Tien-Shan and Pamir. In: *Human Adaptation,* edited by Z. I. Barbashova. Leningrad: Nauka, 1972, p. 125–131.

109. Mirrakhimov, M. M., and T. S. Meimanaliev. Prevalence and diagnostic peculiarities of ischemic heart disease in population of highlanders. *Epidemiol. News.* 30: 125–126, 1981.

110. Mirrakhimov, M. M., and T. S. Meimanaliev. *High-Altitude Cardiology.* Kyrgyzstan: Frunze, 1984.

111. Mirrakhimov, M. M., T. S. Meimanaliev, and K. D. Abdurasulov. Correlation between right ventricular hypertrophy and cardiac arrhythmias in prevalence and diagnostic in population of highlanders. *Jpn. Heart J.* 23: 501–503, 1982.

112. Mirrakhimov, M. M., R. I. Rudenko, T. M. Murataliev, and R. O. Khamzamulin. Circulatory adaptation in man. Proc. clinical and instrumental characteristics of primary high-altitude pulmonary arterial hypertension. *Kardiologiya* 10: 56–61, 1976.

113. Moldotashev, I. K., A. A. Aldashev, T. A. Batyraliev, U. M. Borbugulov, Y. O. Titov, and A. E. Tashpolotov. The reactivity of pulmonary arteries and activity of neurotransmitter receptors in permanent residents of high altitude [Abstract]. *Const. Cong. Ing. Soc. Pathophis.* Moscow, May 28–June 1, 1991.

114. Monge, C. *Les Erythremies de l'Altitude: Les Rapports avec la Maladie de Váquez. Etude Physiologique et Pathologique.* Paris: Masson et Cie, 1929.

115. Monge, C. M., and C. C. Monge. *High-Altitude Diseases. Mechanism and Management.* Springfield: Thomas, 1966.

116. Moret, P. Myocardial metabolism; acute and chronic adaptation to hypoxia. *Med. Sci. Sports Exerc.* 19: 48–63, 1985.

117. Moret, P. R. Coronary blood flow and myocardial metabolism in man at high altitude. In: *High Altitude Physiology: Cardiac and Respiratory Aspects,* edited by R. Porter and J. Knight. Edinburgh: Churchill Livingstone, 1971, p. 131–148.

118. Morpurgo, M., R. Tramarin, C. Rampulla, C. Fracchia, and F. Cobelli. *Pathophysiology and Treatment of Pulmonary Circulation.* Verona: Bi and Gi, 1988.

119. Motley, H. L., A. Cournand, L. Werko, A. Himmelstein, and D. Dresdsle. The influence of short periods of induced acute anoxia upon pulmonary artery pressure in man. *Am. J. Physiol.* 150: 315, 1947.

120. Mustafin, K. S. Morphofunctional alterations in the pulmonary circulation in man and animals at high altitude. *Int. Symp., June 28–30.* Moscow: Frunze, 1982, p. 104–105.

121. Naeije, R., C. Melot, P. Mols, and R. Hallemans. Effects of vasodilators on hypoxic pulmonary vasoconstriction in normal man. *Chest* 82: 404–410, 1982.

122. Nakazava, K., and K. Amaha. Effect of nicardipine hydrochloride on regional hypoxic pulmonary vasoconstriction. *Br J Anaesth* 60: 547–554, 1988.

123. Orbeli, L. A. *The Effect of Decreased Barometrical Pressure on the Central Nervous System.* Leningrad: Nauka, 1940.

124. Panin, L. E. *Energy Aspects of Adaptation.* Leningrad: Medicine, 1978.

125. Parin, V. V. Effect of pulmonary ventilation on pulmonary circulation. *Patologich. Phisiol. Eks. Ter.* 4: 7–12, 1960.

126. Pauli, H. G., B. Truniger, J. K. Larsen, and R. O. Mulhausen.

Renal function during prolonged exposure to hypoxia and carbon monoxide. *Scand. J. Clin. Lab. Invest.* 22: 5, 1968.

127. Peake, M. D., A. L. Harabin, N. J. Breunan, and J. T. Sylvester. Steady-state vascular responses to graded hypoxia in isolated lungs of five species. *J. Appl. Physiol.* 1: 1214–1219, 1981.

128. Peñaloza, D., F. Sime, N. Bachero, R. Gamboa, J. Cruz, and E. Marticorena. Pulmonary hypertension in healthy men born and living at high altitude. *Am. J. Cardiol.* 11: 150–157, 1963.

129. Peñaloza, D., F. Sime, and L. Ruiz. Cor pulmonale in chronic mountain sickness: present concept of Monge's disease. In: *Ciba Foundation Symposium on High-Altitude Physiology: Cardiac and Respiratory Aspects,* edited by R. Porter and J. Knight. Edinburgh: Churchill Livingstone, 1971, p. 41–60.

130. Plotnikov, I. P. Cardiovascular System during Human Adaptation to High Altitude. Dushanbe, 1963. Synopsis of Thesis.

131. Pohl, U., and R. Busse. Hypoxia stimulates release of endothelium-derived relaxant factor. *Am. J. Physiol.* 256 (*Heart Circ. Physiol.* 27): H1595–H1600, 1989.

132. Poupa, O., K. Krofta, J. Prochazka, and Z. Turek. Acclimation to simulated high altitude and acute cardiac necrosis. *Federation Proc.* 25: 1243–1246, 1966.

133. Pugh, L. G. C. E. Cardiac output in muscular exercise at 5800 m (19,000 feet). *J. Appl. Physiol.* 19: 441–447, 1964.

134. Ramos, A., H. Kruger, M. Muro, and J. Arias-Stella. Untitled. *Bol. San. Pan. Am.* 62: 496–502, 1967.

135. Reane, P. M., J. M. Kay, K. L. Suyama, D. Gauthier, and K. Andrew. Lung angiotensin converting enzyme activity in rats with pulmonary hypertension. *Thorax* 37: 198–204, 1982.

136. Redding, G. J., R. Tuck, and P. Escourou. Nifedipine attenuates acute hypoxic pulmonary vasoconstriction in awake piglets. *Am. Rev. Respir. Dis.* 129: 785–789, 1984.

137. Reeves, J. T., E. G. Grover, and R. F. Grover. Pulmonary circulation and oxygen transport in lambs at high altitude. *J. Appl. Physiol.* 18: 560–566, 1963.

138. Reeves, J. T., R. S. Mazzeo, E. E. Wolfel, and A. J. Yong. Increased arterial pressure after acclimatization to 4,300 m: possible role of norepinephrine. *Int. J. Sports Med.* 13: 518–521, 1992.

139. Reid, L. M. Vascular remodelling in pulmonary circulation. In: *Pulmonary circulation: Normal and abnormal,* edited by A. P. Fishman. Philadelphia: University of Pennsylvania Press, 1990, p. 259–282.

140. Replogle, R. L., H. J. Meiselmann, and E. W. Merrill. Clinical implication of rheology studies. *Circulation* 36: 148–160, 1967.

141. Richalet, J., H. Mehdioui, C. Rathat, P. Vignon, A. Keromes, J. P. Herry, C. Sabatler, M. Tauche, and F. Lhoste. Acute hypoxia decreases in cardiac response to catecholamines in exercising humans. *J. Sports Med.* 9: 157–162, 1988.

142. Richalet, J. P., R. Kacimi, and A. M. Antezana. The control of cardiac chronotropic function in hypobaric hypoxia. *Int. J. Sports Med.* 13: S22–S24, 1992.

143. Richardson, T. Q., and A. C. Guyton. Effects of polycythemia and anemia on cardiac output and other circulatory factors. *Am. J. Physiol.* 197: 1167–1170, 1959.

144. Rotta, A. Physiologic condition of the heart in the natives of high altitude. *Am. Heart J.* 33: 669–676, 1947.

145. Rotta, A., A. Canepa, A. Hurtado, T. Velasquez, and R. Chavez. Pulmonary circulation at sea level and high altitude. *J. Appl. Physiol.* 9: 328–336, 1956.

146. Rudenko, R. I. Vectorcardiographic studies in highlanders in problems of adaptation to high altitude. *Proc. Kyrgyz St. Med. Inst.* 69: 162–172, 1971.

147. Ruiz, L., M. Figueroa, C. Horna, and D. Peñaloza. Systemic blood pressure in high altitude residents. In: *Progress Report to the World Health Organization,* 1968.

148. Ruiz, L., C. Figueroa, C. Horna, and D. Peñaloza. Prevalencia

de la hipertensión arterial y cardiopatía isquémica en las grandes alturas. *Am. J. Cardiol.* 39: 474–89, 1969.

149. Ruiz, L. and D. Peñaloza. Altitude and cardiovascular diseases. *Progress Report to the World Health Organization,* 1970.

150. Ruiz, L., and D. Peñaloza. Altitude and hypertension. *Mayo Clin. Proc.* 52: 442, 1977.

151. Rushmer, R. F. Constancy of stroke volume in ventricular responses to exertion. *Am. J. Physiol.* 196: 745, 1959.

152. Saltin, B., and P. Astrand. Maximal oxygen uptake in athletes. *J. Appl. Physiol.* 23: 353, 1967.

153. Sarkisov, D. S. Structural bases of homeostasis. In: *Homeostasis,* edited by P. D. Gorizontov. Moscow: Meditsina, 1976, p. 133–177.

154. Sibley, D. R., J. L. J. Benovic, M. G. Caron, and R. J. Lefkowitz. Regulation of transmembrane signaling by receptor phosphorylation. *Cell* 48: 913–922, 1987.

155. Snyder, G. K., R. T. Byers, and S. R. Kayar. Effects of hypoxia on tissue capilarity in geese. *Respir. Physiol.* 58: 151–160, 1984.

156. Starling, E. H. *The Linacre Lecture on the Law of the Heart.* London: Longmans, Green, 1918.

157. Stenberg, J., B. Ekblom, and R. Messin. Hemodynamic response to work at simulated altitude, 4,000 m. *J. Appl. Physiol.* 21: 1589–1594, 1966.

158. Stenmark, R., A. Aldashev, E. C. J. Orton, A. G. J. Durmowicz, W. C. J. Badesch, R. P. Mecham, N. F. Voelkel, and J. T. Reeves. Cellular adaptation during chronic neonatal hypoxic pulmonary hypertension. *Am. J. Physiol.* 261(suppl.): 97–104, 1991.

159. Stockmann, P. T., D. H. Will, S. D. Sides, S. Brunner, G. D. Wilner, K. M. Leahy, R. C. Wiegand, and P. H. Needelman. Reversible induction of right ventricular atriopeptin synthesis in hypertrophy due to hypoxia. *Circ. Res.* 63: 207–213, 1988.

160. Sun, S. F., T. S. Droma, J. G. Zhang, J. K. Tao, S. Y. Huang, R. McCullough, R. E. McCullough, C. S. Reeves, J. T. Reeves, and L. G. Moore. Greater maximal O$_2$ uptaxes and vital capacities in Tibetan than han residents of Lhasa. *Respir. Physiol.* 79: 151–162, 1990.

161. Swigart, R. H. Polycythemia and right ventricular hypertrophy. *Circ. Res.* 17: 30, 1965.

162. Terrados, N., E. Jansson, C. Sylvin, and L. Kajser. Is hypoxia a stimulus for synthesis of oxidative enzymes and myoglobin? *J. Appl. Physiol.: Respir. Environ. Exerc. Physiol.* 68: 2369–2372, 1990.

163. Theilen, E. O., M. H. Paul, and D. E. Gregg. A comparison of effects of intra-arterial and intravenous transfusion in hemorrhagic hypotensions on coronary blood flow, systemic blood pressure, and ventricular and diastolic pressure. *J. Appl. Physiol.* 7: 248–252, 1954.

164. Tozzi, C. A., G. Y. Poiani, N. H. Edelman, and D. J. Riley. Vascular collagen affects reactivity of hypertensive pulmonary arteries of the rat. *J. Appl. Physiol.: Respir. Environ. Exerc. Physiol.* 66: 1730–1735, 1989.

165. Tucker, A., and S. M. Horvath. Regional blood flow responses to hypoxia and exercise in altitude adapted rats. *Eur. J. Appl. Physiol.* 33: 139–150, 1974.

166. VanLiere, E., and K. Stikney. *Hypoxia.* Moscow: Meditsina, 1967.

167. Vatner, S. F., and E. Braunwald. Cardiovascular control mechanisms in the conscious state. *N. Engl. J. Med.* 293: 970–976, 1975.

168. Vlsdimirov, G. I., I. M. Defolin, and K. Smirnov. The effect of caffeine taking on cardiac output at high altitude. In: *Proceedings Dedicated to London, ES,* Leningrad: Medghiz, 1947, p. 66–78.

169. Voelkel, N., J. Hegstrand, J. T. Reeves, I. E. McMurtry, and P. B. Malinoff. Effects of hypoxia on density of β-adrenergic recep-

tors. *J. Appl. Physiol.: Respir. Environ. Exerc. Physiol.* 50: 313–366, 1981.

170. Vogel, J. H. K. Importance of mild hypoxia on abnormal pulmonary vascular beds. *Adv. Cardiol.* 5: 159–165, 1970.

171. Vogel, J. H. K., J. E. Goss, and H. L. Brammel. Pulmonary circulation in normal man with acute exposure to high altitude. *Circulation* 34: 233, 1966.

172. Vogel, J. H. K., G. Jameeson, M. Delivoria-Papadopoulos, R. D. Luecker, H. L. Brammell, and D. Brane. Coronary blood flow during short term exposure to high altitude. *Adv. Cardiol.* 5: 80–85, 1970.

173. Vogel, J. H. K., W. F. Weaver, R. L. Rose, S. G. Y. R. Blount, and R. F. Grover. Pulmonary hypertension on exertion in normal man living at 10,500 feet (Leadville, Colorado). *Med. Thorac.* 19: 461–477, 1962.

174. Von Euler, U. S., and G. Liljestrand. Observations on the pulmonary arterial blood pressure in the cat. *Acta Physiol. Scand.* 12: 301–324, 1946.

175. Wagenvoort, C. A., and H. Denolin. *Pulmonary Circulation—Advances and Contraversions.* New York: Elsevier, 1989.

176. Wagenvoort, C. A., and N. Wagenvoort. *Pathology of Pulmonary Hypertension.* New York: Wiley, 1977.

177. Walker, B. R., N. F. Voelkel, I. F. J. McMurtry, and E. M. Adams. Evidence for diminished sensitivity of the hamster pulmonary vasculature to hypoxia. *J. Appl. Physiol.: Respir. Environ. Exerc. Physiol.* 52: 1571–1574, 1982.

178. Weir, E. K., D. H. Will, A. F. Alexander, I. F. McMurtry, R. Looda, J. T. Reeves, and R. F. Grover. Vascular hypertrophy in cattle susceptible to hypoxic pulmonary hypertension. *J. Appl. Physiol.: Respir. Environ. Exerc. Physiol.* 46: 517–521, 1979.

179. Will, D. H., J. L. Hicks, C. S. Card, and A. F. Alexander. Inherited susceptibility of cattle to high-altitude pulmonary hypertension. *J. Appl. Physiol.* 38: 491–495, 1970.

180. Winslow, R. M., and C. C. Monge. *Hypoxia, Polycythemia, and Chronic Mountain Sickness.* Baltimore: Johns Hopkins University Press, 1986.

181. Winslow, R. M., C. C. Monge, E. G. Brown, H. G. Klein, F. Sarnquist, and N. J. Winslow. The effect of hemodilution on O_2 transport in high-altitude polycythemia. *J. Appl. Physiol.: Respir. Environ. Exerc. Physiol.* 59: 1495–1502, 1985.

182. Wu, T., Q. Zhang, B. Jin, F. Xu, Q. Cheng, and X. Wan. Chronic mountain sickness (Monge's disease): an observation in Qinhai, Tibet Plateau. In: *High-Altitude Medicine,* edited by G. Ueda. Matsumoto: Shinshu University Press, 1992, p. 314–324.

183. Zhaparov, B. Comparative myocardial morphology of non-adapted, temporary adapted, as well as permanently living at high altitude animals (experimental morphological study). Moscow:, 1987. Synopsis of Thesis (MD).

184. Zierhut, W., and H. Zimmer. Effect of calcium antagonists and other drugs on the hypoxia-induced increase in rat right ventricular pressure. *J. Cardiovasc. Pharmacol.* 14: 311–318, 1989.

185. Zverkova, E. E. Myocardial blood supply and organism resistance to hypoxia during hypoxic and hypercapnic training. Alma-Ata, 1982. Synopsis of Thesis.

54. Endocrine adaptation to hypoxia

HERSHEL RAFF | *Departments of Medicine and Physiology, Medical College of Wisconsin, Endocrine Research Laboratory, St. Luke's Medical Center, Milwaukee, Wisconsin*

ADAPTATION TO ACUTE AND CHRONIC HYPOXIA involves an integrated response of the cardiopulmonary, renal, hematopoietic, and metabolic control systems. This is orchestrated in part by changes in the hormonal milieu.

HYPOTHALAMIC–PITUITARY–ADRENAL AXIS AND ENDORPHINS

Acute Hypoxia

Acute exposure to real or simulated altitude of less than 5,000 m does not appear to increase cortisol in humans (28, 44, 77, 110, 187, 222). Mild acute hypoxia induced by discontinuation of supplemental oxygen in patients with chronic obstructive pulmonary disease (COPD) also does not activate adrenocorticotropic hormone (ACTH) or cortisol secretion (173). Bouissou et al. (28) demonstrated an increase in ACTH during mild acute hypoxic exposure (3,000 m) without a change in plasma cortisol. This suggests that acute hypoxia might decrease adrenal sensitivity to ACTH or alter the clearance of cortisol.

More severe acute hypoxia [>5,000 m/partial pressure of arterial oxygen (PaO_2) < 50 torr] does result in an increase in ACTH and/or corticosteroids in humans (161), conscious and anesthetized dogs (84, 129, 183), pregnant sheep (25), piglets (153), calves (22), and rats (91, 130, 171, 175). Several studies have demonstrated that significant increases in cortisol can occur during acute hypoxia at about 5,000 m in humans with symptoms of acute mountain sickness (AMS) (83, 222). Therefore, these studies suggest that there is a correlation of the severity of the hypoxic exposure with the degree of increase in cortisol but that other influences, such as nausea, can increase corticosteroids even at moderate altitude.

Several studies have examined the cortisol response to exercise under hypoxic conditions. The ACTH and cortisol responses to exercise have been found to be augmented (221), unaltered (32, 110, 117), or decreased (28) under mild hypoxia (<5,000 m) in humans.

Since the steroidogenic enzymes are cytochrome P450 in nature (36), it is possible that exposure to hypoxia could interfere with the synthesis of corticosteroids in the adrenal zona fasciculata and, therefore, the relationship between ACTH and cortisol (that is, adrenal sensitivity). Bouissou et al. (28) found that acute hypoxia disrupted the correlation of ACTH and cortisol in humans at rest and exercise, which implies that hypoxia decreases adrenal sensitivity to ACTH. However, the cortisol response to exogenous ACTH (44, 186) and the correlation of endogenous ACTH and cortisol (117, 178) are not altered by acute hypoxia (Fig. 54.1). Furthermore, basal and stimulated cortisol release from bovine adrenal cells in vitro is not inhibited by decreases in partial pressure of oxygen (PO_2) (168, 172) (Fig. 54.2). Therefore, the production of cortisol by the zona

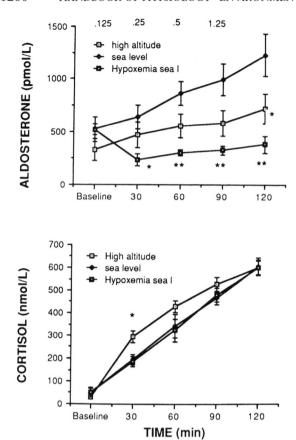

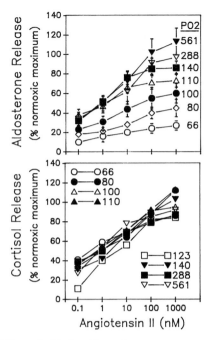

FIG. 54.2. Effect of buffer P_{O_2} (shown in torr) on angiotensin II–stimulated aldosterone and cortisol release from acutely dispersed bovine adrenal zona glomerulosa and fasciculata cells. Aldosterone release was proportional to P_{O_2}, whereas cortisol release was unaffected by changing P_{O_2}. (Adapted from ref. 172 with permission.)

FIG. 54.1. Plasma aldosterone and cortisol concentrations during bolus infusion of ACTH (dose at top in μg) in progressively increasing doses in subjects exposed to normoxemia *(sea level)* and acute hypoxemia at sea level *(Hypoxemia sea I)* and in moderately high altitude *(high altitude)*–adapted subjects. Plasma renin activity was decreased in the sea-level group compared to the other two groups (data not shown). *$P < 0.05$ between subjects at sea level during normoxemia and hypoxemia and between high-altitude subjects. **$P < 0.05$ between subjects at sea level during hypoxemia and between high-altitude subjects. Graph demonstrates that acute and chronic hypoxia decrease aldosterone but not cortisol response to physiological doses of ACTH. (From ref. 186 with permission.)

fasciculata does not appear to be inhibited directly by low O_2.

Changes in adrenal blood flow may alter the adrenal response to ACTH. Hypoxia has been shown to increase adrenocortical blood flow (34, 35, 160). This would tend to increase rather than decrease sensitivity to ACTH and would not explain the apparent dissociation found previously (28).

Mechanisms

Adrenal secretion of corticosteroids in conscious calves is inversely related to the decrease in Pa_{O_2}, with a relationship very similar to that between P_{O_2} and carotid body output (22). Jacobson and Dallman (91) documented an increase in ACTH at an arterial P_{O_2} similar

to that at which ventilation increased, also suggesting a chemoreceptor-mediated process.

As with chemoreceptor response characteristics, there appears to be an interaction between hypercapnia and hypoxia on the adrenocortical control system (113). Anesthetized, paralyzed dogs exhibit a hyperbolic relationship between P_{O_2} and ACTH, which is augmented by hypercapnia in a manner similar to the neural output characteristics of the arterial chemoreceptors (181).

Several studies have demonstrated that denervation of the carotid and aortic bodies attenuated or completely eliminated adrenal secretion of cortisol in response to acute hypoxia (114, 115, 128, 184). These studies were performed after acute denervation, and it is known that surgery alters the cortisol response to hypoxia (178). A subsequent study demonstrated that chronic carotid denervation also attenuated ACTH and cortisol responses to acute hypoxia (179). Therefore, it seems likely that the peripheral arterial chemoreceptors are the source of afferent information to the hypothalamus (116). Nitroprusside-induced vasodilation augments the ACTH response to acute hypoxia in dogs (180), suggesting an interaction between chemoreceptor and baroreceptor afferents.

Although the acute ACTH response to hypoxia is suppressible by exogenous glucocorticoids (130, 182), few in-depth studies have examined the hypothalamic factors [for example, corticotropin-releasing hormone

(CRH)] which activate ACTH secretion under hypoxic conditions. The cortisol response to acute hypoxia in conscious lambs was attenuated by converting enzyme inhibition, suggesting the involvement of angiotensin II (232).

Chronic Hypoxia

In early studies on humans subjects Hornbein (89) could not demonstrate an increase in urinary cortisol at very high altitude (6,400 m), whereas Frayser et al. (61) demonstrated an increase in plasma cortisol at 5,300 m. Moncloa et al. (146) found an inverse correlation between PO_2 ranging from 65 to 40 torr and plasma cortisol ranging from 2 to >20 µg/dl in human subjects. The relationship between PO_2 and cortisol resembled that found in dogs and calves exposed to acute hypoxia (22, 181) and is reminiscent of the ventilatory response to altitude exposure also known to be driven by peripheral arterial chemoreceptors.

Plasma and urinary cortisol responses to altitude [2,000 m (90); 3,500 m (206); 4,300 m (147); 4,600 m (12)] correlate with the time of exposure and/or the development of symptoms of AMS. Furthermore, Maresh et al. (126) demonstrated a greater cortisol response to altitude in low-altitude natives as opposed to moderate-altitude natives, suggesting some adaptation of the system to chronic hypoxia.

Richalet et al. (193) demonstrated that the circadian rhythm is maintained at altitude, albeit at higher average levels of cortisol. A lack of significant rhythm in ACTH was found, which is consistent with an amplification of adrenal sensitivity to ACTH during the circadian rhythm. Furthermore, the system seems to respond normally to other stimuli, such as exercise (136) and hypoglycemia (203), though there may be an alteration of this effect if the system is not studied under basal conditions (60). Finally, adrenal sensitivity to exogenous ACTH is not reduced (Fig. 54.1), nor is the clearance of cortisol significantly accelerated, in humans (145, 186).

As in humans, very mild chronic exposure to hypoxia does not seem to activate cortisol secretion in sheep (223). However, Lau and Timiras (116) found increases in corticosterone in rats exposed for 1 wk to relatively mild hypoxia (3,800 m), whereas Groza et al. (74) did not detect a change at 2,200 m. Several studies have demonstrated a large increase in corticosterone in rats early during hypoxic exposure but dissipation of the response by 7 days (116, 144). Furthermore, an increase in ACTH appears to be the main driving force in this response (127, 177). Rattner et al. (189, 190) carefully examined the relationship between degree of hypoxic exposure and corticosterone levels in conscious mice

and found that the threshold for activation was between 4,600 and 6,100 m.

Pituitary and adrenal hypertrophy have been demonstrated in rats exposed to 5,500 m simulated altitude (69, 163). More specifically, Marks et al. (127) demonstrated an increase in pituitary ACTH content and Gosney (68) demonstrated an increase in the number of corticotrophs in the pituitary. These data suggest chronic activation of the hypothalamic–pituitary–adrenal system during exposure to long-term hypoxia. The increase in corticosteroids during exposure to chronic hypoxia has been suggested to be one of the primary mechanisms for the increase in gluconeogenesis (143).

Endorphins/Lipotropic Pituitary Hormone

Steinbrook et al. (212), studying normobaric hypoxia [oxygen saturation (SaO_2) down to 75%], did not find any change in plasma beta-endorphin levels, nor did the antagonist naloxone induce any change in the relationship between SaO_2 and ventilation. Kraemer et al. (110), studying humans at a simulated altitude of 4,300 m [barometric pressure = 443 mm Hg; partial pressure of inspiratory oxygen (PIO_2) = 83 mm Hg), could not detect an effect of acute hypoxia with or without high-intensity exercise on plasma β-endorphin levels. Freedman et al. (62), studying conscious goats, also found that neither moderate (PaO_2 30–50 torr) nor severe (PaO_2 < 30 torr) hypoxia resulted in any measurable change in plasma or cerebrospinal fluid (CSF) beta-endorphin. Finally, Rochat et al. (194) could not find an effect of acute hypoxia on either plasma beta-endorphin or met-enkephalin concentrations nor a correlation of these peptides with the ventilatory response to hypoxia.

Whereas moderate hypoxia in pregnant ewes (PO_2 ~ 40 torr) did not result in a change in beta-lipotropic pituitary hormone (LPH) or endorphins, severe hypoxia (PaO_2 < 30 torr) did increase these peptides in nonpregnant ewes (231). The same pattern seemed to hold for newborn lambs. Moss et al. (151, 152) found that newborn (~2 days old) piglets demonstrated a larger plasma beta-LPH and endorphin response to hypoxia than did older (38 days old) piglets.

CATECHOLAMINES

Acute Hypoxia

Moderate hypoxia does not increase circulating epinephrine or norepinephrine levels in humans (30–32, 187), conscious (58) and anesthetized (82) dogs, and conscious calves (22). Furthermore, the increase in plasma catecholamines induced by exercise is not altered by concomitant acute hypoxia (30–32, 38, 58). Hypoxia of greater severity ($PaO_2 < 40$ torr) increases plasma catecholamines in dogs (196), conscious calves (22), and rats (40). These data suggest a dose–response relationship between PaO_2 and adrenal catecholamine secretion and/or release of norepinephrine from nerve terminals that spills over into the systemic circulation.

Conscious calves (22), anesthetized dogs (82), and rats (21) increase the adrenal secretion rate of catecholamines even when peripheral levels are unchanged. This apparent conflict may have been resolved by the demonstration of increased catecholamine clearance during acute hypoxia (120). Therefore, adrenal secretion of catecholamines does seem to be inversely correlated with PaO_2, suggesting a possible role of peripheral chemoreceptor afferents in generating the response.

Mechanisms

The chemoreceptor afferent mechanism has been demonstrated by Biesold et al. (21), who found that hypoxia-induced increases in sympathetic nerve activity and epinephrine and norepinephrine secretion measured directly from the adrenal medulla were eliminated by section of the carotid sinus nerve. The efferent segment of the reflex arc was demonstrated by Bloom et al. (22), who showed that resection of the splanchnic nerve eliminated the adrenal medullary response to hypoxia. Finally, Lee et al. (119) demonstrated a direct inhibition of potassium-stimulated catecholamine release from adrenal medullary cells in vitro, also suggesting an explanation for increased sympathetic nerve activity without a dramatic increase in medullary secretion during mild hypoxia.

Chronic Hypoxia

Chronic exposure to altitude or normobaric hypoxia may increase plasma and urinary catecholamines in humans (49, 134, 187) and rats (74, 97), though no increase was found at 4,300 m for 5 days (32). Adrenal medullary hyperplasia has also been demonstrated, suggesting chronic stimulation (69). Finally, it has been suggested that, much like cortisol, plasma catecholamines are elevated more in human subjects with high-altitude pulmonary and cerebral edema, which are symptoms of AMS (12, 109).

ATRIAL NATRIURETIC PEPTIDE

Acute hypoxic exposure has been shown to increase plasma atrial natriuretic peptide (ANP) in humans (53, 118; Fig. 54.3), awake lambs (8), and anesthetized rabbits (5) and pigs (7). However, many studies have been

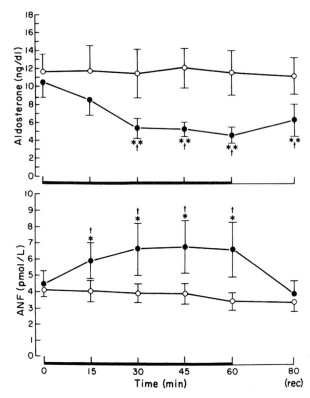

FIG. 54.3. Plasma aldosterone and atrial natriuretic factor (ANF) levels during acute hypoxia (closed circles; $SaO_2 \sim 68 \pm 1\%$) and normoxia (open circles). *$P < 0.05$, **$P < 0.01$ compared with control. †$P < 0.05$, hypoxemia vs. normoxemia. Graph indicates that aldosterone decreases while ANF increases, suggesting that ANF may be a cause of the decreased aldosteronogenesis shown in Figure 54.1. (From ref. 118 with permission.)

unable to detect a change in plasma ANP in humans (101, 187, 201, 216, 228), conscious dogs (43), or conscious rats (233). Furthermore, hypoxia has been shown to blunt (201) or augment (117) the ANP response to exercise. The reasons for these discrepancies cannot by accounted for by the type of subject, diet, or the degree of hypoxemia. Maintenance of isocapnia by artificial ventilation (5, 7) or increased inspired CO_2 (53) may allow the effect of hypoxia per se to be revealed. The increase in ANP during 2 h of acute hypoxia may correlate with diastolic blood pressure rather than arterial PO_2 (53).

Mechanism

Hypoxia appears to directly increase ANP release from isolated perfused hearts (6, 121); this direct effect is not due to direct damage to the myocytes. Interestingly, vagosympathectomy or adrenergic blockade seem to significantly attenuate the ANP response to acute hypoxia (7, 121), indicating a significant direct neural input to ANP release independent of changes in physical forces within the heart. There does not seem to be a significant, direct chemoreceptor afferent component of the system (106), though this aspect of the reflex "arc" requires further investigation.

Chronic Hypoxia

The ANP response to chronic hypoxia is difficult to interpret since chronic hypoxia alters pulmonary hemodynamics and end-diastolic pressures within the heart. Relatively short-term chronic hypoxia (< 2 wk exposure) does not appear to significantly increase plasma ANP in humans (32, 187) and rats (135, 185), though Winter et al. (233) found an increase within 24 h. It is clear, however, that the increase in ANP correlates with the development of hypoxic pulmonary hypertension in rats (93, 135, 185, 215, 233). Longer-term exposure does appear to increase plasma ANP in humans (12, 139, 225).

Several studies have found that exogenous administration of ANP lowers pulmonary artery pressure during hypoxic pulmonary hypertension (1, 94, 95, 123). Treatment of rats with monoclonal ANP antibody exacerbates hypoxic pulmonary hypertension (93). These data suggest that ANP may be a useful pulmonary vasodilator in subjects with hypoxic pulmonary hypertension.

However, Cosby et al. (48) found elevated plasma ANP in subjects with high-altitude pulmonary edema, which normalized during recovery. Furthermore, they suggested that elevated ANP may contribute to the development of pulmonary edema, a notion consistent with Bartsch et al.'s (12) study of subjects with AMS. Milledge et al. (139) found an inverse correlation between ANP levels at low altitude (preexposure) and the symptoms score for AMS. This may represent an index of cardiopulmonary fitness in general rather than a role for ANP in the development of, or protection from, AMS.

Stewart et al. (213) demonstrated elevated ANP in patients with hypoxemic COPD and even higher levels in COPD patients with cor pulmonale. Although patients with cor pulmonale had decreased excretion of a saline load, their ANP responses to saline loading were the same as those in nonedemetous patients. This study questioned the role of ANP in the control of natriuresis during chronic hypoxemia due to lung disease.

ARGININE VASOPRESSIN

Acute Hypoxia

In humans there are several studies that failed to find an increase in plasma arginine vasopressin (AVP) (4, 20, 53, 83) unless some of the symptoms of AMS, including nausea, were present (4, 83). Nausea per se is a stimulus to AVP secretion and is a major confounder in the interpretation of these studies. Large increases in plasma vasopressin have been reported in humans at very high altitude (>6,000 m; 161), suggesting a correlation with PO_2 and AVP as seen with other hormonal systems. One study actually reported a decrease in plasma AVP with mild acute hypoxia (41). This led to the notion that diuresis, at least partly allowed by a decrease or lack of an increase in AVP, is a normal response to hypoxia as long as altitude is tolerated (105). As soon as hypoxia begins to become intolerable (that is, nausea or extreme hypoxemia), AVP is increased, antidiuresis ensues, and edema worsens (169).

Acute severe hypoxia has been shown to increase plasma AVP in rats (59), though two other studies with moderate hypoxia failed to find an increase (73, 98). Anesthetized (3, 180, 181) or conscious (195, 229) dogs have a significant response to severe hypoxia as do conscious sheep (210).

Many of these studies have led to the idea that the AVP response to hypoxia is related to the severity of the stimulus. This appears to be so in rats, where the threshold is an arterial PO_2 of approximately 50 torr (59), and in sheep, where the threshold appears to be between 40 and 30 torr (210, 211). Increased ventilation, which accompanies hypoxemia, is a commonly identified confounding factor because it causes significant hypocapnia, alkalosis, and changes in intrathoracic cardiac dynamics, all of which may influence AVP release. To avoid some of these problems, isocapnic hypoxia has been administered in conscious humans by increased fractional concentration of carbon dioxide in inspired gas ($FICO_2$) with no effect on AVP (42, 53). In contrast, anesthetized, paralyzed, artificially ventilated dogs increased AVP in response to hypoxia, albeit only when PO_2 was very low (181, 230).

Large and sustained increases in AVP in CSF have been found during hypoxia in anesthetized dogs (230) and conscious sheep (210) in a manner which suggests that its release into CSF is controlled independently of release into plasma.

Mechanisms

In a classic study, Share and Levy (204) found that stimulation of the isolated, perfused carotid sinuses with hypoxemic blood resulted in an increase in plasma vasopressin but only when increases in ventilation were prevented. This illustrates the confounding influence of increased vagal afferent activity as well as hypocapnia from both increased tidal volume and minute ventilation.

Several studies have recorded directly from magnocellular (presumably AVP-secreting) cell bodies in the supraoptic nucleus (SON) and the paraventricular nucleus (PVN) during chemoreceptor stimulation and found activation of the SON (81, 234, 235). Interestingly, one study found activation of the arcuate nucleus without an effect on cell bodies within the PVN (10). These data suggest that AVP release during chemoreceptor stimulation may be derived primarily from the SON rather than the PVN. Although many studies have been done with peripheral chemoreceptor denervation and other hormonal systems, only Hanley et al. (78) have convincingly demonstrated that peripheral chemodenervation attenuates the AVP response to acute hypoxia. Another earlier study (3), using antidiuresis as an index of AVP release, found that hypophysectomy and baroreceptor denervation, but not chemoreceptor denervation, attenuated the response.

Acute almitrine bismesylate infusion, which is hypothesized to directly stimulate peripheral chemoreceptors, has been used in one study finding no change in AVP in humans (106), one study finding a decrease in humans (86), and one preliminary study finding an increase in rats (52). Other studies using pharmacological interventions have suggested that central dopamine (blocked with metoclopramide), opioids (blocked with naloxone), and peripheral prostaglandins (blocked with meclofenamate) are involved in the generation of the AVP response to acute hypoxia (59, 229), though their precise roles are as yet unknown.

Interactions

Hypercapnia appears to augment the vasopressin response to hypoxia (181, 195, 230) reminiscent of the effect of hypercapnia on the carotid body response to hypoxia. There also appears to be an interaction of blood pressure (baroreceptor input) with the AVP response to hypoxia; decreased arterial pressure increased the AVP response to hypoxia and vice versa (180, 234, 235). Finally, water restriction, which increases AVP by itself by increasing osmolality and decreasing plasma volume, augments the vasopressin response to acute hypoxemia in rats (73). No interaction was found between 6 h of mild exercise and hypoxia (137).

Chronic Hypoxia

When high-altitude pulmonary edema or AMS are not present, plasma AVP does not appear to increase significantly during exposure to chronic hypoxia in humans (12, 24, 76), sheep (223), or rats (73, 176, 218). Correlations of AVP with high-altitude pulmonary edema (48) and AMS (76) have been reported, whereas several studies could not find correlations with these factors (12, 80, 206).

Since high-altitude pulmonary and cerebral edema are potentially life-threatening, it seems reasonable to expect that a beneficial response would be to decrease plasma vasopressin, which would allow a diuresis as well as, perhaps, decreased pulmonary vascular resistance. In fact, one study did find a decrease in AVP and hypothesized such an enhanced diuresis (165). Another study found an increase in AVP and urine volume at 3,500 m and hypothesized that the increase in cortisol prevented the antidiuretic effect of AVP (33).

Another way to approach the problem is to analyze the concept of an osmotic threshold and/or sensitivity. One study in humans at >5,200 m for more than 3 wk found significant hyperosmolality but no change in vasopressin (24; Fig. 54.4). This suggested not only a lack of stimulation of vasopressin but a decrease in osmosensitivity. A subsequent study found a decreased AVP response to a salt load after 4 days at 3,000 m, confirming a decrease in osmosensitivity (187). Chronic hypoxemia (with or without hypercapnia) in patients with COPD does not appear to alter the vasopressin response to water-loading or hypertonic saline infusion (56). One study in rats has suggested that vasopressin actually has nothing to do with the control of fluid balance during chronic hypoxia (98). It may be that an increase in vasopressin is detrimental and a decrease is beneficial during chronic hypoxia; otherwise, fluid balance is controlled by other factors.

The relationship between fluid intake and vasopressin during hypoxia has been investigated (73, 176). Chronic hypoxia (1–24 h) did not increase vasopressin nor did hypoxia and fluid restriction interact to increase vasopressin. Furthermore, although chronic hypoxia augmented the vasopressin response to acute hemorrhage, this augmentation was lost when fluid was restricted. Since high altitudes often result in a voluntary decrease in fluid intake (99), which might increase vasopressin, the decrease in osmosensitivity appears to be a protective mechanism (187).

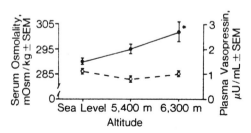

FIG. 54.4. Serum osmolality *(solid line)* and plasma arginine vasopressin concentrations *(dashed line)* at sea level, 5,400 m, and 6,300 m for 7–17 days (n = 13). Graph demonstrates no change in vasopressin despite a significant increase in osmolality. (From ref. 24 with permission.)

RENIN–ANGIOTENSIN–ALDOSTERONE SYSTEM

Since the observation that exposure to altitude or normobaric hypoxia can cause significant disturbances in fluid and electrolyte balance, the renin–angiotensin–aldosterone system has been extensively studied (102, 138, 166, 169, 207, 208). A major point of debate is the existence of a hypoxia-induced dissociation of renin and aldosterone.

Acute Hypoxia

Acute hypoxia without exercise appears to have little effect on plasma renin activity in humans (4, 44, 54, 118), rats (175), and dogs (43), though some experiments have reported increases (179, 224). One study found a hypoxia-induced decrease in renin from elevated levels induced by a decrease in sodium intake (44). Decreased oxygen content due to acute anemia also has little effect on basal and hemorrhage-stimulated renin release (167).

Early studies described a dissociation of renin and aldosterone during hypoxia (207, 208). Subsequent studies tried to assign this dissociation to a direct inhibitory effect of alveolar hypoxia on the activity of angiotensin-converting enzyme (162) such that renin could increase or not change but angiotensin II would decrease (140, 141). Studies using relatively nonspecific angiotensin II assays have failed to find a dissociation of renin and angiotensin II (12, 117, 125, 173, 174, 232, 236). This led to the questioning of the hypothesis that the dissociation of renin and aldosterone was due to an inhibition of converting enzyme and, therefore, angiotensin II generation. However, one study using a highly specific angiotensin II assay with minimal cross-reactivity with other angiotensin peptides did detect a dissociation of renin and angiotensin II (228).

Acute hypoxia interferes with basal and exercise-induced (renin-mediated) aldosterone release (29, 31, 44, 140, 141, 205, 216, 221, 222), though some studies have failed to find consistent suppression (4, 83). The aldosterone response to angiotensin II infusion is normal (45), whereas the aldosterone response to physiological levels of ACTH is inhibited by hypoxia (169, 186; Fig. 54.1). This is consistent with lower ambient angiotensin II levels. However, very severe altitude exposure (6,000 m) resulted in an increase in aldosterone, perhaps due to an increase in ACTH (161).

Chronic Hypoxia

Many studies have found an inhibition of plasma renin activity during long-term exposure to altitude (12, 32, 85), whereas other studies have found no change or an increase (39, 61, 71, 92). Several studies have also failed

to find a dissociation of renin and angiotensin II levels (12, 125, 236). However, a dissociation of renin and aldosterone has been documented in some studies (126, 187, 207, 208) but not in others (39, 61, 90, 131, 132).

Aldosterone levels have been found to decrease (12, 32, 85, 136, 177) or increase (39, 90) during chronic hypoxia. The inhibitory effect may be accentuated in older subjects (100). Furthermore, the aldosterone response to ACTH has been found to be suppressed (170, 186; Fig. 54.1). There are so many confounding factors in the interpretation of these results that the inconsistencies are difficult to resolve.

The renin–angiotensin system is activated in patients with hypoxemic lung disease and in subjects with high-altitude pulmonary edema (48, 173), though this may not account for the disorders of fluid balance observed in these patients (56). These results have been confirmed in patients with COPD (172). Several studies have failed to find a correlation between basal ANP and either renin or aldosterone levels in patients with hypoxic COPD (1, 47). However, ANP infusion inhibited aldosterone but not renin, suggesting it as a possible acute controller of aldosterone secretion in patients with hypoxic COPD (1). Converting enzyme inhibition has been suggested as a treatment for disorders of fluid and electrolyte balance in patients with hypoxic COPD (57) and may alleviate hypoxemic pulmonary hypertension (26), though one study found it to be of little benefit (164). The mineralocorticoid antagonist spironoloctone used prophylactically may prevent AMS (112), the development of which may be correlated with renin and aldosterone levels (139). These data suggest that suppression or antagonism of aldosterone may be beneficial during chronic hypoxic exposure.

Mechanisms

The increase in renin observed in some experiments appears to be mediated by afferent input from peripheral chemoreceptors (179) and efferent sympathetic nerve input. The decrease in renin observed by some is not mediated by beta-adrenergic receptors (32).

The factors influencing aldosterone release provoke even greater debate. As stated above, early studies suggested that the dissociation of renin and aldosterone may be due to inhibition of the conversion of angiotensin I to angiotensin II (140, 141). Several studies measuring angiotensin II failed to find this phenomenon in vivo (12, 118, 125, 173, 174, 232).

Probably the most meticulous study to date was published by Vonmoos et al. (228) in which the factors involved in the control of aldosterone release during acute hypoxia were examined. It was concluded that neither dopaminergic inhibition, ANP, nor hypokalemia were factors in the inhibition of aldosterone release.

Using a very specific assay for angiotensin II, inhibition was detected, thus reactivating the hypothesis that hypoxia significantly inhibits converting enzyme activity in a physiologically significant manner. The lack of specificity of previous assays which failed to find inhibition (12, 118, 125, 173, 174, 232) may account for the failure to distinguish a decrease in angiotensin II from changes in the cross-reacting fragments of angiotensin II.

Other investigators have tried to correlate the dissociation of renin and aldosterone with elevated ANP levels. Several studies using ANP infusion and correlation of endogenous hormone levels during acute hypoxia suggest that ANP may be involved (117, 118; Fig. 54.3). However, many studies have failed to find an increase in ANP (see ATRIAL NATRIURETIC PEPTIDE, above) or a correlation of ANP with the magnitude of the decrease in aldosterone (187). Therefore, ANP may be a factor but not necessarily a unifying mechanism.

Several studies have suggested that stimulation of the peripheral chemoreceptors may, by some unknown mechanism, result in an inhibition of aldosterone release (86, 87, 200), though probably not via dopamine input to the adrenal (228). This interesting hypothesis is still undeveloped.

We have suggested that decreased intraadrenal oxygen tension may directly inhibit basal, cAMP, angiotensin II, and potassium-stimulated aldosteronogenesis (37, 168, 172; Fig. 54.2). Neither cortisol release (168, 172) nor pregnenolone synthesis (36) was inhibited, suggesting an effect specific to the aldosteronogenic late pathway. Brickner et al. (36) have suggested that the conversion of corticosterone to aldosterone, the 18-hydroxylase enzymatic step, may be very sensitive to changes in oxygen tension within the physiological range. Although there is a striking similarity between the in vivo data with ACTH (Fig. 54.1) and the in vitro data with angiotensin II (Fig. 54.2), there are several inconsistencies [for example, with angiotensin infusion in vivo (45)], suggesting that the direct intraadrenal effect may not be the only mechanism governing aldosterone release in vivo.

A decrease in aldosterone may minimize potassium loss, thus helping to maintain plasma potassium (51, 207, 208). However, changes in sodium balance during hypoxia may not be attributable to the renin–angiotensin–aldosterone system (54). Limitation of an increase or an actual decrease in aldosterone may allow sufficient sodium excretion such that edema is minimized and negative potassium balance is maintained (169).

HYPOTHALAMIC–PITUITARY–THYROID AXIS

One might expect a relative hypothyroidism when exposed to hypoxia to reduce the need for oxygen. In fact, this has not been clearly demonstrated in humans.

An early study at 4,300 m for 2 wk found an increase in circulating thyroid stimulating hormone (TSH) activity assessed by thyroid uptake of ^{131}I without a change in basal metabolic rate (BMR) (148). Another early study with only a 3 day stay at 3,700 m found an increase in total and free thyroxine (T_4) without a change in TSH (108). Early studies also described an increase in total T_4 and triiodothyronine (T_3) with no change in thyroid-binding globulin (TBG) or basal TSH and a normal TSH response to thyrotropin-releasing hormone (TRH) during exposure to 3,700 m for 16 days (188). These data suggest an alteration in the set point of thyroid hormone feedback on the pituitary (perhaps due to hypothalamic TRH). Another study described an increase in T_3 and T_4 with a decrease in TSH at 3,700 m for 3 wk, suggesting a scenario consistent with primary hyperthyroxinemia (214). These early studies on thyroid function in humans suggest that circulating total and free thyroid hormone actually increase on exposure to moderate altitude.

One study characterized thyroid adaptation to moderate altitude (3,500 m) in sea-level natives exposed to altitude for 3 wk, in acclimatized sea-level natives, and in high-altitude natives (198). T_4 and T_3 increased within 4 h of exposure to hypoxia without a change in reverse T_3, TBG, or total T_4 binding. Furthermore, TSH was unchanged as was its response to exogenous TRH. This comprehensive study agreed with the older findings and suggested that T_4 and T_3 increase independently of TSH or TBG. Many mechanisms are possible, including a change in the volume of distribution and/or metabolism of thyroid hormone. Furthermore, no symptoms of hyperthyroidism ensued, suggesting no change in free thyroid hormone. This would also explain the lack of an effect on TSH.

One study has been performed in human subjects for a few days to several weeks at very high altitude (5,400–6,300 m) on Mount Everest (150). Total T_4 and T_3 and free T_4 index all increased, as did TSH. Interestingly, serum thyroglobulin did not change, indicating no change in actual thyroid secretion. It was suggested that TSH increased due to a change in feedback sensitivity.

Normal thyroid hormone, resin uptake, and plasma TSH have been documented in patients with chronic moderate hypoxemia due to chronic lung disease (11, 203). However, a delayed or blunted TSH response to TRH has been described in several patients with chronic hypoxemia due to lung disease, suggesting mild secondary hypothyroidism (203). It is likely that thyroid function is only altered in end-stage lung disease when hypoxemia is severe.

Rats

Studies in rats have found quite opposite results. Very early studies using indirect methods suggested that rats may actually become hypothyroid when exposed to simulated altitude or normobaric hypoxia (66, 96). Primary hypothyroidism was subsequently demonstrated in rats exposed to altitudes from 5,500 to 7,600 m (46, 65, 72, 133, 155, 219, 220). However, thyroidal iodide uptake has been found to increase (133), not to change (50, 155), or to decrease (65, 157, 227). Other studies using direct measurements or indirect anatomical evidence have suggested that the decrease in thyroid hormone might be secondary to a decrease in circulating TSH (that is, secondary hypothyroidism) (50, 70). In general, however, most studies have consistently suggested that hypoxia induces primary hypothyroidism.

Connors and Martin (46) summarized the possible mechanisms by which hypoxia might directly inhibit thyroid function. They suggested that decreased oxygen delivery to the thyroid may lead to a decrease in the iodide trap, a decrease in the secretory process, or a decrease in the oxidative steps required to iodinate thyroglobulin within the thyroid. All of these processes are energy-requiring and, therefore, may be oxygen-sensitive.

HORMONES OF REPRODUCTION

Acute hypoxia does not seem to alter gonadotropin or gonadal steroid levels in men with or without concomitant exercise (30). Several studies in men with longer exposures (4–14 days at 4,300 m) did not detect a change in plasma follicle-stimulating hormone (FSH), luteinizing hormone, testosterone, or urinary gonadotropins (209, 226). However, several other studies in humans did find changes in gonadotropins and gonadal steroids. Plasma testosterone has been found to increase [2,000 m (90)] or decrease [3,500 m (197)] with 7–18 days of exposure. Although FSH is not altered during exposure to altitude (197, 226), LH has been reported to decrease (90, 197). There is no obvious explanation for the discrepancy between these studies.

One study found that the urinary testosterone response to 4 days of human chorionic gonadotropin administration to men at 4,250 m was decreased (75). Furthermore, the LH and FSH responses to the gonadotropin-releasing hormone analog appear to be decreased in humans at 2,940 m (217). These data suggest that hypogonadotropic hypogonadism might be a feature of long-term hypoxic exposure. This phenomenon has been demonstrated in men who have decreased fertility and hypogonadotropic hypogonadism [Sherpas living at high altitude (9) and patients with chronic hypoxemia due to lung disease (202, 203)] exposed to chronic hypoxia. Another study in male patients with chronic moderate hypoxemia due to COPD failed to detect a decrease in testosterone (11).

It has been known since the 1940s that rats exposed to 7,620 m for only 4 h/day starting at 14 days of age showed delayed puberty and decreased spermatogenesis and seminiferous tubule dysfunction (males) and decreased gonad/body weight (males and females) (2). These studies have been confirmed by the demonstration of decreased ovarian and testicular size (67, 103, 159), Leydig cell number (67), and viability of sperm (55).

Decreased pituitary contents of FSH and LH with no change in plasma levels have been demonstrated in pregnant and non pregnant female rats exposed to 3,800 m (159). This is not consistent with the findings of increased gonadotroph number in the pituitary of rats exposed to 3,200 m for 2 months (103). In a dose (PIO_2)–response study in male mice, it was found that plasma LH was decreased at only 1,524 m, whereas FSH was unaltered even at very low barometric pressure (189). Another study from the same group found a decrease in FSH at 6,700 m (190).

Hypoxic ventilatory responsiveness is increased in pregnancy, though this was not found to correlate with estradiol or progesterone levels (149). It may be that generalized increases in ovarian steroids during pregnancy increase hypoxic ventilatory responsiveness by both a central and a peripheral (carotid body) action (79). It also appears that the decrease in testicular function may be a cause of decreased weight gain in male rats exposed to very high altitude (142).

Steroidogenesis is an oxygen-requiring synthetic pathway which utilizes cytochrome P450 enzymes to hydroxylate the steroid molecule. One study examined the effect of hypoxia on the conversion of pregnenolone to progesterone in placental cells in culture (64). Only very low O_2 levels (1%) consistently decreased the synthesis of progesterone. This is similar to the relative resistance of the early steroidogenic pathway and of cortisol synthesis to low oxygen and reinforces the selective oxygen sensitivity of the late aldosteronogenic pathway (36, 172; Fig. 54.2).

HORMONES OF THE ISLETS OF LANGERHANS

Baum and co-workers have evaluated the control of intermediate metabolism during hypoxia in dogs (reviewed in ref. 18). Acute hypoxia resulted in an increase in glucose, which did not lead to a sustained increase in insulin levels as would be predicted from glucose infusions in normoxemic dogs (15), suggesting that hypoxia inhibited insulin release possibly via an increase in circulating catecholamines. Alpha-receptor input inhibited insulin per se as well as arginine-mediated and beta-receptor-mediated insulin release during hypoxia (16, 17, 63). Dogs with surgically induced chronic hypoxemia have attenuated norepinephrine-induced lipolysis and glucose-induced insulin release

(13, 14). Interestingly, striking effects were found with severe acute hypoxia (PaO$_2$ < 30 torr) but minimal effects with moderate hypoxia (PaO$_2$ ~ 40 torr). These studies indicate that hypoxia induces hyperglycemia, which is sustained by an increase in catecholamines and glucagon (19) and a direct inhibition of insulin release.

The lack of an effect of moderate hypoxemia and the significant effect of severe hypoxia were confirmed. Bloom et al. (22) demonstrated the temporal sequence by showing that severe (not moderate) hypoxia in conscious calves resulted first in an increase in glucagon and catecholamines, followed by an increase in glucose, and then in an increase in insulin. The increase in insulin secondary to hyperglycemia was attenuated by autonomic blockade.

The possible direct effect of hypoxia on the pancreas was studied in vitro (156). It was found that hypoxia directly inhibited insulin and increased glucagon release. This study suggested that oxygen limitation of intracellular beta cell cAMP production was responsible for the effects found in vitro and could explain the results of Baum and co-workers.

Moderate acute hypoxia (3,000 m) had no effect on glucose, insulin, or glucagon in humans (30). Another study at higher altitude (4,550 m) found an increase in glucose, probably due to increased gluconeogenesis and glycogenolysis (221). Despite the increase in glucose, insulin decreased, in agreement with previous in vivo studies in dogs and in vitro.

Exposure to hypoxia of longer duration in rats (0.5 atmospheres for 24 h) had no effect on glucagon and insulin (143). Increased glucose was thought to be due to an increase in circulating corticosterone, which can both increase gluconeogenesis and decrease insulin sensitivity. Another report studied exposure to 3,500 m for 3 wk in high-altitude natives (199). Although basal glucose was unaltered, the increase in plasma glucose after an oral glucose load was greater at altitude. Although insulin increased, it was suggested that the sustained increase in glucose was due to an increase in counterregulatory hormones. Finally, there does not appear to be a correlation between glucagon and hypoxic survival time in rats (124).

SOMATOMAMMOTROPINS

Prolactin

Acute hypoxia in men has been reported to have no effect (83) or to increase plasma prolactin (42). In contrast, acute hypoxia decreases exercise-induced prolactin release (27). Chronic moderate or severe hypoxia has been reported to have no effect on basal prolactin or the prolactin response to exogenous TRH in men (9, 150).

One study in women has demonstrated a decrease in basal prolactin related to the degree of hypoxemia without an effect on the prolactin response to exercise (104).

One report has suggested a possible association of abnormalities in prolactin secretion in obese women with nocturnal hypoxemia (107). However, in men with hypoxemic chronic lung disease, prolactin responses to hypoglycemia and TRH were normal (202, 203). These studies suggest that, although prolactin appears to be normal in men, it may be inhibited in women during hypoxemia. A study in male mice failed to find an effect of severe hypoxemia (5,500–6,700 m) on prolactin levels (189).

Growth Hormone

As growth hormone is counterregulatory, one might expect its secretion to be increased during hypoxia like the other counterregulatory hormones (cortisol, catecholamines, glucagon). Two studies have suggested that hypoxia increases the growth hormone response to exercise (192, 221). The growth hormone response to hypoglycemia appears to be normal in humans with hypoxemic chronic lung disease (203). Whereas rats appear to have normal growth hormone secretion (158), there has been a suggestion of a decrease in mice (191).

OTHER CIRCULATING HORMONES AND FACTORS

Six days of hypoxia (3,450 m) did not induce a change in parathyroid hormone (PTH) or 1,25(OH)2D despite a decrease in plasma calcium, suggesting a possible decrease in parathyroid function (111). Another study failed to find an effect of hypoxia per se on PTH or on total or ionized calcium in steers (23). However, when exercise was added, PTH increased without a change in calcium, probably due to an increase in circulating catecholamines.

Gastrin inhibitory peptide is increased by acute hypoxemia in neonatal calves without altering gastrin levels (154); the mechanism of its stimulation and function has yet to be determined.

Endothelin-1 is a potent vasoconstrictor released from vascular endothelium. Rats clearly exhibit an endothelin response to acute hypoxia which correlates with the magnitude of the decrease in PO$_2$ (88). Bradykinin, another potent vasoactive substance, is not altered by acute hypoxia (4).

REFERENCES

1. Adnot, S., P. Andrivet, P. E. Chabrier, J. Piquet, P. Plas, P. Braquet, F. Roudot-Thoraval, and C. Brun-Buisson. Atrial natriuretic factor in chronic obstructive lung disease with pulmonary hypertension. *J. Clin. Invest.* 83: 986–993, 1989.

2. Altland, P. D. Effect of discontinuous exposure to 25,000 feet simulated altitude on growth and reproduction of the albino rat. *J. Exp. Zool.* 110: 1–17, 1949.

3. Anderson, R. J., R. G. Pluss, A. S. Berns, J. T. Jackson, P. E. Arnold, R. W. Schrier, and K. M. McDonald. Mechanism of effect of hypoxia on renal water excretion. *J. Clin. Invest.* 62: 769–777, 1978.

4. Ashack, R., M. O. Farber, M. H. Weinberger, G. L. Robertson, N. S. Fineberg, and F. Manfredi. Renal and hormonal responses to acute hypoxia in normal individuals. *J. Lab. Clin. Med.* 106: 12–16, 1985.

5. Baertschi, A. J., J. M. Adams, and M. P. Sullivan. Acute hypoxemia stimulates atrial natriuretic factor secretion in vivo. *Am. J. Physiol.* 255 (*Heart Circ. Physiol.* 26): H295–H300, 1988.

6. Baertschi, A. J., C. Hausmaninger, R. S. Walsh, R. M. Mentzer, Jr., D. A. Wyatt, and R. A. Pence. Hypoxia-induced release of atrial natriuretic factor (ANF) from the isolated rat and rabbit heart. *Biochem. Biophys. Res. Commun.* 140: 427–433, 1986.

7. Baertschi, A. J., J.-H. Jiao, D. E. Carlson, R. W. Campbell, W. G. Teague, D. Willson, and D. S. Gann. Neural control of ANF release in hypoxia and pulmonary hypertension. *Am. J. Physiol.* 259 (*Heart Circ. Physiol.* 30): H735–H744, 1990.

8. Baertschi, A. J., and G. Teague. Alveolar hypoxia is a powerful stimulus for ANF release in conscious lambs. *Am. J. Physiol.* 256 (*Heart Circ. Physiol.* 27): H990–H998, 1989.

9. Bangham, C. R. M., and P. H. Hackett. Effects of high altitude on endocrine function in the Sherpas of Nepal. *J. Endocrinol.* 79: 147–148, 1978.

10. Banks, D., and M. C. Harris. Activation of hypothalamic arcuate but not paraventricular neurons following carotid body stimulation in the rat. *Neuroscience* 24: 967–976, 1988.

11. Banks, W. A., and J. Cooper. Hypoxia and hypercarbia of chronic lung disease: minimal effects on anterior pituitary function. *South. Med. J.* 83: 290–293, 1990.

12. Bartsch, P., S. Shaw, M. Franciolli, M. P. Gnadinger, and P. Weidmann. Atrial natriuretic peptide in acute mountain sickness. *J. Appl. Physiol.: Respir. Environ. Exerc. Physiol.* 65: 1939–1937, 1988.

13. Baum, D., R. Griepp, and D. Porte. Glucose-induced insulin release during acute and chronic hypoxia. *Am. J. Physiol.* 237 (*Endocrinol. Metab. Gastrointest. Physiol.* 6): E45–E50, 1979.

14. Baum, D., and P. Oyer. Norepinephrine-stimulated lipolysis in acute and chronic hypoxemia. *Am. J. Physiol.* 241 (*Endocrinol. Metab.* 4): E28–E34, 1981.

15. Baum, D., and D. Porte. Effect of acute hypoxia on circulating insulin levels. *J. Clin. Endocrinol. Metab.* 29: 991–994, 1969.

16. Baum, D., and D. Porte. A mechanism for regulation of insulin release in hypoxia. *Am. J. Physiol.* 222: 695–699, 1972.

17. Baum, D., and D. Porte. Beta adrenergic receptor dysfunction in hypoxic inhibition of insulin release. *Endocrinology* 98: 359–366, 1976.

18. Baum, D., and D. Porte. Stress hyperglycemia and the adrenergic regulation of pancreatic hormones in hypoxia. *Metabolism* 29: 1176–1185, 1980.

19. Baum, D., D. Porte, and J. Ensinck. Hyperglucagonemia and alpha-adrenergic receptor in acute hypoxia. *Am. J. Physiol.* 237 (*Endocrinol. Metab. Gastrointest. Physiol.* 6): E404–E408, 1979.

20. Baylis, P. H., R. A. Stockley, and D. A. Heath. Effect of acute hypoxaemia on plasma arginine vasopressin in conscious man. *Clin. Sci. Mol. Med.* 53: 401–404, 1977.

21. Biesold, D., M. Kurosawa, A. Sato, and A. Trzebski. Hypoxia and hypercapnia increase the sympathoadrenal medullary functions in anesthetized, artificially ventilated rats. *Jpn. J. Physiol.* 39: 511–522, 1989.

22. Bloom, S. R., A. V. Edwards, and R. N. Hardy. Adrenal and pancreatic endocrine responses to hypoxia and hypercapnia in the calf. *J. Physiol.* 269: 131–143, 1977.

23. Blum, J. W., W. Bianca, F. Naf, P. Kunz, J. A. Fischer, and M. DaPrada. Plasma catecholamine and parathyroid hormone responses in cattle during treadmill exercise at simulated high altitude. *Horm. Metab. Res.* 11: 246–251, 1979.

24. Blume, F. D., S. J. Boyer, L. E. Braverman, A. Cohen, J. Dirkse, and J. P. Mordes. Impaired osmoregulation at high altitude. *JAMA* 252: 524–526, 1984.

25. Boddy, K., C. T. Jones, C. Mantell, J. G. Ratcliffe, and J. S. Robinson. Changes in plasma ACTH and corticosteroid of the maternal and fetal sheep during hypoxia. *Endocrinology* 94: 588–590, 1974.

26. Boschetti, E., C. Tantucci, M. Cocchieri, G. Fornari, V. Grassi, and C. A. Sorbini. Acute effects of captopril in hypoxic pulmonary hypertension. *Respiration* 48: 296–302, 1985.

27. Bouissou, P., G. R. Brisson, F. Peronnet, R. Helie, and M. Ledoux. Inhibition of exercise-induced blood prolactin response by acute hypoxia. *Can. J. Sports Sci.* 12: 49–50, 1987.

28. Bouissou, P., J. Fiet, C. Y. Guezennec, and P. C. Pesquies. Plasma adrenocorticotrophin and cortisol responses to acute hypoxia at rest and during exercise. *Eur. J. Appl. Physiol.* 57: 110–113, 1988.

29. Bouissou, P., C. Y. Geuzennec, F. X. Galen, G. Defer, J. Fiet, and P. C. Pesquies. Dissociated response of aldosterone from plasma renin activity during prolonged exercise under hypoxia. *Horm. Metab. Res.* 20: 517–521, 1988.

30. Bouissou, P., F. Peronnet, G. Brisson, R. Helie, and M. Ledoux. Metabolic and endocrine responses to graded exercise under acute hypoxia. *Eur. J. Appl. Physiol.* 55: 290–294, 1986.

31. Bouissou, P., F. Peronnet, G. Brisson, R. Helie, and M. Ledoux. Fluid–electrolyte shift and renin–aldosterone responses to exercise under hypoxia. *Horm. Metab. Res.* 19: 331–332, 1987.

32. Bouissou, P., J.-P. Richalet, F. X. Galen, M. Lartigue, P. Larmignat, F. Devaux, C. Dubray, and A. Keromes. Effect of β-adrenoreceptor blockade on renin–aldosterone and α-ANF during exercise at altitude. *J. Appl. Physiol.: Respir. Environ. Exerc. Physiol.* 67: 141–146, 1989.

33. Brahmachari, H. D., M. S. Malhotra, K. Ramachandran, and U. Radhakrishnan. Progressive changes in plasma cortisol, antidiuretic hormone, and urinary volume of normal lowlanders during short stay at high altitude. *Indian J. Exp. Biol.* 11: 454–455, 1973.

34. Breslow, M. J., T. D. Ball, C. F. Miller, H. Raff, and R. J. Traystman. Relationship between regional adrenal blood flow and secretory activity during hypoxia in anesthetized, ventilated dogs. *Am. J. Physiol.* 257 (*Heart Circ. Physiol.* 28): H1458–H1465, 1989.

35. Breslow, M. J., J. R. Robin, T. D. Mandress, L. C. Racusen, H. Raff, and R. J. Traystman. Changes in adrenal O_2 consumption during catecholamine secretion in anesthetized dogs. *Am. J. Physiol.* 259 (*Heart Circ. Physiol.* 30): H681–H688, 1990.

36. Brickner, R. C., B. Jankowski, and H. Raff. The conversion of corticosterone to aldosterone is the site of the oxygen sensitivity of the bovine adrenal zona glomerulosa. *Endocrinology* 130: 88–92, 1992.

37. Brickner, R. C., and H. Raff. Oxygen sensitivity of potassium- and angiotensin II–stimulated aldosterone release by bovine adrenal cells. *J. Endocrinol.* 129: 43–48, 1991.

38. Bubb, W. J., E. T. Howley, and R. H. Cox. Effects of various levels of hypoxia on plasma catecholamines at rest and during exercise. *Aviat. Space Environ. Med.* 54: 637–640, 1983.

39. Chakraborti, S., and S. K. Batabyal. Study of moderate and high altitude stress on plasma renin, aldosterone and electrolyte levels in humans. *Indian J. Exp. Biol.* 23: 706–707, 1985.

40. Claustre, J., R. Favre, J. M. Cottet-Emard, and L. Peyrin. Free, glucuronide, and sulfate catecholamines in the rat: effect of hypoxia. *J. Appl. Physiol.: Respir. Environ. Exerc. Physiol.* 59: 12–17, 1985.

41. Claybaugh, J. R., J. E. Hansen, and D. B. Wozniak. Response of antidiuretic hormone to acute exposure to mild and severe hypoxia in man. *J. Endocrinol.* 77: 157–160, 1978.

42. Claybaugh, J. R., C. E. Wade, A. K. Sato, S. A. Cucinell, J. C. Lane, and J. T. Maher. Antidiuretic hormone responses to eucapnic and hypocapnic hypoxia in humans. *J. Appl. Physiol.: Respir. Environ. Exerc. Physiol.* 53: 815–823, 1982.

43. Clozel, J.-P., C. Saunier, D. Hartemann, M. Allam, and W. Fischli. Effect of hypoxia and hypercapnia on atrial natriuretic factor and plasma renin activity in conscious dogs. *Clin. Sci. (Colch.)* 76: 249–254, 1989.

44. Colice, G. L., and G. Ramirez. Effect of hypoxemia on the renin–angiotensin–aldosterone system in humans. *J. Appl. Physiol.: Respir. Environ. Exerc. Physiol.* 58: 724–730, 1985.

45. Colice, G. L., and G. Ramirez. Aldosterone response to angiotensin II during hypoxemia. *J. Appl. Physiol.: Respir. Environ. Exerc. Physiol.* 61: 150–154, 1986.

46. Connors, J. M., and L. G. Martin. Altitude-induced changes in plasma thyroxine, 3, 5, 3'-triiodothyronine, and thyrotropin in rats. *J. Appl. Physiol.: Respir. Environ. Exerc. Physiol.* 53: 313–315, 1982.

47. Corlone, S., P. Palange, E. T. Mannix, M. P. Salatto, P. Serra, M. H. Weinberger, G. R. Aronogg, E. M. Cockerill, and M. O. Farber. Atrial natriuretic peptide, renin, and aldosterone in obstructive lung disease and heart failure. *Am. J. Med. Sci.* 298: 243–248, 1989.

48. Cosby, R. L., A. M. Sophocles, J. A. Durr, C. L. Peerinjacquet, B. Yee, and R. W. Schrier. Elevated plasma atrial natriuretic factor and vasopressin in high-altitude pulmonary edema. *Ann. Intern. Med.* 109: 796–799, 1988.

49. Cunningham, W. L., E. J. Becker, and F. Kreuzer. Catecholamines in plasma and urine at high altitude. *J. Appl. Physiol.* 20: 607–610, 1965.

50. Curbelo, H. M., E. C. Karliner, and A. B. Houssay. Effect of acute hypoxia on blood TSH levels. *Horm. Metab. Res.* 11: 155–157, 1979.

51. Curran-Everett, D. C., J. R. Claybaugh, K. Miki, S. K. Hong, and J. A. Krasney. Hormonal and electrolyte responses of conscious sheep to 96 h of hypoxia. *Am. J. Physiol.* 255 (*Regulatory Integrative Comp. Physiol.* 26): R274–R283, 1988.

52. Doepker, S. K., B. Jankowski, and H. Raff. Vasopressin response to almitrine in normoxic and hypoxic conscious rats. *FASEB J.* 5: A373, 1991.

53. du Souich, P., C. Saunier, D. Hartemann, A. Sautegeau, H. Ong, P. Larose, and R. Babini. Effect of moderate hypoxemia on atrial natriuretic factor and arginine vasopressine (sic) in normal man. *Biochem. Biophys. Res. Commun.* 148: 906–912, 1987.

54. Epstein, M., and T. Saruta. Effects of simulated high altitude on renin–aldosterone and Na homeostasis in normal man. *J. Appl. Physiol.* 33: 204–210, 1972.

55. Fahim, M. S., F. S. Messiha, and S. M. Girgis. Effect of acute and chronic simulated high altitude on male reproduction and testosterone level. *Arch. Androl.* 4: 217–219, 1980.

56. Farber, M. O., S. S. O. Kiblawi, R. A. Strawbridge, G. L. Robertson, M. H. Weinberger, and F. Manfredi. Studies on plasma vasopressin and the renin–angiotensin–aldosterone system in chronic obstructive lung disease. *J. Lab. Clin. Med.* 90: 373–380, 1977.

57. Farber, M. O., M. H. Weinberger, G. L. Robertson, and N. S. Fineberg. The effects of angiotensin-converting enzyme inhibition on sodium handling in patients with advanced chronic obstructive pulmonary disease. *Am. Rev. Respir. Dis.* 136: 862–866, 1987.

58. Favier, R. J., D. Desplanches, J. M. Pequignot, L. Peyrin, and R. Flandrois. Effects of hypoxia on catecholamine and cardiorespiratory responses in exercising dogs. *Respir. Physiol.* 61: 167–177, 1985.

59. Forsling, M. L., and L. A. Aziz. Release of vasopressin in response to hypoxia and the effect of aminergic and opioid antagonists. *J. Endocrinol.* 99: 77–86, 1983.

60. Francesconi, R., and A. Cymerman. Adrenocortical activity and urinary cyclic AMP levels: effects of hypobaric hypoxia. *Aviat. Space Environ. Med.* 46: 50–54, 1975.

61. Frayser, R., I. D. Rennie, G. W. Gray, and C. S. Houston. Hormonal and electrolyte response to exposure to 17,500 ft. *J. Appl. Physiol.* 38: 636–642, 1975.

62. Freedman, A., A. T. Scardella, N. H. Edelman, and T. V. Santiago. Hypoxia does not increase CSF or plasma β-endorphin activity. *J. Appl. Physiol.: Respir. Environ. Exerc. Physiol.* 64: 966–971, 1988.

63. French, J. W., D. Porte, and D. Baum. The effect of hypoxia and epinephrine on arginine induced insulin release. *Proc. Soc. Exp. Biol. Med.* 144: 288–290, 1973.

64. Gabbe, S. G., and C. A. Villee. The effect of hypoxia on progesterone synthesis by placental villi in organ culture. *Am. J. Obstet. Gynecol.* 111: 31–37, 1971.

65. Galton, V. A. Some effects of altitude on thyroid function. *Endocrinology* 91: 1393–1403, 1972.

66. Gordon, A. S., F. J. Tornetta, S. A. D'Angelo, and H. A. Charipper. Effects of low atmospheric pressure on the activity of the thyroid, reproductive system and anterior pituitary in the rat. *Endocrinology* 33: 366–383, 1943.

67. Gosney, J. R. Effects of hypobaric hypoxia on the Leydig cell population of the testis of the rat. *J. Endocrinol.* 103: 59–62, 1984.

68. Gosney, J. R. The effects of hypobaric hypoxia on the corticotrophic population of the adenohypophysis of the male rat. *J. Pathol.* 142: 163–168, 1984.

69. Gosney, J. R. Adrenal corticomedullary hyperplasia in hypobaric hypoxia. *J. Pathol.* 146: 59–64, 1985.

70. Gosney, J. R. Morphological changes in the pituitary and thyroid of the rat in hypobaric hypoxia. *J. Endocrinol.* 109: 119–124, 1986.

71. Gould, A. B., and S. A. Goodman. The effect of hypoxia on the renin–angiotensinogen system. *Lab. Invest.* 22: 443–447, 1970.

72. Gradwell, E. Histological changes in the thyroid gland in rats on acclimatization to simulated high altitude. *J. Pathol.* 125: 33–37, 1978.

73. Griffen, S. C., and H. Raff. Vasopressin responses to hypoxia in conscious rats: interaction with water restriction. *J. Endocrinol.* 125: 61–66, 1990.

74. Groza, P., C. Vladescu, and J. Boerescu. Effects of mild hypoxia (2200 m) on catecholamine and corticosterone secretion. *Physiologie* 12: 165–168, 1975.

75. Guerra-Garcia, R., A. Velasquez, and J. Coyotupa. A test of endocrine gonadal function in men: urinary testosterone after injection of HCG. II. A different response of the high altitude native. *J. Clin. Endocrinol.* 29: 179–182, 1969.

76. Hackett, P. H., M. L. Forsling, J. Milledge, and D. Rennie. Release of vasopressin in man at altitude. *Horm. Metab. Res.* 10: 571–572, 1978.

77. Hale, H. B., G. Sayers, K. L. Sydnor, M. L. Sweat, and D. D. Van Fossan. Blood adrenocorticotrophic hormone and plasma corticosteroids in men exposed to adverse environmental conditions. *J. Clin. Invest.* 36: 1642–1646, 1957.

78. Hanley, D. F., D. A. Wilson, M. A. Feldman, and R. J. Trayst-

man. Peripheral chemoreceptor control of neurohypophyseal blood flow. *Am. J. Physiol.* 254 (*Heart Circ. Physiol.* 25): H742–H750, 1988.

79. Hannhart, B., C. K. Pickett, and L. G. Moore. Effects of estrogen and progesterone on carotid body neural output responsiveness to hypoxia. *J. Appl. Physiol.: Respir. Environ. Exerc. Physiol.* 68: 1909–1916, 1990.

80. Harber, M. J., J. D. Williams, and J. J. Morton. Antidiuretic hormone excretion at high altitude. *Aviat. Space Environ. Med.* 52: 38–40, 1981.

81. Harris, M. C., A. V. Ferguson, and D. Banks. The afferent pathway for carotid body chemoreceptor input to the hypothalamic supraoptic nucleus in the rat. *Pflugers Arch.* 400: 80–81, 1984.

82. Harrison, T. S., and J. Seaton. The relative effects of hypoxia and hypercarbia on adrenal medullary secretion in anesthetized dogs. *J. Surg. Res.* 5: 560–564, 1965.

83. Heyes, M. P., M. O. Farber, F. Manfredi, D. Robertshaw, M. Weinberger, N. Fineberg, and G. Robertson. Acute effects of hypoxia on renal and endocrine function in normal humans. *Am. J. Physiol.* 243 (*Regulatory Integrative Comp. Physiol.* 14): R265–R270, 1982.

84. Hirai, K., G. Atkins, and S. F. Marotta. 17-Hydroxycorticosteroid secretion during hypoxia in anesthetized dogs. *Aerospace Med.* 34: 814–816, 1963.

85. Hogan, R. P., III, T. A. Kotchen, A. E. Boyd III, and L. H. Hartley. Effect of altitude on renin–aldosterone system and metabolism of water and electrolytes. *J. Appl. Physiol.* 35: 385–390, 1973.

86. Honig, A., R. Landgaf, C. Ledderhos, and W. Quies. Plasma vasopressin levels in healthy young men in response to stimulation of the peripheral arterial chemoreceptors by almitrine bismesylate. *Biomed. Biochim. Acta* 46: 1043–1049, 1987.

87. Honig, A., B. Wedler, M. Schmidt, S. Gruska, and A. Twal. Suppression of the plasma aldosterone to renin activity ratio in anaesthetized cats after pharmacological stimulation of the peripheral arterial chemoreceptors with almitrine bismesylate. *Biomed. Biochim. Acta* 46: 1055–1059, 1987.

88. Horio, T., M. Kohno, K. Yokokawa, K.-I. Murakawa, K. Yasunari, H. Fujiwara, N. Kurihara, and T. Takeda. Effect of hypoxia on plasma immunoreactive endothelin-1 concentration in anesthetized rats. *Metabolism* 40: 999–1001, 1991.

89. Hornbein, T. F. Adrenal cortical response to chronic hypoxia. *J. Appl. Physiol.* 17: 246–248, 1962.

90. Humpeler, E., F. Skrabal, and G. Bartsch. Influence of exposure to moderate altitude on the plasma concentration of cortisol, aldosterone, renin, testosterone, and gonadotropins. *Eur. J. Appl. Physiol.* 45: 167–176, 1980.

91. Jacobson, L., and M. F. Dallman. ACTH secretion and ventilation increase at similar arterial P_{O_2} in conscious rats. *J. Appl. Physiol.: Respir. Environ. Exerc. Physiol.* 66: 2245–2250, 1989.

92. Jain, S, W. L. Wilke, and A. Tucker. Age-dependent effects of chronic hypoxia on renin-angiotensin and urinary excretions. *J. Appl. Physiol.: Respir. Environ. Exerc. Physiol.* 69: 141–146, 1990.

93. Jin, H., R.-H. Yang, Y.-F. Chen, R. M. Jackson, H. Itoh, M. Mukoyama, K. Nakao, H. Imura, and S. Oparil. Atrial natriuretic peptide in acute hypoxia-induced pulmonary hypertension in rats. *J. Appl. Physiol.: Respir. Environ. Exerc. Physiol.* 71: 807–814, 1991.

94. Jin, H., R.-H. Yang, Y.-F. Chen, R. M. Jackson, and S. Oparil. Atrial natriuretic peptide attenuates the development of pulmonary hypertension in rats adapted to chronic hypoxia. *J. Clin. Invest.* 85: 115–120, 1990.

95. Jin, H., R.-H. Yang, R. M. Thornton, Y.-F. Chen, R. Jackson, and S. Oparil. Atrial natriuretic peptide lowers pulmonary artery pressure in hypoxia-adapted rats. *J. Appl. Physiol.: Respir. Environ. Exerc. Physiol.* 65: 1729–1735, 1988.

96. Johnson, C. L., and G. LaRoche. Simulated altitude and iodine metabolism in rats: II. Effects of chronic exposure on serum and thyroid iodinated components; effects of blood fractions and some organ weights. *Aerospace Med.* 39: 365–375, 1968.

97. Johnson, T. S., J. B. Young, and L. Landsberg. Sympathoadrenal responses to acute and chronic hypoxia in the rat. *J. Clin. Invest.* 71: 1263–1271, 1983.

98. Jones, R. M., F. T. LaRochelle, Jr., and S. M. Tenney. Role of arginine vasopressin on fluid and electrolyte balance in rats exposed to high altitude. *Am. J. Physiol.* 240 (*Regulatory Integrative Comp. Physiol.* 11) R182–R186, 1981.

99. Jones, R. M., C. Terhaard, J. Zullo, and S. M. Tenney. Mechanism of reduced water intake in rats at high altitude. *Am. J. Physiol.* 240 (*Regulatory Integrative Comp. Physiol.* 11): R187–R191, 1981.

100. Jung, R. C., D. B. Dill, R. Horton, and S. M. Horvath. Effects of age on plasma aldosterone levels and hemoconcentration at altitude. *J. Appl. Physiol.* 31: 593–597, 1971.

101. Kawashima, A., K. Kubo, K. Hirai, S. Yoshikawa, Y. Matsuzawa, and T. Kobayashi. Plasma levels of atrial natriuretic peptide under acute hypoxia in normal subjects. *Respir. Physiol.* 76: 79–92, 1989.

102. Keynes, R. J., G. W. Smith, J. D. H. Slater, M. M. Brown, S. E. Brown, N. N. Payne, T. P. Jowett, and C. C. Monge. Renin and aldosterone at high altitude in man. *J. Endocrinol.* 92: 131–140, 1982.

103. Khmel'nitskii, O. K., and T. Y. Tararek. Effect of exposure to high-altitude hypoxia on morphology of the pituitary-gonads system. *Byull. Eks. Biol. Med.* 111: 432–436, 1991.

104. Knudtzon, J., A. Bogsnes, and N. Norman. Changes in prolactin and growth hormone levels during hypoxia and exercise. *Horm. Metab. Res.* 21: 453–454, 1989.

105. Koller, E. A., A. Buhrer, L. Felder, M. Schopen, and M. B. Vallotton. Altitude diuresis: endocrine and renal responses to acute hypoxia of acclimatized and non-acclimatized subjects. *Eur. J. Appl. Physiol.* 62: 228–234, 1991.

106. Koller, E. A., M. Schopen, M. Keller, R. E. Lang, and M. B. Vallotton. Ventilatory, circulatory, endocrine, and renal effects of almitrine infusion in man: a contribution to high altitude physiology. *Eur. J. Appl. Physiol.* 58: 419–425, 1989.

107. Kopelman, P. G., M. C. P. Apps, T. Cope, and D. W. Empey. Nocturnal hypoxia and prolactin secretion in obese women. *Br. Med. J.* 287: 859–861, 1983.

108. Kotchen, T. A., E. H. Mougey, R. P. Hogan, A. E. Boyd III, L. L. Pennington, and J. W. Mason. Thyroid responses to simulated altitude. *J. Appl. Physiol.* 34: 165–168, 1973.

109. Koyama, S., T. Kobayashi, K. Kubo, M. Fukushima, K. Yoshimura, T. Shibamato, and S. Kusama. The increased sympathoadrenal activity in patients with high altitude pulmonary edema is centrally mediated. *Jpn. J. Med.* 27: 10–16, 1988.

110. Kraemer, W. J., A. J. Hamilton, S. E. Gordon, L. A. Trad, J. T. Reeves, D. W. Zahn, and A. Cymerman. Plasma changes in beta-endorphin to acute hypobaric hypoxia and high intensity exercise. *Aviat. Space Environ. Med.* 62: 754–758, 1991.

111. Krapf, R., P. Jaeger, and H. N. Hulter. Chronic respiratory alkalosis induces renal PTH-resistance, hyperphosphatemia and hypocalcemia in humans. *Kidney Int.* 42: 727–734, 1992.

112. Larsen, R. F., P. B. Rock, C. S. Fulco, B. Edelman, A. J. Young, and A. Cymerman. Effects of sprinolactone on acute mountain sickness. *Aviat. Space Environ. Med.* 57: 543–547, 1986.

113. Lau, C. Effects of O_2-CO_2 changes on hypothalamo–hypophyseal–adrenocortical activation. *Am. J. Physiol.* 221: 607–612, 1971.

114. Lau, C. Role of respiratory chemoreceptors in adrenocortical activation. *Am. J. Physiol.* 221: 602–606, 1971.

115. Lau, C., and S. F. Marotta. Role of peripheral chemoreceptors on adrenocortical secretory rates during hypoxia. *Aerospace Med.* 40: 1065–1068, 1969.

116. Lau, C., and P. S. Timiras. Adrenocortical function in hypothalamic deafferented rats maintained at high altitude. *Am. J. Physiol.* 222: 1040–1042, 1972.

117. Lawrence, D. L., and Y. Shenker. Effect of hypoxic exercise on atrial natriuretic factor and aldosterone regulation. *Am. J. Hypertens.* 4: 341–347, 1991.

118. Lawrence, D. L., J. B. Skatrud, and Y. Shenker. Effect of hypoxia on atrial natriuretic factor and aldosterone regulation in humans. *Am. J. Physiol.* 258 (*Endocrinol. Metab.* 21): E243–E248, 1990.

119. Lee, K., S. Miwa, K. Koshimura, H. Hasegawa, K. Hamahata, and M. Fujiwara. Effects of hypoxia on the catecholamine release, Ca^{2+} uptake, and cytosolic free Ca^{2+} concentration in cultured bovine adrenal chromaffin cells. *J. Neurochem.* 55: 1131–1137, 1990.

120. Leuenberger, U., K. Gleeson, K. Wroblewski, S. Prophet, R. Zelis, C. Zwillich, and L. Sinoway. Norepinephrine clearance is increased during acute hypoxemia in humans. *Am. J. Physiol.* 261(*Heart Circ. Physiol.* 32): H1659–H1664, 1991.

121. Lew, R. A., and A. J. Baertschi. Mechanisms of hypoxia-induced atrial natriuretic factor release from rat hearts. *Am. J. Physiol.* 257(*Heart Circ. Physiol.* 28): H147–H156, 1989.

122. Lewis, R. A., G. W. Thorn, G. F. Koepf, and S. S. Dorrance. The role of the adrenal cortex in acute anoxia. *J. Clin. Invest.* 21: 33–46, 1942.

123. Liu, L., H. Cheng, W. Chin, H. Jin, and S. Oparil. Atrial natriuretic peptide lowers pulmonary artery pressure in patients with high altitude disease. *Am. J. Med. Sci.* 298: 397–401, 1989.

124. Lundy, E. F., L. D. Klima, T. S. Huber, G. B. Zelenock, and L. G. D'Alecy. Elevated blood ketone and glucagon levels cannot account for 1,3-butanediol induced cerebral protection in the Levine rat. *Stroke* 18: 217–222, 1987.

125. Maher, J. T., L. G. Jones, L. H. Hartley, G. H. Williams, and L. I. Rose. Aldosterone dynamics during graded exercise at sea level and high altitude. *J. Appl. Physiol.* 39: 18–22, 1975.

126. Maresh, C. M., B. J. Noble, K. L. Robertson, and J. S. Harvey. Aldosterone, cortisol, and electrolyte responses to hypobaric hypoxia in moderate-altitude natives. *Aviat. Space Environ. Med.* 56: 1078–1084, 1985.

127. Marks, B. H., A. N. Bhattacharya, and J. Vernikos-Danellis. Effect of hypoxia on secretion of ACTH in the rat. *Am. J. Physiol.* 208: 1021–1025, 1965.

128. Marotta, S. F. Roles of aortic and carotid chemoreceptors in activating the hypothalamo–hypophyseal–adrenocortical system during hypoxia. *Proc. Soc. Exp. Biol. Med.* 141: 915–922, 1972.

129. Marotta, S. F., K. Hirai, and G. Atkins. Secretion of 17-hydroxycorticosteroids in conscious and anesthetized dogs exposed to simulated altitude. *Proc. Soc. Exp. Biol. Med.* 114: 403–405, 1963.

130. Marotta, S. F., L. J. Malasanos, and U. Boonayathap. Inhibition of the adrenocortical response to hypoxia by dexamethasone. *Aerospace Med.* 44: 1–4, 1973.

131. Martin, I. H., N. Basso, M. I. Sarchi, and A. C. Taquini. Changes in the renin–angiotensin–aldosterone system in rats of both sexes submitted to chronic hypobaric hypoxia. *Arch. Int. Physiol. Biochem.* 95: 255–262, 1987.

132. Martin, I. H., D. Baulan, N. Basso, and A. C. Taquini. The renin–angiotensin–aldosterone system in rats of both sexes subjected to chronic hypobaric hypoxia. *Arch. Int. Physiol. Biochem.* 90: 129–133, 1982.

133. Martin, L. G., G. E. Wertenberger, and R. W. Bullard. Thyroidal changes in the rat during acclimation to simulated altitude. *Am. J. Physiol.* 221: 1057–1063, 1971.

134. Mazzeo, R. S., P. R. Bender, G. A. Brooks, G. E. Butterfield, B. M. Groves, J. R. Sutton, E. E. Wolfel, and J. T. Reeves. Arterial catecholamine responses during exercise with acute and chronic high-altitude exposure. *Am. J. Physiol.* 261(*Endocrinol. Metab.* 24): E419–D424, 1991.

135. McKenzie, J. C., I. Tanaka, K. Inagami, K. S. Misoni, and R. M. Klein. Alterations in atrial and plasma atrial natriuretic factor (ANF) content during development of hypoxia-induced pulmonary hypertension in the rat. *Proc. Soc. Exp. Biol. Med.* 181: 459–463, 1986.

136. McLean, C. J., C. W. Booth, T. Tattersall, and J. D. Few. The effect of high altitude on saliva aldosterone and glucocorticoid concentrations. *Eur. J. Appl. Physiol.* 58: 341–347, 1989.

137. Meehan, R. T. Renin, aldosterone, and vasopressin responses to hypoxia during 6 hours of mild exercise. *Aviat. Space Environ. Med.* 57: 960–965, 1986.

138. Milledge, J. S. Renin–aldosterone system. In: *High Altitude and Man*, edited by J. B. West and S. Lahiri. Bethesda, MD: Am. Physiol. Soc., 1984, p. 47–57.

139. Milledge, J. S., J. M. Beeley, S. McArthur, and A. H. Morice. Atrial natriuretic peptide, altitude and acute mountain sickness. *Clin. Sci. (Colch.)* 77: 509–514, 1989.

140. Milledge, J. S., and D. M. Catley. Renin, aldosterone, and converting enzyme during exercise and acute hypoxia in humans. *J. Appl. Physiol.: Respir. Environ. Exerc. Physiol.* 52: 320–323, 1982.

141. Milledge, J. S., D. M. Catley, E. S. Williams, W. R. Withey, and B. D. Minty. Effect of prolonged exercise at altitude on the renin–aldosterone system. *J. Appl. Physiol.: Respir. Environ. Exerc. Physiol.* 55: 413–418, 1983.

142. Miranda, de, I. M., J. C. Macome, L. E. Costa, and A. C. Taquini. Adaptation to chronic hypobaric hypoxia and sexual hormones. *Acta Physiol. Pharmacol. Ther. Latinoam.* 27: 65–71, 1977.

143. Mlekusch, W., B. Paletta, W. Truppe, E. Paschke, and R. Grimus. Plasma concentrations of glucose, corticosterone, glucagon, and insulin and liver content of metabolic substrates and enzymes during starvation and additional hypoxia in the rat. *Horm. Metab. Res.* 13: 612–614, 1981.

144. Mohri, M., K. Seto, M. Nagase, K. Tsunashima, and M. Kawakami. Changes in pituitary–adrenal function under continuous exposure to hypoxia in male rats. *Exp. Clin. Endocrinol.* 81: 65–70, 1983.

145. Moncloa, F., L. Beteta, I. Velazco, and C. Gonez. ACTH stimulation and dexamethasone inhibition in newcomers to high altitude. *Proc. Soc. Exp. Biol. Med.* 122: 1029–1031, 1966.

146. Moncloa, F., A. Carcelen, and L. Beteta. Physical exercise, acid-base balance, and adrenal function in newcomers to high altitude. *J. Appl. Physiol.* 28: 151–155, 1970.

147. Moncloa, F., J. Donayre, L. A. Sobrevilla, and R. Guerra-Garcia. Endocrine studies at high altitude. II. Adrenal cortical function in sea level natives exposed to high altitudes (4300 meters) for two weeks. *J. Clin. Endocrinol.* 25: 1640–1642, 1965.

148. Moncloa, F., R. Guerra-Garcia, C. Subauste, L. A. Sobrevilla, and J. Donayre. Endocrine studies at high altitude. I. Thyroid function in sea level natives exposed for two weeks to an altitude of 4300 meters. *J. Clin. Endocrinol.* 26: 1237–1239, 1966.

149. Moore, L. G., R. E. McCullough, and J. V. Weil. Increased HVR in pregnancy: relationship to hormonal and metabolic changes. *J. Appl. Physiol.: Respir. Environ. Exerc. Physiol.* 62: 158–163, 1987.

150. Mordes, J. P., F. D. Blume, S. Boyer, M.-R. Zheng, and L. E.

Braverman. High-altitude pituitary–thyroid dysfunction on Mount Everest. *N. Engl. J. Med.* 308: 1135–1138, 1983.

151. Moss, I. R., and J. G. Inman. Proopiomelanocortin opioids in brain, CSF, and plasma of piglets during hypoxia. *J. Appl. Physiol.: Respir. Environ. Exerc. Physiol.* 66: 2280–2286, 1989.

152. Moss, I. R., and J. D. G. Inman. Effects of pentobarbital on proopiomelanocortin opioid products of neonatal piglets during normoxia and hypoxia. *J. Neuroendocrinol.* 3: 455–460, 1991.

153. Moss, I. R., J. G. Inman, J. C. Porter, and D. J. Faucher. Ontogeny of plasma, CSF and brainstem ACTH in piglets: effects of hypoxia and anesthesia. *Neuroendocrinology* 51: 586–591, 1990.

154. Mouats, A., P. Guilloteau, J. A. Chayvialle, R. Toullec, C. Bernard, J. F. Grongnet, and G. T. Dos Santos. Effet de l'hypoxie sur les concentrations plasmatiques de gastrine et de polypeptide inhibiteur gastrique (GIP) chex le veau nouveau-ne. *Reprod. Nutr. Dev.* 2 (suppl.): 219s–220s, 1990.

155. Mulvey, P. F., and J. M. R. Macaione. Thyroidal dysfunction during simulated altitude conditions. *Federation Proc.* 23: 1243–1246, 1969.

156. Naramiya, M., H. Yamada, I. Matsuba, Y. Ikeda, T. Tanese, and M. Abe. The effect of hypoxia on insulin and glucagon secretion in the perfused pancreas of the rat. *Endocrinology* 111: 1010–1014, 1982.

157. Nelson, B. D., and A. Anthony. Thyroxine biosynthesis and thyroidal uptake of I^{131} in rats at the onset of hypoxia exposure. *Proc. Soc. Exp. Biol. Med.* 121: 1256–1260, 1966.

158. Nelson, M. L., and J. M. Cons. Pituitary hormones and growth retardation in rats raised at simulated high altitude (3800 m). *Environ. Physiol. Biochem.* 5: 273–282, 1975.

159. Nelson, M. L., J. M. Cons, and G. E. Hodgdon. Effects of simulated high altitude (3800 m) on reproductive function in the pregnant rat. *Environ. Physiol. Biochem.* 5: 65–72, 1975.

160. Nishijima, M. K., M. J. Breslow, H. Raff, and R. J. Traystman. Regional adrenal blood flow during hypoxia. *Am. J. Physiol.* 256(*Heart Circ. Physiol.* 27): H94–H100, 1989.

161. Okazaki, S., Y. Tamura, T. Hatano, and N. Matsui. Hormonal disturbances of fluid–electrolyte metabolism under altitude exposure in man. *Aviat. Space Environ. Med.* 55: 200–205, 1984.

162. Oparil, S., A. J. Narkates, R. M. Jackson, and H. S. Ann. Altered angiotensin-converting enzyme in lungs and extrapulmonary tissues of hypoxia-adapted rats. *J. Appl. Physiol.: Respir. Environ. Exerc. Physiol.* 65: 218–227, 1988.

163. Ou, L. C., and S. M. Tenney. Adrenocortical function in rats chronically exposed to high altitude. *J. Appl. Physiol.: Respir. Environ. Exerc. Physiol.* 47: 1185–1187, 1979.

164. Pison, C. M., J. E. Wolf, P. A. Levy, F. Dubois, C. G. Brambilla, and B. Paramelle. Effects of captopril combined with oxygen therapy at rest and on exercise in patients with chronic bronchitis and pulmonary hypertension. *Respiration* 58: 9–14, 1991.

165. Porchet, M., H. Contat, B. Waeber, J. Nussberger, and H. R. Brunner. Response of plasma arginine vasopressin levels to rapid changes in altitude. *Clin. Physiol.* 4: 435–438, 1984.

166. Raff, H. The renin–angiotensin–aldosterone system during hypoxia. In: *Response and Adaptation to Hypoxia—Organ to Organelle*, edited by S. Lahiri, N. Cherniak, and R. S. Fitzgerald. New York: Oxford University Press, 1991, p. 211–222.

167. Raff, H. Renin response to hemorrhage in conscious rats: effect of acute reductions in hematocrit. *Am. J. Physiol.* 258(*Regulatory Integrative Comp. Physiol.* 29): R487–R491, 1990.

168. Raff, H., D. L. Ball, and T. L. Goodfriend. Low oxygen selectively inhibits aldosterone secretion from bovine adrenocortical cells in vitro. *Am. J. Physiol.* 256(*Endocrinol. Metab.* 19): E640–E644, 1989.

169. Raff, H., R. C. Brickner, and B. Jankowski. The renin–angio-

tensin–aldosterone system during hypoxia: is the adrenal an oxygen sensor? In: *Man and Mountain Medicine*, edited by J. R. Sutton and G. Coates. New York: Pergamon, 1992 *Adv. Biosci.* vol. 84, p. 42–49.

170. Raff, H., and K. J. Chadwick. Aldosterone responses to ACTH during hypoxia in conscious rats. *Clin. Exp. Pharmacol. Physiol.* 13: 827–830, 1986.

171. Raff, H., and K. D. Fagin. Measurement of hormones and blood gases during hypoxia in conscious, cannulated rats. *J. Appl. Physiol.: Respir. Environ. Exerc. Physiol.* 56: 1426–1430, 1984.

172. Raff, H., and S. Kohandarvish. The effect of oxygen on aldosterone release from bovine adrenocortical cells in vitro: PO_2 vs. steroidogenesis. *Endocrinology* 127: 682–687, 1990.

173. Raff, H., and S. A. Levy. Renin–angiotensin–aldosterone and ACTH–cortisol control during acute hypoxemia and exercise in patients with chronic obstructive lung disease. *Am. Rev. Respir. Dis.* 133: 396–399, 1986.

174. Raff, H., J. Maselli, and I. A. Reid. Correlation of plasma angiotensin II concentration and plasma renin activity during acute hypoxia in dogs. *Clin. Exp. Pharmacol. Physiol.* 12: 91–94, 1985.

175. Raff, H., and T. P. Roarty. Renin, ACTH, and aldosterone during acute hypercapnia and hypoxia in conscious rats. *Am. J. Physiol.* 254(*Regulatory Integrative Comp. Physiol.* 25): R431–R435, 1988.

176. Raff, H., M. H. Rossing, S. K. Doepker, and S. C. Griffen. Vasopressin response to hemorrhage in the rat: effect of hypoxia and water restriction. *Clin. Exp. Pharmacol. Physiol.* 18: 725–729, 1991.

177. Raff, H., R. B. Sandri, and T. P. Segerson. Renin, ACTH, and adrenocortical function during hypoxia and hemorrhage in conscious rats. *Am. J. Physiol.* 250(*Regulatory Integrative Comp. Physiol.* 21): R240–R244, 1986.

178. Raff, H., J. Shinsako, and M. F. Dallman. Surgery potentiates adrenocortical responses to hypoxia. *Proc. Soc. Exp. Biol. Med.* 172: 400–406, 1983.

179. Raff, H., J. Shinsako, and M. F. Dallman. Renin and ACTH responses to hypercapnia and hypoxia after chronic carotid denervation. *Am. J. Physiol.* 247(*Regulatory Integrative Comp. Physiol.* 18): R412–R417, 1984.

180. Raff, H., J. Shinsako, L. C. Keil, and M. F. Dallman. Vasopressin, ACTH, and blood pressure during hypoxia induced at different rates. *Am. J. Physiol.* 245(*Endocrinol. Metab.* 8): E489–E493, 1983.

181. Raff, H., J. Shinsako, L. C. Keil, and M. F. Dallman. Vasopressin, ACTH, and corticosteroids during hypercapnia and graded hypoxia in dogs. *Am. J. Physiol.* 244(*Endocrinol. Metab.* 7): E453–E458, 1983.

182. Raff, H., J. Shinsako, L. C. Keil, and M. F. Dallman. Feedback inhibition of ACTH and vasopressin responses to hypoxia by physiological increases in endogenous plasma corticosteroids in dogs. *Endocrinology* 114: 1245–1249, 1984.

183. Raff, H., S. P. Tzankoff, and R. S. Fitzgerald. ACTH and cortisol responses to hypoxia in dogs. *J. Appl. Physiol.: Respir. Environ. Exerc. Physiol.* 51: 1257–1260, 1981.

184. Raff, H., S. P. Tzankoff, and R. S. Fitzgerald. Chemoreceptor involvement in cortisol responses to hypoxia in ventilated dogs. *J. Appl. Physiol.: Respir. Environ. Exerc. Physiol.* 52: 1092–1096, 1982.

185. Raffestin, B., S. Adnot, J. J. Mercadier, P. Levame, P. Duc, P. Braquet, I. Viossat, and P. E. Chabrier. Synthesis and secretion of atrial natriuretic factor during chronic hypoxia: a study in the conscious instrumented rat. *Clin. Sci. (Colch.)* 78: 597–603, 1990.

186. Ramirez, G., P. A. Bittle, M. Hammon, C. W. Ayers, J. R. Dietz, and G. L. Colice. Regulation of aldosterone secretion during

hypoxemia at sea level and moderately high altitude. *J. Clin. Endocrinol. Metab.* 67: 1162–1165, 1988.

187. Ramirez, G., M. Hammon, S. J. Agosti, P. A. Bittle, J. R. Dietz, and G. L. Colice. Effects of hypoxemia at sea level and high altitude on sodium excretion and hormonal levels. *Aviat. Space Environ. Med.* 63: 891–898, 1992.

188. Rastogi, G. K., M. S. Malhotra, M. C. Srivastava, R. C. Sawhney, G. I. Dua, K. Sridharan, R. S. Hoon, and I. Singh. Study of the pituitary–thyroid functions at high altitude in man. *J. Clin. Endocrinol. Metab.* 44: 447–452, 1977.

189. Rattner, B. A., B. T. Macmillan, S. D. Michael, and P. D. Altland. Plasma gonadotrophins, prolactin, and corticosterone concentrations in male mice exposed to high altitude. *J. Reprod. Fertil.* 60: 431–436, 1980.

190. Rattner, B. A., S. D. Michael, and P. D. Altland. Plasma concentrations of hypophyseal hormones and corticosterone in male mice acutely exposed to simulated high altitude. *Proc. Soc. Exp. Biol. Med.* 163: 367–371, 1980.

191. Rattner, B. A., S. D. Michael, and H. J. Brinkley. Embryonic implantation, dietary intake, and plasma GH concentration in pregnant mice exposed to hypoxia. *Aviat. Space Environ. Med.* 49: 687–691, 1978.

192. Raynaud, J., L. Drouet, J. P. Martineaud, J. Bordachar, J. Coudert, and J. Durand. Time course of plasma growth hormone during exercise in humans at altitude. *J. Appl. Physiol.: Respir. Environ. Exerc. Physiol.* 50: 229–233, 1981.

193. Richalet, J.-P., V. Rutgers, P. Bouchet, J.-C. Rymer, A. Keromes, G. Duval-Arnould, and C. Rathat. Diurnal variations of acute mountain sickness, colour vision, and plasma cortisol and ACTH at high altitude. *Aviat. Space Environ. Med.* 60: 105–111, 1989.

194. Rochat, T., A. F. Junod, and R. C. Gaillard. Circulating endogenous opioids and ventilatory response to CO_2 and hypoxia. *Respir. Physiol.* 61: 85–93, 1985.

195. Rose, C. E., R. J. Anderson, and R. M. Carey. Antidiuresis and vasopressin release with hypoxemia and hypercapnia in conscious dogs. *Am. J. Physiol.* 247(*Regulatory Integrative Comp. Physiol.* 18): R127–R134, 1984.

196. Rose, C. E., L. B. Latham, V. L. Brashers, K. Y. Rose, M. P. Sandridge, R. M. Carey, J. S. Althaus, and E. D. Miller. Hypoxemia and hypercapnia in conscious dogs: opioid modulation of catecholamines. *Am. J. Physiol.* 254(*Heart Circ. Physiol.* 25): H72–H80, 1988.

197. Sawhney, R. C., P. C. Chhabra, A. S. Malhotra, T. Singh, S. S. Riar, and R. M. Rai. Hormone profiles at high altitude in man. *Andrologia* 17: 178–184, 1985.

198. Sawhney, R. C., and A. S. Malhotra. Thyroid function in sojourners and acclimatised low landers at high altitude in man. *Horm. Metab. Res.* 23: 81–84, 1991.

199. Sawhney, R. C., A. S. Malhotra, T. Singh, R. M. Rai, and K. C. Sinha. Insulin secretion at high altitude in man. *Int. J. Biometeorol.* 30: 231–238, 1986.

200. Schmidt, M., B. Wedler, C. Zingler, C. Ledderhos, and A. Honig. Kidney function during arterial chemoreceptor stimulation. II. Suppression of plasma aldosterone concentration due to hypoxic–hypercapnic perfusion of the carotid bodies in anaesthetized cats. *Biomed. Biochim. Acta* 44: 711–722, 1985.

201. Schmidt, W., G. Brabant, C. Kroger, S. Strauch, and A. Hilgendorf. Atrial natriuretic peptide during and after maximal and submaximal exercise under normoxic and hypoxic conditions. *Eur. J. Appl. Physiol.* 61: 398–407, 1990.

202. Semple, P. d'A., G. H. Beastall, T. M. Brown, K. W. Stirling, R. J. Mills, and W. S. Watson. Sex hormone suppression and sexual impotence in hypoxic pulmonary fibrosis. *Thorax* 39: 46–51, 1984.

203. Semple, P. d'A., G. H. Beastall, W. S. Watson, and R. Hume.

Hypothalamic–pituitary dysfunction in respiratory hypoxia. *Thorax* 36: 605–609, 1981.

204. Share, L., and M. N. Levy. Effect of carotid chemoreceptor stimulation on plasma antidiuretic hormone titer. *Am. J. Physiol.* 210: 157–161, 1966.

205. Shigeoka, J. W., G. L. Colice, and G. Ramirez. Effect of normoxemic and hypoxemic exercise on renin and aldosterone. *J. Appl. Physiol.: Respir. Environ. Exerc. Physiol.* 59: 142–148, 1985.

206. Singh, I., M. S. Malhotra, P. K. Khanna, R. B. Nanda, T. Purshottan, T. N. Upadhyay, U. Radhakrishnan, and H. D. Brahmachari. Changes in plasma cortisol, blood antidiuretic hormone and urinary catecholamine in high-altitude pulmonary oedema. *Int. J. Biometeorol.* 18: 211–221, 1974.

207. Slater, J. D. H., R. E. Tuffley, E. S. Williams, C. H. Beresford, P. H. Sonksen, R. H. T. Edwards, R. P. Ekins, and M. McLaughlin. Control of aldosterone secretion during acclimatization to hypoxia in man. *Clin. Sci. (Colch.)* 37: 327–341, 1969.

208. Slater, J. D. H., E. S. Williams, R. H. T. Edwards, R. P. Ekins, P. H. Sonksen, C. H. Beresford, and M. McLaughlin. Potassium retention during the respiratory alkalosis of mild hypoxia in man: its relationship to aldosterone secretion and other metabolic changes. *Clin. Sci. (Colch.)* 37: 311–326, 1969.

209. Sobrevilla, L. A., I. Romero, F. Moncloa, J. Donayre, and R. Guerra-Garcia. Endocrine studies at high altitude. III. Urinary gonadotropins in subjects native to and living at 14,000 feet and during acute exposure of men living at sea level to high altitudes. *Acta Endocrinol.* 56: 369–375, 1967.

210. Stark, R. I., S. S. Daniel, M. K. Husain, A. B. Zubrow, and L. S. James. Effects of hypoxia on vasopressin concentrations in cerebrospinal fluid and plasma of sheep. *Neuroendocrinology* 38: 453–460, 1984.

211. Stark, R. I., S. L. Wardlaw, S. S. Daniel, M. K. Jusain, U. M. Sanocka, L. S. James, and R. L. Vande Wiele. Vasopressin secretion induced by hypoxia in sheep: developmental changes and relationship to β-endorphin release. *Am. J. Obstet. Gynecol.* 143: 204–215, 1982.

212. Steinbrook, R. A., S. E. Weinberger, D. B. Carr, E. Von Gal, J. Fisher, D. E. Leith, V. Fencl, and M. Rosenblatt. Endogenous opioids and ventilatory responses to hypoxia in normal humans. *Am. Rev. Respir. Dis.* 131: 588–591, 1985.

213. Stewart, A. G., P. A. Bardsley, S. V. Baudouin, J. C. Waterhouse, J. S. Thompson, A. H. Morice, and P. Howard. Changes in atrial natriuretic peptide concentrations during intravenous saline infusion in hypoxic cor pulmonale. *Thorax* 46: 829–834, 1991.

214. Stock, M. J., C. Chapman, J. L. Stirling, and I. T. Campbell. Effects of exercise, altitude, and food on blood hormone and metabolite levels. *J. Appl. Physiol.: Respir. Environ. Exerc. Physiol.* 45: 350–354, 1978.

215. Stockmann, P. T., D. H. Will, S. D. Sides, S. R. Brunnert, G. D. Wilner, K. M. Leahy, R. C. Wiegand, and P. Needleman. Reversible induction of right ventricular atriopeptin synthesis in hypertrophy due to hypoxia. *Circ. Res.* 63: 207–213, 1988.

216. Story, D. A., B. R. Miller, C. M. Shield, and G. Bowes. Atrial natriuretic factor during hypoxia and mild exercise. *Aviat. Space Environ. Med.* 62: 287–290, 1991.

217. Suarez, M. P., J. R. Varea Teran, G. Garces, C. Avila, D. H. Coy, and A. V. Schally. Pituitary response to luteinizing hormone–releasing hormone analog at sea level and high altitudes. *Obstet. Gynecol.* 59: 52–57, 1982.

218. Subramanian, R., B. Bhatia, and H. H. Siddiqui. Urine output and blood ADH in rats under different grades of hypoxia. In: *Selected Topics in Environmental Biology*, edited by B. Bhatia, G. S. Chinna, and B. Singh. New Delhi: Interprint, 1975, chapt. 50, p. 325–332.

219. Surks, M. I. Effect of hypoxia and high altitude on thyroidal iodine metabolism in the rat. *Endocrinology* 78: 307–315, 1966.

220. Surks, M. I. Effect of thyrotropin on thyroidal iodine metabolism during hypoxia. *Am. J. Physiol.* 216: 436–439, 1969.

221. Sutton, J. R. Effect of acute hypoxia on the hormonal response to exercise. *J. Appl. Physiol.: Respir. Environ. Exerc. Physiol.* 42: 587–592, 1977.

222. Sutton, J. R., G. W. Viol, G. W. Gray, M. McFadden, and P. Keane. Renin, aldosterone, electrolyte, and cortisol responses to hypoxic decompression. *J. Appl. Physiol.: Respir. Environ. Exerc. Physiol.* 43: 421–424, 1977.

223. Towell, M. E., J. Figueroa, S. Markowitz, B. Elias, and P. Nathanielsz. The effect of mild hypoxemia maintained for twenty-four hours on maternal and fetal glucose, lactate, cortisol, and arginine vasopressin in pregnant sheep at 122 to 139 days gestation. *Am. J. Obstet. Gynecol.* 157: 1550–1557, 1987.

224. Tuffley, R. E., D. Rubenstein, J. D. H. Slater, and E. S. Williams. Serum renin activity during exposure to hypoxia. *J. Endocrinol.* 48: 497–510, 1970.

225. Tunny, T. J., J. van Gelder, R. D. Gordon, S. A. Klemm, S. M. Hamlet, W. L. Finn, G. M. Carney, and C. Brand-Maher. Effects of altitude on atrial natriuretic peptide: the Bicentennial Mount Everest Expedition. *Clin. Exp. Pharmacol. Physiol.* 16: 287–291, 1989.

226. Vander, A. J., L. G. Moore, G. Brewer, K. M. J. Menon, and B. G. England. Effects of high altitude on plasma concentrations of testosterone and pituitary gonadotropins in man. *Aviat. Space Environ. Med.* 49: 356–357, 1978.

227. Verzar, F., E. Sailer, and V. Vidovic. Changes in thyroid activity at low atmospheric pressures and at high altitudes, as tested with ^{131}I. *J. Endocrinol.* 8: 308–320, 1952.

228. Vonmoos, S., J. Nussberger, B. Waeber, J. Biollax, H. R. Brunner, and P. Leuenberger. Effect of metoclopramide on angiotensins, aldosterone, and atrial peptide during hypoxia. *J. Appl. Physiol.: Respir. Environ. Exerc. Physiol.* 69: 2072–2077, 1990.

229. Walker, B. R. Inhibition of hypoxia-induced ADH release by meclofenamate in the conscious dog. *J. Appl. Physiol.: Respir. Environ. Exerc. Physiol.* 54: 1624–1629, 1983.

230. Wang, B. C., W. D. Sundet, and K. L. Goetz. Vasopressin in plasma and cerebrospinal fluid of dogs during hypoxia or acidosis. *Am. J. Physiol.* 247(*Endocrinol. Metab.* 10): E449–E455, 1984.

231. Wardlaw, S. L., R. I. Stark, S. Daniel, and A. G. Frantz. Effects of hypoxia on β-endorphin and β-lipotropin release in fetal, newborn, and maternal sheep. *Endocrinology* 108: 1710–1715, 1981.

232. Weismann, D. N., J. E. Herrig, O. J. McWeeny, N. A. Ayres, and J. E. Robillard. Renal and adrenal responses to hypoxemia during angiotensin-converting enzyme inhibition in lambs. *Circ. Res.* 52: 179–187, 1983.

233. Winter, R. J. D., L. Meleagros, S. Pervex., H. Jamal, T. Krausz, J. M. Polak, and S. R. Bloom. Atrial natriuretic peptide levels in plasma and in cardiac tissues after chronic hypoxia in rats. *Clin. Sci. (Colch.)* 76: 95–101, 1989.

234. Yamashita, H. Effect of baro- and chemoreceptor activation on supraoptic nuclei neurons in the hypothalamus. *Brain Res.* 126: 551–556, 1977.

235. Yamashita, H., and K. Koizumi. Influence of carotid and aortic baroreceptors on neurosecretory neurons in the supraoptic nuclei. *Brain Res.* 170: 259–277, 1979.

236. Zakheim, R. M., A. Molteni, L. Mattioli, and M. Park. Plasma angiotensin II levels in hypoxic and hypovolemic stress in unanesthetized rabbits. *J. Appl. Physiol.* 41: 462–465, 1976.

55. Body fluid and energy metabolism at high altitude

REED W. HOYT | Environmental Physiology and Medicine Directorate
U. S. Army Research Institute of Environmental Medicine, Natick, Massachusetts

ARNOLD HONIG | Institute of Physiology, Ernst-Moritz-Arndt-University of Greifswald, Greifswald, Germany

BODY FLUID METABOLISM AT HIGH ALTITUDE

Acute Hypoxia

Body Fluid Volume and Sodium and Water Metabolism.
Nonadapted young and healthy mammals exposed to moderate high-altitude hypoxia (3,000–5,000 m) under resting conditions have an increase in blood protein and hemoglobin concentration within the first 1–3 days of exposure (28, 55, 56, 74, 90). In most species this results mainly from a reduced plasma volume (28, 46, 55, 56, 68, 74, 90, 100, 116), though in some animals, such as dogs, the spleen might contribute to this initial hemoconcentration by delivering blood with a high erythrocyte count to the circulation. This rapid increase of hemoglobin concentration in response to acute high-altitude hypoxia probably increases arterial oxygen content and improves delivery of oxygen to tissues (56). Simultaneously, reflex stimulation of the peripheral arterial chemoreceptors increases the tone of venous capacitance vessels (110) and maintains adequate mean circulatory and cardiac filling pressures despite the decreased plasma volume (Fig. 55.1).

As indicated by a rapid decrease in body weight,

young and resting mammals with moderate acute arterial hypoxemia actually lose total body fluid (46, 68, 71, 83). Decreases in extracellular fluid and plasma volumes (9, 10, 19, 36, 72, 73) and intracellular water also have been observed (63, 67, 68, 83, 101, 117).

The natriuresis and diuresis underlying this loss of body fluid occur regardless of the method (ascent into mountains, low-pressure chambers, gas mixtures with low oxygen content, etc.) by which the oxygen deficit is induced. The only essential prerequisite is a decrease of the partial pressure of oxygen in arterial blood. Hypobaria without hypoxia induces neither natriuresis nor a negative sodium balance (30, 31). Thus hypoxic hypoxia resulting in hypoxemia will be referred to simply as *hypoxia* in the following discussion.

Reductions of extracellular fluid volume and total body water require a negative sodium and water balance either by reduced intake or by enhanced excretion. It is well known that in both humans and animals spontaneous water intake is suppressed in acute hypoxia despite the reductions of extracellular fluid and plasma volumes (9, 10, 19, 36, 72, 73); that is, mammals develop a certain state of voluntary dehydration. Although osmolality and sodium concentration in extracellular fluid are precisely controlled, their changes with hypoxia are variable (19, 36, 46, 71, 130). With respect to voluntary (food-independent) sodium intake in spontaneously hypertensive (SHR) and normotensive rats, moderate hypoxia (4,000 m) was accompanied by a sustained suppression of hypertonic saline consumption. This reaction was more pronounced and sustained in SHR than in normotensive rats (10).

In addition, moderate high-altitude hypoxia reportedly reduces sodium and chloride reabsorption in the intestine (126). This could be an additional mechanism contributing to the negative sodium and water balance frequently observed in the first days of hypoxia (19, 47, 71).

Mountaineers have long maintained that a copious urine flow in acute high-altitude hypoxia *(Höhendi-*

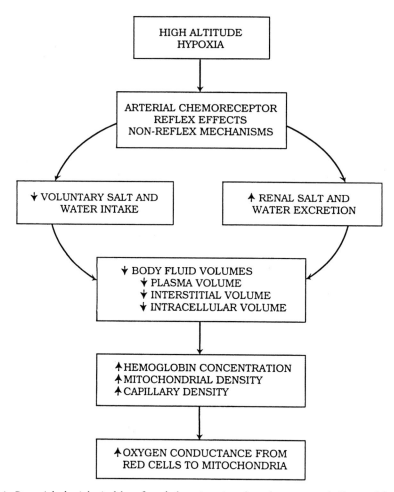

FIG. 55.1. Potential physiological benefits of alterations in salt and water metabolism and body fluid volumes in response to high-altitude hypoxia.

urese) indicates good adaptation to altitude, whereas those who react with antidiuresis are likely to develop acute mountain sickness. Indeed, in acute hypoxia of moderate degree (3,000–5,000 m), nonadapted mammals usually experience increased renal sodium and water excretion (7, 19, 23, 34, 47, 55, 77, 80, 99, 120, 125). In response to acute whole-body hypoxia, innervated kidneys exhibit a moderate increase in renal vascular resistance, sometimes accompanied by an increased filtration fraction, whereas renal blood flow and glomerular filtration rate remain unchanged or decrease slightly (34, 40, 55, 56, 77, 98).

A contrasting pattern is evident in severe acute hypoxia, corresponding to altitudes above 5,000 m or to arterial oxygen tensions below 40 mm Hg, and in poorly tolerated hypoxia. Under these conditions renal blood flow and glomerular filtration rate, as well as urinary sodium and water excretion, usually decrease in nonadapted humans and animals (55, 56, 77, 120).

Control of Body Fluid Volume and Sodium and Water Metabolism. Acute hypoxia has been reported to inhibit the voluntary intake of both sodium and water (9, 10,

19, 36, 72, 73). Hypoxia-induced suppression of voluntary salt intake in SHR was strikingly attenuated by carotid body nerve section (9), indicating that reflexes from the peripheral arterial chemoreceptors have a contributing role. Denervation of the carotid bodies, however, failed to influence inhibition of spontaneous water intake in acute hypoxia (9). Thus one must look for factors other than arterial chemoreceptors to explain reduced drinking. This problem is discussed in detail elsewhere (9, 10, 36, 55, 72, 73).

Arterial chemoreceptor stimulation increases efferent renal nerve activity. This is related at least in part to the control of renal hemodynamics in hypoxia, where the kidneys are in a potentially conflicting situation. On the one hand, vasoconstriction, a response that per se would decrease sodium and water excretion, is necessary to support redistribution of cardiac output toward the brain and cerebral circulation. On the other hand, increases in sodium and water excretion are required to help reduce plasma volume and establish a higher hemoglobin concentration. Apparently, mammalian kidneys cope with this problem by constricting the vasa efferentia, that is, by increasing not only the overall flow resistance of the renal vascular bed but also the filtration fraction, as observed repeatedly during whole-body hypoxia (8, 92, 96, 98, 124). Such increases of filtration fraction were also elicited, but only in innervated kidneys, by stimulating the arterial chemoreceptors in normoxic, narcotized, spontaneously breathing, nonvagotomized cats with unilaterally sectioned renal nerves (58, 60). However, it remains to be confirmed whether the peripheral arterial chemoreceptors, by way of the efferent renal nerves, actually independently control the diameter of the preglomerular and postglomerular arterioles.

A review of the literature showed that the natriuresis and diuresis in moderate acute hypoxia do not result from kidney or brain hypoxia or from acute respiratory and cardiovascular reactions to hypoxia (55, 56). Thus the alternative assumption that the peripheral arterial chemoreceptors might reflexly influence kidney sodium and water excretion in acute moderate hypoxia was tested experimentally. It has been shown that hypoxic perfusion of the vascularly isolated carotid bodies in normoxic anesthetized animals, as well as pharmacological stimulation of the chemoreceptors in normoxic humans and animals, increased absolute and/or fractional renal sodium excretion (5, 6, 55, 58, 60, 76, 81, 85, 113, 131). This response was not abolished by curare administration, relaxation, or constant artificial ventilation of animals, and it is not a pressure natriuresis (5, 6, 55, 58, 76, 113, 131). In addition, it does not depend on an increase in central venous or right atrial pressures and is not prevented by bilateral cervical section of the vagal nerves (that is, exclusion of the vagal low-pressure cardiovascular stretch or volume receptors) (6, 56, 60, 76, 81, 113). In normoxic, anesthetized, vagotomized cats, dogs, and rats, denervation of the carotid bifurcations or inactivation of the carotid body chemoreceptors prevented natriuretic and diuretic responses to chemoreceptor stimulation (6, 56, 76, 113). Altogether, these data show that stimulation of the arterial chemoreceptors in normoxic animals increases renal sodium excretion by specific reflex mechanisms.

Only a few studies have demonstrated the role of the arterial chemoreceptors in the control of renal excretory function in humans. Swenson and coworkers (123) have demonstrated that the isocapnic hypoxic ventilatory drive (HVR) correlated positively to hypoxic natriuresis and diuresis but not to bicarbonate excretion. This not only demonstrates a role for the peripheral arterial chemoreceptors in the control of renal salt and water excretion in humans but also shows that the respiratory alkalosis is very probably not the cause of high-altitude natriuresis and diuresis.

Natriuresis due to arterial chemoreceptor stimulation results from inhibited reabsorption of renal tubular sodium, which may occur despite renal vasoconstriction that often leads to decreases in glomerular filtration rate and renal blood flow (56). Section of the renal nerves prevented vasoconstriction but facilitated the onset of saluresis in response to arterial chemoreceptor stimulation, showing that this type of natriuresis is efferently mediated by hormonal mechanisms (5, 56, 58, 60, 76).

Based on clearance data from experiments in which the arterial chemoreceptors were stimulated by almitrine, it has been thought that arterial chemoreceptor stimulation inhibits sodium reabsorption in the renal proximal tubules (56, 85, 113). However, in recent experiments in young normotensive men, this finding was confirmed when almitrine was used to excite the arterial chemoreceptors, but when whole-body hypoxic hypoxia was used to simulate the arterial chemoreceptors an inhibition of sodium reabsorption in the distal tubules was found (C. Ledderhos, personal communication). Thus, additional studies are needed to clarify the responses of renal tubular sodium reabsorption in acute high-altitude hypoxia and/or during arterial chemoreceptor stimulation.

The question arises as to which hormones might be involved in controlling renal excretory function in hypoxia and during stimulation of the peripheral arterial chemoreceptors in normoxic mammals. Data on plasma renin activity (PRA) and angiotensin II concentration in hypoxia are inconsistent. Depending on the state of sodium balance, as well as on physical activity and age, increases or no changes have been observed (7, 20, 27, 70, 74, 84, 91, 105, 108, 125, 140, 141). In young and healthy humans and animals briefly exposed to moderate hypoxia the secretion, plasma level, and urinary excretion of aldosterone frequently decreased, as did the ratio of plasma aldosterone to renin activity (20, 27, 91, 108). Hypobaria without hypoxia did not induce such changes (30, 31).

Shifting of the renin–aldosterone axis in high-altitude hypoxia is not yet fully explained. The existence of a specific reflex influence of the peripheral arterial chemoreceptors on kidney renin release and the renin–angiotensin–aldosterone system (RAAS) is uncertain (56, 59, 106, 131). With respect to angiotensin II and aldosterone, regardless of the reflex effects of the arterial chemoreceptors, pulmonary angiotensin–converting enzyme activity can be suppressed in arterial hypoxia, and low oxygen pressure selectively inhibits aldosterone secretion from adrenocortical cells in vitro (17, 70, 104).

In contrast, whole-body hypoxia reportedly increased plasma adrenocorticotropic hormone (ACTH) and glucocorticoid levels, and since these responses are abolished or attenuated by arterial chemoreceptor deafferentation, it may be assumed that they are at least partly the reflex results of the stimulation of these sensors (59, 97, 105, 106). Glucocorticoid hormones are known to induce natriuresis at certain doses, and they may influence the control of renal excretory function in hypoxia.

There are reports of no changes or decreases in plasma and urinary antidiuretic hormone (ADH) levels during moderate acute hypoxia (20, 27, 80). Plasma ADH concentrations were increased consistently only in severe or poorly tolerated hypoxia (55, 56). It is still an open question whether the arterial chemoreceptors are involved in the specific reflex control of plasma ADH in acute hypoxia. Chemoreceptor stimulation of the carotid body resulted in an increase in plasma vasopressin, but only in vagotomized, open-chest, artificially ventilated animals (115). The level did not change in healthy, normoxic, spontaneously breathing humans on pharmacologic chemoreceptor stimulation with almitrine (81, 131).

When interpreting arterial chemoreceptor reflex effects on hormone systems, one should take into account the fact that renin and vasopressin release are inhibited by activation of both high- and low-pressure cardiovascular stretch receptors. Arterial chemoreceptor stimulation in spontaneously breathing mammals is always accompanied by enforced oscillations of intrathoracic transmural pressure. This could enhance activity of cardiovascular stretch receptors and secondarily inhibit renin and vasopressin release.

A number of reports suggest that acute whole-body hypoxia increases the plasma concentration of atrial natriuretic factor(s) (ANF) or atrial natriuretic peptide(s) (ANP) (4, 7, 23, 80, 84, 86, 89, 133). It is unresolved whether a specific reflex influence of the arterial chemoreceptors on ANF release exists. Stimulation of the chemoreceptors by almitrine in healthy, conscious, normoxic men elicited only a small and insignificant rise in plasma ANF (81). It is also possible that the elevated concentrations result secondarily from the respiratory and cardiovascular reactions occurring in arterial hypoxia (3, 4). In addition, low oxygen tension in cardiac tissue may directly facilitate the synthesis and release of ANF (86, 89).

In acute hypoxia, ANFs may be involved not only in initiating natriuresis and diuresis but also in inhibiting voluntary sodium intake, and, by influencing the Starling forces in the systemic capillaries, the distribution of fluids between intravascular and extravascular spaces. Moreover, hypoxia is associated with a decrease of the plasma aldosterone to PRA ratio and sometimes of vasopressin concentration. Since ANFs are reported to have all of these effects, it is tempting to speculate that they might be a key for understanding both hormonal and sodium/water responses in acute hypoxia.

In addition, natriuresis and negative sodium balance might develop at high altitude by certain chemoreceptor-independent and nonreflex mechanisms. Animals with denervated arterial chemoreceptors do not show renal vasoconstriction in hypoxia, and since tissue hypoxia might dilate kidney vessels directly, this per se could increase urinary sodium excretion (55, 56). Furthermore, hypoxia is known to release ANF from isolated hearts independently of reflex mechanisms (86) and can also directly inhibit pulmonary angiotensin-converting enzyme activity (17, 70). These changes could result in a decreased plasma aldosterone concentration and thus natriuresis independent of the arterial chemoreceptors.

Such chemoreceptor-independent factors might be phylogenetically and ontogenetically old mechanisms, necessary to control body fluid volume during intrauterine life at a time when no central nervous system and no reflex mechanisms exist. In contrast, control of voluntary sodium intake requires involvement of the central nervous system and reflex mechanisms (tasting and looking for salt). Thus it is reasonable to expect arterial chemoreceptor deafferentation to completely abolish the suppression of spontaneous salt intake in acute hypoxia (9).

Special Conditions Influencing Body Fluid Homeostasis.

Natriuresis, together with inhibition of voluntary sodium intake and perhaps reduced salt reabsorption in the intestine, could explain the negative sodium balance repeatedly observed in well-tolerated acute hypoxia (19, 47, 71). In the simplest case a negative sodium balance (that is, loss of osmotically active particles from the extracellular space) without a change in intracellular concentrations would be followed by diuresis and by water moving into the cells. The result would be a decrease in extracellular fluid and plasma volumes and a brief, transient increase in the intracellular fluid space. Studies of the effects of hypoxia on organ fluid changes suggest that in tissues other than skeletal muscle hypoxia may result in transient (< 12 h) fluid shifts into the cellular compartment (19, 69). Stimulation of arterial chemoreceptors is reported to change the filtration–reabsorption balance in the systemic capillaries (94), but the direction of this response is different in different vascular beds. The contribution of this mechanism to whole-body hypohydration and to differences in the state of hydration in different parts of the body remains to be elucidated.

The normal responses to well-tolerated hypoxia are natriuresis; diuresis; and decreases in total body water (21), extracellular fluid volume, and plasma volume (9, 10, 19, 36, 72, 73). From a whole-body perspective, it also appears that calculated intracellular water generally decreases (63, 67, 68, 83, 101, 117). However, experimental variables can influence the effects of hypoxia on total body water and body fluid distribution. Complicating factors include maturity (growing vs.

adult animals) (48, 114), exercise (physical exercise may influence body fluid distribution by increasing extracellular osmolality) (130), age (28, 74, 90), and possibly gender (26). Steroid hormones can influence body fluid volume distribution (25, 66, 93), but the mechanisms and factors controlling the release of these compounds in acute hypoxia are poorly understood.

Enhanced activity of the arterial chemoreceptors is not always accompanied by natriuresis and diuresis. The natriuretic effect seems to depend on normal filling of the circulation and thus normal activity of the cardiovascular stretch and pressure receptors (3, 56). All facts taken together lead to the view that natriuresis in acute hypoxia (evoked by chemoreceptor stimulation and efferently mediated by hormonal mechanisms) is counteracted by cardiovascular stretch or pressure receptors, by way of efferent renal nerves, when the filling of the circulation is decreased. Such an interaction between chemoreceptor and stretch receptor effects could explain why this natriuresis is transient and why hypovolemia attenuates or prevents it. The interaction may also be effective in controlling voluntary sodium and water intake and could defend against excessive intravascular volume loss in acute hypoxia (56). For instance, the peripheral arterial chemoreceptors are stimulated after severe hemorrhage with low arterial pressures, but under such conditions renal sodium excretion is reduced, not enhanced (56).

In two studies of anesthetized, constantly ventilated, bilaterally vagotomized dogs with both kidneys innervated it was found that the natriuresis due to chemoreceptor stimulation was converted into antinatriuresis when the rise of mean aortic blood pressure during brief carotid body chemoreceptor stimulation was prevented by bleeding (2, 75). The authors concluded that chemoreceptor natriuresis must be a pressure natriuresis (2, 75). However, maintenance of aortic blood pressure during chemoreceptor stimulation in these experiments was simultaneously associated with blood loss, a strong decrease in renal blood flow, and a pronounced fall in glomerular filtration rate (2, 75). In other words, the specific chemoreceptor natriuresis was prevented by three unspecific but massive antinatriuretic stimuli. Furthermore, other studies have shown that at the same arterial pressure only innervated kidneys show an antinatriuresis in response to chemoreceptor stimulation, whereas the contralateral denervated kidneys simultaneously respond with natriuresis (58, 76). This clearly demonstrates that chemoreceptor-mediated natriuresis cannot be a simple pressure natriuresis. Moreover, normoxic rats do not respond to chemoreceptor stimulation with an increase in arterial blood pressure but show a clear natriuretic response of both innervated and denervated kidneys (5, 6). The studies of Al-Obaidi and Karim (2, 75) present direct evidence that chemorecep-

tor and cardiovascular distension receptor reflex effects interact to control renal function. In this respect they are in agreement with hypotheses presented here and elsewhere (55, 56).

When mammals have free access to food and water (that is, when sodium intake is determined by food intake), the usual responses to acute arterial hypoxia are natriuresis and diuresis. In contrast, when SHR have free access not only to food but also to hypertonic saline, they react to acute hypoxia with a reduction in spontaneous intake of hypertonic saline; renal sodium and water excretions either do not change or decrease (10). This suggests that mammals can respond to acute hypoxia either with suppression of voluntary sodium and water intake or with an increase in renal sodium and water excretion, or perhaps by both mechanisms. Electrolytes and water intake should be considered when interpreting renal excretory function in conscious mammals in acute hypoxia.

In severe hypoxia decreases in renal blood flow, glomerular filtration rate, and urine and sodium excretion have been repeatedly found (55, 56). Because renal denervation attenuates these antinatriuretic and antidiuretic responses, it is assumed that they are mediated by the efferent renal sympathetic nerves activated by an oxygen deficiency in the brain (55, 56). Thus it can be speculated that all factors that impair oxygen transport, such as age, anemia, physical exercise, hypovolemia, blood loss, and cardiovascular and respiratory insufficiency, facilitate the onset of general sympathetic activation in arterial hypoxia. This would prevent natriuresis and diuresis, and limit plasma volume reduction, body fluid redistribution, and hemoconcentration (55, 56). Apparently, the state of sodium balance is also significant because sodium restriction can prevent natriuresis due to hypoxia (55, 56).

Young, healthy humans respond to acute hypoxia with decreases in plasma volume and hemoconcentration. Such reactions are not evident or are delayed in the elderly (28, 74, 90) for reasons that remain unclear. It should be taken into account, however, that attenuated ventilatory and cardiovascular responses to acute hypoxia have also been seen in the elderly (82), suggesting weak chemoreceptor reflex effects. Altered endocrine reactions are likewise to be expected (70).

Chronic Hypoxia

Body Fluid Volume and Sodium and Water Metabolism. Most studies of hypoxia lasting longer than 3–4 days have found both lower body weight and lower total body water (21). In a carefully designed study using carcass analysis of full-grown mice, chronic hypoxia (4,300 m and 6,300 m) significantly decreased total body water (48). Humans exposed to elevations of 3,500–5,334 m for 6–14 days experienced reductions in both variables (7, 67, 68, 71, 83, 117). In these studies decreases in body mass were primarily due to the 3%–5% decrease in total body water (67, 68, 83, 117). Total body water volume changes were estimated from either water balance (7, 71) or equilibrium tracer concentrations after administration of D_2O (83) or ^3H_2O (25, 63, 67, 68, 117). The large loss of total body water volume relative to the change in extracellular fluid volume resulted in significant 3%–6% decreases in calculated intracellular fluid volume (67, 68, 83, 117).

Some authors report that both intracellular and extracellular fluid volumes decrease in chronic hypoxia (63, 68, 117), whereas others report decreases only in intracellular fluid volume, with either no change (67, 83) or an increase (63, 101) in extracellular fluid volume. Extracellular fluid volume generally decreases in humans exposed to hypoxia (35, 45, 46, 68, 117), in contrast to the unusual increase seen in herbivorous goats and sheep (63, 101), where potassium, rather than sodium, might play a more important metabolic role. Compared with sea-level values, no consistent differences between extracellular and plasma ion and osmotic concentrations were observed in mammals living permanently at high altitudes (19, 36, 63, 71, 96, 130).

There are conflicting reports that in humans prolonged hypoxia (8–14 days at 4,300–5,334 m) results in only minor (−2% to +3%) changes in total body water (35, 45, 46, 121). The combination of little or no change in total body water with a decrease in extracellular fluid volume resulted in an apparent calculated shift of fluid from the extracellular to the intracellular compartment (35, 45, 46). However, these studies estimated total body water volume from the pattern of nonequilibrium tracer elimination after bolus injections of 4-aminoantipyrine (35, 45, 46) and/or D_2O (45, 121). It may be difficult to accurately estimate total body water by the bolus D_2O method (45, 121). In addition, the 4-aminoantipyrine method used in some early studies apparently overestimated total body water at high altitude (J. P. Hannon, personal communication). The complex effects of hypoxia on the dynamic of water distribution (22, 50) may have compromised the accuracy of total body water estimates made by nonequilibrium methods. An underestimation of the loss in total body water would explain the unusual findings of 3%–13% increases in calculated intracellular fluid volume (35, 45, 46).

Other factors that may influence body fluid volumes in both acute and chronic high-altitude hypoxia include gender (25), age (74), high-altitude tolerance (7), nutrition (71), physical exercise (128), and increased insensible water loss (21).

Humans exposed to 3,500–4,300 m for 8–12 days had decreased plasma and blood volumes (45, 67, 68, 74, 103, 117, 121). Apparently, old age and physical exercise can be accompanied by increased plasma volume in long-lasting hypoxia (74, 130). Compared with sea-level values, the volume of red blood cells in young, healthy humans and animals increases with time at altitude, whereas plasma volume remains reduced or normal (63, 68, 74, 100, 103, 112, 116, 130, 134). Total blood volume may rise after more prolonged residence at altitude due to an increase in red cell mass (103, 112, 130, 135). For example, in humans after 18 days at 4,300 m the decrease in plasma volume was offset by an increase in red cell volume, and blood volume returned to sea-level values (135). A similar pattern was evident in goats after 16 days of hypoxia (5,500 m) (35).

Control of Body Fluid Volume and Sodium and Water Metabolism.

It is well documented that under resting conditions spontaneous food and water intake of mammals is suppressed in the first week of high-altitude exposure but then tends to increase (10, 19, 27, 70–73). Even in long-lasting hypoxia in rats the threshold for osmotic stimulation of drinking increased (73). Mammals apparently try to remain in a state of voluntary dehydration in chronic hypoxia.

In contrast to the transient suppression of spontaneous food and water intake in acute hypoxia, voluntary consumption of hypertonic saline remained reduced in SHR for 20 days at 4,000 m (10), suggesting that food, water, and salt intakes can vary independently in hypoxia. The mechanisms that control these intakes in hypoxia are still a matter of speculation; more detailed discussions are published elsewhere (9, 36, 47, 70, 72, 73). No data are available concerning electrolyte and water absorption in the intestine of chronically hypoxic humans or animals.

In any case, natriuresis and diuresis due to hypoxia must lead to new steady-state balances for body mass (16), for sodium and water, and for volume and distribution of body fluids. This implies that natriuresis must be transient even if high-altitude conditions persist. In this context an interesting observation was that chemoreceptor stimulation by almitrine did not induce a natriuretic effect in chronically hypoxic rats (6). It may be either that the response of the animals' arterial chemoreceptors to the drug was blunted (6) or that these rats had already established a new steady-state balance for sodium and water according to the high altitude at which they lived (56). In the latter case, it would not have been necessary for them to react to chemoreceptor stimulation with additional natriuresis and diuresis. This assumption is supported by the fact that renal responses to acute hypoxia were less pronounced in altitude-acclimatized than in nonacclimatized men (80).

Studies concerning kidney function in hypoxia-adapted mammals revealed normal or enhanced renal blood flow accompanied by low plasma flow (due to high hematocrit) but a strikingly high filtration fraction (8, 92, 96, 124). The high filtration fraction is apparently needed to ensure adequate glomerular filtration despite the low renal plasma flow. The mechanism that induces the high filtration fraction is a matter of speculation; it might be a by-product of the high hematocrit and viscosity of the blood (92), but, as mentioned, it could also be a reflex effect of the peripheral arterial chemoreceptors (56, 58).

After the striking but transient natriuresis and diuresis over 1–3 days in nonacclimatized young and healthy mammals under resting conditions, renal salt and water balance seem to be dominated by intake (19, 47, 70, 72). Renal tubular function is apparently normal in chronic hypoxia (27, 92, 124), but no studies have investigated the reflex control of kidney excretory function in chronically hypoxic humans and animals.

With respect to hormonal systems affecting electrolyte and water metabolism in chronic hypoxia, there is good agreement in the literature that plasma concentrations of ANF are enhanced in both humans and animals (89, 107, 108, 133). In rats exposed to high altitude for 3 months pituitary and adrenal glands hypertrophied and plasma levels of corticosterone were increased compared to a control group at sea level (97). In chronically hypoxic humans and animals the response of the adrenal glucocorticoid hormones to ACTH remains normal (97, 108), whereas the aldosterone response to ACTH seems to be inhibited (108).

Resting levels for components of the RAAS—renal renin granulation, angiotensin I, and PRA—seem to be normal in high-altitude-adapted humans and animals (108, 140). The initial reactions to altitude exposure of PRA, angiotensin II, and aldosterone concentration (PAC) mentioned above return to control values within 10–20 days (91, 141), but the PAC response apparently remains blunted compared with that at sea level (91). The altered reaction pattern of the RAAS in chronic high-altitude hypoxia may be explained by high plasma levels of ANF (108). Age and physical exercise also might modify the responses of the RAAS (74, 131).

In long-lasting and well-tolerated high-altitude hypoxia under resting conditions plasma vasopressin levels are similar to those at sea level (21, 72).

ENERGY METABOLISM AT HIGH ALTITUDE

Cardiovascular Responses to Exercise at Altitude

Cardiorespiratory responses to exercise at moderate altitude have been reviewed (37, 42, 138) but are briefly mentioned here because of their important influence on energy metabolism. Starting at about 2,000 m, maximum aerobic power ($\dot{V}_{O_2max}$) measured during strenuous short-term exercise to exhaustion decreases significantly. Although variable among individuals, there is roughly a 10% decrease in $\dot{V}_{O_2max}$ for every 1,000 m increase in elevation above 2,000 m (129). The $\dot{V}_{O_2}$ at any given submaximal power output, however, remains unchanged from sea level to altitude. Consequently, exercise at altitude at a given power output requires a greater percent of $\dot{V}_{O_2max}$, that is, a higher relative exercise intensity.

The relative intensity of exercise naturally has important physiological consequences (39) independent of those of hypoxia. Under either normoxic or hypoxic conditions hormonal, sympathetic nervous system, and metabolic responses to exercise at a given relative work intensity (%$\dot{V}_{O_2max}$) are similar and highly interrelated (12, 32, 79). Thus when the power output is kept constant during exercise testing at sea level and at altitude, the physiological differences between conditions are due to both hypoxia and the divergence in relative exercise intensity. The confounding effects of differences in percent of environment-specific $\dot{V}_{O_2max}$ can be avoided by reducing submaximal power output at altitude such that the relative exercise intensity remains constant.

Metabolic Responses to Exercise at Altitude

Acute Hypoxia. With the exception of body fluid metabolism, most metabolic responses of unacclimatized lowlanders to exercise at a constant relative work intensity are similar at sea level and on acute exposure to hypoxia (37). Changes in blood concentrations of many substrates (for example, glucose, lactate, free fatty acids) and hormones (for example, norepinephrine, epinephrine, cortisol, testosterone) during exercise at a given %$\dot{V}_{O_2max}$ under normoxic and acutely hypoxic conditions are comparable (12, 32, 87). In addition, muscle lactate accumulation (79) and muscle glycogen use (137) during exercise are similar under normoxic and acutely hypoxic conditions. Although absolute physical work capacity is reduced with ascent to altitude, acute hypoxia does not significantly alter endurance exercise capacity at a given percent of $\dot{V}_{O_2max}$ (62, 87).

Chronic Hypoxia: Acclimatization to Altitude. Sea-level improvements in the capacity for sustained strenuous exercise are usually associated with endurance training and increases in $\dot{V}_{O_2max}$ (53). In contrast, acclimatization to moderate altitude dramatically increases endurance exercise capacity (time to exhaustion during sustained exercise at a constant exercise intensity) (62, 87) but has little or no effect on $\dot{V}_{O_2max}$ (13, 62, 137), at least in the absence of training or detraining effects (139). Increased endurance capacity in high-altitude natives and in sojourners after altitude acclimatization has been reported anecdotally (15, 132) and in systematic studies (62, 87, 88). Compared to sea level, cycle ergometer endurance time at 75% of $\dot{V}O_{2max}$ increased 61% after 12 days at 4,300 m (87), while time to exhaustion with treadmill running at 85% of $\dot{V}O_{2max}$ increased 41% after 16 days at 4,300 m (62).

Another key characteristic of altitude acclimatization, in addition to increased endurance exercise capacity without a commensurate increase in $\dot{V}O_{2max}$, is a decrease in blood lactate accumulation during exercise. In 1936 Edwards (29) reported that blood lactate accumulation during maximal exercise decreased as the altitude to which subjects were acclimatized increased. This phenomenon, verified by many investigators (26, 41, 49, 78, 127), has been named the "lactate paradox" because the magnitude of the increase in lactate accumulation at any given power output with acute exposure to hypoxia is lessened after altitude acclimatization. That is, with acute hypoxia, but not after altitude acclimatization, an increase in relative exercise intensity elicits the expected increase in lactate accumulation. In contrast, blood lactate accumulation with maximal exercise, or with submaximal exercise at a given relative exercise intensity, is depressed after altitude acclimatization. Exercise after altitude acclimatization also results in less muscle glycogen utilization (137) and less muscle lactate accumulation (41) than exercise of the same relative intensity at sea level.

Prolonged exercise in the field or laboratory can result in high rates of energy expenditure and large energy deficits (64, 119) and an acute shift from a carbohydrate- to a fat-predominant fuel metabolism (33, 118). The time course of this transition in energy substrate utilization depends on the subject's $\dot{V}O_{2max}$, physical activity, and diet (1, 33, 118). However, exercise- or diet-induced transitions to a more fat-predominant metabolism differ from that seen with altitude acclimatization in that they are usually associated with muscle glycogen depletion and a reduction in maximum sustainable exercise intensity (65, 102).

The basic mechanism of the lactate paradox remains unclear, though a variety of theories have been proposed (18, 41, 122). Suggestions that the lactate paradox is due to muscle glycogen depletion, glycolytic inhibition as a result of an exaggerated increase in hydrogen ion concentration for a given lactate production caused by a reduced bicarbonate buffer capacity (18), muscle lactate retention (122), or a reduced ability to activate the muscle's contractile apparatus (41) have been dismissed or discounted (88). In addition, studies of the effects of acclimatization to 4,300 m on oxygen delivery to exercising leg muscle found that the positive effects of hemoconcentration and increased arterial saturation on systemic oxygen transport were counterbalanced by

decreased leg blood flow (11). Thus the lactate paradox cannot be ascribed to increased convective oxygen delivery to exercising muscle (109).

One hypothesis attributes the lactate paradox to a closer coupling between skeletal muscle adenosine triphosphate (ATP) demand and supply (51). Closer coupling means that increases in work rate are accompanied by smaller than expected changes in cell phosphorylation potential, phosphocreatine, and pH and consequently less stimulation of glycolysis and less lactate accumulation. This hypothesis is supported indirectly by the finding that plasma ammonia concentration during endurance exercise at a constant relative intensity decreases after acclimatization to moderate altitude, presumably reflecting decreased adenylate breakdown, increased tricarboxylic acid cycle activity, and reduced glycolytic flux (139). A study of Andean Quechua Indian highlanders also supports the hypothesis of a more closely coupled muscle metabolism (88). The Andean highlanders, in spite of low whole-body anaerobic capacities and only moderate $\dot{V}O_{2max}$ levels, were capable of absolute calf muscle work rates equal to those of power- and endurance-trained athletes. In addition, ^{31}P nuclear magnetic resonance spectroscopy revealed that exercise-induced changes in muscle phosphocreatine, inorganic phosphate, and ATP in the Andean highlanders were less pronounced than those of power-trained athletes with equivalent maximum aerobic power but similar to those of endurance-trained athletes with much higher $\dot{V}O_{2max}$ levels (88).

Endurance training and acclimatization to altitude are similar in that both result in an improved capacity for endurance exercise, a tighter coupling of ATP supply and demand, decreased lactate production and accumulation during exercise, and a transition from a carbohydrate- to a fat-predominant fuel metabolism (42, 52–54, 136). However, endurance training increases $\dot{V}O_{2max}$ (53, 54), while altitude acclimatization has little or no effect (13, 62, 137). In addition, peripheral skeletal muscle adaptations with endurance training differ significantly from those associated with altitude acclimatization (127).

With endurance exercise training, muscle fiber size is unchanged or increased, myoglobin concentrations are unchanged, and mitochondrial and capillary volume densities increase due to proliferation (54, 111). In contrast, acclimatization to altitude increased capillary volume density through a reduction in fiber size rather than by capillary proliferation (61, 95, 127), possibly through a loss of interstitial and intracellular fluids. Increases in mitochondrial volume density and cellular enzyme and myoglobin concentrations would be expected to accompany any loss of intracellular fluid with altitude acclimatization. Evidence of such changes comes primarily from studies at moderate altitudes, where acclimatization and eventual body weight stabilization are possible (16, 24, 127). Conflicting findings of decreases in mitochondrial volume density and oxidative enzyme activities come from studies of extreme altitudes (> 6,000 m) (41, 61, 127), where persistent body weight loss, detraining, and negative nitrogen balance are more likely (14, 43, 127).

A loss of fluid from the plasma and the interstitial and intracellular spaces with acclimatization to moderate altitude may result in peripheral metabolic adaptations (for example, increases in muscle capillary or mitochondrial volume densities) that increase O_2 conductance from red cells to skeletal muscle mitochondria. An increase in skeletal muscle O_2 conductance would improve endurance exercise performance and limit lactate accumulation by permitting a higher rate of ATP turnover before adenylate breakdown and stimulation of lactate production. The concept that a loss of body fluid with altitude acclimatization increases tissue O_2 conductance and thereby improves endurance capacity is compatible with studies suggesting that capillary O_2 diffusion limits aerobic metabolism during hypoxemia (44) and that the principal site of resistance to O_2 diffusion from blood to tissue is the "carrier-free" region between the red cell and the sarcolemma that is devoid of hemoglobin or myoglobin (57). The lack of change in $\dot{V}_{O_2max}$ with acclimatization, in spite of increased arterial oxygen content and a postulated improvement in tissue O_2 conductance, is consistent with evidence that $\dot{V}_{O_2max}$ is limited by an altitude-induced decrease in maximal cardiac output (38).

In summary, nonacclimatized, healthy, young mammals respond to acute exposure to moderate high-altitude hypoxia with decreases in plasma and extracellular and total body fluid volumes. The loss of body fluid is caused by a suppression of voluntary sodium and water intake as well as increased insensible water loss and hormonally mediated natriuresis and diuresis. Peripheral arterial chemoreceptor reflex mechanisms play an important role in altering sodium and water homeostasis. The loss of body fluid may result in increased capillary density and other peripheral metabolic adaptations that improve O_2 conductance from red cells to skeletal muscle mitochondria. An increase in tissue O_2 conductance would explain the reported increase in endurance exercise capacity and the decrease in lactate accumulation during exercise following acclimatization to moderate altitude.

The opinions or assertions contained herein are the private views of the authors and are not to be construed as official or reflecting the views of the Department of the Army or the Department of Defense.

REFERENCES

1. Ahlborg, G., and P. Felig. Lactate and glucose exchange across the forearm, legs and splanchnic bed during and after prolonged leg exercise. *J. Clin. Invest.* 69: 45–54, 1982.
2. Al-Obaidi, M., and F. Karim. Primary effects of carotid chemoreceptor stimulation on gracilis muscle and renal blood flow and renal function in dogs. *J. Physiol.* 455: 73–88, 1992.
3. Al-Obaidi, M., E. M. Whitaker, and F. Karim. The effect of discrete stimulation of carotid body chemoreceptors on atrial natriuretic peptide in anaesthetized dogs. *J. Physiol.* 443: 519–531, 1991.
4. Baertschi, A. J., J.-H. Jiao, D. E. Carlson, R. W. Campbell, W. G. Teague, D. Willson, and D. S. Gann. Neural control of ANF release in hypoxia and pulmonary hypertension. *Am. J. Physiol.* 259(*Heart Circ. Physiol.* 28): H735–H744, 1990.
5. Bardsley, P. A., B. F. Johnson, A. G. Stewart, and G. R. Barer. Natriuresis secondary to carotid chemoreceptor stimulation with almitrine bismesylate in the rat: the effect on kidney function and the responses to renal denervation and deficiency of antidiuretic hormone. *Biomed. Biochim. Acta* 50: 175–182, 1991.
6. Bardsley, P. A., and A. J. Suggett. The carotid body and natriuresis: effect of almitrine bismesylate. *Biomed. Biochim. Acta* 46: 1017–1022, 1987.
7. Bärtsch, P., N. Pluger, M. Audetat, S. Shaw, P. Weidmann, P. Vock, W. Vetter, D. Rennie, and O. Oelz. Effects of slow ascent to 4559 m on fluid homeostasis. *Aviat. Space Environ. Med.* 62: 105–110, 1991.
8. Becker, E. L., J. A. Schilling, and R. B. Harvey. Renal function in man acclimatized to high altitude. *J. Appl. Physiol.* 10: 79–80, 1957.
9. Behm, R., B. Gerber, J.-O. Habeck, C. Huckstorf, and K. Ruckborn. Effect of hypobaric hypoxia and almitrine on voluntary salt and water intake in carotid body denervated spontaneously hypertensive rats. *Biomed. Biochim. Acta* 48: 689–695, 1989.
10. Behm, R., A. Honig, M. Griethe, M. Schmidt, and P. Schneider. Sustained suppression of voluntary sodium intake of spontaneously hypertensive rats (SHR) in hypobaric hypoxia. *Biomed. Biochim. Acta* 43: 975–985, 1984.
11. Bender, P. R., B. M. Groves, R. E. McCullough, R. G. McCullough, S. Huang, A. J. Hamilton, P. D. Wagner, A. Cymerman, and J. T. Reeves. Oxygen transport to exercising leg in chronic hypoxia. *J. Appl. Physiol.: Respir. Environ. Exerc. Physiol.* 65: 2592–2597, 1988.
12. Bouissou, P., F. Peronnet, G. Brisson, R. Relie, and M. Ledoux. Metabolic and endocrine responses to graded exercise under acute hypoxia. *Eur. J. Appl. Physiol.* 55: 290–294, 1986.
13. Boutellier, U., D. Dériaz, P. E. Di Prampero, and P. Cerretelli. Aerobic performance at altitude: effects of acclimatization and hematocrit with reference to training. *Int. J. Sports Med.* 11(suppl. 1): S21–S26, 1990.
14. Boyer, S. J., and F. D. Blume. Weight loss and changes in body composition at high altitude. *J. Appl. Physiol.: Respir. Environ. Exerc. Physiol.* 57: 1580–1585, 1984.
15. Buskirk, E. R., J. Kollias, R. F. Akers, E. K. Prokop, and E. R. Reategui. Maximal performance at altitude and on return from altitude in conditioned runners. *J. Appl. Physiol.* 23: 259–266, 1967.
16. Butterfield, G. E., J. Gates, S. Fleming, G. A. Brooks, J. R. Sutton, and J. T. Reeves. Increased energy intake minimizes weight loss in men at high altitude. *J. Appl. Physiol.: Respir. Environ. Exerc. Physiol.* 72: 1741–1748, 1992.
17. Caldwell, R. W., and C. M. Blatteis. Effect of chronic hypoxia on angiotensin-induced pulmonary vasoconstriction and con-

verting enzyme activity in the rat. *Proc. Soc. Exp. Biol. Med.* 172: 346–350, 1983.

18. Cerretelli, P., A. Veicsteinas, and C. Marconi. Anaerobic metabolism at high altitude: the lactacid mechanism. In: *High Altitude Physiology and Medicine,* edited by W. Brendel and R. A. Zink. New York: Springer-Verlag, 1982, chapt. 13, p. 94–102.

19. Christensen, B. M., H. L. Johnson, and A. V. Ross. Organ fluid changes and electrolyte excretion of rats exposed to high altitude. *Aviat. Space Environ. Med.* 46: 16–20, 1975.

20. Claybaugh, J. R., J. E. Hansen, and D. B. Wozniak. Response of antidiuretic hormone to acute exposure to mild and severe hypoxia in man. *J. Endocrinol.* 77: 157–160, 1978.

21. Claybaugh, J. R., C. E. Wade, and S. A. Cucinell. Fluid and electrolyte balance and hormonal response to the hypoxic environment. In: *Hormonal Regulation of Fluid and Electrolytes,* edited by J. R. Claybaugh and C. E. Wade. New York: Plenum, 1989, p. 187–214.

22. Coleman, T. G., D. Manning, Jr., R. A. Norman, Jr., and A. C. Guyton. Dynamics of water isotope distribution. *Am. J. Physiol.* 223: 1371–1375, 1972.

23. Colice, G., S. Yen, G. Ramirez, J. Dietz, and L.-C. Ou. Acute hypoxia-induced diuresis in rats. *Aviat. Space Environ. Med.* 62: 551–554, 1991.

24. Consolazio, C. F., L. O. Matoush, H. L. Johnson, and T. A. Daws. Protein and water balances of young adults during prolonged exposure to high altitude (4300 m). *Am. J. Clin. Nutr.* 21: 154–161, 1968.

25. Costa, L. E., I. H. Martin, J. C. Macome, and A. C. Taquini. Efecto de la hipoxia hipobarica cronica sobre el crecimiento y la composicion corporal en la rata. *Medicina* 39: 604–610, 1979.

26. Cunningham, D. A., and J. R. Magel. The effect of moderate altitude on post-exercise blood lactate. *Int. Z. Angew. Physiol.* 29: 94–100, 1970.

27. Curran-Everett, D. C., J. R. Claybaugh, K. Miki, S. K. Hong, and J. A. Krasney. Hormonal and electrolyte responses of conscious sheep to 96 hours of hypoxia. *Am. J. Physiol.* 255(*Regulatory Integrative Comp. Physiol.* 24): R274–R283, 1988.

28. Dill, D. B., F. G. Hall, K. D. Hall, C. Dawson, and J. L. Newton. Blood, plasma and red cell volumes: age, exercise and environment. *J. Appl. Physiol.* 21: 597–602, 1966.

29. Edwards, H. T. Lactic acid in rest and work at high altitude. *Am. J. Physiol.* 116: 367–375, 1936.

30. Epstein, M., and T. Saruta. Effects of simulated high altitude on renin–aldosterone and Na homeostasis in normal man. *J. Appl. Physiol.* 33: 204–210, 1972.

31. Epstein, M., and T. Saruta. Effects of an hyperoxic hypobaric environment on renin–aldosterone in normal man. *J. Appl. Physiol.* 34: 49–52, 1973.

32. Escourrou, P., D. G. Johnson, and L. B. Rowell. Hypoxemia increases plasma catecholamine concentrations in exercising humans. *J. Appl. Physiol.: Respir. Environ. Exerc. Physiol.* 57: 1507–1511, 1984.

33. Felig, P., and J. Wahren. Fuel homeostasis in exercise. *N. Engl. J. Med.* 293: 1078–1084, 1975.

34. Fishman, A. P., M. H. Maxwell, C. H. Crowder, and P. Morales. Kidney function in cor pulmonale. Particular consideration of changes in renal hemodynamics and sodium excretion during variation in level of oxygenation. *Circulation* 3: 703–721, 1951.

35. Frayser, R., I. D. Rennie, G. W. Gray, and C. S. Houston. Hormonal and electrolyte response to exposure to 17,500 ft. *J. Appl. Physiol.* 38: 636–642, 1975.

36. Fregly, M. J., E. L. Nelson, and P. E. Tyler. Water exchange in rats exposed to cold, hypoxia, and both combined. *Aviat. Space Environ. Med.* 47: 600–607, 1976.

37. Fulco, C. S., and A. Cymerman. Human performance and acute hypoxia. In: *Human Performance and Environmental Medicine at Terrestrial Extremes,* edited by K. B. Pandolf, M. N. Sawka, and R. R. Gonzalez. Indianapolis, In: Benchmark, 1988, p. 467–495.

38. Fulco, C. S., P. B. Rock, L. Trad, V. Forte, Jr., and A. Cymerman. Maximal cardiorespiratory responses to one- and two-legged cycling during acute and long-term exposure to altitude. *Eur. J. Appl. Physiol.* 57: 761–766, 1988.

39. Galbo, H. *Hormonal and Metabolic Adaptation to Exercise.* New York: Thieme-Stratton, 1982.

40. Gotshall, R. W., D. S. Miles, and W. R. Sexson. The combined effects of hypoxemia and mechanical ventilation on renal function. *Aviat. Space Environ. Med.* 57: 782–786, 1986.

41. Green, H. J., J. R. Sutton, P. M. Young, A. Cymerman, and C. S. Houston. Operation Everest II: muscle energetics during maximal exhaustive exercise. *J. Appl. Physiol.: Respir. Environ. Exerc. Physiol.* 66: 142–150, 1989.

42. Grover, R. F., J. V. Weil, and J. T. Reeves. Cardiovascular adaptation to exercise at high altitude. In: *Exercise and Sport Science Reviews,* edited by K. B. Pandolf. New York: Macmillan, 1986, p. 269–302.

43. Guilland, J. C., and J. Klepping. Nutritional alterations at high altitude in man. *Eur. J. Appl. Physiol.* 54: 517–523, 1985.

44. Gutierrez, G., C. Marini, A. L. Acero, and N. Lund. Skeletal muscle P_{O_2} during hypoxemia and isovolemic anemia. *J. Appl. Physiol.: Respir. Environ. Exerc. Physiol.* 68: 2047–2053, 1990.

45. Hannon, J. P., K. S. K. Chinn, and J. L. Shields. Effects of acute high altitude exposure on body fluids. *Federation Proc.* 38: 1178–1184, 1969.

46. Hannon, J. P., K. S. K. Chinn, and J. L. Shields. Alterations in serum and extracellular electrolytes during high-altitude exposure. *J. Appl. Physiol.* 31: 266–273, 1971.

47. Hannon, J. P., L. F. Krabill, T. A. Wooldridge, and D. D. Schnakenberg. Effects of high-altitude and hypophagia on mineral metabolism in rats. *J. Nutr.* 105: 278–287, 1975.

48. Hannon, J. P., and G. B. Rogers. Body composition of mice following exposure to 4300 and 6100 meters. *Aviat. Space Environ. Med.* 46: 1232–1235, 1975.

49. Hansen, J. E., G. P. Stelter, and J. A. Vogel. Arterial pyruvate, lactate, pH, and P_{CO_2} during work at sea level and high altitude. *J. Appl. Physiol.* 23: 523–530, 1967.

50. Heistad, D. D., and F. M. Abboud. Circulatory adjustments to hypoxia. *Circulation* 61: 463–470, 1980.

51. Hochachka, P. W. The lactate paradox: analysis of underlying mechanisms. *Ann. Sports Med.* 4: 184–188, 1988.

52. Hochachka, P. W., C. Stanley, J. Merkt, and J. Sumar-Kalinowski. Metabolic meaning of elevated levels of oxidative enzymes in high altitude adapted animals: an interpretive hypothesis. *Respir. Physiol.* 52: 303–313, 1982.

53. Holloszy, J. O., and F. W. Booth. Biochemical adaptations to endurance exercise in muscle. *Annu. Rev. Physiol.* 38: 273–291, 1976.

54. Holloszy, J. O., and E. F. Coyle. Adaptations of skeletal muscle to endurance exercise and their metabolic consequences. *J. Appl. Physiol.: Respir. Environ. Exerc. Physiol.* 56: 831–838, 1984.

55. Honig, A. Role of the arterial chemoreceptors in the reflex control of renal function and body fluid volumes in acute arterial hypoxia. In: *Physiology of the Peripheral Arterial Chemoreceptors,* edited by H. Acker and R. G. O'Regan. Amsterdam: Elsevier, 1983, p. 395–429.

56. Honig, A. Peripheral arterial chemoreceptors and reflex control of sodium and water homeostasis. *Am. J. Physiol.* 257 (*Regulatory Integrative Comp. Physiol.* 26): R1282–R1302, 1989.

57. Honig, C. R., R. J. Connett, and T. E. J. Gayeski. O_2 transport and its interaction with metabolism; a systems view of aerobic capacity. *Med. Sci. Sports Exerc.* 24: 47–53, 1992.

58. Honig, A., M. Schmidt, and E. J. Freyse. Influence of unilateral renal nerve section on the kidney effects of carotid chemoreceptor stimulation. *Biomed. Biochim. Acta* 38: 1647–1650, 1979.

59. Honig, A., B. Wedler, H. Oppermann, S. Gruska, and M. Schmidt. Influence of arterial chemoreceptor stimulation by almitrine on the plasma levels of ACTH, cortisol, PRA and aldosterone in anaesthetized cats. *Pflugers Arch.* 419(suppl. 1): R78, 1991.

60. Honig, A., B. Wedler, C. Zingler, C. Ledderhos, and M. Schmidt. Kidney function during arterial chemoreceptor stimulation. III. Long-lasting inhibition of renal tubular sodium reabsorption due to pharmacological stimulation of the peripheral arterial chemoreceptors with almitrine bismesylate. *Biomed. Biochim. Acta* 44: 1659–1672, 1985.

61. Hoppeler, H., E. Kleinert, C. Schlegel, H. Claassen, H. Howald, S. R. Kayar, and P. Cerretelli. Morphological adaptations of human skeletal muscle to chronic hypoxia. *Int. J. Sports Med.* 11(suppl. 1): S3–S9, 1990.

62. Horstman, D., R. Weiskopf, and R. E. Jackson. Work capacity during a 3-week sojourn at 4,300 m: effects of relative polycythemia. *J. Appl. Physiol.: Respir. Environ. Exerc. Physiol.* 49: 311–318, 1980.

63. Hoyt, R. W., M. J. Durkot, V. A. Forte, Jr., L. J. Hubbard, L. A. Trad, and A. Cymerman. Hypobaric hypoxia (380 torr) decreases intracellular and total body water in goats. *J. Appl. Physiol.: Respir. Environ. Exerc. Physiol.* 71: 509–513, 1991.

64. Hoyt, R. W., T. E. Jones, T. P. Stein, G. McAninch, H. R. Lieberman, E. W. Askew, and A. Cymerman. Doubly labeled water measurement of human energy expenditure during strenuous exercise. *J. Appl. Physiol.: Respir. Environ. Exerc. Physiol.* 71: 16–22, 1991.

65. Hultman, E. Physiological role of muscle glycogen in man, with special reference to exercise, *Circ. Res.* 20/21(suppl. 1): 99–114, 1967.

66. Ishii, M., M. Yamakado, and S. Murao. The extrarenal effects of aldosterone on the distribution of extracellular fluid in conscious adrenalectomized–nephrectomized rats. *Pflugers Arch.* 404: 273–277, 1985.

67. Jain, S. C., J. Bardhan, Y. V. Swamy, A. Grover, and H. S. Nayar. Body water metabolism in high altitude natives during and after a stay at sea level. *Int. J. Biometeorol.* 25: 47–52, 1981.

68. Jain, S. C., J. Bardhan, Y. Y. Swamy, B. Krishna, and H. S. Nayar. Body fluid compartments in humans during acute high-altitude exposure. *Aviat. Space Environ. Med.* 51: 234–236, 1980.

69. Jain, S. C., S. B. Rawal, H. M. Divekar, and H. S. Nayar. Organ fluid compartments in rats exposed to high altitude. *Indian J. Physiol. Pharmacol.* 24: 177–182, 1980.

70. Jain, S. C., W. L. Wilke, and A. Tucker. Age-dependent effects of chronic hypoxia on renin–angiotensin and urinary excretions. *J. Appl. Physiol.: Respir. Environ. Exerc. Physiol.* 69: 141–146, 1990.

71. Johnson, H. L., C. F. Consolazio, L. O. Matoush, and H. J. Krzywicki. Nitrogen and mineral metabolism at altitude. *Federation Proc.* 28: 1195–1198, 1969.

72. Jones, R. M., F. T. LaRochelle, Jr., and S. M. Tenney. Role of arginine vasopressin on fluid and electrolyte balance in rats exposed to high altitude. *Am. J. Physiol.* 240(*Regulatory Integrative Comp. Physiol.* 9): R182–R186, 1981.

73. Jones, R. M., C. Terhaard, and S. M. Tenney. Mechanism of reduced water intake in rats at high altitude. *Am. J. Physiol.* 240(*Regulatory Integrative Comp. Physiol.* 9): R187–R191, 1981.

74. Jung, R. C., D. B. Dill, R. Horton, and S. M. Horvath. Effects of age on plasma aldosterone levels and hemoconcentration at altitude. *J. Appl. Physiol.* 31: 593–597, 1971.

75. Karim, F., and M. Al-Obaidi. Effects of left atrial receptor stimulation on carotid chemoreceptor–induced renal responses in dogs. *J. Physiol. (Lond.)* 456: 529–539, 1992.

76. Karim, F., S. M. Poucher, and R. A. Summerill. The effects of stimulating carotid chemoreceptors on renal haemodynamics and function in dogs. *J. Physiol. (Lond.)* 392: 451–462, 1987.

77. Kilburn, K. H., and A. R. Dowell. Renal function in respiratory failure. Effects of hypoxia, hyperoxia, and hypercapnia. *Arch. Intern. Med.* 127: 754–762, 1971.

78. Klausen, K. J., D. B. Dill, and S. M. Horvath. Exercise at ambient and high oxygen pressure at high altitude and at sea level. *J. Appl. Physiol.* 29: 456–463, 1970.

79. Knuttgen, H. G., and B. Saltin. Oxygen uptake, muscle high-energy phosphates, and lactate in exercise under acute hypoxic conditions in man. *Acta Physiol. Scand.* 87: 368–376, 1973.

80. Koller, E. A., A. Buhrer, L. Felder, M. Schopen, and M. B. Valloton. Altitude diuresis: endocrine and renal responses to acute hypoxia of acclimatized and non-acclimatized subjects. *Eur. J. Appl. Physiol.* 62: 228–234, 1991.

81. Koller, E. A., M. Schopen, M. Keller, R. E. Lang, and M. B. Valloton. Ventilatory, circulatory, endocrine, and renal effects of almitrine infusion in man: a contribution to high altitude physiology. *Eur. J. Appl. Physiol.* 58: 419–425, 1989.

82. Kronenberg, R. S., and C. W. Drage. Alteration of the ventilatory and heart rate responses to hypoxia and hypercapnia with aging in normal man. *J. Clin. Invest.* 52: 1812–1819, 1973.

83. Krzywicki, H. J., C. F. Consolazio, H. L. Johnson, W. C. Nielsen, and R. A. Barnhart. Water metabolism in humans during acute high-altitude exposure (4300 m). *J. Appl. Physiol.* 30: 806–809, 1971.

84. Lawrence, D. L., J. B. Skatrud, and Y. Shenker. Effect of hypoxia on atrial natriuretic factor and aldosterone regulation in humans. *Am. J. Physiol.* 258(*Endocrinol. Metab.* 21): E243–E248, 1990.

85. Ledderhos, C., R. Sanchez, W. Quies, and R. Schuster. Does the stimulation of peripheral arterial chemoreceptors in humans by almitrine bismesylate inhibit proximal reabsorption. In: *Chemoreceptors and Chemoreceptor Reflexes,* edited by H. Acker, A. Trzebski, and R. G. O'Regan. New York: Plenum, 1990, p. 293–302.

86. Lew, R. A., and A. J. Baertschi. Mechanisms of hypoxia-induced atrial natriuretic factor release from rat hearts. *Am. J. Physiol.* 257(*Heart Circ. Physiol.* 26): H147–H156, 1989.

87. Maher, J. T., L. G. Jones, and L. H. Hartley. Effects of high-altitude exposure on submaximal endurance capacity of men. *J. Appl. Physiol.* 37: 895–898, 1974.

88. Matheson, G. O., P. S. Allen, D. C. Ellinger, C. C. Hanstock, D. Gheorghiu, D. C. McKenzie, C. Stanley, W. S. Parkhouse, and P. W. Hochachka. Skeletal muscle metabolism and work capacity: a ^{31}P-NMR study of Andean native and lowlanders. *J. Appl. Physiol.: Respir. Environ. Exerc. Physiol.* 70: 1963–1976, 1991.

89. McKenzie, J. C., I. Tanaka, T. Inagami, K. S. Misono, and R. M. Klein. Alterations in atrial and plasma atrial natriuretic factor (ANF) content during development of hypoxia-induced pulmonary hypertension in the rat. *Proc. Soc. Exp. Biol. Med.* 181: 459–463, 1986.

90. Mhyre, L. G., D. B. Dill, F. G. Hall, and D. K. Brown. Blood volume changes during three-week residence at high altitude. *Clin. Chem.* 16: 7–14, 1970.

91. Milledge, J. S., D. M. Catley, M. P. Ward, E. S. Williams, and C. R. A. Clarke. Renin–aldosterone and angiotensin-converting enzyme during prolonged altitude exposure. *J. Appl. Physiol.: Respir. Environ. Exerc. Physiol.* 55: 699–702, 1983.

92. Monge, C. C., R. Lozano, C. Marchena, J. Whittembury, and

C. Torres. Kidney function in the high-altitude native. *Federation Proc.* 28: 1199–1203, 1969.

93. Moses, A. M. Influence of adrenal cortex on body water distribution in rats. *Am. J. Physiol.* 208: 662–665, 1965.

94. Öberg, B. Effects of cardiovascular reflexes on net capillary fluid transfer. *Acta Physiol. Scand.* 62(suppl. 229): 43–47, 1964.

95. Oelz, O., H. Howald, P. E. Di Prampero, H. Hoppeler, H. Claassen, R. Jenni, A. Bühlmann, G. Ferretti, J. Brückner, A. Veicsteinas, M. Gussoni, and P. Cerretelli. Physiological profile of world-class high-altitude climbers. *J. Appl. Physiol.: Respir. Environ. Exerc. Physiol.* 60: 1734–1742, 1986.

96. Ou, L. C., J. Silverstein, and B. R. Edwards. Renal function in rats chronically exposed to high altitude. *Am. J. Physiol.* 247(*Renal Fluid Electrolyte Physiol.* 16): F45–F49, 1983.

97. Ou, L. C., and S. M. Tenney. Adrenocortical function in rats chronically exposed to high altitude. *J. Appl. Physiol.: Respir. Environ. Exerc. Physiol.* 47: 1185–1187, 1979.

98. Pauli, H. G., B. Truniger, J. Klarsen, and R. O. Mulhausen. Renal function during prolonged exposure to hypoxia and carbon monoxide. I. Glomerular filtration rate and plasma flow. *Scand. J. Clin. Lab. Invest.* 22(suppl. 103): 55–60, 1968.

99. Pauli, H. G., B. Truniger, J. Klarsen, and R. O. Mulhausen. Renal function during prolonged exposure to hypoxia and carbon monoxide. II. Electrolyte handling. *Scand. J. Clin. Lab. Invest.* 22(suppl. 103): 61–67, 1968.

100. Pepelko, W. E. Effect of hypoxia and hypercapnia and in combination upon the circulating red cell volume of rats. *Proc. Soc. Exp. Biol. Med.* 136: 967–971, 1971.

101. Phillips, R. W., K. L. Knox, W. A. House, and H. N. Jordan. Metabolic responses in sheep chronically exposed to 6,200 m simulated altitude. *Federation Proc.* 28: 974–977, 1969.

102. Phinney, S. D., E. S. Horton, E. A. H. Sims, J. S. Hanson, E. Danforth, Jr., and B. M. LaGrange. Capacity for moderate exercise in obese subjects after adaptation to a hypocaloric, ketogenic diet. *J. Clin. Invest.* 66: 1152–1161, 1980.

103. Pugh, L. G. C. E. Blood volume and hemoglobin concentration at altitudes above 18,000 ft (5,500 m). *J. Physiol. (Lond.)* 170: 344–354, 1964.

104. Raff, H., D. L. Ball, and T. L. Goodfriend. Low oxygen selectively inhibits aldosterone secretion from bovine adrenocortical cells in vitro. *Am. J. Physiol.* 256 (*Endocrinol. Metab.* 19): E640–E644, 1989.

105. Raff, H., J. Shinsako, and M. F. Dallman. Renin and ACTH responses to hypercapnia and hypoxia after chronic carotid chemodenervation. *Am. J. Physiol.* 247 (*Regulatory Integrative Comp. Physiol.* 16): R412–R417, 1984.

106. Raff, H., S. P. Tzankoff, and R. S. Fitzgerald. Chemoreceptor involvement in cortisol responses to hypoxia in ventilated dogs. *J. Appl. Physiol.: Respir. Environ. Exerc. Physiol.* 52: 1092–1096, 1982.

107. Raffestin, B., S. Adnot, J. J. Mercadier, M. Levame, P. Duc, P. Braquet, I. Viossat, and P. E. Chabrier. Synthesis and secretion of atrial natriuretic factor during chronic hypoxia: a study in the conscious instrumented rat. *Clin. Sci. (Colch.)* 78: 597–603, 1990.

108. Ramirez, G., P. A. Bittle, M. Hammond, C. W. Ayers, J. R. Dietz, and G. L. Colice. Regulation of aldosterone secretion during hypoxemia at sea level and moderately high altitude. *J. Clin. Endocrinol. Metab.* 67: 1162–1165, 1988.

109. Reeves, J. T., E. E. Wolfel, H. J. Green, R. S. Mazzeo, A. J. Young, J. R. Sutton, and G. A. Brooks. Oxygen transport during exercise at altitude and the lactate paradox: lessons from Operation Everest II and Pike's Peak. *Exerc. Sports Sci. Rev.* 20: 275–296, 1992.

110. Rothe, C. F. Reflex control of veins and vascular capacitance. *Physiol. Rev.* 63: 1281–1342, 1983.

111. Saltin, B., and P. D. Gollnick. Skeletal muscle adaptability: significance for metabolism and performance. In: *Handbook of Physiology. Skeletal Muscle,* edited by L. D. Peachey. Bethesda, MD: Am. Physiol. Soc., 1983, sect. 10, p. 555–631.

112. Sanchez, C., C. Merino, and M. Figallo. Simultaneous measurement of plasma volume and cell mass in polycythemia of high altitude. *J. Appl. Physiol.* 28: 775–778, 1970.

113. Schmidt, M., B. Kretschmann, R. Schuster, and A. Honig. Influence of sorbitol and mannitol on the reactions of renal excretory function evoked by arterial chemoreceptor stimulation with almitrine in anaesthetized and artificially ventilated cats. *Biomed. Biochim. Acta* 49: 1067–1080, 1990.

114. Schnakenberg, D. D., L. F. Krabill, and P. C. Weiser. The anorectic effect of high altitude on weight gain, nitrogen retention and body composition of rats. *J. Nutr.* 101: 787–795, 1971.

115. Share, L., and M. N. Levy. Effect of carotid chemoreceptor stimulation on plasma antidiuretic hormone titer. *Am. J. Physiol.* 210: 157–161, 1966.

116. Siggaard-Andersen, J., F. B. Petersen, T. I. Hansen, and K. Mellemgaard. Plasma volume and vascular permeability during hypoxia and carbon monoxide exposure. *Scand. J. Clin. Lab. Invest.* 22 (suppl. 103): 39–48, 1968.

117. Singh, M. V., S. C. Jain, S. B. Rawal, H. M. Divekar, R. Parshad, A. K. Tyagi, and K. C. Sinha. Comparative study of acetazolamide and spironolactone on body fluid compartments on induction to high altitude. *Int. J. Biometeorol.* 30: 33–41, 1986.

118. Stein, T. P., R. W. Hoyt, M. O'Toole, M. J. Leskiw, M. D. Schluter, R. R. Wolfe, and W. D. B. Hiller. Protein and energy metabolism during prolonged exercise in trained athletes. *Int. J. Sports Med.* 10: 311–316, 1989.

119. Stein, T. P., R. W. Hoyt, R. G. Settle, M. O'Toole, and W. D. B. Hiller. Determination of energy expenditure during heavy exercise, normal daily activity, and sleep using the doubly-labelled-water ($^2H_2^{18}O$) method. *Am. J. Clin. Nutr.* 45: 534–539, 1987.

120. Stickney, J. C., D. W. Northup, and E. J. Van Liere. The effect of anoxic anoxia on urine secretion in anaesthetized dogs. *Am. J. Physiol.* 147: 616–621, 1946.

121. Surks, M. I., K. S. K. Chinn, and L. O. Matoush. Alterations in body composition in man after acute exposure to high altitude. *J. Appl. Physiol.* 21: 1741–1746, 1966.

122. Sutton, J. R., and G. J. F. Heigenhauser. Lactate at altitude. In: *Hypoxia: the Adaptations,* edited by J. R. Sutton, G. Coates, and J. E. Remmers. Toronto: Decker, 1990, chapt. 17, p. 94–97.

123. Swenson, E. R., T. B. Duncan, S. V. Goldberg, G. Ramirez, S. Ahmad, and R. B. Schoene. The diuretic effect of acute hypoxia in humans: relationship to hypoxic ventilatory responsiveness and renal hormones. *J. Appl. Physiol.* 78: 377–383, 1995.

124. Torres, C., R. Lozano, J. Whittembury, and C. Monge. Effect of angiotensin on the kidney of the high altitude native. *Nephron* 7: 489–498, 1970.

125. Tunny, T. J., J. Van Geldern, R. D. Gordon, S. A. Klemm, S. M. Hamlet, W. L. Finn, G. M. Carney, and C. Brand-Maher. Effects of altitude on atrial natriuretic peptide: the bicentennial Mount Everest expedition. *Clin. Exp. Pharmacol. Physiol.* 16: 287–291, 1989.

126. Van Liere, E. J. The effect of anoxia on the alimentary tract. *Physiol. Rev.* 21: 307–322, 1941.

127. Ward, M. P., J. S. Milledge, and J. B. West. Peripheral tissues. In: *High Altitude Medicine and Physiology,* edited by M. P. Ward, J. S. Milledge, and J. B. West. Philadelphia: University of Pennsylvania Press, 1989, chapt. 11, p. 201–218.

128. West, J. B. Lactate during exercise at extreme altitude. *Federation Proc.* 45: 2953–2957, 1986.

129. West, J. B., S. J. Boyer, D. J. Graber, P. H. Hackett, K. H. Maret, J. S. Milledge, R. M. Peters, Jr., C. J. Pizzo, M. Samaja. F. H.

Sarnquist, R. B. Schoene, and R. M. Winslow. Maximal exercise at extreme altitudes on Mount Everest. *J. Appl. Physiol. Respir. Environ. Exerc. Physiol.* 55: 688–698, 1983.

130. Whithey, W. R., J. S. Milledge, E. S. Williams, B. D. Minty, E. I. Bryson, N. P. Luff, M. W. J. Older, and J. M. Beeley. Fluid and electrolyte homeostasis during prolonged exercise at altitude. *J. Appl. Physiol.: Respir. Environ. Exerc. Physiol.* 55: 409–412, 1983.

131. Wiersbitzky, M., R. Landgraf, S. Gruska, H. Oppermann, R. Schuster, F. Balke, B. Wedler, and A. Honig. Hormonal and renal responses to arterial chemoreceptor stimulation by almitrine in healthy and normotensive men. *Biomed. Biochim. Acta* 49: 1155–1163, 1990.

132. Winslow, R. M., and C. Monge. Exercise capacity. In: *Hypoxia, Polycythemia, and Chronic Mountain Sickness,* edited by R. M. Winslow and C. Monge. Baltimore, MD: Johns Hopkins University Press, 1987, chapt. 8, p. 142–161.

133. Winter, R. J. D., L. Meleagros, S. Pervez, H. Jamal, T. Krausz, J. M. Polak, and S. R. Bloom. Atrial natriuretic peptide levels in plasma and in cardiac tissues after chronic hypoxia in rats. *Clin. Sci. (Colch.)* 76: 95–101, 1989.

134. Wolfe, R. R., and S. M. Horvath. Blood volume responses of rats adapted to different barometric pressures. *Proc. Soc. Exp. Biol. Med.* 148: 89–93, 1975.

135. Wolfel, E. E., B. M. Groves, G. A. Brooks, G. E. Butterfield, R. S. Mazzeo, L. G. Moore, J. R. Sutton, P. R. Bender, T. E. Dahms, R. E. McCullough, R. G. McCullough, S. Huang, S. Sun, R. F. Grover, H. N. Hultgren, and J. T. Reeves. Oxygen transport during steady state, submaximal exercise in chronic hypoxia. *J. Appl. Physiol.: Respir. Environ. Exerc. Physiol.* 70: 1129–1136, 1991.

136. Yoshino, M., K. Kato, K. Murakami, Y. Katsumata, M. Tanaka, and S. Mori. Shift of anaerobic to aerobic metabolism in the rats acclimatized to hypoxia. *Comp. Biochem. Physiol.* 97A: 341–344, 1990.

137. Young, A. J., W. J. Evans, A. Cymerman, K. B. Pandolf, J. J. Knapik, and J. T. Maher. Sparing effect of chronic high-altitude exposure on muscle glycogen utilization. *J. Appl. Physiol.: Respir. Environ. Exerc. Physiol.* 52: 857–862, 1982.

138. Young, A. J., and P. M. Young. Human acclimatization to high terrestrial altitude. In: *Human Performance and Environmental Medicine at Terrestrial Extremes,* edited by K. B. Pandolf, M. N. Sawka, and R. R. Gonzalez. Indianapolis, IN: Benchmark, 1988, p. 467–495.

139. Young, P. M., P. B. Rock, C. S. Fulco, L. A. Trad, V. A. Forte, Jr., and A. Cymerman. Altitude acclimatization attenuates plasma ammonia accumulation during submaximal exercise. *J. Appl. Physiol.: Respir. Environ. Exerc. Physiol.* 63: 758–764, 1987.

140. Zakheim, R. M., F. Bodola, M. K. Park, A. Molteni, and L. Mattioli. Renin–angiotensin system in the llama. *Comp. Biochem. Physiol.* 59A: 375–378, 1978.

141. Zakheim, R. M., A. Molteni, L. Mattioli, and M. Park. Plasma angiotensin II levels in hypoxic and hypovolemic stress in unanaesthetized rabbits. *J. Appl. Physiol.* 41: 462–465, 1976.

56. Brain hypoxia: metabolic and ventilatory depression

P. L. LUTZ | *Department of Biological Sciences, Florida Atlantic University, Boca Raton, Florida*
N. S. CHERNIACK | *School of Medicine, Case Western Reserve University, Cleveland, Ohio*

THE INITIAL RESPONSE OF THE BODY to protect the central nervous system (CNS) from even mild hypoxia is to enhance O_2 delivery (hyperventilation, increased cerebral blood flow), increase substrate supply (cerebral vasodilatation), and, if needed, to "kick start" an emergency energy supply (Pasteur's effect). If the hypoxia is so severe that these measures do not suffice and the brain starts going into energy failure, the next response to prolong survival is to down-regulate the energy demands of the brain (decreased neuronal activity) with a coordinated reduction in energy supply (the anti-Pasteur effect, depression of ventilation). For most vertebrate species the obligatory energy requirements of the brain are so high that the "window of survival" under conditions of severe hypoxia is very narrow, but a few species are able to suppress brain energy costs to such an extent that even prolonged brain anoxia is no longer a survival challenge. The mechanisms by which these processes are achieved are discussed in this chapter.

METABOLIC ASPECTS

Mechanisms of Hypoxic Failure

Most of the brain cells' energy expenditure goes to processes that are essential for neuronal functioning, such as ion transport across cell membranes to restore ion gradients disturbed by the transmission of action potentials and the synthesis, transport, release, and uptake of neurotransmitters (161). The brains of most vertebrate species, in consequence, have little ability to withstand loss of energy supply, and, indeed, even a short exposure to severe hypoxia (minutes) can be sufficient to reduce the cells' energy status and to disturb ionic gradients, while longer exposure will result in irreversible damage (161, 162). The recognition that oxygen insufficiency is a major factor in many human neurological disorders has stimulated intense study into the mechanisms behind this hypoxic vulnerability, and a picture of the basic events is beginning to emerge.

Initially, anoxia produces a brief (< 30 s) period of depolarization in rat hippocampal neurons, characterized by an increase in excitability and no change in membrane input resistance (86). A short (2–3 min for the rat) hyperpolarization phase follows, caused by an increase in K^+ channel conductance (53, 86), during which membrane input resistance decreases, action potential threshold increases, and synaptic transmission is inhibited (86). Potassium channels activated by the decrease in adenosine triphosphate (ATP) (K_{ATP}^+ channels) that occurs during anoxia are thought to mediate hyperpolarization in such regions as the hippocampus and brain stem (50, 65, 114). There is evidence, however, that in the hippocampus K_{ATP}^+ channels are mainly present in glutamate-releasing terminals and serve to reduce anoxic glutamate release (74, 85). Here, an anoxia-induced rise in Ca_i^{2+} (164) is thought to be the major cause of hyperpolarization by activating Ca^{2+}-sensitive K^+ channels (73, 74, 86). During this period membrane ion pumps decrease activity, and there is a net outward leakage of K^+ from the neurons and a slow increase in K_o^+ levels (53). In the mammal when K_o^+ reaches about 10 mM, after about 3–5 min of anoxia, an abrupt transition takes place and K_o^+ rapidly rises to 50–70 mM, producing complete depolarization (53).

Depolarization also facilitates a massive release of the excitatory amino acids (EAA) glutamate and aspartate from presynaptic terminals (103, 148, 162), and at the postsynaptic membrane extracellular glutamate activates kainate and quisqualate glutamate receptors, producing an increase in membrane permeabilities to monovalent cations, which further propagates depolarization (162, 163). Depolarization also allows glutamate to activate the N-methyl-D-asparate (NMDA) glutamate receptor by removing the voltage-dependent Mg^{2+} block, thereby opening Ca^{2+}-permeable channels and producing a flood of Ca^{2+} ions into the cell from the extracellular fluid (163). At this time a fall in intracellular ATP produces a failure of Ca^{2+}-regulating mechanisms, including energy-dependent Ca^{2+}-extrusion mechanisms, and causes the release of bound calcium from the endoplasmic reticulum (9). The resultant uncontrolled and explosive rise in Ca_i^{2+} acts as a trigger for multiple dysfunctional effects, including phospholipid hydrolysis and a rise in harmful free fatty acids, particularly arachidonic acid, that lead to the formation of damaging free radicals (162). In the hypoxic–ischemic brain there are also massive releases of dopamine and norepinephrine, which cause tissue injury (43), and dopamine and serotonin levels may remain elevated even after normoxia is reestablished due to impaired uptake mechanisms (6). Not surprisingly, the uncontrolled release of excitatory neurotransmitters is believed to be an important cause of anoxic/ischemic brain damage (148).

However, different regions of the brain show different degrees of vulnerability; for example, the neocortex is much more resistant to hypoxia than the hippocampus, and within the hippocampus the CA_1 neurons are more sensitive than those of the CA_3 subregion (143). Excitatory amino acids may be a factor here. For example, the greater vulnerability of the striatum to anoxic/ischemic damage has been related to the slower postischemic decline (recovery) in extracellular levels of glycine in the striatum (44).

Clearly, knowledge of the mechanisms that account for the comparatively slight differences in anoxic tolerance of different (sub)regions of the vertebrate brain may help to discover the primary causes of hypoxic damage, but further insights may be obtained from the exceptions to the general rule. Newborn mammals are particularly interesting in that they are more resistant to hypoxia than adults, being able to survive a 100% N_2 atmosphere as much as 10–30 times longer than adults (26). More remarkably, there are even a few species of reptile [some freshwater turtles, *Chrysemys picta*, *Trachemys* (formerly *Pseudemys*) *picta*] and fish (crucian carp, *Carassiuss carassius*, and the goldfish, *C.auratus*) with brains that can withstand complete anoxia for days to months (93). This chapter focuses on how these "exceptions" avoid the catastrophes of brain anoxic failure and compares the mechanisms that account for anoxic tolerance in the turtle (the most widely studied species) with those in the mammalian neonate for possible insights into common adaptive strategies.

Mechanisms of Brain Hypoxic Tolerance

Metabolic Rate Factors. A reduction in oxygen consumption occurs in newborn mammals breathing 10% O_2 (112), and a decrease in O_2 consumption and CO_2 production also occurs in members of many small mammalian species exposed to hypoxia (35). The extraordinary anoxic tolerance of freshwater turtles, however, is only in small part due to the lesser metabolic requirements of reptiles compared to mammals. Although after compensating for body temperature differences the whole-body standard metabolic rate of the reptile is only one-fourth to one-sixth that of the mammal (59, 60), the oxygen consumption rates of turtle and rat brain slices (8) and isolated synaptosomes (30) are almost identical. Indeed, Mink et al. (107) showed that when brain O_2 consumptions for mammals, reptiles, amphibians, and fish were normalized to 37°C results were remarkably similar. Also, while the protein content of reptilian brain tissue is about half that of the mammal, on a per milligram protein basis reptilian brain cytochrome oxidase activity is 1.4 times greater (59). At the same temperature, rat and turtle brains have equivalent activities of citrate synthase, while the turtle has much higher levels of lactate dehydrogenase and hexokinase (171). Furthermore, the brains of lizards *(Anolis sagrei)* are much less anoxia-tolerant than those of *T.scripta* (128).

The anoxia-tolerant brain is not, therefore, a simple consequence of a lower vertebrate level of organization; it is the result of specific adaptations in those few species that survive prolonged anoxia.

Transition to the Anoxic State. As the oxygen supply falls, the glycolytic rate of the turtle brain increases, as indicated by an elevated lactate production (15, 95) and activation of pyruvate kinase (67). During this early hypoxic period heat production remains constant despite lower oxygen consumption (137), indicating that the enhanced glycolysis makes up all of the reduction in oxidative phosphorylation brought about by the decline in partial pressure of oxygen in tissue (PO_2).

The Anoxic State. As the anoxic state becomes established in the turtle brain, protein kinase is inactivated (67, 169), glycogen phosphorylase activity is reduced by 70% (67), and ATP production from glycolysis is reduced to about 10%–20% of initial values (15, 97). Since glycolysis is the only important source of energy for the anoxic brain (138, 158), it can be concluded that the transition to the anoxic state is accompanied by a drastic fall in the manufacture rate of ATP. However, during prolonged anoxia K_o^+ rises only slightly, from 3 to 6 mM in 48 h (158), ATP levels are maintained for greater than 6 h (15), and action potentials can be generated for at least 3 h (24).

It is clear that to remain viable and maintain ionic and energy balance the very substantial reduction in brain cell energy supply during anoxia must be matched by a corresponding reduction in energy use.

Mechanisms of Brain Anoxic Tolerance

Signal for Energy Failure. Although brain levels of ATP, adenosine 5′-diphosphate (ADP), and adenosine monophosphate (AMP) are maintained throughout many hours of anoxia, a limited but significant temporary fall in ATP is seen early in the transition period (67, 95). Nilsson and Lutz (127) have suggested that this fall in ATP acts as a signal of energy insufficiency and that it is involved in initiating the changes that result in the drastic reduction of metabolic rate (see Adenosine, below). When the full hypometabolic state is established, anaerobic glycolysis, albeit at a depressed rate, is sufficient to meet the energy needs of the cell and its energy status is restored.

Electrical Activity. In the turtle brain, anoxia quickly depresses electrical activity and abolishes synaptic transmission (24, 25, 33, 140). This substantial reduction in electrical activity can produce a correspondingly large savings in energy expenditure. The electrical silence seen in isolated mammalian brain slices (89) or produced by deep anesthesia (31) accounts for much of the 50% or more depression of metabolic rate (31, 89).

Channel Arrest. Because such a large proportion of cellular energy is used to maintain ion gradients or to restore them after the firing of action potentials, a major reduction in ion flux should yield important energy savings. This may be achieved by a decrease in plasma membrane ion permeability (97) via "channel arrest" (56, 92). Channel arrest includes two distinct changes (159): (1) "leakage arrest," which could reduce ion leakage by decreasing ion conductance in quiescent (nonspiking) cells, and (2) "spike arrest," which may be brought about by inhibiting channel activity associated with action potentials or by reducing activity by suppressing synaptic transmission.

That some channel arrest changes do occur in the turtle brain is indicated by the finding of Chih and colleagues (16) of a decreased rate of potassium leakage during anoxia. In leakage arrest the decrease in ion conductance would be expected to produce an increase in membrane input resistance. However, Perez-Pinzon and co-workers (140) found a decrease in input resistance in anoxic Purkinje cells of isolated turtle cerebellum, and Doll and colleagues (24) reported no change in whole-cell input resistance of cortical pyramidal neurones during anoxia. These studies indicate that leakage arrest makes no significant contribution toward anoxic tolerance.

Unlike the arrest of leakage channels, spike arrest may be achieved by many mechanisms, including changes in voltage-gated or receptor-gated ion channels. For example, voltage-gated Na^+ channels are responsible for both initiation and conduction of action potentials, and modulation of their functional properties would be important determinants of brain electrical activity. In the isolated turtle cerebellum 4 h of anoxia produced a 42% decline in the density of voltage-gated Na^+ channels (139). Such a mechanism may underlie the anoxia-induced 14 mV increase in sodium action potential thresholds seen in the isolated turtle cerebellum (140), and if it occurs in vivo, it would be an important contributing factor toward the reduction in electrical activity in the anoxic brain by increasing the difficulty for a given synaptic input to provoke action potentials (139). In the isolated turtle synaptosome inhibiting the voltage-dependent Na^+ channels with tetrodotoxin and increasing Na^+ channel activation with veratridine caused substantial decreases and increases in synaptosome energy consumption, respectively (30).

Down-regulation of voltage-dependent sodium channels has been described in mammalian neurons. In fetal neurons increased Na^+ influx into the cell triggers a rapid ($t_{1/2} = 15$ min) disappearance of surface Na^+ channels, thought to be the result of a dissociation of α and β_1 subunits of the Na^+ channel (19). Neumann and co-workers (120) present evidence that Na^+ channel activity in intact rat brain neurons is modulated by protein kinase C and speculate that changing Na^+ channel activity may affect the frequency of action potential generation and alter neurotransmitter release.

Neurotransmitters and Related Substances. Since neurotransmitters are of paramount importance in the regulation of brain activity, it is likely that they and their related metabolites play a major role in the initiation and coordination of the processes that provide anoxic tolerance.

Adenosine. In the mammalian brain hypoxia or ischemia produces an increase in adenosine, which is released into the extracellular space (150, 179), where it increases cerebral blood flow (110), stimulates glycogenolysis (99), and decreases neuronal excitability as well as excitatory neurotransmitter release (142, 168).

Adenosine may even be involved in regulating the whole-body metabolic depression characteristic of anoxia-tolerant species. This is suggested by observations (124) that blocking adenosine receptors by administration of aminophylline caused the anaerobic crucian carp to increase ethanol production (the principal end-product of anaerobic glycolysis for this species). Indeed, adenosine acts as a "retaliatory metabolite" for several tissues, including the heart and brain, during conditions of energy insufficiency by reducing energy consumption while increasing energy supply (121).

In the turtle brain shortly after the onset of brain anoxia there is a marked but temporary rise in extracellular adenosine, probably linked to the contemporaneous fall in ATP (127). This release of adenosine may produce similar compensatory changes while oxidative phosphorylation is becoming increasingly depressed and result in the restoration of ATP levels as the brain goes into a hypometabolic state (93, 96). Infusing the specific adenosine A_1 blocker 8-cyclopentyltheophylline causes the anoxic isolated turtle cerebellum to depolarize (141_a), indicating that at least part of the protective action of adenosine during anoxia is via the A_1 receptor. Adenosine acts through the A_1 receptor to inhibit adenylate cyclase (168) and thereby, perhaps, to decrease the sensitivity of the postsynaptic membrane (142).

Inhibitory amino acids. Inhibitory neurotransmitters are likely to play a key role in anoxia survival since they act to depress neuronal activity and, therefore, energy consumption.

Gamma-aminobutyric acid (GABA) is the most widely studied inhibitory neurotransmitter in the vertebrate nervous system. Increases in tissue concentrations of GABA in the turtle (55, 125) and crucian carp (123) brain are seen over several hours of anoxia, but, more importantly, after about 100 min of anoxia, that is, at about the time when extracellular adenosine is declining (see Adenosine above), there is a large and sustained release of GABA in the turtle striatum, which slowly continues to increase, reaching about 90 times the normoxic level after 240 min of anoxia (126). Kriegstein and Connors (72) have demonstrated that the in vitro dorsal cortex of the turtle shows the classical characteristics of interneuronal inhibition by GABA receptors. The early part of the inhibitory response was mediated by a Cl^- conductance, presumably via the $GABA_A$

receptor; however, the later phase of inhibition appeared to be mediated by a K^+ conductance, possibly coupled to $GABA_B$ receptors.

At the same time GABA is released there are also slow, sustained increases in extracellular concentrations of glycine and taurine, other putative inhibitory amino acids, in the anoxic turtle brain (126). Glycine is a well-established inhibitory neurotransmitter in lower brain areas and the spinal cord of mammals (102). In higher brain areas, however, glycine is a potent allosteric activator of glutamate NMDA receptors (34), acting, therefore, as an excitatory factor. Whether or not taurine can be classified as a true neurotransmitter is still vigorously debated (61), but there is evidence that it is involved in the hyperpolarization of Purkinje cell neurons (132) and in protecting hypoxic neurons from calcium influx (106). Elevated extracellular levels of inhibitory amino acids are likely, therefore, to maintain and even consolidate the depressed state of synaptic excitability that follows the initial metabolic down-regulation (96).

Excitatory amino acids. In contrast to GABA, tissue concentrations of the EAA glutamate decline in anoxic brains of both the turtle (94, 125) and the crucian carp (123). Perhaps more importantly, while a few minutes of anoxia causes a massive release of EAAs in the mammalian brain with devastating effects, no such increase occurs in the turtle brain for at least 4 h of anoxia (126).

However, when depolarized the anoxic turtle brain releases a mammalian-like surge of extracellular glutamate (126), and, interestingly, such brains do not recover evoked potential activity following reoxygenation (33), indicating that functional damage has been incurred. It would appear, therefore, that the turtle brain may be susceptible to the excitotoxic effects described for the mammalian brain and that its primary defense is simply to avoid depolarization. The kinetics, pharmacology, and regulation of the glutamate NMDA receptor complex are similar in rat and turtle brain (151), and both glutamate and aspartate evoke changes in the turtle cerebellum similar to those seen in the mammal, namely, causing K_o^+ to increase and Ca^{2+} to decrease (145) and triggering spreading depression (84). However, embryonic turtle cortical neurons, which also have NMDA and non-NMDA glutamate receptors that show characteristic glutamate-induced excitatory currents, are much more resistant to glutamate neurotoxicity than those of the mammal (172, 187).

Monoamines. Since monoamines require molecular oxygen for synthesis and degradation (160), it is likely that anoxia influences monoamine neurotransmitters and, perhaps, that the transmitters influence anoxic tolerance. Interestingly, while brain concentrations of the putative inhibitory monoamines serotonin, norepinephrine, and epinephrine (159) are preserved in the anoxic turtle and crucian carp brain (123, 125), those of the putative excitatory monoamine dopamine decline in the turtle brain (125). Maintenance of brain levels of the inhibitory monoamines may be related to the higher (four to ten times) concentrations found in turtles compared to mammals and to decreased turnover rates (125).

In summary, the pattern of release of neurotransmitters in the anoxic turtle brain suggests that the early rise in extracellular adenosine is the result of a fall in ATP levels and possibly serves as a signal for compensatory responses. Adenosine acts to reduce energy demand and increase supply and, in the turtle brain at least, to initiate the down-regulatory processes that suppress energy demand and result in the recovery of cell energy status. Adenosine levels in consequence decline, but the down-regulated state is consolidated by sustained release of inhibitory neurotransmitters. By these means anoxic depolarization is avoided as are the consequences of uncontrolled EAA release.

Enhanced Tolerance—The Mammalian Neonate

The greater tolerance of developing mammals compared to adults is due in large measure to the increased resistance of the immature nervous system, which declines to adult levels of susceptibility during the first few weeks after birth (178). Like the turtle, hypoxia produces much slower rates of increase in K_o^+ in the neonatal compared to the adult mammalian brain and slower rates of ATP decline (51, 53, 176). Anoxia also results in smaller changes in synaptic transmission in the neonatal rat hippocampus (13).

One important early response to acute hypoxemia in the fetal sheep is a redistribution of cardiac output, which doubles the blood supply to the brain at the cost of a reduction of blood flow to much of the carcass (136). Interestingly, these changes, which ensure a continued supply of oxygen and substrate to the brain and heart, are similar to the very substantial circulatory adjustments seen in the "classical" diving response of the marine mammal, suggesting that the marine mammal may have retained the fetal response to hypoxia into adulthood, a possible example of physiological neoteny (93).

Without doubt, like the turtle brain, part of the enhanced hypoxic tolerance of the neonatal mammalian brain can be attributed to its lesser energy requirements. Cerebral oxygen and aerobic glucose consumption for the 7 day old rat is about one-tenth that of the adult (178), and the activity of creatine phosphokinase is three- to sixfold less in the neonatal brain (131). Concentrations of mitochondrial enzymes for ATP production are also reduced (49). In the neonatal dog brain the lower rate of oxygen consumption also results in a more favorable (reduced) blood tissue O_2 gradient compared

to the adult (129), which would also contribute to its (systemic) hypoxic tolerance. However, neonatal and adult brains show a critical decline in mitochondrial phosphorylation potential [fall of 50% in brain phosphocreatine/inorganic phosphate (PCr/Pi) ratio] at the same brain vascular percentage of oxygen saturation, though this occurs at a lower arterial PO_2 in the neonate due to the higher oxygen affinity of neonatal blood (129).

Undoubtedly, ontogenetic changes are important factors. Evidence suggests that major changes in ion channel distribution and receptor properties occur early in development that may influence hypoxic tolerance. For example, the newborn rat brain has a much lower density of K_{ATP}^+ channels than the adult, which may contribute to the slower increase in K_o^+ during oxygen deprivation (65). After birth there is a very substantial increase in brain sodium channel density and a change in the ratio of the R_I and R_{II} sodium channel subtypes, which are thought to have distinct functional properties (46). In sharp contrast to adults, benzodiazepines do not potentiate GABA-induced inward currents in neonatal rat hippocampal cells (149). Indeed, in the early postnatal period GABA acts as an excitatory transmitter in hippocampal neurons (14).

While the release of adenosine in the hypoxic neonatal brain may play a role in hypoxic survival (83, 135), the immaturity of the adenosine receptors (22) and the comparatively small increases in extracellular adenosine concentrations (141) indicate that the protective effects may be minor. Indeed, xanthine and hydroxyxanthine accumulate to such an extent in hypoxic fetal sheep cerebral cortex (70) and in the extracellular fluid of neonatal rat striatum (141) that they set the stage for the release of damaging free radicals generated when O_2 is restored (70). Also, unlike turtles, in 7 day old pups moderate hypoxia causes an increase in striatal levels of dopamine and a possible release of dopamine into the extracellular fluid (47), an important putative cause of cellular damage in adults (43).

There is evidence, however, that, although the neonatal brain is particularly sensitive to excitotoxic agents (101), neonates do not suffer as much EAA damage as adults after hypoxia or ischemia (10). This is not apparently due to the absence of functional NMDA and quisqualate receptors (62). The answer may lie in the circumstances that cause EAA release. The anoxic 5 day old rat brain is protected from EAA damage at least in part because EAAs are not released as long as the brain does not depolarize (141). There may be differences in the pattern of release associated with development. Gordon et al. (48) report that hypoxia/ischemia in the striatum of 7 day old rats produces only transient and inconsistent changes in extracellular glutamate and that the rise in glutamate is very much smaller than that seen in the adult. Indeed, Cherici and colleagues (10) report that even depolarization of hippocampal slices from 4–9 day old newborn rats does not result in the release of glutamate or aspartate. It would appear that the most important causes of increased neonatal hypoxic tolerance may be simple consequences of the comparatively undifferentiated state of the brain of the newborn, with its lower energy requirements, slower decline in ATP, lower excitability levels, delayed depolarization, and lesser scope to release harmful EAA.

However, the neonatal brain may also possess some "add-on" features to protect against hypoxia. It is possible that one of the suggested protective mechanisms of survival in turtle brain, namely Na^+ channel down-regulation (139), might be involved in the enhanced neonatal hypoxic tolerance. Interestingly, Dargent and Courand (19) found a rapid down-regulation of sodium channels whenever there was an increase in Na^+ flux in cultured brain cells from the rat fetus. In the anoxic neonatal rat brain an influx in Na^+ probably accompanies the observed slow rise in K_o^+ (141). It has been proposed by Schurr (156), though disputed by Lehman et al. (87), that the higher levels of taurine in the neonatal brain serve to protect it from damage by attenuating Ca^{2+} flux.

HYPOXIC DEPRESSION OF VENTILATION

At the whole-body level, the most immediately noticeable response to hypoxia is a dramatic change in breathing patterns. In this section we concentrate on the more acute effects of hypoxia on ventilation, recognizing that precise patterns of change produced by hypoxia will depend to a degree on the rapidity with which the reduction in oxygen is experienced, as well as its absolute level.

As far as ventilation is concerned, hypoxia excites peripheral chemoreceptors, which in turn, by action on the respiratory neurons of the brain, stimulate breathing (11, 32, 79, 81, 116, 122, 130, 184, 185). Although at least some mammalian species have oxygen-sensitive receptors scattered throughout the body, the excitatory effects on ventilation mainly result from stimulation of the carotid chemoreceptors (the carotid bodies) (32, 79, 81, 184, 185). This ventilation-augmenting action is modified by the effects of hypoxia on other parts of the body. Hypoxia, for example, heightens the secretion of catecholamines, which, in general, has a stimulating effect on breathing. Hypoxia tends to relax some but not all smooth muscles and, overall, reduces resistance to the flow of air to alveoli through the smooth muscle–encased bronchial tree (184, 185).

By contrast, the complex effects of hypoxia on the brain, discussed earlier, which include changes in neurotransmitter release and in central chemoreceptors, act to reduce ventilation levels (111, 175, 180, 186).

Since the respiratory system acts in concert with the circulation to deliver oxygen to the tissues, circulatory effects must be considered in any examination of the actions of hypoxia on breathing (79). In general, environmental hypoxia stimulates the circulation by increasing sympathetic activity to the vasomotor neurons innervating the heart and blood vessels, and this in turn leads to increases in heart rate, blood pressure, and cardiac output (111, 180, 186).

With continued exposure to hypoxia, these cardiorespiratory responses are further altered in a complex fashion as more slowly developing reactions to hypoxia take place. These slower effects include an increase in red blood cell mass, so that the O_2 carrying capacity of the blood is increased, and an increase in the vascularity of tissues (185).

The cardiorespiratory responses to hypoxia, the first line of defense to decreased oxygen in the environment, are altered by these more chronic effects so that these systems can more appropriately respond to any additional acute changes in oxygen level. In this section we first describe the depressive effects of hypoxia on ventilation, then the possible sites at which these effects might be expected and the cellular mechanisms involved, particularly in humans, and finally some of the clinical implications of hypoxic depression.

Cardiorespiratory Responses to Hypoxia

In humans an abrupt exposure to moderate levels of hypoxia generally causes an initial rapid increase in ventilation (within seconds) due to carotid body excitation, which peaks in 3–6 min, but then ventilation declines over the subsequent 20–30 min until it reaches a steady level greater than it was before hypoxic exposure (28, 41, 186). Some of this rise and subsequent fall in ventilation is caused by a reduction in levels of arterial P_{CO_2} arising from the initial increase in ventilation. The lower arterial P_{CO_2} tends to alkalinize the immediate environment of brain CO_2 chemoreceptors and to lessen their excitability (90, 154). These changes in local P_{CO_2} of chemoreceptors are believed to occur more slowly than the changes in arterial P_{O_2} that excite the carotid body, but even when P_{CO_2} levels in the arterial body are maintained constant by experimental manipulations, the decline in ventilation from initial levels still occurs. This decline in ventilation has been termed *hypoxic ventilatory depression* (68).

In adults hypoxic ventilatory depression is caused, in most reports, by a reduction in tidal volume, but a slowing of respiratory frequency has been observed by some (116). Hypoxic depression has been reported to occur not just in sea-level inhabitants but also in individuals born at or acclimatized to altitude (17, 80).

Hypoxic depression of ventilation is more prominent in newborns than in adults. In the newborn, exposure to hypoxia, particularly in cool environments, fails to produce any sustained rise in ventilation and breathing returns to control or even to subnormal levels despite unchanged environmental levels of hypoxia (17, 18, 85).

As ventilation declines, so does the activity of the diaphragm and the phrenic nerve, as well as that of the upper airway muscles (such as the genioglossus) and the expiratory muscles (such as the triangularis sterni) (36, 165, 177). Even during the initial increase in inspiratory activity, there may be no change in the activity of the expiratory muscles, though the activity of expiratory and inspiratory muscles does increase in parallel when breathing is stimulated by CO_2 rather than by hypoxia (36). It is believed that the weak or absent response of the abdominal muscles is caused by their marked sensitivity to the lowering of P_{CO_2}.

Subsequent studies indicate that the depression of ventilation by hypoxia is related to its initial excitatory effect (29, 40, 41, 60). For example, when the response to hypoxia is enhanced in normal adults by prior administration of the agent almitrine, which excites the carotid body, the initial peak of the ventilatory response is enhanced, as is its subsequent decline. Similarly, acute administration of a gas containing less oxygen than the ambient air in animals exposed to CO_2 exaggerates both the initial excitatory response to hypoxia and its later reduction.

When acute exposure to hypoxia is terminated in humans by a return to normoxic conditions, ventilation may continue to be depressed, that is, remain at less than control levels, for up to 60 min. In experimental protocols with healthy adult humans where exposure to acute hypoxia is repetitive and separated by periods of breathing air with normal sea-level percents of O_2, the acute peak of excitatory ventilatory response may be absent if the normoxic intervals between periods of hypoxia are too short (less than 15 min) (29).

A number of different protocols have been used in anesthetized animals to better describe the effects of hypoxia on ventilation (2, 104, 105, 111, 116, 118, 177, 180). After carotid sinus nerve section, the excitatory effects of hypoxia on breathing are usually absent or greatly diminished (2, 116, 180). If severe hypoxia is maintained, apnea eventually occurs. In artificially ventilated animals breathing low concentrations of O_2 (8% or less) normal phrenic nerve discharge disappears, and for a time no phrenic activity is discernible at all (116). This is often followed by a pattern of gasping, and each phrenic nerve discharge peaks abruptly. If artificial ventilation is maintained, systemic blood pressure will fall

to life-threatening levels (116). A similar sequence of phrenic nerve activity changes occurs in decerebrate animals or in animals artificially ventilated with carbon monoxide to produce hypoxia, even though in the reported studies using this technique the development of systemic hypoxia is much slower (116). Both in carbon monoxide–ventilated animals and in those exposed to hypoxia, ventilation will continue to be stimulated by carbon dioxide even after the onset of hypoxic ventilatory depression but apparently not when gasping has occurred (104, 105, 118). In carbon monoxide–ventilated cats where multiple levels of CO_2 can be tested the slope of the phrenic nerve response to hypercapnia remains at control levels during hypoxic depression, even though the absolute levels of activity for a given level of P_{CO_2} are much less (104). A similar observation has been made in humans studied after carotid body resection (57).

Studies also show that respiration is depressed in adult animals when blood equilibrated with low O_2 mixture or sodium cyanide is injected in the vertebral artery, confirming observations in decerebrate animals that higher brain centers are not necessary for hypoxic depression (108).

Studies of the changes in sympathetic nerve activity with hypoxia also have been carried out in anesthetized animals (146, 147, 183). Sympathetic discharge usually rises and falls nearly in synchrony with inspiration and expiration. During hypoxic respiratory depression, these phasic changes in sympathetic activity decline and may eventually vanish, but tonic sympathetic activity increases. This augmented sympathetic activity occurs even when only the medulla is exposed to hypoxia but also in spinalized animals, suggesting both a spinal and a medullary effect of hypoxia on vasomotor discharge.

Thus descriptive studies of hypoxic ventilatory depression in humans and other mammals indicate that multiple systems are involved and that long-lasting effects are produced. The effects seem to depend on at least three factors: the severity of hypoxic exposure, its rate of development, and its duration. Because of the different rates at which physiological systems respond to hypoxia, it is possible that the observed changes in cardiorespiration may not have a single cause.

Potential Sites of Action of Hypoxic Ventilatory Depression

Where hypoxia produces ventilatory depression is still argued. Implicated sites include the pulmonary system itself, respiratory muscles, cerebral blood vessels, peripheral chemoreceptors, and the brain. The predominance of evidence supports the central nervous system as the main anatomical location for hypoxic depression, and, at least in adult animals, data indicate that the medulla itself is involved. However, it is likely that the severity of hypoxic depression is influenced by all of the effects of low levels of oxygen at these sites. Moreover, there is evidence that the influence of hypoxia at these different locations may not be the same in immature mammals as in adults.

Lung and Respiratory Muscles. LaFramboise et al. (76–78) found that the relationship of occlusion pressure and diaphragm electromyogram to ventilation increased during hypoxic ventilatory depression. This finding suggested that greater force was needed to generate a given level of ventilation when breathing was depressed by hypoxia and, hence, that mechanical changes in the lung or the respiratory muscles might contribute to the reduced ventilation observed with hypoxic depression. Functional residual capacity has been reported to be enlarged during hypoxia, probably due to the relative inactivity of the expiratory muscles (153). This would tend to lower the force that could be generated by the inspiratory muscles when stimulated by motor nerves.

Muscle fatigue can occur when hypoxia compromises oxygen delivery (98), but this would require severe hypoxia, while hypoxic ventilatory depression can occur even with mild to moderate hypoxia. Since hypoxic respiratory depression can occur even in artificially ventilated animals, it seems unlikely that the effects of hypoxia on either the lungs or the respiratory muscles account in a significant way for respiratory depression.

Cerebral Blood Vessels. With hypoxia, cerebral blood flow rises so that, even if arterial P_{CO_2} is kept constant, brain tissue P_{CO_2} tends to fall (21). This increase in cerebral blood may be mediated in part by the increase in brain adenosine that occurs with hypoxic exposure (83, 110, 157). The central chemoreceptors which drive breathing as brain P_{CO_2} or H^+ is increased are located in the medulla, even though their precise sites are unknown (90, 154). An increase in cerebral blood flow with hypoxia tends to decrease activity of the central chemoreceptors and, hence, to lessen ventilation.

It has been suggested that the effects of hypoxia occur earliest in higher brain centers. This, in part, may be related to the fact that hypoxia tends to increase blood flow more in caudal areas of the brain (the sites of central chemoreceptors) than in rostral areas (115). This differential effect of hypoxia may be mediated directly by sympathetic activity or by the carotid body, both of which are stimulated by low levels of oxygen (182).

However, in most studies time courses of changes of cerebral blood flow and ventilation do not coincide (3, 130, 173). That is, ventilatory depression decreases while cerebral blood flow is steady. In addition, lactic acid accumulation in the brain during hypoxia tends to offset the decrease in H^+ that might be produced by a cerebral blood flow increase (64, 119). Several investigators report an increase rather than a decrease in brain acidity during hypoxia. Increased acidity at chemoreceptors should stimulate, not reduce, breathing.

Edelman and co-workers (116) proposed that increased cerebral blood may account for most of the ventilatory reduction seen during mild hypoxia in cats breathing CO.

Peripheral Chemoreceptors. Peripheral chemoreceptors contain both excitatory and inhibitory neurotransmitters, and with prolonged exposure to hypoxia (day and weeks) synthesis and release of these agents may change (4, 32, 81, 174). Thus changes in the peripheral chemoreceptors themselves seem to contribute to the increase in ventilation observed in altitude acclimatization. Moreover, the carotid body is subjected to some inhibitory control by the CNS through efferent nerves (82, 133). It is possible that a decrease in carotid body response to hypoxia might participate in the decrease in ventilation from peak values observed in hypoxic depression (7, 155).

Measurements of constant carotid sinus nerve activity by Andronikou and colleagues (2) show conclusively that changes in peripheral chemoreceptor (decreases) do not contribute to hypoxic ventilatory depression in adult anesthetized animals. However, indirect evidence suggests that a decline in carotid body activity plays some role in hypoxic depression in immature animals and perhaps in unanesthetized humans (5, 7, 69, 155).

Schramm and Grunstein (155) studied rabbit pups and found greater hypoxic depression when carotid bodies were functional than when they were not. In addition, they observed that carotid body stimulation by cyanide injection increased ventilation when pups breathed normoxic air but led to depressed breathing when animals were exposed to hypoxic gas. These studies need to be confirmed.

Investigators at Oxford have used the transient changes in ventilation observed ("off transients") when oxygen replaces a hypoxic gas to study hypoxic depression (5, 68, 69). They assumed that oxygen breathing instantaneously shuts down carotid body activity. They examined the time course of ventilation changes when humans breathed oxygen, then a hypoxic gas mixture, and then oxygen once again. They found that switching from oxygen to the hypoxic gas produced an initial peak in ventilation, which then declined, as reported by others. They reasoned that if reduced breathing was caused by CNS depression, a return to O_2 breathing would lead to a level of ventilation less than the prehypoxia level. However, they could discern no significant difference between ventilation levels pre- and posthypoxic exposure and, hence, concluded that in conscious humans adaptation of peripheral chemoreceptors plays a major role in hypoxic depression (68). These intriguing investigations also need to be confirmed by more direct techniques.

Central Nervous System. Most investigators have concluded that in individuals exposed to hypoxia ventilation is the product of a rapidly acting excitatory action on the carotid body and a more slowly developing respiratory inhibitory effect occurring in the brain.

The most convincing evidence in this regard comes from studies demonstrating depression of respiration by hypoxia confined to the medulla (108, 109). Nonetheless, other areas of the brain may influence the severity and time course of hypoxic depression and the temporal pattern of recovery (23, 52).

Different areas of the brain have different effects on breathing which could influence hypoxic depression. In studies of hypoxic responsiveness it appears that the cortex exerts an inhibitory effect on the ventilatory effects of hypoxia, while more caudal areas, like the diencephalon, have a stimulating effect (134, 175).

Gallman and Millhorn (37) have carried out a series of studies suggesting that different areas of the brain may affect the nature of the aftereffects of hypoxia. In intact adult cats exposure to moderate hypoxia produced a long-lasting facilitatory effect, which was converted to a long-lasting inhibitory effect by high decerebration. With low decerebration aftereffects were observed following moderate hypoxia. With severe hypoxic exposure in cats, as in conscious humans, a long-lasting posthypoxic depression was always seen. The ventral medulla may play an important role in hypoxic depression since Mitra et al. (108, 109) have shown that microinjections of cyanide into the rostral half of the ventral medulla of adult cats leads to respiratory depression and sympathetic excitation characteristic of hypoxic depression.

Some investigators have reported that hypoxic ventilatory depression does not occur in immature animals after decerebration, even though it does in adult animals treated the same way (23, 37, 100). Srinivasan et al. (166) found that antagonists of the D_2 receptor of dopamine, an inhibitory neurotransmitter, prevented hypoxic depression. They also showed that the secretion of dopamine into the nucleus tractus solitarius (NTS), where respiratory neurons are concentrated, was increased in immature rabbits by hypoxia when both the mesencephalon and the carotid body were intact. Their finding supports the idea that hypoxic depression of breathing in immature animals depends on the hypoxic effect on both the brain and the carotid body. Studies of immature animals are complicated by the lability of their metabolic rate and by the inhibitory action of hypoxia on body temperature and body metabolism, both actions exerting a respiratory inhibitory action.

Also different from observations in adult animals are the respiratory-stimulating effects of cerebroventricular injection of cyanide in fetal sheep reported by Jansen and Chernick (63).

Mechanisms of Hypoxic Ventilatory Depression

Metabolic Effects. Even though the brain is a major site at which hypoxia acts to inhibit breathing, the mechanisms which cause this depression are unclear (116).

In humans hypoxic ventilatory depression can occur before substantial changes in brain ATP appear. Bacon and co-workers (3) exposed eight healthy adults to isocapnic hypoxia sufficient to reduce arterial oxygen saturation to 80% for 12 min. Cerebral cortical oxygenation (monitored by changes in cytochrome a_1,a_3) as well as cortical blood volume were assessed noninvasively with near infrared spectroscopy. Ventilation decreased, even though cortical oxygenation and blood flow had already changed maximally. However, in many areas of the brain oxygenation is marginal, and it remains possible that local changes in oxygenation and energy metabolism in critical areas (for example, near synapses) contribute to hypoxic depression of ventilation.

It has been suggested that the acidosis produced during severe hypoxia plays an important role in hypoxic ventilatory depression (75, 144). Studies by Neubauer and colleagues (119) demonstrated that dichloroacetic acid (an agent that decreases lactic acid formation) delayed respiratory depression in carbon monoxide–ventilated cats. However, they believed that the effects of dichloroacetic acid on hypoxic depression were indirect. Synthesis of GABA is promoted by acidosis, and by preventing acidosis the accumulation of the inhibitory neurotransmitter GABA is retarded. No effect of dichloroacetic acid on hypoxic ventilatory depression was seen in conscious humans by Georgopoulos et al. (38).

Neurotransmitters. Investigators believe that hypoxic ventilatory depression is mediated by an increase in the synthesis or release of one or more inhibitory neurotransmitters (dopamine, GABA, adenosine, opioids, etc.) or a decrease in excitatory neurotransmitters (glutamate, acetylcholine, aspartate, and, possibly, substance P) (20, 32, 58, 81, 88, 189, 190). Human studies support a role of adenosine in hypoxic depression, though insufficient experiments have been done to rule out the role of other neurotransmitters (20, 39).

Glutamate, the major excitatory neurotransmitter in the brain, stimulates breathing upon injection into the medulla. Its metabolism is decreased under hypoxic conditions (58). Substance P applied to the dorsal medulla of cats increased breathing in decerebrate pups, and substance P antagonists blocked the respiratory response to hypoxia (88, 190). Microdialysis studies have shown increased release of substance P into the NTS in the dorsal medulla during hypoxia.

Dopamine released into the NTS also increased during hypoxia if the pons was intact in immature rabbits (45). Decerebration also abolishes hypoxic depression in these animals but is restored by the injection of apomorphine, a dopamine agonist, into the fourth ventricle. Haloperidol or dopamine antagonists attenuate or block hypoxic ventilatory depression in mature cats (104).

In cats and dogs GABA inhibits respiration (71, 189), and GABA antagonists, such as bicuculline, reverse hypoxic depression (104). In addition, binding to GABA receptors is increased in rats with exposure to hypoxia (71), but there are no studies as yet in conscious animals or humans.

Endorphins depress respiration by acting on the mu-2 and delta receptors of neurons (71). In immature and fetal animals naloxone, an endorphin antagonist, reverses or diminishes hypoxic ventilatory depression (12, 54, 117, 152). Amelioration of depression in immature animals may be indirect and may depend on the effect of naloxone on the action of catecholamines (167). In addition, in some of the animal experiments in which the effects of naloxone have been examined, preparatory surgical procedures may have caused the release of endogenous opioids, which then contributed artifactually to hypoxic depression. Kagawa et al. (66) found no effect of naloxone on hypoxic depression in healthy humans.

Adenosine levels in the brain rise with hypoxia (93, 179, 188). Administration of adenosine reduces both the tidal volume and the frequency of breathing. Success in preventing hypoxic depression in the newborn has been reported by Darnall (20) using aminophylline, an adenosine receptor antagonist. Easton and Anthonisen (27) found that raising aminophylline to 16 mg/liter in eight healthy adults diminished hypoxic ventilatory depression and eliminated posthypoxic respiratory depression. However, restoration of ventilation was due to an increase in breathing frequency, while tidal volume remained depressed. In addition, adenosine has multiple effects which might contribute to this action, including its carotid body–stimulating action (32, 81).

Clinical Implications of Hypoxic Respiratory Depression

Although regulation of P_{CO_2} levels has long been the focus of studies on the control of breathing, naturally occurring stresses to the respiratory system are much more likely to involve changes in the ability to deliver O_2 to the tissues than in the ability to remove CO_2 (52, 91). Aging and diseases of the lung first affect the ability to maintain normal levels of arterial O_2 tension (1).

Depressed ventilatory responses to hypoxia have been reported to be risk factors in sudden infant death syndrome (SIDS) and in the development of acute mountain sickness (91, 113, 170, 181).

It is generally accepted that diseases that attack the respiratory pump (the lungs, the airways, and the respiratory muscles) exert their adverse effects by interfering with gas exchange (91). Although the degree of interference is correlated with the extent and severity of the changes in the mechanical performance of the respiratory pump, the correlation is only rough, and individuals with apparently equal degrees of disease may have quite different levels of arterial oxygen and CO_2 tension, as well as different degrees of impairment in terms of their ability to function (185). For many years it has been suspected that differences in the regulation of breathing, either inherent or acquired in response to the disease process, may significantly influence the ultimate systemic outcome.

Changes in respiratory system compliance and resistance make the energy cost of breathing greater (91). Further increases in ventilatory work occur because of the decreased efficiency in gas exchange caused by lung diseases. As breathing work increases, maximal levels of ventilation fall and blood oxygenation becomes poorer.

In sleep apnea syndrome episodes there is no gas exchange and effective breathing ceases intermittently and repeatedly during the night (170). As a consequence, severe transient hypoxia occurs, which may ultimately lead to pulmonary hypertension and cardiac failure. Sleep apnea is particularly common in the elderly, who already have compromised cardiac function. Moreover, studies have demonstrated that sleep apnea is more likely to occur in patients with heart failure. The intermittent hypoxia of sleep apnea further worsens cardiac function.

In addition to conditions that cause hypoxia generally, there are diseases which affect the adequacy of the oxygen supply to the brain as a whole or to restricted brain regions. These include the cerebral vascular diseases as well as diseases, such as epilepsy, which produce hypoxia by increasing the consumption of oxygen in local brain areas. Although hypoxic ventilatory depression has not been studied in patients with these diseases, it is likely that the degree to which depression occurs will depend on the brain area affected.

Studies of hypoxic respiratory depression have made it clear that how ventilation responds to a hypoxic challenge depends upon at least two different processes: the sensitivity of the carotid body and the sensitivity of the brain to hypoxic depression (113, 184). It is possible, for example, that mysterious clinical events, such as the development of hypoxemia and hypercapnia in some patients with chronic obstructive lung disease of moderate severity, may be the result of differences in the ability of the brain to withstand the adverse effects of hypoxia on respiratory activity. Moreover, it now seems clear that the sensitivity to hypoxic respiratory depression changes with maturation and perhaps with aging, though the studies of Ahmed et al. (1) suggest otherwise. Clues to the mechanism of this maturation process may be important not just in diseases of the newborn, like SIDS, but in problems of adult life. Nonetheless, despite the information that has been gathered, the only agent that appears to be useful in mitigating the effects of hypoxic depression, and then mainly in the newborn, is theophylline.

Neubauer and co-workers (116) have attempted to develop a coherent picture of hypoxic respiratory depression. They believe that hypoxic depression affects the brain in a rostral to caudal direction (the cortex first, the medulla last) and involves three different processes. With mild hypoxia, the increase in cerebral blood flow is the main mechanism depressing breathing; with more severe and prolonged hypoxia, the effects of inhibitory neurotransmitters dominate; and, finally, as hypoxia becomes more severe, respiration is depressed by an inadequate supply of substrate needed to maintain energy levels in neurons. Neubauer and co-workers (116) have also suggested that hypoxic respiratory depression may have some beneficial effects in reducing the energy usage of neurons.

This hypothesis is quite useful but may be too simple because of the multiple interactions, physical and functional, that impinge on respiration and the heterogeneity of brain neurons and their biochemical characteristics. The possible functional significance of brain-mediated hypoxic ventilatory depression also deserves attention. Is it related to the "last ditch" survival strategy, discussed above, of hypoxia-induced hypometabolism when energy failure is threatened? Finally, from a clinical perspective, it is likely that hypoxic depression involves many different processes, which have varying importance in different regions of the brain for different individuals, and there is unlikely, therefore, to be a single agent that will universally reverse the adverse effects of hypoxic depression. Rather, one might expect that as further information is obtained we will be better able to characterize the precise defects inhibiting breathing in individuals and to devise specific interventions that might allow depression to be reversed.

REFERENCES

1. Ahmed, M., G. G. Giesbrecht, C. Serrette, D. Georgopoulos, and N. R. Anthonisen. Ventilatory response to hypoxia in elderly humans. *Respir. Physiol.* 83: 343–352, 1991.
2. Andronikou, S., M. Shirahata, A. Mokashi, and S. Lahiri. Carotid body chemoreceptors and ventilatory response to sus-

tained hypoxia and hypercapnia in the cat. *Respir. Physiol.* 72: 361–374, 1988.
3. Bacon, D. S., S. Afifi, J. A. Griebel, and E. M. Camporesi. Cerebrocortical oxygenation and ventilatory response during sustained hypoxia. *Respir. Physiol.* 80: 245–258, 1990.
4. Barnard, P., S. Andronikou, M. Pokorski, N. Smatresk, A. Mokashi, and S. Lahiri. Time-dependent effect of hypoxia on carotid body chemosensory function. *J. Appl. Physiol.: Respir. Environ. Exerc. Physiol.* 63: 685–691, 1987.
5. Bascom, D. A., I. D. Clement, D. A. Cunningham, R. Painter, and P. A. Robbins. Changes in peripheral chemoreflex sensitivity during sustained hypoxia. *Respir. Physiol.* 82: 161–176, 1990.
6. Broderick, P. A., and G. E. Gibson. Dopamine and serotonin in rat striatum during in vivo hypoxic hypoxia. *Metab. Brain Dis.* 4: 143–153, 1989.
7. Bureau, M. A., J. Lamarche, P. Fouton, and D. Dalle. The ventilatory response to hypoxia in the newborn lamb after carotid body denervation. *Respir. Physiol.* 60: 109–119, 1985.
8. Caliguri, M. A., and E. D. Robins. Prolonged diving and recovery in the freshwater turtle *Pseudemys scripta*—IV. Effects of profound acidosis on O_2 consumption in turtle vs. rat (mammalian) brain and heart slices. *Comp. Biochem. Physiol.* 81A: 603–605, 1985.
9. Carafoli, E. Intracellular calcium homeostasis. *Annu. Rev. Biochem.* 56: 395–433, 1987.
10. Cherici, G., M. Alesiani, D. E. Pellegrini-Gaimpietro, and F. Moroni. Ischemia does not induce the release of excitotoxic amino acids from the hippocampus of newborn rats. *Dev. Brain. Res.* 60: 235–240, 1991.
11. Cherniack, N. S., N. H. Edelman, and S. Lahiri. Hypoxia and hypercapnia as respiratory stimulants and depressants. *Respir. Physiol.* 11: 113–126, 10/71.
12. Chernick, V., and R. J. Craig. Naloxone reverses neonatal depression caused by fetal asphyxia. *Science* 216: 1252–1253, 1982.
13. Cherubini, E., Y. Ben-Ari, and K. Krnjevic. Anoxia produces smaller changes in synaptic transmission, membrane potential and input resistance in immature rat hippocampus. *J. Neurophysiol.* 62: 882–895, 1989.
14. Cherubini, E., J. L. Gaiarsa, and Y. Ben-Ari. GABA: an excitatory transmitter in early postnatal life. *Trends Neurosci.* 14: 515–519, 1991.
15. Chih, C. P., Z. C. Feng, M. Rosenthal, P. L. Lutz, and T. J. Sick. Energy metabolism, ion homeostasis, and evoked potentials in anoxic turtle brain. *Am. J. Physiol.* 257(*Regulatory Integrative Comp. Physiol.* 28): R854–R860, 1989.
16. Chih, C. P., M. Rosenthal, and T. J. Sick. Ion leakage is reduced during anoxia in turtle brain: a potential survival strategy. *Am. J. Physiol.* 255(*Regulatory Integrative Comp. Physiol.* 26): R338–R343, 1989.
17. Cotton, E. K., and M. M. Grunstein. Effect of hypoxia on respiratory control in neonates at high altitude. *J. Appl. Physiol.: Respir. Environ. Exerc. Physiol.* 48: 587–595, 1986.
18. Cross, K. W., J. P. M. Tizzard, and D. A. H. Trythall. The gaseous metabolism of the newborn infant breathing 15% O_2. *Acta Pediatr. Scand.* 47: 217–237, 1958.
19. Dargent, B., and F. Courand. Down-regulation of voltage-dependent sodium channels initiated by sodium influx in developing neurones. *Proc. Natl. Acad. Sci. USA* 87: 5907–5911, 1989.
20. Darnall, R. A., Jr. Aminophylline reduces hypoxic ventilatory depression: possible role of adenosine. *Pediatr. Res.* 19: 706–710, 1985.
21. Darnall, R. A., G. Green, L. Pinto, and N. Hart. Effect of acute hypoxia on respiration and brainstem blood flow in the piglet.

J. Appl. Physiol.: Respir. Environ. Exerc. Physiol. 70: 251–259, 1991.

22. Daval, J., and M. Werck Autoradiography changes in brain adenosine A₁ receptors and their coupling to G proteins following seizures in the developing rat. *Dev. Brain Res.* 59: 237–247, 1991.

23. Dawes, G. S., W. N. Gardner, B. M. Johnston, and D. W. Walker. Breathing in fetal lambs: the effect of brain stem section. *J. Physiol. (Lond.)* 335: 535–553, 1983.

24. Doll, C. J., P. W. Hochachka, and P. B. Reiner. Channel arrest: implications from membrane resistance in turtle neurons. *Am. J. Physiol.* 261(*Regulatory Integrative Comp. Physiol.* 32): R1321–R1324, 1991.

25. Doll, C. J., P. W. Hochachka, and P. B. Reiner. Effects of anoxia and metabolic arrest on turtle and rat cortical neurons. *Am. J. Physiol.* 260(*Regulatory Integrative Comp. Physiol.* 31): R747–R755, 1991.

26. Duffy, T. E., S. J. Kohle, and R. C. Vannucci. Carbohydrate and energy metabolism in perinatal rat brain: relation to survival in anoxia. *J. Neurochem.* 24: 271–276, 1975.

27. Easton, P. A., and N. R. Anthonisen. Ventilatory response to sustained hypoxia after pretreatment with aminophylline. *J. Appl. Physiol.: Respir. Environ. Exerc. Physiol.* 64: 1445–1456, 1988.

28. Easton, P. A., L. J. Slykerman, and N. R. Anthonisen. Ventilatory response to sustained hypoxia in normal adults. *J. Appl. Physiol.: Respir. Environ. Exerc. Physiol.* 61: 906–911, 1986.

29. Easton, P. A., L. J. Slykerman, and N. R. Anthonisen. Recovery of the ventilatory response to hypoxia in normal adults. *J. Appl. Physiol.: Respir. Environ. Exerc. Physiol.* 64: 521–528, 1988.

30. Edwards, R., P. L. Lutz, and D. Baden. Relationship between energy expenditure and ion channel function in the rat and turtle brain. *Am. J. Physiol.* 255(*Regulatory Integrative Comp. Physiol.* 26): R1345–R1359, 1989.

31. Erecinska, M., and I. A. Silver. ATP and brain function. *J. Cereb. Blood Flow Metab.* 9: 2–19, 1989.

32. Eyzaguirre, C., and P. Zapata. Perspectives in carotid body research. *J. Appl. Physiol.: Respir. Environ. Exerc. Physiol.* 57: 931–957, 1984.

33. Feng, Z. C., T. J. Sick, and M. Rosenthal. Orthodromic field potentials and recurrent inhibition during anoxia in turtle brain. *Am. J. Physiol.* 255(*Regulatory Integrative Comp. Physiol.* 26): R484–R491, 1988.

34. Foster, A. C., and J. A. Kemp. Glycine maintains excitement. *Nature* 338: 377–378, 1989.

35. Frappell, P. C., C. Lanthier, R. V. Baudinette, and J. P. Mortola. Metabolism and ventilation in acute hypoxia; a comparative analysis in small mammalian species. *Am. J. Physiol.* 262(*Regulatory Integrative Comp. Physiol.* 33): R1040–R1046, 1992.

36. Fregosi, R. F., S. L. Knuth, D. K. Ward, and D. Bartlett, Jr. Hypoxia inhibits abdominal expiratory nerve activity. *J. Appl. Physiol.: Respir. Environ. Exerc. Physiol.* 63: 211–220, 1987.

37. Gallman, E. A., and D. E. Millhorn. Two long-lasting central respiratory responses following acute hypoxia in glomectomized cats. *J. Physiol. (Lond.)* 395: 333–347, 1988.

38. Georgopoulos, D., D. Berezanski, and N. R. Anthonisen. Effect of dichloroacetate on ventilatory response to sustained hypoxia in normal adults. *Respir. Physiol.* 82: 115–122, 1990.

39. Georgopoulos, D., S. G. Holtby, D. J. Berezanski, and N. R. Anthonisen. Aminophylline effects on ventilatory response to hypoxia and hyperoxia in normal adults. *J. Appl. Physiol.: Respir. Environ. Exerc. Physiol.* 67: 1150–1156, 1989.

40. Georgopoulos, D., S. G. Holtby, D. J. Berezanski, and N. R. Anthonisen. Effects of CO₂ breathing on ventilatory response to sustained hypoxia in normal adults. *J. Appl. Physiol.: Respir. Environ. Exerc. Physiol.* 66: 1071–1078, 1989.

41. Georgopoulos, D., S. Walker, and N. R. Anthonisen. Increased chemoreceptor output and the ventilatory response to sustained hypoxia. *J. Appl. Physiol.: Respir. Environ. Exerc. Physiol.* 67: 1159–1163, 1983.

42. Georgopoulos, D., S. Walker, and N. R. Anthonisen. Effect of sustained hypoxia on ventilatory response to CO₂ in normal adults. *J. Appl. Physiol.: Respir. Environ. Exerc. Physiol.* 68: 891–896, 1990.

43. Globus, M. Y., R. Busto, W. D. Dietrich, E. Martinex, I. Valdes, and M. D. Ginsberg. Effect of ischemia on the in vivo release of striatal dopamine, glutamate and γ-aminobutyric acid studied by intracerebral microdialysis. *J. Neurochem.* 1455–1464, 1988.

44. Globus, M. Y., M. D. Ginsberg, and R. Busto. Excitotoxic index—a biochemical marker of selective vulnerability. *Neurosci. Lett.* 127: 39–42, 1991.

45. Goiny, M., H. Lagercrantz, M. Srinivasan, U. Understedt, and Y. Yamamoto. Hypoxia-mediated in vivo release of dopamine in the nucleus tractus solitarii of the rabbit. *J. Appl. Physiol.: Respir. Environ. Exerc. Physiol.* 70: 2395–2400, 1991.

46. Gordon, D., D. Merrick, V. Auld, R. Dunn, A. L. Goldin, N. Davidson, and W. A. Catterall. Tissue-specific expression of the R₁ and R_{II} sodium channel subtypes. *Proc. Natl. Acad. Sci. USA* 84: 8682–8686, 1987.

47. Gordon, K., D. Statman, M. V. Johnston, T. E. Robinson, J. B. Becker, and F. S. Silverstein. Transient hypoxia alters striatal catecholamine metabolism in immature brain: an in vivo microdialysis study *J. Neurochem.* 54: 605–611, 1990.

48. Gordon, K. E., J. Simpson, D. Statman, and F. S. Silverstein. Effects of perinatal stroke on striatal amino acid efflux in rats studied with in vivo microdialysis. *Stroke* 22: 928–932, 1991.

49. Gregson, N. A., and P. L. Williams. A comparative study of brain and liver mitochondria from new-born and adult rats. *J. Neurochem.* 16: 617–626, 1969.

50. Grigg, J. J., and E. G. Anderson. Glucose and sulfonylureas modify different phases of the membrane potential changes during hypoxia in rat hippocampal slices. *Brain Res.* 409: 302–310, 1989.

51. Haddad, G. G., and D. F. Donnelly. O₂ deprivation induces a major depolarization in brainstem neurons in the adult but not in the neonatal rat. *J. Physiol. (Lond.)* 429: 411–428, 1990.

52. Haddad, G. G., and R. B. Mellins. Hypoxia and respiratory control in early life. *Annu. Rev. Physiol.* 46: 629–643, 1984.

53. Hansen, A. J. Effect of anoxia on ion distribution in the brain. *Physiol. Rev.* 65: 101–148, 1985.

54. Hazinski, T. A., M. M. Grunstein, M. A. Schlueuter, and W. H. Tooley. Effect of naloxone on ventilation in newborn rabbit. *J. Appl. Physiol.: Respir. Environ. Exerc. Physiol.* 50: 713–717, 1981.

55. Hitzig, B. M., V. Kneussl, V. Shih, R. D. Brandstetter, and H. Kazemi. Brain amino acid concentrations during diving and acid-base stress in turtles. *J. Appl. Physiol.: Respir. Environ. Exerc. Physiol.* 58: 1751–1754, 1985.

56. Hochachka, P. W. Defense strategies against hypoxia and hypothermia. *Science* 231: 234–241, 1986.

57. Honda, Y., and I. Hashizume. Evidence for hypoxic depression of CO₂-ventilation response in carotid body–resected humans. *J. Appl. Physiol.: Respir. Environ. Exerc. Physiol.* 70: 590–593, 1991.

58. Hoop, B., D. M. Systrom, V. E. Shih, and H. Kazemi. Central respiratory effects of glutamine synthesis inhibition in dogs. *J. Appl. Physiol.: Respir. Environ. Exerc. Physiol.* 65: 1099–1109, 1988.

59. Hulbert, A. J., and P. L. Else. Evolution of mammalian endo-

thermic metabolism: mitochondrial activity and cell composition. *Am. J. Physiol.* 256 (*Regulatory Integrative Comp. Physiol.* 27): R63–R69, 1989.

60. Hulbert, A. J., and P. L. Else. The cellular basis of endothermic metabolism: a role for "leaky" membranes? *News Physiol. Sci.* 5: 25–28, 1990.

61. Huxtable, R. H. Physiological actions of taurine. *Physiol. Rev.* 72: 101–163, 1992.

62. Insel, T. R., L. P. Miller, and R. E. Gelhard. The ontogeny of excitatory amino acid receptors in the rat forebrain. I. NMDA and quisqualate receptors. *Neuroscience* 35: 31–43, 1990.

63. Jansen, A. H., and V. Chernick. Cardiorespiratory response to central cyanide in fetal sheep. *J. Appl. Physiol.: Respir. Environ. Exerc. Physiol.* 37: 18–21, 1979.

64. Javaheri, S., and L. J. Teppema. Ventral medullary extra medullary pH during hypoxemia. *J. Appl. Physiol.: Respir. Environ. Exerc. Physiol.* 63: 1567–1571, 1987.

65. Jiang, C., Y. Xia, and G. G. Haddad. Role of ATP-sensitive K^+ channels during anoxia: major differences between rat (newborn and adult) and turtle neurons. *J. Physiol. (Lond.)* 448: 599–612, 1992.

66. Kagawa, S., M. J. Stafford, T. B. Waggener, and J. W. Severinghaus. No effect of naloxone on hypoxia induced ventilatory depression in adults. *J. Appl. Physiol.: Respir. Environ. Exerc. Physiol.* 52: 1031–1034, 1982.

67. Kelley, D. A., and K. B. Storey. Organ specific control of glycolysis in anoxic turtles. *Am. J. Physiol.* 255 (*Regulatory Integrative Comp. Physiol.* 26): R774–R779, 1988.

68. Khamnei, S., and P. A. Robbins. Hypoxic depression of ventilation in human: alternative models for the chemoreflexes. *Respir. Physiol.* 81: 117–134, 1990.

69. Khamnei, S., and P. A. Robbins. The transients in ventilation arising from a period of hypoxia at near normal and raised levels of end-tidal P_{CO_2} in man. In: *Respiratory Control: A Modeling Perspective,* edited by G. D. Swanson. New York: Plenum, 1990, p. 207–216.

70. Kjellmer, I., P. Andine, H. Hagberg, and K. Thiringer. Extracellular increases of hypoxanthine and xanthine in the cortex and basal ganglia of fetal lambs during hypoxia–ischemia. *Brain Res.* 478: 241–247, 1989.

71. Kneussl, M. P., P. Pappagianopoulos, B. Hoop, and H. Kazemi. Reversible depression of ventilation and cardiovascular function by ventriculocisternal perfusion with gamma-aminobutyric acid in dogs. *Am. Rev. Respir. Dis.* 133: 1024–1028, 1986.

72. Kriegstein, A. R., and B. W. Connors. Cellular physiology of the turtle visual cortex: synaptic properties and intrinsic circuitry. *J. Neurosci.* 6: 178–191, 1986.

73. Krnjevic, K. Membrane current activation during hypoxia in hippocampal neurones. In: *Surviving Hypoxia: Mechanisms of Control and Adaptation,* edited by P. Hochachka, P. L. Lutz, T. Sick, M. Rosenthal, and G. van den Thilart. Boca Raton, FL: CRC, 1993, p. 365–388.

74. Krynjevic, M. B. Adenosine triphosphate–sensitive potassium channels in anoxia. *Stroke* 21 (suppl. III): 190–193, 1990.

75. Krynjevick, R. M., and B. K. Siesjo. Cortical CO_2 tension and neuronal excitability *J. Physiol. (Lond.)* 176: 105–122, 1965.

76. LaFramboise, W. A., T. A. Standaert, and D. E. Woodrum. Ventilatory depression without neural depression during neonatal hypoxemia [Abstract]. *Federation Proc.* 43: 1008, 1984.

77. LaFramboise, W. A., T. A. Standaert, D. E. Woodrum, and R. G. Guthrie. Occlusion pressure during the ventilatory response to hypoxemia in the newborn monkey. *J. Appl. Physiol.: Respir. Environ. Exerc. Physiol.* 51: 1169–1174, 1981.

78. LaFramboise, W. A., and D. E. Woodrum. Elevated diaphragm electromyogram during neonatal hypoxia ventilatory depression. *J. Appl. Physiol.: Respir. Environ. Exerc. Physiol.* 59: 1040–1045, 1985.

79. Lahiri, S., P. Bernard, and R. Zhang. Initiation and control of ventilatory adaptation to chronic hypoxia of high altitude. In: *Control of Respiration,* edited by D. J. Pallot. London: Croom Helm, 1983, p. 298–325.

80. Lahiri, S., J. S. Brady, E. K. Motoyama, and T. M. Velesquez. Regulation of breathing in the newborn at high altitude. *J. Appl. Physiol.: Respir. Environ. Exerc. Physiol.* 44: 673–678, 1978.

81. Lahiri, S., N. J. Smatresk, and E. Mulligan. Responses of peripheral chemoreceptors to natural stimuli. In: *Physiology of the Peripheral Arterial Chemoreceptors,* edited by H. Acker and R. O'Regan. Amsterdam: Elsevier, 1983, p. 221–256.

82. Lahiri, S., N. Smatresk, M. Pokorski, P. Barnard, and A. Mokashi. Efferent inhibition of carotid body chemoreception in chronically hypoxic cats. *Am. J. Physiol.* 247(*Respir. Environ. Exerc. Physiol.* 18): R24–R28, 1984.

83. Laudignon, N., E. Farri, K. Beharry, J. Rex, and J. V. Aranda. Influence of adenosine on cerebral blood flow during hypoxic hypoxia in the newborn piglet. *J. Appl. Physiol.: Respir. Environ. Exerc. Physiol.* 68: 1534–1541, 1990.

84. Lauritzen, M., M. E. Rice, Y. Iokada, and C. Nicholson. Quisqualate, kainate, and NMDA can initiate spreading depression in the turtle cerebellum. *Brain Res.* 475: 317–327, 1988.

85. Lawson, E. E., and W. W. Long. Central origin of biphasic breathing pattern during hypoxia in newborns. *J. Appl. Physiol.: Respir. Environ. Exerc. Physiol.* 55: 483–488, 1983.

86. Leblond, J., and K. Krnjevic. Hypoxic changes in hippocampal neurones. *J. Neurophysiol.* 62: 1–14, 1989.

87. Lehmann, A., H. Hagberg, P. Andine, and K. Ellern. Taurine and neuronal resistance to hypoxia. *FEBS Lett.* 233: 437–438, 1988.

88. Lindefors, N., Y. Yamamoto, T. Pantaleo, H. Lagercrantz, E. Brodin, and U. Ungerstedt. In vivo release of substance P in the nucleus tractus solitarii increases during hypoxia. *Neurosci. Lett.* 69: 94–97, 1986.

89. Lipton, P., and T. S. Whittingham. Energy metabolism and brain slice function. In: *Brain Slices,* edited by R. Dingledine. New York: Plenum, 1984, p. 113–153.

90. Loeschcke, H. H. Central chemosensitivity and the reaction theory. *J. Physiol. (Lond.)* 332: 1–26, 1982.

91. Longobardo, G. S., N. S. Cherniack, and B. Gothe. Factors affecting respiratory system stability. *Ann. Biomed. Eng.* 17: 377–396, 1989.

92. Lutz, P. L. Interaction between hypometabolism and acid-base balance. *Can. J. Zool.* 67: 3018–3023, 1989.

93. Lutz, P. L. Anoxic defense mechanisms in the vertebrate brain. *Annu. Rev. Physiol.* 54: 601–618, 1992.

94. Lutz, P. L., R. Edwards, and P. McMahon. GABA concentrations are maintained in the anoxic turtle brain. *Am. J. Physiol.* 249(*Regulatory Integrative Comp. Physiol.* 20): R372–R374, 1985.

95. Lutz, P. L., P. McMahon, M. Rosenthal, and T. J. Sick. Relationships between aerobic and anaerobic energy production in turtle brain in situ. *Am. J. Physiol.* 247(*Regulatory Integrative Comp. Physiol.* 18): R740–R744, 1984.

96. Lutz, P. L., and G. Nilsson. Metabolic transitions to anoxia in the turtle brain: the role of neurotransmitters. In: *The Vertebrate Gas Transport Cascade,* edited by E. Bicudo and M. Glass. Boca Raton, FL: CRC, 1993, p. 323–329.

97. Lutz, P. L., M. Rosenthal, and T. J. Sick. Living without oxygen: turtle brain as a model of anaerobic metabolism. *Mol. Physiol.* 8: 411–425, 1985.

98. Macklem, P. T., and C. S. Roussos. Respiratory muscle fatigue:

a cause of respiratory failure. *Clin. Sci. Mol. Med.* 53: 419–422, 1977.

99. Magistretti, P. J., P. R. Hof, and J. L. Martin. Adenosine stimulates glycogenolysis in mouse cerebral cortex: a possible coupling mechanism between neuronal activity and energy metabolism. *J. Neurosci.* 6: 2553–2562, 1986.

100. Martin-Body, R. L., and B. M. Johnston. Central origin of the hypoxic depression of breathing in the newborn. *Respir. Physiol.* 71: 25–32, 1988.

101. McDonald, J. W., F. S. Silverstein, and M. V. Johnston. Neurotoxicity of NDMA is markedly enhanced in developing rat central nervous system. *Brain Res.* 459: 200–203, 1988.

102. McGeer, P. L., and E. G. McGeer. Amino acid neurotransmitters. In: *Basic Neurochemistry: Molecular, Cellular, and Medical Aspects* (4th ed.), edited by G. J. Siegel, B. Agranoff, R. W. Alberts, and P. Molinoff. New York: Raven, 1989, p. 311–332.

103. Meldrum, B. Excitatory amino acids and anoxic–ichemic brain damage. *Trends Neurosci.* 8: 47–48, 1985.

104. Melton, J. E., J. A. Neubauer, and N. H. Edelman. CO_2 sensitivity of cat phrenic neurograms during hypoxic respiratory depression. *J. Appl. Physiol.: Respir. Environ. Exerc. Physiol.* 65: 736–743, 1988.

105. Melton, J. E., J. A. Neubauer, and N. H. Edelman. GABA antagonism reverses hypoxic respiratory depression in the cat. *J. Appl. Physiol.: Respir. Environ. Exerc. Physiol.* 69: 1296–1301, 1990.

106. Menendez, N., J. M. Solis, O. Herreras, A. S. Herranz, and R. Martin Del Rio. Role of endogenous taurine on the glutamate analogue induced neurotoxicity in the rat hippocampus. *J. Neurochem.* 55: 714–717, 1990.

107. Mink, J. W., R. J. Blumenschine, and D. B. Adams. Ratio of central nervous system to body metabolism in vertebrates: its constancy and functional basis. *Am. J. Physiol.* 240(*Regulatory Integrative Comp. Physiol.* 11): R203–R212, 1981.

108. Mitra, J., N. B. Dev, J. R. Romaniuk, R. Trivedi, N. R. Prabhakar, and N. S. Cherniack. Cardiorespiratory changes induced by vertebral artery injection of sodium cyanide in cats. *Respir. Physiol.* 87: 49–61, 1992.

109. Mitra, J., N. B. Dev, R. Trivedi, and N. S. Cherniack. Role of adenosine in central hypoxia produced by vertebral artery injection of NaCN and hypoxic saline in cats. *FASEB J.* 6: A1828, 1992.

110. Mori, S., A. C. Ngai, K. R. Ko, and H. R. Winn. Role of adenosine in regulation of cerebral blood flow: effects of theophylline during normoxia and hypoxia. *Am. J. Physiol.* 253(*Heart Circ. Physiol.* 24): H165–H175, 1987.

111. Morrill, C. G., J. R. Meyer, and J. V. Weil. Hypoxic ventilatory depression in dogs. *J. Appl. Physiol.: Respir. Environ. Exerc. Physiol.* 38: 143–146, 1975.

112. Mortola, J. P., and A. Dotta. Effect of hypoxia and ambient temperature on the gaseous metabolism of newborn rats. *Am. J. Physiol.* 263(*Regulatory Integrative Comp. Physiol.* 34): R262–R272, 1992.

113. Mountain, R., C. Zwillich, and J. V. Weil. Hypoventilation in obstructive lung disease: the role of familial factors. *N. Engl. J. Med.* 298: 521–525, 1978.

114. Mourre, C., Y. Ben Ari, H. Bernardi, M. Fosset, and M. Lazdunski. Antidiabetic sulfonylurea: location of binding sites in the brain and effects on the hyperpolarization induced by anoxia in hippocampal slices. *Brain Res.* 486: 159–164, 1989.

115. Neubauer, J. A., and N. H. Edelman. Nonuniform brain blood flow response to hypoxia. *J. Appl. Physiol.: Respir. Environ. Exerc. Physiol.* 57: 1803–1808, 1984.

116. Neubauer, J. A., J. E. Melton, and N. H. Edelman. Modulation of respiration during brain hypoxia. *J. Appl. Physiol.: Respir. Environ. Exerc. Physiol.* 68: 441–451, 1990.

117. Neubauer, J. A., M. A. Posner, T. V. Santiago, and N. H. Edelman. Naloxone reduces ventilatory depression of brain hypoxia. *J. Appl. Physiol.: Respir. Environ. Exerc. Physiol.* 63: 699–706, 1987.

118. Neubauer, J. A., T. V. Santiago, M. A. Posner, and N. H. Edelman. Ventral medullary pH and ventilatory responses to hyperperfusion and hypoxia. *J. Appl. Physiol.: Respir. Environ. Exerc. Physiol.* 58: 1659–1668, 1985.

119. Neubauer, J. A., A. Simone, and N. H. Edelman. Role of brain lactate acidosis in hypoxic depression of respiration. *J. Appl. Physiol.: Respir. Environ. Exerc. Physiol.* 63: 1324–1331, 1987.

120. Neumann, R., W. A. Catterall, and T. Scheuer. Functional modification of brain sodium channels by protein kinase C phosphorylation. *Science* 245: 115–118, 1991.

121. Newby, A. C., Y. Worku, P. Meghji, M. Nakazawa, and A. C. Skladanowski. Adenosine: a retaliatory metabolite or not? *News Physiol. Sci.* 5: 67–70, 1990.

122. Nielsen, M., and H. Smith. Studies of the regulation of respiration in acute hypoxia. *Acta Physiol. Scand.* 24: 293–313, 1951.

123. Nilsson, G. E. Long-term anoxia in crucian carp: changes in the levels of amino acid and monoamine neurotransmitters in the brain, catecholamines in chromaffin tissue, and liver glycogen. *J. Exp. Biol.* 150: 295–320, 1990.

124. Nilsson, G. E. The adenosine receptor blocker aminophylline increases anoxic ethanol excretion in the crucian carp. *Am. J. Physiol.,* 261(*Regulatory Integrative Comp. Physiol.* 32): R1057–R1060, 1991.

125. Nilsson, G. E., A. A. Alfaro, and P. L. Lutz. The effects of anoxia on turtle brain neurotransmitters and related substances. *Am. J. Physiol.* 259(*Regulatory Integrative Comp. Physiol.* 30): R376–R384, 1990.

126. Nilsson, G. E., and P. L. Lutz. Release of inhibitory neurotransmitters in response to anoxia in turtle brain. *Am. J. Physiol.* 261(*Regulatory Integrative Comp. Physiol.* 32): R32–R37, 1991.

127. Nilsson, G. E., and P. L. Lutz. Adenosine release in the anoxic turtle brain as a mechanism for anoxic survival. *J. Exp. Biol.* 162: 345–351, 1992.

128. Nilsson, G. E., P. L. Lutz, and T. L. Jackson. Neurotransmitters and anoxic survival in the brain: a comparison between anoxia tolerant and anoxia intolerant vertebrates. *Physiol. Zool.* 64: 638–652, 1991.

129. Nioka, S., B. Chance, D. S. Smith, A. Mayevsky, M. P. Reilly, C. Alter, and T. Asakura. Cerebral energy metabolism and oxygen state during hypoxia in neonate and adult dogs. *Pediatr. Res.* 28: 54–62, 1990.

130. Nishimura, M., A. Suzuki, Y. Nishiura, H. Yamamoto, K. Miyamoto, F. Kishi, and Y. Kawakami. Effect of brain blood flow on hypoxic ventilatory response in humans. *J. Appl. Physiol.: Respir. Environ. Exerc. Physiol.* 63: 1100–1106, 1987.

131. Norwood, W. I., J. S. Ingwall, C. Norwood, and E. T. Fossel. Developmental changes in creatine kinase metabolism in rat brain. *Am. J. Physiol.* 244(*Cell Physiol.* 13): C205–C210, 1983.

132. Okamato, K., H. Kimura, and Y. Sakai. Taurine-induced effects of the Cl-conductance of cerebellar Purkinje cell dendrites in vitro. *Brain Res.* 259: 319–323, 1983.

133. O'Regan, R. G., and S. Majcherczyk. Control of peripheral chemoreceptors by efferent fibers. In: *Physiology of the Peripheral Arterial Chemoreceptors*, edited by H. Acker and R. G. O'Regan. Amsterdam: Elsevier, 1983, p. 257–298.

134. Ou, L. C., M. J. Miller, and S. M. Tenney. Hypoxia and CO_2 as separate and interactive depressants of ventilation. *Respir. Physiol.* 28: 347–358, 1976.

135. Park, T. S., D. G. Van Wylen, R. Rubio, and R. M. Berne. Increased brain interstitial fluid adenosine concentration during hypoxia in newborn piglet. *J. Cereb. Blood Flow Metab.* 7: 178–183, 1987.

136. Perez, R., M. Espinoza, R. Riquelme, J. T. Parer, and A. J. Llanos. Arginine vasopressin mediates cardiovascular responses to hypoxemia in fetal sheep. *Am. J. Physiol.* 256(*Regulatory Integrative Comp. Physiol.* 27): R1011–R1018, 1989.

137. Perez-Pinzon, M., J. Bedford, M. Rosenthal, P. L. Lutz, and T. J. Sick. Metabolic adaptations to anoxia in the isolated turtle cerebellum [Abstract]. *Soc. Neurosci.* 17: 1269, 1991.

138. Perez-Pinzon, M., M. Rosenthal, P. L. Lutz, and T. Sick. Anoxic survival in the isolated turtle cerebellum. *J. Comp. Physiol.* 16: 345–351, 1992.

139. Perez-Pinzon, M., M. Rosenthal, T. Sick, P. L. Lutz, and D. Marsh. Down-regulation of sodium channels during anoxia: a putative survival strategy of turtle brain. *Am. J. Physiol.* 262(*Regulatory Integrative Comp. Physiol.* 31): R712–R715, 1992.

140. Perez-Pinzon, M. A., C. Y. Chan, M. Rosenthal, and T. J. Sick. Membrane and synaptic activity during anoxia in the isolated turtle cerebellum. *Am. J. Physiol.* 263(*Regulatory Integrative Comp. Physiol.* 32): R1057–R1063, 1993.

140a. Perez-Pinzon, M., P. L. Lutz, T. Sick, and M. Rosenthal. Adenosine, a "retaliatory" metabolite, promotes anoxia tolerance in turtle brain. *J. Cereb. Blood Flow Metab.* 13: 728–732, 1993.

141. Perez-Pinzon, M. A., G. Nilsson, and P. L. Lutz. Relationships between ion gradients and neurotransmitter release in the newborn rat striatum during anoxia. *Brain Res.* 602: 228–233, 1993.

142. Prince, D. A., and C. F. Stevens. Adenosine decreases neurotransmitter release at central synapses. *Proc. Natl. Acad. Sci. USA* 89: 8585–8590, 1992.

143. Pulsinelli, W. A. Selective neuronal vulnerability: morphological and molecular characteristics. In: *Molecular Mechanisms of Ischemic Brain Damage. Progress in Brain Research,* edited by K. Kogure, K. A. Hossmann, B. K. Siesjö, and F. A. Welsh. Amsterdam: Elsevier, 1985, vol. 63, p. 29–37.

144. Rehncrona, S., I. Rosen, and B. K. Siesjo. Excessive cellular acidosis: an important mechanism of neuronal damage in the brain. *Acta Physiol. Scand.* 110: 435–437, 1980.

145. Rice, M. E., and C. Nicolson. Glutamate and aspartate-induced extracellular potassium and calcium shifts and their relationship to those of kianate, quisqualate, and N-methyl-D-aspartate in the isolated turtle cerebellum. *Neuroscience* 38: 295–310, 1990.

146. Rohlicek, C. V., and C. Polosa. Hypoxic responses of sympathetic preganglionic neurons in the acute spinal cat. *Am. J. Physiol.* 241(*Heart Circ. Physiol.* 12): H679–H683, 1981.

147. Rohlicek, C. V., and C. Polosa. Observations on the hypoxic depression of sympathetic discharge in sinoaortic-denervated cats. *Can. J. Physiol. Pharmacol.* 66: 413–418, 1988.

148. Rothman, S. M., and J. W. Olney. Glutamate and the pathophysiology of hypoxic–ischemic brain damage. *Ann. Neurol.* 19: 105–111, 1986.

149. Rovira, C., and Y. Ben-Ari. Benzodiazapines do not potentiate GABA responses in neonatal hippocampal neurones. *Neurosci. Lett.* 130: 157–161, 1991.

150. Rubio, R., R. M. Berne, E. L. Bockman, and R. R. Curnish. Relationship between adenosine concentration and oxygen supply in rat brain. *Am. J. Physiol.* 228: 1896–1902, 1975.

151. Sakurai, S. NMDA Binding Sites in the Vertebrate CNS: Characterisation using Quantitative Autoradiography of Tritiated MK 801 Binding. University of Michigan, 1991. PhD Thesis.

152. Santiago, T. V., and N. H. Edelman. Opioids and breathing. *J.*

153. Saunders, N. A., M. F. Betts, L. D. Pengelly, and A. S. Rebuck. Changes in lung mechanics induced by acute isocapnia hypoxia. *J. Appl. Physiol.: Respir. Environ. Exerc. Physiol.* 42: 413–419, 1977.

154. Schlaefke, M. Central chemosensitivity: a respiratory drive. *Rev. Physiol. Biochem. Pharmacol.* 90: 171–244, 1981.

155. Schramm, C. M., and M. Grunstein. Respiratory influence of peripheral chemoreceptor stimulation in maturing rabbits. *J. Appl. Physiol.: Respir. Environ. Exerc. Physiol.* 63: 1671–1680, 1987.

156. Schurr, A., M. T. Tseng, C. A. West, and B. M. Rigor. Taurine improves the recovery of neuronal function following cerebral hypoxia: an in vitro study. *Life Sci.* 40: 2059–2066, 1987.

157. Shinozuka, T., E. Nemoto, and P. M. Winter. Mechanism of cerebrovascular O_2 sensitivity from hyperoxia to moderate hypoxia in the rat. *J. Cereb. Blood Flow Metab.* 9: 187–195, 1989.

158. Sick, T. J., M. Rosenthal, J. C. Lemanna, and P. L. Lutz. Brain potassium ion homeostasis, anoxia, and metabolic inhibition in turtles and rats. *Am. J. Physiol.* 243(*Regulatory Integrative Comp. Physiol.* 14): R281–R288, 1982.

159. Sick, T. J., M. Perez-Pinzon, P. L. Lutz, and M. Rosenthal. Maintaining coupled metabolism and membrane function in anoxic brain: a comparison between the turtle and rat. In: *Surviving Hypoxia: Mechanisms of Control and Adaptation,* edited by P. W. Hochachka, P. L. Lutz, T. Sick, and M. R. G. van den Thilart. Boca Raton, FL: CRC, 1993, p. 351–364.

160. Siegel, G., B. Agranoff, R. W. Albers, and P. Molinoff. *Basic Neurochemistry: Molecular, Cellular and Medical Aspects* (4th ed.). New York: Wiley, 1989.

161. Siesjö, B. K. *Brain Energy Metabolism.* New York: Wiley, 1978.

162. Siesjö, B. K. Calcium, excitotoxins, and brain damage. *News Physiol. Sci.* 5: 120–125, 1990.

163. Siesjö, B. K., and F. Bengtsson. Calcium fluxes, calcium antagonists and calcium-related pathology in brain ischemia, and spreading depression: a unifying hypothesis. *J. Cereb. Blood Flow Metab.* 9: 127–140, 1989.

164. Silver, I. A., and M. Enecinska. Intracellular and extracellular changes of $[Ca^{2+}]$ in hypoxia and ischemia in rat brain in vivo. *J. Gen. Physiol.* 95: 837–865, 1990.

165. Smith, C. A., D. M. Ainsworth, K. S. Henderson, and J. A. Dempsey. Differential responses of expiratory muscles to chemical stimuli in awake dogs. *J. Appl. Physiol.: Respir. Environ. Exerc. Physiol.* 66: 384–391, 1989.

166. Srinivasan, M., H. Lagercrantz, and Y. Yamamoto. A possible dopaminergic pathway mediating hypoxic depression in neonatal rabbits. *J. Appl. Physiol.: Respir. Environ. Exerc. Physiol.* 67: 1271–1276, 1989.

167. Srinivasan, M., Y. Yamamoto, and H. Lagercrantz. Ventilatory effects of naloxone on the sympathetic adrenal system in the neonate. *Neurosci. Lett.* 90: 159–164, 1988.

168. Stone, T. W. Receptors for adenosine and adeninine nucleotides. *Gen. Pharmacol.* 22: 25–31, 1991.

169. Storey, J. B., and J. M. Storey. Metabolic rate depression and biochemical adaptation in anaerobiosis, hibernation and estivation. *Q. Rev. Biol.* 65: 145–174, 1990.

170. Strohl, K. P., N. S. Cherniack, and B. Gothe. Physiologic basis of therapy for sleep apnea. *Am. Rev. Respir. Dis.* 134: 791–802, 1986.

171. Suárez, R. K., C. J. Doll, A. E. Buie, T. G. West, G. D. Funk, and P. W. Hochachka. Turtles and rats: a biochemical comparison of anoxia-tolerant and anoxia-sensitive brains. *Am. J.*

Physiol. 257(*Regulatory Integrative Comp. Physiol.* 26): R1083–R1088, 1989.

172. Sullivan, H. C., D. W. Choi, and A. R. Kriegstein. Turtle cortical neurons are resistant to glutamate neurotoxicity [Abstract]. *Neurology* 40: 348, 1990.

173. Suzuki, A., M. Nishimura, H. Yamamoto, K. Miyamoto, F. Kishi, and Y. Kawakami. No effect of brain blood flow on ventilatory depression during sustained hypoxia. *J. Appl. Physiol.: Respir. Environ. Exerc. Physiol.* 66: 1674–1678, 1989.

174. Tatsumi, K., C. K. Pickett, and J. V. Weil. Attenuated carotid body hypoxic sensitivity after prolonged hypoxic exposure. *J. Appl. Physiol.: Respir. Environ. Exerc. Physiol.* 70: 748–755, 1991.

175. Tenney, S. M., and L. C. Ou. Hypoxic ventilatory response of cats at high altitude: an interpretational "blunting." *Respir. Physiol.* 30: 185–199, 1977.

176. Trippenbach, T., D. W. Richter, and H. Acker. Hypoxia and ion activities within the brain stem of newborn rabbits. *J. Appl. Physiol.: Respir. Environ. Exerc. Physiol.* 68: 2494–2503, 1990.

177. van Lunteren, E., R. J. Martin, M. A. Haxhiu, and W. A. Carlo. Diaphragm genioglossus and triangularis sterni responses to poikilocapnic hypoxia. *J. Appl. Physiol.: Respir. Environ. Exerc. Physiol.* 67: 2303–2310, 1989.

178. Vannucci, R. C. Experimental biology of cerebral hypoxia-ischemia: relation to perinatal brain damage. *Pediatr. Res.* 27: 317–326, 1990.

179. Van Wylen, D. G. L., T. S. Park, R. Rubio, and R. M. Berne. Increases in cerebral interstitial fluid adenosine concentration during hypoxia, potassium infusion, and ischemia. *J. Cereb. Blood Flow Metab.* 6: 522–528, 1986.

180. Vizek, M., C. K. Pickett, and J. V. Weil. Biphasic ventilatory response of adult cats to sustained hypoxia has central origin. *J. Appl. Physiol.: Respir. Environ. Exerc. Physiol.* 63: 1658–1664, 1987.

181. Vizek, M., C. K. Pickett, and J. V. Weil. Individual variation of ventilatory response to hypoxia is associated with variation in sensitivity of peripheral chemoreceptors. *J. Appl. Physiol.: Respir. Environ. Exerc. Physiol.* 63: 1884–1889, 1987.

182. Wagerle, L. C., T. M. Heffernon, L. M. Sacks, and M. Delivoria-Papadapoulos. Sympathetic effects on cerebral blood flow regulation in hypoxic new lambs. *Am. J. Physiol.* 245(*Heart Circ. Physiol.* 16): H487–H494, 1983.

183. Wasicko, M. J., J. E. Melton, J. A. Neubauer, N. Krawciw, and N. H. Edelman. Cervical sympathetic and phrenic nerve responses to progressive brain hypoxia. *J. Appl. Physiol.: Respir. Environ. Exerc. Physiol.* 68: 53–58, 1990.

184. Weil, J. V. Ventilatory control at high altitude. In: *Handbook of Physiology. Control of Breathing,* edited by N. S. Cherniack and J. G. Widdicombe. Bethesda, MD: *Am. Physiol. Soc.,* 1986, vol. 2, p. 703–728.

185. Weil, J. V. Lessons from high altitude. Aspen Lung Conference: Chronic Respiratory Failure. *Chest* 97: 70S–76S, 1990.

186. Weiskopf, R. B., and R. A. Gabel. Depression of ventilation during hypoxia in man. *J. Appl. Physiol.* 39: 911–915, 1975.

187. Wilson, A. M., and A. R. Krieggstein. Turtle cortical neurones survive glutamate exposures that are lethal to mammalian neurones. *Brain Res.* 540: 297–301, 1991.

188. Winn, H. R., R. Rubio, and R. M. Berne. Brain adenosine concentration during hypoxia in cats. *Am. J. Physiol.* 241(*Heart Circ. Physiol.* 12): H235–H242, 1981.

189. Yamada, K., P. Homosh, and R. A. Gillis. Respiratory depression produced by activation of GABA receptors in hindbrain of cats. *J. Appl. Physiol.: Respir. Environ. Exerc. Physiol.* 51: 1278–1286, 1981.

190. Yamamoto, Y., and H. Lagercrantz. Some effects of substance P on central respiratory control in rabbit pups. *Acta Physiol. Scand.* 124: 449–455, 1985.

57. Physiology of extreme altitude

JOHN B. WEST | *Department of Medicine, University of California, San Diego, La Jolla, California*

THE STUDY OF THE PHYSIOLOGY OF EXTREME ALTITUDE could be said to have begun with the publication of *La Pression Barométrique* by Paul Bert in 1878 (Table 57.1). In this landmark volume, Bert (4) correctly attributed the physiological effects of high altitude to the low partial pressure of oxygen (PO_2). Other highlights included an extensive and scholarly historical introduction, a very nearly correct prediction of the barometric pressure on the summit of Mt. Everest, and a graphic description of an experiment in which Bert decompressed himself to this pressure in his low-pressure chamber while breathing supplementary oxygen. His pressure for Mt. Everest (24.8 cm Hg; ref. 4, p. 1040) is noteworthy because later physiologists inappropriately used the Standard Atmosphere, thus predicting much lower pressures and causing considerable confusion, which still surfaces from time to time. The deleterious (and occasionally lethal) effects of extreme altitude were known before Bert from the exploits of the balloonists (33), but Bert was the first physiologist to put the effects of high-altitude hypoxia on a solid scientific footing.

In the late nineteenth century Angelo Mosso built the first high-altitude station, the Capanna Regina Margherita, on the Monte Rosa at an altitude of 4,559 m. He attributed many of the deleterious effects of high altitude to a reduced carbon dioxide level in the blood *(acapnia)*. At about this time it was argued that climbers would never be able to ascend above about 6,500 m because of the extreme debilitating effects of very high altitude (32).

However, in 1909 the aristocratic Duke of the Abruzzi reached the remarkable altitude of 7,500 m in the Karakoram Mountains without supplementary oxygen, and the Duke made light of the difficulties associated with this great altitude (23). The climb influenced the physiological community because Douglas, Haldane and co-workers made some calculations indicating that oxygen secretion by the lung must have occurred to maintain the arterial PO_2 at a viable value. Subsequently, an Anglo-American expedition to Pikes Peak, Colorado (4,300 m), was organized by Douglas and Haldane, and experimental evidence for oxygen secretion was apparently obtained (21).

The first comprehensive analysis of the physiological problems at the highest point on earth was made in 1921 by A. M. Kellas, whose contributions have been almost completely overlooked (106). His paper was unfortunately only published in French in a very obscure place (41). On the basis of an extensive study he concluded that "Mt. Everest could be ascended by a man of excellent physical and mental constitution in first-rate training, without adventitious aids [supplementary oxygen] if the physical difficulties of the mountain are not too great." It was not until almost 60 yr later that this prediction was fulfilled.

Norton (59) reached 8,570 m on the north side of Everest without supplementary oxygen in 1924, and this record remained for some 60 yr, strongly supporting the view that the last 300 m is very near the limit of human tolerance to hypoxia. In 1935 the International High-Altitude Expedition to Chile visited a sulfur mine at Aucanquilcha, which they believed to be at 5,800 m, and were impressed by the high level of physical activity of the miners. They reported that the miners lived at an altitude of 5,330 m and walked up to the mine and back each day rather than live at the higher altitude, suggesting that 5,800 m was too high for humans to survive indefinitely (42). However, the mine was visited again later and found to be at an even higher altitude of 5,950 m, with four caretakers living indefinitely at that altitude with occasional excursions to lower altitude. These are probably the highest inhabitants of the world (104).

Mt. Everest was finally climbed by Hillary and Tensing in 1953, though they used supplementary oxygen. The first people to reach the summit without supplementary oxygen were Messner and Habeler in 1978. However, they reported that the last 100 m took more than an hour to climb (53). Since that historic event in

TABLE 57.1. *Some Major Events in the Physiology of Extreme Altitude*

1878	Publication of *La Pression Barométrique* by P. Bert (4), including correct prediction of the barometric pressure on Mt. Everest; Bert exposed himself to this pressure in a chamber while breathing supplementary oxygen.
1893	A. Mosso completed the first high-altitude station, Capanna Regina Margherita, on a summit of Monte Rosa at 4,559 m.
1909	The Duke of the Abruzzi reached 7,500 m in the Karakoram Mountains without supplementary oxygen.
1911	Anglo-American Pikes Peak expedition (4,300 m) during which J. S. Haldane obtained evidence for oxygen secretion at high altitude.
1921	A. M. Kellas published "Sur les possibilités de faire l'ascension du Mount Everest (On the possibility of making an ascent of Mount Everest)" (ref. 41).
1921–1922	International High-Altitude Expedition to Cerro de Pasco, Peru, led by J. Barcroft, including descriptions of the remarkable athletic abilities of high-altitude residents at 4,540 m.
1924	E. F. Norton ascends to 8,570 m on Mt. Everest without supplementary oxygen.
1935	International High-Altitude Expedition to Chile, including a visit to the Aucanquilcha mine at an altitude reported to be 5,800 m (actually 5,950 m).
1946	Operation Everest I carried out by C. S. Houston and R. L. Riley.
1953	First ascent of Mt. Everest by Hillary and Tensing (with supplementary oxygen).
1960–1961	Himalayan Scientific and Mountaineering Expedition in the Everest region. Laboratory at 5,800 m, physiological measurements up to 7,440 m. Scientific leader, L. G. C. E. Pugh.
1978	First ascent of Everest without supplementary oxygen by R. Messner and P. Habeler.
1981	American Medical Research Expedition to Everest, including first physiological measurements on the summit.
1985	Operation Everest II, principal investigators C. S. Houston, J. R. Sutton, and A. Cymerman (see ref. 35).
1987	First ascent of Everest in winter (December) without supplementary oxygen by Sherpa Ang Rita.

the physiology of extreme altitude, many other climbers have repeated the feat.

Several physiological field expeditions and low-pressure chamber studies have contributed greatly to our knowledge of extreme altitude. The first was Operation Everest I (OEI) carried out by Houston and Riley in 1946 in which four volunteers lived in a low-pressure chamber for 32 days, eventually being exposed to a barometric pressure even lower than the summit of Mt. Everest. The results strongly suggested diffusion limitation of oxygen transfer during exercise (36). In 1960–1961, the Himalayan Scientific and Mountaineering Expedition (HSME), led by Hillary with Pugh as scientific leader, set up a laboratory at 5,800 m and made extensive measurements over the course of several months (67). Studies of maximal oxygen consumption were made as high as 7,440 m (70). An Italian Mt. Everest expedition in 1973 with Cerretelli as scientific leader established a laboratory at 5,350 m and looked particularly at the physiology of exercise (11). In 1981 the American Medical Research Expedition to Everest (AMREE) placed laboratories at 5,400 and 6,300 m, and a number of physiological measurements were made above 8,000 m, including the first measurements on the Everest summit itself. Finally, in 1985 Operation Everest II (OEII), principal investigators C. S. Houston, J. R. Sutton, and A. Cymerman, was another low-pressure chamber experiment with an ambitious program, including cardiac catheterization, muscle biopsies, and other invasive procedures not practical in the field (35,

37). Much of what follows is based on the results of AMREE and OEII.

There is no general agreement on what constitutes extreme altitude. One reasonable definition would be altitudes above 6,000 m because, as indicated above, human beings can apparently live more or less indefinitely at an altitude of 5,950 m, as evidenced by the caretakers of the Aucanquilcha mine in north Chile (104). Most of the information in this chapter refers to altitudes of about 6,000 m and above, but there is some emphasis on altitudes above 8,000 m because of the first successful ascent of Mt. Everest without supplementary oxygen and two recent studies designed to elucidate the physiology of human beings at these great altitudes, AMREE and OEII.

BAROMETRIC PRESSURE

Barometric pressure at extreme altitude has been discussed in detail elsewhere (100), but the topic will be summarized here because it is still a subject of confusion and incorrect interpretation of data. As pointed out above, Bert gave a figure of 248 mm Hg for the barometric pressure on the summit of Mt. Everest in 1878. This, of course, was extrapolated from measurements made at much lower altitudes. However, it was essentially correct, as was his figure for the altitude of Mt. Everest, which he gave as 8,840 m. The presently accepted value is 8,848 m, though some measurements, as yet unconfirmed, suggest a somewhat higher or lower value.

Zuntz et al. (118) recognized the important effect of temperature on the altitude–pressure relationship and gave the following formula for determining barometric pressure at any altitude:

$$\log b = \log B - \frac{h}{72\,(256.4 + t)}$$

where h is the altitude difference in meters, t is the mean temperature (°C) of the air column of height h, B is barometric pressure (mm Hg) at the lower altitude, and b is barometric pressure at the higher altitude. This formula was used by early high-altitude physiologists, including FitzGerald (24) and Kellas (40). For example, Kellas predicted a barometric pressure on the summit of Mt. Everest of 251 mm Hg (correct), assuming a mean temperature of 0°C. However, the formula is of limited usefulness because the mean temperature of the air column is not accurately known. For example, when Kellas assumed that the mean temperature was 15°C, he obtained a pressure of 267 mm Hg for the Everest summit, which was much too high. Since the pressure–altitude relationship is so sensitive to temperature, it is clear that it will vary according to latitude. In particular, at the lower latitudes near the equator the barometric pressure for a given altitude will be relatively high.

With the development of the aviation industry after World War I, it became advantageous to develop a "standard atmosphere" with a barometric pressure–altitude relationship that could be universally used for calibrating altimeters, low-pressure chambers, and other devices. This is often referred to as the ICAO Standard Atmosphere (38) or the US Standard Atmosphere (58), which are identical up to altitudes of interest to physiologists. The standard atmosphere assumes a sea-level pressure of 760 mm Hg (1,013 millibars), sea-level temperature of 15°C, and a linear decrease in temperature with altitude (lapse rate) of 6.5°C per kilometer up to an altitude of 11 km.

It should be emphasized that the Standard Atmosphere was never intended to be used to predict the actual barometric pressure at a particular location. Nevertheless, when respiratory physiologists started using low-pressure chambers in World War II, they incorrectly applied the standard atmosphere to mountainous regions of interest, such as the Himalayas and Mt. Everest in particular (36, 37, 71, 80). The result was that the barometric pressure for the Everest summit (8,848 m) was predicted to be 236 mm Hg, which is far too low.

This error became apparent when expeditions to high altitude, especially the Himalayas, resumed after World War II. For example, Pugh (66, 68) noted that the barometric pressures in the Himalayas considerably exceeded those predicted from the standard atmo-

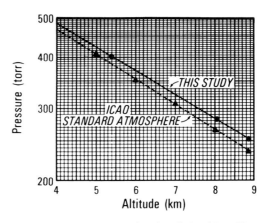

FIG. 57.1. Barometric pressure–altitude relationships. *Upper line* shows measurements made on Mt. Everest during AMREE. *Lower line* shows the relationship for the Standard Atmosphere (38). From (114).

sphere. Figure 57.1 shows a comparison between the Standard Atmosphere pressure and the actual barometric pressure measured on Mt. Everest during the 1981 AMREE. The difference in pressure on the summit was found to be approximately 17 mm Hg, or 7% of the actual pressure. This is sufficient to cause a large change in $\dot{V}_{O_{2}max}$ (see MAXIMAL OXYGEN CONSUMPTION, below).

As indicated earlier, the higher barometric pressure for a given altitude can be ascribed to the low latitude of the Himalayas (and the Andes for the most part). It is remarkable that physiologists took so long to become aware of the importance of latitude on the barometric pressure–altitude relationship, a topic that was well understood by meteorologists in the early part of the century. It also became clear that the barometric pressure at a given altitude varied with season. For example, Figure 57.2 shows the mean monthly pressures for the altitude of Mt. Everest as determined from weather balloons (114). Note that the pressure is highest in the summer months and that the variation between midsummer and midwinter is about 12 mm Hg. This change in barometric pressure is certainly sufficient to affect the maximum work capacity and means that climbing Mt. Everest without supplementary oxygen is a substantially greater challenge in midwinter than in midsummer (see later, under MAXIMAL OXYGEN CONSUMPTION).

Meteorological data obtained from weather balloons also show that the effects of latitude and season on the barometric pressure–altitude relationship are additive. For example, at the equator, barometric pressure at an altitude of 8,848 m is about 253 mm Hg, irrespective of the season. However, at a latitude of 60°N in midwinter barometric pressure is only about 217 mm Hg (103).

It should be added that the effects of latitude and season on barometric pressure are exaggerated to some

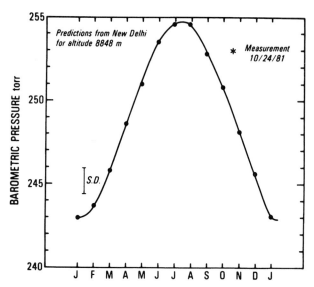

FIG. 57.2. Mean monthly pressures for 8,848 m altitude obtained from weather balloons released from New Delhi, India. Note increase during summer months. Mean monthly standard deviation (SD) also shown. Barometric pressure measured on the Everest summit on October 24, 1981 (*) was unusually high. From (114).

extent by the presence of water vapor in the moist inspired gas. The P_{O_2} of moist inspired gas is given by the equation:

$$P_{O_2} = 0.2094 \ (P_B - P_{H_2O})$$

where P_B is barometric pressure and P_{H_2O} is the water vapor pressure. The latter equals 47 mm Hg at 37°C

and is independent of barometric pressure. The result is that a 10% reduction in barometric pressure at the summit of Mt. Everest results in a 12% reduction in the P_{O_2} of moist inspired gas.

PULMONARY GAS EXCHANGE

Alveolar Gas Composition

One of the most important features of acclimatization to high altitude is an increase in alveolar ventilation, and this is critically important at extreme altitude. One of the best ways to demonstrate this is to portray the alveolar gas composition on an oxygen–carbon dioxide diagram (72). Figure 57.3 shows the alveolar P_{O_2} and P_{CO_2} of acclimatized subjects from sea level (top right) to the Everest summit (bottom left). The solid line was drawn by Rahn and Otis (72) based on data from a number of investigators at increasing altitudes. Note that both alveolar P_{O_2} and P_{CO_2} fall as altitude increases. The P_{O_2} falls because of the reduction of inspired P_{O_2} as altitude increases, while alveolar P_{CO_2} falls because of increasing hyperventilation. The three triangles show mean data obtained by the AMREE expedition, including the Everest summit (bottom triangle). The data fit well with the extrapolation of the line.

Note that the bottom part of the line and its extrapolation are almost vertical, indicating that above a certain altitude (about 7,000 m or barometric pressure 325 mm Hg) alveolar P_{O_2} is essentially constant at about 35

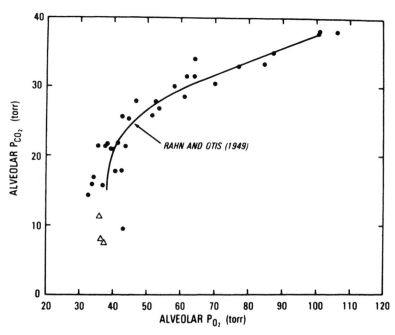

FIG. 57.3. Oxygen–carbon dioxide diagram showing alveolar gas values collated by Rahn and Otis (72) together with values obtained at extreme altitudes by the AMREE expedition (triangles). From (112).

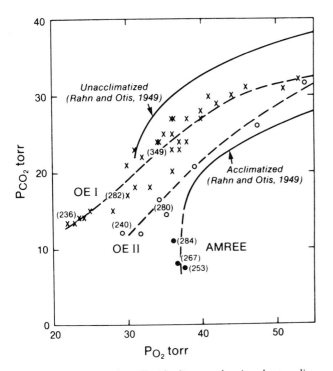

FIG. 57.4. Oxygen–carbon dioxide diagram showing the two lines drawn by Rahn and Otis (72) for unacclimatized and acclimatized subjects at high altitude (compare Fig. 57.3). Alveolar gas failures for OEI, OEII, and AMREE are shown together with corresponding barometric pressures. Note that OEI subjects appeared to be poorly acclimatized at extreme altitudes, whereas OEII subjects had intermediate values. From (107).

AMREE (107). This was particularly the case for OEI, where the chamber was small and the subjects had little opportunity to exercise. In general, a field expedition allows a much longer period of time for acclimatization. However, a chamber experiment enables investigators to carry out much more ambitious and invasive procedures than would be possible in the field. The two types of experiment are therefore complementary.

Blood Gases and Acid-Base Status

Although the alveolar gas values shown in Figures 57.3 and 57.4 are of great interest, in terms of the delivery of oxygen to the peripheral tissues it is clearly desirable to know the arterial blood gases. Arterial P_{O_2}, P_{CO_2}, and pH were measured during OEII at inspired P_{O_2} values of 63, 49, and 43 mm Hg (Table 57.2) (93). With subjects breathing air, these correspond to barometric pressures of 349, 281, and 252 mm Hg, and on Mt. Everest the corresponding altitudes are 6,450, 8,100 and 8,848 m, respectively. The last altitude is the Everest summit. At each altitude, arterial blood was taken at rest and at two or three levels of exercise obtained by means of a bicycle ergometer.

Note that both arterial P_{O_2} and P_{CO_2} declined with altitude and, at any given altitude, with increasing work level. Mean arterial P_{O_2} on the Everest summit was 30 mm Hg at rest, and this fell to 28 mm Hg during exercise levels of 60 and 120 W. Resting arterial P_{CO_2} fell from 20 mm Hg at an inspired P_{O_2} of 63 down to 11 mm Hg on the Everest summit. Exercise on the summit reduced arterial P_{CO_2} to 10 mm Hg. The very low P_{CO_2} values were associated with high pH levels, indicating respiratory alkalosis. The highest arterial pH was 7.56, which was the resting value on the Everest summit.

It is interesting to compare these data with calculated

mm Hg. This means that successful climbers are able to defend their alveolar P_{O_2} by extreme hyperventilation. Another way of saying this is that they insulate the P_{O_2} of alveolar gas from the falling value in the atmosphere around them. This appears to be one of the most important features of acclimatization to extreme altitude. Note that the alveolar P_{CO_2} values on the summit were between 7 and 8 mm Hg (112). To generate the enormous ventilation required for these extremely low P_{CO_2} values, the successful climber needs an adequate hypoxic ventilatory response (see later, under Blood Gases and Acid-Base Status).

Rahn and Otis (72) showed that the alveolar composition of unacclimatized subjects was different from that of fully acclimatized subjects because the former had lower alveolar ventilations at a given altitude. This resulted in alveolar gas compositions being displaced upward and to the left of the fully acclimatized subjects. Figure 57.4 shows solid lines for acclimatized and unacclimatized subjects at high altitude, where alveolar P_{O_2} is less than 55 mm Hg (72). The figure also shows alveolar gas data for OEI (80) and OEII (49a). Note that the degree of acclimatization for these chamber experiments appeared to be less than in the field experiment

TABLE 57.2. *Arterial* P_{O_2} *and* P_{CO_2} *and pH at Extreme Altitude Measured during Operation Everest II*

	Work Level (watts)	P_{O_2}	P_{CO_2}	pH
$P_{IO_2} = 63$, 6,450 m	0	41	20	7.50
	60	35	21	7.49
	120	34	19	7.48
	180–210	34	17	7.44
$P_{IO_2} = 49$, 8,100 m	0	37	13	7.53
	60	33	13	7.51
	120	33	11	7.49
$P_{IO_2} = 43$, 8,848 m	0	30	11	7.56
	60	28	11	7.55
	120	28	10	7.52

P_{O_2} and P_{CO_2} in mm Hg. Altitudes correspond to barometric pressures on Mt. Everest.

P_{IO_2} is partial pressure of inspired oxygen.

From Sutton et al. (93).

arterial blood gases from AMREE (112). These gave an arterial P_{O_2} of 28 mm Hg on the summit, which agrees well with OEII measurements. However, mean alveolar P_{CO_2} was 7.5 mm Hg on the summit, and it is reasonable to assume little if any difference between end-capillary and alveolar values. Thus the P_{CO_2} obtained on AMREE was 3–4 mm Hg lower (about 30%) than that on OEII. This lower P_{CO_2} also caused a much higher calculated arterial pH based on base excess measurements made on venous blood from two climbers on the morning after a successful summit climb. Calculated arterial pH was between 7.7 and 7.8, an extraordinary degree of respiratory alkalosis. Most of the difference in pH between AMREE and OEII measurements can be attributed to the difference in arterial P_{CO_2}.

There are two possible reasons for the much lower P_{CO_2} values found on the field expedition. One is that the climbers were better acclimatized, as discussed in relation to Figure 57.4. In fact, Pizzo, who collected the summit alveolar gas samples on himself, had been exposed to high altitude for some 77 days compared with 36 days for the first summit measurements on OEII. This reflects the fact that logistically it is much easier to spend long periods at high altitude in the field compared with being confined to a low-pressure chamber.

An additional reason for the much lower P_{CO_2} values in the field experiment may have been the unusually high hypoxic ventilatory response of Pizzo. However, against this, his alveolar P_{CO_2} values were in line with those measured on four other climbers at the slightly lower altitude of 8,050 m at Camp 5 (112). Thus it is likely that the main reason for the lower P_{CO_2} values was the longer period of acclimatization.

An interesting feature of the base excess measurements on the two climbers the morning after they reached the summit of Mt. Everest is that they were not different from those measured in 14 subjects living for several weeks at an altitude of 6,300 m (barometric pressure 351 mm Hg), where the mean value was 8.7 ± 1.7 mmol/liter. This suggests that base excess was changing extremely slowly above an altitude of 6,300 m. The reason for this slow change is not known but may be related to the chronic volume depletion observed in climbers living at 6,300 m. At this altitude serum osmolality was 302 mOsm/kg, which was significantly higher than in the same subjects at sea level, where the value was 290 mOsm/kg (6). It is known that the kidney gives a higher priority to correcting dehydration than to acid-base disturbances, and to excrete more bicarbonate to reduce the base excess it would be necessary to lose corresponding cations, which would aggravate volume depletion. This may be the basis for the slow excretion of bicarbonate by the kidney.

These acid-base changes may partly explain why climbers can spend only a relatively short time at extreme altitudes, say above 8,000 m. The marked respiratory alkalosis which increases the oxygen affinity of the hemoglobin at extreme altitudes is beneficial because it enhances the loading of oxygen by the pulmonary capillaries (109). If a climber remains at extreme altitude for several days, presumably there is some renal excretion of bicarbonate, even though this appears to be very slow, and the resulting metabolic compensation would move the pH of the blood back toward 7.4. Thus the advantage of the left-shifted dissociation curve would tend to be lost.

This sequence of events suggests that a good strategy to climb Mt. Everest without supplementary oxygen would be to set up the high camps and then return to base camp at a lower altitude for several days. This period at medium altitude would then allow the body to adjust again to this more moderate oxygen deprivation and enable the blood pH to stabilize nearer normal. The final summit assault would then be as rapid as possible to take advantage of the essentially uncompensated respiratory alkalosis. In fact, this was the pattern adopted by Messner and Habeler (53) in their first ascent of Mt. Everest without supplementary oxygen in 1978.

The very low P_{CO_2} values measured on both AMREE and OEII emphasize the enormous physiological value of extreme hyperventilation at these great altitudes. It is not surprising, therefore, that there is some evidence that climbers with a well-developed hypoxic ventilatory response (HVR) tolerate extreme altitude better than those with a low HVR. For example, Schoene and colleagues (87) showed that among the participants of AMREE, those with the highest HVR values reached higher altitudes than those with low HVR values. Other studies have reported similar results (50, 85). However, this is by no means a universal finding. Some investigators have found that the HVR values of successful climbers to extreme altitudes are intermediate or even occasionally blunted (56, 60, 86). In this regard, it should be noted that high-altitude natives usually have a blunted HVR but often perform better than lowlanders at extreme altitude.

There are additional reasons why a high HVR may be a handicap at extreme altitude. Lahiri et al. (45, 47) showed that a brisk HVR is associated with periodic breathing at high altitude, and since this phenomenon results in very low arterial P_{O_2} levels following the apneic periods, this may reduce tolerance to extreme altitude. Finally, Hornbein et al. (34) showed that climbers with a brisk HVR suffered a greater impairment of mental performance at high altitude, presumably because of the more severe cerebral vasoconstriction as a result of the lower arterial P_{CO_2}.

The very low arterial P_{O_2} values at extreme altitude

are partly the result of diffusion limitation of oxygen transfer across the blood–gas barrier. This possibility has been recognized at least since the 1921–1922 International High-Altitude Expedition to Peru (2). These investigators found that arterial oxygen saturation fell during exercise at an altitude of 4,300 m and correctly concluded that the cause was diffusion-limited oxygen transfer across the blood–gas barrier. Houston and Riley (36) in OEI showed that the alveolar–arterial PO_2 difference increased during exercise at simulated high altitude and ascribed this to diffusion limitation. Measurements during HSME showed that arterial oxygen saturation fell during exercise in the face of a progressive rise in alveolar PO_2. This was strong evidence for diffusion limitation of oxygen transfer (113). Similar results were found on AMREE during exercise at an altitude of 6,300 m (111).

A possible criticism of all these studies is that no account was taken of ventilation–perfusion inequalities within the lung, and that these may have contributed to the observed fall in arterial oxygen saturation and the increased alveolar–arterial PO_2 difference during exercise. An important advance was made by Wagner and colleagues (99), who separately measured ventilation–perfusion inequality during OEII and were able to show that with increasing altitude, its contribution to the fall in arterial PO_2 during exercise decreased while the contribution of diffusion impairment increased.

The reason why diffusion limitation of oxygen transfer occurs at extreme altitude, particularly during exercise, is that the effective slope of the oxygen dissociation curve is very high. Two factors contribute to this. One is that, because alveolar PO_2 is so low, oxygen exchange is occurring on a very steep part of the oxygen dissociation curve. The second factor is that the polycythemia of high altitude increases the change in oxygen concentration of the blood per unit change of PO_2. In effect, oxygen is beginning to behave like carbon monoxide, which has an extremely steep slope of its blood dissociation curve in the working range. For an analysis of the factors predisposing to diffusion limitation across the blood–gas barrier, see Piiper and Scheid (64).

MAXIMAL OXYGEN CONSUMPTION

Maximal work rate (power) and maximal oxygen consumption ($\dot{V}O_{2max}$) fall precipitously with increasing altitude at extreme heights, and it is a remarkable coincidence that the $\dot{V}O_{2max}$ of a climber near the summit of Mt. Everest is just sufficient to allow him to reach the highest point on earth without supplementary oxygen. As pointed out earlier, this is suggested by the history of attempts on the Everest summit. For example, Norton (59) got to within 300 m of the summit in 1924,

but the mountain was not climbed without supplementary oxygen until 1978, that is 54 years later.

Figure 57.5 shows measurements of $\dot{V}O_{2max}$ on both AMREE and OEII down to the inspired PO_2 of the summit of Mt. Everest. The AMREE measurements were made on extremely well-acclimatized subjects at an altitude of 6,300 m with subjects breathing 16% and 14% oxygen (111). The last gave an inspired PO_2 of 42.5 mm Hg, equivalent to that on the Everest summit, where $\dot{V}O_{2max}$ was 15.3 ml/min/kg, equivalent to 1.07 liters/min. Data from OEII were obtained from measurements made after 40 days of exposure to high altitude (20), and, as noted earlier, subjects were apparently not as well acclimatized as those on AMREE. Summit measurements were made at a barometric pressure of 240 mm Hg, with subjects breathing a slightly enriched oxygen mixture giving the same inspired PO_2 of about 43 mm Hg. In spite of this difference between the two studies, the agreement on $\dot{V}O_{2max}$ is extraordinarily close. Notice also that the AMREE subjects had substantially higher $\dot{V}O_{2max}$ values at sea level but nevertheless the data points on the Everest Summit were extremely close.

An oxygen consumption of about 1 liter/min is equivalent to that produced by walking slowly on level ground. It therefore only permits a very slow climbing rate, and calculations show that it fits reasonably well with the maximal climbing rate near the summit reported by Messner and Habeler (102). Note that climbers in the field never work at $\dot{V}O_{2max}$ because their physical activity has to extend over long periods.

An interesting question is what limits maximal oxygen uptake at extreme altitude. People who ask this

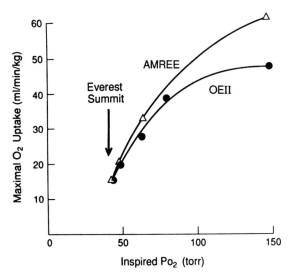

FIG. 57.5. Maximum oxygen consumptions plotted against inspired PO_2 for OEII and AMREE. Note that, although AMREE values were higher at sea level, values measured at the summit PO_2 were essentially identical. From (109).

question often seem to expect a simple answer, such as cardiac output or total ventilation. However, this is naive. Many factors contribute to the limitations of maximal exercise.

One way of approaching this question is to consider the oxygen cascade from the atmosphere to the mitochondria, which includes the processes of convective ventilation of oxygen to the alveoli, diffusion of oxygen across the blood–gas barrier, uptake of oxygen by the hemoglobin in the pulmonary capillaries, convective flow of blood to the peripheral capillaries, unloading of oxygen from the hemoglobin, diffusion of oxygen to the mitochondria, and utilization of oxygen by the electron transport system. Using a theoretical approach, it is possible to calculate the percentage change in total oxygen flow for a given (say 5%) change in resistance (or conductance) at any point along the cascade, assuming that all other factors remain unchanged (102). A number of assumptions are necessary, and the original analysis was carried out with data obtained from AMREE. Figure 57.6 shows the results.

It can be seen that the variable most influencing $\dot{V}O_{2max}$ is barometric pressure. After that, alveolar ventilation, membrane-diffusing capacity of the lung, cardiac output, and blood hemoglobin concentration play approximately equal roles. Note that a reduction in P_{50} (PO_2 for half-saturation of hemoglobin) of the blood is advantageous because this steepens the oxygen dissociation curve, thus enhancing loading of oxygen in the pulmonary capillary and reducing the diffusion limitation. One of the assumptions in this analysis is that the PO_2 of mixed venous blood has to be maintained at 15 mm Hg or above to insure adequate diffusion of oxygen

to the muscle mitochondria. Subsequent measurement on OEII showed that this assumption is reasonable; the PO_2 values of mixed venous blood on the Everest summit for the two highest work levels had mean values of 14.8 and 13.8 mm Hg.

The exquisite sensitivity of $\dot{V}O_{2max}$ to barometric pressure at extreme altitude leads to some interesting predictions. One is that the variation in barometric pressure caused by weather or season will have a significant effect on a climber's ability to reach the summit. For example, Figure 57.2 shows that during the winter, barometric pressure on the Everest summit can be as low as 243 mm Hg, a reduction of nearly 5% compared with the highest midsummer value. Figure 57.6 shows that this would reduce $\dot{V}O_{2max}$ by about 25%, or to less than 800 ml/min. Thus the difficulties of reaching the summit without supplementary oxygen during the winter would be enormously increased by the low barometric pressure, let alone other factors such as the cold temperature. Table 57.3 shows barometric pressures on the summit for some key ascents of Mt. Everest without supplementary oxygen, derived from meteorological data (110).

Another interesting question is how much higher climbers could go above the Everest summit. The answer from Figure 57.6 seems to be very little. For example, a 5% decrease in barometric pressure, which would reduce $\dot{V}O_{2max}$ by about 25%, occurs at an altitude of about 9,250 m at the latitude of Everest, that is about 400 m above the summit. Note that this gives a pressure of about 240 mm Hg, which is still above that predicted for the summit of Mt. Everest by the Standard Atmosphere (see BAROMETRIC PRESSURE, above). Thus it is only the equatorial bulge in barometric pressure which allows humans to reach the highest summit of the world without supplementary oxygen.

It has been suggested that the reduced maximal power at extreme altitude might be related to a reduced muscle force output or increased fatigability. This hypothesis was tested on OEII at barometric pressures of 760, 335, and 282 mm Hg using electrical stimulation of ankle dorsiflexor muscles (25). No impairment of neuromuscular transmission was demonstrated, but there was evidence of increased muscle fatigue apparently related to chronic hypoxia.

The possibility of adaptation of skeletal muscle to prolonged, severe hypoxia was also studied in OEII by taking biopsies from the vastus lateralis muscle at 760, 380, and 282 mm Hg (28, 48). Reductions were found in succinic dehydrogenase, citrate synthetase, and hexokinase at the highest simulated altitude. The number of capillaries per cross-sectional area was increased as a result of a reduction in muscle fiber area. This finding had previously been reported on high-altitude climbers (8, 14, 60). It was concluded that there were no adap-

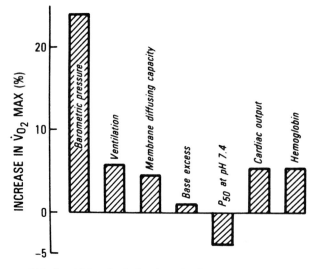

FIG. 57.6. Sensitivity of calculated maximal oxygen consumption to changes in several variables for a climber on the summit of Mt. Everest. Each variable was increased by 5%, leaving all others constant. From (102).

TABLE 57.3. *Barometric Pressures at 8,848 m Altitude for Some Key Ascents of Mt. Everest*

		Altitude (m) at Pressure			Pressure at 8,848 m (mm Hg)
Date	Event	500 mbar	300 mbar	200 mbar	
May 8, 1978	First ascent without supplementary O₂ (by Messner and Habeler)	5,850	9,640	12,390	251
August 20, 1980	First solo ascent without supplementary O₂ (by Messner)	5,830	9,800	12,630	256
October 24, 1981	First direct measurement of pressure on summit (by Pizzo) (value was 253 mm Hg)	5,860	9,650	12,380	252
December 22, 1987	First winter ascent without supplementary O₂ (by Sherpa Ang Rita)	5,810	9,500	12,170	247

Data are from weather balloons released at 1200 h UTC (5:40 PM Nepal time) from New Delhi, India, or Lhasa, Tibet. Altitude readings shown are heights of balloon for the given pressure (500 mbar etc.). Pressures at 8,848 m altitude are by interpolation (110).

tations that would enhance increased oxidative function in the muscles.

Maximal exercise at great altitude is associated with extremely high levels of exercise ventilation. Indeed, this was one of the most obvious features of climbing at extreme altitudes on the early Everest expeditions. For example, Somervell, who attained an altitude of approximately 8,500 m on the north side of Everest in 1924, complained dramatically of extreme breathlessness at these great altitudes, stating that "For every step forward and upward, 7 to 10 complete respirations were required" (59).

The highest levels of exercise ventilation so far recorded were obtained on AMREE at an altitude of 6,300 m (barometric pressure 351 mm Hg). In eight subjects exercising at a work rate of 1,200 kg/min mean total ventilation (body temperature, ambient pressure, saturated with water vapor; BTPS) was 207.2 l/min, with a mean respiratory frequency of 62 breaths/min (111). These values were for a mean oxygen consumption of 2.3 liters/min. These levels of ventilation far exceed anything ever seen at sea level and are approaching the *maximal voluntary ventilation* (MVV), that is, the maximal amount of air that can be moved per minute by breathing in and out as rapidly and deeply as possible. Some measurements of exercise ventilation exceeding 200 liters/min were also obtained during the Himalayan Scientific and Mountaineering Expedition (HSME) at an altitude of 5,800 m (70).

It is interesting that the highest levels of ventilation are not seen at the highest altitudes. For example, when inspired PO₂ was lowered to that equivalent to the Everest summit on AMREE, maximal exercise ventilation was only 162 l/min (111). A reasonable explanation for the lower maximal ventilation is that the work rate was very much lower, being only 450 kg/min as opposed to 1,200 kg/min at 6,300 m during air breathing. A similar pattern was found during the HSME. At an altitude of 5,800 m maximal exercise ventilation had a mean value

of 173 l/min. However, at the higher altitude of 6,400 m this had fallen to 161 liters/min, while at the highest altitude at which measurements were made at that time (7,440 m), the value was only 122 liters/min. The corresponding maximal work levels for the three altitudes were 1,200, 900, and 600 kg/min, respectively. The extremely high exercise ventilations at great altitudes are facilitated by the reduced work of breathing as a result of the lowered density of air (63). It has been shown that MVV (or maximum breathing capacity) increases at simulated high altitudes in a low-pressure chamber (18).

Although maximal exercise ventilation decreases at the highest altitudes near the summit of Mt. Everest, respiratory frequency continues to increase. This means that at the highest altitudes breathing becomes very rapid and shallow. For example, during AMREE, Pizzo measured his ventilation while breathing air and climbing at about 8,300 m, and during the middle 4 min of this period his mean respiratory frequency was 86 breaths/min. No wonder that Messner stated when he reached the Everest summit for the first time with Habeler, "I am nothing more than a single, narrow gasping lung, floating over the mists and the summits" (53).

It is noteworthy that the measurement made during OEII did not show the same decline in maximal exercise ventilation at the highest altitudes. Why these data should differ from those obtained during the two field expeditions, AMREE and HSME, is unclear but may be related to the smaller degree of acclimatization of the OEII subjects.

An interesting question is whether these enormous ventilations at extreme altitude incur such an oxygen cost that they limit external work carried out by the climber. The alveolar PCO₂ of about 7.5 mm Hg measured by Pizzo on the summit indicates that alveolar ventilation was about five times the resting value because it is known that carbon dioxide production both at rest and for a given work level is independent of altitude. Thus if we take the normal resting ventilation to be 7–8 l/min, this means that the resting ventilation on the summit is at least 40 l/min. There is anecdotal confirmation of this high ventilation from a tape made by Pizzo on which he recorded barometric pressure and other information; he had to stop to catch his breath between every two or three words!

It is likely that the oxygen costs of this hyperventilation become a significant proportion of resting oxygen consumption. Sea-level measurements suggest that increasing the ventilation from 8 to 40 liters/min increases oxygen consumption by about 20–40 ml/min (61), though there is considerable variability in the data. As pointed out earlier, the reduced density at great altitudes lessens the work of breathing (63). Nevertheless, it appears that the oxygen cost of ventilation may be 10% of the total oxygen uptake at rest, and it may contribute to a higher proportion of the total oxygen uptake during moderate exercise.

A remarkable feature of maximal exercise at extreme altitude is that, although the aerobic capacity is severely reduced (as shown in Fig. 57.5), anaerobic glycolysis is also greatly restricted. This is in sharp contrast to the situation in acute hypoxia, where a reduction in aerobic capacity is partly compensated for by a large increase in anaerobic glycolysis (13, 105).

The observation that blood lactate levels are low in acclimatized subjects at high altitudes even during maximal work was initially made by Edwards (22) during the 1935 International High-Altitude Expedition to Chile. He plotted blood lactate against work rate (power) at various altitudes up to 5,340 m and showed that when subjects were well acclimatized all points fell on the same line. This means that blood lactate for a given work level was independent of tissue P_{O_2}. Since maximum work capacity declines markedly with increasing altitude (Fig. 57.5), this result also implies that maximal blood lactate falls in acclimatized subjects as altitude increases.

Cerretelli (11–13) has done extensive studies on this topic, and a summary of the results is shown in Figure 57.7 together with additional data obtained at an altitude of 6,300 m on AMREE (111). A provocative finding from these data is that during maximal exercise at altitudes exceeding 7,500 m there will be no increase in blood lactate at all, in spite of the extreme oxygen deprivation. This is a paradox.

The reasons for the low blood lactate levels during maximal exercise in acclimatized subjects at high altitude are not clear and remain the subject of much investigation. It should be pointed out that a low blood lactate per se does not necessarily mean reduced lactate production because the blood level depends on a balance between production and removal. However, stud-

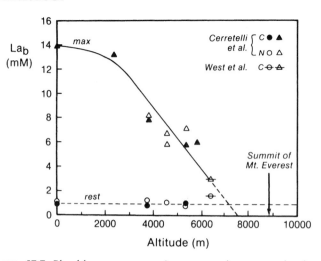

FIG. 57.7. Blood lactate concentrations measured at rest or shortly after maximal exercise at various altitudes. Most of the data are redrawn from Cerretelli (13). *Filled circles* and *triangles* show data for acclimatized Caucasians (C); *open circles* and *triangles* are for high-altitude natives (N). Data for 6,300 m are from AMREE for acclimatized lowlanders. From (105).

ies by muscle biopsy in OEII showed that lactate levels in the muscle were indeed reduced. For example, at a barometric pressure of 282 mm Hg, muscle lactate concentration (mmol/kg dry weight) was 39.2 compared with 113 at sea level after exhaustive exercise (28). An early hypothesis was that plasma bicarbonate depletion resulting from acclimatization interferes with the buffering of released lactate and hydrogen ions, and the resulting fall in local pH inhibits the enzyme phosphofructokinase in the glycolytic cycle. However, there is now evidence against this mechanism. This topic is considered in more detail in the chapter by Lutz and Cherniak in this *Handbook*.

CARDIOVASCULAR SYSTEM

In spite of the great importance of the cardiovascular system in the transport of oxygen to peripheral tissues at high altitude, relatively little information is available on this topic at extreme altitude. The reason for this is the technical difficulty of measuring cardiac output, myocardial function, and pulmonary vascular pressures, which often require complicated equipment or invasive procedures. For example, until the 1980s, only one series of measurements of cardiac output had been made above an altitude of 5,400 m (69). Fortunately, an extensive series of measurements of the cardiovascular system was made during OEII, and most of what we know about this topic comes from this important experiment.

Cardiac output is substantially increased during exposure to acute hypoxia (44, 97). The increase can be considered a useful response because it enhances the delivery of oxygen to the peripheral tissues. However, it is a remarkable fact that in acclimatized subjects, the relationship between cardiac output and work level returns to its sea-level value. This surprising finding has been confirmed on many occasions at moderate altitudes (16, 29, 96). Pugh (69) found the same result at 5,800 m during HSME, where he showed that cardiac output, measured by acetylene rebreathing, fell on the same line when related to work rate as sea-level measurements on the same subjects. A similar finding was reported by Cerretelli at 5,350 m (11).

Essentially the same results were found during OEII (74). These are shown in Figure 57.8, where the top left panel shows that the relationship between cardiac output and oxygen uptake was almost independent of altitude. The only possible deviation is in the measurements made on the Everest summit, but these also lie very close to the sea-level line. High-altitude natives at Cerro de Pasco, Peru (4,305 m), showed these same relationships between cardiac output and oxygen consumption during exercise at sea level in Lima (98).

It is surprising that cardiac output bears the same relationship to work rate (or power) in well-acclimatized lowlanders and high-altitude natives as it does at

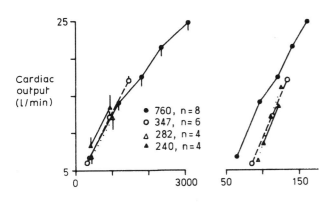

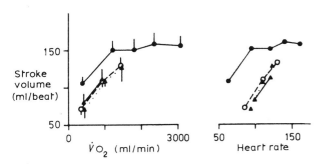

FIG. 57.8. Cardiac output (by thermodilution) and stroke volume plotted against oxygen uptake and heart rate at barometric pressures of 760, 347, 282, and 240 mm Hg during OEII. For measurements at 240 mm Hg, subjects breathed an oxygen mixture to give an inspired PO_2 of 43 mm Hg. From (74).

sea level. A theoretical analysis such as that shown in Figure 57.6 indicates that an increase in cardiac output would improve oxygen delivery to the peripheral tissues. It should be pointed out, however, that in acclimatized subjects at high altitude and in high-altitude natives the hemoglobin concentration of the blood is increased as a result of polycythemia. Therefore, although the cardiac output in relation to work level is unchanged, hemoglobin flow is appreciably increased. Grollman (29) actually suggested back in 1930 that the return of cardiac output to its sea-level value in acclimatized subjects was related in some way to the increase in hemoglobin concentration of the blood. The whole topic emphasizes how little we know about the control of cardiac output.

Heart rate is increased at a given work level during both acute and chronic hypoxia. With acclimatization, resting heart rates may return to approximately sea-level values up to altitudes of about 4,500 m, though there is some individual variation (62, 82). However, heart rates during exercise always exceed sea-level values when related to the same work level (69, 74). Maximal heart rate, that is heart rate at maximal exercise, is reduced in acclimatized subjects at extreme altitude. For example, in OEII maximal heart rates decreased from 160 beats per minute (bpm) at sea level to 137 bpm at a simulated altitude of 6,100 m, 123 bpm at 7,620 m, and 118 bpm at 8,848 m (74). Of course, maximal work rates and oxygen consumptions fall with increasing altitude (Fig. 57.5), and this is a reasonable explanation for the reduced maximal heart rates. As indicated above, when heart rate is related to work level, values are increased at high altitudes.

It is interesting that oxygen breathing in acclimatized subjects at high altitude reduces heart rate for a given work level (69). In fact, heart rate can be actually lower than that corresponding to the same work rate at sea level. A possible explanation is that arterial PO_2 at 5,800 m during 100% oxygen breathing was higher than at sea level and that the subjects had much higher hemoglobin levels than at sea level because of high-altitude polycythemia. It is known that heart rate for a given work level is inversely related to hemoglobin concentration (79).

Stroke volume is reduced at high altitude (including extreme altitude) for a given work level (69). This finding can be deduced from the results already presented because cardiac output is the same as at sea level (Fig. 57.8), whereas heart rate is increased. The reduction in stroke volume was also confirmed in OEII (74). In addition, it was shown that oxygen breathing did not increase stroke volume for a given pulmonary arterial wedge pressure. This suggested that the decline in stroke volume was not caused by hypoxic depression of myocardial contractility (74).

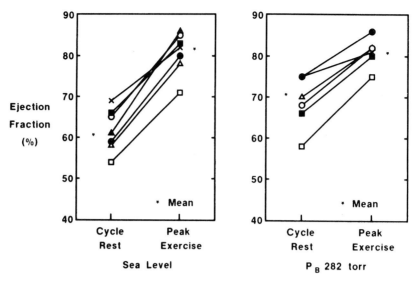

FIG. 57.9. Left ventricular ejection fractions at rest on the cycle ergometer and during peak exercise at sea level and at a simulated altitude of about 8,010 m. Note that ejection fraction was well preserved. From (91).

Myocardial contractility was further measured during OEII with very interesting results. First, it was argued that the reduced stroke volume at high altitude could be caused by either reduced cardiac filling or impaired myocardial contractility. A fall in filling pressures could result from either an increased heart rate or a reduction of circulating blood volume or both.

Both right atrial mean pressure (filling pressure for the right ventricle) and pulmonary arterial wedge pressure (as an index of filling pressure of the left ventricle) were measured. Both values tended to fall as simulated altitude was increased (74). An interesting finding was that right atrial pressures tended to be low in spite of the pulmonary hypertension (see below, this section). In general, the relationship between stroke volume and right atrial pressure was maintained, suggesting maintenance of contractile function. In addition, as stated earlier, oxygen breathing did not increase stroke volume for a given filling pressure, suggesting that the reduced stroke volume was not caused by hypoxic depression of contractility.

Additional evidence for the preservation of myocardial contractility in spite of the severe hypoxemia came from a two-dimensional echocardiography study during OEII (91). It was found that ventricular ejection fraction, the ratio of peak systolic pressure to end-systolic volume, and mean normalized systolic volume at rest were sustained at a barometric pressure of 282 mm Hg, corresponding to an altitude of about 8,010 m. Indeed, the remarkable observation was made that during exercise at a level of 60 W, the ejection fraction was actually higher (70% ± 2% compared with 69% ± 8%) at a barometric pressure of 282 mm Hg compared with sea level (Fig. 57.9). The conclusion was that, despite decreased cardiac volumes, severe hypoxemia, and pulmonary hypertension, cardiac contractile function appeared to be well maintained.

It should be pointed out that this maintenance of normal myocardial contractility occurred in the face of severe hypoxemia. Table 57.2 shows that at a work level of 60 watts for an inspired PO_2 of 49 mm Hg, altitude about 8,010 m, arterial PO_2 was only 33 mm Hg. Thus in spite of this extreme degree of arterial hypoxemia there was no impairment of myocardial function.

This provocative result emphasizes the difference between myocardial hypoxemia and ischemia. The former appears to be well tolerated, whereas the latter leads to severe impairment of myocardial function. It should be emphasized that these results can only be applied to the normal myocardium of the young fit volunteers of OEII. Occasionally, physicians have argued that the high-altitude environment poses no threat for patients with coronary artery disease (77), but this seems an extreme view (108).

Abnormal cardiac rhythms are uncommon at extreme altitude. For example, in an electrocardiographic study of 19 subjects during AMREE, only one subject had premature ventricular contractions, and these were recorded at an altitude of 5,300 m. Another climber showed premature atrial contractions at 6,300 m (39). Occasional premature ventricular and atrial contractions have been observed by others (19, 49, 55). However, marked sinus arrythmia accompanying the periodic breathing of sleep is extremely common at high altitude (see sleep, below).

The pulmonary circulation shows some of the most dramatic changes at extreme altitude. Acute hypoxia causes pulmonary hypertension as a result of hypoxic pulmonary vasoconstriction, predominantly in the small pulmonary arteries. Hypertension is maintained during the chronic hypoxia of exposure to high altitude, and the initial contraction of vascular smooth muscle is supplemented by hypertrophy of the muscle and an increase in collagen and elastin in the wall of the blood vessel, a process known as remodeling (54). In a study on acclimatized lowlanders who spent about 1 yr at an altitude of 4,540 m, mean pulmonary arterial pressure increased from its sea-level value of about 12 to about 18 mm Hg (82, 90).

However pulmonary arterial pressure increases considerably during exercise at high altitude. Figure 57.10 shows the relationship between the mean pulmonary vascular pressure gradient across the lung (mean pulmonary arterial pressure − mean pulmonary wedge pressure) plotted against cardiac output in the subjects of OEII (30). Note that the resting values of the gradient (primarily determined by mean pulmonary arterial pressure) increased, but the most dramatic change was in the slope of the pressure gradient with respect to cardiac output. Mean pulmonary arterial pressure at rest at sea level was 15.0 ± 0.9 mm Hg, and this increased to 33 ± 3 mm Hg at a barometric pressure of 240 mm Hg. Electrocardiographic changes consistent with severe pulmonary hypertension have been seen at extreme altitudes (39, 49, 55).

Inhalation of 100% oxygen in OEII lowered cardiac output and pulmonary artery pressure, but there was no significant fall in pulmonary vascular resistance (30). In interpreting this result it should be noted that a fall in cardiac output normally results in an increase in pulmonary vascular resistance because the reduction in capillary pressure causes derecruitment of capillaries

and a reduction in caliber of those which remain open (27). Thus the fact that pulmonary vascular resistance did not change when it was expected to rise indicated that oxygen breathing probably reduced vascular resistance to some extent. Nevertheless, the degree of irreversibility of the increased pulmonary vascular resistance is remarkable in the subjects who were hypoxic for only 2 or 3 wk when the measurements were made. This indicates that rapid structural changes (remodeling) take place in the pulmonary blood vessels in addition to simple contraction of vascular smooth muscle. Studies in rats and calves exposed to low oxygen mixtures demonstrate that remodeling begins only days or even hours after exposure (51, 65, 95).

Polycythemia occurs at extreme altitude, but whether it is beneficial is arguable. On AMREE, mean hemoglobin concentration at the highest altitude was 18.4 g/dl, and the corresponding value on OEII was 16.9 g/dl (30). These figures are less than the average of 20.5 g/dl reported by Pugh (68), who reviewed the results from five previous field expeditions.

Although high hemoglobin levels increase the amount of oxygen in the arterial blood, they also raise the viscosity of the blood and may cause uneven blood flow and sludging in muscle capillaries. Sarnquist et al. (84) on AMREE found that hemodilution did not impair physical performance; in fact, there was a small improvement in psychometer tests. Winslow et al. (116) reduced the hematocrit of Andean high-altitude residents from 62% to 42% (mean values) and found an increased cardiac output and PO_2 of mixed venous blood.

It should be remembered that the evolutionary pressure for the erythropoietin control system developed at sea level, where it is of value in replacing blood lost by trauma, parasites, and malnutrition. The tissue hypoxia of high altitude is a very different situation, and it may

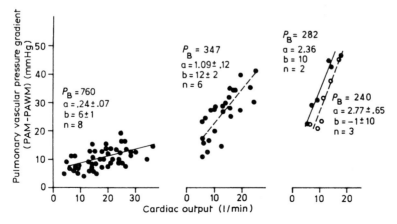

FIG. 57.10. Mean pulmonary artery pressure minus mean pulmonary wedge pressure plotted against cardiac output by thermodilution at various barometric pressures (P_B) during OEII. For measurements at 240 mm Hg, subjects breathed an oxygen mixture to give an inspired PO_2 of 43 mm Hg. From (30).

be that the marked degrees of polycythemia at extreme altitude represent an inappropriate response. This would put polycythemia in the same category as pulmonary hypertension, which also appears to have no physiological value in this setting.

Thromboembolic episodes have been described at very high altitudes from time to time and are presumably related to polycythemia. An extensive study of the clotting cascade during OEII showed no significant differences in clotting factors (1).

OTHER FEATURES

Central Nervous System

In view of the known sensitivity of the central nervous system (CNS) to hypoxia, it is not surprising that changes in neuropsychological function have been described at extreme altitude. These include alterations in special senses, such as vision; higher functions, such as memory; and affective behavior, such as mood. In addition, climbers at extreme altitudes on Mt. Everest have described visual hallucinations, periods of aphasia, and tunnel vision (loss of peripheral vision). Impairment of psychomotor function in Indian troops rapidly airlifted to high altitude during the war between China and India in the early 1960s has been documented (88, 89). Psychiatric disturbances in mountaineers have also been described, including symptoms similar to an organic brain syndrome, that persisted for several weeks after an expedition to 5,500 m (83). During HSME when several normal subjects spent up to 3 months at 5,800 m, mental efficiency was tested by asking the subjects to sort playing cards into bins (26). It was found that the efficiency of sorting cards was less at the high altitude than at sea level and that the inefficiency took the form of a delay in placing the cards into the correct bins rather than errors of sorting. These results reinforce the well-accepted notion that accurate work can be done at extreme altitude but that more time and more effort in concentration are required.

A particularly interesting question is whether there is residual impairment of CNS function after periods at very high altitudes. A careful study was carried out by Townes et al. (94) during AMREE on 21 subjects. A very extensive battery of psychological tests was administered prior to and after the expedition, and a subset of measurements were made at high altitude during the expedition. It was found that there were significant differences between preexpedition, postexpedition, and follow-up performance on neuropsychological tests. Verbal learning and memory declined significantly from the beginning to the end of the expedition as measured by the Wechsler memory scale. In the Halstead-Wep-

man aphasia screening test the number of expressive language errors increased significantly between pre- and post-expedition. The number of aphasic errors was significantly related to the altitude attained by the subject. In addition, finger-tapping speed decreased significantly over the course of the expedition. Of 20 subjects 15 showed impairment of function at Kathmandu immediately after the expedition, and 13 of 16 subjects still had abnormal responses compared with preexpedition measurements when tested 1 yr later. Other investigators have also reported persistent impaired CNS function after exposure to extreme altitude (10, 75), though this has not been a universal finding (17).

Extensive measurement of neuropsychological function were also made during OEII, and these together with results from other field expeditions to high altitude (34) generally confirmed the earlier study during AMREE (94). However, an additional interesting finding was reported, that is that subjects with a high hypoxic ventilatory response (HVR) showed the most impairment of CNS function. For example, a high HVR correlated with a reduction in verbal learning, poor long-term verbal memory, and number of aphasic errors. This is an intriguing finding because subjects with a high HVR would be expected to have an increased arterial P_{O_2} and therefore an increased cerebral P_{O_2}, other things being equal. The investigators suggested that increased ventilation, through its reduction in arterial P_{CO_2}, caused additional cerebral vasoconstriction and reduced cerebral blood flow so that, in spite of the higher arterial P_{O_2}, the degree of cerebral hypoxia was greater. It is well known that a reduced arterial P_{O_2} increases cerebral blood flow (7) and a reduced arterial P_{CO_2} decreases blood flow (31). Climbers at extreme altitude have both severe arterial hypoxemia and hypocapnia (Table 57.2), but the net result of these derangements of blood gases on cerebral blood flow is not known.

Sleep

Anyone who has been at high altitude knows that sleeping is often impaired. Climbers at high altitude frequently complain that they cannot get to sleep for a long period, they wake up frequently during the night, and when they finally get up in the morning they do not feel refreshed. The high frequency of arousals at high altitude has been documented (76, 101).

One of the most striking features of sleep at high and extreme altitudes is periodic breathing. This was first extensively studied by Mosso in his Capanna Regina Margherita at an altitude of 4,559 m (57) and has been confirmed by many investigators since then (3, 46, 49, 92, 101, 115). Figure 57.11 shows an example of periodic breathing at an altitude of 6,300 m (barometric

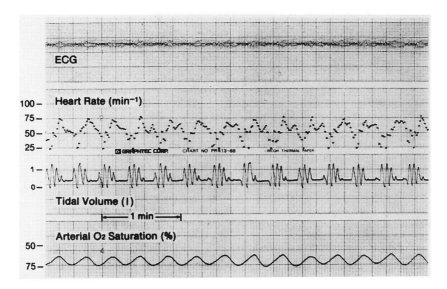

FIG. 57.11. Example of periodic breathing at altitude 6,300 m (barometric pressure 351 mm Hg). From (115).

pressure 351 mm Hg) during AMREE. Marked cyclic variation of heart rate presumably caused by periodic breathing has also been documented at an altitude of 8,050 m (barometric pressure 282 mm Hg) (115).

Some features of the mechanism of periodic breathing were studied during AMREE at an altitude of 5,400 m (45). It was shown that the effect of adding oxygen to inspired gas greatly reduced the strength of periodic breathing, but the periodicity did not totally disappear. Adding carbon dioxide to the inspired gas did not abolish periodic breathing, though it did eliminate the periods of apnea. The investigators also a reported a relationship between the strength of the periodic breathing and HVR. An important finding was that, although lowlanders showed marked periodic breathing at 5,400 m, Sherpas generally did not. The only exception was one Sherpa who had spent long periods of time at low altitudes. In general, Sherpas show low ventilatory responses to hypoxia, though the Sherpa who had some periodic breathing had an intermediate HVR value.

These findings led to the hypothesis that periodic breathing at high altitude is related to the very large hypoxic ventilatory drive and the consequent instability of the control system. This hypothesis has been discussed (115) in the context of the mechanism of periodic breathing presented by Khoo et al. (43). This would explain the correlation between HVR and the strength of periodic breathing and in particular why Sherpas, who tend to have a blunted HVR, generally do not show periodic breathing at high altitude, whereas lowlanders invariably do. Avoidance of periodic breathing by the Sherpas can be considered a useful adaptation to high altitude because there is evidence that arterial PO_2 fol-

lowing the apneic periods in periodic breathing falls to extremely low levels, possibly the lowest levels encountered during a 24 h period (115).

Metabolic Changes

Weight loss is very common at altitudes of 6,000 m and above. During HSME most subjects living for several months at 5,800 m lost between 0.5 and 1.5 kg/wk (67). Boyer and Blume (9) found an average weight loss of 4 kg in thirteen subjects over a mean of 47 days at an altitude of 6,300 m on AMREE. On OEII, subjects lost 7.4 ± 2.2 (SD) kg body weight over a period of 40 days (81).

The causes of the relentless weight loss are unclear. Hypoxia certainly diminishes appetite, and on OEII, where careful measurements of caloric intake were made, this decreased by a mean of 43% over the period of decompression. However, on field expeditions, where living conditions are more agreeable, there is evidence that caloric intake is more or less maintained. On HSME, Pugh (67) estimated a daily intake of 3,000–3,200 kcal/day, which should have been sufficient to maintain body weight. On AMREE, four subjects had a mean intake of 2,224 kcal at 6,300 m over 3 days compared with 2,976 kcal at sea level. Considering the rather sedentary life-style that most climbers have while in camp, this should be sufficient to maintain body weight.

There is some evidence of impaired intestinal absorption at very high altitude. Boyer and Blume (9) on AMREE found that xylose absorption decreased by 24% in six of seven subjects at 6,300 m compared with sea-level controls, and fat absorption was decreased by 49%. Subjects on HSME reported greasy, bulky stools, suggesting fat malabsorption at 5,800 m. However, other studies at somewhat lower altitudes have not found impaired xylose or fat absorption (15, 73).

Climbers returning from extreme altitude have obvious muscle wasting, and tomographic muscle scans on OEII confirmed the reduction in muscle volume (81). This raises the question of whether severe hypoxia affects protein metabolism. However, there is little information on this point. Rennie et al. (78) studied the effect of acute hypoxia on leucine metabolism in forearm muscles at an altitude of 4,550 m in a chamber and found a net loss of amino acids probably due to a fall in protein synthesis. More studies are needed in this area.

Other aspects of metabolism were studied in OEII. During the 40 day decompression, total blood cholesterol concentration decreased 32%, with no change in the total cholesterol/high density lipoprotein ratio. Fasting plasma insulin levels increased approximately twofold, and plasma norepinephrine concentrations increased threefold (117). Increased fasting insulin levels were also found by Blume (5) at 6,300 m on AMREE, though there was a reduced insulin response to glucose loading.

The immune system was studied in OEII at barometric pressures down to 282 mm Hg. The results suggested that T-cell activation is blunted by severe hypoxia, whereas B-cell function and mucosal immunity are preserved (52).

REFERENCES

1. Andrew, M., H. O'Brodovich, and J. Sutton. Operation Everest II: coagulation system during prolonged decompression to 282 torr. *J. Appl. Physiol.: Respir. Environ. Exerc. Physiol.* 63: 1262–1267, 1987.
2. Barcroft, J., C. A. Binger, A. V. Bock, J. H. Doggart, H. S. Forbes, and G. Harrop. Observations upon the effect of high altitude on the physiological processes of the human body, carried out in the Peruvian Andes, chiefly at Cerro de Pasco. *Phil. Trans. R. Soc. Lond. B* 211: 351–480, 1923.
3. Berssenbrugge, A., J. Dempsey, C. Iber, J. Skatrud, and P. Wilson. Mechanisms of hypoxia-induced periodic breathing during sleep in humans. *J. Physiol. (Lond.)* 343: 507–524, 1983.
4. Bert, P. *La Pression Barométrique.* Paris: Masson, 1878. (English translation by M. A. Hitchcock and F. A. Hitchcock, College Book, 1943.)
5. Blume, F. D. Metabolic and endocrine changes at altitude. In: *High Altitude and Man,* edited by J. B. West and S. Lahiri. Bethesda, MD: Am. Physiol. Soc., 1984, p. 37–45.
6. Blume, F. D., S. J. Boyer, L. E. Braverman, and A. Cohen. Impaired osmoregulation at high altitude. *JAMA* 252: 1580–1585, 1984.
7. Borgström, L., H. Johannsson, and B. K. Siesjö. The relationship between arterial PO_2 and cerebral blood flow in hypoxic hypoxia. *Acta Physiol. Scand.* 93: 423–243, 1975.
8. Boutellier, U., H. Howald, P. E. de Prampero, D. Giezendan-

ner, and P. Cerretelli. Human muscle adaptations to chronic hypoxia. *Prog. Clin. Biol. Res.* 136: 273–281, 1983.
9. Boyer, S. J., and F. D. Blume. Weight loss and changes in body composition at high altitude. *J. Appl. Physiol.: Respir. Environ. Exerc. Physiol.* 57: 1580–1585, 1984.
10. Cavaletti, G., R. Moroni, P. Garavaglia, and G. Tredici. Brain damage after high-altitude climbs without oxygen. *Lancet* i: 101, 1987.
11. Cerretelli, P. Limiting factors to oxygen transport on Mount Everest. *J. Appl. Physiol.* 40: 658–667, 1976.
12. Cerretelli, P. Metabolismo ossidativo ed anaerobico nel soggetto acclimatato alla altitudine. *Minerva Aerospace* 67: 11–26, 1976.
13. Cerretelli, P. Gas exchange at high altitude. In: *Pulmonary Gas Exchange,* edited by J. B. West. New York: Academic, 1980, vol. II, p. 97–147.
14. Cerretelli, P., C. Marconi, O. Dériaz, and D. Giezendanner. After effects of chronic hypoxia on cardiac output and muscle blood flow at rest and exercise. *Eur. J. Appl. Physiol.* 53: 92–96, 1984.
15. Chesner, I. M., N. A. Small, and P. W. Dykes. Intestinal absorption at high altitude. *Postgrad. Med. J.* 63: 173–175, 1987.
16. Christensen, C. H., and W. H. Forbes. Der Kreislauf in grossen Hohen. *Skand. Arch. Physiol.* 76: 75–100, 1937.
17. Clark, C. F., R. K. Heaton, and A. N. Wiens. Neuropsychological functioning after prolonged high altitude exposure in mountaineering. *Aviat. Space Environ. Med.* 54: 202–207, 1983.
18. Cotes, J. E. Ventilatory capacity at altitude and its relation to mask design. *Proc. R. Soc. Lond. B* 143: 32–39, 1954.
19. Cummings, P., and M. Lysgaard. Cardiac arrythmia at high altitude. *West. J. Med.* 135: 66–68, 1981.
20. Cymerman, A., J. T. Reeves, J. R. Sutton, P. B. Rock, B. M. Groves, M. K. Malconian, P. M. Young, P. D. Wagner, and C. S. Houston. Operation Everest II: maximal oxygen uptake at extreme altitude. *J. Appl. Physiol.: Respir. Environ. Exerc. Physiol.* 66: 2446–2453, 1989.
21. Douglas, C. G., J. A. Haldane, Y. Henderson, and E. C. Schneider. Physiological observations made on Pike's Peak, Colorado, with special reference to adaptation to low barometric pressures. *Phil. Trans. R. Soc. Lond. B* 203: 185–381, 1913.
22. Edwards, H. T. Lactic acid in rest and work at high altitude. *Am. J. Physiol.* 116: 367–375, 1936.
23. Filippi, de, F. *Karakorum and Western Himalaya.* London: Constable, 1912.
24. FitzGerald, M. P. The changes in the breathing and the blood of various altitudes. *Phil. Trans. R. Soc. Lond. B* 203: 351–371, 1913.
25. Garner, S. H., J. R. Sutton, R. L. Burse, A. J. McComas, A. Cymerman, and C. S. Houston. Operation Everest II: neuromuscular performance under conditions of extreme simulated altitude. *J. Appl. Physiol.: Respir. Environ. Exerc. Physiol.* 68: 1167–1172, 1990.
26. Gill, M. B., E. C. Poulton, A. Carpenter, M. M. Woodhead, and M. H. P. Gregory. Falling efficiency at sorting cards during acclimatization at 19,000 ft. *Nature* 203: 436, 1964.
27. Glazier, J. B., J. M. B. Hughes, J. E. Maloney, and J. B. West. Measurements of capillary dimensions and blood volume in rapidly frozen lungs. *J. Appl. Physiol.* 26: 65–76, 1969.
28. Green, H. J., J. Sutton, P. Young, A. Cymerman, and C. S. Houston. Operation Everest II: muscle energetics during maximal exhaustive exercise. *J. Appl. Physiol.: Respir. Environ. Exerc. Physiol.* 66: 142–150, 1989.
29. Grollman, A. Physiological variations of the cardiac output of man. VII. The effect of high altitude on the cardiac output and

its related functions: an account of experiments conducted on the summit of Pikes Peak, Colorado. *Am. J. Physiol.* 93: 19–40, 1930.

30. Groves, B. M., J. T. Reeves, J. R. Sutton, P. D. Wagner, A. Cymerman, M. K. Malconian, P. B. Rock, P. M. Young, and C. S. Houston. Operation Everest II: elevated high-altitude pulmonary resistance unresponsive to oxygen. *J. Appl. Physiol.: Respir. Environ. Exerc. Physiol.* 63: 521–530, 1987.

31. Harper, A. M., and H. I. Glass. Effect of alterations in the arterial carbon dioxide tension on the blood flow through the cerebral cortex at normal and low arterial blood pressures. *J. Neurol. Neurosurg. Psychiatr.* 28: 449–452, 1965.

32. Hinchliff, T. W. *Over the Sea and Far Away.* London: Longmans Green, 1876.

33. Hitchcock, F. A. Animals in high altitudes: early balloon flights. In: *Handbook of Physiology. Adaptation to the Environment,* edited by D. B. Dill and E. F. Adolph. Washington, DC: Am. Physiol. Soc., 1964.

34. Hornbein, T. F., B. D. Townes, R. B. Schoene, J. R. Sutton, and C. S. Houston. The cost to the central nervous system of climbing to extremely high altitude. *N. Engl. J. Med.* 321: 1714–1719, 1989.

35. Houston, C. S., A. Cymerman, and J. R. Sutton. *Operation Everest II: Biomedical Studies during a Simulated Ascent of Mt. Everest.* Natick, MA: U. S. Army Res. Inst. Exp. Med., 1991.

36. Houston, C. S., and R. L. Riley. Respiratory and circulatory changes during acclimatization to high altitude. *Am. J. Physiol.* 149: 565–588, 1947.

37. Houston, C. S., J. R. Sutton, A. Cymerman, and J. T. Reeves. Operation Everest II: man at extreme altitude. *J. Appl. Physiol.: Respir. Environ. Exerc. Physiol.* 63: 877–882, 1987.

38. International Civil Aviation Organization. *Manual of the ICAO Standard Atmosphere* (2nd ed.) Montreal: Int. Civil Aviation Org., 1964.

39. Karliner, J. S., F. F. Sarnquist, D. M. Graber, Jr., R. M. Peters, and J. B. West. The electrocardiogram at extreme altitude: experience on Mt. Everest. *Am. Heart J.* 109: 505–513, 1985.

40. Kellas, A. M. A consideration of the possibility of ascending the loftier Himalaya. *Geogr. J.* 49: 26–47, 1917.

41. Kellas, A. M. Sur les possibilités de faire l'ascension du Mount Everest. *Comptes Rendus Seances* 1: 451–521, 1921.

42. Keys, A. The physiology of life at high altitude: the International High Altitude Expedition to Chile 1935. *Sci. Monthly* 43: 289–312, 1936.

43. Khoo, M. C. K., R. E. Kronauer, K. P. Strohl, and A. S. Slutsky. Factors inducing periodic breathing in humans: a general model. *J. Appl. Physiol.: Respir. Environ. Exerc. Physiol.* 53: 644–659, 1982.

44. Kontos, H. A., J. E. Levasseur, D. W. Richardson, H. P. Mauck, Jr., and J. L. Patterson, Jr. Comparative circulatory responses to systemic hypoxia in man and in unanesthetized dog. *J. Appl. Physiol.* 23: 381–386, 1967.

45. Lahiri, S., and P. Barnard. Role of arterial chemoreflex in breathing during sleep at high altitude. In: *Hypoxia, Exercise and Altitude,* edited by J. R. Sutton, C. S. Houston, and N. L. Jones. New York: Liss, 1983, p. 75–85.

46. Lahiri, S., K. Maret, and M. G. Sherpa. Dependence of high altitude sleep apnea on ventilatory sensitivity to hypoxia. *Respir. Physiol.* 52: 281–301, 1983.

47. Lahiri, S., K. H. Maret, M. G. Sherpa, and R. M. Peters, Jr. Sleep and periodic breathing at high altitude: Sherpa natives versus sojourners. In: *High Altitude and Man,* edited by J. B. West and S. Lahiri. Bethesda, MD: Am. Physiol. Soc., 1984, p. 73–90.

48. MacDougall, J. D., H. J. Green, J. R. Sutton, G. Coates, A. Cymerman, P. Young, and C. S. Houston. Operation Everest II: structural adaptations in skeletal muscle in response to extreme simulated altitude. *Acta Physiol. Scand.* 142: 421–427, 1991.

49. Malconian, M., H. Hultgren, M. Nitta, J. Anholm, C. Houston, and H. Fails. The sleep electrocardiogram at extreme altitudes (Operation Everest II). *Am. J. Cardiol.* 65: 1014–1020, 1990.

49a. Malconian, M. K., P. B. Rock, J. T. Reeves, and C. S. Houston. Operation Everest II: gas tensions in expired air and arterial blood at extreme altitude. *Aviat. Space Environ. Med.* 64: 37–42, 1993.

50. Masuyama, S., H. Kimura, T. Sugita, T. Kuriyama, K. Tatsumi, F. Kunitomo, S. Okita, H. Tojima, Y. Yuguchi, S. Watanabe, and Y. Honda. Control of ventilation in extreme-altitude climbers. *J. Appl. Physiol.: Respir. Environ. Exerc. Physiol.* 61: 500–506, 1986.

51. Mecham, R. P., L. A. Whitehouse, D. S. Wrenn, W. C. Parks, G. L. Griffin, R. M. Senior, E. C. Crouch, K. R. Stenmark, and N. F. Voelkel. Smooth muscle–mediated connective tissue remodeling in pulmonary hypertension. *Science* 237: 423–426, 1987.

52. Meehan, R., U. Duncan, L. Neale, G. Taylor, H. Muchmore, N. Scott, K. Ramsey, E. Smith, P. Rock, R. Goldblum, and C. Houston. Operation Everest II: alterations in the immune system at high altitudes. *J. Clin. Immunol.* 8: 397–406, 1988.

53. Messner, R. *Everest: Expedition to the Ultimate.* London: Kaye and Ward, 1979.

54. Meyrick, B., and L. Reid. Hypoxia-induced structural changes in the media and adventitia of the rat hilar pulmonary artery and their regression. *Am. J. Pathol.* 100: 151–169, 1980.

55. Milledge, J. S. Electrocardiographic changes at high altitude. *Br. Heart J.* 25: 291–298, 1962.

56. Milledge, J. S., M. P. Ward, E. S. Williams, and C. R. A. Clarke. Cardiorespiratory response to exercise in men repeatedly exposed to extreme altitude. *J. Appl. Physiol.: Respir. Environ. Exerc. Physiol.* 55: 1379–1385, 1983.

57. Mosso, A. *Life of Man on the High Alps.* London: Fisher Unwin, 1898, p. 42–47.

58. National Oceanic and Atmospheric Administration. *U.S. Standard Atmosphere.* Washington, DC: Nat. Oceanic Atmos. Admin., 1976.

59. Norton, E. F. Norton and Somervell's attempt in: *The Fight for Everest, 1924,* edited by E. F. Norton. London: Arnold, 1925, p. 99–119.

60. Oelz, O., H. Howald, P. E. di Prampero, H. Hoppeler, H. Claassen, R. Jenni, A. Bühlmann, G. Ferretti, J.-C. Brückner, A. Veicsteinas, M. Gussoni, and P. Cerretelli. Physiological profile of world-class high-altitude climbers. *J. Appl. Physiol.: Respir. Environ. Exerc. Physiol.* 60: 1734–1742, 1986.

61. Otis, A. B. The work of breathing. In: *Handbook of Physiology. Respiration,* edited by W. O. Fenn and H. Rahn. Washington DC: Am. Physiol. Soc., 1964, vol. I, p. 463–476.

62. Peñaloza, D., F. Sime, N. Banchero, R. Gamboa, J. Cruz, and E. Marticorena. Pulmonary hypertension in healthy men born and living at high altitudes. *Am. J. Cardiol.* 11: 150–157, 1963.

63. Petit, J. M., G. Milie-Emili, and J. Troquet. Travail dynamique pulmonaire et altitude. *Rev. Med. Aeronaut.* 2: 276–279, 1963.

64. Piiper, J., and P. Scheid. Blood-gas equilibration in lungs. In: *Pulmonary Gas Exchange. Ventilation, Blood Flow, and Diffusion,* edited by J. B. West. New York: Academic, 1980, vol. 1, p. 131–171.

65. Poiani, G. J., C. A. Tozzi, S. E. Yohn, R. A. Pierce, S. A. Belsky, R. A. Berg, S. Y. Yu, S. B. Deak, and D. J. Riley. Collagen and elastin metabolism in hypertensive pulmonary arteries of rats. *Circ. Res.* 66: 968–978, 1990.

66. Pugh, L. G. C. E. Resting ventilation and alveolar air on Mount Everest: with remarks on the relation of barometric pressure to altitude in mountains. *J. Physiol. (Lond.)* 135: 590–610, 1957.

67. Pugh, L. G. C. E. Physiological and medical aspects of the Himalayan Scientific and Mountaineering Expedition, 1960–61. *Br. Med. J.* 2: 621–633, 1962.

68. Pugh, L. G. C. E. Animals in high altitudes: man above 5000 m—mountain exploration. In: *Handbook of Physiology. Adaptation to the Environment,* edited by D. B. Dill and E. F. Adolph. Washington, DC: Am. Physiol. Soc., 1964, sect. 4.

69. Pugh, L. G. C. E. Cardiac output in muscular exercise at 5800 m (19,000 ft.). *J. Appl. Physiol.* 19: 441–447, 1964.

70. Pugh, L. G. C. E., M. B. Gill, S. Lahiri, J. S. Milledge, M. P. Ward, and J. B. West. Muscular exercise at great altitudes. *J. Appl. Physiol.* 19: 431–440, 1964.

71. Rahn, H., and W. O. Fenn. *A Graphical Analysis of the Respiratory Gas Exchange.* Washington, DC: Am. Physiol. Soc., 1955.

72. Rahn, H., and A. B. Otis. Man's respiratory response during and after acclimatization to high altitude. *Am. J. Physiol.* 157: 445–462, 1949.

73. Rai, R. M., M. S. Malhotra, G. P. Dimri, and T. Sampathkumar. Utilization of different quantities of fat at high altitude. *Am. J. Clin. Nutr.* 28: 242–245, 1975.

74. Reeves, J. T., B. M. Groves, J. R. Sutton, P. D. Wagner, A. Cymerman, M. K. Malconian, P. B. Rock, P. M. Young, and C. S. Houston. Operation Everest II: preservation of cardiac function at extreme altitude. *J. Appl. Physiol.: Respir. Environ. Exerc. Physiol.* 63: 531–539, 1987.

75. Regard, M., O. Oelz, P. Brugger, and T. Landis. Persistent cognitive impairment in climbers after repeated exposure to extreme altitude. *Neurology* 39: 210–213, 1989.

76. Reite, M., D. Jackson, R. L. Cahoon, and J. V. Weil. Sleep physiology at high altitude. *Electroencephalogr. Clin. Neurophysiol.* 38: 463–471, 1975.

77. Rennie, D. Will mountain trekkers have heart attacks? *JAMA* 261: 1045–1046, 1989.

78. Rennie, M. J., P. Babij, J. R. Sutton, W. J. Tonkins, W. W. Read, C. Ford, and D. Halliday. Effects of acute hypoxia on forearm leucine metabolism. In: *Hypoxia, Exercise and Altitude,* edited by J. R. Sutton, C. S. Houston, and N. L. Jones. New York: Liss, 1983, p. 317–323.

79. Richardson, T. Q., and A. C. Guyton. Effects of polycythemia and anemia on cardiac output and other circulatory factors. *Am. J. Physiol.* 197: 1167–1170, 1959.

80. Riley, R. L., and C. S. Houston. Composition of alveolar air and volume of pulmonary ventilation during long exposure to high altitude. *J. Appl. Physiol.* 3: 526–534, 1951.

81. Rose, M. S., C. S. Houston, C. S. Fulco, G. Coates, J. R. Sutton, and A. Cymerman. Operation Everest II: nutrition and body composition. *J. Appl. Physiol.: Respir. Environ. Exerc. Physiol.* 65: 2545–2551, 1988.

82. Rotta, A., A. Cánepa, A. Hurtado, T. Velásquez, and R. Chávez. Pulmonary circulation at sea level and at high altitudes. *J. Appl. Physiol.* 9: 328–336, 1956.

83. Ryn, Z. Psychopathology in alpinism. *Acta Med. Pol.* 12: 453–467, 1971.

84. Sarnquist, F. H., R. B. Schoene, P. H. Hackett, and B. D. Townes. Hemodilution of polycythemic mountaineers: effects on exercise and mental function. *Aviat. Space Environ. Med.* 57: 313–317, 1986.

85. Schoene, R. B. Control of ventilation in climbers to extreme altitude. *J. Appl. Physiol.: Respir. Environ. Exerc. Physiol.* 53: 886–890, 1982.

86. Schoene, R. B., P. H. Hackett, and R. C. Roach. Blunted hyp-

oxic chemosensitivity at altitude and sea level in an elite high altitude climber. In: *Hypoxia and Cold,* edited by J. R. Sutton, C. S. Houston, and G. Coates. New York: Praeger, 1987, p. 532.

87. Schoene, R. B., S. Lahiri, P. H. Hackett, R. M. Peters, Jr., J. S. Milledge, C. J. Pizzo, F. H. Sarnquist, S. J. Boyer, D. J. Graber, K. H. Maret, and J. B. West. Relationship of hypoxic ventilatory response to exercise performance on Mount Everest. *J. Appl. Physiol.: Respir. Environ. Exerc. Physiol.* 56: 1478–1483, 1984.

88. Sharma, V. M., and M. S. Malhotra. Ethnic variations in psychological performance under altitude stress. *Aviat. Space Environ. Med.* 47: 248–251, 1976.

89. Sharma, V. M., M. S. Malhotra, and A. S. Baskaran. Variations in psychometer efficiency during prolonged stay at high altitude. *Ergonomics* 18: 511–516, 1975.

90. Sime, F., D. Peñaloza, L. Ruiz, N. Gonzales, E. Covarrubias, and R. Postigo. Hypoxemia, pulmonary hypertension, and low cardiac output in newcomers at low altitude. *J. Appl. Physiol.* 36: 561–565, 1974.

91. Suarez, J., J. K. Alexander, and C. S. Houston. Enhanced left ventricular systolic performance at high altitude during Operation Everest II. *Am. J. Cardiol.* 60: 137–142, 1987.

92. Sutton, J. R., C. S. Houston, A. L. Mansell, M. McFadden, P. Hackett, J. R. A. Rigg, and C. P. Powles. Effect of acetazolamide on hypoxemia during sleep at high altitude. *N. Engl. J. Med.* 301: 1329–1331, 1979.

93. Sutton, J. R., J. T. Reeves, P. D. Wagner, B. M. Groves, A. Cymerman, M. K. Malconian, P. B. Rock, P. M. Young, S. D. Walter, and C. S. Houston. Operation Everest II: oxygen transport during exercise at extreme simulated altitude. *J. Appl. Physiol.: Respir. Environ. Exerc. Physiol.* 64: 1309–1321, 1988.

94. Townes, B. D., T. F. Hornbein, R. B. Schoene, F. H. Sarnquist, and I. Grant. Human cerebral function at extreme altitude. In: *High Altitude and Man,* edited by J. B. West and S. Lahiri. Bethesda, MD: Am. Physiol. Soc., 1984, p. 31–36.

95. Tozzi, C. A., G. J. Poiani, A. M. Harangozo, C. D. Boyd, and D. J. Riley. Pressure-induced connective tissue synthesis in pulmonary artery segments is dependent on intact endothelium. *J. Clin. Invest.* 84: 1005–1012, 1989.

96. Vogel, J. A., J. E. Hansen, and C. W. Harris. Cardiovascular responses in man during exhaustive work at sea level and high altitude. *J. Appl. Physiol.* 23: 531–539, 1967.

97. Vogel, J. A., and C. W. Harris. Cardiopulmonary responses of resting man during early exposure to high altitude. *J. Appl. Physiol.* 22: 1124–1128, 1967.

98. Vogel, J. A., L. H. Hartley, and J. C. Cruz. Cardiac output during exercise in altitude natives at sea level and high altitude. *J. Appl. Physiol.* 36: 173–176, 1974.

99. Wagner, P. D., J. R. Sutton, J. T. Reeves, A. Cymerman, B. M. Groves, and M. K. Malconian. Operation Everest II. Pulmonary gas exchange during a simulated ascent of Mt. Everest. *J. Appl. Physiol.: Respir. Environ. Exerc. Physiol.* 63: 2348–2359, 1987.

100. Ward, M. P., J. S. Milledge, and J. B. West. *High Altitude Medicine and Physiology.* London: Chapman and Hall, 1989.

101. Weil, J. V., M. H. Kryger, and C. H. Scoggin. Sleep and breathing at high altitude. In: *Sleep Apnea Syndromes,* edited by C. Guilleminault and W. Dement. New York: Liss, 1978, p. 119–136.

102. West, J. B. Climbing Mt. Everest without oxygen: an analysis of maximal exercise during extreme hypoxia. *Respir. Physiol.* 52: 265–279, 1983.

103. West, J. B. "Oxygenless" climbs and barometric pressure. *Am. Alpine J.* 58: 126–132, 1984.

104. West, J. B. Highest inhabitants in the world. *Nature* 324: 517, 1986.

105. West, J. B. Lactate during exercise at extreme altitude. *Federation Proc.* 45: 2953–2957, 1986.

106. West, J. B. Alexander M. Kellas and the physiological challenge of Mt. Everest. *J. Appl. Physiol.: Respir. Environ. Exerc. Physiol.* 63: 3–11, 1987.

107. West, J. B. Rate of ventilatory acclimatization to extreme altitude. *Respir. Physiol.* 74: 323–333, 1988.

108. West, J. B. The safety of trekking at high altitude after coronary bypass surgery. *JAMA* 260: 2218–2219, 1988.

109. West, J. B. Limiting factors for exercise at extreme altitude. *Clin. Physiol.* 10: 265–272, 1990.

110. West, J. B. Acclimatization and tolerance to extreme altitude. *J. Wild. Med.* 4: 17–26, 1993.

111. West, J. B., S. J. Boyer, D. J. Graber, P. H. Hackett, K. H. Maret, J. S. Milledge, R. M. Peters, Jr., C. J. Pizzo, M. Samaja, F. H. Sarnquist, R. B. Schoene, and R. M. Winslow. Maximal exercise at extreme altitudes on Mount Everest. *J. Appl. Physiol.: Respir. Environ. Exerc. Physiol.* 55: 688–698, 1983.

112. West, J. B., P. H. Hackett, K. H. Maret, J. S. Milledge, R. M. Peters, Jr., C. J. Pizzo, and R. M. Winslow. Pulmonary gas exchange on the summit of Mt. Everest. *J. Appl. Physiol.: Respir. Environ. Exerc. Physiol.* 55: 678–687, 1983.

113. West, J. B., S. Lahiri, M. B. Gill, J. S. Milledge, L. G. C. E. Pugh, and M. P. Ward. Arterial oxygen saturation during exercise at high altitude. *J. Appl. Physiol.* 17: 617–621, 1962.

114. West, J. B., S. Lahiri, K. H. Maret, R. M. Peters, Jr., and C. J. Pizzo. Barometric pressures at extreme altitudes on Mt. Everest: physiological significance. *J. Appl. Physiol.: Respir. Environ. Exerc. Physiol.* 54: 1188–1194, 1983.

115. West, J. B., R. M. Peters, Jr., G. Aksnes, K. H. Maret, J. S. Milledge, and R. B. Schoene. Nocturnal periodic breathing at altitudes of 6,300 and 8,050 m. *J. Appl. Physiol.: Respir. Environ. Exerc. Physiol.* 61: 280–287, 1986.

116. Winslow, R. M., C. C. Monge, E. G. Brown, H. G. Klein, F. Sarnquist, N. J. Winslow, and S. S. McKneally. Effects of hemodilution of O_2 transport in high-altitude polycythemia. *J. Appl. Physiol.: Respir. Environ. Exerc. Physiol.* 59: 1495–1502, 1985.

117. Young, P. M., M. S. Rose, J. R. Sutton, H. J. Green, A. Cymerman, and C. S. Houston. Operation Everest II: plasma lipid and hormonal responses during a simulated ascent of Mt. Everest. *J. Appl. Physiol.: Respir. Environ. Exerc. Physiol.* 66: 1430–1435, 1989.

118. Zuntz, N., A. Loewy, F. Muller, and W. Caspari. Atmospheric pressure at high altitudes. In: *Höhenklima und Bergwanderungen in ihrer Wirkung auf den Menschen.* Berlin: Bong, 1906. (Translation of relevant pages in: *High Altitude Physiology,* edited by J. B. West. Stroudsburg, PA: Hutchins Ross, 1981).

VI | THE CHRONOBIOLOGICAL ENVIRONMENT

Associate Editor   C. M. Blatteis

58. Circadian rhythms

FRED W. TUREK
OLIVIER VAN REETH

Department of Neurobiology and Physiology, and Center for Circadian Biology and Medicine, Northwestern University, Evanston, Illinois

ONE OF THE MOST OBVIOUS ADAPTIVE FEATURES OF LIVING ORGANISMS ON EARTH is the ability to change behavior on a daily or 24 h basis. Daily changes in lifestyle, of course, are correlated with the dramatic changes that take place in the physical environment as the Earth rotates on its axis. While not as readily apparent as behavioral changes, just about every aspect of the internal environment of the organism undergoes pronounced fluctuations over the course of the 24 h day. Claude Bernard's concept of homeostasis, of a *milieu interieur* that remains constant in living organisms, has been modified over the last 30–40 yr to take into account the fact that regular and predictable 24 h changes in the internal milieu are a hallmark of most living organisms.

A remarkable feature of the daily rhythms observed in organisms as diverse as algae, fruit flies, and humans is that they are not simply a response to the 24 h changes in the physical environment imposed by celestial mechanics but instead arise from an internal time-keeping system (132). This time-keeping system, or biological clock(s), allows the organism to predict and prepare for the changes in the physical environment that are associated with night and day. Thus the organism adapts, both behaviorally and physiologically, to meet the challenges associated with the daily changes in the external environment, demonstrating temporal synchronization between the organism and the external environment. The most obvious example of such an adaptation is that many animals are only active during either the light period (diurnal species) or the dark period (nocturnal species) and are inactive during the other part of the day. Such "external synchronization" is of obvious importance for the survival of the species and insures that the organism does the right thing at the right time of day. Of equal, but perhaps less appreciated, importance is the fact that the biological clock, like a conductor of a symphony orchestra, provides internal temporal organization and insures that internal changes take place in coordination with one another. Just as living organisms are organized spatially, they are also organized temporally to insure that there is "internal synchronization" between the myriad biochemical and physiological systems in the body. While lack of synchrony between the organism and the external environment may lead to death, as would be expected if a nocturnal rodent attempted to navigate the hazards of the diurnal world, lack of synchrony within the internal environment may lead to chronic difficulties with equally severe consequences for the health and well-being of the organism.

In addition to reviewing the adaptive significance of diurnal rhythms, this chapter also describes the characteristics and basic properties of these rhythms, the underlying physiological mechanisms that allow diurnal rhythms to be generated and synchronized to the environment, the genetic basis of these rhythms, and the developmental changes that take place in the expression of diurnal rhythms during the lifetime of the organism. The central role played by diurnal rhythm-generating systems in the expression of non-24 h rhythms, is also discussed. The main emphasis of this chapter is on diurnal rhythms in mammals, with brief mention of rhythms in other classes of vertebrates or in invertebrates. For more details on diurnal rhythms in nonmammalian species, the reader is directed to other reviews (37, 64, 161, 189, 206).

GENERATION AND ENTRAINMENT OF CIRCADIAN RHYTHMS

Endogenous Self-Sustained Clocks Drive Circadian Rhythms

Under laboratory conditions devoid of any external time-giving cues from the physical environment, it has been found that just about all diurnal rhythms present under natural conditions continue to be expressed in the laboratory (5, 108, 135). Under constant environmental conditions, however, the period of the rhythm rarely remains exactly 24 h but instead is "about" 24 h. Because the period of diurnal rhythms is close to but not exactly 24 h in duration, they are referred to as *circadian* rhythms, from the Latin *circa diem,* meaning about a day. When a circadian rhythm is expressed in the absence of any 24 h signals in the external environment, it is said to be *free-running;* that is, the rhythm is not synchronized or entrained by any cyclic change in the physical environment. For a population of animals within a given species, the period of a given rhythm (for example, drinking, body temperature, or locomotor activity) will be different between animals, but in general, all will lie in close proximity to 24 h (for example, from 23–24 h in duration). Strictly speaking, a diurnal rhythm should not be referred to as "circadian" until it has been demonstrated that such a rhythm persists under constant environment conditions. The purpose of this distinction is to separate out those rhythms which are simply a response to 24 h changes in the physical environment from those which are driven by some internal time-keeping system. However, for practical purposes, there is little reason to make a distinction between diurnal and circadian rhythms since almost all diurnal rhythms expressed under natural conditions are found to persist under constant environmental condi-

tions in the laboratory. Consequently, the term "circadian" is often used to refer to diurnal rhythms observed under either natural or laboratory conditions.

Genetic, physiological, and behavioral experiments have established that the timing system which underlies the generation of circadian rhythms is endogenous to the organism itself. That is, there is a circadian clock, or clocks, within the organism which somehow regulates the 24 h fluctuations in diverse physiological and behavioral systems. The physiological nature of circadian clocks is reviewed below under CIRCADIAN PACEMAKERS: THE EXECUTIVE CLOCK IN THE SUPRACHIASMATIC NUCLEUS. While literally thousands of rhythms have been monitored in plants and animals, it has not been possible to assay the state of a circadian clock directly in any experimental model to date. Thus attempts to understand the properties of circadian clocks focus on the "hands" of the clock, that is, the expression of overt rhythms regulated by the clock. While the list of biochemical and physiological processes that show circadian fluctuations is enormous, a few select behavioral rhythms (for example, locomotor activity, drinking) are most often utilized to characterize the basic features of the clock system in mammals. Behavioral rhythms are utilized because of their ease of measurement for many cycles without disturbing the animal. Figure 58.1 provides an example of the circadian rhythm of locomotor activity in a male golden hamster held under free-running conditions for 100 days. Not only can this rhythm be monitored for essentially the lifetime of the animal without any interference of the sampling procedure on the rhythm itself, but automated sampling systems allow one to access the state of this rhythm continuously on essentially a minute-to-minute basis. For practical and economic reasons, such long-term and frequent sampling of biochemical and physiological rhythms is often not possible. Nevertheless, data obtained in mammals indicate that behavioral rhythms represent the hands of the same circadian clock system that underlies most, if not all, biochemical and physiological rhythms.

Entrainment of Circadian Clocks: Control of Period and Phase

The fact that the period of circadian rhythms is not exactly equal to the period of the rotation of the earth on its axis demands that 24 h changes in the physical environment must somehow synchronize or entrain the internal clock system regulating circadian rhythms. Otherwise, even a clock with a period only a few minutes shorter or longer than 24 h would soon be totally out of synchrony with the environmental day. An endogenous circadian clock that could not be reset by environmental signals would be of little use to organ-

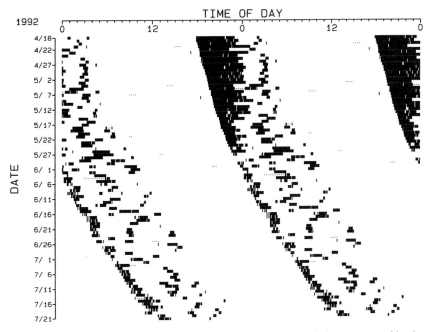

FIG. 58.1. Continuous record of the circadian rhythm of wheel-running behavior in a golden hamster maintained in constant darkness for 100 days. Each revolution of the running wheel was recorded on-line via a personal computer. Black bars represent periods of activity. Successive days are plotted from top to bottom, and the activity record has been double-mounted on a 48 h time interval to aid in the visualization of data. This record shows the remarkable precision of the activity rhythm under free-running conditions devoid of any timing cues. The onset of activity shows a period slightly greater than 24 h throughout the 100 days, with deviations from the mean period being only a few minutes between any 2 days. The decrease in testicular size (and its associated decrease in serum testosterone levels), due to exposure to constant darkness, leads to a decrease in the total amount of activity per day during the latter half of this record.

isms that need to time specific activities to particular times of the day. As discussed below, the light–dark cycle is clearly the major environmental entraining agent of circadian rhythms, though there is substantial evidence that other internal and external factors can influence how circadian rhythms are synchronized to the physical environment. It should be noted that while other stimuli might influence the expression of particular circadian rhythms by influencing a process between the endogenous clock and the effector systems (that is, masking), the focus in this section is on those agents which can control the phase of the circadian clock itself.

Entrainment by Light–Dark Cycles. Except in a few exotic species living in unusual environments (for example, blind cave fish or eyeless mole rats), the light–dark cycle appears to be the primary environmental agent which synchronizes circadian rhythms to the 24 h cycle (134, 191). Thus the period of circadian rhythms exactly matches the period of the 24 h light–dark cycle (Fig. 58.2). From one day to the next, the time between successive recurrences of specific phase points of a rhythm (for example, the onset of locomotor activity, the minimum in body temperature) is the same as the

period or duration of the light–dark cycle. In addition to establishing "period control," an entraining light–dark cycle establishes "phase control" such that specific phases of the circadian rhythm occur at the same times in each cycle relative to the entraining agent. For example, in a hamster entrained to a 14:10 light–dark cycle (that is, 14 h of light followed by 10 h of darkness every 24 h), the onset of the main bout of daily activity always occurs within a few minutes after lights-off, day after day (Fig. 58.2). Following a phase shift in the light–dark cycle, the rhythm reentrains (Fig. 58.2), though the development of a steady-state phase relationship between the circadian rhythm and the entraining light–dark cycle often takes many days due to the limitation on the number of hours per day that the internal circadian clock can be phase shifted by light.

Although circadian rhythms can be entrained to light–dark cycles that are not exactly 24 h in duration, entrainment is restricted to cycles that are close to 24 h (134). The range of entrainment can vary from species to species and is dependent on the experimental conditions (for example, intensity of light–dark cycle, whether period of light–dark cycle is changed gradually or rapidly), but in general animals do not entrain readily

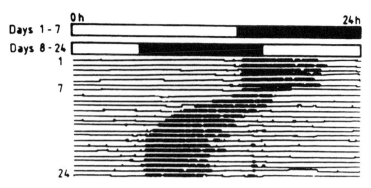

FIG. 58.2. Daily rhythm of wheel-running behavior in a golden hamster exposed to a 14:10 light–dark cycle, which was phase-advanced by 8 h on day 8 of this record. Timing of light–dark cycles before and after the advance in the light cycle is diagrammed at the top of the record. The animal was entrained to the initial cycle such that the onset of activity occurred within a few minutes of lights-off each day. Following the shift in the cycle, it took about 10 days for the animal to become reentrained to the new lighting schedule.

to light–dark cycles that are more than a few hours shorter or longer than the period of the endogenous free-running circadian rhythm. If the period of the light–dark cycle is too short or long for entrainment to occur, the circadian rhythm will free-run with a period close to 24 h.

Since about 1960, one of the most widely used methods to examine how the light–dark cycle influences the circadian system has been to expose animals maintained in constant darkness to a brief pulse of light (for example, 1–60 min in duration) and then return them to darkness (35, 132). The effects of the light pulse on a phase reference point of a circadian rhythm (for example, onset of locomotor activity, minimum of body temperature) in subsequent cycles is then determined. This approach has demonstrated that light pulses can induce phase advances or phase delays or have no effect on free-running circadian rhythms. The direction and magnitude of the shifts are strongly dependent on the circadian time at which the light pulse occurs (Fig. 58.3). A plot of the phase shift induced by an environmental perturbation as a function of the circadian time at which the perturbation is given is called a *"phase response curve"* (PRC). Light pulse PRCs for all organisms share certain characteristics including the fact that light pulses presented near the onset of the subjective night (subjective night and subjective day refer to those parts of the circadian cycle which would occur during the dark or light time, respectively, when the organism is exposed to a light–dark cycle) induce phase delays in the rhythm, while light pulses presented in the late subjective night/early subjective day induce phase advances. In contrast, light pulses presented during most of the subjective day induce no phase shifts in the activity rhythm. Entrainment of the circadian clock to the light–dark cycle is thought to occur by light-inducing phase advances

and/or delays in the clock each day that equal the difference between the free-running period of the clock and the period of the daily light–dark cycle (134, 189).

The importance of the daily light–dark cycle for the entrainment of circadian rhythms has long been recognized in most plant and animal species and is now appreciated in humans. Early studies in humans under conditions of temporal isolation indicated that the light–dark cycle was a very weak synchronizer of human circadian rhythms and that social environmental cues were more important for entrainment (225). Since the late 1970s, Czeisler and colleagues (22, 24) have carried out extensive studies demonstrating that the light–dark cycle could entrain human rhythms and that light could be used to reset human rhythms under a variety of experimental conditions. Indeed, as in other animals, exposure to pulses of bright light can induce phase shifts in free-running human circadian rhythms. A major difference in the entrainment by light of circadian rhythms in humans and most other animal species is the need for apparently much brighter light to synchronize human rhythms, raising questions of the adequacy of normal indoor lighting for the entrainment of rhythms in humans who see very little natural light during the day (22, 24).

Entrainment by Nonphotic Signals. A fundamental assumption in the early development of circadian rhythm research was that, except for the light–dark cycle, endogenous circadian clocks were independent from most changes in the internal and external environment (204). Pittendrigh's (131) early finding that circadian clocks are temperature-compensated (that is, that there is very little change in the period of the clock following an increase or decrease in temperature) led to

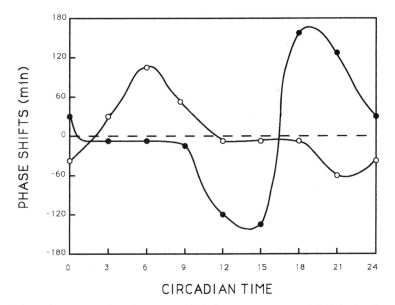

FIG. 58.3. Schematic representation of two generalized phase response curves (PRC) for the phase-shifting effects on the circadian clock regulating the rhythm of locomotor activity following the presentation of either single pulses of light (5–60 min in duration) or stimuli which induce an acute increase in activity (for example, exposure to a novel running wheel or injection of a short-acting benzodiazepine) in golden hamsters free-running in constant darkness. *Closed circles:* light pulse PRC; *open circles:* activity-induced PRC. Circadian time 12 refers to the time of activity onset in this nocturnal species. While exposure to light during the subjective daytime has little or no effect on the phase of the activity rhythm, exposure to light near the time of activity onset (that is, near the time of sunset) induces a delay in the rhythm, while an equivalent light pulse given near the end of the subjective night (that is, near the time of sunrise) induces an advance in the rhythm. While the amplitudes of the phase advances and phase delays that occur in response to the presentation of activity-inducing stimuli are similar to those observed in response to light pulses, the circadian time for the phase delay, phase advance, and unresponsive regions of the two PRCs are dramatically different.

the generalization that to keep accurate time they need to be buffered from most external and internal factors. Since the 1980s, however, a number of internal and external stimuli have been found to influence circadian clocks in a variety of vertebrate species (101, 115, 198). Although the importance of nonphotic factors in the entrainment of circadian rhythms under natural conditions in mammals is not clear, under experimental laboratory conditions changes in ambient temperature, periodic presentation of food, and agents which alter the sleep–wake cycle can all alter the clock regulating overt circadian rhythms (101, 115, 198, 202).

A great deal of attention has been focused on the possibility that changes in the activity–rest state can alter the circadian clock which regulates the rhythm of activity as well as most other circadian rhythms in mammals. The acute presentation of a variety of pharmacological (for example, injections of benzodiazepines) and non-pharmacological (for example, exposing hamsters to a novel running wheel or a pulse of darkness on a background of constant light) stimuli, which can induce phase advances or phase delays in the free-running circadian rhythm of activity, also induces an acute increase in activity (115, 147, 198, 227). These agents induce phase shifts in the circadian clock regulating the activity rhythm, as well as other behavioral and physiological rhythms (200, 226, 227), when hamsters are normally inactive (see Fig. 58.9 for an example of a phase shift induced by a pulse of darkness). The PRCs generated by activity-inducing stimuli are about 180° out of phase with the PRC to light pulses in the hamster (Fig. 58.3). The hypothesis that the increase in locomotor activity is itself somehow responsible for phase shifts in the circadian clock is supported by experiments demonstrating that phase shifts induced by dark pulses or injections of short-acting benzodiazepines can be blocked by confining hamsters to a small nest box or restraining tube during, or for a period of time after, the stimulus is presented (148, 218). Immobilizing hamsters when they are normally very active (that is, during the early part of the subjective night) can also induce phase shifts in the circadian clock underlying the activity rhythm (217). Other experiments have demonstrated that induced activity can accelerate the rate of reentrainment of hamsters following a phase shift in the light–dark cycle (116) and that chronic exposure to a free or locked running wheel can influence the period of the circadian rhythm of locomotor activity in mice, hamsters, and rats (6, 44,

232). While the overall implications of these findings for the normal entrainment and expression of circadian rhythms remain to be determined, it is clear that changes in the behavioral state of the animal can influence the circadian time-keeping system.

From a historical perspective, it is interesting to note that, while early studies in humans minimized the importance of the light–dark cycle and focused on the role of social factors for the entrainment of human circadian rhythms, early studies in animals minimized the importance of behavioral changes and focused on the almost exclusive role of the light–dark cycle in the regulation of circadian rhythms. The relative importance of photic and nonphotic signals for the entrainment of circadian rhythms may well vary between species and is undoubtedly dependent on the evolutionary pressures faced by individual species adapting to the daily changes in the physical environment.

CIRCADIAN PACEMAKERS: THE EXECUTIVE CLOCK IN THE SUPRACHIASMATIC NUCLEUS

In diverse animal species there is substantial evidence that one or only a few specific structures serve as "master circadian pacemakers" or "executive clocks" that can be entrained by the light–dark cycle and regulate most circadian rhythms in the organism. The search for circadian pacemakers in both invertebrate and vertebrate species has focused mainly on the nervous system. A variety of experimental approaches have been used to localize circadian clocks in the brain of silkmoths, the optic lobes of cockroaches, the eyes of molluscs, the pineal gland and eyes of lower vertebrates, and the hypothalamus in mammals and perhaps other vertebrates as well (96, 105, 125, 161, 168, 189, 196). Despite the diversity of systems under investigation, similar questions are being addressed. How is neural tissue organized to generate circadian signals? What are the molecular and cellular processes that underlie rhythm generation? How are circadian clocks coupled to the tissues and organs that express the rhythms? What are the physiological mechanisms by which environmental agents, particularly the light–dark cycle, influence and entrain the circadian clock? In mammals, it appears that two discrete nuclei in the anterior hypothalamus serve as the location of the executive clock(s) that regulates most, if not all, circadian rhythms. The first part of this section reviews the evidence that a central pacemaker lies in these nuclei, as well as some of the characteristics of this structure. The last two parts of this section focus on how this circadian pacemaker is integrated into the overall organization of the circadian system.

The Suprachiasmatic Nucleus of the Hypothalamus

For many behaviors, such as feeding, drinking, sleeping, or sexual activity, it is not possible to point to one area of the brain and say, "there is the control center which regulates this behavior." Instead, many complex behaviors involve a network of brain areas whose total activity underlies the expression of the behavior. In contrast, it appears that in mammals a single anatomical locus involving two small bilaterally paired nuclei are responsible for regulating all the diverse 24 h rhythms of the body. The suprachiasmatic nuclei (SCN) are located in the anterior hypothalamus immediately above the optic chiasm and lateral to the third ventricle in all mammals (106). While each SCN contains only about 8,000 neurons in rodents, from both ultrastructural and immunocytochemical studies the SCN appears to be a complex structure. The SCN clearly functions as the executive clock in mammals, and there is no convincing evidence that any other area of the brain can function as a master circadian pacemaker or that a timing system anywhere else in the brain can be entrained by the light–dark cycle. An entire book on the SCN has been published referring to the SCN in its title as "the mind's clock" (80). Our understanding of the central role of the SCN in the circadian organization of mammals and the wealth of information now available on the SCN is particularly impressive when one takes into account the fact that until the early 1970s this area of the brain was essentially unknown except to perhaps a handful of anatomists. Certainly, a milestone in the field was achieved when the SCN were first destroyed in rodent experiments and the subsequent effects on behavior were observed.

Effects of Destroying the SCN on Circadian Rhythmicity.
Following the pioneering studies of Moore and Eichler (107) and Stephan and Zucker (184) published in 1972, numerous investigators have determined the effects of bilateral destruction of the SCN on the expression of overt circadian rhythms. Under both free-running and entrained conditions, destruction of the SCN in a variety of mammalian species, including primates, leads to the abolishment (that is, the induction of arrhythmicity) or the severe disruption of many behavioral and physiological rhythms, including those of feeding, drinking, locomotor activity, body temperature, sleep–wake, cortisol, pineal melatonin, and growth hormone secretion (92, 101, 105, 162, 168, 196). Neonatal ablation of the SCN in rats permanently eliminates the circadian rhythms of locomotor activity and drinking behavior, suggesting that other regions of the brain do not have the capacity to reorganize and take over the function of the SCN (114). It should be noted that following com-

plete SCN lesions animals are still active, still eat, and still secrete many hormones. The SCN does not regulate the total amount of food or water consumed, nor does it regulate the total amount of sleep. Indeed, many regulatory systems continue to function normally except for one feature: they lose temporal organization.

A few controversial studies indicate that some circadian rhythms may persist after SCN lesions (79), and there is good evidence to suggest that some timing system, which can be entrained by the daily presentation of food at a restricted time, is still present after the abolishment of the SCN (182, 183). While some components of the overall circadian timing system may lie outside of the anatomically defined SCN, the clear role of this structure as the control center for the circadian system, first suggested by the lesion studies, has been confirmed following transplantation of the SCN from one animal to another.

Effects of Transplanting Fetal SCN Tissue on Circadian Rhythmicity.

A major approach in examining structure–function relationships within the central nervous system (CNS) involves the transplantation of fetal tissue from select brain regions into adult animals with the aim of restoring or augmenting neural activity that is specific to the donor tissue (12, 55). Since the mid-1980s, a number of circadian biologists have restored circadian rhythmicity in adult arrhythmic SCN-lesioned rodents by transplanting fetal SCN tissue (for reviews see refs. 85, 142). These findings clearly established the central role of the SCN in the expression of overt circadian rhythms but left open the question of whether the SCN was simply necessary for the expression of a circadian pacemaker located elsewhere or whether the SCN itself was the site of the circadian pacemaker cells. The discovery of a period mutation (referred to as the tau mutant since "tau" is used in referring to the free-running period of a circadian rhythm) in golden hamsters, in which the free-running period of the activity rhythm is shortened to about 20 h in animals homozygous for the mutation (see below), provided Ralph and Menaker (144) with the opportunity to test directly whether the SCN actually contained a circadian clock which drives the expression of overt circadian rhythms. In a series of truly remarkable and elegant experiments Ralph, Menaker, and their colleagues (141, 142) performed a number of reciprocal transplants whereby SCN-lesioned, arrhythmic, wild-type, and tau mutant animals were implanted with fetal SCN tissue from animals with a different genotype. In all cases in which rhythmicity was restored the periods of the restored rhythms were similar to those of the donor genotype (Fig. 58.4), and there was no indication that the host brain significantly

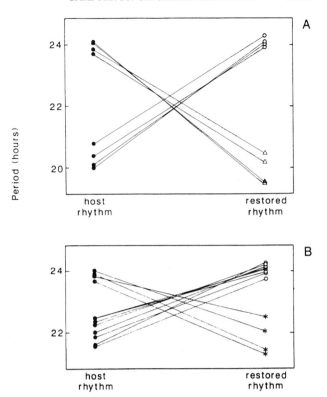

FIG. 58.4. Period of circadian rhythm of locomotor activity before the destruction of the SCN (host rhythm) and following the transplantation of fetal SCN tissue (restored rhythm) in wild-type hamsters and in hamsters homozygous or heterozygous for the tau mutation, which shortens the endogenous free-running period of the activity rhythm. *A*: Reciprocal transplants between homozygous mutant and wild-type animals. *B*: Reciprocal transplants between heterozygous mutant and wild-type hamsters. *Closed circles*: periods of the intact hosts; *open circles*: period of rhythm following transplantation of wild-type tissue; *triangles*: period of rhythm following transplantation of homozygous mutant tissue; *stars*: period of rhythm following transplantation of heterozygous mutant tissue. Reproduced with permission from reference 141.

affected either the period or the long-term stability of the restored rhythm.

There is currently great interest in using SCN grafts to restore rhythmicity and to determine which SCN cells and connections are necessary for the expression of circadian rhythms (143). No specific SCN cell type has been identified as containing a circadian pacemaker, though it has been suggested that specific regions of the SCN may be more important for the generation of circadian signals than other regions (57). It has not been possible to entrain SCN-grafted animals to normal light intensities, presumably because of insufficient retinal innervation to the grafted tissue (85). While SCN grafts clearly make neural connections with the host brain, it is still not clear if the restored rhythms are due to neural or hormonal circadian outputs from the grafted tissue. Although not a consistent finding between laboratories, it appears that some grafts situated far from the site of the host SCN can restore rhythmicity, suggesting a hormonal signal. The rhythms that have been restored following SCN grafting are behavioral ones (for example, locomotor activity, drinking), which probably involve many different areas of the brain. In contrast, it has not been possible to restore the pineal melatonin rhythm, which is known to depend on specific neural connections between the SCN and the pineal gland (86). Very little is known about the nature of the circadian signal emitted by the SCN to other areas of the brain and body, and it may well be that the SCN utilizes hormonal and/or neurohormonal signals when it is sending circadian information to many different areas of the brain at once, whereas neural pathways are used to convey other circadian signals to specific brain regions.

While the SCN transplant studies involving mutant hamsters demonstrate that this nucleus controls the phenotypic expression of circadian rhythms, they do not address the question of whether the SCN itself can generate circadian signals. However, other studies have demonstrated that the SCN expresses circadian rhythms and that even isolated SCN tissue can produce circadian signals.

Intrinsic Oscillations within the SCN. Various experimental approaches have been taken to determine whether the SCN itself can express circadian rhythms, a prerequisite for a circadian pacemaker which generates and relays temporal information to other physiological and behavioral systems. The first overt rhythm to be measured in the SCN was energy metabolism (174). By measuring ^{14}C-labeled 2-deoxyglucose uptake in vivo, Schwartz (173), as well as other workers, demonstrated pronounced diurnal fluctuations in glucose utilization in the SCN of diverse mammalian species, including primates. In all species examined to date, whether nocturnal or diurnal in behavior, the SCN shows higher metabolic activity during the light compared to the dark phase of the light–dark cycle. Like most diurnal rhythms, the SCN metabolic rhythm is not dependent on the presence of a light–dark cycle and persists under constant light.

Consistent with the finding that SCN metabolic activity is higher during the day than during the night is the finding that multiple-unit firing activity is also higher during the light phase, and this rhythm also persists under constant light (73). Importantly, even after surgical isolation of the SCN region from the rest of the brain, a procedure which abolishes the rhythm of locomotor activity and neural firing rhythms in other brain regions, the rhythm in multiple-unit activity persists within the hypothalamic island containing the SCN (73).

Vasopressin levels show pronounced diurnal fluctu-

ations in the cerebrospinal fluid (CSF) in a variety of mammalian species that are circadian in nature (154). The results of numerous studies indicate not only that the CSF rhythm in vasopression content is dependent on circadian signals from the SCN but that the source of the vasopression peptide in the CSF is the SCN (154). Indeed, it has been established that there is a rhythm in vasopressin mRNA in the SCN (89, 205). While this was the first demonstration of a specific gene product showing circadian fluctuations in the SCN, subsequent studies have indicated that many gene products are probably produced on a rhythmic basis within the SCN (see later, under Gene Expression and Protein Synthesis in the SCN).

Perhaps the most convincing evidence that the SCN contains a circadian clock is the finding that in vitro a number of rhythms persist (Fig. 58.5). In both hypothalamic slice and organ culture preparations a variety of rhythms have been observed, including neural firing, vasopressin release, and glucose metabolism (39, 40,

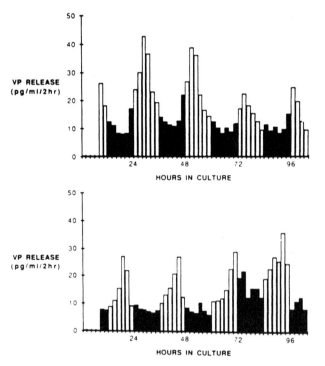

FIG. 58.5. Representative patterns of vasopressin *(VP)* release from two individual SCN explants studied in constant conditions using a perifusion culture technique. Both explants were obtained from rats maintained on an 12:12 light–dark cycle, but preparation of the cultures occurred at two different times. The *top panel* shows the VP rhythm from an explant prepared at the onset of the 12 h light phase, while the *bottom panel* shows the rhythm from an explant prepared near the end of the light phase. Open and closed bars denote determinations of the amount of VP released into the medium over 2 h sampling intervals during the subjective day (that is, the portion of the circadian cycle coinciding with the projected light phase of the cycle) and the subjective night, respectively. Reproduced with permission from reference 40.

57). Being able to investigate the clock "in a dish" provides circadian investigators with a valuable system for determining how SCN cells are regulated and how they can generate circadian rhythms. Of particular interest is determining the functional role of the various cell types and regions within the SCN for the entrainment, generation, and output of circadian signals.

Entrainment Pathways. The eyes are involved in relaying entraining information from the light–dark cycle to the circadian timing system in mammals. However, this seemingly simple and intuitively obvious statement does not convey the uniqueness of the entrainment pathway by which light reaches the circadian pacemaker in the SCN. The expectation that a unique pathway exists from the eye to the circadian clock, one that is separate from the visual system, was undoubtedly suspected by early comparative biologists who knew that in lower vertebrates light influences circadian and reproductive processes via "extraretinal" photoreceptors (124). Although the nature of these photoreceptors remains to be determined, it is clear that in birds, reptiles, amphibians, and fish photoreceptive structures within the brain are responsive to changes in light and dark and that these structures convey this information to other physiological systems, including the circadian system (94, 124, 207). While earlier experiments suggested that in mammals there might be projections from the retina to the hypothalamus, it was not until the early 1970s that Moore and colleagues (17, 104) demonstrated the existence of a direct projection from the retina into the SCN of the hypothalamus, now referred to as the retinohypothalamic tract (RHT). Indeed, it was the finding of this projection which led Moore to suspect that the SCN may be playing a functional role in the generation of circadian rhythms.

At the level of the optic chiasm, retinal projections enter the brain in the region of the SCN and surrounding hypothalamic areas (17). If the primary optic tracts are severed posterior to this innervation of the SCN, entrainment of the various circadian rhythms to the light–dark cycle still occurs, demonstrating that the RHT is sufficient for the entrainment of circadian rhythms (164). The primary visual centers of the brain and/or the perception of light are not necessary for entrainment of circadian rhythms by the light–dark cycle. In addition to the RHT, the SCN also receives retinal information indirectly from the lateral geniculate nucleus (LGN), which receives a direct projection from the retina (17). A geniculohypothalamic tract (GHT) arises from a distinct subdivision of the LGN, referred to as the intergeniculate leaflet (IGL), and gives rise to a dense terminal projection coextensive with the termination of the RHT in the SCN. While the GHT projection to the SCN is not necessary for entrainment to the

light–dark cycle, this tract appears to have a functional photic input to the SCN since destruction of the GHT can modulate the phase angle of entrainment, the circadian period during exposure to constant light, and the rate of reentrainment following a shift in the timing of the light–dark cycle (112).

Early studies suggested that acetylcholine (Ach) may be a neurotransmitter in the circadian system that mediates the effects of light since the cholinergic agonist carbachol was found to mimic many of the effects of light on the circadian clock (41, 235). However, the lack of pharmacological and anatomical support for Ach as an active transmitter in the SCN has put into question the importance of Ach in the photic entrainment of circadian rhythms (92, 163). Increasing attention has focused on the role of excitatory amino acids (EAAs) in the photic response because of a number of findings, including, that (1) some of the effects of light on the circadian system can be mimicked by treatment with glutamate, (2) treatment with EAA antagonists can block the phase-shifting effects of light on the circadian system, (3) glutamate immunoreactivity has been localized to retinal ganglion cells, and (4) optic nerve stimulation can induce the release of glutamate (and aspartate) (2, 19, 92). However, other findings are not consistent with a role for glutamate in the light-input pathway to the SCN, and identification of the neurotransmitter(s) mediating the effects of light on the circadian system is still an open question. Studies on the molecular events that occur in the SCN in response to photic signals raise the possibility that more than one transmitter may be involved in mediating the effects of light (see later, under Gene Expression and Protein Synthesis in the SCN).

Interestingly, while the neural pathways from the retina to the SCN have been well defined, identification of the photoreceptors in the eye that are involved in entrainment is still to be established. Surprisingly, a substantial body of evidence indicates that circadian photoreceptors may be different from the image-forming rods of the retina. In mutant mice with degenerate retinas the photosensitivity of the circadian system over a large range of irradiances is not very different from the sensitivity of mice with normal retinas (42, 53). Studies on the spectral sensitivity of the photoreceptors involved in relaying light to the circadian system as well as the reproductive system in the Djungarian hamster indicate that a short wavelength, cone-like photoreceptor may be the primary receptor relaying nonvisual light information to the brain (69, 98). This hypothesis is particularly attractive in view of studies indicating that blue cones may be more widespread among rodents than previously thought (74). However, the discovery of a class of photoreceptors within the mammalian eye which relay light information to the circadian system cannot be ruled out (53). In any case, it is clear that many properties of the circadian system's response to light are vastly different from the response of the visual system to light (119), highlighting the evolutionary divergence of these two systems. This evolutionary divergence is not surprising given the fact that in all nonmammalian vertebrate classes specialized extraretinal photoreceptors mediate the effects of light on the circadian system (94, 97, 207).

In addition to light, other stimuli, such as induced increases in locomotor activity or restricted feeding periods, can also influence the circadian clock system. How nonphotic information reaches the SCN is still not known, though there is evidence to suggest that the LGN/IGL may be involved in mediating the effects of activity on the clock (110, 112). Phase shifts in the activity rhythm, which are induced by an increase in activity in response to exposure to a novel running wheel for a few hours or injections of the benzodiazepine triazolam, are blocked by lesions of the LGN (75) (Wickland and Turek, unpublished results). Furthermore, the IGL is the source of neuropeptide Y (NPY) innervation of the SCN, and the administration of NPY into the SCN area as well as electrical stimulation of the GHT both induce phase shifts in the hamster locomotor activity rhythm similar to those resulting from activity-inducing stimuli (2, 110). The LGN/IGL may be a common pathway by which information about the lighting environment and the activity–rest state reaches the circadian clock in the SCN, and it may be involved in integrating information in the circadian time-keeping system from the exterior and interior environments.

Gene Expression and Protein Synthesis in the SCN. In view of the pronounced rhythmic changes demonstrated to occur within the SCN itself, it is anticipated that pronounced daily rhythms in the synthesis of specific mRNAs and proteins will eventually be found. Indeed, as noted earlier (see Intrinsic Oscillations within the SCN), a rhythm in vasopressin mRNA, as well as the poly (A) tail length of vasopressin mRNA, within the SCN have been established (89, 205). Diurnal rhythms in mRNA for other peptides, including vasoactive intestinal peptide/peptide histidine isoleucine (VIP/PHI), gastrin-releasing peptide, and somatostatin, have also been observed (2, 192). However, there is no evidence to indicate that any of these gene products plays a fundamental role in the generation of circadian signals, but instead these products appear to represent rhythmic outputs from the clock. Interestingly, Takeuchi et al. (192) reported that in the absence of a light–dark cycle there is still a clear circadian rhythm of somatostatin mRNA levels in the SCN, while the rhythm in VIP/PHI mRNA is not present. This finding raises the possibility that certain peptide rhythms in the SCN are a response to the light–dark cycle, while the

timing of other rhythms depends upon the endogenous oscillator within the SCN.

While no specific gene products that are part of the circadian clock itself have been identified in the SCN, a role for protein synthesis in the generation of circadian rhythms has been established. Acute administration of either of two protein synthesis inhibitors (anisomycin or cycloheximide), with two different mechanisms of action, induces pronounced phase shifts in the circadian clock of hamsters, and the effects of these inhibitors appear to be on cells within the SCN region (72, 190, 231). Interestingly, the PRCs generated by either peripheral or central injections of anisomycin or cycloheximide in hamsters are similar to those measured for protein synthesis inhibitors in microorganisms and invertebrates, suggesting that the biochemical mechanisms generating circadian oscillations in mammals may share common features with those found in very distantly related phylogenetic groups (190). Protein synthesis may be involved in at least some light-induced phase shifts in the circadian pacemaker since treatment with cycloheximide can block phase advances, but apparently not phase delays, in response to a brief pulse of light (Takahashi and Turek, unpublished results).

A number of groups are presently using two-dimensional gel electrophoresis to identify specific proteins that are (1) unique to the SCN, (2) produced at certain times of the circadian cycle when protein synthesis is known to be involved in clock function, and/or (3) responses to entraining signals. One promising lead for the identification of clock-specific proteins is the finding that one of the proteins in the SCN of wild-type hamsters is not present in the SCN of homozygous mutant animals with an abnormally short circadian period (76). Attempts to identify proteins specific to circadian rhythm–generating structures in different invertebrate and vertebrate species represent another promising approach.

A number of laboratories have demonstrated that light can induce the expression of the protooncogene c-fos within the rodent SCN (3, 38, 83, 145, 166). c-fos, as well as other immediate early gene (IEG) products, appears to function by coupling transient stimuli to the regulation of specific genes in the nucleus. The fos protein dimerizes with products of the Jun family of proteins to form a transcriptional regulatory complex, activating protein 1 (AP-1), which can bind to specific regions of the DNA to regulate the transcription of specific genes. Studies indicate that light also regulates jun-B activity in the SCN (84). Of particular interest are findings that the effectiveness of light in inducing c-fos mRNA in the SCN at a particular circadian phase is quantitatively correlated with the magnitude of the light-induced phase shifts in the activity rhythm (Fig. 58.6) (83). Furthermore, the photic induction of c-fos and jun-B in the SCN is gated by the circadian clock such that the synthesis of these genes is induced only in response to light pulses presented at circadian phases where light induces phase shifts in the circadian rhythm of locomotor activity (Fig. 58.7) (83, 84).

Administration of the EAA antagonist MK-801 has been found to block light-induced phase shifts in the activity rhythm of hamsters and also the photic induction of Fos-like immunoreactivity (Fos-lir) in the rostral SCN and the ventrolateral region of the caudal SCN (1). Interestingly, administration of the cholinergic antagonist mecamylamine, which also can block light-induced phase shifts in the activity rhythm, blocks the induction of Fos-lir in the dorsomedial, but not in the ventrolateral, hamster SCN (Zhang and Van Reeth, unpublished results). While these results suggest a possible segrega-

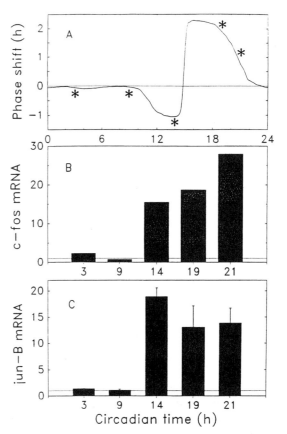

FIG. 58.7. Dependence of light-induced behavioral phase-shifting and jun-B and c-fos mRNA levels in the SCN region as a function of circadian phase. Hamsters were maintained in darkness before being exposed to a 5 min pulse of light. *A:* Asterisks superimposed on the phase response curve to light pulses in hamsters denote the time of the light pulses in *B* and *C. B:* jun-B mRNA in the SCN after light pulses of 5 min duration at circadian time 3, 9, 14, 19, and 21. Animals were returned to darkness for 25 min and brains were prepared for in situ hybridization. Values represent the mean signal in the SCN of light-pulsed hamsters relative to the mean signal at the same circadian time in animals receiving no light. Hamsters receiving no light exhibited no significant jun-B mRNA hybridization in the SCN at all circadian times examined. *C:* c-fos mRNA in the SCN region induced by light following the procedures described in *B.* Reproduced with permission from reference 84.

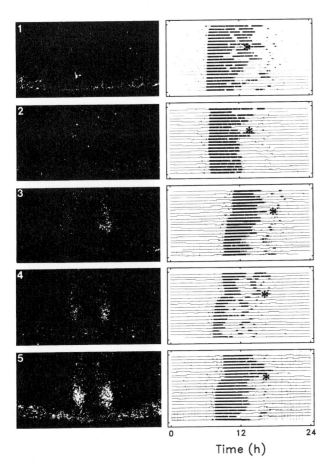

FIG. 58.6. Effects of pulses of light with varying radiance on c-fos mRNA in the SCN region and the phase of the activity rhythm in hamsters maintained in constant darkness. All light pulses were 5 min in duration and were presented at circadian time 19. *Left panels:* Following exposure to the light pulse, animals were returned to constant darkness for 25 min before the brains were prepared for c-fos analysis by in situ hybridization procedures. *Panel 1* shows c-fos levels in an animal receiving no light pulse, while *panels 2–5* show levels in animals exposed to increasing levels of illumination. *Right panels:* Representative activity records of hamsters exposed to light stimuli of similar irradiance to the corresponding left panels. Asterisks mark the time of the light pulse, after which the animals continued to free-run in constant darkness. Reproduced with permission from reference 83.

tion of the distribution of the EAA and cholinergic inputs to the SCN, problems with pharmacological specificity of the antagonists prevent any definitive interpretation. Nevertheless, the identification of early gene products in response to light stimuli opens up a variety of new approaches for studying both the physiological and the molecular events associated with the entrainment and generation of circadian signals.

The Multioscillatory Nature of the Mammalian Circadian Clock System

While there is substantial evidence to indicate that the SCN represents a master circadian pacemaker that provides the temporal signals for most, if not all, circadian rhythms in mammals, there is nevertheless also evidence that the circadian time-keeping system is itself multioscillatory in nature (162). This hypothesis has arisen because of evidence that more than one oscillatory process exists within, as well as outside of, the SCN. Thus the term "multioscillatory," in the context of circadian organization, is often used to describe quite different processes, which may have little to do with one another.

The SCN as a Multioscillatory System.

There are two general findings which indicate that the SCN is made up of more than one circadian oscillator. First, under certain experimental conditions behavioral and endocrine rhythms can dissociate, or "split," into two distinct components which initially free-run with distinctly different circadian periods, resulting in a series of changing phase relationships between the two components (92, 188, 199). Usually these components become recoupled some 12 h out of phase with each other and thereafter assume an identical free-running period. Despite the fact that splitting can occur under certain free-running conditions in the majority of animals (for example, during exposure of hamsters to constant light) and has been documented in different vertebrate species, including hamsters, rats, monkeys, tree shrews, mice squirrels, starlings, and lizards (162, 199), no satisfactory physiological explanation for this intriguing property of the circadian clock has been forthcoming. The etiology of splitting appears to involve pacemakers within the SCN itself since lesioning one of the two SCN can abolish splitting (128). However, it is unlikely that splitting is solely due to each SCN acting as an independent pacemaker since it has been observed in hamsters with only a single SCN (30). Splitting appears to be a characteristic of the inner structure of the central pacemaking system in the SCN region itself since studies indicating that electrical activity within the SCN obtained from animals showing a split rhythm of activity also show a bimodal firing frequency in vitro (237). Any complete explanation of how the SCN can generate circadian signals must take into account this property of the mammalian clock.

A second general finding indicates that the SCN is comprised of more than one circadian oscillator: it has been observed, both in vitro and in vivo, that pieces of the SCN are capable of sustaining circadian oscillations. Thus brain slices containing only a portion of the SCN continue to show rhythmicity in vitro (57), and following lesions of different regions of the SCN rhythmicity is maintained in the whole animal (129). There is even evidence to suggest that dispersed SCN cells, both in vitro and in vivo, can still generate circadian signals, demonstrating that the ability to generate circadian rhythms does not depend on the structural integrity of the SCN (85, 118). Taken together with data from other

models of circadian pacemakers (for example, the isolated avian pineal gland or molluscan eye; 189), it would appear that the ability to generate circadian signals is a cellular property and is not dependent on specific neural networks, though conclusive data in support of this hypothesis have not been obtained.

Oscillators Outside the SCN.

The circadian system in mammals is also thought of as being multioscillatory in nature because of experimental evidence indicating that both free-running and entrained rhythms persist even after total destruction of SCN tissue. However, there is not convincing consistent evidence in mammals that any rhythm can persist for a prolonged period of time under constant environmental conditions following total destruction of the SCN (79, 101, 196). Nevertheless, following large SCN lesions, unstable ultradian rhythms are often observed in the activity patterns of rodents, suggesting that oscillators outside of the SCN, which are normally synchronized by the master pacemaker in the SCN, fail to couple or achieve only weak or unstable coupling after SCN lesions (101).

There are a number of situations in which rhythms persist in SCN-lesioned animals exposed to an external environment with periodic fluctuations. Abnormal but persistent entrainment of the locomotor activity rhythm to a light–dark cycle in hamsters with complete SCN lesions is often observed (101). The presence of direct retinal projections to hypothalamic regions outside the SCN is consistent with the hypothesis that light–dark information may influence other hypothalamic areas, which can provide some temporal information to the animal (17). A number of investigators have demonstrated that periodic food availability can entrain circadian rhythms in mammals. For example, if food is provided to rats for only a few hours per day at a fixed time, an increase in locomotor activity and body temperature anticipates the daily mealtime (101, 182). The anticipatory activity associated with mealtime appears to involve a circadian oscillatory mechanism since such a response does not occur if food is presented at intervals that are much less or greater than 24 h (181). Furthermore, this food-entrainable oscillator appears to lie outside of the SCN region since periodic food presentation can entrain rhythms in SCN-lesioned rats (181, 183).

Early studies in humans indicated that different circadian rhythms might be controlled by different circadian pacemakers (11). The primary evidence for this was the finding that in temporal isolation the rhythms of activity–rest and body temperature were sometimes found to free-run with dramatically different periods (for example, 25 vs. 33 h), a phenomenon referred to as "internal dyschronization." While models have been developed to indicate that a single pacemaking system can explain internal dyschronization in humans (13, 27), the role of non-SCN oscillators in the organization of the circadian system is still an open question.

There are no convincing data to indicate that any structure outside of the SCN acts as a master circadian pacemaker. There is occasional confusion in the literature on this, particularly with respect to the role of the mammalian pineal gland as a possible circadian pacemaker in mammals. One review referred to "our internal clock" as the "pineal system," suggesting some sort of central role for the pineal gland in the clock mechanism (130). In mammals there is no evidence that the pineal gland itself is capable of generating circadian rhythms. Although the pineal gland expresses pronounced circadian rhythms, particularly in the synthesis and release of melatonin, the regulation of this rhythm appears to be totally under the control of circadian neural signals from the SCN (70, 81). The pineal melatonin rhythm may regulate other rhythms, and it is possible that melatonin may have some feedback effects on the circadian pacemaker in the SCN since injections of melatonin under certain experimental conditions can phase shift the circadian clock of rats (18, 146), and melatonin

receptors have been localized to the SCN in diverse mammalian species, including humans (220). Some of the confusion over the role of the pineal gland as a central circadian oscillator in mammals arises from the fact that in lower vertebrates the pineal gland does function as a self-sustained circadian pacemaker regulating other circadian rhythms (93). Indeed, treatment with melatonin can have dramatic effects on the circadian rhythm of locomotor activity in birds (201). Furthermore, the pineal gland of birds and reptiles expresses clear circadian rhythms even in culture (189). No convincing evidence is available to indicate that the mammalian pineal gland can sustain rhythmicity in culture, and the pineal gland is clearly not necessary for the expression of circadian rhythms in rodents (7).

One conceptual way of viewing the overall organization of the circadian system in mammals is provided in Figure 58.8. In this model various entraining signals provide information about environmental time to a hierarchically superior central pacemaker, which is probably located exclusively in the SCN region (211). This central pacemaker drives a large variety of "slave" systems, and whereas some of these systems may not

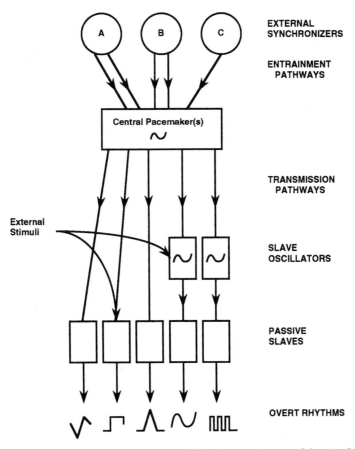

FIG. 58.8. Schematic representation of a putative model for the organization of the circadian system in complex organisms. See text for further details. Reproduced with permission from reference 203.

have the capability to generate intrinsic oscillations *(passive slaves)*, others may be able to show persistence of oscillatory behavior when disconnected from the central pacemaker *(slave oscillations)*. It is important to recognize that many entraining agents may provide temporal information both directly to the central pacemaker and to processes driven by the pacemaker and that at least some driven rhythms (for example, feeding or activity–rest cycles) may have feedback effects on the central pacemaking system.

Although diverse circadian rhythms within a given individual may share a common period and be regulated by the same circadian clock (see Nature of Circadian Code from SCN, below), there is a remarkably wide variation in the shape and amplitude of the measurable overt rhythms. Even within a given physiological system, such as the endocrine system, the characteristics of various rhythms can vary dramatically in the same individual. The diverse shapes of circadian output rhythms have in general been attributed to the fact that the wave-shape of the overt rhythm is highly sensitive to factors that are downstream from the clock actually driving the rhythm. However, little is known about the physiological processes that underlie the variability in waveform and amplitude of circadian rhythms or about the nature of the information relayed from the circadian clock in the SCN to the rest of the brain and body.

Nature of Circadian Code from SCN

Because the physiological nature of the circadian pacemaker in the SCN remains unknown (at both the cellular and the systems level), all studies of circadian outputs have, by necessity, involved the monitoring of some rhythmic variable that is downstream from the clock itself, that is, between the clock and the final events associated with the expression of a particular rhythm. Even circadian rhythms that have been measured within the SCN (for example, electrical activity or glucose metabolism) may simply represent rhythmic processes downstream from the clock mechanism, and such rhythms may or may not be involved in the regulation of circadian rhythms outside of the SCN. Since no direct output of the clock in the SCN that may actually convey circadian information to the rest of the organism has been measured with certainty, we know essentially nothing about the nature of the actual circadian signals, or the "code," by which the SCN communicates with the rest of the organism. In addition, it is not possible to determine which features of the waveform of the measurable overt rhythms are a reflection of circadian outputs per se and which are due to modifications along the pathway between the clock and the final expression of the rhythm.

There is no reason to assume that all circadian infor-

mation leaving the SCN is neural in nature. The circadian clock in the SCN regulates many diverse and complex functions (for example, sleep, eating, drinking, sexual behavior, body temperature) that depend on the coordination of many different regions of the brain. One could argue that the most efficient way of communicating with many different regions of the brain might involve signal transmission that was neurohormonal in nature. The close proximity of the SCN to the ventricular system provides it with a ready-made transportation route to the rest of the brain. There is certainly precedence for circadian pacemakers to use nonneural signals to regulate rhythmicity. In lower vertebrates, there is substantial evidence that a circadian clock in the pineal gland regulates diverse rhythms through the release of melatonin (189). The ability of transplanted fetal SCN tissue to restore rhythmicity in arrhythmic SCN-lesioned animals (85, 142) should provide experimental opportunities to determine if SCN tissue can relay circadian signals to output systems via hormonal and/or neurohormonal mechanisms. It is of great interest to determine if SCN transplants can restore rhythmicity when the transplants are positioned in areas that do not allow for normal neural connections to be made with the rest of the brain (for example, the anterior chamber of the eye) or when physical barriers prevent the transplants from developing neural connections with the host tissue. It may well be that multiple pathways carry circadian signals via both neural and hormonal processes to the variety of output systems under the temporal control of the SCN. As noted by Menaker (93), pathways connecting the "master" pacemaker to its various "slaves" have probably little in common except great length and complexity.

Regardless of whether signal transduction from the SCN relies on neural or neurohormonal factors, the question remains: What are the characteristics of the circadian signal? For example, is the circadian clock providing continuous information to its output systems or are there discrete "on" and "off" signals emitted from the pacemaker? Since most circadian rhythms are usually analyzed by one or two specific phase reference points of the rhythm, such as the onset or offset of locomotor activity, it is easy to assume that the clock sends signals to the activity center(s) of the brain only at specific circadian times. Surely, however, the clock is communicating with activity-generating centers more than just during the time that selected phase reference points are being expressed.

The diverse shapes of circadian output rhythms may reflect the fact that there are multiple circadian codes. For example, while the circadian rhythm in serum cortisol levels in humans is somewhat sinusoidal in shape, the rhythm in preovulatory luteinizing hormone release in rodents is represented by a brief surge of release on a background of low levels (203). The circadian information leaving the SCN may be diverse not only in anatomical terms but also in the characteristics of the signals themselves. To paraphrase Menaker (93), the circadian signals that emanate from the SCN may have little in common except great diversity and complexity.

Not all features of a given overt rhythm represent a property of the circadian pacemaker underlying the generation of that rhythm. Two properties of circadian rhythms that are considered to be parameters representative of the clock itself are the steady-state phase of the oscillation and its period length under nonentrained conditions (197). Thus a change in phase or period of an overt rhythm indicates that the circadian output from the clock has itself been altered. In contrast, since the amplitude of a rhythm is highly sensitive to factors downstream from the clock, changes in rhythm amplitude are not necessarily due to an associated change in the circadian output signal from the SCN. Nevertheless, changes in rhythm amplitude could also be due to changes in the output signal from the clock itself.

Our ignorance of how circadian signals from the SCN are relayed to target tissues may be an important factor in our inability to precisely define the role of circadian abnormalities in various disease states. While circadian abnormalities have been associated with a number of mental and physical disorders (208, 211), the etiologies of the circadian disorders remain unknown. Problems in signal transduction from the clock to its effector systems may be the underlying cause of many diseases associated with disorders of biological time-keeping. A better understanding of the nature of circadian outputs from the SCN and how these output signals may be modified as they wind their way to effector systems is of great importance for elucidating how circadian rhythms are regulated and for delineating the importance of normal temporal organization for the health and well-being of the organism.

GENETIC BASIS

Early studies in a variety of species demonstrated that circadian rhythms were an inherited property of the organism and that environmental factors played little, if any, role in determining the basic properties of the circadian clock system, such as period or phase (4, 131, 132). Circadian rhythms develop normally in diverse species, such as fruit flies, lizards, and mice, even if the animals are never exposed to light–dark cycles (34). Indeed, even when mice are raised under light–dark cycles with periods of 20 or 28 h in duration, upon exposure to constant darkness the period of the free-running rhythm returns immediately to around 24 h (34).

The first clear demonstration that specific genes regulated the period of the circadian cycle was reported in 1971 when Konopka and Benzer (82) established that mutagenesis in *Drosophila* could lead to mutant alleles inducing very short period, long period, or arrhythmic phenotypes. Mutagenesis has uncovered single gene mutations in plants and invertebrates that alter such basic clock properties as the period of free-running rhythms, entrainment to light–dark cycles, and temperature compensation (37, 63, 233). Multiple alleles have been identified at the two most well-studied clock genetic loci in *Drosophila* (the "per" gene) and *Neurospora* (the "frq" gene), with different alleles inducing a variety of circadian phenotypes (37, 63, 88, 157, 233, 234). Mutations in at least six other loci that can alter various circadian parameters have also been identified in these species. In addition to being useful in analyzing the formal properties and the physiological mechanisms that underlie rhythmicity, genetic analysis of these mutations has led to the cloning of clock-related genes and the ability to study their molecular structure and expression (88, 157, 234).

Unfortunately, no such detailed genetic or molecular studies have been performed on clock-related genes in vertebrates. A single gene mutation in a clock-related gene has been identified, and the characteristics of this mutant are described below. Most genetic studies of circadian rhythms in mammals have focused on inbred strains. In addition, studies on human twins are beginning to provide information on the characteristics of genetically influenced human rhythms.

Inbred Strains

Natural genetic differences in various circadian rhythm parameters have been observed between various inbred strains of rodents, indicating that genetic background influences the expression of circadian rhythms in mammals. Small differences in period, phase angle of entrainment, amplitude of rhythms, and PRC to light pulses have been noted between various inbred strains of mouse, rat, and hamster (15, 21, 43, 68, 88, 139, 157, 234). Schwartz and Zimmerman (175) found clear between-strain differences in the free-running period of the locomotor activity rhythms of 12 different inbred strains of mouse. Interestingly, the difference in period between some strains was associated with an interstrain difference in the PRC to light pulses. The interstrain difference in PRC shape appears to compensate for the difference in period because both strains show similar phase relationships between activity onset and the light–dark cycle (175).

In an examination of four inbred strains of golden hamster, it was found that significant differences in the free-running period of the activity rhythm between strains were correlated with the phase angle of entrainment to an entraining light–dark cycle (Vitaterna and Turek, unpublished results). In these same strains there were no apparent differences in the PRC for light-induced phase shifts in the activity rhythm; however, there were clear differences in the phase-shifting effects of an activity-inducing stimulus, triazolam. Indeed, in two of the four inbred strains there were no clear phase shifts in the activity rhythm in response to triazolam and its associated acute increase in activity at any circadian time, while the PRCs to triazolam for the other two strains were similar to the PRCs to triazolam that have been generated in wild-type animals.

While interstrain differences in circadian parameters point to the importance of the genome in the regulation of circadian properties, they do not provide much information on specific genes involved in circadian timing. Nevertheless, examining the physiological basis for differences in the circadian clock system between inbred strains may prove important for elucidating the physiological and cellular events which mediate the entrainment, generation, and expression of circadian rhythms.

Twin Studies

Due to the inherent difficulties in studying human circadian rhythms under well-controlled laboratory conditions, very little is known about the contributions of heredity and the environment for their expression. However, in a study designed to evaluate the possible genetic influences on the expression of circadian rhythmicity, a detailed quantitative analysis of the 24 h profile of plasma cortisol in monozygotic and dizygotic pairs of normal male twins under entrained conditions was performed (87). While there were significant effects of the environment on the 24 h mean levels of cortisol, the nocturnal nadir in cortisol levels was found to be under genetic control. Although very few studies of this type have been performed, these interesting preliminary findings indicate that, while some features of overt measurable circadian rhythms are undoubtedly influenced by behavior (for example, dietary intake, exercise), other features may be under genetic control. It will probably be very difficult to determine in humans whether this genetic control is at the level of the circadian clock itself or in the processing of information into and/or from the circadian clock. Studies in animals should more clearly define the level at which genetic factors influence circadian rhythmicity.

A Mammalian Clock Mutation

The first evidence that the period of a circadian rhythm in mammals could be influenced by a single gene was reported for hamsters in 1988 (144). One male hamster from a commercial supplier was found to have a free-running period in constant darkness of 22 h. Subsequent breeding experiments revealed that this serendipitously discovered short period male was heterozygous for a single autosomal mutant gene, which was called "tau." Further crosses established that the mutant allele was codominant and that the homozygote phenotype of the free-running period was about 20 h. Tau is expressed equally in males and females, and the ranges of the circadian periods between the three genotypes do not overlap (144). As for both the per and frq mutations in *Drosophila* and *Neurospora* (see under GENETIC BASIS, above), it appears that each mutant allele in some way removes 2 h from the course of the oscillation that underlies the expressed circadian rhythm. In addition to altering the period of the circadian clock, the tau mutation alters the shape of the PRC to both photic and nonphotic stimuli (95, 117), indicating that the tau mutation alters the circadian clock system in a number of different ways.

That the tau mutation acts on circadian timing through a product of the SCN has been demonstrated through studies involving transplantation of fetal SCN between wild-type and tau mutant hamsters (141, 142). Experiments in which donor SCN tissue of one genotype is implanted into an SCN-lesioned host animal of a different genotype have demonstrated conclusively that the behavioral rhythms that are restored always have the period of the donor tissue (Fig. 58.4). As noted earlier, the ability to transplant specific circadian characteristics into the host organism by transplanting SCN tissue represents strong evidence that the SCN is functioning as the master circadian pacemaker in mammals.

While the discovery of the tau mutant has led to many interesting observations about the genetic control of circadian rhythms in mammals, as well as to new insight into the physiology of the circadian clock system, use of this mutant for mapping the location and identifying genes that may be involved in the generation of circadian rhythms is unlikely due to the lack of background genetic information in the hamster. Isolation of genetic clock mutants in mice would, in contrast, be very important for mapping clock genes since many genetic markers are available in the mouse.

ONTOGENY OF CIRCADIAN RHYTHMS

While the vast majority of studies on mammalian circadian rhythms have been on adult animals, studies on young (both pre- and postnatal) and old animals have led to a number of surprising findings about how the clock develops and how it ages.

Development of Clock and Expression of Circadian Rhythms

Three different general approaches have been taken to studying the early development of the circadian clock system and its outputs. Because of the central role played by the SCN in circadian organization, some investigators have examined in detail the development of the SCN from a neuroanatomical point of view, while others have attempted to determine when rhythms are first expressed within the SCN itself. A third approach involves determining when various circadian rhythms are first expressed and relating their expression to the development of the SCN. As noted below, the results of various studies have demonstrated that the SCN is functioning as a circadian clock and as a circadian pacemaker for the expression of overt rhythms long before the rhythms are actually present.

Anatomical Development of the SCN.
A comprehensive review on the development of the SCN in rodents has been provided by Moore (106). The SCN is formed in the rat between embryonic day 14 (E14) and E17 (with birth on E21 or E22) and is clearly evident as a distinct nucleus in Nissl stain on E17. Between E17 and postnatal day 10 (P10) the nucleus enlarges and assumes its adult appearance. Synaptogenesis occurs primarily after birth, and while it has been estimated that each adult SCN neuron has 300–1,200 synaptic contacts, at E19 there appears to be less than one synapse per SCN neuron. This is particularly interesting in view of the fact that rhythmicity within the SCN is present before this time (see below). Innervation of the SCN by the RHT is a postnatal event and is clearly present by P4.

Functional Development of the SCN as a Circadian Clock.
In several mammalian species circadian oscillations are present in the fetal SCN (155). By monitoring ^{14}C-labeled deoxyglucose uptake as a functional marker, Reppert and Schwartz (152) demonstrated a circadian rhythm of metabolic activity within the fetal SCN as early as day 19 of gestation, and day–night oscillations in vasopressin mRNA levels were observed in the fetal rat SCN on day 21 of gestation (155). The fetal SCN is entrained by circadian signals from the mother, and although rhythms develop normally in pups born to SCN-lesioned mothers, fetal SCN rhythms are not entrained without the presence of the maternal SCN (155). Studies in both rats and hamsters indicate that the fetal SCN can be entrained by the mother at around the time of neurogenesis of the SCN (31, 155).

It appears that the fetal SCN may be entrained by multiple circadian signals from the mother. Thus while maternal pinealectomy does not disrupt the communi-cation of circadian phase from the mother to the fetus, injections of melatonin into SCN-lesioned hamsters can restore fetal synchrony (32, 153). Restricted feeding of SCN-lesioned pregnant rats can also synchronize rhythms within a litter (153). Whatever the nature of the maternal entraining signal, it is clear that the circadian clock is set by the mother, and that, at least in rats, the postnatal maternal influence on rhythmicity persists for about a week after birth when the pups become capable of responding to the entraining effects of the ambient light–dark cycle.

Development of Overt Rhythms.
Despite the early development of circadian rhythms within the SCN, the ability of the SCN to function as a pacemaker for the regulation of overt behavioral and physiological rhythms does not occur until much later. For example, in rats most behavioral and endocrine rhythms do not appear until the second or third week of life. Thus the circadian clock in the SCN becomes a circadian pacemaker when it develops sufficient afferent, intrinsic, and efferent neural connections to be entrained and to regulate effector systems (106). However, through a series of ingenious experiments, Deguchi (36), Davis (29), and Reppert and Weaver (155) have demonstrated that even before specific rhythms are expressed eventual expression is already being influenced by a central circadian pacemaker. Through experiments involving SCN lesions of the mother during fetal development and/or cross-fostering of pups between dams entrained to different light cycles, it has been demonstrated that the phase of specific rhythms can be set long before the rhythms appear.

Moore (106) has divided the development of circadian function regulated by the SCN into four components. First, the SCN cells are formed, grow, and mature, and rhythmic function within the nucleus is established well before major synaptic contacts either between SCN cells or with the rest of the brain are established. Second, entrainment pathways to the SCN develop (primarily the RHT), with the resultant ability of the SCN to respond to environmental information. Third, SCN projections develop, resulting in the coupling of the SCN to effector systems, often before those systems can express rhythms. Fourth, output systems mature to the point where they can express circadian function.

Aging of the Circadian Clock System

Age-related changes have been documented in endocrine, metabolic, and behavioral circadian rhythms in a variety of animal species, including humans (14, 22, 91, 156, 178, 214–216, 224). Most studies have been largely limited to measurements of the effects of age on the amplitude of circadian rhythms (71, 216). For example, circadian rhythms of temperature (62, 169), corticosterone (123), serum testosterone (100, 194), and melatonin (151, 193) are all dampened in old rats, while reduced light–dark differences in the rhythm of locomotor activity have been reported in mice and rats (90, 113, 126, 169). Old rats also show a reduction in the amplitude of the circadian fluctuations of slow-wave sleep and wakefulness (159, 215), while in some old rats drinking and locomotor activity rhythms have been reported to become arrhythmic (127). Compared to young mice, old mice spend more time asleep during the normal active period while being awake more during the normal sleep period (224).

Alterations of the light–dark fluctuations or amplitude of circadian rhythms do not necessarily imply a change that is intrinsic to the circadian pacemaker system. These changes can also be explained by age-related changes that are either upstream or downstream from the circadian clock. For example, age-related changes in amplitude or entrainment of circadian rhythms could be due to a decrease in the sensory perception of light or in an alteration in the mechanisms regulating various physiological processes that lie between the clock and the overt rhythmic output. Nevertheless, there is convincing evidence in rodents that the circadian clock itself is altered in senescence since the free-running period of the circadian rhythms of locomotor activity (33, 109, 137) and sleep–wakefulness (216) shorten with age in hamsters and rats, as measured under constant lighting conditions. Changes in the intrinsic period of the circadian clock with age may underlie changes in the phase angle of entrainment to light–dark cycles and changes in the rate of reentrainment following a phase shift in the light–dark cycle, which have been observed in old hamsters (158, 236). Support for age-related changes in the clock itself has been provided by Wise and colleagues (223, 228, 229), who found that aging alters the circadian rhythms of glucose utilization and alpha-1 adrenergic receptor levels in the SCN and that these changes are correlated with changes in the circadian rhythm of luteinizing hormone release (that is, the preovulatory "luteinizing hormone surge") that are observed with aging in female rats. In addition, changes in neuropeptide activity within the human SCN have been associated with senility and Alzheimer's disease (187), and patients suffering from Alzheimer's disease show marked disturbances in the levels and timing of daily locomotor activity (171), as well as fragmentation of the daily sleep–wake cycle (140).

In addition to intrinsic changes which may take place in the circadian pacemaker with advanced age, there are changes in the response of the pacemaker to environmental stimuli. Old hamsters appear to be less sensitive to the phase-shifting effects of light pulses (Zhang and

Turek, unpublished results), and it appears that in advanced age there can be a total loss of responsiveness to the phase-shifting effects of activity-inducing stimuli, such as dark pulses or benzodiazepine injections (219) (Fig. 58.9). Interestingly, transplantation of fetal SCN tissue into intact SCN regions of old hamsters can restore the response to the phase-shifting effects of triazolam on the activity rhythm (Van Reeth, unpublished results). However, it is not possible to determine how the donor tissue is contributing to the circadian system to restore responsiveness. The role of the donor tissue could range from that of simply facilitating the aged host SCN (either by restoring some input, intra-SCN,

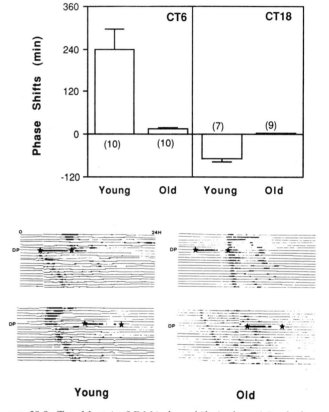

FIG. 58.9. *Top:* Mean (± S.E.M.) phase shifts in the activity rhythm of young and old hamsters maintained in constant light that were subjected, beginning at either circadian time 6 *(left panel)* or circadian time 18 *(right panel)* to a 6 h dark pulse. A value above the solid line indicates an advance in the onset of locomotor activity; a value below indicates a delay. Values in parentheses indicate the number of trials for each group of animals. The mean phase shift in the activity rhythm in response to the dark pulse was significantly greater (P < 0.001) in young than in old hamsters at both circadian times tested. *Bottom:* Representative sections from the wheel-running activity records of two young *(left panels)* and two old *(right panels)* hamsters housed in constant light before and after they were subjected to a 6 h dark pulse beginning at circadian time 6 *(top panels)* or circadian time 18 *(bottom panels).* The day of the dark pulse is designated by *DP* at the left of each record with the exact beginning and end of the dark pulse designated by two stars. Reproduced with permission from reference 219.

or output activity) so that it is capable of expressing a response to an acute increase in activity to that of providing a new pacemaker in the circadian system, which is itself able to regulate the circadian behavior of the organism.

ADAPTIVE SIGNIFICANCE

The fact that circadian rhythms are a central characteristic of all eukaryotic life and the observation that most behavioral and physiological processes fluctuate on a 24 h basis emphasize the evolutionary importance of circadian rhythms for the survival of the species. From a conceptual point of view, it is perhaps useful to consider the adaptive significance of circadian rhythms in two general ways. Circadian rhythms enable organisms to (1) synchronize activities to the dramatic periodic fluctuations in the external environment and (2) integrate and organize the internal milieu so that there is coordination and synchronization of internal processes. However, this view of temporal organization should not obscure the fact that the adaptive significances for external and internal synchronization are intimately related to each other.

It is interesting to note that, while it is accepted as dogma that circadian rhythms are important for the survival of the organism, there have been few direct attempts to demonstrate this importance. In contrast to the hundreds of papers that have been published on the underlying physiological mechanisms for the entrainment and generation of circadian rhythms, only a few studies have been aimed at determining the importance of circadian rhythmicity for the survival of the organism/species. A number of years ago very valuable papers were written on the ecological significance of circadian rhythms (25, 48, 167), and the obvious importance of "doing the right thing at the right time of day" has perhaps discouraged researchers from performing experiments in which the answers were obvious from the beginning. While the importance of external synchronization may be obvious, the importance of internal synchronization is less so, and it is undoubtedly in this area that future research will uncover new insight into the adaptive significance of temporal organization for the health and well-being of the organism.

External Synchronization

As noted a number of years ago by Enright (49), biological clocks have adaptive significance because they enable an organism to get into a "behavioral rut." That is, since a great deal of an animal's environment is predictable from one day to the next, it may be advantageous to repeat on any given day the behavior that was successful the day before. Daily routines in behavior represent a strategy in which animals cope with the time structure of their environment (25). The predictability of the daily changes in the physical environment has provided natural selection with an opportunity to favor those organisms with an internal temporal program of biological function that is in synchrony with the periodicity of environmental change (136). Indeed, as phrased by Pittendrigh (136), there is a "day within" in the life of organisms that matches and copes with the "day outside." Natural selection will favor those genes that enable an organism to develop the most appropriate circadian strategy for coping with the dramatic 24 h changes in the physical environment.

Rusak and Zucker (167) have emphasized that it is difficult to overestimate the importance of temporal organization of behavior since in nature a premium is placed on an animal's ability to perform certain behaviors at the appropriate time of the day (as well as season of the year). Many obvious behaviors, such as feeding, drinking, sleep, exploration, and reproductive activity, that change on a daily basis are in response to the daily changes in the physical environment (for example, temperature, illumination, and humidity) as well as changes in the biological environment (for example, food availability, presence of predators, parasites, competitors, and reproductive mates). Associated with these obvious behavioral changes are changes in perception, sensation, learning, and performance which occur on a daily basis in many animal species, including humans (25, 103, 167). Zucker (238) has suggested that the response characteristics of sensory and motor systems have evolved under selective pressure to provide the organism with high or low responsivity to stimuli depending on the time of day when the stimuli are present. He emphasized that the principal source of rhythmicity is within the organism and not of the external environment.

Underlying the behavioral adaptations to the daily changes in the external environment are a multitude of metabolic, hormonal, and biochemical rhythms. For example, changes in the digestive system occur before behavioral changes lead to food intake, and the rise in body temperature occurs before animals wake up in anticipation of increased metabolic demands (20, 22, 102). Well before female rodents come into behavioral heat, a cascade of neural and endocrine events involved in ovulation have occurred which induce mating behavior (16, 111). It is likely that every organ/tissue in the body is changing in preparation of the behavioral changes which will take place in response to changes in the physical environment. An example of the broad physiological effects of circadian rhythmicity was provided by Aschoff and Pohl (9), who demonstrated that the basal metabolic rate (BMR) measured in darkness, at rest, in a thermoneutral environment, and without food digestion was reduced during the circadian rest phase compared to the circadian active phase. As discussed by Daan and colleagues (28), the reduction in nocturnal core temperature is insufficient to account for the excessive reduction in BMR, and diurnal variations in BMR in homeotherms are primarily due to changes in metabolism of those tissues and organs that are used to provide and transfer energy needed for work and to remove metabolic waste.

While it is generally assumed that circadian changes in behavior that are correlated with the 24 h changes in the physical environment are important for the survival of the organism, there have been very few experimental tests of this hypothesis. That is, few studies have been carried out to determine how a disturbance of normal rhythmicity would affect the survival of the organism on a day-to-day basis. Rusak (see ref. 238) has noted that under laboratory conditions hamsters with SCN lesions would leave their burrows during the daylight, which would presumably increase the risk of predation were such a behavior to occur in the natural environment. Eskes (51) found that the circadian rhythm of copulatory behavior is disrupted in SCN-lesioned hamsters and speculated that such arrhythmic hamsters would probably be at a disadvantage relative to normal clock-intact hamsters in attempting to inseminate females during the relatively brief period of sexual receptivity in the female, which is timed to occur at a specific time of day. However, how a hamster (or any other animal) without a circadian clock would fare under natural conditions still remains a matter of conjecture, with very little experimental evidence to draw upon.

An important adaptive feature of the entrainment of circadian rhythms to the external environment is that the phase relationship of any particular rhythm to the external day is not locked to a particular phase of the light–dark cycle throughout the year. Instead, as the ratio of light to dark changes on a seasonal basis, circadian rhythms take on different phase relationships to the external environment (46, 47, 67, 134). Indeed, it is inherent in the phase resetting mechanisms of biological clocks to light–dark signals that such phase changes will occur for light-induced phase advances and phase delays in the clock to add together to equal the difference between the endogenous period of the clock and the period of the entraining light–dark cycle (136). Maintaining adaptive phase relationships with the external environment is obviously one of the primary functions of circadian clocks. As noted by Pittendrigh (136) the question of why natural selection has used a self-sustaining oscillation in cells as the temporal framework rather than a simpler "hourglass" timer lies, at least in part, in the ability of self-sustained oscillations to be entrained with different phase relationships (that

is, adaptively significant phase relationships) to the entraining agent (that is, the light–dark cycle) as a function of periodic energy input from the external cycle, which varies predictably on a seasonal basis. Thus if the timing of a particular behavior depended on an hourglass timer such that the event always occurred x h after a particular phase point of the light–dark cycle (for example, sunrise), then that behavior would occur at this phase of the light–dark cycle regardless of the length of the day. As the seasons and the length of the day change, there is a need to alter the phase relationship between circadian rhythms and the environmental cycle, and an endogenous circadian oscillator permits this to occur. As day length changes, light falls at different phase advance and phase delay regions of the circadian clock such that stable entrainment still occurs. The coupling of the clock to the light–dark cycle and of the driven rhythms to the clock allow for adaptively significant phase relationships for natural selection to act upon. The precise timing of specific day time behaviors during different seasons will depend on natural selection acting on the circadian clock's period, response to light, and the coupling of driven rhythms to the endogenous pacemaker.

Internal Synchronization

Importance of Maintaining Internal Temporal Order. In addition to having specific phase relationships with the external environment, circadian rhythms have specific phase relationships with each other. A basic tenet of the field of circadian rhythms is that internal temporal organization is central to the health and well-being of the organism and that organisms function most effectively when circadian systems are driven close to their natural frequencies (133). The physical and mental maladies associated with rapid travel across time zones (for example, the jet-lag syndrome) and in having to work and sleep at abnormal circadian times (as in shift workers) are assumed to be due in part to an alteration in the normal phase relationships between various internal rhythms (23, 211). In addition, it has been speculated that it is the alteration of phase relationships between internal rhythms which underlies various mental and physical disorders (211, 222).

Unfortunately, this article of faith, that internal temporal organization is of fundamental importance for the health of the organism, has received very little experimental support due mainly to the lack of good animal models for studying the adverse effects of disrupted circadian rhythms. Early studies demonstrated that fruit flies reared on a 12:12 light–dark cycle live longer than flies exposed to light–dark cycles significantly shorter (21 h) or longer (27 h) than the normal 24 h (138). Likewise, in the blowfly repeated exposure to 6 h phase shifts in the 12:12 light–dark cycle reduced longevity (10, 170). These deleterious effects on life span have been attributed to the disruption of the organism's normal internal temporal organization (137). Other studies in invertebrates have confirmed or failed to confirm these early studies, and, as noted by Brock (14), there is probably a complex interplay of light intensity, frequency, and direction of phase-shifting which must be taken into account in interpreting such longevity studies. While there is some evidence in mammals (121) to indicate that a similar disruption of normal circadian organization decreases longevity, other studies have failed to find any such effect (14). Most of the longevity studies in mammals have focused on repeated phase shifts in the light–dark cycle (see, for example, ref. 121), and little is known about the effects of changes in either the endogenous period of the clock or the period of the entraining agent on longevity.

In any case, determining the effects of disrupted circadian rhythms on longevity under controlled laboratory conditions may tell very little about the importance of internal temporal organization for the health and survival of the organism. The environment in animal care quarters (devoid of predators, with constant food availability, regulated temperature, and a cleanliness never found in the natural environment) protects laboratory animals from the many stresses and challenges of the real world. Determining the effects of temporal disorganization in an animal struggling to survive in a hostile environment might enable us to clearly demonstrate the importance of internal temporal organization for survival. Limited support for this hypothesis comes from studies in squirrel monkeys in which exposure to constant light resulted in impaired thermoregulation in response to mild cold temperatures when compared to monkeys entrained to a 12:12 light–dark cycle (54). Further studies along these lines are clearly needed to define the circumstances in which impaired temporal organization decreases the fitness of the organism. Determining which physiological systems are particularly vulnerable to disrupted timing of internal rhythms in animal models may lead to a better understanding of how either imposed or spontaneous internal dyschronization of rhythms affects human health, performance, and productivity. It is noteworthy that the one species likely to be affected the most by dysynchronized circadian rhythms is *Homo sapiens,* since this is the only species that regularly chooses a life-style which leads to a disruption of the phasing of internal rhythms.

As we know, there are pronounced changes in the phase relationship between circadian rhythms and the external environment that take place in response to the changing day lengths throughout the year. In addition, the phase relationships of internal rhythms to one another are also expected to change on a seasonal basis (136). A study in humans found a seasonal change in the phase relationship between sleep and body temperature (67), raising the possibility that changes in the phase relationship of internal rhythms may underlie some of the mental disorders which have been associated with the change of seasons (160).

Circadian Control of Sleep and Sleep-Associated Rhythms: Implications for Disturbances of Normal Sleep Time. While many circadian rhythms appear to be under the direct control of the circadian clock in the mammalian SCN, the timing of many behavioral, physiological, and biochemical rhythms is indirect and involves the circadian control of the sleep–wake cycle. That is, the expression of many rhythms is dependent on the sleep–wake state of the animal. This is true not only for the obvious behavioral rhythms that require a waking animal for the behavior to occur but also for many internal rhythms. The role of sleep in the expression of many endocrine rhythms has been particularly well studied in humans (210). Thus while some endocrine rhythms are relatively independent of the timing of sleep or wakefulness, other rhythms, such as pituitary prolactin and growth hormone release, are stimulated (or inhibited) by sleep itself. Thus one simple way to look at the circadian clock control of 24 h rhythms is:

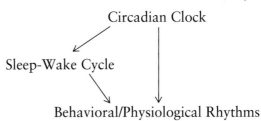

For essentially all animals living in nature the timing of the sleep–wake cycle is usually in the proper phase relationship with the circadian clock since it is the clock which times when the animal is awake or asleep. Thus rhythms regulated by the clock, relatively independently of sleep, as well as rhythms regulated by the clock through its regulation of the sleep–wake cycle are almost always in a normal, and presumably adaptively significant, phase relationship with one another. An exception to this natural order often occurs in humans who can override the circadian control of the sleep–wake cycle for social or work-related reasons. In addition, there may be a dissociation of the circadian and sleep-related rhythms due to mental processes which prevent sleep at the normal circadian time. Regardless of why such a dissociation between the clock and the timing of sleep may occur, the consequences are the same: there is a high degree of internal temporal disorganization since those rhythms gated by the circadian clock are now in different phase relationships with those rhythms gated by the sleep–wake cycle. The situation is made even more complicated since many rhythms are gated by both the timing of the circadian pacemaker and the timing of sleep or wakefulness (210). This discussion of mismatched clock-time and sleep-time has focused on dyschronization between the circadian clock and the sleep–wake cycle in humans since we are the one species which routinely and voluntarily selects a life-style which leads to this dyschronization. However, there may well be cases in other species in which such disassociations also occur. Indeed, in advanced age the normal control of the circadian sleep–wake cycle may break down, leading to a spontaneous dyschronization between rhythms driven by the clock and the sleep cycle. While the importance of disrupted phase relationships of internal rhythms for age-related impairments of mental and physical processes remains to be determined, there is substantial evidence that disrupted rhythms are a hallmark of aging (see Aging of The Circadian Clock System, above).

ROLE OF CIRCADIAN CLOCK IN EXPRESSION OF OTHER BIOLOGICAL RHYTHMS

The circadian clock system plays a fundamental role in the expression of other biological rhythms with periods which are both shorter (*ultradian*) and longer (*infradian*) than 24 h. In addition, many annual rhythms in temperate zone species are regulated by the seasonal change in day length, and the circadian system plays a central role in the measurement of day length and the transfer of this information to other physiological systems. This review focuses only on the importance of circadian rhythms in the expression of ultradian, infradian, and annual rhythms. For more extensive coverage of the adaptive significance and the physiological mechanisms underlying the generation of these other biological rhythms, the reader is directed to other reviews (65, 120, 172, 203, 206, 211).

Ultradian Rhythms

Ultradian rhythms ranging in period from a few minutes to several hours have been well documented in many mammalian species (26, 172, 209, 211). Theoretically, ultradian rhythms may interrelate with the circadian system on two general levels: first, ultradian rhythms may be modulated by circadian signals and, second, common biological processes may underlie the generation of both circadian and ultradian rhythms. Perhaps the best examples of circadian modulation of ultradian rhythms are found in the human endocrine system. There is evidence that the phase, amplitude, and frequency of ultradian oscillations can be modulated by the circadian clock. Phase-setting effects of circadian

time-keeping occur for a wide variety of ultradian oscillations (8). For example, in humans the onset of sleep (itself gated by a circadian clock) has a phase-setting effect on the episodic variations of all hormonal secretions influenced by sleep, including prolactin, growth hormone, and pubertal luteinizing hormone (LH) (211). The concept of amplitude modulation of ultradian hormonal variations by circadian rhythmicity is also supported by strong experimental evidence. The magnitude of the sleep-associated increase in growth hormone (GH) and pubertal LH release depends on the circadian time at which sleep occurs (213). Computer simulations have suggested that the jagged 24 h profile of plasma cortisol results from a succession of pulses of adrenal secretion in which the magnitude is modulated by a circadian rhythm (212). In children approaching puberty the circadian rhythm of LH secretion is at least partially due to an amplification of the magnitude of nocturnal secretory pulses (77). There is less evidence to suggest a role for the circadian system in modulating the frequency of ultradian oscillations. However, in women studied at the beginning of the follicular phase there is a nocturnal slowing of LH pulsatility, indicating that at a specific phase point of the menstrual cycle an interaction between the ultradian rhythm of hormonal release and the circadian sleep–wake cycle may occur (180). While few studies have examined the effects of the circadian clock on ultradian hormonal release in rodents, circadian modulation of ultradian rhythms in locomotor activity has been observed in some rodents (56, 230).

There is little evidence in support of the hypothesis that common neural centers are involved in generating ultradian and circadian rhythms. Although SCN lesions disrupt or abolish most, if not all, circadian rhythms in rodents, such lesions do not disrupt the pulsatile pattern of pituitary LH release in rats (179) or ultradian rhythmicity in locomotor activity in voles (56). Although one study in rats did find that both ultradian and circadian rhythms of locomotor activity were abolished following lesioning of the SCN (230), most of the evidence suggests that different neural centers are responsible for generating circadian and ultradian signals.

Studies in invertebrates indicate that the same gene products may underlie the generation of rhythms, with frequencies ranging from one cycle in less than a minute to one cycle in 24 h (63), and it is tempting to speculate that the cellular and molecular mechanisms underlying the generation of circadian and ultradian rhythms may share common features. Whether pharmacological or genetic (for example, the tau mutant) changes in the period of the circadian clock will also induce similar changes in the period of ultradian oscillations remains to be determined. A major problem in addressing this question is the imprecise nature of the period of most ultradian rhythms (209, 211). Although it is often easy to detect period changes as small as 1% to 5% for circadian rhythms, the detection of similar changes in most ultradian rhythms would be difficult because of greater variability.

Infradian Rhythms

Rhythms with periods ranging from 24 h to 1 yr usually fall into two categories. First, in a variety of marine organisms there are pronounced semilunar (that is, about 14–15 days) and lunar (about 28–30 days) rhythms in both developmental processes and behavioral activities which are programmed to tidal conditions that are associated with the lunar cycle (122). While the circadian system is involved in the expression of lunar cycles, the lack of convincing evidence for such rhythms in mammals puts a discussion of these infradian rhythms outside the scope of this review. In contrast, infradian ovarian cycles with periods ranging from 4 to 28 days are very common in mammals, and, as described below, the circadian clock plays a role in these rhythms. While there have been some unusual claims for the appearance of rhythms with a period of about 7 days and even semiweekly rhythms (61, 176), these appear to be an artifact of the statistics used to extract such rhythms from the raw data (50).

Although specific events in the ovarian cycles of many mammals recur at intervals in the infradian range, there are substantial data from different rodent species demonstrating that various estrus-related events are linked to the circadian system. During the 1960s and 1970s many laboratories working primarily with rats, mice, and hamsters demonstrated that the timing of the proestrous surge in pituitary LH release, ovulation, the increase in progesterone secretion following ovulation, and the onset of sexual receptivity all occur at specific times of day (16, 111, 203). During exposure to constant darkness or constant light, these behavioral and endocrine rhythms continue to occur at specific circadian times. Even though the circadian-timed preovulatory LH surge occurs only once every 4 or 5 days in rats and hamsters, the neural signal for the LH surge can actually be generated every day since ovariectomized, estrogen-primed rats and hamsters show a daily release of LH at a time similar to that observed in proestrus intact animals.

Although there are no conclusive data demonstrating that the timing of the LH surge in non-rodent species is regulated by the circadian system, it should be noted that few comprehensive studies have actually addressed this question. Indeed, there is some evidence to suggest that women the preovulatory LH surge occurs in the majority of women in the early morning, indicating that circadian involvement in the ovarian cycle may be more widespread than previously thought (45, 177, 195), though contrary evidence has also been presented (66).

Annual Rhythms

For the vast majority of species, a multitude of adaptive mechanisms have evolved to cope with the often dramatic changes which occur in the physical environment on an annual basis. Some organisms make use of internal "circannual" clocks, which, like circadian clocks, can free-run under constant environmental conditions but normally are entrained by periodic signals in the environment (59, 60, 239). Other species rely exclusively on periodic changes in the physical environment to time their seasonal cycles, and for many species it is the seasonal change in day length which synchronizes seasonal rhythms to the annual change in the external environment (203). There is little evidence to implicate the circadian clock system in the regulation of endogenous circannual rhythms. Indeed, the circannual rhythm in hibernation in golden-mantled ground squirrels persists even after the SCN is destroyed (239). However, the circadian clock is involved in the timing of seasonal rhythms in many species by way of its central role in measuring the seasonal change in day length.

One of the most pronounced seasonal rhythms observed in the majority of vertebrate species inhabiting the temperate zones of the world is the annual rhythm in reproduction (52, 58, 65). For many species the young are born during specific times of the year when the probability of survival for both parents and offspring is maximal. The primary environmental cue for stimulating gonadal activity and reproductive behavior during the appropriate time of the year is the seasonal change in day length (52, 58, 203). For example, in many small rodents and birds the long days of spring and summer are stimulatory to reproductive activity, while the short days of fall and winter are inhibitory. The value of using the seasonal change in day length for the timing of the reproductive season is undoubtedly due to its reliability as a marker of the phase of the seasonal environmental cycle. While most studies on seasonal rhythms have focused on reproductive activity, a variety of other physiological and behavioral annual cycles have been well documented, including milk production, growth, pelage color, molting, migration, and hibernation. Similar to reproductive cycles, these rhythms are also often regulated by the seasonal change in day length.

Since it was first discovered that day length was a primary environmental signal for the regulation of annual rhythms (that is, the *photoperiodic response*), a great deal of attention has been directed toward the question: "How do living organisms measure the length of the day?" In addressing this question, many different species have been maintained under a variety of unusual light–dark cycles to determine whether the animals (or plants) would show a "long-day" or a "short-day" response to the light cycle. Studies on most photoperiodic vertebrate species exposed to cycles in which light was only present for a short period of time relative to the total amount of darkness have led to the inescapable conclusion that the circadian clock must somehow be involved in photoperiodic time measurement (52, 65, 186, 203). An example of one such experimental finding is shown in Figure 58.10. In this study Djungarian hamsters were exposed to only two 10 min pulses of light per 24 h day (99). The light pulses were separated by dark periods of 7 h 50 min and 15 h 50 min. Some of these nocturnal animals showed a short-day entrainment pattern (that is, locomotor activity was confined to the long period of darkness), whereas other animals showed a long-day entrainment pattern (that is, locomotor activity was confined to the short period of darkness). In this species long days are stimulatory and short days are inhibitory to testicular function. Testicular growth was induced in all animals showing a long-day

entrainment pattern, whereas the testes of animals showing a short-day entrainment pattern remained regressed and were similar in size to the testes of the control animals maintained on non-stimulatory short days. Thus, depending on the way the circadian clock entrains to a "skeleton" photoperiod, the same light cycle can be photostimulatory or inhibitory to testicular growth. The results of this and many other studies point to the importance of the phase relationship between the light–dark cycle and the circadian system in determining whether a given photoperiod will be interpreted as a long or short day. Thus there appears to be a circadian rhythm of sensitivity to light, and whether light is present or absent during those phases will determine if the animal responds as if it is summer or winter.

Soon after the SCN was implicated in the entrainment and generation of mammalian circadian rhythms, studies were performed to determine if the SCN was also involved in the measurement of day length. Early studies in the golden hamster demonstrated that bilateral destruction of the SCN abolishes the photoperiodic gonadal response and that lesioned animals are no

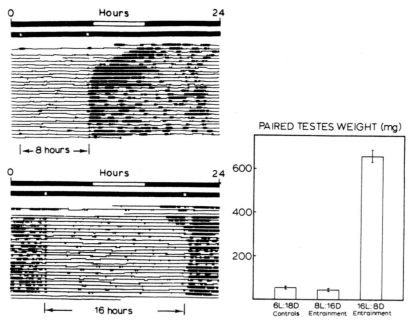

FIG. 58.10. Running wheel activity records from two Djungarian hamsters *(left)* and the mean paired testes weight *(right)* of groups of nine to ten hamsters exposed for 28 days to two 10 min pulses of light per 24 h. The bar at the top of each activity record indicates the light–dark 6:18 photoperiod (unfilled bar represents lights-on) to which the hamsters were exposed for about 21 days prior to transfer to the "skeleton" photoperiods. The second bar at the top of each activity record indicates the time at which the animals were exposed to the two 10 min pulses of light. Depending on the initial placement of the 10 min pulse of light relative to the previous 6:18 light–dark cycle, the activity rhythm of some of the animals was advanced for a few cycles until an 8:16 light–dark entrainment pattern was established (that is, activity confined to the 16 h interval of darkness) *(top left panel)*, whereas in other animals the activity rhythm was delayed for a few cycles until a 16L:8D entrainment pattern was established (that is, activity confined to the 8 h interval of darkness) *(lower left panel)*. Testes weight for control animals maintained on 6:18 light–dark are shown for comparative purposes. Reproduced with permission from reference 99.

longer capable of responding to the inhibitory effects of short days on gonadal function (165, 185). While the role of the SCN in the photoperiodic control of reproduction has been examined in only a few species, in all cases disruption of SCN function disrupts the effects of day length on reproductive activity. The circadian clock in the SCN mediates the effects of day length on reproductive function, at least in part, through its regulation of the diurnal rhythm in pineal melatonin production (70, 149, 150, 186). The synthesis and release of pineal melatonin are tightly coupled to the light–dark cycle such that circulating melatonin levels are high during the night and low during the day, and if the night is interrupted by even a brief period of light, circulating melatonin levels are rapidly depressed (70). Studies originally carried out in the Djungarian hamster and the sheep have established that it is the duration of the uninterrupted nighttime melatonin release that determines whether the photoperiod will be interpreted as a long or short day (58, 78). Studies indicate that the duration of the nighttime secretion of melatonin is also influenced by the seasonal change in day length in humans, though the physiological significance of these changes remains to be determined (221).

It should be noted that while a day length consisting of a short night (for example, 16 h light–8 h dark) and its associated short period of elevated melatonin levels is stimulatory to neuroendocrine–gonadal activity in hamsters, the same day length is inhibitory to reproductive function in short-day breeding sheep (58, 78). It is not known how the circadian melatonin signal is decoded by the brain for measuring day length, nor is it known how a melatonin duration signal is transduced into a change in neuroendocrine–gonadal activity. The finding that the SCN itself contains melatonin receptors (220) raises the possibility of an intricate feedback relationship between the SCN and the pineal gland in the photoperiodic control of seasonal cycles. However, this and other hypotheses on the mechanisms of melatonin's action on the reproductive system remain highly speculative since melatonin's action in the brain is so poorly understood.

SUMMARY

Changes in behavior that occur on a daily or circadian basis represent one of the most ubiquitous strategies by which most living organisms have adapted to the environment. Underlying the daily changes in behavior are a multitude of physiological and biochemical rhythms that provide adaptively significant temporal organization within the organism. In mammals there appears to be a central circadian clock in the hypothalamic SCN that is responsible for generating and coordinating the entire 24 h temporal organization of the animal. While the impact of disturbed circadian rhythms on the survival of the species has received very little attention, the almost universal presence of circadian rhythms in most physiological and biochemical systems argues in support of the hypothesis that a disruption of the normal circadian organization can have serious consequences for the health and well-being of the organism. It is particularly noteworthy that in advanced age various alterations in circadian rhythms have been observed and that these alterations may impair the ability of the animal to adapt normally to the environment. The study of circadian rhythms is a relatively new field of biology, and as a result, much remains to be discovered about the physiological mechanisms that underlie rhythmicity as well as the functional significance of 24 h temporal organization for the survival of the species.

NOTE ADDED IN PROOF

Since this chapter was written, we have reported a circadian clock single gene mutation in the mouse which lengthens the period of the circadian activity rhythm and which disrupts the ability of the mouse to maintain rhythmicity under conditions of constant darkness (219a). Efforts are under way to map the mouse gene, which has been named *clock*.

F.W.T. was supported by a Guggenheim Memorial Fellowship and an NIH Senior Fogarty International Fellowship, and O.V.R. is a fellow of the FNRS.

REFERENCES

1. Abe, H., B. Rusak, and H. A. Robertson. Photic induction of fos protein in the suprachiasmatic nucleus is inhibited by the NMDA receptor antagonist MK-801. *Neurosci. Lett.* 127: 9–12, 1991.
2. Albers, H. E., S.-Y. Liou, C. F. Ferris, E. G. Stopa, and R. T. Zoeller. Neurochemistry of circadian timing. In: *Suprachiasmatic Nucleus: The Mind's Clock,* edited by D. C. Klein, R. Y. Moore, and S. M. Reppert. New York: Oxford University Press, 1991, p. 263–288.
3. Aronin, N., S. M. Sagar, F. R. Sharp, and W. J. Schwartz. Light regulates expression of a fos-related protein in rat suprachiasmatic nuclei. *Proc. Natl. Acad. Sci. USA* 87: 5959–5962, 1990.
4. Aschoff, J. Exogenous and endogenous components in circadian rhythms. *Cold Spring Harb. Symp. Quant. Biol.* 25 11–18, 1960.
5. Aschoff, J. (Ed). *Biological Rhythms. Handbook of Behavioral Neurobiology.* New York: Plenum, 1981, vol. 4.
6. Aschoff, J., J. Figala, and E. Poppel. Circadian rhythms of locomotor activity in the golden hamster (*Mesocricetus auratus*) measured with two different techniques. *J. Comp. Physiol. Psychol.* 85: 20–28, 1973.
7. Aschoff, J., U. Gerecke, C. Von Goetz, G. A. Groos, and F. W. Turek. Phase responses and characteristics of free-running activ-

ity rhythms in the golden hamster: independence of the pineal gland. In: *Vertebrate Circadian Systems: Structure and Physiology,* edited by J. Aschoff, S. Daan, and G. Groos. Berlin: Springer-Verlag, 1982, p. 129–140.

8. Aschoff, J., and M. Gerkema. On diversity and uniformity of ultradian rhythms. In: *Ultradian Rhythms in Physiology and Behavior,* edited by H. Schulz and P. Lavie. Berlin: Springer-Verlag, 1985, p. 321–334.

9. Aschoff, J., and H. Pohl. Rhythmic variations in energy metabolism. *Federation Proc.* 29: 1541–1552, 1970.

10. Aschoff, J., U. V. Saint Paul, and R. Wever. Die lebensdauer von fliegen unter dem enifluss von zeit-verschiebungen. *Naturwissenschaften* 58: 574, 1971.

11. Aschoff, J., and R. Wever. Human circadian rhythms: a multioscillatory system. *Federation Proc.* 35: 2326–2332, 1976.

12. Bjorklund, A., and U. Stenevi. *Neural Grafting in the Mammalian CNS.* Amsterdam: Elsevier, 1985.

13. Borbely, A. A., P. Achermann, L. Trachsel, and I. Tobler. Sleep initiation and initial sleep intensity: interactions of homeostatic and circadian mechnisms. *J. Biol. Rhythms* 4: 149–160, 1989.

14. Brock, M. A. Chronobiology and aging. *J. Am. Geriatr. Soc.* 39: 74–91, 1991.

15. Büttner, D., and F. Wollnik. Strain-differentiated circadian and ultradian rhythms in locomotor activity of the laboratory rat. *Behav. Genet.* 14: 137–152, 1984.

16. Campbell, C. S., and F. W. Turek. Cyclic function in the mammalian ovary. In: *Biological Rhythms,* edited by J. Aschoff. New York: Plenum, 1981, p. 523–545.

17. Card, J. P., and R. Y. Moore. The organization of visual circuits influencing the circadian activity of suprachiasmatic nucleus. In: *Suprachiasmatic Nucleus: The Mind's Clock,* edited by D. C. Klein, R. Y. Moore, and S. M. Reppert. New York: Oxford University Press, 1991, p. 51–76.

18. Cassone, V. M. Melatonin and suprachiasmatic nucleus. In: *Suprachiasmatic Nucleus: The Mind's Clock,* edited by D. C. Klein, R. Y. Moore, and S. M. Reppert. New York: Oxford University Press, 1991, p. 309–323.

19. Colwell, C. S., R. G. Foster, and M. Menaker. NMDA receptor antagonists block the effects of light on circadian behavior in the mouse. *Brain Res.* 554: 105–110, 1991.

20. Comperatore, C. A., and F. K. Stephan. Entrainment of duodenal activity to periodic feeding. *J. Biol. Rhythms* 2: 227–242, 1987.

21. Connolly, M. S., and C. B. Lynch. Circadian variation of strain differences in body temperature and activity in mice. *Physiol. Behav.* 27: 1045–1049, 1981.

22. Czeisler, C. A., A. J. Chiasera, and J. F. Duffy. Research on sleep, circadian rhythms and aging: applications to manned spaceflight. *Exp. Gerontol.* 26: 217–232, 1991.

23. Czeisler, C. A., M. P. Johnson, J. F. Duffy, E. N. Brown, J. M. Ronda, and R. E. Kronauer. Exposure to bright light and darkness to treat physiologic maladaptation to night work. *N. Engl. J. Med.* 322: 1253–1259, 1990.

24. Czeisler, C. A., R. E. Kronauer, J. S. Allan, J. F. Duffy, M. E. Jewett, E. N. Brown, and J. M. Ronda. Bright light induction of strong (type 0) resetting of the human circadian pacemaker. *Science* 244: 1328–1333, 1989.

25. Daan, S. Adaptive daily strategies in behavior. In: *Biological Rhythms,* edited by J. Aschoff. New York: Plenum, 1981, p. 275–298.

26. Daan, S., and J. Aschoff. Short-term rhythms in activity. In: *Biological Rhythms,* edited by J. Aschoff. New York: Plenum, 1981, p. 491–498.

27. Daan, S., G. M. Beersma, and A. A. Borbely. Timing of human sleep: recovery process gated by a circadian pacemaker. *Am. J. Physiol.* 246(*Regulatory Integrative Comp. Physiol.* 17): R161–R178, 1984.

28. Daan, S., D. Masman, A. Strijkstra, and S. Verhulst. Intraspecific allometry of basal metabolic rate: relations with body size, temperature, composition, and circadian phase in the Kerstrel, *Falco tinnunculus. J. Biol. Rhythms* 4: 267–284, 1989.

29. Davis, F. C. Use of postnatal behavioral rhythms to monitor prenatal circadian function. In: *Development of Circadian Rhythmicity and Photoperiodism in Mammals,* edited by S. M. Reppert. Ithaca: Perinatology, 1989, p. 45–65.

30. Davis, F. C., and R. A. Gorski. Unilateral lesions of the hamster suprachiasmatic nuclei: evidence for redundant control of circadian rhythms. *J. Comp. Physiol.* 154: 221–232, 1984.

31. Davis, F. C., and R. A. Gorski. Development of hamster circadian rhythms: role of the maternal suprachiasmatic nucleus. *J. Comp. Physiol. A* 162: 601–610, 1988.

32. Davis, F. C., and J. Mannion. Entrainment of hamster pup circadian rhythms by prenatal melatonin injections to the mother. *Am. J. Physiol.* 255(*Regulatory Integrative Comp. Physiol.* 26): R439–R448, 1988.

33. Davis, F. C., and M. Menaker. Hamsters through time's window: temporal structure of hamster locomotor rhythmicity. *Am. J. Physiol.* 239(*Regulatory Integrative Comp. Physiol.* 10): R149–R155, 1980.

34. Davis, F. C., and M. Menaker. Development of the mouse circadian pacemaker: independence from environmental cycles. *J. Comp. Physiol.* 143: 527–539, 1981.

35. DeCoursey, P. J. Phase control of activity in a rodent. *Cold Spring Harb. Symp. Quant. Biol.* 25: 49–55, 1960.

36. Deguchi, T. Ontogenesis of a biological clock for serotonin acetyl coenzyme A N-acetyltransferase in pineal gland of rat. *Proc. Natl. Acad. Sci. USA* 72: 2914–2920, 1975.

37. Dunlap, J. C. Closely watched clocks. *Trends Genet.* 6: 135–143, 1990.

38. Earnest, D. J., M. Iadarola, H. H. Yeh, and J. A. Olschowka. Photic regulation of c-fos expression in neural components governing the entrainment of circadian rhythms. *Exp. Neurol.* 109: 353–361, 1990.

39. Earnest, D. J., and C. D. Sladek. Circadian rhythms of vasopressin release from individual rat suprachiasmatic explants in vitro. *Brain Res.* 382: 129–133, 1986.

40. Earnest, D. J., and C. D. Sladek. Circadian vasopressin release from perifused rat suprachiasmatic explants in vitro: effects of acute stimulation. *Brain Res.* 422: 398–402, 1987.

41. Earnest, D. J., and F. W. Turek. Neurochemical basis for the photic control of circadian rhythms and seasonal reproductive cycles: role for acetylcholine. *Proc. Natl. Aca. Sci. USA* 82: 4277–4281, 1985.

42. Ebihara, S., and K. Tsuji. Entrainment of the circadian activity rhythm to the light cycle: effective light intensity for Zeitgeber in the retinal degenerate C3H mouse and the normal C57BL mouse. *Physiol. Behav.* 24: 523–527, 1980.

43. Ebihara, S., K. Tsuji, and K. Kondo. Strain differences of the mouse's free-running circadian rhythm in continuous darkness. *Physiol. Behav.* 20: 795–799, 1978.

44. Edgar, D. M., T. S. Kilduff, C. E. Martin, and W. C. Dement. Influence of running wheel activity on free-running sleep/wake and drinking circadian rhythms in mice. *Physiol. Behav.* 50: 373–378, 1991.

45. Edwards, R. G. Test-tube babies. *Nature* 293: 253–256, 1981.

46. Elliott, J. A. Circadian rhythms and photoperiodic time measurement in mammals. *Federation Proc.* 35: 2339–2346, 1976.

47. Ellis, G. B., and F. W. Turek. Changes in locomotor activity associated with the photoperiodic response of the testes in male golden hamsters. *J. Comp. Physiol. A* 132: 277–284, 1979.

48. Enright, J. T. Ecological aspects of endogenous rhythmicity. *Annu. Rev. Ecol. Systemat.* 1: 221–238, 1970.

49. Enright, J. T. The circadian tape recorder and its entrainment. In: *Physiological Adaptation to the Environment,* edited by F. J. Verberg. New York: Intext, 1975, p. 465–476.

50. Enright, J. T. The parallactic view, statistical testing, and circular reasoning. *J. Biol. Rhythms* 4: 295–304, 1989.

51. Eskes, G. A. Functional significance of daily cycles in sexual behavior of the male golden hamster. In: *Vertebrate Circadian Systems: Structure and Physiology,* edited by J. Aschoff, S. Daan, and G. Groos. Berlin: Springer-Verlag, 1982, p. 347–353.

52. Follett, B. K., and D. E. Follett (Eds). *Biological Clocks in Seasonal Reproductive Cycles.* Bristol: Wright, 1981.

53. Foster, R. G., I. Provencio, D. Hudson, S. Fiske, W. De Grip, and M. Menaker. Circadian photoreception in the retinally degenerate mouse (rd/rd). *J. Comp. Physiol. [A]* 169: 39–50, 1991.

54. Fuller, C. A., F. M. Sulzman, and M. C. Moore-Ede. Thermoregulation is impaired in an environment without circadian time cues. *Science* 199: 794–796, 1978.

55. Gash, D. M., and J. R. Sladek, Jr. Functional and non-functional transplants: studies with grafted hypothalamic and preoptic neurons. *Trends Neurosci.* 7: 391–394, 1984.

56. Gerkema, M. P., and S. Daan. Ultradian rhythms in behavior: the case of the common vole *(Microtus arvalis).* In: *Ultradian Rhythms in Physiology and Behavior,* edited by H. Schulz and P. Lavie. Berlin: Springer-Verlag, 1985, p. 11–31.

57. Gillette, M. U. SCN electrophysiology in vitro: rhythmic activity and endogenous clock properties. In: *Suprachiasmatic Nucleus: The Mind's Clock,* edited by D. C. Klein, R. Y. Moore, and S. M. Reppert. New York: Oxford University Press, 1991, p. 125–143.

58. Goldman, B. D., and J. A. Elliott. Photoperiodism and seasonality in hamsters: role of the pineal gland. In: *Processing of Environmental Information in Vertebrates,* edited by M. H. Stetson. New York: Springer-Verlag, 1988, p. 203–218.

59. Gwinner, E. *Circannual Rhythms.* Berlin: Springer-Verglag, 1986.

60. Gwinner, E. Photoperiod as a modifying and limiting factor in the expression of avian circannual rhythms. *J. Biol. Rhythms* 4: 237–266, 1989.

61. Halberg, F., E. Halberg, F. Halberg, and F. Halberg. Circaseptan (about 7-day) and circasemiseptan (about 3.5-day) rhythms and contributions by Ladislav Derer. 2. Examples from botony, zoology and medicine. *Biologia (Bratislava)* 41: 233–252, 1986.

62. Halberg, J., E. Halberg, P. Regal, and F. Halberg. Changes with age characterize circadian rhythm in telemetered core temperature of stroke prone rats. *J. Gerontol.* 36: 28–30, 1981.

63. Hall, J. C. Genetics of circadian rhythms. *Annu. Rev. Genet.* 24: 659–697, 1990.

64. Hall, J. C., and M. Rosbach. Mutations and molecules influencing biological rhythms. *Annu. Rev. Neurosci.* 11: 373–393, 1988.

65. Hastings M. H. Neuroendocrine rhythms. *Pharmacol. Ther.* 50: 35–71, 1991.

66. Hoff, J. D., M. E. Quigley, and S. S. C. Yen. Hormonal dynamics at midcycle: a reevaluation. *J. Clin. Endocrinol. Metab.* 57: 792–796, 1983.

67. Honma, K., S. Honma, M. Kohsaka, and N. Fukuda. Seasonal variation in the human circadian rhythm: dissociation between sleep and temperature rhythm. *Am. J. Physiol.* 262(*Regulatory Integrative Comp. Physiol.* 33): R885–R891, 1992.

68. Hotz, M. M., M. S. Connolly, and C. B. Lynch. Adaptation to daily meal-timing and its effect on circadian temperature rhythms in two inbred strain of mice. *Behav. Genet.* 17: 37–51, 1987.

69. Hotz, M. M., J. J. Milette, J. S. Takahashi, and F. W. Turek. Spectral sensitivity of the circadian clock's response to light in Djungarian hamsters. *Soc. Res. Biol. Rhythms Abstr.* 1: 18, 1990.

70. Illnerova, H. The suprachiasmatic nucleus and rhythmic pineal melatonin production. In: *Suprachiasmatic Nucleus: The Mind's Clock,* edited by D. C. Klein, R. Y. Moore, and S. M. Reppert. New York: Oxford University Press, 1991, p. 197–216.

71. Ingram, D. K., E. D. London, and M. A. Reynolds. Circadian rhythms and sleep: effects of aging in laboratory rats. *Neurobiol. Aging* 3: 287–297, 1982.

72. Inouye, S. I. T., J. S. Takahashi, F. Wollnik, and F. W. Turek. Inhibitor of protein synthesis phase shifts a circadian pacemaker in the mammalian SCN. *Am. J. Physiol.* 255(*Regulatory Integrative Comp. Physiol.* 26): R1055–R1058, 1988.

73. Inouye, S. T., and H. Kawamura. Persistence of circadian rhythmicity in a mammalian hypothalamic island containing the suprachiasmatic nucleus. *Proc. Natl. Acad. Sci. USA* 76: 5962–5966, 1979.

74. Jacobs, G. H., J. Neitz, and J. F. Deegan. Retinal receptors in rodents maximally sensitive to ultraviolet light. *Nature* 353: 655–656, 1991.

75. Johnson, R., L. Smale, R. Y. Moore, and L. P. Morin. Lateral geniculate lesions block circadian phase-shifts responses to a benzodiazepine. *Proc. Natl. Acad. Sci. USA* 85: 5301–5304, 1988.

76. Joy, J. E. Protein differences in tau mutant hamsters: circadian clock proteins. *Soc. Neurosci. Abstr.* 17: 730, 1991.

77. Kapen, S., R. M. Boyar, S. Kapen, L. Hellman, and E. D. Weitzman. Effect of sleep–wake cycle reversal on luteinizing hormone secretory pattern in puberty. *J. Clin. Endocrinol. Metab.* 39: 293–299, 1974.

78. Karsch, F. J., C. J. I. Woodfill, B. Malpaux, J. E. Robinson, and N. L. Wayne. Melatonin and mammalian photoperiodism: synchronization of annual reproductive cycles. In: *Suprachiasmatic Nucleus: The Mind's Clock,* edited by D. C. Klein, R. Y. Moore, and S. M. Reppert. New York: Oxford University Press, 1991, p. 217–232.

79. Kittrell, E. M. W. The suprachiasmatic nucleus and temperature rhythms. In: *Suprachiasmatic Nucleus: The Mind's Clock,* edited by D. C. Klein, R. Y. Moore, and S. M. Reppert. New York: Oxford University Press, 1991, p. 233–245.

80. Klein, D. C., R. Y. Moore, and S. M. Reppert (Eds). *Suprachiasmatic Nucleus: The Mind's Clock.* New York: Oxford University Press, 1991.

81. Klein, D. C., R. Smoot, J. L. Weller, E. Higa, G. J. Creed, and D. M. Jacobwitz. Lesions of the paraventricular nucleus area of the hypothalamus disrupt the suprachiasmatic–spinal cord circuit in the melatonin rhythm generating system. *Brain Res. Bull.* 10: 647–652, 1983.

82. Konopka, R. J., and S. Benzer. Clock mutants of *Drosophila melanogaster. Proc. Natl. Acad. Sci. USA* 68: 2112–2116, 1971.

83. Kornhauser, J. M., D. E. Nelson, K. E. Mayo, and J. S. Takahashi. Photic and circadian regulation of c-Fos gene expression in the hamster suprachiasmatic nucleus. *Neuron* 5: 127–134, 1990.

84. Kornhauser, J. M., D. E. Nelson, K. E. Mayo, and J. S. Takahashi. Regulation of jun-B messenger RNA and AP-1 activity by light and a circadian clock. *Science* 255: 1581–1584, 1992.

85. Lehman, M. N., R. Silver, and E. L. Bittman. Anatomy of suprachiasmatic nucleus grafts. In: *Suprachiasmatic Nucleus: The Mind's Clock,* edited by D. C. Klein, R. Y. Moore, and S. M. Reppert. New York: Oxford University Press, 1991, p. 349–374.

86. Lehman, M. N., R. Silver, W. R. Gladstone, R. M. Kahn, M. Gibson, and E. L. Bittman. Circadian rhythmicity restored by

neural transplant: immunocytochemical characterization of the graft and its integration with the host brain. *J. Neurosci.* 7: 1626–1638, 1987.

87. Linkowski, P., A. Van Onderbergen, M. Kerkofs, D. Bosson, J. Mendlewicz, and E. Van Cauter. A twin study of the circadian and pulsatile variations of plasma cortisol: evidence for genetic control of the human circadian clock. *Am. J. Physiol.* 264(Endocrinol. Metab. 27): E173–E181, 1993.

88. Loros, J. J., S. A. Denome, and J. C. Dunlap. Molecular cloning of genes under control of the circadian clock in *Neurospora. Science* 243: 385–388, 1989.

89. Majzoub, J. A., B. G. Robinson, and R. L. Emanuel. Suprachiasmatic nuclear rhythms of vasopressin mRNA in vivo. In: *Suprachiasmatic Nucleus: The Mind's Clock,* edited by D. C. Klein, R. Y. Moore, and S. M. Reppert. New York: Oxford University Press, 1991, p. 177–190.

90. Martin, J. R., A. Fuchs, R. Bender, and J. Harting. Altered light–dark activity difference with aging in two rat strains. *J. Gerontol.* 44: 2–7, 1985.

91. McGinty, D., and N. Stern. Circadian and sleep-related modulation of hormone levels: changes with aging. In: *Endocrinology of Aging,* edited by J. R. Gower and J. L. Felicetta. New York: Raven, 1988, p. 75–111.

92. Meijer, J. H., and W. J. Rietveld. Neurophysiology of the suprachiasmatic circadian pacemaker in rodents. *Physiol. Rev.* 69: 671–707, 1989.

93. Menaker, M. The search for principals of physiological organization in vertebrate systems. In: *Vertebrate Circadian Systems,* edited by J. Aschoff, S. Daan, and G. Groos. Berlin: Springer-Verlag, 1982, p. 1–12.

94. Menaker, M. Extraretinal photoreception. In: *The Science of Photobiology,* edited by K. Smith. New York: Plenum, 1989, chapt. 8.

95. Menaker, M., and R. Refinetti. The tau mutation in golden hamsters. In: *Molecular Genetics of Biological Rhythms,* edited by M. Young. New York: Marcel-Dekker, p. 255–270, 1993.

96. Menaker, M., J. S. Takahashi, and A. Eskin. The physiology of circadian pacemakers. *Annu. Rev. Physiol.* 59: 501–526, 1978.

97. Menaker, M., and H. Underwood. Extraretinal photoreception in birds. *Photochem. Photobiol.* 23: 299–306, 1978.

98. Milette, J. J., M. M. Hotz, J. S. Takahashi, and F. W. Turek. Characterization of the wavelength of light necessary for initiation of neuroendocrine-gonadal activity in male Djungarian hamsters [Abstracts]. *Biol. Reprod.* 36 (suppl. 1): 110, 1987.

99. Milette, J. J., and F. W. Turek. Circadian and photoperiodic effects of brief light pulses in male Djungarian hamsters. *Biol. Reprod.* 35: 327–335, 1986.

100. Miller, A. E., and G. D. Riegle. Temporal patterns of serum luteinizing hormone and testosterone and endocrine response to luteinizing hormone in aging male rats. *J. Gerontol.* 37: 522–528, 1982.

101. Mistlberger, R., and B. Rusak. Mechanisms and models of the circadian timekeeping system. In: *Principles and Practice of Sleep Medicine,* edited by M. H. Kryger, T. Roth, and W. C. Dement. Philadelphia: Saunders, 1989, p. 141–152.

102. Monk, T. H. Sleep and circadian rhythms. *Exp. Gerontol.* 26: 233–243, 1991.

103. Monk, T. H., J. E. Fookson, M. Moline L., and C. P. Pollak. Diurnal variation in mood and performance in a time-isolated environment. *Chronobiol. Int.* 2: 185–193, 1985.

104. Moore, R. Y. Retinohypothalamic projection in mammals: a comparative study. *Brain Res.* 49: 403–409, 1973.

105. Moore, R. Y. Organization and function of a central nervous system circadian oscillator: the suprachiasmatic hypothalamic nucleus. *Federation Proc.* 42: 2783–2789, 1983.

106. Moore, R. Y. Development of the suprachiasmatic nucleus. In:

Suprachiasmatic Nucleus: The Mind's Clock, edited by D. C. Klein, R. Y. Moore, and S. M. Reppert. New York: Oxford University Press, 1991, p. 391–404.

107. Moore, R. Y., and V. B. Eichler. Loss of a circadian adrenal corticosterone rhythm following suprachiasmatic lesions in the rat. *Brain Res.* 42: 201–206, 1972.

108. Moore-Ede, M. C., F. M. Sulzman, and C. A. Fuller. *The Clocks That Time Us.* Cambridge: Harvard University Press, 1982.

109. Morin, L. P. Age-related changes in hamster circadian period, entrainment and rhythm splitting. *J. Biol. Rhythms* 3: 237–248, 1988.

110. Morin, L. P. Neural control of circadian rhythms as revealed through the use of benzodiazepines. In: *Suprachiasmatic Nucleus: The Mind's Clock,* edited by D. C. Klein, R. Y. Moore, and S. M. Reppert. New York: Oxford University Press, 1991, p. 324–338.

111. Morin, L. P., K. M. Fitzgerald, B. Rusak, and I. Zucker. Circadian organization and neural mediation of hamster reproductive rhythms. *Psychoneuroendocrinology* 2: 73–98, 1977.

112. Morin, L. P., K. M. Michels, L. Smale, and R. Y. Moore. Serotonin regulation of circadian rhythmicity. *Ann. N. Y. Acad. Sci.* 600: 418–426, 1990.

113. Mosko, S. S., G. F. Erickson, and R. Y. Moore. Dampened circadian rhythms in reproductively senescent female rats. *Behav. Neurol. Biol.* 28: 1–14, 1980.

114. Mosko, S. S., and R. Y. Moore. Neonatal suprachiasmatic nucleus lesions: effects on the development of circadian rhythms in the rat. *Brain Res.* 164: 17–38, 1979.

115. Mrosovsky, N. Phase response curves for social entrainment. *J. Comp. Physiol. [A].* 162: 35–46, 1988.

116. Mrosovsky, N., and P. A. Salmon. A behavioral method for accelerating re-entrainment of rhythms to new light–dark cycles. *Nature* 330: 372–373, 1987.

117. Mrosovsky, N., P. A. Salmon, M. Menaker, and M. R. Ralph. Non-photic phase-shifting in hamster clock mutants. *J. Biol. Rhythms* 7: 41–49, 1992.

118. Murakami, N. Long-term cultured neurons from rat SCN retain the capacity for circadian oscillation of vasopressin release. *Brain Res.* 545: 347–350, 1991.

119. Nelson, D. E., and J. S. Takahashi. Sensitivity and integration in a visual pathway for circadian entrainment in the hamster (*Mesocricetus auratus*). *J. Physiol. (Lond.)* 439: 1991.

120. Nelson, R. J., L. L. Badura, and B. D. Goldman. Mechanisms of seasonal cycles of behavior. *Annu. Rev. Psychol.* 41: 81–108, 1990.

121. Nelson, W., and F. Halberg. Schedule-shifts, circadian rhythms and lifespan of freely feeding and meal-fed mice. *Physiol. Behav.* 38: 781, 1986.

122. Neumann, D. Circadian components of semilunar and lunar timing mechanisms. *J. Biol. Rhythms* 4: 285–294, 1989.

123. Nicolau, G. Y., and S. Milcu. Circadian rhythm of corticosterone and nucleic acids in the rat adrenals in relation to age. *Chronobiologia* 4: 136, 1977.

124. Oksche, A. The development of the concept of photoneuroendocrine systems: historical perspective. In *Suprachiasmatic Nucleus: The Mind's Clock,* edited by D. C. Klein, R. Y. Moore, and S. M. Reppert. New York: Oxford University Press, 1991, p. 5–14.

125. Page, T. Neural and endocrine control of circadian rhythmicity in invertebrates. In: *Handbook of Behavioral Neurobiology: Biological Rhythms,* edited by J. Aschoff. New York: Plenum, 1981, p. 145–172.

126. Peng, M. T., M. J. Jiang, and H. K. Hsu. Changes in running-wheel activity, eating and drinking and their day/night distribution throughout the life span of the rat. *J. Gerontol.* 35: 339–347, 1980.

127. Peng, M. T., and M. Kang. Circadian rhythms and patterns of running-wheel activity, feeding and drinking behaviors of old male rats. *Physiol. Behav.* 33: 615–620, 1984.

128. Pickard, G. E., and F. W. Turek. Splitting of the circadian rhythm of activity is abolished by unilateral lesions of the suprachiasmatic nuclei. *Science* 215: 1119–1121, 1982.

129. Pickard, G. E., and F. W. Turek. The suprachiasmatic nuclei: two circadian clocks? *Brain Res.* 268: 201–210, 1983.

130. Pierpaoli, W., and G. J. M. Maestroni. Melatonin: a principal neuroimmunoregulatory and anti-stress hormone: its anti-aging effects. *Immunol. Lett.* 16: 355–362, 1987.

131. Pittendrigh, C. S. On temperature independence in the clock system controlling emergence in *Drosophila. Proc. Natl. Acad. Sci. USA* 40: 2697–2701, 1954.

132. Pittendrigh, C. S. Circadian rhythms and the circadian organization of living organisms. *Cold Spring Harb. Symp. Quant. Biol.* 25: 159–184, 1960.

133. Pittendrigh, C. S. Circadian oscillations in cells and the circadian organization of multicellular systems. In: *The Neurosciences Third Study Program,* edited by F. C. Schmitt and F. G. Worden. Cambridge: MIT, 1974, p. 437–458.

134. Pittendrigh, C. S. Circadian systems: entrainment. In: *Biological Rhythms,* edited by J. Aschoff. New York: Plenum, 1981, p. 95–124.

135. Pittendrigh, C. S. Circadian systems: general perspective. In: *Biological Rhythms,* edited by J. Aschoff. New York: Plenum, 1981, p. 57–80.

136. Pittendrigh, C. S. The photoperiodic phenomena: seasonal modulation of the day within. *J. Biol. Rhythms* 3: 173–188, 1988.

137. Pittendrigh, C. S., and S. Daan. Circadian oscillations in rodents: a systematic increase of their frequency with age. *Science* 186: 548–550, 1974.

138. Pittendrigh, C. S., and D. H. Minis. Circadian systems: longevity as a function of circadian resonance in *Drosophila melanogaster. Proc. Natl. Acad. Sci. USA* 69: 1537–1539, 1972.

139. Possidente, B., and J. P. Hegmann. Circadian complexes: circadian rhythms under common gene control. *J. Comp. Physiol. [B]* 139: 121–125, 1980.

140. Prinz, P. N., E. R. Peskind, P. P. Vitaliano, M. A. Raskind, C. Eisdorfer, N. Zemcuznikov, and C. J. Gerber. Changes in the sleep and waking EEGs of nondemented and demented elderly subjects. *J. Am. Geriatr. Soc.* 30: 86–93, 1982.

141. Ralph, M., R. G. Foster, F. C. Davis, and M. Menaker. Transplanted suprachiasmatic nucleus determines circadian period. *Science* 247: 975–978, 1990.

142. Ralph, M. R. Suprachiasmatic nucleus transplant studies using the tau mutation in golden hamsters. In: *Suprachiasmatic Nucleus: The Mind's Clock,* edited by D. C. Klein, R. Y. Moore, and S. M. Reppert. New York: Oxford University Press, 1991, p. 341–348.

143. Ralph, M. R., and M. N. Lehman. Transplantation: a new tool in the analysis of the mammalian hypothalamic circadian pacemaker. *TINS* 14: 362–366, 1991.

144. Ralph, M. R., and M. Menaker. A mutation of the circadian system in golden hamster. *Science* 241: 1225–1227, 1988.

145. Rea, M. A. Light increases fos-related protein immunoreactivity in the rat suprachiasmatic nuclei. *Brain Res.* 23: 577–581, 1989.

146. Redman, J., S. Armstrong, and K. T. Ng. Free-running activity rhythms in the rat: entrainment by melatonin. *Science* 219: 1089–1091, 1983.

147. Reebs, S., and N. Mrosovsky. Effects of induced wheel running on the circadian activity rhythms of Syrian hamsters: entrainment and phase response curve. *J. Biol. Rhythms* 4: 39–48, 1989.

148. Reebs, S., and N. Mrosovsky. Running activity mediates the phase-advancing effects of dark pulses on hamster circadian rhythms. *J. Comp. Physiol. [A].* 165: 811–818, 1989.

149. Reiter, R. J. Neuroendocrine effects of the pineal gland and of melatonin. In: *Frontiers in Neuroendocrinology,* edited by W. Ganong and L. Martini. New York: Raven, 1982, p. 287–316.

150. Reiter, R. J. Pineal melatonin: cell biology of its synthesis and of its physiological interactions. *Endocr. Rev.* 12: 151–180, 1991.

151. Reiter, R. J., C. M. Craft, J. E. Johnson, Jr., T. S. King, B. A. Richardson, G. M. Vaughan, and M. K. Vaughan. Age associated reduction in nocturnal pineal melatonin levels in female rats. *Endocrinology* 109: 1295–1297, 1981.

152. Reppert, S. M., and W. J. Schwartz. Maternal coordination of the fetal biological clock in utero. *Science* 220: 969–971, 1983.

153. Reppert, S. M., and W. J. Schwartz. Maternal extirpations do not abolish maternal coordination of the fetal circadian clock. *Endocrinology* 119: 1763–1767, 1986.

154. Reppert, S. M., and W. Schwartz. Arginine vasopressin: a novel peptide rhythm in cerebrospinal fluid. *Trends Neurosci.* 10: 76–80, 1987.

155. Reppert, S. M., and D. R. Weaver. A biological clock is oscillating in the fetal suprachiasmatic nuclei. In: *Suprachiasmatic Nucleus: The Mind's Clock,* edited by D. C. Klein, R. Y. Moore, and S. M. Reppert. New York: Oxford University Press, 1991, p. 405–418.

156. Richardson, G. S. Circadian rhythms and aging. In: *Handbook of the Biology of Aging* (3rd ed.), edited by Edward L. Schneider, John W. Rowe. San Diego: Academic Press, 1990, p. 275–305.

157. Rosbash, M., and J. C. Hall. The molecular biology of circadian rhythms. *Neuron* 3: 387–398, 1989.

158. Rosenberg, R., P. Zee, and F. W. Turek. Phase response curve to light in young and old hamsters. *Am. J. Physiol.* 261(*Regulatory Integrative Comp. Physiol.* 32): R491–R495, 1991.

159. Rosenberg, R. S., H. Zepelin, and A. Rechtschaffen. Sleep in young and old rats. *J. Gerontol.* 34: 525–532, 1979.

160. Rosenthal, N. E., D. A. Sack, C. J. Carpenter, B. L. Parry, W. B. Mendelson, and T. A. Wehr. Seasonal affective disorder. *Arch. Gen. Psychiatry* 41: 72–80, 1984.

161. Rosenwasser, A. M. Behavioral neurobiology of circadian pacemakers: a comparative perspective. In: *Progress in Psychobiology and Physiological Psychology,* edited by A. N. Epstein and A. R. Morrison. New York: Academic, 1988, p. 155–226.

162. Rosenwasser, A. M., and N. T. Adler. Structure and function in circadian timing systems: evidence for multiple coupled circadian oscillators. *Neurosci. Behav. Rev.* 10: 431–448, 1986.

163. Rusak, B., and K. G. Bina. Neurotransmitters in the mammalian circadian system. *Annu. Rev. Neurosci.* 13: 387–401, 1990.

164. Rusak, B., and Z. Boulos. Pathways for the photic entrainment of mammalian circadian rhythms. *Photochem. Photobiol.* 34: 267–273, 1981.

165. Rusak, B., and L. P. Morin. Testicular response to photoperiod are blocked by lesions of the suprachiasmatic nuclei in golden hamsters. *Biol. Reprod.* 15: 366–374, 1976.

166. Rusak, B., H. A. Robertson, W. Wisden, and S. P. Hunt. Light pulses that shift rhythms induce gene expression in the suprachiasmatic nucleus. *Science* 248: 1237–1240, 1990.

167. Rusak, B., and I. Zucker. Biological rhythms and animal behavior. *Annu. Rev. Psychol.* 26: 137–171, 1975.

168. Rusak, B., and I. Zucker. Neural regulation of circadian rhythms. *Physiol. Rev.* 59: 449–526, 1979.

169. Sacher, G. A., and P. H. Duffy. Age changes in rhythms of energy metabolism, activity and body core temperature in *Mus musculus* and *Peromyscus.* In: *Aging and Biological Rhythms,* edited by H. V. Samis and A. Capobianio. New York: Plenum Press, 1978, p. 105–124.

170. Saint Paul, U. V., and J. Aschoff. Longevity among blowflies

Phormia terraenovae R. D. kept in non-24 hour light–dark cycles. *J. Comp. Physiol. A* 127: 191, 1978.

171. Satlin, A., M. H. Teicher, H. R. Lieberman, R. J. Baldessarini, L. Volicer, and Y. Rheaume. Circadian locomotor activity rhythms in Alzheimer's disease. *Neuropsychopharmacology* 5: 115–126, 1991.

172. Schulz, H., and P. Lavie (Eds). *Ultradian Rhythms in Physiology and Behavior.* Berlin: Springer-Verlag, 1985.

173. Schwartz, W. J. SCN metabolic activity in vivo. In: *Suprachiasmatic Nucleus: The Mind's Clock,* edited by D. C. Klein, R. Y. Moore, and S. M. Reppert. New York: Oxford University Press, 1991, p. 144–156.

174. Schwartz, W. J., and H. Gainer. Suprachiasmatic nucleus: use of ^{14}C-labeled deoxyglucose uptake as a functional marker. *Science* 197: 1089–1091, 1977.

175. Schwartz, W. J., and P. Zimmerman. Circadian timekeeping in BALB/c and C57 BL/6 inbred mouse strains. *J. Neurosci.* 10: 3685–3694, 1990.

176. Schweiger, H.-G., S. Berger, H. Kretschmer, H. Moerler, E. Halberg, R. B. Southern, and F. Halberg. Evidence for a circaseptan and a circasemisptan growth response to light/dark cycle shifts in nucleated and enucleated Acetabularia cells, respectively. *Proc. Natl. Acad. Sci. USA* 83: 8619–8623, 1989.

177. Seibel, M. M., W. Shine, D. M. Smith, and M. L. Taymor. Biological rhythm of the luteinizing hormone surge. *Fertil. Steril.* 37: 709–711, 1982.

178. Sharma, M., J. Palacios-Bios, G. Schwartz, H. Iskandan, J. Thakus, R. Quinion, and N. Niv. Circadian rhythms of melatonin and cortisol in aging. *Biol. Psychol. Psychiatry* 25: 305–319, 1989.

179. Soper, B. D., and R. F. Weick. Hypothalamic and extrahypothalamic mediation of pulsatile discharges of luteinizing hormone in the ovariectomized rat. *Endocrinology* 106: 348–355, 1980.

180. Soules, M. R., R. A. Steiner, N. L. Cohen, W. J. Bremner, and D. K. Clifton. Nocturnal slowing of pulsatile luteinizing hormone secretion in women during the follicular phase of the menstrual cycle. *J. Clin. Endocrinol. Metab.* 61: 43–49, 1985.

181. Stephan, F. K. Limits of entrainment to periodic feeding in rats with suprachiasmatic lesions. *J. Comp. Physiol. A* 143: 401–410, 1981.

182. Stephan, F. K. Forced dissociation of activity entrained to T cycles of food access in rats with suprachiasmatic lesions. *J. Biol. Rhythms* 4: 467–480, 1989.

183. Stephan, F. K., J. M. Swann, and C. L. Sisk. Anticipation of 24-hr feeding schedules in rats with lesions of the suprachiasmatic nucleus. *Behav. Biol.* 25: 346–363, 1979.

184. Stephan, F. K., and I. Zucker. Circadian rhythm in drinking behavior and locomotor activity of rats are eliminated by hypothalamic lesions. *Proc. Natl. Acad. Sci. USA.* 69: 1583–1586, 1972.

185. Stetson, M. H., and M. Watson-Witmyre. Nucleus suprachiasmaticus: the biological clock of the hamster? *Science* 191: 197–199, 1976.

186. Stetson, M. H., and M. Watson-Whitmyre. Physiology of the pineal and is hormone melatonin in annual reproduction in rodents. In: *The Pineal Gland,* edited by R. J. Reiter. New York: Raven, 1984, p. 109–154.

187. Swaab, D. F., B. Fisser, W. Kamphorst, and D. Troust. The human suprachiasmatic nucleus: neuropeptide changes in senium and Alzheimer's disease. *Basic Appl. Histochem.* 32: 43–54, 1988.

188. Swann, J. M., and F. W. Turek. Multiple circadian oscillators regulate the timing of behavioral and endocrine rhythms in female golden hamsters. *Science* 228: 898–900, 1985.

189. Takahashi, J. S., N. Murakami, S. S. Nlkaido, B. L. Pratt, and L. M. Robertson. The avian pineal, a vertebrate model system of the circadian oscillator: cellular regulation of circadian rhythms by light, second messengers, and macromolecular synthesis. *Recent Prog. Horm. Res.* 45: 279–352, 1989.

190. Takahashi, J. S., and F. W. Turek. Anisomycin, an inhibitor of protein synthesis, perturbs the phase of a mammalian circadian pacemaker. *Brain Res.* 405: 199–203, 1987.

191. Takahashi, J. S., and M. Zatz. Regulation of circadian rhythmicity. *Science* 217: 1102–1111, 1982.

192. Takeuchi, J., H. Nagasake, K. Shinohara, and S. T. Inouye. A circadian rhythm of somatostatin messenger RNA levels, but not vasoactive intestinal polypeptide/peptide histidine isoleucine messenger RNA levels in rat suprachiasmatic nucleus. *Mol. Cell. Neurosci.* 3: 29–35, 1992.

193. Tang, F., M. Hadjiconstantinov, and S. F. Pang. Aging and diurnal rhythms of pineal serotonin, 5-hydroxy-indoleacetic acid, norepinephrine, dopamine and serum melatonin in the rat. *Neuroendocrinology* 40: 160–164, 1985.

194. Tenover, J. S., A. M. Matsumoto, D. K. Clifton, and W. J. Bremner. Age-related alterations in the circadian rhythm of pulsatile luteinizing hormone and testosterone secretion in healthy men. *J. Gerontol.* 43: 163–169, 1988.

195. Testart, J., R. Frydman, and M. Roger. Seasonal influence of diurnal rhythms in the onset of the plasma leutinizing hormone surge in women. *J. Clin. Endocrinol. Metab.* 55: 374–377, 1982.

196. Turek, F. W. Circadian neural rhythms in mammals. *Annu. Rev. Physiol.* 47: 49–64, 1985.

197. Turek, F. W. Pharmacological probes of the mammalian circadian clock: use of the phase response curve approach. *Trends Pharmacol. Sci.* 8: 212–217, 1987.

198. Turek, F. W. Effects of stimulated activity on the circadian pacemaker of vertebrates. *J. Biol. Rhythms* 4: 135–147, 1989.

199. Turek, F. W., D. J. Earnest, and J. Swann. Splitting of the circadian rhythm of activity in hamsters. In: *Vertebrate Circadian Systems,* edited by J. Aschoff, S. Daan, and G. Groos. Berlin: Springer-Verlag, 1982, p. 203–214.

200. Turek, F. W., and S. Losee-Olson. The circadian rhythm of LH release can be shifted by injections of a benzodiazepine in female golden hamsters. *Endocrinology* 122: 756–758, 1988.

201. Turek, F. W., J. P. McMillan, and M. Menaker. Melatonin: effects on the circadian locomotor rhythm of sparrows. *Science* 194: 1441–1443, 1976.

202. Turek, F. W., R. Smith, O. Van Reeth, and C. Wickland. Disturbances of the activity rest cycle alter the circadian clock of mammals. In: *Endogenous Sleep Factors,* edited by S. Inouye and J. M. Krieger. The Hague: SPB Academic, 1990, p. 277–283.

203. Turek, F. W., and E. Van Cauter. Rhythms in reproduction. In: *The Physiology of Reproduction,* edited by E. Knobil and J. Neill. New York: Raven, 1988, p. 1789–1830.

204. Turek, F. W., and O. Van Reeth. Neural and pharmacological control of circadian rhythms. In: *Hormone and Cell Regulation,* edited by J. Nunez, J. E. Dumont, and R. Denton. Libbey, 1989, p. 95–101.

205. Uhl, G. R., and S. M. Reppert. Suprachiasmatic nucleus vasopression messenger RNA: circadian variation in normal and Brattleboro rats. *Science* 232: 390–393, 1986.

206. Underwood, H., and B. D. Goldman. Vertebrate circadian and photoperiodic systems: role of the pineal gland and melatonin. *J. Biol. Rhythms* 2: 279–315, 1987.

207. Underwood, H., and G. A. Groos. Vertebrate circadian rhythms: retinal and extraretinal photoreception. *Experientia* 38: 1113–1121, 1982.

208. Van Cauter, E. Physiology and pathology of circadian rhythms. In: *Recent Advances in Endocrinology and Metabolism,* edited

by C. W. Edwards and D. W. Lincoln. Edinburgh: Churchill Livingstone, 1989, p. 109–134.

209. Van Cauter, E. Diurnal and ultradian rhythms in human endocrine function: a minireview. *Horm. Res.* 34: 45–53, 1990.

210. Van Cauter, E. Hormonal and metabolic changes during sleep. *Int. Acad. Biomed. Drug Res.* 3: 95–108, 1992.

211. Van Cauter, E., and F. W. Turek. *Endocrine and Other Biological Rhythms*, edited by L. J. DeGroot. Philadelphia: W. B. Saunders, 1995, p. 2487–2548.

212. Van Cauter, E., and E. Honickx. The pulsatility of pituitary hormones. In: *Ultradian Rhythms in Physiology and Behavior,* edited by H. Schulz and P. Lavie. Berlin: Springer-Verlag, 1985, p. 41–60.

213. Van Cauter, E., and S. Refetoff. Multifactorial control of the 24-hour secretory profiles of pituitary hormones. *J. Endocrinol. Rev.* 8: 381–391, 1985.

214. Van Coevorden, A., E. Laurent, G. C. Decoster, M. Kerkhoffs, P. Neve, E. Van Cauter, and J. Mockel. Decreased basal circadian and TRH stimulated thyrotropin secretion in healthy aging men with conserved acute secretory response of the thyroid to endogenous TSH. *J. Clin. Endocrinol. Metab.* 69: 177–185, 1989.

215. Van Gool, W. A., and M. Mirmiran. Effects of aging and housing in an enriched environment upon sleep–wake patterns in rats. *Sleep* 9: 335–347, 1986.

216. Van Gool, W. A., W. Witting, and M. Mirmirian. Age-related changes in circadian sleep–wakefulness rhythms in male rats isolated from time cues. *Brain Res.* 413: 384–387, 1987.

217. Van Reeth, O., D. Hinch, J. M. Tecco, and F. W. Turek. The effects of short periods of immobilization on the hamster circadian clock. *Brain Res.* 545: 208–214, 1991.

218. Van Reeth, O., and F. W. Turek. Stimulated activity mediates phase shifts in the hamster circadian clock induced by dark pulses or benzodiazepines. *Nature* 339: 49–51, 1989.

219. Van Reeth, O., Zhang, Y., Zee, P. C., and F. W. Turek. Aging alters the feedback effects of the activity–rest cycle on the circadian clock. *Am. J. Physiol.* 263(*Regulatory Integrative Comp. Physiol.* 32): R981–R986, 1992.

220. Weaver, D. R., S. A. Rivkees, L. L. Carlson, and S. M. Reppert. Localization of melatonin receptors in mammalian brain. In: *Suprachiasmatic Nucleus: The Mind's Clock,* edited by D. C. Klein, R. Y. Moore, and S. M. Reppert. New York: Oxford University Press, 1991, p. 289–308.

221. Wehr, T. A. The duration of human melatonin secretion and sleep respond to changes in daylength (photoperiod). *J. Clin. Endocrinol. Metab.* 73: 1276–1280, 1991.

222. Wehr, T. A., and A. A. Wirz-Justice. Circadian rhythm mechanisms in affective illness and in antidepressant drug action. *Pharmacopsychiatry* 15: 31–39, 1982.

223. Weiland, N. G., and P. M. Wise. Aging progressively decreases the densities and alters the diurnal rhythms of alpha-1 adrenergic receptors in selected hypothalamic regions. *Endocrinology* 126: 2392–2397, 1990.

224. Welsh, D. K., G. S. Richardson, and W. C. Dement. Effect of age on the circadian pattern of sleep and wakefulness in the mouse. *J. Gerontol.* 41: 579–586, 1986.

225. Wever, R. Zur zeitgeber-staerke eines licht-dunkel-wechels fuer die circadiane periodik des menschen. *Eur. J. Physiol.* 321: 133–142, 1970.

226. Wickland, C., and F. W. Turek. Phase-shifting effect of triazolam on the hamster's circadian rhythm of activity is not mediated by a change in body temperature. *Brain Res.* 560: 12–16, 1991.

227. Wickland, C., and F. W. Turek. Phase-shifting effects of acute increases in activity on circadian locomotor rhythms in hamsters. *Am. J. Physiol.* 261(*Regulatory Integrative Comp. Physiol.* 32): R1109–R1117, 1991.

228. Wise, P. M., I. R. Cohen, N. G. Weiland, and D. E. London. Aging alters the circadian rhythm of glucose utilization in the suprachiasmatic nucleus. *Proc. Natl. Acad. Sci. USA.* 85: 5305–5309, 1988.

229. Wise, P. M., R. C. Walovitch, I. R. Cohen, N. G. Weiland, and D. E. London. Diurnal rhythmicity and hypothalamic deficits in glucose utilization in aged ovariectomized rats. *J. Neurosci.* 7: 3469–3473, 1987.

230. Wollnik, F., and F. W. Turek. SCN lesions abolish ultradian and circadian components of activity rhythms in LEW/Ztm rats. *Am. J. Physiol.* 256(*Regulatory Integrative Comp. Physiol.* 27): R1025–R1039, 1989.

231. Wollnik, F., F. W. Turek, P. Majewski, and J. S. Takahashi. Phase shifting the circadian clock with cycloheximide: response of hamsters with an intact or split rhythm of locomotor activity. *Brain Res.* 496: 82–88, 1989.

232. Yamada, N., K. Shimoda, K. Ohi, K. Takahashi, and S. Takahashi. Free-access to a running wheel shortens the period of the free-running rhythm in blinded rats. *Physiol. Behav.* 42: 87–91, 1988.

233. Young, M. W., T. A. Bargiello, M. K. Baylies, L. Saez, and D. C. Spray. Molecular biology of the *Drosophila* clock. In: *Neuronal and Cellular Oscillators,* edited by J. W. Jacklet. New York: Dekker, 1989, p. 529–542.

234. Young, M. W., F. R. Jackson, H. S. Shin, and T. A. Bargiello. A biological clock in *Drosophila*. *Cold Spring Harb. Symp. Quant. Biol.* 50: 865–875, 1985.

235. Zatz, M. Photoentrainment, pharmacology and phase shifts of the circadian rhythm in the rat pineal. *Federation Proc.* 38: 2596–2601, 1979.

236. Zee, P. C., R. S. Rosenberg, and F. W. Turek. Effects of aging on entrainment and rate of resynchronization of the circadian rhythm of locomotor activity in hamsters. *Am. J. Physiol.* 263(*Regulatory Integrative Comp. Physiol.* 32): R1099–R1103, 1992.

237. Zlomanczuk, P., R. R. Margraf, and G. R. Lynch. In vitro electrical activity in the suprachiasmatic nucleus following splitting and masking of wheel-running behavior. *Brain Res.* 559: 94–99, 1991.

238. Zucker, I. Motivation, biological clocks, and temporal organization of behavior. In: *Handbook of Behavioral Neurobiology,* edited by E. Satinoff and P. Teitelbaum. New York: Plenum, 1983, p. 3–21.

239. Zucker, I., T. M. Lee, and J. Dark. The suprachiasmatic nucleus and annual rhythms of mammals. In: *Suprachiasmatic Nucleus: The Mind's Clock,* edited by D. C. Klein, R. Y. Moore, and S. M. Reppert. New York: Oxford University, 1991, p. 246–260.

59. Sleep, thermoregulation, and circadian rhythms

H. CRAIG HELLER — Department of Biological Sciences, Stanford University, Stanford, California

DALE M. EDGAR — Sleep Research Center, Stanford University School of Medicine, Stanford, California

DENNIS A. GRAHN — Department of Biological Sciences, Stanford University, Stanford, California

STEVEN F. GLOTZBACH — Department of Pediatrics, Stanford University School of Medicine, Stanford, California.

CHAPTER CONTENTS

STRONG INTERACTIONS BETWEEN SLEEP, regulation of body temperature, and circadian rhythmicity are known to occur in birds and mammals (38). In assessing any one of these systems as an adaptation to the environment, it is necessary to consider those interactions. For example, most endotherms have a daily cycle of rest and activity, with body temperature regulated at a lower level during the rest phase. Sleep is also more prevalent during the rest phase of the cycle, and body temperature is regulated at a lower level during sleep than wakefulness. Therefore, if we wish to investigate daily lowering of body temperature (that is, shallow torpor) as a possible energy-conserving adaptation, we must examine not only the thermoregulatory system of the species but also the circadian and sleep-control systems. In shallow torpor, for example, has selection modified in this species the thermoregulatory system, the sleep influence on the thermoregulatory system, the circadian influence on the thermoregulatory system, and/or the circadian influence on sleep?

Is sleep an adaptation to the environment? Daily cycles of activity and inactivity as seen in most species are adaptive in that they organize the behavior and physiology of the animal according to the environmental cycle of light and dark (see the chapter by Turek and van Reeth in this *Handbook*). Sleep mostly occurs during the inactive phase of the cycle. Is sleep an environmental adaptation beyond the associated inactivity? We

are limited to considering sleep in birds and mammals only because we do not know if the daily rest periods of other animals are homologous with electrophysiologically defined mammalian and avian sleep. Sleep seems to be essential to endotherms regardless of its temporal organization. The guinea pig, a crepuscular animal with a very weak daily rhythm of rest and activity, has a total daily amount of sleep not very different from mammals with strong diurnal rest–wake cycles (115). Since we do not know the functions of sleep, whether or not electrophysiologically defined states of sleep are adaptations to the environment is a difficult question to answer. The predominant sleep state, non-rapid eye movement sleep (NREMS), exists in a homeostatic relationship with wakefulness (6, 7). This suggests that NREMS is a restorative process that prepares an organism for the subsequent period of wakefulness. Thus we might expect that the timing of NREMS, at least, could be an adaptation to the environment and should show differences related to the activity patterns of the species.

Several hypotheses suggest that sleep states may have evolved as direct adaptations to the environment. Berger (2) has proposed that NREMS evolved in parallel with endothermy for the adaptive purpose of energy conservation. This hypothesis derives support from studies of the thermoregulatory changes that occur during NREMS and the fact that shallow and deep torpor appear to be adaptive extensions of NREMS mechanisms. Other hypotheses have ascribed specific thermoregulatory functions to NREMS or to rapid eye movement sleep (REMS). McGinty and Szymusiak (76) proposed that an important function of NREMS is brain cooling and that the heat load resulting from waking metabolism serves as a feedback signal controlling the expression of NREMS. The evidence supporting this hypothesis is indirect and derived largely from observed interactions between sleep and thermoregulation. The hypothesis is not supported by correlations between body size, thermal niche, and sleep quantity. Small endotherms can lose heat much more rapidly than can large ones, and species in cold habitats can dissipate heat more readily than species in hot habitats. One might expect from the brain cooling hypothesis that large animals should sleep more than small animals and that animals in cold environments should sleep less than animals in warm environments. In contrast, the correlation between sleep duration and body size is negative (131), and animals that show torpor as an energy-conserving adaptation in cold climates have increased sleep time during the cold season (46, 122).

Wehr (127) suggested that REMS serves to heat the brain. Much circumstantial evidence derived from interactions of sleep and thermoregulation has been mustered in support of this idea, but it also has vulnerable points. As will be discussed below (see INFLUENCES OF SLEEP ON THERMOREGULATION), thermoregulation is severely inhibited during REMS. When animals sleep under conditions most likely to decrease brain and body temperatures (namely, at cold ambient temperatures), the amount of REMS decreases. In fact, the amount of REMS is maximal at the upper region of the thermoneutral zone (98, 110) and increases with mild brain warming (98). These are not the observations that might be expected if the function of REMS were to warm the brain. Thermoregulatory hypotheses about the functions of sleep have been reviewed by Horne (56), who concurs that they remain debatable. Nevertheless, research spawned by these hypotheses continues to contribute valuable information about interactions between sleep and autonomic functions.

Without resolving the question of the adaptive nature of sleep, we can recognize that the thermal and chronobiological environments have strong influences on sleep and that sleep profoundly impacts an animal's responses to its environment. Answers to questions of sleep function may come from investigations of the brain mechanisms underlying sleep and sleep control. For example, if we discover the feedback signal that relates NREM sleep to preceding wakefulness, it may tell us something about the restorative function of NREM sleep. Insights into mechanisms frequently are advanced from studies of systems interactions and of the adaptive differences between species. This chapter is not a comprehensive review of these subjects but presents a discussion of some of the general concepts of systems interactions from a variety of studies on different species.

INFLUENCES OF SLEEP ON THERMOREGULATION

Even before polygraphic differentiation of sleep states, changes in body temperatures and metabolic rate at sleep onset in humans suggested a decrease in regulated body temperature during sleep. The fact that these changes were observed in subjects who fasted, were confined to bed, or were paralyzed argued that they were not simple consequences of decreased digestive and/or motor activity. A large number of studies that measured thermoregulatory responses (such as vasomotor activity, sweating, panting, and shivering) as a function of the arousal state in humans and other mammals support the general conclusion that body temperature is regulated at a lower level during NREMS than during wakefulness (for extensive reviews see refs. 38, 48, 88). Decreases in brain temperature always accompany electroencephalogram (EEG)-defined transitions from wakefulness to NREMS in the rat (32, 82). Additionally, Franken et al. (33) correlated brain temperature with percent of time spent awake over 24 h periods in rats and concluded that the wake–sleep distribution accounted for 84% of the variance in brain temperature. The residual variance showed a 24 h periodicity presumably due to a circadian influence. This study may overestimate the sleep influence because of the impact of activity-related thermogenesis. During wakefulness rats are active, which increases their metabolic heat production and heat storage, but sleep is associated with consolidated inactivity, which results in decreased heat production and the opportunity to dissipate stored heat. Thus in the analysis of Franken et al. (33), the influence of activity on body temperature may inflate the estimates of changes in body temperature due to sleep, but clearly sleep is a major influence on body temperature.

Direct approaches to understanding central nervous system (CNS) regulation of body temperature have confirmed that down-regulation of body temperature occurs in NREMS in comparison to wakefulness. In mammals the dominant neural circuits responsible for balancing heat loss and heat gain through activation of thermoregulatory responses reside in the preoptic/anterior hypothalamus (POAH). The major feedback signal to this neural thermostatic system is its local temperature, T_{poah}, which can be independently manipulated with chronically implanted, water-perfused thermodes. It is possible to define a hypothalamic temperature threshold (T_{set}) for a thermoregulatory response and to determine a proportionality constant that relates the magnitude of the response to the difference between T_{poah} and T_{set}. When this approach was applied to sleeping kangaroo rats (36) and marmots (31), both T_{set} and the gain for the metabolic heat production response to hypothalamic cooling were lower during NREMS than during wakefulness. In the pigeon, where the dominant feedback signal for thermoregulation is the temperature of the spinal cord, T_{set} and the gain for heat production by shivering were also lower during NREMS than during wakefulness; similarly, the T_{set} for panting was lower during NREMS than during wakefulness (49). These studies reveal active CNS mechanisms that account for the down-regulation of body temperature in birds and mammals at the transition to NREMS from wakefulness.

The influence of REMS on thermoregulation appeared complex when viewed simply from the perspective of changes in brain, body, and skin temperatures (see ref. 88 for an extensive review). However, the discovery that thermoregulatory responses were actually suppressed during REMS made it easier to understand the variability in the temperature changes associated with REMS. One of the first convincing documentations of inhibition of ongoing thermoregulatory responses during REMS was that of Parmeggiani and Rabini (92), who showed that when cats slept at low ambient temperatures shivering ceased at the onset of REMS, and conversely, at high ambient temperatures panting ceased at the onset of REMS. The resulting conclusion that ongoing thermoregulatory responses are inhibited during REMS has been supported by studies of thermoregulatory vasomotor activity (35, 95) and thermoregulatory sweating (52, 83, 101, 106, 112).

Tissue temperature changes in sleep depend largely on regional blood flow. Studies using radioactive microspheres have shown that changes in regional blood flow during REMS do not indicate thermoregulatory responses (34). The direction of change in blood flow at the onset of REMS depends solely on the prevailing vasomotor state in NREMS. Vasodilated vascular beds showed decreased blood flow due to the slight fall in blood pressure at the NREMS–REMS transition, and constricted beds showed increased flow in REMS compared to NREMS due to smooth muscle relaxation (95). Thus changes in heat content and heat distribution in specific tissues during REMS are highly dependent on the ambient temperature and the thermoregulatory status of the animal just prior to the REMS episode. For example, when sleep recordings are done under thermoneutral conditions, there is little, if any, active peripheral vasoconstriction associated with NREMS. Under these conditions, the major influence on the diameter of arterioles and blood flow through them is transmural pressure. At the onset of REMS, the slight drop in central blood pressure that occurs in some species (130) will cause a decrease in the diameter of peripheral arterioles and a decrease in blood flow and heat transport to the skin, resulting in a decrease in heat loss. This sequence of events, along with increased neural metabolism during REMS, could explain the general observation that brain temperature tends to increase during REMS. However, a different result might be expected when an experiment is conducted at an ambient temperature below thermoneutrality. At the onset of REMS under these conditions there is a relaxation of

vasoconstriction so that blood flow to the skin and heat loss increase (96). In a very small mammal, the pocket mouse, brain temperature fell dramatically during REMS at ambient temperatures below thermoneutrality (124). Even in a larger mammal, the cat, decreases in brain temperature during REMS were observed when ambient temperature was sufficiently low (96).

The most definitive evidence for a centrally mediated disruption of thermoregulatory homeostasis during REMS derives from studies that show a lack of responsiveness to the major feedback signal of the thermoregulatory system, T_{poah}, during REMS (31, 36, 91). There are three possibilities to explain the REMS-related disruption of thermal homeostasis: inhibition of thermal afferents, inhibition of central integration, and/or inhibition of motor output. Inhibition of motor output is a logical candidate because of the general hyperpolarization of skeletal muscle motorneurons during REMS (13). However, this possibility was ruled out as the general mechanism of REMS-related thermoregulatory inhibition by experiments in which the muscle atonia of REMS was eliminated by lesions of the dorsal pontine tegmentum (53). Lesioned cats displayed an increase in muscle tone and motor activity during REMS, but when sleeping in a cold environment, they still ceased shivering at the transition from NREMS to REMS. There is strong evidence, however, for inhibition of thermosensitive and integrative neurons in the POAH during REMS. In both cats and kangaroo rats the temperature sensitivities of both cold- and warm-sensitive POAH neurons decrease in NREMS relative to wakefulness and decrease further at the transition from NREMS to REMS (37, 89, 90). The influence of sleep state transitions on thermoafferent information processing remains unclear (41, 43). In summary, the loss of normal thermoregulatory responses during REMS appears to be due, at least in part, to inhibition of central integrative neurons.

How do we interpret the influence of sleep states on thermoregulation from the standpoint of adaptations to the environment? It has been proposed that NREMS evolved in parallel with endothermy as an adaptation to the elevated energetic costs of maintaining a high metabolic rate and body temperature (2). The rationale is that for early endotherms that were small and poorly insulated it would have been adaptive to lower the level of regulated body temperature during daily periods of inactivity, thus conserving energy without reverting to poikilothermy. Support for this hypothesis is circumstantial. In altricial species NREMS seems to develop in parallel with endothermy (119), and the most extreme mammalian adaptations for energy conservation, shallow torpor and hibernation, appear to be extensions of NREMS (51, 68). The energy conservation hypothesis of NREMS has been criticized, however, on the basis that NREMS is highly conserved in mammals, but the significance of the associated energy conservation is small for many species (for review see ref. 47). Another hypothesis is that NREMS serves a neural restorative function. An aspect of this restorative function is a reduction in both neuronal activity and sensitivity to inputs (109). Thus it is possible that the reduction of thermoregulatory set points and POAH thermosensitivity during NREMS is an indirect consequence of a neuronal restorative process. Nevertheless, such a reduction in thermoregulatory activity resulting in a decline in regulated body temperature could have been a preadaptation enhanced through selection to result in the energy-conserving adaptations of torpor and hibernation.

We can say little about possible adaptive functions of REMS; in fact, the inhibition of homeostatic processes associated with this state of arousal may seem maladaptive. Two speculative viewpoints can be mentioned, however. If it is true that NREMS evolved in parallel with endothermy, then does REMS reflect the sleep processes of ectothermic vertebrates? The homologies between reptilian and mammalian sleep have been debated without resolution because the electrophysiological correlates of mammalian sleep cannot easily be applied to reptiles. If, however, mammalian REMS does reflect primitive vertebrate sleep, its original function could have been the inhibition of motivated behaviors during the inactive phase of the daily cycle. Another viewpoint, from experiments on rats, is that REMS serves a homeostatic function, intrinsic to the CNS, relative to NREMS (1). Periodically, the neuronal events of NREMS must be reversed, and REMS permits that to occur without a reversion to wakefulness.

INFLUENCES OF TEMPERATURE ON SLEEP

Studies of the effects of the thermal environment on mammals and birds must take into account the powerful influence of temperature on sleep. Both environmental and body temperatures affect the amount and distribution of arousal states. Thus for comparative studies of sleep, it is essential to have information on the thermoregulatory status of the animals under study, knowledge of their thermoneutral zones, and measures of the thermal environment. In a number of mammalian species it has been shown that total sleep time (TST) is maximal within the thermoneutral zone (TNZ) and decreases above and below the TNZ. Moreover, as ambient temperature deviates from thermoneutrality, the NREMS/REMS ratio increases due primarily to a reduction in the number of epochs of REMS (93, 94, 98, 102, 107, 118). Environmental temperature even within thermoneutrality affects REMS. The amount of REMS in rats varies significantly within the TNZ (25°–

31°C), peaking at $T_a = 29°C$ (110). Changes in ambient temperature during NREMS can also influence the subsequent distribution of wakefulness and sleep in animals (111) and humans (12). For example, rats sleeping at ambient temperatures above or below thermoneutrality showed increased transitions into REMS if T_a was changed toward thermoneutrality during sleep. In contrast, if T_a was changed away from thermoneutrality, transitions into REMS were fewer while transitions to wakefulness increased compared to control rats sleeping at $T_a = 29°C$ (111).

The influence of ambient temperature on sleep in humans has been studied by a number of investigators (8, 9, 45, 62, 74, 87, 105). In a cold environment wakefulness increased due primarily to decreased REMS and stage 2 NREMS (8, 9). However, subjects selected for their ability to sleep in the cold showed no decrease in REMS when tested at a cold (21°C) vs. a neutral (29°C) T_a (87). In a warm environment the nocturnal sleep period was characterized by increased wakefulness and by reductions in both REMS and NREMS (62, 74). When the same subjects slept at different ambient temperatures (21°–37°C), TST, NREMS stages 3 and 4 (delta sleep), and REMS were maximal at thermoneutrality ($T_a = 29°C$) and progressively decreased as ambient temperatures deviated from thermoneutrality. Duration of REMS also peaked in the TNZ, decreasing outside of thermoneutrality (45).

Adaptation to environmental conditions modifies the influences of ambient temperature on sleep. Libert et al. (75) recorded changes in sleep before, during, and after continuous, multiday exposure of human subjects to a warm (35°C) environment. Although sleep was more fragmented (decreased TST and increased wakefulness) at the warm T_a, there were no differences in the amounts of delta sleep or REMS between baseline (20°C) and experimental conditions. However, during recovery sleep TST and stage 3 NREMS increased, while the number of epochs of wakefulness and the number of transient arousals in REMS decreased. Thus acute and chronic exposures to heat have different effects on sleep organization, which may in part be due to nonspecific stress imposed during acute studies.

To examine the influence of elevated body temperature on sleep in humans, exercise and passive heating of subjects have been used. Both passive heating and high-intensity exercise in trained subjects increased delta sleep and had no effect on REMS (60). When the normal rise in core temperature of about 2°C during running was offset by increasing evaporative heat loss, nocturnal sleep parameters did not differ from baseline recordings, suggesting that the exercise effects appear to be due primarily to the influence of core temperature on sleep (57).

Further studies have investigated the temporal relationship of imposed core temperature changes on sleep parameters in adult humans. Horne and Shackell (59) speculated that as the time between heating and bedtime increases, larger "doses" of heating are needed to increase NREMS. Circadian influences may play a role, however, since heatings given at 7.5 vs. 2.5 h before bedtime are at different phases of the circadian temperature rhythm. Pretreatment of subjects with aspirin (600 mg) at the time of a "late" heating (2.5 h before bedtime) neutralized the increase in NREMS, even though body temperature rise during heating was the same in aspirin-treated and experimental groups (59). Daytime administration of aspirin to healthy subjects resulted in a 12% reduction in stage 4 NREMS (58); this effect appears to be linked to a small decrease in oral temperature seen in the interval 3 h before sleep onset (55).

Bunnell et al. (10) also measured the relationship between proximity of passive body heating (41°C for 1 h) in the morning, afternoon, early evening, and late evening and subsequent sleep changes. Increases in NREMS were seen only following the evening heating periods. Heating during late evening periods also resulted in a decrease in the duration of the first REMS period, but it is unclear whether this is due to increased NREMS drive or to a primary suppression of REMS.

Jordan et al. (61) repeated earlier passive heating protocols and measured rectal temperature during the sleep period to try and resolve the controversy as to whether the amount of NREMS is related more to body temperature at sleep onset (3) or to the rate of fall in T_b following sleep onset (104). Passive heating resulted in a sustained elevation of T_{re} of about 0.2°C compared to controls throughout the night. Both REMS and NREMS were significantly increased during the first part of the night, and thus increases in NREMS were not due to a suppression of REMS. Finally, the increase in NREMS during the first 2.5 h of sleep could not be accounted for by the rate of change of T_b, suggesting that the amount of delta sleep is a function of the level of T_b at sleep onset.

To determine if imposed changes in body temperature influence sleep, body temperature was passively elevated during the circadian T_b nadir (when delta sleep propensity is normally low); subsequent NREMS stage 2–4 were increased significantly in the fourth NREMS cycle (11). The mechanisms by which increases in core temperature increase delta sleep are unknown, but this could result from an acceleration of waking processes that regulate delta sleep (60).

Studies investigating the relationship between sleep and fever in the rat (63) and rabbit (70, 72, 125) indicate that fever produces complex changes in the diurnal organization of sleep. Fevers resulted in a short-term attenuation of the diurnal NREMS rhythm and a suppression of the REMS and T_b rhythms during multiday recording periods in the rat (63). At both 20° and 30°C T_a, NREMS increased more during lights-off, due primarily to an increase in the number of NREMS bouts. Although REMS decreased during lights-on, the total amount of REMS increased at $T_a = 20°C$ and decreased at $T_a = 30°C$, contrary to the usual relationship between T_a and REMS seen in previous studies.

Studies of changes in sleep in rabbits in response to the administration of putative sleep-promoting factors, which are also pyrogens, indicate that the somnogenic effects are not simply a consequence of pyrogenic activity (70, 72, 125). For example, human endogenous pyrogen (interleukin-1 or IL-1) leads to a dose-dependent increase in NREMS concomitant with a rise in brain temperature (T_{br}), but IL-1 still increases sleep after administration of the antipyretic anisomycin, which prevents IL-1 from elevating T_{br}. Many studies of the effects of putative somnogens and pyrogens on sleep and temperature have shown that administration of these agents results in an increase in NREMS, a decrease in REMS, and an increase in core temperature (69, 71). Despite the close links between sleep and thermoregulation, the responses of these two systems to various biochemical agents can be easily uncoupled (69, 84, 86, 126). For example, some cytokines are pyrogenic without increasing sleep (85).

Studies that attempt to determine the efficacy of putative somnogenic agents or that involve pharmacological manipulations of sleep must consider changes in sleep parameters resulting from temperature variations secondary to drug treatment. An example that clearly illustrates this principle is a study of the effects of phentolamine on sleep in rats (64); administration of this α-adrenoreceptor antagonist resulted in decreases in REMS and body temperature. When changes in body temperature were minimized by testing rats in a warm rather than a cool T_a, "drug" effects disappeared. A comprehensive listing of the effects of drugs and a wide variety of biochemical agents on body temperature is available (14–21).

Studies on the effects of reduced body temperatures on sleep are difficult because the resulting thermoregulatory drive is arousing (98). Perhaps the only exception is the hypothermia that occurs naturally in animals during hibernation and shallow torpor. During the entrance into hibernation (121) and during shallow torpor (120) in ground squirrels, NREMS increased and REMS dramatically decreased compared to euthermic sleep distributions.

In summary, the thermal environment has strong influences on sleep structure in a variety of species. Therefore, any attempts to interpret differences in sleep parameters between different species or between similar species in different environments must take into account the thermal environment in which the recordings were made, the prior thermal history of the animal, and the thermoregulatory properties of the animal under study.

INTERACTIONS OF SLEEP, THERMOREGULATION, AND THE CHRONOBIOLOGICAL ENVIRONMENT

The chronobiological environment of an animal includes external and internal components. The internal component consists of the systems that generate endogenous rhythmicity and the external of environmental oscillations that entrain the circadian timing system. Environmental events can also invoke passive oscillations in physiological and behavioral variables that mask the underlying circadian oscillation. Therefore, observed (for example, overt) circadian rhythms normally reflect the sum of influences by external and internal factors. The details of circadian rhythms and their entrainment is the subject of the chapter by Turek and van Reeth in this *Handbook*. We first consider information on how the circadian system of mammals interacts with mechanisms of sleep homeostasis (for example, the processes underlying increased sleepiness and recovery sleep in response to prior sleep loss). Then we consider how the chronobiological environment influences thermoregulation.

The master controller for the circadian system in mammals is localized in the suprachiasmatic nuclei (SCN) of the hypothalamus (79, 97, 108). It has therefore been possible to eliminate the circadian influence on sleep–wakefulness and virtually all other physiological and behavioral parameters by lesioning the SCN (77). Sleep in SCN-lesioned rats maintained in constant conditions (either constant light or constant dark) was equally distributed across the day (24, 78), but the total amount of sleep per 24 h period was essentially the same as that amassed by intact animals (78). If SCN-lesioned rats were sleep-deprived, however, they responded with compensatory sleep rebound similar to control animals, suggesting that a sleep homeostatic system was separate from, but perhaps gated by, the circadian system (78, 114). This concept was formalized in the two-process model of sleep–wake control (4, 5, 23) and is an important theoretical platform upon which many studies have been based. In this model the sleep homeostatic process (Process S) is considered to be the principal determinant of sleepiness/alertness, and the circadian clock modulated upper and lower thresholds through which Process S passed to invoke manifest states of arousal. A large body of data supports the conceptual notion of Process S, but the two-process model does not offer a mechanistic insight into how the circadian system organizes sleep and wakefulness on a daily basis.

Such insight into how the circadian system interacts with the sleep homeostatic process has come from work on SCN-lesioned squirrel monkeys (25, 26). The squirrel monkey is a diurnal new-world primate with consolidated sleep–wake patterns (for example, the prototypical pattern of 16 h awake and 8 h asleep) more similar to humans than is that of any rodent species. Suprachiasmatic nucleus–lesioned squirrel monkeys kept in constant light displayed no circadian rhythms in sleep–wake, sleep states (for example, the daily timing of NREMS and REMS stages), drinking, or brain temperature. Unlike the rat, however, the total amount of sleep per 24 h period was increased considerably (about 4 h) over that of intact animals. This increase was accompanied by a dramatic decrease in wakefulness bout consolidation during the subjective day. Sleep propensity (sleep bout length and frequency of occurrence) was closer to the prelesion nocturnal level than to diurnal levels (Fig. 59.1). The conclusion was that the circadian system interacts with the sleep homeostatic process through the active promotion of wakefulness.

Although SCN lesions eliminate circadian rhythms in mammals, SCN-lesioned animals are not truly arrhythmic in the strictest sense. In the absence of circadian control, rats and monkeys exhibit ultradian frequency oscillations in sleep–wakefulness, body temperature, and behavioral activities, with periods ranging from 1–4 h. Lesions of the SCN do not alter the fundamental cycling of sleep stages (for example, transitions between NREMS and REMS), nor do they impair the essential mechanisms responsible for cortical and behavioral arousal (26). Indeed, it is possible that SCN-lesioned animals arouse more readily than intact animals because, given the increased frequency of short sleep episodes, SCN-lesioned animals experience relatively little sleep loss at any time of day. This view is supported by studies comparing sedative hypnotic drug effects in intact and SCN-lesioned rats. Sleeping aids, such as the benzodiazepine triazolam, interact with sleep loss, and their effects are augmented by sleep deprivation (28, 116). Suprachiasmatic nucleus–lesioned rats, which do not exhibit extended bouts of wakefulness, show little or no increase in sleep following treatment with benzodiazepines (29); however, if the same animals are medicated after being maintained awake for 6 h, benzodiazepine soporific effects are restored (30, 116).

That the SCN promotes wakefulness in diurnal primates raises interesting questions about the circadian control of sleep–wakefulness in nocturnal mammalian species. The electrophysiological and metabolic activity of the SCN peaks during the subjective day in both diurnal and nocturnal species (44, 81, 103). If activity of the SCN is a reflection of SCN effector function, then the output of the circadian pacemaker must somehow be inverted in nocturnal animals. Evidence that the electrical activities within and outside of the SCN are in phase in diurnal mammals but 180° out of phase in nocturnal

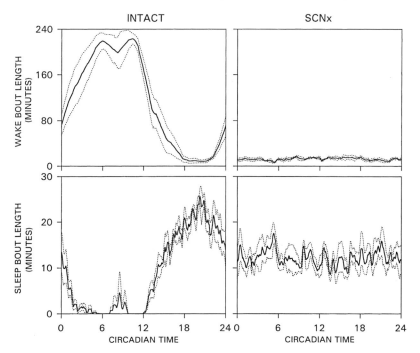

FIG. 59.1. Temporal variation in the longest wake and sleep bout lengths in intact and SCN-lesioned squirrel monkeys. Note the reduction in wake bout lengths during what would otherwise be the subjective day (circadian time 0–12) in the SCN-lesioned population. Data are shown as population means *(solid lines)* ± SEM *(broken lines)*. N = 5 in each group. Reprinted from reference 26.

species (73, 100) seems to support this notion. An important question then remains as to whether the circadian system of nocturnal mammals also promotes wakefulness at specific times of day. Wakefulness bout lengths during the subjective night (activity phase) are significantly greater in intact than in SCN-lesioned rats, suggesting that SCN-dependent alerting could be a common strategy in sleep–wake regulation despite the animal's usual temporal niche (26). In fact, actively promoting vigilance and behavior at times of day that maximize an animal's competitiveness may underscore the importance of such a physiological mechanism.

Apart from its influence on sleep organization and expression, the circadian system modulates the regulation of body temperature, resulting in the well-studied phenomenon of daily rhythms of body temperature. The independence of the circadian rhythm of body temperature from changes due to arousal state variation or activity level is demonstrated by a number of studies on a variety of species. For example, in humans the circadian cycle of body temperature persists during sleep deprivation (65–67, 113). More convincing, however, is the quantification of threshold core temperatures for thermoregulatory responses in exercising (hence, awake) subjects (128). The esophageal temperature at which forearm blood flow increased and sweating was initiated was measured in subjects exercising on a bicycle ergometer. These thresholds were clearly lower at

night than during the day, indicating that active thermoregulatory responses were called into play to maintain the lower nighttime body temperature even in completely awake subjects exercising under controlled conditions.

A daily rhythm of body temperature partially independent of sleep and level of activity has been demonstrated in some animal species. An early approach was to correlate T_b with level of activity and show that the regression line was different at different phases of the daily cycle (54, 80). The most direct evidence for the circadian system having an influence on the regulation of body temperature via the CNS comes from experiments in which internal temperature thresholds for active thermoregulatory responses were measured as a function of time of day. When pigeons were held at an ambient temperature at the high end of thermoneutrality during the day, they had to pant to lower their core body temperatures to the usual nighttime level (40). The birds, therefore, maintained costly, active thermoregulatory responses all night to achieve the low temperature portion of their daily rhythms of body temperature. This experiment, however, did not include an analysis of arousal states, so it was not possible to ascribe the activation of the heat loss response to circadian influences alone.

In subsequent experiments pigeons were instrumented with fine thermodes in the spinal canal, since

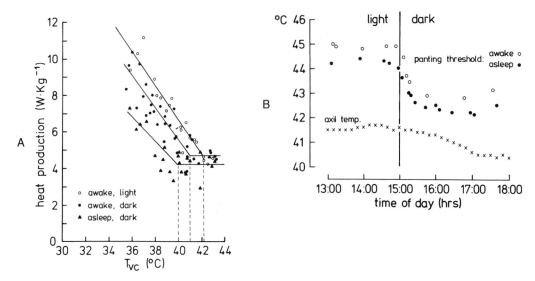

FIG. 59.2. (a) Relationship of metabolic responses to temperature manipulations of the spinal cord in a pigeon plotted as a function of light or dark phase of the daily cycle and for the dark phase, whether the bird was awake or asleep. Time of day accounted for about a 1°C shift in threshold. (b) Axillary and spinal temperature thresholds during the transition from light to dark phase of the daily cycle. In each case panting threshold was determined first during sleep. The bird was then awakened by acoustic stimulation, and panting threshold was again determined. Reprinted from reference 40 with permission.

spinal temperature is the main feedback signal in the avian thermoregulatory system. The birds also received chronically implanted EEG, electrooculographic (EOG), and electromyelographic (EMG) electrodes so that arousal states could be electrographically defined. These studies demonstrated that changes in the threshold spinal temperatures for both heat production and heat loss responses correlated with the daily cycle of body temperature independently of arousal state changes (49) (Fig. 59.2).

There are no studies showing circadian changes in hypothalamic temperature thresholds in mammals, but there is indirect evidence that the circadian system can influence body temperature regulation independently of arousal state changes. When the SCN are partially lesioned, rhythms of body temperature can persist after all rhythmicity in drinking, feeding, and activity has been abolished (99). Conversely, in rats with complete SCN lesions that abolish all rhythms rhythmicity can be restored with transplants of fetal SCN tissue (27). In some of these experiments the circadian rhythm in arousal states was restored before the circadian rhythm in body temperature. Thus arousal state organization is not a sufficient explanation for the circadian rhythm of body temperature.

The general conclusion is that the chronobiological environment has very strong influences on both arousal state expression and the regulation of body temperature. The circadian system in mammals influences sleep expression by actively maintaining wakefulness and thus the accumulation of sleep pressure over a portion

of the daily cycle. When the alerting influence of the circadian system ebbs, accumulated sleep pressure results in the induction and maintenance of sleep states. We do not know if the avian circadian system operates in the same way. In terms of the thermoregulatory system, the circadian system modulates apparent internal temperatures thresholds for thermoregulatory responses, resulting in a daily fluctuation of core body temperature independent of arousal state changes.

INTERACTIVE SYSTEMS IN TORPOR AND HIBERNATION

At the beginning of this chapter, we posed the question of whether the energy-conserving adaptation of torpor evolved as a modification of thermoregulation, circadian influences on thermoregulation, sleep influences on thermoregulation, or the circadian influence on sleep. Hibernation, and presumably shallow torpor, involve regulated declines in body temperature, so clearly a modification in thermoregulatory homeostasis is involved (51). These modifications enable continuous POAH thermosensitivity over a very broad range of POAH temperatures. One study has reported apparent differences in the properties of POAH neurons between hibernators and nonhibernators that correlate with the functional differences in their thermoregulatory systems (129). It is still unclear, however, how these differences evolved and what selection operated on to produce the broad-band thermostat of the hibernator.

NREMS occurs in all mammals and is associated with a decline in the regulated body temperature. Selection could have favored enhanced energy conservation associated with longer durations of NREMS and lower minimum body temperatures during NREMS. Electroencephalographic studies have confirmed that hibernation is entered predominantly through NREMS (121) and that bouts of shallow torpor consist almost entirely of NREMS (120, 123). Since the EEG characteristics change as T_{br} declines, homologies between NREMS and deep hibernation cannot be based on EEG comparisons. The arousal state selectivity of single units, however, remains at brain temperatures below which the EEG is no longer useful to differentiate NREMS, REMS, and wakefulness; in fact, continuous recordings of single units during entire bouts of torpor support the idea that deep hibernation consists mostly of a state homologous with NREMS (68).

What is the possible involvement of the circadian system in the control of torpor and deep hibernation? Since the circadian rhythm of body temperature is a general mammalian phenomenon, it is conceivable that torpor evolved through selection favoring lower and lower

minima of the circadian influence on the regulation of body temperature. The circadian system could be the major modifier of thermoregulation in animals that enter torpor, and the coincidence of sleep with the circadian dip in body temperature might be incidental. Clearly, shallow daily torpor is tightly controlled by the circadian system, and it has been suggested that multiday bouts of deep hibernation could have evolved through a relaxation of the temperature compensation of the circadian clock (50). The consequence of a temperature-sensitive circadian clock in an animal that underwent daily reductions in regulated body temperature would be that the clock would run slower. Thus the lower the temperature drop, the longer the torpor period. This hypothesis seemed plausible in light of the inverse relationship between body temperature and torpor bout length in hibernators (117). We now know, however, that the circadian system continues to run during deep hibernation, with a periodicity of about 24 h (42) (Fig. 59.3), and that bouts of torpor are integer multiples of that period (Fig. 59.4). It is therefore unlikely that the main determinant of the down-regulation of body temperature during torpor is the circa-

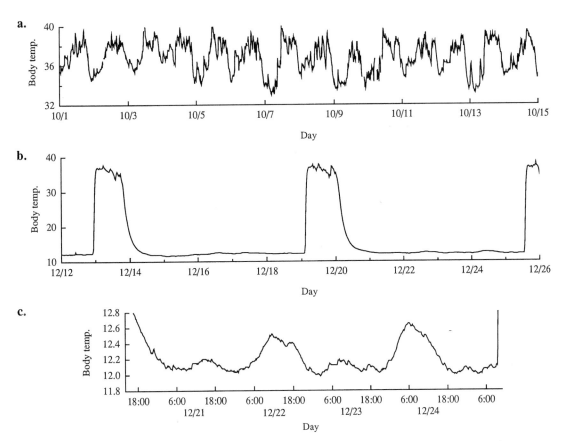

FIG. 59.3. Body temperature (T_b) vs. time of a golden-mantled ground squirrel during euthermia and T_b vs. time of a golden-mantled ground squirrel during hibernation and periodic arousals under constant dim red light at 10°C ambient temperature (a = euthermia; b = hibernation and periodic arousals; c = within a hibernation bout).

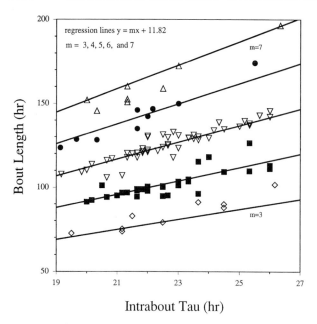

FIG. 59.4. Hibernation bout length for golden-mantled ground squirrels plotted as a function of the intrabout circadian period determined from autocorrelation analysis of telemetered body temperature records. Whereas the circadian periods vary greatly from bout to bout even in the same animal and for contiguous bouts of hibernation, bout length is always a whole integer multiple of the intrabout circadian period. Reprinted from reference 42 with permission.

dian influence on the thermoregulatory system; however, the circadian system appears to time arousals from hibernation.

An enduring question about hibernation concerns the nature of the periodic arousals, which must serve an important function, given the major expenditure of energy utilized during each return to euthermia. While the function of arousal has been quite elusive, a new idea has emerged that focuses on the restorative function of NREMS. The relevant finding is that animals arousing from hibernation exhibit very high levels of EEG delta power (22, 116) and that these levels are proportional to the length of the preceding bout of torpor (116) (Fig. 59.5). The conclusion has been drawn that, even though a bout of torpor is primarily homologous with NREMS, animals must arouse from torpor to gain some restorative benefits of euthermic NREMS. Possibly, the low T_{br} associated with torpor inhibits the biochemical processes of sleep restoration. Until we know what those processes are, this view of the control of hibernation must remain hypothetical.

CONCLUSIONS

This chapter has focused on the strong interactions between the physiological systems controlling sleep, body temperature, and circadian rhythms as well as the responses of these systems to the thermal and chronobiological environments. It is important to realize that it is difficult, if not impossible, to study one component of these interacting systems without considering the others. This caveat applies to mechanistic as well as to adaptational studies. When we look at these systems for adaptations to the environment, we find that selection frequently produces adaptations by shaping the interactions between systems. For example, shallow or daily torpor may be a product of selection favoring an exaggeration of a circadian modulation of body temperature regulation. In contrast, deep torpor or hibernation may be a product of selection favoring an exaggeration of the modulation of thermoregulation by sleep control mechanisms. It may be important to appreciate this difference to understand the mechanisms behind single-day vs. multiday organization of adaptive hypothermia. Similarly, when we look for the mechanism of timing of multiday bouts of torpor, we find the likely explanation not only in the circadian system but possibly in the effects of temperature on processes involved in sleep homeostasis.

A paper by Gould and Lewontin (39) continues to stir up controversy and discussion among students of evolution. The focus of the paper is summarized by the authors: "An adaptationist programme . . . is based on faith in the power of natural selection as an optimizing agent. It proceeds by breaking an organism into unitary 'traits' and proposing an adaptive story for each considered separately." The authors point out that the present utility of a trait of an organism may have nothing to do with its origin and that the existence of a trait may have more to do with developmental and phyletic (and we would add physiological) constraints than with selection. Physiologists, particularly those interested in comparative and environmental studies, would benefit by keeping Gould and Lewontin's admonitions in mind. It is tempting to interpret every physiological measurement in adaptationist terms. For example, very small increases in brain temperature during REMS have been interpreted to support a hypothesis that the function of REMS is brain warming, when the observed changes in temperature may simply be a consequence of thermal gradients, changes in blood flow, and local tissue metabolism. Similarly, apparent declines in regulated body temperature during NREMS have been interpreted to support a hypothesis that the function of NREMS is energy conservation, when the observed changes in body temperature regulation may simply be a consequence of decreasing neuronal activity. Either hypothesis may be correct, but each must be critically tested with experiments that have the power of disproof. Otherwise, it remains an adaptationist story.

A physiological mechanism or phenomenon may not

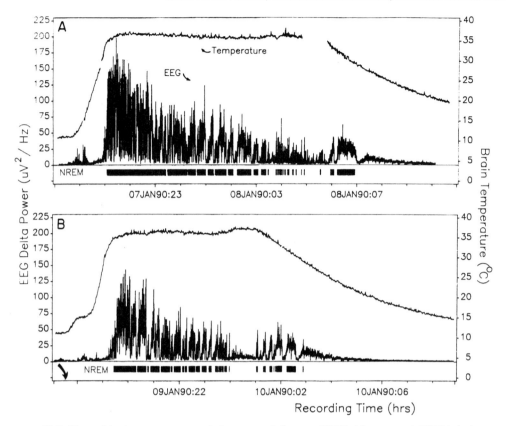

FIG. 59.5. Plots of brain temperature and electroencephalogram *(EEG)* delta power (μV²/Hz) during arousal from hibernation, interbout euthermia, and reentry into hibernation for one animal following a long bout (4.25 days, *A*) and a short bout (1.5 days, *B*) of hibernation. Bold arrow in *B* indicates start of sensory stimulation that initiated the arousal from hibernation after 1.5 days. Arousal in *A* was spontaneous. For each 10 s epoch EEG delta power is depicted, and brain temperature curve connects averages of six consecutive 10 s epochs. Solid vertical bars at the bottom of each panel show NREM sleep. Reprinted from reference 116 with permission.

be an adaptation to the environment per se. Circadian rhythmicity is clearly an adaptation to the fact that the earth turns on its axis once every 24 h. Circadian rhythms give organisms the ability to anticipate environmental changes and are highly adaptive. Sleep, however, is most likely not an adaptation to the environment but serves a physiological need imposed upon the animal as a consequence of the neuronal/metabolic activity associated with waking. Sleep timing, however, can be shaped by selection as an adaptation to specific ecological niches, so we have diurnal sleepers, nocturnal sleepers, and crepuscular animals that distribute their sleep more evenly over the daily cycle. Sleep mechanisms can also be shaped by selection as an adaptation to the environment, hence the hypersomnia and exaggerated lowering of body temperature seen in hibernators, but these specializations do not necessarily tell us anything about the original and basic function of sleep. By investigating mechanisms, physiologists may resolve problematical questions about function and, ultimately, adaptive significance.

REFERENCES

1. Benington, J. H., and H. C. Heller. Evidence that REM sleep propensity accumulates during non-REM sleep rather than waking. *Am. J. Physiol.* 266(*Regulatory Integrative Comp. Physiol.* 35): R1992–R2000.

2. Berger, R. J. Slow wave sleep, shallow torpor and hibernation: homologous states of diminished metabolism and body temperature. *Biol. Psychol.* 19: 305–326, 1984.

3. Berger, R. J., J. W. Palca, J. M. Walker, and N. H. Phillips. Correlations between body temperatures, metabolic rate and slow wave sleep in humans. *Neurosci. Lett.* 86: 230–234, 1988.

4. Borbély, A. A. A two-process model of sleep regulation. *Hum. Neurobiol.* 1: 195–204, 1982.

5. Borbély, A. A., P. Achermann, L. Trachsel, and I. Tobler. Sleep initiation and initial sleep intensity: interactions of homeostatic and circadian mechanisms. *J. Biol. Rhythms* 4: 149–160, 1989.

6. Borbély, A. A., F. Baumann, D. Brandeis, I. Strauch, and D. Lehmann. Sleep deprivation: effect on sleep stages and EEG power density in man. *EEG Clin. Neurophysiol.* 51: 483–495, 1981.

7. Borbély, A. A., and H. U. Neuhaus. Sleep-deprivation: effects on sleep in the rat. *J. Comp. Physiol.* 133: 71–87, 1979.

8. Buguet, A. G., S. D. Livingstone, and L. D. Reed. Skin temperature changes in paradoxical sleep in man in the cold. *Aviat. Space Environ. Med.* 50: 567–570, 1979.

9. Buguet, A. G., S. D. Livingstone, L. D. Reed, and R. E. Limmer. EEG patterns and body temperatures in man during sleep in arctic winter nights. *Int. J. Biometeorol.* 20: 61–69, 1976.

10. Bunnell, D. E., J. A. Agnew, S. M. Horvath, L. Jopson, and M. Wills. Passive body heating and sleep: influence of proximity to sleep. *Sleep* 11: 210–219, 1988.

11. Bunnell, D. E., and S. M. Horvath. Effects of body heating during sleep interruption. *Sleep* 8: 274–282, 1985.

12. Candas, V., J. P. Libert, and A. Muzet. Heating and cooling stimulations during SWS and REM sleep in man. *J. Therm. Biol.* 7: 155–158, 1982.

13. Chase, M. H., and F. R. Morales. The control of motoneurons during sleep. In: *Principles and Practice of Sleep Medicine,* edited by M. H. Kryger, T. Roth, and W. C. Dement. Philadelphia: Saunders, 1989, p. 74–85.

14. Clark, W. G. Changes in body temperature after administration of amino acids, peptides, dopamine, neuroleptics and related agents. *Neurosci. Biobehav. Rev.* 3: 179–231, 1979.

15. Clark, W. G. Changes in body temperature after administration of antipyretics, LSD, delta 9-THC and related agents: II. *Neurosci. Biobehav. Rev.* 11: 35–96, 1987.

16. Clark, W. G., and Y. L. Clark. Changes in body temperature after administration of acetylcholine, histamine, morphine, prostaglandins and related agents. *Neurosci. Biobehav. Rev.* 4: 175–240, 1980.

17. Clark, W. G., and Y. L. Clark. Changes in body temperature after administration of adrenergic and serotonergic agents and related drugs including antidepressants. *Neurosci. Biobehav. Rev.* 4: 281–375, 1980.

18. Clark, W. G., and Y. L. Clark. Changes in body temperature after administration of antipyretics, LSD, delta 9-THC, CNS depressants and stimulants, hormones, inorganic ions, gases, 2,4-DNP and miscellaneous agents. *Neurosci. Biobehav. Rev.* 5: 1–136, 1981.

19. Clark, W. G., and J. M. Lipton. Changes in body temperature after administration of acetylcholine, histamine, morphine, prostaglandins and related agents: II. *Neurosci. Biobehav. Rev.* 9: 479–552, 1985.

20. Clark, W. G., and J. M. Lipton. Changes in body temperature after administration of amino acids, peptides, dopamine, neuroleptics and related agents: II. *Neurosci. Biobehav. Rev.* 9: 299–371, 1985.

21. Clark, W. G., and J. M. Lipton. Changes in body temperature after administration of adrenergic and serotonergic agents and related drugs including antidepressants: II. *Neurosci. Biobehav. Rev.* 10: 153–220, 1986.

22. Daan, S., B. M. Barnes, and A. M. Strijkstra. Warming up for sleep?—Ground squirrels sleep during arousals from hibernation. *Neurosci. Lett.* 128: 265–268, 1991.

23. Daan, S., D. G. M. Beersma, and A. A. Borbély. Timing of human sleep: recovery process gated by a circadian pacemaker. *Am. J. Physiol.* 246(*Regulatory Integrative Comp. Physiol.* 17): R161–R178, 1984.

24. Eastman, C. I., R. E. Mistleberger, and A. Rechtschaffen. Suprachiasmatic nuclei lesions eliminate circadian temperature and sleep rhythms in the rat. *Physiol. Behav.* 32: 357–368, 1984.

25. Edgar, D. M. Circadian Timekeeping in the Squirrel Monkey: Neural and Photic Control of Sleep, Brain Temperature and Drinking. Riverside: University of California, 1986. Dissertation.

26. Edgar, D. M., W. C. Dement, and C. A. Fuller. Effect of SCN lesions on sleep in squirrel monkeys: evidence for opponent processes in sleep–wake regulation. *J. Neurosci.* 13: 1065–1079, 1993.

27. Edgar, D. M., M. R. Ralph, W. F. Seidel, L. K. Lee, and W. C. Dement. Fetal SCN-transplants restore sleep–wake and body

temperature circadian rhythms in SCN-lesioned rats. *Sleep Res.* 21: 371, 1992.

28. Edgar, D. M., W. F. Seidel, and W. C. Dement. Triazolam-induced sleep in the rat: influence of prior sleep, circadian time, and light/dark cycles. *Psychopharmacology (Berl.)* 105: 374–380, 1991.

29. Edgar, D. M., W. F. Seidel, C. E. Martin, P. P. Sayeski, and W. C. Dement. Triazolam fails to induce sleep in suprachiasmatic nucleus–lesioned rats. *Neurosci. Lett.* 125: 125–128, 1991.

30. Edgar, D. M., L. Trachsel, W. F. Seidel, H. C. Heller, and W. C. Dement. The efficacy of triazolam is restored in sleep-deprived SCN-lesioned rats. *Sleep Res.* 19: 352, 1990.

31. Florant, G. L., B. M. Turner, and H. C. Heller. Temperature regulation during wakefulness, sleep, and hibernation in marmots. *Am. J. Physiol.* 235(*Regulatory Integrative Comp. Physiol.* 6): R82–R88, 1978.

32. Franken, P., I. Tobler, and A. A. Borbély. Cortical temperature and EEG slow-wave activity in the rat: analysis of vigilance state related changes. *Eur. J. Physiol.* 420: 500–507, 1992.

33. Franken, P., I. Tobler, and A. A. Borbély. Sleep and waking have a major effect on the 24h rhythm of cortical temperature in the rat. *J. Biol. Rhythms* 7: 341–352, 1992.

34. Franzini, C. The control of peripheral circulation during sleep. In: *The Diencephalon and Sleep,* edited by M. Mancia, and G. Marini. New York: Raven, 1990, p. 343–353.

35. Franzini, C., T. Cianci, P. Lenzi, and P. L. Guidalotti. Neural control of vasomotion in rabbit ear is impaired during desynchronized sleep. *Am. J. Physiol.* 243(*Regulatory Integrative Comp. Physiol.* 14): R142–R146, 1982.

36. Glotzbach, S. F., and H. C. Heller. Central nervous regulation of body temperature during sleep. *Science* 94: 537–539, 1976.

37. Glotzbach, S. F., and H. C. Heller. Changes in the thermal characteristics of hypothalamic neurons during sleep and wakefulness. *Brain Res.* 309: 17–26, 1984.

38. Glotzbach, S. F., and H. C. Heller. Temperature regulation. In: *Principles and Practice of Sleep Medicine* (2nd ed.), edited by M. H. Kryger, T. Roth, and W. C. Dement. Philadelphia: Saunders, 1993, p. 260–275.

39. Gould, S. J., and R. C. Lewontin. The spandrels of San Marco and the Panglossian paradigm: a critique of the adaptationist programme. *Proc. R. Soc. Lond. B.* 205: 581–598, 1979.

40. Graf, R. Diurnal changes of thermoregulatory functions in pigeons, I. Effector mechanisms. *Pflugers Arch.* 386: 173–179, 1980.

41. Grahn, D. A., and H. C. Heller. Activity of most rostral ventromedial medulla neurons reflect EEG/EMG pattern changes. *Am. J. Physiol.* 257(*Regulatory Integrative Comp. Physiol.* 28): R1496–R1505, 1989.

42. Grahn, D. A., J. D. Miller, V. S. Houng, and H. C. Heller. Persistence of circadian rhythmicity in hibernating ground squirrels. *Am. J. Physiol.* 266(*Regulatory Integrative Comp. Physiol.* 35): R1251–R1258, 1994.

43. Grahn, D. A., C. M. Radeke, and H. C. Heller. Arousal state vs. temperature effects on neuronal activity in the subcoeruleus area. *Am. J. Physiol.* 256(*Regulatory Integrative Comp. Physiol.* 27): R840–R849, 1989.

44. Green, D. J., and M. Gillete. Circadian rhythm of firing rate recorded from single cells in the rat suprachiasmatic slice. *Brain Res.* 245: 198–200, 1982.

45. Haskell, E. H., J. W. Palca, J. M. Walker, R. J. Berger, and H. C. Heller. The effects of high and low ambient temperatures on human sleep stages. *EEG Clin. Neurophysiol.* 51: 494–501, 1981.

46. Haskell, E. H., J. M. Walker, and R. J. Berger. Effects of cold stress on sleep of an hibernator, the golden-mantled ground squirrel *(C. lateralis). Physiol. Behav.* 23: 1119–1121, 1979.

47. Heller, H. C. Sleep and hypometabolism. *Can. J. Zool.* 66: 61–69, 1988.

48. Heller, H. C., and S. F. Glotzbach. Thermoregulation and sleep. In: *Heat Transfer in Medicine and Biology—Analysis and Applications,* edited by A. Shitzer and R. C. Eberhart. New York: Plenum, 1985, p. 107–134.

49. Heller, H. C., R. Graf, and W. Rautenberg. Circadian and arousal state influences on thermoregulation in the pigeon. *Am. J. Physiol.* 245(*Regulatory Integrative Comp. Physiol.* 16): R321–328, 1983.

50. Heller, H. C., B. L. Krilowicz, and T. S. Kilduff. Neural mechanisms controlling hibernation. In: *Living in the Cold II, Colloques INSERM,* edited by A. Malan and B. Canguilhem. London: Libbey, 1989, vol. 193, p. 447–458.

51. Heller, H. C., J. M. Walker, G. L. Florant, S. F. Glotzbach, and R. J. Berger. Sleep and hibernation: electrophysiological and thermoregulatory homologies. In: *Strategies in Cold: Natural Torpidity and Thermogenesis,* edited by L. C. H. Wang and J. W. Hudson. New York: Academic, 1978, p. 225–265.

52. Henane, R., A. Buguet, B. Roussel, and J. Bittel. Variations in evaporation and body temperatures during sleep in man. *J. Appl. Physiol.: Respir. Environ. Exerc. Physiol.* 42: 50–55, 1977.

53. Hendricks, J. C. Absence of shivering in the cat during paradoxical sleep without atonia. *Exp. Neurol.* 75: 700–710, 1982.

54. Heusner, A. Mise en évidence d'une variation nycthémérale de la calorification indépendante di cycle de l'activité chez le rat. *C. R. Soc. Biol. (Paris)* 150: 1246, 1957.

55. Horne, J. A. Aspirin and nonfebrile waking oral temperature in healthy men and women: links with SWS changes? *Sleep* 12: 516–521, 1989.

56. Horne, J. A. Human slow wave sleep: a review and appraisal of recent findings, with implications for sleep functions, and psychiatric illness. *Experientia* 48: 941–954, 1992.

57. Horne, J. A., and V. J. Moore. Sleep EEG effects of exercise with and without additional body cooling. *EEG Clin. Neurophysiol.* 60: 33–38, 1985.

58. Horne, J. A., J. E. Percival, and J. R. Traynor. Aspirin and human sleep. *EEG Clin. Neurophysiol.* 49: 409–413, 1980.

59. Horne, J. A., and B. S. Shackell. Slow wave sleep elevations after body heating: proximity to sleep and effects of aspirin. *Sleep* 10: 383–392, 1987.

60. Horne, J. A., and L. H. Staff. Exercise and sleep: body-heating effects. *Sleep* 6: 36–46, 1983.

61. Jordan, J., I. Montgomery, and J. Trinder. The effect of afternoon body heating on body temperature and slow wave sleep. *Psychophysiology* 27: 560–566, 1990.

62. Karacan, I., J. I. Thornby, A. M. Anch, R. L. Williams, and H. M. Perkins. Effects of high ambient temperature on sleep in young men. *Aviat. Space Environ. Med.* 49: 855–860, 1978.

63. Kent, S., M. Price, and E. Satinoff. Fever alters characteristics of sleep in rats. *Physiol. Behav.* 44: 709–715, 1988.

64. Kent, S., and E. Satinoff. Influence of ambient temperature on sleep and body temperature after phentolamine in rats. *Brain Res.* 511: 227–233, 1990.

65. Kleitman, N. Studies of the physiology of sleep. I. The effects of prolonged sleeplessness on man. *Am. J. Physiol.* 66: 67–72, 1923.

66. Kleitman, N. *Sleep and Wakefulness* (2nd ed.). Chicago: University of Chicago Press, 1963.

67. Kreider, M. B. Effects of sleep deprivation on body temperature. *Federation Proc.* 20: 214, 1961.

68. Krilowicz, B. L., S. F. Glotzbach, and H. C. Heller. Neuronal activity during sleep and complete bouts of hibernation. *Am. J. Physiol.* 255(*Regulatory Integrative Comp. Physiol.* 24): R1008–R1019, 1988.

69. Krueger, J. M., and M. L. Karnovsky. Sleep and the immune response. *Ann. N. Y. Acad. Sci.* 496: 510–516, 1987.

70. Krueger, J. M., S. Kubillus, S. Shoham, and D. Davenne. Enhancement of slow-wave sleep by endotoxin and lipid A. *Am. J. Physiol.* 251(*Regulatory Integrative Comp. Physiol.* 22): R591–R597, 1986.

71. Krueger, J. M., and J. A. Majde. Sleep as a host defense: its regulation by microbial products and cytokines. *Clin. Immunol. Immunopathol.* 57: 188–199, 1990.

72. Krueger, J. M., J. Walter, C. A. Dinarello, S. M. Wolff, and L. Chedid. Sleep-promoting effects of endogenous pyrogen (interleukin-1). *Am. J. Physiol.* 246(*Regulatory Integrative Comp. Physiol.* 17): R994–R999, 1984.

73. Kubota, A, S. T. Inouye, and H. Kawamura. Reversal of multiunit activity within and outside the suprachiasmatic nucleus in the rat. *Neurosci. Lett.* 27: 303–308, 1981.

74. Lenzi, P., J. P. Libert, T. Cianci, and C. Franzini. Comparative aspects of the interaction between sleep and thermoregulation. In: *Sleep '90,* edited by J. Horne. Bochum, Germany: Pontenagel, 1990, p. 388–390.

75. Libert, J. P., N. J. Di, H. Fukuda, A. Muzet, J. Ehrhart, and C. Amoros. Effect of continuous heat exposure on sleep stages in humans. *Sleep* 11: 195–209, 1988.

76. McGinty, D., and R. Szymusiak. Keeping cool: a hypothesis about the mechanisms and functions of slow-wave sleep. *Trends Neurosci.* 13: 480–487, 1990.

77. Meijer, J. H., and W. J. Reitveld. Neurophysiology of the suprachiasmatic circadian pacemaker in rodents. *Physiol. Rev.* 69: 671–707, 1989.

78. Mistlberger, R. E., B. M. Bergmann, W. Waldenar, and A. Rechtschaffen. Recovery sleep following sleep deprivation in intact and suprachiasmatic nuclei–lesioned rats. *Sleep* 6: 217–233, 1983.

79. Moore, R. Y., and V. B. Eichler. Loss of a circadian adrenal corticosterone rhythm following suprachiasmatic lesions in the rat. *Brain Res.* 42: 201–206, 1972.

80. Morrison, P. Modification of body temperature by activity in Brazilian hummingbirds. *Condor* 64: 315, 1962.

81. Newman, G. C., and F. E. Hospod. Rhythm of suprachiasmatic nucleus 2-deoxyglucose uptake in vitro. *Brain Res.* 381: 345–350, 1986.

82. Obal, F. J., G. Rubicsek, P. Alfoldi, G. Sary, and F. Obal. Changes in the brain and core temperatures in relation to the various arousal states in rats in the light and dark periods of the day. *Pflugers Arch.* 404: 73–79, 1985.

83. Ogawa, T., T. Satoh, and K. Takagi. Sweating during night sleep. *Jpn. J. Physiol.* 17: 135–148, 1967.

84. Onoe, H., R. Ueno, I. Fujita, H. Nishino, Y. Oomura, and O. Hayaishi. Prostaglandin D2, a cerebral sleep-inducing substance in monkeys. *Proc. Natl. Acad. Sci. USA* 85: 4082–4086, 1988.

85. Opp, M., F. J. Obal, A. B. Cady, L. Johannsen, and J. M. Krueger. Interleukin-6 is pyrogenic but not somnogenic. *Physiol. Behav.* 45: 1069–1072, 1989.

86. Opp, M. R., F. J. Obal, and J. M. Krueger. Interleukin 1 alters rat sleep: temporal and dose-related effects. *Am. J. Physiol.* 260(*Regulatory Integrative Comp. Physiol.* 31): R52–R58, 1991.

87. Palca, J. W., J. M. Walker, and R. J. Berger. Thermoregulation, metabolism, and stages of sleep in cold-exposed men. *J. Appl. Physiol.: Respir. Environ. Exerc. Physiol.* 61: 940–947, 1986.

88. Parmeggiani, P. L. Temperature regulation during sleep: a study in homeostasis. In: *Physiology in Sleep,* edited by J. Orem and C. D. Barnes. New York: Academic, 1980, p. 97–143.

89. Parmeggiani, P. L., A. Azzaroni, D. Devolani, and G. Ferrari. Polygraphic study of anterior hypothalamic-preoptic neuron thermosensitivity during sleep. *EEG Clin. Neurophysiol.* 63: 289–295, 1986.

90. Parmeggiani, P. L., D. Cevolani, A. Azzaroni, and G. Ferrari. Thermosensitivity of anterior hypothalamic-preoptic neurons

during the waking–sleeping cycle: a study in brain functional states. *Brain Res.* 415: 79–89, 1987.

91. Parmeggiani, P. L., C. Franzini, and P. Lenzi. Respiratory frequency as a function of preoptic temperature during sleep. *Brain Res.* 111: 253–260, 1976.

92. Parmeggiani, P. L., and C. Rabini. Shivering and panting during sleep. *Brain Res.* 6: 789–791, 1967.

93. Parmeggiani, P. L., and C. Rabini. Sleep and environmental temperature. *Arch. Ital. Biol.* 108: 369–387, 1970.

94. Parmeggiani, P. L., C. Rabini, and M. Cattalani. Sleep phases at low environmental temperature. *Arch. Sci. Biol. (Bologna)* 53: 277–290, 1969.

95. Parmeggiani, P. L., G. Zamboni, T. Cianci, and M. Calasso. Absence of thermoregulatory vasomotor responses during fast wave sleep in cats. *EEG Clin. Neurophysiol.* 42: 372–380, 1977.

96. Parmeggiani, P. L., G. Zamboni, E. Perez, and P. Lenzi. Hypothalamic temperature during desynchronized sleep. *Exp. Brain Res.* 54: 315–320, 1984.

97. Ralph, M. M., R. D. Foster, F. C. Davis, and M. Menaker. Transplanted suprachiasmatic nucleus determines circadian period. *Science* 247: 975–978, 1990.

98. Sakaguchi, S., S. F. Glotzbach, and H. C. Heller. Influence of hypothalamic and ambient temperatures on sleep in kangaroo rats. *Am. J. Physiol.* 237(*Regulatory Integrative Comp. Physiol.* 8): R80–R88, 1979.

99. Satinoff, E., and R. A. Prosser. Suprachiasmatic nuclear lesions eliminate circadian rhythms of drinking and activity, but not of body temperature in rats. *J. Biol. Rhythms* 3: 1–22, 1988.

100. Sato, T., and H. Kawamura. Circadian rhythms in multiple unit activity inside and outside the suprachiasmatic nucleus in the diurnal chipmunk *(Eutamias sibiricus)*. *Neurosci. Res.* 1: 45–52, 1984.

101. Satoh, T., T. Ogawa, and K. Takagi. Sweating during daytime sleep. *Jpn. J. Physiol.* 15: 523–531, 1965.

102. Schmidek, W. R., K. Hoshino, M. Schmidek, and C. Timo-Iaria. Influence of environmental temperature on the sleep–wakefulness cycle in the rat. *Physiol. Behav.* 8: 363–371, 1972.

103. Schwartz, W. J., and H. Gainer. Suprachiasmatic nucleus: use of ^{14}C-labeled deoxyglucose uptake as a functional marker. *Science* 197: 1089–1091, 1977.

104. Sewitch, D. E. Slow wave sleep deficiency insomnia: a problem in thermo-downregulation at sleep onset. *Psychophysiology* 24: 200–215, 1987.

105. Sewitch, D. E., E. M. Kittrell, D. J. Kupfer, and C. Reynolds. Body temperature and sleep architecture in response to a mild cold stress in women. *Physiol. Behav.* 36: 951–957, 1986.

106. Shapiro, C. M., A. T. Moore, D. Mitchell, and M. L. Yodaiken. How well does man thermoregulate during sleep? *Experientia* 30: 1279–1281, 1974.

107. Sichieri, R., and W. R. Schmidek. Influence of ambient temperature on the sleep–wakefulness cycle in the golden hamster. *Physiol. Behav.* 33: 871–877, 1984.

108. Stephan, F. K., and I. Zucker. Circadian rhythms in drinking behavior and locomotor activity of rats are eliminated by hypothalamic lesions. *Proc. Natl. Acad. Sci. USA* 69: 1583–1586, 1972.

109. Steriade, M. Brain electrical activity and sensory processing during waking and sleep states. In: *Principles and Practice of Sleep Medicine*, edited by M. H. Kryger, T. Roth, and W. C. Dement. Philadelphia: Saunders, 1989, p. 86–103.

110. Szymusiak, R., and E. Satinoff. Maximal REM sleep time defines a narrower thermoneutral zone than does minimal metabolic rate. *Physiol. Behav.* 26: 687–690, 1981.

111. Szymusiak, R., E. Satinoff, T. Schallert, and I. Q. Whishaw. Brief skin temperature changes towards thermoneutrality trigger REM sleep in rats. *Physiol. Behav.* 25: 305–311, 1980.

112. Takagi, K. Sweating during sleep. In: *Physiological and Behavioral Temperature Regulation,* edited by J. D. Hardy, A. P. Gaggeand, and J. A. J. Stolwijk. Springfield, IL: Thomas, 1970, p. 669–675.

113. Timbal, J., J. Colin, C. Boutelier, and J. D. Guieu. Bilan thermique en ambience controlee pendent 24 heures. *Pflugers Arch.* 335: 97–108, 1972.

114. Tobler, I., A. A. Borbély, and G. Groos. The effect of sleep deprivation on sleep in rats with suprachiasmatic lesions. *Neurosci. Lett.* 42: 49–54, 1983.

115. Tobler, I., P. Franken, and K. Jaggi. Vigilance states, EEG spectra, and cortical temperature in the guinea pig. *Am. J. Physiol.* 264(*Regulatory Integrative Comp. Physiol.* 35): R1119–R1132, 1993.

116. Trachsel, L., D. M. Edgar, and H. C. Heller. Are ground squirrels sleep deprived during hibernation? *Am. J. Physiol.* 260(*Regulatory Integrative Comp. Physiol.* 29): R1123–R1129, 1991.

117. Twente, J. W., and J. A. Twente. Regulation of hibernating periods by temperature. *Proc. Natl. Acad. Sci. USA* 54: 1058–1061, 1965.

118. Valatx, J. L., B. Roussel, and M. Cure. Sleep and cerebral temperature in rat during chronic heat exposure. *Brain Res.* 55: 107–122, 1973.

119. Walker, J. M., and R. J. Berger. The ontogenesis of sleep states, thermogenesis, and thermoregulation in the Virginia opossum. *Dev. Psychobiol.* 13: 443–454, 1980.

120. Walker, J. M., A. Garber, R. J. Berger, and H. C. Heller. Sleep and estivation (shallow torpor): continuous processes of energy conservation. *Science* 204: 1098–1100, 1979.

121. Walker, J. M., S. F. Glotzbach, R. J. Berger, and H. C. Heller. Sleep and hibernation in ground squirrels (*Citellus* spp): electrophysiological observations. *Am. J. Physiol.* 233(*Regulatory Integrative Comp. Physiol.* 4): R213–R221, 1977.

122. Walker, J. M., E. H. Haskell, R. J. Berger, and H. C. Heller. Hibernation and circannual rhythms of sleep. *Physiol. Zool.* 53: 8–11, 1980.

123. Walker, J. M., E. H. Haskell, R. J. Berger, and H. C. Heller. Hibernation at moderate temperatures: a continuation of slow wave sleep. *Experientia* 37: 726–728, 1981.

124. Walker, J. M., L. E. Walker, D. V. Harris, and R. J. Berger. Cessation of thermoregulation during REM sleep in the pocket mouse. *Am. J. Physiol.* 244(*Regulatory Integrative Comp. Physiol.* 15): R114–R118, 1983.

125. Walter, J., D. Davenne, S. Shoham, C. A. Dinarello, and J. M. Krueger. Brain temperature changes coupled to sleep states persist during interleukin 1–enhanced sleep. *Am. J. Physiol.* 250(*Regulatory Integrative Comp. Physiol.* 21): R96–R103, 1986.

126. Walter, J. S., P. Meyers, and J. M. Krueger. Microinjection of interleukin-1 into brain: separation of sleep and fever responses. *Physiol. Behav.* 45: 169–176, 1989.

127. Wehr, T. A. A brain-warming function for REM sleep. *Neurosci. Biobehav. Rev.* 16: 379–397, 1992.

128. Wenger, C. B., M. F. Roberts, J. A. J. Stolwijk, and E. R. Nadel. Nocturnal lowering of thresholds for sweating and vasodilation. *J. Appl. Physiol.* 41: 15–19, 1976.

129. Wunnenberg, W., G. Merker, and E. Speulda. Thermosensitivity of preoptic neurones in a hibernator (golden hamster) and a nonhibernator (guinea pig). *Pflugers Arch.* 363: 113–123, 1976.

130. Zanchetti, A., G. Baccelli, and G. Mancia. Cardiovascular regulation during sleep. *Arch. Ital. Biol.* 120: 120–137, 1982.

131. Zepplin, H. Mammalian sleep. In: *Principles and Practice of Sleep Medicine*, edited by M. H. Kryger, T. Roth, and W. C. Dement. Philadelphia: Saunders, 1989, p. 86–103.

60. The relationship between food and sleep

JABER DANGUIR | *Institute of Nutrition, Department of Experimental Nutrition, Tunis, Tunisia*

CHAPTER CONTENTS

SLEEP AND INGESTION-RELATED BEHAVIORS are the most time-consuming and life-supporting activities of animals. For this reason, and because they can be easily measured, these behaviors have been intensively studied by psychologists and physiologists. Sleep and ingestion occur in cyclic patterns and are in competition in the sense that they cannot be accomplished simultaneously. Hitherto, most studies have dealt with only one of these behaviors to the exclusion of the other despite numerous reports about their interrelationship. Among these, early hypotheses concerning the relation between sleep and energy expenditure considered sleep only as an energy-conservation state resulting from the inhibition of muscular activity (46, 51). Subsequently, a comparative study of sleep in various mammalian species showed positive correlations between sleep and energy metabolism (79). Contrary to the viewpoint of energy conservation, sleep may instead play a direct role in the regulation of energy expenditure.

As far as sleep and brain metabolism are concerned, there are contradictory data. Although several authors have found no changes in the cerebral metabolism of sleeping animals (74), other investigators have reported an increased uptake of inorganic phosphates (61), glycogen (43), or adenosine triphosphate (ATP) (61) within the brain during sleep. Furthermore, numerous studies have shown an increase in neuronal activity (28), cerebral blood flow (62), hypothalamic temperature (59), and glucose utilization (36, 60) during various phases of sleep.

A relation between sleep and metabolism can also be inferred from studies showing that several hormones involved in metabolic processes, such as growth hormone (61, 72), prolactin (68), luteinizing hormone (LH) (65), and testosterone (27), exhibit either increased or nearly exclusive secretion during sleep.

In all of these studies one factor was taken into consideration: metabolism. Surprisingly, however, investigators have usually considered the relation between sleep and metabolism only from the viewpoint of what happens to metabolism during sleep and the function of such a state, the idea being that sleep is essentially a process, if not the cause, of metabolic restoration (1). The possibility that the metabolic rate could be one of the primary stimuli for sleep has infrequently been raised, even though everyone has experienced the tendency toward sleep following a big meal.

Since the early 1980s, the relation between feeding and its metabolic consequences and the sleep–waking cycle has been investigated, particularly focusing on the repercussions of peripheral and/or central metabolic fluctuations on sleep.

INTERACTION BETWEEN FEEDING AND SLEEP

Quantitative Aspects

In an initial series of studies (17) we showed a highly significant correlation between the number of calories ingested during a meal and the duration of subsequent sleep in rats. Thus the larger the amount of food ingested during a meal the longer the rat will sleep until the next meal. Any spontaneously or experimentally increased caloric intake should be followed by a proportional increase in the duration of sleep. Conversely, any decrease in food intake should be accompanied by a proportional decrease in the duration of sleep.

Effects of Hyperphagia on Sleep.

In the rat there are two ways of inducing hyperphagia. The first is the presentation of a mixed, palatable, energy-dense choice of foods, referred to as a "cafeteria" diet (13). The second is the classic surgical method of lesioning the ventromedial hypothalamus (VMH).

As shown in Figure 60.1, the cafeteria diet resulted in a significant increase in the daily duration of both slow-wave sleep (SWS) and paradoxical sleep (PS) during the 10 days when a typical cafeteria diet (white bread, chocolate, savory biscuits) was offered. When the cafeteria diet was withdrawn, SWS and PS remained elevated for several days before returning to normal levels (13).

Similar effects on sleep were observed in rats with VMH lesions (18). Such lesions produce a complex syndrome characterized by disturbances of feeding behavior (7). Two distinct phases have been described in this syndrome. The initial dynamic phase is characterized by rapid body weight gain and hyperphagia. It is followed by a static phase, characterized by a relatively stable body weight and feeding comparable to that of normal intact rats. We found that SWS and PS were significantly increased during the dynamic phase and that both stages of sleep returned, as did food intake, to normal values during the static phase (Table 60.1) (18).

Effects of Food Deprivation on Sleep.

The effect of starvation on sleep has been considered in studies of food deprivation (6, 38, 49) and of alteration of the feeding schedule (52). Results have not been consistent. In addition to differences in procedure, none of these experiments took into account the possible importance of the availability of endogenous metabolites as a factor expressing energy deprivation at the cellular level. We have shown that this factor is significant in several

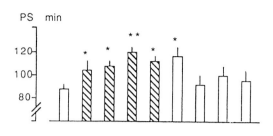
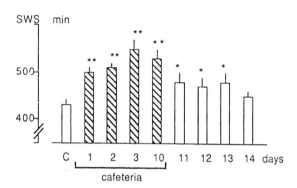

FIG. 60.1. Daily durations of slow-wave sleep *(SWS)* and paradoxical sleep *(PS)* expressed in minutes (mean ± SEM) during the control (C) day, the first 3 days *(1, 2, 3)*, and the tenth day *(10)* during a cafeteria diet presentation and during the 4 days *(11, 12, 13, 14)* when the cafeteria diet was withdrawn.
*$P < 0.01$; **$P < 0.001$ (paired *t* test, experimental vs. control days).

respects. We (19) observed that, although the daily amounts of SWS and PS were progressively decreased in lean (body weight = 200 g) rats during 4 days of food deprivation (Fig. 60.2), sleep was unaffected in large (380 g) rats and also in VMH-lesioned obese (500 g) rats. For VMH-lesioned rats, a much longer food deprivation (more than 6 days) was needed to decrease sleep.

Furthermore, oral restitution of food or direct intravenous (i.v.) infusion of glucose following the period of starvation resulted in an immediate increase in both stages of sleep (19). From these observations, sleep is

TABLE 60.1. *Proportion of Time Feeding and Sleeping in VMH-Lesioned Rats*

Time	Phase	%SWS	%PS	%Waking for Feeding
Light	Normal	48.5 + 1.7	10.4 + 0.4	2.9
	VMH-lesioned dynamic phase	43.4 + 1.1	8.4 + 0.3	11.4
	VMH-lesioned static phase	36.6 + 0.6	5.6 + 0.4	6.3
Dark	Normal	33.0 + 2.3	5.2 + 0.3	10.7
	VMH-lesioned dynamic phase	42.5 + 0.6*	9.5 + 0.3*	14.4
	VMH-lesioned static phase	36.7 + 0.8	7.9 + 0.3	7.9

*$P < 0.001$ (VMH-lesioned dynamic vs. normal).
Slow-wave sleep, PS, and feeding activity expressed as percentages (mean + SEM) of the recording time in VMH hyperphagic rats in the dynamic phase and in the static phase of the syndrome and in normal rats during 12 h light and 12 h dark.

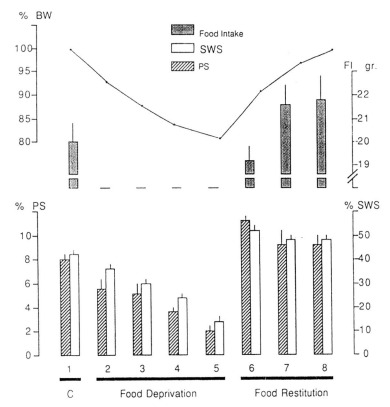

FIG. 60.2. Slow-wave sleep *(SWS)* and paradoxical sleep *(PS)* expressed as percentages (mean ± SEM) of the recording time and body weight *(BW)* expressed as percentages of the control (C) day, during the control day, the 4 days of food deprivation, and the 3 days following restitution of food. Daily food intake *(FI)* was compared between control and refeeding days.

reduced only when the availability of endogenous substrates at the cellular level falls.

Qualitative Aspects

As shown above, the duration of sleep is dependent on the number of calories ingested. In addition, we believe that sleep is also related to the particular macronutrient source of the calories. This conclusion was made possible by studies in which the daily caloric intake was provided through chronic i.v. infusions of equicaloric amounts of glucose, amino acids, or fat (22).

Chronic i.v. infusions of glucose brought about a significant and specific increase of SWS in ad libitum fed rats, while PS remained unchanged (Table 60.2) (22). On the contrary, when a solution of amino acids was infused, there was a selective increase of the daily duration of PS, and SWS remained unchanged. The i.v. infusion of the daily caloric needs of the rat in the form of a lipid solution had no significant effect on the daily amounts of either SWS or PS. Interestingly, the i.v. infusion of a solution containing sugars, lipids, and amino acids resulted in a significant increase in the daily duration of both SWS and PS (Table 60.2) (22).

The exclusive amino-acid infusion, though covering only 60% of the caloric needs (because of the limit of their tolerance by the rat), increased the number of PS episodes (60%–70%), and therefore its total duration, and had no effects on SWS. The finding that amino-acid supply favors specifically PS is in agreement with previous studies suggesting that during PS protein synthesis is enhanced (34, 57). This view is also supported by

TABLE 60.2. *Effect of Specific Macronutrient Infusions on Sleep*

	SWS (%)	PS (%)
Control	40.7 + 1.5	7.2 + 0.2
Glucose	47.2 + 2.1*	6.3 + 0.7
Fat	35.6 + 1.2	6.5 + 0.4
Amino acids	36.8 + 1.4	9.3 + 0.2*
Composite solution	48.5 + 1.4*	9.5 + 0.4*

Slow-wave sleep and PS expressed as percentages (mean + SEM) of the 24 h day in four groups that received their daily caloric needs through i.v. infusions of either glucose, fat, amino acids, or a composite solution. Results were compared to amounts of sleep observed during the control saline infusion days.   *P < 0.001.

observations that large amounts of PS occur during early postnatal ontogeny (40, 63) when increased protein synthesis is occurring and, to some extent, following learning (48), a process which may be related to formation of new proteins (32).

Ischymetric Modulation of Sleep

The above observations may support the hypothesis that sleep can be potentiated by a factor proportional to the rate of utilization of the energy-supplying metabolites. Interestingly, a similar hypothesis was also proposed to explain the onset of hunger and meal taking as a response to the diminution of the metabolic rate (55) and is now known as the ischymetric theory (from Greek, *ischys* = power). At the core of this hypothesis is the view that there are cells that monitor their own power production or the turnover of metabolic substrates as a function of both their effective concentrations and the endogenous cofactors of their processing.

It is tempting to extend the ischymetric hypothesis to sleep. According to this hypothesis, the sensing of increased ischymetry would favor a subsequent augmentation of sleep and satiety. Conversely, the sensing of decreased ischymetry would result in sleep loss and in proportionally specific hunger arousal.

Several experiments support the concept of an ischymetric modulation of sleep (22). For example, replacement of the daily oral caloric needs by a continuous i.v. infusion of glucose did not affect the duration of sleep of normal food-deprived rats. On the contrary, infusion of a similar quantity of glucose but paralleled by an infusion of insulin brought about a significant increase of daily sleep duration (Table 60.3). Similarly, if the glucose infusion was made discontinuous (10 min on, 10 min off), thus increasing the endogenous insulin secretory responsiveness (55), sleep duration was also enhanced (Table 60.3).

These experiments verify the fact that the systemic

TABLE 60.3. *Amounts of Sleep in Rats Receiving Discontinuous Infusion and Continuous Coinfusion of Glucose Plus Insulin*

	SWS (%)	PS (%)
Control (saline)	40.7 + 1.5	7.2 + 0.2
Discontinuous glucose	46.1 + 4.4*	9.3 + 0.6†
Glucose and Insulin	47.8 + 1.4†	8.8 + 0.4*

Slow-wave sleep (SWS) and paradoxical sleep (PS) expressed as percentages (mean + SEM) of the 24 h day in two groups receiving daily caloric needs through either discontinuous glucose infusion or continuous coinfusion of glucose plus insulin for 3 consecutive days. Results were compared to amounts of sleep observed during the control saline infusion days when rats had standard food available ad libitum.
*$P < 0.05$; †$P < 0.01$.

availability of nutrients, particularly when they are furnished parenterally, does not imply a proportional cellular uptake and metabolism. The latter depends also on the concentration of hormones that favor the cellular penetration and the utilization of the circulating substrates. This is particularly established for the insulin-dependent metabolic events because the discharge of this hormone depends not only on systemic but also on orogastric stimulations (53, 54). This could explain why the systemic supply of glucose alone (when continuously administered) in the absence of oral food had no effect on sleep and also why parallel insulin infusion increased sleep in the same condition.

The ischymetric sleep modulation hypothesis could therefore explain the different types of correlation between meal size and subsequent sleep mentioned above. It is supported by the findings that the hyperphagia induced by VMH lesions is also accompanied by hypersomnia (18) and that there are sleep deficits in the lateral hypothalamic syndrome (21), which is characterized by aphagia and adipsia. The ischymetric sleep modulation hypothesis could also explain the decreased sleep following metabolic blockade by means of 2-deoxyglucose injection (58).

Circadian Sleep Patterns: Possible Dependence on Lipolysis and Lipogenesis

In the rat the 12 h/12 h dark–light daily period comprises two metabolically distinct periods. The first is characterized by a high potential for insulin secretion (31, 39) and, therefore, a tendency for lipogenesis. The behavioral consequences of this metabolic state are a transitory physiological hyperphagia and body weight gain. During the same dark period, sleep displays specific properties in terms of durations of SWS and PS (52). The second metabolically distinct period is characterized by a low potential for insulin secretion (31, 39) and, therefore, a tendency for lipolysis, resulting in physiological hypophagia and body weight loss. During the same light period, sleep displays a characteristic pattern (52).

Therefore, it could be hypothesized that the circadian variations of related behaviors, such as sleep and feeding, are due, at least in part, to fluctuations in the lipogenic vs. lipolytic rate resulting from light-dependent sequestration vs. release of circulating metabolites (38). In other words, if the sequence of events influencing sleep and feeding were the light–dark triggering of oscillators favoring either lipolysis or lipogenesis, one would expect to be able to reverse these cyclic behaviors by acting only on the intermediary link, that is, enhancement of lipogenesis or lipolysis.

Under normal lighting conditions (light from 0800 h to 2000 h), an attempt to counteract the natural endo-

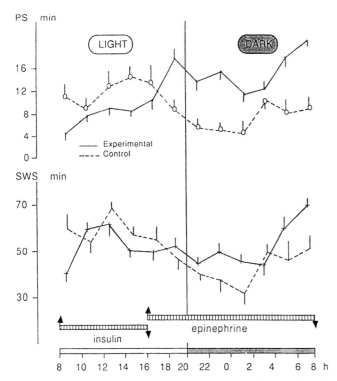

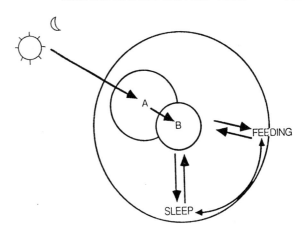

FIG. 60.4. Schematic representation of the possible mechanisms underlying the relation between feeding and sleep. These behaviors are directly influenced by metabolic rate (*B*) fluctuations (ischymetric control), which are dependent on the hypothalamic control of peripheral secretion of metabolic hormones (*A*). The latter are in turn triggered by the light–dark cycle.

FIG. 60.3. Time spent in slow-wave sleep *(SWS)* and in paradoxical sleep *(PS)* (mean ± SEM), expressed in minutes on a 2 h period base throughout day, in rats under control saline infusion *(dashed lines)* and following alternating insulin–epinephrine infusions *(solid lines)*. Insulin was infused from 0800 to 1600 and replaced by epinephrine from 1600 to 0800. Light cycle, 0800 to 2000.

crine command for lipogenesis and lipolysis was made by means of exogenous hormonal supply, that is, a continuous i.v. infusion of insulin (lipogenic hormone) during the normally lipolytic light period that was then replaced by epinephrine (lipolytic hormone) during the normally lipogenic dark period. Such an experimental reversal of the lipogenic vs. the lipolytic rate resulted in an inversion of both meal and sleep patterns (Fig. 60.3) (20). In other words, despite the fact that the normal lighting cycle was maintained, the rat slept more and ate less during the night, whereas hyperphagia and hyposomnia were observed during the light period.

Obviously, these results support the hypothesis that the lighting cycle affects feeding and sleep indirectly through fluctuations of the metabolic rate (and therefore of the ischymetric signal) that are themselves influenced by the hypothalamic control (light-dependent) of the secretion of metabolic hormones by peripheral organs, as illustrated in Figure 60.4.

Nutritional States and Sleep Cycles

As far as both stages of sleep are concerned, most of the experimental manipulations of the metabolism seem to affect SWS and PS nonselectively. In some situations PS seems to be more affected than SWS by nutritional states. However, according to our observations, it seems premature to conclude that SWS is related to tissue restoration and that PS is implicated in memory functions, as suggested (1). Such an assumption was based solely on findings showing that the secretion of various anabolic hormones (growth hormone, prolactin, etc.) is related, at least in human beings, to SWS.

Most of the previously mentioned results express the effect on sleep in terms of total duration. In fact, the effects of nutritional states on sleep seem to be even better reflected in the number and/or the duration of SWS and PS episodes. Indeed, we have observed that in situations in which total sleep duration was unchanged compared to controls SWS and/or PS episodes were dramatically affected. For example, replacement of oral caloric intake by an exclusive i.v. infusion of lipid solution had no effect on the daily duration of sleep. However, under the same lipid infusion the number of PS episodes was higher during the light phase, whereas the mean duration of SWS episodes was significantly shortened (50%) during the dark phase. Interestingly, these results are in agreement with the idea of an ischymetric modulation of sleep states. The increased number of PS episodes during the light phase under i.v. lipid infusion could be interpreted as reflecting facilitated metabolic processing of these substrates since lipolysis prevails during that phase. Conversely, the fact that the dark phase coincides with predominant lipogenesis would not allow an appropriate cellular utilization of the infused lipids, which could explain the shortening of SWS episodes during this period of the nycthemeron.

CENTRAL CONTROL OF FOOD–SLEEP INTERACTION

Serotonin as a Possible Link

Although peripheral metabolic manipulations seem of primary importance in sleep, it is reasonable to hypothesize that the end factor(s) which may trigger sleep or waking (and satiety or hunger) should be at the central levels. Such an assumption is strengthened by numerous studies showing that the same neurochemical mechanisms modulate both sleep and feeding. For example, this is so in the case of monoamines, principally serotonin (5, 37, 41). Interestingly, all neurotransmitters whose syntheses are now known to be influenced by availability of precursors are produced from compounds that must be obtained, in whole or in part, from the diet (78). As far as serotonin is concerned, its precursor, namely tryptophan, cannot be synthesized at all by mammalian cells. Consumption of a meal was shown to increase plasma tryptophan concentration and, therefore, brain serotonin synthesis (30). Accordingly, one could reasonably attribute the relation between food intake and the postprandial duration of sleep to changes in serotonin synthesis and/or release resulting from an increased peripheral tryptophan concentration.

If serotonin had some role in the relation between feeding and sleep, one would expect that any alteration in the metabolism of this monoamine would result in parallel effects on both behaviors. One way of altering serotonin metabolism is by pharmacological manipulation. We have performed experiments using p-chlorophenylalanine (PCPA, an inhibitor of serotonin synthesis) administration in electroencephalogram (EEG) recorded rats. This resulted in an almost total suppression of SWS and PS, but, interestingly, this was preceded by a dramatic reduction of food intake and a consequent body weight loss (24). These results might support the hypothesis of a common serotonergic mechanism underlying the relation between sleep and feeding.

However, according to the findings concerning reduced sleep under food deprivation conditions, one would argue that the PCPA-induced reduction of food intake could be responsible for the inhibition of sleep. This seems not to be the case, however, since i.v. glucose loadings in PCPA-treated rats did not affect sleep.

Therefore, the inhibitory action of PCPA on feeding must be due to some other mechanisms. As a matter of fact, besides its well known inhibitory action on serotonin synthesis, PCPA has been shown to inhibit insulin secretion also, and several authors have reported an active role for serotonin in the regulation of pancreatic islet insulin secretion (4). The possibility that insulin deficiency is directly responsible for PCPA-induced sleep alteration could reasonably be considered. To examine

such a hypothesis, we have administered insulin at nonconvulsive doses to rats made insomniac with PCPA. The acute intraperitoneal (i.p.) injection of insulin to rats made insomniac with PCPA resulted in the reappearance of normal sleep cycles lasting 8 h (presumably the period of effectiveness of the injected dose) followed by the reappearance of insomnia. When insulin was continuously supplied through chronic i.v. infusions, sleep reappeared immediately and lasted as long as the hormone was administered (24).

Of course, these results can be interpreted in terms of the peripheral action of insulin. Indeed, it has been shown that insulin stimulated the transport of neutral amino acids, including tryptophan, from blood to brain (29). Should this be the case, the enhancement of sleep in these particular experiments would not be due to an increased serotonin turnover since the latter was precisely inhibited by PCPA. A more plausible explanation is the insulin-induced increase of peripheral and/or central metabolic rate since, as we have argued in this review, metabolism is closely related to sleep.

Insulin and Slow-Wave Sleep

Effect of Peripheral Administration of Insulin on Sleep. We have observed that the continuous i.v. infusion of insulin in rats resulted in an increase of the daily duration of SWS in spite of the large increase in feeding induced by this hormone (22). Sangiah et al. (67) obtained similar results following an acute i.p. injection of a subconvulsive dose of insulin. We initially suspected that insulin produced sleep through peripheral actions, that is, by reducing blood sugar and affecting other metabolic aspects which might then indirectly alter brain neurochemical functions. However, when insulin (75) and insulin receptors (35) were discovered in the rodent brain, a direct action of this blood-borne polypeptide on neuronal function was considered. Experiments in which insulin was administered through the intracerebroventricular (i.c.v.) route tested this hypothesis.

Effect of Central Infusion of Insulin on Sleep. Chronic i.c.v. infusion of exogenous insulin caused a selective increase of the daily duration of SWS, whereas the duration of PS remained unchanged (23). The increase of SWS was observed immediately following onset of the infusion and lasted as long as insulin administration continued (Fig. 60.5). The insulin-induced increase of SWS was observed during both the light and the dark phases of the circadian cycle (Fig. 60.5).

The mechanism of action of insulin on sleep and the origin of brain insulin are not clear. The existence of brain insulin, together with the effectiveness of insulin

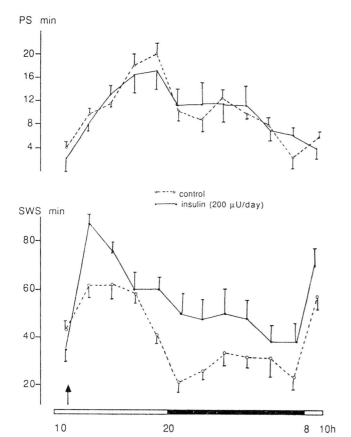

FIG. 60.5. Time spent in slow-wave sleep *(SWS)* and in paradoxical sleep *(PS)*, expressed in minutes (mean ± SEM) on a 2 h period base throughout the nychthemeron, in rats under control conditions and following continuous i.c.v. infusion of insulin. Arrow indicates beginning of infusion.

after peripheral administration, suggest a direct central effect of the circulating hormone. Such reasoning requires that circulating insulin be capable of reaching brain sites. Insulin has not been shown to cross the blood–brain barrier, but the high cerebrospinal fluid (CSF)/plasma ratio for insulin (77) indicates that this peptide may be selectively transported into the brain, perhaps through one or more circumventricular organs, which are characterized by the absence of a blood–brain barrier.

If such a hypothesis were correct, one would expect that the inactivation of peripheral endogenous insulin would result in a selective suppression of SWS. We tested the effects of this manipulation.

Effect of Inactivation of Endogenous Insulin on Sleep.
Specific insulin antiserum (IAS) was continuously infused i.v. or i.c.v. With both routes of infusion, IAS produced a significant decrease in the daily duration of SWS but no change in the duration of PS (23). Twenty-four hours after the end of IAS administration, the dura-

tion of SWS remained lower than the level on control days. No secondary rebound in SWS was observed 48 h after the termination of treatment.

These results are strong evidence against a nonspecific action of insulin. Furthermore, they suggest that the brain insulin responsible for shifting SWS originates from the periphery.

Sleep in Diabetes.
The involvement of insulin in producing SWS was confirmed by an investigation of sleep in diabetic rats (11). Diabetes mellitus was induced in rats by an i.v. injection of streptozotocin, a drug which acts specifically on the pancreatic B (insulin-producing) cells in mammals. As shown in Figure 60.6, the daily durations of both SWS and PS were decreased on the third day following the injection of streptozotocin. However, only SWS remained lower than control levels 2 wk after the induction of diabetes. The suppression of SWS in diabetic rats was due to the absence of endogenous insulin because i.v. administration of exogenous insulin to the diabetic animals restored SWS. Once again, this procedure did not affect PS. Furthermore, the deficit in SWS in diabetic rats was not the consequence of numerous peripherally induced metabolic disturbances caused by the absence of insulin, since a dose-dependent restoration of SWS was also observed when exogenous insulin was infused, not peripherally but directly, into the cerebral ventricle (Fig. 60.6).

The way insulin acts on the brain to induce SWS remains unknown, and only tentative hypotheses can be proposed at this point. The sleep-inducing effect of insulin could be the consequence of some metabolic action in the brain; the peptide might function as a neurotransmitter or a neuromodulator, or some combination of mechanisms may be involved.

Whether the effects of insulin on sleep are physiologically significant remains an open question. What is its role in the rat's normal wake–sleep cycle? To answer this question, we might recall that in the rat the 12 h/12 h dark–light periods comprise two metabolically distinct periods. The dark phase is characterized by increased insulin secretion (31) and, therefore, by hyperphagia and hyposomnia (20). The light phase is characterized by decreased insulin secretion and, therefore, by hypophagia and hypersomnia. Accordingly, if peripheral insulin is active on sleep once it reaches cerebral tissue, one might expect that it would be progressively stored within the brain throughout the dark phase and then would act to induce sleep during the light phase. Although no conclusive evidence is available, several arguments favor such a hypothesis. For example, it has been shown in the rat that when food is offered only during the light period (to produce a reversal of the normal pattern of insulin secretion) sleep is signifi-

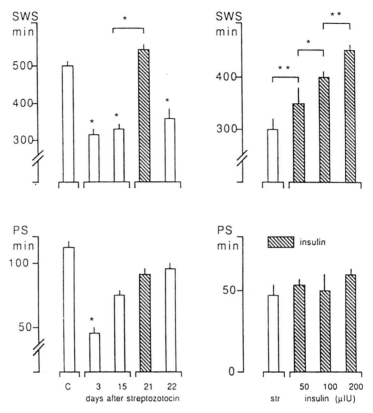

FIG. 60.6. *Left:* Daily durations of slow-wave sleep *(SWS)* and paradoxical sleep *(PS)* (mean ± SEM) on the third day after the injection of streptozotocin and 15 days later, compared to control values *(C)* and following i.v. infusion of exogenous insulin (day 21) and its removal (day 22). *Right:* Specific dose-dependent restoration of *SWS* following chronic i.c.v. infusion of three different doses of insulin during days 4, 5, and 6 after administration of streptozotocin, compared to values obtained on day 3 poststreptozotocin *(str).*
*$P < 0.05$; **$P < 0.01$ (paired t test).

cantly increased during the dark period (52). Similarly, continuous i.v. administration of insulin during the 12 h light period was shown to bring about a significant increase of sleep during the night (20).

Somatostatin and Paradoxical Sleep

Effect of Central Administration of Somatostatin on Sleep. When chronically infused into the third ventricle, somatostatin (SRIF) elicited a specific increase in the daily duration of PS; SWS remained unaffected (Fig. 60.7) (12). These results suggest that SRIF mediates dietary protein–induced increments in PS (69). In addition, they make suspect previous findings relating growth hormone and PS.

A number of studies in rats (26), cats (71), and humans (50) indicate that growth hormone administration results in an elevation of PS. These findings have led to speculations that growth hormone secretion may be involved in PS regulation, but since growth hormone administration has been reported to induce stimulation of SRIF release both in vivo (8) and in vitro (70), the

effects on PS could be the result of increased SRIF release rather than being solely due to an increase in growth hormone.

A report by Olivo et al. (56) strengthens the case for the involvement of SRIF in PS. These authors found that PS was increased in adult rats treated at birth with monosodium glutamate (MSG). Neonatal treatment with MSG is known to destroy the cell bodies located within the arcuate nucleus of the hypothalamus and the periventricular system (44); in addition, Utsumi et al. (73) observed a large increase in plasma SRIF in this animal preparation. Therefore, it is possible that the increased PS in these rats is not due to the lesion per se but to its consequences on the release of SRIF.

As was the case for insulin, a role for SRIF in PS was confirmed in experiments in which endogenous SRIF was inactivated.

Effect of Inactivation of Somatostatin on Sleep. The neutralization of central SRIF by chronic i.c.v. infusion of a specific SRIF antiserum resulted in a selective suppression of PS, whereas SWS remained unchanged. This

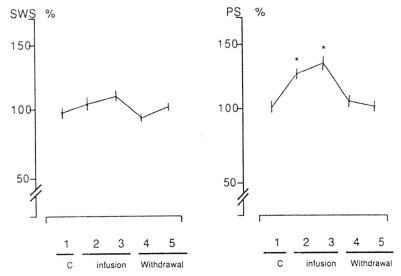

FIG. 60.7. Slow-wave sleep *(SWS)* and paradoxical sleep *(PS)* expressed as percentages (mean ± SEM) of the level on the control (C) day, during the 2 days rats received chronic i.c.v. administration of somatostatin (20 μg/24 h) and during the 2 days which followed its withdrawal. *P < 0.01 (paired *t* test).

result confirmed the role of SRIF in the generation of PS. Moreover, the fact that no secondary rebound in PS was observed after the termination of SRIF antiserum infusion suggests a true hypnogenic role for SRIF, in accordance with the criteria proposed for the definition of sleep-promoting substances (42). These results suggest that the triggering of PS by dietary protein may be due to an increase in SRIF secretion caused by these macronutrients. This possibility was tested in an experiment in which rats received both an i.v. infusion of proteins and a subcutaneous injection of cysteamine, a drug which depletes peripheral and central SRIF. The results showed that the increased PS which follows protein administration was prevented; in fact, a significant decrease occurred compared to control values.

Thus we suggest that SRIF secretion during the waking period, occurring under the influence of protein ingestion, would allow its storage until it triggers PS during sleep. The finding that the daily rhythm of SRIF in the CSF of the rhesus monkey varies, with high levels occurring during darkness (3) (the period during which the animal sleeps), favors such an hypothesis.

Somatostatin and the Cholinergic Theory of Paradoxical Sleep. The case for a role of SRIF in the regulation of PS was strengthened by our (16) demonstration that the increase of PS which follows the injection of cholinergic agonists such as carbachol could be abolished by the administration of SRIF antiserum. In addition, we showed that the well-known suppression of PS by the muscarinic receptor blocker scopolamine was reversed by the administration of a long-lasting and potent

somatostatin analog (octreotide) (15). These data suggest that, contrary to the prevailing concept that the relationship between cholinergic mechanisms and PS is direct (2, 33), acetylcholine may influence the generation of PS through some intervening action on endogenous SRIF.

SLEEP IMPAIRMENTS DURING ADVANCED AGE: POSSIBLE INVOLVEMENT OF PERIPHERAL HORMONAL DEFICITS

The duration of sleep, particularly of PS, declines gradually and substantially from youth to old age (64). The reasons for such age-related sleep loss are still unknown. Also, no substances that improve sleep specifically in old animals or humans are known as yet.

We (14) showed that either i.p. or oral intake of a long-lasting somatostatin analog (octreotide) resulted in a dose-dependent and selective increase of the daily duration of PS in 800 day old rats (Fig. 60.8). These findings are important for several reasons. First, they confirm the role of SRIF in the generation of PS. Second, they favor the involvement of peripheral rather than central dysfunctions in the hyposomnia which develops with age.

The prevailing theories have suggested that sleep loss during aging might be the consequence of only central degenerative processes. This view was essentially based on evidence that several neurotransmitter systems, among them the cholinergic system, undergo reliable changes with advancing age.

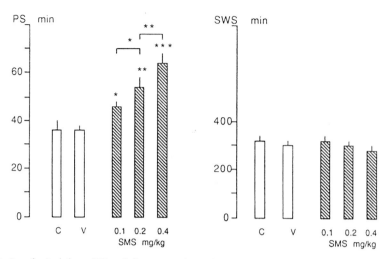

FIG. 60.8. Paradoxical sleep *(PS)* and slow-wave sleep *(SWS)* (mean ± SEM) in rats which received i.p. injections of three different doses of the somatostatin analogue *(SMS)*, compared to values observed under control *(C,* no infusion) or vehicle *(V)* administration. *$P < 0.05$; **$P < 0.01$; ***$P\ 0.001$ (paired t test).

Although no data are available on the levels of peripheral SRIF in aged animals, our results support the hypothesis of an age-related alteration in SRIF secretion. Furthermore, our results raise the question of the possible involvement of PS loss in memory impairments observed in aged rats. Evidence suggests that PS may play a key role in memory processing (48). However, it was shown that SRIF influenced both consolidation and retrieval processess (76). Should this be the case, the restoration of normal amounts of PS by SRIF in aged rats could open a promising outlook in suggesting therapeutic interventions.

WATER INTAKE AND SLEEP: THE CASE OF DIABETES INSIPIDUS

The role of water balance in sleep has never been studied. The only report referring to the incidence of thirst and water intake on sleep cycles concerned Brattleboro diabetes insipidus rats (10), which exhibit hereditary lack of vasopressin. These rats also show sleep deficits, particularly a decrease in PS, which were not due to the lack of vasopressin but to the necessity for these rats to continuously rehydrate themselves. This results in an elevated number of drinking episodes (once every 5–10 min), which often interrupt sleep. Therefore, it seems that water (and probably hydromineral balance) may affect sleep but only indirectly and in extreme deficit situations, as was the case for diabetes insipidus.

CONCLUSION

The studies reviewed above suggest that insulin and SRIF are directly involved in the control of SWS and PS, respectively.

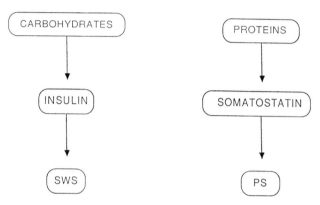

FIG. 60.9. Hypothetical schema of the respective effects of carbohydrates and proteins upon slow-wave sleep *(SWS)* and paradoxical sleep *(PS)*.

The secretions of these hormones are closely related to both the quantity and the quality of food. Carbohydrates, which are known to be particularly potent in producing insulin secretion, seem to affect SWS preferentially. Protein diets, which stimulate SRIF secretion, trigger PS (Fig. 60.9).

These findings allow some speculation on the respective amounts of these stages of sleep in different mammalian species. For example, elevated amounts of PS are observed in species ingesting diets high in protein, such as cats or dogs. Conversely, in species eating diets high in carbohydrates, such as herbivores, a decreased amount of PS and an increased amount of SWS are observed. The relative amounts of the two stages of sleep in species eating more balanced protein/carbohydrate ratios are intermediate. Although no comparative study of respective amounts of insulin and SRIF in mammalian species is available, one would expect that modifications of the protein/carbohydrate ratio in a given species would shift SWS and PS in the expected directions. This prediction was in fact observed by Ruckebusch and Gaujoux (66), who allowed sheep access to a high-protein diet and found a large increase of PS (this species normally exhibits very low levels of PS). Whether the increased PS was accompanied by an increased SRIF secretion was not determined by these investigators.

The role of carbohydrate vs. protein diets in SWS and PS was confirmed in humans (25). Sleep stages were recorded in athletes submitted to a "Scandinavian carbohydrate loading regimen," which consisted of 3 days of a diet rich in fat and protein followed by 3 days of a high-carbohydrate diet. The results showed that PS was significantly and exclusively increased under the protein and fat diets, while a significant increase of SWS was found after the carbohydrate diet. These results might be interpreted as the consecutive change in the hormonal level of insulin and SRIF, but the latter were not investigated in this study.

In conclusion, the above findings suggest that feeding events initiate fundamental sleep processes. Furthermore, they raise practical questions about the possible involvement of metabolic diseases in certain human sleep pathologies. Two types of pathology are at least partly due to a disturbance of feeding. These are the insomnia observed in anorexia nervosa patients, which is partially reversed by forced feeding (9), and the hypersomnia of the Kleine-Levin syndrome, which is accompanied by hyperphagia (45, 47). The studies reviewed here indicate that investigations of some of the feeding dysfunctions of patients suffering from disorders of sleep would be fruitful.

The author thanks J. Elati for helpful revision of the manuscript and S. Chebbi for typing it.

REFERENCES

1. Adam, K., and I. Oswald. Sleep is for tissue restoration. C. R. Coll. Physicians (Lond.) 11: 376–388, 1977.
2. Amatruda, T. A., D. A. Black, T. A. McKenna, R. W. McCarley, and J. A. Hobson. Sleep cycle control and cholinergic mechanisms: differential effects of carbachol injections at pontine brainstem sites. Brain Res. 98: 501–515, 1975.
3. Arnold, M. A., M. J. Perlow, S. M. Reppert, O. P. Rorstad, and J. B. Martin. Daily pattern of somatostatin in the cerebrospinal fluid of the rhesus monkey: effect of environmental lighting. Ann. Neurol. 8: 104, 1980.
4. Bird, J. L., E. E. Wright, and M. Feldman. Pancreatic islets: a tissue rich in serotonin. Diabetes 4: 304–308, 1980.
5. Blundell, J. E. Is there a role of serotonin (5 hydroxytryptamine) in feeding? Int. J. Obes. 1: 15–42, 1977.
6. Borbely, A. A. Sleep in the rat during food deprivation and subsequent restitution of food. Brain Res. 124: 457–471, 1977.
7. Brobeck, J. R., J. Tepperman, and C. N. H. Long. Experimental hypothalamic hyperphagia in the albino rat. Yale J. Biol. Med. 15: 831–853, 1943.

8. Chihora, K. Stimulation by rat growth hormone of somatostatin release into hypophyseal portal blood [Abstract]. In: 61st Annual Meeting of the Endocrine Society, Anaheim, 1979, p. 290.
9. Crisp, A. H. Sleep, activity and mood. Br. J. Psychiatry 137: 1–7, 1980.
10. Danguir, J. Sleep deficits in rats with hereditary diabetes insipidus. Nature 304: 163–164, 1983.
11. Danguir, J. Sleep deficits in diabetic rats: restoration following chronic intravenous or intracerebroventricular infusions of insulin. Brain Res. Bull. 12: 641–645, 1984.
12. Danguir, J. Intracerebroventricular infusion of somatostatin selectively increases paradoxical sleep in rats. Brain Res. 367: 26–30, 1986.
13. Danguir, J. Cafeteria-diet promotes sleep in rats. Appetite 8: 49–53, 1987.
14. Danguir, J. The somatostatin analogue SMS 201–995 promotes paradoxical sleep in aged rats. Neurobiol. Aging 10: 367–369, 1989.
15. Danguir, J., and S. De Saint-Hilaire-Kafi. Scopolamine-induced suppression of paradoxical sleep is reversed by the somatostatin analogue SMS 201–995 in rats. Pharmacol. Biochem. Behav. 30: 295–297, 1988.
16. Danguir, J., and S. De Saint-Hilaire-Kafi. Somatostatin antiserum blocks carbachol-induced increase of paradoxical sleep in the rat. Brain Res. 20: 9–12, 1988.
17. Danguir, J., H. Gerard, and S. Nicolaidis. Relations between sleep and feeding patterns in the rat. J. Comp. Physiol. Psychol. 93: 820–830, 1979.
18. Danguir, J., and S. Nicolaidis. Sleep and feeding patterns in the ventromedial hypothalamic lesioned rat. Physiol. Behav. 21: 769–777, 1978.
19. Danguir, J., and S. Nicolaidis. Dependence of sleep on nutrients' availability. Physiol. Behav. 22: 735–740, 1979.
20. Danguir, J., and S. Nicolaidis. Circadian sleep and feeding patterns in the rat: possible dependence on lipogenesis and lipolysis. Am. J. Physiol. 238(Endocrinol. Metab. 1): E223–E230, 1980.
21. Danguir, J., and S. Nicolaidis. Cortical activity and sleep in the rat lateral hypothalamic syndrome. Brain Res. 185: 305–321, 1980.
22. Danguir, J., and S. Nicolaidis. Intravenous infusion of nutrients and sleep in the rat: an ischymetric sleep regulation hypothesis. Am. J. Physiol. 238(Endocrinol. Metab. 1): E307–E312, 1980.
23. Danguir, J., and S. Nicolaidis. Chronic intracerebroventricular infusion of insulin causes selective increase of slow wave sleep in rats. Brain Res. 306: 97–103, 1984.
24. Danguir, J., and S. Nicolaidis. Feeding, metabolism and sleep: peripheral and central mechanisms of their interaction. In: Brain Mechanisms of Sleep, edited by D. J. McGinty, R. Drucker-Colin, A. Morrison and P. L. Parmeggiani. New York: Raven, 1985, p 321–340.
25. Davenne, D., A. L. Francart, and A. Renaud. The effects of two types of diet on subsequent sleep. Proc. 10th Congr. Eur. Sleep Res. Soc. Gustan Fischer Verlag: Stuttgart, 1990, p. 106.
26. Drucker-Colin, R. R., C. W. Spanis, J. F. Sassin, and J. F. McGaugh. Growth hormone effects on sleep and wakefulness in the rat. Neuroendocrinology 18: 1–8, 1975.
27. Evans, J. I., A. W. McLean, A. A. A. Ismail, and D. Love. Concentrations of plasma testosterone in normal men during sleep. Nature 229: 261–262, 1971.
28. Evarts, E. V. Activity of neurons in visual cortex of the cat during sleep with low voltage fast EEG activity. J. Neurophysiol. 25: 812–816, 1962.
29. Fernstrom, J. D., and R. J. Wurtman. Elevation of plasma tryptophan by insulin in rat. Metabolism 21: 337–342, 1972.
30. Fernstrom, J. D., and R. J. Wurtman. Control of brain serotonin levels by the diet. In: Advances in Psychopharmacology, II, edited

by B. Costa and S. Sandler. Raven Press: New York: 1974, p. 133–142.

31. Gagliardino, J. J., and R. R. Hernandez. Circadian variation of the serum glucose and immunoreactive insulin levels. *Endocrinology* 88: 1529–1531, 1971.

32. Gaito, J., and K. Bonnett. Quantitative versus qualitative RNA and protein changes in the brain during behavior. *Psychol. Bull.* 75: 109–127, 1971.

33. Gnadt, J. W., and G. V. Pegram. Cholinergic brainstem mechanisms of REM sleep in the rat. *Brain Res.* 384: 29–41, 1986.

34. Haider, I., and I. Oswald. Late brain recovery processes after drug overdosage. *Br. Med. J.* 2: 318–322, 1970.

35. Havrankova, J., J. Roth, and M. Browstein. Insulin receptors are widely distributed in the central nervous system. *Nature* 272: 827–829, 1978.

36. Heiss, W. D., G. Pawlik, K. Herholz, R. Wagner, and K. Wienhard. Regional cerebral glucose metabolisme in man during wakefulness, sleep and dreaming. *Brain Res.* 327: 362–366, 1985.

37. Jacobs, B. L., and B. E. Jones. The role of central monoamine and acetylcholine systems in sleep–wakefulness states: mediation or modulation? In: *Cholinergic–Monoaminergic Interactions in the Brain,* edited by L. L. Butcher. New York: Academic, 1978, p. 271–290.

38. Jacobs, B. L., and D. J. McGinty. Effects of food deprivation on sleep and wakefulness. *Exp. Neurol.* 30: 212–222, 1971.

39. Jolin, T., and A. Montes. Daily rhythm of plasma glucose and insulin level in rats. *Horm. Res.* 4: 153–156, 1973.

40. Jouvet-Mounier, D., L. Astic, and D. Lacote. Ontogenesis of the states of sleep in rat, cat and guinea pig during the first postnatal month. *Dev. Psychol.* 2: 216–239, 1970.

41. Jouvet, M. Biogenic amines and the states of sleep. *Science* 163: 32–41, 1969.

42. Jouvet, M. Hypnogenic indolamine-dependent factors and paradoxical sleep rebound. In: *Sleep,* edited by W. P. Koella. Basel: Karger, 1983, p. 2–18.

43. Karadzic, B., and B. Mrsulja. Deprivation of paradoxical sleep and brain glycogen. *J. Neurochem.* 16: 29–34, 1969.

44. Kizer, J. S., C. B. Nemeroff, and N. W. Youngblood. Neurotoxic amino acids and structurally related analogs. *Endocrinology* 29: 301–318, 1978.

45. Kleine, N. Periodische schlafsucht. *Psychiatr. Neurol.* 57: 285–320, 1925.

46. Kleitman, N. *Sleep and Wakefulness* (2nd ed.), Chicago: University of Chicago Press, 1963.

47. Levin, M. Periodic somnolence and morbid hunger: a new syndrome. *Brain* 59: 494–504, 1936.

48. Lucero, M. A. Lengthening of REM sleep duration consecutive to learning in the rat. *Brain Res.* 20: 319–322, 1970.

49. McFayden, U. M., I. Oswald, and V. A. Lewis. Starvation and human slow-wave sleep. *J. Appl. Physiol.* 35: 391–394, 1973.

50. Mendelson, W. B., S. Slater, P. Gold, and J. C. Gillin. The effect of growth hormone administration on sleep: a dose–response study. *Biol. Psychiatry* 15: 613–618, 1980.

51. Moruzzi, G. The functional significance of sleep with particular regard to the brain mechanisms underlying consciousness. In: *Brain and Conscious Experience,* edited by J. Eccles. New York: Springer, 1966, p. 345–388.

52. Mouret, J. R., and P. Bobillier. Diurnal rhythms of sleep in the rat: augmentation of paradoxical sleep following alternation of the feeding schedule. *Int. J. Neurosci.* 2: 265–270, 1971.

53. Nicolaidis, S. Early systemic responses to orogastric stimulation in the regulation of food and water balance. Functional and electrophysiological data. *Ann. N. Y. Acad. Sci.* 167: 1176–1203, 1969.

54. Nicolaidis, S. Rôle des reflexes anticipateurs oro-végétatifs dans la régulation hydrominérale et énergétique. *J. Physiol. (Paris)* 74: 1–19, 1978.

55. Nicolaidis, S., and N. Rowland. Metering of intravenous versus oral nutrients and regulation and energy balance. *Am. J. Physiol.* 231: 661–668, 1976.

56. Olivo, M., K. Kitahama, J. L. Valatx, and M. Jouvet. Neonatal monosodium glutamate dosing alters the sleep–wake cycle of the mature rat. *Neurosci. Lett.* 67: 186–190, 1986.

57. Oswald, I. Human brain protein, drugs and dreams. *Nature* 223: 893–897, 1969.

58. Panksepp, J., J. E. Jalowiec, A. Z. Zolovick, W. C. Stern, and P. J. Morgane. Inhibition of glycolytic metabolism and sleep–waking states in cats. *Pharmacol. Biochem. Behav.* 1: 117–119, 1973.

59. Parmeggiani, P. L., L. F. Agnati, G. Zamboni, and T. Cianci. Hypothalamic temperature during the sleep cycle at different ambient temperatures. *Electroencephalogr. Clin. Neurophysiol.* 38: 589–596, 1975.

60. Petitjean, F., S. Seguin, M. H. Des Rosiers, D. Salvert, C. Buda, M. Janin, G. Debilly, M. Jouvet, and P. Bobillier. Consommation cérébrale locale du glucose au cours de l'éveil et du sommeil lent chez le chat. *C. R. Acad. Sci. III* 292: 1211–1214, 1981.

61. Reich, P., J. K. Driver, and M. L. Karnovsky. Sleep: effects on incorporation of inorganic phosphate into brain fractions. *Science* 157: 336–338, 1967.

62. Reivich, M., G. Isaacs, E. Evarts, and S. Kety. The effect of slow-wave sleep and REM sleep on regional cerebral blood flow in cats. *J. Neurochem.* 15: 301–306, 1968.

63. Roffwarg, H. P., J. Muzio, and W. Dement. The ontogenetic development of the sleep–dream cycle in humans. *Science* 152: 604–619, 1966.

64. Rosenberg, R. S., H. Zepelin, and A. Rechtschaffen. Sleep in young and old rats. *J. Gerontol.* 34: 525–532, 1979.

65. Rubin, R. T., R. E. Poland, and B. B. Tower. Prolactin-related testosterone secretion in normal adult men. *J. Clin. Endocrinol. Metab.* 42: 112–116, 1976.

66. Ruckebusch, Y., and M. Gaujoux. Sleep-inducing effect of a high-protein diet in sheep. *Physiol. Behav.* 17: 9–12, 1976.

67. Sangiah, S., D. F. Caldwell, M. J. Villeneuse, and J. J. Clancy. Sleep: sequential reduction of paradoxical sleep (REM) and elevation of slow-sleep (NREM) by non convulsive dose of insulin in rats. *Life Sci.* 31: 763–769, 1982.

68. Sassin, J. F., A. G. Frantz, S. Kapen, and E. D. Weitzman. The nocturnal rise of human prolactin is dependent on sleep. *J. Clin. Endocrinol. Metab.* 37: 436–440, 1973.

69. Schusdziarra, V. Role of somatostatin in nutrient regulation. In: *Somatostatin,* edited by Y. C. Patel and G. S. Tannenbaum. New York: Plenum, 1985, p. 425–445.

70. Sheppard, M. C., S. Kronheim, and B. L. Pimstone. Stimulation by growth hormone of somatostatin release from the rat hypothalamus in vitro. *Clin. Endocrinol.* 9: 583–587, 1978.

71. Stern, W. C., J. E. Jalowiec, H. Shabshelowitz, and P. J. Morgane. Effect of growth hormone on sleep waking patterns in cats. *Horm. Behav.* 6: 189–196, 1975.

72. Takahashi, S., D. M. Kipnis, and W. H. Daughaday. Growth hormone secretion during sleep. *J. Clin. Invest.* 47: 2079–2090, 1968.

73. Utsumi, M., Y. Hirose, K. Ishihara, H. Makimura, and S. Baba. Hyperinsulinemia and hypersomatostanemia in hypothalamic obese rats induced by monosodium glutamate. *Biomed. Res.* 1(suppl.): 154–158, 1980.

74. Van den Noort, S., and K. Brine. Effect of sleep on brain labile phosphates and metabolic rate. *Am. J. Physiol.* 218: 1434–1439, 1970.

75. Van Houten, M., B. I. Posner, B. M. Kopriwa, and J. R. Brawer. Insulin-binding sites in the rat brain: in vivo localization to the

circumventricular organs by quantitative radioautography. *Endocrinology* 105: 666–673, 1979.

76. Veosei, L., M. Balazs, and G. Telegdy. Action of somatostatin in the central nervous system. *Front. Horm. Res.* 15: 36–57, 1987.

77. Woods, S. C., and D. Porte, Jr. Relationship between plasma and cerebrospinal fluid levels in dogs. *Am. J. Physiol.* 233(*Endocrinol. Metab. Gastrointest. Physiol.* 2) E331–E334, 1977.

78. Wurtman, R. J., and J. D. Fernstrom. Control of brain neurotransmitter synthesis by precursor availability and nutritional state. *Biochem. Pharmacol.* 25: 1691–1696, 1976.

79. Zepelin, H., and A. Rechtschaffen. Mammalian sleep, longevity and energy metabolism. *Brain Behav. Evol.* 10: 425–470, 1974.

VII | THE NUTRITIONAL ENVIRONMENT

Associate Editor   C. M. Blatteis

61. Human adaptation to energy undernutrition

N. G. NORGAN | *Department of Human Sciences, University of Technology, Loughborough, United Kingdom*

A. FERRO-LUZZI | *Istituto Nazionale della Nutrizione, Rome, Italy*

CHAPTER CONTENTS

Energy Undernutrition and Adaptation
    Energy undernutrition
    Adaptation to energy undernutrition
Physiological Adaptation
    Biochemical changes
    Body weight and composition changes
    Energy expenditure
        Basal metabolism
        Thermogenesis
        Physical activity
        Total daily energy expenditure
Functional Consequences
    Reproduction and fertility
    Pregnancy and lactation
    Growth and adolescence
    Infection and disease
    Physical working capacity
        Aerobic working capacity
        Strength and motor performance
    Work output and productivity
    Behavioral and cognitive development
Behavioral Strategies: Household and Community Responses and
    Strategies
Genetic Adaptation to Undernutrition
Conclusions

ENERGY UNDERNUTRITION AND ADAPTATION

Energy Undernutrition

ENERGY UNDERNUTRITION (EUN) is the most common nutritional disorder in the world today. Estimates put the number of people affected at upwards of 780 million (24). The origins of the problem lie in the general poverty and deprivation suffered by many of the world's inhabitants. Energy undernutrition has been a feature of human societies throughout the historical period and perhaps for much longer. This experience has led to adaptive processes and strategies to ameliorate the effects and to promote functioning and survival.

Energy undernutrition, often referred to simply as undernutrition, can be defined as energy intakes below the energy requirement. The energy requirement of an individual is the level of energy intake from food that will balance energy expenditure when the individual has a body size and composition and level of physical activity consistent with long-term good health and that will allow for the maintenance of economically necessary and socially desirable physical activity (25). It is immediately obvious that outcomes such as long-term good health are difficult to define objectively. Inevitably, there are value judgments about whether the adaptive changes are consistent with long-term good health. Further, the adaptive responses to energy undernutrition involve changes in the same variables that the energy requirement seeks to maintain.

Energy undernutrition exists in a variety of forms depending on the severity, duration, and frequency of the undernutrition. Adaptive responses vary according to each of these and are modulated by the initial status of the individual or population; by accompanying nutritional deficits; by other health problems or deprivations, whether environmental, social, or psychological; and by age and sex. Undernutrition may be continuous and of long standing, in which form it is commonly referred to as *chronic energy deficiency* (CED), or intermittent, as may occur in seasonal cycles of plenty and want. It may be intense, as in fasting for therapeutic, religious, or political reasons, or mild, as perhaps existed before the secular increase in height experienced by most developed countries. Chronic energy deficiency is usually of mild to moderate intensity, though its effects may be severe, as severe energy deficiency cannot be withstood indefinitely. Acute energy deficiency may be mild to severe and includes the complete absence of food, referred to in this chapter as *fasting*. However, these terms are relative; there is no consensus on their use and there are spectra of duration and intensities of EUN rather than binary categories.

Chronic energy deficiency affects body size and is commonly defined anthropometrically as deficits in weight and height in children, by comparison with international reference data (144), and in adults by the body

1391

mass index (BMI: weight/height^2, kg/m^2) or Quetelet's index. Cut-off points for three grades of CED based on BMI have been proposed (54) and confirmed (34). Acute energy deficiency affects body weight but not height. However, it must be recognized that anthropometric deficits can arise from the limitation of almost any nutrient and many other environmental factors and that simple single dietary deficiencies, in this case energy, may be rare. Weight, height, and weight-for-height deficits are not conclusive evidence of short- or long-term EUN.

In contrast, in the laboratory EUN may occur as a single, controlled challenge to homeostasis. This is often short-term and of moderate to severe intensity and is referred to here as *energy restriction*. The best known examples are found in the studies of Benedict et al. (10) and Keys et al. (58); in the latter, energy intakes of 32 young men were approximately halved for a period of 24 wk. Energy restriction can be identified as a negative energy balance by measurements of energy intake and expenditure and falling body weight and energy stores. The research findings on the responses to energy restriction are likely to be more applicable to acute than to chronic energy deficiency.

Adaptation to Energy Undernutrition

The concepts and definitions of adaptation have been treated extensively in the opening chapters of this *Handbook*. Adaptation is usually seen as attempting to resist or accommodate changes; in other words, adaptation is a process by which a new or different steady state is reached in response to a change or difference in environment. In classical physiology, adaptation and homeostasis are discussed in relation to particular variables, for example, blood pH or body temperature. In nutrition there is rarely a single, unambiguous variable but rather a vague criterion, such as long-term good health or maintenance of an acceptable level of functions. Hence, adaptation cannot be defined unambiguously or separated easily from such terms as homeostasis, accommodation, acclimation, etc. (139, 142). There is, however, some agreement on the characteristics of nutritional adaptation (143).

An adapted state is a steady state that is sustainable and reversible. To this might be added that adaptation is not a short-lived response but a coordinated state of responses maintained over time. Constancy is never absolute. There is a preferred range of acceptable, sustainable states that can be objectively observed, within which adaptation functions. Every adaptation incurs a cost. An adaptive process seeks to preserve a particular state or function, though one function may be maintained at the expense of another.

Grande (44) emphasized that adaptation is normally considered to include adjustments in behavior and bio-

chemical mechanisms that do not impair body functions. Beaton (7) considered adaptation and accommodation to be processes that have a successful outcome, depending on the function being examined. Scrimshaw and Young (108), however, distinguished between adaptation and accommodation. They defined the adapted state as a long-term steady state achieved while function is maintained within an "acceptable" or "preferred" range. This does not require or imply that all body functions remain unaffected. Accommodation includes the responses that favor the survival of the individual but with impairment or significant losses of important functions. Waterlow (143) considered that as adaptation carries gains and losses and as it is difficult to avoid subjective value judgments, the word should be put aside and replaced by the question, how does function A respond to this stress?

Gould and Lewontin (43) have described three types of adaptation occurring at different hierarchical but overlapping levels. There is physiological adaptation in phenotypic plasticity. Adaptations to high altitude are of this type. These are not heritable, though the capacity to develop them is. There is behavioral adaptation determined by the nature of the society and the status and personality of the individual, part of which may be inherited in a non-Darwinian manner through culture. Finally, there is genetic adaptation arising from the conventional Darwinian mechanism of selection upon genetic variation. The three types operate over different time scales but are interconnected. Genetic makeup determines the type and extent of physiological responses that can be made. This will determine what behavioral adaptations are required. As the rate of change in the environment is increasing, behavioral responses, which can be taken up and implemented quickly, have become more and more important.

Each of the three types of adaptation has been reported or considered in relation to EUN. The adaptive responses to energy restriction act to regain energy homeostasis or balance at the new plane of nutrition. The immediate response is the utilization of the body tissues as a source of energy for metabolism. Energy expenditure falls as a result of the reduced body weight, and there is a reduction in the amount of physical activity and a fall in metabolic activity, which represent an increased efficiency of energy utilization. In CED, reduced growth and adult body size are the major responses. The possibility of genetic adaptation is encapsulated in the thrifty genotype hypothesis.

PHYSIOLOGICAL ADAPTATION

The fundamental adaptation of the body to energy restriction is the utilization of the body tissues as a source of energy for metabolism, with a consequent reduction in body weight and energy expenditure.

Biochemical Changes

There is a coordinated metabolic response to energy restriction which serves to (1) mobilize and utilize the energy stored in the body as fat, (2) maintain glucose production by glycogenolysis and gluconeogenesis for use by the brain and nervous tissue but at the same time minimize protein breakdown and loss, and (3) reduce metabolism with the effect of conserving energy.

The transition from the fed to the fasted state is characterized by a shift in the fuel for metabolism toward fat as glucose and glycogen stores are depleted. Lipolysis is stimulated by hormone-sensitive lipase in adipocytes, bringing about an increase in plasma free fatty acids (FFA) and a reduction in lipogenesis. This is facilitated by falling blood glucose and insulin and rising glucagon levels. After a 12–18 h fast, hepatic gluconeogenesis maintains blood glucose at levels that allow the function of nervous tissue as the brain adjusts to utilizing increasing amounts of keto acids in the plasma. Glucose is derived at first mainly from the amino acids alanine and glutamine. Protein catabolism is determined by glucose need and falls as the dependence of the previously obligate glucose-consuming tissues falls.

Ketosis is common in energy restriction and starvation, reflecting the shortage of reduced nicotinamide adenine dinucleotide phosphate (NADP). However, as ketone bodies can be oxidized by many tissues (98), particularly the brain (84), the ketosis of starvation represents an appropriate physiological response for preserving muscle tissue. Ketone bodies provide 10% of the energy for muscles after an overnight fast and may provide up to 30%–40% of the total energy requirement in the first few days of starvation. This rises to 50%–80% after 3–7 days of energy restriction but falls to 10% after 6 wk, by which time the brain and kidneys are totally dependent on ketones, which, being water-soluble, are able to pass through the blood–brain barrier.

In contrast, in CED respiratory quotients are high and blood glucose levels are unaltered (112). The high carbohydrate content of third world diets leads to a 3% more efficient capture of energy in ATP production than diets of mixed composition. Information on substrate utilization over long periods of time can be gained by serial measurements of body composition. This indicates that although the composition of tissue lost is on average 75% adipose and 25% lean tissue by weight, the actual composition and hence fuel for metabolism varies according to the initial level of energy stores (36, 37). Although there is little information available, protein turnover appears not to be depressed in CED (116).

In energy restriction turnover falls, particularly pro-

tein turnover, through the lowered actions of the thyroid. The contribution of protein turnover to the basal metabolic rate (BMR) may be 15%–20% according to stoichiometry (35) or up to 40% according to isotope studies (116). Therefore, a fall of a quarter would lead to a 5%–10% reduction in BMR. Sodium, potassium, and calcium ion pumping are also reduced in energy restriction as a result of lowered thyroid action. Ion pumping has been variously estimated to contribute 10%–40% of BMR (85), so the potential savings are not insignificant. Futile cycling, superfluous metabolic activity of anabolism and catabolism, is thought to contribute 10%–15% of BMR, and its disappearance would represent another important saving. As the responses to undernutrition are integrated, a number of small savings might add up to an appreciable total.

Body Weight and Composition Changes

Reductions in energy intake invariably cause a fall in body weight best described by an exponential equation, the rate being proportional to the extent of the energy deficit and initial body weight. There is an initial fast component of as much as 5 kg over the first few days of total starvation, representing the emptying of glycogen reserves, the utilization of protein as a substrate for gluconeogenesis, and the shedding of water that accompanies them. This lessens to about 0.5 kg/day as adipose tissue loss replaces lean tissue loss. Thus the rate of weight loss decreases as a result, in the first instance, of the changing composition of the tissue lost. Subsequently, as energy expenditure falls, owing to reductions in basal metabolism and physical activity as a consequence of smaller body weight and to metabolic adaptations, the degree of negative energy balance lessens, as does weight loss.

Energy restriction results in losses of fat-free mass (FFM) as well as of fat. (FFM is used here for all measures of whole-body lean or active tissue.) Forbes (36) describes this as the "companionship of lean and fat tissue." He has shown that the composition of weight lost in energy restrictions greater than 4 wk depends on the initial level of fatness of the individual.

Lean individuals have a higher proportion of FFM in the tissue lost. It is well known that females, who have a higher body fat content, are better able to withstand food shortages than males (147). In rats fasted for 6 days, protein contributed more of the total energy expended in males (18%) than in females (16%). Protein conservation may be more important than the total energy store in withstanding the challenge of energy undernutrition. The degree of energy restriction is a further important factor influencing the composition of tissue lost.

Not all components of FFM are affected to the same extent. Liver and intestinal losses are greater than those of body weight as a whole, heart and kidney being about the same, and brain and nervous tissue show very little loss (44). Muscle tissue losses may be substantial. These have consequences for resting metabolism to be discussed later (Ethnic differences in BMR).

Edema in famine victims is well known. In the Minnesota study (58), plasma and extracellular fluid (ECF) volume remained the same in spite of weight loss. The high ratios of ECF/nitrogen, an index of hydration of the body, and reduced cell mass in severe malnutrition are well established. A low body potassium/nitrogen ratio in malnutrition reflects the fact that collagen protein of muscle and viscera is less affected than noncollagen proteins. In anorexia nervosa lean tissue is lost not only from muscle but also the viscera, including liver, spleen, and kidneys.

In CED, FFM may be well maintained. Lactating British and Gambian women differ in weight and fatness (28% and 18%, respectively) but have similar FFM (41 and 43 kg, respectively) (90). This raises the question as to whether the Gambian women with a BMI of 21 kg/m² are really energy-deficient. In undernourished Guatemalan agriculturists the main change in FFM is a deficiency of muscle cell mass. Some of the changes in body composition, such as edema or loss of muscle mass, may limit the accuracy of the methods used, for example, skin folds, total body water. However, there is general agreement that moderate malnutrition is associated with very low levels of subcutaneous fat and a reduction in FFM, particularly of muscle (111).

Energy Expenditure

Total daily energy expenditure (TDEE) is conveniently divided into three components: (1) resting metabolism, which depends mainly on the FFM; (2) thermogenesis; and (3) physical activity, which potentially is the most variable of the three. Resting metabolism, or resting metabolic rate (RMR), is the largest component (60%–75% of the total) and, as such, has prompted the most enquiry. Basal metabolism, or BMR, is RMR measured under strictly controlled conditions: on waking, in a thermoneutral environment, at complete physical rest, 12–18 h after a meal. The RMR has less stringent measurement conditions and is more representative of the minimal rate of energy expenditure during the day. Food, cold, and drugs elevate BMR and RMR, and this represents the thermogenesis component, which can be separated into two further components: obligate and facultative, or regulatory, thermogenesis.

Basal Metabolism. Basal metabolism has been a key variable in the examination of the effects of undernutrition on energetics. It is a large proportion of the total energy expenditure, it is comparatively easy to measure reproducibly because of the controlled conditions of measurement, and it proves to be responsive to varying planes of nutrition. Basal metabolism is the energy cost of maintaining an integrated metabolism and an intact body. It includes the cost of cell maintenance, ion pumping, anabolism to balance catabolism, and essential heat production for thermostability.

Basal metabolism in energy restriction. In energy restriction, BMR falls. There are two components to this: an early rapid fall of 10%–15% due to reduced metabolic activity and a later sustained mass effect arising from the smaller body weight. The metabolic component has been deduced from the larger falls in BMR than in body weight or FFM. For example, in the Minnesota study (58), BMR (MJ/day) fell by 39% but by 19%/kg body weight and by 16%/kg FFM. Most of the fall/kg FFM was in the first 2 wk. The second component, weight, arises as a consequence of body fat and protein being used as a fuel for metabolism. The relative importance of the two processes depends on the severity and duration of the energy restriction. In short periods of severe restriction or during the first 2–3 wk, reduced metabolic activity ensues quickly, independently of the duration, until weight begins to fall. Grande and coworkers (45) reported 17% falls in 14 day and 12% in 19–20 day energy restriction. Seventy percent of the reduction in BMR was attributed to reduced metabolic activity. However, in long duration energy restriction, the fall in body weight becomes the major factor.

Basal metabolism in CED. The question of adaptation in BMR to CED has been approached in two ways: by examining individuals with anthropometric deficits as evidence of CED and by examining populations where CED is thought to be common. These latter are found mainly in the tropics, so differences may have a nonnutritional basis.

In CED low BMR can be attributed mainly to smaller body weight. The BMR of 11 prisoners of war with 25% weight loss over 12 months was normal per kilogram body weight (9). Here, CED of longer duration but with the same degree of weight loss as in the energy restriction of the Minnesota study (58) had different metabolic effects. However, the BMR/kg FFM was found to be 14% lower in a group of 14 underweight but otherwise fit and healthy Indian laborers with BMI of 16.6 kg/m² compared with 14 controls with BMI of 20.7 kg/m² (112). McNeill et al. (75) were unable to confirm these findings in a group of 12 rural South Indian villagers. They regarded the control BMRs as being high compared to their own control group and other Indian data and found that the significant difference between laborers and controls could owe as much to the high BMR in the controls as to the low BMR in the laborers. In their data BMR/kg FFM was negatively related to weight and BMI, suggesting no metabolic adaptation. Furthermore, although the BMIs of the laborers were the same as in the energy-restricted Minnesota subjects, the BMRs were higher. Thus evidence for metabolic adaptation of BMR in CED is contradictory. Some of this uncertainty, and that concerning ethnic differences in BMR considered below (Ethnic differences in BMR), arises because adjusting BMR for differences in size and composition by dividing by body weight or FFM is not entirely valid. The relationship between BMR and FFM is linear but does not pass through zero (95). There is a positive intercept, which means that smaller individuals will have higher BMR/kg FFM.

In developing countries there are often annual alternations of shortage and abundance of food. The hungry season usually coincides with peak demand for agricultural effort, with the result that there is a seasonality in energy balance, as is well illustrated by seasonal variations in body weight (31, 32). Different strategies to cope with seasonal energy deficits may be adopted in different populations. In Ethiopian women, mean BMI was found to be 18.4 kg/m^2, and BMR/kg fluctuated by 13% and weight by 1.6 kg. In Indian women of slightly lower BMI (mean = 18.0) BMR fluctuated by 5% and weight by 0.5 kg. In contrast, in Beninese women of higher BMI (mean = 20.8) there was no change in BMR with weight fluctuations of 1.2 kg (29). This illustrates three different strategies to meet seasonal energy deficits: falls in BMR and weight (Ethiopians), falls in BMR (Indians), and falls in weight (Beninese). As in most studies, individuals with low BMI lost the least weight. However, body weight changes rarely exceed 5%, and there is some evidence that this is a diminishing phenomenon.

Ethnic differences in BMR. People of the tropics and the Indian subcontinent are generally agreed to have BMR/kg some 10% lower than Westerners (48, 76, 106). This could be interpreted as a metabolic adaptation, since with a lower percentage of body fat in these populations a higher BMR/kg would be expected. A more probable explanation is that the actual relationships between BMR and body weight beyond the range of weights used to derive the reference equations are not linear, as was assumed in analyzing the reference data, but curvilinear (75). This may lead to the BMR of small individuals being overestimated. It has been suggested that CED is not a contributing factor to the lower BMR of adults in the tropics (48). If the several thousand measurements are considered to have been made on healthy subjects, then temperature-induced changes in thyroid gland activity are the likely explanation. However, it is not known whether subjects were exposed to food constraints or not (7).

Significant between-group differences in BMR MJ/day, /kg, and /kg FFM were reported in Scottish, Gam-

bian, and Thai women, but these too were due to size effects (66). The 15% difference in BMR/kg FFM between the heaviest and lightest individuals could be explained on the basis of different compositions of FFM. Some 60% of BMR arises from the activities of the brain, heart, liver, and kidneys. These make up 6% of FFM, but their metabolic rate is 30 times that of the rest of the body. For a 15% difference in BMR the proportions would need to be 8% of the FFM in the lightest and 5% in the heaviest. This is plausible as the FFM of the heaviest has a larger proportion of the nonfat component of adipose tissue and muscle. At the same FFM, there was no difference between the groups, suggesting no apparent effect of race, climate, diet, or nutritional status, even though Gambian women experienced regular seasonal food shortages. This subject needs more research.

As discussed earlier (Basal metabolism in CED), BMR/kg FFM without reference to size is not the optimal basis for comparison. The data require appropriate standardization or an analysis of covariance (87a).

Metabolic adaptation to CED. In CED, the loss of body tissue contributes most to the fall in BMR. There is little information on adaptive metabolic changes. It is usually assumed that the physiological and metabolic responses to CED are the same as those in energy restriction. The hormonal picture shows similarities, low basal noradrenaline and triiodothyronine, (T$_3$), unaffected thyroxine (T$_4$) and free T$_3$, variable insulin, lower thermogenic response to noradrenaline, and a lower fasting T$_3$ and sympathetic drive (113).

A low BMR/kg FFM has usually been accepted as evidence for metabolic adaptation, in spite of the problems with this method of standardization. Indian data show no reduction in BMR/kg FFM (117) even when functional changes in sympathetic nervous system (SNS) activity and thyroid status were observed (63). Indeed, BMR/kg FFM may be significantly higher than in controls (33, 75, 113). This too could arise from a changing composition of FFM such that low-activity muscle mass falls while visceral mass is maintained or increased (3, 27, 116).

In animals of the same age, sex, and weight, BMR can differ up to 40% as a result of immediate prior nutritional experience. Those animals that have been losing weight have lower BMR than those that have been gaining weight. This arises through variations in weight of metabolically active organs that account for a significant amount of basal metabolism, far in excess of the proportional weights of these tissues (60). The BMR has been shown to be directly related to the proportion of FFM made up of the metabolically active organs (38a). Hence, changes or differences in BMR/kg FFM may not reflect changes in metabolic efficiency but changes in the composition of FFM. Support for this conclusion comes from the observation that, although there is a twofold difference in BMR/kg between the infant and the adult, when expressed per kilogram organ weight (sum of the liver, kidney, heart, and brain) BMR remains constant during growth and development (22). Furthermore, the accuracy of the rather indirect estimates of FFM in the chronically energy-deficient, based on estimation equations drawn up on well-nourished Europeans, must be questioned.

Garby (39), in one of the most detailed reviews on metabolic adaptation to decreases in energy intake, concluded that the proposal by the Fifth World Food Survey (24) that a metabolic adaptation of 15% is a realistic possibility in populations of developing countries cannot be supported or rejected by the available experimental data. A tentative conclusion was that energy expenditure at rest decreases following a reduction in energy intake. The magnitude of the effect could not be well established but it is likely to be over 5%. Others have put the range at 10% (55, 82).

Metabolic adaptation to energy restriction: mechanisms. In short-term energy restriction or in the early days of long-term energy restriction decreases in metabolic activity are the major cause of the fall in BMR. The physiological processes have been the subject of much investigation but are not well understood. Energy deficit reduces SNS activity, alters peripheral thyroid metabolism, and lowers insulin secretion (17). The coordinated changes have the twin effects of mobilizing the appropriate substrates and fuels and lowering metabolism by reducing activities such as substrate recycling and protein turnover. The mechanisms have been reviewed (113). Hormones are responsible for the mobilization of the body stores of energy during energy restriction. They also have a role in the metabolic response.

Catecholamines appear to play the key role in modulating the adaptive component of RMR as a metabolic response to energy restriction. Catecholamines from the adrenal medulla have been shown to increase the rate of cellular thermogenesis and the mobilization of substrates which may themselves have a thermogenic effect. The supposition is that changes in SNS activity releasing catecholamines from nerve endings may have the same effect, contributing to the reduced energy expenditure (64). The responses are specific to the carbohydrate content of the diet. Thus there is a similarity in the response of SNS activity and insulin secretion. However, the evidence is not entirely consistent. No difference in metabolic rate and plasma noradrenaline following a meal was found before and after 7 days of underfeeding (70).

Administration of L-dopa, a precursor of the sympathetomimetic agent dopamine, prevents the reduction of RMR associated with a fall in noradrenaline in the slimming obese (114). β-Adrenergic blockade by the administration of propranolol reduces RMR by about

10%, similar to that seen after a 2 wk energy restriction. The major action is the direct inhibition of SNS-mediated stimulation, though it also affects peripheral thyroid metabolism. This is the catecholamine-mediated component to BMR, which falls in energy restriction.

Thyroid hormones are the principal regulators of BMR (49). They are important in the response to energy restriction, influencing both carbohydrate and lipid metabolism. They stimulate metabolism by increasing the membrane Na^+K^+ ATPase activity, hence increasing ion pumping. Energy restriction causes serum T_3 levels to fall within 24 h due to a reduction in synthesis at the periphery but has no effect on circulating T_4. This may protect FFM and conserve energy. Thyroid hormones and catecholamines act synergistically to potentiate each other's actions in the tissues, which may influence their role in the regulation of thermogenesis. Low doses of T_3 during energy restriction can prevent the fall in metabolic rate (146). Thermogenic responses to carbohydrate mediated through the SNS require thyroid hormones. Similarly, the thermogenic actions of thyroid hormones are mediated by Na^+K^+ ATPase, the target for catecholamines via β-adrenergic receptors.

Insulin affects thermogenesis by increasing both glucose uptake (obligate) and catecholamine activity (facultative) (65). It also increases Na pumping across the cell membrane. Falls in insulin levels are a sign of change from the fed to the fasted state. Other hormones, such as glucagon and growth hormone, have potential metabolic and thermogenic effects in energy restriction.

Thermogenesis. Thermogenesis, the second component of the energy expenditure, is inappropriately named as most of the energy expenditure produces heat. By convention, it refers to the effects of food, cold, and thermogenic agents. Diet-induced thermogenesis (DIT) is the main component. It is regarded as having two components, the major one a specific obligatory response to a meal, its thermic effect (TEM), and a facultative-adaptive component that may arise from the recent nutritional history; TEM is the obligatory energy expenditure from the digestion, absorption, net resynthesis, and storage of food, and the facultative component might be dispensed with if the need arose. The saving in energy might not be large. Cold-induced thermogenesis is subdivided into shivering and nonshivering thermogenesis (NST); DIT and NST involve similar processes (100).

The Minnesota study (58) did not measure TEM, but the investigators concluded that the literature may be cited to show that TEM in energy restriction and CED is less than, equal to, or greater than that in the normal state of nutrition. This situation persists. The TEM in Gambian men was found to be lower than that in Europeans (6.3% vs. 12.1%) (77). These men were described as not malnourished but were measured during the nutritionally unfavorable rainy season. The energy saving would be some 0.25 MJ (60 kcal)/day. Others have found or concluded there was no effect in semistarvation (55, 56, 70), but an increased TEM/kg FFM (8.8% vs. 7.0%) has been reported in nine CED Indian men (87).

Physical Activity. A reduction in the amount of physical activity can be the key response to energy restriction and undernutrition in terms of energy saved. However, its occurrence and magnitude are not well documented (30) but the findings are concordant (135).

In the energy restriction of the Minnesota study (58), almost 60% of the fall in energy expenditure was from reduced expenditure in physical activity, and of this most was a result of reduced activity rather than reduced body weight. The savings could have been higher since the volunteers had an exercise program such that their physical activity level (PAL = TDEE/BMR) was 1.6, a level higher than most individuals in Western societies.

A stimulus to the investigation of this topic was the observation that 20 1–3 yr old Ugandan children with energy intakes 70% or less of recommended intakes were gaining height and weight at reference rates but were less active than five control European children (101). Total daily energy expenditures were 326 kJ/kg and 410 kJ/kg, respectively. Further studies have also shown that restriction of energy intakes of about 10% reduced physical activity in infants, toddlers (136), and adults (42). Hence, falls in activity seem the first line of defence in the marginal situation. In contrast, in marginally malnourished Colombian school children peer pressure maintained activity levels and there was no deficit in cognitive or motor development because of reduced activity associated with school or play. Here, slower growth and reduced weight were the first line of defence (126), though the energy savings would be small.

It is not clear if physical activity is lower in areas where CED might be expected. In developing countries PAL of 1.8 have been calculated, not appreciably less than those in developed countries, 1.9 (28). However, current levels in developed countries may well be much lower than this.

Many populations appear to have very low energy intakes, with little energy available for activity. What there is may have to be used for economic productivity at the expense of discretionary activities (those activities leading to the fulfilment of personal needs, to personal interactions between the household and society, to development, and to change) (6). Here, the costs of accommodation to low CED are on social function and psychological development, behavioral factors rather than physiological factors but no less important because of that (85b). Evidence to support the notion of accommodation by reducing discretionary activities is largely anecdotal. Some doubt attaches to low energy intakes, which rarely stand up to independent assessment. Gambian men of low BMI have similar levels of free-living energy expenditure when normalized for weight and FFM to those of average BMI, thus providing no evidence of behavioral accommodation to save energy (18a, 18b)

Mechanical efficiency and economy of effort. The efficiency of muscular work is determined by the combined efficiencies of the coupling of oxidation to phosphorylation and of phosphate-based energy to muscular contraction. These efficiencies are some 30% and are independent of age, sex, acute weight gain, and training status (89). There is evidence that in mice energy restriction may enhance the coupling efficiency of oxidation and phosphorylation in certain tissues (145). Similarly, repeated observations of individuals and groups apparently existing on very low energy intakes yet apparently working moderately hard have led to suggestions of an increased efficiency in human work, but the evidence is tenuous.

"Efficiency" has many meanings, but it can be defined precisely in physical terms. In the energy restriction of the Minnesota study (58) net mechanical efficiency (NME) was unchanged. An earlier claim of an increased efficiency in walking in energy restriction (10) must be regarded as tentative since no initial measurements were made.

There have been two approaches to investigating mechanical efficiency in CED. The first has been to study individuals with some evidence of CED, low dietary intakes, or anthropometric deficits. The second has been to review the work and exercise physiology literature for raised efficiency in groups that might be exposed to food shortages or CED of varying degrees.

The energy cost of a standardized stepping task was found to be reduced in ten Jamaican men reported to subsist on low energy intakes compared to control subjects (1). However, half the subjects lost weight when fed their putative habitual energy intake under controlled conditions in a metabolic ward, a situation found by others (20). Reports of low BMR and high work efficiencies in five high and six low intake Indonesians matched for height and weight (21) do not stand up to close scrutiny (80). The BMR appear to have extreme ranges, and the significant difference in work metabolism appears to arise from a low NME in the high intake group rather than a high efficiency in the low intake group.

Spurr and co-workers found no significant difference in gross (GME) and delta mechanical efficiencies (DME) of treadmill walking either in groups of Colombian men classified as normal, mild, intermediate, and severely nutritionally compromised on the basis of anthropometric and biochemical indices (124) or in marginally malnourished boys (125). A two year longitudinal study of rural Beninese women with several variations in food intake of up to 15% found no seasonal changes in RMR on DME of cycling (1a).

There is evidence from calorimetric studies for a 12% higher NME in six healthy CED men, 20.9%, SD 1.7%, vs. 18.5%, SD 1.3%, in normal-weight controls (62). The intensity of the stepping task was very high, requiring 1.5 and 2.5 liters/min Vo$_2$ in the CED and normal controls, respectively. This is some 40 ml/kg in the CED. The task is likely to have a substantial anaerobic component, which may explain the apparently high NME. The savings from such small differences in NME are quite small. If the energy cost of work were 4.2 MJ (1,000 kcal)/day, the saving could be some 0.1 MJ (25 kcal).

In the second approach, reviews of the appropriate literature have failed to show increased NME in population groups most likely to be exposed to CED. Indeed, most studies show a lower efficiency, perhaps because of the unfamiliarity of the subjects with the ergometer. Where increased efficiency has been found, the intensity of the work was usually so high as to suspect that there is an anaerobic component to energy expenditure and that the total cost is not being represented by oxygen consumption and indirect calorimetry. Higher NME and DME were observed in Gambian than in Swiss men walking on a treadmill at 2 mph at 0% and 10% slopes (NME = 23.2, SEM 0.3% vs. 20.1, SEM 0.4%, respectively) (77). This occurred during the wet, "hungry" season. At such low walking speeds, style and economy of movement would have an important effect on energy expenditure, as was suggested by the investigators. Subsequent studies of Gambian men of BMI <18.5 have found no evidence of raised net work efficiencies compared with men of BMI >22 (18b).

Kenyan women of the Luo and Kikuyu tribes can carry loads of up to 20% of body weight on the head or with a yoke without increasing energy expenditure (69). This remarkable adaptation is thought to arise from maintaining the head or shoulders and the load in a horizontal plane and allowing the spine to act as a spring to accommodate the up and down motion of the body in bipedal locomotion. Such movements do not increase energy expenditure because there is energy recovery in the interconversions of kinetic and potential energy (142). This work has been extended in Gambian women by considering body fat as part of a load (57). There was no increase in energy expenditure provided the body fat plus load was less than 40% of FFM. Lean women would be at an advantage. However, the CED

status of these individuals is not known, loads of 5%–10% of body weight are carried without increased expenditure in most individuals (102), and the normal response to loading, where individuals have a choice, is to slow down to keep the energy cost constant (83). Therefore, the significance of these observations to adaptation in EUN is unclear.

A mechanism for the putative adaptive increases of NME in CED could be a higher proportion of slow twitch muscle fibers with their more efficient aerobic metabolism than fast twitch fibers in terms of mechanical force development per unit of ATP used (47a). Both malnourished patients and hypothyroid patients have higher proportions of slow twitch fibers, mainly due to a reduction of fast twitch fibers (141), an effect which may arise from the lower thyroid status of the CED (62). Alternatively, it may be just an economy of effort through fewer unnecessary body movements.

Information on economy of effort as an adaptation to CED is sparse. Energy expenditure in freestyle, unstandardized activities and hence efficiency is influenced by extraneous body movements and by the demeanor of the individual. The energy cost of fidgeting is substantial and can have an effect on TDEE equivalent to or greater than the observed changes of metabolic adaptation, as well as on mechanical efficiency. Fidgeting could increase the energy cost at rest by 30% and TDEE by 10% (40). Conversely, the quiet demeanor and smooth movements free of unnecessary action of many third world groups have been noticed. An example is the lower proportional increase in energy expenditure when changing from lying to sitting or standing in Africans, Asians, Guatemalans, and Papua New Guineans compared to Europeans. A slow speed of movement has been suggested as a key response to CED (142).

In conclusion, there is little evidence of increased mechanical efficiency in CED. Differences are small and inconsistent. Even in normal subjects, there is unexplained variation in NME and DME.

Total Daily Energy Expenditure. The three general strategies for reducing or economizing energy expenditure in response to EUN are lowered body weight, reduced amount and cost of physical activity, and increased metabolic efficiency. The extent of variation in each of these has been reviewed (30, 90), and the effects have been modeled to show their quantitative significance (27, 28).

A 6 month energy restriction in the Minnesota study (58) caused body weights to fall by 24% and TDEE to fall by 55% (8.0 MJ). One-third of the saving arose from the reduced BMR, but the majority of the remainder was from a reduced energy expenditure in physical activity. Of this, 60% could be attributed to a fall in volitional activity and 40% to the reduced energy cost of activity due to the smaller body weight. Smaller body weight made a major contribution to the reduced basal metabolism (65%), the remainder being metabolic adaptation. Metabolic adaptation can be calculated to have contributed 10% of the total energy saved (0.33 × 0.35 × 100). Although subjects had regained energy balance on the low plane of nutrition, they showed the classical signs of famine victims: edema, anemia, polyuria, bradycardia, weakness, and depression.

Another approach to the investigation of TDEE in CED has been to look at groups with low energy intake. Measurement of TDEE is difficult, and the supposition is that low intake is matched by a compensatory low expenditure. The difficulty here is that measurement of energy intake is neither easy nor apparently often accurate. Individuals on low intakes often lose weight (1). Some of the intakes are less than the basal metabolism (56). Similar problems emerge from studies on small and large eaters. Do small eaters exhibit a hypometabolic state? They are usually of lower weight than large eaters. When matched for size, there is no difference in 24 h energy expenditures and BMR (74). This casts doubt on the representativeness of the intake data. In the original study of Rose and Williams (99) on large and small eaters no differences were found in BMR or the energy cost of resting or standardized work activities of six large eaters (19 MJ/day) and six small eaters (10 MJ/day). The pace of freestyle activities was 10%–15% higher in the large eaters, which is presumably the mechanism for utilization of the extra energy.

Calorimetric studies have confirmed low BMR, high TEM, and NME leading to a lower TDEE in Gambian compared to Swiss men of a similar percentage of fat (77). This is significant because PAL in the calorimeter were low, which may have limited the repertoire of behavioral responses. Measurements were made in the wet, hungry season, so the subjects may have been in negative energy balance. However, they were unable to eat more than 9.5 MJ/day on average and were of good anthropometric status (BMI >20 kg/m²). Thus the differences seem less related to plane of nutrition than to body composition and behavior.

Costless adaptation? Sukhatme and Margen (130) have described an autoregulation in energy intakes and proposed that this was achieved by adjustments in the metabolic efficiency of energy utilization. They suggested that adults can alter their requirements by 30% without altering weight or physical activity and without detriment to health or productive capacity. The implication is that food aid and rural development could be reduced substantially as the numbers of undernourished have been overemphasized.

The ideas of Sukhatme and Margen have been strongly criticized (7, 53). The hypothesis is based mainly on statistical precepts, and there is no physiological evidence for variability up to 30% (55, 82). Evidence adduced against it include the low variability in TDEE under controlled calorimetric conditions (2%), limits of metabolic adaptation of about 10%–15%, and that populations under CED as opposed to energy restriction show no evidence of metabolic adaptation (55).

There are limits to the extent of adaptations to low intakes and there is substantial evidence of widespread losses in function.

FUNCTIONAL CONSEQUENCES

Every adaptation has a cost, and there has been excited debate over the adaptive significance of small body size in adults and children. It is well recognized that severe energy deficiency impairs human functions. The major issue is the significance of milder levels of malnutrition. This has been debated at length (41a, 85a) without resolution (83a). In the context of natural selection, size reduction reduces the pressure on food available to the population, allowing more individuals to survive and to maintain the population over time. Seckler (109), an economist with the Ford Foundation in India, has suggested that growth can be seen as either a genetic potential (and anything less is a deprivation) or a range below the genetic potential, the potential growth space, where function can be maintained until a threshold is reached below which function is compromised. These are described in value-laden terms as *deprivation* and *homeostatic theories of growth*. In the latter, it is possible for individuals to be small but healthy. For example, although mortality is raised in severe malnutrition in childhood, it is not in mild-to-moderate malnutrition (16). The relationship between anthropometric deficits and prospective mortality is nonlinear. The significance of the debate is clear. According to the hypothesis, small children and adults are not stunted but are well adapted and healthy and the problem of undernutrition involves only the much smaller numbers of severely malnourished. It echoes and complements the Sukhatme and Margen hypothesis of costless adaptation. The concept of "preferred ranges" of many body variables as opposed to "set points" is well established (140), but if the hypothesis is substantiated, the question arises, where is the threshold? Do thresholds vary for different functions? Before the former hypothesis can be accepted, however, a detailed accounting of the short- and long-term costs is required.

Some of the functional consequences of undernutrition and small size are considered in the next section to assess the extent to which adaptation to energy undernutrition is successful.

Reproduction and Fertility

Reproduction is an energetically expensive undertaking for a woman, although less so than for other mammals (93). Successful reproduction is likely to depend on adequate energy stores, and natural selection is likely to have fashioned reproduction so that pregnancy is avoided unless success is high. Frisch and Revelle (38) proposed that menarche, the onset of menstruation, was related to nutritional status, in particular to a certain body weight (47 kg) and to fatness (17% for menarche and 22% for continued menstruation). The mechanism was suggested to be a conversion of androgens to estrogens in the adipocyte. The evidence adduced was that body weight at menarche is independent of menarchal age, early maturers usually weigh more than their peers, and that women with low fat contents, such as ballet dancers and anorectics, have a high incidence of amenorrhea. Also, many obese men have increased serum estradiol and decreased serum testosterone.

The hypothesis has been widely criticized (107, 137). The arguments against it are that weight at menarche has been observed to vary between 28 and 97 kg and that the original estimates of body fatness were tenuous, being based on height and weight. Also anorectics stop menstruating before weight loss and ballet students resume menstruating in vacations without radical changes in weight or composition. Many athletes have normal menses, with fat contents below Frisch and Revelle's trigger levels. The hypothesis of critical levels of fatness for menstruation has mixed support.

Fertility declines in times of severe undernutrition. In the Dutch famine of 1944–1945, where good records were kept, there was a marked and immediate reduction in births, the lowest birth rate corresponding with conceptions at the worst time of food shortage (128). Loss of libido has been a common observation in concentration camps and in energy restriction studies. Starvation affects spermatogenesis, sperm motility, and survival (58).

Delayed menarche and early menopause in developing countries have the potential for reducing fertility, but CED, as opposed to acute famine, has not been shown to limit fertility directly. There is, however, usually a positive relationship between a woman's height and/or weight and fertility (23), but in some reports the relationship is an inverted U. Bigger mothers have bigger babies on average and the incidence of low birth weight is higher in small mothers. Although shorter Mayan women have a higher parity, there is a greater infant mortality and fewer surviving children than in taller or middle tertile women (72).

Pregnancy and Lactation

Pregnancy and lactation are events in the life cycle which impose extra demands for energy. Poor preconception nutrition, which may go back to the mother's childhood, has long-term effects on pregnancy. Undernutrition in pregnancy may result in spontaneous abortion if the stress is severe or in fetal growth retardation and low birth weight if dietary intakes are marginal. Babies born to mothers in the siege of Leningrad in World War II were 540 g lighter than those born at other times.

Although the energy cost of pregnancy is high, the evidence is that energy intakes do not increase appreciably. In Gambian women, falls in BMR in the first two trimesters of pregnancy have been reported instead of the expected increase. No significant increases were observed in cost of treadmill walking, 24h energy expenditure, activity or diet-induced thermogenesis in spite of weight gain (89a). In British women such energy-sparing metabolic adaptations have been observed too, particularly in those that tended to be thin and of low initial energy status (91). Changes in TDEE closely paralleled those in BMR. Decreases in net cost of stepping and cycling were observed. However, the key note was interindividual variability.

Many third world populations, such as the Gambians, appear to have very low energy intakes (6.25 MJ/day) yet produce infants of only slightly lower birth weight (2.9 kg) than is thought satisfactory and breast-feed to support acceptable growth rates (94). There is, however, some doubt over the validity of these reported intakes. Gambian women show no evidence of handicapped gestational or lactational performance in the dry season (90). Supplementation had no effect on birth weight, and breast milk volume and energy content were unaffected. Even when seasonal weight changes of almost 1.5 kg occur in the mother, there is little impact on the infant. In contrast, in Indians the incidence of low birth weight was higher in low BMI women than in others of the same height (78).

Human lactation is remarkably resistant to acute energy deficiency and appears only to be compromised by severe undernutrition (92). Lactation in women in developing countries with widespread CED is on average adequate in quantity and quality (91a), but some children may not get enough. During breast-feeding, infants grow well compared with Western reference data.

Growth and Adolescence

That undernutrition affects growth and sexual maturation of animals and humans has been known for many years. Growth can be accelerated and decelerated to a remarkable degree. Whether animals achieve their genetically determined growth potential depends on the stage of development when undernutrition begins, as well as its intensity and duration. Catch-up growth can occur; in humans the growth period may be extended to the early twenties, but the same adult size of well-fed groups is rarely achieved. The effects of undernutrition are overwhelmingly in the first 3 yr of life and have their origins in the troublesome periods of weaning (72). There appear to be critical periods when the deficits that arise are not made up (4–7 months of fetal life in humans—a key period for the establishment of brain structures), and children seem to remain in the growth channel established by 18–24 months of age. The possibilities for catch-up growth are very limited once the child reaches 3–5 yr (71), but this may reflect an unchanged environment rather than a physiological inability. Whatever the cause, by 5 yr, many children are destined to be short adults.

The importance of fetal nutrition is being reaffirmed by epidemiological studies linking birth weight to diabetes and other degenerative diseases in middle age. Poor nutrition in early life increases the child's susceptibility to an affluent diet. Systolic blood pressure at 10 yr and in adult life have been found to be inversely related to birth weight, and in areas with high cardiovascular disease mortality children were shorter and had higher resting pulse rates (4).

The dramatic effects of CED on growth and mortality in infants are not seen in adolescents. It appears that poor nutrition does not hinder height gain during the pubertal spurt in a population (105). The effect on height that results in small adults is apparent before puberty. However, disadvantaged populations have delayed sexual and skeletal maturation compared to the advantaged.

Children in many parts of the world, particularly the tropics, suffer from the effects of chronic hunger, high rates of infection, and general environmental deprivation. As undernutrition and infection go hand-in-hand, it has been difficult to separate out the individual effects on growth. However, the evidence suggests that acute undernutrition plays the major role. In Guatemalan children only 10% of the weight deficit could be attributed to disease, in Gambian children only diarrheal disease had a significant effect on height, and in Sudanese children only a small proportion of the growth deficit could be accounted for by infection.

Infection and Disease

Infection and malnutrition are responsible for two-thirds of all deaths under 5 yr of age in developing countries. The literature is in accord that in an unsanitary environment the smallest children are the most vulner-able, though the situation is far from straightforward. The synergistic effects of nutrition and infection allow infection to contribute to undernutrition by increasing catabolism and causing anorexia. In Bangladesh, the mortality rates over a 2 yr period of the lowest decile for weight and height for age were four times those in the highest decile (16). Also, mortality of small children was higher if the mothers were small. Over the shorter period of 1 month, mid-upper arm circumference was a better predictor of mortality (12). Cultural, behavioral, social, economic, and environmental factors affect the prevalence and outcome of diseases (15, 134).

The usual explanation of raised morbidity and mortality in acute undernutrition is diminished immune competence (59). Small neonates and infants have poor immunological capabilities and are more liable to infectious diseases. There is evidence that the immunological deficit might persist until late childhood even if catch-up in physical growth is achieved. However, older stunted children may not have an impaired immune response (97) nor are they more susceptible to infection (133). The immune system is an important host defence mechanism against not only infectious disease but also a variety of other disorders, such as cancer. Severe undernutrition impairs immune responses. However, not all immune responses are affected to the same extent in varying degrees of undernutrition. Further, with poor nutritional status the integrity of the skin and mucous membranes, normally a highly effective first line of protection, are affected.

Martorell (71) has suggested that nutrition has little to do with who gets sick but may influence the severity of the infection, as evidenced by hospitalization of duration of sickness. Weight-for-height, a measure of wasting, tends to be a stronger predictor of the severity of infections than height, a measure of reduced growth rates not caught up later.

Physical Working Capacity

In much of the world, work is minimally mechanized and economic productivity depends on human energy and labor. It is these same parts where the highest incidence and prevalence of CED are found. The poorly nourished have a lower work capacity and a lower work output, due mainly to their smaller body size. Undernutrition may lead to qualitative changes in skeletal muscle, such as a reduction in muscle glycogen content, a decreased ATP and creatine phosphate content, and reduced activity of oxidative enzymes, as well as a reduction in FFM; but the capacity to work and the productivity of groups depends on socioeconomic, educational, and environmental factors, as well as nutritional and physiological factors (138).

The relationship between nutritional status and physical working capacity has been the subject of several excellent reviews (2, 110, 118, 120, 122, 129, 138).

Aerobic Work Capacity. The effect of energy restriction on work performance depends on several factors, including the type and intensity of the work, the composition of the diet, and the length of time allowed to adapt to the diet (50). For high-intensity exercises (>70% $\dot{V}_{O_{2}max}$), where muscle glycogen is a limiting factor, sufficient carbohydrate must be present in the diet to prevent ketosis. At lower intensities, lipids provide the major substrates for energy production, glucose homeostasis is maintained, and endurance is not impaired.

There was a marked and progressive reduction in $\dot{V}_{O_{2}max}$ in 49 Colombian men with varying degrees of chronic undernutrition, being 21% and 52% lower in the moderate and severely undernourished compared to the mildly malnourished (3). These differences persisted on a per kilogram or a per kilogram muscle mass basis. Eighty percent of the difference was related to differences in muscle cell mass. The remainder may have been due to variations in hemoglobin content or oxidative enzyme levels in muscle. In eight CED Indian men there was a 30% lower $\dot{V}_{O_{2}max}$ but no substantial difference when corrected for body weight or FFM difference (61). Men with CED had a quicker recovery and a lower oxygen debt.

In contrast, endurance (time to exhaustion working at 80% $\dot{V}_{O_{2}max}$) in mildly, moderately, and severely chronically undernourished men were not significantly different (3), but endurance at a given absolute workload was markedly lower in severe undernutrition (1.5 h vs. 8 h). Although a given task imposes a higher strain on the smaller individual, there is no reduction in work capacity or work level when the task is typical of subsistence activities (131).

The potential to perform physical work is maintained in children with mild or moderate malnutrition, but their small size limits their maximum effort (135). In 1,013 Colombian boys 6–16 years of age, there was a 15% lower $\dot{V}_{O_{2}max}$ in the marginally malnourished (low weight for age and height), associated with the lower body weights (127). Similar observations and conclusions have been made on migrant Brazilian boys (19).

Strength and Motor Performance. A review of the literature in 1964 suggested that acute starvation caused marked deterioration in limb and body movement and a slight reduction in coordination (44). Conversely, in prolonged energy restriction, as in the Minnesota experiment (58), speed was only slightly affected but coordination more so. Malnourished patients have a marked impairment in muscle function. Fatigability is increased

in static contraction and there is a changed pattern of contraction and relaxation (122).

The development and refinement of skillful performance in movement abilities is a major developmental task of childhood and youth. The cross-cultural literature on motor development in the first 2 yr of life in CED populations indicates satisfactory motor development in the first year in most groups. Some African groups even show precocity. Toward the end of the first and during the second year there is a developmental lag. This may be related to a discontinuity of nutrition during weaning, the effects of marginal malnutrition, and reduced levels of activity (67, 68). Grantham-McGregor and colleagues (46), however, could find no causal evidence linking the low activity of chronic undernutrition to poor behavioral development. Whereas low activity and exploration are readily improved with rehabilitation, developmental levels are not.

Marginal malnutrition in Mexican boys was associated with reduced grip strength and motor performance (35 yard dash, standing long jump, and softball throw for distance) compared with Mexican–American boys (68). Static strength per unit body size is generally similar in mild–moderate undernutrition and the well nourished. African and Colombian children have lower working capacities than their European peers but these are appropriate for their size. As with adults though, the important criterion is likely to be absolute working capacity. A further consideration is that small children become small adults, which has implications for their productive capacity in heavy physical work as adults (121).

Work Output and Productivity

The importance of energy undernutrition to work output and productivity depends on the extent to which physical effort is required for the common needs of everyday life and survival. Some data illustrate the importance of body size on productivity in strenuous work but less is known of the effects in low-intensity work (26, 51, 73, 119).

During World War II, coal production in Germany closely followed changes in estimated energy intakes. If rations were reduced, output fell and body weight was protected. The introduction of incentives to production, such as cigarettes, increased output but at the cost of a reduced body weight.

Sugar cane cutting is strenuous work too and lends itself to feasible measures of output. In Tanzania, atypically, there were no significant differences in anthropometry of high, medium, and low producers. In those under 35 y of age, there were significant correlations of output with weight, FFM, and leg volume. Fitter individuals were able to produce a given output in less time and took fewer rest days (18). In Jamaica, cutters with weight-for-height less than 85% of reference values had lower productivity than those 95% and over. The self-paced, continuous work of cane cutting in Colombia is associated with higher height, weight, and FFM but not percent fat in high producers (123), but in Guatemalans there was no relationship of productivity to upper arm muscle ratio adjusted for height or weight-for-height (52). Tall workers were more productive and took fewer rest days.

Output in road construction in Kenya was significantly related to weight-for-height (13). In agricultural work in rural southern India, weight-for-height was a significant predictor of wages and productivity. Increasing the weight-for-height to median values in developed countries would increase wages by an estimated 17% (73). In 140 adolescent Indian boys working as farm laborers, wages that were fixed by the employers were directly related to body weight ($r = 0.60$, $P < 0.001$) (104), demonstrating how poor childhood nutritional status influences earning capacity.

In Indian industrial work of low intensity, such as detonator fuse production, output was correlated with weight after allowing for height but not vice versa. Mean weights-for-height were 83% of reference values, similar to other data on adult men in India (103).

Thus with the exception of the industrial Indian data, weight-for-height, a measure of present nutritional status, is a better predictor of productivity than height-for-age, a measure of past status, particularly in less demanding work (73). Height appears to play a role in Latin American sugar cane cutters. Functional impairment in CED is more in relation to capacity than expressed function (57a). If there is no opportunity or incentive to use capacity, because of unemployment or learning, etc., deficits will not be apparent.

Behavioral and Cognitive Development. There is now much evidence that mental development is associated with nutritional status. This comes from (1) animal work on the relation of poor diets to neurological development, activity levels, and learning performance; (2) cross-sectional epidemiological studies where physical growth is correlated with psychological test performance; (3) comparisons of activity levels and intellectual status in previously malnourished and normal children; and (4) the effects of supplementation on behavior and test performance. The topic has been reviewed extensively (14, 115), particularly the methodological issues that arise in the need to separate nutrition from other environmental or social affectors (5, 88).

Long-term severe malnutrition is clearly associated with serious deficits in cognitive performance, particularly if it occurs in the first 6 months of life. Poor developmental levels may be largely explained by factors associated with stunting, rather than with acute episodes (47). What needs to be ascertained is whether those effects are direct (on brain development) or indirect (through lessened play and exploration) and the effects of mild–moderate malnutrition. Severe undernutrition is associated with reduced motor nerve conduction velocity, muscle fiber size, and energy metabolism. Similarly, motor development, such as the ages at sitting up, standing, and walking, is delayed. Reversibility depends on the timing, severity, and duration of the nutritional stress. Motor nerve conduction velocities show normalization with adequate feeding, but persistent deficits in muscle tissue and motor development have been noted (113). The environment that leads to malnutrition is also a poor learning environment.

It is recognized that an understanding of the controversial and unresolved issue of the developmental outcomes of undernourished children requires a consideration of the environmental circumstances as well as the nutritional history. The notion of critical periods in brain development, after which deficits cannot be reversed, is no longer tenable. Environmental conditions can do much to modify the developmental effects of risk factors in early life (88).

BEHAVIORAL STRATEGIES: HOUSEHOLD AND COMMUNITY RESPONSES AND STRATEGIES

The behavioral responses of the individual to energy undernutrition have been described. There is also a considerable repertoire of non-physiological responses at the household and community levels (85b). Small-scale producers rely on the control and coordination of diverse resources and flexibility in choosing options (132). The light–moderate work may be allocated to the low-energy expenders, such as children or women. Some individuals may be more favored than others; Andean children are protected from seasonal stress, as intakes and fatness fall less than in other family members. In the Cameroons, selected young men are reported to be fattened to survive the hungry season. However, the scope for behavioral adaptive strategies is being eroded (86). Much adaptation is coping, which impairs flexibility and potential for change. Survivors may become suited to a passive life of unemployment in poor rural communities with a stereotyped culture. They may be ill equipped for change and disadvantaged in the urban environment.

GENETIC ADAPTATION TO UNDERNUTRITION

One of the most frequently discussed and intuitively attractive hypotheses of adaptation to energy undernutrition, but also the one with the least supportive data, is that of the thrifty genotype (79). This hypothesis suggests that food has often been in short supply and that this favored those with efficient or variable energy utilization. Through natural selection this genotype became common in many populations. When food supplies were limited these individuals survived while others perished.

The evidence for food shortages in the past is equivocal. Grande (44) regarded the history of humans as largely the history of their search for food. There have been over 500 famines recorded (58), but historical times are relatively recent and follow the comparatively recent introduction of agriculture, which has revolutionized diets and ways of life. Traditional hunter–gatherers are said to have rarely suffered food shortages, even under extremely unfavorable circumstances (11). Small population studies suggest that hunter–gatherers and horticulturists present little evidence of nutritional stress by conventional biochemical or clinical criteria, though nonspecific indices of growth performance may suggest otherwise (128).

The evidence for efficient energy utilization centers on a propensity for obesity and diabetes mellitus in traditional economies on modernization or exposure to outside influences. Studies in Pima Indians in America, who have a propensity to obesity and rates of diabetes of nearly 50%, show that those who gain weight have a lower energy expenditure (96). However, other common traits, such as alcoholism or homicide, suggest that much more is happening than an expression of efficient energy utilization. It has been suggested that the reduced thermic response to cold seen in, for example, Kalahari Bushmen and Australian Aborigines, where core temperatures may fall at night, reflects a conservative energy metabolism advantageous in food shortages (56).

Waterlow (141) has pointed out that a group of survivors able to exist on intakes at the lower end of the range of energy requirements should have a lower coefficient of variation of intakes. There is no evidence for this in groups in areas with high rates of CED or in groups proposed as having a thrifty genotype.

CONCLUSIONS

The immediate and prime adaptation to energy restriction is the utilization of body tissue as a source of energy, with a consequent reduction in body weight and energy expenditure. Energy expenditure is further reduced by a lowered physical activity and energy cost of physical activity, through the smaller body weight. Metabolic adaptation may come into play but without affecting the mechanical efficiency of work.

The evidence for metabolic adaptation is not well founded. It is based on a low metabolic rate per kilogram FFM, but the estimates of FFM are tenuous. Further, reductions in some component masses of FFM, for example, the liver, without reductions in component metabolic rate would result in reduced BMR/kg FFM without metabolic adaptation.

The effects of energy undernutrition, more so than with other nutrient deficiencies, involve both physiological and psychosocial aspects of human activity and behavior. It is no longer sufficient to measure undernutrition in terms of conventional biological parameters, such as birth weight, morbidity, and mortality, but to consider the psychosocial effects too if we are interested in the potential for human development rather than mere survival (6). We need to know what activities are being affected and the effects these have on the individual, the household, and the community. What are the true costs of living at a low energy balance and what benefits in what order would accrue from increasing food intakes? How might the existing constraints on intake be removed and at what cost? The key question is, what are the human costs of accommodation to low energy intakes? These cannot as yet be measured accurately (128a).

The "small but healthy" hypothesis fails to distinguish between what is good for the individual and what is good for the population. To be small would be good for a population but not for an individual or a family. Beaton (8) finds the debate is the result of misinterpretations and distracts from the main issues: attained size is a measure of past and present growth failure not nutritional status or functional capacity. He distinguishes between the process of becoming small and the state of being small. In the former there may be common factors giving rise to the process and other impairments, rather than a direct effect. In the latter, the effect is direct, for example, size and work capacity: "Achieved growth is a marker of the environment and not a result of an adaptive process intended to cope with the environment" (8). Anthropometry is a marker of the process not a diagnosis. Smallness may have nonnutritional reasons, and this may explain the lack of effects of many feeding programs (81). Thus it is not being small that matters but becoming small. The objective then is not to make people bigger but to attack the biological and social constraints to growth and development.

Whether smallness is a successful adaptation or not, the conditions imposing the adaptation cannot be regarded as acceptable (41, 143). To say small size is desirable is to affirm the causes of smallness, that is poor diet and infection in childhood. Growth retardation is not innocuous; it is a risk indicator, a warning sign of morbidity and mortality. Good growth means good health. Conditions which give rise to stunted children affect other aspects: cognitive and motor development and immunocompetence.

The most telling argument against the virtues of smallness is that to keep people small they must be kept in poverty, ill health, undernutrition, and low socioeconomic status. Stunted adults are the survivors of a generation of children subjected to inadequate diets and frequent infections. Smallness is an adaptation that incurs a heavy cost.

REFERENCES

1. Ashworth, A. An investigation of very low calorie intakes reported in Jamaica. *Br. J. Nutr.* 22: 341–355, 1968.

1a. Ategbo, E-A. D., J. M. A. van Raaij, F. L. H. A. de Koning, and J. G. A. J. Hautvast. Resting metabolic rate and work efficiency of rural Beninese women: A 2-y longitudinal study. *Am. J. Clin. Nutr.* 61: 466–472,1995.

2. Barac-Nieto, M. Physical work determinants and undernutrition. *World Rev. Nutr. Diet.* 49: 22–65, 1987.

3. Barac-Nieto, M., G. B. Spurr, M. G. Maksud, and H. Lotero. Aerobic work capacity in chronically undernourished adult men. *Am. J. Clin. Nutr.* 32: 981–991, 1979.

4. Barker, D. J. P. The effect of nutrition of the foetus and neonate on cardiovascular disease in adult life. *Proc. Nutr. Soc.* 51: 135–144, 1992.

5. Barrett, D. E., and D. A. Frank. *The Effects of Undernutrition on Children's Behavior.* New York: Gordon and Breach, 1987.

6. Beaton, G. H. Adaptation to and accommodation of long term low energy intake: a commentary on the conference on energy intake and activity. In: *Energy Intake and Activity,* edited by E. Pollitt and P. Amante. New York: Liss, 1984, p. 395–403.

7. Beaton, G. H. The significance of adaptation in the definition of nutrient requirements and for nutrition policy. In: *Nutritional Adaptation in Man,* edited by Sir K. Blaxter and J. C. Waterlow. London: Libbey, 1985, p. 219–232.

8. Beaton, G. H. Small but healthy? Are we asking the right questions? *Hum. Org.* 48: 30–39, 1989.

9. Beattie, J., and P. H. Herbert. The estimation of metabolic rate in the starvation state. *Br. J. Nutr.* 1: 185–191, 1947.

10. Benedict, F. G., W. R. Miles, P. Roth, and H. M. Smith. *Human Vitality and Efficiency Under Prolonged Restricted Diet.* Washington, DC: Carnegie Inst., Publ. 280, 1919.

11. Bradley, P. J. Is obesity an advantageous adaptation? *Int. J. Obes.* 6: 43–52, 1982.

12. Briend, A., B. Wojtyniak, and M. G. M. Rowland. Arm circumference and other factors in children at high risk of death in rural Bangladesh. *Lancet* ii: 725–728, 1987.

13. Brooks, R. M., M. C. Latham, and D. W. T. Crompton. The relationship of nutrition and health to worker productivity in Kenya. *East Afr. Med. J.* 56: 413–421, 1979.

14. Brozek, J., and B. Schürch. Malnutrition and behavior: critical assessment of key issues. Lausanne, Switzerland: Nestlé Foundation, 1984.

15. Chandra, R. K. Nutrition and immunity in the elderly. *Nutr. Res. Rev.* 4: 83–95, 1991.

16. Chen, L. C., A. K. M. A. Chowdhury, and S. L. Huffman. Anthropometric assessment of protein—energy malnutrition and subsequent risk of mortality among preschool aged children. *Am. J. Clin. Nutr.* 33: 1836–1845, 1984.

17. Danforth, E. The role of thyroid hormones and insulin in the regulation of energy metabolism. *Am. J. Clin. Nutr.* 38: 1006–1017, 1983.

18. Davies, C. T. M. Relationship of maximum aerobic power output to productivity and absenteeism of East African sugar cane workers. *Br. J. Indust. Med.* 30: 146–154, 1973.

18a. Della Bianca, P., E. Jequier, and Y. Schutz. High level of free-living energy expenditure in rural Gambian men: lack of behavioural adaptation between low and normal BMI groups. *Eur. J. Clin. Nutr.* 48: 273–278,1994.

18b. Della Bianca, P., E. Jequier, and Y. Schutz. Lack of metabolic and behavioral adaptations in rural Gambian men with low body mass index. *Am.J. Clin. Nutr.* 60: 37–42, 1994.

19. Desai, I. D., C. Wadell, S. Dutra, E. Duarte, M. L. Robazzi, L. S. Cevallos Romero, M. I. Desai, F. L. Vichi, R. B. Bradfield, and J. E. Dutra de Olivreira. Marginal malnutrition and reduced physical work capacity in migrant adolescent boys in Southern Brazil. *Am. J. Clin. Nutr.* 40: 135–145, 1984.

20. Durnin, J. V. G. A. Energy balance in man with particular reference to low intakes. *Bibl. Nutr. Dieta.* 27: 1–10, 1979.

21. Edmundson, W. Adaptation to undernutrition: how much food does man need? *Soc. Sci. Med.* 14D: 119–126, 1980.

22. Elia, M. Organ and tissue contribution to metabolic rate. In: *Energy Metabolism: Tissue Determinants and Cellular Corollaries,* edited by J. M. Kinney and H. N. Tucker. New York: Raven, 1992, p. 61–77.

23. Eveleth, P. B. Nutritional implications of differences in adolescent growth and maturation and in adult body size. In: *Nutritional Adaptation in Man,* edited by Sir K. Blaxter and J. C. Waterlow. London: Libbey, 1985, p. 31–43.

24. FAO/WHO. *Nutrition and Development—A Global Assessment.* Rome: F. A. O., 1992.

25. FAO/WHO/UNU. *Energy and Protein Requirements.* World Health Org. Tech. Rep. Ser. 724. Geneva: World Health Organisation, 1985, p. 12.

26. Ferro-Luzzi, A. Work capacity and productivity in long-term adaptation to low energy intakes. In: *Nutritional Adaptation in Man,* edited by Sir K. Blaxter and J. C. Waterlow. London: Libbey, 1985, p. 61–68.

27. Ferro-Luzzi, A. Range of variation in energy expenditure and scope for regulation. In: *Proc. XIII Int. Congr. Nutr.* edited by T. G. Taylor and N. K. Jenkins. London: Libbey, 1986, p. 393–399.

28. Ferro-Luzzi, A. Marginal energy malnutrition: some speculations on primary energy sparing mechanism. In: *Capacity for Work in the Tropics,* edited by K. J. Collins and D. F. Roberts. Cambridge: Cambridge University Press, 1988, p. 140–164.

29. Ferro-Luzzi, A. Seasonal energy stress in marginally nourished rural women: interpretation and integrated conclusions of a multicentre study in three developing countries. *Eur. J. Clin. Nutr.* 44(suppl. 1): 41–46, 1990.

30. Ferro-Luzzi, A. Social and public health issues in adaptation to low energy intakes. *Am. J. Clin. Nutr.* 51: 309–315, 1990.

31. Ferro-Luzzi, A., F. Branca, and G. Pastore. Body mass index defines risk of seasonal energy stress in the Third World. *Eur. J. Clin. Nutr.* 48, Supp 13: S165–S178, 1994.

32. Ferro-Luzzi, A., G. Pastore, and S. Sette. Seasonality in energy metabolism. In: *Chronic Energy Deficiency: Consequences and Related Issues,* edited by B. Schürch and N. S. Scrimshaw. Lausanne, Switzerland: IDEGG, 1988, p. 37–58.

33. Ferro-Luzzi, A., C. Scaccini, S. Taffese, B. Aberra, and T. Demeke. Seasonal energy deficiency in Ethiopian rural women. *Eur. J. Clin. Nutr.* 44(suppl. 1): 7–18, 1990.

34. Ferro-Luzzi, A., S. Sette, M. Franklin, and W. P. T. James. A

simplified approach of assessing adult chronic energy deficiency. *Eur. J. Clin. Nutr.* 46: 173–176, 1992.

35. Flatt, J. P. The biochemistry of energy expenditure. In: *Recent Advances in Obesity Research II*, edited by G. Bray. Westport, CT: Food Nutrition, 1978, p. 211–228.

36. Forbes, G. B. Composition of weight gains and losses. *Human Body Composition*. New York: Springer-Verlag, 1987, p. 225–247.

37. Forbes, G. B. Lean body mass–body fat interrelationships in humans. *Nutr. Rev.* 45: 225–231, 1987.

38. Frisch, R. E., and R. Revelle. Height and weight at menarche and a hypothesis of critical body weight and adolescent events. *Science* 169: 397–399, 1970.

39. Garby, L. Metabolic adaptation to decreases in energy intake due to changes in the energy cost of low energy expenditure regimen. *World Rev. Nutr. Diet.* 61: 173–208, 1990.

39a. Garby, L., and O. Lammert. An explanation for the non-linearity of the relation between energy expenditure and fat-free mass. *Eur. J. Clin. Nutr.* 46: 235–236, 1992.

40. Garrow, J. S., and J. D. Webster. Thermogenesis to small stimuli. In: *Human Energy Metabolism*, edited by A. J. H. van Es. Wageningen, Netherlands: Euro-Nut., 1985, p. 215–224.

41. Gopalan, C. Stunting: significance and implications for public health policy. In: *Linear Growth Retardation in Less Developed Countries*, edited by J. C. Waterlow. New York: Raven, 1988, p. 265–279.

41a. Gopalan, C. Undernutrition: Measurement and implications. In: *Nutrition and Poverty*, edited by S. R. Osmani. Oxford: Clarendon Press, 1992, p.17–47.

42. Gorsky, R. D., and D. H. Calloway. Activity pattern changes with decreases in food energy intake. *Hum. Biol.* 55: 577–586, 1983.

43. Gould, S. J., and R. C. Lewontin. The spandrels of San Marco and the Panglossian paradigm: a critique of the adaptationist programme. In: *The Evolution of Adaptation by Natural Selection*, edited by J. Maynard Smith and R. Holliday. London: Royal Society, 1979, p. 147–164.

44. Grande, F. Man under caloric deficiency. In: *Handbook of Physiology. Adaptation to the Environment*, edited by D. B. Dill and E. F. Adolph. Washington, DC: Am. Physiol. Soc., 1964, sect. 4, p. 911–936.

45. Grande, F., J. T. Anderson, and A. Keys. Changes of basal metabolic rate in man in semistarvation and refeeding. *J. Appl. Physiol.* 12: 230–238, 1958.

46. Grantham-McGregor, S., J. M. Meeks Gardener, S. Walker, and C. Powell. The relationship between undernutrition, activity levels and development in young children. In: *Activity, Energy Expenditure and Energy Requirements of Infants and Children*, edited by B. Schürch and N. S. Scrimshaw. Lausanne, Switzerland: IDECG, 1991, p. 361–383.

47. Grantham-McGregor, S., C. Powell, and P. Fletcher. Stunting, severe malnutrition and mental development in young children. *Eur. J. Clin. Nutr.* 43: 403–409, 1989.

47a. Henriksson, J. Energy metabolism in muscle: Its possible role in the adaptation to energy deficiency. In: *Energy Metabolism: Tissue Determinants and Cellular Corollaries*, edited by J. M. Kinney and H. N.Tucker. New York: Raven Press, 1992, p. 345–363.

48. Henry, C. J. K., and D. G. Rees. New predictive equations for the estimation of basal metabolic rate in tropical peoples. *Eur. J. Clin. Nutr.* 45: 177–185, 1991.

49. Himms-Hagen, J. Thyroid hormones and thermogenesis. In: *Mammalian Thermogenesis*, edited by L. Girardier and M. J. Stock. London: Chapman Hall, 1983, p. 141–177.

50. Horton, E. S. Effects of low energy diets on work performance. *Am. J. Clin. Nutr.* 35: 1228–1233, 1982.

51. Immink, M. D. C. Economic effects of chronic energy deficiency. In: *Chronic Energy Deficiency: Consequences and Related Issues*, edited by B. Schürch and N. S. Scrimshaw. Lausanne, Switzerland: IDECG, 1988, p. 153–174.

52. Immink, M. D. C., and F. E. Viteri. Energy intake and productivity of Guatemalan sugar-cane cutters: an empirical test of the efficiency wage hypothesis. *J. Dev. Econ.* 9: 251–287, 1981.

53. James, W. P. T. Research relating to energy adaptation in man. In: *Chronic Energy Deficiency: Consequences and Related Issues*, edited by B. Schürch and N. S. Scrimshaw. Lausanne, Switzerland: IDECG, 1988, p. 7–36.

54. James, W. P. T., A. Ferro-Luzzi, and J. C. Waterlow. Definition of chronic energy deficiency in adults. *Eur. J. Clin. Nutr.* 42: 969–981, 1988.

55. James, W. P. T., G. McNeill, and A. Ralph. Metabolism and nutritional adaptation to altered intakes of energy substrates. *Am. J. Clin. Nutr.* 51: 264–269, 1990.

56. James, W. P. T., and P. S. Shetty. Metabolic adaptation and energy requirements in developing countries. *Hum. Nutr. Clin. Nutr.* 36C: 331–336, 1982.

57. Jones, C. D. R., M. S. Jarjou, R. G. Whitehead, and E. Jequier. Fatness and the energy cost of carrying loads in African women. *Lancet* ii: 1331–1332, 1987.

57a. Kennedy, E., and M. Garcia. Body mass index and economic productivity. *Eur. J. Clin. Nutr.* 48, Suppl 3: S45–S55, 1994.

58. Keys, A., J. Brozek, A. Henschell, O. Mickelsen, and H. L. Taylor. *The Biology of Human Starvation*. Minneapolis: University of Minnesota Press, 1950.

59. Kielmann, A. A., I. S. Uberoi, R. K. Chandra, and V. L. Mehra. The effect of nutritional status on immune capacity and immune response in pre-school children in a rural community in India. *Bull. WHO* 54: 477–483, 1976.

60. Koong, L.-J., J. A. Nienaber, J. C. Pekas, and J.-T. Yen. Effects of plane of nutrition on organ size and fasting heat production in pigs. *J. Nutr.* 112: 1638–1642, 1982.

61. Kulkarni, R. N., A. V. Kurpad, and P. S. Shetty. Aerobic capacity and post-exercise recovery characteristics of chronically energy-deficient labourers. *Proc. Nutr. Soc.* 50: 27A, 1991.

62. Kulkarni, R. N., and P. S. Shetty. Increased net mechanical efficiency during stepping in chronically energy deficient human subjects. *Ann. Hum. Biol.* 19: 421–425, 1992.

63. Kurpad, A. V., R. N. Kulkarni, and P. S. Shetty. Reduced thermo-regulatory thermogenesis in undernutrition. *Eur. J. Clin. Nutr.* 43: 27–33, 1989.

64. Landsberg, L., and J. B. Young. Effects of nutritional status on autonomic nervous system function. *Am. J. Clin. Nutr.* 35: 1234–1240, 1982.

65. Landsberg, L., and J. B. Young. Insulin-mediated glucose metabolism in the relationship between dietary intake and sympathetic nervous system activity. *Int. J. Obes.* 9(suppl. 2): 63–68, 1985.

66. Lawrence, M., K. Thongprasert, and J. V. G. A. Durnin. Between-group differences in basal metabolic rates: an analysis of data collected in Scotland, The Gambia and Thailand. *Eur. J. Clin. Nutr.* 42: 877–891, 1988.

67. Malina, R. M. Physical activity and motor development/performance in populations nutritionally at risk. In: *Energy Intake and Activity*, edited by E. Pollitt and P. Amante. New York: Liss, 1984, p. 285–302.

68. Malina, R. M. Motor development and performance of children and youth in undernourished populations. In: *Sport, Health and Nutrition*, edited by F. I. Katch. Champaign, II: Human Kinetics, 1985, p. 213–226.

69. Maloiy, G. M. O., N. C. Hegland, L. M. Prager, G. Cavagna, and C. R. Taylor. Energetic cost of carrying loads: have African

women discovered an economic way? *Nature* 319: 668–669, 1986.

70. Mansell, P. I., and I. A. MacDonald. The effect of underfeeding on the physiological response to food ingestion in normal weight women. *Br. J. Nutr.* 60: 39–48, 1988.

71. Martorell, R. Child growth retardation: a discussion of its causes and its relation to health. In: *Nutritional Adaptation in Man,* edited by Sir K. Blaxter and J. C. Waterlow. London: Libbey, 1985, p. 13–29.

72. Martorell, R. Body size, adaptation and function. *Hum. Org.* 48: 15–20, 1989.

73. Martorell, R., and G. Arroyave. Malnutrition, work output and energy needs. In: *Capacity for Work in the Tropics,* edited by K. J. Collins and D. F. Roberts. Cambridge: Cambridge University Press, 1988, p. 57–75.

74. McNeill, G., A. McBride, and W. P. T. James. The energy expenditure of large and small eaters. *Proc. Nutr. Soc.* 47: 138A, 1988.

75. McNeill, G., P. Payne, and J. Rivers. Patterns of Adult Energy Expenditure in a South Indian Village. Occasional paper 11. London: London School Hyg. Trop. Med. 1988

76. McNeill, G., J. P. W. Rivers, P. R. Payne, J. J. de Britto, and R. Abel. Basal metabolic rate of Indian men: no evidence of metabolic adaptation to a low plane of nutrition. *Hum. Nutr. Clin. Nutr.* 41C: 473–483, 1987.

77. Minghelli, G., Y. Schutz, A. Charbonnier, R. Whitehead, and E. Jequier. Twenty-four-hour energy expenditure and basal metabolic rate measured in a whole-body indirect calorimeter in Gambian men. *Am. J. Clin. Nutr.* 51: 563–570, 1990.

78. Naidu, A. N., and N. P. Rao. Body mass index: a measure of the nutritional status in Indian populations. *Eur. J. Clin. Nutr.* 48, Suppl 3: S131–S160. 1991.

79. Neel, J. M. Diabetes mellitus: a thrifty genotype rendered detrimental by "progress." *Ann. Hum. Genet.* 14: 354–362, 1962.

80. Norgan, N. G. Adaptation of energy balance to level of energy intake. In: *Energy Expenditure Under Field Conditions,* edited by J. Parizkova. Prague: Charles University Press, 1983, p. 56–64.

81. Norgan, N. G. Chronic energy deficiency and the effects of energy supplementation. In: *Chronic Energy Deficiency: Consequences and Related Issues,* edited by B. Schürch and N. S. Scrimshaw. Lausanne, Switzerland: IDECG, 1988, p. 59–76.

82. Norgan, N. G. Thermogenesis above maintenance. *Proc. Nutr. Soc.* 49: 217–226, 1990.

83. Norgan, N. G., A. Ferro-Luzzi, and J. V. G. A. Durnin. The energy and nutrient intake and the energy expenditure of 204 New Guinean adults. *Phil. Trans. R. Soc. Lond.* B 268: 309–348, 1974.

83a. Osmani, S. R. On some controversies in the measurement of undernutrition. In: *Nutrition and Poverty,* edited by S. R. Osmani. Oxford: Clarendon Press, 1992, p. 121–164.

84. Owen, O. E., A. P. Morgan, H. G. Kemp, J. M. Sullivan, M. G. Herrera, and G. F. Cahill. Brain metabolism during fasting. *J. Clin. Invest.* 46: 1589–1595, 1967.

85. Park, H. S., J. M. Kelly, and L. P. Milligan. Energetics and cell membranes. In: *Energy Metabolism: Tissue Determinants and Cellular Corollaries,* edited by J. M. Kinney and H. N. Tucker. New York: Raven, 1992, p. 411–435.

85a. Payne, P. R. Assessing undernutrition: The need for a reconceptualisation. In: *Nutrition and Poverty,* edited by S. R. Osmani. Oxford: Clarendon Press, 1992, p. 49–96.

85b. Payne, P. R., and M. Lipton. How Third World households adapt to dietary energy stress. The evidence and the issues. *Food Policy Research Review, 2.* Washington, DC: International Food Policy Research Institute, 1994.

86. Pelto, G. H., and P. J. Pelto. Small but healthy? An anthropological perspective. *Hum. Org.* 48: 11–15, 1989.

87. Piers, L. S., M. J. Soares, and P. S. Shetty. Thermic effect of a meal 2. Role in chronic undernutrition. *Br. J. Nutr.* 67: 177–185, 1992.

87a. Poehlman, E. T., and M. J. Toth. Mathematical ratios lead to spurious conclusions regarding age- and sex-related differences in resting metabolic rate. *Am. J. Clin. Nutr.* 61: 482–485, 1995.

88. Pollitt, E. A critical view of three decades of research on the effects of chronic energy malnutrition on behavioural development. In: *Chronic Energy Deficiency: Consequences and Related Issues,* edited by B. Schürch and N. S. Scrimshaw. Lausanne, Switzerland: IDECG, 1988, p. 77–93.

89. Poole, D. C., and L. C. Henson. Effect of acute caloric restriction on work efficiency. *Am. J. Clin. Nutr.* 47: 15–18, 1988.

89a. Poppitt, S. D., A. M. Prentice, E. Jequier, E. Schutz, and R. G. Whitehead. Evidence of energy sparing in Gambian women during pregnancy: A longitudinal study using whole-body calorimetry. *Am. J. Clin. Nutr.* 57: 353–364, 1993.

90. Prentice, A. M. Adaptations to long-term low energy intake. In: *Energy Intake and Activity,* edited by E. Pollitt and P Amante. New York: Liss, 1984, p. 3–31.

91. Prentice, A. M., G. R. Goldberg, H. L. Davies, P. R. Murgatroyd, and W. Scott. Energy-sparing adaptations in human pregnancy assessed by whole-body calorimetry. *Br. J. Nutr.* 62: 5–22, 1989.

91a. Prentice, A. M., G. R. Goldberg, and A. Prentice. Body mass index and lactation performance. *Eur. J. Clin. Nutr.* 48, Suppl 3: S78–S89, 1994.

92. Prentice, A. M., and A. Prentice. Energy costs of lactation. *Annu. Rev. Nutr.* 8: 63–79, 1988.

93. Prentice, A. M., and R. G. Whitehead. The energetics of human reproduction. *Symp. Zool. Soc. Lond.* 57: 275–304, 1987.

94. Prentice, A. M., R. G. Whitehead, S. B. Roberts, and A. Paul. Long-term energy balance in child-bearing Gambian women. *Am. J. Clin. Nutr.* 34: 2790–2799, 1981.

95. Ravussin, E., and C. Bogardus. Relationship of genetics, age, and physical-fitness to daily energy-expenditure and fuel utilization. *Am. J. Clin. Nutr.* 49: 968–975, 1989.

96. Ravussin, E., S. Lillioja, W. C. Knowler, L. Christin, D. Freymond, W. G. H. Abbott, V. Boyce, B. V. Howard, and C. Bogardus. Reduced rate of energy expenditure as a risk factor for body-weight gain. *N. Engl. J. Med.* 318: 467–472, 1988.

97. Reddy, V. Malnutrition and immune response. *Ind. J. Nutr. Diet.* 16: 165–169, 1979.

98. Robinson, A. M., and D. H. Williamson. Physiological roles of ketone bodies as substrates and signals in mammalian tissues. *Physiol. Rev.* 60: 143–187, 1980.

99. Rose, G. A., and R. T. Williams. Metabolic studies on large and small eaters. *Br. J. Nutr.* 15: 1–9, 1961.

100. Rothwell, N. J., and M. J. Stock. A role for brown adipose tissue in diet-induced thermogenesis. *Nature* 281: 31–35, 1979.

101. Rutishauser, I. H. E., and R. G. Whitehead. Energy intake and expenditure in 1–3 year old Ugandan children living in a rural environment. *Br. J. Nutr.* 28: 145–152, 1972.

102. Sabiene, F. The mechanisms for minimizing energy expenditure in human locomotion. *Eur. J. Clin. Nutr.* 44(suppl. 1): 65–71, 1990.

103. Satyanarayana, K., A. D. Naidu, B. Chatterjee, and B. S. N. Rao. Body size and work output. *Am. J. Clin. Nutr.* 30: 322–325, 1977.

104. Satyanarayana, K., A. N. Naidu, and B. S. N. Rao. Agricultural employment, wage earnings and nutritional status of teenage rural Hyderabad boys. *Ind. J. Nutr. Diet.* 17: 281–286, 1980.

105. Satyanarayana, K., G. Radhaiah, K. R. Munali-Mohan, B. S. Thimmayanmma, N. P. Rao, and B. S. N. Rao. The adolescent growth spurt and height among rural Indian boys in relation to childhood nutrition background; an 18-year longitudinal study. *Ann. Hum. Biol.* 16: 289–300, 1989.

106. Schofield, W. N., E. C. Schofield, and W. P. T. James. Basal metabolic rates: review and prediction. *Hum. Nutr. Clin. Nutr.* 39C(suppl. 1): 1–96, 1985.

107. Scott, E. C., and F. E. Johnston. Science, nutrition, fat, and policy: tests of the critical-fat hypothesis. *Curr. Anthropol.* 26: 463–473, 1985.

108. Scrimshaw, N. S., and V. R. Young. Adaptation to low protein and energy intakes. *Hum. Org.* 48: 20–30, 1989.

109. Seckler, D. "Small but healthy": a basic hypothesis in the theory, measurement and policy of malnutrition. In: *Newer Concepts in Nutrition and Their Implications for Policy,* edited by P. V. Sukhatme. Pune, India: Maharashtra Assoc., 1982, p. 127–137.

110. Shephard, R. J. Factors associated with population variation in physiological working capacity. *Yearbook Phys. Anthropol.* 28: 97–122, 1985.

111. Shephard, R. J. *Body Composition in Biological Anthropology.* Cambridge: Cambridge University Press, 1991, p. 215–216.

112. Shetty, P. S. Adaptive changes in basal metabolic rate and lean body mass in chronic undernutrition. *Hum. Nutr. Clin. Nutr.* 38C: 443–451, 1984.

113. Shetty, P. S. Physiological mechanisms in the adaptive response of metabolic rates to energy restriction. *Nutr. Res. Rev.* 3: 49–74, 1990.

114. Shetty, P. S., R. T. Yung, and W. P. T. James. Effect of catecholamine replacement with levodopa on the metabolic response to semistarvation. *Lancet* i: 77–79, 1979.

115. Simeon, D. T., and S. M. Grantham-McGregor. Nutritional deficiencies and children's behaviour and mental development. *Nutr. Res. Rev.* 3: 1–24, 1990.

116. Soares, M. J., L. S. Piers, P. S. Shetty, S. Robinson, A. A. Jackson, and J. C. Waterlow. Basal metabolic rate, body composition and whole-body protein turnover in Indian men with differing nutritional status. *Clin. Sci.* 81: 419–425, 1991.

117. Soares, M. J., and P. S. Shetty. Basal metabolics rates and metabolic economy in chronic undernutrition. *Eur. J. Clin. Nutr.* 45: 363–373, 1991.

118. Spurr, G. B. Nutritional status and physical working capacity. *Yearbook Phys. Anthropol.* 26: 1–35, 1983.

119. Spurr, G. B. Physical activity, nutritional status, and physical work capacity in relation to agricultural productivity. In: *Energy Intake and Activity,* edited by E. Pollitt and P Amante. New York: Liss, 1984, p. 207–261.

120. Spurr, G. B. Effects of chronic energy deficiency on stature, work capacity and productivity. In: *Chronic Energy Deficiency: Consequences and Related Issues,* edited by B. Schürch and N. S. Scrimshaw. Lausanne, Switzerland: IDECG, 1988, p. 95–134.

121. Spurr, G. B. Marginal malnutrition in childhood: implications for adult work capacity and productivity. In: *Capacity for Work in the Tropics,* edited by K. J. Collins and D. F. Roberts. Cambridge: Cambridge University Press, 1988, p. 107–140.

122. Spurr, G. B. The impact of chronic undernutrition on physical work capacity and daily energy expenditure. In: *Diet and Disease in Traditional and Developing Societies,* edited by G. A. Harrison and J. C. Waterlow. Cambridge: Cambridge University Press, 1990, p. 24–61.

123. Spurr, G. B., M. Barac-Nieto, and M. G. Maksud. Productivity and maximal oxygen consumption in sugar cane cutters. *Am. J. Clin. Nutr.* 30: 316–321, 1977.

124. Spurr, G. B., M. Barac-Nieto, and M. G. Maksud. Functional assessment of nutritional status: heart rate response to submaximal work. *Am. J. Clin. Nutr.* 32: 767–778, 1979.

125. Spurr, G. B., M. Barac-Nieto, J. C. Reina, and R. Ramirez. Marginal malnutrition in school-aged Colombian boys: efficiency of treadmill walking in submaximal exercise. *Am. J. Clin. Nutr.* 39: 452–459, 1984.

126. Spurr, G. B., and J. C. Reina. Marginal malnutrition in school-aged Colombian girls: dietary intervention and daily energy expenditure. *Hum. Nutr. Clin. Nutr.* 41C: 93–104, 1987.

127. Spurr, G. B., J. C. Reina, H. W. Dahners, and M. Barac-Nieto. Marginal malnutrition in school-aged Colombian boys: functional consequences in maximum exercise. *Am. J. Clin. Nutr.* 37: 834–847, 1983.

128. Stein, Z., M. Susser, G. Saenger, and F. Marolla. Famine and human development. New York: Oxford University Press, 1975, p. 71–86.

128a. Stinson, S. Nutritional adaptation. *Ann. Rev. Anthropology* 21: 143–170, 1992.

129. Strickland, S. S. Traditional economies and patterns of disease. In: *Diet and Disease in Traditional and Developing Societies,* edited by G. A. Harrison and J. C. Waterlow. Cambridge: Cambridge University Press, 1990, 209–239.

130. Sukhatme, P. V., and S. Margen. Auto-regulatory homeostatic nature of energy balance. In: *Newer Concepts in Nutrition and Their Implications for Policy,* edited by P. V. Sukhatme. Pune, India: Maharashtra Assoc. 1982, p. 101–114.

131. Thomas, R. B. Energy flow at high altitude. In: *Man in the Andes,* edited by P. T. Baker and M. A. Little. Stroudsburg, PA: Dowden, Hutchinson and Ross, 1976, p. 379–404.

132. Thomas, R. B., and T. Leatherman. Household coping strategies and contradictions in response to seasonal food shortage. *Eur. J. Clin. Nutr.* 44(suppl 1): 103–112, 1990.

133. Tomkins, A. M. Nutritonal status and severity of diarrhoea among preschool children in rural Nigeria. *Lancet* i: 860–862, 1981.

134. Tomkins, A. M. Protein energy malnutrition and risk of infection. *Proc. Nutr. Soc.* 45: 289–304, 1986.

135. Torun, B. Short- and long-term effects of low or restricted energy intakes on the activity of infants and children. In: *Activity, Energy Expenditure and Energy Requirements of Infants and Children,* edited by B. Schürch and N. S. Scrimshaw. Lausanne, Switzerland: IDECG, 1990, p. 335–358.

136. Torun, B., and F. E. Viteri. Energy requirements of pre-school children and effects of varying energy intakes on protein metabolism. *U. N. Univ. Food Nutr. Bull.* (suppl. 5): 210–228.

137. Trussell, J. Statistical flaws in evidence for the Frisch hypothesis that fatness triggers menarche. *Hum. Biol.* 52: 711–720, 1980.

138. Viteri, F. E. Considerations on the effects of nutrition on body composition and physical working capacity of young Guatemalan adults. In: *Amino-Acid Fortification of Protein Foods,* edited by N. S. Scrimshaw and A. M. Altschul. Cambridge, MA: MIT, 1971, p. 350–375.

139. Waterlow, J. C. Postscript. In: *Nutritional Adaptation in Man,* edited by Sir K. Blaxter and J. C. Waterlow. London: Libbey, 1985, p. 233–235.

140. Waterlow, J. C. What do we mean by adaptation? In: *Nutritional Adaptation in Man,* edited by Sir K. Blaxter and J. C. Waterlow. London: Libbey, 1985, p. 1–11.

141. Waterlow, J. C. Metabolic adaptation to low intakes of energy and protein. *Annu. Rev. Nutr.* 6: 495–526, 1986.

142. Waterlow, J. C. Mechanisms of adaptation to low energy intakes. In: *Diet and Disease in Traditional and Developing Societies,* edited by G. A. Harrison and J. C. Waterlow. Cambridge: Cambridge University Press, 1990, p. 5–23.

143. Waterlow, J. C. Nutritional adaptation in man; general introduction and concepts. *Am. J. Clin. Nutr.* 51: 259–263, 1990.

144. Waterlow, J. C., R. Buzina, W. Keller, J. M. Lane, M. Z. Nichaman, and J. M. Tanner. The presentation and use of height and weight data for comparing the nutritional status of groups of children under the age of 10 years. *Bull. WHO* 55: 489–498, 1977.

145. Weindruch, R. H., M. K. Cheung, A. Verity, and R. L. Walford. Modification of mitochondrial respiration by aging and dietary restriction. *Mech. Ageing Dev.* 12: 375–392, 1980.

146. Welle, S. and R. G. Campbell. Lack of catecholamine mediated thermogenesis during carbohydrate overfeeding in man. *J. Clin. Invest.* 71: 916–925, 1983.

147. Widdowson, E. M. The responses of the sexes to nutritional stress. *Proc. Nutr. Soc.* 35: 175–180, 1976.

62. Physiological responses of mammals to overnutrition

VANESSA H. ROUTH

JUDITH S. STERN

BARBARA A. HORWITZ

Section of Neurobiology, Physiology, and Behavior, Division of Biological Sciences, University of California, Davis, California
Departments of Nutrition and Internal Medicine, University of California, Davis, Davis, California
Section of Neurobiology, Physiology, and Behavior, Division of Biological Sciences, University of California, Davis, Davis, California

CHAPTER CONTENTS

OVERNUTRITION IS DEFINED in this chapter, as positive energy balance leading to increased energy storage, primarily in adipose tissue. Although obesity is the most obvious manifestation of overnutrition, other physiological alterations also occur. The adaptive response to overnutrition differs from responses to other stressors such as cold and altitude in that, although the initial alterations may help the organism to deal with the stressor, the long-term consequences are often detrimental.

CONSEQUENCES OF OVERNUTRITION

There are at least three different types of response to overnutrition. The first is obesity resulting from increased food intake (hyperphagia). The second is independent of hyperphagia and involves alterations in metabolic efficiency. This is seen in many forms of genetic and hypothalamic obesity, where even though animals may be hyperphagic during the development of obesity, this hyperphagia is not necessary for the condition's development. The third involves responses to specific diets. For example, certain animals may become obese on a high-fat but not on a low-fat diet. This chapter will examine the physiological alterations of mammals in response to overnutrition, factors that influence these responses, and some of the mechanisms that underlie them.

When does excess adiposity impair health? The answer is not clear-cut, in part because the indices used to quantify obesity as a health hazard are variable. One reason for this is that the same degree of adiposity may pose a health risk in one individual but not in another. Another reason is that individuals with similar body weights may have different amounts of body fat (97, 246, 248). Despite the fact that it is important to differentiate between overweight and obesity, many so-called quantitative measurements of obesity in humans use weight and height as indices. For example, body mass index (BMI) is equal to weight (kg) divided by height squared (m²). A BMI of 27.8 for men and 27.3 for women is generally considered overweight, whereas a BMI of 31.1 for men and 32.3 for women is considered obese (187). However, the use of BMI as the sole criterion for obesity can be misleading, especially in athletes, because muscle weighs more than fat. Percent body fat is the best measure of obesity. At 18 yrs of age, men whose body fat exceeds 20% and women whose body fat exceeds 28% are considered obese (33).

There are several ways of measuring percent body fat in humans. Among the more traditional are skin-fold thickness, densitometry, and total body water (see ref. 168 for detailed review). However, each of these methods has drawbacks. Tricep skin-fold measurements can be variable and are highly dependent on the skill of the investigator. Nonetheless, if done consistently, the sum of tricep and subscapular or the sum of five skin folds correlates well with hydrometric and densitometric measures of percent body fat (170). Densitometry requires the submersion of the subject in water and the accurate estimate of lung volume. Measurement of total body water by dilution using radioactive tracers or stable isotopes is expensive and requires special equipment. Newer methods include measurement of muscle metabolites, bioelectrical impedance, and photon absorptiometry (168). In laboratory animals, chemical analyses of body composition or measurements of individual fat depots can be performed after death (22).

Thrifty Gene: Altered Lipid Metabolism

A "thrifty genotype" refers to the tendency of some individuals to be more efficient at storing energy as fat than others (188). This thrifty genotype may have been introduced into the human genome via the process of natural selection. One proposed scenario is that selection for a thrifty gene resulted from physiological adaptations that occurred during migration across the land bridge from Asia to North America and finally to the southwestern desert (for a detailed review see ref. 211). Ritenbaugh and Goodby (211) speculate that as prehistoric humans passed over the land bridge they had to adapt to survive on a high-protein, moderate-carbohydrate, low-fat diet (approximately 34%, 45%, and 21%, respectively), as well as intermittent famine. High energy demands for activity and warmth, combined with limited carbohydrate sources, intermittent lipid shortage, and a high availability of protein set the stage for adaptations with interesting physiological consequences. The need to maintain blood glucose above 50 mg/dl for proper function of key organs, such as brain and kidney, was a constraint. Limited dietary carbohydrate would be accompanied by low insulin levels, and fatty acids would most likely be the primary fuel for physical activity and heat production. Thus the ability to store fat provided survival advantages by supplying fuel, sparing glucose, and limiting the amount of

dietary protein needed for energy. As the hunters moved south and developed agriculture, limitations on dietary carbohydrate decreased, and carbohydrate became more readily available as fuel, allowing fat stores to be spared. Moreover, food availability increased in general. Insulin levels increased due to the elevated glucose, further sparing fat. This was especially important during times of famine, particularly for pregnant and lactating women. Those individuals with a greater capacity to store fat were better able to survive a famine (211).

This proposed scenario is not unique to New World populations and, in fact, is consistent with observations of primitive societies all over the world [for example, the Kung San in the Kalahari Desert (37) and the Maori in Polynesia (37, 228)]. It is even reflected in the cultural standards for feminine beauty, which only recently have emphasized thinness. In primitive societies even today, plumpness to the point of obesity is equated with attractiveness and desirability. Thus there was selection for individuals who were better able to store fat (that is, contained a "thrifty gene").

As societies industrialized, dietary fat and carbohydrate increased and protein decreased (37). Moreover, in affluent societies, the threat of famine was eliminated, at least in the upper social classes, and the thrifty gene resulted in an increased probability of obesity. This was exacerbated by decreased physical activity (37, 208, 211, 228). Thus what was once a beneficial adaptation for survival has become one of the industrial world's most widespread health problems, with obesity ranging between 25% and 40% of the population in affluent Western civilizations (100, 246).

Animal Models

Although this section is not inclusive, it will provide a brief overview of three common types of animal models used to study physiological responses to overnutrition: namely, animals whose obesity is induced by diet, by genetic makeup, and by experimental alteration of the hypothalamus. The physiological mechanisms underlying dietary and genetic obesity will be discussed in more depth in the section FACTORS INFLUENCING THE RESPONSE TO OVERNUTRITION, below. The hypothalamic models will be covered below in the section Neural Mechanisms Mediating the Response to Overnutrition. A review by Bray et al. (35) discusses the characteristics of these models in detail.

Diet-Induced Overnutrition. Four examples of diets used to produce obesity in experimental animals include high fat, high sucrose, combined high fat and high sucrose, and "cafeteria." All of these diets can result in long-term metabolic alterations even when a stock diet is resumed.

High-fat diet. As mentioned previously, the increased fat content of the Western diet in combination with decreased physical activity has resulted in the increased occurrence of obesity in humans (211). When fat makes up more than 30% of dietary intake, most animals become fatter and increase their energetic efficiency (33). The degree of obesity appears to be proportional to the amount of fat in the diet (230). For example, the Osborne-Mendel rat strain is very susceptible to this kind of dietary manipulation and will greatly increase its percent body fat, body weight, weight of a variety of organs, and size and number of white adipocytes (193, 261). Similarly, genetically lean Zucker rats fed a high-fat diet will increase adiposity even though brown fat thermogenic capacity may be increased (225). Brown fat contributes significantly to nonshivering heat production in mammals and may be related to dietary thermogenesis in rodents and humans. Fats containing saturated fatty acids and long-chain triglycerides result in greater increases in adiposity than do fats with polyunsaturated fatty acids and medium-chain triglycerides (177). High-fat diets can cause obesity even in the absence of hyperphagia (178, 179). Notably, even elderly rats fed high-fat diets increase the number of fat cells (220).

Diets high in simple sugars. Feeding rats a diet high in sugars (especially in liquid form) induces hyperphagia and obesity, which is more pronounced in females than in males (239). Interestingly, the degree of hyperphagia in rats is dependent on the form of the sugar, with sugar in solid form being least preferred and sugar in liquid form most preferred (239). Moreover, taste preference in rats increases with saccharide length up to 8 glucose units, after which it declines (239). Obese humans also appear to be more responsive to a highly palatable diet (215). As with the high-fat diet, animals can become obese on a high-sucrose diet without overeating (35). Carbohydrate-induced overeating and obesity are discussed in considerable detail by Sclafani (239).

Combined high-fat/high-sugar. Levin and colleagues (163) have shown that a diet of 47% stock, 8% corn oil, and 44% sweetened condensed milk results in hyperphagia and obesity in one-third to two-thirds of Sprague-Dawley rats. The fact that "diet-resistant" rats can be identified prior to becoming obese by their response to a glucose tolerance test (162) or by catecholamine-induced lipolysis of adipose tissue (155) makes them a potentially useful model for studying diet-induced obesity.

Cafeteria diet. Rats will eat more of a highly palatable, varied diet consisting of items such as chocolate chip cookies, salami, cheese, bananas, marshmallows, peanut butter, etc. (239). This results in increased adiposity despite increased thermogenic capacity of brown fat (166, 224). Humans exhibit "sensory-specific satiety;" that is, they will eat more of a variety of foods compared to these same food items presented singly (216).

Genetic Models. Many animal models of genetic obesity are hyperphagic, glucose-intolerant, hyperinsulinemic, or hyperglycemic; have a decreased thermogenic response; and respond to adrenalectomy (ADX) by normalizing their weight gain and most of the accompanying metabolic alterations [see reviews by Johnson et al. (137) and Bray et al. (35)]. Since these variables are under the control of the autonomic nervous system, it has been hypothesized that some forms of genetic obesity result from a defect in the central nervous system (CNS) (289). In general, all genetically obese rodents are hyperphagic in the dynamic phase of the development of obesity. However, their obesity is independent of hyperphagia. In many cases, excess adiposity precedes hyperphagia (137). Thus these models show deficits in the regulation of both energy intake and energy expenditure. Commonly used rodent models are the Zucker obese (*fa/fa*) rat, the Wistar diabetic fatty rat (WDF), the hyperglycemic obese (*ob/ob*) mouse, and the diabetic obese (*db/db*) mouse (35, 137). A comparison of some of the characteristics of genetic animal models is shown in Table 62.1.

There is growing evidence for a genetic component to some human obesities. This evidence comes from studies of isolated populations (85, 132, 145, 208, 228, 263, 271), as well as from studies of twins (25, 27, 205). For example, the Pima Indians are a relatively inbred tribe with an extremely high prevalence of obesity and non-insulin-dependent diabetes (NIDDM) (75% and 45%, respectively). Their metabolic alterations show a very high degree of familial aggregation and are therefore frequently used to study genetic obesity in humans (208). Similarly, studies with monozygotic twins indicate that obesity and the accompanying metabolic alterations are much more highly correlated than in dizygotic twins (206, 271). Moreover, adoption studies have shown that the tendency toward obesity is related more to the biological parents than to the adoptive parents (that is, life-style) (27–29). However, genetic changes require long periods of time to be introduced. Thus the post–World War II increase in obesity is probably a combination of increased dietary fat and decreased physical activity, allowing for the expression of a genetic factor already in the human genome (206).

Hypothalamic Obesity. In humans, hypothalamic obesity is extremely rare. Most cases have resulted from injury or disease (for example, tumors) (35). In the laboratory rat, obesity can be induced from electrolytic lesions of the ventromedial nucleus (VMN) (113, 114), the paraventricular nucleus (PVN) (266), or the arcuate

TABLE 62.1. *Characteristics of Several Rodent Models of Genetic Obesity*

Measured Variable	Zucker *fa/fa* rat	WDF *fa/fa* rat	*ob/ob* mouse	*db/db* mouse
Hyperphagia	+	+	+	+
Lean body mass	−		−	
Protein deposition	−			
Urinary 3-methylhistidine	−			
Glucose intolerance	+	+	+	+
Sympathetic activity	−			
Parasympathetic activity	+			
Testicular size	−			
Response to female pheromones				
Thermogenic capacity	−		−	−
Normalization by ADX	+	+	+	+
Hyperinsulinemia	+	+	+	transient
Hyperglycemia	−	+	+	+
Hypertriglyceridemia	+	+	nd	+
Insulin/glucagon ratio	+		−	var
Corticosterone turnover	+		+	+
Circulating estrogen	−			
Thyrotropin concentration	nd		−	
Circulating thyroid hormone	−		−	−
WAT mass	+	+	+	+
WAT hyperplastic	+	+	+	+
WAT hypertrophic	+	+	+	+
WAT lipoprotein lipase	+	+	+	+
WAT lipogenesis	+		+	+
Basal lipolysis	+		+	+
Isolated WAT insulin binding	+		−	
BAT mass	+		+	
BAT lipoprotein lipase	nd			
GDP-binding/mitochondrial	−		−	−
Pancreatic islet hypertrophy	+	+	+	+
Pancreatic islet hyperplasia	+		+	+
Pancreatic islet glucagon release	−	−	+	+
Pancreatic somatostatin	+		−	−
Pancreatic somatostatin release	+	+	nd	nd
Hepatic lipogenesis	+		+	+
Hepatic ketogenesis	−		−	−
Hepatic Na$^+$/K$^+$ ATPase		nd	−	−
Fructose-2,6 biphosphate	+		+	
Glucose-6PDH activity	+	+		
Fatty acid desaturase	+		+	
Hepatic insulin resistance		+	+	
Isolated hepatic insulin binding	+		−	
Hepatic insulin receptors	var		−	−
Muscle protein	−		−	
Muscle DNA	−			
Muscle RNA	−			
Muscle lipoprotein lipase	nd		−	
Muscle protein synthesis	−		−	
In vitro insulin resistance	+		−	
Skeletal muscle glucose utilization	−			
Muscle insulin receptors	−		−	
CNS insulin content	−			
CNS insulin binding	−	−		
CSF insulin content	+	+		
Gonadotropin secretion	−		−	
Hypothalamic somatostatin secretion	+		nd	nd
Growth hormone secretion	−		−	−
Somatomedin activity	−			
Cholecystokinin activity	+		nd	
Catecholamines/adrenergic system	−		var	
Dopaminergic system	−		var	
Serotonergic system	−		+	
Opioid activity	+		var	

Variables compared to the lean phenotypes; + = obese greater than lean; − = obese less than lean; nd = no differences; var = variable results reported in the literature. [Adapted from Johnson et al. (137).]

nucleus (233, 262). These hypothalamic injuries alter both food intake and metabolic efficiency. Lesions of the VMN result in an obesity syndrome that includes, but is not dependent on, hyperphagia and hyperinsulinemia as well as decreased sympathetic and increased parasympathetic nervous system activity (104, 105, 204, 257). Lesions of the PVN result in a hyperphagia-dependent obesity and will exaggerate dietary obesity (266). Lesions of the arcuate result in obesity and hyperinsulinemia without hyperphagia (233, 262).

Physiological Consequences

Changes in Body Composition. In general, overnutrition results in increased fat depot size due to increased size and/or number of white adipocytes. The distribution of this increased adiposity in humans has an effect on the severity of the accompanying physiological alterations (100, 147, 272). If relatively more fat is distributed in the upper body region than in the lower body region (apple vs. pear), obesity is referred to as *abdominal* or *android*. If the adiposity is relatively more prevalent in the hips and thighs, obesity is referred to as *gluteal-femoral, lower body,* or *gynoid* (33). Obese individuals

with abdominal fat distribution have greater morbidity and mortality than those with gluteal–femoral fat distribution, where fat is especially important as an energy source during pregnancy and lactation (33, 100, 147). Abdominal fat is more readily and rapidly mobilized, contributing significantly to increased blood triglycerides and cholesterol. Abdominal adipocytes are larger than those in gluteal–femoral stores and have greater glucose uptake, but they are more insulin-resistant (246). Upper-body obesity is also associated with decreased hepatic insulin extraction, leading to increased peripheral insulin levels (that is, hyperinsulinemia) (100). Moreover, it tends to correlate more highly with coronary heart disease, stroke, diabetes mellitus, hypertension, and breast cancer (33, 100, 147). Increased levels of free testosterone also accompany upper-body obesity (75).

Obesity in humans results in increased lean body mass to support a heavier frame (63, 240, 282). In addition, organ weights (heart, kidney, liver, pancreas) are increased by overnutrition (Fig. 62.1; 186). In cafeteria-fed rats, both liver and kidney weights are greater than in controls due to increased numbers of cells (DNA) and enlarged cell size (more stored lipid). Hyperplasia and

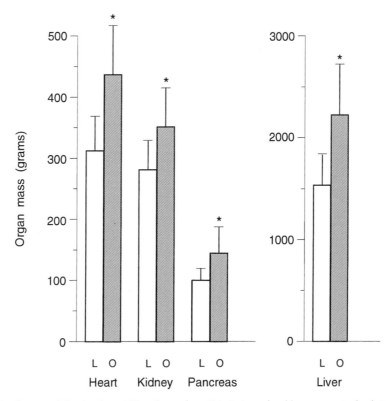

FIG. 62.1. Organ weights in obese (O) and nonobese (L) victims of sudden traumatic death (controlled for sex and body length). Organ weights for heart, liver, and kidney were significantly greater in the obese accident victims (n = 26) than in the nonobese (n = 41) (*P < 0.05). No significant differences in spleen or brain. Obesity in these subjects was defined as body weights 20% greater than upper limits of desirable body weight for individuals with large frames (186).

hypertrophy also occur in brown fat depots (166). Additionally, Zucker obese (fa/fa) rats have heavier livers, as well as hypertrophy and hyperplasia of the pancreatic islets. In contrast to humans, muscle protein and DNA are decreased in these animals and may reflect hypopituitary function (137).

Changes in Whole-Body Metabolism.

It is unclear if the metabolic changes seen in overnutrition result from or cause the maladaptive response to overnutrition. Low resting metabolic rates, including low 24 h energy expenditures, are associated with increased risk of body weight gain (26, 208). Obese individuals have a lower 24 h energy expenditure when expressed per kilogram body weight but not per fat-free mass, suggesting that the reduced metabolism can be explained by the difference in stored fat (63). (Similar results were obtained when data were expressed in terms of mass$^{0.67}$.) When the primary, but not the sole, fuel being burned is fat (for example, after an overnight fast), the respiratory quotient (the ratio between carbon dioxide production and oxygen consumption) is approximately 0.8. When glucose is the primary substrate (for example, after a high-carbohydrate meal), the respiratory quotient is approximately 1.0 (1). Thus the lower the respiratory quotient, the higher the ratio of fat to carbohydrate oxidation. Twenty-four hour respiratory quotient values have been correlated to percent body fat (293). Obesity-prone individuals have increased respiratory quotient values compared to individuals who "resist" becoming obese. For example, nondiabetic Pima Indians with higher quotients are at much greater risk of weight gain independent of their reduced energy expenditure (293).

Altered whole-body metabolism is also correlated with body fat distribution. That is, individuals with gluteal–femoral obesity have lower resting metabolic rates per kilogram fat-free mass than do those with abdominal obesity or those who are not obese. This attenuated metabolism occurs not only at thermoneutrality but in response to cold (282), contributing to greater hypothermia in obese vs. lean humans (158). Additionally, energy expenditure during sleep, as measured in metabolic chambers, is less in obese vs. lean individuals (158).

Genetically obese rodents show similar metabolic alterations. The Zucker obese (fa/fa) rat exhibits blunted thermogenesis at and below thermoneutrality as early as 2 days of age (183). This occurs prior to the accumulation of excess adipose tissue stores, and it appears to be related to decreased brown fat thermogenesis (183). It has also been reported that oxygen consumption in response to a meal is diminished and/or lacking in the (ob/ob) obese hyperglycemic mouse and the Zucker obese (fa/fa) rat relative to lean controls

(117, 124, 268, 269). Thus reduced energy expenditure and fat oxidation are maladaptive metabolic responses to overnutrition that lead to increased energy storage.

Changes in Tissue Utilization (Fuel Partitioning).

According to Ritenbaugh and Goodby's (211) discussion of the thrifty gene hypothesis, when carbohydrate was limited in the diet of prehistoric humans, tissues burned fat to spare glucose. However, as dietary carbohydrate increased, this situation reversed. Weight maintenance is achieved through the balance of energy intake and expenditure, as well as the balance of carbohydrate, fat, and protein. Flatt (80) has proposed a model in which carbohydrate and fat storage and oxidation are regulated to maintain glycogen stores, and imbalanced intake of either one of these nutrients can result in obesity (see ref. 80 for details on the ramifications of this model).

Many obese people are hypertriglyceridemic, a condition alleviated by weight reduction (100, 248). This could be due to overproduction of very low-density lipoprotein (VLDL) triglycerides and/or defective hydrolysis of triglyceride-rich lipoproteins, both situations being seen in obese patients. Obese individuals also tend to have elevated low-density lipoprotein (LDL) cholesterol and decreased high-density lipoprotein (HDL) cholesterol (100, 147, 176, 282, 283). One reason for this is that elevated food intake, especially of saturated fatty acids, raises hepatic VLDL synthesis, which leads to enhanced biliary secretion of cholesterol (100).

Cholesterol is held in solution by phospholipids and bile acids. When levels of cholesterol increase, bile can become supersaturated and cholesterol crystals precipitate out. This leads to increased risk of gallstones (100). Pima Indians, who exhibit elevated biliary secretion of cholesterol, have an extremely high incidence of gallstones (80%) (100, 231).

In the Zucker obese (fa/fa) rat, the synthesis of whole-body fatty acids per hour is four to five times greater than in its lean (Fa/Fa) counterpart (103). The obese rats also have altered tissue phospholipids. For example, arachidonic acid levels are lower in heart, liver, serum, and platelets. Supplementation with a black currant oil concentrate (high in gamma linolenic acid) restores hepatic phospholipid arachidonic acid and decreases food intake and weight gain (200). Omega-3 fatty acids are also lower in heart but elevated in liver, kidney, and adipocytes; n-6 fatty acids are depressed in liver, kidney, and heart. Total lipid weight is elevated in liver from obese Zucker rats (101). Weight of brown fat is greater in the obese (fa/fa) Zucker rat due to increased lipid weight but is functionally less active (268). Finally, obesity-resistant rats are able to oxidize more fat than obesity-prone rats (48).

Changes in Physiological Systems. As discussed above, overnutrition leads to numerous abnormalities other than increased adiposity, many of which are detrimental to the organism (39, 100, 147, 248). The hyperinsulinemic response to overnutrition may initially be advantageous because the resulting insulin resistance would result in less glucose stored as fat. However, the consequence is maladaptive, favoring fat storage (214). One of the most common disorders associated with overnutrition and obesity is NIDDM (100). Increased caloric intake ultimately results in altered regulation of glucose metabolism in humans and genetically obese animals (100, 214). In obese Rhesus monkeys, the insulin secretagogue plasma cell β tropin is highly correlated with body weight and percent body fat (up to 40%) (184). Persistent hyperinsulinemia results in down-regulation of insulin receptors, leading to insulin resistance in some individuals. Substrate overload in the obese could lead to impaired glucose tolerance (100). Moreover, obese subjects with NIDDM show increased post-absorptive endogenous glucose production (26). This may reflect the observed increased flux through the Cori cycle, a glucose recycling process by which hepatic glucose is converted to three carbon units by skeletal muscle and returned to the liver to be made into glucose again (291). The incidence of NIDDM is greater in individuals with upper-body obesity than in those with lower-body obesity. Both male and female Rhesus monkeys that have excess abdominal fat are hyperinsulinemic (143). Steady-state plasma glucose levels correlate with waist to hip ratio. Furthermore, percent of insulin-stimulated glycogen synthetase activity in skeletal muscle decreases as waist to hip ratio increases (147). In obese females, the greater the prehepatic insulin production, the less the hepatic insulin extraction (147).

Another ultimately maladaptive response that may initially be advantageous is sympathetic stimulation resulting from the hyperinsulinemia that accompanies overnutrition in humans (129). Initially, this would protect the individual from increased fat storage. However, this could also contribute to the development of hypertension. Current estimates indicate that 30%–50% of hypertensive patients are obese (100). In addition to sympathetic stimulation, obesity contributes to hypertension by increasing cardiac output due to increased blood volume needed to supply the added mass, both adipose and lean body mass (281). This increased blood volume may also be related in part to elevated sodium consumption accompanying the higher caloric intake or the increased sodium retention by the kidney associated with hyperinsulinemia (281). Overnutrition may also raise total peripheral resistance by a mechanism that is still unknown but is thought to involve circulating androgens, hyperinsulinemia, body fat distribution, and

liver extraction of insulin (147). Moreover, obesity decreases the effectiveness of antihypertensive drugs, exacerbating the situation (181). Loss of some of the excess weight ameliorates hypertension (202). Interestingly, Contreras et al. (53) found no correlation between weight cycling (alternately losing and regaining weight) and hypertension, though this has been reported in some experimental animals and may reflect the special dietary conditions used (74). Whatever the mechanisms, obese individuals are at increased risk for hypertension and coronary heart disease (100, 138, 284).

In addition to responses that may initially have some value, overnutrition causes many changes which are detrimental to health. For example, overnutrition and obesity have been closely correlated to various forms of cancer, such as breast, colon, and prostate (100, 147, 248). Since most obese people also have increased lean body mass, it is unclear if the adipose or lean body mass is the cause. Overnutrition in the young leads to increased body size in adulthood and may play a role in tumor development (100). The food component most often associated with cancer is fat, and obese people and laboratory rats tend to prefer foods high in fat (99, 215). Obese dogs exposed to insecticides appear to be more likely to develop bladder cancer than nonobese animals (93). This may be because hydrophobic insecticides are stored in adipose tissue, prolonging carcinogenic effects (93). In The Netherlands, the high incidence of breast cancer is linked to high intake of fatty foods (39). Breast cancer is more frequent among obese than lean individuals, possibly due to peripheral aromatization of adrenal androstenedione to estrogen. Obese female adolescents tend to reach puberty earlier and are thus exposed to elevated estrogen for longer periods of time (39).

Finally, overnutrition decreases the ability to survive trauma. In fact, obese patients in a trauma ward show more rapid deterioration and decreased survival rate than do lean individuals (49). Metabolic disturbances in the obese interfere with healing. That is, injury results in a catabolic state and increased substrate mobilization (131). Paradoxically, obese individuals do not use lipid from adipose stores but instead decrease muscle mass to provide substrate (131).

Thus although some of the physiological responses to overnutrition may initially have some positive value, the long-term response is maladaptive.

Changes in Behavior. In a society where thinness is stressed, it is difficult to separate behavioral changes resulting from obesity from behaviors that contribute to obesity. Preoccupation with thinness has led more women than men to be dissatisfied with their weight and to undergo dieting cycles, that is, weight loss and regain (243). In industrialized societies, obesity, especially among women, is negatively correlated with social class

(133). In contrast, more primitive cultures, like that of the New Zealand Maori, are much less concerned with obesity (228). In fact, many of these societies still equate plumpness in their women with beauty, for the reasons mentioned above under Thrifty Gene: Altered Lipid Metabolism. Dissatisfaction with body weight leading to dieting may be followed by increased consumption of palatable foods and a high incidence of impulsive eating, which is uncompensated for in the next meal (215). It is also hypothesized that obese people have a lower threshold for physiological arousal coupled with a heightened response to external food cues (215, 236).

FACTORS INFLUENCING THE RESPONSE TO OVERNUTRITION

Gender

In humans, normal-weight young adult women have approximately 20%–25% body fat, whereas 15%–18% is normal for men (97). Thus the definition of obesity differs between genders, over 30% body fat for women and over 25% for men being considered obese (97). Obesity tends to be more prevalent in women than men, especially black women (33).

The pattern of fat distribution in response to overnutrition differs strikingly with gender. The majority of men tend to increase adiposity in the abdominal region (android obesity), while many, but not all, premenopausal women tend to store more fat in the hips and thighs (gluteal–femoral obesity) (33, 41, 44, 97, 147). As mentioned previously, it is the android pattern of fat distribution, in both men and women, that is associated with health risks such as diabetes, certain cancers, hypertension, and heart disease (33, 39, 44, 66, 147). This does not mean that individuals with gynoid obesity do not suffer from these diseases. However, they are at lower risk.

The reason this difference in fat distribution is often sex-linked is unknown. However, in healthy premenopausal women, levels of free testosterone are directly related to the amount of abdominal fat, whereas levels of sex-binding hormone globulin are inversely related to abdominal fat (44, 75). This relationship holds for insulin resistance as well (44, 75), though insulin resistance appears to be more related to body fatness than to hormone status (241). This gender difference may also be related to changes in lipoprotein lipase (LPL) activity associated with the menstrual cycle, pregnancy, or lactation. That is, except during lactation (when there is net mobilization of fats), gluteal–femoral stores exhibit higher LPL activity than do abdominal stores, thus facilitating increased fat deposition (39). This greater LPL activity appears to be correlated with 17β-estradiol lev-

els (209). These studies clearly indicate that gender differences in fat distribution are related to levels of sex hormones [see review by Campaigne (44) for discussion of these data].

Beta-endorphins have been implicated in the regulation of glucose metabolism. Gender-related differences in β-endorphin levels have been described by Ritter et al. (212). Normal-weight women were found to have lower basal β-endorphin levels than men, this difference being abolished by obesity. Thus only women increased endorphin levels in response to obesity. Moreover, men decreased β-endorphin levels during an oral glucose tolerance test but women did not (212). These studies did not control for fat distribution. Leiter (160) has described a Y chromosome–expressed gene in mice that may be linked with glucose intolerance. It is not known how these factors affect the response to overnutrition.

Society plays a large role in the attitudes of men and women toward food intake and adiposity [see, for example, the review by Rothblum (222)]. The fact that many modern societies stress thinness in women has led to increased concern about weight (37), increased guilt about eating (243), more dietary restraint (149), and increased incidence of eating disorders in women (37).

Age

Age tends to result in a greater responsiveness to overnutrition. That is, increasing age is associated with increased BMI in humans (133, 218) and increased adiposity in laboratory animals (189). The exact mechanism by which this occurs is unknown. However, there is considerable evidence indicating that metabolic processes, including thermogenesis and glucose uptake, are impaired in some elderly individuals. Moreover, adult lean body mass (and therefore the requirement for "maintenance" energy) is reduced with increasing age (82).

Thermogenesis in response to cold is depressed in older rats, mice, and humans (52, 58, 172, 173). In rats, this is linked, in part, to a dysfunction in brown adipose tissue (172, 234, 235). The amount of brown fat is reduced in older animals, as are β-adrenergic responsiveness, the number of β_1-adrenergic receptors, and adenylate cyclase activity in this tissue (172, 235). Moreover, oxygen consumption is less in older vs. younger rats in response to cold, norepinephrine, and the putative brown fat β_3-adrenergic agonist CPG 12177A (172, 234). In addition, the thermogenic response of brown fat, as indicated by guanosine diphosphate (GDP) binding to brown fat mitochondria, is attenuated in older vs. younger rats both when given CPG 12177A (234) and during cold exposure (173).

Along with impaired thermogenesis, glucose tolerance in rats and humans decreases with age (89, 245). This is independent of disease, medication, obesity, or decreased physical activity (174, 245) and is not due to a decreased capacity of β cells to produce insulin. Rather, the sensitivity of β cells to glucose is impaired (89). Moreover, the enlarged adipocytes of aged obese rats have reduced levels of the glucose transporter protein Glut-4 (76).

There are changes in other body tissues as well. In mouse skeletal muscle (soleus), oxygen consumption and Na^+/K^+ pump–mediated K^+ uptake are reduced as age increases (5–35 wk) (65). Adrenocorticotropic hormone (ACTH) is less effective in stimulating lipolysis in white adipocytes in older rats (119). Thyroid function is impaired with age. Older rats have lower levels of triiodothyronine and thyroxine, and they do not increase levels of thyroxine in response to cold (127). All of these alterations in older animals contribute to decreased energy expenditure and facilitate fat storage. Thus age tends to increase susceptibility to the effects of overnutrition.

Childhood obesity is becoming much more prevalent and can have unpleasant side effects both physiologically and socially. Obese children have increased blood pressure (125) and/or increased resistance in peripheral vessels, which can predispose them to hypertension (213). Socially, obese children are less popular than their lean counterparts (51).

Past the age of 12–13 yr, self-control with respect to eating decreases (185). Locus of control refers to whether a person believes that any event is contingent upon behavior *(internal control)* or whether the event is dependent on some other force, such as chance or another person *(external control)*. Paradoxically, the actual degree of control decreases but the degree to which subjects believe that body weight is controlled internally increases (185). In general, as children get older, the locus of control becomes more internal, and this has been shown to be true for dietary habits as well. Locus of control is very important for children since parents strongly influence and/or dictate their eating habits (185). Moreover, there is a tendency for obese and/or physically inactive adults to raise obese and/or physically inactive children (217).

Several studies have examined the effects of overfeeding young animals on the induction of obesity in adults, with conflicting results (136, 151, 165, 225). Knittle and Hirsch (151) and Johnson et al. (137) report that overfeeding prior to weaning induced obesity in adults. However, Routh et al. (225) and Lewis et al. (165) found that overfeeding neonates results in larger but not fatter adults. These differences may be due to the fact that the overnutrition was much greater in the studies of Knittle and Hirsch and Johnson et al. Alternatively,

different studies may have dealt with obesity-prone and obesity-resistant individuals within strains (162, 163). Landerholm and Stern (155) found a wide range of weight gain in 3 month old female Sprague-Dawley rats on a high-fat diet, which was inversely correlated to their adipose lipolytic response to epinephrine measured prior to high fat feeding.

Genetics

A number of extensive reviews have been written regarding the role of genetics in the response to overnutrition (26, 28, 29, 137, 255). The genetics of human obesity are very complex according to Bouchard because "the various phenotypes for obesities are generally not of the simple Mendelian kind . . . variation in human body fat is caused by a complex network of genetic, nutritional, energy expenditure, psychological, and social variables" (29). Bouchard proposed a classification for four types of human obesity based on fat distribution and the associated metabolic characteristics. These categories are as follows: Type 1: excess adiposity evenly distributed over the body, Type 2: excess abdominal (visceral or subcutaneous) adiposity (android obesity), Type 3: excess visceral adiposity, and Type 4: excess adiposity in the gluteal–femoral region (gynoid obesity). For a detailed analysis of the effectors, correlates, and heritability of Type 1 obesity the reader is referred to two reviews by Bouchard (28, 29).

The genetic component of obesity has been established in humans using adoption studies of mono- and dizygotic twins separated at birth (255). These studies indicate that genetics is better correlated than environment to the following variables: BMI (206, 256, 271), energy expenditure, resting metabolic rate, thermic effect of food, energy cost of low to moderate intensity exercise, level of spontaneous physical activity (28, 29), skin-fold thickness (25), and body weight (206). Studies of family groups and isolated or inbred cultures have reported similar findings (26, 85, 145, 263).

Animal models are helpful in unraveling the complex genetic factors resulting in obesity. General characteristics of these animal models were discussed earlier in this chapter, and an in-depth discussion of these characteristics, as well as the current status of investigation of genes potentially involved in obesity, has been reviewed by Johnson et al. (137).

The response to overnutrition is an interaction between genetic and environmental effects. For example, in a study of Mexican–American families, exercise suppressed the heritability estimates for obesity; that is, the correlation between family and BMI was decreased in individuals who exercised (140). Also, genetic makeup may predispose an individual to obesity only under a specific environmental condition or type of overnutrition (29).

Exercise

There are several reviews on exercise (30, 144, 150, 242). Briefly, exercise blunts the detrimental responses to overnutrition via a variety of mechanisms. The most obvious is by increasing energy expenditure. Evidence gathered in the 1960s indicated that oxygen consumption may be elevated for up to 24 h postexercise (68, 198). However, further data suggest that this may not be the case (36, 83). This discrepancy may be due to the fact that caloric intake was not controlled in the earlier studies. Although current data point away from a long-term increase in metabolic rate, oxygen consumption does appear to be increased for at least 40 min postexercise (242). This has been hypothesized to be related to the increased rates of substrate cycling seen during this time (190). Exercise also enhances metabolism by increasing catecholamine release, which in turn stimulates lipolysis and glycogenolysis (30). The catecholamine response to exercise can double in exercise-trained individuals (148).

An important effect of exercise is to decrease serum insulin levels while increasing insulin sensitivity of peripheral tissues so that less insulin is needed for the same amount of glucose transport (30, 280). Lower insulin also allows lipolysis to increase (30, 118). Exercise also potentiates diet-induced thermogenesis in lean but not in obese women (242). Although aerobic exercise does not increase the resting metabolic rate in the obese, it does lead to increased work capacity, allowing obese individuals to expend more energy while exercising (242). Moreover, resistance training in combination with dieting will help preserve or even increase lean body mass during weight loss and can lead to elevated resting metabolic rates (14, 15, 70, 161, 242). There is some evidence that, while exercise is helpful in the maintenance of reduced body weight (141), cessation of exercise may result in overnutrition (8, 111, 259).

Stress

The effects of stress on overnutrition are dependent on the individual's response to stress. Two different types of response to stress have been defined by Silvestrini (247). The orthodox response results in increased physical and mental performance under emergency conditions. The paradoxical response results in the opposite behavior, that is, a submissive response. Paradoxical responses, reported in a large percentage of the population, may aid in dealing with stressors for which we are unable to cope and are thus better dealt with through submission. Moreover, although the orthodox response to stress involves increased substrate oxidation, under some situations (such as famine) it is more useful to decrease substrate oxidation to conserve storage. Similarly, food seeking is normally an orthodox behavior which would be beneficial during famine. However, under certain conditions this may not be beneficial and may even be life-threatening. Chronically food-restricted young rats will literally run themselves to death when given access to a running wheel (260). The paradoxical stress response may result in conditions such as hyperphagia, bulimia, depression, panic, etc. (247).

Other investigators have reported that various stressors will result in hyperphagia and obesity. For example, overnutrition can be induced in hamsters by housing them five to a cage. Increased body weight could be seen in the group-housed hamsters after 2 wk. At the end of 10 wk, group-housed animals were fatter and had significantly larger adrenals (175). This could also be related to decreased physical activity or decreased heat loss due to overcrowding. In rats, an acute tail pinch can induce overeating, and chronic tail pinching will lead to increased weight gain (227).

In humans, socioeconomic status plays a role in the development of obesity. This is discussed in great depth in a review by Sobal and Stunkard (254). At very low and high levels of education and income BMI tends to be lower than at the middle ranges (133). This inverse relationship between socioeconomic status and body weight at the higher ranges of education and income is more prevalent in women than in men and children (254). These results are complicated by the fact that smoking (which is associated with decreased body weight) is also inversely related to socioeconomic status, especially in men. Moreover, obese women are often downwardly mobile with respect to socioeconomic status (254).

Restraint describes the state of dietary concern associated with eating seen in people who are anxious about their weight (112). For example, a restrained eater is one who would give up completely on a diet following a single deviation from that diet. Restraint influences the response to overnutrition, especially among women (237). Individuals with high restraint scores tend to develop eating disorders such as binging, bulimia, and anorexia nervosa (154). Moreover, many obese individuals rate high on restraint scales, indicating that they may feel deprived and overeat as a result (149). Interestingly, lean and obese humans respond differently to a glucose load following acute psychological stress. That is, stress delays the glucose response in the lean but not in the obese (285). These data indicate that certain types of stress may result in an increased responsiveness to overnutrition.

Diet Composition

The composition of the diet can have a profound effect on the response to overnutrition (180). There are several types of diet that readily promote weight gain in experimental animals, including high-fat, high-carbohydrate (high in simple sugars), high-fat/high-sugar, and cafeteria diets.

A high-fat diet will produce obesity even when isocalorically matched to a low-fat diet using Atwater factors (13) of 4 kcal/g carbohydrate and 9 kcal/g fat (71, 180, 195, 196, 225, 258). [However, when animals are depositing fat, the efficiency of converting carbohydrate to fat is considerably lower than that of converting fat to fat, leading to the view that under such conditions the caloric equivalents for fat maybe closer to 11 than 9 (69).] For example, lean adult men required a much greater caloric excess to gain the same amount of weight on a mixed diet compared to a high-fat diet (see Fig. 62.2). Moreover, rats fed a high-fat diet will become hyperphagic (207). Along with increased body fat and hyperphagia, high-fat diets result in a number of other physiological alterations. Although Flatt et al. (81) have proposed that fat oxidation is independent of dietary fat intake, Hill et al. (116) reported that a high-fat diet reduced the respiratory quotient in humans, indicating oxidation of fat. This may be related to the finding that brown adipose tissue thermogenic capacity (as indicated by increased mitochondrial GDP binding and amount of uncoupling protein) in lean rats is increased by a high-fat diet (67, 95, 224, 225); interestingly, brown fat

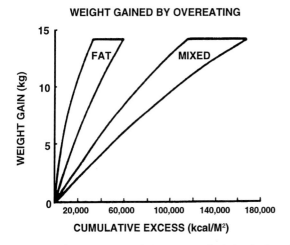

FIG. 62.2. Weight gain vs. cumulative excess caloric intake in male volunteers. Four were overfed a mixed diet for 7 months while in prison with access to caffeine-containing beverages and cigarette smoking; four were overfed fat for 3 months in a clinical research center with restricted activity, smoking, and caffeine-containing beverages. Reprinted with permission from Danforth (60).

mitochondrial GDP binding is unchanged in obese (fa/fa) Zucker rats fed a high-fat diet (95).

High-fat diets also have significant effects on glucose uptake and oxidation. Insulin resistance is increased in high-fat fed rats, while insulin-stimulated glucose transport is decreased. This is associated with decreased amount (protein) and transcription (mRNA) of the glucose transporter Glut 4 (199). Moreover, pyruvate dehydrogenase (a major determinant of the rate of glucose oxidation) is decreased in the heart of both lean and gold thioglucose obese mice in response to a high-fat diet (258).

Finally, a high-fat diet can have effects on the CNS. Neuropeptide Y (NPY) injected into the PVN is a potent stimulator of food intake (159). In the parvocellular PVN, NPY is higher in high-fat-fed vs. high-carbohydrate-fed rats (20, 21).

Diets high in carbohydrates, in the form of simple sugars, also may cause hyperphagia and obesity (2, 71, 196, 258), while the intake of complex carbohydrates has been associated with decreased body fat (180). As in the high-fat diet, cardiac pyruvate dehydrogenase activity is reduced by high carbohydrate (sucrose) (258). Carbohydrate feeding increases both serum glucose and insulin (214, 258). The response to a high-sucrose challenge can be modified through a self-selection diet. Rats that were able to choose between fat, carbohydrate, and protein decreased their intake of fat to compensate for their increased consumption of a high-sucrose formula, thus gaining less weight than rats fed a homogenous mixture of nutrients (2). Interestingly, a high-carbohydrate meal increases blood flow to brown fat to a greater extent than does a high-fat meal (92).

Rats and humans eat more when presented with a highly palatable diet (such as a cafeteria diet), and this leads to increased energy intake and obesity. Additionally, the increased energy intake due to the increased energy density of a high-energy diet also leads to obesity (71, 163, 207, 224). There are conflicting data on the response of brown fat to diet. Rothwell et al. (224) showed increased GDP binding in brown fat of rats fed cafeteria diets. In contrast, Llado et al. (166) reported decreased brown fat mass and protein. Work from our laboratory (95, 225) also demonstrated increased GDP binding in response to a high-fat diet, whereas Fisler et al. (79) reported a decrease. These discrepancies could be due to the percent of fat in the diet. Those diets that stimulate brown fat GDP binding had greater than 45% calories from fat, whereas those diets that did not stimulate brown fat had less than 45%. The greatest increase in energy intake and body weight is seen in a diet high in fats and carbohydrates (that is, a high-energy diet) (207).

Season/Food Availability

In the wild, food availability is primarily dependent on season. During winter, when availability is low, animals resort to several different strategies to survive. Some species hibernate, reducing the body's need for energy. Others store food from times of plenty for subsequent use. This has led to several different metabolic adaptations to "feast or famine."

For many animals, photoperiod and temperature are the main determinants of season, with hours of light decreasing as winter approaches. Most photoperiod-sensitive species show a circannual cycle for both food intake and body weight, both of these variables decreasing as hours of light decrease. Evaluations have been made both in the field [European hamsters (46), ground squirrels (62)] and in the laboratory [meadow voles (61), Siberian hamsters (278, 279)]. In at least one of these species (the Siberian hamster) decreased body weight occurs prior to decreased food intake. Thus it is not simply a lack of available food during an unfavorable season that underlies the photoperiodic response (279). Increased brown fat thermogenic capacity accompanies the weight loss seen in the Siberian hamster under short photoperiodic conditions (110). In contrast, Syrian (Golden) hamsters (also laboratory raised) gain weight (primarily due to increased lipid) during short photoperiods (17, 18, 119). The basis for these species-specific differences is unknown. However, it is clear that the response to overnutrition will be affected by season in photoperiodic species.

Although rats are not normally considered to be photoperiod-sensitive, Larkin et al. (157) report changes in lean (*Fa/Fa*) and obese (*fa/fa*) Zucker rats exposed to a long (14 h light:10 h dark) vs. a short (10 h light:14 h dark) photoperiod. A short photoperiod increased brown fat mitochondrial protein in both genotypes; however, GDP binding tended to be higher only in the obese. Moreover, in the short photoperiod animals, plasma insulin and internal fat stores were lower (that is, the ratios of epididymal, retroperitoneal, and mesenteric fat to total carcass fat were lower).

The increased adiposity of the Syrian hamster in response to a short photoperiod would appear to be a positive response in that it involves storing energy for a time of need. The decreased body weight of other species may decrease energy requirements during the winter (61). Thus in some cases, a short photoperiod enhances the response to overnutrition; in other cases, it attenuates it.

For most humans in affluent societies, food availability is not a problem. Most differences in the amount eaten result from self-imposed restrictions. Cycles of overnutrition and subsequent dieting are increasing in developed countries (100, 290). There is some evidence

that such weight cycling may result in progressively lower basal metabolism and higher energy efficiency, making it more difficult to lose weight (38, 210, 221). However, this view is not universally shared (64, 96, 98, 130, 274). Thus the issue of whether cycles of under- and normal nutrition affect the response to subsequent overnutrition needs further evaluation.

PHYSIOLOGICAL MECHANISMS UNDERLYING RESPONSES TO OVERNUTRITION

Cellular

White Adipose Tissue. Overnutrition perturbs the balance between lipogenesis and lipolysis by increasing the former and/or decreasing the latter, both of which contribute to increased fat deposition. Lipogenesis and lipolysis, which are regulated by neural and hormonal factors, are affected by overnutrition in adipose, as well as in other, tissues. Major features of this regulation in humans are summarized in Figure 62.3. Briefly, triglyceride circulates in the blood in the form of lipoproteins, which are transported into white adipocytes via a process facilitated by LPL. Lipoprotein lipase is located on the plasma membrane of capillary endothelial cells, where it hydrolyzes triglyceride to free fatty acids and glycerol. Free fatty acids are taken up into the cell, re-esterified to form triglyceride, and stored within the large triglyceride droplet in the cell. The situation differs somewhat in rats and mice, where significant de novo fatty acid synthesis occurs in the adipocyte itself (77).

Control of lipolysis is more complex because its rate can be modulated by a variety of signals, including norepinephrine, epinephrine, insulin, and adenosine. These ligands either activate or inhibit the cAMP-mediated cascade, which results in the phosphorylation of hormone-sensitive lipase (HSL), an enzyme that hydrolyzes triglyceride back to free fatty acids and glycerol.

Overnutrition can affect the balance between lipogenesis and lipolysis via several mechanisms. The hyperinsulinemia that accompanies overnutrition enhances lipogenesis by increasing LPL activity (251) and LPL synthesis (137). It also decreases glucose entry into cells by down-regulating insulin receptors (promoting fat storage) (214), as well as by decreasing levels of the Glut-4 glucose transporter protein (76). In addition to enhanced lipogenesis, hyperinsulinemia inhibits lipolysis by inhibiting HSL (86, 118). Moreover, activities of fatty acid synthetase (FAS) and acetyl CoA-carboxylase, as well as levels of FAS mRNA, are increased in the liver of obese (*fa/fa*) Zucker rats (102, 167). Thus the overall effect of insulin is increased fat storage. This is discussed in more detail later in this chapter in the section Insulin.

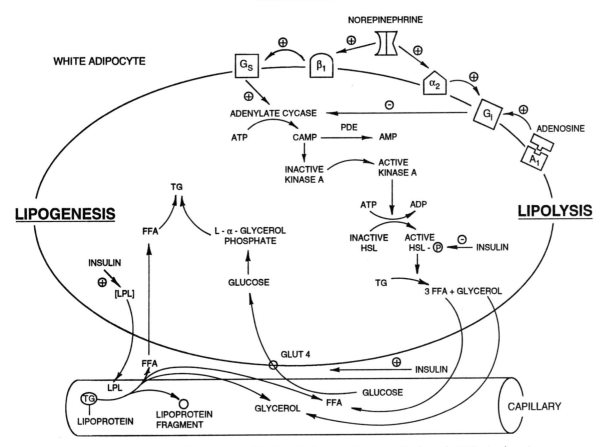

FIG. 62.3. Basic steps in lipogenesis and lipolysis in human adipocytes. Triglyceride *(TG)* circulates in blood in the form of lipoproteins, where it is broken down into free fatty acids *(FFA)* by lipoprotein lipase *(LPL)*, which is synthesized in the white adipocyte and located on the plasma membrane of the capillary endothelial cell. The *FFA* diffuse into the white adipocyte where they are re-esterified with glycerol to form *TG*. Insulin enhances the storage of fat as *TG* by increasing synthesis of *LPL* and facilitating the transport of glucose into the white adipocyte by stimulating the *Glut 4* glucose transporter protein. Lipolysis occurs via a cAMP-mediated cascade, which results in the phosphorylation of hormone-sensitive lipase *(HSL)*, an enzyme which hydrolyzes *TG* into *FFA* and glycerol. These *FFA* are then free to diffuse into the blood. Lipolysis is stimulated by the β_1-noradrenergic receptor [via a G-stimulatory (G_s) protein] and inhibited by the α_2-noradrenergic and A_1-adenosine receptors [via a G-inhibitory (G_i) protein]. Insulin inhibits lipolysis by inhibiting *HSL*. *PDE* is phosphodiesterase. [Expanded from Hirsch et al. (118)].

In addition to its effects of hyperinsulinemia, overnutrition also alters adenosine receptors in adipose tissue. Lipolysis is under both stimulatory (via G_s protein) and inhibitory (via G_i protein) control. Adenosine, acting through its receptor, activates a G_i protein (277). Vannucci et al. (277) demonstrated that regulation of the A_1-adenosine receptor was altered in the obese *(fa/fa)* Zucker rat. In the lean Zucker rat, the effect of guanosine triphosphate (GTP) on adenylate cyclase is biphasic; at low concentrations, adenylate cyclase is stimulated and at high concentrations it is inhibited. In the obese *(fa/fa)* rat the stimulatory effect of low GTP is blunted. Moreover, the A_1-adenosine receptor in the obese *(fa/fa)* rat shows decreased sensitivity to inhibition by the nonhydrolyzable GTP analog GppNHp. This results in greater inhibition of lipolysis in these ani-

mals. There is evidence that decreased cAMP production is responsible for the decreased lipolytic response of older rats (119). Moreover, whole-body fatty acid synthesis is elevated in the young Zucker obese *(fa/fa)* rat (103).

Finally, overnutrition affects adipocyte differentiation. Cultures of preadipocytes from massively obese humans have a greater number of rapidly dividing clones, greater mitogenic capacity, and a greater number of clones susceptible to differentiation (219). Clonal lines derived from mouse fibroblast 3T3 cells that accumulate and mobilize lipid are being used to determine the factors regulating fat cell differentiation (120). Insulin and insulin-like growth factor 1 (IGF-1), both of which are altered by overnutrition, appear to be essential for differentiation (23, 24, 273).

Brown Adipose Tissue. Brown adipose tissue is considered to be a primary mediator of nonshivering thermogenesis and has been implicated in diet-induced thermogenesis (169, 224, 267). Both acute and chronic environmental and nutritional changes can alter the thermogenic capacity of brown fat, and, in general, overnutrition increases the thermogenic capacity of brown fat in lean animals. That is, many investigators have reported that high-energy, high-fat, high-sucrose, or cafeteria diets increase brown fat thermogenic capacity in normal animals (67, 95, 139, 224, 225). However, this response is blunted in both genetic (95, 122, 123, 267) and hypothalamic (35, 123) obesity. Moreover, during the development of gold thioglucose–induced obesity, lipogenesis in brown fat is increased prior to increases in liver and white adipose tissue (54).

This decreased thermogenic capacity reflects changes in the tissue itself, as well as in the sympathetic regulation of thermogenesis. Brown fat in obese animals has less uncoupling protein and more lipid storage (267), and sympathetic stimulation of brown fat is depressed. Holt and York (123) showed that the firing rate of sympathetic nerves to brown fat was reduced in young (6 wk old) obese (fa/fa) Zucker rats. Thus in lean animals, brown fat thermogenesis increases in response to overnutrition, blunting some of its adverse effects. However, this protective mechanism may be dysfunctional in obese animals.

Other Tissues. Brown fat is not the only thermogenically active tissue. The liver has been proposed as a major site of diet-induced thermogenesis. In a study by Ma and Foster (169) partial hepatectomy significantly reduced the increased oxygen consumption of cafeteria-fed rats in comparison to chow-fed controls. Asayama et al. (9) reported that liver and muscle also appear to be involved in an "overflow" oxidation of fatty acids. Peroxisomal β-oxidation is increased in liver and skeletal muscle in the gold thioglucose obese mouse, and this is highly correlated with body weight (9). Thus the liver also appears to protect against adverse responses to overnutrition.

In humans the liver is the primary site of fatty acid synthesis, and hepatocytes from obese humans synthesize more fatty acids than those from lean humans (7). Moreover, insulin stimulates hepatic lipogenesis by 33% in obese humans compared to 8% in lean humans (16). In VMN-lesioned rats the activity of acetyl-CoA carboxylase (the rate-limiting enzyme in fatty acid synthesis) was 1.8 times higher than in controls, as was VLDL synthesis (128). These data indicate that altered regulation of hepatic fatty acid synthesis may result in increased responsiveness to overnutrition.

Membrane phospholipids have been implicated in signal transduction, and there is evidence that they are also altered in obesity. The Zucker obese (fa/fa) rat has decreased 5'-desaturase activity, resulting in altered membrane phospholipid composition in many tissues, including liver, kidney, heart, and brain (101). Moreover, supplementation with black currant oil (high in gamma-linolenic acid) normalizes hepatic phospholipid arachidonic acid and decreases food intake and body weight in the obese (fa/fa) Zucker rat (200). The role of membrane phospholipids in the response to overnutrition is unknown.

Endocrine Mechanisms

This section highlights the endocrine response to overnutrition and the resulting obesity syndrome [see reviews by Glass (91), Kopelman (153), Mohs et al. (182), Rodin (214), Xu and Bjorntorp (288), and Woods et al. (286)].

Insulin. The hyperinsulinemia that results from overnutrition mediates many of the physiological adaptations to overnutrition. Insulin, produced by pancreatic β cells, facilitates glucose transport into cells and the regulation of carbohydrate metabolism. It is also involved in amino acid metabolism as well as fatty acid mobilization and storage. As mentioned previously (see White Adipose Tissue), insulin inhibits HSL and promotes net re-esterification of newly synthesized fatty acids (118). Insulin also stimulates synthesis of LPL, thereby increasing LPL levels and further promoting energy storage in adipose tissue (137). A third mechanism whereby hyperinsulinemia may promote fat storage involves insulin resistance, which develops, in part, due to down-regulation of insulin receptors on target cells (for example, muscle). This can become so severe that glucose intolerance develops. Glucose entry and thus storage of glycogen is reduced, and energy is stored as fat (214).

In addition to the down-regulation of insulin receptor number due to hyperinsulinemia, there are other alterations in insulin dynamics that accompany overnutrition. The Glut 4 glucose transporter, the mediator of insulin-stimulated glucose uptake in rat adipocytes (76), is reduced by a high-fat diet (199). Hypertriglyceridemia is more strongly correlated with insulin resistance than obesity (194). In the Zucker obese rat, pancreatic β cells show decreased sensitivity to glucose (89). Rats susceptible to dietary obesity have greater plasma insulin values on a high-calorie, high-fat diet than do rats that are diet-resistant (163). It has been proposed that insulin enters the brain directly as a signal for the regulation of body weight (286). Insulin receptors are located in all areas of the brain, with the highest concentrations in olfactory bulb, cerebral cortex, and hypothalamus (108, 109).

Growth Hormone. Overnutrition is accompanied by changes in growth hormone levels and thus growth rates of lean body mass (91). Some investigators have reported that growth hormone reduces body fat in the absence of caloric restriction in obese women (250), whereas during caloric restriction, exogenous growth hormone conserved lean body mass without accelerating fat loss (50, 252). These anabolic effects are believed to be mediated by the growth hormone–dependent peptide IGF-1 (253), which has been reported to be reduced in obese men (56). Additionally, growth hormone secretion is reduced in obese patients (91), and this reduction becomes more pronounced with age (262). Moreover, pituitary mRNA levels and secretion of growth hormone is depressed in the obese (fa/fa) Zucker rat (3, 78). This decreased growth hormone secretion may have a permissive effect on the development of adiposity.

Prolactin. Prolactin, an anterior pituitary hormone whose secretion stimulates lactation, may stimulate food intake in the nonlactating female (90). Prolactin secretion is under tonic hypothalamic inhibition by dopamine (55). Although basal prolactin levels are normal in obese subjects, secretion in response to insulin hypoglycemia is impaired in some subjects (153). During lactation, mammary LPL activity is high and adipose LPL activity is low, whereas hypophysectomy reverses this effect (292). Prolactin injections restore mammary LPL and block adipose LPL (292). Prolactin assists the body in changing to a state favoring energy mobilization during lactation. Although prolactin may stimulate food intake, fat storage in adipose may be blocked, thus attenuating the detrimental response to overnutrition. The role of prolactin in the response to overnutrition in the male or nonlactating female is unclear.

Gonadal Steroids. Overnutrition and subsequent obesity are often associated with altered sex steroid secretion (153) and increased extraovarian synthesis of estrogen from androgens in adipose tissue (39, 126). This is particularly evident in postmenopausal obese women, whose elevated estrogen levels derive from the metabolism of androgens (91) and are proportional to excess body weight. Obese men tend to have decreased testosterone and increased estrogen (91).

Androgens are linked to upper body obesity in premenopausal women (75). Increased free testosterone and decreased sex hormone binding globulin are accompanied by increased waist to hip ratios, increased abdominal fat depots, elevated plasma insulin and glucose, and lower insulin sensitivity (75). Diet may also play a role in sex steroid metabolism. In obese women, dietary lipids were negatively correlated with sex hormone binding globulin and dietary carbohydrates were positively correlated with estrone (197). Progesterone

appears to be able to prevent corticosteroid inhibition of adipocyte differentiation by competitive binding in adipocyte precursor cells (288).

Thyroid Hormone. There is mixed evidence for altered thyroid hormone levels in response to overnutrition (91, 153), though hypothyroidism is associated with decreased metabolic rate and increased adiposity (35). Dubuc (72) has reported increased plasma thyroid-stimulating hormone (TSH), triiodothyronine (T_3); and thyroxine (T_4) in the hyperglycemic obese (ob/ob) mouse, while Wu et al. (287) showed that serum T_3 was reduced in the cold-exposed Zucker obese (fa/fa) rat. Diet alters thyroid hormone levels. Obese women on very low-calorie diets show decreased resting metabolic rates and corresponding decreases in T_3 and increases in reverse T_3 (47). As the energy density of the diet increases, the disposal rate of plasma T_3 increases. However, Cavallo (47) found no indication of changes in metabolic pathways. In contrast, 5′-monoiodination accounts for 21% of the plasma disposal of reverse T_3 on a low-calorie diet (22% fat), whereas on a high-calorie diet (36% fat) it accounts for 45%, indicating a diet-induced change in metabolic pathway (40). The relationship between overnutrition and pathways for thyroid hormone clearance is unclear. However, decreased thyroid hormone levels in response to overnutrition would be consistent with the decreased metabolic rate seen in many obese subjects.

Adrenal Corticosteroids. Corticotropin releasing factor (CRF) is manufactured primarily in the parvocellular PVN (232). It is released into the median eminence and enters the hypophysial portal circulation to stimulate the pituitary to release ACTH. This in turn stimulates the release of glucocorticoids from the adrenal gland. Corticotropin releasing factor appears to be released predominantly in response to stress. An excellent review by Rothwell (223) summarizes the central effects of CRF, which have been shown to modify the autonomic nervous system. Also, CRF appears to ameliorate many of the abnormalities associated with obesity, as does adrenalectomy (which raises brain CRF). Thus the adrenal corticosteroid system is believed to play a large role in regulating the response to overnutrition (35).

Humans with Cushing's disease are both obese and hypercortisolemic (33). The obese (fa/fa) Zucker rat has been shown to have elevated plasma levels of corticosterone (34). Since ADX tends to normalize many of the metabolic abnormalities of genetically obese animal models, it has been hypothesized that there is some dysfunction of the hypothalamic pituitary adrenal axis in these animals. Plotsky et al. (203) showed no differences in median eminence CRF or hypothalamic mRNA for CRF. Moreover, there was no genotypic difference in either basal CRF or CRF release in response to restraint stress or hypotension. Secretion into the hypophysial portal circulation was diminished in the obese (fa/fa) rat, and this was increased by ADX. Thus these investigators propose that the obese (fa/fa) rat exhibits a dysfunction in CRF tone proximal to the hypothalamic CRF neurons (203). The sympathoadrenal system is believed to be a dominant efferent control regulating nutrient balance (31, 32, 156).

Neural Mechanisms Mediating the Response to Overnutrition

Since the work of Anand and Brobeck in the 1950s (5), hypothalamic dysfunction has been associated with increased food intake and obesity. In this early work, lesions of the VMN produced obesity and hyperphagia, while lesions of the lateral hypothalamus resulted in aphagia (5). At that time the VMN was believed to be a satiety center and the lateral hypothalamus a feeding center. This view was altered in the 1970s by Han and colleagues, who found that discrete VMN lesions in pair-fed adult (104, 105) and weanling (106, 107) rats produced obesity and hyperinsulinemia independent of hyperphagia. Hyperphagia resulted from lesioning the ventral noradrenergic bundle, which runs lateral to the VMN (10). This fiber tract sends projections to the PVN, a nucleus associated with the regulation of feeding (10). Leibowitz (159) found a variety of substances (monoamines, neuropeptides, hormones, etc.) that influence food intake when injected into the PVN.

The brain is no longer considered to be a collection of centers but of interacting pathways (257). There are several reviews detailing the anatomy of hypothalamic regulation of food intake and metabolism (146, 204, 257). Hypothalamic nuclei exert these effects primarily through the autonomic nervous system. Steffens and colleagues (257) have reviewed the interconnections of hypothalamic nuclei, their relationship to brain stem autonomic nuclei, and their effects on the regulation of food intake. The current view is that the VMN is stimulatory to the sympathetic nervous system and inhibitory to the parasympathetic, while the lateral hypothalamus has the opposite effect (35, 257). The PVN receives input from higher limbic centers as well as circumventricular organs (which lack a blood–brain barrier), keeping it informed of the status of the animal. It also modulates the autonomic nervous system both directly and indirectly (through CRF) (257). Lesions of the VMN result in a hyperphagia-independent obesity, while the obesity resulting from PVN lesions is dependent on hyperphagia. However, in VMN, but not in PVN lesions, meal patterns are disrupted (87, 266).

Electrical stimulation of the VMN increases interscapular brown fat temperature in both lean and obese Zucker rats (121) but not in Sprague-Dawley rats (84). Ventromedial nucleus lesions abolish the cold-induced increased firing rate of sympathetic nerves to brown fat in Sprague-Dawley rats (192). Stimulation of the VMN increases norepinephrine turnover in brown fat in these animals (229). Injection of insulin into the VMN decreases brown fat thermogenesis (4). Lateral hypothalamic lesions decrease insulin levels and food intake, while increasing brown fat GDP binding in obese rats (123). Thus an imbalance in autonomic regulation between the VMN and the lateral hypothalamus could result in many of the physiological responses associated with overnutrition. It is difficult to determine if overnutrition affects the CNS or if dysfunction in these nuclei causes increased responsiveness to overnutrition.

Brain monoamine systems play a role in the control of food intake and energy balance. We have reported that dopaminergic activity is depressed in the PVN, serotonergic activity is depressed in the VMN, and noradrenergic activity is depressed in the preoptic area of obese vs. lean Zucker rats (226). Norepinephrine injected into the PVN stimulates feeding, whereas serotonin inhibits it (159). Moreover, α_2-adrenergic receptors appear to mediate the norepinephrine-induced feeding in the PVN (94), and circulating glucose levels are directly related to α_2-adrenergic receptor density (134). Chronic infusion of norepinephrine into the VMN will produce obesity (244), and VMN noradrenergic receptors have been shown to change in response to diet (164).

In contrast to norepinephrine, serotonin is anorectic when injected into the PVN (159). Flouxetine, a serotonin re-uptake inhibitor, decreases food intake and body weight by decreasing carbohydrate and increasing protein intake (201), and feeding increases lateral hypothalamus serotonin levels (238). Reducing brain levels of serotonin resulted in decreased brown fat thermogenesis (88), and reduced hypothalamic serotonin turnover has been observed in obese animals (35, 159, 226).

Dopamine also plays a role in VMN-lesioned obesity. That is, when dopamine is depleted, VMN-lesioned rats do not become obese (57). Moreover, the increased responsiveness to overnutrition that is associated with increased age may be related to dopaminergic activity. In humans, a decreased ability to cope with heat or cold has been linked to central dopamine dysfunction in the elderly (58). Finally, hypothalamic monoamines have been shown to stimulate CRF release, as well as the converse (42, 43, 59, 73, 134, 135, 171, 244).

Neuropeptides have also been implicated in the control of food intake and obesity. Beck et al. (19) have shown neurotensin to be lower in the anterior pituitary, lateral preoptic area, PVN, VMN, suprachiasmatic nucleus (SCN), and supraoptic nucleus of obese Zucker rats, while NPY, a potent stimulator of food intake, is elevated in the arcuate, PVN, and SCN of these animals. Leu- and met-enkephalin and β-endorphin are elevated in the pituitary of the *ob/ob* mouse (264, 265). Gene expression of NPY is positively regulated by glucocorticoids (115). Peptides that stimulate food intake include galanin, β-endorphin, NPY, dynorphin, and growth hormone releasing hormone. Peptides that inhibit food intake include anorectin, neurotensin, cholecystokinin, glucagon, CRF, and calcitonin (35).

Thus responses to overnutrition include alterations in the neurochemistry of the CNS. Although it is difficult to determine whether overnutrition causes or results from these alterations, one view is that overnutrition/obesity results from a dysfunction of the sympathoadrenal system (11, 12, 32, 156, 223, 270). Bray (31, 32) has described nutrient intake, storage, and metabolism as a controlled system with a controller in the brain, a set of afferent feedback elements, and efferent control elements.

The afferent limb of this system involves nutrient, hormonal, and neural signals providing the central controller with information about the nutritional status of the organism (31, 32). There are a number of reviews dealing with this limb, which will be summarized below (45, 80, 86, 142, 214, 249, 275, 276, 286).

The widely different storage capabilities for carbohydrates and lipids support the concept that intake of these macronutrients is independently regulated. That is, storage capacity for carbohydrates is very low, and daily intake has a major influence on total body carbohydrate. However, storage capacity for lipids is very large, and daily lipid intake has little effect on storage, though small daily increases in intake can yield a large cumulative increase in body fat over time (35, 80).

There is evidence that both glucose and body fat stores regulate food intake (276). Glucose-sensitive neurons exist in the hypothalamus (6), as do insulin receptors (109). Campfield and Smith (45) propose that the transient declines in blood glucose that precede meals actually signal meal initiation. These declines in blood glucose are themselves preceded by a transient insulin spike. Giving intravenous glucose prior to a meal delays meal onset, and experimentally decreasing glucose initiates a meal. Moreover, when food is unavailable during a glucose decline and is then restored, the meal is not initiated until the next decline. This glucose response appears to be at least partly neurally mediated since vagotomy abolished the response 55% of the time. This is consistent with the data indicating that the cephalic phase of a meal is associated with an insulin peak and that in both humans and animals the higher the cephalic phase insulin response, the greater is the tendency to gain weight. Rodin et al. (215) propose that hyperinsulinemia may be associated with a heightened perceived pleasure response to food.

Food intake may also be indirectly regulated by the storage and mobilization of fat and its effect on substrate oxidation (86). That is, increased fuel oxidation (increased ATP) inhibits food intake, and decreased oxidation stimulates it. According to this model, body fat influences food intake via its balance between storage as adipose and its oxidation. This is consistent with the ischymetric hypothesis discussed by Nicolaidis and Even (191), in which feeding is believed to be regulated by metabolic need apart from physical activity ("Metabolism de fond"). Parabiotic studies indicating that endogenous gut signals, plasma levels of nutrients, or signals generated from the metabolism of nutrients influence food intake support the idea of metabolic or nutrient control of food intake (152).

The hypothalamus is the most likely site of the central integrator. As mentioned earlier in this section, the VMN is believed to be stimulatory to the sympathetic and inhibitory to the parasympathetic nervous system, while the lateral hypothalamus has just the opposite effect (35, 257). The PVN receives input from many brain regions (including circumventricular organs which lack a blood–brain barrier) and is directly involved with the regulation of food intake (257).

The primary efferent pathways for the hypothalamus are the autonomic nervous system and the pituitary. Both branches of the autonomic nervous system have been implicated in the regulation of nutrient balance (35, 257). Adrenalectomy normalizes food intake, feeding patterns, and sympathetic activity of obese animals. Thus the adrenal steroids are believed to have a regulatory role as well (31, 32, 156, 223). Moreover, CRF appears to act as a modulator in the brain (223). Since the actions of CRF persist in hypothalamic obesity, the site of action of the adrenal system probably differs from that of the genetic lesion (31, 32, 223). Bray (32) proposes that obesity is accompanied by decreased sympathetic nervous system activity that is dependent on at least normal levels of circulating glucocorticoids. We and others (11, 12, 156, 223, 270) argue that the sympathoadrenal system is a dominant efferent control regulating nutrient balance.

SUMMARY

The physiological responses to overnutrition include increased adiposity, hyperinsulinemia, decreased energy expenditure, and numerous other health risks. The severity of these risks is, in part, dependent on the distribution of fat. Individuals with upper body or android distributions are at greater risk. However, there are numerous other variables that affect the response to overnutrition. Increasing age, stress, and diets high in fat or sugar increase responsiveness to overnutrition, whereas exercise attenuates the response. Fat distribution associated with premenopausal women tends to show fewer maladaptations. Genetic makeup greatly influences individual responses, both to overnutrition and to the variables that affect it.

Regulation of nutrient intake, storage, and metabolism has been described as a controlled system consisting of afferent information concerning nutritional status, a controller probably located in the hypothalamus, and a set of efferent controls, one of which is the sympathoadrenal system. It has been postulated that the detrimental response to overnutrition is due to a dysfunction in the sympathoadrenal system. It is unclear whether the disturbance in this system causes or results from the response to overnutrition. However, it is certain that, unlike responses to other stressors (for example, cold, altitude), the physiological changes occurring in response to overnutrition are ultimately maladaptive in the postreproductive animal.

REFERENCES

1. Acheson, K. J., V. Schultz, T. Bessard, E. Ravussin, E. Jequier, and J. P. Flatt. Nutritional influences on lipogenesis and thermogenesis after a carbohydrate meal. *Am. J. Physiol.* 246 (*Endocrinol. Metab. 9*): E62–E70, 1984.

2. Ackroff, K., and A. Sclafani. Sucrose-induced hyperphagia and obesity in rats fed a macronutrient self-selection diet. *Physiol. Behav.* 44: 181–187, 1988.

3. Ahmad, I., A. W. Steggles, A. J. Carrillo, and J. A. Finklestein. Obesity and sex-related alterations in growth hormone mRNA levels. *Mol. Cell. Endocrinol.* 65: 103–109, 1989.

4. Amir, S., M. Lagiorgia, and R. Pollock. Intra-ventromedial hypothalamic injection of insulin suppresses brown fat thermogenesis in the anaesthetized rat. *Brain Res.* 480: 340–343, 1989.

5. Anand, B. K., and J. R. Brobeck. Hypothalamic control of food intake in rats and cats. *Yale J. Biol. Med.* 24: 123–140, 1951–1952.

6. Anand, B. K., G. S. China, K. N. Sharma, S. Dua, and B. Singh. Activity of single neurons in the hypothalamic feeding centers: effect of glucose. *Am. J. Physiol.* 207: 1146–1154, 1964.

7. Anonymous. Lipogenesis in diabetes and obesity. *Nutr. Rev.* 49: 255–256, 1991.

8. Applegate, E. A., D. E. Upton, and J. S. Stern. Exercise and detraining: effect on food intake, adiposity and lipogenesis in Osborne-Mendel rats made obese by a high fat diet. *J. Nutr.* 114: 447–459, 1984.

9. Asayama, K., Y. Okada, and K. Kato. Peroxisomal beta-oxidation in liver and muscles of gold-thioglucose-induced obese mice: correlation with bodyweight. *Int. J. Obes.* 15: 45–49, 1991.

10. Ashlog, J., P. Randall, and B. Hoebel. Hypothalamic hyperphagia: dissociation from hyperphagia following destruction of noradrenergic neurons. *Science* 190: 399–401, 1975.

11. Astrup, A., T. Andersen, O. Henriksen, N. J. Christensen, J. Bulow, J. Madsen, and F. Quaade. Impaired glucose-induced thermogenesis in skeletal muscle in obesity. The role of the sympathoadrenal system. *Int. J. Obes.* 11: 51–56, 1987.

12. Astrup, A., N. J. Christensen, L. Simonsen, and J. Bulow. Effects of nutrient intake on sympathoadrenal activity and thermogenic mechanisms. *J. Neurosci. Methods* 34: 187–192, 1990.

13. Atwater, W. O., and A. P. Bryand. Results of experiments on the metabolism of matter and energy in the human body. In: *Connecticut (Storrs) Agricultural Experiment Station 12th Annual Report* Storrs, CT: Office of Experimental Stations, 1900, p. 73–123.

14. Ballor, D. Exercise training elevates RMR during moderate but not severe dietary restriction in obese male rats. *J. Appl. Physiol.: Respir. Environ. Exerc. Physiol.* 70: 2303–2310, 1991.

15. Ballor, D. L., V. L. Katch, M. D. Becque, and C. R. Marks. Resistance weight training during caloric restriction enhances lean body weight maintenance. *Am. J. Clin. Nutr.* 47: 19–25, 1988.

16. Barakat, H. A., V. D. McLendon, J. W. Carpenter, R. H. L. Marks, N. Legett, K. O'Brien, and J. F. Caro. Lipogenic potential of liver from morbidly obese patients with and without non-insulin-dependent diabetes. *Metabolism* 40: 280–285, 1991.

17. Bartness, T. J. Species-specific changes in the metabolic control of food intake: integrating the animal with its environment. *Int. J. Obes.* 14(suppl. 3): 115–124, 1990.

18. Bartness, T. J., and G. N. Wade. Photoperiodic control of body weight and energy metabolism in Syrian hamsters (*Mesocricetus auratus*): role of pineal gland, melatonin, gonads, and diet. *Endocrinology* 114: 492–498, 1984.

19. Beck, B., A. Burlet, J.-P. Nicolas, and C. Burlet. Neurotensin in microdissected brain nuclei and in the pituitary of the lean and obese Zucker rats. *Neuropeptides* 13: 1–7, 1989.

20. Beck, B., A. Burlet, J. P. Nicolas, and C. Burlet. Hyperphagia in obesity is associated with a central peptidergic dysregulation in rats. *J. Nutr.* 120: 806–811, 1990.

21. Beck, B., A. Stricker-Krongrad, A. Burlet, J.-P. Nicolas, and C. Burlet. Influence of diet composition on food intake and hypothalamic neuropeptide Y (NPY) in the rat. *Neuropeptides* 17: 197–203, 1990.

22. Bell, G. E., and J. S. Stern. Evaluation of body composition of young obese and lean Zucker rats. *Growth* 41: 63–80, 1977.

23. Bjorntrop, P., M. Karlsson, H. Pertoft, P. Pettersson, L. Sjostrom, and U. Smith. Isolation and characterization of cells from rat adipose tissue developing into adipocytes. *J. Lipid Res.* 19: 316–324, 1978.

24. Bjorntrop, P., M. Karlsson, P. Pettersson, and G. Sypniewska. Differentiation and function of rat adipocyte precursor cells in primary culture. *J. Lipid Res.* 21: 714–723, 1980.

25. Bodurtha, J. N., M. Mosteller, J. K. Hewitt, W. E. Nance, L. J. Eaves, W. B. Moskowitz, S. Katz, and R. M. Schieken. Genetic analysis of anthropometric measures in 11-year-old twins: the Medical College of Virginia twin study. *Pediatr. Res.* 28: 1–4, 1990.

26. Bogardus, C., S. Lillioja, and E. Ravussin. The pathogenesis of obesity in man: result of studies on Pima Indians. *Int. J. Obes.* 14(suppl. 1): 5–15, 1990.

27. Bouchard, C. Genetic factors in obesity. *Med. Clin. North Am.* 73: 67–81, 1989.

28. Bouchard, C. Current understanding of the etiology of obesity: genetic and nongenetic factors. *Am. J. Clin. Nutr.* 53: 1562S–1565S, 1991.

29. Bouchard, C. Heredity and the path to overweight and obesity. *Med. Sci. Sports. Exerc.* 23: 285–291, 1991.

30. Bouchard, C., R. J. Shephard, T. Stephens, J. R. Sutton, and B. D. McPherson (Eds). *Exercise, Fitness and Health: A Concensus of Current Knowledge.* Champaign, IL: Human Kinetics, 1990, p. 217–259, 293–306, 315–320, 497–511.

31. Bray, G. Nutrient balance and obesity: an approach to control of food intake in humans. *Med. Clin. North Am.* 73: 29–44, 1989.

32. Bray, G. Obesity—a state of reduced sympathetic activity and normal or high adrenal activity (the autonomic and adrenal hypothesis revisited). *Int. J. Obes.* 14(suppl. 3): 77–92, 1990.

33. Bray, G. A. Classification and evaluation of the obesities. *Med. Clin. North Am.* 73: 161–184, 1989.

34. Bray, G. A., and L. A. Campfield. Metabolic factors in the control of energy stores. *Metab. Clin. Exp.* 24: 99–117, 1975.

35. Bray, G. A., D. A. York, and J. S. Fisler. Experimental obesity: a homeostatic failure due to defective nutrient stimulation of the sympathetic nervous system. *Vitam. Horm.* 45: 1–125, 1989.

36. Brehm, B. A., and B. Gutin. Recovery energy expenditure for steady state exercise in runners and nonexercisers. *Med. Sci. Sports Exerc.* 18: 205–210, 1986.

37. Brown, P. J., and M. Konner. An anthropological perspective on obesity. *Ann. N. Y. Acad. Sci.* 499: 29–46, 1987.

38. Brownell, K. D., M. R. C. Greenwood, E. Stellar, and E. E. Shrager. The effects of repeated cycles of weight loss and regain in rats. *Physiol. Behav.* 38: 459–464, 1986.

39. Bruning, P. F., J. M. G. Bonfrer, A. Ansink, N. S. Russell, and M. de Jong-Bakker. Why is breast cancer so frequent in The Netherlands? *Eur. J. Surg. Oncol.* 14: 115–122, 1988.

40. Burger, A. G., M. O'Connell, K. Scheidegger, R. Woo, and E. Danforth, Jr. Monodeiodination of triiodothyronine and reverse triiodothyronine during low and high calorie diets. *J. Clin. Endocrinol. Metab.* 65: 829–835, 1987.

41. Callaway, C. W. Obesity. In: *Public Health Reports*, July–Aug (suppl.), Washington, D.C.: Public Health Service, 1987, p. 26–29.

42. Calogero, A. E., R. Bernardini, A. M. Margioris, G. Bagdy, W. T. Galluccini, P. J. Munson, L. Tamarkin, T. P. Tomai, L. Brady, P. W. Gold, and G. P. Chrousos. Effect of serotonergic agonists and antagonists on corticotropin-releasing hormone secretion by explanted rat hypothalami. *Peptides* 10: 189–200, 1989.

43. Calogero, A. E., W. T. Gallucci, G. P. Chrousos, and P. W. Gold. Catecholamine effects upon rat hypothalamic corticotropin–releasing hormone secretion in vitro. *J. Clin. Invest.* 82: 839–846, 1988.

44. Campaigne, B. N. Body fat distribution in females: metabolic consequences and implications for weight loss. *Med. Sci. Sports Exerc.* 22: 291–297, 1990.

45. Campfield, L. A., and F. J. Smith. Transient declines in blood glucose signal meal initiation. *Int. J. Obes.* 14(suppl. 3): 15–33, 1990.

46. Canguilhem, B., J. P. Vaultier, B. Pevet, G. Coumaros, M. Masson-Pevet, and I. Bantz. Photoperiodic regulation of body mass, food intake, hibernation, and reproduction in intact and castrated male European hamsters, *Cricetus cricetus. J. Comp. Physiol. [A]* 163: 549–557, 1988.

47. Cavallo, E., F. Armellini, M. Zamboni, R. Vicnetini, M. P. Milani, and O. Bosello. Resting metabolic rate, body composition and thyroid hormones. Short term effects of very low calorie diet. *Horm. Metab. Res.* 22: 632–635, 1990.

48. Chang, S., B. Graham, F. Yakubu, D. Lin, J. C. Peters, and J. O. Hill. Metabolic differences between obesity-prone and obesity-resistant rats. *Am. J. Physiol.* 259(*Regulatory Integrative Comp. Physiol.* 28): R1103–R1110, 1990.

49. Choban, P. S., L. J. Weireter, Jr., and C. Maynes. Obesity and increased mortality in blunt trauma. *J. Trauma* 31: 1253–1257, 1991.

50. Clemmons, D. R., D. K. Snyder, R. Williams, and L. E. Underwood. Growth hormone administration conserves lean body mass during dietary restriction in obese subjects. *J. Clin. Endocrinol. Metab.* 64: 878–883, 1987.

51. Cohen, R., R. C. Klesges, M. Summerville, and A. W. Meyers. A developmental analysis of the influence of body weight on the sociometry of children. *Addict. Behav.* 14: 473–476, 1989.

52. Collins, K. J. Low indoor temperatures and morbidity in the elderly. *Age Ageing* 15: 212–220, 1987.

53. Contreras, R. J., S. King, L. Rives, A. Williams, and T. Wattleton. Dietary obesity and weight cycling in rats: a model of stress-induced hypertension? *Am. J. Physiol.* 261(*Regulatory Integrative Comp. Physiol.* 30): R848–R857, 1991.

54. Cooney, G. J., M. A. Vanner, J. L. Nicks, P. F. Williams, and I. D. Caterson. Changes in the lipogenic response to feeding of liver, white adipose tissue and brown adipose tissue during the development of obesity in the gold-thioglucose-injected mouse. *Biochem. J.* 259: 652–657, 1989.

55. Cooper, J. R., F. E. Bloom, and R. H. Roth. *The Biochemical Basis of Neuropharmocology* (6th ed.). New York: Oxford University Press, 1991, p. 320.

56. Copeland, K. C., R. B. Colletti, J. T. Devlin, and T. L. McAuliffe. The relationship between insulin-like growth factor-I, adiposity, and aging. *Metabolism* 39: 584–587, 1990.

57. Coscina, D. V., and J. N. Nobrega. 6-Hydroxydopamine-induced blockade of hypothalamic obesity: critical role of brain dopamine–norepinephrine interaction. *Prog. Neuropsychopharmacol. Biol. Psychiatry* 6: 369–372, 1982.

58. Cox, B., T. Lee, and J. Parkes. Decreased ability to cope with heat and cold linked to a dysfunction in a central dopaminergic pathway in elderly rats. *Life Sci.* 28: 2039–2044, 1981.

59. Cummings, S., and V. Seybold. Relationship of alpha-1- and alpha-2-adrenergic binding sites to regions of the paraventricular nucleus of the hypothalamus containing corticotropin-releasing factor and vasopressin neurons. *Neuroendocrinology* 47: 523–532, 1988.

60. Danforth, E., Jr. Diet and obesity. *Am. J. Clin. Nutr.* 41: 1132–1145, 1985.

61. Dark, J., I. Zucker, and G. N. Wade. Photoperiodic regulation of body mass, food intake, and reproduction in meadow voles. *Am. J. Physiol.* 245(*Regulatory Integrative Comp. Physiol.* 16): R334–R338, 1983.

62. Davis, D. E. Hibernation and circannual rhythms of food consumption in marmots and ground squirrels. *Q. Rev. Biol.* 51: 477–514, 1976.

63. De Boer, J. O., A. J. H. Van Es, J. M. A. Van Raaij, and J. G. A. J. Hautvast. Energy requirements and energy expenditure of lean and overweight women, measured by indirect calorimetry. *Am. J. Clin. Nutr.* 46: 13–21, 1987.

64. De Groot, L. C. P. G. M., A. J. H. Van Es, J. M. A. Van Raaij, J. E. Vogt, and G. A. J. Hautvast. Adaptation of energy metabolism of overweight women to alternating and continuous low energy intake. *Am. J. Clin. Nutr.* 50: 1314–1323, 1989.

65. De Luise, M., and M. Harker. Skeletal muscle metabolism: effect of age, obesity, thyroid and nutritional status. *Horm. Metabol. Res.* 21: 410–415, 1989.

66. Den Besten, C., G. Vansant, J. A. Weststrate, and P. Deurenberg. Resting metabolic rate and diet-induced thermogenesis in abdominal and gluteal–femoral obese women before and after weight reduction. *Am. J. Clin. Nutr.* 47: 840–847, 1988.

67. Desautels, M., and R. A. Dulos. Weight gain and brown fat composition of mice selected for high body weight fed a high-fat diet. *Am. J. Physiol.* 258(*Regulatory Integrative Comp. Physiol.* 29): R608–R615, 1990.

68. DeVries, H. A., and D. E. Gray. After effects of exercise upon resting metabolic rate. *Res. Q.* 34: 315–321, 1962.

69. Donato, K., and D. M. Hegsted. Efficiency of utilization of various sources of energy for growth. *Proc. Natl. Acad. Sci. U.S.A.* 82: 4866–4870, 1985.

70. Donnelly, J. E., N. P. Pronk, D. J. Jacobsen, S. J. Pronk, and J. M. Jakicic. Effects of a very-low-calorie diet and physical-training regimens on body composition and resting metabolic rate in obese females. *Am. J. Clin. Nutr.* 54: 56–61, 1991.

71. Dreon, D. M., B. Frey-Hewitt, N. Elsworth, P. T. Williams, R. B. Terry, and P. D. Wood. Dietary fat: carbohydrate ratio and obesity in middle-aged men. *Am. J. Clin. Nutr.* 47: 995–1000, 1988.

72. Dubuc, P. U. Effects of phenotype, feeding condition and cold exposure on thyrotropin and thyroid hormones of obese and lean mice. *Endocr. Regul.* 25: 171–175, 1991.

73. Dunn, A., and C. Berridge. Corticotropin-releasing factor administration elicits a stress-like activation of cerebral catecholaminergic systems. *Pharmacol. Biochem. Behav.* 27: 685–691, 1987.

74. Ernsberger, P., and D. O. Nelson. Refeeding hypertension in dietary obesity. *Am. J. Physiol.* 254(*Regulatory Integrative Comp. Physiol.* 25): R47–R55, 1988.

75. Evans, D. J., R. G. Hoffmann, R. K. Kalkhjof, and A. H. Kissebah. Relationship of androgenic activity to body fat topography, fat cell morphology, and metabolic aberrations in premenopausal women. *J. Clin. Endocrinol. Metab.* 57: 304–310, 1983.

76. Ezaki, O., N. Fukuda, and H. Itakura. Role of two types of glucose transporters in enlarged adipocytes from aged obese rats. *Diabetes* 39: 1543–1549, 1990.

77. Favarger, P., and J. Gerlach. Studies on the synthesis of fats from acetate or glucose. II. The relative roles of adipose tissue and other tissues in lipogenesis in mice. *Helv. Physiol. Pharmacol. Acta* 13: 96–105, 1955.

78. Finklestein, J. A., P. Jervois, M. Menadue, and J. O. Willough. Growth hormone and prolactin secretion in genetically obese Zucker rats. *Endocrinology* 118: 1233–1236, 1986.

79. Fisler, J. S., J. R. Lupien, R. D. Wood, G. A. Bray, and R. A. Schemmel. Brown fat thermogenesis in a rat model of dietary obesity. *Am. J. Physiol.* 253(*Regulatory Integrative Comp. Physiol.* 24): R756–R762, 1987.

80. Flatt, J. P. The difference in the storage capacities for carbohydrate and for fat, and its implications in the regulation of body weight. *Ann. N. Y. Acad. Sci.* 499: 104–123, 1987.

81. Flatt, J. P., E. Ravussin, K. J. Acheson, and E. Jequier. Effects of dietary fat on postprandial substrate oxidation and on carbohydrate and fat balances. *J. Clin. Invest.* 76: 1019–1024, 1985.

82. Forbes, G. B., and J. C. Reina. Adult lean body mass declines with age: some longitudinal observations. *Metabolism* 19: 653–663, 1970.

83. Freedman-Akabas, S., E. Colt, H. R. Kissileff, and F. X. Pi-Sunyer. Lack of sustained increase in V_{O_2} following exercise in fit and unfit subjects. *Am. J. Clin. Nutr.* 41: 545–549, 1985.

84. Freeman, P. H., and P. J. Wellman. Brown adipose tissue thermogenesis induced by low level electrical stimulation of hypothalamus in rats. *Brain Res. Bull.* 18: 7–11, 1987.

85. Friedlander, Y., J. D. Kark, N. A. Kaufmann, E. M. Berry, and Y. Stein. Familial aggregation of body mass index in ethnically diverse families in Jerusalem. The Jerusalem Lipid Research Clinic. *Int. J. Obes.* 12: 237–247, 1988.

86. Friedman, M. I. Body fat and the metabolic control of food intake. *Int. J. Obes.* 14(suppl. 3): 53–67, 1990.

87. Fukushima, M., K. Tokunaga, J. Lupien, J. W. Kemnitz, and G. A. Bray. Dynamic and static phases of obesity following lesions in PVN and VMH. *Am. J. Physiol.* 253(*Regulatory Integrative Comp. Physiol.* 24): R523–R529, 1987.

88. Fuller, N. J., D. M. Stirling, S. Dunnett, G. P. Reynolds, and M. Ashwell. Decreased brown adipose tissue thermogenic activity following a reduction in brain serotonin by intraventricular p-chlorophenylalanine. *Biosci. Rep.* 7: 121–127, 1987.

89. Furnsinn, C., M. Komjati, O. D. Madsen, B. Schneider, and W. Waldhausl. Lifelong sequential changes in glucose tolerance and insulin secretion in genetically obese Zucker rats (fa/fa) fed a diabetogenic diet. *Endocrinology* 128: 1093–1099, 1991.

90. Gerardo-Gettens, T., B. J. Moore, J. S. Stern, and B. A. Horwitz. Prolactin stimulates food intake in the absence of ovarian progesterone. *Am. J. Physiol.* 256(*Regulatory Integrative Comp. Physiol.* 27): R701–R706, 1989.

91. Glass, A. R. Endocrine aspects of obesity. *Med. Clin. North Am.* 73: 139–160, 1989.

92. Glick, Z., S. J. Wickler, J. S. Stern, and B. A. Horwitz. Blood flow into brown fat of rats is greater after high carbohydrate than after a high fat test meal. *J. Nutr.* 114: 1934–1939, 1984.

93. Glickman, L. T., F. S. Schofer, and L. J. McKee. Epidemiologic study of insecticide exposures, obesity, and risk of bladder cancer in household dogs. *J. Toxicol. Environ. Health* 28: 407–414, 1989.

94. Goldman, C. K., L. Marino, and S. F. Leibowitz. Postsynaptic α2-noradrenergic receptors mediate feeding induced by paraventricular nucleus injection of norepinephrine and clondine. *Eur. J. Pharmacol.* 115: 11–19, 1985.

95. Gong, L. T. W., J. S. Stern, and B. A. Horwitz. High fat feeding increases brown fat GDP binding in lean but not obese Zucker rats. *J. Nutr.* 120: 786–792, 1990.

96. Graham, B., S. Chang, D. Lin, F. Yakubu, and J. O. Hill. Effect of weight cycling on susceptibility to dietary obesity. *Am. J. Physiol.* 259(*Regulatory Integrative Comp. Physiol.* 30): R1096–R1102, 1990.

97. Gray, D. S. Diagnosis and prevalence of obesity. *Med. Clin. North Am.* 73: 1–13, 1989.

98. Gray, D. S., J. S. Fisler, and G. A. Bray. Effects of repeated weight loss and regain on body composition in obese rats. *Am. J. Nutr.* 47: 393–399, 1988.

99. Greenberg, D., and S. C. Weatherford. Obese and lean Zucker rats differ in preferences for sham-fed corn oil or sucrose. *Am. J. Physiol.* 259(*Regulatory Integrative Comp. Physiol.* 30): R1093–R1095, 1990.

100. Grundy, S. M., and J. P. Barnett. Metabolic and health complications of obesity. *Dis. Mon.* 36: 643–696, 1990.

101. Guesnet, Ph., J.-M. Bourre, M. Guerre-Millo, G. Pascal, and G. Durand. Tissue phospholipid fatty acid composition in genetically lean (Fa/−) or obese (fa/fa) Zucker female rats on the same diet. *Lipids* 25: 517–522, 1990.

102. Guichard, C., I. Dugail, X. Le Liepvre, and M. Lavau. Genetic regulation of fatty acid synthetase expression in adipose tissue: overtranscription of the gene in genetically obese rats. *J. Lipid Res.* 33: 679–687, 1992.

103. Haggarty, P., K. W. J. Wahle, Peter J. Reeds, and J. M. Fletcher. Whole body fatty acid synthesis and fatty acid intake in young rats of the Zucker strain (*Fa/−* and *fa/fa*). *Int. J. Obes.* 11: 41–50, 1986.

104. Han, P., and L. Frohman. Hyperinsulinemia in tube-fed hypophysectomized rats bearing hypothalamic lesions. *Am. J. Physiol.* 219: 1632–2636, 1970.

105. Han, P., and A. Lui. Obesity and impaired growth of rats force fed 40 days after hypothalamic lesions. *Am. J. Physiol.* 211: 229–231, 1966.

106. Han, P. W. Hypothalamic obesity in rats without hyperphagia. *Trans. N. Y. Acad. Sci.* 30: 229–242, 1967.

107. Han, P. W., C.-H. Lin, K.-C. Chu, J.-Y. Mu, and A.-C. Liu. Hypothalamic obesity in weanling rats. *Am. J. Physiol.* 209: 627–631, 1965.

108. Havrankova, J., M. Brownstein, and J. Roth. Insulin and insulin receptors in rodent brain. *Diabetologia* 20: 268–273, 1981.

109. Havrankova, J., and J. Roth. Insulin receptors are widely distributed in the central nervous system of the rat. *Nature* 272: 827–829, 1978.

110. Heldmaier, G., S. Steinlechner, J. Rafael, and P. Vsiansky. Photoperiodic control and effects of melatonin on nonshivering thermogenesis and brown adipose tissue. *Science* 212: 917–919, 1981.

111. Hering, J. L., P. A. Mole, C. N. Meredith, and J. S. Stern. Effect of suspending exercise training on resting metabolic rate in women. *Med. Sci. Sports Exerc.* 24: 59–65, 1992.

112. Herman, C. P., and J. Polivy. Anxiety, restraint, and eating behavior. *J. Abnorm. Psychol.* 84: 666–672, 1975.

113. Hetherington, A. W. Non-production of hypothalamic obesity in the rat by lesions rostral or dorsal to the ventro-medial hypothalamic nuclei. *J. Comp. Neurol.* 80: 33–45, 1945.

114. Hetherington, A. W., and S. W. Ranson. Hypothalamic lesions and adiposity in the rat. *Anat. Rec.* 78: 149–172, 1940.

115. Higuchi, H., H.-Y. T. Yang, and S. L. Sabol. Rat neuropeptide Y precursor gene expression. *J. Biol. Chem.* 263: 6288–6293, 1987.

116. Hill, J. O., J. C. Peters, G. W. Reed, D. G. Schlundt, T. Sharp, and H. L. Greene. Nutrient balance in humans: effects of diet composition. *Am. J. Clin. Nutr.* 54: 10–17, 1991.

117. Himms-Hagen, J. Defective brown adipose tissue thermogenesis in obese mice. *Int. J. Obes.* 9(suppl. 2): 17–24, 1985.

118. Hirsch, J., S. K. Fried, N. K. Edens, and R. L. Leibel. The fat cell. *Med. Clin. North Am.* 73: 83–96, 1989.

119. Hoffman, R. A., K. Davidson, and K. Steinberg. Influence of photoperiod and temperature on weight gain, food consumption, fat pads and thyroxine in male golden hamsters. *Growth* 46: 150–162, 1982.

120. Hollenberg, C. H. Perspectives in adipose tissue physiology. *Int. J. Obes.* 14(suppl. 3): 135–152, 1990.

121. Holt, S. J., H. V. Wheal, and D. A. York. Hypothalamic control of brown adipose tissue in Zucker lean and obese rats. Effect of electrical stimulation of the ventromedial nucleus and other hypothalamic centres. *Brain Res.* 405: 227–233, 1987.

122. Holt, S. J., H. V. Wheal, and D. A. York. Response of brown adipose tissue to electrical stimulation of hypothalamic centres in intact and adrenalectomized Zucker rats. *Neurosci. Lett.* 84: 63–67, 1988.

123. Holt, S. J., and D. A. York. Effect of lateral hypothalamic lesion on brown adipose tissue of Zucker lean and obese rats. *Physiol. Behav.* 43: 293–299, 1988.

124. Holt, S. J., D. A. York, and J. T. R. Fitzsimons. The effects of corticosterone, cold exposure and overfeeding with sucrose on brown adipose tissue of obese Zucker rats (*fa/fa*). *Biochem. J.* 214: 215–223, 1983.

125. Horswill, C. A., and W. B. Zipf. Elevated blood pressure in obese children: influence of gender, age, weight and serum insulin levels. *Int. J. Obes.* 15: 453–459, 1991.

126. Horton, R., and J. F. Tait. Androstenedione production and interconversion rates measured in peripheral blood and studies on the possible site of its conversion to testosterone. *J. Clin. Invest.* 45: 301–313, 1966.

127. Huang, H. H., R. W. Steger, and J. Meites. Capacity of old versus young male rats to release thyrotropin (TSH), thyroxine (T$_4$) and triiodothyronine (T$_3$) in response to different stimuli. *Exp. Aging Res.* 6: 3–12, 1980.

128. Inui, Y., S. Kawata, Y. Matsuzawa, K. Tokunaga, S. Fujioka, S. Tamura, T. Kobatake, Y. Keno, and S. Tarui. Increased level of apolipoprotein B mRNA in the liver of ventromedial hypothalamus lesioned obese rats. *Biochem. Biophys. Res. Commun.* 163: 1107–1112, 1989.

129. Istfan, N. W., C. S. Plaisted, B. R. Bistrian, and G. L. Blackburn. Insulin resistance versus insulin secretion in the hypertension of obesity. *Hypertension* 19: 385–392, 1992.

130. Jebb, S. A., G. R. Goldberg, W. A. Coward, P. R. Murgatroyd, and A. M. Prentice. Effects of weight cycling caused by intermittent dieting on metabolic rate and body composition in obese women. *Int. J. Obes.* 15: 367–374, 1991.

131. Jeevanandam, M., D. H. Young, and W. R. Schiller. Obesity and the metabolic response to severe multiple trauma in man. *J. Clin. Invest.* 87: 262–269, 1991.

132. Jeffery, R. W. Population perspectives on the prevention and treatment of obesity in minority populations. *Am. J. Clin. Nutr.* 53: 1621S–1624S, 1991.

133. Jeffery, R. W., J. L. Forster, A. R. Folsom, R. V. Luepker, D. R. Jacobs, Jr., and H. Blackburn. The relationship between social status and body mass index in the Minnesota Heart Health Program. *Int. J. Obes.* 13: 59–67, 1989.

134. Jhanwar-Uniyal, M., M. H. Papamichael, and S. F. Leibowitz. Glucose-dependent changes in α_2-noradrenergic receptors in hypothalamic nuclei. *Physiol. Behav.* 44: 611–617, 1988.

135. Joanny, O., J. Steinberg, A. Zamora, C. Conte-Devolx, Y. Millet, and C. Oliver. Corticotropin-releasing factor release from in vitro superfused and incubated rat hypothalamus. Effect of potassium, norepinephrine and dopamine. *Peptides* 10: 903–911, 1989.

136. Johnson, P., J. Stern, M. Greenwood, L. Zucker, and J. Hirsch. Effect of early nutrition on adipose cellularity and pancreatic insulin release in the Zucker rat. *J. Nutr.* 103: 738–743, 1973.

137. Johnson, P. R., M. R. C. Greenwood, B. A. Horwitz, and J. S. Stern. Animal models of obesity: genetic aspects. *Annu. Rev. Nutr.* 11: 325–353, 1991.

138. Jones, A., D. H. Davies, J. R. Dove, M. A. Collinson, and P. M. R. Brown. Identification and treatment of risk factors for coronary heart disease in general practice: a possible screening model. *BMJ* 296: 1711–1714, 1988.

139. Kanarek, R. B., J. R. Aprille, E. Hirsch, L. Gualtiere, and C. A. Brown. Sucrose-induced obesity: effect of diet on obesity and brown adipose tissue. *Am. J. Physiol.* 253(*Regulatory Integrative Comp. Physiol.* 24): R158–R166, 1987.

140. Kaplan, R. M., T. L. Patterson, J. F. Sallis, Jr., and P. R. Nader. Exercise suppresses heritability estimates for obesity in Mexican–American families. *Addict. Behav.* 14: 581–588, 1989.

141. Kayman, S., W. Bruvold, and J. S. Stern. Maintenance and relapse after weight loss in women: behavioral aspects. *Am. J. Clin. Nutr.* 52: 800–807, 1990.

142. Keesey, R. E. Physiological regulation of body weight and the issue of obesity. *Med. Clin. North Am.* 73: 15–26, 1989.

143. Kemnitz, J. W., R. W. Goy, T. J. Flitsch, J. J. Lohmiller, and J. A. Robinson. Obesity in male and female Rhesus monkeys: fat distribution, glucoregulation, and serum androgen. *J. Clin. Endocrinol. Metab.* 69: 287–293, 1989.

144. King, A. C., and D. L. Tribble. The role of exercise in weight regulation in nonathletes. *Sports Med.* 11: 331–349, 1991.

145. King, H., D. C. Rao, K. Bhatia, G. Koki, A. Collins, and P. Zimmet. Family resemblance for glucose tolerance in a Melanesian population, the Tolai. *Hum. Hered.* 39: 212–217, 1989.

146. Kirchgessner, A. L., and A. Sclafani. Histochemical identification of a PVN-hindbrain feeding pathway. *Physiol. Behav.* 42: 529–543, 1988.

147. Kissebah, A. H., D. S. Freedman, and A. N. Peiris. Health risks of obesity. *Med. Clin. North Am.* 73: 111–138, 1989.

148. Kjaer, M., N. J. Christensen, B. Sonne, E. A. Richter, and H. Galbo. Effect of exercise on epinephrine turnover in trained and untrained male subjects. *J. Appl. Physiol.: Respir. Environ. Exerc. Physiol.* 59: 1061–1067, 1985.

149. Klem, M. L., R. C. Klesges, C. R. Bene, and M. W. Mellon. A psychometric study of restraint: the impact of race, gender, weight and marital status. *Addict. Behav.* 15: 147–153, 1990.

150. Klesges, R. C., and L. H. Eck. Effects of obesity, social interactions, and physical environment on physical activity in preschoolers. *Health Psychol.* 9: 435–449, 1990.

151. Knittle, J., and J. Hirsch. Effect of early nutrition on the development of rat epididymal fat pads: cellularity and metabolism. *J. Clin. Invest.* 47: 2091–2098, 1968.

152. Koopmans, H. S. Endogenous gut signals and metabolites control daily food intake. *Int. J. Obes.* 14(suppl. 3): 93–104, 1990.

153. Kopelman, P. G. Neuroendocrine function in obesity. *Clin. Endocrinol.* 28: 675–689, 1988.

154. Kristeller, J. L., and J. Rodin. Identifying eating patterns in male and female undergraduates using cluster analysis. *Addict. Behav.* 14: 631–642, 1989.

155. Landerholm, T. E., and J. S. Stern. Adipose tissue lipolysis in vitro: a predictor of diet-induced obesity in female rats. *Am. J. Physiol.* 263 (*Regulatory, Integrative Comp. Physiol.* 32): R1248–R1253, 1992.

156. Landsberg, L. The sympathoadrenal system, obesity and hypertension. *J. Neurosci. Methods* 34: 179–186, 1990.

157. Larkin, L. M., B. J. Moore, J. S. Stern, and B. A. Horwitz. Effect of photoperiod on body weight and food intake of obese and lean Zucker rats. *Life Sci.* 49: 735–745, 1991.

158. Lean, M. E. J., P. R. Murgatroyd, I. Rothnie, I. W. Reid, and R. Harvey. Metabolic and thyroidal responses to mild cold are abnormal in obese diabetic women. *Clin. Endocrinol.* 28: 665–673, 1988.

159. Leibowitz, S. F. Brain monoamines and peptides: role in the control of eating behavior. *Federation Proc.* 45: 1396–1403, 1986.

160. Leiter, E. H. Control of spontaneous glucose intolerance, hyperinsulinemia, and islet hyperplasia in nonobese C3H.SW male mice by Y-linked locus and adrenal gland. *Metabolism* 37: 689–696, 1988.

161. Lemons, A. D., S. N. Kreitzman, A. Coxon, and A. Howard. Selection of appropriate exercise regimes for weight reduction during VLCD and maintenance. *Int. J. Obes.* 13(suppl. 2): 119–123, 1989.

162. Levin, B. E., and A. C. Sullivan. Glucose-induced norepinephrine and obesity resistance. *Am. J. Physiol.* 253 (*Regulatory Integrative Comp. Physiol.* 23): R475–R481, 1987.

163. Levin, B. E., J. Triscari, S. Hogan, and A. C. Sullivan. Resistance to diet-induced obesity: food intake, pancreatic sympathetic tone, and insulin. *Am. J. Physiol.* 252(*Regulatory Integrative Comp. Physiol.* 23): R471–R478, 1987.

164. Levin, B. E., J. Triscari, and A. C. Sullivan. The effect of diet and chronic obesity on brain catecholamine turnover in the rat. *Pharmacol. Biochem. Behav.* 24: 299–304, 1986.

165. Lewis, D. S., H. A. Bertrand, C. A. McMahan, H. C. McGill, Jr., K. D. Carey, and E. J. Masoro. Influence of preweaning food intake on body composition of young adult baboons. *Am. J. Physiol.* 257(*Regulatory Integrative Comp. Physiol.* 28): R1128–R1135, 1989.

166. Llado, I., A. M. Proenza, F. Serra, A. Palou, and A. Pons. Dietary-induced permanent changes in brown and white adipose tissue composition in rats. *Int. J. Obes.* 15: 415–419, 1990.

167. Lopez-Casillas, F., M. V. Ponce-Castaneda, and K.-H. Kim. In vivo regulation of the activity of the two promoters of the rat acetyl coenzyme-A carboxylase gene. *Endocrinology* 129: 1049–1058, 1991.

168. Lukaski, H. C. Methods for the assessment of human body composition: traditional and new. *Am. J. Clin. Nutr.* 46: 537–556, 1987.

169. Ma, S. W. Y., and D. O. Foster. Brown adipose tissue, liver, and diet-induced thermogenesis in cafeteria diet-fed rats. *Can. J. Physiol. Pharmacol.* 67: 376–381, 1988.

170. Marshall, J. D., D. W. Hazlett, D. W. Spady, and H. A. Quinney. Validity of convenient indicators of obesity. *Hum. Biol.* 63: 137–153, 1991.

171. Martire, M., G. Pistritto, and P. Preziosi. Different regulation of serotonin receptors following adrenal hormone imbalance in the rat hippocampus and hypothalamus. *J. Neural Transm.* 78: 109–120, 1989.

172. McDonald, R. B., B. A. Horwitz, J. S. Hamilton, and J. S. Stern. Cold- and norepinephrine-induced thermogenesis in younger and older Fischer 344 rats. *Am. J. Physiol.* 254(*Regulatory Integrative Comp. Physiol.* 25): R457–R462, 1988.

173. McDonald, R. B., J. S. Stern, and B. A. Horwitz. Thermogenic responses of younger and older rats to cold exposure: comparison of two strains. *J. Gerontol.* 44: B37–B42, 1989.

174. McPhillips, J. B., K. M. Pellettera, E. Barrett-Connor, D. L. Wingard, and M. H. Criqui. Exercise patterns in a population of older adults. *Am. J. Prev. Med.* 2: 65–72, 1989.

175. Meisel, R. L., T. C. Hays, S. N. Del Paine, and V. R. Luttrell. Induction of obesity by group housing in female Syrian hamsters. *Physiol. Behav.* 47: 815–817, 1990.

176. Mela, D. J., R. S. Cohen, and P. M. Kris-Etherton. Lipoprotein metabolism in a rat model of diet-induced adiposity. *J. Nutr.* 117: 1655–1662, 1987.

177. Mercer, S. W., and P. Trayhurn. Effect of high fat diets on energy balance and thermogenesis in brown adipose tissue of lean and genetically obese (*ob/ob*) mice. *J. Nutr.* 117: 2147–2153, 1987.

178. Miller, W. C. Diet composition, energy intake, and nutritional status in relation to obesity in men and women. *Med. Sci. Sports Exerc.* 23: 280–284, 1991.

179. Miller, W. C. Introduction: obesity: diet composition, energy expenditure, and treatment of the obese patient. *Med. Sci. Sports Exerc.* 23: 273–274, 1991.

180. Miller, W. C., A. K. Lindeman, J. Wallace, and M. Niederpruem. Diet composition, energy intake, and exercise in relation to body fat in men and women. *Am. J. Clin. Nutr.* 52: 426–430, 1990.

181. Modan, M., S. Almog, Z. Fuchs, A. Chetrit, A. Lusky, and H. Halkin. Obesity, glucose intolerance, hyperinsulinemia, and response to antihypertensive drugs. *Hypertension* 17: 565–573, 1991.

182. Mohs, M. E., T. K. Leonard, and R. R. Watson. Interrelationships among alcohol abuse, obesity, and type II diabetes mellitus: focus on Native Americans. *World Rev. Nutr. Diet.* 56: 93–172, 1988.

183. Moore, B. J., J. S. Stern, and B. A. Horwitz. Brown fat mediates increased energy expenditure of cold-exposed overfed neonatal rats. *Am. J. Physiol.* 251(*Regulatory Integrative Comp. Physiol.* 22): R518–R524, 1986.

184. Morton, J. L., M. Davenport, A. Beloff-Chain, N. L. Bodkin, and B. C. Hansen. Correlation between β-cell tropin concentrations and body weight in obese rhesus monkeys. *Am. J. Physiol.* 262(*Endocrinol. Metab.* 25): E963–E967, 1992.

185. Moss, N. D., and M. R. Dadds. Body weight attributions and eating self-efficacy in adolescence. *Addict. Behav.* 16: 71–78, 1991.

186. Naeye, R. L., and P. Roode. The sizes and numbers of cells in visceral organs in human obesity. *Am. J. Clin. Pathol.* 251–253, 1970.

187. National Center for Health Statistics. *Plan and Operation of the National Health and Nutrition Examination Survey. 1976–1980.* Hyattsville, MD: U.S. Dept. of Health and Human Services, 1981.

188. Neel, J. V. Diabetes mellitus: a "thrifty" genotype rendered detrimental by "progress." *Am. J. Hum. Genet.* 14: 353–362, 1962.

189. Newby, F. D., M. DiGirolamo, G. A. Cotsonis, and M. H. Kunter. Model of spontaneous obesity in aging male Wistar rats. *Am. J. Physiol.* 259(*Regulatory Integrative Comp. Physiol.* 30): R1117–R1125, 1990.

190. Newsholme, E. A., and B. Crabtree. Substrate cycles in meta-

bolic regulation and heat generation. *Biochem. Soc. Symp.* 41: 61–110, 1976.

191. Nicolaidis, S., and P. C. Even. The ischymetric control of feeding. *Int. J. Obes.* 14 (suppl. 3): 35–52, 1990.

192. Niijima, A., F. Rohner-Jeanrenaud, and B. Jeanrenaud. Role of ventromedial hypothalamus on sympathetic efferents of brown adipose tissue. *Am. J. Physiol.* 247(*Regulatory Integrative Comp. Physiol.* 18): R650–R654, 1984.

193. Obst, B. E., R. A. Schemmel, D. Czajka-Narins, and R. Merkel. Adipocyte size and number in dietary obesity resistant and susceptible rats. *Am. J. Physiol.* 240(*Endocrinol. Metab.* 3): E47–E53, 1981.

194. Olefsky, J. M., J. W. Farquhar, and G. M. Reaven. Reappraisal of the role of insulin in hypertriglyceridemia. *Am. J. Med.* 57: 551–560, 1974.

195. Oscai, L. B., M. M. Brown, and W. C. Miller. Effect of dietary fat on food intake, growth and body composition in rats. *Growth* 48: 415–424, 1984.

196. Oscai, L. B., W. C. Miller, and D. A. Arnall. Effects on dietary sugar and of dietary fat on food intake and body fat content in rats. *Growth* 51: 64–73, 1987.

197. Pasquali, R., D. Antenucci, N. Melchionda, R. Fabbri, S. Venturoli, D. Patrono, and M. Capelli. Sex hormones in obese premenopausal women and their relationships to body fat mass and distribution, β cell function and diet composition. *J. Endocrinol. Invest.* 10: 345–350, 1987.

198. Passmore, R., and R. E. Johnson. Some metabolic changes following prolonged moderate exercise. *Metabol.* 9: 452–455, 1960.

199. Pedersen, O., C. R. Kahn, J. S. Flier, and B. B. Kahn. High fat feeding causes insulin resistance and a marked decrease in the expression of glucose transporters (Glut 4) in fat cells of rats. *Endocrinology* 129: 771–777, 1991.

200. Phinney, S. D., A. B. Tang, D. C. Thurmond, M. T. Makamura, and J. S. Stern. Abnormal polyunsaturated lipid metabolism in the obese Zucker rat, with partial metabolic correction by γ-linolenic acid administration. *Metabolism* 42: 1127–1140, 1993.

201. Pijl, H., H. P. F. Koppeshcaar, F. L. A. Willekens, I. O. de Kamp, H. D. Veldhuis, and A. E. Meinders. Effect of serotonin reuptake inhibition by fluoxetine on body weight and spontaneous food choice in obesity. *Int. J. Obes.* 15: 237–242, 1991.

202. Plaisted, C. S., L. Landsburg, J. Young, G. L. Blackburn, B. R. Bistrian, and N. W. Istfan. Insulin resistance and urinary catecholamine excretion in obese hypertensive patients undergoing weight loss [Abstract]. *Int. J. Obes.* 14(suppl. 2): 74, 1990.

203. Plotsky, P. M., K. V. Thrivikraman, A. G. Watts, and R. L. Hauger. Hypothalamic–pituitary–adrenal axis function in the Zucker obese rat. *Endocrinology* 130: 1931–1941, 1992.

204. Powley, T. L., and W. Laughton. Neural pathways involved in the hypothalamic integration of autonomic responses. *Diabetologia* 20: 378–387, 1981.

205. Price, R. A., T. I. A. Sorensen, and A. J. Stunkard. Component distributions of body mass index defining moderate and extreme overweight in Danish women and men. *Am. J. Epidemiol.* 130: 193–201, 1989.

206. Price, R. A., and A. J. Stunkard. Comingling analysis of obesity in twins. *Hum. Hered.* 39: 121–135, 1989.

207. Ramirez, I., and M. I. Friedman. Dietary hyperphagia in rats: role of fat, carbohydrate, and energy content. *Physiol. Behav.* 47: 1157–1163, 1990.

208. Ravussin, E., and C. Bogardus. Energy expenditure in the obese: is there a thrifty gene? *Infusionstherapie* 17: 108–112, 1990.

209. Rebuffe-Scrive, M., L. Enk, N. Crona, P. Lonnroth, L. Abrahamsson, U. Smith, and P. Bjorntorp. Fat cell metabolism in different regions in women. Effect of menstrual cycle, pregnancy, and lactation. *J. Clin. Invest.* 75: 1973–1976, 1985.

210. Reed, D. R., R. J. Contreras, C. Maggio, M. R. C. Greenwood, and J. Rodin. Weight cycling in female rats increases dietary fat selection and adiposity. *Physiol. Behav.* 42: 389–395, 1988.

211. Ritenbaugh, C., and C. S. Goodby. Beyond the thrifty gene: metabolic implications of prehistoric migration into the New World. *Med. Anthropol.* 11: 227–236, 1989.

212. Ritter, M. M., A. C. Sonnichsen, W. Mohrle, W. O. Richter, and P. Schwandt. β-Endorphin plasma levels and their dependence on gender during an enteral glucose load in lean subjects as well as in obese patients before and after weight reduction. *Int. J. Obes.* 15: 421–427, 1991.

213. Rocchini, A. P., V. Katch, J. Anderson, J. Hinderliter, D. Becque, M. Martin, and C. Marks. Blood pressure in obese adolescents: effect of weight loss. *Pediatrics* 82: 16–23, 1988.

214. Rodin, J. Insulin levels, hunger, and food intake: an example of feedback loops in body weight regulation. *Health Psychol.* 4: 1–24, 1985.

215. Rodin, J., D. Schank, and R. Striegel-Moore. Psychological features of obesity. *Med. Clin. North Am.* 73: 47–66, 1989.

216. Rolls, B. J., M. Hetherington, and V. J. Burley. The specificity of satiety: the influence of foods of different macronutrient content on the development of satiety. *Physiol. Behav.* 43: 145–153, 1988.

217. Romanella, N. E., D. K. Wakat, B. H. Loyd, and L. E. Kelly. Physical activity and attitudes in lean and obese children and their mothers. *Int. J. Obes.* 15: 407–414, 1991.

218. Romieu, I., W. C. Willett, M. J. Stampfer, G. A. Colditz, L. Sampson, B. Rosner, C. H. Hennekens, and G. E. Speizer. Energy intake and other determinants of relative weight. *Am. J. Clin. Nutr.* 47: 406–412, 1988.

219. Roncari, D. A. K. Abnormalities of adipose cells in massive obesity. *Int. J. Obes.* 14(suppl. 3): 187–192, 1990.

220. Rose-Ellis, J. R., R. B. McDonald, and J. S. Stern. A diet high in fat stimulates adipocyte proliferation in older (22 month) rats. *Exp. Gerontol.* 25: 141–148, 1990.

221. Rossner, S., G. Walldus, and H. Bjorvell. Fatty acid composition in serum lipids and adipose tissue in severe obesity before and after 6 weeks of weight loss. *Int. J. Obes.* 13: 603–612, 1989.

222. Rothblum, E. D. Women and weight: fad and fiction. *J. Psychol.* 124: 5–24, 1990.

223. Rothwell, N. J. Central effects of CRF on metabolism and energy balance. *Neurosci. Biobehav. Rev.* 14: 263–271, 1990.

224. Rothwell, N. J., M. J. Stock, and B. P. Warwick. Energy balance and brown fat activity in rats fed cafeteria diets or high fat semisynthetic diets at several levels of intake. *Metab. Clin. Exp.* 34: 474–480, 1985.

225. Routh, V. H., J. S. Hamilton, J. S. Stern, and B. A. Horwitz. Litter size, adrenalectomy and high fat diet alter hypothalamic monoamines in genetically lean (*Fa/Fa*) Zucker rats. *J. Nutr.* 123: 74–84, 1993.

226. Routh, V. H., D. M. Murakami, J. S. Stern, C. A. Fuller, and B. A. Horwitz. Neuronal activity in hypothalamic nuclei of obese and lean Zucker rats. *Int. J. Obes.* 14: 879–891, 1990.

227. Rowland, N. E., and S. M. Antelman. Stress-induced hyperphagia and obesity in rats: a possible model for understanding human obesity. *Science* 191: 310–311, 1975.

228. Sachdev, P. S. Behavioural factors affecting physical health of the New Zealand Maori. *Soc. Sci. Med.* 30: 431–440, 1990.

229. Saito, M., Y. Minokoshi, and T. Shimazu. Ventromedial hypothalamic stimulation accelerates norepinephrine turnover in brown adipose tissue of rats. *Life Sci.* 41: 193–197, 1987.

230. Salmon, D. M. W., and J. P. Flatt. Effect of dietary fat content among ad libitum fed mice. *Int. J. Obes.* 9: 443–449, 1985.

231. Sampliner, R. E., P. H. Bennett, L. J. Comess, et al. Gallbladder

disease in Pima Indians: demonstration of high prevalence and early onset by cholecytography. *N. Engl. J. Med.* 283: 1358–1364, 1977.

232. Sawchenko, P., and L. Swanson. The organization of forebrain afferents to the paraventricular and supraoptic nuclei of the rat. *J. Comp. Neurol.* 218: 121–144, 1983.

233. Scallet, A. C., and J. W. Olney. Components of hypothalamic obesity: bipiperidyl-mustard lesions add hyperphagia to monosodium glutamate-induced hyperinsulinemia. *Brain Res.* 374: 380–384, 1986.

234. Scarpace, P. J., M. Matheny, and S. E. Borst. Thermogenesis and mitochondrial GDP binding with age in response to the novel agonist CGP-12177A. *Am. J. Physiol.* 262(*Endocrinol. Metab.* 33): E185–E190, 1992.

235. Scarpace, P. J., A. D. Mooradian, and J. E. Morley. Age-associated decrease in β-adrenergic receptors and adenylate cyclase activity in rat brown adipose tissue. *J. Gerontol.* 43: B65–B70, 1988.

236. Schlundt, D. G., J. O. Hill, T. Sbrocco, J. Pope-Cordle, and T. Kasser. Obesity: a biogenetic or biobehavioral problem. *Int. J. Obes.* 14: 815–828, 1990.

237. Schlundt, D. G., D. Taylor, J. O. Hill, T. Sbrocco, J. Pope-Cordle, T. Kasser, and D. Arnold. A behavioral taxonomy of obese female participants in a weight-loss program. *Am. J. Clin. Nutr.* 53: 1151–1158, 1991.

238. Schwartz, D. H., S. McClane, L. Hernandez, and B. G. Hoebel. Feeding increases extracellular serotonin in the lateral hypothalamus of the rat as measured by microdialysis. *Brain Res.* 479: 349–354, 1989.

239. Sclafani, A. Carbohydrate taste, appetite, and obesity: an overview. *Neurosci. Behav. Rev.* 11: 131–153, 1987.

240. Scotellaro, P. A., L. L. Ji, J. Gorski, and L. B. Oscai. Body fat accretion: a rat model. *Med. Sci. Sports Exerc.* 23: 275–279, 1991.

241. Segal, K. R., A. Dunaif, B. Gutin, J. Albu, A. Nyuman, and F. X. Pi-Sunyer. Body composition, not body weight, is related to cardiovascular disease risk factors and sex hormone levels in man. *J. Clin. Invest.* 80: 1050–1055, 1987.

242. Segal, K. R., and F. X. Pi-Sunyer. Exercise and obesity. *Med. Clin. North Am.* 73: 217–236, 1989.

243. Seim, H. C., and J. A. Fiola. A comparison of attitudes and behaviors of men and women toward food and dieting. *Fam. Pract. Res. J.* 10: 57–63, 1990.

244. Shimazu, T., M. Noma, and M. Saito. Chronic infusion of norepinephrine into the ventromedial hypothalamus induces obesity in rats. *Brain Res.* 369: 215–223, 1986.

245. Shimokata, Y., D. C. Muller, J. L. Fleg, J. Sorkin, A. W. Ziemba, and R. Andres. Age as an independent determinant of glucose tolerance. *Diabetes* 40: 44–51, 1991.

246. Silverman, A. G. Overnutrition in the diabetic patient. *Mount Sinai J. Med.* 54: 211–216, 1987.

247. Silvestrini, B. The paradoxical stress response: a possible common basis for depression and other conditions. *J. Clin. Psychiatry* 51(suppl. 9): 6–8, 1990.

248. Simopoulos, A. P. Characteristics of obesity: an overview. *Ann. N. Y. Acad. Sci.* 499: 4–13, 1987.

249. Sims, E. A. H. Storage and expenditure of energy in obesity and their implications for management. *Med. Clin. North Am.* 73: 97–110, 1989.

250. Skaggs, S. R., and D. M. Crist. Exogenous human growth hormone reduces body fat in obese women. *Horm. Res.* 35: 19–24, 1991.

251. Smolin, L. A., M. B. Grosvenor, D. J. Handelsmand, and J. Brasel. Diet composition and lipoprotein lipase activity in human obesity. *Br. J. Nutr.* 58: 13–21, 1987.

252. Snyder, D. K., D. R. Clemmons, and L. E. Underwood. Dietary

carbohydrate content determines responsiveness to growth hormone in energy-restricted humans. *J. Clin. Endocrinol. Metab.* 69: 745–752, 1989.

253. Snyder, D. K., L. E. Underwood, and D. R. Clemmons. Anabolic effects of growth hormone in obese diet-restricted subjects are dose dependent. *Am. J. Clin. Nutr.* 52: 431–437, 1990.

254. Sobal, J., and A. J. Stunkard. Socioeconomic status and obesity: a review of the literature. *Psychol. Bull.* 105: 260–275, 1989.

255. Sorensen, T. I. A. Genetic epidemiology utilizing the adoption method: studies of obesity and of premature deaths in adults. *Scand. J. Soc. Med.* 19: 14–19, 1991.

256. Sorensen, T. I. A., and R. A. Price. Secular trends in body mass index among Danish young men. *Int. J. Obes.* 14: 411–419, 1990.

257. Steffens, A. B., A. J. W. Scheurink, P. G. M. Luiten, and B. Bohus. Hypothalamic food intake regulating areas are involved in the homeostasis of blood glucose and plasma FFA levels. *Physiol. Behav.* 44: 581–589, 1988.

258. Steinbeck, K., I. D. Caterson, L. Astbury, and J. R. Turtle. The effect of diet composition on weight gain and pyruvate dehydrogenase activity in heart muscle in the gold thioglucose obese mouse. *Int. J. Obes.* 11: 507–518, 1987.

259. Stern, J. S., and P. R. Johnson. Size and number of adipocytes and their implications. In: *Advances in Modern Nutrition, Diabetes, Obesity and Vascular Disease.* edited by H. M. Katzen and R. J. Mahler. Washington, DC: Hemisphere, 1977, vol. 2, p. 303–341.

260. Stevenson, J. A. F., and R. H. Rixon. Environmental temperature and deprivation of food and water on spontaneous activity of rats. *Yale J. Biol. Med.* 29: 575–584, 1957.

261. Stone, M., R. A. Schemmel, and D. M. Czajka-Narins. Growth and development of kidneys, heart and liver in S 5B/P1 and Osborne-Mendel rats fed high or low-fat diets. *Int. J. Obes.* 4: 65–78, 1980.

262. Tanaka, K., S. Inoue, J. Shiraki, T. Shishido, M. Saito, K. Numata, and Y. Takamura. Age-related decrease in plasma growth hormone: response to growth hormone–releasing hormone, arginine, and L-dopa in obesity. *Metabolism* 40: 1257–1262, 1991.

263. Thomas, J. D., D. C. Thomas, M. M. Doucette, and J. D. Stoeckle. Disease, lifestyle, and consanguinity in 58 American gypsies. *Lancet* 2: 377–379, 1987.

264. Timmers, K., D. L. Coleman, N. R. Voyles, A. M. Powell, A. Rokaeus, and L. Recant. Neuropeptide content in pancreas and pituitary of obese and diabetes mutant mice: strain and sex differences. *Metabolism* 39: 378–383, 1990.

265. Timmers, K., N. R. Voyles, C. Zalenski, S. Wilkins, and L. Recant. Altered β-endorphin, met- and leu-enkephalins, and enkephalin-containing peptides in pancreas and pituitary of genetically obese diabetic (db/db) mice during development of diabetic syndrome. *Diabetes* 35: 1143–1151, 1986.

266. Tokunaga, K., M. Fukushima, J. W. Kemnitz, and G. A. Bray. Comparison of ventromedial and paraventricular lesions in rats that become obese. *Am. J. Physiol.* 251(*Regulatory Integrative Comp. Physiol.* 22): R1221–R1227, 1986.

267. Trayhurn, P. Energy expenditure and thermogenesis: animal studies on brown adipose tissue. *Int. J. Obes.* 14(suppl. 1): 17–29, 1990.

268. Trayhurn, P., and W. P. T. James. Thermoregulation and nonshivering thermogenesis in the genetically obese (*ob/ob*) mouse. *Pflugers Arch.* 373: 189–193, 1978.

269. Triandafillou, J., and J. Himms-Hagen. Brown adipose tissue in genetically obese (*fa/fa*) rats: response to cold and diet. *Am. J. Physiol.* 244(*Endocrinol. Metab.* 7): E145–E150, 1983.

270. Troisi, R. J., S. T. Weiss, D. R. Parker, D. Sparrow, J. B. Young,

and L. Landsberg. Relation of obesity and diet to sympathetic nervous system activity. *Hypertension* 17: 669–677, 1991.

271. Turula, M., J. Kaprio, A. Rissanen, and M. Koskenvuo. Body weight in the Finnish twin cohort. *Diabetes Res. Clin. Pract.* 10: S33–S36, 1990.

272. Vague, J. The degree of masculine differentiation of obesities: a factor determining predisposition to diabetes, atherosclerosis, gout, and uric calculous disease. *Am. J. Clin. Nutr.* 4: 20–34, 1956.

273. Van, R. L. R., C. E. Bayliss, and D. A. K. Roncari. Cytological and enzymological characterization of adult human adipocyte precursors in culture. *J. Clin. Invest.* 58: 699–704, 1976.

274. Van Dale, D., and V. H. M. Saris. Repetitive weight loss and weight regain: effects on weight reduction, resting metabolic rate, and lipolytic activity before and after exercise and/or diet treatment. *Am. J. Clin. Nutr.* 49: 409–416, 1989.

275. Vander Tuig, J. G., J. Kerner, and D. R. Romsos. Hypothalamic obesity, brown adipose tissue, and sympathoadrenal activity in rats. *Am. J. Physiol.* 248(*Endocrinol. Metab.* 11): E607–E617, 1985.

276. Van Itallie, T. B. The glucostatic theory 1953–1988: roots and branches. *Int. J. Obes.* 14(suppl. 3): 1–10, 1990.

277. Vannucci, S. J., C. M. Klim, K. F. LaNoue, and L. F. Martin. Regulation of fat cell adenylate cyclase in young Zucker (fa/fa) rats: alterations in GTP sensitivity of adenosine A_1 mediated inhibition. *Int. J. Obes.* 14(suppl. 3): 125–134, 1990.

278. Vitale, P. M., J. M. Darrow, M. J. Duncan, C. A. Shustak, and B. D. Goldman. Effects of photoperiod, pinealectomy and castration on body weight and daily torpor in Djungarian hamsters. *J. Endocrinol.* 106: 367–375, 1985.

279. Wade, G. N., and T. J. Bartness. Effects of photoperiod and gonadectomy on food intake, body weight, and body composition in Siberian hamsters. *Am. J. Physiol.* 246(*Regulatory Integrative Comp. Physiol.* 17): R26–R30, 1984.

280. Walberg, J. L., D. Upton, and J. S. Stern. Exercise training improves sensitivity in the obese Zucker rat. *Metabolism* 33: 1075–1079, 1984.

281. Weinsier, R. L., L. D. James, B. E. Darnell, H. P. Dustan, R. Birch, and G. R. Hunter. Obesity-related hypertension: evaluation of the separate effects of energy restriction and weight reduction on hemodynamic and neuroendocrine status. *Am. J. Med.* 90: 460–468, 1991.

282. Weststrate, J. A., J. Dekker, M. Stoel, L. Begheijn, P. Deurenberg, and J. G. A. J. Hautvast. Resting energy expenditure in women: impact of obesity and body-fat distribution. *Metabolism* 39: 11–17, 1990.

283. Williams, P. T. Weight set-point theory predicts HDL-cholesterol levels in previously obese long-distance runners. *Int. J. Obes.* 14: 421–427, 1990.

284. Williams, R. R., S. C. Hunt, S. J. Hasstedt, P. N. Hopkins, L. L. Wu, T. D. Berry, B. M. Stults, G. K. Barlow, M. C. Schumacher, R. P. Lifton, and J. M. Lalouel. Are there interactions and relations between genetic and environmental factors predisposing to high blood pressure? *Hypertension* 18(suppl. I): I29–I37, 1991.

285. Wing, R. R., E. H. Blair, L. H. Epstein, and M. D. McDermott. Psychological stress and glucose metabolism in obese and normal-weight subjects: a possible mechanism for differences in stress-induced eating. *Health Psychol.* 9: 693–700, 1990.

286. Woods, S. C., D. P. F. Lattemann, M. W. Schwartz, and D. Porte, Jr. A re-assessment of the regulation of adiposity and appetite by the brain insulin system. *Int. J. Obes.* 14(suppl. 3): 69–76, 1990.

287. Wu, S. Y., J. S. Stern, D. A. Fisher, and Z. Glick. Cold-induced increase in brown fat thyroxine 5-monodeiodinase is attenuated in Zucker obese rat. *Am. J. Physiol.* 252(*Endocrinol. Metab.* 15): E63–E67, 1987.

288. Xu, X., and P. Bjorntorp. The effects of steroid hormones on adipocyte development. *Int. J. Obes.* 14(suppl. 3): 159–163, 1990.

289. York, D. A. Neural activity in hypothalamic and genetic obesity. *Proc. Nutr. Soc.* 46: 105–117, 1987.

290. Zador, D. A., and A. S. Truswell. Nutritional status on admission to a general surgical ward in a Sydney hospital. *Aust. N. Z. J. Med.* 17: 234–240, 1987.

291. Zawadzki, J. K., R. R. Wolfe, D. M. Mott, S. Lillioja, B. V. Howard, and C. Bogardus. Increased rate of Cori Cycle in obese subjects with NIDDM and effect of weight reduction. *Diabetes* 37: 154–159, 1988.

292. Zinder, O., M. Hamosh, T. R. C. Fleck, and R. O. Scow. Effect of prolactin on lipoprotein lipase in mammary gland and adipose tissue of rats. *Am. J. Physiol.* 226: 744–748, 1974.

293. Zurlo, F., S. Lillioja, A. Esposito-Del Puente, B. L. Nyomba, I. Raz, M. F. Saad, B. A. Swinburn, W. C. Knowler, C. Bogardus, and E. Ravussin. Low ratio of fat to carbohydrate oxidation as predictor of weight gain: study of 24-h RQ. *Am. J. Physiol.* 259(*Regulatory Integrative Comp. Physiol.* 30): E650–E657, 1990.

63. Effects of altered vitamin and mineral nutritional status on temperature regulation and thermogenesis in the cold

HENRY C. LUKASKI
SCOTT M. SMITH

United States Department of Agriculture, Agricultural Research Service, Grand Forks Human Nutrition Research Center, Grand Forks, North Dakota

CHAPTER CONTENTS

SURVIVAL DURING EXPOSURE to adverse environmental conditions depends on the capacity of an organism to adapt with integrated thermoregulatory responses. Homeostatic maintenance of body temperature during exposure to a cold environment requires a coordinated increase in endogenous heat production and a decrease in peripheral heat loss.

Two mechanisms to increase heat production are shivering and nonshivering thermogenesis (75). Heat produced by involuntary muscle contraction is termed *shivering thermogenesis*, an inefficient response to maintain body temperature (43). *Nonshivering thermogenesis*, in contrast, is a relatively efficient and prolonged response achieved by the stimulation of meta-bolic processes independent of muscle contraction (88). Nonshivering thermogenesis, also known as facultative thermogenesis, refers to the nonobligatory component of energy expenditure that is responsive to environmental stimuli, such as cold. It involves, to varying degrees, an uncoupling of oxidative metabolism from the production of ATP; the efficiency of oxidative metabolism is adjusted to increase or decrease cellular heat production in response to environmental stimuli (87). Nonshivering thermogenesis can be sustained for prolonged periods and is regulated by the sympathetic nervous system (SNS) through the action of norepinephrine, a neurotransmitter produced within the neurons of the SNS, and triiodothyronine, which is produced in peripheral tissues by the deiodination of thyroxine (46, 47, 54, 55, 75, 87, 124, 134).

The site of cold-induced or nonshivering thermogenesis appears to be species-dependent (88). In laboratory rodents, brown adipose tissue (BAT) is the principal site of regulation of nonshivering thermogenesis (75). In humans, however, the role of BAT is not well understood, but it, as well as skeletal muscle, has the potential to be a major site of facultative thermogenesis (87).

Cooperative adjustments of both the SNS and the thyroid hormone system during cold exposure are required for the appropriate thermogenic responses in BAT (19, 26, 46, 47, 74, 127, 128). An augmentation in SNS activity is indicated by an increase in plasma and urinary norepinephrine concentrations, an increase in the rate of norepinephrine turnover, and an increase in heat production (87). Similarly, the pituitary increases thyroid-stimulating hormone (TSH) release, which stimulates the release of thyroxine (T_4) from the thyroid (46). In addition, peripheral conversion of T_4 to T_3 (triiodothyronine) the physiologically active form of thyroid hormone, is stimulated by the influence of increased SNS activity (26, 127). Importantly, the adrenergic input into BAT activates and stimulates the synthesis of a unique protein, uncoupling protein (UCP), which, by uncoupling oxidative phosphorylation, increases BAT respiration and dissipates the resulting energy as heat (25, 108). The increase in thyroid hormones, particularly T_3, is an important signal to increase oxygen consumption and heat production by chemical thermogenesis in the form of increased substrate flux through futile cycle pathways and increased rates of Na^+–K^+ ATPase activity that are regulated at the level of gene transcription (40). Increases in intracellular T_3 content, particularly in BAT, also amplify the adrenergic signals for heat production (75, 126). Increases in intracellular T_3 also lead to augmented synthesis of many hormones, receptor proteins, and basic metabolic functions needed for resting and facultative metabolic processes (20, 126).

Two extrathyroidal enzymes appear to be principally responsible for the conversion of T_4 to T_3 (85). Type I-5′-deiodinase (5′-D-I), an outer ring deiodinase located primarily in the kidney, liver, and skeletal muscle, is responsible for the production of about 70%–80% of the plasma pool of T_3 in the euthyroid state (89). It is a selenocysteine-containing protein and requires reduced sulphydryl groups for optimal activity (17). Hepatic thyroxine 5%-D-I activity decreases in hypothyroidism and increases in hyperthyroidism (85). These changes are associated with pretranslational regulation of enzyme synthesis (15, 58).

The second enzyme is type II deiodinase (5′-D-II), an inner ring deiodinase located principally in the cerebral cortex, anterior pituitary, placenta, and BAT (89). This enzyme is not selenium-dependent and produces T_3 primarily for intracellular utilization (16, 85). Hypothyroidism, or cold exposure, however, can induce this enzyme to increase T_3 production so that it provides about 50% of the plasma pool (127).

When the gradient of body temperature from the core to the periphery increases, physiological mechanisms are activated not only to increase heat production but also to reduce heat loss. A central mechanism for conserving body heat is peripheral vasoconstriction, which decreases blood flow to the skin or body surface. Thus blood is shunted to the body core to preserve internal body temperature. This mechanism is regulated by the SNS through the action of norepinephrine on blood vessels in the skin (47).

This chapter describes the effect of altered status of some vitamins and minerals on thermogenesis and temperature regulation in the cold and discusses proposed mechanisms of action regarding SNS activity and thyroid hormone status in thermoregulatory function. Information is presented only for nutrients for which experimental data indicate that impaired temperature regulation is related to decreased nutritional status.

VITAMINS

Thiamin

Thiamin, or vitamin B_1, is a relatively simple, water-soluble compound. It consists of a pyrimidine ring and a thiazole ring linked by a methylene bridge. The essentiality of thiamin was demonstrated by its ability to cure beriberi, a disease characterized by damage to the cardiovascular and nervous systems and sometimes accompanied by muscle wasting or edema.

The principal biochemical role of thiamin is to serve as a precursor of thiamin pyrophosphate, which acts as a coenzyme for many dehydrogenase enzyme complexes necessary for the oxidative decarboxylation of α-keto acids to carboxylic acids (for example, pyruvate to acetyl CoA). These reactions are widely distributed in intermediary metabolism, particularly carbohydrate metabolism. In addition, thiamin pyrophosphate serves as the coenzyme for the transketolase reaction of the pentose phosphate shunt (64).

Thiamin deficiency is associated with alterations in carbohydrate metabolism related to an overall decrease in oxidative decarboxylation. Plasma and tissue concentrations of pyruvate are increased during thiamin deficiency. Reduced saturation of thiamin pyrophosphate in erythrocyte transketolase has been found in animals and humans fed diets low in thiamin (65, 71).

Animal Studies. Diet-induced thiamin deficiency in animals has been characterized by the classical signs of malnutrition: anorexia, impaired growth, and reduced oxygen consumption and heat production. Because of the semistarvation associated with thiamin deficiency, it was necessary to determine if the vitamin deficiency per se or reduced energy intake was responsible for the observed reduction in heat production.

Veen et al. (142) delineated the effects of food intake and thiamin restriction on rectal temperature in male rats. Adult male rats were fed a thiamin-adequate diet (5 μg thiamin/g diet), a thiamin-deficient diet (no added thiamin), or pair-fed the control diet in amounts consumed by matched animals fed the deficient diet. Periodically, the animals fed the deficient diet were administered a thiamin supplement by gavage.

The effects of food intake and thiamin status on rectal temperature are shown in Figure 63.1. Voluntary food intake decreased progressively during thiamin restriction and increased from 2.5 g·day^{-1} to 10 g·day^{-1} within 24 h of thiamin supplementation. After withdrawal of supplementation, food intake decreased in the deficient animals to about 2.5 g·day^{-1}. Thiamin-deficient rats demonstrated a progressive decline in rectal temperature during restriction. However, rectal temperature increased to near control values after supplementation.

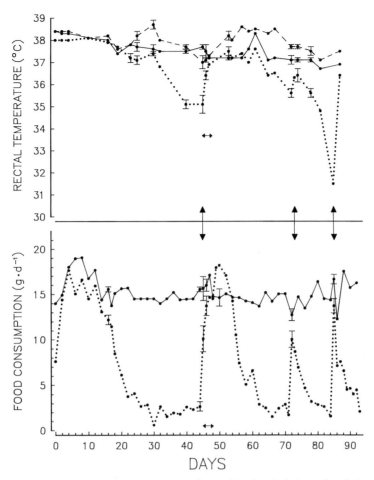

FIG. 63.1. Food consumption and rectal temperature during thiamine depletion and repletion. *Solid line,* control; *dashed line,* pair-fed, *dotted line,* thiamine-deficient. Horizontal arrows indicate feeding control diet to thiamine-deficient rats; vertical arrows indicate treatment of thiamine-deficient rats with 50 μg thiamine. Reproduced by permission of the National Research Council of Canada (135).

Despite restricted food intake, pair-fed animals had rectal temperatures similar to the animals fed a thiamin-adequate diet. These findings indicate that hypothermia associated with thiamin deficiency is not dependent on food intake.

Diet-induced thiamin deficiency results in a reversible impairment of cerebral DNA synthesis (72). During the symptomatic stage of thiamin deficiency, when hypothermic animals (rectal temperature about 33°C) were warmed to 38°C, DNA synthesis was reduced 22%, 37%, 31%, and 19%, respectively, of pair-fed control values in the cortex, brain stem, cerebellum, and subcortical structures. The degree of depressed DNA synthesis increased with the duration of thiamin deficiency. After thiamin supplementation, DNA synthesis in all brain areas increased significantly to and above control values.

Protein synthesis, both visceral and cerebral, is also depressed in thiamin-deficient rats (61). Although some of this reduction in protein synthesis may be attributed

to reduced food intake, the majority is dependent on the concomitant hypothermia associated with vitamin deficiency. When body temperature was increased to 38°C in thiamin-deficient animals, protein synthesis increased significantly more in the brain than in the viscera (60%–70% vs. 30%–35%), which suggests that hypothermia per se is an important factor in regulating protein synthesis in thiamin deficiency.

Human Studies. Human thiamin deficiency, which may be caused by inadequate thiamin intake, most commonly occurs as a result of chronic alcoholism. Alcohol consumption, either acute or chronic, results in decreased thiamin absorption, even in healthy individuals (77). Moderately severe thiamin deficiency in humans is identified by a group of symptoms characteristic of the condition known as Wernicke's encephalopathy or Wernicke-Korsakoff syndrome, a neuropsychiatric disorder. Wernicke's encephalopathy is an acute disorder consisting of variable degrees of ataxia, disordered ocular motility, and disturbances of consciousness. The Wernicke-Korsakoff syndrome or psychosis typically has loss of recent memory and progressive mental impairment as prominent features.

There is increasing evidence from case study reports (2, 24, 76, 81, 84, 91, 97, 117) that patients with Wernicke-Korsakoff syndrome may exhibit hypothermia, independent of cold exposure. Some of these patients, who generally are chronic alcoholics and may be malnourished, present the classical symptom of low body temperature (oral or rectal temperature 36°C or less) and no evidence of primary hypothyroidism. After parenteral administration of thiamin, body temperature increases and, in most cases, neurological deficits are relieved.

An example of the effect of thiamin supplementation on hypothermia and neurological symptoms is presented in Figure 63.2. At admission, the patient, a 67 year old man described as an alcoholic with erratic eating habits, was semicomatose and responsive only to painful stimuli; his skin temperature was described as "very cold." Rectal temperature was 33.3°C, hemoglobin concentration was 144 g·l^{-1}, and hematocrit was 42%, but blood pyruvic acid concentration was abnormal (576 μmol·l^{-1}; normal range: 35–100 μmol·l^{-1}). Rectal temperature declined to 32.2°C during hospitalization, even with the use of a warming blanket. After body temperature increased to 35°C the next morning, the patient was more responsive; this permitted observation of the classical neurological symptoms associated with Wernicke's encephalopathy. The warming blanket was discontinued and thiamin was administered parenterally as three 200 mg doses during the next 12 h.

Ten hours after the initiation of thiamin supplementation, rectal temperature increased to 36.7°C and neurological symptoms subsided.

Inadequate temperature control function has also been reported in Wernicke's encephalopathy (91). Thermolability to exogenous cooling and heating and a lack of normal subjective reactions to changes in temperature were observed. Also, a poor febrile response to an intravenous infusion of a pyrogen was reported. After 2 months of thiamin therapy, the patient regained normal physiological and affective responses to challenges with cold and heat, as well as pyrogen.

Mechanism of Action. The mechanism of action of thiamin deficiency on observed hypothermia in animals and humans is not well understood. In thiamin-deficient, adrenalectomized mice oxygen consumption remains constant with thyroxine administration. However, supplements of thiamin and of thiamin plus thyroxine increase oxygen uptake by 40% and 110%, respectively (18). Thus thiamin is required for basal heat production and thermogenesis.

Chronic thiamin deficiency results in selective destruction of certain cells in the central nervous system (CNS), with sparing of adjacent cells (65). Our understanding of the relationship of thiamin deficiency to specific brain lesions in humans is derived mainly from cases of Wernicke's encephalopathy of a nutritional etiology. Lesions are consistently found in mammillary bodies, thalamus, hypothalamus, and pons. Evidence of CNS lesions in Wernicke-Korsakoff syndrome patients with hypothermia include glial and vascular proliferation in the posterior hypothalamus and mammillary

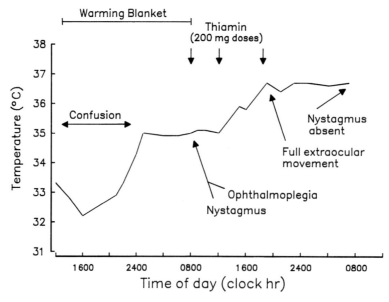

FIG. 63.2. Effects of thiamine supplementation on hypothermia and neurological symptoms in a patient with Wernicke's encephalopathy. From Macaron et al. (91), with permission.

bodies (117, 143). When the central thiamin antagonist, pyrithiamin, is used to induce thiamin deficiency in the rat, structures such as the mammillary bodies, thalamus, hypothalamus, and cerebellum, in addition to the pons, are affected (138). These observations suggest that the hypothermia associated with thiamin deficiency may be explained by lesions in the cold-sensitive portion of the hypothalamus, damage to which is known to cause hypothermia (105).

Because parenteral administration of thiamin to thiamin-deficient animals and humans is known to result in a relatively rapid amelioration of hypothermia, it seems that a biochemical, rather than an anatomical, lesion may have been repaired. Although it may be postulated that peripheral glucose metabolism may be restored with thiamin supplementation, there is no experimental evidence to support this supposition. An alternate hypothesis is that thiamin supplementation improves the activity of enzymes requiring thiamin in the CNS.

Alterations in some CNS neurotransmitter systems have been reported in thiamin deficiency, which suggests that thiamin is important for normal neurotransmitter function (65). Acetylcholine, γ-aminobutyrate, glutamate, and aspartate, which are produced primarily through the oxidation of glucose via the pyruvate dehydrogenase complex and α-ketoglutarate dehydrogenase, are reduced in the brains of thiamin-deficient rats (56, 65). These perturbations have been related to postural abnormalities and seizures (23). Supplementation of thiamin-deficient rats results in an increase in the activities of pyruvate dehydrogenase complex and α-ketoglutarate dehydrogenase in some brain structures but no change in transketolase activity (23).

Changes in serotonin or 5-hydroxytryptamine metabolism, however, are more pronounced than those in other neurotransmitter systems during thiamin deficiency. Studies in animals with symptoms of memory loss similar to Wernicke-Korsakoff syndrome suggest that a decrease in 5-hydroxytryptamine-containing neurons may be a common trait (65). It also has been proposed that the increased piloerection and hypothermia accompanying thiamin deficiency may be caused by serotoninergic neurotransmitter changes (119, 151). Evidence indicates that depletion of serotonin-containing neurons in the hypothalamus is associated with hypothermia in animals (70, 106).

MINERALS

Iron

Iron is one of the most intensively studied nutrients, and thus its physiological function is becoming well understood. This broad understanding of the importance of iron nutriture has been stimulated by the evidence that iron comprises the most common single nutrient deficiency in the United States (35, 118) and in the world (39). Moreover, iron deficiency can be prevented by supplementation (67).

It is estimated that men have more body iron than women (3.8 vs. 2.3 g). Body iron is distributed as iron-containing compounds between two general pools, a functional pool and a storage pool. The functional pool consists of iron-containing compounds that serve a metabolic or an enzymatic role, and the storage pool provides a reservoir for tissue needs and iron transport.

The functional pool includes the essential iron compounds consisting primarily of heme proteins, that is, proteins with an iron–porphyrin prosthetic group. Hemoglobin or erythrocyte iron is the most abundant of the heme proteins and accounts for more than 90% of the functional pool and more than 65% of total body iron. Its function is to transport oxygen via the blood from the lungs to the tissues. Myoglobin is the heme protein found in muscle that transports and stores oxygen for use during muscle contraction; it accounts for about 10% of body iron content.

Heme iron is also a component of some cellular enzyme systems (31, 33). Cytochromes are enzymes located in the mitochondria, involved in the electron transport system, and required for the oxidative production of energy. Other heme enzymes include catalase and peroxidase.

There is another important group of iron-dependent enzymes that do not contain heme and that are involved in oxidative metabolism (33). These compounds include iron–sulfur proteins and metalloflavoproteins that actually account for more iron in the mitochondria than do the cytochromes. These include nicotinamide adenine dinucleotide dehydrogenase and succinic dehydrogenase. The heme and nonheme iron enzymes collectively contribute about 3% of the total body iron.

Another group of enzymes exists in which exogenous iron is required for enzymatic function (33, 34). This group includes aconitase, a component of the tricarboxylic acid cycle; phosphoenolpyruvate carboxykinase, a rate-limiting enzyme in the gluconeogenic pathway; and ribonucleotide reductase, an enzyme needed for DNA synthesis.

Some iron-containing compounds are involved in iron storage and transport (34). The principal iron storage compounds are ferritin and hemosiderin, which are found primarily in the liver, reticuloendothelial cells, and bone marrow. The total amount of storage iron varies considerably, depending on availability and requirements. Ferritin is also found in the plasma, where it can be measured as an index of iron storage (30).

The contribution of storage iron to total body iron varies widely and averages approximately 12% in women and 25% in men. The amount of storage iron indirectly affects iron absorption: as iron stores decrease absorption increases. This autoregulatory, homeostatic mechanism is thought to protect against both overload and deficiency (34).

Iron deficiency remains one of the most prevalent nutritional deficiencies even in industrialized countries. At highest risk are infants and women during their reproductive years (44). Data obtained from the second National Health and Nutritional Examination Survey indicate that almost 6% of women of reproductive age in the United States had anemia and 10%–15% of women in the same group had low concentrations of plasma ferritin (44), indicative of tissue iron deficiency.

Because optimal oxygen transport, delivery, and utilization at the cellular level require iron, manifestations of reduced thermogenesis have been intensively studied in animals and in humans with graded iron deficiency. The findings from these investigations have increased our understanding of the metabolic importance of iron in energy metabolism.

Animal Studies.

Studies in animals have provided the basis for our knowledge of the effects of iron deficiency on thermoregulation and thermogenesis. The first report of altered temperature regulation in iron-deficient animals was by Dillman et al. (42), who noted that, regardless of the degree of anemia, weanling male rats fed diets low in iron ($<8 \ \mu g \cdot g^{-1}$) had increased plasma catecholamine concentrations and urinary catecholamine excretions. The catecholamine response in both iron-deficient and iron-adequate animals was temperature-dependent, with little difference at 30°C but a two- to three fold increase at 25°C and 4°C. The iron-deficient animals were unable to maintain rectal temperatures during a 4 h exposure to 4°C; this impairment was independent of the degree of iron deficiency (that is, magnitude of anemia) and food intake. Importantly, the abnormality in temperature regulation and catecholamine response was reversed following 6 days of iron therapy.

Other investigators have independently determined that, when compared to iron-sufficient rats, iron-deficient rats are unable to maintain core temperature when exposed to cold, which often results in profound hypothermia and occasionally death. Dillmann et al. (41) extended their initial observations of hypothermia and reported a significant reduction in oxygen consumption in iron-deficient rats in comparison to iron-adequate control animals maintained at 4°C for 6 h (Table 63.1). Serum T_4 concentrations were slightly increased and T_3 concentrations were significantly decreased during cold exposure in iron-deficient rats. Interestingly, iron deficiency was associated with a slight increase in oxygen uptake but no difference in thyroid hormone concentrations at 25°C and 30°C.

TABLE 63.1. *Effects of Iron Deficiency on Oxygen Consumption, Serum Thyroxine, (T_4) and Triiodothyronine (T_3) Concentrations in Six Rats at 4°C*

	Oxygen Consumption [ml·(kg·min)$^{-1}$]	T_4 (nmol·liter^{-1})	T_3 (nmol·liter^{-1})
Iron-deficient	39 ± 3*	56 ± 3	0.7 ± 0.1*
Iron-adequate	63 ± 2	46 ± 4	1.1 ± 0.1

*$P < 0.05$.
Values are mean ± SE. Adapted from Dillman et al. (41).

Beard et al. (5) also reported a failure of iron-deficient rats to maintain body temperature and to increase thyroid hormone concentrations when acutely exposed to a cold environment. Importantly, they noted that transfusion, resulting in an increase in hematocrit from 15%–20% to 30%, was associated with improved cold tolerance and an increase in plasma thyroid hormone concentrations.

Virtually all of the studies on the effects of iron status on thermogenesis have utilized male rats. However, because iron deficiency is predominantly a problem for women during their child-bearing years (44), it is appropriate to determine the interactions among reproductive hormones, iron nutriture, and thermogenesis. Two investigations have addressed this important topic.

Smith et al. (129) studied the role of ovarian steroids on metabolic rate and thyroid hormone metabolism in iron-deficient female rats. Indices of thyroid function were examined in ovariectomized rats with high and low iron status and treated with estradiol and progesterone. Relative to sham treatment, exogenous steroid hormones did not affect plasma T_3 concentrations. Estradiol treatment, however, whether alone or in combination with progesterone, significantly increased T_4 concentrations. Iron deficiency reduced plasma T_4 and T_3 concentrations and liver 5′-D-I activity. The BAT 5′-D-II activity was significantly decreased in estradiol-treated animals. Therefore, plasma indices of thyroid function are affected by ovarian steroids.

The effects of the estrous cycle and iron deficiency on thermoregulatory function in the cold have been studied (131). Iron-deficient animals had significantly lower rectal temperatures when exposed to 4°C for 6 h; this effect was most pronounced during proestrus. Ovariectomy resulted in significant hypothermia as well as reduced thyroid hormone concentrations during cold stress. The ability to thermoregulate, as demonstrated by the change in rectal temperature, during cold exposure was affected by the state of the estrous cycle, particularly in iron deficiency (Fig. 63.3).

Human Studies.

Studies in humans (10, 96, 99) confirm the previous findings in animal experiments and clearly

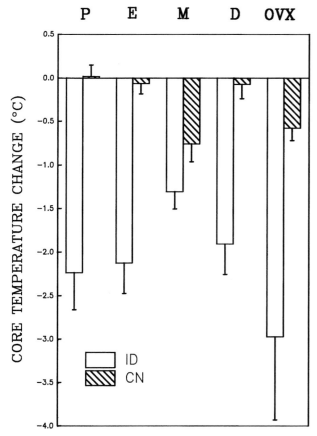

FIG. 63.3. Core temperature change in iron-deficient *(ID)* and control *(CN)* rats exposed to 4°C for 6 h at different stages of the estrous cycle (proestrus, *P*; estrus, *E*; metestrus, *M*; diestrus, *D*) or in ovariectomized *(OVX)* rats. Reprinted with permission of Butterworth-Heinemann (125).

demonstrate the adverse effects of iron deficiency on thermogenesis during acute cold exposure. These studies also provide an integration of biochemical and physical data to support the observations of hypothermia in iron deficiency.

Martinez-Torres et al. (99) studied the effects of graded iron nutritional status on the metabolic and hormonal responses of 21 adults (17 men and 4 women) during immersion in 28°C water for 1 h. Volunteers were classified on the basis of blood biochemical indices of iron status (hemoglobin, serum ferritin, and transferrin saturation) as normal, iron-deficient without anemia, and anemic. Anemia was associated with a significant decrease in oral temperature (-0.9°C), an increase in oxygen uptake, and an increase in plasma norepinephrine concentration. Interestingly, oxygen consumption and plasma norepinephrine were also significantly increased among the non-anemic, iron-deficient adults.

The influence of iron depletion on thermoregulatory function during exposure to cold air was also studied in women (96). Physiological responses during exposure to 16°C were studied after body iron stores were depleted by diet, phlebotomy, and menstruation and then repleted by diet and iron supplementation. The women fasted the night before the cold-exposure studies were conducted. Iron depletion was characterized by a significant decline in hemoglobin, ferritin, and body iron balance. Iron repletion was associated with increased hemoglobin, ferritin, and iron balance. Iron depletion significantly reduced metabolic heat production (49.6 ± 1.1 vs. 53.6 ± 1.2 W·m^{-2}) during cold exposure. The rates of cooling of the body core temperature (-0.43 ± 0.04 vs. -0.14 ± 0.02°C·h^{-1}) and peripheral skin temperature (-3.4 ± 0.1 vs. -2.4 ± 0.1°C·h^{-1}) were significantly greater during iron depletion than repletion. Cold exposure during iron deficiency resulted in blunted postexposure increases in thyroid hormones but significant increases in plasma norepinephrine concentrations.

Examination of the relationships among oxygen consumption and rectal temperature during cold exposure suggests an alteration in the core temperature at which the onset of shivering was observed (Fig. 63.4). These data indicate a blunted oxygen uptake response relative to the faster rate of core cooling in iron deficiency. Such an impaired response in metabolic heat production is consistent with a lower core temperature threshold for the onset of shivering. Thus the increased rate of core cooling and the inability to increase metabolic heat production result in a reduced core temperature threshold for shivering in iron-deficient women.

The effects of graded iron deficiency on thermoregulation during cold water immersion were also assessed in a group of 30 young women (10). Participants were classified according to iron status as normal or control, anemic, and iron-depleted but not anemic. After 100 min of immersion in 28°C water, the anemic women, in comparison with control subjects, had significantly reduced rectal temperatures (36.0 ± 0.2° vs. 36.2 ± 0.1°C) and decreased rates of oxygen consumption (5.28 ± 0.26 vs 5.99 ± 0.29 ml·kg^{-1}·min^{-1}). Plasma T$_4$ and T$_3$ concentrations were significantly reduced in the anemic women both before and after cold exposure. Responses of the iron-depleted but not anemic women were similar to those of the control subjects. Repletion of the anemic women with iron improved thermoregulatory performance but without a complete amelioration of the impairment in thyroid hormone metabolism.

Mechanism of Action. The mechanism through which iron depletion impairs thermogenesis involves the hormone systems regulating energy production, thyroid hormones, and the SNS. Each of these systems has been studied in detail relative to iron deficiency.

Alterations in catecholamine metabolism have been reported in iron-deficient animals and humans. Initial

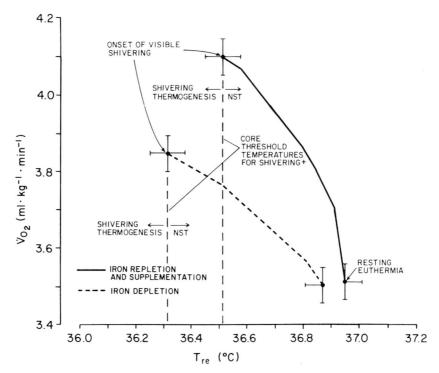

FIG. 63.4. Relationship between mean oxygen uptake ($\dot{V}_{O_2}$) and mean rectal temperature (T_{re}) of women acutely exposed to cold during iron deficiency and after iron repletion and supplementation. *NST* refers to nonshivering thermogenesis. From Lukaski et al. (89), with permission.

studies with iron-deficient children showed a significant elevation in 24 h urinary norepinephrine concentration while epinephrine output was not affected (145). These alterations in urinary monoamine output were corrected rapidly with iron supplementation (Fig. 63.5). Similarly, increased urinary excretion and plasma concentrations of norepinephrine have been noted in iron-deficient rats (5, 41, 42, 60). These observations led to the hypothesis that there was a defect, either direct or indirect, in the

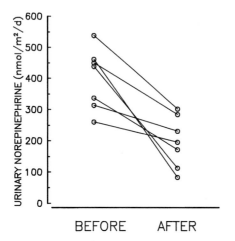

FIG. 63.5. Urinary norepinephrine excretion in iron-deficient children before and after treatment with iron dextran. Adapted from Voorhess et al. (137).

catabolic pathways for catecholamines because iron is known to affect several of the catabolic pathways for these neurotransmitters (33).

One hypothesis to explain the increased plasma and urinary norepinephrine concentrations in iron deficiency was reduced activity of monoamine oxidase (152). This enzyme, which does not contain iron, is responsive to the iron status of the animal. Data from iron-deficient rats indicate that plasma concentrations of norepinephrine are normalized within 24 h, whereas monoamine oxidase activity does not respond for up to 6 days after iron supplementation (147).

Data on tissue-specific metabolism of norepinephrine have been useful in understanding the effects of iron deficiency on SNS activity. The increased turnover of norepinephrine in conjunction with reduced tissue norepinephrine content suggests altered postsynaptic uptake. It is noteworthy that this augmented turnover rate has been associated with increased resting oxygen uptake at room temperature (136). Heart and BAT norepinephrine concentrations, are significantly decreased in iron-deficient rats (9, 12, 21, 136), with increased rates of turnover in each tissue (7, 9). The increased rates of turnover explain the increased norepinephrine concentrations in plasma and urine because there is an increased flux of norepinephrine through the system, with plasma and urine being the final pool in the system.

Alternately, rates of catecholamine synthesis might

influence tissue concentrations of norepinephrine. Experimental data, however, indicate that synthetic rates are not affected by iron deficiency (133).

An important role of thyroid hormone metabolism in thermoregulation was indicated when Dillman et al. (41) observed a prompt increase in the rectal temperatures of iron-deficient rats injected with a bolus of T_3 and when exposed to cold. Subsequently, it was demonstrated that iron-deficient rats administered therapeutic doses of iron improve their capacity to increase body temperature in the cold (12).

Central and peripheral thyroid hormone metabolisms have been investigated in iron deficiency. Plasma TSH concentrations are reduced, and a blunted and delayed response to exogenous thyrotropin-releasing hormone (TRH) challenge in iron-deficient animals has been observed (8, 135). Although endogenous production and release of TRH and TSH have not been studied in detail, it appears that there may be a significant impairment in the central regulation of thyroid hormone system in iron deficiency. It is also possible that reduced circulating T_4 may be the result of an inability of the thyroid to secrete it.

Peripheral T_4 metabolism also has been studied as a limiting factor responsible for impaired thermogenesis. Several studies have documented decreased in vitro 5'-D-I activity in the liver of iron-deficient rats (8, 12, 129, 131, 132), which indicates a reduced conversion rate of T_4 to T_3. These findings are extrapolated to suggest a reduction in the in vivo rate of production of the potent thermogenic hormone T_3 in iron deficiency. This assumption is supported by numerous observations of decreased plasma T_3 concentrations in iron-deficient rats (5, 6, 8, 21, 132, 133, 135) and decreased whole-body production rates of T_3 as calculated from plasma kinetic experiments (8).

An iron-dependent hypothyroidism exerts additional restrictions on thermogenesis at the cellular level. Smith et al. (130) reported that iron-deficient rats have reduced nuclear T_3 receptor occupancy. Also, BAT mitochondrial binding of ^{3}H-GDP, an index of requirement for thermogenesis, was increased at 25°C but decreased at 10°C in iron-deficient rats (11). These observations indicate that iron deficiency impairs thermogenic capacity directly by interrupting the conversion of T_4 to T_3, which restricts the capability of mitochondria to increase heat production.

Data on the metabolic responses of iron-deficient humans during cold exposure seem to conflict. Martinez-Torres et al. (99) reported an increase in oxygen consumption during cold stress, whereas Lukaski et al. (96) and Beard et al. (10) observed a decrease in oxygen uptake. This discrepancy may be reconciled on the basis of body composition. In the study of Martinez-Torres et al. (99) body composition was neither determined nor

controlled in the three groups of participants. Height to weight ratios were reported as 2.61, 2.66, and 3.03 cm·kg^{-1} for control, iron-deficient, and anemic individuals, respectively, which suggests that differences in body composition might be present among the groups. In the studies by Lukaski et al. (96) and Beard et al. (10) body composition did not change in women studied longitudinally. The importance of controlling body composition and, specifically, percent body fat in determining the metabolic response of humans to cold exposure is well established (22).

Copper

The nutritional essentiality of copper was first demonstrated when it was shown to be required, in addition to iron, for prevention of anemia in rats (69). Several other perturbations have been observed in copper-deficient animals, including skeletal demineralization, demyelination and degeneration of the nervous system, defective reproduction, connective tissue abnormalities, myocardial degeneration, lack of arterial elasticity, and depigmentation of hair and wool (38). Diet-induced copper deficiency in humans, principally men, has been shown to produce hypercholesterolemia, hypertension during stress, abnormal glucose metabolism, and abnormal electrocardiograms (82).

In the body, copper exists principally as a complex with some proteins. These proteins are recognized as cuproproteins because copper is a constituent of the molecular structure in which there is a characteristic ratio between moles of protein and atoms of copper. Because of these features, and the fact that the contained copper does not dissociate during the isolation of the protein, these cuproproteins are termed metalloproteins, many of which function as enzymes (100).

O'Dell (110) has summarized the copper-dependent enzymes and the resultant pathologies associated with restricted copper intake in animals and humans. The cuproenzymes are always involved in oxidation-reduction reactions. An important feature of these reactions is the involvement of molecular oxygen or a derivative species, such as the superoxide radical. Because of oxygen consumption, the reaction regulated by the cuproenzyme is usually rate-limiting and irreversible.

After absorption, copper is taken up by the liver and converted into ceruloplasmin, which is then released into the blood. In addition to donating copper to peripheral tissues for synthesis of copper-dependent enzymes, ceruloplasmin facilitates iron utilization. Frieden and Hsieh (50) proposed a unique ferroxidase activity for circulating ceruloplasmin; it is needed for the oxidation of ferrous iron for mobilization and transport of iron from tissue storage sites to peripheral tissues. In copper-deficient animals and humans, serum ceruloplasmin concentration is decreased (100). This defect in copper status and the resultant impairment in iron utilization may be important factors explaining the experimental observations of hypothermia in copper-deficient humans and animals.

Human Studies. In 1962, Menkes and colleagues (101) described in a family of five boys a syndrome characterized by slow growth, progressive cerebral degeneration with death by 2 yr of age, peculiar hair appearance, and X-linked recessive inheritance that was postulated to be the result of a metabolic defect. Other investigators (36, 48, 61, 62, 92, 93) also reported the same signs in male infants. It was noted that these signs were associated with reduced concentrations of copper and ceruloplasmin in the blood and a defect in intestinal absorption of copper (37, 62, 92). These signs have subsequently become known as Menkes' Kinky Hair syndrome because of the sparse, unpigmented, and wiry appearance of body hair (pili torti).

Another unique feature of this disease is hypothermia. Menkes et al. (101) briefly mentioned that "subnormal temperature occurred" in one of the five boys studied. Danks et al. (36), however, systematically reported low body temperatures (29°–35°C) in each of seven afflicted infants and suggested that hypothermia might be an important diagnostic symptom of the disease. The unexplained hypothermia was associated with limited spontaneous activity; hypotonia, drowsiness, and lethargy were presenting symptoms, and blood biochemical studies indicated neither anemia nor hypothyroidism as contributing factors.

Animal Studies. In a preliminary report, Hall et al. (66) examined the effects of a copper-deficient (0.7 $\mu g \cdot g^{-1}$) and a copper-adequate (10 $\mu g \cdot g^{-1}$) diet on rectal temperatures of weanling male rats maintained at 24°C. Hypothermia was observed after 35 days on the copper-deficient diet (Fig. 63.6). Plasma concentrations of T_4 and T_3 were decreased in the copper-deficient rats at approximately the time hypothermia was observed. Hepatic cytochrome oxidase activity, an index of copper status (13), also declined when dietary copper was restricted.

Mechanism of Action. The factors leading to hypothermia in copper deficiency are not well understood. It is known that anemia per se, whether resulting from inadequate dietary iron intake or an inability to utilize body iron stores, can impair thermogenesis by reducing both oxygen-carrying capacity and energy-producing, iron-dependent enzyme activities (33). Anemia is present in animals with diet-induced copper deficiency.

The hypothermia of copper deficiency may be explained in part by altered thyroid hormone metabo-

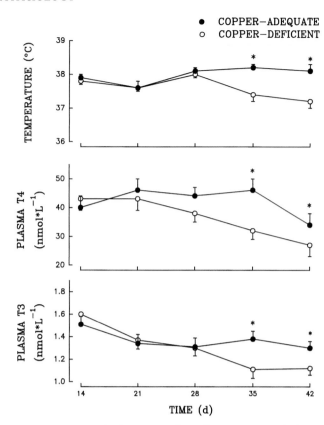

FIG. 63.6. Longitudinal measures of rectal temperature and plasma thyroxine (T_4) and triiodothyronine (T_3) in copper-deficient and copper-adequate male rats. Adapted from Hall et al. (55). *Significantly different ($p < 0.05$) than other treatments.

lism. As shown by Hall et al. (66), plasma T_4 and T_3 concentrations are reduced in copper deficiency. Allen et al. (3) reported no effect of copper status, determined by hepatic copper concentration, on the TSH response to exogenous TRH. However, copper-deficient rats demonstrated a blunted response in circulating T_4 concentrations during the 4 h period after TRH administration. Although not definitive, these findings suggest that copper deficiency may impair T_4 synthesis. The contribution of iron deficiency, which is known to depress plasma T_4 concentrations (8, 12), cannot be eliminated as a modulating factor in the hypothyroidism observed in copper-deficient rats.

In infants with Menkes' disease hypothermia is a prominent symptom in the absence of anemia and hypothyroidism (36, 101). One proposed factor in this hypothermia is reduced energy expenditure. French et al. (48) and Grover and Scrutton (63) observed reduced cytochrome c oxidase activity of skeletal muscles and brains of Menkes' patients. Cytochrome c oxidase is a copper-dependent enzyme located in mitochondria and involved in the utilization of oxygen for the production of energy in the body (13). Reduced cytochrome c oxidase activity has also been found in hypothermic, copper-deficient rats (66).

Altered catecholamine metabolism also may be a factor in the hypothermia of copper deficiency. Dopamine β-hydroxylase, a copper-containing enzyme located in the adrenal medulla and the CNS, catalyzes the conversion of dopamine to norepinephrine (51, 148). Increases in dopamine and concomitant decreases in norepinephrine concentrations of biological fluids and tissues of copper-deficient animals and humans have been reported (61, 83, 93, 102).

In studies of Menkes' patients, investigators have sought to demonstrate the benefit of parenteral copper administration on norepinephrine metabolism. Loyola and Dodson (93) reported an increased concentration of homovanillic acid (HVA) and a decreased concentration of methoxyhydroxyphenylethylene (MHPG) in the cerebral spinal fluid of a patient with Menkes' disease. The low concentration of MHPG and the high concentration of HVA suggest a deficiency of norepinephrine and an excess of dopamine in the CNS. Supplemental copper did not augment epinephrine synthesis, as indicated by the persistent imbalance between HVA and MHPG concentrations. A similar inability of parenteral copper to ameliorate the imbalance in norepinephrine synthesis was reported by Grover et al. (61).

Altered catecholamine metabolism has been observed in the brains of copper-deficient animals. Morgan and O'Dell (103) found that the whole-brain concentrations of both dopamine and norepinephrine were decreased in copper-deficient rats. The catecholamine concentrations correlated with decreased tyrosine hydroxylase activity, which suggests that copper deficiency might affect catecholamine production by reducing substrate availability. Feller and O'Dell (45) found that low copper status in rats decreases the dopamine concentration in the corpus striatum without affecting the norepinephrine concentration of the hypothalamus. Repletion of copper-deficient rats increased brain dopamine and norepinephrine concentrations, which suggests that a catalytic function has been restored (45). Thus apparent alterations in brain catecholamine metabolism associated with copper deficiency are not the cause of hypothermia because hypothalamic norepinephrine is not affected.

In addition to the CNS, copper deficiency influences peripheral tissue catecholamine metabolism. When dietary copper was restricted in weanling rats, cardiac norepinephrine content, synthesis, and turnover decreased significantly (125). Urinary norepinephrine excretion, normalized for creatinine output, was increased about 25% in copper-deficient rats.

A combination of blunted T_4 synthesis and reduced norepinephrine synthesis appears to explain the hypothermia observed at room temperature in copper-deficient animals. Each of these metabolic disruptions may be explained by apparent reductions in the activity of copper-dependent enzymes required for synthesis of thyroid hormone precursors and norepinephrine (51, 98, 148). In addition, reductions in mitochondrial cytochrome c oxidase activity (13), and hence decreased capacity for cellular oxidative processes, contribute to the hypothermia observed in copper deficiency.

Zinc

As an essential nutrient, zinc participates in regulatory, structural, and catalytic functions in many biological systems (150). As a widespread component in all cells, zinc is well suited for regulatory functions because its intracellular concentration can be homeostatically controlled in a tissue-specific manner (32). Structural and functional roles for many metalloenzymes, as well as a stabilizer of macromolecules, particularly receptor molecules and cell membranes, also have been described for zinc (68).

Although zinc is present in virtually all cells, its distribution is tissue-specific (68). Zinc concentrations in soft tissues, such as muscle and heart, are relatively unresponsive to amounts of dietary zinc over most ranges of intake. However, zinc in other tissues, such as bone, prostate, and testes, reflects more precisely dietary zinc. Extracellular fluids represent a small but relatively important pool of metabolically active zinc. These findings indicate that the principal function of zinc is as an intracellular ion.

Manifestations of dietary zinc deficiency are the result of altered zinc metabolism and zinc-dependent biochemical functions. Some classical features of severe zinc deficiency in animals and humans include retarded growth, anorexia, depressed immune function, altered reproductive performance, dermatitis, diarrhea, skeletal abnormalities, and alopecia (68). Many of these manifestations are related to the chronic energy deficits and impaired protein utilization associated with cyclic feeding behavior (28, 149).

Animal Studies.   The first evidence of an influence of zinc status on body temperature came from the observation of reduced rectal temperatures in postpartum zinc-deficient rats. O'Dell et al. (111) fed female rats diets low (<1 $\mu g \cdot g^{-1}$) in or supplemented (100 $\mu g \cdot g^{-1}$) with zinc during gestation. After parturition, rectal temperatures were significantly decreased in the zinc-deficient in comparison to the zinc-supplemented animals ($36.7 \pm 0.2°$ vs. $37.5 \pm 0.1°C$, respectively). Although the investigators concluded that an increased blood loss caused by prolonged labor in the zinc-deficient rats was responsible for the lowered body temperature, they also postulated that the thermogenic defect could be attributed to reduced zinc status.

Topping et al. (137) studied the effects of restricted

zinc intake on rectal temperature during acute cold exposure. Adult male rats were fed diets low (<0.5 μg·g⁻¹) in or supplemented (100 μg·g⁻¹) with zinc. To minimize the effects of reduced food intake, and hence energy restriction, associated with zinc deficiency, each zinc-supplemented animal was pair-fed the amount of diet consumed by the matched zinc-deficient rat. Animals were maintained at 22°C. Rectal temperature at room temperature was not affected after 2 wk of the low-zinc diet. Exposure to 2°C, however, resulted in an initial decrease in rectal temperature in both groups. Whereas the zinc-supplemented animals increased body temperature to 36.8° ± 0.1°C at 60 min, the zinc-deficient rats had a significant 35.8° ± 0.1°C decrease in temperature, which continued to decline during the next 30 min. In other rats fed the same zinc-restricted diet for 3 wk, rectal temperatures were significantly lower at room temperature in comparison to control animals (37.1° ± 0.1°C vs. 38.0° ± 0.1°C). During acute exposure to cold, the zinc-deficient animals had significantly decreased rectal temperatures.

An interesting finding of this study was the influence of nicotine treatment (0.5 mg·kg⁻¹ body weight) on body temperature. Nicotine-treated, zinc-deficient animals initially had lower rectal temperatures at room temperature. After more than 30 min following administration of nicotine, rectal temperatures increased. During cold exposure, rectal temperatures of zinc-deficient rats were significantly depressed until 60 min after nicotine administration, when they exceeded those of control animals.

The effects of dietary zinc and food restriction were also determined in weanling male rats (95). Control rats

were fed diets adequate in all nutrients, including a zinc concentration of 30 μg·g⁻¹. A second group was fed a similar diet but with a low zinc content (<1 μg·g⁻¹). A third group was pair-fed the control diet in amounts equal to that consumed by the matched zinc-deficient group. After 42 days, the animals were fasted, five animals from each group were killed after rectal temperature was measured, and ten animals from each group were exposed to 3°C for 4 h. Rectal temperatures were lower in the zinc-deficient rats at room temperature (23°C) and during cold exposure (Fig. 63.7). Plasma T₄ and T₃ concentrations were reduced in the zinc-deficient rats at room temperature (Table 63.2). Percent body fat was similar among all treatment groups. During cold exposure, TRH and thyroid hormone concentrations were decreased in the zinc-deficient animals. Plasma norepinephrine concentration was increased in the zinc-deficient rats after cold exposure.

Human Studies. The influence of zinc status on thermogenesis in humans has not been intensely studied. Factors, such as the requirement to use either formula diets or conventional foods low in zinc for prolonged periods of time and the difficulty in achieving severe zinc deficiency in adults, have been limitations for investigators.

Wada and King (146) determined the effects of zinc depletion on human energy metabolism. Basal metabolic rate decreased significantly in a group of six young men consuming a diet low in zinc (5.5 mg·day⁻¹) in comparison to a diet containing 16.5 mg·day⁻¹ of zinc (Table 63.3). Serum TSH, T₄, and free T₄ concentrations tended to decrease when zinc intake was low and

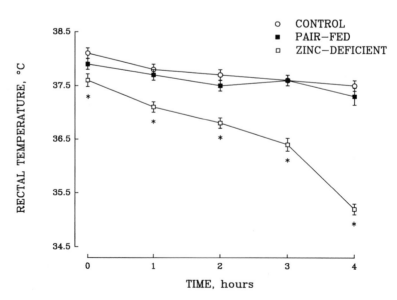

FIG. 63.7. Rectal temperature in zinc-deficient, control (zinc-adequate), and pair-fed rats exposed to a 4°C cold stress. From Lukaski et al. (88), with permission. *Significantly different (p <0.05) than other treatments.

TABLE 63.2. *Effects of Zinc Deficiency (ZD) and Pair-Feeding (PF) on Plasma Concentrations of Thyrotropin (TSH), Total Thyroxine, Total Triiodothyronine, and Norepinephrine of Rats Maintained at Room Temperature (23°C) and Exposed to Cold (3°C) for 4 h.*

Condition	n	Group	TSH (mU·liter^{-1})	Total T$_4$ (nmol·liter^{-1})	Total T$_3$ (nmol·liter^{-1})	Norepinephrine (pmol·liter^{-1})
23°C	5	Control	66 ± 4	33 ± 4	1.2 ± 0.1	—
		ZD	53 ± 4	22 ± 3*	0.9 ± 0.1	—
		PF	59 ± 4	30 ± 3	0.9 ± 0.1	—
3°C	10	Control	76 ± 2	49 ± 3	1.7 ± 0.1	30 ± 0.3
		ZD	58 ± 2*	28 ± 2*	1.1 ± 0.1*	45 ± 0.5*
		PF	68 ± 2	42 ± 3	1.6 ± 0.1	32 ± 0.3

*Significantly ($P < 0.05$) different from other treatment groups at a given temperature.
Values are mean ± SE. Adapted from Lukaski et al. (95).

increase when zinc intake was increased; only the decrease in serum T$_3$ was significant.

Mechanisms of Action. One explanation for the reduced basal and cold-induced heat production in zinc-deficient animals and humans is altered thyroid hormone metabolism. The role of zinc nutriture on thyroid hormone metabolism has been examined in terms of central and peripheral functions.

Initial observations in zinc-deficient weanling rats indicated a potential impairment in the hypothalamic–pituitary–thyroid axis (104). Zinc deficiency and restricted energy intake induced by pair-feeding were characterized by growth depression. Both the pair-fed and zinc-deficient rats had reduced serum T$_3$ and T$_4$ concentrations compared to control animals. However, the T$_3$ concentrations were significantly lower in the zinc-deficient rats than in the energy-restricted animals. Serum TSH concentrations were similar among control, pair-fed, and zinc-deficient animals. Hypothalamic TRH concentration, however, was significantly reduced in the zinc-deficient rats. These initial findings indicate that zinc deficiency reduces circulating T$_3$ concentration more than does semistarvation resulting from pair-feed-

ing and suggest a possible inhibition of peripheral metabolism of T$_4$ in zinc deficiency. They also suggest a reduced hypothalamic biosynthesis of TRH in zinc deficiency.

Attempts to delineate the effects of zinc status on peripheral monodeiodinase activity have produced conflicting results. Fujimoto et al. (52) examined the effects of graded zinc deficiency in weanling male rats fed very low (<2 μg·g^{-1}) and moderate (6.1 μg·g^{-1}) zinc intakes, energy-restricted by pair-feeding, and zinc-supplemented (90 μg·g^{-1}) diets for 5 wk. Serum T$_3$ and T$_4$ concentrations were significantly reduced only in the rats fed the severely zinc-deficient diet. Hepatic 5'-D-I activity was significantly reduced in both groups of rats fed the zinc-deficient diets in comparison to the control animals; zinc-deficient and corresponding pair-fed animals had similar estimates of conversion of T$_4$ to T$_3$.

In contrast, Oliver et al. (114) reported increased hepatic 5'-D-I activity in zinc-deficient rats. Young adult male rats weighing 150–180 g were fed a zinc-deficient diet (5 μg·g^{-1}) and a diet supplemented with zinc (50 μg·g^{-1}). A small but significant effect of dietary zinc was noted on growth. Serum TSH, T$_4$, and T$_3$ con-

TABLE 63.3. *Effects of Zinc Intake on Basal Metabolic Rate (BMR) and Serum Concentrations of TSH, Thyroxine (T$_4$), Free T$_4$ (FT$_4$), and Triiodothyronine (T$_3$) in Six Young Men*[a]

	Zinc Intake (mg·day^{-1})			
	16.5 (11)[b]	5.5 (28)[b]	5.5 (54)[b]	16.5 (9)[b]
BMR, kcal·(kg·h)$^{-1}$	1.00 ± 0.06[c]	0.98 ± 0.07[c]	0.91 ± 0.05[d]	0.95 ± 0.05[c,d]
TSH, mU·liter^{-1}	3.2 ± 0.5	2.4 ± 0.4	2.3 ± 0.4	3.2 ± 0.5
T$_4$, nmol·liter^{-1}	109 ± 11	85 ± 4	90 ± 5	100 ± 12
FT$_4$, nmol·liter^{-1}	19 ± 1[c,e]	15 ± 1[d]	16 ± 1[d,e]	19 ± 14[c]
T$_3$, nmol·liter^{-1}	2.9 ± 0.2	3.0 ± 0.3	3.0 ± 0.2	3.3 ± 0.3

[a]Adapted from Wada and King (146). [b]Values in parentheses indicate number of days on diet. [c-e]Values within a row with different superscripts are significantly different ($P < 0.05$).
Values are mean ± SE.

centrations were not affected by dietary zinc. Hepatic 5'-D-I activity was significantly increased in the zinc-deficient rats (0.23 ± 0.05 vs. 0.61 ± 0.05 ng $T_3 \cdot$mg protein^{-1}).

Differences between the two reported studies indicate that age or biological maturity of the animals fed the zinc-deficient diets and dietary zinc are independent factors affecting thyroid hormone metabolism. For example, zinc deficiency has minimal effects on the hypothalamic–pituitary axis of sexually mature male rats, whereas in sexually immature male rats zinc deprivation leads to impairment of gonadal growth and increases in pituitary TSH and decreases in serum TSH concentrations (123).

Attempts to determine the effect of zinc deficiency on thyroid hormone metabolism have focused on brain TRH metabolism because it has been established that TRH biosynthesis requires the zinc metalloenzyme carboxypeptidase H, also known as carboxypeptidase E or enkaphein convertase (49). The effects of 6 wk of graded zinc intake (<1, 6, and 30 μg$\cdot$g^{-1}) and caloric restriction on TRH biosynthesis in the hypothalamus, brain, and pituitary of young male rats were studied (116). It was determined that severe zinc deficiency selectively reduced the hypothalamic and pituitary total TRH-Gly and other small peptide precursors of TRH-Gly. Pituitary TRH and TRH precursor peptide concentrations were also decreased by zinc deficiency and pair-feeding. These findings indicate that activity of zinc-dependent enzymes, such as carboxypeptidase H, are depressed by severe zinc deficiency and caloric restriction. Thus zinc deficiency influences brain TRH biosynthesis.

Evidence suggests that zinc plays a fundamental role in hormone metabolism at the genomic level. A functionally heterogenous group of DNA-binding proteins, or transcription factors, required for gene replication and translation is known to contain zinc; the function of zinc is to provide conformational or structural stability (141). These chemical structures are termed zinc fingers, zinc clusters, and zinc twists; they fasten onto target genes by binding to specific sequences of DNA base pairs (140). One example is transcriptase factor IIIA, which has been shown to decrease in experimental zinc deficiency, resulting in impaired energy utilization and stunted growth in animals (14, 57).

Zinc also plays a role at the level of the cell membrane, where it acts as a regulator for hormone binding at nuclear receptors. Zinc promotes the binding of androgens to receptors in nuclear membranes (29) and regulates T_3 binding to rc-*erb*A-α and rc-*erb*A-β proteins (94). These findings indicate potential biological mechanisms of zinc deficiency on hormone metabolism and remain to be examined in relation to altered thermogenesis.

Iodine

The principal role of iodine in animal and human nutrition is its absolute requirement for normal growth and development (73). This essential role is evidenced by the incorporation of iodine as a structural component of T_4 and T_3. In addition, restricted dietary iodine has been shown to cause hypothermia in animals challenged by a cold environment.

Animal Studies. Iodine-deficient rats have an increased sensitivity to cold exposure. Riesco et al. (122) reported that five of eight young adult male rats fed diets low in iodine (15–20 ng$\cdot$g^{-1}) for 3–4 months and exposed to 4°–5°C for 25 days experienced severe hypothermia and mortality. This impairment in thermoregulation was associated with undetectable serum T_4 and T_3 concentrations, reduced by more than half of control values. In the three rats that survived the cold stress when fed the low-iodine diet, serum T_3 concentrations were not adversely affected.

Other studies have found two different responses to low-iodine diets: rats which are cold-sensitive (that is, hypothermia and death as a result of cold exposure) and rats which are cold-resistant (107, 122). This effect may not be related to maintenance of plasma T_3 concentrations because there was no correlation between plasma T_3 and cold tolerance and may be the result of an increased activity of BAT 5'-D-II (127). It has been suggested that synthesis of thermogenin, also known as UCP, in BAT was more prominent in cold-resistant rats (115).

An enigma surrounding the apparently conflicting evidence on the effect of dietary iodine deficiency on cold tolerance is the variance in the nutritional adequacy of the experimental diets. The requirement for a very small amount of dietary iodine in a semipurified animal diet poses a difficult problem because the iodine content of macronutrient components (for example, carbohydrate and fat sources) is highly variable based on the geographic site of production. One attempt to standardize an experimental diet low in iodine concentration is the Remington Diet (120). Unfortunately, this diet has been criticized by investigators because of significant batch-to-batch variability in the iodine content (121) and differences in iodine content in diets obtained from different suppliers (112). In addition, the diet may be deficient in essential nutrients (115, 122). This problem is highlighted in a study using the Remington Diet (115), where investigators reported a small, albeit significant, increase in BAT monodeiodinase activity after cold exposure in rats fed the diet supplemented with potassium iodine in the drinking water. In contrast, other investigators using a semipurified or purified diet adequate in iodine observed a more than ten-fold increase in BAT 5′-D-II activity after cold exposure (79, 80, 86, 127). These inconsistent observations in physiological response suggest a possible micronutrient deficiency in the basal diet.

Mechanism of Action. Dietary iodine deficiency causes depletion of thyroid iodine stores with reduced production of T_4. Reduced circulating T_4 stimulates secretion of TSH, which increases thyroid activity and results in thyroid hyperplasia and goiter (1, 53, 107, 109, 122). Plasma concentrations of T_3 are often normal because of both increased efficiency of conversion from T_4 and an increased rate of T_3 secretion from the thyroid gland (1, 59, 115).

SUMMARY

Inadequate intakes or specific genetic diseases resulting in deficient states of some vitamins and minerals can cause impaired thermoregulatory function and hypothermia in the cold. Many of these deficiencies are associated with altered function or metabolism of thyroid hormones or the SNS.

Among the vitamins, only thiamin deficiency has been associated with altered temperature regulation. Thiamin-deficient humans and rodents demonstrate a profound hypothermia at room temperature. The apparent mechanism involves altered neurotransmitter function because of reduced serotonin synthesis in the CNS.

Deficiencies of copper, iron, and zinc elicit hypothermia either at room temperature or during exposure to cold air. One common observation is hypothyroidism. However, the site of the induced metabolic lesion appears to be nutrient-specific. In zinc deficiency, thyroid hormone metabolism is impaired centrally with reduced synthesis of TRH precursors in the hypothalamus and peripherally with altered hepatic 5′-D-I activity. An attenuated T_4 production from the thyroid has been demonstrated in copper-deficient rats. Both thyroidal output of T_4 and peripheral conversion of T_4 to T_3 are decreased in iron-deficient rats. Evidence suggests similar responses in humans whose diets were restricted in zinc and iron, respectively.

Trace element deficiencies also are associated with altered catecholamine metabolism. Zinc-deficient rats exposed to cold air have increased circulating norepinephrine. Tissue norepinephrine contents and fractional turnover rates were decreased in copper-deficient rats. Decreased circulating and tissue norepinephrine are explained by the reduced activity of the copper-dependent enzyme β-hydroxylase. Similarly, cardiac norepinephrine content was reduced in iron-deficient rats. However, cardiac and BAT fractional norepinephrine turnover rates were increased in iron deficiency. Thus

catecholamine metabolism is attenuated in copper deficiency, resulting in hypothermia at room temperature. Adrenergic function is augmented at room temperature and during exposure to cold in an attempt to overcome the hypothyroidism of iron deficiency.

Iodine deficiency is clearly associated with inadequate thyroidal synthesis of T_4.

Findings in the literature clearly indicate that inadequate intake of specific vitamins and minerals results in altered thermogenesis and temperature regulation in animals and humans. Preliminary evidence suggests that alterations in SNS activity and thyroid hormone status may explain these observations. However, additional mechanisms remain to be investigated.

Because the conversion of T_4 to T_3 is impaired in trace element deficiencies, it remains to determined whether zinc, copper, or iron directly influence the activity of the 5′-D-I and 5′-D-II enzymes. In vitro studies are needed to identify the chemical factors (for example, iron, copper, and selenium) regulating the reduced sulphydryl groups, perhaps via a nascent autoreduction system, required for optimal monodeiodinase function (144).

Similarly, additional data are needed to evaluate the effects of therapeutic thyroid hormone supplementation on thyroidal and peripheral deiodinase activity. This approach would provide independent evidence for remediation of impaired T_4 to T_3 conversion in specific nutrient deficiencies and an opportunity to demonstrate amelioration of altered SNS activity.

Although a potential role for altered SNS activity has been demonstrated in copper and iron deficiency, there remains the uncertainty of whether the focus of impaired SNS activity is in the central or the peripheral nervous system. Studies are needed to examine the role of trace mineral deficiencies on the adrenergic regulation of peripheral blood flow during exposures to cold and perhaps hot environments.

In addition to the vitamins and minerals discussed above, other nutrients such as folacin (27, 90), vitamin B_{12} (78, 113), vanadium (139), and selenium (4, 15–17) with reported influences on thyroid hormone or catecholamine metabolism remain to be studied to evaluate their individual impacts on energy metabolism. Determination of the physiological roles of nutrients, and in particular the impairments associated with graded and marginal intakes, is needed to assist in the estimation of human nutrient requirements to optimize human physiological and behavioral performance throughout the life cycle.

REFERENCES

1. Abrams, G. M., and P. R. Larsen. Triiodothyronine and thyroxine in the serum and thyroid glands of iodine deficient rats. *J. Clin. Invest.* 52: 2522–2531, 1973.

2. Ackerman, W. J. Stupor, bradycardia, hypotension and hypothermia: a presentation of Wernicke's encephalopathy with rapid response to thiamine. *West. J. Med.* 121: 428–429, 1974.

3. Allen, D. K., C. A. Hassel, and K. Y. Lei. Function of pituitary–thyroid axis in copper-deficiency rats. *J. Nutr.* 112: 2043–2046, 1982.

4. Arthur, J. R. The role of selenium in thyroid hormone metabolism. *Can. J. Physiol. Pharmacol.* 69: 1648–1652, 1991.

5. Beard, J., C. A. Finch, and W. L. Green. Interactions of iron deficiency, anemia, and thyroid hormone levels in the response of rats to cold exposure. *Life Sci.* 30: 691–697, 1982.

6. Beard, J., W. Green, L. Miller, and C. Finch. Effect of iron-deficiency anemia on hormone levels and thermoregulation during cold exposure. *Am. J. Physiol.* 247(*Regulatory Integrative Comp. Physiol.* 18): R114–R119, 1984.

7. Beard, J., and B. Tobin. Feed efficiency and norepinephrine turnover in iron deficiency. *Proc. Soc. Exp. Biol. Med.* 184: 337–344, 1987.

8. Beard, J., B. Tobin, and W. Green. Evidence for thyroid hormone deficiency in iron-deficient anemic rats. *J. Nutr.* 119: 772–778, 1989.

9. Beard, J., B. Tobin, and S. M. Smith. Norepinephrine turnover in iron deficiency at three environmental temperatures. *Am. J. Physiol.* 255(*Regulatory Integrative Comp. Physiol.* 26): R90–R96, 1988.

10. Beard, J. L., M. J. Borel, and J. Derr. Impaired thermoregulation and thyroid function in iron-deficiency anemia. *Am. J. Clin. Nutr.* 52: 813–819, 1990.

11. Beard, J. L., and S. H. Smith. Re-evaluation of apparent hypometabolism in iron-deficient rats. *J. Nutr. Biochem.* 3: 298–303, 1992.

12. Beard, J. L., B. W. Tobin, and S. M. Smith. Effects of iron repletion and correction of anemia on norepinephrine turnover and thyroid metabolism in iron deficiency. *Proc. Soc. Exp. Biol. Med.* 193: 306–312, 1990.

13. Beinert, H. Cytochrome c oxidase: present knowledge of the state and function of its copper components. In: *The Biochemistry of Copper*, edited by J. Peisach, P. Aisen, and W. E. Blumberg. New York: Academic, 1966, p. 213–234.

14. Berg, J. Potential metal-binding domains in nucleic acid binding proteins. *Science* 252: 485–493, 1986.

15. Berry, M. J., A.-L. Kates, and P. R. Larsen. Thyroid hormone regulates type I deiodinase messenger RNA in rat liver. *Mol. Endocrinol.* 4: 743–748, 1990.

16. Berry, M. J., J. D. Kiefer, and P. R. Larsen. Evidence that cysteine, not selenocysteine, is in the catalytic site of type II iodothyronine deiodinase. *Endocrinology* 129: 550–552, 1991.

17. Berry, M. J., and P. R. Larsen. The role of selenium in thyroid hormone action. *Endocr. Rev.* 13: 207–219, 1992.

18. Bhagat, B., and M. F. Lockett. The failure of thyroxine to influence the oxygen consumption of thiamine-deficient mice. *J. Endocrinol.* 23: 227–230, 1961.

19. Bianco, A. C., and J. E. Silva. Intracellular conversion of thyroxine to triiodothyronine is required for the optimal thermogenic function of brown adipose tissue. *J. Clin. Invest.* 79: 295–300, 1987.

20. Bilezikian, J. P., and J. N. Loeb. The influence of hyperthyroidism and hypothyroidism on α- and β-adrenergic receptor systems and adrenergic responsiveness. *Endocr. Rev.* 4: 378–388, 1987.

21. Borel, M. J., S. H. Smith, D. E. Brigham, and J. L. Beard. The impact of varying degrees of iron deficiency of iron nutriture on several functional consequences of iron deficiency in rats. *J. Nutr.* 121: 729–736, 1991.

22. Buskirk, E. R., R. H. Thompson, and G. D. Whedon. Metabolic response to cold air in men and women in relation to total body fat content. *J. Appl. Physiol.* 18: 603–612, 1963.

23. Butterworth, R. F. Thiamin malnutrition and brain development. In: *Current Topics in Nutrition and Disease*, edited by D. K. Raskin, B. Haber, and B. Drujan. New York: Liss, 1987, vol. 6, p. 287–304.

24. Campbell, A. C. P., and W. R. Russell. Wernicke's encephalopathy: the clinical features and their probable relationship to vitamin B deficiency. *Q. J. Med.* 10: 41–64, 1941.

25. Cannon, B., and J. Nedergaard. The biochemistry of an inefficient tissue: brown adipose tissue. *Essays Biochem.* 2: 110–164, 1985.

26. Carvalho, S. D., E. T. Kimura, A. C. Bianco, and J. E. Silva. Central role of brown adipose tissue thyroxine 5′-deiodinase on thyroid hormone-dependent thermogenic response to cold. *Endocrinology* 128: 2149–2159, 1991.

27. Chan, M. M.-S., and E. L. R. Stokstad. Metabolic responses of folic acid and related compounds to thyroxine in rats. *Biochim. Biophys. Acta* 632: 244–253, 1980.

28. Chesters, J. K., and M. Will. Some factors controlling food intake by zinc-deficient rats. *Br. J. Nutr.* 30: 555–566, 1973.

29. Colvard, D. S., and E. M. Wilson. Zinc potentiation of androgen receptor binding to nuclei in vitro. *Biochemistry* 23: 2371–2378, 1984.

30. Cook, J. D., and C. A. Finch. Assessing iron status of a population. *Am. J. Clin. Nutr.* 32: 2115–2119, 1979.

31. Cook, J. D., and S. R. Lynch. The liabilities of iron deficiency. *Blood* 68: 803–809, 1986.

32. Cousins, R. J. Absorption, transport, and hepatic metabolism of copper and zinc: special reference to metallothionein and ceruloplasmin. *Physiol. Rev.* 65: 238–309, 1985.

33. Dallman, P. R. Biochemical basis for the manifestations of iron deficiency. *Annu. Rev. Nutr.* 6: 13–40, 1986.

34. Dallman, P. R. Iron. In: *Present Knowledge in Nutrition* (6th ed.), edited by M. L. Brown. Washington, DC: Int. Life Sci. Inst., Nutr. Found., 1990, p. 241–250.

35. Dallman, P. R., R. Yip, and C. Johnson. Prevalence and causes of anemia in the United States, 1976–1980. *Am. J. Clin. Nutr.* 39: 437–445, 1984.

36. Danks, D. M., P. E. Campbell, B. J. Stevens, V. Mayne, and E. Cartwright. Menke's kinky hair syndrome: an inherited defect in copper absorption with widespread effects. *Pediatrics* 50: 188–201, 1972.

37. Danks, D. M., B. J. Stevens, P. E. Campbell, J. M. Gillespie, J. Walker-Smith, J. Blomfield, and B. Turner. Menkes' kinky-hair syndrome. *Lancet* 1: 1100–1102, 1972.

38. Davis, G. K., and W. Mertz. Copper. In: *Trace Elements in Human and Animal Nutrition* (5th ed.), edited by W. Mertz. Orlando, FL: Academic, 1987, vol. 2, p. 301–364.

39. DeMaeyer, E., M. Adielis-Tegman, and E. Rayston. The prevalence of anemia in the world. *World Health Stat. Q.* 38: 302–316, 1985.

40. Di Liegro, I., G. Savettieri, and A. Cestelli. Cellular mechanism of action of thyroid hormones. *Differentiation* 35: 165–175, 1987.

41. Dillman, E., C. Gale, W. Green, D. G. Johnson, B. Mackler, and C. Finch. Hypothermia in iron deficiency due to altered triiodothyronine metabolism. *Am. J. Physiol.* 239(*Regulatory Integrative Comp. Physiol.* 10): R377–R381, 1980.

42. Dillman, E., D. G. Johnson, J. Martin, B. Mackler, and C. Finch. Catecholamine elevation in iron deficiency. *Am. J. Physiol.* 237(*Regulatory Integrative Comp. Physiol.* 8): R297–R300, 1979.

43. Dullo, A. G., and D. S. Miller. Obesity: a disorder of the sympathetic nervous system. *World Rev. Nutr. Diet.* 50: 1–56, 1987.

44. Expert Scientific Working Group. Summary of a report on assessment of the iron nutritional status of the United States. *Am. J. Clin. Nutr.* 42: 1318–1330, 1985.

45. Feller, D. J., and B. L. O'Dell. Dopamine and norepinephrine in discrete areas of the copper-deficient rat brain. *J. Neurochem.* 34: 1259–1263, 1980.

46. Fregly, M. J. Activity of the hypothalamic–pituitary–thyroid axis during exposure to cold. In: *Thermoregulation: Physiology and Biochemistry,* edited by E. Schonbaum and P. Lomax. New York: Pergamon, 1990, p. 437–494.

47. Fregly, M. J., F. P. Field, M. J. Katovich, and G. C. Barney. Catecholamine–thyroid hormone interaction in cold-acclimated rats. *Federation Proc.* 38: 2162–2169, 1979.

48. French, J. H., E. S. Sherard, H. Lubell, M. Brotz, and C. L. Moore. Tricho-poliodystrophy. I. Report of a case and biochemical studies. *Arch. Neurol.* 26: 229–244, 1972.

49. Fricker, L. D., C. J. Evans, F. S. Esch, and F. Herbert. Cloning and sequence analysis of cDNA for bovine carboxypeptidase E. *Nature* 323: 461–464, 1986.

50. Frieden, E., and H. S. Hsieh. Ceruloplasmin: the copper transport protein with essential oxidase activity. *Adv. Enzymol.* 44: 187–236, 1976.

51. Friedman, S., and S. Kaufman. 3-4-dihydroxyphenylethylamine β-hydroxylase: a copper protein. *J. Biol. Chem.* 240: 552–554, 1965.

52. Fujimoto, S., Y. Indo, A. Higashi, I. Matsuda, N. Kashiwabara, and I. Nakashima. Conversion of thyroxine into tri-iodothyronine in zinc-deficient rat liver. *J. Pediatr. Gastroenterol. Nutr.* 5: 799–805, 1986.

53. Fukuda, H., N. Yasuda, M. A. Greer, M. Kutas, and S. E. Greer. Changes in plasma thyroxine, triiodothyronine and TSH during adaptation to iodine deficiency in rats. *Endocrinology* 97: 307–314, 1975.

54. Gale, C. C. Neuroendocrine aspects of thermoregulation. *Annu. Rev. Physiol.* 35: 391–430, 1973.

55. Galton, V. A. Thyroid hormone catecholamine interrelationships. *Endocrinology* 77: 278–284, 1965.

56. Gibson, G., L. Barclay, and J. Blass. The role of the cholinergic system in thiamin deficiency. *Ann. N. Y. Acad. Sci.* 305: 382–403, 1982.

57. Giedroc, D. P., K. M. Keating, C. T. Martin, K. R. Williams, and J. E. Coleman. Zinc metalloproteins involved in replication and transcription. *J. Inorg. Biochem.* 28: 155–169, 1986.

58. Goswami, A., and I. N. Rosenberg. Purification and partial characterization of iodothyronine-5'-deiodinase from rat liver microsomes. *Biochem. Biophys. Acta* 173: 6–12, 1990.

59. Greer, M. A., Y. Grimm, and H. Studer. Qualitative changes in the secretion of thyroid hormones induced by iodine deficiency. *Endocrinology* 83: 1193–1198, 1968.

60. Groeneveld, D., H. G. W. Smeets, P. M. Kabra, and P. R. Dallman. Urinary catecholamines in iron-deficient rats at rest and following surgical stress. *Am. J. Clin. Nutr.* 42: 263–269, 1985.

61. Grover, W. D., R. I. Henkin, M. Schwartz, N. Brodsky, E. Hobdell, and J. M. Stolk. A defect in catecholamine metabolism in kinky-hair disease. *Ann. Neurol.* 12: 263–266, 1982.

62. Grover, W. D., W. C. Johnson, and R. I. Henkin. Clinical and biochemical aspects of trichopoliodystrophy. *Ann. Neurol.* 5: 65–71, 1979.

63. Grover, W. D., and M. C. Scrutton. Copper therapy in trichopoliodystrophy. *J. Pediatr.* 86: 216–220, 1975.

64. Gubler, C. J. Thiamin. In: *Handbook of Vitamins* (2nd ed.), edited by L. J. Machlin. New York: Dekker, 1990, p. 233–281.

65. Haas, R. H. Thiamin and the brain. *Annu. Rev. Nutr.* 8: 483–515, 1988.

66. Hall, C. B., C. Perreault, and H. C. Lukaski. A low copper diet influences thermoregulation of the rat. *Proc. N. D. Acad. Sci.* 44: 109, 1990.

67. Hallberg, L., C. Benggston, L. Garby, J. Lennartsson, L. Rossander, and E. Tibblin. An analysis of factors leading to a reduction of iron deficiency in Swedish women. *Bull. WHO* 57: 947–954, 1979.

68. Hambidge, K. M., C. E. Casey, and N. F. Krebs. Zinc. In: *Trace Elements in Human and Animal Nutrition* (5th ed.), edited by W. Mertz. Orlando, FL: Academic, 1986, vol. 2, p. 1–137.

69. Hart, E. B., H. Steenbock, J. Waddell, and C. A. Elvehjem. Iron in nutrition. VII. Copper as a supplement to iron for hemoglobin building in the rat. *J. Biol. Chem.* 77: 797–812, 1928.

70. Hellon, R. F. Monoamine pyrogens and cations: their actions on central control of body temperature. *Pharmacol. Rev.* 26: 381–395, 1975.

71. Henderson, G. I., A. M. Hoyumpa, and S. Schenker. Effects of thiamine deficiency on cerebral and visceral protein synthesis. *Biochem. Pharmacol.* 27: 1677–1683, 1978.

72. Henderson, G. I., and S. Schenker. Reversible impairment of cerebral DNA synthesis in thiamine deficiency. *J. Lab. Clin. Med.* 86: 77–90, 1975.

73. Hetzel, B. S. Iodine deficiency: an international public health problem. In: *Present Knowledge in Nutrition,* (6th ed.), edited by M. L. Brown. Washington, DC: *Int. Life Sci. Inst., Nutr. Found.,* 1990, p. 308–313.

74. Himms-Hagen, J. Brown adipose tissue thermogenesis: role in thermoregulation, energy regulation and obesity. In: *Thermoregulation: Physiology and Biochemistry,* edited by E. Schonbaum and P. Lomax. New York: Pergamon, 1990, p. 327–414.

75. Himms-Hagen, J. Brown adipose tissue and cold-acclimation. In: *Brown Adipose Tissue,* edited by P. Trayhurn and D. G. Nichols. London: Arnold, 1986, p. 214–268.

76. Hunter, J. M. Hypothermia and Wernicke's encephalopathy. *Br. J. Med.* 2: 563–564, 1976.

77. Iber, F. L., J. P. Blass, and M. Brin. Thiamin in the elderly—relation to alcoholism and to neurological degenerative disease. *Am. J. Clin. Nutr.* 36: 1067–1082, 1982.

78. Kasbekan, D. K., W. V. Lavate, D. V. Rege, and A. Sreenivasan. A study of vitamin B$_{12}$ protection in experimental thyrotoxicosis in the rat. *Biochem. J.* 72: 374–383, 1959.

79. Kates, A.-L., and J. Himms-Hagen. Defective cold-induced stimulation of thyroxine 5'-deiodinase in brown adipose tissue of the genetically obese (ob/ob) mouse. *Biochem. Biophys. Res. Commun.* 130: 188–193, 1985.

80. Kates, A.-L., and J. Himms-Hagen. Defective regulation of thyroxine 5'-deiodinase in brown adipose tissue of (ob/ob) mice. *Am. J. Physiol.* 258(*Endocrinol. Metab.* 21): E7–E15, 1990.

81. Kearsley, J. H., and A. F. Musso. Hypothermia and coma in the Wernicke-Korsakoff syndrome. *Med. J. Aust.* 2: 504–506, 1980.

82. Klevay, L. M. Ischemic heart disease as copper deficiency. In: *Copper Bioavailability and Metabolism,* edited by C. Kies. New York: Plenum, 1990, p. 197–208.

83. Klevay, L. M., D. B. Milne, and J. C. Wallwork. Comparison of some indices of copper deficiency in growing rats. *Nutr. Rep. Int.* 31: 963–971, 1985.

84. Koeppen, A. H., J. C. Daniels, and K. D. Barron. Subnormal body temperatures in Wernicke's encephalopathy. *Arch. Neurol.* 21: 493–498, 1969.

85. Kohrle, J., G. Brabant, and R.-D. Hesch. Metabolism of thyroid hormones. *Horm. Res.* 26: 58–78, 1987.

86. Kopecky, J., L. Sigurdson, I. R. A. Park, and J. Himms-Hagen. Thyroxine 5'-deiodinase in brown adipose tissue of myopathic hamsters. *Am. J. Physiol.* 251(*Endocrinol. Metab.* 14): E8–E13, 1986.

87. Landsberg, L., M. E. Saville, and J. B. Young. Sympathoadrenal

system and regulation of thermogenesis. *Am. J. Physiol.* 247(*Endocrinol. Metab.* 10): E181–E189, 1984.

88. Landsberg, L., and J. B. Young. Autonomic regulation of thermogenesis. In: *Mammalian Thermogenesis,* edited by L. Girardier and M. J. Stock. London: Chapman and Hall, 1983, p. 99–140.

89. Leonard, J. L., and T. J. Visser. Biochemistry of deiodination. In: *Thyroid Hormone Metabolism,* edited by G. Hennemann. New York: Dekker, 1986, p. 189–253.

90. Lindenbaum, J., and F. A. Klipstein. Folic acid clearances and basal serum folate levels in patients with thyroid disease. *J. Clin. Pathol.* 17: 666–670, 1964.

91. Lipton, J. M., H. Payne, H. R. Garza, and R. N. Rosenberg. Thermolability in Wernicke's encephalopathy. *Arch. Neurol.* 35: 750–753, 1978.

92. Lott, I. T., R. DiPaolo, D. Schwartz, S. Janowska, and J. N. Kanfer. Copper metabolism in the steely-hair syndrome. *N. Engl. J. Med.* 292: 197–199, 1975.

93. Loyola, M. A., and W. E. Dodson. Metabolic consequences of trichopoliodystrophy. *J. Pediatr.* 98: 588–591, 1981.

94. Lu, C., Y. C. Chan, and P. G. Walfish. Selective effect of zinc compared to other divalent metals on L-triiodothyronine binding to rat c-*erb*A α and β proteins. *Biochem. Int.* 21: 191–198, 1990.

95. Lukaski, H. C., C. B. Hall, and M. J. Marchello. Impaired thyroid hormone status and thermoregulation during cold exposure of zinc-deficient rats. *Horm. Metab. Res.* 24: 363–366, 1992.

96. Lukaski, H. C., C. B. Hall, and F. H. Nielsen. Thermogenesis and thermoregulatory function of iron-deficient women without anemia. *Aviat. Space Environ. Med.* 61: 913–920, 1990.

97. Macaron, C., S. Feero, and M. Goldflies. Hypothermia in Wernicke's encephalopathy. *Postgrad. Med.* 65: 241–243, 1979.

98. Mains, R. E., I. M. Dickerson, D. A. Stoffers, S. N. Perkins, L. H. Ouafik, E. J. Husten, and B. A. Eipper. Cellular and molecular aspects of peptide hormone biosynthesis. *Front. Neuroendocrinol.* 11: 52–89, 1990.

99. Martinez-Torres, C., L. Cubeddu, E. Dillman, G. Brengelmann, I. Leets, M. Layrisse, D. G. Johnson, and C. Finch. Effect of exposure to low temperature on normal and iron-deficient subjects. *Am. J. Physiol.* 246(*Regulatory Integrative Comp. Physiol.* 17): R380–R383, 1984.

100. Mason, K. S. Copper metabolism and requirements of man. *J. Nutr.* 109: 1979–2066, 1979.

101. Menkes, J. H., M. Alter, G. K. Steigleder, D. R. Weakley, and J. H. Sung. A sex-linked recessive disorder with retardation of growth, peculiar hair, and focal cerebral and cerebellar degeneration. *Pediatrics* 29: 764–779, 1962.

102. Missala, K., K. Lloyd, G. Gregoriads, and T. L. Sourkes. Conversion of ^{14}C-dopamine to cardiac ^{14}C-noradrenaline in the copper-deficient rat. *Eur. J. Pharmacol.* 1: 6–10, 1967.

103. Morgan, R. F., and B. L. O'Dell. Effect of copper on the concentration of catecholamines and related enzyme activities in the rat brain. *J. Neurochem.* 28: 207–213, 1977.

104. Morley, J. E., J. Gordon, and J. M. Hershman. Zinc deficiency, chronic starvation, and hypothalamic–pituitary–thyroid function. *Am. J. Clin. Nutr.* 33: 1767–1770, 1980.

105. Myers, R. D. Hypothalamic actions of 5-hydroxytryptamine neurotoxins: feeding, drinking, and body temperature. *Ann. N. Y. Acad. Sci.* 305: 556–575, 1978.

106. Myers, R. D. Hypothalamic control of thermoregulation: neurochemical mechanisms. In: *Behavioral Studies of the Hypothalamus,* edited by P. J. Morgane and J. Panksepp. New York: Dekker, 1980, vol. 3, p. A, p. 83–210.

107. Nakashima, T., A. Taurog, and L. Krulich. Serum thyroxine, triiodothyronine, and TSH levels in iodine-deficient and iodine-

sufficient rats before and after exposure to cold. *Proc. Soc. Exp. Biol. Med.* 167: 45–50, 1981.

108. Nichols, D. G., and R. M. Locke. Thermogenic mechanisms in brown fat. *Physiol. Rev.* 64: 1–64, 1984.

109. Obregon, M. J., P. Santisteban, A. Rodriquex-Peña, A. Paswal, P. Cartagena, A. Ruiz-Marcos, L. Lamas, F. Esgobar Del Rey, and G. Morreale De Esgobar. Cerebral hypothyroidism in rats with adult-onset iodine deficiency. *Endocrinology* 115: 614–624, 1984.

110. O'Dell, B. L. Copper. In: *Present Knowledge in Nutrition* (6th ed.), edited by M. L. Brown. Washington, DC: Int. Life Sci. Inst., Nutr. Found., 1990, p. 261–267.

111. O'Dell, B. L., G. Reynolds, and P. G. Reeves. Analogous effects of zinc deficiency and aspirin toxicity in the pregnant rat. *J. Nutr.* 107: 1222–1228, 1977.

112. Okamura, K., A. Taurog, and L. Krulich. Elevation of serum 3,5,3′-triiodothyronine and thyroxine levels in rats fed Remington diets; opposing effects of nutritional deficiency and iodine deficiency. *Endocrinology* 108: 1247–1256, 1981.

113. Okuda, K., and B. F. Chow. The thyroid and absorption of vitamin B$_{12}$ in rats. *Endocrinology* 68: 607–615, 1961.

114. Oliver, J. W., D. S. Sachan, P. Su, and F. M. Applehans. Effects of zinc deficiency on thyroid function. *Drug Nutr. Interact.* 5: 113–124, 1987.

115. Pazos-Moura, C. C., E. G. Moura, M. L. Dorms, S. Rehnmark, L. Melendez, J. E. Silva, and A. Taurog. Effect of iodine deficiency and cold exposure on thyroxine 5′-deiodinase activity in various rat tissues. *Am. J. Physiol.* 260(*Endocrinol. Metab.* 23): E175–E182, 1991.

116. Pekary, A. E., H. C. Lukaski, I. Mena, and J. M. Hershman. Processing of TRH precursor peptides in rat brain and pituitary is zinc dependent. *Peptides* 12: 1025–1032, 1991.

117. Philip, G., and J. F. Smith. Hypothermia and Wernicke's encephalopathy. *Lancet* II: 122–124, 1973.

118. Pilch, S. M., and F. R. Senti. Assessment of the Iron Nutritional Status of the U.S. Population Based on the Data Collected in the Second National Health and Nutrition Examination Survey, 1976–1980. Bethesda, MD: Fed. Am. Soc. Exp. Biol., 1984, p. 65.

119. Plaitakis, A., E. C. Hwang, M. H. Van Woert, P. I. A. Szilagyi, and S. Beri. Effect of thiamin deficiency on brain neurotransmitter systems. *Ann. N. Y. Acad. Sci.* 378: 367–381, 1982.

120. Remington, R. E. Improved growth in rats on iodine deficient diets. *J. Nutr.* 13: 223–232, 1937.

121. Riesco, G., A. Taurog, P. R. Larsen, and L. Krulich. Variations in the response of the thyroid gland of the rat to different low-iodine diets: correlation with iodine content of diet. *Endocrinology* 99: 270–280, 1976.

122. Riesco, G., A. Taurog, P. R. Larsen, and L. Krulich. Acute and chronic responses to iodine deficiency in rats. *Endocrinology* 100: 303–313, 1977.

123. Root, A. W., G. Duckett, M. Sweetland, and E. O. Reiter. Effects of zinc deficiency upon pituitary function in sexually mature and immature male rats. *J. Nutr.* 109: 958–964, 1979.

124. Sato, T., E. Imura, A. Murata, and N. Igarashi. Thyroid hormone–catecholamine interrelationship during cold acclimation in rats: compensatory role of catecholamines for altered thyroid states. *Acta Endocrinol.* 113: 536–542, 1986.

125. Seidel, K. E., M. L. Failla, and R. W. Rosebrough. Cardiac catecholamine metabolism in copper-deficient rats. *J. Nutr.* 121: 474–483, 1991.

126. Silva, J. E., and A. C. Bianco. Role of local thyroxine 5′-deiodinase on the response of brown adipose tissue to adrenergic stimulation. In: *Obesity: Towards a Molecular Approach,* edited by G. A. Bray, D. Ricquier, and B. M. Spiegelman. New York: Wiley-Liss, 1990, p. 131–146.

127. Silva, J. E., and P. R. Larsen. Adrenergic activation of triiodothyronine production in brown adipose tissue. *Nature* 305: 712–713, 1983.

128. Silva, J. E., and P. R. Larsen. Interrelationships among thyroxine, growth hormone, and the sympathetic nervous system in the regulation of 5'-iodothyronine deiodinase in rat brown adipose tissue. *J. Clin. Invest.* 77: 1214–1223, 1986.

129. Smith, S. M., D. R. Deaver, and J. L. Beard. Metabolic rate and thyroxine monodeiodinase activity in iron-deficient female Sprague-Dawley rats: effects of the ovarian steroids. *J. Nutr. Biochem.* 3: 461–466, 1992.

130. Smith, S. M., J. Finley, L. K. Johnson, and H. C. Lukaski. Indices of in vivo and in vitro thyroid hormone metabolism in iron-deficient rats. *Nutr. Res.* 14: 729–739, 1994.

131. Smith, S. M., and H. C. Lukaski. Estrous cycle and cold stress in iron-deficient rats. *J. Nutr. Biochem.* 3: 23–30, 1992.

132. Smith, S. M., and H. C. Lukaski. Type of dietary carbohydrate affects thyroid hormone deiodination in iron-deficient rats. *J. Nutr.* 122: 1174–1181, 1992.

133. Smith, S. M., S. H. Smith, and J. L. Beard. The roles of sympathetic innervation and catecholamine synthesis in tissue norepinephrine metabolism in iron deficiency anemia in the rat. *J. Nutr. Biochem.* 3: 167–172, 1992.

134. Spaulding, S. W., and R. H. Noth. Thyroid–catecholamine interactions. *Med. Clin. North Am.* 59: 1123–1131, 1975.

135. Tang, F., T. M. Wong, and T. T. Loh. Effects of cold exposure or TRH on the serum TSH levels in the iron-deficient rat. *Horm. Metab. Res.* 20: 616–619, 1988.

136. Tobin, B. W., and J. L. Beard. Interactions of iron deficiency and exercise training relative to tissue norepinephrine turnover, triiodothyronine production and metabolic rate in rats. *J. Nutr.* 120: 900–908, 1990.

137. Topping, D. L., D. G. Clark, and I. E. Dreosti. Impaired thermoregulation in cold-exposed zinc-deficient rats: effect of nicotine. *Nutr. Rep. Int.* 24: 643–648, 1981.

138. Troncoso, S. C., M. V. Johnston, K. M. Hess, J. W. Griffin, and D. L. Price. Model of Wernicke's encephalopathy. *Arch. Neurol.* 38: 350–354, 1981.

139. Uthus, E. O., and F. H. Nielsen. Effect of vanadium, iodine and their interaction on growth, blood variables, liver trace elements and thyroid status indices in rats. *Magn. Trace Elem.* 9: 219–226, 1990.

140. Valle, B. L., J. E. Coleman, and D. S. Auld. Zinc fingers, zinc clusters, and zinc twists in DNA-binding protein domains. *Proc. Natl. Acad. Sci. USA* 88: 999–1003, 1991.

141. Valle, B. L., and K. H. Falchuk. The biochemical basis of zinc physiology. *Physiol. Rev.* 73: 79–118, 1993.

142. Veen, M. J., G. Russell, and G. H. Beaton. Food consumption and rectal temperature in rats during thiamine deficiency and repletion. *Can. J. Biochem. Physiol.* 41: 2463–2471, 1963.

143. Victor, M., R. D. Adams, and G. H. Collins. *The Wernicke-Korsakoff Syndrome*. Philadelphia: Davis, 1971.

144. Visser, T. J. The role of glutathione in the enzymatic deiodination of thyroid hormone. In: *Glutathione: Metabolism and Physiological Functions*, edited by J. Vina. Orlando, FL: CRC, 1991, p. 317–333.

145. Voorhess, M. L., M. J. Stuart, J. A. Stockman, and F. A. Oski. Iron deficiency anemia and increased urinary norepinephrine excretion. *J. Pediatr.* 86: 542–547, 1975.

146. Wada, L., and J. C. King. Effect of low zinc on basal metabolic rate, thyroid hormones and protein utilization. *J. Nutr.* 116: 1045–1053, 1986.

147. Wagner, A., N. Fortier, A. Giroux, J. Lukes, and L. M. Snyder. Catecholamines in adult iron deficiency patients. *Experentia* 35: 681–682, 1978.

148. Weinshilboum, R. M. Serum dopamine β-hydroxylase. *Pharmacol. Rev.* 30: 132–166, 1979.

149. Williams, R. B., and C. F. Mills. The experimental production of zinc deficiency in the rat. *Br. J. Nutr.* 24: 989–1003, 1970.

150. Williams, R. J. P. Zinc: what is its role in biology? *Endeavor* 8: 65–70, 1984.

151. Witt, E. D. Neuroanatomical consequences of thiamin deficiency: a comparative analysis. *Alcohol Alcohol.* 20: 201–221, 1985.

152. Youdim, M. B. H., D. G. Grahame-Smith, and H. F. Woods. Some properties of human platelet monoamine oxidase in iron-deficiency anaemia. *Clin. Sci. Mol. Med.* 50: 479–485, 1976.

64. Nutrition and exercise in adverse environments

E. R. BUSKIRK | *Laboratory for Human Performance Research, The Pennsylvania State University, University Park, Pennsylvania*

CHAPTER CONTENTS

MOST STUDIES OF NUTRITION in those living and working in adverse environments have been done by and for the military organizations of several countries. In the United States all of the military services have been involved, and the many technical reports generated by personnel working in the military-sponsored laboratories (for example, the U.S. Army's Research Institute of Environmental Medicine in Natick, MA) constitute an important source of information. In years past, several U. S. Army laboratories, identified in Table 64.1, were reservoirs of existing knowledge. Some of these laboratories have been closed or moved, but some of their technical reports are still available from libraries established at a few of the bases. Supplementing the technical reports are various reviews that have been published in the open literature, many in physiology journals (3, 4, 9, 16, 34, 37, 39, 40, 49, 58, 67). For the past few years, the Committee on Military Nutrition Research of the Food and Nutrition Board at the Institute of Medicine has paid attention to the nutritional needs of military personnel serving in adverse environments. The Committee's reports serve as an important source of nutritional information.

This chapter briefly explores the general impact on gross nutrition and metabolism of adverse environments involving heat, cold, altitude, and space for those living or working in them. With minor exceptions, no attempt has been made to deal with fluid or salt balance or with micronutrients such as vitamins, minerals, or trace elements.

Although the environmental factors discussed here with respect to nutritional adaptation include heat, cold, altitude, and space, other factors, such as wind, humidity, precipitation, and fog, complicate natural environments. Among other biological factors that interact with environmental factors in modifying nutritional adaptation are pregnancy; lactation; eating disorders such as anorexia nervosa and bulimia; use of pharmaceuticals, socially unacceptable drugs, alcohol, and caffeine; jet-lag; and bed rest. Thus nutritional adaptation is complicated by the status of the organism in relation to the nutrients and other products consumed and the environment in which living and working occurs.

The regulatory mechanisms inherent in intact organisms, including the human body, facilitate adaptability so that essential nutrients are conserved when the diet is inadequate for brief periods. Energy is conserved through reduced physical activity and a lowering of the resting metabolic rate. As certain tissues are modified, their constituents are redistributed and utilized by other tissues. Some nutrients are stored during periods of surfeit, and these stores become available to satisfy immediate needs; for example, sufficient vitamin A (fat-soluble) may be stored in the liver and fat depots to last several months (47), whereas the water-soluble vitamins, such as vitamin C, are stored in quantities that may last several weeks but usually only for several days (54). Compensation for lost stores occurs when a surplus is again consumed. Thus balanced intakes averaged over weekly periods appear adequate to meet long-term needs, including coping with short-term illness or trauma. Relative depletion of stores adversely affects the organism's ability to deal with illness or trauma and reflects overwhelming adaptability.

Food is essential to supply the fuel necessary to satisfy the various energy turnover processes required to sustain life. The necessary physicochemical environment is sustained by these processes as are the electromechanical actions undertaken by the organism. The various processes involved are subject to different efficiencies for energy conversion, which not only leads to interindividual variation but is associated with the ability to adapt to adverse environments and situations. For example, the primary energy store involves ATP and other high-energy phosphate bond–containing compounds, and

TABLE 64.1. *U. S. Army Laboratories that have been Importantly Involved with Nutritional Research*

Food and Container Institute,
Quartermaster Corporation, Chicago, Ill.

Medical Nutrition Laboratory,
Surgeon General's Office, Denver, Co.

Medical Research Laboratory,
Surgeon General's Office, Fort Knox, Ky.

Climatic Research Laboratory,
Quartermaster Corporation, Lawrence, Mass.

Research and Engineering Center,
Quartermaster Corporation, Natick, Mass.

Research Institute of Environmental Medicine,
Surgeon General's Office, Natick, Mass.

Letterman Army Institute for Research,
Surgeon General's Office, Presideo of San Francisco, Cal.

differences in Na^+–K^+ ATPase pump activity in relation to substrate cycling activity are undoubtedly involved in adaptive processes (18).

The higher the quantity of physical activity, the higher the energy intake needs to be to maintain body weight and normal body composition (21). Higher intakes, in general, tend to ensure adequate essential nutrient intake (7, 20). Many of those working in adverse environments (for example, United States military personnel) consume diets fortified with both macro- and micronutrients. High levels of physical activity plus consumption of a fortified diet tend to assure adequate nutrition. Nevertheless, a variety of situations arise, some anticipated—such as winter storms that disrupt logistical supplies or accidents that leave personnel stranded—which modify nutrition by restricting access to food.

Sedentary persons face the problem of decreased caloric needs and the potential for an unbalanced diet. Many elderly face this problem, and they would be well advised to stay physically active to improve the likelihood of their receiving adequate nutrition, assuming their appetites and food selections improve accordingly.

There are large interindividual differences in energy and nutrient requirements; nevertheless, most of the variance can be accounted for in the absence of disease or disability by age, sex, maturation, development, fat-free body or cell mass, physical activity, and environment. The role of genetic factors is undoubtedly important and subject to current investigation but will not be discussed here.

DIETARY DEFICIENCIES

Dietary deficiencies in any environment eventually cause problems, but the speed with which symptoms appear or performance deteriorates may depend on the environment. Johnson originally summarized the more prominent effects of gross nutrient deficiencies, which were cited and modified by Young (70) and are adapted here (see Table 64.2). Most effects are delayed somewhat, but kilocaloric and carbohydrate deficiencies produce symptoms within days. Protein and fat deficiencies produce symptoms much later. Certainly one of the best reviews of undernutrition is the two volume work *The Biology of Human Starvation* (37), which describes the sequential changes that occur with 24 wk of semistarvation and recovery.

The interaction of environmental and nutritional factors in producing compromised physical performance is illustrated in abbreviated form in Table 64.3. Both undernutrition and hypohydration independently and synergistically reduce performance capabilities in any environment.

TABLE 64.2. *Rate of Onset of Deficiency Syndromes in Working Men Exposed to Severe Deficiency of One or More of the Important Gross Nutrients*

Nutrient	Time Before Earliest Effects on Performance Appear After Severe Deficiency	Deficiency Syndrome and End Results
Water	A few hours	Easy fatigue, poorer performance, eventual exhaustion
Kilocalories	2 or 3 days	Easy fatigue, ketosis, poorer performance
Carbohydrate	Several days	Easy fatigue, ketosis, poorer performance, eventual nutritional acidosis
Protein	Probably several weeks	Reduced muscle mass, poorer performance, late result, nutritional edema
Fats	Many months	Earliest effects not known, eventual neurological disorders, poorer performance

Adapted by Young (70) from Johnson (31a) and subsequently modified here.

TABLE 64.3. *Adverse Environments, Physiological Correlates, and Physical Performance*

Environments	
Heat, cold, terrestrial altitude	
Under nutrition	Hypohydration
Poor appetite	Excess water loss
Unavailability of adequate food	Indaequate thirst response
	Inadequate fluid supply
Negative kcal balance	Dehydration
Net result: compromised physical performance	

ENVIRONMENTAL HEAT AND METABOLISM

Various investigations, producing mixed results, have been devoted to the issue of whether environmental heat exposure impacts metabolic rate during either rest or exercise. Several hypothesized factors have been cited for the different observed responses to exercise in the heat and are listed in Table 64.4. Consolazio and colleagues (13, 15, 17) have made the case for a heat-induced elevation in metabolic rate (see Tables 64.5, 64.6) and have cited as the primary cause the increased production and secretion of sweat. In contrast, the observations of Welch et al. (63), Klausen et al. (38), Rowell et al. (53), Sen Gupta et al. (57), Shvartz et al. (59), and Young et al. (69) failed to discern significant differences in energy expenditure when comparable tasks were undertaken in warm or hot as compared to cool or cold environments. The data obtained by Shvartz et al. (59) appear in Table 64.7.

The possibility has been raised by Sen Gupta et al. (57) and Dimri et al. (20), with substantive interpretive comments by Sawka and Wenger (56) and Young (67), that during the performance of submaximal exercise there is lesser aerobic and greater anaerobic metabolism. The hypothesized explanation for such partitioning is the diversion of blood from the muscles and the splanchnic region to the skin to support thermoregulation (52). Further support for this altered partitioning concept comes from the observation that blood lactate accumulation during submaximal exercise is usually greater under hot conditions (69). Nevertheless, for prolonged submaximal work in the heat, there is little evidence to suggest much difference in energy turnover from that experienced in thermoneutral environments.

TABLE 64.4. *Possible Factors for Differences Among Studies: Hypothesized Causes of Different Responses of Metabolic Rate to Exercise in the Heat*

Physical condition of subjects

Extent of heat acclimation/acclimatization

Skill

Frequency of exercise and rest periods

Duration of exercise

Exercise intensity

Environmental heat stress
   Type
   Intensity

Sweating

Hydration state

Febrile state

Clothing worn

Equipment carried

Nutrition
   Meal type, composition and amount
   Spacing of food intake

TABLE 64.5. *Mean Oxygen Uptake (ml $O_2 \cdot min^{-1}$) During Rest and Moderate and Heavy Activity by Young Men (n = 7) in a Room Maintained at Different Temperatures*

Intensity of Activity	Room Temperature		
	21.1°C (70°F)	29.4°C (85°F)	37.7°C (100°F)
Rest	173	282	304*
Moderate	521	525	590*
Heavy	1,422	1,404	1,570

*$P > 0.05$, that is, effect of 37.7°C > 21.1°C or 29.4°C.
Moderate intensity = 50 min on cycle ergometer.
High intensity = 50 min on cycle ergometer at 120 W.
Adapted from Consolazio (15) with permission.

TABLE 64.6. *Oxygen Uptake (ml $O_2 \cdot min^{-1}$) Performing Different Types of Exercise in a Desert Environment—Yuma, Arizona*

Exercise Type	Sun (37.8°C, 100°F)	Shade (37.8°C, 100°F)	Indoors (26.0°C, 78.9°F)
Bicycle I	754*	683*	641
Bicycle II	813*	751*	681
Treadmill	1,156	1,197*	1,110
Resting	340*	322	314

*$P \leq 0.05$, that is, effect of 37.8°C > 26°C.
Adapted from Consolazio et al. (13) with permission.

TABLE 64.7. *Mean Oxygen Uptake (ml $O_2 \cdot m^{-2} \cdot min^{-1}$) Responses to Exercise Before (B) and After (A) Eight Days of Heat Acclimation*

Group	41 W, 23°C		82 W, 23°C		41 W, 39.4°C	
	B	A	B	A	B	A
Trained (n = 7)	623	570*	1,075	960*	634	569*
Untrained (n = 7)	668	577*	1,061	985*	680	586*
Unfit (n = 7)	615	531*	1,050	932*	618	520*
Control (n = 5)	625	578	1,068	1,030	611	615

Heat acclimation regimen = 3 h of bench stepping per day at 41 W.
T_{db} = 39.4°C, T_{wb} = 30.3°C.
*$P < 0.05$. Effect of A < B.
Adapted from Shvartz et al. (59) with permission.

Under prolonged conditions a new balance in muscle, splanchnic, and skin blood flows must be achieved to sustain exercise.

Thus the conclusion as to whether energy turnover is modified during exercise in the heat is dependent on the circumstances. Brief intense exercise in a hot environment may evoke anaerobic processes, but the increment in daily energy turnover is likely quite small and kilocaloric needs relatively unaffected. Possible reasons for a modified metabolic rate in a hot environment are listed in Table 64.8.

TABLE 64.8. *Modified Metabolic Rates in Hot Environments*

Possible reasons for an increase
  Lack of acclimatization
  Psychomotor stress
  Inefficient physical activity
  Q_{10} effect, elevated body temperature
  Greater sweat gland activity
  Tachycardia
  Increased pulmonary ventilation
  Increased anaerobic metabolism
    Increased O_2 debt
    Increased blood lactate
    Increased respiratory quotient
    Increased glycogenolysis
    Increased glycolysis
    Lessened splanchnic blood flow
    Lessened skeletal muscle blood flow
Possible reasons for a decrease
  Complete acclimatization
  Lower basal metabolic rate
  Decreased appetite and associated diet-induced thermogenesis
  Behavioral modifications
    Reduced physical activity, particularly intense activity
    Wearing of lighter weight clothing
    Avoiding solar radiation

HEAT ACCLIMATIZATION/ACCLIMATION

Heat acclimatization (natural environment) or acclimation (laboratory) proceeds quite rapidly, being virtually complete within 10 days (1, 8, 19). During the early period of acclimatization, appetite may be adversely affected, producing undernutrition, a negative water and salt balance, and weight loss. Such an impact may be greater in tropical environments (hot, humid) when sweating rates are high and hyporexia persists (32).

Whether heat acclimatization modifies metabolism has been studied intensively. Sawka et al. (55) and Young et al. (67) drew the conclusion that a modest lowering of metabolism was possibly common. They point to more efficient evaporative cooling, lowering of body temperatures, improved partitioning of blood flow, improved contractile-coupling efficiency, and enhanced mitochondrial phosphorylation as possible factors. Since several of these factors have been incompletely studied, related research is needed. Although heat acclimatization/acclimation constitutes a valuable physiological adaptation, such adaptation plays only a minor role in modifying energy turnover and kilocaloric requirements.

COLD AND CALORIC TURNOVER

A variety of environmental effects are commonly experienced in cold climates which complicate such expo-

sure; some of these are listed in Table 64.9. An appropriately clothed person does not greatly increase energy requirements when exposed to cold, but exercise performed in bulky clothing and accessories plays a major role. Depending on the environmental conditions and, more importantly, the exercise-related variables (see Table 64.10), estimates of the relative kilocaloric increment due to cold have varied from 2% to 15% but are more commonly no more than 5% (63).

An original cross-environmental comparison of average nutrient intake was reported by Johnson and Kark (33). They included data from physically fit, active ground troops who chose their food from the rations provided in arctic and subarctic areas as well as mountain, desert, and Pacific island areas (30, 32). A summary of Johnson and Kark's results, covering 23 different studies, appears in Table 64.11. The respective regression equation that described the linear relation-

TABLE 64.9. *Environmental Effects Commonly Associated with Cold Climates*

Low air temperature
Low humidity
Wind
Precipitation
  Cold rain
  Freezing rain, sleet
  Snow
Ice
Illumination
  Longer darkness
  Glare (reflected light off snow and ice)

TABLE 64.10. *Factors Influencing Energy Needs for Exercise/Work/Sport in Cold Environments*

Body weight
Clothing
  Type
  Weight
  Movement restriction
  Position on body
Mode of movement/locomotion
  Walking, jogging, running, climbing
  Snowshoeing
  Skiing
  Load carrying, hauling
Conditions
  Snow, depth, consistency, water content
  Ice
  Direction and velocity of air movement
  Solar radiation
  Terrain
Physical condition of individual
Available nutrition

Exercise in cold water involves a unique environment that differentiates it from land-based activities. Nevertheless, combinations of cold air/water/land exposure can occur and pose significant adverse physiological consequences.

TABLE 64.11. *Voluntary Average Nutrient Intake of North American Ground Troops Who Remained Healthy, Fit, and Efficient in Different Environments*

Information	U.S. Training Camps	Camp Carson Trials	Exercise "Musk Ox"	Guam Garrison	Luzon 38th Infantry Division
Type of troops	All	Infantry	Motorized	Garrison	Combat
Type of ration	U.S. garrison	U.S. B supplemented	Canadian arctic	U.S. B supplemented	U.S. new C
Duration of time on ration (wk)	Indefinite	8	12	Indefinite	12
Environment	Temperate	Temperate mountain (9,000 ft) summer	Arctic winter	Moist tropics	Moist tropics
Average intake per man per day					
kcal·day^{-1}	3,800	3,900	4,400	3,500	3,200
Protein, g (total)	125	125	120	114	100
Protein, g (anim)	—	75	70	70	60
Fat, g	180	145	190	125	120
Carbohydrate, g	410	520	480	430	
% kcal, fat	43	34	40	32	34
% kcal, carbohydrate	44	53	49	55	54
% kcal, protein	13	13	11	13	12

Adapted from Johnson and Kark (32).

ship between voluntary kilocaloric intake and mean environmental temperature was

$$kcal \cdot day^{-1} = 4.660 - 15.9\ \overline{T}_a(°F)$$
$$r = -0.935,\ r^2 = 0.875$$
$$SEE = 1.25$$

where $\overline{T}_a$ is the mean ambient temperature for the period of study and *SEE* is the standard error of estimate. A consistent increase in voluntary kilocaloric intake was found over the range 37.7°C (100°F) to −28.9°C (−20°F). Unfortunately, Johnson and Kark (32) had few data on actual physical activity to confirm their contention of little difference in physical activity across garrisons. They did state that kilocaloric expenditure for a given task is greater in the cold because of the "hobbling effect" of cold weather clothing. They also recognized that kilocaloric requirements are dependent on the physical activity that troops perform, a more important relationship than the affirmation that more body heat is required in cold than in warm environments to maintain thermal balance. Interestingly, a review of Table 64.11 clearly shows that the percentage of protein intake was essentially the same at each garrison, whereas the percentage of fat intake was higher in the colder areas and lower in the warmer ones.

During intense exercise under adverse conditions involving cold exposure which necessitates the wearing of protective clothing, the accumulation of sweat in the clothing can compromise the insulative capacity of the clothing. Newer clothing that allows some water vapor transfer while preserving insulation reduces the deleterious impact of wetted clothing, but such clothing is expensive and not worn by many who exercise in the cold. In any event, intense exercise usually produces positive heat balance, and no risk is incurred until the activity is moderated or stopped. With lessened activity or cessation, the body cools, both skin and deep body temperatures fall, and shivering is activated to augment resting heat production within the range of one- to threefold (48, 61). Thus intermittent activity in a cold environment can produce rather exceptional caloric turnovers; for example, Hong et al. (29) have estimated that Korean female pearl divers working intermittently in cold water expended approximately 387 kcal per diving shift in summer and 575 kcal in winter. These values include the cumulative heat production from shivering and swimming as well as replacement of the cumulative heat debt. Divers took three shifts in summer and one or two shifts in winter. Thus the cumulative energy expenditure to support the diving was approximately 1,000 kcal·day^{-1}—this in relation to a daily kcal intake of 3,000 kcal·day^{-1} as contrasted to nondiving Korean women of comparable age whose average intake was approximately 2,000 kcal·day^{-1} (28, 34, 35).

COLD AND BODY FUEL

In his thermal model, Wissler (66) assumed that the energy required for thermal homeostasis is derived exclusively from carbohydrates. Nevertheless, the historical evidence invalidates such an assumption, and common sense indicates that glycogen reserves would be depleted by any extended period in the cold without exceptional intakes of carbohydrate. Nevertheless, evidence for increased carbohydrate utilization with acute cold exposure has been documented, albeit by indirect assessment using the respiratory quotient. For example, Vallerand and Jacobs (61) calculated that with a 3°–4°C loss of mean body temperature, carbohydrate oxidation increased more than fivefold and fat oxidation 0.6-fold with no change in protein oxidation. In a follow-up study, they found that adipose tissue and muscle both supply triglycerides to support cold-induced triglyceride clearance, which persisted when nicotinic acid was given (62). The combination of exercise and cold exposure involving shivering was observed to deplete muscle glycogen faster than comparable exercise without shivering (42, 67). Free fatty acids are turned over more rapidly and circulating plasma glycerol is increased in the cold (67). Presumably because muscle glycogen is partially depleted, a second cold exposure within 24 h induces even greater fat metabolism (62). Greater free fatty acid utilization in the cold is importantly related to the increased catecholamine turnover, particularly norepinephrine. Glycogenolysis is stimulated as well. Plasma catecholamines have been demonstrated to increase more if muscles are cooled before exercise (58). Insulin sensitivity in the periphery appears to increase with cold exposure, thus enhancing glucose uptake (60).

Young (67) has speculated about two factors that conceivably could contribute to increased glycolysis and lactate accumulation during exercise in the cold. If muscle force is lessened by muscle cooling, a greater muscle mass might be recruited to perform a given task; that is, muscle efficiency is reduced as muscle temperature decreases. In addition, if plasma catecholamines are increased more when muscle temperatures are reduced, this too could augment glycogenolysis and glycolysis through increased stimulation of glycogen phosphorylase (50). Evidence from submaximal exercise and exposure to cold water has implicated the catecholamine response (22). Relative workload may also increase in the cold by virtue of the fact that aerobic power may be reduced, which would generate a relatively greater catecholamine response (65). Thus it is unclear whether the primary cause of the greater catecholamine response to exercise in the cold emanates from cold stimulation of thermoreceptors and a direct muscle response or the greater relative exercise intensity (67). Shivering and reduced blood flow in the cold are probably ancillary factors associated with accelerated glycolysis.

Shephard (58), in reviewing the literature, suggests that prolonged exposure to cold enhances the normal lipolytic effect of exercise. Shephard cites both animal experiments and the investigations of O'Hara et al. (46), Campbell (11), and Murray et al. (44) among others, implicating ketogenesis and ketonuria, elevated resting metabolism, undetected shivering, nonshivering thermogenesis associated with hypertrophied brown adipose tissue, futile metabolic cycles such as a cold-induced increase in adenylcyclase activity, increased sodium pump activity, and activation of a variable proton leakage pathway in adipose tissue as possible factors in the increased rate of fat utilization with cold exposure. Interestingly, Murray et al. (43) found that fat loss among women was less apparent than among men when comparable exercise was performed in the cold, though the exposure of the women was somewhat less severe. A variety of reasons have been advanced for a possible gender difference in response to cold exposure, including differences in aerobic fitness, rates of fat mobilization, number or sensitivity of catecholamine receptors, structure of the hypothalamic thermoregulatory area, subcutaneous fat thickness, and extent of skeletal muscle involvement in thermoregulation. Obviously, more experimental work needs to be done on cold-induced gender as well as age-related differences. The latter are of concern because of the relatively high incidence of hypothermia among the elderly.

In terms of dietary support for those who must cope with repeated prolonged exercise in the cold (athletes, cross-country trekkers, etc.), the exercise/cold-induced features of possible glycogenolysis, glycolysis with the associated reductions in muscle glycogen, and circulating plasma glucose coupled with increased catecholamine release and elevated circulating free fatty acids pose special nutritional considerations. Not only do glycogen reserves need sustaining but adequate kilocalories need to be supplied, which may only be accomplished by increasing fat intake because of the higher kilocaloric density of fat. Physical conditioning for endurance and cold acclimatization may act synergistically to reduce relative workloads and to sustain enhanced insulative, as well as metabolic, responses, such as lesser rates of glycogen depletion and enhanced free fatty acid utilization (42, 48, 58, 67).

HIGH TERRESTRIAL ALTITUDE

An early effect of an abrupt exposure to terrestrial altitudes above 3,100 m (10,000 ft) is the debilitating influence of acute mountain sickness (AMS) with the associated symptoms of headache, nausea, and hyporexia, among others (24). Such symptoms appear within hours of exposure, peak within 24 to 72 h, and thereafter gradually lessen. Inadequate food intake occurs, which results in a negative energy balance, with interindividual deficits of the order of 10% to 50% (4, 9). Prophylactic pharmacological treatment with the carbonic anhydrase inhibitor acetazolamide has met with some success in preventing AMS or reducing symptoms (5) as has the consumption of a high-carbohydrate diet (70% of kcal) (14). In regard to consuming more carbohydrate, the exact mechanisms by which symptom relief is achieved and physical performance enhanced are unknown, though it has been pointed out that indigenous populations at high altitudes commonly subsist on high-carbohydrate diets. In reviewing the literature, Askew (4) cited experiments implicating relatively elevated arterial blood oxygen tension, greater pulmonary diffusing capacity, and greater energy production per liter of oxygen when higher carbohydrate intakes were consumed. Nevertheless, all of these observations need confirmation because possible mechanisms are not readily apparent.

It is well known that physical endurance is dependent on adequate glycogen stores, and Askew (4) cites experiments designed to impact muscle glycogen stores by having soldiers run as far as possible in 2 h at a 4,100 m (13,500 ft) altitude. Those subjects who supplemented their 45% carbohydrate diet with an additional 200 g of carbohydrate ran significantly farther than those who consumed the common ration. These experiments were conducted each day during the 4 days spent at the higher altitude. Significant differences were found on days 2 and 3 but not 1 and 4, though a trend toward significance occurred.

In reviewing the older literature, Buskirk and Mendez (9) concluded that, despite the physiological effects of hypoxia, energy turnovers at high altitudes are equivalent to those found at sea level for comparable exercise; intakes of 3,200–4,300 kcal·day^{-1} were reported, depending on the respective environments and physical activity levels.

Thus there is little evidence that altitude per se (hypoxia) increases the requirement for any major nutrient, with the possible exception of carbohydrate. As in other stressful environments excess protein intake is counterproductive because of the extra urinary water loss associated with the excretion of nitrogen-containing compounds. An important nutrient, however, is iron because of the elevated erythropoietic response to altitude exposure. Whereas men apparently support adequate hemoglobin synthesis on normal dietary iron intakes, many women may require iron supplementation during long sojourns at high altitudes (25).

An interesting adaptation has been observed by Hochachka et al. (27), who reported that Quechua highlanders exhibit a higher energetic efficiency as demonstrated by a lower net oxygen uptake and presumably a relatively lesser ATP utilization at a given power output, perhaps resulting in a blunting of glycolysis. Among newcomers to altitude, exercise performed soon after ascent results in higher muscle and blood lactate concentrations due to greater glycolysis (6). Following acclimatization, glycolytic fluxes were reduced, resulting in more precise matching to tricarboxylic acid (TCA) cycle turnover during submaximal exercise.

The relative reduction in glycolysis with acclimatization is characterized by a reduced rate of lactate production and release (26). In terms of metabolic fuel, Brooks et al. (6) found no impact of exercise at 4,300 m on the pattern of glycogen utilization but did find enhancement of blood glucose utilization, with dependency on glucose more pronounced following acclimatization. The increase in blood glucose utilization could perhaps reduce glycogen loss from muscle by serving as a fuel in glycolysis or by participating in glycogen synthesis. Increased hexokinase-induced glucose flux has been postulated to provide an efficient fuel (kcal·l O$_2^{-1}$) for working muscle, which serves as a beneficial adaptation to hypoxia (26).

The lessening muscle glycogenolysis during exercise at high terrestrial altitude is probably partially due to the increased mobilization and utilization of free fatty acids. Young et al. (68) have shown that among sojourners acutely exposed in a chamber maintained at 4,300 m, free fatty acids were increased, with a further increment measurable after 18 days of exposure. An elevated plasma glycerol concentration was also apparent after 18 days at 4,300 m. Young et al. (68) concluded that with altitude acclimatization lipolysis is increased but is balanced by fatty acid removal and oxidation. They made this interpretation after observing the considerable increase in plasma glycerol concentration with little change in free fatty acids during exercise on day 18. Direct appraisals of elevated fatty acid oxidation by contracting skeletal muscle remain to be determined.

In reviewing various possible impacting mechanisms on the reduction in glycogenolysis through a norepinephrine-mediated mechanism, Young et al. (67) cited the performance of β-blocking experiments with propranolol and reported no effect on the reduction in glycogen utilization during exercise at high altitudes. Similarly, various oxidative changes failed to explain the response. A reduced ammonia accumulation during submaximal exercise at high altitudes following partial acclimatization suggested that experiments on muscle recruitment patterns should be initiated to ascertain whether they are altered during acclimatization or whether phosphofructokinase stimulation is reduced. Katz and Sahlin (36) have suggested that the lessening of hypoxia in muscle following acclimatization accounts for the reduction in glycogenolysis and lactate accumulation during exercise, but this view also requires confirmation, though observations of improved muscle capillarization per unit fiber area should favor tissue oxygenation. Such an observation has not been observed in the type of experiment showing the acclimatization-induced reduction in glycogenolysis (67).

Although kilocaloric intakes have generally been reported to be somewhat less at high altitudes than at sea level, at least before rather complete acclimation has been established (51), Butterfield et al. (10) found that weight loss at 4,300 m for 21 days was minimized by increasing kilocaloric intake about 300 kcal·day^{-1} over the sea-level intake. Nitrogen balance was also preserved.

NUTRITION IN SPACE

In space, short-term nutrition is little different from that on earth, with the exception of packaging, but long-term flight poses special problems, including the need for food production. To this end, it is likely that plants, algae, or yeast grown on board the spacecraft will constitute an important source of food. West (64), in reviewing long-term needs, described three integrated subsystems. The first involves the food production system, which may well involve hydroponic technology. The second is a food-processing system that provides palatable food from the foodstuffs gleaned from the plants. The third is the system that transforms gaseous, liquid, and solid wastes from the crew or other waste producers into feed for the plants. NASA's calculation of a crew member's nutritional requirements are about 0.5 kg of food and 3 l of water per day. Since about a kilogram of oxygen is also required per day, the importance of a bioregenerative life-support system is apparent.

Personnel associated with the space program have tried to establish the energy requirements of the astronauts in a variety of ways. Based on present knowledge, energy requirements on earth and in space appear roughly equivalent, despite the loss in fat-free body weight that has commonly been observed. The several methods employed have included removal from galley stores, analysis of respiratory gases resident in the spacecraft, metabolic balances, and food intake records. Ostensibly, the best records were obtained from the Skylab missions, using dietary records and metabolic balances, including bomb calorimetry. Available energy was calculated as 2,686 ± 141 (mean ± SD) at 28 days in flight, 2,939 ± 538 at 59 days, and 2,972 ± 78 at 84 days. Thus equivalent amounts of energy were consumed throughout the flight. In addition, there was no difference in kilocaloric intake preflight and in flight. There appeared to be a switch in nutrient proportions in that carbohydrate intake was higher in flight and fat

intake lower. The nonprotein respiratory quotient was elevated in flight. Comparison of rough data from the shuttle and Skylab flights indicates energy turnovers from about 1,950 to 3,600 kcal·day^{-1} (39). Plans to utilize the doubly labeled water technique should provide a more accurate appraisal of in flight energy needs.

A consistent finding to date is that body weight is lost along with a negative nitrogen balance during the first few days of space flight. This has been attributed to the appetite suppression associated with space sickness brought about by the conflict between disturbed otolithic information, acceleration data effects on the semicircular canals, pressure interpretation from proprioceptors, and orientation information provided visually. In addition, a cephalad redistribution of the blood volume is found, which presumably precipitates a diuresis and fluid loss, which may be the main component of the body weight loss (12, 31, 41, 64). Over the long-term, two somewhat conflicting observations have been made: Skylab astronauts showed continued but slowed weight loss over 84 days that was correlated with the reduced kilocaloric intake of the respective crew members, and, in contrast, in the 6 month Soviet missions, three out of four astronauts gained weight (that is, largely stored fat in adipose tissue) (12).

A second consistent finding is the slow decrease in fat-free body weight, perhaps related to adaptations in energy turnover as influenced by modified skeletal muscle activity (2). Analysis of plasma and urine samples obtained in flight suggest hypotrophy of muscle, for example, increased creatinine kinase, nitrogen and potassium concentrations in plasma, as well as increased urinary loss of creatinine, N^{+}-methylhistidine, other amino acids, and potassium. This, along with a negative nitrogen balance, indicated muscle protein catabolism (41). A caution was noted in that preflight nitrogen balance was positive so that a stable baseline was not obtained. Thus there remains a need for careful long-term appraisals of nutritional status that include stable baseline information. Optimal diets in terms of macro- and micronutrients also remain to be determined.

Of related concern are the negative balances of calcium and phosphorus with loss of bone mineral. Combined calcium losses in urine and feces approached 300 mg·day^{-1} by 3 months. It has also been estimated from photon absorptiometry studies that bone mineral losses ranging from 3.2% to 8.0% take place on flights longer than 2 wk and that calcium loss increases in relation to time in space. Because the error of measurement is in the 3%–7% range, more accurate determinations need to be made. Nevertheless, such losses may only partially be regained postflight and represent a serious public health problem. Countermeasures such as exercise have proven only partially effective, and the thought has been expressed that periods of 1g centrifugation in flight may be necessary.

Although glucose/insulin metabolism has been studied during bed rest, only limited information is available during pre- and postflight. The Soviets found lower free fatty acid and higher insulin concentrations in serum 1 day postflight, with insulin remaining elevated for 7 days postflight. No pre- to postflight differences were found for glucose, lactate, or pyruvate concentrations. Speculation as to elevated carbohydrate utilization and lipogenesis reflecting a change in metabolic turnovers awaits definitive investigation (23, 39, 40).

The space shuttle has been supplied with about 70 different food items, most of which are freeze-dried because of a lack of refrigeration (45). About 20 different drinks are also available. Most items qualify as standard American fare. For the space station, a Food Supply and Service System Study has been prepared by NASA Houston so that the industry can develop palatable and nutritious meals. The closed ecological system will hopefully be available for the anticipated 18 month trip to Mars (71).

REFERENCES

1. Adolph, E. F. *Physiology of Man in the Desert.* New York: Interscience, 1947.
2. Altman, P. L., and J. M. Talbot. Nutrition and metabolism in space flight. *J. Nutr.* 117: 421–427, 1987.
3. Askew, E. W. Nutrition and performance under adverse environmental conditions. In: *Nutrition in Exercise and Sport,* edited by J. F. Hickson and I. Wolinsky. Boca Raton, FL: CRC, 1989, p. 367–384.
4. Askew, E. W. Nutrition in a cold environment. *Physician Sportsmed.* 17: 76–89, 1989.
5. Beeckman, D., and E. R. Buskirk. Drug use at high terrestrial altitudes and in cold climates: a brief review. *Hum. Biol.* 60: 663–677, 1988.
6. Brooks, G. A., G. E. Butterfield, R. R. Wolfe, B. M. Groves, R. S. Masseo, J. R. Sutton, E. E. Wolfel, and J. T. Reeves. Increased dependence on blood glucose after acclimatization to 4,300 m. *J. Appl. Physiol.: Respir. Environ. Exerc. Physiol.* 70: 919–927, 1991.
7. Buskirk, E. R. Exercise. In: *Present Knowledge in Nutrition* (6th ed.) edited by M. Brown. Washington, DC: Int. Life Sci. Inst. Nutr. Found., 1990, p. 341–348.
8. Buskirk, E. R., and D. E. Bass. Climate and Exercise. Technical Report EP-61. Natick, MA: Quartermaster Research and Development Center, U.S. Army, 1957.
9. Buskirk, E. R., and J. Mendez. Nutrition, environment and work performance with special reference to altitude. *Federation Proc.* 26: 1760–1767, 1967.
10. Butterfield, G. E., J. Gates, S. Fleming, G. A. Brooks, J. R. Sutton, and J. T. Reeves. Increased energy intake minimizes weight loss in men at high altitude. *J. Appl. Physiol.: Respir. Environ. Exerc. Physiol.* 72: 1741–1748, 1992.
11. Campbell, I. T. Energy intakes on sledding expedition. *Br. J. Nutr.* 45: 89–94, 1981.
12. Committee on Space Biology and Medicine. Human nutrition. In: *A Strategy for Space Biology and Medical Science.* Washington, DC: Space Science Board, National Research Council, 1987, p. 156–159.
13. Consolazio, C. F., H. L. Johnson, and H. J. Krzywicki. Energy Metabolism During Exposure to Extreme Environments. Unnumbered report. Denver, CO: U.S. Army Medical Research and Nutrition Laboratory, Fitzsimmons General Hospital, 1970, p. 16.1–16.9.
14. Consolazio, C. F., L. O. Matoush, H. L. Johnson, H. J. Krzywicki, T. A. Daws, and G. J. Isaac. Effects of high-carbohydrate diets on performance and clinical symptomatology after rapid ascent to high altitude. *Federation Proc.* 28: 937–943, 1969.
15. Consolazio, C. F., L. O. Matoush, R. A. Nelson, J. A. Torres, and G. J. Isaac. Environmental temperature and energy expenditure. *J. Appl. Physiol.* 18: 65–68, 1963.
16. Consolazio, C. F., and D. D. Schnakenberg. Nutrition and the responses to extreme environments. *Federation Proc.* 36: 1673–1678, 1977.
17. Consolazio, C. F., R. Shapiro, J. E. Masterson, and P. S. L. McKinzie. Energy requirements of men in extreme heat. *J. Nutr.* 73: 126–134, 1961.
18. DeLuise, M., G. L. Blackburn, and J. S. Flier. Reduced activity of the red-cell sodium–potassium pump in human obesity. *N. Engl. J. Med.* 303: 1017–1022, 1980.
19. Dill, D. B. *Life, Heat and Altitude.* Cambridge, MA: Harvard University Press, 1938.
20. Dimri, G. P., M. S. Malhotra, J. Sen Gupta, T. S. Kumar, and B. S. Aora. Alterations in aerobic–anerobic proportions of metabolism during work in the heat. *Eur. J. Appl. Physiol.* 45: 43–50, 1980.
21. Durnin, J. V. Muscle in sports medicine—nutrition and muscular performance. *Int. J. Sports Med.* 3(suppl. 1): 52–57, 1982.
22. Galbo, H., M. E. Houston, N. J. Christensen, J. J. Holst, B. Nielsen, E. Nygaard, and J. Suzuki. The effect of water temperature on the hormonal response to prolonged swimming. *Acta Physiol. Scand.* 105: 326–337, 1979.
23. Grigoriev, A. I., I. A. Popova, and A. S. Ushakov. Metabolic and hormonal status of crewmembers in short-term space flights. *Aviat. Space Environ. Med.* 58: A121–A125, 1987.
24. Hackett, P., and D. Rennie. The incidence, importance, and prophylaxis of acute mountain sickness. *Lancet* 2: 1149–1155, 1976.
25. Hannon, J. P. Nutrition at high altitude. In: *Environmental Physiology: Aging, Heat and Altitude.* Edited by S. M. Horvath and M. K. Yousef. Amsterdam: Elsevier, 1980, p. 309–327.
26. Hochachka, P. W. The lactate paradox: analysis of underlying mechanisms. *Ann. Sports Med.* 4: 184–188, 1988.
27. Hochachka, P. W., C. Stanley, G. O. Mathewson, D. C. McKenzie, P. S. Allen, and W. S. Parkhouse. Metabolic and work efficiencies during exercise in Andean natives. *J. Appl. Physiol.: Respir. Environ. Exerc. Physiol.* 70: 1720–1730, 1991.
28. Hong, S. K. Heat exchange and basal metabolism of ama. In: *Physiology of Breath-Hold Diving and Ama of Japan.* Edited by H. Rahn and T. Yokoyama. Washington, DC: NAS-NRC, Publ. 1341, 1965, p. 303–314.
29. Hong, S. K., D. W. Rennie, and Y. S. Park. Cold acclimatization and deacclimatization of Korean women divers. *Exerc. Sport Sci. Rev.* 14: 231–268, 1986.
30. Howe, P. E., and G. H. Berryman. Average food consumption in the training camps of the United States Army (1941–1943). *Am. J. Physiol.* 144: 558–594, 1945.
31. Huntoon, C. L., P. C. Johnson, and N. M. Cintron. Hematology, immunology, endocrinology, and biochemistry. In: *Space Physiology and Medicine.* Edited by A. E. Nicogossian, C. L. Huntoon, and S. L. Pool. Philadelphia: Lea and Febiger, 1989, p. 227–239.

31a. Johnson, R. E. Nutritional standards for men in tropical climates. *Gastroenterology* 1: 832–840, 1943.

32. Johnson, R. E., and R. M. Kark. Feeding Problems in Man as Related to Environment. An Analysis of United States and Canadian Army Ration Trials and Surveys, 1941–1946. Chicago: Quartermaster Food and Container Inst., 1946.

33. Johnson, R. E., and R. M. Kark. Environment and food intake in man. *Science* 10: 378–379, 1947.

34. Kang, D. H., P. K. Kim, B. S. Kang, S. H. Song, and S. K. Hong. Energy metabolism and body temperature in the ama. *J. Appl. Physiol.* 20: 46–50, 1965.

35. Kang, D. H., Y. S. Park, Y. D. Park, I. S. Lee, D. S. Yeon, S. H. Lee, S. Y. Hong, D. W. Rennie, and S. K. Hong. Energetics of wet suit diving in Korean women breath-hold divers. *J. Appl. Physiol.: Respir. Environ. Exerc. Physiol.* 54: 1702–1707, 1983.

36. Katz, A., and K. Sahlin. Effect of decreased oxygen availability on NADH and lactate contents in human skeletal muscle during exercise. *Acta. Physiol. Scand.* 131: 119–127, 1987.

37. Keys, A., J. Brozek, A. Henschel, O. Mickelsen, and H. Taylor. The Biology of Human Starvation. Minneapolis: University of Minnesota Press, 1950, vols. I, II.

38. Klausen, K., D. B. Dill, E. E. Phillips, and D. McGregor. Metabolic reactions to work in the desert. *J. Appl. Physiol.* 22: 292–296, 1967.

39. Lane, H. W. Energy requirements for space flight. *J. Nutr.* 122: 13–18, 1992.

40. Lane, H. W. Nutrition in space: evidence from the U.S. and U.S.S.R. *Nutr. Rev.* 50: 3–6, 1992.

41. Leonard, J. I., C. S. Leach, and P. C. Rambaut. Quantitation of tissue loss during prolonged space flight. *Am. J. Clin. Nutr.* 38: 667–669, 1983.

42. Martineau, L., and I. Jacobs. Muscle glycogen utilization during shivering thermogenesis in humans. *J. Appl. Physiol.* 65: 2046–2050, 1988.

43. Murray, S. J., R. J. Shephard, S. Greaves, C. Allen, and M. Radomski. Effects of cold stress and exercise on fat loss in females. *Eur. J. Appl. Physiol.* 55: 610–618, 1986.

44. Murray, S. M., R. J. Shephard, W. J. Montelpare, and R. C. Goode. Fat loss during moderate exercise in cold environments in relation to fitness level. *Arctic Med. Res.* 47(suppl. 1): 277–279, 1988.

45. NASA. Food for space flight. In: *NASA Facts Report No. NF 133/6-82.* Washington, DC: U.S. Government Printing Office, 1982, p. 1–8.

46. O'Hara, W. J., C. Allen, R. J. Shephard, and G. Allen. Fat loss in the cold: a controlled study. *J. Appl. Physiol.: Respir. Environ. Exerc. Physiol.* 46: 872–877, 1979.

47. Olson, J. A. The storage and metabolism of vitamin A. *Chem. Scripta* 27: 179–183, 1987.

48. Pugh, L. G. C. E. Accidental hypothermia among hill walkers and climbers in Britain. In: *Environmental Effects on Work Performance.* Edited by G. R. Cumming, D. Snidal, and A. W. Taylor. Ottawa: Canadian Assoc. Sports Sci., 1972, p. 41–55.

49. Rambaut, P. C., M. C. Smith, Jr., C. S. Leach, G. D. Whedon, and J. Reid. Nutrition and responses to zero gravity. *Federation Proc.* 36: 1678–1682, 1977.

50. Richter, E. A., N. B. Ruderman, H. Gavras, E. R. Belur, and H. Galbo. Muscle glycogenolysis during exercise: dual control by epinephrine and contractions. *Am. J. Physiol.* 242(*Endocrinol. Metab.* 5): E15–E32, 1982.

51. Rose, M. S., C. S. Houston, C. S. Fulco, G. Coates, J. R. Sutton, and A. Cymerman. Operation Everest II: nutrition and body composition. *J. Appl. Physiol.: Respir. Environ. Exerc. Physiol.* 65: 2545–2551, 1988.

52. Rowell, L. B., G. L. Brengelmann, J. R. Blackmon, T. D. Twiss, and F. Kusumi. Splanchnic blood flow and metabolism in heat-stressed man. *J. Appl. Physiol.* 24: 475–484, 1969.

53. Rowell, L. B., G. L. Brengelmann, J. A. Murray, K. K. Kraning, and F. Fusumi. Human metabolic responses to hyperthermia during mild to maximal exercise. *J. Appl. Physiol.* 26: 395–402, 1969.

54. Sauberlich, H. E. Vitamin C status: methods and findings. *Ann. N. Y. Acad. Sci.* 258: 438–450, 1975.

55. Sawka, M. N., K. B. Pandolf, B. A. Avellini, and Y. Shapiro. Does heat acclimation lower the rate of metabolism elicited by muscular exercise? *Aviat. Space Environ. Med.* 54: 27–31, 1983.

56. Sawka, M. N., and C. B. Wenger. Physiological responses to acute exercise-heat stress. In: *Human Performance Physiology and Environmental Medicine at Terrestrial Extremes.* Edited by K. B. Pandolf, M. N. Sawka, and R. R. Gonzalez. Indianapolis, In: Benchmark, 1988, p. 97–151.

57. Sen Gupta, J., P. Dimri, and M. S. Malhotra. Metabolic responses of Indians during submaximal and maximal work in dry and humid heat. *Ergonomics* 20: 33–40, 1977.

58. Shephard, R. J. Fat metabolism, exercise, and the cold. *Can. J. Sport Sci.* 17: 83–90, 1992.

59. Shvartz, E., Y. Shapiro, A. Magazanik, A. Meroz, H. Bernfeld, A. Mechtinger, and S. Shibolet. Heat acclimation, physical fitness, and responses to exercise in temperate and hot environments. *J. Appl. Physiol.: Respir. Environ. Exerc. Physiol.* 43: 678–683, 1977.

60. Vallerand, A. L., J. Frim, and M. F. Kavanagh. Plasma glucose and insulin responses in cold-exposed humans. *J. Appl. Physiol.: Respir. Environ. Exerc. Physiol.* 65: 2395–2399, 1988.

61. Vallerand, A. L., and I. Jacobs. Rates of energy substrate utilization during cold exposure. *Eur. J. Appl. Physiol.* 58: 873–878, 1989.

62. Vallerand, A. L., and I. Jacobs. Influence of cold exposure and plasma triglyceride clearance in humans. *Metabolism* 39: 1211–1218, 1990.

63. Welch, B. E., E. R. Buskirk, and P. F. Iampietro. Relation of climate and temperature to food and water intake in man. *Metabolism* 7: 141–148, 1958.

64. West, J. B. Life in space. *J. Appl. Physiol.: Respir. Environ. Exerc. Physiol.* 72: 1623–1630, 1992.

65. Winder, W. W., R. C. Hickson, J. M. Hagberg, A. A. Ehsani, and J. A. McLane. Training-induced changes in hormonal responses to submaximal exercise. *J. Appl. Physiol.: Respir. Environ. Exerc. Physiol.* 46: 766–771, 1979.

66. Wissler, E. H. Mathematical simulation of human thermal behavior using whole body models. In: *Heat Transfer in Medicine and Biology.* Edited by A. Shitzer and R. C. Ebhart. New York: Plenum, 1985, p. 325–373.

67. Young, A. J. Energy substrate utilization during exercise in extreme environments. In: *Exercise and Sports Sciences Reviews.* Baltimore, MD: Williams and Wilkins, 1990, vol. 18 p. 65–117.

68. Young, A. J., W. J. Evans, A. Cymerman, K. B. Pandolf, J. J. Knapik, and J. T. Maher. Sparing effect of chronic high altitude exposure on muscle glycogen utilization. *J. Appl. Physiol.: Respir. Environ. Exerc. Physiol.* 52: 857–862, 1982.

69. Young, A. M., M. N. Sawka, L. Levine, B. S. Cadarette, and K. B. Pandolf. Skeletal muscle metabolism during exercise is influenced by heat acclimation. *J. Appl. Physiol.: Respir. Environ. Exerc. Physiol.* 59: 1929–1935, 1985.

70. Young, D. R. *Physiological Performance Fitness and Diet.* Springfield, IL: Thomas, 1977.

71. Ziporyn, T., K. Simmons, C. Marwick, G. A. Raymond, M. F. Goldsmith, B. Merz, and P. Gunby. Aerospace medicine: the first 200 years. Medical news and perspectives. *JAMA* 256: 2010–2052, 1986.

65. Intestinal adaptation to environmental stress

PETER J. HORVATH | *Nutrition Program, State University of New York at Buffalo, Buffalo, New York*

MILTON M. WEISER | *Division of Gastroenterology, Hepatology andNutrition, Department of Medicine, State University of New York at Buffalo and Buffalo General Hospital, Buffalo, New York*

CHAPTER CONTENTS

CHANGES IN SKIN AND IN MUSCULOSKELETAL, cardiovascular, or pulmonary systems usually come to mind when assessing the impact of the environment on humans. For example, chemicals emitted into the atmosphere can produce rashes or breathing difficulties, and high altitude or decreased gravitational effects can lead to muscle weakness and bone loss. What is not usually considered, however, are the effects of environmental stresses on the internal organs, represented by the gastrointestinal system. In this chapter, we will summarize what is known about the effects of the environment on the gastrointestinal system, a system which can be defined embryologically and medically to include the entire gastrointestinal tract, pancreas, biliary system, and liver. A large part of the gastrointestinal system is, in reality, directly exposed to the environment since its tubular lumen mouth to anus is open to the environment and its mucosa, like the skin, functions as the first line of defense against environmental dangers, be they viruses, bacteria, or ingested substances.

Adjustment to environmental food resources requires a digestive system capable of adapting to a particular biota. This interaction affects the psychosocial behavior of a group, which in turn helps form the history of a people. Desert people prepare and eat foods and have dietary taboos different from those who dwell in jungles, polar regions, or mountains. In the desert, water sources are conserved and dates and olives (indigenous to the desert) become staples. Desert nomads cannot maintain cattle but can keep goats, camels, and sheep. The milk from these animals is more often processed, that is, allowed to ferment in the desert heat. Goat milk is used to produce cheese, and very little milk is drunk; thus, lactose intolerance is not a problem unless these desert dwellers move to a different culture and behavioral adaptation occurs.

Similarly, in equatorial Africa a large majority of adults are lactose-intolerant due to a genetically determined lower intestinal lactase activity (50, 245), yet there are Africans who rely on dairy foods for subsistence (198). Adaptation has occurred either by genetic selection (so that the Batutsi nation, for instance, has a high level of intestinal lactase activity similar to lactose-tolerant white Americans; 198) or through food avoidance or alteration (as milk fermentation by the traditional Zulu cattle herders with their buttermilk-like amasi; 198). Milk fermentation appears to be the means by which other populations have adapted to the environmental availability of animal milk as food; most of the world's population have low levels of intestinal lactase activity (Asians, Africans, and the people of the Near East and Mediterranean) and have either limited their intake of milk or resorted to fermentation, as with amasi and yogurt (127, 198).

Thus the study of environmental effects on the gastrointestinal system must take into account the immediate and long-term interactive effects on individuals and their societies using a fairly broad definition of "environment." In this chapter, we include the following under "environment": (1) the geophysical environment, that is, the effects of temperature, gravity, altitude, latitude, and seasonal changes; (2) human-induced environmental changes, that is, industrial pollutants of the air, waters, and earth (besides the by-products of manufacturing such as lead and mercury, this category also includes pesticides, fertilizers, and hormones used to increase food production); (3) cultural and social factors related to the environment that impinge directly on the individual to affect the gastrointestinal system, such as cultural taboos, population density, urbanization, sanitation, poverty, war, public health programs, and racial or class stratifications; (4) personal life-styles that may alter the environment (external and intestinal) through exercise and physical activity, smoking, unusual diets, and alcoholism. We will relate these effects to differences in gender, pregnancy, aging, and genetics.

THE GASTROINTESTINAL SYSTEM AND ITS ABILITY TO RESPOND TO ENVIRONMENTAL STRESS AND INJURY

As indicated above, we define the gastrointestinal system to include the entire gut from mouth to anus and those organs derived from the gut embryologically: liver, pancreas, and biliary tree. The primitive hepatic diverticulum of the embryonic gut near its point of evagination forms the bile duct, and an evagination from it forms the gallbladder. The gut and the organs derived embryologically from the gut are functionally interrelated to expedite the digestion and absorption of nutrients, to process and package these nutrients so that they may be efficiently utilized by all tissues, and to screen and to prevent the absorption of, or to render harmless, nonnutrients that gain access to the gut.

The gut–hepatobiliary system provides a primary antitoxic and immune defense against injurious substances and conditions. The liver, which receives gut-derived substances directly through the portal venous system, is the gateway to the systemic circulation. The liver's ability to remove, detoxify, and excrete harmful organics through the biliary excretory pathway is critical to maintaining health. The particular immune functions of the liver relate to the need to complement the highly specialized gut contribution to humoral immunity (secretory immunoglobulins) and extensive gut-associated lymphoid tissue, making the gut–liver system the largest immune "organ" in the body. It has been estimated that 70%–80% of human B cells are in the intestine: "More IgA is translocated to the gut lumen every day (40 mg/kg body wt) than the total daily pro-

duction of IgG" (24). Surely, this system must be critical to adaptation and survival in encounters with the environment.

Immune Defense Mechanisms

Although there are excellent reviews on immune defense mechanisms (24, 25), a brief discussion here may be helpful in understanding how the environment may affect the gastrointestinal tract through alterations in the gut immune system. Secretion of antibodies by mucosal surfaces entails the synthesis by lamina propria plasma cells of polymeric immunoglobulins (pIgA and pIgM). The uptake of polymeric immunoglobulins by the mucosal epithelial cell occurs at its basal–lateral surface. A secretory component is added by the enterocyte as the polymeric immunoglobulin traverses the mucosal cell and is released at its luminal side as secretory immunoglobulins (sIgA or sIgM). The secretory immunoglobulins are part of a response to foreign antigens presenting at mucosal surfaces, sIgA being the major immunoglobulin secreted by the intestine, liver, and biliary tract.

The specific organs contributing to the synthesis of secretory IgA may differ among species. Secretory immunoglobulins may be the major proteins found in the bile; these are eventually secreted into the intestine (29). In rodents, the liver hepatocyte has been found to bind and translocate circulating pIgA made by intestinal plasma cells upon induction by specific antigens. From the intestine, these pIgAs enter the liver via the portal system, where they bind to the secretory component on the hepatocyte plasma membrane. As in the enterocyte, pIgA is transported through the hepatocyte attached to the secretory component and then secreted into the bile as sIgA (106, 216). In humans, however, most of the biliary sIgA appears to come from biliary epithelial cells, which are also capable of synthesizing the secretory component (29). It is believed that in this manner the biliary system and, by biliary secretion, the upper intestine are thus protected from environmental pathogens entering via the oral route. In the infant an important source of sIgA protection in the intestine is mother's milk. Ductule cells of the breast are capable of binding circulating pIgA and secreting it into milk as sIgA. The importance of this route has been demonstrated by the presence of passive immunization to cholera in breast-fed infants (85).

The mechanism of protection by sIgA is not clear. Secretory IgA does not participate in complement-mediated killing of viruses and bacteria. It is believed that sIgA functions by binding pathogens in ways that prevent their absorption and allow for pathogen destruction by luminal enzymes or other organisms or by their elimination via fecal excretion.

The discovery of the M cell in rodent mucosa was a major contribution to elucidating the mechanism for programming the B cells to produce sIgA specific to an enteric antigen. These mucosal M cells detect and process foreign antigens for presentation to the gut immune system (246). M cells are concentrated in domes overlying lymphoid aggregates with germinal centers called Peyer's patches. The M cells and their precursors occupying these domes differ from the rest of the enterocytes by their lack of secretory components (175). This implies that the concentration of secretory immunoglobulins would be low and the adherence of antigens to these cells high. In effect, this allows pathogens and other potentially harmful antigens (41) to enter an area overlying a concentration of lymphoreticular cells, where the harmful agents may be processed in ways that enhance stimulation of immunologically mediated defenses (153). In addition, Weltzin et al. (237) presented evidence suggesting that M cells preferably absorb sIgA-bound viruses. They postulated that this is a mechanism for controlled, continuous stimulation of specific antibody production needed to maintain adequate protection in the later stages of infection or pathogen exposure. Humans, in contrast to rodents, have fewer M cells covering their Peyer's patches, which are less diffusely distributed, being mainly located in the distal small intestine.

The role of mucosal immunity in protection against enteric viruses is exemplified by the differences between the Salk and Sabin vaccines for polio. The Salk vaccine was administered parenterally, while the Sabin vaccine was taken orally, but both provided effective protection for millions. Many other oral vaccines have been prepared which may prove effective (153) and more efficient. Oral vaccines to cholera and enteric viruses, for example, would be much more amenable to large-scale immunization of populations, as may be required during epidemics.

Cell-mediated defenses are also important to the intestine. Peyer's patches, which underlie the M cells, contain primarily T8 (CD8$^+$) (that is, cytotoxic/suppressor) lymphocytes, whereas the cells in the lamina propria are mainly T4 (CD4$^+$) (that is, helper/inducer) (25). Another group of cytotoxic T8 lymphocytes (intraepithelial, or IEL, cells) is located between villus enterocytes (99). Guy-Grand et al. (99) have suggested that the role of the IEL cells is to "destroy altered epithelial cells." This fits with the findings of Van Kerckhove et al. (226), who have provided evidence that IEL cells are stimulated by "conventional" epithelial cell antigens. That is, they are induced by these epithelial cell antigens to differentiate into a class of T cells with receptors that recognize self-major histocompatibility complex (MHC) antigens on the cell surface of enterocytes, which have been implicated in inducing B lymphocytes to synthesize a specific repertoire of secretory immunoglobulins.

These interactive systems of gut-derived cells and humorally mediated immune defenses appear to be important not only to prevent the entrance of harmful organisms and substances but also to induce tolerance to environmental antigens to which one is constantly exposed. It is an important part of a well-regulated mechanism, which allows a human to adapt to the constant bombardment of foreign, potentially harmful agents in an ever-changing environment. The gastrointestinal system shares this function with the respiratory system, the other environmentally exposed mucosal surface, in a most direct manner, through the process of homing.

Homing allows for lymphocytes to be stimulated, to be induced to differentiate in a clonal manner that is specific for a particular antigen or organism at a mucosal surface, and then to proliferate and redistribute themselves throughout the body to all mucosa-lined surfaces. Selective migration of T lymphocytes to a particular mucosal surface probably involves recognition sites on the surfaces of epithelial cells and lymphocytes. Many lymphocytes originating in the Peyer's patches enter the circulation to return to the intestine, lodging near the crypts as B cells or between villus enterocytes as IEL T8 suppressor lymphocytes.

The immunosuppressive part of this mucosal defense system, which promotes tolerance, has been applied as a form of therapy in diseases thought to be autoimmune-mediated (147). If the antigen responsible for the disease is known, giving it orally appears to induce tolerance and amelioration of the disease. The effectiveness of inducing oral tolerance has been shown in animal models of autoimmune diseases, and currently there are trials in humans involving feeding myelin for multiple sclerosis, collagen for rheumatoid arthritis, and S-antigen for uveitis (73). Failure of inducing oral tolerance to food substances that are structurally and antigenically very close to the equivalent host molecule is, presumably, the postulate invoked in these therapeutic trials.

Failure to induce tolerance, and hence the induction of autoimmune disease, is not the same as food allergies or nonallergic reactions to food contaminants. Eating, it should be emphasized, constitutes the port of entry for many environmental factors along with food, such as pesticides and chemical waste–contaminated soil in which crops were grown or animals grazed, etc. Food allergy as a reaction to the environment may be separated from contaminants of food in that it is an individual's particular immunological reaction to an environmental substance to which most people do not react. This is in contrast to the typical reaction most people may have to an environmental food contaminant, where the reaction depends on the quantity of the contaminant required to produce symptoms. True food hypersensitivities are considered to occur at low frequencies, but it must be admitted that the medically acceptable criteria for defining food allergy are being expanded as knowledge of the intestinal immune mechanisms grows. The presence of antibodies in the sera of individuals is no proof of allergy to the antigen. Many patients with diseases in which the mucosal barrier has been broken have antibodies to many common food proteins but do not exhibit allergic symptoms when ingesting these proteins. Based on strict immunological criteria, the best evidence of an immune reaction to a food antigen is that of immediate type I hypersensitivity, as commonly occurs with peanut and shellfish allergies. The mechanism of this reaction requires IgE-mediated stimulation of mast cells, which are abundant in the intestine. The symptoms are nausea, difficulty in breathing, abdominal cramps, and sweating as signs of mucosal edema of the lungs and intestine develop and urticaria ensues; a generalized anaphylactoid reaction may then ensue with unconsciousness and apnea.

One also has to distinguish between immunologically mediated responses and intolerance to certain foods that may involve unusual metabolic or toxic products. One example of the latter is phenylketonuria, in which a genetic defect in phenylalanine metabolism leads to mental retardation and eczema, and its treatment is the elimination of phenylalanine from the diet. Another more common example is the reaction to monosodiumglutamate, a taste enhancer added to many processed foods and Chinese–American cuisine, which for some individuals causes profound headaches, weakness, flushing, and fainting episodes. Neither disorder has been shown to be immunologically mediated. Other foods or chemical contaminants may involve the intestinal immune system to neutralize or remove the contaminant, but these do not produce clinical symptoms.

Other Defense Mechanisms

There are three features of normal gastrointestinal physiology which also serve as important defense mechanisms to environmental substances: the production of gastric acid, the actions of intestinal motility/secretion, and the extensive adaptative mechanisms of the intestine. The ability to maintain a gastric acid content above 10^{-4} M [H$^+$] (that is, a pH below 4) has been found to destroy many of the bacteria that are commonly ingested. Bacterial resistance to acid is, in part, a determinant of pathogenicity. For example, the *Vibrio cholera* bacterium is very susceptible to acid, and cholera requires the ingestion of over 10^6 more organisms than the acid-resistant *Shigella* bacterium for establishment

of illness. In a cholera epidemic, the elderly are more susceptible partly because a greater number of them have reduced gastric acid production capacity. Similarly, malnourished children may exhibit a reduced capacity to maintain gastric acidity, increasing their susceptibility to cholera (82). Whether viruses are affected by gastric acid has not been studied, and there is little information on gastric acid effects on environmental chemicals and toxins that may be ingested.

Normal intestinal motility and secretion moves harmful substances and pathogens out of the body. Increased motility and enhanced secretion resulting in diarrhea might be viewed as an appropriate defense against harmful substances. Decreased or disorganized motility, however, is often a complication of chronic diseases, such as diabetes or scleroderma, or of surgery. One effect of a chronic alteration in intestinal motility is small bowel bacterial overgrowth. The small intestine is normally kept nearly sterile by peristaltic motility. In the terminal ileum, bacterial counts increase dramatically as the ileal–cecal valve is approached. This represents an area of relative stasis due to the valve. Peristaltic activity in the large bowel is much less frequent and appears designed to maintain a constancy of bacteria since "normal" colonic bacteria fulfill a symbiotic function in stool formation and short-chain fatty acid metabolism. Bacterial overgrowth in the small intestine, as may occur in long-standing diabetes, leads to changes in bile acid absorption, malabsorption, and chronic diarrhea alternating with severe constipation.

The intestine's remarkable ability to adapt is characterized by its large reserve functional capacity and its flexibility in growth (169, 221). Crane (50a) has reported that humans have the capacity to absorb greater than 5 kg of glucose a day. Other studies suggest that the "safety margin" for absorption is more in the range of two to ten times the normal intake (30, 102). In addition, most of what is secreted by the stomach, pancreas, and biliary tract is in excess of what is required to digest and solubilize ingested contents. Many harmful environmental factors are altered or neutralized rapidly by these secretions, which are capable of suspending or solubilizing fairly insoluble compounds, as for example, hydrophobic anilines. The detergent actions of the bile salts and phospholipids are critical since they are mainly responsible for suspending or solubilizing hydrophobic compounds in ways that allow pancreatic and brush-border enzymes to be effective. The enterohepatic circulation of bile salts ensures large quantities where required in the intestinal lumen. This is a very efficient process in which 96% of the luminal bile salts are reabsorbed in the ileum to be used again and again, with minimal requirement for bile salt synthesis by the liver. However, some ingested materials are capable of binding bile acids as may occur with high-fiber diets (38, 120). Such substances will not deplete the capacity of the liver to synthesize more bile acids unless given in pharmacological amounts or with injury or loss of the site of bile salt reabsorption (the ileum).

Intestinal adaptation by growth includes lengthening of the intestine and alterations in enterocyte renewal, the former being largely a feature of the infant and young child (169), though a modest increase in length has been reported in the adult after extensive resection of the small bowel (200, 203) and lactation (233). Enterocyte renewal may lead to shorter or longer villi and crypts, that is, a thickening or thinning of the intestine. The thickening may be to an extent such that the term "hypertrophy" may be used.

Among the unique properties of the intestine are its rapid rate of cell turnover and a histologically defined axis of differentiation from crypt to villus. The proliferative compartment is in the lower crypts, each crypt being the progeny of one stem cell. It has been estimated that approximately 250–300 cells per day emerge from one crypt. A single crypt will contribute to a number of neighboring villi. Moving out of the crypts and up the villi, a cell proceeds in a spiral manner around each villus, with separate "rivers" of cells from different crypts running alongside each other (180). Enterocytes are expelled into the lumen when they approach the tops of villi, having completed what appears to be an apoptotic (programmed) death (223). The life of an enterocyte, from the time it emerges from a stem cell until its loss into the lumen, is approximately 3 days in rats and 5 days in humans. A fully mature villus cell occupies the upper half of the villus, differentiation occurring gradually as the cell traverses the upper crypt and lower villus zones.

What constitutes maturity is defined by the emergence of a most highly polar columnar cell. It is characterized by a plasma membrane domain whose surface area is markedly increased by the formation of prominent microvilli extending into the lumen. Within and upon this luminal membrane are ion channels, unique transporters (for example, Na^+/glucose cotransporter) and many hydrolytic enzymes (for example, sucrase, dipeptidase, and alkaline phosphatase). Within the microvilli are cytoskeletal contractile proteins consisting of actin filaments cross-linked by villin and fibrin (64, 86). These actin bundles are connected to the microvillus membrane by a 110 K calmodulin complex with myosin-like Ca^{2+}-ATPase activity (46, 79, 162). Also characteristic of this highly polarized cell is its lateral plasma membrane domain, whose surface area is also increased by extensive plications (229). Indeed, its surface area may be greater than that attributed to the microvilli. On electron microscopy, this lateral domain is usually seen as two coapted membranes of adjacent cells. During active absorption, however, these membranes may separate to form a potentially large intercellular space filled with fluid and products of digestion and absorption. This space is part of the paracellular route for ion entry or secretion. The lateral plasma membrane is defined by the presence of ouabain-sensitive Na^+/K^+-ATPase activity (sodium pump), the major electrogenic pump of the enterocyte that drives many of the cell's transport functions. As the enterocyte completes its maturation, it also prepares for death, a process termed *apoptosis*.

There are three mechanisms whereby the intestine can respond to direct injury of its mucosal layer, as may occur with high concentrations of alcohol. The first is to have the cells adjacent to the injured or sloughed area assume a more cuboidal or even flattened shape to provide rapid closing of the gap in the mucosal barrier (144). The cells nearest the wound migrate by lamellapodia-like processes to close the denuded or basement membrane–destroyed area. The second response mechanism is a neurally mediated contraction of the villus structure, a shortening that also causes the injury gap to decrease (161). The third is to increase the rate of enterocyte proliferation to meet the demands of increased destruction.

The latter would be the major mechanism for adjusting to chronic injury. This has been shown to be the typical reaction to lectins found in certain foods, such as beans (57, 92, 184), to continuous low-grade injury from bacteria (11), to other presumed microbiological agents (122), to food allergies (98), and to celiac disease. Celiac disease, gluten-sensitive enteropathy, is a human disease induced by the ingestion of gluten-containing foods (ordinary breads) and characterized by "sick," less differentiated enterocytes unable to adequately absorb foods. A major feature of celiac disease is the compensatory increase in mitotic activity leading to a greater depth of the crypt compartment, which is an attempt to replace dysfunctional enterocytes. Renewing the intestine rapidly is an important, critical defense against many environmental injuries.

It is the latter feature, the ability to renew the mucosa, that could be seriously injured in cancer chemotherapy and irradiation. Radiation sickness, as may occur accidently in an overdose during therapy or at sites of atomic bomb testing or as had occurred at Hiroshima, Nagasaki, and Chernobyl, may result in an intestine that cannot renew itself, a condition, until the advent of total parenteral nutrition (TPN), incompatible with life. To what extent other environmental agents or conditions may affect the stem cells or the ability of the enterocytes of the intestine and colon to proliferate and differentiate has not been studied.

GEOPHYSICAL ENVIRONMENT

The geophysical environment may include temperature, gravity, altitude, and latitude, though very few studies have been done on the adaptation by the gastrointestinal tract to latitude (80).

Environmental Temperature

Hyperthermia. External temperature affects the gastrointestinal tract by altering the digestive and absorptive functions and the rate of passage (total time for food to traverse from mouth to anus). As ambient temperature rises, blood flow to the skin increases to maintain body temperature. This results in decreases in blood flow and less volume to the central organs, particularly the gastrointestinal tract. This could occur in hot climates or in hot industrial settings, such as foundries, steel mills, and canning factories. High environment temperatures decrease the rate of gastric emptying, especially if accompanied by lack of water intake (164).

In addition, the gastrointestinal tract responds to the increased loss of salts (Na^+, Cl^-, K^+, and others) in perspiration, as occurs particularly with strenuous activity (work or play) in dry heat. These losses are usually replaced by increased intestinal absorption and reabsorption of these nutrients. That is, salt and water ingested under hyperthermic conditions appear to be more efficiently absorbed in the small and large intestine if the body is in a salt-deficient state (56, 188). In addition, in a response similar to that of the kidney, the intestine also becomes more efficient in absorbing salt and water via the renin–angiotensin system (137, 138). In particular, the large intestine, in response to aldosterone and corticoids (15, 199), increases sodium absorption (75) and actively secretes potassium (77, 152, 179, 244) with excessive perspiration. Chloride is absorbed with sodium by the colon and water follows. It has been estimated that the colon has the capacity to absorb up to six liters of water a day (54). A person becomes aware of these compensatory actions by a decrease in volume and a darkening of the urine and by the more compact, harder, "constipated" stool.

Sodium, chloride, and water are not the only losses that occur during sweating. Sweat also contains other elements, including trace elements whose daily losses can be significant: calcium (especially at low intakes; 47), magnesium (48), zinc (from 0.6 to 0.8 mg/day at moderate levels of physical activity; 9), copper (0.4 mg/day), and iron (0.25 mg/day 9, 111). These represent a substantial portion of the daily requirement. Magnesium loss through sweat in hot environments could account for 25% of the total daily loss (48). With iron the requirement may increase by 25% if absorption by

the intestine remains at 10%. Extreme exercise could result in a loss of up to 2 mg/day of iron (67, 176). The intestine can respond to these losses by increasing absorption of dietary sources and by reabsorption of endogenous losses, especially calcium (70, 72), zinc (74, 76, 208), copper (208), and iron (139, 177), if the diet is adequate. The ability of the small intestine to increase the absorption of iron during a period of deficiency has been well documented (12, 139). Not so commonly appreciated is that up to 2 mg of iron per day can be lost in sweat. When this is accompanied by dietary lack and additional losses, the need for increases in dietary iron may be critical to prevent iron deficiency anemia. For example, in long-distance runners the iron loss in sweat, added to the loss that occurs with increased hemolysis of red cells and exercise-induced trauma of the intestine (see below under Physical Activity and Exercise), can lead to subclinical iron deficiency. Along with its ability to increase iron absorption during increased need, the intestine protects the rest of the body from excess dietary iron by binding iron in the enterocyte for subsequent loss in the fecal material after the cell is sloughed (90, 139). In some cultural settings, dietary iron can be in such excess that the body cannot adapt through cell sloughing (91), as with sub-Saharan African natives where a maize beer-like beverage is made in steel drums (21). In addition, genetic defects in this regulation lead to hemachromatosis and iron-loading anemias (1, 28, 90, 154). This fine-tuned mechanism of iron regulation needs further study.

As emphasized earlier, the intestinal lumen can be considered as part of the external environment. This is especially relevant since food and drink are consumed hot and cold and in large amounts. There is a dramatic circulatory response to ingestion of hot fluids (45°C) with decreased intestinal blood flow (almost to half), arterial hypotension, and increased heart rate (195). This "viscerocirculatory thermoreflex" appears to be mediated by substance P, which can affect organs other than the gastrointestinal tract through afferent C fibers. The response of the rest of the body to food temperature needs to be explored.

Cold Stress. Effects of cold temperatures on the gastrointestinal tract have been studied more extensively than those of hot temperatures. Cold exposure increases caloric requirements, leading to intestinal adaptation. The intestinal tract can increase in length and villus height with external cold exposure (59, 206). This is an indirect response and may be a nonspecific hypertrophic response to the increase in food intake due to the increased metabolic demand required to maintain body temperature in cold exposure (10). The small intestine adapts to this increase in caloric intake by more than just increasing the absorptive area. Absorptive capacity also increases by as much as 25%, as shown for glucose and proline (59). Digestive function also increases as measured by apparent digestibility of dry matter (59), shown by increased food intake but no increase in fecal mass. Apparent digestibility does not take into account endogenous losses (119, 209), and detailed studies need to be done to determine if endogenous losses change with cold exposure.

As the gastrointestinal tract meets these increased demands, it also contributes significantly to the increased need of the body for heat during cold exposure. This has been shown in humans, where blood flow from the splanchnic–visceral region was responsible for almost 40% of the rise in heat production during cold stress (reported in ref. 149). This increase in heat output appears to be due to increased oxygen consumption and/or increased blood flow. In pigs short-term exposure of the whole animal to cold resulted in increased oxygen consumption by the intestine (149). This increase in oxygen consumption occurs even if blood flow decreases (150) or does not change (149). The effect on gastrointestinal blood flow is not uniform, being dependent on the time of the last meal and on the duration and type of cold exposure. Any meal will increase gastrointestinal oxygen consumption, but this is even greater when the animal is exposed to cold. When only extremities are cooled, colonic blood flow increases due to vasodilation in the colon (93–95, 236). Cooling of extremities leads to short-term increases in mesenteric blood flow. This response seems to be mediated through neural mechanisms. The increased blood flow through the intestine passes through the liver, where oxygen consumption also increases upon cold acclimation (234). The size of the liver also increases rapidly (206, 221) and continues to show hyperplasia relative to animals exposed to warm environmental temperatures (100). This increase in liver mass, as in the case of intestinal enlargement, may be due in large part to the increase in caloric intake and metabolic load on the tissue.

Gravity

The physiological effects of space travel may be considered similar in many respects to the effects of long-term bed rest (171), water immersion (96), or head-down tilt experiments (171). Immobilization by bed rest, for example, results in negative balances of potassium, phosphorus, and calcium (97), findings which also occur during space travel. Total body calcium decreased in growing rats after 18 days in space (178). After 14 days in space on COSMOS 2044, however, no loss of bone calcium was observed in adult rats (225). Humans lost almost 0.3 g of body calcium per day during Skylab 4 (186). Gravity-induced bone loss appears to be due to a lack of stress on the bones. This differs from other forms of osteoporosis resulting from a lack of either dietary calcium or active vitamin D hormone. When calcium is lost from bone, increased urinary and fecal calcium excretion results. Space travel osteoporosis does not appear to initiate calcium homeostatic mechanisms. Although one might have assumed that an increase in urinary calcium would be the result of increased calcitonin, no evidence has demonstrated an increase in calcitonin. Calcium absorption by the intestine may be decreased as serum calcium tends to increase during calcium loss from bone. This is similar to what occurs during prolonged bed rest, when parathyroid hormone (PTH) and vitamin D_3 are found not to be elevated, though bone calcium is being lost chronically (132). Corrective measures, such as resistance exercise, have not solved the problem (132). If bone calcium loss can be stopped or reversed, adequate vitamin D is required to maintain calcium absorption and to provide the source of calcium for bone remodeling. This will be especially important during extended space travel. (Data from the long-term experiments of the Soviet space program would be invaluable.) The active metabolite of vitamin D is the hormone 1,25 $(OH)_2$ vitamin D. The vast majority of vitamin D is produced in the skin by ultraviolet light, but the light in current space vehicles does lead to the production of vitamin D. The other source for vitamin D is dietary. Thus dietary vitamin D may become of greater importance during extended space travel. It has been suggested that inclusion of foods higher in vitamin D might be a partial solution (132).

The intestine responds to the loss of muscle mass during space flight by increasing the apparent absorption of amino acids from the diet. Balance studies of nitrogen intake and fecal nitrogen output suggest that increased dietary nitrogen intake did not lead to changes in fecal nitrogen excretion (131, 136). Balance studies, however, do not account for endogenous sources. Fecal nitrogen is predominantly microbial in origin and is proportional to dietary fiber intake (148, 151, 212, 213). Efforts are under way to make the diets consumed during space travel similar to the ones the astronauts consume regularly, but space and meal preparation limit some of the choices. Our analysis of the diet for a Space Shuttle astronaut given by Lane and Schulz (132), showed that dietary fiber intake was about 28 g/day, which is fairly high for the typical American. The fact that fecal nitrogen did not decrease could result from either an increase in "true" amino acid absorption, protein fermentation, or a decrease in absorption, with increasing levels of microbial nitrogen reaching the feces (209, 210). More technically sophisticated studies (for example, triple lumen intubation, stable isotopes, fecal microbial nitrogen, etc.) need to be done if humans are to remain in space for extended periods of time and maintain optimal performance.

Similarly, accurate determination of intestinal motility and colonic function during space travel may require marker pellet studies and dietary fiber balance trials with detailed fecal analysis. Published studies have concluded that little or no change in motility and colonic function occur (131). These studies were done using relatively crude measurements of fecal weight and number of defecations. If the data are reexpressed as percent water in the feces or as fecal output per day, it appears that transit time may have actually decreased and that colonic water reabsorption increased during space travel. This may affect the microflora of the large intestine, and disturbances of the complex microbe/host ecosystem could lead to reduced resistance to infection (157, 217, 218). When the microbial populations are stressed by environmental changes, such as hyperoxic (49), hypobaric, and hyperbaric conditions (81), the normal microbial populations change. This, combined with other hormonal and dietary stresses, has to be considered if long-term health is to be maintained in space and under the sea.

Altitude

The responses of the gastrointestinal tract to high altitude and high hydrostatic pressure appear to be compensatory in nature to the responses of other systems. An example is the increase in iron absorption during increased erythropoiesis and polycythemia with high altitude adaptation (61, 107, 189). Mountain sickness at high altitudes has many gastrointestinal symptoms, including nausea and anorexia. This usually occurs on reaching relatively moderate altitudes (2, 400 m and above). In addition to decreased appetite, utilization of dietary energy (40, 113) and protein (114) appears to be decreased by exposure to high altitude. The nutritional status of many minerals also appears to be affected (114), including that of zinc (58, 205). The intestine may be part of the adaptation to changes in zinc status (58) resulting from the loss of lean body mass associated with high altitude, but specific zinc absorption studies have not been done. Other nutrients, copper and vitamin B_6, were studied but were found not to be affected by high altitude (58).

ENVIRONMENTAL EFFECTS RELATED TO INDUSTRIALIZATION

It is often difficult to separate environmental effects from those related to toxicology, industrial medicine, and pollution. It is also hard to confine the discussion to physiological adaptations rather than the induced

pathology and consequent disease. For example, in the early 1970s a physician noted an increase in the incidence of a rare type of liver tumor in workers assigned to clean out a large tank used in making vinyl plastics; it became a question of exposure to vinyl halides at very low levels in the air over many years (51, 69, 133, 247). It is unclear as to the mechanism by which this low-level exposure could result in a quite rare type of tumor. The way research has tried to answer this and other questions highlights the complexity of the problem. Among the questions raised are: To what extent over these years of environmental exposure was hepatic metabolism changed due to this exposure? Did this chronic long-term environmental/occupational exposure accelerate the detoxification mechanisms of the hepatocyte endoplasmic reticulum, allowing the liver to be exposed to larger concentrations of potentially carcinogenic metabolites? Did alcohol intake, which affects drug metabolism, increase the susceptibility to tumor induction by vinyl halides?

There is no significant bibliography documenting the pathology of industry-related environmental pollutants on the gastrointestinal tract. This may be due to the time required for these pollutants to enter the food chain and the geographic route they often travel. They may enter through run-off soil contamination, chemical fertilizer and pesticide application, or the use of chemicals or sprays to enhance the appearance of an agricultural product. A direct association with illness is not easily apparent due to the ubiquitous occurrence of these compounds.

Industrial Contaminants of the Food Chain

We use the word "contaminants" for those substances that are not normally considered part of a natural agricultural product. Thus the effect of milk lactose on lactase-deficient persons is not the result of lactose being a food contaminant. However, mercury entering the food chain due to industrial pollution of a nearby lake or bay is a food contaminant, even though it could become incorporated into the food through normal biological mechanisms.

Mercury. Mercury compounds are important environmental pollutants (169). Exposure may be by mercury vapor, phenylmercuric compounds, or mercurous and mercuric salts. Often mercury compounds accumulate in lakes, streams, and bays, usually from industrial waste, and enter the food chain via the water supply or by ingestion of fish from contaminated waters. This can lead to health disasters involving large numbers of people (as in Japan in 1956 and 1965, see ref. 44). Other sources are of toxicological or manufacturing origins. Seed wheat treated with a methyl mercury fungicide was used to prepare bread, which led to a disaster of mercury poisoning in Iraq in 1971–1972. Agocs et al. (2) reported a case of mercury intoxication in a child exposed to mercury-containing paint fumes. Their studies demonstrated that latex interior paints, which contained phenylmercuric acetate as a preservative, released mercury fumes on drying and led to increased urinary mercury in those exposed.

These are examples of relatively subacute exposures where the source may be obvious and the symptoms diagnostic. One unanswered question is whether long-term exposure to low levels of contamination might lead to physiological adaptations which could result in chronic disease. Many environmental contaminants are thought to induce permanent changes in the immune system (23). Lead exposure appears to decrease serum immunoglobulins, IgA, and antibody-dependent cellular cytotoxicity (23). Animals chronically exposed to mercury have been shown to develop an autoimmune disease. This disease was similar to a human disease of the kidneys manifested by antibodies to the glomerular basement membrane and membranous glomerulonephritis with proteinuria (35, 142, 183). With time, other organs of the experimental animals were injured by a widespread autoimmune-mediated process including necrotizing vasculitis with intestinal involvement (68). These lesions were induced by repeated parenteral injection of mercuric chloride. An identical process was seen after prolonged oral ingestion of mercuric chloride (123, 124). As previously demonstrated in rat (14), Hultman and Enestrom (109) have shown that the oral dose required to produce disease can be seen with a "body burden similar to that reported in some occupationally exposed humans." Earlier, Knoflach et al. (123, 124) reported that rats given oral mercuric chloride in the drinking water developed immune deposits of IgG and C3 in intestinal vessel walls by 11 wk and later demonstrated similar immune deposits along the intestinal basement membrane. This resulted in increased protein loss through the intestine. Bohme et al. (19, 20) reported that mercury compounds in vitro increased intestinal permeability by loosening tight junctions, a feature, incidentally, that appears to be characteristic of Crohn's disease, even in the uninvolved segment of bowel. Only very select genetic strains of rats (Brown-Norway) and mice (SJL/N) developed immune-complex disease with mercury administration. Could long-term exposure to mercury plus hereditary factors be one cause of Crohn's disease?

Mercuric chloride can inhibit intestinal ion, carbohydrate, and amino acid transport. It has been postulated that this is a result of inhibition of the Na^+, K^+-ATPase pump (5). Bohme et al. (20) have suggested that the secretory diarrhea induced by mercury compounds may be due to increased Cl^- secretion, mediated partly by prostaglandin and partly by both cholinergic and noncholinergic neurons. Thus mercury compounds contaminating the environment usually enter the body through the gastrointestinal tract and, with long-term exposure, can lead to physiological alterations in ion transport, increased intestinal permeability, chronic immune-mediated inflammation, and protein-losing enteropathy.

Pesticides. Pesticides most often affect the liver. Unless it is a relatively acute or subacute effect, the relationship of a particular pesticide to liver disease is difficult to separate from the effects of such agents as alcohol, hepatitis B virus, or aflatoxin. This is especially true in endemic areas of hepatitis B virus where chronic liver disease is common. Aflatoxin, a fungus *(Aspergillus flavus)* product which commonly contaminates food in Asia and Africa, has been implicated as contributing to the induction of hepatoma, that is, liver cancer (126, 174). Hepatitis B, however, is endemic to these populations, making it more difficult to assess the role of aflatoxin.

Pesticides and insecticides have acute and devastating effects on the liver and gastrointestinal tract (141, 215, 219, 235) involving a variety of mechanisms, most of which are related to toxic product formation. Paraquat toxicity is reported to be due to the formation of highly reactive oxygen free radical metabolites (55). A summary of these liver metabolic pathways utilized by pesticides is given by Alvares (4). Drug-metabolizing systems are mainly microsomal, involving cytochrome P-450 systems, but also include cytosolic conjugating enzymes that aid in the excretion of drugs and toxic metabolites. Many of these chemicals lead to an increase in the drug-metabolizing systems, that is, enzyme inducers (222). Depending on the exposure or dose, this increase may lead to the accumulation of metabolites which have effects critical to cell viability or control of transcription. It is postulated that the DNA adducts formed by some metabolites lead eventually to cancer. Although acute or subacute toxic effects of these pesticides are environmental, they are more often categorized as toxicological. Their more pernicious effects on the physiology of the gastrointestinal/liver system leading to pathology should be seen as a result of environmental exposure, but then it becomes more difficult to prove. For example, Dietze et al. (60) used the ^{14}C-aminophenazone breath test to evaluate the hepatic demethylation capacity of 59 workers chronically exposed to pesticides. They found that the values decreased as the years of exposure increased. Lorenz (141), however, did not find any liver pathology characteristic of pesticide exposure in another group of similar workers with a history of long-term exposure. This may merely mean that the chronic long-term effects of many pesticides are to alter liver metabolism, though not to a pathological degree. This functional change, however, may make a person more susceptible to injury from other substances (for example, ethanol) or to malignant transformation. Proof of these associations requires continuing epidemiological and correlative controlled laboratory animal studies.

PERSONAL LIFE-STYLE

There are numerous life stresses that affect the gastrointestinal system. Among these are exercise, starvation, dietary challenges, pregnancy/lactation, aging, etc. Many factors have been implicated in the adaptive response. They include neurological, hormonal [gastrin, glucagon, bombesin, cholecystokinin (CCK), pancreozymin, catecholamines, epidermal growth factor (EGF), etc.] and paracrine (somatostatin, prostaglandins, polyamines, etc.) modulators. For example, polyamines (putrescine, spermine, and spermidine) appear to be a factor in various intestinal adaptations (63), including the growth response of the intestine during refeeding after starvation (63). Catecholamines increase liver glycogen breakdown, gluconeogenesis, amino acid uptake, and lipolysis (13, 89, 168). These are the same changes that occur during cold exposure and strenuous exercise, when increased levels of catecholamines are also observed.

Physical Activity and Exercise

One of the key differences among individuals is the extent of exercise or physical activity performed. Strenuous exercise can have profound effects on the gastrointestinal tract (31, 164). These include motility changes (actual vomiting and increased rate of passage), tissue damage leading to blood loss, and decreased gastric function (164).

Strenuous exercise, such as experienced by marathon runners, may lead to "runner's trots," fecal blood loss, and vomiting. Runner's trots is the urgent need to defecate following strenuous exercise; diarrhea commonly occurs with marathon runners (117, 182, 190). This diarrhea may result from an increased rate of passage (170) and lower fluid absorption by either the small intestine or the colon. This may be due to either an increased load on the small intestine from the stomach or an exaggerated gastrocolic reflex, since strenuous exercise also increases gastric emptying, or both (27, 167). Extensive training may result in adaptation, with a decrease in diarrhea (248). In addition, there are long-term effects, since resting gastric emptying is also more rapid in marathon runners (36). Reports have shown conflicting results, but some studies indicate that small

bowel rate of passage may be increased (170). The small intestine rate of passage, or mouth to cecum transit time (MCTT) may be affected by other mechanisms besides increased bulk from the stomach into the intestine. There is the well-known, yet not fully explained, phenomenon of the gastrocolic reflex. This reflex appears to be mediated through extrinsic hormonal and neural systems. Most of the studies on MCTT and exercise have used the breath hydrogen test, which requires fasting and low fluid intake. This is not what commonly occurs during exercise. This methodological problem may explain the conflicting data of case reports and controlled experiments, since during exercise most individuals consume fluids. Studies have shown that, in men, low-grade exercise decreased MCTT (34, 118), but heavy running for 90 min did not change MCTT (36). In a study that included women, low-level exercise increased MCTT (156), as was also the case in men tested on treadmills (166). Further studies need to be done using techniques that do not require fasting and involve individuals of different ages, genders, and training experience. Studies have suggested that physical activity and exercise may increase large bowel transit time (16, 118, 172). If it is true that the incidences of colon cancer, diverticular disease, and hemorrhoids are negatively related to transit time through the large intestine, then exercise and physical activity may be beneficial.

Exercise could also have more direct harmful effects on the intestine. During exercise, there is a decreased blood flow to the intestine as blood is diverted to the skeletal muscle and peripheral tissues for body cooling (45, 185). This may lead to actual ischemia, especially when combined with jarring damage to the intestine during running. The colon is probably more sensitive to ischemia than the small intestine due to the lack of collateral circulation, especially around the splenic flexure (197). In addition, this decreased blood flow to the intestine may be exacerbated in the older runner because of arteriosclerotic lesions of the mesenteric vessels. The mucosa can regenerate rapidly if the period of ischemia is short, but longer periods can lead to necrosis (197) and catastrophe. If necrosis does not develop, the area of ischemia in both the small and large intestine responds by increasing proliferation to replace the injured enterocytes (155, 191, 192).

Ischemia that may occur during either extreme exercise or running may also result in iron loss through the intestine from the damaged tissues. In addition, intestinal bleeding, especially gastric hemorrhage, appears to be common in runners. Increased fecal occult blood loss has been associated with extreme exercise (165, 214). This blood loss through the intestine, combined with increased iron loss through sweat, could be the underlying basis for the subacute anemia commonly seen in runners (67). Other organs of the gastrointestinal system are also involved in the response to exercise, especially the liver (17, 193), but that is beyond the scope of this chapter.

Starvation

The gastrointestinal tract has evolved to respond to short-term starvation, undernutrition, and changes in dietary components. In the early phase of starvation, the metabolic needs of the gastrointestinal system decrease. The small intestine responds to the lack of lumen nutrients by decreasing the rate of enterocyte proliferation (87, 128). Secretions from the pancreas and biliary systems are also decreased. In addition to the decrease in cell number, there is a decrease in the activities of most brush-border enzymes (3), such as sucrase, maltase (66), and aminopeptidase (78), but some hydrolases, such as lactase, show an increase in activity (78). This response may be mediated at a posttranslational level by thyroid hormones (78).

When the colon is starved for luminally supplied nutrients, as a result of either starvation, intravenous feeding, or a diet low in dietary fiber, the rate of proliferation decreases, with possible atrophy of the colon (112, 163, 196). Aged animals may not be able to respond as well as younger animals (105). With starvation, it was found that older rats had a smaller reduction in crypt cell proliferative rate (105). Crypt depth was shorter in the nonstarved older rats (88), and the crypt cells of the colon of aged rats appeared to proliferate at a faster rate. During starvation, the decrease in proliferation was much less in aged rats, and they overresponded to refeeding (105). Although colon cells grow at the same rate, this suggests that in the older animal the ability of the colon to adapt to starvation stress was lessened. This effect of aging needs to be studied further to understand the underlying mechanisms.

Luminal Environment and Diet

Macronutrients. In a very classical sense, adaptation by the gastrointestinal tract to changes in diet has been well studied. Since the major function of the gastrointestinal tract is to provide nutrients to the rest of the body, it is to be expected that it will adapt to changes in the composition of the diet and the needs of the body for specific nutrients. Excellent reviews are available by Karasov and Diamond (116) and Williamson and Chir (242, 243) on the response of the transport function of the intestine to dietary changes of the macronutrients and hydrolases (3). Up-regulation of glucose transporters can occur within hours after a high-glucose meal (115). In terms of micronutrients and the ability of the intestine to adapt to changes in dietary amounts and nutrient

status, the best examples are calcium (232), iron (139), copper, and zinc (208).

The pancreas is well known for its adaptability to changes in the diet (207). The adaptation occurs during a meal, following daily dietary changes and adjustments to long-term dietary patterns. That is, during a meal pancreatic secretions are controlled through gastrointestinal hormonal and neural reflexes; a salad followed by a steak would elicit high enzyme secretion after the steak. The pancreas also adapts to a previous meal by changing pancreatic protein synthesis in preparation for the coming meal (26, 53, 187, 239). Within hours after the steak, the pancreas increases the synthesis of enzymes, particularly proteases and within 24 h has established a new level of enzyme synthesis (26, 53, 83, 84, 240, 241). If large amounts of protein are not continually consumed, the levels of pancreatic protease will slowly decrease along with their rates of synthesis, and they may take as much as a week to return to the original levels (26, 108). The physiological significance of these late changes is not clear, since it appears that the pancreas secretes many times the amount of enzymes needed for proper digestion. With adaptation, digestion, as measured by intake and fecal output, does increase from 94% to 97.5% (12, 134).

In general, a high-protein diet (above 65% casein) increases pancreatic proteases but decreases amylase secretion. High-carbohydrate diets (between 60% and 75% of calories as starch or glucose and to a lesser extent sucrose and fructose) increase amylase levels and decrease both lipase and protease activities, but a high-lipid diet (40%–75% of the calories) increases lipase and decreases only amylase secretion. The quality of the protein (that is, its biological value) also influences the response of the pancreas, the response being greater with higher quality protein (146). The effect on the pancreas requires the presence of proteins or peptides in the lumen; one cannot get an effect with intravenous amino acids. This is different from the response of the pancreas to elevated plasma glucose levels, which can result in increased pancreatic amylase levels. On a very long-term basis, it would be expected that cultural differences in dietary composition should alter the profiles of pancreatic secretion. Studies on dogs fed diets different in fat capacity found no differences in resting pancreatic secretion profiles, but during a meal the relative concentrations of amylase, lipase, and protease were different (10). These researchers suggested that the difference in the stimulation of pancreatic secretion may be due to the altered stimulation of CCK release in response to the diets. Other studies, in which protein and carbohydrate composition were changed, suggest that the responsiveness of the pancreas to CCK and secretin may be altered under these conditions (65).

Dietary Fiber. Dietary fiber is another variable that depends on cultural and environmental influences. There are wide variations in the amount and type of fiber consumed among cultures; the cultural preferences may depend on availability, as influenced by economics and agriculture. High cassava consumption in tropical areas, high sorghum consumption in arid climates, high rice consumption in Asia, high wheat consumption in North America, and high bean consumption in Central America represent different fiber diets. The composition and amount of dietary fiber affect intestinal microflora metabolism and tissue structure and function. The intestine and liver increase in size and in heat production when dietary fiber levels increase. Even the lungs have been reported to show increased heat production (194). The most pronounced effects occur in the intestine, with increases in weight and changes in structure/function (18, 80) due to the increased bulk of luminal contents and fermentation end-products. This is especially true of the stomach and small intestine. The proliferation rate of colonic mucosa (18) increases along with changes in the cellular makeup of the mucosa, with more goblet cells and mucin production (224). Mucosal enzyme levels increase (201) and there is an increase in the secretion of pancreatic enzymes (202). Also, key intestinal transport processes may be altered, high-fiber diets having been shown to increase Na^+/K^+-ATPase activity in the basolateral membrane of the enterocyte (194). In the small intestine other structural changes are probably due to the intestinal flora and their fermentation end-products, short-chain fatty acids. The villi in humans of Western cultures have a markedly different appearance (finger-like) compared to those of people from the developing tropical countries and of vegetarians (convoluted and feathery with ripples) (37, 173). The former look much like the villi that develop in the rumen upon weaning after the introduction of fiber and fermentation (227). This may be due to the presence of butyrate, a known differentiating agent (101, 129, 135, 181) and an important energy source. The large intestine increases significantly both in size and in microflora mass (249) when dietary fiber is increased.

Gender Differences

The response to pregnancy can be viewed as a unique stress, since the gastrointestinal tract has to respond to the increased nutrient requirement and changes in hormonal balance that are unique to gravid females. The gastric system is sensitive to the increase in estrogens, which may have a protective effect against peptic ulcer disease by increasing mucus production in the stomach and duodenum (42). This may be a partial explanation for the lower incidence of this disease in women that

was so prominent years ago, though smoking and other factors may be more critical. Peptic ulcer disease is uncommon in pregnant women, but the mechanism may be different; it has been postulated that histaminase produced by the placenta degrades histamine, thereby decreasing histamine's stimulation of acid secretion in the stomach (43).

During pregnancy, the length and mass of the small intestine may increase, as does the number of absorptive cells with an increase in villus height (32). Cell proliferation also increases during lactation (33). The underlying mechanism is most likely the hyperphagia required to meet increased energy needs. This process may be mediated by gastrin and other intestinal growth-promoting agents, though studies have failed to provide the needed evidence (140). Fecal weight and water content decrease during pregnancy, reportedly due to the increase in the serum level of progesterone, a weak mineralocorticoid (75). Calcium, sodium, and chloride absorption increase, leading to net water absorption. The longer MCTT that occurs during the luteal phase compared to the follicular phase (230) is about the same as during pregnancy (231). This may be due to higher levels of progesterone, which inhibits small and large intestinal smooth muscle action (130).

Changes in the Gastrointestinal Tract in the Aged

The gastrointestinal system undergoes many changes during the life span of an individual (145, 158), but research in this area is not extensive. These changes influence all of the adaptive mechanisms that have been discussed in this chapter (228). Gastric acid secretion decreases with age, as does the response to the growth hormone properties of gastrin (22, 146, 159, 220). Secretion of intrinsic factor, required for vitamin B_{12} absorption, also decreases (158). Hypochlorhydria occurs commonly with aging (39, 204). Age reduces the production of "adaptative cytoprotective" agents, such as prostaglandins. The lower level of prostaglandins in the elderly has been suggested as a factor for the increase in ulcer disease and nonsteroidal antiinflammatory drug gastropathy in the elderly (52). Gastric emptying is decreased with aging, but it appears that the effect only occurs with the liquid fraction of stomach contents (160). The possible changes that occur in the small intestine with aging include increased rate of proliferation (105) while the mass decreases; increased absorption of vitamin A; and decreased absorption of carbohydrates (71) and lipids (250), vitamin B_{12} (121), calcium (6,211), and zinc (208). The elderly appear to be able to increase calcium absorption when needed, but the degree of response is less when compared to younger subjects (8,110). This lessened ability to adapt to calcium need may be due to a decreased response to vita-min D in effecting active calcium absorption (7). The immunological function of the small intestine may be compromised since the number of Peyer's patches and the underlying lymphatics are decreased (158). The hepatic system shows varied responses with aging. The size and weight of the liver do not change much, but the hepatic blood flow does not decrease with age (158). The enzymes involved with metabolizing xenobiotics (for example, pesticides and drugs) may decrease, but this is not clear (158).

Large intestinal function, in particular motility, decreases in the elderly when assessed using crude mea-surements of bowel habits, such as constipation and stool frequency (62, 158, 238). A decrease in the rate of passage through the large intestine was found in the aged rat when evaluated by the increase in methano-genic bacteria in the colon (143). As mentioned above, exercise can increase small and large intestinal motility. It has been shown that a strength training program can increase whole bowel transit time in both middle-aged and older men (125). With aging, the large intestine may respond differently from the stomach to growth-stimu-lating agents. Epidermal growth factor, which increases proliferation in the young rat's large intestine, appears to have a negative effect in the elderly rat (145). Perhaps this is the mechanism behind the slower proliferation response of the aged intestine to refeeding after starva-tion. The response to refeeding with increased enzyme production is faster in old rats (103). It has been sug-gested that this may be due to a delayed expression of differentiated enterocytes in older intestines (104).

SUMMARY

The gastrointestinal tract is a complex system composed of many organs and tissues. As a major interface with the environment, it digests and absorbs nutrients, excretes lipophilic compounds and heavy metals, metabolizes xenobiotics, and responds to immunologi-cal challenges. Key to its ability to adapt to the environ-ment is the rapid proliferation and cellular modification of the intestinal epithelium. Very little research has been done on the adaptation of the gastrointestinal tract to changes in the geophysical environment, but the poten-tial appears to be great. Many organs of the gastroin-testinal tract have been neglected, such as the salivary glands, esophagus, and gallbladder. In addition, many variables that alter the adaptability of the gastroin-testinal tract have not been well studied, including gender and age. As has been emphasized in this chapter, a great deal of research needs to be done on the relationship between the gastrointestinal tract and the environment.

REFERENCES

1. Adams, P. C., J. V. Frei, C. Bradley, and D. Lam. Hepatic iron and iron absorption in hemochromatosis. Clin. Invest. Med. 13: 256–258, 1990.
2. Agocs, M. M., R. A. Etzel, R. G. Parrish, D. C. Paschal, P. R. Campagna, D. S. Cohen, E. M. Kilbourne, and J. L. Hesse. Mercury exposure from interior latex paint. N. Engl. J. Med. 323: 1096–1100, 1990.
3. Alpers, D. H. Digestion and absorption of carbohydrates and proteins. In: Physiology of the Gastrointestinal Tract, edited by L. R. Johnson. New York: Raven, 1987, p. 1468–1487.
4. Alvares, A. P. Oxidative biotransformation of drugs. In: The Liver. Biology and Pathobiology, edited by I. M. Arias, H. Pop-per, D. Schachter, and D. A. Shafritz. New York: Raven, 1982, p. 267–280.
5. Anner, B. M., and M. Moosmayer. Mercury inhibits Na–K–ATPase primarily at the cytoplasmic side. Am. J. Physiol. 262(Renal Fluid Electrolyte Physiol. 31): F843–F848, 1992.
6. Armbrecht, H. J. Age-related changes in calcium and phospho-rus uptake by the small intestine. Biochem. Biophys. Acta 882: 281–286, 1986.
7. Armbrecht, H. J., T. V. Zenser, M. E. H. Bruns, and B. B. Davis. Effect of age on intestinal calcium absorption and adaptation to dietary calcium. Am. J. Physiol. 236(Endocrinol. Metab. Gas-trointest. Physiol. 5): E769–E774, 1979.
8. Armbrecht, H. J., T. V. Zenser, C. J. Gross, and B. B. Davis. Adaptation to dietary calcium and phosphorus restriction changes with age in the rat. Am. J. Physiol. 239(Endocrinol. Metab. 2): E322–E327, 1980.
9. Baer, M. T., and J. C. King. Tissue zinc levels and zinc excretion during experimental zinc depletion in young men. Am. J. Clin. Nutr. 39: 556–570, 1984.
10. Ballesta, M. C., M. Manas, F. J. Mataix, E. Martinez-Victoria, and I. Seiquer. Long-term adaptation of pancreatic response by dogs to dietary fats of different degrees of saturation. Br. J. Nutr. 64: 487–496, 1990.
11. Banwell, J. G. Pathophysiology of diarrheal disorders. Rev. Infect. Dis. 12: s30–s35, 1990.
12. Baynes, R. D., and T. H. Bothwell. Iron deficiency. Annu. Rev. Nutr. 10: 133–148, 1990.
13. Beisel, W. R. Magnitude of the host nutritional response to infection. Am. J. Clin. Nutr. 30: 1236–1247, 1977.
14. Bernaudin, J. F., E. Druet, P. Druet, and R. Masse. Inhalation or ingestion of organic or inorganic mercurials produces auto-immune disease in rats. Clin. Immunol. Immunopathol. 20: 129–135, 1981.
15. Binder, H. J., F. McGlone, and G. I. Sandle. Effects of cortico-steroid hormones on the electrophysiology of rat distal colon: implications for Na+ and K+ transport. J. Physiol. 410: 425–441, 1989.
16. Bingham, S. A., and J. H. Cummings. Effect of exercise and physical fitness on large intestine function. Gastroenterology 97: 1389–1399, 1989.
17. Boel, J., L. B. Anderson, B. Rasmusin, S. H. Hanson, and M. Dossing. Hepatic drug metabolism and physical fitness. Clin. Pharmacol. Ther. 36: 121–126, 1984.
18. Boffa, L. C., J. R. Lupton, M. R. Mariani, M. Ceppi, H. L. Newmark, A. Scalmati, and M. Lipkin. Modulation of colonic epithelium cell proliferation, histone acetylation, and luminal short chain fatty acids by variation of dietary fiber (wheat bran) in rats. Cancer Res. 52: 5906–5912, 1992.
19. Bohme, M., M. Diener, P. Mestres, and W. Rummel. Direct and indirect actions of HgCl2 and methyl mercury chloride on per-

meability and chloride secretion across the rat colonic mucosa. *Toxicol. Appl. Pharmacol.* 114: 285–294, 1992.

20. Bohme, M., M. Diener, and W. Rummel. Chloride secretion induced by mercury and cadmium: action sites and mechanisms. *Toxicol. Appl. Pharmacol.* 114: 295–301, 1992.

21. Bothwell, T. H., H. Seftel, P. Jacobs, J. D. Torrance, and N. Baumslag. Iron overload in bantu subjects; studies on the availability in bantu beer. *Am. J. Clin. Nutr.* 14: 47–51, 1964.

22. Bowman, B. B., and I. H. Rosenberg. Digestive function and aging. *Hum. Nutr. Clin. Nutr.* 37c: 75–89, 1983.

23. Bozelka, B. E., and J. E. Salvaggio. Immunomodulation by environmental contaminants: asbestos, cadmium, and halogenated biphenyls. A review. *Environ. Carcinogen. Rev.* 3: 1–62, 1985.

24. Brandtzaeg, P., T. S. Halstensen, K. Kett, P. Krajci, D. Kvale, T. O. Rognum, H. Scott, and L. M. Sollid. Immunobiology and immunopathology of human gut mucosa: humoral immunity and intraepithelial lymphocytes. *Gastroenterology* 97: 1562–1584, 1989.

25. Brandtzaeg, P., L. M. Sollid, P. S. Thrane, D. Kvale, K. Bjerke, H. Scott, K. Kett, and T. O. Rognum. Lymphoepithelial interactions in the mucosal immune system. *Gut* 29: 1116–1130, 1988.

26. Brannon, P. M. Adaptation of the exocrine pancreas to diet. *Annu. Rev. Nutr.* 10: 85–105, 1990.

27. Brouns, F., W. H. M. Saris, and N. J. Reher. Abdominal complaints and gastrointestinal function during long-lasting exercise. *Int. J. Sports Med.* 8: 175–189, 1987.

28. Brown, H. M., B. Rydqvist, and H. Moser. Intracellular calcium changes in Balanus photoreceptor. A study with calcium ion-selective electrodes and Arsenazo III. *Cell Calcium* 9: 105–119, 1988.

29. Brown, W., and M. Kloppel. The liver and IgA: immunological, cell biological and clinical implications. *Hepatology* 9: 763–784, 1989.

30. Buddington, R. K., and J. M. Diamond. Ontogenic development of intestinal nutrient transporters. *Annu. Rev. Physiol.* 51: 601–619, 1989.

31. Bunt, J. C. Hormonal alterations due to exercise. *Sports Med.* 3: 331–345, 1986.

32. Burdett, K., and C. Reek. Adaptation of the small intestine during pregnancy. *Biochem. J.* 184: 245–251, 1979.

33. Cairnie, A. B., and R. Bentley. Cell proliferation studies in intestinal epithelium of the rat: hyperplasia during lactation. *Exp. Cell Res.* 46: 428–440, 1967.

34. Cammack, J., N. W. Read, P. A. Cann, D. Greenwood, and A. M. Holgate. Effective prolonged exercise on the passage of a solid meal through the stomach and small intestine. *Gut* 23: 957–961, 1982.

35. Camussi, G., G. Salvidio, G. Biesecker, J. Brentjens, and G. Andres. Heymann antibodies induce complement-dependent injury of rat glomerular visceral epithelial cells. *J. Immunol.* 139: 2906–2914, 1987.

36. Carrio, I., M. Estorch, R. Serra-Grima, M. Ginjaume, R. Notivol, R. Calabuig, and F. Vilardell. Gastric emptying in marathon runners. *Gut* 30: 152–155, 1989.

37. Chacho, C. J. G., K. A. Paulson, V. I. Mathan, and S. J. Bahu. The villus architecture of the small intestine in the tropics. A necroscopy study. *J. Pathol.* 98: 146–151, 1969.

38. Chen, M. L., S. C. Chang, and J. Y. Guoo. Fiber contents of some Chinese vegetables and their in vitro binding capacity of bile acids. *Nutr. Rep. Int.* 26: 1053–1059, 1982.

39. Chernoff, R., and D. A. Lipshitz. Nutrition and aging. In: *Modern Nutrition in Health and Disease,* edited by M. E. Shils and V. R. Young. Philadelphia: Lea and Febiger, 1988, p. 982–1000.

40. Chinn, K. S. K., and J. P. Hannon. Efficiency of food utilization at high altitude. *Federation Proc.* 28: 944–947, 1969.

41. Chu, J. Lupus antiribosomal P antisera contain antibodies to a small fragment of 28S rRNA located in the proposed ribosomal GTPase center. *J. Exp. Med.* 174: 507–514, 1992.

42. Clark, D. H. Peptic ulcer disease in women. *Br. Med. J.* 1: 1254–1257, 1953.

43. Clark, D. H., and H. I. Tankel. Gastric acid and plasma histaminase during pregnancy. *Lancet* ii: 886–887, 1954.

44. Clarkson, T. W. Mercury—an element of mystery. *N. Engl. J. Med.* 323: 1137–1139, 1990.

45. Clausen, J. P. Effective physical training on cardiovascular adjustment to exercise in man. *Physiol. Rev.* 57: 779–815, 1977.

46. Coluccio, L. M., and A. Bretscher. Calcium-regulated cooperative binding of the microvillar 110K-calmodulin complex to F-actin: formation of decorated filaments. *J. Cell Biol.* 105: 325–333, 1987.

47. Consolazio, C. F., L. O. Matoush, R. A. Nelson, L. R. Hackler, and E. E. Preston. Relationship between calcium in sweat, calcium balance and calcium requirements. *J. Nutr.* 78: 78–88, 1962.

48. Consolazio, C. F., L. O. Matoush, R. A. Nelson, R. S. Harding, and J. E. Canham. Excretion of sodium, potassium, magnesium, and iron in human sweat and the relationship of each to balance and requirements. *J. Nutr.* 79: 407–415, 1963.

49. Cordaro, J. T., W. M. Sellers, R. J. Bell, and J. P. Schmidt. Study of man during a 56-day exposure to oxygen–helium atmosphere at 258 mm Hg total pressure. X. Enteric microbial flora. *Aerospace Med.* 37: 594–596, 1966.

50. Cox, J. A., and F. G. Elliott. Primary adult lactose intolerance in the Kivu Lake area: Rwanda and the bushi. *Dig. Dis. Sci.* 19: 714–724, 1974.

50a. Crane, R. K. The physiology of the intestinal absorption of sugars. In: *Physiological Effects of Food Carbohydrates,* edited by A. Jeanes and J. Hodges. Washington, D.C.: American Chemical Society, 1975, p. 2–19.

51. Creech, J. L., and M. N. Johnson. Angiosarcoma of liver in the manufacture of polyvinyl chloride. *J. Occup. Med.* 16: 150–151, 1974.

52. Cryer, B., and M. Feldman. Effects of nonsteroidal anti-inflammatory drugs on endogenous gastrointestinal prostaglandins and therapeutic strategies for prevention and treatment of nonsteroidal anti-inflammatory drug-induced damage. *Arch. Intern. Med.* 152: 1145–1155, 1992.

53. Dagorn, J. C., and R. G. Lahaie. Dietary regulation of pancreatic protein synthesis. I. Rapid and specific modulation of enzyme synthesis by changes in dietary composition. *Biochim. Biophys. Acta* 654: 111–118, 1981.

54. Debongnie, J. C., and S. F. Phillips. Capacity of the human colon to absorb fluid. *Gastroenterology* 74: 698–703, 1978.

55. Degray, J. A., D. N. Rao, and R. P. Mason. Reduction of paraquat and related bipyridylium compounds to free radical metabolites by rat hepatocytes. *Arch. Biochem. Biophys.* 289: 145–152, 1991.

56. Denton, D. A. The study of sheep with permanent unilateral parotid fistulae. *Q. J. Exp. Physiol.* 42: 72–95, 1957.

57. De Oliveira, J. T. A., A. Pusztai, and G. Grant. Changes in organs and tissues induced by feeding of purified kidney bean (*Phaseolus vulgaris*) lectins. *Nutr. Res.* 8: 943–947, 1988.

58. Deuster, P. A., K. L. Gallagher, A. Singh, and R. D. Reynolds. Consumption of a dehydrated ration for 31 days at moderate altitudes: status of zinc, copper and vitamin B-6. *J. Am. Diet. Assoc.* 92: 1372–1375, 1992.

59. Diamond, J. M., and W. H. Karasov. Effect of dietary carbohydrate on monosaccharide uptake by mouse intestine in vitro. *J. Physiol.* 349: 419–440, 1984.

60. Dietze, B., K. O. Haustein, G. Huller, and C. Bruckner. The ^{14}C-

aminophenazone breath test in pesticide workers. *Int. Arch. Occup. Environ. Health* 57: 185–193, 1986.

61. Dill, D. B., J. H. Talbott, and W. V. Conzolazio. Blood as a physiochemical system. *J. Biol. Chem.* 118: 649–666, 1937.

62. Donald, I. P., R. G. Smith, J. G. Cruikshank, R. A. Elton, and M. E. Stoddart. A study of constipation in the elderly living at home. *Gerontology* 31: 112–118, 1985.

63. Dowling, R. H. Polyamines in intestinal adaptation and disease. *Digestion* 46(suppl. 2): 331–334, 1990.

64. Drenckhahn, D., and R. Dermietzel. Organization of the actin-filament cytoskeleton in the intestinal brush border: a quantitative and qualitative immunoelectron microscope study. *J. Cell Biol.* 107: 1037–1048, 1988.

65. Dubick, M. A., A. P. Majumdar, G. A. Kaysen, E. J. Burbige, and M. C. Geokas. Secretagogue-induced enzyme release from the exocrine pancreas of rats following adaptation to a high protein diet. *J. Nutr.* 118: 305–310, 1988.

66. Echnauer, R., and H. Raffler. Effect of starvation on small intestinal enzyme activity in germ-free rats. *Digestion* 18: 45–55, 1978.

67. Ehn, L., B. Calmark, and S. Hoglung. Iron status in athletes involved in intense physical exercise. *Med. Sci. Sports Exerc.* 12: 61–64, 1980.

68. Esnault, V. L. M., P. W. Mathieson, S. Thiru, D. B. G. Oliveira, and C. Martin-Lockwood. Autoantibodies to myeloperoxidase in Brown Norway rats treated with mercuric chloride. *Lab. Invest.* 67: 114–120, 1992.

69. Falk, H., J. L. Creech, C. W. Heath, M. N. Johnson, and M. M. Key. Hepatic disease among workers at a vinyl chloride polymerization plant. *JAMA* 230: 59–63, 1974.

70. Favus, M. J. Factors that influence absorption and secretion of calcium in the small intestine and colon. *Am. J. Physiol.* 248(*Gastrointest. Liver Physiol.* 11): G147–G157, 1985.

71. Feibusch, H. M., and P. R. Holt. Impaired absorption capacity for carbohydrate in the aging human. *Dig. Dis. Sci.* 27: 1095–1100, 1982.

72. Ferraris, R. P., and J. M. Diamond. Specific regulation of intestinal nutrient transporters by their dietary substrates. *Annu. Rev. Physiol.* 51: 125–142, 1989.

73. Filice, G. Antimicrobial properties of Kupffer cells. *Infect. Immunol.* 56: 1430–1435, 1988.

74. Flanagan, P. R., J. Haist, and L. S. Valberg. Alterations in zinc absorption and salivary sediment zinc after a lacto-ovo vegetarian diet. *J. Nutr.* 113: 962–972, 1983.

75. Foster, E. S., T. W. Zimmerman, and J. P. Hayslett. Corticosteroid alteration of active electrolyte transport in the rat distal colon. *Am. J. Physiol.* 245 (*Gastrointest. Liver. Physiol.* 8): G668–G675, 1983.

76. Freeland-Graves, J. H., J. H. Ebangit, and P. J. Hendriksen. Zinc absorption, intraluminal zinc and intestinal metallothionen levels in zinc-deficient and zinc-replete rodents. *Am. J. Clin. Nutr.* 33: 1757–1766, 1980.

77. Frizzell, R. A., and S. G. Schultz. Effect of aldosterone on ion transport by rabbit colon in vitro. *J. Membr. Biol.* 39: 1–26, 1978.

78. Galluser, M., R. Belkhou, J. N. Freund, I. Duluc, N. Trop, M. Danielsen, and F. Raul. Adaptation of intestinal hydrolases to starvation in rats: effect of thyroid function. *J. Comp. Physiol. [B]* 161: 357–361, 1991.

79. Garcia, A., E. Coudrier, J. Carboni, J. Anderson, J. Vandekerkhove, M. Mooseker, D. Louvard, and M. Arpin. Partial deduced sequence of the 110-kD-calmodulin complex of the avian intestinal microvillus shows that this mechanoenzyme is a member of the myosin I family. *J. Cell Biol.* 109: 2895–2903, 1989.

80. Gichev, J. Change of liver function during adaptation of man to northern conditions. *Arctic Med. Res.* 49: 68–73, 1990.

81. Gillmore, J. D., and F. B. Gordon. Effect of exposure to hyperoxic, hyperbaric environments on concentrations of selected aerobic and anaerobic fecal flora of mice. *Appl. Microbiol.* 29: 358–367, 1975.

82. Gilman, R. H., R. Partanen, K. H. Brown, W. M. Spira, S. Khanam, B. Greenberg, S. R. Bloom, and A. Ali. Decreased acid secretion and bacterial colonization of the stomach in severely malnourished Bangladeshi children. *Gastroenterology* 94: 1308–1314, 1988.

83. Giorgi, D., W. Renaud, J. P. Bernard, and J. C. Dagorn. Regulation of amylase messenger RNA concentration in rat pancreas by food content. *EMBO J.* 3: 1521–1524, 1984.

84. Giorgi, D., W. Renaud, J. P. Bernard, and J. C. Dagorn. Regulation of proteolytic enzyme activities and mRNA concentrations in rat pancreas by food content. *Biochem. Biophys. Res. Commun.* 127: 937–942, 1985.

85. Glass, R. I., A.-M. Svennerholm, B. J. Stoll, M. R. Khan, K. M. B. Hossain, M. I. Huq, and J. Holmgren. Protection against cholera in breast-fed children by antibodies in breast milk. *N. Engl. J. Med.* 308: 1389–1392, 1983.

86. Glenney, J., Jr., and P. Glenney. Comparison of Ca^{++}-regulated events in the intestinal brush border. *J. Cell Biol.* 100: 754–763, 1985.

87. Goldsmith, D. P. J. Changes in desquamation rate of jejunal epithelium in cats during fasting. *Digestion* 8: 130–141, 1973.

88. Goodlad, R. A., C. Y. Lee, and N. A. Wright. Colonic cell proliferation and growth fraction in young, adult and old rats. *Virchows Arch. B Cell Pathol.* 61: 415–417, 1992.

89. Gopalan, C. Some recent studies in the nutrition research laboratory, Hyderabad. *Am. J. Clin. Nutr.* 23: 35, 1970.

90. Gordeuk, V. R., B. R. Bacon, and G. M. Brittenham. Iron overload: causes and consequences. *Annu. Rev. Nutr.* 7: 485–508, 1987.

91. Gordeuk, V. R., R. D. Boyd, and G. M. Brittenham. Dietary iron overload persists in rural sub-Saharan Africa. *Lancet* 1(8493): 1310–1313, 1986.

92. Grant, G., W. B. Watt, J. C. Stewart, and A. Pusztai. Local (intestinal) and systemic responses to dietary soyabean lectin. *Lectins Biol. Biochem. Clin. Biochem.* 6: 121–124, 1988.

93. Grayson, J. Responses of the colonic circulation in man to cooling the body. *J. Physiol.* 110: 13P, 1949.

94. Grayson, J. Observations on blood flow in the human intestine. *Br. Med. J.* 2: 1465–1470, 1950.

95. Grayson, J. The measurement of intestinal blood flow in man. *J. Physiol.* 114: 419–434, 1951.

96. Greenleaf, J. E. Physiological responses to prolonged bed rest and fluid immersion in humans. *J. Appl. Physiol.: Respir. Environ. Exerc. Physiol.* 57: 615–633, 1984.

97. Greenleaf, J. E., E. M. Bernauer, L. T. Juhos, W. Stanley, H. L. Young, and J. T. Morse. Effects of exercise on fluid exchange and body composition in man during 14-day bed rest. *J. Appl. Physiol.: Respir. Environ. Exerc. Physiol.* 43: 126–132, 1977.

98. Gryboski, J. D. Gastrointestinal aspects of cow's milk protein intolerance and allergy. *Immunol. Allergy Clin. North Am.* 11: 773–797, 1991.

99. Guy-Grand, D., N. Cerf-Bensussan, B. Malissen, M. Malassis-Seris, C. Briottet, and P. Vassalli. Two gut intraepithelial CD8$^+$ lymphocyte populations with different T cell receptors: a role for the gut epithelium in T cell differentiation. *J. Exp. Med.* 173: 471–481, 1991.

100. Hannon, J. P., and D. A. Vaughan. Effect of exposure duration on selected enzyme indexes of cold acclimatization. *Am. J. Physiol.* 200: 94–98, 1961.

101. Harrison, P. T. C., P. Grasso, and V. Badescu. Early changes in

the forestomach of rats, mice and hamsters exposed to dietary propionic and butyric acid. *Food Chem. Toxicol.* 29: 367–371, 1991.

102. Heroux, O., and N. T. Gridgeman. The effect of cold acclimation on the size of organs and tissues of the rat, with special reference to modes of expression of results. *Can. J. Biochem. Physiol.* 36: 209–216, 1958.

103. Holt, P. R., and D. P. Kotler. Adaptive changes of intestinal enzymes to nutritional intake in the aging rat. *Gastroenterology* 93: 295–300, 1987.

104. Holt, P. R., A. R. Tierney, and D. P. Kotler. Delayed enzyme expression: a defect of the aging rat gut. *Gastroenterology* 89: 1026–1034, 1985.

105. Holt, P. R., and K. Y. Yeh. Colonic proliferation is increased in senescent rats. *Gastroenterology* 95: 1556–1563, 1988.

106. Hoppe, C. A., T. P. Connolly, and A. L. Hubbard. Transcellular transport of polymeric IgA in the rat hepatocyte: biochemical and morphological characterization of the transport pathway. *J. Cell Biol.* 101: 2113–2123, 1985.

107. Hornbein, T. F. Evaluation of iron stores as limiting high altitude polycythemia. *J. Appl. Physiol.* 17: 243–245, 1962.

108. Houghton, M. R., R. G. H. Morgan, and M. Gracey. Effects of long-term dietary modifications on pancreatic enzyme activity. *J. Pediatr. Gastroenterol. Nutr.* 2: 548–554, 1983.

109. Hultman, P., and S. Enestrom. Dose-response studies in murine mercury-induced autoimmunity and immune-complex disease. *Toxicol. Appl. Pharmacol.* 113: 199–208, 1992.

110. Ireland, P., and J. S. Fordtran. Effect of dietary calcium and age on jejunum calcium absorption in humans studied by intestinal perfusion. *J. Clin. Invest.* 52: 2672–2681, 1973.

111. Jacob, R. A., H. H. Sandstead, J. M. Munoz, L. M. Klevay, and D. B. Milne. Whole body surface loss of trace metals in normal males. *Am. J. Clin. Nutr.* 34: 1379–1383, 1981.

112. Janne, P., Y. Carpentier, and G. Willems. Colonic mucosal atrophy induced by a liquid elemental diet in rats. *Am. J. Dig. Dis.* 22: 808–812, 1977.

113. Johnson, H. L., C. F. Consolazio, T. A. Daws, and H. J. Krzywicki. Increased energy requirements of man after abrupt altitude exposure. *Nutr. Rep. Int.* 4: 77–82, 1971.

114. Johnson, H. L., C. F. Consolazio, L. O. Matoush, and H. J. Krzywicki. Nitrogen and mineral metabolism at altitude. *Federation Proc.* 28: 1195–1198, 1969.

115. Karasov, W. H., and J. M. Diamond. Adaptive regulation of sugar and amino acid transport by vertebrate intestine. *Am. J. Physiol.* 245(*Gastrointest. Liver Physiol.* 8): G443–G462, 1983.

116. Karasov, W. H., and J. M. Diamond. Adaptation of intestinal nutrient transport. In: *Physiology of the Gastrointestinal Tract,* edited by L. R. Johnson. New York: Raven, 1986, p. 1489–1498.

117. Keefe, E. B., D. K. Lowe, J. R. Goss, and R. Wayner. Gastrointestinal symptoms of marathon runners. *West. J. Med.* 141: 481–484, 1984.

118. Keeling, W. F., and B. J. Martin. Gastrointestinal transit time during mild exercise. *J. Appl. Physiol.* 63: 978–981, 1987.

119. Kelsay, J. L., K. M. Gehall, and E. S. Prather. Effect of fiber from fruits and vegetables on metabolic responses of human subjects. I. Bowel transit time, number of defecations, fecal weight, urinary excretions of energy and nitrogen and apparent digestibilities of energy, nitrogen and fat. *Am. J. Clin. Nutr.* 31: 1149–1153, 1978.

120. Kern, F., H. Birkner, and V. S. Ostrower. Binding of bile acids by dietary fiber. *Am. J. Clin. Nutr.* 31: S175–S179, 1978.

121. King, C. E., J. Liebach, and P. P. Toskes. Clinically significant vitamin B_{12} deficiency—secondary to malabsorption of protein-bound B_{12}. *Dig. Dis. Sci.* 24: 397–402, 1979.

122. Klipstein, F. A. Tropical sprue in travelers and expatriates living abroad. *Gastroenterology* 80: 590–600, 1981.

123. Knoflach, P., B. Albini, and M. M. Weiser. Autoimmune disease induced by oral administration of mercuric chloride in Brown-Norway rats. *Toxicol. Pathol.* 14: 188–193, 1986.

124. Knoflach, P., M. M. Weiser, and B. Albini. Mercuric chloride-induced protein-losing enteropathy (PLE) in Brown Norway (BN) rats. *Federation Proc.* 45: 952, 1986.

125. Koffler, K. H., A. Menkes, A. Redmond, W. E. Whitehead, R. E. Pratley, and B. F. Hurley. Strength training accelerates gastrointestinal transit time in middle aged and older men. *Med. Sci. Sports Exerc.* 24: 415–419, 1992.

126. Kolars, J. C. Aflatoxin and hepatocellular carcinoma: a useful paradigm for environmentally induced carcinogenesis. *Hepatology* 16: 848–851, 1992.

127. Kolars, J. C., M. D. Levitt, M. Aouji, and D. A. Savaiano. Yogurt—an autodigesting source of lactose. *N. Engl. J. Med.* 310: 1–3, 1984.

128. Komai, M., and S. Kimura. Effects of restricted diet and intestinal flora on the life span of small intestine epithelial cells in mice. *J. Nutr. Sci. Vitaminol* 25: 87–94, 1979.

129. Kruh, J. Effects of sodium butyrate, a new pharmacological agent, on cells in culture. *Mol. Cell Biochem.* 42: 65–82, 1982.

130. Kumar, D. In vitro inhibitory effect on extrauterine human smooth muscle. *Am. J. Obstet. Gynecol.* 84: 1300–1304, 1962.

131. Lane, H. W. Energy requirements for space flight. *J. Nutr.* 122: 13–18, 1992.

132. Lane, H. W., and L. O. Schulz. Nutritional questions relevant to space flight. *Annu. Rev. Nutr.* 12: 257–278, 1992.

133. Laplanche, A., F. Clavel-Chapelon, J. C. Contassot, and C. Lanouziere. Exposure to vinyl chloride monomere: results of a cohort study after a seven year follow up. *Br. J. Ind. Med.* 49: 134–137, 1992.

134. Lee, S. S., Z. Nitsan, and I. E. Liener. Growth, protein utilization and secretion of pancreatic enzymes by rats in response to elevated levels of dietary protein. *Nutr. Res.* 4: 867–876, 1984.

135. Leibovitch, M. P., and J. Kruh. Effect of sodium butyrate on myoblast growth and differentiation. *Biochem. Biophys. Res. Commun.* 87: 896–903, 1979.

136. Leonard, J. I., C. S. Leach, and P. C. Rambaut. Quantitation of tissue loss during prolonged space flight. *Am. J. Clin. Nutr.* 38: 667–679, 1983.

137. Levens, N. R. Modulation of jejunal ion and water absorption by endogenous angiotensin after dehydration. *Am. J. Physiol.* 246(*Gastrointest. Liver Physiol.* 9): G700–G709, 1984.

138. Levens, N. R. Control of intestinal absorption by the renin-angiotensin system. *Am. J. Physiol.* 249(*Gastrointest. Liver Physiol.* 12): G3–G15, 1985.

139. Linder, M. C., and H. N. Munro. The mechanism of iron absorption and its regulation. *Federation Proc.* 36: 2017–2023, 1977.

140. Lipkin, M. Proliferation and differentiation of normal and diseased gastrointestinal cells. In: *Physiology of the Gastrointestinal Tract,* edited by L. R. Johnson. New York: Raven, 1987, p. 255–284.

141. Lorenz, G. Liver damage caused by pesticides. *Zentralbl. Allgemeine Pathol. Pathol. Anat.* 130: 533–538, 1985.

142. Luzzi, G. A., M. Torii, M. Aikawa, and G. Pasvol. Unrestricted growth of plasmodium falciparum in microcytic erythrocytes in iron deficiency and thalassaemia. *Br. J. Haematol.* 74: 519–524, 1990.

143. Maczulak, A. E., M. J. Wolin, and T. L. Miller. Increase in colonic methanogens and total anaerobes in aging rats. *Appl. Environ. Microbiol.* 55: 2468–2473, 1989.

144. Madara, J. L. Loosening tight junctions. Lessons from the intestine. *J. Clin. Invest.* 83: 1089–1094, 1989.

145. Majumdar, A. P., and M. A. Dubick. The aging gastrointestinal tract: cell proliferation and nutritional adaptation. *Prog. Food Nutr. Sci.* 13: 139–160, 1989.

146. Majumdar, A. P. N., E. A. Edgerton, Y. Dayal, and S. N. S. Murthy. Gastrin: levels and trophic action during advancing age. *Am. J. Physiol.* 254(*Gastrointest. Liver Physiol.* 17): G538–G542, 1988.

147. Marx, J. Testing of autoimmune therapy begins. *Science* 252: 27–28, 1991.

148. Mason, V. C. Metabolism of nitrogenous compounds in the large gut. *Proc. Nutr. Soc.* 43: 45–53, 1984.

149. Mayfield, S. R., W. Oh, D. L. Piva, and B. S. Stonestreet. Postprandial gastrointestinal blood flow and oxygen consumption during environmental cold stress. *Am. J. Physiol.* 256(*Gastrointest. Liver Physiol.* 19): G364–G368, 1989.

150. Mayfield, S. R., P. W. Shaul, and B. S. Stonestreet. Gastrointestinal blood flow and oxygen delivery during environmental cold stress: effect of anaemia. *J. Dev. Physiol.* 12: 219–223, 1989.

151. McBurney, M. I., L. U. Thompson, and D. J. A. Jenkins. Comparison of ileal effects, dietary fibers, and whole foods in predicting the physiological importance of colonic fermentation. *Am. J. Gastroenterol.* 83: 536–540, 1988.

152. McCabe, R. D., P. L. Smith, and L. P. Sullivan. Ion transport by rabbit descending colon: mechanisms of transepithelial potassium transport. *Am. J. Physiol.* 246(*Gastrointest. Liver Physiol.* 9): G594–G602, 1984.

153. McGhee, J. R., and J. Mestecky. In defense of mucosal surfaces. In: *Infectious Disease Clinics of North America*, edited by W. Schaffner. Philadelphia: Saunders, 1990, p. 315–341.

154. McLaren, G. D., M. H. Nathanson, A. Jacobs, D. Trevett, and W. Thomson. Control of iron absorption in hemochromatosis: mucosal iron kinetics in vivo. *Ann. N. Y. Acad. Sci.* 526: 185–198, 1988.

155. Menge, H., and J. W. L. Robinson. Early phase of jejunal regeneration after short-term ischemia in the rat. *Lab. Invest.* 40: 25–30, 1979.

156. Meshkinpour, H., C. Kemp, and R. Fairshter. The effect of aerobic exercise on mouth to cecum transit time. *Gastroenterology* 96: 938–941, 1989.

157. Meynell, G. G. Antibacterial mechanisms of the mouse gut II. The role of E_h and volatile fatty acids in the normal gut. *Br. J. Exp. Pathol.* 44: 209–215, 1963.

158. Minaker, K., and J. W. Rowe. Gastrointestinal system. In: *Health and Disease in Old Age*, edited by J. W. Rowe and R. W. Besdine. Boston: Little, Brown, 1982, p. 297–316.

159. Montgomery, R. D., M. R. Heaney, I. N. Ross, H. G. Sammons, A. V. Barford, S. Balakrishan, P. P. Mayer, L. S. Culank, J. Field, and P. Gosling. The aging gut: a study of intestinal absorption in relation to nutrition in the elderly. *Q. J. Med.* 47: 197–211, 1978.

160. Moore, J. G., C. Tweedy, P. E. Christian, and F. L. Datz. The effect of age on gastric emptying of liquid–solid meals in man. *Dig. Dis. Sci.* 28: 340–344, 1983.

161. Moore, R., S. Carlson, and J. L. Madara. Villus contraction aids repair of intestinal epithelium after injury. *Am. J. Physiol.* 257 (*Gastrointest. Liver Physiol.* 20): G274–G283, 1989.

162. Mooseker, M. S., and T. R. Coleman. The 110-kD protein–calmodulin complex of the intestinal microvillus (brush border myosin I) is a mechanoenzyme. *J. Cell Biol.* 108: 2395–2400, 1989.

163. Morin, C. L., V. Ling, and D. Bourassa. Small intestinal and colonic changes induced by a chemically defined diet. *Dig. Dis. Sci.* 25: 123–128, 1980.

164. Moses, F. M. The effect of exercise on the gastrointestinal tract. *Sports Med.* 9: 159–172, 1990.

165. Moses, F. M., R. Baska, G. Graeber, and P. D. Kearny. Gastro-intestinal bleeding during an ultramarathon. *Med. Sci. Sports Exerc.* 21: 578a, 1989.

166. Moses, F. M., C. Ryan, J. Debolt, B. Smoak, A. Hoffman, V. Villanueva, and P. Deuster. Oral cecal transit time during a 2 hr run with injestion of water or glucose polymer. *Am. J. Gastroenterol.* 83: 1055, 1988.

167. Murray, R. The effects of consuming carbohydrate–electrolyte beverages on gastric emptying and fluid absorption during and following exercise. *Sports Med.* 4: 322–351, 1987.

168. Myrvik, Q. N. Nutrition and immunology. In: *Modern Nutrition in Health and Disease*, edited by M. E. Shils and V. R. Young. Philadelphia: Lea and Febiger, 1988, p. 585–616.

169. Nater, E. A., and D. F. Grigal. Regional trends in mercury distribution across the Great Lakes states, north central USA. *Nature* 358: 139–141, 1992.

170. Neufer, P. D., A. J. Young, and M. N. Sawka. Gastric emptying during walking and running: effects of varied exercise intensity. *Eur. J. Appl. Physiol.* 38: 440–445, 1989.

171. Nicogossian, A. E., and L. F. Dietlein. Microgravity: simulation and analogs. In: *Space Physiology and Medicine*, edited by A. E. Nicogossian, C. L. Huntoon, and S. L. Pool. Philadelphia: Lea and Febiger, 1989, p. 240–248.

172. Oettle, G. J. Effect of moderate exercise on bowel habit. *Gut* 32: 941–944, 1991.

173. Owen, R. L., and L. L. Brandborg. Jejunal morphologic consequences of vegetarian diet in humans. *Gastroenterology* 72: A88, 1977.

174. Ozturk, M. p53 mutation in hepatocellular carcinoma after aflatoxin exposure. *Lancet* 338: 1356–1359, 1991.

175. Pappo, J., and R. L. Owen. Absence of secretory component expression by epithelial cells overlying rabbit gut-associated lymphoid tissue. *Gastroenterology* 95: 1173–1177, 1988.

176. Paulev, P., R. Jordal, and N. S. Pederson. Dermal excretion of iron in intensely training athletes. *Clin. Chim. Acta* 127: 19–27, 1983.

177. Peters, T. J., K. B. Raja, R. J. Simpson, and S. Snape. Mechanisms and regulation of intestinal iron absorption. *Ann. N. Y. Acad. Sci.* 526: 141–147, 1988.

178. Pitts, G. C., A. S. Ushakov, N. Pase, T. A. Smirnova, D. F. Rahlmann, and A. H. Smith. Effects of weightlessness on body composition in the rat. *Am. J. Physiol.* 244(*Regulatory Integrative Comp. Physiol.* 15): R332–R337, 1983.

179. Plass, H., A. Gridl, and K. Turnheim. Absorption and secretion of potassium by rabbit descending colon. *Pflugers Arch.* 406: 509–519, 1986.

180. Potten, C. S., and M. Loeffler. A comprehensive model of the crypts of the small intestine of the mouse provides insight into the mechanisms of cell migration and the proliferation hierarchy. *J. Theor. Biol.* 127: 381–391, 1987.

181. Prasad, K. N., and P. K. Sinha. Effect of sodium butyrate on mammalian cells in culture: a review. *In Vitro* 12: 125–132, 1976.

182. Priebe, W. M., and J. A. Priebe. Runners diarrhea—prevalence and clinical symptomatology. *Am. J. Gastroenterol.* 79: 827–828, 1984.

183. Pusey, C. D., A. Dash, M. J. Kershaw, A. Morgan, A. Reilly, A. J. Rees, and C. M. Lockwood. A single autoantigen in Goodpasture's syndrome identified by a monoclonal antibody to human glomerular basement membrane. *Lab. Invest.* 56: 23, 1987.

184. Pusztai, A., G. Grant, D. S. Brown, S. W. B. Ewen, and S. Bardocz. Phaseolus vulgaris lectin induces growth and increases the polyamine content of rat small intestine in vivo. *Med. Sci. Res.* 16: 1283–1284, 1988.

185. Qamar, M., and A. Reed. Effects of exercise on mesenteric blood flow in man. *Gut* 28: 583–587, 1987.

186. Rambaut, P. C., and R. S. Johnston. Prolonged weightless and calcium loss in man. *Acta Astronaut* 6: 1113–1122, 1979.

187. Reboud, J. P., G. Marchis-Mouren, A. Cozzone, and P. Desnuelle. Variations in the biosynthesis rate of pancreatic amylase and chymotrypsin in response to a starch-rich or protein-rich diet. *Biochem. Biophys. Res. Commun.* 22: 94–99, 1966.

188. Renkema, J. A., A. T. Senshu, B. D. E. Gallard, and E. Brouwer. Regulation of sodium excretion and retention by the intestine in cows. *Nature* 195: 389–390, 1962.

189. Reynafarje, C., R. Lozano, and J. Valdivieso. The polycythemia of high altitudes: iron metabolism and related aspects. *Blood* 14: 433–455, 1959.

190. Riddoch, C., and T. Trinick. Gastrointestinal disturbances in marathon runners. *Br. J. Sp. Med.* 22: 71–74, 1988.

191. Rijke, R. P. C., and R. Gart. Epithelial cell kinetics in the descending colon of the rat. I. The effect of ischaemia-induced epithelial cell loss. *Virchows Arch. B Cell Pathol.* 31: 15–22, 1979.

192. Rijke, R. P. C., W. R. Hanson, H. M. Plaisier, and J. W. Osborne. The effect of ischemic villus cell damage on crypt cell proliferation in the small intestine. Evidence for a feedback control mechanism. *Gastroenterology* 71: 786–792, 1976.

193. Ritland, S., N. E. Foss, and E. Gjone. Physical activity and liver disease and liver function in sportsmen. *Scand. J. Soc. Med.* 29 (suppl.): 221–226, 1982.

194. Rompala, R. E., T. A. Hoagland, and J. A. Meister. Effect of dietary bulk on organ mass, fasting heat production and metabolism of the small and large intestines of sheep. *J. Nutr.* 118: 1553–1557, 1988.

195. Rozsa, Z., J. Mattila, and E. D. Jacobson. Substance P mediates a gastrointestinal thermoreflex in rats. *Gastroenterology* 95: 265–276, 1988.

196. Ryan, G. P., S. J. Dudrick, E. M. Copeland, and L. R. Johnson. Effects of various diets on colonic growth in rats. *Gastroenterology* 77: 658–663, 1979.

197. Saegesser, F., U. Roenspies, and J. W. L. Robinson. Ischemic diseases of the large intestine. *Pathobiol. Annu.* 9: 303–337, 1979.

198. Sagel, I., P. P. Gagjee, A. R. Essop, and A. M. Noormohamed. Lactase deficiency in the South African black population. *Am. J. Clin. Nutr.* 38: 901–905, 1983.

199. Sandle, G. I., and H. J. Binder. Corticosteroids and intestinal ion transport. *Gastroenterology* 93: 188–196, 1987.

200. Scheflan, M., S. J. Galli, J. Perrotto, and J. E. Fischer. Intestinal adaptation after extensive resection of the small intestine and prolonged administration of parenteral nutrition. *Surg. Gynecol. Obstet.* 143: 757–762, 1976.

201. Schneeman, B. O. Pancreatic and digestive function. In: *Dietary Fiber in Health and Disease,* edited by G. V. Vahouny and D. Kritchevsky. New York: Plenum, 1982, p. 73–83.

202. Schneeman, B. O., B. D. Richter, and L. R. Jacobs. Response to dietary wheat bran in the exocrine pancreas and intestine of rats. *J. Nutr.* 112: 283–286, 1982.

203. Shin, C. S., A. G. Chaudhry, M. H. Khaddam, P. D. Penha, and R. Dooner. Early morphological changes in the intestine following massive resection of the small intestine and parenteral nutrition therapy. *Surg. Gynecol. Obstet.* 151: 246–250, 1980.

204. Shock, N. The physiology of aging. *Sci. Am.* 206: 110–116, 1962.

205. Singh, A., B. L. Smoak, K. Y. Patterson, L. G. Lemay, C. Veillon, and P. A. Deuster. Biochemical indices of selected trace minerals in men: effect of stress. *Am. J. Clin. Nutr.* 53: 126–131, 1991.

206. Smith, R. E., and D. J. Hoijer. Metabolism and cellular function in cold acclimation. *Physiol. Rev.* 42: 60–142, 1962.

207. Solomon, T. E. Regulation of exocrine pancreatic cell proliferation and enzyme synthesis. In: *Physiology of the Gastrointes-*

tinal Tract, edited by L. R. Johnson. New York: Raven, 1981, p. 873–892.

208. Solomons, N. W. Zinc and copper. In: *Modern Nutrition in Health and Disease,* edited by M. E. Shils and V. R. Young. Philadelphia: Lea and Febiger, 1988, p. 238–262.

209. Southgate, D. A. T. Digestion and absorption of nutrients. In: *Dietary Fiber in Health and Disease,* edited by G. V. Vahouny and D. Kritchevsky. New York: Plenum, 1982, p. 1–7.

210. Southgate, D. A. T., and J. V. G. A. Durin. Calorie conversion factors: an experimental reassessment of the factors used in the calculation of the energy value of human diets. *Br. J. Nutr.* 24: 517–535, 1970.

211. Spencer, H., L. Kramer, and D. Osis. Factors contributing to calcium loss in aging. *Am. J. Clin. Nutr.* 36: 776–787, 1982.

212. Stephen, A. M., and J. H. Cummings. The microbial contribution to human faecal mass. *J. Med. Microbiol.* 13: 45–56, 1980.

213. Stevens, C. E. Comparative physiology of the digestive tract. In: *Duke's Physiology of Domestic Animals,* edited by M. J. Swenson. Ithaca, NY: Comstock, 1977, p. 216–232.

214. Stewart, J. F., D. A. Ahlquist, D. B. McGill, D. M. Ilstrup, S. Schwartz, and R. A. Owen. Gastrointestinal blood loss and anemia in runners. *Ann. Intern. Med.* 100: 843–845, 1984.

215. Sugar, J., K. Toth, O. Csuka, E. Gati, and S. Somfai-Relle. Role of pesticides in hepatocarcinogenesis. *J. Toxicol. Environ. Health* 5: 183–191, 1979.

216. Sztul, E. S., K. E. Howell, and G. E. Palade. Biogenesis of the polymeric IgA receptor in rat hepatocytes. II. Localization of its intracellular forms by cell fractionation studies. *J. Cell Biol.* 100: 1255–1261, 1985.

217. Tannock, G. W. Effect of dietary and environmental stress on the gastrointestinal microbiota. In: *Human Intestinal Microflora in Health and Disease,* edited by D. J. Hentges. New York: Academic, 1983, p. 517–539.

218. Tannock, G. W., and D. C. Savage. Influences of dietary and environmental stress on microbial populations in the marine gastrointestinal tract. *Infect. Immunol.* 9: 591–598, 1974.

219. Tashev, T. S., and D. Markov. Stomach and duodenal lesions in patients with acute organophosphorus pesticide poisonings. *Vutreshni Bolesti* 30: 61–65, 1991.

220. Thomson, A. B. R., and M. Keelan. The aging gut. *Can. J. Physiol. Pharmacol.* 64: 30–38, 1986.

221. Toloza, E. M., M. Lam, and J. Diamond. Nutrient extraction by cold-exposed mice: a test of digestive safety margins. *Am. J. Physiol.* 261(*Gastrointest. Liver Physiol.* 24): G608–G620, 1991.

222. Traber, P. G., J. Chianale, R. Florence, K. Kim, E. Wojcik, and J. J. Gumucio. Expression of cytochrome P450b and P450e genes in small intestinal mucosa of rats following treatment with phenobarbital, polyhalogenated biphenyls, and organochlorine pesticides. *J. Biol. Chem.* 263: 9449–9455, 1988.

223. Uddin, M., G. G. Altmann, and C. P. Leblond. Radioautographic visualization of differences in the patterns of [^{3}H] uridine and [^{3}H] orotic acid incorporation into the RNA of migrating columnar cells in the rat small intestine. *J. Cell Biol.* 98: 1619–1629, 1984.

224. Vahouny, G. V., T. LE, I. Ifrim, S. Satchithanandam, and M. M. Cassidy. Stimulation of intestinal cytokines and mucin turnover in rats fed wheat bran or cellulose. *Am. J. Clin. Nutr.* 41: 895–900, 1985.

225. Vailas, A. C., R. Vanderby, Jr., D. A. Martinez, R. B. Ashman, M. J. Ulm, R. E. Grindeland, G. N. Durnova, and A. Kaplansky. Adaptations of young adult rat cortical bone to 14 days of spaceflight. *J. Appl. Physiol.: Respir. Environ. Exerc. Physiol.* 73(suppl. 2): 4S–9S, 1992.

226. Van Kerckhove, C., G. J. Russell, K. Deusch, K. Reich, A. K. Bhan, H. Dersimonian, and M. B. Brenner. Oligoclonality of

human intestinal intraepithelial T cells. *J. Exp. Med.* 175: 57–63, 1992.

227. Van Soest, P. J. *Nutritional Ecology of the Ruminant.* Corvallis, OR: O and B, 1982.

228. Vellas, B. J., D. Balas, C. Lafont, F. Senegas-Balas, J. L. Albarede, and A. Ribet. Adaptive response of pancreatic and intestinal function to nutritional intake in the aged. *J. Am. Geriatr. Soc.* 38: 254–258, 1990.

229. Vial, J., and K. R. Porter. Scanning microscopy of dissociated tissue cells. *J. Cell Biol.* 67: 345–360, 1975.

230. Wald, A., D. H. Van Theil, L. Hoechstetter, J. S. Gavaler, K. M. Egler, R. Verm, L. Scott, and R. Lester. Gastrointestinal transit: the effect of the menstrual cycle. *Gastroenterology* 80: 1497–1500, 1981.

231. Wald, A., D. H. Van Thiel, X. X. Hoechstetter, J. S. Gavaler, K. M. Egler, R. Verm, L. Scott, and R. Lester. Effect of pregnancy on gastrointestinal transit. *Dig. Dis. Sci.* 27: 1015–1018, 1982.

232. Wasserman, R. H. Gastrointestinal absorption of calcium and phosphorus. In: *Handbook of Physiology. Endocrinology,* edited by G. A. Aurbach. Washington, DC: Am. Physiol. Soc., 1976, vol. VII, sect. 7, p. 137–155.

233. Weaver, L. T., S. Austin, and T. J. Cole. Small intestinal length: a factor essential for gut adaptation. *Gut* 32: 1321–1323, 1991.

234. Weiss, A. K. Adaptation of rats to cold air and effects on tissue oxygen consumption. *Am. J. Physiol.* 177: 201–206, 1954.

235. Weizman, Z., and S. Sofer. Acute pancreatitis in children with anticholinesterase insecticide intoxication. *Pediatrics* 90: 204–206, 1992.

236. Welsh, J. D., and S. N. Wolf. Vascular responses in the exposed human colon. *Am. J. Dig. Dis.* 5: 579–602, 1960.

237. Weltzin, R., P. Lucia-Jandris, P. Michetti, B. N. Fields, J. P. Kraehenbuhl, and M. R. Neutra. Binding and transepithelial transport of immunoglobulins by intestinal M cells: demonstration using monoclonal IgA antibodies against enteric viral proteins. *J. Cell Biol.* 108: 1673–1685, 1989.

238. Whitehead, W. E., D. Drinkwater, J. Cheskin, B. R. Heller, and M. M. Schuster. Constipation in the elderly living at home: def-inition, prevalence, and relationship to lifestyle and health status. *J. Am. Geriatr. Soc.* 37: 423–429, 1989.

239. Wicker, C., and A. Puigserver. Effects of inverse changes in dietary lipid and carbohydrate on the synthesis of some pancreatic secretory proteins. *Eur. J. Biochem.* 162: 25–30, 1987.

240. Wicker, C., A. Puigserver, and G. Scheele. Dietary regulation of levels of active mRNA coding for amylase and serine protease zymogens in the rat pancreas. *Eur. J. Biochem.* 139: 381–387, 1984.

241. Wicker, C., G. A. Scheele, and A. Puigserver. Pancreatic adaptation to dietary lipids is mediated by changes in lipase mRNA. *Biochimie* 70: 1277–1283, 1988.

242. Williamson, R. C., and M. Chir. Intestinal adaptation: structural, functional and cytokinetic changes. *N. Engl. J. Med.* 298: 1393–1402, 1978.

243. Williamson, R. C. N., and M. Chir. Intestinal adaptation: mechanisms of control. *N. Engl. J. Med.* 298: 1444–1450, 1978.

244. Wills, N. K., and B. Biagi. Active potassium transport by rabbit descending colon epithelium. *J. Membr. Biol.* 64: 195–203, 1982.

245. Witte, J., M. Lloyd, V. Lorenzsonn, H. Korsmo, and W. Olsen. The biosynthetic basis of adult lactase deficiency. *J. Clin. Invest.* 86: 1338–1342, 1990.

246. Wolf, J. L. Why antigens are attracted to the dome epithelium: another clue. *Gastroenterology* 95: 1419–1421, 1988.

247. Wong, O., M. D. Whorton, D. E. Foliart, and D. Ragland. An industry-wide epidemiologic study of vinyl chloride workers. *Am. J. Ind. Med.* 20: 317–334, 1991.

248. Worobetz, L. J., and D. F. Gerrard. Gastrointestinal symptoms during exercise and endurance athletes: prevalence and speculations of the etiology. *N. Z. Med. J.* 98: 644–646, 1985.

249. Wyatt, G. M., N. Horn, J. M. Gee, and I. T. Johnson. Intestinal microflora and gastrointestinal adaptation in the rat in response to non-digestible dietary polysaccharides. *Br. J. Nutr.* 60: 197–207, 1988.

250. Young, V. R. Nutrition. In: *Health and Disease in Old Age,* edited by J. W. Rowe and R. W. Besdine. Boston: Little, Brown, 1993, p. 317–333.

THE MICROBIAL ENVIRONMENT

Associate Editor   M. J. Fregly

66. Adaptation to the microbial environment

LINDA A. TOTH | Department of Infectious Diseases, St. Jude Children's Research Hospital, Memphis, Tennessee

CLARK M. BLATTEIS | Department of Physiology and Biophysics, University of Tennessee, Memphis, Tennessee

CHAPTER CONTENTS

THE STUDY OF ADAPTATION generally focuses on mechanisms by which organisms maintain homeostasis during external or internal physiological perturbations. For example, animals subjected to thermal stress as a result of elevated ambient temperatures initiate a wide variety of compensatory physiological and behavioral mechanisms to prevent significant fluctuations in body temperature. Indeed, many of the physiological perturbations to which animals must adapt involve variations in the physical environment per se (for example, ambient temperature, oxygen availability, humidity) or in the availability of essential substances, such as food or water. In addition to adapting to alterations in the physical environment, however, animals must adapt to the presence and actions of other organisms in the environment. Such interactions are often viewed in terms of predator/prey relationships or ecological segregation into competitive or noncompetitive niches. These relationships occur not only among vertebrate species but also between vertebrates and microbial organisms. This chapter will review the mechanisms by which vertebrates adapt to environmental microorganisms.

RELATIONSHIPS BETWEEN MICROBES AND VERTEBRATES

Biological relationships in which two different organisms live in close association are termed symbiotic, and are classified as commensal, mutual, or parasitic. *Commensal* relationships occur when organisms coexist without consequential effects on their mutual viability or well-being; these probably comprise the majority of vertebrate/microbial interactions. *Mutual* symbiosis occurs when both organisms benefit from the relationship. For example, the microbial flora of the mammalian gastrointestinal tract are provided readily available nutrients and a well-modulated environment. They in turn synthesize vitamin K, which cannot be produced by the host but is essential to the process of blood coagulation. In contrast, *parasitic* relationships occur when one organism benefits to the detriment of the other, as exemplified by microbial disease. Thus the term "parasite" can be used broadly to refer to all classes of organism that induce disease in vertebrates, including bacteria, viruses, fungi, protozoa, helminths, and arthropods.

Commensal and Mutual Interactions: The Normal Flora

The wide variety of microorganisms that normally inhabit the epithelial surfaces of vertebrate animals are collectively known as the normal flora. The specific species and strains of these microorganisms vary with and are specific for the species of the vertebrate host. For example, indigenous lactobacilli and staphylococci isolated from rats will adhere to keratinized rat stomach epithelium, whereas isolates from human, swine, or chicken stomach will not (374). Such observations emphasize the specificity with which endogenous microorganisms interact with their hosts. Within a given host species, physiological and environmental factors such as diet, sanitation, and age influence the composition of the normal flora (341). Within a given host, an organism can be commensal at one site but pathogenic elsewhere. *Escherichia coli (E. coli),* for example, is a normal inhabitant of the gastrointestinal tract of many species but is found in the lungs only during disease states. Similarly, the nasopharynx contains many organisms that could be pathogenic if present in bronchi or lungs.

The diversity of microorganisms that normally interact with vertebrate hosts is exemplified by the human flora (138). Human skin is inhabited by organisms ranging from mites to bacteria, most of which inhabit the stratum corneum and the hair follicles. The flora of the mouth also exhibit remarkable segregation into ecological niches. For example, *Streptococcus mitior* is found primarily on the buccal mucosa, *Streptococcus salivarius* on the tongue, and *Streptococcus mutans* and *Streptococcus sanguis* on the teeth and dental plaque (144), whereas the gingival crevices, which have oxygen concentrations of less than 0.5%, support many obligate anaerobes. Compared to other parts of the gastrointestinal tract, the stomach and upper intestinal tract are relatively free of endogenous flora though they may become colonized by organisms indigenous to more proximal or distal gastrointestinal regions if gastric pH increases. In contrast, the colon contains such large numbers of microorganisms, most of which are obligate anaerobes, that they comprise approximately 20% of the fecal mass (138). Commensal organisms from the mouth, nasopharynx, and intestinal tract are occasionally found in lymph nodes or in blood as a transient bacteremia secondary to activities such as chewing or tooth-brushing. Commensal organisms are also present in the male and female urethra. In women the normal urogenital flora is subject to hormonal modulation and varies with age, phase of the menstrual cycle, contraceptive methods, and pregnancy and labor (72, 149, 312, 337, 379). Estrogen, by altering the vaginal pH and the deposition of glycogen in the vaginal epithelium, favors the growth of lactobacilli, which are less prevalent in prepubertal or postmenopausal flora (138).

Normal flora are probably not essential to life (250), as suggested by the long life spans of axenic (microbe-free) animals maintained on appropriate diets under strict conditions (135). Nonetheless, axenic animals demonstrate marked physiological differences from conventional animals that possess intestinal flora (84, 135, 341). For example, conventional rats and mice have significantly higher body temperatures than do axenic rats at all phases of the circadian cycle, suggesting that gut flora provide a tonic stimulatory influence on body temperature (217). Moreover, normal flora can provide clear benefits to the host. For example, bioluminescent fish harbor light-producing bacteria in specialized organs that generate signals used to attract prey, escape predators, and communicate with conspecifics (287). In mammals, intestinal flora synthesize vitamin K, thereby supplementing dietary sources of this vital coagulation factor. The importance of this microbial source is illustrated by the observation that axenic rats maintained on a diet lacking vitamin K develop hemorrhagic diathesis, while rats with conventional microflora remain healthy on the same diet (84, 168, 169). Intestinal flora also hydrolyze conjugated bile acids, thereby promoting their reabsorption and reuse (127, 251, 416). However, intestinal microbes can also convert harmless materials into potential toxins or carcinogens (365).

Perhaps the most striking example of mutual symbiosis between vertebrates and microorganisms occurs in ruminants, such as cattle, sheep, and deer. In these animals, the stomach is anatomically and physiologically complex, with an initial compartment called the rumen. Because food undergoes fermentation in the rumen prior to normal gastric and intestinal digestion, relatively little glucose is available for absorption (5, 248). Unlike other mammals, ruminants use microbially produced short-chain fatty acids as primary energy sources, permitting unusually low normal blood glucose levels (45–80 mg/dl) (201). Ruminants are also highly dependent on proteins and amino acids produced by bacterial, protozoal, and fungal digestion of cellulose and hemicellulose, plant carbohydrates that are largely unavailable as nutrients for nonruminant species (3, 5, 248). Rumen microorganisms also degrade a variety of potentially toxic compounds, including phytotoxins and mycotoxins (1, 248). A microbial relationship similar to that seen in ruminants has also been postulated to occur in the forestomachs of hamsters, leaf-eating Colobus monkeys, and several other species (5, 410).

A less substantiated but nonetheless intriguing suggestion of a mutualistic relationship is that bacterial muramyl peptides, which are the basic monomeric building blocks of bacterial cell wall peptidoglycan, may serve a vitamin-like role in mammals (2). Although enzymatic systems capable of synthesizing muramyl peptides have not been detected in mammalian systems (206, 216), macrophages can process bacterial cell walls and peptidoglycan into biologically active peptide fragments (197, 401). Similarly, enzymes that hydrolyze peptidoglycan and muramyl peptides are present in mammalian serum (174, 393, 398), permitting the rapid in vivo metabolism of exogenously administered peptidoglycan (136, 300). Such microbially derived substances may have a physiological function, as suggested by observations that the administration of exogenous muramyl peptide induces effects ranging from nonspecific immune stimulation to excessive sleep (76, 224).

Parasitic Interactions

Organisms that induce disease in normal hosts are called *pathogens;* in contrast, organisms that induce disease only in abnormal (for example, immunosuppressed) animals are termed *opportunists*. The degree of pathogenicity of a given microorganism is termed *virulence* and is related to the rate or amount of mortality or morbidity induced by the organism. Another feature of pathogenic microorganism is *contagiousness* or *communicability,* which can be defined as the ability to disseminate to new hosts. If a pathogen is of low virulence, host defense mechanisms will usually suppress its replication and thereby secondarily reduce transmission (9).

Microbial properties that influence pathogenicity are known as *virulence factors;* they include adherence, resistance, and toxin production. Because virulence factors may be genetically expressed only in vivo (122, 145, 361), microbial pathogenicity can truly be evaluated only in vivo. For example, physiological sequestration of iron during infection limits iron availability and thereby retards the in vivo growth of many bacteria; alternatively, iron restriction can induce the expression of iron-binding proteins and thereby facilitate multiplication of some gram-negative bacteria (64).

The physiological manifestations of infectious disease in vertebrates can reflect one of three possible adaptive strategies: (1) alterations that confer survival benefit to the host in terms of defense against the pathogen or repair of pathogen-induced damage, (2) alterations that confer survival advantage to the pathogen in terms of improved reproductive capacity or enhanced dispersal to new hosts, or (3) alterations that do not benefit host or pathogen but simply occur as a by-product of the infectious process (126). Disease manifestations included in the last category would reflect either lack of appropriate experimental evaluation or deficiencies in our understanding of the disease process. Examples of the first alternative would include fever and iron sequestration, which have been demonstrated to retard microbial replication (159, 185, 219). Rabies, as an example of the second alternative, is a neurotropic virus that induces behavioral changes in the canine host that promote transfer of the viral pathogen to other hosts. These alternatives are not mutually exclusive. Diarrhea, for

example, can be envisioned to rid the host of the enteric pathogen, yet may also serve to disperse the microbe to uninfected individuals. In general, exposure to microorganisms has one of three outcomes: infection with disease, infection without disease, or no infection (377). The following sections will describe the strategies by which vertebrates prevent or limit invasion by pathogenic microorganisms and adapt to the presence of microbes in the environment, thereby maintaining homeostasis.

ANTIMICROBIAL HOST DEFENSE

Just as vertebrate animals physiologically and behaviorally adapt to external and internal physical perturbations, they have also evolved a repertoire of responses to potential microbial threats. This repertoire is collectively known as the immune response. The evolutionary basis of the immune response is considered to be the organism's need to distinguish "normal self" from "infectious nonself" or "altered self" (for example, neoplastic or damaged cells) (196). The strategies of immune protection include a variety of mechanistic and functional systems.

In general, host defense systems comprise two categories: innate (nonspecific) and acquired (specific or adaptive) immunity (Table 66.1). Despite these seemingly contrasting designations, the two systems are highly complementary. For example, while the innate system affords primary protection from invading microbes, its efficiency is greatly enhanced by interaction with the acquired or specific system. The innate immune system becomes activated and effectively operational within minutes to hours after exposure to infectious agents, while specific systems become fully effective one or more weeks later. Upon reexposure to the same microorganism, the specific system exhibits an anamnestic response ("memory") that develops within a few days and is of greater magnitude than the initial response (153). In contrast, innate mechanisms are generally unaffected by previous exposure except as they

TABLE 66.1. *Comparison of Mechanisms of Innate and Acquired Immunity*

Characteristic	Innate	Acquired
Specificity	Indiscriminate, nonspecific	Discriminating, specific
Induction	Constitutive	Induced by exposure (adaptive)
Onset	Rapid	Delayed (initial exposure) Rapid (secondary exposure)
Mechanical factors	Epithelium	Fibrosis, granulomas
Soluble factors	Lysozyme, complement, acute-phase proteins	Antibody
Cellular factors	Phagocytes, natural killer cells	B and T lymphocytes
Repeated exposure	No memory; latency and efficacy not altered, except by facilitation from acquired immunity	Memory reduces latency and enhances efficacy

are amplified via the actions of the specific system. The innate system is broadly effective against many types and species of microorganism, whereas acquired immunity is highly specific not only for microbial species but in many cases for individual strains. This specificity is largely based on the generation of antibody and the production of antigen/antibody interactions. The innate system is not dependent on such interactions for its activity. However, because the adaptive system is capable of responding to at least 10^8 different antigenic determinants, this specificity does not appear to significantly limit immune response capabilities. Clearly, the innate and acquired mechanisms of host defense interact at many levels. Details of these mechanisms will be presented in the following sections, but a few examples are mentioned here. Antibody, for example, enhances phagocytosis by aggregating and opsonizing pathogens (phagocytes have receptors for the Fc portion of antibody), activates complement on microbial surfaces (the classical pathway of complement activation), and facilitates complement-dependent cytotoxicity. T lymphocytes interact with the innate immune defenses by activating macrophages and secreting lymphokines like interferon-γ (IFNγ), which then stimulate macrophages to release immunomodulators such as interleukin-1 (IL 1). By serving as antigen-presenting cells, macrophages stimulate lymphocyte antibody production and cytotoxicity.

The following sections provide an overview of physiological strategies for host defense. They are intended to familiarize the reader with basic terminology and concepts and to provide examples of key phenomena. A comprehensive discussion of the immune system is beyond the scope of this chapter, and the reader is referred to immunology textbooks or to reviews of specific topics for more detailed information.

Innate or Nonspecific Immunity

Physical Barriers. The primary innate defenses against microbial invasion are the anatomical barriers formed by epithelial surfaces. Four surfaces form an interface between the animal and the environment: skin and conjunctiva, respiratory tract epithelium, gastrointestinal tract epithelium, and urogenital epithelium. These surfaces retard invasion by pathogenic microorganisms via functional properties that include motility, superficial desquamation, and the secretion of antimicrobial substances (395). The skin, for example, undergoes frequent desquamation, shedding potentially infective microorganisms before they penetrate farther. Skin secretions, including oils and fatty acids, deter microbial attachment and penetration and induce a hostile environment by altering the skin pH. Some vertebrate species also secrete specialized antibacterial substances

from the skin. For example, the African clawed toad *Xenopus laevis* secretes broad-spectrum antibacterial substances known as magainins that are postulated to retard microbial invasion in its aquatic habitat (424).

The respiratory tract possesses similar nonspecific defense mechanisms. Respiratory epithelial cells provide a dense protective lining, secrete mucus that traps invading organisms, and contain cilia that mediate the transport of trapped microbes to the oral cavity where they can be expectorated or swallowed. Nasal secretions and tears contain powerful antibacterial lytic enzymes called lysozymes. In addition, the cooler temperatures in the nasal passages relative to those in the deeper respiratory tract retard the growth of microorganisms requiring warmer environments. Interestingly, some microorganisms have evolved to take advantage of this temperature differential. Rhinoviruses, for example, which cause many cases of the common cold, grow better at the lower temperatures found in the nasal passages than at normal core body temperatures (373).

The gastrointestinal tract also has a complement of barriers to microbial invasion. Saliva contains lysozyme and glycolipids that impair the attachment of microbes to epithelial surfaces. Saliva also demonstrates bactericidal properties that are postulated to confer survival benefits to the behaviors of licking of wounds and neonates (176). After microbes are swallowed and enter the stomach, they encounter additional nonspecific host defense mechanisms, such as the highly acidic gastric pH, the digestive enzymes, and the expulsive properties of normal gastrointestinal motility.

The epithelial surface of the urogenital system is similarly protected by the continual flow of urine, which effectively retards the attachment of pathogenic organisms. Spermine, a polyamine present in prostatic secretions, inhibits the growth of gram-positive bacteria in semen.

Normal human milk has antibacterial properties and can kill two common intestinal pathogens, *Entamoeba histolyticum* and *Giardia lamblia*, thereby presumably conferring some protection to the nursing child (146).

Antimicrobial Properties of the Normal Flora. The normal flora provides nonspecific defense against microbial pathogens. The presence of antibiotic-producing commensal bacteria in human skin is associated with a reduced incidence of secondary infections of skin lesions and with a relatively low rate of wound infections after surgery (349). Studies in mice have indicated that a competitive bacterial flora can be more effective than an intact immune system in preventing some types of infection (179). Absent or reduced normal flora in the mouth and intestines lowers the oral infectious dose of pathogens in mice (395) and permits the overgrowth of potential pathogens (367). Normal flora and exogenous organisms compete for nutrients and for attachment sites within a given ecological niche in the host organism. For example, cell wall fragments, lipoteichoic acid, and peptidoglycan from commensal *Lactobacillus* species sterically inhibit the attachment of uropathogens to uroepithelial cells (73). *Lactobacillus* species also control the growth of exogenous organisms by inducing local pH changes, generating superoxides, and producing broad-spectrum antimicrobial substances (312). The endocervical presence of some species of *Lactobacillus* may reduce the risk of *Nisseria gonorrheae* infection following exposure to an infected partner (337).

The importance of microbial competition in the control of potential pathogens is perhaps best illustrated by the fact that antibiotics can alter the normal gastrointestinal flora, permitting the growth of toxin-producing *Clostridium difficile* and other pathogens that induce secondary colitis (59, 230, 232, 313, 359). Systemic treatment of mice with antibiotics allows abnormal intestinal colonization by oral flora and microbial penetration of mesenteric lymph nodes and spleen (396) and alters the flora and endotoxin content of feces (324). Patients with antibiotic-induced disruptions of the normal flora may be more susceptible to colonization by gram-negative enteric pathogens commonly present on raw vegetables (314). When axenic (microbe-free) mice and mice with a pathogen-free normal flora of known composition (gnotobiotic mice) are inoculated with *E. coli*, fewer *E. coli* are found in the gastrointestinal tract and mesenteric lymph nodes of the gnotobiotic mice, indicating that the bacterial antagonism provided by the normal flora helps to confine *E. coli* to the lumen of the intestines (31, 32). Both *Lactobacilli* and *Clostridia* limit *E. coli* multiplication in gnotobiotic mice, but they are maximally suppressive in different regions of the gastrointestinal tract (195). Several commensal pharyngeal organisms, including viridans streptococci, *Neisseria* species, and *Staphyloccocus epidermidis*, inhibit oral colonization by the pathogens *Staphylococcus aureus* and *Neisseria meningitidis*. In some cases, endogenous flora may actively produce antimicrobial products. The skin commensal *Propionibacterium acnes*, for example, produces lipids that suppress colonization by *Staphyloccocus aureus* and other microorganisms. Similarly, free bile salts, which are produced by the hydrolytic actions of gut anaerobes, inhibit the growth of a number of potential pathogens, including *Clostridium difficile* (39).

It is postulated that the normal flora provides continual, low-level immune stimulation that promotes nonspecific immunity, the expression of class II histocompatibility molecules on cells, and the generation of so-called natural antibodies (65). Natural antibodies occur in the absence of prior infection and are postulated to be directed against conserved structures common to many groups of microbes [for example, peptidoglycan, lipopolysaccharides (LPS), and capsular antigens], thereby conferring broad immunity (65, 267, 303, 322, 426, 429). Translocation of intestinal bacteria from the gut to the regional mesenteric lymph nodes and other sites could contribute to this process (412). For example, rats fed killed *Pseudomonas* organisms developed increased production of anti-*Pseudomonas* immunoglobulins (257). Similarly, human volunteers fed nonpathogenic strains of *E. coli* developed anticapsular and bactericidal serum antibodies that cross-reacted with the pathogen *Hemophilus influenzae* (344). The development of natural antibodies is influenced by the microbial flora. For example, axenic rats have low levels of serum antibodies and small lymph nodes; their Peyer's patches lack germinal centers and contain few, if any, antibody-producing cells (15). However, serum immunoglobulins increase in axenic rats exposed to common gut bacteria via association with inoculated cagemates (15). Conventional mice have 3–12 times more potential antibody-producing cells than their axenic counterparts (293). Similarly, antibody titers directed against the microbial antigen hemagglutinin are higher in conventionally raised pigs than in specific-pathogen-free pigs and are very low in axenic pigs (342). Such immune-facilitative processes presumably occur secondary to lysozymal hydrolysis of intestinal flora and the subsequent absorption of immunogenic bacterial fragments (198). Indeed, oral administration of bacteriolytic enzymes has been demonstrated to induce enhanced cellular and humoral immune responses in guinea pigs (286).

Local Inflammatory Response. As we know, animals are constantly exposed to potentially harmful microorganisms in the environment, but they are normally protected by anatomical barriers and by the normal microbial flora. However, this biological balance is very delicate, and the primary host defense system is frequently breached. Cellular damage and/or the entry of foreign material then evokes a concatenation of local and systemic reactions designed to protect the host and to preserve or restore normal function. The general term for these responses is "inflammation." Although the term generally connotes a destructive process, inflammation is a normal, physiological, homeostatic response that becomes pathological only when it escapes the host's regulatory control and/or becomes chronic. This section reviews the mechanisms of the acute inflammatory reaction; chronic inflammation will not be discussed.

The skin is an animal's most conspicuous anatomical barrier against microorganisms in the external environment. However, the outside world also interacts at the epithelia of the conjunctiva and of the gastrointestinal, urogenital, and pulmonary systems. Wherever and however microbial entry may occur, the initial stages of the inflammatory response are local and relatively similar and the mediators involved are generally the same. The response extends over a temporal and regional spectrum determined by the location and severity of the challenge and the efficacy of the host's defenses.

After penetration of the host epithelial cell layer and basement membrane, invading microorganisms pass into the subepithelial tissue fluid, where they encounter an array of physiological defense systems that interact in complex synergistic and antagonistic networks. The initial mediators of microbial inflammation are the neuropeptides substance P, substance K, calcitonin-gene-related peptide, and neurotensin, which are concurrently released by stimulated sensory nerve terminals at the site of inflammation (111). Simultaneously, mast cells normally present in the connective tissue and lining the blood vessel walls release histamine, serotonin, platelet-activating factor (PAF), leukotrienes (particularly C_4), cytokines, and prostaglandins (especially I_2) in response to the irritating stimulus (155, 370). These mediators cause the coincident dilation of arterioles and contraction of endothelial cells of nonmuscular postcapillary venules in and around the inflammatory focus. The latter effect creates endothelial gaps that allow plasma and soluble circulating factors to enter the extravascular space (259). This process results in a serous exudate containing proteins that counterbalance those in blood and thereby slow fluid reabsorption; hence, edema develops. Included among these proteins are immunoglobulins and components of the kinin, complement, coagulation, and fibrinolytic systems that become activated by contact with microbial products (for example, LPS, antigen/antibody complexes) at the inflammatory site. Their active derivatives (for example, bradykinin, complement proteins, thrombin, fibrinopeptides) in turn intensify local vascular permeability and attract neutrophils that emigrate into the inflammatory site by diapedesis through the venular walls (334). This process is described in more detail later in this section.

An important, early event in inflammation is the activation of complement, which is a collective term for approximately 20 functionally related proteins. When activated in a sequential manner, known as the *complement cascade,* the components can induce cellular lysis of the invading organisms. In addition, individual proteins of the complement series possess biological properties such as opsonization and chemoattraction. For example, complement proteins C3a, C4a and C5a are anaphylatoxins. They alter blood flow to the inflamed or damaged area, increase vascular permeability, stimulate the migration of inflammatory cells into the area, and promote the release of lysozyme granules from attracted neutrophils and macrophages. The complement cascade can be activated by either the alternate or the classical pathway. The alternate pathway operates in nonimmune animals and is activated by agents with characteristic, repeating chemical structures, such as sugar moieties in yeast, bacterial cell walls, and endotoxin; antibody is not required. In contrast, antigen/antibody complexes activate the classical complement pathway.

Phagocytosis, or the engulfment and killing of potential pathogens by host cells, is a major nonspecific effector system. The basic physiological mechanisms of phagocytosis include chemotaxis (chemical attraction of the phagocyte to the site of inflammation), adherence or attachment of the phagocyte to the pathogen, ingestion of the microbe, and dissolution or killing of the internalized microbe (161). The two basic types of mammalian phagocytic cell are neutrophils and macrophages. Neutrophils are continuously present in the circulation, though their numbers can increase dramatically during certain types of microbial disease as a result of the acute-phase response (APR; see below, Systemic Inflammation: The Acute-Phase Response). Neutrophils are short-lived, with a circulating half-life of 6–8 h. They predominate early in the course of inflammation. After being recruited by chemoattractants released at the site of tissue damage [for example, N-formyl peptides, leukotrienes, the C5a component of complement (70, 256, 344)], neutrophils release secretory products that loosen endothelial cell junctions and create gaps in the basement membrane for leukocytic emigration (417). These chemoattractants also stimulate neutrophils to develop increased numbers of surface receptors for chemoattractants, immunoglobulins, and complement and an increased capacity for adherence to endothelial cells and particulate matter (132, 331, 398).

Neutrophils contain two types of functional granules. The secondary or specific granules contain lactoferrin, receptors for chemoattractants and adhesins, and complement activators. These products are released extracellularly onto the cell surface and contribute to the initiation and amplification of the inflammatory response (132, 417). The primary or azurophil granules contain acid hydrolases, myeloperoxidases, proteases, and antimicrobial cationic proteins [for example, defensins (139) and bactericidal/permeability-increasing protein (123, 411)]. After phagocytosis, these granules fuse with phagosomes to form phagolysosomal vacuoles and participate in the destruction of vacuolar contents. Activated neutrophils also produce large amounts of oxidants via the *phagocytic respiratory burst.* This process reduces molecular oxygen to superoxide, which is rapidly converted to hydrogen peroxide and hydroxyl radical. Myeloperoxidase in azurophil granules, in the presence of hydrogen peroxide and halide, catalyzes the formation of additional oxidants like hypochlorous acid and free chlorine. These reactive substances provide broad antimicrobial oxidative activity in the phagosome and in the extracellular environment (123) and play an important role in the antimicrobial efficacy of inflammation. As a result of the inflammatory process, neutrophils die and contribute to the formation of pus or abscesses.

Macrophages, in contrast to neutrophils, are long-lived phagocytes. Their tissue turnover time is up to 60 days in the mouse, and they can perform multiple phagocytoses prior to death (150). In sites of inflammation, macrophages are probably derived from circulating monocytes that migrate there. Because the macrophage response generally takes longer than the neutrophil response, neutrophil invasion (suppurative inflammation) is generally associated with acute infection, whereas macrophage invasion (granulomatous inflammation) reflects a more chronic process in which the foreign organisms have not been effectively eliminated. Many tissues possess a fixed macrophage population that provides for rapid clearance of microorganisms (150, 242). Examples include macrophages in lymphoid tissue, Kupffer cells in the liver, microglial cells in the brain, Langerhans cells in skin, alveolar macrophages in lung, and synovial type A cells in the joints. In the basal state, circulating monocytes and tissue macrophages are active in removing small numbers of microbes and damaged cells via receptors for immunoglobulin, complement, and cell surface antigens.

Certain processes can enhance the efficiency of the phagocytic process. Neutrophils, for example, produce a protease that cleaves leukokinin in the circulation to produce tuftsin, which greatly enhances the rate of phagocytosis by macrophages (285). Probably the most important means of enhancing phagocytosis is *opsonization*, a process in which specific substances (opsonins) bind foreign materials to be phagocytized. Important opsonins include antibodies, fibronectin, and the complement C3b. Many microorganisms, such as nonencapsulated bacteria, do not require opsonization but can be bound by phagocytic cells via their intrinsic surface ligands (161). Although suspended organisms are thought to be most commonly phagocytized, the efficiency of the process is improved when neutrophils interact with microbes on cellular surfaces (394). The benefit of this "surface phagocytosis" is obvious, given that attachment of bacteria to epithelial surfaces is often the first phase of microbial invasion and that surface phagocytosis, unlike suspension phagocytosis, does not require opsonization. Also, because levels of opsonizing antibodies are low early in infection and some surfaces (for example, the cerebrospinal canal and alveoli) have relatively low opsonin levels, the lack of dependence on opsonization for surface phagocytosis provides an important defense advantage (235, 354, 394). Activation by cytokines also enhances the phagocytic activity of macrophages against many microbes (see later in this section).

Natural killer cells are also participants in the early inflammatory response (182). These cells are large granular lymphocytes that recognize and kill virus-infected cells and tumor cells (369). The ability of natural killer cells to lyse their targets is not antigen-specific, does not require presensitization, and is not restricted by major histocompatibility complex (MHC) antigens; indeed, absent or abnormal surface MHC antigens are associated with susceptibility to natural killer lysis (172, 369). Because of these properties, natural killer cells are valuable, early effectors in nonspecific host defense.

The phagocytic cells that enter the inflamed area promote and amplify the local inflammatory response via their release of certain cytokines [IL1 and -8, tumor necrosis factor-α (TNFα)], the mRNAs of which they transcribe constitutively and translate in response to a triggering signal, such as microbial products (Tables 66.2 and 66.3). They simultaneously release prostaglan-

TABLE 66.2. *Cell Sources of Cytokines*

Hemopoietic stem cell-derived
  Myeloid lineage
    Mononuclear phagocytes
    Monocytes
    Macrophages
      Kupffer cells
      Alveolar macrophages
      Glomerular mesangial macrophages
      Serosal/peritoneal macrophages
      Splenic macrophages
      Lymph node sinus macrophages
      Brain microglia/astrocytes
      Synovial macrophages
      Placental macrophages
      Neoplastic cell lines
  Polymorphonuclear granulocytes
    Neutrophils
    Eosinophils
    Basophils
    Mast cells
  Platelets
  Lymphoid lineage
    Lymphocytes
      T cells
      B cells
      Natural killer cells
Other cells
  Epithelial cells (keratinocytes, Langerhans cells, corneal, gingival, thymic)
  Fibroblasts (skin, lung, mammary gland)
  Endothelial cells
  Vascular smooth muscle cells
  Uterine stromal cells
  Anterior pituitary cells
  Chondrocytes
  Dendritic cells
  Neurons
  Tumor cell lines

TABLE 66.3. *Microbial Organisms, Their Products, and Exogenous Agents That Induce Inflammatory Cytokines*

Microorganisms
    Bacteria
    Virus
    Fungi
    Parasites
    Protozoa
Microbial Products
    Endotoxins
    Exotoxins
    Peptidoglycans
    Double-stranded RNA
    Specific polysaccharides (for example, yeast)
Antigens
    Microbial (for example, tuberculin)
    Nonmicrobial (for example, ovalbumin, HSA*)
    Immune complexes
Inflammatory agents
    Urate
    Asbestos
    Silica
Synthetic immunoadjuvants
    Polynucleotides (for example, poly I:C)
    Muramyl peptides (for example, MDP*)

*HSA, horse serum albumin; MDP, muramyl dipeptide.

dins (E_1, E_2, and I_2), various lysosomal components, lactoferrin, and other mediators (97, 181, 261). Injured vascular wall cells release additional cytokines and prostaglandins, while platelets and other antigen-sensitized cells in the inflammatory focus also release prostaglandins, histamine, PAF, and other compounds (295, 296, 309, 419). These substances amplify the effects of mediators derived from plasma and induce the production of additional inflammatory factors in plasma and in the exudate that in turn mediate the activation and recruitment of further local host defense functions. Some of these substances also have direct protective effects; for example, neutral proteases degrade digestible particu-

late matter like dead bacteria and cellular debris. Paradoxically, these enzymes also degrade normal tissue elements such as collagen and elastin. As we have seen, phagocytizing neutrophils also produce cytotoxic microbicidal oxidants, such as lactic acid, hydrogen peroxide, and reactive oxygen species (214, 252, 363, 409). These damaging effects are limited, however, by counterregulatory mechanisms, such as the induction of acute-phase proteins (see below, Acute-Phase Proteinemia) and heat-shock proteins of the hsp70 family (180, 311).

IL1, IL8, and TNFα that enter the circulation bind to specific endothelial cell and vascular smooth muscle receptors, inducing their own further production and the production of IL6 (254). These cytokines and histamine in the inflamed site further stimulate the de novo synthesis of the cellular adhesion molecules P-selectin (GMP-140), ICAM-1, VCAM-1, and E-selectin (ELAM-1) by endothelial and smooth muscle cells. These molecules then serve as receptors for the adhesion molecules L-selectin, LFA-1, VLA-4, CR3, and CR4 on the surface of leukocytes (24, 38, 118, 154, 212, 233, 362, 418) (Table 66.4). This process is enhanced by PAF and IL8 but is limited by cAMP in endothelial cells; by IL4 and IFNγ, which are released by T lymphocytes; and by IL6, IL10, and transforming growth factor-β (TGF-β), which are released by monocytic and vascular cells, in the inflamed area (102, 108, 177, 189, 255, 308, 310, 343). Binding of the neutrophils to the inflamed endothelium initiates an organized series of events in which the neutrophils shed L-selectin and engage a second set of adhesion molecules, the leukocyte β_2 integrins (4, 366), which trigger endothelial cell changes that facilitate the margination and further transvenular emigration of leukocytes (30). Simultaneously, IL1β and TNFα stimulate endothelial cells, T lymphocytes, platelets, and histiocytes in the inflamed area to secrete IL8 and the leukotrienes A_4, B_4, C_4, and

TABLE 66.4. *Interactions between Endothelial Cell Surface Molecules and Adhesion Molecules on Leukocytes*

Molecule	Receptor	Ligand	Function
Selectins			
L-selectin (LeCAM-1)	PMN*, monocytes	?	
P-selectin (GMP-140)	Endothelial cells, platelets	Sialyl Lewis X	Rolling
E-selectin (ELAM-1)	Endothelial cells	Sialyl Lewis X	Rolling
Immunoglobulin supergene			
ICAM-1	Endothelial cells	CD11a, b	Adherence, emigration
ICAM-2	Endothelial cells	CD11a	Emigration
Integrins (α/β_2)			
CD11a/CD18 (LFA-1)	All leukocytes	ICAM-1, 2	Adherence, emigration
CD11b/CD18 (CR3)	PMN, monocytes, some lymphocytes	ICAM-1, 2	Adherence, emigration
CD11c/CD18 (CR4)	Granulocytes, monocytes	ICAM-1, 2	Adherence, emigration
VLA-4	Lymphocytes, monocytes, eosinophils	VCAM-1	Adherence

*PMN, polymorphonuclear neutrophils.

D_4. These chemotactic agents in turn attract and activate neutrophils, T lymphocytes, and eosinophils and mediate the passage of these cells through the endothelium (81, 399). These effects are potentiated by complement proteins C3a and C5a and by bradykinin and thrombin derived from their plasma precursors. Endothelial cells activated by IL1, IFNγ, and TNFα also produce the vasodilating substances PGI_2 (prostacylin) and endothelium-dependent relaxing factor. The activity of the latter is mediated by nitric oxide, which is generated from the catalysis of L-arginine by nitric oxide synthetase and liberated into the abluminal space, where it stimulates cGMP production in the underlying smooth muscle cells (191). Macrophages similarly activated also produce nitric oxide, which then inhibits the further production of cytokines and prostaglandins by the macrophages (288, 368), thereby controlling their excessive elevation. Interleukin-1 also induces endothelial cells to produce endothelins (endothelial cell contracting factors 1 and 2), which are potent vasoconstrictors (332). Presumably, the balance between these factors modulates vascular tone at the site of inflammation. In general, however, vasodilation predominates. As a whole, the affected tissue thus displays the classical manifestations of local inflammation: heat and redness (vasodilation), swelling (cell and fluid exudate), and pain [prostaglandin E_2 (PGE_2), histamine, bradykinin, substance P, tissue distention]. Increased blood supply and temperature are thought to maximize the metabolic activity of the leukocytes, and local acidity may inhibit microbial multiplication.

Transudation into the inflamed area eventually subsides as platelets aggregate in the endothelial gaps and tissue pressure approximates intravascular pressure. The exudate and cellular debris drain off via lymphatic vessels. Foreign or microbial material in the lymphatics are subsequently phagocytized by macrophages lining the lymph nodes' marginal sinuses; this process filters the lymph and stimulates the synthesis of additional cytokines by the macrophages. An antigen-mediated immune response may also be initiated here, as discussed below (Acquired or Specific Immunity). However, some microbes or microbial materials may escape phagocytosis or seep through the porous lymphatic vessels and be reabsorbed locally by capillaries and venules. As a consequence of this escape, the inflammatory effects are extended to the whole body (see below, Systemic Inflammation: The Acute-Phase Response) and are augmented locally.

Thus the local inflammatory response is generated and maintained by a complex array of endogenous mediators organized in cascading, interactive circuits that recruit different pro- and anti-inflammatory mediators at different stages of the response. Unless the infectious stimulus persists, the acute response is self-limiting. Debris is removed by macrophages, lost tissue is replaced by fibroblasts, and new blood and lymphatic vessels are formed in the process of healing (86). Hence, the local inflammatory response can be viewed as necessary for recovery, that is, adaptive, because it focuses antimicrobial factors at the infected site and evokes the balance of destructive and restorative effects that contribute to maintaining optimal function.

Systemic Inflammation: The Acute-Phase Response. As we have seen, microorganisms, their products, and various mediators of the acute local inflammatory response can enter the bloodstream and thus evoke a panoply of systemic reactions. These highly organized and carefully regulated events develop relatively quickly after invasion. At their onset, they are nonspecific, are activated by common signals, and function synergistically, comprising a major systemic defense response that combats infection, preserves bodily function, and ultimately promotes survival. Collectively, these reactions are termed the *acute-phase response* (APR) (154). The best recognized and most obvious sign of APR is fever. However, the full range of responses is extensive, including changes in the plasma concentrations of certain proteins and trace metals, mobilization of amino acids from skeletal muscle, neutrophilic leukocytosis, changes in circulating hormone levels, alterations in glucose and lipid metabolism, reduced appetite, and increased sleepiness.

The APR is mediated by numerous cytokines (101, 225). Monocytes/macrophages are probably the principal sources, but other cells, including T and B lymphocytes, natural killer cells, eosinophils, and neural cells, can also produce cytokines (Table 66.2). Bacteria, viruses, fungi, parasites, and their products are potent inducers (Table 66.3), indicating that the synthesis of cytokines by immunocompetent cells represents a critical host defense response against invading pathogens. Cytokine-mediated host defenses depend on the secondary release of many other mediators that enhance or diminish the effects of cytokines, including epinephrine, norepinephrine, serotonin, cortisol, glucagon, PAF, eicosanoids, reactive oxygen species, nitric oxide, and acute-phase proteins. Defining the biological roles of individual factors in this complex cascade, therefore, is exceedingly difficult. Some presumptive interactions were discussed above, under Local Inflammatory Response.

Fever. Development of an abnormally high core temperature is one of the most conserved host defense reactions, being exhibited by many phylogenetically primitive species (215). However, pyrogenic sensitivity varies across species; humans and rabbits are highly sensitive to exogenous pyrogens, whereas rats are relatively insensitive (216). Fever is not synonymous with hyperthermia (371) but is regarded as a regulated, upward adjustment of the thermoregulatory set point (404). In a thermoneutral environment the febrile rise seldom exceeds 3°–4°C (115). Functionally, this increase in core temperature is accomplished by conservation of core heat via constriction of the cutaneous vasculature, cessation of sweating or panting, and increased production of metabolic heat (298). These physiological adjustments are accompanied by sensations of cold that prompt the subject to seek warmer surroundings and by the most visible sign of fever, shivering. The relative contributions of decreased heat loss and increased heat production vary depending on the type and dose of pyrogen; the ambient and initial body temperatures; and the age, size, and species of the host (43, 199, 215, 258, 298, 375). Various cardiovascular, respiratory, neuroendocrine, and metabolic changes accompany the febrile response and primarily reflect physiological processes mediating the enhanced heat production underlying fever (17, 26, 35, 157). Opposite reactions occur during defervescence, in which the thermoregulatory set point returns to its normal level (404). During this period, physiological and behavioral responses resemble those of hyperthermic subjects. The skin vessels become dilated, the skin becomes flushed and hot, perspiration or panting begins, and the defervescing subject seeks a cooler environment.

The febrile response is mediated by cytokines that enter the circulation from a local inflammatory focus or are produced by circulating and fixed mononuclear phagocytes activated by microorganisms or their products. At present, four cytokines, IL1 (α and β), IL6, TNFα, and IFNα_2, have been characterized as endogenous pyrogens. Others (for example, IL2, IL8, IL11, MIP-1, TNFβ, and IFNγ) are also pyrogenic, but they may not act directly (46, 49, 110, 203, 216, 271, 276, 330). Because these mediators are induced by the same stimuli and can promote each other's production and/ or activities (Tables 66.3, 66.5), their individual roles in the modulation of fever are still unresolved. Various cytokines appear to be released at different intervals after microbial challenge (77, 113, 133, 183, 434), and their amounts and combinations may depend on the infectious stimulus (69, 134, 236, 268, 297). The pyrogenic efficacy of specific cytokines also differs across species; for example, IFNα_2 is 100 times more pyrogenic in humans than in rabbits (112), and IL1α is less potent than IL1β in rats but more potent in rabbits (67, 289). In addition, the minimal pyrogenic intravenous doses of different cytokines as well as the doses required to produce a 1°C increase in core temperature range from very low (IL1) to very high (IL6). These doses approximate plasma levels after the bolus injection of an exogenous pyrogen (69, 268), though a temporal correlation between plasma levels of cytokines and the development of fever has been established only for IL6 (216, 236,

TABLE 66.5. Acute-Phase Actions of Some Cytokines

Actions*	IL1α,β	IL6	TNFα,β	IFNα,β,γ
Fever	+	+	+	+
Slow-wave sleep	+	−	+	+
Acute-phase proteinemia	+	+	+	+
Leukocytosis	−/+	+	−/+	−
ACTH secretion	+	+	+	+
PRL secretion	+	0	+	0
LH secretion	−	0	?	?
FSH secretion	0	0	?	?
TSH secretion	−	−	−	−
GH secretion	+	−	+	−
Neutrophil–endothelial cell adhesion	+	−	+	+
T-, B-, and NK-cell activation	+	+	+	?
Anorexia	+	?	+	+

+ = increase; − = decrease; −/+ = initial decrease followed by increase; 0 = no effect; ? = uncertain.
*ACTH, adrenocorticortropic hormone; PRL, prolactin; LH, luteinizing hormone; FSH, follicle-stimulating hormone; TSH, thyroid-stimulating hormone; GH, growth hormone.

237). The magnitudes, patterns, and durations of the fevers induced by these cytokines are also distinct (51, 56). The different temporal patterns may be related to differing neuronal substrates and/or second messengers (351), as well as to the secondary induction of other cytokines (106, 107). In addition, other endogenous factors released concomitantly can reduce the production of cytokines by various mechanisms (for example, corticosteroids, PGE$_2$), prevent interaction with their receptors (for example, receptor antagonists, various soluble receptors), or counteract pyrogenicity [for example, arginine vasopressin, α-melanocyte-stimulating hormone (αMSH)], thereby altering the febrile course (110, 113). Thus the pattern of circulating cytokines is complex, multiple interactions are possible, and no conclusion is yet possible as to their respective contributions to the febrile response.

The febrile response is almost certainly modulated by the brain. Fever can be induced via the microinjection of pyrogenic factors into the cerebral ventricles and into various discrete brain loci (49, 87). The electrical activity of thermosensitive neurons (see below) is similarly affected by peripheral or local application of pyrogens (121, 188, 351), and IL1, IL6, IFNα, and TNFα receptors have been directly visualized in the brain (128, 345). IL1α, IL1β, IL6, TNFα, IFNα_2, and MIP-1 evoke the most rapid and intense febrile responses when microinjected into the preoptic area (POA) of the anterior hypothalamus (49), the brain region that putatively contains the primary thermoregulatory controller (60).

Data have implicated a region in the anteroventral wall of the third ventricle (in the medial POA) as the most responsive site in rats (260, 372). Weaker, more slowly developing fevers have been elicited from the midbrain reticular formation of rabbits (327) and the lateral hypothalamus, pons, and medulla oblongata of guinea pigs (54). In contrast, the pons and medulla of rats and monkeys are not cytokine-sensitive (244, 245). The posterior hypothalamus of several species is also unaffected by the direct microinjection of cytokines (53, 88, 327), but in rabbits the integrity of this region appears to be necessary for fever production (89). The ventromedial hypothalamus appears to be necessary for the febrile response of guinea pigs to systemically injected LPS (350). Other sites, such as the septum, may be involved in counterregulating the pyrogenic action of cytokines (207, 243, 427). All of these regions contain, in differing densities, neurons that are responsive to increases (warm-sensitive) or decreases (cold-sensitive) in the local temperature; stimulation of these neurons evokes systemic heat loss and heat production responses, respectively (60). The response to cytokines is associated with the inhibition of warm-sensitive and the excitation of cold-sensitive neurons (351), consistent with the diminished heat loss and increased heat production that attend fever development (see above). These cells are therefore presumed to constitute the neuronal substrates of pyrogenic cytokine activity. The possibility that these cells are direct targets of IL1β is supported by observations that an IL1 receptor antagonist blocks the IL1β effects on neuronal activity in guinea pig POA slices (420) and that intracerebroventricular (ICV) injection of this antagonist attenuates the febrile response to ICV-injected IL1β in rabbits (193, 294). Furthermore, cells in all of the brain regions implicated in the febrile response express FOS protein following a pyrogenic challenge (75, 292, 408). However, the mechanisms by which the cytokines activate these neurons have not yet been characterized. Both facilitatory and inhibitory factors clearly operate in the further mediation and modulation of fever (for example, various monoamines, eicosanoids, and peptides), but the precise mechanisms are not yet certain. The manner in which circulating cytokine signals are transduced into the brain is also unknown. Reviews of these processes have been published (48, 49, 52, 57, 85, 99, 158, 207, 216, 243, 269, 318, 329, 427).

Despite its phylogenetic ubiquity, fever does not occur under all conditions, such as at the extremes of age. Neonates of various species often exhibit minimal or no fever following an endotoxic challenge (45, 375). This absence has been attributed to arginine vasopressin, a putative endogenous antipyretic that is most active around the time of birth (208, 428). The elderly, particularly those with age-associated functional dis-

abilities, may also have a blunted or absent febrile response to infection (199). In this case, the impaired febrile response appears to be related to a general loss of thermoregulatory capability; however, decreased production of cytokines and increased sensitivity to endogenous antipyretics may also play a role. Exposure to certain environmental stressors may also attenuate pyrogenic responsiveness. For example, the response of guinea pigs to pyrogens is reduced at high altitudes. This effect is thought to be due to a reduced capacity to produce heat under hypoxic conditions (114); depressed pyrogenic responsiveness of POA-thermosensitive neurons may also play a role (376). Febrile sensitivity is also influenced by the circadian cycle. Fevers are lower or absent in rabbits during nighttime sleep (352). Finally, conditions that evoke distress reactions may also reduce fever; for example, restraint attenuates the febrile response of rabbits and guinea pigs to LPS (55, 156). Stress-associated endogenous antipyretics released peripherally and/or within brain regions controlling core temperature could account for this effect (218).

The conservation of the febrile response in evolution is often regarded as evidence for its survival value. Lizards infected with live bacteria voluntarily move to warmer areas of their enclosure and show reduced mortality if permitted to make this behavioral response (215). Immune responsiveness may be increased at high temperatures (117), providing a possible explanation of this benefit. Conversely, the widespread use of antipyretics without apparent adverse effects on the outcome of infection would argue against an essential protective role for fever. Although the importance of fever as a diagnostic and prognostic sign of infectious disease cannot be minimized, its specific benefits in endothermic hosts are not clear (12, 16, 44).

Acute-phase proteinemia. Acute-phase response commonly refers to changes in the plasma concentrations of acute-phase proteins. These changes reflect cytokine-induced alterations in hepatocytic gene expression. The phrase "acute-phase" was introduced in 1941 by Abernethy and Avery (1) to describe human and monkey sera collected during the acute stage of a generalized infection. The sera contained a substance, C-reactive protein (CRP), that was not present in sera from healthy subjects. Like fever, the hepatic acute-phase protein response has also been strongly conserved through evolution in vertebrate species, suggesting that it too plays a major role in systemic host defense (140). However, the profile of plasma proteins during APR differs markedly across species (Table 66.6). For example, serum amyloid A (SAA) is a major acute-phase protein in mice but not in rabbits; conversely, mice do not have a significant CRP response. In addition, the pattern of acute-phase protein responses to different agents may vary (346). Nevertheless, common pathways are used to gen-

TABLE 66.6. *Plasma Levels and Physiological Roles of Some Acute-Phase Proteins*

Acute-Phase Proteins	Δ	Species	Homeostatic functions
CRP	↑↑↑ ↑	Human, rabbit, rat, mouse	1, 2, 6
SAA	↑↑↑ ↑	Human, mouse, rat, rabbit	1, 6
Fibrinogen (Fbg)	↑↑ ↑	Human, rat, mouse, rabbit	1, 2, 6
Haptoglobin (Hpt)	↑↑	Most species	3, 4, 5, 6
Ceruloplasmin (Cer)	↑	Most species	3, 4, 5
α_2-Macroglobulin (α_2M)	↑↑↑ 0	Rat, human, rabbit	3, 4, 5, 6
α_1-Proteinase inhibitor (α_1Pi)	↑↑ ↑	Human, rat, mouse	4, 5, 6
α_1-Acid glycoprotein (α_1AgP)	↑↑	Most species	2, 3, 6
α_1-Acute-phase globulin (α_1AgG)	↑↑ ↑	Rat, human, rabbit, mouse	4, 5
α_1-Antichymotrypsin (α_1ACH)	↑↑ ↑	Human, rat, mouse	4, 5
Albumin (Alb), transferrin	↓	Most species	3

Δ = change (↑↑↑ = increase 100–1,000 ×; ↑↑ = increase 2–5 ×; ↑ = increase 40%–60%; 0 = no change; ↓ = decrease).

1 = binds and removes foreign materials (opsonin); 2 = activates coagulation and fibrinolysis; 3 = transports protein; 4 = antiinflammatory; 5 = protease inhibitor; 6 = modulates immune responses.

erate the patterns of acute-phase proteins unique to different animal species.

At present, the cytokines IL1, IL6, and TNFα are known to directly regulate acute-phase protein synthesis (228). As discussed under Fever, high levels of IL6 and TNFα are rapidly detectable in plasma following microbial challenge, while circulating levels of IL1 and IFNγ are low. These cytokines are delivered to the hepatocytes during the inflammatory response, causing hepatocytic gene regulation at picomolar levels via specific high-affinity receptors (278). During cytokine stimulation, IL6 receptors are up-regulated on hepatocytes but down-regulated on monocytes (19). In rat and human hepatocytes, IL6 is the major inducer, IL1 induces selected acute-phase proteins, and TNFα is less potent than IL1 (140). However, IL1 enhances or diminishes the effect of IL6 on the synthesis of certain proteins (22, 228, 279). All three cytokines induce increased hepatic uptake of amino acids; accumulation of iron and zinc; increased transcription and translation of hepatic mRNA; decreased expression of negative acute-phase proteins, and increased expression of inducible acute-phase proteins and constitutive glycoproteins (47) (Table 66.5). The detailed mechanisms by which hepatocytes transduce these cytokine messages are incompletely understood. Data obtained in vitro suggest that these mechanisms may differ for different proteins, cytokines, and cell systems (that is, primary hepatocyte cultures, cell lines, etc.) (228). C-reactive protein and SAA are inducible acute-phase proteins that are quickly

produced in proportion to the magnitude of the inciting stimulus; they are also rapidly cleared from the circulation after the stimulus has abated. Their concentrations thus correlate temporally with disease severity, and they have become the prototypic acute-phase reactants, used both diagnostically and prognostically (154, 220, 357). Constitutive plasma glycoproteins synthesized in the liver (for example, CRP), as opposed to those synthesized in the intestine and macrophages [for example, complement protein C3, α_1-proteinase inhibitor (α_1Pi), α_1-antichymotrypsin (α_1ACH)], also increase severalfold in concentration during infection, though at a slower rate than do the inducible proteins; due to their relatively long half-lives, they persist in the circulation for days after the infection has resolved (154).

Cytokines other than IL1, IL6, and TNFα can also induce acute-phase proteins (for example, leukemia inhibitory factor released by human keratinocytes) or modulate the expression of acute-phase protein genes stimulated by the primary cytokines [for example, IFNγ, TGF-β, and epidermal growth factor (EGF) (20, 140)]. These substances may thus determine the final level of expression of the individual acute-phase proteins elicited under different inflammatory conditions (20). Other possible variables include posttranslational events, such as altered glycosylation (249), and factors such as glucocorticoids, which are required for the optimal induction of most acute-phase proteins (21). IL1, IL6, and TNFα stimulate the hypothalamus–pituitary–

adrenal axis, thus providing the mechanism for the generation of glucocorticoids (see below, Neuroendocrine Responses). Neuromodulatory influences on acute-phase protein synthesis have also been demonstrated, based on increased plasma protein concentrations following ICV or intrahypothalamic injection of IL1 and IL6 (50, 51, 58). The central action of IL1 may be mediated by peripheral IL6, the concentration of which increases in plasma in response to ICV-injected IL1 (106, 107). However, the mechanism of this effect is unknown.

To support the protein anabolic activity of hepatocytes during APR, IL6 and TNFα induce the mobilization of peripheral proteins, largely from skeletal muscle and collagen; whole-body amino acid oxidation is also augmented (227). It is not certain whether IL1 similarly induces muscle proteolysis (273). Negative nitrogen balance is thus a characteristic sign of infection (26). Cytokine-induced uptake of amino acids by the liver is facilitated by glucagon and glucocorticoids (20, 21) and is modulated by TGF-β and EGF (325). Some of the amino acids are also utilized for the synthesis of immunoglobulins by B lymphocytes. At the same time, gluconeogenesis is interrupted by the IL6-mediated inhibition of the induction of phosphoenolpyruvate carboxykinase (185), the rate-limiting enzyme in gluconeogenesis, thereby contributing to the changes in glucose homeostasis seen during APR (see below, Effects on Intermediary Metabolism).

As discussed earlier, neutrophils and other leukocytes recruited to the site of microbial invasion release hydrolases and other agents that digest and remove inflammatory debris; these substances, if not controlled, can cause severe tissue damage. One function of the acute-phase proteins is to control proteolytic activity and to mediate the clearance of injurious agents. For example, α_1Pi (formerly α_1-antitrypsin), which increases three- to fivefold in plasma, down-regulates neutral proteinases, while ceruloplasmin scavenges reactive oxygen species in blood and tissue fluids; α_2-HS-glycoprotein and fibrinopeptides promote phagocytosis and opsonization; and CRP functions as an opsonin, fixes complement, enhances phagocytosis, and binds chromatin (220). The interested reader is referred to more specialized reviews for further details on the functions of these compounds (141, 151, 185, 227, 229, 239, 302, 347, 357). These proteins thus play a protective role during the early stages of inflammation. The erythrocyte sedimentation rate is an indirect measure of the concentration of these mediators; it characteristically increases during inflammation due to increased plasma levels of acute-phase proteins, in particular fibrinogen. Because they elicit largely proinflammatory actions, IL1 and TNFα are considered proinflammatory cytokines in this context, whereas IL6, which regulates the synthesis of proteinase

inhibitors and fibrinogen, is considered an antiinflammatory cytokine that mediates the homeostatic response of the body.

Despite species-specific differences in the acute-phase proteins, these substances perform common functions in all species. Because many cytokines are elicited following injury, the final pattern of acute-phase proteins is probably determined by the ability of the liver to interpret the complex mixture of signals it receives. In addition, the order of appearance of specific cytokines and their additive or inhibitory actions on each other may modulate the patterns of gene regulation and expression.

Neuroendocrine responses. Although less obvious, the endocrine responses evoked during APR are very important, since, in large measure, they underlie the changes already discussed. For example, elevated levels of plasma glucocorticoids have long been recognized as a nonspecific systemic manifestation of inflammation (27, 80) and are now thought to be an important signal that links the nervous and immune systems via the inhibition of IL1, IL2, and TNFα; the down-regulation of TNF receptors; and the upregulation of IL1 receptors (37, 42). Interestingly, the efficacy of glucocorticoids in this regard may be species-dependent (433, 434). Glucocorticoids represent only one of several humoral signals during acute infection; because their release into the circulation is generally abolished by hypophysectomy, the pituitary gland is known to be involved in APR (18, 34, 37, 148, 340, 353). Indeed, adrenocorticotropic hormone (ACTH), αMSH, β-endorphin, somatostatin, growth hormone, vasopressin, and oxytocin are all secreted in response to systemic LPS. However, secretions of thyroid-stimulating hormone (TSH), prolactin (PRL), chorionic gonadotropin (ChG), luteinizing hormone, and follicle-stimulating hormone (FSH) are inhibited. The peripheral sympathetic nervous system is also activated (33, 131, 321, 338). The precise pattern of these effects varies with the severity and chronicity of the infectious stimulus and with the species. Systemic injection of LPS rapidly stimulates ACTH secretion at doses lower than those that produce fever; hence, this response is independent of fever induction (105).

Many of the cytokines produced during APR have been implicated in the modulation of endocrine responses (18, 33, 37, 171, 194, 263, 320, 339). IL1 and IL6 induce ACTH and, hence, glucocorticoid release. Also stimulatory for ACTH are TNFα, IL2, and IFNγ, but their effectiveness depends on the route and dose of administration (263). Intravenous or central administration of low doses of IL1β also stimulates the release of somatostatin, PRL, growth and hormone, pancreatic glucagon and insulin (103, 315, 340, 421); however, high doses of IL1 inhibit growth hormone and PRL release, probably via corticotropin releasing hormone (CRH) (301). Similarly, TNFα induces secretion of PRL and growth hormone; the doses required, however, are relatively high (98, 316, 364). Interleukins 1 and 6, TNFα, and IFNγ suppress the secretion of TSH and thyroid hormone; IFNγ and IL6 also suppress growth hormone release but have no effect on PRL, luteinizing hormone, FSH, and ChG release (200, 291, 299). However, IL1 inhibits luteinizing hormone release but does not affect FSH release (116, 263). Also, TNFα and IFNγ inhibit pituitary hormone release in vitro (137, 400, 407). Some of these effects can be blocked by αMSH (68, 319). It is probable that different signaling systems are involved in these varied responses. Thus these cytokines generally induce changes in the plasma levels of pituitary hormones similar to those occurring during infection. Indeed, they are the likely mediators of these changes (34).

Whether cytokines act directly on pituitary cells or indirectly via actions on hypothalamic releasing factors is still unknown (18, 37, 339). The available in vivo evidence strongly supports cytokine action at the hypothalamic level, but direct pituitary effects clearly occur in vitro. Both sites are likely to be cytokine targets, depending on the cytokine and other factors, thereby providing redundant systems for modulating components of APR.

Effects on intermediary metabolism. The metabolic effects of acute inflammation are multiple and complex and, like other defense responses, induced by circulating IL1, IL6, TNFα and -β, and IFNα and -γ. The principal actors appear to be ILβ and TNFα, while IL6 and IFN have supportive roles (41, 165, 209, 384). Specific effects are not correlated with circulating cytokine levels, however, but rather with their paracrine actions in various organs. For example, large quantities of TNFα are present in blood within 90 min after endotoxin administration, but these levels decline rapidly as the compound is cleared by the kidneys or binds to its tissue or soluble receptors (183); however, its metabolic effects persist for many more hours (384, 385). In addition to protein metabolism [see Acute-phase proteinemia, above; also ref. 416], these cytokines alter lipid and glucose metabolism. They act in adipose tissue to decrease the expression of lipoprotein lipase (LPL) and the lipogenic enzymes acetyl-CoA-carboxylase and fatty acid synthetase and to increase the expression of hormone-sensitive lipase in adipocytes (130, 209); in skeletal muscle they stimulate glycogenolysis, lactic acid production, and expression of hexose transporters (387); in the liver they increase the rate of lipogenesis (129, 163); and in the gastrointestinal tract they decrease gastric secretion and intestinal motility and absorption (390).

The mechanisms underlying the changes in lipid metabolism are interesting. The cytokines IL1, IL6, and TNFα induce an increase in very low density lipoproteins (VLDL) that are subsequently processed to triglyceride-rich low-density lipoproteins (LDL) (222). The resulting hypertriglyceridemia is not due simply to decreased LPL activity, because plasma triglyceride levels rise within 45 min after cytokine administration, but LPL suppression requires several hours (129); moreover, LPL activity is not depressed in all fat pads (164). Decreased clearance of VLDL also fails to account for the hypertriglyceridemia. Instead, the cytokines appear to increase hepatic production of fatty acids and, consequently, of VLDL. This increased synthesis develops within 30 min after the administration of pyrogenic doses of cytokines and continues for several hours. The free fatty acids mobilized under these conditions are not oxidized but are reesterified into triglycerides and resecreted as VLDL (129, 231); hence, hepatic ketogenesis is reduced (125). Cytokines induce hepatic lipogenesis by increasing hepatic levels of citrate, the activator of acetyl-CoA-carboxylase, the rate-limiting lipogenic enzyme (167); this increase is blocked by IL4 (166). The primary mediator of lipogenesis appears to be TNFα, with other cytokines exerting supplementary actions (130, 209). Interestingly, centrally administered TNFα induces responses similar to or greater than those of systemic TNFα (386). Also, IFNα stimulates fatty acid synthesis by an unknown mechanism. These changes in lipid metabolism presumptively benefit the host, as VLDL and LDL bind LPS and reduce its pyrogenic and other inflammatory effects (173, 391); they also inactivate various viruses (79, 348).

Acute, generalized infection is usually accompanied by an early hyperglycemia followed by a persistent hypoglycemia (246, 415). Hyperglycemia is mediated by skeletal muscle glycogenolysis stimulated by IL1 and TNFα, in combination with glucagon, glucocorticoids, and epinephrine (123). In mice, the subsequent hypoglycemia becomes maximal within 2 h after injection of subpyrogenic doses of IL1 and persists for more than 8 h; it is not the result of glucosuria (103). It develops despite the presence of counterregulatory glucocorticoids, glucagon, and epinephrine (142). At equivalent doses, IL6 and TNFα do not induce hypoglycemia (37). Although IL1 administration is associated with increased blood insulin levels (36, 93, 103, 143, 335, 425), the hypoglycemic effect of IL1 is observed in insulin-resistant diabetic mice, suggesting that hypoglycemia is induced by a mechanism other than the secretagogue action of insulin (35, 36). Indeed, the increase in insulin level may be species- and strain-dependent. For example, in C3H/HeJ mice the hypoglycemic response to IL1 is associated with increased plasma insulin, glucagon, and corticosterone concentrations, while in C57B1/6J mice hyperinsulinemia does not develop (33, 103). Wistar rats may (389) or may not (103) develop hyperinsulinemia in response to IL1 administration. Thus the mechanism of the hypoglycemic effect is still unclear. Possible central nervous system (CNS) mediation is suggested by demonstrations that glucose-responsive neurons in the ventromedial hypothalamus are excited and glucose-sensitive neurons in the lateral hypothalamus are depressed by IL1β, TNFα, and IFNα; IFNγ, IL8, and MIP-1 may also exert effects (226, 270, 305–307). Similarly, 50% of the thermosensitive neurons in the POA are glucose-responsive (61, 226). Direct inhibition of gluconeogenic enzymes is also possible. Thus it has been reported that IL1 antagonizes the induction of phosphoenol pyruvate carboxykinase by glucocorticoids in vitro (184), but in vivo adrenalectomy actually enhances hypoglycemia (103). However, the activities of other gluconeogenic enzymes are also depressed (422). Another possibility is that glucose transport and utilization by peripheral tissues may be enhanced in excess of the initial, infection-induced, high hepatic glucose output. Thus it has been reported that LPS, IL1, and TNFα increase glucose clearance rate by adipocytes, skin, lungs, intestine, liver, spleen, and other lymphoid tissues (14, 92, 265, 266). If hypoglycemia reflects reduced gluconeogenesis, it may benefit the infected host by increasing amino acid availability for acute-phase protein and immunoglobulin syntheses; if due to increased glucose utilization, hypoglycemia may simply reflect the body's augmented functional activities in response to the infection. It should be noted that the kinetics of glucose metabolism described here differ from those in severe sepsis, where blood glucose levels progressively increase (246).

Neutrophilia. Another characteristic of infection is an abnormally high number of circulating neutrophils (356). The initial neutrophilia is transient, occurring within 2–3 h of bacterial and fungal and within 5–8 h of viral invasion. This initial neutrophilia is attributed to epinephrine-stimulated release of cells from the bone marrow reserve and the marginal pools; a modest leukopenia may also occur at this time (211, 317, 382). Subsequently, a slower but sustained increase in numbers of neutrophils, monocytes, and eosinophils, which peaks within 1–2 days, is induced by specific factors that stimulate the proliferation and differentiation of their precursor cells. The cells released are often relatively immature because the increased demand stimulates their rapid production in, and release from, the bone marrow to provide abundant leukocytes for participation in the inflammatory response. The most proximate of the inducing agents are various colony-stimulating factors (CSFs), which are circulating cytokines synthesized by activated endothelial and smooth muscle cells within minutes of the influx of microbial products. These CSFs include G-CSF, M-CSF, and IL5, which act in synergy with IL3 on the precursors of neutrophils, monocytes,

and eosinophils, respectively. The effects of IL3 and M-CSF are potentiated by IL1 and IL6. Also, IL1 and TNFα can induce the synthesis of M-CSF (82, 192, 403). However, other cytokines and mediators released by macrophages and platelets during the inflammatory response, including IFNγ, TGF-β, MIP-1, and PGE$_2$ (432), limit and control these stimulatory effects. In this manner, the response remains within bounds appropriate to the needs of the body. It is probable that these adaptations are modulated in part by the CNS (275, 277).

Other effects. Acute-phase responses in addition to those reviewed above may also be mediated by cytokines. For example, the natriuresis that often occurs early during an infection (28) is mediated by IL1-induced PGE$_2$ in the kidneys (23). Whether the IL1 is derived from the blood or originates in renal mesangial cells is not yet clear, nor are the benefits to the host of sodium and fluid reduction. Systemic IL1α, IL6, IL8, and TNFα also elevate the pain threshold of animals (95, 96, 336). However, central IL1 may be antinociceptive (213).

Some of the effects exerted by cytokines during inflammation (for example, changes in bone metabolism) develop over longer periods of time and may not be relevant to APR. Such effects, therefore, are not considered here. The interested reader is referred to appropriate reviews (247, 283).

Behavioral manifestations of APR. In addition to physiological manifestations, APR includes behavioral changes, such as anorexia (175) and altered sleep (202, 380–382). Anorexia may provide survival advantage to the animal by complementing the reduction in serum iron (an important microbial growth factor) induced during APR and by conserving energy normally used to search for food. Similarly, reduced activity due to increased somnolence would promote conservation of energy reserves. Evolutionary advantages of decreased activity as a means of avoiding predators could also contribute to the survival benefits of enhanced sleep during microbial illness. Studies with mice have demonstrated that forced feeding exacerbates the mortality induced by bacterial infections, whereas food deprivation prior to exposure is associated with increased survival (284, 414). Similarly, a reduced phase of enhanced sleep is associated with a poor prognosis in microbially infected rabbits (383).

Thus taken together the mechanisms that underlie inflammation are very complex, involving multiple, overlapping, concurrent, and parallel pathways. Although cytokines are clearly important in these events, their roles can as yet only be tentatively described, and many other mediators also influence the response. Together they constitute redundant effectors of a generalized host defense reaction that minimizes the deleterious effects of external invaders and mediates the repair of injury. Thus APR can properly be viewed as a homeostatic response to the microbial environment.

Acquired or Specific Immunity

The innate host defense mechanisms are complemented by specific, antibody-mediated mechanisms. A specific immune response begins with the recognition by surface receptors on T and B lymphocytes of antigenic determinants presented by macrophages and other cells (402). This recognition triggers the multiplication and differentiation of immature lymphocytes to produce clones of T lymphocyte–derived cytotoxic effector cells (cellular immune response) and B lymphocyte–derived, antibody-producing cells (humoral immune response) targeted to antigenic determinants. Once established, the specific immune response demonstrates the important property of memory, which enhances the speed, sensitivity, and efficacy of responses to previously encountered antigens.

Humoral Immunity. The humoral immune response refers to the generation of antibodies by B lymphocytes which reside primarily in peripheral lymphoid organs such as the spleen and lymph nodes. After their surface receptors recognize antigenic determinants (epitopes) on soluble multivalent molecules, B lymphocytes internalize and partially digest the triggering antigens, reexpressing them on the cellular surface in association with class II MHC antigens for presentation to T lymphocytes (402). This process triggers the T cells to elaborate and release cytokines that promote the clonal selection, proliferation, and differentiation of the antigen-presenting B cells into plasma cells (terminally differentiated, short-lived cells producing antibody specific for the triggering epitopes) or memory cells, which undergo limited division and then become quiescent for long periods, retaining high antigen affinity (90).

The ability of animals to respond to a seemingly unlimited number of antigens depends on the existence of large numbers of lymphocytic clones, each bearing receptors for distinct antigens. These receptors are immunoglobulin molecules located in the cellular membrane. Following differentiation into plasma cells, B lymphocytes secrete large amounts of soluble specific immunoglobulin known as antibodies. Immunoglobulins are structurally composed of molecular subunits called heavy and light chains. The five main classes of heavy chains characterize the immunoglobulin subtypes: IgM, IgG, IgA, IgE, and IgD. The region closest to the amino terminus, the antigen-binding fragment (Fab), exhibits tremendous variability in amino acid sequence. In contrast, the carboxy terminus (the Fc portion) is a highly conserved sequence that binds to immunoglobulin receptors on cells involved in immune and inflammatory responses.

The multiple interactive potential of immunoglobulin has important host defense functions (178, 274). For example, cytotoxic effector cells bind to and destroy IgG-coated target cells in the process known as antibody-dependent cell-mediated cytotoxicity (ADCC). Binding of IgG also induces the discharge of lysosomal contents and the generation of superoxide anions by neutrophils and facilitates the removal of circulating immune complexes by phagocytic cells. The Fc portion of IgG interacts synergistically at the neutrophil CR3 receptor with complement components on microbial surfaces, enhancing adherence and promoting phagocytosis (253). The Fc portion of IgG and complement component C3b exert both independent and interactive effects on oxidative metabolism and degranulation of neutrophils (152). Exposure to IFNγ increases the number of Fc receptors on neutrophils, thereby facilitating ADCC (304). Like neutrophils, natural killer cells also possess IgG–Fc receptors and lyse cells bound with IgG antibodies. Binding of IgE triggers anaphylactic responses by mast cells and basophils and directs antiparasitic cytotoxicity by eosinophils.

Antibodies have several distinct antimicrobial mechanisms. Perhaps their best-known defensive function is neutralization (109). By binding to microbes, antibodies in some cases can inhibit their biological activities. For example, antibody binding may prevent microbial infection of target cells by blocking structures such as pili or fimbria that mediate attachment to cellular receptors, or they may interfere with the cellular entry or uncoating of viruses. Similar mechanisms enable antibodies to inactivate microbial toxins.

In addition to neutralizing the biological potency of infecting microbes, antibodies also facilitate their destruction and removal via opsonization, in which microorganisms coated with antibody become better targets for phagocytic cells. Microbial targets for opsonizing antibodies include capsular polysaccharides, surface proteins, and peptidoglycans. Opsonic antibodies are most effective against extracellular organisms that are rapidly destroyed after phagocytosis. Factors other than antibodies, such as complement components, fibronectin, and CRP, can also serve as opsonins. Both complement and antibody are necessary for opsonization of some organisms, and opsonic cooperation may be required for optimal host defense. For example, complement component C3b alone may enhance adherence and permit ingestion of microbes by phagocytic cells; alternatively, the Fc component of immunoglobulin may promote ingestion following C3b-facilitated binding (160, 161).

An important result of antigen/antibody complex formation is activation of the classical complement pathway. While free immunoglobulin binds weakly to the first component of complement (C1) and cannot activate the cascade, antigen/antibody complexes bind C1 with high affinity via the Fc component. Possibly because IgM circulates as a pentamer that presents multiple sites for C1 binding, it is much more efficient than IgG in initiating the cascade. If the antibody is bound to an invading microbe, activation of the cascade results in the generation of enzymatic complexes on the microbial surface. Although antibody and complement systems both represent soluble components of serum and are closely related historically and functionally, important differences between these systems include the specificity of antibody-mediated events relative to the general cytotoxic activity of complement and the rapidity of complement activation relative to the delay required for antibody generation. In addition, complement-mediated lysis by serum relies heavily on heat-labile enzymatic components, while antibodies are heat-stable components of serum. An interesting question concerns the need for antibody/antigen complex activation via the classical pathway of the complement system, given that a wide variety of microbial components can directly activate this system via the alternate complement pathway. One proposal suggests that the classical pathway may have evolved to mediate clearance of antigen/antibody complexes and that the merger of this process with the lytic arm of the cascade capitalized on the inherent specificity and adaptability of antibody-mediated processes (153). Clearly, however, the classical pathway permits rapid and efficient activation of the complement system and facilitates the appropriate deposition of complement onto the surfaces of invading pathogens (104).

Cellular Immunity. The cellular immune response is mediated primarily by T lymphocytes, whose functions include direct cytotoxic effects (cytotoxic T lymphocytes or CTL), activation of macrophages, and induction or regulation of antibody production by B lymphocytes (helper T lymphocytes). Unlike B cells, which recognize soluble antigens, T lymphocytes recognize antigens on the surface of antigen-presenting cells (APC) via a molecule known as the T-cell receptor (TCR) (100). Antigen-presenting cells, which include monocytes, tissue macrophages, dendritic cells, and B lymphocytes, present antigens to T lymphocytes as a complex of antigenic peptide fragments and a host MHC antigen (204, 205, 392, 430). The two basic subtypes of T lymphocyte are characterized by different accessory molecules thought to enhance the affinity of the TCR for the antigen–MHC (366). Helper or regulatory T cells express the CD4 receptor, whereas suppressor or cytotoxic T cells express the CD8 molecule. Binding of TCR to the antigen–MHC activates T cells, resulting in lymphokine production, T-cell proliferation, and cytolytic activity.

The MHC is a cluster of genetic loci that encode polymorphic cell membrane molecules involved in antigen binding and T-cell recognition. The T-cell requirement for the presentation of antigen in association with MHC proteins, which is known as MHC restriction, applies to two general systems of antigen processing (Table 66.7). In the exogenous system foreign proteins accumulate in APC after endocytosis, are fragmented in intracellular vesicles, and are then expressed on the cell surface in combination with class II MHC molecules (333); these complexes are recognized by T cells expressing the CD4 receptor (helper or regulatory T cells). Class II MHC proteins are found mainly on APC (for example, B lymphocytes, macrophages, thymic epithelium), which are generally prevalent at lymphoid sites of microbial entrapment; their expression is regulated by IFNγ in many cell types and by IL4 in B lym-

TABLE 66.7. *Comparison of Class I and Class II MHC Restriction*

Characteristic	Class I	Class II
MHC glycoprotein designations		
Mouse H-2 (chromosome 17)	K, D, L	I-A, I-E
Human HLA (chromosome 6)	A, B, C	DR, DQ, DP
Structure	1 polymorphic chain + β_2-microglobulin	Two polymorphic chains
Antigen recognized	Endogenous (intracellular)	Exogenous (extracellular)
Cellular distribution	Virtually all cells	Immune-competent APC (B lymphocytes, macrophages, monocytes, thymic epithelium)
Regulatory vectors	IFN, TNF, prostaglandins, glucocorticoids	IFNγ, IL4
Lymphocyte recognition		
CD phenotype	CD8$^+$ T lymphocytes	CD4$^+$ T lymphocytes
Function	Cytotoxic or suppressor	Helper
Functional role of lymphocytes	Virus surveillance Tumor surveillance Graft rejection	Regulate interactions between immune-competent cells Bacteria, parasite surveillance Graft vs. host rejection

phocytes. Thus the cellular localization and regulation of class II MHC proteins reflect their physiological role in the regulation of the immune response. B lymphocytes, which recognize unprocessed antigen via monomeric IgM and IgD receptors on their cell surfaces and bind and concentrate small amounts of antigen from the environment for processing and presentation to T-lymphocytes, probably serve as major APC via the exogenous system (78, 402). The endogenous system of antigen processing comprises synthesis of foreign protein within host cells (for example, by viral pathogens), fragmentation of the protein in the cytoplasm, and expression of the antigen in association with class I MHC molecules; these complexes are recognized by T cells expressing the CD8 receptor (cytotoxic or suppressor T cells). In addition to genes coding for MHC molecules, the endogenous process also requires genes for transporting and degrading viral proteins; these genes are transcriptionally up-regulated by IFNγ (63, 147, 323). Class I MHC proteins can be found on virtually all cell types; expression is enhanced by TNF and by all three classes of IFN and is suppressed by glucocorticoids. The broad cellular distribution of class I MHC molecules reflects their role in response to viral and neoplastic processes, which can affect virtually any cell population.

Cytokine production is vital to the initiation, amplification, and modulation of the cellular immune response. For example, IL2 promotes the expansion of some T-cell clones, whereas IFNγ stimulates MHC expression and activates the antigen-presenting capabilities of macrophages. Cytokine sensitivity and production have been characterized in two T-helper-cell subtypes known as Th1 and Th2 cells (280, 281); Th1 cells produce IL2 and IFNγ but not IL4 and IL5, and the reverse applies to Th2 cells (186). These relationships reflect the proposed primary roles of Th1 cells in viral and bacterial infections and of Th2 cells in helminthic and allergic conditions (326).

T cell–mediated cytotoxicity occurs when the TCR of CTL engages antigen and class I MHC molecules on target cells (430, 431). T-cell surface antigens such as CD8 and LFA-1 increase the avidity of cell–cell interactions (223, 366). Perforins and granzymes (serine proteases) may mediate the cellular lysis induced by CTL and natural killer cells; expression of these proteins is clearly correlated with in vivo cytotoxic activity (162, 282). Degranulation releases perforin, which is structurally similar to the ninth component of complement. Perforin induces pores in target cell membranes and may mediate the lethality of CTL and natural killer cells (221, 423). However, the demonstration of dissociability between degranulation and cytolysis suggests that other mechanisms are also involved (83).

To efficiently prevent infection, the immune system must respond to low concentrations of antigens; however, lymphocytes have relatively low affinity for antigen prior to antigen-induced generation of memory cells with high antigen affinity. The presence of the protein CD19 in B-cell membranes reduces the threshold for antigen-receptor-dependent stimulation by 100 fold (71). Similarly, in T lymphocytes CD4 and CD8 proteins reduce the number of TCRs that must be engaged to induce cellular activation (7, 234).

The above discussion illustrates the complex nature of interactions between B and T lymphocytes (355, 402). The internalized hydrolyzed antigen, in association with the B-cell class II MHC molecule, is recognized as an antigen epitope by the T-helper TCR. B–T cell interactions are characterized by migration of TCR, CD4, and LFA-1 molecules on T cells, and ICAM and other molecules on B cells, to the membrane site of interaction, where they promote stable B–T adhesion. Binding of TCR and CD4 to the antigen/class II MHC complex induces T-cell proliferation, while stimulated T-helper cells secrete interleukins that promote the proliferation and differentiation of the antigen-presenting B cells into plasma cells. T cells also promote B-cell responsiveness to antigen, the transformation from IgM to IgG production (isotype switching), affinity maturation of the antibody response, and generation of memory (402). Indeed, B–T cell interactions are essential for the ultimate differentiation of B cells into plasmacytes.

Other Factors in Host Defense

In addition to their immunoregulatory role (29), the MHC antigens, by virtue of their extreme polymorphism, limit the ability of parasites to camouflage themselves as host proteins. Viruses, for example, can incorporate host MHC antigens into the viral envelope. When these are transported into secondary hosts, however, MHC polymorphism can trigger a response similar to allograft rejection, reducing the number of primarily infected cells (10). Indeed, CTLs directed against allogeneic MHC antigens are more numerous than those responding to foreign antigens presented with self MHC (241). Expression of different MHC antigens on the cell membranes could also alter receptor sites for pathogen attachment.

The genetic complement of the animal is an important determinant of the type and magnitude of some host responses to microbial challenge. Indeed, the large variability in susceptibility to infectious disease in normal, heterozygous populations, plus the key role of genetic makeup in determining the onset and outcome of infectious conditions, suggest that infections may be causally related to genetic defects (40). Host genetics can influence disease resistance at three levels: microbial attachment and/or penetration, innate defense mechanisms (for example, phagocytosis), and specific host defenses (for example, MHC antigens) (62). Several examples illustrate these points. In mice a variety of genes, such as *Lps, Ity*, and *xid*, modulate the susceptibility to gram-negative bacterial infections (119, 290, 328). Similarly, the susceptibility of mice to mycobacterial infections is controlled by the *Bcg* gene, which regulates antimyco-bacterial T cell–independent macrophage activation (358). Mice also demonstrate a close relationship between MHC type and susceptibility to virus-induced leukemia (240). In humans a reduced or absent cellular immune response associated with genetic markers pro-motes the development of lepromatous as opposed to tuberculoid leprosy (272, 397). Host/pathogen relation-ships may be an important evolutionary factor favoring genetic diversity and protein polymorphism (including MHC diversity) within species (9). In wild populations ranging from plants to mammals approximately 30%–40% of all genetic loci are polymorphic (62). This diversity is postulated to reflect the accumulation of mutations over long evolutionary periods, resulting in multiple specific alleles that survive speciation (11, 120, 264). Epidemiological and genetic models of host/par-asite interactions indicate that host polymorphism is a key factor in maintaining immune defenses against par-asites, which generally evolve at a more rapid rate than host species by virtue of their smaller genome and shorter life cycle (9, 62, 170). The host can counter the rapid adaptability of pathogens by maintaining variety across conspecifics, especially at loci involved in immu-nological defense and self-recognition (62), making it difficult for parasites to camouflage as host or to cir-cumvent immune defenses. Thus MHC polymorphism and immune restriction reduce the risk that a popula-tion will be extinguished by an epidemic infection (238). In fact, catastrophic disease epidemics are more likely to develop in genetically homogeneous populations (62). Host/parasite interactions are proposed as a major evolutionary impetus for sexual reproduction (as opposed to parthenogenesis), which provides recombi-nations that maintain polymorphisms (9).

Factors other than genetic also influence the host response to microbial pathogens. Age, for example, affects the magnitude of the immune response in humans and other vertebrates (124, 378). Although newborn animals receive maternal antibodies placen-tally and/or via colostrum and milk, the neonates of many species undergo a period of increased susceptibil-ity to infectious disease when the maternal antibody has declined but the offspring has not yet generated its own antibodies as a result of immunization or natural expo-sure. In addition, the ability to generate an antibody response matures with age. Humans under 2 yr of age are unable to generate robust antibody responses to polysaccharide antigens, thus rendering them highly susceptible to agents such as group B streptococci and

Hemophilus influenzae (94, 360). The high susceptibil-ity of infants to infection with herpes viruses and intra-cellular microbes, such as *Listeria* and *Toxoplasma*, may be related to immature cellular immunity (124). Neonatal T lymphocytes, for example, produce less of the antiviral mediator IFNγ than do adult cells, and neonatal IFNγ is less effective in enhancing cytotoxic cell function and macrophage production of TNF (66, 405, 406, 413).

Many infectious processes are influenced by nutri-tional status (74). Nonspecific responses, such as phago-cytosis and the complement system, as well as humoral and cellular immunity, are susceptible to nutritional deficiencies; thus nutritional status influences the risk and morbidity of infection, just as infection alters the intake and utilization of food (25, 91). Calorie–protein malnutrition reduces the production and/or efficacy of complement, IgA, and cellular immune processes (74). Protein-malnourished patients also demonstrate reduced ability to synthesize endogenous pyrogens (for example, IL1) (187, 210). Immune function can be altered by a deficiency or excess of calories or of a spe-cific nutrient, by faulty metabolism resulting in absence of key metabolites, or by ingestion of chemicals causing allergic or toxic responses (31). Moreover, age and nutritional factors can interact (74). In infants, for example, age is a primary risk factor for acute respira-tory infection, and malnutrition is a secondary risk fac-tor for mortality as a result of such an infection (388).

The immune competence of the host also is a key determinant in the pathogenicity or relative virulence of many microorganisms. Gastrectomy, for example, can predispose individuals to pulmonary tuberculosis sec-ondary to malnutrition and its effects on immune func-tion or to bacterial gastroenteritis and cholera second-ary to loss of protective gastric acid. In some cases immune competence is impaired as a result of develop-mental or genetic abnormalities. For example, congen-ital hypogammaglobulinemia or genetic conditions associated with granulocytopenia are accompanied by increased susceptibility to pyogenic extracellular bac-teria and to some viruses. Some infectious or noninfec-tious diseases can also secondarily induce immune sup-pression. Perhaps the best-known example is that of the immunodeficiency viruses, which predispose humans and animals to a variety of opportunistic infections. Similarly, Hodgkin's disease induces suppression of cell-mediated immunity, thereby increasing susceptibility to intracellular pathogens like *Mycobacterium tuberculo-sis* and *Histoplasma capsulatum*.

The interplay between parasites and hosts can regu-late host populations, and a critical host density is the-oretically necessary for the persistence of infectious dis-eases in a population (8). Cyclic patterns of disease prevalence could reflect a scenario in which increased population density induces malnutrition in the population as a whole, resulting in impaired immune competence and increased disease incidence; a reduction in populations due to infectious disease would then reduce competition for food, as well as reduce the number of nonimmune organisms available to transmit the pathogen (262).

SUMMARY

In summary, interactions with microorganisms can be advantageous, detrimental, or inconsequential for vertebrate hosts. Advantageous or mutually symbiotic interactions are likely to reflect complex evolutionary integration between organisms. Similarly, the detrimental effects of microbial invasion have provided evolutionary pressures for the development of complex defense systems that include both nonspecific and pathogen-specific mechanisms. Thus the physiological manifestations of infectious disease confer homeostatic benefits to the host via defense reactions against the invading pathogen and/or repair of pathogen-induced damage, though, ironically, the host defense response itself accounts for much of the dysfunction that accompanies infectious processes.

The authors thank Dr. Maryna Eichelberger, Dr. Jerry Rehg, and Ms. Sharon Naron for critical review of this manuscript. Supported in part by NIH Grants NS26429 (to L.A.T.), NS22716 and HL47650 (to C.M.B.) and by CA21765 (core), the American–Lebanese Syrian Associated Charities (to St. Jude Children's Research Hospital).

REFERENCES

1. Abernethy, T. J., and O. T. Avery. The occurrence during acute injection of a protein not normally present in blood. I. Distribution of the reactive protein in patients' sera and the effect of calcium in the flocculation reaction with the C polysaccharide of pneumococcus. *J. Exp. Med.* 73: 173–182, 1941.
2. Adam, A., and E. Lederer. Muramyl peptides: immunomodulators, sleep factors, and vitamins. *Med. Res. Rev.* 4: 111–152, 1984.
3. Akin, D. E., and W. S. Borneman. Role of rumen fungi in fiber degradation. *J. Dairy Sci.* 73: 3023–3032, 1990.
4. Albelda, S. M., and C. A. Buck. Integrins and other cell adhesion molecules. *FASEB J.* 4: 2868–2880, 1990.
5. Allison, M. J. Microbiology of the rumen and small and large intestines. In: *Dukes' Physiology of Domestic Animals* (10th ed.), edited by M. J. Swenson. Ithaca, NY: Cornell University Press, 1984, p. 340–349.
6. Allison, M. J., S. E. Maloy, and R. R. Matson. Inactivation of *Clostridium botulinum* toxin by ruminal microbes from cattle and 2sheep. *Appl. Environ. Microbiol.* 32: 685–688, 1976.
7. Anderson, P., M. L. Blue, C. Moromoto, and S. F. Schlossman. Cross-linking of T3 (CD3) with T4 (CD4) enhances the proliferation of resting T-lymphocytes. *J. Immunol.* 139: 678–682, 1987.
8. Anderson, R. M., and R. M. May. Population biology of infectious diseases: part I. *Nature* 280: 361–367, 1979.
9. Anderson, R. M., and R. M. May. Coevolution of hosts and parasites. *Parasitology* 85: 411–426, 1982.
10. Andersson, L., S. Paabo, and L. Rask. Is allograft rejection a clue to the mechanism promoting MHC polymorphism? *Immunol. Today* 8: 206–209, 1987.
11. Arden, B., and J. Klein. Biochemical comparison of major histocompatibility complex molecules from different subspecies of *Mus musculus*: evidence for trans-specific evolution of alleles. *Proc. Natl. Acad. Sci. USA* 79: 2342–2346, 1982.
12. Atkins, E. Evolving notions on the cause and purpose of fever. In: *Neuro-Immunology of Fever*, edited by T. Bartfai and D. Ottoson. Oxford: Pergamon, 1992, p. 1–10.
13. Bagley, G. J., C. H. Lang, D. M. Hargrove, J. J. Thompson, C. A. Wilson, and J. J. Spitzer. Glucose kinetics in rats infused with endotoxin-induced monokines or tumor necrosis factor. *Circ. Shock* 24: 111–121, 1988.
14. Bagley, G. J., C. H. Lang, N. Skrepnik, and J. J. Spitzer. Attenuation of glucose metabolic changes resulting from TNF-α administration by adrenergic blockade. *Am. J. Physiol.* 262(*Regulatory Integrative Comp. Physiol.* 33): R628–R635, 1992.
15. Balish, E., C. E. Yale, and R. Hong. Serum proteins of gnotobiotic rats. *Infect. Immun.* 6: 112–118, 1972.
16. Banet, M. Fever in mammals; is it beneficial? *Yale J. Biol. Med.* 59: 117–124, 1986.
17. Baracos, V. E., W. T. Whitmore, and R. Gale. The metabolic cost of fever. *Can. J. Physiol. Pharmacol.* 65: 1248–1254, 1987.
18. Bateman, A., A. Singh, T. Kral, and S. Solomon. The immune-hypothalamic–pituitary–adrenal axis. *Endocr. Rev.* 10: 92–112, 1982.
19. Bauer, J., C. Lengyel, T. M. Bauer, G. Acs, and W. Gerok. Regulation of interleukin-6 receptor expression in human monocytes and hepatocytes. *FEBS Lett.* 249: 27–30, 1989.
20. Baumann, H., and J. Gauldie. Regulation of hepatic acute phase plasma protein genes by hepatocyte-stimulating factors and other mediators of inflammation. *Mol. Biol. Med.* 7: 147–160, 1990.
21. Baumann, H., K. R. Prowse, S. Marinkovic, K.-A. Won, and G. P. Jahreis. Stimulation of hepatic acute phase response by cytokines and glucocorticoids. *Ann. N. Y. Acad. Sci.* 557: 280–296, 1989.
22. Baumann, H., C. Richards, and J. Gauldie. Interaction among hepatocyte-stimulating factors, interleukin-1, and glucocorticoids for regulation of acute-phase plasma proteins in human hepatoma (Hep G2) cells. *J. Immunol.* 139: 4122–4128, 1987.
23. Beasley, D., C. A. Dinarello, and J. G. Cannon. Interleukin-1 induces natriuresis in conscious rats: role of renal prostaglandins. *Kidney Int.* 33: 1059–1065, 1988.
24. Beekhuizen, H., and R. van Furth. Monocyte adherence to human vascular endothelium. *J. Leukoc. Biol.* 54: 363–378, 1993.
25. Beisel, W. R. Impact of infection on nutritional status: concluding comments and summary. *Am. J. Clin. Nutr.* 30: 1564–1566, 1977.
26. Beisel, W. R. Metabolic effects of infection. *Prog. Food Nutr. Sci.* 8: 43–75, 1984.
27. Beisel, W. R., and M. I. Rapoport. Inter-relations between adrenocortical functions and infectious illness. *N. Engl. J. Med.* 280: 541–546, 1969.
28. Beisel, W. R., W. D. Sawyer, E. D. Ryll, and D. Crozier. Metabolic effects of intracellular infections in man. *Ann. Intern. Med.* 67: 744–779, 1969.
29. Benacerraf, B. Role of MHC gene products in immune regulation. *Science* 212: 1229–1238, 1981.

30. Berg, E. L. Homing receptors and vascular addressins: cell adhesion molecules that direct lymphocyte traffic. *Immunol. Rev.* 108: 5–18, 1989.

31. Berg, R. D., and A. W. Garlington. Translocation of certain indigenous bacteria from the gastrointestinal tract to the mesenteric lymph nodes and other organs in a gnotobiotic mouse model. *Infect. Immun.* 23: 403–411, 1979.

32. Berg, R. D., and W. E. Owens. Inhibition of translocation of viable *Escherichia coli* from gastrointestinal tract of mice by bacterial antagonism. *Infect. Immun.* 25: 820–827, 1979.

33. Berkenbosch, F., D. E. C. de Goeij, A. del Rey, and H. Besedovsky. Neuroendocrine, sympathetic and metabolic responses induced by interleukin-1. *Neuroendocrinology* 50: 570–576, 1989.

34. Berkenbosch, F., R. De Rijk, K. Schotanus, D. Wolvers, and A.-M. van Dam. The immune–hypothalamo–pituitary–adrenal axis: its role in immunoregulation and tolerance to self-antigens. In: *Interleukin-1 in the Brain*, edited by N. J. Rothwell and R. D. Dantzer. Oxford: Pergamon, 1992, p. 75–91.

35. Besedovsky, H. O., and A. del Rey. Interleukin-1 and glucose homeostasis: an example of the biological relevance of immuno-neuroendocrine interactions. *Horm. Res.* 31: 94–99, 1989.

36. Besedovsky, H. O., and A. del Rey. Metabolic and endocrine actions of interleukin-1. Effects on insulin-resistant animals. *Ann. N. Y. Acad. Sci.* 594: 214–221, 1990.

37. Besedovsky, H. O., and A. del Rey. Immune-neuroendocrine circuits: integrative role of cytokines. In: *Frontiers in Neuroendocrinology*. New York: Raven, 1992, vol. 13, p. 61–94.

38. Bevilacqua, M. P., J. S. Pober, M. E. Wheeler, R. S. Cotran, and M. A. Gimbrone. Interleukin-1 acts on cultured human vascular endothelium to increase the adhesion of polymorphonuclear leukocytes, monocytes, and related leukocyte cell lines. *J. Clin. Invest.* 26: 2003–2011, 1985.

39. Binder, H. J., B. Filburn, and M. Floch. Bile acid inhibition of intestinal anaerobic organisms. *Am. J. Clin. Nutr.* 28: 119–125, 1975.

40. Biozzi, G., D. Mouton, C. Stiffel, and Y. Bouthillier. A major role of the macrophage in quantitative genetic regulation of immunoresponsiveness and antiinfectious immunity. *Adv. Immunol.* 36: 189–234, 1984.

41. Bistrian, B. R., J. Schwartz, and N. W. Istfan. Cytokines, muscle proteolysis, and the catabolic response to infection and inflammation. *Proc. Soc. Exp. Biol. Med.* 200: 220–223, 1992.

42. Blalock, J. E. (Ed). *Neuroimmunoendocrinology* (2nd rev. ed.), *Chemical Immunol.*, Basel: Karger, 1992, vol. 52.

43. Blatteis, C. M. Influence of body weight and temperature on the pyrogenic effect of endotoxin in guinea pigs. *Toxicol. Appl. Pharmacol.* 29: 249–258, 1974.

44. Blatteis, C. M. Fever: is it beneficial? *Yale J. Biol. Med.* 59: 107–116, 1986.

45. Blatteis, C. M. Thermal homeostasis during the neonatal period. *Prog. Biometeor.* 7: 37–42, 1989.

46. Blatteis, C. M. Cytokines as endogenous pyrogens. In: *Cellular and Molecular Aspects of Endotoxin Reactions*, edited by A. Nowotny, J. J. Spitzer, and E. J. Ziegler. Amsterdam: Elsevier, 1990, p. 447–454.

47. Blatteis, C. M. Neuromodulative actions of cytokines. *Yale J. Biol. Med.* 63: 133–146, 1990.

48. Blatteis, C. M. Role of the OVLT in the febrile response to circulating pyrogens. *Prog. Brain Res.* 91: 409–412, 1992.

49. Blatteis, C. M. The pyrogenic action of cytokines. In: *Interleukin-1 in the Brain*, edited by N. J. Rothwell and R. D. Dantzer. Oxford: Pergamon, 1992, p. 93–114.

50. Blatteis, C. M., W. S. Hunter, J. Llanos-Q., R. A. Ahokas, and T. A. Mashburn, Jr. Activation of acute-phase responses by intrapreoptic injections of endogenous pyrogen in guinea pigs. *Brain Res. Bull.* 12: 689–695, 1984.

51. Blatteis, C. M., N. Quan, L. Xin, and A. Ungar. Neuromodulation of acute-phase responses to interleukin-6 in guinea pigs. *Brain Res. Bull.* 25: 895–901, 1990.

52. Blatteis, C. M., and A. A. Romanovsky. Endogenous opioids and fever. In: *Pharmacology of Thermoregulation*, edited by E. Zeisberger, P. Lomax, and E. Schönbaum. Basel: Birkhäuser, 1994, p. 435–441.

53. Blatteis, C. M., and K. A. Smith. Hypothalamic sensitivity to leukocytic pyrogen of adult and newborn guinea-pigs. *J. Physiol. (Lond.)* 296: 177–192, 1979.

54. Blatteis, C. M., A. Ungar, and R. D. Howell. Thermal effects of interleukin-1 (IL1) and various prostaglandins (PG) microinjected into different brain regions. *FASEB J.* 2: A1531, 1988.

55. Blatteis, C. M., L. Xin, and N. Quan. Attenuation of fever by distress. *FASEB J.* 5: A1402, 1991.

56. Blatteis, C. M., L. Xin, and N. Quan. Neuromodulation of fever: apparent involvement of opioids. *Brain Res. Bull.* 26: 219–223, 1991.

57. Blatteis, C. M., L. Xin, and N. Quan. Neuromodulation of fever: a possible role for substance P. *Ann. N. Y. Acad. Sci.* 741: 162–173, 1994.

58. Bornstein, D. L. Leukocytic pyrogen: a major mediator of the acute-phase reaction. *Ann. N. Y. Acad. Sci.* 389: 323–337, 1982.

59. Borriello, S. P. The influence of the normal flora on *Clostridium difficile* colonisation of the gut. *Ann. Med.* 22: 61–67, 1990.

60. Boulant, J. A. Hypothalamic neurons regulating body temperature. In: *Handbook of Physiology. Environmental Physiology*, edited by M. J. Fregly. New York: Oxford University Press, 1995, sect. 11, chapt. 6 (this volume).

61. Boulant, J. A., and N. L. Silva. Interactions of reproductive steroids, osmotic pressure and glucose on thermosensitive neurons in preoptic tissue slices. *Can. J. Physiol. Pharmacol.* 65: 1267–1273, 1987.

62. Bremermann, H. J. Sex and polymorphism as strategies in host–pathogen interactions. *J. Theor. Biol.* 87: 671–702, 1980.

63. Brown, M. G., J. Driscoll, and J. J. Monaco. Structural and serological similarity of MHC-linked LMP and proteasome (multicatalytic proteinase) complexes. *Nature* 353: 355–357, 1991.

64. Brown, M. R. W., and P. Williams. The influence of environment on envelope properties affecting survival of bacteria in infections. *Annu. Rev. Microbiol.* 39: 527–556, 1985.

65. Brubaker, R. R. Mechanisms of bacterial virulence. *Annu. Rev. Microbiol.* 39: 21–50, 1985.

66. Bryson, Y. J., H. S. Winter, S. E. Gard, T. J. Fischer, and E. R. Stiehm. Deficiency of immune interferon production by leukocytes of normal newborns. *Cell. Immunol.* 55: 191–200, 1980.

67. Busbridge, N. J., M. J. Dascombe, F. J. H. Tilders, J. W. A. M. van Oers, E. A. Linton, and N. J. Rothwell. Central activation of thermogenesis and fever by interleukin-1β and interleukin-1α involves different mechanisms. *Biochem. Biophys. Res. Commun.* 62: 591–593, 1989.

68. Cannon, J. G., J. B. Tatro, S. Reichlin, and C. A. Dinarello. α-Melanocyte stimulating hormone inhibits immunostimulatory and inflammatory actions of interleukin-1 *J. Immunol.* 137: 2232–2236, 1986.

69. Cannon, J. G., R. G. Tompkins, J. A. Gelfand, M. R. Michie, G. G. Stanford, J. W. M. van der Meer, S. Endres, G. Lonnemann, J. Corsetti, B. Chernow, D. W. Wilmore, S. M. Wolff, J. F. Burke, and C. A. Dinarello. Circulating interleukin-1 and tumor necrosis factors in humans during sepsis and endotoxin fever. *J. Infect. Dis.* 161: 79–84, 1990.

70. Carp, H. Mitochondrial N-formylmethionyl proteins as chemoattractants for neutrophils. *J. Exp. Med.* 155: 264–275, 1982.

71. Carter, R. H., and D. T. Fearon. CD19: lowering the threshold for antigen receptor stimulation of B lymphocytes. *Science* 256: 105–107, 1992.

72. Chan, R. C. Y., A. W. Bruce, and G. Reid. Adherence of cervical, vaginal and distal urethral normal microbial flora to human uroepithelial cells and the inhibition of adherence of gram-negative uropathogens by competitive exclusion. *J. Urol.* 131: 596–601, 1984.

73. Chan, R. C. Y., G. Reid, R. T. Irvin, A. W. Bruce, and J. W. Costerton. Competitive exclusion of uropathogens from human uroepithelial cells by *Lactobacillus* whole cells and cell wall fragments. *Infect. Immun.* 47: 84–89, 1985.

74. Chandra, R. K. Nutrition, immunity, and infection: present knowledge and future directions. *Lancet* i: 688–691, 1983.

75. Chang, S. L., T. Ren, and J. E. Zadina. Interleukin-1 activation of FOS proto-oncogene protein in the rat hypothalamus. *Brain Res.* 617: 123–130, 1993.

76. Chedid, L., M. Parant, F. Parant, P. Lefrancier, J. Choay, and E. Lederer. Enhancement of nonspecific immunity to *Klebsiella pneumoniae* infection by a synthetic immunoadjuvant (N-acetylmuramyl-L-alanyl-D-isoglutamine) and several analogs. *Proc. Natl. Acad. Sci. USA* 74: 2089–2093, 1977.

77. Chensue, S. W., P. D. Terebuh, D. G. Remick, W. E. Scales, and S. L. Kunkel. In vivo biologic and immunohistochemical analysis of interleukin-1 alpha, beta and tumor necrosis factor during experimental endotoxemia. *Am. J. Pathol.* 138: 395–402, 1991.

78. Chestnut, R. W., and H. M. Grey. Studies of the capacity of B-cells to serve as antigen-presenting cells. *J. Immunol.* 126: 1075–1079, 1981.

79. Chisari, F. V., L. K. Curtiss, and F. C. Jensen. Physiologic concentrations of normal human plasma lipoproteins inhibit the immortalization of peripheral B-lymphocytes by the Epstein Barr virus. *J. Clin. Invest.* 68: 329–336, 1981.

80. Chowers, I., H. T. Hammel, J. Eisenman, R. M. Abrams, and S. M. McCann. A comparison of the effects of environmental and preoptic heating and pyrogen on plasma cortisol levels. *Am. J. Physiol.* 210: 606–610, 1966.

81. Claesson, H. E., and J. Haeggstrom. Metabolism of leukotriene A$_4$ by human endothelial cells: evidence for leukotriene C$_4$ and D$_4$ formation by leukocyte–endothelial cell interaction. In: *Advances in Prostaglandins, Thromboxane and Leukotriene Research*, edited by B. Samuelsson, R. Paoletti, and P. W. Ramwell. New York: Raven, 1989, vol. 17, p. 115–123.

82. Clark, S. C. Interleukin-6. Multiple activities in regulation of the hematopoietic and immune systems. *Ann. N. Y. Acad. Sci.* 557: 438–443, 1989.

83. Clark, W., H. Ostergaard, K. Gorman, and B. Torbett. Molecular mechanisms of CTL-mediated lysis: a cellular perspective. *Immunol. Rev.* 103: 37–51, 1988.

84. Coates, M. E. Gnotobiotic animals in nutrition research. *Proc. Nutr. Soc.* 32: 53–58, 1973.

85. Coceani, F. Prostaglandins and fever. Facts and controversies. In: *Fever: Basic Mechanisms and Management*, edited by P. Mackowiak. New York: Raven, 1991, p. 59–70.

86. Cohen, K., R. F. Diegelmann, and W. J. Lindblad. *Wound Healing: Biochemical and Clinical Aspects*. Philadelphia: Saunders, 1992.

87. Cooper, K. E. The neurobiology of fever: thoughts on recent developments. *Annu. Rev. Neurosci.* 10: 297–324, 1987.

88. Cooper, K. E., W. I. Cranston, and A. J. Honour. Observations on the site and mode of action of pyrogens in the rabbit brain. *J. Physiol. (Lond.)* 191: 325–337, 1967.

89. Cooper, K. E., W. L. Veale, and Q. J. Pittman. Pathogenesis of fever. In: *Brain Dysfunction in Infantile Febrile Convulsions*, edited by M. A. B. Brazier and F. Coceani. New York: Raven, 1976, p. 107–115.

90. Cooper, M. D. B-lymphocytes: normal development and function. *N. Engl. J. Med.* 317: 1452–1457, 1987.

91. Corman, L. C. The relationship between nutrition, infection, and immunity. *Med. Clin. North Am.* 69: 519–531, 1985.

92. Cornelius, P., M. Marlowe, M. D. Lee, and P. H. Pekala. The growth factor-like effects of tumor necrosis factor-α. Stimulation of glucose transport activity and induction of glucose transporter and immediate early gene expression in 3T3-L1 preadipocytes. *J. Biol. Chem.* 265: 20506–20516, 1990.

93. Cornell, R. P., and D. B. Schwartz. Central administration of interleukin-1 elicits hyperinsulinemia in rats. *Am. J. Physiol.* 256(*Regulatory Integrative Comp. Physiol.* 27): R772–R777, 1989.

94. Cowan, J. J., A. J. Ammann, D. W. Wara, V. M. Howie, L. Schultz, H. Doyle, and M. Kaplan. Pneumococcal polysaccharide immunization in infants and children. *Pediatrics* 62: 721–727, 1978.

95. Cunha, F. Q., B. B. Lorenzetti, S. Poole, and S. H. Ferreira. Interleukin-8 as a mediator of sympathetic pain. *Br. J. Pharmacol.* 104: 765–767, 1991.

96. Cunha, F. Q., S. Poole, B. B. Lorenzetti, and S. H. Ferreira. The pivotal role of tumor necrosis factor α in the development of inflammatory hyperalgesia. *Br. J. Pharmacol.* 107: 660–664, 1992.

97. Cybulsky, M. I., M. K. W. Chan, and H. Z. Movat. Acute inflammatory and microthrombosis induced by endotoxin, interleukin-1, and tumor necrosis factor and their implication in gram-negative infection. *Lab. Invest.* 58: 365–378, 1988.

98. Darling, G., D. S. Goldstein, R. Stull, C. M. Gorschboth, and J. A. Norton. Tumor necrosis factor: immune-endocrine interaction. *Surgery* 106: 1155–1160, 1989.

99. Dascombe, M. J. The pharmacology of fever. *Prog. Neurobiol.* 25: 327–373, 1985.

100. Davis, M. M., and P. J. Bjorkman. T-cell antigen receptor genes and T-cell recognition. *Nature* 334: 395–402, 1988.

101. Dawson, M. M. *Lymphokines and Interleukins*. Boca Raton, FL: CRC, 1991.

102. Dejana, E., F. Breviario, A. Erroi, F. Bussolino, M. Gramse, G. Pintucci, B. Casali, C. A. Dinarello, J. Van Damme, and A. Mantovani. Modulation of endothelial cell junction by different molecular species of interleukin-1. *Blood* 69: 695–699, 1987.

103. del Rey, A., and H. Besedovsky. Interleukin-1 affects glucose homeostasis. *Am. J. Physiol.* 253(*Regulatory Integrative Comp. Physiol.* 24): R794–R798, 1987.

104. Densen, P. Complement. In: *Principles and Practice of Infectious Diseases* (3rd ed.), edited by G. L. Mandell, R. G. Douglas, and J. E. Bennett. New York: Churchill-Livingstone, 1990, p. 62–81.

105. DeRijk, R., N. Van Rooyen, F. J. H. Tilders, H. O. Besedovsky, A. del Rey, and F. Berkenbosch. Selective depletion of macrophages prevents pituitary–adrenal activation in response to subpyrogenic but not to pyrogenic doses of bacterial endotoxin in rats. *Endocrinology* 128: 330–338, 1991.

106. De Simoni, M. G., A. De Luigi, L. Gemma, M. Sironi, A. Manfridi, and P. Ghezzi. Modulation of systemic interleukin-6 induction by central interleukin-1. *Am. J. Physiol.* 265(*Regulatory Integrative Comp. Physiol.* 36): R739–R742, 1993.

107. De Simoni, M. G., M. Sironi, A. De Luigi, A. Manfridi, A. Mantovani, and P. Ghezzi. Intracerebroventricular injection of interleukin 1 induces high circulating levels of interleukin 6. *J. Exp. Med.* 171: 1773–1778, 1990.

108. deWaal Malefyt, R., J. Abrams, B. Bennett, C. G. Fidgor, and J. E. deVries. Interleukin-10 (IL-10) inhibits cytokine synthesis by human monocytes: an autoregulatory role of IL-10 produced by monocytes. *J. Exp. Med.* 174: 1209–1220, 1991.

109. Dimmock, N. J. Mechanisms of neutralization of animal viruses. *J. Gen. Virol.* 65: 1015–1022, 1984.

110. Dinarello, C. A. Endogenous pyrogens. The role of cytokines in the pathogenesis of fever. In: *Fever: Basic Mechanisms and Management,* edited by P. Mackowiak. New York: Raven, 1991, p. 23–47.

111. Dinarello, C. A. Role of interleukin-1 in systemic responses to LPS. In: *Bacterial Endotoxic Lipopolysaccharides,* edited by J. L. Ryan and D. C. Morrison. Boca Raton, FL: CRC, 1992, vol. II, p. 105–144.

112. Dinarello, C. A., H. A. Bernheim, G. S. Duff, H. V. Le, T. L. Nagabhushan, N. C. Hamilton, and F. Coceani. Mechanisms of fever induced by recombinant human interferon. *J. Clin. Invest.* 74: 906–918, 1984.

113. Di Padova, F., C. Pozzi, M. J. Tondre, and R. Tritapepe. Selective and early increase of IL-1 inhibitors, IL-6 and cortisol after elective surgery. *Clin. Exp. Immunol.* 85: 137–142, 1991.

114. Doherty, D. W., Jr., and C. M. Blatteis. Hypoxic reduction of endotoxic fever in guinea pigs. *J. Appl. Physiol.* 49: 294–299, 1980.

115. DuBois, E. F. *Fever.* Springfield, IL: Thomas, 1948.

116. Dubuis, J.-M., J.-M. Dayer, C. A. Siegrist-Kaiser, and A. G. Burger. Human recombinant interleukin-1β decreases plasma thyroid hormone and thyroid stimulating hormone levels in rats. *Endocrinology* 123: 2175–2181, 1988.

117. Duff, G. W., and S. K. Durum. Fever and immunoregulation: hyperthermia, interleukin 1 and 2, and T-cell proliferation. *Yale J. Biol. Med.* 55: 437–442, 1982.

118. Dustin, M. L., R. Rothlein, A. K. Bhan, C. A. Dinarello, and T. A. Springer. Induction by IL 1 and interferon-γ: tissue distribution, biochemistry, and function of a natural adherence molecule (ICAM-1). *J. Immunol.* 137: 245–254, 1986.

119. Eden, C. S., R. Shahin, and D. Briles. Host resistance to mucosal gram-negative infection: susceptibility of lipopolysaccharide nonresponder mice. *J. Immunol.* 140: 3180–3185, 1988.

120. Ehrlich, H. A., and U. B. Gyllensten. Shared epitopes among HLA class II alleles: gene conversion, common ancestry and balancing selection. *Immunol. Today* 12: 411–414, 1991.

121. Eisenman, J. S. Electrophysiology of the anterior hypothalamus. Thermoregulation and fever. In: *Pyretics and Antipyretics,* edited by A. S. Milton. Berlin: Springer, 1982, p. 187–217.

122. Ellwood, D. C., and D. W. Tempest. Effects of environment on bacterial wall content and composition. *Adv. Microbiol. Physiol.* 7: 83–118, 1972.

123. Elsbach, P., and J. Weiss. A reevaluation of the roles of the O_2-dependent and O_2-independent microbicidal system of phagocytes. *Rev. Infect. Dis.* 5: 843–853, 1983.

124. English, B. K., and C. B. Wilson. Neonate as an immunocompromised host. In: *Infections of Immunocompromised Infants and Children,* edited by C. C. Patrick. New York: Churchill-Livingstone, 1992, p. 95–118.

125. Evans, R. D., V. Ilic, and D. H. Williamson. Acute administration of tumor necrosis factor-α or interleukin-1-α does not mimic the hypoketonaemia associated with sepsis and inflammatory stress in the rat. *Clin. Sci.* 82: 205–209, 1992.

126. Ewald, P. W. Evolutionary biology and the treatment of signs and symptoms of infectious disease. *J. Theor. Biol.* 86: 169–176, 1980.

127. Eyssen, H. J., G. G. Parmentier, and J. A. Mertens. Sulfated bile acids in germ-free and conventional mice. *Eur. J. Biochem.* 66: 507–514, 1976.

128. Farrar, W. L., P. L. Killian, M. R. Ruff, J. M. Hill, and C. B. Pert. Visualization and characterization of interleukin-1 receptors in brain. *J. Immunol.* 139: 459–463, 1987.

129. Feingold, K. R., and C. Grunfeld. Tumor necrosis factor-alpha stimulates hepatic lipogenesis in the rat in vivo. *J. Clin. Invest.* 80: 184–190, 1987.

130. Feingold, K. R., S. Mounzer, M. K. Serio, A. H. Moser, C. A. Dinarello, and C. Grunfeld. Multiple cytokines stimulate hepatic lipid synthesis in vivo. *Endocrinology* 125: 267–274, 1989.

131. Felten, D. L., S. Y. Felten, D. L. Bellinger, S. L. Carlson, K. D. Ackerman, K. S. Madden, J. A. Olschowka, and S. Livnat. Noradrenergic sympathetic neural interactions with the immune system: structure and function. *Immunol. Rev.* 100: 225–260, 1987.

132. Fletcher, M. P., B. E. Seligmann, and J. I. Gallin. Correlation of human neutrophil secretion, chemoattractant receptor mobilization, and enhanced functional capacity. *J. Immunol.* 128: 941–948, 1982.

133. Flohé, S., P. C. Heinrich, J. Schneider, A. Wendel, and L. Flohé. Time course of IL-6 and TNFα release during endotoxin-induced endotoxin tolerance in rats. *Biochem. Pharmacol.* 41: 1607–1614, 1991.

134. Fong, Y., L. L. Moldawer, M. Marano, H. Wei, S. B. Tatter, R. M. Clarick, U. Santhanam, D. Sherris, L. T. May, P. B. Sehgal, and S. F. Lowry. Endotoxemia elicits increased circulating β₂-IFN/IL-6 in man. *J. Immunol.* 142: 2321–2324, 1989.

135. Foster, H. L. Gnotobiology. In: *The Laboratory Rat,* edited by H. L. Baker, J. R. Lindsey, and S. H. Weisbroth. New York: Academic, 1980, vol. II, p. 43–57.

136. Fox, A., and K. Fox. Rapid elimination of a synthetic adjuvant peptide from the circulation after systemic administration and absence of detectable natural muramyl peptides in normal serum at current analytical levels. *Infect. Immun.* 59: 1202–1205, 1991.

137. Gaillard, R. C., D. Turnill, P. Sappino, and A. F. Muller. Tumor necrosis factor-α inhibits the hormonal response of the pituitary gland to hypothalamic releasing factors. *Endocrinology* 127: 101–106, 1990.

138. Gallis, H. A. Normal flora and opportunistic infections. In: *Zinsser's Microbiology* (19th ed.), edited by W. K. Joklik, H. P. Willett, D. B. Amos, and C. M. Wilfert. Norwalk, CT: Appleton and Lange, 1988, p. 337–342.

139. Ganz, T., M. E. Selsted, D. Szklarek, S. S. L. Harwig, K. Daher, D. F. Bainton, and R. I. Lehrer. Defensins: natural peptide antibiotics of human neutrophils. *J. Clin. Invest.* 76: 1427–1435, 1985.

140. Gauldie, J., and H. Baumann. Cytokines and acute-phase protein expression. In: *Cytokines and Inflammation,* edited by E. S. Kimball. Boca Raton, FL: CRC, 1991, p. 275–305.

141. Gauldie, J., L. Lamontagne, and A. Stadnyk. The acute-phase response in infectious disease. *Surv. Syn. Pathol. Res.* 4: 126–151, 1985.

142. Gelfand, R. A., D. E. Matthews, D. M. Bier, and R. S. Sherwin. Role of counterregulatory hormones in the catabolic response to stress. *J. Clin. Invest.* 74: 2238–2248, 1984.

143. George, D. T., F. B. Abeles, C. A. Mapes, P. E. Sobocinski, T. V. Zenser, and M. C. Powanda. Effect of leukocyte endogenous mediator on endocrine pancreas secretory responses. *Am. J. Physiol.* 233(*Endocrinol. Metab. Gastrointest. Physiol.* 2): E240–E275, 1977.

144. Gibbons, R. J., and J. van Houte. Bacterial adherence in oral microbial ecology. *Annu. Rev. Microbiol.* 29: 19–44, 1975.

145. Gilbert, P., and M. R. W. Brown. Effect of R-plasmid RP1 and nutrient depletion on the gross cellular composition of *Escherichia coli* and its resistance to some uncoupling phenols. *J. Bacteriol.* 133: 1062–1065, 1978.

146. Gillin, F. D., and D. S. Reiner. Human milk kills parasitic intestinal protozoa. *Science* 221: 1290–1292, 1983.

147. Glynne, R., S. H. Powis, S. Beck, A. Kelly, L. A. Kerr, and J. Trowsdale. A proteasome-related gene between the two ABC transporter loci in the class II region of the human MHC. *Nature* 353: 357–360, 1991.

148. Goetzl, E. J., and S. P. Sreedharan. Mediators of communication

and adaptation in the neuroendocrine and immune systems. *FASEB J.* 6: 2646–2652, 1992.

149. Goldacre, M. J., B. Watt, N. Loudon, L. J. R. Milne, J. D. O. Loudon, and M. P. Vessey. Vaginal microbial flora in normal young women. *BMJ* 1: 1450–1453, 1979.

150. Goldstein, E., W. Lippert, and D. Warshauer. Pulmonary alveolar macrophage: defender against bacterial infection of the lung. *J. Clin. Invest.* 54: 519–528, 1974.

151. Goldstein, I. M., H. B. Kaplan, H. S. Edelson, and G. Weissmann. Ceruloplasmin: an acute-phase reactant that scavenges oxygen-derived free radicals. *Ann. N. Y. Acad. Sci.* 389: 368–379, 1982.

152. Goldstein, I. M., H. B. Kaplan, A. Radin, and M. Frosch. Independent effects of IgG and complement upon human polymorphonuclear leukocyte function. *J. Immunol.* 117: 1282–1287, 1976.

153. Golub, E. S., and D. R. Green. *Immunology: A Synthesis.* Sunderlund: Sinauer, 1991, p. 117–132, 352–359.

154. Gordon, A. H., and A. Koj. *The Acute-Phase Response to Injury and Infection.* Amsterdam: Elsevier, 1985.

155. Gordon, J. R., P. R. Burd, and S. J. Galli. Mast cells as a source of multifunctional cytokines. *Immunol. Today* 11: 458–464, 1990.

156. Grant, R. Emotional hypothermia in rabbits. *Am. J. Physiol.* 160: 285–290, 1950.

157. Greisman, S. E. Cardiovascular alterations during fever. In: *Fever: Basic Mechanisms and Management,* edited by P. Mackowiak, New York: Raven, 1991, p. 143–165.

158. Gridland, R. A., and N. W. Kasting. A critical role for central vasopressin in regulation of fever during bacterial infection. *Am. J. Physiol.* 263(*Regulatory Integrative Comp. Physiol.* 34): R1235–R1240, 1992.

159. Griegy, T. A., and M. J. Kluger. Fever and survival: the role of serum iron. *J. Physiol. (Lond.)* 279: 187–196, 1978.

160. Griffin, F. M. Roles of macrophage Fc and C3b receptors in phagocytosis of immunologically coated *Cryptococcus neoformans. Proc. Natl. Acad. Sci. USA* 78: 3853–3857, 1981.

161. Griffin, F. M. Mononuclear cell phagocytic mechanisms and host defense. In: *Advances in Host Defense Mechanisms,* edited by J. I. Gallin and A. S. Fauci. New York: Raven, 1982, vol. 1, p. 31–55.

162. Griffiths, G. M., and C. Mueller. Expression of perforin and granzymes in vivo: potential diagnostic markers for activated cytotoxic cells. *Immunol. Today* 12: 415–419, 1991.

163. Grunfeld, C., and K. R. Feingold. Tumor necrosis factor, interleukin, and interferon induced changes in lipid metabolism as part of host defense. *Proc. Soc. Exp. Biol. Med.* 200: 224–227, 1992.

164. Grunfeld, C., R. Gulli, A. H. Moser, L. A. Gavin, and K. R. Feingold. Effect of tumor necrosis factor administration in vivo on lipoprotein lipase activity in various tissues of the rat. *J. Lipid Res.* 30: 579–586, 1989.

165. Grunfeld, C., and M. A. Palladino, Jr. Tumor necrosis factor: immunologic, antitumor, metabolic, and cardiovascular activities. *Arch. Intern. Med.* 35: 45–72, 1990.

166. Grunfeld, C., M. Soued, S. Adi, A. H. Moser, C. A. Dinarello, and K. R. Feingold. Interleukin-4 inhibits stimulation of hepatic lipogenesis by tumor necrosis factor, interleukin-1 and interleukin-6 but not by interferon-alpha. *Cancer Res.* 51: 2803–2807, 1991.

167. Grunfeld, C., J. A. Verdier, R. A. Neese, A. H. Moser, and F. R. Feingold. Mechanisms by which tumor necrosis factor stimulates hepatic fatty acid synthesis in vivo. *J. Lipid Res.* 29: 1327–1335, 1988.

168. Gustafsson, B. E. Vitamin K deficiency in germfree rats. *Ann. N. Y. Acad. Sci.* 78: 166–174, 1959.

169. Gustafsson, B. E., F. S. Daft, E. G. McDaniel, J. C. Smith, and R. J. Fitzgerald. Effects of vitamin K–active compounds and intestinal microorganisms in vitamin K–deficient rats. *J. Nutr.* 78: 461–468, 1962.

170. Hamilton, W. D., R. Axelrod, and R. Tanese. Sexual reproduction as an adaptation to resist parasites (a review). *Proc. Natl. Acad. Sci. USA* 87: 3566–3573, 1991.

171. Harbuz, M. S., A. Stephanou, R. A. Knight, A. J. Chover-Gonzalez, and S. L. Lightman. Action of interleukin-2 and interleukin-4 on CRF mRNA in the hypothalamus and POMC mRNA in the anterior pituitary. *Brain Behav. Immun.* 6: 214–222, 1992.

172. Harel-Bellan, A., A. Quillet, C. Marchol, R. DeMars, T. Tursz, and D. Fradelizi. Natural killer susceptibility of human cells may be regulated by genes in the HLA region on chromosome 6. *Proc. Natl. Acad. Sci. USA* 83: 5688–5692, 1985.

173. Harris, H. W., C. Grunfeld, K. R. Feingold, and J. R. Rapp. Human VLDL and chylomicrons can protect against endotoxin-induced death in mice. *J. Clin. Invest.* 86: 696–702, 1990.

174. Harrison, J., and A. Fox. Degradation of muramyl dipeptide by mammalian serum. *Infect. Immun.* 50: 320–321, 1985.

175. Hart, B. L. Biological basis of the behavior of sick animals. *Neurosci. Biobehav. Rev.* 12: 123–137, 1988.

176. Hart, B. L., and K. L. Powell. Antibacterial properties of saliva: role in maternal periparturient grooming and in licking wounds. *Physiol. Behav.* 48: 383–386, 1990.

177. Hart, P. H., G. F. Vitti, D. R. Burgess, G. A. Whitty, D. S. Piccoli, and J. A. Hamilton. Potential anti-inflammatory effects of interleukin 4: suppression of human monocyte tumor necrosis factor α, interleukin 1, and prostaglandin E$_2$. *Proc. Natl. Acad. Sci. USA* 86: 3803–3807, 1989.

178. Heinzel, F. P., and R. K. Root. Antibodies. In: *Principles and Practice of Infectious Diseases* (3rd ed.), edited by G. L. Mandell, R. G. Douglas, and J. E. Bennett. New York: Churchill-Livingstone, 1990, p. 41–61.

179. Helstrom, P. B., and E. Balish. Effect of oral tetracycline, the microbial flora, and the athymic state on gastrointestinal colonization and infection of BALB/c mice with *Candida albicans. Infect. Immun.* 23: 764–774, 1979.

180. Hensler T., M. Köller, J. E. Alouf, and W. König. Bacterial toxins induce heat shock proteins in human neutrophils. *Biochem. Biophys. Res. Commun.* 179: 872–879, 1991.

181. Henson, P. M., J. E. Henson, C. Fittschen, D. L. Bratton, and D. W. H. Riches. Degranulation and secretion by phagocytic cells. In: *Inflammation: Basic Principles and Clinical Correlates* (2nd ed.), edited by J. I. Gallin, I. M. Goldstein, and R. Snyderman. New York: Raven, 1992, p. 511–539.

182. Herbermann, R. B., and J. R. Ortaldo. Natural killer cells: their role in defenses against disease. *Science* 214: 24–30, 1981.

183. Hesse, D. G., K. J. Tracey, Y. Fong, K. R. Manogue, M. A. Palladino, Jr., A. Cerami, G. T. Shires, and S. F. Lowry. Cytokine appearance in human endotoxin and primate bacteremia. *Surg. Gynecol. Obstet.* 166: 147–153, 1988.

184. Hill, M. R., R. D. Stith, and R. E. McCallum. Human recombinant IL-1 alters glucocorticoid receptor function in Renfer hepatoma cells. *J. Immunol.* 141: 1522–1528, 1988.

185. Hill, M. R., R. D. Stith, and R. E. McCallum. Mechanims of action of interferon-β2/interleukin-6 on induction of hepatic glucose enzymes. *Ann. N. Y. Acad. Sci.* 557: 502–505, 1989.

186. Hodes, R. J. T-cell-mediated regulation: help and suppression. In: *Fundamental Immunology* (2nd ed.), edited by W. E. Paul. New York: Raven, 1989, p. 587–620.

187. Hoffman-Goetz, L., D. McFarlane, B. R. Bistrian, and G. L. Blackburn. Febrile and plasma iron responses of rabbits injected with endogenous pyrogen from malnourished patients. *Am. J. Clin. Nutr.* 34: 1109–1116, 1981.

188. Hori, T. An update on thermosensitive neurons in the brain: from cellular biology to thermal and nonthermal homeostatic functions. *Jpn. J. Physiol.* 41: 1–22, 1991.

189. Huber, A. R., S. L. Kunkel, R. F. Todd, III, and S. J. Weiss. Regulation of transendothelial neutrophil migration by endogenous interleukin-8. *Science* 254: 99–102, 1991.

190. Husseini, R. H., C. Sweet, M. H. Collie, and H. Smith. Elevation of nasal viral levels by suppression of fever in ferrets infected with influenza viruses of differing virulence. *J. Infect. Dis.* 145: 520–524, 1982.

191. Ignarro, L. J. Endothelium-derived nitric oxide: actions and properties. *FASEB J.* 3: 31–36, 1989.

192. Ihle, J. N. Lymphokine regulation of hematopoietic cell development. In: *Immunophysiology. The Role of Cells and Cytokines in Immunity and Inflammation,* edited by J. J. Oppenheim and E. M. Shevach. New York: Oxford University Press, 1991, p. 166–193.

193. Imeri, L., M. R. Opp, and J. M. Krueger. An IL-1 receptor and an IL-1 receptor antagonist attenuate muramyl dipeptide and IL-1-induced sleep and fever. *Am. J. Physiol.* 265(*Regulatory Integrative Comp. Physiol.* 36): R907–R913, 1993.

194. Imura, H., J. Fukata, Y. Nakai, T. Usui, Y. Naito, T. Tominaga, N. Murakami, H. Kobayashi, O. Ebisui, and H. Segawa. Effects of cytokines on the hypothalamo–pituitary system. In: *New Trends in Autonomic Nervous System Research,* edited by M. Yoshikawa. Amsterdam: Elsevier, 1990, p. 51–54.

195. Itoh, K., and R. Freter. Control of *Escherichia coli* populations by a combination of indigenous clostridia and lactobacilli in gnotobiotic mice and continuous-flow cultures. *Infect. Immun.* 57: 559–565, 1989.

196. Janeway, C. A. The immune system evolved to discriminate infectious nonself from noninfectious self. *Immunol. Today* 13: 11–16, 1992.

197. Johannsen, L., J. Wecke, F. Obal, and J. M. Krueger. Macrophages produce somnogenic and pyrogenic muramyl peptides during digestion of staphylococci. *Am. J. Physiol.* 260(*Regulatory Integrative Comp. Physiol.* 31): R126–R133, 1991.

198. Jolles, P. A possible physiological function of lysozyme. *Biomedicine* 25: 275–276, 1976.

199. Jones, S. R. Fever in the elderly. In: *Fever: Basic Mechanisms and Clinical Management,* edited by P. Mackowiak. New York: Raven, 1991, p. 233–242.

200. Kalra, P. S., A. Sahn, and S. P. Kalra. Interleukin-1 inhibits the ovarian steroid–induced luteinizing hormone surge and release of hypothalamic luteinizing hormone-releasing hormone in rats. *Endocrinology* 126: 2145–2152, 1990.

201. Kaneko, J. J. Carbohydrate metabolism and its diseases. In: *Clinical Biochemistry of Domestic Animals* (4th ed.), edited by J. J. Kaneko. New York: Academic, 1989, p. 44–85.

202. Kapás, L., L. Hong, A. B. Cady, M. R. Opp, A. E. Postlethwaite, J. M. Seyer, and J. M. Krueger. Somnogenic, pyrogenic, and anorectic activities of tumor necrosis factor-α and TNF-α fragments. *Am. J. Physiol.* 263(*Regulatory Integrative Comp. Physiol.* 34): R708–R715, 1992.

203. Kapás, L., and J. M. Krueger. Tumor necrosis factor-β induces sleep, fever, and anorexia. *Am. J. Physiol.* 263(*Regulatory Integrative Comp. Physiol.* 34): R703–R707, 1992.

204. Kappler, J. W., and P. C. Marrack. Helper T cells recognize antigen and macrophage surface components simultaneously. *Nature* 262: 797–799, 1976.

205. Kappler, J. W., B. Skidmore, J. White, and P. Marrack. Antigen-inducible, H-2-restricted, interleukin-2-producing T cell hybridomas: lack of independent antigen and H-2 recognition. *J. Exp. Med.* 153: 1198–1214, 1981.

206. Karnovsky, M. L. Muramyl peptides in mammalian tissues and their effects at the cellular level. *Federation Proc.* 42: 2556–2560, 1986.

207. Kasting, N. W. Criteria for establishing a physiological role of vasopressin in thermoregulation during fever and antipyresis. *Brain Res. Rev.* 14: 143–153, 1989.

208. Kasting, N. W., and M. F. Wilkinson. Vasopressin functions as endogenous antipyretic in the newborn. *Biol. Neonate* 51: 249–254, 1987.

209. Kawakami, M., N. Watanabe, H. Ogawa, T. Murase, N. Yamada, H. Sando, S. Shibata, T. Oda, and F. Takaku. Specificity in metabolic effects of cachectin/TNF and other related cytokines. *Ann. N. Y. Acad. Sci.* 587: 339–350, 1990.

210. Keenan, R. A., L. L. Moldawer, R. D. Yang, I. Kawamura, G. L. Blackburn, and B. R. Bistrian. An altered response by peripheral leukocytes to synthesize or release leukocyte endogenous mediator in critically ill, protein-malnourished patients. *J. Lab. Clin. Med.* 100: 844–857, 1982.

211. Kimura-Takeuchi, M., J. A. Majde, L. A. Toth, and J. M. Krueger. Influenza virus–induced changes in rabbit sleep and acute-phase responses. *Am. J. Physiol.* 263(*Regulatory Integrative Comp. Physiol.* 34): R1115–R1121, 1992.

212. Kishimoto, T. K., and D. C. Anderson. The role of integrins in inflammation. In: *Inflammation: Basic Principles and Clinical Correlates* (2nd ed.), edited by J. I. Gallin, I. M. Goldstein, and R. Snyderman. New York: Raven, 1992, p. 353–406.

213. Kita, A., K. Imano, and H. Nakamura. Involvement of corticotropin-releasing factor in the antinociception produced by interleukin-1 in mice. *Eur. J. Pharmacol.* 237: 317–322, 1993.

214. Klebanoff, S. J. Oxygen metabolites from phagocytes. In: *Inflammation: Basic Principles and Clinical Correlates* (2nd ed.), edited by J. I. Gallin, I. M. Goldstein, and R. Snyderman. New York: Raven, 1992, p. 541–588.

215. Kluger, M. J. *Fever. Its Biology, Evolution, and Function.* Princeton, NJ: Princeton University Press, 1979.

216. Kluger, M. J. Fever: role of pyrogens and cryogens. *Physiol. Rev.* 71: 93–127, 1991.

217. Kluger, M. J., C. A. Conn, B. Franklin, R. Freter, and G. D. Abrams. Effect of gastrointestinal flora on body temperature of rats and mice. *Am. J. Physiol.* 258(*Regulatory Integrative Comp. Physiol.* 29): R552–R557, 1990.

218. Kluger, M. J., R. O'Reilly, T. R. Shope, and A. J. Vander. Further evidence that stress hyperthermia is a fever. *Physiol. Behav.* 39: 763–766, 1987.

219. Kluger, M. J., and B. A. Rothenburg. Fever and reduced iron: their interaction as a host defense response to bacterial infection. *Science* 203: 374–376, 1979.

220. Koj, A. Biological functions of acute-phase proteins. In: *The Acute-Phase Response to Injury and Infection,* edited by A. H. Gordon and A. Koj. Amsterdam: Elsevier, 1985, p. 145–160.

221. Krahenbuhl, O., and J. Tschopp. Perforin-induced pore formation. *Immunol. Today* 12: 399–402, 1991.

222. Krauss, R. M., C. Grunfeld, W. T. Doerrler, and K. R. Feingold. Tumor necrosis factor acutely increases plasma levels of very low density lipoproteins of normal size and composition. *Endocrinology* 127: 1016–1021, 1990.

223. Krensky, A. M., F. Sanchez-Madrid, E. Robbins, J. A. Nagy, T. A. Springer, and S. J. Burakoff. The functional significance, distribution, and structure of LFA-1, LFA-2, and LFA-3: cell surface antigens associated with CTL-target interactions. *J. Immunol.* 131: 611–616, 1983.

224. Krueger, J. M., J. R. Pappenheimer, and M. L. Karnovsky. Sleep-promoting effects of muramyl peptides. *Proc. Natl. Acad. Sci. USA* 79: 6102–6106, 1982.

225. Kunkel, S. L., and D. G. Remick. *Cytokines in Health and Disease.* New York: Dekker, 1992.

226. Kuriyama, K., T. Hori, T. Mori, and T. Nakashima. Actions of

interferon-α and interleukin-1β on the glucose-responsive neurons in the ventromedial hypothalamus. *Brain Res. Bull.* 24: 803–810, 1990.

227. Kushner, I. The phenomenon of the acute-phase response. *Ann. N. Y. Acad. Sci.* 389: 39–48, 1982.

228. Kushner, I., M. Ganapathi, and D. Schultz. The acute-phase response is mediated by heterogeneous mechanisms. *Ann. N. Y. Acad. Sci.* 557: 19–30, 1989.

229. Kushner, I., and A. Mackiewicz. Acute-phase proteins as disease markers. *Dis. Markers* 5: 1–11, 1987.

230. LaMont, J. T., E. B. Sonnenblick, and S. Rothman. Role of clostridial toxin in the pathogenesis of clindamycin colitis in rabbits. *Gastroenterology* 76: 356–361, 1979.

231. Lanza-Jacoby, S., and A. Tabares. Triglyceride kinetics, tissue lipoprotein lipase and liver lipogenesis in septic rats. *Am. J. Physiol.* 258(*Endocrinol. Metab.* 21): E678–E685, 1990.

232. Larson, H. E., A. B. Price, P. Honour, and S. P. Borriello. *Clostridium difficile* and the aetiology of pseudomembranous colitis. *Lancet* i: 1063–1066, 1978.

233. Lasky, L. A., and S. D. Rosen. The selectins. In: *Inflammation: Basic Principles and Clinical Correlates* (2nd ed.), edited by J. I. Gallin, I. M. Goldstein, and R. Snyderman. New York: Raven, 1992, p. 407–419.

234. Ledbetter, J. A., C. H. June, P. S. Rabinovitch, A. Grossman, T. T. Tsu, and J. B. Imboden. Signal transduction through CD4 receptors: stimulatory vs. inhibitory activity is regulated by CD4 proximity to the CD3/T cell receptor. *Eur. J. Immunol.* 18: 525–532, 1988.

235. Lee, D. A., J. R. Hoidal, D. J. Garlich, C. C. Clawson, P. G. Quie, and P. K. Peterson. Opsonin-independent phagocytosis of surface-adherent bacteria by human alveolar macrophages. *J. Leukocyte Biol.* 36: 689–701, 1984.

236. LeMay, D. R., L. G. LeMay, M. J. Kluger, and L. G. D'Alecy. Plasma profiles of IL6 and TNF with fever-inducing doses of lipopolysaccharides in dogs. *Am. J. Physiol.* 259(*Regulatory Integrative Comp. Physiol.* 30): R126–R132, 1990.

237. LeMay, L. G., A. J. Vander, and M. J. Kluger. Role of interleukin-6 fever in rats. *Am. J. Physiol.* 258(*Regulatory Integrative Comp. Physiol.* 29): R798–R803, 1990.

238. Lewin, H. A. Disease resistance and immune response genes in cattle: strategies for their detection and evidence of their existence. *J. Dairy Sci.* 72: 1334–1348, 1989.

239. Li, J. J., K. P. W. J. McAdam, and L. L. Bausserman. The regulatory role of acute-phase reactants in human immune responses. *Ann. N. Y. Acad. Sci.* 389: 456, 1982.

240. Lilly, F., E. A. Boyse, and L. J. Old. Genetic basis of susceptibility to viral leukaemogenesis. *Lancet* ii: 1207–1209, 1964.

241. Lindahl, K. F., and D. B. Wilson. Histocompatibility antigen-activated cytotoxic T lymphocytes. II. Estimates of the frequency and specificity of precursors. *J. Exp. Med.* 145: 508–522, 1977.

242. Lipscomb, M. F., J. M. Onofrio, E. J. Nash, A. K. Pierce, and G. B. Toews. A morphological study of the role of phagocytes in the clearance of *Staphylococcus aureus* from the lung. *J. Reticuloendothel. Soc.* 33: 429–442, 1983.

243. Lipton, J. M., and A. Catania. α-MSH peptides modulate fever and inflammation. In: *Neuro-Immunology of Fever,* edited by T. Bartfai and D. Ottoson. Oxford: Pergamon, 1992, p. 123–136.

244. Lipton, J. M., and G. P. Trczinka. Persistence of febrile responses to pyrogens after PO/AH lesions in squirrel monkeys. *Am. J. Physiol.* 231: 1638–1648, 1976.

245. Lipton, J. M., J. P. Welch, and W. G. Clark. Changes in body temperature produced by injecting prostaglandin E₁, EGTA and bacterial endotoxins into the PO/AH region and the medulla oblongata of the rat. *Experientia* 29: 806–808, 1973.

246. Long, C. L. Energy balance and carbohydrate metabolism in

infection and sepsis. *Am. J. Clin. Nutr.* 30: 1301–1310, 1977.

247. Lorenzo, J. A. Cytokines and bone metabolism: resorption and formation. In: *Cytokines and Inflammation,* edited by E. S. Kimball. Boca Raton, FL: CRC, 1991, p. 145–168.

248. Mackie, R. I., and B. A. White. Recent advances in rumen microbial ecology and metabolism: potential impact of nutrient output. *J. Dairy Sci.* 73: 2971–2995, 1990.

249. Mackiewicz, A., and I. Kushner. Role of IL6 in acute-phase protein glycosylation. *Ann. N. Y. Acad. Sci.* 557: 515–517, 1989.

250. Mackowiak, P. A. The normal microbial flora. *N. Engl. J. Med.* 307: 83–93, 1982.

251. Madsen, D., M. Beaver, L. Chang, E. Bruckner-Kardoss, and B. Wostmann. Analysis of bile acids in conventional and germfree rats. *J. Lipid Res.* 17: 107–111, 1976.

252. Maeda, H., and T. Akaike. Oxygen free radicals as pathogenic molecules in viral diseases. *Proc. Soc. Exp. Biol. Med.* 198: 721–727, 1991.

253. Malech, H. L., and J. I. Gallin. Neutrophils in human diseases. *N. Engl. J. Med.* 317: 687–694, 1987.

254. Mantovani, A., F. Bussolino, and E. Dejana. Cytokine regulation of endothelial cell function. *FASEB J.* 6: 2591–2599, 1992.

255. Mantovani, A., and E. Dejana. Functional responses elicited in endothelial cells by cytokines. In: *Cytokines in Health and Disease,* edited by S. L. Kunkel and D. G. Remick. New York: Dekker, 1992, p. 297–307.

256. Marasco, W. A., S. H. Phan, H. Krutzsch, H. J. Showell, D. E. Feltner, R. Nairn, E. L. Becker, and P. A. Ward. Purification and identification of formyl-methionyl-leucyl-phenylalanine as the major peptide neutrophil chemotactic factor produced by *Escherichia coli. J. Biol. Chem.* 259: 5430–5439, 1984.

257. Marshall, J. C., N. V. Christou, and J. L. Meakins. Immunomodulation by altered gastrointestinal tract flora. *Arch. Surg.* 123: 1465–1469, 1988.

258. Martin, L. W., L. B. Deeter, and J. M. Lipton. Acute-phase response to endogenous pyrogen in rabbit: effects of age and route of administration. *Am. J. Physiol.* 257(*Regulatory Integrative Comp. Physiol.* 28): R189–R193, 1989.

259. Martin, S., K. Maruta, V. Burkart, S. Gillis, and H. Kolb. IL-1 and IFN-γ increase vascular permeability. *Immunology* 64: 301–305, 1988.

260. Mastuda, T., T. Hori, and T. Nakashima. Thermal and PGE₂ sensitivity of the *organum vasculosum lamina terminalis* region and preoptic area in rat brain slices. *J. Physiol. (Lond.)* 454: 197–212, 1992.

261. Matsushima, K., and J. J. Oppenheim. Interleukin-8 and MCAF: novel inflammatory cytokines inducible by IL1 and TNF. *Cytokine* 1: 2–13, 1989.

262. May, R. M., and R. M. Anderson. Population biology of infectious diseases: part II. *Nature* 280: 455–461, 1979.

263. McCann, S. M., S. Karanth, A. Kamat, W. L. Dees, K. Lyson, M. Gimeno, and V. Rettori. Induction by cytokines of the pattern of pituitary hormone secretion in infection. *Neuroimmunomodulation* 1: 2–13, 1994.

264. McConnel, T. J., W. S. Talbot, R. A. McIndoe, and E. K. Wakeland. The origin of MHC class II gene polymorphism within the genus *Mus. Nature* 332: 651–654, 1988.

265. Meszaros, K., J. Bojta, A. P. Bautista, C. H. Lang, and J. J. Spitzer. Glucose utilization by Kupfer cells, endothelial cells, and granulocytes in endotoxemic rat liver. *Am. J. Physiol.* 260(*Gastrointest. Liver Physiol.* 23): G7–G12, 1991.

266. Meszaros, K., C. H. Lang, G. J. Bagley, and J. J. Spitzer. Contribution of different organs to increased glucose consumption after endotoxin administration. *J. Biol. Chem.* 262: 10965–10970, 1987.

267. Michael, J. G., and F. S. Rosen. Association of "natural" anti-

bodies to gram-negative bacteria with the γ_1-macroglobulins. *J. Exp. Med.* 118: 619–626, 1963.

268. Michie, H. R., K. R. Manogue, D. R. Spriggs, A. Revhaug, S. O'Dwyer, C. A. Dinarello, A. Cerami, S. M. Wolff, and D. W. Wilmore. Detection of circulating tumor necrosis factor after endotoxin administration. *N. Engl. J. Med.* 318: 1481–1486, 1988.

269. Milton, A. S. Thermoregulatory actions of eicosanoids in the central nervous system with particular regard to the pathogenesis of fever. *Ann. N. Y. Acad. Sci.* 539: 392–410, 1989.

270. Miñano, F. J., and R. D. Myers. Anorexia and adipsia: dissociation from fever after MIP-1 injection into ventromedial hypothalamus and preoptic area of rats. *Brain Res. Bull.* 27: 273–278, 1991.

271. Miñano, F. J., M. Sancibrian, M. Vizcaino, X. Paez, G. Davatelis, T. Fahey, B. Sherry, A. Cerami, and R. D. Myers. Macrophage inflammatory protein-1: unique action in the hypothalamus to evoke fever. *Brain Res. Bull.* 24: 849–852, 1990.

272. Modlin, R. L., J. Melancon-Kaplan, S. M. M. Young, C. Pirmez, H. Kino, J. Convit, T. H. Rea, and B. R. Bloom. Learning from lesions: patterns of tissue inflammation in leprosy. *Proc. Natl. Acad. Sci. USA* 85: 1213–1217, 1988.

273. Moldawer, L. L., C. Andersen, J. Gerlin, and K. G. Lundholm. Regulation of food intake and hepatic protein synthesis by recombinant derived cytokines. *Am. J. Physiol.* 254(*Gastrointest. Liver Physiol.* 17): G450–G456, 1988.

274. Morgan, E. L., and W. O. Weigle. Biological activities residing in the Fc region of immunoglobulin. *Adv. Immunol.* 40: 61–134, 1987.

275. Morimoto, A., N. Murakami, T. Nakamori, and T. Watanabe. Evidence for separate mechanisms of induction of biphasic fever inside and outside the blood–brain barrier in rabbits. *J. Physiol. (Lond.)* 383: 629–637, 1987.

276. Morimoto, A., N. Murakami, M. Takada, S. Teshigori, and T. Watanabe. Fever and acute-phase responses induced in rabbits by human recombinant interferon-γ. *J. Physiol. (Lond.)* 391: 209–218, 1987.

277. Morimoto, A., Y. Sakata, T. Watanabe, and N. Murakami. Leukocytosis induced in rabbits by intravenous or central injection of granulocyte colony stimulating factor. *J. Physiol. (Lond.)* 426: 117–126, 1990.

278. Morrone, G., G. Ciliberto, S. Oliviero, R. Arcones, L. Dente, J. Content, and E. Cortese. Recombinant interleukin-6 regulates the transcriptional activation of a set of human acute-phase genes. *J. Biol. Chem.* 263: 12554–12558, 1988.

279. Moshage, H. J., H. M. J. Roelofs, J. F. van Pelt, B. P. C. Hazenberg, M. A. van Leeuwen, P. C. Limburg, L. A. Arden, and S. H. Yap. The effect of interleukin-1, interleukin-6 and its interrelationship on the synthesis of serum amyloid A and C-reactive protein in primary cultures of adult human hepatocytes. *Biochem. Biophys. Res. Commun.* 155: 112–117, 1988.

280. Mosmann, T. R., H. Cherwinski, M. W. Bond, M. A. Giedlin, and R. L. Coffman. Two types of murine helper T cell clones. I. Definition according to profiles of lymphokine activities and secreted proteins. *J. Immunol.* 136: 2348–2357, 1986.

281. Mosmann, T. R., and R. L. Coffman. Two types of mouse helper T-cell clones: implications for immune regulation. *Immunol. Today* 8: 233–236, 1987.

282. Muller, C., D. Kagi, T. Aebischer, B. Odermatt, W. Held, E. R. Podack, R. M. Zinkernagel, and H. Hengartner. Detection of perforin and granzyme A mRNA in infiltrating cells during infection of mice with lymphocytic choriomeningitis virus. *Eur. J. Immunol.* 19: 1253–1259, 1989.

283. Mundy, G., and L. F. Bonewald. Effects of immune cell products on bone. In: *Macrophage-Derived Cell Regulatory Factors,* edited by C. Sorg. Basel: Karger, 1989, p. 38–53.

284. Murray, M. J., and A. B. Murray. Anorexia of infection as a mechanism of host defense. *Am. J. Clin. Nutr.* 32: 593–596, 1979.

285. Najjar, V. A. Tuftsin, a natural activator of phagocyte cells: an overview. *Ann. N. Y. Acad. Sci.* 419: 1–11, 1983.

286. Namba, Y., Y. Hidaka, K. Taki, and T. Morimoto. Effect of oral administration of lysozyme or digested bacterial cell walls on immunostimulation in guinea pigs. *Infect. Immun.* 31: 580–583, 1981.

287. Nealson, K. H., and J. W. Hastings. Bacterial bioluminescence: its control and ecological significance. *Microbiol. Rev.* 43: 496–518, 1979.

288. Nussler, A. K., and T. R. Billiar. Inflammation, immunoregulation and inducible nitric oxide synthase. *J. Leukoc. Biol.* 54: 171–178, 1993.

289. Obál, F., Jr., M. Opp, A. B. Cady, L. Johannsen, A. E. Postlethwaite, H. M. Poppleton, J. M. Seyer, and J. M. Krueger. Interleukin-1α and interleukin-1β fragments are somnogenic. *Am. J. Physiol.* 259(*Regulatory Integrative Comp. Physiol.* 30): R439–R446, 1990.

290. O'Brien, A. D., D. L. Rosenstreich, I. Scher, G. H. Campbell, R. P. MacDermott, and S. B. Formal. Genetic control of susceptibility to *Salmonella typhimurium* in mice: role of the LPS gene. *J. Immunol.* 124: 20–24, 1980.

291. Ohashi, K., F. Saji, M. Kato, A. Wakimoto, and O. Tanisawa. Tumor necrosis factor-α inhibits human chorionic gonadotropin secretion. *J. Clin. Endocrinol. Metab.* 74: 130–134, 1992.

292. Oladehin, A., J. A. Barriga-Calle, and C. M. Blatteis. Lipopolysaccharide (LPS)-induced FOS expression in the brains of febrile rats. In: *Thermal Physiology 1993,* edited by A. S. Milton. Basel: Birkhäuser, 1994, p. 81–85.

293. Olson, G. B., and B. S. Wostmann. Lymphocytopoiesis, plasmacytopoiesis and cellular proliferation in nonantigenically stimulated germfree mice. *J. Immunol.* 97: 267–274, 1966.

294. Opp, M., and J. M. Krueger. Interleukin-1 receptor antagonist blocks interleukin-1-induced sleep and fever. *Am. J. Physiol.* 260(*Regulatory Integrative Comp. Physiol.* 31): R453–R457, 1991.

295. Packham, M. A., E. Nishizawa, and J. F. Mustard. Response of platelets to tissue injury. *Biochem. Pharmacol.* 17 (suppl.): 171–184, 1968.

296. Page, C. P. Platelets as inflammatory cells. *Immunopharmacology* 17: 51–59, 1989.

297. Palma, C., A. Cassone, D. Serbousek, C. A. Pearson, and J. Y. Djeu. Lactoferrin release and interleukin-1, interleukin-6, and tumor necrosis factor production by human polymorphonuclear cells stimulated by various lipopolysaccharides: relationship to growth inhibition of *Candida albicans. Infect. Immun.* 60: 4604–4611, 1992.

298. Palmes, E. D., and C. R. Park. The regulation of body temperature during fever. *Arch. Environ. Health* 11: 749–759, 1965.

299. Pang, X.-P., J. M. Hershman, C. J. Mirell, and A. E. Pekary. Impairment of hypothalamic–pituitary–thyroid function in rats treated with human recombinant tumor necrosis factor-α (cachectin). *Endocrinology* 125: 76–84, 1989.

300. Parant, M., C. Damais, F. Audibert, F. Parant, L. Chedid, E. Sache, P. Lefrancier, J. Choay, and E. Lederer. In vivo and in vitro stimulation of nonspecific immunity by the b-D-p-aminophenyl glycoside of N-acetylmuramyl-L-alanyl-D-isoglutamine and an oligomer prepared by cross-linking with glutaraldehyde. *J. Infect. Dis.* 138: 378–386, 1978.

301. Payne, L. C., F. Obál, Jr., M. R. Opp, and J. M. Krueger. Stimulation and inhibition of growth hormone secretion by interleukin-1β: the involvement of growth hormone releasing hormone. *Neuroendocrinology* 56: 118–123, 1992.

302. Pepys, M. B., and M. L. Baltz. Acute phase proteins with special

reference to C-reactive protein and related proteins (pentaxins) and serum amyloid A protein. *Adv. Immunol.* 34: 141–212, 1983.

303. Peterson, P. K., B. J. Wilkinson, Y. Kim, D. Schmeling, S. D. Douglas, P. G. Quie, and J. Verhoef. The key role of peptidoglycan in the opsonization of *Staphylococcus aureus. J. Clin. Invest.* 61: 597–609, 1978.

304. Petroni, K. C., L. Shen, and P. M. Guyre. Modulation of human polymorphonuclear leukocyte IgG Fc receptors and Fc receptor–mediated functions by IFN-γ and glucocorticoids. *J. Immunol.* 140: 3467–3472, 1988.

305. Plata-Salamán, C. R. Interferons and central regulation of feeding. *Am. J. Physiol.* 263(*Regulatory Integrative Comp. Physiol.* 34): R1222–R1227, 1992.

306. Plata-Salamán, C. R., and J. P. Borkoski. Interleukin-8 modulates feeding by direct action in the central nervous system. *Am. J. Physiol.* 265(*Regulatory Integrative Comp. Physiol.* 36): R877–R882, 1993.

307. Plata-Salamán, C. R., Y. Oomura, and Y. Kai. Tumor necrosis factor and interleukin-1β: suppression of food intake by direct action in the central nervous system. *Brain Res.* 448: 106–114, 1988.

308. Pober, J. S., and R. S. Cotran. Cytokines and endothelial cell biology. *Physiol. Rev.* 70: 427–451, 1990.

309. Pober, J. S., and R. S. Cotran. Overview: the role of endothelial cells in inflammation. *Transplantation* 50: 537–544, 1990.

310. Pober, J. S., M. R. Slowik, L. G. de Luca, and A. J. Ritchie. Elevated cyclic AMP inhibits endothelial cell synthesis and expression of TNF-induced endothelial leukocyte adhesion molecule-1 and vascular cell adhesion molecule-1, but not intercellular adhesion molecule-1. *J. Immunol.* 150: 5114–5123, 1993.

311. Polla, B. S. A role for heat shock proteins in inflammation? *Immunol. Today* 9: 134–137, 1988.

312. Redondo-Lopez, V., R. L. Cook, and J. D. Sobel. Emerging role of lactobacilli in the control and maintenance of vaginal bacterial microflora. *Rev. Infect. Dis.* 12: 856–872, 1990.

313. Rehg, J. E., and S. P. Pakes. *Clostridium difficile* antitoxin neutralization of cecal toxin(s) from guinea pigs with penicillin-associated colitis. *Lab. Anim. Sci.* 31: 156–160, 1981.

314. Remington, J. S., and S. C. Schimpff. Please don't eat the salads. *N. Engl. J. Med.* 304: 433–434, 1981.

315. Rettori, V., J. Jurcovicova, and S. M. McCann. Central action of interleukin-1 in altering the release of TSH, growth hormone, and prolactin in the male rat. *J. Neurosci. Res.* 18: 179–183, 1987.

316. Rettori, V., L. Milenkovic, B. A. Beutler, and S. M. McCann. Hypothalamic action of cachectin to alter pituitary hormone release. *Brain Res. Bull.* 23: 471–475, 1989.

317. Richardson, R. P., C. D. Rhyne, Y. Fong, D. G. Hesse, K. J. Tracey, M. A. Marano, S. F. Lowry, A. C. Antonacci, and S. E. Calvano. Peripheral blood leukocyte kinetics following in vivo lipopolysaccharide (LPS) administration to normal human subjects. *Ann. Surg.* 210: 239–245, 1989.

318. Riedel, W. Mechanics of fever. *J. Basic Clin. Physiol. Pharmacol.* 1: 291–322, 1990.

319. Rivier, C., R. Chizzonite, and W. Vale. In the mouse, the activation of the hypothalamic–pituitary–adrenal axis by a lipopolysaccharide (endotoxin) is mediated through interleukin-1. *Endocrinology* 125: 2800–2805, 1989.

320. Rivier, C., and W. Vale. In the rat, interleukin-1α acts at the level of the brain and the gonads to interfere with gonadotropin and sex steroid secretion. *Endocrinology* 124: 2105–2109, 1989.

321. Rivier, C., W. Vale, and M. Brown. In the rat, interleukin-1α and β stimulate adrenocorticotropin and catecholamine release. *Endocrinology* 125: 3096–3102, 1989.

322. Robbins, J. B., R. L. Myerowitz, J. K. Whisnant, M. Argaman, R. Schneerson, Z. T. Handzel, and E. C. Gotschlich. Enteric bacteria cross-reactive with *Neisseria meningitidis* groups A and C and *Diplococcus pneumoniae* types I and III. *Infect. Immun.* 6: 651–656, 1972.

323. Robertson, M. Proteasomes in the pathway. *Nature* 353: 300–301, 1991.

324. Rogers, M. J., R. Moore, and J. Cohen. The relationship between faecal endotoxin and faecal microflora of the C57BL mouse. *J. Hyg.* 95: 397–402, 1985.

325. Rokita, H., J. Bereta, A. Koj, A. H. Gordon, and J. Gauldie. Epidermal growth factor and transferring growth factor-β differently modulate the acute phase response elicited by interleukin-6 in cultured liver cells from man, rat and mouse. *Comp. Biochem. Physiol.* 95A: 41–45, 1990.

326. Romagnani, S. Induction of T$_H$1 and T$_H$2 responses: a key role for the "natural" immune response? *Immunol. Today* 13: 379–381, 1992.

327. Rosendorff, C., and J. J. Mooney. Central nervous system sites of action of a purified leukocyte pyrogen. *Am. J. Physiol.* 220: 597–603, 1971.

328. Rosenstreich, D. L., A. C. Weinblatt, and A. D. O'Brien. Genetic control of resistance to infection in mice. *Crit. Rev. Immunol.* 3: 263–330, 1982.

329. Rothwell, N. J. Metabolic responses to interleukin-1. In: *Interleukin-1 in the Brain,* edited by N. J. Rothwell and R. D. Dantzer. Oxford: Pergamon, 1992, p. 115–134.

330. Rothwell, N. J., A. Hardwick, and L. Lindley. Central actions of interleukin-8 in the rat are independent of prostaglandins. *Horm. Metabol. Res.* 22: 595–596, 1990.

331. Rotrosen, D., H. L. Malech, and J. I. Gallin. Formyl peptide leukocyte chemoattractant uptake and release by cultured human umbilical vein endothelial cells. *J. Immunol.* 139: 3034–3040, 1987.

332. Rubanyi, G. M., and L. H. Butelho. Endothelins. *FASEB J.* 5: 2713–2720, 1991.

333. Rudensky, A. Y., P. Preston-Hurlburt, S. C. Hong, A. Barlow, and C. A. Janeway. Sequence analysis of peptides bound to MHC class II molecules. *Nature* 353: 622–627, 1991.

334. Ryan, G. B., and G. Majno. Acute inflammation. *Am. J. Pathol.* 86: 183–276, 1977.

335. Sacco-Gibson, N., and J. P. Filkins. Glucoregulatory effects of interleukin-1: implications to the carbohydrate dyshomeostasis of septic shock. *Prog. Clin. Biol. Res.* 264: 355–360, 1988.

336. Sacerdote, P., M. Bianchi, P. Ricciardi-Castagnoli, and A. E. Panerai. Tumor necrosis factor alpha and interleukin-1 alpha increase pain thresholds in the rat. *Ann. N. Y. Acad. Sci.* 650: 197–201, 1992.

337. Saigh, J. H., C. C. Sanders, and W. E. Sanders. Inhibition of *Neisseria gonorrhoeae* by aerobic and facultatively anaerobic components of the endocervical flora: evidence for a protective effect against infection. *Infect. Immun.* 19: 704–710, 1978.

338. Saigusa, T. Participation of interleukin-1 and tumor necrosis factor in the responses of the sympathetic nervous system during lipopolysaccharide-induced fever. *Pflügers Arch.* 416: 225–229, 1989.

339. Scarborough, D. E. Cytokine modulation of pituitary hormone secretion. *Ann. N. Y. Acad. Sci.* 594: 169–187, 1990.

340. Scarborough, D. E. Somatostatin regulation by cytokines. *Metabolism* 39(suppl. 2): 108–111, 1990.

341. Schaedler, R. W., and R. P. Orcutt. Gastrointestinal microflora. In: *The Mouse in Biomedical Research,* edited by H. L. Foster, J. D. Small, and J. G. Fox. New York: Academic, 1983, vol. III, p. 327–345.

342. Scheffel, J. W., and Y. B. Kim. Role of environment in the devel-

opment of "natural" hemagglutinins in Minnesota miniature swine. *Infect. Immun.* 26: 202–210, 1979.

343. Schindler, R., J. Mancilla, S. Endres, R. Ghorbani, S. C. Clark, and C. A. Dinarello. Correlations and interactions in the production of interleukin-6 (IL-6), IL-1, and tumor necrosis factor (TNF) in human blood mononuclear cells: IL-6 suppresses IL-1 and TNF. *Blood* 75: 40–47, 1990.

344. Schneerson, R., and J. B. Robbins. Induction of serum *Haemophilus influenzae* type B capsular antibodies in adult volunteers fed cross-reacting *Escherichia coli* 075:K100:H5. *N. Engl. J. Med.* 292: 1093–1096, 1975.

345. Schöbitz, B., D. A. M. Voorhuis, and E. R. DeKloet. Localization of interleukin-6 mRNA and interleukin-6 receptor mRNA in rat brain. *Neurosci. Lett.* 136: 189–192, 1992.

346. Schreiber, G. A., A. Tsykin, A. R. Aldred, T. Thomas, W. P. Fung, P. W. Dickson, T. Cole, H. Birch, F. A. De Jong, and J. Milland. The acute-phase response in the rodent. *Ann. N. Y. Acad. Sci.* 557: 61–85, 1989.

347. Segal, P. B., G. Grieninger, and G. Tosato (Eds). Regulation of the acute phase and immune responses: interleukin-6. *Ann. N. Y. Acad. Sci.* 557: 1–583, 1989.

348. Seganti, L., M. Grassi, P. Mastromarino, A. Pan'a, F. Superti, and N. Orsi. Activity of human serum lipoproteins on the infectivity of rhabdoviruses. *Microbiology* 6: 91–99, 1983.

349. Selwyn, S. Natural antibiosis among skin bacteria as a primary defense against infection. *Br. J. Dermatol.* 93: 487–493, 1975.

350. Seydoux, J., J. Llanos-Q., and C. M. Blatteis. Possible involvement of the ventromedial hypothalamus (VMH) in fever production. *Federation Proc.* 44: 438, 1985.

351. Shibata, M., and C. M. Blatteis. Differential effects of cytokines on thermosensitive neurons in guinea pig preoptic area slices. *Am. J. Physiol.* 261(*Regulatory Integrative Comp. Physiol.* 32): R1096–R1103, 1991.

352. Shoham, S., and J. M. Krueger. Muramyl dipeptide–induced sleep and fever: effects of ambient temperature and time of injections. *Am. J. Physiol.* 255(*Regulatory Integrative Comp. Physiol.* 26): R157–R165, 1988.

353. Silverstein, R. The endocrine response to endotoxin. In: *Bacterial Endotoxic Lipopolysaccharides,* edited by J. L. Ryan and D. C. Morrison. Boca Raton, FL: CRC, 1992, p. 295–309.

354. Simberkoff, M. S., N. H. Moldover, and J. J. Rahal. Absence of detectable bacteriocidal and opsonic activities in normal and infected human cerebrospinal fluids. *J. Lab. Clin. Med.* 95: 362–372, 1980.

355. Singer, S. J. Intercellular communication and cell–cell adhesion. *Science* 255: 1671–1677, 1992.

356. Sipe, J. D. Cellular and humoral components of the early inflammatory reaction. In: *The Acute-Phase Response to Injury and Infection,* edited by A. H. Gordon and A. Koj. Amsterdam: Elsevier, 1985, p. 3–21.

357. Sipe, J. D. The acute-phase response. In: *Immunophysiology. The Role of Cells and Cytokines in Immunity and Inflammation,* edited by J. J. Oppenheim and E. M. Shevach. New York: Oxford University Press, 1990, p. 259–273.

358. Skamene, E. Genetic control of susceptibility to mybobacterial infection. *Rev. Infect. Dis.* 11: S394–S399, 1989.

359. Small, J. D. Fatal enterocolitis in hamsters given lincomycin hydrochloride. *Lab. Anim. Care* 18: 411–420, 1968.

360. Smith, D. H., G. Peter, D. L. Ingram, A. L. Harding, and P. Anderson. Responses of children immunized with capsular polysaccharide of *Haemophilus influenzae. Pediatrics* 52: 637–665, 1973.

361. Smith, H. Microbial surfaces in relation to pathogenicity. *Bact. Rev.* 41: 475–500, 1977.

362. Smith, W., T. K. Kishimoto, O. Abbass, B. Hughes, R. Rothlein, L. V. McIntire, E. Butcher, and D. C. Anderson. Chemotactic

factors regulate lectin adhesion molecule 1 (LECAM-1)–dependent neutrophil adhesion to cytokine-stimulated endothelial cells in vitro. *J. Clin. Invest.* 87: 609–618, 1991.

363. Southorn, P. A., and G. Powis. Free radicals in medicine. *Mayo Clin. Proc.* 63: 381–408, 1988.

364. Spangelo, B. L., and R. M. MacLeod. Regulation of the acute-phase response and neuroendocrine function by interleukin-6. *Prog. Neuroendocrinimmunol.* 3: 167–175, 1990.

365. Spatz, M., D. W. E. Smith, E. G. McDaniel, and G. L. Laqueur. Role of intestinal microorganisms in determining cycasin toxicity. *Proc. Soc. Exp. Biol. Med.* 124: 691–697, 1967.

366. Springer, T. A. Adhesion receptors of the immune system. *Nature* 346: 425–434, 1990.

367. Sprunt, K., and W. Redman. Evidence suggesting importance of role of interbacterial inhibition in maintaining balance of normal flora. *Ann. Intern. Med.* 68: 579–590, 1968.

368. Stedler, J., B. G. Harbrecht, M. DiSilvio, R. D. Curran, M. L. Jordan, R. L. Simmonds, and T. R. Billar. Endogenous nitric oxide inhibits the synthesis of cyclooxygenase products and interleukin-6 by rat Kupffer cells. *J. Leukoc. Biol.* 53: 165–172, 1993.

369. Stern, P., M. Gidlund, A. Orn, and H. Wigzell. Natural killer cells mediate lysis of embryonal carcinoma cells lacking MHC. *Nature* 285: 341–342, 1980.

370. Stevens, R. L., and K. F. Austen. Recent advances in the cellular and molecular biology of mast cells. *Immunol. Today* 10: 381–386, 1989.

371. Stitt, J. T. Fever versus hyperthermia. *Federation Proc.* 38: 39–43, 1979.

372. Stitt, J. T. Differential sensitivity in the sites of fever production by prostaglandin E₁ within the hypothalamus of the rat. *J. Physiol. (Lond.)* 432: 99–110, 1991.

373. Stott, E. J., and R. A. Killington. Rhinoviruses. *Annu. Rev. Microbiol.* 26: 503–524, 1972.

374. Suegara, N., M. Morotomi, T. Watanabe, Y. Kawai, and M. Mutai. Behavior of microflora in the rat stomach: adhesion of lactobacilli to the keratinized epithelial cells of rat stomach in vitro. *Infect. Immun.* 12: 173–179, 1975.

375. Székely, M., and Z. Szelényi. The pathophysiology of fever in the neonate. In: *Pyretics and Antipyretics,* edited by A. S. Milton. Berlin: Springer, 1982, p. 479–528.

376. Tamaki, Y., and T. Nakayama. Effects of air constituents on thermosensitivities of preoptic neurons: hypoxia versus hypercapnia. *Pflügers Arch.* 409: 1–6, 1987.

377. Templeton, J. W., R. Smith, and G. Adams. Natural disease resistance in domestic animals. *J. Am. Vet. Med. Assoc.* 192: 1306–1315, 1988.

378. Tizard, I. Immunity in the fetus and newborn animal. In: *An Introduction to Veterinary Immunology.* Philadelphia: Saunders, 1974, p. 165–177.

379. Thadepalli, H., W. H. Chan, J. E. Maidman, and E. C. Davidson. Microflora of the cervix during normal labor and the puerperium. *J. Infect. Dis.* 137: 568–572, 1978.

380. Toth, L. A., and J. M. Krueger. Alterations in sleep during *Staphylococcus aureus* infection in rabbits. *Infect. Immun.* 56: 1785–1791, 1988.

381. Toth, L. A., and J. M. Krueger. Effects of microbial challenge on sleep in rabbits. *FASEB J.* 3: 2062–2066, 1989.

382. Toth, L. A., and J. M. Krueger. Somnogenic, pyrogenic, and hematologic effects of experimental pasteurellosis in rabbits. *Am. J. Physiol.* 258(*Regulatory Integrative Comp. Physiol.* 29): R536–R542, 1990.

383. Toth, L. A., E. A. Tolley, and J. M. Krueger. Sleep as a prognostic indicator during infectious disease in rabbits. *Proc. Soc. Exp. Biol. Med.* 203: 179–192, 1993.

384. Tracey, K. J., and A. Cerami. Metabolic responses to cachectin/

TNF. A brief review. *Ann. N. Y. Acad. Sci.* 587: 325–331, 1990.

385. Tracey, K. J., and A. Cerami. Tumor necrosis factor and regulation of metabolism in infection: role of systemic versus tissue levels. *Proc. Exp. Soc. Biol. Med.* 200: 233–239, 1992.

386. Tracey, K. J., S. Morgello, B. Koplin, T. J. Fahey III, J. Fox, and A. Aledo. Metabolic effects of cachectin/tumor necrosis factor are modified by site of production. *J. Clin. Invest.* 86: 2014–2024, 1990.

387. Tredget, E. E., Y. M. Yu, S. Zhong, R. Burini, S. Okusawa, J. A. Gelfand, C. A. Dinarello, V. R. Young, and J. F. Burke. Role of interleukin-1 and tumor necrosis factor on energy metabolism in rabbits. *Am. J. Physiol.* 255(*Endocrinol. Metab.* 18): E760–E768, 1988.

388. Tupasi, T. E., M. A. Velmonte, M. E. G. Sanvictores, L. Abraham, L. E. De Leon, S. A. Tan, C. A. Miguel, and M. C. Saniel. Determinants of morbidity and mortality due to acute respiratory infections: implications for intervention. *J. Infect. Dis.* 157: 615–623, 1988.

389. Uehara, A., T. Okumura, Y. Kumei, Y. Takasugi, and M. Namiki. Indomethacin reverses interleukin-1-induced hyperinsulinemia in conscious and freely moving rats. *Eur. J. Pharmacol.* 192: 185–187, 1992.

390. Uehara, A., T. Okumura, C. Sekiya, K. Okamura, Y. Takasugi, and M. Namiki. Interleukin-1 inhibits the secretion of gastric acid in rats: possible involvement of prostaglandin. *Biochem. Biophys. Res. Commun.* 162: 1578–1581, 1989.

391. Ulevitch, R. J., A. R. Johnston, and D. B. Weinstein. New function for high density lipoproteins. Their participation in intravascular reactions of bacterial lipopolysaccharides. *J. Clin. Invest.* 64: 1516–1524, 1979.

392. Unanue, E. R., and P. M. Allen. The basis for the immunoregulatory role of macrophages and other accessory cells. *Science* 236: 551–557, 1987.

393. Valinger, Z., B. Ladesic, and J. Tomasic. Partial purification and characterization of N-acetylmuramyl-L-alanine amidase from human and mouse serum. *Biochim. Biophys. Acta* 701: 63–71, 1982.

394. Vandenbroucke-Grauls, C. M. J. E., and J. Verhoef. Bacteria, phagocytes, and bystander cells: interactions between *Staphylococcus aureus,* polymorphonuclear leukocytes, and endothelial cells. *Pathol. Immunopathol. Res.* 7: 149–161, 1988.

395. van der Waaij, D. Colonization pattern of the digestive tract by potentially pathogenic microorganisms: colonization-controlling mechanisms and consequences for antibiotic treatment. *Infection* 11: S90–S92, 1983.

396. van der Waaij, D., J. M. Berghuis, and J. E. C. Lekkerkerk. Colonization resistance of the digestive tract of mice during systemic antibiotic treatment. *J. Hyg.* 70: 605–610, 1972.

397. van Eden, W., N. M. Gonzalez, R. R. P. de Vries, J. Convit, and J. J. van Rood. HLA-linked control of predisposition to lepromatous leprosy. *J. Infect. Dis.* 151: 9–14, 1985.

398. Van Epps, D. E., J. G. Bender, S. J. Simpson, and D. E. Chenoweth. Relationship of chemotactic receptors for formyl peptide and C5a to CR1, CR3, and Fc receptors on human neutrophils. *J. Leukocyte Biol.* 47: 519–527, 1990.

399. Vanhoutte, P. M. Platelet-derived serotonin, the endothelium and cardiovascular disease. *J. Cardiovasc. Pharmacol.* 17(suppl. 5): S6–S12, 1991.

400. Vankelecom, H., P. Carmeliet, H. Heremans, J. van Damme, R. Dijkmans, A. Billiau, and C. Denef. Interferon-γ inhibits stimulated adrenocorticotropin, prolactin, and growth hormone secretion in normal rat anterior pituitary cell cultures. *Endocrinology* 126: 2919–2926, 1990.

401. Vermeulon, M. W., and G. R. Grey. Processing of *Bacillus subtilis* peptidoglycan by a mouse macrophage cell line. *Infect. Immun.* 46: 476–483, 1984.

402. Vitetta, E. S., R. Fernandez-Botran, C. D. Myers, and V. M. Sanders. Cellular interactions in the humoral immune response. *Adv. Immunol.* 45: 1–105, 1992.

403. Vogel, S. N., S. S. Douches, E. N. Kaufman, and R. Neta. Induction of colony stimulating factor in vivo by recombinant interleukin-1 and tumor necrosis factor. *J. Immunol.* 138: 2143–2148, 1987.

404. von Liebermeister, C. *Handbuch der Pathologie und Therapie des Fiebers.* Leipzig: Vogel, 1875.

405. Wakasugi, N., and J. L. Virelizier. Defective IFN-γ production in the human neonate. I. Dysregulation rather than intrinsic abnormality. *J. Immunol.* 134: 167–171, 1985.

406. Wakasugi, N., J. L. Virelizier, F. Arenzana-Seisdedos, B. Rothgut, J. M. M. Huerta, F. Russo-Marie, and W. Fiers. Defective IFN-γ production in the human neonate. II. Role of increased sensitivity to the suppressive effects of prostaglandin E. *J. Immunol.* 134: 172–176, 1985.

407. Walton, P. E., and M. J. Cronin. Tumor necrosis factor-α inhibits growth hormone secretion from cultured anterior pituitary cells. *Endocrinology* 125: 925–929, 1989.

408. Wan, W., L. Janz, C. Y. Vriend, C. M. Sorensen, A. H. Greenberg, and D. M. Nance. Differential induction of *c-fos* immunoreactivity in hypothalamus and brain stem nuclei following central and peripheral administration of endotoxin. *Brain Res. Bull.* 32: 581–587, 1993.

409. Ward, P. A., J. S. Warren, J. Varani, and K. J. Johnson. PAF, cytokines, toxic oxygen products and cell injury. *Mol. Aspects Med.* 12: 169–174, 1991.

410. Warner, R. G., and F. R. Ehle. Nutritional idiosyncrasies of the golden hamster (*Mesocricetus auratus*). *Lab. Anim. Sci.* 26: 670–673, 1976.

411. Weiss, J., M. Victor, and P. Elsbach. Role of charge and hydrophobic interactions in the action of bactericidal/permeability-increasing protein of neutrophils on gram-negative bacteria. *J. Clin. Invest.* 71: 540–549, 1983.

412. Wells, C. L., M. A. Maddaus, and R. L. Simmons. Proposed mechanisms for the translocation of intestinal bacteria. *Rev. Infect. Dis.* 10: 958–979, 1988.

413. Wilson, C. B., and D. B. Lewis. Basis and implications of selectively diminished cytokine production in neonatal susceptibility to infection. *Rev. Infect. Dis.* 12: S410–S420, 1990.

414. Wing, E. J., and J. R. Young. Acute starvation protects mice against *Listeria monocytogenes. Infect. Immun.* 28: 771–776, 1980.

415. Wolfe, R. R. Glucose metabolism in sepsis and endotoxicosis. In: *Infection: The Physiologic and Metabolic Responses of the Host,* edited by M. C. Powanda and P. G. Canonico. Amsterdam: Elsevier, 1987, p. 213–243.

416. Wostmann, B. S. Intestinal bile acids and cholesterol absorption in the germfree rat. *J. Nutr.* 103: 982–990, 1973.

417. Wright, D. G. The neutrophil as a secretory organ of host defense. In: *Advances in Host Defense Mechanisms,* edited by J. I. Gallin and A. S. Fauci. New York: Raven, 1982, vol. 1, p. 75–110.

418. Wright, S. D. Receptors for complement and the biology of phagocytosis. In: *Inflammation: Basic Principles and Clinical Correlates* (2nd ed.), edited by J. I. Gallin, I. M. Goldstein, and R. Snyderman. New York: Raven, 1992, p. 477–495.

419. Wu, K. K. Endothelial cells in hemostasis, thrombosis, and inflammation. *Hosp. Pract.* 27: 145–166, 1992.

420. Xin, L., and C. M. Blatteis. Blockade of interleukin-1 receptor antagonist (IL1ra) of IL1β-induced neuronal activity in guinea pig preoptic area slices. *Brain Res.* 569: 348–352, 1992.

421. Yamaguchi, M., K. Koike, Y. Yoshimoto, H. Ikegami, A. Miyake, and O. Tanizawa. Effect of TNF-α on prolactin secretion from rat anterior pituitary and dopamine release from the

hypothalamus: comparison with the effect of interleukin-1β. *Endocrinol. Japan.* 38: 357–361, 1991.

422. Yasmineh, W. G., and A. Theologides. Effects of tumor necrosis factor on enzymes of gluconeogenesis in the rat. *Proc. Soc. Exp. Biol. Med.* 199: 97–103, 1992.

423. Young, J. D., H. Hengartner, E. R. Podack, and Z. A. Cohn. Purification and characterization of a cytolytic pore-forming protein from granules of cloned lymphocytes with natural killer activity. *Cell* 44: 849–859, 1988.

424. Zaslof, M. Magainins, a class of antimicrobial peptides from *Xenopus* skin: isolation, characterization of two active forms, and partial cDNA sequence of a precursor. *Proc. Natl. Acad. Sci. USA* 84: 5449–5453, 1987.

425. Zawalich, W. S., and K. C. Zawalich. Interleukin-1 is a potent stimulus of islet insulin secretion and phosphoinositide hydrolysis. *Am. J. Physiol.* 256(*Endocrinol. Metab.* 19): E19–E24, 1989.

426. Zeigler, A. R., C. U. Tuazon, and J. N. Sheagren. Antibody levels to bacterial peptidoglycan in human sera during the time course of endocarditis and bacteremic infections caused by *Staphylococcus aureus*. *Infect. Immun.* 33: 795–800, 1981.

427. Zeisberger, E. Antipyretic action of vasopressin in the ventral septal area of the guinea pig brain. In: *Thermoregulation: Research and Clinical Applications*, edited by P. Lomax and E. Schönbaum. Basel: Karger, 1989, p. 65–68.

428. Zeisberger, E. The role of septal peptides in thermoregulation and fever. In: *Thermoreception and Temperature Regulation*, edited by J. Bligh and K. Voight. Berlin: Springer, 1990, p. 274–283.

429. Ziegler, E. J., H. Douglas, J. E. Sherman, C. E. Davis, and A. I. Braude. Treatment of *E. coli* and *Klebsiella* bacteremia in agranulocytic animals with antiserum to a UDP-gal epimerase–deficient mutant. *J. Immunol.* 111: 433–438, 1973.

430. Zinkernagel, R. M., and P. C. Doherty. H-2 compatibility requirement for T-cell-mediated lysis of target cells infected with lymphocytic choriomeningitis virus. *J. Exp. Med.* 141: 1427–1436, 1975.

431. Zinkernagel, R. M., and P. C. Doherty. MHC-restricted cytotoxic T cells: studies on the biological role of polymorphic major transplantation antigens determining T-cell restriction-specificity, function, and responsiveness. *Adv. Immunol.* 27: 51–177, 1979.

432. Zipori, D. Regulation of hemopoiesis by cytokines that restrict options for growth and differentiation. *Cancer Cells* 2: 205–211, 1990.

433. Zuckerman, S. H., and A. M. Bendele. Regulation of serum tumor necrosis factor in glucocorticoid-sensitive and -resistant rodent endotoxin shock models. *Infect. Immun.* 57: 3009–3013, 1989.

434. Zuckerman, S. H., J. Shellhaas, and L. D. Butler. Differential regulation of lipopolysaccharide-induced interleukin 1 and tumor necrosis factor synthesis: effects of endogenous and exogenous glucocorticoids and the role of the pituitary–adrenal axis. *Eur. J. Immunol.* 19: 301–305, 1989.

IX | THE PSYCHOSOCIAL ENVIRONMENT

Associate Editor   C. M. Blatteis

67. The place of behavior in physiology

MICHEL CABANAC | *Department of Physiology, Faculty of Medicine, Laval University, Québec, Canada*

CONSIDERING BEHAVIOR AS A PART OF PHYSIOLOGY has been a tradition from the time of the Greek philosophers, for whom behavior was biology. The first students of what we now consider to be the science of physiology were, in fact, concerned with behavior, motivation, and cognition. Thus Johannes Müller (68) studied sensations, while Claude Bernard (5) studied the determinants of thirst and its behavioral response, water intake. Pavlov (74) discovered conditioning and conditioned reflexes as a development of his studies on digestive secretions. This trend was also followed by Cannon (15), who studied hunger and emotion in humans and animals, and by Hess (39), who was the first to investigate coordinated behavioral responses after either electrical stimulation or lesion in the diencephalon of animals.

BEHAVIOR AS A PHYSIOLOGICAL RESPONSE

Although the study of behavior as an integrated physiological response has persisted, behavior has remained a minor branch of physiology, very few reviews being published since the early 1980s (1, 3, 55, 47, 50). The vast majority of contemporary physiologists have tended to focus their attention more and more on cellular mechanisms. This reductionist trend has also occurred among neurophysiologists. Behavior is used by them mostly as a convenient window through which to probe the structure and function of the brain. In the past, a good reason for this development of neurophysiology may well have been the constraints imposed by the use of anesthetized animals and stereotaxic apparatus that did not permit animals to behave. Meanwhile, behavior was studied by natural scientists and psychologists.

The entry of behavior into natural science is usually attributed to Darwin (21). Later, behavior was studied in Europe by zoologists who founded the discipline of *ethology,* which is the scientific study of the behavior of animals integrated in their natural environment. In 1951, Tinbergen (91) introduced the term "ethophysiology." Ethology is largely devoted to the study of animal fitness, which includes not only reproductive success but also how behavior optimizes physiological function, that is, whole-body physiology. An important branch of ethology is behavioral ecology (49), where the integration of animals into their environment is studied and measured quantitatively.

Until the end of the eighteenth century psychology was the branch of philosophy that studied the human mind. Psychology became a science, that is, experimental psychology, only when the study of behavior became experimental, under the influences of psychophysicists such as Helmoltz (37) and psychologists such as Thorndike (90) and Watson (97). Psychology started to diversify; one important branch of that science is physiological psychology, in which behavior is studied for itself or as a means of exploring motivation. As a result, behaviors with clear physiological aims, such as food and water intake, have been studied mostly by psychologists (85). The interest of psychologists in physiology has materialized in a vast bulk of physiological data on food intake (6, 29, 48, 80, 102), on general physiological adaptation (22, 23, 77, 105), on optimality (65), and in innumerable works on sensory inputs. Further, in 1991 63% of the authors of research articles published in the journal *Physiology and Behavior* identified themselves clearly as psychologists or as physiologists; of these, 69% were from departments of psychology and 31% from departments of physiology.

How the body satisfies its needs is left mostly, therefore, to psychologists and how it quantitatively adjusts to the environment is left mostly to ethologists. Traditional physiology deals with the functioning of the various systems within the body. The modern trend is toward a finer level of analysis of the systems whose sum constitutes the living organism. Modern physiology has become so analytical that its borderline with biochemistry hardly exists anymore. Thus physiologists may be losing sight of an aspect of paramount importance: the living being is not a closed system, isolated from the environment, but is open and exchanges energy and matter with the environment. Short-sighted physiologists take for granted the permanent inflow from the environment into the body, and outflow from the body to the environment of particular variables. In that regard they are similar to electricians who are most interested in a properly functioning television set, less concerned with the implied heat production, and unconcerned with the energy needed to run the device.

The return of behavior to physiology has occurred largely under the influence of environmental physiologists. Environmental physiology, by its very nature, is integrative and deals in most cases with the functioning of the entire organism. Environmental physiologists, of course, have always been aware of the environmental constraints that influence life processes, especially in the case of energy balance. Early studies showed that behavior was the only response used by reptiles to maintain a stable core temperature (52). However, the majority of physiologists have acknowledged that behavior is a powerful way to cope with environmental constraints. Temperature regulation is mostly behavioral and was therefore important in heightening the awareness of behavior among physiologists. Hardy (34; see Fig. 67.1) was among the first of the contemporary physiologists to consider behavior as deserving to be studied, and Bligh (7, 44) pointed out that one should not oppose physiological and behavioral temperature regulation. Since the behavioral response is also physiological, he suggested the terms "autonomic" and "behavioral" temperature regulation.

The concept that behavior belongs as much to physiology as to psychology and zoology (ethology) is not yet commonplace, but there are signs that the pendulum is swinging back and that behavior again is a subject of challenge for curious and thoughtful physiologists. Behavioral research takes place in several physiological journals. The notion of emergence has become accepted by physiologists, and even mentation is of interest to them (10).

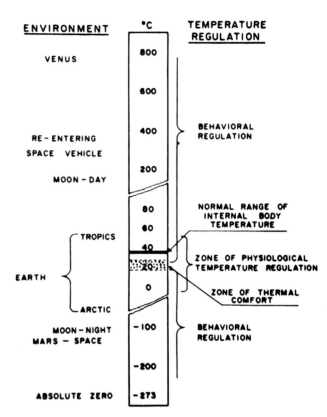

FIG. 67.1. Relative contributions of autonomic and behavioral responses in reducing threats of body temperature variation when humans are exposed to potentially hazardous conditions of heat and cold. *Left:* Temperatures of environments that humans could possibly encounter. *Right:* Thermoregulatory responses in relation to their capabilities and ranges of internal body temperature and thermal comfort. Human freedom to move about on the earth would be limited without the behavioral response [from Hardy (34)].

THE LIVING BEING IS AN OPEN SYSTEM

Living beings are not closed systems but are open to the environment. At each instant they receive from and lose to the environment equal flows of energy and matter. The very process of life entails this exchange, and, in fact, without these flows life is not possible. The water tank of Figure 67.2 represents any one of the flows that traverse the living body.

In this figure, a water tank receives a permanent inflow and loses an equal outflow of water. The figure is a model of each of the flows that traverse the body and its subcompartments, that is, water, energy, or mass of any of the constituents of the body, such as nitrogen, sodium, calcium, etc. In the model, the mass of water in the tank is analogous to the amount of heat, water, nitrogen, glucose, sodium, etc. received from the environment and stored in the body, and, *h*, the level of the water in the tank, indicates the tension reached by the

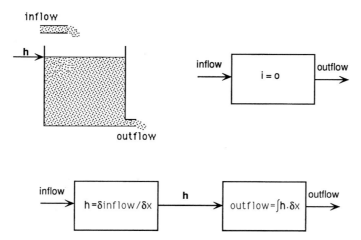

FIG. 67.2. Living animals are analogous to the above tank. They are open systems in steady-state equilibrium receiving a continuous inflow and losing an equal outflow of matter and energy. Black box in upper right describes the system. Block diagram below analyzes the system with its input and output faucets related by the level h, which is a derivative of inflow and an integral of outflow [from Cabanac and Russek (13)].

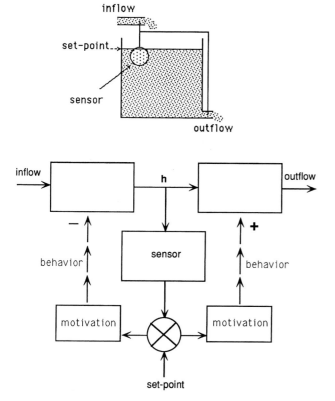

FIG. 67.3. Steady state is now equipped with a sensor (float) of the regulated variable (h), a negative feedback loop on the input faucet, and a positive feed-forward loop on the output faucet. Both loops are behavioral [from Cabanac and Russek (13)].

variable in question. Thus the level of the water in the tank is analogous to the concentration of solutes, the temperature, the pressure, etc. achieved in the body or in its subcompartments.

Proper functioning of the water tank implies modulation of the inflow and/or of the outflow faucet, as, for example, is achieved in Figure 67.3. The water tank is now equipped with a sensor of the water level, the float, a negative feedback loop modulating the inflow, and a positive feed-forward loop modulating the outflow. Any increase of h will tend to retroact negatively and reduce the inflow (negative feedback) and to anteact positively and increase the outflow (positive feed-forward). Reciprocally, any decrease of h will retroact negatively to increase the inflow and anteact positively to decrease the outflow. The water tank now regulates its level through modulation of the interface with the environment and maintains the *milieu intérieur* nearly constant. Modulation of these flows is not physiological, therefore, in the classical definition of the word. It is essentially behavioral because it involves the environment.

Behavioral Control of Inflows

The body receives permanent or intermittent vital inflows of chemical potential energy, vitamins, metabolites, oligoelements, and water from the environment. All of those enter the body through the mouth and the digestive tract; that is, they are behaviorally modulated. The body also receives occasional inflows of heat either from the environment or from the muscles, and these are also behaviorally modulated. Finally, the body receives a quasicontinuous inflow of oxygen through the airways and lungs. This inflow is modulated autonomically but can, on occasion, also be modulated behaviorally for short periods. Everything that enters the body is under either total or occasional behavioral control. Therefore, the process of living depends upon behavior in animals.

Behavioral Control of Outflows

The body returns heat to the environment in a permanent flow which is under both autonomic and behavioral modulation. It also returns water in sweat, which is autonomically modulated; in respiratory gas, which is usually not modulated but can occasionally become behaviorally modulated; and in urine with behavioral relaxation-type modulated micturition. Finally, the body returns unabsorbed mass in feces, a mass of catabolites in urine under behavioral modulation, CO_2 in respiratory gas under occasional behavioral modulation, and various bodily secretions such as milk or sperm under mostly behavioral modulation. The loss of exfoliates from skin and phaneres is practically independent of behavior. The part played by behavior in the control of outflows is, therefore, more modest than that in the control of inflows. Yet the control of outflows is important from a theoretical point of view. It shows the ubiquity of behavior even when these behaviors are more subtle and definitely not vital. In addition, propositional control of ventilation eventually permits phonation and oral language.

Almost the total exchange of the body's fluid, electrolytes, mass, energy, etc. with its environment is, therefore, under behavioral modulation. Thus animal life would not be possible without behavior. These behavioral responses modify either the environment or the subject itself.

THE VARIOUS BEHAVIORS

All behaviors with physiological purpose consist of adjustments of the subject's relationship with its environment. Behaviors may be ranked by order of complexity.

Postural Adaptation

Postural adaptation, which is among the simplest of behaviors, is quite efficacious when applied to the flux of energy from solar radiation. All terrestrial animals position their bodies parallel to the solar beams when body temperature is high and perpendicular when body temperature is low. In dry climates the sun irradiates about $1 \text{ kW} \cdot \text{m}^{-2}$. Such a heat flux when applied to the wings of a butterfly is sufficient to warm the animal in seconds. Even when the mass/area ratio is larger, as in large mammals, postural adaptation can modulate radiative heat gain by a factor of 3.

Posture is also adapted to the satisfaction of other physiological aims, such as food and water intakes, sleep, and hibernation.

Migration

Migration can be understood in a classical sense, that is, seasonal continental travel by groups of birds and mammals caused by the prospect of food scarcity in the approaching winter. Examples are the migrations of passerine birds, geese, and caribous. Migration, however, can also be understood as more modest travel when animals seek a water hole, forage for food, or reach a screen to protect them from a cold gale or from solar radiation. In all cases of migration locomotion is a behavioral means to satisfy a physiological need. We are not accustomed to looking at locomotion in this way, but each time an animal moves to satisfy a physiological need it can be considered as migrating. The main difference between the classical migration of geese flying south from the 75th to the 38th parallel and the upward dive to the surface of a cachalot foraging at a 2,000 m depth in the ocean lies in the time constant of the physiological variables involved. Geese, which migrate for food, possess energy storage within the body that decays over days, whereas the cachalot, which returns to the surface for oxygen at the end of a 30 min dive, must fill its oxygen storage within minutes. The difference between goose and cachalot behaviors lies in the time constant for the depletion of the regulated variable that will benefit from the migration. They are similar in the principle that a variable is depleted and that locomotion satisfies a physiological need.

Building Microenvironments

The building of microclimates spares locomotion and at the same time satisfies physiological needs. This is commonplace in the human species, which lives in a primarily artificial environment. In our species, the natural environment is artifactual. The human species has been able to invade all continents by building thermal microclimates. Heated houses, of course, provide favorable temperatures, but clothes (82) and beds (14, 31) also slow body heat loss and provide warmer surface temperatures. As a result, humans live permanently in tropical conditions, even under extreme latitudes.

One can find many examples of artificial microclimates in animals. Nests, burrows, and dens are seen usually as protection against predators, but they also provide a favorable thermal microclimate. This behavior is not limited to homeotherms, since social insects also regulate environmental temperature. Oxygen consumption within the beehive increases, maintaining a safe internal temperature when outside ambient temperature decreases (58). On warm days bees behaviorally increase the internal convection of the hive and bring water inside to be evaporated (36).

Microenvironments are not limited to the function of temperature regulation. Several animals build respiratory microenvironments. The aquatic spider *Argyroneta aquatica* builds a web under water and stores under it a reserve of air that it fetches from the surface (32). A respiratory microenvironment is maintained by bees in the hive (88). Diving suits, aqualungs, space suits, underwater houses, and space stations also depend on respiratory microenvironments.

Hoarding food (94) in the nest, the den, a cache, a hive, or a home is also an enrichment of the immediate environment in response to an anticipated physiological need of nutrients and energy.

Finally, a microenvironment may be arranged to solve several environmental problems at one time. A human house and a beehive provide at the same time favorable temperature and hoarded food. Sometimes the microenvironment needed for a given purpose may be noxious from another point of view. Prairie dogs build burrows that protect them against predators but at the risk of enriching their microclimate in CO_2. They solve the problem by building the extremities of the burrows in such a way as to be ventilated by a venturi effect when wind is blowing outside (96).

Operant Behavior

A variation of building microclimates is the use of operant behavior, which most often modifies the subject's immediate environment. This method has been used extensively for experimental purposes because it simplifies the quantification of the behavior displayed by the animal. Bar pressing has been used mostly for the study of food intake, but also for temperature regulation, since it was first described by Weiss and Laties (98). Figures 67.4 and 67.5 give examples of such behavioral responses. Figure 67.4 illustrates a thermoregulatory behavior achieved by a bird embryo. The behavioral response was proportional to the thermal need on both sides of a set point near 37.8°C. The correcting response in nature is the parents' behavior. Figure 67.5 illustrates a nutritional behavior leading to food intake.

Parental Behavior

When parents feed their offspring or provide them with thermal protection (8), they behaviorally ensure the environmental conditions necessary for survival. Sometimes the protection consists of providing oxygen to the

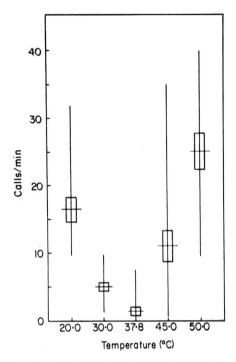

FIG. 67.4. Calls of American white pelican embryos at the pipped egg stage when exposed in vitro to various temperatures for 10 min. *Horizontal line:* mean; *open box:* ± 1 SE; *vertical line:* range; N = 15 eggs, each tested under all conditions. Embryos called for parental rescue proportionally to the deviation of temperature above and below 37.8°C [from Evans (26)].

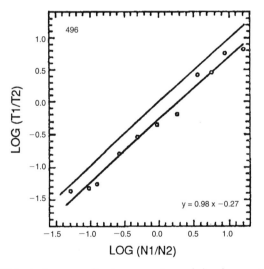

FIG. 67.5. A pigeon standing in an experimental chamber receives a grain when it stays on the correct side of the chamber as indicated by colored lights. The rhythm of lights on and off was variable. The figure gives the logarithm of the ratio of time spent on the left of the chamber plotted as a function of the logarithm of the ratio of number of grains received on the left to number of grains received on the right during an experimental session. Symbols give the animal's behavior. The *solid line* is the linear regression (equation in lower right corner). The *dashed line* has a slope of one and passes through the origin; it represents the performance of perfect matching. It can be seen that the bird's behavior was almost perfect [from Baum and Rachlin (4)].

offspring (20, 95). Figure 67.4 illustrates an example in which the corrective regulatory response to the pipping egg (26) should be parental. In neonatal humans thermoregulatory behavior consists of an audible request of parental protection, even by premature infants (9).

Behavioral Self-Adjustment

Behavior may be focused not only on the modulation of the environment, of course, but on the subject itself. In the latter case behavior is quite similar to an autonomic response, and a separation of behavioral from autonomic responses may not be easy since the response lies in the body. Examples of such responses are thermoregulatory heat production from muscular exercise (12) and intravenous self-drinking (70). Micturition and defecation are initiated reflexively, but they are completed by behaviors that imply self-control by the subject. Food and water intakes start as behaviors and end as reflexes that modify the subject. Hibernation may be considered a reflex, but it is accompanied by a whole sequence of behaviors that render hibernation possible and safe for the animal. Breath-holding and vocalization are behavioral responses rendered possible by the control of the subject on itself. Finally, sleep enters into the category of behavioral self-adjustment, though the physiological result of sleep is still unknown.

Experimentally in animals (66, 93) and clinically in humans (25, 40, 51, 86, 99) it has been demonstrated that subjects are able to modulate visceral function. The principle is that of operant conditioning. Subjects lower or enhance a given function, for example, heart rate, for the sake of obtaining a reward, such as intracranial electrical stimulation in the case of animals or the satisfaction of improving one's health in the case of humans.

An interesting response of some endothermic species to hypoxia and low oxygen supply is the seeking of cooler environments to lower body temperature and in turn metabolism (101). This behavior seems to be common in ectothermic vertebrates and can be described as *behavioral anapyrexia* (44).

The behaviors described in this section modify the internal state of the subjects, but they are not necessarily accomplished with the goal of satisfying a physiological need. Subjects simply satisfy their motivation to behave, and the result is physiological. However, when human subjects purposely modify food intake or increase aerobic heat production to lower blood cholesterol or body weight, they intend to modify behaviorally their physiological state. These efforts are somewhat different since the behavior does not fulfill a physiological need but rather opposes the subject's normal physiological functioning.

Ultimate Behavioral Self-Adjustment: Autostimulation of the Central Nervous System

When a subject displays any of the behaviors listed above, one may question how the subject knows that it behaves appropriately. The answer to that question is beyond the scope of the present chapter, but we may state that the subject's nervous system receives messages from sensors located in either the periphery or the core of the body and that behavior is adjusted as a function of that information. There are behaviors, though, that modify neither the environment nor the *milieu intérieur* but that shunt the step of afferent inputs and actuate directly brain circuits. The intracranial stimulus may be nonspecific, as discovered by Olds (30, 71) with intracranial electrical self-stimulation; pharmacological (72, 100), which is exemplified by immemorial human drug addiction; or specific, as is the case with intracranial temperature self-stimulation (19).

EFFICACY AND PRECISION OF BEHAVIOR

The few examples provided above show that behavior is both proteiform and ubiquitous in its service to physiological aims. Behavior is the obligatory step in the life process of animals. This section will show that this servo-control is not only qualitative to permit life but also precisely quantitative. The adjustment of behavior to the physiological needs of the body occurs for short-term as well as long-term needs. It is in this field that the most remarkable developments have occurred. (Interested readers will find a review of these developments in ref. 79.)

Short-Term Adjustments

The best example of short-term adjustment is found in temperature regulation (11). Figure 67.4 provides an example of temperature regulation mediated by a behavioral response proportional to the deviation of core temperature from its set point near 37.8°C. Following are equations describing thermoregulatory corrective responses as functions of body temperatures where sensors have been identified are strikingly similar when the response is either behavioral (18; equation 1) or autonomic (33; equation 2):

$$R = a\,(T_{hypothalamus} - T_{set}) \qquad (1)$$

$$R = a(T_{body} - T_{set}) \qquad (2)$$

Other short-term behavioral responses occur in the adjustment of water and sodium balances (87). The example provided in Figure 67.5 shows how perfectly an animal will match the amount of food obtained with the cost of that food in terms of energy and time. Behavior is quantitatively optimized; that is, energy cost vs. energy gain is maximized at each step in the sequence of actions in the procurement of food: search, procurement, and handling (16, 17).

Ethologists have developed the concept that what an animal does at a given time depends on the benefits and costs of its actions. The theory was developed specifically to characterize the utilization of food resources and foraging behavior (24, 59), but McFarland (62) has applied the concept of optimality to all mechanisms of decision making.

Long-Term Adjustments

During the course of a lifetime, regulations using behavioral responses can be extremely precise, as shown, for example, by Hervey (38), who estimated that the average English woman ingests 20 tons of food between the ages of 25 and 65 while she gains 11 kg; yet during this period her body weight change corresponds to an average error of only 100 mg per meal. If body weight is regulated, as suggested by this extremely small error, then the drift of 11 kg is part of the regulatory process and is due to a drift of the set point for body weight with aging. Hence, the behavioral response may be considered perfect.

The hoarding of food provides another example of long-term regulation. The amount of food stored by a rat is proportional to the decrease of its body weight below set point (27). The application of behavior to long-term physiological regulations has been developed by Mrosovsky (67).

THE RANKING OF PRIORITIES

The examples of behaviors adapted to physiological aims provided in the above sections were similar to autonomic responses in that they modulated the body's inflow or outflow of matter and energy. The similarity of behavioral and autonomic responses will be found also in other chapters of this *Handbook* dealing with peripheral signals (Chapter 5), afferent inputs, nervous centers, and laws relating signals and responses of various physiological regulations (Chapters 7, 13, 26, and 27). Autonomic and behavioral responses are therefore complementary and can be substituted for one another (Chapter 3). Thus an organism may have several possible responses to a given environmental challenge and may be able to play with possible substitutions of responses. This flexibility finds its usefulness when several physiological functions compete for a given response and when several motivations compete for the behavioral final common path (64). Figure 67.6 is a chart of the main physiological functions ranked in decreasing order of priority, that is, in increasing order of time constant tolerated before satisfying a need. A

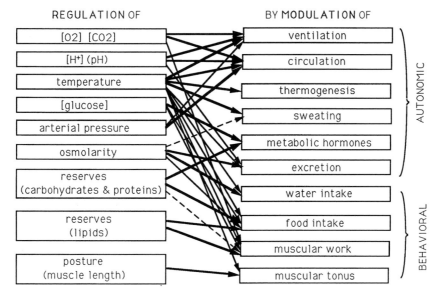

FIG. 67.6. *Left:* Some regulated variables of the organism ranked from top to bottom in order of priority, defined by the time constant tolerated before correcting a perturbation. *Right:* Non-exhaustive list of somatic functions. Regulation of variables on left achieved by control (modulation) of functions on right. *Heavy arrows:* tight control; *light arrows:* loose control; *dashed arrows:* hypothetical control. Behavioral responses participate in defense of most regulated variables of the organism [from Cabanac and Russek (13)].

given regulation may modulate several systems, and several regulations may converge on a given system. Some of the modulated functions are autonomic, others are behavioral. Therefore, the organism has at its disposal a keyboard made of various behavioral and autonomic responses.

Complementarity of Autonomic and Behavioral Responses: The Gain of Freedom

Figure 67.7 provides an example of the complementarity of autonomic and behavioral responses. In that experiment pigeons used their operant response to counterbalance behaviorally the ambient temperature stress and to maintain both a stable body temperature and a low evaporative heat loss.

In another experiment pigeons behaviorally compensated for a food scarcity by selecting a higher ambient temperature (73). These examples demonstrate that pigeons can substitute a behavioral thermoregulatory response for an autonomic one. The opposite is also true, and pigeons can use their autonomic response to free their behavior for another purpose, for example, flight from a predator. The gain in freedom is larger when adaptation improves the autonomic performance and especially in the case of cross-adaptation (54). The upward drift in the body weight of aging humans might be considered an adaptation of the somatic nature to declining behavioral capacity to forage.

Ectothermic animals must rely on behavior to regulate body temperature. This is also true for some mammals. The mole rat, a poikilothermic mammal, does not possess an autonomic defense against cold stress and must use behavior to prevent hypothermia when ambient temperature decreases (103). Similarly, the pig does not sweat, is a poor panter, and responds autonomically only by skin vasomotion. It must, therefore, use its behavior to prevent hyperthermia in a warm environment (42). This makes it seem reasonable that pigs should cover themselves with wet mud in the summer.

Some species have developed particular autonomic capacities that make them far better performers in a given environment than other species or breeds. For example, the black bedouin goat is able to store water in its rumen, maintain an extremely low urine flow, recycle in its gut up to 90% of its urea production, and digest fibers that are indigestible to other goats (83). These capacities make the black bedouin goat better suited than other goats, and most other mammals including their predators, to survival in a desert environment. The camel thrives in the desert because it also recycles its urea; can depress its thyroid and insulin activities; tolerates a high core temperature, glycemia, and osmotic pressure when needed; and is able to rehydrate within minutes (104). The mole rat tolerates high

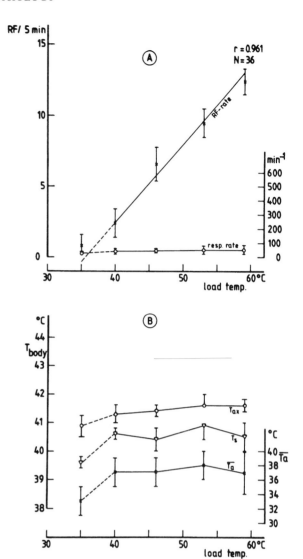

FIG. 67.7. A pigeon stands in a climatic chamber, the temperature of which (load temperature) is imposed by the experimenter. The bird can peck at a key and thus obtain a burst of cool air. The figure gives the mean resulting body temperature and behavior of three pigeons performing in 43 sessions. A: Rate of pecking: RF, respiratory frequency. The animals' behavior was proportional to ambient temperature and therefore directly thermoregulatory. As a result, the birds saved evaporative heat loss and did not hyperventilate while maintaining stable body temperatures (B); $T_{a,x}$, axillary temperature; T_s, dorsal skin temperature; T_a temporal mean ambient temperature [from Schmidt (81)].

concentrations of carbon dioxide and low oxygen concentration in its respiratory environment, the underground collective burrows; hence, it is able to maintain normal activity and metabolism (2). The emperor penguin is able to store 40% of its body weight as fat. This allows almost 4 months of fasting during the reproductive season in the Antarctic continent, 120 km from the seashore (56).

These extraordinary adaptations of autonomic functions free the behavior and allow the dwelling of these species in areas where predators cannot follow them. One may wonder what started first: the autonomic properties that allow long stays in hostile environments, such as desert, underground burrows rich in CO_2, and the Antarctic, or the behavioral trend to stay free from predators in environments that demand a particular autonomic adaptation. The behavioral trend to seek security from predators and the extreme efficacy of autonomic responses to resist environmental constraints probably result from the coevolution of autonomic and behavioral traits. There must be a cost for these exceptional autonomic adaptations, if not by the individual, at least by the species that is able to survive in these extreme environments. The cost for these species will be found probably in low reproductive rates due to scarcity of food.

The complementarity of behavioral and autonomic functions can be found in pathological situations. When the autonomic response to environmental stress is hindered by disease or by the experimenter, the behavioral response takes control over the regulatory process. This may arouse another cost, behavioral in this instance, which is to be paid by all species. Psychological or behavioral aggressions may have somatic impacts. Indeed, psychological stimuli are among the most potent to affect the pituitary adrenal cortical system, and they lead to a stress reaction only if the subject shows an emotional response (61). Stress is now viewed mainly as a general biological response to environmental demands, and experimental studies of stress always consist of behavioral hindrance.

Conflicts of Motivation: The Behavioral Final Common Path

In natural environments a motivation is rarely present alone. Most often, several motivations are present simultaneously but cannot be satisfied simultaneously because the behaviors involved are mutually incompatible; for example, it is not possible to eat and sleep at the same time. How behavior succeeds in satisfying both physiological and nonphysiological motivations is one of the most fascinating problems of modern physiology. It was seen in the preceding section that animals are able to substitute autonomic for behavioral responses, thus gaining one degree of freedom. Optimization of behavior and maximization of fitness is obtained by trade-offs among the various competing motivations.

Thus behavior is a final common path on which all motivations converge (64). As a result, motivations compete to have access to behavior as a final common path, and it may be stated that at any time the motivation being satisfied by behavior is the most important for the subject at that very instant. Behavior is therefore

used sequentially for the service of various functions. When a cachalot dives to forage for food, the need for energy and nutrients ranks first; when it rushes to the surface to breathe, the need of oxygen ranks above the need of food, though the latter may be still present. The situation of conflict has been studied theoretically (62, 63) and experimentally. An example of the systematic study of such an alternative is provided by the fulfillment of the needs for both water and sodium by ingestion of more or less concentrated solutions of sodium chloride in water (87). Other examples can be found of situations where animals alternated their choice between food and a shelter against excessive ambient temperature (41, 45, 46, 60, 69), between food and water (57, 84), between food and the urge to reproduce (75), and between intracranial electrical self-stimulation and various physiological needs (28, 43, 76, 89). In the most complex situation studied experimentally three motivations were combined: the needs for food, water, and temperature regulation (78).

OPTIMAL BEHAVIOR

When animal or human subjects are placed experimentally in situations where two motivations clash, measurement of their behavioral choice shows that they tend to optimize their behavior. The word *optimality* applied to behavior can be ambiguous because it bears somewhat different meanings when used by either ethologists, economists, or physiologists (53). Ethologists differentiate between goal and cost. Economists differentiate between utility and cost. The goal of a subject, as well as utility, is some entity that an optimal behavior will tend to maximize, a definition of which may appear tautological to the physiologist. The cost is a characteristic of the environment that optimal behavior will tend to minimize. All would agree that an optimal behavior gives the maximal net benefit (or fitness) to the behaving individual. Specialists diverge in their definition of benefit (or fitness). Benefit can be defined in terms of reproductive efficacy (49) as well as financial profit and physiological function (64). We are concerned here with the latter aspect.

Figures 67.8 and 67.9 provide two examples of experiments in which subjects were placed in conflicts of motivation and where their behavioral response was shown to be clearly optimal, as seen from the point of view of physiology. In the first case hunger was pitted against thirst and in the second cold discomfort against muscular fatigue.

When the distance between food and water was enlarged by the experimenter by up to several hours of walking, to increase the cost of obtaining both, sheep reduced the number of shuttle trips between food and

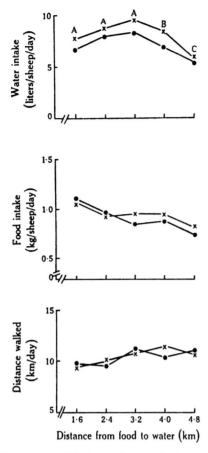

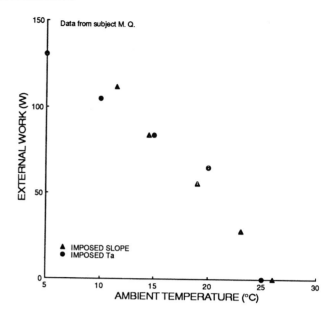

FIG. 67.8. Two groups of Merino wethers can feed and drink at two locations separated by the various distances indicated on the abscissa; water intake, food intake, number of shuttle trips between food and water, and resulting distance walked were monitored. A: Two drinks per day; B: Three drinks every 2 days; C: one drink per day. It can be seen that the animals modulated behavior to maintain approximately constant water intake, food intake, and cost measured by walking distance [from Squires and Wilson (84)].

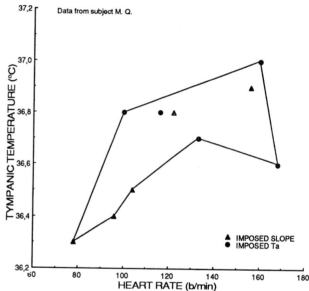

water from two to one per day then to two per 3 days, etc. At the same time, they managed to keep energy expenditure, food intake, and water intake remarkably constant (Fig. 67.8).

When human subjects were placed in a cold room on a treadmill and could warm themselves at the cost of fatigue, they increased heat production proportionally to the decrease of ambient temperature and, at the same time, limited the increase of heart rate (12). This was achieved when they could manipulate the treadmill slope and ambient temperature was fixed, as well as when they could manipulate ambient temperature and the treadmill slope was imposed (Fig. 67.9).

In cases such as the behaviors shown in Figs. 67.8 and 67.9 the physiologist possesses strong reasons to consider as optimal the behavior displayed by the subjects and measured by the experiments. Fitness here does not refer to the subjects' reproductive efficiency, which is the

FIG. 67.9. A subject walks at a steady rate of 3 km·h⁻¹ on a treadmill in a climactic chamber. He can adjust either the slope of the treadmill or the ambient temperature. When the subject adjusts the slope, ambient temperature is imposed by the experimenter. When the subject adjusts ambient temperature, the slope is imposed by the experimenter. Each symbol shows the subject's state at the end of a 1 h session. Results obtained when slope was imposed were not different from results when ambient temperature was imposed. This means that the subject used the degree of freedom left to him to reach, with his behavioral choice, the same physiological state. In addition, results show that with one behavioral degree of freedom the subject managed to produce external work, that is, heat production, that was inversely proportional to ambient temperature and to curb increases in both heart rate and core temperature. Behavior was therefore optimal as seen from the point of view of temperature regulation [from Cabanac and Leblanc (12)].

ethologist's criterion of fitness. The most efficacious behaviors are those which do not require use of the behavioral final common path. In the same way as we saw the complementarity of autonomic and behavioral responses that led to the acquisition of one degree of freedom, various behaviors can also be complementary and result in freeing the behavioral final common path. This is achieved by improving the microenvironment. The human species thus is able to modulate artificially, and even to regulate, the environment to suppress most environmental stresses. A diving suit, for example, suppresses each of the thermal, respiratory, and buoyancy challenges posed to a human in a water environment and thus frees the behavior for other purposes.

CONCLUSIONS

This chapter is a plea to physiologists that behavior is extremely important as a major homeostatic mechanism that they cannot afford to ignore. When the investment in research is judged from the point of view of rentability in terms of financial return and when the sciences are pitted against one another to compete for research funds, physiologists cannot afford to ignore that all human industry is behavior. Yet there are more important and challenging reasons for physiologists to study behavior: from the origins of the science of physiology, behavior has found its place in physiological laboratories. Short- as well as long-term behaviors are similar qualitatively to autonomic responses in this regard in that they modulate the flow of energy and matter

exchanged with the environment. Yet quantitatively behavior is far more powerful than the autonomic response to that modulation. Modulation takes place by control not only of the immediate environment of the subject but also of the subject itself since behavior is also able to modulate the internal functions of the body. The span of potential behavioral responses is, therefore, most extensive. When appropriately measured, behavioral responses can be used equally as well as autonomic responses in experiments probing the functioning of regulatory functions. The behavioral final common path is used in a time-sharing pattern for the service of various functions. Some freedom is gained by the complementarity of autonomic and behavioral responses. Additional freedom is gained by the ranking of emergencies. The first motivations to be satisfied are the most urgent. The most urgent needs can be defined as those with the shortest time constant.

With the extreme flexibility, variety, and power of behavioral responses in mind, one may look at phylogenesis with a new respect for the so-called lower phyla as compared to mammals and birds. Behavioral responses have been able to efficiently secure the survival of ectothermic vertebrates. In the long tinkering process of phylogeny towards freedom, autonomic and behavioral responses have complemented one another. The few examples provided illustrate how closely behavior can be used by animals and humans to fulfill physiological goals. This raises the question of how the brain "knows" when, how, and how much to behave. The understanding of this process is an immense field open to research by physiologists.

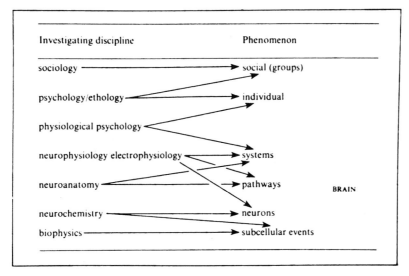

FIG. 67.10. Diagram illustrating the phenomena and disciplines of the brain and behavioral sciences. The sciences *(left column)* are ranked from bottom to top in order of increasing complexity of the phenomena studied *(center column)*. The understanding of behavior is the understanding of whole organisms [from Toates (92)].

A system can be defined as a set of elements with an input and an output related by a law (35). When two of these elements are known, it is also possible to know the third element. Many problems in research are characterized by situations in which the inputs and outputs of a system are known but not the laws relating them. Research, in any science consists of identifying the laws. Research in behavioral physiology is, therefore, not different in principle from research in any other field of science. Behavior merges into physiology as soon as responses involve the whole organism rather than subcellular, cellular, tissue, and organ systems. Behavioral physiology is concerned with systems that include the whole organism and its environment (Fig. 67.10). This is to say that behavioral physiology is integrative by essence and implies knowledge and understanding of environmental biology in addition to cellular and systemic biology. The larger number of variables involved makes behavioral physiology an extremely difficult area of research, where the excitement lies not in the application of technology but in the intelligence of systems.

REFERENCES

1. Amit, Z., and Z. H. Galina. Stress-induced analgesia: adaptive pain suppression. *Physiol. Rev.* 66: 1091–1120, 1986.
2. Arieli, R., A. Ar, and A. Shkolnik. Metabolic responses of a fossorial rodent *(Spalax ehrenbergi)* to simulated burrow conditions. *Physiol. Zool.* 50: 61–75, 1977.
3. Baile, C. A., C. L. McLaughlin, and M. A. Della-Fera. Role of cholecystokinin and opioid peptides in control of food intake. *Physiol. Rev.* 66: 172–234, 1986.
4. Baum, W. M., and H. C. Rachlin. Choice as time allocation. *J. Exp. Anal. Behav.* 12: 861–874, 1969.
5. Bernard, C. *Leçons de Physiologie Expérimentale Appliquée à la Médecine. Cours du Semestre d'Été, 1855.* Paris: Baillière, 1855, vol. II. p. 49–52.
6. Berstein, I. L., and S. Borson. Learned food aversion: a component of anorexia syndromes. *Psychol. Rev.* 93: 462–472, 1986.
7. Bligh, J. *Temperature Regulation in Mammals and Other Vertebrates.* Amsterdam: North Holland, 1973, p. 436.
8. Blix, A. S., and J. B. Steen. Temperature regulation in newborn polar homeotherms. *Physiol. Rev.* 59: 285–304, 1979.
9. Brück, K. Which environmental temperature does the premature infant prefer? *Pediatrics* 41: 1027–1030, 1968.
10. Bunge, M. From neuron to mind. *News in Physiological Science* 4: 206–209, 1989.
11. Cabanac, M. Thermoregulatory behavior. In: *Essays on Temperature Regulation,* edited by J. Bligh and R. Moore. Amsterdam: North Holland, 1972, p. 19–36.
12. Cabanac, M., and J. Leblanc. Physiological conflict in humans: fatigue vs. cold discomfort. *Am. J. Physiol.* 244(*Regulatory Integrative Comp. Physiol.* 15): R621–R628, 1983.
13. Cabanac, M., and M. Russek. *Régulation et Controle en Biologie.* Québec: Presses de l'Université Laval, 1982, p. 242.
14. Candas, V., J. P. Libert, J. J. Vogt, J. Ehrhart, and A. Muzet. Body temperature during sleep under different conditions. In: *Proc. Int. Indoor Climate Symp.,* edited by O. Fanger. København, 1978.
15. Cannon, W. B. *The Wisdom of the Body.* New York: Norton, 1932, p. 101.
16. Collier, G. The economics of hunger, thirst, satiety, and regulation. *Ann. N. Y. Acad. Sci.* 575: 136–154, 1989.
17. Collier, G., and C. K. Rovee-Collier. A comparative analysis of optimal foraging behavior: laboratory simulations. In: *Foraging Behavior,* edited by A. C. Kamil and T. D. Sargent. New York: Garland STPM 1981, p. 39–76.
18. Corbit, J. D. Behavioral regulation of body temperature. In: *Physiological and Behavioral Temperature Regulation,* edited by J. D. Hardy, A. P. Gagge, and J. A. J. Stolwijk. Springfield, IL: Thomas, 1970, p. 777–801.
19. Corbit, J. D. Voluntary control of hypothalamic temperature. *J. Comp. Physiol. Psychol.* 83: 394–411, 1973.
20. Courtenay, S. C., and M. H. A. Keeleyside. Wriggler-hanging: a response to hypoxia by brood-rearing *Herotilapia multispinosa* (Teleostei, Cichlidae). *Behaviour* 85: 183–197, 1983.
21. Darwin, C. *The Expression of the Emotions in Man and the Animals,* Chicago: University of Chicago Press, 1965, p. 372.
22. Dienstbier, R. A. Arousal and physiological toughness: implications for mental and physical health. *Psychol. Rev.* 96: 84–100, 1989.
23. Eikelboom, R., and J. Stewart. Conditioning of drug-induced physiological responses. *Psychol. Rev.* 89: 507–528, 1982.
24. Emlen, J. M. The role of time and energy in food preference. *Am. Natr.* 100: 611–617, 1966.
25. Engel, B. T., and N. Schneiderman. Operant conditioning and the modulation of cardiovascular function. *Annu. Rev. Physiol.* 46: 199–210, 1984.
26. Evans, R. M. Vocal regulation of temperature by avian embryos: a laboratory study with pipped eggs of American white pelican. *Anim. Behav.* 40: 969–979, 1990.
27. Fantino, M., and M. Cabanac. Body weight regulation with a proportional hoarding response in the rat. *Physiol. Behav.* 24: 939–942, 1980.
28. Frank, R. A., W. S. Pritchard, and R. M. Stutz. Food versus intracranial self-stimulation: failure of limited access self depriving rats to self-deprive in a continuous access paradigm. *Behav. Neural Biol.* 33: 503–508, 1981.
29. Galef, B. G., Jr. A contrarian view of the wisdom of the body as it relates to dietary self-selection. *Psychol. Rev.* 98: 218–223, 1991.
30. Gallistel, C. R. Determining the quantitative characteristics of a reward pathway. In: *Quantitative Analyses of Behavior. Biological Determinants of Reinforcement,* edited by M. L. Commons, R. M. Chuch, J. R. Stellar, and A. R. Wagner. Hillsdale, NJ: Laurence Erlbaum, 1988, vol. 7, p. 1–30.
31. Goldsmith, R., R. Hampton, and I. F. G. Hampton. Nocturnal microclimate of man. *J. Physiol. (Lond.)* 194: 32–33, 1968.
32. Grassé, P. P., and R. Poisson. *Précis de Sciences Biologiques. Zoologie I.* Paris: Masson, 1961, p. 919.
33. Hardy, J. D. The set-point concept in physiological temperature regulation. In: *Physiological Controls and Regulations,* edited by W. S. Yamamoto and J. R. Brobeck. Philadelphia: Saunders, 1965, p. 98–116.
34. Hardy, J. D. Thermal comfort and health. *A.S.H.R.A.E. J.* 43–51, February 1971.
35. Hardy, J. D., and J. A. J. Stolwijk. Regulation and control in physiology. In: *Medical Physiology* (12th ed.), edited by V. B. Mountcastle. St. Louis: Mosby, 1968, p. 591–610.
36. Hazelhoff, E. H. Ventilation in a bee-hive during summer. *Physiol. Comp. Œcol.* 3: 343–364, 1954.
37. Helmoltz, H. *Die Tatsachen in der Wahrnemung.* Berlin: Hirschwald, 1879, p. 112.
38. Hervey, G. R. Regulation of energy balance. *Nature* 223: 629–631, 1969.

39. Hess, W. R. Hypothalamus und die Zentren des autonomen Nervensystems: Physiologie. *Arch. Psychiatr. Nervenkr.* 104: 548–557, 1936.

40. Huang, M. H., J. Ebey, and S. Wolf. Manipulating the QT interval of the ECG by cognitive effort. *Pavl. J. Biol. Sci.* 24: 102–108, 1989.

41. Ingram, D. L., and K. F. Legge. The thermoregulatory behavior of young pigs in a natural environment. *Physiol. Behav.* 5: 981–987, 1970.

42. Ingram, D. L., and L. E. Mount. *Man and Animals in Hot Environments.* Berlin: Springer-Verlag, 1975, p. 185.

43. Ishikawa, Y. H., T. Tanaka, T. Nakayama, and K. Kanosue. Competition between lever pressing behavior and thermoregulatory behavior on exposure to heat in intracranial self stimulating rats. *Physiol. Behav.* 42: 599–603, 1988.

44. I.U.P.S. Commission for thermal physiology. Glossary of terms for thermal physiology. *Pflügers Arch.* 410: 410–587, 1987.

45. Johnson, K. G., and M. Cabanac. Homeostatic competition between food intake and temperature regulation in rats. *Physiol. Behav.* 28: 675–679, 1982.

46. Johnson, K. G., and R. Strack. Adaptive behaviour of laboratory rats feeding in hot conditions. *Comp. Biochem. Physiol.* 94A: 69–72, 1989.

47. Koob, G. F., and F. E. Bloom. Behavioral effects of neuropeptides: endorphins and vasopressin. *Annu. Rev. Physiol.* 44: 571–582, 1982.

48. Kraly, F. S. Physiology of drinking elicited by eating. *Psychol. Rev.* 91: 478–490, 1984.

49. Krebs, J. R., and N. B. Davies. *An Introduction to Behavioural Ecology.* Sunderland, MA: Sinauer 1981, p. 292.

50. Kung, C., and Y. Saimi. The physiological basis of taxes in *Paramecium. Annu. Rev. Physiol.* 44: 519–534, 1982.

51. Lang, P. J. Learned control of human heart rate in a computer directed environment. In: *Cardiovascular Psychophysiology: Current Issues in Response Mechanisms, Biofeedback, and Methodology.* edited by P. A. Obrist, A. H. Black, J. Brener, and L. V. Dicara. Chicago: Aldipe, 1974, p. 392–405.

52. Langlois, P. La régulation thermique des poikilothermes. *J. Physiol. (Paris)* 2: 249–256, 1902.

53. Lea, S. E. G., R. M. Tarpy, and P. Webley. *The Individual in the Economy, a Survey of Economic Psychology.* Cambridge: Cambridge University Press, 1987, p. 627.

54. Leblanc, J. Interactions between adaptation to cold and to altitude. In: *Matsumoto Int. Symp. High-Altitude Med. Sci.,* edited by G. Ueda, Matsumoto, Japan: Shinshu University Press, p. 475–481, 1992.

55. Lemagnen, J. Body energy balance and food intake: a neuroendocrine regulatory mechanism. *Physiol. Rev.* 63: 314–386, 1983.

56. Lemaho, Y. The emperor penguin: a strategy to live and breed in the cold. *Am. Sci.* 65: 680–693, 1977.

57. Lester, N. P. The "feed drink" decision. *Behaviour* 89: 200–219, 1984.

58. Lindauer, M. Die Temperaturregulierung der Bienen bei Stocküberhitzung. *Naturwissenschaften* 38: 308–309, 1951.

59. MacArthur, R. H., and E. R. Pianka. On the optimal use of a patching environment. *Am. Nat.* 100: 603–609, 1966.

60. Malechek, J. C., and B. S. Smith. Behavior of range cows in response to winter weather. *J. Range Manage.* 29: 9–12, 1976.

61. Mason, J. W., J. T. Maher, L. H. Hartley, E. H. Mougey, M. J. Perlow, and G. J. Jones. Selectivity of corticosteroid and catecholamine response to various natural stimuli. In: *Psychopathology of Human Adaptation.* edited by G. Sarban. New York: Plenum, 1976.

62. McFarland, D. J. Decision making in animals. *Nature* 269: 15–21, 1977.

63. McFarland, D. J., and A. Houston. *Quantitative Ethology, the State Space Approach.* Boston: Pitman, 1981, p. 204.

64. McFarland, D. J., and R. M. Sibly. The behavioural final common path. *Phil. Trans. R. Soc. Lond.* 270: 265–293, 1975.

65. Meyer, D. E., R. A. Abrams, S. Kornblum, C. E. Wright, and J. E. K. Smith. Optimality in human motor performance: ideal control of rapid aimed movements. *Psychol. Rev.* 95: 340–370, 1988.

66. Miller, N. E. Learning of visceral and glandular responses. *Science* 163: 434–445, 1969.

67. Mrosovsky, N. *Rheostasis, the Physiology of Change.* New York: Oxford University Press, 1990, p. 183.

68. Müller, J. *Handbuch der Physiologie des Menschen für Vorlesungen.* Coblenz: Hölscher, 1834–1840, 2 vol.

69. Murakami, N., and K. Kinoshita. Spontaneous activity and heat avoidance of mice. *J. Appl. Physiol.: Respir. Environ. Exerc. Physiol.* 43: 573–576, 1977.

70. Nicolaidis, S., and N. Rowland. Long-term self intravenous drinking in the rat. *J. Comp. Physiol. Psychol.* 87: 1–15, 1974.

71. Olds, J. Physiological mechanism of reward. In: *Nebraska Symp. Motivation.* edited by M. R. Jones. Lincoln: University of Nebraska Press, 1955.

72. Olds, J., A. Yuwiler, M. E. Olds, and C. Yun. Neurohumors in hypothalamic substrates of reward. *Am. J. Physiol.* 207: 242–254, 1964.

73. Ostheim, J. Coping with food-limited conditions: feeding behavior, temperature preference, and nocturnal hypothermia in pigeons. *Physiol. Behav.* 51: 353–361, 1992.

74. Pavlov, I. P. *Essential Works of Pavlov,* edited by M. Kaplan. Toronto: Bantam, 1966, p. 392.

75. Perrigo, G. Breeding and feeding strategies in deer mice and house mice when females are challenged to work for their food. *Anim. Behav.* 35: 1298–1316, 1987.

76. Phillips, A. G., C. W. Morgan, and G. J. Mogenson. Changes in self-stimulation as a function of incentive of alternative rewards. *Can. J. Psychol.* 24: 289–297, 1970.

77. Poulos, C. X., and H. Cappel. Homeostatic theory of drug tolerance: a general model of physiological adaptation. *Psychol. Rev.* 98: 390–408, 1991.

78. Rautenberg, W., B. May, and G. Arabin. Behavioral and autonomic temperature regulation in competition with food intake and water balance of pigeons. *Pflügers Arch.* 384: 253–260, 1980.

79. Rowland, N. Interplay of behavioral and physiological mechanisms in adaptation. *Handbook of Physiology. Adaptation to the Environment,* edited by D. B. Dill and E. F. Adolph. Washington, DC: Am. Physiol. Soc., 1964, sect. 4.

80. Rozin, P., and A. E. Fallon. A perspective on disgust. *Psychol. Rev.* 94: 23–41, 1987.

81. Schmidt, I. Interactions of behavioral and autonomic thermoregulation in heat stressed pigeons. *Pflügers Arch.* 374: 47–55, 1978.

82. Scholander, P. F., N. Anderson, J. Krog, F. V. Lorentzen, and J. Steen. Critical temperature in Lapps. *J. Appl. Physiol.* 10: 231–234, 1957.

83. Shkolnik, A. The black bedouin goat. *Bielefelder Ökol. Beitr.* 6: 53–60, 1992.

84. Squires, V. R., and A. D. Wilson. Distance between food and water supply and its effects on drinking frequency, and food and water intake of Merino and Border Leicester sheep. *Aust. J. Agricult. Res.* 22: 283–290, 1971.

85. Stellar, E. The physiology of motivation. *Psychol. Rev.* 61: 5–22, 1954.

86. Sterman, M. B., and L. Friar. Suppression of seizures in an epileptic following sensorimotor EEG feedback training. *EEG Clin. Neurophysiol.* 33: 89–95, 1972.

87. Stricker, E. M., and J. G. Verbalis. Hormones and behavior: the biology of thirst and sodium appetite. *Am. Sci.* 76: 261–267, 1988.

88. Stussi, T. Thermogénèse de l'Abeille et Ses Rapports avec le Niveau Thermique de la Ruche. Lyon: Univ. of Lyon, 1967. Thèse Doctorat d'État.

89. Stutz, R. M., R. R. Rossi, and A. M. Bowring. Competition between food and rewarding brain shock. *Physiol. Behav.* 7: 753–757, 1971.

90. Thorndike, E. L. The fundamentals of learning. New York: AMS Press, 1971, p. 638.

91. Tinbergen, N. *The Study of Instinct.* Oxford: Clarendon Press, 1951, p. 228.

92. Toates, F. *Biological Foundations of Behaviour.* Milton Keynes, UK: Open University Press, 1986, p. 130.

93. Trowill, J. A. Instrumental conditioning of the heart rate in the curarized rat. *J. Comp. Physiol. Psychol.* 63: 7–11, 1967.

94. Vander Wall, S. B. *Food Hoarding in Animals.* Chicago: University of Chicago Press, 1990, p. 445.

95. Van Iersel, J. J. A. An analysis of the parental behavior of the male three-spined stickleback (*Gastrosteus aculeatus* L.). *Behav. Suppl.* 3: 1–159, 1953.

96. Vogel, S., C. P. Elington, Jr., and D. L. Kilgore. Wind-induced ventilation of the burrow of the prairie dog *Cynomys ludovicianus. J. Comp. Physiol.* 85: 1–14, 1973.

97. Watson, J. B. Psychology as the behaviorist views it. *Psychol. Rev.* 20: 158–177, 1913.

98. Weiss, B., and V. G. Laties. Behavioral thermoregulation. *Science* 133: 1338–1344, 1961.

99. Weiss, T., and B. T. Engel. Operant conditioning of heart rate in patients with premature ventricular contractions. *Psychosom. Med.* 33: 301–321, 1971.

100. Wise, R. A., and D. C. Hoffman. Localization of drug reward mechanisms by intracranial injections. *Synapse* 10: 247–263, 1992.

101. Wood, S. C. Interactions between hypoxia and hypothermia. *Annu. Rev. Physiol.* 53: 71–85, 1991.

102. Woods, S. C. The eating paradox: how we tolerate food. *Psychol. Rev.* 98: 488–505, 1991.

103. Yagil, R. The desert camel. Krager: Comparative animal nutrition series No. 5, 1985.

104. Yahav, S., and R. Buffenstein. Huddling behavior facilitates homeothermy in the naked mole rat *Heterocephalus glaber. Physiol. Zool.* 64: 871–884, 1991.

105. Zajonc, R. B., S. T. Murphy, and M. Inglehart. Feeling and facial efference: implication of the vascular theory of emotion. *Psychol. Rev.* 96: 395–416, 1989.

68. Neuroimmunomodulation: neuroimmune interactions with the environment

NOVERA HERBERT SPECTOR

Division of Fundamental Neurosciences, NINDS, NIH, Bethesda, Maryland, and The American Institute for Neuroimmunomodulation Research, Phoenix, Arizona

SVETLANA DOLINA

San Francisco, California and Rehovot, Israel

GERMAINE CORNELISSEN
FRANZ HALBERG

Chronobiology Laboratories, University of Minnesota, Roseville, Minnesota

BRANISLAV M. MARKOVIĆ

Immunology Research Center, Belgrade, Yugoslavia

BRANISLAV D. JANKOVIĆ

Immunology Research Center, Belgrade, Yugoslavia

CHAPTER CONTENTS

PREFACE

ALL MAMMALS, AND HUMANS in particular, in addition to the physical factors of an environment, must cope also with a variety of other external stressors that act upon the senses (for example, excessive, continuous noise, as well as mental stressors that can arise from within as a result of external psychosocial pressures—disruptions of society by war, floods, earthquakes, isolation, job-related stresses, etc.). It is fair to say that the maintenance of health depends to a significant extent on the ability of the exposed host to respond appropriately and, eventually, to adapt to these stressors. Indeed, it is now well established that inappropriate or maladaptive responses to such stressors weaken the body's resistance to other stimuli from the environment. For example, the stress of bereavement reduces the ability of the immune system to adequately combat invading pathogenic organisms, leading to a greater susceptibility to infections (65). It is thus proper to consider the psychosocial environment as an integral part of the general environment, disturbances of which evoke homeostatic responses. Recognition of this fact over the last 100 years has generated an ever-growing body of evidence, which is clearly demonstrating that the social effects of the environment indeed have an impact on the rest of the body via redundant and reciprocal interactions between the body and the brain. These are linked by the nervous, endocrine, and immune systems (Figs. 68.1, 68.2) and utilize a large array of chemical messengers, including hormones (137), cytokines, neurotransmitters, neuro- and immunomodulators, and other transducers and signaling pathways (42–44, 50, 52, 65, 66, 77, 86, 113, 128, 139, 144, 148, 149, 151, 153, 154).

Clark M. Blatteis, Ed.

INTEGRATIVE PHYSIOLOGY OF NEUROIMMUNOMODULATION (NIM)

Neuroimmunomodulation refers to the continuous interactions between the nervous and immune systems. Table 68.1 lists examples of stimuli from the external and internal environments that can trigger the complex

Integrative Physiology of NIM
Interactions with the External Environment

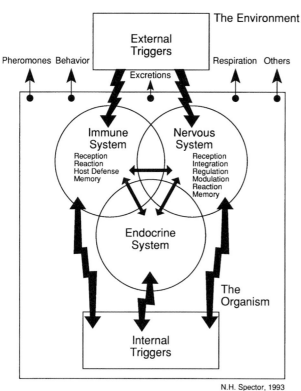

FIG. 68.1. Continuously interacting spheres of physiological functions: all are interacting continuously with the external and internal environments.

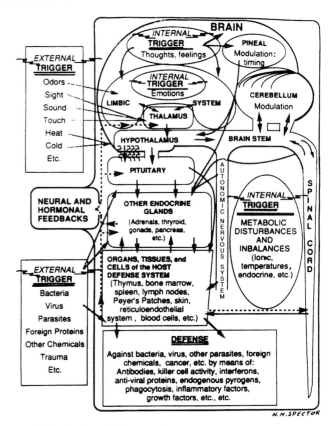

FIG. 68.2. Outline of the internal circuitry in response to and interaction with triggers (stimuli that reach physiological thresholds) from the external and internal environments. Among the factors not shown in this cartoon are the conditions of the substrate (nutrition, chronome: genome, etc.) and the ever-changing conditions of the external and internal environments. (Modified from Spector [151], with permission.)

machinery of NIM. Not all such stimuli need be consistently harmful. A given stressor may have different (and occasionally even opposite) effects at different times. Thus neuroimmunologic reactions to the environment, at the microscopic or the macroscopic level, may be useful (*benetensive*) or harmful (*maltensive*). Organisms, moreover, can volitionally intervene to alter, in some degree, their own relations with their physical, biological, and social environments, thereby minimizing their harmful, "distressful," or maltensive aspects or, alternatively, maximizing their beneficial, "eustressful," or benetensive aspects.

Among the maltensive stimuli that have produced immunosuppression in laboratory animals are high-level noise, unpredictable electric shocks, exposure to unusual lighting or unusual temperatures, disruption of usual housing or feeding conditions, moving to strange surroundings, isolation, crowding, prolonged physical restraint, early separation from maternal care, handling by strangers, the presence of predators, disrupted biorhythms, and many others. Similar results were obtained in experiments with humans. Immunosuppression was measured in students faced with difficult examinations and in people who have lost the love (or

life) of a partner. Other distressful events can trigger various illnesses, as shown in epidemiological comparisons of matched human populations.

We now have abundant evidence that there are functional afferent nerve endings in the tissues of the immune system arising from both the sympathetic and parasympathetic systems (27, 28, 42–45, 50, 66, 71, 72, 84, 87, 116, 118, 153, 162). It is also clear that many neurotransmitters and neuromodulators (including neuroendocrine substances and neuroactive peptides) can drastically change many immune functions (70, 120, 121). However, little is known about the direct afferent connections; the vagus may be a pathway. Indirect evidence for these influences has appeared in the form of unit electrical recordings from cells of the hypothalamus. These changes in frequency and rhythm have been recorded during the course of a peripheral immune challenge (16, 17, 37, 84, 86, 153). Additional evidence for afferent links, but still without clearly delineated wiring networks, comes from experiments showing changes in brain chemistry paralleling or following peripheral immune responses (15, 16, 65–68, 71, 117, 136).

Table 68.1. *Integrative Physiology of NIM*

External Environment External Triggers (Partial List)		Internal Environment Internal Triggers (Partial List)	
Heat	Φ	Abstract thoughts	Φ-Ψ
Cold	Φ	Heat, cold	Φ
Light	Φ	Illness	Φ
Darkness	Φ	Hormonal shifts	Φ-Ψ
Isolation	Σ-Ψ	Ionic shifts	Φ
Crowding	Σ-Ψ	Fear	Ψ-Φ
Hostile gestures	Ψ-Φ	Muscle tension	Φ-Ψ
Friendly gestures	Ψ-Φ	Anxiety	Ψ-Φ
Shock	Φ-Ψ	Joy	Ψ-Φ
Electric	Φ	Lesions, wounds	Φ
Vascular	Φ	Bleeding	Φ
Trauma, injury	Σ-Ψ-Φ	Fight	Σ-Ψ-Φ
Fight	Φ-Ψ-Σ	Flight	Φ
Flight	Σ-Ψ-Φ	Sexual activity	Σ-Ψ-Φ
Sexual activity	Φ	Hunger	Φ-Ψ
Food	Φ	Thirst	Φ-Ψ
Water	Φ	Satiation	Φ-Ψ
Seasonal changes (external clocks)	Ψ-Σ	"Zeitgebers" (internal clocks)	Φ
Changes in family relationships	Ψ-Φ	Intense emotions	Φ-Ψ
Pheromones		Infection	Φ
		Sleep	Φ-Ψ
		Arousal	Φ-Ψ

Interactions between the external and internal environments that characterize NIM. Key: Ψ, psychic factors; Φ, physiological factors; Σ, social factors. There is, of course, considerable overlap among these three categories. Reactions to these triggers are continuously modified by the physiological–chronobiological–nutritional state of the substrate (the internal environment). See also Figure 68.2. (Modified from Spector [151], with permission.)

Environmental Stimuli ("Stressors") and Immunity

As mentioned above, cold, heat, noise, electrical current, restraint, disruption of housing conditions, early separation from maternal care, and magnetic fields are a few of the environmental influences that may affect susceptibility to infections, development of tumors, specific humoral and cell-mediated immune responses, and functions of accessory cells engaged in nonspecific defense mechanisms. The effects of these factors on susceptibility to infections (82) and on tumor development (78) have been reviewed extensively elsewhere. Physiologic effects of changes in ambient temperature are discussed extensively in other chapters of this *Handbook*; in this chapter our comments on temperature are limited to the effects of heat and cold on neuroimmune functions.

Cold. Exposure to low temperatures has been widely used experimentally as a "stressor." Its effect on delayed-type hypersensitivity responses was found to vary as a function of the interval between immunization and cold exposure. Mice kept at the ambient temperature of 5°C for 2 days before immunization showed a decreased delayed-type hypersensitivity response (21). Opposite effects of cold stress on the same type of immune response were found in different animal species. For example, contact sensitivity reactions to dinitrofluorobenzene were enhanced in mice (21) and suppressed in calves (83) following exposure to low ambient temperature. Suppression of the secondary antibody response to human immunoglobulin was found in mice kept at 4°C for 10 days before immunization and throughout the experiment (134). Exposure of mice to 8°C for 1–3 days resulted in a transient increase in antigen uptake by peritoneal cells and a persistent decrease in spleen and liver uptake (30). Altered functions of macrophages, as manifested by increased prostaglandin E_2 secretion, increased interleukin-1 (IL-1) production, decreased Ia-antigen expression, decreased T-lymphocytic mitogenic response, and decreased natural killer cell activity, were described by Jiang and colleagues (75) following exposure of mice to cold water for 1 min, twice daily, for 4 days.

Heat. Exposure to high temperatures has been used as a "stressor" (*e.g.* 159), and again inconsistent results have been reported. Morgan and colleagues (108) found that heat suppressed systemic anaphylaxis in chickens. Acute heat either suppressed (158), enhanced (58), or had no effect (128) on the antibody response in chickens to sheep red blood cells. The same type of immune response reportedly was elevated or suppressed after exposure to heat, depending on the animal species used in the experiment. For example, exposure to an ambient temperature of 35°C for 8–10 days enhanced both delayed-type hypersensitivity and contact sensitivity reactions in mice (21), while in calves these responses were reduced (83).

Heat shock proteins, induced under these stressful

conditions, are often also associated with changes in immune responses, but little detailed information is available as yet.

Noise. There have been numerous studies of the effect of high-level sound on immunity. (Not entirely relevant to the present context are those strong auditory stimuli that induce convulsions in some strains of mice and rats.) Thus mice subjected to subconvulsive audiogenic stimulation showed suppression of turpentine-induced inflammation (140). Jensen and Rasmussen (74) observed that daily 3 h exposures to high-intensity sound increased resistance to passive anaphylaxis. Monjan and Collector (107) reported that high-intensity sound could induce either immunosuppression or immunopotentiation, depending on the length of exposure to the stimulation. They found that 40-day nocturnal exposure to an intermittent high-amplitude auditory stimulus induced depression of T- and B-cell proliferative responses to mitogens during the first 25 days and enhancement of the same response during the last 15 days. This biphasic response may have been due to a simple rebound phenomenon, common in physiologic reactions. A more complex but ubiquitous reason for bi- or multiphasic changes is the influence of rhythmic changes (chronobiology, especially chronoNIM). Adult rats exposed to high-intensity sound showed depression of antibody production to bovine serum albumin (65). Rats subjected to sound stimuli on postnatal days 15, 18, and 21 exhibited decreased Arthus and delayed-type hypersensitivity skin reactions when immunized as adults (143). Exposure to high-intensity sound during the last week of pregnancy suppressed delayed-type hypersensitivity reactions in the offspring (142). An increased incidence of collagen-induced arthritis was found in rats subjected to a sound stimulus, as compared to unexposed controls (133). Bernard and colleagues (14) reported that subjecting rats to a sound stimulus provoked a longer duration of experimental allergic encephalomyelitis with more intense clinical symptoms. Another group, however, reported that the same type of stimulus did not affect the incidence of experimental allergic encephalomyelitis or the severity of clinical signs and histological lesions, except that the disease appeared later in rats that were sound-stimulated after immunization (26).

Electric Current. An early experimental procedure using electrical current as a stimulus involved avoidance learning. In electric footshock experiments, a signal (light or sound) preceded the shock and the animal could learn to avoid actively or passively the aversive stimulus. Avoidance learning as a stress reduced the susceptibility to passive anaphylaxis (127, 161). This type of stimulus was shown to induce prolonged skin allo-graft survival time (168) and to reduce interferon (IFN) synthesis (73) in mice. Croiset and colleagues (33) reported that one-trial passive avoidance learning led to an increase in the proliferative response of rat splenocytes stimulated with concanavalin A (Con A). An immunopotentiating effect, observed as an increased plaque-forming cell response in the spleen of rats subjected to passive avoidance learning, was reported by the same group (32). Rats subjected to avoidance learning, however, exhibited reduced lymphoproliferative responses to mitogens and reduced natural killer (NK) cell activity (65).

Escapable electrical footshock was used in the experiments cited above. Another experimental procedure consists of administering inescapable electrical footshocks, signaled or unsignaled, to animals (e.g. 165). Mettrop and Visser (105) showed that contact sensitivity reactions were more severe in guinea pigs subjected to inescapable footshocks administered before the application of chloro-dinitrobenzine. Further studies demonstrated that inescapable footshock reduced plaque-forming cell responses (at 72 h postshock) and antibody responses to sheep red blood cells in mice (170, 171), reduced NK cell cytotoxicity in rats (138), and reduced secondary antibody response to keyhole limpet hemocyanin and serum immunoglobulin-G (IgG) concentrations in mice (65). Cunnick and colleagues (35) demonstrated that unsignaled electrical footshock suppressed the T lymphocytic mitogenic response to Con A in rats and potentiated this response in mice. The same group reported that signaled footshock induced suppression of NK cell activity in rats. Administering naltrexone before the shock session prevented the suppression of NK activity, suggesting that this effect was mediated by endogenous opioids.

A third experimental procedure using electrical current as a stimulus involves the application of painful electrical shocks to the tail of the animal. It should be noted that this kind of administration has usually been accompanied by another stressor, namely restraint of the animal, and that tailshocks could also be signaled or unsignaled. Tailshock administered to a restrained animal appeared to be an intensive stress. This procedure resulted in suppression of lymphocytic proliferation induced by phytohemagglutinin in rats (80). Similar results were obtained in adrenalectomized rats (81), suggesting that this suppression of immune function was mediated by corticosteroid-independent mechanisms. Suppression of the humoral immune response to keyhole limpet hemocyanin was reported in rats after one 2 h tailshock session (90), while suppression of T-cell mitogenesis, NK cell activity, and production of IL-2 and IFN were demonstrated only after a 19 h tailshock session (166). Decreased proliferative responses to T-cell mitogens, decreased numbers of T- and B-peripheral lymphocytes, and decreased IL-2 production were observed in rats following single or repeated tailshock sessions combined with restraint (12).

The effect of inescapable electric tailshocks on anaphylaxis also has been investigated. The intensity of systemic anaphylaxis in the rat decreased after one or four tailshock sessions (101). The antianaphylactic effect of acute (1 day) tailshock stimulation was dose-dependent; the greatest protective effect was found in animals that received the highest shocking dose of antigen. Tailshock applied for 19 days after immunization suppressed the appearance and development of experimental allergic encephalomyelitis in the rat (26), while the same tailshock performed before immunization had no effect.

Signaled, yet escapable, shock is another experimental model in which electrical current is used as a stimulus. In this model physical stimulation (electrical current) can be separated from psychologic stimulation (inability to predict and control the stimulus). Inescapable electric shock enhanced tumor growth in mice (138) and rats (116), while escapable electrical shock had no effect, although both groups of animals received the same amount of stimulation in physical terms. An experiment dealing more directly with measures of immunity revealed that rats receiving inescapable shocks showed suppression of mitogen-induced proliferation of splenic T lymphocytes, while the mitogenic response of the animals receiving escapable shocks did not differ from controls (92). It seems, therefore, that predictability and controllability of a stimulus are important factors in determining its effect on immunity. In in vitro assays in which mitogen-induced proliferation of lymphocytes and activity of NK cells are investigated, neuronal and/or endocrine mediation of immune functions is restricted to the effects persisting at the moment of killing the animal. Thus for an elucidation of the roles of the psychological vs. the physical components of stressors on the humoral and cellular immune reactions, further in vivo investigations are required.

Restraint (Immobilization).

Immobilization (for example, the physical restraint of an experimental animal) is a strongly distressful and maltensive procedure. Marsh and Rasmussen (102), among others, reported that restraint induced adrenal hypertrophy, leucopenia, and involution of the thymus and the spleen. The timing of immobilization was found to be an important factor in determining the effects.

Physical restraint can be construed to represent a change in the habitual environment of an organism. Indeed, it results in many changes in neural, endocrine, and immune responses (13, 21–23, 46, 50–52, 94, 114, 146, 172). Changes were reported in responses to foreign red blood cells, expression of antigens on macrophages, thymic cells, contact sensitivity, cytotoxic T-cell activity, NK-cell activity, expression of allergic

encephalomyelitis, tumor growth, and survival times in cancer. A number of investigators have observed that forced restraint in rodents leads to a rapid increase in adrenal size; an increase in circulating adrenocorticosteroids; a reduction of various immune responses, including NK-cell activity; increases in brain concentrations of indolealkylamines; a decrease in mitogen responses; and, upon repeated applications, a decreased ability to produce IFN in response to an IFN inducer, poly I:C (151, 153, Spector, unpublished observations). In two different murine cancer models, it was observed that, in the case of osteocarcinoma, restrained mice survived significantly longer than unrestrained controls and that in mice injected with MOPC-104E (myeloma) tumor cells, the growth of tumor cells proceeded more rapidly at first (up to day 43) and then much more slowly so that by day 50 control animals had almost twice as many tumor cells as the restrained ones (46, 47, 146). In another experiment (Spector, unpublished observation) mice under repeated daily restraint retained their ability to produce IFN in response to poly I:C injections, in contrast to several nonrestrained control groups, all of which lost the response to poly I:C within 10 days. Apparently, a given stimulus, which may be maltensive under some conditions, may, under others, become benetensive, depending on the response and when it is measured.

Restraint also may produce biphasic (or multiphasic) responses. For example, restraint in mice initially induced increased blood levels of corticosterone, norepinephrine, and plaque-forming cells in spleen and bone marrow, while reverse effects on all of these responses were observed after 3 days (85).

Psychosocial Factors.

Psychosocial stimuli are ubiquitous in human experience, and they may lead to immunosuppression or immunoenhancement, depending on the nature of the stimulus. For example, mitogen-induced lymphocytic proliferation was found to be lower during the first 2 months postbereavement and to return to prebereavement levels in most of the (human) subjects within a year. In a group of patients hospitalized with acute major depressive disorders, a decrease in total number of T and B cells was found (156).

Experiments on monkeys were designed to simulate events that may occur in human life. Monthly repeated group reorganizations altered the stable social conditions in group-housed cynomolgus macaques (*Macaca fascicularis*). An increase in an antibody titer with a lower total IgG concentration was found in monkeys exposed to chronic social reorganization, compared to the socially stable group. In addition, subordinate animals (those who were defeated by all other animals of the group) had a greater primary antibody response to tetanus toxoid, along with a lower total serum IgG. A lack of habituation to chronic changes was demonstrated: ten monthly repeated reorganizations still affected immunity (34). The formation of a new social group of eight unfamiliar female rhesus monkeys with an adult male led to immunosuppression as the predominant response to this potent psychosocial stimulus. A dominance hierarchy was established within 48 h. An increase in circulating cortisol and a decrease in the total number of lymphocytes and CD4+ and CD8+ T cells were found at 24 h after formation of the hierarchy. These indices were restored to "normal" values much faster in high-ranking than in low-ranking monkeys. Low-ranking animals showed lower values of CD4+ and CD8+ T cells for up to 9 weeks (53).

Housing Conditions: Isolation and Overcrowding.

Housing conditions have been shown to influence immune responsiveness. Vessey (164) reported an increased antibody response to protein antigen in socially isolated mice. However, mice housed singly exhibited decreased antibody production to bovine serum albumin compared with group-housed mice (49). It was found that overcrowding for 1 week prior to immunization with flagellin lowered antibody production in rats (136). Suppressed splenic and peripheral blood NK activity was reported in rats after 1 week of isolation. Suppression was more pronounced in 12-month-old than in 3-month-old animals (48). Edwards et al. (41) showed that overcrowding of mice lowered humoral antibody formation in response to typhoid–paratyphoid vaccine. A change in housing conditions from isolation to group housing decreased the antibody response to bovine serum albumin. Rabin and Salvin (124) reported that male mice kept in isolation showed an increased plaque-forming cell response, an increased T-cell response to mitogens, and an increased IL-2 production by comparison with group-housed animals.

The lymphoproliferative response to thyroglobulin was increased in female rats subjected to overcrowding for 5 weeks before immunization. This response was decreased in rats kept in isolation (76). Rabin and Salvin (124) reported that isolation of male mice induced a transient potentiation of lymphocytic mitogenic responses to Con A and an increase in plaque-forming cell responses. Peritoneal macrophages from individually housed mice showed increased activity, as measured by phagocytosis and production of IL-1 (136). Isolation of experimental animals, apart from affecting immune reactivity, markedly influenced behavior (57).

The relationship between housing conditions and autoimmunity was the subject of two reports. Overcrowding of male rats increased the intensity and accelerated the peak of adjuvant-induced arthritis (8). In a subsequent study overcrowding was followed by an increase in the production of antimyelin basic protein autoantibodies (38).

Handling.

Even gentle handling of experimental animals has been shown to act as a stress. Handling and transportation of rats beginning 3 days before immunization with type II collagen decreased the incidence of arthritis (133). Experimental manipulations performed early in life, such as preweaning handling, preweaning maternal deprivation, and premature weaning, affected immune functions of animals. Daily handling of rats from birth to weaning enhanced primary and secondary responses to flagellin at 10–12 weeks of age (145). However, early handling of mice led to a decreased plaque-forming cell response at 50–90 days of age (128). Daily handling during the preweaning period significantly enhanced the mitogenic response of splenic T and B lymphocytes of mice at 60 days of age but did not have an effect when mice were tested at 21 days of age; early maternal deprivation and premature weaning decreased the number of plaque-forming cells at 10 weeks of age (106). Separation of infant monkeys from their biological mothers suppressed lymphocytic proliferation in response to mitogens (91). Premature separation of rat pups from their mothers induced a decreased response of peripheral blood lymphocytes to phytohemagglutinin at 40 days of age (1).

In various laboratories, under varying conditions and depending on a number of other factors, such as chronome stage (time of day, week, month, etc.), prior experiences, nutritional status, etc., handling may produce a variety of neuroendocrine-immune changes. As with other stimuli, depending on such circumstances, handling can produce even opposite effects. For example, "gentle" handling resulted in progressively higher corticosterone concentrations as the procedure was repeated day after day (141).

Transportation and Vibration.

Moving animals from one housing situation to another can also modify their behavioral and physiological responses to such an extent that failure to take into account the environmental history of an animal can confound the results of any experiment. While this is well known anecdotally, it has been a more difficult task to demonstrate scientifically. In the preceding section, a few examples of demonstration of one type of environmental–social manipulation were given. The economic consequences of transportation effects upon the health of farm animals are of great importance (for example, "shipping fever" in calves), yet there are only a few examples in the literature clearly showing such effects. In Czechoslovakia and some other European countries, in the 1970s and 1980s, livestock was kept in very large, mostly indoor, holding barns. Calves at the very early age of 8–20 days were transported to other facilities, often mixed in trucks with calves from other farms. Many animals died of respiratory infections and/or diarrhea. Rasková and colleagues (125) noted that, following transport, these animals appeared more agitated than the nontransported ones. They walked, moved their tails, and kicked considerably more than did their nontransported cohorts. When metipranol (trimepranol; Sofia, Prague), a beta-adrenergic blocking agent, was administered, 2 mg/kg orally, the signs of agitation were significantly reduced. Subsequently, turguride was employed for the same purpose. Turguride, an ergoleine derivative, is an adrenolytic, serotoninolytic, and dopamine-antagonistic drug. These successful treatments for transported calves were followed by attempts to develop parallel models of "transportation stress" in mice and rabbits. Vibration could produce similar symptoms and could be alleviated with the same type of treatment. They also measured changes in various immune responses. In both untreated calves and vibrated mice leukopenia and changes in phagocytic activity were observed. The same adrenolytic drugs, given prophylactically, had an acute sedative effect on behavior and ameliorated changes in delayed-type hypersensitivity even when tested weeks after exposure to the stresses. The improvement in immunological functions was correlated with a reduction in the incidence of respiratory infections in the calves (126). Both behavioral and immunological effects were confirmed by others (93). Among other effects of vibration in the laboratory animals, Rasková and colleagues (unpublished) noted analgesia, suggesting a possible role for endogenous opioids. The significance of opioids in NIM has been reviewed elsewhere (121, 122).

The formation of free radicals, *peroxidation*, is considered, in most cases, to be damaging to the immune system. In 1991, Lavicky and Rasková (93) compared the consequences of i.v. injections of endotoxin or peptidoglycan with those of vibration in mice and rabbits. Both toxins induced an increase in lipid peroxidation, while the vibration was associated with some scavenging of free radicals. This perhaps unexpected result suggests that caution must be observed and oversimplification avoided in predicting the physiological effects of any environmental stimulus.

Gravitational and Positional Stimuli

The interaction of gravitational fields and physiological systems has been treated in detail in other chapters in this *Handbook*. Here, we shall discuss only effects upon the immune system. Our cartoon of environmental triggers that set off cascades of NIM reactions within the body (Fig 68.2) is, of course, an oversimplification and, in the interest of clarity, understates the roles of the cerebellum and the vestibular nervous system (central and peripheral) in immunity. It is safe to assume that some of the effects of transportation stress and vibration stress, discussed in the preceding section, are mediated,

at least in part, by stimuli to proprioceptors in muscles, the semicircular canals, and baroreceptors in the vascular system and elsewhere and are relayed via vestibular and other afferents to the central nervous system (CNS) and via neural efferents and hormones, communicated to the organs, tissues, and cells of the immune system.

Changes in position, acceleration, deceleration, rocking, swinging, and turning motions may influence immune status. These effects have been documented chiefly by the experiments of Maiti (98–100), who states:

> we designed a benign stress-reducing device . . . inducing non-invasive vestibular stimulation in animals, which would allow the animal to rotate or swing in a circular or perpendicular horizontal motion with varying speeds while the maximum gravitational force . . . was less than one g. . . . The preliminary observation . . . showed . . . some suppressive influences . . . in tumor development. (98)

It is interesting to note here that rotation of mice by Riley (132) at a considerably higher speed resulted in immunosuppression, elevation of adrenal steroids, increases in viral infections, and other deleterious effects. Indeed, unlike most investigators of "stress," Riley used a definition of stress that was quantitative; that is, in his use of the term, stress could be measured by the number and speed of rotations. If we look at Maiti's data together with Riley's, we can see, as in so many other physiological–immunological responses to environmental factors, a biphasic response curve: very slow rotational (vestibular) stimulation can be beneficial, higher rates of rotation harmful. Interestingly, Maiti (personal communication, 1992) found that, while vestibular stimulation in medulla-lesioned rats accelerated the growth of chemically induced mammary tumors, the same stimulus in nonlesioned rats "showed suppressive influence to DMBA-induced tumorigenesis." In addition, rotation of normal animals in one direction only resulted in leukopenia, while rats rotated in alternating directions developed leukocytosis.

Maiti extended his early studies of the physiological effects of rocking, swinging, balancing, and rotation in rats and cats to later clinical experiments with humans. His many papers on this subject will not be reviewed here, but two examples of vestibular–postural gravity effects on human immunity are worth mentioning. Adult human volunteers, aged 24–36, male and female, were given vestibular stimulation by rotation in alternating directions for 15 days, 15 min/day. Total and differential leukocyte counts from arterial blood were made, and IgA from saliva was measured. At the end of the 15 days, lymphocyte numbers increased 25%–35% and IgA was elevated from 56 mg/dl to 68 mg/dl (99). Similar results were obtained in studies of autistic and normal children after vestibular stimulation by swinging (100). Chronically ill cancer patients, after vestibular stimulation, reported "notable relief" of the pain, nausea, vertigo, and anorexia of the chemotherapeutic toxic syndrome (99).

Magnetic Fields.

Are magnetic fields maltensive, benetensive, or neither? Under natural conditions, all living beings on Earth are constantly and inescapably exposed to magnetic waves. Since the Earth's magnetic forces of 0.5 Gauss are normal constituents of life, many regard them as not harmful to biological systems. Yet experiments with the whole body, organs, and cells of animals and plants have demonstrated that biological processes are magnetosensitive (e.g. 112, 134). However, most experiments have used magnetic forces of a strength much higher than that coming from the natural environment.

At the biochemical level, it has been shown that magnetic fields can induce polymerization of macromolecules (157), alterations in DNA synthesis (95), disregulation of intracellular concentrations of ions (147), changes of membrane fluidity (88), etc. At the cellular level, magnetic fields were found to induce changes in membranes (111), cytoskeleton, mitochondria, and nuclei (18) and at the cellular level in bacteria (19), flatworms (24), muscle fibers (9), retinal cells (31), and erythrocytes (111). At the functional level, magnetic fields modified circadian rhythms, body temperature, food consumption, and survival rate, among other variables (25, 54). The effects of environmental magnetic and electromagnetic fields on the immune, nervous, endocrine, hemopoietic, cardiovascular, and reproductive systems have been described (7, 123, 157). Of particular interest, within the scope of this article, are chemical, physiological, and behavioral changes associated with exposure of the CNS to magnetic waves and magnetosensitivity of the cerebral cortex (10, 131), as well as changes of immune functions such as resistance to infection (163), elevated antibody production (76), increased number of antibody-producing cells (36), augmented T- and B-cell mitogenic responses, and others. Most of these experiments dealt with repeated exposure of the whole animal to static or electromagnetic fields at different time intervals.

The sensitivity of the CNS to magnetic fields (7) and its participation in mechanisms underlying immune functions (71, 72) motivated experiments in which static magnetic forces were investigated in the context of neuroimmune interactions (70). For that purpose, micromagnets, with influx density of 60 mT (600 Gauss), were fixed to the frontal, parietal, or occipital regions of the rat skull. Animals bearing micromagnets showed a significant increase in several immune reactions: plaque-forming cell response, hemagglutinin production, elaboration of antibodies against bovine serum

albumin, Arthus and delayed-type hypersensitivity skin reactions to bovine serum albumin, and experimental allergic encephalomyelitis. In addition, there was a marked increase in CD4+ and a decrease in CD8+ lymphocytes in the peripheral blood. The greatest improvements were observed after exposure of the occipital region to micromagnets for 24 days. Thus these immunoenhancing effects of magnetic fields were dependent on the duration of exposure. They varied also as a function of placement, that is, which brain region received the maximal exposure.

Several explanations can be offered for these results. First, a magnetic field may directly affect circulating lymphocytes since these cells are magnetosensitive (64). However, this explanation does not fit the finding that micromagnets, applied frontally, parietally, and occipitally, exerted a differential influence on peripheral blood lymphocytes. Second, magnetic fields may show an indirect effect on lymphocytes by acting on certain brain structures (hypothalamus, reticular formation, limbic system) and the hypopohysis, and thereby on lymphocytes. Third, since both the cell populations (131) and melatonin synthesis (115) in the pineal gland are sensitive to magnetic fields, it may well be that micromagnets accomplish their immunopotentiating activity via the pineal gland. Fourth, magnetic stimuli have been shown to influence the opioid system and its relevant receptors (79), and opioid peptides play an important role in the regulation/modulation of the immune system (69–72, 121). Therefore, magnetic forces may exert their activity throughout the CNS–opioid system. Thus although the mechanisms of magnet-induced immunopotentiation are unclear, the excitatory immune phenomena produced by prolonged exposure of parts of the brain to micromagnetic fields provide further evidence for the complex interactions between the CNS and the immune system and for the constant impingement of environmental changes upon these complex functions of living organisms.

ENDO- AND EXOENVIRONMENTAL RHYTHMS

Rhythms, once viewed as a source of error or as "noise" in data, are now recognized as valuable tools. Another chapter in this volume deals with biorhythms having a frequency of approximately 24 h (circadian). There are, in addition, rhythms of many diverse frequencies: infradian (less than 24 h periods), ultradian (periods greater than 25 h), circaseptan (7-day), circannual, life-span, cosmic, etc. These rhythms may originate in the internal (endo-) or external (exo-) environments and may be biological or nonbiological (that is, physicochemical).

As already discussed, NIM is a feature of both integration within the organism and adaptation to its environment (Fig. 68.1), with the changes modulated from within and from without and occurring mostly within the range of usual (that is, "normal") values. Table 68.1 does not stress chronobiological factors. One should distinguish, however, among benetensive, maltensive, and seemingly neutral (*transtensive*) stimuli and be aware that a stimulus that may be benetensive during one phase of a time cycle (chronome) may very well be maltensive or transtensive during another phase. In the context of NIM, chronomes that characterize behavioral, biochemical, and biophysical (including electrophysiological) variables, some yet to be elucidated, may modulate the nervous, endocrine, circulatory, and immune systems, as well as the very difficult-to-assess mind and psyche. Knowledge of the temporal details of a condition or treatment, therefore, can be essential (55). For example, it was shown (11) that melatonin, the major secretory product of the pineal gland, inhibits, stimulates, or has no effect upon tumor growth and survival as a function of the timing of its administration. The possible time-dependence of melatonin effects was investigated on 115 CD_2F_1 female mice, housed four per cage, on a staggered schedule of light and darkness alternating at 12 h intervals. All animals were injected intraperitoneally with 5,000 live L1210 leukemia cells. Starting 72 h later and continuing until death, 1 mg/kg of melatonin or saline was injected subcutaneously daily at one of six circadian stages. A third group of mice was handled for weighing only every other day, rather than daily, as were the other two groups. The effect of melatonin was shown to be circadian stage-dependent (56, 89). Thus melatonin injected in the evening had a statistically significant beneficial effect, but injected in the morning it either had no effect or was slightly deleterious, as compared to the saline controls. The importance of circaseptan and circannual rhythm stages was also suggested by another, similar study (168). These findings are relevant to NIM in that the pineal gland is affected externally, for example, by light (130) and magnetic fields (70), and internally by several hormones (29) and sympathetic inputs (160, 169) and is classically viewed as coordinating influences on several aspects of endocrine, immune, and metabolic activities (20, 97, 119).

As a result of more than 20 years of study of the pineal and its hormone, W. Pierpaoli and his colleagues have called melatonin "the aging clock," perhaps the ultimate example of NIM and the internal and external environments.

CONDITIONING AND NIM

All living things are continually conditioned by their environments. In the 1920s and 1930s, at the Pasteur Institute in Paris, a series of ingenious experiments was carried out by Metal'nikov and his students (103, 104,

148) on rabbits and guinea pigs. Bacteria or bacteria-derived substances were injected while the animals' skin was scratched. Tests of the animals' blood showed that they were producing antibodies to the bacteria. After repeated pairings of the two stimuli, it was demonstrated that the skin-scratching alone, with no bacterial injection, could elicit the antibody response. Successes were reported later in the conditioning of both specific and nonspecific reactions. In Russia, Dolin and colleagues (39, 40) demonstrated both immunosuppression and immunoenhancement. Among the experimental animals were mice, rats, guinea pigs, rabbits, dogs, oxen, monkeys, and humans. Among the antigenic substances employed were viruses; red blood cells from other species; albumin; malarial parasites and other whole organisms; various vaccines; extracts from salmonellae; dysentery amoebae; and bacilli, including typhoid, paratyphoid, diphtheria, *Escherichia coli*, and staphylococci (96). In the earlier experiments (39) immunoenhancement was obtained by pairing vaccine of a 24 h culture of Garther's paratyphoid (*B. enteritis* Garthneri) as the unconditioned stimulus with confinement in a box as the conditioned stimulus. For rabbits, the vaccine was injected in gradually increasing doses, first after 30 min, then in gradual steps after 3 h of confinement in a box. The animal remained in the box for another 3 h after the vaccine injection. Twelve such conditioned–unconditioned stimuli trials were done in 1 month. After a 1-month rest, rabbits were reexposed to the conditioned stimuli and injected with saline rather than the vaccine. Consistent elevation in agglutinin titer in response to the saline was obtained. Conditioned immune responses (CIR) were stable and resistant to extinction for more than 240 days. Later analysis by others showed that the results of some of these studies were highly statistically significant (2).

Subsequently, new immune-conditioning techniques have been demonstrated. One series of experiments showed that NK-cell activity could be either enhanced greatly or depressed, depending on the design of the experiment (46, 47, 146, 148, 150, 152). It was also shown that prior conditioning could be employed to prolong significantly the life of mice injected with cancer cells. Indeed, several mice thus treated showed a reversal of cancer growth and some remained free of the otherwise lethal cancer for as long as they were observed (47, 150, 152). Other experiments showed that by conditioned immunosuppression mice with an autoimmune disease could enjoy longer life spans than their untreated controls (4). Experiments in Italy have demonstrated that very old mice can increase their NK activity (and thus anticancer and antiviral activities) fourfold or more, after conditioning by the "Spector technique" (150) in response to the conditioned stimulus alone (155).

There are now numerous published reports of CIR,

which have confirmed and broadened the pioneering experiments of Metal'nikov and Chorine (103, 104) and Dolin et al. (39, 40, 96). Newer and much more elegant models than those initial, old-fashioned ones have been developed. These new studies have filled gaps and shortcomings in the initial studies by more elaborate controls and modern statistics. Conditioning stimuli such as odors or even the odor of disturbed animals can elicit conditioned reflexes (46, 146, 148).

Conditioned immunosuppression has been demonstrated also by Ader and colleagues (2–6, 109). Using a taste-aversion model, these workers reported that immunosuppression could be conditioned to saccharine (SAC) as the stimulus that had previously, even once, been paired with cyclophosphamide (CY, an immunosuppressive drug) as the unconditioned stimulus. Lowered antibody reactions to thymus-dependent and thymus-independent antigenic stimulation were found in animals reexposed to SAC after SAC was paired with CY. Conditioned immunosuppression was not observed if lithium chloride, an illness-inducing but not immunosuppressive drug, was used as the unconditioned stimulus. A diminished IgG antibody forming the cellular response to red blood cells, using the same protocol, was also reported.

Although the mechanisms by which the CNS may regulate and/or modulate the immune process remain partially unknown, it is evident that associative processes play an important role in immune functioning in the natural environment. When comparing the data obtained both in early and later research, it further becomes evident that CIRs are characterized by rapid acquisition, intensity, stability, and partial resistance to extinction and are influenced by multilinked regulatory and/or coordinating circuits connecting the neuroendocrine and immune systems, that is, by those circuits which are activated by immunological challenge. The details of these circuits and their mechanisms are under investigation. Involved in the afferent loop is IFN; various neurotransmitters are required for integration within the CNS; and endogenous opioids are essential for the efferent loop—that is, the conditioned response is blocked by naloxone or naltrexone (59–63).

CONCLUSIONS

Any environmental stimulus to the nervous system will have resultant effects upon the immune system and vice versa. In most of these reactions the endocrine system plays an indispensable role. As each cell interacts with other cells in other systems, so does each cell interact with its microenvironment; and each organism is continually interacting with the macroenvironment around it. This macroenvironment is physical, chemical, biolog-

ical, social, and cosmic. Both macro- and microenvironments are constantly changing, usually in rhythmic patterns. The vast and rapid accumulation of scientific evidence pointing to the prevalence of NIM has emphasized the integrative nature of physiology (44, 65, 148, 153, 154). In this chapter, we have scanned only a portion of this evidence, illustrating only some of the aspects of NIM that deal with the environment; but, despite the already enormous contributions to the understanding of NIM from every branch of biomedical science, most of the research still lies ahead. We invite the reader to join in the fun.

We thank Joan D. Levin and Helena Rašková for their critical reading of the manuscript, Cleveland Cooper III and Joan D. Levin for their assistance in preparing this paper, Ione Auston for bibliographic help, and the editors of this *Handbook* for their patient encouragement and support.

REFERENCES

1. Ackerman, S. H., S. E. Keller, S. J. Schleifer, R. D. Shindledecker, M. Cameribo, M. A. Hofer, H. Weiner, and M. Stein. Premature maternal separation and lymphocyte function. *Brain Behav. Immun.* 2: 161–165, 1988.
2. Ader, R. A historical account of conditioned immunobiologic responses. In: *Psychoneuroimmunology*, edited by R. Ader. New York: Academic, 1981, p. 321–352.
3. Ader, R., and N. Cohen. Behaviorally conditioned immunosuppresion. *Psychosom. Med.* 37: 333–340, 1975.
4. Ader, R., and N. Cohen. Behaviorally conditioned immunosuppression and systemic lupus erythematosis. *Science* 124: 1534–1536, 1982.
5. Ader, R., and N. Cohen. CNS-immune interactions: conditioning phenomenon. *Behav. Brain Sci.* 8: 379–426, 1985.
6. Ader, R., N. Cohen, and L. J. Grota. Adrenal involvement in conditioned immunosuppression. *Int. J. Immunopharmacol.* 1: 141–145, 1979.
7. Adey, W. R. Tissue interaction with nonionizing electromagnetic fields. *Physiol. Rev.* 61: 435–514, 1981.
8. Amkraut, A. A., G. F. Solomon, and H. C. Kramer. Stress, early experience and adjuvant-induced arthritis in the rat. *Psychosom. Med.* 33: 203–214, 1971.
9. Arnold, W., R. Steele, and H. Mueller. On the magnetic asymmetry of muscle fibers. *Proc. Natl. Acad. Sci. USA* 44: 1–4, 1958.
10. Barker, A. T., and R. Jalinous. Non-invasive magnetic stimulation of human motor cortex. *Lancet* 1: 1106–1107, 1985.
11. Bartsch, H., and C. Bartsch. Effect of melatonin on experimental tumors under different photoperiods and times of administration. *J. Neur. Transm. Gen.* 52: 269–279, 1981.
12. Bautman, O. A., D. Sajewski, J. E. Ottenweller, D. L. Pitman, and B. H. Natelson. Effects of repeated stress on T cell numbers and function in rats. *Brain Behav. Immun.* 4: 105–117, 1990.
13. Ben-Eliyahu, S., R. Yirmiya, J. C. Liebeskind, A. N. Taylor, and R. P. Gale. Stress increases metastic spread of a mammary tumor in rats: evidence for mediation by the immune system. *Brain Behav. Immun.* 5: 193–205, 1991.
14. Bernard, C. C. A., E. Grgacic, and G. Singer. The effect of stress, corticosterone and adrenalectomy on the development of exper-

imental autoimmune encephalomyelitis. *J. Neuroimmunol.* 17: 254–258, 1988.

15. Besedovsky, H., A. del Rey, E. Sorkin, M. da Prada, R. Burri, and C. Honegger. The immune response evokes changes in brain noradrenergic neurons. *Science* 221: 564–566, 1983.

16. Besedovsky, H., E. Sorkin, D. Felix, and H. Haas. Hypothalamic changes during the immune response. *Eur. J. Immunol.* 7: 325–328, 1977.

17. Birmanns, B., D. Saphier, and O. Abramsky. β-interferon modifies cortical EEG activity: dose-dependency and antagonism by naloxone. *J. Neurol.* 100: 22–26, 1990.

18. Bistolti, F. Classification of possible targets of interaction of magnetic fields with living matter. *Panminerva Med.* 29: 71–73, 1987.

19. Blakemore, R. P. Magnetotactic bacteria. *Science* 190: 377–379, 1975.

20. Blask, D. E. The pineal: an oncostatic gland? In: *The Pineal Gland*, edited by R. J. Reiter. New York: Raven, 1984, p. 253–284.

21. Blecha, F., A. Barry, and K. W. Kelley. Stress-induced alternations in delayed type hypersensitivity to SRBC and contact sensitivity to DNFB in mice. *Proc. Soc. Exp. Biol. Med.* 169: 239–246, 1982.

22. Bonneau, R. H., J. F. Sheridan, N. Feng, and R. Glaser. Stress-induced effects on cell-mediated innate and adaptive memory components of the murine immune response to herpes simplex virus infection. *Brain Behav. Immun.* 5: 274–295, 1991.

23. Bonneau, R. H., J. F. Sheridan, N. Feng, and R. Glaser. Stress-induced suppression of herpes simplex virus (HSV)-specific cytotoxic T lymphocyte and natural killer cell activity: an enhancement of acute pathogenesis following local HSV infection. *Brain Behav. Immun.* 5: 170–192, 1991.

24. Brown, F. A., and Y. H. Park. Duration of an after-effect in planarians following a reversed horizontal magnetic vector. *Biol. Bull.* 128: 347–355, 1965.

25. Brown, F. A., and K. M. Scow. Magnetic induction of a circadian cycle in hamsters. *J. Interdisc. Cycle Res.* 9: 137–145, 1978.

26. Bukilica, M., S. Djordjević, I. Maric, M. Dimitrijević, B. M. Marković, and B. D. Janković. Stress-induced suppression of experimental allergic encephalomyelitis in the rat. *Int. J. Neurosci.* 59: 167–175, 1991.

27. Bulloch, K. Neuroanatomy of lymphoid tissue: a review. In: *Neural Modulation of Immunity*, edited by R. Guillemin, M. Cohn, and T. Melnechuk. New York: Raven, 1985, p. 111–141.

28. Calvo, W. The innervation of the bone marrow in laboratory animals. *Am. J. Anat.* 123: 315–328, 1968.

29. Cardinali, D. P., and M. I. Vacas. Mechanism underlying hormone effects on pineal function: a model for the study of integrative neuroendocrine processes. *J. Endocrinol. Invest.* 1: 89–96, 1976.

30. Casey, F. B., J. Eisenberg, D. Peterson, and D. Pieper. Altered antigen uptake and distribution due to exposure to extreme environmental temperatures or sleep deprivation. *J. Reticuloendothel. Soc.* 15: 87–95, 1974.

31. Changeux, R., H. Changeux, and N. Chalazonitis. Disease in magnetic anisotropy of external segments of the retinal rods after total photolysis. *Biophys. J.* 18: 125–128, 1977.

32. Croiset, G., R. E. Ballieux, D. De Wied, and C. J. Heijnen. Effects of environmental stimuli on immuno-reactivity: further studies on passive avoidance behavior. *Brain Behav. Immun.* 3: 138–146, 1989.

33. Croiset, G., H. D. Veldhuis, R. E. Ballieux, D. De Wied, and C. J. Heijnen. The impact of mild emotional stress induced by the passive avoidance procedure on immune reactivity. *Ann. N. Y. Acad. Sci.* 496: 477–484, 1987.

34. Cunnick, J. E., S. Cohen, B. S. Rabin, A. B. Carpenter, S. B. Manuck, and J. R. Kaplan. Alterations in specific antibody production due to rank and social instability. *Brain Behav. Immun.* 5: 357–369, 1991.

35. Cunnick, J. E., D. T. Lysle, A. Armfield, and B. S. Rabin. Shock-induced modulation of lymphocyte responsiveness and natural killer activity: differential mechanisms of induction. *Brain Behav. Immun.* 2: 102–113, 1988.

36. Czerski, P. Microwave effects on the blood-forming system with particular reference to the lymphocyte. *Ann. N. Y. Acad. Sci.* 2476: 232–242, 1975.

37. Dafny, N., R. Prieto-Gomez, and C. Reyes-Vazquez. Does the immune system communicate with the central nervous system? Interferon modifies central nervous activity. *Neuroimmunology* 9: 1–12, 1985.

38. Djordjević, S., M. Bukilica, M. Dimitrijević, B. M. Marković, and B. D. Janković. Anti-myelin basic antibodies in rats stressed by overcrowding. *Ann. N. Y. Acad. Sci.* 650: 302–306, 1992.

39. Dolin, A. O., and V. N. Krylov. Role of the brain in the organism's immune responses. *Zh. Vyssh. Nerv. Deiat.* 2: 547–560, 1952.

40. Dolin, A. O., B. N. Krylov, V. I. Lukyanenko, and B. A. Flerov. New experimental data on conditioned reflex reproduction and suppression of immune and allergic reactions. *Zh. Vyssh. Nerv. Deiat.* 10: 832–841, 1960.

41. Edwards, E. A., R. H. Rahe, P. M. Stephens, and J. P. Henry. Antibody response to bovine serum albumin in mice: the effects of psychosocial environmental change. *Proc. Soc. Exp. Biol. Med.* 164: 478–481, 1980.

42. Fabris, N. *Immunoregulation.* New York: Plenum, 1983.

43. Fabris, N., B. D. Janković, B. N. Marković, and N. H. Spector. Ontogenetic and phylogenetic mechanisms of neuroimmunomodulation: from molecular biology to psychosocial sciences. *Ann. N. Y. Acad. Sci.* 650: 1–369, 1992.

44. Fabris, N., B. M. Marković, N. H. Spector, and B. D. Janković. Neuroimmunomodulation: the state of the art. *Ann. N. Y. Acad. Sci.* 741: 1–372, 1994.

45. Felten, S. Y., and D. L. Felten. Innervation of lymphoid tissue. In: *Psychoneuroimmunology* (2nd ed.), edited by R. Ader, D. L. Felten, and N. Cohen. New York: Academic, 1991, p. 27–69.

46. Ghanta, V., R. Hiramoto, H. Solvason, and N. H. Spector. Neural and environmental influences on neoplasia and conditioning of NK activity. *J. Immunol.* 135: 848s–852s, 1985.

47. Ghanta, V., R. N. Hiramoto, B. Solvason, and N. H. Spector. Influence of conditioned natural immunity on tumor growth. *Ann. N. Y. Acad. Sci.* 496: 637–646, 1987.

48. Ghoneum, M., G. Gill, P. Assanah, and W. Stevens. Susceptibility of natural killer cell activity of old rats to stress. *Immunology* 60: 461–465, 1987.

49. Glenn, W. G., and R. E. Becker. Individual versus group housing in mice: immunological response to time and phase injection. *Physiol. Zool.* 42: 411–416, 1969.

50. Goetzl, E., and N. H. Spector. *Neuroimmune Networks: Physiology and Disease.* New York: Liss, 1989.

51. Griffin, A. C., and C. C. Whitacre. Differential effects of stress on the disease outcome in experimental autoimmune encephalomyelitis. *FASEB J.* 4: A2038, 1990.

52. Guillemin, R., M. Cohn, and T. Melnechuk. *Neural Modulation of Immunity.* New York: Raven, 1983.

53. Gust, D. A., T. P. Gordon, M. E. Wilson, A. Ahmed-Ansari, A. R. Brodie, and H. M. McClure. Formation of a new social group of unfamiliar female rhesus monkeys affects the immune and pituitary adrenocortical systems. *Brain Behav. Immun.* 5: 296–307, 1991.

54. Halberg, F. Chronobiology. *Annu. Rev. Physiol.* 31: 675–725, 1969.

55. Halberg, F., G. Cornelissen, R. B. Sothern, L. A. Wallach, E.

Halberg, A. Ahlgren, M. Kuzel, A. Radke, J. Barbosa, F. Goetz, J. Buckley, J. Mandel, L. Schuman, E. Haus, D. Lakatua, L. Sackett, H. Berg, H. W. Wendt, T. Kawasaki, M. Ueno, K. Uezono, M. Matsuoka, T. Omae, B. Tarquini, M. Cagnoni, M. Garcia Sainz, E. Perez-Vega, D. Wilson, K. Griffiths, L. Donati, P. Tatti, M. Vasta, I. Locatelli, A. Camagna, R. Lauro, G. Tritsch, and L. Wetterberg. International geographic studies of oncological interest on chronobiological variables. In: *Neoplasms—Comparative Pathology of Growth in Animals, Plants, and Man*, edited by H. Kaiser. Baltimore: Williams and Wilkins, 1981, p. 553–596.

56. Halberg, F., S. Sanchez de la Peña, L. Wetterberg, J. Halberg, F. Halberg, H. Wrba, A. Dutter, and R. C. Hermida-Dominguez. Chronobiology as a tool for research, notably on melatonin and tumor development. In: *Biorhythms and Stress in the Physiopathology of Reproduction*, edited by P. Pancheri and L. Zichella. New York: Hemisphere, 1988, p. 131–175.

57. Hatch, A., T. Balazs, G. S. Wiberg, and H. C. Grice. Long-term isolation stress in rats. *Science* 142: 507–508, 1963.

58. Heller, E. D., D. B. Nathan, and M. Perek. Short heat stress as an immunostimulant in chicks. *Avian Pathol.* 8: 195–203, 1979.

59. Hiramoto, R. N., V. K. Ghanta, J. F. Lorden, H. B. Solvason, S-J. Soong, C. F. Rogers, C.-M. Hsueh, and N. S. Hiramoto. Conditioning of the NK cell response: effect of interstimulus intervals and evidence for long delayed learning. *Prog. Neuroendocrinimmunol.* 5: 13–20, 1992.

60. Hiramoto, R. N., V. K. Ghanta, C. F. Rogers, and N. S. Hiramoto. Conditioned fever: a host defensive reflex response. *Life Sci.* 49: 93–99, 1991.

61. Hiramoto, R. N., V. K. Ghanta, H. B. Solvason, J. F. Lorden, C.-M. Hsueh, C. F. Rogers, S. Demissie, and N. S. Hiramoto. Identification of specific pathways of communication between the CNS and NK cell system. *Life Sci.* 53: 527–540, 1993.

62. Hiramoto, R. N., H. B. Solvason, V. K. Ghanta, J. Lorden, and N. S. Hiramoto. Effect of reserpine on retention of the conditioned NK cell response. *Pharmacol. Biochem. Behav.* 36: 51–56, 1990.

63. Hsueh, C.-F., R. N. Hiramoto, and V. K. Ghanta. The central effect of methionine-enkephalin on NK cell activity. *Brain Res.* 578: 142–148, 1992.

64. Huang, A. T., M. E. Engle, A. Elder, J. B. Kinn, and T. R. Ward. The effect of microwave radiation (2450 MHz) on the morphology and chromosomes of lymphocytes. *Radio Sci.* 12: 173–177, 1977.

65. Janković, B. D. The relationship between the immune system and the nervous system: old and new strategies. In: *Immunology 1930–1980: Essays on the History of Immunology*, edited by P. M. H. Mazumdar. Toronto: Wall and Thompson, 1988, p. 203–220.

66. Janković, B. D. Neuroimmunomodulation: facts and dilemmas. *Immunol. Lett.* 21: 101–118, 1989.

67. Janković, B. D. Neuroendocrine–immune interactions during ontogeny. *Int. J. Neurosci.* 5: 377–379, 1990.

68. Janković, B. D., K. Isaković, M. Mičić, and Z. Knezević. The embryonic lympho-neuro-endocrine relationship. *Clin. Immunol. Immunopathol.* 18: 108–120, 1981.

69. Janković, B. D., and D. Marić. Modulation of in vivo immune response by enkephalins. *Clin. Neuropharmacol.* 9: 447–476, 1986.

70. Janković, B. D., D. Marić, J. Ranin, and J. Velić. Magnetic fields, brain and immunity: effect on humoral and cell mediated immune response. *Int. J. Neurosci.* 59: 25–43, 1991.

71. Janković, B. D., B. M. Marković, and N. H. Spector. Neuroimmune interactions. Proceedings of the Second International Workshop on Neuroimmunmodulation. *Ann. N. Y. Acad. Sci.* 496: 1–756, 1987.

72. Janković, B. D., and N. H. Spector. Effects on the immune system of lesioning and stimulation of the nervous system: neuroimmunomodulation. In: *Enkephalins and Endorphins: Stress and the Immune System*, edited by N. K. Plotnikoff, R. E. Faith, A. J. Murgo, and R. A. Good. New York: Plenum, 1986, p. 189–220.

73. Jensen, M. M. Transitory impairment of interferon production in stressed mice. *J. Infect. Dis.* 118: 230–234, 1968.

74. Jensen, M. M., and A. F. Rasmussen. Stress and susceptibility of viral infection. I. Response of adrenals, liver, thymus, spleen and peripheral leukocyte counts to sound stress. *J. Immunol.* 90: 17–20, 1963.

75. Jiang, C. G., J. L. Morrow-Tesch, D. I. Beller, E. M. Levy, and P. H. Black. Immunosuppression in mice induced by cold water stress. *Brain Behav. Immun.* 4: 278–291, 1990.

76. Jitariu, P., N. Lascu, N. Topala, and M. Lazar. The influence of magnetic fields on the antitoxic-antitetanic immunity in guinea pigs. *Rev. Rom. Biol. Zool.* 10: 33–38, 1965.

77. Joasoo, A., and J. M. McKenzie. Stress and the immune response in rats. *Int. Arch. Allergy Appl. Immunol.* 50: 659–663, 1976.

78. Justice, A. Review of the effects of stress on cancer in laboratory animals: importance of time of stress application and type of tumor. *Psychol. Bull.* 98: 108–138, 1985.

79. Kavaliers, M., and K. P. Ossenkopp. Magnetic fields differentially inhibit mu, delta, kappa and sigma opiate-induced analgesia in mice. *Peptides* 7: 449–453, 1985.

80. Keller, S. E., J. M. Weiss, S. J. Schleifer, N. E. Miller, and M. Stein. Suppression of immunity by stress: effect of a graded series of stressors on lymphocyte stimulation in the rat. *Science* 213: 1397–1399, 1981.

81. Keller, S. E., J. M. Weiss, S. J. Schleifer, N. E. Miller, and M. Stein. Stress induced suppression of immunity in adrenalectomized rats. *Science* 221: 1301–1304, 1983.

82. Kelley, K. W. Stress and immune function: a bibliographic review. *Annu. Rev. Vet.* 11: 445–478, 1980.

83. Kelley, K. W., R. E. Greenfield, J. F. Evermann, S. M. Parish, and L. E. Perryman. Delayed-type hypersensitivity, contact sensitivity, and phytohemagglutinin skin-test responses of heat- and cold-stressed calves. *Am. J. Vet. Res.* 43: 775–779, 1982.

84. Klimenko, V. M. The study of some neuronal mechanisms of hypothalamic regulation of immune reactions in rabbits. *Avtoref. Kand. Diss. Inst. Exp. Med.* I. P. Pavlova, 1972.

85. Komori, T., R. Fujivara, J. Nomura, and M. M. Yokoyama. Effects of immobilization and fasting on plaque-forming cell response in mice [Abstract]. *Int. J. Neurosci.* 73: 1993.

86. Korneva, E. A., and V. M. Klimenko. Neuronale hypothalamusaktivität und homostatische reaktionen. *Ergeb. Exp. Med.* 23: 373–382, 1976.

87. Kudoh, G., K. Hoski, and T. Murakami. Fluorescence, microscopic and enzyme histochemical studies of the innervation of the human spleen. *Arch. Histol. Japon.* 42: 169–180, 1979.

88. Labes, M. M. A possible explanation for the effect of magnetic fields on biological systems. *Nature* 211: 968, 1966.

89. Langevin, T., W. Hrushesky, S. Sanchez, and F. Halberg. Melatonin modulates survival of CD2F1 mice with L1210 leukemia. *Chronobiologia* 10: 173–174, 1983.

90. Laudenslager, M. L., M. Fleshner, P. Hofstadter, P. E. Held, L. Simons, and S. F. Maier. Suppression of specific antibody production by inescapable shock: stability under varying conditions. *Brain Behav. Immun.* 2: 92–101, 1988.

91. Laudenslager, M., M. Reite, and R. Harbeck. Suppressed immune response in infant monkeys associated with maternal separation. *Behav. Neural Biol.* 36: 40–48, 1982.

92. Laudenslager, M. L., S. M. Ryan, R. C. Drugen, R. L. Hyson, and S. F. Maier. Coping and immunosuppression: inescapable but not escapable shock suppresses lymphocyte proliferation. *Science* 221: 568–570, 1983.

93. Lavicky, J., and H. Rasková. Different stressors and blood lipid peroxidation. *Arzneimittelforschug* 41: 793–796, 1991.

94. Levine, S., R. Strebel, E. J. Wenk, and P. J. Harman. Suppression of experimental allergic encephalomyelitis by stress. *Proc. Soc. Exp. Biol. Med.* 109: 294–298, 1962.

95. Liboff, A. R., T. Williams, D. M. Strong, and R. Wistar. Time-varying magnetic fields: effect on DNA synthesis. *Science* 233: 818–819, 1984.

96. Lukyanenko, V. I. The problem of conditioned-reflex regulation of immunologic reactions. *Usp. Sovrem. Biol.* 51: 170–187, 1961.

97. Maestroni, G. J. M., A. Conti, and W. Pierpaoli. Pineal melatonin, its fundamental immunoregulatory role in aging and cancer. In: *Neuroimmunomodulation: Interventions in Aging and Cancer*, edited by W. Pierpaoli and N. H. Spector. New York: Ann. N. Y. Acad. Sci. 1988, vol. 521, p. 140–148.

98. Maiti, A. J. Conditioned immunogenesis by vestibular stimulation? In: *Proc. 13th Ann. Conf. Indian Immunol. Soc.* Karala, India: Trichur, 1985.

99. Maiti, A. J. Vestibular–cerebellar interactions and immunity. (Lecture given at National Institutes of Health, Bethesda, MD. Synopsis available from A. J. M., Director, University of Calcutta Centre for Neuroscience, 2448 A. J. Chandra Bose Road, Calcutta, 700020, India), 1986.

100. Maiti, A. J. Vestibular stimulation to improve immune function in autistic children [Abstract]. In: *Proc. 2nd World Congr. Neurosci.*, IBRO, 1987.

101. Marković, B. M., I. L. Djordević, M. Lazarević, Z. Sporcić, V. J. Djuric, and B. D. Janković. Chronic stress and anaphylactic shock in the rat. *Period. Biol.* 92: 69–70, 1990.

102. Marsh, J. T., and A. F. Rasmussen, Jr. Response of adrenals, thymus, spleen and leucocytes to shuttle box and confinement stress. *Proc. Soc. Exp. Biol. Med.* 104: 180–183, 1960.

103. Metal'nikov, S. *Role du Système Nerveux et des Facteurs Biologiques et Psychiques dans l'Immunité.* Masson: Paris, 1934.

104. Metal'nikov, S., and V. Chorine. Role des réflexes conditionnels dans l'immunité. *Ann. Inst. Pasteur* 40: 893–900, 1926.

105. Mettrop, P. J., and P. Visser. Exteroceptive stimulation as a contingent factor in induction and elicitation of delayed-type hypersensitivity reactions to 1-chloro-, 2–4, dinitrobenzene in guinea pigs. *Psychophysiology* 5: 385–388, 1969.

106. Michaut, R. L., R. P. Dechambre, S. Doumero, B. Lesourd, A. DeVillechabroile, and R. Moulias. Influence of early maternal deprivation on adult humoral immune response in mice. *Physiol. Behav.* 26: 189–191, 1981.

107. Monjan, A., and M. I. Collector. Stress induced modulation of the immune response. *Science* 196: 307–308, 1977.

108. Morgan, G. W., P. Thaxton, and F. W. Edens. Reduced symptoms of anaphylaxis in chickens by ACTH or heat. *Poultry Sci.* 55: 1498–1504, 1976.

109. Moynihan, J. A., R. Ader, L. J. Grota, T. R. Schachtman, and N. Cohen. The effects of stress on the development of immunological memory following low-dose antigen priming in mice. *Brain Behav. Immun.* 4: 1–12, 1990.

110. Moynihan, J., D. Koota, G. Brenner, N. Cohen, and R. Ader. Repeated intraperitoneal injections of saline attenuate the antibody response to a subsequent intraperitoneal injection of antigen. *Brain Behav. Immun.* 3: 90–96, 1989.

111. Murayama, M. Orientation of sickled erythrocytes in a magnetic field. *Nature* 206: 420–422, 1965.

112. Neugebauer, D. C., A. E. Blaurock, and D. L. Worcester. Magnetic orientation of purple membranes demonstrated by optical measurements and neutron scattering. *FEBS Lett.* 78: 31–35, 1977.

113. Odio, M., A. Brodish, and M. L. Ricardo, Jr. Effects on immune responses by chronic stress are modulated by aging. *Brain Behav. Immun.* 1: 204–215, 1987.

114. Okimura, T., and Y. Nigo. Stress and immune responses. I. Suppression of T cell function in restraint-stressed mice. *Jpn. J. Pharmacol.* 40: 505–511, 1986.

115. Olesce, J., and S. Reuss. Magnetic field effects on pineal gland melatonin synthesis: comparative studies on albino and pigmented rodents. *Brain Res.* 396: 365–368, 1986.

116. Pavlidis, N., and M. Chirigos. Stress-induced impairment of macrophage tumoricidal function. *Psychosom. Med.* 42: 47–54, 1980.

117. Pierpaoli, W., and N. Fabris. Physiological senescence and its postponement: theoretical approaches and rational interventions. *Ann. N. Y. Acad. Sci.* 621: 1–454, 1991.

118. Pierpaoli, W., N. Fabris, and E. Sorkin. Developmental hormones and immunological maturation. In: *Hormones and the Immune Response*, edited by G. E. W. Wolstenholme and J. Knight. London: Churchill, 1970, p. 126–143.

119. Pierpaoli, W., W. Regelson, and N. Fabris. The aging clock. *Ann. N. Y. Acad. Sci.* 791: 1–558, 1993.

120. Pierpaoli, W., and N. H. Spector. Neuroimmunomodulation: interventions in aging and cancer. First Stromboli Conference on Aging and Cancer. *Ann. N. Y. Acad. Sci.* 521: 1–361, 1988.

121. Plotnikoff, N. P., R. E. Faith, A. J. Murgo, and R. A. Good. *Enkephalins and Endorphins: Stress and the Immune System.* New York: Plenum, 1986.

122. Plotnikoff, N. P., R. E. Faith, A. J. Murgo, and J. Wybran. *Stress and Immunity.* Boca Raton, FL: CRC, 1991.

123. Polk, C., and E. Postow. *CRC Handbook of Biological Effects of Electromagnetic Fields.* Boca Raton, FL: CRC, 1986.

124. Rabin, B. S., and S. B. Salvin. Effect of differential housing and time on immune reactivity to sheep erythrocytes and *Candida.* *Brain Behav. Immun.* 1: 267–275, 1987.

125. Rasková, H., Z. Urbanova, L. Celeda, V. Trcka, J. Cerny, and A. Rubicek. Some drug induced modifications of stress in calves. In: *Neuropharmacology '85*, edited by K. Kelemen, K. Magyar, and E. S. Vizi. Budapest: Akademiai Kiado, 1985, p. 249–254.

126. Rasková, H., J. Vanecek, D. Celeda, and Z. Urbanova. Pharmacology of nonspecific resistance revisited. In: *Trends in the Pharmacology of Neurotransmission*, edited by T. Stoitchev, D. Staneva-Stoitcheva, R. Rodomirov, S. Todorov, and R. Ovcharov. Sofia: Bulg. Acad. Sci., 1987, p. 229–236.

127. Rasmussen, A. F., Jr., E. S. Spencer, and J. T. Marsh. Decrease in susceptibility of mice to passive anaphylaxis following avoidance-learning stress. *Proc. Soc. Exp. Biol. Med.* 100: 878–879, 1959.

128. Raymond, L. N., E. Reyes, S. Tokuda, and B. C. Jones. Differential immune response in two handled inbred strains of mice. *Physiol. Behav.* 37: 295–297, 1986.

129. Regnier, J. A., K. W. Kelley, and C. T. Gaskins. Acute thermal stressors and synthesis of antibodies in chickens. *Poultry Sci.* 985–990, 1980.

130. Reiter, R. J. Interaction of photoperiod, pineal and seasonal reproduction as exemplified by findings in the hamster. In: *The Pineal in Reproduction. Progress in Reproductive Biology*, edited by R. J. Reiter. Basel: Karger, 1978, vol. 4, p. 169–190.

131. Reuss, S., P. Semm, and L. Volrath. Different types of magnetically sensitive cells in the rat pineal gland. *Neurosci. Lett.* 40: 23–26, 1983.

132. Riley, V. Psychoneuroendocrine influences on immunocompetence and neoplasia. *Science* 212: 1100–1109, 1981.

133. Rogers, M. P., D. E. Trentham, and P. Reich. Modulation of collagen-induced arthritis by different stress protocols. *Psychosom. Med.* 42: 72, 1980.

134. Rosen, A. D., and J. Lubowsky. Magnetic field influence on the central nervous system function. *Exp. Neurol.* 95: 679–687, 1987.

135. Sabiston, B. H., J. E. Rose, and B. Cinader. Temperature stress and immunity in mice. Effects of environmental temperature on the antibody response to human immunoglobulin of mice, differing in age and strain. *J. Immunogenet.* 5: 197–212, 1978.

136. Salvin, S. B., B. S. Rabin, and R. Neta. Evaluation of immunologic assays to determine the effects of differential housing on immune reactivity. *Brain Behav. Immun.* 4: 180–188, 1990.

137. Saphier, D. Neurophysiological and endocrine consequences of immune activity. *Psychoneuroendocrinology* 14: 63–87, 1989.

138. Shavit, Y., F. C. Martin, A. R. Yirmiy, S. Ben-Eliyahu, G. W. Terman, H. Weiner, R. P. Gale, and J. C. Liebeskind. Effects of single administration of morphine or footshock stress on natural killer cell cytotoxicity. *Brain Behav. Immun.* 1: 318–328, 1987.

139. Sklar, L. S., and H. Anisman. Stress and coping factors influence tumor growth. *Science* 205: 513–515, 1979.

140. Smith, L. W., N. Molomut, and B. Gottfried. Effect of subconvulsive audiogenic stress in mice on turpentine induced inflammation. *Proc. Soc. Exp. Biol. Med.* 103: 370–372, 1960.

141. Smolensky, M. H., F. Halberg, J. Harter, B. Hsi, and W. Nelson. Higher corticosterone values at a fixed single timepoint in serum from mice "trained" by prior handling. *Chronobiologia* 5: 1–13, 1978.

142. Sobrian, S. K., B. Marković, V. Djurić, M. Lazarević, and B. D. Janković. Alterations in the immune function following prenatal sound stress. *FASEB J.* 2: A1261, 1988.

143. Sobrian, S. K., B. M. Marković, M. Lazarević, V. J. Djurić, and B. D. Janković. Parameters influencing the effect of sound stress on immune function. *J. Neuroimmunol.* 16: 164, 1987.

144. Solomon, G. F. Stress and antibody response in rats. *Int. Arch. Allergy Appl. Immunol.* 35: 97–104, 1969.

145. Solomon, G. F., S. Levine, and J. K. Kraft. Early experience and immunity. *Nature* 220: 821–822, 1968.

146. Solvason, H. B., V. Ghanta, R. Hiramoto, and N. H. Spector. Natural killer cell activity augmented by classical (Pavlovian) conditioning. In: *Proc. First International Workshop on Neuroimmunomodulation.* Bethesda: International Working Group on Neuroimmunomodulation (IWGN, Bethesda, MD), 1985, p. 188.

147. Speber, D., K. Dransfeld, G. Maret, and M. H. Weisenseel. Oriented growth of pollen tubes in a strong magnetic field. *Naturwissenschaften* 68: 40–44, 1981.

148. Spector, N. H. The central state of the hypothalamus in health and disease: old and new concepts. In: *Physiology of the Hypothalamus,* edited by P. Morgane and J. Panksepp. New York: Dekker, 1980, p. 453–517.

149. Spector, N. H. *Neuroimmunomodulation. Proc. First Int. Workshop* (2nd ed.), New York: Gordon and Breach, 1988, p. 1–306.

150. Spector, N. H. Old and new strategies in the conditioning of immune responses. *Ann. N. Y. Acad. Sci.* 496: 522–531, 1989.

151. Spector, N. H. Basic mechanisms and pathways of neuroimmunomodulation: triggers. *Int. J. Neurosci.* 51: 335–337, 1990.

152. Spector, N. H. Strategien bei der Konditionierung der Immunantwort. In: *Psychoimmunologie,* edited by W. P. Kaschka and H. N. Aschauer. Stuttgart: Georg Thieme, 1990, p. 2–9.

153. Spector, N. H. New information explosions in NIM research. *Immunol. Today* 11: 381–383, 1991.

154. Spector, N. H., and E. A. Korneva. Neurophysiology, immunophysiology and neuroimmunomodulation. In: *Psychoneuroimmunology,* edited by R. Ader. New York: Academic, 1981, p. 449–473.

155. Spector, N. H., M. Provinciali, G. DiStefano, M. Muzzioli, D. Bulian, C. Vitticchi, F. Rossano, and N. Fabris. Immune enhancement by conditioning of senescent mice: comparison of old and young mice in learning ability and inability to increase natural killer cell activity and other host-defense reactions in response to a conditioned stimulus. *Ann. N. Y. Acad. Sci.* 741: 283–292, 1994.

156. Stein, M., S. E. Keller, and S. J. Schleifer. Stress and immunomodulation: the role of depression and neuroendocrine function. *J. Immunol.* 135: 827s–833s, 1985.

157. Tenforde, T. S. *Magnetic Field Effects on Biological Systems.* New York: Plenum, 1979.

158. Teshima, H., H. Sogawa, H. Kihara, S. Nagata, Y. Ago, and T. Nakagawa. Changes in populations of T-cell subsets due to stress. *Ann. N. Y. Acad. Sci.* 496: 459–466, 1987.

159. Thaxton, P., C. R. Sadler, and B. Glick. Immune response of chickens following heat exposure or injections with ACTH. *Poultry Sci.* 47: 264–266, 1968.

160. Torbet, J., J. M. Freyssinet, and G. Hudry-Clergeon. Oriented fibrin gels formed by polymerization in strong magnetic fields. *Nature* 289: 91–93, 1981.

161. Treadwell, P. E., and A. F. Rasmussen, Jr. Role of the adrenals in stress-induced resistance to anaphylactic shock. *J. Immunol.* 87: 492–497, 1961.

162. Ueck, M. Innervation of the vertebrate pineal. *Prog. Brain Res.* 52: 45–88, 1979.

163. Vasilyev, N. V., and L. P. Boginitch. *Influence of Magnetic Fields on Processes in Infection and Immunity.* Tomsk: Univ. of Tomsk, 1973.

164. Vessey, S. H. Effects of grouping on levels of circulating antibodies in mice. *Proc. Soc. Exp. Biol. Med.* 115: 252–255, 1964.

165. Visintainer, M. A., J. R. Volpicelli, and M. E. P. Seligman. Tumor rejection in rats after inescapable and escapable shock. *Science* 216: 437–439, 1982.

166. Weiss, J. M., S. K. Sundar, and K. J. Brecker. Stress induced immunosuppression and immunoenhancement: cellular immune changes and mechanisms. In: *Neuroimmune Networks: Physiology and Diseases,* edited by E. J. Goetzl and N. H. Spector. New York: Liss, 1989, p. 193–206.

167. Wistar, R., Jr., and W. H. Hildemann. Effect of stress on skin transplantation immunity in mice. *Science* 131: 159–160, 1960.

168. Wrba, H., A. Dutter, S. Sanchez de la Peña, J. Wu, F. Carandente, G. Cornelissen, and F. Halberg. Secular or circannual effects of placebo and melatonin on murine breast cancer. In: *Chronobiology: Its Role in Clinical Medicine, General Biology, and Agriculture,* edited by D. K. Hayes, J. E. Pauly, and R. J. Reiter. New York: Wiley-Liss, 1990, pt. A, p. 31–40.

169. Wurtman, R. J., J. Axelrod, and D. E. Kelly. *The Pineal.* New York: Academic, 1968.

170. Zalcman, S., L. Kerr, and H. Anisman. Immunosuppression elicited by stressors and stressor-related odors. *Brain Behav. Immun.* 5: 262–273, 1991.

171. Zalcman, S., A. Minkiewicz-Janda, M. Richter, and H. Anisman. Critical periods associated with stressor effects on antibody titers and on the plaque-forming cell responses to sheep red blood cells. *Brain Behav. Immun.* 2: 254–266, 1988.

172. Zwilling, B. S., M. Dinkins, R. Christner, M. Faris, A. Griffin, M. Hilburger, M. McPeek, and D. Pearl. Restraint stress induces suppression of MHC class II expression by murine peritoneal macrophages. *J. Neuroimmunol.* 29: 125–130, 1990.

Index